3. *Sum of two cubes:*
$$A^3 + B^3 = (A + B)(A^2 - AB + B^2)$$

4. *Difference of two cubes:*
$$A^3 - B^3 = (A - B)(A^2 + AB + B^2)$$

Variation

English Statement	Equation
y varies directly as x.	$y = kx$
y varies directly as x^n.	$y = kx^n$
y varies inversely as x.	$y = \dfrac{k}{x}$
y varies inversely as x^n.	$y = \dfrac{k}{x^n}$
y varies jointly as x and z.	$y = kxz$

Exponents
Definitions of Rational Exponents

1. $a^{\frac{1}{n}} = \sqrt[n]{a}$

2. $a^{\frac{m}{n}} = \left(\sqrt[n]{a}\right)^m$ or $\sqrt[n]{a^m}$

3. $a^{-\frac{m}{n}} = \dfrac{1}{a^{\frac{m}{n}}}$

Properties of Rational Exponents
If m and n are rational exponents, and a and b are real numbers for which the following expressions are defined, then

1. $b^m \cdot b^n = b^{m+n}$

2. $\dfrac{b^m}{b^n} = b^{m-n}$

3. $\left(b^m\right)^n = b^{mn}$

4. $(ab)^n = a^n b^n$

5. $\left(\dfrac{a}{b}\right)^n = \dfrac{a^n}{b^n}$

Radicals

1. If n is even, then $\sqrt[n]{a^n} = |a|$.
2. If n is odd, then $\sqrt[n]{a^n} = a$.
3. The product rule: $\sqrt[n]{a} \cdot \sqrt[n]{b} = \sqrt[n]{ab}$
4. The quotient rule: $\dfrac{\sqrt[n]{a}}{\sqrt[n]{b}} = \sqrt[n]{\dfrac{a}{b}}$

Complex Numbers

1. The imaginary unit i is defined as
$$i = \sqrt{-1}, \quad \text{where} \quad i^2 = -1.$$

The set of numbers in the form $a + bi$ is called the set of complex numbers. If $b = 0$, the complex number is a real number. If $b \neq 0$, the complex number is an imaginary number.

2. The complex numbers $a + bi$ and $a - bi$ are conjugates. Conjugates can be multiplied using the formula
$$(A + B)(A - B) = A^2 - B^2.$$

The multiplication of conjugates results in a real number.

3. To simplify powers of i, rewrite the expression in terms of i^2. Then replace i^2 by -1 and simplify.

Quadratic Equations and Functions

1. The solutions of a quadratic equation in standard form
$$ax^2 + bx + c = 0, \quad a \neq 0,$$
are given by the quadratic formula
$$x = \dfrac{-b \pm \sqrt{b^2 - 4ac}}{2a}.$$

2. The discriminant, $b^2 - 4ac$, of the quadratic equation $ax^2 + bx + c = 0$ determines the number and type of solutions.

Discriminant	Solutions
Positive perfect square with a, b, and c rational numbers	2 rational solutions
Positive and not a perfect square	2 irrational solutions
Zero, with a, b, and c rational numbers	1 rational solution
Negative	2 imaginary solutions

3. The graph of the quadratic function
$$f(x) = a(x - h)^2 + k, \quad a \neq 0,$$
is called a parabola. The vertex, or turning point, is (h, k). The graph opens upward if a is positive and downward if a negative. The axis of symmetry is a vertical line passing through the vertex. The graph can be obtained using the vertex, x-intercepts, if any, [set $f(x)$ equal to zero], and the y-intercept (set $x = 0$).

4. A parabola whose equation is in the form
$$f(x) = ax^2 + bx + c, \quad a \neq 0,$$
has its vertex at
$$\left(-\dfrac{b}{2a}, f\left(-\dfrac{b}{2a}\right)\right).$$

If $a > 0$, then f has a minimum that occurs at $x = -\dfrac{b}{2a}$. If $a < 0$, then f has a maximum that occurs at $x = -\dfrac{b}{2a}$.

(*continued on inside back cover*)

Introductory and Intermediate Algebra

FOR COLLEGE STUDENTS

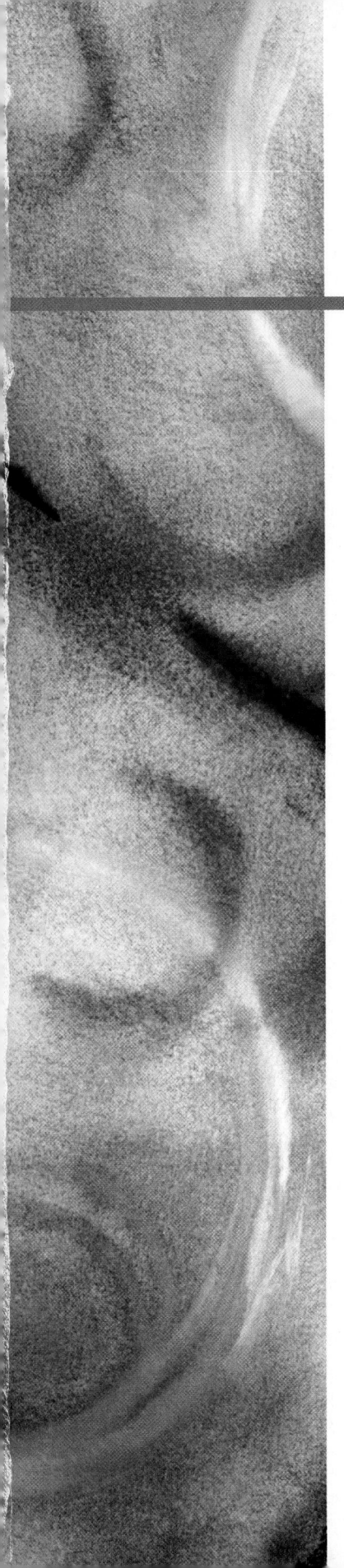

INTRODUCTORY AND INTERMEDIATE ALGEBRA
FOR COLLEGE STUDENTS

Robert Blitzer
Miami-Dade Community College

PRENTICE HALL, Upper Saddle River, New Jersey 07458

Library of Congress Cataloging-in-Publication Data

Blitzer, Robert.
  Introductory and intermediate algebra for college students /Robert Blitzer.-1st ed.
    p. cm.
  Introductory and intermediate algebra for college students is part of a series of three
  texts that include Introductory algebra for college students, third ed., and Intermediate
  algebra for college students, third ed.—Pref.
  Includes index.
  ISBN 0-13-032842-1
    1. Algebra. I. Title.

  QA152.3.B65 2002
  512.9—dc21   2001055484

Executive Acquisition Editor: *Karin E. Wagner*
Editor-in-Chief: *Christine Hoag*
Senior Managing Editor: *Linda Mihatov Behrens*
Executive Managing Editor: *Kathleen Schiaparelli*
Vice President/Director of Production and Manufacturing: *David W. Riccardi*
Production Editor/Assistant Managing Editor: *Bayani Mendoza de Leon*
Manufacturing Buyer: *Alan Fischer*
Manufacturing Manager: *Trudy Pisciotti*
Executive Marketing Manager: *Eilish Collins Main*
Development Editor: *Don Gecewicz*
Editor-in-Chief, Development: *Carol Trueheart*
Media Project Manager, Developmental Math: *Audra J. Walsh*
Art Director: *Maureen Eide*
Assistant to the Art Director: *John Christiana*
Interior Designer: *Donna Wickes*
Cover Designer: *Joseph Sengotta*
Managing Editor, Audio/Video Assets: *Grace Hazeldine*
Creative Director: *Carole Anson*
Director of Creative Services: *Paul Belfanti*
Photo Researcher: *Melinda Alexander*
Photo Editor: *Beth Boyd*
Cover Image: *Richard Blair* (www.richardblair.com)
Art Studio: *Scientific Illustrators*
Compositor: *Prepare, Inc.*

©2002 by Prentice-Hall, Inc.
Upper Saddle River, New Jersey 07458

Printed in the United States of America
10 9 8 7 6 5

ISBN 0-13-032842-1

Pearson Education Ltd., *London*
Pearson Education Australia Pty., Limited, *Sydney*
Pearson Education *Singapore*, Pte. Ltd
Pearson Education North Asia Ltd., *Hong Kong*
Pearson Education Canada, Ltd., *Toronto*
Pearson Education de Mexico, S.A. de C.V.
Pearson Education—Japan, *Tokyo*
Pearson Education Malaysia, Pte. Ltd.

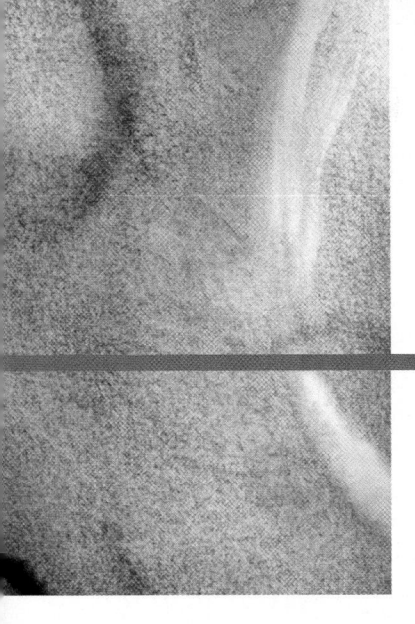

Contents

CHAPTER 1 The Real Number System   3

CHAPTER 2 Linear Equations and Inequalities in One Variable   101

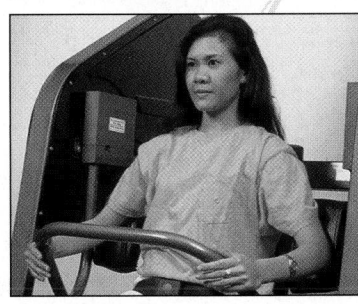

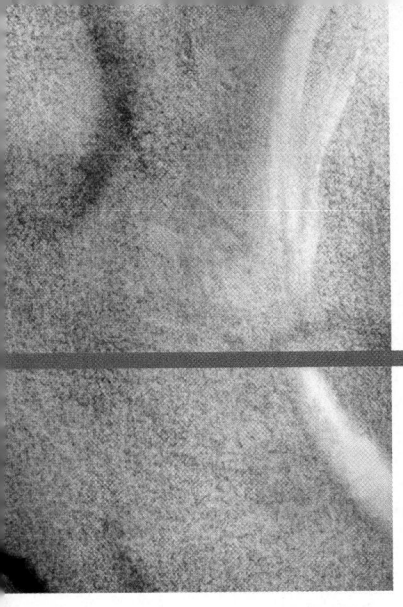

Preface

Introductory and Intermediate Algebra for College Students provides comprehensive, in- depth coverage of the topics required in a course combining the study of introductory and intermediate algebra. The primary goals of the book are to help students acquire a solid foundation in introductory and intermediate algebra, without the repetition of topics in two separate texts, and to show how algebra can model and solve authentic real-world problems.

A source of frustration for me and my colleagues is that very few students read their textbook. When I ask students why they do not take full advantage of the text, their responses generally fall into two categories:

- "I cannot follow the explanations."
- "The applications are not interesting."

I thought about both of these objections in writing every page of this book.

"I can't follow the explanations." For many of my students, textbook explanations are too compressed. The chapters in *Introductory and Intermediate Algebra for College Students* have been written to make them extremely accessible. Every section contains a range of simple, intermediate, and challenging examples. Voice balloons allow for specific annotations in examples, further clarifying procedures and concepts.

"The applications are not interesting." One of the things I enjoy most about teaching in a large urban community college is the diversity of who my students are and what interests them. Real-world data that celebrate this variety are used to bring relevance to examples, discussions, and applications. I selected all up-to-date real-world data to be interesting and intriguing to students. By connecting algebra to the whole spectrum of their interests, it is my intent to show students that their world is profoundly mathematical and, indeed, pi is in the sky.

Key Pedagogical Features

Introductory and Intermediate Algebra for College Students is part of a series of four texts that include *Introductory Algebra for College Students*, Third Edition, *Intermediate Algebra for College Students*, Third Edition, and *Algebra for College Students*, Fourth Edition. The following features are found throughout the series.

- **Chapter-Opening and Section-Opening Scenarios.** Every chapter *and every section* opens with a compelling image that supports a scenario presenting a unique application of algebra in students' lives outside the classroom. Each scenario is revisited later in the chapter or section.

- **Section Objectives.** Learning objectives open every section. The objectives are stated in the margin at their point of use.

- **Detailed Illustrative Examples**. Each illustrative example is titled, making clear the purpose of the example. Examples are clearly written and provide students with detailed step-by-step solutions. No steps are omitted and each step is explained.

- **Check Point Examples.** Each worked example is followed by a similar matched problem for the student to work while reading the material. This actively involves the student in the learning process and gives students the opportunity to work with a concept as soon as they have learned it. Answers to all Check Points are given in the answer section.

- **Graphing**. Chapter 1 contains an introduction to graphing, a topic that is integrated throughout the book. Line, bar, circle, and rectangular coordinate graphs that use real data appear in nearly every section and exercise set. Many examples and exercises use graphs to explore relationships between data and to provide ways of visualizing a problem's solution.

- **Geometric Problem Solving.** Section 2.6 on problem solving in geometry teaches geometric concepts that are important to a student's understanding of algebra. There is frequent emphasis on problem solving in geometric situations, as well as on geometric models that allow students to visualize algebraic formulas.

- **Functions.** Functions are introduced in Chapter 8, with functions emphasized throughout the second half of the book.

- **Thorough, Yet Optional Technology.** Although the use of graphing utilities is optional, they are utilized in Using Technology boxes to enable students to visualize algebraic concepts. The use of graphing utilities is also reinforced in the technology exercises appearing in the exercise sets for those who want this option. With the book's early introduction to graphing, students can look at the calculator screens in the Using Technology boxes and gain an increased understanding of an example's solution even if they are not using a graphing utility in the course.

- **Enrichment Essays.** Enrichment essays provide historical, interdisciplinary, and otherwise interesting connections throughout the text.

- **Study Tips.** Study Tip boxes offer suggestions for problem solving, point out common student errors, and provide informal tips and suggestions. These invaluable hints appear in abundance throughout the book.

- **Discovery.** Discover for Yourself boxes, found throughout the text, encourage students to further explore algebraic concepts. These explorations are optional and their omission does not interfere with the continuity of the topic under consideration.

- **Exercise Sets.** An extensive collection of exercises is included in an exercise set at the end of each section. The text organizes exercises by level within six category types: Practice Exercises, Application Exercises, Writing in Mathematics, Technology Exercises, Critical Thinking Exercises, and

Review Exercises. This format makes it easy to create well-rounded home-work assignments. Writing exercises offer students the opportunity to write about every objective covered in each section, as well as to discuss, interpret, and give opinions about data. Each review exercise contains the section number and example number of a similar worked-out example.

- **Chapter Projects.** At the end of each chapter are collaborative activities that give students the opportunity to work cooperatively as they think and talk about mathematics. Many of these exercises should result in interesting group discussions.

- **Chapter Review Grids.** Each chapter contains a review chart that summarizes the definitions and concepts in every section of the chapter. Examples that illustrate these key concepts are also included in the chart. Like the summary grid, review exercises are organized by each section of the chapter.

- **End-of-Chapter Materials.** The review grids provide a focused summary and illustrative examples for each section in the chapter. A comprehensive collection of review exercises for each of the chapter's sections follows the review grid. This is followed by a chapter test. Beginning with Chapter 2, each chapter concludes with a comprehensive collection of cumulative review exercises.

- **A Review of Introductory Algebra.** Appendix A, entitled *Are You Prepared for Intermediate Algebra?*, provides students with a fast way to review introductory algebra topics before starting the intermediate algebra portion of the book.

- **Supplements Package.** This text is supported by a wealth of supplements designed for added effectiveness and efficiency. These items are described on pages xiii through xv.

Supplements for the Instructor

Printed Resources

Annotated Instructor's Edition (0–13–032843–X)
- Answers to exercises on the same text page or in Graphing Answer Section.
- Graphing Answer Section contains answers to exercises requiring graphical solutions.

Instructor's Solutions Manual (0–13–034328–5)
- Step-by-step solutions for every even-numbered section exercise.
- Step-by-step solutions for every (even and odd) Check Point exercise, Chapter Review exercise, Chapter Test and Cumulative Review exercise.

Instructor's Resource Manual (0–13–034319–6)
- Notes to the Instructor
- Eight Chapter Tests per chapter (5 free response, 3 multiple choice)
- Eight Final Exams (4 free response, 4 multiple choice)
- Twenty additional exercises per section for added test exercises or worksheets.
- Answers to all items

Media Resources

TestGen-EQ with QuizMaster-EQ (CD-ROM for IBM and Macintosh 0–13–034324–2)
- Algorithmically driven, text specific testing program.
- Networkable for administering tests and capturing grades on-line.
- Edit or add your own questions to create a nearly unlimited number of tests and worksheets.
- Use the new "Function Plotter" to create graphs.
- Tests can be easily exported to HTML so they can be posted to the Web.

Computerized Tutorial Software Course Management System

MathPro Explorer 4.0
- Network version for IBM and Macintosh
- Enables instructors to create either customized or algorithmically generated practice quizzes from any section of a chapter.
- Includes an e-mail function for networked users, enabling instructors to send a message to a specific student or to an entire group.
- Network based reports and summaries for a class or student and for cumulative or selected scores are available.

MathPro 5
- Anytime. Anywhere.
- Online tutorial with enhanced class and student management features.
- Integration of TestGen-EQ allows for testing to operate within the tutorial environment.
- Course management tracking of both tutorial and testing activity.

Online Options for Distance Learning

WebCT/Blackboard/CourseCompass
- Prentice Hall offers three different on-line interactivity and delivery options for a variety of distance learning needs. Instructors may access or adopt these in conjunction with this text.

Supplements for the Student

Printed Resources

Student Solutions Manual (0–13–034327–7)
- Step-by-step solutions for every odd-numbered section exercise.
- Step-by-step solutions for every (even and odd) Check Point exercise, Chapter Review exercise, Chapter Test and Cumulative Review exercise.

How to Study Mathematics
- Have your instructor contact the local Prentice Hall sales representative.

Math on the Internet: A Student's Guide
- Have your instructor contact the local Prentice Hall sales representative.

Media Resources

Computerized Tutorial Software

MathPro Explorer 4.0
- Keyed to each section of the text for text-specific tutorial exercises and instruction.
- Warm-up exercises and graded Practice Problems.
- Video clips show a problem being explained and worked out on the board.
- Algorithmically generated exercises. On-line help, glossary and summary of scores.

MathPro 5–Anytime.　Anywhere.
- Enhanced, Internet-based version of Prentice Hall's popular tutorial software.

Lecture Videos
- Keyed to each section of the text.

Digitized Lecture Videos on CD.
- Have your instructor contact the local Prentice Hall sales representative.

Prentice Hall Tutoring Center
- Provides one-on-one tutorial assistance by phone, e-mail, or fax.

Companion Website
- Offers Warm-ups, Real World Activities and Chapter Quizzes.
- E-mail results to your instructor.
- Destination links provide additional opportunities to explore other related sites.

Acknowledgments

I wish to express my appreciation to all the reviewers whose helpful criticisms and suggestions, frequently transmitted with wit, humor and intelligence, contributed to this text. In particular, I would like to thank:

Thomas B. Clark	*Trident Technical College*
Bettyann Daley	*University of Delaware*
Marion K. Glasby	*Anne Arundel Community College*
Andrea Hendricks	*Georgia Perimeter College*
Sandee House	*Georgia Perimeter College*
Laura Hoye	*Trident Technical College*
Marcella Jones	*Minneapolis Community and Technical College*
Shelbra B. Jones	*Wake Technical Community College*
Sharon Keenee	*Georgia Perimter College*
Kristi Laird	*Jackson Community College*
R$obert Leibman	*Austin Community College*
Ann M. Loving	*J. Sargent Reynolds Community College*
Kent MacDougall	*Temple College*

Jean P. Millen	*Georgia Perimeter College*
Terri Moser	*Austin Community College*
Lois Jean Niemi	*Minneapolis Community and Technical College*
Jeff Parent	*Oakland Community College*
Robert Patenaude	*College of the Canyons*
Christopher Reisch	*The State University of New York at Buffalo*
Haazim Sabree	*Georgia Perimeter College*
Margaret Williamson	*Milwaukee Area Technical College*

Special thanks go to the faculty at College of the Canyons, specifically Cherie Choate, Marlene Demerjian, Charles Johnson, Dennis Morrow, and Bob Patenaude, who were so supportive of my initial efforts at writing this text and gave me some invaluable feedback for this new book which I have incorporated.

Additional acknowledgments are extended to Jacquelyn White for creating the dynamic video tape series covering every section of the book; Penny Arnold for solving all the book's exercises; Lea Rosenberry for the Herculean task of preparing the solutions manuals, as well as serving as accuracy checker; the team at Laurel Technical Services for preparing the answer section and also serving as accuracy checker; Henry Small, research assistant, for obtaining much of the book's data and information for the new section openers; Melinda Alexander, photo researcher, for obtaining the book's photographs; Richard Blair for granting us permission to use one of his outstanding photographs from *Point Reyes Visions* for the book's cover; Brian Morris and the team of graphic artists and mathematicians at Scientific Illustrators, whose superb illustrations and graphs provide visual support to the verbal portions of the text; Prepare Inc., the book's compositor, for inputting hundreds of pages with hardly an error; and Bayani Mendoza de Leon, whose talents as production editor contributed to the book's wonderful look.

I would like to thank my editor at Prentice Hall, Karin Wagner, who guided and coordinated the many details of this project, and my development editor, Don Gecewicz. My thanks also to Eilish Main, Executive Marketing Manager, for her innovative marketing efforts, as well as to the entire Prentice Hall team for their confidence in and enthusiasm for my book.

Robert Blitzer

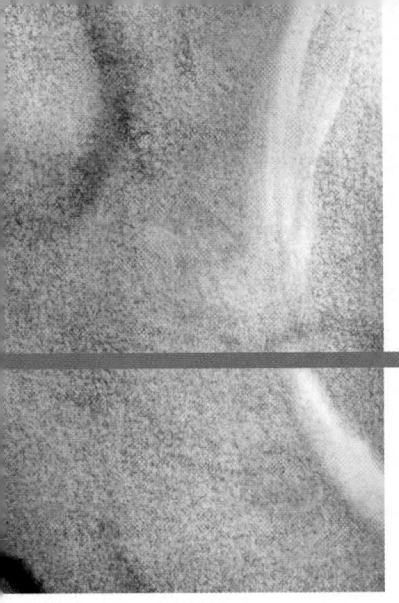

To the Student

I've written this book so that you can learn about the power of algebra and how it relates directly to your life outside the classroom. All concepts are carefully explained, important definitions and procedures are set off in boxes, and worked-out examples that present solutions in a step-by-step manner appear in every section. Each example is followed by a similar matched problem, called a Check Point, for you to try so that you can actively participate in the learning process as you read the book. (Answers to all Check Points appear in the back of the book.) Study Tips offer hints and suggestions and often point out common errors to avoid. A great deal of attention has been given to applying algebra to your life to make your learning experience both interesting and relevant.

As you begin your studies, I would like to offer some specific suggestions for using this book and for being successful in this course:

1. Attend all lectures. No book is intended to be a substitute for valuable insights and interactions that occur in the classroom. In addition to arriving for lecture on time and being prepared, you will find it useful to read the section before it is covered in lecture. This will give you a clear idea of the new material that will be discussed.

2. Read the book. Read each section with pen (or pencil) in hand. Move through the illustrative examples with great care. These worked-out examples provide a model for doing exercises in the exercise sets. As you proceed through the reading, do not give up if you do not understand every single word. Things will become clearer as you read on and see how various procedures are applied to specific worked-out examples.

3. Work problems every day and check your answers. The way to learn mathematics is by doing mathematics, which means working the Check Points and assigned exercises in the exercise sets. The more exercises you work, the better you will understand the material.

4. Prepare for chapter exams. After completing a chapter, study the summary chart, work the exercises in the Chapter Review, and work the exercises in the Chapter Test. Answers to all these exercises are given in the back of the book.

5. Use the supplements available with this book. A solutions manual containing worked-out solutions to the book's odd-numbered exercises, all review exercises, and all Check Points, a dynamic web page, and video tapes created for every section of the book are among the supplements created to help you tap into the power of mathematics. Ask your instructor or bookstore what supplements are available and where you can find them.

I wrote this book in Point Reyes National Seashore, 40 miles north of San Francisco. The park consists of 75,000 acres with miles of pristine surf-washed beaches, forested ridges, and bays bordered by white cliffs. It was my hope to convey the beauty and excitement of mathematics using nature's unspoiled beauty as a source of inspiration and creativity. Enjoy the pages that follow as you empower yourself with the algebra needed to succeed in college, your career, and in your life.

Regards,
Bob
Robert Blitzer

About the Author

Bob Blitzer is a native of Manhattan and received a Bachelor of Arts degree with dual majors in mathematics and psychology (minor: English literature) from the City College of New York. His unusual combination of academic interests led him toward a Master of Arts in mathematics from the University of Miami and a doctorate in behavioral sciences from Nova University. Bob is most energized by teaching mathematics and has taught a variety of mathematics courses at Miami-Dade Community College for nearly 30 years. He has received numerous teaching awards, including Innovator of the Year from the League for Innovations in the Community College, and was among the first group of recipients at Miami-Dade Community College for an endowed chair based on excellence in the classroom. In addition to *Introductory and Intermediate Algebra for College Students*, Bob has written *Introductory Algebra for College Students, Intermediate Algebra for College Students, Algebra for College Students, Thinking Mathematically, College Algebra, Algebra and Trigonometry*, and *Precalculus*, all published by Prentice Hall.

*M*ost things in life depend on many variables. Temperature and precipitation are two variables that affect whether regions are forests, grasslands, or deserts. In this chapter, you will learn methods for modeling your world with inequalities. You will even see how inequalities are used to describe some of the most magnificent places in our nation's landscape.

You are in Yosemite National Park in California, surrounded by evergreen forests, alpine meadows, and sheer walls of granite. The beauty of soaring cliffs, plunging waterfalls, gigantic trees, rugged canyons, mountains, and valleys is overwhelming. This is so different from where you live and attend college, a region in which grasslands predominate.

▶ **SECTION 9.1**   *Interval Notation and Business Applications Using Linear Inequalities*

Objectives

1 Use interval notation.

2 Review how to solve linear inequalities.

3 Use linear inequalities to solve problems involving revenue, cost, and profit.

SSM   PH Tutor   CD-   Video
     Center   ROM

Driving through your neighborhood, you see kids selling lemonade. Would it surprise you to know that this activity can be analyzed using linear inequalities? By doing so, you will view profit and loss in the business world in a new way. In this section, we use linear inequalities to solve problems and model business ventures.

1 Use interval notation.

Interval Notation

Recall from Chapter 2 that any inequality in the form $ax + b \leq c$ is called a **linear inequality in one variable**. The symbol between $ax + b$ and can be $\leq$

page 588

A Strategy for Solving Word Problems Using Equations

Problem solving is an important part of algebra. The problems in this book are presented in English. We must translate from the ordinary language of English into the language of algebraic equations. To translate, however, we must understand the English prose and be familiar with the forms of algebraic language. Here are some general steps we will follow in solving word problems.

Strategy for Solving Word Problems

Step 1 Read the problem carefully. Attempt to state the problem in your own words and state what the problem is looking for. Let x (or any variable) represent one of the quantities in the problem.

Step 2 If necessary, write expressions for any other unknown quantities in the problem in terms of x.

Step 3 Write an equation in x that describes the verbal conditions of the problem.

Step 4 Solve the equation and answer the problem's question.

Step 5 Check the solution *in the original wording* of the problem, not in the equation obtained from the words.

---

page 136

page 136

page 533

page 303

Emphasis on Problem Solving

The text has been organized to emphasize problem solving.

Section 2.5 is "An Introduction to Problem Solving," and Chapter 4, "Systems of Linear Equations," fully explores problem-solving strategies.

For example,

$$(T - S)(2000) = T(2000) - S(2000) = \$10{,}500 - \$3400 = \$7100.$$

> In 2000, the difference between total tax and state and local tax was $7100. This is the per capita federal tax.

Figure 8.9 illustrates that information involving differences of functions often appears in graphs seen in newspapers and magazines. Like numbers and algebraic expressions, two functions can be added, subtracted, multiplied, or divided as long as there are numbers common to the domains of both functions. The common domain for functions T and S in Figure 8.9 is

$$\{1900, 1901, 1902, 1903, \ldots, 2000\}.$$

Voice Balloons

Voice Balloons call out key problem-solving tips and observations to clarify the problem-solving process.

Relevant Applications

Over 90% of the *applications* and many of the *examples* have been written to incorporate current, real-world data drawn from familiar sources, such as the Statistical Abstract published by the U.S. Census Bureau.

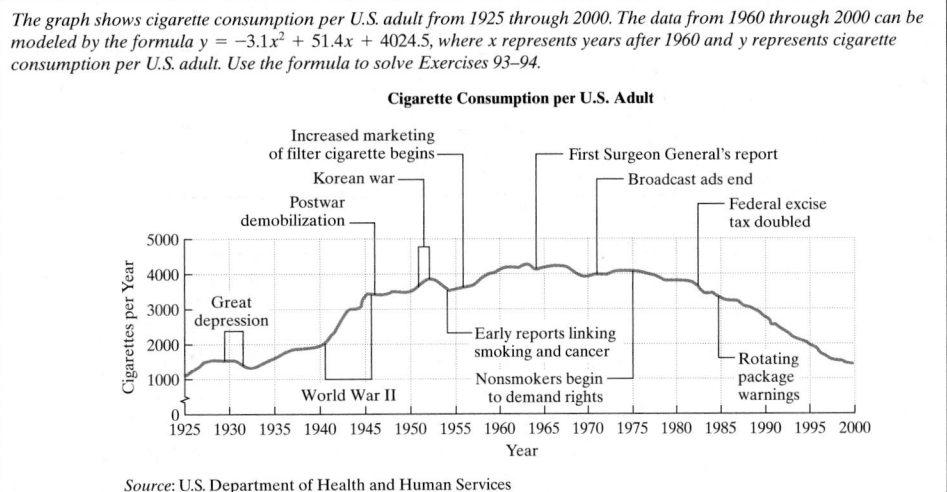

The graph shows cigarette consumption per U.S. adult from 1925 through 2000. The data from 1960 through 2000 can be modeled by the formula $y = -3.1x^2 + 51.4x + 4024.5$, where x represents years after 1960 and y represents cigarette consumption per U.S. adult. Use the formula to solve Exercises 93–94.

Cigarette Consumption per U.S. Adult

Source: U.S. Department of Health and Human Services

Critical Thinking Exercises

A wide array of exercises help students check concept mastery.

Critical Thinking Exercises

69. Which one of the following statements is true?
   a. The equation $3(x + 4) = 3(4 + x)$ has precisely one solution.
   b. The equation $2y + 5 = 0$ is equivalent to $2y = 5$.
   c. If $2 - 3y = 11$ and the solution to the equation is substituted into $y^2 + 2y - 3$, a number results that is neither positive nor negative.

page 125

Writing in Mathematics Exercises

This text provides numerous exercises that ask students to explain concepts in their own words—providing reinforcement and encouraging acquisition of a mathematical vocabulary.

Writing in Mathematics

58. How do you determine whether an ordered pair is a solution of an equation in two variables, x and y?
59. Explain how to find ordered pairs that are solutions of an equation in two variables, x and y.

page 197

Discover for Yourself

Obtain a second point in Example 4 by writing the slope as follows:

$$m = \frac{2}{-5} = \frac{\text{Rise}}{\text{Run}}.$$

$-\dfrac{2}{5}$ can be expressed as $\dfrac{-2}{5}$ or $\dfrac{2}{-5}$.

Obtain a second point in Figure 3.27 by moving *up* 2 units and to the *left* 5 units, starting at $(0,0)$. What do you observe once you graph the line?

Discover for Yourself Boxes

Discover for Yourself boxes encourage students to actively participate in the learning process as they read the book.

page 222

Study Tip

Try to avoid the following common errors that can occur when simplifying exponential expressions.

Correct	Incorrect	Description of Error
$\dfrac{2^{20}}{2^4} = 2^{20-4} = 2^{16}$	$\dfrac{2^{20}}{2^4} = 2^5$	Exponents should be subtracted, not divided.
$-8^0 = -1$	$-8^0 = 1$	Only 8 is raised to the 0 power.
$\left(\dfrac{x}{5}\right)^2 = \dfrac{x^2}{5^2} = \dfrac{x^2}{25}$	$\left(\dfrac{x}{5}\right)^2 = \dfrac{x^2}{5}$	The numerator and denominator must both be squared.

Study Tips

Study Tips make mathematical content more accessible to the student.

page 335

Check Points

Each example is followed by an exercise that provides for a more interactive text and gives students the opportunity to work with a concept as soon as they have learned it.

EXAMPLE 6   An Application of Subtraction Using the Word "Difference"

The bar graph in Figure 1.17 shows that in 1995, Social Security had an annual cash surplus of $233 billion. By 2020, this amount is expected to be a negative number—a deficit of $244 billion. What is the difference between the 1995 surplus and the projected 2020 deficit?

Solution

The difference	is	the 1995 surplus	minus	the 2020 deficit.
=		233	–	(−244)

$$= 233 + 244 = 477$$

The difference between the 1995 surplus and the projected 2020 deficit is $477 billion. ∎

✔ **CHECK POINT 6**   The peak of Mount Everest is 8848 meters above sea level. The Marianas Trench, on the floor of the Pacific Ocean, is 10,915 meters below sea level. What is the difference in elevation between the peak of Mount Everest and the Marianas Trench?

page 59

CHAPTER 8 GROUP PROJECTS

1. Consult an almanac, newspaper, magazine, or the Internet to find data displayed in the style of Figure 8.9 on page 532 or in Exercises 53–56 in Exercise Set 8.2 on page 539. Using the two graphs that group members find most interesting, introduce two or more functions that are related to the graphs. Then write and solve a problem involving function addition and function subtraction for each selected graph.

2. Group members should consult an almanac, newspaper, magazine, or the Internet and return to the group with data that have a variable that is first decreasing and then increasing over time, or vice versa. The group should select the two most interesting data sets. For each set selected:
   a. Identify three data points and use the function $y = ax^2 + bx + c$ to model the data. Let x represent the number of years after the first year in the data set.
   b. Use the quadratic function to make a prediction about what might occur in the future.

3. The group should write three original word problems that can be solved using a system of linear equations in three variables. Each problem should be on a different topic. The group should turn in the three problems and their algebraic solutions.

4. Turn on your computer and read your e-mail or write a paper. When you need to do research, use the Internet to browse through art museums and photography exhibits. When you need a break, load a flight simulator program and fly through a photorealistic computer world. As different as these experiences may be, thay all share one thing —you're looking at images based on matrices. Matrices have applications in numerous fields, including the new technology of digital photography in which pictures are represented by numbers rather than film. Members of the group should research

Chapter Projects and Group Activities

Extended applications conclude each chapter. Some activities feature related Websites for student research and exploration.

page 574

Review Exercises

In Exercises 75–76, insert either $<$ or $>$ in the box between each pair of numbers to make a true statement.

75. $-24 \; \square \; -20$ (Section 1.2, Example 6)

76. $-\dfrac{1}{3} \; \square \; -\dfrac{1}{5}$ (Section 1.2, Example 6)

77. Simplify: $-9 - 11 + 7 - (-3)$. (Section 1.6, Example 3)

page 125

Review Exercises

Review Exercises in section exercise sets are cross-referenced back to an example in the text—giving students a pattern for problem solving.

CHAPTER SUMMARY, REVIEW, AND TEST

SUMMARY

DEFINITIONS AND CONCEPTS	EXAMPLES

Section 8.1 Introduction to Functions

A relation is any set of ordered pairs. The set of first components of the ordered pairs is the domain and the set of second components is the range. A function is a relation in which each member of the domain corresponds to exactly one member of the range. No two ordered pairs of a function can have the same first components and different second components.

The domain of the relation $\{(1, 2), (3, 4), (3, 7)\}$ is $\{1, 3\}$. The range is $\{2, 4, 7\}$. The relation is not a function: 3, in the domain, corresponds to both 4 and 7 in the range.

If a function is defined as an equation, the notation $f(x)$, read "f of x" or "f at x," describes the value of the function at the number, or input, x.

$$\text{If } f(x) = 7x - 5, \text{ then}$$
$$f(a + 2) = 7(a + 2) - 5$$
$$= 7a + 14 - 5$$
$$= 7a + 9.$$

The graph of a function is the graph of its ordered pairs.

The Vertical Line Test for Functions
If any vertical line intersects a graph in more than one point, the graph does not define y as a function of x.

At the left or right of a function's graph, you will often find closed dots, open dots, or arrows. A closed dot shows the graph ends and the point belongs to the graph. An open dot shows the graph ends and the point does not belong to the graph. An arrow indicates the graph extends indefinitely.

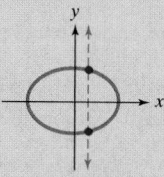

Not the graph of a function

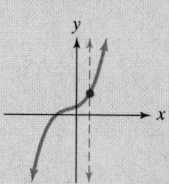

The graph of a function

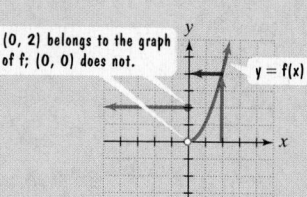

(0, 2) belongs to the graph of f; (0, 0) does not.

$y = f(x)$

To find $f(2)$, locate 2 on the x-axis. The graph shows $f(2) = 4$.

page 143

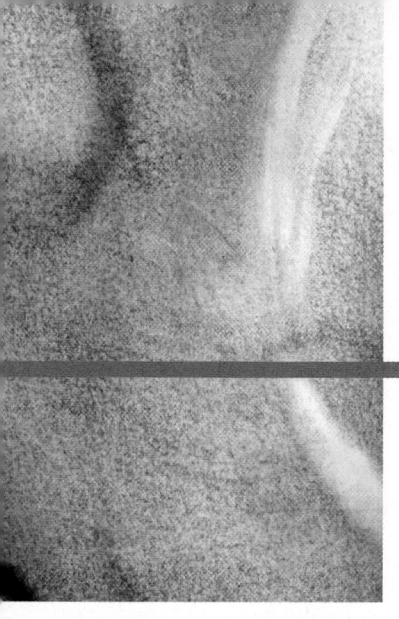

Applications Index

Introductory and Intermediate Algebra

FOR COLLEGE STUDENTS

When you encounter formulas with symbols such as those that appear in the estimation of a child's adult height, don't panic! You are already familiar with many symbols—the smiley face, the peace symbol, the heart symbol, the dollar sign, and even symbols on your calculator or computer. In this chapter, you will become familiar with the special symbolic notation of algebra. You will see that the language of algebra describes your world and holds the power to solve many of its problems.

Sitting in the biology department office, you overhear two of the professors discussing the possible adult heights of their respective children. Looking at the blackboard that they've been writing on, you see that there are formulas that can estimate the height a child will attain as an adult. If the child is x years old and h inches tall, that child's adult height, H, in inches, is approximated by one of the following formulas.

Girl:

$$H = \frac{h}{0.00028x^3 - 0.0071x^2 + 0.0926x + 0.3524}$$

Boys:

$$H = \frac{h}{0.00011x^3 - 0.0032x^2 + 0.0604x + 0.3796}$$

The Real Number System

▶ SECTION 1.1   *Fractions*

Objectives

1 Reduce or simplify fractions.

2 Multiply fractions.

3 Divide fractions.

4 Add and subtract fractions with identical denominators.

5 Add and subtract fractions with unlike denominators.

6 Solve problems involving fractions.

SSM   PH Tutor   CD-   Video
         Center   ROM

Study Tip

Mathematics is based on a few fundamental assumptions from which everything else follows. To understand algebra, a solid foundation is essential. By devoting special attention to the basic skills explained in this chapter, you'll be well on your way to mastering algebra.

" ...I have a dream that my four little children will one day live in a nation where they will not be judged by the color of their skin but by the content of their character."
- Dr. Martin Luther King, Jr. (1929–1968)

A Time/CNN poll asked African American teenagers and African American adults the following question:

> Have you ever been a victim of discrimination because you are black?

Here are the poll's results:

Black Teens	Black Adults
Yes: 23%	Yes: 55%
No: 77%	No: 45%

Source: Time/CNN

The 23% means $\frac{23}{100}$. We can say that the *fraction* of black teens who have experienced discrimination is $\frac{23}{100}$. Because polls have a margin of error, we can conclude that in every group of 100 black teens, *approximately* 23 people have experienced discrimination.

In a fraction, the number that is written above the fraction bar is called the **numerator**. The number below the fraction bar is called the **denominator**.

Fraction bar ⟶ $\dfrac{23}{100}$ ⟵ 23 is the numerator.
                    ⟵ 100 is the denominator.

Fractions appear throughout algebra. We are frequently given numerical information that involves fractions. In this section, we present a brief review of operations with fractions that we will use in algebra.

Reducing Fractions

1 Reduce or simplify fractions.

To *reduce*, or *simplify* a fraction, we need to factor the numerator and the denominator. To **factor** a number means to write it as a multiplication. For example, 21 can be factored as 7 · 3. In the statement 7 · 3 = 21, 7 and 3 are called the **factors** and 21 is the **product**.

$$7 \cdot 3 = 21$$

7 is a factor of 21.   3 is a factor of 21.   The product of 7 and 3 is 21.

Two fractions are **equivalent** if they represent the same value. Writing a fraction as an equivalent fraction with a smaller denominator is called **reducing a fraction**. A fraction is **reduced to its lowest terms** when the numerator and denominator have no common factors other than 1.

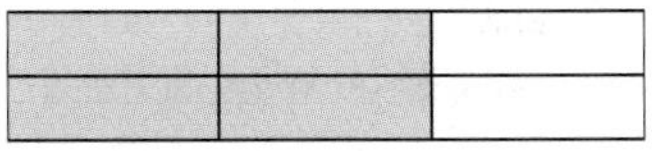

Look at the rectangle in Figure 1.1. Can you see that it is divided into 6 equal parts? Of these 6 parts, 4 of the parts are red. Thus, $\frac{4}{6}$ of the rectangle is red.

The rectangle in Figure 1.1 is also divided into 3 equal stacks and 2 of the stacks are red. Thus, $\frac{2}{3}$ of the rectangle is red. Because both $\frac{4}{6}$ and $\frac{2}{3}$ of the rectangle are red, we can conclude that $\frac{4}{6}$ and $\frac{2}{3}$ are equivalent fractions.

We reduce $\frac{4}{6}$ to lowest terms and obtain $\frac{2}{3}$ by removing a factor of 1 as follows:

$$\frac{4}{6} = \frac{2 \cdot 2}{3 \cdot 2} = \frac{2}{3} \cdot \frac{2}{2} = \frac{2}{3} \cdot 1 = \frac{2}{3}.$$

Multiplying a number by 1 does not change that number.

We can speed up this process by dividing the numerator and the denominator of $\frac{4}{6}$ by 2, the **greatest common factor** of 4 and 6. The **greatest common factor** of two or more numbers is the largest counting number that divides all the numbers.

$$\frac{4}{6} = \frac{2 \cdot 2}{3 \cdot 2} = \frac{2}{3}$$

Simplifying a Fraction

To reduce a fraction to its lowest terms, divide both the numerator and the denominator by their greatest common factor.

EXAMPLE 1   Reducing Fractions

Reduce each fraction to its lowest terms:

a. $\dfrac{6}{8}$ **b.** $\dfrac{45}{27}$ **c.** $\dfrac{11}{25}$ **d.** $\dfrac{11}{33}$.

Solution   For each fraction, divide the numerator and the denominator by their greatest common factor.

a. $\dfrac{6}{8} = \dfrac{2 \cdot 3}{2 \cdot 4} = \dfrac{3}{4}$ 2 is the greatest common factor of 6 and 8. Divide the numerator and denominator by 2.

b. $\dfrac{45}{27} = \dfrac{9 \cdot 5}{9 \cdot 3} = \dfrac{5}{3}$ 9 is the greatest common factor of 45 and 27. Divide the numerator and denominator by 9.

c. Because 11 and 25 share no common factor (other than 1), $\frac{11}{25}$ is already reduced to its lowest terms.

d. $\dfrac{11}{33} = \dfrac{11 \cdot 1}{11 \cdot 3} = \dfrac{1}{3}$ 11 is the greatest common factor of 11 and 33. Divide the numerator and denominator by 11. ∎

✔ **CHECK POINT 1**   Reduce each fraction to its lowest terms:

a. $\dfrac{10}{15}$ **b.** $\dfrac{42}{24}$ **c.** $\dfrac{13}{15}$ **d.** $\dfrac{9}{45}$.

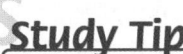

Figure 1.1

Study Tip

After working each Check Point, check your answer in the answer section before continuing your reading.

2 Multiply fractions.

Multiplying Fractions

The result of multiplying two fractions is called their **product**.

> ### Multiplying Fractions
> The product of two or more fractions is the product of their numerators divided by the product of their denominators.

Here is an example that illustrates the rule in the box.

$$\frac{3}{8} \cdot \frac{5}{11} = \frac{3 \cdot 5}{8 \cdot 11} = \frac{15}{88}$$

The product of $\frac{3}{8}$ and $\frac{5}{11}$ is $\frac{15}{88}$.

Multiply numerators and denominators.

EXAMPLE 2   Multiplying Fractions

Multiply. If possible, reduce the product to its lowest terms:

a. $\frac{3}{7} \cdot \frac{2}{5}$ **b.** $5 \cdot \frac{7}{12}$ **c.** $\frac{2}{3} \cdot \frac{9}{4}$.

Solution

a. $\frac{3}{7} \cdot \frac{2}{5} = \frac{3 \cdot 2}{7 \cdot 5} = \frac{6}{35}$ Multiply numerators and denominators.

b. $5 \cdot \frac{7}{12} = \frac{5}{1} \cdot \frac{7}{12} = \frac{5 \cdot 7}{1 \cdot 12} = \frac{35}{12}$ Write 5 as $\frac{5}{1}$. Then multiply numerators and denominators.

c. $\frac{2}{3} \cdot \frac{9}{4} = \frac{2 \cdot 9}{3 \cdot 4} = \frac{18}{12} = \frac{\cancel{6} \cdot 3}{\cancel{6} \cdot 2} = \frac{3}{2}$

Simplify $\frac{18}{12}$; 6 is the greatest common factor of 18 and 12.

✔ **CHECK POINT 2**   Multiply. If possible, reduce the product to its lowest terms:

a. $\frac{4}{11} \cdot \frac{2}{3}$ **b.** $6 \cdot \frac{3}{5}$ **c.** $\frac{3}{7} \cdot \frac{2}{3}$.

3 Divide fractions.

Dividing Fractions

The result of dividing two fractions is called their **quotient**. A geometric figure is useful for developing a process for determining the quotient of two fractions.

Consider the division

$$\frac{4}{5} \div \frac{1}{10}.$$

We want to know how many $\frac{1}{10}$'s are in $\frac{4}{5}$. We can use Figure 1.2 to find this quotient. The rectangle is divided into fifths. The dashed lines further divide the rectangle into tenths.

Figure 1.2 shows that $\frac{4}{5}$ of the rectangle is red. How many $\frac{1}{10}$'s of the rectangle does this include? Can you see that this includes eight of the $\frac{1}{10}$ pieces? Thus, there are eight $\frac{1}{10}$'s in $\frac{4}{5}$:

$$\frac{4}{5} \div \frac{1}{10} = 8.$$

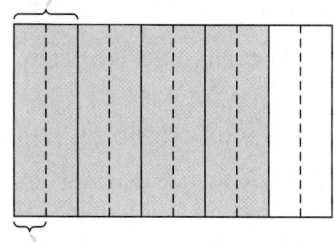

This is $\frac{1}{5}$ of the figure.

This is $\frac{1}{10}$ of the figure.

Figure 1.2

We can obtain the quotient 8 in the following way:

$$\frac{4}{5} \div \frac{1}{10} = \frac{4}{5} \cdot \frac{10}{1} = \frac{4 \cdot 10}{5 \cdot 1} = \frac{40}{5} = 8.$$

Change the division to multiplication.

Invert the divisor, $\frac{1}{10}$.

Generalizing from this result gives us the following rule.

Dividing Fractions

To find the quotient of two fractions, invert the divisor and multiply. (The divisor is the second fraction if the division is written with ÷.)

EXAMPLE 3   Dividing Fractions

Divide:

a. $\dfrac{2}{3} \div \dfrac{7}{15}$

b. $\dfrac{3}{4} \div 5.$

Solution

a. $\dfrac{2}{3} \div \dfrac{7}{15} = \dfrac{2}{3} \cdot \dfrac{15}{7} = \dfrac{2 \cdot 15}{3 \cdot 7} = \dfrac{30}{21} = \dfrac{\cancel{3} \cdot 10}{\cancel{3} \cdot 7} = \dfrac{10}{7}$

Change division to multiplication.

Invert the divisor.

Simplify. 3 is the greatest common factor of 30 and 21.

b. $\dfrac{3}{4} \div 5 = \dfrac{3}{4} \div \dfrac{5}{1} = \dfrac{3}{4} \cdot \dfrac{1}{5} = \dfrac{3 \cdot 1}{4 \cdot 5} = \dfrac{3}{20}$

Change division to multiplication.

Invert the divisor.

■

✔ **CHECK POINT 3**   Divide:

a. $\dfrac{5}{4} \div \dfrac{3}{8}$

b. $\dfrac{2}{3} \div 3.$

4 Add and subtract fractions with identical denominators.

Adding and Subtracting Fractions with Identical Denominators

The result of adding two fractions is called their **sum**. The result of subtracting two fractions is called their **difference**. A geometric figure is useful for developing a process for determining the sum or difference of two fractions with identical denominators.

Consider the addition

$$\frac{3}{7} + \frac{2}{7}.$$

We can use Figure 1.3 to find this sum. The rectangle is divided into sevenths. On the left, $\frac{3}{7}$ of the rectangle is red. On the right, $\frac{2}{7}$ of the rectangle is red. Including both the left and the right, a total of $\frac{5}{7}$ of the rectangle is red. Thus,

$$\frac{3}{7} + \frac{2}{7} = \frac{5}{7}.$$

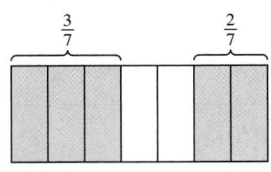

$\frac{3}{7}$ $\frac{2}{7}$

Figure 1.3

We can obtain the sum $\frac{5}{7}$ in the following way:

$$\frac{3}{7} + \frac{2}{7} = \frac{3+2}{7} = \frac{5}{7}.$$

Add numerators and put this result over the common denominator.

Generalizing from this result gives us the following rule.

Adding and Subtracting Fractions with Identical Denominators

Add or subtract numerators. Put this result over the common denominator.

EXAMPLE 4   Adding and Subtracting Fractions with Identical Denominators

Perform the indicated operations:

a. $\dfrac{3}{11} + \dfrac{4}{11}$ **b.** $\dfrac{11}{12} - \dfrac{5}{12}.$

Solution

a. $\dfrac{3}{11} + \dfrac{4}{11} = \dfrac{3+4}{11}$ Add the numerators. Put this sum over the common denominator.

$= \dfrac{7}{11}$ Perform the addition.

b. $\dfrac{11}{12} - \dfrac{5}{12} = \dfrac{11-5}{12}$ Subtract the numerators. Put this difference over the common denominator.

$= \dfrac{6}{12}$ Perform the subtraction.

$= \dfrac{\cancel{6} \cdot 1}{\cancel{6} \cdot 2}$ Now simplify: 6 is the greatest common factor of 6 and 12.

$= \dfrac{1}{2}$ Divide the numerator and denominator by 6.   ∎

✔ **CHECK POINT 4**   Perform the indicated operations:

a. $\dfrac{2}{11} + \dfrac{3}{11}$ **b.** $\dfrac{5}{6} - \dfrac{1}{6}.$

5 Add and subtract fractions with unlike denominators.

Adding and Subtracting Fractions with Unlike Denominators

How do we add or subtract fractions with different denominators? We must first rewrite them as equivalent fractions with the same denominator. We do this by multiplying fractions by 1, as shown in the next example. Multiplication by 1 does not change the value of a number.

EXAMPLE 5   Writing an Equivalent Fraction

Write $\frac{3}{4}$ as an equivalent fraction with a denominator of 16.

Solution To obtain a denominator of 16, we must multiply the denominator of the given fraction, $\frac{3}{4}$, by 4. So that we do not change the value of the fraction, we also multiply the numerator by 4. Multiplying by $\frac{4}{4} = 1$ does not change the given fraction's value.

$$\frac{3}{4} = \frac{3}{4} \cdot \frac{4}{4} = \frac{3 \cdot 4}{4 \cdot 4} = \frac{12}{16}$$ ∎

✔ **CHECK POINT 5** Write $\frac{2}{3}$ as an equivalent fraction with a denominator of 21.

Equivalent fractions can be used to add fractions with different denominators, such as $\frac{1}{2}$ and $\frac{1}{3}$. Figure 1.4 indicates that the sum of half the whole figure and one-third of the whole figure results in 5 parts out of 6, or $\frac{5}{6}$, of the figure. Thus,

$$\frac{1}{2} + \frac{1}{3} = \frac{5}{6}.$$

We can obtain the sum $\frac{5}{6}$ if we rewrite each fraction as an equivalent fraction with a denominator of 6.

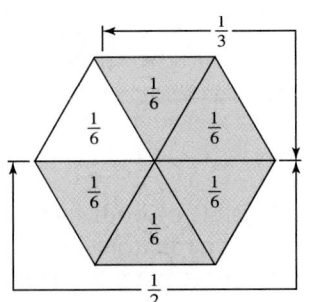

Figure 1.4 $\frac{1}{2} + \frac{1}{3} = \frac{5}{6}$

$$\frac{1}{2} + \frac{1}{3} = \frac{1}{2} \cdot \frac{3}{3} + \frac{1}{3} \cdot \frac{2}{2}$$ Rewrite each fraction as an equivalent fraction with a denominator of 6. $\frac{3}{3} = 1$ and $\frac{2}{2} = 1$, and multiplying by 1 does not change a number's value.

$$= \frac{3}{6} + \frac{2}{6}$$ Multiply. We now have a common denominator.

$$= \frac{3 + 2}{6}$$ Add the numerators and place this sum over the common denominator.

$$= \frac{5}{6}$$ Perform the addition.

When adding $\frac{1}{2}$ and $\frac{1}{3}$, there are many common denominators that we can use, such as 6, 12, 18, and so on. The given denominators, 2 and 3, divide into all of these numbers. However, the denominator 6 is the smallest number that 2 and 3 divide into. For this reason, 6 is called the *least common denominator,* abbreviated LCD.

Adding and Subtracting Fractions with Unlike Denominators

1. Rewrite the fractions as equivalent fractions with the least common denominator.

2. Add or subtract the numerators, putting this result over the common denominator.

EXAMPLE 6 Adding and Subtracting Fractions with Unlike Denominators

Perform the indicated operation:

a. $\dfrac{1}{5} + \dfrac{3}{4}$
b. $\dfrac{3}{4} - \dfrac{1}{6}$
c. $\dfrac{7}{10} - \dfrac{1}{5}$.

Discover for Yourself

Try Example 6a, $\frac{1}{5} + \frac{3}{4}$, using a common denominator of 40. Because both 5 and 4 divide into 40, 40 is a common denominator, although not the *least* common denominator. Describe what happens. What is the advantage of using the least common denominator?

Solution

a. Just by looking, you can tell that the smallest number divisible by both 5 and 4 is 20. Thus, the least common denominator for the denominators 5 and 4 is 20. We rewrite both fractions as equivalent fractions with the least common denominator, 20.

$$\frac{1}{5} + \frac{3}{4} = \frac{1}{5} \cdot \frac{4}{4} + \frac{3}{4} \cdot \frac{5}{5}$$

Multiply each fraction by 1. Because $5 \cdot 4 = 20$, multiply the first fraction by $\frac{4}{4}$. Because $4 \cdot 5 = 20$, multiply the second fraction by $\frac{5}{5}$.

$$= \frac{4}{20} + \frac{15}{20}$$

Perform the multiplications.

$$= \frac{4 + 15}{20}$$

Add the numerators and put this sum over the least common denominator.

$$= \frac{19}{20}$$

Perform the addition.

b. By looking, can you tell that the smallest number divisible by both 4 and 6 is 12? Thus, the least common denominator for the denominators 4 and 6 is 12. We rewrite both fractions as equivalent fractions with the least common denominator, 12.

$$\frac{3}{4} - \frac{1}{6} = \frac{3}{4} \cdot \frac{3}{3} - \frac{1}{6} \cdot \frac{2}{2}$$

Rewrite each fraction as an equivalent fraction with a denominator of 12.

$$= \frac{9}{12} - \frac{2}{12}$$

Multiply: $3 \cdot 3 = 9$ and $1 \cdot 2 = 2$.

$$= \frac{9 - 2}{12}$$

Subtract the numerators and put this difference over the least common denominator.

$$= \frac{7}{12}$$

Perform the subtraction.

c. The smallest number divisible by both 10 and 5 is 10. Thus, the least common denominator for the denominators 10 and 5 is 10. Because one of the fractions already has a denominator of 10, we only have to rewrite one of the fractions.

$$\frac{7}{10} - \frac{1}{5} = \frac{7}{10} - \frac{1}{5} \cdot \frac{2}{2}$$

Rewrite the second fraction as an equivalent fraction with a denominator of 10.

$$= \frac{7}{10} - \frac{2}{10}$$

Multiply: $1 \cdot 2 = 2$.

$$= \frac{7 - 2}{10}$$

Subtract numerators, putting this difference over the least common denominator.

$$= \frac{5}{10}$$

Perform the subtraction.

$$= \frac{\not{5} \cdot 1}{\not{5} \cdot 2}$$

Simplify: 5 is the greatest common factor of 5 and 10.

$$= \frac{1}{2}$$

Divide the numerator and denominator by 5. ∎

✔ CHECK POINT 6 Perform the indicated operation:

a. $\dfrac{1}{2} + \dfrac{3}{5}$ **b.** $\dfrac{4}{3} - \dfrac{3}{4}$ **c.** $\dfrac{13}{18} - \dfrac{2}{9}$.

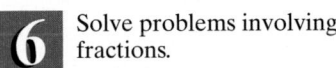

6 Solve problems involving fractions.

Applications

Because numerical information frequently involves fractions, many practical problems can be solved using operations with fractions.

EXAMPLE 7   Problem Solving with Fractions

At a workshop on enhancing creativity, $\frac{1}{4}$ of the participants are musicians, $\frac{2}{5}$ are artists, $\frac{1}{10}$ are actors, and the remaining participants are writers. Find the fraction of people at the workshop who are writers.

Solution   We begin by finding the fraction of people at the workshop who are *not* writers. Add the three given fractions.

$$\frac{1}{4} + \frac{2}{5} + \frac{1}{10}$$
The least common denominator is 20.

$$= \frac{1}{4} \cdot \frac{5}{5} + \frac{2}{5} \cdot \frac{4}{4} + \frac{1}{10} \cdot \frac{2}{2}$$
Rewrite each fraction as an equivalent fraction with a denominator of 20.

$$= \frac{5}{20} + \frac{8}{20} + \frac{2}{20}$$
Multiply: $1 \cdot 5 = 5, 2 \cdot 4 = 8$, and $1 \cdot 2 = 2$.

$$= \frac{5 + 8 + 2}{20}$$
Add the numerators and put this sum over the least common denominator.

$$= \frac{15}{20}$$

$$= \frac{\cancel{5} \cdot 3}{\cancel{5} \cdot 4}$$
Simplify; 5 is the greatest common factor of 15 and 20.

$$= \frac{3}{4}$$
Divide the numerator and denominator by 5.

Thus, $\frac{3}{4}$ of the participants are musicians, artists, and actors. This does not include the writers, the only other group at the workshop. The fractions representing all four groups at the workshop must add up to 1. Therefore, the fraction of writers at the workshop can be found by subtracting $\frac{3}{4}$ from 1. The fraction of the people at the workshop who are writers is

$$1 - \frac{3}{4} = \frac{4}{4} - \frac{3}{4} = \frac{1}{4}.$$

We see that $\frac{1}{4}$ of the participants are writers. ∎

✔ **CHECK POINT 7**   The graph shows how a particular student spends her weekdays. What fraction of the day is spent in class?

The Weekdays of a Student

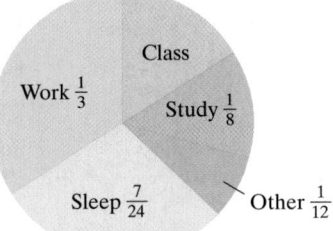

Class

Work $\frac{1}{3}$

Study $\frac{1}{8}$

Sleep $\frac{7}{24}$

Other $\frac{1}{12}$

EXERCISE SET 1.1

Practice Exercises

In Exercises 1–12, simplify each fraction by reducing it to its lowest terms.

1. $\dfrac{10}{16}$ **2.** $\dfrac{8}{14}$ **3.** $\dfrac{15}{18}$

4. $\dfrac{18}{45}$ **5.** $\dfrac{35}{50}$ **6.** $\dfrac{45}{50}$

7. $\dfrac{32}{80}$ **8.** $\dfrac{75}{80}$ **9.** $\dfrac{44}{50}$

10. $\dfrac{38}{50}$ **11.** $\dfrac{120}{86}$ **12.** $\dfrac{116}{86}$

In Exercises 13–56, perform the indicated operation. Where possible, reduce the answer to its lowest terms.

13. $\dfrac{2}{5} \cdot \dfrac{1}{3}$ **14.** $\dfrac{3}{7} \cdot \dfrac{1}{4}$

15. $\dfrac{3}{8} \cdot \dfrac{7}{11}$ **16.** $\dfrac{5}{8} \cdot \dfrac{3}{11}$

17. $9 \cdot \dfrac{4}{7}$ **18.** $8 \cdot \dfrac{3}{7}$

19. $\dfrac{1}{10} \cdot \dfrac{5}{6}$ **20.** $\dfrac{1}{8} \cdot \dfrac{2}{3}$

21. $\dfrac{5}{4} \cdot \dfrac{6}{7}$ **22.** $\dfrac{7}{4} \cdot \dfrac{6}{11}$

23. $\dfrac{5}{4} \div \dfrac{4}{3}$ **24.** $\dfrac{7}{8} \div \dfrac{2}{3}$

25. $\dfrac{18}{5} \div 2$ **26.** $\dfrac{12}{7} \div 3$

27. $2 \div \dfrac{18}{5}$ **28.** $3 \div \dfrac{12}{7}$

29. $\dfrac{3}{4} \div \dfrac{1}{4}$ **30.** $\dfrac{3}{7} \div \dfrac{1}{7}$

31. $\dfrac{7}{6} \div \dfrac{5}{3}$ **32.** $\dfrac{7}{4} \div \dfrac{3}{8}$

33. $\dfrac{1}{14} \div \dfrac{1}{7}$ **34.** $\dfrac{1}{8} \div \dfrac{1}{4}$

35. $\dfrac{2}{11} + \dfrac{4}{11}$ **36.** $\dfrac{5}{13} + \dfrac{2}{13}$

37. $\dfrac{7}{12} + \dfrac{1}{12}$ **38.** $\dfrac{5}{16} + \dfrac{1}{16}$

39. $\dfrac{5}{8} + \dfrac{5}{8}$ **40.** $\dfrac{3}{8} + \dfrac{3}{8}$

41. $\dfrac{7}{12} - \dfrac{5}{12}$ **42.** $\dfrac{13}{18} - \dfrac{5}{18}$

43. $\dfrac{16}{7} - \dfrac{2}{7}$ **44.** $\dfrac{17}{5} - \dfrac{2}{5}$

45. $\dfrac{1}{2} + \dfrac{1}{5}$ **46.** $\dfrac{1}{3} + \dfrac{1}{5}$

47. $\dfrac{3}{4} + \dfrac{3}{20}$ **48.** $\dfrac{2}{5} + \dfrac{2}{15}$

49. $\dfrac{3}{8} + \dfrac{5}{12}$ **50.** $\dfrac{3}{10} + \dfrac{2}{15}$

51. $\dfrac{11}{18} - \dfrac{2}{9}$ **52.** $\dfrac{17}{18} - \dfrac{4}{9}$

53. $\dfrac{4}{3} - \dfrac{3}{4}$ **54.** $\dfrac{3}{2} - \dfrac{2}{3}$

55. $\dfrac{7}{10} - \dfrac{3}{16}$ **56.** $\dfrac{9}{10} - \dfrac{5}{16}$

Different operations with the same fractions usually result in different answers. Exercises 57–58 illustrate some curious exceptions.

57. Show that $\dfrac{13}{4} + \dfrac{13}{9}$ and $\dfrac{13}{4} \times \dfrac{13}{9}$ give the same answer.

58. Show that $\dfrac{169}{30} + \dfrac{13}{15}$ and $\dfrac{169}{30} \div \dfrac{13}{15}$ give the same answer.

Application Exercises

Exercises 59–64 are based on a Time/CNN poll focusing on teenagers and adults and their opinions on racial issues.

59. Of the 800 white adults surveyed, 512 replied that racism is a big problem. What fractional part of the white adults, expressed in lowest terms, thought that racism is a big problem?

60. Of the 400 black adults surveyed, 312 replied that racism is a big problem. What fractional part of the black adults, expressed in lowest terms, thought that racism is a big problem?

61. Of the 300 black teenagers surveyed, 186 replied that racism is a big problem. What fractional part of the black teenagers, expressed in lowest terms, thought that racism is *not* a big problem?

62. Of the 300 white teenagers surveyed, 174 replied that racism is a big problem. What fractional part of the white teenagers, expressed in lowest terms, thought that racism is *not* a big problem?

63. $\frac{19}{20}$ of the 300 black teenagers surveyed said that they were planning to go to college. How many planned on college?

64. $\frac{23}{25}$ of the 300 white teenagers surveyed said that they were planning to go to college. How many planned on college?

65. A recipe calls for $\frac{3}{4}$ cup of sugar. How much is needed to make half of the recipe?

66. A recipe calls for $\frac{3}{4}$ cup of shortening. How much is needed to triple the recipe?

67. A franchise is owned by three people. The first owns $\frac{5}{12}$ of the business and the second owns $\frac{1}{4}$ of the business. What fractional part of the business is owned by the third person?

68. If you walk $\frac{3}{4}$ mile and then jog $\frac{2}{5}$ mile, what is the total distance covered? How much farther did you walk than jog?

69. Some companies pay people extra when they work more than a regular 40-hour work week. The overtime pay is often $1\frac{1}{2}$, or $\frac{3}{2}$, times the regular hourly rate. This is called time and a half. A summer job for students pays $12 an hour and offers time and a half for the hours worked over 40. If a student works 46 hours during one week, what is the student's total pay before taxes?

70. A will states that $\frac{3}{5}$ of the estate is to be divided among relatives. Of the remaining estate, $\frac{1}{4}$ goes to the National Foundation for AIDS Research. What fraction of the estate goes to the National Foundation for AIDS Research?

Writing in Mathematics

Writing about mathematics will help you to learn mathematics. For all writing exercises in this book, use complete sentences to respond to the questions. Some writing exercises can be answered in a sentence; others require a paragraph or two. You can decide how much you need to write as long as your writing clearly and directly answers the question in the exercise. Standard references such as a dictionary and a thesaurus should be helpful.

71. Explain how to reduce a fraction to its lowest terms. Give an example with your explanation.

72. Explain how to multiply fractions and give an example.

73. Explain how to divide fractions and give an example.

74. Describe how to add or subtract fractions with identical denominators. Provide an example with your description.

75. Explain how to add fractions with different denominators. Use $\frac{5}{6} + \frac{1}{2}$ as an example.

76. Explain what is wrong with this statement. "If you'd like to save some money, I'll be happy to sell you my computer system for only $\frac{3}{2}$ of the price I originally paid for it."

Critical Thinking Exercises

77. Which one of the following is true?

a. $\frac{1}{2} + \frac{1}{5} = \frac{2}{7}$

b. $\frac{2+6}{2} = \frac{\cancel{2}+6}{\cancel{2}} = 6$

c. $\frac{1}{2} \div 4 = 2$

d. Every fraction has infinitely many equivalent fractions.

78. If 2 servings of rice call for $\frac{1}{4}$ teaspoon of salt, and 4 servings require $\frac{1}{2}$ teaspoon of salt, how much salt is needed for 3 servings?

79. Shown below is a short excerpt from "The Star-Spangled Banner." The time is $\frac{3}{4}$, which means that each measure must contain notes that add up to $\frac{3}{4}$. The values of the different notes tell musicians how long to hold each note.

$$\mathbf{o} = 1 \qquad \text{\musDblWhole} = \frac{1}{2} \qquad \text{\musQuarter} = \frac{1}{4} \qquad \text{\musEighth} = \frac{1}{8}$$

Use vertical lines to divide this line of "The Star-Spangled Banner" into measures.

Technology Exercises

80. Some calculators have a fraction feature. This feature allows you to perform operations with fractions and displays the answer as a fraction reduced to its lowest terms. If your calculator has this feature, use it to verify any five of the answers that you obtained in Exercises 13–56.

In Exercises 81–83, use a calculator with a fraction feature to perform the indicated operation.

81. $\frac{5}{24} + \frac{7}{30}$

82. $\frac{7}{108} + \frac{55}{144}$

83. $\frac{7}{24} - \frac{1}{15}$

▶ **SECTION 1.2**   *The Real Numbers*

Objectives

1 Define the sets that make up the real numbers.

2 Graph numbers on a number line.

3 Express rational numbers as decimals.

4 Classify numbers as belonging to one or more sets of the real numbers.

5 Understand and use inequality symbols.

6 Find the absolute value of a real number.

SSM   PH Tutor   CD-   Video
        Center   ROM

The U.N. building is designed with three golden rectangles.

The United Nations Building in New York was designed to represent its mission of promoting world harmony. Viewed from the front, the building looks like three rectangles stacked upon each other. In each rectangle, the ratio of the width to height is $\sqrt{5} + 1$ to 2, approximately 1.618 to 1. The ancient Greeks believed that such a rectangle, called a **golden rectangle**, was the most visually pleasing of all rectangles.

The ratio 1.618 to 1 is approximate because $\sqrt{5}$ is an irrational number, a special kind of real number. Irrational? Real? Let's make sense of all this by describing the kinds of numbers you will encounter in this course.

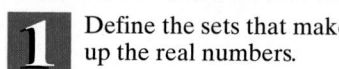

1 Define the sets that make up the real numbers.

Natural Numbers and Whole Numbers

Before we describe the set of real numbers, let's be sure you are familiar with some basic ideas about sets. A **set** is a collection of objects whose contents can be clearly determined. The objects in a set are called the **elements** of the set. For example, the set of numbers used for counting can be represented by

$$\{1, 2, 3, 4, 5, \ldots\}.$$

The braces, { }, indicate that we are representing a set. This form of representing a set uses commas to separate the elements of the set. The three dots after the 5 indicate that there is no final element and that the listing goes on forever.

The set of numbers used for counting is called the set of **natural numbers**. When we combine the number 0 with the natural numbers, we obtain the set of **whole numbers**.

Natural Numbers and Whole Numbers

The set of **natural numbers** is $\{1, 2, 3, 4, 5, \ldots\}$.
The set of **whole numbers** is $\{0, 1, 2, 3, 4, 5, \ldots\}$.

Integers and the Number Line

The whole numbers do not allow us to describe certain everyday situations. For example, if the balance in your checking account is $30 and you write a check for $35, your checking account is overdrawn by $5. We can write this as −5, read *negative* 5. The set consisting of the natural numbers, 0, and the negatives of the natural numbers is called the set of **integers**.

Integers

The set of **integers** is

$$\{\ldots, -4, -3, -2, -1, 0, 1, 2, 3, 4, \ldots\}.$$

$\underbrace{}_{\text{Negative integers}}$ $\underbrace{}_{\text{Positive integers}}$

Notice that the term **positive integers** is another name for the natural numbers. The positive integers can be written in two ways:

1. Use a "+" sign. For example, +4 is "positive four."
2. Do not write any sign. For example, 4 is assumed to be "positive four."

EXAMPLE 1 Practical Examples of Negative Integers

Write a negative integer that describes each of the following situations:

a. A debt of $10
b. The shore surrounding the Dead Sea is 1312 feet below sea level.

Solution

a. A debt of $10 can be expressed by the negative integer −10 (negative ten).
b. The shore surrounding the Dead Sea is 1312 feet below sea level, expressed as −1312. ∎

✔ **CHECK POINT 1** Write a negative integer that describes each of the following situations:

a. A debt of $500
b. Death Valley, the lowest point in North America, is 282 feet below sea level.

2 Graph numbers on a number line.

The **number line** is a graph we use to visualize the set of integers, as well as sets of other numbers. The number line is shown in Figure 1.5.

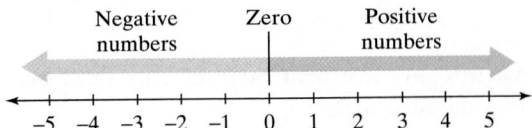

Figure 1.5 The number line

The number line extends indefinitely in both directions, shown by the arrows on the left and the right. Zero separates the positive numbers from the negative numbers on the number line. The positive integers are located to the right of 0, and the negative integers are located to the left of 0. Zero is neither positive nor negative. For every positive integer on a number line, there is a corresponding negative integer on the opposite side of 0.

Integers are graphed on a number line by placing a dot at the correct location for each number.

EXAMPLE 2 Graphing Integers on a Number Line

Graph: **a.** −3   **b.** 4   **c.** 0.

Solution Place a dot at the correct location for each integer.

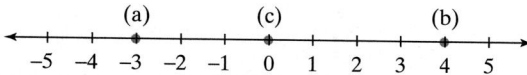

CHECK POINT 2 Graph: **a.** −4   **b.** 0   **c.** 3.

Rational Numbers

If two integers are added, subtracted, or multiplied, the result is always another integer. This, however, is not always the case with division. For example, 10 divided by 5 is the integer 2. By contrast, 5 divided by 10 is $\frac{1}{2}$, and $\frac{1}{2}$ is not an integer. To permit divisions such as $\frac{5}{10}$, we enlarge the set of integers, calling the new collection the *rational numbers*. The set of **rational numbers** consists of all the numbers that can be expressed as a quotient of two integers, with the denominator not 0.

> ### The Rational Numbers
>
> The set of **rational numbers** is the set of all numbers which can be expressed in the form $\frac{a}{b}$, where a and b are integers and b is not equal to 0, written $b \neq 0$. The integer a is called the **numerator** and the integer b is called the **denominator**.

Study Tip

In Section 1.7, you will learn that a negative number divided by a positive number gives a negative result. Thus, $\frac{-3}{4}$ can also be written as $-\frac{3}{4}$.

Here are two examples of rational numbers:

- $\dfrac{1}{2}$ *a, the integer in the numerator, is 1.*
  b, the integer in the denominator, is 2.

- $\dfrac{-3}{4}$ *a, the integer in the numerator, is −3.*
  b, the integer in the denominator, is 4.

Is the integer 5 another example of a rational number? Yes. The integer can be written with a denominator of 1.

$$5 = \frac{5}{1}$$

a, the integer in the numerator, is 5.
b, the integer in the denominator, is 1.

All integers are also rational numbers because they can be written with a denominator of 1.

Rational numbers are graphed on a number line by placing a dot at the correct location for each number.

EXAMPLE 3 Graphing Rational Numbers on a Number Line

Graph: **a.** $\dfrac{7}{2}$ **b.** −4.6.

Solution Place a dot at the correct location for each rational number.

a. Because $\frac{7}{2} = 3\frac{1}{2}$, its graph is midway between 3 and 4.

b. Because $-4.6 = -4\frac{6}{10}$, its graph is $\frac{6}{10}$ of a unit to the left of -4.

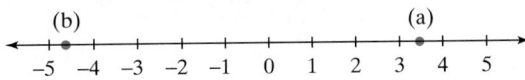

CHECK POINT 3   Graph: **a.** $\frac{9}{2}$ **b.** -1.2.

Every rational number can be expressed as a fraction or a decimal. To express the fraction $\frac{a}{b}$ as a decimal, divide the denominator, b, into the numerator, a.

③ Express rational numbers as decimals.

EXAMPLE 4   Expressing Rational Numbers as Decimals

Express each rational number as a decimal:

a. $\frac{5}{8}$ **b.** $\frac{7}{11}$.

Solution   In each case, divide the denominator into the numerator.

a.

$$
\begin{array}{r}
0.625 \\
8\overline{)5.000} \\
\underline{4\,8} \\
20 \\
\underline{16} \\
40 \\
\underline{40} \\
0
\end{array}
$$

b.

$$
\begin{array}{r}
0.6363\ldots \\
11\overline{)7.0000\ldots} \\
\underline{6\,6} \\
40 \\
\underline{33} \\
70 \\
\underline{66} \\
40 \\
\underline{33} \\
70 \\
\vdots
\end{array}
$$

In Example 4, the decimal for $\frac{5}{8}$, namely 0.625, stops and is called a **terminating decimal**. Other examples of terminating decimals are:

$$\frac{1}{4} = 0.25, \quad \frac{2}{5} = 0.4, \quad \frac{7}{8} = 0.875.$$

By contrast, the division process for $\frac{7}{11}$ results in $0.6363\ldots$, with the digits 63 repeating over and over indefinitely. To indicate this, write a bar over the digits that repeat. Thus,

$$\frac{7}{11} = 0.\overline{63}.$$

The decimal for $\frac{7}{11}$, $0.\overline{63}$, is called a **repeating decimal**. Other examples of repeating decimals are:

$$\frac{1}{3} = 0.333\ldots = 0.\overline{3} \quad \text{and} \quad \frac{2}{3} = 0.666\ldots = 0.\overline{6}.$$

Rational Numbers and Decimals

Any rational number can be expressed as a decimal. The resulting decimal will either terminate (stop), or it will have a digit that repeats or a block of digits that repeat.

✔ CHECK POINT 4   Express each rational number as a decimal:

a. $\frac{3}{8}$ **b.** $\frac{5}{11}$.

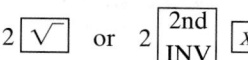

ENRICHMENT ESSAY

The Best and Worst of π

In 1997, two Japanese mathematicians used a Hitachi SR2201 computer to calculate π to over 51 billion digits. The calculations took the computer 29 hours!

The most inaccurate version of π came from the 1897 General Assembly of Indiana. Bill No. 246 stated that "π was by law 4."

 Classify numbers as belonging to one or more sets of the real numbers.

Irrational Numbers

Can you think of a number that, when written in decimal form, neither terminates nor repeats? An example of such a number is $\sqrt{2}$ (read: "the square root of 2"). The number $\sqrt{2}$ is a number that can be multiplied by itself to obtain 2. No terminating or repeating decimal can be multiplied by itself to get 2. However, some approximations come close to 2.

- 1.4 is an approximation of $\sqrt{2}$:
$$1.4 \times 1.4 = 1.96.$$
- 1.41 is an approximation of $\sqrt{2}$:
$$1.41 \times 1.41 = 1.9881.$$
- 1.4142 is an approximation of $\sqrt{2}$:
$$1.4142 \times 1.4142 = 1.99996164.$$

Can you see how each approximation in the list is getting better? This is because the products are getting closer and closer to 2.

The number $\sqrt{2}$, whose decimal representation does not come to an end and does not have a block of repeating digits, is an example of an **irrational number**.

The Irrational Numbers

Any number that can be represented on the number line that is not a rational number is called an **irrational number**. Thus, the set of irrational numbers is the set of numbers whose decimal representations are neither terminating nor repeating.

Perhaps the best known of all the irrational numbers is π (pi). This irrational number represents the distance around a circle (its circumference) divided by the diameter of the circle. In the *Star Trek* episode "Wolf in the Fold," Spock foils an evil computer by telling it to "compute the last digit in the value of π." Because π is an irrational number, there is no last digit in its decimal representation:

$$\pi = 3.14159265358979323846264338332795\ldots.$$

Because irrational numbers cannot be represented by decimals that come to an end, mathematicians use symbols such as $\sqrt{2}$, $\sqrt{3}$, and π to represent these numbers. However, **not all square roots are irrational**. For example, $\sqrt{25} = 5$ because 5 multiplied by itself is 25. Thus, $\sqrt{25}$ is a natural number, a whole number, an integer, and a rational number $\left(\sqrt{25} = \frac{5}{1}\right)$.

The Set of Real Numbers

All numbers that can be represented by points on the number line are called **real numbers**. Thus, the set of real numbers is formed by combining the rational numbers and the irrational numbers. Every real number is either rational or irrational.

The sets that make up the real numbers are summarized in Table 1.1. Notice the use of the symbol $\approx$ in the examples of irrational numbers. The symbol $\approx$ means "is approximately equal to."

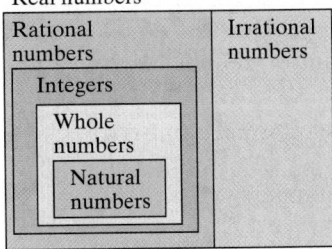

Real numbers

Rational numbers	Irrational numbers
Integers	
Whole numbers	
Natural numbers	

This diagram shows that every real number is rational or irrational.

Table 1.1   The Sets that Make Up the Real Numbers

Name	Description	Examples
Natural numbers	$\{1, 2, 3, 4, 5, \dots\}$ These numbers are used for counting.	$2, 3, 5, 17$
Whole numbers	$\{0, 1, 2, 3, 4, 5, \dots\}$ The whole numbers add 0 to the set of natural numbers.	$0, 2, 3, 5, 17$
Integers	$\{\dots, -5, -4, -3, -2, -1, 0, 1, 2, 3, 4, 5, \dots\}$ The integers add the negatives of the natural numbers to the set of whole numbers.	$-17, -5, -3, -2, 0,$ $2, 3, 5, 17$
Rational numbers	These numbers can be expressed as an integer divided by a nonzero integer: $\frac{a}{b}$: a and b are integers: $b \neq 0$. Rational numbers can be expressed as terminating or repeating decimals.	$-17 = \frac{-17}{1}, -5 = \frac{-5}{1}, -3, -2,$ $0, 2, 3, 5, 17,$ $\frac{2}{5} = 0.4,$ $\frac{-2}{3} = -0.6666\dots = -0.\overline{6}$
Irrational numbers	This is the set of numbers whose decimal representations are neither terminating nor repeating. Irrational numbers cannot be expressed as a quotient of integers.	$\sqrt{2} \approx 1.414214$ $-\sqrt{3} \approx -1.73205$ $\pi \approx 3.142$ $-\dfrac{\pi}{2} \approx -1.571$

EXAMPLE 5   Classifying Real Numbers

Consider the following set of numbers:

$$\left\{-7, -\frac{3}{4}, 0, 0.\overline{6}, \sqrt{5}, \pi, 7.3, \sqrt{81}\right\}.$$

List the numbers in the set that are:

a. natural numbers.   **b.** whole numbers.

c. integers.   **d.** rational numbers.

e. irrational numbers.   **f.** real numbers.

Solution

a. Natural numbers: The natural numbers are the numbers used for counting. The only natural number in the set is $\sqrt{81}$ because $\sqrt{81} = 9$. (9 multiplied by itself is 81.)

b. Whole numbers: The whole numbers consist of the natural numbers and 0. The elements of the set that are whole numbers are 0 and $\sqrt{81}$.

c. Integers: The integers consist of the natural numbers, 0, and the negatives of the natural numbers. The elements of the set that are integers are $\sqrt{81}$, 0, and -7.

d. Rational numbers: All numbers in the set that can be expressed as the quotient of integers are rational numbers. These include $-7\left(-7 = -\frac{7}{1}\right), -\frac{3}{4}$, $0\left(0 = \frac{0}{1}\right)$, and $\sqrt{81}\left(\sqrt{81} = \frac{9}{1}\right)$. Furthermore, all numbers in the set that are terminating or repeating decimals are also rational numbers. These include $0.\overline{6}$ and 7.3.

e. Irrational numbers: The irrational numbers in the set are $\sqrt{5}\,(\sqrt{5} \approx 2.236)$ and $\pi\,(\pi \approx 3.14)$. Both $\sqrt{5}$ and π are only approximately equal to 2.236 and 3.14, respectively. In decimal form, $\sqrt{5}$ and π neither terminate nor have blocks of repeating digits.

f. Real numbers: All the numbers in the given set are real numbers. ∎

✔ **CHECK POINT 5**   Consider the following set of numbers:

$$\left\{-9, -1.3, 0, 0.\overline{3}, \frac{\pi}{2}, \sqrt{9}, \sqrt{10}\right\}.$$

List the numbers in the set that are:

a. natural numbers.   **b.** whole numbers.

c. integers.   **d.** rational numbers.

e. irrational numbers.   **f.** real numbers.

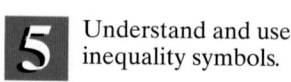

5   Understand and use inequality symbols.

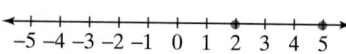

Figure 1.6

Ordering the Real Numbers

On the real number line, the real numbers increase from left to right. The lesser of two real numbers is the one farther to the left on a number line. The greater of two real numbers is the one farther to the right on a number line.

Look at the number line in Figure 1.6. The integers 2 and 5 are graphed. Observe that 2 is to the left of 5 on the number line. This means that 2 is less than 5:

$2 < 5$:   2 is less than 5 because 2 is to the *left* of 5 on the number line.

In Figure 1.6, we can also observe that 5 is to the right of 2 on the number line. This means that 5 is greater than 2.

$5 > 2$:   5 is greater than 2 because 5 is to the *right* of 2 on the number line.

The symbols $<$ and $>$ are called **inequality symbols**. These symbols always point to the lesser of the two real numbers when the inequality is true.

2 is less than 5. $2 < 5$ The symbol points to 2, the lesser number.

5 is greater than 2. $5 > 2$ The symbol points to 2, the lesser number.

EXAMPLE 6   Using Inequality Symbols

Insert either $<$ or $>$ in the shaded area between each pair of numbers to make a true statement:

a. 3 ▢ 17   **b.** -4.5 ▢ 1.2   **c.** -5 ▢ -83 **d.** $\dfrac{4}{5}$ ▢ $\dfrac{2}{3}$.

Solution   In each case, mentally compare the graph of the first number to the graph of the second number. If the first number is to the left of the second number, insert the symbol $<$ for "is less than." If the first number is to the right of the second number, insert the symbol $>$ for "is greater than."

a. Compare the graphs of 3 and 17 on the number line. Because 3 is to the left of 17, this means that 3 is less than 17: $3 < 17$.

b. Compare the graphs of -4.5 and 1.2. Because -4.5 is to the left of 1.2, this means that -4.5 is less than 1.2: $-4.5 < 1.2$.

c. Compare the graphs of -5 and -83. Because -5 is to the right of -83, this means that -5 is greater than -83: $-5 > -83$.

d. Compare the graphs of $\frac{4}{5}$ and $\frac{2}{3}$. To do so, convert to decimal notation or use a common denominator. Using decimal notation, $\frac{4}{5} = 0.8$ and $\frac{2}{3} = 0.\overline{6}$. Because 0.8 is to the right of $0.\overline{6}$, this means that $\frac{4}{5}$ is greater than $\frac{2}{3}$: $\frac{4}{5} > \frac{2}{3}$. ∎

✔ **CHECK POINT 6** Insert either $<$ or $>$ in the shaded area between each pair of numbers to make a true statement:

a. $14 \quad 5$ **b.** $-5.4 \quad 2.3$

c. $-19 \quad -6$ **d.** $\dfrac{1}{4} \quad \dfrac{1}{2}$.

The symbols $<$ and $>$ may be combined with an equal sign, as shown in the table.

Symbol	Meaning	Example	Explanation
$a \le b$	a is less than or equal to b.	$3 \le 7$	Because $3 < 7$
		$7 \le 7$	Because $7 = 7$
$b \ge a$	b is greater than or equal to a.	$7 \ge 3$	Because $7 > 3$
		$-5 \ge -5$	Because $-5 = -5$

When using the symbol $\le$ (is less than or equal to), the inequality is a true statement if either the $<$ part or the $=$ part is true. When using the symbol $\ge$ (is greater than or equal to), the inequality is a true statement if either the $>$ part or the $=$ part is true.

EXAMPLE 7   Using Inequality Symbols

Determine whether each inequality is true or false:

a. $-7 \le 4$ **b.** $-7 \le -7$ **c.** $-9 \ge 6$.

Solution

a. $-7 \le 4$ is true because $-7 < 4$ is true.

b. $-7 \le -7$ is true because $-7 = -7$ is true.

c. $-9 \ge 6$ is false because neither $-9 > 6$ nor $-9 = 6$ is true. ∎

✔ **CHECK POINT 7** Determine whether each inequality is true or false:

a. $-2 \le 3$ **b.** $-2 \ge -2$ **c.** $-4 \ge 1$.

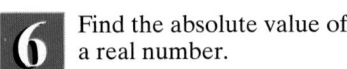

6   Find the absolute value of a real number.

Absolute Value

Suppose you are feeling a bit lazy. Instead of your usual jog, you decide to use your finger to stroll along the number line. You start at 0 and end the finger-walk at -5. You have covered a distance of 5 units.

Absolute value describes distance from 0 on a number line. If a represents a real number, the symbol $|a|$ represents its absolute value, read "the absolute value of a." In terms of your walk, we can write

$$|-5| = 5.$$

The absolute value of −5 is 5 because −5 is 5 units from 0 on the number line.

> **Absolute Value**
>
> The absolute value of a real number a, denoted by $|a|$, is the distance from 0 to a on the number line. Because absolute value describes a distance, it is never negative.

EXAMPLE 8 Finding Absolute Value

Find the absolute value:

a. $|-3|$ **b.** $|5|$ **c.** $|0|$.

Solution

a. $|-3| = 3$ The absolute value of −3 is 3 because −3 is 3 units from 0.

b. $|5| = 5$ 5 is 5 units from 0.

c. $|0| = 0$ 0 is 0 units from itself.

Absolute value describes distance from 0 on a number line.

Can you see that the absolute value of a real number is either positive or zero? Zero is the only real number whose absolute value is 0: $|0| = 0$. **The absolute value of any other real number is always positive**.

✔ **CHECK POINT 8** Find the absolute value:

a. $|-4|$ **b.** $|6|$ **c.** $|-\sqrt{2}|$.

EXERCISE SET 1.2

Practice Exercises

In Exercises 1–8, write a positive or negative integer that describes each situation.

1. Meteorology: 20° below zero
2. Navigation: 65 feet above sea level
3. Health: A gain of 8 pounds
4. Economics: A loss of $12,500.00
5. Banking: A withdrawal of $3000.00
6. Physics: An automobile slowing down at a rate of 3 meters per second each second.
7. Economics: A budget deficit of 4 billion dollars
8. Football: A 14-yard loss

In Exercises 9–20, start by drawing a number line that shows integers from −5 to 5. Then graph each of the following real numbers on your number line.

9. 2   10. 5   11. −5

12. −2   13. $2\frac{1}{2}$ 14. $3\frac{1}{4}$

15. $\dfrac{13}{3}$ 16. $\dfrac{7}{3}$ 17. −2.8

18. −4.8   19. $-\dfrac{16}{5}$ 20. $-\dfrac{11}{5}$

In Exercises 21–32, express each rational number as a decimal.

21. $\dfrac{3}{4}$ 22. $\dfrac{3}{5}$ 23. $\dfrac{7}{20}$

24. $\dfrac{3}{20}$ 25. $\dfrac{7}{8}$ 26. $\dfrac{5}{16}$

27. $\dfrac{9}{11}$ 28. $\dfrac{3}{11}$ 29. $-\dfrac{1}{2}$

30. $-\dfrac{1}{4}$ 31. $-\dfrac{5}{6}$ 32. $-\dfrac{7}{6}$

In Exercises 33–36, list all numbers from the given set that are: **a.** *natural numbers,* **b.** *whole numbers,* **c.** *integers,* **d.** *rational numbers,* **e.** *irrational numbers,* **f.** *real numbers.*

33. $\{-9, -\frac{4}{5}, 0, 0.25, \sqrt{3}, 9.2, \sqrt{100}\}$

34. $\{-7, -0.\overline{6}, 0, \sqrt{49}, \sqrt{50}\}$

35. $\left\{-11, -\dfrac{5}{6}, 0, 0.75, \sqrt{5}, \pi, \sqrt{64}\right\}$

36. $\{-5, -0.\overline{3}, 0, \sqrt{2}, \sqrt{4}\}$

37. Give an example of a whole number that is not a natural number.

38. Give an example of an integer that is not a whole number.

39. Give an example of a rational number that is not an integer.

40. Give an example of a rational number that is not a natural number.

41. Give an example of a number that is an integer, a whole number, and a natural number.

42. Give an example of a number that is a rational number, an integer, and a real number.

43. Give an example of a number that is an irrational number and a real number.

44. Give an example of a number that is a real number, but not an irrational number.

In Exercises 45–62, insert either $<$ or $>$ in the shaded area between each pair of numbers to make a true statement.

45. $\dfrac{1}{2}$ ▢ 2

46. 4 ▢ -3

47. 3 ▢ $-\dfrac{5}{2}$

48. 3 ▢ $\dfrac{3}{2}$

49. -4 ▢ -6

50. $-\dfrac{5}{2}$ ▢ $-\dfrac{5}{3}$

51. -2.5 ▢ 1.5

52. -1.25 ▢ -0.5

53. $-\dfrac{3}{4}$ ▢ $-\dfrac{5}{4}$

54. 0 ▢ $-\dfrac{1}{2}$

55. -4.5 ▢ 3

56. -5.5 ▢ 2.5

57. $\sqrt{2}$ ▢ 1.5

58. $\sqrt{3}$ ▢ 2

59. $0.\overline{3}$ ▢ 0.3

60. 0.6 ▢ $0.\overline{6}$

61. $-\pi$ ▢ -3.5

62. $-\dfrac{\pi}{2}$ ▢ -2.3

In Exercises 63–70, determine whether each inequality is true or false.

63. $-5 \geq -13$

64. $-5 \leq -8$

65. $-9 \geq -9$

66. $-14 \leq -14$

67. $0 \geq -6$

68. $0 \geq -13$

69. $-17 \geq 6$

70. $-14 \geq 8$

In Exercises 71–78, find each absolute value.

71. $|6|$

72. $|3|$

73. $|-7|$

74. $|-9|$

75. $\left|\dfrac{2}{3}\right|$

76. $\left|\dfrac{1}{2}\right|$

77. $|-\sqrt{13}|$

78. $|-\sqrt{17}|$

Application Exercises

Temperatures sometimes fall below zero. A combination of low temperature and wind makes it feel colder than the actual temperature. The table shows how cold it feels when low temperatures are combined with different wind speeds.

Windchill

Wind (mph)	Temperature (°F)											
	35	30	25	20	15	10	5	0	−5	−10	−15	−20
5	33	27	21	16	12	7	0	−5	−10	−15	−21	−26
10	22	16	10	3	−3	−9	−15	−22	−27	−34	−40	−46
15	16	9	2	−5	−11	−18	−25	−31	−38	−45	−51	−58
20	12	4	−3	−10	−17	−24	−31	−39	−46	−53	−60	−67
25	8	1	−7	−15	−22	−29	−36	−44	−51	−59	−66	−74

Use the information from the table to solve Exercises 79–80.

79. Write a negative integer that indicates how cold the temperature feels when the temperature is 25° Fahrenheit and the wind is blowing at 20 miles per hour.

80. Write a negative integer that indicates how cold the temperature feels when the temperature is 20° Fahrenheit and the wind is blowing at 15 miles per hour.

The following table shows the amount of money, in millions of dollars, collected and spent by the U.S. government from 1996 through 2000.

Year	Money Collected	Money Spent
1996	1,453,100	1,560,500
1997	1,579,300	1,601,200
1998	1,721,800	1,652,600
1999	1,827,500	1,703,000
2000	1,956,300	1,789,600

Money is expressed in millions of dollars.
Source: Office of Management and Budget

Use the information from the table to solve Exercises 81–82.

81. List the years for which money collected < money spent. Was there a budget surplus or deficit in these years?

82. List the years for which money collected > money spent. Was there a budget surplus or deficit in these years?

Writing in Mathematics

Writing about mathematics will help you to learn mathematics. For all writing exercises in this book, use complete sentences to respond to the questions. Some writing exercises can be answered in a sentence; others require a paragraph or two. You can decide how much you need to write as long as your writing clearly and directly answers the question in the exercise. Standard references such as a dictionary and a thesaurus should be helpful.

83. What is a set?
84. What are the natural numbers?
85. What are the whole numbers?
86. What are the integers?
87. How does the set of integers differ from the set of whole numbers?

88. Describe how to graph a number on the number line.
89. What is a rational number?
90. Explain how to express $\frac{3}{8}$ as a decimal.
91. Describe the difference between a rational number and an irrational number.
92. If you are given two real numbers, explain how to determine which one is the lesser.
93. Describe what is meant by the absolute value of a number. Give an example with your explanation.
94. Give an example of an everyday situation that can be described using integers but not using whole numbers.

Critical Thinking Exercises

95. Which one of the following statements is true?
  a. Every rational number is an integer.
  b. Some whole numbers are not integers.
  c. Some rational numbers are not positive.
  d. Irrational numbers cannot be negative.

96. Which one of the following statements is true?
  a. $\sqrt{36}$ is an irrational number.
  b. Some real numbers are not rational numbers.
  c. Some integers are not rational numbers.
  d. All whole numbers are positive.

97. Answer this question without using a calculator. Between which two consecutive integers is $-\sqrt{47}$?

98. We have used a code to describe an activity that is important to your success in school and at work. Here is the coded message:

$$(20, 18) \quad (-8, -12) \quad (0, 9) \quad (-14, -17)$$
$$(0, 11) \quad (-17, 9) \quad (-16, 14) \quad (-22, 7).$$

Read across the two rows. Each number pair represents one letter of the coded message. The first number pair represents the first letter, the second pair, the second letter, and so on. Use the following table to decode the message.

First Decoding	Second Decoding							
Use the greater of the two numbers in each pair. If the greater number is negative, then use its absolute value.	0 = blank space		1 = A	2 = B		3 = C	4 = D	
	5 = E	6 = F	7 = G	8 = H		9 = I	10 = J	
	11 = K	12 = L	13 = M	14 = N		15 = O	16 = P	
	17 = Q	18 = R	19 = S	20 = T		21 = U	22 = V	
	23 = W	24 = X	25 = Y	26 = Z				

Technology Exercises

In Exercises 99–102, use a calculator to find a decimal approximation for each irrational number, correct to three decimal places. Between which two integers should you graph each of these numbers on the number line?

99. $\sqrt{3}$ **100.** $-\sqrt{12}$ **101.** $1 - \sqrt{2}$ **102.** $2 - \sqrt{5}$

▶ SECTION 1.3 *Ordered Pairs and Graphs*

Objectives

1 Plot ordered pairs in the rectangular coordinate system.

2 Find coordinates of points in the rectangular coordinate system.

3 Interpret information given by line graphs.

4 Interpret information given by bar graphs.

5 Interpret information given by circle graphs.

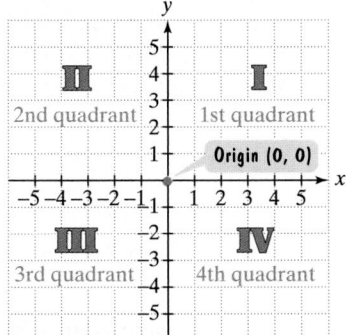

SSM
PH Tutor   CD-   Video
Center   ROM

The beginning of the seventeenth century was a time of innovative ideas and enormous intellectual progress in Europe. English theatergoers enjoyed a succession of exciting new plays by Shakespeare. William Harvey proposed the radical notion that the heart was a pump for blood rather than the center of emotion. Galileo, with his new-fangled invention called the telescope, supported the theory of Polish astronomer Copernicus that the sun, not the Earth, was the center of the solar system. Monteverdi was writing the world's first grand operas. French mathematicians Pascal and Fermat invented a new field of mathematics called probability theory.

Into this arena of intellectual electricity stepped French aristocrat René Descartes (1596–1650). Descartes, propelled by the creativity surrounding him, developed a new branch of mathematics that brought together algebra and geometry in a unified way—a way that visualized numbers as points on a graph, equations as geometric figures, and geometric figures as equations. This new branch of mathematics, called *analytic geometry*, established Descartes as one of the founders of modern thought and among the most original mathematicians and philosophers of any age. We begin this section by looking at Descartes's deceptively simple idea, called the **rectangular coordinate system** or (in his honor) the **Cartesian coordinate system**.

Points and Ordered Pairs

Descartes used two number lines that intersect at right angles at their zero points, as shown in Figure 1.7. The horizontal number line is the **x-axis**. The vertical number line is the **y-axis**. The point of intersection of these axes is their zero points, called the **origin**. Positive numbers are shown to the right and above the origin. Negative numbers are shown to the left and below the origin. The axes divide the plane into four quarters, called **quadrants**. The points located on the axes are not in any quadrant.

Each point in the rectangular coordinate system corresponds to an **ordered pair** of real numbers, (x, y). Examples of such pairs are $(4, 2)$ and $(-5, -3)$. The first number in each pair, called the **x-coordinate**, denotes the distance and direction from the origin along the x-axis. The second number, called the **y-coordinate**, denotes vertical distance and direction along a line parallel to the y-axis or along the y-axis itself.

1 Plot ordered pairs in the rectangular coordinate system.

Figure 1.7 The rectangular coordinate system

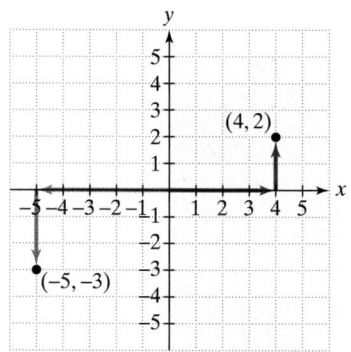

Figure 1.8 Plotting (4, 2) and (−5, −3)

Figure 1.8 shows how we **plot**, or locate, the points corresponding to the ordered pairs (4, 2) and (−5, −3). We plot (4, 2) by going 4 units from 0 to the right along the *x*-axis. Then we go 2 units up parallel to the *y*-axis. We plot (−5, −3) by going 5 units from 0 to the left along the *x*-axis and 3 units down parallel to the *y*-axis. The phrase "the point corresponding to the ordered pair (−5, −3)" is often abbreviated as "the point (−5, −3)."

EXAMPLE 1 Plotting Points in the Rectangular Coordinate System

Plot the points: $A(-3, 5)$, $B(2 -4)$, $C(5, 0)$, $D(-5, -3)$, $E(0, 4)$, and $F(0, 0)$.

Solution   See Figure 1.9. We plot the points in the following way:

$A(-3, 5)$:   3 units left, 5 units up (in quadrant II)

$B(2, -4)$:   2 units right, 4 units down (in quadrant IV)

$C(5, 0)$:   5 units right, 0 units up or down (on the *x*-axis)

$D(-5, -3)$:   5 units left, 3 units down (in quadrant III)

$E(0, 4)$:   0 units right or left, 4 units up (on the *y*-axis)

$F(0, 0)$:   0 units right or left, 0 units up or down (at the origin).

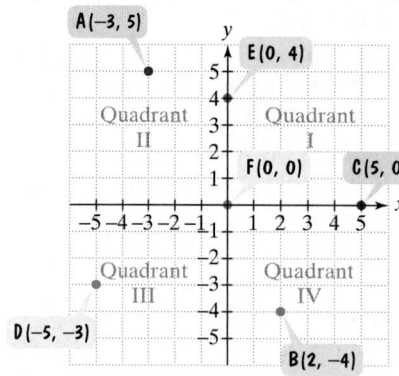

Figure 1.9 Plotting points

The phrase *ordered pair* is used because **order is important**. For example, the points (2, 5) and (5, 2) are not the same. To plot (2, 5), move 2 units right and 5 units up. To plot (5, 2), move 5 units right and 2 units up. The points (2, 5) and (5, 2) are in different locations. **The order in which coordinates appear makes a difference in a point's location.**

✔ **CHECK POINT 1**   Plot the points:

$A(-2, 4)$, $B(4, -2)$, $C(-3, 0)$, and $D(0, -3)$.

2 Find coordinates of points in the rectangular coordinate system.

In the rectangular coordinate system, each ordered pair corresponds to exactly one point. Example 2 illustrates that each point in the rectangular coordinate system corresponds to exactly one ordered pair.

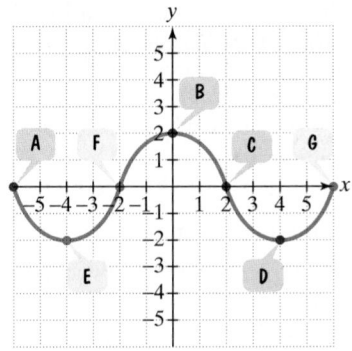

Figure 1.10 Finding coordinates of points

EXAMPLE 2   Finding Coordinates of Points

Determine the coordinates of points A, B, C, and D shown in Figure 1.10.

Solution

Point	Position	Coordinates
A	6 units left, 0 units up or down	$(-6, 0)$
B	0 units right or left, 2 units up	$(0, 2)$
C	2 units right, 0 units up or down	$(2, 0)$
D	4 units right, 2 units down	$(4, -2)$

■

✔ **CHECK POINT 2**   Determine the coordinates of points E, F, and G shown in Figure 1.10.

The rectangular coordinate system lets us visualize relationships between two quantities, as shown in the next example.

EXAMPLE 3   An Application of the Rectangular Coordinate System

Comedian David Letterman has been known to drop watermelons out of windows and watch them fall. Suppose that he drops a watermelon from the observation deck at the top of New York's Empire State Building. (Yes, the sidewalk on the ground below the building has been cleared!) Letterman drops the melon and watches it glide past the side of the building, smiling at the instant it splatters on the ground. The points in Figure 1.11 show the height of the melon above the ground at different times. Find the coordinates of point A. Then interpret the coordinates in terms of the information given.

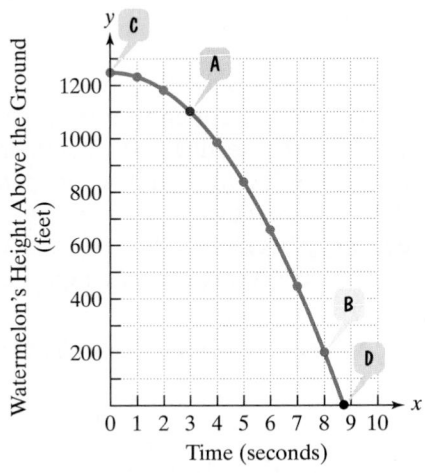

Figure 1.11

Solution   Let's take a few minutes to look at the rectangular coordinate system. The x-axis represents the time, in seconds, that the watermelon is in the air. Each mark on the x-axis represents one second. The y-axis represents the melon's height, in feet, above the ground. Each mark on the y-axis represents 100 feet. Because time and height are not negative, Figure 1.11 shows only the first quadrant of the rectangular coordinate system and its boundary.

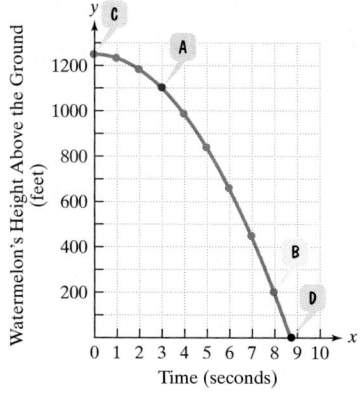

Figure 1.11 repeated

Study Tip

Any point on the *x*-axis has a *y*-coordinate of 0. Any point on the *y*-axis has an *x*-coordinate of 0.

3 Interpret information given by line graphs.

Now let us find the coordinates of point *A*. Point *A* is 3 units right and 1100 units up. Thus, the coordinates of point *A* are (3, 1100). This means that after 3 seconds, the watermelon is 1100 feet above the ground. ∎

✔ **CHECK POINT 3**   Use Figure 1.11 to find the coordinates of point *B*. Then interpret the coordinates in terms of the information given.

Take another look at Figure 1.11. How can we find the coordinates of points *C* and *D*? It appears that we must estimate one or more of these coordinates and arrive at reasonable approximations.

EXAMPLE 4   Estimating Coordinates of a Point

Use Figure 1.11 to estimate the coordinates of point *C*. Then interpret the coordinates in terms of the information given.

Solution   Point *C* is 0 units right. Thus, its *x*-coordinate is 0. How far up is point *C*? Its distance up is approximately midway between 1200 and 1300. Thus, its *y*-coordinate is approximately 1250. A reasonable estimate is that the coordinates of point *C* are (0, 1250). This means that after 0 seconds, or at the very instant Letterman dropped the watermelon, it was approximately 1250 feet above the ground. Equivalently, the melon was dropped from a height of 1250 feet. ∎

✔ **CHECK POINT 4**   Use Figure 1.11 to estimate the coordinates of point *D*. Then interpret the coordinates in terms of the information given.

Throughout your study of algebra, you will see how the rectangular coordinate system is used to create graphs that show relationships between quantities. Magazines and newspapers also use graphs to display information. In the remainder of this section, we will discuss how to interpret line, bar, and circle graphs.

Line Graphs

Line graphs are often used to illustrate trends over time. Some measure of time, such as months or years, frequently appears on the horizontal axis. Amounts are generally listed on the vertical axis. Points are drawn to represent the given information. The graph is formed by connecting the points with line segments.

Figure 1.12 is an example of a typical line graph. The graph shows the average age at which women in the United States married for the first time over a 110-year period. The years are listed on the horizontal axis and the ages are listed on the vertical axis.

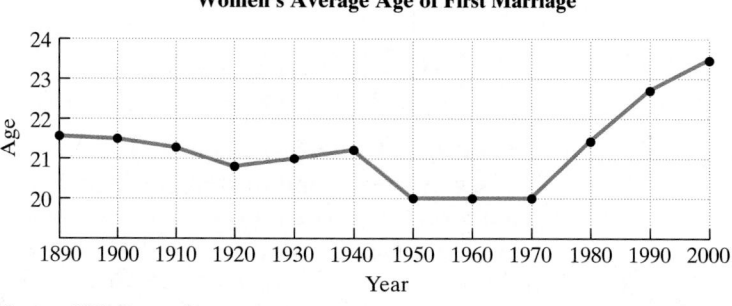

Figure 1.12 Average age at which U.S. women married for the first time

Source: U.S. Census Bureau

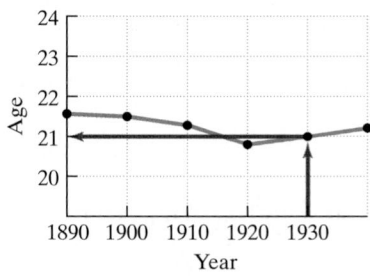

Figure 1.13 In 1930, women were 21 on average when they married for the first time.

Figure 1.13 shows how to find the average age at which women married for the first time in 1930. (Only the part of the graph that reveals what occurred through about 1940 is shown in the margin because we are interested in 1930.)

Step 1 Locate 1930 on the horizontal axis.

Step 2 Locate the point above 1930.

Step 3 Read across to the corresponding age on the vertical axis.

The age is 21. Thus, in 1930, women in the United States married for the first time at an average age of 21.

EXAMPLE 5   Using a Line Graph

Use the line graph in Figure 1.12 to estimate the maximum average age at which U.S. women married for the first time. When did this occur?

Solution   The maximum average age at which U.S. women married for the first time can be found by locating the highest point on the graph. This point lies above the number 2000 on the horizontal axis. Read across to the corresponding age on the vertical axis. The age falls approximately midway between 23 and 24, at $23\frac{1}{2}$. Thus, according to the graph, the maximum average age at which U.S. women married for the first time is about $23\frac{1}{2}$. This occurred in 2000. ■

✔ **CHECK POINT 5**   Use the line graph in Figure 1.12 to find the minimum average age at which U.S. women married for the first time. State one year in which this occurred.

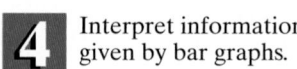 Interpret information given by bar graphs.

Bar Graphs

Bar graphs are convenient for showing comparisons among items. The bars may be either horizontal or vertical, and they are used to show the amounts of different items. Figure 1.14 is an example of a typical bar graph. The graph shows the percentage of teenagers naming the brand listed on the vertical axis as one of three "coolest."

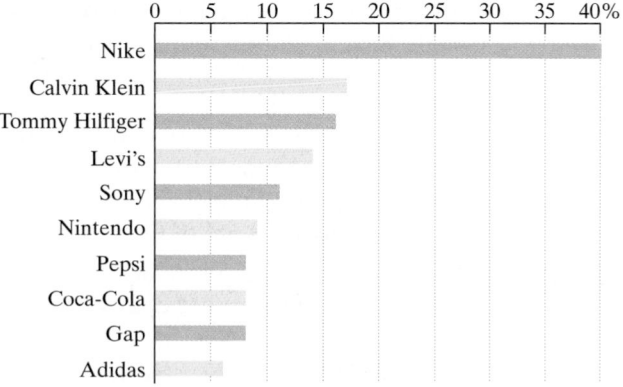

"Coolest" Brands for U.S. Teenagers

Figure 1.14

Source: Teenage Research Unlimited Inc.

"Coolest" Brands for U.S. Teenagers

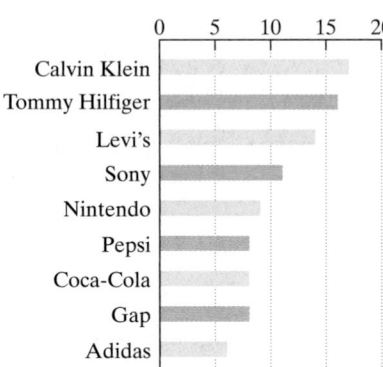

Figure 1.14 repeated, with Nike omitted

EXAMPLE 6   Using a Bar Graph

Using Figure 1.14, answer the following:

a. Estimate the percentage of teenagers who named Calvin Klein as one of the three "coolest."

b. Which brands were rated "coolest" by fewer than 10% of teenagers?

Solution

a. We look at the right edge of the bar representing Calvin Klein and then read the percent scale. The bar extends to 15% plus approximately $\frac{2}{5}$ of the distance between the 15% lines and the 20% lines. $\frac{2}{5}$ of this distance is $\frac{2}{5}$ of 5%, or 2%. Thus, approximately 15% + 2%, or 17%, of teenagers named Calvin Klein as one of the three "coolest."

b. We locate the 10% mark on the percent scale and then look for bars ending before 10%. There are five such bars, namely the five bars located in the bottom half of the graph. The brand names on these bars show that the brands rated "coolest" by fewer than 10% of teenagers are Nintendo, Pepsi, Coca-Cola, Gap, and Adidas. ∎

✔ **CHECK POINT 6** Using the Figure 1.14 complete on page 29, answer the following:

a. Estimate the percentage of teenagers who named Levi's as one of the "coolest."

b. Which brands were rated "coolest" by more than 15% of the teenagers?

Circle Graphs

Circle graphs, also called **pie charts**, show how a whole quantity is divided into parts. Circle graphs are divided into pieces called **sectors**. Figure 1.15 is an example of a typical circle graph. The graph shows the expected U.S. population by race and Hispanic origin for the year 2050.

5 Interpret information given by circle graphs.

U.S. Population in 2050

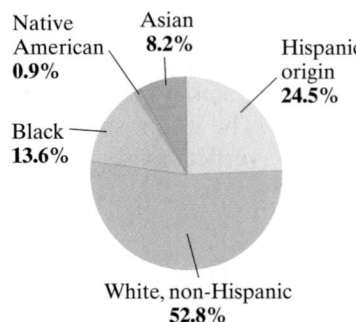

Source: U.S. Census Bureau

Figure 1.15 U.S. population in 2050

Study Tip

To convert from percent form to decimal form, move the decimal point two places to the left and drop the percent sign. For example,

$$24.5\% = 0.245.$$

EXAMPLE 7   Using a Circle Graph

Using Figure 1.15, answer the following:

a. What information is shown by the sector labeled "White, non-Hispanic"?

b. If the U.S. population for the year 2050 is expected to be 393,931,000, estimate the population for persons of Hispanic origin.

Solution

a. The sector labeled "White, non-Hispanic" shows that 52.8% of the U.S. population for the year 2050 will be white, non-Hispanic. Whites will be the largest racial group in 2050. (To put this number into perspective, in 1999 whites made up 71.8% of the U.S. population.)

b. The circle graph indicates that 24.5% of the population will be Hispanic in 2050. The actual number of Hispanics is 24.5% of the total 393,931,000.

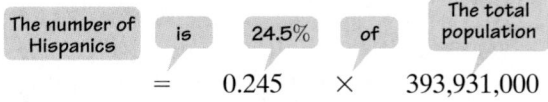

$$= \quad 0.245 \quad \times \quad 393,931,000$$

Because 24.5% can be rounded to 25%, which is $\frac{1}{4}$ of the total, we may obtain an estimate as follows:

$$\begin{array}{ccc} 393{,}931{,}000 & \rightarrow & 400{,}000{,}000 \\ \times \quad 0.245 & \rightarrow & \times \quad \frac{1}{4} \end{array} : \frac{400{,}000{,}000}{4} = 100{,}000{,}000.$$

An estimate of the Hispanic population for the year 2050 is 100 million. The actual number can be found by multiplying 0.245 and 393,931,000 with a calculator. The product is 96,513,095, or almost 97 million. Our estimate is close to the actual number. ∎

 CHECK POINT 7 Using Figure 1.15, answer the following:

a. What information is shown by the sector labeled "Black"?

b. If the U.S. population for the year 2050 is expected to be approximately 400,000,000, estimate the population for Asian Americans.

EXERCISE SET 1.3

Practice Exercises

In Exercises 1–8, plot the given point in a rectangular coordinate system. Indicate in which quadrant each point lies.

1. $(3, 4)$ **2.** $(4, 3)$
3. $(-4, 1)$ **4.** $(1, -4)$
5. $(-2, -5)$ **6.** $(-5, -2)$
7. $(4, -3)$ **8.** $(-3, 4)$

In Exercises 9–24, plot the given point in a rectangular coordinate system.

9. $(-3, -3)$ **10.** $(-5, -5)$ **11.** $(-2, 0)$
12. $(-5, 0)$ **13.** $(0, 2)$ **14.** $(0, 5)$
15. $(0, -3)$ **16.** $(0, -5)$ **17.** $\left(\frac{5}{2}, \frac{7}{2}\right)$
18. $\left(\frac{7}{2}, \frac{5}{2}\right)$ **19.** $\left(-5, \frac{3}{2}\right)$ **20.** $\left(-\frac{9}{2}, -4\right)$
21. $(0, 0)$ **22.** $\left(-\frac{5}{2}, 0\right)$ **23.** $\left(0, -\frac{5}{2}\right)$ **24.** $\left(0, \frac{7}{2}\right)$

In Exercises 25–32, give the ordered pairs that correspond to the points labeled in the figure.

25. A
26. B
27. C
28. D
29. E
30. F
31. G
32. H

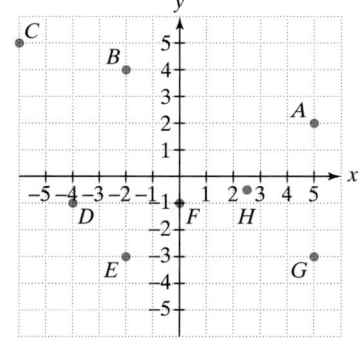

Application Exercises

A football is thrown by a quarterback to a receiver. The points in the Figure show the height of the football, in feet, above the ground in terms of its distance, in yards, from the quarterback. Use this information to solve Exercises 33–38.

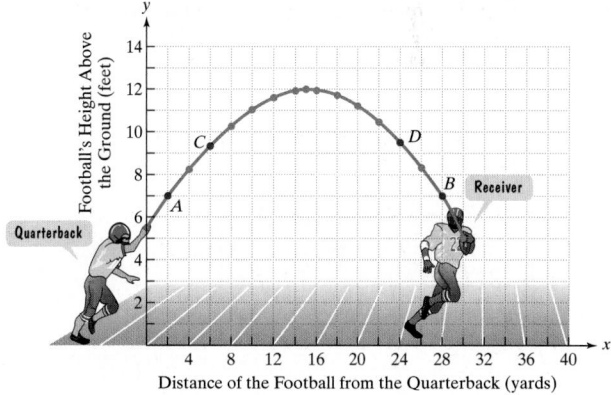

33. Find the coordinates of point A. Then interpret the coordinates in terms of the information given.

34. Find the coordinates of point B. Then interpret the coordinates in terms of the information given.

35. Estimate the coordinates of point C.

36. Estimate the coordinates of point D.

37. What is the football's maximum height? What is its distance from the quarterback when it reaches its maximum height?

38. What is the football's height when it is caught by the receiver? What is the receiver's distance from the quarterback when he catches the football?

The line graph shows the population of the United States from 1970 through 2000 for people under 16. In Exercises 39–42, find (or estimate) the coordinates of the given point. Then interpret the coordinates in terms of the information given by the graph.

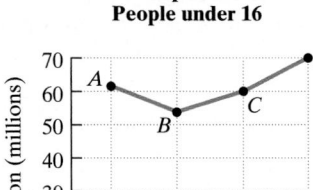

U.S. Population of People under 16

Source: U.S. Census Bureau

39. A   **40.** B

41. C   **42.** D

The line graph shows the U.S. unemployment rate from 1965 through 2000. Use the graph to solve Exercises 43–46.

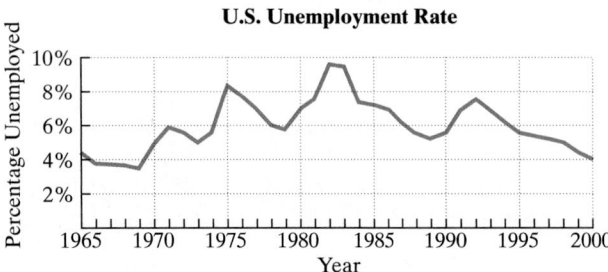

U.S. Unemployment Rate

Source: U.S. Bureau of Labor Statistics

43. Find an estimate for the unemployment rate in 1970.

44. Find an estimate for the unemployment rate in 1980.

45. For the period shown, when did the unemployment rate reach a maximum? What is a reasonable estimate for the rate during that year?

46. For the period shown, when did the unemployment rate reach a minimum? What is a reasonable estimate for the rate during that year?

The bar graph shows the percentage of vacations that included the activity listed on the left. Use the graph to solve Exercises 47–50.

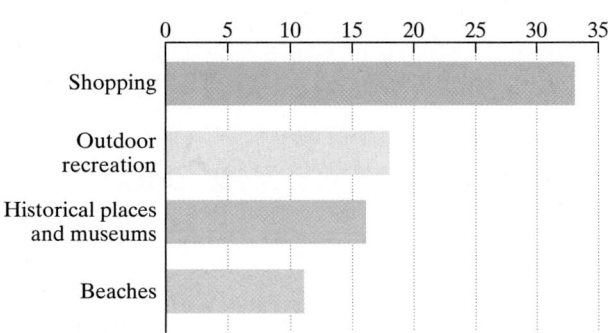

Percentage of U.S. Vacations That Included the Activity

Source: Travel Industry Association of America

47. Estimate the percentage of vacations that include shopping.

48. Estimate the percentage of vacations that include beaches.

49. Which activities are included on more than 15% of vacations?

50. Which activities are included on at least 14% of vacations and at most 30%?

The bar graph shows life expectancy in the United States by year of birth. Use the graph to solve Exercises 51–54.

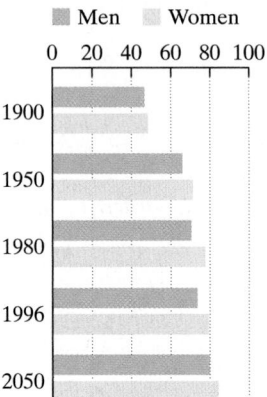

Life Expectancy in the U.S. by Birth Year

Source: National Center for Health Statistics, U.S. Census Bureau

51. Estimate the life expectancy for men born in 1900.

52. Estimate the life expectancy for women born in 2050.

53. By approximately how many more years can women born in 1996 expect to live as compared to men born in 1950?

54. For which genders and for which birth years does life expectancy exceed 40 years but is at most 60 years?

The bar graph shows the average yearly amount spent in U.S. households buying sports gear for the sports listed on the horizontal axis. Use this graph to solve Exercises 55–58.

The Most and Least Expensive Sports in the U.S.

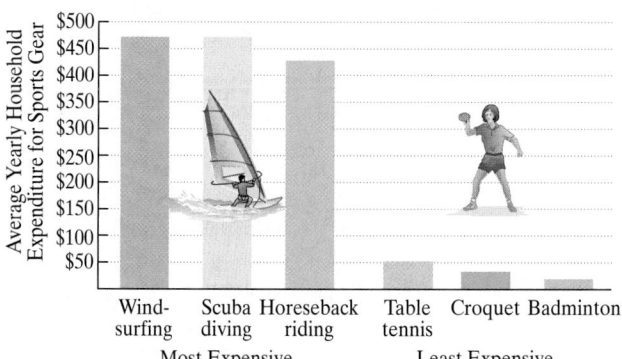

Source: National Sporting Goods Association

55. Estimate the amount spent buying sports gear for scuba diving.

56. Estimate the amount spent buying sports gear for horseback riding.

57. Approximately how much more is spent buying sports gear for table tennis than for croquet?

58. Approximately how much more is spent buying sports gear for table tennis than for badminton?

The circle graph shows the percentage of Americans who belong to various religious groups and who are members of a congregation. The population of the United States in 1999 was 272,878,000. Use the graph to solve Exercises 59–60.

Religious Affiliation

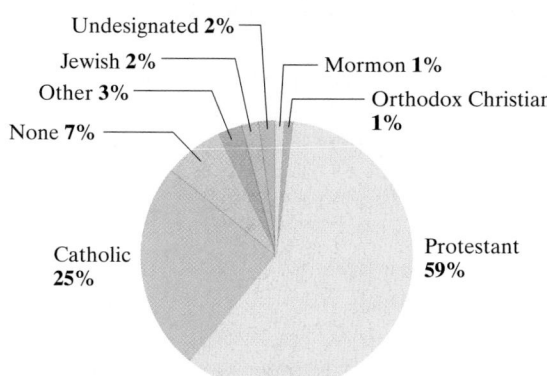

Source: The Gallup Organization

59. Estimate the number of Protestants.

60. Estimate the number of Catholics.

In 1999, there were 10,021 hate crimes reported in the United States. The circle graph shows the motivation for these reported incidents. Use the graph to solve Exercises 61–62.

Motivation For U.S. Hate-Crime Incidents

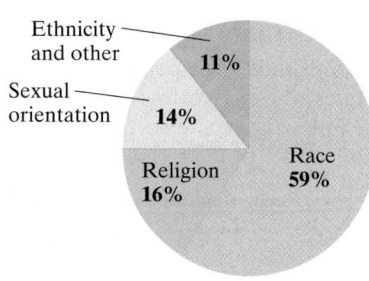

Source: FBI

61. Estimate the number of hate-crime incidents that were motivated by race.

62. Estimate the number of hate-crime incidents that were motivated by sexual orientation.

Writing in Mathematics

63. What is the rectangular coordinate system?

64. Explain how to plot a point in the rectangular coordinate system. Give an example with your explanation.

65. Explain why $(5, -2)$ and $(-2, 5)$ do not represent the same point.

66. Explain how to find the coordinates of a point in the rectangular coordinate system.

67. Describe a line graph.

68. Describe a bar graph.

69. Describe a circle graph.

70. What is a sector?

71. The circle graphs compare class attendance of successful and unsuccessful students. Write one sentence that summarizes the information conveyed by the graphs.

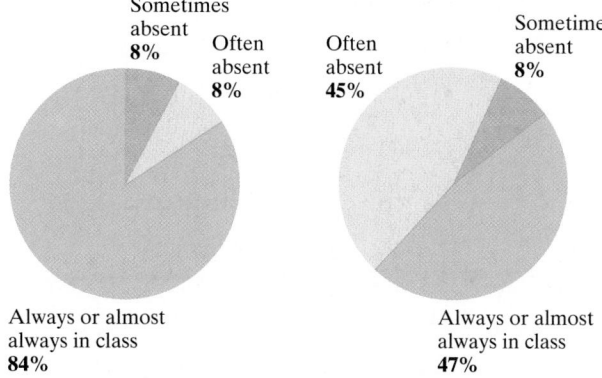

Source: *The Psychology of College Success: A Dynamic Approach*, by permission of H. C. Lindgren, 1969

72. Find a graph in a newspaper, magazine, or almanac and describe what the graph illustrates.

Critical Thinking Exercises

73. The bar graph shows the percentage of people in the United States who smoke, by level of education. Which one of the following is false according to the graph?

Percentage of U.S. Smokers by Level of Education

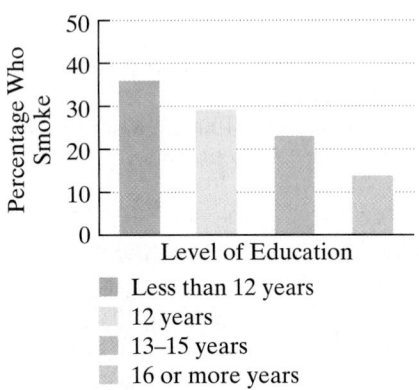

Percentage Who Smoke

Level of Education

- ■ Less than 12 years
- ▨ 12 years
- ▨ 13–15 years
- ▨ 16 or more years

Source: Centers for Disease Control and Prevention

a. The percentage of people who smoke decreases as level of education increases.

b. Less than 16% of college graduates smoke.

c. More than 39% of people with less than a high school education smoke.

d. There is only one education level shown in the graph for which the percentage of smokers exceeds 20% but is not more than 25%.

74. The circle graph shows the regional breakdown for the millions of households around the world that are online. Estimate the percentage of households in each of the four sectors.

World Online Households by Region

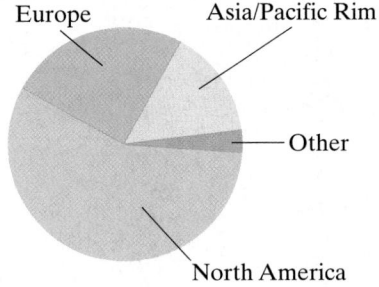

Europe   Asia/Pacific Rim

Other

North America

Source: Jupiter Communications

Review Exercises

From here on, each exercise set will contain three review exercises. It is essential to review previously covered topics to improve your understanding of the topics and to help maintain your mastery of the material. If you are not certain how to solve a review exercise, turn to the section and the illustrative example given in parentheses at the end of each exercise.

75. Add: $\frac{3}{4} + \frac{2}{5}$. (Section 1.1; Example 6)

76. Insert $<$ or $>$ in the shaded area to make a true statement: $-\frac{1}{4}$ ▨ 0. (Section 1.2, Example 6)

77. Find the absolute value: $|-5.83|$. (Section 1.2, Example 8)

▶ SECTION 1.4   *Basic Rules of Algebra*

Objectives

1 Evaluate algebraic expressions.

2 Understand and use the vocabulary of algebraic expressions.

3 Use commutative properties.

4 Use associative properties.

5 Use distributive properties.

6 Combine like terms.

7 Simplify algebraic expressions.

8 Use algebraic expressions that model reality.

SSM
PH Tutor   CD-   Video
Center   ROM

Algebraic Expressions

Feeling attractive with a suntan that gives you a "healthy glow"? Think again. Direct sunlight is known to promote skin cancer. Although sunscreens protect you from burning, skin doctors are concerned with the long-term damage that results from the sun even without sunburn.

Algebra uses letters, such as x and y, to represent numbers. Such letters are called **variables**. For example, we can let x represent the number of minutes that a person can stay in the sun without burning with no sunscreen. With a number 6 sunscreen, exposure time without burning is six times as long, or 6 times x. This can be written $6 \cdot x$, but it is usually expressed as $6x$. Placing a number and a letter next to one another indicates multiplication.

Notice that $6x$ combines the number 6 and the variable x using the operation of multiplication. A combination of variables and numbers using the operations of addition, subtraction, multiplication, or division, as well as powers or roots, is called an **algebraic expression**. Here are some examples of algebraic expressions:

$$x + 6, \quad x - 6, \quad 6x, \quad \frac{x}{6}, \quad 3x + 5, \quad \sqrt{x} + 7.$$

1 Evaluate algebraic expressions.

Evaluating Algebraic Expressions

We can replace a variable that appears in an algebraic expression by a number. We are **substituting** the number for the variable. The process is called **evaluating the expression**. For example, we can evaluate $6x$ (from the sunscreen example) when $x = 15$. We substitute 15 for x. We obtain $6 \cdot 15$, or 90. This means if you can stay in the sun for 15 minutes without burning when you don't put on any lotion, then with a number 6 lotion, you can "cook" for 90 minutes without burning.

Many algebraic expressions involve more than one operation. The order in which we add, subtract, multiply, and divide is important. In Section 1.8, we will discuss the rules for the order in which operations should be done. For now, follow this order:

1. Perform calculations within parentheses first.

2. Perform multiplication before addition.

EXAMPLE 1   Evaluating an Algebraic Expression

The algebraic expression $2.35x + 179.5$ describes the population of the United States, in millions, x years after 1960. Evaluate the expression when $x = 40$. Describe what the answer means in practical terms.

Solution   We begin by substituting 40 for x. Because $x = 40$, we will be finding the U.S. population 40 years after 1960, in the year 2000.

$$2.35x + 179.5$$

Replace x by 40.

$$= 2.35(40) + 179.5$$

$$= 94 + 179.5 \qquad \text{Perform the multiplication: } 2.35(40) = 94$$

$$= 273.5 \qquad \text{Perform the addition.}$$

Thus, in 2000, the population of the United States was 273.5 million.   ∎

✔ **CHECK POINT 1**   Evaluate: $2.35x + 179.5$ when $x = 20$.
Describe what your answer means in practical terms.

In Example 1 and Check Point 1, we used an algebraic expression that describes population. Many algebraic expressions describe some aspect of reality, and we say that they *model* reality.

The Vocabulary of Algebraic Expressions

2 Understand and use the vocabulary of algebraic expressions.

We have seen that an algebraic expression combines numbers and variables. Here is another example of an algebraic expression:

$$7x + 3.$$

The **terms** of an algebraic expression are those parts that are separated by addition. For example, the algebraic expression $7x + 3$ contains two terms, namely $7x$ and 3. Notice that a term is a number, a variable, or a number multiplied by one or more variables.

The numerical part of a term is called its **numerical coefficient**. In the term $7x$, the 7 is the numerical coefficient. If a term containing one or more variables is written without a numerical coefficient, the numerical coefficient is understood to be 1. Thus, x means $1x$ and ab means $1ab$.

A term that consists of just a number is called a **constant term**. The constant term of $7x + 3$ is 3.

The parts of each term that are multiplied are called the **factors of the term**. The factors of the term $7x$ are 7 and x.

Like terms are terms that have exactly the same variable factors. Here are two examples of like terms.

$$7x \quad \text{and} \quad 3x \qquad \text{These terms have the same variable factor, x.}$$

$$4y \quad \text{and} \quad 9y \qquad \text{These terms have the same variable factor, y.}$$

By contrast, here are some examples of terms that are not like terms. These terms do not have the same variable factor.

$$7x \quad \text{and} \quad 3$$

The variable factor of the first term is x.
The second term has no variable factor.

$$7x \quad \text{and} \quad 3y$$

The variable factor of the first term is x. The variable factor of the second term is y.

Constant terms are like terms. Thus, the constant terms 7 and −12 are like terms.

EXAMPLE 2   Using the Vocabulary of Algebraic Expressions

Use the algebraic expression

$$4x + 7 + 5x$$

to answer the following questions:

a. How many terms are there in the algebraic expression?
b. What is the numerical coefficient of the first term?
c. What is the constant term?
d. What are the like terms in the algebraic expression?

Solution

a. Because terms are separated by addition, the algebraic expression $4x + 7 + 5x$ contains three terms.

$$4x \quad + \quad 7 \quad + \quad 5x$$

| First term | Second term | Third term |

b. The numerical coefficient of the first term, $4x$, is 4.
c. The constant term in $4x + 7 + 5x$ is 7.
d. The like terms in $4x + 7 + 5x$ are $4x$ and $5x$. These terms have the same variable factor, x. ∎

✔ **CHECK POINT 2**   Use the algebraic expression $6x + 2x + 11$ to answer each of the four questions in Example 2.

Equivalent Algebraic Expressions

In Example 2, we considered the algebraic expression

$$4x + 7 + 5x.$$

Let's compare this expression with a second algebraic expression

$$9x + 7.$$

Evaluate each expression for some choice for x. We will select $x = 2$.

$4x + 7 + 5x$	$9x + 7$
Replace x by 2.	Replace x by 2.
$= 4 \cdot 2 + 7 + 5 \cdot 2$	$= 9 \cdot 2 + 7$
$= 8 + 7 + 10$	$= 18 + 7$
$= 25$	$= 25$

Discover for Yourself

Show that $4x + 7 + 5x$ and $9x + 7$ have the same value when $x = 10$.

Both algebraic expressions have the same value when $x = 2$. Regardless of what number you select for x, the algebraic expressions $4x + 7 + 5x$ and $9x + 7$ will have the same value. These expressions are called *equivalent algebraic expressions*. Two algebraic expressions that have the same value for all replacements are called **equivalent algebraic expressions**. Because $4x + 7 + 5x$ and $9x + 7$ are equivalent algebraic expressions, we write

$$4x + 7 + 5x = 9x + 7.$$

Properties of Real Numbers and Algebraic Expressions

We now turn to basic properties or rules that you know from past experiences in working with whole numbers and fractions. These properties will be extended to include all real numbers and algebraic expressions. We will give each property a name so that we can refer to it throughout the study of algebra.

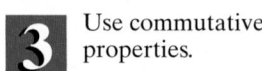

Use commutative properties.

The Commutative Properties

The addition or multiplication of two real numbers can be done in any order. For example, $3 + 5 = 5 + 3$ and $3 \cdot 5 = 5 \cdot 3$. Changing the order does not change the answer of a sum or a product. These facts are called **commutative properties**.

Study Tip

The commutative property does not hold for subtraction or division.

$$6 - 1 \neq 1 - 6$$

$$8 \div 4 \neq 4 \div 8$$

The Commutative Properties

Let a, b, and c represent real numbers, variables, or algebraic expressions.

Commutative Property of Addition

$$a + b = b + a$$

Changing order when adding does not affect the sum.

Commutative Property of Multiplication

$$ab = ba$$

Changing order when multiplying does not affect the product.

EXAMPLE 3   Using the Commutative Properties

Use the commutative properties to write an algebraic expression equivalent to each of the following:

    a. $y + 6$ **b.** $5x$.

Solution

    a. By the commutative property of addition, an algebraic expression equivalent to $y + 6$ is $6 + y$. Thus,

$$y + 6 = 6 + y.$$

    b. By the commutative property of multiplication, an algebraic expression equivalent to $5x$ is $x5$. Thus,

$$5x = x5.$$ ■

CHECK POINT 3 Use the commutative properties to write an algebraic expression equivalent to each of the following:

    a. $x + 14$ **b.** $7y$.

EXAMPLE 4 Using the Commutative Properties

Write an algebraic expression equivalent to $13x + 8$ using:

    a. the commutative property of addition.

    b. the commutative property of multiplication.

Solution

    a. By the commutative property of addition, we change the order of the terms being added. This means that an algebraic expression equivalent to $13x + 8$ is $8 + 13x$:

$$13x + 8 = 8 + 13x.$$

    b. By the commutative property of multiplication, we change the order of the factors being multiplied. This means that an algebraic expression equivalent to $13x + 8$ is $x13 + 8$:

$$13x + 8 = x13 + 8. \qquad \blacksquare$$

 CHECK POINT 4 Write an algebraic expression equivalent to $5x + 17$ using:

    a. the commutative property of addition.

    b. the commutative property of multiplication.

ENRICHMENT ESSAY

Commutative Words and Sentences

The commutative property states that a change in order produces no change in the answer. The words and sentences listed here are commutative; they read the same from left to right and from right to left!

dad	Draw, o coward!	Revolting is error. Resign it, lover.
repaper	Dennis sinned.	Naomi, did I moan?
never odd or even	Ma is a nun, as I am.	Al lets Della call Ed Stella.

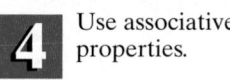 Use associative properties.

The Associative Properties

Parentheses indicate groupings. We perform operations within the parentheses first. For example,

$$(2 + 5) + 10 = 7 + 10 = 17$$

and

$$2 + (5 + 10) = 2 + 15 = 17.$$

In general, the way in which three numbers are grouped does not change their sum. It also does not change their product. These facts are called the **associative properties**.

Study Tip

The associative property does not hold for subtraction or division.

$$(6 - 1) - 3 \neq 6 - (1 - 3)$$

$$(8 \div 4) \div 2 \neq 8 \div (4 \div 2)$$

The Associative Properties

Let a, b, and c represent real numbers, variables, or algebraic expressions.

Associative Property of Addition

$$(a + b) + c = a + (b + c)$$

Changing grouping when adding does not affect the sum.

Associative Property of Multiplication

$$(ab)\,c = a\,(bc)$$

Changing grouping when multiplying does not affect the product.

The associative properties can be used to simplify some algebraic expressions by removing the parentheses.

EXAMPLE 5   Simplifying Using the Associative Properties

Simplify:

   a. $3 + (8 + x)$ **b.** $8(4x)$.

Solution

  a. $3 + (8 + x)$ This is the given algebraic expression.

      $= (3 + 8) + x$ Use the associative property of addition to group the first two numbers.

      $= 11 + x$ Add within parentheses.

Using the commutative property of addition, this simplified algebraic expression can also be written as $x + 11$.

  b. $8(4x)$ This is the given algebraic expression.

      $= (8 \cdot 4)x$ Use the associative property of multiplication to group the first two numbers.

      $= 32x$ Multiply within parentheses.

We can use the commutative property of multiplication and write this simplified algebraic expression as $x32$ or $x \cdot 32$. However, it is customary to express a term with its numerical coefficient on the left. Thus, we use $32x$ as the simplified form of the algebraic expression. ∎

 **CHECK POINT 5** Simplify:

    a. $8 + (12 + x)$ **b.** $6(5x)$.

ENRICHMENT ESSAY

The Associative Property and the English Language

In the English language, sentences can take on different meanings depending on the way the words are associated with commas. Here are two examples.

- *Do not break your bread or roll in your soup.*
  Do not break your bread, or roll in your soup.

- *Woman, without her man, is nothing.*
  Woman, without her, man is nothing.

The next example involves the use of both basic properties to simplify an algebraic expression.

EXAMPLE 6 Using the Commutative and Associative Properties

Simplify: $7 + (x + 2)$.

Solution

$$7 + (x + 2) \quad \text{This is the given algebraic expression.}$$
$$= 7 + (2 + x) \quad \text{Use the commutative property to change the order of the addition.}$$
$$= (7 + 2) + x \quad \text{Use the associative property to group the first two numbers.}$$
$$= 9 + x \quad \text{Add within parentheses.}$$

Using the commutative property of addition, an equivalent algebraic expression is $x + 9$. ∎

✔ **CHECK POINT 6**   Simplify: $8 + (x + 4)$.

5 Use distributive properties.

The Distributive Properties

The **distributive property** involves both multiplication and addition. The property shows how to multiply the sum of two numbers by a third number. Consider, for example, $4(7 + 3)$, which can be calculated in two ways. One way is to perform the addition within the grouping symbols and then multiply.

$$4(7 + 3) = 4(10) = 40$$

The other way is to *distribute* the multiplication by 4 over the addition by first multiplying each number within the parentheses by 4 and then adding.

$$4(7 + 3) = 4 \cdot 7 + 4 \cdot 3 = 28 + 12 = 40$$

The result in both cases is 40. Thus,

$$4(7 + 3) = 4 \cdot 7 + 4 \cdot 3 \quad \text{Multiplication } \textit{distributes over addition.}$$

The distributive property allows us to rewrite the product of a number and a sum as the sum of two products.

The Distributive Property

Let a, b, and c represent real numbers, variables, or algebraic expressions.

$$a(b + c) = ab + ac$$

Multiplication distributes over addition.

EXAMPLE 7 Using the Distributive Property

Multiply: $6(x + 4)$.

Solution   Multiply *each term* inside the parentheses, x and 4, by the multiplier outside, 6.

$$6(x + 4) = 6x + 6 \cdot 4 \quad \text{Use the distributive property to remove parentheses.}$$
$$= 6x + 24 \quad \text{Multiply: } 6 \cdot 4 = 24.$$ ∎

✔ **CHECK POINT 7**   Multiply: $5(x + 3)$.

Study Tip

When using a distributive property to remove parentheses, be sure to multiply *each term* inside the parentheses by the multiplier outside.

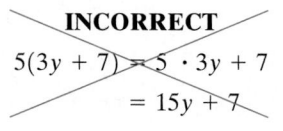

INCORRECT

$5(3y + 7) = 5 \cdot 3y + 7$
$= 15y + 7$

EXAMPLE 8 Using the Distributive Property

Multiply: $5(3y + 7)$.

Solution Multiply *each term* inside the parentheses, $3y$ and 7, by the multiplier outside, 5.

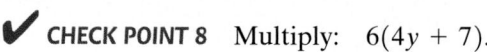

$5(3y + 7) = 5 \cdot 3y + 5 \cdot 7$ Use the distributive property to remove parentheses.

$\qquad = 15y + 35$ Multiply. Use the associative property of multiplication to find $5 \cdot 3y$: $5 \cdot 3y = (5 \cdot 3)y = 15y$. ∎

✔ **CHECK POINT 8** Multiply: $6(4y + 7)$.

Table 1.2 shows a number of other forms of the distributive property.

Table 1.2   Other Forms of the Distributive Property

Property	Meaning	Example
$a(b - c) = ab - ac$	Multiplication distributes over subtraction.	$5(4x - 3) = 5 \cdot 4x - 5 \cdot 3$ $= 20x - 15$
$a(b + c + d) = ab + ac + ad$	Multiplication distributes over three or more terms in parentheses.	$4(x + 10 + 3y)$ $= 4x + 4 \cdot 10 + 4 \cdot 3y$ $= 4x + 40 + 12y$
$(b + c)a = ba + ca$	Multiplication on the right distributes over addition (or subtraction).	$(x + 7)9 = x \cdot 9 + 7 \cdot 9$ $= 9x + 63$

6 Combine like terms.

Combining Like Terms

The distributive property

$$a(b + c) = ab + ac$$

lets us add and subtract like terms. To do this, we will usually apply the property in the form

$$ax + bx = (a + b)x$$

and then combine a and b. For example,

$$3x + 7x = (3 + 7)x = 10x.$$

This process is called **combining like terms**.

Discover for Yourself

Can you think of a fast method that will immediately give each result in Example 9? Describe the method.

EXAMPLE 9   Combining Like Terms

Combine like terms:

a. $4x + 15x$ **b.** $7a - 2a$.

Solution

a. $4x + 15x$ These are like terms because $4x$ and $15x$ have identical variable factors.

$\quad = (4 + 15)x$ Apply the distributive property.

$\quad = 19x$ Add within parentheses.

b. $7a - 2a$ These are like terms because $7a$ and $2a$ have identical variable factors.

     $= (7 - 2)a$ Apply the distributive property.

     $= 5a$ Subtract within parentheses. ∎

✔ **CHECK POINT 9**   Combine like terms:

     a. $7x + 3x$ **b.** $9a - 4a$.

When combining like terms, you may find yourself leaving out the details of the distributive property. For example, you may simply write

$$7x + 3x = 10x.$$

It might be useful to think along these lines: Seven things plus three of the (same) things give ten of those things. To add like terms, add the numerical coefficients and copy the common variable.

Combining Like Terms Mentally

1. Add or subtract the numerical coefficients of the terms.

2. Use the result of step 1 as the numerical coefficient of the term's variable factor.

When an expression contains three or more terms, use the commutative and associative properties to group like terms. Then combine the like terms.

EXAMPLE 10   Grouping and Combining Like Terms

Simplify:

     a. $7x + 5 + 3x + 8$ **b.** $4x + 7y - 2x - 3y$.

Solution

     a. $7x + 5 + 3x + 8$

         $= (7x + 3x) + (5 + 8)$ Rearrange terms and group the like terms using the commutative and associative properties. This step is often done mentally.

         $= 10x + 13$ Combine like terms: $7x + 3x = 10x$. Combine constant terms: $5 + 8 = 13$.

     b. $4x + 7y - 2x - 3y$

         $= (4x - 2x) + (7y - 3y)$ Group like terms.

         $= 2x + 4y$ Combine like terms by subtracting coefficients and keeping the variable factor. ∎

✔ **CHECK POINT 10**   Simplify:

     a. $8x + 7 + 10x + 3$ **b.** $9x + 6y - 5x - 2y$.

7   Simplify algebraic expressions.

Simplifying Algebraic Expressions

An algebraic expression is **simplified** when parentheses have been removed and like terms have been combined.

> **Simplifying Algebraic Expressions**
>
> **1.** Use the distributive property to remove parentheses.
> **2.** Rearrange terms and group like terms using commutative and associative properties. This step may be done mentally.
> **3.** Combine like terms by combining the coefficients of the terms and keeping the same variable factor.

EXAMPLE 11   Simplifying an Algebraic Expression

Simplify: $5(3x - 7) - 6x$.

Solution

$$5(3x - 7) - 6x$$

$$= 5 \cdot 3x - 5 \cdot 7 - 6x \qquad \text{Use the distributive property to remove the parentheses.}$$

$$= 15x - 35 - 6x \qquad \text{Multiply.}$$

$$= (15x - 6x) - 35 \qquad \text{Group like terms.}$$

$$= 9x - 35 \qquad \text{Combine like terms.} \qquad \blacksquare$$

✔ **CHECK POINT 11**   Simplify: $7(2x - 3) - 11x$.

Discover for Yourself

> **Discover**
> **for Yourself**
>
> Substitute 10 for x in both $5(3x - 7) - 6x$ and $9x - 35$. Do you get the same answer in each case? Which form of the expression is easier to work with?

EXAMPLE 12   Simplifying an Algebraic Expression

Simplify: $6(2x - 4y) + 10(4x + 3y)$.

Solution

$$6(2x - 4y) + 10(4x + 3y)$$

$$= 6 \cdot 2x - 6 \cdot 4y + 10 \cdot 4x + 10 \cdot 3y \qquad \text{Use the distributive property to remove the parentheses.}$$

$$= 12x - 24y + 40x + 30y \qquad \text{Multiply.}$$

$$= (12x + 40x) + (30y - 24y) \qquad \text{Group like terms.}$$

$$= 52x + 6y \qquad \text{Combine like terms.} \qquad \blacksquare$$

✔ **CHECK POINT 12**   Simplify: $7(4x + 3y) + 2(5x - y)$.

8 Use algebraic expressions that model reality.

Applications

Have you had a good workout lately? The next example involves an algebraic expression that describes, or models, your optimum heart rate during exercise.

EXAMPLE 13   Modeling Optimum Heart Rate

The optimum heart rate is the rate that a person should achieve during exercise for the exercise to be most beneficial. The algebraic expression

$$0.6(220 - a)$$

describes a person's optimum heart rate, in beats per minute, where a represents the age of the person in years.

a. Use the distributive property to rewrite the algebraic expression without parentheses.

b. Use each form of the algebraic expression to determine the optimum heart rate for a 20-year-old runner.

Solution

a. $0.6(220 - a) = 0.6(220) - 0.6a$ *Use the distributive property to remove parentheses.*

$= 132 - 0.6a$ *Multiply: $0.6(220) = 132$*

b. To determine the optimum heart rate for a 20-year-old runner, substitute 20 for a in each form of the algebraic expression.

Using $0.6(220 - a)$:	**Using $132 - 0.6a$:**
$0.6(220 - 20)$	$132 - 0.6(20)$
$= 0.6(200)$	$= 132 - 12$
$= 120$	$= 120$

Both forms of the algebraic expression indicate that the optimum heart rate for a 20-year-old runner is 120 beats per minute. ∎

✔ **CHECK POINT 13**   Use each form of the algebraic expression in Example 13 to determine the optimum heart rate for a 40-year-old runner.

EXERCISE SET 1.4

Practice Exercises

In Exercises 1–10, evaluate each algebraic expression for the given value of the variable.

1. $x + 13$; $x = 5$
2. $x - 9$; $x = 12$
3. $7x$; $x = 10$
4. $9x$; $x = 11$
5. $5x + 7$; $x = 4$
6. $9x + 6$; $x = 5$
7. $4(x + 3)$; $x = 2$
8. $6(x + 4)$; $x = 3$
9. $\frac{5}{9}(F - 32)$; $F = 77$
10. $\frac{5}{9}(F - 32)$; $F = 50$

In Exercises 11–16, an algebraic expression is given. Use each expression to answer the following questions.

a. *How many terms are there in the algebraic expression?*
b. *What is the numerical coefficient of the first term?*
c. *What is the constant term?*
d. *Does the algebraic expression contain like terms? If so, what are the like terms?*

11. $3x + 5$
12. $9x + 4$

13. $x + 2 + 5x$
14. $x + 6 + 7x$

15. $4y + 1 + 3x$
16. $8y + 1 + 10x$

In Exercises 17–24, use the commutative property of addition to write an equivalent algebraic expression.

17. $y + 4$
18. $x + 7$

19. $5 + 3x$
20. $4 + 9x$
21. $4x + 5y$
22. $10x + 9y$
23. $5(x + 3)$
24. $6(x + 4)$

In Exercises 25–32, use the commutative property of multiplication to write an equivalent algebraic expression.

25. $9x$
26. $8x$
27. $x + y6$
28. $x + y7$
29. $7x + 23$
30. $13x + 11$
31. $5(x + 3)$
32. $6(x + 4)$

In Exercises 33–36, use an associative property to rewrite each algebraic expression. Once the grouping has been changed, simplify the resulting algebraic expression.

33. $7 + (5 + x)$
34. $9 + (3 + x)$
35. $7(4x)$
36. $8(5x)$

In Exercises 37–56, use a distributive property to rewrite each algebraic expression without parentheses.

37. $3(x + 5)$
38. $4(x + 6)$
39. $8(2x + 3)$
40. $9(2x + 5)$
41. $\frac{1}{3}(12 + 6r)$
42. $\frac{1}{4}(12 + 8r)$
43. $5(x + y)$
44. $7(x + y)$
45. $3(x - 2)$
46. $4(x - 5)$

47. $2(4x - 5)$
48. $6(3x - 2)$
49. $\frac{1}{2}(5x - 12)$
50. $\frac{1}{3}(7x - 21)$
51. $(2x + 7)4$
52. $(5x + 3)6$
53. $6(x + 3 + 2y)$
54. $7(2x + 4 + y)$
55. $5(3x - 2 + 4y)$
56. $4(5x - 3 + 7y)$

In Exercises 57–74, simplify each algebraic expression.

57. $7x + 10x$
58. $5x + 13x$
59. $11a - 3a$
60. $14b - 5b$
61. $3 + (x + 11)$
62. $7 + (x + 10)$
63. $5y - 3 + 6y$
64. $8y - 7 + 10y$
65. $2x + 5 + 7x - 4$
66. $7x + 8 + 2x - 3$
67. $11a + 12 - 3a - 2$
68. $13a + 15 - 2a - 11$
69. $5(3x + 2) - 4$
70. $2(5x + 4) - 3$
71. $12 + 5(3x - 2)$
72. $14 + 2(5x - 1)$
73. $7(3a + 2b) + 5(4a - 2b)$
74. $11(6a + 3b) + 4(12a - 5b)$

Application Exercises

75. Suppose you can stay in the sun for x minutes without burning when you don't put on any lotion. The algebraic expression $15x$ describes how long you can tan without burning with a number 15 lotion. Evaluate the algebraic expression when $x = 20$. Describe what the answer means in practical terms.

76. Suppose that the cost of an item, excluding tax, is x dollars. The algebraic expression $0.06x$ describes the sales tax on that item. Evaluate the algebraic expression when $x = 400$. Describe what the answer means in practical terms.

77. The algebraic expression $1527x + 31{,}290$ describes average yearly earnings for elementary and secondary teachers in the United States x years after 1990. Evaluate the algebraic expression when $x = 10$. Describe what the answer means in practical terms.

78. The algebraic expression $3{,}184{,}000x + 48{,}264{,}000$ describes the number of cable television subscribers in the United States x years after 1990. Evaluate the algebraic expression when $x = 10$. Describe what the answer means in practical terms.

79. The equivalent algebraic expressions
$$\frac{DA + D}{24} \quad \text{and} \quad \frac{D(A + 1)}{24}$$

describe the drug dosage for children between the ages of 2 and 13. In each algebraic expression, D stands for an adult dose and A represents the child's age. If an adult dose of ibuprofen is 200 milligrams, what is the proper dose for a 12-year-old child? Use both forms of the algebraic expressions to answer the question. Which form is easier to use?

80. The algebraic expression
$$2(x + 100) + 0.35x - 20.5$$

describes the population of the United States, in millions, x years after 1960.
a. Simplify the algebraic expression.
b. Use the simplified algebraic expression to find the U.S. population for 2000.

Writing in Mathematics

81. What is an algebraic expression? Provide an example with your description.
82. What does it mean to evaluate an algebraic expression? Provide an example with your description.
83. What is a term? Provide an example with your description.
84. What are like terms? Provide an example with your description.
85. What are equivalent algebraic expressions?
86. State a commutative property and give an example.
87. State an associative property and give an example.
88. State a distributive property and give an example.
89. Explain how to add like terms. Give an example.
90. What does it mean to simplify an algebraic expression?
91. An algebra student incorrectly used the distributive property and wrote $3(5x + 7) = 15x + 7$. If you were that student's teacher, what would you say to help the student avoid this kind of error?
92. You can transpose the letters in the word "conversation" to form the phrase "voices rant on." From "total abstainers" we can form "sit not at ale bars." What two algebraic properties do each of these transpositions (called anagrams) remind you of? Explain your answer.

Critical Thinking Exercises

93. Which one of the following statements is true?
a. Subtraction is a commutative operation.
b. $(24 \div 6) \div 2 = 24 \div (6 \div 2)$
c. $7y + 3y = (7 + 3)y$ for any value of y.
d. $2x + 5 = 5x + 2$
94. Which one of the following statements is true?
a. $a + (bc) = (a + b)(a + c)$ In words, addition can be distributed over multiplication.

b. $4(x + 3) = 4x + 3$

c. Not every algebraic expression can be simplified.

d. Like terms contain the same numerical coefficients.

95. A business that manufactures small alarm clocks has weekly fixed costs of $5000. The average cost per clock for the business to manufacture x clocks is described by

$$\frac{0.5x + 5000}{x}.$$

a. Find the average cost when $x = 100, 1000,$ and $10,000$.

b. Like all other businesses, the alarm clock manufacturer must make a profit. To do this, each clock must be sold for at least 50¢ more than what it costs to man-

ufacture. Due to competition from a larger company, the clocks can be sold for $1.50 each and no more. Our small manufacturer can only produce 2000 clocks weekly. Does this business have much of a future? Explain.

 Review Exercises

96. Express $\frac{4}{9}$ as a decimal. (Section 1.2; Example 4)

97. Plot $(-3, -1)$ in a rectangular coordinate system. (Section 1.3; Example 1)

98. Divide: $\frac{3}{7} \div \frac{15}{7}$. (Section 1.1; Example 3)

▶ SECTION 1.5   *Addition of Real Numbers*

Objectives

1 Add numbers with a number line.

2 Find sums using identity and inverse properties.

3 Add numbers without a number line.

4 Use addition rules to simplify algebraic expressions.

5 Solve applied problems using a series of additions.

SSM
PH Tutor   CD-   Video
Center   ROM

It has not been a good day! First, you lost your wallet with $30 in it. Then, to get through the day, you borrowed $10, which you somehow misplaced. Your loss of $30 followed by a loss of $10 is an overall loss of $40. This can be written

$$-30 + (-10) = -40.$$

The result of adding two or more numbers is called the **sum** of the numbers. The sum of -30 and -10 is -40. You can think of gains and losses of money to find sums. For example, to find $17 + (-13)$, think of a gain of $17 followed by a loss of $13. There is an overall gain of $4. Thus, $17 + (-13) = 4$. In the same way, to find $-17 + 13$, think of a loss of $17 followed by a gain of $13. There is an overall loss of $4, so $-17 + 13 = -4$.

1 Add numbers with a number line.

Adding with a Number Line

We use the number line to help picture the addition of real numbers. Here is the procedure for finding $a + b$, the sum of a and b, using the number line.

ENRICHMENT ESSAY

Debts and Negative Numbers

Wall Street in New York City is the financial capital of the United States. In 1929, the financial crash on Wall Street saw stock values fall, leaving many people who had invested in stocks in debt. In accounting, a debt is usually written in parentheses. For example, ($2500) shows a debt of $2500, indicating that $2500 is owed. In mathematics, we write −2500.

Using the Number Line to Find a Sum

Let a and b represent real numbers. To find $a + b$ using a number line,

1. Start at a.
2. **a.** If b is **positive**, move b units to the **right**.
   b. If b is **negative**, move b units to the **left**.
   c. If b is **0**, **stay** at a.
3. The number where we finish on the number line represents the sum of a and b.

This procedure is illustrated in Examples 1 and 2. Think of moving to the right as a gain and moving to the left as a loss.

EXAMPLE 1   Adding Real Numbers Using a Number Line

Find the sum using a number line:

$$3 + (-5).$$

Solution   The solution is illustrated by the number line in Figure 1.16.

Step 1 We consider 3 to be the first number, represented by a in the preceding box. We start at a, or 3.

Step 2 We consider −5 to be the second number, represented by b. Because this number is negative, we move 5 units to the left.

Step 3 We finish at −2 on the number line. The number where we finish represents the sum of 3 and −5. Thus,

$$3 + (-5) = -2.$$

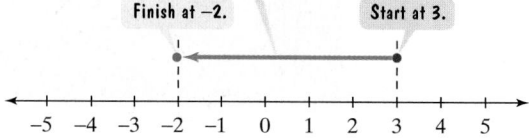

Figure 1.16 $3 + (-5) = -2$

Observe that if there is a gain of $3 followed by a loss of $5, there is an overall loss of $2. ∎

✔ **CHECK POINT 1**   Find the sum using a number line:

$$4 + (-7).$$

EXAMPLE 2   Adding Real Numbers Using a Number Line

Find each sum using a number line:

a. $-3 + (-4)$ **b.** $-6 + 2.$

Solution

a. To find $-3 + (-4)$, start at −3. Move 4 units to the left. We finish at −7. Thus,

$$-3 + (-4) = -7.$$

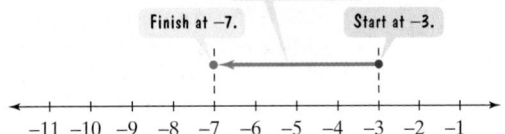

Observe that if there is a loss of $3 followed by a loss of $4, there is an overall loss of $7.

b. To find $-6 + 2$, start at -6. Move 2 units to the right because 2 is positive. We finish at -4. Thus,

$$-6 + 2 = -4.$$

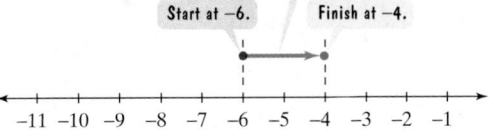

Observe that if there is a loss of $6 followed by a gain of $2, there is an overall loss of $4. ■

✔ **CHECK POINT 2** Find each sum using a number line:

a. $-1 + (-3)$ **b.** $-5 + 3.$

2 Find sums using identity and inverse properties.

The Number Line and Properties of Addition

The number line can be used to picture some useful properties of addition. For example, let's see what happens if we add two numbers with different signs but the same absolute value. Two such numbers are 3 and -3. To find $3 + (-3)$ on a number line, we start at 3 and move 3 units to the left. We finish at 0. Thus,

$$3 + (-3) = 0.$$

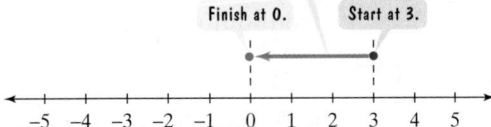

Numbers that are opposites, such as 3 and -3, are called *additive inverses*. **Additive inverses** are pairs of real numbers that are the same number of units from zero on the number line, but are on opposite sides of zero. Thus, -3 is the additive inverse of 3, and 5 is the additive inverse of -5. The additive inverse of 0 is 0. Other additive inverses come in pairs.

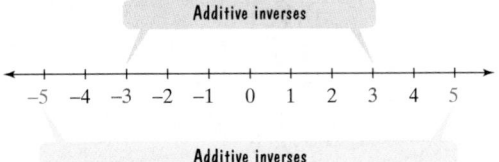

In general, the sum of any real number, denoted by a, and its additive inverse, denoted by $-a$, is zero:

$$a + (-a) = 0.$$

This property is called the **inverse property of addition**. In Section 1.4, we discussed the commutative and associative properties of addition. We now add two additional properties to our previous list, shown in Table 1.3.

Table 1.3   Identity and Inverse Properties of Addition

Let a be a real number, a variable, or an algebraic expression.		
Property	**Meaning**	**Examples**
Identity Property of Addition	Zero can be deleted from a sum. $a + 0 = a$ $0 + a = a$	• $4 + 0 = 4$ • $-3x + 0 = -3x$ • $0 + (5a + b) = 5a + b$
Inverse Property of Addition	The sum of a real number and its additive inverse is 0, the additive identity. $a + (-a) = 0$ $(-a) + a = 0$	• $6 + (-6) = 0$ • $3x + (-3x) = 0$ • $[-(2y + 1)] + (2y + 1) = 0$

3 Add numbers without a number line.

Adding without a Number Line

Now that we can picture the addition of real numbers, we look at two rules for using absolute value to add signed numbers.

Adding Two Numbers with the Same Sign

1. Add the absolute values.

2. Use the common sign as the sign of the sum.

EXAMPLE 3   Adding Real Numbers

Add without using a number line:

a. $-11 + (-15)$ **b.** $-0.2 + (-0.8)$ **c.** $-\dfrac{3}{4} + \left(-\dfrac{1}{2}\right).$

Solution   In each part of this example, we are adding numbers with the same sign.

a. $-11 + (-15) = -26$ Add absolute values: $11 + 15 = 26$.

Use the common sign.

b. $-0.2 + (-0.8) = -1$ Add absolute values: $0.2 + 0.8 = 1.0$, or 1.

Use the common sign.

c. $-\dfrac{3}{4} + \left(-\dfrac{1}{2}\right) = -\dfrac{5}{4}$ Add absolute values: $\dfrac{3}{4} + \dfrac{1}{2} = \dfrac{3}{4} + \dfrac{2}{4} = \dfrac{5}{4}$.

Use the common sign. ∎

> **Study Tip**
>
> The sum of two positive numbers is always positive. The sum of two negative numbers is always negative.

✔ **CHECK POINT 3**   Add without using a number line:

a. $-10 + (-25)$ **b.** $-0.3 + (-1.2)$ **c.** $-\dfrac{2}{3} + \left(-\dfrac{1}{6}\right).$

We also use absolute value to add two real numbers with different signs.

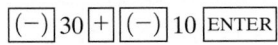
Study Tip

The sum of two numbers with different signs may be positive or negative. Keep in mind that the sign of the sum is the sign of the number with the greater absolute value.

Adding Two Numbers with Different Signs

1. Subtract the smaller absolute value from the greater absolute value.
2. Use the sign of the number with the greater absolute value as the sign of the sum.

EXAMPLE 4   Adding Real Numbers

Add without using a number line:

a. $-13 + 4$ **b.** $-0.2 + 0.8$ **c.** $-\dfrac{3}{4} + \dfrac{1}{2}$.

Solution   In each part of this example, we are adding numbers with different signs.

a. $-13 + 4 = -9$ Subtract absolute values: $13 - 4 = 9$.

Use the sign of the number with the greater absolute value.

b. $-0.2 + 0.8 = 0.6$ Subtract absolute values: $0.8 - 0.2 = 0.6$.

Use the sign of the number with the greater absolute value. The sign is assumed to be positive.

c. $-\dfrac{3}{4} + \dfrac{1}{2} = -\dfrac{1}{4}$ Subtract absolute values: $\dfrac{3}{4} - \dfrac{1}{2} = \dfrac{3}{4} - \dfrac{2}{4} = \dfrac{1}{4}$.

Use the sign of the number with the greater absolute value.

✔ **CHECK POINT 4**   Add without using a number line:

a. $-15 + 2$ **b.** $-0.4 + 1.6$ **c.** $-\dfrac{2}{3} + \dfrac{1}{6}$.

Algebraic Expressions

The rules for adding real numbers can be used to simplify certain algebraic expressions.

 Use addition rules to simplify algebraic expressions.

EXAMPLE 5   Simplifying Algebraic Expressions

Simplify:

a. $-11x + 7x$
b. $7y + (-12z) + (-9y) + 15z$
c. $3(8 - 7x) + 5(2x - 4)$.

Solution

a. $-11x + 7x$ The given algebraic expression has two like terms. $-11x$ and $7x$ have identical variable factors.

$\quad = (-11 + 7)x$ Apply the distributive property.

$\quad = -4x$ Add within parentheses.

b. $7y + (-12z) + (-9y) + 15z$

The colors indicate that there are two pairs of like terms.

$= 7y + (-9y) + (-12z) + 15z$

Arrange like terms so that they are next to one another.

$= [7 + (-9)]y + [(-12) + 15]z$

Apply the distributive property.

$= -2y + 3z$

Add within the grouping symbols.

c. $3(8 - 7x) + 5(2x - 4)$

$= 3 \cdot 8 - 3 \cdot 7x + 5 \cdot 2x - 5 \cdot 4$

Use the distributive property to remove the parentheses.

$= 24 - 21x + 10x - 20$

Multiply.

$= (-21x + 10x) + (24 - 20)$

Group like terms.

$= (-21 + 10)x + (24 - 20)$

Apply the distributive property.

$= -11x + 4$

Perform operations within parentheses. ∎

✔ **CHECK POINT 5**   Simplify:

a. $-20x + 3x$ **b.** $3y + (-10z) + (-10y) + 16z$

c. $4(10 - 8x) + 4(3x - 5)$.

---

5 Solve applied problems using a series of additions.

Applications

Positive and negative numbers are used in everyday life to represent such things as gains and losses in the stock market, rising and falling temperatures, deposits and withdrawals on bank statements, and ascending and descending motion. Positive and negative numbers are used to solve applied problems involving a series of additions.

One way to add a series of positive and negative numbers is to use the commutative and associative properties.

• Add all the positive numbers.

• Add all the negative numbers.

• Add the sums obtained in the first two steps.

The next example illustrates this idea.

EXAMPLE 6   An Application of Adding Signed Numbers

A glider was towed 1000 meters into the air and then let go. It descended 70 meters into a thermal (rising bubble of warm air), which took it up 2100 meters. At this point it dropped 230 meters into a second thermal. Then it rose 1200 meters. What was its altitude at that point?

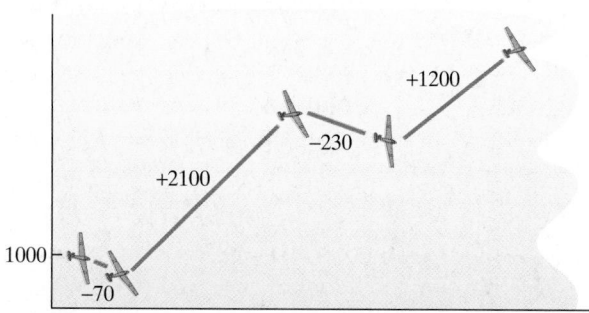

Discover for Yourself

Try working Example 6 by adding from left to right. You should still obtain 4000 for the sum. Which method do you find easier?

Solution   We use the problem's conditions to write a sum. The altitude of the glider is expressed by the following sum.

Towed to 1000 meters	then	Descended 70 meters	then	Taken up 2100 meters	then	Dropped 230 meters	then	Rose 1200 meters.
1000	+	(-70)	+	2100	+	(-230)	+	1200

$$1000 + (-70) + 2100 + (-230) + 1200$$

This is the sum arising from the problem's conditions

$$= (1000 + 2100 + 1200) + \left[(-70) + (-230)\right]$$

Use the commutative and associative properties to group the positive and negative numbers.

$$= 4300 + (-300)$$

Add the positive numbers.
Add the negative numbers.

$$= 4000$$

Add the results.

The altitude of the glider is 4000 meters. ∎

✔ **CHECK POINT 6**   The water level of a reservoir is measured over a five-month period. During this time, the level rose 2 feet, then fell 4 feet, then rose 1 foot, then fell 5 feet, and then rose 3 feet. What was the change in the water level at the end of the five months?

EXERCISE SET 1.5

Practice Exercises

In Exercises 1–8, find each sum using a number line.

1. $3 + (-7)$
2. $2 + (-7)$
3. $-4 + (-5)$
4. $-1 + (-6)$
5. $-8 + 2$
6. $-7 + 3$
7. $2 + (-2)$
8. $4 + (-4)$

In Exercises 9–46, find each sum without the use of a number line.

9. $-4 + 0$
10. $-6 + 0$
11. $9 + (-9)$
12. $13 + (-13)$
13. $-9 + (-9)$
14. $-13 + (-13)$
15. $-7 + (-5)$
16. $-3 + (-4)$
17. $-0.4 + (-0.9)$
18. $-1.5 + (-5.3)$
19. $-\dfrac{7}{10} + \left(-\dfrac{3}{10}\right)$
20. $-\dfrac{7}{8} + \left(-\dfrac{1}{8}\right)$
21. $-9 + 4$
22. $-7 + 3$
23. $12 + (-8)$
24. $13 + (-5)$
25. $6 + (-9)$
26. $3 + (-11)$
27. $-3.6 + 2.1$
28. $-6.3 + 5.2$
29. $-3.6 + (-2.1)$
30. $-6.3 + (-5.2)$
31. $\dfrac{9}{10} + \left(-\dfrac{3}{5}\right)$
32. $\dfrac{7}{10} + \left(-\dfrac{2}{5}\right)$

33. $-\dfrac{5}{8} + \dfrac{3}{4}$
34. $-\dfrac{5}{6} + \dfrac{1}{3}$
35. $-\dfrac{3}{7} + \left(-\dfrac{4}{5}\right)$
36. $-\dfrac{3}{8} + \left(-\dfrac{2}{3}\right)$
37. $4 + (-7) + (-5)$
38. $10 + (-3) + (-8)$
39. $85 + (-15) + (-20) + 12$
40. $60 + (-50) + (-30) + 25$
41. $17 + (-4) + 2 + 3 + (-10)$
42. $19 + (-5) + 1 + 8 + (-13)$
43. $-45 + \left(-\dfrac{3}{7}\right) + 25 + \left(-\dfrac{4}{7}\right)$
44. $-50 + \left(-\dfrac{7}{9}\right) + 35 + \left(-\dfrac{11}{9}\right)$
45. $3.5 + (-45) + (-8.4) + 72$
46. $6.4 + (-35) + (-2.6) + 14$

In Exercises 47–64, simplify each algebraic expression.

47. $-8x + 5x$
48. $-17x + 10x$
49. $15y + (-12y)$
50. $16y + (-14y)$
51. $-7a + (-10a)$
52. $-8a + (-12a)$
53. $-4 + 7x + 5 + (-13x)$
54. $-5 + 8x + 3 + (-16x)$
55. $7b + 2 + (-b) + (-6)$
56. $10b + 7 + (-b) + (-15)$

57. $7x + (-5y) + (-9x) + 2y$
58. $13x + (-9y) + (-11x) + 3y$
59. $4(5x - 3) + 6$
60. $5(2x - 3) + 4$
61. $8(3 - 4y) + 35y$
62. $7(5 - 3y) + 25y$
63. $6(2 - 9a) + 7(3a + 5)$
64. $8(3 - 7a) + 4(2a + 3)$

Application Exercises

Solve Exercises 65–74 by writing a sum of signed numbers and adding.

65. The greatest temperature variation recorded in a day is 100 degrees in Browning, Montana, on January 23, 1916. The low temperature was $-56°$ F. What was the high temperature?

66. In Spearfish, South Dakota, on January 22, 1943, the temperature rose 49 degrees in two minutes. If the initial temperature was $-4°$ F, what was the high temperature?

67. The Dead Sea is the lowest elevation on earth, 1312 feet below sea level. What is the elevation of a person standing 712 feet above the Dead Sea?

68. Lake Assal in Africa is 512 feet below sea level. What is the elevation of a person standing 642 feet above Lake Assal?

69. The temperature at 8:00 A.M. was $-7°$F. By noon it had risen 15°F, but by 4:00 P.M. it had fallen 5°F. What was the temperature at 4:00 P.M.?

70. On three successive plays, a football team lost 15 yards, gained 13 yards, and then lost 4 yards. What was the team's total gain or loss for the three plays?

71. A football team started with the football at the 27-yard line, advancing toward the center of the field (the 50-yard line). Four successive plays resulted in a 4-yard gain, a 2-yard loss, an 8-yard gain, and a 12-yard loss. What was the location of the football at the end of the fourth play?

72. The water level of a reservoir is measured over a five-month period. At the beginning, the level is 20 feet. During this time, the level rose 3 feet, then fell 2 feet, then fell 1 foot, then fell 4 feet, and then rose 2 feet. What is the reservoir's water level at the end of the five months?

73. The bar graph at the top of the next column shows the number of women athletes in the Olympics from 1976 through 2000. In 1976, there were 1274 women in the Olympics. The number of women athletes then changed as follows:

1980: decreased by 82; 1984: increased by 428; 1988: increased by 818; 1992: increased by 570; 1996: increased by 676; 2000: increased by 716.

How many women athletes participated in the 2000 Olympics?

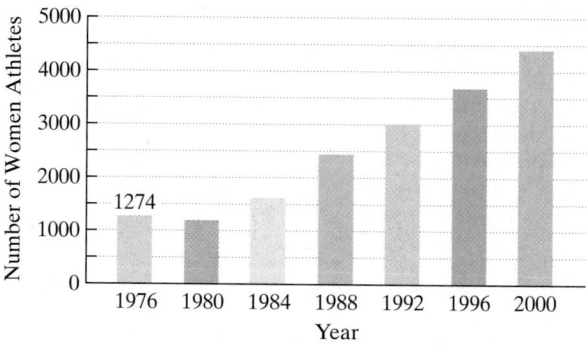

Number of Women in the Olympics

Source: International Olympic Committee

74. The bar graph shows the number of married Americans, in millions, by age. There are 15.7 million married Americans in the 35–39 age group. The number of married Americans then changes by age group as follows:

40–44: decreases by 0.1 million; 45–54: increases by 9.4 million; 55–64: decreases by 8.6 million; 65–74: decreases by 4.7 million; 75–84: decreases by 6.1 million.

How many married Americans, in millions, are there in the 75–84 age group?

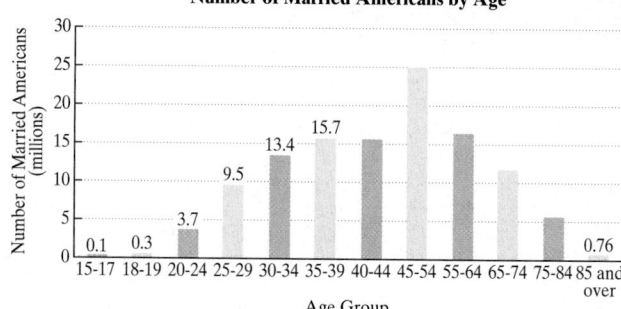

Number of Married Americans by Age

Source: National Center for Health Statistics

Writing in Mathematics

75. Explain how to add two numbers with a number line. Provide an example with your explanation.

76. What are additive inverses?

77. Describe how the inverse property of addition
$$a + (-a) = 0$$
can be shown on a number line.

78. Without using a number line, describe how to add two numbers with the same sign. Give an example.

79. Without using a number line, describe how to add two numbers with different signs. Give an example.

80. Write a problem that can be solved by finding the sum of at least three numbers, some positive and some negative. Then explain how to solve the problem.

81. Without a calculator, you can add numbers using a number line, using absolute value, or using gains and losses. Which method do you find most helpful? Why is this so?

Critical Thinking Exercises

82. Which one of the following statements is true?
   a. The sum of a positive number and a negative number is a negative number.
   b. $|-9 + 2| = 9 + 2$
   c. If two numbers are both positive or both negative, then the absolute value of their sum equals the sum of their absolute values.
   d. $\dfrac{3}{4} + \left(-\dfrac{3}{5}\right) = -\dfrac{3}{20}$

83. Which one of the following statements is true?
   a. The sum of a positive number and a negative number is a positive number.
   b. If one number is positive and the other negative, then the absolute value of their sum equals the sum of their absolute values.
   c. $\dfrac{3}{4} + \left(-\dfrac{2}{3}\right) = -\dfrac{1}{12}$
   d. The sum of zero and a negative number is always a negative number.

In Exercises 84–85, find the missing term.

84. $5x + \underline{\quad} + (-11x) + (-6y) = -6x + 2y$

85. $\underline{\quad} + 11x + (-3y) + 3x = 7(2x - 3y)$

Technology Exercises

86. Use a calculator to verify any five of the sums that you found in Exercises 17–46.

87. Use a calculator to verify any three of the answers that you obtained in Application Exercises 65–74.

In Exercises 88–89, use a calculator to estimate each sum to four decimal places.

88. $-\sqrt{2} + \sqrt{5} - \sqrt{7} + \sqrt{3}$

89. $3\sqrt{5} - 2\sqrt{7} - \sqrt{11} + 4\sqrt{3}$

Review Exercises

90. Determine whether this inequality is true or false: $19 \ge -18$. (Section 1.2, Example 7)

91. Consider the set
$$\{-6, -\pi, 0, 0.\overline{7}, \sqrt{3}, \sqrt{4}\}.$$
List all numbers from the set that are **a.** natural numbers, **b.** whole numbers, **c.** integers, **d.** rational numbers, **e.** irrational numbers, **f.** real numbers. (Section 1.2, Example 5)

92. Plot $\left(\frac{7}{2}, -\frac{5}{2}\right)$ in a rectangular coordinate system. In which quadrant does the point lie? (Section 1.3, Example 1)

▶ SECTION 1.6   *Subtraction of Real Numbers*

Objectives

1 Subtract real numbers.

2 Simplify a series of additions and subtractions.

3 Use the definition of subtraction to identify terms.

4 Use the subtraction definition to simplify algebraic expressions.

5 Solve problems involving subtraction.

SSM
PH Tutor   CD-   Video
  Center   ROM

People are going to live longer in the 21st century. This will put added pressure on the Social Security and Medicare systems. How insecure is Social Security's future? In this section, we use subtraction of real numbers to numerically describe one aspect of the insecurity.

The Meaning of Subtraction

Time for a new computer! Your favorite model, which normally sells for $1500, has an incredible price reduction of $600. The computer's reduced price, $900, can be expressed in two ways:

$$1500 - 600 = 900 \quad \text{or} \quad 1500 + (-600) = 900.$$

This means that

$$1500 - 600 = 1500 + (-600).$$

To subtract 600 from 1500, we add 1500 and the additive inverse of 600. Generalizing from this situation, we define subtraction as follows:

Definition of Subtraction

For all real numbers a and b,

$$a - b = a + (-b).$$

In words: To subtract b from a, add the additive inverse of b to a. The result of subtraction is called the **difference**.

1 Subtract real numbers.

A Procedure for Subtracting Real Numbers

The definition of subtraction gives us a procedure for subtracting real numbers.

Subtracting Real Numbers

1. Change the subtraction operation to addition.
2. Change the sign of the number being subtracted.
3. Add, using one of the rules for adding numbers with the same sign or different signs.

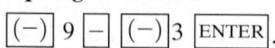
EXAMPLE 1   Using the Definition of Subtraction

Subtract:

a. $7 - 10$ **b.** $5 - (-6)$ **c.** $-9 - (-3)$.

Solution

a. $7 - 10 = 7 + (-10) = -3$

> Change the subtraction to addition.   Replace 10 with its additive inverse.

b. $5 - (-6) = 5 + 6 = 11$

> Change the subtraction to addition.   Replace -6 with its additive inverse.

c. $-9 - (-3) = -9 + 3 = -6$

> Change the subtraction to addition.   Replace -3 with its additive inverse.

✔ **CHECK POINT 1**   Subtract:

a. $3 - 11$ **b.** $4 - (-5)$ **c.** $-7 - (-2)$.

The definition of subtraction can be applied to real numbers that are not integers.

EXAMPLE 2   Using the Definition of Subtraction

Subtract:

a. $-5.2 - (-11.4)$ **b.** $-\dfrac{3}{4} - \dfrac{2}{3}$ **c.** $4\pi - (-9\pi)$.

Solution

a. $-5.2 - (-11.4) = -5.2 + 11.4 = 6.2$

Change the subtraction to addition.

Replace -11.4 with its additive inverse.

b. $-\dfrac{3}{4} - \dfrac{2}{3} = -\dfrac{3}{4} + \left(-\dfrac{2}{3}\right) = -\dfrac{9}{12} + \left(-\dfrac{8}{12}\right) = -\dfrac{17}{12}$

Change the subtraction to addition.

Replace $\frac{2}{3}$ with its additive inverse.

c. $4\pi - (-9\pi) = 4\pi + 9\pi = (4 + 9)\pi = 13\pi$

Change the subtraction to addition.

Replace -9π with its additive inverse.

∎

Reading the symbol "−" can be a bit tricky. The way you read it depends on where it appears. For example,

$$-5.2 - (-11.4)$$

is read "negative five point two minus negative eleven point four." Read parts (b) and (c) of Example 2 aloud. When is "−" read "negative" and when is it read "minus"?

✔ **CHECK POINT 2** Subtract:

a. $-3.4 - (-12.6)$ **b.** $-\dfrac{3}{5} - \dfrac{1}{3}$ **c.** $5\pi - (-2\pi)$.

2 Simplify a series of additions and subtractions.

Problems Containing a Series of Additions and Subtractions

In some problems, several additions and subtractions occur together. We begin by converting all subtractions to additions of additive inverses, or opposites.

Simplifying a Series of Additions and Subtractions

1. Change all subtractions to additions of additive inverses.
2. Group and then add all the positive numbers.
3. Group and then add all the negative numbers.
4. Add the results of steps 2 and 3.

EXAMPLE 3   Simplifying a Series of Additions and Subtractions

Simplify: $7 - (-5) - 11 - (-6) - 19$.

Solution

$$7 - (-5) - 11 - (-6) - 19$$

$= 7 + 5 + (-11) + 6 + (-19)$ Write subtractions as additions of additive inverses.

$= (7 + 5 + 6) + [(-11) + (-19)]$ Group the positive numbers. Group the negative numbers.

$= 18 + (-30)$ Add the positive numbers. Add the negative numbers.

$= -12$ Add the results.   ∎

✔ **CHECK POINT 3** Simplify: $10 - (-12) - 4 - (-3) - 6$.

3 Use the definition of subtraction to identify terms.

Subtraction and Algebraic Expressions

We know that the terms of an algebraic expression are separated by addition signs. How can we use this idea to identify the terms of the algebraic expression

$$9x - 4y - 5?$$

Because terms are separated by addition, we rewrite the algebraic expression as additions of additive inverses, or opposites. Thus,

$$9x - 4y - 5 = 9x + (-4y) + (-5).$$

The three terms of the algebraic expression are $9x$, $-4y$, and -5.

EXAMPLE 4   Using the Definition of Subtraction to Identify Terms

Identify the terms of the algebraic expression:

$$2xy - 13y - 6.$$

Solution   Rewrite the algebraic expression as additions of additive inverses.

$$2xy - 13y - 6 = 2xy + (-13y) + (-6)$$

First term   Second term   Third term

Because terms are separated by addition, the terms are $2xy$, $-13y$, and -6.   ∎

 CHECK POINT 4   Identify the terms of the algebraic expression:

$$-6 + 4a - 7ab.$$

4 Use the subtraction definition to simplify algebraic expressions.

The procedure for subtracting real numbers can be used to simplify certain algebraic expressions that involve subtraction.

EXAMPLE 5   Simplifying Algebraic Expressions

Simplify:

    a. $2 + 3x - 8x$ **b.** $-4x - 9y - 2x + 12y.$

Solution

    a. $2 + 3x - 8x$ This is the given algebraic expression.

      $= 2 + 3x + (-8x)$ Write the subtraction as the addition of an additive inverse.

      $= 2 + [3 + (-8)]x$ Apply the distributive property.

      $= 2 + (-5x)$ Add within the grouping symbols.

      $= 2 - 5x$ Be concise and express as subtraction.

    b. $-4x - 9y - 2x + 12y$

      $= -4x + (-9y) + (-2x) + 12y$ Write the subtractions as the additions of additive inverses.

      $= -4x + (-2x) + (-9y) + 12y$ Arrange like terms so that they are next to one another.

      $= [-4 + (-2)]x + (-9 + 12)y$ Apply the distributive property.

      $= -6x + 3y$ Add within the grouping symbols.   ∎

Study Tip

You can think of gains and losses of money to work the distributive property mentally.

- $3x - 8x = -5x$ A gain of 3 dollars followed by a loss of 8 of those dollars is a net loss of 5 of those dollars.
- $-9y + 12y = 3y$ A loss of 9 dollars followed by a gain of 12 of those dollars is a net gain of 3 of those dollars.

✔ **CHECK POINT 5** Simplify:

a. $4 + 2x - 9x$ **b.** $-3x - 10y - 6x + 14y.$

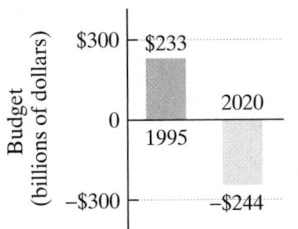

5 Solve problems involving subtraction.

Applications

Subtraction is used to solve problems in which the word "difference" appears. The difference between real numbers a and b is expressed as $a - b$.

EXAMPLE 6 An Application of Subtraction Using the Word "Difference"

The bar graph in Figure 1.17 shows that in 1995, Social Security had an annual cash surplus of $233 billion. By 2020, this amount is expected to be a negative number—a deficit of $244 billion. What is the difference between the 1995 surplus and the projected 2020 deficit?

Solution

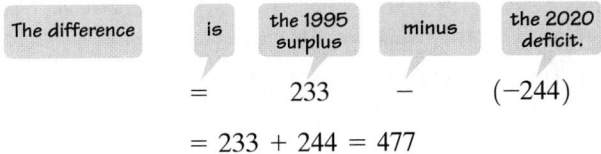

$$= \quad 233 \quad - \quad (-244)$$

$$= 233 + 244 = 477$$

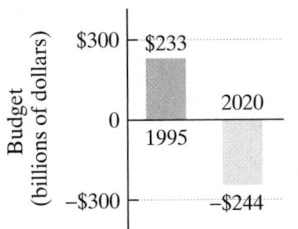

Figure 1.17 Social Security Annual Operating Budget
Source: U.S. Office of Management and Budget

The difference between the 1995 surplus and the projected 2020 deficit is $477 billion. ∎

✔ **CHECK POINT 6** The peak of Mount Everest is 8848 meters above sea level. The Marianas Trench, on the floor of the Pacific Ocean, is 10,915 meters below sea level. What is the difference in elevation between the peak of Mount Everest and the Marianas Trench?

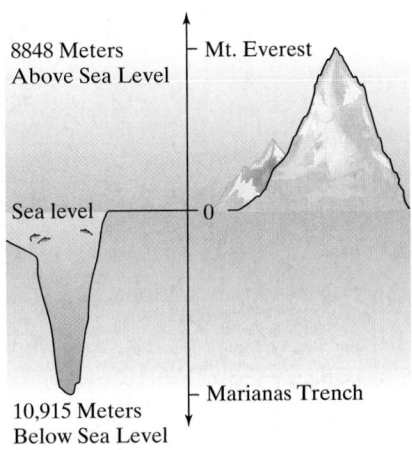

EXERCISE SET 1.6

Practice Exercises

1. Consider the subtraction $5 - 12$.
  a. Find the additive inverse, or opposite, of 12.
  b. Rewrite the subtraction as the addition of the additive inverse of 12.

2. Consider the subtraction $4 - 10$.
  a. Find the additive inverse, or opposite, of 10.
  b. Rewrite the subtraction as the addition of the additive inverse of 10.

3. Consider the subtraction $5 - (-7)$.
  a. Find the additive inverse, or opposite, of -7.
  b. Rewrite the subtraction as the addition of the additive inverse of -7.

4. Consider the subtraction $2 - (-8)$.
  a. Find the additive inverse, or opposite, of -8.
  b. Rewrite the subtraction as the addition of the additive inverse of -8.

In Exercises 5–50, perform the indicated subtraction.

5. $13 - 8$ **6.** $14 - 3$

7. $8 - 15$ **8.** $9 - 20$

9. $4 - (-10)$ **10.** $3 - (-17)$

11. $-6 - (-17)$ **12.** $-4 - (-19)$

13. $-12 - (-3)$ **14.** $-19 - (-2)$

15. $-11 - 17$ **16.** $-19 - 21$

17. $-25 - (-25)$ **18.** $-50 - (-50)$

19. $13 - 13$ **20.** $18 - 18$

21. $7 - (-7)$ **22.** $10 - (-10)$

23. $0 - 8$ **24.** $0 - 9$

25. $0 - (-3)$ **26.** $0 - (-7)$

27. $\dfrac{3}{7} - \dfrac{5}{7}$ **28.** $\dfrac{4}{9} - \dfrac{7}{9}$

29. $\dfrac{1}{5} - \left(-\dfrac{3}{5}\right)$ **30.** $\dfrac{1}{7} - \left(-\dfrac{3}{7}\right)$

31. $-\dfrac{4}{5} - \dfrac{1}{5}$ **32.** $-\dfrac{4}{9} - \dfrac{1}{9}$

33. $-\dfrac{4}{5} - \left(-\dfrac{1}{5}\right)$ **34.** $-\dfrac{4}{9} - \left(-\dfrac{1}{9}\right)$

35. $\dfrac{1}{2} - \left(-\dfrac{1}{4}\right)$ **36.** $\dfrac{2}{5} - \left(-\dfrac{1}{10}\right)$

37. $\dfrac{1}{2} - \dfrac{1}{4}$ **38.** $\dfrac{2}{5} - \dfrac{1}{10}$

39. $9.8 - 2.2$ **40.** $5.7 - 3.3$

41. $-3.1 - (-1.1)$ **42.** $-4.6 - (-1.1)$

43. $1.3 - (-1.3)$ **44.** $1.4 - (-1.4)$

45. $-2.06 - (-2.06)$ **46.** $-3.47 - (-3.47)$

47. $5\pi - 2\pi$ **48.** $9\pi - 7\pi$

49. $3\pi - (-10\pi)$ **50.** $4\pi - (-12\pi)$

In Exercises 51–68, simplify each series of additions and subtractions.

51. $13 - 2 - (-8)$ **52.** $14 - 3 - (-7)$

53. $9 - 8 + 3 - 7$ **54.** $8 - 2 + 5 - 13$

55. $-6 - 2 + 3 - 10$ **56.** $-9 - 5 + 4 - 17$

57. $-10 - (-5) + 7 - 2$ **58.** $-6 - (-3) + 8 - 11$

59. $-23 - 11 - (-7) + (-25)$

60. $-19 - 8 - (-6) + (-21)$

61. $-823 - 146 - 50 - (-832)$

62. $-726 - 422 - 921 - (-816)$

63. $1 - \dfrac{2}{3} - \left(-\dfrac{5}{6}\right)$ **64.** $2 - \dfrac{3}{4} - \left(-\dfrac{7}{8}\right)$

65. $-0.16 - 5.2 - (-0.87)$

66. $-1.9 - 3 - (-0.26)$

67. $-\dfrac{3}{4} - \dfrac{1}{4} - \left(-\dfrac{5}{8}\right)$ **68.** $-\dfrac{1}{2} - \dfrac{2}{3} - \left(-\dfrac{1}{3}\right)$

In Exercises 69–72, identify the terms in each algebraic expression.

69. $-3x - 8y$ **70.** $-9a - 4b$

71. $12x - 5xy - 4$

72. $8a - 7ab - 13$

In Exercises 73–84, simplify each algebraic expression.

73. $3x - 9x$ **74.** $2x - 10x$

75. $4 + 7y - 17y$ **76.** $5 + 9y - 29y$

77. $2a + 5 - 9a$

78. $3a + 7 - 11a$

79. $4 - 6b - 8 - 3b$

80. $5 - 7b - 13 - 4b$

81. $13 - (-7x) + 4x - (-11)$

82. $15 - (-3x) + 8x - (-10)$

83. $-5x - 10y - 3x + 13y$

84. $-6x - 9y - 4x + 15y$

Application Exercises

85. The peak of Mount Kilimanjaro, the highest point in Africa, is 19,321 feet above sea level. Qattara Depression, Egypt, the lowest point in Africa, is 436 feet below sea level. What is the difference in elevation between the peak of Mount Kilimanjaro and the Qattara Depression?

86. The peak of Mount Whitney is 14,494 feet above sea level. Mount Whitney can be seen directly above Death Valley, which is 282 feet below sea level. What is the difference in elevation between these geographic locations?

It seems that Phideau's medical bills are costing us an arm and a paw. The bar graph shows veterinary costs, in billions of dollars, for dogs and cats in five selected years. Use the graph to solve Exercises 87–88.

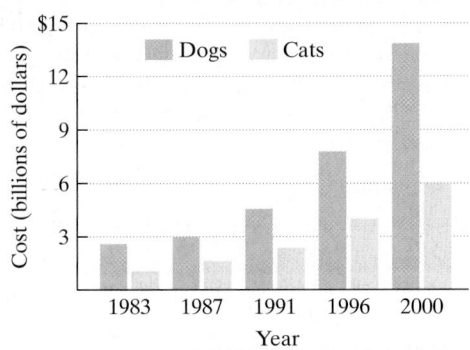

Veterinary Costs in the U.S.

Source: American Veterinary Medical Association

87. Estimate the difference between veterinary costs for dogs and cats in 2000.

88. Estimate the difference between veterinary costs for dogs in 2000 and in 1996.

Do you enjoy cold weather? If so, try Fairbanks, Alaska. The average daily low temperature for each month in Fairbanks is shown in the bar graph. Use the graph to solve Exercises 89–92.

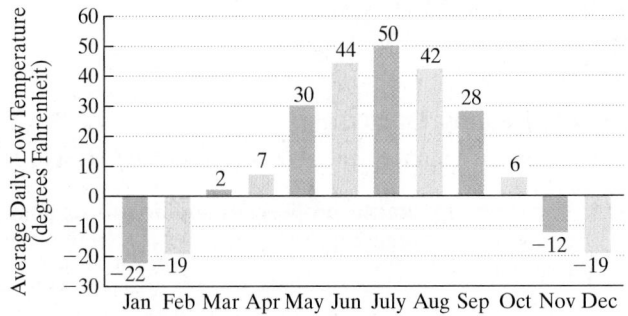

Each Month's Average Daily Low Temperature in Fairbanks, Alaska

Source: The Weather Channel Enterprises, Inc.

89. What is the difference between the average daily low temperatures for March and February?

90. What is the difference between the average daily low temperatures for October and November?

91. How many degrees warmer is February's average low temperature than January's average low temperature?

92. How many degrees warmer is November's average low temperature than December's average low temperature?

When a person receives a drug injected into a muscle, the concentration of the drug in the body depends on the time elapsed since the injection. The points in the rectangular system show the concentration of the drug, measured in milligrams per 100 milliliters, from the time the drug was injected until 13 hours later. Use this information to solve Exercises 93–98.

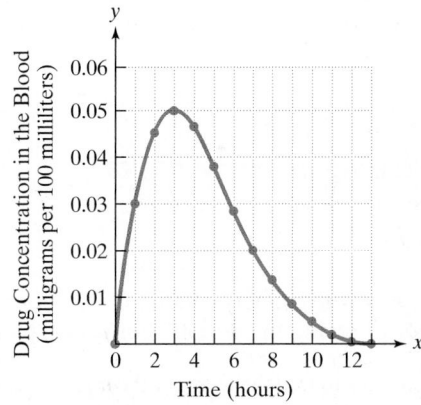

93. What is the drug's maximum concentration and when does this occur?

94. What happens by the end of 13 hours?

95. What is the difference between the drug's concentration 4 hours after it was injected and 1 hour after it was injected?

96. What is the difference between the drug's concentration 4 hours after it was injected and 7 hours after it was injected?

97. When is the drug's concentration increasing?

98. When is the drug's concentration decreasing?

The graph shows projections for the five occupations with the greatest increase and greatest decrease from 1996 through 2006. Use the graph to solve Exercises 99–100.

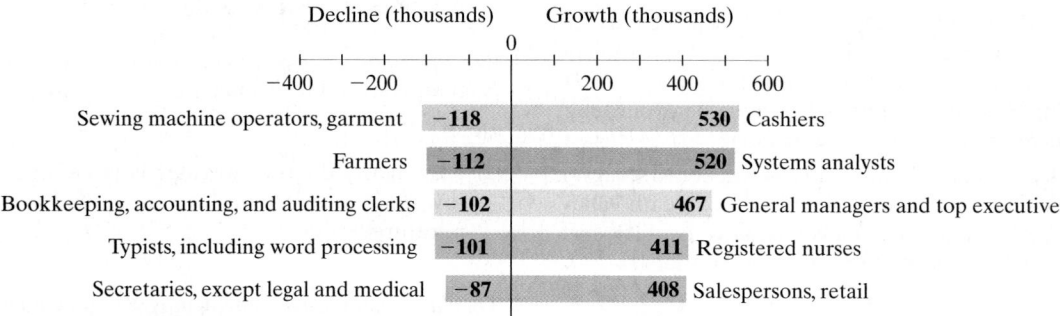

Projected Changes in Jobs from 1996–2006

| | Decline (thousands) | | Growth (thousands) | |

Sewing machine operators, garment **−118** **530** Cashiers
Farmers **−112** **520** Systems analysts
Bookkeeping, accounting, and auditing clerks **−102** **467** General managers and top executives
Typists, including word processing **−101** **411** Registered nurses
Secretaries, except legal and medical **−87** **408** Salespersons, retail

Projected changes in Jobs from 1996–2006
Source: U.S. Bureau of Labor Statistics

99. Find the difference in growth between systems analysts and farmers.

100. Find the difference in growth between registered nurses and typists.

Writing in Mathematics

101. Explain how to subtract real numbers.

102. How is $4 - (-2)$ read?

103. Explain how to simplify a series of additions and subtractions. Provide an example with your explanation.

104. Explain how to find the terms of the algebraic expression $5x - 2y - 7$.

105. Write a problem that can be solved by finding the difference between two numbers. At least one of the numbers should be negative. Then explain how to solve the problem.

Critical Thinking Exercises

106. Which one of the following statements is true?
   a. If a and b are negative numbers, then $a - b$ is a negative number.
   b. $7 - (-2) = 5$
   c. The difference between 0 and a negative number is always a positive number.
   d. None of the given statements is true.

107. The golden age of Athens culminated in 212 B.C. and the golden age of India culminated in A.D. 500. Determine the number of years that elapsed between these dates. (*Note:* When the calendar was reformed, the number 0 had not been invented. There was no year 0 and the year A.D. 1 followed the year 1 B.C. Calculate the difference between the years in the usual way and then use this added bit of information to modify your answer.)

108. Find the value:
$$-1 + 2 - 3 + 4 - 5 + 6 - \cdots - 99 + 100.$$

Technology Exercises

109. Use a calculator to verify any five of the differences that you found in Exercises 5–46.

110. Use a calculator to verify any three of the answers that you found in Exercises 51–68.

Use a calculator to estimate the value of each expression in Exercises 111–112. Round to four decimal places.

111. $4\sqrt{2} - (-3\sqrt{5}) - (-\sqrt{7}) + \sqrt{3}$

112. $-5\sqrt{3} - (-2\sqrt{11}) - (-\sqrt{17}) + \sqrt{6}$

Review Exercises

113. Graph on a number line: -4.5. (Section 1.2; Example 3)

114. Use the commutative property of addition to write an equivalent algebraic expression: $10(a + 4)$. (Section 1.4; Example 4)

115. Give an example of an integer that is not a natural number (Section 1.2; Example 5)

▶ SECTION 1.7   Multiplication and Division of Real Numbers

Objectives

1 Multiply real numbers.

2 Multiply more than two real numbers.

3 Find multiplicative inverses.

4 Use the definition of division.

5 Divide real numbers.

6 Simplify algebraic expressions involving multiplication.

7 Use algebraic expressions that model reality.

SSM   PH Tutor   CD-   Video
        Center   ROM

Technology is now promising to bring light, fast, and beautiful wheelchairs to millions of disabled people. The cost of manufacturing these radically different wheelchairs can be modeled by an algebraic expression containing division. In this section, we will see how this algebraic expression illustrates that low prices are possible with high production levels, urgently needed in this situation. There are more than half a billion people with disabilities in developing countries; an estimated 20 million need wheelchairs right now.

Multiplying Real Numbers

1 Multiply real numbers.

Suppose that things go from bad to worse for Social Security, and the projected $244 billion deficit in 2020 triples by the end of the twenty-first century. The new deficit is

$$3(-244) = (-244) + (-244) + (-244) = -732$$

or $732 billion. Thus,

$$3(-244) = -732.$$

The result of the multiplication, -732, is called the **product** of 3 and -244. The numbers being multiplied, 3 and -244, are called the **factors** of the product.

Rules for multiplying real numbers are described in terms of absolute value. For example, $3(-244) = -732$ illustrates that the product of numbers with different signs is found by multiplying their absolute values. The product is negative.

$$3(-244) = -732$$

Multiply absolute values:
$|3| \cdot |-244| = 3 \cdot 244 = 732.$

Factors have different signs and the product is negative.

The following rules are used to determine the sign of the product of two numbers.

The Product of Two Real Numbers

- The product of two real numbers with **different signs** is found by multiplying their absolute values. The product is **negative**.
- The product of two real numbers with the **same sign** is found by multiplying their absolute values. The product is **positive**.
- The product of 0 and any real number is 0. Thus, for any real number a,

$$a \cdot 0 = 0 \quad \text{and} \quad 0 \cdot a = 0.$$

EXAMPLE 1 Multiplying Real Numbers

Multiply:

a. $6(-3)$ **b.** $-\dfrac{1}{5} \cdot \dfrac{2}{3}$ **c.** $(-9)(-10)$ **d.** $(-1.4)(-2)$ **e.** $(-372)(0)$.

Solution

a. $6(-3) = -18$ Multiply absolute values: $6 \cdot 3 = 18$.

Different signs: negative product

b. $-\dfrac{1}{5} \cdot \dfrac{2}{3} = -\dfrac{2}{15}$ Multiply absolute values: $\frac{1}{5} \cdot \frac{2}{3} = \frac{1 \cdot 2}{5 \cdot 3} = \frac{2}{15}$.

Different signs: negative product

c. $(-9)(-10) = 90$ Multiply absolute values: $9 \cdot 10 = 90$.

Same sign: positive product

d. $(-1.4)(-2) = 2.8$ Multiply absolute values: $(1.4)(2) = 2.8$.

Same sign: positive product

e. $(-372)(0) = 0$ The product of 0 and any real number is 0: $a \cdot 0 = 0$.

✔ **CHECK POINT 1**   Multiply:

a. $8(-5)$ **b.** $-\dfrac{1}{3} \cdot \dfrac{4}{7}$

c. $(-12)(-3)$ **d.** $(-1.1)(-5)$

e. $(-543)(0)$.

2 Multiply more than two real numbers.

Multiplying More Than Two Numbers

How do we perform more than one multiplication, such as

$$-4(-3)(-2)?$$

Because of the associative and commutative properties, we can order and group the numbers in any manner. Each pair of negative numbers will produce a positive product. Thus, the product of an even number of negative numbers is always positive. By contrast, the product of an odd number of negative numbers is always negative.

$$-4(-3)(-2) = -24$$ Multiply absolute values: $4 \cdot 3 \cdot 2 = 24$.

Odd number of negative numbers (three): negative product

Multiplying More Than Two Numbers

1. Assuming that no factor is zero,
- The product of an **even** number of **negative numbers** is **positive**.
- The product of an **odd** number of **negative numbers** is **negative**.

The multiplication is performed by multiplying the absolute values of the given numbers.

2. If any factor is 0, the product is 0.

EXAMPLE 2   Multiplying More Than Two Numbers

Multiply:

    a. $(-3)(-1)(2)(-2)$ **b.** $(-1)(-2)(-2)(3)(-4)$.

Solution

    a. $(-3)(-1)(2)(-2) = -12$ Multiply absolute values: $3 \cdot 1 \cdot 2 \cdot 2 = 12$.

    Odd number of negative numbers (three): negative product

    b. $(-1)(-2)(-2)(3)(-4) = 48$ Multiply absolute values: $1 \cdot 2 \cdot 2 \cdot 3 \cdot 4 = 48$.

Even number of negative numbers (four): positive product

■

✔ **CHECK POINT 2** Multiply:

    a. $(-2)(3)(-1)(4)$ **b.** $(-1)(-3)(2)(-1)(5)$.

Is it always necessary to count the number of negative factors when multiplying more than two numbers? No. If any factor is 0, you can immediately write 0 for the product. For example,

$$(-37)(423)(0)(-55)(-3.7) = 0.$$

If any factor is 0, the product is 0.

3 Find multiplicative inverses.

The Meaning of Division

The result of dividing the real number a by the nonzero real number b is called the **quotient** of a and b. We can write this quotient as $a \div b$ or $\dfrac{a}{b}$.

We know that subtraction is defined in terms of addition of an additive inverse, or opposite:

$$a - b = a + (-b).$$

In a similar way, we can define division in terms of multiplication. For example, the quotient of 8 and 2 can be written as multiplication:

$$8 \div 2 = 8 \cdot \frac{1}{2}.$$

We call $\frac{1}{2}$ the *multiplicative inverse*, or *reciprocal*, of 2. Two numbers whose product is 1 are called **multiplicative inverses** or **reciprocals** of each other. Thus, the multiplicative inverse of 2 is $\frac{1}{2}$ and the multiplicative inverse of $\frac{1}{2}$ is 2 because $2 \cdot \frac{1}{2} = 1$.

EXAMPLE 3   Find Multiplicative Inverses

Find the multiplicative inverse of each number:

    a. 5   **b.** $\dfrac{1}{3}$ **c.** -4 **d.** $-\dfrac{4}{5}$.

Solution

a. The multiplicative inverse of 5 is $\frac{1}{5}$ because $5 \cdot \frac{1}{5} = 1$.

b. The multiplicative inverse of $\frac{1}{3}$ is 3 because $\frac{1}{3} \cdot 3 = 1$.

c. The multiplicative inverse of -4 is $-\frac{1}{4}$ because $(-4)\left(-\frac{1}{4}\right) = 1$.

d. The multiplicative inverse of $-\frac{4}{5}$ is $-\frac{5}{4}$ because $\left(-\frac{4}{5}\right)\left(-\frac{5}{4}\right) = 1$. ∎

✔ **CHECK POINT 3** Find the multiplicative inverse of each number:

a. 7   b. $\frac{1}{8}$ c. -6 d. $-\frac{7}{13}$.

Can you think of a real number that has no multiplicative inverse? The number **0 has no multiplicative inverse** because 0 multiplied by any number is never 1, but always 0.

We now define division in terms of multiplication by a multiplicative inverse.

4 Use the definition of division.

Definition of Division

If a and b are real numbers and b is not 0, then the quotient of a and b is defined as

$$a \div b = a \cdot \frac{1}{b}.$$

In words: The quotient of two real numbers is the product of the first number and the multiplicative inverse of the second number.

Using Technology

You can use a calculator to multiply and divide signed numbers. Here are the keystrokes for finding

$(-173)(-256)$:

Scientific Calculator

173 $\boxed{+/-}$ $\boxed{\times}$ 256 $\boxed{+/-}$ $\boxed{=}$

Graphing Calculator

$\boxed{(-)}$ 173 $\boxed{\times}$ $\boxed{(-)}$ 256 $\boxed{ENTER}$.

The number 44288 should be displayed.

Division is performed in the same manner, using $\boxed{\div}$ instead of $\boxed{\times}$. What happens when you divide by 0? Try entering

8 $\boxed{\div}$ 0

and pressing $\boxed{=}$ or $\boxed{ENTER}$.

EXAMPLE 4 Using the Definition of Division

Use the definition of division to find each quotient:

a. $-15 \div 3$ b. $\frac{-20}{-4}$.

Solution

a. $-15 \div 3 = -15 \cdot \frac{1}{3} = -5$

Change the division to multiplication.   Replace 3 with its multiplicative inverse.

b. $\frac{-20}{-4} = -20 \cdot \left(-\frac{1}{4}\right) = 5$

Change the division to multiplication.   Replace -4 with its multiplicative inverse. ∎

✔ **CHECK POINT 4** Use the definition of division to find each quotient:

a. $-28 \div 7$ b. $\frac{-16}{-2}$.

5   Divide real numbers.

A Procedure for Dividing Real Numbers

Because the quotient $a \div b$ is defined as the product $a \cdot \dfrac{1}{b}$, the sign rules for dividing numbers are the same as the sign rules for multiplying them.

The Quotient of Two Real Numbers

- The quotient of two real numbers with **different signs** is found by dividing their absolute values. The quotient is **negative**.
- The quotient of two real numbers with the **same sign** is found by dividing their absolute values. The quotient is **positive**.
- Division of a nonzero number by zero is undefined.
- Any nonzero number divided into 0 is 0.

EXAMPLE 5   Dividing Real Numbers

Divide:

   a. $\dfrac{8}{-2}$ **b.** $-\dfrac{3}{4} \div \left(-\dfrac{5}{9}\right)$ **c.** $\dfrac{-20.8}{4}$ **d.** $\dfrac{0}{-7}$.

Solution

   a. $\dfrac{8}{-2} = -4$ Divide absolute values: $\frac{8}{2} = 4$.

Different signs: negative quotient

   b. $-\dfrac{3}{4} \div \left(-\dfrac{5}{9}\right) = \dfrac{27}{20}$ Divide absolute values: $\frac{3}{4} \div \frac{5}{9} = \frac{3}{4} \cdot \frac{9}{5} = \frac{27}{20}$.

Same sign: positive quotient

   c. $\dfrac{-20.8}{4} = -5.2$ Divide absolute values: $4\overline{)20.8}$ → 5.2.

Different signs: negative quotient

   d. $\dfrac{0}{-7} = 0$ Any nonzero number divided into 0 is 0.

Can you see why $\frac{0}{-7}$ must be 0? The definition of division tells us that

$$\frac{0}{-7} = 0 \cdot \left(-\frac{1}{7}\right)$$

and the product of 0 and any real number is 0. By contrast, the definition of division does not allow for division by 0 because 0 does not have a multiplicative inverse. It is incorrect to write

$$\frac{-7}{0} = -7 \cdot \frac{1}{0}.$$

0 does not have a multiplicative inverse.

Division by zero is not allowed or not defined. Thus, $\frac{-7}{0}$ does not represent a real number. A real number can never have a denominator of 0.

 **CHECK POINT 5** Divide:

a. $\dfrac{-32}{-4}$ **b.** $-\dfrac{2}{3} \div \dfrac{5}{4}$ **c.** $\dfrac{21.9}{-3}$ **d.** $\dfrac{0}{-5}$.

6 Simplify algebraic expressions involving multiplication.

Multiplication and Algebraic Expressions

In Section 1.4, we discussed the commutative and associative properties of multiplication. We also know that multiplication distributes over addition and subtraction. We now add some additional properties to our previous list (Table 1.4). These properties are frequently helpful in simplifying algebraic expressions.

Table 1.4 Additional Properties of Multiplication

Let a be a real number, a variable, or an algebraic expression.		
Property	**Meaning**	**Examples**
Identity Property of Multiplication	1 can be deleted from a product. $a \cdot 1 = a$ $1 \cdot a = a$	• $\sqrt{3} \cdot 1 = \sqrt{3}$ • $1x = x$ • $1(2x + 3) = 2x + 3$
Inverse Property of Multiplication	If a is not 0: $a \cdot \dfrac{1}{a} = 1$ $\dfrac{1}{a} \cdot a = 1$ The product of a nonzero number and its multiplicative inverse, or reciprocal, gives 1, the multiplicative identity.	• $6 \cdot \dfrac{1}{6} = 1$ • $3x \cdot \dfrac{1}{3x} = 1 \ (x \text{ is not } 0.)$ • $\dfrac{1}{(y-2)} \cdot (y-2) = 1 \ (y \text{ is not } 2.)$
Multiplication Property of -1	Negative 1 times a is the additive inverse, or opposite, of a. $-1 \cdot a = -a$ $a(-1) = -a$	• $-1 \cdot \sqrt{3} = -\sqrt{3}$ • $-1(-\frac{3}{4}) = \frac{3}{4}$ • $-1x = -x$ • $-(x + 4) = -1(x + 4)$ $\quad = -x - 4$
Double Negative Property	The additive inverse of $-a$ is a. $-(-a) = a$	• $-(-4) = 4$ • $-(-6y) = 6y$

In the preceding table, we used three steps to remove the parentheses from $-(x + 4)$. First, we used the multiplication property of -1.

$$-(x + 4) = -1(x + 4)$$

Then we used the distributive property, distributing -1 to each term in parentheses.

$$-1(x + 4) = (-1)x + (-1)4 = -x + (-4) = -x - 4$$

There is a fast way to obtain $-(x + 4) = -x - 4$ in just one step.

Negative Signs and Parentheses

If a negative sign precedes parentheses, remove the parentheses and change the sign of every term within the parentheses.

Here are some examples that illustrate this method.

$$-(11x + 5) = -11x - 5$$
$$-(11x - 5) = -11x + 5$$
$$-(-11x + 5) = 11x - 5$$
$$-(-11x - 5) = 11x + 5$$

EXAMPLE 6   Simplifying Algebraic Expressions

Simplify:

a. $-2(3x)$ **b.** $6x + x$ **c.** $8a - 9a$

d. $-3(2x - 5)$ **e.** $-(3y - 8)$.

Solution   We will show all steps in the solution process. However, you probably are working many of these steps mentally.

a. $-2(3x)$ This is the given algebraic expression.

  $= (-2 \cdot 3)x$ Use the associative property and group the first two numbers.

  $= -6x$ Numbers with opposite signs have a negative product.

b. $6x + x$ This is the given algebraic expression.

  $= 6x + 1x$ Use the multiplication property of 1.

  $= (6 + 1)x$ Apply the distributive property.

  $= 7x$ Add within parentheses.

c. $8a - 9a$ This is the given algebraic expression.

  $= (8 - 9)a$ Apply the distributive property.

  $= -1a$ Subtract within parentheses: $8 - 9 = 8 + (-9) = -1$.

  $= -a$ Apply the multiplication property of -1.

d. $-3(2x - 5)$ This is the given algebraic expression.

  $= -3(2x) - (-3) \cdot (5)$ Apply the distributive property.

  $= -6x - (-15)$ Multiply.

  $= -6x + 15$ Subtraction is the addition of an additive inverse.

e. $-(3y - 8)$ This is the given algebraic expression.

  $= -3y + 8$ Remove parentheses by changing the sign of every term inside the parentheses. ■

✔ **CHECK POINT 6**   Simplify:

a. $-4(5x)$ **b.** $9x + x$ **c.** $13b - 14b$

d. $-7(3x - 4)$ **e.** $-(7y - 6)$.

Before turning to applications, let's try one additional example involving simplification.

Discover for Yourself

Verify each simplification in Example 6 by substituting 5 for the variable. The value of the given expression should be the same as the value of the simplified expression. Which expression is easier to evaluate?

EXAMPLE 7   Simplifying an Algebraic Expression

Simplify: $5(2y - 9) - (9y - 8)$.

Solution

$5(2y - 9) - (9y - 8)$	This is the given algebraic expression.
$= 5 \cdot 2y - 5 \cdot 9 - (9y - 8)$	Apply the distributive property in the first parentheses.
$= 10y - 45 - (9y - 8)$	Multiply.
$= 10y - 45 - 9y + 8$	Remove the second parentheses by changing the sign, of each term within parentheses.
$= (10y - 9y) + (-45 + 8)$	Group like terms.
$= 1y + (-37)$	Combine like terms. For the variable terms, $10y - 9y = 10y + (-9y) = \left[10 + (-9)\right]y = 1y$.
$= y + (-37)$	Use the multiplication property of 1: $1y = y$.
$= y - 37$	Express addition of an additive inverse as subtraction. ∎

✔ **CHECK POINT 7**   Simplify: $4(3y - 7) - (13y - 2)$.

A Summary of Operations with Real Numbers

Operations with real numbers are summarized in Table 1.5.

Table 1.5   Summary of Operations with Real Numbers

Signs of Numbers	Addition	Subtraction	Multiplication	Division
Both Numbers Are Positive Examples 8 and 2 2 and 8	Sum Is Always Positive $8 + 2 = 10$ $2 + 8 = 10$	Difference May Be Either Positive or Negative $8 - 2 = 6$ $2 - 8 = -6$	Product Is Always Positive $8 \cdot 2 = 16$ $2 \cdot 8 = 16$	Quotient Is Always Positive $8 \div 2 = 4$ $2 \div 8 = \frac{1}{4}$
One Number Is Positive and the Other Number Is Negative Examples 8 and -2 -8 and 2	Sum May Be Either Positive or Negative $8 + (-2) = 6$ $-8 + 2 = -6$	Difference May Be Either Positive or Negative $8 - (-2) = 10$ $-8 - 2 = -10$	Product Is Always Negative $8(-2) = -16$ $-8(2) = -16$	Quotient Is Always Negative $8 \div (-2) = -4$ $-8 \div 2 = -4$
Both Numbers Are Negative Examples -8 and -2 -2 and -8	Sum Is Always Negative $-8 + (-2) = -10$ $-2 + (-8) = -10$	Difference May Be Either Positive or Negative $-8 - (-2) = -6$ $-2 - (-8) = 6$	Product Is Always Positive $-8(-2) = 16$ $-2(-8) = 16$	Quotient Is Always Positive $-8 \div (-2) = 4$ $-2 \div (-8) = \frac{1}{4}$

7 Use algebraic expressions that model reality.

Applications

Algebraic expressions that model reality frequently contain division.

EXAMPLE 8   Average Cost of Producing a Wheelchair

A company that manufactures wheelchairs has monthly fixed costs of $500,000. The average cost per wheelchair for the company to manufacture x wheelchairs per month is modeled by the algebraic expression

$$\frac{400x + 500,000}{x}.$$

Find the average cost per wheelchair for the company to manufacture:

a. 10,000 wheelchairs per month.

b. 50,000 wheelchairs per month.

c. 100,000 wheelchairs per month.

What happens to the average cost per wheelchair as the production level increases?

Solution

a. We are interested in the average cost per wheelchair for the company if 10,000 wheelchairs are manufactured per month. Because x represents the number of wheelchairs manufactured per month, we substitute 10,000 for x in the given algebraic expression.

$$\frac{400x + 500,000}{x} = \frac{400(10,000) + 500,000}{10,000} = \frac{4,000,000 + 500,000}{10,000}$$

$$= \frac{4,500,000}{10,000} = 450$$

The average cost per wheelchair of producing 10,000 wheelchairs per month is $450.

b. Now, 50,000 wheelchairs are manufactured per month. We find the average cost per wheelchair by substituting 50,000 for x in the given algebraic expression.

$$\frac{400x + 500,000}{x} = \frac{400(50,000) + 500,000}{50,000} = \frac{20,000,000 + 500,000}{50,000}$$

$$= \frac{20,500,000}{50,000} = 410$$

The average cost per wheelchair of producing 50,000 wheelchairs per month is $410.

c. Finally, the production level has increased to 100,000 wheelchairs per month. We find the average cost per wheelchair for the company by substituting 100,000 for x in the given algebraic expression.

$$\frac{400x + 500,000}{x} = \frac{400(100,000) + 500,000}{100,000} = \frac{40,000,000 + 500,000}{100,000}$$

$$= \frac{40,500,000}{100,000} = 405$$

The average cost per wheelchair of producing 100,000 wheelchairs per month is $405.

As the production level increases, the average cost of producing each wheelchair decreases. This illustrates the difficulty with small businesses. It is nearly impossible to have competitively low prices when production levels are low. ■

The points in the rectangular coordinate system in Figure 1.18 show the relationship between production level and cost. The x-axis represents the number of wheelchairs produced per month. The y-axis represents the average cost per wheelchair for the company. The symbol $\frac{1}{7}$ on the y-axis indicates that there is a break in the values of y between 0 and 400. Thus, the values of y begin at 400.

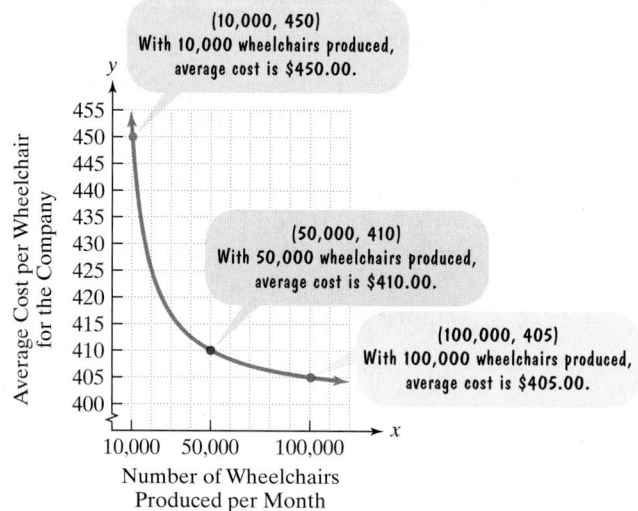

Figure 1.18 As production level increases, the average cost per wheelchair for the company decreases.

The three points with the voice balloons illustrate our computations in Example 8. The points in Figure 1.18 are falling from left to right. Can you see how this shows that the company's cost per wheelchair is decreasing as their production level increases?

✔ **CHECK POINT 8** A company that manufactures running shoes has weekly fixed costs of $300,000. The average cost per pair of running shoes for the company to manufacture x pairs per week is modeled by the algebraic expression

$$\frac{30x + 300,000}{x}.$$

Find the average cost per pair of running shoes for the company to manufacture:

    a. 1000 pairs per week.

    b. 10,000 pairs per week.

    c. 100,000 pairs per week.

EXERCISE SET 1.7

Practice Exercises

In Exercises 1–34, perform the indicated multiplication.

1. $6(-9)$ **2.** $5(-7)$

3. $(-7)(-3)$ **4.** $(-8)(-5)$

5. $(-2)(6)$ **6.** $(-3)(10)$

7. $(-13)(-1)$ **8.** $(-17)(-1)$

9. $0(-5)$ **10.** $0(-8)$

11. $\frac{1}{2}(-14)$ **12.** $\frac{1}{3}(-15)$

13. $\left(-\frac{3}{4}\right)(-20)$ **14.** $\left(-\frac{4}{5}\right)(-25)$

15. $-\frac{3}{5} \cdot \left(-\frac{4}{7}\right)$ **16.** $-\frac{5}{7} \cdot \left(-\frac{3}{8}\right)$

17. $-\frac{7}{9} \cdot \frac{2}{3}$ **18.** $-\frac{5}{11} \cdot \frac{2}{7}$

19. $3(-1.2)$ **20.** $4(-1.2)$

21. $-0.2(-0.6)$ **22.** $-0.3(-0.7)$

23. $(-5)(-2)(3)$ **24.** $(-6)(-3)(10)$

25. $(-4)(-3)(-1)(6)$ **26.** $(-2)(-7)(-1)(3)$

27. $-2(-3)(-4)(-1)$ **28.** $-3(-2)(-5)(-1)$

29. $(-3)(-3)(-3)$ **30.** $(-4)(-4)(-4)$

31. $5(-3)(-1)(2)(3)$ **32.** $2(-5)(-2)(3)(1)$

33. $(-8)(-4)(0)(-17)(-6)$ **34.** $(-9)(-12)(-18)(0)(-3)$

In Exercises 35–42, find the multiplicative inverse of each number.

35. 4

36. 3

37. $\dfrac{1}{5}$

38. $\dfrac{1}{7}$

39. -10

40. -12

41. $-\dfrac{2}{5}$

42. $-\dfrac{4}{9}$

In Exercises 43–46,

  a. *Rewrite the division as multiplication involving a multiplicative inverse.*

  b. *Use the multiplication from part (a) to find the given quotient.*

43. $-32 \div 4$

44. $-18 \div 6$

45. $\dfrac{-60}{-5}$

46. $\dfrac{-30}{-5}$

In Exercises 47–76, perform the indicated division or state that the expression is undefined.

47. $\dfrac{12}{-4}$

48. $\dfrac{40}{-5}$

49. $\dfrac{-21}{3}$

50. $\dfrac{-60}{6}$

51. $\dfrac{-90}{-3}$

52. $\dfrac{-66}{-6}$

53. $\dfrac{0}{-7}$

54. $\dfrac{0}{-8}$

55. $\dfrac{7}{0}$

56. $\dfrac{-8}{0}$

57. $-15 \div 3$

58. $-80 \div 8$

59. $120 \div (-10)$

60. $130 \div (-10)$

61. $(-180) \div (-30)$

62. $(-150) \div (-25)$

63. $0 \div (-4)$

64. $0 \div (-10)$

65. $-4 \div 0$

66. $-10 \div 0$

67. $\dfrac{-12.9}{3}$

68. $\dfrac{-21.6}{3}$

69. $-\dfrac{1}{2} \div \left(-\dfrac{3}{5}\right)$

70. $-\dfrac{1}{2} \div \left(-\dfrac{7}{9}\right)$

71. $-\dfrac{14}{9} \div \dfrac{7}{8}$

72. $-\dfrac{5}{16} \div \dfrac{25}{8}$

73. $\dfrac{1}{3} \div \left(-\dfrac{1}{3}\right)$

74. $\dfrac{1}{5} \div \left(-\dfrac{1}{5}\right)$

75. $6 \div \left(-\dfrac{2}{5}\right)$

76. $8 \div \left(-\dfrac{2}{9}\right)$

In Exercises 77–96, simplify each algebraic expression.

77. $-5(2x)$

78. $-9(3x)$

79. $-4\left(-\dfrac{3}{4}y\right)$

80. $-5\left(-\dfrac{3}{5}y\right)$

81. $8x + x$

82. $12x + x$

83. $-5x + x$

84. $-6x + x$

85. $6b - 7b$

86. $12b - 13b$

87. $-y + 4y$

88. $-y + 9y$

89. $-4(2x - 3)$

90. $-3(4x - 5)$

91. $-3(-2x + 4)$

92. $-4(-3x + 2)$

93. $-(2y - 5)$

94. $-(3y - 1)$

95. $4(2y - 3) - (7y + 2)$

96. $5(3y - 1) - (14y - 2)$

☆ **Application Exercises**

    The number of liposuctions in the United States, in thousands, can be approximated by the algebraic expression

$$30x + 113$$

where x is the number of years after 1996. Use this algebraic expression to solve Exercises 97–98.

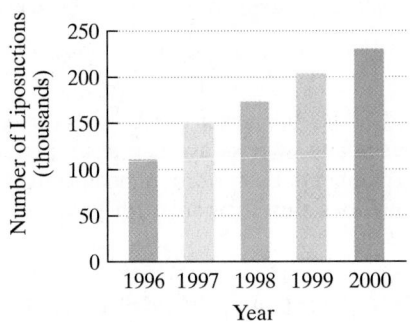

Liposuctions: U.S. Men and Women

Source: U.S. Department of Health and Human Services

97. How many liposuctions were there in the United States in 2000?

98. How many liposuctions were there in the United States in 1998?

In an experiment on memory, students in a language class are asked to memorize 40 vocabulary words in Latin, a language with which the students are not familiar. After studying the words for one day, students are tested each day after to see how many words they remember. The class average is taken and the results are graphed as points in the rectangular coordinate system. Use the points shown to solve Exercises 99–100.

**Average Number of Words
Remembered over Time**

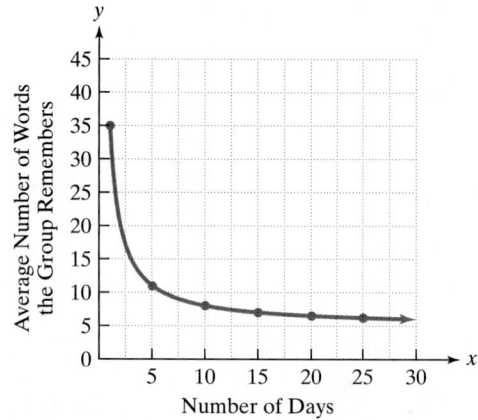

99. a. Find a reasonable estimate of the number of Latin words remembered after 5 days.
   b. The algebraic expression

$$\frac{5x + 30}{x}$$

   models the number of Latin words remembered by the students after x days. Use this expression to find the number of Latin words remembered after 5 days. How does this compare with your estimate from part (a)?

100. a. Find a reasonable estimate of the number of Latin words remembered after 15 days.
   b. The algebraic expression

$$\frac{5x + 30}{x}$$

   models the number of Latin words remembered by the students after x days. Use this expression to find the number of Latin words remembered after 15 days. How does this compare with your estimate from part (a)?

In Palo Alto, California, a government agency ordered computer-related companies to contribute to a pool of money to clean up underground water supplies. (The companies had stored toxic chemicals in leaking underground containers.) The algebraic expression

$$\frac{200x}{100 - x}$$

models the cost, in tens of thousands of dollars for removing x percent of the contaminants. Use this algebraic expression to solve Exercises 101–102.

101. a. Substitute 50 for x and find the cost, in tens of thousands of dollars, for removing 50% of the con- taminants.
   b. Find the cost, in tens of thousands of dollars, for removing 80% of the contaminants.
   c. Describe what is happening to the cost of the cleanup as the percentage of contaminant removed increases.

102. a. Substitute 60 for x and find the cost, in tens of thousands of dollars, for removing 60% of the contaminants.
   b. Find the cost, in tens of thousands of dollars, for re- moving 90% of the contaminants.
   c. Describe what is happening to the cost of the cleanup as the percentage of contaminant removed increases.

Writing in Mathematics

103. Explain how to multiply two real numbers. Provide ex- amples with your explanation.
104. Explain how to determine the sign of a product that involves more than two numbers.
105. Explain how to find the multiplicative inverse of a number.
106. Why is it that 0 has no multiplicative inverse?
107. Explain how to divide real numbers.
108. Why is division by zero undefined?
109. Explain how to simplify an algebraic expression in which a negative sign precedes parentheses.
110. A politician promises to do "whatever it takes" to seize "one hundred percent" of all illegal drugs that enter the country. Suppose that the cost, in millions of dollars, for seizing x percent of the illegal drugs entering the country is modeled by the algebraic expression

$$\frac{130x}{100 - x}.$$

According to this algebraic expression, can the politician keep his or her promise? Explain your answer.

Critical Thinking Exercises

111. Which one of the following statements is true?
   a. Multiplying a negative number by a nonnegative number will always give a negative number.
   b. The product of two negative numbers is always a positive number.
   c. The product of -3 and 4 is 12.
   d. The product of real numbers a and b is not always equal to the product of real numbers b and a.

112. Which one of the following statements is true?
   a. The product of two negative numbers is sometimes a negative number.
   b. Both the addition and the multiplication of two negative numbers result in a positive number.
   c. $\left(-\frac{1}{2}\right)\left(-\frac{1}{2}\right) = \frac{1}{4}$
   d. Reversing the order of the two factors in a product results in a different answer.

In Exercises 113–114, write an algebraic expression for the given English phrase.

113. The value, in cents, of x nickels
114. The monthly salary, in dollars, for a person earning x dollars per year

Technology Exercises

115. Use a calculator to verify any five of the products that you found in Exercises 1–34.

116. Use a calculator to verify any five of the quotients that you found in Exercises 47–76.

117. Simplify using a calculator:

$$0.3(4.7x - 5.9) - 0.07(3.8x - 61).$$

118. Use your calculator to attempt to find the quotient of -3 and 0. Describe what happens. Does the same thing occur when finding the quotient of 0 and -3? Explain the difference. Finally, what happens when you enter the quotient of 0 and itself?

Review Exercises

In Exercises 119–121, perform the indicated operation.

119. $-6 + (-3)$ (Section 1.5; Example 3)
120. $-6 - (-3)$ (Section 1.6; Example 1)
121. $-6 \div (-3)$ (Section 1.7; Example 4)

▶ **SECTION 1.8**   *Exponents, Order of Operations, and Mathematical Models*

Objectives

1 Evaluate exponential expressions.
2 Simplify algebraic expressions with exponents.
3 Use the order of operations agreement.
4 Evaluate formulas.

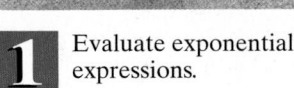

SSM   PH Tutor   CD-   Video
        Center   ROM

It's been another one of those days! Traffic is really backed up on the highway. Finally, you see the source of the traffic jam—a minor fender-bender. Still stuck in traffic, you notice that the driver appears to be quite young. This might seem like a strange observation. After all, what does a driver's age have to do with his or her chance of getting into an accident? In this section, we see how algebra describes your world, including a relationship between age and numbers of car accidents.

1 Evaluate exponential expressions.

Powers of Ten

$$10 = 10^1$$
$$100 = 10^2$$
$$1000 = 10^3$$
$$10,000 = 10^4$$
$$100,000 = 10^5$$
$$1,000,000 = 10^6 \quad \text{million}$$
$$10,000,000 = 10^7$$
$$100,000,000 = 10^8$$
$$1,000,000,000 = 10^9 \quad \text{billion}$$

Natural Number Exponents

Although people do a great deal of talking, the total output since the beginning of gabble to the present day, including all baby talk, love songs, and congressional debates, only amounts to about 10 million billion words. This can be expressed as 16 factors of 10, or 10^{16} words.

Exponents such as 2, 3, 4, and so on are used to indicate repeated multiplication. For example,

$$10^2 = 10 \cdot 10 = 100,$$
$$10^3 = 10 \cdot 10 \cdot 10 = 1000, \quad 10^4 = 10 \cdot 10 \cdot 10 \cdot 10 = 10,000.$$

The 10 that is repeated when multiplying is called the **base**. The small numbers above and to the right of the base are called **exponents**. The exponent tells the

Using Technology

You can use a calculator to evaluate exponential expressions. For example, to evaluate 5^3, press the following keys:

Scientific Calculator

$5 \boxed{x^y} 3 \boxed{=}$

Graphing Calculator

$5 \boxed{\wedge} 3 \boxed{\text{ENTER}}$.

Although calculators have special keys to evaluate powers of ten and squaring bases, you can always use one of the sequences shown here.

number of times the base is to be used when multiplying. In 10^3, the base is 10 and the exponent is 3.

Any number with an exponent of 1 is the number itself. Thus, $10^1 = 10$.

Multiplications that are expressed in exponential notation are read as follows:

10^1: "ten to the first power"

10^2: "ten to the second power" or "ten squared"

10^3: "ten to the third power" or "ten cubed"

10^4: "ten to the fourth power"

10^5: "ten to the fifth power"

etc.

Any real number can be used as the base. Thus,

$$7^2 = 7 \cdot 7 = 49 \quad \text{and} \quad (-3)^4 = (-3)(-3)(-3)(-3) = 81.$$

The bases are 7 and -3, respectively. Do not confuse $(-3)^4$ and -3^4.

$$-3^4 = -3 \cdot 3 \cdot 3 \cdot 3 = -81$$

> The negative is not taken to the power because it is not inside parentheses.

An exponent applies only to a base. A negative sign is not part of a base unless it appears in parentheses.

EXAMPLE 1 Evaluating Exponential Expressions

Evaluate:

  a. 4^2 **b.** $(-5)^3$ **c.** $(-2)^4$ **d.** -2^4.

Solution

> Exponent is 2.

a. $4^2 = 4 \cdot 4 = 16$ The exponent indicates that the base is used as a factor two times.

> Base is 4.

We read $4^2 = 16$ as "4 to the second power is 16" or "4 squared is 16."

> Exponent is 3.

b. $(-5)^3 = (-5)(-5)(-5)$ The exponent indicates that the base is used as a factor three times.

> Base is -5.

$$= -125$$ An odd number of negative factors yields a negative product.

We read $(-5)^3 = -125$ as "the number negative 5 to the third power is negative 125" or "negative 5 cubed is negative 125."

> Exponent is 4.

c. $(-2)^4 = (-2)(-2)(-2)(-2)$ The exponent indicates the base is used as a factor four times.

> Base is -2.

$$= 16$$ An even number of negative factors yields a positive product.

We read $(-2)^4 = 16$ as "the number negative 2 to the fourth power is 16."

> Exponent is 4.

d. $-2^4 = -2 \cdot 2 \cdot 2 \cdot 2$ The negative is not inside parentheses and is not taken to the fourth power.

> Base is 2.

$$= -16$$ Multiply the twos and copy the negative.

We read $-2^4 = -16$ as "the negative of 2 raised to the fourth power is negative 16" or "the opposite, or additive inverse, of 2 raised to the fourth power is negative 16." ∎

✔ **CHECK POINT 1** Evaluate:

    a. 6^2 **b.** $(-4)^3$ **c.** $(-1)^4$ **d.** -1^4.

The formal algebraic definition of a natural number exponent summarizes our discussion.

Definition of a Natural Number Exponent

If b is a real number and n is a natural number,

> Exponent

$$b^n = \underbrace{b \cdot b \cdot b \cdot \cdots \cdot b}.$$

> Base b appears as a factor n times.

b^n is read "the nth power of b" or "b to the nth power." Thus, the nth power of b is defined as the product of n factors of b. The expression b^n is called an **exponential expression**.

Furthermore, $b^1 = b$.

---

② Simplify algebraic expressions with exponents.

Exponents and Algebraic Expressions

The distributive property can be used to simplify certain algebraic expressions that contain exponents. For example, we can use the distributive property to combine like terms in the algebraic expression $4x^2 + 6x^2$:

$$4x^2 + 6x^2 \qquad = \qquad (4 + 6)x^2 = 10x^2.$$

> First term with variable factor x^2 Second term with variable factor x^2 The common variable factor is x^2.

EXAMPLE 2   Simplifying Algebraic Expressions

Simplify, if possible:

    a. $7x^3 + 2x^3$ **b.** $5x^2 + x^2$ **c.** $3x^2 + 4x^3$.

Solution

    a. $7x^3 + 2x^3$ There are two like terms with the same variable factor, namely x^3.

        $= (7 + 2)x^3$ Apply the distributive property.

        $= 9x^3$ Add within parentheses.

Study Tip

When adding algebraic expressions, if you have like terms you add only the numerical coefficients—not the exponents. **Exponents are never added when the operation is addition.** Avoid these common errors.

- $7x^3 + 2x^3 = 9x^6$
- $5x^2 + x^2 = 6x^4$
- $3x^2 + 4x^3 = 7x^5$

INCORRECT!

Show that all three results are incorrect by substituting 2 for x. You should get a different number on both sides of the equal sign.

b. $5x^2 + x^2$ There are two like terms with the same variable factor, namely x^2.

$\quad = 5x^2 + 1x^2$ Use the multiplication property of 1.

$\quad = (5 + 1)x^2$ Apply the distributive property.

$\quad = 6x^2$ Add within parentheses.

c. $3x^2 + 4x^3$ cannot be simplified. The terms $3x^2$ and $4x^3$ are not like terms because they have different variable factors, namely x^2 and x^3. ■

✔ **CHECK POINT 2**   Simplify, if possible:

   a. $16x^2 + 5x^2$ **b.** $7x^3 + x^3$ **c.** $10x^2 + 8x^3$.

3   Use the order of operations agreement.

Order of Operations

Suppose that you want to find the value of $3 + 7 \cdot 5$. Which procedure shown is correct?

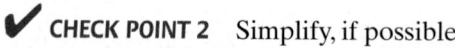

$$3 + 7 \cdot 5 = 3 + 35 = 38 \quad \text{or} \quad 3 + 7 \cdot 5 = 10 \cdot 5 = 50$$

If you know the answer, you probably know certain rules, called the **order of operations**, to make sure that there is only one correct answer. One of these rules states that if a problem contains no parentheses or other grouping symbols, perform multiplication before addition. Thus, the procedure on the left is correct because the multiplication of 7 and 5 is done first. Then the addition is performed. The correct answer is 38.

     Some problems contain grouping symbols, such as parentheses, ( ), brackets, [ ], braces, { } absolute value symbols, | |, or fractions bars. These grouping symbols tell us what to do first. Here are two examples:

• $(3 + 7) \cdot 5 = 10 \cdot 5 = 50$

> First, perform operations in grouping symbols.

• $8|6 - 16| = 8|-10| = 8 \cdot 10 = 80.$

Here are the rules for determining the order in which operations should be performed.

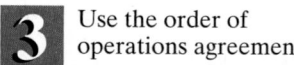
Order of Operations

1. Perform all operations within grouping symbols.
2. Evaluate all exponential expressions.
3. Do all multiplications and divisions in the order in which they occur, working from left to right.
4. Finally, do all additions and subtractions in the order in which they occur, working from left to right.

In the third step, be sure to do all multiplications and divisions *as they occur* from left to right. For example,

$$8 \div 4 \cdot 2 = 2 \cdot 2 = 4$$

     Do the division first because it occurs first.

$$8 \cdot 4 \div 2 = 32 \div 2 = 16.$$

     Do the multiplication first because it occurs first.

EXAMPLE 3   Using the Order of Operations

Simplify: $18 + 2 \cdot 3 - 10$.

Solution   There are no grouping symbols or exponential expressions. In cases like this, we multiply and divide before adding and subtracting.

$$18 + 2 \cdot 3 - 10 = 18 + 6 - 10 \qquad \text{Multiply: } 2 \cdot 3 = 6.$$
$$= 24 - 10 \qquad \text{Add and subtract from left to right: } 18 + 6 = 24.$$
$$= 14 \qquad \text{Subtract: } 24 - 10 = 14. \quad \blacksquare$$

✔ **CHECK POINT 3**   Simplify: $20 + 4 \cdot 3 - 17$.

EXAMPLE 4   Using the Order of Operations

Simplify: $6^2 - 24 \div 2^2 \cdot 3 - 1$.

Solution   There are no grouping symbols. Thus, we begin by evaluating exponential expressions. Then we multiply or divide. Finally, we add or subtract.

$$6^2 - 24 \div 2^2 \cdot 3 - 1$$
$$= 36 - 24 \div 4 \cdot 3 - 1 \qquad \text{Evaluate exponential expressions:}$$
$$6^2 = 6 \cdot 6 = 36 \text{ and}$$
$$2^2 = 2 \cdot 2 = 4.$$
$$= 36 - 6 \cdot 3 - 1 \qquad \text{Perform the multiplications and divisions from left to right. Start with } 24 \div 4 = 6.$$
$$= 36 - 18 - 1 \qquad \text{Now do the multiplication: } 6 \cdot 3 = 18.$$
$$= 18 - 1 \qquad \text{Finally, perform the subtraction from left to right: } 36 - 18 = 18.$$
$$= 17 \qquad \text{Complete the subtraction: } 18 - 1 = 17. \quad \blacksquare$$

✔ **CHECK POINT 4**   Simplify: $7^2 - 48 \div 4^2 \cdot 5 - 2$.

EXAMPLE 5   Using the Order of Operations

Simplify:
   a. $(2 \cdot 5)^2$
   b. $2 \cdot 5^2$.

Solution

   a. Because $(2 \cdot 5)^2$ contains grouping symbols, namely parentheses, we perform the operation within parentheses first.

$$(2 \cdot 5)^2 = 10^2 \qquad \text{Multiply within parentheses: } 2 \cdot 5 = 10.$$
$$= 100 \qquad \text{Evaluate the exponential expression: } 10^2 = 10 \cdot 10 = 100.$$

b. Because $2 \cdot 5^2$ does not contain grouping symbols, we begin by evaluating the exponential expression.

$$2 \cdot 5^2 = 2 \cdot 25 \qquad \text{Evaluate the exponential}$$
$$\text{expression: } 5^2 = 5 \cdot 5 = 25.$$

$$= 50 \qquad \text{Now do the multiplication:}$$
$$2 \cdot 25 = 50. \qquad ■$$

✔ **CHECK POINT 5** Simplify:

    a. $(3 \cdot 2)^2$ **b.** $3 \cdot 2^2$.

EXAMPLE 6   Using the Order of Operations

Simplify: $(-6)^2 - (5 - 7)^2(-3)$.

Solution Because grouping symbols appear, we perform the operation within parentheses first.

$$(-6)^2 - (5 - 7)^2(-3)$$
$$= (-6)^2 - (-2)^2(-3) \qquad \text{Work inside parentheses first:}$$
$$5 - 7 = 5 + (-7) = -2.$$

$$= 36 - 4(-3) \qquad \text{Evaluate exponential expressions:}$$
$$(-6)^2 = (-6)(-6) = 36$$
$$\text{and } (-2)^2 = (-2)(-2) = 4.$$

$$= 36 - (-12) \qquad \text{Multiply: } 4(-3) = -12.$$
$$= 48 \qquad \text{Subtract: } 36 - (-12) = 36 + 12 = 48.$$

$$■$$

✔ **CHECK POINT 6** Simplify: $(-8)^2 - (10 - 13)^2(-2)$.

Some expressions contain many grouping symbols. An example of such an expression is $2[5(4 - 7) + 9]$. The grouping symbols are the parentheses and the brackets.

> The parentheses, the innermost grouping symbols, group $4 - 7$.

$$2\big[5(4 - 7) + 9\big]$$

> The brackets, the outermost grouping symbols, group $5(4 - 7) + 9$.

When combinations of grouping symbols appear, perform operations within the innermost grouping symbols first. Then work to the outside, performing operations within the outermost grouping symbols.

EXAMPLE 7   Using the Order of Operations

Simplify: $2[5(4 - 7) + 9]$.

Solution

$$2\big[5(4 - 7) + 9\big]$$
$$= 2\big[5(-3) + 9\big] \qquad \text{Work inside parentheses first:}$$
$$4 - 7 = 4 + (-7) = -3.$$

$$= 2[-15 + 9] \qquad$$ Work inside brackets and multiply: $5(-3) = -15.$

$$= 2[-6] \qquad$$ Add inside brackets: $-15 + 9 = -6.$ The resulting problem can also be expressed as $2(-6).$

$$= -12 \qquad$$ Multiply: $2[-6] = -12.$ ∎

Parentheses can be used for both innermost and outermost grouping symbols. For example, the expression $2[5(4 - 7) + 9]$ can also be written $2(5(4 - 7) + 9)$. However, too many parentheses can be confusing. The use of both parentheses and brackets makes it easier to identify inner and outer groupings.

✔ **CHECK POINT 7** Simplify: $4[3(6 - 11) + 5].$

EXAMPLE 8 Using the Order of Operations

Simplify: $18 \div 6 + 4[5 + 2(8 - 10)^3].$

Solution

$$18 \div 6 + 4[5 + 2(8 - 10)^3]$$

$$= 18 \div 6 + 4[5 + 2(-2)^3] \qquad$$ Work inside parentheses first: $8 - 10 = 8 + (-10) = -2.$

$$= 18 \div 6 + 4[5 + 2(-8)] \qquad$$ Work inside brackets and evaluate the exponential expression: $(-2)^3 = (-2)(-2)(-2) = -8.$

$$= 18 \div 6 + 4[5 + (-16)] \qquad$$ Work inside brackets and multiply: $2(-8) = -16.$

$$= 18 \div 6 + 4[-11] \qquad$$ Work inside brackets and add: $5 + (-16) = -11.$

$$= 3 + 4[-11] \qquad$$ Perform the multiplications and divisions from left to right. Start with $18 \div 6 = 3.$

$$= 3 + (-44) \qquad$$ Now do the multiplication: $4(-11) = -44.$

$$= -41 \qquad$$ Finally, perform the addition: $3 + (-44) = -41.$ ∎

✔ **CHECK POINT 8** Simplify: $25 \div 5 + 3[4 + 2(7 - 9)^3].$

Fraction bars are grouping symbols that separate expressions into two parts, the numerator and the denominator. Consider, for example,

The numerator is one part of the expression.

Fraction bar is the grouping symbol.

$$\frac{2(3 - 12) + 6 \cdot 4}{2^4 + 1}$$

The denominator is the other part of the expression.

We can use brackets instead of the fraction bar. An equivalent expression is

$$[2(3 - 12) + 6 \cdot 4] \div [2^4 + 1].$$

The grouping at the bottom of page 81 suggests a method for simplifying expressions with fraction bars as grouping symbols:

- Simplify the numerator.
- Simplify the denominator.
- If possible, simplify the fraction.

EXAMPLE 9   Using the Order of Operations

Simplify: $\dfrac{2(3 - 12) + 6 \cdot 4}{2^4 + 1}$.

Solution

$$\frac{2(3 - 12) + 6 \cdot 4}{2^4 + 1}$$

$$= \frac{2(-9) + 6 \cdot 4}{16 + 1}$$

Work inside parentheses in the numerator: $3 - 12 = 3 + (-12) = -9$. Evaluate the exponential expression in the denominator: $2^4 = 2 \cdot 2 \cdot 2 \cdot 2 = 16$.

$$= \frac{-18 + 24}{16 + 1}$$

Multiply in the numerator: $2(-9) = -18$ and $6 \cdot 4 = 24$.

$$= \frac{6}{17}$$

Perform the addition in the numerator and the denominator. ∎

✔ **CHECK POINT 9**   Simplify: $\dfrac{5(4 - 9) + 10 \cdot 3}{2^3 - 1}$.

EXAMPLE 10   Using the Order of Operations

Evaluate: $-x^2 - 7x$ for $x = -2$.

Solution   We begin by substituting -2 for each occurrence of x in the algebraic expression. Then we use the order of operations to evaluate the expression.

$$-x^2 - 7x$$

Replace x by −2.

$$= -(-2)^2 - 7(-2)$$

$$= -4 - 7(-2)$$

Evaluate the exponential expression: $(-2)^2 = (-2)(-2) = 4$.

$$= -4 - (-14)$$

Multiply: $7(-2) = -14$.

$$= 10$$

Subtract:
$-4 - (-14) = -4 + 14 = 10$. ∎

✔ **CHECK POINT 10**   Evaluate: $-x^2 - 4x$ for $x = -5$.

Some algebraic expressions contain two sets of grouping symbols. Using the order of operations, grouping symbols are removed from innermost (parentheses) to outermost (brackets).

EXAMPLE 11   Simplifying an Algebraic Expression

Simplify: $18x^2 + 4 - [6(x^2 - 2) + 5]$.

Solution

$18x^2 + 4 - [6(x^2 - 2) + 5]$

$= 18x^2 + 4 - [6x^2 - 12 + 5]$ Use the distributive property to remove parentheses:
$6(x^2 - 2) = 6x^2 - 6 \cdot 2 = 6x^2 - 12$.

$= 18x^2 + 4 - [6x^2 - 7]$ Add inside brackets: $-12 + 5 = -7$.

$= 18x^2 + 4 - 6x^2 + 7$ Remove brackets by changing the sign of each term within brackets.

$= (18x^2 - 6x^2) + 4 + 7$ Group like terms.

$= 12x^2 + 11$ Combine like terms.   ■

✔ **CHECK POINT 11**   Simplify: $14x^2 + 5 - [7(x^2 - 2) + 4]$.

4 Evaluate formulas.

Applications: Formulas and Mathematical Models

One aim of algebra is to provide a compact, symbolic description of the world. These descriptions involve the use of **formulas**, statements of equality expressing a relationship among two or more variables. For example, one variety of crickets chirp faster as the temperature rises. You can calculate the temperature by counting the number of times a cricket chirps per minute and applying the following formula:

$$T = 0.3n + 40.$$

In the formula, T is the temperature in degrees Fahrenheit and n is the number of cricket chirps per minute. We can use this formula to determine the temperature if you are sitting on your porch and count 80 chirps per minute. Here is how to do so.

$T = 0.3n + 40$ This is the given formula.

$T = 0.3(80) + 40$ Substitute 80 for n.

$T = 24 + 40$ Multiply: $0.3(80) = 24$.

$T = 64$ Add.

When there are 80 cricket chirps per minute, the temperature is 64 degrees.

In developing formulas that describe, or model, variables such as the number of cricket chirps and temperature, mathematicians strive for both accuracy and simplicity. The formula $T = 0.3n + 40$ is relatively simple to use. However, you should not get upset if you count 80 cricket chirps per minute and the actual temperature is 62 degrees, rather than 64 degrees, as predicted by the formula. Many formulas give an approximate, rather than an exact, relationship between variables. Such formulas are called **mathematical models**.

Formulas express relationships between quantities. People use formulas in almost all academic disciplines, as well as in everyday life. For example, there are mathematical models that estimate what height, H, a child who is x years old and is h inches tall will attain as an adult. The formulas for girls and boys are

Girls: $H = \dfrac{h}{0.00028x^3 - 0.0071x^2 + 0.0926x + 0.3524}$

Boys: $H = \dfrac{h}{0.00011x^3 - 0.0032x^2 + 0.0604x + 0.3796}.$

In both formulas, H represents adult height, in inches.

EXAMPLE 12   Car Accidents and Age

The mathematical model

$$N = 0.4x^2 - 36x + 1000$$

approximates the number of accidents, N, per 50 million miles driven, for drivers who are x years old. The formula applies to drivers ages 16 to 74, inclusive. How many accidents, per 50 million miles driven, are there for 20-year-old drivers?

Solution   In the mathematical model, x represents the age of the driver. We are interested in 20-year-old drivers. Thus, we substitute 20 for each occurrence of x. Then we use the order of operations to find N, the number of accidents per 50 million miles driven.

$N = 0.4x^2 - 36x + 1000$	This is the given mathematical model.
$N = 0.4(20)^2 - 36(20) + 1000$	Replace each occurrence of x by 20.
$N = 0.4(400) - 36(20) + 1000$	Evaluate the exponential expression: $(20)^2 = (20)(20) = 400.$
$N = 160 - 720 + 1000$	Multiply from left to right: $0.4(400) = 160$ and $36(20) = 720.$
$N = -560 + 1000$	Perform additions and subtractions from left to right. Subtract: $160 - 720 = 160 + (-720) = -560.$
$N = 440$	Add: $-560 + 1000 = 440.$

Study Tip

To find

$$160 - 720 + 1000$$
$$= 160 + (-720) + 1000,$$

you can also first add the positive numbers

$$160 + 1000 = 1160$$

and then add the negative number to that result:

$$1160 + (-720) = 440.$$

Thus, 20-year-old drivers have 440 accidents per 50 million miles driven.   ■

How do the number of accidents for 20-year-old drivers compare to, say, the number for 45-year-old drivers? The answer is given by the points in the rectangular coordinate system in Figure 1.19. The x-coordinate of each point represents the driver's age. The y-coordinate represents the number of accidents per 50 million miles driven.

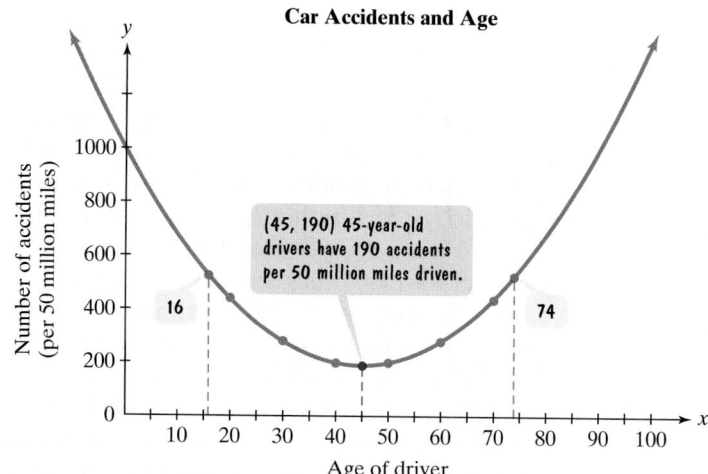

Figure 1.19

Can you see that the point (45,190) is lower than any of the other points? In practical terms, this indicates that 45-year-olds have the least number of car accidents, 190 per 50 million miles driven. Drivers both younger and older than 45 have more accidents per 50 million miles driven.

✔ **CHECK POINT 12**   Use the mathematical model described in Example 12, $N = 0.4x^2 - 36x + 1000$, to answer this question: How many accidents, per 50 million miles driven, are there for 40-year-old drivers?

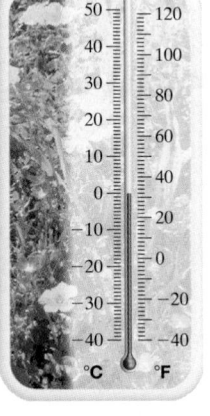

Figure 1.20 The Celsius scale is on the left and the Fahrenheit scale is on the right.

Some formulas give an exact, rather than an approximate, relationship between variables. For example, Figure 1.20 shows temperatures on the Celsius scale and the Fahrenheit scale. The formula

$$C = \frac{5}{9}(F - 32)$$

expresses an exact relationship between Fahrenheit temperature, F, and Celsius temperature, C.

The Formula	What the Formula Tells Us
$C = \dfrac{5}{9}(F - 32)$	If 32 is subtracted from the Fahrenheit temperature, $F - 32$, and this difference is multiplied by $\frac{5}{9}$, the resulting product, $\frac{5}{9}(F - 32)$, gives the Celsius temperature.

EXAMPLE 13   Converting from Fahrenheit to Celsius

The temperature on a warm spring day is 77°F. Use the formula $C = \frac{5}{9}(F - 32)$ to find the equivalent temperature on the Celsius scale.

Solution   Because the temperature is 77°F, we substitute 77 for F in the given formula. Then we use the order of operations to find the value of C.

$$C = \frac{5}{9}(F - 32) \qquad \text{This is the given formula.}$$

$$C = \frac{5}{9}(77 - 32) \qquad \text{Replace F by 77.}$$

$$C = \frac{5}{9}(45) \qquad \begin{array}{l}\text{Work inside parentheses}\\ \text{first: } 77 - 32 = 45.\end{array}$$

$$C = 25 \qquad \text{Multiply: } \frac{5}{9}(45) = \frac{5}{\cancel{9}_1} \cdot \frac{\overset{5}{\cancel{45}}}{1} = \frac{25}{1} = 25.$$

Thus, 77°F is equivalent to 25°C.   ∎

✔ **CHECK POINT 13**   The temperature on a warm summer day is 86°F. Use the formula $C = \frac{5}{9}(F - 32)$ to find the equivalent temperature on the Celsius scale.

EXERCISE SET 1.8

Practice Exercises

In Exercises 1–14, evaluate each exponential expression.

1. 9^2
2. 3^2
3. 4^3
4. 6^3
5. $(-4)^2$
6. $(-10)^2$
7. $(-4)^3$
8. $(-10)^3$
9. $(-5)^4$
10. $(-1)^6$
11. -5^4
12. -1^6
13. -10^2
14. -8^2

In Exercises 15–28, simplify each algebraic expression, or explain why the expression cannot be simplified.

15. $6x^2 + 11x^2$
16. $5x^2 + 17x^2$
17. $9x^3 + 4x^3$
18. $13x^3 + 7x^3$
19. $7x^4 + x^4$
20. $13x^4 + x^4$
21. $16x^2 - 17x^2$
22. $19x^2 - 20x^2$
23. $17x^3 - 16x^3$
24. $20x^3 - 19x^3$
25. $2x^2 + 2x^3$
26. $3x^2 + 3x^3$
27. $6x^2 - 6x^2$
28. $7x^2 - x^2$

In Exercises 29–66, use the order of operations to simplify each expression.

29. $7 + 6 \cdot 3$
30. $3 + 4 \cdot 5$
31. $45 \div 5 \cdot 3$
32. $40 \div 4 \cdot 2$
33. $6 \cdot 8 \div 4$
34. $8 \cdot 6 \div 2$
35. $14 - 2 \cdot 6 + 3$
36. $36 - 12 \div 4 + 2$
37. $8^2 - 16 \div 2^2 \cdot 4 - 3$
38. $10^2 - 100 \div 5^2 \cdot 2 - 1$
39. $3(-2)^2 - 4(-3)^2$
40. $5(-3)^2 - 2(-4)^2$
41. $(4 \cdot 5)^2 - 4 \cdot 5^2$
42. $(3 \cdot 5)^2 - 3 \cdot 5^2$
43. $(2 - 6)^2 - (3 - 7)^2$
44. $(4 - 6)^2 - (5 - 9)^2$
45. $6(3 - 5)^3 - 2(1 - 3)^3$
46. $-3(-6 + 8)^3 - 5(-3 + 5)^3$
47. $[2(6 - 2)]^2$
48. $[3(4 - 6)]^3$
49. $2[5 + 2(9 - 4)]$
50. $3[4 + 3(10 - 8)]$
51. $[7 + 3(2^3 - 1)] \div 21$
52. $[11 - 4(2 - 3^3)] \div 37$
53. $\dfrac{10 + 8}{5^2 - 4^2}$
54. $\dfrac{6^2 - 4^2}{2 - (-8)}$
55. $\dfrac{37 + 15 \div (-3)}{2^4}$
56. $\dfrac{22 + 20 \div (-5)}{3^2}$
57. $\dfrac{(-11)(-4) + 2(-7)}{7 - (-3)}$
58. $\dfrac{-5(7 - 2) - 3(4 - 7)}{-13 - (-5)}$
59. $4|10 - (8 - 20)|$
60. $6|7 - 4 \cdot 3|$

61. $8(-10) + |4(-5)|$
62. $4(-15) + |3(-10)|$
63. $-2^2 + 4[16 \div (3 - 5)]$
64. $-3^2 + 2[20 \div (7 - 11)]$
65. $24 \div \dfrac{3^2}{8 - 5} - (-6)$
66. $30 \div \dfrac{5^2}{7 - 12} - (-9)$

In Exercises 67–74, evaluate each algebraic expression for the given value of the variable.

67. $x^2 + 5x$; $x = 3$
68. $x^2 - 2x$; $x = 6$
69. $3x^2 - 8x$; $x = -2$
70. $4x^2 - 2x$; $x = -3$
71. $-x^2 - 10x$; $x = -1$
72. $-x^2 - 14x$; $x = -1$
73. $\dfrac{6y - 4y^2}{y^2 - 15}$; $y = 5$
74. $\dfrac{3y - 2y^2}{y(y - 2)}$; $y = 5$

In Exercises 75–82, simplify each algebraic expression by removing parentheses and brackets.

75. $3[5(x - 2) + 1]$
76. $4[6(x - 3) + 1]$
77. $3[6 - (y + 1)]$
78. $5[2 - (y + 3)]$
79. $7 - 4[3 - (4y - 5)]$
80. $6 - 5[8 - (2y - 4)]$
81. $2(3x^2 - 5) - [4(2x^2 - 1) + 3]$
82. $4(6x^2 - 3) - [2(5x^2 - 1) + 1]$

Application Exercises

On the average, infant girls weigh 7 pounds at birth and gain 1.5 pounds for each month for the first six months. The formula

$$W = 1.5x + 7$$

models a baby girl's weight, W, in pounds, after x months, where x is less than or equal to 6. Use the formula to solve Exercises 83–84.

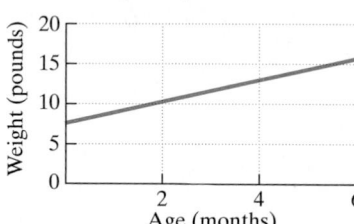

Average Weight for Infant Girls

83. What does an infant girl weigh after four months? Identify your computation as an appropriate point on the line graph.

84. What does an infant girl weigh after six months? Identify your computation as an appropriate point on the line graph.

Medical researchers have found that the desirable heart rate, R, in beats per minute, for beneficial exercises is approximated by the mathematical models

$$R = 165 - 0.75A \qquad \text{for men}$$
$$R = 143 - 0.65A \qquad \text{for women}$$

where A is the person's age. Use these mathematical models to solve Exercises 85–86.

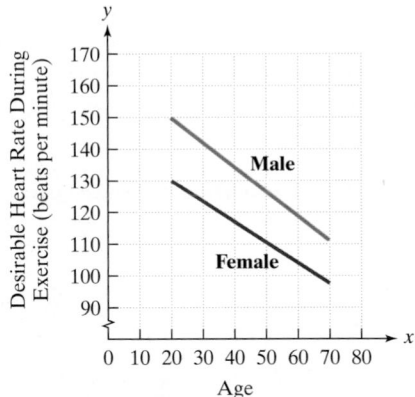

85. What is the desirable heart rate during exercise for a 40-year-old man? Identify your computation as an appropriate point in the rectangular coordinate system.

86. What is the desirable heart rate during exercise for a 40-year-old woman? Identify your computation as an appropriate point in the rectangular coordinate system.

The Internet is the world's largest communications network. Although millions of computer owners access the Internet using phone lines, the process of downloading files can be slow and tedious. Cable TV modems dramatically speed up this process. By contrast to phone modems, which transmit 56,000 bits per second, cable modems are capable of transmitting 10 million bits per second. The graph shows the millions of Internet users in the United States with this new technology. The data can be modeled by the formula $N = 0.4x^2 + 0.5$, where N represents the millions of people in the United States using cable modems x years after 1996. Use this formula to solve Exercises 87–88.

Number of People in the United States Using Cable TV Modems

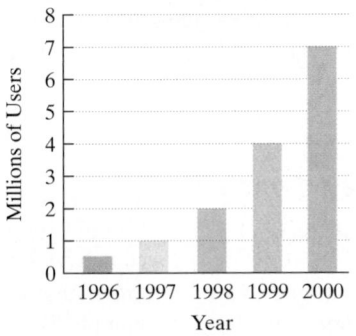

Source: The New York Times

87. According to the formula, how many millions of Americans used cable TV modems in 2000? How well does the formula describe the actual number of users for that year shown by the bar graph?

88. According to the formula, how many millions of Americans used cable TV modems in 1999? How well does the formula describe the actual number of users for that year shown by the bar graph?

The bar graph shows the cost of Medicare, in billions of dollars, projected through 2005. The data can be modeled by the formula

$$N = 1.2x^2 + 15.2x + 181.4,$$

where N represents Medicare spending, in billions of dollars, x years after 1995. Use this formula to solve Exercises 89–90.

Medicare Spending

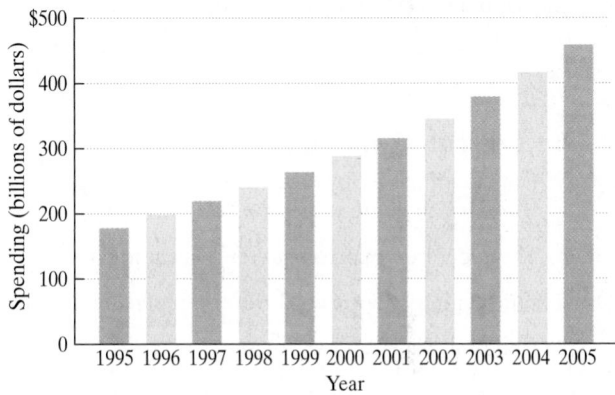

Source: Congressional Budget Office

89. According to the formula, what is the cost of Medicare, in billions of dollars, in 2000? How well does the formula describe the cost for that year shown by the bar graph?

90. According to the formula, what is the cost of Medicare, in billions of dollars, in 2005? How well does the formula describe the cost for that year shown by the bar graph?

The formula

$$C = \frac{5}{9}(F - 32)$$

expresses the relationship between Fahrenheit temperature, F, and Celsius temperature, C. In Exercises 91–94, use the formula to convert the given Fahrenheit temperature to its equivalent temperature on the Celsius scale.

91. 68°F

92. 41°F

93. −22°F

94. −31°F

Writing in Mathematics

95. Describe what it means to raise a number to a power. In your description, include a discussion of the difference between -5^2 and $(-5)^2$.

96. Explain how to simplify $4x^2 + 6x^2$. Why is the sum not equal to $10x^4$?

97. Why is the order of operations agreement needed?

98. What is a formula?

99. The formula $F = \frac{9}{5}C + 32$ expresses the relationship between Celsius temperature, C, and Fahrenheit temperature, F. You'll be leaving the cold of winter for a Hawaii vacation. CNN International reports a temperature in Hawaii of 30°C. Should you cancel the trip? Use the formula to explain your answer.

Critical Thinking Exercises

100. Which one of the following is true?
   a. If x is -3, then the value of $-3x - 9$ is -18.
   b. The algebraic expression $\dfrac{6x + 6}{x + 1}$ cannot have the same value when two different replacements are made for x such as $x = -3$ and $x = 2$.
   c. A miniature version of a space shuttle is an example of a mathematical model.
   d. The value of $\dfrac{|3 - 7| - 2^3}{(-2)(-3)}$ is the fraction that results when $\dfrac{1}{3}$ is subtracted from $-\dfrac{1}{3}$.

101. Simplify: $\dfrac{1}{4} - 6(2 + 8) \div \left(-\dfrac{1}{3}\right)\left(-\dfrac{1}{9}\right)$.

Grouping symbols can be inserted into $4 + 3 \cdot 7 - 4$ so that the resulting value is 45. By placing parentheses around the addition we obtain

$$(4 + 3) \cdot 7 - 4 = 7 \cdot 7 - 4 = 49 - 4 = 45.$$

In Exercises 102–103, insert parentheses in each expression so that the resulting value is 45.

102. $2 \cdot 3 + 3 \cdot 5$

103. $2 \cdot 5 - \dfrac{1}{2} \cdot 10 \cdot 9$

Technology Exercises

Use a calculator to solve Exercises 104–106.
The bar graph shows the cumulative number of deaths from AIDS in the United States from 1992 through 2000. The data can be modeled by the formula

$$N = -1.65x^2 + 51.8x + 111.44,$$

where N represents the cumulative number of U.S. AIDS deaths, in thousands, x years after 1990. Use this formula to solve Exercises 104–105.

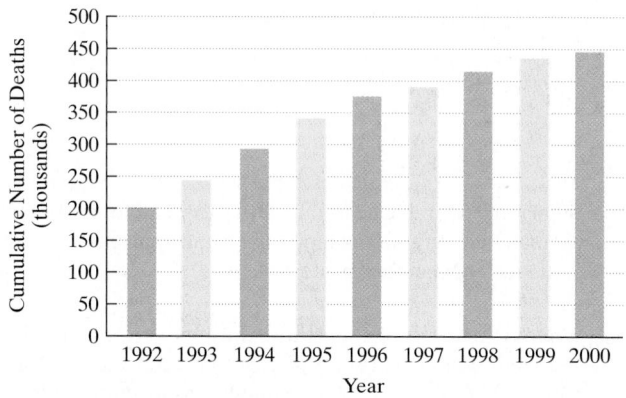

Source: Centers for Disease Control

104. According to the formula, what is the cumulative number of U.S. AIDS deaths in 1998? How well does the formula describe the cumulative number for that year shown by the bar graph?

105. According to the formula, what is the cumulative number of U.S. AIDS deaths in 1999? How well does the formula describe the cumulative number for that year shown by the bar graph?

106. The formula

$$P = \dfrac{72,900}{100x^2 + 729}$$

models the percentage of people in the United States, P, with x years of education who are unemployed. What percentage of college graduates, with 16 years of education, are unemployed? Round to the nearest tenth of a percent.

Review Exercises

107. Simplify: $-8 - 2 - (-5) + 11$ (Section 1.6, Example 3)

108. Multiply: $-4(-1)(-3)(2)$. (Section 1.7, Example 2)

109. Give an example of a real number that is not an irrational number. (Section 1.2, Example 5)

CHAPTER 1 GROUP PROJECTS

1. The points in the rectangular coordinate system in Figure 1.19 on page 84 show a relationship between a driver's age and the number of car accidents. For this activity, group members should list as many characteristics as possible that vary with a person's age. For each characteristic, group members should show points in the rectangular coordinate system. The *x*-axis should be labeled age and the *y*-axis should be labeled the characteristic. Plot points that show what happens to the characteristic from childhood to old age. Are there similarities in some of the graphs even though they describe different characteristics?

2. Group members should write and solve a problem similar to Example 12 on page 84. However, do not use the "car accidents and age" formula. Instead, use one of the formulas that estimates the adult height of a child given in the chapter introduction. Make your problem realistic by using the actual height of a child whose age, *x*, and height in inches, *h*, you know. A calculator will be helpful.

3. One measure of physical fitness is your *resting heart rate*. Generally speaking, the more fit you are, the lower your resting heart rate. The best time to take this measurement is when you first awaken in the morning, before you get out of bed. Lie on your back with no body parts crossed and take your pulse in your neck or wrist. Use your index and second fingers and count your pulse beat for one full minute to get your resting heart rate. A resting heart rate under 48 to 57 indicates high fitness, 58 to 62, above average fitness, 63 to 70, average fitness, 71 to 82, below average fitness, and 83 or more, low fitness.

   Another measure of physical fitness is your percentage of body fat. You can estimate your body fat using the following formulas:

   For men: $\quad$ Bodyfat $= -98.42 + 4.15w - 0.082b$

   For women: $\quad$ Bodyfat $= -76.76 + 4.15w - 0.082b$

   where w = waist measurement, in inches, and b = total body weight, in pounds. Then divide your body fat by your total weight to get your body fat percentage. For men, less than 15% is considered athletic, 25% about average. For women, less than 22% is considered athletic, 30% about average.

   Each group member should bring his or her age, resting heart rate, and body fat percentage to the group. Using the data, the group should create three graphs.

   a. Create a graph that shows age and resting heart rate for group members.
   b. Create a graph that shows age and body fat percentage for group members.
   c. Create a graph that shows resting heart rate and body fat percentage for group members.

   For each graph, select the style (line, bar, or circle) that is most appropriate.

CHAPTER SUMMARY, REVIEW, AND TEST

SUMMARY

DEFINITIONS AND CONCEPTS	EXAMPLES
Section 1.1 Fractions	
A fraction is reduced to its lowest terms when the numerator and denominator have no common factors other than 1. To reduce a fraction to its lowest terms, divide both the numerator and the denominator by their greatest common factor.	Reduce to lowest terms: $$\frac{8}{14} = \frac{\cancel{2} \cdot 4}{\cancel{2} \cdot 7} = \frac{4}{7}$$
Multiplying Fractions The product of two or more fractions is the product of their numerators divided by the product of their denominators.	Multiply: $$\frac{2}{7} \cdot \frac{5}{9} = \frac{2 \cdot 5}{7 \cdot 9} = \frac{10}{63}$$
Dividing Fractions To find the quotient of two fractions, invert the divisor and multiply.	Divide: $$\frac{4}{9} \div \frac{3}{7} = \frac{4}{9} \cdot \frac{7}{3} = \frac{4 \cdot 7}{9 \cdot 3} = \frac{28}{27}$$
Adding and Subtracting Fractions with Identical Denominators Add or subtract numerators. Put this result over the common denominator.	Subtract: $$\frac{5}{8} - \frac{3}{8} = \frac{5-3}{8} = \frac{2}{8} = \frac{\cancel{2} \cdot 1}{\cancel{2} \cdot 4} = \frac{1}{4}$$
Adding and Subtracting Fractions with Unlike Denominators Rewrite the fractions as equivalent fractions with the least common denominator. Then add or subtract numerators, putting this result over the common denominator.	Add: $$\frac{3}{8} + \frac{5}{12} = \frac{3}{8} \cdot \frac{3}{3} + \frac{5}{12} \cdot \frac{2}{2}$$ $$= \frac{9}{24} + \frac{10}{24} = \frac{19}{24}$$
Section 1.2 The Real Numbers	
A set is a collection of objects, called elements, whose contents can be cleary determined .	$\{a, b, c\}$
A line used to visualize numbers is called a number line.	 $-4\ -3\ -2\ -1\ \ 0\ \ 1\ \ 2\ \ 3\ \ 4$
Real Numbers: the set of all numbers that can be represented by points on the number line **The Sets That Make Up the Real Numbers** • Natural Numbers: $\{1, 2, 3, 4, \dots\}$ • Whole Numbers: $\{0, 1, 2, 3, 4, \dots\}$ • Integers: $\{\dots, -3, -2, -1, 0, 1, 2, 3, \dots\}$ • Rational Numbers: the set of numbers that can be expressed as the quotient of an integer and a nonzero integer; can be expressed as terminating or repeating decimals • Irrational Numbers: the set of numbers that cannot be expressed as the quotient of integers; decimal representations neither terminate nor repeat.	Given the set $$\left\{-1.4, 0, 0.\overline{7}, \frac{9}{10}, \sqrt{2}, \sqrt{4}\right\}$$ list the • natural numbers: $\sqrt{4}$, or 2 • whole numbers: $0, \sqrt{4}$ • rational numbers: $-1.4, 0, 0.\overline{7}, \frac{9}{10}, \sqrt{4}$ • irrational numbers: $\sqrt{2}$ • real numbers: $$-1.4, 0, 0.\overline{7}, \frac{9}{10}, \sqrt{2}, \sqrt{4}$$

DEFINITIONS AND CONCEPTS	EXAMPLES

Section 1.2 The Real Numbers (cont.)

For any two real numbers, a and b, a is less than b if a is to the left of b on the number line.

Inequality Symbols

$<$: is less than $>$: is greater than
$\leq$: is less than or equal to $\geq$: is greater than or equal to

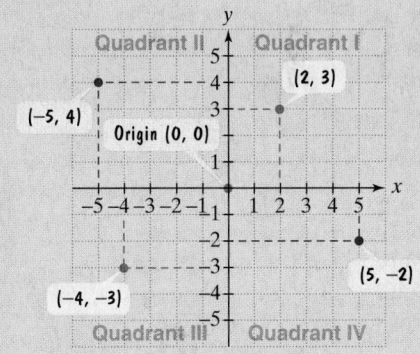

$-2 < 0$ $0 > -2$
$0 < 2.5$ $2.5 > 0$

The absolute value of a, written $|a|$, is the distance from 0 to a on the number line.

$|4| = 4$ $|0| = 0$ $|-6| = 6$

Section 1.3 Ordered Pairs and Graphs

The rectangular coordinate system consists of a horizontal number line, the x-axis, and a vertical number line, the y-axis, intersecting at their zero points, the origin. Each point in the system corresponds to an ordered pair of real numbers (x, y). The first number in the pair is the x-coordinate; the second number is the y-coordinate.

Plot: $(2, 3), (-5, 4), (-4, -3), (5, -2)$.

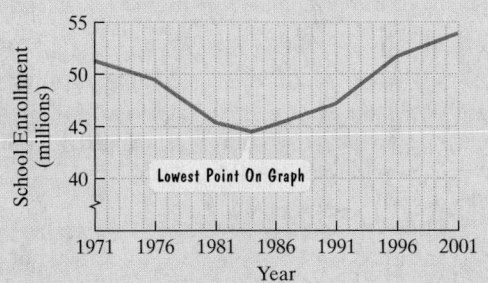

Information is often displayed using line graphs, bar graphs, and circle graphs. Line graphs are often used to illustrate trends over time. Bar graphs are convenient for showing comparisons among items. Circle graphs, divided into pieces called sectors, show how a whole quantity is divided into parts.

Millions of Students Enrolled in U.S. Schools

Source: National Education Association

In which year did enrollment reach a minimum? 1984
Estimate enrollment for that year: 44.5 million.

Section 1.4 Basic Rules of Algebra

A letter used to represent a number is called a variable. An algebraic expression is a combination of variables, numbers, and operation symbols. Terms are separated by addition. The parts of each term that are multiplied are its factors. Like terms have the same variable factors raised to the same powers. To evaluate an algebraic expression, substitute a given number for the variable and simplify.

Evaluate: $6(x + 3) + 4x$ when $x = 5$.

Replace x by 5.

$= 6(5 + 3) + 4 \cdot 5$
$= 6(8) + 4 \cdot 5$
$= 48 + 20$
$= 68$

DEFINITIONS AND CONCEPTS	EXAMPLES

Section 1.4 Basic Rules of Algebra (cont.)

Properties of Real Numbers and Algebraic Expressions

- Commutative Properties:

  $$a + b = b + a \quad ab = ba$$

- Associative Properties

  $$(a + b) + c = a + (b + c) \quad (ab)c = a(bc)$$

- Distributive Properties:

  $$a(b + c) = ab + ac \quad (b + c)a = ba + ca$$
  $$a(b - c) = ab - ac \quad (b - c)a = ba - ca$$
  $$a(b + c + d) = ab + ac + ad$$

Commutative of Addition:
$$5x + 4 = 4 + 5x$$
Commutative of Multiplication:
$$5x + 4 = x5 + 4$$
Associative of Addition:
$$6 + (4 + x) = (6 + 4) + x = 10 + x$$
Associative of Multiplication:
$$7(10x) = (7 \cdot 10)x = 70x$$
Distributive:
$$8(x + 5 + 4y) = 8x + 40 + 32y$$
Distributive to Combine Like Terms:
$$8x + 12x = (8 + 12)x = 20x$$

Simplifying Algebraic Expressions

Use the distributive property to remove grouping symbols. Then combine like terms.

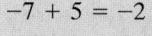
$$4(5x - 7) - 13x$$
$$= 20x - 28 - 13x$$
$$= (20x - 13x) - 28$$
$$= 7x - 28$$

Section 1.5 Addition of Real Numbers

Sums on a Number Line

To find $a + b$, the sum of a and b, on a number line, start at a. If b is positive, move b units to the right. If b is negative, move b units to the left. If b is 0, stay at a. The number where we finish on the number line represents $a + b$.

$$-7 + 5 = -2$$

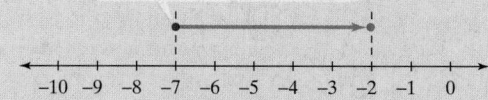

Additive inverses are pairs of real numbers that are the same number of units from zero on the number line, but on opposite sides of zero.

- Identity Property of Addition

  $$a + 0 = 0 \quad 0 + a = a$$

- Inverse Property of Addition

  $$a + (-a) = 0 \quad (-a) + a = 0$$

The additive inverse (or opposite) of 4 is -4. The additive inverse of -1.7 is 1.7.

Identity Property of Addition:
$$4x + 0 = 4x$$

Inverse Property of Addition:
$$4x + (-4x) = 0$$

Addition without a Number Line

To add two numbers with the same sign, add their absolute values and use their common sign. To add two numbers with different signs, subtract the smaller absolute value from the greater absolute value and use the sign of the number with the greater absolute value.

Add:
$$10 + 4 = 14$$
$$-4 + (-6) = -10$$
$$-30 + 5 = -25$$
$$12 + (-8) = 4$$

To add a series of positive and negative numbers, add all the positive numbers and add all the negative numbers. Then add the resulting positive and negative sums.

$$5 + (-3) + (-7) + 2$$
$$= (5 + 2) + [(-3) + (-7)]$$
$$= 7 + (-10)$$
$$= -3$$

DEFINITIONS AND CONCEPTS	EXAMPLES

Section 1.6 Subtraction of Real Numbers

To subtract b from a, add the additive inverse of b to a: $\qquad a - b = a + (-b)$. The result is called the difference between a and b.	Subtract: $-7 - (-5) = -7 + 5 = -2$ $-\dfrac{3}{4} - \dfrac{1}{2} = -\dfrac{3}{4} + \left(-\dfrac{1}{2}\right)$ $\qquad = -\dfrac{3}{4} + \left(-\dfrac{2}{4}\right) = -\dfrac{5}{4}$
To simplify a series of additions and subtractions, change all subtractions to additions of additive inverses. Then use the procedure for adding a series of positive and negative numbers.	Simplify: $\quad -6 - 2 - (-3) + 10$ $\qquad = -6 + (-2) + 3 + 10$ $\qquad = -8 + 13$ $\qquad = 5$

Section 1.7 Multiplication and Division of Real Numbers

The result of multiplying a and b, ab, is called the product of a and b. If the two numbers have different signs, the product is negative. If the two numbers have the same sign, the product is positive. If either number is 0, the product is 0.	Multiply: $-5(-10) = 50$ $\dfrac{3}{4}\left(-\dfrac{5}{7}\right) = -\dfrac{3}{4} \cdot \dfrac{5}{7} = -\dfrac{15}{28}$
Assuming that no number is 0, the product of an even number of negative numbers is positive. The product of an odd number of negative numbers is negative. If any number is 0, the product is 0.	Multiply: $(-3)(-2)(-1)(-4) = 24$ $(-3)(2)(-1)(-4) = -24$
The result of dividing the real number a by the nonzero real number b is called the quotient of a and b. If two numbers have different signs, their quotient is negative. If two numbers have the same sign, their quotient is positive. Division by zero is undefined.	Divide: $\dfrac{21}{-3} = -7$ $-\dfrac{1}{3} \div (-3) = \dfrac{1}{3} \cdot \dfrac{1}{3} = \dfrac{1}{9}$
Two numbers whose product is 1 are called multiplicative inverses or reciprocals of each other. The number 0 has no multiplicative inverse. • Identity Property of Multiplication $\qquad a \cdot 1 = a \qquad 1 \cdot a = a$ • Inverse Property of Multiplication If a is not 0: $\qquad a \cdot \dfrac{1}{a} = 1 \qquad \dfrac{1}{a} \cdot a = 1$ • Multiplication Property of -1 $\qquad -1a = -a \qquad a(-1) = -a$ • Double Negative Property $\qquad -(-a) = a$	The multiplicative inverse of 4 is $\frac{1}{4}$. The multiplicative inverse of $-\frac{1}{3}$ is -3. Simplify: $1x = x$ $7x \cdot \dfrac{1}{7x} = 1$ $4x - 5x = -1x = -x$ $-(-7y) = 7y$
If a negative sign precedes parentheses, remove parentheses and change the sign of every term within parentheses.	Simplify: $-(7x - 3y + 2) = -7x + 3y - 2$

DEFINITIONS AND CONCEPTS	EXAMPLES
Section 1.8 Exponents, Order of Operations, and Mathematical Models	

If b is a real number and n is a natural number, b^n, the nth power of b, is the product of n factors of b. Furthermore, $b^1 = b$.	Evaluate: $8^2 = 8 \cdot 8 = 64$ $(-5)^3 = (-5)(-5)(-5) = -125$
Order of Operations 1. Perform operations within grouping symbols, starting with the innermost grouping symbols. 2. Evaluate exponential expressions. 3. Multiply and divide in order from left to right. 4. Add and subtract in order from left to right.	Simplify: $5(4-6)^2 - 2(1-3)^3$ $= 5(-2)^2 - 2(-2)^3$ $= 5(4) - 2(-8)$ $= 20 - (-16)$ $= 20 + 16 = 36$
Some algebraic expressions contain two sets of grouping symbols: parentheses, the inner grouping symbols, and brackets, the outer grouping symbols. To simplify such expressions, use the order of operations and remove grouping symbols from innermost (parentheses) to outermost (brackets).	Simplify: $5 - 3[2(x+1) - 7]$ $= 5 - 3[2x + 2 - 7]$ $= 5 - 3[2x - 5]$ $= 5 - 6x + 15$ $= -6x + 20$
A formula is a statement of equality expressing a relationship among two or more variables. Many formulas describe, or model, real-world variables. Formulas that give an approximate, rather than an exact, relationship among variables are called mathematical models.	The formula $N = 0.4x^2 + 0.5$ models the millions of people, N, in the United States using cable modems x years after 1996.

Review Exercises

1.1 *In Exercises 1–2, simplify each fraction by reducing it to its lowest terms.*

1. $\dfrac{15}{33}$

2. $\dfrac{40}{75}$

In Exercises 3–7, perform the indicated operation. Where possible, reduce the answer to its lowest terms.

3. $\dfrac{3}{5} \cdot \dfrac{7}{10}$

4. $\dfrac{4}{5} \div \dfrac{3}{10}$

5. $\dfrac{2}{9} + \dfrac{4}{9}$

6. $\dfrac{5}{6} + \dfrac{7}{9}$

7. $\dfrac{3}{4} - \dfrac{2}{15}$

8. The gas tank of a car is filled to its capacity. The first day, $\frac{1}{4}$ of the tank's gas is used for travel. The second day, $\frac{1}{3}$ of the tank's original amount of gas is used for travel. What fraction of the tank is filled with gas at the end of the second day?

1.2 *In Exercises 9–10, graph each real number on a number line.*

9. -2.5

10. $4\dfrac{3}{4}$

In Exercises 11–12, express each rational number as a decimal.

11. $\dfrac{5}{8}$

12. $\dfrac{3}{11}$

13. Consider the set

$$\left\{ -17, -\dfrac{9}{13}, 0, 0.75, \sqrt{2}, \pi, \sqrt{81} \right\}.$$

List all numbers from the set that are: **a.** natural numbers, **b.** whole numbers, **c.** integers, **d.** rational numbers, **e.** irrational numbers, **f.** real numbers.

14. Give an example of an integer that is not a natural number.

15. Give an example of a rational number that is not an integer.

16. Give an example of a real number that is not a rational number.

In Exercises 17–20, insert either $<$ or $>$ in the shaded area between each pair of numbers to make a true statement.

17. -93 ▨ 17

18. -2 ▨ -200

19. 0 ▨ $-\dfrac{1}{3}$

20. $-\dfrac{1}{4}$ ▨ $-\dfrac{1}{5}$

In Exercises 21–22, determine whether each inequality is true or false.

21. $-13 \geq -11$

22. $-126 \leq -126$

In Exercises 23–24, find each absolute value.

23. $|-58|$

24. $|2.75|$

1.3 *In Exercises 25–28, plot the given point in a rectangular coordinate system. Indicate in which quadrant each point lies.*

25. $(1, -5)$

26. $(4, -3)$

27. $\left(\dfrac{7}{2}, \dfrac{3}{2}\right)$

28. $(-5, 2)$

29. Give the ordered pairs that correspond to the points labeled in the figure.

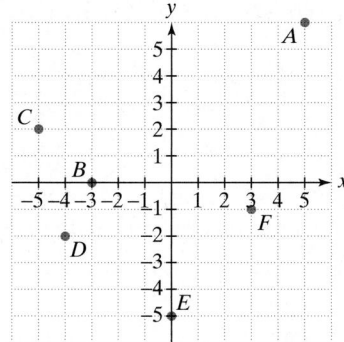

The line graph shows the number of murders per 100,000 people in the United States from 1900 through 2000. Use the graph to solve Exercises 30–31.

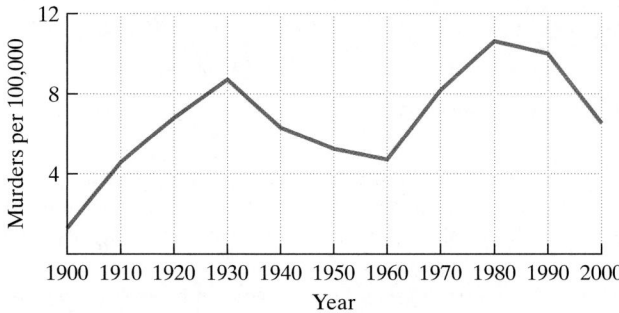

Murders Per 100,000 People in the United States

Source: National Center for Health Statistics

30. Find an estimate for the number of murders per 100,000 people in 2000.

31. For the period shown, when did the murder rate reach a maximum? What is a reasonable estimate for the number of murders per 100,000 people for that year?

The bar graph shows the percentage of households in the United States using various new technologies. Use the graph to solve Exercises 32–33.

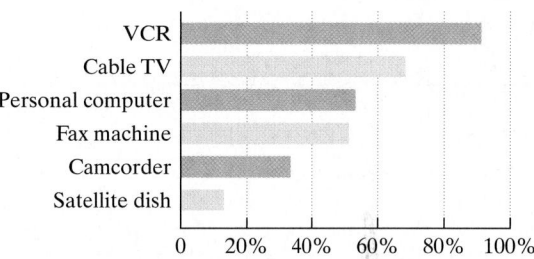

Percentage of Households in the U.S. Using New Technologies

Source: U.S. Consumer Electronics Industry

32. Estimate the percentage of households in the United States with VCRs.

33. Which technologies are used by fewer than 40% of households in the United States?

34. Approximately 24 million Americans are runners. Use the circle graph to determine the number of female runners.

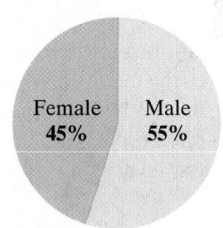

Runners in the U.S.

Source: U.S. Census Bureau

1.4 *In Exercises 35–36, evaluate each algebraic expression for the given value of the variable.*

35. $7x + 3; x = 10$

36. $5(x - 4); x = 12$

37. Use the commutative property of addition to write an equivalent algebraic expression: $7 + 13y$.

38. Use the commutative property of multiplication to write an equivalent algebraic expression: $9(x + 7)$.

In Exercises 39–40, use an associative property to rewrite each algebraic expression. Then simplify the resulting algebraic expression.

39. $6 + (4 + y)$

40. $7(10x)$

41. Use the distributive property to rewrite without parentheses: $6(4x - 2 + 5y)$.

In Exercises 42–43, simplify each algebraic expression.

42. $4a + 9 + 3a - 7$

43. $6(3x + 4) + 5(2x - 1)$

44. Suppose that a store is selling all computers at 25% off the regular price. If x is the regular price, the algebraic expression $x - 0.25x$ describes the sale price. Evaluate the expression when $x = 2400$. Describe what the answer means in practical terms.

1.5

45. Use a number line to find the sum: $-6 + 8$.

In Exercises 46–48, find each sum without the use of a number line.

46. $8 + (-11)$

47. $-\dfrac{3}{4} + \dfrac{1}{5}$

48. $7 + (-5) + (-13) + 4$

In Exercises 49–50, simplify each algebraic expression.

49. $8x + (-6y) + (-12x) + 11y$

50. $10(4 - 3y) + 28y$

51. The Dead Sea is the lowest elevation on Earth, 1312 feet below sea level. If a person is standing 512 feet above the Dead Sea, what is that person's elevation?

52. The water level of a reservoir is measured over a five-month period. At the beginning, the level is 25 feet. During this time, the level fell 3 feet, then rose 2 feet, then rose 1 foot, then fell 4 feet, and then rose 2 feet. What is the reservoir's water level at the end of the five months?

1.6

53. Rewrite $9 - 13$ as the addition of an additive inverse.

In Exercises 54–56, perform the indicated subtraction.

54. $-9 - (-13)$

55. $-\dfrac{7}{10} - \dfrac{1}{2}$

56. $-3.6 - (-2.1)$

In Exercises 57–58, simplify each series of additions and subtractions.

57. $-7 - (-5) + 11 - 16$

58. $-25 - 4 - (-10) + 16$

59. Simplify: $3 - 6a - 8 - 2a$.

60. What is the difference in elevation between a plane flying 26,500 feet above sea level and a submarine traveling 650 feet below sea level?

1.7 *In Exercises 61–63, perform the indicated multiplication.*

61. $-7(-12)$

62. $\dfrac{3}{5}\left(-\dfrac{5}{11}\right)$

63. $5(-3)(-2)(-4)$

In Exercises 64–66, perform the indicated division or state that the expression is undefined.

64. $\dfrac{45}{-5}$

65. $-17 \div 0$

66. $-\dfrac{4}{5} \div \left(-\dfrac{2}{5}\right)$

In Exercises 67–68, simplify each algebraic expression.

67. $-4\left(-\dfrac{3}{4}x\right)$

68. $-3(2x - 1) - (4 - 5x)$

1.8 *In Exercises 69–71, evaluate each exponential expression.*

69. $(-6)^2$

70. -6^2

71. $(-2)^5$

In Exercises 72–73, simplify each algebraic expression, or explain why the expression cannot be simplified.

72. $4x^3 + 2x^3$

73. $4x^3 + 4x^2$

In Exercises 74–80, use the order of operations to simplify each expression.

74. $-40 \div 5 \cdot 2$

75. $-6 + (-2) \cdot 5$

76. $6 - 4(-3 + 2)$

77. $28 \div (2 - 4^2)$

78. $36 - 24 \div 4 \cdot 3 - 1$

79. $-8[-4 - 5(-3)]$

80. $\dfrac{6(-10 + 3)}{2(-15) - 9(-3)}$

In Exercises 81–82, evaluate each algebraic expression for the given value of the variable.

81. $x^2 - 2x + 3; x = -1$

82. $-x^2 - 7x; x = -2$

In Exercises 83–84, simplify each algebraic expression.

83. $4[7(a - 1) + 2]$

84. $-6[4 - (y + 2)]$

The line graph shows the number of women and the number of men, in millions, enrolled in U.S. colleges, projected through 2008. The data can be modeled by the formulas

$$N = 0.07x + 4.1 \qquad \text{for women}$$
$$N = 0.01x + 3.9 \qquad \text{for men}$$

where N represents enrollment, in millions, x years after 1984. Use this information to solve Exercises 85–86.

Enrollment in U.S. Colleges

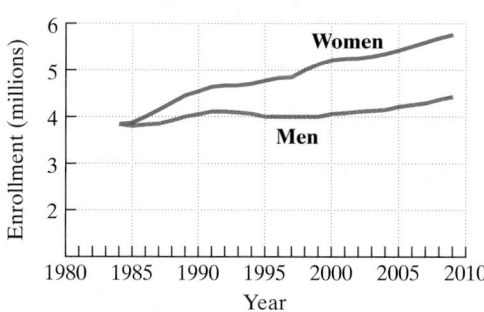

Source: Department of Education

85. According to the formula, what is the projected enrollment for women, in millions, in 2004? How well does the formula describe enrollment for that year shown by the line graph?

86. According to the formula, what is the projected enrollment for men, in millions, in 2004? How well does the formula describe enrollment for that year shown by the line graph?

The line graph shows the number of inmates, in thousands, in U.S. state and federal prisons from 1980 through 2000. The data can be modeled by the formula

$$N = 2x^2 + 22x + 320$$

where N represents the number of inmates, in thousands, in state and federal prisons x years after 1980. Use this information to solve Exercises 87–88.

Number of Inmates in U.S. State and Federal Prisons

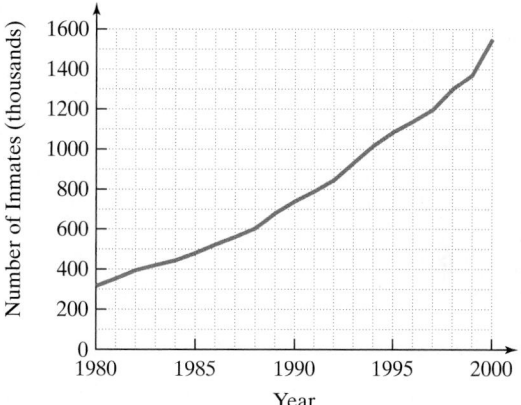

Source: U.S. Justice Department

87. According to the formula, what was the prison population, in thousands, in 1990? How well does the formula describe the number of inmates for that year shown by the line graph?

88. What do you notice about the prison population from the graph that is not obvious by looking at the formula?

Chapter 1 Test

In Exercises 1–9, perform the indicated operation or operations.

1. $1.4 - (-2.6)$

2. $-9 + 3 + (-11) + 6$

3. $3(-17)$

4. $\left(-\dfrac{3}{7}\right) \div \left(-\dfrac{15}{7}\right)$

5. $-50 \div 10$

6. $-6 - (5 - 12)$

7. $(-3)(-4) \div (7 - 10)$

8. $(6 - 8)^2(5 - 7)^3$

9. $\dfrac{3(-2) - 2(2)}{-2(8 - 3)}$

In Exercises 10–12, simplify each algebraic expression.

10. $11x - (7x - 4)$

11. $5(3x - 4y) - (2x - y)$

12. $6 - 2[3(x + 1) - 5]$

13. List all the rational numbers in this set.

$$\left\{-7, -\dfrac{4}{5}, 0, 0.25, \sqrt{3}, \sqrt{4}, \dfrac{22}{7}, \pi\right\}$$

14. Insert either $<$ or $>$ in the shaded area to make a true statement: -1 ▓ -100.

15. Find the absolute value: $|-12.8|$.

16. Plot $(-4, 3)$ in a rectangular coordinate system. In which quadrant does the point lie?

17. Find the coordinates of point A in the figure.

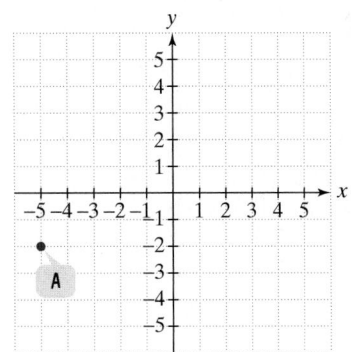

In Exercises 18–19, evaluate each algebraic expression for the given value of the variable.

18. $5(x - 7); x = 4$

19. $x^2 - 5x; x = -10$

20. Use the commutative property of addition to write an equivalent algebraic expression: $2(x + 3)$.

21. Use the associative property of multiplication to rewrite $-6(4x)$. Then simplify the expression.

22. Use the distributive property to rewrite without parentheses: $7(5x - 1 + 2y)$.

A number of elk are placed into a newly acquired habitat. The points in the rectangular coordinate system show the elk population every ten years. Use this information to solve Exercises 23–24.

Change in Elk Population over Time

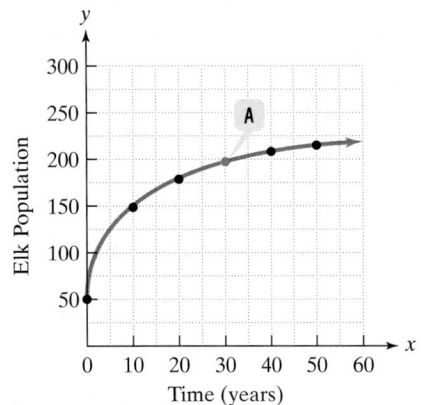

23. Find the coordinates of point A. Then interpret the coordinates in terms of the information given.

24. How many elk were introduced into the habitat?

25. Use the bar graph to find a reasonable estimate for the projected number of U.S. households, in millions, investing online in 2003.

Projected Number of U.S. Households, in Millions, Investing Online

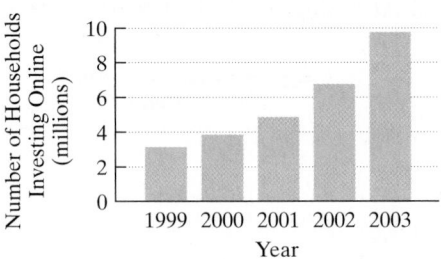

Source: Tradeline/IDD Information Services

26. The circle graph shown is based on a survey of 17.1 million acres of lakes in the United States. Using this survey by the Environmental Protection Agency, estimate the number of acres of impaired lakes.

Water Quality in U.S. Lakes

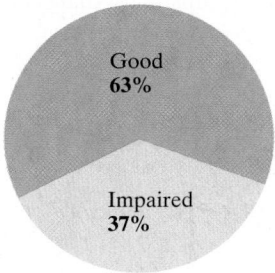

Source: U.S. Environmental Protection Agency

27. The formula

$$T = 3(A - 20)^2 \div 50 + 10$$

describes the average running time, T, in seconds, for a person who is A years old to run the 100-yard dash. How long does it take a 30-year-old runner to run the 100-yard dash?

28. Use the graph to find a reasonable estimate for the average mortgage loan in 1990.

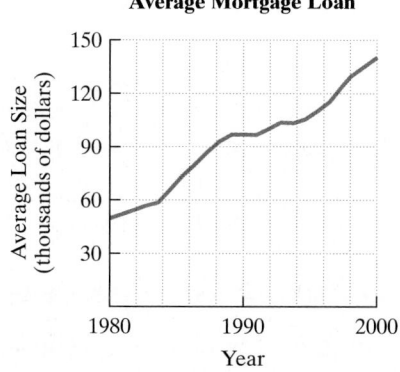

Average Mortgage Loan

Source: Mortgage Bankers Association of America

29. The data in Exercise 28 can be modeled by the formula

$$N = 3.5x + 58$$

where N represents the average mortgage loan, in thousands of dollars, x years after 1980. Use this formula to find the average mortgage loan in 1990.

30. What is the difference in elevation between a plane flying 16,200 feet above sea level and a submarine traveling 830 feet below sea level?

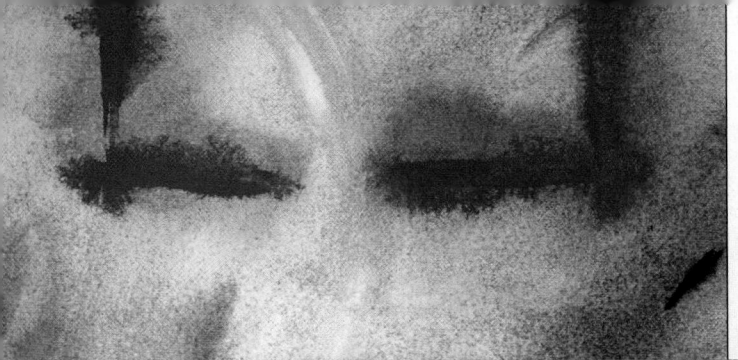

Listening to the radio on the way to work, you hear candidates in the upcoming election discussing the problem of the rising cost of prescription drugs. With millions of baby boomers pushing 60 and new drugs coming on the market, our nation's drug tab is headed for the stratosphere.

*I*n 1999, prescription drug spending in the United States was $125 billion. The bitter cost of better pills can be modeled by a formula that predicts spending will exceed $250 billion by 2008. Formulas can be used to explain what is happening in the present and to make predictions about what might occur in the future. In this chapter, you will learn to use formulas in new ways that will help you to recognize patterns, logic, and order in a world that can appear chaotic to the untrained eye.

Chapter 2

Linear Equations and Inequalities in One Variable

▶ SECTION 2.1 The Addition Property of Equality

Objectives

1 Check whether a number is a solution to an equation.

2 Use the addition property of equality to solve equations.

3 Solve applied problems using formulas.

SSM
PH Tutor CD- Video
Center ROM

Unfortunately, many of us have been fined for driving over the speed limit. The amount of the fine depends on how fast we are speeding. Suppose that a highway has a speed limit of 60 miles per hour. The amount that speeders are fined, F, is described by the formula

$$F = 10x - 600,$$

where x is the speed in miles per hour. A friend, whom we shall call Leadfoot, borrows your car and returns a few hours later with a $400 speeding fine. Leadfoot is furious, protesting that the car was barely driven over the speed limit. Should you believe Leadfoot?

In order to decide if Leadfoot is telling the truth, use the formula

$$F = 10x - 600.$$

Leadfoot was fined $400, so substitute 400 for F:

$$400 = 10x - 600.$$

Now you need to find the value for x. This variable represents Leadfoot's speed, which resulted in the $400 fine.

The use of algebra in a variety of everyday applications often leads to an *equation*. An **equation** is a statement that two algebraic expressions are equal. Thus, $400 = 10x - 600$ is an example of an equation. The equal sign divides the equation into two parts, the left side and the right side:

$$\boxed{400} = \boxed{10x - 600}$$

Left side   Right side

The two sides of an equation can be reversed, so we can express this equation as

$$10x - 600 = 400.$$

The form of this equation is $ax + b = c$, with $a = 10$, $b = -600$, and $c = 400$. Any equation in this form is called a **linear equation in one variable**. The exponent on the variable in such an equation is 1.

In the next three sections, we will study how to solve linear equations. **Solving an equation** is the process of finding the number (or numbers) that make the equation a true statement. These numbers are called the **solutions**, or **roots**, of the equation, and we say that they **satisfy** the equation.

1 Check whether a number is a solution to an equation.

Checking Whether a Number Is a Solution to an Equation

A proposed solution to an equation can be checked by substituting that number for each occurrence of the variable in the equation. If the substitution results in a true statement, the number is a solution. If the substitution results in a false statement, the number is not a solution.

EXAMPLE 1   Checking Proposed Solutions (Is Leadfoot Telling the Truth?)

Consider the equation

$$10x - 600 = 400.$$

(Remember that x represents Leadfoot's speed that resulted in the $400 fine.) Determine whether:

    a. 60 is a solution.   **b.** 100 is a solution.

Solution

    a. To determine whether 60 is a solution to the equation, we substitute 60 for x.

$$10x - 600 = 400 \quad \text{This is the given equation.}$$
$$10(60) - 600 \stackrel{?}{=} 400 \quad \text{Substitute 60 for } x. \text{ The question mark over the equal sign indicates that we do not know yet if the statement is true.}$$
$$600 - 600 \stackrel{?}{=} 400 \quad \text{Multiply: } 10(60) = 600.$$
$$\boxed{\text{This statement is false.}} \quad 0 = 400 \quad \text{Subtract: } 600 - 600 = 0.$$

Because the check results in a false statement, we conclude that 60 is not a solution to the given equation. (Leadfoot was not doing 60 miles per hour in your borrowed car.)

    b. To determine whether 100 is a solution to the equation, we substitute 100 for x.

$$10x - 600 = 400 \quad \text{This is the given equation.}$$
$$10(100) - 600 \stackrel{?}{=} 400 \quad \text{Substitute 100 for } x.$$
$$1000 - 600 \stackrel{?}{=} 400 \quad \text{Multiply: } 10(100) = 1000.$$
$$\boxed{\text{This statement is true.}} \quad 400 = 400 \quad \text{Subtract: } 1000 - 600 = 400.$$

Because the check results in a true statement, we conclude that 100 is a solution to the given equation. Thus, 100 satisfies the equation. (Leadfoot was doing an outrageous 100 miles per hour, and lied with the claim that your car was barely driven over the speed limit.) ■

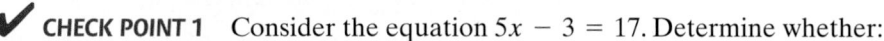 **CHECK POINT 1** Consider the equation $5x - 3 = 17$. Determine whether:
    a. 3 is a solution.
    b. 4 is a solution.

2 Use the addition property of equality to solve equations.

Using the Addition Property of Equality to Solve Equations

Consider the equation

$$x = 11.$$

By inspection, we can see that the solution to this equation is 11. If we substitute 11 for x, we obtain the true statement $11 = 11$.

Study Tip

An equation is like a balanced scale—balanced because its two sides are equal. To maintain this balance, whatever is done to one side must also be done to the other side.

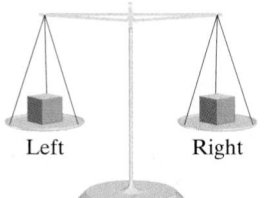

Consider $x - 3 = 8$.

The scale is balanced if the left and right sides are equal.

Add 3 to the left side.

Keep the scale balanced by adding 3 to the right side.

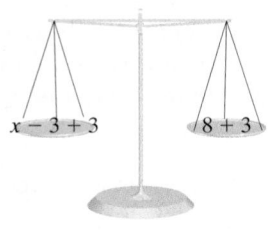

Thus, $x = 11$.

Now consider the equation

$$x - 3 = 8.$$

If we substitute 11 for x, we obtain $11 - 3 \overset{?}{=} 8$. Subtracting on the left, we get the true statement $8 = 8$.

The equations $x - 3 = 8$ and $x = 11$ both have the same solution, namely 11, and are called *equivalent equations*. **Equivalent equations** are equations that have the same solution.

The idea in solving a linear equation is to get an equivalent equation with the variable (the letter) by itself on one side of the equal sign and a number by itself on the other side. For example, consider the equation $x - 3 = 8$. To get x by itself on the left side, add 3 to the left side, because $x - 3 + 3$ gives $x + 0$, or just x. You must then add 3 to the right side also. By doing this, we are using the **addition property of equality**.

The Addition Property of Equality

The same real number (or algebraic expression) may be added to both sides of an equation without changing the equation's solution. This can be expressed symbolically as follows:

$$\text{If } a = b, \text{ then } a + c = b + c.$$

EXAMPLE 2   Solving an Equation Using the Addition Property

Solve the equation: $x - 3 = 8$.

Solution   We can isolate the variable, x, by adding 3 to both sides of the equation.

$$x - 3 = 8 \qquad \text{This is the given equation.}$$
$$x - 3 + 3 = 8 + 3 \qquad \text{Add 3 to both sides.}$$
$$x + 0 = 11 \qquad \text{This step is often done mentally and not listed.}$$
$$x = 11$$

By inspection, we can see that the solution to $x = 11$ is 11. To check this proposed solution, replace x with 11 in the original equation.

$$\textbf{Check} \quad x - 3 = 8 \qquad \text{This is the original equation.}$$
$$11 - 3 \overset{?}{=} 8 \qquad \text{Substitute 11 for x.}$$
$$\boxed{\text{This statement is true.}} \quad 8 = 8 \qquad \text{Subtract: } 11 - 3 = 8.$$

Because the check results in a true statement, we conclude that the solution to the given equation is 11.   ∎

The set of an equation's solutions is called its **solution set**. Thus, the solution set of the equation in Example 2 is $\{11\}$. The solution can be expressed as 11 or, using set notation, $\{11\}$. However, do not write the solution as $x = 11$. **The solution of an equation should not be given as an equivalent equation.**

✔ **CHECK POINT 2** Solve the equation and check your proposed solution:
$$x - 5 = 12.$$

When we use the addition property of equality, we add the same number to both sides of an equation. We know that subtraction is the addition of an additive inverse. Thus, the addition property also lets us subtract the same number from both sides of an equation without changing the equation's solution.

EXAMPLE 3   Subtracting the Same Number from Both Sides

Solve and check: $z + 1.4 = 2.06$.

Solution

$$z + 1.4 = 2.06 \qquad \text{This is the given equation.}$$
$$z + 1.4 - 1.4 = 2.06 - 1.4 \qquad \text{Subtract 1.4 from both sides. This is equivalent to adding −1.4 to both sides.}$$
$$z = 0.66 \qquad \text{Subtracting 1.4 from both sides eliminates 1.4 on the left.}$$

Can you see that the solution to $z = 0.66$ is 0.66? To check this proposed solution, replace z with 0.66 in the original equation.

$$\textbf{Check} \qquad z + 1.4 = 2.06 \qquad \text{This is the original equation.}$$
$$0.66 + 1.4 \stackrel{?}{=} 2.06 \qquad \text{Substitute 0.66 for z.}$$
$$2.06 = 2.06 \qquad \text{True}$$

This true statement indicates that the solution is 0.66, or the solution set is $\{0.66\}$. ∎

✔ **CHECK POINT 3**   Solve and check: $z + 2.8 = 5.09$.

When isolating the variable, it can be isolated on either the left or right side of an equation.

EXAMPLE 4   Isolating the Variable on the Right

Solve and check: $-\dfrac{1}{2} = x - \dfrac{2}{3}$.

Solution   We can isolate the variable, x, on the right side by adding $\frac{2}{3}$ to both sides of the equation.

$$-\frac{1}{2} = x - \frac{2}{3} \qquad \text{This is the given equation.}$$
$$-\frac{1}{2} + \frac{2}{3} = x - \frac{2}{3} + \frac{2}{3} \qquad \text{Add } \tfrac{2}{3} \text{ to both sides, isolating x on the right.}$$
$$-\frac{3}{6} + \frac{4}{6} = x \qquad \text{Rewrite fractions as equivalent fractions with a denominator of 6: } -\frac{1}{2} + \frac{2}{3} = -\frac{1}{2} \cdot \frac{3}{3} + \frac{2}{3} \cdot \frac{2}{2} = -\frac{3}{6} + \frac{4}{6}.$$
$$\frac{1}{6} = x \qquad \text{Add on the left side: } -\frac{3}{6} + \frac{4}{6} = \frac{-3+4}{6} = \frac{1}{6}.$$

Take a few minutes to check the proposed solution, $\frac{1}{6}$. Substitute $\frac{1}{6}$ for x in the original equation. You should obtain $-\frac{1}{2} = -\frac{1}{2}$. This true statement indicates that the solution is $\frac{1}{6}$, or the solution set is $\{\frac{1}{6}\}$. ∎

✔ **CHECK POINT 4**   Solve and check: $-\dfrac{1}{2} = x - \dfrac{3}{4}$.

Study Tip

The equations $a = b$ and $b = a$ have the same meaning. If you prefer, you can solve

$$-\frac{1}{2} = x - \frac{2}{3}$$

by reversing both sides and solving

$$x - \frac{2}{3} = -\frac{1}{2}.$$

In Example 5, we combine like terms before using the addition property.

EXAMPLE 5 Combining Like Terms Before Using the Addition Property

Solve and check: $5y + 3 - 4y - 8 = 6 + 9$.

Solution

$$5y + 3 - 4y - 8 = 6 + 9 \qquad \text{This is the given equation.}$$

$$y - 5 = 15 \qquad \text{Combine like terms: } 5y - 4y = y, 3 - 8 = -5,$$
$$\text{and } 6 + 9 = 15.$$

$$y - 5 + 5 = 15 + 5 \qquad \text{Add 5 to both sides.}$$

$$y = 20$$

To check the proposed solution, 20, replace y with 20 in the original equation.

Check $\qquad 5y + 3 - 4y - 8 = 6 + 9 \qquad$ Be sure to use the original equation and not the simplified form in the second step. (Why?)

$$5(20) + 3 - 4(20) - 8 \stackrel{?}{=} 6 + 9 \qquad \text{Substitute 20 for } y.$$

$$100 + 3 - 80 - 8 \stackrel{?}{=} 6 + 9 \qquad \text{Multiply on the left.}$$

$$103 - 88 \stackrel{?}{=} 6 + 9 \qquad \text{Combine positive and negative numbers on the left.}$$

$$15 = 15 \qquad \text{True}$$

This true statement verifies that the solution is 20, or the solution set is {20}. ∎

✔ **CHECK POINT 5** Solve and check: $8y + 7 - 7y - 10 = 6 + 4$.

Adding and Subtracting Variable Terms On Both Sides of an Equation

In some equations, variable terms appear on both sides. Here is an example:

$$4x = 7 + 3x.$$

A variable term, 4x, is on the left side.   A variable term, 3x, is on the right side.

Our goal is to isolate all the variable terms on one side of the equation. We can use the addition property of equality to do this. The property allows us to add or subtract the same variable term on both sides of an equation without changing the solution. Let's see how we can use this idea to solve $4x = 7 + 3x$.

EXAMPLE 6 Using the Addition Property to Isolate Variable Terms

Solve and check: $4x = 7 + 3x$.

Solution In the given equation, variable terms appear on both sides. We can isolate them on one side by subtracting $3x$ from both sides of the equation.

$$4x = 7 + 3x \qquad \text{This is the given equation.}$$

$$4x - 3x = 7 + 3x - 3x \qquad \text{Subtract 3x from both sides and isolate variable terms on the left.}$$

$$x = 7 \qquad \text{Subtracting 3x from both sides eliminates 3x on the right. On the left, 4x − 3x = 1x = x.}$$

To check the proposed solution, 7, replace x with 7 in the original equation.

Check $\quad 4x = 7 + 3x \qquad \text{Use the original equation.}$

$$4(7) \stackrel{?}{=} 7 + 3(7) \qquad \text{Substitute 7 for x.}$$

$$28 \stackrel{?}{=} 7 + 21 \qquad \text{Multiply: } 4(7) = 28 \text{ and } 3(7) = 21.$$

$$28 = 28 \qquad \text{True}$$

This true statement verifies that the solution is 7, or the solution set is {7}. ∎

✔ **CHECK POINT 6** Solve and check: $7x = 12 + 6x$.

EXAMPLE 7 Solving an Equation by Isolating the Variable

Solve and check: $3y - 9 = 2y + 6$.

Solution Our goal is to isolate variable terms on one side and constant terms on the other side. Let's begin by isolating the variable on the left.

$$3y - 9 = 2y + 6 \qquad \text{This is the given equation.}$$

$$3y - 2y - 9 = 2y - 2y + 6 \qquad \text{Isolate the variable terms on the left by subtracting 2y from both sides.}$$

$$y - 9 = 6 \qquad \text{Subtracting 2y from both sides eliminates 2y on the right. On the left, 3y − 2y = 1y = y.}$$

Now we isolate the constant terms on the right by adding 9 to both sides.

$$y - 9 + 9 = 6 + 9 \qquad \text{Add 9 to both sides.}$$

$$y = 15$$

Check $\quad 3y - 9 = 2y + 6 \qquad \text{Use the original equation.}$

$$3(15) - 9 \stackrel{?}{=} 2(15) + 6 \qquad \text{Substitute 15 for y.}$$

$$45 - 9 \stackrel{?}{=} 30 + 6 \qquad \text{Multiply: } 3(15) = 45 \text{ and } 2(15) = 30.$$

$$36 = 36 \qquad \text{True}$$

The solution is 15, or the solution set is {15}. ∎

✔ **CHECK POINT 7** Solve and check: $3x - 6 = 2x + 5$.

3 Solve applied problems using formulas.

Applications

Our next example shows how the addition property of equality can be used to find the value of a variable in a mathematical model.

EXAMPLE 8   An Application: Vocabulary and Age

There is a relationship between a child's vocabulary, V, and the child's age, A, in months. This relationship can be modeled by the formula

$$V + 900 = 60A.$$

Use the formula to find the vocabulary of a child at the age of 30 months.

Solution   In the formula, A represents the child's age, in months. We are interested in a 30-month-old child. Thus, we substitute 30 for A. Then we use the addition property of equality to find V, the number of words in the child's vocabulary.

$V + 900 = 60A$	This is the given formula.
$V + 900 = 60(30)$	Substitute 30 for A.
$V + 900 = 1800$	Multiply: $60(30) = 1800$.
$V + 900 - 900 = 1800 - 900$	Subtract 900 from both sides and solve for V.
$V = 900$	

At the age of 30 months, a child has a vocabulary of 900 words.   ■

The points in the rectangular coordinate system in Figure 2.1 allow us to "see" the formula $V + 900 = 60A$. The x-coordinate of each point represents the child's age, in months. The y-coordinate represents the child's vocabulary. The points are rising steadily from left to right. This shows that a typical child's vocabulary is steadily increasing with age.

✔ **CHECK POINT 8**   Use the formula $V + 900 = 60A$ to find the vocabulary of a child at the age of 50 months.

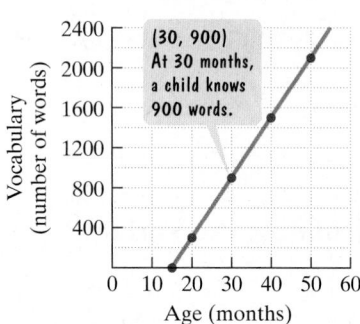

Vocabulary and Age

(30, 900)
At 30 months, a child knows 900 words.

Figure 2.1

EXERCISE SET 2.1

Practice Exercises

Solve each equation in Exercises 1–42 using the addition property of equality. Be sure to check your proposed solutions.

1. $x - 7 = 13$

2. $y - 3 = -17$

3. $z + 5 = -12$

4. $z + 12 = -14$

5. $-3 = x + 14$

6. $-12 = x + 17$

7. $-18 = y - 5$

8. $-20 = y - 6$

9. $7 + z = 13$

10. $18 + z = 11$

11. $-3 + y = -17$

12. $-5 + y = -19$

13. $x + \dfrac{1}{3} = \dfrac{7}{3}$

14. $x + \dfrac{7}{8} = \dfrac{9}{8}$

15. $t + \dfrac{5}{6} = -\dfrac{7}{12}$

16. $t + \dfrac{2}{3} = -\dfrac{7}{6}$

17. $x - \dfrac{3}{4} = \dfrac{9}{2}$

18. $x - \dfrac{3}{5} = \dfrac{7}{10}$

19. $-\dfrac{1}{5} + y = -\dfrac{3}{4}$

20. $-\dfrac{1}{8} + y = -\dfrac{1}{4}$

21. $3.2 + x = 7.5$

22. $-2.7 + w = -5.3$

23. $x + \dfrac{3}{4} = -\dfrac{9}{2}$

24. $r + \dfrac{3}{5} = -\dfrac{7}{10}$

25. $5 = -13 + y$

26. $-11 = 8 + x$

27. $-\dfrac{3}{5} = -\dfrac{3}{2} + s$

28. $\dfrac{7}{3} = -\dfrac{5}{2} + z$

29. $830 + y = 520$

30. $-90 + t = -35$

31. $r + 3.7 = 8$

32. $x + 10.6 = -9$

33. $-3.7 + m = -3.7$

34. $y + \dfrac{7}{11} = \dfrac{7}{11}$

35. $6y + 3 - 5y = 14$

36. $-3x - 5 + 4x = 9$

37. $7 - 5x + 8 + 2x + 4x - 3 = 2 + 3 \cdot 5$

38. $13 - 3r + 2 + 6r - 2r - 2r - 1 = 3 + 2 \cdot 9$

39. $7y + 4 = 6y - 9$

40. $4r - 3 = 5 + 3r$

41. $18 - 7x = 12 - 6x$

42. $26 - 8s = 20 - 7s$

 ### Application Exercises

With more Americans using more prescription drugs for more conditions, the nation's drug bill is on the rise. The data shown in the line graph can be modeled by the formula

$$D - 15x = 62,$$

where D is the amount, in billions, spent on prescription drugs x years after 1995. Use this formula to solve Exercises 43–44.

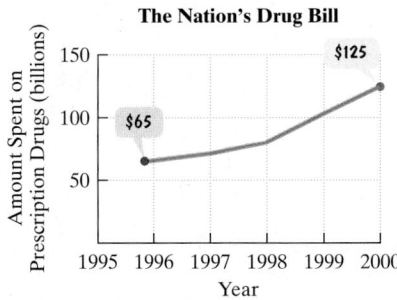

The Nation's Drug Bill

Source: Health Care Financing Administration

43. Predict the amount spent on prescription drugs in 2005.

44. Predict the amount spent on prescription drugs in 2002.

Formulas frequently appear in the business world. For example, the cost, C, of an item (the price paid by a retailer) plus the markup, M, on that item (the retailer's profit) equals the selling price, S, of the item. The formula is

$$C + M = S.$$

Use the formula to solve Exercises 45–46.

45. The selling price of a computer is $1850. If the markup on the computer is $150, find the cost to the retailer for the computer.

46. The selling price of a television is $650. If the cost to the retailer for the television is $520, find the markup.

The formula

$$d + 525,000 = 5000c$$

models the relationship between the annual number of deaths in the United States from heart disease, d, and the average cholesterol level, c, of blood. (Cholesterol level, c, is expressed in milligrams per deciliter of blood.) Use this formula to solve Exercises 47–48.

47. The average cholesterol level for people in the United States is 210. According to the formula, how many deaths per year from heart disease can be expected with this cholesterol level?

48. Suppose that the average cholesterol level for people in the United States could be reduced to 180. Determine the number of annual deaths from heart disease that can be expected with this reduced cholesterol level.

 ### Writing in Mathematics

49. What is an equation?

50. Explain how to determine whether a number is a solution to an equation.

51. State the addition property of equality and give an example.

52. Explain why $x + 2 = 9$ and $x + 2 = -6$ are not equivalent equations.

53. What is the difference between solving an equation such as

$$5y + 3 - 4y - 8 = 6 + 9$$

and simplifying an algebraic expression such as

$$5y + 3 - 4y - 8?$$

If there is a difference, which topic should be taught first? Why?

 ### Critical Thinking Exercises

54. Which one of the following statements is true?
   a. If $y - a = -b$, then $y = a + b$.
   b. If $y + 7 = 0$, then $y = 7$.
   c. The solution to $4 - x = -3x$ is -2.
   d. If 7 is added on one side of an equation, then it should be subtracted on the other side.

55. Solve for x: $|x| + 4 = 10$.

Technology Exercises

Use a calculator to solve each equation in Exercises 56–57.

56. $x - 7.0463 = -9.2714$

57. $6.9825 = 4.2296 + y$

 ### Review Exercises

58. Plot $(-3, 1)$ in rectangular coordinates. In which quadrant does the point lie? (Section 1.3, Example 1)

59. Simplify: $-16 - 8 \div 4 \cdot (-2)$. (Section 1.8, Example 4)

60. Simplify: $3[7x - 2(5x - 1)]$. (Section 1.8, Example 11)

▶ SECTION 2.2   The Multiplication Property of Equality

Objectives

1 Use the multiplication property of equality to solve equations.

2 Solve equations in the form $-x = c$.

3 Use the addition and multiplication properties to solve equations.

4 Solve applied problems using formulas.

SSM
PH Tutor   CD-   Video
Center   ROM

Could you live to be 125? The number of Americans aged 100 or older could approach 850,000 by 2050. Some scientists predict that by 2100, our descendants could live to be 200 years of age. In this section, we will see how a formula can be used to make these kinds of predictions as we turn to a new property for solving linear equations.

Using the Multiplication Property of Equality to Solve Equations

Can the addition property of equality be used to solve every linear equation in one variable? No. For example, consider the equation

$$\frac{x}{5} = 9.$$

We cannot isolate the variable x by adding or subtracting 5 on both sides. To get x by itself on the left side, multiply the left side by 5:

$$5 \cdot \frac{x}{5} = \left(5 \cdot \frac{1}{5}\right)x = 1\,x = x.$$

5 is the multiplicative inverse of $\frac{1}{5}$.

You must then multiply the right side by 5 also. By doing this, we are using the **multiplication property of equality**.

---

1 Use the multiplication property of equality to solve equations.

> ### The Multiplication Property of Equality
> The same nonzero real number (or algebraic expression) may multiply both sides of an equation without changing the solution. This can be expressed symbolically as follows:
> $$\text{If } a = b \text{ and } c \neq 0, \text{ then } ac = bc.$$

**EXAMPLE 1   Solving an Equation
                Using the Multiplication Property**

Solve the equation: $\dfrac{x}{5} = 9.$

Solution   We can isolate the variable, x, by multiplying both sides of the equation by 5.

$$\frac{x}{5} = 9 \qquad \text{This is the given equation.}$$

$$5 \cdot \frac{x}{5} = 5 \cdot 9 \qquad \text{Multiply both sides by 5.}$$

$$1x = 45 \qquad \text{Simplify.}$$

$$x = 45 \qquad \text{1x = x}$$

By substituting 45 for x in the given equation, we obtain the true statement $9 = 9$. This verifies that the solution is 45, or the solution set is $\{45\}$. ∎

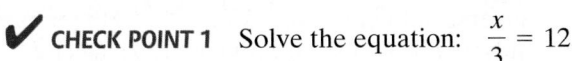 **CHECK POINT 1**   Solve the equation: $\frac{x}{3} = 12$.

When we use the multiplication property of equality, we multiply both sides of an equation by the same nonzero number. We know that division is multiplication by a multiplicative inverse. Thus, the multiplication property also lets us divide both sides of an equation by a nonzero number without changing the solution set.

EXAMPLE 2   Dividing Both Sides by the Same Number

Solve:

    a. $6x = 30$ **b.** $-7y = 56$ **c.** $-18.9 = 3z$.

Solution   In each equation, the variable is multiplied by a number. We can isolate the variable by dividing both sides of the equation by that number.

  a. $6x = 30$ This is the given equation.

$$\frac{6x}{6} = \frac{30}{6} \qquad \text{Divide both sides by 6.}$$

$$1x = 5 \qquad \text{Simplify.}$$

$$x = 5 \qquad \text{1x = x}$$

By substituting 5 for x in the given equation, we obtain the true statement $30 = 30$. The solution is 5, or the solution set is $\{5\}$.

  b. $-7y = 56$ This is the given equation.

$$\frac{-7y}{-7} = \frac{56}{-7} \qquad \text{Divide both sides by } -7.$$

$$1y = -8 \qquad \text{Simplify.}$$

$$y = -8 \qquad \text{1y = y}$$

By substituting -8 for y in the given equation, we obtain the true statement $56 = 56$. The solution is -8, or the solution set is $\{-8\}$.

  c. $-18.9 = 3z$ This is the given equation.

$$\frac{-18.9}{3} = \frac{3z}{3} \qquad \text{Divide both sides by 3.}$$

$$-6.3 = 1z \qquad \text{Simplify.}$$

$$-6.3 = z \qquad \text{1z = z}$$

By substituting -6.3 for z in the given equation, we obtain the true statement $-18.9 = -18.9$. The solution is -6.3, or the solution set is $\{-6.3\}$. ∎

✔ **CHECK POINT 2**   Solve:

a. $4x = 84$

b. $-11y = 44$

c. $-15.5 = 5z$.

Some equations have a variable term with a fractional coefficient. Here is an example:

$$\frac{3}{4}y = 12.$$

> The numeral coefficient of the term $\frac{3}{4}y$ is $\frac{3}{4}$.

To isolate the variable, multiply both sides of the equation by the multiplicative inverse of the fraction. For example, the multiplicative inverse of $\frac{3}{4}$ is $\frac{4}{3}$. Thus, we solve $\frac{3}{4}y = 12$ by multiplying both sides by $\frac{4}{3}$.

EXAMPLE 3   Using the Multiplication Property to Eliminate a Fractional Coefficient

Solve:

a. $\dfrac{3}{4}y = 12$

b. $9 = -\dfrac{3}{5}x$.

Solution

a. $\dfrac{3}{4}y = 12$ This is the given equation.

$\dfrac{4}{3}\left(\dfrac{3}{4}y\right) = \dfrac{4}{3} \cdot 12$ Multiply both sides by $\dfrac{4}{3}$, the multiplicative inverse of $\dfrac{3}{4}$.

$1y = 16$ On the left, $\dfrac{4}{3}\left(\dfrac{3}{4}y\right) = \left(\dfrac{4}{3}\cdot\dfrac{3}{4}\right)y = 1y$.

On the right, $\dfrac{4}{3}\cdot\dfrac{12}{1} = \dfrac{48}{3} = 16$.

$y = 16$ $1y = y$

By substituting 16 for y in the given equation, we obtain the true statement $12 = 12$. The solution is 16, or the solution set is $\{16\}$.

b. $9 = -\dfrac{3}{5}x$ This is the given equation.

$-\dfrac{5}{3}\cdot 9 = -\dfrac{5}{3}\left(-\dfrac{3}{5}x\right)$ Multiply both sides by $-\frac{5}{3}$, the multiplicative inverse of $-\frac{3}{5}$.

$-15 = 1x$ Simplify.

$-15 = x$ $1x = x$

By substituting -15 for x in the given equation, we obtain the true statement $9 = 9$. The solution is -15, or the solution set is $\{-15\}$. ■

> **Study Tip**
>
> The equation
> $$9 = -\frac{3}{5}x$$
> can be expressed as
> $$9 = -\frac{3x}{5}$$
> or
> $$9 = \frac{-3x}{5}.$$

✔ **CHECK POINT 3**   Solve:

a. $\dfrac{2}{3}y = 16$

b. $28 = -\dfrac{7}{4}x$.

2 Solve equations in the form $-x = c$.

Equations and Coefficients of −1

How do we solve an equation in the form $-x = c$, such as $-x = 4$? Because the equation means $-1x = 4$, we have not yet obtained a solution. The solution of an equation is obtained from the form $x =$ some number. The equation $-x = 4$ is not yet in this form. We still need to isolate x. We can do this by multiplying or dividing both sides of the equation by −1. We will multiply by −1.

EXAMPLE 4 Solving Equations in the Form $-x = c$

Solve:

    a. $-x = 4$ **b.** $-x = -7$.

Solution We multiply both sides of each equation by −1. This will isolate x on the left side.

    a.

$-x = 4$	This is the given equation.
$-1x = 4$	Rewrite $-x$ as $-1x$.
$(-1)(-1x) = (-1)(4)$	Multiply both sides by −1.
$1x = -4$	On the left, $(-1)(-1) = 1$. On the right, $(-1)(4) = -4$.
$x = -4$	$1x = x$

Check	$-x = 4$	This is the original equation.
	$-(-4) \overset{?}{=} 4$	Substitute −4 for x.
	$4 = 4$	$-(-a) = a$, so $-(-4) = 4$.

This true statement indicates that the solution is −4, or the solution set is $\{-4\}$.

    b.

$-x = -7$	This is the given equation.
$-1x = -7$	Rewrite $-x$ as $-1x$.
$(-1)(-1x) = (-1)(-7)$	Multiply both sides by −1.
$1x = 7$	$(-1)(-1) = 1$ and $(-1)(-7) = 7$.
$x = 7$	$1x = x$

By substituting 7 for x in the given equation, we obtain the true statement $-7 = -7$. The solution is 7, or the solution set is $\{7\}$. ∎

 **CHECK POINT 4** Solve:

    a. $-x = 5$ **b.** $-x = -3$.

3 Use the addition and multiplication properties to solve equations.

Equations Requiring Both the Addition and Multiplication Properties

When an equation does not contain fractions, we will often use the addition property of equality before the multiplication property of equality. Our overall goal is to isolate the variable with a coefficient of 1 on either the left or right side of the equation.

Here is the procedure that we will be using to solve the equations in the next three examples.

- Use the addition property of equality to isolate the variable term.
- Use the multiplication property of equality to isolate the variable.

EXAMPLE 5   Using Both the Addition and Multiplication Properties

Solve: $3x + 1 = 7$.

Solution   We begin by isolating the variable term, $3x$, subtracting 1 from both sides. Then we isolate the variable, x, by dividing both sides by 3.

- **Use the addition property of equality to isolate the variable term.**

$3x + 1 = 7$ This is the given equation

$3x + 1 - 1 = 7 - 1$ Use the addition property, subtracting 1 from both sides.

$3x = 6$ Simplify.

- **Use the multiplication property of equality to isolate the variable.**

$$\frac{3x}{3} = \frac{6}{3}$$ Divide both sides by 3.

$x = 2$ Simplify.

By substituting 2 for x in the given equation, we obtain the true statement $7 = 7$. The solution is 2, or the solution set is $\{2\}$. ∎

✔ **CHECK POINT 5**   Solve: $4x + 3 = 27$.

EXAMPLE 6   Using Both the Addition and Multiplication Properties

Solve: $-2y - 28 = 4$.

Solution   We begin by isolating the variable term, $-2y$, adding 28 to both sides. Then we isolate the variable, y, by dividing both sides by -2.

- **Use the addition property of equality to isolate the variable term.**

$-2y - 28 = 4$ This is the given equation.

$-2y - 28 + 28 = 4 + 28$ Use the addition property, adding 28 to both sides.

$-2y = 32$ Simplify.

- **Use the multiplication property of equality to isolate the variable.**

$$\frac{-2y}{-2} = \frac{32}{-2}$$ Divide both sides by -2.

$y = -16$ Simplify.

Take a moment to substitute -16 for y in the given equation. Do you obtain the true statement $4 = 4$? The solution is -16, or the solution set is $\{-16\}$. ∎

✔ **CHECK POINT 6**   Solve: $-4y - 15 = 25$.

EXAMPLE 7   Using Both the Addition and Multiplication Properties

Solve: $3x - 14 = -2x + 6$.

Solution   We will use the addition property to collect all terms involving x on the left and all numerical terms on the right. Then we will isolate the variable, x, by dividing both sides by its numerical coefficient.

- **Use the addition property of equality to isolate the variable term.**

$$3x - 14 = -2x + 6 \qquad \text{This is the given equation.}$$
$$3x + 2x - 14 = -2x + 2x + 6 \qquad \text{Add 2x to both sides.}$$
$$5x - 14 = 6 \qquad \text{Simplify.}$$
$$5x - 14 + 14 = 6 + 14 \qquad \text{Add 14 to both sides.}$$
$$5x = 20 \qquad \text{Simplify. The variable, 5x, is isolated on the left.}$$
$$\text{The numerical term, 20, is isolated on the right.}$$

- **Use the multiplication property of equality to isolate the variable.**

$$\frac{5x}{5} = \frac{20}{5} \qquad \text{Divide both sides by 5.}$$
$$x = 4 \qquad \text{Simplify.}$$

Check $\quad 3x - 14 = -2x + 6 \qquad \text{Use the original equation.}$

$$3(4) - 14 \stackrel{?}{=} -2(4) + 6 \qquad \text{Substitute the proposed solution, 4, for x.}$$
$$12 - 14 \stackrel{?}{=} -8 + 6 \qquad \text{Multiply.}$$
$$-2 = -2 \qquad \text{Simplify.}$$

The true statement $-2 = -2$ verifies that the solution is 4, or the solution set is $\{4\}$. ∎

✔ **CHECK POINT 7**   Solve: $2x - 15 = -4x + 21$.

4 Solve applied problems using formulas.

Applications

Your life expectancy is related to the year you were born. The graph in Figure 2.2 shows life expectancy in the United States by year of birth. The data for U.S. women shown in the bar graph can be modeled by the formula

$$E = 0.215t + 71.05,$$

where E is the life expectancy for women born t years after 1950.

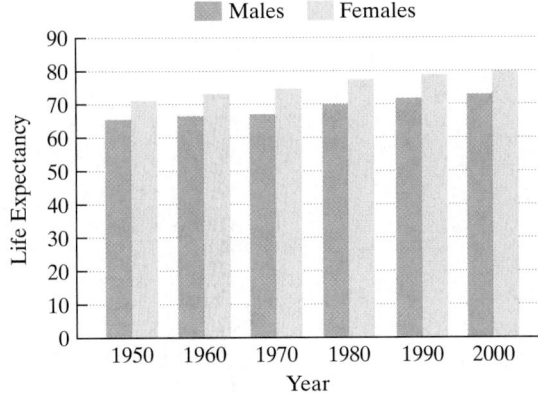

Life Expectancy by Year of Birth

Source: U.S. Bureau of the Census

Figure 2.2 Life expectancy by year of birth

EXAMPLE 8   Using the Formula For Life Expectancy

Use the formula

$$E = 0.215t + 71.05$$

to determine the year of birth for which U.S. women can expect to live 77.5 years.

Solution   We are given that the life expectancy for women is 77.5 years, so substitute 77.5 for E in the formula and solve for t.

$$E = 0.215t + 71.05 \qquad \text{This is the given formula}$$

$$77.5 = 0.215t + 71.05 \qquad \text{Replace E by 77.5 and solve for t.}$$

$$77.5 - 71.05 = 0.215t + 71.05 - 71.05 \qquad \text{Isolate the term containing t by subtracting 71.05 from both sides.}$$

$$6.45 = 0.215t \qquad \text{Simplify.}$$

$$\frac{6.45}{0.215} = \frac{0.215t}{0.215} \qquad \text{Divide both sides by 0.215.}$$

$$30 = t \qquad \text{Simplify: } 0.215\overline{)6.450}^{\;30.}$$

The formula indicates that U.S. women born 30 years after 1950, or in 1980, can expect to live 77.5 years. ◾

✔ **CHECK POINT 8**   The formula $D = 0.2F - 1$ models death rate from breast cancer per 100,000 women, D, and daily fat intake, F, in grams. The death rate of American women from breast cancer is 19 women per 100,000. What is the daily fat intake for women in America?

EXERCISE SET 2.2

Practice Exercises

Solve each equation in Exercises 1–28 using the multiplication property of equality. Be sure to check your proposed solutions.

1. $\dfrac{x}{3} = 4$

2. $\dfrac{x}{5} = 3$

3. $\dfrac{x}{-5} = 11$

4. $\dfrac{x}{-7} = 2$

5. $5y = 45$

6. $6y = 18$

7. $-7y = 56$

8. $-4y = 48$

9. $-24 = 8z$

10. $-25 = 5z$

11. $-15 = -3z$

12. $-45 = -9z$

13. $-8x = 2$

14. $-6x = 3$

15. $7y = 0$

16. $-3y = 0$

17. $\dfrac{2}{3}y = 8$

18. $\dfrac{3}{4}y = 12$

19. $21 = -\dfrac{7}{2}x$

20. $25 = -\dfrac{5}{8}x$

21. $-x = 7$

22. $-x = 10$

23. $-15 = -y$

24. $-22 = -y$

25. $-\dfrac{x}{5} = -10$

26. $-\dfrac{x}{7} = -1$

27. $2x - 8x = 24$

28. $8x - 5x = -21$

Solve each equation in Exercises 29–54 using both the addition and multiplication properties of equality. Check proposed solutions.

29. $2x + 1 = 11$

30. $2x + 5 = 13$

31. $2x - 3 = 9$

32. $3x - 2 = 9$

33. $-2y + 5 = 7$

34. $-3y + 4 = 13$

35. $-3y - 7 = -1$

36. $-2y - 5 = 7$

37. $12 = 4z + 3$

38. $14 = 5z - 21$

39. $-x - 3 = 3$

40. $-x - 5 = 5$

41. $6y = 2y - 12$

42. $8y = 3y - 10$

43. $3z = -2z - 15$

44. $2z = -4z + 18$

45. $-5x = -2x - 12$

46. $-7x = -3x - 8$

47. $8y + 4 = 2y - 5$

48. $5y + 6 = 3y - 6$

49. $6z - 5 = z + 5$

50. $6z - 3 = z + 2$

51. $6x + 14 = 2x - 2$

52. $9x + 2 = 6x - 4$

53. $-3y - 1 = 5 - 2y$

54. $-3y - 2 = -5 - 4y$

Application Exercises

The formula

$$M = \frac{n}{5}$$

models your distance, M, in miles, from a lightning strike in a thunderstorm if it takes n seconds to hear thunder after seeing lightning. Use this formula to solve Exercises 55–56.

55. If you are 2 miles away from the lightning flash, how long will it take the sound of thunder to reach you?

56. If you are 3 miles away from the lightning flash, how long will it take the sound of thunder to reach you?

The Mach number is a measurement of speed, named after the man who suggested it, Ernst Mach (1838–1916). The formula

$$M = \frac{A}{740}$$

indicates that the speed of an aircraft, A, in miles per hour, divided by the speed of sound, approximately 740 miles per hour, results in the Mach number, M. Use the formula to determine the speed, in miles per hour, of the aircrafts in Exercises 57–58. (Note: When an aircraft's speed increases beyond Mach 1, it is said to have broken the sound barrier.)

57.

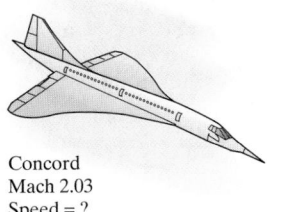

Concord
Mach 2.03
Speed = ?

58.

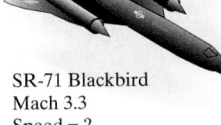

SR-71 Blackbird
Mach 3.3
Speed = ?

59. The formula $P = -0.5d + 100$ models the percentage, P, of lost hikers found in search and rescue missions when members of the search team walk parallel to one another separated by a distance of d yards. If a search and rescue team finds 70% of lost hikers, substitute 70 for P in the

formula and find the parallel distance of separation between members of the search party.

60. The formula $M = 420x + 720$ models the data for the amount of money lost to credit card fraud worldwide, M, in millions of dollars, x years after 1989. In which year did losses amount to 4080 million dollars?

Writing in Mathematics

61. State the multiplication property of equality and give an example.

62. Explain how to solve the equation $-x = -50$.

63. Explain how to solve the equation $2x + 8 = 5x - 3$.

64. In this section, we used a formula to find that U.S. women born in 1980 can expect to live 77.5 years. According to the Bureau of the Census, U.S. women born in 1980 can expect to live 77.4 years. Is there something wrong with the formula? Explain.

Critical Thinking Exercises

65. Which one of the following statements is true?
a. If $7x = 21$, then $x = 21 - 7$.
b. If $3x - 4 = 16$, then $3x = 12$.
c. If $3x + 7 = 0$, then $x = \frac{7}{3}$.
d. The solution to $6x = 0$ is not a natural number.

66. Write three equations whose solution is 5.

67. Take another look at the formula in Exercise 59. It indicates that the chance of lost hikers being found increases as searchers in any area get closer to one another. If this is the case, why not simply have almost no distance separating members of the search team to obtain near certainty of finding the lost hikers?

Technology Exercises

Solve each equation in Exercises 68–69. Use a calculator to help with the arithmetic. Check your solution using the calculator.

68. $3.7x - 19.46 = -9.988$
69. $-72.8y - 14.6 = -455.43 - 4.98y$

Review Exercises

70. Evaluate: $(-10)^2$. (Section 1.8, Example 1)
71. Evaluate: -10^2. (Section 1.8, Example 1)
72. Evaluate $x^3 - 4x$ if $x = -1$. (Section 1.8, Example 10)

▶ **SECTION 2.3**   *Solving Linear Equations*

Objectives

1. Solve linear equations.

2. Solve linear equations containing fractions.

3. Identify equations with no solution or infinitely many solutions.

4. Solve applied problems using formulas.

SSM PH Tutor CD- Video
Center ROM

Yes we overindulged, but it was delicious. Now if we could just grow a few inches, we'd be back in line with our recommended weights.

In this section, we will see how algebra models weight and height. To use this mathematical model, it would be helpful to have a systematic procedure for solving linear equations. We open the section with such a procedure.

1   Solve linear equations.

A Step-By-Step Procedure for Solving Linear Equations

Here is a step-by-step procedure for solving a linear equation in one variable. Not all of these steps are necessary to solve every equation.

> **Solving a Linear Equation**
>
> 1. Simplify the algebraic expression on each side.
> 2. Collect all the variable terms on one side and all the constant terms on the other side.
> 3. Isolate the variable and solve.
> 4. Check the proposed solution in the original equation.

EXAMPLE 1   Solving a Linear Equation

Solve and check: $2x - 8x + 40 = 13 - 3x - 3$.

Solution

Step 1   Simplify the algebraic expression on each side.

$$2x - 8x + 40 = 13 - 3x - 3 \quad \text{This is the given equation.}$$
$$-6x + 40 = 10 - 3x \quad \text{Combine like terms: } 2x - 8x = -6x \text{ and } 13 - 3 = 10.$$

Step 2   Collect variable terms on one side and constant terms on the other side.
We will collect variable terms on the left by adding $3x$ to both sides. We will collect the numbers on the right by subtracting 40 from both sides.

$$-6x + 40 + 3x = 10 - 3x + 3x \quad \text{Add } 3x \text{ to both sides.}$$
$$-3x + 40 = 10 \quad \text{Simplify: } -6x + 3x = -3x.$$
$$-3x + 40 - 40 = 10 - 40 \quad \text{Subtract 40 from both sides.}$$
$$-3x = -30 \quad \text{Simplify.}$$

Discover for Yourself

Solve the equation in Example 1 by collecting terms with the variable on the right and numbers on the left. What do you observe?

Step 3   Isolate the variable and solve. We isolate the variable, x, by dividing both sides by -3.

$$\frac{-3x}{-3} = \frac{-30}{-3} \quad \text{Divide both sides by } -3.$$

$$x = 10 \quad \text{Simplify.}$$

Step 4   Check the proposed solution in the original equation. Substitute 10 for x in the original equation.

$$2x - 8x + 40 = 13 - 3x - 3 \quad \text{This is the original equation.}$$

$$2 \cdot 10 - 8 \cdot 10 + 40 \overset{?}{=} 13 - 3 \cdot 10 - 3 \quad \text{Substitute 10 for x.}$$

$$20 - 80 + 40 \overset{?}{=} 13 - 30 - 3 \quad \text{Perform the indicated multiplications.}$$

$$-60 + 40 \overset{?}{=} -17 - 3 \quad \text{Subtract: } 20 - 80 = -60 \text{ and } 13 - 30 = -17.$$

$$-20 = -20 \quad \text{Simplify.}$$

By substituting 10 for x in the given equation, we obtain the true statement $-20 = -20$. This verifies that the solution is 10, or the solution set is $\{10\}$. ■

✔ **CHECK POINT 1**   Solve and check: $-7x + 25 + 3x = 16 - 2x - 3$.

EXAMPLE 2   Solving a Linear Equation

Solve and check: $5x = 8(x + 3)$.

Solution

Step 1   Simplify the algebraic expression on each side. Use the distributive property to remove parentheses on the right.

$$5x = 8(x + 3) \quad \text{This is the given equation.}$$

$$5x = 8x + 24 \quad \text{Use the distributive property.}$$

Step 2   Collect variable terms on one side and constant terms on the other side. We will collect variable terms on the left by subtracting $8x$ from both sides. The only constant term, 24, is already on the right.

$$5x - 8x = 8x + 24 - 8x \quad \text{Subtract 8x from both sides.}$$

$$-3x = 24 \quad \text{Simplify: } 5x - 8x = -3x.$$

Step 3   Isolate the variable and solve. We isolate the variable, x, by dividing both sides by -3.

$$\frac{-3x}{-3} = \frac{24}{-3} \quad \text{Divide both sides by } -3.$$

$$x = -8 \quad \text{Simplify.}$$

Step 4   Check the proposed solution in the original equation. Substitute -8 for x in the original equation.

$$5x = 8(x + 3) \quad \text{This is the original equation.}$$

$$5(-8) \overset{?}{=} 8(-8 + 3) \quad \text{Substitute } -8 \text{ for x.}$$

$$5(-8) \overset{?}{=} 8(-5) \quad \text{Perform the addition in parentheses: } -8 + 3 = -5.$$

$$-40 = -40 \quad \text{Multiply.}$$

The true statement $-40 = -40$ verifies that -8 is the solution, or the solution set is $\{-8\}$. ■

The compact, symbolic notation of algebra enables us to use a clear step-by-step method for solving equations, designed to avoid the confusion shown in the painting. Squeak Carnwath "Equations" 1981, oil on cotton canvas 96 in. h × 72 in. w.

✔ **CHECK POINT 2** Solve and check: $8x = 2(x + 6)$.

EXAMPLE 3 Solving a Linear Equation

Solve and check: $2(x - 3) - 17 = 13 - 3(x + 2)$.

Solution

Step 1 **Simplify the algebraic expression on each side.**

$$2(x - 3) - 17 = 13 - 3(x + 2)$$ This is the given equation.
$$2x - 6 - 17 = 13 - 3x - 6$$ Use the distributive property.
$$2x - 23 = -3x + 7$$ Combine like terms.

Step 2 **Collect variable terms on one side and constant terms on the other side.** We will collect variable terms on the left by adding $3x$ to both sides. We will collect the numbers on the right by adding 23 to both sides.

$$2x - 23 + 3x = -3x + 7 + 3x$$ Add 3x to both sides.
$$5x - 23 = 7$$ Simplify: 2x + 3x = 5x.
$$5x - 23 + 23 = 7 + 23$$ Add 23 to both sides.
$$5x = 30$$ Simplify.

Step 3 **Isolate the variable and solve.** We isolate the variable, x, by dividing both sides by 5.

$$\frac{5x}{5} = \frac{30}{5}$$ Divide both sides by 5.
$$x = 6$$ Simplify.

Step 4 **Check the proposed solution in the original equation.** Substitute 6 for x in the original equation.

$$2(x - 3) - 17 = 13 - 3(x + 2)$$ This is the original equation.
$$2(6 - 3) - 17 \stackrel{?}{=} 13 - 3(6 + 2)$$ Substitute 6 for x.
$$2(3) - 17 \stackrel{?}{=} 13 - 3(8)$$ Simplify inside parentheses.
$$6 - 17 \stackrel{?}{=} 13 - 24$$ Multiply.
$$-11 = -11$$

The true statement $-11 = -11$ verifies that 6 is the solution, or the solution set is {6}. ∎

✔ **CHECK POINT 3** Solve and check: $4(2x + 1) - 29 = 3(2x - 5)$.

2 Solve linear equations containing fractions.

Linear Equations with Fractions

Equations are easier to solve when they do not contain fractions. How do we solve equations involving fractions? We begin by multiplying both sides of the equation by the least common denominator. The least common denominator is the smallest number that all denominators will divide into. Multiplying every term on both sides of the equation by the least common denominator will eliminate the fractions in the equation. Example 4 shows how we "clear an equation of fractions."

EXAMPLE 4 Solving a Linear Equation Involving Fractions

Solve and check: $\dfrac{3x}{2} = \dfrac{x}{5} - \dfrac{39}{5}$.

Solution The denominators are 2, 5, and 5. The smallest number that is divisible by 2, 5, and 5 is 10. We begin by multiplying both sides of the equation by 10, the least common denominator.

$$\frac{3x}{2} = \frac{x}{5} - \frac{39}{5}$$ This is the given equation.

$$10 \cdot \frac{3x}{2} = 10\left(\frac{x}{5} - \frac{39}{5}\right)$$ Multiply both sides by 10.

$$10 \cdot \frac{3x}{2} = 10 \cdot \frac{x}{5} - 10 \cdot \frac{39}{5}$$ Use the distributive property. Be sure to multiply all terms by 10.

$$\overset{5}{\cancel{10}} \cdot \frac{3x}{\underset{1}{\cancel{2}}} = \overset{2}{\cancel{10}} \cdot \frac{x}{\underset{1}{\cancel{5}}} - \overset{2}{\cancel{10}} \cdot \frac{39}{\underset{1}{\cancel{5}}}$$ Divide out common factors in the multiplications.

$$15x = 2x - 78$$ Complete the multiplication. The fractions are now cleared.

At this point, we have an equation similar to those we have previously solved. Collect the variable terms on one side and the constant terms on the other side.

$$15x - 2x = 2x - 2x - 78$$ Subtract 2x to get the variable terms on the left.

$$13x = -78$$ Simplify.

Isolate x by dividing both sides by 13.

$$\frac{13x}{13} = \frac{-78}{13}$$ Divide both sides by 13.

$$x = -6$$ Simplify.

Check the proposed solution. Substitute -6 for x in the original equation. You should obtain $-9 = -9$. This true statement verifies that the solution is -6, or the solution set is $\{-6\}$. ∎

 CHECK POINT 4 Solve and check: $\dfrac{x}{4} = \dfrac{2x}{3} + \dfrac{5}{6}$.

③ Identify equations with no solution or infinitely many solutions.

Linear Equations with No Solution or Infinitely Many Solutions

Thus far, each equation that we have solved has had a single solution. However, some equations are not true for even one real number. Such an equation is called an **inconsistent equation**, or a **contradiction**. Here is an example of such an equation:

$$x = x + 4.$$

There is no number that is equal to itself plus 4. This equation has no solution. Its solution set is written either as

$$\{\ \} \text{ or } \varnothing.$$

These symbols stand for the empty set, a set with no elements.

An equation that is true for all real numbers is called an **identity**. An example of an identity is

$$x + 3 = x + 2 + 1.$$

Every number plus 3 is equal to that number plus 2 plus 1. Every real number is a solution to this equation. Thus, the solution set to this equation is the set of all real numbers. This set is written as $\mathbb{R}$. In section 2.7, you will learn a second notation for representing the set of all real numbers.

If you attempt to solve an equation with no solution, you will eliminate the variable and obtain a false statement, such as $2 = 5$. If you attempt to solve an equation that is true for every real number, you will eliminate the variable and obtain a true statement, such as $4 = 4$.

EXAMPLE 5 Solving a Linear Equation

Solve: $2x + 6 = 2(x + 4)$.

Solution

$2x + 6 = 2(x + 4)$	This is the given equation.
$2x + 6 = 2x + 8$	Use the distributive property.
$2x + 6 - 2x = 2x + 8 - 2x$	Subtract 2x from both sides.
$6 = 8$	Simplify.

The original equation is equivalent to the false statement $6 = 8$, which is false for every value of x. The equation is inconsistent and has no solution. You can express this by writing "no solution" or using one of the symbols for the empty set, $\{\ \}$ or $\varnothing$. ∎

✔ **CHECK POINT 5** Solve: $3x + 7 = 3(x + 1)$.

EXAMPLE 6 Solving a Linear Equation

Solve: $-3x + 5 + 5x = 4x - 2x + 5$.

Solution

$-3x + 5 + 5x = 4x - 2x + 5$	This is the given equation.
$2x + 5 = 2x + 5$	Combine like terms: $-3x + 5x = 2x$ and $4x - 2x = 2x$.
$2x + 5 - 2x = 2x + 5 - 2x$	Subtract 2x from both sides.
$5 = 5$	Simplify.

The original equation is equivalent to the true statement $5 = 5$, which is true for every value of x. The equation is an identity and all real numbers are solutions. You can express this by writing "all real numbers" or using the notation $\mathbb{R}$ for the set of all real numbers. ∎

✔ **CHECK POINT 6** Solve: $3(x - 1) + 9 = 8x + 6 - 5x$.

An equation that is not an identity, but that is true for at least one real number, is called a **conditional equation**. The equation $2x + 3 = 17$ is an example of a conditional equation. The equation is not an identity and is true only if x is 7.

4 Solve applied problems using formulas.

Applications

The next example shows how our procedure for solving equations with fractions can be used to find the value of a variable in a mathematical model.

EXAMPLE 7 Modeling Weight and Height

The formula

$$\frac{W}{2} - 3H = 53$$

models the recommended weight, W, in pounds, for a male, where H represents the man's height, in inches, over 5 feet. What is the recommended weight for a man who is 6 feet, 3 inches tall?

Solution Keep in mind that H represents height in inches *above 5 feet*. A man who is 6 feet, 3 inches tall is 1 foot, 3 inches above 5 feet. Because 12 inches = 1 foot, he is $12 + 3$, or 15 inches, above 5 feet tall. To find his recommended weight, we substitute 15 for H in the formula and solve for W:

$$\frac{W}{2} - 3H = 53 \quad \text{This is the given formula.}$$

$$\frac{W}{2} - 3 \cdot 15 = 53 \quad \text{Substitute 15 for H.}$$

$$\frac{W}{2} - 45 = 53 \quad \text{Multiply: } 3 \cdot 15 = 45.$$

Multiply both sides of the equation by 2, the least common denominator:

$$2\left(\frac{W}{2} - 45\right) = 2 \cdot 53$$

$$2 \cdot \frac{W}{2} - 2 \cdot 45 = 2 \cdot 53 \quad \text{Use the distributive property.}$$

$$W - 90 = 106 \quad \text{Multiply: } 2 \cdot \frac{W}{2} = \frac{2}{1} \cdot \frac{W}{2} = \frac{2W}{2} = 1W = W.$$

$$W - 90 + 90 = 106 + 90 \quad \text{Add 90 to both sides.}$$

$$W = 196$$

The recommended weight for a man whose height is 6 feet, 3 inches is 196 pounds. ∎

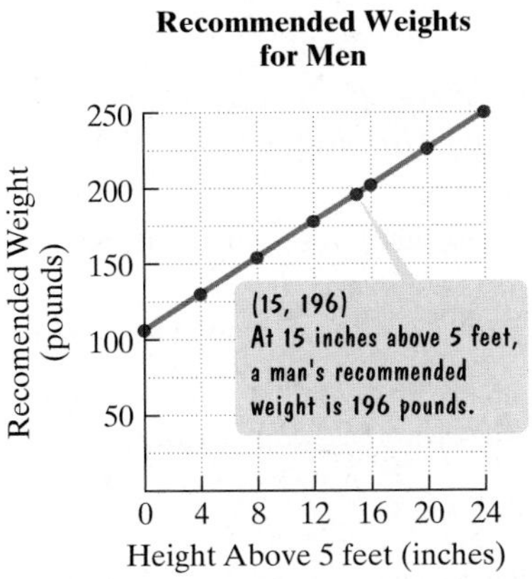

The points in the rectangular coordinate system in Figure 2.3 allow us to "see" the formula $\frac{W}{2} - 3H = 53$. The x-coordinate of each point represents a man's height, in inches, above 5 feet. The y-coordinate represents that man's recommended weight, in pounds. The points are rising steadily from left to right. This shows the increasing recommended weights as men's heights increase.

✔ **CHECK POINT 7** Use the formula $\frac{W}{2} - 3H = 53$ to find the recommended weight for a man who is 5 feet, 3 inches tall.

Recommended Weights for Men

(15, 196) At 15 inches above 5 feet, a man's recommended weight is 196 pounds.

Figure 2.3

EXERCISE SET 2.3

Practice Exercises

In Exercises 1–30, solve each equation. Be sure to check your proposed solution by substituting it for the variable in the given equation.

1. $5x + 3x - 4x = 10 + 2$
2. $4x + 8x - 2x = 20 - 15$
3. $3x - 7x + 30 = 10 - 2x$
4. $2x - 8x + 35 = 5 - 3x$
5. $3x + 6 - x = 8 + 3x - 6$
6. $3x + 2 - x = 6 + 3x - 8$
7. $3(x - 2) = 12$
8. $3(x + 2) = 21$
9. $7(2x - 1) = 21$
10. $4(2x - 3) = 36$
11. $25 = 5(3y + 4)$
12. $21 = 3(5y + 2)$
13. $2(4z + 3) - 8 = 46$
14. $3(3z + 5) - 7 = 89$
15. $6x - (3x + 10) = 14$
16. $5x - (2x + 14) = 10$
17. $14(y - 2) = 10(y + 4)$
18. $6(y + 2) = 2(y - 2)$
19. $3(5 - x) = 4(2x + 1)$
20. $3(3x - 1) = 4(3 + 3x)$
21. $8(y + 2) = 2(3y + 4)$
22. $8(y + 3) = 3(2y + 12)$
23. $3(x + 1) = 7(x - 2) - 3$
24. $5x - 4(x + 9) = 2x - 3$
25. $5(2x - 8) - 2 = 5(x - 3) + 3$
26. $7(3x - 2) + 5 = 6(2x - 1) + 24$
27. $6 = -4(1 - x) + 3(x + 1)$
28. $100 = -(x - 1) + 4(x - 6)$
29. $10(z + 4) - 4(z - 2) = 3(z - 1) + 2(z - 3)$
30. $-2(z - 4) - (3z - 2) = -2 - (6z - 2)$

Solve and check each equation in Exercises 31–46. Begin your work by rewriting each equation without fractions.

31. $\dfrac{x}{5} - 4 = -6$
32. $\dfrac{x}{2} + 13 = -22$
33. $\dfrac{2x}{3} - 5 = 7$
34. $\dfrac{3x}{4} - 9 = -6$
35. $\dfrac{2y}{3} - \dfrac{3}{4} = \dfrac{5}{12}$
36. $\dfrac{3y}{4} - \dfrac{2}{3} = \dfrac{7}{12}$
37. $\dfrac{x}{3} + \dfrac{x}{2} = \dfrac{5}{6}$
38. $\dfrac{x}{4} - \dfrac{x}{5} = 1$
39. $20 - \dfrac{z}{3} = \dfrac{z}{2}$
40. $\dfrac{z}{5} - \dfrac{1}{2} = \dfrac{z}{6}$

41. $\dfrac{y}{3} + \dfrac{2}{5} = \dfrac{y}{5} - \dfrac{2}{5}$
42. $\dfrac{y}{12} + \dfrac{1}{6} = \dfrac{y}{2} - \dfrac{1}{4}$
43. $\dfrac{3x}{4} - 3 = \dfrac{x}{2} + 2$
44. $\dfrac{3x}{5} - \dfrac{6}{15} = \dfrac{x}{3} + \dfrac{2}{5}$
45. $\dfrac{3x}{5} - x = \dfrac{x}{10} - \dfrac{5}{2}$
46. $2x - \dfrac{2x}{7} = \dfrac{x}{2} + \dfrac{17}{2}$

In Exercises 47–58, solve each equation. Identify equations that have no solution, or equations that are true for all real numbers.

47. $3x - 7 = 3(x + 1)$
48. $2(x - 5) = 2x + 10$
49. $2(x + 4) = 4x + 5 - 2x + 3$
50. $3(x - 1) = 8x + 6 - 5x - 9$
51. $4x + 1 - 5x = 5 - (x + 4)$
52. $5x - 5 = 3x - 7 + 2(x + 1)$
53. $4(x + 2) + 1 = 7x - 3(x - 2)$
54. $5x - 3(x + 1) = 2(x + 3) - 5$
55. $\dfrac{x}{3} + 2 = \dfrac{x}{3}$
56. $\dfrac{x}{4} + 3 = \dfrac{x}{4}$
57. $3 - x = 2x + 3$
58. $5 - x = 4x + 5$

Application Exercises

In Massachusetts, speeding fines are determined by the formula

$$F = 10(x - 65) + 50$$

where F is the cost, in dollars, of the fine if a person is caught driving x miles per hour. Use this formula to solve Exercises 59–60.

59. If a fine comes to $250, how fast was that person speeding?
60. If a fine comes to $400, how fast was that person speeding?

The formula

$$\dfrac{c}{2} + 80 = 2F$$

models the relationship between temperature, F, in degrees Fahrenheit, and the number of cricket chirps per minute, c, for the snow tree cricket. Use this mathematical model to solve Exercises 61–62.

61. Find the number of chirps per minute at a temperature of 70°F.

62. Find the number of chirps per minute at a temperature of 80°F.

The formula

$$p = 15 + \frac{5d}{11}$$

describes the pressure of sea water, p, in pounds per square foot, at a depth of d feet below the surface. Use the formula to solve Exercises 63–64.

63. The record depth for breath-held diving, by Francisco Ferreras (Cuba) off Grand Bahama Island, on November 14, 1993, involved pressure of 201 pounds per square foot. To what depth did Ferreras descend on this ill-advised venture? (He was underwater for 2 minutes and 9 seconds!)

64. At what depth is the pressure 20 pounds per square foot?

Writing in Mathematics

65. In your own words, describe how to solve a linear equation.

66. Explain how to solve a linear equation containing fractions.

67. Suppose that you solve $\frac{x}{5} - \frac{x}{2} = 1$ by multiplying both sides by 20, rather than the least common denominator of 5 and 2 (namely, 10). Describe what happens. If you get the correct solution, why do you think we clear the equation of fractions by multiplying by the *least* common denominator?

68. Suppose you are an algebra teacher grading the following solution on an examination:

Solve: $-3(x - 6) = 2 - x$
Solution: $-3x - 18 = 2 - x$
            $-2x - 18 = 2$
            $-2x = -16$
            $x = 8.$

You should note that 8 checks, and the solution is 8. The student who worked the problem therefore wants full credit. Can you find any errors in the solution? If full credit is 10 points, how many points should you give the student? Justify your position.

Critical Thinking Exercises

69. Which one of the following statements is true?
   a. The equation $3(x + 4) = 3(4 + x)$ has precisely one solution.
   b. The equation $2y + 5 = 0$ is equivalent to $2y = 5$.
   c. If $2 - 3y = 11$ and the solution to the equation is substituted into $y^2 + 2y - 3$, a number results that is neither positive nor negative.
   d. The equation $x + \frac{1}{3} = \frac{1}{2}$ is equivalent to $x + 2 = 3$.

70. A woman's height, h, is related to the length of the femur, f (the bone from the knee to the hip socket), by the formula $f = 0.432h - 10.44$. Both h and f are measured in inches. A partial skeleton is found of a woman in which the femur is 16 inches long. Police find the skeleton in an area where a woman slightly over 5 feet tall has been missing for over a year. Can the partial skeleton be that of the missing woman? Explain.

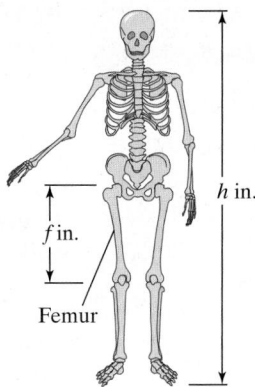

h in.

f in.

Femur

Solve each equation in Exercises 71–72.

71. $\frac{2x - 3}{9} + \frac{x - 3}{2} = \frac{x + 5}{6} - 1$

72. $2(3x + 4) = 3x + 2[3(x - 1) + 2]$

Technology Exercises

Solve each equation in Exercises 73–74. Use a calculator to help with the arithmetic. Check your solution using the calculator.

73. $2.24y - 9.28 = 5.74y + 5.42$

74. $4.8y + 32.5 = 124.8 - 9.4y$

Review Exercises

In Exercises 75–76, insert either $<$ or $>$ in the box between each pair of numbers to make a true statement.

75. $-24 \ \square \ -20$ (Section 1.2, Example 6)

76. $-\frac{1}{3} \ \square \ -\frac{1}{5}$ (Section 1.2, Example 6)

77. Simplify: $-9 - 11 + 7 - (-3)$. (Section 1.6, Example 3)

▶ SECTION 2.4 *Formulas and Percents*

Objectives

1 Solve for a variable in a formula.

2 Express a decimal as a percent.

3 Express a percent as a decimal.

4 Use the percent formula.

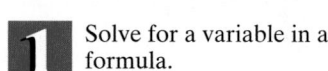

SSM   PH Tutor   CD-   Video
      Center   ROM

Is an early retirement awaiting you?

Dreaming of being a millionaire and an early retirement? Your dream is shared by a majority of college students. Numerical information about this shared dream is given with percents. In this section, you will learn to use a formula that will help you understand percents and their applications.

1 Solve for a variable in a formula.

Solving for a Variable in a Formula

We have seen how the addition and multiplication properties of equality are used to find the value of a variable in a formula. These properties can also be used to solve formulas for a specified variable. For example, consider the formula:

$$C + M = S.$$

This formula states that the cost, C, of an item (the price paid by the retailer) plus the markup, M, on that item (the retailer's profit) equals the selling price, S, of the item. How do we solve this formula for C? Use the addition property to isolate C by subtracting M from both sides.

> We need to isolate C.

$$C + M = S \qquad \text{This is the given formula.}$$
$$C + M - M = S - M \qquad \text{Subtract M from both sides.}$$
$$C = S - M \qquad \text{Simplify.}$$

Our final result, $C = S - M$, tells us that the cost of an item for a retailer is the item's selling price minus its markup.

You can solve for a variable in a formula by isolating the variable on one side of the equation. Use the addition property of equality to isolate all terms with the specified variable on one side of the equation and all terms without the specified variable on the other side. Then use the multiplication property of equality to get the specified variable alone.

Our first example involves the formula for the area of a rectangle. The **area of a two-dimensional figure** is the number of square units it takes to fill the interior of the figure. A **square unit** is a square, each of whose sides is one unit in length, illustrated in Figure 2.4. The figure shows that there are 12 square units contained within the rectangle. The area of the rectangle is 12 square units. Notice that the area can be determined in the following manner:

Square unit of measure

Figure 2.4 The area of the region on the left is 12 square units.

Across   Down

$$4 \text{ units} \cdot 3 \text{ units} = 4 \cdot 3 \cdot \text{units} \cdot \text{units} = 12 \text{ square units.}$$

The area of a rectangle is the product of the distance across, its length, and the distance down, its width.

Area of a Rectangle

The area, A, of a rectangle with length l and width w is given by the formula

$$A = lw.$$

EXAMPLE 1 Solving for a Variable in a Formula

Solve the formula $A = lw$ for w.

Solution Our goal is to get w by itself on one side of the formula. There is only one term with w, lw, and it is already isolated on the right side. We isolate w on the right by using the multiplication property of equality and dividing both sides by l.

We need to isolate w.

$$A = lw \qquad \text{This is the given formula.}$$

$$\frac{A}{l} = \frac{lw}{l} \qquad \text{Isolate } w \text{ by dividing both sides by } l.$$

$$\frac{A}{l} = w \qquad \text{Simplify: } \frac{lw}{l} = 1w = w.$$

The formula solved for w is $\dfrac{A}{l} = w$ or $w = \dfrac{A}{l}$. Thus, the area of a rectangle divided by its length is equal to its width. ∎

✔ **CHECK POINT 1** Solve the formula $A = lw$ for l.

The perimeter, P, of a two-dimensional figure is the sum of the lengths of its sides. Perimeter is measured in linear units, such as inches, feet, yards, meters, or kilometers.

Example 2 involves the perimeter of a rectangle. Because perimeter is the sum of the lengths of the sides, the perimeter of the rectangle shown in Figure 2.5 is $l + w + l + w$. This can be expressed as

$$P = 2l + 2w.$$

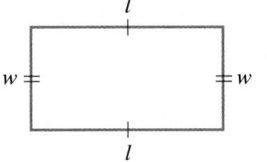

Figure 2.5 A rectangle with length l and width w

Perimeter of a Rectangle

The perimeter, P, of a rectangle with length l and width w is given by the formula

$$P = 2l + 2w.$$

The perimeter of a rectangle is the sum of twice the length and twice the width.

EXAMPLE 2   Solving for a Variable in a Formula

Solve the formula $2l + 2w = P$ for w.

Solution   First, isolate $2w$ on the left by subtracting $2l$ from both sides. Then solve for w by dividing both sides by 2.

> We need to isolate w.

$$2l + 2w = P \qquad \text{This is the given formula.}$$

$$2l - 2l + 2w = P - 2l \qquad \text{Isolate 2w by subtracting 2l from both sides.}$$

$$2w = P - 2l \qquad \text{Simplify.}$$

$$\frac{2w}{2} = \frac{P - 2l}{2} \qquad \text{Isolate w by dividing both sides by 2.}$$

$$w = \frac{P - 2l}{2} \qquad \text{Simplify.} \qquad \blacksquare$$

✔ **CHECK POINT 2**   Solve the formula $2l + 2w = P$ for l.

EXAMPLE 3   Solving for a Variable in a Formula

The total price of an article purchased on a monthly deferred payment plan is described by the following formula:

$$T = D + pm.$$

In this formula, T is the total price, D is the down payment, p is the monthly payment, and m is the number of months one pays. Solve the formula for p.

Solution   First, isolate pm on the right by subtracting D from both sides. Then isolate p from pm by dividing both sides of the formula by m.

> We need to isolate p.

$$T = D + pm \qquad \text{This is the given formula. We want p alone.}$$

$$T - D = D - D + pm \qquad \text{Isolate pm by subtracting D from both sides.}$$

$$T - D = pm \qquad \text{Simplify.}$$

$$\frac{T - D}{m} = \frac{pm}{m} \qquad \text{Now isolate p by dividing both sides by m.}$$

$$\frac{T - D}{m} = p \qquad \text{Simplify: } \frac{pm}{m} = \frac{p \cdot \overset{1}{\cancel{m}}}{\underset{1}{\cancel{m}}} = p \cdot 1 = p. \qquad \blacksquare$$

✔ **CHECK POINT 3**   Solve the formula $T = D + pm$ for m.

The next example has a formula that contains a fraction. When formulas contain fractions, we first eliminate the fractions by multiplying both sides of the formula by the least common denominator. Then we solve for the specified variable.

EXAMPLE 4   Solving for a Variable in a Formula Containing a Fraction

Solve the formula $\dfrac{W}{2} - 3H = 53$ for W.

Solution   Do you remember seeing this formula in the last section? It models the recommended weight, W, for a male, where H represents the man's height, in inches, over 5 feet. We begin by multiplying both sides of the formula by 2 to eliminate the fraction. Then we isolate the variable W.

$$\frac{W}{2} - 3H = 53 \qquad \text{This is the given formula.}$$

$$2\left(\frac{W}{2} - 3H\right) = 2 \cdot 53 \qquad \text{Multiply both sides by 2.}$$

$$2 \cdot \frac{W}{2} - 2 \cdot 3H = 2 \cdot 53 \qquad \text{Use the distributive property.}$$

We need to isolate W.

$$W - 6H = 106 \qquad \text{Simplify.}$$

$$W - 6H + 6H = 106 + 6H \qquad \text{Isolate } W \text{ by adding } 6H \text{ to both sides.}$$

$$W = 106 + 6H \qquad \text{Simplify.}$$

This form of the formula makes it easy to find a man's recommended weight, W, if we know his height, H, in inches, over 5 feet. ∎

✔ **CHECK POINT 4**   Solve for x: $\dfrac{x}{3} - 4y = 5$.

Before turning to a formula involving percent, let's review some of the basics of percent.

Basics of Percent

Did you know that one of the most common ways that you are given numerical information is with percents? **Percents** are the result of expressing numbers as a part of 100. The word *percent* means *per hundred*. For example, the circle graph in Figure 2.6 shows that 57 out of every 100 one-family houses have three bedrooms. Thus, $\frac{57}{100} = 57\%$, indicating that 57% of the houses have three bedrooms. The percent sign, %, is used to indicate the number of parts out of one hundred parts.

By definition, $57\% = \frac{57}{100}$. We can express the fraction $\frac{57}{100}$ in decimal notation as 0.57. Thus, $0.57 = 57\%$. Here is a general procedure for expressing a decimal number as a percent.

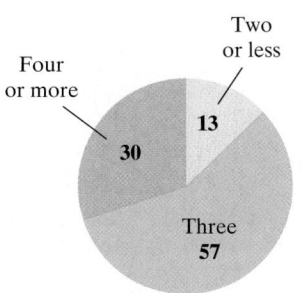

Figure 2.6 Number of bedrooms in privately owned one-family U.S. houses per 100 houses. *Source:* U.S. Census Bureau and HUD

Expressing a Decimal Number as a Percent

1. Move the decimal point two places to the right.
2. Attach a percent sign.

2 Express a decimal as a percent.

EXAMPLE 5 Expressing a Decimal as a Percent

Express 0.47 as a percent.

Solution

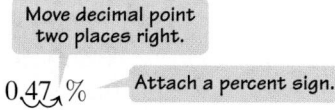

Thus, $0.47 = 47\%$. ∎

✔ **CHECK POINT 5** Express 0.023 as a percent.

3 Express a percent as a decimal.

We reverse the procedure of Example 5 to express a percent as a decimal number.

Expressing a Percent as a Decimal Number

1. Move the decimal point two places to the left.
2. Remove the percent sign.

EXAMPLE 6 Expressing Percents as Decimals

Express each percent as a decimal.

a. 19% **b.** 180%

Solution Use the two steps in the box.

a. $19\% = 19.\% = 0.19$ ⚡

> The percent sign is removed.
> The decimal point starts at the far right.
> The decimal point is moved two places to the left.

Thus, $19\% = 0.19$.

b. $180\% = 1.80$ ⚡ $= 1.80$ or 1.8. ∎

✔ **CHECK POINT 6** Express each percent as a decimal.
a. 67% **b.** 250%

4 Use the percent formula.

A Formula Involving Percent

Percents are useful in comparing two numbers. To compare the number A to the number B using a percent P, the following formula is used.

> A   is   P percent   of   B.
$$A = P \cdot B$$

In the formula

$$A = PB$$

B = the base number, P = the percent (in decimal form), and A = the number compared to B.

There are three basic types of percent problems that can be solved using the percent formula

$$A = PB. \quad \text{A is P percent of B.}$$

Question	Given	Percent Formula
What is P percent of B?	P and B	Solve for A.
A is P percent of what?	A and P	Solve for B.
A is what percent of B?	A and B	Solve for P.

Let's look at an example of each type of problem.

EXAMPLE 7   **Using the Percent Formula: What is P Percent of B?**

What is 8% of 20?

Solution   We use the formula $A = PB$: A is P percent of B. We are interested in finding the quantity A in this formula.

What   is   8%   of   20?

$$A = 0.08 \cdot 20 \quad \text{Express 8\% as 0.08.}$$

$$A = 1.6 \quad \text{Multiply:} \quad \begin{array}{r} .08 \\ \times\ 20 \\ \hline 1.60 \end{array}$$

Thus, 1.6 is 8% of 20. The answer is 1.6. ∎

✔ **CHECK POINT 7**   What is 9% of 50?

EXAMPLE 8   **Using the Percent Formula: A is P Percent of What?**

4 is 25% of what?

Solution   We use the formula $A = PB$: A is P percent of B. We are interested in finding the quantity B in this formula.

4   is   25%   of   what?

$$4 = 0.25 \cdot B \quad \text{Express 25\% as 0.25.}$$

$$\frac{4}{0.25} = \frac{0.25\,B}{0.25} \quad \text{Divide both sides by 0.25.}$$

$$16 = B \quad \text{Simplify:} \quad 0.25\,\overline{)4.00}^{\ 16.}$$

Thus, 4 is 25% of 16. The answer is 16. ∎

✔ **CHECK POINT 8**   9 is 60% of what?

EXAMPLE 9 Using the Percent Formula: A is What Percent of B?

1.3 is what percent of 26?

Solution We use the formula $A = PB$: A is P percent of B. We are interested in finding the quantity P in this formula.

| 1.3 | is | what percent | of | 26? |

$$1.3 = P \cdot 26$$

$$\frac{1.3}{26} = \frac{P \cdot 26}{26} \qquad \text{Divide both sides by 26.}$$

$$0.05 = P \qquad \text{Simplify: } 26\overline{)1.30} \;\; .05$$

We change 0.05 to a percent by moving the decimal point two places to the right and adding a percent sign: $0.05 = 5\%$. Thus, 1.3 is 5% of 26. The answer is 5%. ∎

✔ **CHECK POINT 9** 18 is what percent of 50?

Applications

There are many kinds of applied problems that can be solved using the percent formula.

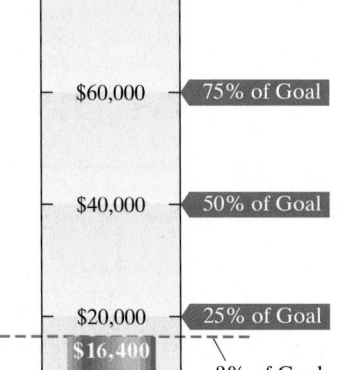

Raising money for a charity

EXAMPLE 10 Using the Percent Formula

A charity has raised \$16,400, with a goal of raising \$80,000. What percent of the goal has been raised?

Solution The question that we must address is "\$16,400 is what percent of \$80,000?"

We use the formula $A = PB$: A is P percent of B. We are interested in finding the quantity P in this formula.

| \$16,4000 | is | what percent | of | 80,000? |

$$16,400 = P \cdot 80,000$$

$$\frac{16,400}{80,000} = \frac{P \cdot 80,000}{80,000} \qquad \text{Divide both sides by 80,000.}$$

$$0.205 = P \qquad \text{Simplify: } \frac{16,400}{80,000} = \frac{4 \cdot 41}{4 \cdot 200}$$

$$200\overline{)41.000} \;\; .205$$

We change 0.205 to a percent by moving the decimal point two places to the right and adding a percent sign: $0.205 = 20.5\%$. Thus, 20.5% of the charity's goal has been raised. ∎

✔ **CHECK POINT 10** In England, the average price per gallon of premium gasoline is \$4.30. Of this amount, \$3.44 is for taxes. What percent of fuel cost is for taxes?

EXERCISE SET 2.4

Practice Exercises

In Exercises 1–26, solve each formula for the specified variable. Do you recognize the formula? If so, what does it describe?

1. $d = rt$ for r

2. $d = rt$ for t

3. $I = Prt$ for P

4. $I = Prt$ for r

5. $C = 2\pi r$ for r

6. $C = \pi d$ for d

7. $E = mc^2$ for m

8. $V = \pi r^2 h$ for h

9. $y = mx + b$ for m

10. $y = mx + b$ for x

11. $T = D + pm$ for p

12. $P = C + MC$ for M

13. $A = \dfrac{1}{2}bh$ for b

14. $A = \dfrac{1}{2}bh$ for h

15. $M = \dfrac{n}{5}$ for n

16. $M = \dfrac{A}{740}$ for A

17. $\dfrac{c}{2} + 80 = 2F$ for c

18. $p = 15 + \dfrac{5d}{11}$ for d

19. $A = \dfrac{1}{2}(a + b)$ for a

20. $A = \dfrac{1}{2}(a + b)$ for b

21. $S = P + Prt$ for r

22. $S = P + Prt$ for t

23. $A = \dfrac{1}{2}h(a + b)$ for b

24. $A = \dfrac{1}{2}h(a + b)$ for a

25. $Ax + By = C$ for x

26. $Ax + By = C$ for y

In Exercises 27–34, express each decimal as a percent.

27. 0.59

28. 0.96

29. 0.003

30. 0.007

31. 2.87

32. 9.83

33. 100

34. 95

In Exercises 35–44, express each percent as a decimal.

35. 72%

36. 38%

37. 43.6%

38. 6.25%

39. 130%

40. 260%

41. 2%

42. 6%

43. 62.5%

44. 87.5%

Use the percent formula, $A = PB$: A is P percent of B, to solve Exercises 45–56.

45. What is 3% of 200?

46. What is 8% of 300?

47. What is 18% of 40?

48. What is 16% of 90?

49. 3 is 60% of what?

50. 8 is 40% of what?

51. 24% of what number is 40.8?

52. 32% of what number is 51.2?

53. 3 is what percent of 15?

54. 18 is what percent of 90?

55. What percent of 2.5 is 0.3?

56. What percent of 7.5 is 0.6?

Application Exercises

57. The average, or mean, A, of three exam grades, x, y, and z, is given by the formula
$$A = \frac{x + y + z}{3}.$$

a. Solve the formula for z.

b. Use the formula in part (a) to solve this problem. On your first two exams, your grades are 86% and 88%: $x = 86$ and $y = 88$. What must you get on the third exam to have an average of 90%?

58. The average, or mean, A, of four exam grades, x, y, z, and w, is given by the formula
$$A = \frac{x + y + z + w}{4}.$$

a. Solve the formula for w.

b. Use the formula in part (a) to solve this problem. On your first three exams, your grades are 76%, 78%, and 79%: $x = 76$, $y = 78$, and $z = 79$. What must you get on the fourth exam to have an average of 80%?

59. If you are traveling in your car at an average rate of r miles per hour for t hours, then the distance, d, in miles, that you travel is described by the formula $d = rt$: distance equals rate times time.

   a. Solve the formula for t.

   b. Use the formula in part (a) to find the time that you travel if you cover a distance of 100 miles at an average rate of 40 miles per hour.

60. The formula $F = \dfrac{9}{5}C + 32$ gives the relationship between Celsius temperature, C, and Fahrenheit temperature, F.

   a. Solve the formula for C.

   b. Use the formula from part (a) to find the equivalent Celsius temperature for a Fahrenheit temperature of 59°.

The circle graph shows where your dollar goes when you purchase prescription drugs. Use the graph to solve Exercises 61–62.

Where Your Dollar Goes Buying Prescription Drugs

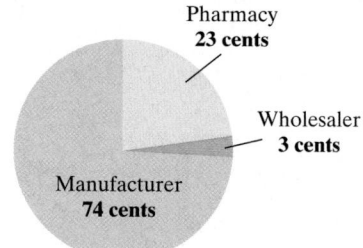

Pharmacy 23 cents

Wholesaler 3 cents

Manufacturer 74 cents

Source: U.S. Department of Health and Human Services

61. What percent of the cost of prescription drugs goes to the pharmacy?

62. What percent of the cost of prescription drugs goes to the drug's manufacturer?

Because of the nationwide crackdown on violent crime, the number of executions in the United States continues to rise sharply. The circle graph shows the total number of executions in the United States from 1976 to 2000. Use this information to solve Exercises 63–64.

Total Execution in U.S., 1976–2000

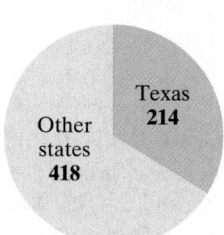

Texas 214

Other states 418

Source: Death Penalty Information Center

63. What percent of the total executions are in Texas? Round to the nearest percent.

64. What percent of the total executions are not in Texas? Round to the nearest percent.

A recent Time/CNN telephone poll included never-married single women between the ages of 18 and 49 and never-married single men between the ages of 18 and 49. The circle graphs show the results for one of the questions in the poll. Use this information to solve Exercises 65–66.

If You Couldn't Find the Perfect Mate, Would You Marry Someone Else?

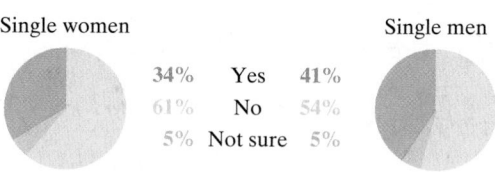

Single women			Single men
34%	Yes	41%	
61%	No	54%	
5%	Not sure	5%	

Source: Time, August 28, 2000

65. There were 122 women in the poll who would marry someone other than the perfect mate. How many single women participated in the poll?

66. There were 135 men in the poll who would marry someone other than the perfect mate. How many single men participated in the poll?

67. A charity has raised $7500, with a goal of raising $60,000. What percent of the goal has been raised?

68. A charity has raised $225,000, with a goal of raising $500,000. What percent of the goal has been raised?

69. A restaurant bill came to $60. If 15% of this bill is left as a tip, how much was the tip?

70. If income tax is $3502 plus 28% of taxable income over $23,000, how much is the income tax on a taxable income of $35,000?

71. Suppose that the local sales tax rate is 6% and you buy a car for $16,800.
   a. How much tax is due?
   b. What is the car's total cost?

72. Suppose that the local sales tax rate is 7% and you buy a graphing calculator for $96.
   a. How much tax is due?
   b. What is the calculator's total cost?

73. An exercise machine with an original price of $860 is on sale at 12% off.
   a. What is the discount amount?
   b. What is the exercise machine's sale price?

74. A dictionary that normally sells for $16.50 is on sale at 40% off.
   a. What is the discount amount?
   b. What is the dictionary's sale price?

Writing in Mathematics

75. Explain what it means to solve for a variable in a formula.

76. What is a percent?

77. Describe how to express a decimal number as a percent and give an example.

78. Describe how to express a percent as a decimal number and give an example.

79. What does the percent formula, $A = PB$, describe? Give an example of how the formula is used.

80. Describe one way in which you use percents in your daily life.

Critical Thinking Exercises

81. Which one of the following statements is true?

a. If $ax + b = 0$, then $x = \dfrac{b}{a}$.

b. If $A = lw$, then $w = \dfrac{l}{A}$.

c. If $A = \dfrac{1}{2}bh$, then $b = \dfrac{A}{2h}$.

d. Solving $x - y = -7$ for y gives $y = x + 7$.

82. In psychology, an intelligence quotient, Q, also called IQ, is measured by the formula

$$Q = \frac{100M}{C}$$

where $M = $ mental age and $C = $ chronological age. Solve the formula for C.

83. The height, h, in feet, of water in a fountain is described by the formula

$$h = -16t^2 + 64t$$

and the velocity, v, in feet per second, of water in the fountain is described by $v = -32t + 64$. Find the time when the water's velocity is 16 feet per second, and then find the water's height at that time.

Review Exercises

84. Solve and check: $5x + 20 = 8x - 16$. (Section 2.2, Example 7)

85. Solve and check: $5(2y - 3) - 1 = 4(6 + 2y)$. (Section 2.3, Example 3)

86. Simplify: $x - 0.3x$. (Section 1.4, Example 9)

▶ SECTION 2.5   An Introduction to Problem Solving

Objectives

1 Translate English phrases into algebraic expressions.

2 Solve algebraic word problems using linear equations.

SSM
PH Tutor   CD-   Video
Center   ROM

You started college with your best friend. This semester, you and your friend are taking two classes together. However, your friend frequently misses class and is not doing the necessary homework between classes to succeed. What can you say to your friend, who values your advice and who is in danger of flunking out of college if things continue on their present course?

Some problems have many plans for finding an answer. To solve your friend's problem, or any problem for that matter, we need to understand the problem fully, devise a plan for solving it, and then carry out the plan. However, problem solving in algebra is easier than solving the many problematic situations encountered in everyday life. Why? Algebra provides a step-by-step strategy for solving problems. As you become familiar with this strategy, you will learn to solve a wide variety of problems.

A Strategy for Solving Word Problems Using Equations

Problem solving is an important part of algebra. The problems in this book are presented in English. We must translate from the ordinary language of English into the language of algebraic equations. To translate, however, we must understand the English prose and be familiar with the forms of algebraic language. Here are some general steps we will follow in solving word problems.

Strategy for Solving Word Problems

Step 1 Read the problem carefully. Attempt to state the problem in your own words and state what the problem is looking for. Let x (or any variable) represent one of the quantities in the problem.

Step 2 If necessary, write expressions for any other unknown quantities in the problem in terms of x.

Step 3 Write an equation in x that describes the verbal conditions of the problem.

Step 4 Solve the equation and answer the problem's question.

Step 5 Check the solution *in the original wording* of the problem, not in the equation obtained from the words.

Take great care with step 1. Reading a word problem is not the same as reading a newspaper. Reading the problem involves slowly working your way through its parts, making notes on what is given, and perhaps rereading the problem a few times. Only at this point should you let x represent one of the quantities.

The most difficult step in this process is step 3 because it involves translating verbal conditions into an algebraic equation. Translations of some commonly used English phrases are listed in Table 2.1. We choose to use x to represent the variable, but we can use any letter.

1 Translate English phrases into algebraic expressions.

Study Tip

Cover the right column in Table 2.1 with a sheet of paper and attempt to formulate the algebraic expression in the column on your own. Then slide the paper down and check your answer. Work through the entire table in this manner.

Table 2.1   Algebraic Translations of English Phrases

English Phrase	Algebraic Expression
Addition	
The sum of a number and 7	$x + 7$
Five more than a number; a number plus 5	$x + 5$
A number increased by 6; 6 added to a number	$x + 6$
Subtraction	
A number minus 4	$x - 4$
A number decreased by 5	$x - 5$
A number subtracted from 8	$8 - x$
The difference between a number and 6	$x - 6$
The difference between 6 and a number	$6 - x$
Seven less than a number	$x - 7$
Seven minus a number	$7 - x$
Nine fewer than a number	$x - 9$
Multiplication	
Five times a number	$5x$
The product of 3 and a number	$3x$
Two-thirds of a number (used with fractions)	$\frac{2}{3}x$
Seventy-five percent of a number (used with decimals)	$0.75x$
Thirteen multiplied by a number	$13x$

(continues)

Table 2.1 *(cont.)*   **Algebraic Translations of English Phrases**

English Phrase	Algebraic Expression
Multiplication (cont.)	
A number multiplied by 13	$13x$
Twice a number	$2x$
Division	
A number divided by 3	$\dfrac{x}{3}$
The quotient of 7 and a number	$\dfrac{7}{x}$
The quotient of a number and 7	$\dfrac{x}{7}$
The reciprocal of a number	$\dfrac{1}{x}$
More than one operation	
The sum of twice a number and 7	$2x + 7$
Twice the sum of a number and 7	$2(x + 7)$
Three times the sum of 1 and twice a number	$3(1 + 2x)$
Nine subtracted from 8 times a number	$8x - 9$
Twenty-five percent of the sum of 3 times a number and 14	$0.25(3x + 14)$
Seven times a number, increased by 24	$7x + 24$
Seven times the sum of a number and 24	$7(x + 24)$

Study Tip

Here are three similar English phrases that have very different translations:

7 minus 10: $7 - 10$ 7 less than 10: $10 - 7$ 7 is less than 10: $7 < 10$

Think carefully about what is expressed in English before you translate into the language of algebra.

EXAMPLE 1   Translating English Phrases into Algebraic Expressions

Write each English phrase as an algebraic expression. Let x represent the number.

a. Six subtracted from 5 times a number
b. The quotient of 9 and a number, decreased by 4 times the number

Solution

a.

Six subtracted from   5 times a number

$5x$ -6

The algebraic expression for "six subtracted from 5 times a number" is $5x - 6$.

b.

The quotient of 9 and a number,   decreased by   4 times the number

$\dfrac{9}{x}$ $-$ $4x$

The algebraic expression for "the quotient of 9 and a number, decreased by 4 times the number" is $\dfrac{9}{x} - 4x$. ∎

✔ **CHECK POINT 1** Write each English phrase as an algebraic expression. Let x represent the number.

    a. Four times a number, increased by 6

    b. The quotient of a number decreased by 4 and 9

2 Solve algebraic word problems using linear equations.

Applying the Strategy for Solving Word Problems

Now that we've practiced writing algebraic expessions for English phrases, let's apply our five step strategy for solving word problems.

EXAMPLE 2 Solving a Word Problem

Nine subtracted from eight times a number is 39. Find the number.

Solution

Step 1 Let x represent one of the quantities. Because we are asked to find a number, let

$$x = \text{the number.}$$

Step 2 Represent other quantities in terms of x. There are no other unknown quantities to find, so we can skip this step.

Step 3 Write an equation in x that describes the conditions.

Nine subtracted from	eight times a number	is	39.

$$8x \qquad -9 \qquad = \qquad 39$$

Step 4 Solve the equation and answer the question.

$$
\begin{aligned}
8x - 9 &= 39 &&\text{This is the equation for the problem's conditions.}\\
8x - 9 + 9 &= 39 + 9 &&\text{Add 9 to both sides.}\\
8x &= 48 &&\text{Simplify.}\\
\frac{8x}{8} &= \frac{48}{8} &&\text{Divide both sides by 8.}\\
x &= 6 &&\text{Simplify.}
\end{aligned}
$$

The number is 6.

Step 5 Check the proposed solution in the original wording of the problem.
"Nine subtracted from eight times a number is 39." The proposed number is 6. Eight times 6 is $8 \cdot 6$, or 48. Nine subtracted from 48 is $48 - 9$, or 39. The proposed solution checks in the problem's wording, verifying that the number is 6. ∎

✔ **CHECK POINT 2** Four subtracted from six times a number is 68. Find the number.

EXAMPLE 3 Pet Population

Americans love their pets. The number of cats in the United States exceeds the number of dogs by 7.5 million. The number of cats and dogs combined is 114.7 million. Determine the number of dogs and cats in the United States.

Solution

Step 1 Let x represent one of the quantities. We know something about the number of cats: the cat population exceeds the dog population by 7.5 million.

U.S. Pet Population

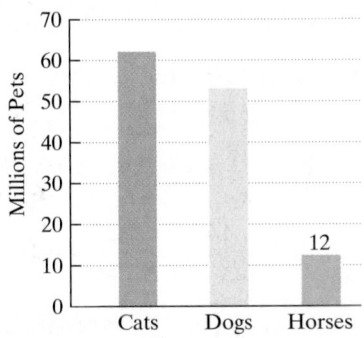

Source: American Veterinary Medical Association

Americans spend more than $21 billion a year on their pets. 31.4% of households have cats and 34.3% have dogs.

This means that there are 7.5 million more cats than dogs. We will let

$$x = \text{the number (in millions) of dogs in the United States.}$$

Step 2 Represent other quantities in terms of x. The other unknown quantity is the number of cats. Because there are 7.5 million more cats than dogs, let

$$x + 7.5 = \text{the number (in millions) of cats in the United States.}$$

Step 3 Write an equation in x that describes the conditions. The number of cats and dogs combined is 114.7 million

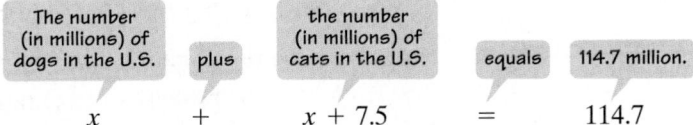

The number (in millions) of dogs in the U.S.	plus	the number (in millions) of cats in the U.S.	equals	114.7 million.
x	$+$	$x + 7.5$	$=$	114.7

Step 4 Solve the equation and answer the question.

$$x + x + 7.5 = 114.7 \qquad \text{This is the equation specified by the conditions of the problem.}$$

$$2x + 7.5 = 114.7 \qquad \text{Combine like terms on the left side.}$$

$$2x + 7.5 - 7.5 = 114.7 - 7.5 \qquad \text{Subtract 7.5 from both sides.}$$

$$2x = 107.2 \qquad \text{Simplify.}$$

$$\frac{2x}{2} = \frac{107.2}{2} \qquad \text{Divide both sides by 2.}$$

$$x = 53.6 \qquad \text{Simplify.}$$

Because x represents the number (in millions) of dogs, there are 53.6 million dogs in the United States. Because $x + 7.5$ represents the number (in millions) of cats, there are $53.6 + 7.5$, or 61.1 million cats in the United States.

Step 5 Check the proposed solution in the original wording of the problem. The problem states that the number of cats and dogs combined is 114.7 million. By adding 53.6 million, the dog population, and 61.1 million, the cat population, we do, indeed, obtain a sum of 114.7 million. ■

✔ **CHECK POINT 3** Two of the top-selling music albums of all time are *Jagged Little Pill* (Alanis Morissette) and *Saturday Night Fever* (Bee Gees). The Morissette album sold 5 million more copies than that of the Bee Gees. Combined, the two albums sold 27 million copies. Determine the number of sales for each of the albums.

EXAMPLE 4   Consecutive Integers

Two pages that face each other in a book have 145 as the sum of their page numbers. What are the page numbers?

Solution

Step 1 Let x represent one of the quantities. We will let

$$x = \text{the page number of the page on the left.}$$

Step 2   Represent other quantities in terms of *x*. The other unknown quantity is the page number of the facing page on the right. Page numbers on facing pages are consecutive integers. Thus,

$$x + 1 = \text{the page number of the page on the right.}$$

Step 3   Write an equation in *x* that describes the conditions. The two facing pages have 145 as the sum of their page numbers.

The page number on the left	plus	the page number on the right	equals	145.
x	$+$	$(x + 1)$	$=$	145

Step 4   Solve the equation and answer the question.

$x + (x + 1) = 145$	This is the equation for the problem's conditions.
$2x + 1 = 145$	Regroup and combine like terms.
$2x + 1 - 1 = 145 - 1$	Subtract 1 from both sides.
$2x = 144$	Simplify.
$\dfrac{2x}{2} = \dfrac{144}{2}$	Divide both sides by 2.
$x = 72$	Simplify.

Thus,

$$\text{the page number on the left} = x = 72$$

and

$$\text{the page number on the right} = x + 1 = 72 + 1 = 73.$$

The page numbers are 72 and 73.

Step 5   Check the proposed solution in the original wording of the problem. The problem states that the sum of the page numbers on the facing pages is 145. By adding 72, the page number on the left, and 73, the page number on the right, we do, indeed, obtain a sum of 145. ■

✔ **CHECK POINT 4**   Two pages that face each other in a book have 193 as the sum of their page numbers. What are the page numbers?

Example 4 and Check Point 4 involved consecutive integers. By contrast, some word problems involve consecutive odd integers, such as 5, 7, and 9. Other word problems involve consecutive even integers, such as 6, 8, and 10. When working with consecutive even or consecutive odd integers, we must continuously add 2 to move from one integer to the next successive integer in the list.

Table 2.2 should be helpful in solving consecutive integer problems.

Table 2.2   Consecutive Integers

English Phrase	Algebraic Expressions	Example
Two consecutive integers	$x, x + 1$	$13, 14$
Three consecutive integers	$x, x + 1, x + 2$	$-8, -7, -6$
Two consecutive even integers	$x, x + 2$	$40, 42$
Two consecutive odd integers	$x, x + 2$	$-37, -35$
Three consecutive even integers	$x, x + 2, x + 4$	$30, 32, 34$
Three consecutive odd integers	$x, x + 2, x + 4$	$9, 11, 13$

EXAMPLE 5   Renting a Car

Rent-a-Heap Agency charges $125 per week plus $0.20 per mile to rent a small car. How many miles can you travel for $335?

Solution

Step 1   Let x represent one of the quantities. Because we are asked to find the number of miles we can travel for $335, let

$$x = \text{the number of miles.}$$

Step 2   Represent other quantities in terms of x. There are no other unknown quantities to find, so we can skip this step.

Step 3   Write an equation in x that describes the conditions. Before writing the equation, let us consider a few specific values for the number of miles traveled. The rental charge is $125 plus $0.20 for each mile.

3 miles: The rental charge is $125 + $0.20(3).

30 miles: The rental charge is $125 + $0.20(30).

100 miles: The rental charge is $125 + $0.20(100).

x miles: The rental charge is $125 + $0.20x$.

The weekly charge of $125	plus	the charge of $0.20 per mile for x miles	equals	the total $335 rental charge.
125	+	0.20x	=	335

Step 4   Solve the equation and answer the question.

$$125 + 0.20x = 335 \qquad \text{This is the equation specified by the conditions of the problem.}$$

$$125 + 0.20x - 125 = 335 - 125 \qquad \text{Subtract 125 from both sides.}$$

$$0.20x = 210 \qquad \text{Simplify.}$$

$$\frac{0.20x}{0.20} = \frac{210}{0.20} \qquad \text{Divide both sides by 0.20.}$$

$$x = 1050 \qquad \text{Simplify.}$$

You can travel 1050 miles for $335.

Step 5   Check the proposed solution in the original wording of the problem. Traveling 1050 miles should result in a total rental charge of $335. The mileage charge of $0.20 per mile is

$$\$0.20(1050) = \$210.$$

Adding this to the $125 weekly charge gives a total rental charge of

$$\$125 + \$210 = \$335.$$

Because this results in the given rental charge of $335, this verifies that you can travel 1050 miles.   ■

✔ **CHECK POINT 5**   A taxi charges $2.00 to turn on the meter plus $0.25 each eighth of a mile. If you have $10.00, how many eighths of a mile can you go? How many miles is that?

We will be using the formula for the perimeter of a rectangle, $P = 2l + 2w$, in our next example. Twice the rectangle's length plus twice the rectangle's width is its perimeter.

EXAMPLE 6 Finding the Dimensions of a Soccer Field

A rectangular soccer field is twice as long as it is wide. If the perimeter of a soccer field is 300 yards, what are the field's dimensions?

Solution

Step 1 Let x represent one of the quantities. We know something about the length; the field is twice as long as it is wide. We will let

$$x = \text{the width.}$$

Step 2 Represent other quantities in terms of x. Because the field is twice as long as it is wide, let

$$2x = \text{the length.}$$

Figure 2.7 illustrates the soccer field and its dimensions.

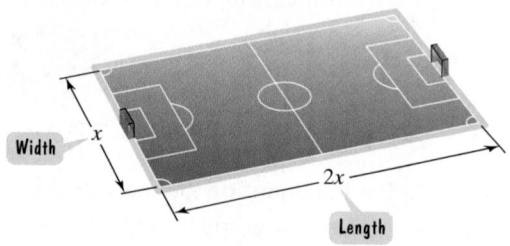

Figure 2.7

Step 3 Write an equation in x that describes the conditions. Because the perimeter of a soccer field is 300 yards,

Twice the length	plus	twice the width	is	the perimeter.
$2 \cdot 2x$	$+$	$2 \cdot x$	$=$	$300.$

Step 4 Solve the equation and answer the question.

$$2 \cdot 2x + 2 \cdot x = 300 \quad \text{This is the equation for the problem's conditions.}$$
$$4x + 2x = 300 \quad \text{Multiply.}$$
$$6x = 300 \quad \text{Combine like terms.}$$
$$\frac{6x}{6} = \frac{300}{6} \quad \text{Divide both sides by 6.}$$
$$x = 50 \quad \text{Simplify.}$$

Thus,

$$\text{Width} = x = 50$$
$$\text{Length} = 2x = 2(50) = 100.$$

The dimensions of a soccer field are 50 yards by 100 yards.

Step 5 Check the proposed solution in the original wording of the problem. The perimeter of the soccer field using the dimensions that we found is

$2(50 \text{ feet}) + 2(100 \text{ feet}) = 100 \text{ feet} + 200 \text{ feet}$, or 300 feet. Because the problem's wording tells us that the perimeter is 300 feet, our dimensions are correct. ■

✔ **CHECK POINT 6** A rectangular swimming pool is three times as long as it is wide. If the perimeter of the pool is 320 feet, what are the pool's dimensions?

EXAMPLE 7 A Price Reduction

Your local computer store is having a sale. After a 30% price reduction, you purchase a new computer for $980. What was the computer's price before the reduction?

Solution

Step 1 Let x represent one of the quantities. We will let

$x =$ the original price of the computer prior to the reduction.

Step 2 Represent other quantities in terms of x. There are no other unknown quantities to find, so we can skip this step.

Step 3 Write an equation in x that describes the conditions. The computer's original price minus the 30% reduction is the reduced price, $980.

Original price	minus	the reduction (30% of the original price)	is	the reduced price, $980.
x	$-$	$0.3x$	$=$	980

Step 4 Solve the equation and answer the question.

$x - 0.3x = 980$ This is the equation for the problem's conditions.

$0.7x = 980$ Combine like terms: $x - 0.3x = 1x - 0.3x = 0.7x$.

$\dfrac{0.7x}{0.7} = \dfrac{980}{0.7}$ Divide both sides by 0.7

$x = 1400$ Simplify: $0.7\overline{)980.0}$ 1400

The computer's price before the reduction was $1400.

Step 5 Check the proposed solution in the original wording of the problem. The price before the reduction, $1400, minus the reduction in price should equal the reduced price given in the original wording, $980. The reduction in price is equal to 30% of the price before the reduction, $1400. To find the reduction, we multiply the decimal equivalent of 30%, 0.30 or 0.3, by the original price, $1400:

$$30\% \text{ of } \$1400 = (0.3)(\$1400) = \$420.$$

Now we can determine whether the calculation for the price before the reduction, $1400, minus the reduction, $420, is equal to the reduced price given in the problem, $980. We subtract:

$$\$1400 - \$420 = \$980.$$

This verifies that the price of the computer before the reduction was $1400. ■

✔ **CHECK POINT 7** After a 40% price reduction, an exercise machine sold for $564. What was the exercise machine's price before this reduction?

EXERCISE SET 2.5

Practice Exercises

In Exercises 1–14, let x represent the number. Write each English phrase as an algebraic expression.

1. The sum of a number and 9
2. A number increased by 13
3. A number subtracted from 20
4. 13 less than a number
5. 8 decreased by 5 times a number
6. 14 less than the product of 6 and a number
7. The quotient of 15 and a number
8. The quotient of a number and 15
9. The sum of twice a number and 20
10. Twice the sum of a number and 20
11. 30 subtracted from 7 times a number
12. The quotient of 12 and a number, decreased by 3 times the number
13. Four times the sum of a number and 12
14. Five times the difference of a number and 6

In Exercises 15–34, let x represent the number. Use the given conditions to write an equation. Solve the equation and find the number.

15. A number increased by 40 is equal to 450. Find the number.
16. The sum of a number and 29 is 54. Find the number.
17. A number decreased by 13 is equal to 123. Find the number.
18. The difference between a number and 14 is 28. Find the number.
19. The product of 7 and a number is 91. Find the number.
20. The product of 8 and a number is 184. Find the number.
21. The quotient of a number and 18 is 6. Find the number.
22. The quotient of a number and 13 is 9. Find the number.
23. The sum of four and twice a number is 36. Find the number.
24. The sum of five and three times a number is 29. Find the number.
25. Seven subtracted from five times a number is 123. Find the number.
26. Eight subtracted from six times a number is 184. Find the number.

27. A number increased by 5 is two times the number. Find the number.
28. A number increased by 12 is four times the number. Find the number.
29. Twice the sum of four and a number is 36. Find the number.
30. Three times the sum of five and a number is 48. Find the number.
31. Nine times a number is 30 more than three times that number. Find the number.
32. Five more than four times a number is that number increased by 35. Find the number.
33. If the quotient of three times a number and five is increased by four, the result is 34. Find the number.
34. If the quotient of three times a number and 4 is decreased by three, the result is 9. Find the number.

Application Exercises

In Exercises 35–62, use the five-step strategy to solve each problem.

35. Two of the most expensive movies ever made were *Titanic* and *Waterworld*. The cost to make *Titanic* exceeded the cost to make *Waterwold* by $40 million. The combined cost to make the two movies was $360 million. Find the cost of making each of these movies.

Paramount Pictures Corporation, Inc.

36. The 1998 baseball season set a record for the most home runs. During that year, Mark McGwire hit four more home runs than Sammy Sosa. Combined, the two athletes hit 136 home runs. Determine the number of home runs hit by McGwire and Sosa.

37. Each year, Americans in 68 urban areas waste almost 7 billion gallons of fuel sitting in traffic. The bar graph shows the number of hours in traffic per year for the average motorist in ten cities. The average motorist in Los Angeles spends 32 hours less than twice that of the average motorist in Miami stuck in traffic each year. In the two cities combined, 139 hours are spent by the average motorist per year in traffic. How many hours are wasted in traffic by the average motorist in Los Angeles and Miami?

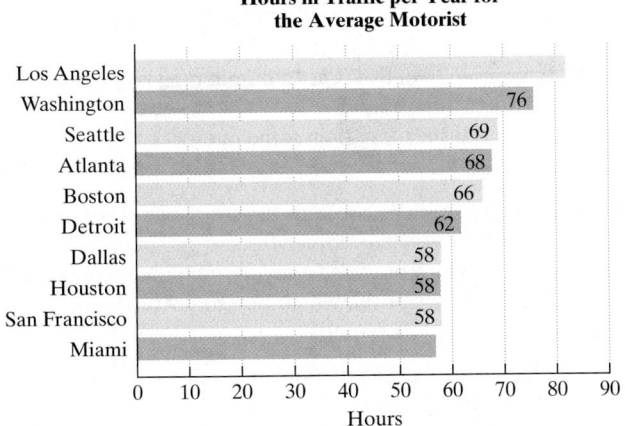

Hours in Traffic per Year for the Average Motorist

Source: Texas Transportation Institute

38. The graph reflects fear of crime in ten selected countries. The percentage of people feeling unsafe in their neighborhoods after dark in the United States exceeds twice that of Sweden by 14%. Futhermore, the U.S. percentage exceeds that of Sweden by 27.5%. Find the percentage of people feeling unsafe after dark for Sweden and the United States.

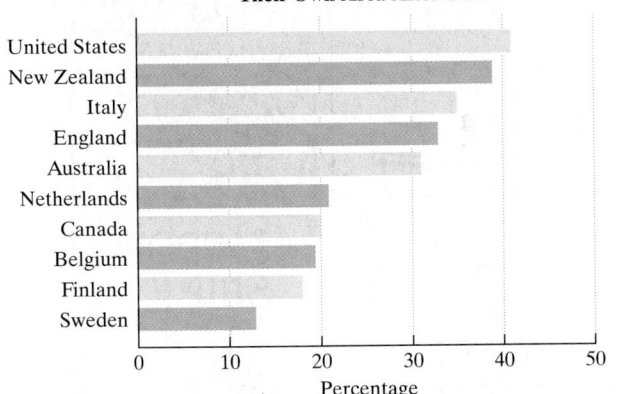

Percentage of the Public Feeling Unsafe When Walking in Their Own Area After Dark

Source: Ministry of Justice, The Netherlands

39. The sum of the page numbers on the facing pages of a book is 629. What are the page numbers?

40. The sum of the page numbers on the facing pages of a book is 525. What are the page numbers?

41. The first Super Bowl was played between the Green Bay Packers and the Kansas City Chiefs in 1967. Only once, in 1991, were the winning and losing scores in the Super Bowl consecutive integers. If the sum of the scores was 39, what were the scores?

42. Before Mark McGwire and Sammy Sosa, Roger Maris held the record for the most home runs in one season. Just behind Maris was Babe Ruth. The numbers of home runs hit by these two athletes in their record-breaking seasons form consecutive integers. Combined, the two athletes hit 121 home runs. Determine the number of home runs hit by Maris and Ruth in their record-breaking seasons.

43. Find two consecutive even integers whose sum is 66.

44. Find two consecutive odd integers whose sum is 72.

45. A car rental agency charges $200 per week plus $0.15 per mile to rent a car. How many miles can you travel in one week for $320?

46. A car rental agency charges $180 per week plus $0.25 per mile to rent a car. How many miles can you travel in one week for $395?

47. The average weight for female infants at birth is 7 pounds, with a monthly weight gain of 1.5 pounds. After how many months does a baby girl weigh 16 pounds?

48. In 1995, the average yearly salary for teachers in the United States was $38,556. If the salary increases by $1496 per year, in which year will the salary reach $56,508?

49. A rectangular field is four times as long as it is wide. If the perimeter of the field is 500 yards, what are the field's dimensions?

50. A rectangular field is five times as long as it is wide. If the perimeter of the field is 288 yards, what are the field's dimensions?

51. An American football field is a rectangle with a perimeter of 1040 feet. The length is 200 feet more than the width. Find the width and length of the rectangular field.

52. A basketball court is a rectangle with a perimeter of 86 meters. The length is 13 meters more than the width. Find the width and length of the basketball court.

53. A bookcase is to be constructed as shown in the figure. The length is to be 3 times the height. If 60 feet of lumber is available for the entire unit, find the length and height of the bookcase.

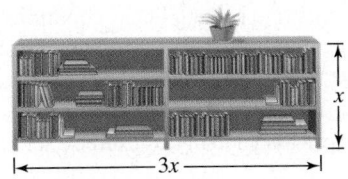

54. The height of the bookcase in the figure is 3 feet longer than the length of a shelf. If 18 feet of lumber is available for the entire unit, find the length and height of the unit.

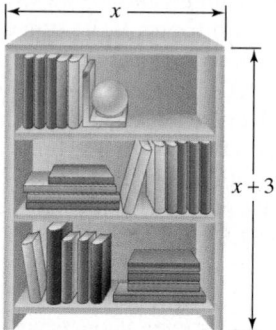

55. After a 20% reduction, you purchase a television for $320. What was the television's price before the reduction?

56. After a 30% reduction, you purchase a VCR for $98. What was the VCR's price before the reduction?

57. After a dictionary's price is reduced by $\frac{1}{4}$ of its original price, you purchase it for $21. What was the dictionary's price before the reduction?

58. After a graphing calculator's price is reduced by $\frac{1}{3}$ of its original price, you purchase it for $64. What was the graphing calculator's price before the reduction?

59. Including 6% sales tax, a car sold for $15,370. Find the price of the car before the tax was added.

60. Including 8% sales tax, a bed-and-breakfast inn charges $172.80 per night. Find the inn's nightly cost before the tax is added.

61. An automobile repair shop charged a customer $448, listing $63 for parts and the remainder for labor. If the cost of labor is $35 per hour, how many hours of labor did it take to repair the car?

62. A repair bill on a sailboat came to $1603, including $532 for parts and the remainder for labor. If the cost of labor is $63 per hour, how many hours of labor did it take to repair the sailboat?

Writing in Mathematics

63. In your own words, describe a step-by-step approach for solving algebraic word problems.

64. Many students find solving linear equations much easier than solving algebraic word problems. Discuss some of the reasons why this is the case.

65. Did you have some difficulties solving some of the problems that were assigned in this exercise set? Discuss what you did if this happened to you. Did your course of action enhance your ability to solve algebraic word problems?

66. Write an original word problem that can be solved using a linear equation. Then solve the problem.

Critical Thinking Exercises

67. Which English statement given below is correctly translated into an algebraic equation?
   a. Ten pounds less than Bill's weight (x) equals 160 pounds: $10 - x = 160$.
   b. Four more than five times a number (x) is one less than six times that number: $5x + 4 = 1 - 6x$.
   c. Seven is three more than some number (x): $7 + 3 = x$.
   d. None of the above is correctly translated.

68. An HMO pamphlet contains the following recommended weight for women: "Give yourself 100 pounds for the first 5 feet plus 5 pounds for every inch over 5 feet tall." Using this description, which height corresponds to an ideal weight of 135 pounds?

69. The rate for a particular international telephone call is $0.55 for the first minute and $0.40 for each additional minute. Determine the length of a call that costs $6.95.

70. In a film, the actor Charles Coburn plays an elderly "uncle" character criticized for marrying a woman when he is 3 times her age. He wittily replies, "Ah, but in 20 years time I shall only be twice her age." How old is the "uncle" and the woman?

71. Answer the question in the following *Peanuts* cartoon strip. (*Note:* You may not use the answer given in the cartoon!)

PEANUTS reprinted by permission of United Features Syndicate, Inc.

Review Exercises

72. Solve and check: $\frac{4}{5}x = -16$. (Section 2.2, Example 3)

73. Solve and check: $6(y - 1) + 7 = 9y - y + 1$. (Section 2.3, Example 3)

74. Solve for w: $V = \frac{1}{3}lwh$. (Section 2.4, Example 4)

▶ SECTION 2.6   *Problem Solving in Geometry*

Objectives

1 Solve problems using formulas for perimeter and area.

2 Solve problems using formulas for a circle's area and circumference.

3 Solve problems using formulas for volume.

4 Solve problems involving the angles of a triangle.

5 Solve problems involving complementary and supplementary angles.

SSM PH Tutor CD- Video
     Center   ROM

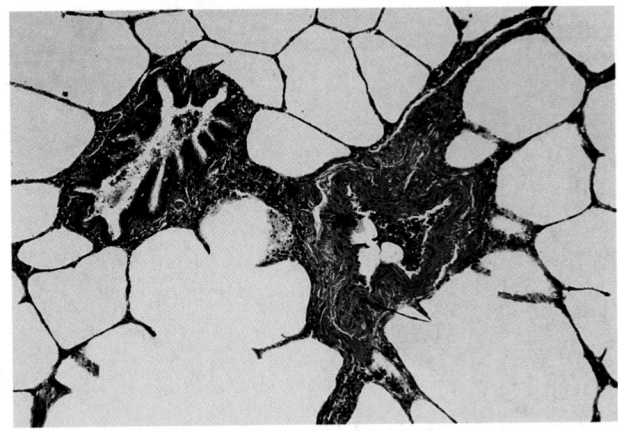

Human lung, normal tissue, with two blood vessels and alveoli magnified 160 times

Geometry is about the space you live in and the shapes that surround you. You're even made of it. The human lung consists of nearly 300 spherical air sacs, geometrically designed to provide the greatest surface area within the limited volume of our bodies. Viewed in this way, geometry becomes an intimate experience.

For thousands of years, people have studied geometry in some form to obtain a better understanding of the world in which they live. A study of the shape of your world will provide you with many practical applications that will help to increase your problem-solving skills.

Geometric Formulas for Perimeter and Area

1 Solve problems using formulas for perimeter and area.

Solving geometry problems often requires using basic geometric formulas. Formulas for perimeter and area are summarized in Table 2.3. Remember that perimeter is measured in linear units, such as feet or meters, and area is measured in square units, such as square feet, ft², or square meters, m².

Table 2.3   Common Formulas for Perimeter and Area

Square	Rectangle	Triangle	Trapezoid
$A = s^2$	$A = lw$	$A = \frac{1}{2}bh$	$A = \frac{1}{2}h(a + b)$
$P = 4s$	$P = 2l + 2w$		

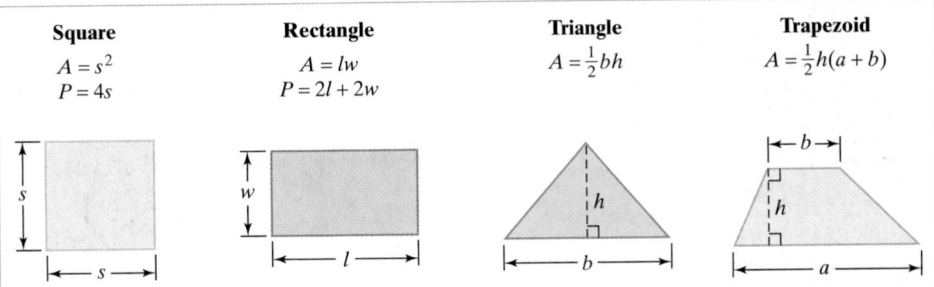

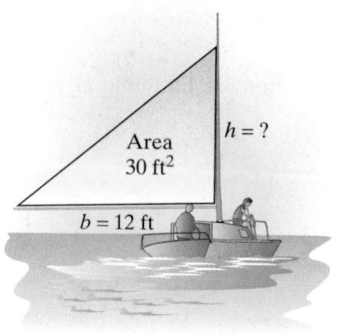

Figure 2.8 Finding the height of a triangular sail

EXAMPLE 1   Using the Formula for the Area of a Triangle

A sailboat has a triangular sail with an area of 30 square feet and a base that is 12 feet long. (See Figure 2.8.) Find the height of the sail.

Solution   We begin with the formula for the area of a triangle given in Table 2.3.

$$A = \frac{1}{2}bh \qquad \text{The area of a triangle is } \frac{1}{2} \text{ the product of its base and height.}$$

$$30 = \frac{1}{2}(12)h \qquad \text{Substitute 30 for A and 12 for b.}$$

$$30 = 6h \qquad \text{Simplify.}$$

$$\frac{30}{6} = \frac{6b}{6} \qquad \text{Divide both sides by 6.}$$

$$5 = h \qquad \text{Simplify.}$$

The height of the sail is 5 feet.

Check

The area is $A = \frac{1}{2}bh = \frac{1}{2}(12 \text{ feet})(5 \text{ feet}) = 30 \text{ feet}^2$. ∎

✔ **CHECK POINT 1**   A sailboat has a triangular sail with an area of 24 square feet and a base that is 4 feet long. Find the height of the sail.

2 Solve problems using formulas for a circle's area and circumference.

Geometric Formulas for Circumference and Area of a Circle

It's a good idea to know your way around a circle. Clocks, angles, maps, and compasses are based on circles. Circles occur everywhere in nature: in ripples on water, patterns on a butterfly's wings, and cross sections of trees. Some consider the circle to be the most pleasing of all shapes.

A **circle** is actually a set of points in the plane equally distant from a given point, its **center**. Figure 2.9 shows three circles. Each point on the circle on the left is 3 inches away from the center. Each 3-inch line segment is called a **radius** (plural: radii). The radius, r, is a line segment from the center to any point on the circle. For a given circle, all radii have the same length. Each 6-inch line segment is called a **diameter**. The diameter, d, is a line segment through the center whose endpoints both lie on the circle. In any circle, the **length of the diameter is twice the length of the radius**.

The point at which a pebble hits a flat surface of water becomes the center of a number of circular ripples.

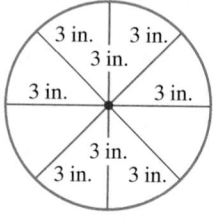
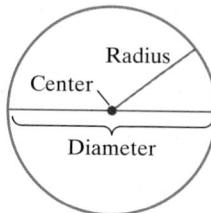
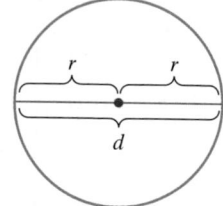

Figure 2.9

The words *radius* and *diameter* refer to both the line segments in Figure 2.9 as well as their linear measures. The distance around a circle (its perimeter) is called its **circumference**. Formulas for the area and circumference of a circle are given in terms of π and appear in Table 2.4. We have seen that π is an irrational number and is approximately equal to 3.14.

Table 2.4 Formulas for Circles

Circle	Area	Circumference
	$A = \pi r^2$	$C = 2\pi r$

When computing a circle's circumference by hand, round π to 3.14. When using a calculator, use the $\boxed{\pi}$ key, which gives the value of π rounded to approximately 11 decimal places. In either case, calculations involving π give approximate answers. These answers can vary slightly depending on how π is rounded. The symbol $\approx$ (is approximately equal to) will be written in these calculations.

EXAMPLE 2   Finding the Area and Circumference of a Circle

Find the area and circumference of a circle whose diameter measures 20 inches.

Solution   The radius is half the diameter, so $r = \frac{20}{2} = 10$ inches.

$A = \pi r^2 \qquad C = 2\pi r$ Use the formulas for area and circumference of a circle.

$A = \pi(10)^2 \quad C = 2\pi(10)$ Substitute 10 for r.

$A = 100\pi \qquad C = 20\pi$

The area of the circle is 100π square inches and the circumference is 20π inches. Using the fact that $\pi \approx 3.14$, the area is approximately $100(3.14)$, or 314 square inches. The circumference is approximately $20(3.14)$, or 62.8 inches. ■

✔ **CHECK POINT 2**   The diameter of a circular landing pad for helicopters is 40 feet. Find the area and circumference of the landing pad. Express answers in terms of π. Then round answers to the nearest square foot and foot, respectively.

EXAMPLE 3   Problem Solving Using the Formula for a Circle's Area

Which one of the following is the better buy: a large pizza with a 16-inch diameter for $15.00 or a medium pizza with an 8-inch diameter for $7.50?

Solution   The better buy is the pizza with the cheaper price per square inch. The radius of the large pizza is $\frac{1}{2} \cdot 16$ inches, or 8 inches, and the radius of the medium pizza is $\frac{1}{2} \cdot 8$ inches, or 4 inches. The area of the surface of each circular pizza is determined using the formula for the area of a circle.

Large pizza: $A = \pi r^2 = \pi(8 \text{ in.})^2 = 64\pi \text{ in.}^2 \approx 201 \text{ in.}^2$

Medium pizza: $A = \pi r^2 = \pi(4 \text{ in.})^2 = 16\pi \text{ in.}^2 \approx 50 \text{ in.}^2$

For each pizza, the price per square inch is found by dividing the price by the area:

$$\text{Price per square inch for large pizza} = \frac{\$15.00}{64\pi} \approx \frac{\$15.00}{201 \text{ in.}^2} \approx \frac{\$0.07}{\text{in.}^2}$$

$$\text{Price per square inch for medium pizza} = \frac{\$7.50}{16\pi} \approx \frac{\$7.50}{50 \text{ in.}^2} = \frac{\$0.15}{\text{in.}^2}$$

The large pizza is the better buy. ∎

Using Technology

You can use your calculator to obtain the price per square inch for each pizza in Example 3. The price per square inch for the large pizza,

$$\frac{\$15}{64\pi}$$

is approximated by one of the following keystrokes:

Scientific Calculator   **Display**

15 ÷ ((64 × π)) =   0.0746039

Graphing Calculator

15 ÷ ((64 π)) ENTER   0.074603879574

In Example 3, did you at first think that the price per square inch would be the same for the large and the medium pizzas? After all, the radius of the large pizza is twice that of the medium pizza, and the cost of the large is twice that of the medium. However, the large pizza's area, 64π square inches, is *four times the area* of the medium pizza's, 16π square inches. Doubling the radius of a circle increases its area by four times the original amount.

3 Solve problems using formulas for volume.

✔ **CHECK POINT 3**   Which one of the following is the better buy: a large pizza with an 18-inch diameter for $20.00 or a medium pizza with a 14-inch diameter for $14.00?

Geometric Formulas for Volume

A shoe box and a basketball are examples of three-dimensional figures. **Volume** refers to the amount of space occupied by such a figure. In order to measure this space, we begin by selecting a cubic unit. One such cubic unit is shown in Figure 2.10.

The edges of a cube all have the same length. Other cubic units used to measure volume include 1 cubic inch (in.^{3}) and 1 cubic foot (ft^3). The volume of a solid is the number of cubic units that can be contained in the solid.

Formulas for volumes of three-dimensional figures are given in Table 2.5.

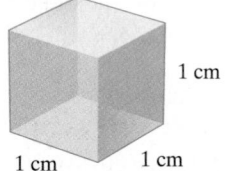

Figure 2.10 A cubic unit for measuring volume

1 cm, 1 cm, 1 cm

Table 2.5   Common Formulas for Volume

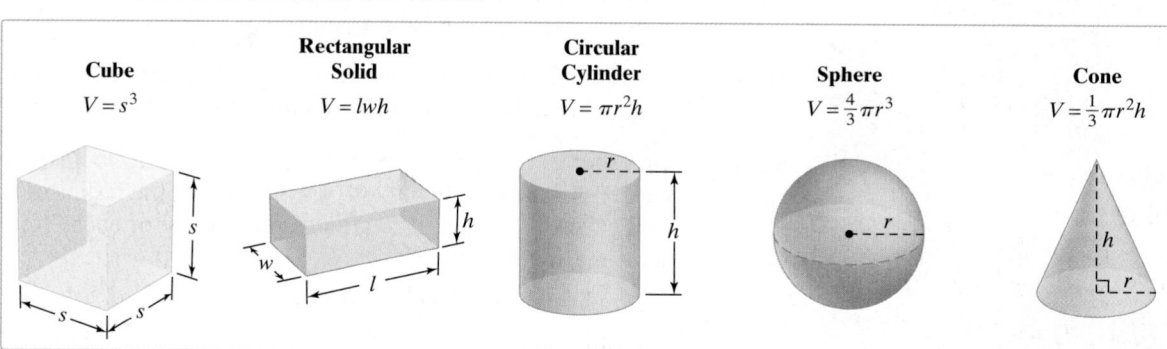

Cube	Rectangular Solid	Circular Cylinder	Sphere	Cone
$V = s^3$	$V = lwh$	$V = \pi r^2 h$	$V = \frac{4}{3}\pi r^3$	$V = \frac{1}{3}\pi r^2 h$

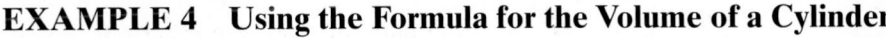

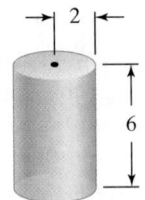

Radius: 2 inches
Height: 6 inches

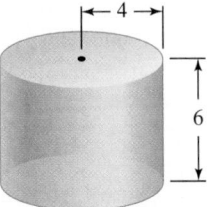

Radius: 4 inches
Height: 6 inches

Figure 2.11 Doubling a cylinder's radius

EXAMPLE 4   Using the Formula for the Volume of a Cylinder

A cylinder with a radius of 2 inches and a height of 6 inches has its radius doubled. (See Figure 2.11.) How many times greater is the volume of the larger cylinder than the volume of the smaller cylinder?

Solution   We begin with the formula for the volume of a cylinder given in Table 2.5. Find the volume of the smaller cylinder and the volume of the larger cylinder. To compare the volumes, divide the volume of the larger cylinder by the volume of the smaller cylinder.

$$V = \pi r^2 h$$
Use the formula for the volume of a cylinder.

Radius is doubled.

$$V_{\text{Smaller}} = \pi(2)^2(6) \quad V_{\text{Larger}} = \pi(4)^2(6)$$
Substitute the given values.

$$V_{\text{Smaller}} = \pi(4)(6) \quad V_{\text{Larger}} = \pi(16)(6)$$

$$V_{\text{Smaller}} = 24\pi \quad V_{\text{Larger}} = 96\pi$$

The volume of the smaller cylinder is 24π cubic inches. The volume of the larger cylinder is 96π cubic inches. We use division to compare the volumes:

$$\frac{V_{\text{Larger}}}{V_{\text{Smaller}}} = \frac{96\pi}{24\pi} = \frac{4}{1}.$$

Thus, the volume of the larger cylinder is 4 times the volume of the smaller cylinder. ∎

✔ **CHECK POINT 4**   A cylinder with a radius of 3 inches and a height of 5 inches has its height doubled. How many times greater is the volume of the larger cylinder than the volume of the smaller cylinder?

EXAMPLE 5   Applying Volume Formulas

An ice cream cone is 5 inches deep and has a radius of 1 inch. A spherical scoop of ice cream also has a radius of 1 inch. (See Figure 2.12.) If the ice cream melts into the cone, will it overflow?

Solution   The ice cream will overflow if the volume of the ice cream, a sphere, is greater than the volume of the cone. Find the volume of each.

$$V_{\text{cone}} = \frac{1}{3}\pi r^2 h = \frac{1}{3}\pi(1 \text{ in.})^2 \cdot 5 \text{ in.} = \frac{5\pi}{3} \text{ in.}^3 \approx 5 \text{ in.}^3$$

$$V_{\text{sphere}} = \frac{4}{3}\pi r^3 = \frac{4}{3}\pi(1 \text{ in.})^3 = \frac{4\pi}{3} \text{ in.}^3 \approx 4 \text{ in.}^3$$

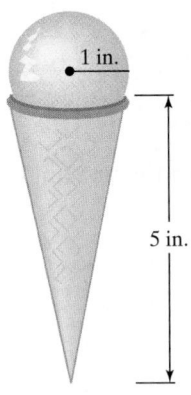

1 in.

5 in.

Figure 2.12

The volume of the spherical ice cream is less than the volume of the cone, so there will be no overflow. ∎

✔ **CHECK POINT 5**   A basketball has a radius of 4.5 inches. If the ball is filled with 350 cubic inches of air, is this enough air to fill it completely?

4 Solve problems involving the angles of a triangle.

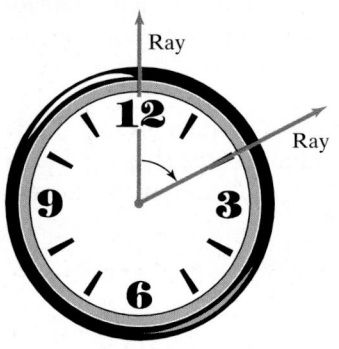

Figure 2.13 Clock with hands forming an angle

The Angles of a Triangle

The hour hand of a clock moves from 12 to 2 o'clock. The hour hand suggests a **ray**, a part of a line that has only one endpoint and extends forever in the opposite direction. An *angle* is formed as the ray in Figure 2.13 rotates from 12 to 2.

An **angle**, symbolized ∡, is made up of two rays that have a common endpoint. Figure 2.14 shows an angle. The common endpoint, *B* in the figure, is called the **vertex** of the angle. The two rays that form the angle are called its **sides**. The four ways of naming the angle are shown to the right of Figure 2.14.

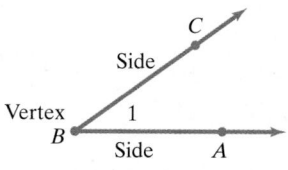

Naming the Angle

∡ 1   ∡ *B*   ∡ *ABC*   ∡ *CBA*

Vertex alone   Vertex letter in the middle

Figure 2.14 An angle: two rays with a common endpoint

One way to measure angles is in **degrees**, symbolized by a small, raised circle °. Think of the hour hand of a clock. From 12 noon to 12 midnight, the hour hand moves around in a complete circle. By definition, the ray has rotated through 360 degrees, or 360°. Using 360° as the amount of rotation of a ray back onto itself, a degree, 1°, is $\frac{1}{360}$ of a complete rotation.

Our next problem is based on the relationship among the three angles of any triangle.

The Angles of a Triangle

The sum of the measures of the three angles of any triangle is 180°.

EXAMPLE 6   Angles of a Triangle

In a triangle, the measure of the first angle is twice the measure of the second angle. The measure of the third angle is 20° less than the second angle. What is the measure of each angle?

Solution

Step 1 Let *x* represent one of the quantities. Let

$$x = \text{the measure of the second angle.}$$

Step 2 Represent other quantities in terms of *x*. The measure of the first angle is twice the measure of the second angle. Thus, let

$$2x = \text{the measure of the first angle.}$$

The measure of the third angle is 20° less than the second angle. Thus, let

$$x - 20 = \text{the measure of the third angle.}$$

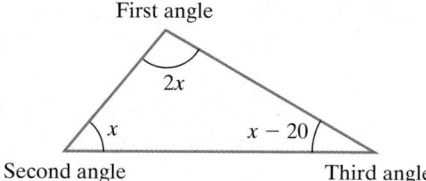

Step 3 Write an equation in x that describes the conditions. Because we are working with a triangle, the sum of the measures of its three angles is 180°.

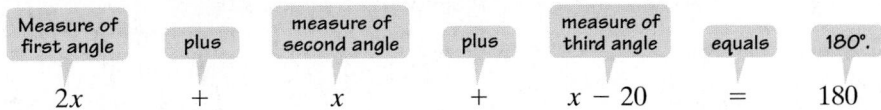

Measure of first angle	plus	measure of second angle	plus	measure of third angle	equals	180°.
$2x$	$+$	x	$+$	$x - 20$	$=$	180

Step 4 Solve the equation and answer the question.

$$2x + x + x - 20 = 180 \qquad \text{This is the equation that describes the sum of the measures of the angles.}$$

$$4x - 20 = 180 \qquad \text{Combine like terms.}$$

$$4x - 20 + 20 = 180 + 20 \qquad \text{Add 20 to both sides.}$$

$$4x = 200 \qquad \text{Simplify.}$$

$$\frac{4x}{4} = \frac{200}{4} \qquad \text{Divide both sides by 4.}$$

$$x = 50 \qquad \text{Simplify.}$$

Measure of first angle $= 2x = 2 \cdot 50° = 100°$

Measure of second angle $= x = 50°$

Measure of third angle $= x - 20 = 50 - 20 = 30°$

The angles measure 100°, 50°, and 30°.

Step 5 Check the proposed solution in the original wording of the problem. The problem tells us that we are working with a triangle's angles. Thus, the sum of the measures should be 180°. Adding the three measures, we obtain $100° + 50° + 30°$, giving the required sum of 180°. ∎

✔ **CHECK POINT 6** In a triangle, the measure of the first angle is three times the measure of the second angle. The measure of the third angle is 20° less than the second angle. What is the measure of each angle?

5 Solve problems involving complementary and supplementary angles.

Complementary and Supplementary Angles

Two angles whose measures have a sum of 90° are called **complementary angles**. For example, angles measuring 70° and 20° are complementary angles because $70° + 20° = 90°$. For angles such as those measuring 70° and 20°, each angle is a **complement** of the other: the 70° angle is the complement of the 20° angle and the 20° angle is the complement of the 70° angle. The measure of the complement can be found by subtracting the angle's measure from 90°. For example, we can find the complement of a 25° angle by subtracting 25° from 90°: $90° - 25° = 65°$. Thus, an angle measuring 65° is the complement of one measuring 25°.

Two angles whose measures have a sum of 180° are called **supplementary angles**. For example, angles measuring 110° and 70° are supplementary angles because $110° + 70° = 180°$. For angles such as those measuring 110° and 70°, each angle is a **supplement** of the other: the 110° angle is the supplement of the 70° angle, and the 70° angle is the supplement of the 110° angle. The measure of the supplement can be found by subtracting the angle's measure from 180°. For example, we can find the supplement of a 25° angle by subtracting 25° from 180°: $180° - 25° = 155°$. Thus, an angle measuring 155° is the supplement of one measuring 25°.

Algebraic Expressions for Complements and Supplements

Measure of an angle: $x°$

Measure of the angle's complement: $90° - x°$

Measure of the angle's supplement: $180° - x°$

EXAMPLE 7   Angle Measures and Complements

The measure of an angle is 40° less than four times the measure of its complement. What is the angle's measure?

Solution

Step 1 Let x represent one of the quantities. Let

$$x = \text{the measure of the angle.}$$

Step 2 Represent other unknown quantities in terms of x. Because this problem involves an angle and its complement, let

$$90 - x = \text{the measure of the complement.}$$

Step 3 Write an equation in x that describes the conditions.

The angle's measure	is	40° less than	four times the measure of its complement.
x	$=$	$4(90 - x)$	$- 40$

Step 4 Solve the equation and answer the question.

$x = 4(90 - x) - 40$ This is the equation that describes the problem's conditions.

$x = 360 - 4x - 40$ Use the distributive property.

$x = 320 - 4x$ Simplify: $360 - 40 = 320$.

$x + 4x = 320 - 4x + 4x$ Add $4x$ to both sides.

$5x = 320$ Simplify.

$\dfrac{5x}{5} = \dfrac{320}{5}$ Divide both sides by 5.

$x = 64$ Simplify.

The angle measures 64°.

Step 5 Check the proposed solution in the original wording of the problem. The measure of the complement is $90° - 64° = 26°$. Four times the measure of the complement is $4 \cdot 26°$, or 104°. The angle's measure, 64°, is 40° less than 104°: $104° - 40° = 64°$. As specified by the problem's wording, the angle's measure is 40° less than four times the measure of its complement.   ∎

✔ **CHECK POINT 7**   The measure of an angle is twice the measure of its complement. What is the angle's measure?

EXERCISE SET 2.6

Practice Exercises

Use the formulas for perimeter and area in Table 2.3 on page 147 to solve Exercises 1–12.

In Exercises 1–2, find the perimeter and area of each rectangle.

1.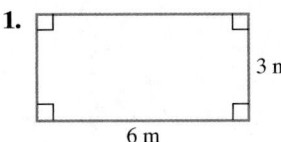

3 m

6 m

2.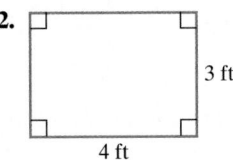

3 ft

4 ft

In Exercises 3–4, find the area of each triangle.

3.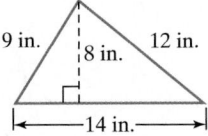

9 in. 8 in. 12 in.

14 in.

4.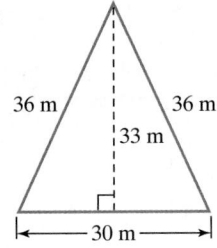

36 m 36 m

33 m

30 m

In Exercises 5–6, find the area of each trapezoid.

5.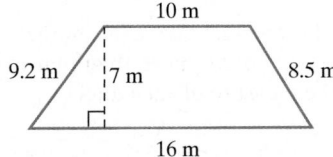

10 m

9.2 m 7 m 8.5 m

16 m

6.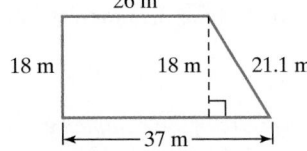

26 m

18 m 18 m 21.1 m

37 m

7. A rectangular swimming pool has a width of 25 feet and an area of 1250 square feet. What is the pool's length?

8. A rectangular swimming pool has a width of 35 feet and an area of 2450 square feet. What is the pool's length?

9. A triangle has a base of 5 feet and an area of 20 square feet. Find the triangle's height.

10. A triangle has a base of 6 feet and an area of 30 square feet. Find the triangle's height.

11. A rectangle has a width of 44 centimeters and a perimeter of 188 centimeters. What is the rectangle's length?

12. A rectangle has a width of 46 centimeters and a perimeter of 208 centimeters. What is the rectangle's length?

Use the formulas for the area and a circumference of a circle in Table 2.4 on page 149 to solve Exercises 13–18.

In Exercises 13–16, find the area and circumference of each circle. Express answers in terms of π. Then round to the nearest whole number.

13.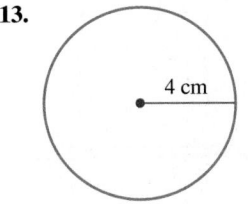

4 cm

14.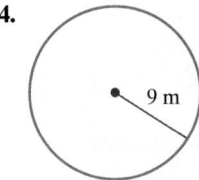

9 m

15.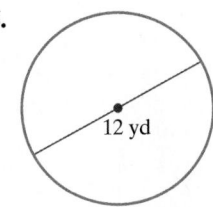

12 yd

16.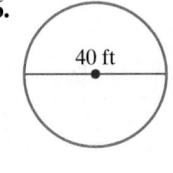

40 ft

17. The circumference of a circle is 14π inches. Find the circle's radius and diameter.

18. The circumference of a circle is 16π inches. Find the circle's radius and diameter.

Use the formulas for volume in Table 2.5 on page 150 to solve Exercises 19–30.

In Exercises 19–26, find the volume of each figure. Where applicable, express answers in terms of π. Then round to the nearest whole number.

19.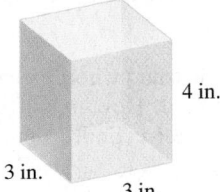

4 in.

3 in.

3 in.

20.

3 cm

3 cm 5 cm

21.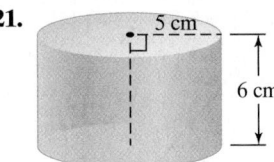

5 cm

6 cm

22.
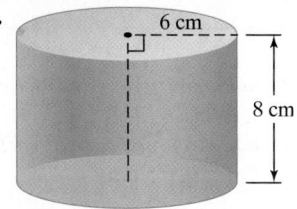

23.
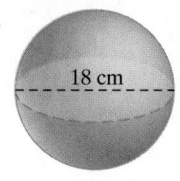

24.
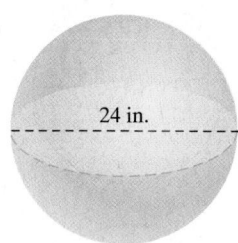

25.
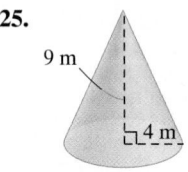

26.
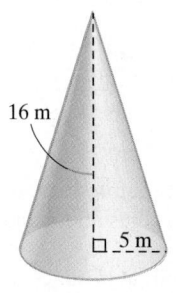

27. Solve the formula for the volume of a circular cylinder for h.

28. Solve the formula for the volume of a cone for h.

29. A cylinder whose radius is 3 inches and whose height is 4 inches has its radius tripled. How many times greater is the volume of the larger cylinder than the smaller cylinder?

30. A cylinder whose radius is 2 inches and whose height is 3 inches has its radius quadrupled. How many times greater is the volume of the larger cylinder than the smaller cylinder?

Use the relationship among the three angles of any triangle to solve Exercises 31–36.

31. Two angles of a triangle have the same measure and the third angle is 30° greater than the measure of the other two. Find the measure of each angle.

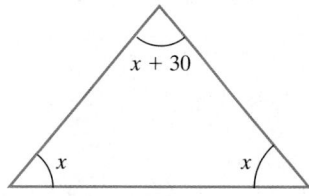

32. One angle of a triangle is three times as large as another. The measure of the third angle is 40° more than that of the smallest angle. Find the measure of each angle.

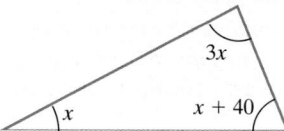

Find the measure of each angle in the triangles in Exercises 33–34.

33.
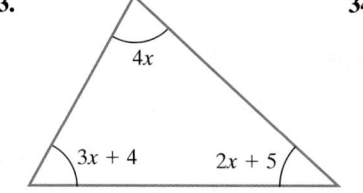

34.
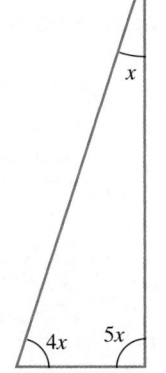

35. One angle of a triangle is twice as large as another. The measure of the third angle is 20° more than that of the smallest angle. Find the measure of each angle.

36. One angle of a triangle is three times as large as another. The measure of the third angle is 30° greater than that of the smallest angle. Find the measure of each angle.

In Exercises 37–40, find the measure of the complement of each angle.

37. 48°

38. 52°

39. 89°

40. 1°

In Exercises 41–44, find the measure of the supplement of each angle.

41. 111°

42. 95°

43. 90°

44. 179°

In Exercises 45–50, use the five-step problem-solving strategy to find the measure of the angle described.

45. The angle's measure is 60° more than that of its complement.

46. The angle's measure is 78° less than that of its complement.

47. The angle's measure is three times that of its supplement.

48. The angle's measure is 16° more than triple that of its supplement.

49. The measure of the angle's supplement is 10° more than three times that of its complement.

50. The measure of the angle's supplement is 52° more than twice that of its complement.

 Application Exercises

Use the formulas for perimeter and area in Table 2.3 on page 147 to solve Exercises 51–52.

51. Taxpayers with an office in their home may deduct a percentage of their home-related expenses. This percentage is based on the ratio of the office's area to the area of the home. A taxpayer with an office in a 2200 square foot home maintains a 20 foot by 16 foot office. If the yearly electric bills for the home come to $4800, how much of this is deductible?

52. The lot in the figure shown, except for the house, shed, and driveway, is lawn. One bag of lawn fertilizer costs $25.00 and covers 4000 square feet.
 a. Determine the minimum number of bags of fertilizer needed for the lawn.
 b. Find the total cost of the fertilizer.

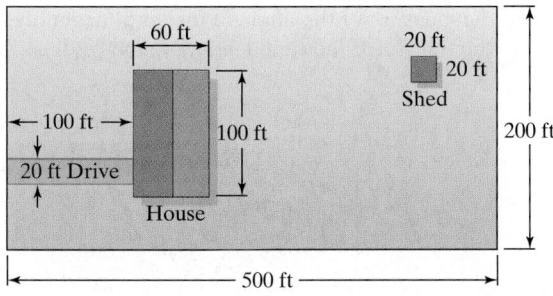

Use the formulas for the area and the circumference of a circle in Table 2.4 on page 149 to solve Exercises 53–58. Round all circumference and area calculations to the nearest whole number.

53. Which one of the following is a better buy: a large pizza with a 14-inch diameter for $12.00 or a medium pizza with a 7-inch diameter for $5.00?

54. Which one of the following is a better buy: a large pizza with a 16-inch diameter for $12.00 or two small pizzas, each with a 10-inch diameter, for $12.00?

55. If asphalt pavement costs $0.80 per square foot, find the cost to pave the circular road in the figure shown.

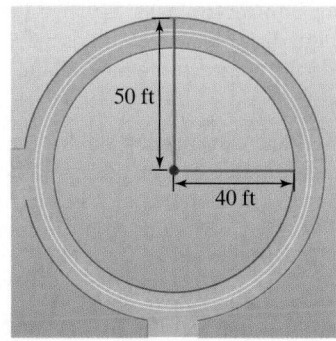

56. Hardwood flooring costs $10.00 per square foot. How much will it cost (to the nearest dollar) to cover the dance floor shown in the figure with hardwood flooring?

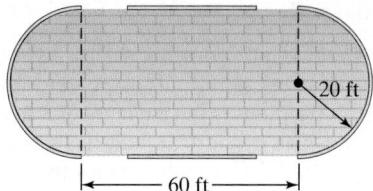

57. A glass window is to be placed in a house. The window consists of a rectangle, 6 feet high by 3 feet wide, with a semicircle at the top. Approximately how many feet of stripping will be needed to frame the window?

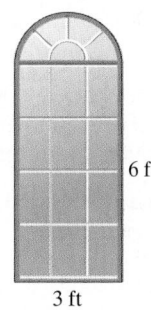

58. How many plants spaced every 6 inches are needed to surround a circular garden with a 30-foot radius?

Use the formulas for volume in Table 2.5 on page 150 to solve Exercises 59–63. When necessary, round all volume calculations to the nearest whole number.

59. A water reservoir is shaped like a rectangular solid with a base that is 50 yards by 30 yards, and a vertical height of 20 yards. At the start of a three-month period of no rain, the reservoir was completely full. At the end of this period, the height of the water was down to 6 yards. How much water was used in the three-month period?

60. A building contractor is to dig a foundation 4 yards long, 3 yards wide, and 2 yards deep for a toll booth's foundation. The contractor pays $10 per load for trucks to remove the dirt. Each truck holds 6 cubic yards. What is the cost to the contractor to have all the dirt hauled away?

61. Two cylindrical cans of soup sell for the same price. One can has a diameter of 6 inches and a height of 5 inches. The other has a diameter of 5 inches and a height of 6 inches. Which can contains more soup and, therefore, is the better buy?

62. The tunnel under the English Channel that connects England and France is the world's longest tunnel. There are actually three separate tunnels built side by side. Each is a half-cylinder that is 50,000 meters long and 4 meters high. How many cubic meters of dirt had to be removed to build the tunnel?

63. You are about to sue your contractor who promised to install a water tank that holds 500 gallons of water. You know that 500 gallons is the capacity of a tank that holds 67 cubic feet. The cylindrical tank has a radius of 3 feet and a height of 2 feet 4 inches. Does the evidence indicate you can win the case against the contractor if it goes to court?

Writing in Mathematics

64. Using words only, describe how to find the area of a triangle.

65. Describe the difference between the following problems: How much fencing is needed to enclose a garden? How much fertilizer is needed for the garden?

66. Describe how volume is measured. Explain why linear or square units cannot be used.

67. What is an angle?

68. If the measures of two angles of a triangle are known, explain how to find the measure of the third angle.

69. Can a triangle contain two 90° angles? Explain your answer.

70. What are complementary angles? Describe how to find the measure of an angle's complement.

71. What are supplementary angles? Describe how to find the measure of an angle's supplement?

72. Describe an application of a geometric formula involving area or volume.

73. Write and solve an original problem involving the measures of the three angles of a triangle.

Critical Thinking Exercises

74. Which one of the following is true?
  a. It is not possible to have a circle whose circumference is numerically equal to its area.
  b. When the measure of a given angle is added to three times the measure of its complement, the sum equals the sum of the measures of the complement and supplement of the angle.
  c. The complement of an angle that measures less than 90° is an angle that measures more than 90°.
  d. Two complementary angles cannot be equal in measure.

75. Suppose you know the cost for building a rectangular deck measuring 8 feet by 10 feet. If you decide to increase the dimensions to 12 feet by 15 feet, by how many times will the cost increase?

76. A rectangular swimming pool measures 14 feet by 30 feet. The pool is surrounded on all four sides by a path that is 3 feet wide. If the cost to resurface the path is $2 per square foot, what is the total cost of resurfacing the path?

77. What happens to the volume of a sphere if its radius is doubled?

78. A scale model of a car is constructed so that its length, width, and height are each $\frac{1}{10}$ the length, width, and height of the actual car. By how many times does the volume of the car exceed its scale model?

79. Find the measure of the angle of inclination, denoted by $x°$ in the figure, for the road leading to the bridge.

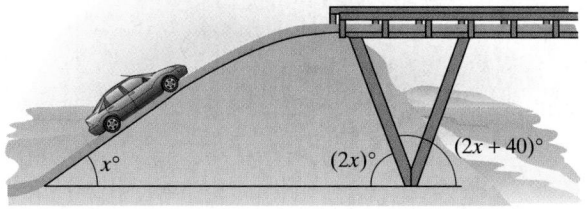

Review Exercises

80. Solve for s: $P = 2s + b$. (Section 2.4, Example 3)

81. Solve for x: $\frac{x}{2} + 7 = 13 - \frac{x}{4}$ (Section 2.3, Example 4)

82. Simplify: $\left[3(12 \div 2^2 - 3)^2\right]^2$. (Section 1.8, Example 8)

▸ SECTION 2.7 *Solving Linear Inequalities*

Objectives

1. Graph the solutions of an inequality on a number line.

2. Use set-builder notation.

3. Understand properties used to solve linear inequalities.

4. Solve linear inequalities.

5. Identify inequalities with no solution or infinitely many solutions.

6. Solve problems using linear inequalities.

SSM
PH Tutor CD- Video
Center ROM

Do you remember Rent-a-Heap, the car rental company that charged $125 per week plus $0.20 per mile to rent a small car? In Example 5 on page 141 we asked the question: How many miles can you travel for $335? We let x represent the number of miles and set up a linear equation as follows:

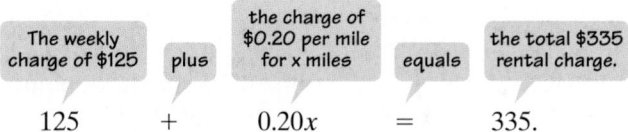

$$125 \quad + \quad 0.20x \quad = \quad 335.$$

Because we are limited by how much money we can spend on everything from buying clothing to renting a car, it is also possible to ask: How many miles can you travel if you can spend *at most* $335? We again let x represent the number of miles. Spending *at most* $335 means that the amount spent on the weekly rental must be *less than or equal to* $335:

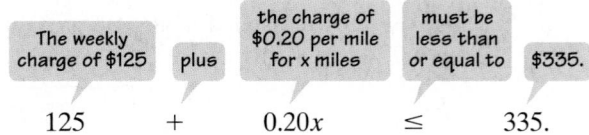

$$125 \quad + \quad 0.20x \quad \leq \quad 335.$$

Using the commutative property of addition, we can express this inequality as $0.20x + 125 \leq 335$. The form of this inequality is $ax + b \leq c$, with $a = 0.20$, $b = 125$, and $c = 335$. Any inequality in this form is called a **linear inequality in one variable**. The symbol between $ax + b$ and c can be $\leq$ (is less than or equal to), $<$ (is less than), $\geq$ (is greater than or equal to), or $>$ (is greater than). The greatest exponent on the variable in such an inequality is 1.

In this section, we will study how to solve linear inequalities such as $0.20x + 125 \leq 335$. **Solving an inequality** is the process of finding the set of numbers that will make the inequality a true statement. These numbers are called the **solutions** of the inequality, and we say that they **satisfy** the inequality. The set of all solutions is called the **solution set** of the inequality.

1 Graph the solutions of an inequality on a number line.

Graphs of Inequalities

There are infinitely many solutions to the inequality $x < 3$, namely, all real numbers that are less than 3. Although we cannot list all the solutions, we can make a drawing on a number line that represents these solutions. Such a drawing is called the **graph of the inequality**.

Graphs of linear inequalities are shown on a number line by shading all points representing numbers that are solutions. *Open dots* (○) indicate endpoints that are *not solutions* and *closed dots* (●) indicate endpoints that *are solutions*.

EXAMPLE 1   Graphing Inequalities

Graph the solutions of:

  a. $x < 3$ **b.** $x \geq -1$ **c.** $-1 < x \leq 3.$

Solution

a. The solutions of $x < 3$ are all real numbers that are less than 3. They are graphed on a number line by shading all points to the left of 3. The open dot at 3 indicates that 3 is not a solution, but numbers such as 2.9999 and 2.6 are.

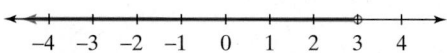

b. The solutions of $x \geq -1$ are all real numbers that are greater than or equal to -1. We shade all points to the right of -1 and the point for -1 itself. The closed dot at -1 shows that -1 is a solution of the given inequality.

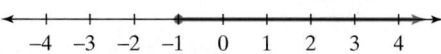

c. The inequality $-1 < x \leq 3$ is read "-1 is less than x and x is less than or equal to 3," or "x is greater than -1 and less than or equal to 3." The solutions of $-1 < x \leq 3$ are all real numbers between -1 and 3, not including -1 but including 3. In the graph for all real numbers between -1, exclusive, and 3, inclusive, the open dot at -1 indicates that -1 is not a solution. The closed dot at 3 shows that 3 is a solution. Shading indicates the other solutions.

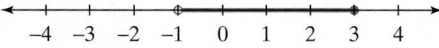

■

✔ **CHECK POINT 1**   Graph the solutions of:

  a. $x < 4$ **b.** $x \geq -2$ **c.** $-4 \leq x < 1.$

2 Use set-builder notation.

Solution Sets

The solutions of $x < 3$ are all real numbers that are less than 3. We can use the set concept introduced in Chapter 1 and state that the solution is the *set of all real numbers less than 3*. We use **set-builder notation** to write the solution set of $x < 3$ as

$$\{x \mid x < 3\}.$$

We read this as "the set of all x such that x is less than 3." Solutions of inequalities should be expressed in set-builder notation.

3 Understand properties used to solve linear inequalities.

Properties Used to Solve Linear Inequalities

Back to our question: How many miles can you drive on your Rent-a-Heap car if you can spend at most $335 per week? We answer the question by solving

$$0.20x + 125 \le 335$$

for x. The solution procedure is nearly identical to that for solving

$$0.20x + 125 = 335.$$

Our goal is to get x by itself on the left side. We do this by isolating $0.20x$, subtracting 125 from both sides:

$0.20x + 125 \le 335$	This is the given inequality.
$0.20x + 125 - 125 \le 335 - 125$	Subtract 125 from both sides.
$0.20x \le 210$	Simplify.

Finally, we isolate x from $0.20x$ by dividing both sides of the inequality by 0.20:

$\dfrac{0.20x}{0.20} \le \dfrac{210}{0.20}$	Divide both sides by 0.20.
$x \le 1050$	Simplify.

With at most $335 per week to spend, you can travel at most 1050 miles.

We started with the inequality $0.20x + 125 \le 335$ and obtained the inequality $x \le 1050$ in the final step. Both of these inequalities have the same solution set, namely $\{x \,|\, x \le 1050\}$. Inequalities such as these, with the same solution set, are said to be **equivalent**.

We isolated x from $0.20x$ by dividing both sides of $0.20x \le 210$ by 0.20, a positive number. Let's see what happens if we divide both sides of an inequality by a negative number. Consider the inequality $10 < 14$. Divide 10 and 14 by -2:

$$\frac{10}{-2} = -5 \quad \text{and} \quad \frac{14}{-2} = -7.$$

Because -5 lies to the right of -7 on the number line, -5 is greater than -7:

$$-5 > -7.$$

Notice that the direction of the inequality symbol is reversed:

$$10 < 14$$
$$-5 > -7$$

In general, **when we multiply or divide both sides of an inequality by a negative number, the direction of the inequality symbol is reversed**. When we reverse the direction of the inequality symbol, we say that we change the *sense* of the inequality.

We can isolate a variable in a linear inequality the same way we can isolate a variable in a linear equation. The following properties are used to create equivalent inequalities.

Study Tip

We can express $x \le 1050$ in English in at least three ways:
- You can travel at most 1050 miles.
- You can travel no more than 1050 miles.
- Your travel cannot exceed 1050 miles.

Properties of Inequalities

Property	The Property in Words	Example
The Addition Property of Inequality If $a < b$, then $a + c < b + c$. If $a < b$, then $a - c < b - c$.	If the same quantity is added to or subtracted from both sides of an inequality, the resulting inequality is equivalent to the original one.	$2x + 3 < 7$ Subtract 3: $2x + 3 - 3 < 7 - 3$ Simplify: $2x < 4$
The Positive Multiplication Property of Inequality If $a < b$ and c is positive, then $ac < bc$. If $a < b$ and c is positive, then $\dfrac{a}{c} < \dfrac{b}{c}$.	If we multiply or divide both sides of an inequality by the same positive quantity, the resulting inequality is equivalent to the original one.	$2x < 4$ Divide by 2: $\dfrac{2x}{2} < \dfrac{4}{2}$ Simplify: $x < 2$
The Negative Multiplication Property of Inequality If $a < b$ and c is negative, then $ac > bc$. If $a < b$ and c is negative, then $\dfrac{a}{c} > \dfrac{b}{c}$.	If we multiply or divide both sides of an inequality by the same negative quantity and reverse the direction of the inequality symbol, the result is an equivalent inequality.	$-4x < 20$ Divide by -4 and reverse the sense of the inequality: $\dfrac{-4x}{-4} > \dfrac{20}{-4}$ Simplify: $x > -5$

 Solve linear inequalities.

Solving Linear Inequalities Involving Only One Property of Inequality

If you can solve a linear equation, it is likely that you can solve a linear inequality. Why? The procedure for solving linear inequalities is nearly the same as the procedure for solving linear equations, with one important exception: When multiplying or dividing by a negative number, reverse the direction of the inequality symbol, changing the sense of the inequality.

EXAMPLE 2   Solving a Linear Inequality

Solve and graph the solution set on a number line:

$$x + 3 < 8.$$

Solution   Our goal is to isolate x. We can do this by using the addition property, subtracting 3 from both sides.

$$x + 3 < 8 \qquad \text{This is the given inequality.}$$
$$x + 3 - 3 < 8 - 3 \qquad \text{Subtract 3 from both sides.}$$
$$x < 5 \qquad \text{Simplify.}$$

The solution set consists of all real numbers that are less than 5. We express this in set-builder notation as

$$\{x \mid x < 5\}. \quad \text{This is read "the set of all x such that x is less than 5."}$$

The graph of the solution set is shown as follows:

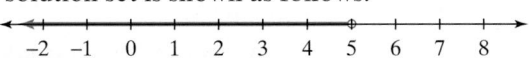

■

Discover for Yourself

Can you check all solutions to Example 2 in the given inequality? Is a partial check possible? Select a real number that is less than 5 and show that it satisfies $x + 3 < 8$.

✔ **CHECK POINT 2**   Solve and graph the solution set on a number line:

$$x + 6 < 9.$$

EXAMPLE 3   Solving a Linear Inequality

Solve and graph the solution set on a number line:

$$4x - 1 \geq 3x - 6.$$

Solution   Our goal is to isolate all terms involving x on one side and all numerical terms on the other side, exactly as we did when solving equations. Let's begin by using the addition property to isolate variable terms on the left.

$$4x - 1 \geq 3x - 6 \qquad \text{This is the given inequality.}$$
$$4x - 3x - 1 \geq 3x - 3x - 6 \qquad \text{Subtract 3x from both sides.}$$
$$x - 1 \geq -6 \qquad \text{Simplify.}$$

Now we isolate the numerical terms on the right. Use the addition property and add 1 to both sides.

$$x - 1 + 1 \geq -6 + 1 \qquad \text{Add 1 to both sides.}$$
$$x \geq -5 \qquad \text{Simplify.}$$

The solution set consists of all real numbers that are greater than or equal to -5. We express this in set-builder notation as

$$\{x \mid x \geq -5\}. \qquad \text{This is read "the set of all x such that x is greater}$$
$$\text{than or equal to } -5."$$

The graph of the solution set is shown as follows:

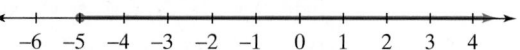

■

✔ **CHECK POINT 3**   Solve and graph the solution set on a number line:

$$8x - 2 \geq 7x - 4.$$

We solved the inequalities in Examples 2 and 3 using the addition property of inequality. Now let's practice using the multiplication property of inequality. Do not forget to reverse the direction of the inequality symbol when multiplying or dividing both sides by a negative number.

EXAMPLE 4   Solving Linear Inequalities

Solve and graph the solution set on a number line:

a. $\dfrac{1}{3}x < 5$ **b.** $-3x < 21.$

Solution   In each case, our goal is to isolate x. In the first inequality, this is accomplished by multiplying both sides by 3. In the second inequality, we can do this by dividing both sides by -3.

a. $\dfrac{1}{3}x < 5 \qquad \text{This is the given inequality.}$

$3 \cdot \dfrac{1}{3}x < 3 \cdot 5 \qquad$ Isolate x by multiplying by 3 on both sides.
$\qquad\qquad\qquad$ The symbol $<$ stays the same because we are multiplying by a positive number.

$\qquad x < 15 \qquad \text{Simplify.}$

The solution set is $\{x \,|\, x < 15\}$. The graph of the solution set is shown as follows:

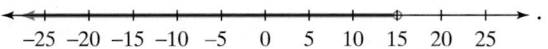

b. $-3x < 21$ This is the given inequality.

$$\frac{-3x}{-3} > \frac{21}{-3}$$ Isolate x by dividing by −3 on both sides.
The symbol $<$ must be reversed because we are dividing by a negative number.

$x > -7$ Simplify.

The solution set is $\{x \,|\, x > -7\}$. The graph of the solution set is shown as follows:

■

✔ **CHECK POINT 4**   Solve and graph the solution set on a number line:

a. $\dfrac{1}{4}x < 2$ **b.** $-6x < 18$.

Inequalities Requiring Both the Addition and Multiplication Properties

If an equality does not contain fractions, it can be solved using the following procedure. Notice, again, how similar this procedure is to the procedure for solving an equation.

> **Solving a Linear Inequality**
>
> 1. Simplify the algebraic expression on each side.
> 2. Use the addition property of inequality to collect all the variable terms on one side and all the constant terms on the other side.
> 3. Use the multiplication property of inequality to isolate the variable and solve. Reverse the sense of the inequality when multiplying or dividing both sides by a negative number.
> 4. Express the solution set in set-builder notation and graph the solution set on a number line.

EXAMPLE 5   Solving a Linear Inequality

Solve and graph the solution set on a number line:

$$4y - 7 \geq 5.$$

Solution

Step 1   Simplify each side. Because each side is already simplified, we can skip this step.

Step 2 Collect variable terms on one side and constant terms on the other side. The variable term, $4y$, is already on the left. We will collect constant terms on the right by adding 7 to both sides.

$$4y - 7 \geq 5 \qquad \text{This is the given inequality.}$$
$$4y - 7 + 7 \geq 5 + 7 \qquad \text{Add 7 to both sides.}$$
$$4y \geq 12 \qquad \text{Simplify.}$$

Step 3 Isolate the variable and solve. We isolate the variable, y, by dividing both sides by 4. Because we are dividing by a positive number, we do not reverse the inequality symbol.

$$\frac{4y}{4} \geq \frac{12}{4} \qquad \text{Divide both sides by 4.}$$
$$y \geq 3 \qquad \text{Simplify.}$$

Step 4 Express the solution set in set-builder notation and graph the set on a number line. The solution set consists of all real numbers that are greater than or equal to 3, expressed in set-builder notation as $\{y \mid y \geq 3\}$. The graph of the solution set is shown as follows:

✔ **CHECK POINT 5** Solve and graph the solution set on a number line:
$$5y - 3 \geq 17.$$

Discover for Yourself

As a partial check, select one number from the solution set for the inequality in Example 5. Substitute that number into the original inequality. Perform the resulting computations. You should obtain a true statement.

Is it possible to perform a partial check using a number that is not in the solution set? What should happen in this case? Try doing this.

EXAMPLE 6 Solving a Linear Inequality

Solve and graph the solution set on a number line:
$$7x + 15 \geq 13x + 51.$$

Solution

Step 1 Simplify each side. Because each side is already simplified, we can skip this step.

Step 2 Collect variable terms on one side and constant terms on the other side. We will collect variable terms on the left and constant terms on the right.

$$7x + 15 \geq 13x + 51 \qquad \text{This is the given inequality.}$$
$$7x + 15 - 13x \geq 13x + 51 - 13x \qquad \text{Subtract 13x from both sides.}$$
$$-6x + 15 \geq 51 \qquad \text{Simplify.}$$
$$-6x + 15 - 15 \geq 51 - 15 \qquad \text{Subtract 15 from both sides.}$$
$$-6x \geq 36 \qquad \text{Simplify.}$$

Step 3 Isolate the variable and solve. We isolate the variable, x, by dividing both sides by -6. Because we are dividing by a negative number, we must reverse the inequality symbol.

$$\frac{-6x}{-6} \leq \frac{36}{-6} \qquad \text{Divide both sides by 6 and reverse the sense of the inequality.}$$
$$x \leq -6 \qquad \text{Simplify.}$$

Study Tip

You can solve
$$7x + 15 \geq 13x + 51$$
by isolating x on the right side. Subtract $7x$ from both sides:
$$7x + 15 - 7x$$
$$\geq 13x + 51 - 7x$$
$$15 \geq 6x + 51.$$
Now subtract 51 from both sides:
$$15 - 51 \geq 6x + 51 - 51$$
$$-36 \geq 6x.$$
Finally, divide both sides by 6:
$$\frac{-36}{6} \geq \frac{6x}{6}$$
$$-6 \geq x.$$
This last inequality means the same thing as
$$x \leq -6.$$

Step 4   Express the solution set in set-builder notation and graph the set on a number line. The solution set consists of all real numbers that are less than or equal to −6, expressed in set-builder notation as $\{x \mid x \leq -6\}$. The graph of the solution set is shown as follows:

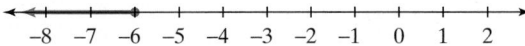

✔ **CHECK POINT 6**   Solve and graph the solution set: $6 - 3x \leq 5x - 2$.

EXAMPLE 7   Solving a Linear Inequality

Solve and graph the solution set on a number line:

$$2(x - 3) + 5x \leq 8(x - 1).$$

Solution

Step 1   Simplify each side. We use the distributive property to remove parentheses. Then we combine like terms.

$$2(x - 3) + 5x \leq 8(x - 1) \qquad \text{This is the given inequality.}$$
$$2x - 6 + 5x \leq 8x - 8 \qquad \text{Use the distributive property.}$$
$$7x - 6 \leq 8x - 8 \qquad \text{Add like terms on the left.}$$

Step 2   Collect variable terms on one side and constant terms on the other side. We will collect variable terms on the left and constant terms on the right.

$$7x - 8x - 6 \leq 8x - 8x - 8 \qquad \text{Subtract 8x from both sides.}$$
$$-x - 6 \leq -8 \qquad \text{Simplify.}$$
$$-x - 6 + 6 \leq -8 + 6 \qquad \text{Add 6 to both sides.}$$
$$-x \leq -2 \qquad \text{Simplify.}$$

Step 3   Isolate the variable and solve. To isolate x, we must eliminate the negative sign in front of the x. Because $-x$ means $-1x$, we can do this by multiplying (or dividing) both sides of the inequality by −1. We are multiplying by a negative number. Thus, we must reverse the inequality symbol.

$$(-1)(-x) \geq (-1)(-2) \qquad \text{Multiply both sides by −1 and reverse the sense of the inequality.}$$
$$x \geq 2 \qquad \text{Simplify.}$$

Step 4   Express the solution set in set-builder notation and graph the set on a number line. The solution set consists of all real numbers that are greater than or equal to 2, expressed in set-builder notation as $\{x \mid x \geq 2\}$. The graph of the solution set is shown as follows:

✔ **CHECK POINT 7**   Solve and graph the solution set:
$$2(x - 3) - 1 \leq 3(x + 2) - 14.$$

5 Identify inequalities with no solution or infinitely many solutions.

Inequalities with Unusual Solution Sets

We have seen that some equations have no solution. This is also true for some inequalities. An example of such an inequality is

$$x > x + 1.$$

There is no number that is greater than itself plus 1. This inequality has no solution. Its solution set is $\emptyset$, the empty set.

By contrast, some inequalities are true for all real numbers. An example of such an inequality is

$$x < x + 1.$$

Every real number is less than itself plus 1. The solution set is $\{x \mid x \text{ is a real number}\}$ or $\mathbb{R}$.

If you attempt to solve an inequality that has no solution, you will eliminate the variable and obtain a false statement, such as $0 > 1$. If you attempt to solve an inequality that is true for all real numbers, you will eliminate the variable and obtain a true statement, such as $0 < 1$.

EXAMPLE 8   Solving a Linear Inequality

Solve: $3(x + 1) > 3x + 5$.

Solution

$3(x + 1) > 3x + 5$	This is the given inequality.
$3x + 3 > 3x + 5$	Apply the distributive property.
$3x + 3 - 3x > 3x + 5 - 3x$	Subtract 3x from both sides.
$3 > 5$	Simplify.

The original inequality is equivalent to the statement $3 > 5$, which is false for every value of x. The inequality has no solution. The solution set is $\emptyset$, the empty set. ∎

 **CHECK POINT 8**   Solve: $4(x + 2) > 4x + 15$

EXAMPLE 9   Solving a Linear Inequality

Solve: $2(x + 5) \le 5x - 3x + 14$.

Solution

$2(x + 5) \le 5x - 3x + 14$	This is the given inequality.
$2x + 10 \le 5x - 3x + 14$	Apply the distributive property.
$2x + 10 \le 2x + 14$	Combine like terms.
$2x + 10 - 2x \le 2x + 14 - 2x$	Subtract 2x from both sides.
$10 \le 14$	Simplify.

The original inequality is equivalent to the true statement $10 \le 14$, which is true for every value of x. The solution is the set of all real numbers, written

$$\{x \mid x \text{ is a real number}\} \text{ or } \mathbb{R}.$$ ∎

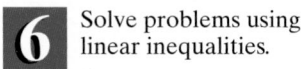

6 Solve problems using linear inequalities.

✔ **CHECK POINT 9** Solve: $3(x + 1) \geq 2x + 1 + x$.

Applications

As you know, different professors may use different grading systems to determine your final course grade. Some professors require a final examination; others do not. In our next example, not only is a final exam required, but it also counts as two grades.

EXAMPLE 10   An Application: Final Course Grade

To earn an A in a course, you must have a final average of at least 90%. On the first four examinations, you have grades of 86%, 88%, 92%, and 84%. If the final examination counts as two grades, what must you get on the final to earn an A in the course?

Solution   We will use our five-step strategy for solving algebraic word problems.

Steps 1 and 2   Represent unknown quantities in terms of x. Let x = your grade on the final examination.

Step 3   Write an inequality in x that describes the conditions. The average of the six grades is found by adding the grades and dividing the sum by 6.

$$\text{Average} = \frac{86 + 88 + 92 + 84 + x + x}{6}$$

Because the final counts as two grades, the x (your grade on the final examination) is added twice. This is also why the sum is divided by 6.

In order to get an A, your average must be at least 90. This means that your average must be greater than or equal to 90.

| Your average | must be greater than or equal to | 90. |

$$\frac{86 + 88 + 92 + 84 + x + x}{6} \qquad \geq \qquad 90$$

Step 4   Solve the inequality and answer the problem's question.

$$\frac{86 + 88 + 92 + 84 + x + x}{6} \geq 90 \qquad \text{This is the inequality for the given conditions.}$$

$$\frac{350 + 2x}{6} \geq 90 \qquad \text{Combine like terms in the numerator.}$$

$$6\left(\frac{350 + 2x}{6}\right) \geq 6(90) \qquad \text{Multiply both sides by 6, clearing fractions.}$$

$$350 + 2x \geq 540 \qquad \text{Multiply.}$$

$$350 + 2x - 350 \geq 540 - 350 \qquad \text{Subtract 350 from both sides.}$$

$$2x \geq 190 \qquad \text{Simplify.}$$

$$\frac{2x}{2} \geq \frac{190}{2} \qquad \text{Divide both sides by 2.}$$

$$x \geq 95$$

You must get at least 95% on the final examination to earn an A in the course.

Step 5   Check. We can perform a partial check by computing the average with any grade that is at least 95. We will use 96. If you get 96% on the final examination, your average is

$$\frac{86 + 88 + 92 + 84 + 96 + 96}{6} = \frac{542}{6} = 90\frac{1}{3}.$$

Because $90\frac{1}{3} > 90$, you earn an A in the course.   ■

✔ **CHECK POINT 10**   To earn a B in a course, you must have a final average of at least 80%. On the first three examinations, you have grades of 82%, 74%, and 78%. If the final examination counts as two grades, what must you get on the final to earn a B in the course?

EXERCISE SET 2.7

 Practice Exercises

In Exercises 1–12, graph the solutions of each inequality on a number line.

1. $x > 6$ 　　　　　　 **2.** $x > -2$

3. $x < -4$ 　　　　　　 **4.** $x < 0$

5. $x \geq -3$ 　　　　　 **6.** $x \geq -5$

7. $x \leq 4$ 　　　　　　 **8.** $x \leq 7$

9. $-2 < x \leq 5$ 　　　 **10.** $-3 \leq x < 7$

11. $-1 < x < 4$ 　　　 **12.** $-7 \leq x \leq 0$

Describe each graph in Exercises 13–18 using set-builder notation.

13.
14.
15.
16.
17.
18.

Use the addition property of inequality to solve each inequality in Exercises 19–36. Express the solution set in set-builder notation and graph the set on a number line.

19. $x - 3 > 2$ 　　　　　 **20.** $x + 1 < 5$

21. $x + 4 \leq 9$ 　　　　 **22.** $x - 5 \geq 1$

23. $y - 3 < 0$ 　　　　　 **24.** $y + 4 \geq 0$

25. $3x + 4 \leq 2x + 7$

26. $2x + 9 \leq x + 2$

27. $5x - 9 < 4x + 7$

28. $3x - 8 < 2x + 11$

29. $7x - 7 > 6x - 3$

30. $8x - 9 > 7x - 3$

31. $x - \dfrac{2}{3} > \dfrac{1}{2}$ 　　　 **32.** $x - \dfrac{1}{3} \geq \dfrac{5}{6}$

33. $y + \dfrac{7}{8} \leq \dfrac{1}{2}$ 　　　 **34.** $y + \dfrac{1}{3} \leq \dfrac{3}{4}$

35. $-15y + 13 > 13 - 16y$

36. $-12y + 17 > 20 - 13y$

Use the multiplication property of inequality to solve each inequality in Exercises 37–54. Express the solution set in set-builder notation and graph the set on a number line.

37. $\dfrac{1}{2}x < 4$ 　　　　 **38.** $\dfrac{1}{2}x > 3$

39. $\dfrac{x}{3} > -2$ 　　　　 **40.** $\dfrac{x}{4} < -1$

41. $4x < 20$ 　　　　　 **42.** $6x < 18$

43. $3x \geq -21$ 　　　　 **44.** $7x \geq -56$

45. $-3x < 15$ 　　　　 **46.** $-7x > 21$

47. $-3x \geq 15$ 　　　　 **48.** $-7x \leq 21$

49. $-16x > -48$ 　　　 **50.** $-20x > -140$

51. $-4y \le \dfrac{1}{2}$ **52.** $-2y \le \dfrac{1}{2}$

53. $-x < 4$ **54.** $-x > -3$

Use both the addition and multiplication properties of inequality to solve each inequality in Exercises 55–78. Express the solution set in set-builder notation and graph the set on a number line.

55. $2x - 3 > 7$ **56.** $3x + 2 \le 14$

57. $3x + 3 < 18$ **58.** $8x - 4 > 12$

59. $3 - 7x \le 17$

60. $5 - 3x \ge 20$

61. $-2x - 3 < 3$

62. $-3x + 14 < 5$

63. $5 - x \le 1$

64. $3 - x \ge -3$

65. $2x - 5 > -x + 6$

66. $6x - 2 \ge 4x + 6$

67. $2y - 5 < 5y - 11$

68. $4y - 7 > 9y - 2$

69. $3(2y - 1) < 9$

70. $4(2y - 1) > 12$

71. $3(x + 1) - 5 < 2x + 1$

72. $4(x + 1) + 2 \ge 3x + 6$

73. $8x + 3 > 3(2x + 1) - x + 5$

74. $7 - 2(x - 4) < 5(1 - 2x)$

75. $\dfrac{x}{3} - 2 \ge 1$ **76.** $\dfrac{x}{4} - 3 \ge 1$

77. $1 - \dfrac{x}{2} > 4$ **78.** $1 - \dfrac{x}{2} < 5$

In Exercises 79–88, solve each inequality. Identify inequalities that have no solution, or inequalities that are true for all real numbers.

79. $4x - 4 < 4(x - 5)$

80. $3x - 5 < 3(x - 2)$

81. $x + 3 < x + 7$

82. $x + 4 < x + 10$

83. $7x \le 7(x - 2)$

84. $3x + 1 \le 3(x - 2)$

85. $2(x + 3) > 2x + 1$

86. $5(x + 4) > 5x + 10$

87. $5x - 4 \le 4(x - 1)$

88. $6x - 3 \le 3(x - 1)$

Application Exercises

The list shown ranks the ten best-educated cities in the United States measured by the percentage of the population ages 25 and older with 16 or more years of education. Let x represent the percentage of the population with 16 or more years of education. In Exercises 89–94, write the name or names of the city or cities described by the given inequality.

Most Educated

City	% with 16 + Years of Education
1. Raleigh, NC	40.6%
2. Seattle, WA	37.9
3. San Francisco, CA	35.0
4. Austin, TX	34.4
5. Washington, DC	33.3
6. Lexington-Fayette, KY	30.6
7. Minneapolis, MN	30.3
8. Boston, MA	30.0
Arlington, TX	30.0
10. San Diego, CA	29.8

Source: U.S. Census Bureau

89. $x \ge 34.4\%$

90. $x > 35.0\%$

91. $x < 30.0\%$

92. $x \le 30.3\%$

93. $30.0\% \le x < 34.4\%$

94. $30.6\% < x \le 35.0\%$

The bar graph shows the number of people in the United States with various disorders of mental illness. Use the information in the graph to solve Exercises 95–96.

Mental Illness in America, by Disorder

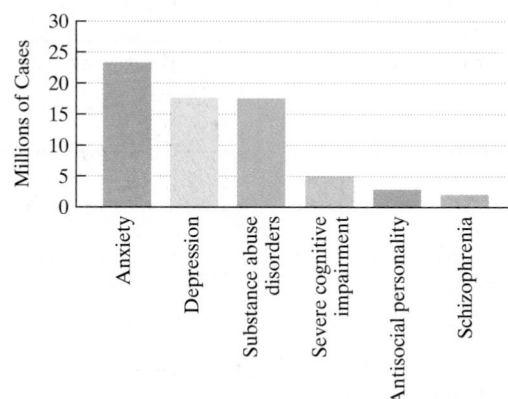

Source: National Institute of Mental Health.

95. Let x represent the millions of people with the various illnesses shown by the graph. Which illnesses are described by $3x - 4 < 11$ cases? Begin by solving the inequality.

96. Let x represent the millions of people with the various illnesses shown by the graph. Which illness or illnesses are described by $3x - 4 > 56$ cases? Begin by solving the inequality.

The line graph shows the declining consumption of cigarettes in the United States. The data shown by the graph can be modeled by

$$N = 550 - 9x$$

where N is the number of cigarettes consumed, in billions, x years after 1988. Use this formula to solve Exercises 97–98.

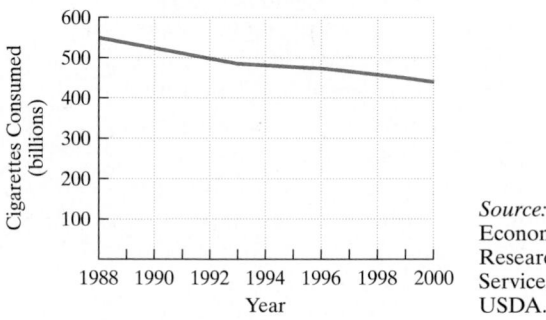

Consumption of Cigarettes in the U.S.

Source: Economic Research Service, USDA.

97. Describe how many years after 1988 cigarette consumption will be less than 370 billion cigarettes each year. Which years are included in your description?

98. Describe how many years after 1988 cigarette consumption will be less than 325 billion cigarettes each year. Which years are included in your description?

99. On two examinations, you have grades of 86 and 88. There is an optional final examination, which counts as one grade. You decide to take the final in order to get a course grade of A, meaning a final average of at least 90.
  a. What must you get on the final to earn an A in the course?
  b. By taking the final, if you do poorly, you might risk the B that you have in the course based on the first two exam grades. If your final average is less than 80, you will lose your B in the course. Describe the grades on the final that will cause this to happen.

100. On three examinations, you have grades of 88, 78, and 86. There is still a final examination, which counts as one grade.
  a. In order to get an A, your average must be at least 90. If you get 100 on the final, compute your average and determine if an A in the course is possible.
  b. To earn a B in the course, you must have a final average of at least 80. What must you get on the final to earn a B in the course?

101. A car can be rented from Continental Rental for $80 per week plus 25 cents for each mile driven. How many miles can you travel if you can spend at most $400 for the week?

102. A car can be rented from Basic Rental for $60 per week plus 50 cents for each mile driven. How many miles can you travel if you can spend at most $600 for the week?

103. An elevator at a construction site has a maximum capacity of 3000 pounds. If the elevator operator weighs 245 pounds and each cement bag weighs 95 pounds, how many bags of cement can be safely lifted on the elevator in one trip?

104. An elevator at a construction site has a maximum capacity of 2800 pounds. If the elevator operator weighs 265 pounds and each cement bag weighs 65 pounds, how many bags of cement can be safely lifted on the elevator in one trip?

Writing in Mathematics

105. When graphing the solutions of an inequality, what is the difference between an open dot and a closed dot?

106. When solving an inequality, when is it necessary to change the direction of the inequality symbol? Give an example.

107. Describe ways in which solving a linear inequality is similar to solving a linear equation.

108. Describe ways in which solving a linear inequality is different from solving a linear equation.

109. Using current trends, future costs of Medicare can be modeled by $C = 18x + 250$, where x represents the number of years after 2000 and C represents the cost of Medicare, in billions of dollars. Use the formula to write a word problem that can be solved using a linear inequality. Then solve the problem.

Critical Thinking Exercises

110. Which one of the following statements is true?
  a. The inequality $x - 3 > 0$ is equivalent to $x < 3$.
  b. The statement "x is at most 5" is written $x < 5$.
  c. The inequality $-4x < -20$ is equivalent to $x > -5$.
  d. The statement "the sum of x and 6% of x is at least 80" is written $x + 0.06x \geq 80$.

111. A car can be rented from Basic Rental for $260 week with no extra charge for mileage. Continental charges $80 per week plus 25 cents for each mile driven to rent the same car. How many miles should be driven in a week to make the rental cost for Basic Rental a better deal than Continental's?

112. Membership in a fitness club costs $500 yearly plus $1 per hour spent working out. A competing club charges $440 yearly plus $1.75 per hour for use of their equipment. How many hours must a person work out yearly to make membership in the first club cheaper than membership in the second club?

Technology Exercises

Solve each inequality in Exercises 113–114. Use a calculator to help with the arithmetic.

113. $1.45 - 7.23x > -1.442$

114. $126.8 - 9.4y \leq 4.8y + 34.5$

Review Exercises

115. 8 is 40% of what number? (Section 2.4, Example 8)

116. The length of a rectangle exceeds the width by 5 inches. The perimeter is 34 inches. What are the rectangle's dimensions? (Section 2.5, Example 6)

117. Solve and check: $5x + 16 = 3(x + 8)$. (Section 2.3, Example 2)

CHAPTER 2 GROUP PROJECTS

1. One of the best ways to learn how to *solve* a word problem in algebra is to *design* word problems of your own. Creating a word problem makes you very aware of precisely how much information is needed to solve the problem. You must also focus on the best way to present information to a reader and on how much information to give. As you write your problem, you gain skills that will help you solve problems created by others.

The group should design five different word problems that can be solved using an algebraic equation. All of the problems should be on different topics. For example, the group should not have more than one problem on finding a number. The group should turn in both the problems and their algebraic solutions.

2. Group members are to write a helpful list of items for a pamphlet called "A Student's Guide to Solving Linear Equations, Inequalities, and Word Problems." The pamphlet will be used primarily by students who sit, stare, and get nervous every time they are asked to solve an algebraic word problem. It will also be used by students who make errors such as these when solving equations and inequalities:

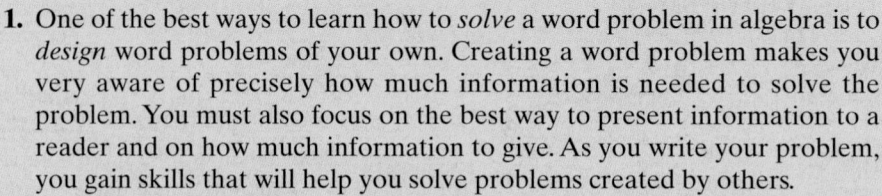

$$4x + 24 = 32 \qquad 7x = 21 \qquad 5 - 3(x + 1) = -4$$
$$4x = 56 \qquad 7x - 7 = 21 - 7 \qquad 2(x + 1) = -4$$
$$x = -14 \qquad x = 14 \qquad 2x + 2 = -4$$
$$2x = -6$$
$$x = -3$$

$$\frac{3}{4} = \frac{x}{2} + 1 \qquad -2x \leq -6$$
$$4 \cdot \frac{3x}{4} = 4 \cdot \frac{x}{2} + 1 \qquad \frac{-2x}{-2} \leq \frac{-6}{-2}$$
$$3x = 2x + 1 \qquad x \leq 3$$
$$x = 1$$

INCORRECT!

In your pamphlet, list common errors that can arise in solving equations and inequalities. Include ways of helping students to avoid them. What helpful guidelines for solving word problems can you offer from the perspective of a student that you probably won't find in math books? If you have your own strategies that work particularly well, include them in the pamphlet.

CHAPTER SUMMARY, REVIEW, AND TEST

SUMMARY

DEFINITIONS AND CONCEPTS	EXAMPLES
Section 2.1 The Addition Property of Equality	
A linear equation in one variable can be written in the form $ax + b = c$, where a is not zero.	$3x + 7 = 9$ is a linear equation.
Equivalent equations have the same solution.	$2x - 4 = 6$, $2x = 10$, and $x = 5$ are equivalent equations.
The Addition Property of Equality Adding the same number (or algebraic expression) to or subtracting the same number (or algebraic expression) from both sides of an equation does not change its solution.	$\bullet$ $\quad x - 3 = 8$ $x - 3 + 3 = 8 + 3$ $x = 11$ $\bullet$ $\quad x + 4 = 10$ $x + 4 - 4 = 10 - 4$ $x = 6$
Section 2.2 The Multiplication Property of Equality	
The Multiplication Property of Equality Multiplying both sides or dividing both sides of an equation by the same nonzero real number (or algebraic expression) does not change the solution.	$\bullet$ $\quad \dfrac{x}{-5} = 6$ $-5\left(\dfrac{x}{-5}\right) = -5(6)$ $x = -30$ $\bullet$ $\quad -50 = -5y$ $\dfrac{-50}{-5} = \dfrac{-5y}{-5}$ $10 = y$
Equations and Coefficients of -1 If $-x = c$, multiply both sides by -1 to solve for x. The solution is the additive inverse of c.	$-x = -12$ $(-1)(-x) = (-1)(-12)$ $x = 12$
Using the Addition and Multiplication Properties If an equation does not contain fractions, • Use the addition property to isolate the variable term. • Use the multiplication property to isolate the variable.	$-2x - 5 = 11$ $-2x - 5 + 5 = 11 + 5$ $-2x = 16$ $\dfrac{-2x}{-2} = \dfrac{16}{-2}$ $x = -8$
Section 2.3 Solving Linear Equations	
Solving a Linear Equation **1.** Simplify each side.	Solve: $\quad 7 - 4(x - 1) = x + 1.$ $7 - 4x + 4 = x + 1$ $-4x + 11 = x + 1$
2. Collect all the variable terms on one side and all the constant terms on the other side.	$-4x - x + 11 = x - x + 1$ $-5x + 11 = 1$ $-5x + 11 - 11 = 1 - 11$ $-5x = -10$

DEFINITIONS AND CONCEPTS	EXAMPLES

Section 2.3 Solving Linear Equations (cont.)

3. Isolate the variable and solve. (If the variable is eliminated and a false statement results, the inconsistent equation has no solution. If a true statement results, all real numbers are solutions of the identity.)	$$\frac{-5x}{-5} = \frac{-10}{-5}$$ $$x = 2$$
4. Check the proposed solution in the original equation.	$$7 - 4(2 - 1) \overset{?}{=} 2 + 1$$ $$7 - 4(1) \overset{?}{=} 2 + 1$$ $$7 - 4 \overset{?}{=} 2 + 1$$ $$3 = 3, \text{ true}$$ The solution is 2, or the solution set is $\{2\}$.
Equations Containing Fractions Multiply both sides (all terms) by the least common denominator. This clears the equation of fractions.	$$\frac{x}{5} + \frac{1}{2} = \frac{x}{2} - 1$$ $$10\left(\frac{x}{5} + \frac{1}{2}\right) = 10\left(\frac{x}{2} - 1\right)$$ $$10 \cdot \frac{x}{5} + 10 \cdot \frac{1}{2} = 10 \cdot \frac{x}{2} - 10 \cdot 1$$ $$2x + 5 = 5x - 10$$ $$-3x = -15$$ $$x = 5$$ The solution is 5, or the solution set is $\{5\}$.
Types of Equations An equation that is true for all real numbers, $\mathbb{R}$, is called an identity. When solving an identity, the variable is eliminated and a true statement, such as $3 = 3$, results. An equation that is not true for even one real number is called an inconsistent equation. A false statement, such as $3 = 7$, results when solving such an equation, whose solution set is $\varnothing$, the empty set. A conditional equation is not an identity, but is true for at least one real number.	Solve: $4x + 5 = 4(x + 2)$. $$4x + 5 = 4x + 8$$ $$5 = 8, \quad \text{false}$$ The inconsistent equation has no solution: $\varnothing$. Solve: $5x - 4 = 5(x + 1) - 9$ $$5x - 4 = 5x + 5 - 9$$ $$5x - 4 = 5x - 4$$ $$-4 = -4, \quad \text{true}$$ All real numbers satisfy the identity: $\mathbb{R}$.

Section 2.4 Formulas and Percents

To solve for a variable in a formula, use the steps for solving a linear equation and isolate the specified variable on one side of the equation.	Solve for l: $\quad w = \dfrac{P - 2l}{2}$. $$2w = 2\left(\frac{P - 2l}{2}\right)$$ $$2w = P - 2l$$ $$2w - P = P - P - 2l$$ $$2w - P = -2l$$ $$\frac{2w - P}{-2} = \frac{-2l}{-2}$$ $$\frac{2w - P}{-2} = l$$

DEFINITIONS AND CONCEPTS	EXAMPLES
Section 2.4 Formulas and Percents (cont.)	
The word *percent* means *per hundred*. The symbol % denotes percent.	$47\% = \dfrac{47}{100}$ $3\% = \dfrac{3}{100}$
To express a decimal as a percent, move the decimal point two places to the right and add a percent sign.	$0.37 = 37\%$ $0.006 = 0.6\%$
To express a percent as a decimal, move the decimal point two places to the left and remove the percent sign.	$250\% = 250\% = 2.5$ $4\% = 04.\% = 0.04$
A Formula Involving Percent $\boxed{A}$ $\boxed{\text{is}}$ $\boxed{P \text{ percent}}$ $\boxed{\text{of}}$ $\boxed{B.}$ $A \quad = \quad P \quad \cdot \quad B$ In the formula $A = PB$, P is expressed as a decimal.	• $\boxed{\text{What}}$ $\boxed{\text{is}}$ $\boxed{5\%}$ $\boxed{\text{of}}$ $\boxed{20?}$ $A \ = \ 0.05 \ \cdot \ 20$ $A = 1$ • $\boxed{6}$ $\boxed{\text{is}}$ $\boxed{30\%}$ $\boxed{\text{of}}$ $\boxed{\text{what?}}$ $6 = \ 0.3 \ \cdot \ B$ $\dfrac{6}{0.3} = B$ $20 = B$ • $\boxed{33}$ $\boxed{\text{is}}$ $\boxed{\text{what percent}}$ $\boxed{\text{of}}$ $\boxed{75?}$ $33 = \quad P \quad \cdot \quad 75$ $\dfrac{33}{75} = P$ $P = 0.44 = 44\%$
Section 2.5 An Introduction to Problem Solving	
Strategy for Solving Word Problems **Step 1** Let x represent one of the quantities.	The length of a rectangle exceeds the width by 3 inches. The perimeter is 26 inches. What are the rectangle's dimensions? Let x = the width.
Step 2 Represent other quantities in terms of x.	$x + 3$ = the length
Step 3 Write an equation that describes the conditions.	$\boxed{\substack{\text{Twice} \\ \text{length}}}$ $\boxed{\text{plus}}$ $\boxed{\substack{\text{twice} \\ \text{width}}}$ $\boxed{\text{is}}$ $\boxed{\text{perimeter.}}$ $2(x + 3) \ + \ 2x \ = \ 26$
Step 4 Solve the equation and answer the question.	$2x + 6 + 2x = 26$ $4x + 6 = 26$ $4x = 20$ $x = 5$ The width (x) is 5 inches and the length $(x + 3)$ is $5 + 3$, or 8 inches.
Step 5 Check the proposed solution in the original wording of the problem.	Perimeter $= 2(5 \text{ in.}) + 2(8 \text{ in.})$ $= 10 \text{ in.} + 16 \text{ in.} = 26 \text{ in.}$ This checks with the given perimeter.

DEFINITION AND CONCEPTS	EXAMPLES

Section 2.6 Problem Solving in Geometry

Solving geometry problems often requires using basic geometric formulas. Formulas for perimeter, area, circumference, and volume are given in Tables 2.3 (page 147), 2.4 (page 149), and 2.5 (page 150) in Section 2.6.	A sailboat's triangular sail has an area of 24 ft² and a base of 8 ft. Find its height. $$A = \frac{1}{2} bh$$ $$24 = \frac{1}{2}(8) h$$ $$24 = 4h$$ $$6 = h$$ The sail's height is 6 ft.
The sum of the measures of the three angles of any triangle is 180°.	In a triangle, the first angle measures 3 times the second and the third measures 40° less than the second. Find each angle's measure. $$\text{Second angle} = x$$ $$\text{First angle} = 3x$$ $$\text{Third angle} = x - 40$$ Sum of measures is 180°. $$x + 3x + x - 40 = 180$$ $$5x - 40 = 180$$ $$5x = 220$$ $$x = 44$$ The angles measure $x = 44°$, $3x = 3 \cdot 44° = 132°$, and $x - 40 = 44° - 40° = 4°$.
Two complementary angles have measures whose sum is 90°. Two supplementary angles have measures whose sum is 180°. If an angle measures $x°$, its complement measures $90° - x°$, and its supplement measures $180° - x°$.	An angle measures five times its complement. Find the angle's measure. $$x = \text{angle's measure}$$ $$90 - x = \text{measure of complement}$$ $$x = 5(90 - x)$$ $$x = 450 - 5x$$ $$6x = 450$$ $$x = 75$$ The angle measures 75°.

DEFINITIONS AND CONCEPTS	EXAMPLES

Section 2.7 Solving Linear Inequalities

A linear inequality in one variable can be written in one of these forms:

$$ax + b < c \qquad ax + b \le c$$
$$ax + b > c \qquad ax + b \ge c$$

where a is not 0.

$3x + 6 > 12$ is a linear inequality.

Set-Builder Notation and Graphs

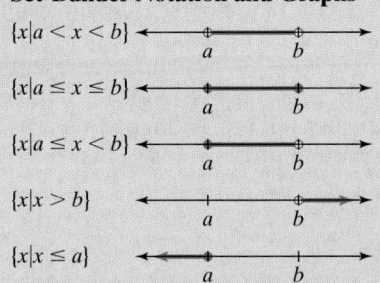

$\{x | a < x < b\}$

$\{x | a \le x \le b\}$

$\{x | a \le x < b\}$

$\{x | x > b\}$

$\{x | x \le a\}$

- Graph the solution of $x < 4$.

  $-3\ -2\ -1\ \ 0\ \ 1\ \ 2\ \ 3\ \ 4\ \ 5$

- Graph the solution of $-2 < x \le 1$.

  $-4\ -3\ -2\ -1\ \ 0\ \ 1\ \ 2\ \ 3\ \ 4$

The Addition Property of Inequality

Adding the same number to or subtracting the same number from both sides of an inequality does not change the solutions.

$$x + 3 < 8$$
$$x + 3 - 3 < 8 - 3$$
$$x < 5$$

The Positive Multiplication Property of Inequality

Multiplying or dividing both sides of an inequality by the same positive number does not change the solutions.

$$\frac{x}{6} \ge 5$$
$$6 \cdot \frac{x}{6} \ge 6 \cdot 5$$
$$x \ge 30$$

The Negative Multiplication Property of Inequality

Multiplying or dividing both sides of an inequality by the same negative number and reversing the direction of the inequality sign does not change the solutions.

$$-3x \le 12$$
$$\frac{-3x}{-3} \ge \frac{12}{-3}$$
$$x \ge -4$$

Solving Linear Inequalities

Use the procedure for solving linear equations. When multiplying or dividing by a negative number, reverse the direction of the inequality symbol. Express the solution set in set-builder notation and graph the set on a number line. If the variable is eliminated and a false statement results, the inequality has no solution. The solution set is $\varnothing$, the empty set. If a true statement results, the solution is the set of all real numbers: $\{x | x$ is a real number$\}$ or $\mathbb{R}$.

Solve:
$$x + 4 \ge 6x - 16$$
$$x + 4 - 6x \ge 6x - 16 - 6x$$
$$-5x + 4 \ge -16$$
$$-5x + 4 - 4 \ge -16 - 4$$
$$-5x \ge -20$$
$$\frac{-5x}{-5} \le \frac{-20}{-5}$$
$$x \le 4$$
$$\{x | x \le 4\}$$

$-3\ -2\ -1\ \ 0\ \ 1\ \ 2\ \ 3\ \ 4\ \ 5$

Review Exercises

2.1 *Solve each equation in Exercises 1–5 using the addition property of equality. Be sure to check proposed solutions.*

1. $x - 10 = 32$

2. $-14 = y + 6$

3. $7z - 3 = 6z + 9$

4. $4(x + 3) = 3x - 10$

5. $6x - 3x - 9 + 1 = -5x + 7x - 3$

2.2 *Solve each equation in Exercises 6–13 using the multiplication property of equality. Be sure to check proposed solutions.*

6. $\dfrac{x}{7} = 10$

7. $\dfrac{y}{-8} = 4$

8. $7z = 77$

9. $-36 = -9y$

10. $\dfrac{3}{5}x = -9$

11. $30 = -\dfrac{5}{2}y$

12. $-x = 14$

13. $\dfrac{-x}{3} = -1$

Solve each equation in Exercises 14–18 using both the addition and multiplication properties of equality. Check proposed solutions.

14. $4x + 9 = 33$

15. $-3y - 2 = 13$

16. $5z + 20 = 3z$

17. $5x - 3 = x + 5$

18. $3 - 2x = 9 - 8x$

19. The formula $N = 0.07x + 4.1$ models the number of women, N, in millions, enrolled in U.S. colleges x years after 1984. How many years after 1984 is the projected enrollment for women expected to reach 6.2 million? In which year is this expected to occur?

2.3 *Solve and check each equation in Exercises 20–28.*

20. $5x + 9 - 7x + 6 = x + 18$

21. $3(x + 4) = 5x - 12$

22. $1 - 2(6 - y) = 3y + 2$

23. $2(x - 4) + 3(x + 5) = 2x - 2$

24. $-2(y - 4) - (3y - 2) = -2 - (6y - 2)$

25. $\dfrac{2x}{3} = \dfrac{x}{6} + 1$

26. $\dfrac{x}{2} - \dfrac{1}{10} = \dfrac{x}{5} + \dfrac{1}{2}$

27. $3(8x - 1) = 6(5 + 4x)$

28. $4(2x - 3) + 4 = 8x - 8$

29. The optimum heart rate that a person should achieve during exercise for the exercise to be most beneficial is modeled by $r = 0.6(220 - a)$, where a represents a person's age and r represents that person's optimum heart rate, in beats per minute. If the optimum heart rate is 120 beats per minute, how old is that person?

2.4 *In Exercises 30–34, solve each formula for the specified variable.*

30. $I = Pr$ for r

31. $V = \dfrac{1}{3}Bh$ for h

32. $P = 2l + 2w$ for w

33. $A = \dfrac{B + C}{2}$ for B

34. $T = D + pm$ for m

In Exercises 35–36, express each decimal as a percent.

35. 0.72

36. 0.0035

In Exercises 37–39, express each percent as a decimal.

37. 65%

38. 150%

39. 3%

40. What is 8% of 120?

41. 90 is 45% of what?

42. 36 is what percent of 75?

43. The radius is one of two bones that connect the elbow and the wrist. The formula $r = \dfrac{h}{7}$ models the length of a woman's radius, r, in inches, and her height, h, in inches.
   a. Solve the formula for h.
   b. Use the formula in part (a) to find a woman's height if her radius is 9 inches long.

44. Every year, approximately 1760 Americans suffer spinal cord injuries due to falls. This represents 22% of the total number of Americans who suffer spinal cord injuries yearly. Determine the number of Americans who suffer spinal cord injuries each year.

Causes of U.S. Spinal Cord Injuries

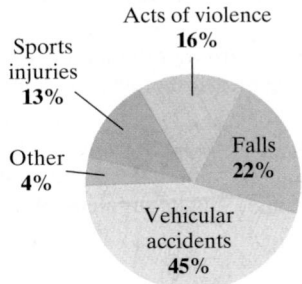

Source: U.S. News and World Report

2.5 *In Exercises 45–52, use the five-step strategy to solve each problem.*

45. Six times a number, decreased by 20, is four times the number. Find the number.

46. On average, the number of unhealthy air days per year in Los Angeles exceeds three times that of New York City by 48 days. If Los Angeles and New York combined have 268 unhealthy air days per year, determine the number of unhealthy days for each city. (*Source*: Enviromental Protection Agency)

47. Two pages that face each other in a book have 93 as the sum of their page numbers. What are the page numbers?

48. The two female artists in the United States with the most platinum albums are Barbra Streisand followed by Madonna. (A platinum album represents one album sold per 266 people.) The number of platinum albums by these two singers form consecutive odd integers. Combined, they have 96 platinum albums. Determine the number of platinum albums by Streisand and Madonna.

49. In 2000, the average weekly salary for workers in the United States was $567. If this amount is increasing by $15 yearly, in how many years after 2000 will the average salary reach $702. In which year will that be?

50. A bank's total monthly charge for a checking account is $6 plus $0.05 per check. If your total monthly charge is $6.90, how many checks did you write during that month?

51. A rectangular field is three times as long as it is wide. If the perimeter of the field is 400 yards, what are the field's dimensions?

52. After a 25% reduction, you purchase a table for $180. What was the table's price before the reduction?

2.6

Use a formula for area to find the area of each figure in Exercises 53–55.

53.

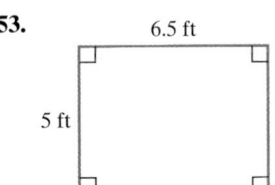

54.

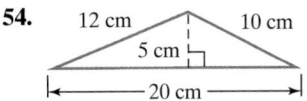

55.

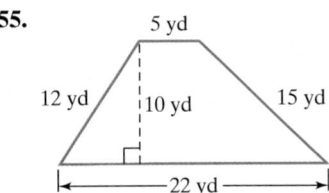

56. Find the circumference and the area of a circle with a diameter of 20 meters. Round answers to the nearest whole number.

57. A sailboat has a triangular sail with an area of 42 square feet and a base that measures 14 feet. Find the height of the sail.

58. A rectangular kitchen floor measures 12 feet by 15 feet. A stove on the floor has a rectangular base measuring 3 feet by 4 feet, and a refrigerator covers a rectangular area of the floor measuring 3 feet by 4 feet. How many square feet of tile will be needed to cover the kitchen floor not counting the area used by the stove and the refrigerator?

59. A yard that is to be covered with mats of grass is shaped like a trapezoid. The bases are 80 feet and 100 feet, and the height is 60 feet. What is the cost of putting the grass mats on the yard if the landscaper charges $0.35 per square foot?

60. Which one of the following is a better buy: a medium pizza with a 14-inch diameter for $6.00 or two small pizzas, each with an 8-inch diameter, for $6.00?

Use a formula for volume to find the volume of each figure in Exercises 61–63. Where applicable, express answers in terms of π. Then round to the nearest whole number.

61.

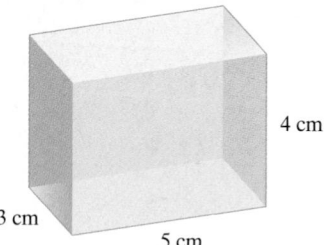

4 cm

3 cm

5 cm

62.

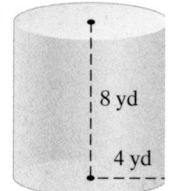

8 yd

4 yd

63.

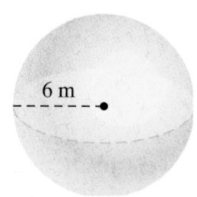

6 m

64. A train is being loaded with freight containers. Each box is 8 meters long, 4 meters wide, and 3 meters high. If there are 50 freight containers, how much space is needed?

65. A cylindrical fish tank has a diameter of 6 feet and a height of 3 feet. How many tropical fish can be put in the tank if each fish needs 5 cubic feet of water?

66. Find the measure of each angle of the triangle shown in the figure.

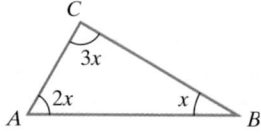

C

$3x$

$2x$

x

A

B

67. In a triangle, the measure of the first angle is 15° more than twice the measure of the second angle. The measure of the third angle exceeds that of the second angle by 25°. What is the measure of each angle?

68. Find the measure of the complement of a 57° angle.

69. Find the measure of the supplement of a 75° angle.

70. How many degrees are there in an angle that measures 25° more than the measure of its complement?

71. The measure of the supplement of an angle is 45° less than four times the measure of the angle. Find the measure of the angle and its supplement.

2.7 *In Exercises 72–73, graph the solution of each inequality on a number line.*

72. $x < -1$

73. $-2 < x \le 4$

Describe each graph in Exercises 74–75 using set-builder notation.

74.

```
←--+---+---+---+---+---+---+---○---+---+---+--→
   -3  -2  -1   0   1   2   3   4   5   6   7
```

75.

```
←--+---+---+---●---+---+---+---+---+---+---+--→
   -5  -4  -3  -2  -1   0   1   2   3   4   5
```

Solve each inequality in Exercises 76–83. Express the solution set in set-builder notation and graph the set on a number line. If the inequality has no solution or is true for all real numbers, so state. It is not necessary to graph solution sets for these inequalities.

76. $2x - 5 < 3$

77. $\dfrac{x}{2} > -4$

78. $3 - 5x \le 18$

79. $4x + 6 < 5x$

80. $6x - 10 \ge 2(x + 3)$

81. $4x + 3(2x - 7) \le x - 3$

82. $2(2x + 4) > 4(x + 2) - 6$

83. $-2(x - 4) \le 3x + 1 - 5x$

84. To pass a course, a student must have an average on three examinations of at least 60. If a student scores 42 and 74 on the first two tests, what must be earned on the third test to pass the course?

85. A long distance telephone service charges 10¢ for the first minute and 5¢ for each minute thereafter. The cost, C, in cents, for a call lasting x minutes is modeled by the formula

$$C = 10 + 5(x - 1).$$

How many minutes can you talk on the phone if you do not want the cost to exceed $5, or 500¢?

Chapter 2 Test

In Exercises 1–6, solve each equation.

1. $4x - 5 = 13$

2. $12x + 4 = 7x - 21$

3. $8 - 5(x - 2) = x + 26$

4. $3(2y - 4) = 9 - 3(y + 1)$

5. $\dfrac{3}{4}x = -15$

6. $\dfrac{x}{10} + \dfrac{1}{3} = \dfrac{x}{5} + \dfrac{1}{2}$

7. The formula $N = 2.4x + 180$ models U.S. population, N, in millions x years after 1960. How many years after 1960 is the U.S. population expected to reach 324 million? In which year is this expected to occur?

In Exercises 8–9, solve each formula for the specified variable.

8. $V = \pi r^2 h$ for h

9. $l = \dfrac{P - 2w}{2}$ for w

10. What is 6% of 140?

11. 120 is 80% of what?

12. 12 is what percent of 240?

In Exercises 13–17, solve each problem.

13. The product of 5 and a number, decreased by 9, is 310. What is the number?

14. At the time they took office, Ronald Reagan and James Buchanan were among the oldest U.S. presidents. Reagan was 4 years older than Buchanan. The sum of their ages was 134. Determine Reagan's age and Buchanan's age at the time each man took office.

15. A long distance telephone plan has a monthly fee of $15.00 and a rate of $0.05 per minute. How many minutes can you chat long distance in a month for a total cost, including the $15.00, of $45.00?

16. A rectangular field is twice as long as it is wide. If the perimeter of the field is 450 yards, what are the field's dimensions?

17. After a 20% reduction, you purchase a new Stephen King novel for $28. What was the book's price before the reduction?

In Exercises 18–19, find the area of each figure.

18.

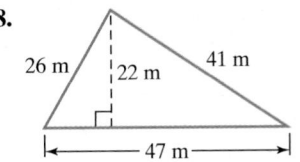

19.

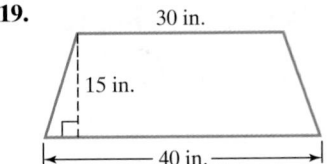

In Exercises 20–21, find the volume of each figure. Where applicable, express answers in terms of π. Then round to the nearest whole number.

20.

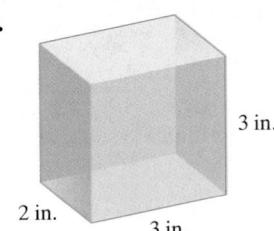

21.

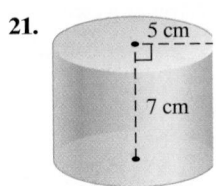

22. What will it cost to cover a rectangular floor measuring 40 feet by 50 feet with square tiles that measure 2 feet on each side if a package of 10 tiles costs $13 per package?

23. A sailboat has a triangular sail with an area of 56 square feet and a base that measures 8 feet. Find the height of the sail.

24. In a triangle, the measure of the first angle is three times that of the second angle. The measure of the third angle is 30° less than the measure of the second angle. What is the measure of each angle?

25. How many degrees are there in an angle that measures 16° more than the measure of its complement?

In Exercises 26–27, graph the solution of each inequality on a number line.

26. $x > -2$

27. $-4 \le x < 1$

28. Use set-builder notation to describe the following graph.

```
←──┼──┼──┼──┼──┼──●──┼──┼──┼──┼──┼──→
  -5  -4  -3  -2  -1  0  1  2  3  4  5
```

Solve each inequality in Exercises 29–31. Express the solution set in set-builder notation and graph the set on a number line.

29. $\dfrac{x}{2} < -3$

30. $6 - 9x \ge 33$

31. $4x - 2 > 2(x + 6)$

32. A student has grades on three examinations of 76, 80, and 72. What must the student earn on a fourth examination in order to have an average of at least 80?

33. The length of a rectangle is 20 inches. For what widths is the perimeter greater than 56 inches?

Cumulative Review Exercises (Chapters 1–2)

In Exercises 1–3, perform the indicated operation or operations.

1. $-8 - (12 - 16)$

2. $(-3)(-2) + (-2)(4)$

3. $(8 - 10)^3(7 - 11)^2$

4. Simplify: $2 - 5[x + 3(x + 7)]$.

5. List all the rational numbers in this set:

$$\left\{-4, -\frac{1}{3}, 0, \sqrt{2}, \sqrt{4}, \frac{\pi}{2}, 1063\right\}.$$

6. Plot $(-2, -1)$ in a rectangular coordinate system. In which quadrant does the point lie?

7. Insert either $<$ or $>$ in the box to make a true statement: $-10,000 \,\square\, -2$.

8. Use the distributive property to rewrite without parentheses:

$$6(4x - 1 - 5y).$$

The graph shows the unemployment rate in the United States from 1990 through 2000. Use the information in the graph to solve Exercises 9–10.

U.S. Unemployment Rate

Source: Bureau of Labor Statistics

9. For the period shown, in which year was the unemployment rate at a minimum? What percentage of the work force was unemployed in in that year?

10. For the period shown, during which year did the unemployment rate reach a maximum? Estimate the percentage of the work force unemployed, to the nearest tenth of a percent, at that time.

In Exercises 11–12, solve each equation.

11. $5 - 6(x + 2) = x - 14$

12. $\dfrac{x}{5} - 2 = \dfrac{x}{3}$

13. Solve for A: $V = \dfrac{1}{3} Ah$.

14. 48 is 30% of what?

15. The length of a rectangular parking lot is 10 yards less than twice its width. If the perimeter of the lot is 400 yards, what are its dimensions?

16. A gas station owner makes a profit of 40 cents per gallon of gasoline sold. How many gallons of gasoline must be sold in a year to make a profit of $30,000 from gasoline sales?

17. Graph the solution set of $-2 < x \leq 3$ on a number line.

Solve each inequality in Exercises 18–19. Express the solution set in set-builder notation and graph the set on a number line.

18. $3 - 3x > 12$

19. $5 - 2(3 - x) \leq 2(2x + 5) + 1$

20. You take a summer job selling medical supplies. You are paid $600 per month plus 4% of the sales price of all the supplies you sell. If you want to earn more than $2500 per month, what value of medical supplies must you sell?

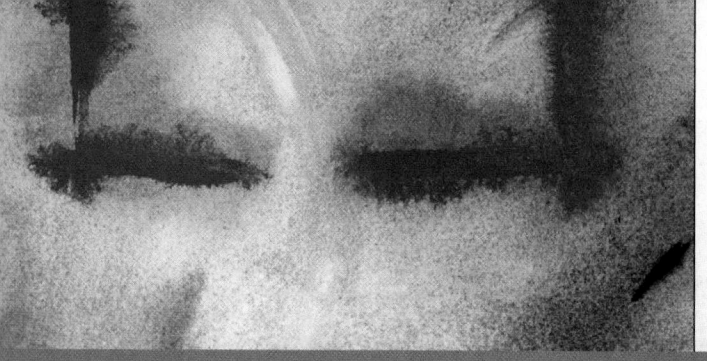

The salmon fishing industry is one of the major industries of the states along the Pacific coast. It provides 60,000 jobs and more than $1 billion in income per year for the region. The equation in two variables

$$y = -0.22x + 9.6$$

in which y is the Pacific salmon population, in millions, x years after 1960 tells of a severely threatened population. In this chapter, you will learn graphing and modeling methods for situtations involving two variables. With these methods, you will be able to create formulas that model the data of your world.

For fish lovers, the Pacific salmon is in a class by itself. However, the drastic decline in their population has raised alarm and become one of the most important conservation issues in the Pacific Northwest.

Chapter 3

Linear Equations in Two Variables

▶ **SECTION 3.1** *Graphing Linear Equations*

Objectives

1 Determine whether an ordered pair is a solution of an equation.

2 Find solutions of an equation in two variables.

3 Use point plotting to graph linear equations.

4 Use point plotting to graph other kinds of equations.

5 Use graphs of linear equations to solve problems.

SSM   PH Tutor   CD-   Video
       Center   ROM

A picture, as they say, is worth a thousand words. Have you seen pictures of gas-guzzling cars from the 1950s, with their huge fins and overstated designs? The worst year for automobile fuel efficiency was 1958, when U.S. cars averaged a dismal 12.4 miles per gallon.

There is a formula that describes fuel efficiency of U.S. cars over time. The formula is

$$y = 0.0075x^2 - 0.2672x + 14.8.$$

The variable x represents the number of years after 1940. The variable y represents the average number of miles per gallon for U.S. automobiles. Looking at the formula does not make it obvious that 1958, 18 years after 1940, was the worst year for fuel efficiency. However, if we could somehow make a picture of the formula, such as the one shown in Figure 3.1, the lowest point on the picture would reveal approximately 1958 as the year in which gas-guzzling cars averaged less than 13 miles per gallon. The shape of the graph also shows decreasing fuel efficiency from 1940 through 1958 and increasing fuel efficiency after 1958. In this chapter we will be making pictures of equations. We can use these pictures to visualize the behavior of the variables in the equation.

Solutions of Equations

The rectangular coordinate system allows us to visualize relationships between two variables by connecting any equation in two variables with a geometric figure. Consider, for example, the following equation in two variables:

$$x + y = 10.$$

The sum of two numbers, x and y, is 10.

Many pairs of numbers fit the description in the voice balloon, such as $x = 1$ and $y = 9$, or $x = 3$ and $y = 7$. The phrase "$x = 1$ and $y = 9$" is abbreviated using the ordered pair $(1, 9)$. Similarly, the phrase "$x = 3$ and $y = 7$" is abbreviated using the ordered pair $(3, 7)$.

A **solution of an equation in two variables**, x and y, is an ordered pair of real numbers with the following property: When the x-coordinate is substituted for x and the y-coordinate is substituted for y in the equation, we obtain a true statement. For example, $(1, 9)$ is a solution of the equation $x + y = 10$. When 1 is substituted for x and 9 is substituted for y, we obtain the true statement

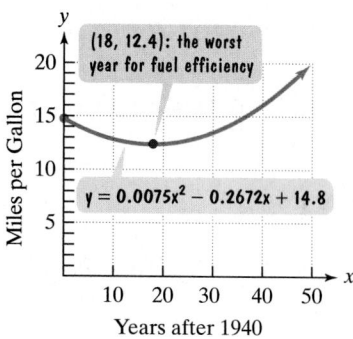

Figure 3.1 A picture of a formula

1 Determine whether an ordered pair is a solution of an equation.

$1 + 9 = 10$, or $10 = 10$. Because there are infinitely many pairs of numbers that have a sum of 10, the equation $x + y = 10$ has infinitely many solutions. Each ordered-pair solution is said to **satisfy** the equation. Thus, $(1, 9)$ satisfies the equation $x + y = 10$.

EXAMPLE 1 Deciding Whether an Ordered Pair Satisfies an Equation

Determine whether each ordered pair is a solution of the equation

$$x - 4y = 14:$$

a. $(2, -3)$ **b.** $(12, 1)$.

Solution

a. To determine whether $(2, -3)$ is a solution of the equation, we substitute 2 for x and -3 for y.

$$x - 4y = 14 \qquad \text{This is the given equation.}$$
$$2 - 4(-3) \stackrel{?}{=} 14 \qquad \text{Substitute 2 for x and } -3 \text{ for y.}$$
$$2 - (-12) \stackrel{?}{=} 14 \qquad \text{Multiply: } 4(-3) = -12.$$
$$14 = 14 \qquad \text{Subtract: } 2 - (-12) = 2 + 12 = 14.$$

This statement is true.

Because we obtain a true statement, we conclude that $(2, -3)$ is a solution of the equation $x - 4y = 14$. Thus, $(2, -3)$ satisfies the equation.

b. To determine whether $(12, 1)$ is a solution of the equation, we substitute 12 for x and 1 for y.

$$x - 4y = 14 \qquad \text{This is the given equation.}$$
$$12 - 4(1) \stackrel{?}{=} 14 \qquad \text{Substitute 12 for x and 1 for y.}$$
$$12 - 4 \stackrel{?}{=} 14 \qquad \text{Multiply: } 4(1) = 4$$
$$8 = 14 \qquad \text{Subtract: } 12 - 4 = 8.$$

This statement is false.

Because we obtain a false statement, we conclude that $(12, 1)$ is not a solution of $x - 4y = 14$. The ordered pair $(12, 1)$ does not satisfy the equation. ∎

✔ **CHECK POINT 1** Determine whether each ordered pair is a solution of the equation $x - 3y = 9$:

a. $(3, -2)$ **b.** $(-2, 3)$.

In this chapter, we will use x and y to represent the variables in an equation in two variables. However, any two letters can be used. Solutions are still ordered pairs. The first number in an ordered pair usually replaces the variable that occurs first alphabetically. The second number in an ordered pair usually replaces the variable that occurs last alphabetically.

2 Find solutions of an equation in two variables.

How do we find ordered pairs that are solutions of an equation in two variables, x and y?

- Select a value for one of the variables.
- Substitute that value in the equation and find the value of the other variable.
- Use the values of the two variables to form an ordered pair (x, y). This pair is a solution of the equation.

EXAMPLE 2 Finding Solutions of an Equation

Find five solutions of

$$y = 2x - 1.$$

Select integers for x, starting with -2 and ending with 2.

Solution We organize the process of finding solutions in the following table of values.

Start with these values of x.	Substitute x in y = 2x − 1 and compute y.	Use values for x and y to form an ordered-pair solution.

x	$y = 2x - 1$	(x, y)
-2	$y = 2(-2) - 1 = -4 - 1 = -5$	$(-2, -5)$
-1	$y = 2(-1) - 1 = -2 - 1 = -3$	$(-1, -3)$
0	$y = 2 \cdot 0 - 1 = 0 - 1 = -1$	$(0, -1)$
1	$y = 2 \cdot 1 - 1 = 2 - 1 = 1$	$(1, 1)$
2	$y = 2 \cdot 2 - 1 = 4 - 1 = 3$	$(2, 3)$

Look at the ordered pairs in the last column. Five solutions of $y = 2x - 1$ are $(-2, -5), (-1, -3), (0, -1), (1, 1),$ and $(2, 3)$. ∎

✔ **CHECK POINT 2** Find five solutions of $y = 3x + 2$. Select integers for x, starting with -2 and ending with 2.

3 Use point plotting to graph linear equations.

Graphing Linear Equations in the Form $y = mx + b$

In Example 2, we found five solutions of $y = 2x - 1$. We can generate as many ordered-pair solutions as desired to $y = 2x - 1$ by substituting numbers for x and then finding the values for y. The **graph of the equation** is the set of all points whose coordinates satisfy the equation.

One method for graphing an equation such as $y = 2x - 1$ is the **point-plotting method**.

The Point-Plotting Method for Graphing an Equation in Two Variables

1. Find several ordered pairs that are solutions of the equation.
2. Plot these ordered pairs as points in the rectangular coordinate system.
3. Connect the points with a smooth curve or line.

EXAMPLE 3 Graphing an Equation Using the Point-Plotting Method

Graph the equation: $y = 3x$.

Solution

Step 1 Find several ordered pairs that are solutions of the equation. Because there are infinitely many solutions, we cannot list them all. To find some solutions of the equation, we select integers for x, starting with -2 and ending with 2.

Start with these values of x.	Substitute x in $y = 3x$ and compute y.	These are some solutions of $y = 3x$.
x	$y = 3x$	(x, y)
-2	$y = 3(-2) = -6$	$(-2, -6)$
-1	$y = 3(-1) = -3$	$(-1, -3)$
0	$y = 3 \cdot 0 = 0$	$(0, 0)$
1	$y = 3 \cdot 1 = 3$	$(1, 3)$
2	$y = 3 \cdot 2 = 6$	$(2, 6)$

Step 2 Plot these ordered pairs as points in the rectangular coordinate system. The five ordered pairs in the table of values are plotted in Figure 3.2(a).

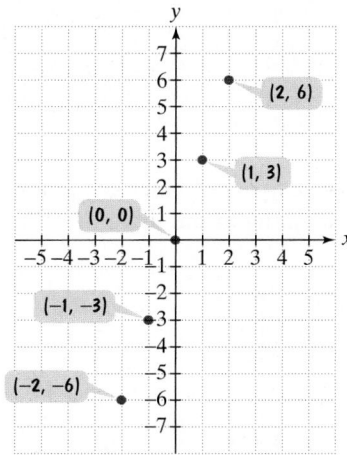

Figure 3.2(a) Some solutions of $y = 3x$ plotted as points

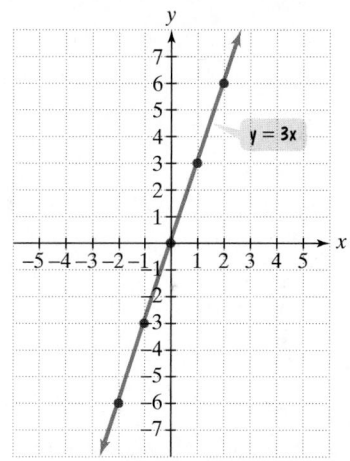

Figure 3.2(b) The graph of $y = 3x$

Step 3 Connect the points with a smooth curve or line. The points lie along a straight line. The graph of $y = 3x$ is shown in Figure 3.2(b). The arrows on both ends of the line indicate that it extends indefinitely in both directions. ∎

✔ **CHECK POINT 3** Graph the equation: $y = 2x$.

Equations like $y = 3x$ and $y = 2x$ are called **linear equations in two variables** because the graph of each equation is a line. Any equation that can be written in the form $y = mx + b$, where m and b are constants, is a linear equation in two variables. Here are examples of linear equations in two variables:

$$y = 3x \qquad\qquad y = 3x - 2$$

$$\text{or} \quad y = 3x + 0 \quad\quad \text{or} \quad y = 3x + (-2).$$

This is in the form $y = mx + b$ with $m = 3$ and $b = 0$.

This is in the form $y = mx + b$ with $m = 3$ and $b = -2$.

Can you guess how the graph of the linear equation $y = 3x - 2$ compares with the graph of $y = 3x$? In Example 3, we graphed $y = 3x$. Now, let's graph the equation $y = 3x - 2$.

EXAMPLE 4 Graphing a Linear Equation in Two Variables

Graph the equation: $y = 3x - 2$.

Solution

Step 1 Find several ordered pairs that are solutions of the equation. To find some solutions, we select integers for x, starting with -2 and ending with 2.

x	$y = 3x - 2$	(x, y)
-2	$y = 3(-2) - 2 = -6 - 2 = -8$	$(-2, -8)$
-1	$y = 3(-1) - 2 = -3 - 2 = -5$	$(-1, -5)$
0	$y = 3 \cdot 0 - 2 = 0 - 2 = -2$	$(0, -2)$
1	$y = 3 \cdot 1 - 2 = 3 - 2 = 1$	$(1, 1)$
2	$y = 3 \cdot 2 - 2 = 6 - 2 = 4$	$(2, 4)$

Step 2 Plot these ordered pairs as points in the rectangular coordinate system. The five ordered pairs in the table of values are plotted in Figure 3.3(a).

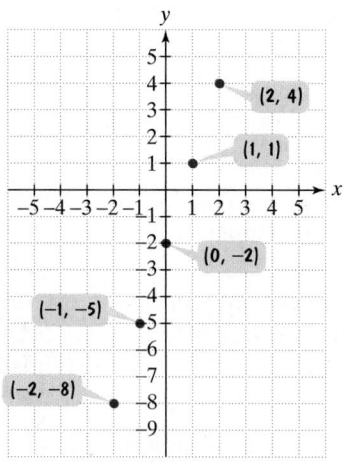

Figure 3.3(a) Some solutions of $y = 3x - 2$ plotted as points

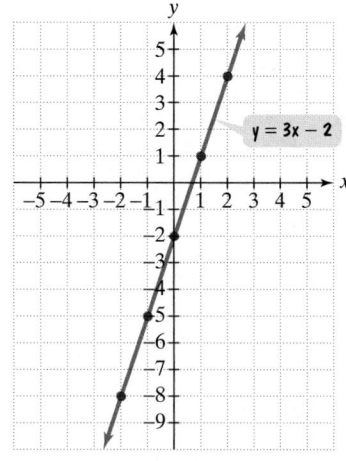

Figure 3.3(b) The graph of $y = 3x - 2$

Step 3 Connect the points with a smooth curve or line. The points lie along a straight line. The graph of $y = 3x - 2$ is shown in Figure 3.3(b). ∎

Now we are ready to compare the graphs of $y = 3x - 2$ and $y = 3x$. The graphs of both linear equations are shown in the same rectangular coordinate system in Figure 3.4. Can you see that the graph of $y = 3x - 2$ looks exactly like the graph of $y = 3x$, but shifted 2 units down? Instead of crossing the y-axis at $(0, 0)$, the graph now crosses the y-axis at $(0, -2)$.

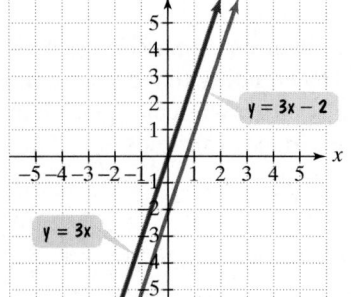

Figure 3.4

Comparing Graphs of Linear Equations

If the value of m does not change:

- The graph of $y = mx + b$ is the graph of $y = mx$ shifted b units up when b is a positive number.
- The graph of $y = mx + b$ is the graph of $y = mx$ shifted b units down when b is a negative number.

✔ **CHECK POINT 4**   Graph the equation: $y = 2x - 2$.

EXAMPLE 5   Graphing a Linear Equation in Two Variables

Graph the equation: $y = \dfrac{2}{3}x + 1$.

Solution

Step 1 Find several ordered pairs that are solutions of the equation. Notice that m, the coefficient of x, is $\frac{2}{3}$. When m is a fraction, we will select values of x that are multiples of the denominator. In this way, we can avoid values of y that are fractions. Because the denominator of $\frac{2}{3}$ is 3, we select multiples of 3 for x. Let's use $-6, -3, 0, 3,$ and 6.

x	$y = \dfrac{2}{3}x + 1$	(x, y)
-6	$y = \dfrac{2}{3}(-6) + 1 = -4 + 1 = -3$	$(-6, -3)$
-3	$y = \dfrac{2}{3}(-3) + 1 = -2 + 1 = -1$	$(-3, -1)$
0	$y = \dfrac{2}{3} \cdot 0 + 1 = 0 + 1 = 1$	$(0, 1)$
3	$y = \dfrac{2}{3} \cdot 3 + 1 = 2 + 1 = 3$	$(3, 3)$
6	$y = \dfrac{2}{3} \cdot 6 + 1 = 4 + 1 = 5$	$(6, 5)$

Step 2 Plot these ordered pairs as points in the rectangular coordinate system. The five ordered pairs in the table of values are plotted in Figure 3.5.

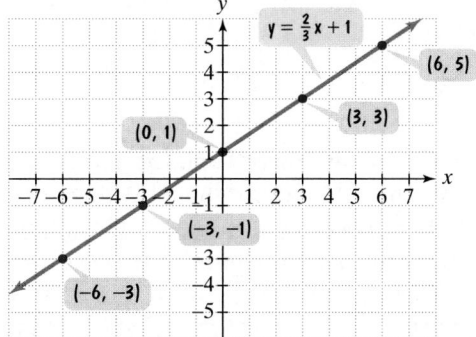

Figure 3.5 The graph of $y = \frac{2}{3}x + 1$

Step 3 Connect the points with a smooth curve or line. The points lie along a straight line. The graph of $y = \frac{2}{3}x + 1$ is shown in Figure 3.5. ■

✔ **CHECK POINT 5** Graph the equation: $y = \dfrac{1}{3}x + 2$.

4 Use point plotting to graph other kinds of equations.

Graphing Other Kinds of Equations in Two Variables

Look at the picture of this gymnast. He has created a perfect balance in which the two halves of his body are mirror images of each other. Is it possible for graphs to have mirrorlike qualities? Yes. Although our next graph is not a straight line, we can obtain its cuplike shape using the point-plotting method for graphing an equation in two variables.

EXAMPLE 6   Graphing an Equation in Two Variables

Graph the equation: $y = x^2 - 4$.

Solution   The given equation involves two variables, x and y. However, because the variable x is squared, it is not a linear equation in two variables.

$$y = x^2 - 4$$

This is not in the form $y = mx + b$ because x is squared.

Although the graph is not a line, it is still a picture of all the ordered-pair solutions of $y = x^2 - 4$. Thus, we can use the point-plotting method to obtain the graph.

Step 1 Find several ordered pairs that are solutions of the equation. To find some solutions, we select integers for x, starting with -3 and ending with 3.

x	$y = x^2 - 4$	(x, y)
-3	$y = (-3)^2 - 4 = 9 - 4 = 5$	$(-3, 5)$
-2	$y = (-2)^2 - 4 = 4 - 4 = 0$	$(-2, 0)$
-1	$y = (-1)^2 - 4 = 1 - 4 = -3$	$(-1, -3)$
0	$y = 0^2 - 4 = 0 - 4 = -4$	$(0, -4)$
1	$y = 1^2 - 4 = 1 - 4 = -3$	$(1, -3)$
2	$y = 2^2 - 4 = 4 - 4 = 0$	$(2, 0)$
3	$y = 3^2 - 4 = 9 - 4 = 5$	$(3, 5)$

Step 2 Plot these ordered pairs as points in the rectangular coordinate system. The seven ordered pairs in the table of values are plotted in Figure 3.6(a).

Step 3 Connect the points with a smooth curve. The seven points are joined with a smooth curve in Figure 3.6(b). The graph of $y = x^2 - 4$ is a curve where the part of the graph to the right of the y-axis is a reflection of the part to the left of it, and vice versa. The arrows on both ends of the curve indicate that it extends indefinitely in both directions.

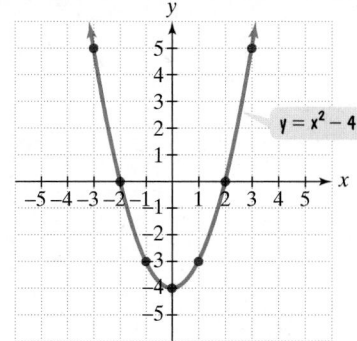

Figure 3.6(a) Some solutions of $y = x^2 - 4$ plotted as points

Figure 3.6(b) The graph of $y = x^2 - 4$

✔ **CHECK POINT 6** Graph the equation: $y = x^2 - 1$. Select integers for x, starting with -3 and ending with 3.

5 Use graphs of linear equations to solve problems.

Applications

Part of the beauty of the rectangular coordinate system is that it allows us to "see" mathematical formulas and visualize the solution to a problem. This idea is demonstrated in Example 7.

EXAMPLE 7   An Application Using Graphs of Linear Equations

The toll to a bridge costs $2.50. Commuters who use the bridge frequently have the option of purchasing a monthly coupon book for $21.00. With the coupon book, the toll is reduced to $1.00. The monthly cost, y, of using the bridge x times can be described by the following formulas.

Without the coupon book:

$$y = 2.50x$$

> The monthly cost, y, is $2.50 times the number of times, x, that the bridge is used.

With the coupon book:

$$y = 21 + 1 \cdot x$$
$$y = 21 + x$$

> The monthly cost, y, is $21 for the book plus $1 times the number of times, x, that the bridge is used.

a. Let $x = 0, 2, 4, 10, 12, 14$, and 16. Make a table of values showing seven solutions for each of the linear equations.

b. Graph the equations in the same rectangular coordinate system.

c. What are the coordinates of the intersection point for the two graphs? Interpret the coordinates in practical terms.

Solution

a. A table of values showing seven solutions for each equation follows.

Without the Coupon Book

x	$y = 2.5x$	(x, y)
0	$y = 2.5(0) = 0$	$(0, 0)$
2	$y = 2.5(2) = 5$	$(2, 5)$
4	$y = 2.5(4) = 10$	$(4, 10)$
10	$y = 2.5(10) = 25$	$(10, 25)$
12	$y = 2.5(12) = 30$	$(12, 30)$
14	$y = 2.5(14) = 35$	$(14, 35)$
16	$y = 2.5(16) = 40$	$(16, 40)$

With the Coupon Book

x	$y = 21 + x$	(x, y)
0	$y = 21 + 0 = 21$	$(0, 21)$
2	$y = 21 + 2 = 23$	$(2, 23)$
4	$y = 21 + 4 = 25$	$(4, 25)$
10	$y = 21 + 10 = 31$	$(10, 31)$
12	$y = 21 + 12 = 33$	$(12, 33)$
14	$y = 21 + 14 = 35$	$(14, 35)$
16	$y = 21 + 16 = 37$	$(16, 37)$

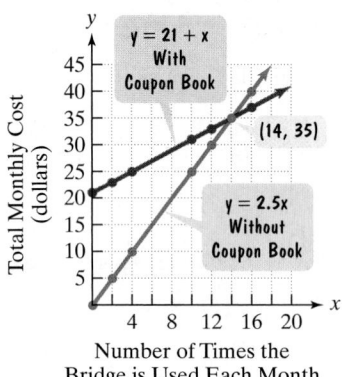

Figure 3.7 Options for a toll

b. Now we are ready to graph the two equations. Because the x- and y-coordinates are nonnegative, it is only necessary to use the origin, the positive portions of the x- and y-axes, and the first quadrant of the rectangular coordinate system. The x-coordinates begin at 0 and end at 16. We will let each tick-mark on the x-axis represent two units. However, the y-coordinates begin at 0 and get as large as 40 in the formula without the coupon book. So that our y-axis does not get too long, we will let each tick mark on the y-axis represent five units. Using this setup and the two tables of values, we construct the graphs of $y = 2.5x$ and $y = 21 + x$, shown in Figure 3.7.

c. The graphs intersect at $(14, 35)$. This means that if the bridge is used 14 times in a month, the total monthly cost without the coupon book is the same as the total monthly cost with the coupon book, namely $35. ∎

In Figure 3.7, look at the two graphs to the right of the intersection point $(14, 35)$. The red graph of $y = 21 + x$ lies below the blue graph of $y = 2.5x$. This means that if the bridge is used more than 14 times in a month $(x > 14)$, the monthly cost, y, with the coupon book is cheaper than the monthly cost, y, without the coupon book.

✔ **CHECK POINT 7** The toll to a bridge costs $2.00. If you use the bridge x times in a month, the monthly cost, y, is $y = 2x$. With a $10 coupon book, the toll is reduced to $1.00. The monthly cost, y, of using the bridge x times in a month with the coupon book is $y = 10 + x$.

a. Let $x = 0, 2, 4, 6, 8, 10$, and 12. Make a table of values showing seven solutions of $y = 2x$ and seven solutions of $y = 10 + x$.

b. Graph the equations in the same rectangular coordinate system.

c. What are the coordinates of the intersection point for the two graphs? Interpret the coordinates in practical terms.

Mathematical Blossom

Graph of an equation in a three-dimensional Cartesian plane

This picture is a graph in a three-dimensional rectangular coordinate system. For many, the picture is more interesting than its equation:

$$z = (|x| - |y|)^2 + \frac{2|xy|}{\sqrt{x^2 + y^2}}.$$

Sometimes it's pleasant to simply get the "feel" of an equation by seeing its picture. Turning equations into visual images hints at the beauty that can be derived from mathematics.

Using Technology

Graphing calculators or graphing software packages for computers are referred to as **graphing utilities** or graphers. A graphing utility is a powerful tool that quickly generates the graph of an equation in two variables. Figure 3.8 shows two such graphs for the equations in Examples 3 and 4.

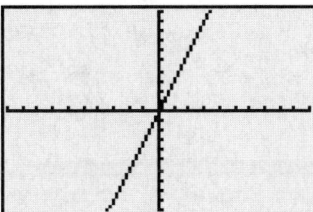

Figure 3.8(a) The graph of $y = 3x$

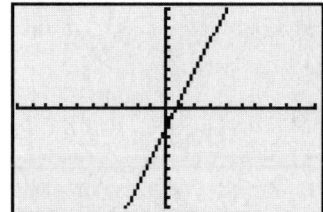

Figure 3.8(b) The graph of $y = 3x - 2$

What differences do you notice between these graphs and the graphs that we drew by hand? They do seem a bit "jittery." Arrows do not appear on both ends of the graphs. Furthermore, numbers are not given along the axes. For both graphs in Figure 3.8, the x-axis extends from −10 to 10 and the y-axis also extends from −10 to 10. The distance represented by each consecutive tick mark is one unit. We say that the **viewing rectangle** is $[-10, 10, 1]$ by $[-10, 10, 1]$.

$$[-10, \qquad 10, \qquad 1] \quad \text{by} \quad [-10, \qquad 10, \qquad 1].$$

The minimum x-value along the x-axis is −10.	The maximum x-value along the x-axis is 10.	The scale on the x-axis is 1 unit per tick mark.	The minimum y-value along the y-axis is −10.	The maximum y-value along the y-axis is 10.	The scale on the y-axis is 1 unit per tick mark.

To graph an equation in x and y using a graphing utility, enter the equation and specify the size of the viewing rectangle. The size of the viewing rectangle sets minimum and maximum values for both the x- and y-axes. Enter these values, as well as the values between consecutive tick marks, on the respective axes. The $[-10, 10, 1]$ by $[-10, 10, 1]$ viewing rectangle used in Figure 3.8 is called the **standard viewing rectangle**.

On most graphing utilities, the display screen is two-thirds as high as it is wide. By using a square setting, you can make the x and y tick marks be equally spaced. (This does not occur in the standard viewing rectangle.) Graphing utilities can also *zoom in* and *zoom out*. When you zoom in, you see a smaller portion of the graph, but you do so in greater detail. When you zoom out, you see a larger portion of the graph. Thus, zooming out may help you to develop a better understanding of the overall character of the graph. With practice, you will become more comfortable with graphing equations in two variables using your graphing utility. You will also develop a better sense of the size of the viewing rectangle that will reveal needed information about a particular graph.

EXERCISE SET 3.1

Practice Exercises

In Exercises 1–12, determine whether each ordered pair is a solution of the given equation.

1. $y = 3x$ $(2, 3), (3, 2), (-4, -12)$

2. $y = 4x$ $(3, 12), (12, 3), (-5, -20)$.

3. $y = -4x$ $(-5, -20), (0, 0), (9, -36)$

4. $y = -3x$ $(-5, 15), (0, 0), (7, -21)$

5. $y = 2x + 6$ $(0, 6), (-3, 0), (2, -2)$

6. $y = 8 - 4x$ $(8, 0), (16, -2), (3, -4)$

7. $3x + 5y = 15$ $(-5, 6), (0, 5), (10, -3)$

8. $2x - 5y = 0$ $(-2, 0), (-10, 6), (5, 0)$

9. $x + 3y = 0$ $(0, 0), \left(1, \dfrac{1}{3}\right), \left(2, -\dfrac{2}{3}\right)$

10. $4x - y = 0$ $(1, 4), (-2, -8), (-3, 12)$

11. $x - 4 = 0$ $(4, 7), (3, 4), (0, -4)$

12. $y + 2 = 0$ $(0, 2), (2, 0), (0, -2)$

In Exercises 13–20, find five solutions of each equation. Select integers for x, starting with −2 and ending with 2. Organize your work in a table of values.

13. $y = 10x$ **14.** $y = 20x$

15. $y = -6x$ **16.** $y = -8x$

17. $y = 5x - 8$ **18.** $y = 4x - 6$

19. $y = -7x + 3$ **20.** $y = -7x + 5$

In Exercises 21–44, graph each linear equation in two variables. Find at least five solutions in your table of values for each equation.

21. $y = x$ **22.** $y = x + 1$

23. $y = x - 1$ **24.** $y = x - 2$

25. $y = 2x + 1$ **26.** $y = 2x - 1$

27. $y = -x + 2$ **28.** $y = -x + 3$

29. $y = -3x - 1$ **30.** $y = -3x - 2$

31. $y = \dfrac{1}{2}x$ **32.** $y = -\dfrac{1}{2}x$

33. $y = -\dfrac{1}{4}x$ **34.** $y = \dfrac{1}{4}x$

35. $y = \dfrac{1}{3}x + 1$ **36.** $y = \dfrac{1}{3}x - 1$

37. $y = -\dfrac{3}{2}x + 1$ **38.** $y = -\dfrac{3}{2}x + 2$

39. $y = -\dfrac{5}{2}x - 1$ **40.** $y = -\dfrac{5}{2}x + 1$

41. $y = x + \dfrac{1}{2}$ **42.** $y = x - \dfrac{1}{2}$

43. $y = 4$, or $y = 0x + 4$ **44.** $y = 3$, or $y = 0x + 3$

Graph each equation in Exercises 45–50. Find seven solutions in your table of values for each equation by using integers for x, starting with −3 and ending with 3.

45. $y = x^2$ **46.** $y = x^2 - 2$

47. $y = x^2 + 1$ **48.** $y = x^2 + 2$

49. $y = 4 - x^2$ **50.** $y = 9 - x^2$

Application Exercises

The graph shows U.S. population projections from 2000 through 2050. These projections, as well as the actual U.S. population from 1960 through 2000, can be modeled by the linear equation in two variables $y = 2.4x + 180$, in which x is the number of years after 1960 and y is the U.S. population, in millions. Use this equation to solve Exercises 51–52.

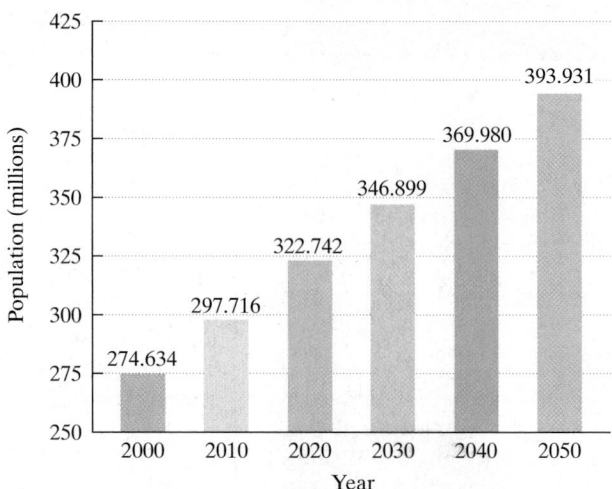

U.S. Population Projections: 2000–2050

Source: U.S. Census Bureau

51. a. Find three solutions of $y = 2.4x + 180$. Use 40, 50, and 60 for x. Organize your work in a table of values.
   b. How well does the linear equation model the projections for each of the first three years shown in the bar graph?

52. a. Find three solutions of $y = 2.4x + 180$. Use 70, 80, and 90 for x. Organize your work in a table of values.
   b. How well does the linear equation model the projections for each of the last three years shown in the bar graph?

The data show the number of registered automatic weapons, in thousands, and the murder rate, in murders per 100,000, for eight randomly selected states.

Automatic weapons, x	11.6	8.3	6.9	3.6	2.6	2.5	2.4	0.6
Murder rate, y	13.1	10.6	11.5	10.1	5.3	6.6	3.6	4.4

Source: FBI and Bureau of Alcohol, Tobacco, and Firearms

The data in the table can be modeled by the linear equation in two variables $y = 0.85x + 4.05$. Use this equation to solve Exercises 53–54.

53. a. Find two solutions of $y = 0.85x + 4.05$. Use 11.6 and 8.3 for x. Organize your work in a table of values.
   b. How well does the linear equation model the actual data shown in the table?

54. a. Find two solutions of $y = 0.85x + 4.05$. Use 6.9 and 3.6 for x. Organize your work in a table of values.
   b. How well does the linear equation model the actual data shown in the table?

55. A rental company charges $40.00 a day plus $0.35 per mile to rent a moving truck. The total cost, y, for a day's rental if x miles are driven is described by $y = 40 + 0.35x$. A second company charges $36.00 a day plus $0.45 per mile, so the daily cost, y, if x miles are driven is described by $y = 36 + 0.45x$. The graphs of the two equations are shown in the same rectangular coordinate system.

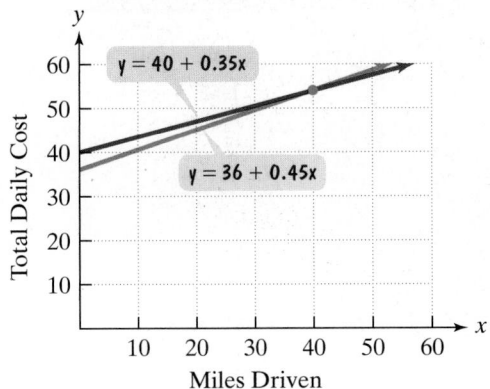

a. What is the x-coordinate of the intersection point of the graphs? Describe what this x-coordinate means in practical terms.

b. What is a reasonable estimate for the y-coordinate of the intersection point?

c. Substitute the x-coordinate of the intersection point into each of the equations and find the corresponding value for y. Describe what this value represents in practical terms. How close is this value to your estimate from part (b)?

56. The linear equation in two variables $y = 166x + 1781$ models the cost, y, in tuition and fees per year, of a four-year public college x years after 1990.
   a. Find five solutions of $y = 166x + 1781$. Use 0, 5, 10, 15, and 20 for x. Organize your work in a table of values.
   b. Use the solutions in part (a) to graph $y = 166x + 1781$. What does the shape of the graph indicate about the cost of a four-year public college?

57. The linear equation in two variables $y = 50x + 30,000$ models the total weekly cost, y, in dollars, for a business that manufactures x racing bicycles each week. The equation indicates that the business has weekly fixed costs of $30,000 plus a cost of $50 to manufacture each bicycle.
   a. Find five solutions of $y = 50x + 30,000$. Use 0, 10, 20, 30, and 40 for x. Organize your work in a table of values.
   b. Use the solutions in part (a) to graph $y = 50x + 30,000$.

Writing in Mathematics

58. How do you determine whether an ordered pair is a solution of an equation in two variables, x and y?

59. Explain how to find ordered pairs that are solutions of an equation in two variables, x and y.

60. What is the graph of an equation?

61. Explain how to graph an equation in two variables in the rectangular coordinate system.

Critical Thinking Exercises

62. Which one of the following is true?
   a. The graph of $y = 3x + 1$ looks exactly like the graph of $y = 2x$, but shifted up 1 unit.
   b. The graph of any equation in the form $y = mx + b$ passes through the point $(0, b)$.
   c. The ordered pair $(3, 4)$ satisfies the equation
   $$2y - 3x = -6.$$
   d. If $(2, 5)$ satisfies an equation, then $(5, 2)$ also satisfies the equation.

Graph each equation in Exercises 63–64. Find seven solutions in your table of values by using integers for x, starting with −3 and ending with 3.

63. $y = |x|$ **64.** $y = |x| + 1$

65. Although the level of air pollution varies from day to day and from hour to hour, during the summer the level of air pollution depends on the time of day. The equation in two variables
   $$y = 0.1x^2 - 0.4x + 0.6$$
   describes the level of air pollution (in parts per million [ppm]), where x corresponds to the number of hours after 9 A.M.
   a. Find six solutions of the equation. Select integers for x, starting with 0 and ending with 5.
   b. Researchers have determined that a level of 0.3 ppm or more of pollutants in the air can be hazardous to your health. Based on the six solutions in part (a), at what time of day should runners exercise to avoid unsafe air?

Technology Exercises

Use a graphing utility to graph each equation in Exercises 66–69 in a standard viewing rectangle. Then use the TRACE *feature to trace along the line and find the coordinates of two points.*

66. $y = 2x - 1$ **67.** $y = -3x + 2$
68. $y = \frac{1}{2}x$ **69.** $y = \frac{3}{4}x - 2$

70. Use a graphing utility to verify any five of your hand-drawn graphs in Exercises 21–44. Use an appropriate range setting and the ZOOM SQUARE feature to make the graph look like the one you drew by hand.

71. The linear equation $y = 2.4x + 180$ models U.S. population, y, in millions, x years after 1960. Use a graphing utility to graph the equation in a [0, 90, 10] by [0, 500, 100] viewing rectangle. What does the shape of the graph indicate about changing U.S. population over time?

 **Review Exercises**

72. Solve: $3x + 5 = 4(2x - 3) + 7$. (Section 2.3, Example 3)

73. Simplify: $3(1 - 2 \cdot 5) - (-28)$. (Section 1.8, Example 7)

74. Solve for h: $V = \frac{1}{3}Ah$. (Section 2.4, Example 4)

▶ **SECTION 3.2**   *Graphing Linear Equations Using Intercepts*

Objectives

1 Use a graph to identify intercepts.

2 Graph a linear equation in two variables using intercepts.

3 Graph horizontal or vertical lines.

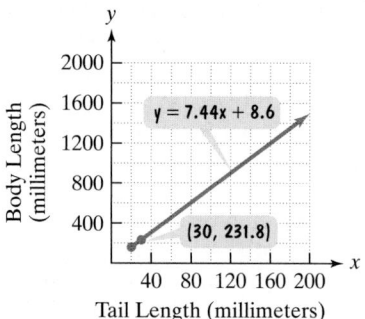

SSM PH Tutor CD- Video
      Center   ROM

y

2000
1600 $y = 7.44x + 8.6$
1200
800
400   (30, 231.8)

Body Length (millimeters)

40 80 120 160 200   x
Tail Length (millimeters)

Figure 3.9 A graph showing the relationship between a snake's tail length and its body length

We hope that you are not watching the horror film *Anaconda* as you glance at the image above. If you are, think of smaller snakes whose body lengths range from 160 millimeters (about 6.25 inches) to 1800 millimeters (about 5.8 feet, which is large enough if you are phobic about snakes). The graph in Figure 3.9 shows a relationship between a snake's tail length and its body length. The point (30, 231.8) indicates that a snake with a 30-millimeter tail length has a body length of 231.8 millimeters.

The graph in Figure 3.9 is a straight line. The equation of this line is

$$y = 7.44x + 8.6.$$

The variable x represents the snake's tail length, in millimeters. The variable y represents the snake's body length, in millimeters.

There is another way that we can write the equation $y = 7.44x + 8.6$. We will collect the x- and y-terms on the left side. This is done by subtracting $7.44x$ from both sides:

$$-7.44x + y = 8.6.$$

The form of this equation is $Ax + By = C$.

$$-7.44x + y = 8.6$$

A, the coefficient of x, is −7.44.   B, the coefficient of y, is 1.   C, the constant on the right, is 8.6.

All equations of the form $Ax + By = C$ are straight lines when graphed as long as A and B are not both zero. To graph linear equations of this form, we will use two important points: the *intercepts*.

1 Use a graph to identify intercepts.

Intercepts

An **x-intercept** of a graph is the *x*-coordinate of a point where a graph intersects the *x*-axis. For example, look at the graph of $2x - 4y = 8$ in Figure 3.10. The graph crosses the *x*-axis at $(4, 0)$. Thus, the *x*-intercept is 4. **The y-coordinate corresponding to an x-intercept is always zero.**

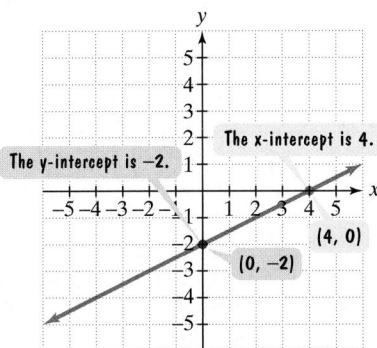

Figure 3.10 The graph of $2x - 4y = 8$

A **y-intercept** of a graph is the *y*-coordinate of a point where a graph intersects the *y*-axis. The graph of $2x - 4y = 8$ in Figure 3.10 shows that the graph crosses the *y*-axis at $(0, -2)$. Thus, the *y*-intercept is -2. **The x-coordinate corresponding to a y-intercept is always zero.**

EXAMPLE 1   Identifying Intercepts

Identify the *x*- and *y*-intercepts.

a.

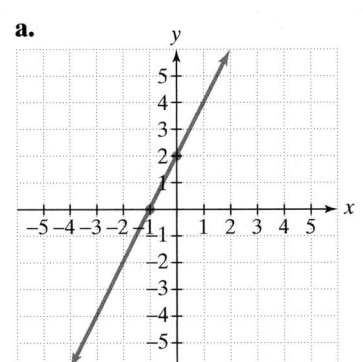

b.

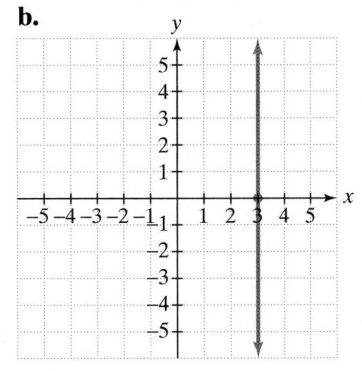

c.

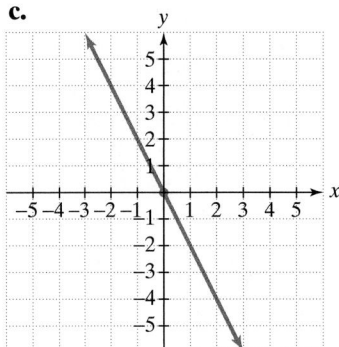

Solution

a. The graph crosses the *x*-axis at $(-1, 0)$. Thus, the *x*-intercept is -1. The graph crosses the *y*-axis at $(0, 2)$. Thus, the *y*-intercept is 2.

b. The graph crosses the *x*-axis at $(3, 0)$, so the *x*-intercept is 3. This vertical line does not cross the *y*-axis. Thus, there is no *y*-intercept.

c. This graph crosses the *x*- and *y*-axes at the same point, the origin. Because the graph crosses both axes at $(0, 0)$, the *x*-intercept is 0 and the *y*-intercept is 0.   ■

✔ **CHECK POINT 1**   Identify the x- and y-intercepts.

a.

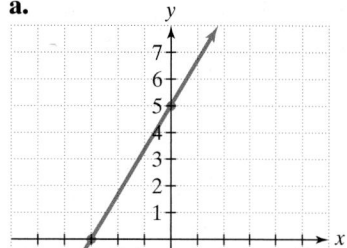

b.

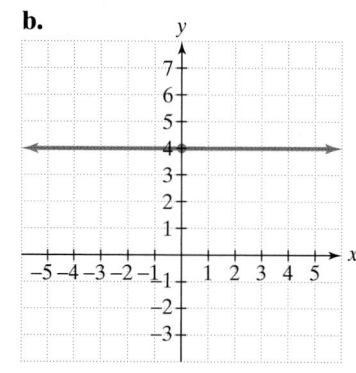

c.

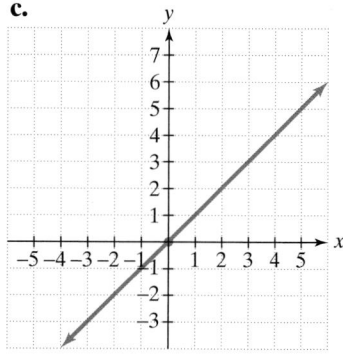

2 Graph a linear equation in two variables using intercepts.

Graphing Using Intercepts

An equation of the form $Ax + By = C$, where A, B, and C are integers, is called the **standard form** of the equation of a line. The equation can be graphed by finding the x- and y-intercepts, plotting the intercepts, and drawing a straight line through these points. How do we find the intercepts of a line, given its equation? Because the y-coordinate of the x-intercept is 0, to find the x-intercept:

• Substitute 0 for y in the equation.
• Solve for x.

EXAMPLE 2   Finding the x-Intercept

Find the x-intercept of the graph of $3x - 4y = 24$.

Solution   To find the x-intercept, let $y = 0$ and solve for x.

$$3x - 4y = 24 \qquad \text{This is the given equation.}$$
$$3x - 4 \cdot 0 = 24 \qquad \text{Let } y = 0.$$
$$3x = 24 \qquad \text{Simplify: } 4 \cdot 0 = 0 \text{ and } 3x - 0 = 3x.$$
$$x = 8 \qquad \text{Divide both sides by 3.}$$

The x-intercept is 8. The graph of $3x - 4y = 24$ passes through the point $(8, 0)$. ■

✔ **CHECK POINT 2**   Find the x-intercept of the graph of $4x - 3y = 12$.

Because the x-coordinate of the y-intercept is 0, to find the y-intercept:

• Substitute 0 for x in the equation.
• Solve for y.

EXAMPLE 3   Finding the y-Intercept

Find the y-intercept of the graph of $3x - 4y = 24$.

Solution   To find the y-intercept, let $x = 0$ and solve for y.

$$3x - 4y = 24 \qquad \text{This is the given equation.}$$
$$3 \cdot 0 - 4y = 24 \qquad \text{Let } x = 0.$$
$$-4y = 24 \qquad \text{Simplify: } 3 \cdot 0 = 0 \text{ and } 0 - 4y = -4y.$$
$$y = -6 \qquad \text{Divide both sides by } -4.$$

The y-intercept is -6. The graph of $3x - 4y = 24$ passes through the point $(0, -6)$. ■

✔ **CHECK POINT 3**   Find the *y*-intercept of the graph of $4x - 3y = 12$.

When graphing using intercepts, it is a good idea to use a third point, a checkpoint, before drawing the line. A checkpoint can be obtained by selecting a value for either variable, other than 0, and finding the corresponding value for the other variable. The checkpoint should lie on the same line as the *x*- and *y*-intercepts. If it does not, recheck your work and find the error.

> **Using Intercepts to Graph $Ax + By = C$**
>
> **1.** Find the *x*-intercept. Let $y = 0$ and solve for *x*.
>
> **2.** Find the *y*-intercept. Let $x = 0$ and solve for *y*.
>
> **3.** Find a checkpoint, a third ordered-pair solution.
>
> **4.** Graph the equation by drawing a line through the three points.

EXAMPLE 4   Using Intercepts to Graph a Linear Equation

Graph: $3x + 2y = 6$.

Solution

Step 1 Find the *x*-intercept. Let $y = 0$ and solve for *x*.

$$3x + 2 \cdot 0 = 6$$
$$3x = 6$$
$$x = 2$$

The *x*-intercept is 2, so the line passes through $(2, 0)$.

Step 2 Find the *y*-intercept. Let $x = 0$ and solve for *y*.

$$3 \cdot 0 + 2y = 6$$
$$2y = 6$$
$$y = 3$$

The *y*-intercept is 3, so the line passes through $(0, 3)$.

Step 3 Find a checkpoint, a third ordered-pair solution. For our checkpoint, we will let $x = 1$ and find the corresponding value for *y*.

$3x + 2y = 6$	This is the given equation.
$3 \cdot 1 + 2y = 6$	Substitute 1 for x.
$3 + 2y = 6$	Simplify.
$2y = 3$	Subtract 3 from both sides.
$y = \dfrac{3}{2}$	Divide both sides by 2.

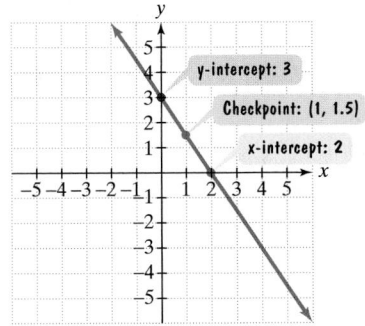

Figure 3.11 The graph of $3x + 2y = 6$

The checkpoint is the ordered pair $\left(1, \dfrac{3}{2}\right)$, or $(1, 1.5)$.

Step 4 Graph the equation by drawing a line through the three points. The three points in Figure 3.11 lie along the same line. Drawing a line through the three points results in the graph of $3x + 2y = 6$. ∎

Using Technology

You can use a graphing utility to graph equations of the form $Ax + By = C$. Begin by solving the equation for y. For example, to graph $3x + 2y = 6$, solve the equation for y.

$$3x + 2y = 6 \qquad \text{This is the equation to be graphed.}$$
$$3x - 3x + 2y = -3x + 6 \qquad \text{Subtract 3x from both sides.}$$
$$2y = -3x + 6 \qquad \text{Simplify.}$$
$$\frac{2y}{2} = \frac{-3x + 6}{2} \qquad \text{Divide both sides by 2.}$$
$$y = -\frac{3}{2}x + 3 \qquad \text{Simplify.}$$

This is the equation to enter in your graphing utility. The graph of $y = -\frac{3}{2}x + 3$ or, equivalently, $3x + 2y = 6$ is shown below in a $[-6, 6, 1]$ by $[-6, 6, 1]$ viewing rectangle.

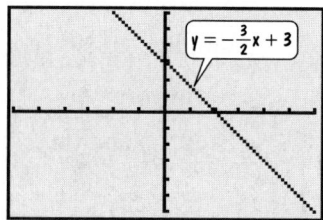

✔ **CHECK POINT 4** Graph: $2x + 3y = 6$.

EXAMPLE 5 Using Intercepts to Graph a Linear Equation

Graph: $2x - y = 4$.

Solution

Step 1 Find the x-intercept. Let $y = 0$ and solve for x.
$$2x - 0 = 4$$
$$2x = 4$$
$$x = 2$$

The x-intercept is 2, so the line passes through $(2, 0)$.

Step 2 Find the y-intercept. Let $x = 0$ and solve for y.
$$2 \cdot 0 - y = 4$$
$$-y = 4$$
$$y = -4$$

The y-intercept is -4, so the line passes through $(0, -4)$.

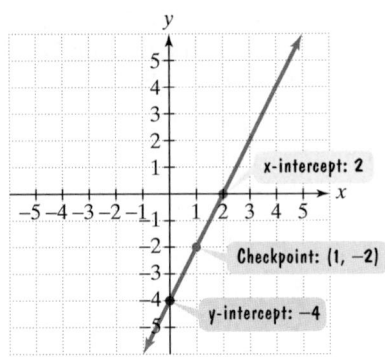

Figure 3.12 The graph of $2x - y = 4$

Step 3 Find a checkpoint, a third ordered-pair solution. For our checkpoint, we will let $x = 1$ and find the corresponding value for y.

$2x - y = 4$	This is the given equation.
$2 \cdot 1 - y = 4$	Substitute 1 for x.
$2 - y = 4$	Simplify.
$-y = 2$	Subtract 2 from both sides.
$y = -2$	Multiply (or divide) both sides by −1.

The checkpoint is $(1, -2)$.

Step 4 Graph the equation by drawing a line through the three points. The three points in Figure 3.12 lie along the same line. Drawing a line through the three points results in the graph of $2x - y = 4$. ∎

✔ **CHECK POINT 5** Graph: $x - 2y = 4$.

We have seen that not all lines have two different intercepts. Some lines pass through the origin. Thus, they have an x-intercept of 0 and a y-intercept of 0. Is it possible to recognize these lines by their equations? Yes. **The graph of the linear equation $Ax + By = 0$ passes through the origin.** Notice that the constant on the right side of this equation is 0.

An equation of the form $Ax + By = 0$ can be graphed by using the origin as one point on the line. Find two other points by finding two other solutions of the equation. Select values for either variable, other than 0, and find the corresponding values for the other variable.

EXAMPLE 6 Graphing a Linear Equation of the Form $Ax + By = 0$

Graph: $x + 2y = 0$.

Solution Because the constant on the right is 0, the graph passes through the origin. The x- and y-intercepts are both 0. Remember that we are using two points and a checkpoint to determine a line. Thus, we still want to find two other points. Let $y = -1$ to find a second ordered-pair solution. Let $y = 1$ to find a third ordered-pair (checkpoint) solution.

$x + 2y = 0$	$x + 2y = 0$
Let y = −1.	Let y = 1.
$x + 2(-1) = 0$	$x + 2 \cdot 1 = 0$
$x + (-2) = 0$	$x + 2 = 0$
$x = 2$	$x = -2$

The solutions are $(2, -1)$ and $(-2, 1)$. Plot these two points, as well as the origin—that is, $(0, 0)$. The three points in Figure 3.13 lie along the same line. Drawing a line through the three points results in the graph of $x + 2y = 0$. ∎

✔ **CHECK POINT 6** Graph: $x + 3y = 0$.

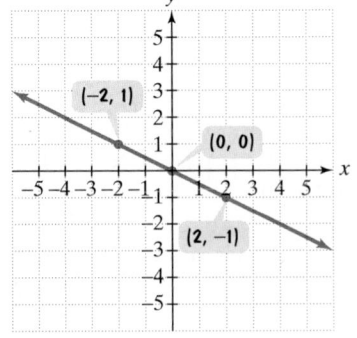

Figure 3.13 The graph of $x + 2y = 0$

3 Graph horizontal or vertical lines.

Equations of Horizontal and Vertical Lines

Some things change very little. For example, from 1985 to the present, the number of Americans participating in downhill skiing has remained relatively constant, indicated by the graph shown in Figure 3.14. Shown in the figure is a horizontal line that passes through or near most of the data points.

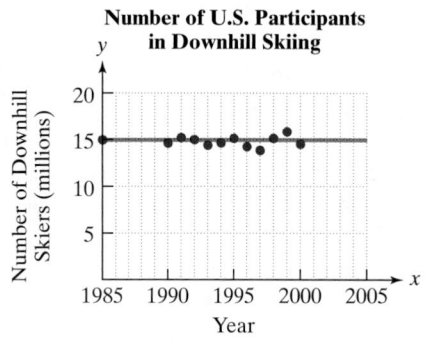

Source: National Ski Areas Association

Figure 3.14

We can use the horizontal line in Figure 3.14 to write an equation that reasonably models the data. The *y*-intercept of the line is 15. Furthermore, all points on the line have a value of *y* that is always 15. Thus, an equation that models the number of participants in downhill skiing for the period shown is

$$y = 15.$$

The popularity of downhill skiing has remained relatively constant in the United States at approximately 15 million participants each year.

The equation $y = 15$ can be expressed as $0x + 1y = 15$. We know that the graph of any equation of the form $Ax + By = C$ is a line as long as A and B are not both zero. The graph of $y = 15$ suggests that when A is zero, the line is horizontal.

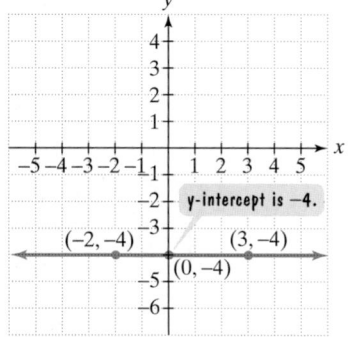

Figure 3.15 The graph of $y = -4$

EXAMPLE 7   Graphing a Horizontal Line

Graph the linear equation: $y = -4$.

Solution   All ordered pairs that are solutions of $y = -4$ have a value of y that is always -4. Any value can be used for x. Let us select three of the possible values for x: $-2, 0$, and 3. So, three ordered pairs that are solutions of $y = -4$ are $(-2, -4), (0, -4)$, and $(3, -4)$. Plot each of these points. Drawing a line that passes through the three points gives the horizontal line shown in Figure 3.15. ∎

✔ **CHECK POINT 7**   Graph the linear equation: $y = 3$.

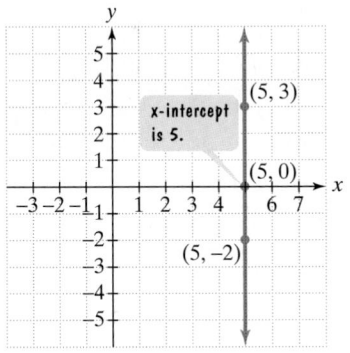

Figure 3.16 The graph $x = 5$

EXAMPLE 8   Graphing a Vertical Line

Graph the linear equation: $x = 5$.

Solution   All ordered pairs that are solutions of $x = 5$ have a value of x that is always 5. Any value can be used for y. Let us select three of the possible values for y: $-2, 0$, and 3. So, three ordered pairs that are solutions of $x = 5$ are $(5, -2)$, $(5, 0)$, and $(5, 3)$. Drawing a line that passes through the three points gives the vertical line shown in Figure 3.16.   ■

✔ **CHECK POINT 8**   Graph the linear equation: $x = -2$.

Horizontal and Vertical Lines

The graph of a linear equation in one variable is a horizontal or vertical line.

The graph of $y = b$ is a horizontal line. The y-intercept is b.

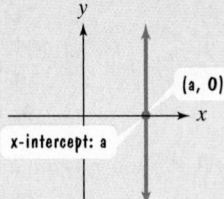

The graph of $x = a$ is a vertical line. The x-intercept is a.

EXERCISE SET 3.2

Practice Exercises

In Exercises 1–8, use the graph to identify the:
a. *x-intercept, or state that there is no x-intercept;*
b. *y-intercept, or state that there is no y-intercept.*

1.

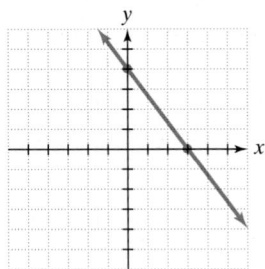

2.

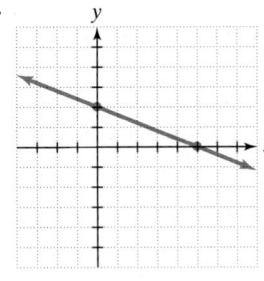

3.

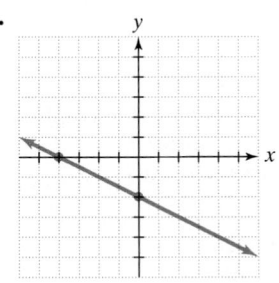

4.

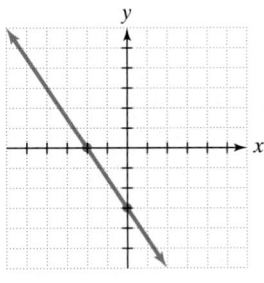

5.

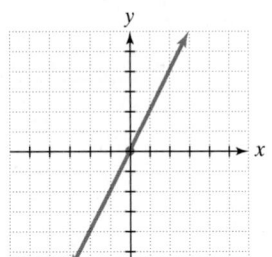

6.

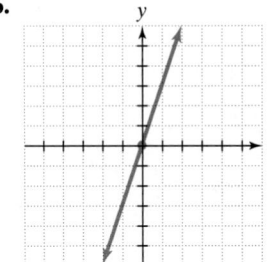

7.

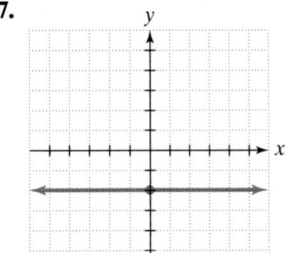

8.

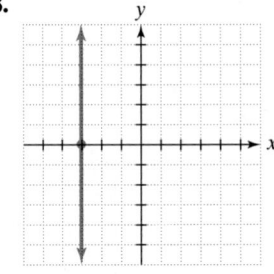

In Exercises 9–18, find the x-intercept and the y-intercept of the graph of each equation. Do not graph the equation.

9. $4x + 5y = 20$

10. $3x + 7y = 21$

11. $7x - 3y = 42$

12. $5x - 3y = 30$

13. $-x + 4y = -8$

14. $-x + 3y = -9$

15. $3x - 5y = 0$

16. $2x - 3y = 0$

17. $2x = 3y - 6$

18. $3x = 2y - 12$

In Exercises 19–40, use intercepts and a checkpoint to graph each equation.

19. $x + y = 3$

20. $x + y = 4$

21. $3x + y = 6$

22. $x + 2y = 4$

23. $9x - 6y = 18$

24. $2x - 6y = 12$

25. $-x + 4y = 8$

26. $-x + 3y = 12$

27. $2x - y = 6$

28. $3x - y = 9$

29. $5x = 3y - 15$

30. $3x = 2y + 6$

31. $50y = 100 - 25x$

32. $40y = 60 - 10x$

33. $8x - 2y = 12$

34. $6x - 3y = 15$

35. $x + y = 0$

36. $x - y = 0$

37. $2x + y = 0$

38. $3x + y = 0$

39. $y - 2x = 0$

40. $y - 3x = 0$

In Exercises 41–46, write an equation for each graph.

41.

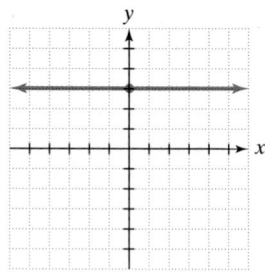

42.

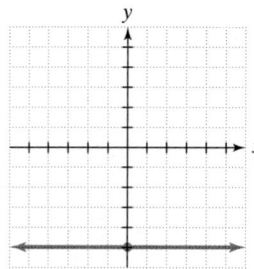

43.

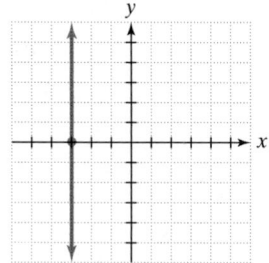

44.

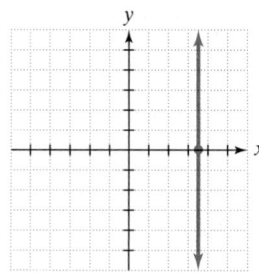

45.

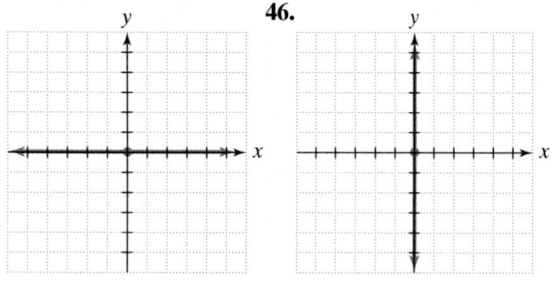

46.

In Exercises 47–62, graph each equation.

47. $y = 4$

48. $y = 2$

49. $y = -2$

50. $y = -3$

51. $x = 2$

52. $x = 4$

53. $x + 1 = 0$

54. $x + 5 = 0$

55. $y - 3.5 = 0$

56. $y - 2.5 = 0$

57. $x = 0$

58. $y = 0$

59. $3y = 9$

60. $5y = 20$

61. $12 - 3x = 0$

62. $12 - 4x = 0$

Application Exercises

The flight of a vulture is observed for 30 seconds. The graph shows the vulture's height, in meters, during this period of time. Use the graph to solve Exercises 63–67.

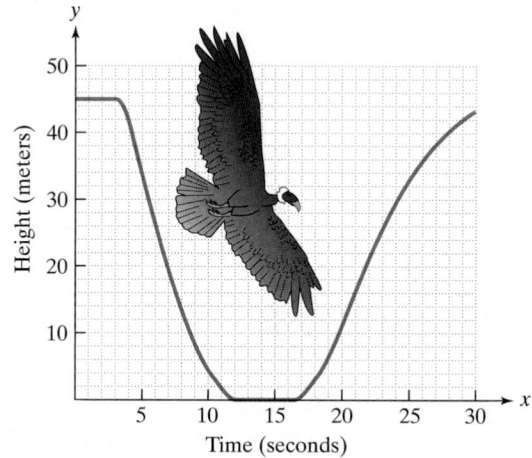

63. During which period of time is the vulture's height decreasing?

64. During which period of time is the vulture's height increasing?

65. What is the y-intercept? What does this mean about the vulture's height at the beginning of the observation?

66. During the first three seconds of observation, the vulture's flight is graphed as a horizontal line. Write the equation of the line. What does this mean about the vulture's flight pattern during this time?

67. Use integers to write five x-intercepts of the graph. What is the vulture doing during these times?

Too late for that flu shot now! It's only 8 A.M. and you're feeling lousy. Fascinated by the way that algebra models the world (your author is projecting a bit here), you decide to construct a graph showing your body temperature from 8 A.M. through 3 P.M. You decide to let x represent the number of hours after 8 A.M. and y your temperature at time x. The graph is shown. Use it to solve Exercises 68–72.

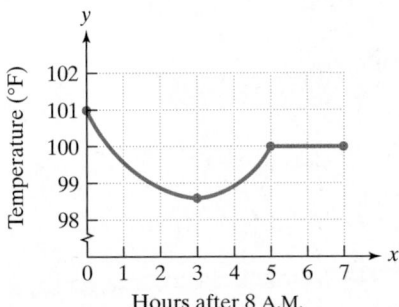

68. What is the y-intercept? What does this mean about your temperature at 8 A.M.?

69. During which period of time is your temperature decreasing?

70. Estimate your minimum temperature during the time period shown. How many hours after 8 A.M. does this occur? At what time does this occur?

71. During which period of time is your temperature increasing?

72. From five hours after 8 A.M. until seven hours after 8 A.M., your temperature is graphed as a portion of a horizontal line. Write the equation of the line. What does this mean about your temperature over this period of time?

73. As shown in the bar graph, the percentage of people in the United States satisfied with their lives remains relatively constant for all age groups. If x represents a person's age and y represents the percentage of people satisfied with their lives at that age, write an equation that reasonably models the data.

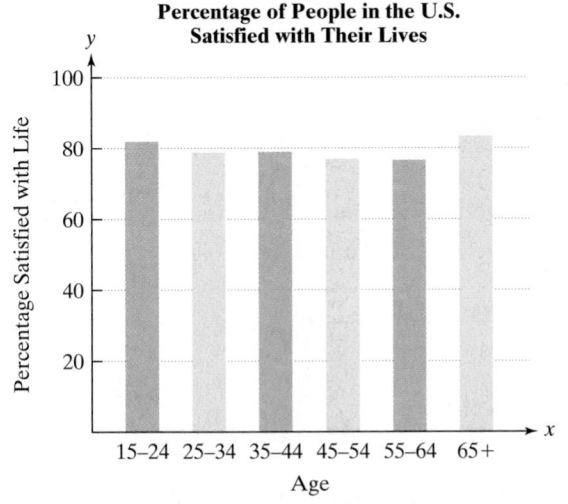

Percentage of People in the U.S. Satisfied with Their Lives

Source: Culture Shift in Advanced Industrial Society, Princeton University Press

Writing in Mathematics

74. What is an x-intercept of a graph?

75. What is a y-intercept of a graph?

76. If you are given an equation of the form $Ax + By = C$, explain how to find the x-intercept.

77. If you are given an equation of the form $Ax + By = C$, explain how to find the y-intercept.

78. Explain how to graph $Ax + By = C$ if C is not equal to zero.

79. Explain how to graph a linear equation of the form $Ax + By = 0$.

80. How many points are needed to graph a line? How many should actually be used? Explain.

81. Describe the graph of $y = 200$.

82. Describe the graph of $x = -100$.

83. We saw that the number of skiers in the United States has remained constant over time. Exercise 73 showed that the percentage of people satisfied with their lives remains constant for all age groups. Give another example of a real-world phenomenon that has remained relatively constant. Try writing an equation that models this phenomenon.

Critical Thinking Exercises

84. Write the equation of a line passing through the point $(5, 6)$ and parallel to the line whose equation is $y = -1$.

In Exercises 85–86, find the coefficients that must be placed in each shaded area so that the equation's graph will be a line with the specified intercepts.

85. ▢$x +$ ▢$y = 10$;   x-intercept $= 5$; y-intercept $= 2$

86. ▢$x +$ ▢$y = 12$;   x-intercept $= -2$; y-intercept $= 4$

Technology Exercises

87. Use a graphing utility to verify any five of your hand-drawn graphs in Exercises 19–40. Solve the equation for y before entering it.

In Exercises 88–91, use a graphing utility to graph each equation. You will need to solve the equation for y before entering it. Use the equation displayed on the screen to identify the x-intercept and the y-intercept.

88. $2x + y = 4$ **89.** $3x - y = 9$

90. $2x + 3y = 30$ **91.** $4x - 2y = -40$

Review Exercises

92. Find the absolute value: $|-13.4|$. (Section 1.2, Example 8)

93. Simplify: $7x - (3x - 5)$. (Section 1.7, Example 7)

94. Graph: $-2 \leq x < 4$. (Section 2.7, Example 1)

▶ SECTION 3.3   *Slope*

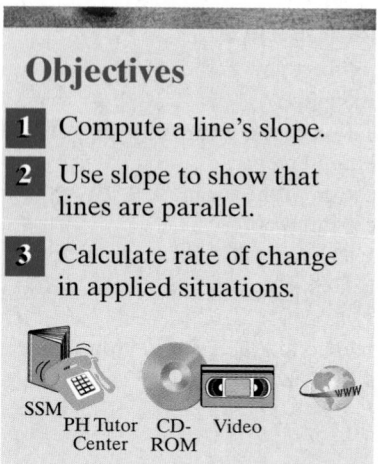

Objectives

1 Compute a line's slope.

2 Use slope to show that lines are parallel.

3 Calculate rate of change in applied situations.

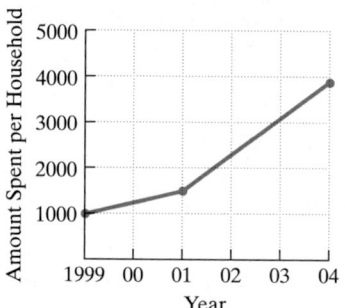

Good news: Projections indicate that in the next decades we'll live longer and move somewhere warmer where we'll shop online and chat on our tiny video cell phones. Figure 3.17 shows projected online shopping per U.S. online household through 2004. The graph is composed of two line segments. The segment on the right is steeper than the one on the left. This shows that online shopping is expected to increase more per year in 2001–2004 than in 1999–2001.

Data often fall on or near a line. In this section, we will study the idea of a line's steepness from a mathematical perspective.

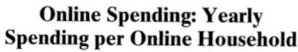

Online Spending: Yearly Spending per Online Household

Source: Forrester Research

Figure 3.17

The Slope of a Line

Mathematicians have developed a useful measure of the steepness of a line, called the **slope** of the line. Slope compares the vertical change (the **rise**) to the horizontal change (the **run**) when moving from one fixed point to another along the line. To calculate the slope of a line, mathematicians use a ratio comparing the change in y (the rise) to the change in x (the run).

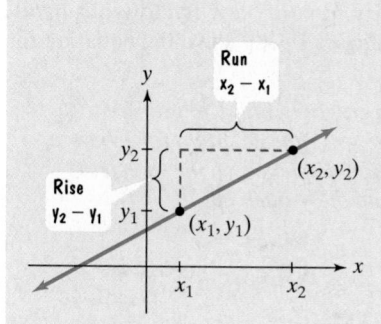

Definition of Slope

The **slope** of the line through the distinct points (x_1, y_1) and (x_2, y_2) is

$$\frac{\text{Change in } y}{\text{Change in } x} = \frac{\text{Rise}}{\text{Run}}$$

$$= \frac{y_2 - y_1}{x_2 - x_1}$$

where $x_2 - x_1 \neq 0$.

It is common notation to let the letter m represent the slope of a line. The letter m is used because it is the first letter of the French verb *monter*, meaning to rise, or to ascend.

1 Compute a line's slope.

EXAMPLE 1 Using the Definition of Slope

Find the slope of the line passing through each pair of points:

a. $(-3, -1)$ and $(-2, 4)$ **b.** $(-3, 4)$ and $(2, -2)$.

Solution

a. Let $(x_1, y_1) = (-3, -1)$ and $(x_2, y_2) = (-2, 4)$. We obtain a slope of

$$m = \frac{\text{Change in } y}{\text{Change in } x} = \frac{y_2 - y_1}{x_2 - x_1} = \frac{4 - (-1)}{-2 - (-3)} = \frac{5}{1} = 5.$$

The situation is illustrated in Figure 3.18(a). The slope of the line is 5, indicating that there is a vertical change, a rise, of 5 units for each horizontal change, a run, of 1 unit. The slope is positive, and the line rises from left to right.

Study Tip

When computing slope, it makes no difference which point you call (x_1, y_1) and which point you call (x_2, y_2). If we let $(x_1, y_1) = (-2, 4)$ and $(x_2, y_2) = (-3, -1)$, the slope is still 5:

$$m = \frac{\text{Change in } y}{\text{Change in } x} = \frac{y_2 - y_1}{x_2 - x_1} = \frac{-1 - 4}{-3 - (-2)} = \frac{-5}{-1} = 5.$$

However, you should not subtract in one order in the numerator $(y_2 - y_1)$ and then in a different order in the denominator $(x_1 - x_2)$. The slope is *not*

$$\frac{-1 - 4}{-2 - (-3)} \quad \frac{-5}{1} = -5. \quad \text{Incorrect}$$

b. We can let $(x_1, y_1) = (-3, 4)$ and $(x_2, y_2) = (2, -2)$. The slope of the line shown in Figure 3.18(b) is computed as follows:

$$m = \frac{\text{Change in } y}{\text{Change in } x} = \frac{y_2 - y_1}{x_2 - x_1} = \frac{-2 - 4}{2 - (-3)} = \frac{-6}{5} = -\frac{6}{5}.$$

The slope of the line is $-\frac{6}{5}$. For every vertical change of -6 units (6 units down), there is a corresponding horizontal change of 5 units. The slope is negative and the line falls from left to right.

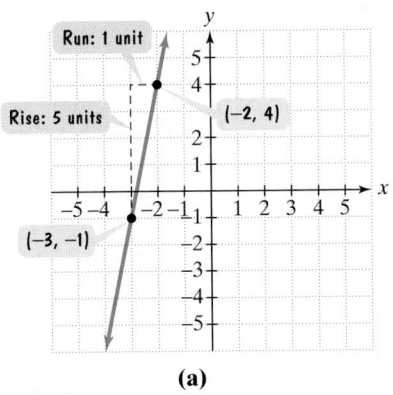

(a)

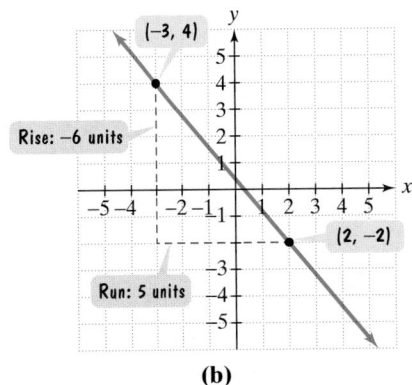

(b)

Figure 3.18 Visualizing slope

✔ **CHECK POINT 1**   Find the slope of the line passing through each pair of points:

a. $(-3, 4)$ and $(-4, -2)$ **b.** $(4, -2)$ and $(-1, 5)$.

EXAMPLE 2   Using the Definition of Slope for Horizontal and Vertical Lines

Find the slope of the line passing through each pair of points:

a. $(5, 4)$ and $(3, 4)$ **b.** $(2, 5)$ and $(2, 1)$.

Solution

a. Let $(x_1, y_1) = (5, 4)$ and $(x_2, y_2) = (3, 4)$. We obtain a slope of

$$m = \frac{\text{Change in } y}{\text{Change in } x} = \frac{y_2 - y_1}{x_2 - x_1} = \frac{4 - 4}{3 - 5} = \frac{0}{-2} = 0.$$

The situation is illustrated in Figure 3.19(a). Can you see that the line is horizontal? Because any two points on a horizontal line have the same y-coordinate, these lines neither rise nor fall from left to right. The change in y, $y_2 - y_1$, is always zero. Thus, **the slope of any horizontal line is zero**.

b. We can let $(x_1, y_1) = (2, 5)$ and $(x_2, y_2) = (2, 1)$. Figure 3.19(b) shows that these points are on a vertical line. We attempt to compute the slope as follows:

$$m = \frac{\text{Change in } y}{\text{Change in } x} = \frac{1 - 5}{2 - 2} = \frac{-4}{0}. \quad \boxed{\text{Division by zero is undefined.}}$$

Because division by zero is undefined, the slope of the vertical line in Figure 3.19(b) is undefined. In general, **the slope of any vertical line is undefined**.

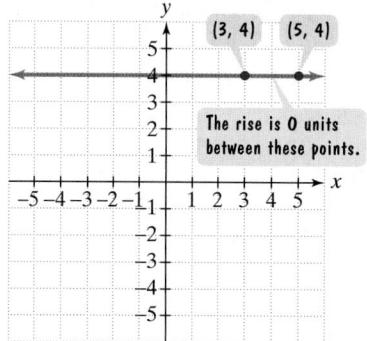

(a) Horizontal lines have no vertical change.   (b) Vertical lines have no horizontal change.

Figure 3.19 Visualizing Slope

■

Table 3.1 summarizes the four possibilities for the slope of a line.

Table 3.1   Possibilities for a Line's Slope

Positive Slope	Negative Slope	Zero Slope	Undefined Slope
$m > 0$	$m < 0$	$m = 0$	m is undefined.
Line rises from left to right.	Line falls from left to right.	Line is horizontal.	Line is vertical.

✔ **CHECK POINT 2**   Find the slope of the line passing through each pair of points or state that the slope is undefined:

a. $(6, 5)$ and $(2, 5)$ **b.** $(1, 6)$ and $(1, 4)$.

Slope and Parallel Lines

2 Use slope to show that lines are parallel.

Number of People in the U.S. Living Alone

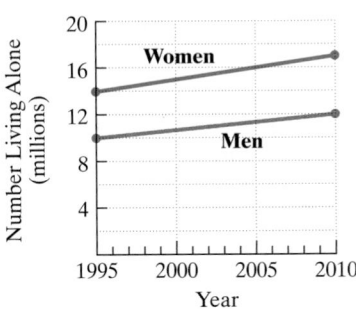

Source: Forrester Research

Figure 3.20

A best guess at the look of our nation in the next decades indicates that the number of men and women living alone will increase each year. Figure 3.20 shows that by 2010, approximately 12 million men and 17 million women will be living alone. Can you tell by the line graphs in the figure if the yearly increase for women is the same as the yearly increase for men? We can use a relationship between slope and parallel lines to answer this question.

Two nonintersecting lines that lie in the same plane are **parallel**. If two lines do not intersect, the ratio of the vertical change to the horizontal change is the same for each line. Because two parallel lines have the same "steepness," they must have the same slope.

Slope and Parallel Lines

1. If two nonvertical lines are parallel, then they have the same slope.
2. If two distinct nonvertical lines have the same slope, then they are parallel.
3. Two distinct vertical lines, each with undefined slope, are parallel.

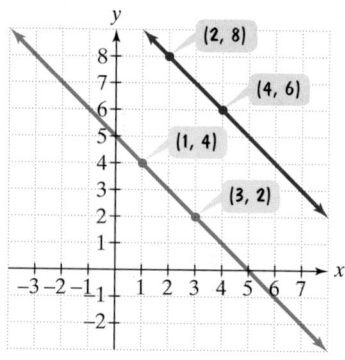

Figure 3.21 Using slope to show that lines are parallel

EXAMPLE 3   Using Slope to Show That Lines Are Parallel

Show that the line passing through $(1, 4)$ and $(3, 2)$ is parallel to the line passing through $(2, 8)$ and $(4, 6)$.

Solution   The situation is illustrated in Figure 3.21. The lines certainly look like they are parallel. Let's use equal slopes to confirm this fact. For each line, we compute the ratio of the difference in y-coordinates to the difference in x-coordinates. Be sure to subtract the coordinates in the same order.

Slope of the line through $(1, 4)$ and $(3, 2)$ is:

$$\frac{4 - 2}{1 - 3} = \frac{2}{-2} = -1.$$

Slope of the line through $(2, 8)$ and $(4, 6)$ is:

$$\frac{8 - 6}{2 - 4} = \frac{2}{-2} = -1.$$

With equal slopes, the lines are parallel.   ∎

✔ **CHECK POINT 3**   Show that the line passing through $(4, 2)$ and $(6, 6)$ is parallel to the line passing through $(0, -2)$ and $(1, 0)$.

3 Calculate rate of change in applied situations.

Applications

Slope is defined as the ratio of a change in y to a corresponding change in x. Our next example shows how slope can be interpreted as a rate of change in an applied situation.

EXAMPLE 4   Slope as a Rate of Change

The line graphs for the number of women and men living alone are shown again in Figure 3.22. Find the slope of the line segment for the women. Describe what this slope represents.

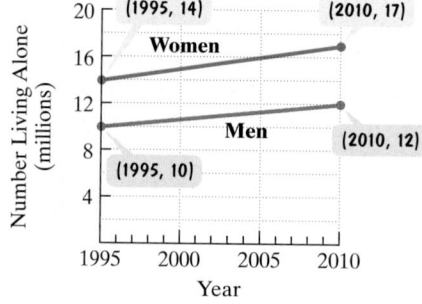

Source: Forrester Research   **Figure 3.22**

Solution   We let x represent a year and y the number of women living alone in that year. The two points shown on the line segment for women have the following coordinates:

$$(1995, 14) \quad \text{and} \quad (2010, 17).$$

In 1995, 14 million U.S. women lived alone.

In 2010, 17 million U.S. women are projected to live alone.

Now we compute the slope:

$$m = \frac{\text{Change in } y}{\text{Change in } x} = \frac{17 - 14}{2010 - 1995}$$

The unit in the numerator is million people.

$$= \frac{3}{15} = \frac{1}{5} = \frac{0.2 \text{ million people}}{\text{year}}.$$

The unit in the denominator is year.

The slope indicates that the number of U.S. women living alone is projected to increase by 0.2 million each year. The rate of change is 0.2 million women per year. ∎

✔ **CHECK POINT 4**   Use the graph in Example 4 to find the slope of the line segment for the men. Express the slope correct to two decimal places and describe what it represents.

In Check Point 4, did you find that the slope of the line segment for the men is different from that of the women? The rate of change for men living alone is not equal to the rate of change for women living alone. Because of these different slopes, if you extend the line segments in Figure 3.22, the resulting lines will intersect. They are not parallel.

ENRICHMENT ESSAY

Railroads and Highways

The steepest part of Mt. Washington Cog Railway in New Hampshire has a 37% grade. This is equivalent to a slope of $\frac{37}{100}$. For every horizontal change of 100 feet, the railroad ascends 37 feet vertically. Engineers denote slope by grade, expressing slope as a percentage.

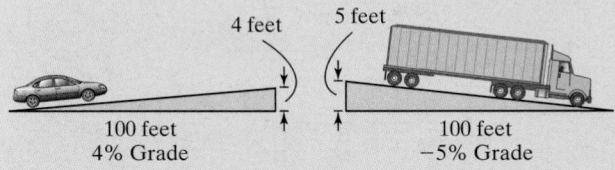

4 feet

5 feet

100 feet
4% Grade

100 feet
−5% Grade

Railroad grades are usually less than 2%, although in the mountains they may go as high as 4%. The grade of the Mt. Washington Cog Railway is phenomenal, making it necessary for locomotives to *push* single cars up its steepest part.

A Mount Washington Cog Railway locomotive pushing a single car up the steepest part of the railroad. The locomotive is about 120 years old.

EXERCISE SET 3.3

Practice Exercises

In Exercises 1–10, find the slope of the line passing through each pair of points or state that the slope is undefined. Then indicate whether the line through the points rises, falls, is horizontal, or is vertical.

1. $(4, 7)$ and $(8, 10)$

2. $(2, 1)$ and $(3, 4)$

3. $(-2, 1)$ and $(2, 2)$

4. $(-1, 3)$ and $(2, 4)$

5. $(4, -2)$ and $(3, -2)$

6. $(4, -1)$ and $(3, -1)$

7. $(-2, 4)$ and $(-1, -1)$

8. $(6, -4)$ and $(4, -2)$

9. $(5, 3)$ and $(5, -2)$

10. $(3, -4)$ and $(3, 5)$

In Exercises 11–22, find the slope of each line, or state that the slope is undefined.

11.

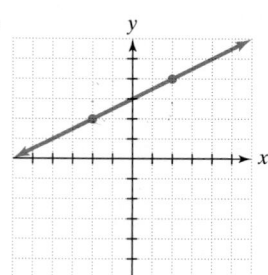

12.

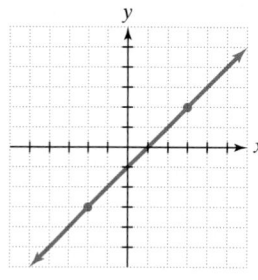

21.

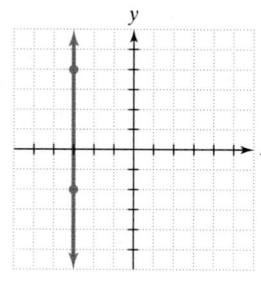

22.

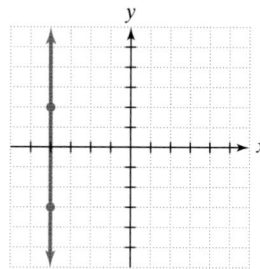

13.

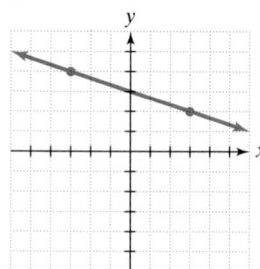

14.

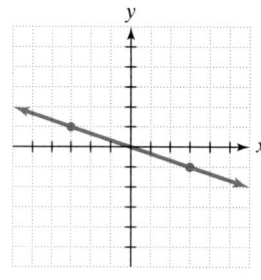

In Exercises 23–26, determine whether the lines through each pair of points are parallel.

23. $(-2, 0)$ and $(0, 6)$
$(1, 8)$ and $(0, 5)$

24. $(2, 4)$ and $(6, 1)$
$(-3, 1)$ and $(1, -2)$

25. $(0, 3)$ and $(1, 5)$
$(-1, 7)$ and $(1, 10)$

26. $(-7, 6)$ and $(0, 4)$
$(-9, -3)$ and $(1, 5)$

15.

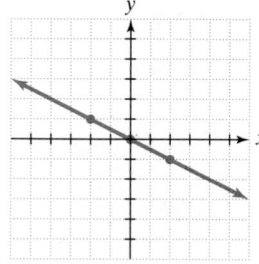

16.

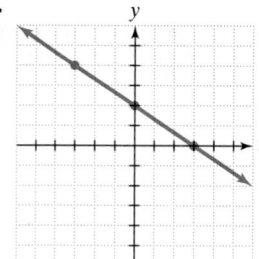

Application Exercises

The graph shows the projected online shopping per U.S. online household through 2004. Use the information provided by the graph to solve Exercises 27–28.

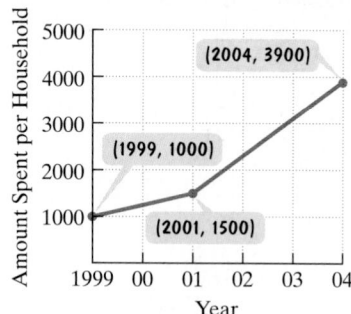

Online Spending: Yearly Spending per Online Household

Source: Forrester Research

17.

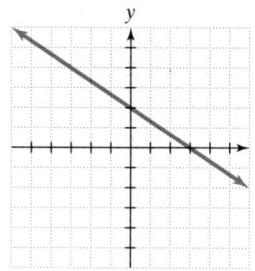

18.

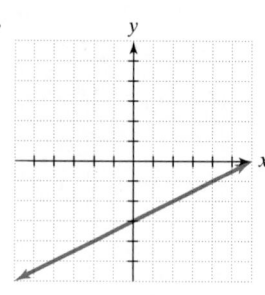

27. Find the slope of the line passing through (1999, 1000) and (2001, 1500). What does this represent in terms of the increase in online shopping per year?

19.

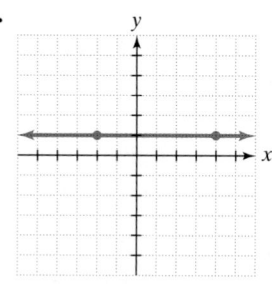

20.

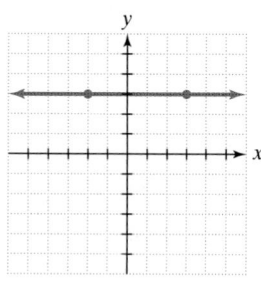

28. Find the slope of the line passing through (2001, 1500) and (2004, 3900). What does this represent in terms of the increase in online shopping per year?

If talk about the federal budget surplus sounds too good to be true, that's because it might be. The Congressional Budget Office's estimates for 2010 range from a $1.2 trillion budget surplus to a $286 billion deficit. Use the information provided by the Congressional Budget Office graphs to solve Exercises 29–30.

Federal Budget Projections

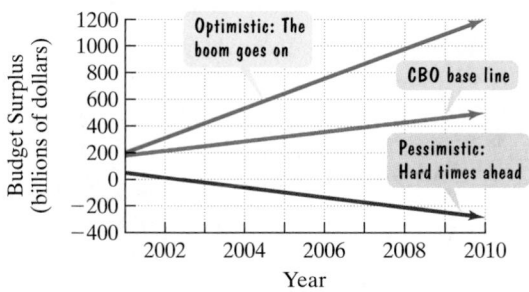

Source: Congressional Budget Office

29. Look at the line that indicates hard times ahead. Find the slope of this line using (2001, 50) and (2010, −286). What does this represent in terms of the rate of decrease in budget surplus per year?

30. Look at the line that indicates the boom goes on. Find the slope of this line using (2001, 200) and (2010, 1200). What does this represent in terms of the rate of increase in the budget surplus per year?

From 1993 through 2000, the number of books sold by book publishers in the United States has not changed from year to year. By contrast, dollar sales have increased steadily. Use the information provided by the graphs to solve Exercises 31–32.

Book Publishers' Dollar Sales and Unit Sales

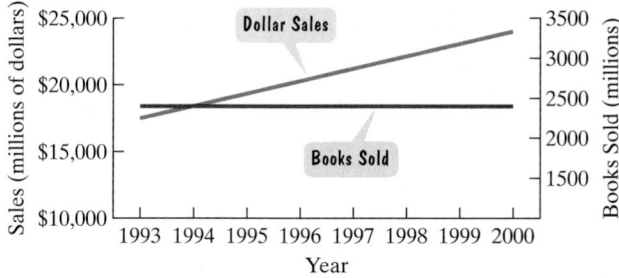

Source: Book Industry Study Group

31. Find the slope of the line segment representing books sold. Describe what this slope represents.

32. Find the slope of the line segment representing dollar sales. (First, you will need to estimate the coordinates of two points that lie on the line segment.) Describe what this slope represents.

Use slope to solve Exercises 33–34.

33. How much does it cost per mile to own and operate a full-size pickup truck?

Cost to Own and Operate a Full-Size Truck

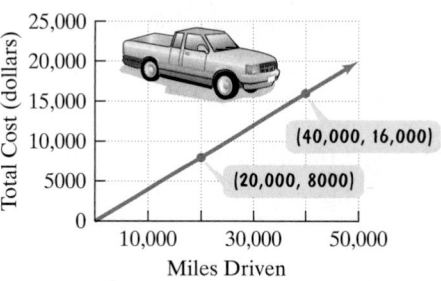

Source: Federal Highway Administration

34. How much does it cost per mile to own and operate a compact car?

Cost to Own and Operate a Compact Car

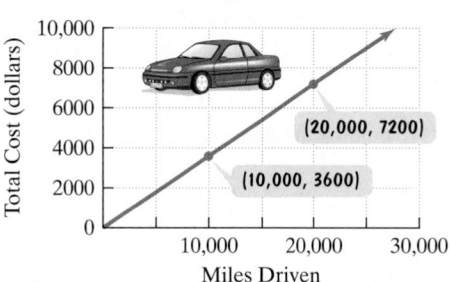

Source: Federal Highway Administration

The pitch of a roof refers to the absolute value of its slope. In Exercises 35–36, find the pitch of each roof shown.

35.

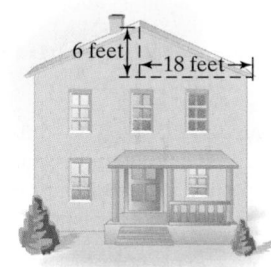

36.

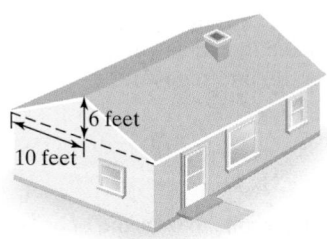

6 feet

10 feet

The grade of a road or ramp refers to its slope expressed as a percent. Use this information to solve Exercises 37–38.

37. Construction laws are very specific when it comes to access ramps for the disabled. Every vertical rise of 1 foot requires a horizontal run of 12 feet. What is the grade of such a ramp? Round to the nearest tenth of a percent.

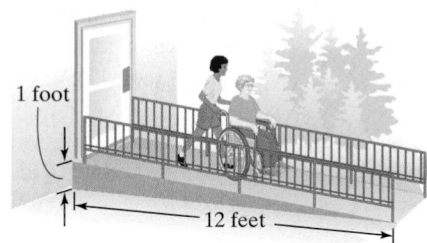

1 foot

12 feet

38. A college campus goes beyond the standards described in Exercise 37. All wheelchair ramps on campus are designed so that every vertical rise of 1 foot is accompanied by a horizontal run of 14 feet. What is the grade of such a ramp? Round to the nearest tenth of a percent.

Writing in Mathematics

39. What is the slope of a line?

40. Describe how to calculate the slope of a line passing through two points.

41. What does it mean if the slope of a line is zero?

42. What does it mean if the slope of a line is undefined?

43. If two lines are parallel, describe the relationship between their slopes.

44. Look back at Figure 3.17 on page 208. Do you think that the line through the points corresponding to the year 2001 and the year 2004 will model online spending per online household in the year 2040? Explain your answer.

Critical Thinking Exercises

45. Which one of the following is true?
   a. Slope is run divided by rise.
   b. The line through $(2, 2)$ and the origin has slope 1.
   c. A line with slope 3 can be parallel to a line with slope -3.
   d. The line through $(3, 1)$ and $(3, -5)$ has zero slope.

In Exercises 46–47, use the figure shown to make the indicated list.

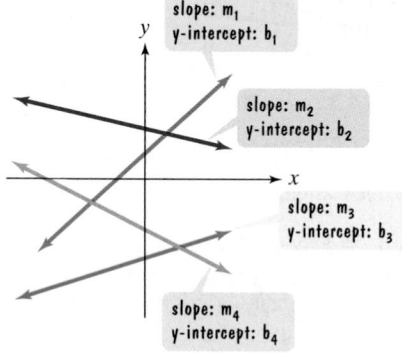

slope: m_1
y-intercept: b_1

slope: m_2
y-intercept: b_2

slope: m_3
y-intercept: b_3

slope: m_4
y-intercept: b_4

46. List the slopes $m_1, m_2, m_3,$ and m_4 in order of decreasing size.

47. List the y-intercepts $b_1, b_2, b_3,$ and b_4 in order of decreasing size.

Technology Exercises

Use a graphing utility to graph each equation in Exercises 48–51. Then use the TRACE *feature to trace along the line and find the coordinates of two points. Use these points to compute the line's slope.*

48. $y = 2x + 4$

49. $y = -3x + 6$

50. $y = -\frac{1}{2}x - 5$

51. $y = \frac{3}{4}x - 2$

52. In Exercises 48–51, compare the slope that you found with the line's equation. What relationship do you observe between the line's slope and one of the constants in the equation?

Review Exercises

53. A 36-inch board is cut into two pieces. One piece is twice as long as the other. How long are the pieces? (Section 2.5, Example 3)

54. Simplify: $-10 + 16 \div 2(-4)$. (Section 1.8, Example 4)

55. Solve and graph the solution set on a number line: $2x - 3 \le 5$. (Section 2.7, Example 5)

▶ SECTION 3.4   *The Slope-Intercept Form of the Equation of a Line*

Objectives

1 Find a line's slope and *y*-intercept from its equation.

2 Graph lines in slope-intercept form.

3 Use slope and *y*-intercept to graph $Ax + By = C$.

4 Interpret slope and *y*-intercept in linear mathematical models.

SSM
PH Tutor CD- Video
Center ROM

Can the same form of an equation model the cost of a four-year college, the percentage of runners injured in the Boston Marathon, the rising cost of prescription drugs, and the number of multiple births in the United States? Yes. Understanding this equation will help you to interpret data in everyday situations and perhaps even view these situations that appear to be different in a new, and unified, way.

1 Find a line's slope and *y*-intercept from its equation.

The Slope-Intercept Form of the Equation of a Line

Let's begin with an example that shows how easy it is to find a line's slope and *y*-intercept from its equation.

Figure 3.23 shows the graph of $y = 2x + 4$. Verify that the *x*-intercept is -2 by setting *y* equal to 0 and solving for *x*. Similarly, verify that the *y*-intercept is 4 by setting *x* equal to 0 and solving for *y*.

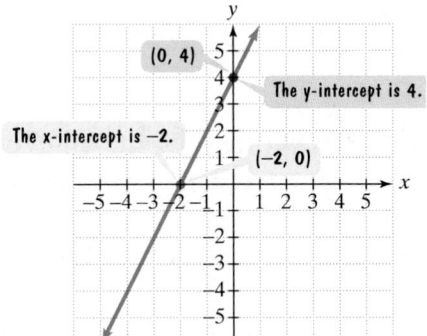

Figure 3.23 The graph of $y = 2x + 4$

Now that we have two points on the line, we can calculate the slope of the graph of $y = 2x + 4$.

$$\text{Slope} = \frac{\text{Change in } y}{\text{Change in } x}$$

$$= \frac{4 - 0}{0 - (-2)} = \frac{4}{2} = 2$$

We see that the slope of the line is 2, the same as the coefficient of x in the equation $y = 2x + 4$. The y-intercept is 4, the same as the constant in the equation $y = 2x + 4$.

$$y = 2x + 4$$

The slope is 2.   The y-intercept is 4.

It is not merely a coincidence that the x-coefficient is the line's slope and the constant term is the y-intercept. Let's find the equation of any nonvertical line with slope m and y-intercept b. Because the y-intercept is b, the point $(0, b)$ lies on the line. Now, let (x, y) represent any other point on the line, shown in Figure 3.24. Keep in mind that the point (x, y) is arbitrary and is not in one fixed position. By contrast, the point $(0, b)$ is fixed.

Regardless of where the point (x, y) is located, the steepness of the line in Figure 3.24 remains the same. Thus, the ratio for slope stays a constant m. This means that for all points along the line,

$$m = \frac{\text{Change in } y}{\text{Change in } x} = \frac{y - b}{x - 0} = \frac{y - b}{x}.$$

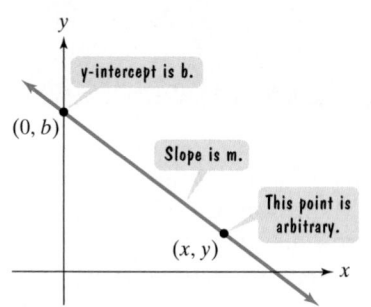

Figure 3.24 A line with slope m and y-intercept b

We can clear the fraction by multiplying both sides by x, the least common denominator.

$$m = \frac{y - b}{x} \qquad \text{This is the slope of the line in Figure 4.24.}$$

$$mx = \frac{y - b}{x} \cdot x \qquad \text{Multiply both sides by x.}$$

$$mx = y - b \qquad \text{Simplify: } \frac{y - b}{\cancel{x}} \cdot \cancel{x} = y - b.$$

$$mx + b = y - b + b \qquad \text{Add b to both sides and solve for y.}$$

$$mx + b = y \qquad \text{Simplify.}$$

Now, if we reverse the two sides, we obtain the **slope-intercept form** of the equation of a line.

Slope-Intercept Form of the Equation of a Line

The **slope-intercept equation** of a nonvertical line with slope m and y-intercept b is

$$y = mx + b.$$

Thus, if a line's equation is written with y isolated on one side, the x-coefficient is the line's slope and the constant term is the y-intercept.

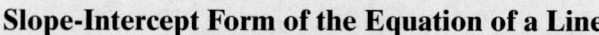

EXAMPLE 1   Finding a Line's Slope and y-Intercept from Its Equation

Find the slope and the y-intercept of the line with the given equation:

a. $y = 2x - 4$ **b.** $y = \frac{1}{2}x + 2$ **c.** $5x + y = 4$.

Solution

a. We write $y = 2x - 4$ as $y = 2x + (-4)$. The slope is the x-coefficient and the y-intercept is the constant term.

$$y = 2x + (-4)$$

The slope is 2.   The y-intercept is -4.

b. The equation $y = \frac{1}{2}x + 2$ is in the form $y = mx + b$. We can find the slope, m, by identifying the coefficient of x. We can find the y-intercept, b, by identifying the constant term.

$$y = \frac{1}{2}x + 2$$

The slope is $\frac{1}{2}$.   The y-intercept is 2.

c. The equation $5x + y = 4$ is not in the form $y = mx + b$. We can obtain this form by isolating y on one side. We isolate y on the left side by subtracting $5x$ from both sides.

$$5x + y = 4 \qquad \text{This is the given equation.}$$
$$5x - 5x + y = -5x + 4 \qquad \text{Subtract 5x from both sides.}$$
$$y = -5x + 4 \qquad \text{Simplify.}$$

Now, the equation is in the form $y = mx + b$. The slope is the coefficient of x and the y-intercept is the constant term.

$$y = -5x + 4$$

The slope is -5.   The y-intercept is 4.

✔ **CHECK POINT 1**   Find the slope and the y-intercept of the line with the given equation:

a. $y = 5x - 3$

b. $y = \frac{2}{3}x + 4$

c. $7x + y = 6$.

---

2 Graph lines in slope-intercept form.

Graphing $y = mx + b$ by Using the Slope and y-Intercept

If a line's equation is written with y isolated on one side, we can use the y-intercept and the slope to obtain its graph.

Graphing $y = mx + b$ by Using the Slope and y-Intercept

1. Plot the point containing the y-intercept on the y-axis. This is the point $(0, b)$.
2. Obtain a second point using the slope, m. Write m as a fraction, and use rise over run, starting at the point containing the y-intercept, to plot this point.
3. Use a straightedge to draw a line through the two points. Draw arrowheads at the ends of the line to show that the line continues indefinitely in both directions.

EXAMPLE 2   Graphing by Using the Slope and *y*-Intercept

Graph the line whose equation is $y = 4x - 3$.

Solution   We write $y = 4x - 3$ in the form $y = mx + b$.

$$y = 4x + (-3)$$

The slope is 4.   The *y*-intercept is −3.

Now that we have identified the slope and the *y*-intercept, we use the three steps in the box to graph the equation.

Step 1 Plot the point containing the *y*-intercept on the *y*-axis. The *y*-intercept is −3. We plot the point $(0, -3)$, shown in Figure 3.25(a).

Step 2 Obtain a second point using the slope, *m*. Write *m* as a fraction, and use rise over run, starting at the point containing the *y*-intercept, to plot this point. The slope, 4, can be written as a fraction:

$$m = \frac{4}{1} = \frac{\text{Rise}}{\text{Run}}.$$

We plot the second point on the line by starting at $(0, -3)$, the first point. Based on the slope, we move 4 units *up* (the rise) and 1 unit to the *right* (the run). This puts us at a second point on the line, $(1, 1)$, shown in Figure 3.25(b).

Step 3 Use a straightedge to draw a line through the two points. The graph of $y = 4x - 3$ is shown in Figure 3.25(c).

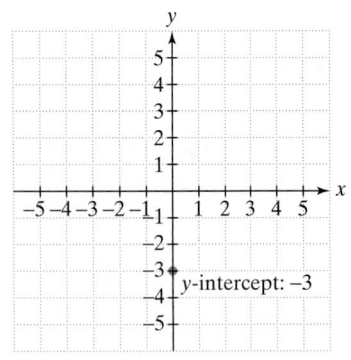

(a) The *y*-intercept is −3, so $(0, -3)$ is a point on the line.

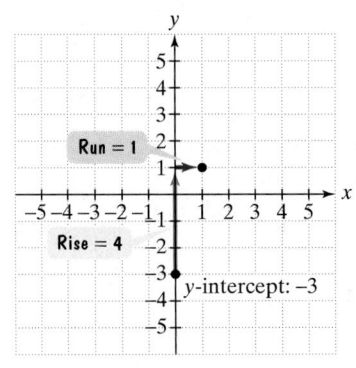

(b) The slope is 4.

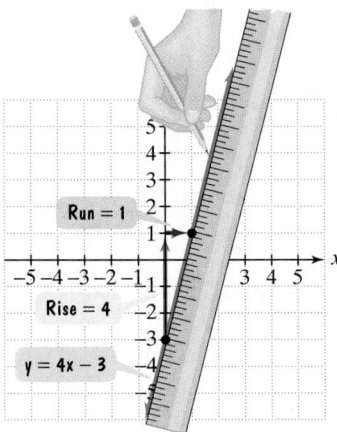

(c) The graph of $y = 4x - 3$

Figure 3.25 Graphing $y = 4x - 3$ using the *y*-intercept and slope

✔ **CHECK POINT 2**   Graph the line whose equation is $y = 3x - 2$.

EXAMPLE 3   Graphing by Using the Slope and *y*-Intercept

Graph the line whose equation is $y = \frac{2}{3}x + 2$.

Solution   The equation of the line is in the form $y = mx + b$. We can find the slope, m, by identifying the coefficient of x. We can find the y-intercept, b, by identifying the constant term.

$$y = \frac{2}{3}x + 2$$

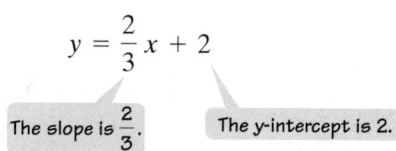

The slope is $\frac{2}{3}$.   The y-intercept is 2.

Now that we have identified the slope and the y-intercept, we use the three-step procedure to graph the equation.

Step 1 Plot the point containing the y-intercept on the y-axis. The y-intercept is 2. We plot $(0, 2)$, shown in Figure 3.26.

Step 2 Obtain a second point using the slope, m. Write m as a fraction, and use rise over run, starting at the point containing the y-intercept, to plot this point. The slope, $\frac{2}{3}$, is already written as a fraction.

$$m = \frac{2}{3} = \frac{\text{Rise}}{\text{Run}}$$

We plot the second point on the line by starting at $(0, 2)$, the first point. Based on the slope, we move 2 units *up* (the rise) and 3 units to the *right* (the run). This puts us at a second point on the line, $(3, 4)$, shown in Figure 3.26.

Step 3 Use a straightedge to draw a line through the two points. The graph of $y = \frac{2}{3}x + 2$ is shown in Figure 3.26. ∎

✔ **CHECK POINT 3**   Graph the line whose equation is $y = \dfrac{3}{5}x + 1$.

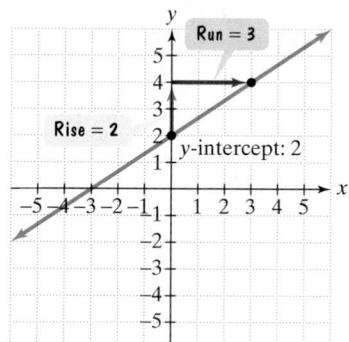

Figure 3.26 The graph of $y = \frac{2}{3}x + 2$

3 Use slope and y-intercept to graph $Ax + By = C$.

Graphing $Ax + By = C$ by Using the Slope and y-Intercept

Earlier in this chapter, we considered linear equations of the form $Ax + By = C$. We used x- and y-intercepts, as well as checkpoints, to graph these equations. It is also possible to obtain the graphs by using the slope and y-intercept. To do this, begin by solving $Ax + By = C$ for y. This will put the equation in slope-intercept form. Then use the three-step procedure to graph the equation. This is illustrated in Example 4.

EXAMPLE 4   Graphing by Using the Slope and y-Intercept

Graph the linear equation $2x + 5y = 0$ by using the slope and y-intercept.

Solution   We put the equation in slope-intercept form by solving for y.

$2x + 5y = 0$	This is the given equation.
$2x - 2x + 5y = -2x + 0$	Subtract 2x from both sides.
$5y = -2x + 0$	Simplify.
$\dfrac{5y}{5} = \dfrac{-2x + 0}{5}$	Divide both sides by 5.
$y = \dfrac{-2x}{5} + \dfrac{0}{5}$	Divide each term in the numerator by 5.
$y = -\dfrac{2}{5}x + 0$	Simplify.

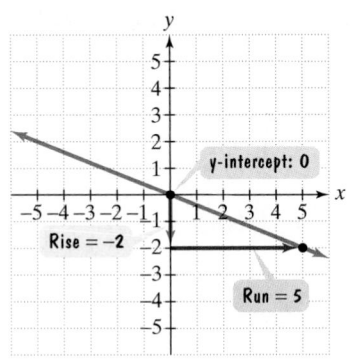

Figure 3.27 The graph of $2x + 5y = 0$, or $y = -\frac{2}{5}x + 0$

Now that the equation is in slope-intercept form, we can use the slope and y-intercept to obtain its graph. Examine the slope-intercept form:

$$y = -\frac{2}{5}x + 0.$$

slope: $-\frac{2}{5}$ y-intercept: 0

Note that the slope is $-\frac{2}{5}$ and the y-intercept is 0. Use the y-intercept to plot $(0, 0)$ on the y-axis. Then locate a second point by using the slope.

$$m = -\frac{2}{5} = \frac{-2}{5} = \frac{\text{Rise}}{\text{Run}}$$

Because the rise is -2 and the run is 5, move *down* 2 units and to the *right* 5 units, starting at the point $(0, 0)$. This puts us at a second point on the line, $(5, -2)$. The graph of $2x + 5y = 0$ is the line drawn through these points, shown in Figure 3.27.   ∎

Discover for Yourself

Obtain a second point in Example 4 by writing the slope as follows:

$$m = \frac{2}{-5} = \frac{\text{Rise}}{\text{Run}}.$$

$-\frac{2}{5}$ can be expressed as $\frac{-2}{5}$ or $\frac{2}{-5}$.

Obtain a second point in Figure 3.27 by moving *up* 2 units and to the *left* 5 units, starting at $(0, 0)$. What do you observe once you graph the line?

✔ **CHECK POINT 4**   Graph the linear equation $3x + 4y = 0$ by using the slope and y-intercept.

Modeling with the Slope-Intercept Form of the Equation of a Line

4 Interpret slope and y-intercept in linear mathematical models.

If an equation in slope-intercept form models some physical situation, then the slope and y-intercept have physical interpretations. For the equation $y = mx + b$, the y-intercept, b, tells us what is happening to y when x is 0. If x represents time, the y-intercept describes the value of y at the beginning, or when time equals 0. The slope represents the rate of change in y per unit change in x.

These ideas are illustrated in Table 3.2.

Table 3.2   Interpreting Slope and y-Intercept

Linear Equation	What the Equation Models	Interpretation
$y = 166x + 1781$ Slope y-intercept	The approximate cost, y, in dollars, (tuition and fees per year) of a four-year public college x years after 1990	1781 is the y-intercept. At the beginning (in 1990), the cost of college was $1781 per year. 166 is the slope. The cost of college is increasing $166 per year.
$y = -0.22x + 9.6$ Slope y-intercept	The Pacific salmon population, y, in millions, x years after 1960	9.6 is the y-intercept. At the beginning (in 1960), the Pacific salmon population was 9.6 million. -0.22 is the slope. The Pacific salmon population is decreasing by 0.22 million per year.

EXERCISE SET 3.4

Practice Exercises

In Exercises 1–12, find the slope and the y-intercept of the line with the given equation.

1. $y = 3x + 2$

2. $y = 9x + 4$

3. $y = 3x - 5$

4. $y = 4x - 2$

5. $y = -\frac{1}{2}x + 5$

6. $y = -\frac{3}{4}x + 6$

7. $y = 7x$

8. $y = 10x$

9. $y = 10$

10. $y = 7$

11. $y = 4 - x$

12. $y = 5 - x$

In Exercises 13–26, begin by solving the linear equation for y. This will put the equation in slope-intercept form. Then find the slope and the y-intercept of the line with this equation.

13. $-5x + y = 7$

14. $-9x + y = 5$

15. $x + y = 6$

16. $x + y = 8$

17. $6x + y = 0$

18. $8x + y = 0$

19. $3y = 6x$

20. $3y = -9x$

21. $2x + 7y = 0$

22. $2x + 9y = 0$

23. $3x + 2y = 3$

24. $4x + 3y = 4$

25. $3x - 4y = 12$

26. $5x - 2y = 10$

In Exercises 27–38, graph each linear equation using the slope and y-intercept.

27. $y = 2x + 3$

28. $y = 2x + 1$

29. $y = -2x + 4$

30. $y = -2x + 3$

31. $y = \frac{1}{2}x + 3$

32. $y = \frac{1}{2}x + 2$

33. $y = \frac{2}{3}x - 4$

34. $y = \frac{3}{4}x - 5$

35. $y = -\frac{3}{4}x + 4$

36. $y = -\frac{2}{3}x + 5$

37. $y = -\frac{5}{3}x$

38. $y = -\frac{4}{3}x$

In Exercises 39–46:
a. *Put the equation in slope-intercept form by solving for y.*
b. *Identify the slope and the y-intercept.*
c. *Use the slope and y-intercept to graph the equation.*

39. $3x + y = 0$

40. $2x + y = 0$

41. $3y = 4x$

42. $4y = 5x$

43. $2x + y = 3$

44. $3x + y = 4$

45. $7x + 2y = 14$

46. $5x + 3y = 15$

In Exercises 47–50, graph both linear equations in the same rectangular coordinate system. If the lines are parallel, explain why.

47. $y = 3x + 1$
$y = 3x - 3$

48. $y = 2x + 4$
$y = 2x - 3$

49. $y = -3x + 2$
$y = 3x + 2$

50. $y = -2x + 1$
$y = 2x + 1$

Application Exercises

51. The formula $y = -0.4x + 38$ models the percentage of U.S. men, y, smoking cigarettes x years after 1980.
  a. Use the formula to find the percentage of men who smoked in 1980, 1981, 1982, 1983, 1990, and 2000.

  b. What is the slope of this model? What does it represent in this situation?

  c. What is the y-intercept of this model? What does it represent in this situation?

52. A salesperson receives a fixed salary plus a percentage of all sales. The linear equation $y = 0.05x + 500$ describes the weekly salary, y, in dollars, in terms of weekly sales, x, also in dollars.
  a. Use the formula to find the weekly salary for sales of $0, $1, $2, $3, $4, $5, $100, and $1000.

  b. What is the slope of this equation? What does it represent in this situation?

  c. What is the y-intercept of this equation? What does it represent in this situation?

53. Horrified at the cost the last time you needed a prescription drug? The graph shows that the cost of the average retail prescription has been rising steadily since 1991.

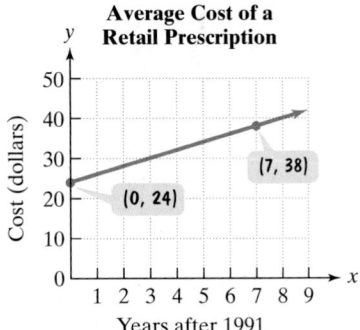

Average Cost of a Retail Prescription

Cost (dollars) / Years after 1991

(0, 24) (7, 38)

Source: Newsweek

a. According to the graph, what is the y-intercept? Describe what this represents in this situation.

b. Use the coordinates of the two points shown to compute the slope. What does this mean about the cost of the average retail prescription?

c. Use the y-intercept from part (a) and the slope from part (b) to write an equation that models the cost of the average retail prescription, y, x years after 1991. Write your model in the form $y = mx + b$.

d. Use your model from part (c) to predict the cost of the average retail prescription in 2005.

54. For 61 years, Social Security has been a huge success. It is the primary source of income for 66% of Americans over 65 and the only thing that keeps 42% of the elderly from poverty. However, the number of workers per Social Security beneficiary has been declining steadily since 1950.

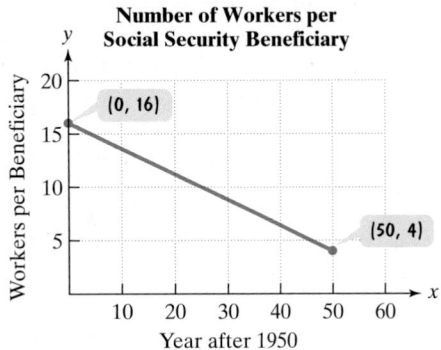

Number of Workers per Social Security Beneficiary

Workers per Beneficiary / Year after 1950

(0, 16) (50, 4)

Source: Social Security Administration

a. According to the graph, what is the y-intercept? Describe what this represents in this situation.

b. Use the coordinates of the two points shown to compute the slope. What does this mean about the number of workers per beneficiary?

c. Use the y-intercept from part (a) and the slope from part (b) to write an equation that models the number of workers per beneficiary, y, x years after 1950. Write your model in $y = mx + b$ form.

d. Use your model from part (c) to predict the number of workers per beneficiary in 2010. For every 8 workers, how many beneficiaries will there be?

Writing in Mathematics

55. Describe how to find the slope and the y-intercept of a line whose equation is given.

56. Describe how to graph a line using the slope and y-intercept. Provide an original example with your description.

57. A formula in the form $y = mx + b$ models the cost, y, of a four-year college x years after 2003. Would you expect m to be positive, negative, or zero? Explain your answer.

Critical Thinking Exercises

58. Which one of the following is true?
a. The equation $y = mx + b$ shows that no line can have a y-intercept that is numerically equal to its slope.
b. Every line in the rectangular coordinate system has an equation that can be expressed in slope-intercept form.
c. The line $3x + 2y = 5$ has slope $-\frac{3}{2}$.
d. The line $2y = 3x + 7$ has a y-intercept of 7.

59. The number of multiple births in the United States (twins, triplets, etc.) per 1000 live births can be modeled by $y = 0.463x + 18.888$, where x represents the number of years since 1980 and y represents multiple births per 1000 live births. Explain why the equation of this line cannot be parallel to the line representing the number of births in the United States since 1980.

60. The following graph indicates that lower fertility rates (the number of births per woman) are related to the percentage of the population using contraceptives. A line that best fits the data is shown. Estimate the y-intercept and the slope of this line. Then write the line's slope-intercept equation. Use the equation to find the number of births per woman if 90% of married women of child-bearing age used contraceptives.

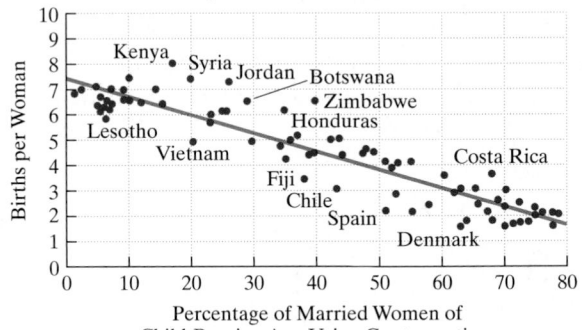

Contraceptive Prevalence and Average Number of Births per Woman, Selected Countries

Births per Woman / Percentage of Married Women of Child-Bearing Age Using Contraceptives

Kenya, Syria, Jordan, Botswana, Zimbabwe, Honduras, Lesotho, Vietnam, Fiji, Chile, Spain, Costa Rica, Denmark

Source: Population Reference Bureau

Review Exercises

61. Solve: $\dfrac{x}{2} + 7 = 13 - \dfrac{x}{4}$. (Section 2.3, Example 4)

62. Simplify: $3(12 \div 2^2 - 3)^2$. (Section 1.8, Example 6)

63. 14 is 25% of what? (Section 2.4, Example 8)

▶ **SECTION 3.5**   *The Point-Slope Form of the Equation of a Line*

Objectives

1 Use the point-slope form to write equations of a line.

2 Find slopes and equations of parallel and perpendicular lines.

3 Write linear equations that model data and make predictions.

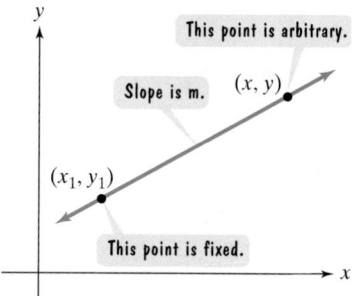

SSM
PH Tutor   CD-   Video
Center   ROM

Surprised by the number of people smoking cigarettes in movies and television shows made in the 1940s and 1950s? At that time, there was no awareness of the relationship between tobacco use and numerous diseases. Cigarette smoking was seen as a healthy way to relax and help digest a hearty meal. Then, in 1964, a linear equation changed everything. To understand the mathematics behind this turning point in public health, we explore another form of a line's equation.

Point-Slope Form

We can use the slope of a line to obtain another useful form of the line's equation. Consider a nonvertical line that has a slope of m and contains the point (x_1, y_1). Now, let (x, y) represent any other point on the line, shown in Figure 3.28. Keep in mind that the point (x, y) is arbitrary and is not in one fixed position. By contrast, the point (x_1, y_1) is fixed.

Regardless of where the point (x, y) is located, the steepness of the line in Figure 3.28 remains the same. Thus, the ratio for slope stays a constant m. This means that for all points along the line,

$$m = \frac{\text{Change in } y}{\text{Change in } x} = \frac{y - y_1}{x - x_1}.$$

We can clear the fraction by multiplying both sides by $x - x_1$, the least common denominator.

$$m = \frac{y - y_1}{x - x_1} \qquad \text{\small This is the slope of the line in Figure 3.28.}$$

$$m(x - x_1) = \frac{y - y_1}{x - x_1} \cdot (x - x_1) \qquad \text{\small Multiply both sides by } x - x_1.$$

$$m(x - x_1) = y - y_1 \qquad \text{\small Simplify: } \frac{y - y_1}{x - x_1} \cdot x - x_1 = y - y_1$$

Figure 3.28 A line passing through (x_1, y_1) with slope m

This point is arbitrary.
Slope is m. (x, y)
(x_1, y_1)
This point is fixed.

Now, if we reverse the two sides, we obtain the **point-slope form** of the equation of a line.

Point-Slope Form of the Equation of a Line

The **point-slope equation** of a nonvertical line with slope m that passes through the point (x_1, y_1) is

$$y - y_1 = m(x - x_1).$$

For example, the point-slope equation of the line passing through $(1, 1)$ with a slope of $2(m = 2)$ is

$$y - 1 = 2(x - 1).$$

1 Use the point-slope form to write equations of a line.

Using the Point-Slope Form to Write a Line's Equation

If we know the slope of a line and a point through which the line passes, the point-slope form is the equation that we should use. Once we have obtained this equation, it is customary to solve for y and write the equation in slope-intercept form. Examples 1 and 2 illustrate these ideas.

EXAMPLE 1   Writing the Point-Slope Form and the Slope-Intercept Form

Write the point-slope form and the slope-intercept form of the equation of the line with slope 4 that passes through the point $(-1, 3)$.

Solution   We begin with the point-slope equation of a line with $m = 4$, $x_1 = -1$, and $y_1 = 3$.

$$y - y_1 = m(x - x_1) \qquad \text{This is the point-slope form of the equation.}$$

$$y - 3 = 4[x - (-1)] \qquad \text{Substitute the given values.}$$

$$y - 3 = 4(x + 1) \qquad \text{We now have the point-slope form of the equation of the given line.}$$

Now we solve this equation for y and write an equivalent equation in slope-intercept form $(y = mx + b)$.

$$y - 3 = 4(x + 1) \qquad \text{This is the point-slope equation.}$$

$$y - 3 = 4x + 4 \qquad \text{Use the distributive property.}$$

$$y = 4x + 7 \qquad \text{Add 3 to both sides.}$$

The slope-intercept form of the line's equation is $y = 4x + 7$.   ■

✔ **CHECK POINT 1**   Write the point-slope form and the slope-intercept form of the equation of the line with slope 6 that passes through the point $(2, -5)$.

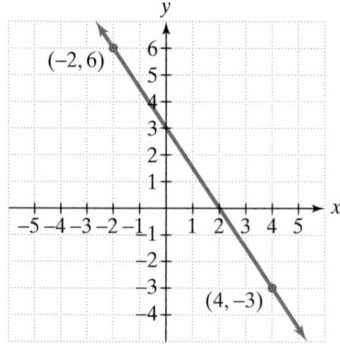

Figure 3.29

EXAMPLE 2 Writing the Point-Slope Form and the Slope-Intercept Form

A line passes through the points $(4, -3)$ and $(-2, 6)$. (See Figure 3.29.) Find the equation of the line:

a. in point-slope form. **b.** in slope-intercept form.

Solution

a. To use the point-slope form, we need to find the slope. The slope is the change in the y-coordinates divided by the corresponding change in the x-coordinates.

$$m = \frac{6 - (-3)}{-2 - 4} = \frac{9}{-6} = -\frac{3}{2} \qquad \text{This is the definition of slope using } (4, -3) \text{ and } (-2, 6).$$

We can take either point on the line to be (x_1, y_1). Let's use $(x_1, y_1) = (4, -3)$. Now, we are ready to write the point-slope equation.

$$y - y_1 = m(x - x_1) \qquad \text{This is the point-slope form of the equation.}$$

$$y - (-3) = -\frac{3}{2}(x - 4) \qquad \text{Substitute: } (x_1, y_1) = (4, -3) \text{ and } m = -\frac{3}{2}.$$

$$y + 3 = -\frac{3}{2}(x - 4) \qquad \text{Simplify.}$$

This equation is the point-slope form of the equation of the line shown in Figure 3.29.

b. Now, we solve this equation for y and write an equivalent equation in slope-intercept form ($y = mx + b$).

$$y + 3 = -\frac{3}{2}(x - 4) \qquad \text{This is the point-slope equation.}$$

$$y + 3 = -\frac{3}{2}x + 6 \qquad \text{Use the distributive property.}$$

$$y = -\frac{3}{2}x + 3 \qquad \text{Subtract 3 from both sides.}$$

This equation is the slope-intercept form of the equation of the line shown in Figure 3.29. ∎

Discover for Yourself

If you are given two points on a line, you can use either point for (x_1, y_1) when you write its point-slope equation. Rework Example 2 using $(-2, 6)$ for (x_1, y_1). Once you solve for y, you should obtain the same slope-intercept equation as the one shown in the last line of the solution to Example 2.

✔ **CHECK POINT 2** A line passes through the points $(-2, -1)$ and $(-1, -6)$. Find the equation of the line:

a. in point-slope form. **b.** in slope-intercept form.

In Examples 1 and 2, we eventually write a line's equation in slope-intercept form? But where do we start our work?

Starting with $y = mx + b$	Starting with $y - y_1 = m(x - x_1)$
Begin with the slope-intercept form if you know:	Begin with the point-slope form if you know:
1. The slope of the line and the y-intercept.	**1.** The slope of the line and a point on the line or **2.** Two points on the line.

2 Find slopes and equations of parallel and perpendicular lines.

Parallel and Perpendicular Lines

The next example uses the fact parallel lines have the same slope.

EXAMPLE 3   Writing Equations of a Line Parallel to a Given Line

Write an equation of the line passing through $(-3, 2)$ and parallel to the line whose equation is $y = 2x + 1$. Express the equation in point-slope form and slope-intercept form.

Solution   The situation is illustrated in Figure 3.30. We are looking for the equation of the line shown on the left. How do we obtain this equation? Notice that the line passes through the point $(-3, 2)$. Using the point-slope form of the line's equation, we have $x_1 = -3$ and $y_1 = 2$.

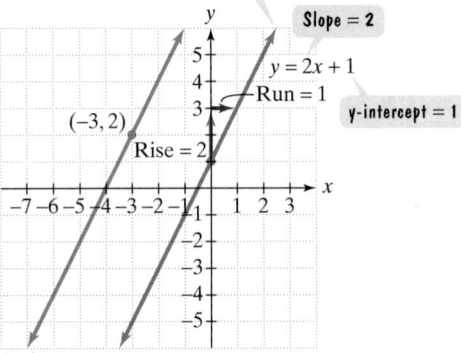

The equation of this line is given: $y = 2x + 1$.

Slope = 2

$y = 2x + 1$

Run = 1

y-intercept = 1

$(-3, 2)$

Rise = 2

We must write the equation of this line.

Figure 3.30

$$y - y_1 = m(x - x_1)$$

$y_1 = 2$ $x_1 = -3$

Now the only thing missing from the equation is m, the slope of the line on the left. Do we know anything about the slope of either line in Figure 3.30? The answer is yes; we know the slope of the line on the right, whose equation is given.

$$y = 2x + 1$$

The slope of the line on the right in Figure 3.30 is 2.

Parallel lines have the same slope. Because the slope of the line with the given equation is 2, $m = 2$ for the line whose equation we must write.

$$y - y_1 = m(x - x_1)$$

$y_1 = 2$ $m = 2$ $x_1 = -3$

The point-slope form of the line's equation is

$$y - 2 = 2[x - (-3)] \text{ or}$$
$$y - 2 = 2(x + 3).$$

Solving for y, we obtain the slope-intercept form of the equation.

$$y - 2 = 2x + 6 \quad \text{Apply the distributive property.}$$

$$y = 2x + 8 \quad \text{Add 2 to both sides. This is the slope-intercept form,}$$
$$y = mx + b, \text{ of the equation.} \quad\blacksquare$$

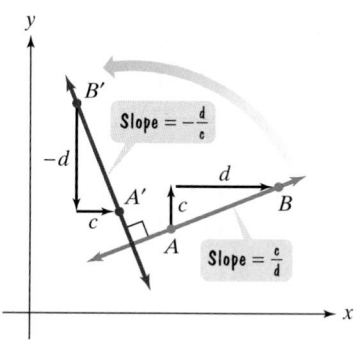

Figure 3.31 Slopes of perpendicular lines

✔ **CHECK POINT 3** Write an equation of the line passing through $(-2, 5)$ and parallel to the line whose equation is $y = 3x + 1$. Express the equation in point-slope form and slope-intercept form.

Two lines that intersect at a right angle (90°) are said to be **perpendicular**, shown in Figure 3.31. The relationship between the slopes of perpendicular lines is not as obvious as the relationship between parallel lines. Figure 3.31 shows line AB, with a slope of $\dfrac{c}{d}$. Rotate line AB through 90° to the left to obtain line $A'B'$, perpendicular to line AB. The figure indicates that the rise and the run of the new line are reversed from the original line, but the rise is now negative. This means that the slope of the new line is $-\dfrac{d}{c}$. Notice that the product of the slopes of the two perpendicular lines is -1:

$$\left(\frac{c}{d}\right)\left(-\frac{d}{c}\right) = -1.$$

This relationship holds for all perpendicular lines and is summarized in the following box.

Slope and Perpendicular Lines

1. If two nonvertical lines are perpendicular, then the product of their slopes is -1.
2. If the product of the slopes of two lines is -1, then the lines are perpendicular.
3. A horizontal line having zero slope is perpendicular to a vertical line having undefined slope.

An equivalent way of stating this relationship is to say that one line is perpendicular to another line if its slope is the *negative reciprocal* of the slope of the other. For example, if a line has slope 5, any line having slope $-\frac{1}{5}$ is perpendicular to it. Similarly, if a line has slope $-\frac{3}{4}$, any line having slope $\frac{4}{3}$ is perpendicular to it.

EXAMPLE 4   Finding the Slope of a Line Perpendicular to a Given Line

Find the slope of any line that is perpendicular to the line whose equation is $x + 4y - 8 = 0$.

Solution   We begin by writing the equation of the given line in slope-intercept form. Solve for y.

$$x + 4y - 8 = 0 \qquad \text{This is the given equation.}$$
$$4y = -x + 8 \qquad \text{To isolate the y-term, subtract x and add 8 on both sides.}$$
$$y = -\frac{1}{4}x + 2 \qquad \text{Divide both sides by 4.}$$

$\boxed{\text{Slope is } -\dfrac{1}{4}.}$

The given line has slope $-\frac{1}{4}$. Any line perpendicular to this line has a slope that is the negative reciprocal of $-\frac{1}{4}$. Thus, the slope of any perpendicular line is 4. ∎

✔ **CHECK POINT 4** Find the slope of any line that is perpendicular to the line whose equation is $x + 3y - 12 = 0$.

3 Write linear equations that model data and make predictions.

Applications

Linear equations are useful for modeling data that fall on or near a line. For example, Table 3.3 gives the population of the United States, in millions, in the indicated year. The data are displayed as a set of six points in Figure 3.32.

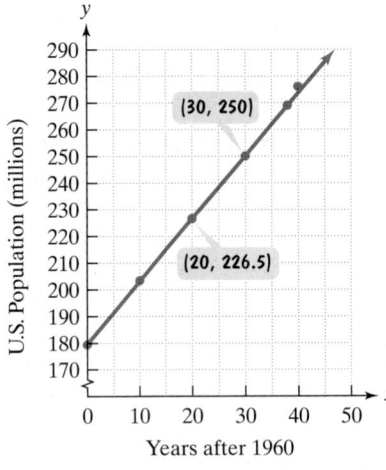

Figure 3.32

Table 3.3

Year	x (Years after 1960)	y (U.S. Population) (in millions)
1960	0	179.3
1970	10	203.3
1980	20	226.5
1990	30	250.0
1998	38	268.9
2000	40	276.0

A set of points representing data is called a **scatter plot**. Also shown in Figure 3.32 is a line that passes through or near the six points. By writing the equation of this line, we can obtain a model of the data and make predictions about the population of the United States in the future.

EXAMPLE 5 Modeling U.S. Population

Write the slope-intercept equation of the line shown in Figure 3.32. Use the equation to predict U.S. population in 2010.

Solution The line in Figure 3.32 passes through (20, 226.5) and (30, 250). We start by finding the slope.

$$m = \frac{\text{Change in } y}{\text{Change in } x} = \frac{250 - 226.5}{30 - 20} = \frac{23.5}{10} = 2.35$$

Now, we write the line's slope-intercept equation.

$$y - y_1 = m(x - x_1)$$ Begin with the point-slope form.

$$y - 250 = 2.35(x - 30)$$ Either ordered pair can be (x_1, y_1).
Let $(x_1, y_1) = (30, 250)$. From above, $m = 2.35$.

$$y - 250 = 2.35x - 70.5$$ Apply the distributive property on the right.

$$y = 2.35x + 179.5$$ Add 250 to both sides and solve for y.

A linear equation that models U.S. population, y, in millions, x years after 1960 is

$$y = 2.35x + 179.5.$$

Now, let's use this equation to predict U.S. population in 2010. Because 2010 is 50 years after 1960, substitute 50 for x and compute y.

$$y = 2.35(50) + 179.5 = 297$$

Our equation predicts that the population of the United States in the year 2010 will be 297 million. (The projected figure from the U.S. Census Bureau is 297.716 million.) ∎

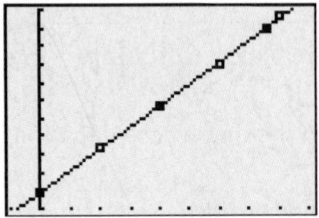

✔ **CHECK POINT 5** Use the data points (10, 203.3) and (20, 226.5) from Table 3.3 to write the slope-intercept equation that models U.S. population *x* years after 1960. Use the equation to predict U.S. population in 2020.

Why is the study of population important? Population growth drives poverty, environmental destruction, migration, and conflict. Recognizing the connections between population and critical issues enables us to work more effectively for positive change.

ENRICHMENT ESSAY

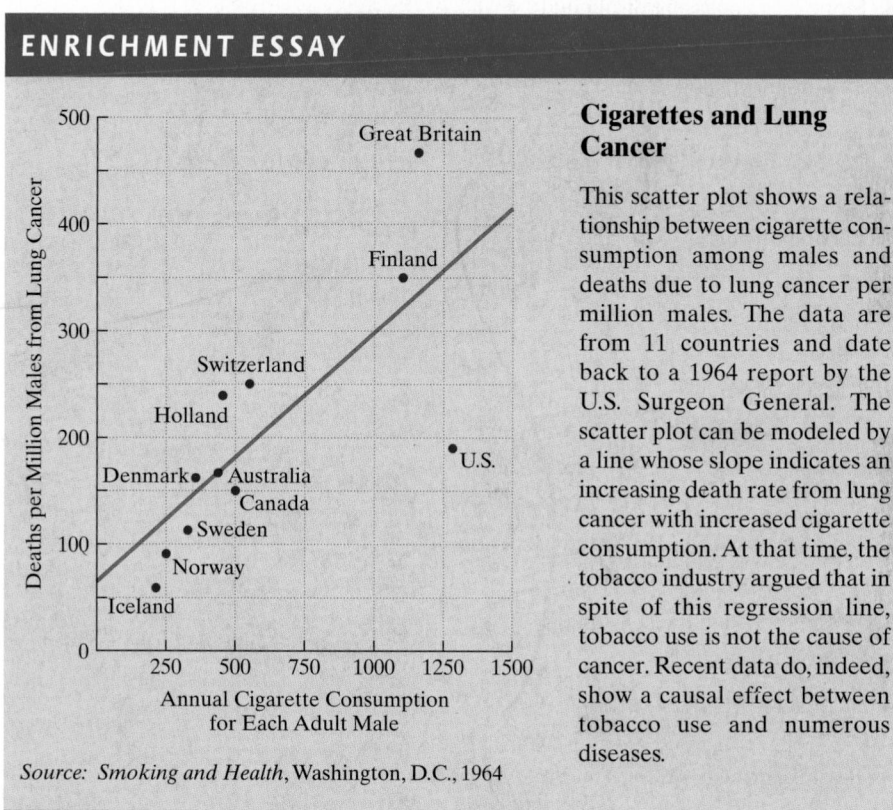

Cigarettes and Lung Cancer

This scatter plot shows a relationship between cigarette consumption among males and deaths due to lung cancer per million males. The data are from 11 countries and date back to a 1964 report by the U.S. Surgeon General. The scatter plot can be modeled by a line whose slope indicates an increasing death rate from lung cancer with increased cigarette consumption. At that time, the tobacco industry argued that in spite of this regression line, tobacco use is not the cause of cancer. Recent data do, indeed, show a causal effect between tobacco use and numerous diseases.

Source: Smoking and Health, Washington, D.C., 1964

EXERCISE SET 3.5

Practice Exercises

✔ *Write the point-slope form of the line satisfying each of the conditions in Exercises 1–28. Then use the point-slope form of the equation to write the slope-intercept form of the equation.*

1. Slope = 2, passing through (3, 5)

2. Slope = 4, passing through (1, 3)

3. Slope = 6, passing through (−2, 5)

4. Slope = 8, passing through (4, −1)

5. Slope = −3, passing through (−2, −3)

6. Slope = −5, passing through (−4, −2)

7. Slope = −4, passing through (−4, 0)

8. Slope = −2, passing through (0, −3)

9. Slope = −1, passing through $\left(-\frac{1}{2}, -2\right)$

10. Slope $= -1$, passing through $\left(-4, -\frac{1}{4}\right)$

11. Slope $= \frac{1}{2}$, passing through the origin

12. Slope $= \frac{1}{3}$, passing through the origin

13. Slope $= -\frac{2}{3}$, passing through $(6, -2)$

14. Slope $= -\frac{3}{5}$, passing through $(10, -4)$

15. Passing through $(1, 2)$ and $(5, 10)$

16. Passing through $(3, 5)$ and $(8, 15)$

17. Passing through $(-3, 0)$ and $(0, 3)$

18. Passing through $(-2, 0)$ and $(0, 2)$

19. Passing through $(-3, -1)$ and $(2, 4)$

20. Passing through $(-2, -4)$ and $(1, -1)$

21. Passing through $(-3, -2)$ and $(3, 6)$

22. Passing through $(-3, 6)$ and $(3, -2)$

23. Passing through $(-3, -1)$ and $(4, -1)$

24. Passing through $(-2, -5)$ and $(6, -5)$

25. Passing through $(2, 4)$ with x-intercept $= -2$

26. Passing through $(1, -3)$ with x-intercept $= -1$

27. x-intercept $= -\frac{1}{2}$ and y-intercept $= 4$

28. x-intercept $= 4$ and y-intercept $= -2$

In Exercises 29–44, the equation of a line is given. Find the slope of a line that is (a) parallel to the line with the given equation; and (b) perpendicular to the line with the given equation.

29. $y = 5x$

30. $y = 3x$

31. $y = -7x$

32. $y = -9x$

33. $y = \dfrac{1}{2}x + 3$

34. $y = \dfrac{1}{4}x - 5$

35. $y = -\dfrac{2}{5}x - 1$

36. $y = -\dfrac{3}{7}x - 2$

37. $4x + y = 7$

38. $8x + y = 11$

39. $2x + 4y - 8 = 0$

40. $3x + 2y - 6 = 0$

41. $2x - 3y - 5 = 0$

42. $3x - 4y + 7 = 0$

43. $x = 6$

44. $y = 9$

In Exercises 45–48, write an equation for line L in point-slope form and slope-intercept form.

45.

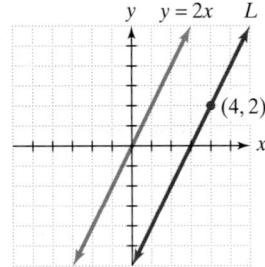

L is parallel to $y = 2x$.

46.

L is parallel to $y = -2x$.

47.

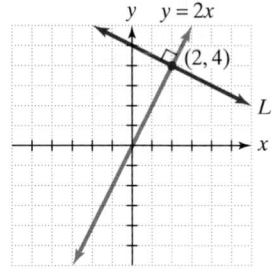

L is perpendicular to $y = 2x$.

48.

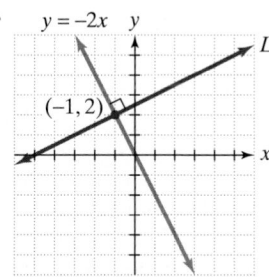

L is perpendicular to $y = -2x$.

In Exercises 49–56, use the given conditions to write an equation for each line in point-slope form and slope-intercept form.

49. Passing through $(-8, -10)$ and parallel to the line whose equation is $y = -4x + 3$

50. Passing through $(-2, -7)$ and parallel to the line whose equation is $y = -5x + 4$

51. Passing through $(2, -3)$ and perpendicular to the line whose equation is $y = \frac{1}{5}x + 6$

52. Passing through $(-4, 2)$ and perpendicular to the line whose equation is $y = \frac{1}{3}x + 7$

53. Passing through $(-2, 2)$ and parallel to the line whose equation is $2x - 3y - 7 = 0$

54. Passing through $(-1, 3)$ and parallel to the line whose equation is $3x - 2y - 5 = 0$

55. Passing through $(4, -7)$ and perpendicular to the line whose equation is $x - 2y - 3 = 0$

56. Passing through $(5, -9)$ and perpendicular to the line whose equation is $x + 7y - 12 = 0$

Application Exercises

57. We seem to be fed up with being lectured about our waistlines. The points in the graph show the average weight of American adults from 1990 through 2000. Also shown is a line that passes through or near the points.

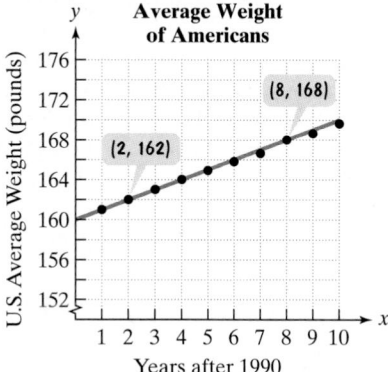

Source: Diabetes Care

a. Use the two points whose coordinates are shown by the voice balloons to find the point-slope equation of the line that models average weight of Americans, y, in pounds, x years after 1990.

b. Write the equation in part (a) in slope-intercept form.

c. Use the slope-intercept equation to predict the average weight of Americans in 2005.

58. Films may not be getting any better, but in this era of moviegoing, the number of screens available for new films and the classics has exploded. The points in the graph show the number of screens in the United States from 1995 through 2000. Also shown is a line that passes through or near the points.

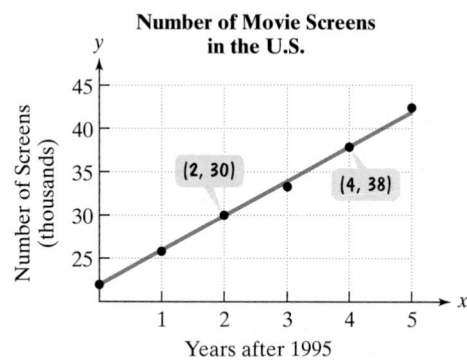

Source: Motion Picture Association of America

a. Use the two points whose coordinates are shown by the voice balloons to find the point-slope equation of the line that models the number of screens, y, in thousands, x years after 1995.

b. Write the equation in part (a) in slope-intercept form.

c. Use the slope-intercept equation to predict the number of screens, in thousands, in 2005.

59. Is there a relationship between education and prejudice? With increased education, does a person's level of prejudice tend to decrease? The scatter plot shows ten data points, each representing the number of years of school completed and the score on a test measuring prejudice for each subject. Higher scores on this 1-to-10 test indicate greater prejudice. Also shown is the regression line, the line that best fits the data. Use two points on this line to write both its point-slope and slope-intercept equations. Then use the slope-intercept equation to predict the score on the prejudice test for a person with seven years of education.

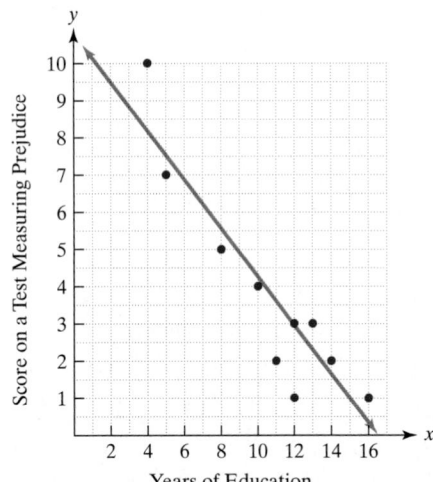

Years of Education

60. A business discovers a linear relationship between the number of shirts it can sell and the price per shirt. In particular, 20,000 shirts can be sold at $19 each, and 2000 of the same shirts can be sold at $55 each. Write the point-slope and slope-intercept equations of the *demand line* through the ordered pairs

(20,000 shirts, $19) and (2000 shirts, $55).

Then determine the number of shirts that can be sold at $50 each.

61. The scatter plot shows the average number of minutes each that 16 people exercise per week and the average number of headaches per month each person experiences.

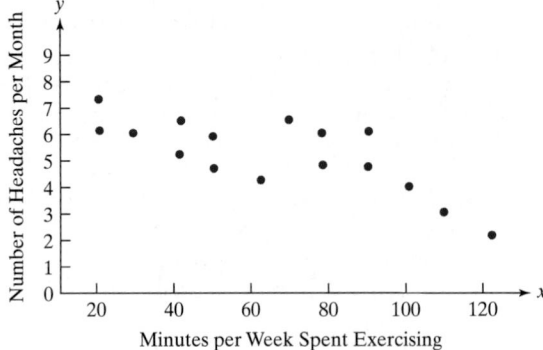

Minutes per Week Spent Exercising

a. Draw a line that fits the data so that the spread of the data points around the line is as small as possible.
b. Use the coordinates of two points along your line to write its point-slope and slope-intercept equations.
c. Use the equation in part (b) to predict the number of headaches per month for a person exercising 130 minutes per week.

Writing in Mathematics

62. Describe how to write the equation of a line if its slope and a point along the line are known.

63. Describe how to write the equation of a line if two points along the line are known.

64. Take a second look at the scatter plot in Exercise 59. Although there is a relationship between education and prejudice, we cannot necessarily conclude that increased education causes a person's level of prejudice to decrease. Offer two or more possible explanations for the data in the scatter plot.

Critical Thinking Exercises

65. Which one of the following is true?
   a. If a line has undefined slope, then it has no equation.
   b. The line whose equation is $y - 3 = 7(x + 2)$ passes through $(-3, 2)$.
   c. The point-slope form will not work for the line through the points $(2, -5)$ and $(2, 6)$.
   d. The slope of the line whose equation is $3x + y = 7$ is 3.

66. Write the point-slope form of the equation of a line having an x-intercept of -3 and perpendicular to the line passing throuhg $(0, 0)$ and $(6, -2)$. Then use the point-slope form of the equation to write the slope-intercept form.

67. Excited about the success of celebrity stamps, post office officials were rumored to have put forth a plan to institute two new types of thermometers. On these new scales, $°E$ represents degrees Elvis and $°M$ represents degrees Madonna. If it is known that $40°E = 25°M$, $280°E = 125°M$, and degrees Elvis is linearly related to degrees Madonna, write an equation expressing E in terms of M.

Technology Exercises

68. Use a graphing utility to graph $y = 1.75x - 2$. Select the best viewing rectangle possible by experimenting with the range settings to show that the line's slope is $\frac{7}{4}$.

69. Use a graphing utility to graph the slope-intercept equation that you wrote in Exercise 60. Then select an appropriate range setting and use the $\boxed{\text{TRACE}}$ feature to graphically show the number of shirts that can be sold at $50 each.

70. a. Use the statistical menu of a graphing utility to enter the ten data points shown in the scatter plot in Exercise 59.

b. Use the $\boxed{\text{DRAW}}$ menu and the scatter plot capability to draw a scatter plot of the data points like the one shown in Exercise 59.

c. Select the linear regression option. Use your utility to obtain values for a and b for the equation of the regression line, $y = ax + b$. You may also be given a *correlation coefficient, r.* Values of r close to 1 indicate that the points can be described by a linear relationship and the regression line has a positive slope. Values of r close to -1 indicate that the points can be described by a linear relationship and the regression line has a

negative slope. Values of r close to 0 indicate no linear relationship between the variables.

d. Use the appropriate sequence (consult your manual) to graph the regression equation on top of the points in the scatter plot.

Review Exercises

71. How many sheets of paper, weighing 2 grams each, can be put in an envelope weighing 4 grams if the total weight must not exceed 29 grams? (Section 2.7, Example 8, and Section 2.5, Example 5)

72 List all the natural numbers in this set:

$$\left\{-2, 0, \frac{1}{2}, 1, \sqrt{3}, \sqrt{4}\right\}.$$

(Section 1.2, Example 5)

73. Use intercepts to graph $3x - 5y = 15$. (Section 3.2, Example 4)

CHAPTER 3 GROUP PROJECTS

1. a. Researchers at Yale University have suggested that levels of passion and commitment in human relations change over time. Based on the shapes of the graphs shown, group members should determine which depicts passion and which represents commitment. Explain how you arrived at your decision.

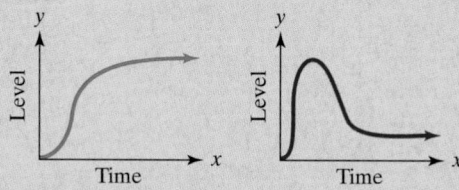

b. For the second part of this activity, each group member will be creating a graph of a particular experience that involved feelings of love, anger, sadness, or any other emotion you choose. The horizontal axis should be labeled time and the vertical axis the emotion you are graphing. You will not be using your algebra skills to create your graph; however, you should try to make the graph as precise as possible. You may use negative numbers on the vertical axis, if appropriate. After each group member has created a graph, pool together all of the graphs and study them to see if there are any similarities in the graphs for a particular emotion or for all emotions.

2. Group members should consult an almanac, newspaper, magazine, or the Internet to find data that lie approximately on or near a straight line. Working by hand or using a graphing utility, construct a scatter plot for the data. If working by hand, draw a line that approximately fits the data and then write its equation. If using a graphing utility, obtain the equation of the regression line. Then use the equation of the line to make a prediction about what might happen in the future. Are there circumstances that might affect the accuracy of this prediction? List some of these circumstances.

CHAPTER SUMMARY, REVIEW, AND TEST

SUMMARY

DEFINITIONS AND CONCEPTS	EXAMPLES
Section 3.1 Graphing Linear Equations	
An ordered pair is a solution of an equation in two variables if replacing the variables by the coordinates of the ordered pair results in a true statement.	Is $(-1, 4)$ a solution of $2x + 5y = 18$? $2(-1) + 5 \cdot 4 \stackrel{?}{=} 18$ $-2 + 20 \stackrel{?}{=} 18$ $18 = 18$, true Thus, $(-1, 4)$ is a solution.

One method for graphing an equation in two variables is point plotting. Find several ordered-pair solutions, plot them as points, and connect the points with a smooth curve or line.	Graph: $y = 2x + 1$.		

x	$y = 2x + 1$	(x, y)
-2	$y = 2(-2) + 1 = -3$	$(-2, -3)$
-1	$y = 2(-1) + 1 = -1$	$(-1, -1)$
0	$y = 2 \cdot 0 + 1 = 1$	$(0, 1)$
1	$y = 2 \cdot 1 + 1 = 3$	$(1, 3)$
2	$y = 2 \cdot 2 + 1 = 5$	$(2, 5)$

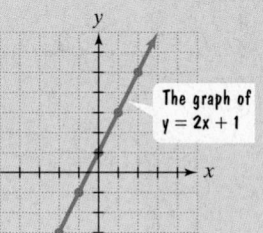

The graph of $y = 2x + 1$

Section 3.2 Graphing Linear Equations Using Intercepts

If a graph intersects the x-axis at $(a, 0)$, then a is an x-intercept. If a graph intersects the y-axis at $(0, b)$, then b is a y-intercept.	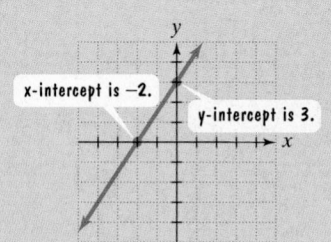 x-intercept is −2. y-intercept is 3.
An equation of the form $Ax + By = C$, where A, B, and C are integers, is called the standard form of the equation of a line. The graph of $Ax + By = C$ is a line that can be obtained using intercepts. To find the x-intercept, let $y = 0$ and solve for x. To find the y-intercept, let $x = 0$ and solve for y. Find a checkpoint, a third ordered-pair solution. Graph the equation by drawing a line through the three points.	Graph using intercepts: $4x + 3y = 12$. x-intercept: $\quad 4x = 12$ $\qquad\qquad\quad x = 3$ y-intercept: $\quad 3y = 12$ $\qquad\qquad\quad y = 4$ Checkpoint: $\quad$ Let $x = 2$. $\qquad\qquad 8 + 3y = 12$ $\qquad\qquad\quad 3y = 4$ $\qquad\qquad\quad\ y = \frac{4}{3}$

DEFINITIONS AND CONCEPTS	**EXAMPLES**

Section 3.2 Graphing Linear Equations Using Intercepts (cont.)

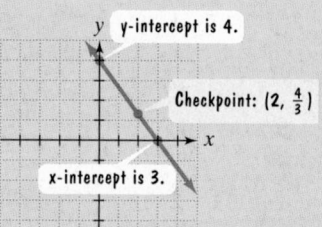

The graph of $Ax + By = 0$ is a line that passes through the origin. Find two other points by finding two other solutions of the equation. Graph the equation by drawing a line through the origin and these two points.

Graph: $x + 2y = 0$.

$x = 2$: $2 + 2y = 0$

$2y = -2$

$y = -1$

$y = 1$: $x + 2(1) = 0$

$x = -2$

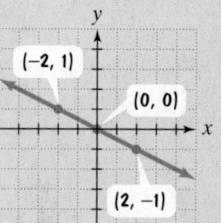

Horizontal and Vertical Lines
The graph of $y = b$ is a horizontal line. The y-intercept is b.
The graph of $x = a$ is a vertical line. The x-intercept is a.

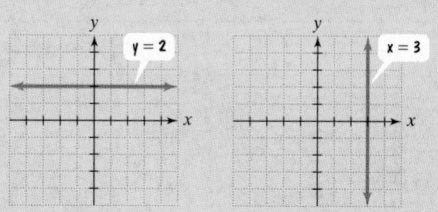

Section 3.3 Slope

The slope, m, of the line through the points (x_1, y_1) and (x_2, y_2) is

$$m = \frac{y_2 - y_1}{x_2 - x_1}, \quad x_2 - x_1 \neq 0.$$

If the slope is positive, the line rises from left to right. If the slope is negative, the line falls from left to right.

The slope of a horizontal line is 0. The slope of a vertical line is undefined.

If two distinct nonvertical lines have the same slope, then they are parallel.

Find the slope of the line passing through the points shown.

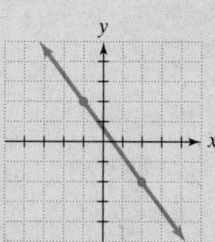

Let $(x_1, y_1) = (-1, 2)$ and $(x_2, y_2) = (2, -2)$.

$$m = \frac{y_2 - y_1}{x_2 - x_1} = \frac{-2 - 2}{2 - (-1)} = \frac{-4}{3} = -\frac{4}{3}$$

DEFINITIONS AND CONCEPTS	EXAMPLES

Section 3.4 The Slope-Intercept Form of the Equation of a Line

The slope-intercept equation of a nonvertical line with slope m and y-intercept b is

$$y = mx + b.$$

Find the slope and the y-intercept of the line with the given equation.

• $y = -2x + 5$

Slope is -2. y-intercept is 5.

• $2x + 3y = 9$ (Solve for y.)

$3y = -2x + 9$ Subtract $2x$.

$y = -\frac{2}{3}x + 3$ Divide by 3.

Slope is $-\frac{2}{3}$. y-intercept is 3.

Graphing $y = mx + b$ Using the Slope and y-Intercept

1. Plot the point containing the y-intercept on the y-axis. This is the point $(0, b)$.
2. Use the slope, m, to obtain a second point. Write m as a fraction, and use rise over run, starting at the point containing the y-intercept, to plot this point.
3. Graph the equation by drawing a line through the two points.

Graph: $y = -\frac{3}{4}x + 1$.

Slope is $-\frac{3}{4}$. y-intercept is 1.

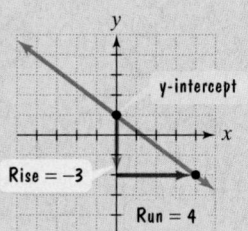

Section 3.5 The Point-Slope Form of the Equation of a Line

The point-slope equation of a nonvertical line of slope m that passes through the point (x_1, y_1) is

$$y - y_1 = m(x - x_1).$$

Slope $= -3$, passing through $(-1, 5)$

$m = -3$ $x_1 = -1$ $y_1 = 5$

The line's point-slope equation is

$$y - 5 = -3[x - (-1)].$$

Simplify:

$$y - 5 = -3(x + 1).$$

To write the point-slope form of the line passing through two points, begin by using the points to compute the slope, m. Use either given point as (x_1, y_1) and write the point-slope equation:

$$y - y_1 = m(x - x_1).$$

Solving this equation for y gives the slope-intercept form of the line's equation.

Write an equation in point-slope form and slope-intercept form of the line passing through $(-1, -3)$ and $(4, 2)$.

$$m = \frac{2 - (-3)}{4 - (-1)} = \frac{2 + 3}{4 + 1} = \frac{5}{5} = 1$$

Using $(4, 2)$ as (x_1, y_1), the point-slope equation is

$$y - 2 = 1(x - 4).$$

Solve for y to obtain the slope-intercept form.

$$y = x - 2 \quad \text{Add 2 to both sides.}$$

DEFINITIONS AND CONCEPTS	EXAMPLES

Section 3.5 The Point-Slope Form of the Equation of a Line (cont.)

Nonvertical parallel lines have the same slope. If the product of the slopes of two lines is −1, then the lines are perpendicular. One line is perpendicular to another line if its slope is the negative reciprocal of the slope of the other.

Write point-slope and slope-intercept equations of the line passing through $(2, -1)$

x_1 y_1

and perpendicular to $y = -\dfrac{1}{5}x + 6$.

slope

Slope, m, of a perpendicular line is 5, the negative reciprocal of $-\frac{1}{5}$.

$$y - (-1) = 5(x - 2)$$ Point-slope equation

$$y + 1 = 5(x - 2)$$

$$y + 1 = 5x - 10$$ Slope-intercept equation

$$y = 5x - 11$$

Review Exercises

3.1 *In Exercises 1–2, determine whether each ordered pair is a solution of the given equation.*

1. $y = 3x + 6$ $(-3, 3), (0, 6), (1, 9)$

2. $3x - y = 12$ $(0, 4), (4, 0), (-1, 15)$

In Exercises 3–4:
a. *Find five solutions of each equation. Organize your work in a table of values.*
b. *Use the five solutions in the table to graph each equation.*

3. $y = 2x - 3$

4. $y = \frac{1}{2}x + 1$

5. Graph the equation: $y = x^2 - 3$. Select integers for x, starting with −3 and ending with 3.

6. The linear equation in two variables $y = 5x - 41$ models the percentage of U.S. adults, y, with x years of education who are doing volunteer work.

   a. Find four solutions of the equation. Use 10, 12, 14, and 16 for x. Organize your work in a table of values.

   b. How well does the given equation model the data shown in the following table? Explain your answer.

Years of Education	10	12	14	16
Percentage Doing Volunteer Work	8.3%	18.8%	28.1%	38.4%

Source: U.S. Bureau of Labor

3.2 *In Exercises 7–9, use the graph to identify the*
a. *x-intercept, or state that there is no x-intercept.*
b. *y-intercept, or state that there is no y-intercept.*

7.

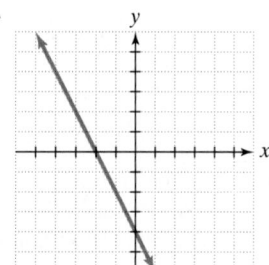

8.

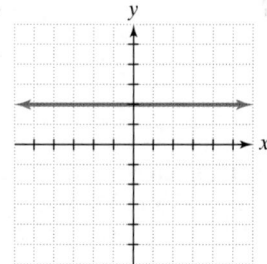

9.

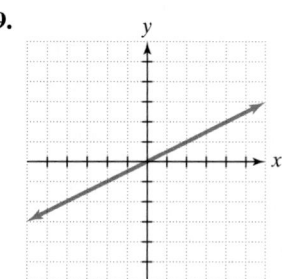

In Exercises 10–13, use intercepts to graph each equation.

10. $2x + y = 4$

11. $3x - 2y = 12$

12. $3x = 6 - 2y$

13. $3x - y = 0$

In Exercises 14–17, graph each equation.

14. $x = 3$ **15.** $y = -5$

16. $y + 3 = 5$ **17.** $2x = -8$

18. The graph shows the Fahrenheit temperature, y, x hours after noon.

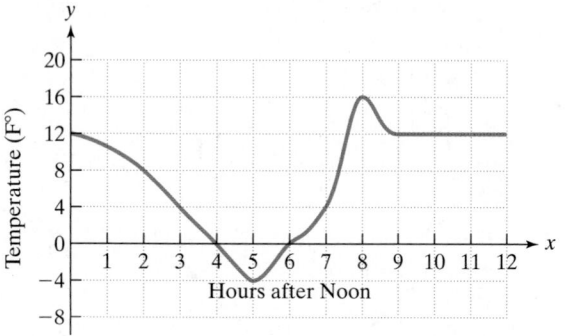

a. At what time did the minimum temperature occur? What is the minimum temperature?

b. At what time did the maximum temperature occur? What is the maximum temperature?

c. What are the x-intercepts? In terms of time and temperature, interpret the meaning of these intercepts.

d. What is the y-intercept? What does this mean in terms of time and temperature?

e. From 9 P.M. until midnight, the graph is shown as a horizontal line. What does this mean about the temperature over this period of time?

3.3 *In Exercises 19–22, calculate the slope of the line passing through the given points. If the slope is undefined, so state. Then indicate whether the line rises, falls, is horizontal, or is vertical.*

19. $(3, 2)$ and $(5, 1)$ **20.** $(-1, 2)$ and $(-3, -4)$

21. $(-3, 4)$ and $(6, 4)$ **22.** $(5, 3)$ and $(5, -3)$

In Exercises 23–26, find the slope of each line, or state that the slope is undefined.

23.

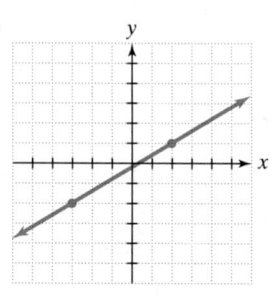

24.

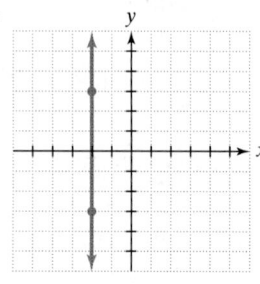

25.

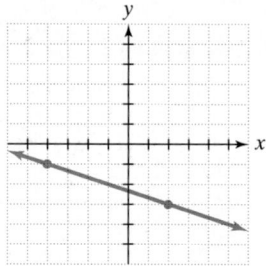

26.

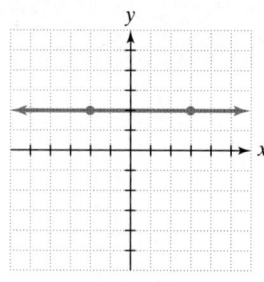

In Exercises 27–28, determine whether the lines through each pair of points are parallel.

27. $(-1, -3)$ and $(2, -8)$ **28.** $(5, 4)$ and $(9, 7)$
$(8, -7)$ and $(9, 10)$ $(-6, 0)$ and $(-2, 3)$

29. The graph shows the number of lawyers in the United States from 1950 through 2000.

Number of U.S. Lawyers

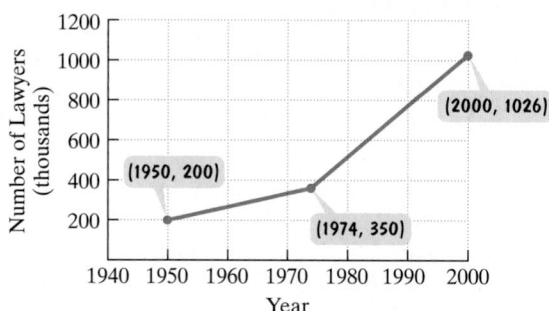

Source: Hudson Institute

a. Find the slope of the line passing through (1974, 350) and (2000, 1026). What does this represent in terms of the increase in the thousands of U.S. lawyers per year?

b. Find and interpret the slope of the line passing through (1950, 200) and (1974, 350).

3.4 *In Exercises 30–33, find the slope and the y-intercept of the line with the given equation.*

30. $y = 5x - 7$ **31.** $y = 6 - 4x$

32. $y = 3$ **33.** $2x + 3y = 6$

In Exercises 34–36, graph each linear equation using the slope and y-intercept.

34. $y = 2x - 4$

35. $y = \dfrac{1}{2}x - 1$

36. $y = -\dfrac{2}{3}x + 5$

In Exercises 37–38, write each equation in slope-intercept form. Then use the slope and y-intercept to graph the equation.

37. $y - 2x = 0$

38. $\dfrac{1}{3}x + y = 2$

39. Graph $y = -\dfrac{1}{2}x + 4$ and $y = -\dfrac{1}{2}x - 1$ in the same rectangular coordinate system. Are the lines parallel? If so, explain why.

40. The graph shows the average age of U.S. whites, African Americans, and Americans of Hispanic origin from 1990 through 2000.

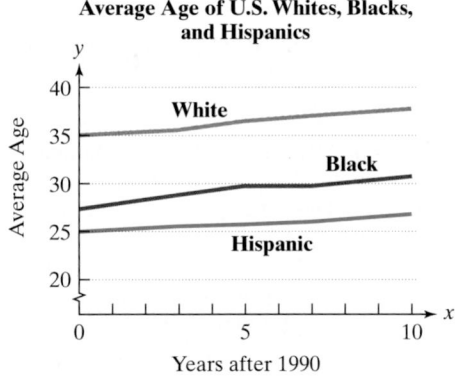

Average Age of U.S. Whites, Blacks, and Hispanics

Source: U.S. Census Bureau

a. What is the smallest y-intercept? Describe what this represents in this situation.

b. The average age of the group with the greatest y-intercept was approximately 38 in 2000. Use the points $(0, 35)$ and $(10, 38)$ to compute the slope for this group. What does this mean about their average age for the period shown?

c. Use the slope from part (b) and the y-intercept for this group shown by the graph to write an equation that models the group's average age, y, x years after 1990. Write your model in the form $y = mx + b$.

d. Use your model from part (c) to predict the average age for the group in 2010.

4.5 *Write the point-slope form of the line satisfying the conditions in Exercises 41–42. Then use the point-slope form of the equation to write the slope-intercept form.*

41. Slope $= 6$, passing through $(-4, 7)$

42. Passing through $(3, 4)$ and $(2, 1)$

43. Passing through $(4, -7)$ and parallel to the line whose equation is $3x + y - 9 = 0$

44. Passing through $(-2, 6)$ and perpendicular to the line whose equation is $y = \frac{1}{3}x + 4$.

45. In 1900, the typical surfboard was 16 feet long. Since then, they have become shorter and shorter. Here are two data measurements for a typical surfboard's length. (A scatter plot of all such data measurements through 1980 would show all data points on or near a straight line.)

x (Years since 1900)	y (Average Surfboard Length, in Feet)
0	16
30	12.1

Tom Blake with six of his surfboards, 1930. (*Source:* Bishop Museum)

a. Write the point-slope form of the equation of the line on which these measurements fall.

b. Use the point-slope form of the equation to write the slope-intercept form of the equation.

c. Use the equation in part (b) to find average surfboard length in 1970 and 1980.

d. Does the equation in part (b) reasonably describe reality in 2000?

Chapter 3 Test

1. Determine whether each ordered pair is a solution of $4x - 2y = 10$:

$$(0, -5), \quad (-2, 1), \quad (4, 3).$$

2. Find five solutions of $y = 3x + 1$. Organize your work in a table of values. Then use the five solutions in the table to graph the equation.

3. Graph: $y = x^2 - 1$. Select integers for x, starting with -3 and ending with 3.

4. Use the graph to identify the:

a. x-intercept, or state that there is no x-intercept.

b. y-intercept, or state that there is no y-intercept.

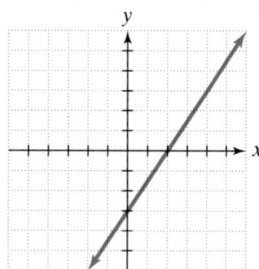

5. Use intercepts to graph $4x - 2y = -8$.

6. Graph $y = 4$ in a rectangular coordinate system.

In Exercises 7–8, calculate the slope of the line passing through the given points. If the slope is undefined, so state. Then indicate whether the line rises, falls, is horizontal, or is vertical.

7. $(-3, 4)$ and $(-5, -2)$

8. $(6, -1)$ and $(6, 3)$

9. Find the slope of the line in the figure shown or state that the slope is undefined.

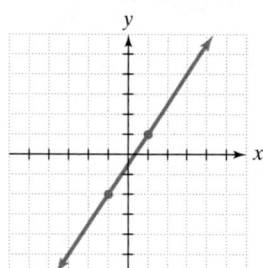

10. Determine whether the line through $(2, 4)$ and $(6, 1)$ is parallel to the line through $(-3, 1)$ and $(1, -2)$.

In Exercises 11–12, find the slope and the y-intercept of the line with the given equation.

11. $y = -x + 10$

12. $2x + y = 6$

In Exercises 13–14, graph each linear equation using the slope and y-intercept.

13. $y = \dfrac{2}{3}x - 1$

14. $y = -2x + 3$

In Exercises 15–16, use the given conditions to write an equation for each line in point-slope form and slope-intercept form.

15. Slope $= -2$, passing through $(-1, 4)$

16. Passing through $(2, 1)$ and $(-1, -8)$

17. Passing through $(-2, 3)$ and perpendicular to the line whose equation is $y = -\frac{1}{2}x - 4$.

18. The graph shows spending per pupil in public schools. Find the slope of the line passing through $(1970, 2100)$ and $(2000, 5280)$. Describe what the slope represents in this situation.

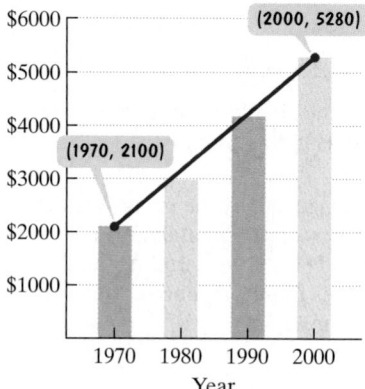

Spending Per Pupil in America's Public School

Source: National Education Association

19. Strong demand plus higher fuel and labor costs are driving up the price of flying. The graph shows the national averages for one-way fares. Also shown is a line that models the data.

National Averages for One-Way Airline Fares: Business Travel

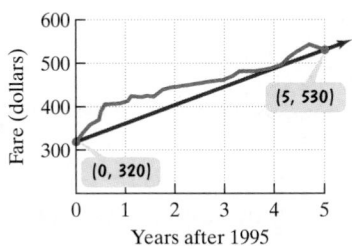

Years after 1995

Source: American Express

a. Use the two points whose coordinates are shown by the voice balloons to find the point-slope equation of the line that models the average one-way fare, y, in dollars, x years after 1995.

b. Write the equation in part (a) in slope-intercept form.

c. Use the equation in part (b) to predict the national average for one-way fares in 2008.

Cumulative Review Exercises (Chapters 1–3)

1. Perform the indicated operations:
$$\frac{10 - (-6)}{3^2 - (4 - 3)}.$$

2. Simplify: $6 - 2[3(x - 1) + 4]$.

3. List all the irrational numbers in this set: $\{-3, 0, 1, \sqrt{4}, \sqrt{5}, \frac{11}{2}\}$.

In Exercises 4–5, solve each equation.

4. $6(2x - 1) - 6 = 11x + 7$

5. $x - \dfrac{3}{4} = \dfrac{1}{2}$

6. Solve for x: $y = mx + b$.

7. 120 is 15% of what?

8. The formula $y = 4.5x - 46.7$ models the stopping distance, y, in feet, for a car traveling x miles per hour. If the stopping distance is 133.3 feet, how fast was the car traveling?

In Exercises 9–10, solve each inequality. Express the solution set in set-builder notation and graph the set on a number line.

9. $2 - 6x \geq 2(5 - x)$

10. $6(2 - x) > 12$

11. Evaluate $x^2 - 10x$ if $x = -3$.

12. Insert either $<$ or $>$ in the shaded area to make a true statement:
$$-2000 \quad\rule{0.5cm}{0.3cm}\quad -3.$$

13. On February 8, the temperature in Manhattan at 10 P.M. was $-4°$F. By 3 A.M. the next day, the temperature had fallen $11°$, but by noon the temperature increased by $21°$. What was the temperature at noon?

14. The amount of money owed by doctors graduating from medical school is on the rise. The formula
$$D = 4x + 30$$
models the average debt, D, in thousands of dollars, of indebted medical-school graduates x years after 1985. Use the formula to determine in which year this debt will reach $150 thousand.

Average Debt of Indebted Medical School Graduates

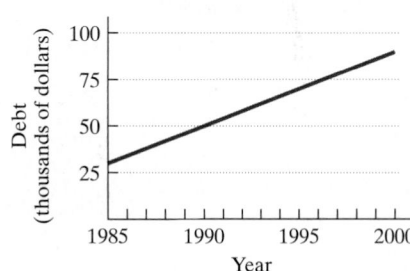

Year

Source: American Medical Association

15. The length of a rectangular football field is 14 meters more than twice the width. If the perimeter is 346 meters, find the field's dimensions.

16. After a 10% weight loss, a person weighed 180 pounds. What was the weight before the loss?

17. A plumber charged a customer $228, listing $18 for parts and the remainder for labor. If the cost of the labor is $35 per hour, how many hours did the plumber work?

18. In a triangle, the measure of the second angle is $20°$ greater than the measure of the first angle. The measure of the third angle is twice that of the first. What is the measure of each angle?

In Exercises 19–20, graph each equation in the rectangular coordinate system.

19. $2x - y = 4$

20. $y = -4x + 3$

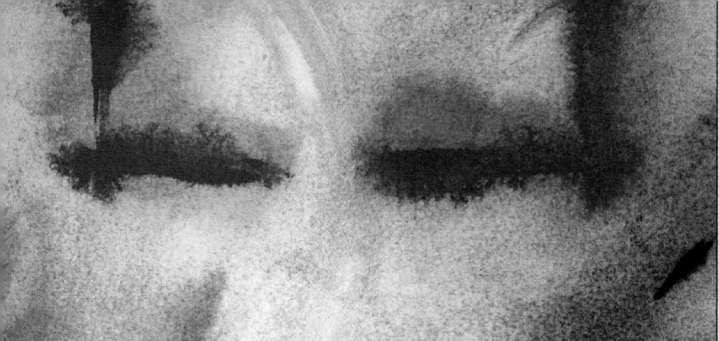

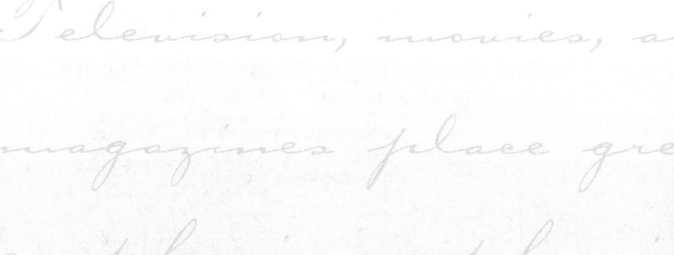

Mary Katherine Campbell,
Miss America 1922

Angela Perez Baraquio,
Miss America 2000

You are not a great fan of beauty pageants. However, as you were channel surfing, you tuned into the Miss America festivities. Is it your imagination, or does the icon of American beauty have that lean and hungry look?

Television, movies, and magazines place great emphasis on physical beauty. Our culture emphasizes physical appearance to such an extent that it is a central factor in the perception and judgment of others. The modern emphasis on thinness as the ideal body shape has been suggested as a major cause of eating disorders among adolescent women. In this chapter, you will learn how systems of linear equations in two variables reveal the hidden patterns of your world, including a relationship between our changing cultural values of physical attractiveness and undernutrition.

Chapter 4

Systems of Linear Equations

► SECTION 4.1 *Solving Systems of Linear Equations by Graphing*

Objectives

1 Decide whether an ordered pair is a solution of a linear system.

2 Solve systems of linear equations by graphing.

3 Use graphing to identify systems with no solution or infinitely many solutions.

SSM
PH Tutor   CD-   Video
Center   ROM

Key West residents Brian Goss (left), George Wallace, and Michael Mooney (right) hold on to each other as they battle 90 mph winds along Houseboat Row in Key West, Fla., on Friday, Sept. 25, 1998. The three had sought shelter behind a Key West hotel as Hurricane Georges descended on the Florida Keys but were forced to seek other shelter when the storm conditions became too rough. Hundreds of people were killed by the storm when it swept through the Caribbean.

Real-world problems often involve solving thousands of equations, sometimes containing a million variables. Problems ranging from scheduling airline flights to controlling traffic flow to routing phone calls over the nation's communication network often require solutions in a matter of moments. AT&T's domestic long distance network involves 800,000 variables! Meteorologists describing atmospheric conditions surrounding a hurricane must solve problems involving thousands of equations rapidly and efficiently. The difference between a two-hour warning and a two-day warning is a life-and-death issue for thousands of people in the path of one of nature's most destructive forces.

Although we will not be solving 800,000 equations with 800,000 variables, we will turn our attention to two equations with two variables, such as

$$2x - 3y = -4$$
$$2x + y = 4.$$

The methods that we consider for solving such problems provide the foundation for solving far more complex systems with many variables.

Systems of Linear Equations and Their Solutions

1 Decide whether an ordered pair is a solution of a linear system.

We have seen that all equations in the form $Ax + By = C$ are straight lines when graphed. Two such equations, such as those listed above, are called a **system of linear equations** or a **linear system**. A **solution to a system of linear equations** is an ordered pair that satisfies all equations in the system. For example, (3, 4) satisfies the system

$$x + y = 7 \qquad \text{(3 + 4 is, indeed, 7.)}$$
$$x - y = -1. \qquad \text{(3 - 4 is, indeed, -1.)}$$

Thus, $(3, 4)$ satisfies both equations and is a solution of the system. The solution can be described by saying that $x = 3$ and $y = 4$. The solution can also be described using set notation. The solution set to the system is $\{(3, 4)\}$—that is, the set consisting of the ordered pair $(3, 4)$.

A system of linear equations can have exactly one solution, no solution, or infinitely many solutions. We begin with systems having exactly one solution.

EXAMPLE 1   **Determining Whether Ordered Pairs Are Solutions of a System**

Consider the system:

$$x + 2y = 2$$
$$x - 2y = 6.$$

Determine if each ordered pair is a solution of the system:

   a. $(4, -1)$ **b.** $(-4, 3)$.

Solution

a. We begin by determining whether $(4, -1)$ is a solution. Because 4 is the x-coordinate and -1 is the y-coordinate of $(4, -1)$, we replace x by 4 and y by -1.

$$
\begin{array}{ll}
x + 2y = 2 & x - 2y = 6 \\
4 + 2(-1) \overset{?}{=} 2 & 4 - 2(-1) \overset{?}{=} 6 \\
4 + (-2) \overset{?}{=} 2 & 4 - (-2) \overset{?}{=} 6 \\
2 = 2, \ \text{true} & 4 + 2 \overset{?}{=} 6 \\
& 6 = 6, \ \text{true}
\end{array}
$$

The pair $(4, -1)$ satisfies both equations: It makes each equation true. Thus, the orderd pair is a solution of the system.

b. To determine whether $(-4, 3)$ is a solution, we replace x by -4 and y by 3.

$$
\begin{array}{ll}
x + 2y = 2 & x - 2y = 6 \\
-4 + 2 \cdot 3 \overset{?}{=} 2 & -4 - 2 \cdot 3 \overset{?}{=} 6 \\
-4 + 6 \overset{?}{=} 2 & -4 - 6 \overset{?}{=} 6 \\
2 = 2, \ \text{true} & -10 = 6, \ \text{false}
\end{array}
$$

The pair $(-4, 3)$ fails to satisfy *both* equations: It does not make both equations true. Thus, the ordered pair is not a solution of the system. ■

✔ **CHECK POINT 1**   Consider the system:

$$2x - 3y = -4$$
$$2x + \ y = \ \ 4.$$

Determine if each ordered pair is a solution of the system:

   a. $(1, 2)$ **b.** $(7, 6)$.

2 Solve systems of linear equations by graphing.

Solving Linear Systems by Graphing

The solution of a system of linear equations can be found by graphing both of the equations in the same rectangular coordinate system. For a system with one solution, **the coordinates of the point of intersection give the system's solution.** For example, the system in Example 1,

$$x + 2y = 2$$
$$x - 2y = 6$$

is graphed in Figure 4.1. The solution of the system, $(4, -1)$, corresponds to the point of intersection of the lines.

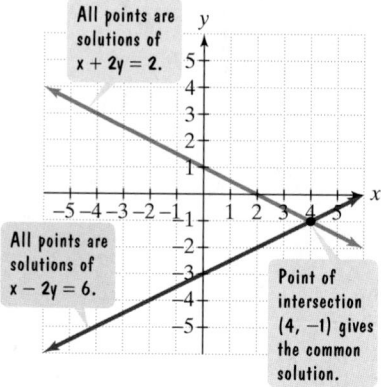

Figure 4.1 Visualizing a system's solution

Solving Systems of Two Linear Equations in Two Variables, x and y, by Graphing

1. Graph the first equation.
2. Graph the second equation on the same axes.
3. If the lines representing the two equations intersect at a point, determine the coordinates of this point of intersection. The ordered pair is the solution to the system.
4. Check the solution in both equations.

EXAMPLE 2   Solving a Linear System by Graphing

Solve by graphing:

$$2x + 3y = 6$$
$$2x + y = -2.$$

Solution

Step 1 Graph the first equation. We use intercepts to graph $2x + 3y = 6$.

x-intercept (Set $y = 0$.)	**y-intercept** (Set $x = 0$.)
$2x + 3 \cdot 0 = 6$	$2 \cdot 0 + 3y = 6$
$2x = 6$	$3y = 6$
$x = 3$	$y = 2$

The x-intercept is 3, so the line passes through $(3, 0)$. The y-intercept is 2, so the line passes through $(0, 2)$. The graph of $2x + 3y = 6$ is shown in Figure 4.2.

Step 2 Graph the second equation on the same axes. We use intercepts to graph $2x + y = -2$.

x-intercept (Set $y = 0$.)	**y-intercept** (Set $x = 0$.)
$2x + 0 = -2$	$2 \cdot 0 + y = -2$
$2x = -2$	$y = -2$
$x = -1$	

The x-intercept is -1, so the line passes through $(-1, 0)$. The y-intercept is -2, so the line passes through $(0, -2)$. The graph of $2x + y = -2$ is shown in Figure 4.2.

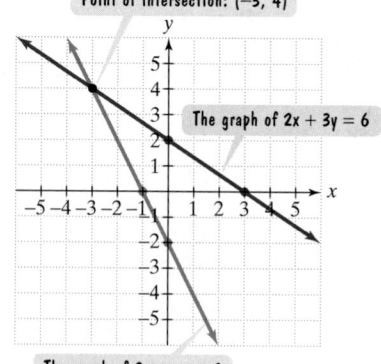

Figure 4.2

Step 3 Determine the coordinates of the intersection point. This ordered pair is the system's solution. Using Figure 4.2, it appears that the lines intersect at $(-3, 4)$. The "apparent" solution of the system is $(-3, 4)$.

Step 4 Check the solution in both equations.

<table>
<tr><td align="center">Check $(-3, 4)$ in
$2x + 3y = 6$:</td><td align="center">Check $(-3, 4)$ in
$2x + y = -2$:</td></tr>
<tr><td align="center">$2(-3) + 3 \cdot 4 \overset{?}{=} 6$</td><td align="center">$2(-3) + 4 \overset{?}{=} -2$</td></tr>
<tr><td align="center">$-6 + 12 \overset{?}{=} 6$</td><td align="center">$-6 + 4 \overset{?}{=} -2$</td></tr>
<tr><td align="center">$6 = 6,$ true</td><td align="center">$-2 = -2,$ true</td></tr>
</table>

Because both equations are satisfied, $(-3, 4)$ is the solution of the system and $\{(-3, 4)\}$ is the solution set. ∎

✔ **CHECK POINT 2**   Solve by graphing:

$$2x + y = \ \ 6$$
$$2x - y = -2.$$

Discover for Yourself

Must two lines intersect at exactly one point? Sketch two lines that have less than one intersection point. Now sketch two lines that have more than one intersection point. What does this say about each of these systems?

EXAMPLE 3   Solving a Linear System by Graphing

Solve by graphing:

$$y = -3x + 2$$
$$y = \ \ 5x - 6.$$

Solution   Each equation is in the form $y = mx + b$. Thus, we use the y-intercept, b, and the slope, m, to graph each line.

Step 1 Graph the first equation.

$$y = -3x + 2$$

The slope is -3.   The y-intercept is 2.

The y-intercept is 2, so the line passes through $(0, 2)$. The slope is $-\frac{3}{1}$. Start at the y-intercept and move 3 units down (the rise) and 1 unit to the right (the run). The graph of $y = -3x + 2$ is shown in Figure 4.3.

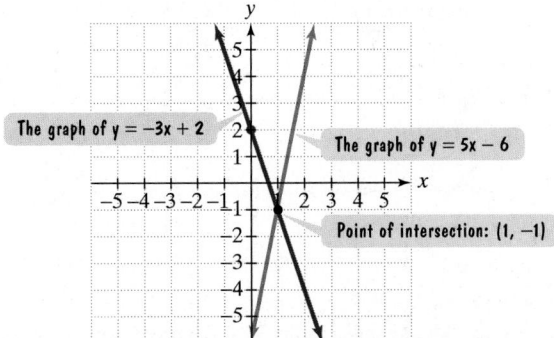

The graph of $y = -3x + 2$

The graph of $y = 5x - 6$

Point of intersection: $(1, -1)$

Figure 4.3

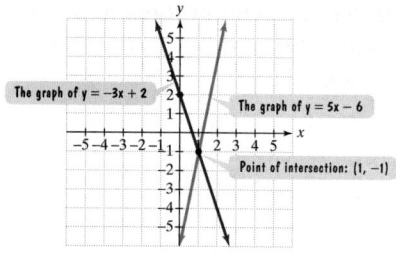

Figure 4.3, repeated

Step 2 Graph the second equation on the same axes.

$$y = 5x - 6$$

The slope is 5.   The y-intercept is −6.

The y-intercept is -6, so the line passes through $(0, -6)$. The slope is $\frac{5}{1}$. Start at the y-intercept and move 5 units up (the rise) and 1 unit to the right (the run). The graph of $y = 5x - 6$ is shown in Figure 4.3.

Step 3 Determine the coordinates of the intersection point. This ordered pair is the system's solution. Using Figure 4.3, it appears that the lines intersect at $(1, -1)$. The "apparent" solution of the system is $(1, -1)$.

Step 4 Check the solution in both equations.

Check $(1, -1)$ in
$y = -3x + 2$:
$-1 \overset{?}{=} -3 \cdot 1 + 2$
$-1 \overset{?}{=} -3 + 2$
$-1 = -1$, true

Check $(1, -1)$ in
$y = 5x - 6$:
$-1 \overset{?}{=} 5 \cdot 1 - 6$
$-1 \overset{?}{=} 5 - 6$
$-1 = -1$, true

Because both equations are satisfied, $(1, -1)$ is the solution and the system's solution set is $\{(1, -1)\}$. ∎

✔ **CHECK POINT 3** Solve by graphing:

$$y = -x + 6$$
$$y = 3x - 6.$$

3 Use graphing to identify systems with no solution or infinitely many solutions.

Linear Systems Having No Solution or Infinitely Many Solutions

We have seen that a system of linear equations in two variables represents a pair of lines. The lines either intersect, are parallel, or are identical. Thus, there are three possibilities for the number of solutions to a system of two linear equations.

The Number of Solutions to a System of Two Linear Equations

The number of solutions to a system of two linear equations in two variables is given by one of the following. (See Figure 4.4.)

Number of Solutions	What This Means Graphically
Exactly one ordered-pair solution	The two lines intersect at one point.
No solution	The two lines are parallel.
Infinitely many solutions	The two lines are identical.

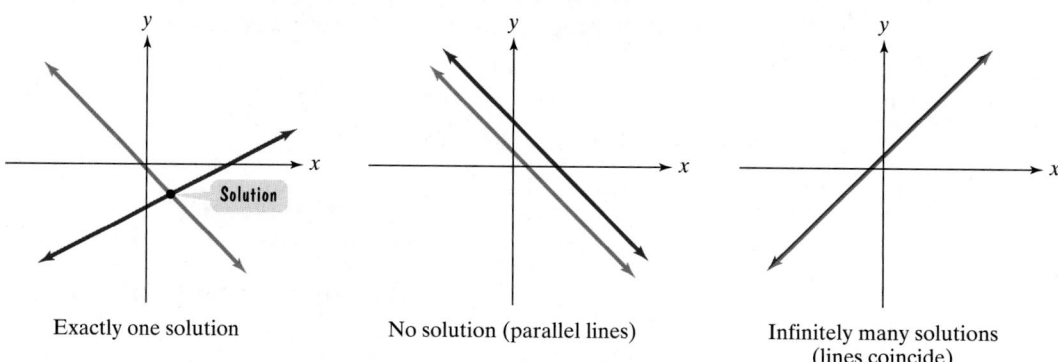

Exactly one solution   No solution (parallel lines)   Infinitely many solutions (lines coincide)

Figure 4.4 Possible graphs for a system of two linear equations in two variables

A linear system with no solution is called an **inconsistent system**. If you attempt to solve such a system by graphing, you will obtain two parallel lines. The solution set is the empty set, $\varnothing$.

EXAMPLE 4 A System with No Solution

Solve by graphing:

$$y = 2x - 1$$
$$y = 2x + 3.$$

Solution Compare the slopes and y-intercepts in the two equations.

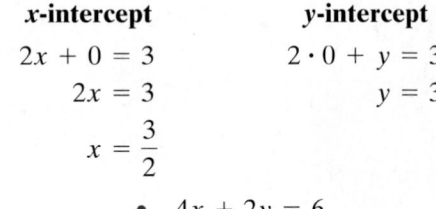

Figure 4.5 shows the graphs of the two equations. Because both equations have the same slope, 2, but different y-intercepts, the lines are parallel. The system is inconsistent and has no solution. The solution set is the empty set, $\varnothing$. ∎

✔ **CHECK POINT 4** Solve by graphing:

$$y = 3x - 2$$
$$y = 3x + 1.$$

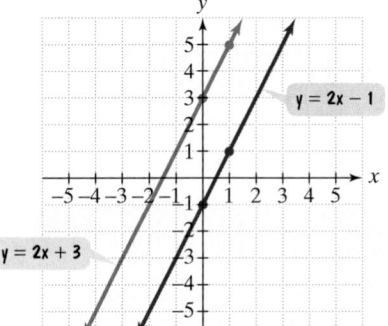

Figure 4.5 The graph of an inconsistent system

EXAMPLE 5 A System with Infinitely Many Solutions

Solve by graphing:

$$2x + y = 3$$
$$4x + 2y = 6.$$

Solution We use intercepts to graph each equation.

• $2x + y = 3$

x-intercept	**y-intercept**
$2x + 0 = 3$	$2 \cdot 0 + y = 3$
$2x = 3$	$y = 3$
$x = \dfrac{3}{2}$	

• $4x + 2y = 6$

x-intercept	**y-intercept**
$4x + 2 \cdot 0 = 6$	$4 \cdot 0 + 2y = 6$
$4x = 6$	$2y = 6$
$x = \dfrac{6}{4} = \dfrac{3}{2}$	$y = 3$

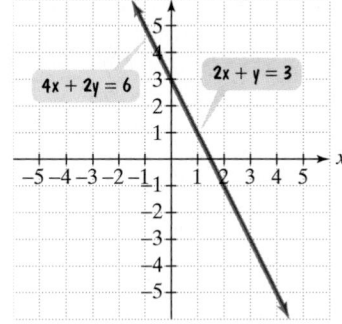

Figure 4.6 The graph of a system with infinitely many solutions

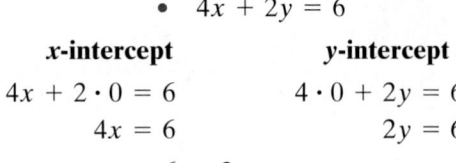

Intersecting lines and the repetition of simple forms play a role in modern architecture.

Both lines have the same x-intercept, $\frac{3}{2}$, or 1.5, and the same y-intercept, 3. Thus, the graphs of the two equations in the system are the same line, shown in Figure 4.6. The two equations have the same solutions. Any ordered pair that is a solution of one equation is a solution of the other, and, consequently, a solution of the system. The system has an infinite number of solutions, namely all points that are solutions of either equation.

We express the solution set for the system in one of two equivalent ways:

$$\{(x, y) \mid 2x + y = 3\}$$ The set of all ordered pairs (x, y) such that $2x + y = 3$

$$\{(x, y) \mid 4x + 2y = 6\}.$$ The set of all ordered pairs (x, y) such that $4x + 2y = 6$ ■

Take a second look at the two equations in Example 5. If you multiply both sides of the first equation, $2x + y = 3$, by 2, you will obtain the second equation, $4x + 2y = 6$.

$2x + y = 3$	This is the first equation in the system.
$2(2x + y) = 2 \cdot 3$	Multiply both sides by 2.
$2 \cdot 2x + 2y = 2 \cdot 3$	Use the distributive property.
$4x + 2y = 6$	Simplify.

> This is the second equation in the system.

Because $2x + y = 3$ and $4x + 2y = 6$ are different forms of the same equation, these equations are called *dependent equations*. In general, the equations in a linear system with infinitely many solutions are called **dependent equations**.

✔ **CHECK POINT 5**   Solve by graphing:

$$x + y = 3$$
$$2x + 2y = 6.$$

ENRICHMENT ESSAY

Missing America

Here she is, Miss America, the icon of American beauty. Always thin, she is becoming more so. The scatter plot in the figure shows Miss America's body-mass index, a ratio comparing weight divided by the square of height. Two lines are also shown: a line that passes near the data points, called the regression line, and a horizontal line representing the World Health Organization's cutoff point for undernutrition. The intersection point indicates that in approximately 1978, Miss America reached this cutoff. There she goes: If the trend continues, Miss America's body-mass index could reach zero in about 320 years.

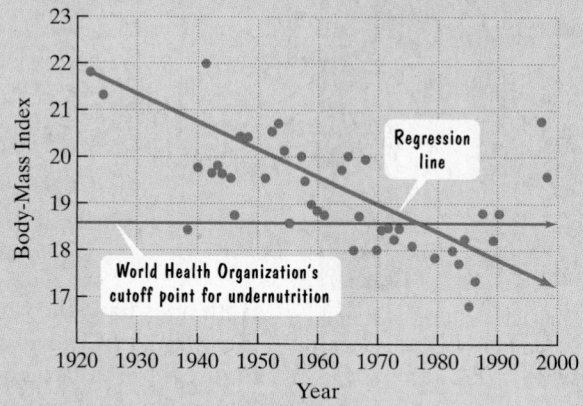

Body-Mass Index of Miss America

Source: John Hopkins School of Public Health

EXERCISE SET 4.1

Practice Exercises

In Exercises 1–10, determine whether the given ordered pair is a solution of the system.

1. $(2, 3)$
$x + 3y = 11$
$x - 5y = -13$

2. $(-3, 5)$
$9x + 7y = 8$
$8x - 9y = -69$

3. $(-3, -1)$
$5x - 11y = -4$
$6x - 8y = -10$

4. $(-2, 6)$
$7x + 3y = 4$
$8x + 7y = 26$

5. $(2, 5)$
$2x + 3y = 17$
$x + 4y = 16$

6. $(3, -1)$
$2x - y = 7$
$3x = 6$

7. $\left(\dfrac{1}{3}, 1\right)$
$6x - 9y = -7$
$9x + 5y = 8$

8. $\left(\dfrac{1}{3}, \dfrac{1}{2}\right)$
$15x + 4y = 7$
$6x + 14y = 9$

9. $(8, 5)$
$5x - 4y = 20$
$3y = 2x + 1$

10. $(5, -2)$
$4x - 3y = 26$
$x = 15 - 5y$

In Exercises 11–42, solve each system by graphing. If there is no solution or an infinite number of solutions, so state, and use set notation to express solution sets.

11. $x + y = 6$
$x - y = 2$

12. $x + y = 2$
$x - y = 4$

13. $x + y = 1$
$y - x = 3$

14. $x + y = 4$
$y - x = 4$

15. $2x - 3y = 6$
$4x + 3y = 12$

16. $x + 2y = 2$
$x - y = 2$

17. $4x + y = 4$
$3x - y = 3$

18. $5x - y = 10$
$2x + y = 4$

19. $y = x + 5$
$y = -x + 3$

20. $y = x + 1$
$y = 3x - 1$

21. $y = 2x$
$y = -x + 6$

22. $y = 2x + 1$
$y = -2x - 3$

23. $y = -2x + 3$
$y = -x + 1$

24. $y = 3x - 4$
$y = -2x + 1$

25. $y = 2x - 1$
$y = 2x + 1$

26. $y = 3x - 1$
$y = 3x + 2$

27. $x + y = 4$
$x = -2$

28. $x + y = 6$
$y = -3$

29. $x - 2y = 4$
$2x - 4y = 8$

30. $2x + 3y = 6$
$4x + 6y = 12$

31. $y = 2x - 1$
$x - 2y = -4$

32. $y = -2x - 4$
$4x - 2y = 8$

33. $x + y = 5$
$2x + 2y = 12$

34. $x - y = 2$
$3x - 3y = -6$

35. $x - y = 0$
$y = x$

36. $2x - y = 0$
$y = 2x$

37. $x = 2$
$y = 4$

38. $x = 3$
$y = 5$

39. $x = 2$
$x = -1$

40. $x = 3$
$x = -2$

41. $y = 0$
$y = 4$

42. $y = 0$
$y = 5$

Application Exercises

43. The graph shows the number of births in Massachusetts for women under 30 years old and women 30 years or older.

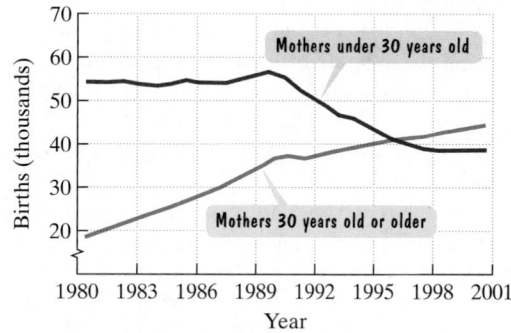

Number of Births in Massachusetts

Source: Massachusetts Department of Public Health

a. Estimate the coordinates of the point of intersection. What does this mean in terms of the number of babies born to older mothers?

b. Describe what is happening to the number of births in Massachusetts to the right of the intersection point.

44. The figure shows scatter plots for the men's and women's winning times, in seconds, in the Olympic 100-meter freestyle swimming race. Also shown are lines that best fit the data, one for the men and one for the women.

Winning Times in the Olympic 100-Meter Freestyle Swimming Race

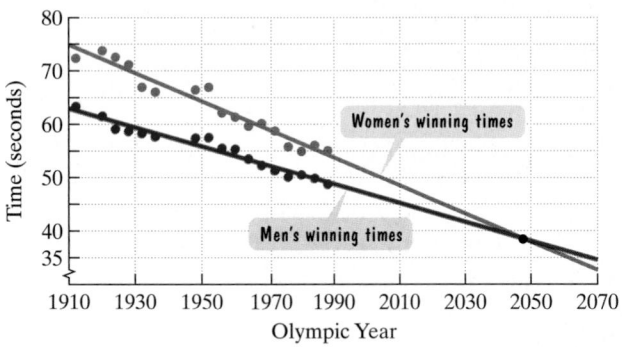

a. Estimate the coordinates of the point of intersection. What does this mean in terms of the women's time and the men's time?

b. Make a prediction about the swimming race in the Olympic years to the right of the intersection point.

Writing in Mathematics

45. What is a system of linear equations? Provide an example with your description.
46. What is a solution to a system of linear equations?
47. Explain how to determine if an ordered pair is a solution of a system of linear equations.
48. Explain how to solve a system of linear equations by graphing.
49. What is an inconsistent system? What happens if you attempt to solve such a system by graphing?
50. Explain how a linear system can have infinitely many solutions.
51. What are dependent equations? Provide an example with your description.
52. Suppose that two lines are shown in the same rectangular coordinate system. One line is the graph of a formula that models the number of corded phones sold in the United States from 1990 to the present. The other line is the graph of a formula that models the number of cordless phones (not including cellular phones) sold in the United States from 1990 to the present. Describe whether there is an intersection point. What does this point represent? What happens to the right of the intersection point?

Critical Thinking Exercises

53. Which one of the following statements is true?

a. If a linear system has graphs with equal slopes, the system must be inconsistent.
b. If a linear system has graphs with equal y-intercepts, the system must have infinitely many solutions.
c. If a linear system has two points that are solutions, then the graphs of the system's equations have equal slopes and equal y-intercepts.
d. It is possible for a linear system with one solution to have graphs with equal slopes.

54. Write a system of linear equations whose solution is $(5, 1)$. How many different systems are possible? Explain.

55. Write a system of equations with one solution, a system of equations with no solution, and a system of equations with infinitely many solutions. Explain how you were able to think of these systems.

56. Graph $y = x^2$ and $y = x + 2$ on the same axes. Find two ordered pairs that satisfy the system. Check that your answers satisfy both equations in the system.

Technology Exercises

57. Verify your solutions to any five exercises from 11 through 36 by using a graphing utility to graph the two equations in the system in the same viewing rectangle. After entering the two equations, one as y_1 and the other as y_2, and graphing them, use the ⌜TRACE⌝ and ⌜ZOOM⌝ features to find the coordinates of the intersection point. (It may first be necessary to solve the equation for y before entering it.) Many graphing utilities have a special ⌜INTERSECTION⌝ feature that displays the coordinates of the intersection point once the equations are graphed. Consult your manual.

Read Exercise 57. Then use a graphing utility to solve the systems in Exercises 58–65.

58. $y = 2x + 2$
$y = -2x + 6$

59. $y = -x + 5$
$y = x - 7$

60. $x + 2y = 2$
$x - y = 2$

61. $2x - 3y = 6$
$4x + 3y = 12$

62. $3x - y = 5$
$-5x + 2y = -10$

63. $2x - 3y = 7$
$3x + 5y = 1$

64. $y = \frac{1}{3}x + \frac{2}{3}$
$y = \frac{5}{7}x - 2$

65. $y = -\frac{1}{2}x + 2$
$y = \frac{3}{4}x + 7$

Review Exercises

In Exercises 66–68, perform the indicated operation.

66. $-3 + (-9)$ (Section 1.7, Table 1.5)
67. $-3 - (-9)$ (Section 1.7, Table 1.5)
68. $-3(-9)$ (Section 1.7, Table 1.5)

▶ **SECTION 4.2** *Solving Systems of Linear Equations by the Substitution Method*

Objectives

1 Solve linear systems by the substitution method.

2 Use the substitution method to identify systems with no solution or infinitely many solutions.

3 Solve problems using the substitution method.

SSM
PH Tutor CD- Video
Center ROM

1 Solve linear systems by the substitution method.

Other than outrage, what is going on at the gas pumps? With surging demand, are gas shortages created as a way to increase oil prices? Like all things in a free market economy, the price of a commodity is based on supply and demand. In this section, we use a second method for solving linear systems, the *substitution method*, to understand this economic phenomenon.

Eliminating a Variable Using the Substitution Method

Finding the solution to a linear system by graphing equations may not be easy to do. For example, a solution of $\left(-\frac{2}{3}, \frac{157}{29}\right)$ would be difficult to "see" as an intersection point on a graph.

Let's consider a method that does not depend on finding a system's solution visually: the substitution method. This method involves converting the system to one equation in one variable by an appropriate substitution.

EXAMPLE 1 Solving a System by Substitution

Solve by the substitution method:

$$y = -x - 1$$
$$4x - 3y = 24.$$

Solution

Step 1 Solve either of the equations for one variable in terms of the other. This step has already been done for us. The first equation, $y = -x - 1$, has y solved in terms of x.

Step 2 Substitute the expression from step 1 into the other equation. We substitute the expression $-x - 1$ for y in the other equation:

$$y = \boxed{-x - 1} \qquad 4x - 3\boxed{y} = 24. \quad \text{Substitute } -x - 1 \text{ for } y.$$

This gives us an equation in one variable, namely

$$4x - 3(-x - 1) = 24.$$

The variable y has been eliminated.

Step 3 Solve the resulting equation containing one variable.

$$4x - 3(-x - 1) = 24 \qquad \text{This is the equation containing one variable.}$$
$$4x + 3x + 3 = 24 \qquad \text{Apply the distributive property.}$$
$$7x + 3 = 24 \qquad \text{Combine like terms.}$$
$$7x = 21 \qquad \text{Subtract 3 from both sides.}$$
$$x = 3 \qquad \text{Divide both sides by 7.}$$

Using Technology

A graphing utility can be used to solve the system in Example 1. Graph each equation and use the intersection feature. The utility displays the solution $(3, -4)$ as $x = 3, y = -4$.

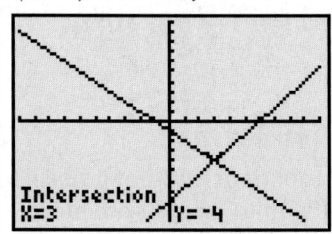

Step 4 Back-substitute the obtained value into the equation from step 1. We now know that the x-coordinate of the solution is 3. To find the y-coordinate, we back-substitute the x-value into the equation from step 1,

$$y = -x - 1.$$

Substitute 3 for x.

$$y = -3 - 1 = -4$$

With $x = 3$ and $y = -4$, the proposed solution is $(3, -4)$.

Step 5 Check the proposed solution in both of the system's given equations. Replace x with 3 and y with -4.

$$y = -x - 1 \qquad\qquad 4x - 3y = 24$$
$$-4 \overset{?}{=} -3 - 1 \qquad\qquad 4(3) - 3(-4) \overset{?}{=} 24$$
$$-4 = -4 \quad \text{true} \qquad\qquad 12 + 12 \overset{?}{=} 24$$
$$24 = 24 \quad \text{true}$$

The pair $(3, -4)$ satisfies both equations. The system's solution is $(3, -4)$ and the solution set is $\{(3, -4)\}$. ∎

✔ CHECK POINT 1 Solve by the substitution method:

$$y = 5x - 13$$
$$2x + 3y = 12.$$

Before considering additional examples, let's summarize the steps used in the substitution method.

Study Tip

In step 1, if possible, solve for a variable whose coefficient is 1 or −1 to avoid working with fractions.

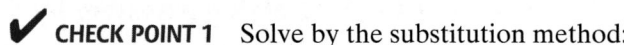

Solving Linear Systems by Substitution

1. Solve either of the equations for one variable in terms of the other. (If one of the equations is already in this form, you can skip this step.)
2. Substitute the expression found in step 1 into the other equation. This will result in an equation in one variable.
3. Solve the equation obtained in step 2.
4. Back-substitute the value found in step 3 into the equation from step 1. Simplify and find the value of the remaining variable.
5. Check the proposed solution in both of the system's given equations.

EXAMPLE 2   Solving a System by Substitution

Solve by the substitution method:

$$5x - 4y = 9$$
$$x - 2y = -3.$$

Solution

Step 1 Solve either of the equations for one variable in terms of the other. We begin by isolating one of the variables in either of the equations. By solving for x in the second equation, which has a coefficient of 1, we can avoid fractions.

$$x - 2y = -3 \qquad \text{This is the second equation in the given system.}$$
$$x = 2y - 3 \qquad \text{Solve for x by adding 2y to both sides.}$$

Step 2 Substitute the expression from step 1 into the other equation. We substitute $2y - 3$ for x in the first equation.

$$x = \boxed{2y - 3} \qquad 5\boxed{x} - 4y = 9$$

This gives us an equation in one variable, namely

$$5(2y - 3) - 4y = 9.$$

The variable x has been eliminated.

Step 3 Solve the resulting equation containing one variable.

$$5(2y - 3) - 4y = 9 \qquad \text{This is the equation containing one variable.}$$
$$10y - 15 - 4y = 9 \qquad \text{Apply the distributive property.}$$
$$6y - 15 = 9 \qquad \text{Combine like terms.}$$
$$6y = 24 \qquad \text{Add 15 to both sides.}$$
$$y = 4 \qquad \text{Divide both sides by 6.}$$

Step 4 Back-substitute the obtained value into the equation from step 1. Now that we have the y-coordinate of the solution, we back-substitute 4 for y in the equation $x = 2y - 3$.

$$x = 2y - 3 \qquad \text{Use the equation obtained in step 1.}$$
$$x = 2(4) - 3 \qquad \text{Substitute 4 for y.}$$
$$x = 8 - 3 \qquad \text{Multiply.}$$
$$x = 5 \qquad \text{Subtract.}$$

With $x = 5$ and $y = 4$, the proposed solution is $(5, 4)$.

Step 5 Check. Take a moment to show that $(5, 4)$ satisfies both given equations. The solution is $(5, 4)$ and the solution set is $\{(5, 4)\}$. ∎

Study Tip

Get into the habit of checking ordered-pair solutions in *both* equations of the system.

✔ **CHECK POINT 2**   Solve by the substitution method:

$$3x + 2y = -1$$
$$x - y = 3.$$

2 Use the substitution method to identify systems with no solution or infinitely many solutions.

The Substitution Method with Linear Systems Having No Solution or Infinitely Many Solutions

Recall that a linear system with no solution is called an **inconsistent system**. If you attempt to solve such a system by substitution, you will eliminate both variables. A false statement such as $0 = 17$ will be the result.

Using Technology

A graphing utility was used to graph the equations in Example 3. The lines are parallel and have no point of intersection. This verifies that the system is inconsistent.

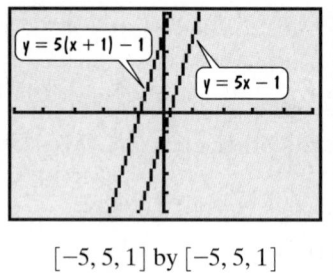

$[-5, 5, 1]$ by $[-5, 5, 1]$

EXAMPLE 3 Using the Substitution Method On an Inconsistent System

Solve the system:

$$y + 1 = 5(x + 1)$$
$$y = 5x - 1.$$

Solution The variable y is isolated in the second equation. We use the substitution method and substitute the expression for y in the first equation.

$$\boxed{y} + 1 = 5(x + 1) \qquad y = \boxed{5x - 1} \qquad \text{Substitute } 5x - 1 \text{ for } y.$$

$$5x - 1 + 1 = 5(x + 1) \qquad \text{This substitution gives an equation in one variable.}$$

$$5x = 5x + 5 \qquad \text{Simplify on the left side. Use the distributive property on the right side.}$$

$$0 = 5, \quad \text{false} \qquad \text{Subtract } 5x \text{ from both sides.}$$

The false statement $0 = 5$ indicates that the system is inconsistent and has no solution. The solution set is the empty set, $\varnothing$. ∎

✔ **CHECK POINT 3** Solve the system:

$$3x + y = -5$$
$$y = -3x + 3.$$

Do you remember that the equations in a linear system with infinitely many solutions are called **dependent**? If you attempt to solve such a system by substitution, you will eliminate both variables. However, a true statement such as $5 = 5$ will be the result.

EXAMPLE 4 Using the Substitution Method On a System with Infinitely Many Solutions

Solve the system:

$$y = 3 - 2x$$
$$4x + 2y = 6.$$

Solution The variable y is isolated in the first equation. We use the substitution method and substitute the expression for y in the second equation.

$$y = \boxed{3 - 2x} \quad 4x + 2\boxed{y} = 6 \qquad \text{Substitute } 3 - 2x \text{ for } y.$$

$$4x + 2(3 - 2x) = 6 \qquad \text{This substitution gives an equation in one variable.}$$

$$4x + 6 - 4x = 6 \qquad \text{Apply the distributive property:}$$

$$2(3 - 2x) = 2 \cdot 3 - 2 \cdot 2x = 6 - 4x.$$

$$6 = 6, \quad \text{true} \qquad \text{Simplify: } 4x - 4x = 0.$$

This true statement indicates that the system contains dependent equations and has infinitely many solutions.

We express the solution set for the system in one of two equivalent ways:

$$\{(x, y) \mid y = 3 - 2x\} \quad \text{The set of all ordered pairs } (x, y) \text{ such that } y = 3 - 2x$$
$$\text{or } \{(x, y) \mid 4x + 2y = 6\}. \quad \text{The set of all ordered pairs } (x, y) \text{ such that } 4x + 2y = 6$$

■

✔ **CHECK POINT 4** Solve the system:

$$y = 3x - 4$$
$$9x - 3y = 12.$$

3 Solve problems using the substitution method.

Applications

An important application of systems of equations arises in connection with supply and demand. As the price of a product increases, the demand for that product decreases. However, at higher prices suppliers are willing to produce greater quantities of the product.

EXAMPLE 5   Supply and Demand Models

A chain of video stores specializes in cult films. The weekly demand and supply models for *The Rocky Horror Picture Show* are given by

$$N = -13p + 760 \quad \text{Demand model}$$
$$N = 2p + 430 \quad \text{Supply model}$$

in which p is the price of the video and N is the number of copies of the video sold or supplied each week to the chain of stores.

a. How many copies of the video can be sold and supplied at $18 per copy?

b. Find the price at which supply and demand are equal. At this price, how many copies of *Rocky Horror* can be supplied and sold each week?

Solution

a. To find how many copies of the video can be sold and supplied at $18 per copy, we substitute 18 for p in the demand and supply models.

Demand Model	**Supply Model**
$N = -13p + 760$	$N = 2p + 430$
Substitute 18 for p.	Substitute 18 for p.
$N = -13 \cdot 18 + 760 = 526$	$N = 2 \cdot 18 + 430 = 466$

At $18 per video, the chain can sell 526 copies of *Rocky Horror* in a week. The manufacturer is willing to supply 466 copies per week. This will result in a shortage of copies of the video. Under these conditions, the retail chain is likely to raise the price of the video.

b. We can find the price at which supply and demand are equal by solving the demand-supply linear system. We will use substitution, substituting $-13p + 760$ for N in the second equation.

$$N = \boxed{-13p + 760} \quad \boxed{N} = 2p + 430 \qquad \text{Substitute } -13p + 760 \text{ for } N.$$

$$-13p + 760 = 2p + 430 \qquad \begin{array}{l}\text{The resulting equation contains} \\ \text{only one variable.}\end{array}$$

$$-15p + 760 = 430 \qquad \text{Subtract } 2p \text{ from both sides.}$$

$$-15p = -330 \qquad \text{Subtract 760 from both sides.}$$

$$p = 22 \qquad \text{Divide both sides by } -15.$$

The price at which supply and demand are equal is $22 per video. To find the value of N, the number of videos supplied and sold weekly at this price, we back-substitute 22 for p into either the demand or the supply model. We'll use both models to make sure we get the same number in each case.

Demand Model

$N = -13p + 760$

Substitute 22 for p.

$N = -13 \cdot 22 + 760 = 474$

Supply Model

$N = 2p + 430$

Substitute 22 for p.

$N = 2 \cdot 22 + 430 = 474$

At a price of $22, 474 units of the video can be supplied and sold weekly. The intersection point, $(22, 474)$, is shown in Figure 4.7.

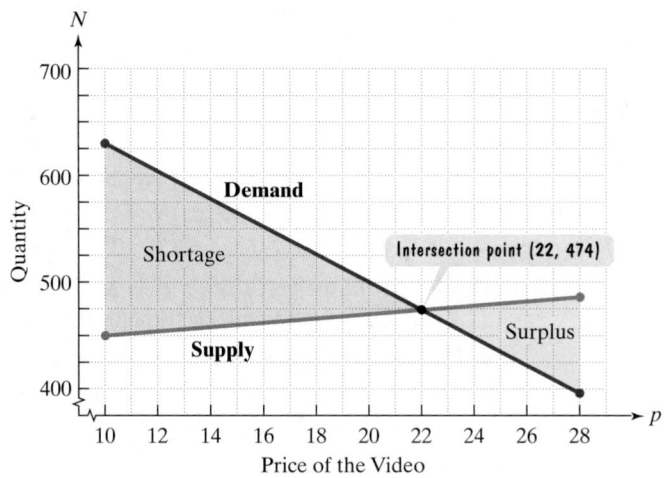

Figure 4.7 Priced at $22, 474 copies of the video can be supplied and sold weekly.

✔ **CHECK POINT 5** The demand for a product is modeled by $N = -20p + 1000$ and the supply for the product by $N = 5p + 250$. In these models, p is the price of the product and N is the number supplied or sold weekly. At what price will supply equal demand? At that price, how many units of the product will be supplied and sold each week?

EXERCISE SET 4.2

Practice Exercises

In Exercises 1–32, solve each system by the substitution method. If there is no solution or an infinite number of solutions, so state, and use set notation to express solution sets.

1. $x + y = 4$
   $y = 3x$

2. $x + y = 6$
   $y = 2x$

3. $x + 3y = 8$
   $y = 2x - 9$

4. $2x - 3y = -13$
   $y = 2x + 7$

5. $x + 3y = 5$
   $4x + 5y = 13$

6. $x + 2y = 5$
   $2x - y = -15$

7. $2x - y = -5$
   $x + 5y = 14$

8. $2x + 3y = 11$
   $x - 4y = 0$

9. $2x - y = 3$
   $5x - 2y = 10$

10. $-x + 3y = 10$
   $2x + 8y = -6$

11. $x + 8y = 6$
   $2x + 4y = -3$

12. $-4x + y = -11$
   $2x - 3y = 5$

13. $x = 9 - 2y$
   $x + 2y = 13$

14. $6x + 2y = 7$
   $y = 2 - 3x$

15. $y = 3x - 5$
   $21x - 35 = 7y$

16. $9x - 3y = 12$
   $y = 3x - 4$

17. $5x + 2y = 0$
$x - 3y = 0$
18. $4x + 3y = 0$
$2x - y = 0$
19. $2x + 5y = -4$
$3x - y = 11$
20. $2x + 5y = 1$
$-x + 6y = 8$
21. $2(x - 1) - y = -3$
$y = 2x + 3$
22. $x + y - 1 = 2(y - x)$
$y = 3x - 1$

23. $x = 4y - 2$
$x = 6y + 8$
24. $x = 3y + 7$
$x = 2y - 1$
25. $y = 2x - 8$
$y = 3x - 13$
26. $y = -3x - 1$
$y = -4x + 2$
27. $y = \dfrac{1}{3}x + \dfrac{2}{3}$

$y = \dfrac{5}{7}x - 2$

28. $y = -\dfrac{1}{2}x + 2$

$y = \dfrac{3}{4}x + 7$

29. $\dfrac{x}{6} - \dfrac{y}{2} = \dfrac{1}{3}$

$x + 2y = -3$
30. $\dfrac{x}{4} - \dfrac{y}{4} = -1$

$x + 4y = -9$
31. $2x - 3y = 8 - 2x$
$3x + 4y = x + 3y + 14$
32. $3x - 4y = x - y + 4$
$2x + 6y = 5y - 4$

Application Exercises

33. At a price of p dollars per ticket, the number of tickets to a rock concert that can be sold is given by the demand model $N = -25p + 7500$. At a price of p dollars per ticket, the number of tickets that the concert's promoters are willing to make available is given by the supply model $N = 5p + 6000$.
  a. How many tickets can be sold and supplied for $40 per ticket?
  b. Find the ticket price at which supply and demand are equal. At this price, how many tickets will be supplied and sold?

34. The weekly demand and supply models for a particular brand of scientific calculator for a chain of stores are given by the demand model $N = -53p + 1600$, and the supply model $N = 75p + 320$. In these models, p is the price of the calculator and N is the number of calculators sold or supplied each week to the stores.
  a. How many calculators can be sold and supplied at $12 per calculator?
  b. Find the price at which supply and demand are equal. At this price, how many calculators of this type can be supplied and sold each week?

A business breaks even when the cost for running the business is equal to the money taken in by the business. In Exercises 35–36, determine how many units must be sold so that a business breaks even, experiencing neither loss nor profit.

35. A gasoline station has weekly costs and revenue (the money taken in by the station) that depend on the number of gallons of gasoline purchased and sold. If x gallons are purchased and sold, weekly costs are given by $y = 1.2x + 1080$ and weekly revenue by $y = 1.6x$. How many gallons of gasoline must be sold weekly for the station to break even?

36. An artist has monthly costs and revenue (the money taken in by the artist) that depend on the number of ceramic pieces produced and sold. If x ceramic pieces are produced and sold, monthly costs are given by $y = 4x + 2000$ and monthly revenue by $y = 9x$. How many ceramic pieces must be sold monthly for the artist to break even?

37. The June 7, 1999 issue of *Newsweek* presents statistics showing progress African Americans have made in education, health, and finance. Infant mortality for blacks is decreasing at a faster rate than it is for whites, shown by the graphs below. Infant mortality for blacks can be modeled by $M = -0.41x + 22$ and for whites by $M = -0.18x + 10$. In both models, x is the number of years since 1980 and M is infant mortality, measured in deaths per 1000 live births. Use these models to project when infant mortality for blacks and whites will be the same. What is infant mortality for both groups at that time?

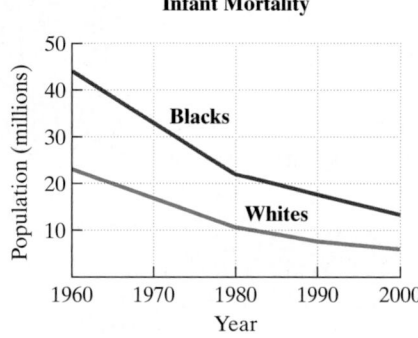

Infant Mortality

Source: National Center for Health Statistics

38. The equation $x + 10y = 2120$ models deaths from gunfire in the United States, y, in deaths per hundred thousand Americans, in year x. The equation

$$7x + 8y = 14,065$$

models deaths from car accidents in the United States, y, in deaths per hundred thousand Americans, in year x. Solve the linear system formed by the two models. Then describe what the solution means in terms of the variables in the given models.

Writing in Mathematics

39. Describe a problem that might arise when solving a system of equations using graphing. Assume that both equations in the system have been graphed correctly and the system has exactly one solution.

40. Explain how to solve a system of equations using the substitution method. Use $y = 3 - 3x$ and $3x + 4y = 6$ to illustrate your explanation.

41. When using the substitution method, how can you tell if a system of linear equations has no solution?

42. When using the substitution method, how can you tell if a system of linear equations has infinitely many solutions?

43. The law of supply and demand states that, in a free market economy, a commodity tends to be sold at its equilibrium price. At this price, the amount that the seller will supply is the same amount that the consumer will buy. Explain how systems of equations can be used to determine the equilibrium price.

Critical Thinking Exercises

44. Which one of the following is true?
   a. Solving an inconsistent system by substitution results in a true statement.
   b. The line passing through the intersection of the graphs of $x + y = 4$ and $x - y = 0$ with slope $= 3$ has an equation given by $y - 2 = 3(x - 2)$.
   c. Unlike the graphing method, where solutions cannot be seen, the substitution method provides a way to visualize solutions as intersection points.
   d. To solve the system

$$2x - y = 5$$
$$3x + 4y = 7$$

   by substitution, replace y in the second equation by $5 - 2x$.

45. If $x = 3 - y - z, 2x + y - z = -6$, and $3x - y + z = 11$, find the values for x, y, and z.

46. Find the value of m that makes

$$y = mx + 3$$
$$5x - 2y = 7$$

an inconsistent system.

Review Exercises

47. Graph: $4x + 6y = 12$. (Section 3.2, Example 4)

48. Solve: $4(x + 1) = 25 + 3(x - 3)$. (Section 2.3, Example 3)

49. List all the integers in this set: $\left\{-73, -\dfrac{2}{3}, 0, \dfrac{3}{1}, \dfrac{3}{2}, \dfrac{\pi}{1}\right\}$.

(Section 1.2, Example 5)

▶ **SECTION 4.3**   *Solving Systems of Linear Equations by the Addition Method*

Objectives

1 Solve linear systems by the addition method.

2 Use the addition method to identify systems with no solution or infinitely many solutions.

3 Determine the most efficient method for solving a linear system.

SSM   PH Tutor   CD-   Video
      Center   ROM

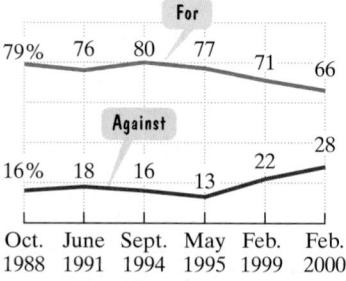

Are You in Favor of the Death Penalty for a Person Convicted of Murder?

The graphs shown are based on 543 adults polled nationally by *Newsweek*. If these trends continue, when will the percentage of Americans in favor of the death penalty be the same as the percentage of those who oppose it? The question can be answered by modeling the data with a system of linear equations and solving the system. However, the substitution method is not always the easiest way to solve linear systems. In this section we consider a third method for solving these systems.

1 Solve linear systems by the addition method.

Eliminating a Variable Using the Addition Method

The substitution method is most useful if one of the given equations has an isolated variable. A third, and frequently the easiest, method for solving a linear system is the addition method. Like the substitution method, the addition method involves eliminating a variable and ultimately solving an equation containing only one variable. However, this time we eliminate a variable by adding the equations.

For example, consider the following equations:

$$3x - 4y = 11$$
$$-3x + 2y = -7.$$

When we add these two equations, the x-terms are eliminated. This occurs because the coefficients of the x-terms, 3 and -3, are opposites (additive inverses) of each other.

$$3x - 4y = 11$$
$$\underline{-3x + 2y = -7}$$
Add: $\qquad -2y = 4$
$$y = -2 \qquad \text{Solve for } y, \text{ dividing both sides by } -2.$$

Now we can back-substitute -2 for y into one of the original equations to find x. It does not matter which equation you use; you will obtain the same value for x in either case. If we use either equation, we can show that $x = 1$ and the solution $(1, -2)$ satisfies both equations in the system.

When we use the addition method, we want to obtain two equations whose sum is an equation containing only one variable. The key step is to obtain, for one of the variables, coefficients that differ only in sign. To do this, we may need to multiply one or both equations by some nonzero number so that the coefficients of one of the variables, x or y, become opposites. Then when the two equations are added, this variable is eliminated.

EXAMPLE 1   Solving a System by the Addition Method

Solve by the addition method:

$$x + y = 4$$
$$x - y = 6.$$

Solution   The coefficients of y in the two equations, 1 and -1, differ only in sign. Therefore, by adding the two left sides and the two right sides, we can eliminate the y-terms.

$$x + y = 4$$
$$\underline{x - y = 6}$$
Add: $2x + 0y = 10$
$$2x = 10 \qquad \text{Simplify.}$$

Now y is eliminated and we can solve $2x = 10$ for x.

$$2x = 10$$
$$x = 5 \qquad \text{Divide both sides by 2 and solve for } x.$$

We back-substitute 5 for x into one of the original equations to find y. We will use both equations to show that we obtain the same value for y in either case.

Use the first equation:	Use the second equation:	
$x + y = 4$	$x - y = 6$	
$5 + y = 4$	$5 - y = 6$	Replace x with 5.
$y = -1$	$-y = 1$	Solve for y.
	$y = -1$	

Thus, $x = 5$ and $y = -1$. The proposed solution, $(5, -1)$, can be shown to satisfy both equations in the system. Consequently, the solution is $(5, -1)$ and the solution set is $\{(5, -1)\}$. ∎

✔ **CHECK POINT 1** Solve by the addition method:
$$x + y = 5$$
$$x - y = 9.$$

EXAMPLE 2   Solving a System by the Addition Method

Solve by the addition method:
$$3x - \ y = 11$$
$$2x + 5y = 13.$$

Solution   We must rewrite one or both equations in equivalent forms so that the coefficients of the same variable (either x or y) differ only in sign. Consider the terms in y in each equation, that is, $-1y$ and $5y$. To eliminate y, we can multiply each term of the first equation by 5 and then add the equations.

$$3x - \ y = 11 \xrightarrow{\text{Multiply by 5.}} 15x - 5y = 55$$
$$2x + 5y = 13 \xrightarrow{\text{No change}} 2x + 5y = 13$$

$$\text{Add:} \qquad 17x + 0y = 68$$
$$17x = 68 \qquad \text{Simplify.}$$
$$x = 4 \qquad \text{Divide both sides by 17 and solve for x.}$$

Thus, $x = 4$. To find y, we back-substitute 4 for x into either one of the given equations. We'll use the second one.

$2x + 5y = 13$	This is the second equation in the given system.
$2 \cdot 4 + 5y = 13$	Substitute 4 for x.
$8 + 5y = 13$	Multiply: $2 \cdot 4 = 8$.
$5y = 5$	Subtract 8 from both sides.
$y = 1$	Divide both sides by 5.

The solution is $(4, 1)$. Check to see that it satisfies both of the original equations in the system. The solution set is $\{(4, 1)\}$ ∎

✔ **CHECK POINT 2** Solve by the addition method:
$$4x - \ y = 22$$
$$3x + 4y = 26.$$

Before considering additional examples, let's summarize the steps for solving linear systems by the addition method.

Solving Linear Systems by Addition

1. If necessary, rewrite both equations in the form $Ax + By = C$.
2. If necessary, multiply either equation or both equations by appropriate nonzero numbers so that the sum of the x-coefficients or the sum of the y-coefficients is 0.
3. Add the equations in step 2. The sum is an equation in one variable.
4. Solve the equation in one variable.
5. Back-substitute the value obtained in step 4 into either of the given equations and solve for the other variable.
6. Check the solution in both of the original equations.

EXAMPLE 3   Solving a System by the Addition Method

Solve by the addition method:

$$3x + 2y = 48$$
$$9x - 8y = -24.$$

Solution

Step 1 Rewrite both equations in the form $Ax + By = C$. Both equations are already in this form. Variable terms appear on the left and constants appear on the right.

Step 2 If necessary, multiply either equation or both equations by appropriate numbers so that the sum of the x-coefficients or the sum of the y-coefficients is 0. We can eliminate x or y. Let's eliminate x. Consider the terms in x in each equation, that is, $3x$ and $9x$. To eliminate x, we can multiply each term of the first equation by -3 and then add the equations.

$$3x + 2y = 48 \quad \xrightarrow{\text{Multiply by } -3.} \quad -9x - 6y = -144$$
$$9x - 8y = -24 \quad \xrightarrow{\text{No change}} \quad 9x - 8y = -24$$

Step 3 Add the equations. $\qquad\qquad$ Add: $\qquad -14y = -168$

Step 4 Solve the equation in one variable. We solve $-14y = -168$ by dividing both sides by -14.

$$\frac{-14y}{-14} = \frac{-168}{-14} \qquad \text{Divide both sides by } -14.$$
$$y = 12 \qquad \text{Simplify.}$$

Step 5 Back-substitute and find the value for the other variable. We can back-substitute 12 for y into either one of the given equations. We'll use the first one.

$$3x + 2y = 48 \qquad \text{This the first equation in the given system.}$$
$$3x + 2(12) = 48 \qquad \text{Substitute 12 for } y.$$
$$3x + 24 = 48 \qquad \text{Multiply.}$$
$$3x = 24 \qquad \text{Subtract 24 from both sides.}$$
$$x = 8 \qquad \text{Divide both sides by 3.}$$

The solution is $(8, 12)$.

Step 6 Check. Take a few minutes to show that $(8, 12)$ satisfies both of the original equations in the system. The solution set is $\{(8, 12)\}$. ∎

✔ **CHECK POINT 3**   Solve by the addition method:

$$4x + 5y = 3$$
$$2x - 3y = 7.$$

Some linear systems have solutions that are not integers. If the value of one variable turns out to be a "messy" fraction, back-substitution might lead to cumbersome arithmetic. If this happens, you can return to the original system and use addition to find the value of the other variable.

EXAMPLE 4   Solving a System by the Addition Method

Solve by the addition method:

$$2x = 7y - 17$$
$$5y = 17 - 3x.$$

Solution

Step 1 Rewrite both equations in the form $Ax + By = C$. We first arrange the system so that variable terms appear on the left and constants appear on the right. We obtain:

$$2x - 7y = -17 \qquad \text{Subtract 7y from both sides of the first equation.}$$
$$3x + 5y = 17. \qquad \text{Add 3x to both sides of the second equation.}$$

Step 2 If necessary, multiply either equation or both equations by appropriate numbers so that the sum of the x-coefficients or the sum of the y-coefficients is 0. We can eliminate x or y. Let's eliminate x by multiplying the first equation by 3 and the second equation by -2.

$$2x - 7y = -17 \xrightarrow{\text{Multiply by 3.}} 3 \cdot 2x - 3 \cdot 7y = 3(-17) \longrightarrow 6x - 21y = -51$$

$$3x + 5y = 17 \xrightarrow{\text{Multiply by } -2.} -2 \cdot 3x + (-2) \cdot 5y = -2(17) \longrightarrow \underline{-6x - 10y = -34}$$

Step 3 Add the equations.   Add: $-31y = -85$

Step 4 Solve the equation in one variable. We solve $-31y = -85$ by dividing both sides by -31.

$$\frac{-31y}{-31} = \frac{-85}{-31} \qquad \text{Divide both sides by } -31.$$

$$y = \frac{85}{31} \qquad \text{Simplify.}$$

Step 5 Back-substitute and find the value for the other variable. Back-substitution of $\frac{85}{31}$ for y into either of the given equations results in cumbersome arithmetic. Instead, let's use the addition method on the given system in the form

$Ax + By = C$ to find the value for x. Thus, we eliminate y by multiplying the first equation by 5 and the second equation by 7.

$$2x - 7y = -17 \xrightarrow{\text{Multiply by 5.}} 10x - 35y = -85$$
$$3x + 5y = 17 \xrightarrow{\text{Multiply by 7.}} 21x + 35y = 119$$
$$\text{Add:} \qquad 31x = 34$$
$$x = \frac{34}{31} \qquad \text{Divide both sides by 31.}$$

The solution is $\left(\dfrac{34}{31}, \dfrac{85}{31} \right)$.

Step 6 Check. For this system, a calculator is helpful in showing that $\left(\frac{34}{31}, \frac{85}{31} \right)$ satisfies both of the original equations in the system. The solution set is $\left\{ \left(\frac{34}{31}, \frac{85}{31} \right) \right\}$. ∎

✔ **CHECK POINT 4** Solve by the addition method:
$$2x = 9 + 3y$$
$$4y = 8 - 3x.$$

2 Use the addition method to identify systems with no solution or infinitely many solutions.

The Addition Method with Linear Systems Having No Solution or Infinitely Many Solutions

As with the substitution method, if the addition method results in a false statement, the linear system is inconsistent and has no solution.

EXAMPLE 5 Using the Addition Method on an Inconsistent System

Solve the system:
$$4x + 6y = 12$$
$$6x + 9y = 12.$$

Solution We can eliminate x or y. Let's eliminate x by multiplying the first equation by 3 and the second equation by -2.

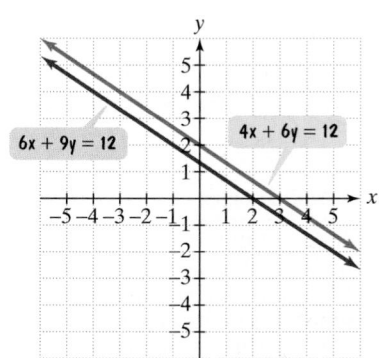

$$4x + 6y = 12 \xrightarrow{\text{Multiply by 3.}} 12x + 18y = 36$$
$$6x + 9y = 12 \xrightarrow{\text{Multiply by -2.}} -12x - 18y = -24$$
$$\text{Add:} \qquad 0 = 12$$

There are no values of x and y for which $0 = 12$.

Figure 4.8 Visualizing the inconsistent system in Example 5

The false statement $0 = 12$ indicates that the system is inconsistent and has no solution. The solution set is the empty set, $\varnothing$. The graphs of the system's equations are shown in Figure 4.8. The lines are parallel and have no point of intersection. ∎

✔ **CHECK POINT 5** Solve the system:
$$x + 2y = 4$$
$$3x + 6y = 13.$$

If you use the addition method, how can you tell if a system has infinitely many solutions? As with the substitution method, you will eliminate both variables and obtain a true statement.

EXAMPLE 6 Using the Addition Method on a System with Infinitely Many Solutions

Solve by the addition method:

$$2x - y = 3$$
$$-4x + 2y = -6.$$

Solution We can eliminate y by multiplying the first equation by 2.

$$
\begin{array}{rcl}
2x - y = 3 & \xrightarrow{\text{Multiply by 2.}} & 4x - 2y = 6 \\
-4x + 2y = -6 & \xrightarrow{\text{No change}} & -4x + 2y = -6 \\
& \text{Add:} & 0 = 0
\end{array}
$$

The true statement $0 = 0$ indicates that the system contains dependent equations and has infinitely many solutions. Any ordered pair that satisfies the first equation also satisfies the second equation. We express the solution set for the system in one of two equivalent ways:

$$\{(x, y) \mid 2x - y = 3\} \text{ or } \{(x, y) \mid -4x + 2y = -6\}. \qquad \blacksquare$$

✔ **CHECK POINT 6** Solve by the addition method:

$$x - 5y = 7$$
$$3x - 15y = 21.$$

3 Determine the most efficient method for solving a linear system.

Comparing the Three Solution Methods

The following chart compares the graphing, substitution, and addition methods for solving systems of linear equations. With increased practice, it becomes easier for you to select the best method for solving a particular linear system.

Method	Advantages	Disadvantages
Graphing	You can see the solutions.	If the solution does not involve integers or is too large to be seen on the graph, it's impossible to tell exactly what the solutions are.
Substitution	Gives exact solutions.	Solutions cannot be seen.
	Easy to use if a variable is on one side by itself.	Can introduce extensive work with fractions when no variable has a coefficient of 1 or −1.
Addition	Gives exact solutions.	Solutions cannot be seen.
	Easy to use if no variable has a coefficient of 1 or −1.	

EXERCISE SET 4.3

Practice Exercises

In Exercises 1–44, solve each system by the addition method. If there is no solution or an infinite number of solutions, so state, and use set notation to express solution sets.

1. $x + y = 1$
$x - y = 3$

2. $x + y = 6$
$x - y = -2$

3. $2x + 3y = 6$
$2x - 3y = 6$

4. $3x + 2y = 14$
$3x - 2y = 10$

5. $x + 2y = 7$
$-x + 3y = 18$

6. $2x + y = -2$
$-2x - 3y = -6$

7. $5x - y = 14$
$-5x + 2y = -13$

8. $7x - 4y = 13$
$-7x + 6y = -11$

9. $x + 2y = 2$
$-4x + 3y = 25$

10. $2x - y = -7$
$3x + 2y = 0$

11. $2x - 7y = 2$
$3x + y = -20$

12. $5x + 2y = -7$
$x + 3y = 9$

13. $x + 5y = -1$
$2x + 7y = 1$

14. $2x + y = 1$
$6x + 5y = 13$

15. $4x + 3y = 15$
$2x - 5y = 1$

16. $3x - 7y = 13$
$6x + 5y = 7$

17. $3x - 4y = 11$
$2x + 3y = -4$

18. $2x + 3y = -16$
$5x - 10y = 30$

19. $3x + 2y = -1$
$-2x + 7y = 9$

20. $5x + 3y = 27$
$7x - 2y = 13$

21. $3x = 2y + 7$
$5x = 2y + 13$

22. $9x = 25 + y$
$2y = 4 - 9x$

23. $2x = 3y - 4$
$-6x + 12y = 6$

24. $5x = 4y - 8$
$3x + 7y = 14$

25. $2x - y = 3$
$4x + 4y = -1$

26. $3x - y = 22$
$4x + 5y = -21$

27. $4x = 5 + 2y$
$2x + 3y = 4$

28. $3x = 4y + 1$
$4x + 3y = 1$

29. $3x - y = 1$
$3x - y = 2$

30. $4x - 9y = -2$
$-4x + 9y = -2$

31. $x + 3y = 2$
$3x + 9y = 6$

32. $4x - 2y = 2$
$2x - y = 1$

33. $7x - 3y = 4$
$-14x + 6y = -7$

34. $2x + 4y = 5$
$3x + 6y = 6$

35. $5x + y = 2$
$3x + y = 1$

36. $2x - 5y = -1$
$2x - y = 1$

37. $x = 5 - 3y$
$2x + 6y = 10$

38. $4x = 36 + 8y$
$3x - 6y = 27$

39. $4(3x - y) = 0$
$3(x + 3) = 10y$

40. $2(2x + 3y) = 0$
$7x = 3(2y + 3) + 2$

41. $x + y = 11$
$\dfrac{x}{5} + \dfrac{y}{7} = 1$

42. $x - y = -3$

$\dfrac{x}{9} - \dfrac{y}{7} = -1$

43. $\dfrac{4}{5}x - y = -1$

$\dfrac{2}{5}x + y = 1$

44. $\dfrac{x}{3} + y = 3$

$\dfrac{x}{2} - \dfrac{y}{4} = 1$

In Exercises 45–56, solve each system by the method of your choice. If there is no solution or an infinite number of solutions, so state, and use set notation to express solution sets. Explain why you selected one method over the other two.

45. $3x - 2y = 8$

$x = -2y$

46. $2x - y = 10$

$y = 3x$

47. $3x + 2y = -3$

$2x - 5y = 17$

48. $2x - 7y = 17$

$4x - 5y = 25$

49. $3x - 2y = 6$

$y = 3$

50. $2x + 3y = 7$

$x = 2$

51. $y = 2x + 1$

$y = 2x - 3$

52. $y = 2x + 4$

$y = 2x - 1$

53. $2(x + 2y) = 6$

$3(x + 2y - 3) = 0$

54. $2(x + y) = 4x + 1$

$3(x - y) = x + y - 3$

55. $3y = 2x$

$2x + 9y = 24$

56. $4y = -5x$

$5x + 8y = 20$

Writing in Mathematics

57. Explain how to solve a system of equations using the addition method. Use $3x + 5y = -2$ and $2x + 3y = 0$ to illustrate your explanation.

58. When using the addition method, how can you tell if a system of linear equations has no solution?

59. When using the addition method, how can you tell if a system of linear equations has infinitely many solutions?

60. Take a second look at the data about the death penalty shown on page 296. Do you think that these trends will continue? Explain your answer.

61. The formula $3239x + 96y = 134{,}014$ models the number of daily evening newspapers, y, x years after 1980. The formula $-665x + 36y = 13{,}800$ models the number of daily morning newspapers, y, x years after 1980. What is the most efficient method for solving this system? Explain why. What does the solution mean in terms of the variables in the formulas? (It is not necessary to actually solve the system.)

Critical Thinking Exercises

62. Which one of the following statements is true?
  a. Once x is eliminated by the addition method, y cannot be eliminated by using the original equations of the system.
  b. If $Ax + 2y = 2$ and $2x + By = 10$ have graphs that intersect at $(2, -2)$, then $A = -3$ and $B = 3$.
  c. The equations $y = x - 1$ and $x = y + 1$ are dependent.
  d. If the two equations in a linear system are $5x - 3y = 7$ and $4x + 9y = 11$, multiplying the first equation by 4, the second by 5, and then adding equations will eliminate x.

In Exercises 63–64, solve each system using the addition method. First clear fractions by multiplying both sides of each equation by the least common denominator.

63. $\dfrac{3x}{5} + \dfrac{4y}{5} = 1$

$\dfrac{x}{4} - \dfrac{3y}{8} = -1$

64. $\dfrac{x}{3} - \dfrac{y}{2} = \dfrac{2}{3}$

$\dfrac{2x}{3} + y = \dfrac{4}{3}$

In Exercises 65–66, solve each system using the addition method.

65. $0.5x - 0.2y = 0.5$

$0.4x + 0.7y = 0.4$

66. $0.02x - 0.04y = 0.26$

$0.07x - 0.09y = 0.66$

Technology Exercises

67. Some graphing utilities can give the solution to a linear system of equations. (Consult your manual for details.) This capability is usually accessed with the $\boxed{\text{SIMULT}}$ (simultaneous equations) feature. First, you will enter 2, for two equations in two variables. With each equation in $Ax + By = C$ form, you will then enter the coefficients for x and y and the constant term, one equation at a time. After entering all six numbers, press $\boxed{\text{SOLVE}}$. The solution will be displayed on the screen. (The x-value may be displayed as $x_1 =$ and the y-value as $x_2 = $.) Use this capability to verify the solution to any five of the exercises you solved in the practice exercises of this exercise set. Describe what happens when you use your graphing utility on a system with no solution or infinitely many solutions.

If your graphing utility has the feature described in Exercise 67, use it to solve each system in Exercises 68–70.

68. $\dfrac{1}{4} x - \dfrac{1}{4} y = -1$

$-3x + 7y = 8$

69. $x = 5y$

$2x - 3y = 7$

70. $0.6x + 0.08y = 4$

$3x + 2y = 4$

Review Exercises

71. For which number is 5 times the number equal to the number increased by 40? (Section 2.5, Example 2)

72. In which quadrant is $\left(-\frac{3}{2}, 15\right)$ located? (Section 1.3, Example 1)

73. Solve: $29,700 + 150x = 5000 + 1100x$. (Section 2.2, Example 7)

▶ SECTION 4.4   *Problem Solving Using Systems of Equations*

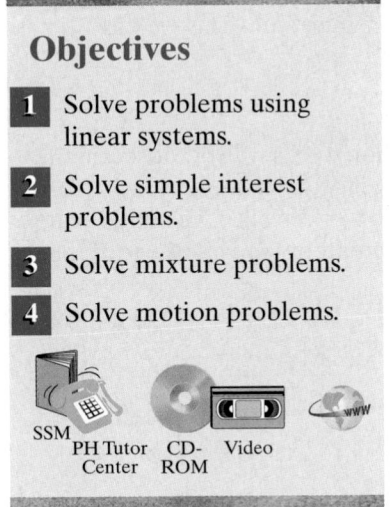

Objectives

1 Solve problems using linear systems.

2 Solve simple interest problems.

3 Solve mixture problems.

4 Solve motion problems.

SSM
PH Tutor   CD-   Video
Center   ROM

Enjoy chatting long distance on the phone? Telecommunication companies want your business. You can go online and get a list of their plans. Is there a monthly fee? What is the rate per minute? Does the plan involve a monthly minimum? In this section, you will learn to use systems of equations to select a plan that will save you the most money.

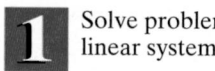

Solve problems using linear systems.

A Strategy for Solving Word Problems Using Systems of Equations

When we solved problems in Chapter 2, we let x represent a quantity that was unknown. Problems in this section involve two unknown quantities. We will let x and y represent these quantities. We then translate from the verbal conditions of the problem to a *system* of linear equations.

EXAMPLE 1   The World's Longest Snakes

The royal python and the anaconda are the world's longest snakes. The maximum length for each of these snakes is implied by the following description:

> Three royal pythons and two anacondas measure 161 feet. The royal python's length increased by triple the anaconda's length is 119 feet. Find the maximum length for each of these snakes.

Solution

Step 1 Use variables to represent unknown quantities. Let x represent the royal python's length, in feet. Let y represent the anaconda's length, in feet.

Step 2 Write a system of equations describing the problem's conditions.

Three royal pythons	and	two anacondas	measure	161 feet.
$3x$	$+$	$2y$	$=$	161

The royal python's length	increased by	triple the anaconda's length	is	119 feet.
x	$+$	$3y$	$=$	119

Step 3 Solve the system and answer the problem's question. The system

$$3x + 2y = 161$$
$$x + 3y = 119$$

can be solved by substitution or addition. Substitution works well because x in the second equation has a coefficient of 1. We can solve for x by subtracting $3y$ from both sides, thereby avoiding fractions. Addition also works well; if we multiply the second equation by -3, adding equations will eliminate x. We will use addition.

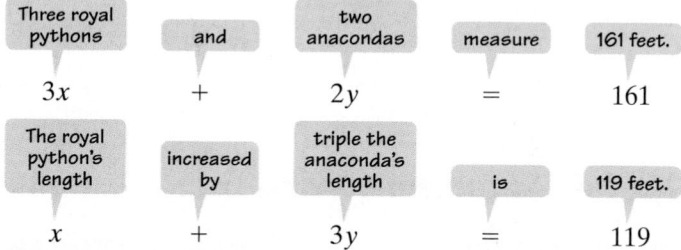

$$
\begin{array}{ll}
3x + 2y = 161 & \xrightarrow{\text{No change}} \\
x + 3y = 119 & \xrightarrow{\text{Multiply by } -3.}
\end{array}
\quad
\begin{array}{l}
3x + 2y = 161 \\
-3x - 9y = -357 \\
\hline
{-7y} = -196 \\
y = \dfrac{-196}{-7} = 28
\end{array}
$$

Add:

Because y represents the anaconda's length, we see that the anaconda is 28 feet long. Now we can find x, the royal python's length. We do so by back-substituting 28 for y in either of the system's equations.

$$x + 3y = 119 \qquad \text{We'll use the second equation.}$$
$$x + 3 \cdot 28 = 119 \qquad \text{Back-substitute 28 for } y.$$
$$x + 84 = 119 \qquad \text{Multiply: } 3 \cdot 28 = 84.$$
$$x = 35 \qquad \text{Subtract 84 from both sides.}$$

Because $x = 35$ and $y = 28$, a royal python is 35 feet long and the anaconda is 28 feet long.

Step 4 Check the proposed answers in the original wording of the problem. Three royal pythons and two anacondas should measure 161 feet:

$$3(35 \text{ ft}) + 2(28 \text{ ft}) = 105 \text{ ft} + 56 \text{ ft} = 161 \text{ ft}.$$

The royal python's length increased by triple the anaconda's length should be 119 feet:

$$35 \text{ ft} + 3(28 \text{ ft}) = 35 \text{ ft} + 84 \text{ ft} = 119 \text{ ft}.$$

This verifies that the royal python's and the anaconda's length are 35 feet and 28 feet, respectively. ∎

✔ **CHECK POINT 1** The bustard and the condor are the world's heaviest flying bird and the world's heaviest bird of prey, respectively. The maximum weight for each of these birds is implied by the following description:

> Two bustards and three condors weigh 173 pounds. The bustard's weight increased by double the condor's weight is 100 pounds. What is the maximum weight for each of these birds?

EXAMPLE 2 Cholesterol and Heart Disease

The verdict is in. After years of research, the nation's health experts agree that high cholesterol in the blood is a major contributor to heart disease. Thus, cholesterol intake should be limited to 300 milligrams or less each day. Fast foods provide a cholesterol carnival. All together, two McDonald's Quarter Pounders and three Burger King Whoppers with cheese contain 520 milligrams of cholesterol. Three Quarter Pounders and one Whopper with cheese exceed the suggested daily cholesterol intake by 53 milligrams. Determine the cholesterol content in each item.

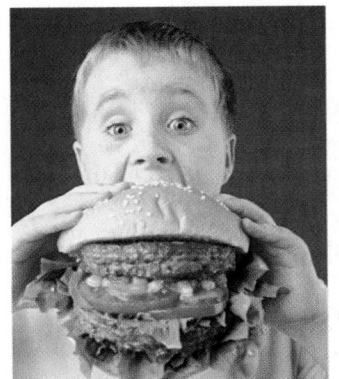

About 15 million hamburgers are eaten every day in the United States.

Solution

Step 1 Use variables to represent the unknown quantities. Let x represent the cholesterol content, in milligrams, of a Quarter Pounder and y the cholesterol content, in milligrams, of a Whopper with cheese.

Step 2 Write a system of equations describing the problem's conditions.

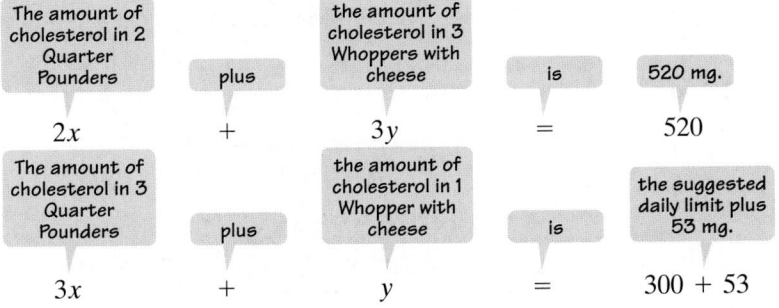

The amount of cholesterol in 2 Quarter Pounders	plus	the amount of cholesterol in 3 Whoppers with cheese	is	520 mg.
$2x$	$+$	$3y$	$=$	520

The amount of cholesterol in 3 Quarter Pounders	plus	the amount of cholesterol in 1 Whopper with cheese	is	the suggested daily limit plus 53 mg.
$3x$	$+$	y	$=$	$300 + 53$

Step 3 Solve the system and answer the problem's question. The system

$$2x + 3y = 520$$
$$3x + y = 353$$

can be solved by substitution or addition. We will use addition; if we multiply the second equation by -3, adding equations will eliminate y.

$$
\begin{array}{lll}
2x + 3y = 520 & \xrightarrow{\text{No change}} & 2x + 3y = 520 \\
3x + y = 353 & \xrightarrow{\text{Multiply by }-3.} & -9x - 3y = -1059 \\
& \text{Add:} & \overline{-7x = -539} \\
& & x = \dfrac{-539}{-7} = 77
\end{array}
$$

Because x represents the cholesterol content of a Quarter Pounder, we see that a Quarter Pounder contains 77 milligrams of cholesterol. Now we can find y, the cholesterol content of a Whopper with cheese. We do so by back-substituting 77 for x in either of the system's equations.

$$3x + y = 353 \qquad \text{We'll use the second equation.}$$
$$3(77) + y = 353 \qquad \text{Back-substitute 77 for x.}$$
$$231 + y = 353 \qquad \text{Multiply.}$$
$$y = 122 \qquad \text{Subtract 231 from both sides.}$$

Because $x = 77$ and $y = 122$, a Quarter Pounder contains 77 milligrams of cholesterol and a Whopper with cheese contains 122 milligrams of cholesterol.

Step 4 Check the proposed answers in the original wording of the problem. Two Quarter Pounders and three Whoppers with cheese contain

$$2(77 \text{ mg}) + 3(122 \text{ mg}) = 520 \text{ mg,}$$

which checks with the given conditions. Furthermore, three Quarter Pounders and one Whopper with cheese contain

$$3(77 \text{ mg}) + 1(122 \text{ mg}) = 353 \text{ mg,}$$

which does exceed the daily limit of 300 milligrams by 53 milligrams. ∎

Top Meat Eaters

Country	Pounds Consumed per Person per Year
United States	257
Australia	229
New Zealand	222

Source: International Food Policy Research Institute

✔ **CHECK POINT 2** How do the Quarter Pounder and Whopper with cheese measure up in the calorie department? Actually, not too well. Two Quarter Pounders and three Whoppers with cheese provide 2607 calories. Even one of each provides enough calories to bring tears to Jenny Craig's eyes—9 calories in excess of what is allowed on a 1000 calorie-a-day diet. Find the caloric content of each item.

EXAMPLE 3   Solar and Electric Heating Systems

The costs for two different kinds of heating systems for a three-bedroom home are given in the following table.

System	Cost to Install	Operating Cost/Year
Solar	$29,700	$150
Electric	$5000	$1100

After how many years will total costs for solar heating and electric heating be the same? What will be the cost at that time?

Solution

Step 1 Use variables to represent unknown quantities. Let x represent the number of years the heating system is used. Let y represent the total cost for the heating system.

Step 2 Write a system of equations describing the problem's conditions.

System	Cost to Install	Operating Cost/Year
Solar	$29,700	$150
Electric	$5000	$1100

Costs for heating systems, repeated

Total cost for the solar system	equals	installation cost	plus	yearly operating cost	times	the number of years the system is used.
y	$=$	29,700	$+$	150	$\cdot$	x

Total cost for the electric system	equals	installation cost	plus	yearly operating cost	times	the number of years the system is used.
y	$=$	5000	$+$	1100	$\cdot$	x

Step 3 Solve the system and answer the problem's question. We want to know after how many years total costs for the two systems will be the same. We must solve the system

$$y = 29{,}700 + 150x$$

$$y = 5000 + 1100x.$$

Substitution works well because y is isolated in each equation.

$y = \boxed{29{,}700 + 150x} \quad \boxed{y} = 5000 + 1100x$ Substitute 29,700 + 150x for y.

$29{,}700 + 150x = 5000 + 1100x$ This substitution gives an equation in one variable.

$29{,}700 = 5000 + 950x$ Subtract 150x from both sides.

$24{,}700 = 950x$ Subtract 5000 from both sides.

$26 = x$ Divide both sides by 950: $\dfrac{24{,}700}{950} = 26.$

Because x represents the number of years the heating system is used, we see that after 26 years, the total costs for the two systems will be the same. Now we can find y, the total cost. Back-substitute 26 for x in either of the system's equations. We will use the second equation, $y = 5000 + 1100x$.

$$y = 5000 + 1100 \cdot 26 = 5000 + 28{,}600 = 33{,}600$$

Because $x = 26$ and $y = 33{,}600$, after 26 years, the total costs for the two systems will be the same. The cost for each system at that time will be $33,600.

Step 4 Check the proposed answers in the original wording of the problem.
Let's verify that after 26 years, the two systems will cost the same amount.
The installation cost for the solar system is $29,700 and the yearly operating cost
is $150. Thus, the total cost after 26 years is:

$$\$29,700 + \$150(26) = \$29,700 + \$3900 = \$33,600.$$

The installation cost for the electric system is $5000 and the yearly operating cost
is $1100. Thus, the total cost after 26 years is:

$$\$5000 + \$1100(26) = \$5000 + \$28,600 = \$33,600.$$

This verifies that after 26 years the two systems will cost the same amount, namely $33,600.   ■

✔ **CHECK POINT 3**   Costs for two different kinds of heating systems for a three-
bedroom house are given in the following table.

System	Cost to Install	Operating Cost/Year
Electric	$5000	$1100
Gas	$12,000	$700

After how long will total costs for electric heating and gas heating be the same?
What will be the cost at that time?

Next, we will solve problems involving investments, mixtures, and motion
with systems of equations. We will continue using our four-step problem-solving
strategy. We will also use tables to help organize the information in the problems.

2 Solve simple interest problems.

Dual Investments with Simple Interest

Simple interest involves interest calculated only on the amount of money that we
invest, called the **principal**. The formula $I = Pr$ is used to find the simple interest, I, earned for one year when the principal, P, is invested at an annual interest
rate, r. Dual investment problems involve different amounts of money in two or
more investments, each paying a different rate.

EXAMPLE 4   Solving a Dual Investment Problem

Your grandmother needs your help. She has $50,000 to invest. Part of this money
is to be invested in noninsured bonds paying 15% annual interest. The rest of this
money is to be invested in a government-insured certificate of deposit paying
7% annual interest. She told you that she requires $6000 per year in extra income
from both of these investments. How much money should be placed in each
investment?

Solution

Step 1 Use variables to represent unknown quantities.
Let x = the amount invested in the 15% noninsured bonds.
Let y = the amount invested in the 7% certificate of deposit.

Step 2 Write a system of equations describing the problem's conditions. Because Grandma has $50,000 to invest,

The amount invested at 15%	plus	the amount invested at 7%	equals	$50,000.
x	$+$	y	$=$	50,000

Furthermore, Grandma requires $6000 in total interest. We can use a table to organize the information in the problem and obtain a second equation.

	Principal (amount invested)	$\times$	Interest rate	$=$	Interest earned
15% Investment	x		0.15		$0.15x$
7% Investment	y		0.07		$0.07y$

The interest for the two investments combined must be $6000.

Interest from the 15% investment	plus	interest from the 7% investment	is	$6000.
$0.15x$	$+$	$0.07y$	$=$	6000

Step 3 Solve the system and answer the problem's question. The system

$$x + y = 50{,}000$$

$$0.15x + 0.07y = 6000$$

can be solved by substitution or addition. Substitution works well because both variables in the first equation have coefficients of 1. Addition also works well; if we multiply the first equation by -0.15 or -0.07, adding equations will eliminate a variable. We will use addition.

$$
\begin{array}{l}
x + y = 50{,}000 \xrightarrow{\text{Multiply by } -0.07.} -0.07x - 0.07y = -3500 \\
0.15x + 0.07y = 6000 \xrightarrow{\quad \text{No change} \quad} 0.15x + 0.07y = 6000 \\
\hline
\qquad\qquad\qquad\qquad\qquad \text{Add:} \quad 0.08x \qquad\qquad = 2500
\end{array}
$$

$$x = \frac{2500}{0.08} = 31{,}250$$

Because x represents the amount that should be invested at 15%, Grandma should place $31,250 in 15% noninsured bonds. Now we can find y, the amount that she should place in the 7% certificate of deposit. We do so by back-substituting 31,250 for x in either of the system's equations.

$$x + y = 50{,}000 \qquad \text{We'll use the first equation.}$$

$$31{,}250 + y = 50{,}000 \qquad \text{Back-substitute 31,250 for x.}$$

$$y = 18{,}750 \qquad \text{Subtract 31,250 from both sides.}$$

Because $x = 31{,}250$ and $y = 18{,}750$, Grandma should invest $31,250 at 15% and $18,750 at 7%.

Step 4 Check the proposed answers in the original wording of the problem. Has Grandma invested $50,000?

$$\$31{,}250 + \$18{,}750 = \$50{,}000$$

Yes, all her money was placed in the dual investments. Can she count on $6000 interest? The interest earned on $31,250 at 15% is ($31,250)(0.15), or $4687.50. The interest earned on $18,750 at 7% is ($18,750)(0.07), or $1312.50. The total interest is $4687.50 + $1312.50, or $6000, exactly as it should be. You've made your grandmother happy. (Now if you would just visit her more often…) ∎

✔ **CHECK POINT 4**   You inherited $16,000 with the stipulation that for the first year the money had to be invested in two accounts paying 6% and 8% annual interest. How much did you invest at each rate if the total interest earned for the year was $1180?

3 Solve mixture problems.

Problems Involving Mixtures

Chemists and pharmacists often have to change the concentration of solutions and other mixtures. In these situations, the amount of a particular ingredient in the solution or mixture is expressed as a percentage of the total.

EXAMPLE 5   Solving a Mixture Problem

A chemist working on a flu vaccine needs to mix a 10% sodium-iodine solution with a 60% sodium-iodine solution to obtain 50 milliliters of a 30% sodium-iodine solution. How many milliliters of the 10% solution and of the 60% solution should be mixed?

Solution

Step 1 Use variables to represent unknown quantities.
Let x = the number of milliliters of the 10% solution to be used in the mixture.
Let y = the number of milliliters of the 60% solution to be used in the mixture.

Step 2 Write a system of equations describing the problem's conditions. The situation is illustrated in Figure 4.9.

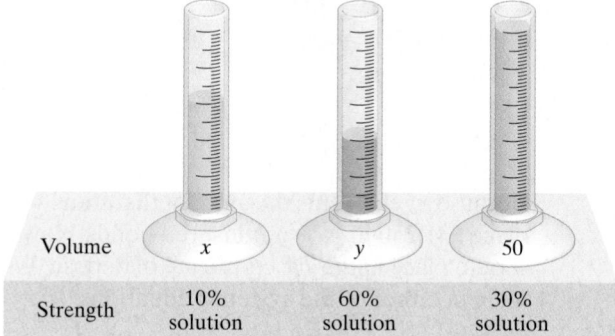

Figure 4.9

The chemist needs 50 milliliters of a 30% sodium-iodine solution. We form a table that shows the amount of sodium-iodine in each of the three solutions.

Solution	Number of milliliters	×	Percent of Sodium-Iodine	=	Amount of Sodium-Iodine
10% Solution	x		10% = 0.1		$0.1x$
60% Solution	y		60% = 0.6		$0.6y$
30% Mixture	50		30% = 0.3		$0.3(50) = 15$

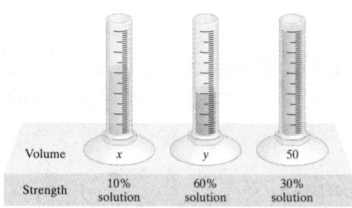

Figure 4.9, repeated

The chemist needs to obtain a 50-milliliter mixture.

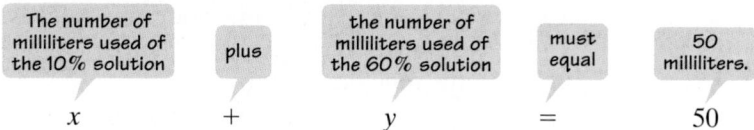

$$x \quad + \quad y \quad = \quad 50$$

The 50-milliliter mixture must be 30% sodium-iodine. The amount of sodium-iodine must be 30% of 50, or $(0.3)(50) = 15$ milliliters.

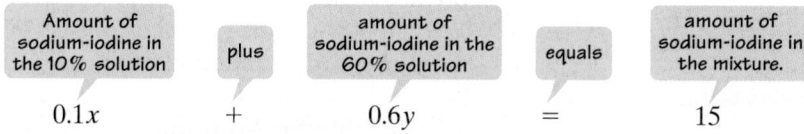

$$0.1x \quad + \quad 0.6y \quad = \quad 15$$

Step 3 Solve the system and answer the problem's question. The system

$$x + y = 50$$
$$0.1x + 0.6y = 15$$

can be solved by substitution or addition. Let's use substitution. Solving the first equation for y, we obtain $y = 50 - x$.

$$y = \boxed{50 - x} \qquad 0.1x + 0.6\boxed{y} = 15$$

We substitute $50 - x$ for y in the second equation. This gives us an equation in one variable.

$0.1x + 0.6(50 - x) = 15$	This equation contains one variable, x.
$0.1x + 30 - 0.6x = 15$	Apply the distributive property.
$-0.5x + 30 = 15$	Combine like terms.
$-0.5x = -15$	Subtract 30 from both sides.
$x = \dfrac{-15}{-0.5} = 30$	Divide both sides by −0.5.

Back-substituting 30 for x in either of the system's equations ($x + y = 50$ is easier to use) gives $y = 20$. Because x represents the number of milliliters of the 10% solution and y the number of milliliters of the 60% solution, the chemist should mix 30 milliliters of the 10% solution with 20 milliliters of the 60% solution.

Step 4 Check the proposed solution in the original wording of the problem.
The problem states that the chemist needs 50 milliliters of a 30% sodium-iodine solution. The amount of sodium-iodine in this mixture is 0.3(50), or 15 milliliters. The amount of sodium-iodine in 30 milliliters of the 10% solution is 0.1(30), or 3 milliliters. The amount of sodium-iodine in 20 milliliters of the 60% solution is $0.6(20) = 12$ milliliters. The amount of sodium-iodine in the two solutions used in the mixture is 3 milliliters + 12 milliliters, or 15 milliliters, exactly as it should be. ∎

✔ **CHECK POINT 5**   A chemist needs to mix an 18% acid solution with a 45% acid solution to obtain 12 liters of a 36% acid solution. How many liters of each of the acid solutions must be used?

4 Solve motion problems.

Problems Involving Motion

Suppose that you ride your bike at an average speed of 12 miles per hour. What distance do you cover in 2 hours? Your distance is the product of your speed and the time that you travel:

$$\frac{12 \text{ miles}}{\cancel{\text{hour}}} \times 2 \; \cancel{\text{hours}} = 24 \text{ miles.}$$

Your distance is 24 miles. Notice how the hour units cancel. The distance is expressed in miles.

In general, the distance covered by any moving body is the product of its average speed, or rate, and its time in motion.

A Formula for Motion

$$d = rt$$

Distance equals rate times time.

Wind speed and water current have the effect of increasing or decreasing a traveler's rate.

EXAMPLE 6   Solving a Motion Problem

When a small airplane flies with the wind, it can travel 450 miles in 3 hours. When the same airplane flies in the opposite direction against the wind, it takes 5 hours to fly the same distance. Find the speed of the plane in still air and the speed of the wind.

Solution

Step 1 Use variables to represent unknown quantities.
Let $x =$ the speed of the plane in still air.
Let $y =$ the speed of the wind.

Step 2 Write a system of equations describing the problem's conditions. As it travels with the wind, the plane's speed is increased. The net rate is its speed in still air, x, plus the speed of the wind, y, given by the expression $x + y$. As it travels against the wind, the plane's speed is decreased. The net rate is its speed in still air, x, minus the speed of the wind, y, given by the expression $x - y$. Here is a chart that summarizes the problem's information and includes the increased and decreased rates.

	Rate	×	Time	=	Distance
Trip with the Wind	$x + y$		3		$3(x + y)$
Trip against the Wind	$x - y$		5		$5(x - y)$

The problem states that the distance in each direction is 450 miles. We use this information to write our system of equations.

The distance of the trip with the wind	is	450 miles.
$3(x + y)$	$=$	450

The distance of the trip against the wind	is	450 miles.
$5(x - y)$	$=$	450

Step 3 Solve the system and answer the problem's question. We can simplify the system by dividing both sides of the equations by 3 and 5, respectively.

$$3(x + y) = 450 \xrightarrow[\text{Divide by 5.}]{\text{Divide by 3.}} x + y = 150$$
$$5(x - y) = 450 \xrightarrow{} x - y = 90$$

Solve the system on the right by the addition method.

$$x + y = 150$$
$$x - y = 90$$
$$\text{Add:} \quad \overline{2x = 240}$$
$$x = 120 \quad \text{Divide both sides by 2.}$$

Back-substituting 120 for x in either of the system's equations gives $y = 30$. Because $x = 120$ and $y = 30$, the speed of the plane in still air is 120 miles per hour and the speed of the wind is 30 miles per hour.

Step 4 Check the proposed solution in the original wording of the problem. The problem states that the distance in each direction is 450 miles. The speed of the plane with the wind is $120 + 30 = 150$ miles per hour. In 3 hours, it travels $150 \cdot 3$, or 450 miles, which checks with the stated condition. Furthermore, the speed of the plane against the wind is $120 - 30 = 90$ miles per hour. In 5 hours, it travels $90 \cdot 5 = 450$ miles, which is the stated distance. ■

✔ **CHECK POINT 6** With the current, a motorboat can travel 84 miles in 2 hours. Against the current, the same trip takes 3 hours. Find the speed of the boat in still water and the speed of the current.

EXERCISE SET 4.4

Practice Exercises

In Exercises 1–4, let x represent one number and let y represent the other number. Use the given conditions to write a system of equations. Solve the system and find the numbers.

1. The sum of two numbers is 7. If one number is subtracted from the other, their difference is −1. Find the numbers.

2. The sum of two numbers is 2. If one number is subtracted from the other, their difference is 8. Find the numbers.

3. Three times a first number decreased by a second number is 1. The first number increased by twice the second number is 12. Find the numbers.

4. The sum of three times a first number and twice a second number is 8. If the second number is subtracted from twice the first number, the result is 3. Find the numbers.

Application Exercises

5. The graph makes Super Bowl Sunday look like a day of snack food bingeing in the United States. Combined, we wolf down 10.4 million pounds of potato chips and tortilla chips. The difference between consumption of potato chips and tortilla chips is 1.2 million pounds. How many millions of pounds of potato chips and tortilla chips are consumed on Super Bowl Sunday?

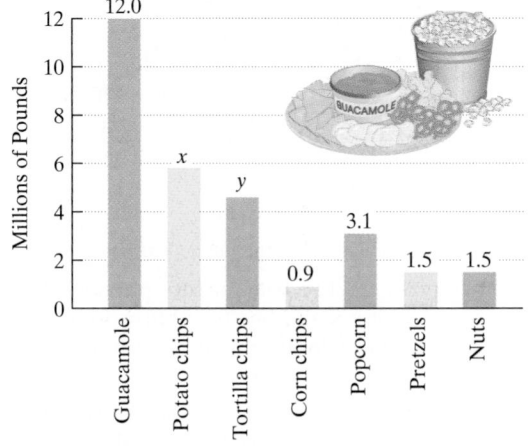

Millions of Pounds of Snack Food Consumed on Super Bowl Sunday

Source: Association of American Snack Foods

6. The graph shows the top six registered U.S. dog breeds. Combined, there are 169,000 rottweilers and German shepherds. The difference between the number of rottweilers and German shepherds is 10,800. How many rottweilers and German shepherds are registered in the United States?

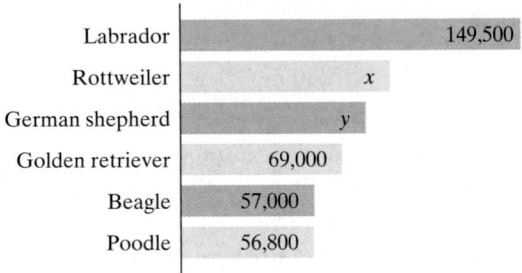

Top Six Registered U.S. Dog Breeds

Labrador — 149,500
Rottweiler — x
German shepherd — y
Golden retriever — 69,000
Beagle — 57,000
Poodle — 56,800

Source: American Kennel Club

The graph shows the calories in some favorite fast foods. Use the information in Exercises 7–8 to find the exact caloric content of the specified foods.

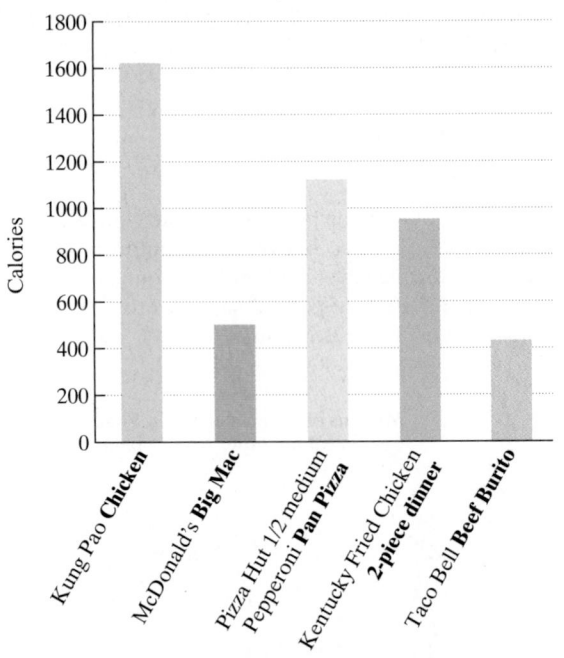

Calories in Some Favorite Fast Foods

Fast Food

Source: Center for Science in the Public Interest

7. One pan pizza and two beef burritos provide 1980 calories. Two pan pizzas and one beef burrito provide 2670 calories. Find the caloric content of each item.

8. One Kung Pao chicken and two Big Macs provide 2620 calories. Two Kung Pao chickens and one Big Mac provide 3740 calories. Find the caloric content of each item.

9. Cholesterol intake should be limited to 300 mg or less each day. One serving of scrambled eggs from McDonalds and one Double Beef Whopper from Burger King exceed this intake by 241 mg. Two servings of scrambled eggs and three Double Beef Whoppers provide 1257 mg of cholesterol. Determine the cholesterol content in each item.

10. Two medium eggs and three cups of ice cream contain 701 milligrams of cholesterol. One medium egg and one cup of ice cream exceed the suggested daily cholesterol intake of 300 milligrams by 25 milligrams. Determine the cholesterol content in each item.

11. In a discount clothing store, all sweaters are sold at one fixed price, and all shirts are sold at another fixed price. If one sweater and three shirts cost $42, while three sweaters and two shirts cost $56, find the price of one sweater and one shirt.

12. A restaurant purchased eight tablecloths and five napkins for $106. A week later, a tablecloth and six napkins were bought for $24. Find the cost of one tablecloth and one napkin, assuming the same prices for both purchases.

13. You are choosing between two long distance telephone plans. Plan A has a monthly fee of $20 with a charge of $0.05 per minute for all long distance calls. Plan B has a monthly fee of $5 with a charge of $0.10 per minute for all long distance calls.
   a. For what number of minutes of long-distance calls will the costs for the two plans be the same? What will be the cost for each plan?
   b. If you make approximately 10 long-distance calls per month, each averaging 20 minutes, which plan should you select? Explain your answer.

14. You are choosing between two long-distance telephone plans. Plan A has a monthly fee of $15 with a charge of $0.08 per minute for all long distance calls. Plan B has a monthly fee of $3 with a charge of $0.12 per minute for all long distance calls.
   a. For what number of minutes of long-distance calls will the costs for the two plans be the same? What will be the cost for each plan?
   b. If you make approximately 15 long-distance calls per month, each averaging 30 minutes, which plan should you select? Explain your answer.

15. You are choosing between two plans at a discount warehouse. Plan A offers an annual membership fee of $100 and you pay 80% of the manufacturer's recommended list price. Plan B offers an annual membership fee of $40 and you pay 90% of the manufacturer's recommended list price. How many dollars of merchandise would you have to purchase in a year to pay the same amount under both plans? What will be the cost for each plan?

16. You are choosing between two plans at a discount warehouse. Plan A offers an annual membership fee of $300 and you pay 70% of the manufacturer's recommended list price. Plan B offers an annual membership fee of $40 and you pay 90% of the manufacturer's recommended list price. How many dollars of merchandise would you have to purchase in a year to pay the same amount under both plans? What will be the cost for each plan?

The graphs show average weekly earnings of full-time wage and salary workers 25 and older, by educational attainment. Exercises 17–18 involve the information in these graphs.

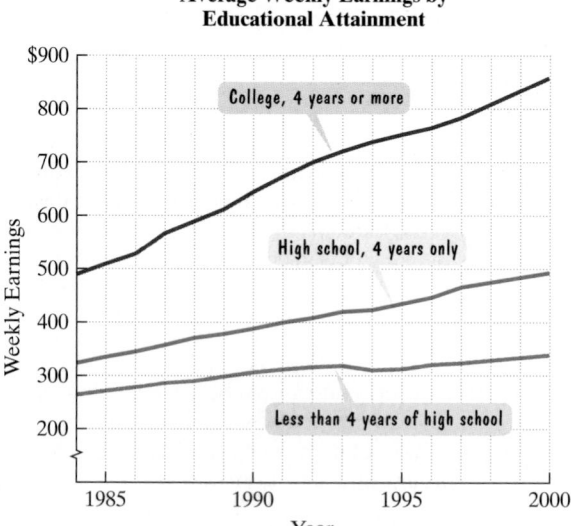

Average Weekly Earnings by Educational Attainment

Source: U.S. Bureau of Labor Statistics

17. In 1985, college graduates averaged $508 in weekly earnings. This amount has increased by approximately $25 in weekly earnings per year. By contrast, in 1985, high school graduates averaged $345 in weekly earnings. This amount has only increased by approximately $9 in weekly earnings per year. How many years after 1985 will college graduates be earning twice the amount per week of high school graduates? In which year will this occur? What will be the weekly earnings for each group at that time?

18. In 1985, college graduates averaged $508 in weekly earnings. This amount has increased by approximately $25 in weekly earnings per year. By contrast, in 1985, people with less than four years of high school averaged $270 in weekly earnings. This amount has only increased by approximately $4 in weekly earnings per year. How many years after 1985 will college graduates be earning three times the amount per week of people with less than four years of high school? (Round to the nearest whole number.) In which year will this occur? What will be the weekly earnings for each group at that time?

19. Nutritional information for macaroni and broccoli is given in the table. How many servings of each would it take to get exactly 14 grams of protein and 48 grams of carbohydrates?

	Macaroni	Broccoli
Protein (grams/serving)	3	2
Carbohydrates (grams/serving)	16	4

20. The calorie-nutrient information for an apple and an avocado is given in the table. How many of each should be eaten to get exactly 1000 calories and 100 grams of carbohydrates?

	One Apple	One Avocado
Calories	100	350
Carbohydrates (grams)	24	14

Exercises 21–26 involve simple interest. Use the four-step strategy to solve each problem.

21. You invest $20,000 in two accounts paying 7% and 9% annual interest, respectively. If the total interest earned for the year is $1550, how much was invested at each rate?

22. You invest $20,000 in two accounts paying 7% and 8% annual interest, respectively. If the total interest earned for the year is $1520, how much was invested at each rate?

23. A bank loaned out $250,000, part of it at the rate of 8% annual mortgage interest and the rest at the rate of 18% annual credit card interest. The interest received on both loans totaled $23,000. How much was loaned at each rate?

24. A bank loaned out $120,000, part of it at the rate of 8% annual mortgage interest and the rest at the rate of 18% annual credit card interest. The interest received on both loans totaled $10,000. How much was loaned at each rate?

25. Things did not go quite as planned. You invested $8000, part of it in a fund that paid 12% annual interest. However, the rest of the money suffered a 5% loss. If the total annual income from both investments was $620, how much was invested at each rate?

26. Things did not go quite as planned. You invested $12,000, part of it in a fund that paid 14% annual interest. However, the rest of the money suffered a 6% loss. If the total annual income from both investments was $680, how much was invested at each rate?

Exercises 27–32 involve mixtures. Use the four-step strategy to solve each problem.

27. A chef needs to mix a 45% fat content cheese with a 20% fat content cheese to obtain 30 grams of a cheese mixture that is 30% fat. How many grams of each kind of cheese must be used?

28. A candy company needs to mix a 30% fat content chocolate with a 12% fat content chocolate to obtain 50 pounds of a 20% fat content chocolate. How many pounds of each kind of chocolate must be used?

29. At the north campus of a small liberal arts college, 10% of the students are women. At the south campus, 50% of the students are women. The campuses are merged into one east campus. If 40% of the 1200 students at the east campus are women, how many students did the north and south campuses have before the merger?

30. At the north campus of a performing arts school, 10% of the students are music majors. At the south campus, 90% of the students are music majors. The campuses are merged into one east campus. If 42% of the 1000 students at the east campus are music majors, how many students did the north and south campuses have before the merger?

31. A grocer needs to mix tea worth $6.00 per pound with tea worth $8.00 per pound to obtain 144 pounds of a tea mixture worth $7.50 per pound. How many pounds of each kind of tea must be used?

32. A grocer needs to mix cashews worth $6 per pound with peanuts worth $2 per pound to obtain 10 pounds of a mixture worth $3 per pound. How many pounds of each kind of nut must be used?

Exercises 33–38 involve motion. Use the four-step strategy to solve each problem.

33. When a small plane flies with the wind, it can travel 800 miles in 5 hours. When the plane flies in the opposite direction, against the wind, it takes 8 hours to fly the same

distance. Find the speed of the plane in still air and the speed of the wind.

34. When a plane flies with the wind, it can travel 4200 miles in 6 hours. When the plane flies in the opposite direction, against the wind, it takes 7 hours to fly the same distance. Find the speed of the plane in still air and the speed of the wind.

35. A boat's crew rowed 16 kilometers downstream, with the current, in 2 hours. The return trip upstream, against the current, covered the same distance, but took 4 hours. Find the crew's rowing rate in still water and the rate of the current.

36. A motorboat traveled 36 miles downstream, with the current, in 1.5 hours. The return trip upstream, against the current, covered the same distance, but took 2 hours. Find the boat's rate in still water and the rate of the current.

37. With the current, you can canoe 24 miles in 4 hours. Against the same current, you can canoe only $\frac{3}{4}$ of this distance in 6 hours. Find your speed in still water and the speed of the current.

38. With the current, you can row 24 miles in 3 hours. Against the same current, you can row only $\frac{2}{3}$ of this distance in 4 hours. Find your rowing speed in still water and the speed of the current.

Writing in Mathematics

39. Describe the conditions in a problem that enable it to be solved using a system of linear equations.

40. Write a word problem that can be solved by translating to a system of linear equations. Then solve the problem.

41. Exercises 13–16 involve using systems of linear equations to compare costs of long distance telephone plans and plans at a discount warehouse. Describe another situation that involves choosing between two options that can be modeled and solved with a linear system.

Critical Thinking Exercises

42. A set of identical twins can only be recognized by the characteristic that one always tells the truth and the other always lies. One twin tells you of a lucky number pair: "When I multiply my first lucky number by 3 and my second lucky number by 6, the addition of the resulting numbers produces a sum of 12. When I add my first lucky number and twice my second lucky number, the sum is 5." Which twin is talking?

43. Tourist: "How many birds and lions do you have in your zoo?" Zookeeper: "There are 30 heads and 100 feet." Tourist: "I can't tell from that." Zookeeper: "Oh, yes, you can!" Can you? Find the number of each.

44. Find the measures of the angles marked x and y in the figure.

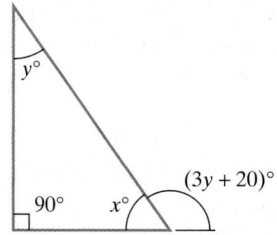

45. One apartment is directly above a second apartment. The resident living downstairs calls his neighbor living above him and states, "If one of you is willing to come downstairs, we'll have the same number of people in both apartments." The upstairs resident responds, "We're all too tired to move. Why don't one of you come up here? Then we'll have twice as many people up here as you've got down there." How many people are in each apartment?

46. In Lewis Carroll's *Through the Looking Glass*, the following dialogue takes place:

> **Tweedledum (to Tweedledee):** The sum of your weight and twice mine is 361 pounds.

> **Tweedledee (to Tweedledum):** Contrawise, the sum of your weight and twice mine is 362 pounds.

Find the weight of each of the two characters.

Technology Exercise

47. Select any two problems that you solved from Exercises 5–20. Use a graphing utility to graph the system of equations that you wrote for that problem. Then use the [TRACE] or [INTERSECTION] feature to show the point on the graphs that corresponds to the problem's solution.

Review Exercises

48. Find the slope of the line containing the points $(-6, 1)$ and $(2, -1)$. (Section 3.3, Example 1)

49. Add: $\frac{1}{5} + \left(-\frac{3}{4}\right)$. (Section 1.5, Example 4)

50. Graph: $y = x^2$. (Section 3.1, Example 6)

CHAPTER 4 GROUP PROJECTS

1. Group members should go online and obtain a list of telecommunication companies that provide residential long-distance service. The list should contain the monthly fee, the monthly minimum, and the rate per minute for each service provider.

   a. For each provider in the list, write an equation that describes the total monthly cost, y, of x minutes of long-distance phone calls.

   b. Compare two of the plans. After how many minutes of long-distance calls will the costs for the two plans be the same? Solve a linear system to obtain your answer. What will be the cost for each plan?

   c. Repeat part (b) for another two of the plans.

   d. Each person should estimate the number of minutes he or she spends talking long distance each month. Group members should assist that person in selecting the plan that will save the most amount of money. Be sure to factor in the monthly minimum, if any, when choosing a plan. Whenever possible, use the equations that you wrote in part (a).

   Caution: We've left something out! Your comparisons do not take into account the in-state rates of the plan. Furthermore, if you make international calls, international rates should also be one of your criteria in choosing a plan. Problem solving in real life can get fairly complicated.

2. Group members should turn to the graphs on page 296 that show the percentage of people in favor of and the percentage of people opposed to the death penalty. Let x = the number of years after 1988. Let y = the percentage of people in favor of or opposed to the death penalty. For each graph, draw a line that approximately fits the data and then write its equation. Solve the resulting system of equations and, assuming these trends continue, find the year when the percentage of Americans in favor of the death penalty will be the same as the percentage of those who oppose it. Are there circumstances that might affect the accuracy of this prediction? Group members should list some of these circumstances.

3. The group should write five different word problems that can be solved using a system of linear equations in two variables. Each problem should be on a different topic. Select from the following topics: a problem using supply and demand models (see Example 5 on page 259), a problem based on two missing numbers in a graph (see Exercises 5 and 6 on pages 281–282), a problem based on calories in fast foods (see Exercises 7, 8, and the bar graph on page 282), a problem involving choosing between two plans, such as telephone plans (see Exercises 13 and 14 on page 282) or discount warehouse plans (see Exercises 15 and 16 on pages 282–283), and problems involving interest (see Example 4 on page 276), mixtures (see Example 5 on page 278), and motion (see Example 6 on page 280). Of course, you can also base the problem on any topic of interest, but remember—only one problem per topic. The group should turn in the five problems and their algebraic solutions.

CHAPTER SUMMARY, REVIEW, AND TEST

SUMMARY

DEFINITIONS AND CONCEPTS	EXAMPLES

Section 4.1 Solving Systems of Linear Equations by Graphing

A system of linear equations in two variables, x and y, consists of two equations of the form $Ax + By = C$. A solution is an ordered pair of numbers that satisfies both equations.

Determine whether $(3, -1)$ is a solution of

$$2x + 5y = 1$$
$$4x + y = 11.$$

Replace x by 3 and y by -1 in both equations.

$$2x + 5y = 1 \qquad\qquad 4x + y = 11$$
$$2 \cdot 3 + 5(-1) \stackrel{?}{=} 1 \qquad 4 \cdot 3 + (-1) \stackrel{?}{=} 11$$
$$6 + (-5) \stackrel{?}{=} 1 \qquad 12 + (-1) \stackrel{?}{=} 11$$
$$1 = 1, \text{ true} \qquad\qquad 11 = 11, \text{ true}$$

Thus, $(3, -1)$ is a solution of the system.

Using the graphing method, a solution of a linear system is a point common to the graphs of both equations in the system. If the graphs are parallel lines, the system has no solution and is called inconsistent. If the graphs are the same line, the system has infinitely many solutions. The equations are called dependent.

Solve by graphing: $\quad 2x + y = 4$
$\qquad\qquad\qquad\qquad x + y = 2.$

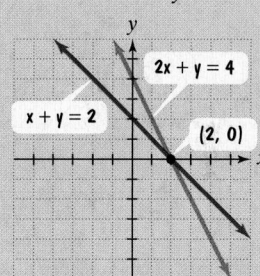

The solution is $(2, 0)$ and the solution set is $\{(2, 0)\}$.

Section 4.2 Solving Systems of Linear Equations by the Substitution Method

To solve a linear system by the substitution method,

1. Solve one equation for a variable.
2. Substitute the expression for that variable into the other equation.
3. Solve the equation in one variable from step 2.
4. Back-substitute the value of the variable found in step 3 in the equation from step 1 and find the value of the remaining variable.
5. Check the proposed solution in both equations.

If both variables are eliminated and a false statement results, the system has no solution. If both variables are eliminated and a true statement results, the system has infinitely many solutions.

Solve by the substitution method:

$$y = 2x + 3$$
$$7x - 5y = -18.$$

Substitute $2x + 3$ for y in the second equation.

$$7x - 5(2x + 3) = -18$$
$$7x - 10x - 15 = -18$$
$$-3x - 15 = -18$$
$$-3x = -3$$
$$x = 1$$

Find y. Substitute 1 for x in $y = 2x + 3$.

$$y = 2 \cdot 1 + 3 = 2 + 3 = 5$$

The solution, $(1, 5)$, checks and $\{(1, 5)\}$ is the solution set.

DEFINITIONS AND CONCEPTS	EXAMPLES

Section 4.3 Solving Systems of Linear Equations by the Addition Method

To solve a linear system by the addition method,

1. Write equations in $Ax + By = C$ form.
2. Multiply one or both equations by nonzero numbers so that coefficients of a variable are opposites.
3. Add equations.
4. Solve the resulting equation in one variable.
5. Back-substitute the value of the variable in either original equation and find the value of the remaining variable.
6. Check the proposed solution in both equations.

If both variables are eliminated and a false statement results, the system has no solution. If both variables are eliminated and a true statement results, the system has infinitely many solutions.

Solve by the addition method:

$$3x + y = -11$$
$$6x - 2y = -2.$$

Eliminate y. Multiply both sides of the first equation by 2.

$$6x + 2y = -22$$
$$\underline{6x - 2y = -2}$$

Add: $12x \qquad = -24$

$$x = -2$$

Find y. Back-substitute -2 for x. Using the first equation:

$$3(-2) + y = -11$$
$$-6 + y = -11$$
$$y = -5$$

The solution, $(-2, -5)$, checks. The solution set is $\{(-2, -5)\}$

Section 4.4 Problem Solving Using Systems of Equations

A Problem-Solving Strategy

1. Use variables, usually x and y, to represent unknown quantities.
2. Write a system of equations describing the problem's conditions.
3. Solve the system and answer the problem's question.
4. Check proposed answers in the problem's wording.

You invested $14,000 in two funds paying 7% and 9% interest. Total year-end interest was $1180. How much was invested at each rate?

Let x = amount invested at 7% and
y = amount invested at 9%.

amount invested at 7%		amount invested at 9%

$$x + y = 14,000$$

Interest from 7% investment		Interest from 9% investment

$$0.07x + 0.09y = 1180$$

Solving by substitution or addition, $x = 4000$ and $y = 10,000$. Thus, $4000 was invested at 7% and $10,000 at 9%.

Review Exercises

4.1 *In Exercises 1–2, determine whether the given ordered pair is a solution of the system.*

1. $(1, -5)$
$$4x - y = 9$$
$$2x + 3y = -13$$

2. $(-5, 2)$
$$2x + 3y = -4$$
$$x - 4y = -10$$

3. Can the graphing-utility-generated screen be the solution for the following system? Explain.

$$x + y = 2$$
$$2x + y = -5?$$

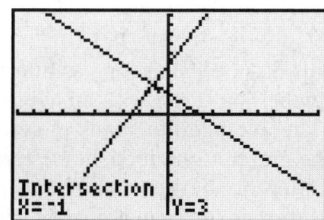

In Exercises 4–14, solve each system by graphing. If there is no solution or an infinite number of solutions, so state, and use set notation to express solution sets.

4. $x + y = 2$
$x - y = 6$

5. $2x - 3y = 12$
$-2x + y = -8$

6. $3x + 2y = 6$
$3x - 2y = 6$

7. $y = \dfrac{1}{2}x$
$y = 2x - 3$

8. $x + 2y = 2$
$y = x - 5$

9. $x + 2y = 8$
$3x + 6y = 12$

10. $2x - 4y = 8$
$x - 2y = 4$

11. $y = 3x - 1$
$y = 3x + 2$

12. $x - y = 4$
$x = -2$

13. $x = 2$
$y = 5$

14. $x = 2$
$x = 5$

4.2 *In Exercises 15–23, solve each system by the substitution method. If there is no solution or an infinite number of solutions, so state, and use set notation to express solution sets.*

15. $2x - 3y = 7$
$y = 3x - 7$

16. $2x - y = 6$
$x = 13 - 2y$

17. $2x - 5y = 1$
$3x + y = -7$

18. $3x + 4y = -13$
$5y - x = -21$

19. $y = 39 - 3x$
$y = 2x - 61$

20. $4x + y = 5$
$12x + 3y = 15$

21. $4x - 2y = 10$
$y = 2x + 3$

22. $x - 4 = 0$
$9x - 2y = 0$

23. $8y = 4x$
$7x + 2y = -8$

24. The weekly demand and supply models for the video *Titanic* at a chain of stores that sells videos are given by the demand model $N = -60p + 1000$ and the supply model $N = 4p + 200$, in which p is the price of the video and N is the number of videos sold or supplied each week to the chain of stores. Find the price at which supply and demand are equal. At this price, how many copies of *Titanic* can be supplied and sold each week?

4.3 *In Exercises 25–35, solve each system by the addition method. If there is no solution or an infinite number of solutions, so state, and use set notation to express solution sets.*

25. $x + y = 6$
$2x + y = 8$

26. $3x - 4y = 1$
$12x - y = -11$

27. $3x - 7y = 13$
$6x + 5y = 7$

28. $8x - 4y = 16$
$4x + 5y = 22$

29. $5x - 2y = 8$
$3x - 5y = 1$

30. $2x + 7y = 0$
$7x + 2y = 0$

31. $x + 3y = -4$
$3x + 2y = 3$

32. $2x + y = 5$
$2x + y = 7$

33. $3x - 4y = -1$
$-6x + 8y = 2$

34. $2x = 8y + 24$
$3x + 5y = 2$

35. $5x - 7y = 2$
$3x = 4y$

In Exercises 36–41, solve each system by the method of your choice. If there is no solution or an infinite number of solutions, so state, and use set notation to express solution sets..

36. $3x + 4y = -8$
$2x + 3y = -5$

37. $6x + 8y = 39$
$y = 2x - 2$

38. $x + 2y = 7$
$2x + y = 8$

39. $y = 2x - 3$
$y = -2x - 1$

40. $3x - 6y = 7$
$3x = 6y$

41. $y - 7 = 0$
$7x - 3y = 0$

4.4

42. Most animals do not live as long as human beings. One exception is the box turtle with an average life span of 100 years. The average life spans of some animals are shown in the graph. Combined, horses and lions live 35 years. The difference between a horse's average life span and a lion's average life span is 5 years. Find the average life span for each of these animals.

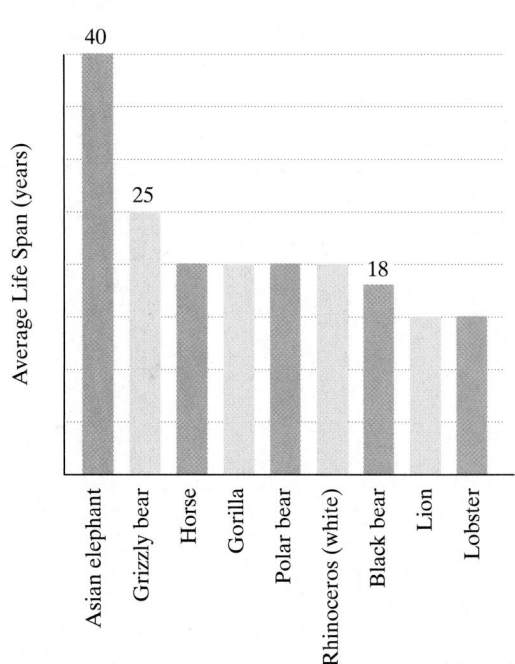

Source: The World Almanac

43. The gorilla and orangutan are the heaviest of the world's apes. Two gorillas and three orangutans weigh 1465 pounds. A gorilla's weight increased by twice an orangutan's weight is 815 pounds. Find the weight for each of these primates.

44. Health experts agree that cholesterol intake should be limited to 300 milligrams or less each day. Three ounces of shrimp and 2 ounces of scallops contain 156 milligrams of cholesterol. Five ounces of shrimp and 3 ounces of scallops contain 45 milligrams of cholesterol less than the suggested maximum daily intake. Determine the cholesterol content in an ounce of each item.

45. The perimeter of a table tennis top is 28 feet. The difference between 4 times the length and 3 times the width is 21 feet. Find the dimensions.

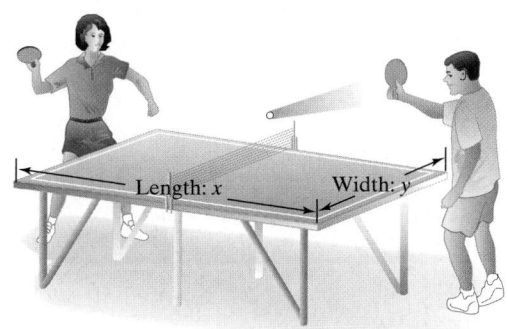

46. A travel agent offers two package vacation plans. The first plan costs $360 and includes 3 days at a hotel and a rental car for 2 days. The second plan costs $500 and includes 4 days at a hotel and a rental car for 3 days. The daily charge for the room is the same under each plan, as is the daily charge for the car. Find the cost per day for the room and for the car.

47. You are choosing between two long-distance telephone plans. One plan has a monthly fee of $15 with a charge of $0.05 per minute for all long distance calls. The other plan has a monthly fee of $10 with a charge of $0.075 per minute for all long-distance calls. For what number of minutes of long-distance calls will the costs for the two plans be the same? What will be the cost for each plan?

48. You invested $10,000 in two funds paying 8% and 10% annual interest, respectively. At the end of the year, the total interest from these investments was $940. How much was invested at each rate?

49. A chemist needs to mix a solution that is 34% silver nitrate with one that is 4% silver nitrate to obtain 100 milliliters of a mixture that is 7% silver nitrate. How many milliliters of each of the solutions must be used?

50. When a plane flies with the wind, it can travel 2160 miles in 3 hours. When the plane flies in the opposite direction, against the wind, it takes 4 hours to fly the same distance. Find the speed of the plane in still air and the speed of the wind.

Chapter 4 Test

In Exercises 1–2, determine whether the given ordered pair is a solution of the system.

1. $(5, -5)$
$2x + y = 5$
$x + 3y = -10$

2. $(-3, 2)$
$x + 5y = 7$
$3x - 4y = 1$

In Exercises 3–4, solve each system by graphing. If there is no solution or an infinite number of solutions, so state, and use set notation to express solution sets.

3. $x + y = 6$
$4x - y = 4$

4. $2x + y = 8$
$y = 3x - 2$

In Exercises 5–7, solve each system by the substitution method. If there is no solution or an infinite number of solutions, so state, and use set notation to express solution sets.

5. $x = y + 4$
$3x + 7y = -18$

6. $2x - y = 7$
$3x + 2y = 0$

7. $2x - 4y = 3$
$x = 2y + 4$

In Exercises 8–10, solve each system by the addition method. If there is no solution or an infinite number of solutions, so state, and use set notation to express solution sets.

8. $2x + y = 2$
$4x - y = -8$

9. $2x + 3y = 1$
$3x + 2y = -6$

10. $3x - 2y = 2$
$-9x + 6y = -6$

11. World War II and the Vietnam War were America's costliest wars. In current dollars, the two wars combined cost $500 billion. The difference between the cost of World War II and the Vietnam War was $120 billion. What was the cost of each of these wars in current dollars?

12. You are choosing between two long-distance telephone plans. One plan has a monthly fee of $15 with a charge of $0.05 per minute. The other plan has a monthly fee of $5 with a charge of $0.07 per minute. For how many minutes of long-distance calls will the costs for the two plans be the same? What will be the cost for each plan?

13. You invested $6000 in two stocks paying 9% and 6% annual interest, respectively. At the end of the year, the total interest from these investments was $480. How much was invested at each rate?

14. A chemist needs to mix a 20% acid solution with a 50% acid solution to obtain 60 ounces of a 30% acid solution. How many ounces of each of the solutions must be used?

15. When a plane flies with the wind, it can travel 1600 kilometers in 2 hours. When the plane flies in the opposite direction, against the wind, it takes 3 hours to travel 1950 kilometers. Find the speed of the plane in still air and the speed of the wind.

Cumulative Review Exercises (Chapters 1–4)

1. Perform the indicated operations:

$$-14 - [18 - (6 - 10)].$$

2. Simplify: $6(3x - 2) - (x - 1)$.

In Exercises 3–4, solve each equation.

3. $17(x + 3) = 13 + 4(x - 10)$

4. $\dfrac{x}{4} - 1 = \dfrac{x}{5}$

5. Solve for t: $A = P + Prt$.

6. Solve and graph the solution set on a number line: $2x - 5 < 5x - 11$.

In Exercises 7–9, graph each equation in the rectangular coordinate system.

7. $x - 3y = 6$

8. $y = 4 - x^2$

9. $y = -\dfrac{3}{5}x + 2$

In Exercises 10–11, solve each linear system.

10. $3x - 4y = 8$
$4x + 5y = -10$

11. $2x - 3y = 9$
$y = 4x - 8$

12. Find the slope of the line passing through $(5, -6)$ and $(6, -5)$.

13. Write the point-slope form and the slope-intercept form of the equation of the line passing through $(-1, 6)$ with slope $= -4$.

14. The area of a triangle is 80 square feet. Find the height if the base is 16 feet.

15. If 10 pens and 15 pads cost $26, and 5 of the same pens and 10 of the same pads cost $16, find the cost of a pen and a pad.

16. List all the integers in this set:

$$\left\{-93, -\frac{7}{3}, 0, \sqrt{3}, \frac{7}{1}, \sqrt{100}\right\}.$$

The graphs show the percentage of U.S. households with one computer and multiple computers. Use the information provided by the graphs to solve Exercises 17–20.

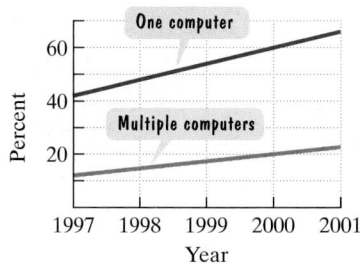

Source: Forrester Research. Inc.

17. What percentage of U.S. households had multiple computers in 2000?

18. Which graph has the greater slope? What does this mean in terms of the variables in this situation?

19. In 1997, 42% of U.S. households had one computer. This is increasing by approximately 6% per year. If this trend continues, in how many years after 1997 will 84% of U.S. households have one computer? In which year will that be?

20. The formula $y = \frac{8}{3}x + 12$ models the percentage of U.S. households, y, with multiple computers x years after 1997. Use the formula to find in which year 52% of U.S. households will have multiple computers.

There is a formula that models the age in human years, y, of a dog that is x years old:

$$y = -0.001618x^4 + 0.077326x^3 - 1.2367x^2 + 11.460x + 2.914.$$

The algebraic expression on the right side of the formula contains variables to powers that are whole numbers and is an example of a **polynomial**. Much of what we do in algebra involves operations with polynomials. In this chapter, we study these operations, as well as the many applications of polynomials.

One of the joys of your life is your dog, your very special buddy. Lately, however, you've noticed that your companion is slowing down a bit. He is now 8 years old and you wonder how this translates into human years. You remember something about every year of a dog's life being equal to seven years for a human. Is there a more accurate description?

Chapter 5

Exponents and Polynomials

▶ SECTION 5.1 *Adding and Subtracting Polynomials*

Objectives

1 Understand the vocabulary used to describe polynomials.

2 Add polynomials.

3 Subtract polynomials.

4 Use mathematical models that contain polynomials.

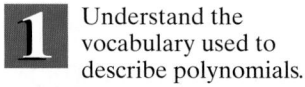

SSM
PH Tutor Center
CD-ROM
Video

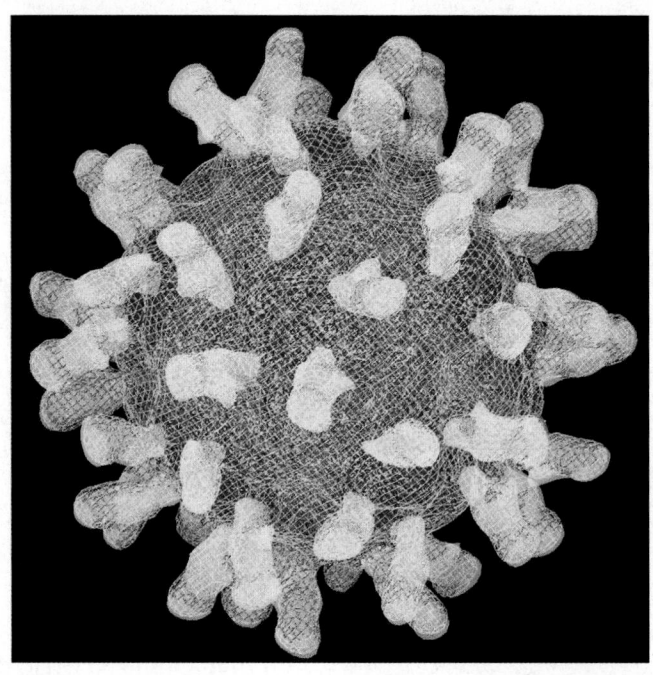

This computer-simulated model of the common cold virus was developed by researchers at Purdue University. Their discovery of how the virus infects human cells could lead to more effective treatment for the illness.

Runny nose? Sneezing? You are probably familiar with the unpleasant onset of a cold. We "catch cold" when the cold virus enters our bodies, where it multiplies. Fortunately, at a certain point the virus begins to die. The algebraic expression $-0.75x^4 + 3x^3 + 5$ describes the billions of viral particles in our bodies after x days of invasion. The expression enables mathematicians to determine the day on which there is a maximum number of viral particles and, consequently, the day we feel sickest.

The algebraic expression $-0.75x^4 + 3x^3 + 5$ is an example of a polynomial. A **polynomial** is a single term or the sum of two or more terms containing variables in the numerator with whole number exponents. This particular polynomial contains three terms. Equations containing polynomials are used in such diverse areas as science, business, medicine, psychology, and sociology. In this section, we present basic ideas about polynomials. We then use our knowledge of combining like terms to find sums and differences of polynomials.

1 Understand the vocabulary used to describe polynomials.

How We Describe Polynomials

Consider the polynomial

$$7x^3 - 9x^2 + 13x - 6.$$

We can express this polynomial as

$$7x^3 + (-9x^2) + 13x + (-6).$$

The polynomial contains four terms. It is customary to write the terms in the order of descending powers of the variables. This is the **standard form** of a polynomial.

We begin this chapter by limiting our discussion to polynomials containing only one variable. Each term of a polynomial in x is of the form ax^n. The **degree** of ax^n is n. For example, the degree of the term $7x^3$ is 3.

Study Tip

We can express 0 in many ways, including $0x$, $0x^2$, and $0x^3$. It is impossible to assign a unique exponent to the variable. This is why 0 has no defined degree.

The Degree of ax^n

If $a \neq 0$, the degree of ax^n is n. The degree of a nonzero constant is 0. The constant 0 has no defined degree.

Here is an example of a polynomial and the degree of each of its four terms:

$$6x^4 - 3x^3 + 2x - 5.$$

degree 4 | degree 3 | degree 1 | degree of non-zero constant: 0

Notice that the exponent on x for the term $2x$ is understood to be 1: $2x^1$. For this reason, the degree of $2x$ is 1.

A polynomial with exactly one term is called a **monomial**. A **binomial** is a polynomial that has two terms, each with a different exponent. A **trinomial** is a polynomial with three terms, each with a different exponent. Polynomials with four or more terms have no special names.

The degree of a polynomial is the highest degree of all the terms of the polynomial. For example, $4x^2 + 3x$ is a binomial of degree 2 because the degree of the first term is 2, and the degree of the other term is less than 2. Also, $7x^5 - 2x^2 + 4$ is a trinomial of degree 5 because the degree of the first term is 5, and the degrees of the other terms are less than 5.

Up to now, we have used x to represent the variable in a polynomial. However, any letter can be used. For example,

- $7x^5 - 3x^3 + 8$ is a polynomial (in x) of degree 5. Because there are three terms, the polynomial is a trinomial.

- $6y^3 + 4y^2 - y + 3$ is a polynomial (in y) of degree 3. Because there are four terms, the polynomial has no special name.

- $z^7 + \sqrt{2}$ is a polynomial (in z) of degree 7. Because there are two terms, the polynomial is a binomial.

2 Add polynomials.

Adding Polynomials

Polynomials are added by combining like terms. For example, we can add the monomials $-9x^3$ and $13x^3$ as follows:

$$-9x^3 + 13x^3 = (-9 + 13)x^3 = 4x^3.$$

Add coefficients and keep the same variable factor, x^3.

EXAMPLE 1   Adding Polynomials

Add: $(-9x^3 + 7x^2 - 5x + 3) + (13x^3 + 2x^2 - 8x - 6)$.

Solution   The like terms are $-9x^3$ and $13x^3$, containing the same variable to the same power (x^3), as well as $7x^2$ and $2x^2$ (both contain x^2), $-5x$ and $-8x$ (both contain x) and the constant terms 3 and -6. We begin by grouping these pairs of like terms.

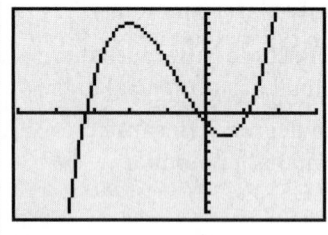
$$(-9x^3 + 7x^2 - 5x + 3) + (13x^3 + 2x^2 - 8x - 6)$$
$$= (-9x^3 + 13x^3) + (7x^2 + 2x^2) \qquad \text{Group like terms.}$$
$$+ (-5x - 8x) + (3 - 6)$$
$$= 4x^3 + 9x^2 + (-13x) + (-3) \qquad \text{Combine like terms.}$$
$$= 4x^3 + 9x^2 - 13x - 3 \qquad \blacksquare$$

✔ **CHECK POINT 1** Add: $(-11x^3 + 7x^2 - 11x - 5) + (16x^3 - 3x^2 + 3x - 15)$.

Polynomials can be added by arranging like terms in columns. Then combine like terms, column by column.

EXAMPLE 2 Adding Polynomials Vertically

Add: $(-9x^3 + 7x^2 - 5x + 3) + (13x^3 + 2x^2 - 8x - 6)$.

Solution

$$
\begin{array}{rrrr}
-9x^3 & 7x^2 & -5x & 3 \\
13x^3 & 2x^2 & -8x & -6 \\
\hline
4x^3 & 9x^2 & -13x & -3
\end{array}
$$

We consider each term separately and write like terms in columns.

Add, column by column.

Now add the four sums together:

$$4x^3 + 9x^2 + (-13x) + (-3) = 4x^3 + 9x^2 - 13x - 3.$$

This is the same answer found in Example 1. $\blacksquare$

✔ **CHECK POINT 2** Add the polynomials in Check Point 1 using a vertical format. Begin by arranging like terms in columns.

3 Subtract polynomials.

Subtracting Polynomials

We subtract real numbers by adding the opposite, or additive inverse, of the number being subtracted. For example,

$$8 - 3 = 8 + (-3) = 5.$$

Similarly, we subtract one polynomial from another by adding the opposite of the polynomial being subtracted.

> **Subtracting Polynomials**
>
> To subtract two polynomials, change the sign of every term of the second polynomial. Add this result to the first polynomial.

EXAMPLE 3   Subtracting Polynomials

Subtract: $(7x^2 + 3x - 4) - (4x^2 - 6x - 7)$.

Solution

$$(7x^2 + 3x - 4) - (4x^2 - 6x - 7)$$
$$= (7x^2 + 3x - 4) + (-4x^2 + 6x + 7)$$ Change the sign of each term of the second polynomial and add the two polynomials.

$$= (7x^2 - 4x^2) + (3x + 6x) + (-4 + 7)$$ Group like terms.
$$= 3x^2 + 9x + 3$$ Combine like terms.   ∎

✔ **CHECK POINT 3**   Subtract: $(9x^2 + 7x - 2) - (2x^2 - 4x - 6)$.

Study Tip

Be careful of the order in Example 4. For example, subtracting 2 from 5 is equivalent to $5 - 2$. In general, subtracting B from A becomes $A - B$. The order of the resulting problem is not the same as the order in English.

EXAMPLE 4   Subtracting Polynomials

Subtract $2x^3 - 6x^2 - 3x + 9$ from $7x^3 - 8x^2 + 9x - 6$.

Solution

$$(7x^3 - 8x^2 + 9x - 6) - (2x^3 - 6x^2 - 3x + 9)$$
$$= (7x^3 - 8x^2 + 9x - 6) + (-2x^3 + 6x^2 + 3x - 9)$$ Change the sign of each term of the second polynomial and add the two polynomials.

$$= (7x^3 - 2x^3) + (-8x^2 + 6x^2)$$
$$+ (9x + 3x) + (-6 - 9)$$ Group like terms.

$$= 5x^3 + (-2x^2) + 12x + (-15)$$ Combine like terms.

$$= 5x^3 - 2x^2 + 12x - 15$$ ∎

✔ **CHECK POINT 4**   Subtract $3x^3 - 8x^2 - 5x + 6$ from $10x^3 - 5x^2 + 7x - 2$.

Subtraction can also be performed in vertical columns.

EXAMPLE 5   Subtracting Polynomials Vertically

Use the method of subtracting by columns to find

$$(12y^3 - 9y^2 - 11y - 3) - (4y^3 - 5y + 8).$$

Solution   Arrange like terms in columns.

$$\begin{array}{r} 12y^3 - 9y^2 - 11y - 3 \\ -(4y^3 \qquad - 5y + 8) \\ \hline \end{array}$$ Leave space for the missing term.

Change the sign of each term in the second row, and combine like terms.

$$12y^3 - 9y^2 - 11y - 3$$
$$\underline{+ \ -4y^3 \qquad\quad + \ 5y - 8} \qquad \text{\small Change the sign of each term.}$$
$$8y^3 - 9y^2 - 6y - 11 \qquad \text{\small Combine like terms.} \qquad \blacksquare$$

✔ **CHECK POINT 5** Use the method of subtracting by columns to find

$$\left(8y^3 - 10y^2 - 14y - 2\right) - \left(5y^3 - 3y + 6\right).$$

4 Use mathematical models that contain polynomials.

Applications

The wage gap is used to compare the status of women's earnings relative to men's. The wage gap is expressed as a percent and is calculated by dividing the average annual earnings for women by the average annual earnings for men. Based on data provided by the U.S. Women's Bureau, the formula

$$y = 0.022x^2 - 0.4x + 60.07$$

models women's earnings, y, as a percentage of men's, x years after 1960. For example, to calculate the wage gap in 1998, substitute 38 for x because 1998 is 38 years after 1960:

$$y = 0.022(38)^2 - 0.4(38) + 60.07 \approx 76.6.$$

Thus, in 1998 women earned 76.6% as much as men.

The formula $y = 0.022x^2 - 0.4x + 60.07$ contains the polynomial $0.022x^2 - 0.4x + 60.07$. Can you see that this polynomial has three terms and is, therefore, a trinomial? Polynomials often appear in formulas that model real-world situations.

EXAMPLE 6   An Application: Death Rate

The formula

$$y = 0.036x^2 - 2.8x + 58.14$$

models the number of deaths per year per thousand people, y, for people who are x years old, $40 \leq x \leq 60$. Approximately how many people per thousand who are 50 years old die each year?

Solution   Because we are interested in people who are 50 years old, substitute 50 for x in the formula's polynomial.

$y = 0.036x^2 - 2.8x + 58.14$	This is the given formula.
$y = 0.036(50)^2 - 2.8(50) + 58.14$	Substitute 50 for x.
$y = 0.036(2500) - 2.8(50) + 58.14$	Evaluate the exponential expression: $50^2 = 50 \cdot 50 = 2500$.
$y = 90 - 140 + 58.14$	Perform the multiplications.
$y = 8.14 \approx 8$	Simplify.

Approximately 8 people per thousand who are 50 years old die each year. ■

We can use point plotting or a graphing utility to graph formulas that contain polynomials. These graphs contain only rounded curves with no sharp corners. For example, the graph of $y = 0.036x^2 - 2.8x + 58.14$, the formula from Example 6 that models number of deaths per thousand, is shown in Figure 5.1. Our work in Example 6 can be visualized as a point on the curve.

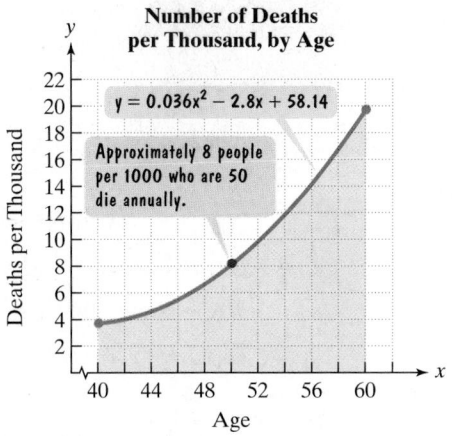

Figure 5.1 The graph of a formula containing a polynomial

✔ **CHECK POINT 6**   Use the formula $y = 0.036x^2 - 2.8x + 58.14$ to answer this question: Approximately how many people per thousand who are 40 years old die annually? Identify your solution as a point on the curve in Figure 5.1.

EXERCISE SET 5.1

Practice Exercises

 *In Exercises 1–16, identify each polynomial as a monomial, binomial, or trinomial. Give the degree of the polynomial.*

1. $3x + 7$
2. $5x - 2$
3. $x^3 - 2x$
4. $x^5 - 7x$
5. $8x^2$
6. $10x^2$
7. 5
8. 9
9. $x^2 - 3x + 4$
10. $x^2 - 9x + 2$
11. $7y^2 - 9y^4 + 5$
12. $3y^2 - 14y^5 + 6$
13. $15x - 7x^3$
14. $9x - 5x^3$
15. $-9y^{23}$
16. $-11y^{26}$

In Exercises 17–38, add the polynomials.

17. $(5x + 7) + (-8x + 3)$
18. $(7x - 3) + (-9x + 11)$
19. $(3x^2 + 7x - 9) + (7x^2 + 8x - 2)$
20. $(8x^2 + 5x - 3) + (12x^2 + 7x - 14)$
21. $(5x^2 - 3x) + (2x^2 - x)$
22. $(-2x^2 + x) + (4x^2 + 7x)$
23. $(3x^2 - 7x + 10) + (x^2 + 6x + 8)$
24. $(-5x^2 + 7x + 4) + (2x^2 + x + 3)$

25. $(4y^3 + 7y - 5) + (10y^2 - 6y + 3)$
26. $(2y^3 + 3y + 10) + (3y^2 + 5y - 22)$
27. $(2x^2 - 6x + 7) + (3x^3 - 3x)$
28. $(4x^3 + 5x + 13) + (-4x^2 + 22)$
29. $(4y^2 + 8y + 11) + (-2y^3 + 5y + 2)$
30. $(7y^3 + 5y - 1) + (2y^2 - 6y + 3)$
31. $(-2y^6 + 3y^4 - y^2) + (-y^6 + 5y^4 + 2y^2)$
32. $(7r^4 + 5r^2 + 2r) + (-18r^4 - 5r^2 - r)$
33. $\left(9x^3 - x^2 - x - \dfrac{1}{3}\right) + \left(x^3 + x^2 + x + \dfrac{4}{3}\right)$
34. $\left(12x^3 - x^2 - x + \dfrac{4}{3}\right) + \left(x^3 + x^2 + x - \dfrac{1}{3}\right)$
35. $\left(\dfrac{1}{5}x^4 + \dfrac{1}{3}x^3 + \dfrac{3}{8}x^2 + 6\right) +$
$\left(-\dfrac{3}{5}x^4 + \dfrac{2}{3}x^3 - \dfrac{1}{2}x^2 - 6\right)$
36. $\left(\dfrac{2}{5}x^4 + \dfrac{2}{3}x^3 + \dfrac{5}{8}x^2 + 7\right) +$
$\left(-\dfrac{4}{5}x^4 + \dfrac{1}{3}x^3 - \dfrac{1}{4}x^2 - 7\right)$

37. $(0.03x^5 - 0.1x^3 + x + 0.03) +$
$(-0.02x^5 + x^4 - 0.7x + 0.3)$

38. $(0.06x^5 - 0.2x^3 + x + 0.05) +$
$(-0.04x^5 + 2x^4 - 0.8x + 0.5)$

In Exercises 39–54, use a vertical format to add the polynomials.

39. $5y^3 - 7y^2$
$6y^3 + 4y^2$

40. $13x^4 - x^2$
$7x^4 + 2x^2$

41. $3x^2 - 7x + 4$
$-5x^2 + 6x - 3$

42. $7x^2 - 5x - 6$
$-9x^2 + 4x + 6$

43. $\frac{1}{4}x^4 - \frac{2}{3}x^3 - 5$
$-\frac{1}{2}x^4 + \frac{1}{5}x^3 + 4.7$

44. $\frac{1}{3}x^9 - \frac{1}{5}x^5 - 2.7$
$-\frac{3}{4}x^9 + \frac{2}{3}x^5 + 1$

45. $y^3 + 5y^2 - 7y - 3$
$-2y^3 + 3y^2 + 4y - 11$

46. $y^3 + y^2 - 7y + 9$
$-y^3 - 6y^2 - 8y + 11$

47. $4x^3 - 6x^2 + 5x - 7$
$-9x^3 - 4x + 3$

48. $-4y^3 + 6y^2 - 8y + 11$
$2y^3 + 9y - 3$

49. $7x^4 - 3x^3 + x^2$
$ x^3 - x^2 + 4x - 2$

50. $7y^5 - 3y^3 + y^2$
$ 2y^3 - y^2 - 4y - 3$

51. $7x^2 - 9x + 3$
$4x^2 + 11x - 2$
$-3x^2 + 5x - 6$

52. $7y^2 - 11y - 6$
$8y^2 + 3y + 4$
$-9y^2 - 5y + 2$

53. $1.2x^3 - 3x^2 + 9.1$
$7.8x^3 - 3.1x^2 + 8$
$ 1.2x^2 - 6$

54. $7.9x^3 - 6.8x^2 + 3.3$
$6.1x^3 - 2.2x^2 + 7$
$ 4.3x^2 - 5$

In Exercises 55–74, subtract the polynomials.

55. $(x - 8) - (3x + 2)$

56. $(x - 2) - (7x + 9)$

57. $(x^2 - 5x - 3) - (6x^2 + 4x + 9)$

58. $(3x^2 - 8x - 2) - (11x^2 + 5x + 4)$

59. $(x^2 - 5x) - (6x^2 - 4x)$

60. $(3x^2 - 2x) - (5x^2 - 6x)$

61. $(x^2 - 8x - 9) - (5x^2 - 4x - 3)$

62. $(x^2 - 5x + 3) - (x^2 - 6x - 8)$

63. $(y - 8) - (3y - 2)$

64. $(y - 2) - (7y - 9)$

65. $(6y^3 + 2y^2 - y - 11) - (y^2 - 8y + 9)$

66. $(5y^3 + y^2 - 3y - 8) - (y^2 - 8y + 11)$

67. $(7n^3 - n^7 - 8) - (6n^3 - n^7 - 10)$

68. $(2n^2 - n^7 - 6) - (2n^3 - n^7 - 8)$

69. $(y^6 - y^3) - (y^2 - y)$

70. $(y^5 - y^3) - (y^4 - y^2)$

71. $(7x^4 + 4x^2 + 5x) - (-19x^4 - 5x^2 - x)$

72. $(-3x^6 + 3x^4 - x^2) - (-x^6 + 2x^4 + 2x^2)$

73. $\left(\frac{3}{7}x^3 - \frac{1}{5}x - \frac{1}{3}\right) - \left(-\frac{2}{7}x^3 + \frac{1}{4}x - \frac{1}{3}\right)$

74. $\left(\frac{3}{8}x^2 - \frac{1}{3}x - \frac{1}{4}\right) - \left(-\frac{1}{8}x^2 + \frac{1}{2}x - \frac{1}{4}\right)$

In Exercises 75–88, use a vertical format to subtract the polynomials.

75. $7x + 1$
$-(3x - 5)$

76. $4x + 2$
$-(3x - 5)$

77. $7x^2 - 3$
$-(-3x^2 + 4)$

78. $9y^2 - 6$
$-(-5y^2 + 2)$

79. $7y^2 - 5y + 2$
$-(11y^2 + 2y - 3)$

80. $3x^5 - 5x^3 + 6$
$-(7x^5 + 4x^3 - 2)$

81. $7x^3 + 5x^2 - 3$
$-(-2x^3 - 6x^2 + 5)$

82. $3y^4 - 4y^2 + 7$
$-(-5y^4 - 6y^2 - 13)$

83. $5y^3 + 6y^2 - 3y + 10$
$-(6y^3 - 2y^2 - 4y - 4)$

84. $4y^3 + 5y^2 + 7y + 11$
$-(-5y^3 + 6y^2 - 9y - 3)$

85. $7x^4 - 3x^3 + 2x^2$
$-(- x^3 - x^2 + x - 2)$

86. $5x^6 - 3y^3 + 2y^2$
$-(- y^3 - y^2 - y - 1)$

87. $0.07x^3 - 0.01x^2 + 0.02x$
$-(0.02x^3 - 0.03x^2 - x)$

88. $0.04x^3 - 0.03x^2 + 0.05x$
$-(0.02x^3 - 0.06x^2 - x)$

Application Exercises

89. The polynomial $-0.02A^2 + 2A + 22$ is used by coaches to get athletes fired up so that they can perform well. The polynomial represents the performance level related to various levels of enthusiasm, from $A = 1$ (almost no enthusiasm) to $A = 100$ (maximum level of enthusiasm). Evaluate the polynomial when $A = 20$, $A = 50$, and $A = 80$. Describe what happens to performance as we get more and more fired up.

90. The common cold is caused by a rhinovirus. The polynomial

$$-0.75x^4 + 3x^3 + 5$$

describes the billions of viral particles in our bodies after x days of invasion. Find the number of viral particles, in billions, after 0 days (the time of the cold's onset), 1 day, 2 days, 3 days, and 4 days. After how many days is the number of viral particles at a maximum and consequently the day we feel the sickest? By when should we feel completely better?

The formula $y = 0.022x^2 - 0.4x + 60.07$ models women's earnings as a percentage of men's, y, x years after 1960. Use the formula to solve Exercises 91–92. Round to the nearest tenth of a percent.

91. What were women's earnings as a percentage of men's in 2000?

92. What were women's earnings as a percentage of men's in 1990?

The graph shows cigarette consumption per U.S. adult from 1925 through 2000. The data from 1960 through 2000 can be modeled by the formula $y = -3.1x^2 + 51.4x + 4024.5$, where x represents years after 1960 and y represents cigarette consumption per U.S. adult. Use the formula to solve Exercises 93–94.

Cigarette Consumption per U.S. Adult

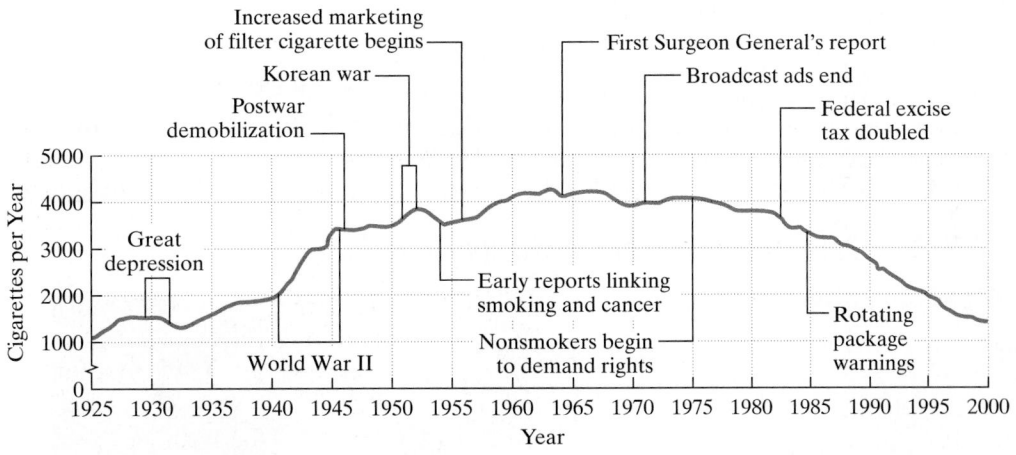

Source: U.S. Department of Health and Human Services

93. What was cigarette consumption per adult in 2000? How well does the formula model the information given by the graph?

94. What was cigarette consumption per adult in 1980? How well does the formula model the information given by the graph?

The formula

$$y = -0.001618x^4 + 0.077326x^3 - 1.2367x^2 + 11.460x + 2.914$$

models the age in human years, y, of a dog that is x years old, where $x > 1$. The coefficients make it difficult to use this formula when computing by hand. However, a graph of the formula makes approximations possible. Use the graph shown to solve Exercises 95–98.

Dog's Age in Human Years

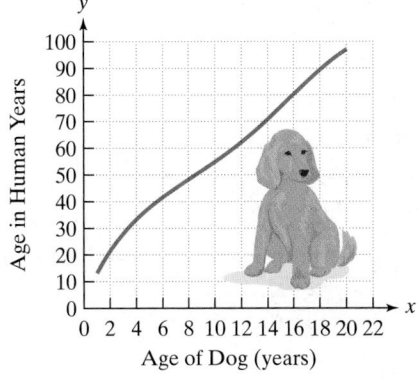

Source: U.C. Davis

95. If your dog is 6 years old, estimate the equivalent age in human years.

96. If your dog is 16 years old, estimate the equivalent age in human years.

97. If you are 25, what is the equivalent age for dogs?

98. If you are 45, what is the equivalent age for dogs?

Writing in Mathematics

99. What is a polynomial?

100. What is a monomial? Give an example with your explanation.

101. What is a binomial? Give an example with your explanation.

102. What is a trinomial? Give an example with your explanation.

103. What is the degree of a polynomial? Provide an example with your explanation.

104. Explain how to add polynomials.

105. Explain how to subtract polynomials.

106. A friend who is blind is having difficulty visualizing the relationship between age and deaths per thousand. Describe this relationship for your friend as age increases from 40 through 60. Use Figure 5.1 on page 301.

107. For Exercise 89, explain why performance levels do what they do as we get more and more fired up. If possible, describe an example of a time when you were too enthused and thus did poorly at something you were hoping to do well.

Critical Thinking Exercises

108. Which one of the following is true?

a. In the polynomial $3x^2 - 5x + 13$, the coefficient of x is 5.

b. The degree of $3x^2 - 7x + 9x^3 + 5$ is 2.

c. $\dfrac{1}{5x^2} + \dfrac{1}{3x}$ is a binomial.

d. $(2x^2 - 8x + 6) - (x^2 - 3x + 5) = x^2 - 5x + 1$ for any value of x.

109. What polynomial must be subtracted from $5x^2 - 2x + 1$ so that the difference is $8x^2 - x + 3$?

110. The number of people who catch a cold t weeks after January 1 is $5t - 3t^2 + t^3$. The number of people who recover t weeks after January 1 is $t - t^2 + \frac{1}{3}t^3$. Write a polynomial in standard form for the number of people who are still ill with a cold t weeks after January 1.

111. Explain why it is not possible to add two polynomials of degree 3 and get a polynomial of degree 4.

Review Exercises

112. Simplify: $(-10)(-7) \div (1 - 8)$. (Section 1.8, Example 8)

113. Subtract: $-4.6 - (-10.2)$. (Section 1.6, Example 2)

114. Solve: $3(x - 2) = 9(x + 2)$. (Section 2.3, Example 3)

▶ **SECTION 5.2**   *Multiplying Polynomials*

Objectives

1 Use the product rule for exponents.

2 Use the power rule for exponents.

3 Use the products-to-powers rule.

4 Multiply monomials.

5 Multiply a monomial and a polynomial.

6 Multiply polynomials when neither is a monomial.

SSM   PH Tutor   CD-   Video
       Center   ROM

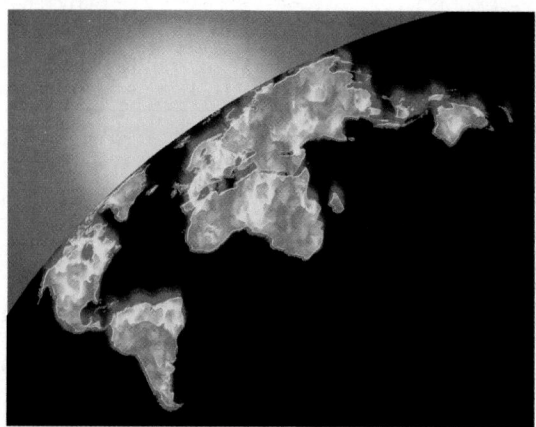

Recent advances in our understanding of climate have changed global warming from a subject for a disaster movie (the Statue of Liberty up to its chin in water) to a serious but manageable scientific and policy issue. Global warming may be related to the burning of fossil fuels, which adds carbon dioxide to the atmosphere. In the new millennium, we will see whether our use of fossil fuels will add enough carbon dioxide to the atmosphere to change it (and our climate) in significant ways. In this section's essay, you will see how a polynomial models trends in global warming through 2040. In the section itself, you will learn to multiply these algebraic expressions that play a significant role in modeling your world.

Before studying how polynomials are multiplied, we must develop some rules for working with exponents.

1 Use the product rule for exponents.

The Product Rule for Exponents

We have seen that exponents are used to indicate repeated multiplication. For example, 2^4, where 2 is the base and 4 is the exponent, indicates that 2 occurs as a factor four times:

$$2^4 = 2 \cdot 2 \cdot 2 \cdot 2.$$

Now consider the multiplication of two exponential expressions, such as $2^4 \cdot 2^3$. We are multiplying 4 factors of 2 and 3 factors of 2. We have a total of 7 factors of 2:

4 factors of 2   3 factors of 2

$$2^4 \cdot 2^3 = (2 \cdot 2 \cdot 2 \cdot 2) \cdot (2 \cdot 2 \cdot 2).$$

Total: 7 factors of 2

Thus,

$$2^4 \cdot 2^3 = 2^7.$$

We can quickly find the exponent, 7, of the product by adding 4 and 3, the original exponents:

$$2^4 \cdot 2^3 = 2^{4+3} = 2^7.$$

This suggests the following rule:

The Product Rule

$$b^m \cdot b^n = b^{m+n}$$

When multiplying exponential expressions with the same base, add the exponents. Use this sum as the exponent of the common base.

EXAMPLE 1 Using the Product Rule

Multiply each expression using the product rule:

    a. $2^2 \cdot 2^3$ **b.** $x^7 \cdot x^9$ **c.** $y \cdot y^5$ **c.** $y^3 \cdot y^2 \cdot y^5$.

Solution

    a. $2^2 \cdot 2^3 = 2^{2+3} = 2^5$ or 32

    b. $x^7 \cdot x^9 = x^{7+9} = x^{16}$

    c. $y \cdot y^5 = y^1 \cdot y^5 = y^{1+5} = y^6$

    d. $y^3 \cdot y^2 \cdot y^5 = y^{3+2+5} = y^{10}$ ■

✔ **CHECK POINT 1** Multiply each expression using the product rule:

    a. $2^2 \cdot 2^4$ **b.** $x^6 \cdot x^4$ **c.** $y \cdot y^7$ **d.** $y^4 \cdot y^3 \cdot y^2$.

2 Use the power rule for exponents.

The Power Rule for Exponents

The next property of exponents applies when an exponential expression is raised to a power. Here is an example:

$$\left(3^2\right)^4.$$

> The exponential expression 3^2 is raised to the fourth power.

There are 4 factors of 3^2. Thus,

$$\left(3^2\right)^4 = 3^2 \cdot 3^2 \cdot 3^2 \cdot 3^2 = 3^{2+2+2+2} = 3^8.$$

> Add exponents when multiplying with the same base.

We can obtain the answer, 3^8, by multiplying the exponents:

$$\left(3^2\right)^4 = 3^{2 \cdot 4} = 3^8.$$

This suggests the following rule:

> ### The Power Rule (Powers to Powers)
>
> $$\left(b^m\right)^n = b^{mn}$$
>
> When an exponential expression is raised to a power, multiply the exponents. Place the product of the exponents on the base and remove the parentheses.

EXAMPLE 2 Using the Power Rule

Simplify each expression using the power rule:

    a. $\left(2^3\right)^5$ **b.** $\left(x^6\right)^4$ **c.** $\left[(-3)^7\right]^5.$

Solution

    a. $\left(2^3\right)^5 = 2^{3 \cdot 5} = 2^{15}$

    b. $\left(x^6\right)^4 = x^{6 \cdot 4} = x^{24}$

    c. $\left[(-3)^7\right]^5 = (-3)^{7 \cdot 5} = (-3)^{35}$ ■

✔ **CHECK POINT 2** Simplify each expression using the power rule:

    a. $\left(3^4\right)^5$ **b.** $\left(x^9\right)^{10}$ **c.** $\left[(-5)^7\right]^3.$

3 Use the products-to-powers rule.

The Products-to-Powers Rule for Exponents

The next property of exponents applies when we are raising a product to a power. Here is an example:

$$\left(2x\right)^4.$$

> The product $2x$ is raised to the fourth power.

There are four factors of $2x$. Thus,

$$\left(2x\right)^4 = 2x \cdot 2x \cdot 2x \cdot 2x = 2 \cdot 2 \cdot 2 \cdot 2 \cdot x \cdot x \cdot x \cdot x = 2^4 x^4.$$

We can obtain the answer, 2^4x^4, by raising each factor within the parentheses to the fourth power:

$$(2x)^4 = 2^4x^4.$$

This suggests the following rule:

Products to Powers

$$(ab)^n = a^nb^n$$

When a product is raised to a power, raise each factor to the power.

EXAMPLE 3 Using the Products-to-Powers Rule

Simplify each expression using the products-to-powers rule:

a. $(5x)^3$
b. $(-2y^4)^5.$

Solution

a. $(5x)^3 = 5^3x^3$ Raise each factor to the third power.
$\qquad = 125x^3$ $5^3 = 5 \cdot 5 \cdot 5 = 125$

b. $(-2y^4)^5 = (-2)^5(y^4)^5$ Raise each factor to the fifth power.
$\qquad\qquad = (-2)^5y^{4\cdot5}$ To raise an exponential expression to a power, multiply exponents: $(b^m)^n = b^{mn}$.

$\qquad\qquad = -32y^{20}$ $(-2)^5 = (-2)(-2)(-2)(-2)(-2) = -32$ ■

✔ **CHECK POINT 3** Simplify each expression using the products-to-powers rule:

a. $(2x)^4$
b. $(-4y^2)^3.$

Study Tip

Try to avoid the following common errors that can occur when simplifying exponential expressions.

Correct	Incorrect	Description of Error
$b^3 \cdot b^4 = b^{3+4} = b^7$	$b^3 \cdot b^4 = b^{12}$	Exponents should be added, not multiplied.
$3^2 \cdot 3^4 = 3^{2+4} = 3^6$	$3^2 \cdot 3^4 = 9^{2+4} = 9^6$	The common base should be retained, not multiplied.
$(x^5)^3 = x^{5\cdot3} = x^{15}$	$(x^5)^3 = x^{5+3} = x^8$	Exponents should be multiplied, not added, when raising a power to a power.
$(4x)^3 = 4^3x^3$	$(4x)^3 = 4x^3$	Both factors should be cubed.
$\qquad = 64x^3$		

4 Multiply monomials.

Multiplying Monomials

Now that we have developed three properties of exponents, we are ready to turn to polynomial multiplication. We begin with the product of two monomials, such as $-8x^6$ and $5x^3$. This product is obtained by multiplying the coefficients, -8 and 5, and then multiplying the variables using the product rule for exponents.

$$(-8x^6)(5x^3) = -8 \cdot 5x^{6+3} = -40x^9$$

Multiply coefficients and add exponents.

Multiplying Monomials

To multiply monomials, multiply the coefficients and then multiply the variables. Use the product rule for exponents to multiply the variables: Keep the variable and add the exponents.

EXAMPLE 4   Multiplying Monomials

Multiply:

a. $(2x)(4x^2)$ **b.** $(-10x^6)(6x^{10})$.

Solution

a. $(2x)(4x^2) = (2 \cdot 4)(x \cdot x^2)$ Multiply the coefficients and multiply the variables.

$\qquad\qquad\quad = 8x^{1+2}$ Add exponents: $b^m \cdot b^n = b^{m+n}$.

$\qquad\qquad\quad = 8x^3$ Simplify.

b. $(-10x^6)(6x^{10}) = (-10 \cdot 6)(x^6 \cdot x^{10})$ Multiply the coefficients and multiply the variables.

$\qquad\qquad\qquad\quad = -60x^{6+10}$ Add exponents: $b^m \cdot b^n = b^{m+n}$.

$\qquad\qquad\qquad\quad = -60x^{16}$ Simplify. ∎

✔ **CHECK POINT 4**   Multiply:

a. $(7x^2)(10x)$ **b.** $(-5x^4)(4x^5)$.

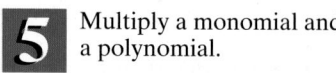 **5** Multiply a monomial and a polynomial.

Multiplying a Monomial and a Polynomial That Is Not a Monomial

We use the distributive property to multiply a monomial and a polynomial that is not a monomial. For example,

$$3x^2(2x^3 + 5x) = 3x^2 \cdot 2x^3 + 3x^2 \cdot 5x = 3 \cdot 2x^{2+3} + 3 \cdot 5x^{2+1} = 6x^5 + 15x^3.$$

Monomial   Binomial   Multiply coefficients and add exponents.

Multiplying a Monomial and a Polynomial That Is Not a Monomial

To multiply a monomial and a polynomial, multiply each term of the polynomial by the monomial.

EXAMPLE 5 Multiplying a Monomial and a Polynomial

Multiply:

    a. $2x(x + 4)$ **b.** $3x^2(4x^3 - 5x + 2)$.

Solution

    a. $2x(x + 4) = 2x \cdot x + 2x \cdot 4$ Use the distributive property.

                      $= 2 \cdot 1x^{1+1} + 2 \cdot 4x$ To multiply the monomials, multiply coefficients and add exponents.

                      $= 2x^2 + 8x$ Simplify.

    b. $3x^2(4x^3 - 5x + 2)$

          $= 3x^2 \cdot 4x^3 - 3x^2 \cdot 5x + 3x^2 \cdot 2$ Use the distributive property.

          $= 3 \cdot 4x^{2+3} - 3 \cdot 5x^{2+1} + 3 \cdot 2x^2$ To multiply the monomials, multiply coefficients and add exponents.

          $= 12x^5 - 15x^3 + 6x^2$ Simplify.   ■

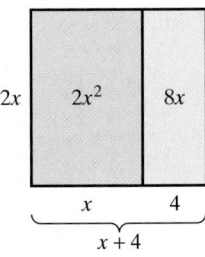

Rectangles often make it possible to visualize polynomial multiplication. For example, Figure 5.2 shows a rectangle with length $2x$ and width $x + 4$. The area of the large rectangle is

$$2x(x + 4).$$

The sum of the areas of the two smaller rectangles is

$$2x^2 + 8x.$$

Figure 5.2

Conclusion:

$$2x(x + 4) = 2x^2 + 8x.$$

✔ **CHECK POINT 5** Multiply:

    a. $3x(x + 5)$ **b.** $6x^2(5x^3 - 2x + 3)$

⑥   Multiply polynomials when neither is a monomial.

Multiplying Polynomials When Neither Is a Monomial

How do we multiply two polynomials if neither is a monomial? For example, consider

$$(2x + 3)(x^2 + 4x + 5).$$

One way to perform this multiplication is to distribute $2x$ throughout the trinomial

$$2x(x^2 + 4x + 5)$$

and 3 throughout the trinomial

$$3(x^2 + 4x + 5).$$

Then combine the like terms that result. In general, the product of two polynomials is the polynomial obtained by multiplying each term of one polynomial by each term of the other polynomial and then combining like terms.

Multiplying Polynomials When Neither Is a Monomial

Multiply each term of one polynomial by each term of the other polynomial. Then combine like terms.

EXAMPLE 6   Multiplying Binomials

Multiply:

a. $(x + 3)(x + 2)$ **b.** $(3x + 7)(2x - 4)$.

Solution   We begin by multiplying each term of the second binomial by each term of the first binomial.

a. $(x + 3)(x + 2)$

$= x(x + 2) + 3(x + 2)$ Multiply the second binomial by each term of the first binomial.

$= x \cdot x + x \cdot 2 + 3 \cdot x + 3 \cdot 2$ Use the distributive property.

$= x^2 + 2x + 3x + 6$ Multiply. Note that $x \cdot x = x^1 \cdot x^1 = x^{1+1} = x^2$.

$= x^2 + 5x + 6$ Combine like terms.

b. $(3x + 7)(2x - 4)$

$= 3x(2x - 4) + 7(2x - 4)$ Multiply the second binomial by each term of the first binomial.

$= 3x \cdot 2x - 3x \cdot 4 + 7 \cdot 2x - 7 \cdot 4$ Use the distributive property.

$= 6x^2 - 12x + 14x - 28$ Multiply.

$= 6x^2 + 2x - 28$ Combine like terms. ∎

✔ **CHECK POINT 6**   Multiply:

a. $(x + 4)(x + 5)$ **b.** $(5x + 3)(2x - 7)$.

Study Tip

You can visualize the polynomial multiplication in Example 6(a) using the rectangle with dimensions $x + 3$ and $x + 2$.

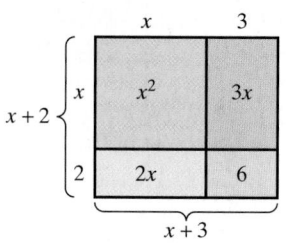

Area of large rectangle

$= (x + 3)(x + 2)$

Sum of areas of smaller rectangles

$= x^2 + 3x + 2x + 6$

$= x^2 + 5x + 6$

Conclusion:

$(x + 3)(x + 2) = x^2 + 5x + 6$.

Using Technology

A graphing utility can be used to see if a polynomial operation has been performed correctly. For example, to check

$$(x + 3)(x + 2) = x^2 + 5x + 6$$

graph the left and right sides on the same screen, using

$$y_1 = (x + 3)(x + 2)$$

and

$$y_2 = x^2 + 5x + 6.$$

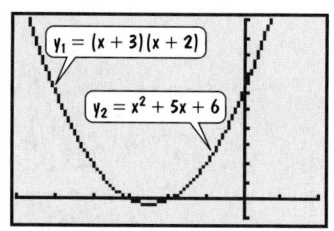

$[-6, 2, 1]$ by $[-1, 10, 1]$

As shown in the figure, both graphs are the same, verifying that the binomial multiplication was performed correctly.

EXAMPLE 7   Multiplying a Binomial and a Trinomial

Multiply: $(2x + 3)(x^2 + 4x + 5)$.

Solution

$(2x + 3)(x^2 + 4x + 5)$

$= 2x(x^2 + 4x + 5) + 3(x^2 + 4x + 5)$ Multiply the trinomial by each term of the binomial.

$= 2x \cdot x^2 + 2x \cdot 4x + 2x \cdot 5 + 3x^2 + 3 \cdot 4x + 3 \cdot 5$ Use the distributive property.

$= 2x^3 + 8x^2 + 10x + 3x^2 + 12x + 15$ Multiply monomials: Multiply coefficients and add exponents.

$= 2x^3 + 11x^2 + 22x + 15$ Combine like terms: $8x^2 + 3x^2 = 11x^2$ and $10x + 12x = 22x$. ∎

✔ **CHECK POINT 7**   Multiply: $(5x + 2)(x^2 - 4x + 3)$.

Another method for solving Example 7 is to use a vertical format similar to that used for multiplying whole numbers.

Write like terms in the same column.

$$
\begin{array}{r}
x^2 + 4x + 5 \\
2x + 3 \\
\hline
3x^2 + 12x + 15 \\
2x^3 + 8x^2 + 10x \\
\hline
2x^3 + 11x^2 + 22x + 15
\end{array}
$$

$3(x^2 + 4x + 5)$
$2x(x^2 + 4x + 5)$
Combine like terms.

EXAMPLE 8   Multiplying Polynomials Using a Vertical Format

Multiply: $(2x^2 - 3x)(5x^3 - 4x^2 + 7x)$.

Solution   To use the vertical format, it is most convenient to write the polynomial with the greatest number of terms in the top row.

$$
\begin{array}{r}
5x^3 - 4x^2 + 7x \\
2x^2 - 3x
\end{array}
$$

We now multiply each term in the top polynomial by the last term in the bottom polynomial.

$$
\begin{array}{r}
5x^3 - 4x^2 + 7x \\
2x^2 - 3x \\
\hline
-15x^4 + 12x^3 - 21x^2
\end{array}
$$

$-3x(5x^3 - 4x^2 + 7x)$

Then we multiply each term in the top polynomial by $2x^2$, the first term in the bottom polynomial. Like terms are placed in columns because the final step involves adding them.

$$5x^3 - 4x^2 + 7x$$
$$2x^2 - 3x$$

Write like terms in the same column.

$$-15x^4 + 12x^3 - 21x^2 \qquad -3x(5x^3 - 4x^2 + 7x)$$
$$10x^5 - 8x^4 + 14x^3 \qquad 2x^2(5x^3 - 4x^2 + 7x)$$
$$10x^5 - 23x^4 + 26x^3 - 21x^2$$

Add like terms, which are lined up in columns.

✔ **CHECK POINT 8** Multiply using a vertical format: $(3x^2 - 2x)(2x^3 - 5x^2 + 4x)$.

ENRICHMENT ESSAY

Is It Hot in Here Or Is It Just Me?

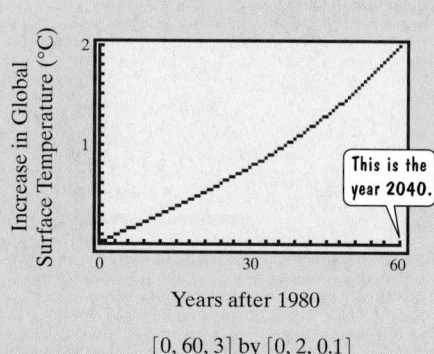

This is the year **2040**.

Years after 1980

$[0, 60, 3]$ by $[0, 2, 0.1]$

In the 1980s, a rising trend in global surface temperature was observed and the term "global warming" was coined. Scientists are more convinced than ever that burning coal, oil, and gas results in a buildup of gases and particles that trap heat and raise the planet's temperature. The average increase in global surface temperature, y, in degrees Centigrade, x years after 1980 can be modeled by the polynomial formula

$$y = \frac{21}{5,000,000}x^3 - \frac{127}{1,000,000}x^2 + \frac{1293}{50,000}x.$$

The graph of this formula is shown above in a $[0, 60, 3]$ by $[0, 2, 0.1]$ viewing rectangle. The graph illustrates that the model predicts global warming will increase through the year 2040. Furthermore, the increasing steepness of the curve shows that global warming will increase at greater rates near the middle of the twenty-first century.

EXERCISE SET 5.2

 Practice Exercises

In Exercises 1–8, multiply each expression using the product rule.

1. $x^{10} \cdot x^5$

2. $x^{12} \cdot x^3$

3. $y \cdot y^7$

4. $y \cdot y^8$

5. $x^2 \cdot x^5 \cdot x^4$

6. $x^3 \cdot x^2 \cdot x^4$

7. $3^9 \cdot 3^{10}$

8. $4^7 \cdot 4^{10}$

In Exercises 9–14, simplify each expression using the power rule.

9. $(3^9)^{10}$

10. $(4^7)^{10}$

11. $(x^4)^5$

12. $(x^5)^8$

13. $[(-2)^3]^3$

14. $[(-5)^3]^3$

In Exercises 15–24, simplify each expression using the products-to-powers rule.

15. $(2x)^3$

16. $(4x)^3$

17. $(-5x)^2$

18. $(-6x)^2$

19. $(4x^3)^2$

20. $(6x^3)^2$

21. $(-2y^6)^4$

22. $(-2y^5)^4$

23. $(-2x^7)^5$

24. $(-2x^{11})^5$

In Exercises 25–34, multiply the monomials.

25. $(7x)(2x)$

26. $(8x)(3x)$

27. $(6x)(4x^2)$

28. $(10x)(3x^2)$

29. $(-5y^4)(3y^3)$

30. $(-6y^4)(2y^3)$

31. $\left(-\dfrac{1}{2}a^3\right)\left(-\dfrac{1}{4}a^2\right)$

32. $\left(-\dfrac{1}{3}a^4\right)\left(-\dfrac{1}{2}a^2\right)$

33. $(2x^2)(-3x)(8x^4)$

34. $(3x^3)(-2x)(5x^6)$

In Exercises 35–54, find each product of the monomial and the polynomial.

35. $4x(x + 3)$

36. $6x(x + 5)$

37. $x(x - 3)$

38. $x(x - 7)$

39. $2x(x - 6)$

40. $3x(x - 5)$

41. $-4y(3y + 5)$

42. $-5y(6y + 7)$

43. $4x^2(x + 2)$

44. $5x^2(x + 6)$

45. $2y^2(y^2 + 3y)$

46. $4y^2(y^2 + 2y)$

47. $2y^2(3y^2 - 4y + 7)$

48. $4y^2(5y^2 - 6y + 3)$

49. $(3x^3 + 4x^2)(2x)$

50. $(4x^3 + 5x^2)(2x)$

51. $(x^2 + 5x - 3)(-2x)$

52. $(x^3 - 2x + 2)(-4x)$

53. $-3x^2(-4x^2 + x - 5)$

54. $-6x^2(3x^2 - 2x - 7)$

In Exercises 55–78, find each product. In each case, neither factor is a monomial.

55. $(x + 3)(x + 5)$

56. $(x + 4)(x + 6)$

57. $(2x + 1)(x + 4)$

58. $(2x + 5)(x + 3)$

59. $(x + 3)(x - 5)$

60. $(x + 4)(x - 6)$

61. $(x - 11)(x + 9)$

62. $(x - 12)(x + 8)$

63. $(2x - 5)(x + 4)$

64. $(3x - 4)(x + 5)$

65. $\left(\dfrac{1}{4}x + 4\right)\left(\dfrac{3}{4}x - 1\right)$

66. $\left(\dfrac{1}{5}x + 5\right)\left(\dfrac{3}{5}x - 1\right)$

67. $(x + 1)(x^2 + 2x + 3)$

68. $(x + 2)(x^2 + x + 5)$

69. $(y - 3)(y^2 - 3y + 4)$

70. $(y - 2)(y^2 - 4y + 3)$

71. $(2a - 3)(a^2 - 3a + 5)$

72. $(2a - 1)(a^2 - 4a + 3)$

73. $(x + 1)(x^3 + 2x^2 + 3x + 4)$

74. $(x + 1)(x^3 + 4x^2 + 7x + 3)$

75. $\left(x - \dfrac{1}{2}\right)(4x^3 - 2x^2 + 5x - 6)$

76. $\left(x - \dfrac{1}{3}\right)(3x^3 - 6x^2 + 5x - 9)$

77. $(x^2 + 2x + 1)(x^2 - x + 2)$

78. $(x^2 + 3x + 1)(x^2 - 2x - 1)$

In Exercises 79–92, use a vertical format to find each product.

79. $x^2 - 5x + 3$
$\underline{\hspace{1.5cm} x + 8}$

80. $x^2 - 7x + 9$
$\underline{\hspace{1.5cm} x + 4}$

81. $x^2 - 3x + 9$
$\underline{\hspace{1.5cm} 2x - 3}$

82. $y^2 - 5y + 3$
$\underline{\hspace{1.5cm} 4y - 5}$

83. $2x^3 + x^2 + 2x + 3$
$\underline{\hspace{3cm} x + 4}$

84. $3y^3 + 2y^2 + y + 4$
$\underline{\hspace{3cm} y + 3}$

85. $4z^3 - 2z^2 + 5z - 4$
$\underline{\hspace{3cm} 3z - 2}$

86. $5z^3 - 3z^2 + 4z - 3$
$\underline{\hspace{3cm} 2z - 4}$

87. $7x^3 - 5x^2 + 6x$
$\underline{\hspace{2cm} 3x^2 - 4x}$

88. $9y^3 - 7y^2 + 5y$
$\underline{\hspace{2cm} 3y^2 + 5y}$

89. $2y^5 - 3y^3 + y^2 - 2y + 3$
$\underline{\hspace{4cm} 2y - 1}$

90. $n^4 - n^3 + n^2 - n + 1$
$\underline{\hspace{4cm} 2n + 3}$

91. $x^2 + 7x - 3$
$\underline{x^2 - x - 1}$

92. $x^2 + 6x - 4$
$\underline{x^2 - x - 2}$

Application Exercises

93. Find a trinomial for the area of the rectangular rug shown below whose sides are $x + 5$ feet and $2x - 3$ feet.

94. The base of a triangular sail is $4x$ feet and its height is $3x + 10$ feet. Write a binomial in terms of x for the area of the sail.

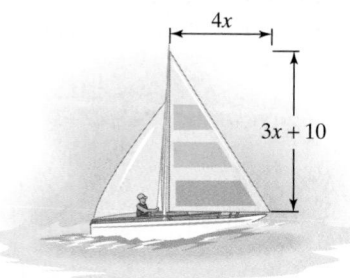

In Exercises 95–96:

   a. *Express the area of the large rectangle as the product of two binomials.*

   b. *Find the sum of the areas of the four smaller rectangles.*

   c. *Use polynomial multiplication to show that your expressions for area in parts (a) and (b) are equal.*

95.

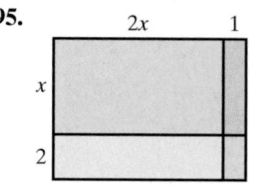

96.

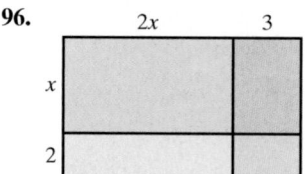

Writing in Mathematics

97. Explain the product rule for exponents. Use $2^3 \cdot 2^5$ in your explanation.

98. Explain the power rule for exponents. Use $(3^2)^4$ in your explanation.

99. Explain how to simplify an expression that involves a product raised to a power. Provide an example with your explanation.

100. Explain how to multiply monomials. Give an example.

101. Explain how to multiply a monomial and a polynomial that is not a monomial. Give an example.

102. Explain how to multiply polynomials when neither is a monomial. Give an example.

103. Explain the difference between performing these two operations:
$$2x^2 + 3x^2 \quad \text{and} \quad (2x^2)(3x^2).$$

104. Discuss situations in which a vertical format, rather than a horizontal format, is useful for multiplying polynomials.

105. Describe one change that might alter the prediction about global warming given by the model and graph on page 000.

Critical Thinking Exercises

106. Which one of the following is true?
  a. $4x^3 \cdot 3x^4 = 12x^{12}$
  b. $5x^2 \cdot 4x^6 = 9x^8$
  c. $(y - 1)(y^2 + y + 1) = y^3 - 1$
  d. Some polynomial multiplications can only be performed by using a vertical format.

107. Find a polynomial in descending powers of x representing the area of the shaded region.

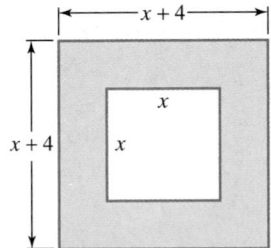

108. Find each of the products in parts (a)–(c).
  a. $(x - 1)(x + 1)$
  b. $(x - 1)(x^2 + x + 1)$
  c. $(x - 1)(x^3 + x^2 + x + 1)$
  d. Using the pattern found in parts (a)–(c), find $(x - 1)(x^4 + x^3 + x^2 + x + 1)$ without actually multiplying.

109. Find the missing factor.

$$(\underline{\hspace{1cm}})\left(-\frac{1}{4}xy^3\right) = 2x^5y^3$$

Review Exercises

110. Solve: $4x - 7 > 9x - 2$. (Section 2.7, Example 6)

111. Graph $3x - 2y = 6$ using intercepts. (Section 3.2, Example 4)

112. Find the slope of the line passing through the points $(-2, 8)$ and $(1, 6)$. (Section 3.3, Example 1)

▶ SECTION 5.3 *Special Products*

Objectives

1 Use FOIL in polynomial multiplication.

2 Multiply the sum and difference of two terms.

3 Find the square of a binomial sum.

4 Find the square of a binomial difference.

SSM PH Tutor CD- Video
Center ROM

Let's cut to the chase. Are there fast methods for finding products of polynomials? Yes. In this section, we use the distributive property to develop patterns that will let you multiply certain binomials quite rapidly.

 1 Use FOIL in polynomial multiplication.

The Product of Two Binomials: FOIL

Frequently we need to find the product of two binomials. We can use a method called FOIL, which is based on the distributive property, to do so. For example, we can find the product of the binomials $3x + 2$ and $4x + 5$ as follows:

$$(3x + 2)(4x + 5) = 3x(4x + 5) + 2(4x + 5)$$

First, distribute $3x$ over $4x + 5$. Then distribute 2.

$$= 3x(4x) + 3x(5) + 2(4x) + 2(5)$$

$$= 12x^2 + 15x + 8x + 10.$$

Two binomials can be quickly multiplied by using the FOIL method, in which F represents the product of the **first** terms in each binomial, O represents the product of the **outside** terms, I represents the product of the two **inside** terms, and L represents the product of the **last**, or second, terms in each binomial.

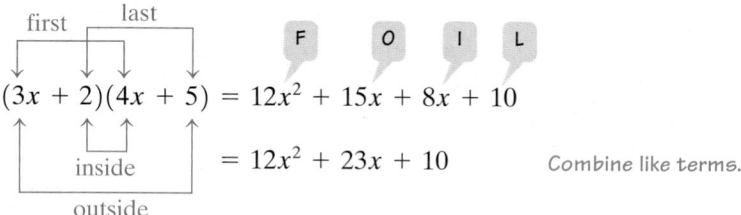

$$= 12x^2 + 23x + 10$$ Combine like terms.

In general, here's how to use the FOIL method to find the product of $ax + b$ and $cx + d$:

Using the FOIL Method to Multiply Binomials

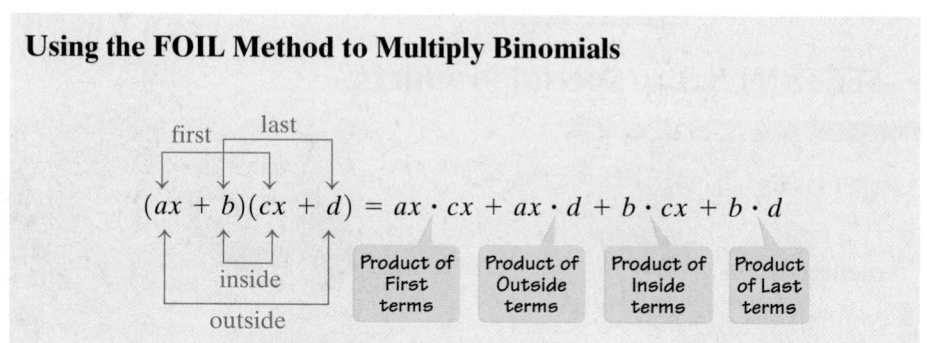

EXAMPLE 1 Using the FOIL Method

Multiply: $(x + 3)(x + 4)$.

Solution

F: First terms $= x \cdot x = x^2$ $(x + 3)(x + 4)$

O: Outside terms $= x \cdot 4 = 4x$ $(x + 3)(x + 4)$

I: Inside terms $= 3 \cdot x = 3x$ $(x + 3)(x + 4)$

L: Last terms $= 3 \cdot 4 = 12$ $(x + 3)(x + 4)$

$$\underbrace{(x + 3)(x + 4)}_{} = \overset{F}{x \cdot x} + \overset{O}{x \cdot 4} + \overset{I}{3 \cdot x} + \overset{L}{3 \cdot 4}$$

$$= x^2 + 4x + 3x + 12$$

$$= x^2 + 7x + 12 \qquad \textit{Combine like terms.} \ \blacksquare$$

✔ **CHECK POINT 1** Multiply: $(x + 5)(x + 6)$.

EXAMPLE 2 Using the FOIL Method

Multiply: $(3x + 4)(5x - 3)$.

Solution

$$(3x + 4)(5x - 3) = \overset{F}{3x \cdot 5x} + \overset{O}{3x(-3)} + \overset{I}{4 \cdot 5x} + \overset{L}{4(-3)}$$

$$= 15x^2 - 9x + 20x - 12$$

$$= 15x^2 + 11x - 12 \qquad \textit{Combine like terms.} \ \blacksquare$$

✔ **CHECK POINT 2** Multiply: $(7x + 5)(4x - 3)$.

EXAMPLE 3 Using the FOIL Method

Multiply: $(2 - 5x)(3 - 4x)$.

Solution

$$(2 - 5x)(3 - 4x) = \overset{F}{2 \cdot 3} + \overset{O}{2(-4x)} + \overset{I}{(-5x)(3)} + \overset{L}{(-5x)(-4x)}$$

$$= 6 - 8x - 15x + 20x^2$$

$$= 6 - 23x + 20x^2 \qquad \textit{Combine like terms.}$$

The product can also be expressed as $20x^2 - 23x + 6$. $\blacksquare$

✔ **CHECK POINT 3** Multiply: $(4 - 2x)(5 - 3x)$.

2 Multiply the sum and difference of two terms.

Multiplying the Sum and Difference of Two Terms

We can use the FOIL method to multiply $A + B$ and $A - B$ as follows:

$$(A + B)(A - B) = \overset{F}{A^2} - \overset{O}{AB} + \overset{I}{AB} - \overset{L}{B^2} = A^2 - B^2.$$

Notice that the outside and inside products have a sum of 0 and the terms cancel. The FOIL multiplication provides us with a quick rule for multiplying the sum and difference of two terms, referred to as a special-product formula.

The Product of the Sum and Difference of Two Terms

$$(A + B)(A - B) = A^2 - B^2$$

| The product of the sum and the difference of the same two terms | is | the square of the first term minus the square of the second term. |

EXAMPLE 4 Finding the Product of the Sum and Difference of Two Terms

Find each product by using the preceding rule:

    a. $(4y + 3)(4y - 3)$ **b.** $(3x - 7)(3x + 7)$ **c.** $(5a^4 + 6)(5a^4 - 6)$.

Solution Use the special-product formula shown.

$$(A + B)(A - B) \quad = \quad A^2 \quad - \quad B^2$$

	First term squared	−	Second term squared	=	Product
a. $(4y + 3)(4y - 3)$ =	$(4y)^2$	−	3^2	=	$16y^2 - 9$
b. $(3x - 7)(3x + 7)$ =	$(3x)^2$	−	7^2	=	$9x^2 - 49$
c. $(5a^4 + 6)(5a^4 - 6)$ =	$(5a^4)^2$	−	6^2	=	$25a^8 - 36$ ■

✔ **CHECK POINT 4** Find each product:

    a. $(7y + 8)(7y - 8)$

    b. $(4x - 5)(4x + 5)$

    c. $(2a^3 + 3)(2a^3 - 3)$.

3 Find the square of a binomial sum.

The Square of a Binomial

Let us find $(A + B)^2$, the square of a binomial sum. To do so, we begin with the FOIL method and look for a general rule.

$$\overset{\text{F}\quad\text{O}\quad\text{I}\quad\text{L}}{(A + B)^2 = (A + B)(A + B) = A \cdot A + A \cdot B + A \cdot B + B \cdot B}$$

$$= A^2 + 2AB + B^2$$

This result implies the following rule, which is another example of a special-product formula.

Study Tip

Caution! The square of a sum is *not* the sum of the squares.

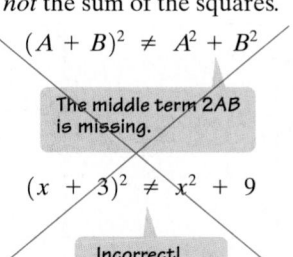

$$(A + B)^2 \neq A^2 + B^2$$

The middle term $2AB$ is missing.

$$(x + 3)^2 \neq x^2 + 9$$

Incorrect!

Show that $(x + 3)^2$ and $x^2 + 9$ are not equal by substituting 5 for x in each expression and simplifying.

The Square of a Binomial Sum

$(A + B)^2$	$=$	A^2	$+$	$2AB$	$+$	B^2
The square of a binomial sum	is	first term squared	plus	2 times the product of the terms	plus	last term squared.

EXAMPLE 5   Finding the Square of a Binomial Sum

Square each binomial using the preceding rule:

a. $(x + 3)^2$ **b.** $(3x + 7)^2$.

Solution   Use the special-product formula shown.

$$(A + B)^2 = A^2 + 2AB + B^2$$

	(First Term)^2 $+$	2 · Product of the Terms $+$	(Last Term)^2	= Product
a. $(x + 3)^2 =$	x^2	$2 \cdot x \cdot 3$	3^2	$= x^2 + 6x + 9$
b. $(3x + 7)^2 =$	$(3x)^2$	$2(3x)(7)$	7^2	$= 9x^2 + 42x + 49$

■

 CHECK POINT 5   Square each binomial:

a. $(x + 10)^2$ **b.** $(5x + 4)^2$.

4 Find the square of a binomial difference.

Using the FOIL method on $(A - B)^2$, the square of a binomial difference, we obtain the following rule.

The Square of a Binomial Difference

$(A - B)^2$	$=$	A^2	$-$	$2AB$	$+$	B^2
The square of a binomial difference	is	first term squared	minus	2 times the product of the terms	plus	last term squared.

EXAMPLE 6   Finding the Square of a Binomial Difference

Square each binomial using the preceding rule:

a. $(x - 4)^2$ **b.** $(5y - 6)^2$.

Solution   Use the special-product formula shown.

$$(A - B)^2 = A^2 - 2AB + B^2$$

	(First Term)^2 $-$	2 · Product of the Terms $+$	(Last Term)^2	= Product
a. $(x - 4)^2 =$	x^2	$2 \cdot x \cdot 4$	4^2	$= x^2 - 8x + 16$
b. $(5y - 6)^2 =$	$(5y)^2$	$2(5y)(6)$	6^2	$= 25y^2 - 60y + 36$

■

✔ **CHECK POINT 6** Square each binomial:

a. $(x - 9)^2$ **b.** $(7x - 3)^2$.

Figure 5.3 makes it possible to visualize the square of a binomial sum. The area of the large square is

$$(A + B)(A + B) \quad \text{or} \quad (A + B)^2.$$

The sum of the areas of the four smaller rectangles that make up the square is

$$A^2 + AB + AB + B^2$$

or

$$A^2 + 2AB + B^2.$$

Conclusion:

$$(A + B)^2 = A^2 + 2AB + B^2.$$

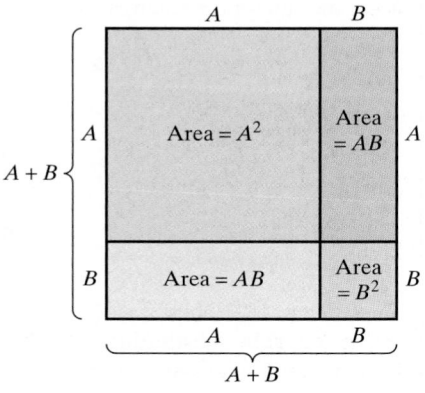

Figure 5.3

The following box summarizes the FOIL method and the two special products. The special products occur so frequently in algebra that it is convenient to memorize the form or pattern of these formulas.

FOIL and Special Products

Let A, B, C, and D be real numbers, variables, or algebraic expressions.

FOIL	*Example*
F O I L	F O I L
$(A + B)(C + D) = AC + AD + BC + BD$	$(2x + 3)(4x + 5) = (2x)(4x) + (2x)(5) + (3)(4x) + (3)(5)$
	$= 8x^2 + 10x + 12x + 15$
	$= 8x^2 + 22x + 15$
Sum and Difference of Two Terms	*Example*
$(A + B)(A - B) = A^2 - B^2$	$(2x + 3)(2x - 3) = (2x)^2 - 3^2$
	$= 4x^2 - 9$
Square of a Binomial	*Example*
$(A + B)^2 = A^2 + 2AB + B^2$	$(2x + 3)^2 = (2x)^2 + 2(2x)(3) + 3^2$
	$= 4x^2 + 12x + 9$
$(A - B)^2 = A^2 - 2AB + B^2$	$(2x - 3)^2 = (2x)^2 - 2(2x)(3) + 3^2$
	$= 4x^2 - 12x + 9$

EXERCISE SET 5.3

Practice Exercises

✔ *In Exercises 1–24, use the FOIL method to find each product. Express the product in descending powers of the variable.*

1. $(x + 3)(x + 5)$

2. $(x + 7)(x + 2)$

3. $(y - 5)(y + 3)$

4. $(y - 1)(y + 2)$

5. $(2x - 1)(x + 2)$

6. $(2x - 5)(x + 3)$

7. $(2y - 3)(y + 1)$

8. $(3y - 5)(y + 4)$

9. $(2x - 3)(5x + 3)$

10. $(2x - 5)(7x + 2)$

11. $(3y - 7)(4y - 5)$

12. $(4y - 5)(7y - 4)$

13. $(7 + 3x)(1 - 5x)$

14. $(2 + 5x)(1 - 4x)$

15. $(5 - 3y)(6 - 2y)$

16. $(7 - 2y)(10 - 3y)$

17. $(5x^2 - 4)(3x^2 - 7)$

18. $(7x^2 - 2)(3x^2 - 5)$

19. $(6x - 5)(2 - x)$

20. $(4x - 3)(2 - x)$

21. $(x + 5)(x^2 + 3)$

22. $(x + 4)(x^2 + 5)$

23. $(8x^3 + 3)(x^2 + 5)$

24. $(7x^3 + 5)(x^2 + 2)$

In Exercises 25–44, multiply by using the rule for finding the product of the sum and difference of two terms.

25. $(x + 3)(x - 3)$

26. $(y + 5)(y - 5)$

27. $(3x + 2)(3x - 2)$

28. $(2x + 5)(2x - 5)$

29. $(3r - 4)(3r + 4)$

30. $(5z - 2)(5z + 2)$

31. $(3 + r)(3 - r)$

32. $(4 + s)(4 - s)$

33. $(5 - 7x)(5 + 7x)$

34. $(4 - 3y)(4 + 3y)$

35. $\left(2x + \dfrac{1}{2}\right)\left(2x - \dfrac{1}{2}\right)$

36. $\left(3y + \dfrac{1}{3}\right)\left(3y - \dfrac{1}{3}\right)$

37. $(y^2 + 1)(y^2 - 1)$

38. $(y^2 + 2)(y^2 - 2)$

39. $(r^3 + 2)(r^3 - 2)$

40. $(m^3 + 4)(m^3 - 4)$

41. $(1 - y^4)(1 + y^4)$

42. $(2 - s^5)(2 + s^5)$

43. $(x^{10} + 5)(x^{10} - 5)$

44. $(x^{12} + 3)(x^{12} - 3)$

In Exercises 45–62, multiply by using the rule for the square of a binomial.

45. $(x + 2)^2$

46. $(x + 5)^2$

47. $(2x + 5)^2$

48. $(5x + 2)^2$

49. $(x - 3)^2$

50. $(x - 6)^2$

51. $(3y - 4)^2$

52. $(4y - 3)^2$

53. $(4x^2 - 1)^2$

54. $(5x^2 - 3)^2$

55. $(7 - 2x)^2$

56. $(9 - 5x)^2$

57. $\left(2x + \dfrac{1}{2}\right)^2$

58. $\left(3x + \dfrac{1}{3}\right)^2$

59. $\left(4y - \dfrac{1}{4}\right)^2$

60. $\left(2y - \dfrac{1}{2}\right)^2$

61. $(x^8 + 3)^2$

62. $(x^8 + 5)^2$

In Exercises 63–82, multiply by the method of your choice.

63. $(x - 1)(x^2 + x + 1)$

64. $(x + 1)(x^2 - x + 1)$

65. $(x - 1)^2$

66. $(x + 1)^2$

67. $(3y + 7)(3y - 7)$

68. $(4y + 9)(4y - 9)$

69. $3x^2(4x^2 + x + 9)$

70. $5x^2(7x^2 + x + 6)$

71. $(7y + 3)(10y - 4)$

72. $(8y + 3)(10y - 5)$

73. $(x^2 + 1)^2$

74. $(x^2 + 2)^2$

75. $(x^2 + 1)(x^2 + 2)$

76. $(x^2 + 2)(x^2 + 3)$

77. $(x^2 + 4)(x^2 - 4)$

78. $(x^2 + 5)(x^2 - 5)$

79. $(2 - 3x^5)^2$

80. $(2 - 3x^6)^2$

81. $\left(\frac{1}{4}x^2 + 12\right)\left(\frac{3}{4}x^2 - 8\right)$

82. $\left(\frac{1}{4}x^2 + 16\right)\left(\frac{3}{4}x^2 - 4\right)$

In Exercises 83–88, find the area of each shaded region. Write the answer as a polynomial in descending powers of x.

83.
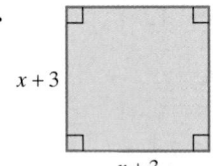
$x + 1$
$x + 1$

84.
$x + 3$
$x + 3$

85.
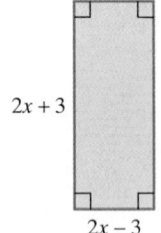
$2x + 3$
$2x - 3$

86.
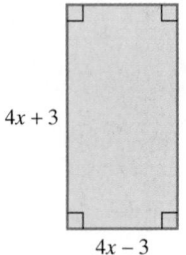
$4x + 3$
$4x - 3$

87.
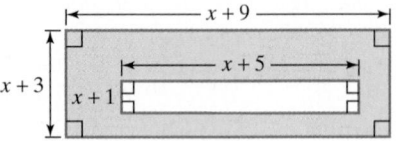
$x + 9$
$x + 5$
$x + 3$
$x + 1$

88.
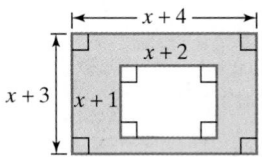
$x + 4$
$x + 2$
$x + 3$ $x + 1$

★ **Application Exercises**

The square garden shown in the figure measures x yards on each side. The garden is to be expanded so that one side is increased by 2 yards and an adjacent side is increased by 1 yard. The graph shows the area of the expanded garden in terms of the length of one of its original square sides. Use this information to solve Exercises 89–92.

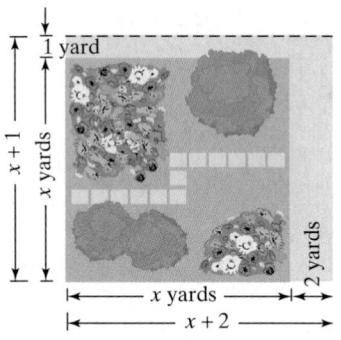

1 yard
$x + 1$
x yards
x yards
$x + 2$
2 yards

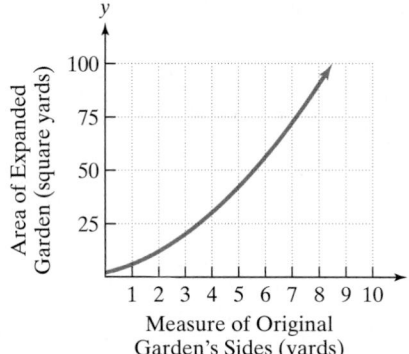

Area of Expanded Garden (square yards)

Measure of Original Garden's Sides (yards)

89. Write a product of two binomials that expresses the area of the larger garden.

90. Write a polynomial in descending powers of x that expresses the area of the larger garden.

91. If the original garden measures 6 yards on a side, use your expression from Exercise 89 to find the area of the larger garden. Then identify your solution as a point on the graph shown.

92. If the original garden measures 8 yards on a side, use your polynomial from Exercise 90 to find the area of the larger garden. Then identify your solution as a point on the graph shown.

The square painting in the figure measures x inches on each side. The painting is uniformly surrounded by a frame that measures 1 inch wide. Use this information to solve Exercises 93–94.

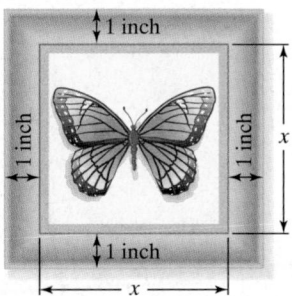

93. Write a polynomial in descending powers of x that expresses the area of the square that includes the painting and the frame.

94. Write an algebraic expression that describes the area of the frame. (*Hint:* The area of the frame is the area of the square that includes the painting and the frame minus the area of the painting.)

Writing in Mathematics

95. Explain how to multiply two binomials using the FOIL method. Give an example with your explanation.

96. Explain how to find the product of the sum and difference of two terms. Give an example with your explanation.

97. Explain how to square a binomial sum. Give an example with your explanation.

98. Explain how to square a binomial difference. Give an example with your explanation.

99. Explain why the graph for Exercises 89–92 is shown only in quadrant I.

Critical Thinking Exercises

100. Which one of the following is true?
 a. $(3 + 4)^2 = 3^2 + 4^2$
 b. $(2y + 7)^2 = 4y^2 + 28y + 49$
 c. $(3x^2 + 2)(3x^2 - 2) = 9x^2 - 4$
 d. $(x - 5)^2 = x^2 - 5x + 25$

101. Which two binomials must be multiplied using the FOIL method to give a product of $x^2 - 8x - 20$?

102. Express the volume of the box as a polynomial in standard form.

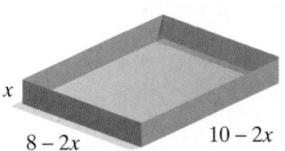

103. Express the area of the plane figure shown as a polynomial in standard form.

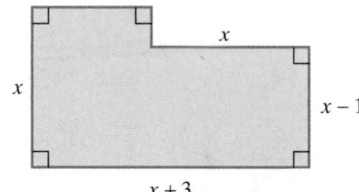

Technology Exercises

 In Exercises 104–107, use a graphing utility to graph each side of the equation in the same viewing rectangle. (Call the left side y_1 and the right side y_2.) If the graphs coincide, verify that the multiplication has been performed correctly. If the graphs do not appear to coincide, this indicates that the multiplication is incorrect. In these exercises, correct the right side of the equation. Then graph the left side and the corrected right side to verify that the graphs coincide.

104. $(x + 1)^2 = x^2 + 1$; Use a $[-5, 5, 1]$ by $[0, 20, 1]$ viewing rectangle.

105. $(x + 2)^2 = x^2 + 2x + 4$; Use a $[-6, 5, 1]$ by $[0, 20, 1]$ viewing rectangle.

106. $(x + 1)(x - 1) = x^2 - 1$; Use a $[-6, 5, 1]$ by $[-2, 18, 1]$ viewing rectangle.

107. $(x - 2)(x + 2) + 4 = x^2$; Use a $[-6, 5, 1]$ by $[-2, 18, 1]$ viewing rectangle.

 ## Review Exercises

In Exercises 108–109, solve each system by the method of your choice.

108. $2x + 3y = 1$
 $y = 3x - 7$
 (Section 4.2, Example 1)

109. $3x + 4y = 7$
 $2x + 7y = 9$
 (Section 4.3, Example 3)

110. Graph: $y = \dfrac{1}{3}x$.
 (Section 3.4, Example 3)

▶ **SECTION 5.4** *Polynomials in Several Variables*

Objectives

1 Evaluate polynomials in several variables.

2 Understand the vocabulary of polynomials in two variables.

3 Add and subtract polynomials in several variables.

4 Multiply polynomials in several variables.

SSM
PH Tutor   CD-   Video
Center   ROM

The next time you visit the lumberyard and go rummaging through piles of wood, think *polynomials*, although polynomials a bit different from those we have encountered so far. The construction industry uses a polynomial in two variables to determine the number of board feet that can be manufactured from a tree with a diameter of x inches and a length of y feet. This polynomial is

$$\frac{1}{4}x^2y - 2xy + 4y.$$

We call a polynomial containing two or more variables a **polynomial in several variables**. These polynomials can be evaluated, added, subtracted, and multiplied just like polynomials that contain only one variable.

Evaluating a Polynomial in Several Variables

Two steps can be used to evaluate a polynomial in several variables.

1 Evaluate polynomials in several variables.

Evaluating a Polynomial in Several Variables

1. Substitute the given value for each variable.
2. Perform the resulting computation using the order of operations.

EXAMPLE 1   Evaluating a Polynomial in Two Variables

Evaluate $2x^3y + xy^2 + 7x - 3$ for $x = -2$ and $y = 3$.

Solution   We begin by substituting -2 for x and 3 for y in the polynomial.

$2x^3y + xy^2 + 7x - 3$	This is the given polynomial.
$= 2(-2)^3 \cdot 3 + (-2) \cdot 3^2 + 7(-2) - 3$	Replace x by -2 and y by 3.
$= 2(-8) \cdot 3 + (-2) \cdot 9 + 7(-2) - 3$	Evaluate exponential expressions: $(-2)^3 = (-2)(-2)(-2) = -8$ and $3^2 = 3 \cdot 3 = 9$.
$= -48 + (-18) + (-14) - 3$	Perform the indicated multiplications.
$= -83$	Add from left to right. ∎

✔ **CHECK POINT 1**   Evaluate $3x^3y + xy^2 + 5y + 6$ for $x = -1$ and $y = 5$.

The Vocabulary of Polynomials in Two Variables

2 Understand the vocabulary of polynomials in two variables.

In this section, we will limit our discussion of polynomials in several variables to two variables.

In general, a **polynomial in two variables**, x and y, contains the sum of one or more monomials in the form $ax^n y^m$. The constant a is the **coefficient**. The exponents, n and m, represent whole numbers. The **degree** of the monomial $ax^n y^m$ is $n + m$. We'll use the polynomial from the construction industry to illustrate these ideas.

The coefficients are $\frac{1}{4}$, -2, and 4.

$$\frac{1}{4}\,x^2 y \qquad -2xy \qquad +4y$$

| Degree of monomial: $2 + 1 = 3$ | Degree of monomial: $1 + 1 = 2$ | Degree of monomial: 1 |

The **degree of a polynomial in two variables** is the highest degree of all its terms. For the preceding polynomial, the degree is 3.

EXAMPLE 2 Using the Vocabulary of Polynomials

Determine the coefficient of each term, the degree of each term, and the degree of the polynomial:

$$7x^2 y^3 - 17x^4 y^2 + xy - 6y^2 + 9.$$

Solution

Think of xy as $1x^1 y^1$.

Term	Coefficient	Degree (Sum of Exponents on the Variables)
$7x^2 y^3$	7	$2 + 3 = 5$
$-17x^4 y^2$	-17	$4 + 2 = 6$
xy	1	$1 + 1 = 2$
$-6y^2$	-6	2
9	9	0

The degree of the polynomial is the highest degree of all its terms, which is 6. ■

✔ **CHECK POINT 2** Determine the coefficient of each term, the degree of each term, and the degree of the polynomial:

$$8x^4 y^5 - 7x^3 y^2 - x^2 y - 5x + 11.$$

3 Add and subtract polynomials in several variables.

Adding and Subtracting Polynomials in Several Variables

Polynomials in several variables are added by combining like terms. For example, we can add the monomials $-7xy^2$ and $13xy^2$ as follows:

$$-7xy^2 + 13xy^2 = (-7 + 13)xy^2 = 6xy^2.$$

Add coefficients and keep the same variable factors, xy^2.

EXAMPLE 3 Adding Polynomials in Two Variables

Add: $(6xy^2 - 5xy + 7) + (9xy^2 + 2xy - 6)$.

Solution

$$(6xy^2 - 5xy + 7) + (9xy^2 + 2xy - 6)$$

$$= (6xy^2 + 9xy^2) + (-5xy + 2xy) + (7 - 6) \qquad \text{Group like terms.}$$

$$= 15xy^2 - 3xy + 1 \qquad \text{Combine like terms by adding coefficients and keeping the same variable factors.} \ \blacksquare$$

✔ **CHECK POINT 3** Add: $(-8x^2y - 3xy + 6) + (10x^2y + 5xy - 10)$.

We subtract polynomials in two variables just as we did when subtracting polynomials in one variable. Change the sign of every term of the polynomial being subtracted. Add this result to the first polynomial.

EXAMPLE 4 Subtracting Polynomials in Two Variables

Subtract:

$$(5x^3 - 9x^2y + 3xy^2 - 4) - (3x^3 - 6x^2y - 2xy^2 + 3).$$

Solution

$$(5x^3 - 9x^2y + 3xy^2 - 4) - (3x^3 - 6x^2y - 2xy^2 + 3)$$

$$= (5x^3 - 9x^2y + 3xy^2 - 4) + (-3x^3 + 6x^2y + 2xy^2 - 3)$$

Change the sign of each term in the second polynomial and add the two polynomials.

$$= (5x^3 - 3x^3) + (-9x^2y + 6x^2y) + (3xy^2 + 2xy^2) + (-4 - 3)$$

Group like terms.

$$= 2x^3 - 3x^2y + 5xy^2 - 7 \quad \text{Combine like terms by adding coefficients and keeping the same variable factors.} \ \blacksquare$$

✔ **CHECK POINT 4** Subtract:
$$(7x^3 - 10x^2y + 2xy^2 - 5) - (4x^3 - 12x^2y - 3xy^2 + 5).$$

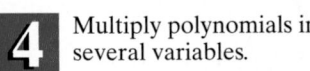
Multiply polynomials in several variables.

Multiplying Polynomials in Several Variables

The product of monomials forms the basis of polynomial multiplication. As with monomials in one variable, multiplication can be done mentally by multiplying coefficients and adding exponents on variables with the same base.

EXAMPLE 5   Multiplying Monomials

Multiply: $(7x^2y)(5x^3y^2)$.

Solution

$$(7x^2y)(5x^3y^2)$$

$$= (7 \cdot 5)(x^2 \cdot x^3)(y \cdot y^2) \qquad \text{This regrouping can be worked mentally.}$$

$$= 35x^{2+3}y^{1+2} \qquad \text{Multiply coefficients and add exponents on variables with same base.}$$

$$= 35x^5y^3 \qquad \text{Simplify.} \qquad \blacksquare$$

✔ **CHECK POINT 5**   Multiply: $(6xy^3)(10x^4y^2)$.

How do we multiply a monomial and a polynomial that is not a monomial? As we did with polynomials in one variable, multiply each term of the polynomial by the monomial.

EXAMPLE 6   Multiplying a Monomial and a Polynomial

Multiply: $3x^2y(4x^3y^2 - 6x^2y + 2)$.

Solution

$$3x^2y(4x^3y^2 - 6x^2y + 2)$$

$$= 3x^2y \cdot 4x^3y^2 - 3x^2y \cdot 6x^2y + 3x^2y \cdot 2 \qquad \text{Use the distributive property.}$$

$$= 12x^{2+3}y^{1+2} - 18x^{2+2}y^{1+1} + 6x^2y \qquad \text{Multiply coefficients and add exponents on variables with the same base.}$$

$$= 12x^5y^3 - 18x^4y^2 + 6x^2y \qquad \text{Simplify.}$$

✔ **CHECK POINT 6**   Multiply: $6xy^2(10x^4y^5 - 2x^2y + 3)$.

FOIL and the special products formulas can be used to multiply polynomials in several variables.

EXAMPLE 7   Multiplying Polynomials in Two Variables

Multiply: **a.** $(x + 4y)(3x - 5y)$ **b.** $(5x + 3y)^2$.

Solution   We will perform the multiplication in part (a) using the FOIL method. We will multiply in part (b) using the formula for the square of a binomial, $(A + B)^2$.

a. $(x + 4y)(3x - 5y)$ Multiply these binomials using the FOIL method.

F   O   I   L

$$= (x)(3x) + (x)(-5y) + (4y)(3x) + (4y)(-5y)$$
$$= 3x^2 - 5xy + 12xy - 20y^2$$
$$= 3x^2 + 7xy - 20y^2 \quad \text{Combine like terms.}$$

$(A + B)^2 = A^2 + 2 \cdot A \cdot B + B^2$

b. $(5x + 3y)^2 = (5x)^2 + 2(5x)(3y) + (3y)^2$
$$= 25x^2 + 30xy + 9y^2$$ ∎

✔ **CHECK POINT 7**   Multiply:
   a. $(7x - 6y)(3x - y)$
   b. $(2x + 4y)^2$.

EXAMPLE 8   Multiplying Polynomials in Two Variables

Multiply: **a.** $(4x^2y + 3y)(4x^2y - 3y)$ **b.** $(x + y)(x^2 - xy + y^2)$.

Solution   We perform the multiplication in part (a) using the formula for the product of the sum and difference of two terms. We perform the multiplication in part (b) by multiplying each term of the trinomial, $x^2 - xy + y^2$, by x and y, respectively, and then adding like terms.

$(A + B) \ (A - B) \ = \ A^2 - B^2$

a. $(4x^2y + 3y)(4x^2y - 3y) = (4x^2y)^2 - (3y)^2$
$$= 16x^4y^2 - 9y^2$$

b. $(x + y)(x^2 - xy + y^2)$

$= x(x^2 - xy + y^2) + y(x^2 - xy + y^2)$ Multiply the trinomial by each term of the binomial.

$= x \cdot x^2 - x \cdot xy + x \cdot y^2 + y \cdot x^2 - y \cdot xy + y \cdot y^2$ Use the distributive property.

$= x^3 - x^2y + xy^2 + x^2y - xy^2 + y^3$ Add exponents on variables with the same base.

$= x^3 + y^3$ Combine like terms:
$-x^2y + x^2y = 0$ and
$xy^2 - xy^2 = 0.$ ∎

✔ **CHECK POINT 8**   Multiply:
   a. $(6xy^2 + 5x)(6xy^2 - 5x)$

   b. $(x - y)(x^2 + xy + y^2)$.

EXERCISE SET 5.4

Practice Exercises

In Exercises 1–6, evaluate each polynomial for $x = 2$ and $y = -3$.

1. $x^2 + 2xy + y^2$ **2.** $x^2 + 3xy + y^2$

3. $xy^3 - xy + 1$ **4.** $x^3y - xy + 2$

5. $2x^2y - 5y + 3$ **6.** $3x^2y - 4y + 5$

In Exercises 7–8, determine the coefficient of each term, the degree of each term, and the degree of the polynomial.

7. $x^3y^2 - 5x^2y^7 + 6y^2 - 3$ **8.** $12x^4y - 5x^3y^7 - x^2 + 4$

In Exercises 9–20, add or subtract as indicated.

9. $(5x^2y - 3xy) + (2x^2y - xy)$

10. $(-2x^2y + xy) + (4x^2y + 7xy)$

11. $(4x^2y + 8xy + 11) + (-2x^2y + 5xy + 2)$

12. $(7x^2y + 5xy + 13) + (-3x^2y + 6xy + 4)$

13. $(7x^4y^2 - 5x^2y^2 + 3xy) + (-18x^4y^2 - 6x^2y^2 - xy)$

14. $(6x^4y^2 - 10x^2y^2 + 7xy) + (-12x^4y^2 - 3x^2y^2 - xy)$

15. $(x^3 + 7xy - 5y^2) - (6x^3 - xy + 4y^2)$

16. $(x^4 - 7xy - 5y^3) - (6x^4 - 3xy + 4y^3)$

17. $(3x^4y^2 + 5x^3y - 3y) - (2x^4y^2 - 3x^3y - 4y + 6x)$

18. $(5x^4y^2 + 6x^3y - 7y) - (3x^4y^2 - 5x^3y - 6y + 8x)$

19. $(x^3 - y^3) - (-4x^3 - x^2y + xy^2 + 3y^3)$

20. $(x^3 - y^3) - (-6x^3 + x^2y - xy^2 + 2y^3)$

21. Add:

$5x^2y^2 - 4xy^2 + 6y^2$
$-8x^2y^2 + 5xy^2 - y^2$

22. Add:

$7a^2b^2 - 5ab^2 + 6b^2$
$-10a^2b^2 + 6ab^2 + 6b^2$

23. Subtract:

$3a^2b^4 - 5ab^2 + 7ab$
$-(-5a^2b^4 - 8ab^2 - ab)$

24. Subtract:

$13x^2y^4 - 17xy^2 + xy$
$-(-7x^2y^4 - 8xy^2 - xy)$

25. Subtract $11x - 5y$ from the sum of $7x + 13y$ and $-26x + 19y$.

26. Subtract $23x - 5y$ from the sum of $6x + 15y$ and $x - 19y$.

In Exercises 27–76, find each product.

27. $(6x^2y)(3xy)$ **28.** $(4x^2y)(5xy)$

29. $(-7x^3y^4)(2x^2y^5)$

30. $(6x^4y^5)(-10x^7y^{11})$

31. $5xy(2x + 3y)$

32. $4xy(5x + 2y)$

33. $3xy^2(6x^2 - 2y)$

34. $5x^2y(4x^2 - 3y)$

35. $3ab^2(6a^2b^3 + 5ab)$

36. $5ab^2(10a^2b^3 + 7ab)$

37. $-b(a^2 - ab + b^2)$

38. $-b(a^3 - ab + b^3)$

39. $(x + 5y)(7x + 3y)$

40. $(x + 9y)(6x + 7y)$

41. $(x - 3y)(2x + 7y)$

42. $(3x - y)(2x + 5y)$

43. $(3xy - 1)(5xy + 2)$

44. $(7xy + 1)(2xy - 3)$

45. $(2x + 3y)^2$

46. $(2x + 5y)^2$

47. $(xy - 3)^2$

48. $(xy - 5)^2$

49. $(x^2 + y^2)^2$

50. $(2x^2 + y^2)^2$

51. $(x^2 - 2y^2)^2$

52. $(x^2 - y^2)^2$

53. $(3x + y)(3x - y)$

54. $(x + 5y)(x - 5y)$

55. $(ab + 1)(ab - 1)$

56. $(ab + 2)(ab - 2)$

57. $(x + y^2)(x - y^2)$

58. $(x^2 + y)(x^2 - y)$

59. $(3a^2b + a)(3a^2b - a)$

60. $(5a^2b + a)(5a^2b - a)$

61. $(3xy^2 - 4y)(3xy^2 + 4y)$

62. $(7xy^2 - 10y)(7xy^2 + 10y)$

63. $(a + b)(a^2 - b^2)$

64. $(a - b)(a^2 + b^2)$

65. $(x + y)(x^2 + 3xy + y^2)$

66. $(x + y)(x^2 + 5xy + y^2)$

67. $(x - y)(x^2 - 3xy + y^2)$

68. $(x - y)(x^2 - 4xy + y^2)$

69. $(xy + ab)(xy - ab)$

70. $(xy + ab^2)(xy - ab^2)$

71. $(x^2 + 1)(x^4y + x^2 + 1)$

72. $(x^2 + 1)(xy^4 + y^2 + 1)$

73. $(x^2y^2 - 3)^2$

74. $(x^2y^2 - 5)^2$

75. $(x + y + 1)(x + y - 1)$

76. $(x + y + 1)(x - y + 1)$

In Exercises 77–80 write a polynomial in two variables that describes the total area of each shaded region. Express each polynomial as the sum or difference of terms.

77. 3x + 5y ... x + y

78. x + 3y ... x + 3y

79.
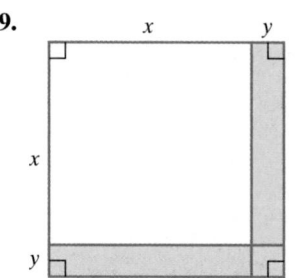

80.
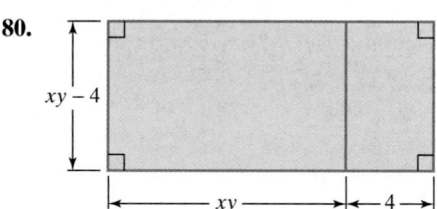

Application Exercises

81. The number of board feet, N, that can be manufactured from a tree with a diameter of x inches and a length of y feet is modeled by the formula

$$N = \frac{1}{4}x^2y - 2xy + 4y.$$

A building contractor estimates that 3000 board feet of lumber is needed for a job. The lumber company has just milled a fresh load of timber from 20 trees that averaged 10 inches in diameter and 16 feet in length. Is this enough to complete the job? If not, how many additional board feet of lumber are needed?

82. The storage shed shown in the figure has a volume given by the polynomial

$$2x^2y + \frac{1}{2}\pi x^2y.$$

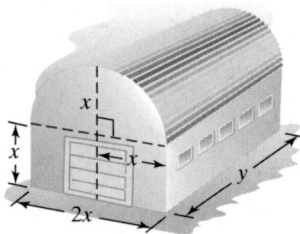

a. A small business is considering having a shed installed like the one shown in the figure. The shed's height, $2x$, is 26 feet and its length, y, is 27 feet. Using $x = 13$ and $y = 27$, find the volume of the storage shed.

b. The business requires at least 18,000 cubic feet of storage space. Should they construct the storage shed described in part (a)?

An object that is falling or vertically projected into the air has its height, in feet, above the ground given by

$$s = -16t^2 + v_0 t + s_0$$

where s is the height, in feet, v_0 is the original velocity of the object, in feet per second, t is the time the object is in motion, in seconds, and s_0 is the height, in feet, from which the object is dropped or projected. The figure shows that a ball is thrown straight up from a rooftop at an original velocity of 80 feet per second from a height of 96 feet. The ball misses the rooftop on its way down and eventually strikes the ground. Use the formula and this information to solve Exercises 83–85.

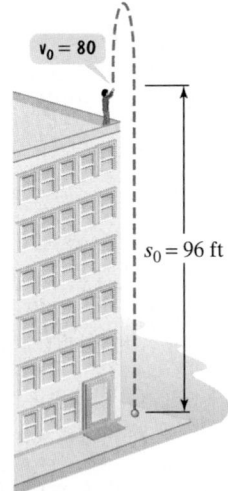

83. How high above the ground will the ball be 2 seconds after being thrown?

84. How high above the ground will the ball be 4 seconds after being thrown?

85. How high above the ground will the ball be 6 seconds after being thrown? Describe what this means in practical terms.

The graph visually displays the information about the thrown ball described in Exercises 83–85. The horizontal axis represents the ball's time in motion, in seconds. The vertical axis represents the ball's height above the ground, in feet. Use the graph to solve Exercises 86–91.

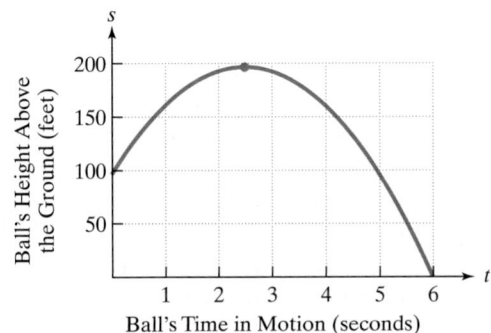

86. During which time period is the ball rising?

87. During which time period is the ball falling?

88. Identify your answer from Exercise 84 as a point on the graph.

89. Identify your answer from Exercise 83 as a point on the graph.

90. After how many seconds does the ball strike the ground?

91. After how many seconds does the ball reach its maximum height above the ground? What is a reasonable estimate of this maximum height?

Writing in Mathematics

92. What is a polynomial in two variables? Provide an example with your description.

93. Explain how to find the degree of a polynomial in two variables.

94. Suppose that you take up sky diving. Explain how to use the formula for Exercises 83–85 to determine your height above the ground at every instant of your fall.

Critical Thinking Exercises

95. Which one of the following is true?
  a. The degree of $5x^{24} - 3x^{16}y^9 - 7xy^2 + 6$ is 24.
  b. In the polynomial $4x^2y + x^3y^2 + 3x^2y^3 + 7y$, the term x^3y^2 has degree 5 and no numerical coefficient.
  c. $(2x + 3 - 5y)(2x + 3 + 5y) = 4x^2 + 12x + 9 - 25y^2$
  d. $(6x^2y - 7xy - 4) - (6x^2y + 7xy - 4) = 0$

In Exercises 96–97, find a polynomial in two variables that describes the area of the shaded region of each figure. Write each polynomial as the sum or difference of terms.

96.

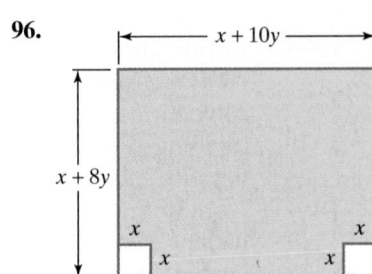

97.

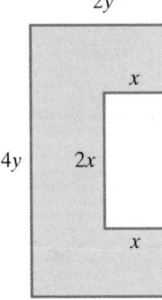

98. Multiply by using the rule for finding the product of the sum and difference of two terms:

$$[5y + (2x + 3)][5y - (2x + 3)].$$

99. Use the formulas for the volume of a rectangular solid and a cylinder to derive the polynomial in Exercise 82 that describes the volume of the storage building.

 Review Exercises

100. Solve for $W: R = \dfrac{L + 3W}{2}$. (Section 2.4, Example 4)

101. Subtract: $-6.4 - (-10.2)$. (Section 1.6, Example 2)

102. Write the point-slope and slope-intercept equations of a line passing through the point $(-2, 5)$ and parallel to the line whose equation is $3x - y = 9$. (Section 3.5, Example 3)

▶ **SECTION 5.5**   *Dividing Polynomials*

Objectives

1 Use the quotient rule for exponents.

2 Use the zero-exponent rule.

3 Use the quotients-to-powers rule.

4 Divide monomials.

5 Check polynomial division.

6 Divide a polynomial by a monomial.

SSM
PH Tutor   CD-   Video
Center   ROM

To play the part of Charlie Chaplin, actor Robert Downey Jr. (1965-) learned to pantomime, speak two British dialects, and play left-handed tennis. His problems with substance abuse have fueled the debate over whether the illness should be handled by our criminal justice system or by health care professionals.

As you learn more mathematics, you will discover new ways to describe your world. Almost anything that you can think of involving variables can be modeled by a formula. For example, a polynomial models the annual number of drug convictions in the United States and another polynomial models drug arrests. By dividing the respective polynomials, we obtain an algebraic expression that describes the conviction rate for drug arrests.

In the next two sections, you will learn how to divide polynomials. Before turning to polynomial division, we must develop some additional rules for working with exponents.

1 Use the quotient rule for exponents.

The Quotient Rule for Exponents

Consider the quotient of two exponential expressions, such as the quotient of 2^7 and 2^3. We are dividing 7 factors of 2 by 3 factors of 2. We are left with 4 factors of 2:

$$\frac{2^7}{2^3} = \frac{2 \cdot 2 \cdot 2 \cdot 2 \cdot 2 \cdot 2 \cdot 2}{2 \cdot 2 \cdot 2} = \frac{\cancel{2} \cdot \cancel{2} \cdot \cancel{2} \cdot 2 \cdot 2 \cdot 2 \cdot 2}{\cancel{2} \cdot \cancel{2} \cdot \cancel{2}} = 2 \cdot 2 \cdot 2 \cdot 2.$$

Thus,

$$\frac{2^7}{2^3} = 2^4.$$

We can quickly find the exponent, 4, on the quotient by subtracting the original exponents:

$$\frac{2^7}{2^3} = 2^{7-3}.$$

This suggests the following rule:

The Quotient Rule

$$\frac{b^m}{b^n} = b^{m-n}, \quad b \neq 0$$

When dividing exponential expressions with the same nonzero base, subtract the exponent in the denominator from the exponent in the numerator. Use this difference as the exponent of the common base.

EXAMPLE 1   Using the Quotient Rule

Divide each expression using the quotient rule:

a. $\dfrac{2^8}{2^4}$ **b.** $\dfrac{x^{13}}{x^3}$ **c.** $\dfrac{y^{15}}{y}$.

Solution

a. $\dfrac{2^8}{2^4} = 2^{8-4} = 2^4$ or 16

b. $\dfrac{x^{13}}{x^3} = x^{13-3} = x^{10}$

c. $\dfrac{y^{15}}{y} = \dfrac{y^{15}}{y^1} = y^{15-1} = y^{14}$ ∎

✔ **CHECK POINT 1**   Divide each expression using the quotient rule:

a. $\dfrac{5^{12}}{5^4}$ **b.** $\dfrac{x^9}{x^2}$ **c.** $\dfrac{y^{20}}{y}$.

2 Use the zero-exponent rule.

Zero as an Exponent

A nonzero base can be raised to the 0 power. The quotient rule can be used to help determine what zero as an exponent should mean. Consider the quotient of b^4 and b^4, where b is not zero. We can determine this quotient in two ways.

$$\frac{b^4}{b^4} = 1 \qquad\qquad \frac{b^4}{b^4} = b^{4-4} = b^0$$

Any nonzero expression divided by itself is 1.

Use the quotient rule and subtract exponents.

This means that b^0 must equal 1.

The Zero-Exponent Rule

If b is any real number other than 0,

$$b^0 = 1.$$

EXAMPLE 2   Using the Zero-Exponent Rule

Use the zero-exponent rule to simplify each expression:

   a. 7^0 **b.** $(-5)^0$ **c.** -5^0 **d.** $10x^0$ **e.** $(10x)^0$.

Solution

   a. $7^0 = 1$ Any nonzero number raised to the 0 power is 1.
   b. $(-5)^0 = 1$ Any nonzero number raised to the 0 power is 1.
   c. $-5^0 = -1$ $-5^0 = -1(5^0) = -1(1) = -1$

Only 5 is raised to the 0 power.

   d. $10x^0 = 10 \cdot 1 = 10$

Only x is raised to the 0 power.

   e. $(10x)^0 = 1$

The entire expression, 10x, is raised to the 0 power.

■

✔ **CHECK POINT 2**   Use the zero-exponent rule to simplify each expression:

   a. 14^0 **b.** $(-10)^0$ **c.** -10^0 **d.** $20x^0$ **e.** $(20x)^0$.

---

3 Use the quotients-to-powers rule.

The Quotients -to-Powers Rule for Exponents

We have seen that when a product is raised to a power, we raise every factor in the product to the power:

$$(ab)^n = a^n b^n.$$

There is a similar property for raising a quotient to a power.

Quotients to Powers

If a and b are real numbers and b is nonzero, then

$$\left(\frac{a}{b}\right)^n = \frac{a^n}{b^n}.$$

When a quotient is raised to a power, raise the numerator to the power and divide by the denominator to the power.

EXAMPLE 3   Using the Quotients-to-Powers Rule

Simplify each expression using the quotients-to-powers rule:

a. $\left(\dfrac{x}{4}\right)^2$ **b.** $\left(\dfrac{x^2}{5}\right)^3$ **c.** $\left(\dfrac{2a^3}{b^4}\right)^5$.

Solution

a. $\left(\dfrac{x}{4}\right)^2 = \dfrac{x^2}{4^2} = \dfrac{x^2}{16}$

Square the numerator and the denominator.

b. $\left(\dfrac{x^2}{5}\right)^3 = \dfrac{\left(x^2\right)^3}{5^3} = \dfrac{x^{2\cdot3}}{5\cdot5\cdot5} = \dfrac{x^6}{125}$

Cube the numerator and the denominator.

c. $\left(\dfrac{2a^3}{b^4}\right)^5 = \dfrac{\left(2a^3\right)^5}{\left(b^4\right)^5}$

Raise the numerator and the denominator to the fifth power.

$= \dfrac{2^5\left(a^3\right)^5}{\left(b^4\right)^5}$

Raise each factor in the numerator to the fifth power.

$= \dfrac{2^5 a^{3\cdot5}}{b^{4\cdot5}}$

To raise exponential expressions to powers, multiply exponents: $\left(b^m\right)^n = b^{mn}$.

$= \dfrac{32a^{15}}{b^{20}}$

Simplify. ∎

✔ **CHECK POINT 3**   Simplify each expression using the quotients-to-powers rule:

a. $\left(\dfrac{x}{5}\right)^2$ **b.** $\left(\dfrac{x^4}{2}\right)^3$ **c.** $\left(\dfrac{2a^{10}}{b^3}\right)^4$.

Study Tip

Try to avoid the following common errors that can occur when simplifying exponential expressions.

Correct	Incorrect	Description of Error
$\dfrac{2^{20}}{2^4} = 2^{20-4} = 2^{16}$	$\dfrac{2^{20}}{2^4} = 2^5$	Exponents should be subtracted, not divided.
$-8^0 = -1$	$-8^0 = 1$	Only 8 is raised to the 0 power.
$\left(\dfrac{x}{5}\right)^2 = \dfrac{x^2}{5^2} = \dfrac{x^2}{25}$	$\left(\dfrac{x}{5}\right)^2 = \dfrac{x^2}{5}$	The numerator and denominator must both be squared.

4 Divide monomials.

Dividing Monomials

Now that we have developed three additional properties of exponents, we are ready to turn to polynomial division. We begin with the quotient of two monomials, such as $16x^{14}$ and $8x^2$. This quotient is obtained by dividing the coefficients, 16 and 8, and then dividing the variables using the quotient rule for exponents.

$$\frac{16x^{14}}{8x^2} = \frac{16}{8} x^{14-2} = 2x^{12}$$

Divide coefficients and subtract exponents.

Dividing Monomials

To divide monomials, divide the coefficients and then divide the variables. Use the quotient rule for exponents to divide the variables: Keep the variable and subtract the exponents.

EXAMPLE 4   Dividing Monomials

Divide:

a. $\dfrac{-12x^8}{4x^2}$ **b.** $\dfrac{2x^3}{8x^3}$ **c.** $\dfrac{15x^5y^4}{3x^2y}$.

Solution

a. $\dfrac{-12x^8}{4x^2} = \dfrac{-12}{4} x^{8-2} = -3x^6$

b. $\dfrac{2x^3}{8x^3} = \dfrac{2}{8} x^{3-3} = \dfrac{1}{4} x^0 = \dfrac{1}{4} \cdot 1 = \dfrac{1}{4}$

c. $\dfrac{15x^5y^4}{3x^2y} = \dfrac{15}{3} x^{5-2}y^{4-1} = 5x^3y^3$

> ### Study Tip
>
> Look at the solution to Example 4(b). Rather than subtracting exponents for division that results in a 0 exponent, you might prefer to cancel.
>
> $$\frac{2\cancel{x^3}}{8\cancel{x^3}} = \frac{2}{8} = \frac{1}{4}$$

✔ **CHECK POINT 4**   Divide:

a. $\dfrac{-20x^{12}}{10x^4}$ **b.** $\dfrac{3x^4}{15x^4}$ **c.** $\dfrac{9x^6y^5}{3xy^2}$.

5 Check polynomial division.

Checking Division of Polynomial Problems

The answer to a division problem can be checked. For example, consider the following problem:

Dividend: the polynomial you are dividing into

$$\frac{15x^5y^4}{3x^2y} = 5x^3y^3.$$

Quotient: the answer to your division problem

Divisor: the polynomial you are dividing by

The quotient is correct if the product of the divisor and the quotient is the dividend. Is the quotient shown in the preceding equation correct?

$$(3x^2y)(5x^3y^3) = 3 \cdot 5x^{2+3}y^{1+3} = 15x^5y^4$$

Divisor   Quotient   This is the dividend.

Because the product of the divisor and the quotient is the dividend, the answer to the division problem is correct.

> ### Checking Division of Polynomials
>
> To check a quotient in a division problem, multiply the divisor and the quotient. If this product is the dividend, the quotient is correct.

6 Divide a polynomial by a monomial.

Dividing a Polynomial That Is Not a Monomial by a Monomial

To divide a polynomial by a monomial, we divide each term of the polynomial by the monomial. For example,

$$\underset{\substack{\text{polynomial}\\\text{dividend}}}{}\frac{10x^8 + 15x^6}{5x^3} = \underset{\substack{\text{monomial}\\\text{divisor}}}{}\frac{10x^8}{5x^3} + \frac{15x^6}{5x^3} = \underset{\substack{\text{Divide the first}\\\text{term by }5x^3.}}{\frac{10}{5}x^{8-3}} + \underset{\substack{\text{Divide the second}\\\text{term by }5x^3.}}{\frac{15}{5}x^{6-3}} = 2x^5 + 3x^3.$$

Is the quotient correct? Multiply the divisor and the quotient.

$$5x^3(2x^5 + 3x^3) = 5x^3 \cdot 2x^5 + 5x^3 \cdot 3x^3$$
$$= 5 \cdot 2x^{3+5} + 5 \cdot 3x^{3+3} = 10x^8 + 15x^6$$

Because this product gives the dividend, the quotient is correct.

> ### Dividing a Polynomial That Is Not a Monomial by a Monomial
>
> To divide a polynomial by a monomial, divide each term of the polynomial by the monomial.

Study Tip

Try to avoid this common error:

Incorrect:

$$\frac{x^4 - \overset{1}{\cancel{x}}}{\cancel{x}} = \frac{x^4 - 1}{1} = x^4 - 1$$

Correct:

$$\frac{x^4 - x}{x} = \frac{x^4}{x} - \frac{x}{x}$$
$$= x^{4-1} - x^{1-1}$$
$$= x^3 - x^0$$
$$= x^3 - 1$$

Don't leave out the 1.

EXAMPLE 5   Dividing a Polynomial by a Monomial

Find the quotient: $(-12x^8 + 4x^6 - 8x^3) \div 4x^2$.

Solution

$$\frac{-12x^8 + 4x^6 - 8x^3}{4x^2}$$ Rewrite the division in a vertical format.

$$= \frac{-12x^8}{4x^2} + \frac{4x^6}{4x^2} - \frac{8x^3}{4x^2}$$ Divide each term of the polynomial by the monomial.

$$= \frac{-12}{4}x^{8-2} + \frac{4}{4}x^{6-2} - \frac{8}{4}x^{3-2}$$ Divide coefficients and subtract exponents.

$$= -3x^6 + x^4 - 2x$$ Simplify.

To check the answer, multiply the divisor and the quotient.

$$\underset{\text{Divisor}}{4x^2}(\underset{\text{Quotient}}{-3x^6 + x^4 - 2x}) = 4x^2(-3x^6) + 4x^2 \cdot x^4 - 4x^2(2x)$$
$$= 4(-3)x^{2+6} + 4x^{2+4} - 4 \cdot 2x^{2+1}$$
$$= -12x^8 + 4x^6 - 8x^3 \quad \text{This is the dividend.}$$

Because the product of the divisor and the quotient is the dividend, the answer—that is, the quotient—is correct. ∎

✔ **CHECK POINT 5**   Find the quotient: $(-15x^9 + 6x^5 - 9x^3) \div 3x^2$.

EXAMPLE 6   Dividing a Polynomial by a Monomial

Divide: $\dfrac{16x^5 - 9x^4 + 8x^3}{2x^3}$.

Solution

$$\dfrac{16x^5 - 9x^4 + 8x^3}{2x^3}$$ *This is the given polynomial division.*

$$= \dfrac{16x^5}{2x^3} - \dfrac{9x^4}{2x^3} + \dfrac{8x^3}{2x^3}$$ *Divide each term by $2x^3$.*

$$= \dfrac{16}{2}x^{5-3} - \dfrac{9}{2}x^{4-3} + \dfrac{8}{2}x^{3-3}$$ *Divide coefficients and subtract exponents. Did you immediately write the last term as 4?*

$$= 8x^2 - \dfrac{9}{2}x + 4x^0$$ *Simplify.*

$$= 8x^2 - \dfrac{9}{2}x + 4$$ *$x^0 = 1$, so $4x^0 = 4 \cdot 1 = 4$.*

Check the answer by showing that the product of the divisor and the quotient is the dividend.   ∎

✔ **CHECK POINT 6**   Divide: $\dfrac{25x^9 - 7x^4 + 10x^3}{5x^3}$.

EXAMPLE 7   Dividing Polynomials in Two Variables

Divide: $(15x^5y^4 - 3x^3y^2 + 9x^2y) \div 3x^2y$.

Solution

$$\dfrac{15x^5y^4 - 3x^3y^2 + 9x^2y}{3x^2y}$$ *Rewrite the division in a vertical format.*

$$= \dfrac{15x^5y^4}{3x^2y} - \dfrac{3x^3y^2}{3x^2y} + \dfrac{9x^2y}{3x^2y}$$ *Divide each term of the polynomial by the monomial.*

$$= \dfrac{15}{3}x^{5-2}y^{4-1} - \dfrac{3}{3}x^{3-2}y^{2-1} + \dfrac{9}{3}x^{2-2}y^{1-1}$$ *Divide coefficients and subtract exponents.*

$$= 5x^3y^3 - xy + 3$$ *Simplify.*

Check the answer by showing that the product of the divisor and the quotient is the dividend.   ∎

✔ **CHECK POINT 7**   Divide: $(18x^7y^6 - 6x^2y^3 + 60xy^2) \div 6xy^2$.

EXERCISE SET 5.5

Practice Exercises

In Exercises 1–10, divide each expression using the quotient rule. Express any, numerical answers in exponential form.

1. $\dfrac{3^{20}}{3^5}$

2. $\dfrac{3^{30}}{3^{10}}$

3. $\dfrac{x^6}{x^2}$

4. $\dfrac{x^8}{x^4}$

5. $\dfrac{y^{13}}{y^5}$

6. $\dfrac{y^{19}}{y^6}$

7. $\dfrac{5^6 \cdot 2^8}{5^3 \cdot 2^4}$

8. $\dfrac{3^6 \cdot 2^8}{3^3 \cdot 2^4}$

9. $\dfrac{x^{100}y^{50}}{x^{25}y^{10}}$

10. $\dfrac{x^{200}y^{40}}{x^{25}y^{10}}$

In Exercises 11–24, use the zero exponent rule to simplify each expression.

11. 2^0

12. 4^0

13. $(-2)^0$

14. $(-4)^0$

15. -2^0

16. -4^0

17. $100y^0$

18. $200y^0$

19. $(100y)^0$

20. $(200y)^0$

21. $-5^0 + (-5)^0$

22. $-6^0 + (-6)^0$

23. $-\pi^0 - (-\pi)^0$

24. $-\sqrt{3}^0 - (-\sqrt{3})^0$

In Exercises 25–36, simplify each expression using the quotients to powers rule. If possible, evaluate exponential expressions.

25. $\left(\dfrac{x}{3}\right)^2$

26. $\left(\dfrac{x}{5}\right)^2$

27. $\left(\dfrac{x^2}{4}\right)^3$

28. $\left(\dfrac{x^2}{3}\right)^3$

29. $\left(\dfrac{2x^3}{5}\right)^2$

30. $\left(\dfrac{3x^4}{7}\right)^2$

31. $\left(\dfrac{-4}{3a^3}\right)^3$

32. $\left(\dfrac{-5}{2a^3}\right)^3$

33. $\left(\dfrac{-2a^7}{b^4}\right)^5$

34. $\left(\dfrac{-2a^8}{b^3}\right)^5$

35. $\left(\dfrac{x^2y^3}{2z}\right)^4$

36. $\left(\dfrac{x^3y^2}{2z}\right)^4$

In Exercises 37–52, divide the monomials. Check each answer by showing that the product of the divisor and the quotient is the dividend.

37. $\dfrac{30x^{10}}{10x^5}$

38. $\dfrac{45x^{12}}{15x^4}$

39. $\dfrac{-8x^{22}}{4x^2}$

40. $\dfrac{-15x^{40}}{3x^4}$

41. $\dfrac{-9y^8}{18y^5}$

42. $\dfrac{-15y^{13}}{45y^9}$

43. $\dfrac{7y^{17}}{5y^5}$

44. $\dfrac{9y^{19}}{7y^{11}}$

45. $\dfrac{30x^7y^5}{5x^2y}$

46. $\dfrac{40x^9y^5}{2x^2y}$

47. $\dfrac{-18x^{14}y^2}{36x^2y^2}$

48. $\dfrac{-15x^{16}y^2}{45x^2y^2}$

49. $\dfrac{9x^{20}y^{20}}{7x^{20}y^{20}}$

50. $\dfrac{7x^{30}y^{30}}{15x^{30}y^{30}}$

51. $\dfrac{-5x^{10}y^{12}z^6}{50x^2y^3z^2}$

52. $\dfrac{-8x^{12}y^{10}z^4}{40x^2y^3z^2}$

In Exercises 53–78, divide the polynomial by the monomial. Check each answer by showing that the product of the divisor and the quotient is the dividend.

53. $\dfrac{6x^4 + 2x^3}{2}$

54. $\dfrac{10x^4 + 5x^3}{5}$

55. $\dfrac{6x^4 - 2x^3}{2x}$

56. $\dfrac{10x^4 - 5x^3}{5x}$

57. $\dfrac{y^5 - 3y^2 + y}{y}$

58. $\dfrac{y^6 - 2y^3 + y}{y}$

59. $\dfrac{15x^3 - 24x^2}{-3x}$

60. $\dfrac{20x^3 - 10x^2}{-5x}$

61. $\dfrac{18x^5 + 6x^4 + 9x^3}{3x^2}$

62. $\dfrac{18x^5 + 24x^4 + 12x^3}{6x^2}$

63. $\dfrac{12x^4 - 8x^3 + 40x^2}{4x}$

64. $\dfrac{49x^4 - 14x^3 + 70x^2}{-7x}$

65. $(4x^2 - 6x) \div x$

66. $(16y^2 - 8y) \div y$

67. $\dfrac{30z^3 + 10z^2}{-5z}$

68. $\dfrac{12y^4 - 42y^2}{-4y}$

69. $\dfrac{8x^3 + 6x^2 - 2x}{2x}$

70. $\dfrac{9x^3 + 12x^2 - 3x}{3x}$

71. $\dfrac{25x^7 - 15x^5 - 5x^4}{5x^3}$

72. $\dfrac{49x^7 - 28x^5 - 7x^4}{7x^3}$

73. $\dfrac{18x^7 - 9x^6 + 20x^5 - 10x^4}{-2x^4}$

74. $\dfrac{25x^8 - 50x^7 + 3x^6 - 40x^5}{-5x^5}$

75. $\dfrac{12x^2y^2 + 6x^2y - 15xy^2}{3xy}$

76. $\dfrac{18a^3b^2 - 9a^2b - 27ab^2}{9ab}$

77. $\dfrac{20x^7y^4 - 15x^3y^2 - 10x^2y}{-5x^2y}$

78. $\dfrac{8x^6y^3 - 12x^8y^2 - 4x^{14}y^6}{-4x^6y^2}$

Writing in Mathematics

79. Explain the quotient rule for exponents. Use $\dfrac{3^6}{3^2}$ in your explanation.

80. Explain how to find any nonzero number to the 0 power.

81. Explain the difference between $(-7)^0$ and -7^0.

82. Explain how to simplify an expression that involves a quotient raised to a power. Provide an example with your explanation.

83. Explain how to divide monomials. Give an example.

84. Explain how to divide a polynomial that is not a monomial by a monomial. Give an example.

85. Are the expressions

$$\dfrac{12x^2 + 6x}{3x} \quad \text{and} \quad 4x + 2$$

equal for every value of x? Explain.

Critical Thinking Exercises

86. Which one of the following is true?
 a. $x^{10} \div x^2 = x^5$ for all nonzero real numbers x.
 b. $\dfrac{12x^3 - 6x}{2x} = 6x^2 - 6x$ **c.** $\dfrac{x^2 + x}{x} = x$
 d. If a polynomial in x of degree 6 is divided by a monomial in x of degree 2, the degree of the quotient is 4.

87. What polynomial, when divided by $3x^2$, yields the trinomial $6x^6 - 9x^4 + 12x^2$ as a quotient?

In Exercises 88–89, find the missing coefficients and exponents designated by question marks.

88. $\dfrac{?x^8 - ?x^6}{3x^?} = 3x^5 - 4x^3$

89. $\dfrac{3x^{14} - 6x^{12} - ?x^7}{?x^?} = -x^7 + 2x^5 + 3$

Review Exercises

90. Find the absolute value: $|-20.3|$. (Section 1.2, Example 8)

91. Express $\frac{7}{8}$ as a decimal. (Section 1.2, Example 4)

92. Graph: $y = \dfrac{1}{3}x + 2$. (Section 3.4, Example 3)

▶ **SECTION 5.6** **Dividing Polynomials by Binomials; Synthetic Division**

Objectives

1 Divide polynomials by binomials.

2 Divide polynomials using synthetic division.

SSM   PH Tutor   CD-   Video
      Center   ROM

For those of you who are dog lovers, you might still be thinking of the polynomial formula that models the age in human years, y, of a dog that is x years old, namely

$$y = -0.001618x^4 + 0.077326x^3 - 1.2367x^2 + 11.460x + 2.914.$$

Suppose that you are in your twenties, say 25. What is Fido's equivalent age? To answer this question, we must substitute 25 for y and solve the resulting polynomial equation for x:

$$25 = -0.001618x^4 + 0.077326x^3 - 1.2367x^2 + 11.460x + 2.914.$$

Don't panic! We won't be solving an equation as complicated as this—yet. You will learn to solve polynomial equations in which the highest power of the variable is 4 in more advanced algebra courses. Part of the method for solving such an equation involves the division of a polynomial by a binomial, such as

$$x + 3 \overline{)x^2 + 10x + 21}.$$

| Divisor has two terms and is a binomial. | The polynomial dividend has three terms and is a trinomial. |

In this section, you will learn how to perform such divisions.

1 Divide polynomials by binomials.

The Steps in Dividing a Polynomial by a Binomial

Dividing a polynomial by a binomial may remind you of long division.

Discover for Yourself

Divide 3983 by 26 without the use of a calculator. Describe the process of the division using the four steps—*divide, multiply, subtract*, and *bring down*. What do you observe about this process? When does it come to an end?

When a divisor is a binomial, the four steps used to divide whole numbers—**divide, multiply, subtract, bring down the next term**—form the repetitive procedure for dividing a polynomial by a binomial.

EXAMPLE 1   Dividing a Polynomial by a Binomial

Divide $x^2 + 10x + 21$ by $x + 3$.

Solution   The following steps illustrate how polynomial division is very similar to numerical division.

$$x + 3 \overline{)x^2 + 10x + 21}$$

Arrange the terms of the dividend $\left(x^2 + 10x + 21\right)$ and the divisor $(x + 3)$ in descending powers of x.

$$\begin{array}{r} x \\ x + 3 \overline{)x^2 + 10x + 21} \end{array}$$

Divide x^2 (the first term in the dividend) by x (the first term in the divisor): $\dfrac{x^2}{x} = x$. Align like terms.

$$\begin{array}{r} x \\ x + 3 \overline{)x^2 + 10x + 21} \\ x^2 + 3x \end{array}$$
times ⟶ equals

Multiply each term in the divisor $(x + 3)$ by x, aligning terms of the product under like terms in the dividend.

$$\begin{array}{r} x \\ x + 3 \overline{)x^2 + 10x + 21} \\ \underline{x^2 + 3x} \\ 7x \end{array}$$

Subtract $x^2 + 3x$ from $x^2 + 10x$ by changing the sign of each term in the lower expression and adding.

$$\begin{array}{r} x \\ x + 3 \overline{)x^2 + 10x + 21} \\ \underline{x^2 + 3x} \\ 7x + 21 \end{array}$$

Bring down 21 from the original dividend and add algebraically to form a new dividend.

$$\begin{array}{r} x + 7 \\ x + 3 \overline{\smash{)}\ x^2 + 10x + 21} \\ \underline{x^2 + 3x} \\ 7x + 21 \end{array}$$

Find the second term of the quotient. **Divide** the first term of $7x + 21$ by x, the first term of the divisor: $\dfrac{7x}{x} = 7.$

$$\begin{array}{r} \text{times} \qquad\ x + 7 \\ x + 3 \overline{\smash{)}\ x^2 + 10x + 21} \\ \underline{x^2 + 3x} \\ 7x + 21 \\ \text{equals} \quad \dfrac{7x + 21}{0} \end{array}$$

Multiply the divisor $(x + 3)$ by 7, aligning under like terms in the new dividend. Then subtract to obtain the remainder of 0.

Remainder

The quotient is $x + 7$ and the remainder is 0. We will not list a remainder of 0 in the answer. Thus,

$$(x^2 + 10x + 21) \div (x + 3) = x + 7.$$ ∎

When dividing polynomials by binomials, the answer can be checked. Find the product of the divisor and the quotient and add the remainder. If the result is the dividend, the answer to the division problem is correct. For example, let's check our work in Example 1.

$$(x^2 + 10x + 21) \div (x + 3) = x + 7$$

Dividend   Divisor   Quotient we wish to check

Multiply the divisor and the quotient and add the remainder, 0:

$$(x + 3)(x + 7) + 0 = x^2 + 7x + 3x + 21 + 0 = x^2 + 10x + 21.$$

Divisor   Quotient   Remainder   This is the dividend.

Because we obtained the dividend, the quotient is correct.

✔ **CHECK POINT 1** Divide $x^2 + 14x + 45$ by $x + 9$.

Before considering additional examples, let's summarize the general procedure for dividing a polynomial by a binomial.

> **Dividing a Polynomial by a Binomial**
>
> 1. **Arrange the terms** of both the dividend and the divisor in descending powers of any variable.
> 2. **Divide** the first term in the dividend by the first term in the divisor. The result is the first term of the quotient.
> 3. **Multiply** every term in the divisor by the first term in the quotient. Write the resulting product beneath the dividend with like terms lined up.
> 4. **Subtract** the product from the dividend.
> 5. **Bring down** the next term in the original dividend and write it next to the remainder to form a new dividend.
> 6. Use this new expression as the dividend and repeat this process until the remainder can no longer be divided. This will occur when the degree of the remainder (the highest exponent on a variable in the remainder) is less than the degree of the divisor.

In our next division, we will obtain a nonzero remainder.

EXAMPLE 2 Dividing a Polynomial by a Binomial

Divide: $\dfrac{7x - 9 - 4x^2 + 4x^3}{2x - 1}$.

Solution We begin by writing the dividend in descending powers of x.

$$7x - 9 - 4x^2 + 4x^3 = 4x^3 - 4x^2 + 7x - 9$$

> Think of 9 as $9x^0$. The powers descend from 3 to 0.

$$2x - 1 \overline{\smash{)}\, 4x^3 - 4x^2 + 7x - 9}$$

This is the problem with the dividend in descending powers of x.

$$\begin{array}{r} 2x^2 \\ 2x - 1 \overline{\smash{)}\, 4x^3 - 4x^2 + 7x - 9} \end{array}$$

Divide: $\dfrac{4x^3}{2x} = 2x^2$.

$$\begin{array}{r} \text{times} \quad 2x^2 \\ 2x - 1 \overline{\smash{)}\, 4x^3 - 4x^2 + 7x - 9} \\ 4x^3 - 2x^2 \\ \text{equals} \end{array}$$

Multiply: $2x^2(2x - 1) = 4x^3 - 2x^2$.

$$\begin{array}{r} 2x^2 \\ 2x - 1 \overline{\smash{)}\, 4x^3 - 4x^2 + 7x - 9} \\ \ominus 4x^3 \oplus 2x^2 \\ \hline - 2x^2 \end{array}$$

Change signs of the polynomial being subtracted

Subtract: $4x^3 - 4x^2 - (4x^3 - 2x^2)$
$= 4x^3 - 4x^2 - 4x^3 + 2x^2$
$= -2x^2$.

$$\begin{array}{r} 2x^2 \\ 2x - 1 \overline{\smash{)}\, 4x^3 - 4x^2 + 7x - 9} \\ 4x^3 - 2x^2 \quad \downarrow \\ \hline - 2x^2 + 7x \end{array}$$

Bring down 7x. The new dividend is $-2x^2 + 7x$.

$$\begin{array}{r} 2x^2 - x \\ 2x - 1 \overline{\smash{)}\, 4x^3 - 4x^2 + 7x - 9} \\ 4x^3 - 2x^2 \\ \hline - 2x^2 + 7x \end{array}$$

Divide: $\dfrac{-2x^2}{2x} = -x$.

$$\begin{array}{r} \text{times} \quad 2x^2 - x \\ 2x - 1 \overline{\smash{)}\, 4x^3 - 4x^2 + 7x - 9} \\ 4x^3 - 2x^2 \\ \hline - 2x^2 + 7x \\ \text{equals} \quad - 2x^2 + x \end{array}$$

Multiply: $-x(2x - 1) = -2x^2 + x$.

$$\begin{array}{r} 2x^2 - x \\ 2x - 1 \overline{\smash{)}\, 4x^3 - 4x^2 + 7x - 9} \\ 4x^3 - 2x^2 \\ \hline - 2x^2 + 7x \\ \oplus - 2x^2 \ominus + x \\ \hline 6x \end{array}$$

Subtract: $-2x^2 + 7x - (-2x^2 + x)$
$= -2x^2 + 7x + 2x^2 - x$
$= 6x$.

$$\begin{array}{r} 2x^2 - x \\ 2x - 1 \overline{\smash{\big)}4x^3 - 4x^2 + 7x - 9} \\ 4x^3 - 2x^2 \\ \hline -2x^2 + 7x \\ -2x^2 + x \\ \hline 6x - 9 \end{array}$$

Bring down -9. **The new dividend is** $6x - 9$.

$$\begin{array}{r} 2x^2 - x + 3 \\ 2x - 1 \overline{\smash{\big)}4x^3 - 4x^2 + 7x - 9} \\ 4x^3 - 2x^2 \\ \hline -2x^2 + 7x \\ -2x^2 + x \\ \hline 6x - 9 \end{array}$$

Divide: $\dfrac{6x}{2x} = 3$.

times

$$\begin{array}{r} 2x^2 - x + 3 \\ 2x - 1 \overline{\smash{\big)}4x^3 - 4x^2 + 7x - 9} \\ 4x^3 - 2x^2 \\ \hline -2x^2 + 7x \\ -2x^2 + x \\ \hline 6x - 9 \end{array}$$

equals $\;\rightarrow\; 6x - 3$

Multiply: $3(2x - 1) = 6x - 3$.

$$\begin{array}{r} 2x^2 - x + 3 \\ 2x - 1 \overline{\smash{\big)}4x^3 - 4x^2 + 7x - 9} \\ 4x^3 - 2x^2 \\ \hline -2x^2 + 7x \\ -2x^2 + x \\ \hline 6x - 9 \\ \ominus\;6x \oplus 3 \\ \hline -6 \end{array}$$

Subtract: $6x - 9 - (6x - 3)$
$= 6x - 9 - 6x + 3 = -6$.

Remainder

The quotient is $2x^2 - x + 3$ and the remainder is -6. When there is a nonzero remainder, as in this example, list the quotient, plus the remainder above the divisor. Thus,

$$\frac{7x - 9 - 4x^2 + 4x^3}{2x - 1} = 2x^2 - x + 3 + \frac{-6}{2x - 1}$$

Remainder above divisor

Quotient

or

$$\frac{7x - 9 - 4x^2 + 4x^3}{2x - 1} = 2x^2 - x + 3 - \frac{6}{2x - 1}.$$

Check this result by showing that the product of the divisor and the quotient,

$$(2x - 1)(2x^2 - x + 3),$$

plus the remainder, -6, is the dividend, $7x - 9 - 4x^2 + 4x^3$. ■

✔ **CHECK POINT 2** Divide: $\dfrac{6x + 8x^2 - 12}{2x + 3}$.

If a power of a variable is missing in a dividend, add that power of the variable with a coefficient of 0 and then divide. In this way, like terms will be aligned as you carry out the division.

EXAMPLE 3   Dividing a Polynomial with Missing Terms

Divide: $\dfrac{8x^3 - 1}{2x - 1}$.

Solution   We write the dividend, $8x^3 - 1$, as

$$8x^3 + 0x^2 + 0x - 1.$$

Use a coefficient of 0 with missing terms.

By doing this, we will keep all like terms aligned.

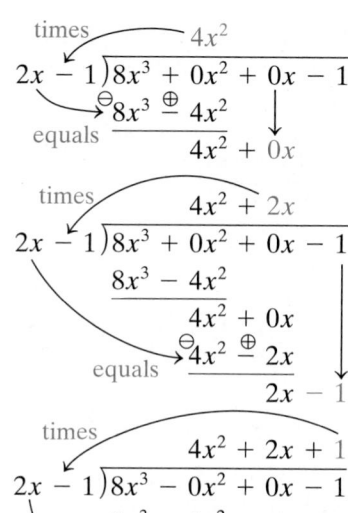

Divide $\left(\dfrac{8x^3}{2x} = 4x^2\right)$, multiply, subtract, and bring down the next term.
The new dividend is $4x^2 + 0x$.

Divide $\left(\dfrac{4x^2}{2x} = 2x\right)$, multiply $\left[2x(2x - 1) = 4x^2 - 2x\right]$, subtract, and bring down the next term.
The new dividend is $2x - 1$.

Divide $\left(\dfrac{2x}{2x} = 1\right)$, multiply $\left[1(2x - 1) = 2x - 1\right]$, and subtract. The remainder is 0.

Thus,

$$\frac{8x^3 - 1}{2x - 1} = 4x^2 + 2x + 1.$$

Check this result by showing that the product of the divisor and the quotient,

$$(2x - 1)(4x^2 + 2x + 1)$$

plus the remainder, 0, is the dividend, $8x^3 - 1$.   ■

✔ **CHECK POINT 3**   Divide: $\dfrac{x^3 - 1}{x - 1}$.

Using Technology

The graphs of $y_1 = \dfrac{8x^3 - 1}{2x - 1}$ and $y_2 = 4x^2 + 2x + 1$ are shown below.

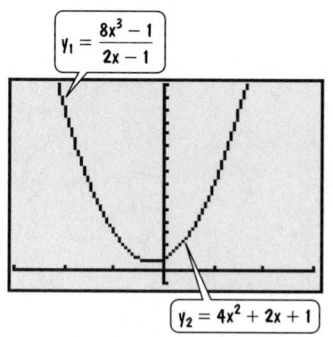

$[-3, 3, 1]$ by $[-1, 15, 1]$

The graphs coincide. Thus,

$$\frac{8x^3 - 1}{2x - 1} = 4x^2 + 2x + 1.$$

② Divide polynomials using synthetic division.

Dividing Polynomials Using Synthetic Division

We can use **synthetic division** to divide polynomials if the divisor is of the form $x - c$. This method provides a quotient more quickly than long division. Let's compare the two methods showing $x^3 + 4x^2 - 5x + 5$ divided by $x - 3$.

Long Division   *Quotient*

$$\begin{array}{r} x^2 + 7x + 16 \\ x - 3\overline{)x^3 + 4x^2 - 5x + 5} \end{array}$$

Divisor $x - c$; $c = 3$

$\underline{x^3 - 3x^2}$ *Dividend*
$7x^2 - 5x$
$\underline{7x^2 - 21x}$
$16x + 5$
$\underline{16x - 48}$ *Remainder*
53

Synthetic Division

$$\begin{array}{r|rrrr} 3 & 1 & 4 & -5 & 5 \\ & & 3 & 21 & 48 \\ \hline & 1 & 7 & 16 & 53 \end{array}$$

Notice the relationship between the polynomials in the long division process and the numbers that appear in synthetic division.

These are the coefficients of the dividend $x^3 + 4x^2 - 5x + 5$.

The divisor is $x - 3$. This is 3, or c in $x - c$.

$$\begin{array}{r|rrrr} 3 & 1 & 4 & -5 & 5 \\ & & 3 & 21 & 48 \\ \hline & 1 & 7 & 16 & 53 \end{array}$$

These are the coefficients of the quotient $x^2 + 7x + 16$.

This is the remainder.

Now let's look at the steps involved in synthetic division.

Synthetic Division

To divide a polynomial by $x - c$:

Example

1. Arrange polynomials in descending powers, with a 0 coefficient for any missing term.

$$x - 3\overline{)x^3 + 4x^2 - 5x + 5}$$

2. Write c for the divisor, $x - c$. To the right, write the coefficients of the dividend.

$$\begin{array}{r|rrrr} 3 & 1 & 4 & -5 & 5 \end{array}$$

3. Write the leading coefficient of the dividend on the bottom row.

$$\begin{array}{r|rrrr} 3 & 1 & 4 & -5 & 5 \\ & \downarrow & & & \text{Bring down.} \\ \hline & 1 & & & \end{array}$$

4. Multiply c (in this case, 3) times the value just written on the bottom row. Write the product in the next column in the second row.

$$\begin{array}{r|rrrr} 3 & 1 & 4 & -5 & 5 \\ & & 3 & & \\ \hline & 1 & & & \end{array}$$

Multiply by 3: $3 \cdot 1 = 3$.

5. Add the values in this new column, writing the sum in the bottom row.

$$
\begin{array}{r|rrrr}
3 & 1 & 4 & -5 & 5 \\
 & & 3 & & \text{Add.} \\
\hline
 & 1 & 7 &
\end{array}
$$

6. Repeat this series of multiplications and additions until all columns are filled in.

$$
\begin{array}{r|rrrr}
3 & 1 & 4 & -5 & 5 \\
 & & 3 & 21 & \text{Add.} \\
\hline
 & 1 & 7 & 16 &
\end{array}
$$

Multiply by 3: $3 \cdot 7 = 21$.

$$
\begin{array}{r|rrrr}
3 & 1 & 4 & -5 & 5 \\
 & & 3 & 21 & 48 \quad \text{Add.} \\
\hline
 & 1 & 7 & 16 & 53
\end{array}
$$

Multiply by 3: $3 \cdot 16 = 48$.

7. Use the numbers in the last row to write the quotient and remainder in fractional form. **The degree of the first term of the quotient is one less than the degree of the first term of the dividend.** The final value in this row is the remainder.

Written from the last row of the synthetic division.

$$
x - 3 \overline{)x^3 + 4x^2 - 5x + 5} \quad \Rightarrow \quad 1x^2 + 7x + 16 + \dfrac{53}{x - 3}
$$

EXAMPLE 4   Using Synthetic Division

Use synthetic division to divide $5x^3 + 6x + 8$ by $x + 2$.

Solution   The divisor must be in the form $x - c$. Thus, we write $x + 2$ as $x - (-2)$. This means that $c = -2$. Writing a 0 coefficient for the missing x^2-term in the dividend, we can express the division as follows:

$$
x - (-2) \,\overline{)5x^3 + 0x^2 + 6x + 8}.
$$

Now we are ready to set up the problem so that we can use synthetic division.

Use the coefficients of the dividend $5x^3 + 0x^2 + 6x + 8$ in descending powers of x.

This is c in $x - (-2)$.

$$
\begin{array}{r|rrrr}
-2 & 5 & 0 & 6 & 8
\end{array}
$$

We begin the synthetic division process by bringing down 5. This is followed by a series of multiplications and additions.

1. Bring down 5.

$$
\begin{array}{r|rrrr}
-2 & 5 & 0 & 6 & 8 \\
 & & & & \\
\hline
 & 5 & & &
\end{array}
$$

2. Multiply: $-2(5) = -10$.

$$
\begin{array}{r|rrrr}
-2 & 5 & 0 & 6 & 8 \\
 & & -10 & & \\
\hline
 & 5 & & &
\end{array}
$$

Multiply by -2.

3. Add: $0 + (-10) = -10$.

$$
\begin{array}{r|rrrr}
-2 & 5 & 0 & 6 & 8 \\
 & & -10 & & \text{Add.} \\
\hline
 & 5 & -10 & &
\end{array}
$$

4. Multiply: $-2(-10) = 20.$

$$
\begin{array}{r|rrrr}
-2 & 5 & 0 & 6 & 8 \\
   &   & -10 & 20 & \\
\hline
   & 5 & -10 & &
\end{array}
$$

Multiply by -2.

5. Add: $6 + 20 = 26.$

$$
\begin{array}{r|rrr|r}
-2 & 5 & 0 & 6 & 8 \\
   &   & -10 & 20 & \\
\hline
   & 5 & -10 & 26 &
\end{array}
$$

Add.

6. Multiply: $-2(26) = -52.$

$$
\begin{array}{r|rrrr}
-2 & 5 & 0 & 6 & 8 \\
   &   & -10 & 20 & -52 \\
\hline
   & 5 & -10 & 26 &
\end{array}
$$

Multiply by -2.

7. Add: $8 + (-52) = -44.$

$$
\begin{array}{r|rrrr|}
-2 & 5 & 0 & 6 & 8 \\
   &   & -10 & 20 & -52 \\
\hline
   & 5 & -10 & 26 & -44
\end{array}
$$

Add.

The numbers in the last row represent the coefficients of the quotient and the remainder. The degree of the first term of the quotient is one less than that of the dividend. Because the degree of the dividend, $5x^3 + 6x + 8$, is 3, the degree of the quotient is 2. This means that the 5 in the last row represents $5x^2$.

$$
\begin{array}{r|rrrr}
-2 & 5 & 0 & 6 & 8 \\
   &   & -10 & 20 & -52 \\
\hline
   & 5 & -10 & 26 & -44
\end{array}
$$

The quotient is $5x^2 - 10x + 26.$ The remainder is $-44.$

Thus,

$$
x + 2 \overline{)5x^3 + 6x + 8} \quad = \quad 5x^2 - 10x + 26 - \frac{44}{x + 2}
$$

✔ **CHECK POINT 4** Use synthetic division to divide $x^3 - 7x - 6$ by $x + 2$.

EXERCISE SET 5.6

Practice Exercises

In Exercises 1–36, divide using long division. Check each answer by showing that the product of the divisor and the quotient, plus the remainder, is the dividend.

1. $\dfrac{x^2 + 6x + 8}{x + 2}$

2. $\dfrac{x^2 + 7x + 10}{x + 5}$

3. $\dfrac{2x^2 + x - 10}{x - 2}$

4. $\dfrac{2x^2 + 13x + 15}{x + 5}$

5. $\dfrac{x^2 - 5x + 6}{x - 3}$

6. $\dfrac{x^2 - 2x - 24}{x + 4}$

7. $\dfrac{2y^2 + 5y + 2}{y + 2}$

8. $\dfrac{2y^2 - 13y + 21}{y - 3}$

9. $\dfrac{x^2 - 5x + 8}{x - 3}$

10. $\dfrac{x^2 + 7x - 8}{x + 3}$

11. $\dfrac{5y + 10 + y^2}{y + 2}$

12. $\dfrac{-8y + y^2 - 9}{y - 3}$

13. $\dfrac{x^3 - 6x^2 + 7x - 2}{x - 1}$

14. $\dfrac{x^3 + 3x^2 + 5x + 3}{x + 1}$

15. $\dfrac{12y^2 - 20y + 3}{2y - 3}$

16. $\dfrac{4y^2 - 8y - 5}{2y + 1}$

17. $\dfrac{4a^2 + 4a - 3}{2a - 1}$

18. $\dfrac{2b^2 - 9b - 5}{2b + 1}$

19. $\dfrac{3y - y^2 + 2y^3 + 2}{2y + 1}$

20. $\dfrac{9y + 18 - 11y^2 + 12y^3}{4y + 3}$

21. $\dfrac{2x^2 - 9x + 8}{2x + 3}$

22. $\dfrac{4y^2 + 8y + 3}{2y - 1}$

23. $\dfrac{x^3 + 4x - 3}{x - 2}$

24. $\dfrac{x^3 + 2x^2 - 3}{x - 2}$

25. $\dfrac{4y^3 + 8y^2 + 5y + 9}{2y + 3}$

26. $\dfrac{2y^3 - y^2 + 3y + 2}{2y + 1}$

27. $\dfrac{6y^3 - 5y^2 + 5}{3y + 2}$

28. $\dfrac{4y^3 - y - 5}{2y + 3}$

29. $\dfrac{27x^3 - 1}{3x - 1}$ **30.** $\dfrac{8x^3 + 27}{2x + 3}$

31. $\dfrac{81 - 12y^3 + 54y^2 + y^4 - 108y}{y - 3}$

32. $\dfrac{8y^3 + y^4 + 16 + 32y + 24y^2}{y + 2}$

33. $\dfrac{4y^2 + 6y}{2y - 1}$

34. $\dfrac{10x^2 - 3x}{x + 3}$

35. $\dfrac{y^4 - 2y^2 + 5}{y - 1}$

36. $\dfrac{y^4 - 6y^2 + 3}{y - 1}$

In Exercises 37–54, divide using synthetic division. In the first two exercises, begin the process as shown.

37. $(2x^2 + x - 10) \div (x - 2)$

$2\rfloor\quad 2 \quad 1 \quad -10$

38. $(x^2 + x - 2) \div (x - 1)$

$1\rfloor\quad 1 \quad 1 \quad -2$

39. $(3x^2 + 7x - 20) \div (x + 5)$

40. $(5x^2 - 12x - 8) \div (x + 3)$

41. $(4x^3 - 3x^2 + 3x - 1) \div (x - 1)$

42. $(5x^3 - 6x^2 + 3x + 11) \div (x - 2)$

43. $(6x^5 - 2x^3 + 4x^2 - 3x + 1) \div (x - 2)$

44. $(x^5 + 4x^4 - 3x^2 + 2x + 3) \div (x - 3)$

45. $(x^2 - 5x - 5x^3 + x^4) \div (5 + x)$

46. $(x^2 - 6x - 6x^3 + x^4) \div (6 + x)$

47. $(3x^3 + 2x^2 - 4x + 1) \div \left(x - \dfrac{1}{3}\right)$

48. $(2x^4 - x^3 + 2x^2 - 3x + 1) \div \left(x - \dfrac{1}{2}\right)$

49. $\dfrac{x^5 + x^3 - 2}{x - 1}$

50. $\dfrac{x^7 + x^5 - 10x^3 + 12}{x + 2}$

51. $\dfrac{x^4 - 256}{x - 4}$

52. $\dfrac{x^7 - 128}{x - 2}$

53. $\dfrac{2x^5 - 3x^4 + x^3 - x^2 + 2x - 1}{x + 2}$

54. $\dfrac{x^5 - 2x^4 - x^3 + 3x^2 - x + 1}{x - 2}$

Application Exercises

55. A rectangle with length $2x - 1$ inches has an area of $2x^2 + 5x - 3$ square inches. Write a binomial that represents its width.

Width = ?	Area = $2x^2 + 5x - 3$ square inches

Length $= 2x - 1$ inches

56. If the distance traveled is $x^3 + 3x^2 + 5x + 3$ miles and the rate is $x + 1$ miles per hour, write a trinomial for the time traveled.

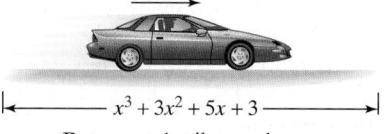

$\overleftarrow{\qquad x^3 + 3x^2 + 5x + 3 \qquad}\overrightarrow{}$

Rate $= x + 1$ miles per hour

57. Two people are 25 years old and 20 years old. In x years from now, their ages can be represented by $x + 25$ and $x + 20$.

a. Use long division to rewrite $\dfrac{x + 25}{x + 20}$, the ratio of the older person's age in x years to the younger person's age in x years.

b. Complete the following table.

x	0	5	10	25	50	75
$\dfrac{x + 25}{x + 20}$						

c. Describe what is happening to the ratio $\dfrac{x + 25}{x + 20}$ as x increases. How can this be verified using the result of the long division in part (a)?

Writing in Mathematics

58. In your own words, explain how to divide a polynomial by a binomial. Use $\dfrac{x^2 + 4}{x + 2}$ in your explanation.

59. When dividing a polynomial by a binomial, explain when to stop dividing.

60. After dividing a polynomial by a binomial, explain how to check the answer.

61. When dividing a binomial into a polynomial with missing terms, explain the advantage of writing the missing terms with zero coefficients.

62. Explain how to perform synthetic division. Use the division problem
$$(2x^3 - 3x^2 - 11x + 7) \div (x - 3)$$
to support your explanation.

Critical Thinking Exercises

63. Which one of the following is true?
  a. If $4x^2 + 25x - 3$ is divided by $4x + 1$, the remainder is 9.
  b. If polynomial division results in a remainder of zero, then the product of the divisor and the quotient is the dividend.
  c. The degree of a polynomial is the power of the term that appears in the first position.
  d. When a polynomial is divided by a binomial, the division process stops when the last term of the dividend is brought down.

64. When a certain polynomial is divided by $2x + 4$, the quotient is
$$x - 3 + \dfrac{17}{2x + 4}.$$
What is the polynomial?

65. Find the number k such that when $16x^2 - 2x + k$ is divided by $2x - 1$, the remainder is 0.

66. Describe the pattern that you observe in the following quotients and remainders.

$$\dfrac{x^3 - 1}{x + 1} = x^2 - x + 1 - \dfrac{2}{x + 1}$$

$$\dfrac{x^5 - 1}{x + 1} = x^4 - x^3 + x^2 - x + 1 - \dfrac{2}{x + 1}$$

Use this pattern to find $\dfrac{x^7 - 1}{x + 1}$. Verify your result by dividing.

Technology Exercises

In Exercises 67–71, use a graphing utility to determine whether the divisions have been performed correctly. Graph each side of the given equation in the same viewing rectangle. The graphs should coincide. If they do not, correct the expression on the right side by using polynomial division. Then use your graphing utility to show that the division has been performed correctly.

67. $\dfrac{x^2 - 4}{x - 2} = x + 2$

68. $\dfrac{x^2 - 25}{x - 5} = x - 5$

69. $\dfrac{2x^2 + 13x + 15}{x - 5} = 2x + 3$

70. $\dfrac{6x^2 + 16x + 8}{3x + 2} = 2x - 4$

71. $\dfrac{x^3 + 3x^2 + 5x + 3}{x + 1} = x^2 - 2x + 3$

Review Exercises

72. Solve the system:
$$7x - 6y = 17$$
$$3x + y = 18.$$
(Section 4.3, Example 2)

73. What is 6% of 20? (Section 2.4, Example 7)

74. Solve: $\dfrac{x}{3} + \dfrac{2}{5} = \dfrac{x}{5} - \dfrac{2}{5}$ (Section 2.3, Example 4)

▶ SECTION 5.7   Negative Exponents and Scientific Notation

Objectives

1 Use the negative exponent rule.

2 Simplify exponential expressions.

3 Convert from scientific notation to decimal notation.

4 Convert from decimal notation to scientific notation.

5 Compute with scientific notation.

6 Solve applied problems with scientific notation.

SSM
PH Tutor   CD-   Video
Center   ROM

You are listening to a discussion of the country's 5.6 trillion dollar deficit. It seems that this is a real problem, but then you realize that you don't really know what this number means. How can you look at this deficit in the proper perspective? If the national debt were evenly divided among all citizens of the country, how much would each citizen have to pay?

In the new millennium, literacy with numbers, called **numeracy**, will be a necessary skill for functioning in a meaningful way personally, professionally, and as a citizen. In this section, you will learn to use exponents to provide a way of putting large and small numbers in perspective.

1 Use the negative exponent rule.

Negative Integers as Exponents

A nonzero base can be raised to a negative power. The quotient rule can be used to help determine what a negative integer as an exponent should mean. Consider the quotient of b^3 and b^5, where b is not zero. We can determine this quotient in two ways.

$$\frac{b^3}{b^5} = \frac{\cancel{b} \cdot \cancel{b} \cdot \cancel{b}}{\cancel{b} \cdot \cancel{b} \cdot \cancel{b} \cdot b \cdot b} = \frac{1}{b^2}$$

After cancelling, we have two factors of b in the denominator.

$$\frac{b^3}{b^5} = b^{3-5} = b^{-2}$$

Use the quotient rule and subtract exponents.

Notice that $\dfrac{b^3}{b^5}$ equals both b^{-2} and $\dfrac{1}{b^2}$. This means that b^{-2} must equal $\dfrac{1}{b^2}$. This example is a special case of the **negative exponent rule**.

The Negative Exponent Rule

If b is any real number other than 0 and n is a natural number, then

$$b^{-n} = \frac{1}{b^n}.$$

Study Tip

A negative exponent does not automatically mean that the value of an expression negative. For example,

$$7^{-2} = \frac{1}{7^2} = \frac{1}{49}$$

is positive. Avoid these common errors:

INCORRECT!

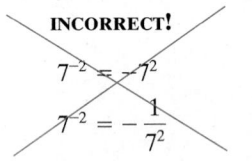

EXAMPLE 1 Using the Negative Exponent Rule

Use the negative exponent rule to write each expression with a positive exponent. Then simplify the expression.

a. 7^{-2} **b.** 4^{-3} **c.** $(-2)^{-3}$ **d.** 5^{-1}

Solution

a. $7^{-2} = \dfrac{1}{7^2} = \dfrac{1}{7 \cdot 7} = \dfrac{1}{49}$

b. $4^{-3} = \dfrac{1}{4^3} = \dfrac{1}{4 \cdot 4 \cdot 4} = \dfrac{1}{64}$

c. $(-2)^{-3} = \dfrac{1}{(-2)^3} = \dfrac{1}{(-2)(-2)(-2)} = \dfrac{1}{-8} = -\dfrac{1}{8}$

d. $5^{-1} = \dfrac{1}{5^1} = \dfrac{1}{5}$

∎

✔ **CHECK POINT 1** Use the negative exponent rule to write each expression with a positive exponent. Then simplify the expression.

a. 6^{-2} **b.** 5^{-3}

c. $(-4)^{-3}$ **d.** 8^{-1}

Negative exponents can appear in denominators. For example,

$$\frac{1}{2^{-10}} = \frac{1}{\dfrac{1}{2^{10}}} = 1 \div \frac{1}{2^{10}} = 1 \cdot \frac{2^{10}}{1} = 2^{10}.$$

In general, if a negative exponent appears in a denominator, an expression can be written with a positive exponent using

$$\frac{1}{b^{-n}} = b^n.$$

For example,

$$\frac{1}{2^{-3}} = 2^3 = 8 \qquad \text{and} \qquad \frac{1}{(-6)^{-2}} = (-6)^2 = 36.$$

Change only the sign of the exponent and not the sign of the base, −6.

Negative Exponents in Numerators and Denominators

If b is any real number other than 0 and n is a natural number, then

$$b^{-n} = \frac{1}{b^n} \qquad \text{and} \qquad \frac{1}{b^{-n}} = b^n.$$

When a negative number appears as an exponent, switch the position of the base (from numerator to denominator or from denominator to numerator) and make the exponent positive. The sign of the base does not change.

EXAMPLE 2 Using Negative Exponents

Write each expression with positive exponents only. Then simplify, if possible.

a. $\dfrac{4^{-3}}{5^{-2}}$

b. $\left(\dfrac{3}{4}\right)^{-2}$

c. $\dfrac{1}{4x^{-3}}$

d. $\dfrac{x^{-5}}{y^{-1}}$

Solution

a. $\dfrac{4^{-3}}{5^{-2}} = \dfrac{5^2}{4^3} = \dfrac{5 \cdot 5}{4 \cdot 4 \cdot 4} = \dfrac{25}{64}$

Switch the position of the bases and make the exponents positive.

b. $\left(\dfrac{3}{4}\right)^{-2} = \dfrac{3^{-2}}{4^{-2}} = \dfrac{4^2}{3^2} = \dfrac{4 \cdot 4}{3 \cdot 3} = \dfrac{16}{9}$

Switch the position of the bases and make the exponents positive.

c. $\dfrac{1}{4x^{-3}} = \dfrac{x^3}{4}$

Switch the position of the base and make the exponent positive. Note that only x is raised to the −3 power

d. $\dfrac{x^{-5}}{y^{-1}} = \dfrac{y^1}{x^5} = \dfrac{y}{x^5}$

✔ **CHECK POINT 2** Write each expression with positive exponents only. Then simplify, if possible.

a. $\dfrac{2^{-3}}{7^{-2}}$

b. $\left(\dfrac{4}{5}\right)^{-2}$

c. $\dfrac{1}{7y^{-2}}$

d. $\dfrac{x^{-1}}{y^{-8}}$

2 Simplify exponential expressions.

Simplifying Exponential Expressions

Properties of exponents are used to simplify exponential expressions. An exponential expression is **simplified** when:

- No parentheses appear.
- No powers are raised to powers.
- Each base occurs only once.
- No negative or zero exponents appear.

Simplifying Exponential Expressions

1. If necessary, remove parentheses by using

$$(ab)^n = a^n b^n \quad \text{or} \quad \left(\frac{a}{b}\right)^n = \frac{a^n}{b^n}.$$

Example

$$(xy)^3 = x^3 y^3$$

2. If necessary, simplify powers to powers by using

$$\left(b^m\right)^n = b^{mn}.$$

$$\left(x^4\right)^3 = x^{4 \cdot 3} = x^{12}$$

3. If necessary, be sure that each base appears only once, by using

$$b^m \cdot b^n = b^{m+n} \quad \text{or} \quad \frac{b^m}{b^n} = b^{m-n}.$$

$$x^4 \cdot x^3 = x^{4+3} = x^7$$

4. If necessary, rewrite exponential expressions with zero powers as 1 $\left(b^0 = 1\right)$. Furthermore, write the answer with positive exponents by using

$$b^{-n} = \frac{1}{b^n} \quad \text{or} \quad \frac{1}{b^{-n}} = b^n.$$

$$\frac{x^5}{x^8} = x^{5-8} = x^{-3} = \frac{1}{x^3}$$

The following examples show how to simplify exponential expressions. In each example, assume that the variable in the denominator is not equal to zero.

Study Tip

There is often more than one method to simplify an exponential expression. For example, you may prefer to simplify Example 3 as follows:

$$x^{-9} \cdot x^4 = \frac{x^4}{x^9} = x^{4-9} = x^{-5} = \frac{1}{x^5}.$$

EXAMPLE 3   Simplifying an Exponential Expression

Simplify: $x^{-9} \cdot x^4$.

Solution

$$x^{-9} \cdot x^4 = x^{-9+4} \qquad b^m \cdot b^n = b^{m+n}$$

$$= x^{-5} \qquad \text{The base, x, now appears only once.}$$

$$= \frac{1}{x^5} \qquad b^{-n} = \frac{1}{b^n}$$

■

✔ **CHECK POINT 3**   Simplify: $x^{-12} \cdot x^2$.

EXAMPLE 4   Simplifying Exponential Expressions

Simplify:

a. $\dfrac{x^4}{x^9}$ **b.** $\dfrac{25x^6}{5x^8}$ **c.** $\dfrac{10y^7}{-2y^{10}}$.

Solution

a. $\dfrac{x^4}{x^9} = x^{4-9} = x^{-5} = \dfrac{1}{x^5}$

b. $\dfrac{25x^6}{5x^8} = \dfrac{25}{5} \cdot \dfrac{x^6}{x^8} = 5x^{6-8} = 5x^{-2} = \dfrac{5}{x^2}$

c. $\dfrac{10y^7}{-2y^{10}} = \dfrac{10}{-2} \cdot \dfrac{y^7}{y^{10}} = -5y^{7-10} = -5y^{-3} = -\dfrac{5}{y^3}$ ■

✔ **CHECK POINT 4** Simplify:

a. $\dfrac{x^2}{x^{10}}$ **b.** $\dfrac{75x^3}{5x^9}$ **c.** $\dfrac{50y^8}{-25y^{14}}.$

EXAMPLE 5 Simplifying an Exponential Expression

Simplify: $\dfrac{(5x^3)^2}{x^{10}}.$

Solution

$$\dfrac{(5x^3)^2}{x^{10}} = \dfrac{5^2(x^3)^2}{x^{10}}$$ Raise each factor in the product to the second power. Parentheses are removed using $(ab)^n = a^n b^n$.

$$= \dfrac{5^2 x^{3\cdot 2}}{x^{10}}$$ Multiply powers to powers using $\left(b^m\right)^n = b^{mn}$.

$$= \dfrac{25x^6}{x^{10}}$$ Simplify.

$$= 25x^{6-10}$$ When dividing with the same base, subtract exponents: $\dfrac{b^m}{b^n} = b^{m-n}$.

$$= 25x^{-4}$$ Simplify. The base, x, now appears only once.

$$= \dfrac{25}{x^4}$$ Rewrite with a positive exponent using $b^{-n} = \dfrac{1}{b^n}$. ■

✔ **CHECK POINT 5** Simplify: $\dfrac{(6x^4)^2}{x^{11}}.$

EXAMPLE 6 Simplifying an Exponential Expression

Simplify: $\left(\dfrac{x^5}{x^2}\right)^{-3}.$

Solution
Method 1. Remove parentheses first by raising the numerator and denominator to the -3 power.

$$\left(\frac{x^5}{x^2}\right)^{-3} = \frac{\left(x^5\right)^{-3}}{\left(x^2\right)^{-3}}$$

Use $\left(\dfrac{a}{b}\right)^n = \dfrac{a^n}{b^n}$ and raise the numerator and denominator to the -3 power.

$$= \frac{x^{5(-3)}}{x^{2(-3)}}$$

Multiply powers to powers using $\left(b^m\right)^n = b^{mn}$.

$$= \frac{x^{-15}}{x^{-6}}$$

Simplify.

$$= x^{-15-(-6)}$$

When dividing with the same base, subtract the exponent in the denominator from the exponent in the numerator: $\dfrac{b^m}{b^n} = b^{m-n}$.

$$= x^{-9}$$

Subtract: $-15 - (-6) = -15 + 6 = -9$. The base, x, now appears only once.

$$= \frac{1}{x^9}$$

Rewrite with a positive exponent using $b^{-n} = \dfrac{1}{b^n}$.

Method 2. First perform the division within the parentheses.

$$\left(\frac{x^5}{x^2}\right)^{-3} = \left(x^{5-2}\right)^{-3}$$

Within parentheses, divide by subtracting exponents: $\dfrac{b^m}{b^n} = b^{m-n}$.

$$= \left(x^3\right)^{-3}$$

Simplify. The base, x, now appears only once.

$$= x^{3(-3)}$$

Multiply powers to powers: $\left(b^m\right)^n = b^{mn}$.

$$= x^{-9}$$

Simplify.

$$= \frac{1}{x^9}$$

Rewrite with a positive exponent using $b^{-n} = \dfrac{1}{b^n}$.

Which method do you prefer? ∎

✔ **CHECK POINT 6** Simplify: $\left(\dfrac{x^8}{x^4}\right)^{-5}$.

Scientific Notation

The national debt of the United States is about \$5.6 trillion. A stack of \$1 bills equaling the national debt would rise to twice the distance from the Earth to the moon. Because a trillion is 10^{12}, the national debt can be expressed as

$$5.6 \times 10^{12}.$$

The number 5.6×10^{12} is written in a form called **scientific notation**. A number in scientific notation is expressed as a number greater than or equal to 1 and less than 10 multiplied by some power of 10. It is customary to use the multiplication symbol, ×, rather than a dot in scientific notation.

Here are two examples of numbers in scientific notation:

- Each day, 2.6×10^7 pounds of dust from the atmosphere settle on Earth.
- The diameter of a hydrogen atom is 1.016×10^{-8} centimeter.

3 Convert from scientific notation to decimal notation.

We can use the exponent on the 10 to change a number in scientific notation to decimal notation. If the exponent is *positive*, move the decimal point in the number to the *right* the same number of places as the exponent. If the exponent is *negative*, move the decimal point in the number to the *left* the same number of places as the exponent.

EXAMPLE 7 Converting from Scientific to Decimal Notation

Write each number in decimal notation:

a. 2.6×10^7 **b.** 1.016×10^{-8}.

Solution

a. We express 2.6×10^7 in decimal notation by moving the decimal point in 2.6 seven places to the right. We need to add six zeros.

$$2.6 \times 10^7 = 26{,}000{,}000$$

b. We express 1.016×10^{-8} in decimal notation by moving the decimal point in 1.016 eight places to the left. We need to add seven zeros to the right of the decimal point.

$$1.016 \times 10^{-8} = 0.00000001016 \qquad \blacksquare$$

✔ **CHECK POINT 7**   Write each number in decimal notation:

a. 7.4×10^9 **b.** 3.017×10^{-6}.

---

4 Convert from decimal notation to scientific notation.

To convert from decimal notation to scientific notation, we reverse the procedure of Example 7.

- Move the decimal point in the given number to obtain a number greater than or equal to 1 and less than 10.
- The number of places the decimal point moves gives the exponent on 10; the exponent is positive if the given number is greater than 10 and negative if the given number is between 0 and 1.

EXAMPLE 8 Converting from Decimal Notation to Scientific Notation

Write each number in scientific notation:

a. 4,600,000   **b.** 0.00023.

Solution

a. $4{,}600{,}000 = 4.6 \times 10^?$ Decimal point moves 6 places. $\longrightarrow = 4.6 \times 10^6$

b. $0.00023 = 2.3 \times 10^{-?}$ Decimal point moves 4 places. $\longrightarrow = 2.3 \times 10^{-4}$ $\qquad \blacksquare$

✔ **CHECK POINT 8**   Write each number in scientific notation:

a. 7,410,000,000   **b.** 0.000000092.

---

5 Compute with scientific notation.

Computations with Scientific Notation

Because numbers in scientific notation are exponential expressions with base 10, multiplication and division can be performed by using special cases of three exponential properties.

$$10^m \cdot 10^n = 10^{m+n} \qquad \frac{10^m}{10^n} = 10^{m-n} \qquad (10^m)^n = 10^{mn}$$

Using Technology

$(4 \times 10^5)(2 \times 10^9)$ On a Calculator:

Many Scientific Calculators

4 [EE] 5 [×] 2 [EE] 9 [=]

Display: 8. 14

Many Graphing Calculators

4 [EE] 5 [×] 2 [EE] 9 [ENTER]

Display: 8E14

EXAMPLE 9   Computations with Scientific Notation

Perform the indicated computations, writing the answers in scientific notation:

a. $(4 \times 10^5)(2 \times 10^9)$ **b.** $\dfrac{1.2 \times 10^6}{4.8 \times 10^{-3}}$ **c.** $(5 \times 10^{-4})^3.$

Solution

a. $(4 \times 10^5)(2 \times 10^9) = 4 \cdot 2 \cdot 10^5 \cdot 10^9$ Regroup factors.

$= 8 \times 10^{5+9}$ $10^m \cdot 10^n = 10^{m+n}$

$= 8 \times 10^{14}$

b. $\dfrac{1.2 \times 10^6}{4.8 \times 10^{-3}} = \dfrac{1.2}{4.8} \times \dfrac{10^6}{10^{-3}}$

$= 0.25 \times 10^{6-(-3)}$ $\dfrac{10^m}{10^n} = 10^{m-n}$

$= 0.25 \times 10^9$ Because 0.25 is not between 1 and 10, it must be written in scientific notation.

$= 2.5 \times 10^{-1} \times 10^9$ $0.25 = 2.5 \times 10^{-1}$

$= 2.5 \times 10^{-1+9}$ $10^m \cdot 10^n = 10^{m+n}$

$= 2.5 \times 10^8$

c. $(5 \times 10^{-4})^3 = 5^3 \times (10^{-4})^3$ $(ab)^n = a^n b^n$. Cube each factor in parentheses.

$= 5^3 \times 10^{-12}$ $(10^m)^n = 10^{mn}$

$= 125 \times 10^{-12}$ 125 must be expressed in scientific notation.

$= 1.25 \times 10^2 \times 10^{-12}$ $125 = 1.25 \times 10^2$

$= 1.25 \times 10^{2+(-12)}$ $10^m \cdot 10^n = 10^{m+n}$

$= 1.25 \times 10^{-10}$ ∎

✔ **CHECK POINT 9**   Perform the indicated computation, writing the answers in scientific notation:

a. $(3 \times 10^8)(2 \times 10^2)$ **b.** $\dfrac{8.4 \times 10^7}{4 \times 10^{-4}}$

c. $(4 \times 10^{-2})^3.$

6 Solve applied problems with scientific notation.

Applications: Putting Numbers in Perspective

Due to tax cuts and spending increases, the United States began accumulating large deficits in the 1980s. To finance the deficit, the government had borrowed $5.6 trillion as of the end of 2000. The graph in Figure 5.4 shows the national debt increasing over time.

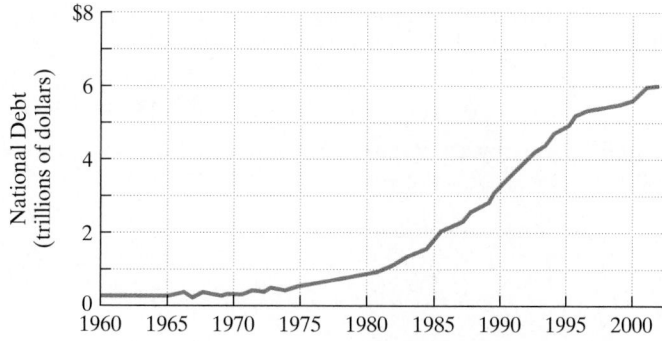

Figure 5.4 The national debt

Source: Office of Management and Budget

Example 10 shows how we can use scientific notation to comprehend the meaning of a number such as 5.6 trillion.

Using Technology

Here is the keystroke sequence for solving Example 10 using a calculator:

5.6 ⌈EE⌉ 12 ⌈÷⌉ 2.8 ⌈EE⌉ 8.

The quotient is displayed by pressing ⌈=⌉ on a scientific calculator and ⌈ENTER⌉ on a graphing calculator. The answer can be displayed in scientific or decimal notation. Consult your manual.

EXAMPLE 10   The National Debt

As of the end of 2000, the national debt was $5.6 trillion, or 5.6×10^{12} dollars. At that time, the U.S. population was approximately 280,000,000 (280 million), or 2.8×10^8. If the national debt was evenly divided among every individual in the United States, how much would each citizen have to pay?

Solution   The amount each citizen must pay is the total debt, 5.6×10^{12} dollars, divided by the number of citizens, 2.8×10^8.

$$\frac{5.6 \times 10^{12}}{2.8 \times 10^8} = \left(\frac{5.6}{2.8}\right) \times \left(\frac{10^{12}}{10^8}\right)$$

$$= 2 \times 10^{12-8}$$

$$= 2 \times 10^4$$

$$= 20,000$$

Every U.S. citizen would have to pay about $20,000 to the federal government to pay off the national debt. A family of three would owe $60,000. ∎

✔ **CHECK POINT 10**   Approximately 2×10^4 people run in the New York City Marathon each year. Each runner runs a distance of 26 miles. Write the total distance covered by all the runners (assuming that each person completes the marathon) in scientific notation.

ENRICHMENT ESSAY

Earthquakes and Exponents

The earthquake that ripped through northern California on October 17, 1989, measured 7.1 on the Richter scale, killed more than 60 people, and injured more than 2400. Shown here is San Francisco's Marina district, where shock waves tossed houses off their foundations and into the street.

The Richter scale is misleading because it is not actually a 1 to 8, but rather a 1 to 10 million scale. Each level indicates a tenfold increase in magnitude from the previous level, making a 7.0 earthquake a million times greater than a 1.0 quake.

The following is a translation of the Richter scale:

Richter number (R)	Magnitude (10^{R-1})
1	$10^{1-1} = 10^0 = 1$
2	$10^{2-1} = 10^1 = 10$
3	$10^{3-1} = 10^2 = 100$
4	$10^{4-1} = 10^3 = 1000$
5	$10^{5-1} = 10^4 = 10{,}000$
6	$10^{6-1} = 10^5 = 100{,}000$
7	$10^{7-1} = 10^6 = 1{,}000{,}000$
8	$10^{8-1} = 10^7 = 10{,}000{,}000$

EXERCISE SET 5.7

Practice Exercises

In Exercises 1–28, write each expression with positive exponents only. Then simplify, if possible.

1. 5^{-2}

2. 4^{-2}

3. 2^{-3}

4. 3^{-3}

5. $(-2)^{-2}$

6. $(-3)^{-2}$

7. $(-3)^{-3}$

8. $(-1)^{-3}$

9. 4^{-1}

10. 6^{-1}

11. $2^{-1} + 3^{-1}$

12. $3^{-1} - 6^{-1}$

13. $\dfrac{1}{3^{-2}}$

14. $\dfrac{1}{4^{-3}}$

15. $\dfrac{1}{(-3)^{-2}}$

16. $\dfrac{1}{(-2)^{-2}}$

17. $\dfrac{2^{-3}}{8^{-2}}$

18. $\dfrac{4^{-3}}{2^{-2}}$

19. $\left(\dfrac{1}{4}\right)^{-2}$

20. $\left(\dfrac{1}{5}\right)^{-2}$

21. $\left(\dfrac{3}{5}\right)^{-3}$

22. $\left(\dfrac{3}{4}\right)^{-3}$

23. $\dfrac{1}{6x^{-5}}$

24. $\dfrac{1}{8x^{-6}}$

25. $\dfrac{x^{-8}}{y^{-1}}$

26. $\dfrac{x^{-12}}{y^{-1}}$

27. $\dfrac{3}{(-5)^{-3}}$

28. $\dfrac{4}{(-3)^{-3}}$

In Exercises 29–78, simplify each exponential expression. Assume that variables in denominators do not equal zero.

29. $x^{-8} \cdot x^3$

30. $x^{-11} \cdot x^5$

31. $(4x^{-5})(2x^2)$

32. $(5x^{-7})(3x^3)$

33. $\dfrac{x^3}{x^9}$

34. $\dfrac{x^5}{x^{12}}$

35. $\dfrac{y}{y^{100}}$

36. $\dfrac{y}{y^{50}}$

37. $\dfrac{30z^5}{10z^{10}}$

38. $\dfrac{45z^4}{15z^{12}}$

39. $\dfrac{-8x^3}{2x^7}$

40. $\dfrac{-15x^4}{3x^9}$

41. $\dfrac{-9a^5}{27a^8}$

42. $\dfrac{-15a^8}{45a^{13}}$

43. $\dfrac{7w^5}{5w^{13}}$

44. $\dfrac{7w^8}{9w^{14}}$

45. $\dfrac{x^3}{\left(x^4\right)^2}$

46. $\dfrac{x^5}{\left(x^3\right)^2}$

47. $\dfrac{y^{-3}}{\left(y^4\right)^2}$

48. $\dfrac{y^{-5}}{\left(y^3\right)^2}$

49. $\dfrac{\left(4x^3\right)^2}{x^8}$

50. $\dfrac{\left(5x^3\right)^2}{x^7}$

51. $\dfrac{\left(6y^4\right)^3}{y^{-5}}$

52. $\dfrac{\left(4y^5\right)^3}{y^{-4}}$

53. $\left(\dfrac{x^4}{x^2}\right)^{-3}$

54. $\left(\dfrac{x^6}{x^2}\right)^{-3}$

55. $\left(\dfrac{4x^5}{2x^2}\right)^{-4}$

56. $\left(\dfrac{6x^7}{2x^2}\right)^{-4}$

57. $\left(3x^{-1}\right)^{-2}$

58. $\left(4x^{-1}\right)^{-2}$

59. $\left(-2y^{-1}\right)^{-3}$

60. $\left(-3y^{-1}\right)^{-3}$

61. $\dfrac{2x^5 \cdot 3x^7}{15x^6}$

62. $\dfrac{3x^3 \cdot 5x^{14}}{20x^{14}}$

63. $\left(x^3\right)^5 \cdot x^{-7}$

64. $\left(x^4\right)^3 \cdot x^{-5}$

65. $\left(2y^3\right)^4 y^{-6}$

66. $\left(3y^4\right)^3 y^{-7}$

67. $\dfrac{\left(y^3\right)^4}{\left(y^2\right)^7}$

68. $\dfrac{\left(y^2\right)^5}{\left(y^3\right)^4}$

69. $\left(y^{10}\right)^{-5}$

70. $\left(y^{20}\right)^{-5}$

71. $\left(a^4b^5\right)^{-3}$

72. $\left(a^5b^3\right)^{-4}$

73. $\left(a^{-2}b^6\right)^{-4}$

74. $\left(a^{-7}b^2\right)^{-5}$

75. $\left(\dfrac{x^2}{2}\right)^{-2}$

76. $\left(\dfrac{x^2}{2}\right)^{-3}$

77. $\left(\dfrac{x^2}{y^3}\right)^{-3}$

78. $\left(\dfrac{x^3}{y^2}\right)^{-4}$

In Exercises 79–90, write each number in decimal notation without the use of exponents.

79. 2.7×10^2

80. 4.75×10^3

81. 9.12×10^5

82. 8.14×10^4

83. 3.4×10^0

84. 9.115×10^0

85. 7.9×10^{-1}

86. 8.6×10^{-1}

87. 2.15×10^{-2}

88. 3.14×10^{-2}

89. 7.86×10^{-4}

90. 4.63×10^{-5}

In Exercises 91–106, write each number in scientific notation.

91. 32,400

92. 327,000

93. 220,000,000

94. 370,000,000,000

95. 713

96. 623

97. 6751

98. 9832

99. 0.0027

100. 0.00083

101. 0.0000202

102. 0.00000103

103. 0.005

104. 0.006

105. 3.14159

106. 2.71828

In Exercises 107–126, perform the indicated computations. Write the answers in scientific notation.

107. $\left(2 \times 10^3\right)\left(3 \times 10^2\right)$

108. $\left(3 \times 10^4\right)\left(3 \times 10^2\right)$

109. $\left(2 \times 10^5\right)\left(8 \times 10^3\right)$

110. $\left(4 \times 10^3\right)\left(5 \times 10^4\right)$

111. $\dfrac{12 \times 10^6}{4 \times 10^2}$

112. $\dfrac{20 \times 10^{20}}{10 \times 10^{10}}$

113. $\dfrac{15 \times 10^4}{5 \times 10^{-2}}$

114. $\dfrac{18 \times 10^2}{9 \times 10^{-3}}$

115. $\dfrac{15 \times 10^{-4}}{5 \times 10^2}$

116. $\dfrac{18 \times 10^{-2}}{9 \times 10^3}$

117. $\dfrac{180 \times 10^6}{2 \times 10^3}$

118. $\dfrac{180 \times 10^8}{2 \times 10^4}$

119. $\dfrac{3 \times 10^4}{12 \times 10^{-3}}$

120. $\dfrac{5 \times 10^2}{20 \times 10^{-3}}$

121. $\left(5 \times 10^2\right)^3$

122. $\left(4 \times 10^3\right)^2$

123. $\left(3 \times 10^{-2}\right)^4$

124. $\left(2 \times 10^{-3}\right)^5$

125. $\left(4 \times 10^6\right)^{-1}$

126. $\left(5 \times 10^4\right)^{-1}$

Application Exercises

In Exercises 127–130, rewrite the number in each statement in scientific notation.

127. King Mongut of Siam (the king in the musical *The King and I*) had 9200 wives.

128. The top-selling music album of all time is Michael Jackson's "Thriller," selling 25,000,000 copies.

129. The volume of a bacterium is 0.00000000000000025 cubic meter.

130. Home computers can perform a multiplication in 0.00000000036 second.

In Exercises 131–134, use 10^6 for one million and 10^9 for one billion to rewrite the number of dollars in each statement in scientific notation.

131. In 1999, the U.S. government collected $1,694,300 million.

132. In 1999, the U.S. government spent $1,751,800 million.

133. The federal government is expected to provide nearly $60 billion in student aid in 2002.

134. In 1998, U.S. consumers spent $5493.7 billion.

135. The United States government spends approximately $1.6 trillion per year. Use 10^{12} for one trillion and the graph to write the amount that it spends on Social Security in scientific notation.

Where the U.S. Government Spends Money

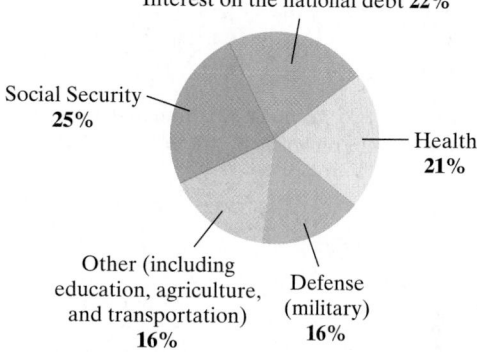

Interest on the national debt **22%**

Social Security **25%**

Health **21%**

Other (including education, agriculture, and transportation) **16%**

Defense (military) **16%**

Source: U.S. Office of Management and Budget

136. Americans say they lead active lives, but for the 205 million of us ages 18 and older, walking is often as strenuous as it gets. Use the graph to write the number of Americans whose lifestyle is not very active. Express the answer in scientific notation.

How Active is Your Lifestyle?

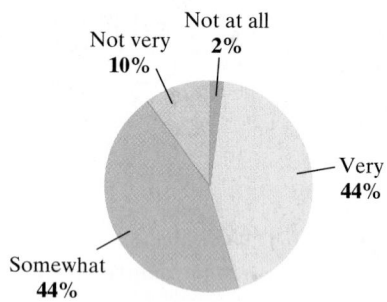

Not at all **2%**

Not very **10%**

Very **44%**

Somewhat **44%**

Source: Discovery Health Media

137. If the population of the United States is 2.7×10^8 and each person spends about $120 per year on ice cream, express the total annual spending on ice cream in scientific notation.

138. A human brain contains 3×10^{10} neurons and a gorilla brain contains 7.5×10^9 neurons. How many times as many neurons are in the brain of a human as in the brain of a gorilla?

Writing in Mathematics

139. Explain the negative exponent rule and give an example.

140. How do you know if an exponential expression is simplified?

141. How do you know if a number is written in scientific notation?

142. Explain how to convert from scientific to decimal notation and give an example.

143. Explain how to convert from decimal to scientific notation and give an example.

144. Describe one advantage of expressing a number in scientific notation over decimal notation.

Critical Thinking Exercises

145. Which one of the following is true?
a. $4^{-2} < 4^{-3}$
b. $5^{-2} > 2^{-5}$
c. $(-2)^4 = 2^{-4}$
d. $5^2 \cdot 5^{-2} > 2^5 \cdot 2^{-5}$

146. Which one of the following is true?
a. $534.7 = 5.347 \times 10^3$
b. $\dfrac{8 \times 10^{30}}{4 \times 10^{-5}} = 2 \times 10^{25}$
c. $(7 \times 10^5) + (2 \times 10^{-3}) = 9 \times 10^2$
d. $(4 \times 10^3) + (3 \times 10^2) = 4.3 \times 10^3$

147. Give an example of a number where there is no advantage to using scientific notation instead of decimal notation. Explain why this is the case.

148. The mad Dr. Frankenstein has gathered enough bits and pieces (so to speak) for $2^{-1} + 2^{-2}$ of his creature-to-be. Write a fraction that represents the amount of his creature that must still be obtained.

Technology Exercises

149. Use a calculator in a fraction mode to check any five of your answers in Exercises 1–22.

150. Use a calculator to check any three of your answers in Exercises 79–90.

151. Use a calculator to check any three of your answers in Exercises 91–106.

152. Use a calculator with an $\boxed{\text{EE}}$ or $\boxed{\text{EXP}}$ key to check any four of your computations in Exercises 107–126. Display the result of the computation in scientific notation.

Review Exercises

153. Solve: $8 - 6x > 4x - 12$. (Section 2.7, Example 6)

154. Simplify: $24 \div 8 \cdot 3 + 28 \div (-7)$. (Section 1.8, Example 8)

155. List the whole numbers in this set:
$$\left\{-4, -\frac{1}{5}, 0, \pi, \sqrt{16}, \sqrt{17}\right\}.$$
(Section 1.2, Example 5)

CHAPTER 5 GROUP PROJECTS

1. **Putting Numbers into Perspective.** A large number can be put into perspective by comparing it with another number. For example, we put the $5.6 trillion national debt into perspective by comparing it to the number of U.S. citizens. The total distance covered by all the runners in the New York City Marathon (Check Point 10 on page 359) can be put into perspective by comparing this distance with, say, the distance from New York to San Francisco.

   For this project, each group member should consult an almanac, a newspaper, or the World Wide Web to find a number greater than one million. Explain to other members of the group the context in which the large number is used. Express the number in scientific notation. Then put the number into perspective by comparing it with another number.

2. **Polynomials and Number Tricks.** Here's a number puzzle that can be verified using polynomial operations.

   Take a number. Add 1. Square the result. Subtract the product of the original number times two more than the original number. The answer will always be 1.

   Verification

   Take a number: x.

   Add one: $x + 1$.

   Square the result: $(x + 1)^2$.

   Subtract the product of the original number times two more than the original number: $(x + 1)^2 - x(x + 2)$.

   Using polynomial operations:

   $$(x + 1)^2 - x(x + 2) = x^2 + 2x + 1 - x^2 - 2x = 1.$$

   The answer is 1, regardless of what number is originally chosen!

   Group members should verify the puzzles in parts (a) and (b) using polynomial operations.

   a. Take a number. Add 2. Square the sum. Add 25 to this result. Subtract the product of the original number times four more than the original number. Add 6 to the difference. The result is always 35.

   b. Take a number and multiply it by five more than the number. Subtract from this the product of seven more than the original number and two less than the original number. Multiply the total by 5 and subtract 50. The result is always 20.

   c. The group should write and then verify two puzzles similar to those in parts (a) and (b).

CHAPTER SUMMARY, REVIEW, AND TEST

SUMMARY

DEFINITIONS AND CONCEPTS	EXAMPLES

Section 5.1 Adding and Subtracting Polynomials

A polynomial is a single term or the sum of two or more terms containing variables in the numerator with whole number exponents. A monomial is a polynomial with exactly one term; a binomial has exactly two terms; a trinomial has exactly three terms. The degree of a polynomial is the highest power of all the terms.	Polynomials Monomial: $2x^5$ Degree is 5. Binomial: $6x^3 + 5x$ Degree is 3. Trinomial: $7x + 4x^2 - 5$ Degree is 2.
To add polynomials, add like terms.	$(6x^3 + 5x^2 - 7x) + (-9x^3 + x^2 + 6x)$ $= (6x^3 - 9x^3) + (5x^2 + x^2) + (-7x + 6x)$ $= -3x^3 + 6x^2 - x$
To subtract two polynomials, change the sign of every term of the second polynomial. Add this result to the first polynomial.	$(5y^3 - 9y^2 - 4) - (3y^3 - 12y^2 - 5)$ $= (5y^3 - 9y^2 - 4) + (-3y^3 + 12y^2 + 5)$ $= (5y^3 - 3y^3) + (-9y^2 + 12y^2) + (-4 + 5)$ $= 2y^3 + 3y^2 + 1$

Section 5.2 Multiplying Polynomials

Properties of Exponents Product Rule: $b^m \cdot b^n = b^{m+n}$ Power Rule: $(b^m)^n = b^{mn}$ Products to Powers: $(ab)^n = a^n b^n$	$x^3 \cdot x^8 = x^{3+8} = x^{11}$ $(x^3)^8 = x^{3 \cdot 8} = x^{24}$ $(-5x^2)^3 = (-5)^3(x^2)^3 = -125x^6$
To multiply monomials, multiply coefficients and add exponents.	$(-6x^4)(3x^{10}) = -6 \cdot 3x^{4+10} = -18x^{14}$
To multiply a monomial and a polynomial, multiply each term of the polynomial by the monomial.	$2x^4(3x^2 - 6x + 5)$ $= 2x^4 \cdot 3x^2 - 2x^4 \cdot 6x + 2x^4 \cdot 5$ $= 6x^6 - 12x^5 + 10x^4$
To multiply polynomials when neither is a monomial, multiply each term of one polynomial by each term of the other polynomial. Then combine like terms.	$(2x + 3)(5x^2 - 4x + 2)$ $= 2x(5x^2 - 4x + 2) + 3(5x^2 - 4x + 2)$ $= 10x^3 - 8x^2 + 4x + 15x^2 - 12x + 6$ $= 10x^3 + 7x^2 - 8x + 6$

DEFINITIONS AND CONCEPTS	EXAMPLES

Section 5.3 Special Products

The FOIL method may be used when multiplying two binomials: First terms multiplied. Outside terms multiplied. Inside terms multiplied. Last terms multiplied.	F O I L $(3x + 7)(2x - 5) = 3x \cdot 2x + 3x(-5) + 7 \cdot 2x + 7(-5)$ $= 6x^2 - 15x + 14x - 35$ $= 6x^2 - x - 35$
The Product of the Sum and Difference of Two Terms $(A + B)(A - B) = A^2 - B^2$	$(4x + 7)(4x - 7) = (4x)^2 - 7^2$ $= 16x^2 - 49$
The Square of a Binomial Sum $(A + B)^2 = A^2 + 2AB + B^2$	$(x^2 + 6)^2 = (x^2)^2 + 2 \cdot x^2 \cdot 6 + 6^2$ $= x^4 + 12x^2 + 36$
The Square of a Binomial Difference $(A - B)^2 = A^2 - 2AB + B^2$	$(9x - 3)^2 = (9x)^2 - 2 \cdot 9x \cdot 3 + 3^2$ $= 81x^2 - 54x + 9$

Section 5.4 Polynomials in Several Variables

To evaluate a polynomial in several variables, substitute the given value for each variable and perform the resulting computation.	Evaluate $4x^2y + 3xy - 2x$ for $x = -1$ and $y = -3$. $4x^2y + 3xy - 2x$ $= 4(-1)^2(-3) + 3(-1)(-3) - 2(-1)$ $= 4(1)(-3) + 3(-1)(-3) - 2(-1)$ $= -12 + 9 + 2 = -1$
For a polynomial in two variables, the degree of a term is the sum of the exponents on its variables. The degree of the polynomial is the highest degree of all its terms.	$7x^2y + 12x^4y^3 - 17x^5 + 6$ degree: $2 + 1 = 3$ degree: $4 + 3 = 7$ degree: 5 degree: 0 Degree of polynomial $= 7$
Polynomials in several variables are added, subtracted, and multiplied using the same rules for polynomials in one variable.	$(5x^2y^3 - xy + 4y^2) - (8x^2y^3 - 6xy - 2y^2)$ $= (5x^2y^3 - xy + 4y^2) + (-8x^2y^3 + 6xy + 2y^2)$ $= -3x^2y^3 + 5xy + 6y^2$ F O I L $(3x - 2y)(x - y) = 3x \cdot x + 3x(-y) + (-2y)x + (-2y)(-y)$ $= 3x^2 - 3xy - 2xy + 2y^2$ $= 3x^2 - 5xy + 2y^2$

DEFINITIONS AND CONCEPTS	EXAMPLES

Section 5.5 Dividing Polynomials

Additional Properties of Exponents

Quotient Rule: $\dfrac{b^m}{b^n} = b^{m-n}, \quad b \neq 0$

Zero Exponent Rule: $b^0 = 1, \quad b \neq 0$

Quotients to Powers: $\left(\dfrac{a}{b}\right)^n = \dfrac{a^n}{b^n}, \quad b \neq 0$

$$\frac{x^{12}}{x^4} = x^{12-4} = x^8$$

$$(-3)^0 = 1 \qquad -3^0 = -1(3^0) = -1 \cdot 1 = -1$$

$$\left(\frac{y^2}{4}\right)^3 = \frac{(y^2)^3}{4^3} = \frac{y^{2\cdot3}}{4\cdot4\cdot4} = \frac{y^6}{64}$$

To divide monomials, divide coefficients and subtract exponents.

$$\frac{-40x^{40}}{20x^{20}} = \frac{-40}{20}x^{40-20} = -2x^{20}$$

To divide a polynomial by a monomial, divide each term of the polynomial by the monomial.

$$\frac{8x^6 - 4x^3 + 10x}{2x}$$

$$= \frac{8x^6}{2x} - \frac{4x^3}{2x} + \frac{10x}{2x}$$

$$= 4x^{6-1} - 2x^{3-1} + 5x^{1-1} = 4x^5 - 2x^2 + 5$$

Section 5.6 Dividing Polynomials by Binomials; Synthetic Division

To divide by a polynomial containing more than one term, use long division. If necessary, arrange the dividend in descending powers of the variable. If a power of a variable is missing, add that power with a coefficient of 0. Repeat the four steps of the long division process—divide, multiply, subtract, bring down the next term—until the degree of the remainder is less than the degree of the divisor.

Divide: $(2x^3 - x^2 - 7) \div (x - 2)$.

$$
\begin{array}{r}
2x^2 + 3x + 6 \\
x - 2 \overline{)\,2x^3 - x^2 + 0x - 7} \\
\underline{2x^3 - 4x^2} \\
3x^2 + 0x \\
\underline{3x^2 - 6x} \\
6x - 7 \\
\underline{6x - 12} \\
5
\end{array}
$$

The answer is $2x^2 + 3x + 6 + \dfrac{5}{x-2}$.

A shortcut to long division, called synthetic division, can be used to divide a polynomial by a binomial of the form $x - c$.

Here is the division problem shown above using synthetic division.

$$(2x^3 - x^2 - 7) \div (x - 2)$$

Coefficients of the dividend,
$2x^3 - x^2 + 0x - 7$

This is c in $x - c$.
For $x - 2$, c is 2.

$$
\begin{array}{r|rrrr}
2 & 2 & -1 & 0 & -7 \\
& \downarrow & 4 & 6 & 12 \\
\hline
& 2 & 3 & 6 & 5
\end{array}
$$

Coefficients of quotient   Remainder

The answer is $2x^2 + 3x + 6 + \dfrac{5}{x-2}$.

DEFINITIONS AND CONCEPTS	EXAMPLES

Section 5.7 Negative Exponents and Scientific Notation

Negative Exponents in Numerators and Denominators

If $b \neq 0$, $b^{-n} = \dfrac{1}{b^n}$ and $\dfrac{1}{b^{-n}} = b^n$.

$$6^{-2} = \frac{1}{6^2} = \frac{1}{36}$$

$$\frac{1}{(-2)^{-4}} = (-2)^4 = 16$$

$$\left(\frac{2}{3}\right)^{-3} = \frac{2^{-3}}{3^{-3}} = \frac{3^3}{2^3} = \frac{27}{8}$$

An exponential expression is simplified when:
- No parentheses appear.
- No powers are raised to powers.
- Each base occurs only once.
- No negative or zero exponents appear.

Simplify: $\dfrac{(2x^4)^3}{x^{18}}$.

$$\frac{(2x^4)^3}{x^{18}} = \frac{2^3(x^4)^3}{x^{18}} = \frac{8x^{12}}{x^{18}} = 8x^{12-18} = 8x^{-6} = \frac{8}{x^6}$$

A number in scientific notation is expressed as a number greater than or equal to 1 and less than 10 multiplied by some power of 10.

Write 2.9×10^{-3} in decimal notation.

$$2.9 \times 10^{-3} = .0029 = 0.0029$$

Write 16,000 in scientific notation.

$$16,000 = 1.6 \times 10^4$$

Use

$$10^m \cdot 10^n = 10^{m+n}, \quad \frac{10^m}{10^n} = 10^{m-n}, \quad \text{and} \quad (10^m)^n = 10^{mn}$$

to perform computations with scientific notation.

$$(5 \times 10^3)(4 \times 10^{-8})$$
$$= 5 \cdot 4 \times 10^{3-8}$$
$$= 20 \times 10^{-5}$$
$$= 2 \times 10^1 \times 10^{-5} = 2 \times 10^{-4}$$

Review Exercises

5.1 In Exercises 1–3, identify each polynomial as a monomial, binomial, or trinomial. Give the degree of the polynomial.

1. $7x^4 + 9x$

2. $3x + 5x^2 - 2$

3. $16x$

In Exercises 4–8, add or subtract as indicated.

4. $(-6x^3 + 7x^2 - 9x + 3) + (14x^3 + 3x^2 - 11x - 7)$

5. $(9y^3 - 7y^2 + 5) + (4y^3 - y^2 + 7y - 10)$

6. $(5y^2 - y - 8) - (-6y^2 + 3y - 4)$

7. $(13x^4 - 8x^3 + 2x^2) - (5x^4 - 3x^3 + 2x^2 - 6)$

8. Subtract $x^4 + 7x^2 - 11x$ from $-13x^4 - 6x^2 + 5x$.

In Exercises 9–11, add or subtract as indicated.

9. Add.
$$7y^4 - 6y^3 + 4y^2 - 4y$$
$$\underline{\qquad y^3 - \quad y^2 + 3y - 4}$$

10. Subtract.
$$7x^2 - 9x + 2$$
$$\underline{-(4x^2 - 2x - 7)}$$

11. Subtract.
$$5x^3 - 6x^2 - \quad 9x + 14$$
$$\underline{-(-5x^3 + 3x^2 - 11x + \quad 3)}$$

12. The polynomial $104.5x^2 - 1501.5x + 6016$ models the death rate per year per 100,000 men for men averaging x hours of sleep each night. Evaluate the polynomial when $x = 10$. Describe what the answer means in practical terms.

5.2 *In Exercises 13–17, simplify each expression.*

13. $x^{20} \cdot x^3$

14. $y \cdot y^5 \cdot y^8$

15. $(x^{20})^5$

16. $(10y)^2$

17. $(-4x^{10})^3$

In Exercises 18–26, find each product.

18. $(5x)(10x^3)$

19. $(-12y^7)(3y^4)$

20. $(-2x^5)(-3x^4)(5x^3)$

21. $7x(3x^2 + 9)$

22. $5x^3(4x^2 - 11x)$

23. $3y^2(-7y^2 + 3y - 6)$

24. $2y^5(8y^3 - 10y^2 + 1)$

25. $(x + 3)(x^2 - 5x + 2)$

26. $(3y - 2)(4y^2 + 3y - 5)$

In Exercises 27–28, use a vertical format to find each product.

27.
$$\begin{array}{r} y^2 - 4y + 7 \\ 3y - 5 \\ \hline \end{array}$$

28.
$$\begin{array}{r} 4x^3 - 2x^2 - 6x - 1 \\ 2x + 3 \\ \hline \end{array}$$

5.3 *In Exercises 29–41, find each product.*

29. $(x + 6)(x + 2)$

30. $(3y - 5)(2y + 1)$

31. $(4x^2 - 2)(x^2 - 3)$

32. $(5x + 4)(5x - 4)$

33. $(7 - 2y)(7 + 2y)$

34. $(y^2 + 1)(y^2 - 1)$

35. $(x + 3)^2$

36. $(3y + 4)^2$

37. $(y - 1)^2$

38. $(5y - 2)^2$

39. $(x^2 + 4)^2$

40. $(x^2 + 4)(x^2 - 4)$

41. $(x^2 + 4)(x^2 - 5)$

42. Write a polynomial in descending powers of x that represents the area of the shaded region.

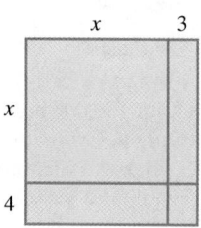

43. The parking garage shown in the figure measures 30 yards by 20 yards. The length and the width are each increased by a fixed amount, x yards. Write a trinomial that describes the area of the expanded garage.

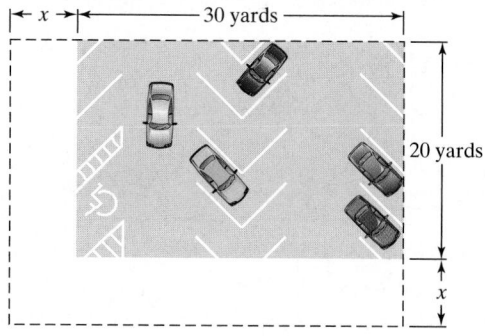

5.4

44. Evaluate $2x^3y - 4xy^2 + 5y + 6$ for $x = -1$ and $y = 2$.

45. Determine the coefficient of each term, the degree of each term, and the degree of the polynomial:
$$4x^2y + 9x^3y^2 - 17x^4 - 12.$$

In Exercises 46–55, perform the indicated operations.

46. $(7x^2 - 8xy + y^2) + (-8x^2 - 9xy + 4y^2)$

47. $(13x^3y^2 - 5x^2y - 9x^2) - (11x^3y^2 - 6x^2y - 3x^2 + 4)$

48. $(-7x^2y^3)(5x^4y^6)$

49. $5ab^2(3a^2b^3 - 4ab)$

50. $(x + 7y)(3x - 5y)$

51. $(4xy - 3)(9xy - 1)$

52. $(3x + 5y)^2$

53. $(xy - 7)^2$

54. $(7x + 4y)(7x - 4y)$

55. $(a - b)(a^2 + ab + b^2)$

5.5 *In Exercises 56–62, simplify each expression.*

56. $\dfrac{6^{40}}{6^{10}}$

57. $\dfrac{x^{18}}{x^3}$

58. $(-10)^0$

59. -10^0

60. $400x^0$

61. $\left(\dfrac{x^4}{2}\right)^3$

62. $\left(\dfrac{-3}{2y^6}\right)^4$

In Exercises 63–67, divide and check each answer.

63. $\dfrac{-15y^8}{3y^2}$

64. $\dfrac{40x^8y^6}{5xy^3}$

65. $\dfrac{18x^4 - 12x^2 + 36x}{6x}$

66. $\dfrac{30x^8 - 25x^7 - 40x^5}{-5x^3}$

67. $\dfrac{27x^3y^2 - 9x^2y - 18xy^2}{3xy}$

5.6 *In Exercises 68–71, divide using long division and check each answer.*

68. $\dfrac{2x^2 + 3x - 14}{x - 2}$

69. $\dfrac{2x^3 - 5x^2 + 7x + 5}{2x + 1}$

70. $\dfrac{x^3 - 2x^2 - 33x - 7}{x - 7}$

71. $\dfrac{y^3 - 27}{y - 3}$

In Exercises 72–74, divide using synthetic division.

72. $(4x^3 - 3x^2 - 2x + 1) \div (x + 1)$

73. $(3x^4 - 2x^2 - 10x - 20) \div (x - 2)$

74. $(x^4 + 16) \div (x + 4)$

5.7 *In Exercises 75–79, write each expression with positive exponents only and then simplify.*

75. 7^{-2}

76. $(-4)^{-3}$

77. $2^{-1} + 4^{-1}$

78. $\dfrac{1}{5^{-2}}$

79. $\left(\dfrac{2}{5}\right)^{-3}$

In Exercises 80–88, simplify each exponential expression. Assume that variables in denominators do not equal zero.

80. $\dfrac{x^3}{x^9}$

81. $\dfrac{30y^6}{5y^8}$

82. $(5x^{-7})(6x^2)$

83. $\dfrac{x^4 \cdot x^{-2}}{x^{-6}}$

84. $\dfrac{(3y^3)^4}{y^{10}}$

85. $\dfrac{y^{-7}}{(y^4)^3}$

86. $(2x^{-1})^{-3}$

87. $\left(\dfrac{x^7}{x^4}\right)^{-2}$

88. $\dfrac{(y^3)^4}{(y^{-2})^4}$

In Exercises 89–91, write each number in decimal notation without the use of exponents.

89. 2.3×10^4

90. 1.76×10^{-3}

91. 9×10^{-1}

In Exercises 92–95, write each number in scientific notation.

92. 73,900,000

93. 0.00062

94. 0.38

95. 3.8

In Exercises 96–98, perform the indicated computation. Write the answers in scientific notation.

96. $(6 \times 10^{-3})(1.5 \times 10^6)$

97. $\dfrac{2 \times 10^2}{4 \times 10^{-3}}$

98. $(4 \times 10^{-2})^2$

99. A microsecond is 10^{-6} second and a nanosecond is 10^{-9} second. How many nanoseconds make a microsecond?

100. The world's population is approximately 6.1×10^9 people. Current projections double this population in 40 years. Write the population 40 years from now in scientific notation.

Chapter 5 Test

1. Identify $9x + 6x^2 - 4$ as a monomial, binomial, or trinomial. Give the degree of the polynomial.

In Exercises 2–3, add or subtract as indicated.

2. $(7x^3 + 3x^2 - 5x - 11) + (6x^3 - 2x^2 + 4x - 13)$

3. $(9x^3 - 6x^2 - 11x - 4) - (4x^3 - 8x^2 - 13x + 5)$

In Exercises 4–10, find each product.

4. $(-7x^3)(5x^8)$

5. $6x^2(8x^3 - 5x - 2)$

6. $(3x + 2)(x^2 - 4x - 3)$

7. $(3y + 7)(2y - 9)$

8. $(7x + 5)(7x - 5)$

9. $(x^2 + 3)^2$

10. $(5x - 3)^2$

11. Evaluate $4x^2y + 5xy - 6x$ for $x = -2$ and $y = 3$.

In Exercises 12–14, perform the indicated operations.

12. $(8x^2y^3 - xy + 2y^2) - (6x^2y^3 - 4xy - 10y^2)$

13. $(3a - 7b)(4a + 5b)$

14. $(2x + 3y)^2$

In Exercises 15–17, divide and check each answer.

15. $\dfrac{-25x^{16}}{5x^4}$

16. $\dfrac{15x^4 - 10x^3 + 25x^2}{5x}$

17. $\dfrac{2x^3 - 3x^2 + 4x + 4}{2x + 1}$

18. Divide using synthetic division:
$$(3x^4 + 11x^3 - 20x^2 + 7x + 35) \div (x + 5).$$

In Exercises 19–20, write each expression with positive exponents only and then simplify.

19. 10^{-2}

20. $\dfrac{1}{4^{-3}}$

In Exercises 21–26, simplify each expression.

21. $(-3x^2)^3$

22. $\dfrac{20x^3}{5x^8}$

23. $(-7x^{-8})(3x^2)$

24. $\dfrac{(2y^3)^4}{y^8}$

25. $(5x^{-4})^{-2}$

26. $\left(\dfrac{x^{10}}{x^5}\right)^{-3}$

27. Write 3.7×10^{-4} in decimal notation.

28. Write 7,600,000 in scientific notation.

In Exercises 29–30, perform the indicated computation. Write the answers in scientific notation.

29. $(4.1 \times 10^2)(3 \times 10^{-5})$

30. $\dfrac{8.4 \times 10^6}{4 \times 10^{-2}}$

Cumulative Review Exercises (Chapters 1–5)

In Exercises 1–2, perform the indicated operation or operations.

1. $(-7)(-5) \div (12 - 3)$

2. $(3 - 7)^2(9 - 11)^3$

3. What is the difference in elevation between a plane flying 14,300 feet above sea level and a submarine traveling 750 feet below sea level?

In Exercises 4–5, solve each equation.

4. $2(x + 3) + 2x = x + 4$

5. $\dfrac{x}{5} - \dfrac{1}{3} = \dfrac{x}{10} - \dfrac{1}{2}$

6. The length of a rectangular sign is 2 feet less than three times its width. If the perimeter of the sign is 28 feet, what are its dimensions?

7. Solve: $7 - 8x \le -6x - 5$. Express the solution set in set-builder notation and graph the solution set on a number line.

8. You invested $6000 in two stocks paying 12% and 14% annual interest, respectively. At the end of the year, the total interest from these investment was $772. How much was intested at each rate?

9. You need to mix a solution that is 70% antifreeze with one that ia 30% antifreeze to obtain 20 liters of a mixture that is 60% antifreeze. How many liters of each of the solutions must be used?

10. Graph $y = -\frac{2}{5}x + 2$ using the slope and y-intercept.

11. Graph $x - 2y = 4$ using intercepts.

12. Find the slope of the line passing through the points $(-3, 2)$ and $(2, -4)$. Is the line rising, falling, horizontal, or vertical?

13. The slope of a line is -2 and the line passes through the point $(3, -1)$. Write the line's equation in point-slope form and slope-intercept form.

In Exercises 14–15, solve each system by the method of your choice.

14. $3x + 2y = 10$
  $4x - 3y = -15$

15. $2x + 3y = -6$
  $y = 3x - 13$

16. You are choosing between two long-distance telephone plans. One has a monthly fee of $15 with a charge of $0.05 per minute for all long-distance calls. The other plan has a monthly fee of $5 with a charge of $0.07 per minute for all long-distance calls. For how many minutes of long-distance calls will the costs for the two plans be the same? What will be the cost for each plan?

17. Subtract: $(9x^5 - 3x^3 + 2x - 7) - (6x^5 + 3x^3 - 7x - 9)$.

18. Divide: $\dfrac{x^3 + 3x^2 + 5x + 3}{x + 1}$.

19. Simplify: $\dfrac{(3x^2)^4}{x^{10}}$.

20. Write 2.4×10^{-3} in decimal notation.

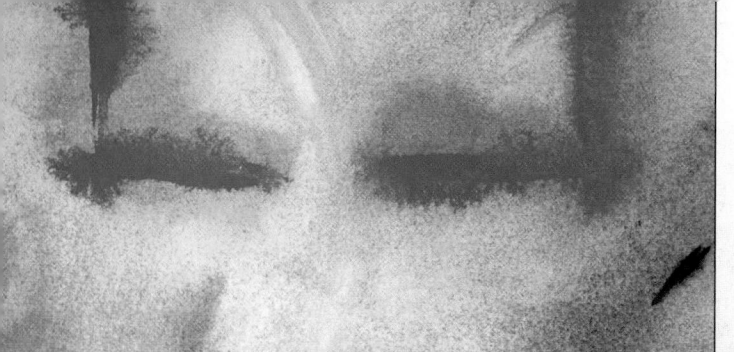

Landscaped parks often form an oasis of greenery and calm in which to escape from the bustle of city life. Flowers, trees, ponds, and fountains provide a natural setting that mirrors the interest that many city dwellers have in their environment.

H ave you ever thought about developing an attractive and inviting home landscape? Algebra and geometry play an important role in landscape design. In this chapter, you will see how rewriting a polynomial sum or difference in terms of multiplication can be used in the creation of horticultural masterpieces.

Chapter 6

Factoring Polynomials

► SECTION 6.1   *The Greatest Common Factor By Grouping*

Objectives

1 Factor monomials.

2 Find the greatest common factor.

3 Factor out the greatest common factor of a polynomial.

4 Factor by grouping.

SSM
PH Tutor CD- Video
Center ROM

A two-year-old boy is asked, "Do you have a brother?" He answers, "Yes." "What is your brother's name?" "Tom." Asked if Tom has a brother, the two-year-old replies, "No." The child can go in the direction from self to brother, but he cannot reverse this direction and move from brother back to self.

As our intellects develop, we learn to reverse the direction of our thinking. Reversibility of thought is found throughout algebra. For example, we can multiply polynomials and show that

$$5x(2x + 3) = 10x^2 + 15x.$$

We can also reverse this process and express the resulting polynomial as

$$10x^2 + 15x = 5x(2x + 3).$$

Factoring is the process of writing a polynomial as the product of two or more polynomials. The **factors** of $10x^2 + 15x$ are $5x$ and $2x + 3$.

In this chapter, we will be factoring over the set of integers, meaning that the coefficients in the factors are integers. Polynomials that cannot be factored using integer coefficients are called **prime polynomials** over the set of integers.

1 Factor monomials.

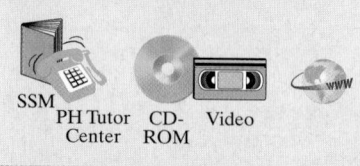

Discover for Yourself

Write three more ways of factoring the monomial $30x^2$.

Factoring Monomials

Factoring a monomial means finding two monomials whose product gives the original monomial. For example, $30x^2$ can be factored in a number of different ways, such as:

$30x^2 = (5x)(6x)$	The factors are 5x and 6x.
$30x^2 = (15x)(2x)$	The factors are 15x and 2x.
$30x^2 = (10x^2)(3)$	The factors are 10x² and 3.
$30x^2 = (-6x)(-5x).$	The factors are −6x and −5x.

Observe that each part of the factorization is called a *factor* of the given monomial.

Factoring Out the Greatest Common Factor

We use the distributive property to multiply a monomial and a polynomial of two or more terms. When we factor, we reverse this process, expressing the polynomial as a product.

Multiplication	**Factoring**
$a(b + c) = ab + ac$	$ab + ac = a(b + c)$

Here is a specific example:

Multiplication	Factoring
$5x(2x + 3)$	$10x^2 + 15x$
$= 5x \cdot 2x + 5x \cdot 3$	$= 5x \cdot 2x + 5x \cdot 3$
$= 10x^2 + 15x$	$= 5x(2x + 3).$

In the process of finding an equivalent expression for $10x^2 + 15x$ that is a product, we used the fact that $5x$ is a factor of both $10x^2$ and $15x$. The factoring on the right shows that $5x$ is a *common factor* for all the terms of the binomial $10x^2 + 15x$.

In any factoring problem, the first step is to look for the *greatest common factor*. The **greatest common factor**, abbreviated GCF, is an expression of the highest degree that divides each term of the polynomial. Can you see that $5x$ is the greatest common factor of $10x^2 + 15x$? 5 is the greatest integer that divides 10 and 15. Furthermore, x is the greatest expression that divides x^2 and x.

The variable part of the greatest common factor always contains the smallest power of a variable that appears in all terms of the polynomial. For example, consider the polynomial

$$10x^2 + 15x.$$

> x^1, or x, is the variable raised to the smallest exponent.

We see that x is the variable part of the greatest common factor, $5x$.

EXAMPLE 1   Finding the Greatest Common Factor

Find the greatest common factor of each list of terms:

a. $6x^3$ and $10x^2$ **b.** $15y^5$, $-9y^4$, and $27y^3$

c. x^5y^3, x^4y^4, and x^3y^2.

Solution   Use numerical coefficients to determine the coefficient of the GCF. Use variable factors to determine the variable factor of the GCF.

> 2 is the greatest integer that divides 6 and 10.

a. $6x^3$ and $10x^2$

> x^2 is the variable raised to the smallest exponent.

We see that 2 is the numerical coefficient of the GCF and x^2 is the variable factor of the GCF. Thus, the GCF of $6x^3$ and $10x^2$ is $2x^2$.

> 3 is the greatest integer that divides 15, -9, and 27.

b. $15y^5$, $-9y^4$,   and $27y^3$

> y^3 is the variable raised to the smallest exponent.

We see that 3 is the numerical coefficient of the GCF and y^3 is the variable factor of the GCF. Thus, the GCF of $15y^5$, $-9y^4$, and $27y^3$ is $3y^3$.

---

2 Find the greatest common factor.

ENRICHMENT ESSAY

Friendly Numbers

You probably do not describe numbers as sad, angry, happy, or friendly. But the ancient Greeks described two numbers as **friendly** if each was the sum of the other's factors, excluding the numbers themselves. The Greeks knew of only one such pair, 220 and 284. Factors of 220 have a sum of 284:

$1 + 2 + 4 + 5 + 10 + 11 + 20 + 22 + 44 + 55 + 110 = 284.$

Factors of 284 have a sum of 220:

$1 + 2 + 4 + 71 + 142 = 220.$

In 1636, the French mathematician Pierre de Fermat discovered a second pair of friendly numbers, 17,296 and 18,416. By the middle of the nineteenth century, the number of known pairs of friendly numbers exceeded 60. We still have unanswered questions about friendly numbers. All known friendly pairs consist of either two odd or two even numbers. Are pairs consisting of an odd and an even number possible? Why are all the known odd friendly numbers multiples of 3?

x^3 is the variable, x, raised to the smallest exponent.

c. $x^5 y^3, \quad x^4 y^4, \quad x^3 y^2$

y^2 is the variable, y, raised to the smallest exponent.

Because all terms have numerical coefficients of 1, 1 is the greatest integer that divides these coefficients. Thus, 1 is the numerical coefficient of the GCF. The voice balloons show that x^3 and y^2 are the variable factors of the GCF. Thus, the GCF of $x^5 y^3$, $x^4 y^4$, and $x^3 y^2$ is $x^3 y^2$. ■

✔ **CHECK POINT 1** Find the greatest common factor of each list of terms:

  a. $18x^3$ and $15x^2$ **b.** $-20x^2$, $12x^4$, and $40x^3$
  c. $x^4 y$, $x^3 y^2$, and $x^2 y$.

3 Factor out the greatest common factor of a polynomial.

When we factor a monomial from a polynomial, we determine the greatest common factor of all terms in the polynomial. Sometimes there may not be a GCF other than 1. When a GCF other than 1 exists, we use the following procedure:

Factoring a Monomial from a Polynomial

1. Determine the greatest common factor of all terms in the polynomial.
2. Express each term as the product of the GCF and its other factor.
3. Use the distributive property to factor out the GCF.

EXAMPLE 2   Factoring Out the Greatest Common Factor

Factor: $5x^2 + 30$.

Solution The GCF of $5x^2$ and 30 is 5.

$$5x^2 + 30$$
$$= 5 \cdot x^2 + 5 \cdot 6 \qquad \text{Express each term as the product of the GCF and its other factor.}$$
$$= 5(x^2 + 6) \qquad \text{Factor out the GCF.}$$

Because factoring reverses the process of multiplication, all factoring results can be checked by multiplying.

$$5(x^2 + 6) = 5 \cdot x^2 + 5 \cdot 6 = 5x^2 + 30$$

The factoring is correct because multiplication gives us the original polynomial. ■

✔ **CHECK POINT 2** Factor: $6x^2 + 18$.

EXAMPLE 3   Factoring Out the Greatest Common Factor

Factor: $18x^3 + 27x^2$.

Solution We begin by determining the greatest common factor.

9 is the greatest integer that divides 18 and 27.

$$18x^3 \qquad \text{and} \qquad 27x^2$$

x^2 is the variable raised to the smallest exponent.

The GCF of the two terms in the polynomial is $9x^2$.

$$18x^3 + 27x^2$$

$$= 9x^2(2x) + 9x^2(3) \qquad \text{Express each term as the product of the GCF and its other factor.}$$

$$= 9x^2(2x + 3) \qquad \text{Factor out the GCF.}$$

We can check this factorization by multiplying $9x^2$ and $2x + 3$, obtaining the original polynomial as the answer. ∎

✔ **CHECK POINT 3** Factor: $25x^2 + 35x^3$.

Discover for Yourself

What happens if you factor out $3x^2$ rather than $9x^2$ from $18x^3 + 27x^2$? Although $3x^2$ is a common factor of the two terms, it is not the *greatest* common factor. Remove $3x^2$ from $18x^3 + 27x^2$ and describe what happens with the second factor. Now factor again. Make the final result look like the factorization in Example 3. What is the advantage of factoring out the greatest common factor rather than just a common factor?

EXAMPLE 4   Factoring Out the Greatest Common Factor

Factor: $16x^5 - 12x^4 + 4x^3$.

Solution   First, determine the greatest common factor.

> 4 is the greatest integer that divides 16, −12, and 4.

$$16x^5, \quad -12x^4, \quad \text{and} \quad 4x^3$$

> x^3 is the variable raised to the smallest exponent.

The GCF of the three terms of the polynomial is $4x^3$.

$$16x^5 - 12x^4 + 4x^3$$

$$= 4x^3 \cdot 4x^2 - 4x^3 \cdot 3x + 4x^3 \cdot 1 \qquad \text{Express each term as the product of the GCF and its other factor.}$$

$$= 4x^3(4x^2 - 3x + 1) \qquad \text{Factor out the GCF.}$$

> Don't leave out the 1.

∎

✔ **CHECK POINT 4** Factor: $15x^5 + 12x^4 - 27x^3$.

EXAMPLE 5   Factoring Out the Greatest Common Factor

Factor: $27x^2y^3 - 9xy^2 + 81xy$.

Solution   First, determine the greatest common factor.

> 9 is the greatest integer that divides 27, −9, and 81.

$$27x^2y^3, \quad -9xy^2, \quad \text{and} \quad 81xy$$

> The variables raised to the smallest exponents are x and y.

The GCF of the three terms of the polynomial is $9xy$.

$$27x^2y^3 - 9xy^2 + 81xy$$

$$= 9xy \cdot 3xy^2 - 9xy \cdot y + 9xy \cdot 9 \qquad \text{Express each term as the product of the GCF and its other factor.}$$

$$= 9xy(3xy^2 - y + 9) \qquad \text{Factor out the GCF.}$$

∎

✔ **CHECK POINT 5** Factor: $8x^3y^2 - 14x^2y + 2xy$.

4 Factor by grouping.

Factoring by Grouping

Up to now, we have factored a monomial from a polynomial. By contrast, in our next example, the greatest common factor of the polynomial is a binomial.

EXAMPLE 6   Factoring Out the Greatest Common Binomial Factor

Factor:

a. $x^2(x + 3) + 5(x + 3)$ **b.** $x(y + 1) - 2(y + 1)$.

Solution   Let's identify the common binomial factor in each part of the problem.

$x^2(x + 3)$ and $5(x + 3)$ $x(y + 1)$ and $-2(y + 1)$

The GCF, a binomial, is x + 3.   The GCF, a binomial, is y + 1.

We factor out this common binomial factor as follows.

a. $x^2(x + 3) + 5(x + 3)$

$= (x + 3)x^2 + (x + 3)5$ Express each term as the product of the GCF and its other factor, in that order. Hereafter, we omit this step.

$= (x + 3)(x^2 + 5)$ Factor out the GCF, x + 3.

b. $x(y + 1) - 2(y + 1)$

$= (y + 1)(x - 2)$ Factor out the GCF, y + 1. ∎

 CHECK POINT 6   Factor:

a. $x^2(x + 1) + 7(x + 1)$ **b.** $x(y + 4) - 7(y + 4)$.

Some polynomials have only a greatest common factor of 1. However, by a suitable grouping of the terms, it still may be possible to factor. This process, called **factoring by grouping**, is illustrated in Example 7.

EXAMPLE 7   Factoring by Grouping

Factor: $x^3 + 4x^2 + 3x + 12$.

Solution   There is no factor other than 1 common to all terms. However, we can group terms that have a common factor:

$$\boxed{x^3 + 4x^2} + \boxed{3x + 12}.$$

Common factor is x^2.   Common factor is 3.

We now factor the given polynomial as follows:

$x^3 + 4x^2 + 3x + 12$

$= (x^3 + 4x^2) + (3x + 12)$ Group terms with common factors.

$= x^2(x + 4) + 3(x + 4)$ Factor out the greatest common factor from the grouped terms. The remaining two terms have x + 4 as a common binomial factor.

$= (x + 4)(x^2 + 3)$ Factor out the GCF, x + 4.

Discover for Yourself

In Example 7, group the terms as follows:

$(x^3 + 3x) + (4x^2 + 12)$.

Factor out the greatest common factor from each group and complete the factoring process. Describe what happens. What can you conclude?

Thus, $x^3 + 4x^2 + 3x + 12 = (x + 4)(x^2 + 3)$. Check the factorization by multiplying the right side of the equation using the FOIL method. If the factorization is correct, you will obtain the original polynomial. ∎

✔ **CHECK POINT 7** Factor: $x^3 + 5x^2 + 2x + 10$.

Factoring by Grouping

1. Group terms that have a common monomial factor. There will usually be two groups. Sometimes the terms must be rearranged.
2. Factor out the common monomial factor from each group.
3. Factor out the remaining binomial factor (if one exists).

EXAMPLE 8 Factoring by Grouping

Factor: $xy + 5x - 4y - 20$.

Solution There is no factor other than 1 common to all terms. However, we can group terms that have a common factor:

$$\boxed{xy + 5x} \quad + \quad \boxed{-4y - 20}.$$

Common factor is x:
$xy + 5x = x(y + 5)$.

Use −4, rather than 4, as the common factor:
$-4y - 20 = -4(y + 5)$. In this way, the common binomial factor, $y + 5$, appears.

The voice balloons illustrate that it is sometimes necessary to factor out a negative number from a grouping to obtain a common binomial factor for the two groupings. We now factor the given polynomial as follows:

$$xy + 5x - 4y - 20$$
$$= x(y + 5) - 4(y + 5) \qquad \text{Factor x and −4, respectively, from each grouping.}$$
$$= (y + 5)(x - 4) \qquad \text{Factor out the GCF, } y + 5.$$

Thus, $xy + 5x - 4y - 20 = (y + 5)(x - 4)$. Using the commutative property of multiplication, the factorization can also be expressed as $(x - 4)(y + 5)$. Multiply these factors using the FOIL method to verify that, regardless of the order, these are the correct factors. ∎

✔ **CHECK POINT 8** Factor: $xy + 3x - 5y - 15$.

EXERCISE SET 6.1

Practice Exercises

In Exercises 1–6, find three factorizations for each monomial.

1. $8x^3$

2. $20x^4$

3. $-12x^5$

4. $-15x^6$

5. $36x^4$

6. $27x^5$

In Exercises 7–18, find the greatest common factor of each list of terms.

7. 4 and $8x$

8. 5 and $15x$

9. $12x^2$ and $8x$

10. $20x^2$ and $15x$

11. $-2x^4$ and $6x^3$

12. $-3x^4$ and $6x^3$

13. $9y^5, 18y^2,$ and $-3y$

14. $10y^5, 20y^2,$ and $-5y$

15. xy, xy^2, and xy^3

16. x^2y, $3x^3y$, and $6x^2$

17. $16x^5y^4$, $8x^6y^3$, and $20x^4y^5$

18. $18x^5y^4$, $6x^6y^3$, and $12x^4y^5$

In Exercises 19–54, factor each polynomial using the greatest common factor. If there is no common factor other than 1 and the polynomial cannot be factored, so state.

19. $5x + 5$

20. $7x + 7$

21. $3y - 3$

22. $6y - 6$

23. $8x + 16$

24. $3x + 12$

25. $25x - 10$

26. $14x - 7$

27. $x^2 + x$

28. $x^2 + 2x$

29. $18y^2 + 24$

30. $7y^2 + 21$

31. $36x^3 + 24x^2$

32. $6x^3 + 2x^2$

33. $25y^2 - 13y$

34. $30y^2 - 11y$

35. $9y^4 + 27y^6$

36. $10y^4 + 15y^6$

37. $8x^2 - 4x^4$

38. $12x^2 - 4x^4$

39. $12y^2 + 16y - 8$

40. $15y^2 - 3y + 9$

41. $9x^4 + 18x^3 + 6x^2$

42. $32x^4 + 2x^3 + 8x^2$

43. $100y^5 - 50y^3 + 100y^2$

44. $26y^5 - 13y^3 + 39y^2$

45. $10x - 20x^2 + 5x^3$

46. $6x - 4x^2 + 2x^3$

47. $11x^2 - 23$

48. $12x^2 - 25$

49. $6x^3y^2 + 9xy$

50. $4x^2y^3 + 6xy$

51. $30x^2y^3 - 10xy^2 + 20xy$

52. $27x^2y^3 - 18xy^2 + 45x^2y$

53. $32x^3y^2 - 24x^3y - 16x^2y$

54. $18x^3y^2 - 12x^3y - 24x^2y$

In Exercises 55–66, factor each polynomial using the greatest common binomial factor.

55. $x(x + 5) + 3(x + 5)$

56. $x(x + 7) + 10(x + 7)$

57. $x(x + 2) - 4(x + 2)$

58. $x(x + 3) - 8(x + 3)$

59. $x(y + 6) - 7(y + 6)$

60. $x(y + 9) - 11(y + 9)$

61. $3x(x + y) - (x + y)$

62. $7x(x + y) - (x + y)$

63. $4x(3x + 1) + 3x + 1$

64. $5x(2x + 1) + 2x + 1$

65. $7x^2(5x + 4) + 5x + 4$

66. $9x^2(7x + 2) + 7x + 2$

In Exercises 67–84, factor by grouping.

67. $x^2 + 2x + 4x + 8$

68. $x^2 + 3x + 5x + 15$

69. $x^2 + 3x - 5x - 15$

70. $x^2 + 7x - 4x - 28$

71. $x^3 - 2x^2 + 5x - 10$

72. $x^3 - 3x^2 + 4x - 12$

73. $x^3 - x^2 + 2x - 2$

74. $x^3 + 6x^2 - 2x - 12$

75. $xy + 5x + 9y + 45$

76. $xy + 6x + 2y + 12$

77. $xy - x + 5y - 5$

78. $xy - x + 7y - 7$

79. $3x^2 - 6xy + 5xy - 10y^2$

80. $10x^2 - 12xy + 35xy - 42y^2$

81. $3x^3 - 2x^2 - 6x + 4$

82. $4x^3 - x^2 - 12x + 3$

83. $x^2 - ax - bx + ab$

84. $x^2 + ax + bx + ab$

Application Exercises

85. The polynomial $8x^2 + 20x + 2488$ models the number, in thousands, of high school graduates in the United States x years after 1993.

   a. According to this model, how many students will graduate from U.S. high schools in 2003?

   b. Factor the polynomial.

   c. Use the factored form of the polynomial in part (b) to find the number of high school graduates in 2003. Do you get the same answer as you did in part (a)? If so, does this prove that your factorization is correct? Explain.

86. A rocket is fired upward from the ground with an initial velocity of 80 feet per second. The polynomial $80x - 16x^2$ describes the height of the rocket, in feet, after x seconds.
  a. Find the rocket's height after 4 seconds.
  b. Factor the polynomial.
  c. Use the factored form of the polynomial in part (b) to find the rocket's height after 4 seconds. Do you get the same answer as you did in part (a)? If so, does this prove that your factorization is correct? Explain.

In Exercises 87–88, write a polynomial for the length of each rectangle.

87.

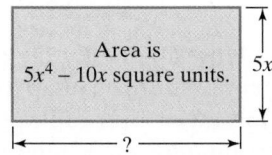

Area is $5x^4 - 10x$ square units. $5x$?

88.

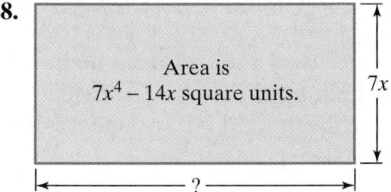

Area is $7x^4 - 14x$ square units. $7x$?

Writing in Mathematics

89. What is factoring?

90. What is a prime polynomial?

91. Explain how to find the greatest common factor of a list of terms. Give an example with your explanation.

92. Use an example and explain how to factor out the greatest common factor of a polynomial.

93. Suppose that a polynomial contains four terms and can be factored by grouping. Explain how to obtain the factorization.

94. Write a sentence that uses the word "factor" as a noun. Then write a sentence that uses the word "factor" as a verb.

Critical Thinking Exercises

95. Which one of the following is true?
  a. Because a monomial contains one term, it follows that a monomial can be factored in precisely one way.

  b. The GCF for $8x^3 - 16x^2$ is $8x$.
  c. The integers 10 and 31 have no GCF.
  d. $-4x^2 + 12x$ can be factored as $-4x(x - 3)$ or $4x(-x + 3)$.

In Exercises 96–97, factor each polynomial.

96. $2x + 4y + 8 + xy$

97. $x^{4n} + x^{2n} + x^{3n}$ (n is a natural number.)

In Exercises 98–99, write a polynomial that fits the given description. Do not use a polynomial that appears in this section or in the exercise set.

98. The polynomial has four terms and can be factored using a greatest common factor that has both a numerical co-efficient and a variable.

99. The polynomial has four terms and can be factored by grouping.

Technology Exercises

In Exercises 100–102, use a graphing utility to graph each side of the equation in the same viewing rectangle. Do the graphs coincide? If so, this means that the polynomial on the left side has been factored correctly. If not, factor the polynomial correctly and then use your graphing utility to verify the factorization.

100. $-3x - 6 = -3(x - 2)$

101. $x^2 - 2x + 5x - 10 = (x - 2)(x - 5)$

102. $x^2 + 2x + x + 2 = x(x + 2) + 1$

Review Exercises

103. Multiply: $(x + 7)(x + 10)$. (Section 5.3, Example 1)

104. Solve the system by graphing:

$$2x - y = -4$$
$$x - 3y = 3.$$

(Section 4.1, Example 2)

105. Write the point-slope form of a line passing through $(-7, 2)$ and $(-4, 5)$. Then use the point-slope equation to write the slope-intercept equation. (Section 3.5, Example 2)

▶ SECTION 6.2 *Factoring Trinomials Whose Leading Coefficient Is One*

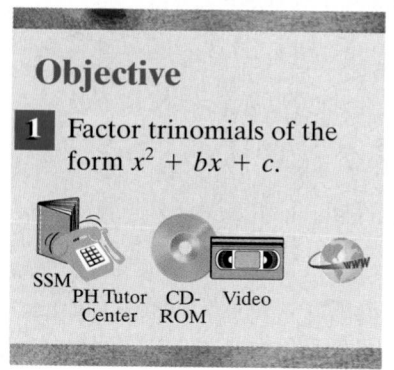

Objective

1 Factor trinomials of the form $x^2 + bx + c$.

SSM
PH Tutor CD- Video
Center ROM

Not afraid of heights and cutting-edge excitement? How about sky diving? Behind your exhilarating experience is the world of algebra. After you jump from the airplane, your height above the ground at every instant of your fall can be described by a formula involving a variable that is squared. At a height of approximately 2000 feet, you'll need to open your parachute. How can you determine when you must do so?

The answer to this critical question involves using the factoring technique presented in this section. In Section 6.6, in which applications are discussed, this technique is applied to models involving the height of any free-falling object—in this case, you.

1 Factor trinomials of the form $x^2 + bx + c$.

A Strategy for Factoring $x^2 + bx + c$

In Section 5.3, we used the FOIL method to multiply two binomials. The product was often a trinomial. The following are some examples.

Factored Form	**F O I L**		**Trinomial Form**

$$(x + 3)(x + 4) = x^2 + 4x + 3x + 12 = x^2 + 7x + 12$$
$$(x - 3)(x - 4) = x^2 - 4x - 3x + 12 = x^2 - 7x + 12$$
$$(x + 3)(x - 5) = x^2 - 5x + 3x - 15 = x^2 - 2x - 15$$

Observe that each trinomial is of the form $x^2 + bx + c$, where the coefficient of the squared term is 1. Our goal in this section is to start with the trinomial form and, assuming that it is factorable, return to the factored form.

Let's start by multiplying $(x + 3)(x + 4)$ using the FOIL method to obtain $x^2 + 7x + 12$.

$$(x + 3)(x + 4) = x^2 + 7x + 12$$

Now, we can make several important observations.

$$x^2 + 7x + 12 = (x + 3)(x + 4) \qquad x^2 + 7x + 12 = (x + 3)(x + 4) \qquad x^2 + 7x + 12 = (x + 3)(x + 4)$$

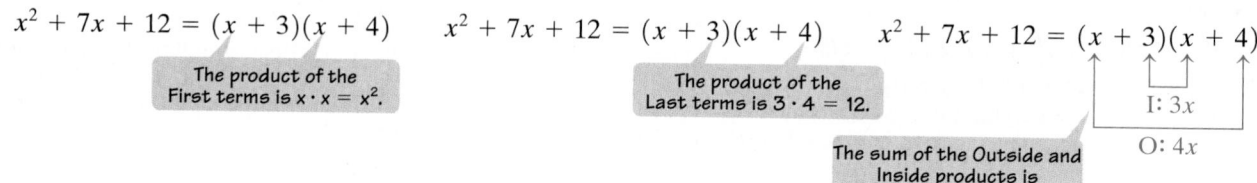

The product of the
First terms is x · x = x².

The product of the
Last terms is 3 · 4 = 12.

The sum of the Outside and
Inside products is
4x + 3x = 7x.

I: 3x

O: 4x

These observations provide us with a procedure for factoring $x^2 + bx + c$.

> **A Strategy for Factoring $x^2 + bx + c$**
>
> **1.** Enter x as the first term of each factor.
> $$(x \quad)(x \quad) = x^2 + bx + c$$
> **2.** List pairs of factors of the constant c.
> **3.** Try various combinations of these factors. Select the combination in which the sum of the *O*utside and *I*nside products is equal to bx.
>
>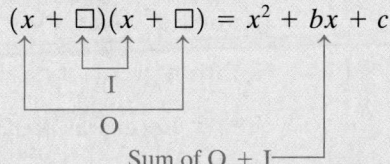
>
> **4.** Check your work by multiplying the factors using the FOIL method. You should obtain the original trinomial.
>
> If none of the possible combinations yield an *O*utside product and an *I*nside product whose sum is equal to bx, the trinomial cannot be factored using integers and is called **prime** over the set of integers.

EXAMPLE 1   Factoring a Trinomial in $x^2 + bx + c$ Form

Factor: $x^2 + 6x + 8$.

Solution

Step 1 Enter x as the first term of each factor.
$$x^2 + 6x + 8 = (x \quad)(x \quad)$$

To find the second term of each factor, we must find two integers whose product is 8 and whose sum is 6.

Step 2 List pairs of factors of the constant, 8.

Factors of 8	8, 1	4, 2	−8, −1	−4, −2

Step 3 Try various combinations of these factors. The correct factorization of $x^2 + 6x + 8$ is the one in which the sum of the *O*utside and *I*nside products is equal to $6x$. Here is a list of the possible factorizations.

Possible Factorizations of $x^2 + 6x + 8$	Sum of Outside and Inside Products (Should Equal $6x$)
$(x + 8)(x + 1)$	$x + 8x = 9x$
$(x + 4)(x + 2)$	$2x + 4x = 6x$
$(x - 8)(x - 1)$	$-x - 8x = -9x$
$(x - 4)(x - 2)$	$-2x - 4x = -6x$

This is the required middle term.

Thus, $x^2 + 6x + 8 = (x + 4)(x + 2)$.

Using Technology

A graphing utility can be used to check factorizations. For example, graph

$$y_1 = x^2 + 6x + 8$$
and
$$y_2 = (x + 4)(x + 2)$$

on the same screen, as shown below. The graphs are identical, so we can conclude that

$x^2 + 6x + 8$
$$= (x + 4)(x + 2).$$

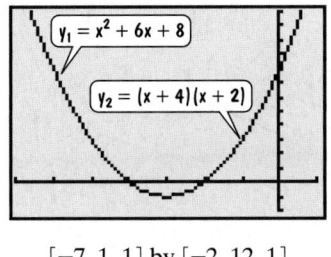

$[-7, 1, 1]$ by $[-2, 12, 1]$

Check this result by multiplying the right side using the FOIL method. You should obtain the original trinomial. Because of the commutative property, we can also say that

$$x^2 + 6x + 8 = (x + 2)(x + 4).$$ ■

✔ **CHECK POINT 1** Factor: $x^2 + 5x + 6$.

EXAMPLE 2 Factoring a Trinomial in $x^2 + bx + c$ Form

Factor: $x^2 - 5x + 6$.

Solution

Step 1 Enter x as the first term of each factor.

$$x^2 - 5x + 6 = (x \quad)(x \quad)$$

To find the second term of each factor, we must find two integers whose product is 6 and whose sum is −5.

Step 2 List pairs of factors of the constant, 6.

Factors of 6	6, 1	3, 2	−6, −1	−3, −2

Step 3 Try various combinations of these factors. The correct factorization of $x^2 - 5x + 6$ is the one in which the sum of the *Outside* and *Inside* products is equal to −5x. Here is a list of the possible factorizations.

Possible Factorizations of $x^2 - 5x + 6$	Sum of Outside and Inside Products (Should Equal −5x)	
$(x + 6)(x + 1)$	$x + 6x = 7x$	
$(x + 3)(x + 2)$	$2x + 3x = 5x$	
$(x - 6)(x - 1)$	$-x - 6x = -7x$	
$(x - 3)(x - 2)$	$-2x - 3x = -5x$	This is the required middle term.

Thus, $x^2 - 5x + 6 = (x - 3)(x - 2)$. Verify this result using the FOIL method. ■

Study Tip

To factor $x^2 + bx + c$ when c is positive, find two numbers with the same sign as the middle term.

$x^2 + 6x + 8 = (x + 2)(x + 4)$

Same signs

$x^2 - 5x + 6 = (x - 3)(x - 2)$

Same signs

In factoring a trinomial of the form $x^2 + bx + c$, you can speed things up by listing the factors of c and then finding their sums. We are interested in a sum of b. For example, in factoring $x^2 - 5x + 6$, we are interested in the factors of 6 whose sum is −5.

Factors of 6	6, 1	3, 2	−6, −1	−3, −2
Sum of Factors	7	5	−7	−5

This is the desired sum.

Thus, $x^2 - 5x + 6 = (x - 3)(x - 2)$.

✔ **CHECK POINT 2** Factor: $x^2 - 6x + 8$.

EXAMPLE 3   Factoring a Trinomial in $x^2 + bx + c$ Form

Factor: $x^2 + 2x - 35$.

Solution

Step 1 Enter x as the first term of each factor.

$$x^2 + 2x - 35 = (x \qquad)(x \qquad).$$

To find the second term of each factor, we must find two integers whose product is -35 and whose sum is 2.

Step 2 List pairs of factors of the constant, -35.

Factors of -35	$-35, 1$	$-7, 5$	$35, -1$	$7, -5$

Step 3 Try various combinations of these factors. We are looking for the pair of factors whose sum is 2.

Factors of -35	$-35, 1$	$-7, 5$	$35, -1$	$7, -5$
Sum of Factors	-34	-2	34	2

This is the desired sum.

Thus, $x^2 + 2x - 35 = (x + 7)(x - 5)$.

Step 4 Verify the factorization using the FOIL method.

$$(x + 7)(x - 5) = x^2 - 5x + 7x - 35 = x^2 + 2x - 35$$

Because the product of the factors is the original polynomial, the factorization is correct. ■

✔ **CHECK POINT 3**   Factor: $x^2 + 3x - 10$.

EXAMPLE 4   Factoring a Trinomial Whose Leading Coefficient Is One

Factor: $y^2 - 2y - 99$.

Solution

Step 1 Enter y as the first term of each factor.

$$y^2 - 2y - 99 = (y \qquad)(y \qquad)$$

To find the second term of each factor, we must find two integers whose product is -99 and whose sum is -2.

Step 2 List pairs of factors of the constant, −99.

Factors of −99	−99, 1	−11, 9	−33, 3	99, −1	11, −9	33, −3

Step 3 Try various combinations of these factors. We are interested in the pair of factors whose sum is −2.

Factors of −99	−99, 1	−11, 9	−33, 3	99, −1	11, −9	33, −3
Sum of Factors	−98	−2	−30	98	2	30

This is the desired sum.

Thus, $y^2 - 2y - 99 = (y - 11)(y + 9)$. Verify this result using the FOIL method. ∎

✔ **CHECK POINT 4** Factor: $y^2 - 6y - 27$.

EXAMPLE 5 Trying to Factor a Trinomial in $x^2 + bx + c$ Form

Factor: $x^2 + x - 5$.

Solution

Step 1 Enter x as the first term of each factor.

$$x^2 + x - 5 = (x \quad)(x \quad)$$

To find the second term of each factor, we must find two integers whose product is −5 and whose sum is 1.

Steps 2 and 3 List pairs of factors of the constant, −5, and try various combinations of these factors. We are interested in the pair of factors whose sum is 1.

Factors of −5	−5, 1	5, −1
Sum of Factors	−4	4

No pair gives the desired sum, 1.

Because neither pair has a sum of 1, $x^2 + x - 5$ cannot be factored using integers. This trinomial is prime. ∎

✔ **CHECK POINT 5** Factor: $x^2 + x - 7$.

EXAMPLE 6 Factoring a Trinomial in Two Variables

Factor: $x^2 - 5xy + 6y^2$.

Solution

Step 1 Enter x as the first term of each factor. Because the last term of the trinomial contains y^2, the second term of each factor must contain y.

$$x^2 - 5xy + 6y^2 = (x \quad ?y)(x \quad ?y)$$

The question marks indicate that we are looking for the coefficients of y in each factor. To find these coefficients, we must find two integers whose product is 6 and whose sum is -5.

Steps 2 and 3 List pairs of factors of the coefficient of the last term, 6, and try various combinations of these factors. We are interested in the pair of factors whose sum is -5.

Factors of 6	6, 1	3, 2	−6, −1	−3, −2
Sum of Factors	7	5	−7	−5

This is the desired sum.

Thus, $x^2 - 5xy + 6y^2 = (x - 3y)(x - 2y)$.

Step 4 Verify the factorization using the FOIL method.

$$(x - 3y)(x - 2y) = x^2 - 2xy - 3xy + 6y^2 = x^2 - 5xy + 6y^2$$

Because the product of the factors is the original polynomial, the factorization is correct. ∎

✔ **CHECK POINT 6**   Factor: $x^2 - 4xy + 3y^2$.

Some polynomials can be factored using more than one technique. **Always begin by trying to factor out the greatest common factor.** A polynomial is **factored completely** when it is written as the product of prime polynomials.

EXAMPLE 7   Factoring Completely

Factor: $3x^3 - 15x^2 - 42x$.

Solution   The GCF of the three terms of the polynomial is $3x$. We begin by factoring out $3x$. Then we factor the remaining trinomial by the methods of this section.

$$3x^3 - 15x^2 - 42x$$
$$= 3x(x^2 - 5x - 14) \qquad \text{Factor out the GCF.}$$
$$= 3x(x \quad)(x \quad) \qquad \text{Begin factoring } x^2 - 5x - 14. \text{ Find two integers whose product is } -14 \text{ and whose sum is } -5.$$
$$= 3x(x - 7)(x + 2) \qquad \text{The integers are } -7 \text{ and 2.}$$

Thus,

$$3x^3 - 15x - 42x = 3x(x - 7)(x + 2).$$

Be sure to include the GCF in the factorization.

How can we check this factorization? We will multiply the binomials using the FOIL method. Then use the distributive property and multiply each term of this product by $3x$. If the factorization is correct, we should obtain the original polynomial.

$$3x(x - 7)(x + 2) = 3x(x^2 + 2x - 7x - 14) = 3x(x^2 - 5x - 14) = 3x^3 - 15x^2 - 42x$$

Use the FOIL method on
$(x - 7)(x + 2)$.

This is the original
polynomial.

The factorization is correct.

✔ **CHECK POINT 7** Factor: $2x^3 + 6x^2 - 56x$.

EXERCISE SET 6.2

Practice Exercises

In Exercises 1–42, factor each trinomial, or state that the trinomial is prime. Check each factorization using FOIL multiplication.

1. $x^2 + 6x + 5$

2. $x^2 + 8x + 7$

3. $x^2 + 8x + 15$

4. $x^2 + 8x + 12$

5. $x^2 + 12x + 11$

6. $x^2 + 15x + 14$

7. $x^2 - 8x + 15$

8. $x^2 - 14x + 45$

9. $x^2 - 14x + 49$

10. $x^2 - 10x + 25$

11. $y^2 - 15y + 36$

12. $y^2 - 10y + 9$

13. $x^2 + 3x - 10$

14. $x^2 + 3x - 28$

15. $y^2 + 10y - 39$

16. $y^2 + 5y - 24$

17. $x^2 - 2x - 15$

18. $x^2 - 4x - 5$

19. $x^2 - 2x - 8$

20. $x^2 - 5x - 6$

21. $x^2 + 4x + 12$

22. $x^2 + 4x + 5$

23. $y^2 - 16y + 48$

24. $y^2 - 10y + 21$

25. $x^2 - 3x + 6$

26. $x^2 + 4x - 10$

27. $w^2 - 30w - 64$

28. $w^2 + 12w - 64$

29. $y^2 - 18y + 65$

30. $y^2 - 22y + 72$

31. $r^2 + 12r + 27$

32. $r^2 - 15r - 16$

33. $y^2 - 7y + 5$

34. $y^2 - 15y + 5$

35. $x^2 + 7xy + 6y^2$

36. $x^2 + 6xy + 8y^2$

37. $x^2 - 8xy + 15y^2$

38. $x^2 - 9xy + 14y^2$

39. $x^2 - 3xy - 18y^2$

40. $x^2 - xy - 30y^2$

41. $a^2 - 18ab + 45b^2$

42. $a^2 - 18ab + 80b^2$

In Exercises 43–62, factor completely.

43. $3x^2 + 15x + 18$

44. $3x^2 + 21x + 36$

45. $4y^2 - 4y - 8$

46. $3y^2 + 3y - 18$

47. $10x^2 - 40x - 600$

48. $2x^2 + 10x - 48$

49. $3x^2 - 33x + 54$

50. $2x^2 - 14x + 24$

51. $2r^3 + 6r^2 + 4r$

52. $2r^3 + 8r^2 + 6r$

53. $4x^3 + 12x^2 - 72x$

54. $3x^3 - 15x^2 + 18x$

55. $2r^3 + 8r^2 - 64r$

56. $3r^3 - 9r^2 - 54r$

57. $y^4 + 2y^3 - 80y^2$

58. $y^4 - 12y^3 + 35y^2$

59. $x^4 - 3x^3 - 10x^2$

60. $x^4 - 22x^3 + 120x^2$

61. $2w^4 - 26w^3 - 96w^2$

62. $3w^4 + 54w^3 + 135w^2$

Application Exercises

63. You dive directly upward from a board that is 32 feet high. After t seconds, your height above the water is described by the polynomial $-16t^2 + 16t + 32$. Factor the polynomial completely. Begin by factoring -16 from each term.

64. If x represents a positive integer, factor $x^3 + 3x^2 + 2x$ to show that the trinomial represents the product of three consecutive integers.

Then use the dimensions given on the box and show that its volume is equivalent to the factorization that you obtain.

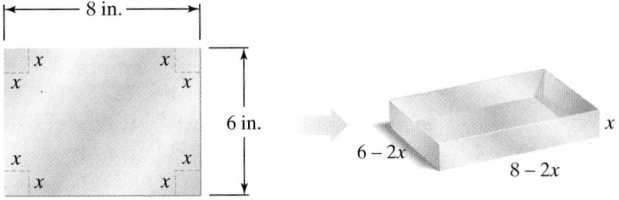

Writing in Mathematics

65. Explain how to factor $x^2 + 8x + 15$.

66. Give two helpful suggestions for factoring $x^2 - 5x + 6$.

67. In factoring $x^2 + bx + c$, describe how the last terms in each factor are related to b and c.

68. Without actually factoring and without multiplying the given factors, explain why the following factorization is not correct:
$$x^2 + 46x + 513 = (x - 27)(x - 19).$$

Critical Thinking Exercises

69. Which one of the following is true?
   a. A factor of $x^2 + x + 20$ is $x + 5$.
   b. A trinomial can never have two identical factors.
   c. A factor of $y^2 + 5y - 24$ is $y - 3$.
   d. $x^2 + 4 = (x + 2)(x + 2)$

In Exercises 70–71, find all positive integers b so that the trinomial can be factored.

70. $x^2 + bx + 15$

71. $x^2 + 4x + b$

72. Factor: $x^{2n} + 20x^n + 99$.

73. A box with no top is to be made from an 8-inch by 6-inch piece of metal by cutting identical squares from each corner and turning up the sides (see the figure). The volume of the box is modeled by the polynomial $4x^3 - 28x^2 + 48x$. Factor the polynomial completely.

Technology Exercises

In Exercises 74–77, use a graphing utility to graph each side of the equation in the same viewing rectangle. Do the graphs coincide? If so, this means that the polynomial on the left side has been correctly factored. If not, factor the trinomial correctly and then use your graphing utility to verify the factorization.

74. $x^2 - 5x + 6 = (x - 2)(x - 3)$

75. $2x^2 + 2x - 12 = 2(x - 3)(x + 2)$

76. $x^2 - 2x + 1 = (x + 1)(x - 1)$

77. $2x^2 + 8x + 6 = (x + 3)(x + 1)$

Review Exercises

78. Multiply: $(2x + 3)(x - 2)$. (Section 5.3, Example 2)

79. Multiply: $(3x + 4)(3x + 1)$. (Section 5.3, Example 2)

80. Solve: $4(x - 2) = 3x + 5$. (Section 2.3, Example 2)

▶ SECTION 6.3 *Factoring Trinomials Whose Leading Coefficient Is Not One*

Objectives

1 Factor trinomials by trial and error.

2 Factor trinomials by grouping.

SSM
PH Tutor   CD-   Video
Center   ROM

Georges Tooker, American, born 1920. "Farewell" 1966, egg tempera on gessoed masonite, 61×60.1 cm. p. 967.76. Hood Museum of Art. Darmouth College, Hanover, New Hampshire; gift of Pennington Haile, Class of 1924.

The special significance of the number 1 is reflected in our language. "One," "an," and "a" mean the same thing. The words "unit," "unity," "union," "unique," and "universal" are derived from the Latin word for "one." For the ancient Greeks, 1 was the indivisible unit from which all other numbers arose.

The Greeks' philosophy of 1 applies to our work in this section. Factoring trinomials whose leading coefficient is 1 is the basic technique from which other methods of factoring $ax^2 + bx + c$, where a is not equal to 1, follow.

Factoring by the Trial-and-Error Method

1 Factor trinomials by trial and error.

How do we factor a trinomial such as $3x^2 - 20x + 28$? Notice that the leading coefficient is 3. We must find two binomials whose product is $3x^2 - 20x + 28$. The product of the *F*irst terms must be $3x^2$:

$$(3x \quad)(x \quad).$$

From this point on, the factoring strategy is exactly the same as the one we use to factor trinomials whose leading coefficient is 1.

A Strategy for Factoring $ax^2 + bx + c$

Assume, for the moment, that there is no greatest common factor.

1. Find two *F*irst terms whose product is ax^2:

$$(\Box x + \quad)(\Box x + \quad) = ax^2 + bx + c.$$

2. Find two *L*ast terms whose product is c:

$$(\Box x + \Box)(\Box x + \Box) = ax^2 + bx + c.$$

3. By trial and error, perform steps 1 and 2 until the sum of the *O*utside product and *I*nside product is bx:

$$(\Box x + \Box)(\Box x + \Box) = ax^2 + bx + c.$$
$$\underset{\text{(sum of O + I)}}{\underset{\text{O}}{\text{I}}}$$

If no such combinations exist, the polynomial is prime.

EXAMPLE 1 Factoring a Trinomial Whose Leading Coefficient Is Not One

Factor: $3x^2 - 20x + 28$.

Solution

Step 1 Find two *First* terms whose product is $3x^2$.

$$3x^2 - 20x + 28 = (3x \quad)(x \quad)$$

Step 2 Find two *Last* terms whose product is 28. The number 28 has pairs of factors that are either both positive or both negative. Because the middle term, $-20x$, is negative, both factors must be negative. The negative factorizations of 28 are $-1(-28), -2(-14)$, and $-4(-7)$.

Study Tip

With practice, you will find that it is not necessary to list every possible factor of the trinomial. As you practice factoring, you will be able to narrow down the list of possible factors to just a few. When it comes to factoring, practice makes perfect. (Sorry about the cliché.)

Step 3 Try various combinations of these factors. The correct factorization of $3x^2 - 20x + 28$ is the one in which the sum of the *O*utside and *I*nside products is equal to $-20x$. Here is a list of the possible factorizations.

Possible Factorizations of $3x^2 - 20x + 28$	Sum of Outside and Inside Products (Should Equal $-20x$)
$(3x - 1)(x - 28)$	$-84x - x = -85x$
$(3x - 28)(x - 1)$	$-3x - 28x = -31x$
$(3x - 2)(x - 14)$	$-42x - 2x = -44x$
$(3x - 14)(x - 2)$	$-6x - 14x = -20x$ ← This is the required middle term.
$(3x - 4)(x - 7)$	$-21x - 4x = -25x$
$(3x - 7)(x - 4)$	$-12x - 7x = -19x$

Thus,

$$3x^2 - 20x + 28 = (3x - 14)(x - 2) \quad \text{or} \quad (x - 2)(3x - 14).$$

Show that this factorization is correct by multiplying the factors using the FOIL method. You should obtain the original trinomial. ∎

✔ **CHECK POINT 1** Factor: $5x^2 - 14x + 8$.

EXAMPLE 2 Factoring a Trinomial Whose Leading Coefficient Is Not One

Factor: $8x^2 - 10x - 3$.

Solution

Step 1 Find two *F*irst terms whose product is $8x^2$.

$$8x^2 - 10x - 3 \overset{?}{=} (8x \quad)(x \quad)$$
$$8x^2 - 10x - 3 \overset{?}{=} (4x \quad)(2x \quad)$$

Step 2 Find two *L*ast terms whose product is -3. The possible factorizations are $1(-3)$ and $-1(3)$.

Step 3 Try various combinations of these factors. The correct factorization of $8x^2 - 10x - 3$ is the one in which the sum of the *O*utside and *I*nside products is equal to $-10x$. Here is a list of the possible factorizations.

Possible Factorizations of $8x^2 - 10x - 3$	Sum of Outside and Inside Products (Should Equal $-10x$)
$(8x + 1)(x - 3)$	$-24x + x = -23x$
$(8x - 3)(x + 1)$	$8x - 3x = 5x$
$(8x - 1)(x + 3)$	$24x - x = 23x$
$(8x + 3)(x - 1)$	$-8x + 3x = -5x$
$(4x + 1)(2x - 3)$	$-12x + 2x = -10x$ ← This is the required middle term.
$(4x - 3)(2x + 1)$	$4x - 6x = -2x$
$(4x - 1)(2x + 3)$	$12x - 2x = 10x$
$(4x + 3)(2x - 1)$	$-4x + 6x = 2x$

Thus,

$$8x^2 - 10x - 3 = (4x + 1)(2x - 3) \quad \text{or} \quad (2x - 3)(4x + 1).$$

Use FOIL multiplication to check either of these factorizations. ∎

Study Tip

Here are some suggestions for reducing the list of possible factors for $ax^2 + bx + c$:

1. If b is relatively small, avoid the larger factors of a.

2. If c is positive, the signs in both binomial factors must match the sign of b.

3. If the trinomial has no common factor, no binomial factor can have a common factor.

4. Reversing the signs in the binomial factors reverses the sign of bx, the middle term.

✔ **CHECK POINT 2** Factor: $6x^2 + 19x - 7$.

EXAMPLE 3 Factoring a Trinomial in Two Variables

Factor: $2x^2 - 7xy + 3y^2$.

Solution

Step 1 Find two First terms whose product is $2x^2$.

$$2x^2 - 7xy + 3y^2 = (2x \quad)(x \quad)$$

Step 2 Find two Last terms whose product is $3y^2$. The possible factorizations are $(y)(3y)$ and $(-y)(-3y)$.

Step 3 Try various combinations of these factors. The correct factorization of $2x^2 - 7xy + 3y^2$ is the one in which the sum of the Outside and Inside products is equal to $-7xy$. Here is a list of possible factorizations.

Possible Factorizations of $2x^2 - 7xy + 3y^2$	Sum of Outside and Inside Products (Should Equal $-7xy$)	
$(2x + 3y)(x + y)$	$2xy + 3xy = 5xy$	
$(2x + y)(x + 3y)$	$6xy + xy = 7xy$	
$(2x - 3y)(x - y)$	$-2xy - 3xy = -5xy$	
$(2x - y)(x - 3y)$	$-6xy - xy = -7xy$	This is the required middle term.

Thus,

$$2x^2 - 7xy + 3y^2 = (2x - y)(x - 3y) \quad \text{or} \quad (x - 3y)(2x - y).$$

Use FOIL multiplication to check either of these factorizations. ■

✔ **CHECK POINT 3** Factor: $3x^2 - 13xy + 4y^2$.

---

2 Factor trinomials by grouping.

Factoring by the Grouping Method

A second method for factoring $ax^2 + bx + c$, $a \neq 0$, is called the **grouping method**. The method involves both trial and error, as well as grouping. The trial and error in factoring $ax^2 + bx + c$ depends on finding two numbers, p and q, for which $p + q = b$. Then we factor $ax^2 + px + qx + c$ using grouping.

Let's see how this works by looking at our factorization in Example 2:

$$8x^2 - 10x - 3 = (2x - 3)(4x + 1).$$

If we multiply using FOIL on the right, we obtain:

$$(2x - 3)(4x + 1) = 8x^2 + 2x - 12x - 3.$$

In this case, the desired numbers, p and q, are $p = 2$ and $q = -12$. Compare these numbers to ac and b in the given polynomial:

$$8x^2 - 10x - 3. \quad ac = 8(-3) = -24$$

$$a = 8 \quad b = -10 \quad c = -3$$

Can you see that p and q, 2 and -12, are factors of ac, or -24? Furthermore, p and q have a sum of b, namely -10. By expressing the middle term, $-10x$, in terms of p and q, we can factor by grouping as follows:

$$8x^2 - 10x - 3$$
$$= 8x^2 + (2x - 12x) - 3 \qquad \text{Rewrite } -10x \text{ as } 2x - 12x.$$
$$= (8x^2 + 2x) + (-12x - 3) \qquad \text{Group terms.}$$
$$= 2x(4x + 1) - 3(4x + 1) \qquad \text{Factor from each group.}$$
$$= (4x + 1)(2x - 3) \qquad \text{Factor out the common binomial factor.}$$

As we obtained in Example 2,
$$8x^2 - 10x - 3 = (4x + 1)(2x - 3).$$

Generalizing from this example, here's how to factor a trinomial by grouping:

> **Factoring $ax^2 + bx + c$ Using Grouping ($a \neq 1$)**
>
> **1.** Multiply the leading coefficient, a, and the constant, c.
> **2.** Find the factors of ac whose sum is b.
> **3.** Rewrite the middle term, bx, as a sum or difference using the factors from step 2.
> **4.** Factor by grouping.

EXAMPLE 4   Factoring by Grouping

Factor by grouping: $2x^2 - x - 6$.

Solution   The trinomial is of the form $ax^2 + bx + c$.

$$2x^2 - x - 6$$

$a = 2$ $b = -1$ $c = -6$

Step 1 Multiply the leading coefficient, a, and the constant, c. Using $a = 2$ and $c = -6$,
$$ac = 2(-6) = -12.$$

Step 2 Find the factors of ac whose sum is b. We want the factors of -12 whose sum is b, or -1. The factors of -12 whose sum is -1 are -4 and 3.

Step 3 Rewrite the middle term, $-x$, as a sum or difference using the factors from step 2, -4 and 3.

$$2x^2 - x - 6 = 2x^2 - 4x + 3x - 6$$

Step 4 Factor by grouping.
$$= (2x^2 - 4x) + (3x - 6) \qquad \text{Group terms.}$$
$$= 2x(x - 2) + 3(x - 2) \qquad \text{Factor from each group.}$$
$$= (x - 2)(2x + 3) \qquad \text{Factor out the common binomial factor.}$$

Thus,
$$2x^2 - x - 6 = (x - 2)(2x + 3) \quad \text{or} \quad (2x + 3)(x - 2). \qquad \blacksquare$$

Discover for Yourself

In step 2 we discovered that the desired numbers were -4 and 3, and we wrote $-x$ as $-4x + 3x$. What happens if we write $-x$ as $3x - 4x$? Use factoring by grouping on

$$2x^2 - x - 6$$
$$= 2x^2 + 3x - 4x - 6.$$

Describe what happens.

✔ **CHECK POINT 4**   Factor by grouping: $3x^2 - x - 10$.

EXAMPLE 5 Factoring by Grouping

Factor by grouping: $8x^2 - 22x + 5$.

Solution The trinomial is of the form $ax^2 + bx + c$.

$$8x^2 - 22x + 5$$

$a = 8 \qquad b = -22 \qquad c = 5$

Step 1 Multiply the leading coefficient, a, and the constant, c. Using $a = 8$ and $b = 5$, $ac = 8 \cdot 5 = 40$.

Step 2 Find the factors of ac whose sum is b. We want the factors of 40 whose sum is b, or -22. The factors of 40 whose sum is -22 are -2 and -20.

Step 3 Rewrite the middle term, $-22x$, as a sum or difference using the factors from step 2, -2 and -20.

$$8x^2 - 22x + 5 = 8x^2 - 2x - 20x + 5$$

Step 4 Factor by grouping.
$$= (8x^2 - 2x) + (-20x + 5) \quad \text{Group terms.}$$
$$= 2x(4x - 1) - 5(4x - 1) \quad \text{Factor from each group.}$$
$$= (4x - 1)(2x - 5) \quad \text{Factor out the common binomial factor.}$$

Thus,

$$8x^2 - 22x + 5 = (4x - 1)(2x - 5) \quad \text{or} \quad (2x - 5)(4x - 1). \quad \blacksquare$$

✔ **CHECK POINT 5** Factor by grouping: $8x^2 - 10x + 3$.

Factoring Completely

If each term of a trinomial has a common factor, always begin your work by factoring out the greatest common factor. After doing this, you should attempt to factor the remaining trinomial by one of the methods presented in this section.

EXAMPLE 6 Factoring Completely

Factor completely: $15y^4 + 26y^3 + 7y^2$.

Solution We will first factor out a common monomial factor from the polynomial and then factor the resulting trinomial by the methods of this section. The GCF of the three terms is y^2.

$$15y^4 + 26y^3 + 7y^2 = y^2(15y^2 + 26y + 7) \quad \text{Factor out the GCF.}$$
$$= y^2(5y + 7)(3y + 1) \quad \text{Factor } 15y^2 + 26y + 7 \text{ using trial and error or grouping.}$$

Thus,

$$15y^4 + 26y^3 + 7y^2 = y^2(5y + 7)(3y + 1) \quad \text{or} \quad y^2(3y + 1)(5y + 7). \quad \blacksquare$$

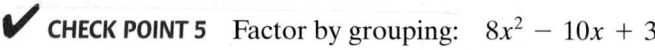
Be sure to include the GCF, y^2, in the factorization.

✔ **CHECK POINT 6** Factor completely: $5y^4 + 13y^3 + 6y^2$.

EXERCISE SET 6.3

Practice Exercises

In Exercises 1–58, use the method of your choice to factor each trinomial, or state that the trinomial is prime. Check each factorization using FOIL multiplication.

1. $2x^2 + 7x + 3$

2. $3x^2 + 7x + 2$

3. $3x^2 + 8x + 4$

4. $2x^2 + 11x + 5$

5. $2x^2 + 13x + 20$

6. $2x^2 + 17x + 35$

7. $5y^2 - 8y + 3$

8. $5y^2 - 13y + 6$

9. $3y^2 - y - 2$

10. $3y^2 - 2y - 5$

11. $3x^2 + x - 10$

12. $2x^2 + 5x - 3$

13. $3x^2 - 22x + 7$

14. $3x^2 - 10x + 7$

15. $5y^2 - 16y + 3$

16. $5y^2 - 8y + 3$

17. $3x^2 - 17x + 10$

18. $3x^2 - 25x - 28$

19. $6w^2 - 11w + 4$

20. $6w^2 - 17w + 12$

21. $8x^2 + 33x + 4$

22. $7x^2 + 43x + 6$

23. $5x^2 + 33x - 14$

24. $3x^2 + 22x - 16$

25. $14y^2 + 15y - 9$

26. $6y^2 + 7y - 24$

27. $6x^2 - 7x + 3$

28. $9x^2 + 3x + 2$

29. $25z^2 - 30z + 9$

30. $9z^2 + 12z + 4$

31. $15y^2 - y - 2$

32. $15y^2 + 13y - 2$

33. $5x^2 + 2x + 9$

34. $3x^2 - 5x + 1$

35. $10y^2 + 43y - 9$

36. $16y^2 - 46y + 15$

37. $8x^2 - 2x - 1$

38. $8x^2 - 22x + 5$

39. $9y^2 - 9y + 2$

40. $9y^2 + 5y - 4$

41. $20x^2 + 27x - 8$

42. $15x^2 - 19x + 6$

43. $2x^2 + 3xy + y^2$

44. $3x^2 + 4xy + y^2$

45. $3x^2 + 5xy + 2y^2$

46. $3x^2 + 11xy + 6y^2$

47. $2x^2 - 9xy + 9y^2$

48. $3x^2 + 5xy - 2y^2$

49. $6x^2 - 5xy - 6y^2$

50. $6x^2 - 7xy - 5y^2$

51. $15x^2 + 11xy - 14y^2$

52. $15x^2 - 31xy + 10y^2$

53. $2a^2 + 7ab + 5b^2$

54. $2a^2 + 5ab + 2b^2$

55. $15a^2 - ab - 6b^2$

56. $3a^2 - ab - 14b^2$

57. $12x^2 - 25xy + 12y^2$

58. $12x^2 + 7xy - 12y^2$

In Exercises 59–86, factor completely.

59. $4x^2 + 26x + 30$

60. $4x^2 - 18x - 10$

61. $9x^2 - 6x - 24$

62. $12x^2 - 33x + 21$

63. $4y^2 + 2y - 30$

64. $36y^2 + 6y - 12$

65. $9y^2 + 33y - 60$

66. $16y^2 - 16y - 12$

67. $3x^3 + 4x^2 + x$

68. $3x^3 + 14x^2 + 8x$

69. $2x^3 - 3x^2 - 5x$

70. $6x^3 + 4x^2 - 10x$

71. $9y^3 - 39y^2 + 12y$

72. $10y^3 + 12y^2 + 2y$

73. $60z^3 + 40z^2 + 5z$

74. $80z^3 + 80z^2 - 60z$

75. $15x^4 - 39x^3 + 18x^2$

76. $24x^4 + 10x^3 - 4x^2$

77. $10x^5 - 17x^4 + 3x^3$

78. $15x^5 - 2x^4 - x^3$

79. $6x^2 - 3xy - 18y^2$

80. $4x^2 + 14xy + 10y^2$

81. $12x^2 + 10xy - 8y^2$

82. $24x^2 + 3xy - 27y^2$

83. $8x^2y + 34xy - 84y$

84. $6x^2y - 2xy - 60y$

85. $12a^2b - 46ab^2 + 14b^3$

86. $12a^2b - 34ab^2 + 14b^3$

Application Exercises

It is possible to construct geometric models for factorizations so that you can see the factoring. This idea is developed in Exercises 87–88.

87. Consider the following figure.

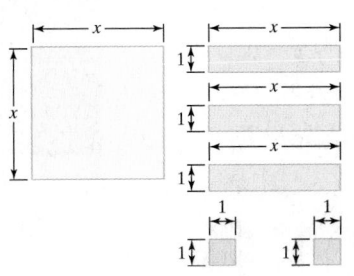

(a)

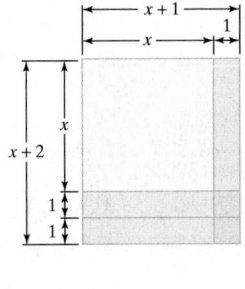

(b)

a. Write a trinomial that expresses the sum of the areas of the six rectangular pieces shown in figure (a).

b. Express the area of the large rectangle in figure (b) as the product of two binomials.

c. Are the pieces in figures (a) and (b) the same? Set the expressions that you wrote in parts (a) and (b) equal to each other. What factorization is illustrated?

88. Copy the figure and cut out the six pieces. Use the pieces to create a geometric model for the factorization

$$2x^2 + 3x + 1 = (2x + 1)(x + 1)$$

by forming a large rectangle using all the pieces.

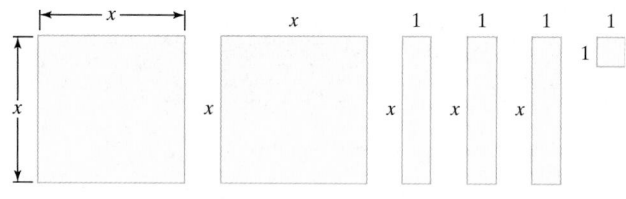

Writing in Mathematics

89. Explain how to factor $2x^2 - x - 1$.

90. Why is it a good idea to factor out the GCF first and then use other methods of factoring? Use $3x^2 - 18x + 15$ as an example. Discuss what happens if one first uses trial and error FOIL rather than first factoring out the GCF.

91. In factoring $3x^2 - 10x - 8$, a student lists $(3x - 2)(x + 4)$ as a possible factorization. Use FOIL multiplication to determine if this factorization is correct. If it is not correct, describe how the correct factorization can quickly be obtained using these factors.

92. Explain why $2x - 10$ cannot be one of the factors in the correct factorization of $6x^2 - 19x + 10$.

Critical Thinking Exercises

93. Which one of the following is true?

a. Once a GCF is factored from $18y^2 - 6y + 6$, the remaining trinomial factor is prime.

b. A factor of $12x^2 - 13x + 3$ is $4x + 3$.

c. A factor of $4y^2 - 11y - 3$ is $y + 3$.

d. The trinomial $3x^2 + 2x + 1$ has relatively small coefficients and therefore can be factored.

In Exercises 94–95, find all integers b so that the trinomial can be factored.

94. $3x^2 + bx + 2$

95. $2x^2 + bx + 3$

96. Factor: $3x^{10} - 4x^5 - 15$.

97. Factor: $2x^{2n} - 7x^n - 4$.

Review Exercises

In Exercises 98–100, perform the indicated operations.

98. $(9x + 10)(9x - 10)$ (Section 5.3, Example 4)

99. $(4x + 5y)^2$ (Section 5.3, Example 5)

100. $(x + 2)(x^2 - 2x + 4)$ (Section 5.2, Example 7)

▶ **SECTION 6.4   *Factoring Special Forms***

Objectives

1 Factor the difference of two squares.

2 Factor perfect square trinomials.

3 Factor the sum or difference of two cubes.

Do you enjoy solving puzzles? The process is a natural way to develop problem-solving skills that are important to every area of our lives. Engaging in problem solving for sheer pleasure releases chemicals in the brain that enhance our feeling of well-being. Perhaps this is why puzzles date back 12,000 years.

In this section, we develop factoring techniques by reversing the formulas for special products discussed in Chapter 5. These factorizations can be visualized by fitting pieces of a puzzle together to form rectangles.

1 Factor the difference of two squares.

Factoring the Difference of Two Squares

A method for factoring the difference of two squares is obtained by reversing the special product for the sum and difference of two terms.

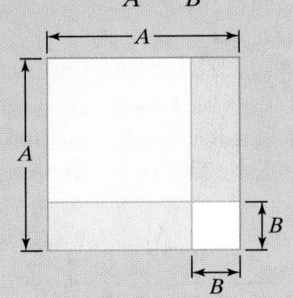
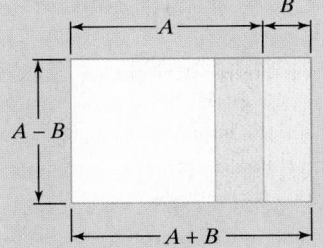
> **The Difference of Two Squares**
>
> If A and B are real numbers, variables, or algebraic expressions, then
> $$A^2 - B^2 = (A + B)(A - B).$$
>
> In words: The difference of the squares of two terms is factored as the product of the sum and the difference of those terms.

EXAMPLE 1 Factoring the Difference of Two Squares

Factor: **a.** $x^2 - 4$ **b.** $81x^2 - 49$.

Solution We must express each term as the square of some monomial. Then we use the formula for factoring $A^2 - B^2$.

a. $x^2 - 4 = x^2 - 2^2 = (x + 2)(x - 2)$
$$A^2 - B^2 = (A + B)(A - B)$$

b. $81x^2 - 49 = (9x)^2 - 7^2 = (9x + 7)(9x - 7)$ ∎

✔ **CHECK POINT 1** Factor:

a. $x^2 - 81$ **b.** $36x^2 - 25$.

Can $x^2 - 5$ be factored using integers and the formula for factoring $A^2 - B^2$? No. The number 5 is not the square of an integer. Thus, $x^2 - 5$ is prime over the set of integers.

EXAMPLE 2 Factoring the Difference of Two Squares

Factor: **a.** $9 - 16x^{10}$ **b.** $25x^2 - 4y^2$.

Solution Begin by expressing each term as the square of some monomial. Then use the formula for factoring $A^2 - B^2$.

a. $9 - 16x^{10} = 3^2 - (4x^5)^2 = (3 + 4x^5)(3 - 4x^5)$
$$A^2 - B^2 = (A + B)(A - B)$$

b. $25x^2 - 4y^2 = (5x)^2 - (2y)^2 = (5x + 2y)(5x - 2y)$ ∎

✔ **CHECK POINT 2** Factor:

a. $25 - 4x^{10}$ **b.** $100x^2 - 9y^2$.

When factoring, always check first for common factors. If there are common factors, factor out the GCF and then factor the resulting polynomial.

EXAMPLE 3   Factoring Out the GCF and Then Factoring the Difference of Two Squares

Factor: **a.** $12x^3 - 3x$ **b.** $80 - 125x^2$.

Solution

a. $12x^3 - 3x = 3x(4x^2 - 1) = 3x[(2x)^2 - 1^2] = 3x(2x + 1)(2x - 1)$

Factor out the GCF. $A^2 \quad - \quad B^2 \quad = \quad (A \quad + \quad B) \quad (A \quad - \quad B)$

b. $80 - 125x^2 = 5(16 - 25x^2) = 5[4^2 - (5x)^2] = 5(4 + 5x)(4 - 5x)$ ∎

✔ **CHECK POINT 3**   Factor:

a. $18x^3 - 2x$ **b.** $72 - 18x^2$.

We have seen that a polynomial is factored completely when it is written as the product of prime polynomials. To be sure that you have factored completely, check to see whether the factors can be factored.

EXAMPLE 4   A Repeated Factorization

Factor completely: $x^4 - 81$.

Solution

$x^4 - 81 = (x^2)^2 - 9^2$ Express as the difference of two squares.

$= (x^2 + 9)(x^2 - 9)$ The factors are the sum and difference of the squared terms.

$= (x^2 + 9)(x^2 - 3^2)$ The factor $x^2 - 9$ is the difference of two squares and can be factored.

$= (x^2 + 9)(x + 3)(x - 3)$ The factors of $x^2 - 9$ are the sum and difference of the expressions being squared.   ∎

Are you tempted to further factor $x^2 + 9$, the sum of two squares, in Example 4? Resist the temptation! **The sum of two squares, $A^2 + B^2$, with no common factor other than 1 is a prime polynomial**.

✔ **CHECK POINT 4**   Factor completely: $81x^4 - 16$.

Study Tip

Factoring $x^4 - 81$ as

$$(x^2 + 9)(x^2 - 9)$$

is not a complete factorization. The second factor, $x^2 - 9$, is itself a difference of two squares and can be factored.

2 Factor perfect square trinomials.

Factoring Perfect Square Trinomials

Our next factoring technique is obtained by reversing the special products for squaring binomials. The trinomials that are factored using this technique are called **perfect square trinomials**.

Visualizing the Factoring for a Perfect Square Trinomial

Area:

$$(A + B)^2$$

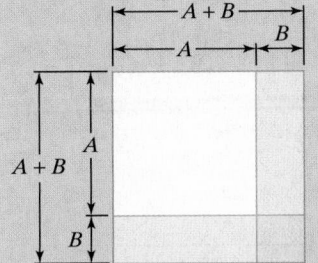

Sum of Areas:

$$A^2 + 2AB + B^2$$

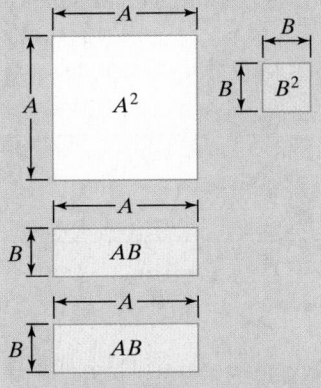

Conclusion:

$$A^2 + 2AB + B^2 = (A + B)^2$$

Factoring Perfect Square Trinomials

Let A and B be real numbers, variables, or algebraic expressions.

1. $A^2 + 2AB + B^2 = (A + B)^2$

Same sign

2. $A^2 - 2AB + B^2 = (A - B)^2$

Same sign

The two items in the box show that perfect square trinomials come in two forms: one in which the middle term is positive and one in which the middle term is negative. Here's how to recognize a perfect square trinomial:

1. The first and last terms are squares of monomials or integers.

2. The middle term is twice the product of the expressions being squared in the first and last terms.

EXAMPLE 5   Factoring Perfect Square Trinomials

Factor: **a.** $x^2 + 6x + 9$ **b.** $x^2 - 16x + 64$ **c.** $25x^2 - 60x + 36$.

Solution

a. $x^2 + 6x + 9 = x^2 + 2 \cdot x \cdot 3 + 3^2 = (x + 3)^2$ *The middle term has a positive sign.*

$A^2 + 2AB + B^2 = (A + B)^2$

b. $x^2 - 16x + 64 = x^2 - 2 \cdot x \cdot 8 + 8^2 = (x - 8)^2$ *The middle term has a negative sign.*

$A^2 - 2AB + B^2 = (A - B)^2$

c. We suspect that $25x^2 - 60x + 36$ is a perfect square trinomial because $25x^2 = (5x)^2$ and $36 = 6^2$. The middle term can be expressed as twice the product of $5x$ and 6.

$$25x^2 - 60x + 36 = (5x)^2 - 2 \cdot 5x \cdot 6 + 6^2 = (5x - 6)^2$$

$A^2 - 2AB + B^2 = (A - B)^2$ ∎

✔ **CHECK POINT 5**   Factor:

a. $x^2 + 14x + 49$ **b.** $x^2 - 6x + 9$
c. $16x^2 - 56x + 49$.

EXAMPLE 6   Factoring a Perfect Square Trinomial in Two Variables

Factor: $16x^2 + 40xy + 25y^2$.

Solution   Observe that $16x^2 = (4x)^2$, $25y^2 = (5y)^2$, and $40xy$ is twice the product of $4x$ and $5y$. Thus, we have a perfect square trinomial.

$$16x^2 + 40xy + 25y^2 = (4x)^2 + 2 \cdot 4x \cdot 5y + (5y)^2 = (4x + 5y)^2$$

$A^2 + 2AB + B^2 = (A + B)^2$ ∎

3 Factor the sum or difference of two cubes.

✔ **CHECK POINT 6** Factor: $4x^2 + 12xy + 9y^2$.

Factoring the Sum or Difference of Two Cubes

We can use the following formulas to factor the sum or the difference of two cubes.

Factoring the Sum or Difference of Two Cubes

1. Factoring the Sum of Two Cubes

$$A^3 + B^3 = (A + B)(A^2 - AB + B^2)$$

Same sign   Opposite signs

2. Factoring the Difference of Two Cubes

$$A^3 - B^3 = (A - B)(A^2 + AB + B^2)$$

Same sign   Opposite signs

Using Technology

You can use a graphing utility to verify the factorization in Example 7. Graphing

$$y_1 = x^3 + 8$$

and

$$y_2 = (x + 2)(x^2 - 2x + 4)$$

results in the same graph, so

$x^3 + 8$

$\quad = (x + 2)(x^2 - 2x + 4)$.

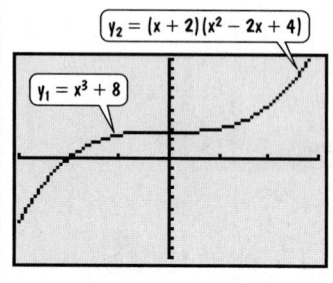

$y_2 = (x + 2)(x^2 - 2x + 4)$

$y_1 = x^3 + 8$

$[-3, 3, 1]$ by $[-30, 30, 3]$

EXAMPLE 7   Factoring the Sum of Two Cubes

Factor: $x^3 + 8$.

Solution   We must express each term as the cube of some monomial. Then we use the formula for factoring $A^3 + B^3$.

$$x^3 + 8 = x^3 + 2^3 = (x + 2)(x^2 - x \cdot 2 + 2^2) = (x + 2)(x^2 - 2x + 4)$$

$$A^3 + B^2 \ = \ (A + B) \ (A^2 - AB \ + B^2)$$

■

✔ **CHECK POINT 7**   Factor: $x^3 + 27$.

EXAMPLE 8   Factoring the Difference of Two Cubes

Factor: $27 - y^3$.

Solution   Express each term as the cube of some monomial. Then use the formula for factoring $A^3 - B^3$.

$$27 - y^3 = 3^3 - y^3 = (3 - y)(3^2 + 3y + y^2) = (3 - y)(9 + 3y + y^2)$$

$$A^3 - B^3 \ = \ (A - B) \ (A^2 + AB \ + B^2)$$

■

✔ **CHECK POINT 8**   Factor: $1 - y^3$.

EXAMPLE 9   Factoring the Sum of Two Cubes

Factor: $64x^3 + 125$.

Solution   Express each term as the cube of some monomial. Then use the formula for factoring $A^3 + B^3$.

$$64x^3 + 125 = (4x)^3 + 5^3 = (4x + 5)[(4x)^2 - (4x)(5) + 5^2]$$

$$A^3 \ + B^3 \ = \ (A \ + B) \ (A^2 \ - \ AB \ + B^2)$$

$$= (4x + 5)(16x^2 - 20x + 25)$$

■

✔ **CHECK POINT 9**   Factor: $125x^3 + 8$.

EXERCISE SET 6.4

 Practice Exercises

In Exercises 1–26, factor each difference of two squares.

1. $x^2 - 25$

2. $x^2 - 16$

3. $y^2 - 1$

4. $y^2 - 9$

5. $4x^2 - 9$

6. $9x^2 - 25$

7. $25 - x^2$

8. $16 - x^2$

9. $1 - 49x^2$

10. $1 - 64x^2$

11. $9 - 25y^2$

12. $16 - 49y^2$

13. $x^4 - 9$

14. $x^4 - 25$

15. $49y^4 - 16$

16. $49y^4 - 25$

17. $x^{10} - 9$

18. $x^{10} - 1$

19. $25x^2 - 16y^2$

20. $9x^2 - 25y^2$

21. $x^4 - y^{10}$

22. $x^{14} - y^4$

23. $x^4 - 16$

24. $x^4 - 1$

25. $16x^4 - 81$

26. $81x^4 - 1$

In Exercises 27–40, factor completely, or state that the polynomial is prime.

27. $2x^2 - 18$

28. $5x^2 - 45$

29. $2x^3 - 72x$

30. $2x^3 - 8x$

31. $x^2 + 36$

32. $x^2 + 4$

33. $2x^3 + 72x$

34. $2x^3 + 8x$

35. $50 - 2y^2$

36. $72 - 2y^2$

37. $8y^3 - 2y$

38. $12y^3 - 48y$

39. $2x^3 - 2x$

40. $3x^3 - 3x$

In Exercises 41–62, factor any perfect square trinomials, or state that the polynomial is prime.

41. $x^2 + 2x + 1$

42. $x^2 + 4x + 4$

43. $x^2 - 14x + 49$

44. $x^2 - 10x + 25$

45. $x^2 - 2x + 1$

46. $x^2 - 4x + 4$

47. $x^2 + 22x + 121$

48. $x^2 + 24x + 144$

49. $4x^2 + 4x + 1$

50. $9x^2 + 6x + 1$

51. $25y^2 - 10y + 1$

52. $64y^2 - 16y + 1$

53. $x^2 - 10x + 100$

54. $x^2 - 7x + 49$

55. $x^2 + 14xy + 49y^2$

56. $x^2 + 16xy + 64y^2$

57. $x^2 - 12xy + 36y^2$

58. $x^2 - 18xy + 81y^2$

59. $x^2 - 8xy + 64y^2$

60. $x^2 + 9xy + 16y^2$

61. $16x^2 - 40xy + 25y^2$

62. $9x^2 + 48xy + 64y^2$

In Exercises 63–70, factor completely.

63. $12x^2 - 12x + 3$

64. $18x^2 + 24x + 8$

65. $9x^3 + 6x^2 + x$

66. $25x^3 - 10x^2 + x$

67. $2y^2 - 4y + 2$

68. $2y^2 - 40y + 200$

69. $2y^3 + 28y^2 + 98y$

70. $50y^3 + 20y^2 + 2y$

In Exercises 71–88, factor using the formula for the sum or difference of two cubes.

71. $x^3 + 1$

72. $x^3 + 64$

73. $x^3 - 27$

74. $x^3 - 64$

75. $8y^3 - 1$

76. $27y^3 - 1$

77. $27x^3 + 8$

78. $125x^3 + 8$

79. $x^3y^3 - 64$

80. $x^3y^3 - 27$

81. $27y^4 + 8y$

82. $64y - y^4$

83. $54 - 16y^3$

84. $128 - 250y^3$

85. $64x^3 + 27y^3$

86. $8x^3 + 27y^3$

87. $125x^3 - 64y^3$

88. $125x^3 - y^3$

Application Exercises

In Exercises 89–92, find the formula for the area of the shaded region and express it in factored form.

89.

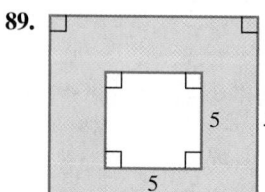

90.

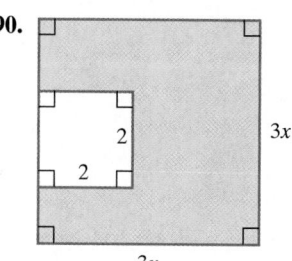

91.

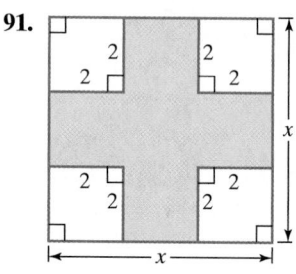

92.

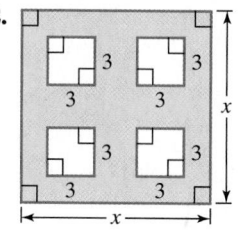

Writing in Mathematics

93. Explain how to factor the difference of two squares. Provide an example with your explanation.

94. What is a perfect square trinomial and how is it factored?

95. Explain why $x^2 - 1$ is factorable, but $x^2 + 1$ is not.

96. Explain how to factor $x^3 + 1$.

Critical Thinking Exercises

97. Which one of the following is true?
  a. Because $x^2 - 25 = (x + 5)(x - 5)$, then $x^2 + 25 = (x - 5)(x + 5)$.
  b. All perfect square trinomials are squares of binomials.
  c. Any polynomial that is the sum of two squares is prime.
  d. The polynomial $16x^2 + 20x + 25$ is a perfect square trinomial.

98. Where is the error in this "proof" that $2 = 0$?

$a = b$	Suppose that a and b are any equal real numbers.
$a^2 = b^2$	Square both sides of the equation.
$a^2 - b^2 = 0$	Subtract b^2 from both sides.

$2(a^2 - b^2) = 2 \cdot 0$	Multiply both sides by 2.
$2(a^2 - b^2) = 0$	On the right side, $2 \cdot 0 = 0$.
$2(a + b)(a - b) = 0$	Factor $a^2 - b^2$.
$2(a + b) = 0$	Divide both sides by $a - b$.
$2 = 0$	Divide both sides by $a + b$.

In Exercises 99–103, factor each polynomial.

99. $(x + 1)^2 - 25$

100. $x^{2n} - 25y^{2n}$

101. $4x^{2n} + 12x^n + 9$

102. $(x + 3)^2 - (x + 2)^2$

103. $(x + 3)^2 - 2(x + 3) + 1$

In Exercises 104–105, find all integers k so that the trinomial is a perfect square trinomial.

104. $9x^2 + kx + 1$

105. $64x^2 - 16x + k$

Technology Exercises

In Exercises 106–109, use a graphing utility to graph each side of the equation in the same viewing rectangle. Do the graphs coincide? If so, this means that the polynomial on the left side has been correctly factored. If not, factor the polynomial correctly and then use your graphing utility to verify the factorization.

106. $4x^2 - 9 = (4x + 3)(4x - 3)$

107. $x^2 - 6x + 9 = (x - 3)^2$

108. $4x^2 - 4x + 1 = (4x - 1)^2$

109. $x^3 - 1 = (x - 1)(x^2 - x + 1)$

Review Exercises

110. Simplify: $(2x^2y^3)^4(5xy^2)$. (Section 5.7, Example 5)

111. Subtract: $(10x^2 - 5x + 2) - (14x^2 - 5x - 1)$. (Section 5.1, Example 3)

112. Divide: $\dfrac{6x^2 + 11x - 10}{3x - 2}$. (Section 5.6, Example 1)

▶ SECTION 6.5   *A General Factoring Strategy*

Objectives

1 Recognize the appropriate method for factoring a polynomial.

2 Use a general strategy for factoring polynomials.

SSM
PH Tutor   CD-   Video
Center   ROM

Successful problem solving involves understanding the problem, devising a plan for solving it, and then carrying out the plan. In this section, you will learn a step-by-step strategy that provides a plan and direction for solving factoring problems.

A Strategy for Factoring Polynomials

It is important to practice factoring a wide variety of polynomials so that you can quickly select the appropriate technique. The polynomial is factored completely when all its polynomial factors, except possibly the monomial factors, are prime. Because of the commutative property, the order of the factors does not matter.

Here is a general strategy for factoring polynomials.

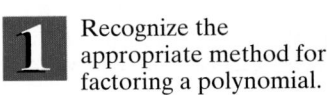

1 Recognize the appropriate method for factoring a polynomial.

A Strategy for Factoring a Polynomial

1. If there is a common factor, factor out the GCF.

2. Determine the number of terms in the polynomial and try factoring as follows.

   a. If there are two terms, can the binomial be factored by one of the following special forms?

      Difference of two squares: $A^2 - B^2 = (A + B)(A - B)$

      Sum of two cubes: $A^3 + B^3 = (A + B)(A^2 - AB + B^2)$

      Difference of two cubes: $A^3 - B^3 = (A - B)(A^2 + AB + B^2)$

   b. If there are three terms, is the trinomial a perfect square trinomial? If so, factor by one of the following special forms:

$$A^2 + 2AB + B^2 = (A + B)^2$$
$$A^2 - 2AB + B^2 = (A - B)^2.$$

     If the trinomial is not a perfect square trinomial, try factoring by trial and error or grouping.

   c. If there are four or more terms, try factoring by grouping.

3. Check to see if any factors with more than one term in the factored polynomial can be factored further. If so, factor completely.

4. Check by multiplying.

2 Use a general strategy for factoring polynomials.

The following examples and those in the exercise set are similar to the previous factoring problems. One difference is that although these polynomials may be factored using the techniques we have studied in this chapter, each must be factored using at least two techniques. Also different is that these factorizations are not all of the same type; they are intentionally mixed to promote the development of a general factoring strategy.

EXAMPLE 1   Factoring a Polynomial

Factor: $4x^4 - 16x^2$.

Solution

Step 1 If there is a common factor, factor out the GCF. Because $4x^2$ is common to both terms, we factor it out.

$$4x^4 - 16x^2 = 4x^2(x^2 - 4) \quad \text{\footnotesize Factor out the GCF.}$$

Step 2 Determine the number of terms and factor accordingly. The factor $x^2 - 4$ has two terms. It is the difference of two squares: $x^2 - 2^2$. We factor using the special form for the difference of two squares and rewrite the GCF.

$$4x^4 - 16x^2 = 4x^2(x + 2)(x - 2) \quad \text{\footnotesize Use } A^2 - B^2 = (A + B)(A - B)$$
$$\text{\footnotesize on } x^2 - 4: A = x \text{ and } B = 2.$$

Step 3 Check to see if any factors with more than one term can be factored further. No factor with more than one term can be factored further, so we have factored completely.

Step 4 Check by multiplying.

$$4x^2(x + 2)(x - 2) = 4x^2(x^2 - 4) = 4x^4 - 16x^2$$

This is the original polynomial, so the factorization is correct.

■

✔ **CHECK POINT 1**   Factor: $5x^4 - 45x^2$.

EXAMPLE 2   Factoring a Polynomial

Factor: $3x^2 - 6x - 45$.

Solution

Step 1 If there is a common factor, factor out the GCF. Because 3 is common to all terms, we factor it out.

$$3x^2 - 6x - 45 = 3(x^2 - 2x - 15) \quad \text{\footnotesize Factor out the GCF.}$$

Step 2 Determine the number of terms and factor accordingly. The factor $x^2 - 2x - 15$ has three terms, but it is not a perfect square trinomial. We factor it using trial and error.

$$3x^2 - 6x - 45 = 3(x^2 - 2x - 15) = 3(x - 5)(x + 3)$$

Step 3 Check to see if factors can be factored further. In this case, they cannot, so we have factored completely.

Step 4 Check by multiplying.

$$3(x - 5)(x + 3) = 3(x^2 - 2x - 15) = 3x^2 - 6x - 45$$

FOIL

This is the original polynomial, so the factorization is correct. ∎

✔ **CHECK POINT 2** Factor: $4x^2 - 16x - 48$.

EXAMPLE 3 Factoring a Polynomial

Factor: $7x^5 - 7x$.

Solution

Step 1 If there is a common factor, factor out the GCF. Because $7x$ is common to both terms, we factor it out.

$$7x^5 - 7x = 7x(x^4 - 1) \qquad \text{Factor out the GCF.}$$

Step 2 Determine the number of terms and factor accordingly. The factor $x^4 - 1$ has two terms. This binomial can be expressed as $(x^2)^2 - 1^2$, so it can be factored as the difference of two squares.

$$7x^5 - 7x = 7x(x^4 - 1) = 7x(x^2 + 1)(x^2 - 1) \qquad \begin{array}{l}\text{Use } A^2 - B^2 = (A + B)(A - B) \\ \text{on } x^4 - 1: A = x^2 \text{ and } B = 1.\end{array}$$

Step 3 Check to see if factors can be factored further. We note that $(x^2 - 1)$ is also the difference of two squares, $x^2 - 1^2$, so we continue factoring.

$$7x^5 - 7x = 7x(x^2 + 1)(x + 1)(x - 1) \qquad \begin{array}{l}\text{Factor } x^2 - 1 \text{ as the dif-}\\ \text{ference of two squares.}\end{array}$$

Step 4 Check by multiplying.

$$7x(x^2 + 1)(x + 1)(x - 1) = 7x(x^2 + 1)(x^2 - 1) = 7x(x^4 - 1) = 7x^5 - 7x$$

We obtain the original polynomial, so the factorization is correct. ∎

✔ **CHECK POINT 3** Factor: $4x^5 - 64x$.

EXAMPLE 4 Factoring a Polynomial

Factor: $x^3 - 5x^2 - 4x + 20$.

Solution

Step 1 If there is a common factor, factor out the GCF. Other than 1, there is no common factor.

Step 2 Determine the number of terms and factor accordingly. There are four terms. We try factoring by grouping.

$$x^3 - 5x^2 - 4x + 20$$

$$= (x^3 - 5x^2) + (-4x + 20) \qquad \text{Group terms with common factors.}$$

$$= x^2(x - 5) - 4(x - 5) \qquad \text{Factor from each group.}$$

$$= (x - 5)(x^2 - 4) \qquad \text{Factor out the common binomial factor, } x - 5.$$

Step 3 Check to see if factors can be factored further. We note that $(x^2 - 4)$ is the difference of two squares, $x^2 - 2^2$, so we continue factoring.

$$x^3 - 5x^2 - 4x + 20 = (x - 5)(x + 2)(x - 2) \qquad \text{Factor } x^2 - 4 \text{ as the difference of two squares.}$$

We have factored completely because no factor with more than one term can be factored further.

Step 4 Check by multiplying.

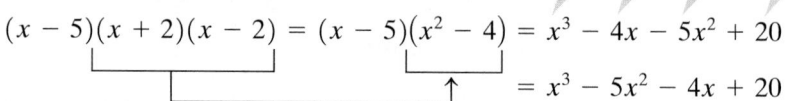

$$(x - 5)(x + 2)(x - 2) = (x - 5)(x^2 - 4) = x^3 - 4x - 5x^2 + 20$$

$$= x^3 - 5x^2 - 4x + 20$$

We obtain the original polynomial, so the factorization is correct. ∎

✔ **CHECK POINT 4** Factor: $x^3 - 4x^2 - 9x + 36$.

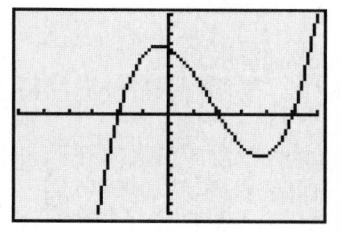
EXAMPLE 5 Factoring a Polynomial

Factor: $2x^3 - 24x^2 + 72x$.

Solution

Step 1 If there is a common factor, factor out the GCF. Because $2x$ is common to all terms, we factor it out.

$$2x^3 - 24x^2 + 72x = 2x(x^2 - 12x + 36) \qquad \text{Factor out the GCF.}$$

Step 2 Determine, the number of terms and factor accordingly. The factor $x^2 - 12x + 36$ has three terms. Is it a perfect square trinomial? Yes. The first term, x^2, is the square of a monomial. The last term, 36, or 6^2, is the square of an integer. The middle term involves twice the product of x and 6. We factor using $A^2 - 2AB + B^2 = (A - B)^2$.

$$2x^3 - 24x^2 + 72x = 2x(x^2 - 12x + 36)$$

$$= 2x(x^2 - 2 \cdot x \cdot 6 + 6^2) \qquad \text{The second factor is a perfect square trinomial.}$$

$$A^2 - 2 \quad A \quad B + B^2$$

$$= 2x(x - 6)^2 \qquad A^2 - 2AB + B^2 = (A - B)^2$$

Step 3 Check to see if factors can be factored further. In this problem, they cannot, so we have factored completely.

Step 4 Check by multiplying.

$$2x(x - 6)^2 = 2x(x^2 - 12x + 36) = 2x^3 - 24x^2 + 72x$$

We obtain the original polynomial, so the factorization is correct. ∎

✔ **CHECK POINT 5** Factor: $3x^3 - 30x^2 + 75x$.

EXAMPLE 6 Factoring a Polynomial

Factor: $3x^5 + 24x^2$.

Solution

Step 1 If there is a common factor, factor out the GCF. Because $3x^2$ is common to all terms, we factor it out.

$$3x^5 + 24x^2 = 3x^2(x^3 + 8) \qquad \text{Factor out the GCF.}$$

Step 2 Determine the number of terms and factor accordingly. The factor $x^3 + 8$ has two terms. This binomial can be expressed as $x^3 + 2^3$, so it can be factored as the sum of two cubes.

$$3x^5 + 24x^2 = 3x^2(\underbrace{x^3 + 2^3}_{A^3 + B^3}) \qquad \begin{array}{l}\text{Express } x^3 + 8 \text{ as the sum}\\ \text{of two cubes.}\end{array}$$

$$= 3x^2\underbrace{(x + 2)(x^2 - 2x + 4)}_{(A + B)(A^2 - AB + B^2)} \qquad \text{Factor the sum of two cubes.}$$

Step 3 Check to see if factors can be factored further. In this problem, they cannot, so we have factored completely.

Step 4 Check by multiplying.

$$3x^2(x + 2)(x^2 - 2x + 4) = 3x^2[x(x^2 - 2x + 4) + 2(x^2 - 2x + 4)]$$

$$= 3x^2(x^3 - 2x^2 + 4x + 2x^2 - 4x + 8)$$

$$= 3x^2(x^3 + 8) = 3x^5 + 24x^2$$

We obtain the original polynomial, so the factorization is correct. ∎

✔ **CHECK POINT 6** Factor: $2x^5 + 54x^2$.

Discover for Yourself

In Examples 1–6, substitute 1 for the variable in both the given polynomial and in its factored form. What do you observe? Do this for a second value of the variable. Is this a complete check or only a partial check of the factorization? Explain.

EXAMPLE 7 Factoring a Polynomial in Two Variables

Factor: $32x^4y - 2y^5$.

Solution

Step 1 If there is a common factor, factor out the GCF. Because $2y$ is common to all terms, we factor it out.

$$32x^4y - 2y^5 = 2y(16x^4 - y^4) \qquad \text{Factor out the GCF.}$$

Step 2 Determine the number of terms and factor accordingly. The factor $16x^4 - y^4$ has two terms. It is the difference of two squares: $(4x^2)^2 - (y^2)^2$. We factor using the special form for the difference of two squares.

$$32x^4y - 2y^5 = 2y\left[(4x^2)^2 - (y^2)^2\right]$$

Express $16x^4 - y^4$ as the difference of two squares.

$$= 2y(4x^2 + y^2)(4x^2 - y^2)$$

$A^2 - B^2 = (A + B)(A - B)$

Step 3 Check to see if factors can be factored further. We note that the last factor, $4x^2 - y^2$, is also the difference of two squares, $(2x)^2 - y^2$, so we continue factoring.

$$32x^4y - 2y^5 = 2y(4x^2 + y^2)(2x + y)(2x - y)$$

Step 4 Check by multiplying. Multiply the factors in the factorization and verify that you obtain the original polynomial. ∎

✔ **CHECK POINT 7** Factor: $3x^4y - 48y^5$.

EXAMPLE 8 Factoring a Polynomial in Two Variables

Factor: $18x^3 + 48x^2y + 32xy^2$.

Solution

Step 1 If there is a common factor, factor out the GCF. Because $2x$ is common to all terms, we factor it out.

$$18x^3 + 48x^2y + 32xy^2 = 2x(9x^2 + 24xy + 16y^2)$$

Step 2 Determine the number of terms and factor accordingly. The factor $9x^2 + 24xy + 16y^2$ has three terms. Is it a perfect square trinomial? Yes. The first term, $9x^2$, or $(3x)^2$, and the last term, $16y^2$, or $(4y)^2$, are squares of monomials. The middle term, $24xy$, is twice the product of $3x$ and $4y$. We factor using $A^2 + 2AB + B^2 = (A + B)^2$.

$$18x^3 + 48x^2y + 32xy^2 = 2x(9x^2 + 24xy + 16y^2)$$

$$= 2x\left[(3x)^2 + 2 \cdot 3x \cdot 4y + (4y)^2\right]$$

The second factor is a perfect square trinomial.

$$= 2x(3x + 4y)^2$$

$A^2 + 2AB + B^2$
$= (A + B)^2$

Step 3 Check to see if factors can be factored further. In this problem, they cannot, so we have factored completely.

Step 4 Check by multiplication. Multiply the factors in the factorization and verify that you obtain the original polynomial. ∎

✔ **CHECK POINT 8** Factor: $12x^3 + 36x^2y + 27xy^2$.

EXERCISE SET 6.5

Practice Exercises

In Exercises 1–62, factor completely, or state that the polynomial is prime. Check factorizations using multiplication or a graphing utility.

1. $3x^3 - 3x$
2. $5x^3 - 45x$

3. $3x^3 + 3x$
4. $5x^3 + 45x$

5. $4x^2 - 4x - 24$
6. $6x^2 - 18x - 60$

7. $2x^4 - 162$
8. $7x^4 - 7$

9. $x^3 + 2x^2 - 9x - 18$
10. $x^3 + 3x^2 - 25x - 75$

11. $3x^3 - 24x^2 + 48x$
12. $5x^3 - 20x^2 + 20x$

13. $2x^5 + 2x^2$
14. $2x^5 + 128x^2$

15. $6x^2 + 8x$
16. $21x^2 - 35x$

17. $2y^2 - 2y - 112$
18. $6x^2 - 6x - 12$

19. $7y^4 + 14y^3 + 7y^2$
20. $2y^4 + 28y^3 + 98y^2$

21. $y^2 + 8y - 16$
22. $y^2 - 18y - 81$

23. $16y^2 - 4y - 2$
24. $32y^2 + 4y - 6$

25. $r^2 - 25r$
26. $3r^2 - 27r$

27. $4w^2 + 8w - 5$
28. $35w^2 - 2w - 1$

29. $x^3 - 4x$
30. $9x^3 - 9x$

31. $x^2 + 64$
32. $y^2 + 36$

33. $9y^2 + 13y + 4$
34. $20y^2 + 12y + 1$

35. $y^3 + 2y^2 - 4y - 8$
36. $y^3 + 2y^2 - y - 2$

37. $9y^2 + 24y + 16$
38. $9y^2 + 6y + 1$

39. $5y^3 - 45y^2 + 70y$
40. $14y^3 + 7y^2 - 7y$

41. $y^5 - 81y$
42. $y^5 - 16y$

43. $20a^4 - 45a^2$
44. $48a^4 - 3a^2$

45. $12y^2 - 11y + 2$
46. $21x^2 - 25x - 4$

47. $9y^2 - 64$
48. $100y^2 - 49$

49. $9y^2 + 64$
50. $100y^2 + 49$

51. $2y^3 + 3y^2 - 50y - 75$
52. $12y^3 + 16y^2 - 3y - 4$

53. $2r^3 + 30r^2 - 68r$
54. $3r^3 - 27r^2 - 210r$

55. $8x^5 - 2x^3$
56. $y^9 - y^5$

57. $3x^2 + 243$
58. $27x^2 + 75$

59. $x^4 + 8x$
60. $x^4 + 27x$

61. $2y^5 - 2y^2$
62. $2y^5 - 128y^2$

Exercises 63–92 contain polynomials in several variables. Factor each polynomial completely and check using multiplication.

63. $6x^2 + 8xy$
64. $21x^2 - 35xy$

65. $xy - 7x + 3y - 21$
66. $xy - 5x + 2y - 10$

67. $x^2 - 3xy - 4y^2$
68. $x^2 - 4xy - 12y^2$

69. $72a^3b^2 + 12a^2 - 24a^4b^2$

70. $24a^4b + 60a^3b^2 + 150a^2b^3$

71. $3a^2 + 27ab + 54b^2$
72. $3a^2 + 15ab + 18b^2$

73. $48x^4y - 3x^2y$
74. $16a^3b^2 - 4ab^2$

75. $6a^2b + ab - 2b$
76. $16a^2 - 32ab + 12b^2$

77. $7x^5y - 7xy^5$
78. $3x^4y^2 - 3x^2y^2$

79. $10x^3y - 14x^2y^2 + 4xy^3$
80. $18x^3y + 57x^2y^2 + 30xy^3$

81. $2bx^2 + 44bx + 242b$
82. $3xz^2 - 72xz + 432x$

83. $15a^2 + 11ab - 14b^2$
84. $25a^2 + 25ab + 6b^2$

85. $36x^3y - 62x^2y^2 + 12xy^3$

86. $10a^4b^2 - 15a^3b^3 - 25a^2b^4$
87. $a^2y - b^2y - a^2x + b^2x$

88. $bx^2 - 4b + ax^2 - 4a$
89. $9ax^3 + 15ax^2 - 14ax$

90. $4ay^3 - 12ay^2 + 9ay$
91. $81x^4y - y^5$

92. $16x^4y - y^5$

Application Exercises

93. A rock is dropped from the top of a 256-foot cliff. The height, in feet, of the rock above the water after t seconds is modeled by the polynomial $256 - 16t^2$. Factor this expression completely.

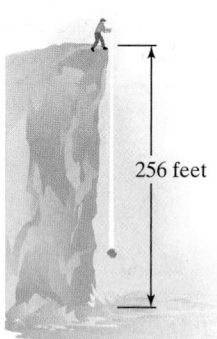

256 feet

94. The building shown in the figure has a height represented by x feet. The building's base is a square and its volume is $x^3 - 60x^2 + 900x$ cubic feet. Express the building's dimensions in terms of x.

x

95. Express the area of the shaded ring shown in the figure in terms of π. Then factor this expression completely.

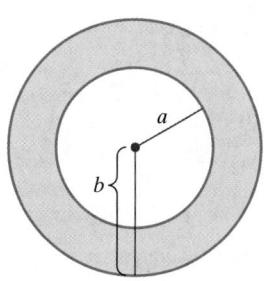

a

b

Writing in Mathematics

96. Describe a strategy that can be used to factor polynomials.

97. Describe some of the difficulties in factoring polynomials. What suggestions can you offer to overcome these difficulties?

98. You are about to take a great picture of fog rolling into San Francisco from the middle of the Golden Gate Bridge, 400 feet above the water. Whoops! You accidently lean too far over the safety rail and drop your camera. The height, in feet, of the camera after t seconds is modeled by the polynomial $400 - 16t^2$. The factored form of the polynomial is $16(5 + t)(5 - t)$. Describe something about your falling camera that is easier to see from the factored form, $16(5 + t)(5 - t)$, than from the form $400 - 16t^2$.

Critical Thinking Exercises

99. Which one of the following is true?
  a. $x^2 - 9 = (x - 3)^2$ for any real number x.
  b. The polynomial $4x^2 + 100$ is the sum of two squares and therefore cannot be factored.
  c. If the general factoring strategy is used to factor a polynomial, at least two factorizations are necessary before the given polynomial is factored completely.
  d. Once a common monomial factor is removed from $3xy^3 + 9xy^2 + 21xy$, the remaining trinomial factor cannot be factored further.

In Exercises 100–104, factor completely.

100. $3x^5 - 21x^3 - 54x$ **101.** $5y^5 - 5y^4 - 20y^3 + 20y^2$

102. $x^2(x + 3) - x(x + 3) - 6(x + 3)$

103. $(x + 5)^2 - 20(x + 5) + 100$

104. $3x^{2n} - 27y^{2n}$

Technology Exercises

In Exercises 105–109, use a graphing utility to graph each side of the equation in the given viewing rectangle. Do the graphs coincide? If so, this means that the polynomial on the left side has been correctly factored. If not, factor the polynomial correctly and then use your graphing utility to verify the factorization.

105. $4x^2 - 12x + 9 = (4x - 3)^2$; $[-5, 5, 1]$ by $[0, 20, 1]$

106. $3x^3 - 12x^2 - 15x = 3x(x + 5)(x - 1)$; $[-5, 7, 1]$ by $[-80, 80, 10]$

107. $6x^2 + 10x - 4 = 2(3x - 1)(x + 2)$;
[−5, 5, 1] by [−20, 20, 2]

108. $x^4 - 16 = (x^2 + 4)(x + 2)(x - 2)$;
[−5, 5, 1] by [−20, 20, 2]

109. $2x^3 + 10x^2 - 2x - 10 = 2(x + 5)(x^2 + 1)$;
[−8, 4, 1] by [−100, 100, 10]

Review Exercises

110. Factor: $9x^2 - 16$. (Section 6.4, Example 1)

111. Graph using intercepts: $5x - 2y = 10$. (Section 3.2, Example 4)

112. The second angle of a triangle measures three times that of the first angle's measure. The third angle measures 80° more than the first. Find the measure of each angle. (Section 2.6, Example 6)

▶ **SECTION 6.6**   *Solving Quadratic Equations by Factoring*

Objectives

1 Use the zero-product principle.

2 Solve quadratic equations by factoring.

3 Solve problems using quadratic equations.

SSM
PH Tutor CD- Video
Center   ROM

The alligator, an endangered species, was the subject of a protection program at Florida's Everglades National Park. Park rangers used the formula

$$P = -10x^2 + 475x + 3500$$

to estimate the alligator population, P, after x years of the protection program. Their goal was to bring the population up to 7250. To find out how long the program had to be continued for this to occur, we need to substitute 7250 for P in the formula and solve for x:

$$7250 = -10x^2 + 475x + 3500.$$

Do you see how this equation differs from a linear equation? The exponent on x is 2. Solving such an equation involves finding the numbers that will make the equation a true statement. In this section, we use factoring for solving equations in the form $ax^2 + bx + c = 0$. We also look at applications of these equations.

The Standard Form of a Quadratic Equation

We begin by defining a quadratic equation.

Definition of a Quadratic Equation

A **quadratic equation** in x is an equation that can be written in the **standard form**

$$ax^2 + bx + c = 0$$

where $a, b,$ and c are real numbers, with $a \neq 0$. A quadratic equation in x is also called a **second-degree polynomial equation** in x.

An example of a quadratic equation in standard form is $x^2 - 7x + 10 = 0$. The coefficient of x^2 is $1 (a = 1)$, the coefficient of x is $-7 (b = -7)$, and the constant term is $10 (c = 10)$.

1 Use the zero-product principle.

Solving Quadratic Equations by Factoring

We can factor the left side of the quadratic equation $x^2 - 7x + 10 = 0$. We obtain $(x - 5)(x - 2) = 0$. If a quadratic equation has zero on one side and a factored expression on the other side, it can be solved using the **zero-product principle**.

The Zero-Product Principle

If the product of two algebraic expressions is zero, then at least one of the factors is equal to zero.

$$\text{If } AB = 0, \text{ then } A = 0 \text{ or } B = 0.$$

For example, consider the equation $(x - 5)(x - 2) = 0$. According to the zero-product principle, this product can be zero only if at least one of the factors is zero. We set each individual factor equal to zero and solve each resulting equation for x.

$$(x - 5)(x - 2) = 0$$

$$x - 5 = 0 \quad \text{or} \quad x - 2 = 0$$
$$x = 5 \qquad\qquad x = 2$$

We can check each of these proposed solutions in the original quadratic equation, $x^2 - 7x + 10 = 0$. Substitute each one separately for x in the equation.

Check 5:	**Check 2:**
$x^2 - 7x + 10 = 0$	$x^2 - 7x + 10 = 0$
$5^2 - 7 \cdot 5 + 10 \overset{?}{=} 0$	$2^2 - 7 \cdot 2 + 10 \overset{?}{=} 0$
$25 - 35 + 10 \overset{?}{=} 0$	$4 - 14 + 10 \overset{?}{=} 0$
$0 = 0, \quad \text{true}$	$0 = 0, \quad \text{true}$

The resulting true statements indicate that the solutions are 5 and 2. Note that with a quadratic equation, we can have two solutions, compared to the linear equation that usually had one.

EXAMPLE 1 Using the Zero-Product Principle

Solve the equation: $(3x - 1)(x + 2) = 0$.

Solution The product $(3x - 1)(x + 2)$ is equal to zero. By the zero-product principle, the only way that this product can be zero is if at least one of the factors is zero. Thus,

$$3x - 1 = 0 \quad \text{or} \quad x + 2 = 0.$$
$$3x = 1 \qquad\qquad x = -2 \qquad \textit{Solve each equation for x.}$$
$$x = \frac{1}{3}$$

Because each linear equation has a solution, the original equation given above, $(3x - 1)(x + 2) = 0$, has two solutions, $\frac{1}{3}$ and -2. Check these solutions by substituting each one separately in the original equation. The equation's solution set is $\left\{-2, \frac{1}{3}\right\}$. ∎

✔ **CHECK POINT 1**   Solve the equation: $(2x + 1)(x - 4) = 0.$

2 Solve quadratic equations by factoring.

In Example 1 and Check Point 1, the given equations were in factored form. Here is a procedure for solving a quadratic equation when we must first do the factoring.

Solving a Quadratic Equation by Factoring

1. If necessary, rewrite the equation in the form $ax^2 + bx + c = 0$, moving all terms to one side, thereby obtaining zero on the other side.
2. Factor.
3. Apply the zero-product principle, setting each factor equal to zero.
4. Solve the equations in step 3.
5. Check the solutions in the original equation.

EXAMPLE 2   Solving a Quadratic Equation by Factoring

Solve: $2x^2 + 7x - 4 = 0.$

Solution

Step 1 Move all terms to one side and obtain zero on the other side. All terms are already on the left and zero is on the other side, so we can skip this step.

Step 2 Factor.
$$2x^2 + 7x - 4 = 0$$
$$(2x - 1)(x + 4) = 0$$

Study Tip

Do not confuse factoring a polynomial with solving a quadratic equation by factoring.

INCORRECT!

~~Factor: $2x^2 + 7x - 4.$~~

~~$(2x - 1)(x + 4)$~~

~~$2x - 1 = 0$ or $x + 4 = 0$~~

~~$x = \dfrac{1}{2}$ $x = -4$~~

Steps 3 and 4 Set each factor equal to zero and solve each resulting equation.

$$2x - 1 = 0 \quad \text{or} \quad x + 4 = 0$$
$$2x = 1 \qquad\qquad x = -4$$
$$x = \frac{1}{2}$$

Step 5 Check the solutions in the original equation.

Check $\dfrac{1}{2}$:

$$2x^2 + 7x - 4 = 0$$
$$2\left(\frac{1}{2}\right)^2 + 7\left(\frac{1}{2}\right) - 4 \stackrel{?}{=} 0$$
$$2\left(\frac{1}{4}\right) + 7\left(\frac{1}{2}\right) - 4 \stackrel{?}{=} 0$$
$$\frac{1}{2} + \frac{7}{2} - 4 \stackrel{?}{=} 0$$
$$4 - 4 \stackrel{?}{=} 0$$
$$0 = 0, \quad \text{true}$$

Check -4:

$$2x^2 + 7x - 4 = 0$$
$$2(-4)^2 + 7(-4) - 4 \stackrel{?}{=} 0$$
$$2(16) + 7(-4) - 4 \stackrel{?}{=} 0$$
$$32 + (-28) - 4 \stackrel{?}{=} 0$$
$$4 - 4 \stackrel{?}{=} 0$$
$$0 = 0, \quad \text{true}$$

The solutions are -4 and $\dfrac{1}{2}$, and the solution set is $\left\{-4, \dfrac{1}{2}\right\}$. ∎

✔ **CHECK POINT 2** Solve: $x^2 - 6x + 5 = 0$.

Using Technology

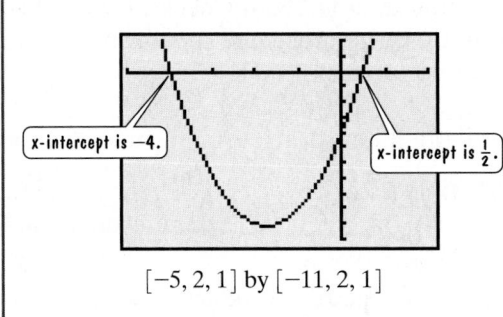

x-intercept is −4.

x-intercept is $\frac{1}{2}$.

$[-5, 2, 1]$ by $[-11, 2, 1]$

You can use a graphing utility to check the real number solutions to a quadratic equation. **The solutions to $ax^2 + bx + c = 0$ correspond to the x-intercepts for the graph of $y = ax^2 + bx + c$.** For example, to check the solutions of $2x^2 + 7x - 4 = 0$, graph $y = 2x^2 + 7x - 4$, as shown on the left. The x-intercepts are -4 and $\frac{1}{2}$, verifying -4 and $\frac{1}{2}$ as the solutions.

EXAMPLE 3   Solving a Quadratic Equation by Factoring

Solve: $3x^2 = 2x$.

Solution

Step 1 Move all terms to one side and obtain zero on the other side. Subtract $2x$ from both sides and write the equation in standard form.

$$3x^2 - 2x = 2x - 2x$$
$$3x^2 - 2x = 0$$

Step 2 Factor. We factor out x from the two terms on the left side.

$$3x^2 - 2x = 0$$
$$x(3x - 2) = 0$$

Steps 3 and 4 Set each factor equal to zero and solve the resulting equations.

$$x = 0 \qquad \text{or} \qquad 3x - 2 = 0$$
$$3x = 2$$
$$x = \frac{2}{3}$$

Step 5 Check the solutions in the original equation.

Check 0:

$$3x^2 = 2x$$
$$3 \cdot 0^2 \stackrel{?}{=} 2 \cdot 0$$
$$0 = 0, \qquad \text{true}$$

Check $\frac{2}{3}$:

$$3x^2 = 2x$$
$$3\left(\frac{2}{3}\right)^2 \stackrel{?}{=} 2\left(\frac{2}{3}\right)$$
$$3\left(\frac{4}{9}\right) \stackrel{?}{=} 2\left(\frac{2}{3}\right)$$
$$\frac{4}{3} = \frac{4}{3}, \qquad \text{true}$$

Study Tip

Avoid dividing both sides of $3x^2 = 2x$ by x. You will obtain $3x = 2$ and, consequently, $x = \frac{2}{3}$. The other solution, 0, is lost. We can divide both sides of an equation by any *nonzero* real number. If x is zero, we lose the second solution.

The solutions are 0 and $\frac{2}{3}$, and the solution set is $\left\{0, \frac{2}{3}\right\}$. ∎

✔ **CHECK POINT 3** Solve: $4x^2 = 2x$.

EXAMPLE 4   Solving a Quadratic Equation by Factoring

Solve: $x^2 = 6x - 9$.

Solution

Step 1 Move all terms to one side and obtain zero on the other side. To obtain zero on the right, we subtract $6x$ and add 9 on both sides.

$$x^2 - 6x + 9 = 6x - 6x - 9 + 9$$
$$x^2 - 6x + 9 = 0$$

Step 2 Factor. The trinomial on the left side is a perfect square trinomial: $x^2 - 6x + 9 = x^2 - 2 \cdot x \cdot 3 + 3^2$. We factor using $A^2 - 2AB + B^2 = (A - B)^2$: $A = x$ and $B = 3$.

$$x^2 - 6x + 9 = 0$$
$$(x - 3)^2 = 0$$

Steps 3 and 4 Set each factor equal to zero and solve the resulting equations. Because both factors are the same, it is only necessary to set one of them equal to zero.

$$x - 3 = 0$$
$$x = 3$$

Step 5 Check the solution in the original equation.

     Check 3:

$$x^2 = 6x - 9$$
$$3^2 \overset{?}{=} 6 \cdot 3 - 9$$
$$9 \overset{?}{=} 18 - 9$$
$$9 = 9, \quad \text{true}$$

The solution is 3 and the solution set is $\{3\}$.   ■

✔ **CHECK POINT 4** Solve: $x^2 = 10x - 25$.

Using Technology

The graph of $y = x^2 - 6x + 9$ is shown below. Notice that there is only one x-intercept, namely 3, verifying that the solution of

$$x^2 - 6x + 9 = 0$$

is 3.

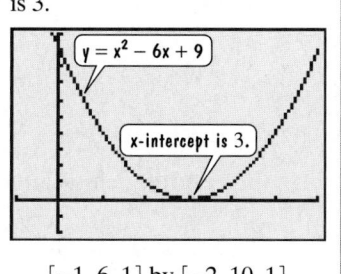

$y = x^2 - 6x + 9$

x-intercept is 3.

$[-1, 6, 1]$ by $[-2, 10, 1]$

EXAMPLE 5   Solving a Quadratic Equation by Factoring

Solve: $9x^2 = 16$.

Solution

Step 1 Move all terms to one side and obtain zero on the other side. Subtract 16 from both sides and write the equation in standard form.

$$9x^2 - 16 = 16 - 16$$
$$9x^2 - 16 = 0$$

Step 2 Factor. The binomial on the left side is the difference of two squares: $9x^2 - 16 = (3x)^2 - 4^2$. We factor using $A^2 - B^2 = (A + B)(A - B)$: $A = 3x$ and $B = 4$.

$$9x^2 - 16 = 0$$
$$(3x + 4)(3x - 4) = 0$$

Steps 3 and 4 Set each factor equal to zero and solve the resulting equations.

$$3x + 4 = 0 \quad \text{or} \quad 3x - 4 = 0$$
$$3x = -4 \quad\quad\quad 3x = 4$$
$$x = -\frac{4}{3} \quad\quad\quad x = \frac{4}{3}$$

Step 5 Check the solutions in the original equation. Do this now and verify that the solutions are $-\frac{4}{3}$ and $\frac{4}{3}$. The equation's solution set is $\left\{-\frac{4}{3}, \frac{4}{3}\right\}$. ■

✔ **CHECK POINT 5** Solve: $16x^2 = 25$.

EXAMPLE 6 Solving a Quadratic Equation by Factoring

Solve: $(x - 2)(x + 3) = 6$.

Solution

Step 1 Move all terms to one side and obtain zero on the other side. We write the equation in standard form by multiplying out the product on the left side and then subtracting 6 from both sides.

$$(x - 2)(x + 3) = 6 \qquad \text{This is the given equation.}$$
$$x^2 + 3x - 2x - 6 = 6 \qquad \text{Use the FOIL method.}$$
$$x^2 + x - 6 = 6 \qquad \text{Simplify.}$$
$$x^2 + x - 6 - 6 = 6 - 6 \qquad \text{Subtract 6 from both sides.}$$
$$x^2 + x - 12 = 0 \qquad \text{Simplify.}$$

Step 2 Factor.

$$x^2 + x - 12 = 0$$
$$(x + 4)(x - 3) = 0$$

Steps 3 and 4 Set each factor equal to zero and solve the resulting equations.

$$x + 4 = 0 \quad \text{or} \quad x - 3 = 0$$
$$x = -4 \quad\quad\quad x = 3$$

Step 5 Check the solutions in the original equation. Do this now and verify that the solutions are -4 and 3. The equation's solution set is $\{-4, 3\}$. ■

✔ **CHECK POINT 6** Solve: $(x - 5)(x - 2) = 28$.

③ Solve problems using quadratic equations.

Applications of Quadratic Equations

Solving quadratic equations by factoring can be used to answer questions about variables contained in mathematical models.

Figure 6.1

EXAMPLE 7   Modeling Motion

You throw a ball straight up from a rooftop 160 feet high with an initial speed of 48 feet per second. The formula

$$h = -16t^2 + 48t + 160$$

describes the ball's height above the ground, h, in feet, t seconds after you threw it. The ball misses the rooftop on its way down and eventually strikes the ground. The situation is illustrated in Figure 6.1. How long will it take for the ball to hit the ground?

Solution   The ball hits the ground when h, its height above the ground, is 0 feet. Thus, we substitute 0 for h in the given formula and solve for t.

$$h = -16t^2 + 48t + 160 \qquad \text{\small{This is the formula that models the ball's height.}}$$
$$0 = -16t^2 + 48t + 160 \qquad \text{\small{Substitute 0 for } h.}$$

It is easier to factor a trinomial with a positive leading coefficient. Thus, if a negative squared term appears in a quadratic equation, we make it positive by multiplying both sides of the equation by -1.

$$-1 \cdot 0 = -1(-16t^2 + 48t + 160)$$
$$0 = 16t^2 - 48t - 160$$

Do you see that each term on the right side of the equation changed sign? The left side of the equation remained zero. Now we continue to solve the equation.

$16t^2 - 48t - 160 = 0$	Reverse the two sides of the equation. This step is optional.
$16(t^2 - 3t - 10) = 0$	Factor out the GCF, 16.
$16(t - 5)(t + 2) = 0$	Factor the trinomial.
$t - 5 = 0 \quad \text{or} \quad t + 2 = 0$	Set each variable factor equal to 0.
$t = 5 \qquad\qquad t = -2$	Solve for t.

> Do not set the constant, 16, equal to zero: $16 \neq 0$.

Because we begin describing the ball's height at $t = 0$, we discard the solution $t = -2$. The ball hits the ground after 5 seconds. ∎

Figure 6.2 shows the graph of the formula $h = -16t^2 + 48t + 160$. The horizontal axis is labeled t, for the ball's time in motion. The vertical axis is labeled h, for the ball's height above the ground. Because time and height are both positive, the model is graphed in quadrant I only.

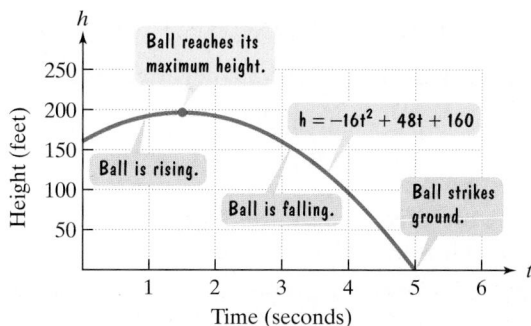

Figure 6.2

The graph visually shows what we discovered algebraically: The ball hits the ground after 5 seconds. The graph also reveals that the ball reaches its maximum height, nearly 200 feet, after 1.5 seconds. Then the ball begins to fall.

✔ **CHECK POINT 7** Use the formula $h = -16t^2 + 48t + 160$ to determine when the ball's height is 192 feet. Identify your solutions as points on the graph in Figure 6.2.

In our next example, we use our five-step strategy for solving word problems.

EXAMPLE 8 Solving a Problem About a Rectangle's Area

An architect is allowed no more than 15 square meters to add a small bedroom to a house. Because of the room's design in relationship to the existing structure, the width of its rectangular floor must be 7 meters less than two times the length. Find the precise length and width of the rectangular floor of maximum area that the architect is permitted.

Solution

Step 1 Let x represent one of the quantities. We know something about the width: It must be 7 meters less than two times the length. We will let

$$x = \text{the length of the floor.}$$

Step 2 Represent other quantities in terms of x. Because the width must be 7 meters less than two times the length, let

$$2x - 7 = \text{the width of the floor.}$$

The problem is illustrated in Figure 6.3.

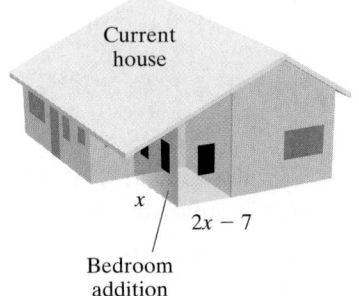

Current house

x

$2x - 7$

Bedroom addition

Figure 6.3

Step 3 Write an equation that describes the conditions. Because the architect is allowed no more than 15 square meters, an area of 15 square meters is the maximum area permitted. The area of a rectangle is the product of its length and its width.

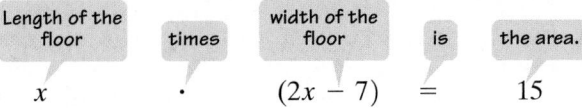

Length of the floor	times	width of the floor	is	the area.
x	$\cdot$	$(2x - 7)$	$=$	15

Step 4 Solve the equation and answer the question.

$$\begin{aligned}
x(2x - 7) &= 15 && \text{This is the equation for the problem's conditions.} \\
2x^2 - 7x &= 15 && \text{Use the distributive property.} \\
2x^2 - 7x - 15 &= 0 && \text{Subtract 15 from both sides.} \\
(2x + 3)(x - 5) &= 0 && \text{Factor.} \\
2x + 3 = 0 \quad \text{or} \quad x - 5 &= 0 && \text{Set each factor equal to zero.} \\
2x = -3 \qquad\qquad x &= 5 && \text{Solve the resulting equations.} \\
x = -\frac{3}{2}
\end{aligned}$$

A rectangle cannot have a negative length. Thus,

$$\text{Length} = x = 5$$
$$\text{Width} = 2x - 7 = 2 \cdot 5 - 7 = 10 - 7 = 3.$$

The architect is permitted a room of maximum area whose length is 5 meters and whose width is 3 meters.

Step 5 Check the proposed solution in the original wording of the problem. The area of the floor using the dimensions that we found is

$$A = lw = (5 \text{ meters})(3 \text{ meters}) = 15 \text{ square meters.}$$

Because the problem's wording tells us that the maximum area permitted is 15 square meters, our dimensions are correct. ∎

✔ **CHECK POINT 8**   The length of a rectangular sign is 3 feet longer than the width. If the sign's area is 54 square feet, find its length and width.

EXERCISE SET 6.6

 Practice Exercises

In Exercises 1–8, solve each equation using the zero-product principle.

1. $x(x + 3) = 0$

2. $x(x - 2) = 0$

3. $(x - 8)(x + 5) = 0$

4. $(x - 5)(x + 11) = 0$

5. $(x - 2)(4x + 5) = 0$

6. $(3x - 1)(x + 9) = 0$

7. $4(x - 3)(2x + 7) = 0$

8. $3(x - 6)(2x + 5) = 0$

In Exercises 9–56, use factoring to solve each quadratic equation. Check by substitution or by using a graphing utility to identify x-intercepts.

9. $x^2 + 8x + 15 = 0$

10. $x^2 + 5x + 6 = 0$

11. $x^2 - 2x - 15 = 0$

12. $x^2 + x - 42 = 0$

13. $x^2 - 4x = 21$

14. $x^2 + 7x = 18$

15. $x^2 + 9x = -8$

16. $x^2 - 11x = -10$

17. $x^2 + 4x = 0$

18. $x^2 - 6x = 0$

19. $x^2 - 5x = 0$

20. $x^2 + 3x = 0$

21. $x^2 = 4x$

22. $x^2 = 8x$

23. $2x^2 = 5x$

24. $3x^2 = 5x$

25. $3x^2 = -5x$

26. $2x^2 = -3x$

27. $x^2 + 4x + 4 = 0$

28. $x^2 + 6x + 9 = 0$

29. $x^2 = 12x - 36$

30. $x^2 = 14x - 49$

31. $4x^2 = 12x - 9$

32. $9x^2 = 30x - 25$

33. $2x^2 = 7x + 4$

34. $3x^2 = x + 4$

35. $5x^2 = 18 - x$

36. $3x^2 = 15 + 4x$

37. $x^2 - 49 = 0$

38. $x^2 - 25 = 0$

39. $4x^2 - 25 = 0$

40. $9x^2 - 100 = 0$

41. $81x^2 = 25$

42. $25x^2 = 49$

43. $x(x - 4) = 21$

44. $x(x - 3) = 18$

45. $4x(x + 1) = 15$

46. $x(3x + 8) = -5$

47. $(x - 1)(x + 4) = 14$

48. $(x - 3)(x + 8) = -30$

49. $(x + 1)(2x + 5) = -1$

50. $(x + 3)(3x + 5) = 7$

51. $y(y + 8) = 16(y - 1)$

52. $y(y + 9) = 4(2y + 5)$

53. $4y^2 + 20y + 25 = 0$

54. $4y^2 + 44y + 121 = 0$

55. $64w^2 = 48w - 9$

56. $25w^2 = 80w - 64$

Application Exercises

A ball is thrown straight up from a rooftop 300 feet high. The formula

$$h = -16t^2 + 20t + 300$$

describes the ball's height above the ground, h, in feet, t seconds after it was thrown. The ball misses the rooftop on its way down and eventually strikes the ground. The graph of the formula is shown, with tick marks omitted along the horizontal axis. Use the formula to solve Exercises 57–59.

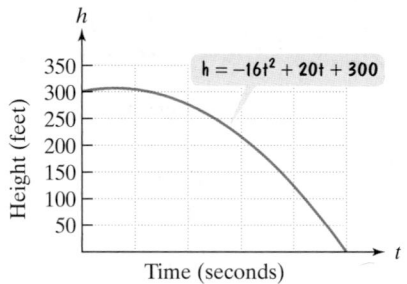

57. How long will it take for the ball to hit the ground? Use this information to provide tick marks with appropriate numbers along the horizontal axis in the figure shown.

58. When will the ball's height be 304 feet? Identify the solution as a point on the graph.

59. When will the ball's height be 276 feet? Identify the solution as a point on the graph.

An explosion causes debris to rise vertically with an initial speed of 72 feet per second. The formula

$$h = -16t^2 + 72t$$

describes the height of the debris above the ground, h, in feet, t seconds after the explosion. Use this information to solve Exercises 60–61.

60. How long will it take for the debris to hit the ground?

61. When will the debris be 32 feet above the ground?

The formula

$$N = 2x^2 + 22x + 320$$

models the number of inmates, in thousands, in U.S. state and federal prisons x years after 1980. The graph of the formula is shown in a [0, 20, 1] by [0, 1600, 100] viewing rectangle. Use the formula to solve Exercises 62–63.

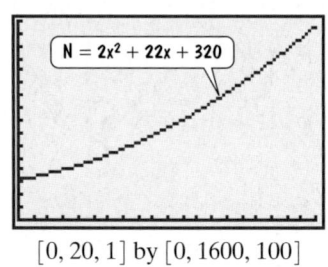

[0, 20, 1] by [0, 1600, 100]

62. In which year were there 740 thousand inmates in U.S. state and federal prisons? Identify the solution as a point on the graph shown.

63. In which year were there 1100 thousand inmates in U.S. state and federal prisons? Identify the solution as a point on the graph shown.

The alligator, an endangered species, is the subject of a protection program. The formula

$$P = -10x^2 + 475x + 3500$$

models the alligator population, P, after x years of the protection program, where $0 \le x \le 12$. Use the formula to solve Exercises 64–65.

64. After how long is the population up to 5990?

65. After how long is the population up to 7250?

The graph of the alligator population is shown over time. Use the graph to solve Exercises 66–67.

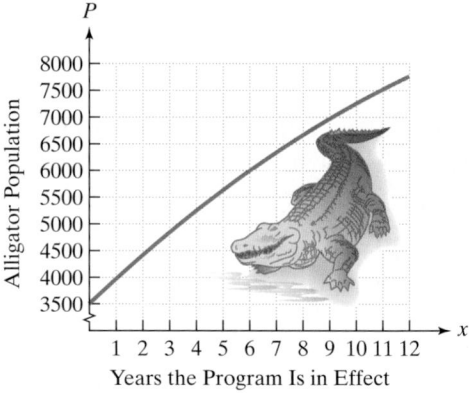

66. Identify your solution in Exercise 64 as a point on the graph.

67. Identify your solution in Exercise 65 as a point on the graph.

The formula

$$N = \frac{t^2 - t}{2}$$

describes the number of football games, N, that must be played in a league with t teams if each team is to play every other team once. Use this information to solve Exercises 68–69.

68. If a league has 36 games scheduled, how many teams belong to the league, assuming that each team plays every other team once?

69. If a league has 45 games scheduled, how many teams belong to the league, assuming that each team plays every other team once?

70. The length of a rectangular garden is 5 feet greater than the width. The area of the rectangle is 300 square feet. Find the length and the width.

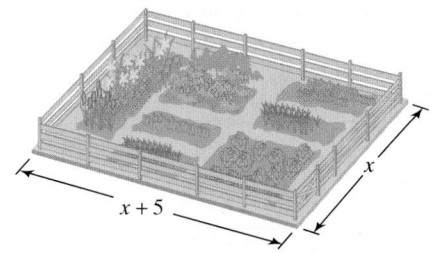

71. A rectangular parking lot has a length that is 3 yards greater than the width. The area of the parking lot is 180 square yards. Find the length and the width.

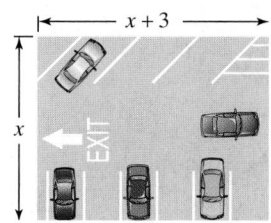

72. Each end of a glass prism is a triangle with a height that is 1 inch shorter than twice the base. If the area of the triangle is 60 square inches, how long are the base and height?

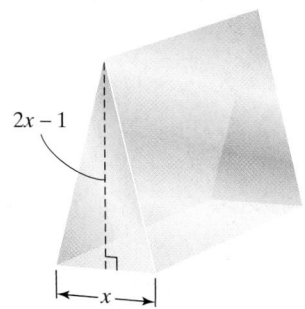

73. Great white sharks have triangular teeth with a height that is 1 centimeter longer than the base. If the area of one tooth is 15 square centimeters, find its base and height.

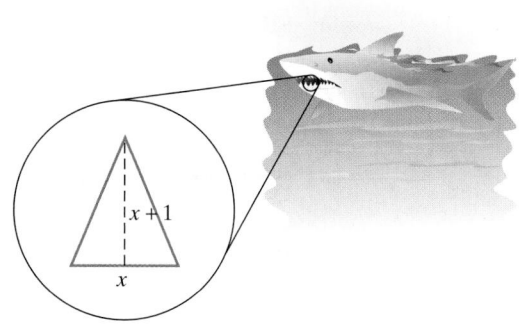

74. A vacant rectangular lot is being turned into a community vegetable garden measuring 15 meters by 12 meters. A path of uniform width is to surround the garden. If the area of the lot is 378 square meters, find the width of the path surrounding the garden.

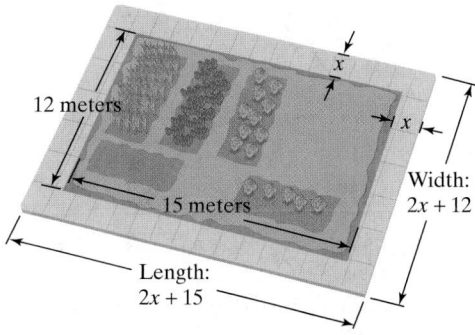

75. As part of a landscaping project, you put in a flower bed measuring 10 feet by 12 feet. You plan to surround the bed with a uniform border of low-growing plants.

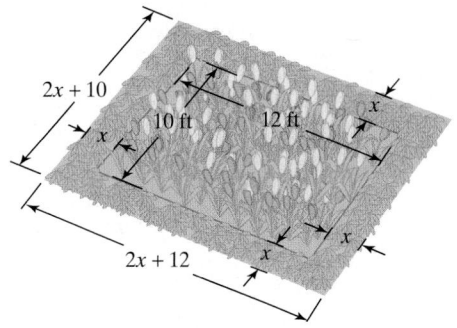

a. Write a polynomial that describes the area of the uniform border that surrounds your flower bed. (*Hint*: The area of the border is the area of the large rectangle shown in the figure minus the area of the flower bed.)

b. The low-growing plants surrounding the flower bed require 1 square foot each when mature. If you have 168 of these plants, how wide a strip around the flower bed should you prepare for the border?

Writing in Mathematics

76. What is a quadratic equation?

77. Explain how to solve $x^2 + 6x + 8 = 0$ using factoring and the zero-product principle.

78. If $(x + 2)(x - 4) = 0$ indicates that $x + 2 = 0$ or $x - 4 = 0$, explain why $(x + 2)(x - 4) = 6$ does not mean $x + 2 = 6$ or $x - 4 = 6$. Could we solve the equation using $x + 2 = 3$ and $x - 4 = 2$ because $3 \cdot 2 = 6$?

79. According to the U.S. Justice Department, in 1990 there were 739,980 inmates in U.S. state and federal prisons. By 1995, this number had increased to 1,085,022. Use this information to describe how well the formula that you used in Exercise 62 or 63 models the actual number of inmates for the year in question.

Critical Thinking Exercises

80. Which one of the following is true?
  a. If $(x + 3)(x - 4) = 2$, then $x + 3 = 0$ or $x - 4 = 0$.
  b. The solutions to the equation $4(x - 5)(x + 3) = 0$ are 4, 5, and −3.
  c. Equations solved by factoring always have two different solutions.
  d. Both 0 and $-\pi$ are solutions of the equation $x(x + \pi) = 0$.

81. Write a quadratic equation in standard form whose solutions are −3 and 5.

In Exercises 82–85, solve each equation.

82. $(x + 5)(x^2 + 13x + 36) = 0$

83. $x^3 + 3x^2 - 10x = 0$

84. $3^{x^2-9x+20} = 1$

85. $(x^2 - 5x + 5)^3 = 1$

In Exercises 86–89, match each equation with its graph. The graphs are labeled (a) through (d).

86. $y = x^2 - x - 2$ **87.** $y = x^2 + x - 2$

88. $y = x^2 - 4$ **89.** $y = x^2 - 4x$

a.

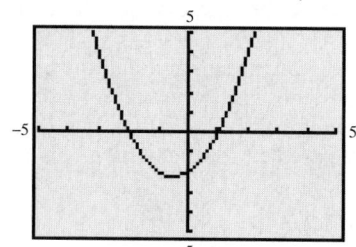

b.

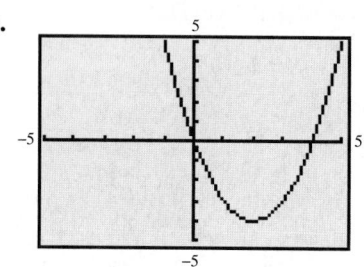

c.

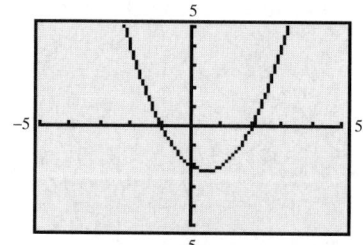

d.

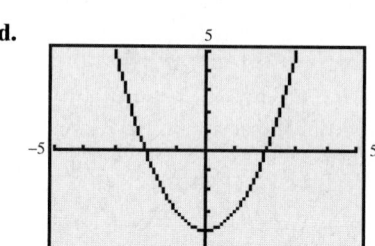

Technology Exercises

 In Exercises 90–93, use the x-intercepts for the graph in a $[-10, 10, 1]$ by $[-13, 10, 1]$ viewing rectangle to solve the quadratic equation. Check by substitution.

90. Use the graph of $y = x^2 + 3x - 4$ to solve
$$x^2 + 3x - 4 = 0.$$

91. Use the graph of $y = x^2 + x - 6$ to solve
$$x^2 + x - 6 = 0.$$

92. Use the graph of $y = (x - 2)(x + 3) - 6$ to solve
$$(x - 2)(x + 3) - 6 = 0.$$

93. Use the graph of $y = x^2 - 2x + 1$ to solve
$$x^2 - 2x + 1 = 0.$$

94. Use the technique of identifying x-intercepts on a graph generated by a graphing utility to check any five equations that you solved in Exercises 9–56.

95. If you have access to a calculator that solves quadratic equations, consult the owner's manual to determine how to use this feature. Then use your calculator to solve any five of the equations in Exercises 9–56.

Review Exercises

96. Graph: $y = -\dfrac{2}{3}x + 1$. (Section 3.4, Example 3)

97. Simplify: $\left(\dfrac{8x^4}{4x^7}\right)^2$. (Section 5.7, Example 6)

98. Solve: $5x + 28 = 6 - 6x$. (Section 2.2, Example 7)

CHAPTER 6 GROUP PROJECTS

1. Divide the group in half. Without looking at any factoring problems in the book, each group should create five factoring problems. Make sure that some of your problems require at least two factoring strategies. Next, exchange problems with the other half of the group. Work to factor the five problems. After completing the factorizations, evaluate the factoring problems that you were given. Are they too easy? Too difficult? Can the polynomials really be factored? Share your responses with the half of the group that wrote the problems. Finally, grade each other's work in factoring the polynomials. Each factoring problem is worth 20 points. You may award partial credit. If you take off points, explain why points are deducted and how you decided to take off a particular number of points for the error(s) that you found.

2. Group members are on the board of a condominium association. The condominium has just installed a small 35-foot-by-30-foot pool. Your job is to choose a material to surround the pool to create a border of uniform width.

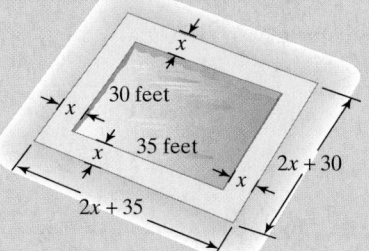

   a. Begin by writing an algebraic expression for the area, in square feet, of the border around the pool. (*Hint*: The border's area is the combined area of the pool and border minus the area of the pool.)
   b. You must select one of the following options for the border. Write an algebraic expression for the cost of installing the border for each of these options.

Options for the Border	Price
Cement	$6 per square foot
Outdoor carpeting	$5 per square foot plus $10 per foot to install edging around the rectangular border
Brick	$8 per square foot plus a $60 charge for delivering the bricks

   c. You would like the border to be 5 feet wide. Use the algebraic expressions in part (b) to find the cost of the border for each of the three options.
   d. You would prefer not to use cement. However, the condominium association is limited by a $5000 budget. Given this limitation, approximately how wide can the border be using outdoor carpeting or brick? Which option should you select and why?

3. **The Oddball's Oddball.** In their March 29, 1999 issue on the century's greatest minds, *Time* magazine called Paul Erdös (1913–1996) "the most prolific and arguably the cleverest mathematician of the century." Mathematicians gathered to hear his fascinating ideas about integers, whole numbers, and primes. Erdös solved problems about the nature of numbers and their factorizations, although he could not figure out how to boil an egg or wash his underwear. Speaking of mathematicians in general and Erdös in particular, the writers of *Time* stated, "In a profession with no shortage of oddballs, he was the strangest." Group members should do research on Paul Erdös, presenting a report on his work with numbers and his eccentricities.

CHAPTER SUMMARY, REVIEW, AND TEST

SUMMARY

DEFINITIONS AND CONCEPTS	EXAMPLES
Section 6.1 The Greatest Common Factor and Factoring By Grouping	
Factoring is the process of writing a polynomial as the product of two or more polynomials. The greatest common factor, GCF, is an expression that divides every term of the polynomial. The variable part of the GCF contains the smallest power of a variable that appears in all terms of the polynomial.	Find the GCF of $16x^2y, 20x^3y^2$, and $8x^2y^3$. The GCF of 16, 20, and 8 is 4. The GCF of x^2, x^3, and x^2 is x^2. The GCF of y, y^2, and y^3 is y. $$\text{GCF} = 4 \cdot x^2 \cdot y = 4x^2y$$
To factor a monomial from a polynomial, express each term as the product of the GCF and its other factor. Then use the distributive property to factor out the GCF.	$16x^2y + 20x^3y^2 + 8x^2y^3$ $= 4x^2y \cdot 4 + 4x^2y \cdot 5xy + 4x^2y \cdot 2y^2$ $= 4x^2y(4 + 5xy + 2y^2)$
To factor by grouping, factor out the GCF from each group. Then factor out the remaining factor.	$xy + 5x - 3y - 15$ $= x(y + 5) - 3(y + 5)$ $= (y + 5)(x - 3)$
Section 6.2 Factoring Trinomials Whose Leading Coefficient Is One	
To factor a trinomial of the form $x^2 + bx + c$, find two numbers whose product is c and whose sum is b. The factorization is $$(x + \text{ one number})(x + \text{ other number}).$$	Factor: $x^2 + 9x + 20$. Find two numbers whose product is 20 and whose sum is 9. The numbers are 4 and 5. $$x^2 + 9x + 20 = (x + 4)(x + 5)$$
Section 6.3 Factoring Trinomials Whose Leading Coefficient Is Not One	
To factor $ax^2 + bx + c$ by trial and error, try various combinations of factors of ax^2 and c until a middle term of bx is obtained for the sum of outside and inside products.	Factor: $3x^2 + 7x - 6$. Factors of $3x^2$: $3x, x$ Factors of -6: 1 and $-6, -1$ and 6, 2 and $-3, -2$ and 3. A possible combination of these factors is $$(3x - 2)(x + 3).$$ Sum of outside and inside products should equal $7x$. $$9x - 2x = 7x$$ Thus, $3x^2 + 7x - 6 = (3x - 2)(x + 3)$.
To factor $ax^2 + bx + c$ by grouping, find the factors of ac whose sum is b. Write bx using these factors. Then factor by grouping.	Factor: $3x^2 + 7x - 6$. Find the factors of $3(-6)$, or -18, whose sum is 7. They are 9 and -2. $3x^2 + 7x - 6$ $= 3x^2 + 9x - 2x - 6$ $= 3x(x + 3) - 2(x + 3) = (x + 3)(3x - 2)$

DEFINITIONS AND CONCEPTS	EXAMPLES

Section 6.4 Factoring Special Forms

The Difference of Two Squares

$$A^2 - B^2 = (A + B)(A - B)$$

$9x^2 - 25y^2$

$$= (3x)^2 - (5y)^2 = (3x + 5y)(3x - 5y)$$

Perfect Square Trinomials

$$A^2 + 2AB + B^2 = (A + B)^2$$
$$A^2 - 2AB + B^2 = (A - B)^2$$

$x^2 + 16x + 64 = x^2 + 2 \cdot x \cdot 8 + 8^2 = (x + 8)^2$

$25x^2 - 30x + 9 = (5x)^2 - 2 \cdot 5x \cdot 3 + 3^2 = (5x - 3)^2$

Sum and Difference of Cubes

$$A^3 + B^3 = (A + B)(A^2 - AB + B^2)$$
$$A^3 - B^3 = (A - B)(A^2 + AB + B^2)$$

$8x^3 - 125 = (2x)^3 - 5^3$

$$= (2x - 5)[(2x)^2 + 2x \cdot 5 + 5^2]$$

$$= (2x - 5)(4x^2 + 10x + 25)$$

Section 6.5 A General Factoring Strategy

A Factoring Strategy
1. Factor out the GCF.
2. a. If two terms, try
   $$A^2 - B^2 = (A + B)(A - B)$$
   $$A^3 + B^3 = (A + B)(A^2 - AB + B^2)$$
   $$A^3 - B^3 = (A - B)(A^2 + AB + B^2).$$
   b. If three terms, try
   $$A^2 + 2AB + B^2 = (A + B)^2$$
   $$A^2 - 2AB + B^2 = (A - B)^2.$$
   If not a perfect square trinomial, try trial and error or grouping.
   c. If four terms, try factoring by grouping.
3. See if any factors can be factored further.
4. Check by multiplying.

Factor: $2x^4 + 10x^3 - 8x^2 - 40x$.
The GCF is $2x$.

$$2x^4 + 10x^3 - 8x^2 - 40x$$
$$= 2x(x^3 + 5x^2 - 4x - 20)$$

Four terms: Try grouping.

$$= 2x[x^2(x + 5) - 4(x + 5)]$$
$$= 2x(x + 5)(x^2 - 4)$$

This can be factored further.

$$= 2x(x + 5)(x + 2)(x - 2)$$

Section 6.6 Solving Quadratic Equations by Factoring

The Zero-Product Principle
If $AB = 0$, then $A = 0$ or $B = 0$.

Solve: $(x - 6)(x + 10) = 0$
$$x - 6 = 0 \quad \text{or} \quad x + 10 = 0$$
$$x = 6 \qquad\qquad x = -10$$
The solution are -10 and 6, and the solution set is $\{-10, 6\}$.

A quadratic equation in x is an equation that can be written in the standard form

$$ax^2 + bx + c = 0, \quad a \neq 0.$$

To solve by factoring, write the equation in standard form, factor, set each factor equal to zero, and solve each resulting equation. Check proposed solutions in the original equation.

Solve: $4x^2 + 9x = 9$.
$$4x^2 + 9x - 9 = 0$$
$$(4x - 3)(x + 3) = 0$$
$$4x - 3 = 0 \quad \text{or} \quad x + 3 = 0$$
$$x = \frac{3}{4} \qquad\qquad x = -3$$
The solution are -3 and $\frac{3}{4}$, and the solution set is $\{-3, \frac{3}{4}\}$.

Review Exercises

6.1 *In Exercises 1–5, factor each polynomial using the greatest common factor. If there is no common factor other than 1 and the polynomial cannot be factored, so state.*

1. $30x - 45$

2. $12x^3 + 16x^2 - 400x$

3. $30x^4y + 15x^3y + 5x^2y$

4. $7(x + 3) - 2(x + 3)$

5. $7x^2(x + y) - (x + y)$

In Exercises 6–9, factor by grouping.

6. $x^3 + 3x^2 + 2x + 6$

7. $xy + y + 4x + 4$

8. $x^3 + 5x + x^2 + 5$

9. $xy + 4x - 2y - 8$

6.2 *In Exercises 10–17, factor completely, or state that the trinomial is prime.*

10. $x^2 - 3x + 2$

11. $x^2 - x - 20$

12. $x^2 + 19x + 48$

13. $x^2 - 6xy + 8y^2$

14. $x^2 + 5x - 9$

15. $x^2 + 16xy - 17y^2$

16. $3x^2 + 6x - 24$

17. $3x^3 - 36x^2 + 33x$

6.3 *In Exercises 18–26, factor completely, or state that the trinomial is prime.*

18. $3x^2 + 17x + 10$

19. $5y^2 - 17y + 6$

20. $4x^2 + 4x - 15$

21. $5y^2 + 11y + 4$

22. $8x^2 + 8x - 6$

23. $2x^3 + 7x^2 - 72x$

24. $12y^3 + 28y^2 + 8y$

25. $2x^2 - 7xy + 3y^2$

26. $5x^2 - 6xy - 8y^2$

6.4 *In Exercises 27–30, factor each difference of two squares completely.*

27. $4x^2 - 1$

28. $81 - 100y^2$

29. $25a^2 - 49b^2$

30. $z^4 - 16$

In Exercises 31–34, factor completely, or state that the polynomial is prime.

31. $2x^2 - 18$

32. $x^2 + 1$

33. $9x^3 - x$

34. $18xy^2 - 8x$

In Exercises 35–41, factor any perfect square trinomials, or state that the polynomial is prime.

35. $x^2 + 22x + 121$

36. $x^2 - 16x + 64$

37. $9y^2 + 48y + 64$

38. $16x^2 - 40x + 25$

39. $25x^2 + 15x + 9$

40. $36x^2 + 60xy + 25y^2$

41. $25x^2 - 40xy + 16y^2$

In Exercises 42–45, factor using the formula for the sum or difference of two cubes.

42. $x^3 - 27$

43. $64x^3 + 1$

44. $54x^3 - 16y^3$

45. $27x^3y + 8y$

In Exercises 46–47, find the formula for the area of the shaded region and express it in factored form.

46.

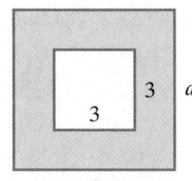

47.

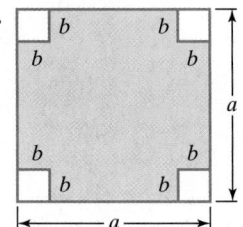

48. The figure shows a geometric interpretation of a factorization. Use the sum of the areas of the four pieces on the left and the area of the square on the right to write the factorization that is illustrated.

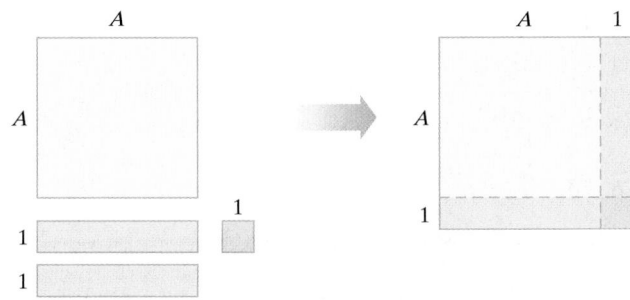

6.5 *In Exercises 49–81, factor completely, or state that the polynomial is prime.*

49. $x^3 - 8x^2 + 7x$

50. $10y^2 + 9y + 2$

51. $128 - 2y^2$

52. $9x^2 + 6x + 1$

53. $20x^7 - 36x^3$

54. $x^3 - 3x^2 - 9x + 27$

55. $y^2 + 16$

56. $2x^3 + 19x^2 + 35x$

57. $3x^3 - 30x^2 + 75x$

58. $3x^5 - 24x^2$

59. $4y^4 - 36y^2$

60. $5x^2 + 20x - 105$

61. $9x^2 + 8x - 3$

62. $10x^5 - 44x^4 + 16x^3$

63. $100y^2 - 49$

64. $9x^5 - 18x^4$

65. $x^4 - 1$

66. $2y^3 - 16$

67. $x^3 + 64$

68. $6x^2 + 11x - 10$

69. $3x^4 - 12x^2$

70. $x^2 - x - 90$

71. $25x^2 + 25xy + 6y^2$

72. $x^4 + 125x$

73. $32y^3 + 32y^2 + 6y$

74. $2y^2 - 16y + 32$

75. $x^2 - 2xy - 35y^2$

76. $x^2 + 7x + xy + 7y$

77. $9x^2 + 24xy + 16y^2$

78. $2x^4y - 2x^2y$

79. $100y^2 - 49z^2$

80. $x^2 + xy + y^2$

81. $3x^4y^2 - 12x^2y^4$

6.6 *In Exercises 82–83, solve each equation using the zero-product principle.*

82. $x(x - 12) = 0$

83. $3(x - 7)(4x + 9) = 0$

In Exercises 84–92, use factoring to solve each quadratic equation.

84. $x^2 + 5x - 14 = 0$

85. $5x^2 + 20x = 0$

86. $2x^2 + 15x = 8$

87. $x(x - 4) = 32$

88. $(x + 3)(x - 2) = 50$

89. $x^2 = 14x - 49$

90. $9x^2 = 100$

91. $3x^2 + 21x + 30 = 0$

92. $3x^2 = 22x - 7$

93. You dive from a board that is 32 feet above the water. The formula

$$h = -16t^2 + 16t + 32$$

describes your height above the water, h, in feet, t seconds after you dive. How long will it take you to hit the water?

94. The length of a rectangular sign is 3 feet longer than the width. If the sign has space for 40 square feet of advertising, find its length and its width.

95. The square lot shown here is being turned into a garden with a 3-meter path at one end. If the area of the garden is 88 square meters, find the dimensions of the square lot.

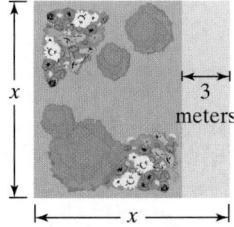

Chapter 6 Test

In Exercises 1–21, factor completely, or state that the polynomial is prime.

1. $x^2 - 9x + 18$

2. $x^2 - 14x + 49$

3. $15y^4 - 35y^3 + 10y^2$

4. $x^3 + 2x^2 + 3x + 6$

5. $x^2 - 9x$

6. $x^3 + 6x^2 - 7x$

7. $14x^2 + 64x - 30$

8. $25x^2 - 9$

9. $x^3 + 8$

10. $x^2 - 4x - 21$

11. $x^2 + 4$

12. $6y^3 + 9y^2 + 3y$

13. $4y^2 - 36$

14. $16x^2 + 48x + 36$

15. $2x^4 - 32$

16. $36x^2 - 84x + 49$

17. $7x^2 - 50x + 7$

18. $x^3 + 2x^2 - 5x - 10$

19. $12y^3 - 12y^2 - 45y$

20. $y^3 - 125$

21. $5x^2 - 5xy - 30y^2$

In Exercises 22–27, solve each quadratic equation.

22. $x^2 + 2x - 24 = 0$

23. $3x^2 - 5x = 2$

24. $x(x - 6) = 16$

25. $6x^2 = 21x$

26. $16x^2 = 81$

27. $(5x + 4)(x - 1) = 2$

28. Find a formula for the area of the shaded region and express it in factored form.

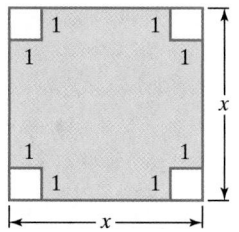

29. A model rocket is launched from a height of 96 feet. The formula

$$h = -16t^2 + 80t + 96$$

describes the rocket's height, h, in feet, t seconds after it was launched. How long will it take the rocket to reach the ground?

30. The length of a rectangular garden is 6 feet longer than its width. If the area of the garden is 55 square feet, find its length and its width.

Cumulative Review Exercises (Chapters 1–6)

1. Simplify: $6[5 + 2(3 - 8) - 3]$.

2. Solve: $4(x - 2) = 2(x - 4) + 3x$.

3. Solve: $\dfrac{x}{2} - 1 = \dfrac{x}{3} + 1$.

4. Solve and express the solution set in set-builder notation. Graph the solution set on a number line.

$$5 - 5x > 2(5 - x) + 1$$

5. Find the measures of the angles of a triangle whose two base angles have equal measure and whose third angle is 10° less than three times the measure of a base angle.

6. A dinner for six people cost $159, including a 6% tax. What was the dinner's cost before tax?

7. Graph using the slope and y-intercept: $y = -\frac{3}{5}x + 3$.

8. Write the point-slope form of the line passing through $(2, -4)$ and $(3, 1)$. Then use the point-slope form of the equation to write the slope-intercept equation.

9. Graph: $5x - 6y = 30$.

10. Solve the system:

$$5x + 2y = 13$$
$$y = 2x - 7.$$

11. Solve the system:

$$2x + 3y = 5$$
$$3x - 2y = -4.$$

12. Subtract: $\dfrac{4}{5} - \dfrac{9}{8}$.

In Exercises 13–15, perform the indicated operations.

13. $\dfrac{6x^5 - 3x^4 + 9x^2 + 27x}{3x}$

14. $(3x - 5y)(2x + 9y)$

15. $\dfrac{6x^3 + 5x^2 - 34x + 13}{3x - 5}$

16. Write 0.0071 in scientific notation.

In Exercises 17–19, factor completely.

17. $3x^2 + 11x + 6$

18. $y^5 - 16y$

19. $4x^2 + 12x + 9$

20. The length of a rectangle is 2 feet greater than its width. If the rectangle's area is 24 square feet, find its dimensions.

At the start of the twenty-first century, we are plagued by questions about the environment. Will we run out of gas? How hot will it get? Will there be neighborhoods where the air is pristine? Can we make garbage disappear? Will there be any wilderness left? Which wild animals will become extinct? How much will it cost to clean up toxic wastes from our rivers so that they can safely provide food, recreation, and enjoyment of wildlife for the millions who live along and visit their shores?

W hen making decisions on public policies dealing with the environment, two important questions are:

- What are the costs?
- What are the benefits?

Algebraic fractions play an important role in modeling these costs. By learning to work with these fractional expressions, you will gain new insights into phenomena as diverse as the dosage of drugs prescribed for children, inventory costs for a business, the cost of environmental cleanup, and even the shape of our heads.

Chapter 7

Rational Expressions

431

▶ SECTION 7.1 *Rational Expressions and Their Simplification*

Objectives

1 Find numbers for which a rational expression is undefined.

2 Simplify rational expressions.

3 Solve applied problems involving rational expressions.

SSM
PH Tutor CD- Video
Center ROM

This photograph was selected prior to September 11, 2001, our new date of infamy. Its inclusion should not be taken as insensitivity to the events of that terrible day.

How do we describe the costs of reducing environmental pollution? We often use algebraic expressions involving quotients of polynomials. For example, the algebraic expression

$$\frac{250x}{100 - x}$$

describes the cost, in millions of dollars, to remove x percent of the pollutants that are discharged into a river. Removing a modest percentage of pollutants, say 40%, is far less costly than removing a substantially greater percentage, such as 95%. We see this by evaluating the algebraic expression for $x = 40$ and $x = 95$.

Evaluating $\dfrac{250x}{100 - x}$ for

$x = 40$:	$x = 95$:
Cost is $\dfrac{250(40)}{100 - 40} \approx 167.$	Cost is $\dfrac{250(95)}{100 - 95} = 4750.$

The cost increases from approximately $167 million to a possibly prohibitive $4750 million, or $4.75 billion. Costs spiral upward as the percentage of removed pollutants increases.

Many algebraic expressions that describe costs of environmental projects are examples of *rational expressions*. In this section, we introduce rational expressions and their simplification.

Discover for Yourself

What happens if you try substituting 100 for x in

$$\frac{250x}{100 - x} \; ?$$

What does this tell you about the cost of cleaning up all of the river's pollutants?

1 Find numbers for which a rational expression is undefined.

Excluding Numbers from Rational Expressions

A **rational expression** is the quotient of two polynomials. Some examples are:

$$\frac{x - 2}{4}, \quad \frac{4}{x - 2}, \quad \frac{x}{x^2 - 1}, \quad \text{and} \quad \frac{x^2 + 1}{x^2 + 2x - 3}.$$

Rational expressions indicate division and division by zero is undefined. This means that **we must exclude any value or values of the variable that make the denominator zero**. For example, consider the rational expression

$$\frac{4}{x - 2}.$$

When x is replaced by 2, the denominator is 0 and the expression is undefined.

If $x = 2$: $\quad \dfrac{4}{x - 2} = \dfrac{4}{2 - 2} = \dfrac{4}{0}.$ Division by zero is undefined.

Notice that if x is replaced by a number other than 2, such as 1, the expression is defined because the denominator is nonzero.

If $x = 1$: $\quad \dfrac{4}{x - 2} = \dfrac{4}{1 - 2} = \dfrac{4}{-1} = -4.$

Thus, only 2 must be excluded as a replacement for x in the rational expression $\dfrac{4}{x - 2}.$

Discover for Yourself

Use a graphing utility to graph the equation

$$y = \frac{4}{x - 2}.$$

Then use the ⌐TRACE⌐ feature to determine what happens as x gets close to 2 and what happens at $x = 2$. Describe what you observe.

Excluding Values From Rational Expressions

If a variable in a rational expression is replaced by a number that causes the denominator to be 0, that number must be excluded as a replacement for the variable. The rational expression is undefined at any value that produces a denominator of 0.

How do we determine the value or values of the variable for which a rational expression is undefined? Set the denominator equal to 0 and then solve the resulting equation for the variable.

EXAMPLE 1 Determining Numbers for Which Rational Expressions Are Undefined

Find all the numbers for which the rational expression is undefined:

a. $\dfrac{6x + 12}{7x - 28}$

b. $\dfrac{2x + 6}{x^2 + 3x - 10}.$

Solution In each case, we set the denominator equal to 0 and solve.

$$\dfrac{6x + 12}{7x - 28} \qquad \text{Exclude values of } x \text{ that make these denominators 0.} \qquad \dfrac{2x + 6}{x^2 + 3x - 10}$$

a. $7x - 28 = 0$ Set the denominator equal to 0.

$\qquad 7x = 28$ Add 28 to both sides.

$\qquad\quad x = 4$ Divide both sides by 7.

Thus, $\dfrac{6x + 12}{7x - 28}$ is undefined for $x = 4.$

b. $x^2 + 3x - 10 = 0$ Set the denominator equal to 0.

$\quad (x + 5)(x - 2) = 0$ Factor.

$\qquad x + 5 = 0 \quad \text{or} \quad x - 2 = 0$ Set each factor equal to 0.

$\qquad\qquad x = -5 \qquad\qquad x = 2$ Solve the resulting equations.

Thus, $\dfrac{2x + 6}{x^2 + 3x - 10}$ is undefined for $x = -5$ and $x = 2.$ ∎

Using Technology

When using a graphing utility to graph an equation containing a rational expression, you might not be pleased with the quality of the display. Compare these two graphs of $y = \dfrac{6x + 12}{7x - 28}$.

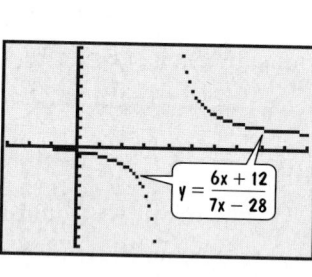
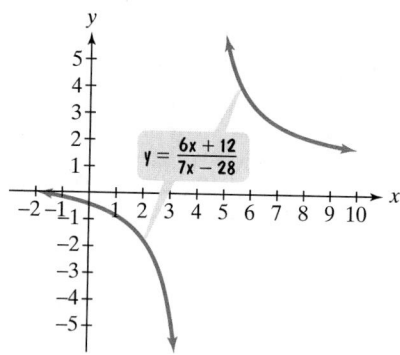

The graph on the left was obtained using the $\boxed{\text{DOT}}$ mode in a $[-3, 10, 1]$ by $[-10, 10, 1]$ viewing rectangle. Examine the behavior of the graph near $x = 4$, the number for which the rational expression is undefined. The values of the rational expression are decreasing as the values of x get closer to 4 on the left and increasing as the values of x get closer to 4 on the right. However, there is no point on the graph corresponding to $x = 4$. Would you agree that this behavior is better illustrated in the hand-drawn graph on the right?

✔ **CHECK POINT 1**   Find all the numbers for which the rational expression is undefined:

a. $\dfrac{7x - 28}{8x - 40}$ **b.** $\dfrac{8x - 40}{x^2 + 3x - 28}.$

Is every rational expression undefined for at least one number? No. Consider

$$\frac{x - 2}{4}.$$

Because the denominator is not zero for any value of x, the rational expression is defined for all real numbers. Thus, it is not necessary to exclude any values for x.

2 Simplify rational expressions.

Simplifying Rational Expressions

A rational expression is **simplified** if its numerator and denominator have no common factors other than 1 or −1. The following principle is used to simplify a rational expression.

Fundamental Principle of Rational Expressions

If P, Q, and R are polynomials, and Q and R are not 0,

$$\frac{PR}{QR} = \frac{P}{Q}.$$

As you read the Fundamental Principle, can you see why $\dfrac{PR}{QR}$ is not simplified? The numerator and denominator have a common factor, the polynomial R. By dividing the numerator and the denominator by the common factor, R, we obtain the simplified form $\dfrac{P}{Q}$. This is often shown as follows:

$$\frac{P\overset{1}{\cancel{R}}}{Q\underset{1}{\cancel{R}}} = \frac{P}{Q}.$$

Observe that
$$\frac{PR}{QR} = \frac{P}{Q} \cdot \frac{R}{R} = \frac{P}{Q} \cdot 1 = \frac{P}{Q}.$$

The following procedure can be used to simplify rational expressions.

Simplifying Rational Expressions

1. Factor the numerator and denominator completely.

2. Divide both the numerator and denominator by the common factors.

EXAMPLE 2   Simplifying a Rational Expression

Simplify: $\dfrac{5x + 35}{20x}$.

Solution

$$\frac{5x + 35}{20x} = \frac{5(x + 7)}{5 \cdot 4x}$$

Factor the numerator and denominator. Because the denominator is $20x$, $x \neq 0$.

$$= \frac{\overset{1}{\cancel{5}}(x + 7)}{\underset{1}{\cancel{5}} \cdot 4x}$$

Divide out the common factor of 5.

$$= \frac{x + 7}{4x}$$

■

✔ **CHECK POINT 2**   Simplify: $\dfrac{7x + 28}{21x}$.

EXAMPLE 3   Simplifying a Rational Expression

Simplify: $\dfrac{x^3 + x^2}{x + 1}$.

Solution

$$\frac{x^3 + x^2}{x + 1} = \frac{x^2(x + 1)}{x + 1}$$

Factor the numerator. Because the denominator is $x + 1$, $x \neq -1$.

$$= \frac{x^2 \overset{1}{\cancel{(x + 1)}}}{\underset{1}{\cancel{x + 1}}}$$

Divide out the common factor of $x + 1$.

$$= x^2$$

■

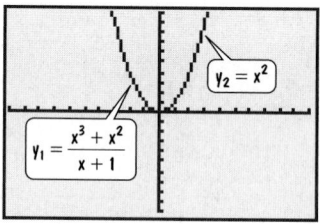
Simplifying a rational expression can change the numbers that make it undefined. For example, we just showed that

$$\frac{x^3 + x^2}{x + 1} = x^2.$$

This is undefined for $x = -1$.

The simplified form is defined for all real numbers.

Thus, to equate the two expressions, we must restrict the values for x in the simplified expression to exclude -1. We can write

$$\frac{x^3 + x^2}{x + 1} = x^2, \quad x \neq -1.$$

Hereafter, we will assume that the simplified rational expression is equal to the original rational expression for all real numbers except those for which either denominator is 0.

✔ **CHECK POINT 3** Simplify: $\dfrac{x^3 - x^2}{7x - 7}$.

EXAMPLE 4 Simplifying a Rational Expression

Simplify: $\dfrac{x^2 + 6x + 5}{x^2 - 25}$.

Solution

$$\frac{x^2 + 6x + 5}{x^2 - 25} = \frac{(x + 5)(x + 1)}{(x + 5)(x - 5)}$$

Factor the numerator and denominator. Because the denominator is $(x + 5)(x - 5)$, $x \neq -5$ and $x \neq 5$.

$$= \frac{\overset{1}{\cancel{(x + 5)}}(x + 1)}{\underset{1}{\cancel{(x + 5)}}(x - 5)}$$

Divide out the common factor of $x + 5$.

$$= \frac{x + 1}{x - 5}$$

✔ **CHECK POINT 4** Simplify: $\dfrac{x^2 - 1}{x^2 + 2x + 1}$.

Factors That Are Opposites

How do we simplify rational expressions that contain factors in the numerator and denominator that are opposites, or additive inverses? Here is an example of such an expression:

$$\frac{x - 3}{3 - x}.$$

The numerator and denominator are opposites. They differ only in their signs.

Study Tip

When simplifying rational expressions, only *factors* that are common to the *entire numerator* and the *entire denominator* can be divided out. **It is incorrect to divide out common terms from the numerator and denominator.**

INCORRECT

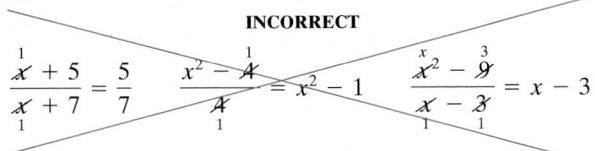

$$\frac{\overset{1}{\cancel{x}} + 5}{\underset{1}{\cancel{x}} + 7} = \frac{5}{7} \qquad \frac{\overset{1}{x^2} - \overset{1}{\cancel{4}}}{\underset{1}{\cancel{4}}} = x^2 - 1 \qquad \frac{\overset{x}{x^2} - \overset{3}{\cancel{9}}}{\underset{1}{\cancel{x}} - \underset{1}{\cancel{3}}} = x - 3$$

The first two expressions, $\dfrac{x + 5}{x + 7}$ and $\dfrac{x^2 - 4}{4}$, have no common factors in their numerators and denominators. Thus, these rational expressions are in simplified form. The rational expression $\dfrac{x^2 - 9}{x - 3}$ can be simplified as follows:

CORRECT

$$\frac{x^2 - 9}{x - 3} = \frac{(x + 3)\overset{1}{\cancel{(x - 3)}}}{\underset{1}{\cancel{x - 3}}} = x + 3.$$

Divide out the common factor, $x - 3$.

Factor out -1 from either the numerator or the denominator. Then divide out the common factor.

$$\frac{x - 3}{3 - x} = \frac{-1(-x + 3)}{3 - x}$$

Factor -1 from the numerator. Notice how the sign of each term in the polynomial changes.

$$= \frac{-1(3 - x)}{3 - x}$$

In the numerator, use the commutative property to rewrite $-x + 3$ as $3 - x$.

$$= \frac{-1\overset{1}{\cancel{(3 - x)}}}{\underset{1}{\cancel{3 - x}}}$$

Divide out the common factor of $3 - x$.

$$= -1$$

Our result, -1, suggests the following useful property.

Simplifying Rational Expressions with Opposite Factors in the Numerator and Denominator

The quotient of two polynomials that have opposite signs and are additive inverses is -1.

EXAMPLE 5 Simplifying a Rational Expression

Simplify: $\dfrac{4x^2 - 25}{15 - 6x}$.

Solution

$$\frac{4x^2 - 25}{15 - 6x} = \frac{(2x + 5)(2x - 5)}{3(5 - 2x)} \qquad \text{Factor the numerator and denominator.}$$

$$= \frac{(2x + 5)\overset{-1}{\cancel{(2x - 5)}}}{3\cancel{(5 - 2x)}} \qquad \text{The quotient of polynomials with opposite signs is } -1.$$

$$= \frac{-(2x + 5)}{3} \quad \text{or} \quad -\frac{2x + 5}{3} \quad \text{or} \quad \frac{-2x - 5}{3}$$

Each of these forms is an acceptable answer.

3 Solve applied problems involving rational expressions.

✔ **CHECK POINT 5** Simplify: $\dfrac{9x^2 - 49}{28 - 12x}$.

Applications

The equation

$$y = \frac{250x}{100 - x}$$

models the cost, in millions of dollars, to remove x percent of the pollutants that are discharged into a river. This equation contains the rational expression that we looked at in the opening to this section. Do you remember how costs were spiraling upward as the percentage of removed pollutants increased?

Is it possible to clean up the river completely? To do this, we must remove 100% of the pollutants. The problem is that the rational expression is undefined for $x = 100$.

$$y = \frac{250x}{100 - x} \qquad \boxed{\text{If } x = 100, \text{ the value of the denominator is } 0.}$$

Notice how the graph of $y = \dfrac{250x}{100 - x}$, shown in Figure 7.1, approaches but never touches the dashed vertical line drawn through $x = 100$, our undefined value. The graph continues to rise more and more steeply, visually showing the escalating costs. By never touching the dashed vertical line, the graph illustrates that no amount of money will be enough to remove all pollutants from the river.

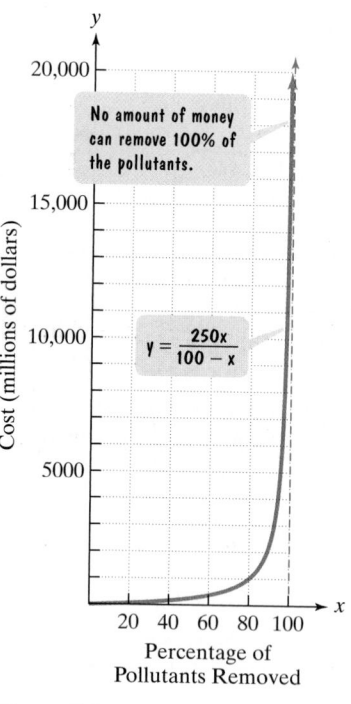

No amount of money can remove 100% of the pollutants.

$y = \dfrac{250x}{100 - x}$

Percentage of Pollutants Removed

Figure 7.1

Practice Exercises

In Exercises 1–20, find all numbers for which each rational expression is undefined. If the rational expression is defined for all real numbers, so state.

1. $\dfrac{7}{2x}$

2. $\dfrac{8}{3x}$

3. $\dfrac{x}{x - 7}$

4. $\dfrac{x}{x - 5}$

5. $\dfrac{7}{5x - 15}$

6. $\dfrac{9}{6x - 18}$

7. $\dfrac{x + 4}{(x + 7)(x - 3)}$

8. $\dfrac{x + 4}{(x + 6)(x - 8)}$

9. $\dfrac{13x}{(3x - 15)(x + 2)}$

10. $\dfrac{7x}{(x - 1)(2x + 6)}$

11. $\dfrac{x + 5}{x^2 + x - 12}$

12. $\dfrac{7x - 14}{x^2 - 9x + 20}$

13. $\dfrac{x + 5}{5}$

14. $\dfrac{x + 7}{7}$

15. $\dfrac{y + 3}{4y^2 + y - 3}$

16. $\dfrac{y + 8}{6y^2 - y - 2}$

17. $\dfrac{y + 5}{y^2 - 25}$

18. $\dfrac{y + 7}{y^2 - 49}$

19. $\dfrac{5}{x^2 + 1}$

20. $\dfrac{8}{x^2 + 4}$

In Exercises 21–76, simplify each rational expression. If the rational expression cannot be simplified, so state.

21. $\dfrac{14x^2}{7x}$

22. $\dfrac{9x^2}{6x}$

23. $\dfrac{5x - 15}{25}$

24. $\dfrac{7x + 21}{49}$

25. $\dfrac{2x - 8}{4x}$

26. $\dfrac{3x - 9}{6x}$

27. $\dfrac{3}{3x - 9}$

28. $\dfrac{12}{6x - 18}$

29. $\dfrac{-15}{3x - 9}$

30. $\dfrac{-21}{7x - 14}$

31. $\dfrac{3x + 9}{x + 3}$

32. $\dfrac{5x - 10}{x - 2}$

33. $\dfrac{x + 5}{x^2 - 25}$

34. $\dfrac{x + 4}{x^2 - 16}$

35. $\dfrac{2y - 10}{3y - 15}$

36. $\dfrac{6y + 18}{11y + 33}$

37. $\dfrac{x + 1}{x^2 - 2x - 3}$

38. $\dfrac{x + 2}{x^2 - x - 6}$

39. $\dfrac{4x - 8}{x^2 - 4x + 4}$

40. $\dfrac{x^2 - 12x + 36}{4x - 24}$

41. $\dfrac{y^2 - 3y + 2}{y^2 + 7y - 18}$

42. $\dfrac{y^2 + 5y + 4}{y^2 - 4y - 5}$

43. $\dfrac{2y^2 - 7y + 3}{2y^2 - 5y + 2}$

44. $\dfrac{3y^2 + 4y - 4}{6y^2 - y - 2}$

45. $\dfrac{2x + 3}{2x + 5}$

46. $\dfrac{3x + 7}{3x + 10}$

47. $\dfrac{x^2 + 12x + 36}{x^2 - 36}$

48. $\dfrac{x^2 - 14x + 49}{x^2 - 49}$

49. $\dfrac{x^3 - 2x^2 + x - 2}{x - 2}$

50. $\dfrac{x^3 + 4x^2 - 3x - 12}{x + 4}$

51. $\dfrac{x^3 - 8}{x - 2}$

52. $\dfrac{x^3 - 125}{x^2 - 25}$

53. $\dfrac{(x - 4)^2}{x^2 - 16}$

54. $\dfrac{(x + 5)^2}{x^2 - 25}$

55. $\dfrac{x}{x + 1}$

56. $\dfrac{x}{x + 7}$

57. $\dfrac{x + 4}{x^2 + 16}$

58. $\dfrac{x + 5}{x^2 + 25}$

59. $\dfrac{x - 5}{5 - x}$

60. $\dfrac{x - 7}{7 - x}$

61. $\dfrac{2x - 3}{3 - 2x}$

62. $\dfrac{5x - 4}{4 - 5x}$

63. $\dfrac{x - 5}{x + 5}$

64. $\dfrac{x - 7}{x + 7}$

65. $\dfrac{4x - 6}{3 - 2x}$

66. $\dfrac{9x - 15}{5 - 3x}$

67. $\dfrac{4 - 6x}{3x^2 - 2x}$

68. $\dfrac{9 - 15x}{5x^2 - 3x}$

69. $\dfrac{x^2 - 1}{1 - x}$

70. $\dfrac{x^2 - 4}{2 - x}$

71. $\dfrac{y^2 - y - 12}{4 - y}$

72. $\dfrac{y^2 - 7y + 12}{3 - y}$

73. $\dfrac{x^2 y - x^2}{x^3 - x^3 y}$

74. $\dfrac{xy - 2x}{3y - 6}$

75. $\dfrac{x^2 + 2xy - 3y^2}{2x^2 + 5xy - 3y^2}$

76. $\dfrac{x^2 + 3xy - 10y^2}{3x^2 - 7xy + 2y^2}$

Application Exercises

77. The polynomial
$$6t^4 - 207t^3 + 2128t^2 - 6622t + 15{,}220$$

describes the annual number of drug convictions in the United States t years after 1984. The polynomial
$$28t^4 - 711t^3 + 5963t^2 - 1695t + 27{,}424$$

describes the annual number of drug arrests in the United States t years after 1984. Write a rational expression that describes the conviction rate for drug arrests in the United States t years after 1984.

78. The polynomial $-0.14t^2 + 0.51t + 31.6$ describes the U.S. population, in millions, ages 65 and older t years after 1990. The polynomial $0.54t^2 + 12.64t + 107.1$ describes the total yearly cost of Medicare, in billions of dollars, t years after 1990. Write a rational expression that describes the average cost, in thousands of dollars, of Medicare per person ages 65 or older t years after 1990.

79. The rational expression

$$\frac{130x}{100 - x}$$

describes the cost, in millions of dollars, to inoculate x percent of the population against a particular strain of flu.

a. Evaluate the expression for $x = 40$, $x = 80$, and $x = 90$. Describe the meaning of each evaluation in terms of percentage inoculated and cost.

b. For what value of x is the expression undefined?

c. What happens to the cost as x approaches 100%? How can you interpret this observation?

80. The rational expression

$$\frac{60,000x}{100 - x}$$

describes the cost, in dollars, to remove x percent of the air pollutants in the smokestack emission of a utility company that burns coal to generate electricity.

a. Evaluate the expression for $x = 20$, $x = 50$, and $x = 80$. Describe the meaning of each evaluation in terms of percentage of pollutants removed and cost.

b. For what value of x is the expression undefined?

c. What happens to the cost as x approaches 100%? How can you interpret this observation?

Doctors use the rational expression

$$\frac{DA}{A + 12}$$

to determine the dosage of a drug prescribed for children. In this expression, A = the child's age and D = the adult dosage. Use the expression to solve Exercises 81–82.

81. If the normal adult dosage of medication is 1000 milligrams, what dosage should an 8-year-old child receive?

82. If the normal adult dosage of medication is 1000 milligrams, what dosage should a 4-year-old child receive?

83. A company that manufactures bicycles has costs given by the equation

$$C = \frac{100x + 100,000}{x}$$

in which x is the number of bicycles manufactured and C is the cost to manufacture each bicycle.

a. Find the cost per bicycle when manufacturing 500 bicycles.

b. Find the cost per bicycle when manufacturing 4000 bicycles.

c. Does the cost per bicycle increase or decrease as more bicycles are manufactured? Explain why this happens.

84. A company that manufactures small canoes has costs given by the equation

$$C = \frac{20x + 20,000}{x}$$

in which x is the number of canoes manufactured and C is the cost to manufacture each canoe.

a. Find the cost per canoe when manufacturing 100 canoes.

b. Find the cost per canoe when manufacturing 10,000 canoes.

c. Does the cost per canoe increase or decrease as more canoes are manufactured? Explain why this happens.

A drug is injected into a patient and the concentration of the drug in the bloodstream is monitored. The drug's concentration, y, in milligrams per liter, after x hours is modeled by

$$y = \frac{5x}{x^2 + 1}.$$

The graph of this equation, obtained with a graphing utility, is shown in the figure in a $[0, 10, 1]$ by $[0, 3, 1]$ viewing rectangle. Use this information to solve Exercises 85–87.

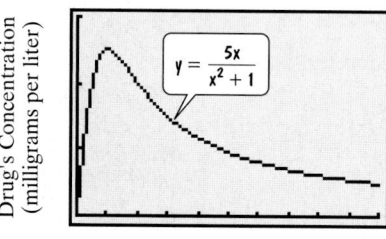

Hours After Injection

85. Use the equation to find the drug's concentration after 3 hours. Then identify the point on the equation's graph that conveys this information.

86. Use the equation to find the drug's concentration after 4 hours. Then identify the point on the equation's graph that conveys this information.

87. Use the graph of the equation to find after how many hours the drug reaches its maximum concentration. Then use the equation to find the drug's concentration at this time.

Writing in Mathematics

88. What is a rational expression? Give an example with your explanation.

89. Explain how to find the number or numbers, if any, for which a rational expression is undefined.

90. Explain how to simplify a rational expression.

91. Explain how to simplify a rational expression with opposite factors in the numerator and denominator.

92. A politician claims that each year the conviction rate for drug arrests in the United States is increasing. Explain how to use the polynomials in Exercise 77 to verify this claim.

93. Use the graph shown for Exercises 85–87 to write a description of the drug's concentration over time. In your description, try to convey as much information as possible that is displayed visually by the graph.

Critical Thinking Exercises

94. Which one of the following is true?

a. $\dfrac{x+5}{x} = 5$

b. $\dfrac{x^2+3}{3} = x^2 + 1$

c. $\dfrac{3x+9}{3x+13} = \dfrac{9}{13}$

d. The expression $\dfrac{-3y-6}{y+2}$ reduces to the consecutive integer that follows -4.

95. Write a rational expression that cannot be simplified.

96. Write a rational expression that is undefined for $x = -4$.

97. Write a rational expression with $x^2 - x - 6$ in the numerator and that can be simplified to $x - 3$.

98. Simplify: $\dfrac{\frac{1}{7}x^5 - \frac{3}{11}x^4}{x^4}$.

Technology Exercises

In Exercises 99–101, use a graphing utility to determine if the rational expression has been correctly simplified by graphing the original and simplified expressions on the same screen. If the simplification is wrong, correct it and then verify your answer using the graphing utility.

99. $\dfrac{3x+15}{x+5} = 3, \quad x \neq -5$

100. $\dfrac{2x^2 - x - 1}{x - 1} = 2x^2 - 1, \quad x \neq 1$

101. $\dfrac{x^2 - x}{x} = x^2 - 1, \quad x \neq 0$

102. Use a graphing utility to verify the graph in Figure 7.1 on page 438. $\boxed{\text{TRACE}}$ along the graph as x approaches 100. What do you observe?

Review Exercises

103. Multiply: $\dfrac{5}{6} \cdot \dfrac{9}{25}$. (Section 1.1, Example 2)

104. Divide: $\dfrac{2}{3} \div 4$. (Section 1.1, Example 3)

105. Solve by the addition method:

$2x - 5y = -2$

$3x + 4y = 20$. (Section 4.3, Example 3)

▶ **SECTION 7.2**   *Multiplying and Dividing Rational Expressions*

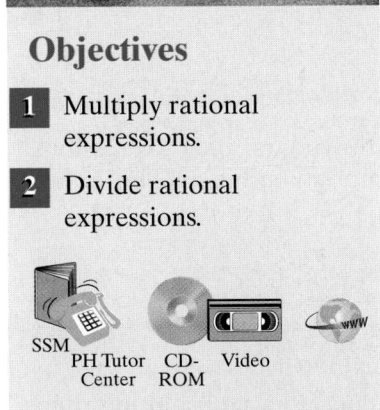

Objectives

1 Multiply rational expressions.

2 Divide rational expressions.

SSM   PH Tutor   CD-   Video
Center   ROM

Your psychology class is learning various techniques to double what we remember over time. At the beginning of the course, students memorize 40 words in Latin, a language with which they are not familiar. The rational expression

$$\frac{5t + 30}{t}$$

models the class average for the number of words remembered after t days, where $t \geq 1$. If the techniques are successful, what will be the new memory model?

The new model can be found by multiplying the given rational expression by 2. In this section, you will see that we multiply rational expressions in the same way that we multiply rational numbers. Thus, we multiply numerators and multiply denominators. The rational expression for doubling what the class remembers over time is

$$\frac{2}{1} \cdot \frac{5t + 30}{t} = \frac{2(5t + 30)}{1 \cdot t} = \frac{2 \cdot 5t + 2 \cdot 30}{t} = \frac{10t + 60}{t}.$$

Here's to all our improved memories: May the distributive property be with you for a long time!

1 Multiply rational expressions.

Multiplying Rational Expressions

The product of two rational expressions is the product of their numerators divided by the product of their denominators.

> **Multiplying Rational Expressions**
>
> If $P, Q, R,$ and S are polynomials, where $Q \neq 0$ and $S \neq 0$, then
>
> $$\frac{P}{Q} \cdot \frac{R}{S} = \frac{PR}{QS}.$$

EXAMPLE 1   Multiplying Rational Expressions

Multiply: $\dfrac{7}{x + 3} \cdot \dfrac{x - 2}{5}$.

Solution

$$\frac{7}{x + 3} \cdot \frac{x - 2}{5} = \frac{7(x - 2)}{(x + 3)5}$$

Multiply numerators. Multiply denominators. $(x \neq -3)$

$$= \frac{7x - 14}{5x + 15}$$

∎

✔ **CHECK POINT 1**   Multiply: $\dfrac{9}{x + 4} \cdot \dfrac{x - 5}{2}$.

Here is a step-by-step procedure for multiplying rational expressions. Before multiplying, divide out any factors common to both a numerator and a denominator.

> **Multiplying Rational Expressions**
>
> 1. Factor all numerators and denominators completely.
> 2. Divide numerators and denominators by common factors.
> 3. Multiply the remaining factors in the numerators and multiply the remaining factors in the denominators.

EXAMPLE 2   Multiplying Rational Expressions

Multiply: $\dfrac{x-3}{x+5} \cdot \dfrac{10x+50}{7x-21}$.

Solution

$$\dfrac{x-3}{x+5} \cdot \dfrac{10x+50}{7x-21}$$

$$= \dfrac{x-3}{x+5} \cdot \dfrac{10(x+5)}{7(x-3)} \qquad \text{Factor as many numerators and denominators as possible.}$$

$$= \dfrac{\overset{1}{\cancel{x-3}}}{\underset{1}{\cancel{x+5}}} \cdot \dfrac{10\overset{1}{\cancel{(x+5)}}}{7\underset{1}{\cancel{(x-3)}}} \qquad \text{Divide numerators and denominators by common factors.}$$

$$= \dfrac{10}{7} \qquad \text{Multiply the remaining factors in the numerators and denominators.} \quad \blacksquare$$

✔ **CHECK POINT 2**   Multiply: $\dfrac{x+4}{x-7} \cdot \dfrac{3x-21}{8x+32}$.

EXAMPLE 3   Multiplying Rational Expressions

Multiply: $\dfrac{x-7}{x-1} \cdot \dfrac{x^2-1}{3x-21}$.

Solution

$$\dfrac{x-7}{x-1} \cdot \dfrac{x^2-1}{3x-21}$$

$$= \dfrac{x-7}{x-1} \cdot \dfrac{(x+1)(x-1)}{3(x-7)} \qquad \text{Factor as many numerators and denominators as possible.}$$

$$= \dfrac{\overset{1}{\cancel{x-7}}}{\underset{1}{\cancel{x-1}}} \cdot \dfrac{(x+1)\overset{1}{\cancel{(x-1)}}}{3\underset{1}{\cancel{(x-7)}}} \qquad \text{Divide numerators and denominators by common factors.}$$

$$= \dfrac{x+1}{3} \qquad \text{Multiply the remaining factors in the numerators and denominators.} \quad \blacksquare$$

✔ **CHECK POINT 3**   Multiply: $\dfrac{x-5}{x-2} \cdot \dfrac{x^2-4}{9x-45}$.

EXAMPLE 4   Multiplying Rational Expressions

Multiply: $\dfrac{4x+8}{6x-3x^2} \cdot \dfrac{3x^2-4x-4}{9x^2-4}$.

Solution

$$\frac{4x + 8}{6x - 3x^2} \cdot \frac{3x^2 - 4x - 4}{9x^2 - 4}$$

$$= \frac{4(x + 2)}{3x(2 - x)} \cdot \frac{(3x + 2)(x - 2)}{(3x + 2)(3x - 2)}$$

Factor as many numerators and denominators as possible.

$$= \frac{4(x + 2)}{3x \overset{1}{(2 - x)}} \cdot \frac{\overset{1}{\cancel{(3x + 2)}} \overset{-1}{\cancel{(x - 2)}}}{\underset{1}{\cancel{(3x + 2)}}(3x - 2)}$$

Divide numerators and denominators by common factors. Because $2 - x$ and $x - 2$ have opposite signs, their quotient is -1.

$$= \frac{-4(x + 2)}{3x(3x - 2)} \quad \text{or} \quad -\frac{4(x + 2)}{3x(3x - 2)}$$

Multiply the remaining factors in the numerators and denominators.

It is not necessary to carry out these multiplications.

✔ **CHECK POINT 4** Multiply: $\dfrac{5x + 5}{7x - 7x^2} \cdot \dfrac{2x^2 + x - 3}{4x^2 - 9}$.

2 Divide rational expressions.

Dividing Rational Expressions

The quotient of two rational expressions is the product of the first expression and the multiplicative inverse, or reciprocal, of the second. The reciprocal is found by interchanging the numerator and the denominator.

Dividing Rational Expressions

If P, Q, R, and S are polynomials, where $Q \neq 0, S \neq 0$, and $R \neq 0$, then

$$\frac{P}{Q} \div \frac{R}{S} = \frac{P}{Q} \cdot \frac{S}{R} = \frac{PS}{QR}.$$

Change division to multiplication.

Replace $\dfrac{R}{S}$ by its reciprocal by interchanging numerator and denominator.

Thus, **we find the quotient of two rational expressions by inverting the divisor and multiplying**. For example,

$$\frac{x}{7} \div \frac{6}{y} = \frac{x}{7} \cdot \frac{y}{6} = \frac{xy}{42}.$$

Change the division to multiplication.

Replace $\dfrac{6}{y}$ with its reciprocal by interchanging numerator and denominator.

Study Tip

When performing operations with rational expressions, if a rational expression is written without a denominator, it is helpful to write the expression with a denominator of 1. In Example 5, we wrote $x + 5$ as

$$\frac{x + 5}{1}.$$

EXAMPLE 5 Dividing Rational Expressions

Divide: $(x + 5) \div \dfrac{x - 2}{x + 9}$.

Solution

$$(x + 5) \div \frac{x - 2}{x + 9} = \frac{x + 5}{1} \cdot \frac{x + 9}{x - 2}$$

Invert the divisor and multiply.

$$= \frac{(x + 5)(x + 9)}{x - 2}$$

Multiply the factors in the numerators and denominators. We need not carry out the multiplication in the numerator. ∎

✔ **CHECK POINT 5** Divide: $(x + 3) \div \dfrac{x - 4}{x + 7}$.

EXAMPLE 6 Dividing Rational Expressions

Divide: $\dfrac{x^2 - 2x - 8}{x^2 - 9} \div \dfrac{x - 4}{x + 3}$.

Solution

$$\dfrac{x^2 - 2x - 8}{x^2 - 9} \div \dfrac{x - 4}{x + 3}$$

$$= \dfrac{x^2 - 2x - 8}{x^2 - 9} \cdot \dfrac{x + 3}{x - 4} \qquad \text{Invert the divisor and multiply.}$$

$$= \dfrac{(x - 4)(x + 2)}{(x + 3)(x - 3)} \cdot \dfrac{x + 3}{x - 4} \qquad \begin{array}{l}\text{Factor as many numerators and} \\ \text{denominators as possible.}\end{array}$$

$$= \dfrac{\overset{1}{\cancel{(x - 4)}}(x + 2)}{\underset{1}{\cancel{(x + 3)}}(x - 3)} \cdot \dfrac{\overset{1}{\cancel{(x + 3)}}}{\underset{1}{\cancel{(x - 4)}}} \qquad \begin{array}{l}\text{Divide numerators and denominators} \\ \text{by common factors.}\end{array}$$

$$= \dfrac{x + 2}{x - 3} \qquad \begin{array}{l}\text{Multiply the remaining factors in the} \\ \text{numerators and the denominators.}\end{array} \quad \blacksquare$$

✔ **CHECK POINT 6** Divide: $\dfrac{x^2 + 5x + 6}{x^2 - 25} \div \dfrac{x + 2}{x + 5}$.

EXAMPLE 7 Dividing Rational Expressions

Divide: $\dfrac{y^2 + 7y + 12}{y^2 + 9} \div (7y^2 + 21y)$.

Solution

$$\dfrac{y^2 + 7y + 12}{y^2 + 9} \div \dfrac{7y^2 + 21y}{1} \qquad \begin{array}{l}\text{It is helpful to write the divisor} \\ \text{with a denominator of 1.}\end{array}$$

$$= \dfrac{y^2 + 7y + 12}{y^2 + 9} \cdot \dfrac{1}{7y^2 + 21y} \qquad \text{Invert the divisor and multiply.}$$

$$= \dfrac{(y + 4)(y + 3)}{y^2 + 9} \cdot \dfrac{1}{7y(y + 3)} \qquad \begin{array}{l}\text{Factor as many numerators and} \\ \text{denominators as possible.}\end{array}$$

$$= \dfrac{(y + 4)\overset{1}{\cancel{(y + 3)}}}{y^2 + 9} \cdot \dfrac{1}{7y\underset{1}{\cancel{(y + 3)}}} \qquad \begin{array}{l}\text{Divide numerators and denominators} \\ \text{by common factors.}\end{array}$$

$$= \dfrac{y + 4}{7y(y^2 + 9)} \qquad \begin{array}{l}\text{Multiply the remaining factors in the} \\ \text{numerators and the denominators.}\end{array} \quad \blacksquare$$

✔ **CHECK POINT 7** Divide: $\dfrac{y^2 + 3y + 2}{y^2 + 1} \div (5y^2 + 10y)$.

EXERCISE SET 7.2

Practice Exercises

In Exercises 1–32, multiply as indicated.

1. $\dfrac{5}{x + 2} \cdot \dfrac{x - 3}{7}$

2. $\dfrac{7}{x - 1} \cdot \dfrac{x + 4}{3}$

3. $\dfrac{x}{2} \cdot \dfrac{4}{x + 1}$

4. $\dfrac{x}{3} \cdot \dfrac{6}{x - 1}$

5. $\dfrac{3}{x} \cdot \dfrac{2x}{9}$

6. $\dfrac{7}{x} \cdot \dfrac{4x}{28}$

7. $\dfrac{x - 2}{x + 3} \cdot \dfrac{2x + 6}{5x - 10}$

8. $\dfrac{x - 3}{x + 7} \cdot \dfrac{3x + 21}{2x - 6}$

9. $\dfrac{x^2 + 7x + 12}{x + 4} \cdot \dfrac{1}{x + 3}$

10. $\dfrac{x^2 + 7x + 10}{x + 5} \cdot \dfrac{1}{x + 2}$

11. $\dfrac{x^2 - 25}{x^2 - 3x - 10} \cdot \dfrac{x + 2}{x}$

12. $\dfrac{x^2 - 49}{x^2 - 4x - 21} \cdot \dfrac{x + 3}{x}$

13. $\dfrac{4y + 30}{y^2 - 3y} \cdot \dfrac{y - 3}{2y + 15}$

14. $\dfrac{9y + 21}{y^2 - 2y} \cdot \dfrac{y - 2}{3y + 7}$

15. $\dfrac{y^2 - 7y - 30}{y^2 - 6y - 40} \cdot \dfrac{2y^2 + 5y + 2}{2y^2 + 7y + 3}$

16. $\dfrac{3y^2 + 17y + 10}{3y^2 - 22y - 16} \cdot \dfrac{y^2 - 4y - 32}{y^2 - 8y - 48}$

17. $\left(y^2 - 9\right) \cdot \dfrac{4}{y - 3}$

18. $\left(y^2 - 16\right) \cdot \dfrac{3}{y - 4}$

19. $\dfrac{x^2 - 5x + 6}{x^2 - 2x - 3} \cdot \dfrac{x^2 - 1}{x^2 - 4}$

20. $\dfrac{x^2 + 5x + 6}{x^2 + x - 6} \cdot \dfrac{x^2 - 9}{x^2 - x - 6}$

21. $\dfrac{x^3 - 8}{x^2 - 4} \cdot \dfrac{x + 2}{3x}$

22. $\dfrac{x^2 + 6x + 9}{x^3 + 27} \cdot \dfrac{1}{x + 3}$

23. $\dfrac{(x - 2)^3}{(x - 1)^3} \cdot \dfrac{x^2 - 2x + 1}{x^2 - 4x + 4}$

24. $\dfrac{(x + 4)^3}{(x + 2)^3} \cdot \dfrac{x^2 + 4x + 4}{x^2 + 8x + 16}$

25. $\dfrac{6x + 2}{x^2 - 1} \cdot \dfrac{1 - x}{3x^2 + x}$

26. $\dfrac{8x + 2}{x^2 - 9} \cdot \dfrac{3 - x}{4x^2 + x}$

27. $\dfrac{25 - y^2}{y^2 - 2y - 35} \cdot \dfrac{y^2 - 8y - 20}{y^2 - 3y - 10}$

28. $\dfrac{2y}{3y - y^2} \cdot \dfrac{2y^2 - 9y + 9}{8y - 12}$

29. $\dfrac{x^2 - y^2}{x} \cdot \dfrac{x^2 + xy}{x + y}$

30. $\dfrac{4x - 4y}{x} \cdot \dfrac{x^2 + xy}{x^2 - y^2}$

31. $\dfrac{x^2 + 2xy + y^2}{x^2 - 2xy + y^2} \cdot \dfrac{4x - 4y}{3x + 3y}$

32. $\dfrac{x^2 - y^2}{x + y} \cdot \dfrac{x + 2y}{2x^2 - xy - y^2}$

In Exercises 33–64, divide as indicated.

33. $\dfrac{x}{7} \div \dfrac{5}{3}$

34. $\dfrac{x}{3} \div \dfrac{3}{8}$

35. $\dfrac{3}{x} \div \dfrac{12}{x}$

36. $\dfrac{x}{5} \div \dfrac{20}{x}$

37. $\dfrac{15}{x} \div \dfrac{3}{2x}$

38. $\dfrac{9}{x} \div \dfrac{3}{4x}$

39. $\dfrac{x + 1}{3} \div \dfrac{3x + 3}{7}$

40. $\dfrac{x + 5}{7} \div \dfrac{4x + 20}{9}$

41. $\dfrac{7}{x - 5} \div \dfrac{28}{3x - 15}$

42. $\dfrac{4}{x - 6} \div \dfrac{40}{7x - 42}$

43. $\dfrac{x^2 - 4}{x} \div \dfrac{x + 2}{x - 2}$

44. $\dfrac{x^2 - 4}{x - 2} \div \dfrac{x + 2}{4x - 8}$

45. $\left(y^2 - 16\right) \div \dfrac{y^2 + 3y - 4}{y^2 + 4}$

46. $\left(y^2 + 4y - 5\right) \div \dfrac{y^2 - 25}{y + 7}$

47. $\dfrac{y^2 - y}{15} \div \dfrac{y - 1}{5}$

48. $\dfrac{y^2 - 2y}{15} \div \dfrac{y - 2}{5}$

49. $\dfrac{4x^2 + 10}{x - 3} \div \dfrac{6x^2 + 15}{x^2 - 9}$

50. $\dfrac{x^2 + x}{x^2 - 4} \div \dfrac{x^2 - 1}{x^2 + 5x + 6}$

51. $\dfrac{x^2 - 25}{2x - 2} \div \dfrac{x^2 + 10x + 25}{x^2 + 4x - 5}$

52. $\dfrac{x^2 - 4}{x^2 + 3x - 10} \div \dfrac{x^2 + 5x + 6}{x^2 + 8x + 15}$

53. $\dfrac{y^3 + y}{y^2 - y} \div \dfrac{y^3 - y^2}{y^2 - 2y + 1}$

54. $\dfrac{3y^2 - 12}{y^2 + 4y + 4} \div \dfrac{y^3 - 2y^2}{y^2 + 2y}$

55. $\dfrac{y^2 + 5y + 4}{y^2 + 12y + 32} \div \dfrac{y^2 - 12y + 35}{y^2 + 3y - 40}$

56. $\dfrac{y^2 + 4y - 21}{y^2 + 3y - 28} \div \dfrac{y^2 + 14y + 48}{y^2 + 4y - 32}$

57. $\dfrac{2y^2 - 128}{y^2 + 16y + 64} \div \dfrac{y^2 - 6y - 16}{3y^2 + 30y + 48}$

58. $\dfrac{3y + 12}{y^2 + 3y} \div \dfrac{y^2 + y - 12}{9y - y^3}$

59. $\dfrac{2x + 2y}{3} \div \dfrac{x^2 - y^2}{x - y}$

60. $\dfrac{5x + 5y}{7} \div \dfrac{x^2 - y^2}{x - y}$

61. $\dfrac{x^2 - y^2}{8x^2 - 16xy + 8y^2} \div \dfrac{4x - 4y}{x + y}$

62. $\dfrac{4x^2 - y^2}{x^2 + 4xy + 4y^2} \div \dfrac{4x - 2y}{3x + 6y}$

63. $\dfrac{xy - y^2}{x^2 + 2x + 1} \div \dfrac{2x^2 + xy - 3y^2}{2x^2 + 5xy + 3y^2}$

64. $\dfrac{x^2 - 4y^2}{x^2 + 3xy + 2y^2} \div \dfrac{x^2 - 4xy + 4y^2}{x + y}$

Application Exercises

65. In the section opener, we used

$$\dfrac{250x}{100 - x}$$

to describe the cost, in millions of dollars, to remove x percent of the pollutants that are discharged into the river. We were wrong. The cost will be half of what we originally anticipated. Write a rational expression that represents the reduced cost.

66. We originally thought that the cost, in dollars, to manufacture each of x bicycles was

$$\dfrac{100x + 100,000}{x}.$$

We were wrong. We can manufacture each bicycle at half of what we originally anticipated. Write a rational expression that represents the reduced cost.

Writing in Mathematics

67. Explain how to multiply rational expressions.

68. Explain how to divide rational expressions.

69. In dividing polynomials

$$\dfrac{P}{Q} \div \dfrac{R}{S},$$

why is it necessary to state that polynomial R is not equal to 0?

Critical Thinking Exercises

70. Which one of the following is true?

a. $5 \div x = \dfrac{1}{5} \cdot x$ for any nonzero number x.

b. $\dfrac{4}{x} \div \dfrac{x - 2}{x} = \dfrac{4}{x - 2}$ if $x \neq 0$ and $x \neq 2$.

c. $\dfrac{x - 5}{6} \cdot \dfrac{3}{5 - x} = \dfrac{1}{2}$ for any value of x except 5.

d. The quotient of two rational expressions can be found by dividing their numerators and dividing their denominators.

71. Find the missing polynomials: $\dfrac{\blacksquare}{\blacksquare} \cdot \dfrac{3x - 12}{2x} = \dfrac{3}{2}$.

72. Find the missing polynomials: $-\dfrac{1}{2x - 3} \div \dfrac{\blacksquare}{\blacksquare} = \dfrac{1}{3}$.

73. Simplify:

$$\left(\dfrac{y - 2}{y^2 - 9y + 18} \cdot \dfrac{y^2 - 4y - 12}{y + 2} \right) \div \dfrac{y^2 - 4}{y^2 + 5y + 6}.$$

Technology Exercises

In Exercises 74–77, use a graphing utility to determine if the multiplication or division has been performed correctly by graphing the expressions on both sides on the same screen. If the answer is wrong, correct it and then verify your correction using the graphing utility.

74. $\dfrac{x^2 + x}{3x} \cdot \dfrac{6x}{x + 1} = 2x$

75. $\dfrac{x^3 - 25x}{x^2 - 3x - 10} \cdot \dfrac{x + 2}{x} = x + 5$

76. $\dfrac{x^2 - 9}{x + 4} \div \dfrac{x - 3}{x + 4} = x - 3$

77. $(x - 5) \div \dfrac{2x^2 - 11x + 5}{4x^2 - 1} = 2x - 1$

Review Exercises

78. Solve: $2x + 3 < 3(x - 5)$. (Section 2.7, Example 6)

79. Factor completely: $3x^2 - 15x - 42$. (Section 6.5, Example 2)

80. Solve: $x(2x + 9) = 5$. (Section 6.6, Example 6)

▶ SECTION 7.3 Adding and Subtracting Rational Expressions with the Same Denominators

Objectives

1 Add rational expressions with the same denominators.

2 Subtract rational expressions with the same denominators.

3 Add and subtract rational expressions with opposite denominators.

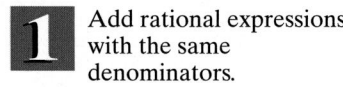

SSM PH Tutor CD- Video
Center ROM

Are you long, medium, or round? Your skull, that is? The varying shapes of the human skull create glorious diversity in the human species. By learning to add and subtract rational expressions with the same denominator, you will obtain an expression that models this diversity.

1 Add rational expressions with the same denominators.

Addition When Denominators Are the Same

To add rational numbers having the same denominators, such as $\frac{2}{9}$ and $\frac{5}{9}$, we add the numerators and place the sum over the common denominator:

$$\frac{2}{9} + \frac{5}{9} = \frac{2 + 5}{9} = \frac{7}{9}.$$

We add rational expressions with the same denominator in an identical manner.

Adding Rational Expressions with Common Denominators

If $\dfrac{P}{R}$ and $\dfrac{Q}{R}$ are rational expressions, then

$$\frac{P}{R} + \frac{Q}{R} = \frac{P + Q}{R}.$$

To add rational expressions with the same denominators, add numerators and place the sum over the common denominator. If possible, simplify the final result.

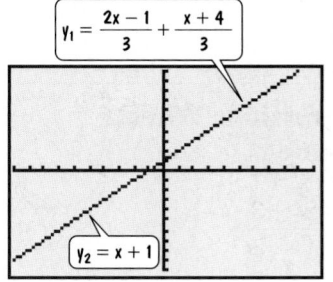
EXAMPLE 1 Adding Rational Expressions When Denominators Are the Same

Add: $\dfrac{2x - 1}{3} + \dfrac{x + 4}{3}$.

Solution

$$\frac{2x - 1}{3} + \frac{x + 4}{3} = \frac{2x - 1 + x + 4}{3}$$

Add numerators. Place this sum over the common denominator.

$$= \frac{3x + 3}{3}$$

Combine like terms.

$$= \frac{\overset{1}{\cancel{3}}(x + 1)}{\underset{1}{\cancel{3}}}$$

Factor and simplify.

$$= x + 1$$ ∎

✔ **CHECK POINT 1** Add: $\dfrac{3x - 2}{5} + \dfrac{2x + 12}{5}$.

EXAMPLE 2 Adding Rational Expressions When Denominators Are the Same

Add: $\dfrac{x^2}{x^2 - 9} + \dfrac{9 - 6x}{x^2 - 9}$.

Solution

$$\frac{x^2}{x^2 - 9} + \frac{9 - 6x}{x^2 - 9} = \frac{x^2 + 9 - 6x}{x^2 - 9}$$

Add numerators. Place this sum over the common denominator.

$$= \frac{x^2 - 6x + 9}{x^2 - 9}$$

Write the numerator in descending powers of x.

$$= \frac{(x - 3)\cancel{(x - 3)}^{1}}{(x + 3)\underset{1}{\cancel{(x - 3)}}}$$

Factor and simplify. What values of x are not permitted?

$$= \frac{x - 3}{x + 3}$$ ∎

✔ **CHECK POINT 2** Add: $\dfrac{x^2}{x^2 - 25} + \dfrac{25 - 10x}{x^2 - 25}$.

2 Subtract rational expressions with the same denominators.

Subtraction When Denominators Are the Same

The following box shows how to subtract rational expressions with the same denominators:

Subtracting Rational Expressions with Common Denominators

If $\dfrac{P}{R}$ and $\dfrac{Q}{R}$ are rational expressions, then

$$\frac{P}{R} - \frac{Q}{R} = \frac{P - Q}{R}.$$

To subtract rational expressions with the same denominators, subtract numerators and place the difference over the common denominator. If possible, simplify the final result.

EXAMPLE 3   Subtracting Rational Expressions When Denominators Are the Same

Subtract: **a.** $\dfrac{2x + 3}{x + 1} - \dfrac{x}{x + 1}$ **b.** $\dfrac{5x + 1}{x^2 - 9} - \dfrac{4x - 2}{x^2 - 9}.$

Solution

a. $\dfrac{2x + 3}{x + 1} - \dfrac{x}{x + 1} = \dfrac{2x + 3 - x}{x + 1}$ Subtract numerators. Place this difference over the common denominator.

$= \dfrac{x + 3}{x + 1}$ Combine like terms.

b. $\dfrac{5x + 1}{x^2 - 9} - \dfrac{4x - 2}{x^2 - 9} = \dfrac{5x + 1 - (4x - 2)}{x^2 - 9}$ Subtract numerators and include parentheses to indicate that both terms are subtracted. Place this difference over the common denominator.

$= \dfrac{5x + 1 - 4x + 2}{x^2 - 9}$ Remove parentheses and then change the sign of each term.

$= \dfrac{x + 3}{x^2 - 9}$ Combine like terms.

$= \dfrac{\overset{1}{\cancel{x + 3}}}{\underset{1}{\cancel{(x + 3)}}(x - 3)}$ Factor and simplify ($x \neq -3$ and $x \neq 3$).

$= \dfrac{1}{x - 3}$ ■

✔ **CHECK POINT 3**   Subtract:

a. $\dfrac{4x + 5}{x + 7} - \dfrac{x}{x + 7}$ **b.** $\dfrac{3x^2 + 4x}{x - 1} - \dfrac{11x - 4}{x - 1}.$

Study Tip

When a numerator is being subtracted, be sure to **subtract every term in that expression**.

The − sign applies to the entire numerator, 4x − 2.

Insert parentheses to indicate this.

The sign of every term of 4x − 2 changes.

$$\frac{5x+1}{x^2-9} - \frac{4x-2}{x^2-9} = \frac{5x+1-(4x-2)}{x^2-9} = \frac{5x+1-4x+2}{x^2-9}$$

The entire numerator of the second rational expression must be subtracted. Avoid the common error of subtracting only the first term.

INCORRECT!

−2 must also be subtracted.

$$\frac{5x+1}{x^2-9} - \frac{4x-2}{x^2-9} = \frac{5x+1-4x-2}{x^2-9}$$

EXAMPLE 4 Subtracting Rational Expressions When Denominators Are the Same

Subtract: $\dfrac{20y^2+5y+1}{6y^2+y-2} - \dfrac{8y^2-12y-5}{6y^2+y-2}$.

Solution

$$\frac{20y^2+5y+1}{6y^2+y-2} - \frac{8y^2-12y-5}{6y^2+y-2}$$

Don't forget the parentheses.

$$=\frac{20y^2+5y+1-(8y^2-12y-5)}{6y^2+y-2}$$

Subtract numerators. Place this difference over the common denominator.

$$=\frac{20y^2+5y+1-8y^2+12y+5}{6y^2+y-2}$$

Remove parentheses and then change the sign of each term.

$$=\frac{(20y^2-8y^2)+(5y+12y)+(1+5)}{6y^2+y-2}$$

Group like terms. This step is usually performed mentally.

$$=\frac{12y^2+17y+6}{6y^2+y-2}$$

Combine like terms.

$$=\frac{\overset{1}{\cancel{(3y+2)}}(4y+3)}{\underset{1}{\cancel{(3y+2)}}(2y-1)}$$

Factor and simplify.

$$=\frac{4y+3}{2y-1}$$

✔ **CHECK POINT 4** Subtract: $\dfrac{y^2+3y-6}{y^2-5y+4} - \dfrac{4y-4-2y^2}{y^2-5y+4}$.

3 Add and subtract rational expressions with opposite denominators.

Addition and Subtraction When Denominators Are Opposites

How do we add or subtract rational expressions when denominators are opposites, or additive inverses? Here is an example of this type of addition problem:

$$\frac{x^2}{x-5} + \frac{4x+5}{5-x}.$$

These denominators are opposites.
They differ only in their signs.

Multiply the numerator and the denominator of either of the rational expressions by -1. Then they will both have the same denominator.

EXAMPLE 5   Adding Rational Expressions When Denominators Are Opposites

Add: $\dfrac{x^2}{x-5} + \dfrac{4x+5}{5-x}.$

Solution

$$\frac{x^2}{x-5} + \frac{4x+5}{5-x}$$

$$= \frac{x^2}{x-5} + \frac{(-1)}{(-1)} \cdot \frac{4x+5}{5-x} \qquad \text{Multiply the numerator and denominator of the second rational expression by } -1.$$

$$= \frac{x^2}{x-5} + \frac{-4x-5}{-5+x} \qquad \text{Perform the multiplications by } -1 \text{ by changing every term's sign.}$$

$$= \frac{x^2}{x-5} + \frac{-4x-5}{x-5} \qquad \text{Rewrite } -5+x \text{ as } x-5. \text{ Both rational expressions have the same denominator.}$$

$$= \frac{x^2+(-4x-5)}{x-5} \qquad \text{Add numerators. Place this sum over the common denominator.}$$

$$= \frac{x^2-4x-5}{x-5} \qquad \text{Remove parentheses.}$$

$$= \frac{\overset{1}{\cancel{(x-5)}}(x+1)}{\underset{1}{\cancel{x-5}}} \qquad \text{Factor and simplify.}$$

$$= x+1 \qquad\qquad\qquad\qquad\qquad \blacksquare$$

✔ **CHECK POINT 5**   Add: $\dfrac{x^2}{x-7} + \dfrac{4x+21}{7-x}.$

Adding and Subtracting Rational Expressions with Opposite Denominators

When one denominator is the additive inverse of the other, first multiply either rational expression by $\frac{-1}{-1}$ to obtain a common denominator.

EXAMPLE 6   Subtracting Rational Expressions When Denominators Are Opposites

Subtract: $\dfrac{5x - x^2}{x^2 - 4x - 3} - \dfrac{3x - x^2}{3 + 4x - x^2}$.

Solution   We note that $x^2 - 4x - 3$ and $3 + 4x - x^2$ are opposites. We multiply the second rational expression by $\frac{-1}{-1}$.

$$\frac{(-1)}{(-1)} \cdot \frac{3x - x^2}{3 + 4x - x^2} = \frac{-3x + x^2}{-3 - 4x + x^2}$$ Multiply the numerator and denominator by −1 by changing every term's sign.

$$= \frac{x^2 - 3x}{x^2 - 4x - 3}$$ Write the numerator and the denominator in descending powers of x.

We now return to the original subtraction problem.

$$\frac{5x - x^2}{x^2 - 4x - 3} - \frac{3x - x^2}{3 + 4x - x^2}$$

$$= \frac{5x - x^2}{x^2 - 4x - 3} - \frac{x^2 - 3x}{x^2 - 4x - 3}$$ Replace the second rational expression by the form obtained through multiplication by $\frac{-1}{-1}$.

$$= \frac{5x - x^2 - \left(x^2 - 3x\right)}{x^2 - 4x - 3}$$ Subtract numerators. Place this difference over the common denominator. Don't forget parentheses!

$$= \frac{5x - x^2 - x^2 + 3x}{x^2 - 4x - 3}$$ Remove parentheses and then change the sign of each term.

$$= \frac{-2x^2 + 8x}{x^2 - 4x - 3}$$ Combine like terms in the numerator. Although the numerator can be factored, further simplification is not possible. ∎

✔ **CHECK POINT 6**   Subtract: $\dfrac{7x - x^2}{x^2 - 2x - 9} - \dfrac{5x - 3x^2}{9 + 2x - x^2}$.

EXERCISE SET 7.3

Practice Exercises

In Exercises 1–38, add or subtract as indicated. Simplify the result, if possible.

1. $\dfrac{4x}{9} + \dfrac{3x}{9}$

2. $\dfrac{5x}{11} + \dfrac{4x}{11}$

3. $\dfrac{7x}{12} + \dfrac{x}{12}$

4. $\dfrac{7x}{14} + \dfrac{x}{14}$

5. $\dfrac{x - 3}{8} + \dfrac{3x + 7}{8}$

6. $\dfrac{x - 4}{6} + \dfrac{5x + 10}{6}$

7. $\dfrac{5}{x} + \dfrac{3}{x}$

8. $\dfrac{7}{x} + \dfrac{4}{x}$

9. $\dfrac{7}{9x} + \dfrac{5}{9x}$

10. $\dfrac{5}{6x} + \dfrac{4}{6x}$

11.  $\dfrac{5}{x + 3} + \dfrac{4}{x + 3}$

12. $\dfrac{8}{x + 6} + \dfrac{10}{x + 6}$

13. $\dfrac{x}{x - 3} + \dfrac{4x + 5}{x - 3}$

14. $\dfrac{x}{x - 4} + \dfrac{9x + 7}{x - 4}$

15. $\dfrac{4x + 1}{6x + 5} + \dfrac{8x + 9}{6x + 5}$

16. $\dfrac{3x + 2}{3x + 4} + \dfrac{3x + 6}{3x + 4}$

17. $\dfrac{y^2 + 7y}{y^2 - 5y} + \dfrac{y^2 - 4y}{y^2 - 5y}$

18. $\dfrac{y^2 - 2y}{y^2 + 3y} + \dfrac{y^2 + y}{y^2 + 3y}$

19. $\dfrac{4y - 1}{5y^2} + \dfrac{3y + 1}{5y^2}$

20. $\dfrac{y + 2}{6y^3} + \dfrac{3y - 2}{6y^3}$

21. $\dfrac{x^2 - 2}{x^2 + x - 2} + \dfrac{2x - x^2}{x^2 + x - 2}$

22. $\dfrac{x^2 + 9x}{4x^2 - 11x - 3} + \dfrac{3x - 5x^2}{4x^2 - 11x - 3}$

23. $\dfrac{x^2 - 4x}{x^2 - x - 6} + \dfrac{4x - 4}{x^2 - x - 6}$

24. $\dfrac{x}{2x + 7} - \dfrac{2}{2x + 7}$

25. $\dfrac{3x}{5x - 4} - \dfrac{4}{5x - 4}$

26. $\dfrac{x}{x - 1} - \dfrac{1}{x - 1}$

27. $\dfrac{4x}{4x - 3} - \dfrac{3}{4x - 3}$

28. $\dfrac{2y + 1}{3y - 7} - \dfrac{y + 8}{3y - 7}$

29. $\dfrac{14y}{7y + 2} - \dfrac{7y - 2}{7y + 2}$

30. $\dfrac{2x + 3}{3x - 6} - \dfrac{3 - x}{3x - 6}$

31. $\dfrac{3x + 1}{4x - 2} - \dfrac{x + 1}{4x - 2}$

32. $\dfrac{x^3 - 3}{2x^4} - \dfrac{7x^3 - 3}{2x^4}$

33. $\dfrac{3y^2 - 1}{3y^3} - \dfrac{6y^2 - 1}{3y^3}$

34. $\dfrac{y^2 + 3y}{y^2 + y - 12} - \dfrac{y^2 - 12}{y^2 + y - 12}$

35. $\dfrac{4y^2 + 5}{9y^2 - 64} - \dfrac{y^2 - y + 29}{9y^2 - 64}$

36. $\dfrac{2y^2 + 6y + 8}{y^2 - 16} - \dfrac{y^2 - 3y - 12}{y^2 - 16}$

37. $\dfrac{6y^2 + y}{2y^2 - 9y + 9} - \dfrac{2y + 9}{2y^2 - 9y + 9} - \dfrac{4y - 3}{2y^2 - 9y + 9}$

38. $\dfrac{3y^2 - 2}{3y^2 + 10y - 8} - \dfrac{y + 10}{3y^2 + 10y - 8} - \dfrac{y^2 - 6y}{3y^2 + 10y - 8}$

In Exercises 39–64, denominators are additive inverses. Add or subtract as indicated. Simplify the result, if possible.

39. $\dfrac{4}{x - 3} + \dfrac{2}{3 - x}$

40. $\dfrac{6}{x - 5} + \dfrac{2}{5 - x}$

41. $\dfrac{6x + 7}{x - 6} + \dfrac{3x}{6 - x}$

42. $\dfrac{6x + 5}{x - 2} + \dfrac{4x}{2 - x}$

43. $\dfrac{5x - 2}{3x - 4} + \dfrac{2x - 3}{4 - 3x}$

44. $\dfrac{9x - 1}{7x - 3} + \dfrac{6x - 2}{3 - 7x}$

45. $\dfrac{x^2}{x - 2} + \dfrac{4}{2 - x}$

46. $\dfrac{x^2}{x - 3} + \dfrac{9}{3 - x}$

47. $\dfrac{y - 3}{y^2 - 25} + \dfrac{y - 3}{25 - y^2}$

48. $\dfrac{y - 7}{y^2 - 16} + \dfrac{7 - y}{16 - y^2}$

49. $\dfrac{6}{x - 1} - \dfrac{5}{1 - x}$

50. $\dfrac{10}{x - 2} - \dfrac{6}{2 - x}$

51. $\dfrac{10}{x + 3} - \dfrac{2}{-x - 3}$

52. $\dfrac{11}{x + 7} - \dfrac{5}{-x - 7}$

53. $\dfrac{y}{y - 1} - \dfrac{1}{1 - y}$

54. $\dfrac{y}{y - 4} - \dfrac{4}{4 - y}$

55. $\dfrac{3 - x}{x - 7} - \dfrac{2x - 5}{7 - x}$

56. $\dfrac{4 - x}{x - 9} - \dfrac{3x - 8}{9 - x}$

57. $\dfrac{x - 2}{x^2 - 25} - \dfrac{x - 2}{25 - x^2}$

58. $\dfrac{x - 8}{x^2 - 16} - \dfrac{x - 8}{16 - x^2}$

59. $\dfrac{x}{x - y} + \dfrac{y}{y - x}$

60. $\dfrac{2x - y}{x - y} + \dfrac{x - 2y}{y - x}$

61. $\dfrac{2x}{x^2 - y^2} + \dfrac{2y}{y^2 - x^2}$

62. $\dfrac{2y}{x^2 - y^2} + \dfrac{2x}{y^2 - x^2}$

63. $\dfrac{x^2 - 2}{x^2 + 6x - 7} + \dfrac{19 - 4x}{7 - 6x - x^2}$

64. $\dfrac{2x + 3}{x^2 - x - 30} + \dfrac{x - 2}{30 + x - x^2}$

Application Exercises

65. Anthropologists and forensic scientists classify skulls using

$$\frac{L + 60W}{L} - \frac{L - 40W}{L}$$

where L is the skull's length and W is its width.

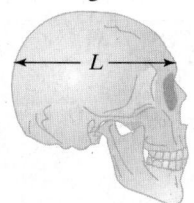

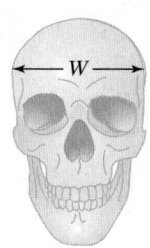

a. Express the classification as a single rational expression.

b. If the value of the rational expression in part (a) is less than 75, a skull is classified as long. A medium skull has a value between 75 and 80, and a round skull has a value over 80. Use your rational expression from part (a) to classify a skull that is 5 inches wide and 6 inches long.

66. The temperature, in degrees Fahrenheit, of a dessert placed in a freezer for t hours is modeled by

$$\frac{t + 30}{t^2 + 4t + 1} - \frac{t - 50}{t^2 + 4t + 1}.$$

a. Express the temperature as a single rational expression.

b. Use your rational expression from part (a) to find the temperature of the dessert after 1 hour and after 2 hours.

In Exercises 67–68, find the perimeter of each rectangle.

67.

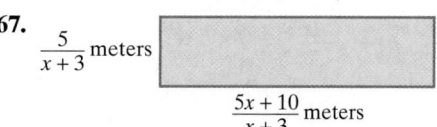

$\dfrac{5}{x + 3}$ meters

$\dfrac{5x + 10}{x + 3}$ meters

68.

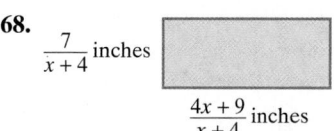

$\dfrac{7}{x + 4}$ inches

$\dfrac{4x + 9}{x + 4}$ inches

Writing in Mathematics

69. Explain how to add rational expressions when the denominators are the same. Give an example with your explanation.

70. Explain how to subtract rational expressions when the denominators are the same. Give an example with your explanation.

71. Describe two similarities between the following problems:

$$\frac{3}{8} + \frac{1}{8} \quad \text{and} \quad \frac{x}{x^2 - 1} + \frac{1}{x^2 - 1}.$$

72. Explain how to add rational expressions when the denominators are opposites. Use an example to support your explanation.

Critical Thinking Exercises

73. Which one of the following is true?

a. The sum of two rational expressions with the same denominator can be found by adding numerators, adding denominators, and then simplifying.

b. $\dfrac{4}{b} - \dfrac{2}{-b} = -\dfrac{2}{b}$

c. The difference between two rational expressions with the same denominator can always be simplified.

d. $\dfrac{2x + 1}{x - 7} + \dfrac{3x + 1}{x - 7} - \dfrac{5x + 2}{x - 7} = 0$

In Exercises 74–76, perform the indicated operations. Simplify the result if possible.

74. $\dfrac{5x - 3}{x^2 - 4} - \dfrac{2x^2 - 6}{x^2 - 4} + \dfrac{10x^2 - 1}{x^2 - 4}$

75. $\left(\dfrac{3x - 1}{x^2 + 5x - 6} - \dfrac{2x - 7}{x^2 + 5x - 6} \right) \div \dfrac{x + 2}{x^2 - 1}$

76. $\left(\dfrac{3x^2 - 4x + 4}{3x^2 + 7x + 2} - \dfrac{10x + 9}{3x^2 + 7x + 2} \right) \div \dfrac{x - 5}{x^2 - 4}$

In Exercises 77–81, find the missing expression.

77. $\dfrac{2x}{x + 3} + \dfrac{\boxed{}}{x + 3} = \dfrac{4x + 1}{x + 3}$

78. $\dfrac{3x}{x + 2} - \dfrac{\boxed{}}{x + 2} = \dfrac{6 - 17x}{x + 2}$

79. $\dfrac{6}{x - 2} + \dfrac{\boxed{}}{2 - x} = \dfrac{13}{x - 2}$

80. $\dfrac{a^2}{a - 4} - \dfrac{\boxed{}}{a - 4} = a + 3$

81. $\dfrac{3x}{x - 5} + \dfrac{\boxed{}}{5 - x} = \dfrac{7x + 1}{x - 5}$

Technology Exercises

In Exercises 82–84, use a graphing utility to determine if the subtraction has been performed correctly by graphing the expressions on both sides on the same screen. If the answer is wrong, correct it and then verify your correction using the graphing utility.

82. $\dfrac{3x + 6}{2} - \dfrac{x}{2} = x + 3$

83. $\dfrac{x^2 + 4x + 3}{x + 2} - \dfrac{5x + 9}{x + 2} = x - 2, x \neq -2$

84. $\dfrac{x^2 - 13}{x + 4} - \dfrac{3}{x + 4} = x + 4, x \neq -4$

Review Exercises

85. Subtract: $\dfrac{13}{15} - \dfrac{8}{45}$. (Section 1.1, Example 6)

86. Factor completely: $81x^4 - 1$. (Section 6.4, Example 4)

87. Divide: $\dfrac{3x^3 + 2x^2 - 26x - 15}{x + 3}$. (Section 5.6, Example 2)

▶ SECTION 7.4 Adding and Subtracting Rational Expressions with Different Denominators

Objectives

1 Find the least common denominator.

2 Add and subtract rational expressions with different denominators.

SSM PH Tutor CD- Video
Center ROM

When my aunt asked how I liked my five-year-old nephew, I replied "medium rare." Unfortunately, my little joke did not get me out of baby sitting for Dennis the Menace of our family. Now the little squirt doesn't want to go to bed because his head hurts. Does my aunt have any aspirin? What is the proper dosage for a child his age?

In this section's exercise set, you will use two formulas that model drug dosage for children. Before working with these models, we continue drawing on your experience from arithmetic to add and subtract rational expressions that have different denominators.

1 Find the least common denominator.

Finding the Least Common Denominator

We can gain insight into adding rational expressions with different denominators by looking closely at what we do when adding fractions with different denominators. For example, suppose that we want to add $\frac{1}{2}$ and $\frac{2}{3}$. We must first write the fractions with the same denominator. We look for the smallest number that contains both 2 and 3 as factors. This number, 6, is then used as the *least common denominator*, or LCD.

The **least common denominator** of a group of rational expressions is a polynomial of least degree whose factors include all the factors of the denominators in the group.

Finding the Least Common Denominator

1. Factor each denominator completely.

2. List the factors of the first denominator.

3. Add to the list in step 2 any factors of the second denominator that do not appear in the list.

4. Form the product of each different factor from the list in step 3. This product is the least common denominator.

EXAMPLE 1   Finding the Least Common Denominator

Find the LCD of $\dfrac{7}{6x^2}$ and $\dfrac{2}{9x}$.

Solution

Step 1 Factor each denominator completely.

$$6x^2 = 3 \cdot 2x^2 \quad (\text{or } 3 \cdot 2 \cdot x \cdot x)$$

$$9x = 3 \cdot 3x$$

Step 2 List the factors of the first denominator.

$$3, 2, x^2 \quad (\text{or } 3, 2, x, x)$$

Step 3 Add any unlisted factors from the second denominator. Two factors from $3 \cdot 3x$ are already in our list. These factors include x and one factor of 3. We add the other factor of 3 to our list. We have

$$3, 3, 2, x^2.$$

Step 4 The least common denominator is the product of all factors in the final list. Thus,

$$3 \cdot 3 \cdot 2x^2$$

or $18x^2$ is the least common denominator.   ∎

✔ **CHECK POINT 1**   Find the LCD of $\dfrac{3}{10x^2}$ and $\dfrac{7}{15x}$.

EXAMPLE 2   Finding the Least Common Denominator

Find the LCD of $\dfrac{3}{x + 1}$ and $\dfrac{5}{x - 1}$.

Solution

Step 1 Factor each denominator completely.
$$x + 1 = 1(x + 1)$$
$$x - 1 = 1(x - 1)$$

Step 2 List the factors of the first denominator.
$$1, x + 1$$

Step 3 Add any unlisted factors from the second denominator. These factors include 1 and $x - 1$. One factor, 1, is already in our list, but the other factor, $x - 1$, is not. We add $x - 1$ to the list. We have
$$1, x + 1, x - 1.$$

Step 4 The least common denominator is the product of all factors in the final list. Thus,
$$1(x + 1)(x - 1)$$
or $(x + 1)(x - 1)$ is the least common denominator. ∎

✔ **CHECK POINT 2**   Find the LCD of $\dfrac{2}{x + 3}$ and $\dfrac{4}{x - 3}$.

EXAMPLE 3   Finding the Least Common Denominator

Find the LCD of

$$\frac{7}{5x^2 + 15x} \quad \text{and} \quad \frac{9}{x^2 + 6x + 9}.$$

Solution

Step 1 Factor each denominator completely.
$$5x^2 + 15x = 5x(x + 3)$$
$$x^2 + 6x + 9 = (x + 3)^2$$

Step 2 List the factors of the first denominator.
$$5, x, (x + 3)$$

Step 3 Add any unlisted factors from the second denominator. The second denominator is $(x + 3)^2$ or $(x + 3)(x + 3)$. One factor of $x + 3$ is already in our list, but the other factor is not. We add $x + 3$ to the list. We have
$$5, x, (x + 3), (x + 3).$$

Step 4 The least common denominator is the product of all factors in the final list. Thus,
$$5x(x + 3)(x + 3) \quad \text{or} \quad 5x(x + 3)^2$$
is the least common denominator. ∎

✔ **CHECK POINT 3**   Find the LCD of $\dfrac{9}{7x^2 + 28x}$ and $\dfrac{11}{x^2 + 8x + 16}$.

2 Add and subtract rational expressions with different denominators.

Adding and Subtracting Rational Expressions with Different Denominators

Finding the least common denominator for two (or more) rational expressions is the first step needed to add or subtract the expressions. For example, to add $\frac{1}{2}$ and $\frac{2}{3}$, we first determine that the LCD is 6. Then we write each fraction in terms of the LCD.

$$\frac{1}{2} + \frac{2}{3} = \frac{1}{2} \cdot \frac{3}{3} + \frac{2}{3} \cdot \frac{2}{2}$$

Multiply the numerator and denominator of each fraction by whatever extra factors are required to form 6, the LCD.

$\frac{3}{3} = 1$ and $\frac{2}{2} = 1$. Multiplying by 1 does not change a fraction's value.

$$= \frac{3}{6} + \frac{4}{6}$$

$$= \frac{3 + 4}{6}$$

Add numerators. Place this sum over the LCD.

$$= \frac{7}{6}$$

We follow the same steps in adding or subtracting rational expressions with different denominators.

Adding and Subtracting Rational Expressions That Have Different Denominators

1. Find the LCD of the rational expressions.
2. Rewrite each rational expression as an equivalent expression whose denominator is the LCD. To do so, multiply the numerator and the denominator of each rational expression by any factor(s) needed to convert the denominator into the LCD.
3. Add or subtract numerators, placing the resulting expression over the LCD.
4. If necessary, simplify the resulting rational expression.

EXAMPLE 4   Adding Rational Expressions with Different Denominators

Add: $\dfrac{7}{6x^2} + \dfrac{2}{9x}$.

Solution

Step 1 Find the least common denominator. In Example 1, we found that the LCD for these rational expressions is $18x^2$.

Step 2 Write equivalent expressions with the LCD as denominators. We must rewrite each rational expression with a denominator of $18x^2$.

$$\frac{7}{6x^2} \cdot \frac{3}{3} = \frac{21}{18x^2} \qquad\qquad \frac{2}{9x} \cdot \frac{2x}{2x} = \frac{4x}{18x^2}$$

Multiply the numerator and denominator by 3 to get $18x^2$, the LCD.

Multiply the numerator and denominator by $2x$ to get $18x^2$, the LCD.

Study Tip

It is incorrect to add rational expressions by adding numerators and denominators. Avoid this common error.

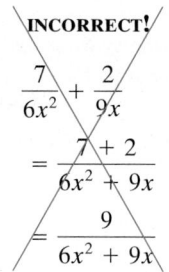

INCORRECT!

$$\frac{7}{6x^2} + \frac{2}{9x}$$

$$= \frac{7+2}{6x^2 + 9x}$$

$$= \frac{9}{6x^2 + 9x}$$

Because $\frac{3}{3} = 1$ and $\frac{2x}{2x} = 1$, we are not changing the value of either rational expression, only its appearance. In summary, we have:

$$\frac{7}{6x^2} + \frac{2}{9x}$$ The LCD is $18x^2$.

$$= \frac{7}{6x^2} \cdot \frac{3}{3} + \frac{2}{9x} \cdot \frac{2x}{2x}$$ Write equivalent expressions with the LCD.

$$= \frac{21}{18x^2} + \frac{4x}{18x^2}.$$

Steps 3 and 4 Add numerators, putting this sum over the LCD. Simplify if possible.

$$= \frac{21}{18x^2} + \frac{4x}{18x^2}$$

$$= \frac{21 + 4x}{18x^2} \quad \text{or} \quad \frac{4x + 21}{18x^2}$$

The numerator is prime and further simplification is not possible. ∎

✔ **CHECK POINT 4** Add: $\dfrac{3}{10x^2} + \dfrac{7}{15x}$.

EXAMPLE 5   Adding Rational Expressions with Different Denominators

Add: $\dfrac{3}{x+1} + \dfrac{5}{x-1}$.

Solution

Step 1 Find the least common denominator. The factors of the denominators are $x + 1$ and $x - 1$. In Example 2, we found that the LCD is $(x + 1)(x - 1)$.

Step 2 Write equivalent expressions with the LCD as denominators.

$$\frac{3}{x+1} + \frac{5}{x-1}$$

$$= \frac{3(x-1)}{(x+1)(x-1)} + \frac{5(x+1)}{(x+1)(x-1)}$$ Multiply each numerator and denominator by the extra factor required to form $(x + 1)(x - 1)$, the LCD.

Steps 3 and 4 Add numerators, putting this sum over the LCD. Simplify if possible.

$$= \frac{3(x-1) + 5(x+1)}{(x+1)(x-1)}$$

$$= \frac{3x - 3 + 5x + 5}{(x+1)(x-1)}$$ Use the distributive property to multiply and remove grouping symbols.

$$= \frac{8x + 2}{(x+1)(x-1)}$$ Combine like terms. ∎

We can factor 2 from the numerator of the answer in Example 5 to obtain

$$\frac{2(4x + 1)}{(x + 1)(x - 1)}.$$

Because the numerator and denominator do not have any common factors, further simplification is not possible. In this section, unless there is a common factor in the numerator and denominator, we will leave an answer's numerator in unfactored form and the denominator in factored form.

✔ **CHECK POINT 5** Add: $\dfrac{2}{x + 3} + \dfrac{4}{x - 3}$.

EXAMPLE 6 Subtracting Rational Expressions with Different Denominators

Subtract: $\dfrac{x}{x + 3} - 1$.

Solution

Step 1 Find the least common denominator. We know that 1 means $\frac{1}{1}$. The factor of the first denominator is $x + 3$. Adding the factor of the second denominator, 1, the LCD is $1(x + 3)$ or $x + 3$.

Step 2 Write equivalent expressions with the LCD as denominators.

$$\frac{x}{x + 3} - 1$$

$$= \frac{x}{x + 3} - \frac{1}{1} \qquad \text{Write 1 as } \tfrac{1}{1}.$$

$$= \frac{x}{x + 3} - \frac{1(x + 3)}{1(x + 3)} \qquad \begin{array}{l}\text{Multiply the numerator and denominator}\\ \text{of } \tfrac{1}{1} \text{ by the extra factor required to form}\\ x + 3, \text{ the LCD.}\end{array}$$

Steps 3 and 4 Subtract numerators, putting this difference over the LCD. Simplify if possible.

$$= \frac{x - (x + 3)}{x + 3}$$

$$= \frac{x - x - 3}{x + 3} \qquad \begin{array}{l}\text{Remove parentheses and then}\\ \text{change the sign of each term.}\end{array}$$

$$= \frac{-3}{x + 3} \quad \text{or} \quad -\frac{3}{x + 3} \qquad \text{Simplify.} \qquad ■$$

✔ **CHECK POINT 6** Subtract: $\dfrac{x}{x + 5} - 1$.

EXAMPLE 7   Subtracting Rational Expressions with Different Denominators

Subtract: $\dfrac{y + 2}{4y + 16} - \dfrac{2}{y^2 + 4y}$.

Solution

Step 1 Find the least common denominator. Start by factoring the denominators.

$$4y + 16 = 4(y + 4)$$
$$y^2 + 4y = y(y + 4)$$

The factors of the first denominator are 4 and $y + 4$. The only factor from the second denominator that is unlisted is y. Thus, the least common denominator is $4y(y + 4)$.

Step 2 Write equivalent expressions with the LCD as denominators.

$$\frac{y + 2}{4y + 16} - \frac{2}{y^2 + 4y}$$

$$= \frac{y + 2}{4(y + 4)} - \frac{2}{y(y + 4)} \qquad \text{Factor denominators.}$$

The LCD is $4y(y + 4)$.

$$= \frac{(y + 2)y}{4y(y + 4)} - \frac{2 \cdot 4}{4y(y + 4)} \qquad \begin{array}{l}\text{Multiply each numerator and}\\\text{denominator by the extra factor}\\\text{required to form } 4y(y + 4)\text{, the LCD.}\end{array}$$

Steps 3 and 4 Subtract numerators, putting this difference over the LCD. Simplify if possible.

$$= \frac{(y + 2)y - 2 \cdot 4}{4y(y + 4)}$$

$$= \frac{y^2 + 2y - 8}{4y(y + 4)} \qquad \begin{array}{l}\text{Use the distributive property:}\\(y + 2)y = y^2 + 2y.\text{ Multiply: } 2 \cdot 4 = 8.\end{array}$$

$$= \frac{\overset{1}{\cancel{(y + 4)}}(y - 2)}{4y\underset{1}{\cancel{(y + 4)}}} \qquad \text{Factor and simplify.}$$

$$= \frac{y - 2}{4y}$$

∎

✔ **CHECK POINT 7**   Subtract: $\dfrac{5}{y^2 - 5y} - \dfrac{y}{5y - 25}$.

In some situations, after factoring denominators, a factor in one denominator is the opposite of a factor in the other denominator. When this happens, we can use the following procedure:

Adding and Subtracting Rational Expressions When Denominators Contain Opposite Factors

When one denominator contains the opposite factor of the other, first multiply either rational expression by $\frac{-1}{-1}$. Then apply the procedure for adding or subtracting rational expressions that have different denominators to the rewritten problem.

EXAMPLE 8   Adding Rational Expressions with Opposite Factors in the Denominators

Add: $\dfrac{x^2 - 2}{2x^2 - x - 3} + \dfrac{x - 2}{3 - 2x}$.

Solution

Step 1 Find the least common denominator. Start by factoring the denominators.

$$2x^2 - x - 3 = (2x - 3)(x + 1)$$

$$3 - 2x = 1(3 - 2x)$$

Do you see that $2x - 3$ and $3 - 2x$ are opposite factors? Thus, we multiply either rational expression by $\frac{-1}{-1}$. We will use the second rational expression, resulting in $2x - 3$ in the denominator.

$$\dfrac{x^2 - 2}{2x^2 - x - 3} + \dfrac{x - 2}{3 - 2x}$$

$$= \dfrac{x^2 - 2}{(2x - 3)(x + 1)} + \dfrac{(-1)}{(-1)} \cdot \dfrac{x - 2}{3 - 2x}$$

Factor the first denominator.
Multiply the second rational
expression by $\frac{-1}{-1}$.

$$= \dfrac{x^2 - 2}{(2x - 3)(x + 1)} + \dfrac{-x + 2}{-3 + 2x}$$

Perform the multiplications by
−1 by changing every term's sign.

$$= \dfrac{x^2 - 2}{(2x - 3)(x + 1)} + \dfrac{2 - x}{2x - 3}$$

The LCD of our rewritten addition problem is $(2x - 3)(x + 1)$.

Step 2 Write equivalent expressions with the LCD as denominators.

$$= \dfrac{x^2 - 2}{(2x - 3)(x + 1)} + \dfrac{(2 - x)(x + 1)}{(2x - 3)(x + 1)}$$

Multiply the numerator and denomina-
tor of the second rational expression
by the extra factor required to form
$(2x - 3)(x + 1)$, the LCD.

Steps 3 and 4 Add numerators, putting this sum over the LCD. Simplify if possible.

$$= \dfrac{x^2 - 2 + (2 - x)(x + 1)}{(2x - 3)(x + 1)}$$

$$= \dfrac{x^2 - 2 + 2x + 2 - x^2 - x}{(2x - 3)(x + 1)}$$

Use the FOIL method to multiply
$(2 - x)(x + 1)$.

$$= \dfrac{(x^2 - x^2) + (2x - x) + (-2 + 2)}{(2x - 3)(x + 1)}$$

Group like terms.

$$= \dfrac{x}{(2x - 3)(x + 1)}$$

Combine like terms. ∎

Discover for Yourself

In Example 8, the denominators can be factored as follows:

$$2x^2 - x - 3 =$$

$$(2x - 3)(x + 1)$$

$$3 - 2x = -1(2x - 3).$$

Using these factorizations, what is the LCD? Solve Example 8 by obtaining this LCD in each rational expression. Then combine the expressions. How does your solution compare with the one shown on the right?

✔ **CHECK POINT 8** Add: $\dfrac{4x}{x^2 - 25} + \dfrac{3}{5 - x}$.

EXERCISE SET 7.4

Practice Exercises

In Exercises 1–16, find the least common denominator of the rational expressions.

1. $\dfrac{7}{15x^2}$ and $\dfrac{13}{24x}$

2. $\dfrac{11}{25x^2}$ and $\dfrac{17}{35x}$

3. $\dfrac{8}{15x^2}$ and $\dfrac{5}{6x^5}$

4. $\dfrac{7}{15x^2}$ and $\dfrac{11}{24x^5}$

5. $\dfrac{4}{x-3}$ and $\dfrac{7}{x+1}$

6. $\dfrac{2}{x-5}$ and $\dfrac{3}{x+7}$

7. $\dfrac{5}{7(y+2)}$ and $\dfrac{10}{y}$

8. $\dfrac{8}{11(y+5)}$ and $\dfrac{12}{y}$

9. $\dfrac{2}{x+3}$ and $\dfrac{5}{x^2-9}$

10. $\dfrac{2}{x-5}$ and $\dfrac{3}{x^2-25}$

11. $\dfrac{7}{y^2-4}$ and $\dfrac{15}{y(y+2)}$

12. $\dfrac{7}{y^2-100}$ and $\dfrac{13}{y(y-10)}$

13. $\dfrac{3}{y^2-25}$ and $\dfrac{y}{y^2-10y+25}$

14. $\dfrac{8}{y^2-16}$ and $\dfrac{y}{y^2-8y+16}$

15. $\dfrac{3}{x^2-x-20}$ and $\dfrac{x}{2x^2+7x-4}$

16. $\dfrac{7}{x^2-5x-6}$ and $\dfrac{x}{x^2-4x-5}$

In Exercises 17–82, add or subtract as indicated. Simplify the result, if possible.

17. $\dfrac{3}{x}+\dfrac{5}{x^2}$

18. $\dfrac{4}{x}+\dfrac{8}{x^2}$

19. $\dfrac{2}{9x}+\dfrac{11}{6x}$

20. $\dfrac{5}{6x}+\dfrac{7}{8x}$

21. $\dfrac{4}{x}+\dfrac{7}{2x^2}$

22. $\dfrac{10}{x}+\dfrac{3}{5x^2}$

23. $1+\dfrac{1}{x}$

24. $4+\dfrac{1}{x}$

25. $\dfrac{3}{x}+5$

26. $\dfrac{5}{x}+3$

27. $\dfrac{x-1}{6}+\dfrac{x+2}{3}$

28. $\dfrac{x+3}{2}+\dfrac{x+5}{4}$

29. $\dfrac{4}{x}+\dfrac{3}{x-5}$

30. $\dfrac{3}{x}+\dfrac{4}{x-6}$

31. $\dfrac{2}{x-1}+\dfrac{3}{x+2}$

32. $\dfrac{3}{x-2}+\dfrac{4}{x+3}$

33. $\dfrac{2}{y+5}+\dfrac{3}{4y}$

34. $\dfrac{3}{y+1}+\dfrac{2}{3y}$

35. $\dfrac{x}{x+7}-1$

36. $\dfrac{x}{x+6}-1$

37. $\dfrac{7}{x+5}-\dfrac{4}{x-5}$

38. $\dfrac{8}{x+6}-\dfrac{2}{x-6}$

39. $\dfrac{2x}{x^2-16}+\dfrac{x}{x-4}$

40. $\dfrac{4x}{x^2-25}+\dfrac{x}{x+5}$

41. $\dfrac{5y}{y^2-9}-\dfrac{4}{y+3}$

42. $\dfrac{8y}{y^2-16}-\dfrac{5}{y+4}$

43. $\dfrac{7}{x-1}-\dfrac{3}{(x-1)^2}$

44. $\dfrac{5}{x+3}-\dfrac{2}{(x+3)^2}$

45. $\dfrac{3y}{4y-20}+\dfrac{9y}{6y-30}$

46. $\dfrac{4y}{5y-10}+\dfrac{3y}{10y-20}$

47. $\dfrac{y+4}{y}-\dfrac{y}{y+4}$

48. $\dfrac{y}{y-5}-\dfrac{y-5}{y}$

49. $\dfrac{2x+9}{x^2-7x+12}-\dfrac{2}{x-3}$

50. $\dfrac{3x+7}{x^2-5x+6}-\dfrac{3}{x-3}$

51. $\dfrac{3}{x^2-1}+\dfrac{4}{(x+1)^2}$

52. $\dfrac{6}{x^2-4}+\dfrac{2}{(x+2)^2}$

53. $\dfrac{3x}{x^2+3x-10}-\dfrac{2x}{x^2+x-6}$

54. $\dfrac{x}{x^2-2x-24}-\dfrac{x}{x^2-7x+6}$

55. $\dfrac{y}{y^2 + 2y + 1} + \dfrac{4}{y^2 + 5y + 4}$

56. $\dfrac{y}{y^2 + 5y + 6} + \dfrac{4}{y^2 - y - 6}$

57. $\dfrac{x - 5}{x + 3} + \dfrac{x + 3}{x - 5}$

58. $\dfrac{x - 7}{x + 4} + \dfrac{x + 4}{x - 7}$

59. $\dfrac{5}{2y^2 - 2y} - \dfrac{3}{2y - 2}$

60. $\dfrac{7}{5y^2 - 5y} - \dfrac{2}{5y - 5}$

61. $\dfrac{4x + 3}{x^2 - 9} - \dfrac{x + 1}{x - 3}$

62. $\dfrac{2x - 1}{x + 6} - \dfrac{6 - 5x}{x^2 - 36}$

63. $\dfrac{y^2 - 39}{y^2 + 3y - 10} - \dfrac{y - 7}{y - 2}$

64. $\dfrac{y^2 - 6}{y^2 + 9y + 18} - \dfrac{y - 4}{y + 6}$

65. $4 + \dfrac{1}{x - 3}$

66. $7 + \dfrac{1}{x - 5}$

67. $3 - \dfrac{3y}{y + 1}$

68. $7 - \dfrac{4y}{y + 5}$

69. $\dfrac{9x + 3}{x^2 - x - 6} + \dfrac{x}{3 - x}$

70. $\dfrac{x^2 + 9x}{x^2 - 2x - 3} + \dfrac{5}{3 - x}$

71. $\dfrac{x + 3}{x^2 + x - 2} - \dfrac{2}{x^2 - 1}$

72. $\dfrac{x}{x^2 - 10x + 25} - \dfrac{x - 4}{2x - 10}$

73. $\dfrac{y + 3}{5y^2} - \dfrac{y - 5}{15y}$

74. $\dfrac{y - 7}{3y^2} - \dfrac{y - 2}{12y}$

75. $\dfrac{x + 3}{3x + 6} + \dfrac{x}{4 - x^2}$

76. $\dfrac{x + 7}{4x + 12} + \dfrac{x}{9 - x^2}$

77. $\dfrac{y}{y^2 - 1} + \dfrac{2y}{y - y^2}$

78. $\dfrac{y}{y^2 - 1} + \dfrac{5y}{y - y^2}$

79. $\dfrac{x - 1}{x} + \dfrac{y + 1}{y}$

80. $\dfrac{x + 2}{y} + \dfrac{y - 2}{x}$

81. $\dfrac{3x}{x^2 - y^2} - \dfrac{2}{y - x}$

82. $\dfrac{7x}{x^2 - y^2} - \dfrac{3}{y - x}$

⭐ **Application Exercises**

Two formulas that approximate the dosage of a drug prescribed for children are

$$\text{Young's Rule: } C = \dfrac{DA}{A + 12}$$

$$\text{and Cowling's Rule: } C = \dfrac{D(A + 1)}{24}.$$

In each formula, A = the child's age, in years, D = an adult dosage, and C = the proper child's dosage. The formulas apply for ages 2 through 13. Use the formulas to solve Exercises 83–86.

83. Use Young's rule to find the difference in a child's dosage for an 8-year-old child and a 3-year-old child. Express the answer as a single rational expression in terms of D. Then describe what your answer means in terms of the variables in the model.

84. Use Young's rule to find the difference in a child's dosage for a 10-year-old child and a 3-year-old child. Express the answer as a single rational expression in terms of D. Then describe what your answer means in terms of the variables in the model.

85. For a 12-year-old child, what is the difference in the dosage given by Cowling's rule and Young's rule? Express the answer as a single rational expression in terms of D. Then describe what your answer means in terms of the variables in the models.

86. Use Cowling's rule to find the difference in a child's dosage for a 12-year-old child and a 10-year-old child. Express the answer as a single rational expression in terms of D. Then describe what your answer means in terms of the variables in the model.

The graphs illustrate Young's rule and Cowling's rule when the dosage of a drug prescribed for an adult is 1000 milligrams. Use the graphs to solve Exercises 87–90.

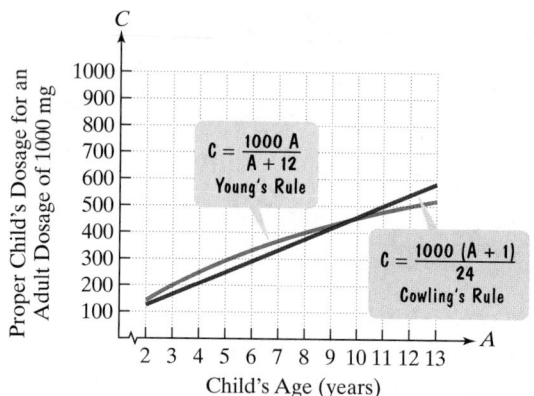

87. Does either formula consistently give a smaller dosage than the other? If so, which one?

88. Is there an age at which the dosage given by one formula becomes greater than the dosage given by the other? If so, what is a reasonable estimate of that age?

89. For what age under 11 is the difference in dosage given by the two formulas the greatest?

90. For what age over 11 is the difference in dosage given by the two formulas the greatest?

In Exercises 91–92, express the perimeter of each rectangle as a single rational expression.

91.

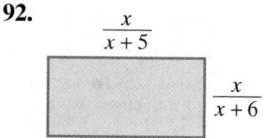

92.

$\dfrac{x}{x+5}$

$\dfrac{x}{x+6}$

Writing in Mathematics

93. Explain how to find the least common denominator for denominators of $x^2 - 100$ and $x^2 - 20x + 100$.

94. Explain how to add rational expressions that have different denominators. Use $\dfrac{3}{x+5} + \dfrac{7}{x+2}$ in your explanation.

Explain the error in Exercises 95–96. Then rewrite the right side of the equation to correct the error that now exists.

95. $\dfrac{1}{x} + \dfrac{2}{5} = \dfrac{3}{x+5}$

96. $\dfrac{1}{x} + 7 = \dfrac{1}{x+7}$

97. The formulas in Exercises 83–86 relate the dosage of a drug prescribed for children to the child's age. Describe another factor that might be used when determining a child's dosage. Is this factor more or less important than age? Explain why.

Critical Thinking Exercises

98. Which one of the following is true?

a. $x - \dfrac{1}{5} = \dfrac{4}{5}x$

b. The LCD of $\dfrac{1}{x}$ and $\dfrac{2x}{x-1}$ is $x^2 - 1$.

c. $\dfrac{1}{x} + \dfrac{x}{1} = \dfrac{1}{\cancel{x}} + \dfrac{\overset{1}{\cancel{x}}}{1} = 1 + 1 = 2$

d. $\dfrac{2}{x} + 1 = \dfrac{2 + x}{x}, x \neq 0$

In Exercises 99–101, perform the indicated operations. Simplify the result, if possible.

99. $\dfrac{x+6}{x^2-4} - \dfrac{x+3}{x+2} + \dfrac{x-3}{x-2}$

100. $\dfrac{1}{x+y} - \dfrac{1}{x-y} + \dfrac{2x}{x^2-y^2}$

101. $\dfrac{y^2+5y+4}{y^2+2y-3} \cdot \dfrac{y^2+y-6}{y^2+2y-3} - \dfrac{2}{y-1}$

In Exercises 102–103, find the missing rational expression.

102. $\dfrac{2}{x-1} + \boxed{} = \dfrac{2x^2+3x-1}{x^2(x-1)}$

103. $\dfrac{4}{x-2} - \boxed{} = \dfrac{2x+8}{(x-2)(x+1)}$

Review Exercises

104. Multiply: $(3x + 5)(2x - 7)$. (Section 5.3, Example 2)

105. Graph: $3x - y = 3$. (Section 3.2, Example 5)

106. Write the slope-intercept form of the equation of the line passing through $(-3, -4)$ and $(1, 0)$. (Section 3.5, Example 2)

▶ SECTION 7.5   *Complex Rational Expressions*

Objectives

1 Simplify complex rational expressions by dividing.

2 Simplify complex rational expressions by multiplying by the LCD.

SSM
PH Tutor   CD-   Video
Center   ROM

Do you drive to and from campus each day? If the one-way distance of your round-trip commute is d, then your average speed is given by the expression

$$\frac{2d}{\dfrac{d}{r_1} + \dfrac{d}{r_2}}$$

in which r_1 and r_2 are your average speeds on the outgoing and return trips, respectively. Do you notice anything unusual about this expression? It has two separate rational expressions in its denominator.

Numerator $\longrightarrow$ $\dfrac{2d}{\dfrac{d}{r_1} + \dfrac{d}{r_2}}$ $\longleftarrow$ Main fraction bar

Denominator $\longrightarrow$

Separate rational expressions occur in the denominator.

Complex rational expressions, also called **complex fractions**, have numerators or denominators containing one or more rational expressions. Here is another example of such an expression:

Numerator $\longrightarrow$

Main fraction bar $\longrightarrow$ $\dfrac{1 + \dfrac{1}{x}}{1 - \dfrac{1}{x}}$. $\longleftarrow$ Separate rational expressions occur in the numerator and denominator.

Denominator $\longrightarrow$

In this section, we study two methods for simplifying complex rational expressions.

1 Simplify complex rational expressions by dividing.

Simplifying by Rewriting Complex Rational Expressions as a Quotient of Two Rational Expressions

One method for simplifying a complex rational expression is to combine its numerator into a single expression and combine its denominator into a single expression. Then perform the division by inverting the denominator and multiplying.

Simplifying a Complex Rational Expression by Dividing

1. If necessary, add or subtract to get a single rational expression in the numerator.
2. If necessary, add or subtract to get a single rational expression in the denominator.
3. Perform the division indicated by the main fraction bar: Invert the denominator of the complex rational expression and multiply.
4. If possible, simplify.

The following examples illustrate the use of this first method.

EXAMPLE 1   Simplifying a Complex Rational Expression

Simplify:

$$\frac{\dfrac{1}{3} + \dfrac{2}{5}}{\dfrac{2}{5} - \dfrac{1}{3}}.$$

Solution   Let's first identify the parts of this complex rational expression.

Numerator —————
Main fraction bar —————
Denominator —————
$$\frac{\dfrac{1}{3} + \dfrac{2}{5}}{\dfrac{2}{5} - \dfrac{1}{3}}$$

Step 1 Add to get a single rational expression in the numerator.

$$\frac{1}{3} + \frac{2}{5} = \frac{1 \cdot 5}{3 \cdot 5} + \frac{2 \cdot 3}{5 \cdot 3} = \frac{5}{15} + \frac{6}{15} = \frac{11}{15}$$

The LCD is 3 · 5, or 15.

Step 2 Subtract to get a single rational expression in the denominator.

$$\frac{2}{5} - \frac{1}{3} = \frac{2 \cdot 3}{5 \cdot 3} - \frac{1 \cdot 5}{3 \cdot 5} = \frac{6}{15} - \frac{5}{15} = \frac{1}{15}$$

The LCD is 15.

Steps 3 and 4 Perform the division indicated by the main fraction bar: Invert and multiply. If possible, simplify.

$$\frac{\dfrac{1}{3} + \dfrac{2}{5}}{\dfrac{2}{5} - \dfrac{1}{3}} = \frac{\dfrac{11}{15}}{\dfrac{1}{15}} = \frac{11}{15} \cdot \frac{15}{1} = \frac{11}{\cancel{15}} \cdot \frac{\overset{1}{\cancel{15}}}{1} = 11$$

Invert and multiply.

■

✔ **CHECK POINT 1**   Simplify: $\dfrac{\dfrac{1}{4} + \dfrac{2}{3}}{\dfrac{2}{3} - \dfrac{1}{4}}.$

EXAMPLE 2   Simplifying a Complex Rational Expression

Simplify:

$$\frac{1 + \dfrac{1}{x}}{1 - \dfrac{1}{x}}.$$

Solution

Step 1 Add to get a single rational expression in the numerator.

$$1 + \frac{1}{x} = \frac{1}{1} + \frac{1}{x} = \frac{1 \cdot x}{1 \cdot x} + \frac{1}{x} = \frac{x}{x} + \frac{1}{x} = \frac{x + 1}{x}$$

The LCD is 1 · x, or x.

Step 2 Subtract to get a single rational expression in the denominator.

$$1 - \frac{1}{x} = \frac{1}{1} - \frac{1}{x} = \frac{1 \cdot x}{1 \cdot x} - \frac{1}{x} = \frac{x}{x} - \frac{1}{x} = \frac{x - 1}{x}$$

The LCD is 1 · x, or x.

Steps 3 and 4 Perform the division indicated by the main fraction bar: Invert and multiply. If possible, simplify.

$$\frac{1 + \dfrac{1}{x}}{1 - \dfrac{1}{x}} = \frac{\dfrac{x + 1}{x}}{\dfrac{x - 1}{x}} = \frac{x + 1}{x} \cdot \frac{x}{x - 1} = \frac{x + 1}{\cancel{x}_1} \cdot \frac{\cancel{x}^1}{x - 1} = \frac{x + 1}{x - 1}$$

Invert and multiply.

∎

✔ **CHECK POINT 2**   Simplify: $\dfrac{2 - \dfrac{1}{x}}{2 + \dfrac{1}{x}}.$

EXAMPLE 3   Simplifying a Complex Rational Expression

Simplify:

$$\frac{\dfrac{1}{xy}}{\dfrac{1}{x} + \dfrac{1}{y}}.$$

Solution

Step 1 Get a single rational expression in the numerator. The numerator, $\dfrac{1}{xy}$, already contains a single rational expression, so we can skip this step.

Step 2 Add to get a single rational expression in the denominator.

$$\frac{1}{x} + \frac{1}{y} = \frac{1 \cdot y}{x \cdot y} + \frac{1 \cdot x}{x \cdot y} = \frac{y}{xy} + \frac{x}{xy} = \frac{y + x}{xy}$$

The LCD is xy.

Steps 3 and 4 Perform the division indicated by the main fraction bar: Invert and multiply. If possible, simplify.

$$\frac{\dfrac{1}{xy}}{\dfrac{1}{x} + \dfrac{1}{y}} = \frac{\dfrac{1}{xy}}{\dfrac{y + x}{xy}} = \frac{1}{xy} \cdot \frac{xy}{y + x} = \frac{1}{\cancel{xy}} \cdot \frac{\overset{1}{\cancel{xy}}}{y + x} = \frac{1}{y + x}$$

Invert and multiply.

■

✔ **CHECK POINT 3** Simplify: $\dfrac{\dfrac{1}{x} - \dfrac{1}{y}}{\dfrac{1}{xy}}$.

2 | Simplify complex rational expressions by multiplying by the LCD.

Simplifying Complex Rational Expressions by Multiplying by the LCD

A second method for simplifying a complex rational expression is to find the least common denominator of all the rational expressions in its numerator and denominator. Then multiply each term in its numerator and denominator by this least common denominator. Because we are multiplying by a form of 1, we will obtain an equivalent expression that does not contain fractions in its numerator or denominator.

> ### Simplifying a Complex Rational Expression by Multiplying by the LCD
>
> 1. Find the LCD of all rational expressions within the complex rational expression.
> 2. Multiply both the numerator and the denominator of the complex rational expression by this LCD.
> 3. Use the distributive property and multiply each term in the numerator and denominator by this LCD. Simplify. No fractional expressions should remain.
> 4. If possible, factor and simplify.

We now rework Examples 1, 2, and 3 using the method of multiplying by the LCD. Compare the two simplification methods to see if there is one method that you prefer.

EXAMPLE 4   Simplifying a Complex Rational Expression by the LCD Method

Simplify:

$$\frac{\dfrac{1}{3} + \dfrac{2}{5}}{\dfrac{2}{5} - \dfrac{1}{3}}.$$

Solution   The denominators in the complex rational expression are $3, 5, 5,$ and 3. The LCD is $3 \cdot 5$, or 15. Multiply both the numerator and denominator of the complex rational expression by 15.

$$\frac{\dfrac{1}{3} + \dfrac{2}{5}}{\dfrac{2}{5} - \dfrac{1}{3}} = \frac{15}{15} \cdot \frac{\left(\dfrac{1}{3} + \dfrac{2}{5}\right)}{\left(\dfrac{2}{5} - \dfrac{1}{3}\right)} = \frac{15 \cdot \dfrac{1}{3} + 15 \cdot \dfrac{2}{5}}{15 \cdot \dfrac{2}{5} - 15 \cdot \dfrac{1}{3}} = \frac{5 + 6}{6 - 5} = \frac{11}{1} = 11$$

$\dfrac{15}{15} = 1$, so we are not changing the complex fraction's value.

■

✔ **CHECK POINT 4**   Simplify by the LCD method: $\dfrac{\dfrac{1}{4} + \dfrac{2}{3}}{\dfrac{2}{3} - \dfrac{1}{4}}.$

EXAMPLE 5   Simplifying a Complex Rational Expression by the LCD Method

Simplify: $\dfrac{1 + \dfrac{1}{x}}{1 - \dfrac{1}{x}}.$

Solution   The denominators in the complex rational expression are $1, x, 1,$ and x.

$$\frac{1 + \dfrac{1}{x}}{1 - \dfrac{1}{x}} = \frac{\dfrac{1}{1} + \dfrac{1}{x}}{\dfrac{1}{1} - \dfrac{1}{x}} \quad \text{Denominators}$$

Denominators

The LCD is $1 \cdot x$, or x. Multiply both the numerator and denominator of the complex rational expression by x.

$$\frac{1 + \dfrac{1}{x}}{1 - \dfrac{1}{x}} = \frac{x}{x} \cdot \frac{\left(1 + \dfrac{1}{x}\right)}{\left(1 - \dfrac{1}{x}\right)} = \frac{x \cdot 1 + x \cdot \dfrac{1}{x}}{x \cdot 1 - x \cdot \dfrac{1}{x}} = \frac{x + 1}{x - 1}$$

■

✔ **CHECK POINT 5**   Simplify by the LCD method: $\dfrac{2 - \dfrac{1}{x}}{2 + \dfrac{1}{x}}.$

EXAMPLE 6   Simplifying a Complex Rational Expression by the LCD Method

Simplify: $\dfrac{\dfrac{1}{xy}}{\dfrac{1}{x}+\dfrac{1}{y}}$.

Solution   The denominators in the complex rational expression are xy, x, and y. The LCD is xy. Multiply both the numerator and denominator of the complex rational expression by xy.

$$\frac{\dfrac{1}{xy}}{\dfrac{1}{x}+\dfrac{1}{y}}=\frac{xy}{xy}\cdot\frac{\left(\dfrac{1}{xy}\right)}{\left(\dfrac{1}{x}+\dfrac{1}{y}\right)}=\frac{xy\cdot\dfrac{1}{xy}}{xy\cdot\dfrac{1}{x}+xy\cdot\dfrac{1}{y}}=\frac{1}{y+x}.$$

■

✔ **CHECK POINT 6**   Simplify by the LCD method: $\dfrac{\dfrac{1}{x}-\dfrac{1}{y}}{\dfrac{1}{xy}}$.

EXERCISE SET 7.5

Practice Exercises

In Exercises 1–40, simplify each complex rational expression by the method of your choice.

1. $\dfrac{\dfrac{1}{2}+\dfrac{1}{4}}{\dfrac{1}{2}+\dfrac{1}{3}}$

2. $\dfrac{\dfrac{1}{3}+\dfrac{1}{4}}{\dfrac{1}{3}+\dfrac{1}{6}}$

3. $\dfrac{3+\dfrac{1}{2}}{4-\dfrac{1}{4}}$

4. $\dfrac{1+\dfrac{3}{5}}{2-\dfrac{1}{4}}$

5. $\dfrac{\dfrac{2}{5}-\dfrac{1}{3}}{\dfrac{2}{3}-\dfrac{3}{4}}$

6. $\dfrac{\dfrac{3}{5}-\dfrac{2}{3}}{\dfrac{2}{3}-\dfrac{5}{6}}$

7. $\dfrac{\dfrac{3}{4}-x}{\dfrac{3}{4}+x}$

8. $\dfrac{\dfrac{2}{3}-x}{\dfrac{2}{3}+x}$

9. $\dfrac{5-\dfrac{2}{x}}{3+\dfrac{1}{x}}$

10. $\dfrac{4+\dfrac{2}{y}}{1-\dfrac{3}{y}}$

11. $\dfrac{2+\dfrac{3}{y}}{1-\dfrac{7}{y}}$

12. $\dfrac{4-\dfrac{7}{y}}{3-\dfrac{2}{y}}$

13. $\dfrac{\dfrac{1}{y}-\dfrac{3}{2}}{\dfrac{1}{y}+\dfrac{3}{4}}$

14. $\dfrac{\dfrac{1}{y}-\dfrac{3}{4}}{\dfrac{1}{y}+\dfrac{2}{3}}$

15. $\dfrac{\dfrac{x}{5}-\dfrac{5}{x}}{\dfrac{1}{5}+\dfrac{1}{x}}$

16. $\dfrac{\dfrac{3}{x}+\dfrac{x}{3}}{\dfrac{x}{3}-\dfrac{3}{x}}$

17. $\dfrac{1+\dfrac{1}{x}}{1-\dfrac{1}{x^2}}$

18. $\dfrac{\dfrac{1}{x^2}-1}{\dfrac{1}{x}+1}$

19. $\dfrac{\dfrac{1}{7}-\dfrac{1}{y}}{\dfrac{7-y}{7}}$

20. $\dfrac{\dfrac{1}{9}-\dfrac{1}{y}}{\dfrac{9-y}{9}}$

21. $\dfrac{x+\dfrac{1}{y}}{\dfrac{x}{y}}$

22. $\dfrac{x-\dfrac{1}{y}}{\dfrac{x}{y}}$

23. $\dfrac{\dfrac{1}{x} + \dfrac{1}{y}}{xy}$

24. $\dfrac{\dfrac{1}{x} + \dfrac{1}{y}}{x + y}$

25. $\dfrac{\dfrac{x}{y} + \dfrac{1}{x}}{\dfrac{y}{x} + \dfrac{1}{x}}$

26. $\dfrac{\dfrac{1}{x} + \dfrac{1}{y}}{\dfrac{1}{x} - \dfrac{1}{y}}$

27. $\dfrac{\dfrac{1}{y} + \dfrac{2}{y^2}}{\dfrac{2}{y} + 1}$

28. $\dfrac{\dfrac{1}{y} + \dfrac{3}{y^2}}{\dfrac{3}{y} + 1}$

29. $\dfrac{\dfrac{12}{x^2} - \dfrac{3}{x}}{\dfrac{15}{x} - \dfrac{9}{x^2}}$

30. $\dfrac{\dfrac{8}{x^2} - \dfrac{2}{x}}{\dfrac{10}{x} - \dfrac{6}{x^2}}$

31. $\dfrac{2 + \dfrac{6}{y}}{1 - \dfrac{9}{y^2}}$

32. $\dfrac{3 + \dfrac{12}{y}}{1 - \dfrac{16}{y^2}}$

33. $\dfrac{\dfrac{1}{x + 2}}{1 + \dfrac{1}{x + 2}}$

34. $\dfrac{\dfrac{1}{x - 2}}{1 - \dfrac{1}{x - 2}}$

35. $\dfrac{x - 5 + \dfrac{3}{x}}{x - 7 + \dfrac{2}{x}}$

36. $\dfrac{x + 9 - \dfrac{7}{x}}{x - 6 + \dfrac{4}{x}}$

37. $\dfrac{\dfrac{3}{xy^2} + \dfrac{2}{x^2y}}{\dfrac{1}{x^2y} + \dfrac{2}{xy^3}}$

38. $\dfrac{\dfrac{2}{x^3y} + \dfrac{5}{xy^4}}{\dfrac{5}{x^3y} - \dfrac{3}{xy}}$

39. $\dfrac{\dfrac{3}{x + 1} - \dfrac{3}{x - 1}}{\dfrac{5}{x^2 - 1}}$

40. $\dfrac{\dfrac{3}{x + 2} - \dfrac{3}{x - 2}}{\dfrac{5}{x^2 - 4}}$

Application Exercises

41. The average speed on a round-trip commute having a one-way distance d is given by the complex rational expression

$$\dfrac{2d}{\dfrac{d}{r_1} + \dfrac{d}{r_2}}$$

in which r_1 and r_2 are the average speeds on the outgoing and return trips, respectively. Simplify the expression. Then find your average speed if you drive to campus averaging 40 miles per hour and return home on the same route averaging 30 miles per hour.

42. If two electrical resistors with resistances R_1 and R_2 are connected in parallel (see the figure), then the total resistance in the circuit is given by the complex rational expression

$$\dfrac{1}{\dfrac{1}{R_1} + \dfrac{1}{R_2}}.$$

Simplify the expression. Then find the total resistance if $R_1 = 10$ ohms and $R_2 = 20$ ohms.

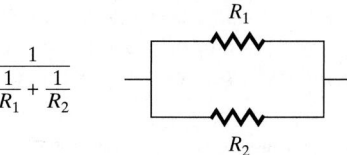

Writing in Mathematics

43. What is a complex rational expression? Give an example with your explanation.

44. Describe two ways to simplify $\dfrac{\dfrac{3}{x} + \dfrac{2}{x^2}}{\dfrac{1}{x^2} + \dfrac{2}{x}}$.

45. Which method do you prefer for simplifying complex rational expressions? Why?

Critical Thinking Exercises

46. Which one of the following is true?

a. The fraction $\dfrac{31{,}729{,}546}{72{,}578{,}112}$ is a complex rational expression.

b. $\dfrac{y - \dfrac{1}{2}}{y + \dfrac{3}{4}} = \dfrac{4y - 2}{4y + 3}$ for any value of y except $-\dfrac{3}{4}$.

c. $\dfrac{\dfrac{1}{4} - \dfrac{1}{3}}{\dfrac{1}{3} + \dfrac{1}{6}} = \dfrac{1}{12} \div \dfrac{3}{6} = \dfrac{1}{6}$

d. Some complex rational expressions cannot be simplified by both methods discussed in this section.

47. In one short sentence, five words or less, explain what

$$\frac{\dfrac{1}{x} + \dfrac{1}{x^2} + \dfrac{1}{x^3}}{\dfrac{1}{x^4} + \dfrac{1}{x^5} + \dfrac{1}{x^6}}$$

does to each number x.

In Exercises 48–49, simplify completely.

48. $\dfrac{1}{1 + \dfrac{1}{1 + \dfrac{1}{x}}}$

49. $\dfrac{1 + \dfrac{1}{y} - \dfrac{6}{y^2}}{1 - \dfrac{5}{y} + \dfrac{6}{y^2}} - \dfrac{1 - \dfrac{1}{y}}{1 - \dfrac{2}{y} - \dfrac{3}{y^2}}$

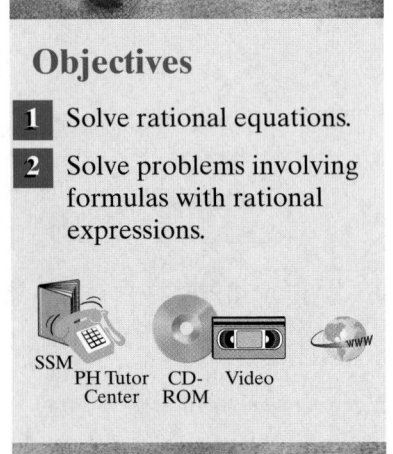
Technology Exercises

In Exercises 50–52, use a graphing utility to determine if the simplification is correct by graphing the expressions on both sides on the same screen. If the

answer is wrong, correct it and then verify your corrected simplification using the graphing utility.

50. $\dfrac{x - \dfrac{1}{2x + 1}}{1 - \dfrac{x}{2x + 1}} = 2x - 1$

51. $\dfrac{\dfrac{1}{x} + 1}{\dfrac{1}{x}} = 2$

52. $\dfrac{\dfrac{1}{x} + \dfrac{1}{3}}{\dfrac{1}{3x}} = x + \dfrac{1}{3}$

Review Exercises

53. Factor completely: $2x^3 - 20x^2 + 50x$. (Section 6.5, Example 2)

54. Solve: $2 - 3(x - 2) = 5(x + 5) - 1$. (Section 2.3, Example 3)

55. Multiply: $(x + y)(x^2 - xy + y^2)$. (Section 5.2, Example 7)

▶ **SECTION 7.6   Solving Rational Equations**

Objectives

1 Solve rational equations.

2 Solve problems involving formulas with rational expressions.

SSM PH Tutor CD- Video www
       Center ROM

The time has come to clean up the river. Suppose that the government has committed $375 million for this project. We know that

$$y = \frac{250x}{100 - x}$$

models the cost, in millions of dollars, to remove x percent of the river's pollutants. What percentage of pollutants can be removed for $375 million?

In order to determine the percentage, we use the given model. The government has committed $375 million, so substitute 375 for y:

$$375 = \frac{250x}{100 - x} \quad \text{or} \quad \frac{250x}{100 - x} = 375.$$

This equation contains a
rational expression.

Now we need to solve the equation and find the value for x. This variable represents the percentage of pollutants that can be removed for $375 million.

A **rational**, or **fractional**, **equation** is an equation containing one or more rational expressions. The preceding equation is an example of a rational equation. Do you see that there is a variable in a denominator? This is a characteristic of many rational equations. In this section, you will learn a procedure for solving such equations.

1 Solve rational equations.

Solving Rational Equations

We have seen that the LCD is used to add and subtract rational expressions. By contrast, when solving rational equations, **the LCD is used as a multiplier that clears an equation of fractions**.

Using Technology

We can use a graphing utility to verify the solution to Example 1. Graph each side of the equation, namely,

$$y_1 = \frac{x}{4}$$

$$y_2 = \frac{1}{4} + \frac{x}{6}.$$

Trace along the lines or use the utility's intersection feature. The solution, as shown below, is the first coordinate of the point of intersection. Thus, the solution is 3.

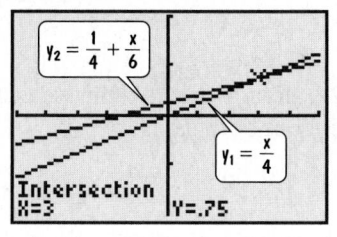

Intersection
X=3   Y=.75

EXAMPLE 1   Solving a Rational Equation

Solve: $\dfrac{x}{4} = \dfrac{1}{4} + \dfrac{x}{6}$.

Solution   The LCD of 4, 4, and 6 is 12. To clear the equation of fractions, we multiply both sides by 12.

$$\frac{x}{4} = \frac{1}{4} + \frac{x}{6}$$

This is the given equation.

$$12\left(\frac{x}{4}\right) = 12\left(\frac{1}{4} + \frac{x}{6}\right)$$

Multiply both sides by 12, the LCD of all the fractions in the equation.

$$12 \cdot \frac{x}{4} = 12 \cdot \frac{1}{4} + 12 \cdot \frac{x}{6}$$

Use the distributive property on the right side.

$$3x = 3 + 2x$$

Simplify: $\dfrac{\overset{3}{\cancel{12}}}{1} \cdot \dfrac{x}{\cancel{4}} = 3x; \ \overset{3}{\cancel{12}} \cdot \dfrac{1}{\cancel{4}} = 3; \ \overset{2}{\cancel{12}} \cdot \dfrac{x}{\cancel{6}} = 2x.$

$$x = 3$$

Subtract 2x from both sides.

Substitute 3 for x in the original equation. You should obtain the true statement $\dfrac{3}{4} = \dfrac{3}{4}$. This verifies that the solution is 3 and the solution set is $\{3\}$. ∎

✔ **CHECK POINT 1**   Solve: $\dfrac{x}{6} = \dfrac{1}{6} + \dfrac{x}{8}$.

In Example 1, we solved a rational equation with constants in denominators. Now, let's consider an equation such as

$$\frac{1}{x} = \frac{1}{5} + \frac{3}{2x}.$$

Can you see how this equation differs from the rational equation that we solved earlier? The variable, x, appears in two of the denominators. The procedure for solving this equation still involves multiplying each side by the least common denominator. However, we must avoid any values of the variable that make a denominator zero. For example, examine the denominators in the equation

$$\frac{1}{x} = \frac{1}{5} + \frac{3}{2x}.$$

> This denominator would equal zero if x = 0.

> This denominator would equal zero if x = 0.

We see that x cannot equal zero. With this in mind, let's solve the equation.

EXAMPLE 2   Solving a Rational Equation

Solve: $\dfrac{1}{x} = \dfrac{1}{5} + \dfrac{3}{2x}.$

Solution   The denominators are x, 5, and $2x$. The least common denominator is $10x$. We begin by multiplying both sides of the equation by $10x$. We will also write the restriction that x cannot equal zero to the right of the equation.

$$\frac{1}{x} = \frac{1}{5} + \frac{3}{2x}, \quad x \neq 0 \qquad \text{This is the given equation.}$$

$$10x \cdot \frac{1}{x} = 10x\left(\frac{1}{5} + \frac{3}{2x}\right) \qquad \text{Multiply both sides by 10x.}$$

$$10x \cdot \frac{1}{x} = 10x \cdot \frac{1}{5} + 10x \cdot \frac{3}{2x} \qquad \begin{array}{l}\text{Use the distributive property. Be}\\\text{sure to multiply all terms by 10x.}\end{array}$$

$$10x \cdot \frac{1}{x} = \overset{2}{\cancel{10}}x \cdot \frac{1}{\underset{1}{\cancel{5}}} + \overset{5}{\cancel{10}}x \cdot \frac{3}{\underset{1}{\cancel{2x}}} \qquad \begin{array}{l}\text{Divide out common factors in the}\\\text{multiplications.}\end{array}$$

$$10 = 2x + 15 \qquad \text{Simplify.}$$

Observe that the resulting equation,

$$10 = 2x + 15,$$

is now cleared of fractions. With the variable term, $2x$, already on the right, we will collect constant terms on the left by subtracting 15 from both sides.

$$-5 = 2x \qquad \text{Subtract 15 from both sides.}$$

$$-\frac{5}{2} = x \qquad \text{Divide both sides by 2.}$$

We check our solution by substituting $-\frac{5}{2}$ into the original equation or by using a calculator. With a calculator, evaluate each side of the equation for $x = -\frac{5}{2}$, or for $x = -2.5$. Note that the original restriction that $x \neq 0$ is met. The solution is $-\frac{5}{2}$ and the solution set is $\left\{-\frac{5}{2}\right\}$. ■

✔ **CHECK POINT 2**   Solve: $\dfrac{5}{2x} = \dfrac{17}{18} - \dfrac{1}{3x}.$

The following steps may be used to solve a rational equation:

> **Solving Rational Equations**
>
> 1. List restrictions on the variable. Avoid any values of the variable that make a denominator zero.
> 2. Clear the equation of fractions by multiplying both sides by the LCD of all rational expressions in the equation.
> 3. Solve the resulting equation.
> 4. Reject any proposed solution that is in the list of restrictions on the variable. Check other proposed solutions in the original equation.

EXAMPLE 3   Solving a Rational Equation

Solve: $x + \dfrac{1}{x} = \dfrac{5}{2}$.

Solution

Step 1 List restrictions on the variable.

$$x + \frac{1}{x} = \frac{5}{2}$$

This denominator would equal zero if $x = 0$.

The restriction is $x \neq 0$.

Step 2 Multiply both sides by the LCD. The denominators are x and 2. Thus, the LCD is $2x$. We multiply both sides by $2x$.

$$x + \frac{1}{x} = \frac{5}{2}, \quad x \neq 0 \qquad \text{This is the given equation.}$$

$$2x\left(x + \frac{1}{x}\right) = 2x\left(\frac{5}{2}\right) \qquad \text{Multiply both sides by the LCD.}$$

$$2x \cdot x + 2x \cdot \frac{1}{x} = 2x \cdot \frac{5}{2} \qquad \begin{array}{l}\text{Use the distributive property} \\ \text{on the left side.}\end{array}$$

$$2x^2 + 2 = 5x \qquad \text{Simplify.}$$

Step 3 Solve the resulting equation. Can you see that we have a quadratic equation? Write the equation in standard form and solve for x.

$$2x^2 - 5x + 2 = 0 \qquad \text{Subtract 5x from both sides.}$$

$$(2x - 1)(x - 2) = 0 \qquad \text{Factor.}$$

$$2x - 1 = 0 \quad \text{or} \quad x - 2 = 0 \qquad \text{Set each factor equal to 0.}$$

$$2x = 1 \qquad\qquad x = 2 \qquad \text{Solve the resulting equations.}$$

$$x = \frac{1}{2}$$

Step 4 Check proposed solutions in the original equation. The proposed solutions, $\frac{1}{2}$ and 2, are not part of the restriction that $x \neq 0$. Neither makes a denominator in the original equation equal to zero.

Check $\dfrac{1}{2}$:   **Check 2:**

$$x + \frac{1}{x} = \frac{5}{2} \qquad\qquad x + \frac{1}{x} = \frac{5}{2}$$

$$\frac{1}{2} + \frac{1}{\frac{1}{2}} \stackrel{?}{=} \frac{5}{2} \qquad\qquad 2 + \frac{1}{2} \stackrel{?}{=} \frac{5}{2}$$

$$\frac{1}{2} + 2 \stackrel{?}{=} \frac{5}{2} \qquad\qquad \frac{4}{2} + \frac{1}{2} \stackrel{?}{=} \frac{5}{2}$$

$$\frac{1}{2} + \frac{4}{2} \stackrel{?}{=} \frac{5}{2} \qquad\qquad \frac{5}{2} = \frac{5}{2}, \text{ true}$$

$$\frac{5}{2} = \frac{5}{2}, \text{ true}$$

The solutions are $\dfrac{1}{2}$ and 2, and $\left\{\dfrac{1}{2}, 2\right\}$ is the solution set.   ■

✔ **CHECK POINT 3**   Solve: $x + \dfrac{6}{x} = -5$.

EXAMPLE 4   Solving a Rational Equation

Solve: $\dfrac{3x}{x^2 - 9} + \dfrac{1}{x - 3} = \dfrac{3}{x + 3}$.

Solution

Step 1 List restrictions on the variable. By factoring denominators, it makes it easier to see values that make denominators zero.

$$\frac{3x}{(x + 3)(x - 3)} + \frac{1}{x - 3} = \frac{3}{x + 3}$$

This denominator is zero if $x = -3$ or $x = 3$.   This denominator is zero if $x = 3$.   This denominator is zero if $x = -3$.

The restrictions are $x \neq -3$ and $x \neq 3$.

Step 2 Multiply both sides by the LCD. The LCD is $(x + 3)(x - 3)$.

$$\frac{3x}{(x + 3)(x - 3)} + \frac{1}{x - 3} = \frac{3}{x + 3}, \quad x \neq -3, x \neq 3 \qquad \text{This is the given equation with a denominator factored.}$$

$$(x + 3)(x - 3)\left[\frac{3x}{(x + 3)(x - 3)} + \frac{1}{x - 3}\right] = (x + 3)(x - 3) \cdot \frac{3}{x + 3} \qquad \text{Multiply both sides by the LCD.}$$

$$(x + 3)(x - 3) \cdot \frac{3x}{(x + 3)(x - 3)} + (x + 3)(x - 3) \cdot \frac{1}{x - 3}$$

$$= (x + 3)(x - 3) \cdot \frac{3}{x + 3} \qquad \text{Use the distributive property on the left side.}$$

$$3x + (x + 3) = 3(x - 3) \qquad \text{Simplify.}$$

Step 3 Solve the resulting equation.

$$3x + (x + 3) = 3(x - 3)$$ This is the equation cleared of fractions.

$$4x + 3 = 3x - 9$$ Combine like terms on the left side. Use the distributive property on the right side.

$$x + 3 = -9$$ Subtract 3x from both sides.

$$x = -12$$ Subtract 3 from both sides.

Step 4 Check proposed solutions in the original equation. The proposed solution, -12, is not part of the restriction that $x \neq -3$ and $x \neq 3$. Substitute -12 for x in the given equation and show that -12 is the solution. The equation's solution set is $\{-12\}$. ∎

✔ **CHECK POINT 4** Solve: $\dfrac{11}{x^2 - 25} + \dfrac{4}{x + 5} = \dfrac{3}{x - 5}$.

EXAMPLE 5   Solving a Rational Equation

Solve: $\dfrac{8x}{x + 1} = 4 - \dfrac{8}{x + 1}$.

Solution

Step 1 List restrictions on the variable.

$$\frac{8x}{x + 1} = 4 - \frac{8}{x + 1}$$

These denominators are zero if x = −1.

The restriction is $x \neq -1$.

Step 2 Multiply both sides by the LCD. The LCD is $x + 1$.

$$\frac{8x}{x + 1} = 4 - \frac{8}{x + 1}, \quad x \neq -1$$ This is the given equation.

$$(x + 1) \cdot \frac{8x}{x + 1} = (x + 1)\left[4 - \frac{8}{x + 1}\right]$$ Multiply both sides by the LCD.

$$\cancel{(x + 1)} \cdot \frac{8x}{\cancel{x + 1}} = (x + 1) \cdot 4 - \cancel{(x + 1)} \cdot \frac{8}{\cancel{x + 1}}$$ Use the distributive property on the right side.

$$8x = 4(x + 1) - 8$$ Simplify.

Step 3 Solve the resulting equation.

$$8x = 4(x + 1) - 8$$ This is the equation cleared of fractions.

$$8x = 4x + 4 - 8$$ Use the distributive property on the right side.

$$8x = 4x - 4$$ Simplify.

$$4x = -4$$ Subtract 4x from both sides.

$$x = -1$$ Divide both sides by 4.

Study Tip

Reject any proposed solution that causes any denominator in a rational equation to equal 0.

Step 4 Check proposed solutions. The proposed solution, -1, is *not* a solution because of the restriction that $x \neq -1$. Notice that -1 makes two of the denominators zero in the original equation. There is *no solution to this equation*. The solution set is $\varnothing$, the empty set. ∎

✔ **CHECK POINT 5** Solve: $\dfrac{x}{x-3} = \dfrac{3}{x-3} + 9$.

Study Tip

It is important to distinguish between adding and subtracting rational expressions and solving rational equations. We *simplify* sums and differences of terms. On the other hand, we *solve* equations. This is shown in the following two problems, both with an LCD of $3x$.

Adding Rational Expressions	Solving Rational Equations
Simplify:	Solve:

$$\frac{5}{3x} + \frac{3}{x}.$$

$$\frac{5}{3x} + \frac{3}{x} = 1.$$

$$= \frac{5}{3x} + \frac{3}{x} \cdot \frac{3}{3}$$

$$3x\left(\frac{5}{3x} + \frac{3}{x}\right) = 3x \cdot 1$$

$$= \frac{5}{3x} + \frac{9}{3x}$$

$$3x \cdot \frac{5}{3x} + 3x \cdot \frac{3}{x} = 3x$$

$$= \frac{5+9}{3x}$$

$$5 + 9 = 3x$$

$$= \frac{14}{3x}$$

$$14 = 3x$$

$$\frac{14}{3} = x$$

Applications of Rational Equations

2 Solve problems involving formulas with rational expressions.

Rational equations can be solved to answer questions about variables contained in mathematical models.

EXAMPLE 6   A Government-Funded Cleanup

The formula

$$y = \frac{250x}{100 - x}$$

models the cost, in millions of dollars, to remove x percent of a river's pollutants. If the government commits \$375 million for this project, what percentage of pollutants can be removed?

Solution   Substitute 375 for y and solve the resulting rational equation for x.

$$375 = \frac{250x}{100 - x}$$ The LCD is $100 - x$.

$$(100 - x)375 = (100-x) \cdot \frac{250x}{100-x}$$ Multiply both sides by the LCD.

$$375(100 - x) = 250x$$ Simplify.

$$37,500 - 375x = 250x$$

Use the distributive property on the left side.

$$37,500 = 625x$$

Add 375x to both sides.

$$\frac{37,500}{625} = \frac{625x}{625}$$

Divide both sides by 625.

$$60 = x$$

Simplify.

If the government spends \$375 million, 60% of the river's pollutants can be removed. ∎

✔ **CHECK POINT 6**   Use the model in Example 6 to answer this question: If government funding is increased to \$750 million, what percentage of pollutants can be removed?

EXERCISE SET 7.6

Practice Exercises

In Exercises 1–44, solve each rational equation. If an equation has no solution, so state.

1. $\dfrac{x}{3} = \dfrac{x}{2} - 2$

2. $\dfrac{x}{5} = \dfrac{x}{6} + 1$

3. $\dfrac{4x}{3} = \dfrac{x}{18} - \dfrac{x}{6}$

4. $\dfrac{5x}{4} = \dfrac{x}{12} - \dfrac{x}{2}$

5. $2 - \dfrac{8}{x} = 6$

6. $1 - \dfrac{9}{x} = 4$

7. $\dfrac{4}{x} + \dfrac{1}{2} = \dfrac{5}{x}$

8. $\dfrac{5}{x} + \dfrac{1}{3} = \dfrac{6}{x}$

9. $\dfrac{2}{x} + 3 = \dfrac{5}{2x} + \dfrac{13}{4}$

10. $\dfrac{7}{2x} = \dfrac{5}{3x} + \dfrac{22}{3}$

11. $\dfrac{2}{3x} + \dfrac{1}{4} = \dfrac{11}{6x} - \dfrac{1}{3}$

12. $\dfrac{5}{2x} - \dfrac{8}{9} = \dfrac{1}{18} - \dfrac{1}{3x}$

13. $\dfrac{6}{x+3} = \dfrac{4}{x-3}$

14. $\dfrac{3}{x+1} = \dfrac{5}{x-1}$

15. $\dfrac{x-2}{2x} + 1 = \dfrac{x+1}{x}$

16. $\dfrac{7x-4}{5x} = \dfrac{9}{5} - \dfrac{4}{x}$

17. $x + \dfrac{6}{x} = -7$

18. $x + \dfrac{7}{x} = -8$

19. $\dfrac{x}{5} - \dfrac{5}{x} = 0$

20. $\dfrac{x}{4} - \dfrac{4}{x} = 0$

21. $x + \dfrac{3}{x} = \dfrac{12}{x}$

22. $x + \dfrac{3}{x} = \dfrac{19}{x}$

23. $\dfrac{4}{y} - \dfrac{y}{2} = \dfrac{7}{2}$

24. $\dfrac{4}{3y} - \dfrac{1}{3} = y$

25. $\dfrac{x-4}{x} = \dfrac{15}{x+4}$

26. $\dfrac{x-4}{x} = \dfrac{6}{x+4}$

27. $\dfrac{1}{x-1} + 5 = \dfrac{11}{x-1}$

28. $\dfrac{3}{x+4} - 7 = \dfrac{-4}{x+4}$

29. $\dfrac{8y}{y+1} = 4 - \dfrac{8}{y+1}$

30. $\dfrac{2}{y-2} = \dfrac{y}{y-2} - 2$

31. $\dfrac{3}{x-1} + \dfrac{8}{x} = 3$

32. $\dfrac{2}{x-2} + \dfrac{4}{x} = 2$

33. $\dfrac{3y}{y-4} - 5 = \dfrac{12}{y-4}$

34. $\dfrac{10}{y+2} = 3 - \dfrac{5y}{y+2}$

35. $\dfrac{1}{x} + \dfrac{1}{x-3} = \dfrac{x-2}{x-3}$

36. $\dfrac{1}{x-1} + \dfrac{2}{x} = \dfrac{x}{x-1}$

37. $\dfrac{x+1}{3x+9} + \dfrac{x}{2x+6} = \dfrac{2}{4x+12}$

38. $\dfrac{3}{2y-2} + \dfrac{1}{2} = \dfrac{2}{y-1}$

39. $\dfrac{4y}{y^2-25} + \dfrac{2}{y-5} = \dfrac{1}{y+5}$

40. $\dfrac{4}{y+5} + \dfrac{2}{y-5} = \dfrac{32}{y^2-25}$

41. $\dfrac{1}{x-4} - \dfrac{5}{x+2} = \dfrac{6}{x^2-2x-8}$

42. $\dfrac{6}{x+3} - \dfrac{5}{x-2} = \dfrac{-20}{x^2+x-6}$

43. $\dfrac{2}{x+3} - \dfrac{2x+3}{x-1} = \dfrac{6x-5}{x^2+2x-3}$

44. $\dfrac{x-3}{x-2} + \dfrac{x+1}{x+3} = \dfrac{2x^2-15}{x^2+x-6}$

⭐ Application Exercises

A company that manufactures wheelchairs has fixed costs of $500,000. The average cost per wheelchair, C, for the company to manufacture x wheelchairs per month is modeled by the formula

$$C = \dfrac{400x + 500,000}{x}.$$

Use this mathematical model to solve Exercises 45–46.

45. How many wheelchairs per month can be produced at an average cost of $450 per wheelchair?

46. How many wheelchairs per month can be produced at an average cost of $405 per wheelchair?

In Palo Alto, California, a government agency ordered computer-related companies to contribute to a pool of money to clean up underground water supplies. (The companies had stored toxic chemicals in leaking underground containers.) The formula

$$C = \dfrac{2x}{100 - x}$$

models the cost, in millions of dollars, for removing x percent of the contaminants. Use this mathematical model to solve Exercises 47–48.

47. What percentage of the contaminants can be removed for $2 million?

48. What percentage of the contaminants can be removed for $8 million?

We have seen that Young's rule

$$C = \dfrac{DA}{A + 12}$$

can be used to approximate the dosage of a drug prescribed for children. In this formula, A = the child's age, in years, D = an adult dosage, and C = the proper child's dosage. Use this formula to solve Exercises 49–50.

49. When the adult dosage is 1000 milligrams, a child is given 300 milligrams. What is that child's age? Round to the nearest year.

50. When the adult dosage is 1000 milligrams, a child is given 500 milligrams. What is that child's age?

A grocery store sells 4000 cases of canned soup per year. By averaging costs to purchase soup and pay storage costs, the owner has determined that if x cases are ordered at a time, the yearly inventory cost, C, can be modeled by

$$C = \dfrac{10,000}{x} + 3x.$$

The graph of this model is shown below. Use this information to solve Exercises 51–52.

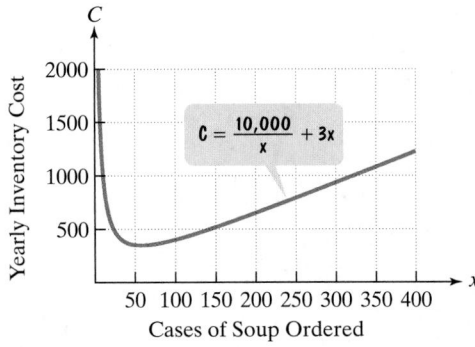

51. How many cases should be ordered at a time for yearly inventory costs to be $350? Identify your solutions as points on the graph.

52. How many cases should be ordered at a time for yearly inventory costs to be $790? Identify your solutions as points on the graph.

In baseball, a player's batting average is the total number of hits divided by the total number of times at bat. Use this information to solve Exercises 53–54.

53. A player has 12 hits after 40 times at bat. How many additional consecutive times must the player hit the ball to achieve a batting average of 0.440?

54. A player has eight hits after 50 times at bat. How many additional consecutive times must the player hit the ball to achieve a batting average of 0.250?

Writing in Mathematics

55. What is a rational equation?

56. Explain how to solve a rational equation.

57. Explain how to find restrictions on the variable in a rational equation.

58. Why should restrictions on the variable in a rational equation be listed before you begin solving the equation?

59. Describe similarities and differences between the procedures needed to solve the following problems:

$$\text{Add: } \frac{2}{x} + \frac{3}{4} \qquad \text{Solve: } \frac{2}{x} + \frac{3}{4} = 1.$$

60. The equation

$$P = \frac{72,900}{100x^2 + 729}$$

models the percentage of people in the United States, P, who have x years of education and are unemployed. Use this model to write a problem that can be solved using a rational equation. It is not necessary to solve the problem.

Critical Thinking Exercises

61. Which one of the following is true?

a. $\frac{1}{x} + \frac{1}{6} = 6x\left(\frac{1}{x} + \frac{1}{6}\right) = 6 + x$

b. If a is any real number, the equation $\frac{a}{x} + 1 = \frac{a}{x}$ has no solution.

c. All real numbers satisfy the equation $\frac{3}{x} - \frac{1}{x} = \frac{2}{x}$.

d. To solve $\frac{5}{3x} + \frac{3}{x} = 1$, we must first add the rational expressions on the left side.

62. Solve for f: $\frac{1}{p} + \frac{1}{q} = \frac{1}{f}$.

63. Solve for f_2: $f = \frac{f_1 f_2}{f_1 + f_2}$.

In Exercises 64–65, solve each rational equation.

64. Solve:

$$\frac{x}{x - 3} - \frac{7x - 6}{x^2 - x - 6} = \frac{2}{x + 2}.$$

65. Solve:

$$\left(\frac{x + 1}{x + 7}\right)^2 \div \left(\frac{x + 1}{x + 7}\right)^4 = 0.$$

66. Find b so that the solution of

$$\frac{7x + 4}{b} + 13 = x$$

is -6.

Technology Exercises

In Exercises 67–69, use a graphing utility to solve each rational equation. Graph each side of the equation in the given viewing rectangle. The solution is the first coordinate of the point(s) of intersection. Check by direct substitution.

67. $\frac{x}{2} + \frac{x}{4} = 6$

$[-5, 10, 1]$ by $[-5, 10, 1]$

68. $\frac{50}{x} = 2x$

$[-10, 10, 1]$ by $[-20, 20, 2]$

69. $x + \frac{6}{x} = -5$

$[-10, 10, 1]$ by $[-10, 10, 1]$

Review Exercises

70. Factor completely: $x^4 + 2x^3 - 3x - 6$. (Section 6.1, Example 7)

71. Simplify: $(3x^2)(-4x^{-10})$. (Section 5.7, Example 3)

72. Simplify: $-5[4(x - 2) - 3]$. (Section 1.8, Example 11)

▶ SECTION 7.7 Applications Using Rational Equations and Proportions

Objectives

1 Solve problems involving motion.

2 Solve problems involving work.

3 Solve problems involving proportions.

4 Solve problems involving similar triangles.

SSM   PH Tutor   CD-   Video
        Center   ROM

The possibility of seeing a blue whale, the largest mammal ever to grace the earth, increases the excitement of gazing out over the ocean's swell of waves. Blue whales were hunted to near extinction in the last half of the nineteenth and the first half of the twentieth centuries. Using a method for estimating wildlife populations that we discuss in this section, by the mid-1960s it was determined that the world population of blue whales was less than 1000. This led the International Whaling Commission to ban the killing of blue whales to prevent their extinction. A dramatic increase in blue whale sightings indicates an ongoing increase in their population and the success of the killing ban.

1 Solve problems involving motion.

Problems Involving Motion

We have seen that the distance, d, covered by any moving body is the product of its average rate, r, and its time in motion, t: $d = rt$. Rational expressions appear in motion problems when the conditions of the problem involve the time traveled. We can obtain an expression for t, the time traveled, by dividing both sides of $d = rt$ by r.

$$d = rt \qquad \text{Distance equals rate times time.}$$

$$\frac{d}{r} = \frac{rt}{r} \qquad \text{Divide both sides by } r.$$

$$\frac{d}{r} = t \qquad \text{Simplify.}$$

Time in Motion

$$t = \frac{d}{r}$$

$$\text{Time traveled} = \frac{\text{Distance traveled}}{\text{Rate of travel}}$$

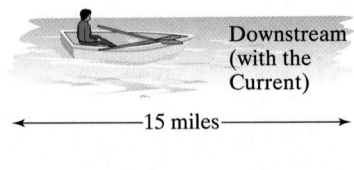

Downstream (with the Current)

◄——————15 miles——————►

Upstream (against the Current)

◄—— 9 miles ——►

EXAMPLE 1   A Motion Problem Involving Time

You were told not to go puttering around in your small boat because of the current. In still water, your boat averages 8 miles per hour. It takes you the same amount of time to travel 15 miles downstream, with the current, as 9 miles upstream, against the current. What is the rate of the water's current?

Solution

Step 1 Let x represent one of the quantities. Let

$$x = \text{the rate of the current.}$$

Step 2 Represent other quantities in terms of x. We still need expressions for the rate of your boat with the current and the rate against the current. Traveling with the current, the boat's rate in still water, 8 miles per hour, is increased by the current's rate, x miles per hour. Thus,

$$8 + x = \text{the boat's rate with the current.}$$

Traveling against the current, the boat's rate in still water, 8 miles per hour, is decreased by the current's rate, x miles per hour. Thus,

$$8 - x = \text{the boat's rate against the current.}$$

Step 3 Write an equation that describes the conditions. By reading the problem again, we discover that the crucial idea is that the time spent going 15 miles with the current equals the time spent going 9 miles against the current. This information is summarized in the following table.

	Distance	Rate	Time $= \dfrac{\text{Distance}}{\text{Rate}}$
With the Current	15	$8 + x$	$\dfrac{15}{8 + x}$
Against the Current	9	$8 - x$	$\dfrac{9}{8 - x}$

These two times are equal.

We are now ready to write an equation that describes the problem's conditions.

The time spent going 15 miles with the current — equals — the time spent going 9 miles against the current.

$$\frac{15}{8 + x} = \frac{9}{8 - x}$$

Step 4 Solve the equation and answer the question.

$$\frac{15}{8 + x} = \frac{9}{8 - x} \qquad \text{This is the equation for the problem's conditions.}$$

$$(8+x)(8-x) \cdot \frac{15}{8+x} = (8+x)(8-x) \cdot \frac{9}{8-x} \qquad \text{Multiply both sides by the LCD, } (8 + x)(8 - x).$$

$$15(8 - x) = 9(8 + x) \qquad \text{Simplify.}$$

$$120 - 15x = 72 + 9x \qquad \text{Use the distributive property.}$$

$$120 = 72 + 24x \qquad \text{Add 15x to both sides.}$$

$$48 = 24x \qquad \text{Subtract 72 from both sides.}$$

$$2 = x \qquad \text{Divide both sides by 24.}$$

The rate of the water's current is 2 miles per hour.

Step 5 Check the proposed solution in the original wording of the problem. Does it take you the same amount of time to travel 15 miles downstream as 9 miles upstream if the current is 2 miles per hour? Keep in mind that your rate in still water is 8 miles per hour.

$$\text{Time required to travel 15 miles with the current} = \frac{\text{Distance}}{\text{Rate}} = \frac{15}{8+2} = \frac{15}{10} = 1\frac{1}{2} \text{ hours}$$

$$\text{Time required to travel 9 miles against the current} = \frac{\text{Distance}}{\text{Rate}} = \frac{9}{8-2} = \frac{9}{6} = 1\frac{1}{2} \text{ hours}$$

These times are the same, which checks with the original conditions of the problem. ∎

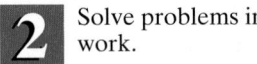 **CHECK POINT 1** Forget the small boat! This time we have you canoeing on the Colorado River. In still water, your average canoeing rate is 3 miles per hour. It takes you the same amount of time to travel 10 miles downstream, with the current, as 2 miles upstream, against the current. What is the rate of the water's current?

② Solve problems involving work.

Problems Involving Work

You are thinking of designing your own Web site. You estimate that it will take 30 hours to do the job. In 1 hour, $\frac{1}{30}$ of the job is completed. In 2 hours, $\frac{2}{30}$, or $\frac{1}{15}$, of the job is completed. In 3 hours, the fractional part of the job done is $\frac{3}{30}$, or $\frac{1}{10}$. In x hours, the fractional part of the job that you can complete is $\frac{x}{30}$.

Your friend, who has experience developing Web sites, took 20 hours working on his own to design an impressive site. You wonder about the possibility of working together. How long would it take both of you to design your Web site?

Problems involving work usually have two people working together to complete a job. The amount of time it takes each person to do the job working alone is frequently known. The question deals with how long it will take both people working together to complete the job.

In work problems, **the number 1 represents one whole job completed.** For example, the completion of your Web site is represented by 1. Equations in work problems are based on the following condition:

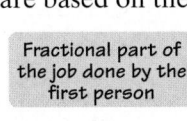

| Fractional part of the job done by the first person | + | Fractional part of the job done by the second person | = | 1 (one whole job completed). |

EXAMPLE 2 Solving a Problem Involving Work

You can design a Web site in 30 hours. Your friend can design the same site in 20 hours. How long will it take to design the Web site if you both work together?

Solution

Step 1 Let x represent one of the quantities. Let

x = the time, in hours, for you and your friend to design the Web site together.

Step 2 Represent other quantities in terms of *x*. Because there are no other unknown quantities, we can skip this step.

Step 3 Write an equation that describes the conditions. We construct a table to help find the fractional part of the task completed by you and your friend in *x* hours.

	Fractional part of job completed in 1 hour	Time working together	Fractional part of job completed in *x* hours
You	$\dfrac{1}{30}$	x	$\dfrac{x}{30}$
Your friend	$\dfrac{1}{20}$	x	$\dfrac{x}{20}$

You can design the site in 30 hours.

Your friend can design the site in 20 hours.

Fractional part of the job done by you	+	fractional part of the job done by your friend	=	one whole job.
$\dfrac{x}{30}$	+	$\dfrac{x}{20}$	=	1

Step 4 Solve the equation and answer the question.

$$\frac{x}{30} + \frac{x}{20} = 1 \qquad \text{This is the equation for the problem's conditions.}$$

$$60\left(\frac{x}{30} + \frac{x}{20}\right) = 60 \cdot 1 \qquad \text{Multiply both sides by 60, the LCD.}$$

$$60 \cdot \frac{x}{30} + 60 \cdot \frac{x}{20} = 60 \qquad \text{Use the distributive property on the left side.}$$

$$2x + 3x = 60 \qquad \text{Simplify: } \frac{\overset{2}{\cancel{60}}}{1} \cdot \frac{x}{\underset{1}{\cancel{30}}} = 2x \text{ and } \frac{\overset{3}{\cancel{60}}}{1} \cdot \frac{x}{\underset{1}{\cancel{20}}} = 3x.$$

$$5x = 60 \qquad \text{Combine like terms.}$$

$$x = 12 \qquad \text{Divide both sides by 5.}$$

If you both work together, you can design your Web site in 12 hours.

Step 5 Check the proposed solution in the original wording of the problem. Will you both complete the job in 12 hours? In 12 hours, you can complete $\frac{12}{30}$, or $\frac{2}{5}$, of the job. In 12 hours, your friend can complete $\frac{12}{20}$, or $\frac{3}{5}$, of the job. Notice that $\frac{2}{5} + \frac{3}{5} = 1$, which represents the completion of the entire job, or one whole job. ■

Study Tip

Let

$$a = \text{the time it takes person } A \text{ to do a job working alone}$$

$$b = \text{the time it takes person } B \text{ to do the same job working alone.}$$

If *x* represents the time it takes for *A* and *B* to complete the entire job working together, then the situation can be modeled by the rational equation

$$\frac{x}{a} + \frac{x}{b} = 1.$$

✔ **CHECK POINT 2** One person can paint the outside of a house in 8 hours. A second person can do it in 4 hours. How long will it take them to do the job if they work together?

③ Solve problems involving proportions.

Problems Involving Proportions

A **ratio** compares quantities by division. For example this year's entering class at a medical school contains 60 women and 30 men. The ratio of women to men is $\frac{60}{30}$. We can express this ratio as a fraction reduced to lowest terms:

$$\frac{60}{30} = \frac{\cancel{30} \cdot 2}{\cancel{30} \cdot 1} = \frac{2}{1}.$$

This ratio can be expressed as 2:1, or 2 to 1.

A **proportion** is a statement that says that two ratios are equal. If the ratios are $\frac{a}{b}$ and $\frac{c}{d}$, then the proportion is

$$\frac{a}{b} = \frac{c}{d}.$$

We can clear this rational equation of fractions by multiplying both sides by bd, the least common denominator:

$$\frac{a}{b} = \frac{c}{d} \qquad \text{This is the given proportion.}$$

$$bd \cdot \frac{a}{b} = bd \cdot \frac{c}{d} \qquad \text{Multiply both sides by } bd (b \ne 0 \text{ and } d \ne 0). \text{ Then simplify. On the}$$
$$\text{left, } \frac{\cancel{b}d}{1} \cdot \frac{a}{\cancel{b}} = da = ad. \text{ On the right, } \frac{b\cancel{d}}{1} \cdot \frac{c}{\cancel{d}} = bc.$$

$$ad = bc$$

We see that the following principle is true for any proportion:

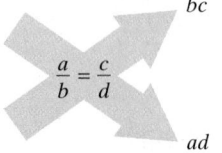

bc

$\frac{a}{b} = \frac{c}{d}$

ad

The cross-products principle: $ad = bc$

The Cross-Products Principle for Proportions

If $\dfrac{a}{b} = \dfrac{c}{d}$, then $ad = bc$. ($b \ne 0$ and $d \ne 0$)

The cross products ad and bc are equal.

For example, if $\frac{2}{3} = \frac{6}{9}$, we see that $2 \cdot 9 = 3 \cdot 6$, or $18 = 18$.

Here is a procedure for solving problems involving proportions:

Solving Applied Problems Using Proportions

1. Read the problem and represent the unknown quantity by x (or any letter).
2. Set up a proportion by listing the given ratio on one side and the ratio with the unknown quantity on the other side.
3. Drop units and apply the cross-products principle.
4. Solve for x and answer the question.

EXAMPLE 3   Applying Proportions: Calculating Taxes

The tax on a house whose assessed value is $65,000 is $825. Determine the tax on a house with an assessed value of $180,000, assuming the same tax rate.

Solution

Step 1 Represent the unknown by x. Let x = the tax on a $180,000 house.

Step 2 Set up a proportion. We will set up a proportion comparing taxes to assessed value.

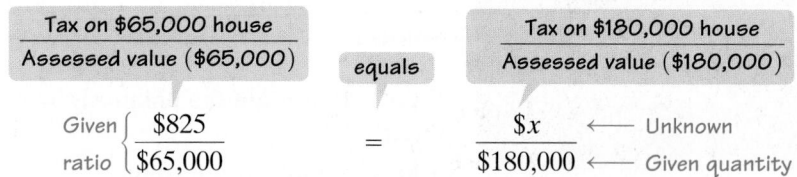

$$\dfrac{\text{Tax on \$65,000 house}}{\text{Assessed value (\$65,000)}} \quad\text{equals}\quad \dfrac{\text{Tax on \$180,000 house}}{\text{Assessed value (\$180,000)}}$$

$$\text{Given ratio}\begin{cases}\\\end{cases}\dfrac{\$825}{\$65,000} = \dfrac{\$x}{\$180,000}\quad\begin{array}{l}\leftarrow\text{ Unknown}\\\leftarrow\text{ Given quantity}\end{array}$$

Step 3 Drop the units and apply the cross products principle. We drop the dollar signs and begin to solve for x.

$$\frac{825}{65,000} = \frac{x}{180,000}$$

$$65,000x = (825)(180,000) \qquad \text{Apply the cross-products principle.}$$

$$65,000x = 148,500,000 \qquad \text{Multiply.}$$

Step 4 Solve for x and answer the question.

$$\frac{65,000x}{65,000} = \frac{148,500,000}{65,000} \qquad \text{Divide both sides by 65,000.}$$

$$x \approx 2284.62 \qquad \text{Round the value of } x \text{ to the nearest cent.}$$

The tax on the $180,000 house is approximately $2284.62. ∎

✔ **CHECK POINT 3**   The tax on a house whose assessed value is $45,000 is $600. Determine the tax on a house with an assessed value of $112,500, assuming the same tax rate.

The method described in the section opener that was used to estimate the blue whale population is called the **capture-recapture method**. Because it is impossible to count each individual animal within a population, wildlife biologists randomly catch and tag a given number of animals. Sometime later they recapture a second sample of animals and count the number of recaptured tagged animals. The total size of the wildlife population is then estimated using the following proportion.

$$\underset{\substack{\text{Initially}\\\text{unknown}\\(x)\,\longrightarrow}}{}\dfrac{\substack{\text{Original number of}\\\text{tagged animals}}}{\substack{\text{Total number}\\\text{of animals in the}\\\text{population}}} = \dfrac{\substack{\text{Number of recaptured}\\\text{tagged animals}}}{\substack{\text{Number of animals}\\\text{in second sample}}}\left.\begin{array}{c}\\\\\\\end{array}\right\}\substack{\text{Known}\\\text{ratio}}$$

Although this is called the capture-recapture method, it is not necessary to recapture animals in order to observe whether or not they are tagged. This could be done from a distance, with binoculars for instance.

EXAMPLE 4   Applying Proportions: Estimating Wildlife Population

Wildlife biologists catch, tag, and then release 135 deer back into a wildlife refuge. Two weeks later they observe a sample of 140 deer, 30 of which are tagged. Assuming the ratio of tagged deer in the sample holds for all deer in the refuge, approximately how many deer are in the refuge?

Solution

Step 1 Represent the unknown by x. Let $x =$ the total number of deer in the refuge.

Step 2 Set up a proportion.

Unknown ⟶ $\dfrac{\text{Original number of tagged deer} \rightarrow 135}{\text{Total number of deer} \rightarrow x}$ equals $\dfrac{\text{Number of tagged deer in the observed sample} \rightarrow 30}{\text{Total number of deer in the observed sample} \rightarrow 140}$ } Known ratio

$$\frac{135}{x} = \frac{30}{140}$$

Steps 3 and 4 Apply the cross products principle, solve, and answer the question.

$$\frac{135}{x} = \frac{30}{140}$$

$$(135)(140) = 30x \qquad \text{Apply the cross-products principle.}$$

$$18{,}900 = 30x \qquad \text{Multiply.}$$

$$\frac{18{,}900}{30} = \frac{30x}{30} \qquad \text{Divide both sizes by 30.}$$

$$630 = x \qquad \text{Simplify.}$$

There are approximately 630 deer in the refuge.   ∎

✔ **CHECK POINT 4**   Wildlife biologists catch, tag, and then release 120 deer back into a wildlife refuge. Two weeks later they observe a sample of 150 deer, 25 of which are tagged. Assuming the ratio of tagged deer in the sample holds for all deer in the refuge, approximately how many deer are in the refuge?

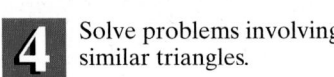

4 Solve problems involving similar triangles.

Similar Triangles

Shown in the margin on page 491 is an international road sign. This sign is shaped just like the actual sign, although its size is smaller. Figures that have the same shape, but not the same size, are used in **scale drawings**. A scale drawing always pictures the exact shape of the object that the drawing represents. Architects, engineers, landscape gardeners, and interior decorators use scale drawings in planning their work.

Pedestrian Crossing

Figures that have the same shape, but not necessarily the same size, are called **similar figures**. In Figure 7.2, triangles ABC and DEF are similar. Angles A and D measure the same number of degrees and are called **corresponding angles**. Angles C and F are corresponding angles, as are angles B and E. Angles with the same number of tick marks in Figure 7.2 are the corresponding angles.

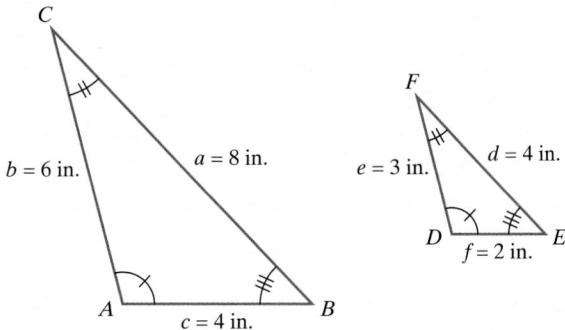

Figure 7.2

The sides opposite the corresponding angles are called **corresponding sides**. In similar triangles, the **measures of corresponding angles are equal**, and **corresponding sides are proportional**. This means that

$$\frac{a}{d} = \frac{b}{e} = \frac{c}{f}.$$

For the triangles in Figure 7.2, each ratio of corresponding sides is equal to 2.

$$\frac{a}{d} = \frac{8}{4} = 2 \qquad \frac{b}{e} = \frac{6}{3} = 2 \qquad \frac{c}{f} = \frac{4}{2} = 2$$

Proportions can be used to solve problems involving similar triangles.

EXAMPLE 5   Using Similar Triangles

The triangles in Figure 7.3 are similar. Find the missing length, x.

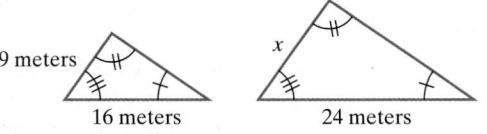

Figure 7.3

9 meters

16 meters   24 meters

Solution   Because the triangles are similar, their corresponding sides are proportional.

Left side of small △ — $\dfrac{9}{x}$ — Corresponding side on left of big △

$= \dfrac{16}{24}$ — Bottom side of small △ — Corresponding side on bottom of big △

We solve this proportion by applying the cross-products principle. (You can also multiply both sides by the LCD, $24x$.)

$$9 \cdot 24 = 16x \qquad \text{Apply the cross-products principle.}$$
$$216 = 16x \qquad \text{Multiply: } 9 \cdot 24 = 216.$$
$$13.5 = x \qquad \text{Divide both sides by 16.}$$

The length of the side marked with an x is 13.5 meters.   ∎

CHECK POINT 5 The similar triangles in the figure are positioned so that they have the same orientation. Find the missing length, x.

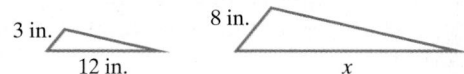

How can we quickly determine if two triangles are similar? If the measures of two angles of one triangle are equal to those of two angles of a second triangle, then the two triangles are similar. If the triangles are similar, then their corresponding sides are proportional.

EXERCISE SET 7.7

Practice and Application Exercises

Use rational equations to solve Exercises 1–10. Each exercise is a problem involving motion.

1. How bad is the heavy traffic? You can walk 10 miles in the same time that it takes to travel 15 miles by car. If the car's rate is 3 miles per hour faster than your walking rate, find the average rate of each.

	Distance	Rate	Time = $\dfrac{\text{Distance}}{\text{Rate}}$
Walking	10	x	$\dfrac{10}{x}$
Car in Heavy Traffic	15	$x + 3$	$\dfrac{15}{x + 3}$

2. You can travel 40 miles on motorcycle in the same time that it takes to travel 15 miles on bicycle. If your motorcycle's rate is 20 miles per hour faster than your bicycle's, find the average rate for each.

	Distance	Rate	Time = $\dfrac{\text{Distance}}{\text{Rate}}$
Motorcycle	40	$x + 20$	$\dfrac{40}{x + 20}$
Bicycle	15	x	$\dfrac{15}{x}$

3. A jogger runs 4 miles per hour faster downhill than uphill. If the jogger can run 5 miles downhill in the same time that it takes to run 3 miles uphill, find the jogging rate in each direction.

4. A truck can travel 120 miles in the same time that it takes a car to travel 180 miles. If the truck's rate is 20 miles per hour slower than the car's, find the average rate for each.

5. In still water, a boat averages 15 miles per hour. It takes the same amount of time to travel 20 miles downstream, with the current, as 10 miles upstream, against the current. What is the rate of the water's current?

6. In still water, a boat averages 18 miles per hour. It takes the same amount of time to travel 33 miles downstream, with the current, as 21 miles upstream, against the current. What is the rate of the water's current?

7. As part of an exercise regimen, you walk 2 miles on an indoor track. Then you jog at twice your walking speed for another 2 miles. If the total time spent walking and jogging is 1 hour, find the walking and jogging rates.

8. The joys of the Pacific Coast! You drive 90 miles along the Pacific Coast Highway and then take a 5-mile walk along a hiking trail in Point Reyes National Seashore. Your driving rate is nine times that of your walking rate. If the total time for driving and hiking is 3 hours, find the average rate driving and the average rate hiking.

9. The water's current is 2 miles per hour. A boat can travel 6 miles downstream, with the current, in the same amount of time it travels 4 miles upstream, against the current. What is the boat's average rate in still water?

10. The water's current is 2 miles per hour. A canoe can travel 6 miles downstream, with the current, in the same amount of time it travels 2 miles upstream, against the current. What is the canoe's average rate in still water?

Use a rational equation to solve Exercises 11–16. Each exercise is a problem involving work.

11. You must leave for campus in 10 minutes, or you will be late for class. Unfortunately, you are snowed in. You can shovel the driveway in 20 minutes, and your brother claims he can do it in 15 minutes. If you shovel together, how long will it take to clear the driveway? Will this give you enough time before you have to leave?

12. You promised your parents that you would wash the family car. You have not started the job, and they are due home in 16 minutes. You can wash the car in 40 minutes, and your sister claims she can do it in 30 minutes. If you work together, how long will it take to do the job? Will this give you enough time before your parents return?

13. The MTV crew will arrive in one week and begin filming the city for *Real World Kalamazoo.* The mayor is desperate to clean the city streets before filming begins. Two teams are available, one that requires 400 hours and one that requires 300 hours. If the teams work together, how long will it take to clean all of Kalamazoo's streets? Is this enough time before the cameras begin rolling?

14. A hurricane strikes and a rural area is without food or water. Three crews arrive. One can dispense needed supplies in 10 hours, a second in 15 hours, and a third in 20 hours. How long will it take all three crews working together to dispense food and water?

15. A pool can be filled by one pipe in 4 hours and by a second pipe in 6 hours. How long will it take using both pipes to fill the pool?

16. A pool can be filled by one pipe in 3 hours and by a second pipe in 6 hours. How long will it take using both pipes to fill the pool?

Use a proportion to solve Exercises 17–24.

17. The tax on a property with an assessed value of $65,000 is $720. Find the tax on a property with an assessed value of $162,500.

18. The maintenance bill for a shopping center containing 180,000 square feet is $45,000. What is the bill for a store in the center that is 4800 square feet?

19. St. Paul Island in Alaska has 12 fur seal rookeries (breeding places). In 1961, to estimate the fur seal pup population in the Gorbath rookery, 4963 fur seal pups were tagged in early August. In late August, a sample of 900 pups was observed and 218 of these were found to have been previously tagged. Estimate the total number of fur seal pups in this rookery.

20. To estimate the number of bass in a lake, wildlife biologists tagged 50 bass and released them in the lake. Later they netted 108 bass and found that 27 of them were tagged. Approximately how many bass are in the lake?

21. The ratio of monthly child support to a father's yearly income is 1:40. How much should a father earning $38,000 annually pay in monthly child support?

22. A 6-foot canoe weighs 75 pounds. Find the weight of a 16-foot canoe of this type.

23. Height is proportional to foot length. A person whose foot length is 10 inches is 67 inches tall. In 1951, photos of large footprints were published. Some believed that these footprints were made by the "Abominable Snowman." Each footprint was 23 inches long. If indeed they belonged to the Abominable Snowman, how tall is the critter?

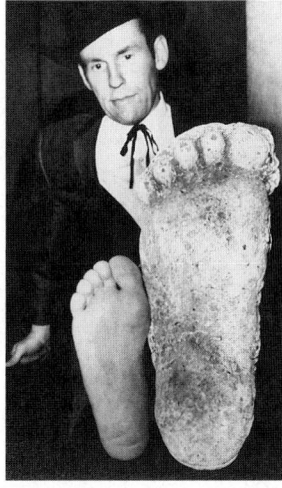

Roger Patterson comparing his foot with a plaster cast of a footprint of the purported "Bigfoot" which Mr. Patterson said he sighted in a California forest in 1967.

24. A person's hair length is proportional to the number of years it has been growing. After 2 years, a person's hair length is 8 inches. The longest moustache on record was grown by Kalyan Sain of India. Sain grew his moustache for 17 years. How long was it?

In Exercises 25–30, use similar triangles and the fact that corresponding sides are proportional to find the length of the side marked with an x.

25.

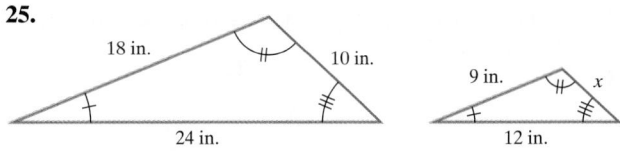

26.

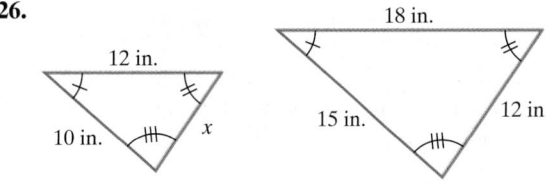

27.

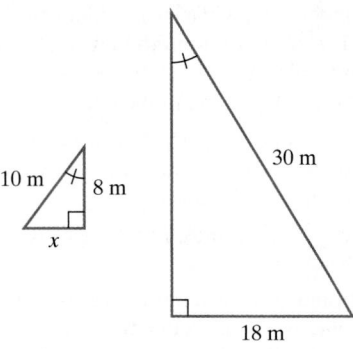

28.

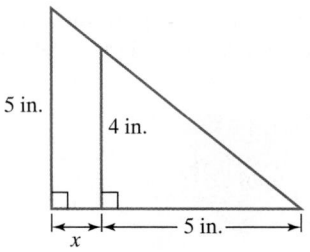

29.

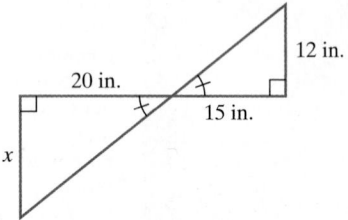

30.

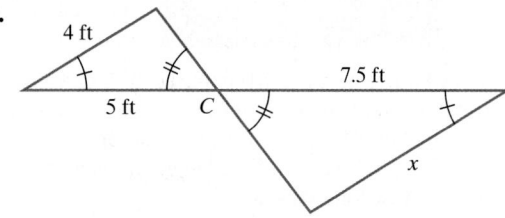

Use similar triangles to solve Exercises 31–32.

31. A tree casts a shadow 12 feet long. At the same time, a vertical rod 8 feet high casts a shadow of 6 feet long. How tall is the tree?

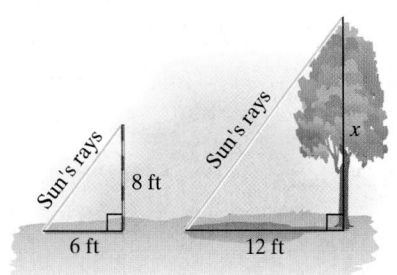

32. A person who is 5 feet tall is standing 80 feet from the base of a tree, and the tree casts an 86-foot shadow. The person's shadow is 6 feet in length. What is the tree's height?

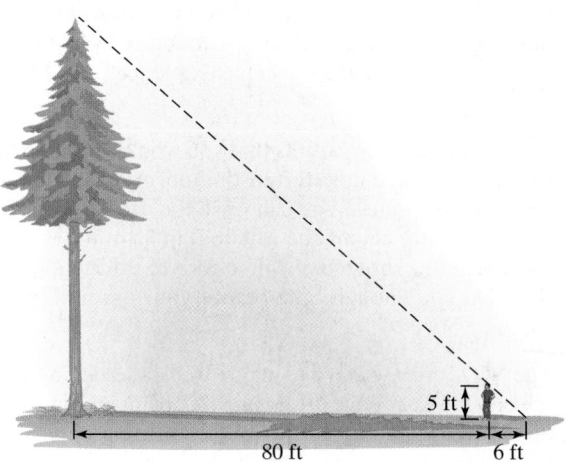

Writing in Mathematics

33. What is the relationship among time traveled, distance traveled, and rate of travel?

34. If you know how many hours it takes for you to do a job, explain how to find the fractional part of the job you can complete in x hours.

35. If you can do a job in 6 hours and your friend can do the same job in 3 hours, explain how to find how long it takes to complete the job working together. It is not necessary to solve the problem.

36. When two people work together to complete a job, describe one factor that can result in more or less time than the time given by the rational equations we have been using.

37. What is a proportion? Give an example with your description.

38. What are similar triangles?

39. If the ratio of the corresponding sides of two similar triangles is 1 to 1 $\left(\frac{1}{1}\right)$, what must be true about the triangles?

40. What are corresponding angles in similar triangles?

41. Describe how to identify the corresponding sides in similar triangles.

Critical Thinking Exercises

42. Two skiers begin skiing along a trail at the same time. The faster skier averages 9 miles per hour and the slower skier averages 6 miles per hour. The faster skier completes the trail $\frac{1}{4}$ hour before the slower skier. How long is the trail?

43. An experienced carpenter can panel a room 3 times faster than an apprentice can. Working together, they can panel the room in 6 hours. How long would it take each one working alone to do the job?

44. It normally takes 2 hours to fill a swimming pool. The pool has developed a slow leak. If the pool were full, it would take 10 hours for all the water to leak out. If the pool is empty, how long will it take to fill it?

45. The front sprocket on a bicycle has 60 teeth, and the rear sprocket has 20 teeth. For mountain biking, an owner needs a 5:1 front:rear ratio. If only one of the sprockets is to be replaced, describe the two ways in which this can be done.

Review Exercises

46. Factor: $25x^2 - 81$. (Section 6.4, Example 1)

47. Solve: $x^2 - 12x + 36 = 0$. (Section 6.6, Example 4)

48. Graph: $y = -\frac{2}{3}x + 4$. (Section 3.4, Example 3)

▶ SECTION 7.8   *Modeling Using Variation*

Objectives

1 Solve direct variation problems.

2 Solve inverse variation problems.

3 Solve combined variation problems.

4 Solve problems involving joint variation.

SSM

PH Tutor CD-   Video
Center   ROM

Have you ever wondered how telecommunication companies estimate the number of phone calls expected per day between two cities? The formula

$$N = \frac{400P_1P_2}{d^2}$$

shows that the daily number of phone calls, N, increases as the populations of the cities, P_1 and P_2, in thousands, increase and decreases as the distance, d, between the cities increases.

Certain formulas occur so frequently in applied situations that they are given special names. Variation formulas show how one quantity changes in relation to other quantities. Quantities can vary *directly, inversely*, or *jointly*. In this section, we look at situations that can be modeled by each of these kinds of variation. And think of this. The next time you get one of those "all-circuits-are-busy" messages, you will be able to use a variation formula to estimate how many other callers you're competing with for those precious 5-cent minutes.

Solve direct variation problems.

Direct Variation

Because light travels faster than sound, during a thunderstorm we see lightning before we hear thunder. The formula

$$d = 1080t$$

describes the distance, in feet, of the storm's center if it takes t seconds to hear thunder after seeing lightning. Thus,

If $t = 1$, $d = 1080 \cdot 1 = 1080$.   If it takes 1 second to hear thunder, the storm's center is 1080 feet away.

If $t = 2$, $d = 1080 \cdot 2 = 2160$.   If it takes 2 seconds to hear thunder, the storm's center is 2160 feet away.

If $t = 3$, $d = 1080 \cdot 3 = 3240$.   If it takes 3 seconds to hear thunder, the storm's center is 3240 feet away.

As the formula $d = 1080t$ illustrates, the distance to the storm's center is a constant multiple of how long it takes to hear the thunder. When the time is doubled, the storm's distance is doubled; when the time is tripled, the storm's distance is tripled; and so on. Because of this, the distance is said to **vary directly** as the time. The **equation of variation** is

$$d = 1080t.$$

Generalizing, we obtain the following statement:

Direct Variation

If a situation is described by an equation in the form

$$y = kx$$

where k is a constant, we say that **y varies directly as x.** The number k is called the **constant of variation**.

Problems involving direct variation can be solved using the following procedure. This procedure applies to direct variation problems as well as to the other kinds of variation problems that we will discuss.

Solving Variation Problems

1. Write an equation that describes the given English statement.
2. Substitute the given pair of values into the equation in step 1 and find the value of k.
3. Substitute the value of k into the equation in step 1.
4. Use the equation from step 3 to answer the problem's question.

The graph shows distance d on the vertical axis labeled "Distance of the Storm's Center (feet)" with values 1080, 2160, 3240, 4320, 5400, 6480, and time t on the horizontal axis labeled "Time (seconds) to Hear Thunder" with values 1, 2, 3, 4, 5, 6.

The graph of $d = 1080t$. Distance to a storm's center varies directly as the time it takes to hear thunder.

EXAMPLE 1   Solving a Direct Variation Problem

The amount of garbage, G, varies directly as the population, P. Allegheny County, Pennsylvania, has a population of 1.3 million and creates 26 million pounds of garbage each week. Find the weekly garbage produced by New York City with a population of 7.3 million.

Solution

Step 1 Write an equation. We know that y varies directly as x is expressed as

$$y = kx.$$

By changing letters, we can write an equation that describes the following English statement: Garbage production, G, varies directly as the population, P.

$$G = kP$$

Step 2 Use the given values to find k. Allegheny County has a population of 1.3 million and creates 26 million pounds of garbage weekly. Substitute 26 for G and 1.3 for P in the direct variation equation. Then solve for k.

$$G = kP$$
$$26 = k \cdot 1.3$$
$$\frac{26}{1.3} = \frac{k \cdot 1.3}{1.3} \qquad \text{Divide both sides by 1.3.}$$
$$20 = k \qquad \text{Simplify.}$$

Step 3 Substitute the value of k into the equation.

$$G = kP \qquad \text{Use the equation from step 1.}$$
$$G = 20P \qquad \text{Replace } k, \text{ the constant of variation, with 20.}$$

Step 4 Answer the problem's question. New York City has a population of 7.3 million. To find its weekly garbage production, substitute 7.3 for P in $G = 20P$ and solve for G.

$$G = 20P \qquad \text{Use the equation from step 3.}$$
$$G = 20(7.3) \qquad \text{Substitute 7.3 for } P.$$
$$G = 146$$

The weekly garbage produced by New York City weighs approximately 146 million pounds.   ■

✔ **CHECK POINT 1**   The pressure, P, of water on an object below the surface varies directly as its distance, D, below the surface. If a submarine experiences a pressure of 25 pounds per square inch 60 feet below the surface, how much pressure will it experience 330 feet below the surface?

2 Solve inverse variation problems.

Inverse Variation

The distance from Atlanta, Georgia, to Orlando, Florida, is 450 miles. The time that it takes to drive from Atlanta to Orlando depends on the rate at which one drives and is given by

$$\text{Time} = \frac{450}{\text{Rate}}.$$

$$\boxed{\text{Time} = \frac{\text{Distance}}{\text{Rate}}}$$

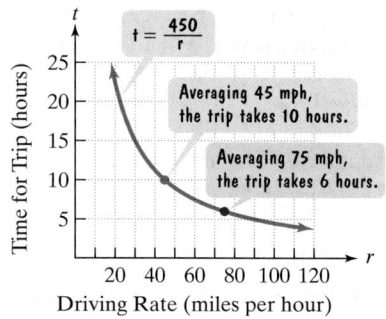

$$t = \frac{450}{r}$$

Averaging 45 mph, the trip takes 10 hours.

Averaging 75 mph, the trip takes 6 hours.

Figure 7.4

For example, if you average 45 miles per hour, the time for the drive is

$$\text{Time} = \frac{450}{45} = 10,$$

or 10 hours. If you ignore speed limits and average 75 miles per hour, the time for the drive is

$$\text{Time} = \frac{450}{75} = 6,$$

or 6 hours. As your rate (or speed) increases, the time for the trip decreases and vice versa. This is illustrated by the graph in Figure 7.4.

We can express the time for the Atlanta-Orlando trip using t for time and r for rate:

$$t = \frac{450}{r}.$$

This equation is an example of an **inverse variation** equation. Time, t, **varies inversely** as rate, r. When two quantities vary inversely, one quantity increases as the other decreases, and vice versa.

Generalizing, we obtain the following statement:

Inverse Variation

If a situation is described by an equation in the form

$$y = \frac{k}{x}$$

where k is a constant, we say that **y varies inversely as x**. The number k is called the **constant of variation**.

We use the same procedure to solve inverse variation problems as we did to solve direct variation problems. Example 2 illustrates this procedure.

EXAMPLE 2 Solving an Inverse Variation Problem

Will the surging oil prices shown in Figure 7.5 put a brake on boom times? The price, P, of oil varies inversely as the supply, S. An OPEC nation sells oil for $19.50 per barrel when its daily production level is 4 million barrels. At what price will it sell oil if the daily production level is decreased to 3 million barrels?

Solution

Step 1 Write an equation. We know that y varies inversely as x is expressed as

$$y = \frac{k}{x}.$$

By changing letters, we can write an equation that describes the following English statement: Price, P, varies inversely as supply, S.

$$P = \frac{k}{S}$$

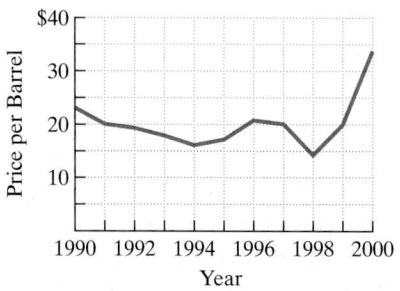

Crude-Oil Price per Barrel

Source: U.S. Department of Commerce

Figure 7.5

Step 2 Use the given values to find k. Oil is sold for \$19.50 at a production level of 4 million barrels. Substitute \$19.50 for P and 4 for S in the inverse variation equation. Then solve for k.

$$P = \frac{k}{S}$$

$$19.50 = \frac{k}{4}$$

$$(19.50)(4) = \frac{k}{4} \cdot 4 \qquad \text{Multiply both sides by 4.}$$

$$78 = k \qquad \text{Simplify.}$$

Step 3 Substitute the value of k into the equation.

$$P = \frac{k}{S} \qquad \text{Use the equation from step 1.}$$

$$P = \frac{78}{S} \qquad \begin{array}{l}\text{Replace } k, \text{ the constant of} \\ \text{variations, with 78.}\end{array}$$

Step 4 Answer the problem's question. We need to find at what price the OPEC nation will sell oil if it reduces daily production to 3 million barrels. Substitute 3 for S in the equation and solve for P.

$$P = \frac{78}{S} = \frac{78}{3} = 26$$

The price will be be \$26 per barrel. ∎

✔ **CHECK POINT 2** When you use a spray can and press the valve at the top, you decrease the pressure of the gas in the can. This decrease of pressure causes the volume of the gas in the can to increase. Because the gas needs more room than is provided in the can, it expands in spray form through the small hole near the valve. In general, if the temperature is constant, the pressure, P, of a gas in a container varies inversely as the volume, V, of the container. The pressure of a gas sample in a container whose volume is 8 cubic inches is 12 pounds per square inch. If the sample expands to a volume of 22 cubic inches, what is the new pressure of the gas?

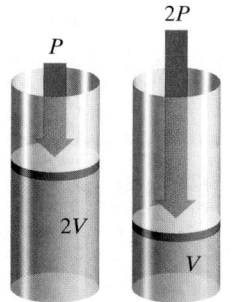

P $2P$

$2V$ V

Doubling the pressure halves the volume.

3 Solve combined variation problems.

Combined Variation

In a **combined variation** situation, direct and inverse variation occur at the same time. For example, as the advertising budget, A, of a company increases, its monthly sales, S, also increase. Monthly sales vary directly as the advertising budget:

$$S = kA.$$

By contrast, as the price of the company's product, P, increases, its monthly sales, S, decrease. Monthly sales vary inversely as the price of the product:

$$S = \frac{k}{P}.$$

We can combine these two variation equations into one combined equation:

$$S = \frac{kA}{P}.$$

The following example illustrates the application of combined variation.

EXAMPLE 3   Solving a Combined Variation Problem

The owners of Rollerblades Now determine that the monthly sales, S, of its skates vary directly as its advertising budget, A, and inversely as the price of the skates, P. When $60,000 is spent on advertising and the price of the skates is $40, the monthly sales are 12,000 pairs of rollerblades.

    a. Write an equation of variation that describes this situation.

    b. Determine monthly sales if the amount of the advertising budget is increased to $70,000.

Solution

    a. Write an equation.

$$S = \frac{kA}{P}$$

Translate "sales vary directly as the advertising budget and inversely as the skates' price."

Use the given values to find k.

$$12,000 = \frac{k(60,000)}{40}$$

When $60,000 is spent on advertising ($A = 60,000$) and the price is $40 ($P = 40$), monthly sales are 12,000 units ($S = 12,000$).

$$12,000 = k \cdot 1500$$

Divide 60,000 by 40.

$$\frac{12,000}{1500} = \frac{k \cdot 1500}{1500}$$

Divide both sides of the equation by 1500.

$$8 = k$$

Simplify.

Therefore, the equation of variation that describes monthly sales is

$$S = \frac{8A}{P}.$$

    b. The advertising budget is increased to $70,000, so $A = 70,000$. The skates' price is still $40, so $P = 40$.

$$S = \frac{8A}{P}$$

This is the equation from part (a).

$$S = \frac{8(70,000)}{40}$$

Substitute 70,000 for A and 40 for P.

$$S = 14,000$$

With a $70,000 advertising budget and $40 price, the company can expect to sell 14,000 pairs of rollerblades in a month (up from 12,000). ■

✔ **CHECK POINT 3**   The number of minutes needed to solve an exercise set of variation problems varies directly as the number of problems and inversely as the number of people working to solve the problems. It takes 4 people 32 minutes to solve 16 problems. How many minutes will it take 8 people to solve 24 problems?

4 Solve problems involving joint variation.

Joint Variation

Joint variation is a variation in which a variable varies directly as the product of two or more other variables. Thus, the equation $y = kxz$ is read "y varies jointly as x and z."

Joint variation plays a critical role in Isaac Newton's formula for gravitation:

$$F = G\frac{m_1 m_2}{d^2}.$$

The formula states that the force of gravitation, F, between two bodies varies jointly as the product of their masses, m_1 and m_2, and inversely as the square of the distance between them, d. (G is the gravitational constant.) The formula indicates that gravitational force exists between any two objects in the universe, increasing as the distance between the bodies decreases. One practical result is that the pull of the moon on the oceans is greater on the side of Earth closer to the moon. This gravitational imbalance is what produces tides.

EXAMPLE 4   Modeling Centrifugal Force

The centrifugal force, C, of a body moving in a circle varies jointly with the radius of the circular path, r, and the body's mass, m, and inversely with the square of the time, t, it takes to move about one full circle. A 6-gram body moving in a circle with radius 100 centimeters at a rate of 1 revolution in 2 seconds has a centrifugal force of 6000 dynes. Find the centrifugal force of an 18-gram body moving in a circle with radius 100 centimeters at a rate of 1 revolution in 3 seconds.

Solution

$$C = \frac{krm}{t^2}$$
Translate "Centrifugal force, C, varies jointly with radius, r, and mass, m, and inversely with the square of time, t."

$$6000 = \frac{k(100)(6)}{2^2}$$
If $r = 100$, $m = 6$, and $t = 2$, then $C = 6000$.

$$6000 = 150k$$
Simplify.

$$40 = k$$
Divide both sides by 150 and solve for k.

$$C = \frac{40rm}{t^2}$$
Substitute 40 for k in the model for centrifugal force.

$$= \frac{40(100)(18)}{3^2}$$
Find C when $r = 100$, $m = 18$, and $t = 3$.

$$= 8000$$
Simplify.

The centrifugal force is 8000 dynes. ∎

✔ **CHECK POINT 4**   The volume of a cone, V, varies jointly as its height, h, and the square of its radius r. A cone with a radius measuring 6 feet and a height measuring 10 feet has a volume of 120π cubic feet. Find the volume of a cone having a radius of 12 feet and a height of 2 feet.

EXERCISE SET 7.8

 Practice Exercises

Use the four-step procedure for solving variation problems given on page 496 to solve Exercises 1–8.

1. y varies directly as x. $y = 35$ when $x = 5$. Find y when $x = 12$.

2. y varies directly as x. $y = 55$ when $x = 5$. Find y when $x = 13$.

3. y varies inversely as x. $y = 10$ when $x = 5$. Find y when $x = 2$.

4. y varies inversely as x. $y = 5$ when $x = 3$. Find y when $x = 9$.

5. y varies directly as x and inversely as the square of z. $y = 20$ when $x = 50$ and $z = 5$. Find y when $x = 3$ and $z = 6$.

6. a varies directly as b and inversely as the square of c. $a = 7$ when $b = 9$ and $c = 6$. Find a when $b = 4$ and $c = 8$.

7. y varies jointly as x and z. $y = 25$ when $x = 2$ and $z = 5$. Find y when $x = 8$ and $z = 12$.

8. C varies jointly as A and T. $C = 175$ when $A = 2100$ and $T = 4$. Find C when $A = 2400$ and $T = 6$.

 Application Exercises

9. A person's fingernail growth, G, in inches, varies directly as the number of weeks it has been growing, W.
   a. Write an equation that expresses this relationship.

   b. Fingernails grow at a rate of about 0.02 inch per week. Substitute 0.02 for k, the constant of variation, in the equation in part (a) and write the equation for fingernail growth.
   c. Substitute 52 for W to determine your fingernail growth at the end of one year if for some bizarre reason you decided not to cut them and they did not break.

10. A person's wages, W, vary directly as the number of hours worked, h.
    a. Write an equation that expresses this relationship.

    b. For a 40-hour work week, Gloria earned $1400. Substitute 1400 for W and 40 for h in the equation from part (a) and find k, the constant of variation.

c. Substitute the value of k into your equation in part (a) and write the equation that describes Gloria's wages in terms of the number of hours she works.
d. Use the equation from part (c) to find Gloria's wages for 25 hours of work.

Use the four-step procedure for solving variation problems given on page 496 to solve Exercises 11–29.

11. The cost, C, of an airplane ticket varies directly as the number of miles, M, in the trip. A 3000-mile trip costs $400. What is the cost of a 450-mile trip?

12. An object's weight on the moon, M, varies directly as its weight on Earth, E. A person who weighs 55 kilograms on Earth weighs 8.8 kilograms on the moon. What is the moon weight of a person who weighs 90 kilograms on Earth?

13. The Mach number is a measurement of speed named after the man who suggested it, Ernst Mach (1838–1916). The speed of an aircraft varies directly as its Mach number. Shown here are two aircraft. Use the figures for the Concorde to determine the Blackbird's speed.

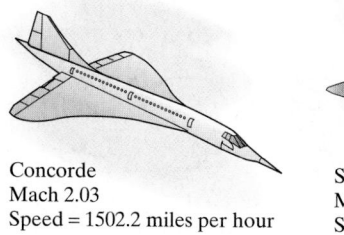

Concorde
Mach 2.03
Speed = 1502.2 miles per hour

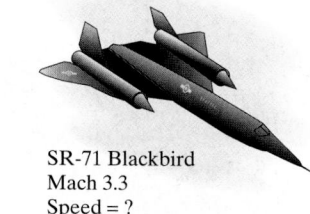

SR-71 Blackbird
Mach 3.3
Speed = ?

14. Do you still own records, or are you strictly a CD person? Record owners claim that the quality of sound on good vinyl surpasses that of a CD, although this is up for debate. This, however, is not debatable: The number of revolutions a record makes as it is being played varies directly as the time that it is on the turntable. A record that lasted 3 minutes made 135 revolutions. If a record takes 2.4 minutes to play, how many revolutions does it make?

15. If all men had identical body types, their weight would vary directly as the cube of their height. Shown is Robert Wadlow, who reached a record height of 8 feet 11 inches (107 inches) before his death at age 22. If a man who is 5 feet 10 inches tall (70 inches) with the same body type as

Mr. Wadlow weighs 170 pounds, what was Robert Wadlow's weight shortly before his death?

16. The distance that an object falls varies directly as the square of the time it has been falling. An object falls 144 feet in 3 seconds. Find how far it will fall in 7 seconds.

17. The time that it takes you to get to campus varies inversely as your driving rate. Averaging 20 miles per hour in terrible traffic, it takes you 1.5 hours to get to campus. How long would the trip take averaging 60 miles per hour?

18. The weight that can be supported by a 2-inch by 4-inch piece of pine (called a 2-by-4) varies inversely as its length. A 10-foot 2-by-4 can support 500 pounds. What weight can be supported by a 5-foot 2-by-4?

19. The volume of a gas in a container at a constant temperature varies inversely as the pressure. If the volume is 32 cubic centimeters at a pressure of 8 pounds, find the pressure when the volume is 40 cubic centimeters.

20. The current in a circuit varies inversely as the resistance. The current is 20 amperes when the resistance is 5 ohms. Find the current for a resistance of 16 ohms.

21. A person's body-mass index is used to assess levels of fatness, with an index from 20 to 26 considered in the desirable range. The index varies directly as one's weight, in pounds, and inversely as one's height, in inches. A person who weighs 150 pounds and is 70 inches tall has an index of 21. What is the body-mass index of a person who weighs 240 pounds and is 74 inches tall? Because the index is rounded to the nearest whole number, do so and then determine if this person's level of fatness is in the desirable range.

22. The volume of a gas varies directly as its temperature and inversely as its pressure. At a temperature of 100 Kelvin and a pressure of 15 kilograms per square meter, the gas occupies a volume of 20 cubic meters. Find the volume at a temperature of 150 Kelvin and a pressure of 30 kilograms per square meter.

23. The intensity of illumination on a surface varies inversely as the square of the distance of the light source from the surface. The illumination from a source is 25 foot-candles at a distance of 4 feet. What is the illumination when the distance is 6 feet?

24. The gravitational force with which Earth attracts an object varies inversely as the square of the distance from the center of Earth. A gravitational force of 160 pounds acts on an object 400 miles from Earth's center. Find the force of attraction on an object 6000 miles from the center of Earth.

25. Kinetic energy varies jointly as the mass and the square of the velocity. A mass of 8 grams and velocity of 3 centimeters per second has a kinetic energy of 36 ergs. Find the kinetic energy for a mass of 4 grams and velocity of 6 centimeters per second.

26. The electrical resistance of a wire varies directly as its length and inversely as the square of its diameter. A wire of 720 feet with $\frac{1}{4}$-inch diameter has a resistance of $1\frac{1}{2}$ ohms. Find the resistance for 960 feet of the same kind of wire if its diameter is doubled.

27. The average number of phone calls between two cities in a day varies jointly as the product of their populations and inversely as the square of the distance between them. The population of Minneapolis is 2538 thousand, and the population of Cincinnati is 1818 thousand. The average number of telephone calls per day between the two cities, which are 608 miles apart, is 158,233. Find the average number of telephone calls per day between Orlando, Florida (population 1225 thousand) and Seattle, Washington (population 2970 thousand), two cities that are 3403 miles apart.

28. The force of attraction between two bodies varies jointly as the product of their masses and inversely as the square of the distance between them. Two 1-kilogram masses separated by 1 meter exert a force of attraction of 6.67×10^{-11} newton. What is the gravitational force exerted by Earth on a 1000-kilogram satellite orbiting at an altitude of 300 kilometers? (The mass of Earth is 5.98×10^{24} kilograms and its radius is 6400 kilometers. Consequently, the distance between Earth and the satellite is 6400 + 300, or 6700 kilometers.)

29. The force of wind blowing on a window positioned at a right angle to the direction of the wind varies jointly as the area of the window and the square of the wind's speed. It is known that a wind of 30 miles per hour blowing on a window measuring 4 feet by 5 feet exerts a force of 150 pounds. During a storm with winds of 60 miles per hour, should hurricane shutters be placed on a window that measures 3 feet by 4 feet and is capable of withstanding 300 pounds of force?

30. The table of values shows the values for the current, I, in an electric circuit and the resistance, R, of the circuit.

I (amperes)	0.5	1.0	1.5	2.0	2.5	3.0	4.0	5.0
R (ohms)	12	6.0	4.0	3.0	2.4	2.0	1.5	1.2

a. Graph the ordered pairs in the table of values, with values of I along the x-axis and values of R along the y-axis. Connect the eight points with a smooth curve.

b. Does current vary directly or inversely as resistance? Use your graph and explain how you arrived at your answer.

c. Write an equation of variation for I and R, using one of the ordered pairs in the table to find the constant of variation. Then use your variation equation to verify the other seven ordered pairs in the table.

Writing in Mathematics

31. What does it mean if two quantities vary directly?

32. In your own words, explain how to solve a variation problem.

33. What does it mean if two quantities vary inversely?

34. Explain what is meant by combined variation. Give an example with your explanation.

35. Explain what is meant by joint variation. Give an example with your explanation.

In Exercises 36–37, describe in words the variation shown by the given equation.

36. $z = \dfrac{k\sqrt{x}}{y^2}$

37. $z = kx^2\sqrt{y}$

38. We have seen that the daily number of phone calls between two cities varies jointly as their populations and inversely as the square of the distance between them. This model, used by telecommunication companies to estimate the line capacities needed among various cities, is called the *gravity model*. Compare the model to Newton's formula for gravitation on page 501 and describe why the name *gravity model* is appropriate.

Critical Thinking Exercises

39. In a hurricane, the wind pressure varies directly as the square of the wind velocity. If wind pressure is a measure of a hurricane's destructive capacity, what happens to this destructive power when the wind speed doubles?

40. The illumination from a light source varies inversely as the square of the distance from the light source. If you raise a lamp from 15 inches to 30 inches over your desk, what happens to the illumination?

41. The heat generated by a stove element varies directly as the square of the voltage and inversely as the resistance. If the voltage remains constant, what needs to be done to triple the amount of heat generated?

42. Galileo's telescope brought about revolutionary changes in astronomy. A comparable leap in our ability to observe the universe took place as a result of the Hubble Space Telescope. The space telescope can see stars and galaxies whose brightness is $\frac{1}{50}$ of the faintest objects now observable using ground-based telescopes. Use the fact that the brightness of a point source, such as a star, varies inversely as the square of its distance from an observer to show that the space telescope can see about seven times farther than a ground-based telescope.

Technology Exercise

43. Use a graphing utility to graph any three of the variation equations in Exercises 11–20. Then TRACE along each curve and identify the point that corresponds to the problem's solution.

Review Exercises

44. Solve:
$$8(2 - x) = -5x.$$
(Section 2.3, Example 2)

45. Divide:
$$\frac{27x^3 - 8}{3x + 2}.$$
(Section 5.6, Example 3)

46. Factor:
$$6x^3 - 6x^2 - 120x.$$
(Section 6.5, Example 2)

CHAPTER 7 GROUP PROJECTS

1. Group members are to write a helpful list of items for a pamphlet called "Errors That Occur When Working with Rational Expressions and How to Avoid Them." This pamphlet will be used by students who use the math lab because they are having problems simplifying, multiplying, dividing, adding, and subtracting rational expressions, as well as simplifying complex rational expressions and solving rational equations. What helpful guidelines for avoiding errors with rational expressions can you offer from the perspective of a student that you won't find in math books?

2. Begin by deciding on a product that interests the group because you are now in charge of advertising this product. Members were told that the demand for the product varies directly as the amount spent on advertising and inversely as the price of the product. However, as more money is spent on advertising, the price of your product rises. Under what conditions would members recommend an increased expense in advertising? Once you've determined what your product is, write formulas for the given conditions and experiment with hypothetical numbers. What other factors might you take into consideration in terms of your recommendation? How do these factors affect the demand for your product?

3. Group members make up the sales team for a company that makes computer video games. It has been determined that the formula

$$y = \frac{200x}{x^2 + 100}$$

models the monthly sales, y, in thousands of games, of a new video game x months after the game is introduced. The figure shows the graph of the formula. What are the team's recommendations to the company in terms of how long the video game should be on the market before another new video game is introduced? What other factors might members want to take into account in terms of the recommendations? What will eventually happen to sales, and how is this indicated by the graph? What could the company do to change the behavior of this model and continue generating sales? Would this be cost effective?

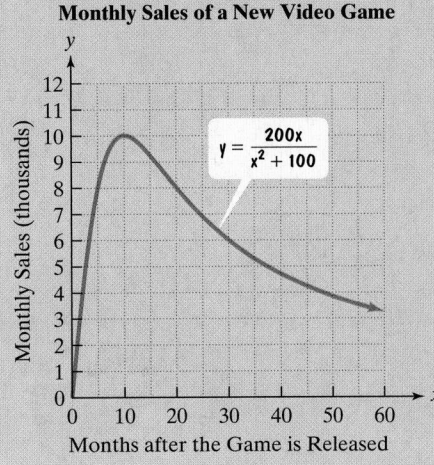

Monthly Sales of a New Video Game

$y = \dfrac{200x}{x^2 + 100}$

Months after the Game is Released

CHAPTER SUMMARY, REVIEW, AND TEST

SUMMARY

DEFINITIONS AND CONCEPTS	EXAMPLES

Section 7.1 Rational Expressions and Their Simplification

A rational expression is the quotient of two polynomials. To find values for which a rational expression is undefined, set the denominator equal to 0 and solve.

Find all numbers for which
$$\frac{7x}{x^2 - 3x - 4}$$
is undefined.

$$x^2 - 3x - 4 = 0$$
$$(x - 4)(x + 1) = 0$$
$$x - 4 = 0 \quad \text{or} \quad x + 1 = 0$$
$$x = 4 \qquad\qquad x = -1$$

Undefined at 4 and -1

To simplify a rational expression:

1. Factor the numerator and denominator completely.
2. Divide the numerator and denominator by the common factors.

If factors in the numerator and denominator are opposites, their quotient is -1.

Simplify: $\dfrac{3x + 18}{x^2 - 36}$.

$$\frac{3x + 18}{x^2 - 36} = \frac{3\overset{1}{\cancel{(x + 6)}}}{\underset{1}{\cancel{(x + 6)}}(x - 6)} = \frac{3}{x - 6}$$

Section 7.2 Multiplying and Dividing Rational Expressions

Multiplying Rational Expressions

1. Factor completely.
2. Divide numerators and denominators by common factors.
3. Multiply remaining factors in the numerators and the denominators.

$$\frac{x^2 + 3x - 10}{x^2 - 2x} \cdot \frac{x^2}{x^2 - 25}$$

$$= \frac{\overset{1}{\cancel{(x + 5)}}\overset{1}{\cancel{(x - 2)}}}{\underset{1}{\cancel{x}}\underset{1}{\cancel{(x - 2)}}} \cdot \frac{\overset{1}{\cancel{x}} \cdot x}{\underset{1}{\cancel{(x + 5)}}(x - 5)}$$

$$= \frac{x}{x - 5}$$

Dividing Rational Expressions
Invert the divisor and multiply.

$$\frac{3y + 3}{(y + 2)^2} \div \frac{y^2 - 1}{y + 2}$$

$$= \frac{3y + 3}{(y + 2)^2} \cdot \frac{y + 2}{y^2 - 1}$$

$$= \frac{3\overset{1}{\cancel{(y + 1)}}}{(y + 2)\underset{1}{\cancel{(y + 2)}}} \cdot \frac{\overset{1}{\cancel{(y + 2)}}}{\underset{1}{\cancel{(y + 1)}}(y - 1)}$$

$$= \frac{3}{(y + 2)(y - 1)}$$

DEFINITIONS AND CONCEPTS	EXAMPLES

Section 7.3 Adding and Subtracting Rational Expressions with the Same Denominators

To add or subtract rational expressions with the same denominators, add or subtract the numerators and place the result over the common denominator. If possible, simplify the resulting expression.	$\dfrac{y^2 - 3y + 4}{y^2 + 8y + 15} - \dfrac{y^2 - 5y - 2}{y^2 + 8y + 15}$ $= \dfrac{y^2 - 3y + 4 - (y^2 - 5y - 2)}{y^2 + 8y + 15}$ $= \dfrac{y^2 - 3y + 4 - y^2 + 5y + 2}{y^2 + 8y + 15}$ $= \dfrac{2y + 6}{(y + 5)(y + 3)}$ $= \dfrac{2\overset{1}{\cancel{(y + 3)}}}{(y + 5)\underset{1}{\cancel{(y + 3)}}} = \dfrac{2}{y + 5}$
To add or subtract rational expressions with opposite denominators, multiply either rational expression by $\frac{-1}{-1}$ to obtain a common denominator.	$\dfrac{7}{x - 6} + \dfrac{x + 4}{6 - x}$ $= \dfrac{7}{x - 6} + \dfrac{(-1)}{(-1)} \cdot \dfrac{x + 4}{6 - x}$ $= \dfrac{7}{x + 6} + \dfrac{-x - 4}{x - 6}$ $= \dfrac{7 - x - 4}{x - 6} = \dfrac{3 - x}{x - 6}$

Section 7.4 Adding and Subtracting Rational Expressions with Different Denominators

Finding the Least Common Denominator (LCD) 1. Factor denominators completely. 2. List factors of the first denominator. 3. Add to the list factors of the second denominator that are not already in the list. 4. The LCD is the product of factors in step 3.	Find the LCD of $\dfrac{x + 1}{2x - 2}$ and $\dfrac{2x}{x^2 + 2x - 3}$. $2x - 2 = 2(x - 1)$ $x^2 + 2x - 3 = (x - 1)(x + 3)$ Factors of first denominator: $2, x - 1$ Factors of second denominator not in the list: $x + 3$ LCD: $2(x - 1)(x + 3)$
Adding and Subtracting Rational Expressions with Different Denominators 1. Find the LCD. 2. Rewrite each rational expression as an equivalent expression with the LCD. 3. Add or subtract numerators, placing the resulting expression over the LCD. 4. If possible, simplify.	$\dfrac{x + 1}{2x - 2} - \dfrac{2x}{x^2 + 2x - 3}$ $= \dfrac{x + 1}{2(x - 1)} - \dfrac{2x}{(x - 1)(x + 3)}$ LCD is $2(x - 1)(x + 3)$. $= \dfrac{(x + 1)(x + 3)}{2(x - 1)(x + 3)} - \dfrac{2x \cdot 2}{2(x - 1)(x + 3)}$ $= \dfrac{x^2 + 4x + 3 - 4x}{2(x - 1)(x + 3)}$ $= \dfrac{x^2 + 3}{2(x - 1)(x + 3)}$

DEFINITIONS AND CONCEPTS	EXAMPLES

Section 7.5 Complex Rational Expressions

Complex rational expressions have numerators or denominators containing one or more rational expressions. Complex rational expressions can be simplified by obtaining single expressions in the numerator and denominator and then dividing. They can also be simplified by multiplying the numerator and denominator by the LCD of all rational expressions within the complex rational expression.

Simplify by dividing: $\dfrac{\dfrac{1}{x} + 5}{\dfrac{1}{x} - \dfrac{1}{3}}$.

$$= \dfrac{\dfrac{1}{x} + \dfrac{5x}{x}}{\dfrac{3}{3x} - \dfrac{x}{3x}} = \dfrac{\dfrac{1+5x}{x}}{\dfrac{3-x}{3x}} = \dfrac{1+5x}{\cancel{x}} \cdot \dfrac{\overset{1}{\cancel{3x}}}{3-x}$$

$$= \dfrac{3(1+5x)}{3-x} \text{ or } \dfrac{3+15x}{3-x}$$

Simplify by the LCD method: $\dfrac{\dfrac{1}{x} + 5}{\dfrac{1}{x} - \dfrac{1}{3}}$.

LCD is $3x$.

$$\dfrac{3x}{3x} \cdot \dfrac{\left(\dfrac{1}{x} + 5\right)}{\left(\dfrac{1}{x} - \dfrac{1}{3}\right)} = \dfrac{3x \cdot \dfrac{1}{x} + 3x \cdot 5}{3x \cdot \dfrac{1}{x} - 3x \cdot \dfrac{1}{3}}$$

$$= \dfrac{3+15x}{3-x}$$

Section 7.6 Solving Rational Equations

A rational equation is an equation containing one or more rational expressions.

Solving Rational Equations

1. List restrictions on the variable.
2. Clear fractions by multiplying both sides by the LCD.
3. Solve the resulting equations.
4. Reject any proposed solution in the list of restrictions. Check other proposed solutions in the original equation.

Solve: $\dfrac{7x}{x^2-4} + \dfrac{5}{x-2} = \dfrac{2x}{x^2-4}$

$$\dfrac{7x}{(x+2)(x-2)} + \dfrac{5}{x-2} = \dfrac{2x}{(x+2)(x-2)}$$

Denominators would equal 0 if $x = -2$ or $x = 2$. Restrictions: $x \neq -2$ and $x \neq 2$.

LCD is $(x+2)(x-2)$.

$$(x+2)(x-2)\left[\dfrac{7x}{(x+2)(x-2)} + \dfrac{5}{x-2}\right]$$
$$= (x+2)(x-2) \cdot \dfrac{2x}{(x+2)(x-2)}$$

$$7x + 5(x+2) = 2x$$
$$7x + 5x + 10 = 2x$$
$$12x + 10 = 2x$$
$$10 = -10x$$
$$-1 = x$$

The proposed solution, -1, is not part of the restriction $x \neq -2$ and $x \neq 2$. It checks. The solution is -1 and the solution set is $\{-1\}$.

DEFINITIONS AND CONCEPTS	EXAMPLES

Section 7.7 Applications Using Rational Equations and Proportions

Motion problems involving time are solved using

$$t = \frac{d}{r}.$$

$$\text{Time traveled} = \frac{\text{Distance traveled}}{\text{Rate of travel}}$$

It takes a cyclist who averages 16 miles per hour in still air the same time to travel 48 miles with the wind as 16 miles against the wind. What is the wind's rate?

$$x = \text{wind's rate}$$
$$16 + x = \text{cyclist's rate with wind}$$
$$16 - x = \text{cyclist's rate against wind}$$

	Distance	Rate	Time = $\dfrac{\text{Distance}}{\text{Rate}}$
With wind	48	$16 + x$	$\dfrac{48}{16 + x}$
Against wind	16	$16 - x$	$\dfrac{16}{16 - x}$

Two times are equal.

$$\frac{48}{16 + x} = \frac{16}{16 - x}$$

$$(16 + x)(16 - x) \cdot \frac{48}{16 + x} = \frac{16}{16 - x} \cdot (16 + x)(16 - x)$$

$$48(16 - x) = 16(16 + x)$$

Solving this equation, $x = 8$.
The wind's rate is 8 miles per hour.

Work problems are solved using the following condition:

$$\boxed{\text{Fraction of job done by first}} + \boxed{\text{fraction of job done by second}} = \boxed{1.}$$

One pipe fills a pool in 20 hours and a second pipe in 15 hours. How long will it take to fill the pool using both pipes?

$$x = \text{time using both pipes}$$

$$\boxed{\begin{array}{c}\text{Fraction of pool}\\\text{filled by pipe 1 in}\\x \text{ hours}\end{array}} + \boxed{\begin{array}{c}\text{fraction of pool}\\\text{filled by pipe 2 in}\\x \text{ hours}\end{array}} = \boxed{1}$$

$$\frac{x}{20} + \frac{x}{15} = 1$$

$$60\left(\frac{x}{20} + \frac{x}{15}\right) = 60 \cdot 1$$

$$3x + 4x = 60$$

$$7x = 60$$

$$x = \frac{60}{7} = 8\frac{4}{7} \text{ hours}$$

It will take $8\frac{4}{7}$ hours for both pipes to fill the pool.

DEFINITIONS AND CONCEPTS	EXAMPLES

Section 7.7 Applications Using Rational Equations and Proportions (continued)

Proportions

The Cross-Products Principle:

If $\dfrac{a}{b} = \dfrac{c}{d}$, then $ad = bc$ ($b \neq 0$ and $d \neq 0$).

Solving Applied Problems Using Proportions

1. Read the problem and represent the unknown quantity by x (or any letter).
2. Set up a proportion by listing the given ratio on one side and the ratio with the unknown quantity on the other side.
3. Drop units and apply the cross-products principle.
4. Solve for x and answer the question.

30 elk are tagged and released. Sometime later, a sample of 80 elk are observed and 10 are tagged. How many elk are there?

$x = $ number of elk

Tagged $\to$
Total $\to$
$$\frac{30}{x} = \frac{10}{80}$$

$$10x = 30 \cdot 80$$

$$10x = 2400$$

$$x = 240$$

There are 240 elk.

Similar triangles have the same shape, but not necessarily the same size. Corresponding angles have the same measure, and corresponding sides are proportional. If the measures of two angles of one triangle are equal to those of two angles of a second triangle, then the two triangles are similar.

Find x for these similar triangles.

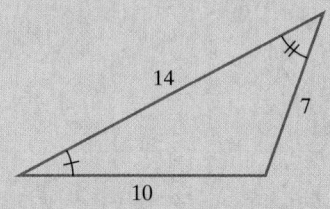

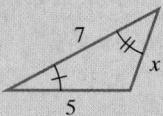

Corresponding sides are proportional:

$$\frac{7}{x} = \frac{10}{5}.$$

$7 \cdot 5 = 10x$ Apply the cross-products principle.

$35 = 10x$ Simplify.

$$x = \frac{35}{10} = 3.5$$

DEFINITIONS AND CONCEPTS	EXAMPLES

Section 7.8 Modeling Using Variation

English Statement	Equation
y varies directly as x.	$y = kx$
y varies directly as x^n.	$y = kx^n$
y varies inversely as x.	$y = \dfrac{k}{x}$
y varies inversely as x^n.	$y = \dfrac{k}{x^n}$
y varies directly as x and inversely as z.	$y = \dfrac{kx}{z}$
y varies jointly as x and z.	$y = kxz$

The number k is called the constant of variation.

The time that it takes you to get to campus varies inversely as your driving rate. Averaging 20 miles per hour in terrible traffic, it takes you 1.5 hours to get to campus. How long would the trip take averaging 60 miles per hour?

1. $t = \dfrac{k}{r}$

2. It takes 1.5 hours at 20 miles per hour:

$$1.5 = \frac{k}{20}$$

$$k = 1.5(20) = 30.$$

3. $t = \dfrac{30}{r}$

4. How long at 60 miles per hour? Substitute 60 for r.

$$t = \frac{30}{60} = \frac{1}{2}$$

It takes $\dfrac{1}{2}$ hour at 60 miles per hour.

Review Exercises

7.1 *In Exercises 1–4, find all numbers for which each rational expression is undefined. If the rational expression is defined for all real numbers, so state.*

1. $\dfrac{5x}{6x - 24}$

2. $\dfrac{x + 3}{(x - 2)(x + 5)}$

3. $\dfrac{x^2 + 3}{x^2 - 3x + 2}$

4. $\dfrac{7}{x^2 + 81}$

In Exercises 5–12, simplify each rational expression. If the rational expression cannot be simplified, so state.

5. $\dfrac{16x^2}{12x}$

6. $\dfrac{x^2 - 4}{x - 2}$

7. $\dfrac{x^3 + 2x^2}{x + 2}$

8. $\dfrac{x^2 + 3x - 18}{x^2 - 36}$

9. $\dfrac{x^2 - 4x - 5}{x^2 + 8x + 7}$

10. $\dfrac{y^2 + 2y}{y^2 + 4y + 4}$

11. $\dfrac{x^2}{x^2 + 4}$

12. $\dfrac{2x^2 - 18y^2}{3y - x}$

7.2 *In Exercises 13–17, multiply as indicated.*

13. $\dfrac{x^2 - 4}{12x} \cdot \dfrac{3x}{x + 2}$

14. $\dfrac{5x + 5}{6} \cdot \dfrac{3x}{x^2 + x}$

15. $\dfrac{x^2 + 6x + 9}{x^2 - 4} \cdot \dfrac{x - 2}{x + 3}$

16. $\dfrac{y^2 - 2y + 1}{y^2 - 1} \cdot \dfrac{2y^2 + y - 1}{5y - 5}$

17. $\dfrac{2y^2 + y - 3}{4y^2 - 9} \cdot \dfrac{3y + 3}{5y - 5y^2}$

In Exercises 18–22, divide as indicated.

18. $\dfrac{x^2 + x - 2}{10} \div \dfrac{2x + 4}{5}$

19. $\dfrac{6x + 2}{x^2 - 1} \div \dfrac{3x^2 + x}{x - 1}$

20. $\dfrac{1}{y^2 + 8y + 15} \div \dfrac{7}{y + 5}$

21. $\dfrac{y^2 + y - 42}{y - 3} \div \dfrac{y + 7}{(y - 3)^2}$

22. $\dfrac{8x + 8y}{x^2} \div \dfrac{x^2 - y^2}{x^2}$

7.3 *In Exercises 23–28, add or subtract as indicated. Simplify the result, if possible.*

23. $\dfrac{4x}{x + 5} + \dfrac{20}{x + 5}$

24. $\dfrac{8x - 5}{3x - 1} + \dfrac{4x + 1}{3x - 1}$

25. $\dfrac{3x^2 + 2x}{x - 1} - \dfrac{10x - 5}{x - 1}$

26. $\dfrac{6y^2 - 4y}{2y - 3} - \dfrac{12 - 3y}{2y - 3}$

27. $\dfrac{x}{x - 2} + \dfrac{x - 4}{2 - x}$

28. $\dfrac{x + 5}{x - 3} - \dfrac{x}{3 - x}$

7.4 *In Exercises 29–31, find the least common denominator of the rational expressions.*

29. $\dfrac{7}{9x^3}$ and $\dfrac{5}{12x}$

30. $\dfrac{3}{x^2(x - 1)}$ and $\dfrac{11}{x(x - 1)^2}$

31. $\dfrac{x}{x^2 + 4x + 3}$ and $\dfrac{17}{x^2 + 10x + 21}$

In Exercises 32–42, add or subtract as indicated. Simplify the result, if possible.

32. $\dfrac{7}{3x} + \dfrac{5}{2x^2}$

33. $\dfrac{5}{x + 1} + \dfrac{2}{x}$

34. $\dfrac{7}{x + 3} + \dfrac{4}{(x + 3)^2}$

35. $\dfrac{6y}{y^2 - 4} - \dfrac{3}{y + 2}$

36. $\dfrac{y - 1}{y^2 - 2y + 1} - \dfrac{y + 1}{y - 1}$

37. $\dfrac{x + y}{y} - \dfrac{x - y}{x}$

38. $\dfrac{2x}{x^2 + 2x + 1} + \dfrac{x}{x^2 - 1}$

39. $\dfrac{5x}{x + 1} - \dfrac{2x}{1 - x^2}$

40. $\dfrac{4}{x^2 - x - 6} - \dfrac{4}{x^2 - 4}$

41. $\dfrac{7}{x + 3} + 2$

42. $\dfrac{2y - 5}{6y + 9} - \dfrac{4}{2y^2 + 3y}$

7.5 *In Exercises 43–47, simplify each complex rational expression.*

43. $\dfrac{\dfrac{1}{2} + \dfrac{3}{8}}{\dfrac{3}{4} - \dfrac{1}{2}}$

44. $\dfrac{\dfrac{1}{x}}{1 - \dfrac{1}{x}}$

45. $\dfrac{\dfrac{1}{x} + \dfrac{1}{y}}{\dfrac{1}{xy}}$

46. $\dfrac{\dfrac{1}{x} - \dfrac{1}{2}}{\dfrac{1}{3} - \dfrac{x}{6}}$

47. $\dfrac{3 + \dfrac{12}{x}}{1 - \dfrac{16}{x^2}}$

7.6 *In Exercises 48–55, solve each rational equation. If an equation has no solution, so state.*

48. $\dfrac{3}{x} - \dfrac{1}{6} = \dfrac{1}{x}$

49. $\dfrac{3}{4x} = \dfrac{1}{x} + \dfrac{1}{4}$

50. $x + 5 = \dfrac{6}{x}$

51. $4 - \dfrac{x}{x + 5} = \dfrac{5}{x + 5}$

52. $\dfrac{2}{x - 3} = \dfrac{4}{x + 3} + \dfrac{8}{x^2 - 9}$

53. $\dfrac{2}{x} = \dfrac{2}{3} + \dfrac{x}{6}$

54. $\dfrac{13}{y - 1} - 3 = \dfrac{1}{y - 1}$

55. $\dfrac{1}{x + 3} - \dfrac{1}{x - 1} = \dfrac{x + 1}{x^2 + 2x - 3}$

56. Park rangers introduce 50 elk into a wildlife preserve. The formula

$$P = \frac{250(3t + 5)}{t + 25}$$

models the elk population, P, after t years. How many years will it take for the population to increase to 125 elk?

57. The formula

$$S = \frac{C}{1 - r}$$

describes the selling price, S, of a product in terms of its cost to the retailer, C, and its markup, r, usually expressed as a percent. A small television cost a retailer $140 and was sold for $200. Find the markup. Express the answer as a percent.

7.7

58. In still water, a paddle boat averages 20 miles per hour. It takes the boat the same amount of time to travel 72 miles downstream, with the current, as 48 miles upstream, against the current. What is the rate of the water's current?

59. A car travels 60 miles in the same time that a car traveling 10 miles per hour faster travels 90 miles. What is the rate of each car?

60. A painter can paint a fence around a house in 6 hours. Working alone, the painter's apprentice can paint the same fence in 12 hours. How many hours would it take them to do the job if they worked together?

61. If a school board determines that there should be 3 teachers for every 50 students, how many teachers are needed for an enrollment of 5400 students?

62. To determine the number of trout in a lake, a conservationist catches 112 trout, tags them, and returns them to the lake. Later, 82 trout are caught, and 32 of them are found to be tagged. How many trout are in the lake?

63. The triangles shown in the figure are similar. Find the length of the side marked with an x.

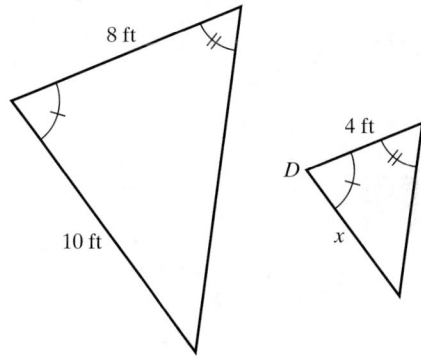

7.8 *Solve the variation problems in Exercises 64–68.*

64. An electric bill varies directly as the amount of electricity used. The bill for 1400 kilowatts of electricity is $98. What is the bill for 2200 kilowatts of electricity?

65. The time it takes to drive a certain distance varies inversely as the rate of travel. If it takes 4 hours at 50 miles per hour to drive the distance, how long will it take at 40 miles per hour?

66. The loudness of a stereo speaker, measured in decibels, varies inversely as the square of your distance from the speaker. When you are 8 feet from the speaker, the loudness is 28 decibels. What is the loudness when you are 4 feet from the speaker?

67. The time required to assemble computers varies directly as the number of computers assembled and inversely as the number of workers. If 30 computers can be assembled by 6 workers in 10 hours, how long would it take 5 workers to assemble 40 computers?

68. The volume of a pyramid varies jointly as its height and the area of its base. A pyramid with a height of 15 feet and a base with an area of 35 square feet has a volume of 175 cubic feet. Find the volume of a pyramid with a height of 20 feet and a base with an area of 120 square feet.

Chapter 7 Test

1. Find all numbers for which

$$\frac{x + 7}{x^2 + 5x - 36}$$

is undefined.

In Exercises 2–3, simplify each rational expression.

2. $\dfrac{x^2 + 2x - 3}{x^2 - 3x + 2}$

3. $\dfrac{4x^2 - 20x}{x^2 - 4x - 5}$

In Exercises 4–16, perform the indicated operations. Simplify the result, if possible.

4. $\dfrac{x^2 - 16}{10} \cdot \dfrac{5}{x + 4}$

5. $\dfrac{x^2 - 7x + 12}{x^2 - 4x} \cdot \dfrac{x^2}{x^2 - 9}$

6. $\dfrac{2x + 8}{x - 3} \div \dfrac{x^2 + 5x + 4}{x^2 - 9}$

7. $\dfrac{5y + 5}{(y - 3)^2} \div \dfrac{y^2 - 1}{y - 3}$

8. $\dfrac{2y^2 + 5}{y + 3} + \dfrac{6y - 5}{y + 3}$

9. $\dfrac{y^2 - 2y + 3}{y^2 + 7y + 12} - \dfrac{y^2 - 4y - 5}{y^2 + 7y + 12}$

10. $\dfrac{x}{x + 3} + \dfrac{5}{x - 3}$

11. $\dfrac{2}{x^2 - 4x + 3} + \dfrac{6}{x^2 + x - 2}$

12. $\dfrac{4}{x - 3} + \dfrac{x + 5}{3 - x}$

13. $1 + \dfrac{3}{x - 1}$

14. $\dfrac{2x + 3}{x^2 - 7x + 12} - \dfrac{2}{x - 3}$

15. $\dfrac{8y}{y^2 - 16} - \dfrac{4}{y - 4}$

16. $\dfrac{(x - y)^2}{x + y} \div \dfrac{x^2 - xy}{3x + 3y}$

In Exercises 17–18, simplify each complex rational expression.

17. $\dfrac{5 + \dfrac{5}{x}}{2 + \dfrac{1}{x}}$

18. $\dfrac{\dfrac{1}{x} - \dfrac{1}{y}}{\dfrac{1}{x}}$

In Exercises 19–21, solve each rational equation.

19. $\dfrac{5}{x} + \dfrac{2}{3} = 2 - \dfrac{2}{x} - \dfrac{1}{6}$

20. $\dfrac{3}{y + 5} - 1 = \dfrac{4 - y}{2y + 10}$

21. $\dfrac{2}{x - 1} = \dfrac{3}{x^2 - 1} + 1$

22. In still water, a boat averages 30 miles per hour. It takes the boat the same amount of time to travel 16 miles downstream, with the current, as 14 miles upstream, against the current. What is the rate of the water's current?

23. One pipe can fill a hot tub in 20 minutes and a second pipe can fill it in 30 minutes. If the hot tub is empty, how long will it take both pipes to fill it?

24. Park rangers catch, tag, and release 200 tule elk back into a wildlife refuge. Two weeks later they observe a sample of 150 elk, of which 5 are tagged. Assuming that the ratio of tagged elk in the sample holds for all elk in the refuge, how many elk are there in the park?

25. The triangles in the figure are similar. Find the length of the side marked with an x.

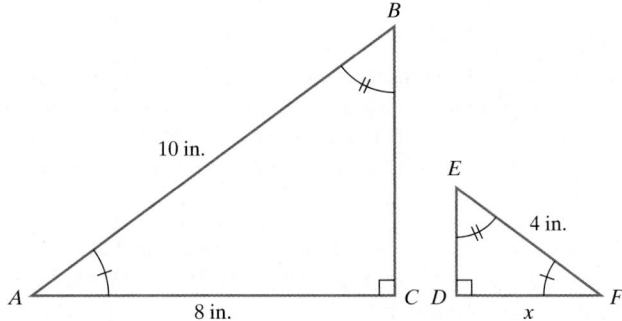

26. The amount of current flowing in an electrical circuit varies inversely as the resistance in the circuit. When the resistance in a particular circuit is 5 ohms, the current is 42 amperes. What is the current when the resistance is 4 ohms?

Cumulative Review Exercises (Chapters 1–7)

In Exercises 1–6, solve each equation, inequality, or system of equations.

1. $2(x - 3) + 5x = 8(x - 1)$

2. $-3(2x - 4) > 2(6x - 12)$

3. $x^2 + 3x = 18$

4. $\dfrac{2x}{x^2 - 4} + \dfrac{1}{x - 2} = \dfrac{2}{x + 2}$

5. $y = 2x - 3$
    $x + 2y = 9$

6. $3x + 2y = -2$
    $-4x + 5y = 18$

In Exercises 7–9, graph each equation in a rectangular coordinate system.

7. $3x - 2y = 6$ **8.** $y = -2x + 3$

9. $y = -3$

In Exercises 10–12, simplify each expression.

10. $-21 - 16 - 3(2 - 8)$

11. $\left(\dfrac{4x^5}{2x^2}\right)^3$ **12.** $\dfrac{\dfrac{1}{x} - 2}{4 - \dfrac{1}{x}}$

In Exercises 13–15, factor completely.

13. $4x^2 - 13x + 3$

14. $4x^2 - 20x + 25$

15. $3x^2 - 75$

In Exercises 16–18, perform the indicated operations.

16. $\left(4x^2 - 3x + 2\right) - \left(5x^2 - 7x - 6\right)$

17. $\dfrac{-8x^6 + 12x^4 - 4x^2}{4x^2}$

18. $\dfrac{x + 6}{x - 2} + \dfrac{2x + 1}{x + 3}$

19. You invest $4000, part at 5% and the remainder at 9% annual interest. At the end of the year, the total interest from these investments was $311. How much was invested at each rate?

20. A 68-inch board is to be cut into two pieces. If one piece must be three times as long as the other, find the length of each piece.

Traffic jams getting you down? Powerful computers, able to solve systems of equations with hundreds of thousands of variables in a single bound, may promise a gridlock-free future. The computer in your car could be linked to a central computer that manages traffic flow by controlling traffic lights, rerouting you away from traffic congestion, issuing weather reports, and selecting the best route to your destination. New technologies could eventually drive your car at a steady 75 miles per hour along automated highways as you comfortably nap.

Real-world problems often involve solving thousands of equations, sometimes containing a million variables. The methods that you will learn in this chapter provide the foundation for solving the kinds of complex systems needed to create a future free of traffic jams.

Chapter 8

Functions; More on Systems of Linear Equations

▶ SECTION 8.1   *Introduction to Functions*

Objectives

1 Find the domain and range of a relation.

2 Determine whether a relation is a function.

3 Evaluate a function.

4 Use the vertical line test to identify functions.

5 Obtain information about a function from its graph.

SSM
PH Tutor   CD-   Video
Center   ROM

The cost of mailing a package depends on its weight. Sales of solar energy systems depend on the number of years after 1995. In both these situations, the relationship between variables can be illustrated with the notion of a *function*. Understanding this concept will give you a new perspective on many ordinary situations. Much of your work in the second half of this book will be devoted to the important topic of functions and how they model your world.

1 Find the domain and range of a relation.

Study Tip

The algebra in the second half of this book is based on the topics in chapters 1 through 7. Are you prepared for intermediate algebra? See the appendix that begins on page A1.

Relations

Concerned about the rising costs of electricity? Perhaps you should consider a solar energy system, which converts sunlight into electricity. Users of these often elaborate and pricey systems estimate that they save, on average, $200 a month on utility bills. The graph in Figure 8.1 shows an increase in sales of solar systems.

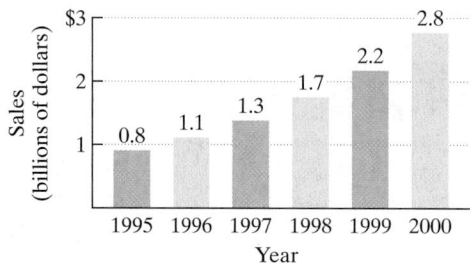

Sales of Solar Energy Systems

Figure 8.1 *Source*: Solar Energy Industries Association

If we let x represent a year and y the sales, in billions of dollars, of solar systems, the graph in Figure 8.1 shows a correspondence between the two variables x and y. We can write this correspondence using a set of ordered pairs:

$$\{(1995, 0.8), (1996, 1.1), (1997, 1.3), (1998, 1.7), (1999, 2.2), (2000, 2.8)\}.$$

The mathematical term for a set of ordered pairs is a **relation**.

Definition of a Relation

A **relation** is any set of ordered pairs. The set of all first components of the ordered pairs is called the **domain** of the relation, and the set of all second components is called the **range** of the relation.

EXAMPLE 1 Analyzing Solar Energy Sales as a Relation

Find the domain and range of the relation

$$\{(1995, 0.8), (1996, 1.1), (1997, 1.3), (1998, 1.7), (1999, 2.2), (2000, 2.8)\}.$$

Solution The domain is the set of all first components. Thus, the domain is

$$\{1995, 1996, 1997, 1998, 1999, 2000\}.$$

The range is the set of all second components. Thus, the range is

$$\{0.8, 1.1, 1.3, 1.7, 2.2, 2.8\}. \qquad \blacksquare$$

✔ **CHECK POINT 1** Find the domain and the range of the relation:

$$\{(5, 12.8), (10, 16.2), (15, 18.9), (20, 20.7), (25, 21.8)\}.$$

As you worked Check Point 1, did you wonder if the numbers in each ordered pair represented anything? Think paid vacation days! The first number in each ordered pair is the number of years for full-time workers at medium to large U.S. companies. The second number is the average number of paid vacation days each year. Consider, for example, the ordered pair (15, 18.9).

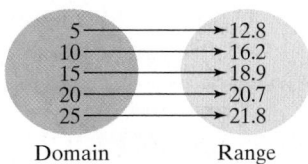

The relation in the vacation-days example can be pictured as follows:

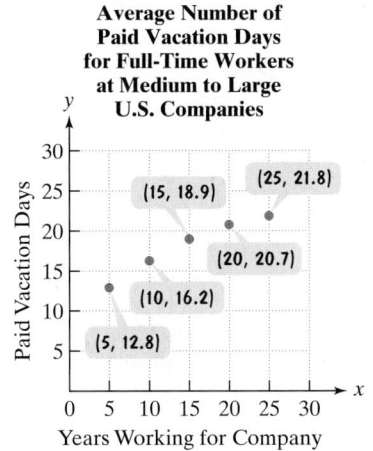

Average Number of Paid Vacation Days for Full-Time Workers at Medium to Large U.S. Companies

Figure 8.2 The graph of a relation showing a correspondence between years with a company and paid vacation days
Source: Bureau of Labor Statistics

The five points in Figure 8.2 are another way to visually represent the relation.

Functions

The SAT is the test that everyone loves to hate. The set of points in Figure 8.3 shows a relation indicating a correspondence between SAT scores and grade point averages for the first year in college for a group of randomly selected college students. The domain is the set of SAT scores for the students. The range is the set of their grade point averages. Is it possible for two students with the same SAT score to have different grade point averages? Look for two or more data points that are aligned vertically. We see that there are two students who have the same SAT score, 700, but their grade point averages are different. One student has a grade point average of approximately 2.4 and the other a grade point average of approximately 3.7. These students are represented by the following ordered pairs:

$$(700, 2.4) \qquad (700, 3.7).$$

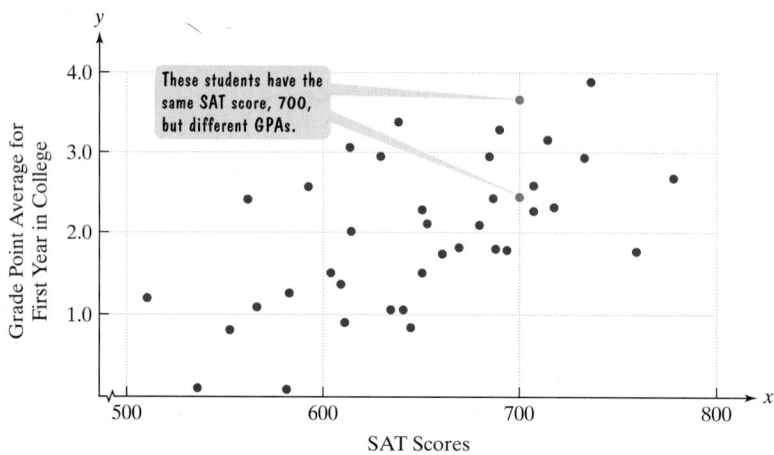

Figure 8.3

A relation in which each member of the domain corresponds to exactly one member of the range is a **function**. The relation in Figure 8.3 is not a function because at least one member of the domain corresponds to two members of the range.

$$(700, 2.4) \qquad (700, 3.7)$$

The member of the domain, 700, corresponds to two members of the range, 2.4 and 3.7. Because a function is a relation in which **no two ordered pairs have the same first component and different second components**, the ordered pairs $(700, 2.4)$ and $(700, 3.7)$ are not ordered pairs of a function.

Same first components

$$(700, 2.4) \qquad (700, 3.7)$$

Different second components

2 Determine whether a relation is a function.

Definition of a Function

A **function** is a relation in which each member of the domain corresponds to exactly one member of the range. No two ordered pairs of a function can have the same first component and different second components.

EXAMPLE 2   Determining Whether a Relation Is a Function

Determine whether each relation is a function:

a. $\{(1, 6), (2, 6), (3, 8), (4, 9)\}$ **b.** $\{(6, 1), (6, 2), (8, 3), (9, 4)\}.$

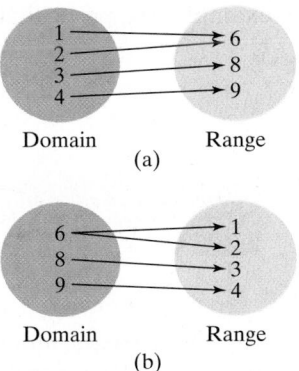

Domain (a) Range

Domain (b) Range

Figure 8.4

Solution We begin by making a figure for each relation that shows the domain and the range (Figure 8.4).

a. Figure 8.4(a) shows that every element in the domain corresponds to exactly one element in the range. The element 1 in the domain corresponds to the element 6 in the range. Furthermore, 2 corresponds to 6, 3 corresponds to 8, and 4 corresponds to 9. No two ordered pairs in the given relation have the same first component and different second components. Thus, the relation is a function.

b. Figure 8.4(b) shows that 6 corresponds to both 1 and 2. If any element in the domain corresponds to more than one element in the range, the relation is not a function. This relation is not a function; two ordered pairs have the same first component and different second components.

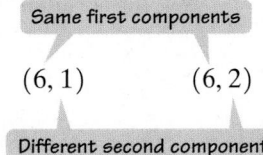

Same first components

$(6, 1)$　　$(6, 2)$

Different second components ∎

Look at Figure 8.4(a) again. The fact that 1 and 2 in the domain correspond to the same number, 6, in the range does not violate the definition of a function. A function can have two different first components with the same second component. By contrast, a relation is not a function when two different ordered pairs have the same first component and different second components. Thus, the relation in Example 2(b) is not a function.

✔ **CHECK POINT 2** Determine whether each relation is a function:

a. $\{(1, 2), (3, 4), (5, 6), (5, 8)\}$

b. $\{(1, 2), (3, 4), (6, 5), (8, 5)\}$.

Functions as Equations and Function Notation

Functions are usually given in terms of equations rather than as sets of ordered pairs. For example, here is an equation that models paid vacation days each year as a function of years working for a company:

$$y = -0.016x^2 + 0.93x + 8.5.$$

The variable x represents years working for a company. The variable y represents the average number of vacation days each year. The variable y is a function of the variable x. For each value of x, there is one and only one value of y. The variable x is called the **independent variable** because it can be assigned any value from the domain. Thus, x can be assigned any positive integer representing years working for a company. The variable y is called the **dependent variable** because its value depends on x. Paid vacation days depend on years working for a company. The value of the dependent variable, y, is calculated after selecting a value for the independent variable, x.

If an equation in x and y gives one and only one value of y for each value of x, then the variable y is a function of the variable x. When an equation represents a function, the function is often named by a letter such as f, g, h, F, G, or H. Any letter can be used to name a function. Suppose that f names a function. Think of the domain as the set of the function's inputs and the range as the set of the function's outputs. As shown in Figure 8.5, the input is represented by x and the output by $f(x)$. The special notation $f(x)$, read "f of x" or "f at x," represents the **value of the function at the number x.**

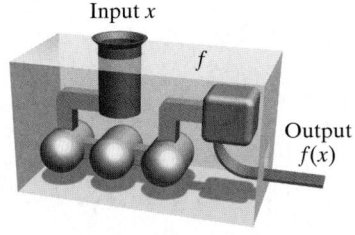

Input x

f

Output $f(x)$

Figure 8.5 A "function machine" with inputs and outputs

3 Evaluate a function.

Let's make this clearer by considering a specific example. We know that the equation

$$y = -0.016x^2 + 0.93x + 8.5$$

defines y as a function of x. We'll name the function f. Now, we can apply our new function notation.

Input	Output	Equation	We read this equation as "f of x equals $-0.016x^2 + 0.93x + 8.5$."
x	$f(x)$	$f(x) = -0.016x^2 + 0.93x + 8.5$	

Suppose we are interested in finding $f(10)$, the function's output when the input is 10. To find the value of the function at 10, we substitute 10 for x. We are **evaluating the function** at 10.

$$f(x) = -0.016x^2 + 0.93x + 8.5 \qquad \text{This is the given function.}$$
$$f(10) = -0.016(10)^2 + 0.93(10) + 8.5 \qquad \text{Replace each occurrence of } x \text{ by 10.}$$
$$= -0.016(100) + 0.93(10) + 8.5 \qquad \text{Evaluate the exponential expression:} \quad 10^2 = 100.$$
$$= -1.6 + 9.3 + 8.5 \qquad \text{Perform the multiplications.}$$
$$= 16.2 \qquad \text{Add from left to right.}$$

The statement $f(10) = 16.2$, read "f of 10 equals 16.2," tells us that the value of the function at 10 is 16.2. When the function's input is 10, its output is 16.2. (After 10 years, workers average 16.2 vacation days each year.) To find other function values, such as $f(15)$, $f(20)$, or $f(23)$, substitute the specified input values for x into the function's equation.

If a function is named f and x represents the independent variable, the notation $f(x)$ corresponds to the y-value for a given x. Thus,

$$f(x) = -0.016x^2 + 0.93x + 8.5 \quad \text{and} \quad y = -0.016x^2 + 0.93x + 8.5$$

define the same function. This function may be written as

$$y = f(x) = -0.016x^2 + 0.93x + 8.5.$$

Study Tip

The notation $f(x)$ does *not* mean "f times x." The notation describes the value of the function at x.

EXAMPLE 3   Using Function Notation

Find the indicated function value:

a. $f(4)$ for $f(x) = 2x + 3$ **b.** $g(-2)$ for $g(x) = 2x^2 - 1$
c. $h(-5)$ for $h(r) = r^3 - 2r^2 + 5$ **d.** $F(a + h)$ for $F(x) = 5x + 7$.

Solution

a. $f(x) = 2x + 3$ This is the given function.
$\quad f(4) = 2 \cdot 4 + 3$ To find f of 4, replace x by 4.
$\quad\quad = 8 + 3$
$\quad\quad = 11$ Thus, f of 4 is 11.

b. $g(x) = 2x^2 - 1$ This is the given function.
$\quad g(-2) = 2(-2)^2 - 1$ To find g of -2, replace x by -2.
$\quad\quad = 2(4) - 1$
$\quad\quad = 8 - 1$
$\quad\quad = 7$ Thus, g of -2 is 7.

c. $h(r) = r^3 - 2r^2 + 5$ — The function's name is h and r represents the independent variable.

$h(-5) = (-5)^3 - 2(-5)^2 + 5$ — To find h of -5, replace each occurrence of r by -5.

$= -125 - 2(25) + 5$ — Evaluate exponential expressions.

$= -125 - 50 + 5$ — Multiply.

$= -170$ — Thus, h of -5 is -170.

d. $F(x) = 5x + 7$ — This is the given function.

$F(a + h) = 5(a + h) + 7$ — Replace x by $a + h$.

$= 5a + 5h + 7$ — Apply the distributive property. Thus, F of $a + h$ is $5a + 5h + 7$. ■

✔ **CHECK POINT 3** Find the indicated function value:

a. $f(6)$ for $f(x) = 4x + 5$

b. $g(-5)$ for $g(x) = 3x^2 - 10$

c. $h(-4)$ for $h(r) = r^2 - 7r + 2$

d. $F(a + h)$ for $F(x) = 6x + 9$.

Study Tip

Equations of the form $y = mx + b$, now expressed as $f(x) = mx + b$, were presented in Section 3.4. You can find a brief review of the slope-intercept form of the equation of a line in the Section 3.4 summary on page 238.

The functions in Check Point 3(a) and 3(d), $f(x) = 4x + 5$ and $F(x) = 6x + 9$, are examples of *linear functions*. **Linear functions** have equations of the form $f(x) = mx + b$. By contrast, the functions in Check Point 3(b) and 3(c), $g(x) = 3x^2 - 10$ and $h(r) = r^2 - 7r + 2$, are examples of *quadratic functions*. **Quadratic functions** have equations of the form $f(x) = ax^2 + bx + c$, $a \neq 0$.

4 Use the vertical line test to identify functions.

Graphs of Functions and the Vertical Line Test

The **graph of a function** is the graph of its ordered pairs. For example, the graph of $f(x) = 3x + 1$ is the set of points (x, y) in the rectangular coordinate system satisfying the equation $y = 3x + 1$. Thus, the graph of f is a line with slope 3 and y-intercept 1. Similarly, the graph of $f(x) = x^2 + 3x + 5$ is the set of points (x, y) in the rectangular coordinate system satisfying the equation $y = x^2 + 3x + 5$. Thus, the graph of g is a parabola.

Not every graph in the rectangular coordinate system is the graph of a function. The definition of a function specifies that no value of x can be paired with two or more different values of y. Consequently, if a graph contains two or more different points with the same first coordinate, the graph cannot represent a function. This is illustrated in Figure 8.6. Observe that points sharing a common first coordinate are vertically above or below each other.

This observation is the basis of a useful test for determining whether a graph defines y as a function of x. The test is called the **vertical line test**.

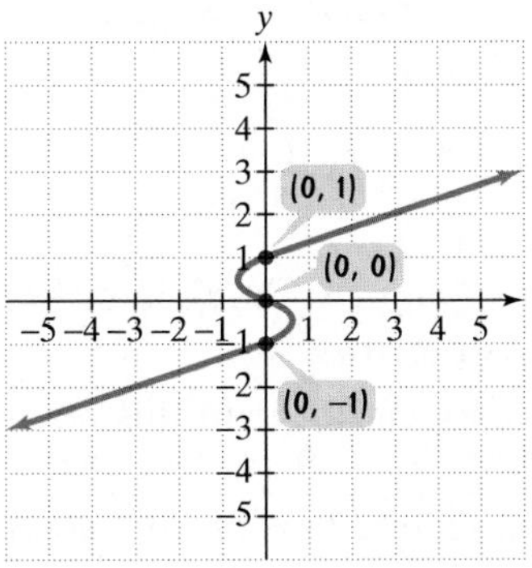

Figure 8.6 y is not a function of x because 0 is paired with three values of y, namely, 1, 0, and −1.

The Vertical Line Test for Functions

If any vertical line intersects a graph in more than one point, the graph does not define y as a function of x.

EXAMPLE 4 Using the Vertical Line Test

Use the vertical line test to identify graphs in which y is a function of x.

a. 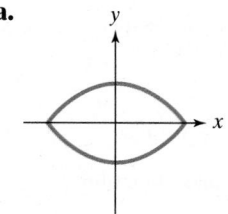 **b.** 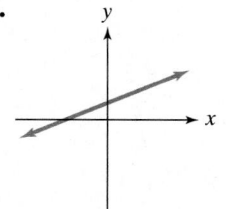 **c.** 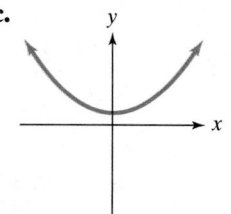 **d.**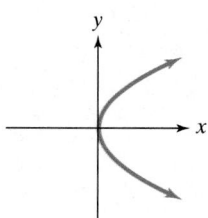

Solution y is a function of x for the graphs in (b) and (c).

a. 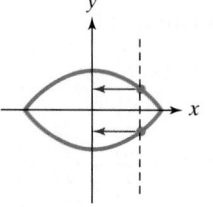 **b.** 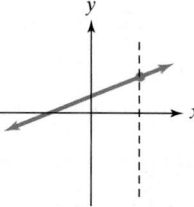 **c.** 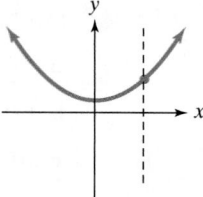 **d.**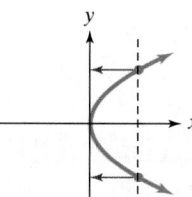

y **is not a function** of x. Two values of y correspond to an x-value.

y **is a function** of x.

y **is a function** of x.

y **is not a function** of x. Two values of y correspond to an x-value. ∎

✔ **CHECK POINT 4** Use the vertical line test to identify graphs in which y is a function of x.

a. 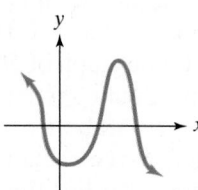 **b.** 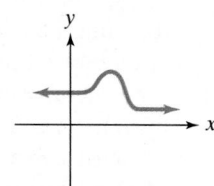 **c.**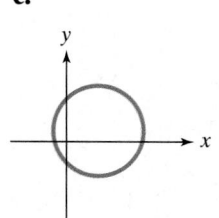

5 Obtain information about a function from its graph.

Obtaining Information from Graphs

You can obtain information about a function from its graph. At the right or left of a graph, you will often find closed dots, open dots, or arrows.

- A closed dot indicates that the graph does not extend beyond this point and the point belongs to the graph.
- An open dot indicates that the graph does not extend beyond this point and the point does not belong to the graph.
- An arrow indicates that the graph extends indefinitely in the direction in which the arrow points.

Average Number of Paid Vacation Days for Full-Time Workers at Medium to Large U.S. Companies

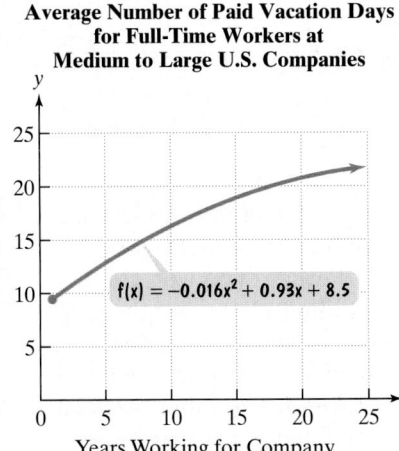

Figure 8.7 *Source:* Bureau of Labor Statistics

EXAMPLE 5 Analyzing the Graph of a Function

The function

$$f(x) = -0.016x^2 + 0.93x + 8.5$$

models the average number of paid vacation days each year, $f(x)$, for full-time workers at medium to large U.S. companies after x years. The graph of f is shown in Figure 8.7.

a. Explain why f represents the graph of a function.
b. Use the graph to find a reasonable estimate of $f(5)$.
c. For what value of x is $f(x) = 20$?
d. Describe the general trend shown by the graph.

Solution

a. No vertical line intersects the graph of f more than once. By the vertical line test, f represents the graph of a function.

b. To find $f(5)$, or f of 5, we locate 5 on the x-axis. The figure shows the point on the graph of f for which 5 is the first coordinate. From this point, we look to the y-axis to find the corresponding y-coordinate. A reasonable estimate of the y-coordinate is 13. Thus, $f(5) \approx 13$. After 5 years, a worker can expect approximately 13 paid vacation days.

Average Number of Paid Vacation Days for Full-Time Workers at Medium to Large U.S. Companies

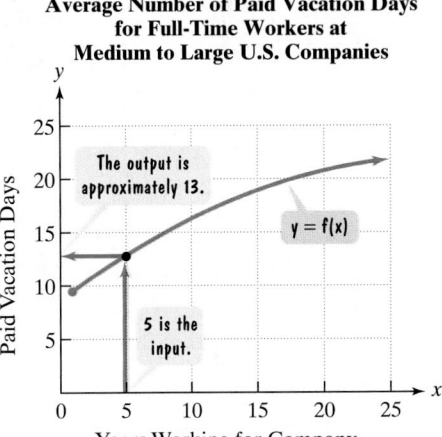

c. To find the value of x for which $f(x) = 20$, we locate 20 on the y-axis. The figure shows that there is one point on the graph of f for which 20 is the second coordinate. From this point, we look to the x-axis to find the corresponding x-coordinate. A reasonable estimate of the x-coordinate is 18. Thus, $f(x) = 20$ for $x \approx 18$. A worker with 20 paid vacation days has been with the company approximately 18 years.

Average Number of Paid Vacation Days for Full-Time Workers at Medium to Large U.S. Companies

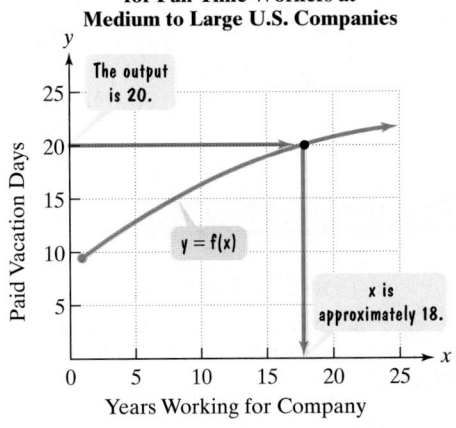

d. The graph of f is rising from left to right. This shows that paid vacation days increase as time with the company increases. However, the rate of increase is slowing down as the graph moves to the right. This means that the increase in paid vacation days takes place more slowly the longer an employee is with the company. ∎

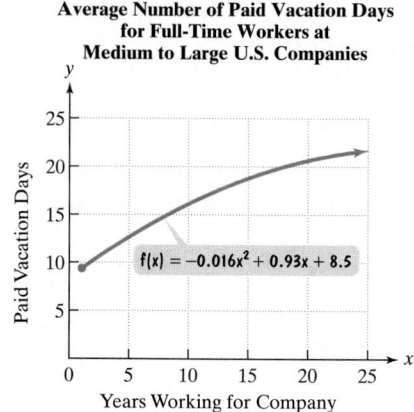

Average Number of Paid Vacation Days for Full-Time Workers at Medium to Large U.S. Companies

$f(x) = -0.016x^2 + 0.93x + 8.5$

Years Working for Company

Figure 8.7, repeated

✔ **CHECK POINT 5**

a. Use the graph of f in Figure 8.7, repeated in the margin, to find a reasonable estimate of $f(10)$.

b. For what value of x is $f(x) = 15$? Round to the nearest whole number.

Relationships between variables that you encounter in many everyday situations can be illustrated by functions and their graphs. For example, have you ever mailed a letter that seemed heavier than usual? Perhaps you worried that the letter would not have enough postage. Costs for mailing a letter weighing up to 5 ounces are given in Table 8.1 below. If your letter weighs an ounce or less, the cost is $0.34. If your letter weighs 1.05 ounces, 1.50 ounces, 1.90 ounces, or 2.00 ounces, the cost "steps" to $0.55. The cost does not take on any value between $0.34 and $0.55. If your letter weighs 2.05 ounces, 2.50 ounces, 2.90 ounces, or 3 ounces, the cost "steps" to $0.76. Cost increases are $0.21 per step.

Table 8.1 Cost of First-Class Mail (Effective January 7, 2001)

Weight Not Over	Cost
1 ounce	$0.34
2 ounces	0.55
3 ounces	0.76
4 ounces	0.97
5 ounces	1.18

Source: U.S. Postal Service

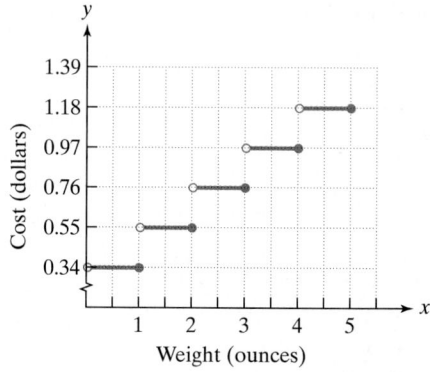

The graph of the function that models this situation is shown to the right of the table. Let

$$x = \text{the weight of the letter, in ounces, and}$$
$$y = f(x) = \text{the cost of mailing a letter weighing } x \text{ ounces.}$$

Notice how the graph of f consists of a series of steps that jump vertically 0.21 unit at each integer. The graph is constant between each pair of consecutive integers. The open dot on the left of each horizontal step shows that the left point is not included on the graph. By contrast, the closed dot on the right shows that the right point of each horizontal step is part of the graph.

EXERCISE SET 8.1

Practice Exercises

In Exercises 1–8, determine whether each relation is a function. Give the domain and range for each relation.

1. $\{(1, 2), (3, 4), (5, 5)\}$

2. $\{(4, 5), (6, 7), (8, 8)\}$

3. $\{(3, 4), (3, 5), (4, 4), (4, 5)\}$

4. $\{(5, 6), (5, 7), (6, 6), (6, 7)\}$

5. $\{(-3, -3), (-2, -2), (-1, -1), (0, 0)\}$

6. $\{(-7, -7), (-5, -5), (-3, -3), (0, 0)\}$

7. $\{(1, 4), (1, 5), (1, 6)\}$

8. $\{(4, 1), (5, 1), (6, 1)\}$

In Exercises 9–18, find the indicated function values.

9. $f(x) = x + 1$
    a. $f(0)$ **b.** $f(5)$ **c.** $f(-8)$
    d. $f(2a)$ **e.** $f(a + 2)$

10. $f(x) = x + 3$
    a. $f(0)$ **b.** $f(5)$ **c.** $f(-8)$
    d. $f(2a)$ **e.** $f(a + 2)$

11. $g(x) = 3x - 2$
    a. $g(0)$ **b.** $g(-5)$ **c.** $g\left(\dfrac{2}{3}\right)$
    d. $g(4b)$ **e.** $g(b + 4)$

12. $g(x) = 4x - 3$
    a. $g(0)$ **b.** $g(-5)$ **c.** $g\left(\dfrac{3}{4}\right)$
    d. $g(5b)$ **e.** $g(b + 5)$

13. $h(x) = 3x^2 + 5$
    a. $h(0)$ **b.** $h(-1)$ **c.** $h(4)$
    d. $h(-3)$ **e.** $h(4b)$

14. $h(x) = 2x^2 - 4$
    a. $h(0)$ **b.** $h(-1)$ **c.** $h(5)$
    d. $h(-3)$ **e.** $h(5b)$

15. $f(x) = 2x^2 + 3x - 1$
    a. $f(0)$ **b.** $f(3)$ **c.** $f(-4)$
    d. $f(b)$
    e. $f(5a)$

16. $f(x) = 3x^2 + 4x - 2$
    a. $f(0)$ **b.** $f(3)$ **c.** $f(-5)$
    d. $f(b)$ **e.** $f(5a)$

17. $f(x) = \dfrac{2x - 3}{x - 4}$

    a. $f(0)$ **b.** $f(3)$ **c.** $f(-4)$

    d. $f(-5)$ **e.** $f(a + h)$

    f. Why must 4 be excluded from the domain of f?

18. $f(x) = \dfrac{3x - 1}{x - 5}$

    a. $f(0)$ **b.** $f(3)$ **c.** $f(-3)$

    d. $f(10)$ **e.** $f(a + h)$

    f. Why must 5 be excluded from the domain of f?

In Exercises 19–26, use the vertical line test to identify graphs in which y is a function of x.

19.
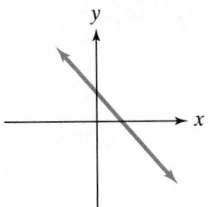

20.
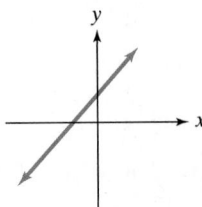

21.
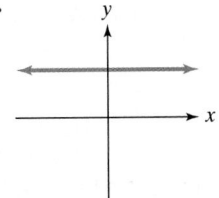

22.
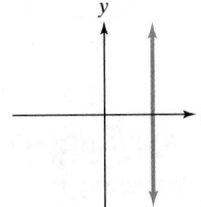

23.
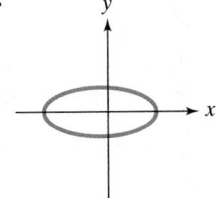

24.
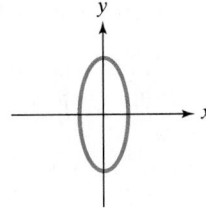

25.
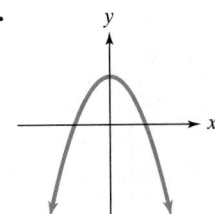

26.
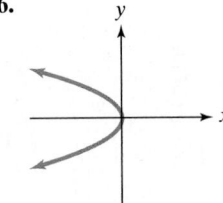

In Exercises 27–32, use the graph of f to find each indicated function value.

27. $f(-2)$

28. $f(2)$

29. $f(4)$

30. $f(-4)$

31. $f(-3)$

32. $f(-1)$

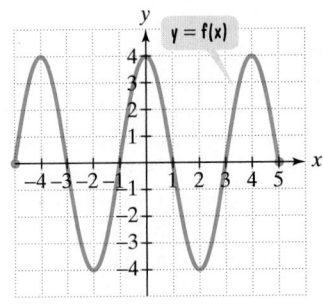

Use the graph of g to solve Exercises 33–38.

33. Find $g(-4)$.

34. Find $g(2)$.

35. Find $g(-10)$.

36. Find $g(10)$.

37. For what value of x is $g(x) = 1$?

38. For what value of x is $g(x) = -1$?

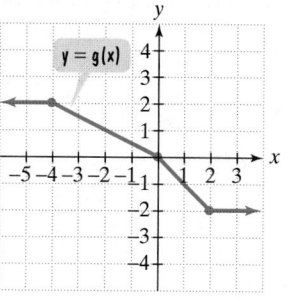

Gun Deaths per 100,000 People, by Country

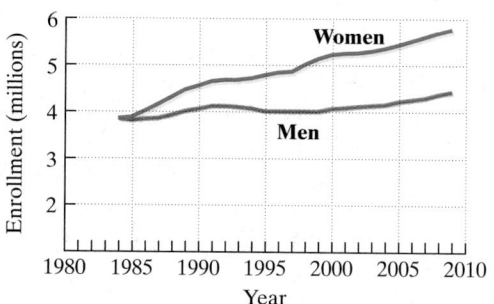

1. United States
2. Brazil
3. Mexico
4. Finland
5. France
6. Australia
7. Germany
8. England/Wales

Source: Centers for Disease Control and Prevention

⭐ ## Application Exercises

39. The bar graph shows the percentage of people in the United States using the Internet by education level. Write five ordered pairs for

(education level, percentage using Internet)

as a relation. Find the domain and the range of the relation. Is this relation a function? Explain your answer.

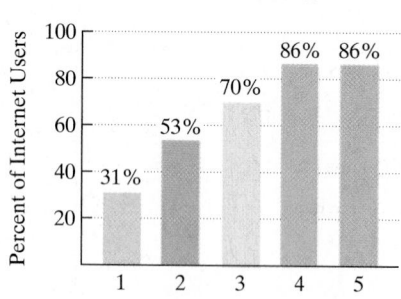

Internet Use in the U.S. by Education Level

31%, 53%, 70%, 86%, 86%

Percent of Internet Users

Education Level
(1 = less than high school, 2 = high school graduate, 3 = some college, 4 = college graduate, 5 = advanced degree)

Source: U.C.L.A. Center for Communication Policy

40. The debate over the place of guns in our society rages on. The graph at the top of the next column shows gun deaths in eight selected countries. Each gun represents one death per 100,000 people. Write eight ordered pairs for

(country number, gun deaths per 100,000 people)

as a relation. Find the domain and the range of the relation. Is this relation a function? Explain your answer.

The male minority? The graphs show enrollment in U.S. colleges, with projections from 2000 to 2009. The trend indicated by the graphs is among the hottest topics of debate among college-admissions officers. Some private liberal arts colleges have quietly begun special efforts to recruit men—including admissions preferences for them.

Enrollment in U.S. Colleges

Women

Men

Enrollment (millions)

Year

Source: Department of Education

The function

$$W(x) = 0.07x + 4.1$$

models the number of women, $W(x)$, in millions, enrolled in U.S. colleges x years after 1984. The function

$$M(x) = 0.01x + 3.9$$

models the number of men, $M(x)$, in millions, enrolled in U.S. colleges x years after 1984. Use these equations to solve Exercises 41–44.

41. Find and interpret $W(16)$. Identify this information as a point on the graph for women.

42. Find and interpret $M(16)$. Identify this information as a point on the graph for men.

43. Find and interpret $W(20) - M(20)$.

44. Find and interpret $W(25) - M(25)$.

The function

$$f(x) = 0.4x^2 - 36x + 1000$$

models the number of accidents, $f(x)$, per 50 million miles driven as a function of the driver's age, x, in years, where x includes drivers from ages 16 through 74. The graph of f is shown. Use the equation for f to solve Exercises 45–48.

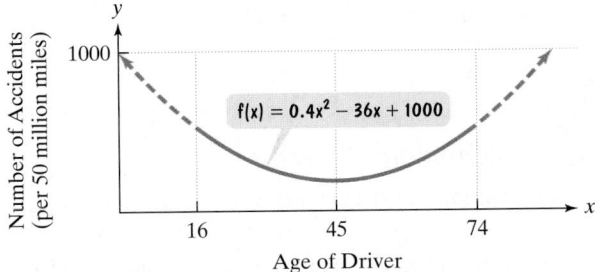

45. Find and interpret $f(20)$. Identify this information as a point on the graph of f.

46. Find and interpret $f(50)$. Identify this information as a point on the graph of f.

47. For what value of x does the graph reach its lowest point? Use the equation for f to find the minimum value of y. Describe the practical significance of this minimum value.

48. Use the graph to identify two different ages for which drivers have the same number of accidents. Use the equation for f to find the number of accidents for drivers at each of these ages.

The figure shows the percentage of the U.S. population, $f(x)$, made up of Jewish Americans as a function of time, x, where x is the number of years after 1900. Use the graph to solve Exercises 49–56.

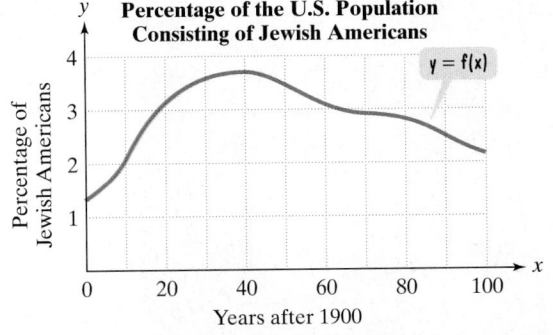

Source: American Jewish Yearbook

49. Use the graph to find a reasonable estimate of $f(60)$. What does this mean in terms of the variables in this situation?

50. Use the graph to find a reasonable estimate of $f(100)$. What does this mean in terms of the variables in this situation?

51. For what value or values of x is $f(x) = 3$? Round to the nearest year. What does this mean in terms of the variables in this situation?

52. For what value or values of x is $f(x) = 2.5$? Round to the nearest year. What does this mean in terms of the variables in this situation?

53. In which year did the percentage of Jewish Americans in the U.S. population reach a maximum? What is a reasonable estimate of the percentage for that year?

54. In which year was the percentage of Jewish Americans in the U.S. population at a minimum? What is a reasonable estimate of the percentage for that year?

55. Explain why f represents the graph of a function.

56. Describe the general trend shown by the graph.

The figure shows the cost of mailing a first-class letter, $f(x)$, as a function of its weight, x, in ounces. Use the graph to solve Exercises 57–60.

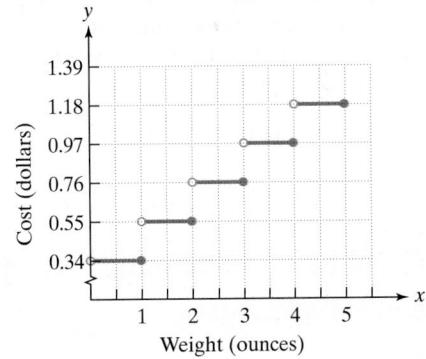

57. Find $f(3)$. What does this mean in terms of the variables in this situation?

58. Find $f(4)$. What does this mean in terms of the variables in this situation?

59. What is the cost of mailing a letter that weighs 1.5 ounces?

60. What is the cost of mailing a letter that weighs 1.8 ounces?

Writing in Mathematics

61. What is a relation? Describe what is meant by its domain and its range.

62. Explain how to determine whether a relation is a function. What is a function?

63. Does $f(x)$ mean f times x when referring to function f? If not, what does $f(x)$ mean? Provide an example with your explanation.

64. What is the graph of a function?

65. Explain how the vertical line test is used to determine whether a graph is a function.

66. For people filing a single return, federal income tax is a function of adjusted gross income because for each value of adjusted gross income there is a specific tax to be paid. By contrast, the price of a house is not a function of the lot size on which the house sits because houses on same-sized lots can sell for many different prices.
   a. Describe an everyday situation between variables that is a function.
   b. Describe an everyday situation between variables that is not a function.

67. Do you believe that the trend shown by the graphs for Exercises 41–44 should be reversed by providing admissions preferences for men? Explain your position on this issue.

Critical Thinking Exercises

68. Which one of the following is true?
   a. All relations are functions.
   b. No two ordered pairs of a function can have the same second component and different first components.
   c. The graph of every line is a function.
   d. A horizontal line can intersect the graph of a function in more than one point.

69. If $f(x) = 3x + 7$, find $\dfrac{f(a + h) - f(a)}{h}$.

70. Take another look at the cost of first-class mail and its graph (Table 8.1 and the figure to its right on page 526). Change the description of the heading in the left column of Table 8.1 so that the graph includes the point on the left of each horizontal step, but does not include the point on the right.

71. If $f(x + y) = f(x) + f(y)$ and $f(1) = 3$, find $f(2)$, $f(3)$, and $f(4)$. Is $f(x + y) = f(x) + f(y)$ for all functions?

 Technology Exercise

72. The function
$$f(x) = -0.00002x^3 + 0.008x^2 - 0.3x + 6.95$$
models the number of annual physician visits, $f(x)$, by a person of age x. Graph the function in a $[0, 100, 5]$ by $[0, 40, 2]$ viewing rectangle. What does the shape of the graph indicate about the relationship between one's age and the number of annual physician visits? Use the $\boxed{\text{TRACE}}$ or minimum function capability to find the coordinates of the minimum point on the graph of the function. What does this mean?

 Review Exercises

73. Simplify: $24 \div 4[2 - (5 - 2)]^2 - 6$. (Section 1.8, Example 8)

74. Simplify: $\left(\dfrac{3x^2y^{-2}}{y^3}\right)^{-2}$. (Section 5.7, Example 6)

75. Solve: $\dfrac{x}{3} = \dfrac{3x}{5} + 4$. (Section 2.3, Example 4)

▶ **SECTION 8.2   The Algebra of Functions**

Objectives

1 Find the domain of a function.

2 Find the sum of two functions.

3 Use the algebra of functions to combine functions and determine domains.

© Bettmann Corbis

America is a nation of immigrants. Since 1820, over 40 million people have immigrated to the United States from all over the world. They chose to come for various reasons, such as to live in freedom, to practice religion freely, to escape poverty or oppression, and to make better lives for themselves and their children. We open this section by looking at America's many faces from the perspective of functions. You will see that functions can be combined using procedures that will remind you of combining algebraic expressions.

The Domain of a Function

The early part of the twentieth century was the golden age of immigration in America. More than 13 million people migrated to the United States between 1900 and 1914. By 1910, foreign-born residents accounted for 15% of the total U.S. population. The graph in Figure 8.8 shows the percentage of Americans who were foreign born throughout the twentieth century.

Although the graph in Figure 8.8 consists of three line segments, no vertical line can be drawn that intersects any one of these line segments more than once. This shows that the percentage of foreign-born Americans is a function of time. The domain of this function is

$$\{1900, 1901, 1902, 1903, \ldots, 2000\}.$$

Functions that model data often have their domains explicitly given on the horizontal axis of a coordinate system, as in Figure 8.8, or along with the function's equation. However, for most functions, only an equation is given, and the domain is not specified. In cases like this, the domain of function f is the largest set of real numbers for which the value of $f(x)$ is a real number. For example, consider the function

$$f(x) = \frac{1}{x - 3}.$$

Because division by 0 is undefined (and not a real number), the denominator $x - 3$ cannot be 0. Thus, x cannot equal 3. The domain of the function consists of all real numbers other than 3, represented by

$$\text{Domain of } f = \{x \mid x \text{ is a real number and } x \neq 3\}.$$

In Chapter 10, we will be studying square root functions such as

$$g(x) = \sqrt{x}.$$

The equation tells us to take the square root of x. Because only nonnegative numbers have square roots that are real numbers, the expression under the square root sign, x, must be nonnegative. Thus,

$$\text{Domain of } g = \{x \mid x \text{ is a nonnegative real number}\}.$$

Finding a Function's Domain

If a function f does not model data or verbal conditions, its domain is the largest set of real numbers for which the value of $f(x)$ is a real number. Exclude from a function's domain real numbers that cause division by zero and real numbers that result in a square root of a negative number.

EXAMPLE 1   Finding the Domain of a Function

Find the domain of each function:

a. $f(x) = 3x + 2$

b. $g(x) = \dfrac{3x + 2}{x + 1}.$

---

1 Find the domain of a function.

Percentage of Americans Who Are Foreign Born

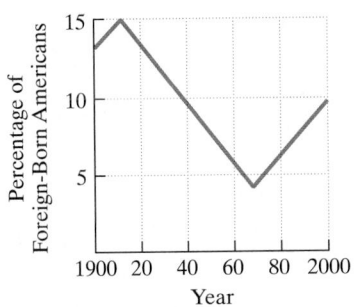

Source: U.S. Census Bureau

Figure 8.8

Using Technology

You can graph a function and visually determine its domain. For example,

$$g(x) = \sqrt{x}, \text{ or } y = \sqrt{x}$$

appears only for x greater than or equal to 0: $x \geq 0$. This verifies that

$$\{x \mid x \text{ is a nonnegative real number}\}$$

is the domain.

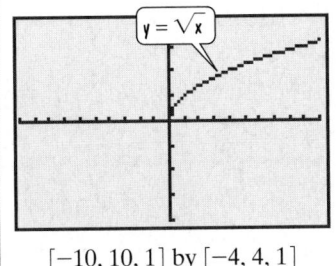

$[-10, 10, 1]$ by $[-4, 4, 1]$

Solution

a. The function $f(x) = 3x + 2$ contains neither division nor a square root. For every real number, x, the algebraic expression $3x + 2$ is a real number. Thus, the domain of f is the set of all real numbers.

$$\text{Domain of } f = \{x \mid x \text{ is a real number}\}$$

b. The function $g(x) = \dfrac{3x + 2}{x + 1}$ contains division. Because division by 0 is undefined, we must exclude from the domain the value of x that causes $x + 1$ to be 0. Thus, x cannot equal -1.

$$\text{Domain of } g = \{x \mid x \text{ is a real number and } x \neq -1\} \qquad \blacksquare$$

✔ **CHECK POINT 1** Find the domain of each function:

a. $f(x) = \dfrac{1}{2}x + 3$

b. $g(x) = \dfrac{7x + 4}{x + 5}$.

The Algebra of Functions

The United States was a low-tax country in the early part of the twentieth century. Figure 8.9 shows how the tax burden has grown since then.

U.S. Per Capita Tax Burden in 2000 Dollars

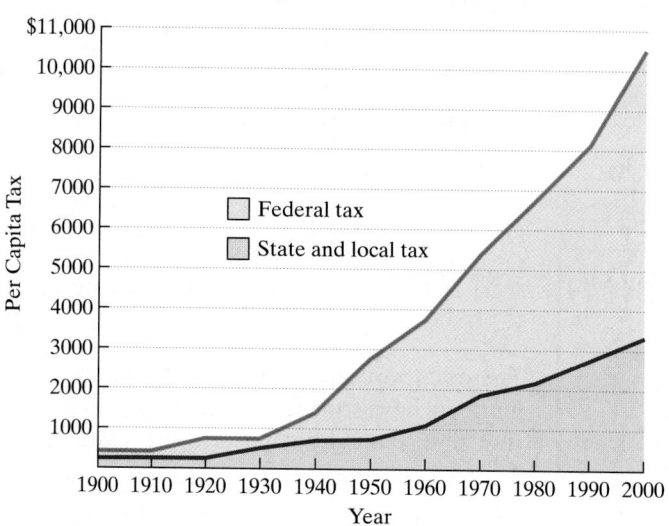

Figure 8.9   *Source:* Tax Foundation

Take a look at the information shown for the year 2000. The total per capita tax burden is approximately $10,500. The per capita state and local tax is approximately $3400. The per capita federal tax is the difference between these amounts:

$$\text{Per capita federal tax} = \$10{,}500 - \$3400 = \$7100.$$

We can think of this subtraction as the subtraction of function values. We do this by introducing the following functions:

Let $T(x) =$ total per capital tax in year x.

Let $S(x) =$ per capita state and local tax in year x.

Using Figure 8.9, we see that

$$T(2000) = \$10{,}500 \quad \text{and} \quad S(2000) = \$3400.$$

We can subtract these function values by introducing a new function, $T - S$, defined by the subtraction of $T(x)$ and $S(x)$. Thus,

$$(T - S)(x) = T(x) - S(x) = \text{total per capita tax in year } x$$
$$\text{minus state and local per}$$
$$\text{capita tax in year } x.$$

For example,

$$(T - S)(2000) = T(2000) - S(2000) = \$10{,}500 - \$3400 = \$7100.$$

> In 2000, the difference between total tax and state and local tax was $7100. This is the per capita federal tax.

Figure 8.9 illustrates that information involving differences of functions often appears in graphs seen in newspapers and magazines. Like numbers and algebraic expressions, two functions can be added, subtracted, multiplied, or divided as long as there are numbers common to the domains of both functions. The common domain for functions T and S in Figure 8.9 is

$$\{1900, 1901, 1902, 1903, \ldots, 2000\}.$$

2 Find the sum of two functions.

Because functions are usually given as equations, we perform operations by carrying out these operations with the algebraic expressions that appear on the right side of the equations. For example, we can combine the following two functions using addition:

$$f(x) = 2x + 1 \quad \text{and} \quad g(x) = x^2 - 4.$$

To do so, we add the terms to the right of the equal sign for $f(x)$ to the terms to the right of the equal sign for $g(x)$. Here is how it's done:

$$(f + g)(x) = f(x) + g(x)$$
$$= (2x + 1) + (x^2 - 4) \qquad \text{Add terms for } f(x) \text{ and } g(x).$$
$$= 2x - 3 + x^2 \qquad \text{Combine like terms.}$$
$$= x^2 + 2x - 3. \qquad \text{Arrange terms in descending powers of } x.$$

The name of this new function is $f + g$. Thus, the sum $f + g$ is the function defined by $(f + g)(x) = x^2 + 2x - 3$. The domain of $f + g$ consists of the numbers x that are in the domain of f and in the domain of g. Because neither f nor g contains division or square roots, the domain of each function is the set of all real numbers. Thus, the domain of $f + g$ is also the set of all real numbers.

EXAMPLE 2   Finding the Sum of Two Functions

Let $f(x) = x^2 - 3$ and $g(x) = 4x + 5$. Find:

a. $(f + g)(x)$ **b.** $(f + g)(3)$.

Solution

a. $(f + g)(x) = f(x) + g(x) = (x^2 - 3) + (4x + 5) = x^2 + 4x + 2$

Thus,

$$(f + g)(x) = x^2 + 4x + 2.$$

b. We find $(f + g)(3)$ by substituting 3 for x in the equation for $f + g$.

$$(f + g)(x) = x^2 + 4x + 2 \qquad \text{This is the equation for } f + g.$$

Substitute 3 for x.

$$(f + g)(3) = 3^2 + 4 \cdot 3 + 2 = 9 + 12 + 2 = 23 \qquad ■$$

✔ **CHECK POINT 2** Let $f(x) = 3x^2 + 4x - 1$ and $g(x) = 2x + 7$. Find:

a. $(f + g)(x)$

b. $(f + g)(4)$.

Here is a general definition for function addition:

The Sum of Functions

Let f and g be two functions. The **sum $f + g$** is the function defined by

$$(f + g)(x) = f(x) + g(x).$$

The domain of $f + g$ is the set of all real numbers that are common to the domain of f and the domain of g.

EXAMPLE 3   Adding Functions and Determining the Domain

Let $f(x) = \dfrac{4}{x}$ and $g(x) = \dfrac{3}{x + 2}$. Find:

a. $(f + g)(x)$ **b.** the domain of $f + g$.

Solution

a. $(f + g)(x) = f(x) + g(x) = \dfrac{4}{x} + \dfrac{3}{x + 2}$

(In Chapter 7, we discussed how to add these rational expressions. In our study of the algebra of functions, we will leave these fractions in the form shown.)

b. The domain of $f + g$ is the set of all real numbers that are common to the domain of f and the domain of g. Thus, we must find the domains of f and g. We will do so for f first.

Note that $f(x) = \dfrac{4}{x}$ is a function involving division. Because division by 0 is undefined, x cannot equal 0.

$$\text{Domain of } f = \{x \mid x \text{ is a real number and } x \neq 0\}$$

The function $g(x) = \dfrac{3}{x + 2}$ is also a function involving division. Because division by 0 is undefined, x cannot equal -2.

$$\text{Domain of } g = \{x \mid x \text{ is a real number and } x \neq -2\}$$

In order to find $f(x) + g(x)$, x must be in both domains listed. Thus,

$$\text{Domain of } f + g = \{x \mid x \text{ is a real number and } x \neq 0 \text{ and } x \neq -2\}. \qquad ■$$

✔ **CHECK POINT 3** Let $f(x) = \dfrac{5}{x}$ and $g(x) = \dfrac{7}{x-8}$. Find:

a. $(f + g)(x)$

b. the domain of $f + g$.

3 Use the algebra of functions to combine functions and determine domains.

We can also combine functions using subtraction, multiplication, and division by performing operations with the algebraic expressions that appear on the right side of the equations. For example, the functions $f(x) = 2x$ and $g(x) = x - 1$ can be combined to form the difference, product, and quotient of f and g. Here's how it's done:

Difference: $f - g$
$$(f - g)(x) = f(x) - g(x)$$
$$= 2x - (x - 1) = 2x - x + 1 = x + 1$$

Product: fg
$$(fg)(x) = f(x) \cdot g(x)$$
$$= 2x(x - 1) = 2x^2 - 2x$$

Quotient: $\dfrac{f}{g}$
$$\left(\dfrac{f}{g}\right)(x) = \dfrac{f(x)}{g(x)} = \dfrac{2x}{x-1}, x \neq 1.$$

Just like the domain for $f + g$, the domain for each of these functions consists of all real numbers that are common to the domains of f and g. In the case of the quotient function $\dfrac{f(x)}{g(x)}$, we must remember not to divide by 0, so we add the further restriction that $g(x) \neq 0$.

The following definitions summarize our discussion.

The Algebra of Functions: Sum, Difference, Product, and Quotient of Functions

Let f and g be two functions. The **sum** $f + g$, the **difference** $f - g$, the **product** fg, and the **quotient** $\dfrac{f}{g}$ are functions whose domains are the set of all real numbers common to the domains of f and g, defined as follows:

1. Sum: $(f + g)(x) = f(x) + g(x)$

2. Difference: $(f - g)(x) = f(x) - g(x)$

3. Product: $(fg)(x) = f(x) \cdot g(x)$

4. Quotient: $\left(\dfrac{f}{g}\right)(x) = \dfrac{f(x)}{g(x)}$, provided $g(x) \neq 0$.

EXAMPLE 4   Using the Algebra of Functions

Let $f(x) = x^2 + x$ and $g(x) = x - 5$. Find:

a. $(f + g)(4)$

b. $(f - g)(x)$ and $(f - g)(-3)$

c. $(fg)(x)$ and $(fg)(-2)$

d. $\left(\dfrac{f}{g}\right)(x)$ and $\left(\dfrac{f}{g}\right)(7)$.

Solution

a. We can find $(f + g)(4)$ using $f(4)$ and $g(4)$.

$$f(x) = x^2 + x \qquad\qquad g(x) = x - 5$$
$$f(4) = 4^2 + 4 = 20 \qquad g(4) = 4 - 5 = -1$$

Thus,

$$(f + g)(4) = f(4) + g(4) = 20 + (-1) = 19.$$

We can also find $(f + g)(4)$ by first finding $(f + g)(x)$ and then substituting 4 for x:

$$(f + g)(x) = f(x) + g(x) \qquad \text{This is the definition of the sum } f + g.$$
$$= (x^2 + x) + (x - 5) \quad \text{Substitute the given functions.}$$
$$= x^2 + 2x - 5. \qquad\quad \text{Simplify.}$$

Using $(f + g)(x) = x^2 + 2x - 5$, we have

$$(f + g)(4) = 4^2 + 2 \cdot 4 - 5 = 16 + 8 - 5 = 19.$$

b. $(f - g)(x) = f(x) - g(x)$ This is the definition of the difference $f - g$.
$$= (x^2 + x) - (x - 5) \quad \text{Substitute the given functions.}$$
$$= x^2 + x - x + 5 \qquad \text{Remove parentheses and change the sign of each term in the second set of parentheses.}$$
$$= x^2 + 5 \qquad\qquad\quad \text{Simplify.}$$

Using $(f - g)(x) = x^2 + 5$, we have

$$(f - g)(-3) = (-3)^2 + 5 = 9 + 5 = 14.$$

Study Tip

Here are the details of the FOIL method we used to multiply the binomials:

$(x^2 + x)(x - 5)$

   F   O   I   L

$= x^2 \cdot x + x^2(-5) + x \cdot x + x(-5)$
$= x^3 - 5x^2 + x^2 - 5x.$

Special product of polynomials are reviewed in the Section 5.3 summary on page 365.

c. $(fg)(x) = f(x) \cdot g(x)$ This is the definition of the product fg.
$$= (x^2 + x)(x - 5) \qquad \text{Substitute the given functions.}$$
$$= x^3 - 5x^2 + x^2 - 5x \quad \text{Multiply using the FOIL method.}$$
$$= x^3 - 4x^2 - 5x \qquad\quad \text{Combine like terms: } -5x^2 + x^2 = -4x^2.$$

Using $(fg)(x) = x^3 - 4x^2 - 5x$, we have
$$(fg)(-2) = (-2)^3 - 4(-2)^2 - 5(-2)$$
$$= -8 - 4(4) - 5(-2) \qquad \text{Evaluate exponential expressions.}$$
$$= -8 - 16 + 10 = -14.$$

We can also find $(fg)(-2)$ using the fact that

$$(fg)(-2) = f(-2) \cdot g(-2).$$

$$f(x) = x^2 + x \qquad\qquad\qquad g(x) = x - 5$$
$$f(-2) = (-2)^2 + (-2) = 4 - 2 = 2 \qquad g(-2) = -2 - 5 = -7$$

Thus,

$$(fg)(-2) = f(-2) \cdot g(-2) = 2(-7) = -14.$$

d. $\left(\dfrac{f}{g}\right)(x) = \dfrac{f(x)}{g(x)}$ *This is the definition of the quotient $\dfrac{f}{g}$.*

$\qquad\quad = \dfrac{x^2 + x}{x - 5}$ *Substitute the given functions.*

Using $\left(\dfrac{f}{g}\right)(x) = \dfrac{x^2 + x}{x - 5}$, we have

$$\left(\dfrac{f}{g}\right)(7) = \dfrac{7^2 + 7}{7 - 5} = \dfrac{56}{2} = 28. \qquad\blacksquare$$

✔ **CHECK POINT 4**   Let $f(x) = x^2 - 2x$ and $g(x) = x + 3$. Find:

a. $(f + g)(5)$

b. $(f - g)(x)$ and $(f - g)(-1)$

c. $(fg)(x)$ and $(fg)(-4)$

d. $\left(\dfrac{f}{g}\right)(x)$ and $\left(\dfrac{f}{g}\right)(7)$.

EXERCISE SET 8.2

Practice Exercises

In Exercises 1–10, find the domain of each function.

1. $f(x) = 3x + 5$

2. $f(x) = 4x + 7$

3. $g(x) = \dfrac{1}{x + 4}$

4. $g(x) = \dfrac{1}{x + 5}$

5. $f(x) = \dfrac{2x}{x - 3}$

6. $f(x) = \dfrac{4x}{x - 2}$

7. $g(x) = x + \dfrac{3}{5 - x}$

8. $g(x) = x + \dfrac{7}{6 - x}$

9. $f(x) = \dfrac{1}{x + 7} + \dfrac{3}{x - 9}$

10. $f(x) = \dfrac{1}{x + 8} + \dfrac{3}{x - 10}$

In Exercises 11–16, find **a.** $(f + g)(x)$ **b.** $(f + g)(5)$.

11. $f(x) = 3x + 1,\ g(x) = 2x - 6$

12. $f(x) = 4x + 2, g(x) = 2x - 9$

13. $f(x) = x - 5, g(x) = 3x^2$

14. $f(x) = x - 6, g(x) = 2x^2$

15. $f(x) = 2x^2 - x - 3, g(x) = x + 1$

16. $f(x) = 4x^2 - x - 3, g(x) = x + 1$

In Exercises 17–28, for each pair of functions, f and g, determine the domain of $f + g$.

17. $f(x) = 3x + 7, \ g(x) = 9x + 10$

18. $f(x) = 7x + 4, \ g(x) = 5x - 2$

19. $f(x) = 3x + 7, \ g(x) = \dfrac{2}{x - 5}$

20. $f(x) = 7x + 4, g(x) = \dfrac{2}{x - 6}$

21. $f(x) = \dfrac{1}{x}, \ g(x) = \dfrac{2}{x - 5}$

22. $f(x) = \dfrac{1}{x}, g(x) = \dfrac{2}{x - 6}$

23. $f(x) = \dfrac{8x}{x - 2}, g(x) = \dfrac{6}{x + 3}$

24. $f(x) = \dfrac{9x}{x - 4}, g(x) = \dfrac{7}{x + 8}$

25. $f(x) = \dfrac{8x}{x - 2}, g(x) = \dfrac{6}{2 - x}$

26. $f(x) = \dfrac{9x}{x - 4}, g(x) = \dfrac{7}{4 - x}$

27. $f(x) = x^2, g(x) = x^3$

28. $f(x) = x^2 + 1, g(x) = x^3 - 1$

In Exercises 29–48, let

$$f(x) = x^2 + 4x \quad \text{and} \quad g(x) = 2 - x.$$

Find each of the following.

29. $(f + g)(x)$ and $(f + g)(3)$

30. $(f + g)(x)$ and $(f + g)(4)$

31. $f(-2) + g(-2)$

32. $f(-3) + g(-3)$

33. $(f - g)(x)$ and $(f - g)(5)$

34. $(f - g)(x)$ and $(f - g)(6)$

35. $f(-2) - g(-2)$

36. $f(-3) - g(-3)$

37. $(fg)(x)$ and $(fg)(2)$

38. $(fg)(x)$ and $(fg)(3)$

39. $(fg)(5)$

40. $(fg)(6)$

41. $\left(\dfrac{f}{g}\right)(x)$ and $\left(\dfrac{f}{g}\right)(1)$

42. $\left(\dfrac{f}{g}\right)(x)$ and $\left(\dfrac{f}{g}\right)(3)$

43. $\left(\dfrac{f}{g}\right)(-1)$

44. $\left(\dfrac{f}{g}\right)(0)$

45. The domain of $f + g$

46. The domain of $f - g$

47. The domain of $\dfrac{f}{g}$

48. The domain of fg

Application Exercises

The graph shows veterinary costs, in billions of dollars, for dogs and cats in five selected years. Let

$D(x)$ = veterinary costs, in billions of dollars, for dogs in year x

$C(x)$ = veterinary costs, in billions of dollars, for cats in year x.

Use the graph to solve Exercises 49–52.

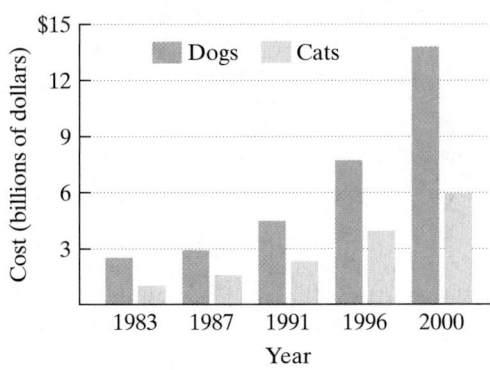

Veterinary Costs in the U.S.

Source: American Veterinary Medical Association

49. Find an estimate of $(D + C)(2000)$. What does this mean in terms of the variables in this situation?

50. Find an estimate of $(D - C)(2000)$. What does this mean in terms of the variables in this situation?

51. Using the information shown in the graph, what is the domain of $D + C$?

52. Using the information shown in the graph, what is the domain of $D - C$?

Consider the following functions:

$f(x)$ = population of the world's more-developed regions in year x

$g(x)$ = the population of the world's less-developed regions in year x

$h(x)$ = total world population in year x.

Use these functions and the graphs shown to answer Exercises 53–56.

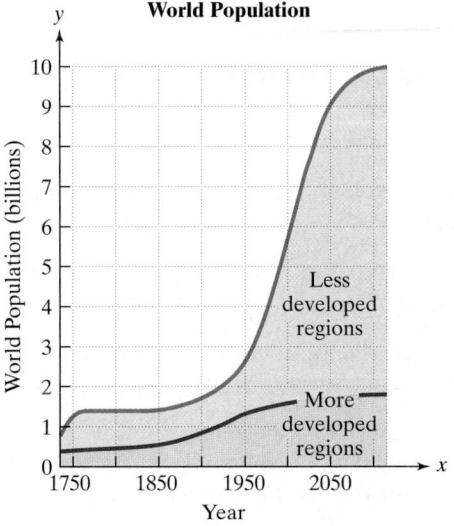

World Population

Source: Population Reference Bureau

53. What does the function $f + g$ represent?

54. What does the function $h - g$ represent?

55. Use the graph to estimate $(f + g)(2000)$.

56. Use the graph to estimate $(h - g)(2000)$.

57. A company that sells radios has yearly fixed costs of $600,000. It costs the company $45 to produce each radio. Each radio will sell for $65. The company's costs and revenue are modeled by the following functions:

$C(x) = 600,000 + 45x$ This function models the company's costs.

$R(x) = 65x$ This function models the company's revenue.

Find and interpret $(R - C)(20,000)$, $(R - C)(30,000)$ and $(R - C)(40,000)$.

58. The function $f(t) = -0.14t^2 + 0.51t + 31.6$ models the U.S. population, $f(t)$, in millions, ages 65 and older t years after 1990. The function $g(t) = 0.54t^2 + 12.64t + 107.1$ models the total yearly cost of Medicare, $g(t)$, in billions of dollars, t years after 1990.

a. What does the function $\dfrac{g}{f}$ represent?

b. Find and interpret $\dfrac{g}{f}(10)$.

Writing in Mathematics

59. If a function is defined by an equation, explain how to find its domain.

60. If equations for functions f and g are given, explain how to find $f + g$.

61. If the equations of two functions are given, explain how to obtain the quotient function and its domain.

62. If equations for functions f and g are given, describe two ways to find $(f - g)(3)$.

63. Explain how to use the graphs in Figure 8.9 on page 532 to estimate the per capita federal tax for any one of the years shown on the horizontal axis.

Critical Thinking Exercises

64. Which one of the following is false?
   a. If $(f + g)(a) = 0$, then $f(a)$ and $g(a)$ must be opposites, or additive inverses.
   b. If $(f - g)(a) = 0$, then $f(a)$ and $g(a)$ must be equal.
   c. If $\left(\dfrac{f}{g}\right)(a) = 0$, then $f(a)$ must be 0.
   d. If $(fg)(a) = 0$, then $f(a)$ must be 0.

65. Use the graphs given in Exercises 53–56 to create a graph that shows the population, in billions, of less-developed regions from 1950 through 2050.

Technology Exercises

66. Graph the function $\dfrac{g}{f}$ from Exercise 58 in a $[0, 15, 1]$ by $[0, 60, 1]$ viewing rectangle. What does the shape of the graph indicate about the per capita costs of Medicare for the U.S. population ages 65 and over with increasing time?

In Exercises 67–70, graph each of the three functions in the same $[-10, 10, 1]$ by $[-10, 10, 1]$ viewing rectangle.

67. $y_1 = 2x + 3$
   $y_2 = 2 - 2x$
   $y_3 = y_1 + y_2$

68. $y_1 = x - 4$
   $y_2 = 2x$
   $y_3 = y_1 - y_2$

69. $y_1 = x$
   $y_2 = x - 4$
   $y_3 = y_1 \cdot y_2$

70. $y_1 = x^2 - 2x$
   $y_2 = x$
   $y_3 = \dfrac{y_1}{y_2}$

71. In Exercise 70, use the $\boxed{\text{TRACE}}$ feature to trace along y_3. What happens at $x = 0$? Explain why this occurs.

Review Exercises

72. Solve by the addition method:

$$11x + 4y = -3$$
$$-13x + y = 15.$$

(Section 4.3, Example 2)

73. Solve: $3(6 - x) = 3 - 2(x - 4)$.

(Section 2.3, Example 3)

74. If $f(x) = 6x - 4$, find $f(b + 2)$.

(Section 8.1, Example 3)

▶ SECTION 8.3   *Systems of Linear Equations in Three Variables*

Objectives

1 Verify the solution of a system of linear equations in three variables.

2 Solve systems of linear equations in three variables.

3 Identify inconsistent and dependent systems.

4 Solve problems using systems in three variables.

SSM
PH Tutor   CD-   Video
Center   ROM

All animals sleep, but the length of time they sleep varies widely: Cattle sleep for only a few minutes at a time. We humans seem to need more sleep than other animals, up to eight hours a day. Without enough sleep, we have difficulty concentrating, make mistakes in routine tasks, lose energy, and feel bad-tempered. There is a relationship between hours of sleep and death rate per year per 100,000 people. How many hours of sleep will put you in the group with the minimum death rate? In this section you will learn how to solve systems of linear equations with more than two variables in order to answer this question.

Study Tip

Our work in this section is based on understanding how to solve systems of linear equations in two variables. The substitution and addition methods for solving such systems are reviewed in the Section 4.2 and 4.3 summaries on pages 287 and 288.

1 Verify the solution of a system of linear equations in three variables.

Systems of Linear Equations in Three Variables and Their Solutions

An equation such as $x + 2y - 3z = 9$ is called a **linear equation in three variables**. In general, any equation of the form

$$Ax + By + Cz = D$$

where A, B, C, and D are real numbers such that A, B, and C are not all 0, is a linear equation in the variables x, y, and z. The graph of this linear equation in three variables is a plane in three-dimensional space.

The process of solving a system of three linear equations in three variables is geometrically equivalent to finding the point of intersection (assuming that there is one) of three planes in space (see Figure 8.10). A **solution** to a system of linear equations in three variables is an ordered triple of real numbers that satisfies all equations of the system. The **solution set** of the system is the set of all its solutions.

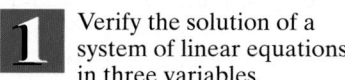

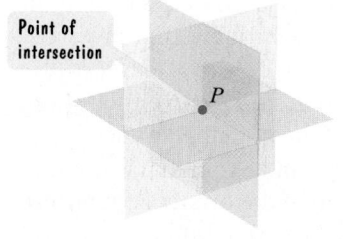

Point of intersection

P

Figure 8.10

EXAMPLE 1 Determining Whether an Ordered Triple Satisfies a System

Show that the ordered triple $(-1, 2, -2)$ is a solution of the system:

$$x + 2y - 3z = 9$$
$$2x - y + 2z = -8$$
$$-x + 3y - 4z = 15.$$

Solution Because -1 is the x-coordinate, 2 is the y-coordinate, and -2 is the z-coordinate of $(-1, 2, -2)$, we replace x by -1, y by 2, and z by -2 in each of the three equations.

$$x + 2y - 3z = 9$$
$$-1 + 2(2) - 3(-2) \stackrel{?}{=} 9$$
$$-1 + 4 + 6 \stackrel{?}{=} 9$$
$$9 = 9, \quad \text{true}$$

$$2x - y + 2z = -8$$
$$2(-1) - 2 + 2(-2) \stackrel{?}{=} -8$$
$$-2 - 2 - 4 \stackrel{?}{=} -8$$
$$-8 = -8, \quad \text{true}$$

$$-x + 3y - 4z = 15$$
$$-(-1) + 3(2) - 4(-2) \stackrel{?}{=} 15$$
$$1 + 6 + 8 \stackrel{?}{=} 15$$
$$15 = 15, \quad \text{true}$$

The ordered triple $(-1, 2, -2)$ satisfies the three equations: It makes each equation true. Thus, the ordered triple is a solution of the system. ■

✔ **CHECK POINT 1** Show that the ordered triple $(-1, -4, 5)$ is a solution of the system:

$$x - 2y + 3z = 22$$
$$2x - 3y - z = 5$$
$$3x + y - 5z = -32.$$

2 Solve systems of linear equations in three variables.

Solving Systems of Linear Equations in Three Variables by Eliminating Variables

The method for solving a system of linear equations in three variables is similar to that used on systems of linear equations in two variables. We use addition to eliminate any variable, reducing the system to two equations in two variables. Once we obtain a system of two equations in two variables, we use addition or substitution to eliminate a variable. The result is a single equation in one variable. We solve this equation to get the value of the remaining variable. Other variable values are found by back-substitution.

Solving Linear Systems in Three Variables by Eliminating Variables

1. Reduce the system to two equations in two variables. This is usually accomplished by taking two different pairs of equations and using the addition method to eliminate the same variable from each pair.
2. Solve the resulting system of two equations in two variables using addition or substitution. The result is an equation in one variable that gives the value of that variable.
3. Back-substitute the value of the variable found in step 2 into either of the equations in two variables to find the value of the second variable.
4. Use the values of the two variables from steps 2 and 3 to find the value of the third variable by back-substituting into one of the original equations.
5. Check the proposed solution in each of the original equations.

EXAMPLE 2 Solving a System in Three Variables

Solve the system:

$$5x - 2y - 4z = 3 \qquad \text{Equation 1}$$
$$3x + 3y + 2z = -3 \qquad \text{Equation 2}$$
$$-2x + 5y + 3z = 3. \qquad \text{Equation 3}$$

Solution There are many ways to proceed. Because our initial goal is to reduce the system to two equations in two variables, **the central idea is to take two different pairs of equations and eliminate the same variable from each pair.**

Step 1 Reduce the system to two equations in two variables. We choose any two equations and use the addition method to eliminate a variable. Let's eliminate z from Equations 1 and 2. We do so by multiplying Equation 2 by 2. Then we add equations.

$$
\begin{array}{llll}
\text{(Equation 1)} & 5x - 2y - 4z = 3 & \xrightarrow{\text{No change}} & 5x - 2y - 4z = 3 \\
\text{(Equation 2)} & 3x + 3y + 2z = -3 & \xrightarrow{\text{Multiply by 2.}} & 6x + 6y + 4z = -6 \\
& & \text{Add:} & \overline{11x + 4y \qquad = -3} \quad \text{(Equation 4)}
\end{array}
$$

Now we must eliminate the *same* variable from another pair of equations. We can eliminate z from Equations 2 and 3. First, we multiply Equation 2 by -3. Next, we multiply Equation 3 by 2. Finally, we add equations.

$$
\begin{array}{llll}
\text{(Equation 2)} & 3x + 3y + 2z = -3 & \xrightarrow{\text{Multiply by } -3.} & -9x - 9y - 6z = 9 \\
\text{(Equation 3)} & -2x + 5y + 3z = 3 & \xrightarrow{\text{Multiply by 2.}} & -4x + 10y + 6z = 6 \\
& & \text{Add:} & \overline{-13x + \quad y \qquad = 15} \quad \text{(Equation 5)}
\end{array}
$$

Equations 4 and 5 give us a system of two equations in two variables.

Step 2 Solve the resulting system of two equations in two variables. We will use the addition method to solve Equations 4 and 5 for x and y. To do so, we multiply Equation 5 on both sides by -4 and add this to Equation 4.

$$
\begin{array}{llll}
\text{(Equation 4)} & 11x + 4y = -3 & \xrightarrow{\text{No change}} & 11x + 4y = -3 \\
\text{(Equation 5)} & -13x + \;\; y = 15 & \xrightarrow{\text{Multiply by } -4.} & 52x - 4y = -60 \\
& & \text{Add:} & \overline{63x \qquad = -63} \\
& & & \;\;\; x \qquad = -1 \quad \text{Divide both sides by 63.}
\end{array}
$$

Step 3 Use back-substitution in one of the equations in two variables to find the value of the second variable. We back-substitute -1 for x in either Equation 4 or 5 to find the value of y.

$$
\begin{array}{ll}
-13x + y = 15 & \text{Equation 5} \\
-13(-1) + y = 15 & \text{Substitute } -1 \text{ for } x. \\
13 + y = 15 & \text{Multiply.} \\
y = 2 & \text{Subtract 13 from both sides.}
\end{array}
$$

Step 4 Back-substitute the values found for two variables into one of the original equations to find the value of the third variable. We can now use any one of the original equations and back-substitute the values of x and y to find the value for z. We will use Equation 2.

$$3x + 3y + 2z = -3 \qquad \text{Equation 2}$$

$$3(-1) + 3(2) + 2z = -3 \qquad \text{Substitute } -1 \text{ for } x \text{ and } 2 \text{ for } y.$$

$$3 + 2z = -3 \qquad \text{Multiply and then add:} \\ 3(-1) + 3(2) = -3 + 6 = 3.$$

$$2z = -6 \qquad \text{Subtract 3 from both sides.}$$

$$z = -3 \qquad \text{Divide both sides by 2.}$$

With $x = -1$, $y = 2$, and $z = -3$, the proposed solution is the ordered triple $(-1, 2, -3)$.

Step 5 Check. Check the proposed solution, $(-1, 2, -3)$, by substituting the values for x, y, and z into each of the three original equations. These substitutions yield three true statements. Thus, the solution is $(-1, 2, -3)$ and the solution set is $\{(-1, 2, -3)\}$. ∎

✔ **CHECK POINT 2** Solve the system:

$$x + 4y - z = 20$$
$$3x + 2y + z = 8$$
$$2x - 3y + 2z = -16.$$

In some examples, one of the variables is already eliminated from a given equation. In this case, the same variable should be eliminated from the other two equations, thereby making it possible to omit one of the elimination steps. We illustrate this idea in Example 3.

EXAMPLE 3 Solving a System of Equations with a Missing Term

Solve the system:

$$x + z = 8 \qquad \text{Equation 1}$$
$$x + y + 2z = 17 \qquad \text{Equation 2}$$
$$x + 2y + z = 16. \qquad \text{Equation 3}$$

Solution

Step 1 Reduce the system to two equations in two variables. Because Equation 1 contains only x and z, we could eliminate y from Equations 2 and 3. This will give us two equations in x and z. To eliminate y from Equations 2 and 3, we multiply Equation 2 by -2 and add Equation 3.

$$\text{(Equation 2)} \quad x + y + 2z = 17 \xrightarrow{\text{Multiply by } -2.} -2x - 2y - 4z = -34$$
$$\text{(Equation 3)} \quad x + 2y + z = 16 \xrightarrow{\text{No change}} \underline{x + 2y + z = 16}$$
$$\text{Add:} \quad -x - 3z = -18 \quad \text{(Equation 4)}$$

Equation 4 and the given Equation 1 provide us with a system of two equations in two variables.

Step 2 Solve the resulting system of two equations in two variables. We will solve Equations 1 and 4 for x and z.

$$
\begin{array}{rl}
x + z = 8 & \quad \text{Equation 1} \\
\underline{-x - 3z = -18} & \quad \text{Equation 4} \\
\text{Add:} \quad -2z = -10 & \\
z = 5 & \quad \text{Divide both sides by } -2.
\end{array}
$$

Step 3 Use back-substitution in one of the equations in two variables to find the value of the second variable. To find x, we back-substitute 5 for z in either Equation 1 or 4. We will use Equation 1.

$$
\begin{array}{rl}
x + z = 8 & \quad \text{Equation 1} \\
x + 5 = 8 & \quad \text{Substitute 5 for z.} \\
x = 3 & \quad \text{Subtract 5 from both sides.}
\end{array}
$$

Step 4 Back-substitute the values found for two variables into one of the original equations to find the value of the third variable. To find y, we back-substitute 3 for x and 5 for z into Equation 2 or 3. We can't use Equation 1 because y is missing in this equation. We will use Equation 2.

$$
\begin{array}{rl}
x + y + 2z = 17 & \quad \text{Equation 2} \\
3 + y + 2(5) = 17 & \quad \text{Substitute 3 for x and 5 for z.} \\
y + 13 = 17 & \quad \text{Multiply and add.} \\
y = 4 & \quad \text{Subtract 13 from both sides.}
\end{array}
$$

We found that $z = 5$, $x = 3$, and $y = 4$. Thus, the proposed solution is the ordered triple $(3, 4, 5)$.

Step 5 Check. Substituting 3 for x, 4 for y, and 5 for z into each of the three original equations yields three true statements. Consequently, the solution is $(3, 4, 5)$ and the solution set is $\{(3, 4, 5)\}$. ∎

✔ **CHECK POINT 3** Solve the system:

$$
\begin{array}{r}
2y - z = 7 \\
x + 2y + z = 17 \\
2x - 3y + 2z = -1.
\end{array}
$$

3 Identify inconsistent and dependent systems.

Inconsistent and Dependent Systems

A system of linear equations in three variables represents three planes. The three planes need not intersect at one point. The planes may have no common point of intersection and represent an **inconsistent system** with no solution. Figure 8.11 illustrates some of the geometric possibilities for inconsistent systems.

Three planes are parallel with no common intersection point.

Two planes are parallel with no common intersection point.

Planes intersect two at a time. There is no intersection point common to all three planes.

Figure 8.11 Three planes may have no common point of intersection.

If you attempt to solve an inconsistent system algebraically, at some point in the solution process you will eliminate all three variables. A false statement such as $0 = 17$ will be the result. For example, consider the system

$$2x + 5y + z = 12 \qquad \text{Equation 1}$$
$$x - 2y + 4z = 10 \qquad \text{Equation 2}$$
$$-3x + 6y - 12z = 20. \qquad \text{Equation 3}$$

Suppose we reduce the system to two equations in two variables by eliminating x. To eliminate x in Equations 2 and 3, we multiply Equation 2 by 3 and add Equation 3:

$$
\begin{array}{ll}
x - 2y + 4z = -10 & \xrightarrow{\text{Multiply by 3.}} \quad 3x - 6y + 12z = -30 \\
-3x + 6y - 12z = 20 & \xrightarrow{\text{No change}} \quad -3x + 6y - 12z = \underline{20} \\
& \qquad\qquad \text{Add:} \qquad\qquad\qquad 0 = -10.
\end{array}
$$

There are no values of x, y, and z for which $0 = -10$. The false statement $0 = -10$ indicates that the system is inconsistent and has no solution. The solution set is the empty set, $\varnothing$.

We have seen that a linear system that has at least one solution is called a **consistent system**. Planes that intersect at one point and planes that intersect at infinitely many points both represent consistent systems. Figure 8.12 illustrates planes that intersect at infinitely many points. The equations in these linear systems with infinitely many solutions are called **dependent**.

The planes coincide.

Figure 8.12 Three planes may intersect at infinitely many points.

The planes intersect along a common line.

If you attempt to solve a system with dependent equations algebraically, at some point in the solution process you will eliminate all three variables. A true statement such as $0 = 0$ will be the result. If this occurs as you are solving a linear system, simply state that the equations are dependent.

4 Solve problems using systems in three variables.

Applications

Systems of equations may allow us to find models for data without using a graphing utility. Three data points that do not lie on or near a line determine the graph of a function of the form

$$y = ax^2 + bx + c, a \neq 0.$$

Recall that such a function is called a **quadratic function**. Its graph has a cuplike shape, making it ideal for modeling situations in which values of y are decreasing and then increasing. In Chapter 11, we'll have lots of interesting things to tell you about quadratic functions and their graphs.

The process of determining a function whose graph contains given points is called **curve fitting**. In our next example, we fit three data points to the curve whose equation is $y = ax^2 + bx + c$. Using a system of equations, we find values for a, b, and c.

EXAMPLE 4   Modeling Data Relating Sleep and Death Rate

In a study relating sleep and death rate, the following data were obtained. Use the function $y = ax^2 + bx + c$ to model the data.

x (Average Number of Hours of Sleep)	y (Death Rate per Year per 100,000 Males)
4	1682
7	626
9	967

Solution   We need to find values for a, b, and c. We can do so by solving a system of three linear equations in a, b, and c. We obtain the three equations by using the values of x and y from the data as follows:

$y = ax^2 + bx + c$ *Use the quadratic function to model the data.*

When x = 4, y = 1682: $1682 = a \cdot 4^2 + b \cdot 4 + c$ or $16a + 4b + c = 1682$

When x = 7, y = 626: $626 = a \cdot 7^2 + b \cdot 7 + c$ or $49a + 7b + c = 626$

When x = 9, y = 967: $967 = a \cdot 9^2 + b \cdot 9 + c$ or $81a + 9b + c = 967.$

The easiest way to solve this system is to eliminate c from two pairs of equations, obtaining two equations in a and b. Solving this system gives $a = 104.5$, $b = -1501.5$, and $c = 6016$. We now substitute the values for a, b, and c into $y = ax^2 + bx + c$. The function that models the given data is

$$y = 104.5x^2 - 1501.5x + 6016.$$ ∎

We can use the model that we obtained in Example 4 to find the death rate of males who average, say, 6 hours of sleep. First, write the model in function notation:

$$f(x) = 104.5x^2 - 1501.5x + 6016.$$

Substitute 6 for x:

$$f(6) = 104.5(6)^2 - 1501.5(6) + 6016 = 769.$$

According to the model, the death rate for males who average 6 hours of sleep is 769 deaths per 100,000 males.

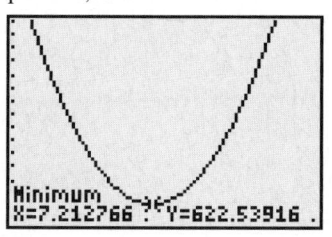

Study Tip

The strategy for solving problems with two unknowns is reviewed in the Section 4.4 summary on page 288.

✔ **CHECK POINT 4** Find the quadratic function $y = ax^2 + bx + c$ whose graph passes through the points $(1, 4)$, $(2, 1)$, and $(3, 4)$.

Problems involving three unknowns can be solved using the same strategy for solving problems with two unknown quantities. You can let x, y, and z represent the unknown quantities. We then translate from the verbal conditions of the problem to a system of three equations in three variables. Problems of this type are included in the exercise set that follows.

EXERCISE SET 8.3

Practice Exercises

In Exercises 1–4, determine if the given ordered triple is a solution of the system.

1.
$$x + y + z = 4$$
$$x - 2y - z = 1$$
$$2x - y - z = -1$$
$$(2, -1, 3)$$

2.
$$x + y + z = 0$$
$$x + 2y - 3z = 5$$
$$3x + 4y + 2z = -1$$
$$(5, -3, -2)$$

3.
$$x - 2y = 2$$
$$2x + 3y = 11$$
$$y - 4z = -7$$
$$(4, 1, 2)$$

4.
$$x - 2z = -5$$
$$y - 3z = -3$$
$$2x - z = -4$$
$$(-1, 3, 2)$$

Solve each system in Exercises 5–22. If there is no solution or if there are infinitely many solutions and a system's equations are dependent, so state.

5.
$$x + y + 2z = 11$$
$$x + y + 3z = 14$$
$$x + 2y - z = 5$$

6.
$$2x + y - 2z = -1$$
$$3x - 3y - z = 5$$
$$x - 2y + 3z = 6$$

7.
$$4x - y + 2z = 11$$
$$x + 2y - z = -1$$
$$2x + 2y - 3z = -1$$

8.
$$x - y + 3z = 8$$
$$3x + y - 2z = -2$$
$$2x + 4y + z = 0$$

9.
$$3x + 2y - 3z = -2$$
$$2x - 5y + 2z = -2$$
$$4x - 3y + 4z = 10$$

10.
$$2x + 3y + 7z = 13$$
$$3x + 2y - 5z = -22$$
$$5x + 7y - 3z = -28$$

11.
$$2x - 4y + 3z = 17$$
$$x + 2y - z = 0$$
$$4x - y - z = 6$$

12.
$$x + z = 3$$
$$x + 2y - z = 1$$
$$2x - y + z = 3$$

13.
$$2x + y = 2$$
$$x + y - z = 4$$
$$3x + 2y + z = 0$$

14.
$$x + 3y + 5z = 20$$
$$y - 4z = -16$$
$$3x - 2y + 9z = 36$$

15.
$$x + y = -4$$
$$y - z = 1$$
$$2x + y + 3z = -21$$

16.
$$x + y = 4$$
$$x + z = 4$$
$$y + z = 4$$

17.
$$2x + y + 2z = 1$$
$$3x - y + z = 2$$
$$x - 2y - z = 0$$

18.
$$3x + 4y + 5z = 8$$
$$x - 2y + 3z = -6$$
$$2x - 4y + 6z = 8$$

19.
$$5x - 2y - 5z = 1$$
$$10x - 4y - 10z = 2$$
$$15x - 6y - 15z = 3$$

20.
$$x + 2y + z = 4$$
$$3x - 4y + z = 4$$
$$6x - 8y + 2z = 8$$

21.
$$3(2x + y) + 5z = -1$$
$$2(x - 3y + 4z) = -9$$
$$4(1 + x) = -3(z - 3y)$$

22.
$$7z - 3 = 2(x - 3y)$$
$$5y + 3z - 7 = 4x$$
$$4 + 5z = 3(2x - y)$$

In Exercises 23–26, find the quadratic function
$y = ax^2 + bx + c$ *whose graph passes through the given points.*

23. $(-1, 6), (1, 4), (2, 9)$

24. $(-2, 7), (1, -2), (2, 3)$

25. $(-1, -4), (1, -2), (2, 5)$

26. $(1, 3), (3, -1), (4, 0)$

In Exercises 27–28, let x represent the first number, y the second number, and z the third number. Use the given conditions to write a system of equations. Solve the system and find the numbers.

27. The sum of three numbers is 16. The sum of twice the first number, 3 times the second number, and 4 times the third number is 46. The difference between 5 times the first number and the second number is 31. Find the three numbers.

28. Three numbers are unknown. Three times the first number plus the second number plus twice the third number is 5. If 3 times the second number is subtracted from the sum of the first number and 3 times the third number, the result is 2. If the third number is subtracted from 2 times the first number and 3 times the second number, the result is 1. Find the numbers.

Application Exercises

29. The bar graph shows that the percentage of the U.S. population that was foreign-born decreased between 1960 and 1970 and then increased between 1970 and 1980.

Percentage of U.S. Population That Is Foreign-Born, 1900-1999

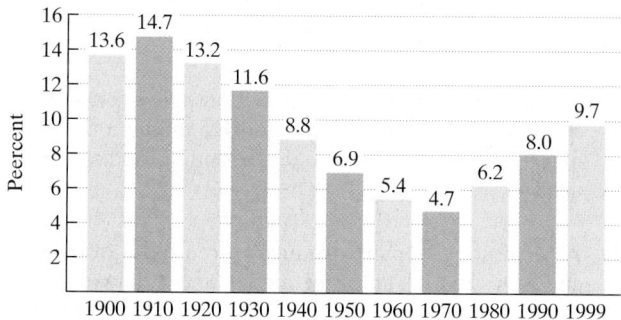

Source: Year U.S. Census Bureau

   a. Write the data for 1960, 1970, and 1980 as ordered pairs (x, y), where x is the number of years after 1960 and y is the percentage of the U.S. population that was foreign-born in that year.

   b. The three data points in part (a) can be modeled by the quadratic function $y = ax^2 + bx + c$. Substitute each ordered pair into this function, one ordered pair at a time, and write a system of linear equations in three variables that can be used to find values for a, b, and c. It is not necessary to solve the system.

30. The bar graph shows that the U.S. divorce rate increased between 1970 and 1985 and then decreased between 1985 and 1998.

U.S. Divorce Rates: Number of Divorces per 1000 People

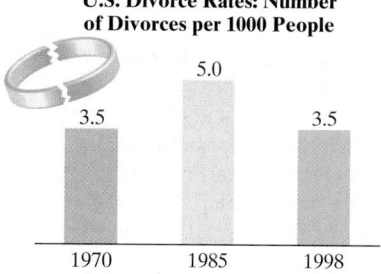

Source: U.S. Census Bureau

   a. Write the data for 1970, 1985, and 1998 as ordered pairs (x, y), where x is the number of years after 1970 and y is that year's divorce rate.

   b. The three data points in part (a) can be modeled by the quadratic function $y = ax^2 + bx + c$. Substitute each ordered pair into this function, one ordered pair at a time, and write a system of linear equations in three variables that can be used to find values for a, b, and c. It is not necessary to solve the system.

31. You throw a ball straight up from a rooftop. The ball misses the rooftop on its way down and eventually strikes the ground. A mathematical model can be used to describe the relationship for the ball's height above the ground, y, after x seconds. Consider the following data.

x, seconds after the ball is thrown	y, ball's height, in feet, above the ground
1	224
3	176
4	104

   a. Find the quadratic function $y = ax^2 + bx + c$ whose graph passes through the given points.

   b. Use the function in part (a) to find the value for y when $x = 5$. Describe what this means.

32. A mathematical model can be used to describe the relationship between the number of feet a car travels once the brakes are applied, y, and the number of seconds the car is in motion after the brakes are applied, x. A research firm collects the following data.

x, seconds in motion after brakes are applied	y, feet car travels once the brakes are applied
1	46
2	84
3	114

   a. Find the quadratic function $y = ax^2 + bx + c$ whose graph passes through the given points.

   b. Use the function in part (a) to find the value for y when $x = 6$. Describe what this means.

In Exercises 33–40, use the four-step strategy to solve each problem. Use x, y, and z to represent unknown quantities. Then translate from the verbal conditions of the problem to a system of three equations in three variables.

33. The bar graph shows the average starting salaries for the five top-paying fields for college graduates. If we add the average starting salaries for college graduates who are chemical, mechanical, and electrical engineers, the total is $121,421. The difference between the starting salaries for chemical and mechanical engineers is $2906. The difference between the starting salaries for mechanical engineers and electrical engineers is $1041. Find the average starting salaries for chemical, mechanical, and electrical engineers.

Average Starting Salaries for the Five Top Paying Fields for College Graduates in 1999

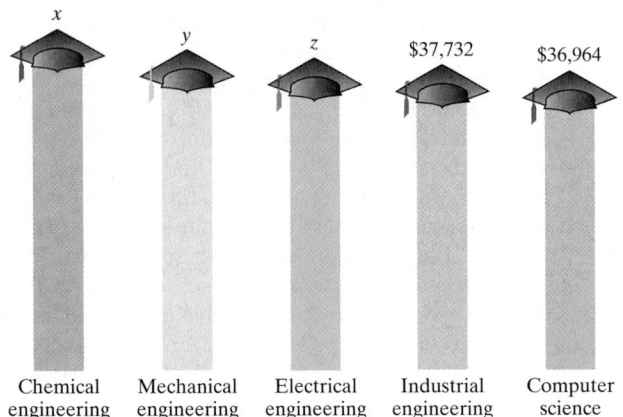

Chemical engineering	Mechanical engineering	Electrical engineering	Industrial engineering	Computer science

Source: Michigan State University

34. The circle graph indicates computers in use for the United States and the rest of the world. The percentage of the world's computers in Europe and Japan combined is 13% less than the percentage of the world's computers in the United States. If the percentage of the world's computers in Europe is doubled, it is only 3% more than the percentage of the world's computers in the United States. Find the percentage of the world's computers in the United States, Europe, and Japan.

Percentage of the World's Computers: U.S. and the World

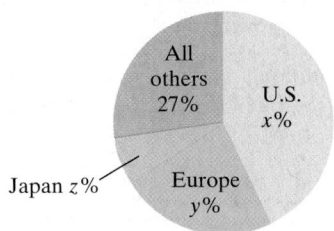

Source: Jupiter Communications

35. A person invested $6700 for one year, part at 8%, part at 10%, and the remainder at 12%. The total annual income from these investments was $716. The amount of money invested at 12% was $300 more than the amount invested at 8% and 10% combined. Find the amount invested at each rate.

36. A person invested $17,000 for one year, part at 10%, part at 12%, and the remainder at 15%. The total annual income from these investments was $2110. The amount of money invested at 12% was $1000 less than the amount invested at 10% and 15% combined. Find the amount invested at each rate.

37. At a college production of *Evita*, 400 tickets were sold. The ticket prices were $8, $10, and $12, and the total income from ticket sales was $3700. How many tickets of each type were sold if the combined number of $8 and $10 tickets sold was 7 times the number of $12 tickets sold?

38. A certain brand of razor blades comes in packages of 6, 12, and 24 blades, costing $2, $3, and $4 per package, respectively. A store sold 12 packages containing a total of 162 razor blades and took in $35. How many packages of each type were sold?

39. Three foods have the following nutritional content per ounce.

	Calories	Protein (in grams)	Vitamin C (in milligrams)
Food *A*	40	5	30
Food *B*	200	2	10
Food *C*	400	4	300

If a meal consisting of the three foods allows exactly 660 calories, 25 grams of protein, and 425 milligrams of vitamin C, how many ounces of each kind of food should be used?

40. A furniture company produces three types of desks: a children's model, an office model, and a deluxe model. Each desk is manufactured in three stages: cutting, construction, and finishing. The time requirements for each model and manufacturing stage are given in the following table.

	Children's model	Office model	Deluxe model
Cutting	2 hr	3 hr	2 hr
Construction	2 hr	1 hr	3 hr
Finishing	1 hr	1 hr	2 hr

Each week the company has available a maximum of 100 hours for cutting, 100 hours for construction, and 65 hours for finishing. If all available time must be used, how many of each type of desk should be produced each week?

Writing in Mathematics

41. What is a system of linear equations in three variables?

42. How do you determine whether a given ordered triple is a solution of a system in three variables?

43. Describe in general terms how to solve a system in three variables.

44. Describe what happens when using algebraic methods to solve an inconsistent system.

45. Describe what happens when using algebraic methods to solve a system with dependent equations.

46. AIDS is taking a deadly toll on southern Africa. Describe how to use the techniques that you learned in this section to obtain a model for African life span using projections with AIDS. Let x represent the number of years after 1985 and let y represent African life span in that year.

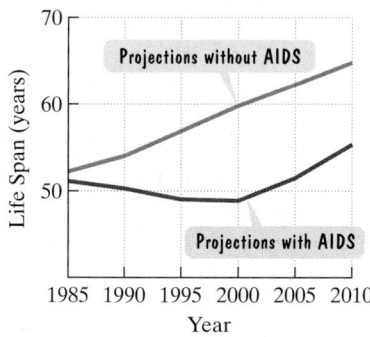

African Life Span

Source: United Nations

Critical Thinking Exercises

47. Which one of the following is true?
   a. The ordered triple $(2, 15, 14)$ is the only solution to the equation $x + y - z = 3$.
   b. The equation $x - y - z = -6$ is satisfied by $(2, -3, 5)$.
   c. If two equations in a system are $x + y - z = 5$ and $2x + 2y - 2z = 7$, then the system must be inconsistent.
   d. An equation with four variables, such as $x + 2y - 3z + 5w = 2$, cannot be satisfied by real numbers.

48. Describe how the system

$$\begin{aligned} x + y - z - 2w &= -8 \\ x - 2y + 3z + w &= 18 \\ 2x + 2y + 2z - 2w &= 10 \\ 2x + y - z + w &= 3 \end{aligned}$$

could be solved. Is it likely that in the near future a graphing utility will be available to provide a geometric solution (using intersecting graphs) to this system? Explain.

49. A modernistic painting consists of triangles, rectangles, and pentagons, all drawn so as to not overlap or share sides. Within each rectangle are drawn 2 red roses, and each pentagon contains 5 carnations. How many triangles, rectangles, and pentagons appear in the painting if the painting contains a total of 40 geometric figures, 153 sides of geometric figures, and 72 flowers?

50. Find the values of the angles marked $x°$, $y°$, and $z°$ in the following triangle.

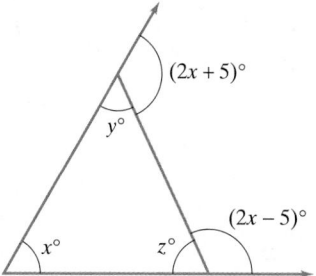

Technology Exercises

51. Does your graphing utility have a feature that allows you to solve linear systems by entering coefficients and constant terms? If so, use this feature to verify the solutions to any five exercises that you worked by hand from Exercises 5–16.

52. Verify your results in Exercises 23–26 by using a graphing utility to graph the quadratic function. Trace along the curve and convince yourself that the three points given in the exercise lie on the function's graph.

Review Exercises

53. Graph: $4x - 5y = 20$. (Section 3.2, Example 4)

54. Solve and graph the solution on a number line: $2(x - 3) > 4x + 10$. (Section 2.7, Example 7)

55. Divide: $\dfrac{1}{x^2 - 17x + 30} \div \dfrac{1}{x^2 + 7x - 18}$. (Section 7.2, Example 6)

▶ SECTION 8.4   *Matrix Solutions to Systems of Linear Equations*

Objectives

1 Write the augmented matrix for a linear system.

2 Perform matrix row operations.

3 Use matrices to solve systems.

4 Use matrices to identify inconsistent and dependent systems.

SSM   PH Tutor   CD-   Video
       Center   ROM

Yes, we overindulged, but it was delicious. Anyway, a few hours of moderate activity and we'll just burn off those extra calories. The following chart should help. We see that the number of calories burned per hour depends on our weight. Four hours of tennis and we'll be as good as new!

How Fast You Burn Off Calories

| Activity | Weight (pounds) | | | | | |
	110	132	154	176	187	209
	Calories Burned per Hour					
Housework	175	210	245	285	300	320
Cycling	190	215	245	270	280	295
Tennis	335	380	425	470	495	520
Watching TV	60	70	80	85	90	95

The 24 numbers inside the red brackets are arranged in four rows and six columns. This rectangular array of 24 numbers, arranged in rows and columns and placed in brackets, is an example of a **matrix** (plural: **matrices**). The numbers inside the brackets are called **elements** of the matrix. Matrices are used to display information and to solve systems of linear equations.

Solving Linear Systems by Using Matrices

1 Write the augmented matrix for a linear system.

A matrix gives us a shortened way of writing a system of equations. The first step in solving a system of linear equations using matrices is to write the augmented matrix. An **augmented matrix** has a vertical bar separating the columns of the matrix into two groups. The coefficients of each variable are placed to the left of the vertical line, and the constants are placed to the right. If any variable is missing, its coefficient is 0. Here are two examples.

System of Linear Equations	**Augmented Matrix**

$$\begin{aligned} x + 3y &= 5 \\ 2x - y &= -4 \end{aligned} \qquad \begin{bmatrix} 1 & 3 & | & 5 \\ 2 & -1 & | & -4 \end{bmatrix}$$

$$\begin{aligned} x + 2y - 5z &= -19 \\ y + 3z &= 9 \\ z &= 4 \end{aligned} \qquad \begin{bmatrix} 1 & 2 & -5 & | & -19 \\ 0 & 1 & 3 & | & 9 \\ 0 & 0 & 1 & | & 4 \end{bmatrix}$$

Notice how the matrix at the bottom of page 552 contains 1s down the diagonal from upper left to lower right and 0s below the 1s. This arrangement makes it easy to find the solution of the system of equations, as Example 1 shows.

EXAMPLE 1   Solving a System Using a Matrix

Write the solution for a system of equations represented by the matrix

$$\left[\begin{array}{ccc|c} 1 & 2 & -5 & -19 \\ 0 & 1 & 3 & 9 \\ 0 & 0 & 1 & 4 \end{array}\right].$$

Solution   The system represented by the given matrix is

$$\left[\begin{array}{ccc|c} 1 & 2 & -5 & -19 \\ 0 & 1 & 3 & 9 \\ 0 & 0 & 1 & 4 \end{array}\right] \rightarrow \begin{array}{l} 1x + 2y - 5z = -19 \\ 0x + 1y + 3z = 9 \\ 0x + 0y + 1z = 4 \end{array}.$$

This system can be simplified as follows.

$$x + 2y - 5z = -19 \qquad \text{Equation 1}$$

$$y + 3z = 9 \qquad \text{Equation 2}$$

$$z = 4 \qquad \text{Equation 3}$$

The value of z is known. We can find y by back-substitution.

$$y + 3z = 9 \qquad \text{Equation 2}$$

$$y + 3(4) = 9 \qquad \text{Substitute 4 for z.}$$

$$y + 12 = 9 \qquad \text{Multiply.}$$

$$y = -3 \qquad \text{Subtract 12 from both sides.}$$

With values for y and z, we can now use back-substitution to find x.

$$x + 2y - 5z = -19 \qquad \text{Equation 1}$$

$$x + 2(-3) - 5(4) = -19 \qquad \text{Substitute -3 for y and 4 for z.}$$

$$x - 6 - 20 = -19 \qquad \text{Multiply.}$$

$$x - 26 = -19 \qquad \text{Add.}$$

$$x = 7 \qquad \text{Add 26 to both sides.}$$

We see that $x = 7$, $y = -3$, and $z = 4$. The solution is $(7, -3, 4)$ and the solution set for the system is $\{(7, -3, 4)\}$. ■

✔ **CHECK POINT 1**   Write the solution for a system of equations represented by the matrix

$$\left[\begin{array}{ccc|c} 1 & -1 & 1 & 8 \\ 0 & 1 & -12 & -15 \\ 0 & 0 & 1 & 1 \end{array}\right].$$

Our goal in solving a linear system using matrices is to produce a matrix with 1s down the main diagonal and 0s below the 1s. In general, the matrix will be one of the following forms.

This is the desired form for systems with two equations.

$$\begin{bmatrix} 1 & a & | & b \\ 0 & 1 & | & c \end{bmatrix}$$

$$\begin{bmatrix} 1 & a & b & | & c \\ 0 & 1 & d & | & e \\ 0 & 0 & 1 & | & f \end{bmatrix}$$

This is the desired form for systems with three equations.

The last row of these matrices gives us the value of one variable. The other variables can then be found by back-substitution.

2 Perform matrix row operations.

A matrix with 1s down the main diagonal and 0s below the 1s is said to be in **triangular form** or **row-echelon form**. How do we produce a matrix in this form? We use **row operations** on the augmented matrix. These row operations are just like what you did when solving a linear system by the addition method. The difference is that we no longer write the variables, usually represented by x, y, and z.

Matrix Row Operations

These row operations produce matrices that represent systems with the same solution.

1. Two rows of a matrix may be interchanged. This is the same as interchanging two equations in the linear system.
2. The elements in any row may be multiplied by a nonzero number. This is the same as multiplying both sides of an equation by a nonzero number.
3. The elements in any row may be multiplied by a nonzero number, and these products may be added to the corresponding elements in any other row. This is the same as multiplying both sides of an equation by a nonzero number and then adding equations to eliminate a variable.

Two matrices are **row equivalent** if one can be obtained from the other by a sequence of row operations.

Study Tip

When performing the row operation

$$kR_i + R_j$$

you use row i to find the products. However, **elements in row i do not change. It is the elements in row j that change:** Add k times the elements in row i to the corresponding elements in row j. Replace elements in row j by these sums.

Each matrix row operation in the preceding box can be expressed symbolically as follows:

1. Interchange the elements in the ith and jth rows: $R_i \leftrightarrow R_j$.
2. Multiply each element in the ith row by k: kR_i.
3. Add k times the elements in row i to the corresponding elements in row j: $kR_i + R_j$.

EXAMPLE 2 Performing Matrix Row Operations

Use the matrix

$$\begin{bmatrix} 3 & 18 & -12 & | & 21 \\ 1 & 2 & -3 & | & 5 \\ -2 & -3 & 4 & | & -6 \end{bmatrix}$$

and perform each indicated row operation:

a. $R_1 \leftrightarrow R_2$ **b.** $\dfrac{1}{3} R_1$ **c.** $2R_2 + R_3$.

Solution

a. The notation $R_1 \leftrightarrow R_2$ means to interchange the elements in row 1 and row 2. This results in the row-equivalent matrix

$$\begin{bmatrix} 1 & 2 & -3 & | & 5 \\ 3 & 18 & -12 & | & 21 \\ -2 & -3 & 4 & | & -6 \end{bmatrix}.$$

This was row 2; now it's row 1.

This was row 1; now it's row 2.

b. The notation $\frac{1}{3} R_1$ means to multiply each element in row 1 by $\frac{1}{3}$. This results in the row-equivalent matrix

$$\begin{bmatrix} \frac{1}{3}(3) & \frac{1}{3}(18) & \frac{1}{3}(-12) & | & \frac{1}{3}(21) \\ 1 & 2 & -3 & | & 5 \\ -2 & -3 & 4 & | & -6 \end{bmatrix} = \begin{bmatrix} 1 & 6 & -4 & | & 7 \\ 1 & 2 & -3 & | & 5 \\ -2 & -3 & 4 & | & -6 \end{bmatrix}.$$

c. The notation $2R_2 + R_3$ means to add 2 times the elements in row 2 to the corresponding elements in row 3. Replace the elements in row 3 by these sums. First, we find 2 times the elements in row 2:

$$2(1) \text{ or } 2, \qquad 2(2) \text{ or } 4, \qquad 2(-3) \text{ or } -6, \qquad 2(5) \text{ or } 10.$$

Now we add these products to the corresponding elements in row 3. Although we use row 2 to find the products, row 2 does not change. It is the elements in row 3 that change, resulting in the row-equivalent matrix

$$\begin{bmatrix} 3 & 18 & -12 & | & 21 \\ 1 & 2 & -3 & | & 5 \\ -2+2=0 & -3+4=1 & 4+(-6)=-2 & | & -6+10=4 \end{bmatrix}$$

$$= \begin{bmatrix} 3 & 18 & -12 & | & 21 \\ 1 & 2 & -3 & | & 5 \\ 0 & 1 & -2 & | & 4 \end{bmatrix}. \quad \blacksquare$$

✔ **CHECK POINT 2** Use the matrix

$$\begin{bmatrix} 4 & 12 & -20 & | & 8 \\ 1 & 6 & -3 & | & 7 \\ -3 & -2 & 1 & | & -9 \end{bmatrix}$$

and perform each indicated row operation:

a. $R_1 \leftrightarrow R_2$

b. $\dfrac{1}{4} R_1$

c. $3R_2 + R_3$.

3 Use matrices to solve systems.

The process that we use to solve linear systems using matrix row operations is often called **Gaussian elimination**, after the German mathematician Carl Friedrich Gauss (1777–1835). Here are the steps used in solving systems with matrices.

Solving Linear Systems Using Matrices

1. Write the augmented matrix for the system.
2. Use matrix row operations to simplify the matrix to one with 1s down the diagonal from upper left to lower right, and 0s below the 1s.
3. Write the system of linear equations corresponding to the matrix in step 2, and use back-substitution to find the system's solution.

EXAMPLE 3   Using Matrices to Solve a Linear System

Use matrices to solve the system:

$$4x - 3y = -15$$
$$x + 2y = -1.$$

Solution

Step 1 Write the augmented matrix for the system.

Linear System	Augmented Matrix
$4x - 3y = -15$	$\begin{bmatrix} 4 & -3 & \mid & -15 \\ 1 & 2 & \mid & -1 \end{bmatrix}$
$x + 2y = -1$	

Step 2 Use matrix row operations to simplify the matrix to one with 1s down the diagonal from upper left to lower right, and 0s below the 1s. Our goal is to obtain a matrix of the form

$$\begin{bmatrix} 1 & a & \mid & b \\ 0 & 1 & \mid & c \end{bmatrix}.$$

Our first step in achieving this goal is to get 1 in the top position of the first column.

We want 1 in this position. $\begin{bmatrix} 4 & -3 & \mid & -15 \\ 1 & 2 & \mid & -1 \end{bmatrix}$

To get 1 in this position, we interchange rows 1 and 2: $R_1 \leftrightarrow R_2$.

$$\begin{bmatrix} 1 & 2 & \mid & -1 \\ 4 & -3 & \mid & -15 \end{bmatrix}$$

This was row 2; now it's row 1.
This was row 1; now it's row 2.

Now we want 0 below the 1 in the first column.

We want 0 in this position. $\begin{bmatrix} 1 & 2 & \mid & -1 \\ 4 & -3 & \mid & -15 \end{bmatrix}$

Let's get a 0 where there is now a 4. If we multiply the top row of numbers by -4 and add these products to the second row of numbers, we will get 0 in this position: $-4R_1 + R_2$. *We change only row 2.*

$$\begin{bmatrix} 1 & 2 & \mid & -1 \\ -4(1) + 4 & -4(2) + (-3) & \mid & -4(-1) + (-15) \end{bmatrix} = \begin{bmatrix} 1 & 2 & \mid & -1 \\ 0 & -11 & \mid & -11 \end{bmatrix}$$

row 1 element   row 2 element   row 1 element   row 2 element   row 1 element   row 2 element

We move on to the second column. We want 1 in the second row, second column.

We want 1 in
this position. $\begin{bmatrix} 1 & 2 & | & -1 \\ 0 & -11 & | & -11 \end{bmatrix}$

To get 1 in the desired position, we multiply -11 by its multiplicative inverse, or reciprocal, $-\frac{1}{11}$. Therefore, we multiply all the numbers in the second row by $-\frac{1}{11} : -\frac{1}{11} R_2$.

$$\begin{bmatrix} 1 & 2 & | & -1 \\ -\dfrac{1}{11}(0) & -\dfrac{1}{11}(-11) & | & -\dfrac{1}{11}(-11) \end{bmatrix} = \begin{bmatrix} 1 & 2 & | & -1 \\ 0 & 1 & | & 1 \end{bmatrix}$$

We now have the desired matrix with 1s down the diagonal and 0 below the 1.

Step 3 Write the system of linear equations corresponding to the matrix in step 2, and use back-substitution to find the system's solution. The system represented by the matrix in step 2 is

$$\begin{bmatrix} 1 & 2 & | & -1 \\ 0 & 1 & | & 1 \end{bmatrix} \rightarrow \begin{array}{c} 1x + 2y = -1 \\ 0x + 1y = 1 \end{array} \quad \text{or} \quad \begin{array}{c} x + 2y = -1 \\ y = 1 \end{array}.$$

We immediately see that the value for y is 1. To find x, we back-substitute 1 for y in the first equation.

$$x + 2y = -1 \quad \text{Equation 1}$$
$$x + 2 \cdot 1 = -1 \quad \text{Substitute 1 for } y.$$
$$x = -3 \quad \text{Subtract 2 from both sides.}$$

With $x = -3$ and $y = 1$, the proposed solution is $(-3, 1)$. Take a moment to show that $(-3, 1)$ satisfies both equations. The solution is $(-3, 1)$ and the solution set is $\{(-3, 1)\}$. ■

✔ **CHECK POINT 3**   Use matrices to solve the system:

$$2x - y = -4$$
$$x + 3y = 5.$$

EXAMPLE 4   Using Matrices to Solve a Linear System

Use matrices to solve the system:

$$3x + y + 2z = 31$$
$$x + y + 2z = 19$$
$$x + 3y + 2z = 25.$$

Solution

Step 1 Write the augmented matrix for the system.

Linear System	Augmented Matrix
$3x + y + 2z = 31$	
$x + y + 2z = 19$	$\begin{bmatrix} 3 & 1 & 2 & | & 31 \\ 1 & 1 & 2 & | & 19 \\ 1 & 3 & 2 & | & 25 \end{bmatrix}$
$x + 3y + 2z = 25$	

You can carry out step 2 on the right using the following order. Start with the augmented matrix.

$$\begin{bmatrix} * & * & * & | & * \\ * & * & * & | & * \\ * & * & * & | & * \end{bmatrix}$$

Get 1 in the upper left-hand corner.

$$\begin{bmatrix} 1 & * & * & | & * \\ * & * & * & | & * \\ * & * & * & | & * \end{bmatrix}$$

Get 0s in the first column beneath the 1.

$$\begin{bmatrix} 1 & * & * & | & * \\ 0 & * & * & | & * \\ 0 & * & * & | & * \end{bmatrix}$$

Get 1 in the second row/second column position.

$$\begin{bmatrix} 1 & * & * & | & * \\ 0 & 1 & * & | & * \\ 0 & * & * & | & * \end{bmatrix}$$

Get 0 below the 1 in the second column.

$$\begin{bmatrix} 1 & * & * & | & * \\ 0 & 1 & * & | & * \\ 0 & 0 & * & | & * \end{bmatrix}$$

Get 1 in the third row/third column position.

$$\begin{bmatrix} 1 & * & * & | & * \\ 0 & 1 & * & | & * \\ 0 & 0 & 1 & | & * \end{bmatrix}$$

Step 2 Use matrix row operations to simplify the matrix to one with 1s down the diagonal from upper left to lower right, and 0s below the 1s. Our goal is to obtain a matrix of the form

$$\begin{bmatrix} 1 & a & b & | & c \\ 0 & 1 & d & | & e \\ 0 & 0 & 1 & | & f \end{bmatrix}.$$

Our first step in achieving this goal is to get 1 in the top position of the first column.

We want 1 in this position.
$$\begin{bmatrix} 3 & 1 & 2 & | & 31 \\ 1 & 1 & 2 & | & 19 \\ 1 & 3 & 2 & | & 25 \end{bmatrix}$$

To get 1 in this position, we interchange rows 1 and 2: $R_1 \leftrightarrow R_2$. (We could also interchange rows 1 and 3 to attain our goal.)

$$\begin{bmatrix} 1 & 1 & 2 & | & 19 \\ 3 & 1 & 2 & | & 31 \\ 1 & 3 & 2 & | & 25 \end{bmatrix}$$

This was row 2; now it's row 1.

This was row 1; now it's row 2.

Now we want to get 0s below the 1 in the first column.

We want 0 in these positions.
$$\begin{bmatrix} 1 & 1 & 2 & | & 19 \\ 3 & 1 & 2 & | & 31 \\ 1 & 3 & 2 & | & 25 \end{bmatrix}$$

To get a 0 where there is now a 3, multiply the top row of numbers by -3 and add these products to the second row of numbers: $-3R_1 + R_2$. To get a 0 where there is now a 1, multiply the top row of numbers by -1 and add these products to the third row of numbers: $-1R_1 + R_3$. Although we are using row 1 to find the products, the numbers in row 1 do not change.

$\begin{matrix} -3R_1 + R_2 \\ -1R_1 + R_3 \end{matrix}$
$$\begin{bmatrix} 1 & 1 & 2 & | & 19 \\ -3(1) + 3 & -3(1) + 1 & -3(2) + 2 & | & -3(19) + 31 \\ -1(1) + 1 & -1(1) + 3 & -1(2) + 2 & | & -1(19) + 25 \end{bmatrix}$$

$$= \begin{bmatrix} 1 & 1 & 2 & | & 19 \\ 0 & -2 & -4 & | & -26 \\ 0 & 2 & 0 & | & 6 \end{bmatrix}$$

We move on to the second column. We want 1 in the second row, second column.

We want 1 in this position.
$$\begin{bmatrix} 1 & 1 & 2 & | & 19 \\ 0 & -2 & -4 & | & -26 \\ 0 & 2 & 0 & | & 6 \end{bmatrix}$$

To get 1 in the desired position, we multiply -2 by its reciprocal, $-\frac{1}{2}$. Therefore, we multiply all the numbers in the second row by $-\frac{1}{2}: -\frac{1}{2}R_2$.

$$\begin{bmatrix} 1 & 1 & 2 & | & 19 \\ -\frac{1}{2}(0) & -\frac{1}{2}(-2) & -\frac{1}{2}(-4) & | & -\frac{1}{2}(-26) \\ 0 & 2 & 0 & | & 6 \end{bmatrix} = \begin{bmatrix} 1 & 1 & 2 & | & 19 \\ 0 & 1 & 2 & | & 13 \\ 0 & 2 & 0 & | & 6 \end{bmatrix}.$$

We want 0 in this position.

We are not yet done with the second column. The voice balloon shows that we want to get a 0 where there is now a 2. If we multiply the second row of numbers by -2 and add these products to the third row of numbers, we will get 0 in this position: $-2R_2 + R_3$. Although we are using the numbers in row 2 to find the products, the numbers in row 2 do not change.

$$-2R_2 + R_3 \quad \begin{bmatrix} 1 & 1 & 2 & | & 19 \\ 0 & 1 & 2 & | & 13 \\ -2(0)+0 & -2(1)+2 & -2(2)+0 & | & -2(13)+6 \end{bmatrix} = \begin{bmatrix} 1 & 1 & 2 & | & 19 \\ 0 & 1 & 2 & | & 13 \\ 0 & 0 & -4 & | & -20 \end{bmatrix}$$

We move on to the third column. We want 1 in the third row, third column.

We want 1 in this position.

$$\begin{bmatrix} 1 & 1 & 2 & | & 19 \\ 0 & 1 & 2 & | & 13 \\ 0 & 0 & \boxed{-4} & | & -20 \end{bmatrix}$$

To get 1 in the desired position, we multiply -4 by its reciprocal, $-\frac{1}{4}$. Therefore, we multiply all the numbers in the third row by $-\frac{1}{4} : -\frac{1}{4}R_3$.

$$\begin{bmatrix} 1 & 1 & 2 & | & 19 \\ 0 & 1 & 2 & | & 13 \\ -\frac{1}{4}(0) & -\frac{1}{4}(0) & -\frac{1}{4}(-4) & | & -\frac{1}{4}(-20) \end{bmatrix} = \begin{bmatrix} 1 & 1 & 2 & | & 19 \\ 0 & 1 & 2 & | & 13 \\ 0 & 0 & 1 & | & 5 \end{bmatrix}.$$

We now have the desired matrix with 1s down the diagonal and 0s below the 1s.

Step 3 **Write the system of linear equations corresponding to the matrix in step 2, and use back-substitution to find the system's solution.** The system represented by the matrix in step 2 is

$$\begin{bmatrix} 1 & 1 & 2 & | & 19 \\ 0 & 1 & 2 & | & 13 \\ 0 & 0 & 1 & | & 5 \end{bmatrix} \rightarrow \begin{array}{l} 1x + 1y + 2z = 19 \\ 0x + 1y + 2z = 13 \\ 0x + 0y + 1z = 5 \end{array} \quad \text{or} \quad \begin{array}{r} x + y + 2z = 19 \\ y + 2z = 13. \\ z = 5 \end{array}$$

We immediately see that the value for z is 5. To find y, we back-substitute 5 for z in the second equation.

$$\begin{array}{ll} y + 2z = 13 & \text{Equation 2} \\ y + 2(5) = 13 & \text{Substitute 5 for z.} \\ y = 3 & \text{Solve for y.} \end{array}$$

Finally, back-substitute 3 for y and 5 for z in the first equation.

$$\begin{array}{ll} x + y + 2z = 19 & \text{Equation 1} \\ x + 3 + 2(5) = 19 & \text{Substitute 3 for y and 5 for z.} \\ x + 13 = 19 & \text{Multiply and add.} \\ x = 6 & \text{Subtract 13 both sides.} \end{array}$$

The solution for the original system is $(6, 3, 5)$ and the solution set is $\{(6, 3, 5)\}$. Check to see that the solution satisfies all three equations in the given system. ∎

✔ **CHECK POINT 4** Use matrices to solve the system

$$\begin{array}{rr} 2x + y + 2z = & 18 \\ x - y + 2z = & 9 \\ x + 2y - z = & 6. \end{array}$$

Modern supercomputers are capable of solving systems with more than 600,000 variables. The augmented matrices for such systems are huge, but the solution using matrices is exactly like what we did in Example 4. Work with the augmented matrix, one column at a time. First, get 1 in the desired position. Then get 0s below the 1.

Inconsistent Systems and Systems with Dependent Equations

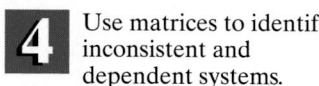

Use matrices to identify inconsistent and dependent systems.

When solving a system using matrices, you might obtain a matrix with a row in which the numbers to the left of the vertical bar are all zeros but a nonzero number number appears on the right. In such a case, the system is inconsistent and has no solution. For example, a system of equations that yields the following matrix is an inconsistent system:

$$\begin{bmatrix} 1 & -2 & | & 3 \\ 0 & 0 & | & -4 \end{bmatrix}.$$

The second row of the matrix represents the equation $0x + 0y = -4$, which is false for all values of x and y.

If you obtain a matrix in which a 0 appears across an entire row, the system contains dependent equations and has infinitely many solutions. This row of zeros represents $0x + 0y = 0$ or $0x + 0y + 0z = 0$. These equations are satisfied by infinitely many ordered pairs or triples.

EXERCISE SET 8.4

Practice Exercises

In Exercises 1–8, write the system of linear equations represented by the augmented matrix. Use x, y, and, if necessary, z, for the variables. Once the system is written, use back-substitution to find its solution.

1. $\begin{bmatrix} 1 & -3 & | & 11 \\ 0 & 1 & | & -3 \end{bmatrix}$

2. $\begin{bmatrix} 1 & 3 & | & 5 \\ 0 & 1 & | & 2 \end{bmatrix}$

3. $\begin{bmatrix} 1 & -3 & | & 1 \\ 0 & 1 & | & -1 \end{bmatrix}$

4. $\begin{bmatrix} 1 & 2 & | & 13 \\ 0 & 1 & | & 4 \end{bmatrix}$

5. $\begin{bmatrix} 1 & 0 & -4 & | & 5 \\ 0 & 1 & -12 & | & 13 \\ 0 & 0 & 1 & | & -\frac{1}{2} \end{bmatrix}$

6. $\begin{bmatrix} 1 & 2 & 1 & | & 0 \\ 0 & 1 & 0 & | & -2 \\ 0 & 0 & 1 & | & 3 \end{bmatrix}$

7. $\begin{bmatrix} 1 & \frac{1}{2} & 1 & | & \frac{11}{2} \\ 0 & 1 & \frac{3}{2} & | & 7 \\ 0 & 0 & 1 & | & 4 \end{bmatrix}$

8. $\begin{bmatrix} 1 & 1 & 0 & | & 3 \\ 0 & 1 & \frac{3}{2} & | & -2 \\ 0 & 0 & 1 & | & 0 \end{bmatrix}$

In Exercises 9–22, perform each matrix row operation and write the new matrix.

9. $\begin{bmatrix} 2 & 2 & | & 5 \\ 1 & -\frac{3}{2} & | & 5 \end{bmatrix} R_1 \leftrightarrow R_2$

10. $\begin{bmatrix} -6 & 9 & | & 4 \\ 1 & -\frac{3}{2} & | & 4 \end{bmatrix} R_1 \leftrightarrow R_2$

11. $\begin{bmatrix} -6 & 8 & | & -12 \\ 3 & 5 & | & -2 \end{bmatrix} -\frac{1}{6} R_1$

12. $\begin{bmatrix} -2 & 3 & | & -10 \\ 4 & 2 & | & 5 \end{bmatrix} -\frac{1}{2} R_1$

13. $\begin{bmatrix} 1 & -3 & | & 5 \\ 2 & 6 & | & 4 \end{bmatrix} -2R_1 + R_2$

14. $\begin{bmatrix} 1 & -3 & | & 1 \\ 2 & 1 & | & -5 \end{bmatrix} -2R_1 + R_2$

15. $\begin{bmatrix} 1 & -\frac{3}{2} & | & \frac{7}{2} \\ 3 & 4 & | & 2 \end{bmatrix} -3R_1 + R_2$

16. $\begin{bmatrix} 1 & -\frac{2}{5} & | & \frac{3}{4} \\ 4 & 2 & | & -1 \end{bmatrix} -4R_1 + R_2$

17. $\begin{bmatrix} 2 & -6 & 4 & | & 10 \\ 1 & 5 & -5 & | & 0 \\ 3 & 0 & 4 & | & 7 \end{bmatrix} \frac{1}{2}R_1$

18. $\begin{bmatrix} 3 & -12 & 6 & | & 9 \\ 1 & -4 & 4 & | & 0 \\ 2 & 0 & 7 & | & 4 \end{bmatrix} \frac{1}{3}R_1$

19. $\begin{bmatrix} 1 & -3 & 2 & | & 0 \\ 3 & 1 & -1 & | & 7 \\ 2 & -2 & 1 & | & 3 \end{bmatrix} -3R_1 + R_2$

20. $\begin{bmatrix} 1 & -1 & 5 & | & -6 \\ 3 & 3 & -1 & | & 10 \\ 1 & 3 & 2 & | & 5 \end{bmatrix} -3R_1 + R_2$

21. $\begin{bmatrix} 1 & 1 & -1 & | & 6 \\ 2 & -1 & 1 & | & -3 \\ 3 & -1 & -1 & | & 4 \end{bmatrix} \begin{matrix} -2R_1 + R_2 \\ \text{and} \\ -3R_1 + R_3 \end{matrix}$

22. $\begin{bmatrix} 1 & 2 & 1 & | & 2 \\ -2 & -1 & 2 & | & 5 \\ 1 & 3 & -2 & | & -8 \end{bmatrix} \begin{matrix} 2R_1 + R_2 \\ \text{and} \\ -1R_1 + R_3 \end{matrix}$

In Exercises 23–46, solve each system using matrices. If there is no solution or if there are infinitely many solutions and a system's equations are dependent, so state.

23. $\begin{aligned} x + y &= 6 \\ x - y &= 2 \end{aligned}$

24. $\begin{aligned} x + 2y &= 11 \\ x - y &= -1 \end{aligned}$

25. $\begin{aligned} 2x + y &= 3 \\ x - 3y &= 12 \end{aligned}$

26. $\begin{aligned} 3x - 5y &= 7 \\ x - y &= 1 \end{aligned}$

27. $\begin{aligned} 5x + 7y &= -25 \\ 11x + 6y &= -8 \end{aligned}$

28. $\begin{aligned} 3x - 5y &= 22 \\ 4x - 2y &= 20 \end{aligned}$

29. $\begin{aligned} 4x - 2y &= 5 \\ -2x + y &= 6 \end{aligned}$

30. $\begin{aligned} -3x + 4y &= 12 \\ 6x - 8y &= 16 \end{aligned}$

31. $\begin{aligned} x - 2y &= 1 \\ -2x + 4y &= -2 \end{aligned}$

32. $\begin{aligned} 3x - 6y &= 1 \\ 2x - 4y &= \frac{2}{3} \end{aligned}$

33. $\begin{aligned} x + y - z &= -2 \\ 2x - y + z &= 5 \\ -x + 2y + 2z &= 1 \end{aligned}$

34. $\begin{aligned} x - 2y - z &= 2 \\ 2x - y + z &= 4 \\ -x + y - 2z &= -4 \end{aligned}$

35. $\begin{aligned} x + 3y &= 0 \\ x + y + z &= 1 \\ 3x - y - z &= 11 \end{aligned}$

36. $\begin{aligned} 3y - z &= -1 \\ x + 5y - z &= -4 \\ -3x + 6y + 2z &= 11 \end{aligned}$

37. $\begin{aligned} 2x + 2y + 7z &= -1 \\ 2x + y + 2z &= 2 \\ 4x + 6y + z &= 15 \end{aligned}$

38. $\begin{aligned} 3x + 2y + 3z &= 3 \\ 4x - 5y + 7z &= 1 \\ 2x + 3y - 2z &= 6 \end{aligned}$

39. $\begin{aligned} x + y + z &= 6 \\ x - z &= -2 \\ y + 3z &= 11 \end{aligned}$

40. $\begin{aligned} x + y + z &= 3 \\ -y + 2z &= 1 \\ -x + z &= 0 \end{aligned}$

41. $\begin{aligned} x - y + 3z &= 4 \\ 2x - 2y + 6z &= 7 \\ 3x - y + 5z &= 14 \end{aligned}$

42. $\begin{aligned} 3x - y + 2z &= 4 \\ -6x + 2y - 4z &= 1 \\ 5x - 3y + 8z &= 0 \end{aligned}$

43.
$$x - 2y + z = 4$$
$$5x - 10y + 5z = 20$$
$$-2x + 4y - 2z = -8$$

44.
$$x - 3y + z = 2$$
$$4x - 12y + 4z = 8$$
$$-2x + 6y - 2z = -4$$

45.
$$x + y = 1$$
$$y + 2z = -2$$
$$2x - z = 0$$

46.
$$x + 3y = 3$$
$$y + 2z = -8$$
$$x - z = 7$$

Application Exercises

47. The table shows the number of inmates in federal and state prisons in the United States for three selected years.

x (Number of Years after 1980)	1	5	10
y (Number of Inmates, in thousands)	344	480	740

  a. Use the quadratic function $y = ax^2 + bx + c$ to model the data. Solve the system of linear equations involving $a, b,$ and c using matrices.

  b. Predict the number of inmates in the year 2010.

  c. List one factor that would change the accuracy of this model for the year 2010.

48. A football is kicked straight upward. The graph shows the ball's height, y, in feet, after x seconds.

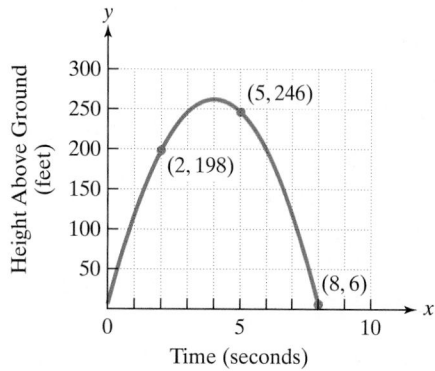

  a. Find the quadratic function

$$y = ax^2 + bx + c$$

  whose graph passes through the three points labeled on the graph. Solve the system of linear equations involving $a, b,$ and c using matrices.

b. Use the function in part (a) to find the value for y when $x = 7$. Describe what this means.

Write a system of linear equations in three variables to solve Exercises 49–50. Then use matrices to solve the system.

49. The circle graph indicates the ages of the 40 million online users in the United States. The percentage of online users in the youngest (under 30) and oldest (50 and over) age groups combined exceeds the percentage in the 30–49 age group by 2%. If the percentage of users in the oldest age group is doubled, it is 3% less than the percentage of users in the youngest age group. Find the percentage of online users in each of the three age groups.

Age of U.S. Online Users

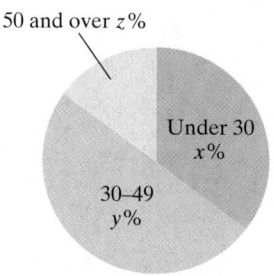

Source: U.S. Census Bureau

50. From *shlemiel* to *shlemazal*, evocative Yiddish words have entered our vocabulary. Yiddish, Thai, and Persian are spoken by 621 thousand people in the United States. The difference between the number of people who speak Yiddish and the number who speak Thai is 7 thousand. The difference between the number of people who speak Thai and the number who speak Persian is 4 thousand. Find the thousands of people in the United States who speak Yiddish, Thai, and Persian.

Writing in Mathematics

51. What is a matrix?

52. Describe what is meant by the augmented matrix of a system of linear equations.

53. In your own words, describe each of the three matrix row operations. Give an example of each of the operations.

54. Describe how to use matrices and row operations to solve a system of linear equations.

55. When solving a system using matrices, how do you know if the system has no solution?

56. When solving a system using matrices, how do you know if the system has infinitely many solutions?

Critical Thinking Exercises

57. Which one of the following is true?
  a. A matrix row operation such as $-\frac{4}{5}R_1 + R_2$ is not permitted because of the negative fraction.
  b. The augmented matrix for the system

$$\begin{array}{rcl} x - 3y & = & 5 \\ y - 2z & = & 7 \\ 2x + z & = & 4 \end{array} \quad \text{is} \quad \left[\begin{array}{ccc|c} 1 & -3 & & 5 \\ 1 & -2 & & 7 \\ 2 & & 1 & 4 \end{array}\right].$$

  c. In solving a linear system of three equations in three variables, we begin with the augmented matrix and use row operations to obtain a row-equivalent matrix with 0s down the diagonal from left to right and 1s below each 0.
  d. None of the above is true.

58. Matrices can be used to solve systems involving more than three equations. Write the augmented matrix for the following system. Then use matrix row operations to obtain a matrix with 1s down the diagonal from upper left to lower right, and 0s below the 1s. Use this matrix and back-substitution to solve the system.

$$\begin{array}{rcl} 2x + y + 3z - w & = & 6 \\ x - y + 2z - 2w & = & -1 \\ x - y - z + w & = & -4 \\ -x + 2y - 2z - w & = & -7 \end{array}$$

59. The vitamin content per ounce for three foods is given in the following table.

	Milligrams per Ounce		
	Thiamin	Riboflavin	Niacin
Food A	3	7	1
Food B	1	5	3
Food C	3	8	2

  a. Use matrices to show that no combination of these foods can provide exactly 14 milligrams of thiamin, 32 milligrams of riboflavin, and 9 milligrams of niacin.

  b. Use matrices to describe in practical terms what happens if the riboflavin requirement is increased by 5 mg and the other requirements stay the same.

Technology Exercises

60. Most graphing utilities can perform row operations on matrices. Consult the owner's manual for your graphing utility to learn proper keystroke sequences for performing these operations. Then duplicate the row operations of any three exercises that you solved from Exercises 11–20.

61. The final augmented matrix that we obtain when using Gaussian elimination is said to be in **row-echelon form**. For systems of linear equations with unique solutions, this form results when each entry in the main diagonal is 1 and all entries below the main diagonal are 0s. Some graphing utilities can transform a matrix to row-echelon form. Consult your owner's manual. If your utility has this capability, use it to verify of any five of the matrices you obtained in Exercises 23–46.

Review Exercises

62. If $f(x) = -3x + 10$, find $f(2a - 1)$. (Section 8.1, Example 3)

63. If $f(x) = 3x$ and $g(x) = 2x - 3$, find $(fg)(x)$ and $(fg)(-1)$. (Section 8.2, Example 4)

64. Factor completely:

$$2x^3 - 16x^2 + 30x.$$

(Section 6.5, Example 2)

▶ **SECTION 8.5** *Determinants and Cramer's Rule*

Objectives

1 Evaluate a second-order determinant.

2 Solve a system of linear equations in two variables using Cramer's rule.

3 Evaluate a third-order determinant.

4 Solve a system of linear equations in three variables using Cramer's rule.

5 Use determinants to identify inconsistent and dependent systems.

SSM
PH Tutor   CD-   Video   www
Center   ROM

A portion of Charles Babbage's unrealized Difference Engine

As cyberspace absorbs more and more of our work, play, shopping, and socializing, where will it all end? Which activities will still be offline in 2025?

Our technologically transformed lives can be traced back to the English inventor Charles Babbage (1792–1871). Babbage knew of a method for solving linear systems called *Cramer's rule*, in honor of the Swiss geometer Gabriel Cramer (1704–1752). Cramer's rule was simple, but involved numerous multiplications for large systems. Babbage designed a machine, called the "difference engine," that consisted of toothed wheels on shafts for performing these multiplications. Despite the fact that only one-seventh of the functions ever worked, Babbage's invention demonstrated how complex calculations could be handled mechanically. In 1944, scientists at IBM used the lessons of the difference engine to create the world's first computer.

Those who invented computers hoped to relegate the drudgery of repeated computation to a machine. In this section, we look at a method for solving linear systems that played a critical role in this process. The method uses arrays of numbers called *determinants*. As with matrix methods, solutions are obtained by writing down the coefficients and constants of a linear system and performing operations with them.

1 Evaluate a second-order determinant.

The Determinant of a 2 × 2 Matrix

A matrix of **order** $m \times n$ has m rows and n columns. If $m = n$, a matrix has the same number of rows as columns and is called a **square matrix**. Associated with every square matrix is a real number, called its **determinant**. The determinant for a 2 × 2 square matrix is defined as follows:

Definition of the Determinant of a 2 × 2 Matrix

The determinant of the matrix $\begin{bmatrix} a_1 & b_1 \\ a_2 & b_2 \end{bmatrix}$ is denoted by $\begin{vmatrix} a_1 & b_1 \\ a_2 & b_2 \end{vmatrix}$ and is defined by

$$= \begin{vmatrix} a_1 & b_1 \\ a_2 & b_2 \end{vmatrix} = a_1 b_2 - a_2 b_1.$$

We also say that the **value** of the **second-order determinant** $\begin{vmatrix} a_1 & b_1 \\ a_2 & b_2 \end{vmatrix}$ is $a_1 b_2 - a_2 b_1$.

Study Tip

To evaluate a 2 × 2 determinant, find the difference of the products of the two diagonals.

$$\begin{vmatrix} a_1 & b_1 \\ a_2 & b_2 \end{vmatrix} = a_1 b_2 - a_2 b_1$$

Example 1 illustrates that the determinant of a matrix may be positive or negative. The determinant can also have 0 as its value.

EXAMPLE 1 Evaluating the Determinant of a 2×2 Matrix

Evaluate the determinant of:

a. $\begin{bmatrix} 5 & 6 \\ 7 & 3 \end{bmatrix}$
b. $\begin{bmatrix} 2 & 4 \\ -3 & -5 \end{bmatrix}$.

Solution We multiply and subtract as indicated.

a. $\begin{vmatrix} 5 & 6 \\ 7 & 3 \end{vmatrix} = 5 \cdot 3 - 7 \cdot 6 = 15 - 42 = -27$ *The value of the second-order determinant is −27.*

b. $\begin{vmatrix} 2 & 4 \\ -3 & -5 \end{vmatrix} = 2(-5) - (-3)(4) = -10 + 12 = 2$ *The value of the second-order determinant is 2.* ∎

✔ **CHECK POINT 1** Evaluate the determinant of:

a. $\begin{bmatrix} 10 & 9 \\ 6 & 5 \end{bmatrix}$

b. $\begin{bmatrix} 4 & 3 \\ -5 & -8 \end{bmatrix}$.

2 Solve a system of linear equations in two variables using Cramer's rule.

Solving Linear Systems of Equations in Two Variables Using Determinants

Determinants can be used to solve a linear system in two variables. In general, such a system appears as

$$a_1 x + b_1 y = c_1$$
$$a_2 x + b_2 y = c_2.$$

Let's first solve this system for x using the addition method. We can solve for x by eliminating y from the equations. Multiply the first equation by b_2 and the second equation by $-b_1$. Then add the two equations.

$$a_1 x + b_1 y = c_1 \xrightarrow{\text{Multiply by } b_2.} a_1 b_2 x + b_1 b_2 y = c_1 b_2$$
$$a_2 x + b_2 y = c_2 \xrightarrow{\text{Multiply by } -b_1.} -a_2 b_1 x - b_1 b_2 y = -c_2 b_1$$

$$\text{Add:} \quad (a_1 b_2 - a_2 b_1)x = c_1 b_2 - c_2 b_1$$

$$x = \frac{c_1 b_2 - c_2 b_1}{a_1 b_2 - a_2 b_1}$$

Because

$$\begin{vmatrix} c_1 & b_1 \\ c_2 & b_2 \end{vmatrix} = c_1 b_2 - c_2 b_1 \quad \text{and} \quad \begin{vmatrix} a_1 & b_1 \\ a_2 & b_2 \end{vmatrix} = a_1 b_2 - a_2 b_1$$

we can express our answer for x as the quotient of two determinants:

$$x = \frac{\begin{vmatrix} c_1 & b_1 \\ c_2 & b_2 \end{vmatrix}}{\begin{vmatrix} a_1 & b_1 \\ a_2 & b_2 \end{vmatrix}}.$$

In a similar way, we could use the addition method to solve our system for y, again expressing y as the quotient of two determinants. This method of using determinants to solve the linear system, called **Cramer's rule**, is summarized in the box.

Solving a Linear System in Two Variables Using Determinants

Cramer's Rule

If

$$a_1 x + b_1 y = c_1$$
$$a_2 x + b_2 y = c_2$$

then

$$x = \frac{\begin{vmatrix} c_1 & b_1 \\ c_2 & b_2 \end{vmatrix}}{\begin{vmatrix} a_1 & b_1 \\ a_2 & b_2 \end{vmatrix}} \quad \text{and} \quad y = \frac{\begin{vmatrix} a_1 & c_1 \\ a_2 & c_2 \end{vmatrix}}{\begin{vmatrix} a_1 & b_1 \\ a_2 & b_2 \end{vmatrix}}$$

where

$$\begin{vmatrix} a_1 & b_1 \\ a_2 & b_2 \end{vmatrix} \neq 0.$$

Here are some helpful tips when solving

$$a_1 x + b_1 y = c_1$$
$$a_2 x + b_2 y = c_2$$

using determinants:

1. Three different determinants are used to find x and y. The determinants in the denominators for x and y are identical. The determinants in the numerators for x and y differ. In abbreviated notation, we write

$$x = \frac{D_x}{D} \quad \text{and} \quad y = \frac{D_y}{D} \text{ where } D \neq 0.$$

2. The elements of D, the determinant in the denominator, are the coefficients of the variables in the system.

$$D = \begin{vmatrix} a_1 & b_1 \\ a_2 & b_2 \end{vmatrix}$$

3. D_x, the determinant in the numerator of x, is obtained by replacing the x-coefficients in D, a_1 and a_2, with the constants on the right side of the equations, c_1 and c_2.

$$D = \begin{vmatrix} a_1 & b_1 \\ a_2 & b_2 \end{vmatrix} \quad \text{and} \quad D_x = \begin{vmatrix} c_1 & b_1 \\ c_2 & b_2 \end{vmatrix} \qquad \text{Replace the column with } a_1 \text{ and } a_2 \text{ with the constants } c_1 \text{ and } c_2 \text{ to get } D_x.$$

4. D_y, the determinant in the numerator for y, is obtained by replacing the y-coefficients in D, b_1 and b_2, with the constants on the right side of the equations, c_1 and c_2.

$$D = \begin{vmatrix} a_1 & b_1 \\ a_2 & b_2 \end{vmatrix} \quad \text{and} \quad D_y = \begin{vmatrix} a_1 & c_1 \\ a_2 & c_2 \end{vmatrix} \qquad \text{Replace the column with } b_1 \text{ and } b_2 \text{ with the constants } c_1 \text{ and } c_2 \text{ to get } D_y.$$

Example 2 illustrates the use of Cramer's rule.

EXAMPLE 2   Using Cramer's Rule to Solve a Linear System

Use Cramer's rule to solve the system:

$$5x - 4y = 2$$
$$6x - 5y = 1.$$

Solution   Because

$$x = \frac{D_x}{D} \quad \text{and} \quad y = \frac{D_y}{D},$$

we will set up and evaluate the three determinants $D, D_x,$ and D_y.

1. D, the determinant in both denominators, consists of the x- and y-coefficients.

$$D = \begin{vmatrix} 5 & -4 \\ 6 & -5 \end{vmatrix} = (5)(-5) - (6)(-4) = -25 + 24 = -1$$

Because this determinant is not zero, we continue to use Cramer's rule to solve the system.

2. D_x, the determinant in the numerator for x, is obtained by replacing the x-coefficients in D, 5 and 6, by the constants on the right side of the equation, 2 and 1.

$$D_x = \begin{vmatrix} 2 & -4 \\ 1 & -5 \end{vmatrix} = (2)(-5) - (1)(-4) = -10 + 4 = -6$$

3. D_y, the determinant in the numerator for y, is obtained by replacing the y-coefficients in D, -4 and -5, by the constants on the right side of the equation, 2 and 1.

$$D_y = \begin{vmatrix} 5 & 2 \\ 6 & 1 \end{vmatrix} = (5)(1) - (6)(2) = 5 - 12 = -7$$

4. Thus,

$$x = \frac{D_x}{D} = \frac{-6}{-1} = 6 \quad \text{and} \quad y = \frac{D_y}{D} = \frac{-7}{-1} = 7.$$

As always, the ordered pair $(6, 7)$ can be checked by substituting these values into the original equations. The solution is $(6, 7)$ and the solution set is $\{(6, 7)\}$.   ∎

✔ **CHECK POINT 2**   Use Cramer's rule to solve the system:

$$5x + 4y = 12$$
$$3x - 6y = 24.$$

3 Evaluate a third-order determinant.

The Determinant of a 3 × 3 Matrix

Associated with every square matrix is a real number called its determinant. The determinant for a 3 × 3 matrix is defined in terms of second-order determinants.

Definition of the Determinant of a 3 × 3 Matrix

A third-order determinant is defined by

Subtract.   Add.

$$\begin{vmatrix} a_1 & b_1 & c_1 \\ a_2 & b_2 & c_2 \\ a_3 & b_3 & c_3 \end{vmatrix} = a_1 \begin{vmatrix} b_2 & c_2 \\ b_3 & c_3 \end{vmatrix} - a_2 \begin{vmatrix} b_1 & c_1 \\ b_3 & c_3 \end{vmatrix} + a_3 \begin{vmatrix} b_1 & c_1 \\ b_2 & c_2 \end{vmatrix}.$$

The a's on the right come
from the first column.

Here are some tips that may be helpful when evaluating the determinant of a 3 × 3 matrix:

Evaluating the Determinant of a 3 × 3 Matrix

1. Each of the three terms in the definition contains two factors—a numerical factor and a second-order determinant.
2. The numerical factor in each term is an element from the first column of the third-order determinant.
3. The minus sign precedes the second term.
4. The second-order determinant that appears in each term is obtained by crossing out the row and the column containing the numerical factor.

$$a_1 \begin{vmatrix} b_2 & c_2 \\ b_3 & c_3 \end{vmatrix} - a_2 \begin{vmatrix} b_1 & c_1 \\ b_3 & c_3 \end{vmatrix} + a_3 \begin{vmatrix} b_1 & c_1 \\ b_2 & c_2 \end{vmatrix}$$

$$\begin{vmatrix} a_1 & b_1 & c_1 \\ a_2 & b_2 & c_2 \\ a_3 & b_3 & c_3 \end{vmatrix} \quad \begin{vmatrix} a_1 & b_1 & c_1 \\ a_2 & b_2 & c_2 \\ a_3 & b_3 & c_3 \end{vmatrix} \quad \begin{vmatrix} a_1 & b_1 & c_1 \\ a_2 & b_2 & c_2 \\ a_3 & b_3 & c_3 \end{vmatrix}$$

The **minor** of an element is the determinant that remains after deleting the row and column of that element. For this reason, we call this method **expansion by minors**.

EXAMPLE 3   Evaluating the Determinant of a 3 × 3 Matrix

Evaluate the determinant of:

$$\begin{bmatrix} 4 & 1 & 0 \\ -9 & 3 & 4 \\ -3 & 8 & 1 \end{bmatrix}.$$

Solution   We know that each of the three terms in the determinant contains a numerical factor and a second-order determinant. The numerical factors are from the first column of the determinant of the given matrix. They are shown in red in the following matrix:

$$\begin{bmatrix} 4 & 1 & 0 \\ -9 & 3 & 4 \\ -3 & 8 & 1 \end{bmatrix}.$$

We find the minor for each numerical factor by deleting the row and column of that element:

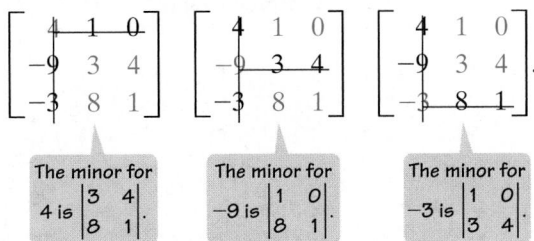

| The minor for 4 is $\begin{vmatrix} 3 & 4 \\ 8 & 1 \end{vmatrix}$. | The minor for -9 is $\begin{vmatrix} 1 & 0 \\ 8 & 1 \end{vmatrix}$. | The minor for -3 is $\begin{vmatrix} 1 & 0 \\ 3 & 4 \end{vmatrix}$. |

Now we have three numerical factors, 4, −9, and −3, and three second-order determinants. We multiply each numerical factor by its second-order determinant to find the three terms of the third-order determinant:

$$4\begin{vmatrix} 3 & 4 \\ 8 & 1 \end{vmatrix}, \quad -9\begin{vmatrix} 1 & 0 \\ 8 & 1 \end{vmatrix}, \quad -3\begin{vmatrix} 1 & 0 \\ 3 & 4 \end{vmatrix}.$$

Based on the preceding definition, we subtract the second term from the first term and add the third term:

Using Technology

Verify the result of Example 3 by using your graphing utility. First enter the given matrix:

$$A = \begin{bmatrix} 4 & 1 & 0 \\ -9 & 3 & 4 \\ -3 & 8 & 1 \end{bmatrix}.$$

Then enter

det [A] ENTER .

The result should be −119.

$$\begin{vmatrix} 4 & 1 & 0 \\ -9 & 3 & 4 \\ -3 & 8 & 1 \end{vmatrix} = 4\begin{vmatrix} 3 & 4 \\ 8 & 1 \end{vmatrix} - (-9)\begin{vmatrix} 1 & 0 \\ 8 & 1 \end{vmatrix} - 3\begin{vmatrix} 1 & 0 \\ 3 & 4 \end{vmatrix}$$

Don't forget to supply the minus sign.

$$= 4(3 \cdot 1 - 8 \cdot 4) + 9(1 \cdot 1 - 8 \cdot 0) - 3(1 \cdot 4 - 3 \cdot 0)$$

$$= 4(3 - 32) + 9(1 - 0) - 3(4 - 0) \qquad \text{Evaluate the three second-order determinants.}$$

$$= 4(-29) + 9(1) - 3(4) \qquad \text{Subtract within parentheses.}$$

$$= -116 + 9 - 12 \qquad \text{Multiply.}$$

$$= -119 \qquad \text{Add and subtract as indicated.} \quad \blacksquare$$

✔ **CHECK POINT 3** Evaluate the determinant of:

$$\begin{bmatrix} 2 & 1 & 7 \\ -5 & 6 & 0 \\ -4 & 3 & 1 \end{bmatrix}.$$

4 Solve a system of linear equations in three variables using Cramer's rule.

Solving Systems of Linear Equations in Three Variables Using Determinants

Cramer's rule can be applied to solving systems of linear equations in three variables. The determinants in the numerator and denominator of all variables are third-order determinants.

Solving Three Equations in Three Variables Using Determinants

Cramer's Rule

If

$$a_1 x + b_1 y + c_1 z = d_1$$
$$a_2 x + b_2 y + c_2 z = d_2$$
$$a_3 x + b_3 y + c_3 z = d_3$$

then

$$x = \frac{D_x}{D}, \ y = \frac{D_y}{D}, \text{ and } z = \frac{D_z}{D}.$$

These four third-order determinants are given by:

$$D = \begin{vmatrix} a_1 & b_1 & c_1 \\ a_2 & b_2 & c_2 \\ a_3 & b_3 & c_3 \end{vmatrix}$$ These are the coefficients of the variables x, y, and z, $D \neq 0$.

$$D_x = \begin{vmatrix} d_1 & b_1 & c_1 \\ d_2 & b_2 & c_2 \\ d_3 & b_3 & c_3 \end{vmatrix}$$ Replace x-coefficients in D with the constants at the right of the three equations.

$$D_y = \begin{vmatrix} a_1 & d_1 & c_1 \\ a_2 & d_2 & c_2 \\ a_3 & d_3 & c_3 \end{vmatrix}$$ Replace y-coefficients in D with the constants at the right of the three equations.

$$D_z = \begin{vmatrix} a_1 & b_1 & d_1 \\ a_2 & b_2 & d_2 \\ a_3 & b_3 & d_3 \end{vmatrix}.$$ Replace z-coefficients in D with the constants at the right of the three equations.

ENRICHMENT ESSAY

Why Software Programs for Solving Linear Systems Do Not Use Cramer's Rule

Supercomputers can perform one trillion (10^{12}) multiplications per second. To solve a linear system with a "mere" 20 equations using Cramer's rule requires over 5.1×10^{19} multiplications. This would take a supercomputer more than 590 days. The cost for this venture? At $2200 per hour, typical of the costs for supercomputer time, the computations required by Cramer's rule would cost more than $30 million!

EXAMPLE 4 Using Cramer's Rule to Solve a Linear System in Three Variables

Use Cramer's rule to solve:

$$x + 2y - \ z = -4$$
$$x + 4y - 2z = -6$$
$$2x + 3y + \ z = 3.$$

Solution Because

$$x = \frac{D_x}{D}, y = \frac{D_y}{D}, \text{ and } z = \frac{D_z}{D},$$

we need to set up and evaluate four determinants.

Step 1 Set up the determinants.

1. D, the determinant in all three denominators, consists of the x-, y-, and z-coefficients.

$$D = \begin{vmatrix} 1 & 2 & -1 \\ 1 & 4 & -2 \\ 2 & 3 & 1 \end{vmatrix}$$

2. D_x, the determinant in the numerator for x, is obtained by replacing the x-coefficients in D, 1, 1, and 2, with the constants on the right side of the equation, -4, -6, and 3.

$$D_x = \begin{vmatrix} -4 & 2 & -1 \\ -6 & 4 & -2 \\ 3 & 3 & 1 \end{vmatrix}$$

3. D_y, the determinant in the numerator for y, is obtained by replacing the y-coefficients in D, 2, 4, and 3, with the constants on the right side of the equation, -4, -6, and 3.

$$D_y = \begin{vmatrix} 1 & -4 & -1 \\ 1 & -6 & -2 \\ 2 & 3 & 1 \end{vmatrix}$$

4. D_z, the determinant in the numerator for z, is obtained by replacing the z-coefficients in D, -1, -2, and 1, with the constants on the right side of the equation, -4, -6, and 3.

$$D_z = \begin{vmatrix} 1 & 2 & -4 \\ 1 & 4 & -6 \\ 2 & 3 & 3 \end{vmatrix}$$

Step 2 Evaluate the four determinants.

$$D = \begin{vmatrix} 1 & 2 & -1 \\ 1 & 4 & -2 \\ 2 & 3 & 1 \end{vmatrix} = 1\begin{vmatrix} 4 & -2 \\ 3 & 1 \end{vmatrix} - 1\begin{vmatrix} 2 & -1 \\ 3 & 1 \end{vmatrix} + 2\begin{vmatrix} 2 & -1 \\ 4 & -2 \end{vmatrix}$$

$$= 1(4 + 6) - 1(2 + 3) + 2(-4 + 4)$$

$$= 1(10) - 1(5) + 2(0) = 5$$

Using the same technique to evaluate each determinant, we obtain
$$D_x = -10, \quad D_y = 5, \quad \text{and} \quad D_z = 20.$$

Step 3 Substitute these four values and solve the system.

$$x = \frac{D_x}{D} = \frac{-10}{5} = -2$$

$$y = \frac{D_y}{D} = \frac{5}{5} = 1$$

$$z = \frac{D_z}{D} = \frac{20}{5} = 4$$

The ordered triple $(-2, 1, 4)$ can be checked by substitution into the original three equations. The solution is $(-2, 1, 4)$ and the solution set is $\{(-2, 1, 4)\}$.

■

✔ **CHECK POINT 4** Use Cramer's rule to solve the system:

$$\begin{aligned} 3x - 2y + z &= 16 \\ 2x + 3y - z &= -9. \\ x + 4y + 3z &= 2 \end{aligned}$$

5 Use determinants to identify inconsistent and dependent systems.

Cramer's Rule with Inconsistent and Dependent Systems

If D, the determinant in the denominator, is 0, the variables described by the quotient of determinants are not real numbers. However, when $D = 0$, this indicates that the system is inconsistent or contains dependent equations. This gives rise to the following two situations:

Determinants: Inconsistent and Dependent Systems

1. If $D = 0$ and at least one of the determinants in the numerator is not 0, then the system is inconsistent. The solution set is $\varnothing$.
2. If $D = 0$ and all the determinants in the numerators are 0, then the equations in the system are dependent. The system has infinitely many solutions.

Although we have focused on applying determinants to solve linear systems, they have other applications, some of which we consider in the exercise set that follows.

EXERCISE SET 8.5

Practice Exercises

Evaluate each determinant in Exercises 1–10.

1. $\begin{vmatrix} 5 & 7 \\ 2 & 3 \end{vmatrix}$

2. $\begin{vmatrix} 4 & 8 \\ 5 & 6 \end{vmatrix}$

3. $\begin{vmatrix} -4 & 1 \\ 5 & 6 \end{vmatrix}$

4. $\begin{vmatrix} 7 & 9 \\ -2 & -5 \end{vmatrix}$

5. $\begin{vmatrix} -7 & 14 \\ 2 & -4 \end{vmatrix}$

6. $\begin{vmatrix} 1 & -3 \\ -8 & 2 \end{vmatrix}$

7. $\begin{vmatrix} -5 & -1 \\ -2 & -7 \end{vmatrix}$

8. $\begin{vmatrix} \frac{1}{5} & \frac{1}{6} \\ -6 & 5 \end{vmatrix}$

9. $\begin{vmatrix} \frac{1}{2} & \frac{1}{2} \\ \frac{1}{8} & -\frac{3}{4} \end{vmatrix}$

10. $\begin{vmatrix} \frac{2}{3} & \frac{1}{3} \\ -\frac{1}{2} & \frac{3}{4} \end{vmatrix}$

For Exercises 11–26, use Cramer's rule to solve each system or to determine that the system is inconsistent or contains dependent equations.

11. $x + y = 7$
    $x - y = 3$
12. $2x + y = 3$
    $x - y = 3$
13. $12x + 3y = 15$
    $2x - 3y = 13$
14. $x - 2y = 5$
    $5x - y = -2$
15. $4x - 5y = 17$
    $2x + 3y = 3$
16. $3x + 2y = 2$
    $2x + 2y = 3$
17. $x - 3y = 4$
    $3x - 4y = 12$
18. $2x - 9y = 5$
    $3x - 3y = 11$
19. $3x - 4y = 4$
    $2x + 2y = 12$

20. $3x = 7y + 1$
    $2x = 3y - 1$
21. $2x = 3y + 2$
    $5x = 51 - 4y$
22. $y = -4x + 2$
    $2x = 3y + 8$
23. $3x = 2 - 3y$
    $2y = 3 - 2x$
24. $x + 2y - 3 = 0$
    $12 = 8y + 4x$
25. $4y = 16 - 3x$
    $6x = 32 - 8y$
26. $2x = 7 + 3y$
    $4x - 6y = 3$

Evaluate each determinant in Exercises 27–32.

27. $\begin{vmatrix} 3 & 0 & 0 \\ 2 & 1 & -5 \\ 2 & 5 & -1 \end{vmatrix}$

28. $\begin{vmatrix} 4 & 0 & 0 \\ 3 & -1 & 4 \\ 2 & -3 & 5 \end{vmatrix}$

29. $\begin{vmatrix} 3 & 1 & 0 \\ -3 & 4 & 0 \\ -1 & 3 & -5 \end{vmatrix}$

30. $\begin{vmatrix} 2 & -4 & 2 \\ -1 & 0 & 5 \\ 3 & 0 & 4 \end{vmatrix}$

31. $\begin{vmatrix} 1 & 1 & 1 \\ 2 & 2 & 2 \\ -3 & 4 & -5 \end{vmatrix}$

32. $\begin{vmatrix} 1 & 2 & 3 \\ 2 & 2 & -3 \\ 3 & 2 & 1 \end{vmatrix}$

In Exercises 33–40, use Cramer's rule to solve each system.

33. $x + y + z = 0$
    $2x - y + z = -1$
    $-x + 3y - z = -8$

34. $x - y + 2z = 3$
    $2x + 3y + z = 9$
    $-x - y + 3z = 11$

35. $4x - 5y - 6z = -1$
    $x - 2y - 5z = -12$
    $2x - y = 7$

36. $x - 3y + z = -2$
    $x + 2y = 8$
    $2x - y = 1$

37. $x + y + z = 4$
$x - 2y + z = 7$
$x + 3y + 2z = 4$

38. $2x + 2y + 3z = 10$
$4x - y + z = -5$
$5x - 2y + 6z = 1$

39. $x + 2z = 4$
$2y - z = 5$
$2x + 3y = 13$

40. $3x + 2z = 4$
$5x - y = -4$
$4y + 3z = 22$

Application Exercises

Determinants are used to find the area of a triangle whose vertices are given by three points in a rectangular coordinate system. The area of a triangle with vertices $(x_1, y_1), (x_2, y_2),$ and (x_3, y_3) is:

$$\text{Area} = \pm \frac{1}{2} \begin{vmatrix} x_1 & y_1 & 1 \\ x_2 & y_2 & 1 \\ x_3 & y_3 & 1 \end{vmatrix}.$$

where the symbol $(\pm)$ indicates that the appropriate sign should be chosen to yield a positive area. Use this information to work Exercises 41–42.

41. Use determinants to find the area of the triangle whose vertices are $(3, -5), (2, 6),$ and $(-3, 5)$.

42. Use determinants to find the area of the triangle whose vertices are $(1, 1), (-2, -3),$ and $(11, -3)$.

Determinants are used to show that three points lie on the same line (are collinear). If

$$\begin{vmatrix} x_1 & y_1 & 1 \\ x_2 & y_2 & 1 \\ x_3 & y_3 & 1 \end{vmatrix} = 0$$

then the points $(x_1, y_1), (x_2, y_2),$ and (x_3, y_3) are collinear. If the determinant does not equal 0, then the points are not collinear. Use this information to work Exercises 43–44.

43. Are the points $(3, -1), (0, -3),$ and $(12, 5)$ collinear?

44. Are the points $(-4, -6), (1, 0),$ and $(11, 12)$ collinear?

Determinants are used to write an equation of a line passing through two points. An equation of the line passing through the distinct points (x_1, y_1) and (x_2, y_2) is given by

$$\begin{vmatrix} x & y & 1 \\ x_1 & y_1 & 1 \\ x_2 & y_2 & 1 \end{vmatrix} = 0.$$

Use this information to work Exercises 45–46.

45. Use the determinant to write an equation for the line passing through $(3, -5)$ and $(-2, 6)$. Then expand the determinant, expressing the line's equation in slope-intercept form.

46. Use the determinant to write an equation for the line passing through $(-1, 3)$ and $(2, 4)$. Then expand the determinant, expressing the line's equation in slope-intercept form.

Writing in Mathematics

47. Explain how to evaluate a second-order determinant.

48. Describe the determinants $D, D_x,$ and D_y in terms of the coefficients and constants in a system of two equations in two variables.

49. Explain how to evaluate a third-order determinant.

50. When expanding a determinant by minors, when is it necessary to supply a minus sign?

51. Without going into too much detail, describe how to solve a linear system in three variables using Cramer's rule.

52. In applying Cramer's rule, what does it mean if $D = 0$?

53. The process of solving a linear system in three variables using Cramer's rule can involve tedious computation. Is there a way of speeding up this process, perhaps using Cramer's rule to find the value for only one of the variables? Describe how this process might work, presenting a specific example with your description. Remember that your goal is still to find the value for each variable in the system.

54. If you could use only one method to solve linear systems in three variables, which method would you select? Explain why this is so.

Critical Thinking Exercises

55. Which one of the following is true?
   a. If $D = 0$, then every variable has a value of 0.
   b. Using Cramer's rule, we use $\dfrac{D}{D_y}$ to get the value of y.
   c. Because there are different determinants in the numerators of x and y, if a system is solved using Cramer's rule, x and y cannot have the same value.
   d. Only one 2×2 determinant is needed to evaluate

$$\begin{vmatrix} 2 & 3 & -2 \\ 0 & 1 & 3 \\ 0 & 4 & -1 \end{vmatrix}.$$

56. Solve for x: $\begin{vmatrix} 2x + 1 & 5 \\ x - 1 & -3 \end{vmatrix} = 3x - 1$.

57. What happens to the value of a second-order determinant if the two columns are interchanged?

58. Consider the system

$$a_1 x + b_1 y = c_1$$
$$a_2 x + b_2 y = c_2.$$

Use Cramer's rule to prove that if the first equation of the system is replaced by the sum of the two equations, the resulting system has the same solution as the original system.

59. Show that the equation of a line through (x_1, y_1) and (x_2, y_2) is given by the determinant

$$\begin{vmatrix} x & y & 1 \\ x_1 & y_1 & 1 \\ x_2 & y_2 & 1 \end{vmatrix} = 0.$$

Technology Exercises

60. Use the feature of your graphing utility that evaluates the determinant of a square matrix to verify any five of the determinants that you evaluated by hand in Exercises 1–10, or 27–32.

61. What is the fastest method for solving a linear system with your graphing utility?

Review Exercises

62. Solve: $2x - 1 = x^2 - 4x + 4.$ (Section 6.6, Example 4)

63. Simplify: $(2x^2)^{-3}$. (Section 5.7, Example 6)

64. Multiply: $\dfrac{x^2 - 6x + 9}{12} \cdot \dfrac{3}{x^2 - 9}$. (Section 7.2, Example 3)

CHAPTER 8 GROUP PROJECTS

1. Consult an almanac, newspaper, magazine, or the Internet to find data displayed in the style of Figure 8.9 on page 532 or in Exercises 53–56 in Exercise Set 8.2 on page 539. Using the two graphs that group members find most interesting, introduce two or more functions that are related to the graphs. Then write and solve a problem involving function addition and function subtraction for each selected graph.

2. Group members should consult an almanac, newspaper, magazine, or the Internet and return to the group with data that have a variable that is first decreasing and then increasing over time, or vice versa. The group should select the two most interesting data sets. For each set selected:
   a. Identify three data points and use the function $y = ax^2 + bx + c$ to model the data. Let x represent the number of years after the first year in the data set.
   b. Use the quadratic function to make a prediction about what might occur in the future.

3. The group should write three original word problems that can be solved using a system of linear equations in three variables. Each problem should be on a different topic. The group should turn in the three problems and their algebraic solutions.

4. Turn on your computer and read your e-mail or write a paper. When you need to do research, use the Internet to browse through art museums and photography exhibits. When you need a break, load a flight simulator program and fly through a photorealistic computer world. As different as these experiences may be, thay all share one thing —you're looking at images based on matrices. Matrices have applications in numerous fields, including the new technology of digital photography in which pictures are represented by numbers rather than film. Members of the group should research applications of matrices that they find intriguing. The group should then present a seminar to the class about these applications.

CHAPTER SUMMARY, REVIEW, AND TEST

SUMMARY

DEFINITIONS AND CONCEPTS	EXAMPLES

Section 8.1 Introduction to Functions

A relation is any set of ordered pairs. The set of first components of the ordered pairs is the domain and the set of second components is the range. A function is a relation in which each member of the domain corresponds to exactly one member of the range. No two ordered pairs of a function can have the same first components and different second components.

The domain of the relation $\{(1, 2), (3, 4), (3, 7)\}$ is $\{1, 3\}$. The range is $\{2, 4, 7\}$. The relation is not a function: 3, in the domain, corresponds to both 4 and 7 in the range.

If a function is defined as an equation, the notation $f(x)$, read "f of x" or "f at x," describes the value of the function at the number, or input, x.

If $f(x) = 7x - 5$, then
$$f(a + 2) = 7(a + 2) - 5$$
$$= 7a + 14 - 5$$
$$= 7a + 9.$$

The graph of a function is the graph of its ordered pairs.

The Vertical Line Test for Functions
If any vertical line intersects a graph in more than one point, the graph does not define y as a function of x.

At the left or right of a function's graph, you will often find closed dots, open dots, or arrows. A closed dot shows the graph ends and the point belongs to the graph. An open dot shows the graph ends and the point does not belong to the graph. An arrow indicates the graph extends indefinitely.

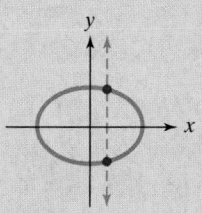

Not the graph of a function

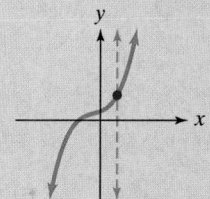

The graph of a function

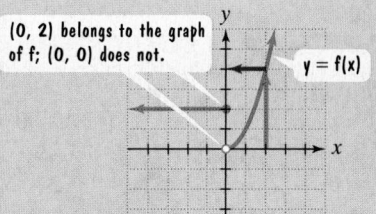

To find $f(2)$, locate 2 on the x-axis. The graph shows $f(2) = 4$.

Section 8.2 The Algebra of Functions

A Function's Domain
If a function f does not model data or verbal conditions, its domain is the largest set of real numbers for which the value of $f(x)$ is a real number. Exclude from a function's domain real numbers that cause division by zero and real numbers that result in a square root of a negative number.

$$f(x) = 7x + 13$$

Domain of $f = \{x \mid x \text{ is a real number}\}$

$$g(x) = \frac{7x}{12 - x}$$

Domain of g
$$= \{x \mid x \text{ is a real number and } x \neq 12\}$$

DEFINITIONS AND CONCEPTS	EXAMPLES

Section 8.2 The Algebra of Functions (continued)

The Algebra of Functions

Let f and g be two functions. The sum $f + g$, the difference $f - g$, the product fg, and the quotient $\dfrac{f}{g}$ are functions whose domains are the set of all real numbers common to the domains of f and g, defined as follows:

1. **Sum:** $(f + g)(x) = f(x) + g(x)$
2. **Difference:** $(f - g)(x) = f(x) - g(x)$
3. **Product:** $(fg)(x) = f(x) \cdot g(x)$
4. **Quotient:** $\left(\dfrac{f}{g}\right)(x) = \dfrac{f(x)}{g(x)}, g(x) \neq 0.$

Let $f(x) = x^2 + 2x$ and $g(x) = 4 - x$.

- $(f + g)(x) = (x^2 + 2x) + (4 - x) = x^2 + x + 4$

  $(f + g)(-2) = (-2)^2 + (-2) + 4 = 4 - 2 + 4 = 6$

- $(f - g)(x) = (x^2 + 2x) - (4 - x) = x^2 + 2x - 4 + x$

  $= x^2 + 3x - 4$

  $(f - g)(5) = 5^2 + 3 \cdot 5 - 4 = 25 + 15 - 4 = 36$

- $(fg)(x) = (x^2 + 2x)(4 - x) = 4x^2 - x^3 + 8x - 2x^2$

  $= -x^3 + 2x^2 + 8x$

  $(fg)(1) = -1^3 + 2 \cdot 1^2 + 8 \cdot 1 = -1 + 2 + 8 = 9$

- $\left(\dfrac{f}{g}\right)(x) = \dfrac{x^2 + 2x}{4 - x}, x \neq 4$

  $\left(\dfrac{f}{g}\right)(3) = \dfrac{3^2 + 2 \cdot 3}{4 - 3} = \dfrac{9 + 6}{1} = 15$

Section 8.3 Systems of Linear Equations in Three Variables

A system of linear equations in three variables, x, y, and z, consists of three equations of the form $Ax + By + Cz = D$. The solution set is the set of ordered triples that satisfy all three equations. The solution represents the point of intersection of three planes in space.

Is $(2, -1, 3)$ a solution of

$$3x + 5y - 2z = -5$$

$$2x + 3y - z = -2$$

$$2x + 4y + 6z = 18?$$

Replace x by 2, y by -1, and z by 3.
Using Equation 1, we obtain:

$$3 \cdot 2 + 5(-1) - 2(3) \overset{?}{=} -5$$

$$6 - 5 - 6 \overset{?}{=} -5$$

$$-5 = -5, \text{ true}$$

The ordered triple $(2, -1, 3)$ satisfies the first equation. In a similar manner, it satisfies the other two equations and is a solution.

DEFINITIONS AND CONCEPTS	EXAMPLES

Section 8.3 Systems of Linear Equations in Three Variables (continued)

To solve a linear system in three variables by eliminating variables:

1. Reduce the system to two equations in two variables.
2. Solve the resulting system of two equations in two variables.
3. Use back-substitution in one of the equations in two variables to find the value of the second variable.
4. Back-substitute the values for two variables into one of the original equations to find the value of the third variable.
5. Check.

If all variables are eliminated and a false statement results, the system is inconsistent and has no solution. If a true statement results, the system contains dependent equations and has infinitely many solutions.

Solve:
$$2x + 3y - 2z = 0 \quad \boxed{1}$$
$$x + 2y - z = 1 \quad \boxed{2}$$
$$3x - y + z = -15. \quad \boxed{3}$$

Add equations $\boxed{2}$ and $\boxed{3}$ to eliminate z.
$$4x + y = -14 \quad \boxed{4}$$

Eliminate z again. Multiply equation $\boxed{3}$ by 2 and add to equation $\boxed{1}$.
$$8x + y = -30 \quad \boxed{5}$$

Multiply equation $\boxed{4}$ by -1 and add to equation $\boxed{5}$.
$$-4x - y = 14$$
$$\underline{8x + y = -30}$$
$$\text{Add:} \quad 4x = -16$$
$$x = -4$$

Substitute -4 for x in $\boxed{4}$.
$$4(-4) + y = -14$$
$$y = 2$$

Substitute -4 for x and 2 for y in $\boxed{3}$.
$$3(-4) - 2 + z = -15$$
$$-14 + z = -15$$
$$z = -1$$

Checking verifies that $(-4, 2, -1)$ is the solution and $\{(-4, 2, -1)\}$ is the solution set.

Curve Fitting
Curve fitting is determining a function whose graph contains given points. Three points that do not lie on a line determine the graph of a quadratic function
$$y = ax^2 + bx + c.$$
Use the three given points to create a system of three equations. Solve the system to find a, b, and c.

Find the quadratic function $y = ax^2 + bx + c$ whose graph passes through the points $(-1, 2)$, $(1, 8)$, and $(2, 14)$.

Use $y = ax^2 + bx + c$.

When $x = -1, y = 2$: $2 = a(-1)^2 + b(-1) + c$
When $x = 1, y = 8$: $8 = a \cdot 1^2 + b \cdot 1 + c$
When $x = 2, y = 14$: $14 = a \cdot 2^2 + b \cdot 2 + c$

Solving,
$$a - b + c = 2$$
$$a + b + c = 8$$
$$4a + 2b + c = 14,$$

$a = 1, b = 3$, and $c = 4$. The quadratic function, $y = ax^2 + bx + c$, is $y = x^2 + 3x + 4$.

DEFINITIONS AND CONCEPTS	EXAMPLES

Section 8.4 Matrix Solutions to Linear Systems

A matrix is a rectangular array of numbers. The augmented matrix of a linear system is obtained by writing the coefficients of each variable, a vertical bar, and the constants of the system.

$$x + 4y = 9$$
$$3x + y = 5$$

Augmented matrix is $\begin{bmatrix} 1 & 4 & | & 9 \\ 3 & 1 & | & 5 \end{bmatrix}$.

The following row operations produce matrices that represent systems with the same solution. Two matrices are row equivalent if one can be obtained from the other by a sequence of these row operations.

1. Interchange the elements in the ith and jth rows: $R_i \leftrightarrow R_j$.
2. Multiply each element in the ith row by k: kR_i.
3. Add k times the elements in row i to the corresponding elements in row j: $kR_i + R_j$.

Find $-4R_1 + R_2$:

$$\begin{bmatrix} 1 & 0 & -2 & | & 5 \\ 4 & -1 & 2 & | & 6 \\ 3 & -7 & 9 & | & 10 \end{bmatrix}.$$

Add -4 times the elements in row 1 to the corresponding elements in row 2.

$$\begin{bmatrix} 1 & 0 & -2 & | & 5 \\ -4(1)+4 & -4(0)+(-1) & -4(-2)+2 & | & -4(5)+6 \\ 3 & -7 & 9 & | & 10 \end{bmatrix}$$

$$= \begin{bmatrix} 1 & 0 & -2 & | & 5 \\ 0 & -1 & 10 & | & -14 \\ 3 & -7 & 9 & | & 10 \end{bmatrix}$$

Solving Linear Systems Using Matrices

1. Write the augmented matrix for the system.
2. Use matrix row operations to simplify the matrix to one with 1s down the diagonal from upper left to lower right, and 0s below the 1s.
3. Write the system of linear equations corresponding to the matrix in step 2, and use back-substitution to find the system's solution.

If you obtain a matrix with a row containing 0s to the left of the vertical bar and a nonzero number on the right, the system is inconsistent. If 0s appear across an entire row, the system contains dependent equations.

Solve using matrices:

$$3x + y = 5$$
$$x + 4y = 9.$$

$$\begin{bmatrix} 3 & 1 & | & 5 \\ 1 & 4 & | & 9 \end{bmatrix} \xrightarrow{R_1 \leftrightarrow R_2} \begin{bmatrix} 1 & 4 & | & 9 \\ 3 & 1 & | & 5 \end{bmatrix}$$

$$\xrightarrow{-3R_1 + R_2} \begin{bmatrix} 1 & 4 & | & 9 \\ 0 & -11 & | & -22 \end{bmatrix} \xrightarrow{-\frac{1}{11}R_2} \begin{bmatrix} 1 & 4 & | & 9 \\ 0 & 1 & | & 2 \end{bmatrix}$$

$$\rightarrow x + 4y = 9$$
$$y = 2$$

When $y = 2$, $x + 4 \cdot 2 = 9$, so $x = 1$. The solution is $(1, 2)$ and the solution set is $\{(1, 2)\}$.

Section 8.5 Determinants and Cramer's Rule

A square matrix has the same number of rows as columns. A determinant is a real number associated with a square matrix. The determinant is denoted by placing vertical bars about the array of numbers. The value of a second-order determinant is

$$\begin{vmatrix} a_1 & b_1 \\ a_2 & b_2 \end{vmatrix} = a_1 b_2 - a_2 b_1.$$

Evaluate:

$$\begin{vmatrix} 2 & -1 \\ 3 & 4 \end{vmatrix} = 2(4) - 3(-1) = 8 + 3 = 11.$$

DEFINITIONS AND CONCEPTS	EXAMPLES

Section 8.5 Determinants and Cramer's Rule (continued)

Cramer's Rule for Two Linear Equations in Two Variables.

If

$$a_1 x + b_1 y = c_1$$
$$a_2 x + b_2 y = c_2$$

then

$$x = \frac{\begin{vmatrix} c_1 & b_1 \\ c_2 & b_2 \end{vmatrix}}{\begin{vmatrix} a_1 & b_1 \\ a_2 & b_2 \end{vmatrix}} = \frac{D_x}{D} \quad \text{and} \quad y = \frac{\begin{vmatrix} a_1 & c_1 \\ a_2 & c_2 \end{vmatrix}}{\begin{vmatrix} a_1 & b_1 \\ a_2 & b_2 \end{vmatrix}} = \frac{D_y}{D}, \quad D \neq 0.$$

If $D = 0$ and any numerator is not zero, the system is inconsistent and has no solution. If all determinants are 0, the system contains dependent equations and has infinitely many solutions.

Solve by Cramer's rule:

$$5x + 3y = 7$$
$$-x + 2y = 9.$$

$$D = \begin{vmatrix} 5 & 3 \\ -1 & 2 \end{vmatrix} = 5(2) - (-1)(3) = 10 + 3 = 13$$

$$D_x = \begin{vmatrix} 7 & 3 \\ 9 & 2 \end{vmatrix} = 7 \cdot 2 - 9 \cdot 3 = 14 - 27 = -13$$

$$D_y = \begin{vmatrix} 5 & 7 \\ -1 & 9 \end{vmatrix} = 5(9) - (-1)(7) = 45 + 7 = 52$$

$$x = \frac{D_x}{D} = \frac{-13}{13} = -1, \, y = \frac{D_y}{D} = \frac{52}{13} = 4$$

The solution is $(-1, 4)$ and the solution set is $\{(-1, 4)\}$.

The value of a third-order determinant is

$$\begin{vmatrix} a_1 & b_1 & c_1 \\ a_2 & b_2 & c_2 \\ a_3 & b_3 & c_3 \end{vmatrix}$$

$$= a_1 \begin{vmatrix} b_2 & c_2 \\ b_3 & c_3 \end{vmatrix} - a_2 \begin{vmatrix} b_1 & c_1 \\ b_3 & c_3 \end{vmatrix} + a_3 \begin{vmatrix} b_1 & c_1 \\ b_2 & c_2 \end{vmatrix}.$$

Each 2×2 determinant is called a minor.

Evaluate:

$$\begin{vmatrix} 1 & -2 & 1 \\ 3 & 1 & -2 \\ 5 & 5 & 3 \end{vmatrix}.$$

$$= 1 \begin{vmatrix} 1 & -2 \\ 5 & 3 \end{vmatrix} - 3 \begin{vmatrix} -2 & 1 \\ 5 & 3 \end{vmatrix} + 5 \begin{vmatrix} -2 & 1 \\ 1 & -2 \end{vmatrix}$$

$$= 1(3 - (-10)) - 3(-6 - 5) + 5(4 - 1)$$

$$= 1(13) - 3(-11) + 5(3)$$

$$= 13 + 33 + 15 = 61$$

Cramer's Rule for Three Linear Equations in Three Variables

If

$$a_1 x + b_1 y + c_1 z = d_1$$
$$a_2 x + b_2 y + c_2 z = d_2$$
$$a_3 x + b_3 y + c_3 z = d_3$$

Solve by Cramer's rule:

$$x - 2y + z = 4$$
$$3x + y - 2z = 3$$
$$5x + 5y + 3z = -8.$$

DEFINITIONS AND CONCEPTS	EXAMPLES

Section 8.5 Determinants and Cramer's Rule (continued)

then

$$x = \frac{D_x}{D}, \qquad y = \frac{D_y}{D}, \qquad z = \frac{D_z}{D}.$$

$$D = \begin{vmatrix} a_1 & b_1 & c_1 \\ a_2 & b_2 & c_2 \\ a_3 & b_3 & c_3 \end{vmatrix} \neq 0, \quad D_x = \begin{vmatrix} d_1 & b_1 & c_1 \\ d_2 & b_2 & c_2 \\ d_3 & b_3 & c_3 \end{vmatrix}$$

$$D_y = \begin{vmatrix} a_1 & d_1 & c_1 \\ a_2 & d_2 & c_2 \\ a_3 & d_3 & c_3 \end{vmatrix}, \qquad D_z = \begin{vmatrix} a_1 & b_1 & d_1 \\ a_2 & b_2 & d_2 \\ a_3 & b_3 & d_3 \end{vmatrix}$$

If $D = 0$ and any numerator is not zero, the system is inconsistent. If all determinants are 0, the system contains dependent equations.

$$D = \begin{vmatrix} 1 & -2 & 1 \\ 3 & 1 & -2 \\ 5 & 5 & 3 \end{vmatrix} = 61 \qquad \textit{worked out on page 579}$$

$$D_x = \begin{vmatrix} 4 & -2 & 1 \\ 3 & 1 & -2 \\ -8 & 5 & 3 \end{vmatrix} = 61$$

$$D_y = \begin{vmatrix} 1 & 4 & 1 \\ 3 & 3 & -2 \\ 5 & -8 & 3 \end{vmatrix} = -122$$

$$D_z = \begin{vmatrix} 1 & -2 & 4 \\ 3 & 1 & 3 \\ 5 & 5 & -8 \end{vmatrix} = -61$$

$$x = \frac{D_x}{D} = \frac{61}{61} = 1, \quad y = \frac{D_y}{D} = \frac{-122}{61} = -2,$$

$$z = \frac{D_z}{D} = \frac{-61}{61} = -1.$$

The solution is $(1, -2, -1)$ and the solultion set is $\{(1, -2, -1)\}$.

Review Exercises

8.1 *In Exercises 1–3, determine whether each relation is a function. Give the domain and range for each relation.*

1. $\{(3, 10), (4, 10), (5, 10)\}$

2. $\{(1, 12), (2, 100), (3, \pi), (4, -6)\}$

3. $\{(13, 14), (15, 16), (13, 17)\}$

In Exercises 4–5, find the indicated function values.

4. $f(x) = 7x - 5$
    a. $f(0)$ **b.** $f(3)$ **c.** $f(-10)$
    d. $f(2a)$ **e.** $f(a + 2)$

5. $g(x) = 3x^2 - 5x + 2$
    a. $g(0)$ **b.** $g(5)$ **c.** $g(-4)$
    d. $g(b)$ **e.** $g(4a)$

In Exercises 6–9, use the vertical line test to identify graphs in which y is a function of x.

6.

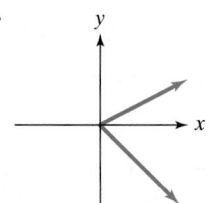

7.

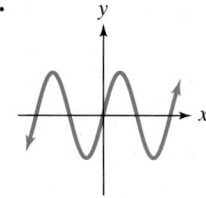

8.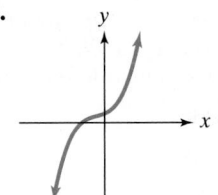

9.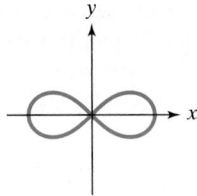

Use the graph of f to solve Exercises 10–12.

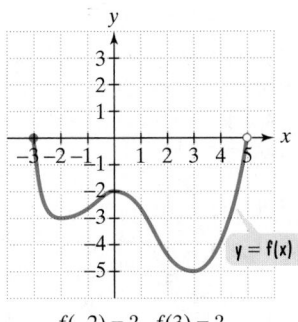

$f(-2) = ?\quad f(3) = ?$

10. Find $f(-2)$.

11. Find $f(0)$.

12. For what value of x is $f(x) = -5$?

13. The graph shows the height, in meters, of a vulture in terms of its time, in seconds, in flight.

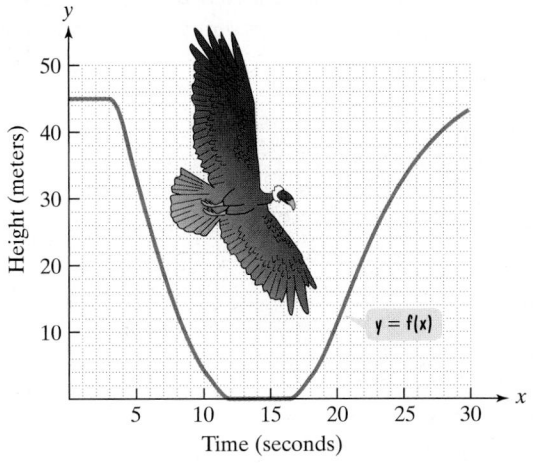

a. Use the graph to explain why the vulture's height is a function of its time in flight.

b. Find $f(15)$. Describe what this means in practical terms.

c. What is a reasonable estimate of the vulture's maximum height?

d. For what values of x is $f(x) = 20$? Describe what this means in practical terms.

e. Use the graph of the function to write a description of the vulture's flight.

8.2 *In Exercises 14–16, find the domain of each function.*

14. $f(x) = 7x - 3$

15. $g(x) = \dfrac{1}{x + 8}$

16. $f(x) = x + \dfrac{3x}{x - 5}$

In Exercises 17–18, find **a.** $(f + g)(x)$ *and* **b.** $(f + g)(3)$.

17. $f(x) = 4x - 5,\quad g(x) = 2x + 1$

18. $f(x) = 5x^2 - x + 4,\quad g(x) = x - 3$

In Exercises 19–20, for each pair of functions, f and g, determine the domain of f + g.

19. $f(x) = 3x + 4,\quad g(x) = \dfrac{5}{4 - x}$

20. $f(x) = \dfrac{7x}{x + 6},\quad g(x) = \dfrac{4}{x + 1}$

In Exercises 21–28, let
$$f(x) = x^2 - 2x \quad \text{and} \quad g(x) = x - 5.$$
Find each of the following.

21. $(f + g)(x)\quad$ and $\quad(f + g)(-2)$

22. $f(3) + g(3)$

23. $(f - g)(x)\quad$ and $\quad(f - g)(1)$

24. $f(4) - g(4)$

25. $(fg)(x)\quad$ and $\quad(fg)(-3)$

26. $\left(\dfrac{f}{g}\right)(x)\quad$ and $\quad\left(\dfrac{f}{g}\right)(4)$

27. The domain of $f - g$

28. The domain of $\dfrac{f}{g}$

8.3

29. Is $(-3, -2, 5)$ a solution of the system

$$\begin{aligned} x + y + z &= 0 \\ 2x - 3y + z &= 5 \\ 4x + 2y + 4z &= 3? \end{aligned}$$

Solve each system in Exercises 30–32 by eliminating variables using the addition method. If there is no solution or if there are infinitely many solutions and a system's equations are dependent, so state.

30. $\begin{aligned} 2x - y + z &= 1 \\ 3x - 3y + 4z &= 5 \\ 4x - 2y + 3z &= 4 \end{aligned}$

31. $\begin{aligned} x + 2y - z &= 5 \\ 2x - y + 3z &= 0 \\ 2y + z &= 1 \end{aligned}$

32. $\begin{aligned} 3x - 4y + 4z &= 7 \\ x - y - 2z &= 2 \\ 2x - 3y + 6z &= 5 \end{aligned}$

33. Find the quadratic function $y = ax^2 + bx + c$ whose graph passes through the points $(1, 4)$, $(3, 20)$, and $(-2, 25)$.

34. The graph shows a low savings rate in the United States compared to that of many industrialized countries. Adding the rates for Japan, Germany, and France gives 45%. The savings rate in Japan exceeds that for Germany by 1% and is 12% less than twice that for France. Find the savings rates for Japan, Germany, and France.

Comparative Savings Rates

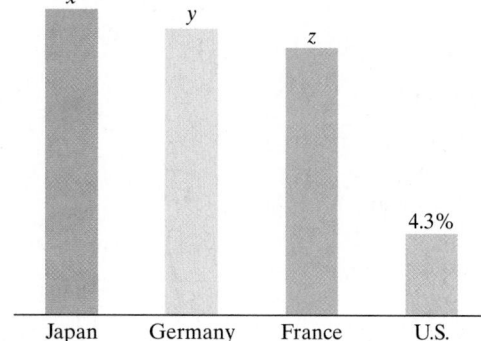

4.3%

Japan   Germany   France   U.S.

Source: Office of Management and Budget

8.4

In Exercises 35–36, write the system of linear equations represented by the augmented matrix. Use x, y, and, if necessary, z, for the variables. Once the system is written, use back-substitution to find its solution.

35. $\left[\begin{array}{cc|c} 2 & 3 & -10 \\ 0 & 1 & -6 \end{array}\right]$

36. $\left[\begin{array}{ccc|c} 1 & 1 & 3 & 12 \\ 0 & 1 & -2 & -4 \\ 0 & 0 & 1 & 3 \end{array}\right]$

In Exercises 37–40, perform each matrix row operation and write the new matrix.

37. $\left[\begin{array}{cc|c} 1 & -8 & 3 \\ 0 & 7 & -14 \end{array}\right] \frac{1}{7} R_2$

38. $\left[\begin{array}{cc|c} 1 & -3 & 1 \\ 2 & 1 & -5 \end{array}\right] -2R_1 + R_2$

39. $\left[\begin{array}{ccc|c} 2 & -2 & 1 & -1 \\ 1 & 2 & -1 & 2 \\ 6 & 4 & 3 & 5 \end{array}\right] \frac{1}{2} R_1$

40. $\left[\begin{array}{ccc|c} 1 & 2 & 2 & 2 \\ 0 & 1 & -1 & 2 \\ 0 & 5 & 4 & 1 \end{array}\right] -5R_2 + R_3$

In Exercises 41–44, solve each system using matrices. If there is no solution or if a system's equations are dependent, so state.

41. $\begin{aligned} x + 4y &= 7 \\ 3x + 5y &= 0 \end{aligned}$

42. $\begin{aligned} 2x - 3y &= 8 \\ -6x + 9y &= 4 \end{aligned}$

43. $\begin{aligned} x + 2y + 3z &= -5 \\ 2x + y + z &= 1 \\ x + y - z &= 8 \end{aligned}$

44. $\begin{aligned} x - 2y + z &= 0 \\ y - 3z &= -1 \\ 2y + 5z &= -2 \end{aligned}$

8.5

In Exercises 45–48, evaluate each determinant.

45. $\begin{vmatrix} 3 & 2 \\ -1 & 5 \end{vmatrix}$

46. $\begin{vmatrix} -2 & -3 \\ -4 & -8 \end{vmatrix}$

47. $\begin{vmatrix} 2 & 4 & -3 \\ 1 & -1 & 5 \\ -2 & 4 & 0 \end{vmatrix}$

48. $\begin{vmatrix} 4 & 7 & 0 \\ -5 & 6 & 0 \\ 3 & 2 & -4 \end{vmatrix}$

In Exercises 49–52, use Cramer's rule to solve each system. If there is no solution or if a system's equations are dependent, so state.

49. $x - 2y = 8$
$3x + 2y = -1$

50. $7x + 2y = 0$
$2x + y = -3$

51. $x + 2y + 2z = 5$
$2x + 4y + 7z = 19$
$-2x - 5y - 2z = 8$

52. $2x + y = -4$
$y - 2z = 0$
$3x - 2z = -11$

53. Use the quadratic function $y = ax^2 + bx + c$ to model the following data:

x (Age of a Driver)	y (Average Number of Automobile Accidents per Day in the United States)
20	400
40	150
60	400

Use Cramer's rule to determine values for a, b, and c. Then use the model to write a statement about the average number of automobile accidents in which 30-year-olds and 50-year-olds are involved daily.

Chapter 8 Test

In Exercises 1–2, determine whether each relation is a function. Give the domain and range for each relation.

1. $\{(1, 2), (3, 4), (5, 6), (6, 6)\}$

2. $\{(2, 1), (4, 3), (6, 5), (6, 6)\}$

3. If $f(x) = 3x - 2$, find $f(a + 4)$.

4. If $f(x) = 4x^2 - 3x + 6$, find $f(-2)$.

In Exercises 5–6, identify the graph or graphs in which y is a function of x.

5.

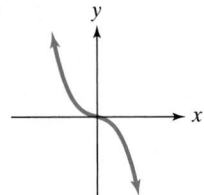

6.

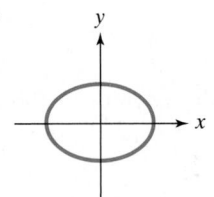

Use the graph of f to solve Exercises 7–8.

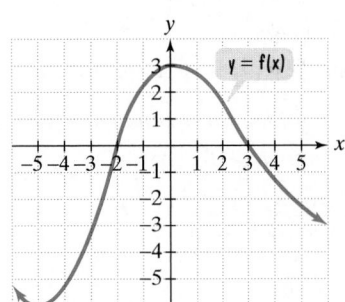

7. Find $f(6)$.

8. List a value of x for which $f(x) = -6$.

9. Find the domain of $f(x) = \dfrac{6}{10 - x}$.

In Exercises 10–14, let

$$f(x) = x^2 + 4x \quad and \quad g(x) = x + 2.$$

Find each of the following.

10. $(f + g)(x)$ and $(f + g)(3)$

11. $(f - g)(x)$ and $(f - g)(-1)$

12. $(fg)(x)$ and $(fg)(-5)$

13. $\left(\dfrac{f}{g}\right)(x)$ and $\left(\dfrac{f}{g}\right)(2)$

14. The domain of $\dfrac{f}{g}$

15. Solve by eliminating variables using the addition method:

$$x + y + z = 6$$

$$3x + 4y - 7z = 1$$

$$2x - y + 3z = 5.$$

16. Perform the indicated matrix row operation and write the new matrix.

$$\begin{bmatrix} 1 & 0 & -4 & | & 5 \\ 6 & -1 & 2 & | & 10 \\ 2 & -1 & 4 & | & -3 \end{bmatrix} -6R_1 + R_2$$

In Exercises 17–18, solve each system using matrices.

17. $2x + y = 6$
$3x - 2y = 16$

18. $x - 4y + 4z = -1$
$2x - y + 5z = 6$
$-x + 3y - z = 5$

In Exercises 19–20, evaluate each determinant.

19. $\begin{vmatrix} -1 & -3 \\ 7 & 4 \end{vmatrix}$

20. $\begin{vmatrix} 3 & 4 & 0 \\ -1 & 0 & -3 \\ 4 & 2 & 5 \end{vmatrix}$

In Exercises 21–22, use Cramer's rule to solve each system.

21. $4x - 3y = 14$
$3x - y = 3$

22. $2x + 3y + z = 2$
$3x + 3y - z = 0$
$x - 2y - 3z = 1$

Cumulative Review Exercises (Chapters 1–8)

In Exercises 1–5, solve each equation or system of equations.

1. $2x + 3x - 5 + 7 = 10x + 3 - 6x - 4$

2. $2x^2 + 5x = 12$

3. $8x - 5y = -4$
$2x + 15y = -66$

4. $\dfrac{15}{x} - 4 = \dfrac{6}{x} + 3$

5. $-3x - 7 = 8$

6. If $f(x) = 2x^2 - 5x + 2$ and $g(x) = x^2 - 2x + 3$, find $(f - g)(x)$ and $(f - g)(3)$.

In Exercises 7–11, simplify each expression.

7. $\dfrac{8x^3}{-4x^7}$

8. $-8 - (-3) \cdot 4$

9. $\dfrac{\dfrac{1}{x} - \dfrac{1}{2}}{\dfrac{1}{3} - \dfrac{x}{6}}$

10. $\dfrac{4 - x^2}{3x^2 - 5x - 2}$

11. $-5 - (-8) - (4 - 6)$

In Exercises 12–13, factor completely.

12. $x^2 - 18x + 77$

13. $x^3 - 25x$

In Exercises 14–17, perform the indicated operations. If possible, simplify the answer.

14. $\dfrac{6x^3 - 19x^2 + 16x - 4}{x - 2}$

15. $(2x - 3)(4x^2 + 6x + 9)$

16. $\dfrac{3x}{x^2 + x - 2} - \dfrac{2}{x + 2}$

17. $\dfrac{5x^2 - 6x + 1}{x^2 - 1} \div \dfrac{16x^2 - 9}{4x^2 + 7x + 3}$

18. Solve by eliminating variables using addition method:
$$x + 3y - z = 5$$
$$-x + 2y + 3z = 13$$
$$2x - 5y - z = -8.$$

In Exercises 19–20, graph each equation in a rectangular coordinate system

19. $2x - y = 4$.

20. $y = -\dfrac{2}{3}x$

21. Evaluate: $\begin{vmatrix} 0 & 1 & -2 \\ -7 & 0 & -4 \\ 3 & 0 & 5 \end{vmatrix}$.

22. Find the slope of the line through $(-1, 5)$ and $(2, -3)$.

23. Write the point-slope form of the equation of the line with slope $= 5$, passing through $(-2, -3)$. Then use the point-slope equation to write the slope-intercept form of the line's equation.

24. Solve using matrices:

$$\begin{aligned} 2x + 3y - z &= -1 \\ x + 2y + 3z &= 2 \\ 3x + 5y - 2z &= -3. \end{aligned}$$

25. Solve using Cramer's rule (determinants):

$$\begin{aligned} 3x + 4y &= -1 \\ -2x + y &= 8. \end{aligned}$$

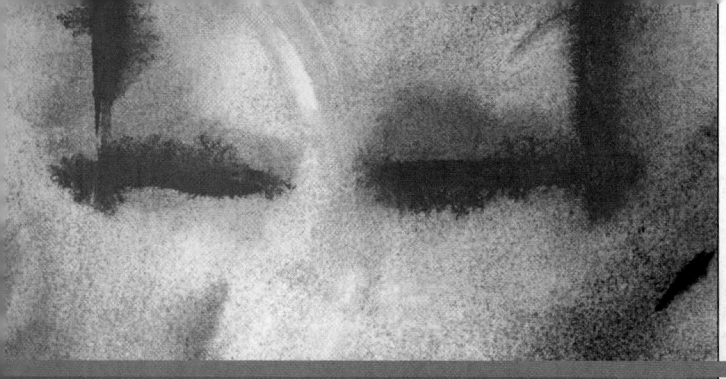

*M*ost things in life depend on many variables. Temperature and precipitation are two variables that affect whether regions are forests, grasslands, or deserts. In this chapter, you will learn methods for modeling your world with inequalities. You will even see how inequalities are used to describe some of the most magnificent places in our nation's landscape.

You are in Yosemite National Park in California, surrounded by evergreen forests, alpine meadows, and sheer walls of granite. The beauty of soaring cliffs, plunging waterfalls, gigantic trees, rugged canyons, mountains, and valleys is overwhelming. This is so different from where you live and attend college, a region in which grasslands predominate.

Inequalities and Problem Solving

▶ SECTION 9.1 *Interval Notation and Business Applications Using Linear Inequalities*

Objectives

1 Use interval notation.

2 Review how to solve linear inequalities.

3 Use linear inequalities to solve problems involving revenue, cost, and profit.

SSM
PH Tutor CD- Video
Center ROM

Driving through your neighborhood, you see kids selling lemonade. Would it surprise you to know that this activity can be analyzed using linear inequalities? By doing so, you will view profit and loss in the business world in a new way. In this section, we use linear inequalities to solve problems and model business ventures.

 Use interval notation.

Interval Notation

Recall from Chapter 2 that any inequality in the form $ax + b \leq c$ is called a **linear inequality in one variable**. The symbol between $ax + b$ and c can be $\leq$ (is less than or equal to), $<$ (is less than), $\geq$ (is greater than or equal to), or $>$ (is greater than). The greatest exponent on the variable in such an inequality is 1.

Solving an inequality is the process of finding the set of numbers that make the inequality a true statement. These numbers are called the **solutions** of the inequality, and we say that they **satisfy** the inequality. The set of all solutions is called the **solution set** of the inequality.

Graphs of solutions to linear inequalities are shown on a number line by shading all points representing numbers that are solutions. For example, the solutions of $x > 4$ are all real numbers that are greater than 4. They are graphed on a number line by shading all points to the right of 4. The open dot at 4 indicates that 4 is not a solution:

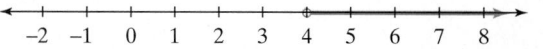

Throughout the remainder of this book, a convenient notation, called *interval notation*, will also be used to represent solution sets of inequalities. To help understand this notation, a different graphing notation will be used. Parentheses, rather than open dots, indicate endpoints that are not solutions. Thus, we graph the solutions of $x > 4$ as

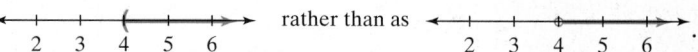

Square brackets, rather than closed dots, indicate endpoints that are solutions. Thus, we graph the solutions of $x \geq 4$ as

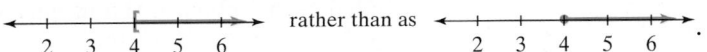

rather than as

EXAMPLE 1 Graphing Inequalities

Graph the solutions of:

a. $x < 3$ **b.** $x \geq -1$ **c.** $-1 < x \leq 3$.

Solution

a. The solutions of $x < 3$ are all real numbers that are less than 3. They are graphed on a number line by shading all points to the left of 3. The parenthesis at 3 indicates that 3 is not a solution, but numbers such as 2.9999 and 2.6 are. The arrow shows that the graph extends indefinitely to the left.

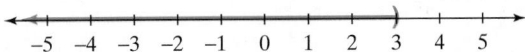

b. The solutions of $x \geq -1$ are all real numbers that are greater than or equal to -1. We shade all points to the right of -1 and the point for -1 itself. The bracket at -1 shows that -1 is a solution for the given inequality. The arrow shows that the graph extends indefinitely to the right.

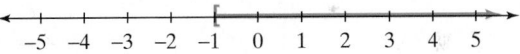

c. The inequality $-1 < x \leq 3$ is read "-1 is less than x *and* x is less than or equal to 3," or "x is greater than -1 *and* less than or equal to 3." The solutions of $-1 < x \leq 3$ are all real numbers between -1 and 3, not including -1 but including 3. The parenthesis at -1 indicates that -1 is not a solution. By contrast, the bracket at 3 shows that 3 is a solution. Shading indicates the other solutions.

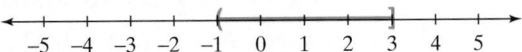

✔ **CHECK POINT 1** Graph the solutions of:

a. $x \leq 2$ **b.** $x > -4$ **c.** $2 \leq x < 6$.

In Chapter 2, we used *set-builder notation* to represent solution sets for inequalities. Using this method, the solution set for $x > -4$ can be expressed as

$$\{x \mid x > -4\}.$$

The set of all x   such that

We read this as "the set of all real numbers x such that x is greater than -4."

As previously mentioned, another method used to represent solution sets for inequalities is **interval notation**. Using this notation, the solution set for $x > -4$ is expressed as $(-4, \infty)$. The parenthesis at -4 indicates that -4 is not included in the interval. The infinity symbol, ∞, does not represent a real number. It indicates that the interval extends indefinitely to the right.

Table 9.1 lists nine possible types of intervals used to describe sets of real numbers.

Table 9.1   Intervals on the Real Number Line

Let a and b be real numbers such that $a < b$.

Interval Notation	Set-Builder Notation	Graph
(a, b)	$\{x \mid a < x < b\}$	
$[a, b]$	$\{x \mid a \leq x \leq b\}$	
$[a, b)$	$\{x \mid a \leq x < b\}$	
$(a, b]$	$\{x \mid a < x \leq b\}$	
(a, ∞)	$\{x \mid x > a\}$	
$[a, \infty)$	$\{x \mid x \geq a\}$	
$(-\infty, b)$	$\{x \mid x < b\}$	
$(-\infty, b]$	$\{x \mid x \leq b\}$	
$(-\infty, \infty)$	$\mathbb{R}$ (set of all real numbers)	

EXAMPLE 2   Intervals and Inequalities

Express the intervals in terms of inequalities and graph:

    a. $(-1, 4]$ **b.** $[2.5, 4]$ **c.** $(-4, \infty)$.

Solution

    a. $(-1, 4] = \{x \mid -1 < x \leq 4\}$

    b. $[2.5, 4] = \{x \mid 2.5 \leq x \leq 4\}$

    c. $(-4, \infty) = \{x \mid x > -4\}$

■

✔ **CHECK POINT 2** Express the intervals in terms of inequalities and graph:

    a. $[-2, 5)$ **b.** $[1, 3.5]$ **c.** $(-\infty, -1)$.

2 Review how to solve linear inequalities.

Solving Linear Inequalities

Recall from Chapter 2 the following procedure for solving a linear inequality. The only difference is that we will now use two notations, set-builder notation and interval notation, to express solution sets.

Study Tip

If you need a more detailed presentation on solving linear inequalities, see Section 2.7, Examples 2 through 7, on pages 162 through 166.

> ### Solving a Linear Inequality
>
> 1. Simplify the algebraic expression on each side.
> 2. Use the addition property of inequality to collect all the variable terms on one side and all the constant terms on the other side.
> 3. Use the multiplication property of inequality to isolate the variable and solve. Reverse the sense of the inequality when multiplying or dividing both sides by a negative number.
> 4. Express the solution set in set-builder or interval notation and graph the solution set on a number line.

EXAMPLE 3   Solving a Linear Inequality

Solve and graph the solution set on a number line:

$$2 - 12x \le 7(1 - x).$$

Solution

Step 1 Simplify each side. We use the distributive property to remove parentheses on the right side.

$$2 - 12x \le 7\overset{\frown}{(1 - x)} \qquad \text{This is the given inequality.}$$
$$2 - 12x \le 7 - 7x \qquad \text{Use the distributive property.}$$

Study Tip

You can solve

$$2 - 12x \le 7 - 7x$$

by isolating x on the right side. Add $12x$ to both sides:

$$2 - 12x + 12x$$
$$\le 7 - 7x + 12x$$
$$2 \le 7 + 5x.$$

Now subtract 7 from both sides:

$$2 - 7 \le 7 + 5x - 7$$
$$-5 \le 5x.$$

Finally, divide both sides by 5:

$$\frac{-5}{5} \le \frac{5x}{5}$$
$$-1 \le x.$$

This last inequality means the same thing as

$$x \ge -1.$$

Step 2 Collect variable terms on one side and constant terms on the other side. We will collect variable terms on the left and constant terms on the right.

$$2 - 12x + 7x \le 7 - 7x + 7x \qquad \text{Add 7x to both sides.}$$
$$2 - 5x \le 7 \qquad \text{Simplify.}$$
$$2 - 5x - 2 \le 7 - 2 \qquad \text{Subtract 2 from both sides.}$$
$$-5x \le 5 \qquad \text{Simplify.}$$

Step 3 Isolate the variable and solve. We isolate the variable, x, by dividing both sides by -5. Because we are dividing by a negative number, we must reverse the inequality symbol.

$$\frac{-5x}{-5} \ge \frac{5}{-5} \qquad \text{Divide both sides by −5 and reverse the sense of the inequality.}$$
$$x \ge -1 \qquad \text{Simplify.}$$

Step 4 Express the solution set in set-builder or interval notation and graph the set on a number line. The solution set consists of all real numbers that are greater than or equal to -1, expressed in set-builder notation as $\{x \mid x \ge -1\}$. The interval notation for this solution set is $[-1, \infty)$. The graph of the solution set is shown as follows:

$$\begin{array}{c} \xleftarrow{\hspace{1cm}} \quad \overset{[}{\rule{0pt}{1em}} \quad \xrightarrow{\hspace{1cm}} \\ -5 \;\; -4 \;\; -3 \;\; -2 \;\; -1 \;\; 0 \;\; 1 \;\; 2 \;\; 3 \;\; 4 \;\; 5 \end{array}.$$ ∎

CHECK POINT 3 Solve and graph the solution set on a number line:
$$-12 - 8x \leq 4(6 - x).$$

If an inequality contains fractions, begin by multiplying both sides by the least common denominator. This will clear the inequality of fractions.

EXAMPLE 4   Solving a Linear Inequality

Solve and graph the solution set on a number line:

$$\frac{x + 3}{4} > \frac{x - 2}{3} + \frac{1}{4}.$$

Solution   The denominators are 4, 3, and 4. The least common denominator is 12. We begin by multiplying both sides of the inequality by 12.

$$\frac{x + 3}{4} > \frac{x - 2}{3} + \frac{1}{4}$$ This is the given inequality.

$$12\left(\frac{x + 3}{4}\right) > 12\left(\frac{x - 2}{3} + \frac{1}{4}\right)$$ Multiply both sides by 12. Multiplying by a positive number preserves the sense of the inequality.

$$\frac{12}{1} \cdot \frac{x + 3}{4} > \frac{12}{1} \cdot \frac{x - 2}{3} + \frac{12}{1} \cdot \frac{1}{4}$$ Use the distributive property and multiply each term by 12.

$$\overset{3}{\cancel{12}} \cdot \frac{x + 3}{\underset{1}{\cancel{4}}} > \overset{4}{\cancel{12}} \cdot \frac{x - 2}{\underset{1}{\cancel{3}}} + \overset{3}{\cancel{12}} \cdot \frac{1}{\underset{1}{\cancel{4}}}$$ Divide out common factors in each multiplication.

$$3(x + 3) > 4(x - 2) + 3$$ The fractions are now cleared.

Now that fractions are cleared, we follow the four steps that we used in the previous example.

Step 1 Simplify each side.

$$3(x + 3) > 4(x - 2) + 3$$ This is the inequality with the fractions cleared.

$$3x + 9 > 4x - 8 + 3$$ Use the distributive property.
$$3x + 9 > 4x - 5$$ Simplify.

Step 2 Collect variable terms on one side and constant terms on the other side.
We will collect variable terms on the left and constant terms on the right.

$$3x + 9 - 4x > 4x - 5 - 4x$$ Subtract 4x from both sides.

$$-x + 9 > -5$$ Simplify.

$$-x + 9 - 9 > -5 - 9$$ Subtract 9 from both sides.

$$-x > -14$$ Simplify.

Step 3 Isolate the variable and solve. To isolate x, we must eliminate the negative sign in front of the x. Because $-x$ means $-1x$, we can do this by multiplying (or dividing) both sides of the inequality by -1. We are multiplying by a negative number. Thus, we must reverse the inequality symbol.

$$(-1)(-x) < (-1)(-14)$$ Multiply both sides by -1 and reverse the sense of the inequality.

$$x < 14$$ Simplify.

Step 4 Express the solution set in set-builder or interval notation and graph the set on a number line. The solution set consists of all real numbers that are less than 14, expressed in set-builder notation as $\{x \mid x < 14\}$. The interval notation for this solution set is $(-\infty, 14)$. The graph of the solution set is shown as follows:

```
 ←+——+——+——+——+——+——+——+——+——)——+——→
    5   6   7   8   9  10  11  12  13  14  15
```

CHECK POINT 4   Solve and graph the solution set on a number line:

$$\frac{x-4}{2} > \frac{x-2}{3} + \frac{5}{6}.$$

③ Use linear inequalities to solve problems involving revenue, cost, and profit.

Functions of Business and Linear Inequalities

As a young entrepreneur, did you ever try selling lemonade in your front yard? Suppose that you charged 55 cents for each cup and you sold 45 cups. Your **revenue** is your income from selling these 45 units, or $\$0.55(45) = \$24.75$. Your *revenue function* from selling x cups is

$$R(x) = 0.55x.$$

This is the unit price: 55¢ for each cup.   This is the number of units sold.

For any business, the **revenue function**, $R(x)$, is the money generated by selling x units of the product:

$$R(x) = px.$$

Price per unit x units sold

Back to selling lemonade and energizing the neighborhood with white sugar: Is your revenue for the afternoon also your profit? No. We need to consider the cost of the business. You estimate that the lemons, white sugar, and bottled water cost 5 cents per cup. Furthermore, mommy dearest is charging you a $\$10$ rental fee for use of your (her?) front yard. Your *cost function* for selling x cups of lemonade is

$$C(x) = 10 + 0.05x.$$

This is your $\$10$ fixed cost.   This is your variable cost: 5¢ for each cup produced.

For any business, the **cost function**, $C(x)$, is the cost of producing x units of the product:

$$C(x) = (\text{fixed cost}) + cx.$$

Cost per unit x units produced

The term on the right, cx, represents **variable cost**, because it varies based on the number of units produced. Thus, the cost function is the sum of the fixed cost and the variable cost.

What does every entrepreneur, from a kid selling lemonade to Donald Trump, want to do? Generate profit, of course. The *profit* made is the money taken in, or the revenue, minus the money spent, or the cost. This relationship between revenue and cost allows us to define the *profit function*, $P(x)$.

The Profit Function

The profit, $P(x)$, generated after producing and selling x units of a product is given by the **profit function**

$$P(x) = R(x) - C(x)$$

where R and C are the revenue and cost functions, respectively.

Figure 9.1 shows the graphs of the revenue and cost functions for the lemonade business. Similar graphs and models apply no matter how small or large a business venture may be.

$$R(x) = 0.55x \qquad\qquad C(x) = 10 + 0.05x$$

Revenue is 55¢ times the number of cups sold.

Cost is $10 plus 5¢ times the number of cups produced.

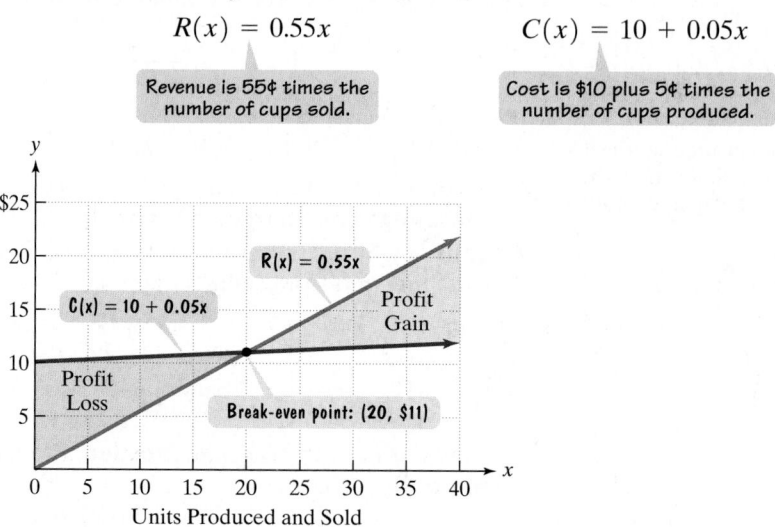

Figure 9.1

The lines intersect at the point $(20, 11)$. This means that when 20 cups are produced and sold, both cost and revenue are $11. In business, this point of intersection is called the **break-even point**. At the break-even point, the money coming in is equal to the money going out. Can you see what happens for x-values less than 20? The cost graph is above the revenue graph: $C(x) > R(x)$. The cost is greater than the revenue and the business is losing money. Thus, if you sell fewer than 20 cups of lemonade, the result is a *profit loss*. By contrast, look at what happens for x-values greater than 20. The revenue graph is above the cost graph: $R(x) > C(x)$. The revenue is greater than the cost and the business is making money. Thus, if you sell more than 20 cups of lemonade, the result is a *profit gain*.

EXAMPLE 5   Writing a Profit Function to Determine a Profit Gain

a. Use the revenue and cost functions shown in Figure 9.1,

$$R(x) = 0.55x \quad \text{and} \quad C(x) = 10 + 0.05x,$$

to write the profit function for producing and selling x cups of lemonade.

b. Use and solve a linear inequality to answer this question: More than how many cups of lemonade must be sold to have a profit gain?

Solution

a. The profit function is the difference between the revenue function and the cost function.

$$P(x) = R(x) - C(x)$$ This is the definition of the profit function.

$$= 0.55x - (10 + 0.05x)$$ Substitute the given functions.

$$= 0.55x - 10 - 0.05x$$ Distribute −1 to each term in parentheses.

$$= 0.50x - 10$$ Simplify.

The profit function is $P(x) = 0.50x - 10$.

b. We are interested in a profit gain. This occurs when $P(x) > 0$.

$$0.50x - 10 > 0$$ This inequality models a profit gain.

$$0.50x > 10$$ Add 10 to both sides.

$$\frac{0.50x}{0.50} > \frac{10}{0.50}$$ Divide both sides by 0.50. Division by a positive number does not reverse the sense of the inequality.

$$x > 20$$ Simplify.

More than 20 cups of lemonade must be sold to have a profit gain.   ∎

The graph of the profit function, $P(x) = 0.50x - 10$, is shown in Figure 9.2. The red portion of the graph lies below the x-axis and shows a profit loss when fewer than 20 units are sold. Thus, if $x < 20$ then $P(x) < 0$. The lemonade business is "in the red." The black portion of the graph lies above the x-axis and shows a profit gain when more than 20 units are sold. Thus, if $x > 20$ then $P(x) > 0$. The lemonade business is "in the black."

✔ **CHECK POINT 5**   **a.** Use the revenue and cost functions

$$R(x) = 200x$$
$$C(x) = 160,000 + 75x$$

to write the profit function for producing and selling x units.

b. More than how many units must be produced and sold to have a profit gain?

Discover for Yourself

A profit gain occurs when revenue exceeds cost:

$$R(x) > C(x).$$

Use $R(x) = 0.55x$ and $C(x) = 10 + 0.05x$ to verify that more than 20 cups of lemonade must be sold to have a profit gain.

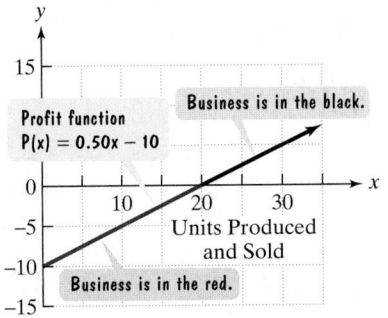

Figure 9.2 In the red and in the black

EXAMPLE 6   Writing a Profit Function to Determine a Profit Gain

Technology is now promising to bring light, fast, and beautiful wheelchairs to millions of disabled people. A company is planning to manufacture these radically different wheelchairs. Fixed cost will be $500,000 and it will cost $400 to produce each wheelchair. Each wheelchair will be sold for $600.

a. Write the cost function, C, of producing x wheelchairs.

b. Write the revenue function, R, from the sale of x wheelchairs.

c. Write the profit function, P, from producing and selling x wheelchairs.

d. More than how many wheelchairs must be produced and sold to have a profit gain?

Summary of Given Information

Fixed cost: $500,000
Cost to produce each wheelchair: $400
Selling price per wheelchair: $600

Solution

a. The cost function is the sum of the fixed cost and variable cost.

> Fixed cost of $500,000   plus   Variable cost: $400 for each chair produced

$$C(x) = 500,000 + 400x$$

b. The revenue function is the money generated from the sale of x wheelchairs.

> Revenue per chair, $600, times   the number of chairs sold

$$R(x) = 600x$$

c. The profit function is the difference between the revenue function and the cost function.

$$P(x) = R(x) - C(x)$$ This is the definition of the profit function.

$$= 600x - (500,000 + 400x)$$ Substitute the given functions.

$$= 600x - 500,000 - 400x$$ Distribute –1 to each term in parentheses.

$$= 200x - 500,000$$ Simplify.

The profit function is $P(x) = 200x - 500,000$.

d. A profit gain occurs when $P(x) > 0$.

$$200x - 500,000 > 0$$ Set the profit function greater than 0.

$$200x > 500,000$$ Add 500,000 to both sides.

$$\frac{200x}{200} > \frac{500,000}{200}$$ Divide both sides by 200.

$$x > 2500$$ Simplify.

More than 2500 wheelchairs must be produced and sold to have a profit gain.   ■

Using Technology

The graphs of the wheelchair company's cost and revenue functions

$$C(x) = 500,000 + 400x$$

and $R(x) = 600x$
are shown in a

$$[0, 5000, 1000]\text{ by}$$

$$[0, 3,000,000, 1,000,000]$$

viewing rectangle. To the right of the intersection point, the graph of the revenue function lies above that of the cost function. This confirms that producing and selling more than 2500 wheelchairs results in a profit gain.

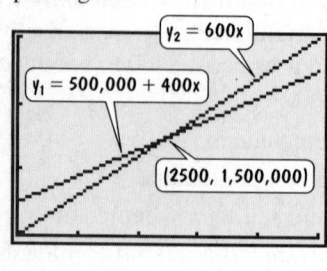

$y_2 = 600x$
$y_1 = 500,000 + 400x$
(2500, 1,500,000)

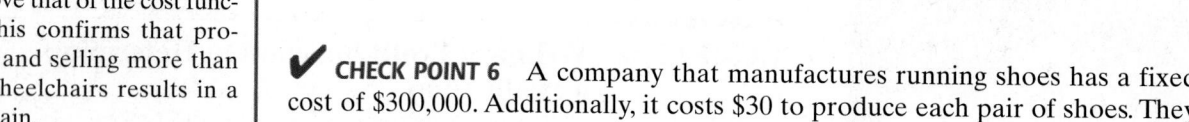

✔ **CHECK POINT 6**   A company that manufactures running shoes has a fixed cost of $300,000. Additionally, it costs $30 to produce each pair of shoes. They are sold at $80 per pair.

a. Write the cost function, C, of producing x pairs of running shoes.

b. Write the revenue function, R, from the sale of x pairs of running shoes.

c. Write the profit function, P, from producing and selling x pairs of running shoes.

d. More than how many pairs of running shoes must be produced and sold to have a profit gain?

EXERCISE SET 9.1

Practice Exercises

In Exercises 1–14, express each interval in terms of an inequality and graph the interval on a number line.

1. $(1, 6]$ **2.** $(-2, 4]$

3. $[-5, 2)$ **4.** $[-4, 3)$

5. $[-3, 1]$ **6.** $[-2, 5]$

7. $(2, \infty)$ **8.** $(3, \infty)$

9. $[-3, \infty)$ **10.** $[-5, \infty)$

11. $(-\infty, 3)$ **12.** $(-\infty, 2)$

13. $(-\infty, 5.5)$ **14.** $(-\infty, 3.5]$

In Exercises 15–34, solve each linear inequality and graph the solution set on a number line.

15. $5x + 11 < 26$ **16.** $2x + 5 < 17$

17. $3x - 8 \geq 13$ **18.** $8x - 2 \geq 14$

19. $-9x \geq 36$ **20.** $-5x \leq 30$

21. $8x - 11 \leq 3x - 13$ **22.** $18x + 45 \leq 12x - 8$

23. $4(x + 1) + 2 \geq 3x + 6$

24. $8x + 3 > 3(2x + 1) + x + 5$

25. $2x - 11 < -3(x + 2)$ **26.** $-4(x + 2) > 3x + 20$

27. $1 - (x + 3) \geq 4 - 2x$ **28.** $5(3 - x) \leq 3x - 1$

29. $\dfrac{x}{4} - \dfrac{1}{2} \leq \dfrac{x}{2} + 1$ **30.** $\dfrac{3x}{10} + 1 \geq \dfrac{1}{5} - \dfrac{x}{10}$

31. $1 - \dfrac{x}{2} > 4$ **32.** $7 - \dfrac{4}{5}x < \dfrac{3}{5}$

33. $\dfrac{x - 4}{6} \geq \dfrac{x - 2}{9} + \dfrac{5}{18}$ **34.** $\dfrac{4x - 3}{6} + 2 \geq \dfrac{2x - 1}{12}$

In Exercises 35–38, cost and revenue functions for producing and selling x units of a product are given. Cost and revenue are expressed in dollars.

a. *Write the profit function from producing and selling x units of the product.*

b. *More than how many units must be produced and sold to have a profit gain?*

35. $C(x) = 25{,}500 + 15x$
$R(x) = 32x$

36. $C(x) = 15{,}000 + 12x$
$R(x) = 32x$

37. $C(x) = 105x + 70{,}000$
$R(x) = 245x$

38. $C(x) = 1.2x + 1500$
$R(x) = 1.7x$

Application Exercises

The bar graph shows how we spend our leisure time. Let x represent the percentage of the population regularly participating in an activity. In Exercises 39–46, write the name or names of the activity described by the given inequality or interval.

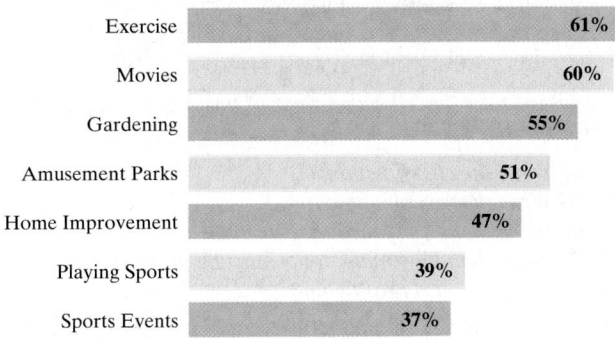

Percentage of U.S. Population Participating in Each Activity on a Regular Basis

Activity	Percentage
Exercise	61%
Movies	60%
Gardening	55%
Amusement Parks	51%
Home Improvement	47%
Playing Sports	39%
Sports Events	37%

Source: U.S. Census Bureau

39. $x < 40\%$

40. $x < 50\%$

41. $[51\%, 61\%]$

42. $[47\%, 60\%]$

43. $(51\%, 61\%)$
44. $(47\%, 60\%)$
45. $(39\%, 55\%]$

46. $(37\%, 47\%]$
47. The percentage, P, of U.S. adults who read the daily newspaper can be modeled by the formula

$$P = -0.7x + 80$$

where x is the number of years after 1965. In which years will fewer than 52% of U.S. adults read the daily newspaper?

48. The formula
$$W = 0.3x + 22.1$$
models the average number of hours, W, that Americans work per week x years after 1900. In which years will Americans average more than 55.1 hours of work per week?

The Olympic 500-meter speed skating times have generally been decreasing over time. The formulas

$$W = -0.19t + 57 \quad \text{and} \quad M = -0.15t + 50$$

model the winning times, in seconds, for women, W, and men, M, t years after 1900. Use these models to solve Exercises 49–50.

49. Find values of t such that $W < M$. Describe what this means in terms of winning times.

50. Find values of t such that $W > M$. Describe what this means in terms of winning times.

Exercises 51–54 describe a number of business ventures. For each exercise:

   a. *Write the cost function, C.*
   b. *Write the revenue function, R.*
   c. *Write the profit function, P.*
   d. *More than how many units must be produced and sold to have a profit gain?*

51. A company that manufactures small canoes has a fixed cost of $18,000. It costs $20 to produce each canoe. The selling price is $80 per canoe. (In solving this exercise, let x represent the number of canoes produced and sold.)

52. A company that manufactures bicycles has a fixed cost of $100,000. It costs $100 to produce each bicycle. The selling price is $300 per bike. (In solving this exercise, let x represent the number of bicycles produced and sold.)

53. You invest in a new play. The cost includes an overhead of $30,000, plus production costs of $2500 per performance. A sold-out performance brings in $3125. (In solving this exercise, let x represent the number of sold-out performances.)

54. You invested $30,000 and started a business writing greeting cards. Supplies cost 2¢ per card and you are selling each card for 50¢. (In solving this exercise, let x represent the number of cards produced and sold.)

55. A company manufactures and sells blank audiocassette tapes. The weekly fixed cost is $10,000 and it costs $0.40 to produce each tape. The selling price is $2.00 per tape. How many tapes must be produced and sold each week for the company to have a profit gain?

56. A company manufactures and sells personalized stationery. The weekly fixed cost is $3000 and it costs $3.00 to produce each package of stationery. The selling price is $5.50 per package. How many packages of stationery must be produced and sold each week for the company to have a profit gain?

57. You are choosing between two long-distance telephone plans. Plan A has a monthly fee of $15 with a charge of $0.08 per minute for all long-distance calls. Plan B has a monthly fee of $3 with a charge of $0.12 per minute for all long-distance calls. How many minutes of long-distance calls in a month make plan A the better deal?

58. A city commission has proposed two tax bills. The first bill requires that a homeowner pay $1800 plus 3% of the assessed home value in taxes. The second bill requires taxes of $200 plus 8% of the assessed home value. What price range of home assessment would make the first bill a better deal?

Writing in Mathematics

59. When graphing the solutions of an inequality, what does a parenthesis signify? What does a bracket signify?

60. Describe a revenue function for a business venture.

61. Describe a cost function for a business venture. What are the two kinds of costs that are modeled by this function?

62. What is the profit function for a business venture and how is it determined?

63. If the profit function for a business venture is known, explain how to find how many units must be produced and sold to have a profit gain.

64. The formula

$$V = 3.5x + 120$$

models Super Bowl viewers, V, in millions, x years after 1990. Use the formula to write a word problem that can be solved using a linear inequality. Then solve the problem.

Critical Thinking Exercises

65. Which one of the following is true?
   a. The inequality $3x > 6$ is equivalent to $2 > x$.
   b. The smallest real number in the solution set of $2x > 6$ is 4.
   c. If x is at least 7, then $x > 7$.
   d. The inequality $-3x > 6$ is equivalent to $-2 > x$.

66. What's wrong with this argument? Suppose x and x represent two real numbers, where $x > y$.

$2 > 1$	This is a true statement.
$2(y - x) > 1(y - x)$	Multiply both sides by $y - x$.
$2y - 2x > y - x$	Use the distributive property.
$y - 2x > -x$	Subtract y from both sides.
$y > x$	Add 2x to both sides.

The final inequality, $y > x$, is impossible because we were initially given $x > y$.

67. The graphs of $y = 6$, $y = 3(-x - 5) - 9$, and $y = 0$ are shown in the figure. The graph was obtained using a graphing utility and a $[-12, 1, 1]$ by $[-2, 8, 1]$ viewing rectangle. Use the graphs to write the solution set for the inequality $0 < 3(-x - 5) - 9 < 6$.

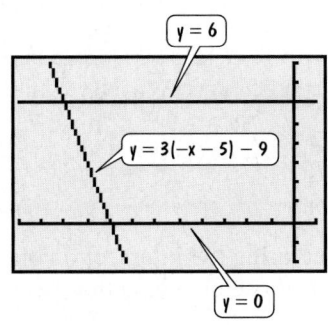

Technology Exercises

In Exercises 68–69, solve each inequality using a graphing utility. Graph each side separately. Then determine the values of x for which the graph on the left side lies above the graph on the right side.

68. $-3(x - 6) > 2x - 2$ **69.** $-2(x + 4) > 6x + 16$

70. A bank offers two checking account plans. Plan A has a base service charge of $4.00 per month plus 10¢ per check. Plan B charges a base service charge of $2.00 per month plus 15¢ per check.
   a. Write models for the total monthly costs for each plan if x checks are written.
   b. Use a graphing utility to graph the models in the same $[0, 50, 1]$ by $[0, 10, 1]$ viewing rectangle.
   c. Use the graphs (and the TRACE or intersection feature) to determine for what number of checks per month plan A will be better than plan B.
   d. Verify the result of part (c) algebraically by solving an inequality.

In Exercises 71–72, graph the revenue and cost functions in the same viewing rectangle. Then use the intersection feature to determine how many units must be produced and sold to have a profit gain.

71. $R(x) = 50x$, $C(x) = 20x + 180$

72. $R(x) = 92.5x$, $C(x) = 52x + 1782$

Review Exercises

73. If $f(x) = x^2 - 2x + 5$, find $f(-4)$. (Section 8.1, Example 3)

74. Solve the system:

$$2x - y - z = -3$$
$$3x - 2y - 2z = -5$$
$$-x + y + 2z = 4.$$

(Section 8.3, Example 2)

75. Factor: $25x^2 - 81$. (Section 6.4, Example 1)

▶ **SECTION 9.2**   *Compound Inequalities*

Objectives

1 Find the intersection of two sets.

2 Solve compound inequalities involving "and."

3 Find the union of two sets.

4 Solve compound inequalities involving "or."

SSM
PH Tutor   CD-   Video   www
Center   ROM

Boys and girls differ in their toy preferences. These differences are functions of our gender stereotypes—that is, our widely shared beliefs about males' and females' abilities, personality traits, and social behavior. Generally, boys have less leeway to play with "feminine" toys than girls do with "masculine" toys.

Which toys are requested by more than 40% of boys *and* more than 10% of girls? Which toys are requested by more than 40% of boys *or* more than 10% of girls? These questions are not the same. One involves inequalities joined by the word "and"; the other involves inequalities joined by the word "or." A **compound inequality** is formed by joining two inequalities with the word *and* or *or*.

Examples of Compound Inequalities

$$x - 3 < 5 \quad \text{and} \quad 2x + 4 < 14$$
$$3x - 5 \le 13 \quad \text{or} \quad 5x + 2 > -3$$

In this section, you will learn to solve compound inequalities. With this skill, you will analyze childrens' toy preferences in the exercise set.

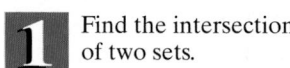

1 Find the intersection of two sets.

Compound Inequalities Involving "And"

You need to determine whether there is sufficient support on campus to have a blood drive. You take a survey to obtain information, asking students the following:

> Would you be willing to donate blood?
>
> Would you be willing to help serve a free breakfast to blood donors?

Part of your report involves the number of students willing to both donate blood *and* serve breakfast.

You must focus on the set of all students who are common to both the set of donors and the set of breakfast servers. This set is called the *intersection* of the two sets.

Definition of the Intersection of Sets

The **intersection** of sets A and B, written $A \cap B$, is the set of elements common to both set A and set B. This definition can be expressed in set-builder notation as follows:

$$A \cap B = \{x \mid x \in A \text{ AND } x \in B\}.$$

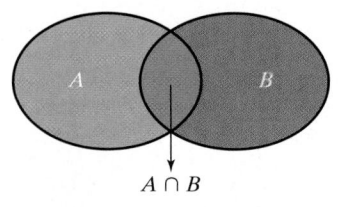

$A \cap B$

Figure 9.3 Picturing the intersection of two sets

Figure 9.3 shows a useful way of picturing the intersection of sets A and B. The figure indicates that $A \cap B$ contains those elements that belong to both A and B at the same time.

EXAMPLE 1 Finding the Intersection of Two Sets

Find the intersection: $\{7, 8, 9, 10, 11\} \cap \{6, 8, 10, 12\}$.

Solution The elements common to $\{7, 8, 9, 10, 11\}$ and $\{6, 8, 10, 12\}$ are 8 and 10. Thus,

$$\{7, 8, 9, 10, 11\} \cap \{6, 8, 10, 12\} = \{8, 10\}. \qquad \blacksquare$$

✔ **CHECK POINT 1** Find the intersection: $\{3, 4, 5, 6, 7\} \cap \{3, 7, 8, 9\}$.

2 Solve compound inequalities involving "and."

A number is a **solution of a compound inequality** formed by the word "and" if it is a solution of both inequalities. For example, the solution set of the compound inequality

$$x \le 6 \quad \text{and} \quad x \ge 2$$

is the set of values of x that satisfy $x \le 6$ and $x \ge 2$. Thus, the solution set is the intersection of the solution sets of the two inequalities.

What are the numbers that satisfy both $x \le 6$ and $x \ge 2$? These numbers are easier to see if we graph the solution set to each inequality on a number line. These graphs are shown in Figure 9.4. The intersection is shown in the third graph.

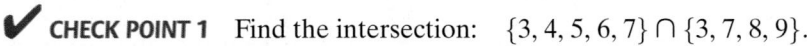

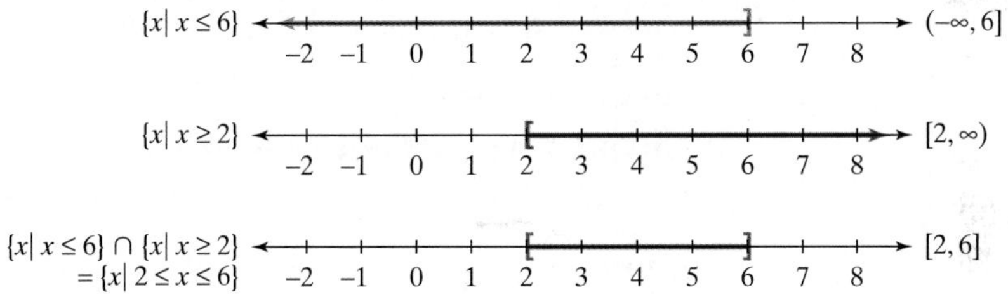

Figure 9.4 Numbers satisfying both $x \le 6$ and $x \ge 2$

The numbers common to both sets are those that are less than or equal to 6 and greater than or equal to 2. This set is $\{x \mid 2 \le x \le 6\}$, or, in interval notation, $[2, 6]$.

Here is a procedure for finding the solution set of a compound inequality containing the word "and."

Solving Compound Inequalities Involving "And"

1. Solve each inequality separately.

2. Graph the solution set to each inequality on a number line and take the intersection of these solution sets.

EXAMPLE 2 Solving a Compound Inequality with "And"

Solve: $x - 3 < 5$ and $2x + 4 < 14$.

Solution

Step 1 Solve each inequality separately.

$$\begin{array}{ccc} x - 3 < 5 & \text{and} & 2x + 4 < 14 \\ x < 8 & & 2x < 10 \\ & & x < 5 \end{array}$$

Step 2 Take the intersection of the solution sets of the two inequalities. We graph the solution sets of $x < 8$ and $x < 5$. The intersection is shown in the third graph.

$\{x \mid x < 8\}$ <img> $(-\infty, 8)$

$\{x \mid x < 5\}$ <img> $(-\infty, 5)$

$\{x \mid x < 8\} \cap \{x \mid x < 5\}$
$= \{x \mid x < 5\}$ <img> $(-\infty, 5)$

The numbers common to both sets are those that are less than 5. The solution set is $\{x \mid x < 5\}$, or, in interval notation, $(-\infty, 5)$. Take a moment to check that any number in $(-\infty, 5)$ satisfies both of the original inequalities. ∎

✔ **CHECK POINT 2** Solve: $x + 2 < 5$ and $2x - 4 < -2$.

EXAMPLE 3   Solving a Compound Inequality with "And"

Solve: $2x - 7 > 3$ and $5x - 4 < 6$.

Solution

Step 1 Solve each inequality separately.

$$2x - 7 > 3 \qquad \text{and} \qquad 5x - 4 < 6$$
$$2x > 10 \qquad\qquad\qquad 5x < 10$$
$$x > 5 \qquad\qquad\qquad x < 2$$

Step 2 Take the intersection of the solution sets of the two inequalities. We graph the solution sets of $x > 5$ and $x < 2$. We use these graphs to find their intersection.

$\{x \mid x > 5\}$ <img> $(5, \infty)$

$\{x \mid x < 2\}$ <img> $(-\infty, 2)$

$\{x \mid x > 5\} \cap \{x \mid x < 2\}$
$= \varnothing$ <img> $\varnothing$

There is no number that is both greater than 5 and less than 2. Thus, the solution set is the empty set, $\varnothing$. ∎

✔ **CHECK POINT 3** Solve: $4x - 5 > 7$ and $5x - 2 < 3$.

If $a < b$, the compound inequality

$$a < x \text{ and } x < b$$

can be written in the shorter form

$$a < x < b.$$

For example, the compound inequality
$$-3 < 2x + 1 \text{ and } 2x + 1 < 3$$
can be abbreviated
$$-3 < 2x + 1 < 3.$$
The word "and" does not appear when the inequality is written in the shorter form, although it is implied. The shorter form enables us to solve both inequalities at once. By performing the same operations on all three parts of the inequality, our goal is to **isolate x in the middle**.

EXAMPLE 4   Solving a Compound Inequality

Solve and graph the solution set:
$$-3 < 2x + 1 \le 3.$$

Solution   We would like to isolate x in the middle. We can do this by first subtracting 1 from all three parts of the compound inequality. Then we isolate x from $2x$ by dividing all three parts of the inequality by 2.

$-3 < 2x + 1 \le 3$	This is the given inequality.
$-3 - 1 < 2x + 1 - 1 \le 3 - 1$	Subtract 1 from all three parts.
$-4 < 2x \le 2$	Simplify.
$\dfrac{-4}{2} < \dfrac{2x}{2} \le \dfrac{2}{2}$	Divide each part by 2.
$-2 < x \le 1$	Simplify.

The solution set consists of all real numbers greater than -2 and less than or equal to 1, represented by $\{x \mid -2 < x \le 1\}$ in set-builder notation and $(-2, 1]$ in interval notation. The graph is shown as follows:

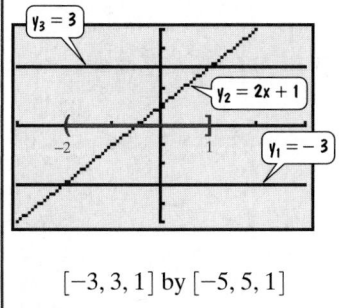

✔ **CHECK POINT 4**   Solve and graph the solution set: $1 \le 2x + 3 < 11$.

Compound Inequalities Involving "Or"

You continue to sort the results of the survey. Part of your report involves the number of students willing to donate blood *or* serve breakfast *or* do both. You must focus on the set containing donors or the set of breakfast servers or both of these sets. This set is called the *union* of two sets.

Definition of the Union of Sets

The **union** of sets A and B, written $A \cup B$, is the set of elements that are members of set A or of set B or of both sets. This definition can be expressed in set-builder notation as follows:
$$A \cup B = \{x \mid x \in A \text{ OR } x \in B\}.$$

Figure 9.5 shows a useful way of picturing the union of sets A and B. The figure indicates that $A \cup B$ is formed by joining the sets together.

We can find the union of set A and set B by listing the elements of set A. Then, we include any elements of set B that have not already been listed. Enclose all elements that are listed with braces. This shows that the union of two sets is also a set.

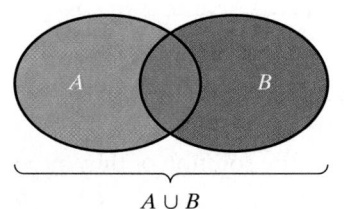
3 Find the union of two sets.

Figure 9.5 Picturing the union of two sets

EXAMPLE 5   Finding the Union of Two Sets

Find the union: $\{7, 8, 9, 10, 11\} \cup \{6, 8, 10, 12\}$.

Solution   To find $\{7, 8, 9, 10, 11\} \cup \{6, 8, 10, 12\}$, start by listing all the elements from the first set, namely 7, 8, 9, 10, and 11. Now list all the elements from the second set that are not in the first set, namely 6 and 12. The union is the set consisting of all these elements. Thus,

$$\{7, 8, 9, 10, 11\} \cup \{6, 8, 10, 12\} = \{6, 7, 8, 9, 10, 11, 12\}.$$ ■

✔ **CHECK POINT 5**   Find the union: $\{3, 4, 5, 6, 7\} \cup \{3, 7, 8, 9\}$.

4 Solve compound inequalities involving "or."

A number is a **solution of a compound inequality** formed by the word "or" if it is a solution of either inequality. Thus, the solution set of a compound inequality formed by the word "or" is the union of the solution sets of the two inequalities.

Solving Compound Inequalities Involving "Or"

1. Solve each inequality separately.
2. Graph the solution set to each inequality on a number line and take the union of these solution sets.

EXAMPLE 6   Solving a Compound Inequality with "Or"

Solve: $2x - 3 < 7$ or $35 - 4x \le 3$.

Solution

Step 1 Solve each inequality separately.

$$
\begin{array}{ccc}
2x - 3 < 7 & \text{or} & 35 - 4x \le 3 \\
2x < 10 & & -4x \le -32 \\
x < 5 & & x \ge 8
\end{array}
$$

Step 2 Take the union of the solution sets of the two inequalities. We graph the solution sets of $x < 5$ and $x \ge 8$. We use these graphs to find their union.

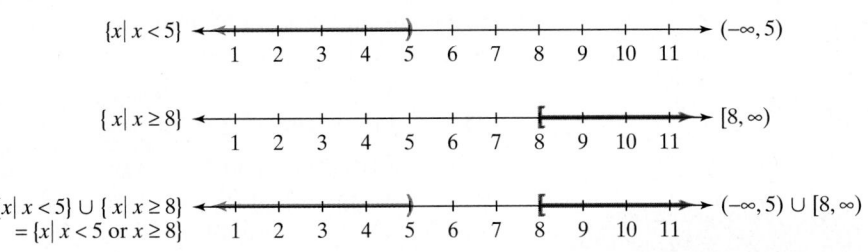

The solution set consists of all numbers that are less than 5 or greater than or equal to 8. The solution set is $\{x \mid x < 5 \text{ or } x \ge 8\}$, or, in interval notation, $(-\infty, 5) \cup [8, \infty)$. There is no shortcut way to express the solution of this union when interval notation is used. ■

✔ **CHECK POINT 6**   Solve: $3x - 5 \le -2$ or $10 - 2x < 4$.

EXAMPLE 7   Solving a Compound Inequality with "Or"

Solve: $3x - 5 \le 13$ or $5x + 2 > -3$.

Solution

Step 1 Solve each inequality separately.

$$3x - 5 \le 13 \qquad \text{or} \qquad 5x + 2 > -3$$
$$3x \le 18 \qquad\qquad\qquad 5x > -5$$
$$x \le 6 \qquad\qquad\qquad\quad x > -1$$

Step 2 Take the union of the solution sets of the two inequalities. We graph the solution sets of $x \le 6$ and $x > -1$. We use these graphs to find their union.

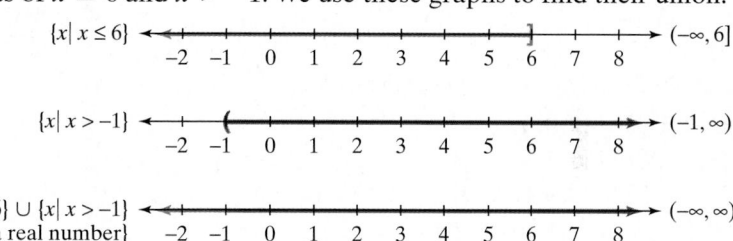

Because all real numbers are less than or equal to 6 or greater than -1, the union of the two sets fills the entire number line. Thus, the solution set is $\{x \mid x \text{ is a real number}\}$, or $\mathbb{R}$. The solution set in interval notation is $(-\infty, \infty)$. Any real number that you select will satisfy either or both of the original inequalities. ■

✔ **CHECK POINT 7**   Solve: $2x + 5 \ge 3$ or $2x + 3 < 3$.

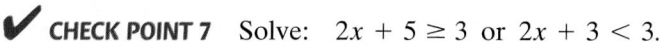

EXERCISE SET 9.2

Practice Exercises

In Exercises 1–4, find the intersection of the sets.

1. $\{1, 2, 3, 4\} \cap \{2, 4, 5\}$

2. $\{1, 3, 7\} \cap \{2, 3, 8\}$

3. $\{1, 3, 5, 7\} \cap \{2, 4, 6, 8, 10\}$

4. $\{0, 1, 3, 5\} \cap \{-5, -3, -1\}$

In Exercises 5–22, solve each compound inequality. Use graphs to show the solution set to each of the two given inequalities, as well as a third graph that shows the solution set of the compound inequality. Except for the empty set, express the solution set in both set-builder and interval notations.

5. $x > 3$ and $x > 6$

6. $x > 2$ and $x > 4$

7. $x \le 5$ and $x \le 1$

8. $x \le 6$ and $x \le 2$

9. $x < 2$ and $x \ge -1$

10. $x < 3$ and $x \ge -1$

11. $x > 2$ and $x < -1$

12. $x > 3$ and $x < -1$

13. $5x < -20$ and $3x > -18$

14. $3x \le 15$ and $2x > -6$

15. $x - 4 \le 2$ and $3x + 1 > -8$

16. $3x + 2 > -4$ and $2x - 1 < 5$

17. $2x > 5x - 15$ and $7x > 2x + 10$

18. $6 - 5x > 1 - 3x$ and $4x - 3 > x - 9$

19. $4(1 - x) < -6$ and $\dfrac{x - 7}{5} \le -2$

20. $5(x - 2) > 15$ and $\dfrac{x - 6}{4} \le -2$

21. $x - 1 \le 7x - 1$ and $4x - 7 < 3 - x$

22. $2x + 1 > 4x - 3$ and $x - 1 \ge 3x + 5$

In Exercises 23–30, solve each inequality and graph the solution set on a number line. Express the solution set in both set-builder and interval notations.

23. $6 < x + 3 < 8$

24. $7 < x + 5 < 11$

25. $-3 \le x - 2 < 1$

26. $-6 < x - 4 \le 1$

27. $-11 < 2x - 1 \le -5$

28. $3 \le 4x - 3 < 19$

29. $-3 \le \dfrac{2x}{3} - 5 < -1$

30. $-6 \le \dfrac{x}{2} - 4 < -3$

In Exercises 31–34, find the union of the sets.

31. $\{1, 2, 3, 4\} \cup \{2, 4, 5\}$

32. $\{1, 3, 7, 8\} \cup \{2, 3, 8\}$

33. $\{1, 3, 5, 7\} \cup \{2, 4, 6, 8, 10\}$

34. $\{0, 1, 3, 5\} \cup \{2, 4, 6\}$

In Exercises 35–50, solve each compound inequality. Use graphs to show the solution set to each of the two given inequalities, as well as a third graph that shows the solution set of the compound inequality. Express the solution set in both set-builder and interval notations.

35. $x > 3$ or $x > 6$

36. $x > 2$ or $x > 4$

37. $x \le 5$ or $x \le 1$

38. $x \le 6$ or $x \le 2$

39. $x < 2$ or $x \ge -1$

40. $x < 3$ or $x \ge -1$

41. $x \ge 2$ or $x < -1$

42. $x \ge 3$ or $x < -1$

43. $3x > 12$ or $2x < -6$

44. $3x < 3$ or $2x > 10$

45. $3x + 2 \le 5$ or $5x - 7 \ge 8$

46. $2x - 5 \le -11$ or $5x + 1 \ge 6$

47. $4x + 3 < -1$ or $2x - 3 \ge -11$

48. $2x + 1 < 15$ or $3x - 4 \ge -1$

49. $-2x + 5 > 7$ or $-3x + 10 > 2x$

50. $16 - 3x \ge -8$ or $13 - x > 4x + 3$

Application Exercises

As a result of cultural expectations about what is appropriate behavior for each gender, boys and girls differ substantially in their toy preferences. The graph shows the percentage of boys and girls asking for various types of toys in letters to Santa Claus. Use the information in the graph to solve Exercises 51–56.

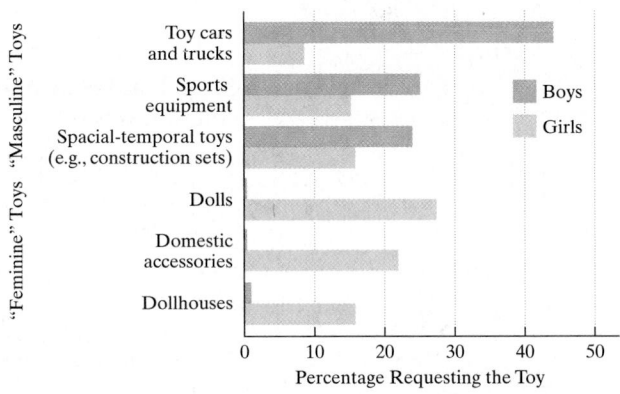

Toys Requested by Children

Source: Richardson, J. G., & Simpson, C. H. (1982). Children, gender and social structure: An analysis of the contents of letters to Santa Claus. *Child Development, 53,* 429–436.

51. Which toys were requested by more than 10% of the boys *and* less than 20% of the girls?

52. Which toys were requested by fewer than 5% of the boys *and* fewer than 20% of the girls?

53. Which toys were requested by more than 10% of the boys *or* less than 20% of the girls?

54. Which toys were requested by fewer than 5% of the boys *or* fewer than 20% of the girls?

55. Which toys were requested by more than 40% of the boys *and* more than 10% of the girls?

56. Which toys were requested by more than 40% of the boys *or* more than 10% of the girls?

57. The formula

$$T = 0.01x + 56.7$$

models the global mean temperature, T, in degrees Fahrenheit, of Earth x years after 1905. For which range of years was the global mean temperature at least 52°F and at most 57.2°F?

58. The formula for converting Fahrenheit temperature, F, to Celsius temperature, C, is

$$C = \frac{5}{9}(F - 32).$$

If Celsius temperature ranges from 15° to 35°, inclusive, what is the range for the Fahrenheit temperature? Use interval notation to express this range.

59. On the first four exams, your grades are 70, 75, 87, and 92. There is still one more exam, and you are hoping to earn a B in the course. This will occur if the average of your five exam grades is greater than or equal to 80 and less than 90. What range of grades on the fifth exam will result in earning a B? Use interval notation to express this range.

60. On the first four exams, your grades are 82, 75, 80, and 90. There is still a final exam, and it counts as two grades. You are hoping to earn a B in the course: This will occur if the average of your six exam grades is greater than or equal to 80 and less than 90. What range of grades on the final exam will result in earning a B? Use interval notation to express this range.

61. The toll to a bridge is $3.00. A three-month pass costs $7.50 and reduces the toll to $0.50. A six-month pass costs $30 and permits crossing the bridge for no additional fee. How many crossings per three-month period does it take for the three-month pass to be the best deal?

62. Parts for an automobile repair cost $175. The mechanic charges $34 per hour. If you receive an estimate for at least $226 and at most $294 for fixing the car, what is the time interval that the mechanic will be working on the job?

Writing in Mathematics

63. Describe what is meant by the intersection of two sets. Give an example.

64. Explain how to solve a compound inequality involving "and."

65. Why is $1 < 2x + 3 < 9$ a compound inequality? Where are the two inequalities and what is the word that joins them?

66. Explain how to solve $1 < 2x + 3 < 9$.

67. Describe what is meant by the union of two sets. Give an example.

68. Explain how to solve a compound inequality involving "or."

69. How many Christmas trees are sold each holiday season? The function $T(x) = 0.9x + 32$ models the number of trees sold, $T(x)$, in millions, x years after 1990. Use the function to write a problem that can be solved using a compound inequality. Then solve the problem.

Critical Thinking Exercises

70. Which one of the following is true?
 a. $(-\infty, -1] \cap [-4, \infty) = [-4, -1]$
 b. $(-\infty, 3) \cup (-\infty, -2) = (-\infty, -2)$
 c. The union of two sets can never give the same result as the intersection of those sets.
 d. The solution set of $x < a$ and $x > a$ is the set of all real numbers excluding a.

In Exercises 71–72, solve each compound inequality. Express solution sets in interval notation.

71. $-7 \le 8 - 3x \le 20$ and $-7 < 6x - 1 < 41$

72. $2x - 3 < 3x + 1 < 6x + 2$

73. Graph the linear function $f(x) = -x + 4$. Then use your graph to solve $2 < 4 - x < 7$.

74. Graph the linear function $f(x) = x + 3$. Then use your graph to solve $x + 3 > 3$ or $x + 3 < 2$.

The graphs of $f(x) = \sqrt{4 - x}$ and $g(x) = \sqrt{x + 1}$ are shown in a $[-3, 10, 1]$ by $[-2, 5, 1]$ viewing rectangle.

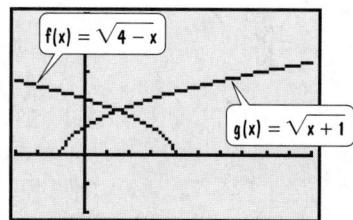

In Exercises 75–78, use the graphs and interval notation to express the domain of the given function.

75. The domain of f

76. The domain of g

77. The domain of $f + g$

78. The domain of $\dfrac{f}{g}$

79. At the end of the day, the change machine at a laundrette contained at least $3.20 and at most $5.45 in nickels, dimes, and quarters. There were 3 fewer dimes than twice the number of nickels and 2 more quarters than twice the number of nickels. What was the least possible number and the greatest possible number of nickels?

Technology Exercises

In Exercises 80–83, solve each inequality using a graphing utility. Graph each of the three parts of the inequality separately in the same viewing rectangle. The solution set consists of all values of x for which the graph of the linear function in the middle lies between the graphs of the constant functions on the left and the right.

80. $1 < x + 3 < 9$

81. $-1 < \dfrac{x + 4}{2} < 3$

82. $1 \leq 4x - 7 \leq 3$

83. $2 \leq 4 - x \leq 7$

Review Exercises

84. If $f(x) = x^2 - 3x + 4$ and $g(x) = 2x - 5$, find $(g - f)(x)$ and $(g - f)(-1)$. (Section 8.2, Example 4)

85. Graph: $y = -\dfrac{2}{3}x + 4$. (Section 3.4, Example 3)

86. Simplify: $4 - [2(x - 4) - 5]$. (Section 1.8, Example 11)

▶ SECTION 9.3   Equations and Inequalities Involving Absolute Value

Objectives

1 Solve equations involving absolute value.

2 Solve inequalities involving absolute value.

3 Recognize absolute value inequalities with no solution or infinitely many solutions.

4 Solve problems using absolute value inequalities.

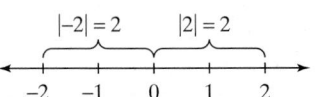

SSM PH Tutor CD- Video
Center ROM

M*A*S*H took place in the early 1950s during the Korean War. By the final episode, the show had lasted four times as long as the Korean War.

At the end of the twentieth century, there were 94 million households in the United States with television sets. The television program viewed by the greatest percentage of such households in that century was the final episode of *M*A*S*H*. Over 50 million American households watched this program.

Numerical information, such as the number of households watching a television program, is often given with a margin of error. Inequalities involving absolute value are used to describe errors in polling, as well as errors of measurement in manufacturing, engineering, science, and other fields. In this section, you will learn to solve equations and inequalities containing absolute value. With these skills, you will analyze the percentage of households that watched the final episode of *M*A*S*H*.

Equations Involving Absolute Value

We have seen that the absolute value of a, $|a|$, is the distance from 0 to a on the number line. Now consider **absolute value equations**, such as

$$|x| = 2.$$

This means that we must determine real numbers whose distance from the origin on the number line is 2. Figure 9.6 shows that there are two numbers such that $|x| = 2$, namely, 2 and -2. We write $x = 2$ or $x = -2$. This observation can be generalized as follows:

1 Solve equations involving absolute value.

$|-2| = 2$ $|2| = 2$

-2 -1   0   1   2

Figure 9.6 If $|x| = 2$, then $x = 2$ or $x = -2$.

Rewriting an Absolute Value Equation without Absolute Value Bars

If c is a positive real number and X represents any algebraic expression, then $|X| = c$ is equivalent to $X = c$ or $X = -c$.

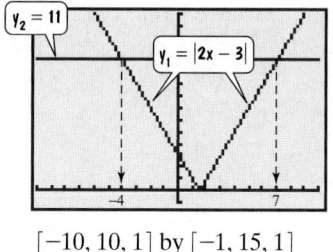
EXAMPLE 1   Solving an Equation Involving Absolute Value

Solve: $|2x - 3| = 11.$

Solution

$\|2x - 3\| = 11$		This is the given equation.
$2x - 3 = 11$ or $2x - 3 = -11$		Rewrite the equation without absolute value bars.
$2x = 14$	$2x = -8$	Add 3 to both sides of each equation.
$x = 7$	$x = -4$	Divide both sides of each equation by 2.

Check 7:	**Check $-$4:**	
$\|2x - 3\| = 11$	$\|2x - 3\| = 11$	This is the original equation.
$\|2(7) - 3\| \overset{?}{=} 11$	$\|2(-4) - 3\| \overset{?}{=} 11$	Substitute the proposed solutions.
$\|14 - 3\| \overset{?}{=} 11$	$\|-8 - 3\| \overset{?}{=} 11$	Perform operations inside the absolute value bars.
$\|11\| \overset{?}{=} 11$	$\|-11\| \overset{?}{=} 11$	
$11 = 11,$ true	$11 = 11,$ true	These true statements indicate that 7 and -4 are solutions.

The solutions are -4 and 7. We can also say that the solution set is $\{-4, 7\}$. ∎

✔ **CHECK POINT 1**   Solve: $|2x - 1| = 5.$

The absolute value of a number is never negative. Thus, if X is an algebraic expression and c is a negative number, then $|X| = c$ has no solution. For example, the equation $|3x - 6| = -2$ has no solution because $|3x - 6|$ cannot be negative. The solution set is $\varnothing$, the empty set.

The absolute value of 0 is 0. Thus, if X is an algebraic expression and $|X| = 0$, the solution is found by solving $X = 0$. For example, the solution of $|x - 2| = 0$ is obtained by solving $x - 2 = 0$. The solution is 2 and the solution set is $\{2\}$.

To solve some absolute value equations, it is necessary to first isolate the expression containing the absolute value symbols. For example, consider the equation

$$3|2x - 3| - 8 = 25.$$

We need to isolate $|2x - 3|$.

How can we isolate $|2x - 3|$? Add 8 to both sides of the equation and then divide both sides by 3.

$3\|2x - 3\| - 8 = 25$	This is the given equation.
$3\|2x - 3\| = 33$	Add 8 to both sides.
$\|2x - 3\| = 11$	Divide both sides by 3.

This results in the equation we solved in Example 1.

Some equations have two absolute value expressions, such as

$$|3x - 1| = |x + 5|.$$

These absolute value expressions are equal when the expressions inside the absolute value bars are equal to or opposites of each other.

Rewriting an Absolute Value Equation with Two Absolute Values without Absolute Value Bars

If $|X_1| = |X_2|$, then $X_1 = X_2$ or $X_1 = -X_2$.

EXAMPLE 2   Solving an Absolute Value Equation with Two Absolute Values

Solve: $|3x - 1| = |x + 5|$.

Solution   We rewrite the equation without absolute value bars.

| $|X_1| = |X_2|$ | means | $X_1 = X_2$ | or | $X_1 = -X_2.$ |

$|3x - 1| = |x + 5|$ means $3x - 1 = x + 5$ or $3x - 1 = -(x + 5)$.

We now solve the two equations that do not contain absolute value bars.

$$3x - 1 = x + 5 \quad \text{or} \quad 3x - 1 = -(x + 5)$$
$$2x - 1 = 5 \qquad\qquad 3x - 1 = -x - 5$$
$$2x = 6 \qquad\qquad 4x - 1 = -5$$
$$x = 3 \qquad\qquad 4x = -4$$
$$\qquad\qquad\qquad x = -1$$

Take a moment to complete the solution process by checking the two proposed solutions in the original equation. The solutions are -1 and 3, and the solution set is $\{-1, 3\}$. ∎

✔ **CHECK POINT 2**   Solve: $|2x - 7| = |x + 3|$.

2 Solve inequalities involving absolute value.

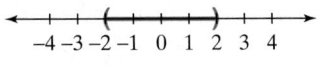

Figure 9.7 $|x| < 2$, so $-2 < x < 2$.

Figure 9.8 $|x| > 2$, so $x < -2$ or $x > 2$.

Inequalities Involving Absolute Value

Absolute value can also arise in inequalities. Consider, for example,

$$|x| < 2.$$

This means that the distance of x from 0 is *less than* 2, as shown in Figure 9.7. The interval shows values of x that lie less than 2 units from 0. Thus, x can lie between -2 and 2. That is, x is greater than -2 and less than 2: $-2 < x < 2$.

Some absolute value inequalities use the "greater than" symbol. For example, $|x| > 2$ means that the distance of x from 0 is *greater than* 2, as shown in Figure 9.8. Thus, x can be less than -2 *or* greater than 2: $x < -2$ or $x > 2$.

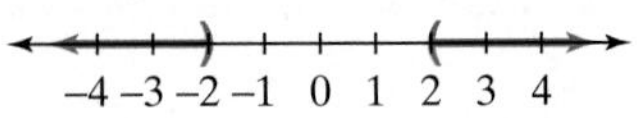

These observations suggest the following principles for solving inequalities with absolute value.

Solving an Absolute Value Inequality

If X is an algebraic expression and c is a positive number:

1. The solutions of $|X| < c$ are the numbers that satisfy $-c < X < c$.
2. The solutions of $|X| > c$ are the numbers that satisfy $X < -c$ or $X > c$.

These rules are valid if $<$ is replaced by $\leq$ and $>$ is replaced by $\geq$.

EXAMPLE 3   Solving an Absolute Value Inequality with <

Solve and graph the solution set on a number line: $|x - 4| < 3$.

Solution   We rewrite the inequality without absolute value bars.

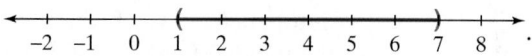

$$|x - 4| < 3 \quad \text{means} \quad -3 < x - 4 < 3.$$

We solve the compound inequality by adding 4 to all three parts.

$$-3 < x - 4 < 3$$
$$-3 + 4 < x - 4 + 4 < 3 + 4$$
$$1 < x < 7$$

The solution set is all real numbers greater than 1 and less than 7, denoted by $\{x \mid 1 < x < 7\}$ or $(1, 7)$. The graph of the solution set is shown as follows:

$$\xrightarrow[\hspace{0.3cm}-2\hspace{0.3cm}-1\hspace{0.3cm}0\hspace{0.3cm}(\hspace{0.15cm}1\hspace{0.3cm}2\hspace{0.3cm}3\hspace{0.3cm}4\hspace{0.3cm}5\hspace{0.3cm}6\hspace{0.3cm})\hspace{0.15cm}7\hspace{0.3cm}8]{}$$

We can use the rectangular coordinate system to visualize the solution set of

$$|x - 4| < 3.$$

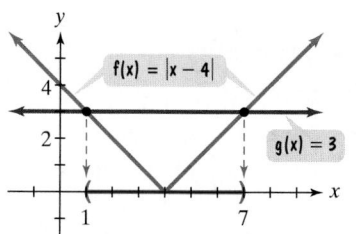

Figure 9.9 The solution set of $|x - 4| < 3$ is $(1, 7)$.

Figure 9.9 shows the graphs of $f(x) = |x - 4|$ and $g(x) = 3$. The solution set of $|x - 4| < 3$ consists of all values of x for which the blue graph of f lies below the red graph of g. These x-values make up the interval $(1, 7)$, which is the solution set.

✔ **CHECK POINT 3**   Solve and graph the solution set on a number line: $|x - 2| < 5$.

EXAMPLE 4   Solving an Absolute Value Inequality with ≥

Solve and graph the solution set on a number line: $|2x + 3| \geq 5$.

Solution   We rewrite the inequality without absolute value bars.

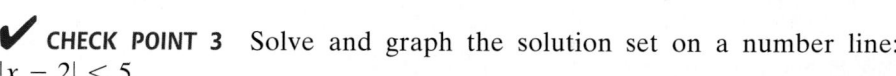

$$|2x + 3| \geq 5 \quad \text{means} \quad 2x + 3 \leq -5 \quad \text{or} \quad 2x + 3 \geq 5.$$

We solve this compound inequality by solving each of these inequalities separately. Then we take the union of their solution sets.

$$2x + 3 \leq -5 \quad \text{or} \quad 2x + 3 \geq 5 \qquad \text{These are the inequalities without absolute value bars.}$$

$$2x \leq -8 \quad \text{or} \quad 2x \geq 2 \qquad \text{Subtract 3 from both sides.}$$

$$x \leq -4 \quad \text{or} \quad x \geq 1 \qquad \text{Divide both sides by 2.}$$

The solution set consists of all numbers that are less than or equal to -4 or greater than or equal to 1. The solution set is $\{x \mid x \leq -4 \text{ or } x \geq 1\}$, or, in interval notation, $(-\infty, -4] \cup [1, \infty)$. The graph of the solution set is shown as follows:

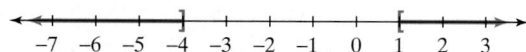

✔ **CHECK POINT 4** Solve and graph the solution set on a number line: $|2x - 5| \geq 3$.

③ Recognize absolute value inequalities with no solution or infinitely many solutions.

Absolute Value Inequalities with Unusual Solution Sets

If c is a positive number, the solutions of $|X| < c$ comprise a single interval. By contrast, the solutions of $|X| > c$ make up two intervals.

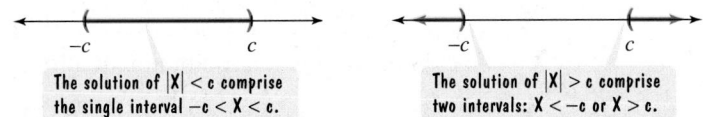

The solution of $|X| < c$ comprise the single interval $-c < X < c$.

The solution of $|X| > c$ comprise two intervals: $X < -c$ or $X > c$.

Now let's see what happens to these inequalities if c is a negative number. Consider, for example, $|x| < -2$. Because $|x|$ always has a value that is greater than or equal to 0, there is no number whose absolute value is less than -2. The inequality $|x| < -2$ has no solution. The solution set is $\varnothing$.

Now consider the inequality $|x| > -2$. Because $|x|$ is never negative, all numbers have an absolute value that is greater than -2. All real numbers satisfy the inequality $|x| > -2$. The solution set is $(-\infty, \infty)$.

Absolute Value Inequalities with Unusual Solution Sets

If X is an algebraic expression and c is a negative number:

1. The inequality $|X| < c$ has no solution.

2. The inequality $|X| > c$ is true for all real numbers for which X is defined.

---

④ Solve problems using absolute value inequalities.

Applications

When you were between the ages of 6 and 14, how would you have responded to this question:

> What is bad about being a kid?

In a random sample of 1172 children ages 6 to 14, 17% of the children responded, "Getting bossed around." The problem is that this is a single random sample. Do 17% of kids in the entire population of children ages 6 to 14 think that getting bossed around as a kid is a bad thing?

If you look at the results of a poll like the one in Table 9.2, you will observe that a **margin of error** is reported. The margin of error is $\pm 2.9\%$. This means that the actual percentage of children who feel getting bossed around is a bad thing is at most 2.9% greater than or less than 17%. If x represents the percentage of children in the population who think that getting bossed around is a bad thing, then the poll's margin of error can be expressed as an absolute value inequality:

$$|x - 17| \leq 2.9.$$

Table 9.2 What Is Bad About Being a Kid?

Kids Say	
Getting bossed around	17%
School, homework	15%
Can't do everything I want	11%
Chores	9%
Being grounded	9%

Source: Penn, Schoen, and Berland using 1172 interviews with children ages 6 to 14 from May 14 to June 1, 1999, Margin of error: $\pm 2.9\%$

Note the margin of error.

EXAMPLE 5   Analyzing a Poll's Margin of Error

The inequality

$$|x - 9| \leq 2.9$$

describes the percentage of children in the population who think that being grounded is a bad thing about being a kid. (See Table 9.2.) Solve the inequality and interpret the solution.

Solution   We rewrite the inequality without absolute value bars.

$$|X| \leq c \qquad \text{means} \qquad -c \leq X \leq c.$$

$$|x - 9| \leq 2.9 \qquad \text{means} \qquad -2.9 \leq x - 9 \leq 2.9.$$

We solve the compound inequality by adding 9 to all three parts.

$$-2.9 \leq x - 9 \leq 2.9$$
$$-2.9 + 9 \leq x - 9 + 9 \leq 2.9 + 9$$
$$6.1 \leq x \leq 11.9$$

The percentage of children in the population who think that being grounded is a bad thing is somewhere between a low of 6.1% and a high of 11.9%. Notice that these percents are always within 2.9%, the poll's margin of error, of 9%. ■

✔ **CHECK POINT 5**   Solve the inequality:

$$|x - 11| \leq 2.9.$$

Interpret the solution in terms of the information in Table 9.2 on page 612.

EXERCISE SET 9.3

Practice Exercises

In Exercises 1–38, find the solution set for each equation.

1. $|x| = 8$

2. $|x| = 6$

3. $|x - 2| = 7$

4. $|x + 1| = 5$

5. $|2x - 1| = 7$

6. $|2x - 3| = 11$

7. $\left|\dfrac{4x - 2}{3}\right| = 2$

8. $\left|\dfrac{3x - 1}{5}\right| = 1$

9. $|x| = -8$

10. $|x| = -6$

11. $|x + 3| = 0$

12. $|x + 2| = 0$

13. $2|y + 6| = 10$

14. $3|y + 5| = 12$

15. $3|2x - 1| = 21$

16. $2|3x - 2| = 14$

17. $|6y - 2| + 4 = 32$

18. $|3y - 1| + 10 = 25$

19. $7|5x| + 2 = 16$

20. $7|3x| + 2 = 16$

21. $|x + 1| + 5 = 3$

22. $|x + 1| + 6 = 2$

23. $|4y + 1| + 10 = 4$

24. $|3y - 2| + 8 = 1$

25. $|2x - 1| + 3 = 3$

26. $|3x - 2| + 4 = 4$

27. $|5x - 8| = |3x + 2|$

28. $|4x - 9| = |2x + 1|$

29. $|2x - 4| = |x - 1|$

30. $|6x| = |3x - 9|$

31. $|2x - 5| = |2x + 5|$

32. $|3x - 5| = |3x + 5|$

33. $|x - 3| = |5 - x|$

34. $|x - 3| = |6 - x|$

35. $|2y - 6| = |10 - 2y|$

36. $|4y + 3| = |4y + 5|$

37. $\left|\dfrac{2x}{3} - 2\right| = \left|\dfrac{x}{3} + 3\right|$

38. $\left|\dfrac{x}{2} - 2\right| = \left|x - \dfrac{1}{2}\right|$

In Exercises 39–70, solve and graph the solution set on a number line.

39. $|x| < 3$

40. $|x| < 5$

41. $|x - 2| < 1$

42. $|x - 1| < 5$

43. $|x + 2| \leq 1$

44. $|x + 1| \leq 5$

45. $|2x - 6| < 8$

46. $|3x + 5| < 17$

47. $|x| > 3$

48. $|x| > 5$

49. $|x + 3| > 1$

50. $|x - 2| > 5$

51. $|x - 4| \geq 2$

52. $|x - 3| \geq 4$

53. $|3x - 8| > 7$

54. $|5x - 2| > 13$

55. $|2(x - 1) + 4| \leq 8$

56. $|3(x - 1) + 2| \leq 20$

57. $\left|\dfrac{2y + 6}{3}\right| < 2$

58. $\left|\dfrac{3x - 3}{4}\right| < 6$

59. $\left|\dfrac{2x + 2}{4}\right| \geq 2$

60. $\left|\dfrac{3x - 3}{9}\right| \geq 1$

61. $\left|3 - \dfrac{2x}{3}\right| > 5$

62. $\left|3 - \dfrac{3x}{4}\right| > 9$

63. $|x - 2| < -1$ **64.** $|x - 3| < -2$

65. $|x + 6| > -10$

66. $|x + 4| > -12$

67. $|x + 2| + 9 \leq 16$

68. $|x - 2| + 4 \leq 5$

69. $2|2x - 3| + 10 > 12$

70. $3|2x - 1| + 2 > 8$

Application Exercises

The three television programs viewed by the greatest percentage of U.S. households in the twentieth century are shown in the table. The data are from a random survey of 4000 TV households by Nielsen Media Research. In Exercises 71–72, let x represent the actual viewing percentage in the U.S. population.

TV Programs with the Greatest U.S. Audience Viewing Percentage of the Twentieth Century

Program	Viewing Percentage in Survey
1. "M*A*S*H" Feb. 28, 1983	60.2%
2. "Dallas" Nov. 21, 1980	53.3%
3. "Roots" Part 8 Jan. 30, 1977	51.1%

Source: Nielsen Media Research

71. Solve the inequality: $|x - 60.2| \leq 1.6$. Interpret the solution in terms of the information in the table. What is the margin of error?

72. Solve the inequality: $|x - 51.1| \leq 1.6$. Interpret the solution in terms of the information in the table. What is the margin of error?

73. The inequality $|T - 57| \leq 7$ describes the range of monthly average temperature, T, in degrees Fahrenheit, for San Francisco, California. Solve the inequality and interpret the solution.

74. The inequality $|T - 50| \leq 22$ describes the range of monthly average temperature, T, in degrees Fahrenheit, for Albany, New York. Solve the inequality and interpret the solution.

The specifications for machine parts are given with tolerance limits that describe a range of measurements for which the part is acceptable. In Exercises 75–76, x represents the length of a machine part, in centimeters. The tolerance limit is 0.01 centimeter.

75. Solve: $|x - 8.6| \leq 0.01$. If the length of the machine part is supposed to be 8.6 centimeters, interpret the solution.

76. Solve: $|x - 9.4| \leq 0.01$. If the length of the machine part is supposed to be 9.4 centimeters, interpret the solution.

77. It a coin is tossed 100 times, we would expect approximately 50 of the outcomes to be heads. It can be demonstrated that a coin is unfair if h, the number of outcomes that result in heads, satisfies $\left|\dfrac{h - 50}{5}\right| \geq 1.645$. Describe the number of outcomes that determine an unfair coin that is tossed 100 times.

Writing in Mathematics

78. Explain how to solve an equation containing one absolute value expression.

79. Explain why the procedure that you described in Exercise 78 does not apply to the equation $|x - 5| = -3$. What is the solution set of this equation?

80. Describe how to solve an absolute value equation with two absolute values.

81. Describe how to solve an absolute value inequality involving the symbol $<$. Give an example.

82. Explain why the procedure that you described in Exercise 81 does not apply to the inequality $|x - 5| < -3$. What is the solution set of this inequality?

83. Describe how to solve an absolute value inequality involving the symbol $>$. Give an example.

84. Explain why the procedure that you described in Exercise 83 does not apply to the inequality $|x - 5| > -3$. What is the solution set of this inequality?

85. The final episode of $M*A*S*H$ was viewed by more than 58% of U.S. television households. Is it likely that a popular television series in the twenty-first century will achieve a 58% market share? Explain your answer.

Critical Thinking Exercises

86. Which one of the following is true?
   a. All absolute value equations have two solutions.
   b. The equation $|x| = -6$ is equivalent to $x = 6$ or $x = -6$.
   c. We can rewrite the inequality $x > 5$ or $x < -5$ more compactly as $-5 < x < 5$.
   d. Absolute value inequalities in the form $|ax + b| < c$ translate into *and* compound statements, which may be written as three-part inequalities.

In Exercises 87–88, use the graph of $f(x) = |4 - x|$ to solve each equation or inequality.

87. $|4 - x| = 1$
88. $|4 - x| < 5$

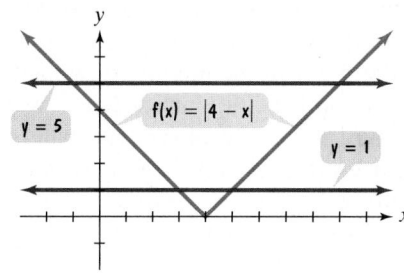

89. The percentage, p, of defective products manufactured by a company is given as $|p - 0.3\%| \leq 0.2\%$. If 100,000 products are manufactured and the company offers a $5 refund for each defective product, describe the company's cost for refunds.

90. Solve: $|2x + 5| = 3x + 4$.

Technology Exercises

In Exercises 91–93, solve each equation using a graphing utility. Graph each side separately in the same viewing rectangle. The solutions are the x-coordinates of the intersection points.

91. $|x + 1| = 5$

92. $|3(x + 4)| = 12$

93. $|2x - 3| = |9 - 4x|$

In Exercises 94–96, solve each inequality using a graphing utility. Graph each side separately in the same viewing rectangle. The solution set consists of all values of x for which the the graph of the left side lies below the graph of the right side.

94. $|2x + 3| < 5$

95. $\left| \dfrac{2x - 1}{3} \right| < \dfrac{5}{3}$

96. $|x + 4| < -1$

In Exercises 97–99, solve each inequality using a graphing utility. Graph each side separately in the same viewing rectangle. The solution set consists of all values of x for which the graph of the left side lies above the graph of the right side.

97. $|2x - 1| > 7$

98. $|0.1x - 0.4| + 0.4 > 0.6$

99. $|x + 4| > -1$

100. Use a graphing utility to verify the solution sets for any five equations or inequalities that you solved by hand in Exercises 1–70.

Review Exercises

In Exercises 101–103, graph each linear function.

101. $3x - 5y = 15$ (Section 3.2, Example 5)

102. $f(x) = -\dfrac{2}{3}x$ or $y = -\dfrac{2}{3}x$ (Section 3.4, Example 4)

103. $f(x) = -2$ or $y = -2$ (Section 3.2, Example 7)

▶ SECTION 9.4   *Linear Inequalities in Two Variables*

Objectives

1 Graph a linear inequality in two variables.

2 Graph a system of linear inequalities.

3 Solve applied problems involving systems of inequalities.

SSM
PH Tutor Center   CD-ROM   Video   www

This book was written in Point Reyes National Seashore, 40 miles north of San Francisco. The park consists of 75,000 acres with miles of pristine surf-washed beaches, forested ridges, and bays bordered by white cliffs.

Like your author, many people are kept inspired and energized surrounded by nature's unspoiled beauty. In this section, you will see how systems of inequalities model whether a region's natural beauty manifests itself in forests, grasslands, or deserts.

Linear Inequalities in Two Variables and Their Solution

We have seen that equations in the form $Ax + By = C$ are straight lines when graphed. If we change the $=$ sign to $>, <, \geq$, or $\leq$, we obtain a **linear inequality in two variables**. Some examples of linear inequalities in two variables are $x + y > 2$, $3x - 5y \leq 15$, and $2x - y < 4$.

A **solution of an inequality in two variables**, x and y, is an ordered pair of real numbers with the following property: When the x-coordinate is substituted for x and the y-coordinate is substituted for y in the inequality, we obtain a true statement. For example, $(3, 2)$ is a solution of the inequality $x + y > 1$. When 3 is substituted for x and 2 is substituted for y, we obtain the true statement $3 + 2 > 1$, or $5 > 1$. Because there are infinitely many pairs of numbers that have a sum greater than 1, the inequality $x + y > 1$ has infinitely many solutions. Each ordered pair solution is said to **satisfy** the inequality. Thus, $(3, 2)$ satisfies the inequality $x + y > 1$.

1 Graph a linear inequality in two variables.

The Graph of a Linear Inequality in Two Variables

We know that the graph of an equation in two variables is the set of all points whose coordinates satisfy the equation. Similarly, the **graph of an inequality in two variables** is the set of all points whose coordinates satisfy the inequality.

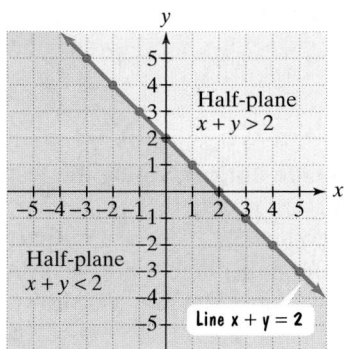

Figure 9.10

Let's use Figure 9.10 to get an idea of what the graph of a linear inequality in two variables looks like. Part of the figure shows the graph of the linear equation $x + y = 2$. The line divides the points in the rectangular coordinate system into three sets. First, there is the set of points along the line, satisfying $x + y = 2$. Next, there is the set of points in the green region above the line. Points in the green region satisfy the linear inequality $x + y > 2$. Finally, there is the set of points in the pink region below the line. Points in the pink region satisfy the linear inequality $x + y < 2$.

A **half-plane** is the set of all the points on one side of a line. In Figure 9.10, the green region is a half-plane. The pink region is also a half-plane. A half-plane is the graph of a linear inequality that involves $>$ or $<$. The graph of an inequality that involves $\geq$ or $\leq$ is a half-plane and a line. A solid line is used to show that the line is part of the graph. A dashed line is used to show that a line is not part of a graph.

Graphing a Linear Inequality in Two Variables

1. Replace the inequality symbol with an equal sign and graph the corresponding linear equation. Draw a solid line if the original inequality contains a $\leq$ or $\geq$ symbol. Draw a dashed line if the original inequality contains a $<$ or $>$ symbol.

2. Choose a test point in one of the half-planes that is not on the line. Substitute the coordinates of the test point into the inequality.

3. If a true statement results, shade the half-plane containing this test point. If a false statement results, shade the half-plane not containing this test point.

EXAMPLE 1   Graphing a Linear Inequality in Two Variables

Graph: $3x - 5y \geq 15$.

Solution

Step 1 Replace the inequality symbol by $=$ and graph the linear equation. We need to graph $3x - 5y = 15$. We can use intercepts to graph this line.

We set $y = 0$ to find the x-intercept:	We set $x = 0$ to find the y-intercept:
$3x - 5y = 15$	$3x - 5y = 15$
$3x - 5 \cdot 0 = 15$	$3 \cdot 0 - 5y = 15$
$3x = 15$	$-5y = 15$
$x = 5.$	$y = -3.$

The x-intercept is 5, so the line passes through $(5, 0)$. The y-intercept is -3, so the line passes through $(0, -3)$. Using the intercepts, the line is shown in Figure 9.11 as a solid line. This is because the inequality $3x - 5y \geq 15$ contains a $\geq$ symbol, in which equality is included.

Step 2 Choose a test point in one of the half-planes that is not on the line. Substitute its coordinates into the inequality. The line $3x - 5y = 15$ divides the plane into three parts —the line itself and two half-planes. The points in one half-plane satisfy $3x - 5y > 15$. The points in the other half-plane satisfy $3x - 5y < 15$. We need to find which half-plane belongs to the solution. To do so, we test a point from either half-plane. The origin, $(0, 0)$, is the easiest point to test.

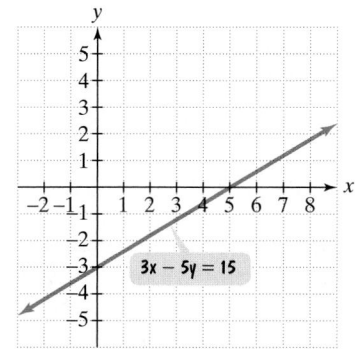

Figure 9.11 Preparing to graph $3x - 5y \geq 15$

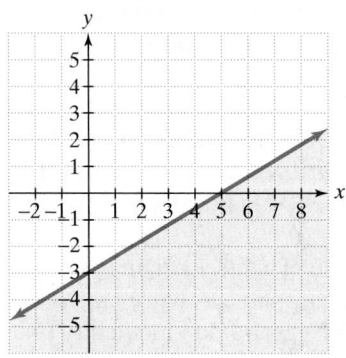

Figure 9.12 The graph of
$3x - 5y \geq 15$

$$3x - 5y \geq 15 \quad \text{This is the given inequality.}$$
$$3 \cdot 0 - 5 \cdot 0 \overset{?}{\geq} 15 \quad \text{Test } (0, 0) \text{ by substituting 0 for } x \text{ and 0 for } y.$$
$$0 - 0 \overset{?}{\geq} 15 \quad \text{Multiply.}$$
$$0 \geq 15 \quad \text{This statement is false.}$$

Step 3 If a false statement results, shade the half-plane not containing the test point. Because 0 is not greater than or equal to 15, the test point, $(0, 0)$, is not part of the solution set. Thus, the half-plane below the solid line $3x - 5y = 15$ is part of the solution set. The solution set is the line and the half-plane that does not contain the point $(0, 0)$, indicated by shading this half-plane. The graph is shown using green shading and a blue line in Figure 9.12. ∎

✔ **CHECK POINT 1** Graph: $2x - 4y \geq 8$.

When graphing a linear inequality, test a point that lies in one of the half-planes and *not on the line dividing the half-planes*. The test point $(0, 0)$ is convenient because it is easy to calculate when 0 is substituted for each variable. However, if $(0, 0)$ lies on the dividing line and not in a half-plane, a different test point must be selected.

EXAMPLE 2 Graphing a Linear Inequality in Two Variables

Graph: $y > -\dfrac{2}{3}x$.

Solution

Step 1 Replace the inequality symbol by = and graph the linear equation. We need to graph $y = -\frac{2}{3}x$. We can use the slope and the y-intercept to graph this linear function.

$$y = -\frac{2}{3}x + 0$$

$$\text{Slope} = \frac{-2}{3} = \frac{\text{rise}}{\text{run}} \qquad \text{y-intercept} = 0$$

The y-intercept is 0, so the line passes through $(0, 0)$. Using the y-intercept and the slope, the line is shown in Figure 9.13 as a dashed line. This is because the inequality $y > -\frac{2}{3}x$ contains a $>$ symbol, in which equality is not included.

Step 2 Choose a test point in one of the half-planes that is not on the line. Substitute its coordinates into the inequality. We cannot use $(0, 0)$ as a test point because it lies on the line and not in a half-plane. Let's use $(1, 1)$, which lies in the half-plane above the line.

$$y > -\frac{2}{3}x \quad \text{This is the given inequality.}$$
$$1 \overset{?}{>} -\frac{2}{3} \cdot 1 \quad \text{Test } (1, 1) \text{ by substituting 1 for } x \text{ and 1 for } y.$$
$$1 > -\frac{2}{3} \quad \text{This statement is true.}$$

Step 3 If a true statement results, shade the half-plane containing the test point. Because 1 is greater than $-\frac{2}{3}$, the test point $(1, 1)$ is part of the solution set. All the points on the same side of the line $y = -\frac{2}{3}x$ as the point $(1, 1)$ are members of the solution set. The solution set is the half-plane that contains the point $(1, 1)$, indicated by shading this half-plane. The graph is shown using green shading and a dashed blue line in Figure 9.13. ∎

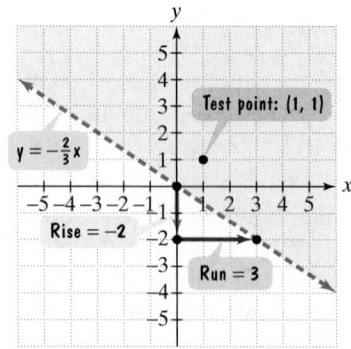

Figure 9.13 The graph of $y > -\frac{2}{3}x$

Using Technology

Most graphing utilities can graph inequalities in two variables with the SHADE feature. The procedure varies by model, so consult your manual. For most graphing utilities, you must first solve for y if it is not already isolated. The figure shows the graph of $y > -\frac{2}{3}x$. Most displays do not distinguish between dashed and solid boundary lines.

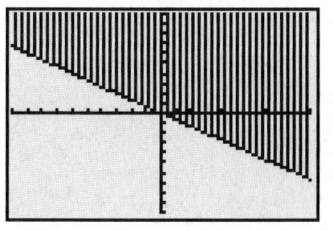

✔ **CHECK POINT 2** Graph: $y > -\dfrac{3}{4}x.$

You can graph inequalities in the form $y > mx + b$ or $y < mx + b$ without using test points. The inequality symbol indicates which half-plane to shade.

- If $y > mx + b$, shade above the line $y = mx + b$.
- If $y < mx + b$, shade below the line $y = mx + b$.

In Section 3.2, we learned that $y = b$ graphs as a horizontal line, where b is the y-intercept. Similarly, the graph of $x = a$ is a vertical line, where a is the x-intercept. Half-planes can be separated by horizontal or vertical lines. For example, Figure 9.14 shows the graph of $y \leq 2$. Because $(0, 0)$ satisfies this inequality ($0 \leq 2$ is true), the graph consists of the half-plane below the line $y = 2$ and the line. Similarly, Figure 9.15 shows the graph of $x < 4$.

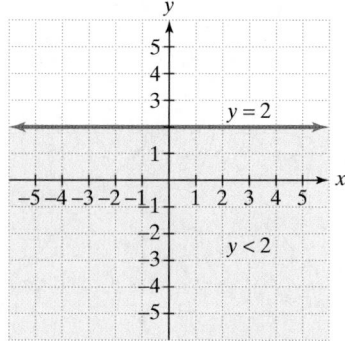

Figure 9.14 The graph of $y \leq 2$

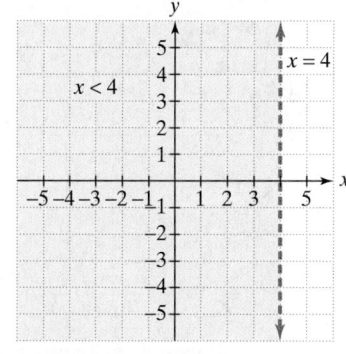

Figure 9.15 The graph of $x < 4$

2 Graph a system of linear inequalities.

Systems of Linear Inequalities

Just as two linear equations make up a system of linear equations, two or more linear inequalities make up a **system of linear inequalities**. Here is an example of a system of linear inequalities:

$$2x - y < 4$$
$$x + y \geq -1.$$

A **solution of a system of linear inequalities** in two variables is an ordered pair that satisfies each inequality in the system. The set of all such ordered pairs is the **solution set** of the system. Thus, to graph a system of inequalities in two variables, begin by graphing each individual inequality in the same rectangular coordinate system. Then find the region, if there is one, that is common to every graph in the system. This region of intersection gives a picture of the system's solution set.

EXAMPLE 3 Graphing a System of Linear Inequalities

Graph the solution set of the system:

$$2x - y < 4$$
$$x + y \geq -1.$$

Solution We begin by graphing $2x - y < 4$. Because the inequality contains a $<$ symbol, rather than $\leq$, we graph $2x - y = 4$ as a dashed line. (If $x = 0$, then $y = -4$, and if $y = 0$, then $x = 2$. The x-intercept is 2 and the y-intercept is -4.) Because $(0, 0)$ makes the inequality $2x - y < 4$ true, we shade the half-plane containing $(0, 0)$, shown in yellow in Figure 9.16.

Now we graph $x + y \geq -1$ in the same rectangular coordinate system. Because the inequality contains a $\geq$ symbol, in which equality is included, we graph $x + y = -1$ as a solid line. (If $x = 0$, then $y = -1$, and if $y = 0$, then $x = -1$. The x-intercept and y-intercept are both -1.) Because $(0, 0)$ makes the inequality true, we shade the half-plane containing $(0, 0)$. This is shown in Figure 9.17 using green vertical shading above the solid red line.

The solution set of the system is shown graphically by the intersection (the overlap) of the two half-planes. This is shown in Figure 9.17 as the region in which the yellow shading and the green vertical shading overlap. The solutions of the system are shown again in Figure 9.18.

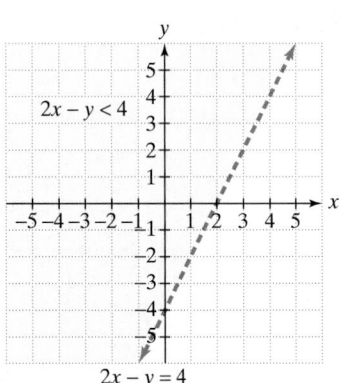

Figure 9.16 The graph of $2x - y < 4$

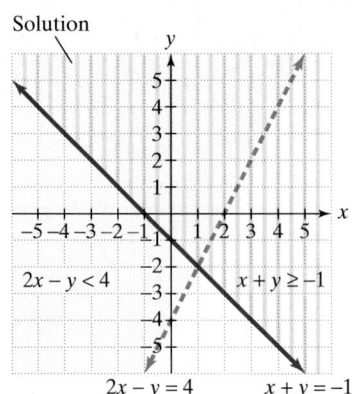

Figure 9.17 Adding the graph of $x + y \geq -1$

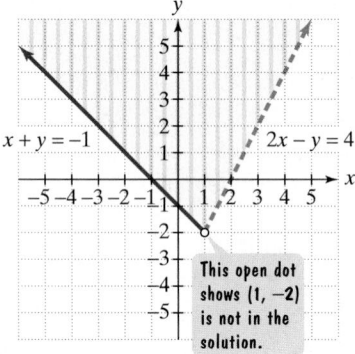

This open dot shows $(1, -2)$ is not in the solution.

Figure 9.18 The graph of $2x - y < 4$ and $x + y \geq -1$ ∎

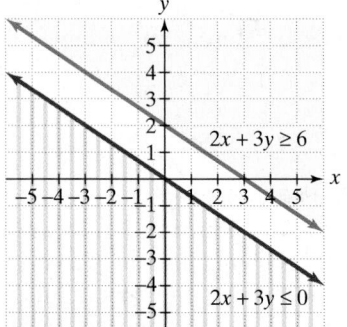

Figure 9.19 A system of inequalities with no solution.

✔ **CHECK POINT 3** Graph the solution set of the system:

$$x + 2y > 4$$
$$2x - 3y \leq -6.$$

A system of inequalities has no solution if there are no points in the rectangular coordinate system that simultaneously satisfy each inequality in the system. For example, the system

$$2x + 3y \geq 6$$
$$2x + 3y \leq 0$$

whose separate graphs are shown in Figure 9.19 has no overlapping region. Thus, the system has no solution. The solution set is $\varnothing$, the empty set.

EXAMPLE 4   Graphing a System of Inequalities

Graph the solution set:

$$x - y < 2$$
$$-2 \le x < 4$$
$$y < 3.$$

Solution   We begin by graphing $x - y < 2$, the first given inequality. The line $x - y = 2$ has an x-intercept of 2 and a y-intercept of -2. The test point $(0, 0)$ makes the inequality $x - y < 2$ true, and its graph is shown in Figure 9.20.

Now, let's consider the second given inequality, $-2 \le x < 4$. Replacing the inequality symbols by $=$, we obtain $x = -2$ and $x = 4$, graphed as vertical lines. The line of $x = 4$ is not included. Using $(0, 0)$ as a test point and substituting the x-coordinate, 0, into $-2 \le x < 4$, we obtain the true statement $-2 \le 0 < 4$. We therefore shade the region between the vertical lines. We've added this region to Figure 9.20, intersecting the region between the vertical lines with the yellow region in Figure 9.20. The resulting region is shown in yellow and green vertical shading in Figure 9.21.

Finally, let's consider the third given inequality, $y < 3$. Replacing the inequality symbol by $=$, we obtain $y = 3$, which graphs as a horizontal line. Because $(0, 0)$ satisfies $y < 3$ ($0 < 3$ is true), the graph consists of the half-plane below the line $y = 3$. We've added this half-plane to the region in Figure 9.21, intersecting the half-plane with this region. The resulting region is shown in yellow and green vertical shading in Figure 9.22. This region represents the graph of the solution set of the given system.   ∎

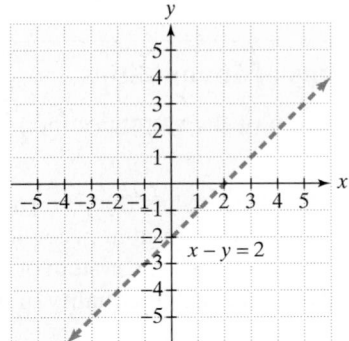

Figure 9.20 The graph of $x - y < 2$

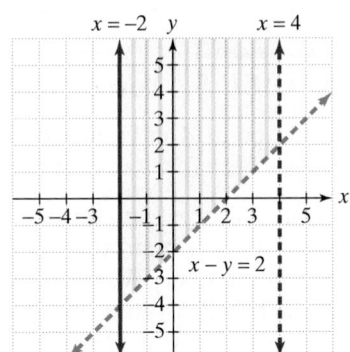

Figure 9.21 The graph of $x - y < 2$ and $-2 \le x < 4$

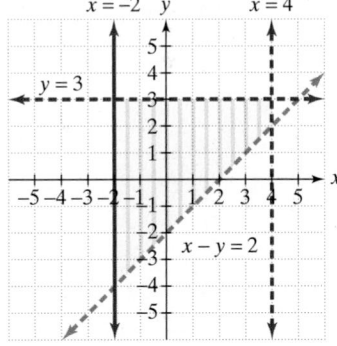

Figure 9.22 The graph of $x - y < 2$ and $-2 \le x < 4$ and $y < 3$

✔ **CHECK POINT 4**   Graph the solution set:

$$x + y < 2$$
$$-2 \le x < 1$$
$$y > -3.$$

Applications

3 Solve applied problems involving systems of inequalities.

Temperature and precipitation affect whether or not trees and forests can grow. At certain levels of precipitation and temperature, only grasslands and deserts will exist. Figure 9.23 on page 622 shows three kinds of regions—deserts, grasslands, and forests—that result from various ranges of temperature, T, and precipitation, P.

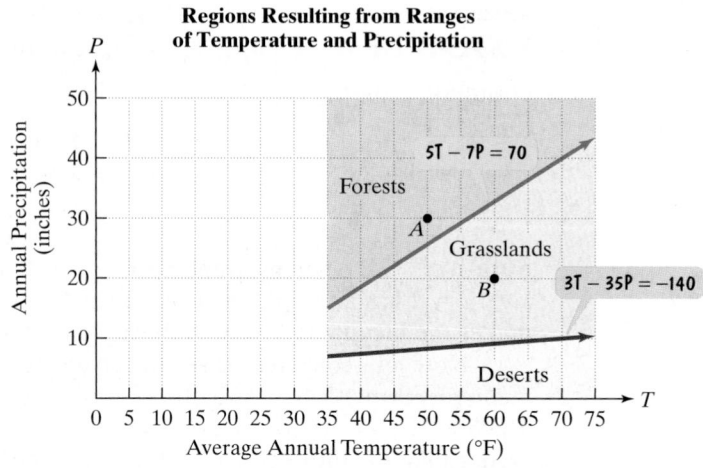

Figure 9.23   *Source*: A. Miller and J. Thompson, *Elements of Meteorology*

Systems of inequalities can be used to describe where forests, grasslands, and deserts occur. Because these regions occur when the average annual temperature, T, is 35°F or greater, each system contains the inequality $T \geq 35$.

Forests occur if	**Grasslands occur if**	**Deserts occur if**
$T \geq 35$	$T \geq 35$	$T \geq 35$
$5T - 7P < 70.$	$5T - 7P \geq 70$	$3T - 35P > -140.$
	$3T - 35P \leq -140.$	

EXAMPLE 5   Forests and Systems of Inequalities

Show that point A in Figure 9.23 is a solution of the system of inequalities that describes where forests occur.

Solution   Point A has coordinates $(50, 30)$. This means that if a region has an average annual temperature of 50°F and an average annual precipitation of 30 inches, a forest occurs. We can show that $(50, 30)$ satisfies the system of inequalities for forests by substituting 50 for T and 30 for P in each inequality in the system.

$$T \geq 35 \qquad\qquad 5T - 7P < 70$$
$$50 \geq 35, \text{ true} \qquad 5 \cdot 50 - 7 \cdot 30 \overset{?}{<} 70$$
$$250 - 210 \overset{?}{<} 70$$
$$40 < 70, \text{ true}$$

The coordinates $(50, 30)$ make each inequality true. Thus, $(50, 30)$ satisfies the system for forests. ∎

✔ **CHECK POINT 5**   Show that point B in Figure 9.23 is a solution of the system of inequalities that describes where grasslands occur.

EXERCISE SET 9.4

Practice Exercises

In Exercises 1–22, graph each inequality.

1. $x + y \geq 3$

2. $x + y \geq 2$

3. $x - y < 5$

4. $x - y < 6$

5. $x + 2y > 4$

6. $2x + y > 6$

7. $3x - y \leq 6$

8. $x - 3y \leq 6$

9. $\dfrac{x}{2} + \dfrac{y}{3} < 1$

10. $\dfrac{x}{4} + \dfrac{y}{2} < 1$

11. $y > \dfrac{1}{3}x$

12. $y > \dfrac{1}{4}x$

13. $y \leq 3x + 2$

14. $y \leq 2x - 1$

15. $y < -\dfrac{1}{4}x$

16. $y < -\dfrac{1}{3}x$

17. $x \leq 2$

18. $x \leq -4$

19. $y > -4$

20. $y > -2$

21. $y \geq 0$

22. $x \leq 0$

In Exercises 23–46 graph the solution set of each system of inequalities or indicate that the system has no solution.

23. $3x + 6y \leq 6$
    $2x + y \leq 8$

24. $x - y \geq 4$
    $x + y \leq 6$

25. $2x - 5y \leq 10$
    $3x - 2y > 6$

26. $2x - y \leq 4$
    $3x + 2y > -6$

27. $y > 2x - 3$
    $y < -x + 6$

28. $y < -2x + 4$
    $y < x - 4$

29. $x + 2y \leq 4$
    $y \geq x - 3$

30. $x + y \leq 4$
    $y \geq 2x - 4$

31. $x \leq 2$
    $y \geq -1$

32. $x \leq 3$
    $y \leq -1$

33. $-2 \leq x < 5$

34. $-2 < y \leq 5$

35. $x - y \leq 1$
    $x \geq 2$

36. $4x - 5y \geq -20$
    $x \geq -3$

37. $x + y > 4$
    $x + y < -1$

38. $x + y > 3$
    $x + y < -2$

39. $x + y > 4$
    $x + y > -1$

40. $x + y > 3$
    $x + y > -2$

41. $x - y \leq 2$
    $x \geq -2$
    $y \leq 3$

42. $3x + y \leq 6$
    $x \geq -2$
    $y \leq 4$

43. $x \geq 0$
    $y \geq 0$
    $2x + 5y \leq 10$
    $3x + 4y \leq 12$

44. $x \geq 0$
    $y \geq 0$
    $2x + y \leq 4$
    $2x - 3y \leq 6$

45. $3x + y \leq 6$
    $2x - y \leq -1$
    $x \geq -2$
    $y \leq 4$

46. $2x + y \leq 6$
    $x + y \geq 2$
    $1 \leq x \leq 2$
    $y \leq 3$

Application Exercises

For people between ages 10 and 70, inclusive, the target zone for aerobic exercise is given by the following system of inequalities in which a represents one's age and p is one's pulse rate.

$$10 \leq a \leq 70$$
$$2a + 3p \geq 450$$
$$a + p \leq 190$$

The graph of this target zone is shown in the figure. Use this information to solve Exercises 47–48.

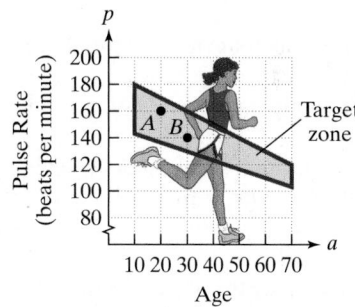

47. **a.** What are the coordinates of point A and what does this mean in terms of age and pulse rate?

    b. Show that point A is a solution of the system of inequalities.

48. **a.** What are the coordinates of point B and what does this mean in terms of age and pulse rate?

    b. Show that point B is a solution of the system of inequalities.

49. Use Figure 9.23 on page 622 to identify the coordinates of a point that lies on the boundary line between grasslands and forests. Then show that this point is a solution of the system of inequalities that describes where grasslands occur.

50. Use Figure 9.23 on page 622 to identify the coordinates of a point that lies on the boundary line between grasslands and deserts. Then show that this point is a solution of the system of inequalities that describes where grasslands occur.

The shaded region in the figure shows recommended weight and height combinations based on information from the Department of Agriculture. Points along the blue and red lines also belong to the healthy weight region. Use this region to solve Exercises 51–55.

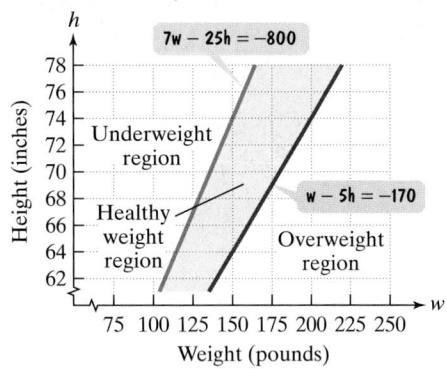

51. Estimate the maximum healthy weight for a person who is 6 feet tall.

52. Estimate the minimum healthy weight for a person who is 6 feet tall.

53. Write an inequality that describes the underweight region. Then identify a point in this region and show that it is a solution of the inequality.

54. Write an inequality that describes the overweight region. Then identify a point in this region and show that it is a solution of the inequality.

55. Write a system of inequalities that describes the healthy weight region. Then identify a point in this region and show that it is a solution of the system.

Writing in Mathematics

56. What is a linear inequality in two variables? Provide an example with your description.

57. How do you determine whether an ordered pair is a solution of an inequality in two variables, x and y?

58. What is a half-plane?

59. What does a solid line mean in the graph of an inequality?

60. What does a dashed line mean in the graph of an inequality?

61. Explain how to graph $x - 2y < 4$.

62. What is a system of linear inequalities?

63. What is a solution of a system of linear inequalities?

64. Explain how to graph the solution set of a system of inequalities.

65. What does it mean if a system of linear inequalities has no solution?

66. Look at the shaded region showing recommended weight and height combinations in the figure for Exercises 51–55. Describe why a system of inequalities, rather than a linear equation in w and h, is better suited to give the recommended combinations.

Critical Thinking Exercises

67. Write a linear inequality in two variables whose graph is shown.

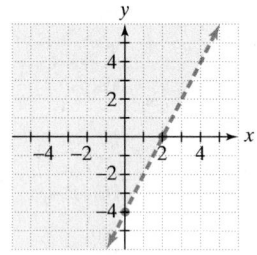

68. Write a system of inequalities whose solution set includes every point in the rectangular coordinate system.

69. A person with no more than $15,000 to invest plans to place the money in two investments. One investment is high risk, high yield; the other is low risk, low yield. At least $2000 is to be placed in the high-risk investment. Furthermore, the amount invested at low risk should be at least three times the amount invested at high risk. Find and graph a system of inequalities that describes all possibilities for placing the money in the high- and low-risk investments.

70. Sketch the graph of the solution set for the following system of inequalities:

$$y \geq nx + b \ (n < 0, b > 0)$$

$$y \leq mx + b \ (m > 0, b > 0).$$

Technology Exercises

Graphing utilities can be used to shade regions in the rectangular coordinate system, thereby graphing an inequality in two variables. Read the section of the user's manual for your graphing utility that describes how to shade a region. Then use your graphing utility to graph the inequalities in Exercises 71–74.

71. $y \leq 4x + 4$

72. $y \geq \dfrac{2}{3}x - 2$

73. $2x + y \leq 6$

74. $3x - 2y \geq 6$

75. Does your graphing utility have any limitations in terms of graphing inequalities? If so, what are they?

76. Use a graphing utility with a $\boxed{\text{SHADE}}$ feature to verify any five of the graphs that you drew by hand in Exercises 1–22.

77. Use a graphing utility with a $\boxed{\text{SHADE}}$ feature to verify any five of the graphs that you drew by hand for the systems in Exercises 23–46.

Review Exercises

78. Solve using matrices:

$$3x - y = 8$$
$$x - 5y = -2.$$

(Section 8.4, Example 3)

79. Solve by graphing:

$$y = 3x - 2$$
$$y = -2x + 8.$$

(Section 4.1, Example 3)

80. Evaluate:
$$\begin{vmatrix} 8 & 2 & -1 \\ 3 & 0 & 5 \\ 6 & -3 & 4 \end{vmatrix}.$$

(Section 8.5, Example 3)

▶ SECTION 9.5 *Linear Programming*

Objectives

1 Write an objective function describing a quantity that must be maximized or minimized.

2 Use inequalities to describe limitations in a situation.

3 Use linear programming to solve problems.

SSM PH Tutor CD- Video
      Center   ROM

West Berlin children at Tempelhof airport watch fleets of U.S. airplanes bringing in supplies to circumvent the Russian blockade. The airlift began June 28, 1948 and continued for 15 months.

The Berlin Airlift (1948–1949) was an operation by the United States and Great Britain. It was a response to military action by the former Soviet Union: Soviet troops closed all roads and rail lines between West Germany and Berlin, cutting off supply routes to the city. The Allies used a mathematical technique developed during World War II to maximize the amount of supplies transported. During the 15-month airlift, 278,228 flights provided basic necessities to blockaded Berlin, saving one of the world's great cities.

In this section, we will look at an important application of systems of linear inequalities. Such systems arise in **linear programming**, a method for solving problems in which a particular quantity that must be maximized or minimized is limited. Linear programming is one of the most widely used tools in management science. It helps businesses allocate resources to manufacture products in a way that will maximize profit. Linear programming accounts for more than 50% and perhaps as much as 90% of all computing time used for management decisions in business. The Allies used linear programming to save Berlin.

1 Write an objective function describing a quantity that must be maximized or minimized.

Objective Functions in Linear Programming

Many problems involve quantities that must be maximized or minimized. Businesses are interested in maximizing profit. An operation in which bottled water and medical kits are shipped to earthquake victims needs to maximize the number of victims helped by this shipment. An **objective function** is an algebraic expression in two or more variables describing a quantity that must be maximized or minimized.

EXAMPLE 1   Writing an Objective Function

Bottled water and medical supplies are to be shipped to victims of an earthquake by plane. Each container of bottled water will serve 10 people and each medical kit will aid 6 people. If x represents the number of bottles of water to be shipped and y represents the number of medical kits, write the objective function that describes the number of people that can be helped.

Solution   Because each bottle of water serves 10 people and each medical kit aids 6 people, we have

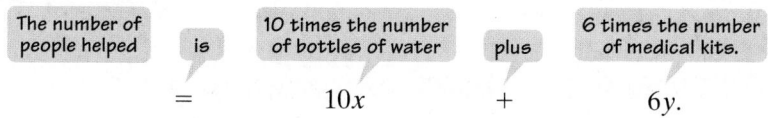

$$= \qquad 10x \qquad + \qquad 6y.$$

Using z to represent the objective function, we have

$$z = 10x + 6y.$$

Unlike the functions that we have seen so far, the objective function is an equation in three variables. For a value of x and a value of y, there is one and only one value of z. Thus, z is a function of x and y. ∎

✔ **CHECK POINT 1**   A company manufactures bookshelves and desks for computers. Let x represent the number of bookshelves manufactured daily and y the number of desks manufactured daily. The company's profits are $25 per bookshelf and $55 per desk. Write the objective function that describes the company's total daily profit, z, from x bookshelves and y desks. (Check Points 1 through 4 are related to this situation, so keep track of your answers.)

Constraints in Linear Programming

2 Use inequalities to describe limitations in a situation.

Ideally, the number of earthquake victims helped in Example 1 should increase without restriction so that every victim receives water and medical kits. However, the planes that ship these supplies are subject to weight and volume restrictions. In linear programming problems, such restrictions are called **constraints**. Each constraint is expressed as a linear inequality. The list of constraints forms a system of linear inequalities.

EXAMPLE 2   Writing a Constraint

Each plane can carry no more than 80,000 pounds. The bottled water weighs 20 pounds per container and each medical kit weighs 10 pounds. If x represents the number of bottles of water to be shipped and y represents the number of medical kits, write an inequality that describes this constraint.

Solution   Because each plane can carry no more than 80,000 pounds, we have

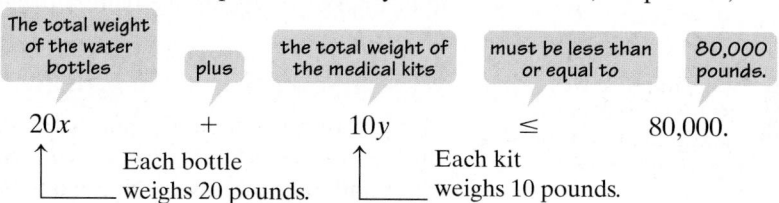

The plane's weight constraint is described by the inequality

$$20x + 10y \le 80{,}000.$$ ∎

✔ **CHECK POINT 2** To maintain high quality, the company in Check Point 1 should not manufacture more than 80 bookshelves and desks per day. Write an inequality that describes this constraint.

In addition to a weight constraint on its cargo, each plane has a limited amount of space in which to carry supplies. Example 3 demonstrates how to express this constraint.

EXAMPLE 3   Writing a Constraint

Planes can carry a total volume for supplies that does not exceed 6000 cubic feet. Each water bottle is 1 cubic foot and each medical kit also has a volume of 1 cubic foot. With x still representing the number of water bottles and y the number of medical kits, write an inequality that describes this second constraint.

Solution   Because each plane can carry a volume of supplies that does not exceed 6000 cubic feet, we have

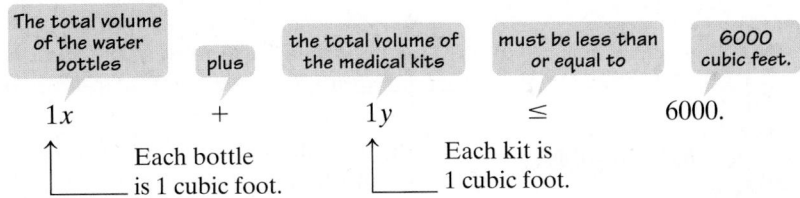

| The total volume of the water bottles | plus | the total volume of the medical kits | must be less than or equal to | 6000 cubic feet. |

$$1x \quad + \quad 1y \quad \leq \quad 6000.$$

Each bottle is 1 cubic foot.   Each kit is 1 cubic foot.

The plane's volume constraint is described by the inequality $x + y \leq 6000$. ∎

In summary, here's what we have described in this aid-to-earthquake-victims situation:

$$z = 10x + 6y$$
This is the objective function describing the number of people helped with x bottles of water and y medical kits.

$$20x + 10y \leq 80{,}000$$
$$x + y \leq 6000.$$
These are the constraints based on each plane's weight and volume limitations.

✔ **CHECK POINT 3** To meet customer demand, the company in Check Point 1 must manufacture between 30 and 80 bookshelves per day, inclusive. Furthermore, the company must manufacture at least 10 and no more than 30 desks per day. Write an inequality that describes each of these sentences. Then summarize what you have described about this company by writing the objective function for its profits, and the three constraints.

3 Use linear programming to solve problems.

Solving Problems with Linear Programming

The problem in the earthquake situation described previously is to maximize the number of victims who can be helped, subject to the planes' weight and volume constraints. The process of solving this problem is called linear programming, based on a theorem that was proven during World War II.

Solving a Linear Programming Problem

Let $z = ax + by$ be an objective function that depends on x and y. Furthermore, z is subject to a number of constraints on x and y. If a maximum or minimum value of z exists, it can be determined as follows:

1. Graph the system of inequalities representing the constraints.
2. Find the value of the objective function at each corner, or **vertex**, of the graphed region. The maximum and minimum of the objective function occur at one or more of the corner points.

EXAMPLE 4 Solving a Linear Programming Problem

Determine how many bottles of water and how many medical kits should be sent on each plane to maximize the number of earthquake victims who can be helped.

Solution   We must maximize $z = 10x + 6y$ subject to the constraints:

$$20x + 10y \le 80{,}000$$
$$x + y \le 6000.$$

Step 1 Graph the system of inequalities representing the constraints. Because x (the number of bottles of water per plane) and y (the number of medical kits per plane) must be nonnegative, we need to graph the system of inequalities in quadrant I and its boundary only. To graph the inequality $20x + 10y \le 80{,}000$, we graph the equation $20x + 10y = 80{,}000$ as a solid blue line (Figure 9.24). Setting $y = 0$, the x-intercept is 4000 and setting $x = 0$, the y-intercept is 8000. Using $(0, 0)$ as a test point, the inequality is satisfied, so we shade below the blue line, as shown in yellow in Figure 9.24. Now we graph $x + y \le 6000$ by first graphing $x + y = 6000$ as a solid line. Setting $y = 0$, the x-intercept is 6000. Setting $x = 0$, the y-intercept is 6000. Using $(0, 0)$ as a test point, the inequality is satisfied, so we shade below the red line, as shown using green vertical shading in Figure 9.24.

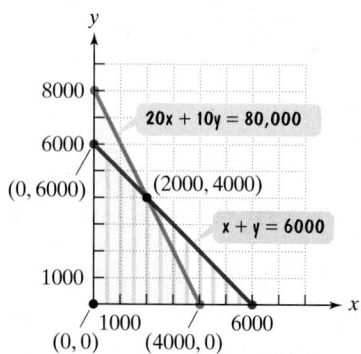

Figure 9.24 The region in quadrant I representing the constraints
$20x + 10y \le 80{,}000$
$x + y \le 6000$

We use the addition method to find where the lines $20x + 10y = 80{,}000$ and $x + y = 6000$ intersect.

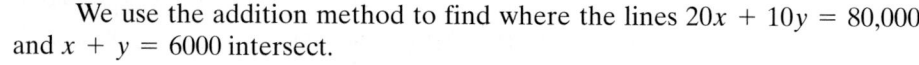

$$
\begin{array}{lll}
20x + 10y = 80{,}000 & \xrightarrow{\text{No change}} & 20x + 10y = 80{,}000 \\
x + y = 6000 & \xrightarrow{\text{Multiply by }-10.} & -10x - 10y = -60{,}000 \\
& \text{Add:} & 10x = 20{,}000 \\
& & x = 2000
\end{array}
$$

Back-substituting 2000 for x in $x + y = 6000$, we find $y = 4000$, so the intersection point is $(2000, 4000)$.

The system of inequalities representing the constraints is shown by the region in which the yellow shading and the green vertical shading overlap in Figure 9.24. The graph of the system of inequalities is shown again in Figure 9.25. The red and blue line segments are included in the graph.

Step 2 Find the value of the objective function at each corner of the graphed region. The maximum and minimum of the objective function occur at one or more of the corner points. We must evaluate the objective function, $z = 10x + 6y$, at the four corners, or vertices, of the region in Figure 9.25.

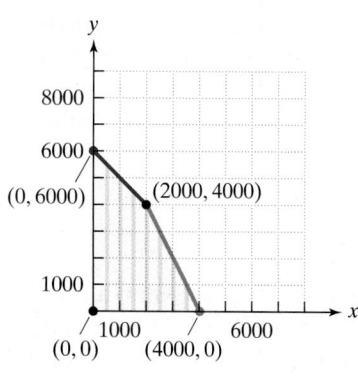

Figure 9.25

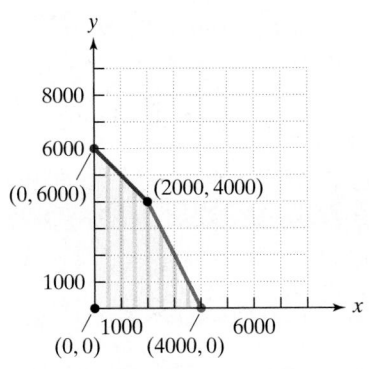

Corner (x, y)	Objective Function $z = 10x + 6y$	
$(0, 0)$	$z = 10(0) + 6(0) = 0$	
$(4000, 0)$	$z = 10(4000) + 6(0) = 40{,}000$	
$(2000, 4000)$	$z = 10(2000) + 6(4000) = 44{,}000$	← maximum

Figure 9.25, repeated

Thus, the maximum value of z is 44,000 and this occurs when $x = 2000$ and $y = 4000$. In practical terms, this means that the maximum number of earthquake victims who can be helped with each plane shipment is 44,000. This can be accomplished by sending 2000 water bottles and 4000 medical kits per plane. ∎

✔ **CHECK POINT 4** For the company in Check Points 1–3, how many bookshelves and how many desks should be manufactured per day to obtain maximum profit? What is the maximum daily profit?

EXAMPLE 5 Solving a Linear Programming Problem

Find the maximum value of the objective function

$$z = 2x + y$$

subject to the constraints:

$$x \geq 0, \ y \geq 0$$
$$x + 2y \leq 5$$
$$x - y \leq 2.$$

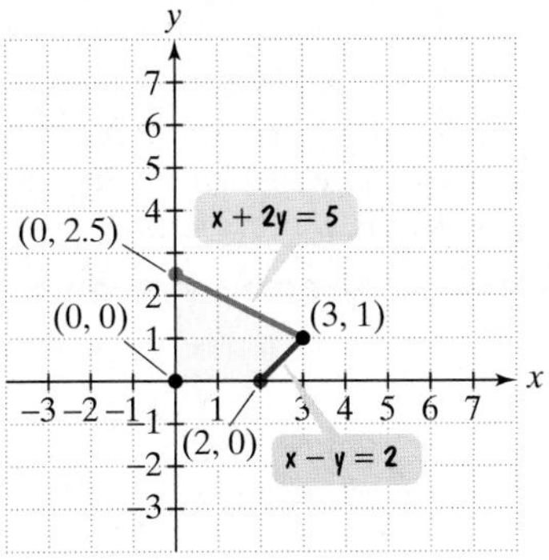

Solution We begin by graphing the region in quadrant I ($x \geq 0$, $y \geq 0$) formed by the constraints. The graph is shown in Figure 9.26.

Now we evaluate the objective function at the four vertices of this region.

Objective function: $z = 2x + y$

At $(0, 0)$: $z = 2 \cdot 0 + 0 = 0$

At $(2, 0)$: $z = 2 \cdot 2 + 0 = 4$

At $(3, 1)$: $z = 2 \cdot 3 + 1 = 7$ Maximum value of z

At $(0, 2.5)$: $z = 2 \cdot 0 + 2.5 = 2.5$

Thus, the maximum value of z is 7, and this occurs when $x = 3$ and $y = 1$. ∎

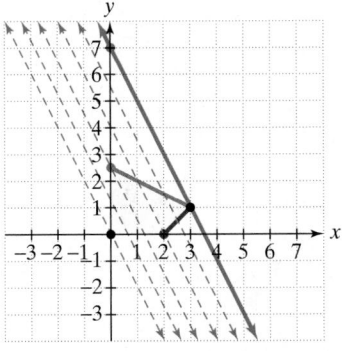

Figure 9.26 The graph of $x + 2y \leq 5$ and $x - y \leq 2$ in quadrant I

We can see why the objective function in Example 5 has a maximum value that occurs at a vertex by solving the equation for y.

$z = 2x + y$ This is the objective function of Example 5.

$y = -2x + z$ Solve for y. Recall that the slope-intercept form of a line is $y = mx + b$.

Slope $= -2$ y-intercept $= z$

In this form, z represents the y-intercept of the objective function. The equation describes infinitely many parallel lines, each with a slope of -2. The process in linear programming involves finding the maximum z-value for all lines that intersect the region determined by the constraints. Of all the lines whose slope is -2, we're looking for the one with the greatest y-intercept that intersects the given region. As we see in Figure 9.27, such a line will pass through one (or possibly more) of the vertices of the region.

Figure 9.27 The line with slope -2 with the greatest y-intercept that intersects the shaded region passes through one of its vertices.

✔ **CHECK POINT 5** Find the maximum value of the objective function $z = 3x + 5y$ subject to the constraints $x \geq 0, y \geq 0, x + y \geq 1, x + y \leq 6$.

ENRICHMENT ESSAY

Faster and Faster

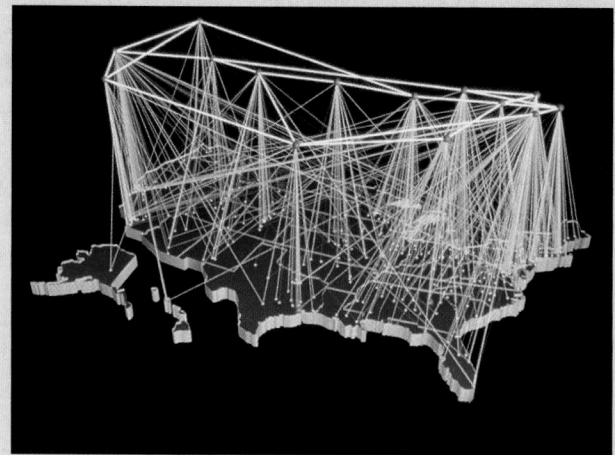

The network of computer linkages in the United States

The problems we solve nowadays have thousands of equations, sometimes a million variables. One of the things that still amazes me is to see a program run on the computer—and to see the answer come out. If we think of the number of combinations of different solutions that we're trying to choose the best of, it's akin to the stars in the heavens. Yet we solve them in a matter of moments. This, to me, is staggering. Not that we can solve them—but that we can solve them so rapidly and efficiently.
—George Dantzig
Inventor of a linear programming method

Problems in linear programming can involve objective functions with thousands of variables subject to thousands of constraints. Several nongeometric linear programming methods are available on software for solving such problems. And we continue to search for faster and faster linear programming methods. This area of applied mathematics has a direct impact on the efficiency and profitability of numerous industries, including telephone and computer communications, and the airlines.

EXERCISE SET 9.5

Practice Exercises

In Exercises 1–4, find the value of the objective function at each corner of the graphed region. What is the maximum value of the objective function? What is the minimum value of the objective function?

1. Objective Function $z = 5x + 6y$

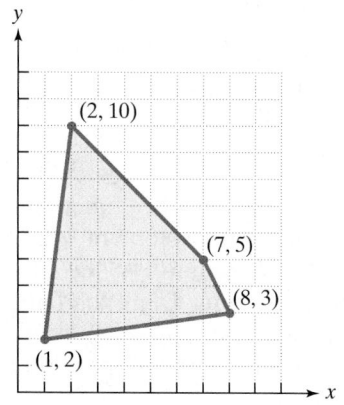

2. Objective Function $z = 3x + 2y$

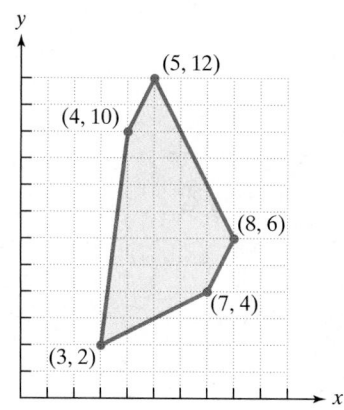

3. Objective Function $z = 40x + 50y$

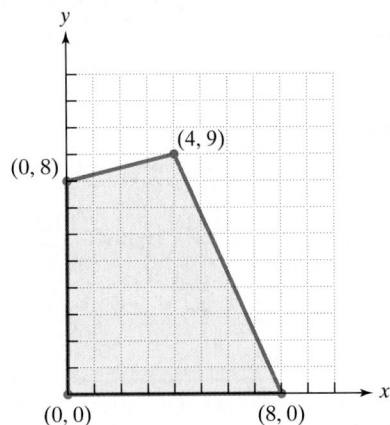

4. Objective Function $z = 30x + 45y$

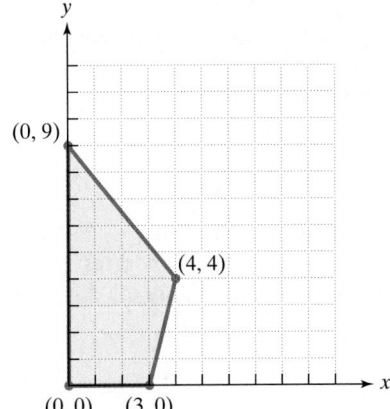

In Exercises 5–14, an objective function and a system of linear inequalities representing constraints are given.

    a. *Graph the system of inequalities representing the constraints.*

    b. *Find the value of the objective function at each corner of the graphed region.*

    c. *Use the values in part (b) to determine the maximum value of the objective function and the values of x and y for which the maximum occurs.*

5. Objective Function $z = 3x + 2y$
   Constraints $x \geq 0, y \geq 0$
                       $2x + y \leq 8$
                       $x + y \geq 4$

6. Objective Function $z = 2x + 3y$
   Constraints $x \geq 0, y \geq 0$
                       $2x + y \leq 8$
                       $2x + 3y \leq 12$

7. Objective Function $z = 4x + y$
   Constraints $x \geq 0, y \geq 0$
                       $2x + 3y \leq 12$
                       $x + y \geq 3$

8. Objective Function $z = x + 6y$
   Constraints $x \geq 0, y \geq 0$
                       $2x + y \leq 10$
                       $x - 2y \geq -10$

9. Objective Function $z = 3x - 2y$
   Constraints $1 \leq x \leq 5$
                       $y \geq 2$
                       $x - y \geq -3$

10. Objective Function $z = 5x - 2y$
   Constraints $0 \leq x \leq 5$
                       $0 \leq y \leq 3$
                       $x + y \geq 2$

11. Objective Function $z = 4x + 2y$
   Constraints $x \geq 0, y \geq 0$
                       $2x + 3y \leq 12$
                       $3x + 2y \leq 12$
                       $x + y \geq 2$

12. Objective Function $z = 2x + 4y$
   Constraints $x \geq 0, y \geq 0$
                       $x + 3y \geq 6$
                       $x + y \geq 3$
                       $x + y \leq 9$

13. Objective Function $z = 10x + 12y$
   Constraints $x \geq 0, y \geq 0$
                       $x + y \leq 7$
                       $2x + y \leq 10$
                       $2x + 3y \leq 18$

14. Objective Function $z = 5x + 6y$
   Constraints $x \geq 0, y \geq 0$
                       $2x + y \geq 10$
                       $x + 2y \geq 10$
                       $x + y \leq 10$

Application Exercises

15. A television manufacturer makes console and wide-screen televisions. The profit per unit is \$125 for the console televisions and \$200 for the wide-screen televisions.

    a. Let x = the number of consoles manufactured in a month and y = the number of wide-screens manufactured in a month. Write the objective function that describes the total monthly profit.

b. The manufacturer is bound by the following constraints:

1. Equipment in the factory allows for making at most 450 console televisions in one month.
2. Equipment in the factory allows for making at most 200 wide-screen televisions in one month.
3. The cost to the manufacturer per unit is $600 for the console televisions and $900 for the wide-screen televisions. Total monthly costs cannot exceed $360,000.

Write a system of three inequalities that describes these constraints.

c. Graph the system of inequalities in part (b). Use only the first quadrant and its boundary, because x and y must both be nonnegative.

d. Evaluate the objective function for total monthly profit at each of the five vertices of the graphed region. (The vertices should occur at $(0, 0)$, $(0, 200)$, $(300, 200)$, $(450, 100)$, and $(450, 0)$.)

e. Complete the missing portions of this statement: The television manufacturer will make the greatest profit by manufacturing _____ console televisions each month and _____ wide-screen televisions each month. The maximum monthly profit is $_____.

16. a. A student earns $10 per hour for tutoring and $7 per hour as a teacher's aid. Let x = the number of hours each week spent tutoring, and y = the number of hours each week spent as a teacher's aid. Write the objective function that describes total weekly earnings.

b. The student is bound by the following constraints:
- To have enough time for studies, the student can work no more than 20 hours a week.
- The tutoring center requires that each tutor spend at least three hours a week tutoring.
- The tutoring center requires that each tutor spend no more than eight hours a week tutoring.

Write a system of three inequalities that describes these constraints.

c. Graph the system of inequalities in part (b). Use only the first quadrant and its boundary, because x and y are nonnegative.

d. Evaluate the objective function for total weekly earnings at each of the four vertices of the graphed region. (The vertices should occur at $(3,0)$, $(8,0)$, $(3,17)$, and $(8,12)$.)

e. Complete the missing portions of this statement: The student can earn the maximum amount per week by tutoring for __ hours per week and working as a teacher's aid for ___ hours per week. The maximum amount that the student can earn each week is $____.

Use the two steps for solving a linear programming problem, given in the box on page 628, to solve the problems in Exercises 17–23.

17. A manufacturer produces two models of mountain bicycles. The times (in hours) required for assembling and painting each model are given in the following table.

	Model *A*	Model *B*
Assembling	5	4
Painting	2	3

The maximum total weekly hours available in the assembly department and the paint department are 200 hours and 108 hours, respectively. The profits per unit are $25 for model A and $15 for model B. How many of each type should be produced to maximize profit?

18. A large institution is preparing lunch menus containing foods A and B. The specifications for the two foods are given in the following table.

Food	Units of Fat per Ounce	Units of Carbohydrates per Ounce	Units of Protein per Ounce
A	1	2	1
B	1	1	1

Each lunch must provide at least 6 units of fat per serving, no more than 7 units of protein, and at least 10 units of carbohydrates. The institution can purchase food A for $0.12 per ounce and food B for $0.08 per ounce. How many ounces of each food should a serving contain to meet the dietary requirements at the least cost?

19. Food and clothing are shipped to victims of a natural disaster. Each carton of food will feed 12 people, while each carton of clothing will help 5 people. Each 20-cubic-foot box of food weighs 50 pounds and each 10-cubic-foot box of clothing weighs 20 pounds. The commercial carriers transporting food and clothing are bound by the following constraints:

1. The total weight per carrier cannot exceed 19,000 pounds.
2. The total volume must be less than 8000 cubic feet.

How many cartons of food and clothing should be sent with each plane shipment to maximize the number of people who can be helped?

20. On June 24, 1948, the former Soviet Union blocked all land and water routes through East Germany to Berlin. A gigantic airlift was organized using American and British planes to bring food, clothing, and other supplies to the more than 2 million people in West Berlin. The cargo capacity was 30,000 cubic feet for an American plane and 20,000 cubic feet for a British plane. To break the Soviet blockade, the Western Allies had to maximize cargo capacity, but were subject to the following restrictions:

1. No more than 44 planes could be used.

2. The larger American planes required 16 personnel per flight, double that of the requirement for the British planes. The total number of personnel available could not exceed 512.

3. The cost of an American flight was $9000 and the cost of a British flight was $5000. Total weekly costs could not exceed $300,000.

Find the number of American and British planes that were used to maximize cargo capacity.

21. A theater is presenting a program on drinking and driving for students and their parents. The proceeds will be donated to a local alcohol information center. Admission is $2.00 for parents and $1.00 for students. However, the situation has two constraints: The theater can hold no more than 150 people and every two parents must bring at least one student. How many parents and students should attend to raise the maximum amount of money?

22. You are about to take a test that contains computation problems worth 6 points each and word problems worth 10 points each. You can do a computation problem in 2 minutes and a word problem in 4 minutes. You have 40 minutes to take the test and may answer no more than 12 problems. Assuming you answer all the problems attempted correctly, how many of each type of problem must you do to maximize your score? What is the maximum score?

23. In 1978, a ruling by the Civil Aeronautics Board allowed Federal Express to purchase larger aircraft. Federal Express's options included 20 Boeing 727s that United Airlines was retiring and/or the French-built Dassault Fanjet Falcon 20. To aid in their decision, executives at Federal Express analyzed the following data:

	Boeing 727	Falcon 20
Direct Operating Cost	$1400 per hour	$500 per hour
Payload	42,000 pounds	6000 pounds

Federal Express was faced with the following constraints:

1. Hourly operating cost was limited to $35,000.

2. Total payload had to be at least 672,000 pounds.

3. Only twenty 727s were available.

Given the constraints, how many of each kind of aircraft should Federal Express have purchased to maximize the number of aircraft?

Writing in Mathematics

24. What kinds of problems are solved using the linear programming method?

25. What is an objective function in a linear programming problem?

26. What is a constraint in a linear programming problem? How is a constraint represented?

27. In your own words, describe how to solve a linear programming problem.

28. Describe a situation in your life in which you would like to maximize something, but are limited by at least two constraints. Can linear programming be used in this situation? Explain your answer.

Critical Thinking Exercises

29. Suppose that you inherit $10,000. The will states how you must invest the money. Some (or all) of the money must be invested in stocks and bonds. The requirements are that at least $3000 be invested in bonds, with expected returns of $0.08 per dollar, and at least $2000 be invested in stocks, with expected returns of $0.12 per dollar. Because the stocks are medium risk, the final stipulation requires that the investment in bonds should never be less than the investment in stocks. How should the money be invested so as to maximize your expected returns?

30. Consider the objective function $z = Ax + By$ ($A > 0$ and $B > 0$) subject to the following constraints: $2x + 3y \leq 9$, $x - y \leq 2$, $x \geq 0$, and $y \geq 0$. Prove that the objective function will have the same maximum value at the vertices $(3, 1)$ and $(0, 3)$ if $A = \frac{2}{3}B$.

Review Exercises

31. Solve: $x^2 - 12x + 36 = 0$.
(Section 6.6, Example 4)

32. Divide: $\dfrac{1}{x^2 - 17x + 30} \div \dfrac{1}{x^2 + 7x - 18}$
(Section 7.2, Example 6)

33. If $f(x) = x^3 + 2x^2 - 5x + 4$, find $f(-1)$.
(Section 8.1, Example 3)

CHAPTER 9 GROUP PROJECTS

1. Each group member should research one situation that provides two different pricing options. These can involve areas such as public transportation options (with or without coupon books) or long-distance telephone plans or anything of interest. Be sure to bring in all the details for each option. At the group meeting, select the two pricing situations that are most interesting and relevant. Using each situation, write a word problem about selecting the better of the two options. The word problem should be one that can be solved using a linear inequality. The group should turn in the two problems and their solutions.

2. For this activity, group members will conduct interviews with a random sample of students on campus. Each student is to be asked, "What is the worst thing about being a student?" One response should be recorded for each student.

   a. Each member should interview enough students so that there are at least 50 randomly selected students in the sample.

   b. After all responses have been recorded, the group should organize the four most common answers. For each answer, compute the percentage of students in the sample who felt that this is the worst thing about being a student.

   c. If there are n students in your sample, the poll's margin of error is given by $\pm\dfrac{1}{\sqrt{n}}$. Compute the margin of error for your poll. Express the margin of error as a percent, correct to the nearest tenth of a percent.

   d. For each of the four most common answers, write a statement about the percentage of all students, x, on your campus who feel that this is the worst thing about being a student. Use the inequality

   $$|x - \text{percentage in the sample}| \leq \text{margin of error}.$$

3. Group members should choose a particular field of interest. Research how linear programming is used to solve problems in that field. If possible, investigate the solution of a specific practical problem. Present a report on your findings, including the contributions of George Dantzig, Narendra Karmarkar, and L.G. Khachion to linear programming.

4. Members of the group should interview a business executive who is in charge of deciding the product mix for a business. How are production policy decisions made? Are other methods used in conjunction with linear programming? What are these methods? What sort of academic background, particularly in mathematics, does this executive have? Present a group report addressing these questions, emphasizing the role of linear programming for the business.

CHAPTER SUMMARY, REVIEW, AND TEST

SUMMARY

DEFINITIONS AND CONCEPTS	EXAMPLES

Section 9.1 Interval Notation and Business Applications Using Linear Inequalities

A linear inequality in one variable can be written in the form $ax + b < c$, $ax + b \leq c$, $ax + b > c$, or $ax + b \geq c$. The set of all numbers that make the inequality a true statement is its solution set. Graphs of solution sets are shown on a number line by shading all points representing numbers that are solutions. Parentheses indicate endpoints that are not solutions. Square brackets indicate endpoints that are solutions.

- $(-2, 1] = \{x \mid -2 < x \leq 1\}$

$$\xleftarrow{\hspace{0.5cm}} \underset{-4\ -3\ -2\ -1\ \ 0\ \ 1\ \ 2\ \ 3\ \ 4}{\rule{4cm}{0.4pt}} \xrightarrow{\hspace{0.5cm}}$$

- $[-2, \infty) = \{x \mid x \geq -2\}$

$$\xleftarrow{\hspace{0.5cm}} \underset{-4\ -3\ -2\ -1\ \ 0\ \ 1\ \ 2\ \ 3\ \ 4}{\rule{4cm}{0.4pt}} \xrightarrow{\hspace{0.5cm}}$$

Solving a Linear Inequality

1. Simplify each side.
2. Collect variable terms on one side and constant terms on the other side.
3. Isolate the variable and solve.

If an inequality is multiplied or divided by a negative number, the inequality symbol must be reversed.

Solve: $2(x + 3) - 5x \leq 15$.

$$2x + 6 - 5x \leq 15$$
$$-3x + 6 \leq 15$$
$$-3x \leq 9$$
$$\frac{-3x}{-3} \geq \frac{9}{-3}$$
$$x \geq -3$$

Solution set: $\{x \mid x \geq -3\}$ or $[-3\ \infty)$

$$\xleftarrow{\hspace{0.5cm}} \underset{-4\ -3\ -2\ -1\ \ 0\ \ 1\ \ 2\ \ 3\ \ 4}{\rule{4cm}{0.4pt}} \xrightarrow{\hspace{0.5cm}}$$

Functions of Business

A company produces and sells x units of a product.

Revenue Function

$$R(x) = (\text{price per unit})x$$

Cost Function

$$C(x) = \text{fixed cost} + (\text{cost per unit produced})x$$

Profit Function

$$P(x) = R(x) - C(x)$$

If $P(x) > 0$, there is a profit gain. If $P(x) < 0$, there is a profit loss.

A company that manufactures lamps has a fixed cost of $80,000 and it costs $20 to produce each lamp. Lamps are sold for $70.

a. Write the cost function.

$$C(x) = 80{,}000 + 20x$$

Fixed cost   Variable cost: $20 per lamp.

b. Write the revenue function.

$$R(x) = 70x$$

Revenue per lamp, $70, times number of lamps sold

c. Write the profit function.

$$P(x) = R(x) - C(x)$$
$$= 70x - (80{,}000 + 20x)$$
$$= 50x - 80{,}000$$

d. More than how many lamps must be produced and sold to have a profit gain?

Solve: $P(x) > 0$.

$$50x - 80{,}000 > 0$$
$$50x > 80{,}000$$
$$x > 1600$$

More than 1600 lamps must be produced and sold to have a profit gain.

DEFINITIONS AND CONCEPTS	EXAMPLES

Section 9.2 Compound Inequalities

Intersection (∩) and Union (∪)
$A \cap B$ is the set of elements common to both set A and set B.
$A \cup B$ is the set of elements that are members of set A or set B or of both sets.

$\{1, 3, 5, 7\} \cap \{5, 7, 9, 11\} = \{5, 7\}$
$\{1, 3, 5, 7\} \cup \{5, 7, 9, 11\} = \{1, 3, 5, 7, 9, 11\}$

A compound inequality is formed by joining two inequalities with the word *and* or *or*. When the connecting word is *and*, graph each inequality separately and take the intersection of their solution sets.

Solve: $x + 1 > 3$ and $x + 4 \le 8$.
$\qquad x > 2$ and $\qquad x \le 4$

$$\xleftarrow{\quad}\underset{-4\,-3\,-2\,-1\ \ 0\ \ 1\ \ 2\ \ 3\ \ 4}{+\ +\ +\ +\ +\ +\ (\ +\ +\ ]}\xrightarrow{\quad}$$

Solution set: $\{x \mid 2 < x \le 4\}$ or $(2, 4]$

The compound inequality $a < x < b$ means $a < x$ and $x < b$. Solve by isolating the variable in the middle.

Solve: $-1 < \dfrac{2x + 1}{3} \le 2$.

$\quad -3 < 2x + 1 \le 6$ Multiply by 3.
$\quad -4 < 2x \le 5$ Subtract 1.
$\quad -2 < x \le \dfrac{5}{2}$ Divide by 2.

Solution set: $\left\{ x \mid -2 < x \le \dfrac{5}{2} \right\}$ or $\left(-2, \dfrac{5}{2} \right]$

$$\xleftarrow{\quad}\underset{-4\,-3\,-2\,-1\ \ 0\ \ 1\ \ 2\ \ 3\ \ 4}{+\ +\ (\ +\ +\ +\ +\ ]\ +\ +}\xrightarrow{\quad}$$

When the connecting word in a compound inequality is *or*, graph each inequality separately and take the union of their solution sets.

Solve: $x - 2 > -3$ or $2x \le -6$.
$\qquad x > -1$ or $\qquad x \le -3$

$$\xleftarrow{\quad}\underset{-6\,-5\,-4\,-3\,-2\,-1\ \ 0\ \ 1\ \ 2}{+\ ]\ +\ +\ +\ +\ (\ +\ +}\xrightarrow{\quad}$$

Solution set: $\{x \mid x \le -3 \text{ or } x > -1\}$ or
$(-\infty, -3] \cup (-1, \infty)$

Section 9.3 Equations and Inequalities Involving Absolute Value

Absolute Value Equations
  1. If $c > 0$, then $|X| = c$ means $X = c$ or $X = -c$.
  2. If $c < 0$, then $|X| = c$ has no solution.
  3. If $c = 0$, then $|X| = 0$ means $X = 0$.

Solve: $|2x - 7| = 3$.
$\quad 2x - 7 = 3$ or $2x - 7 = -3$
$\qquad 2x = 10 \qquad\qquad 2x = 4$
$\qquad\ \ x = 5 \qquad\qquad\ \ x = 2$
The solution set is $\{2, 5\}$.

Absolute Value Equations with Two Absolute Value Bars
If $|X_1| = |X_2|$, then $X_1 = X_2$ or $X_1 = -X_2$.

Solve: $|x - 6| = |2x + 1|$.
$\quad x - 6 = 2x + 1$ or $x - 6 = -(2x + 1)$
$\quad -x - 6 = 1 \qquad\qquad x - 6 = -2x - 1$
$\qquad -x = 7 \qquad\qquad\ 3x - 6 = -1$
$\qquad\ \ x = -7 \qquad\qquad\quad 3x = 5$
$$x = \dfrac{5}{3}$$

The solutions are -7 and $\frac{5}{3}$, and the solution set is $\left\{ -7, \frac{5}{3} \right\}$.

DEFINITIONS AND CONCEPTS	EXAMPLES

Section 9.3 Equations and Inequalities Involving Absolute Value (continued)

If c is a positive number, then to solve $|X| < c$, solve the compound inequality $-c < X < c$. If c is negative, $|X| < c$ has no solution.

Solve: $|x - 4| < 3$.
$$-3 < x - 4 < 3$$
$$1 < x < 7$$
The solution set is $\{x \mid 1 < x < 7\}$ or $(1, 7)$.

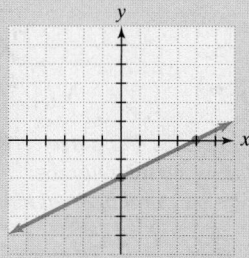
(number line graph: 0 1 2 3 4 5 6 7 8)

If c is a positive number, then to solve $|X| > c$, solve the compound inequality $X < -c \ or \ X > c$. If c is negative, $|X| > c$ is true for all real numbers for which X is defined.

Solve: $\left|\dfrac{x}{3} - 1\right| \geq 2$.

$$\frac{x}{3} - 1 \leq -2 \quad \text{or} \quad \frac{x}{3} - 1 \geq 2$$

$$x - 3 \leq -6 \quad \text{or} \quad x - 3 \geq 6 \quad \text{Multiply by 3.}$$

$$x \leq -3 \quad \text{or} \quad x \geq 9 \quad \text{Add 3.}$$

The solution set is $\{x \mid x \leq -3 \text{ or } x \geq 9\}$ or $(-\infty, -3] \cup [9, \infty)$.

(number line graph: $-6 -4 -2 \ 0 \ 2 \ 4 \ 6 \ 8 \ 10$)

Section 9.4 Linear Inequalities in Two Variables

If the equal sign in $Ax + By = C$ is replaced with an inequality symbol, the result is a linear inequality in two variables. Its graph is the set of all points whose coordinates satisfy the inequality. To obtain the graph:

1. Replace the inequality symbol with an equal sign and graph the boundary line. Use a solid line for $\leq$ or $\geq$ and a dashed line for $<$ or $>$.
2. Choose a test point not on the line and substitute its coordinates into the inequality.
3. If a true statement results, shade the half-plane containing the test point. If a false statement results, shade the half-plane not containing the test point.

Graph: $x - 2y \leq 4$.

1. Graph $x - 2y = 4$. Use a solid line because the inequality symbol is $\leq$.
2. Test $(0, 0)$.
$$0 - 2 \cdot 0 \overset{?}{=} 4$$
$$0 \leq 4, \quad \textbf{true}$$
3. The inequality is true. Shade the half-plane containing $(0, 0)$.

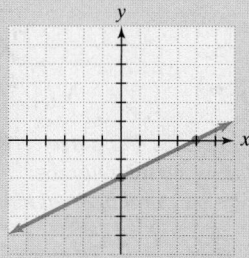

Two or more linear inequalities make up a system of linear inequalities. A solution is an ordered pair satisfying all inequalities in the system. To graph a system of inequalities, graph each inequality in the system. The overlapping region represents the solutions of the system.

Graph the solutions of the system:
$$y \leq -2x$$
$$x - y \geq 3.$$

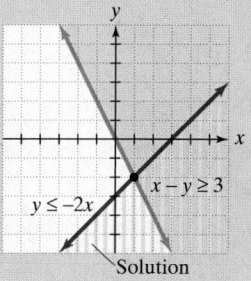
$x - y \geq 3$
$y \leq -2x$
Solution

DEFINITIONS AND CONCEPTS	EXAMPLES

Section 9.5 Linear Programming

Linear programming is a method for solving problems in which a particular quantity that must be maximized or minimized is limited. An objective function is an algebraic expression in three variables describing a quantity that must be maximized or minimized. Constraints are restrictions, expressed as linear inequalities.

Solving a Linear Programming Problem

1. Graph the system of inequalities representing the constraints.
2. Find the value of the objective function at each corner, or vertex, of the graphed region. The maximum and minimum of the objective function occur at one or more vertices.

Find the maximum value of the objective function $z = 3x + 2y$ subject to the following constraints: $x \geq 0, y \geq 0, 2x + 3y \leq 18, 2x + y \leq 10$.

1. Graph the system of inequalities representing the constraints.

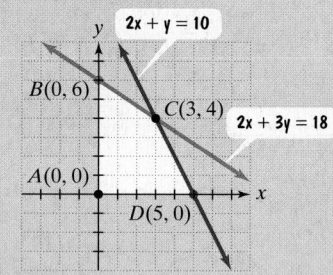

2. Evaluate the objective function at each vertex.

Vertex	$z = 3x + 2y$
$A(0, 0)$	$z = 3(0) + 2(0) = 0$
$B(0, 6)$	$z = 3(0) + 2(6) = 12$
$C(3, 4)$	$z = 3(3) + 2(4) = 17$
$D(5, 0)$	$z = 3(5) + 2(0) = 15$

The maximum value of the objective function is 17.

Review Exercises

9.1 *In Exercises 1–3, express each interval in terms of an inequality and graph the interval on a number line.*

1. $(-2, 3]$
2. $[-1.5, 2]$
3. $(-1, \infty)$

In Exercises 4–8, solve each linear inequality and graph the solution set on a number line. Express the solution set in both set-builder and interval notations.

4. $-6x + 3 \leq 15$
5. $6x - 9 \geq -4x - 3$
6. $\dfrac{x}{3} - \dfrac{3}{4} - 1 > \dfrac{x}{2}$
7. $6x + 5 > -2(x - 3) - 25$
8. $3(2x - 1) - 2(x - 4) \geq 7 + 2(3 + 4x)$

9. The cost and revenue functions for producing and selling x units of a toaster oven are
$$C(x) = 40x + 357{,}000 \quad \text{and} \quad R(x) = 125x.$$

a. Write the profit function, P, from producing and selling x toaster ovens.

b. More than how many tooster ovens must be produced and sold to have profit gain?

Use this information to solve Exercises 10–13: A company is planning to produce and sell a new line of computers. The fixed cost will be $360,000, and it will cost $850 to produce each computer. Each computer will be sold for $1150.

10. Write the cost function, C, of producing x computers.

11. Write the revenue function, R, from the sale of x computers.

12. Write the profit function, P, from producing and selling x computers.

13. More than how many computers must be produced and sold to have a profit gain?

14. A person can choose between two charges on a checking account. The first method involves a fixed cost of $11 per month plus 6¢ for each check written. The second method involves a fixed cost of $4 per month plus 20¢ for each check written. How many checks should be written to make the first method a better deal?

9.2 *In Exercises 15–18, let* $A = \{a, b, c\}$, $B = \{a, c, d, e\}$ *and* $C = \{a, d, f, g\}$. *Find the indicated set.*

15. $A \cap B$

16. $A \cap C$

17. $A \cup B$

18. $A \cup C$

In Exercises 19–29, solve each compound inequality. Except for the empty set, express the solution set in both set-builder and interval notations. Graph the solution set on a number line.

19. $x \leq 3$ and $x < 6$

20. $x \leq 3$ or $x < 6$

21. $-2x < -12$ and $x - 3 < 5$

22. $5x + 3 \leq 18$ and $2x - 7 \leq -5$

23. $2x - 5 > -1$ and $3x < 3$

24. $2x - 5 > -1$ or $3x < 3$

25. $x + 1 \leq -3$ or $-4x + 3 < -5$

26. $5x - 2 \leq -22$ or $-3x - 2 > 4$

27. $5x + 4 \geq -11$ or $1 - 4x \geq 9$

28. $-3 < x + 2 \leq 4$

29. $-1 \leq 4x + 2 \leq 6$

30. On the first of four exams, your grades are 72, 73, 94, and 80. There is still one more exam, and you are hoping to earn a B in the course. This will occur if the average of your five exam grades is greater than or equal to 80 and less than 90. What range of grades on the fifth exam will result in receiving a B? Use interval notation to express this range.

9.3 *In Exercises 31–34, find the solution set for each equation.*

31. $|2x + 1| = 7$

32. $|3x + 2| = -5$

33. $2|x - 3| - 7 = 10$

34. $|4x - 3| = |7x + 9|$

In Exercises 35–38, solve and graph the solution set on a number line. Except for the empty set, express the solution set in both set-builder and interval notations.

35. $|2x + 3| \leq 15$

36. $\left| \dfrac{2x + 6}{3} \right| > 2$

37. $|2x + 5| - 7 < -6$

38. $|2x - 3| + 4 \geq -10$

39. Approximately 90% of the population sleeps h hours daily, where h is modeled by the inequality $|h - 6.5| \leq 1$. Write a sentence describing the range for the number of hours that most people sleep. Do *not* use the phrase "absolute value" in your description.

9.4 *In Exercises 40–45, graph each inequality in a rectangular coordinate system.*

40. $3x - 4y > 12$

41. $x - 3y \leq 6$

42. $y \leq -\dfrac{1}{2}x + 2$

43. $y > \dfrac{3}{5}x$

44. $x \leq 2$

45. $y > -3$

In Exercises 46–54, graph the solution set of each system of inequalities or indicate that the system has no solution.

46. $3x - y \leq 6$
$x + y \geq 2$

47. $y < -x + 4$
$y > x - 4$

48. $-3 \leq x < 5$

49. $-2 < y \leq 6$

50. $x \geq 3$
$y \leq 0$

51. $2x - y > -4$
$x \geq 0$

52. $x + y \leq 6$
$y \geq 2x - 3$

53. $3x + 2y \geq 4$
$x - y \leq 3$
$x \geq 0, y \geq 0$

54. $2x - y > 2$
$2x - y < -2$

9.5

55. Find the value of the objective function $z = 2x + 3y$ at each corner of the graphed region shown. What is the maximum value of the objective function? What is the minimum value of the objective function?

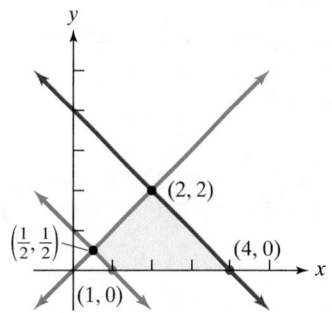

In Exercises 56–58, graph the region determined by the constraints. Then find the maximum value of the given objective function, subject to the constraints.

56. Objective Function $\quad z = 2x + 3y$
Constraints $\quad x \geq 0, y \geq 0$
$x + y \leq 8$
$3x + 2y \geq 6$

57. Objective Function $\quad z = x + 4y$
Constraints $\quad 0 \leq x \leq 5, 0 \leq y \leq 7$
$x + y \geq 3$

58. Objective Function $\quad z = 5x + 6y$
Constraints $\quad x \geq 0, y \geq 0$
$y \leq x$
$2x + y \leq 12$
$2x + 3y \geq 6$

59. A paper manufacturing company converts wood pulp to writing paper and newsprint. The profit on a unit of writing paper is $500 and the profit on a unit of newsprint is $350.

  a. Let x represent the number of units of writing paper produced daily. Let y represent the number of units of newsprint produced daily. Write the objective function that models total daily profit.

  b. The manufacturer is bound by the following constraints:
   1. Equipment in the factory allows for making at most 200 units of paper (writing paper and newsprint) in a day.
   2. Regular customers require at least 10 units of writing paper and at least 80 units of newsprint daily. Write a system of inequalities that models these constraints.

  c. Graph the inequalities in part (b). Use only the first quadrant, because x and y must both be positive. (*Suggestion*: Let each unit along the x- and y-axes represent 20.)

  d. Evaluate the objective profit function at each of the three vertices of the graphed region.

  e. Complete the missing portions of this statement: The company will make the greatest profit by producing ___ units of writing paper and ___ units of newsprint each day. The maximum daily profit is $ _____ .

60. A manufacturer of lightweight tents makes two models whose specifications are given in the following table.

	Cutting Time per Tent	Assembly Time per Tent
Model A	0.9 hour	0.8 hour
Model B	1.8 hours	1.2 hours

On a monthly basis, the manufacturer has no more than 864 hours of labor available in the cutting department and at most 672 hours in the assembly division. The profits come to $25 per tent for model A and $40 per tent for model B. How many of each should be manufactured monthly to maximize the profit?

Chapter 9 Test

In Exercises 1–2, express each interval as an inequality and graph the interval on a number line.

  1. $[-3, 2)$
  2. $(-\infty, -1]$

In Exercises 3–4, solve and graph the solution set on a number line. Express the solution set in both set-builder and interval notations.

  3. $3(x + 4) \geq 5x - 12$
  4. $\dfrac{x}{6} + \dfrac{1}{8} \leq \dfrac{x}{2} - \dfrac{3}{4}$

  5. A company is planning to manufacture computer desks. The fixed cost will be $60,000 and it will cost $200 to produce each desk. Each desk will be sold for $450.

   a. Write the cost function, C, of producing x desks.

   b. Write the revenue function, R, from the sale of x desks.

   c. Write the profit function, P, from producing and selling x desks.

   d. More than how many desks must be produced and sold to have a profit gain?

  6. Find the intersection: $\{2, 4, 6, 8, 10\} \cap \{4, 6, 12, 14\}$.
  7. Find the union: $\{2, 4, 6, 8, 10\} \cup \{4, 6, 12, 14\}$.

In Exercises 8–12, solve each compound inequality. Except for the empty set, express the solution set in both set-builder and interval notations. Graph the solution set on a number line.

  8. $2x + 4 < 2$ and $x - 3 > -5$
  9. $x + 6 \geq 4$ and $2x + 3 \geq -2$
  10. $2x - 3 < 5$ or $3x - 6 \leq 4$
  11. $x + 3 \leq -1$ or $-4x + 3 < -5$

  12. $-3 \leq \dfrac{2x + 5}{3} < 6$

In Exercises 13–14, find the solution set for each equation.

  13. $|5x + 3| = 7$
  14. $|6x + 1| = |4x + 15|$

In Exercises 15–16, solve and graph the solution set on a number line. Express the solution set in both set-builder and interval notations.

  15. $|2x - 1| < 7$
  16. $|2x - 3| \geq 5$

  17. The inequality $|T - 74| \leq 8$ describes the range of monthly average temperature, T, in degrees Fahrenheit, for Miami, Florida. Solve the inequality and interpret the solution.

In Exercises 18–20, graph each inequality in a rectangular coordinate system.

18. $3x - 2y < 6$

19. $y \geq \dfrac{1}{2}x - 1$

20. $y \leq -1$

In Exercises 21–23, graph the solution set of each system of inequalities.

21. $\begin{aligned} x + y &\geq 2 \\ x - y &\geq 4 \end{aligned}$

22. $\begin{aligned} 3x + y &\leq 9 \\ 2x + 3y &\geq 6 \\ x \geq 0, \; y &\geq 0 \end{aligned}$

23. $-2 < x \leq 4$

24. Find the maximum value of the objective function $z = 3x + 5y$ subject to the following constraints: $x \geq 0$, $y \geq 0$, $x + y \leq 6$, $x \geq 2$.

25. A manufacturer makes two types of jet skis, regular and deluxe. The profit on a regular jet ski is $200 and the profit on the deluxe model is $250. To meet customer demand, the company must manufacture at least 50 regular jet skis per week and at least 75 deluxe models. To maintain high quality, the total number of both models of jet skis manufactured by the company should not exceed 150 per week. How many jet skis of each type should be manufactured per week to obtain maximum profit? What is the maximum weekly profit?

Cumulative Review Exercises (Chapters 1–9)

In Exercises 1–2, solve each equation.

1. $5(x + 1) + 2 = x - 3(2x + 1)$

2. $\dfrac{2(x + 6)}{3} = 1 + \dfrac{4x - 7}{3}$

3. Simplify: $\dfrac{-10x^2 y^4}{15x^7 y^{-3}}$.

4. If $f(x) = x^2 - 3x + 4$, find $f(-3)$ and $f(2a)$.

5. If $f(x) = 3x^2 - 4x + 1$ and $g(x) = x^2 - 5x - 1$, find $(f - g)(x)$ and $(f - g)(2)$.

6. Use function notation to write the equation of the line passing through $(2, 3)$ and perpendicular to the line whose equation is $y = 2x - 3$.

In Exercises 7–10, graph each equation or inequality in a rectangular coordinate system.

7. $f(x) = 2x + 1$

8. $y > 2x$

9. $2x - y \geq 6$

10. $f(x) = -1$

11. Solve the system:

$\begin{aligned} 3x - y + z &= -15 \\ x + 2y - z &= 1. \\ 2x + 3y - 2z &= 0 \end{aligned}$

12. Solve using matrices:

$\begin{aligned} 2x - y &= -4 \\ x + 3y &= 5 \end{aligned}$

13. Evaluate: $\begin{vmatrix} 4 & 3 \\ -1 & -5 \end{vmatrix}$.

14. A motel with 60 rooms charges $90 per night for rooms with kitchen facilities and $80 per night for rooms without kitchen facilities. When all rooms are occupied, the nightly revenue is $5260. How many rooms of each kind are there?

15. Which of the following are functions?

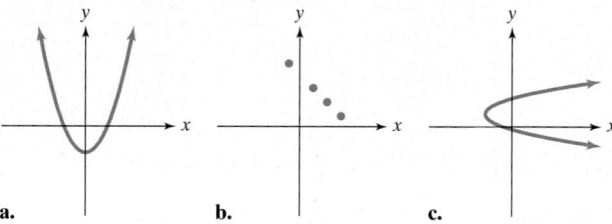

a.   **b.**   **c.**

In Exercises 16–20, solve and graph the solution set on a number line. Express the solution set in both set-builder and interval notations.

16. $\dfrac{x}{4} - \dfrac{3}{4} - 1 \leq \dfrac{x}{2}$

17. $2x + 5 \leq 11$ and $-3x > 18$

18. $x - 4 \geq 1$ or $-3x + 1 \geq -5 - x$

19. $|2x + 3| \leq 17$

20. $|3x - 8| > 7$

An increase in the median age at which Americans first marry and the increased rate of divorce have contributed to the growth of our single population. Data indicate that the number of Americans living alone increased quite rapidly toward the end of the twentieth century. However, now this growth rate is beginning to slow down. In this chapter, you will see why radical functions with square roots are used to describe phenomena that are continuing to grow, but whose growth is leveling off. By learning about radicals and radical functions, you will have new algebraic tools for describing your world.

You enjoy the single life. You have a terrific career, a support group of close friends, and your own home. Still, you feel the pressure to marry. Your parents belong to a generation that believes that you are not complete until you have found your "other half." They see your staying unattached as a "failure to marry." Perhaps you will someday enter into a permanent relationship, but at the moment you are satisfied with your life and career, and plan to stay single into the future.

Chapter 10

Radicals, Radical Functions, and Rational Exponents

643

▶ SECTION 10.1   *Radical Expressions and Functions*

Objectives

1 Evaluate square roots.

2 Evaluate square root functions.

3 Find the domain of a square root function.

4 Use models that are square root functions.

5 Simplify expressions of the form $\sqrt{a^2}$.

6 Evaluate cube root functions.

7 Simplify expressions of the form $\sqrt[3]{a^3}$.

8 Find even and odd roots.

9 Simplify expressions of the form $\sqrt[n]{a^n}$.

SSM   PH Tutor   CD-   Video
      Center   ROM   www

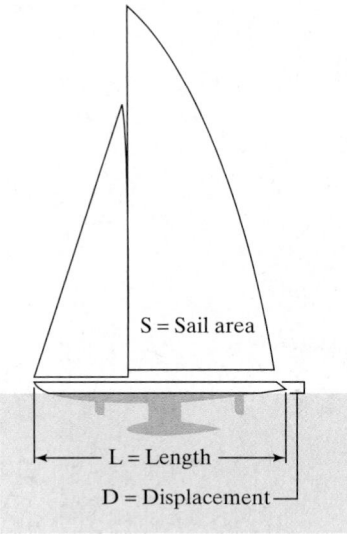

S = Sail area

L = Length

D = Displacement

The America's Cup is the supreme event in ocean sailing. Competition is fierce and the costs are huge. Competitors look to mathematics to provide the critical innovation that can make the difference between winning and losing. The basic dimensions of competitors' yachts must satisfy an inequality containing square roots and cube roots:

$$L + 1.25\sqrt{S} - 9.8\sqrt[3]{D} \le 16.296.$$

In the inequality, L is the yacht's length, in meters, S is its sail area, in square meters, and D is its displacement, in cubic meters.

In this section, we introduce a new category of expressions and functions that contain roots. You will see why square root functions are used to describe phenomena that are continuing to grow but whose growth is leveling off.

1 Evaluate square roots.

Square Roots

From our earlier work with exponents, we are aware that the square of both 5 and −5 is 25:

$$5^2 = 25 \quad \text{and} \quad (-5)^2 = 25.$$

The reverse operation of squaring a number is finding the *square root* of the number. For example,

- A square root of 25 is 5 because $5^2 = 25$.
- A square root of 25 is also −5 because $(-5)^2 = 25$.

In general, **if $b^2 = a$, then b is a square root of a.**

The symbol $\sqrt{\ }$ is used to denote the *positive* or *principal square root* of a number. For example,

- $\sqrt{25} = 5$ because $5^2 = 25$ and 5 is positive.
- $\sqrt{100} = 10$ because $10^2 = 100$ and 10 is positive.

The symbol $\sqrt{}$ that we use to denote the principal square root is called a **radical sign**. The number under the radical sign is called the **radicand**. Together we refer to the radical sign and its radicand as a **radical expression**.

Radical sign — $\sqrt{a}$ — Radicand

Radical expression

> ### Definition of the Principal Square Root
> If a is a nonnegative real number, the nonnegative number b such that $b^2 = a$, denoted by $b = \sqrt{a}$, is the **principal square root** of a.

The symbol $-\sqrt{}$ is used to denote the negative square root of a number. For example,

- $-\sqrt{25} = -5$ because $(-5)^2 = 25$ and -5 is negative.
- $-\sqrt{100} = -10$ because $(-10)^2 = 100$ and -10 is negative.

EXAMPLE 1   Evaluating Square Roots

Evaluate:

    a. $\sqrt{81}$ **b.** $-\sqrt{9}$ **c.** $\sqrt{\dfrac{4}{49}}$

    d. $\sqrt{0.0064}$ **e.** $\sqrt{36 + 64}$ **f.** $\sqrt{36} + \sqrt{64}$.

Solution

Study Tip

In Example 1, parts (e) and (f), observe that $\sqrt{36 + 64}$ is not equal to $\sqrt{36} + \sqrt{64}$. In general,

$$\sqrt{a + b} \neq \sqrt{a} + \sqrt{b}$$

and

$$\sqrt{a - b} \neq \sqrt{a} - \sqrt{b}.$$

a. $\sqrt{81} = 9$ The principal square root of 81 is 9 because $9^2 = 81$.

b. $-\sqrt{9} = -3$ The negative square root of 9 is -3 because $(-3)^2 = 9$.

c. $\sqrt{\dfrac{4}{49}} = \dfrac{2}{7}$ The principal square root of $\dfrac{4}{49}$ is $\dfrac{2}{7}$ because $\left(\dfrac{2}{7}\right)^2 = \dfrac{4}{49}$.

d. $\sqrt{0.0064} = 0.08$ The principal square root of 0.0064 is 0.08 because $(0.08)^2 = (0.08)(0.08) = 0.0064$.

e. $\sqrt{36 + 64} = \sqrt{100}$ Simplify the radicand.

             $= 10$ Take the principal square root of 100, which is 10.

f. $\sqrt{36} + \sqrt{64} = 6 + 8$ $\sqrt{36} = 6$ because $6^2 = 36$. $\sqrt{64} = 8$

                $= 14$ because $8^2 = 64$. ∎

✔ **CHECK POINT 1** Evaluate:

    a. $\sqrt{64}$ **b.** $-\sqrt{49}$ **c.** $\sqrt{\dfrac{16}{25}}$

    d. $\sqrt{0.0081}$ **e.** $\sqrt{9 + 16}$ **f.** $\sqrt{9} + \sqrt{16}$.

Let's see what happens to the radical expression $\sqrt{x}$ if x is a negative number. Is the square root of a negative number a real number? For example, consider $\sqrt{-25}$. Is there a real number whose square is -25? No. Thus, $\sqrt{-25}$ is not a real number. In general, **a square root of a negative number is not a real number**.

2 Evaluate square root functions.

Square Root Functions

Because each nonnegative real number, x, has precisely one principal square root, $\sqrt{x}$, there is a **square root function** defined by

$$f(x) = \sqrt{x}.$$

The domain of this function is $[0, \infty)$. We can graph $f(x) = \sqrt{x}$ by selecting nonnegative real numbers for x. It is easiest to choose perfect squares, numbers that have rational square roots. Table 10.1 shows five such choices for x and the calculations for the corresponding outputs. We plot these ordered pairs as points in the rectangular coordinate system and connect the points with a smooth curve. The graph of $f(x) = \sqrt{x}$ is shown in Figure 10.1.

Table 10.1

x	$f(x) = \sqrt{x}$	(x, y) or $(x, f(x))$
0	$f(0) = \sqrt{0} = 0$	$(0, 0)$
1	$f(1) = \sqrt{1} = 1$	$(1, 1)$
4	$f(4) = \sqrt{4} = 2$	$(4, 2)$
9	$f(9) = \sqrt{9} = 3$	$(9, 3)$
16	$f(16) = \sqrt{16} = 4$	$(16, 4)$

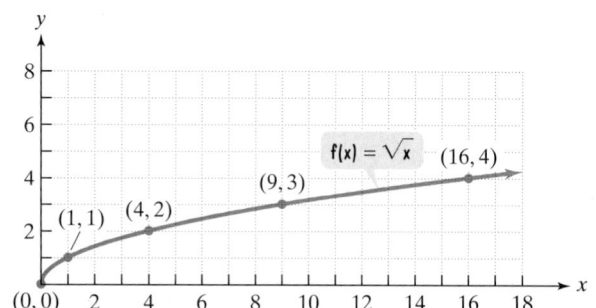

Figure 10.1 The graph of the square root function $f(x) = \sqrt{x}$

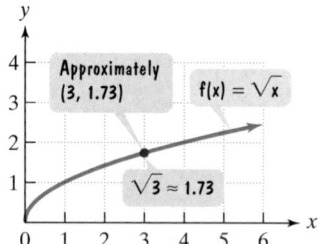

Figure 10.2 Visualizing $\sqrt{3}$ as a point on the graph of $f(x) = \sqrt{x}$

Is it possible to choose values of x in Table 10.1 that are not squares of integers, or perfect squares? Yes. For example, we can let $x = 3$. Thus, $f(3) = \sqrt{3}$. Because 3 is not a perfect square, $\sqrt{3}$ is an irrational number, one that cannot be expressed as a quotient of integers. We can use a calculator to find a decimal approximation of $\sqrt{3}$.

Most Scientific Calculators

$3\boxed{\sqrt{}}$

Most Graphing Calculators

$\boxed{\sqrt{}}\ 3\ \boxed{\text{ENTER}}$

Rounding the displayed number to two decimal places, $\sqrt{3} \approx 1.73$. This information is shown visually as a point, approximately $(3, 1.73)$, on the graph of $f(x) = \sqrt{x}$ in Figure 10.2.

To evaluate a square root function, we use substitution, just as we do to evaluate other functions.

EXAMPLE 2   Evaluating Square Root Functions

For each function, find the indicated function value:

a. $f(x) = \sqrt{5x - 6}$; $f(2)$ **b.** $g(x) = -\sqrt{64 - 8x}$; $g(-3)$.

Solution

a. $f(2) = \sqrt{5 \cdot 2 - 6}$ Substitute 2 for x in $f(x) = \sqrt{5x - 6}$.

$ = \sqrt{4} = 2$ Simplify the radicand and take the square root.

b. $g(-3) = -\sqrt{64 - 8(-3)}$ Substitute -3 for x in $g(x) = -\sqrt{64 - 8x}$.

$ = -\sqrt{88} \approx -9.38$ Simplify the radicand: $64 - 8(-3) = 64 - (-24) = 64 + 24 = 88$. Then use a calculator to approximate $\sqrt{88}$.

■

✔ **CHECK POINT 2** For each function, find the indicated function value:
a. $f(x) = \sqrt{12x - 20}$; $f(3)$
b. $g(x) = -\sqrt{9 - 3x}$; $g(-5)$.

3 Find the domain of a square root function.

We have seen that the domain of a function f is the largest set of real numbers for which the value of $f(x)$ is a real number. Because only nonnegative numbers have real square roots, the domain of a square root function is the set of real numbers for which the radicand is nonnegative.

EXAMPLE 3 Finding the Domain of a Square Root Function

Find the domain of
$$f(x) = \sqrt{3x + 12}.$$

Solution The domain is the set of real numbers, x, for which the radicand, $3x + 12$, is nonnegative. We set the radicand greater than or equal to 0 and solve the resulting inequality.

$$3x + 12 \geq 0$$
$$3x \geq -12$$
$$x \geq -4$$

The domain of f is $\{x \mid x \geq -4\}$ or $[-4, \infty)$. ∎

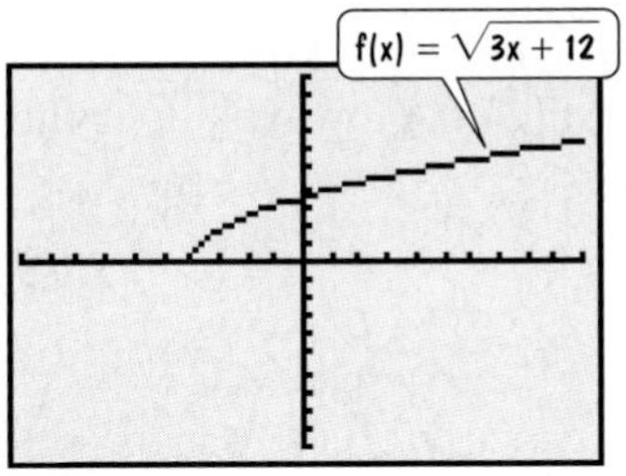

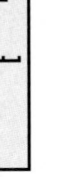

$f(x) = \sqrt{3x + 12}$

Figure 10.3

Figure 10.3 shows the graph of $f(x) = \sqrt{3x + 12}$ in a $[-10, 10, 1]$ by $[-10, 10, 1]$ viewing rectangle. The graph appears only for $x \geq -4$, verifying $[-4, \infty)$ as the domain. Can you see how the graph also illustrates this square root function's range? The graph only appears for nonnegative values of y. Thus, the range is $[0, \infty)$.

✔ **CHECK POINT 3** Find the domain of
$$f(x) = \sqrt{9x - 27}.$$

4 Use models that are square root functions.

The graph of the square root function $f(x) = \sqrt{x}$ is increasing from left to right. However, the rate of increase is slowing down as the graph moves to the right. This is why square root functions are often used to model growing phenomena with growth that is leveling off. The amount of money for new student loans, shown from 1993 through 2000 in Figure 10.4, displays this growth pattern.

Amount of New Student Loans

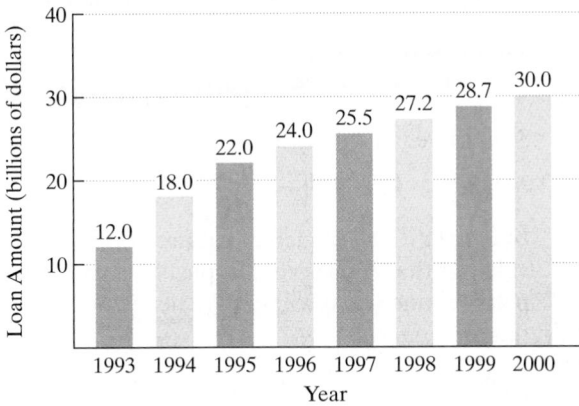

Figure 10.4 *Source:* U.S. Department of Education

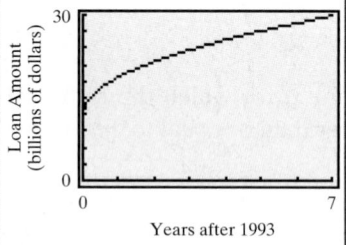
EXAMPLE 4   Modeling with a Square Root Function

The data in Figure 10.4 can be modeled by the function
$$f(x) = 6.75\sqrt{x} + 12$$
where $f(x)$ is the amount, in billions of dollars, of new student loans x years after 1993. According to the model, how much was loaned in 2000? Round to the nearest tenth of a billion. How well does the model describe the actual data?

Solution   Because 2000 is 7 years after 1993, we substitute 7 for x and evaluate the function at 7.

$$f(x) = 6.75\sqrt{x} + 12 \qquad \text{Use the given function.}$$
$$f(7) = 6.75\sqrt{7} + 12 \qquad \text{Substitute 7 for x.}$$
$$\approx 29.9 \qquad \text{Use a calculator.}$$

Rounding to the nearest tenth of a billion, the amount of money in new student loans in 2000 was $29.9 billion. The amount given in the data in Figure 10.4 on the bottom of page 647 is $30.0 billion. Thus, the model describes the actual data fairly well.   ■

✔ **CHECK POINT 4**   Use the square root function in Example 4 to find how much was loaned in 1997. Round to the nearest tenth of a billion. How well does the model describe the actual data?

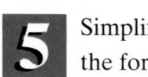 Simplify expressions of the form $\sqrt{a^2}$.

Simplifying Expressions of the Form $\sqrt{a^2}$

You may think that $\sqrt{a^2} = a$. However, this is not necessarily true. Consider the following examples:

$$\sqrt{4^2} = \sqrt{16} = 4$$
$$\sqrt{(-4)^2} = \sqrt{16} = 4.$$

The result is not -4, but rather the absolute value of -4, or 4.

Here is a rule for simplifying expressions of the form $\sqrt{a^2}$:

> **Simplifying $\sqrt{a^2}$**
>
> For any real number a,
> $$\sqrt{a^2} = |a|.$$
>
> In words, the principal square root of a^2 is the absolute value of a.

EXAMPLE 5   Simplifying Radical Expressions

Simplify each expression:

  a. $\sqrt{(-6)^2}$ **b.** $\sqrt{(x+5)^2}$ **c.** $\sqrt{25x^6}$ **d.** $\sqrt{x^2 - 4x + 4}$.

Solution   The principal square root of an expression squared is the absolute value of that expression. In parts (a) and (b), we are given squared radicands. In parts (c) and (d), it will first be necessary to express the radicand as an expression that is squared.

  a. $\sqrt{(-6)^2} = |-6| = 6$

  b. $\sqrt{(x+5)^2} = |x+5|$

c. To simplify $\sqrt{25x^6}$, first write $25x^6$ as an expression that is squared: $25x^6 = (5x^3)^2$. Then simplify.

$$\sqrt{25x^6} = \sqrt{(5x^3)^2} = |5x^3|$$

d. To simplify $\sqrt{x^2 - 4x + 4}$, first write $x^2 - 4x + 4$ as an expression that is squared by factoring the perfect square trinomial: $x^2 - 4x + 4 = (x - 2)^2$. Then simplify.

$$\sqrt{x^2 - 4x + 4} = \sqrt{(x - 2)^2} = |x - 2| \qquad \blacksquare$$

✔ **CHECK POINT 5** Simplify each expression:

a. $\sqrt{(-7)^2}$

b. $\sqrt{(x + 8)^2}$

c. $\sqrt{49x^{10}}$

d. $\sqrt{x^2 - 6x + 9}$.

In some situations we are told that no radicands involve negative quantities to even powers. When the expression being squared is nonnegative, it is not necessary to use absolute value when simplifying $\sqrt{a^2}$. For example, assuming that no radicands contain negative quantities that are squared,

$$\sqrt{x^8} = \sqrt{(x^4)^2} = x^4$$

$$\sqrt{25x^2 + 10x + 1} = \sqrt{(5x + 1)^2} = 5x + 1.$$

6 Evaluate cube root functions.

Cube Roots and Cube Root Functions

Finding the square root of a number reverses the process of squaring a number. Similarly, finding the cube root of a number reverses the process of cubing a number. For example, $2^3 = 8$, and so the cube root of 8 is 2. The notation that we use is $\sqrt[3]{8} = 2$.

Definition of the Cube Root of a Number

The **cube root** of a real number a is written $\sqrt[3]{a}$.

$$\sqrt[3]{a} = b \quad \text{means that} \quad b^3 = a.$$

Study Tip

Some cube roots occur so frequently that you might want to memorize them.

$\sqrt[3]{1} = 1$

$\sqrt[3]{8} = 2$

$\sqrt[3]{27} = 3$

$\sqrt[3]{64} = 4$

$\sqrt[3]{125} = 5$

$\sqrt[3]{216} = 6$

$\sqrt[3]{1000} = 10$

For example,

$$\sqrt[3]{64} = 4 \quad \text{because} \quad 4^3 = 64.$$

$$\sqrt[3]{-27} = -3 \quad \text{because} \quad (-3)^3 = -27.$$

By contrast to square roots, the cube root of a negative number is a real number. All real numbers have cube roots. The cube root of a positive number is positive. The cube root of a negative number is negative.

Because every real number, x, has precisely one cube root, $\sqrt[3]{x}$, there is a **cube root function** defined by

$$f(x) = \sqrt[3]{x}.$$

The domain of this function is the set of all real numbers. We can graph $f(x) = \sqrt[3]{x}$ by selecting perfect cubes, numbers that have rational cube roots, for x. Table 10.2 shows five such choices for x and the calculations for the corresponding outputs. We plot these ordered pairs as points in the rectangular coordinate system and connect the points with a smooth curve. The graph of $f(x) = \sqrt[3]{x}$ is shown in Figure 10.5.

Table 10.2

x	$f(x) = \sqrt[3]{x}$	(x, y) or $(x, f(x))$
-8	$f(-8) = \sqrt[3]{-8} = -2$	$(-8, -2)$
-1	$f(-1) = \sqrt[3]{-1} = -1$	$(-1, -1)$
0	$f(0) = \sqrt[3]{0} = 0$	$(0, 0)$
1	$f(1) = \sqrt[3]{1} = 1$	$(1, 1)$
8	$f(8) = \sqrt[3]{8} = 2$	$(8, 2)$

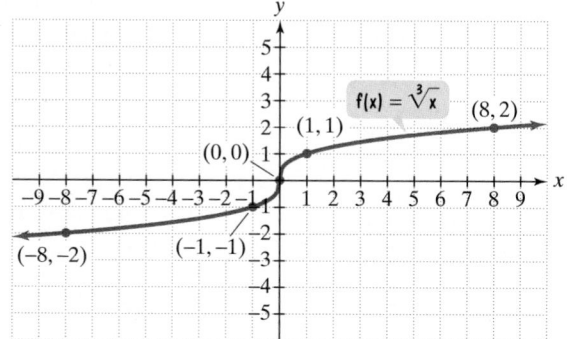

Figure 10.5 The graph of the cube root function $f(x) = \sqrt[3]{x}$

Notice that both the domain and range of $f(x) = \sqrt[3]{x}$ are sets of all real numbers.

EXAMPLE 6 Evaluating Cube Root Functions

For each function, find the indicated function value:

    a. $f(x) = \sqrt[3]{x - 2}$; $f(127)$ **b.** $g(x) = \sqrt[3]{8x - 8}$; $g(-7)$.

Solution

    a. $f(127) = \sqrt[3]{127 - 2}$ Substitute 127 for x in $f(x) = \sqrt[3]{x - 2}$.

               $= \sqrt[3]{125}$ Simplify the radicand.

               $= 5$ $\sqrt[3]{125} = 5$ because $5^3 = 125$.

    b. $g(-7) = \sqrt[3]{8(-7) - 8}$ Substitute -7 for x in $g(x) = \sqrt[3]{8x - 8}$.

               $= \sqrt[3]{-64}$ Simplify the radicand: $8(-7)-8 = -56 - 8 = -64$.

               $= -4$ $\sqrt[3]{-64} = -4$ because $(-4)^3 = -64$.   ■

✔ **CHECK POINT 6**   For each function, find the indicated function value:

    a. $f(x) = \sqrt[3]{x - 6}$; $f(33)$
    b. $g(x) = \sqrt[3]{2x + 2}$; $g(-5)$.

7 Simplify expressions of the form $\sqrt[3]{a^3}$.

Because the cube root of a positive number is positive and the cube root of a negative number is negative, absolute value is not needed to simplify expressions of the form $\sqrt[3]{a^3}$.

Simplifying $\sqrt[3]{a^3}$

For any real number a,

$$\sqrt[3]{a^3} = a.$$

In words, the cube root of any expression cubed is that expression.

EXAMPLE 7   Simplifying a Cube Root

Simplify: $\sqrt[3]{-64x^3}$.

Solution   Begin by expressing the radicand as an expression that is cubed: $-64x^3 = (-4x)^3$. Then simplify.

$$\sqrt[3]{-64x^3} = \sqrt[3]{(-4x)^3} = -4x$$

We can check our answer by cubing $-4x$:

$$(-4x)^3 = (-4)^3 x^3 = -64x^3.$$

By obtaining the radicand, we know that our simplification is correct.   ∎

✔ **CHECK POINT 7**   Simplify: $\sqrt[3]{-27x^3}$.

8   Find even and odd roots.

Even and Odd nth Roots

Up to this point, we have focused on square and cube roots. Other radical expressions have different roots. For example, the fifth root of a, written $\sqrt[5]{a}$, is the number b for which $b^5 = a$. Thus,

$$\sqrt[5]{32} = 2 \quad \text{because} \quad 2^5 = 2 \cdot 2 \cdot 2 \cdot 2 \cdot 2 = 32.$$

The radical expression $\sqrt[n]{a}$ represents the **nth root of a**. The number n is called the **index**. An index of 2 represents a square root and is not written. An index of 3 represents a cube root.

If the index n in $\sqrt[n]{a}$ is an odd number, a root is said to be an **odd root**. A cube root is an odd root. Other odd roots have the same characteristics as cube roots. Every real number has exactly one real root when n is odd. The (odd) nth root of a, $\sqrt[n]{a}$, is the number b for which $b^n = a$. An odd root of a positive number is positive, and an odd root of a negative number is negative. For example,

$$\sqrt[5]{243} = 3 \quad \text{because} \quad 3^5 = 3 \cdot 3 \cdot 3 \cdot 3 \cdot 3 = 243$$

and $\sqrt[5]{-243} = -3$ because $(-3)^5 = (-3)(-3)(-3)(-3)(-3) = -243.$

If the index n in $\sqrt[n]{a}$ is an even number, a root is said to be an **even root**. A square root is an even root. Other even roots have the same characteristics as square roots. Every positive real number has two real roots when n is even. One root is positive and one is negative. The positive root, called the **principal nth root** and represented by $\sqrt[n]{a}$, is the nonnegative number b for which $b^n = a$. The symbol $-\sqrt[n]{a}$ is used to denote the negative nth root. **An even root of a negative number is not a real number.**

Study Tip

Some higher even and odd roots occur so frequently that you might want to memorize them.

Fourth Roots	Fifth Roots
$\sqrt[4]{1} = 1$	$\sqrt[5]{1} = 1$
$\sqrt[4]{16} = 2$	$\sqrt[5]{32} = 2$
$\sqrt[4]{81} = 3$	$\sqrt[5]{243} = 3$
$\sqrt[4]{256} = 4$	
$\sqrt[4]{625} = 5$	

EXAMPLE 8   Finding Even and Odd Roots

Find the indicated root, or state that the expression is not a real number:

   a. $\sqrt[4]{81}$ **b.** $-\sqrt[4]{81}$ **c.** $\sqrt[4]{-81}$ **d.** $\sqrt[5]{-32}$.

Solution

   a. $\sqrt[4]{81} = 3$ The principal fourth root of 81 is 3 because $3^4 = 3 \cdot 3 \cdot 3 \cdot 3 = 81.$

   b. $-\sqrt[4]{81} = -3$ The negative fourth root of 81 is -3 because $(-3)^4 = (-3)(-3)(-3)(-3) = 81.$

c. $\sqrt[4]{-81}$ is not a real number because the index, 4, is even and the radicand, -81, is negative. No real number can be raised to the fourth power to give a negative result such as -81. Real numbers to even powers can only result in nonnegative numbers.

d. $\sqrt[5]{-32} = -2$ because $(-2)^5 = (-2)(-2)(-2)(-2)(-2) = -32$. An odd root of a negative real number is always negative. ∎

✔ **CHECK POINT 8** Find the indicated root, or state that the expression is not a real number:

a. $\sqrt[4]{16}$ **b.** $-\sqrt[4]{16}$

c. $\sqrt[4]{-16}$ **d.** $\sqrt[5]{-1}$.

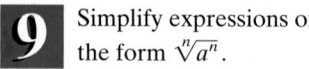 Simplify expressions of the form $\sqrt[n]{a^n}$.

Simplifying Expressions of the Form $\sqrt[n]{a^n}$

We have seen that

$$\sqrt{a^2} = |a| \quad \text{and} \quad \sqrt[3]{a^3} = a.$$

Expressions of the form $\sqrt[n]{a^n}$ can be simplified in the same manner. Unless a is known to be nonnegative, absolute value notation is needed when n is even. When the index is odd, absolute value bars are not necessary.

Simplifying $\sqrt[n]{a^n}$

For any real number a,

1. If n is even, $\sqrt[n]{a^n} = |a|$.

2. If n is odd, $\sqrt[n]{a^n} = a$.

EXAMPLE 9   Simplifying Radical Expressions

Simplify:

     a. $\sqrt[4]{(x-3)^4}$ **b.** $\sqrt[5]{(2x+7)^5}$ **c.** $\sqrt[6]{(-5)^6}$.

Solution   Each expression involves the nth root of a radicand raised to the nth power. Thus, each radical expression can be simplified. Absolute value bars are necessary in parts (a) and (c) because n is even.

     a. $\sqrt[4]{(x-3)^4} = |x-3|$ $\sqrt[n]{a^n} = |a|$ if n is even.

     b. $\sqrt[5]{(2x+7)^5} = 2x+7$ $\sqrt[n]{a^n} = a$ if n is odd.

     c. $\sqrt[6]{(-5)^6} = |-5| = 5$ $\sqrt[n]{a^n} = |a|$ if n is even. ∎

✔ **CHECK POINT 9**   Simplify:

     a. $\sqrt[4]{(x+6)^4}$

     b. $\sqrt[5]{(3x-2)^5}$

     c. $\sqrt[6]{(-8)^6}$.

EXERCISE SET 10.1

Practice Exercises

In Exercises 1–16, evaluate each expression, or state that the expression is not a real number.

1. $\sqrt{36}$

2. $\sqrt{16}$

3. $-\sqrt{36}$

4. $-\sqrt{16}$

5. $\sqrt{-36}$

6. $\sqrt{-16}$

7. $\sqrt{\dfrac{1}{25}}$

8. $\sqrt{\dfrac{1}{49}}$

9. $-\sqrt{0.04}$

10. $-\sqrt{0.64}$

11. $\sqrt{25 - 16}$

12. $\sqrt{144 + 25}$

13. $\sqrt{25} - \sqrt{16}$

14. $\sqrt{144} + \sqrt{25}$

15. $\sqrt{16 - 25}$

16. $\sqrt{25 - 144}$

In Exercises 17–22, find the indicated function values for each function. If necessary, round to two decimal places. If the function value is not a real number and does not exist, so state.

17. $f(x) = \sqrt{x - 2}; \quad f(18), f(3), f(2), f(-2)$

18. $f(x) = \sqrt{x - 3}; \quad f(28), f(4), f(3), f(-1)$

19. $g(x) = -\sqrt{2x + 3}; \quad g(11), g(1), g(-1), g(-2)$

20. $g(x) = -\sqrt{2x + 1}; \quad g(4), g(1), g\left(-\dfrac{1}{2}\right), g(-1)$

21. $h(x) = \sqrt{(x - 1)^2}; \quad h(5), h(3), h(0), h(-5)$

22. $h(x) = \sqrt{(x - 2)^2}; \quad h(5), h(3), h(0), h(-5)$

In Exercises 23–28, find the domain of each square root function. Then use the domain to match the radical function with its graph. [The graphs are labeled (a) through (f) and are shown in $[-10, 10, 1]$ by $[-10, 10, 1]$ viewing rectangles.]

23. $f(x) = \sqrt{x - 3}$

24. $f(x) = \sqrt{x + 2}$

25. $f(x) = \sqrt{3x + 15}$

26. $f(x) = \sqrt{3x - 15}$

27. $f(x) = \sqrt{6 - 2x}$

28. $f(x) = \sqrt{8 - 2x}$

(a)

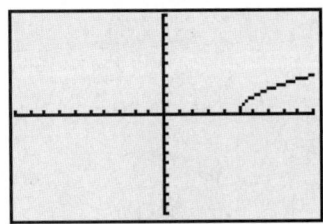

(b)

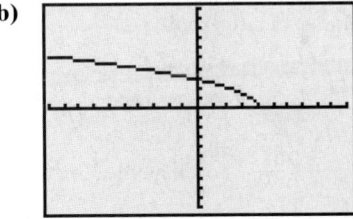

(c)

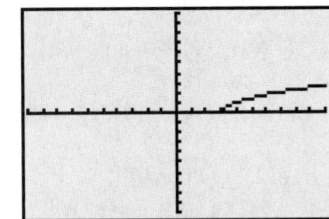

(d)

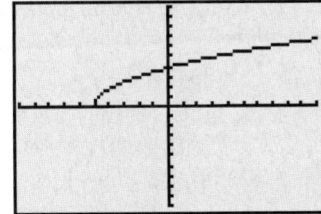

(e)

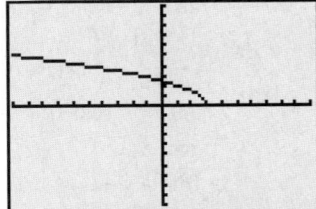

(f)

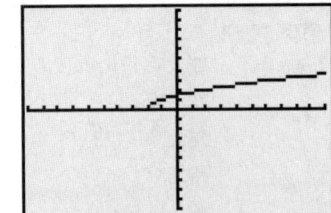

In Exercises 29–42, simplify each expression.

29. $\sqrt{5^2}$

30. $\sqrt{7^2}$

31. $\sqrt{(-4)^2}$

32. $\sqrt{(-10)^2}$

33. $\sqrt{(x-1)^2}$

34. $\sqrt{(x-2)^2}$

35. $\sqrt{36x^4}$

36. $\sqrt{81x^4}$

37. $-\sqrt{100x^6}$

38. $-\sqrt{49x^6}$

39. $\sqrt{x^2+12x+36}$

40. $\sqrt{x^2+14x+49}$

41. $-\sqrt{x^2-8x+16}$

42. $-\sqrt{x^2-10x+25}$

In Exercises 43–48, find each cube root.

43. $\sqrt[3]{27}$

44. $\sqrt[3]{64}$

45. $\sqrt[3]{-27}$

46. $\sqrt[3]{-64}$

47. $\sqrt[3]{\dfrac{1}{125}}$

48. $\sqrt[3]{\dfrac{1}{1000}}$

In Exercises 49–52, find the indicated function values for each function.

49. $f(x) = \sqrt[3]{x-1};\quad f(28), f(9), f(0), f(-63)$

50. $f(x) = \sqrt[3]{x-3};\quad f(30), f(11), f(2), f(-122)$

51. $g(x) = -\sqrt[3]{8x-8}; \; g(2), g(1), g(0)$

52. $g(x) = -\sqrt[3]{2x+1};\quad g(13), g(0), g(-63)$

In Exercises 53–70, find the indicated root, or state that the expression is not a real number.

53. $\sqrt[4]{1}$

54. $\sqrt[5]{1}$

55. $\sqrt[4]{16}$

56. $\sqrt[4]{81}$

57. $-\sqrt[4]{16}$

58. $-\sqrt[4]{81}$

59. $\sqrt[4]{-16}$

60. $\sqrt[4]{-81}$

61. $\sqrt[5]{-1}$

62. $\sqrt[7]{-1}$

63. $\sqrt[6]{-1}$

64. $\sqrt[8]{-1}$

65. $-\sqrt[4]{256}$

66. $-\sqrt[4]{10,000}$

67. $\sqrt[6]{64}$

68. $\sqrt[5]{32}$

69. $-\sqrt[5]{32}$

70. $-\sqrt[6]{64}$

In Exercises 71–84, simplify each expression. Include absolute value bars where necessary.

71. $\sqrt[3]{x^3}$

72. $\sqrt[5]{x^5}$

73. $\sqrt[4]{y^4}$

74. $\sqrt[6]{y^6}$

75. $\sqrt[3]{-8x^3}$

76. $\sqrt[3]{-125x^3}$

77. $\sqrt[3]{(-5)^3}$

78. $\sqrt[3]{(-6)^3}$

79. $\sqrt[4]{(-5)^4}$

80. $\sqrt[6]{(-6)^6}$

81. $\sqrt[4]{(x+3)^4}$

82. $\sqrt[4]{(x+5)^4}$

83. $\sqrt[5]{-32(x-1)^5}$

84. $\sqrt[5]{-32(x-2)^5}$

Application Exercises

The table shows the median, or average, heights for boys of various ages in the United States, from birth through 60 months, or five years. The data can be modeled by the radical function

$$f(x) = 2.9\sqrt{x} + 20.1$$

where $f(x)$ is the median height, in inches, of boys who are x months of age. Use the function to solve Exercises 85–86.

Boys' Median Heights

Age (months)	Height (inches)
0	20.5
6	27.0
12	30.8
18	32.9
24	35.0
36	37.5
48	40.8
60	43.4

Source: The Portable Pediatrician for Parents, by Laura Walther Nathanson, M.D., FAAP

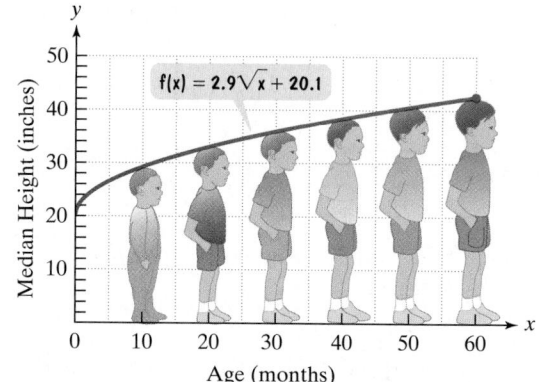

85. According to the model, what is the median height of boys who are 48 months, or four years, old? Use a calculator and round to the nearest tenth of an inch. How well does the model describe the actual median height shown in the table?

86. According to the model, what is the median height of boys who are 60 months, or five years, old? Use a calculator and round to the nearest tenth of an inch. How well does the model describe the actual median height shown in the table?

Police use the function $f(x) = \sqrt{20x}$ to estimate the speed of a car, $f(x)$, in miles per hour, based on the length, x, in feet, of its skid marks upon sudden braking on a dry asphalt road. Use the function to solve Exercises 87–88.

87. A motorist is involved in an accident. A police officer measures the car's skid marks to be 245 feet long. Estimate the speed at which the motorist was traveling before braking. If the posted speed limit is 50 miles per hour and the motorist tells the officer he was not speeding, should the officer believe him? Explain.

88. A motorist is involved in an accident. A police officer measures the car's skid marks to be 45 feet long. Estimate the speed at which the motorist was traveling before braking. If the posted speed limit is 35 miles per hour and the motorist tells the officer she was not speeding, should the officer believe her? Explain.

Writing in Mathematics

89. What are the square roots of 36? Explain why each of these numbers is a square root.

90. What does the symbol $\sqrt{}$ denote? Which of your answers in Exercise 89 is given by this symbol? Write the symbol needed to obtain the other answer.

91. Explain why $\sqrt{-1}$ is not a real number.

92. Explain how to find the domain of a square root function.

93. Explain how to simplify $\sqrt{a^2}$. Give an example with your explanation.

94. Explain why $\sqrt[3]{8}$ is 2. Then describe what is meant by the cube root of a real number.

95. Describe two differences between odd and even roots.

96. Explain how to simplify $\sqrt[n]{a^n}$ if n is even and if n is odd. Give examples with your explanations.

97. Explain the meaning of the words "radical," "radicand," and "index." Give an example with your explanation.

98. Describe the trend in a boy's growth from birth through five years, shown in the table for Exercises 85–86. Why is a square root function a useful model for the data?

99. The function $f(x) = 6.85\sqrt{x} + 19$ models the percentage, $f(x)$, of U.S. households online x years after 1997. Write a word problem using the function. Then solve the problem.

Critical Thinking Exercises

100. Which one of the following is true?
  a. The domain of $f(x) = \sqrt[3]{x} - 4$ is $[4, \infty)$.
  b. If n is odd and b is negative, then $\sqrt[n]{b}$ is not a real number.
  c. The expression $\sqrt[n]{4}$ represents increasingly larger numbers for $n = 2, 3, 4, 5, 6$, and so on.
  d. None of the above is true.

101. Write a function whose domain is $(-\infty, 5]$.

102. Let $f(x) = \sqrt{x - 3}$ and $g(x) = \sqrt{x + 1}$. Find the domain of $f + g$ and $\dfrac{f}{g}$.

103. Simplify: $\sqrt{(2x + 3)^{10}}$.

In Exercises 104–105, graph each function by hand. Then describe the relationship between the function that you graphed and the graph of $f(x) = \sqrt{x}$.

104. $g(x) = \sqrt{x} + 3$

105. $h(x) = \sqrt{x + 3}$

 Technology Exercises

106. Use a graphing utility to graph $y_1 = \sqrt{x}$, $y_2 = \sqrt{x + 4}$, and $y_3 = \sqrt{x - 3}$ in the same $[-5, 10, 1]$ by $[0, 6, 1]$ viewing rectangle. Describe one similarity and one difference that you observe among the graphs. Use the word "shift" in your response.

107. Use a graphing utility to graph $y = \sqrt{x}$, $y = \sqrt{x} + 4$, and $y = \sqrt{x} - 3$ in the same $[-1, 10, 1]$ by $[-10, 10, 1]$ viewing rectangle. Describe one similarity and one difference that you observe among the graphs.

108. Use a graphing utility to graph $f(x) = \sqrt{x}$, $g(x) = -\sqrt{x}$, $h(x) = \sqrt{-x}$, and $k(x) = -\sqrt{-x}$ in the same $[-10, 10, 1]$ by $[-4, 4, 11]$ viewing rectangle. Use the graphs to describe the domain and range of functions f, g, h, and k.

109. Use a graphing utility to graph $y_1 = \sqrt{x^2}$ and $y_2 = -x$ in the same viewing rectangle.
  a. For what values of x is $\sqrt{x^2} = -x$?
  b. For what values of x is $\sqrt{x^2} \neq -x$?

 Review Exercises

110. Simplify: $3x - 2[x - 3(x + 5)]$.
  (Section 1.8, Example 11)

111. Simplify: $(-3x^{-4}y^3)^{-2}$. (Section 5.7, Example 6)

112. Solve: $|3x - 4| > 11$.
  (Section 9.3, Example 4)

▶ SECTION 10.2   *Rational Exponents*

Objectives

1 Use the definition of $a^{\frac{1}{n}}$.

2 Use the definition of $a^{\frac{m}{n}}$.

3 Use the definition of $a^{-\frac{m}{n}}$.

4 Use models that are functions with rational exponents.

5 Simplify expressions with rational exponents.

6 Simplify radical expressions using rational exponents.

SSM PH Tutor CD- Video
      Center   ROM

Marine iguanas of the Galápagos Islands

The Galápagos Islands are a volcanic chain of islands lying 600 miles west of Ecuador. They are famed for over 5000 species of plants and animals, including a rare flightless cormorant, marine iguanas, and giant tortoises weighing more than 600 pounds. Early in 2001, the plants and wildlife that live in the Galápagos chain were at risk from a massive oil spill that flooded 150,000 gallons of toxic fuel into one of the world's most fragile ecosystems. The long-term danger of the accident is that fuel sinking to the ocean floor will destroy algae that is vital to the food chain. Any imbalance in the food chain could threaten the rare Galápagos plant and animal species that have evolved for thousands of years in isolation with little human intervention.

At risk on these ecologically vulnerable islands are unique flora and fauna that helped to inspire Charles Darwin's theory of evolution. Darwin made an enormous collection of the islands' plant species. The function

$$f(x) = 29x^{\frac{1}{3}}$$

models the number of plant species, $f(x)$, on the various islands of the Galápagos chain in terms of the area, x, in square miles, of a particular island. x to the *what* power? How can we interpret the information given by this function? In this section, we turn our attention to rational exponents such as $\frac{1}{3}$ and their relationship to roots of real numbers.

Defining Rational Exponents

We define rational exponents so that their properties are the same as the properties for integer exponents. For example, suppose that $x = 7^{\frac{1}{3}}$. We know that exponents are multiplied when an exponential expression is raised to a power. For this to be true,

$$x^3 = \left(7^{\frac{1}{3}}\right)^3 = 7^{\frac{1}{3} \cdot 3} = 7^1 = 7.$$

We see that $x^3 = 7$. This means that x is the number whose cube is 7. Thus, $x = \sqrt[3]{7}$. Remember that we began with $x = 7^{\frac{1}{3}}$. This means that

$$7^{\frac{1}{3}} = \sqrt[3]{7}.$$

We can generalize this idea with the following definition:

1 Use the definition of $a^{\frac{1}{n}}$.

The Definition of $a^{\frac{1}{n}}$

If $\sqrt[n]{a}$ represents a real number and $n \geq 2$ is an integer, then

$$a^{\frac{1}{n}} = \sqrt[n]{a}.$$

> The denominator of the rational exponent is the radical's index.

If a is negative, n must be odd. If a is nonnegative, n can be any index.

EXAMPLE 1   Using the Definition of $a^{\frac{1}{n}}$

Use radical notation to rewrite each expression. Simplify, if possible:

a. $64^{\frac{1}{2}}$ **b.** $(-125)^{\frac{1}{3}}$ **c.** $(6x^2y)^{\frac{1}{5}}$.

Solution

a. $64^{\frac{1}{2}} = \sqrt{64} = 8$

> The denominator is the index.

b. $(-125)^{\frac{1}{3}} = \sqrt[3]{-125} = -5$ **c.** $(6x^2y)^{\frac{1}{5}} = \sqrt[5]{6x^2y}$ ∎

✔ **CHECK POINT 1**   Use radical notation to rewrite each expression. Simplify, if possible:

a. $25^{\frac{1}{2}}$ **b.** $(-8)^{\frac{1}{3}}$ **c.** $(5xy^2)^{\frac{1}{4}}$.

In our next example, we begin with radical notation and rewrite the expression with rational exponents.

> The radical's index becomes the exponent's denominator.

$$\sqrt[n]{a} = a^{\frac{1}{n}}$$

> The radicand becomes the base.

EXAMPLE 2   Using the Definition of $a^{\frac{1}{n}}$

Rewrite with rational exponents:

a. $\sqrt[5]{13ab}$ **b.** $\sqrt[7]{\dfrac{xy^2}{17}}$.

Solution   Parentheses are needed to show that the entire radicand becomes the base.

a. $\sqrt[5]{13ab} = (13ab)^{\frac{1}{5}}$

> The index is the exponent's denominator.

b. $\sqrt[7]{\dfrac{xy^2}{17}} = \left(\dfrac{xy^2}{17}\right)^{\frac{1}{7}}$ ∎

✔ **CHECK POINT 2** Rewrite with rational exponents:

a. $\sqrt[4]{5xy}$ **b.** $\sqrt[5]{\dfrac{a^3b}{2}}$.

2 Use the definition of $a^{\frac{m}{n}}$.

Can rational exponents have numerators other than 1? The answer is yes. If the numerator is some other integer, we still want to multiply exponents when raising a power to a power. For this reason,

$$a^{\frac{2}{3}} = \left(a^{\frac{1}{3}}\right)^2 \quad \text{and} \quad a^{\frac{2}{3}} = \left(a^2\right)^{\frac{1}{3}}.$$

This means $\left(\sqrt[3]{a}\right)^2$.   This means $\sqrt[3]{a^2}$.

Thus,

$$a^{\frac{2}{3}} = \left(\sqrt[3]{a}\right)^2 = \sqrt[3]{a^2}.$$

Do you see that the denominator, 3, of the rational exponent is the same as the index of the radical? The numerator, 2, of the rational exponent serves as an exponent in each of the two radical forms. We generalize these ideas with the following definition:

The Definition of $a^{\frac{m}{n}}$

If $\sqrt[n]{a}$ represents a real number, and $\dfrac{m}{n}$ is a positive rational number, $n \geq 2$, then

$$a^{\frac{m}{n}} = \left(\sqrt[n]{a}\right)^m$$

and

$$a^{\frac{m}{n}} = \sqrt[n]{a^m}.$$

The first form of the definition shown in the box involves taking the root first. This form is often preferable because smaller numbers are involved. Notice that the rational exponent consists of two parts, indicated by the following voice balloons.

The numerator is the exponent.

$$a^{\frac{m}{n}} = \left(\sqrt[n]{a}\right)^m$$

The denominator is the radical's index.

EXAMPLE 3   Using the Definition of $a^{\frac{m}{n}}$

Use radical notation to rewrite each expression and simplify:

a. $27^{\frac{2}{3}}$ **b.** $9^{\frac{3}{2}}$ **c.** $-32^{\frac{4}{5}}$.

Solution

a. $27^{\frac{2}{3}} = \left(\sqrt[3]{27}\right)^2 = 3^2 = 9$

b. $9^{\frac{3}{2}} = \left(\sqrt{9}\right)^3 = 3^3 = 27$

c. $-32^{\frac{4}{5}} = -\left(\sqrt[5]{32}\right)^4 = -2^4 = -16$

The base is 32 and the negative sign is not affected by the exponent.

■

✔ **CHECK POINT 3**   Use radical notation to rewrite each expression and simplify.

a. $27^{\frac{4}{3}}$ **b.** $4^{\frac{3}{2}}$ **c.** $-16^{\frac{3}{4}}$

In our next example, we begin with radical notation and rewrite the expression with rational exponents. When changing from radical form to exponential form, the index becomes the denominator of the rational exponent.

EXAMPLE 4   Using the Definition of $a^{\frac{m}{n}}$

Rewrite with rational exponents:

a. $\sqrt[3]{7^5}$ **b.** $\left(\sqrt[4]{13xy}\right)^9$.

Solution

a. $\sqrt[3]{7^5} = 7^{\frac{5}{3}}$

> The index is the exponent's denominator.

b. $\left(\sqrt[4]{13xy}\right)^9 = (13xy)^{\frac{9}{4}}$ ∎

✔ **CHECK POINT 4**   Rewrite with rational exponents:

a. $\sqrt[3]{6^4}$ **b.** $\left(\sqrt[5]{2xy}\right)^7$.

3 Use the definition of $a^{-\frac{m}{n}}$.

Can a rational exponent be negative? Yes. The following definition is similar to the way that negative integer exponents are defined.

> ### The Definition of $a^{-\frac{m}{n}}$
>
> If $a^{\frac{m}{n}}$ is a nonzero real number, then
>
> $$a^{-\frac{m}{n}} = \frac{1}{a^{\frac{m}{n}}}.$$

EXAMPLE 5   Using the Definition of $a^{-\frac{m}{n}}$

Rewrite each expression with a positive exponent. Simplify, if possible:

a. $100^{-\frac{1}{2}}$ **b.** $27^{-\frac{1}{3}}$ **c.** $81^{-\frac{3}{4}}$ **d.** $(7xy)^{-\frac{4}{7}}$.

Solution

a. $100^{-\frac{1}{2}} = \dfrac{1}{100^{\frac{1}{2}}} = \dfrac{1}{\sqrt{100}} = \dfrac{1}{10}$

b. $27^{-\frac{1}{3}} = \dfrac{1}{27^{\frac{1}{3}}} = \dfrac{1}{\sqrt[3]{27}} = 3$

c. $81^{-\frac{3}{4}} = \dfrac{1}{81^{\frac{3}{4}}} = \dfrac{1}{\left(\sqrt[4]{81}\right)^3} = \dfrac{1}{3^3} = \dfrac{1}{27}$

d. $(7xy)^{-\frac{4}{7}} = \dfrac{1}{(7xy)^{\frac{4}{7}}}$ ∎

Using Technology

Here are the calculator keystroke sequences for $81^{-\frac{3}{4}}$:

Scientific calculator

81 $\boxed{y^x}$ $\boxed{(}$ 3 $\boxed{+/-}$ $\boxed{\div}$ 4 $\boxed{)}$ $\boxed{=}$

Graphing Calculator

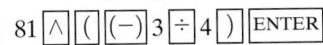
81 $\boxed{\wedge}$ $\boxed{(}$ $\boxed{(-)}$ 3 $\boxed{\div}$ 4 $\boxed{)}$ $\boxed{\text{ENTER}}$

✔ **CHECK POINT 5**   Rewrite each expression with a positive exponent. Simplify, if possible.

a. $25^{-\frac{1}{2}}$

b. $64^{-\frac{1}{3}}$

c. $32^{-\frac{4}{5}}$

d. $(3xy)^{-\frac{5}{9}}$.

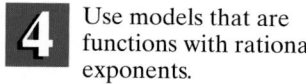

Use models that are functions with rational exponents.

Applications

Now that you know the meaning of rational exponents, you can work with functions that contain these exponents.

EXAMPLE 6   Filing Tax Returns Early

The bar graph in Figure 10.6 shows the percentage of U.S. taxpayers who file their tax returns at least one month before the April 15 deadline. The function

$$P(x) = 152x^{-\frac{1}{5}}$$

models the percentage, $P(x)$, of taxpayers who are x years old who file early. What percentage of 32-year-olds are expected to file early?

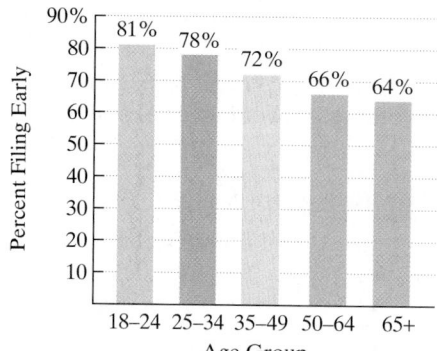

Percentage of Taxpayers Filing at Least One Month Before the Deadline

Source: Bruskin/Goldring Research

Figure 10.6

Solution   Because we are interested in taxpayers of age 32, substitute 32 for x and evaluate the function at 32.

$$P(x) = 152x^{-\frac{1}{5}}$$ This is the given function.

$$P(32) = 152 \cdot 32^{-\frac{1}{5}}$$ Substitute 32 for x.

$$= \frac{152}{32^{\frac{1}{5}}} = \frac{152}{\sqrt[5]{32}} = \frac{152}{2} = 76$$

We see that $P(32) = 76$. This means that 76% of 32-year-olds are expected to file early. ∎

✔ **CHECK POINT 6**   The function $S(x) = 63.25x^{\frac{1}{4}}$ models the average sale price, $S(x)$, in thousands of dollars, of single-family homes in the U.S. Midwest x years after 1981. What was the average sale price in 1997?

Properties of Rational Exponents

The same properties apply to rational exponents as to integer exponents. The following is a summary of these properties:

Properties of Rational Exponents

If m and n are rational exponents, and a and b are real numbers for which the following expressions are defined, then

1. $b^m \cdot b^n = b^{m+n}$ When multiplying exponential expressions with the same base, add the exponents.

2. $\dfrac{b^m}{b^n} = b^{m-n}$ When dividing exponential expressions with the same base, subtract the exponents.

3. $\left(b^m\right)^n = b^{mn}$ When an exponential expression is raised to a power, multiply the exponents.

4. $(ab)^n = a^n b^n$ When a product is raised to a power, raise each factor to that power and multiply.

5. $\left(\dfrac{a}{b}\right)^n = \dfrac{a^n}{b^n}$ When a quotient is raised to a power, raise the numerator and denominator to that power and divide.

5 Simplify expressions with rational exponents.

We can use these properties to simplify exponential expressions with rational exponents. As with integer exponents, an expression with rational exponents is **simplified** when:

- No parentheses appear.
- No powers are raised to powers.
- Each base occurs only once.
- No negative or zero exponents appear.

EXAMPLE 7   Simplifying Expressions with Rational Exponents

Simplify:

a. $6^{\frac{1}{7}} \cdot 6^{\frac{4}{7}}$ **b.** $\dfrac{32x^{\frac{1}{2}}}{16x^{\frac{3}{4}}}$ **c.** $\left(8.3^{\frac{3}{4}}\right)^{\frac{2}{3}}$ **d.** $\left(x^{-\frac{2}{5}}y^{\frac{1}{3}}\right)^{\frac{1}{2}}$

Solution

a. $6^{\frac{1}{7}} \cdot 6^{\frac{4}{7}} = 6^{\frac{1}{7}+\frac{4}{7}}$ To multiply with the same base, add exponents.

$\qquad\qquad = 6^{\frac{5}{7}}$ Simplify: $\dfrac{1}{7} + \dfrac{4}{7} = \dfrac{5}{7}$.

b. $\dfrac{32x^{\frac{1}{2}}}{16x^{\frac{3}{4}}} = \dfrac{32}{16}x^{\frac{1}{2}-\frac{3}{4}}$ Divide coefficients. To divide expressions with the same base, subtract exponents.

$\qquad\qquad = 2x^{\frac{2}{4}-\frac{3}{4}}$ Write exponents in terms of the LCD, 4.

$\qquad\qquad = 2x^{-\frac{1}{4}}$ Subtract: $\dfrac{2}{4} - \dfrac{3}{4} = -\dfrac{1}{4}$.

$\qquad\qquad = \dfrac{2}{x^{\frac{1}{4}}}$ Rewrite with a positive exponent: $a^{-\frac{m}{n}} = \dfrac{1}{a^{\frac{m}{n}}}$.

c. $\left(8.3^{\frac{3}{4}}\right)^{\frac{2}{3}} = 8.3^{\left(\frac{3}{4}\right)\left(\frac{2}{3}\right)}$ To raise a power to a power, multiply exponents.

$= 8.3^{\frac{1}{2}}$ Multiply: $\dfrac{3}{4} \cdot \dfrac{2}{3} = \dfrac{6}{12} = \dfrac{1}{2}$.

d. $\left(x^{-\frac{2}{5}}y^{\frac{1}{3}}\right)^{\frac{1}{2}} = \left(x^{-\frac{2}{5}}\right)^{\frac{1}{2}}\left(y^{\frac{1}{3}}\right)^{\frac{1}{2}}$ To raise a product to a power, raise each factor to the power.

$= x^{-\frac{1}{5}}y^{\frac{1}{6}}$ Multiply: $-\dfrac{2}{5} \cdot \dfrac{1}{2} = -\dfrac{1}{5}$ and $\dfrac{1}{3} \cdot \dfrac{1}{2} = \dfrac{1}{6}$.

$= \dfrac{y^{\frac{1}{6}}}{x^{\frac{1}{5}}}$ Rewrite with a positive exponent.   ■

✔ **CHECK POINT 7**   Simplify:

a. $7^{\frac{1}{2}} \cdot 7^{\frac{1}{3}}$

b. $\dfrac{50x^{\frac{1}{3}}}{10x^{\frac{4}{3}}}$

c. $\left(9.1^{\frac{2}{5}}\right)^{\frac{3}{4}}$

d. $\left(x^{-\frac{3}{5}}y^{\frac{1}{4}}\right)^{\frac{1}{3}}$.

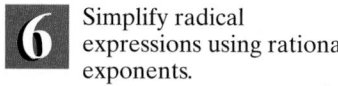 Simplify radical expressions using rational exponents.

Using Rational Exponents to Simplify Radical Expressions

Some radical expressions can be simplified using rational exponents. We will use the following procedure:

> **Simplifying Radical Expressions Using Rational Exponents**
>
> 1. Rewrite each radical expression as an exponential expression with a rational exponent.
> 2. Simplify using properties of rational exponents.
> 3. Rewrite in radical notation when rational exponents still appear.

EXAMPLE 8   Simplifying Radical Expressions Using Rational Exponents

Use rational exponents to simplify:

a. $\sqrt[10]{x^5}$ **b.** $\sqrt[3]{27a^{15}}$ **c.** $\sqrt[4]{x^6y^2}$ **d.** $\sqrt{x} \cdot \sqrt[3]{x}$ **e.** $\sqrt[3]{\sqrt{x}}$.

Solution

a. $\sqrt[10]{x^5} = x^{\frac{5}{10}}$ Rewrite as an exponential expression.

$= x^{\frac{1}{2}}$ Simplify the exponent.

$= \sqrt{x}$ Rewrite in radical notation.

b. $\sqrt[3]{27a^{15}} = \left(27a^{15}\right)^{\frac{1}{3}}$ Rewrite as an exponential expression.

$= 27^{\frac{1}{3}}\left(a^{15}\right)^{\frac{1}{3}}$ Raise each factor in parentheses to the $\frac{1}{3}$ power.

$= \sqrt[3]{27} \cdot a^{15\left(\frac{1}{3}\right)}$ To raise a power to a power, multiply exponents.

$= 3a^5$ $\sqrt[3]{27} = 3$. Multiply exponents: $15 \cdot \frac{1}{3} = 5$.

c. $\sqrt[4]{x^6 y^2} = \left(x^6 y^2\right)^{\frac{1}{4}}$ Rewrite as an exponential expression.

$\qquad = \left(x^6\right)^{\frac{1}{4}} \left(y^2\right)^{\frac{1}{4}}$ Raise each factor in parentheses to the $\frac{1}{4}$ power.

$\qquad = x^{\frac{6}{4}} y^{\frac{2}{4}}$ To raise powers to powers, multiply.

$\qquad = x^{\frac{3}{2}} y^{\frac{1}{2}}$ Simplify.

$\qquad = \left(x^3 y\right)^{\frac{1}{2}}$ $a^n b^n = (ab)^n$

$\qquad = \sqrt{x^3 y}$ Rewrite in radical notation.

d. $\sqrt{x} \cdot \sqrt[3]{x} = x^{\frac{1}{2}} \cdot x^{\frac{1}{3}}$ Rewrite as exponential expressions.

$\qquad = x^{\frac{1}{2} + \frac{1}{3}}$ To multiply with the same base, add exponents.

$\qquad = x^{\frac{3}{6} + \frac{2}{6}}$ Write exponents in terms of the LCD, 6.

$\qquad = x^{\frac{5}{6}}$ Add: $\frac{3}{6} + \frac{2}{6} = \frac{5}{6}$.

$\qquad = \sqrt[6]{x^5}$ Rewrite in radical notation.

e. $\sqrt[3]{\sqrt{x}} = \sqrt[3]{x^{\frac{1}{2}}}$ Write the radicand as an exponential expression.

$\qquad = \left(x^{\frac{1}{2}}\right)^{\frac{1}{3}}$ Write the entire expression in exponential form.

$\qquad = x^{\frac{1}{6}}$ To raise powers to powers, multiply: $\frac{1}{2} \cdot \frac{1}{3} = \frac{1}{6}$.

$\qquad = \sqrt[6]{x}$ Rewrite in radical notation. ∎

✔ **CHECK POINT 8** Use rational exponents to simplify:

a. $\sqrt[6]{x^3}$ **b.** $\sqrt[3]{8a^{12}}$ **c.** $\sqrt[8]{x^4 y^2}$ **d.** $\dfrac{\sqrt{x}}{\sqrt[3]{x}}$ **e.** $\sqrt{\sqrt[3]{x}}$.

ENRICHMENT ESSAY

Rational Exponents and Windchill

The way that we perceive the temperature on a cold day depends on both air temperature and wind speed. The **windchill temperature** is what the air temperature would have to be with no wind to achieve the same chilling effect on the skin. The formula that describes windchill temperature, W, in terms of the velocity of the wind, v, in miles per hour, and the actual air temperature, t, in degrees Fahrenheit, is

$$W = 91.4 - \frac{\left(10.5 + 6.7 v^{\frac{1}{2}} - 0.45v\right)(457 - 5t)}{110}.$$

Use your calculator to describe how cold the air temperature feels (that is, the windchill temperature) when the temperature is 15° Fahrenheit and the velocity of the wind is 5 miles per hour. Contrast this with a temperature of 40° Fahrenheit and a wind blowing at 50 miles per hour.

EXERCISE SET 10.2

Practice Exercises

In Exercises 1–20, use radical notation to rewrite each expression. Simplify, if possible.

1. $49^{\frac{1}{2}}$
2. $100^{\frac{1}{2}}$
3. $(-27)^{\frac{1}{3}}$
4. $(-64)^{\frac{1}{3}}$
5. $-16^{\frac{1}{4}}$
6. $-81^{\frac{1}{4}}$
7. $(xy)^{\frac{1}{3}}$
8. $(xy)^{\frac{1}{4}}$
9. $(2xy^3)^{\frac{1}{5}}$
10. $(3xy^4)^{\frac{1}{5}}$
11. $81^{\frac{3}{2}}$
12. $25^{\frac{3}{2}}$
13. $125^{\frac{2}{3}}$
14. $1000^{\frac{2}{3}}$
15. $(-32)^{\frac{3}{5}}$
16. $(-27)^{\frac{2}{3}}$
17. $27^{\frac{2}{3}} + 16^{\frac{3}{4}}$
18. $4^{\frac{5}{2}} - 8^{\frac{2}{3}}$
19. $(xy)^{\frac{4}{7}}$
20. $(xy)^{\frac{4}{9}}$

In Exercises 21–38, rewrite each expression with rational exponents.

21. $\sqrt{7}$
22. $\sqrt{13}$
23. $\sqrt[3]{5}$
24. $\sqrt[3]{6}$
25. $\sqrt[5]{11x}$
26. $\sqrt[5]{13x}$
27. $\sqrt{x^3}$
28. $\sqrt{x^5}$
29. $\sqrt[5]{x^3}$
30. $\sqrt[7]{x^4}$
31. $\sqrt[5]{x^2y}$
32. $\sqrt[7]{xy^3}$
33. $(\sqrt{19xy})^3$
34. $(\sqrt{11xy})^3$
35. $(\sqrt[6]{7xy^2})^5$
36. $(\sqrt[6]{9x^2y})^5$
37. $2x\sqrt[3]{y^2}$
38. $4x\sqrt[5]{y^2}$

In Exercises 39–54, rewrite each expression with a positive rational exponent. Simplify, if possible.

39. $49^{-\frac{1}{2}}$
40. $9^{-\frac{1}{2}}$
41. $27^{-\frac{1}{3}}$
42. $125^{-\frac{1}{3}}$
43. $16^{-\frac{3}{4}}$
44. $81^{-\frac{5}{4}}$
45. $8^{-\frac{2}{3}}$
46. $32^{-\frac{4}{5}}$
47. $\left(\frac{8}{27}\right)^{-\frac{1}{3}}$
48. $\left(\frac{8}{125}\right)^{-\frac{1}{3}}$
49. $(-64)^{-\frac{2}{3}}$
50. $(-8)^{-\frac{2}{3}}$

51. $(2xy)^{-\frac{7}{10}}$
52. $(4xy)^{-\frac{4}{7}}$
53. $5xz^{-\frac{1}{3}}$
54. $7xz^{-\frac{1}{4}}$

In Exercises 55–78, use properties of rational exponents to simplify each expression. Assume that all variables represent positive numbers.

55. $3^{\frac{3}{4}} \cdot 3^{\frac{1}{4}}$
56. $5^{\frac{2}{3}} \cdot 5^{\frac{1}{3}}$
57. $\dfrac{16^{\frac{3}{4}}}{16^{\frac{1}{4}}}$
58. $\dfrac{100^{\frac{3}{4}}}{100^{\frac{1}{4}}}$
59. $x^{\frac{1}{2}} \cdot x^{\frac{1}{3}}$
60. $x^{\frac{1}{2}} \cdot x^{\frac{2}{3}}$
61. $\dfrac{x^{\frac{4}{5}}}{x^{\frac{1}{5}}}$
62. $\dfrac{x^{\frac{3}{7}}}{x^{\frac{1}{7}}}$
63. $\dfrac{x^{\frac{1}{3}}}{x^{\frac{3}{4}}}$
64. $\dfrac{x^{\frac{1}{4}}}{x^{\frac{3}{5}}}$
65. $(5^{\frac{2}{3}})^3$
66. $(3^{\frac{4}{5}})^5$
67. $(y^{-\frac{2}{3}})^{\frac{1}{4}}$
68. $(y^{-\frac{3}{4}})^{\frac{1}{6}}$
69. $(2x^{\frac{1}{5}})^5$
70. $(2x^{\frac{1}{4}})^4$
71. $(25x^4y^6)^{\frac{1}{2}}$
72. $(125x^9y^6)^{\frac{1}{3}}$
73. $(x^{\frac{1}{2}}y^{-\frac{3}{5}})^{\frac{1}{2}}$
74. $(x^{\frac{1}{4}}y^{-\frac{2}{5}})^{\frac{1}{3}}$
75. $\dfrac{3^{\frac{1}{2}} \cdot 3^{\frac{3}{4}}}{3^{\frac{1}{4}}}$
76. $\dfrac{5^{\frac{3}{4}} \cdot 5^{\frac{1}{2}}}{5^{\frac{1}{4}}}$
77. $\dfrac{(3y^{\frac{1}{4}})^3}{y^{\frac{1}{12}}}$
78. $\dfrac{(2y^{\frac{1}{5}})^4}{y^{\frac{3}{10}}}$

In Exercises 79–106, use rational exponents to simplify each expression. If rational exponents appear after simplifying, write the answer in radical notation. Assume that all variables represent positive numbers.

79. $\sqrt[8]{x^2}$
80. $\sqrt[10]{x^2}$
81. $\sqrt[3]{8a^6}$
82. $\sqrt[3]{27a^{12}}$
83. $\sqrt[5]{x^{10}y^{15}}$
84. $\sqrt[5]{x^{15}y^{20}}$

85. $\left(\sqrt[3]{xy}\right)^{18}$

86. $\left(\sqrt[3]{xy}\right)^{21}$

87. $\sqrt[10]{(3y)^2}$

88. $\sqrt[12]{(3y)^2}$

89. $\left(\sqrt[6]{2a}\right)^4$

90. $\left(\sqrt[8]{2a}\right)^6$

91. $\sqrt[9]{x^6 y^3}$

92. $\sqrt[4]{x^2 y^6}$

93. $\sqrt{2} \cdot \sqrt[3]{2}$

94. $\sqrt{3} \cdot \sqrt[3]{3}$

95. $\dfrac{\sqrt[4]{x}}{\sqrt[5]{x}}$

96. $\dfrac{\sqrt[3]{x}}{\sqrt[4]{x}}$

97. $\dfrac{\sqrt[3]{y^2}}{\sqrt[6]{y}}$

98. $\dfrac{\sqrt[5]{y^2}}{\sqrt[10]{y^3}}$

99. $\sqrt[4]{\sqrt{x}}$

100. $\sqrt[5]{\sqrt{x}}$

101. $\sqrt{\sqrt{x^2 y}}$

102. $\sqrt{\sqrt{xy^2}}$

103. $\sqrt[4]{\sqrt[3]{2x}}$

104. $\sqrt[5]{\sqrt[3]{2x}}$

105. $\left(\sqrt[4]{x^3 y^5}\right)^{12}$

106. $\left(\sqrt[5]{x^4 y^2}\right)^{20}$

Application Exercises

The Galápagos Islands, lying 600 miles west of Ecuador, are famed for their extraordinary wildlife. The function

$$f(x) = 29x^{\frac{1}{3}}$$

models the number of plant species, $f(x)$, on the various islands of the Galápagos chain in terms of the area, x, in square miles, of a particular island. Use the function to solve Exercises 107–108.

107. How many species of plants are on a Galápagos island that has an area of 8 square miles?

108. How many species of plants are on a Galápagos island that has an area of 27 square miles?

The function

$$f(x) = 0.07x^{\frac{3}{2}}$$

models the duration of a storm, $f(x)$, in hours, in terms of the diameter of the storm, x, in miles. Use the function to solve Exercises 109–110. Round answers to the nearest tenth of an hour.

109. What is the duration of a storm with a diameter of 9 miles?

110. What is the duration of a storm with a diameter of 16 miles?

According to the American Management Association, the percentage of potential employees testing positive for illegal drugs is on the decline. The function

$$p(t) = \frac{73t^{\frac{1}{3}} - 28t^{\frac{2}{3}}}{t}$$

models the percentage, $p(t)$, of people applying for jobs who test positive t years after 1985. Use this model to solve Exercises 111–112.

111. What percentage of people applying for jobs tested positive for illegal drugs in 1993?

112. What percentage of people applying for jobs tested positive for illegal drugs in 2001? Use a calculator and round to nearest hundredth of a percent.

The function

$$f(x) = 70x^{\frac{3}{4}}$$

models the number of calories per day, $f(x)$, a person needs to maintain life in terms of that person's weight, x, in kilograms. (1 kilogram is approximately 2.2 pounds.) Use this model and a calculator to solve Exercises 113–114. Round answers to the nearest calorie.

113. How many calories per day does a person who weighs 80 kilograms (approximately 176 pounds) need to maintain life?

114. How many calories per day does a person who weighs 70 kilograms (approximately 154 pounds) need to maintain life?

Your job is to determine whether or not yachts are eligible for the America's Cup, the supreme event in ocean sailing. The basic dimensions of competitors' yachts must satisfy

$$L + 1.25\sqrt{S} - 9.8\sqrt[3]{D} \le 16.296,$$

where L is the yacht's length, in meters; S is its sail area, in square meters, and D is its displacement, in cubic meters. Use this information to solve Exercises 115–116.

115. a. Rewrite the inequality using rational exponents.

b. Use your calculator to determine if a yacht with length 20.85 meters, sail area 276.4 square meters, and displacement 18.55 cubic meters is eligible for the America's Cup.

116. a. Rewrite the inequality using rational exponents.

b. Use your calculator to determine if a yacht with length 22.85 meters, sail area 312.5 square meters, and displacement 22.34 cubic meters is eligible for the America's Cup.

Writing in Mathematics

117. What is the meaning of $a^{\frac{1}{n}}$? Give an example to support your explanation.

118. What is the meaning of $a^{\frac{m}{n}}$? Give an example.

119. What is the meaning of $a^{-\frac{m}{n}}$? Give an example.

120. Explain why $a^{\frac{1}{n}}$ is negative when n is odd and a is negative. What happens if n is even and a is negative? Why?

121. In simplifying $36^{\frac{3}{2}}$, is it better to use $a^{\frac{m}{n}} = \sqrt[n]{a^m}$ or $a^{\frac{m}{n}} = (\sqrt[n]{a})^m$? Explain.

122. How can you tell if an expression with rational exponents is simplified?

123. Explain how to simplify $\sqrt[3]{x} \cdot \sqrt{x}$.

124. Explain how to simplify $\sqrt[3]{\sqrt{x}}$.

125. In Exercises 111–112, you used a model for the percentage of people applying for jobs who test positive for illegal drugs. Drug testing is a controversial issue. Is testing of this kind an infringement of our right to privacy or an appropriate procedure for companies to screen potential new hires? Explain your position on this issue.

Critical Thinking Exercises

126. Which one of the following is true?

a. If n is odd, then $(-b)^{\frac{1}{n}} = -b^{\frac{1}{n}}$.

b. $(a + b)^{\frac{1}{n}} = a^{\frac{1}{n}} + b^{\frac{1}{n}}$

c. $8^{-\frac{2}{3}} = -4$

d. None of the above is true.

127. A mathematics professor recently purchased a birthday cake for her son with the inscription

$$\text{Happy } (2^{\frac{5}{2}} \cdot 2^{\frac{3}{4}} \div 2^{\frac{1}{4}})\text{th Birthday.}$$

How old is the son?

128. The birthday boy in Exercise 127, excited by the inscription on the cake, tried to wolf down the whole thing. Professor Mom, concerned about the possible metamorphosis of her son into a blimp, exclaimed, "Hold on! It is your birthday, so why not take $\dfrac{8^{-\frac{4}{3}} + 2^{-2}}{16^{-\frac{3}{4}} + 2^{-1}}$ of the cake? I'll eat half of what's left over." How much of the cake did the professor eat?

129. Simplify: $\left[3 + (27^{\frac{2}{3}} + 32^{\frac{2}{5}})\right]^{\frac{3}{2}} - 9^{\frac{1}{2}}$.

130. Find the domain of $f(x) = (x - 3)^{\frac{1}{2}}(x + 4)^{-\frac{1}{2}}$.

Technology Exercises

131. Use a scientific or graphing calculator to verify your results in Exercises 15–18.

132. Use a scientific or graphing calculator to verify your results in Exercises 45–50.

133. The graph of $y = x^{\frac{2}{3}}$ is shown in the figure in a $[-10, 10, 1]$ by $[-10, 10, 1]$ viewing rectangle.

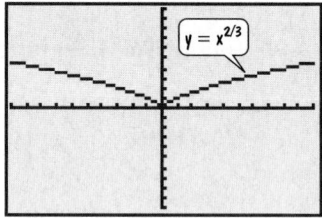

a. Use a graphing utility to obtain this graph by entering

$$Y_1 = X \boxed{\wedge} (2 \boxed{\div} 3).$$

Describe what happens.

b. Repeat part (a), but this time enter

$$Y_1 = (X \boxed{\wedge} (1 \boxed{\div} 3)) \boxed{\wedge} 2.$$

Describe what happens.

134. a. Graph $y = x^{\frac{2}{6}}$ by entering

$$Y_1 = (X \boxed{\wedge} (1 \boxed{\div} 6)) \boxed{\wedge} 2.$$

b. Graph $y = x^{\frac{1}{3}}$ by entering

$$Y_1 = X \boxed{\wedge} (1 \boxed{\div} 3).$$

c. Does your graphing utility produce the same graph in parts (a) and (b)? Can you explain what is happening?

Review Exercises

135. Write the equation of the linear function whose graph passes through $(5, 1)$ and $(4, 3)$.
(Section 3.5, Example 2)

136. Graph, $y \leq -\dfrac{3}{2}x + 3$.
(Section 9.4, Example 2)

137. Solve by Cramer's rule:
$$5x - 3y = 3$$
$$7x + y = 25.$$
(Section 8.5, Example 2)

▶ **SECTION 10.3** *Multiplying and Simplifying Radical Expressions*

Objectives

1 Use the product rule to multiply radicals.

2 Use factoring and the product rule to simplify radicals.

3 Multiply radicals and then simplify.

SSM PH Tutor CD- Video www
Center ROM

A difficult challenge of middle adulthood is confronting the aging process. Middle-aged adults are forced to acknowledge their mortality as they witness the deaths of parents, colleagues, and friends. In addition, there are a number of physical transformations, including changes in vision that require glasses for reading, the onset of wrinkles and sagging skin, and a decrease in heart response. A change in heart response occurs fairly early; after 20, our hearts become less adept at accelerating in response to exercise. In this section, you will see how a radical function models changes in heart function throughout the aging process, as we turn to multiplying and simplifying radical expressions.

1 Use the product rule to multiply radicals.

The Product Rule for Radicals

A rule for multiplying radicals can be generalized by comparing $\sqrt{25} \cdot \sqrt{4}$ and $\sqrt{25 \cdot 4}$. Notice that

$$\sqrt{25} \cdot \sqrt{4} = 5 \cdot 2 = 10 \quad \text{and} \quad \sqrt{25 \cdot 4} = \sqrt{100} = 10.$$

Because we obtain 10 in both situations, the original radical expressions must be equal. That is,

$$\sqrt{25} \cdot \sqrt{4} = \sqrt{25 \cdot 4}.$$

This result is a special case of the **product rule for radicals** that can be generalized as follows:

> ### The Product Rule for Radicals
> If $\sqrt[n]{a}$ and $\sqrt[n]{b}$ are real numbers, then
> $$\sqrt[n]{a} \cdot \sqrt[n]{b} = \sqrt[n]{ab}.$$
> The product of two nth roots is the nth root of the product.

EXAMPLE 1   Using the Product Rule for Radicals

Multiply:

a. $\sqrt{3} \cdot \sqrt{7}$ **b.** $\sqrt{x+7} \cdot \sqrt{x-7}$ **c.** $\sqrt[3]{7} \cdot \sqrt[3]{9}$ **d.** $\sqrt[8]{10x} \cdot \sqrt[8]{8x^4}$.

Solution   In each problem, the indices are the same. Thus, we multiply by multiplying the radicands.

a. $\sqrt{3} \cdot \sqrt{7} = \sqrt{3 \cdot 7} = \sqrt{21}$

b. $\sqrt{x+7} \cdot \sqrt{x-7} = \sqrt{(x+7)(x-7)} = \sqrt{x^2 - 49}$

> This is *not* equal to $\sqrt{x^2} - \sqrt{49}$.

c. $\sqrt[3]{7} \cdot \sqrt[3]{9} = \sqrt[3]{7 \cdot 9} = \sqrt[3]{63}$

d. $\sqrt[8]{10x} \cdot \sqrt[8]{8x^4} = \sqrt[8]{10x \cdot 8x^4} = \sqrt[8]{80x^5}$ ∎

✔ **CHECK POINT 1**   Multiply:

a. $\sqrt{5} \cdot \sqrt{11}$ **b.** $\sqrt{x+4} \cdot \sqrt{x-4}$

c. $\sqrt[3]{6} \cdot \sqrt[3]{10}$ **d.** $\sqrt[7]{2x} \cdot \sqrt[7]{6x^3}$.

2 Use factoring and the product rule to simplify radicals.

Using Factoring and the Product Rule to Simplify Radicals

A number that is the square of an integer is a **perfect square**. For example, 100 is a perfect square because $100 = 10^2$. A number is a **perfect cube** if it is the cube of an integer. Thus, 125 is a perfect cube because $125 = 5^3$. In general, a number is a **perfect nth power** if it is the nth power of an integer. Thus, p is a perfect nth power if there is an integer q such that $p = q^n$.

A radical of index n is **simplified** when its radicand has no factors other than 1 that are perfect nth powers. For example, $\sqrt{300}$ is not simplified because it can be expressed as $\sqrt{100 \cdot 3}$ and 100 is a perfect square. We can use the product rule in the form

$$\sqrt[n]{ab} = \sqrt[n]{a} \cdot \sqrt[n]{b}$$

to simplify $\sqrt[n]{ab}$ when $\sqrt[n]{a}$ or $\sqrt[n]{b}$ is a perfect nth power. Consider $\sqrt{300}$. To simplify, we factor 300 so that one of its factors is the largest perfect square possible.

$$\sqrt{300} = \sqrt{100 \cdot 3}$$ Factor 300. 100 is the largest perfect square factor.

$$= \sqrt{100} \cdot \sqrt{3}$$ Use the product rule: $\sqrt[n]{ab} = \sqrt[n]{a} \cdot \sqrt[n]{b}$.

$$= 10\sqrt{3}$$ Write $\sqrt{100}$ as 10. We read $10\sqrt{3}$ as "ten times the square root of three."

Simplifying Radical Expressions by Factoring

A radical expression whose index is n is **simplified** when its radicand has no factors that are perfect nth powers. To simplify:

1. Write the radicand as the product of two factors, one of which is the largest perfect nth power.
2. Use the product rule to take the nth root of each factor.
3. Find the nth root of the perfect nth power.

EXAMPLE 2   Simplifying Radicals by Factoring

Simplify by factoring:

a. $\sqrt{75}$ **b.** $\sqrt[3]{54}$ **c.** $\sqrt[5]{64}$ **d.** $\sqrt{500xy^2}$.

Solution

a. $\sqrt{75} = \sqrt{25 \cdot 3}$ | 25 is the largest perfect square that is a factor of 75.

$\qquad = \sqrt{25} \cdot \sqrt{3}$ | Take the square root of each factor: $\sqrt[n]{ab} = \sqrt[n]{a} \cdot \sqrt[n]{b}$.

$\qquad = 5\sqrt{3}$ | Write $\sqrt{25}$ as 5.

b. $\sqrt[3]{54} = \sqrt[3]{27 \cdot 2}$ | 27 is the largest perfect cube that is a factor of 54: $27 = 3^3$.

$\qquad = \sqrt[3]{27} \cdot \sqrt[3]{2}$ | Take the cube root of each factor: $\sqrt[n]{ab} = \sqrt[n]{a} \cdot \sqrt[n]{b}$.

$\qquad = 3\sqrt[3]{2}$ | Write $\sqrt[3]{27}$ as 3.

c. $\sqrt[5]{64} = \sqrt[5]{32 \cdot 2}$ | 32 is the largest perfect fifth power that is a factor of 64: $32 = 2^5$.

$\qquad = \sqrt[5]{32} \cdot \sqrt[5]{2}$ | Take the fifth root of each factor: $\sqrt[n]{ab} = \sqrt[n]{a} \cdot \sqrt[n]{b}$.

$\qquad = 2\sqrt[5]{2}$ | Write   as 2.

d. $\sqrt{500xy^2} = \sqrt{100y^2 \cdot 5x}$ | $100y^2$ is the largest perfect square that is a factor of $500xy^2$: $100y^2 = (10y)^2$.

$\qquad = \sqrt{100y^2} \cdot \sqrt{5x}$ | Factor into two radicals.

$\qquad = 10|y|\sqrt{5x}$ | Take the square root of $100y^2$.   ■

✔ **CHECK POINT 2**   Simplify by factoring:

a. $\sqrt{80}$ **b.** $\sqrt[3]{40}$

c. $\sqrt[4]{32}$ **d.** $\sqrt{200x^2y}$.

EXAMPLE 3   Simplifying a Radical Function

If

$$f(x) = \sqrt{2x^2 + 4x + 2},$$

express the function, f, in simplified form.

Solution   Begin by factoring the radicand. The GCF is 2. Simplification is possible if we obtain a factor that is a perfect square.

$f(x) = \sqrt{2x^2 + 4x + 2}$ | This is the given function.

$\qquad = \sqrt{2(x^2 + 2x + 1)}$ | Factor out the GCF.

$\qquad = \sqrt{2(x + 1)^2}$ | Factor the perfect square trinomial: $A^2 + 2AB + B^2 = (A + B)^2$.

$\qquad = \sqrt{2} \cdot \sqrt{(x + 1)^2}$ | Take the square root of each factor. The factor $(x + 1)^2$ is a perfect square.

$\qquad = \sqrt{2}\,|x + 1|$ | Take the square root of $(x + 1)^2$.

In simplified form,

$$f(x) = \sqrt{2}\,|x + 1|.$$ ■

Using Technology

The graphs of

$$f(x) = \sqrt{2x^2 + 4x + 2}, \quad g(x) = \sqrt{2}|x + 1|, \quad \text{and} \quad h(x) = \sqrt{2}(x + 1)$$

are shown in three separate $[-5, 5, 1]$ by $[-5, 5, 1]$ viewing rectangles. The graphs in parts (a) and (b) are identical. This verifies that our simplification in Example 3 is correct. Now compare the graphs in parts (a) and (c). Can you see that they are not the same? This illustrates the importance of not leaving out absolute value bars.

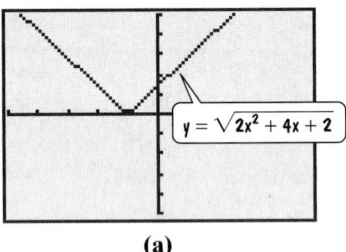

(a)

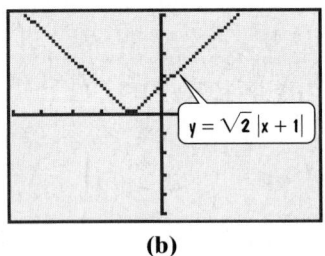

(b)

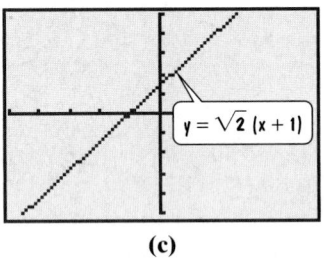

(c)

✔ **CHECK POINT 3** If $f(x) = \sqrt{3x^2 - 12x + 12}$, express the function, f, in simplified form.

For the remainder of this chapter, in situations that do not involve functions, we will **assume that no radicands involve negative quantities raised to even powers. Based upon this assumption, absolute value bars are not necessary when taking even roots.**

Simplifying When Variables to Even Powers in a Radicand Are Nonnegative Quantities

For any nonnegative real number a,
$$\sqrt[n]{a^n} = a.$$

In simplifying an nth root, how do we find variable factors in the radicand that are perfect nth powers? The **perfect nth powers have exponents that are divisible by n.** Simplification is possible by observation or by using rational exponents. Here are some examples:

- $\sqrt{x^6} = \sqrt{(x^3)^2} = x^3$ or $\sqrt{x^6} = (x^6)^{\frac{1}{2}} = x^3$

  6 is divisible by the index, 2. Thus, x^6 is a perfect square.

- $\sqrt[3]{y^{21}} = \sqrt[3]{(y^7)^3} = y^7$ or $\sqrt[3]{y^{21}} = (y^{21})^{\frac{1}{3}} = y^7$

  21 is divisible by the index, 3. Thus, y^{21} is a perfect cube.

- $\sqrt[6]{z^{24}} = \sqrt[6]{(z^4)^6} = z^4$ or $\sqrt[6]{z^{24}} = (z^{24})^{\frac{1}{6}} = z^4$.

  24 is divisible by the index, 6. Thus, z^{24} is a perfect 6th power.

EXAMPLE 4   Simplifying a Radical by Factoring

Simplify: $\sqrt{x^5 y^{13} z^7}$.

Solution We write the radicand as the product of the largest perfect square factor and another factor. Because the index is 2, variables that have exponents

Discover for Yourself

Square the answer in Example 4 and show that it is correct. If it is a square root, you should obtain the given radicand, $x^5y^{13}z^7$.

that are divisible by 2 are part of the perfect square factor. We use the largest exponents that are divisible by 2.

$$\sqrt{x^5y^{13}z^7} = \sqrt{x^4 \cdot x \cdot y^{12} \cdot y \cdot z^6 \cdot z} \quad \text{Use the largest even power of each variable.}$$

$$= \sqrt{(x^4y^{12}z^6)(xyz)} \quad \text{Group the perfect square factors.}$$

$$= \sqrt{x^4y^{12}z^6} \cdot \sqrt{xyz} \quad \text{Factor into two radicals.}$$

$$= x^2y^6z^3\sqrt{xyz} \quad \sqrt{x^4y^{12}z^6} = (x^4y^{12}z^6)^{\frac{1}{2}} = x^2y^6z^3 \quad \blacksquare$$

✔ **CHECK POINT 4** Simplify: $\sqrt{x^9y^{11}z^3}$.

EXAMPLE 5   Simplifying a Radical by Factoring

Simplify: $\sqrt[3]{32x^8y^{16}}$.

Solution   We write the radicand as the product of the largest perfect cube factor and another factor. Because the index is 3, variables that have exponents that are divisible by 3 are part of the perfect cube factor. We use the largest exponents that are divisible by 3.

$$\sqrt[3]{32x^8y^{16}} = \sqrt[3]{8 \cdot 4 \cdot x^6 \cdot x^2 \cdot y^{15} \cdot y} \quad \text{Identify perfect cube factors.}$$

$$= \sqrt[3]{(8x^6y^{15})(4x^2y)} \quad \text{Group the perfect cube factors.}$$

$$= \sqrt[3]{8x^6y^{15}} \cdot \sqrt[3]{4x^2y} \quad \text{Factor into two radicals.}$$

$$= 2x^2y^5\sqrt[3]{4x^2y} \quad \sqrt[3]{8} = 2 \text{ and } \sqrt[3]{x^6y^{15}} = (x^6y^{15})^{\frac{1}{3}} = x^2y^5. \quad \blacksquare$$

✔ **CHECK POINT 5** Simplify: $\sqrt[3]{40x^{10}y^{14}}$.

EXAMPLE 6   Simplifying a Radical by Factoring

Simplify: $\sqrt[5]{64x^3y^7z^{29}}$.

Solution   We write the radicand as the product of the largest perfect 5th power and another factor. Because the index is 5, variables that have exponents that are divisible by 5 are part of the perfect fifth factor. We use the largest exponents that are divisible by 5.

$$\sqrt[5]{64x^3y^7z^{29}} = \sqrt[5]{32 \cdot 2 \cdot x^3 \cdot y^5 \cdot y^2 \cdot z^{25} \cdot z^4} \quad \text{Identify perfect fifth factors.}$$

$$= \sqrt[5]{(32y^5z^{25})(2x^3y^2z^4)} \quad \text{Group the perfect fifth factors.}$$

$$= \sqrt[5]{32y^5z^{25}} \cdot \sqrt[5]{2x^3y^2z^4} \quad \text{Factor into two radicals.}$$

$$= 2yz^5\sqrt[5]{2x^3y^2z^4} \quad \sqrt[5]{32} = 2 \text{ and}$$
$$\sqrt[5]{y^5z^{25}} = (y^5z^{25})^{\frac{1}{5}} = yz^5. \quad \blacksquare$$

✔ **CHECK POINT 6** Simplify: $\sqrt[5]{32x^{12}y^2z^8}$.

3 Multiply radicals and then simplify.

Multiplying and Simplifying Radicals

We have seen how to use the product rule when multiplying radicals with the same index. Sometimes after multiplying, we can simplify the resulting radical.

EXAMPLE 7 Multiplying Radicals and then Simplifying

Multiply and simplify:

a. $\sqrt{15} \cdot \sqrt{3}$ **b.** $7\sqrt[3]{4} \cdot 5\sqrt[3]{6}$ **c.** $\sqrt[4]{8x^3y^2} \cdot \sqrt[4]{8x^5y^3}$.

Solution

a. $\sqrt{15} \cdot \sqrt{3} = \sqrt{15 \cdot 3}$ Use the product rule.

$\qquad = \sqrt{45} = \sqrt{9 \cdot 5}$ 9 is the largest perfect square factor of 45.

$\qquad = \sqrt{9} \cdot \sqrt{5} = 3\sqrt{5}$

b. $7\sqrt[3]{4} \cdot 5\sqrt[3]{6} = 35\sqrt[3]{4 \cdot 6}$ Use the product rule.

$\qquad = 35\sqrt[3]{24} = 35\sqrt[3]{8 \cdot 3}$ 8 is the largest perfect cube factor of 24.

$\qquad = 35\sqrt[3]{8} \cdot \sqrt[3]{3} = 35 \cdot 2 \cdot \sqrt[3]{3}$

$\qquad = 70\sqrt[3]{3}$

c. $\sqrt[4]{8x^3y^2} \cdot \sqrt[4]{8x^5y^3} = \sqrt[4]{8x^3y^2 \cdot 8x^5y^3}$ Use the product rule.

$\qquad = \sqrt[4]{64x^8y^5}$ Multiply.

$\qquad = \sqrt[4]{16 \cdot 4 \cdot x^8 \cdot y^4 \cdot y}$ Identify perfect fourth factors.

$\qquad = \sqrt[4]{(16x^8y^4)(4y)}$ Group the perfect fourth factors.

$\qquad = \sqrt[4]{16x^8y^4} \cdot \sqrt[4]{4y}$ Factor into two radicals.

$\qquad = 2x^2y\sqrt[4]{4y}$ $\sqrt[4]{16} = 2$ and

 CHECK POINT 7   Multiply and simplify:

a. $\sqrt{6} \cdot \sqrt{2}$

b. $10\sqrt[3]{16} \cdot 5\sqrt[3]{2}$

c. $\sqrt[4]{4x^2y} \cdot \sqrt[4]{8x^6y^3}$.

EXERCISE SET 10.3

 Practice Exercises

In Exercises 1–20, use the product rule to multiply.

1. $\sqrt{3} \cdot \sqrt{5}$

2. $\sqrt{7} \cdot \sqrt{5}$

3. $\sqrt[3]{2} \cdot \sqrt[3]{9}$

4. $\sqrt[3]{5} \cdot \sqrt[3]{4}$

5. $\sqrt[4]{11} \cdot \sqrt[4]{3}$

6. $\sqrt[5]{9} \cdot \sqrt[5]{3}$

7. $\sqrt{3x} \cdot \sqrt{11y}$

8. $\sqrt{5x} \cdot \sqrt{11y}$

9. $\sqrt[5]{6x^3} \cdot \sqrt[5]{4x}$

10. $\sqrt[4]{6x^2} \cdot \sqrt[4]{3x}$

11. $\sqrt{x+3} \cdot \sqrt{x-3}$

12. $\sqrt{x+6} \cdot \sqrt{x-6}$

13. $\sqrt[6]{x-4} \cdot \sqrt[6]{(x-4)^4}$

14. $\sqrt[6]{x-5} \cdot \sqrt[6]{(x-5)^4}$

15. $\sqrt{\dfrac{2x}{3}} \cdot \sqrt{\dfrac{3}{2}}$

16. $\sqrt{\dfrac{2x}{5}} \cdot \sqrt{\dfrac{5}{2}}$

17. $\sqrt[4]{\dfrac{x}{7}} \cdot \sqrt[4]{\dfrac{3}{y}}$

18. $\sqrt[4]{\dfrac{x}{3}} \cdot \sqrt[4]{\dfrac{7}{y}}$

19. $\sqrt[7]{7x^2y} \cdot \sqrt[7]{11x^3y^2}$

20. $\sqrt[9]{12x^2y^3} \cdot \sqrt[9]{3x^3y^4}$

In Exercises 21–32, simplify by factoring.

21. $\sqrt{50}$

22. $\sqrt{27}$

23. $\sqrt{45}$

24. $\sqrt{28}$

25. $\sqrt{75x}$

26. $\sqrt{40x}$

27. $\sqrt[3]{16}$

28. $\sqrt[3]{54}$

29. $\sqrt[3]{27x^3}$

30. $\sqrt[3]{250x^3}$

31. $\sqrt[3]{-16x^2y^3}$

32. $\sqrt[3]{-32x^2y^3}$

In Exercises 33–38, express the function, f, in simplified form. Assume that x can be any real number.

33. $f(x) = \sqrt{36(x + 2)^2}$

34. $f(x) = \sqrt{81(x - 2)^2}$

35. $f(x) = \sqrt[3]{32(x + 2)^3}$

36. $f(x) = \sqrt[3]{48(x - 2)^3}$

37. $f(x) = \sqrt{3x^2 - 6x + 3}$

38. $f(x) = \sqrt{5x^2 - 10x + 5}$

In Exercises 39–60, simplify by factoring. Assume that all variables in a radicand represent positive real numbers and no radicands involve negative quantities raised to even powers.

39. $\sqrt{x^7}$

40. $\sqrt{x^5}$

41. $\sqrt{x^8y^9}$

42. $\sqrt{x^6y^7}$

43. $\sqrt{48x^3}$

44. $\sqrt{40x^3}$

45. $\sqrt[3]{y^8}$

46. $\sqrt[3]{y^{11}}$

47. $\sqrt[3]{x^{14}y^3z}$

48. $\sqrt[3]{x^3y^{17}z^2}$

49. $\sqrt[3]{81x^8y^6}$

50. $\sqrt[3]{32x^9y^{17}}$

51. $\sqrt[3]{(x + y)^5}$

52. $\sqrt[3]{(x + y)^4}$

53. $\sqrt[5]{y^{17}}$

54. $\sqrt[5]{y^{18}}$

55. $\sqrt[5]{64x^6y^{17}}$

56. $\sqrt[5]{64x^7y^{16}}$

57. $\sqrt[4]{80x^{10}}$

58. $\sqrt[4]{96x^{11}}$

59. $\sqrt[4]{(x - 3)^{10}}$

60. $\sqrt[4]{(x - 2)^{14}}$

In Exercises 61–82, multiply and simplify. Assume that all variables in a radicand represent positive real numbers and no radicands involve negative quantities raised to even powers.

61. $\sqrt{12} \cdot \sqrt{2}$

62. $\sqrt{3} \cdot \sqrt{6}$

63. $\sqrt{5x} \cdot \sqrt{10y}$

64. $\sqrt{8x} \cdot \sqrt{10y}$

65. $\sqrt{12x} \cdot \sqrt{3x}$

66. $\sqrt{20x} \cdot \sqrt{5x}$

67. $\sqrt{50xy} \cdot \sqrt{4xy^2}$

68. $\sqrt{5xy} \cdot \sqrt{10xy^2}$

69. $2\sqrt{5} \cdot 3\sqrt{40}$

70. $3\sqrt{15} \cdot 5\sqrt{6}$

71. $\sqrt[3]{12} \cdot \sqrt[3]{4}$

72. $\sqrt[4]{4} \cdot \sqrt[4]{8}$

73. $\sqrt{5x^3} \cdot \sqrt{8x^2}$

74. $\sqrt{2x^7} \cdot \sqrt{12x^4}$

75. $\sqrt[3]{25x^4y^2} \cdot \sqrt[3]{5xy^{12}}$

76. $\sqrt[3]{6x^7y} \cdot \sqrt[3]{9x^4y^{12}}$

77. $\sqrt[4]{8x^2y^3z^6} \cdot \sqrt[4]{2x^4yz}$

78. $\sqrt[4]{4x^2y^3z^3} \cdot \sqrt[4]{8x^4yz^6}$

79. $\sqrt[5]{8x^4y^6} \cdot \sqrt[5]{8xy^7}$

80. $\sqrt[5]{8x^4y^3} \cdot \sqrt[5]{8xy^9}$

81. $\sqrt[3]{x - y} \cdot \sqrt[3]{(x - y)^7}$

82. $\sqrt[3]{x - 6} \cdot \sqrt[3]{(x - 6)^7}$

Application Exercises

Racing cyclists use the function $r(x) = 4\sqrt{x}$ to determine the maximum rate, $r(x)$, in miles per hour, to turn a corner of radius x, in feet, without tipping over. Use this function to solve Exercises 83–84.

83. What is the maximum rate that a cyclist should travel around a corner of radius 8 feet without tipping over? Write the answer in simplified radical form. Then use the simplified radical form and a calculator to express the answer to the nearest tenth.

84. What is the maximum rate that a cyclist should travel around a corner of radius 12 feet without tipping over? Write the answer in simplified radical form. Then use the simplified radical form and a calculator to express the answer to the nearest tenth.

The function

$$T(x) = \sqrt{\frac{x}{16}}$$

models the time, $T(x)$, in seconds, that it takes an object to fall x feet. Use this function to solve Exercises 85–86.

85. How long will it take a ball dropped from the top of a 320-foot building to hit the ground? Write the answer in simplified radical form. Then use the simplified radical form and a calculator to express the answer to the nearest tenth of a second.

86. How long will it take a ball dropped from the top of a 640-foot building to hit the ground? Write the answer in simplified radical form. Then use the simplified radical form and a calculator to express the answer to the nearest tenth of a second.

*Your **cardiac index** is your heart's output, in liters of blood per minute, divided by your body's surface area, in square meters. The cardiac index, $C(x)$, can be modeled by*

$$C(x) = \frac{7.644}{\sqrt[4]{x}}, \quad 10 \leq x \leq 80,$$

where x is an individual's age, in years. The graph of the function is shown. Use the function to solve Exercises 87–88.

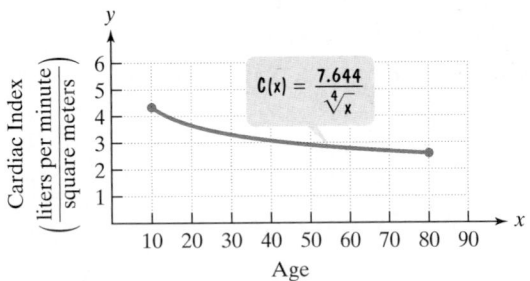

87. a. Find the cardiac index of a 32-year-old. Express the denominator in simplified radical form and reduce the fraction.

   b. Use the form of the answer in part (a) and a calculator to express the cardiac index to the nearest hundredth. Identify your solution as a point on the graph.

88. a. Find the cardiac index of an 80-year-old. Express the denominator in simplified radical form and reduce the fraction.

   b. Use the form of the answer in part (a) and a calculator to express the cardiac index to the nearest hundredth. Identify your solution as a point on the graph.

Writing in Mathematics

89. What is the product rule for radicals? Give an example to show how it is used.

90. Explain why $\sqrt{50}$ is not simplified. What do we mean when we say a radical expression is simplified?

91. In simplifying an nth root, explain how to find variable factors in the radicand that are perfect nth powers.

92. Without showing all the details, explain how to simplify $\sqrt[3]{16x^{14}}$.

93. As you get older, what would you expect to happen to your heart's output? Explain how this is shown in the graph for Exercises 87–88. Is this trend taking place progressively more rapidly or more slowly over the entire interval? What does this mean about this aspect of aging?

Critical Thinking Exercises

94. Which one of the following is true?
   a. $2\sqrt{5} \cdot 6\sqrt{5} = 12\sqrt{5}$
   b. $\sqrt[3]{4} \cdot \sqrt[3]{4} = 4$
   c. $\sqrt{12} = 2\sqrt{6}$
   d. $\sqrt[5]{3^{25}} = 243$

95. If a number is tripled, what happens to its square root?

96. What must be done to a number so that its cube root is tripled?

97. If $f(x) = \sqrt[3]{2x}$ and $(fg)(x) = 2x$, find $g(x)$.

98. Graph $f(x) = \sqrt{(x-1)^2}$ by hand.

Technology Exercises

99. Use a calculator to provide numerical support for your simplifications in Exercises 21–24 and 27–28. In each case, find a decimal approximation for the given expression. Then find a decimal approximation for your simplified expression. The approximations should be the same.

In Exercises 100–103, determine if each simplification is correct by graphing the function on each side of the equation with your graphing utility. Use the given viewing rectangle. The graphs should be the same. If they are not, correct the right side of the equation and then use your graphing utility to verify the simplification.

100. $\sqrt{x^4} = x^2$
   $[0, 5, 1]$ by $[0, 20, 1]$

101. $\sqrt{8x^2} = 4x\sqrt{2}$
   $[-5, 5, 1]$ by $[-5, 20, 1]$

102. $\sqrt{3x^2 - 6x + 3} = (x - 1)\sqrt{3}$
   $[-5, 5, 1]$ by $[-5, 5, 1]$

103. $\sqrt[3]{2x} \cdot \sqrt[3]{4x^2} = 4x$
   $[-10, 10, 1]$ by $[-10, 10, 1]$

Review Exercises

104. Solve: $2x - 1 \leq 21$ and $2x + 2 \geq 12$.
   (Section 9.2, Example 2)

105. Solve by the addition method:
$$5x + 2y = 2$$
$$4x + 3y = -4.$$
   (Section 4.3, Example 4)

106. Factor: $64x^3 - 27$.
   (Section 6.4, Example 8)

▶ **SECTION 10.4**   *Adding, Subtracting, and Dividing Radical Expressions*

Objectives

1 Add and subtract radical expressions.

2 Use the quotient rule to simplify radical expressions.

3 Use the quotient rule to divide radical expressions.

SSM
PH Tutor   CD-   Video
Center   ROM

The future is now: You have the opportunity to explore the cosmos in a starship traveling near the speed of light. The experience will enable you to understand the mysteries of the universe in deeply personal ways, transporting you to unimagined levels of knowing and being. The down side: According to Einstein's theory of relativity, close to the speed of light, your aging rate relative to friends on Earth is nearly zero. You will return from your two-year journey to a futuristic world in which friends and loved ones are long dead. Do you explore space or stay here on Earth? In this section, in addition to learning to perform various operations with radicals, you will see how they model your return to a world in which the people you knew have long departed.

1 Add and subtract radical expressions.

Adding and Subtracting Radical Expressions

Two or more radical expressions that have the same indices and radicands are called **like radicals**. Like radicals are combined in exactly the same way that we combine like terms. For example,

$$7\sqrt{11} + 6\sqrt{11} = (7 + 6)\sqrt{11} = 13\sqrt{11}.$$

7 square roots of 11 plus 6 square roots of 11 result in 13 square roots of 11.

EXAMPLE 1   Adding and Subtracting Like Radicals

Simplify (add or subtract) by combining like radical terms:

a. $7\sqrt{2} + 5\sqrt{2}$ **b.** $\sqrt[3]{5} - 4x\sqrt[3]{5} + 8\sqrt[3]{5}$

c. $8\sqrt[6]{5x} - 5\sqrt[6]{5x} + 4\sqrt[3]{5x}.$

Solution

a. $7\sqrt{2} + 5\sqrt{2}$

$= (7 + 5)\sqrt{2}$ *Apply the distributive property.*

$= 12\sqrt{2}$ *Simplify.*

b. $\sqrt[3]{5} - 4x\sqrt[3]{5} + 8\sqrt[3]{5}$

$= (1 - 4x + 8)\sqrt[3]{5}$ *Apply the distributive property.*

$= (9 - 4x)\sqrt[3]{5}$ *Simplify.*

c. $8\sqrt[6]{5x} - 5\sqrt[6]{5x} + 4\sqrt[3]{5x}$

$= (8 - 5)\sqrt[6]{5x} + 4\sqrt[3]{5x}$ *Apply the distributive property to the two like radicals.*

$= 3\sqrt[6]{5x} + 4\sqrt[3]{5x}$ *The indices, 6 and 3, differ. These are not like terms and cannot be combined.* ■

✔ **CHECK POINT 1** Simplify by combining like radical terms:

a. $8\sqrt{13} + 2\sqrt{13}$
b. $9\sqrt[3]{7} - 6x\sqrt[3]{7} + 12\sqrt[3]{7}$
c. $7\sqrt[4]{3x} - 2\sqrt[4]{3x} + 2\sqrt[3]{3x}$.

In some cases, radical expressions can be combined once they have been simplified. For example, to add $\sqrt{2}$ and $\sqrt{8}$, we can write $\sqrt{8}$ as $\sqrt{4 \cdot 2}$ because 4 is a perfect square factor of 8.

$$\sqrt{2} + \sqrt{8} = \sqrt{2} + \sqrt{4 \cdot 2} = 1\sqrt{2} + 2\sqrt{2} = (1 + 2)\sqrt{2} = 3\sqrt{2}$$

EXAMPLE 2 Combining Radicals That First Require Simplification

Simplify by combining like radical terms, if possible:

a. $7\sqrt{18} + 5\sqrt{8}$ **b.** $4\sqrt{27x} - 8\sqrt{12x}$ **c.** $7\sqrt{3} - 2\sqrt{5}$.

Solution

a. $7\sqrt{18} + 5\sqrt{8}$

$= 7\sqrt{9 \cdot 2} + 5\sqrt{4 \cdot 2}$	Factor the radicands using the largest perfect square factors.
$= 7\sqrt{9} \cdot \sqrt{2} + 5\sqrt{4} \cdot \sqrt{2}$	Take the square root of each factor.
$= 7 \cdot 3 \cdot \sqrt{2} + 5 \cdot 2 \cdot \sqrt{2}$	$\sqrt{9} = 3$ and $\sqrt{4} = 2$.
$= 21\sqrt{2} + 10\sqrt{2}$	Multiply.
$= (21 + 10)\sqrt{2}$	Apply the distributive property.
$= 31\sqrt{2}$	Simplify.

b. $4\sqrt{27x} - 8\sqrt{12x}$

$= 4\sqrt{9 \cdot 3x} - 8\sqrt{4 \cdot 3x}$	Factor the radicands using the largest perfect square factors.
$= 4\sqrt{9} \cdot \sqrt{3x} - 8\sqrt{4} \cdot \sqrt{3x}$	Take the square root of each factor.
$= 4 \cdot 3 \cdot \sqrt{3x} - 8 \cdot 2 \cdot \sqrt{3x}$	$\sqrt{9} = 3$ and $\sqrt{4} = 2$.
$= 12\sqrt{3x} - 16\sqrt{3x}$	Multiply.
$= (12 - 16)\sqrt{3x}$	Apply the distributive property.
$= -4\sqrt{3x}$	Simplify.

c. $7\sqrt{3} - 2\sqrt{5}$ cannot be simplified. The radical expressions have different radicands, namely 3 and 5, and are not like terms. ∎

✔ **CHECK POINT 2** Simplify by combining like radical terms, if possible:

a. $3\sqrt{20} + 5\sqrt{45}$
b. $3\sqrt{12x} - 6\sqrt{27x}$
c. $8\sqrt{5} - 6\sqrt{2}$.

EXAMPLE 3 Adding and Subtracting with Higher Indices

Simplify by combining like radical terms, if possible:

a. $2\sqrt[3]{16} - 4\sqrt[3]{54}$ **b.** $5\sqrt[3]{xy^2} + \sqrt[3]{8x^4y^5}$.

Solution

a. $2\sqrt[3]{16} - 4\sqrt[3]{54}$

$= 2\sqrt[3]{8 \cdot 2} - 4\sqrt[3]{27 \cdot 2}$ Factor the radicands using the largest perfect cube factors.

$= 2\sqrt[3]{8} \cdot \sqrt[3]{2} - 4\sqrt[3]{27} \cdot \sqrt[3]{2}$ Take the cube root of each factor.

$= 2 \cdot 2 \cdot \sqrt[3]{2} - 4 \cdot 3 \cdot \sqrt[3]{2}$ $\sqrt[3]{8} = 2$ and $\sqrt[3]{27} = 3$.

$= 4\sqrt[3]{2} - 12\sqrt[3]{2}$ Multiply.

$= (4 - 12)\sqrt[3]{2}$ Apply the distributive property.

$= -8\sqrt[3]{2}$ Simplify.

b. $5\sqrt[3]{xy^2} + \sqrt[3]{8x^4y^5}$

$= 5\sqrt[3]{xy^2} + \sqrt[3]{(8x^3y^3)xy^2}$ Factor the second radicand using the largest perfect cube factor.

$= 5\sqrt[3]{xy^2} + \sqrt[3]{8x^3y^3} \cdot \sqrt[3]{xy^2}$ Take the cube root of each factor.

$= 5\sqrt[3]{xy^2} + 2xy\sqrt[3]{xy^2}$ $\sqrt[3]{8} = 2$ and $\sqrt[3]{x^3y^3} = (x^3y^3)^{\frac{1}{3}} = xy$.

$= (5 + 2xy)\sqrt[3]{xy^2}$ Apply the distributive property.   ■

✔ **CHECK POINT 3**   Simplify by combining like radical terms, if possible.

a. $3\sqrt[3]{24} - 5\sqrt[3]{81}$

b. $5\sqrt[3]{x^2y} + \sqrt[3]{27x^5y^4}$.

Dividing Radical Expressions

We have seen that the root of a product is the product of the roots. The root of a quotient can also be expressed as the quotient of roots. Here is an example:

$$\sqrt{\frac{64}{4}} = \sqrt{16} = 4 \qquad \text{and} \qquad \frac{\sqrt{64}}{\sqrt{4}} = \frac{8}{2} = 4.$$

> This expression is the square root of a quotient.

> This expression is the quotient of two square roots.

The two procedures produce the same result, 4. This is a special case of the **quotient rule for radicals**.

2   Use the quotient rule to simplify radical expressions.

The Quotient Rule for Radicals

If $\sqrt[n]{a}$ and $\sqrt[n]{b}$ are real numbers and $b \neq 0$, then

$$\sqrt[n]{\frac{a}{b}} = \frac{\sqrt[n]{a}}{\sqrt[n]{b}}.$$

The nth root of a quotient is the quotient of the nth roots.

We know that a radical is simplified when its radicand has no factors other than 1 that are perfect nth powers. The quotient rule can be used to simplify some radicals. Keep in mind that all variables in radicands represent positive real numbers.

EXAMPLE 4 Using the Quotient Rule to Simplify Radicals

Simplify using the quotient rule:

a. $\sqrt[3]{\dfrac{16}{27}}$ **b.** $\sqrt{\dfrac{x^2}{25y^6}}$ **c.** $\sqrt[4]{\dfrac{7y^5}{x^{12}}}.$

Solution We simplify each expression by taking the roots of the numerator and denominator. Then we use factoring to simplify the resulting radicals, if possible.

a. $\sqrt[3]{\dfrac{16}{27}} = \dfrac{\sqrt[3]{16}}{\sqrt[3]{27}} = \dfrac{\sqrt[3]{8 \cdot 2}}{3} = \dfrac{\sqrt[3]{8} \cdot \sqrt[3]{2}}{3} = \dfrac{2\sqrt[3]{2}}{3}$

b. $\sqrt{\dfrac{x^2}{25y^6}} = \dfrac{\sqrt{x^2}}{\sqrt{25y^6}} = \dfrac{x}{5(y^6)^{\frac{1}{2}}} = \dfrac{x}{5y^3}$

Try to do this step mentally.

c. $\sqrt[4]{\dfrac{7y^5}{x^{12}}} = \dfrac{\sqrt[4]{7y^5}}{\sqrt[4]{x^{12}}} = \dfrac{\sqrt[4]{y^4 \cdot 7y}}{\sqrt[4]{x^{12}}} = \dfrac{y\sqrt[4]{7y}}{x^3}$ ∎

✔ **CHECK POINT 4** Simplify using the quotient rule:

a. $\sqrt[3]{\dfrac{24}{125}}$ **b.** $\sqrt{\dfrac{9x^3}{y^{10}}}$

c. $\sqrt[3]{\dfrac{8y^7}{x^{12}}}.$

3 Use the quotient rule to divide radical expressions.

By reversing the two sides of the quotient rule, we obtain a procedure for dividing radical expressions.

Dividing Radical Expressions

If $\sqrt[n]{a}$ and $\sqrt[n]{b}$ are real numbers and $b \neq 0$, then

$$\dfrac{\sqrt[n]{a}}{\sqrt[n]{b}} = \sqrt[n]{\dfrac{a}{b}}.$$

To divide two radical expressions with the same index, divide the radicands and retain the common index.

EXAMPLE 5 Dividing Radical Expressions

Divide and, if possible, simplify:

a. $\dfrac{\sqrt{48x^3}}{\sqrt{6x}}$ **b.** $\dfrac{\sqrt{45xy}}{2\sqrt{5}}$ **c.** $\dfrac{\sqrt[3]{16x^5y^2}}{\sqrt[3]{2xy^{-1}}}.$

Solution In each part of this problem, the indices in the numerator and the denominator are the same. Perform each division by dividing the radicands and retaining the common index.

a. $\dfrac{\sqrt{48x^3}}{\sqrt{6x}} = \sqrt{\dfrac{48x^3}{6x}} = \sqrt{8x^2} = \sqrt{4x^2 \cdot 2} = \sqrt{4x^2} \cdot \sqrt{2} = 2x\sqrt{2}$

b. $\dfrac{\sqrt{45xy}}{2\sqrt{5}} = \dfrac{1}{2} \cdot \sqrt{\dfrac{45xy}{5}} = \dfrac{1}{2} \cdot \sqrt{9xy} = \dfrac{1}{2} \cdot 3\sqrt{xy}$ or $\dfrac{3\sqrt{xy}}{2}$

c. $\dfrac{\sqrt[3]{16x^5y^2}}{\sqrt[3]{2xy^{-1}}} = \sqrt[3]{\dfrac{16x^5y^2}{2xy^{-1}}}$ Divide the radicands and retain the common index.

$= \sqrt[3]{8x^{5-1}y^{2-(-1)}}$ Divide factors in the radicand. Subtract exponents on common bases.

$= \sqrt[3]{8x^4y^3}$ Simplify.

$= \sqrt[3]{(8x^3y^3)x}$ Factor using the largest perfect cube factor.

$= \sqrt[3]{8x^3y^3} \cdot \sqrt[3]{x}$ Factor into two radicals.

$= 2xy\sqrt[3]{x}$ Simplify.   ∎

✔ **CHECK POINT 5**   Divide and, if possible, simplify:

a. $\dfrac{\sqrt{40x^5}}{\sqrt{2x}}$ **b.** $\dfrac{\sqrt{50xy}}{2\sqrt{2}}$

c. $\dfrac{\sqrt[3]{48x^7y}}{\sqrt[3]{6xy^{-2}}}.$

EXERCISE SET 10.4

 Practice Exercises

In this exercise set, assume that all variables represent positive real numbers.

In Exercises 1–10, add or subtract as indicated.

1. $8\sqrt{5} + 3\sqrt{5}$ **2.** $7\sqrt{3} + 2\sqrt{3}$

3. $9\sqrt[3]{6} - 2\sqrt[3]{6}$ **4.** $9\sqrt[3]{7} - 4\sqrt[3]{7}$

5. $4\sqrt[5]{2} + 3\sqrt[5]{2} - 5\sqrt[5]{2}$

6. $6\sqrt[5]{3} + 2\sqrt[5]{3} - 3\sqrt[5]{3}$

7. $3\sqrt{13} - 2\sqrt{5} - 2\sqrt{13} + 4\sqrt{5}$

8. $8\sqrt{17} - 5\sqrt{19} - 6\sqrt{17} + 4\sqrt{19}$

9. $3\sqrt{5} - \sqrt[3]{x} + 4\sqrt{5} + 3\sqrt[3]{x}$

10. $6\sqrt{7} - \sqrt[3]{x} + 2\sqrt{7} + 5\sqrt[3]{x}$

In Exercises 11–28, add or subtract as indicated. You will need to simplify terms to identify the like radicals.

11. $\sqrt{3} + \sqrt{27}$ **12.** $\sqrt{5} + \sqrt{20}$

13. $7\sqrt{12} + \sqrt{75}$ **14.** $5\sqrt{12} + \sqrt{75}$

15. $3\sqrt{32x} - 2\sqrt{18x}$

16. $5\sqrt{45x} - 2\sqrt{20x}$

17. $5\sqrt[3]{16} + \sqrt[3]{54}$

18. $3\sqrt[3]{24} + \sqrt[3]{81}$

19. $3\sqrt{45x^3} + \sqrt{5x}$

20. $8\sqrt{45x^3} + \sqrt{5x}$

21. $\sqrt[3]{54xy^3} + y\sqrt[3]{128x}$

22. $\sqrt[3]{24xy^3} + y\sqrt[3]{81x}$

23. $\sqrt[3]{54x^4} - \sqrt[3]{16x}$

24. $\sqrt[3]{81x^4} - \sqrt[3]{24x}$

25. $\sqrt{9x - 18} + \sqrt{x - 2}$

26. $\sqrt{4x - 12} + \sqrt{x - 3}$

27. $2\sqrt[3]{x^4y^2} + 3x\sqrt[3]{xy^2}$

28. $4\sqrt[3]{x^4y^2} + 5x\sqrt[3]{xy^2}$

In Exercises 29–44, simplify using the quotient rule.

29. $\sqrt{\dfrac{11}{4}}$

30. $\sqrt{\dfrac{19}{25}}$

31. $\sqrt[3]{\dfrac{19}{27}}$

32. $\sqrt[3]{\dfrac{11}{64}}$

33. $\sqrt{\dfrac{x^2}{36y^8}}$

34. $\sqrt{\dfrac{x^2}{144y^{12}}}$

35. $\sqrt{\dfrac{8x^3}{25y^6}}$

36. $\sqrt{\dfrac{50x^3}{81y^8}}$

37. $\sqrt[3]{\dfrac{x^4}{8y^3}}$

38. $\sqrt[3]{\dfrac{x^5}{125y^3}}$

39. $\sqrt[3]{\dfrac{50x^8}{27y^{12}}}$

40. $\sqrt[3]{\dfrac{81x^8}{8y^{15}}}$

41. $\sqrt[4]{\dfrac{9y^6}{x^8}}$

42. $\sqrt[4]{\dfrac{13y^7}{x^{12}}}$

43. $\sqrt[5]{\dfrac{64x^{13}}{y^{20}}}$

44. $\sqrt[5]{\dfrac{64x^{14}}{y^{15}}}$

In Exercises 45–62, divide and, if possible, simplify.

45. $\dfrac{\sqrt{40}}{\sqrt{5}}$

46. $\dfrac{\sqrt{200}}{\sqrt{10}}$

47. $\dfrac{\sqrt[3]{48}}{\sqrt[3]{6}}$

48. $\dfrac{\sqrt[3]{54}}{\sqrt[3]{2}}$

49. $\dfrac{\sqrt{54x^3}}{\sqrt{6x}}$

50. $\dfrac{\sqrt{72x^3}}{\sqrt{2x}}$

51. $\dfrac{\sqrt{x^5y^3}}{\sqrt{xy}}$

52. $\dfrac{\sqrt{x^7y^6}}{\sqrt{x^3y^2}}$

53. $\dfrac{\sqrt{200x^3}}{\sqrt{10x^{-1}}}$

54. $\dfrac{\sqrt{500x^3}}{\sqrt{10x^{-1}}}$

55. $\dfrac{\sqrt{72xy}}{2\sqrt{2}}$

56. $\dfrac{\sqrt{50xy}}{2\sqrt{2}}$

57. $\dfrac{\sqrt[3]{24x^3y^5}}{\sqrt[3]{3y^2}}$

58. $\dfrac{\sqrt[3]{250x^5y^3}}{\sqrt[3]{2x^3}}$

59. $\dfrac{\sqrt[4]{32x^{10}y^8}}{\sqrt[4]{2x^2y^{-2}}}$

60. $\dfrac{\sqrt[5]{96x^{12}y^{11}}}{\sqrt[5]{3x^2y^{-2}}}$

61. $\dfrac{\sqrt[3]{x^2+5x+6}}{\sqrt[3]{x+2}}$

62. $\dfrac{\sqrt[3]{x^2+7x+12}}{\sqrt[3]{x+3}}$

Application Exercises

What does travel in space have to do with radicals? Imagine that in the future we will be able to travel in starships at velocities approaching the speed of light (approximately 186,000 miles per second). According to Einstein's theory of relativity, time would pass more quickly on Earth than it would in the moving starship. The radical expression

$$R_f \dfrac{\sqrt{c^2 - v^2}}{\sqrt{c^2}}$$

gives the aging rate of an astronaut relative to the aging rate of a friend, R_f, on Earth. In the expression, v is the astronaut's velocity and c is the speed of light. Use the expression to solve Exercises 63–64. Imagine that you are the astronaut on the starship.

63. a. Use the quotient rule and simplify the expression that shows your aging rate relative to a friend on Earth. Working in a step-by-step manner, express your aging rate as

$$R_f \sqrt{1 - \left(\dfrac{v}{c}\right)^2}.$$

b. You are moving at velocities approaching the speed of light. Substitute c, the speed of light, for v in the simplified expression from part (a). Simplify completely. Close to the speed of light, what is your aging rate relative to a friend of Earth? What does this mean?

64. a. Use the quotient rule and simplify the expression that shows your aging rate relative to a friend on Earth. Working in a step-by-step manner, express your aging rate as

$$R_f \sqrt{1 - \left(\dfrac{v}{c}\right)^2}.$$

b. You are moving at 90% of the speed of light. Substitute $0.9c$ for v, your velocity, in the simplified expression from part (a). What is your aging rate, correct to two decimal places, relative to a friend on Earth? If you are gone for 44 weeks, approximately how many weeks have passed for your friend?

In Exercises 65–66, find the perimeter and area of each rectangle. Express answers in simplified radical form.

65.

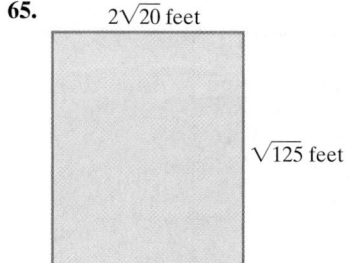

$2\sqrt{20}$ feet

$\sqrt{125}$ feet

66.

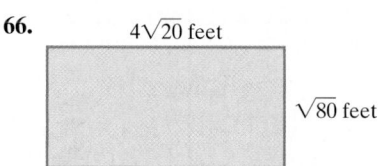

$4\sqrt{20}$ feet

$\sqrt{80}$ feet

Writing in Mathematics

67. What are like radicals? Give an example with your explanation.

68. Explain how to add like radicals. Give an example with your explanation.

69. If only like radicals can be combined, why is it possible to add $\sqrt{2}$ and $\sqrt{8}$?

70. Explain how to simplify a radical expression using the quotient rule. Provide an example.

71. Explain how to divide radical expressions with the same index.

72. Answer the question posed in the section opener on page 675. What will you do: explore space or stay here on Earth? What are the reasons for your choice?

Critical Thinking Exercises

73. Which one of the following is true?
   a. $\sqrt{5} + \sqrt{5} = \sqrt{10}$
   b. $4\sqrt{3} + 5\sqrt{3} = 9\sqrt{6}$

   c. If any two radical expressions are completely simplified, they can then be combined through addition or subtraction.
   d. None of the above is true.

74. If an irrational number is decreased by $2\sqrt{18} - \sqrt{50}$, the result is $\sqrt{2}$. What is the number?

75. Simplify: $\dfrac{\sqrt{20}}{3} + \dfrac{\sqrt{45}}{4} - \sqrt{80}$.

76. Simplify: $\dfrac{6\sqrt{49xy}\sqrt{ab^2}}{7\sqrt{36x^{-3}y^{-5}}\sqrt{a^{-9}b^{-1}}}$.

Technology Exercises

77. Use a calculator to provide numerical support to any four exercises that you worked from Exercises 1–62 that do not contain variables. Begin by finding a decimal approximation for the given expression. Then find a decimal approximation for your answer. The two decimal approximations should be the same.

In Exercises 78–80, determine if each operation is performed correctly by graphing the function on each side of the equation with your graphing utility. Use the given viewing rectangle. The graphs should be the same. If they are not, correct the right side of the equation and then use your graphing utility to verify the correction.

78. $\sqrt{4x} + \sqrt{9x} = 5\sqrt{x}$
   $[0, 5, 1]$ by $[0, 10, 1]$

79. $\sqrt{16x} - \sqrt{9x} = \sqrt{7x}$
   $[0, 5, 1]$ by $[0, 5, 1]$

80. $x\sqrt{8} + x\sqrt{2} = x\sqrt{10}$
   $[-5, 5, 1]$ by $[-15, 15, 1]$

Review Exercises

81. Solve: $2(3x - 1) - 4 = 2x - (6 - x)$.
   (Section 2.3, Example 3)

82. Factor: $x^2 - 8xy + 12y^2$.
   (Section 6.2, Example 6)

83. Add: $\dfrac{2}{x^2 + 5x + 6} + \dfrac{3x}{x^2 + 6x + 9}$.
   (Section 7.4, Example 7)

▶ **SECTION 10.5**   *Multiplying with More Than One Term and Rationalizing Denominators*

Objectives

1 Multiply radicals with more than one term.

2 Use polynomial special products to multiply radicals.

3 Rationalize denominators containing one term.

4 Rationalize denominators containing two terms.

5 Rationalize numerators.

SSM
PH Tutor   CD-   Video
Center   ROM

PEANUTS reprinted by permission of United Feature Syndicate, Inc.

The late Charles Schultz, creator of the "Peanuts" comic strip, transfixed 350 million readers worldwide with the joys and angst of his hapless Charlie Brown and Snoopy, a romantic self-deluded beagle. In 18,250 comic strips that spanned nearly 50 years, mathematics was often featured. Is the discussion of radicals shown above the real thing, or is it just an algebraic scam? By the time you complete this section on multiplying and dividing radicals, you will be able to decide.

1 Multiply radicals with more than one term.

Multiplying Radicals with More Than One Term

Radical expressions with more than one term are multiplied in much the same way that polynomials with more than one term are multiplied. Example 1 uses the distributive property and the FOIL method to perform multiplications.

EXAMPLE 1   Multiplying Radicals

Multiply:

    a. $\sqrt{7}(x + \sqrt{2})$ **b.** $\sqrt[3]{x}(\sqrt[3]{6} - \sqrt[3]{x^2})$ **c.** $(5\sqrt{2} + 2\sqrt{3})(4\sqrt{2} - 3\sqrt{3}).$

Solution

    a. $\sqrt{7}(x + \sqrt{2})$

        $= \sqrt{7} \cdot x + \sqrt{7} \cdot \sqrt{2}$ Use the distributive property.

        $= x\sqrt{7} + \sqrt{14}$ Multiply the radicals.

    b. $\sqrt[3]{x}(\sqrt[3]{6} - \sqrt[3]{x^2})$

        $= \sqrt[3]{x} \cdot \sqrt[3]{6} - \sqrt[3]{x} \cdot \sqrt[3]{x^2}$ Use the distributive property.

        $= \sqrt[3]{6x} - \sqrt[3]{x^3}$ Multiply the radicals: $\sqrt[n]{a} \cdot \sqrt[n]{b} = \sqrt[n]{ab}$.

        $= \sqrt[3]{6x} - x$ Simplify: $\sqrt[3]{x^3} = x.$

c. $\left(5\sqrt{2} + 2\sqrt{3}\right)\left(4\sqrt{2} - 3\sqrt{3}\right)$ Use the FOIL method.

$$= \left(5\sqrt{2}\right)\left(4\sqrt{2}\right) + \left(5\sqrt{2}\right)\left(-3\sqrt{3}\right) + \left(2\sqrt{3}\right)\left(4\sqrt{2}\right) + \left(2\sqrt{3}\right)\left(-3\sqrt{3}\right)$$

$$= 20 \cdot 2 - 15\sqrt{6} + 8\sqrt{6} - 6 \cdot 3$$ Multiply. Note that $\sqrt{2} \cdot \sqrt{2} = \sqrt{4} = 2$ and $\sqrt{3} \cdot \sqrt{3} = \sqrt{9} = 3$.

$$= 40 - 15\sqrt{6} + 8\sqrt{6} - 18$$ Complete the multiplications.

$$= (40 - 18) + \left(-15\sqrt{6} + 8\sqrt{6}\right)$$ Group like terms. Try to do this step mentally.

$$= 22 - 7\sqrt{6}$$ Combine numerical terms and like radicals.

■

✔ **CHECK POINT 1**   Multiply:

   a. $\sqrt{6}\left(x + \sqrt{10}\right)$

   b. $\sqrt[3]{y}\left(\sqrt[3]{y^2} - \sqrt[3]{7}\right)$

   c. $\left(6\sqrt{5} + 3\sqrt{2}\right)\left(2\sqrt{5} - 4\sqrt{2}\right)$.

2 Use polynomial special products to multiply radicals.

Some radicals can be multiplied using the special products for multiplying polynomials.

EXAMPLE 2   Using Special Products to Multiply Radicals

Multiply:

   a. $\left(\sqrt{3} + \sqrt{7}\right)^2$ **b.** $\left(\sqrt{7} + \sqrt{3}\right)\left(\sqrt{7} - \sqrt{3}\right)$ **c.** $\left(\sqrt{a} - \sqrt{b}\right)\left(\sqrt{a} + \sqrt{b}\right)$.

Solution   Use the special-product formulas shown.

a. $(A + B)^2 = A^2 + 2 \cdot A \cdot B + B^2$

$$\left(\sqrt{3} + \sqrt{7}\right)^2 = \left(\sqrt{3}\right)^2 + 2 \cdot \sqrt{3} \cdot \sqrt{7} + \left(\sqrt{7}\right)^2$$ Use the special product for $(A + B)^2$.

$$= 3 + 2\sqrt{21} + 7$$ Multiply the radicals.

$$= 10 + 2\sqrt{21}$$ Simplify.

b. $(A + B) \cdot (A - B) = A^2 - B^2$

$$\left(\sqrt{7} + \sqrt{3}\right)\left(\sqrt{7} - \sqrt{3}\right) = \left(\sqrt{7}\right)^2 - \left(\sqrt{3}\right)^2$$ Use the special product for $(A + B)(A - B)$.

$$= 7 - 3$$ Simplify: $\left(\sqrt{a}\right)^2 = a$.

$$= 4$$

$(A - B) \cdot (A + B) = A^2 - B^2$

c. $\left(\sqrt{a} - \sqrt{b}\right)\left(\sqrt{a} + \sqrt{b}\right) = \left(\sqrt{a}\right)^2 - \left(\sqrt{b}\right)^2 = a - b$ ■

Study Tip

In case you forgot them, special products for multiplying polynomials are summarized as follows:

Square of a Binomial:

$(A + B)^2 = A^2 + 2AB + B^2$

$(A - B)^2 = A^2 - 2AB + B^2$

Product of the Sum and Difference of Two Terms:

$(A + B)(A - B) = A^2 - B^2$.

Radical expressions that involve the sum and difference of the same two terms are called **conjugates**. For example,

$$\sqrt{7} + \sqrt{3} \quad \text{and} \quad \sqrt{7} - \sqrt{3}$$

are conjugates of each other. Parts (b) and (c) of Example 2 illustrate that the product of two radical expressions need not be a radical expression:

$$(\sqrt{7} + \sqrt{3})(\sqrt{7} - \sqrt{3}) = 4$$

$$(\sqrt{a} - \sqrt{b})(\sqrt{a} + \sqrt{b}) = a - b.$$

> The product of conjugates does not contain a radical.

Later in this section, we will use conjugates to simplify quotients.

✔ **CHECK POINT 2** Multiply:

a. $(\sqrt{5} + \sqrt{6})^2$ **b.** $(\sqrt{6} + \sqrt{5})(\sqrt{6} - \sqrt{5})$

c. $(\sqrt{a} - \sqrt{7})(\sqrt{a} + \sqrt{7})$.

3 Rationalize denominators containing one term.

Rationalizing Denominators Containing One Term

You can use a calculator to compare the approximate values for $\dfrac{1}{\sqrt{3}}$ and $\dfrac{\sqrt{3}}{3}$. The two approximations are the same. This is not a coincidence:

$$\frac{1}{\sqrt{3}} = \frac{1}{\sqrt{3}} \cdot \boxed{\frac{\sqrt{3}}{\sqrt{3}}} = \frac{\sqrt{3}}{\sqrt{9}} = \frac{\sqrt{3}}{3}.$$

> Any number divided by itself is 1. Multiplication by 1 does not change the value of $\dfrac{1}{\sqrt{3}}$.

This process involves rewriting a radical expression as an equivalent expression in which the denominator no longer contains a radical. The process is called **rationalizing the denominator**. When the denominator contains a single radical with an nth root, **multiply the numerator and denominator by a radical of index n that produces a perfect nth power in the denominator's radicand.**

EXAMPLE 3   Rationalizing Denominators

Rationalize each denominator:

a. $\dfrac{\sqrt{5}}{\sqrt{6}}$ **b.** $\sqrt[3]{\dfrac{7}{16}}$.

Solution

a. If we multiply the numerator and denominator of $\dfrac{\sqrt{5}}{\sqrt{6}}$ by $\sqrt{6}$, the denominator becomes $\sqrt{6} \cdot \sqrt{6} = \sqrt{36} = 6$. The denominator's radicand, 36, is a perfect square. The denominator no longer contains a radical. Therefore, we multiply by 1, choosing $\dfrac{\sqrt{6}}{\sqrt{6}}$ for 1.

$$\frac{\sqrt{5}}{\sqrt{6}} = \frac{\sqrt{5}}{\sqrt{6}} \cdot \frac{\sqrt{6}}{\sqrt{6}}$$

Multiply the numerator and denominator by $\sqrt{6}$ to remove the radical in the denominator.

$$= \frac{\sqrt{30}}{\sqrt{36}}$$

Multiply numerators and denominators. The denominator's radicand, 36, is a perfect square.

$$= \frac{\sqrt{30}}{6}$$

Simplify: $\sqrt{36} = 6$.

b. Using the quotient rule, we can express $\sqrt[3]{\dfrac{7}{16}}$ as $\dfrac{\sqrt[3]{7}}{\sqrt[3]{16}}$. We have cube roots, so we want the denominator's radicand to be a perfect cube. Right now, the denominator's radicand is 16 or 4^2. We know that $\sqrt[3]{4^3} = 4$. If we multiply the numerator and denominator of $\dfrac{\sqrt[3]{7}}{\sqrt[3]{16}}$ by $\sqrt[3]{4}$, the denominator becomes

$$\sqrt[3]{16} \cdot \sqrt[3]{4} = \sqrt[3]{4^2} \cdot \sqrt[3]{4} = \sqrt[3]{4^3} = 4.$$

The denominator's radicand, 4^3, is a perfect cube. The denominator no longer contains a radical. Therefore, we multiply by 1, choosing $\dfrac{\sqrt[3]{4}}{\sqrt[3]{4}}$ for 1.

$$\sqrt[3]{\frac{7}{16}} = \frac{\sqrt[3]{7}}{\sqrt[3]{16}}$$

Use the quotient rule and rewrite as the quotient of radicals.

$$= \frac{\sqrt[3]{7}}{\sqrt[3]{4^2}}$$

Write the denominators radicand as an exponential expression.

$$= \frac{\sqrt[3]{7}}{\sqrt[3]{4^2}} \cdot \frac{\sqrt[3]{4}}{\sqrt[3]{4}}$$

Multiply the numerator and denominator by $\sqrt[3]{4}$ to remove the radical in the denominator.

$$= \frac{\sqrt[3]{28}}{\sqrt[3]{4^3}}$$

Multiply numerators and denominators. The denominator's radicand, 4^3, is a perfect cube.

$$= \frac{\sqrt[3]{28}}{4}$$

Simplify: $\sqrt[3]{4^3} = 4$. ∎

✔ **CHECK POINT 3**   Rationalize each denominator:

a. $\dfrac{\sqrt{3}}{\sqrt{7}}$

b. $\sqrt[3]{\dfrac{2}{9}}$.

Example 3 showed that it is helpful to express the denominator's radicand using exponents. In this way, we can find the extra factor needed to produce a perfect nth power. For example, suppose that $\sqrt[5]{8}$ appears in the denominator. We want a perfect fifth power. By expressing $\sqrt[5]{8}$ as $\sqrt[5]{2^3}$, we would multiply the numerator and denominator by $\sqrt[5]{2^2}$ because

$$\sqrt[5]{2^3} \cdot \sqrt[5]{2^2} = \sqrt[5]{2^5} = 2.$$

EXAMPLE 4   Rationalizing Denominators

Rationalize each denominator:

a. $\sqrt{\dfrac{3x}{5y}}$ **b.** $\dfrac{\sqrt[3]{x}}{\sqrt[3]{36y}}$ **c.** $\dfrac{10y}{\sqrt[5]{4x^3y}}$.

Solution   By examining each denominator, you can determine how to multiply by 1. Let's first look at the denominators. For the square root, we must produce exponents of 2 in the radicand. For the cube root, we need exponents of 3, and for the fifth root, we want exponents of 5.

- $\sqrt{5y}$
- $\sqrt[3]{36y}$ or $\sqrt[3]{6^2y}$
- $\sqrt[5]{4x^3y}$ or $\sqrt[5]{2^2x^3y}$

Multiply by $\sqrt{5y}$:	Multiply by $\sqrt[3]{6y^2}$:	Multiply by $\sqrt[5]{2^3x^2y^4}$:
$\sqrt{5y} \cdot \sqrt{5y} = \sqrt{25y^2} = 5y.$	$\sqrt[3]{6^2y} \cdot \sqrt[3]{6y^2} = \sqrt[3]{6^3y^3} = 6y.$	$\sqrt[5]{2^2x^3y} \cdot \sqrt[5]{2^3x^2y^4} = \sqrt[5]{2^5x^5y^5} = 2xy.$

a. $\sqrt{\dfrac{3x}{5y}} = \dfrac{\sqrt{3x}}{\sqrt{5y}} = \dfrac{\sqrt{3x}}{\sqrt{5y}} \cdot \dfrac{\sqrt{5y}}{\sqrt{5y}} = \dfrac{\sqrt{15xy}}{\sqrt{25y^2}} = \dfrac{\sqrt{15xy}}{5y}$

Multiply by 1. $25y^2$ is a perfect square.

b. $\dfrac{\sqrt[3]{x}}{\sqrt[3]{36y}} = \dfrac{\sqrt[3]{x}}{\sqrt[3]{6^2y}} = \dfrac{\sqrt[3]{x}}{\sqrt[3]{6^2y}} \cdot \dfrac{\sqrt[3]{6y^2}}{\sqrt[3]{6y^2}} = \dfrac{\sqrt[3]{6xy^2}}{\sqrt[3]{6^3y^3}} = \dfrac{\sqrt[3]{6xy^2}}{6y}$

Multiply by 1. 6^3y^3 is a perfect cube.

c. $\dfrac{10y}{\sqrt[5]{4x^3y}} = \dfrac{10y}{\sqrt[5]{2^2x^3y}} = \dfrac{10y}{\sqrt[5]{2^2x^3y}} \cdot \dfrac{\sqrt[5]{2^3x^2y^4}}{\sqrt[5]{2^3x^2y^4}}$

Multiply by 1.

$= \dfrac{10y\sqrt[5]{2^3x^2y^4}}{\sqrt[5]{2^5x^5y^5}} = \dfrac{10y\sqrt[5]{8x^2y^4}}{2xy} = \dfrac{5\sqrt[5]{8x^2y^4}}{x}$

$2^5x^5y^5$ is a perfect   Simplify: Divide numerator
5th power.   and denominator by 2y.

■

✔ **CHECK POINT 4**   Rationalize each denominator:

a. $\sqrt{\dfrac{2x}{7y}}$

b. $\dfrac{\sqrt[3]{x}}{\sqrt[3]{9y}}$

c. $\dfrac{6x}{\sqrt[5]{8x^2y^4}}$.

4 Rationalize denominators containing two terms.

Rationalizing Denominators Containing Two Terms

How can we rationalize a denominator if the denominator contains two terms with one or more square roots? Multiply the numerator and the denominator by the conjugate of the denominator. Here are three examples of such expressions:

• $\dfrac{8}{3\sqrt{2} + 4}$

The conjugate of the denominator is $3\sqrt{2} - 4$.

• $\dfrac{2 + \sqrt{5}}{\sqrt{6} - \sqrt{3}}$

The conjugate of the denominator is $\sqrt{6} + \sqrt{3}$.

• $\dfrac{h}{\sqrt{x + h} - \sqrt{x}}$

The conjugate of the denominator is $\sqrt{x + h} + \sqrt{x}$.

The product of the denominator and its conjugate is found using the formula

$$(A + B)(A - B) = A^2 - B^2.$$

The product will not contain a radical.

EXAMPLE 5 Rationalizing a Denominator Containing Two Terms

Rationalize the denominator: $\dfrac{8}{3\sqrt{2} + 4}$.

Solution The conjugate of the denominator is $3\sqrt{2} - 4$. If we multiply the numerator and denominator by $3\sqrt{2} - 4$, the denominator will not contain a radical. Therefore, we multiply by 1, choosing $\dfrac{3\sqrt{2} - 4}{3\sqrt{2} - 4}$ for 1.

$$\frac{8}{3\sqrt{2} + 4} = \frac{8}{3\sqrt{2} + 4} \cdot \frac{3\sqrt{2} - 4}{3\sqrt{2} - 4}$$ Multiply by 1.

$$= \frac{8(3\sqrt{2} - 4)}{(3\sqrt{2})^2 - 4^2}$$ $(A + B)(A - B) = A^2 - B^2$

Leave the numerator in factored form. This helps simplify, if possible.

$$= \frac{8(3\sqrt{2} - 4)}{18 - 16}$$ $(3\sqrt{2})^2 = 9 \cdot 2 = 18$

$$= \frac{8(3\sqrt{2} - 4)}{2}$$ This expression can still be simplified.

$$= \frac{\overset{4}{\cancel{8}}(3\sqrt{2} - 4)}{\underset{1}{\cancel{2}}}$$ Divide the numerator and denominator by 2.

$$= 4(3\sqrt{2} - 4) \quad \text{or} \quad 12\sqrt{2} - 16$$ ∎

✔ **CHECK POINT 5** Rationalize the denominator: $\dfrac{18}{2\sqrt{3} + 3}$.

EXAMPLE 6   Rationalizing a Denominator Containing Two Terms

Rationalize the denominator: $\dfrac{2 + \sqrt{5}}{\sqrt{6} - \sqrt{3}}$.

Solution   Multiplication of both the numerator and denominator by $\sqrt{6} + \sqrt{3}$ will rationalize the denominator.

$$\frac{2 + \sqrt{5}}{\sqrt{6} - \sqrt{3}} = \frac{2 + \sqrt{5}}{\sqrt{6} - \sqrt{3}} \cdot \frac{\sqrt{6} + \sqrt{3}}{\sqrt{6} + \sqrt{3}}$$ Multiply by 1.

$$= \frac{\overset{F}{2\sqrt{6}} + \overset{O}{2\sqrt{3}} + \overset{I}{\sqrt{5} \cdot \sqrt{6}} + \overset{L}{\sqrt{5} \cdot \sqrt{3}}}{\left(\sqrt{6}\right)^2 - \left(\sqrt{3}\right)^2}$$ Use FOIL on the numerator and $(A - B)(A + B) = A^2 - B^2$ in the denominator.

$$= \frac{2\sqrt{6} + 2\sqrt{3} + \sqrt{30} + \sqrt{15}}{6 - 3}$$

$$= \frac{2\sqrt{6} + 2\sqrt{3} + \sqrt{30} + \sqrt{15}}{3}$$ Further simplification is not possible. ∎

✔ **CHECK POINT 6**   Rationalize the denominator: $\dfrac{3 + \sqrt{7}}{\sqrt{5} - \sqrt{2}}$.

5 Rationalize numerators.

Rationalizing Numerators

We have seen that square root functions are often used to model growing phenomena with growth that is leveling off. Figure 10.7 shows a male's height as a function of his age. The pattern of his growth suggests modeling with a square root function.

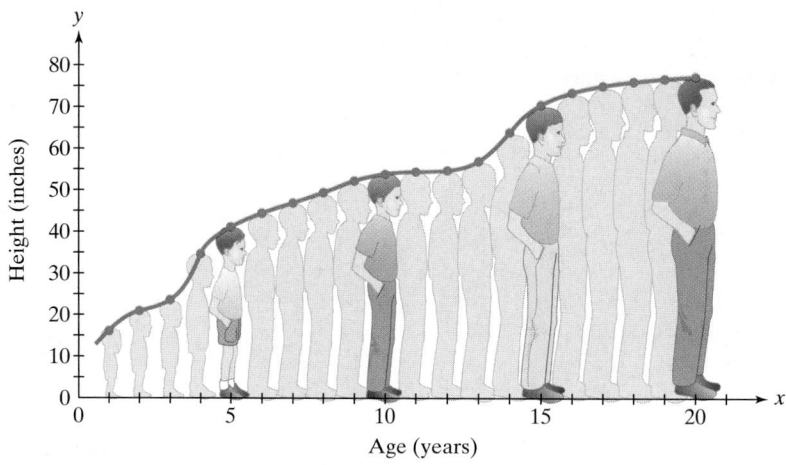

Figure 10.7

If we use $f(x) = \sqrt{x}$ to model height, $f(x)$, at age x, the expression

$$\frac{f(a + h) - f(a)}{h} = \frac{\sqrt{a + h} - \sqrt{a}}{h}$$

describes the man's average growth rate from age a to age $a + h$. Can you see that this expression is not defined if $h = 0$? However, to explore the man's average growth rates for successively shorter periods of time, we need to know what happens to the expression as h takes on values that get closer and closer to 0.

What happens to growth near the instant in time that the man is age a? The question is answered by first **rationalizing the numerator**. The procedure is similar to rationalizing the denominator. **To rationalize a numerator, multiply by 1 to eliminate the radical in the *numerator*.**

EXAMPLE 7   Rationalizing a Numerator

Rationalize the numerator:

$$\frac{\sqrt{a + h} - \sqrt{a}}{h}.$$

Solution   The conjugate of the numerator is $\sqrt{a + h} + \sqrt{a}$. If we multiply the numerator and denominator by $\sqrt{a + h} + \sqrt{a}$, the numerator will not contain radicals. Therefore, we multiply by 1, choosing $\dfrac{\sqrt{a + h} + \sqrt{a}}{\sqrt{a + h} + \sqrt{a}}$ for 1.

$$\frac{\sqrt{a + h} - \sqrt{a}}{h} = \frac{\sqrt{a + h} - \sqrt{a}}{h} \cdot \frac{\sqrt{a + h} + \sqrt{a}}{\sqrt{a + h} + \sqrt{a}} \qquad \text{Multiply by 1.}$$

$$= \frac{(\sqrt{a + h})^2 - (\sqrt{a})^2}{h(\sqrt{a + h} + \sqrt{a})} \qquad \begin{array}{l} (A - B)(A + B) = A^2 - B^2 \\ \text{Leave the denominator in} \\ \text{factored form.} \end{array}$$

$$= \frac{a + h - a}{h(\sqrt{a + h} + \sqrt{a})} \qquad \begin{array}{l} (\sqrt{a + h})^2 = a + h \\ \text{and } (\sqrt{a})^2 = a. \end{array}$$

$$= \frac{h}{h(\sqrt{a + h} + \sqrt{a})} \qquad \text{Simplify the numerator.}$$

$$= \frac{1}{\sqrt{a + h} + \sqrt{a}} \qquad \begin{array}{l} \text{Simplify by dividing the} \\ \text{numerator and the} \\ \text{denominator by } h. \end{array}$$

■

What happens to the rationalized expression in Example 7 as h is close to 0? The expression is close to

$$\frac{1}{\sqrt{a + 0} + \sqrt{a}} = \frac{1}{\sqrt{a} + \sqrt{a}} = \frac{1}{2\sqrt{a}}.$$

The man's growth rate at one instant in time, age a, can be determined with this expression. This expression reveals how the square root function $f(x) = \sqrt{x}$ is changing, instant by instant.

✔ **CHECK POINT 7**   Rationalize the numerator: $\dfrac{\sqrt{x + 3} - \sqrt{x}}{3}.$

EXERCISE SET 10.5

Practice Exercises

In this exercise set, assume that all variables represent positive real numbers.

In Exercises 1–38, multiply as indicated. If possible, simplify any radical expressions that appear in the product.

1. $\sqrt{2}(x + \sqrt{7})$

2. $\sqrt{5}(x + \sqrt{3})$

3. $\sqrt{6}(7 - \sqrt{6})$

4. $\sqrt{3}(5 - \sqrt{3})$

5. $\sqrt{3}(4\sqrt{6} - 2\sqrt{3})$

6. $\sqrt{6}(4\sqrt{6} - 3\sqrt{2})$

7. $\sqrt[3]{2}(\sqrt[3]{6} + 4\sqrt[3]{5})$

8. $\sqrt[3]{3}(\sqrt[3]{6} + 7\sqrt[3]{4})$

9. $\sqrt[3]{x}(\sqrt[3]{16x^2} - \sqrt[3]{x})$

10. $\sqrt[3]{x}(\sqrt[3]{24x^2} - \sqrt[3]{x})$

11. $(5 + \sqrt{2})(6 + \sqrt{2})$

12. $(7 + \sqrt{2})(8 + \sqrt{2})$

13. $(6 + \sqrt{5})(9 - 4\sqrt{5})$

14. $(4 + \sqrt{5})(10 - 3\sqrt{5})$

15. $(6 - 3\sqrt{7})(2 - 5\sqrt{7})$

16. $(7 - 2\sqrt{7})(5 - 3\sqrt{7})$

17. $(\sqrt{2} + \sqrt{7})(\sqrt{3} + \sqrt{5})$

18. $(\sqrt{3} + \sqrt{2})(\sqrt{10} + \sqrt{11})$

19. $(\sqrt{2} - \sqrt{7})(\sqrt{3} - \sqrt{5})$

20. $(\sqrt{3} - \sqrt{2})(\sqrt{10} - \sqrt{11})$

21. $(3\sqrt{2} - 4\sqrt{3})(2\sqrt{2} + 5\sqrt{3})$

22. $(3\sqrt{5} - 2\sqrt{3})(4\sqrt{5} + 5\sqrt{3})$

23. $(\sqrt{3} + \sqrt{5})^2$

24. $(\sqrt{2} + \sqrt{7})^2$

25. $(\sqrt{3x} - \sqrt{y})^2$

26. $(\sqrt{2x} - \sqrt{y})^2$

27. $(\sqrt{5} + 7)(\sqrt{5} - 7)$

28. $(\sqrt{6} + 2)(\sqrt{6} - 2)$

29. $(2 - 5\sqrt{3})(2 + 5\sqrt{3})$

30. $(3 - 5\sqrt{2})(3 + 5\sqrt{2})$

31. $(3\sqrt{2} + 2\sqrt{3})(3\sqrt{2} - 2\sqrt{3})$

32. $(4\sqrt{3} + 3\sqrt{2})(4\sqrt{3} - 3\sqrt{2})$

33. $(3 - \sqrt{x})(2 - \sqrt{x})$

34. $(4 - \sqrt{x})(3 - \sqrt{x})$

35. $(\sqrt[3]{x} - 4)(\sqrt[3]{x} + 5)$

36. $(\sqrt[3]{x} - 3)(\sqrt[3]{x} + 7)$

37. $(x + \sqrt[3]{y^2})(2x - \sqrt[3]{y^2})$

38. $(x - \sqrt[5]{y^3})(2x + \sqrt[5]{y^3})$

In Exercises 39–64, rationalize each denominator.

39. $\dfrac{\sqrt{2}}{\sqrt{5}}$

40. $\dfrac{\sqrt{7}}{\sqrt{3}}$

41. $\sqrt{\dfrac{11}{x}}$

42. $\sqrt{\dfrac{6}{x}}$

43. $\dfrac{9}{\sqrt{3y}}$

44. $\dfrac{12}{\sqrt{3y}}$

45. $\dfrac{1}{\sqrt[3]{2}}$

46. $\dfrac{1}{\sqrt[3]{3}}$

47. $\dfrac{6}{\sqrt[3]{4}}$

48. $\dfrac{10}{\sqrt[3]{5}}$

49. $\sqrt[3]{\dfrac{2}{3}}$

50. $\sqrt[3]{\dfrac{3}{4}}$

51. $\dfrac{4}{\sqrt[3]{x}}$

52. $\dfrac{7}{\sqrt[3]{x}}$

53. $\sqrt[3]{\dfrac{2}{y^2}}$

54. $\sqrt[3]{\dfrac{5}{y^2}}$

55. $\dfrac{7}{\sqrt[3]{2x^2}}$

56. $\dfrac{10}{\sqrt[3]{4x^2}}$

57. $\sqrt[3]{\dfrac{2}{xy^2}}$

58. $\sqrt[3]{\dfrac{3}{xy^2}}$

59. $\dfrac{3}{\sqrt[4]{x}}$

60. $\dfrac{5}{\sqrt[4]{x}}$

61. $\dfrac{6}{\sqrt[5]{8x^3}}$

62. $\dfrac{10}{\sqrt[5]{16x^2}}$

63. $\dfrac{2x^2y}{\sqrt[5]{4x^2y^4}}$

64. $\dfrac{3xy^2}{\sqrt[5]{8xy^3}}$

In Exercises 65–82, rationalize each denominator. Simplify, if possible.

65. $\dfrac{8}{\sqrt{5} + 2}$

66. $\dfrac{15}{\sqrt{6} + 1}$

67. $\dfrac{13}{\sqrt{11} - 3}$

68. $\dfrac{17}{\sqrt{10} - 2}$

69. $\dfrac{6}{\sqrt{5} + \sqrt{3}}$

70. $\dfrac{12}{\sqrt{7} + \sqrt{3}}$

71. $\dfrac{\sqrt{a}}{\sqrt{a} - \sqrt{b}}$

72. $\dfrac{\sqrt{b}}{\sqrt{a} - \sqrt{b}}$

73. $\dfrac{25}{5\sqrt{2} - 3\sqrt{5}}$

74. $\dfrac{35}{5\sqrt{2} - 3\sqrt{5}}$

75. $\dfrac{\sqrt{5} + \sqrt{3}}{\sqrt{5} - \sqrt{3}}$

76. $\dfrac{\sqrt{11} - \sqrt{5}}{\sqrt{11} + \sqrt{5}}$

77. $\dfrac{\sqrt{x} + 1}{\sqrt{x} + 3}$

78. $\dfrac{\sqrt{x} - 2}{\sqrt{x} - 5}$

79. $\dfrac{5\sqrt{3} - 3\sqrt{2}}{3\sqrt{2} - 2\sqrt{3}}$

80. $\dfrac{2\sqrt{6} + \sqrt{5}}{3\sqrt{6} - \sqrt{5}}$

81. $\dfrac{2\sqrt{x} + \sqrt{y}}{\sqrt{y} - 2\sqrt{x}}$

82. $\dfrac{3\sqrt{x} + \sqrt{y}}{\sqrt{y} - 3\sqrt{x}}$

In Exercises 83–94, rationalize each numerator. Simplify, if possible.

83. $\sqrt{\dfrac{3}{2}}$

84. $\sqrt{\dfrac{5}{3}}$

85. $\dfrac{\sqrt[3]{4x}}{\sqrt[3]{y}}$

86. $\dfrac{\sqrt[3]{2x}}{\sqrt[3]{y}}$

87. $\dfrac{\sqrt{x} + 3}{\sqrt{x}}$

88. $\dfrac{\sqrt{x} + 4}{\sqrt{x}}$

89. $\dfrac{\sqrt{a} + \sqrt{b}}{\sqrt{a} - \sqrt{b}}$

90. $\dfrac{\sqrt{a} - \sqrt{b}}{\sqrt{a} + \sqrt{b}}$

91. $\dfrac{\sqrt{x + 5} - \sqrt{x}}{5}$

92. $\dfrac{\sqrt{x + 7} - \sqrt{x}}{7}$

93. $\dfrac{\sqrt{x} + \sqrt{y}}{x^2 - y^2}$

94. $\dfrac{\sqrt{x} - \sqrt{y}}{x^2 - y^2}$

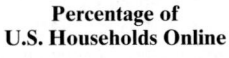

Application Exercises

The bar graph shows the percentage of U.S. households online from 1997 through 2001. The function

$$P(t) = 6.85\sqrt{t} + 19$$

models the percentage, $P(t)$, of U.S. households online t years after 1997. Use the function to solve Exercises 95–96.

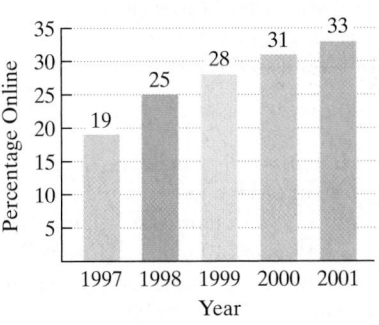

Percentage of U.S. Households Online

Source: Forrester Research

95. Find $P(4)$ and describe what this means. Round to the nearest percent. How well does your answer model the actual data shown in the graph?

96. Find $P(3)$ and describe what this means. Use a calculator and round to the nearest percent. How well does your answer model the actual data shown in the graph?

In Exercises 97–98, use the data shown in the bar graph for the percentage of U.S. households online.

97. The average rate of change in the percentage of households online from 1997 through 2001 is

$$\dfrac{\text{percentage online in 2001} - \text{percentage online in 1997}}{2001 - 1997}.$$

Change in percent   Change in time

Find the average yearly increase in the percentage of online households during this period.

98. The average rate of change in the percentage of households online from 1998 through 2001 is

$$\dfrac{\text{percentage online in 2001} - \text{percentage online in 1998}}{2001 - 1998}.$$

Change in percent   Change in time

Find the average yearly increase in the percentage of online households during this period.

The algebraic expression

$$6.85 \left(\frac{\sqrt{t+h} - \sqrt{t}}{h} \right)$$

models the average rate of change in the percentage of households online from t years after 1997 to t + h years after 1997. Use the expression to solve Exercises 99–102.

99. Let $t = 0$ and $h = 4$. Evaluate the expression to find the average rate of change in the percentage of households online from 1997 through 2001—that is, between 0 years and 4 years after 1997. How well does your answer model the actual yearly increase in the percentage that you found in Exercise 97?

100. Let $t = 1$ and $h = 3$. Evaluate the expression to find the average rate of change in the percentage of households online from 1998 through 2001—that is, between 1 year and $1 + 3$, or 4, years after 1997. How well does your answer model the actual yearly increase in the percentage that you found in Exercise 98?

101. a. Rewrite the average rate of change as an equivalent expression by rationalizing the numerator of the rational expression in parentheses.

b. Substitute 0 for h in your expression from part (a) and simplify. The resulting expression gives the rate of change in the percentage of households online precisely t years after 1997.

c. Use the expression from part (b) to find the rate of change in the percentage of households online in 2001. Round to the nearest tenth of a percent.

102. Solve parts (a) and (b) of Exercise 101. Then find the rate of change in the percentage of households online in 1998. Round to the nearest tenth of a percent.

103. Do you expect to pay more taxes than were withheld? Would you be surprised to know that the percentages of taxpayers who receive a refund and the percentage of taxpayers who pay more taxes varies according to age? The function

$$p(x) = \frac{x(13 + \sqrt{x})}{5\sqrt{x}}$$

models the percentage, $p(x)$, of taxpayers who are x years old who must pay more taxes.

a. What percentage of 25-year-olds must pay more taxes?

b. Rewrite the function by rationalizing the denominator.

104. Did you know that the U.S. infant mortality rate has been declining steadily? The function

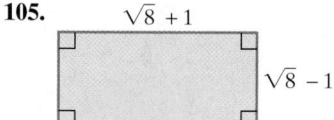

$$f(x) = \frac{28.46}{\sqrt[3]{x^2}}$$

models the infant mortality rate, $f(x)$, in deaths per 1000 live births, x years after 1959.

a. Find and interpret $f(8)$.

b. Rewrite the function by rationalizing the denominator.

In Exercises 105–106, write expressions for the perimeter and area of each figure. Then simplify these expressions. Assume that all measures are given in inches.

105.

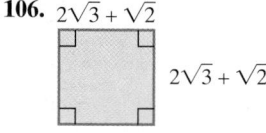

$\sqrt{8} + 1$

$\sqrt{8} - 1$

106. $2\sqrt{3} + \sqrt{2}$

$2\sqrt{3} + \sqrt{2}$

107. In the "Peanuts" cartoon shown in the section opener on page 682, Woodstock appears to be working steps mentally. Fill in the missing steps that show how to go from $\dfrac{7\sqrt{2 \cdot 2 \cdot 3}}{6}$ to $\dfrac{7}{3}\sqrt{3}$.

108. Rationalize the denominator of the golden ratio

$$\frac{2}{\sqrt{5} - 1},$$

discussed in the enrichment essay on page 687. Then use a calculator and find the ratio of width to height, correct to the nearest hundredth, in golden rectangles.

Writing in Mathematics

109. Explain how to perform this multiplication: $\sqrt{2}(\sqrt{7} + \sqrt{10})$.

110. Explain how to perform this multiplication: $(2 + \sqrt{3})(4 + \sqrt{3})$.

111. Explain how to perform this multiplication: $(2 + \sqrt{3})^2$.

112. What are conjugates? Give an example with your explanation.

113. Describe how to multiply conjugates.

114. Describe what it means to rationalize a denominator. Use both $\dfrac{1}{\sqrt{5}}$ and $\dfrac{1}{5 + \sqrt{5}}$ in your explanation.

115. When a radical expression has its denominator rationalized, we change the denominator so that it no longer contains a radical. Doesn't this change the value of the radical expression? Explain.

116. Suppose that you do not have a calculator and wish to obtain a decimal approximation for $\dfrac{5}{\sqrt{7}}$. From a table of square roots, you know that $\sqrt{7} \approx 2.6458$. Describe the advantage of rationalizing the denominator to obtain a decimal approximation for $\dfrac{5}{\sqrt{7}}$.

117. Square the real number $\dfrac{2}{\sqrt{3}}$. Observe that the radical is eliminated from the denominator. Explain whether this process is equivalent to rationalizing the denominator.

118. Describe the trend shown in the bar graph in Exercises 95–96 for the percentage of U.S. households online from 1997 through 2001. What explanation can you offer for this trend?

Critical Thinking Exercises

119. Which one of the following is true?

a. $\dfrac{\sqrt{3} + 7}{\sqrt{3} - 2} = -\dfrac{7}{2}$

b. $\dfrac{4}{\sqrt{x} + y} = \dfrac{4\sqrt{x} - y}{x - y}$

c. $\dfrac{4\sqrt{x}}{\sqrt{x} - y} = \dfrac{4x + 4y\sqrt{x}}{x - y^2}$

d. $(\sqrt{x} - 7)^2 = x - 49$

120. Simplify: $\sqrt{2} + \sqrt{\dfrac{1}{2}}$.

121. Simplify: $(\sqrt{2 + \sqrt{3}} + \sqrt{2 - \sqrt{3}})^2$.

122. Rationalize the denominator: $\dfrac{1}{\sqrt{2} + \sqrt{3} + \sqrt{4}}$.

123. Show that $\dfrac{3 + \sqrt{3}}{3}$ is a solution of $3x^2 + 2 = 6x$.

Technology Exercises

In Exercises 124–127, determine if each operation is performed correctly by graphing the function on each side of the equation with your graphing utility. Use the given viewing rectangle. The graphs should be the same. If they are not, correct the right side of the equation and then use your graphing utility to verify the correction.

124. $(\sqrt{x} - 1)(\sqrt{x} - 1) = x + 1$
$[0, 5, 1]$ by $[-1, 2, 1]$

125. $(\sqrt{x} + 2)(\sqrt{x} - 2) = x^2 - 4$ for $x \geq 0$
$[0, 10, 1]$ by $[-10, 10, 1]$

126. $(\sqrt{x} + 1)^2 = x + 1$
$[0, 8, 1]$ by $[0, 15, 1]$

127. $\dfrac{3}{\sqrt{x + 3} - \sqrt{x}} = \sqrt{x + 3} + \sqrt{x}$
$[0, 8, 1]$ by $[0, 6, 1]$

Review Exercises

128. Solve using Cramer's rule (determinants):

$$4x + 3y = 18$$
$$5x - 9y = 48.$$

(Section 8.5, Example 2)

129. Subtract: $\dfrac{6x}{x^2 - 4} - \dfrac{3}{x + 2}$. (Section 7.4, Example 7)

130. Factor completely: $2x^3 - 16x^2 + 30x$. (Section 6.5, Example 2)

▶ SECTION 10.6　*Radical Equations*

Objectives

1. Solve radical equations.
2. Use models that are radical functions to solve problems.

A best guess at the look of our nation in the next decades indicates that the number of men and women living alone will increase each year. By 2010, approximately 28 million men and women are projected to be living alone. The function

$$f(x) = 2.6\sqrt{x} + 11$$

models the number of Americans living alone, $f(x)$, in millions, x years after 1970. How can we predict the year when, say, 29 million of us will live alone? Substitute 29 for $f(x)$ and solve for x:

$$29 = 2.6\sqrt{x} + 11.$$

The resulting equation contains a variable in the radicand and is called a *radical equation*. A **radical equation** is an equation in which the variable occurs in a square root, cube root, or any higher root. Some examples of radical equations are

$$\sqrt{2x + 3} = 5, \quad \sqrt{3x + 1} - \sqrt{x + 4} = 1, \quad \text{and} \quad \sqrt[3]{3x - 1} + 4 = 0.$$

Variables occur in radicands.

In this section, you will learn how to solve radical equations. Solving such equations will enable you to solve new kinds of problems using radical functions.

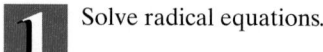　Solve radical equations.

Solving Radical Equations

Solving equations involving radicals involves raising both sides of the equation to a power equal to the radical's index. Unfortunately, if the index is even, all the solutions of the equation raised to the even power may not be solutions of the original equation. Consider, for example, the equation

$$x = 4.$$

If we square both sides, we obtain

$$x^2 = 16.$$

This new equation has two solutions, -4 and 4. By contrast, only 4 is a solution of the original equation, $x = 4$. For this reason, **when raising both sides of an equation to an even power, always check proposed solutions in the original equation.**

Here is a general method for solving radical equations with nth roots:

> **Solving Radical Equations Containing *n*th Roots**
>
> 1. If necessary, arrange terms so that one radical is isolated on one side of the equation.
> 2. Raise both sides of the equation to the *n*th power to eliminate the *n*th root.
> 3. Solve the resulting equation. If this equation still contains radicals, repeat steps 1 and 2.
> 4. Check all proposed solutions in the original equation.

EXAMPLE 1   Solving a Radical Equation

Solve: $\sqrt{2x + 3} = 5$.

Solution

Step 1 Isolate a radical on one side. The radical, $\sqrt{2x + 3}$, is already isolated on the left side of the equation, so we can skip this step.

Step 2 Raise both sides to the *n*th power. Because *n*, the index, is 2, we square both sides.

$$\sqrt{2x + 3} = 5 \qquad \text{This is the given equation.}$$
$$\left(\sqrt{2x + 3}\right)^2 = 5^2 \qquad \text{Square both sides to eliminate the radical.}$$
$$2x + 3 = 25 \qquad \text{Simplify.}$$

Step 3 Solve the resulting equation.

$$2x + 3 = 25 \qquad \text{The resulting equation is a linear equation.}$$
$$2x = 22 \qquad \text{Subtract 3 from both sides.}$$
$$x = 11 \qquad \text{Divide both sides by 2.}$$

Step 4 Check the proposed solution in the original equation. Because both sides were raised to an even power, this check is essential.

CHECK 11:

$$\sqrt{2x + 3} = 5$$
$$\sqrt{2 \cdot 11 + 3} \stackrel{?}{=} 5$$
$$\sqrt{25} \stackrel{?}{=} 5$$
$$5 = 5, \quad \text{true}$$

The solution is 11 and the solution set is $\{11\}$. ∎

✔ **CHECK POINT 1** Solve: $\sqrt{3x + 4} = 8$.

EXAMPLE 2   Solving a Radical Equation

Solve: $\sqrt{x - 3} + 6 = 5$.

Solution

Step 1 Isolate a radical on one side. The radical, $\sqrt{x - 3}$, can be isolated by subtracting 6 from both sides. We obtain

$$\sqrt{x - 3} = -1.$$

Does anything look strange about this equation? Can a principal square root be negative? It appears that this equation has no solution. However, we'll continue the solution procedure to see what happens.

Step 2 Raise both sides to the *n*th power. Because n, the index, is 2, we square both sides.

$$\left(\sqrt{x-3}\right)^2 = (-1)^2$$

$$x - 3 = 1 \qquad \text{Simplify.}$$

Step 3 Solve the resulting equation.

$$x - 3 = 1 \quad \text{The resulting equation is a linear equation.}$$

$$x = 4 \quad \text{Add 3 to both sides.}$$

Step 4 Check the proposed solution in the original equation.

CHECK 4:

$$\sqrt{x-3} + 6 = 5$$
$$\sqrt{4-3} + 6 \overset{?}{=} 5$$
$$\sqrt{1} + 6 \overset{?}{=} 5$$
$$1 + 6 \overset{?}{=} 5$$
$$7 = 5, \quad \text{false}$$

This false statement indicates that 4 is not a solution. Thus, the equation has no solution. The solution set is $\varnothing$, the empty set. ∎

Example 2 illustrates that extra solutions may be introduced when you raise both sides of a radical equation to an even power. Such solutions, which are not solutions of the given equation, are called **extraneous solutions.** Thus, 4 is an extraneous solution of $\sqrt{x-3} + 6 = 5$.

✔ **CHECK POINT 2** Solve: $\sqrt{x-1} + 7 = 2$.

EXAMPLE 3 Solving a Radical Equation

Solve: $x + \sqrt{26 - 11x} = 4$.

Solution

Step 1 Isolate a radical on one side. We isolate the radical by subtracting x from both sides. We obtain

$$\sqrt{26 - 11x} = 4 - x.$$

Step 2 Square both sides.

$$\left(\sqrt{26 - 11x}\right)^2 = (4 - x)^2$$

$$26 - 11x = 16 - 8x + x^2 \quad \begin{array}{l}\text{Simplify. Use the formula}\\ (A - B)^2 = A^2 - 2AB + B^2 \text{ to}\\ \text{square } 4 - x.\end{array}$$

Step 3 Solve the resulting equation. Because of the x^2-term, the resulting equation is a quadratic equation. We need to write this quadratic equation in standard form. We can obtain zero on the left side by subtracting 26 and adding $11x$ on both sides.

$$26 - 26 - 11x + 11x = 16 - 26 - 8x + 11x + x^2$$

$$0 = x^2 + 3x - 10 \qquad \text{Simplify.}$$

$$0 = (x + 5)(x - 2) \qquad \text{Factor.}$$

$$x + 5 = 0 \quad \text{or} \quad x - 2 = 0 \qquad \text{Set each factor equal to zero.}$$

$$x = -5 \qquad\qquad x = 2 \qquad \text{Solve for } x.$$

Step 4 Check the proposed solutions in the original equation.

CHECK −5:

$$x + \sqrt{26 - 11x} = 4$$
$$-5 + \sqrt{26 - 11(-5)} \stackrel{?}{=} 4$$
$$-5 + \sqrt{81} \stackrel{?}{=} 4$$
$$-5 + 9 \stackrel{?}{=} 4$$
$$4 = 4, \quad \text{true}$$

CHECK 2:

$$x + \sqrt{26 - 11x} = 4$$
$$2 + \sqrt{26 - 11 \cdot 2} \stackrel{?}{=} 4$$
$$2 + \sqrt{4} \stackrel{?}{=} 4$$
$$2 + 2 \stackrel{?}{=} 4$$
$$4 = 4, \quad \text{true}$$

The solutions are -5 and 2, and the solution set is $\{-5, 2\}$. ∎

✔ **CHECK POINT 3**   Solve: $\sqrt{6x + 7} - x = 2$.

The solution of radical equations with two or more square root expressions involves isolating a radical, squaring both sides, and then repeating this process. Let's consider an equation containing two square root expressions.

EXAMPLE 4   Solving an Equation Involving Two Radicals

Solve: $\sqrt{3x + 1} - \sqrt{x + 4} = 1$.

Solution

Step 1 Isolate a radical on one side. We can isolate the radical $\sqrt{3x + 1}$ by adding $\sqrt{x + 4}$ to both sides. We obtain

$$\sqrt{3x + 1} = \sqrt{x + 4} + 1.$$

Step 2 Square both sides.

$$\left(\sqrt{3x + 1}\right)^2 = \left(\sqrt{x + 4} + 1\right)^2$$

Squaring the expression on the right side of the equation can be a bit tricky. We need to use the formula

$$(A + B)^2 = A^2 + 2AB + B^2.$$

Focusing on just the right side, here is how the squaring is done.

$$\underbrace{(A + B)^2}_{} = \underbrace{A^2}_{} + \underbrace{2}_{} \cdot \underbrace{A}_{} \cdot \underbrace{B}_{} + \underbrace{B^2}_{}$$

$$\left(\sqrt{x + 4} + 1\right)^2 = \left(\sqrt{x + 4}\right)^2 + 2 \cdot \sqrt{x + 4} \cdot 1 + 1^2$$

This simplifies to $x + 4 + 2\sqrt{x + 4} + 1$. Thus, our equation can be written as follows:

$$3x + 1 = x + 4 + 2\sqrt{x + 4} + 1.$$

$$3x + 1 = x + 5 + 2\sqrt{x + 4} \qquad \text{Combine numerical terms on the right.}$$

Can you see that the resulting equation still contains a radical, namely $\sqrt{x + 4}$? Thus, we need to repeat the first two steps.

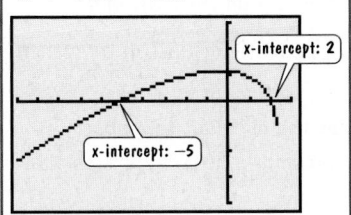

Repeat Step 1 Isolate a radical on one side. We isolate $2\sqrt{x+4}$, the radical term, by subtracting $x + 5$ from both sides. We obtain

$$3x + 1 = x + 5 + 2\sqrt{x+4}$$
This is the equation on the bottom of page 697.

$$2x - 4 = 2\sqrt{x+4}.$$
Subtract x + 5 from both sides.

Although we can simplify the equation by dividing both sides by 2, this sort of simplification is not always possible. Thus, we will work with the equation in this form.

Repeat Step 2 Square both sides.

$$\left(2x - 4\right)^2 = \left(2\sqrt{x+4}\right)^2$$

$$4x^2 - 16x + 16 = 4(x + 4)$$
Square both the 2 and the $\sqrt{x+4}$ on the right side.

Step 3 Solve the resulting equation. We solve this quadratic equation by writing it in standard form.

$$4x^2 - 16x + 16 = 4x + 16$$
Use the distributive property.

$$4x^2 - 20x = 0$$
Subtract 4x + 16 from both sides.

$$4x(x - 5) = 0$$
Factor.

$$4x = 0 \quad \text{or} \quad x - 5 = 0$$
Set each factor equal to zero.

$$x = 0 \qquad\qquad x = 5$$
Solve for x.

Step 4 Check the proposed solutions in the original equation.

CHECK 0:	**CHECK 5:**
$\sqrt{3x+1} - \sqrt{x+4} = 1$	$\sqrt{3x+1} - \sqrt{x+4} = 1$
$\sqrt{3\cdot 0 + 1} - \sqrt{0+4} \overset{?}{=} 1$	$\sqrt{3\cdot 5 + 1} - \sqrt{5+4} \overset{?}{=} 1$
$\sqrt{1} - \sqrt{4} \overset{?}{=} 1$	$\sqrt{16} - \sqrt{9} \overset{?}{=} 1$
$1 - 2 \overset{?}{=} 1$	$4 - 3 \overset{?}{=} 1$
$-1 = 1, \quad \text{false}$	$1 = 1, \quad \text{true}$

The check indicates that 0 is not a solution. It is an extraneous solution brought about by squaring each side of the equation. The only solution is 5, and the solution set is {5}. ∎

✔ **CHECK POINT 4** Solve: $\sqrt{x+5} - \sqrt{x-3} = 2$.

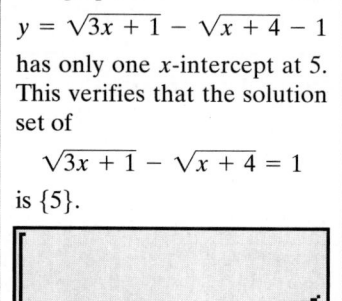

Using Technology

The graph of

$$y = \sqrt{3x+1} - \sqrt{x+4} - 1$$

has only one x-intercept at 5. This verifies that the solution set of

$$\sqrt{3x+1} - \sqrt{x+4} = 1$$

is {5}.

x-intercept: 5

EXAMPLE 5 Solving a Radical Equation

Solve: $(3x - 1)^{\frac{1}{3}} + 4 = 0$.

Solution Although we can rewrite the equation in radical form

$$\sqrt[3]{3x-1} + 4 = 0,$$

it is not necessary to do so. Because the equation involves a cube root, we isolate the radical term—that is, the term with the rational exponent—and cube both sides.

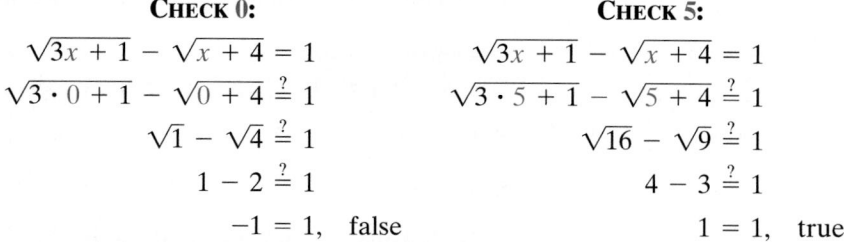

$$(3x - 1)^{\frac{1}{3}} + 4 = 0$$
This is the given equation.

$$(3x - 1)^{\frac{1}{3}} = -4$$
Subtract 4 from both sides.

$$\left[(3x - 1)^{\frac{1}{3}}\right]^3 = (-4)^3$$
Cube both sides.

$$3x - 1 = -64$$
Multiply exponents on the left side and simplify.

$$3x = -63 \qquad \text{Add 1 to both sides.}$$
$$x = -21 \qquad \text{Divide both sides by 3.}$$

Because both sides were raised to an odd power, it is not essential to check the proposed solution, -21. However, checking is always a good idea. Do so now and verify that -21 is the solution and the solution set is $\{-21\}$. ■

Example 5 illustrates that a radical equation with rational exponents can be solved by:

1. isolating the expression with the rational exponent, and
2. raising both sides of the equation to a power that is the reciprocal of the rational exponent.

Keep in mind that it is essential to check when both sides have been raised to even powers. Thus, equations with rational exponents such as $\frac{1}{2}$ and $\frac{1}{4}$ must be checked.

✔ **CHECK POINT 5** Solve: $(2x - 3)^{\frac{1}{3}} + 3 = 0$.

2 Use models that are radical functions to solve problems.

Applications of Radical Equations

Radical equations can be solved to answer questions about variables contained in radical functions.

EXAMPLE 6 Living Alone

The function

$$f(x) = 2.6\sqrt{x} + 11$$

models the number of Americans living alone, $f(x)$, in millions, x years after 1970. The U.S. population in 2000 was approximately 280 million. According to the model, when will the number of people living alone be $\frac{1}{8}$ of the population in 2000?

Solution We are interested in $\frac{1}{8}$ of the total population, 280 million:

$$\frac{1}{8} \cdot 280 \text{ million} = 35 \text{ million.}$$

When will 35 million Americans live alone? Substitute 35 for $f(x)$ in the given function. Then solve for x, the number of years after 1970.

$$f(x) = 2.6\sqrt{x} + 11 \qquad \text{This is the given function.}$$
$$35 = 2.6\sqrt{x} + 11 \qquad \text{Substitute 35 for f(x).}$$
$$24 = 2.6\sqrt{x} \qquad \text{Subtract 11 from both sides.}$$
$$\frac{24}{2.6} = \sqrt{x} \qquad \text{Divide both sides by 2.6.}$$
$$\left(\frac{24}{2.6}\right)^2 = \left(\sqrt{x}\right)^2 \qquad \text{Square both sides.}$$
$$85 \approx x \qquad \text{Use a calculator.}$$

The model indicates that 35 million Americans will live alone approximately 85 years after 1970. Because $1970 + 85 = 2055$, this is predicted to occur in 2055. ■

✔ **CHECK POINT 6** Use the information in Example 6 to find in which year the number of Americans living alone will be $\frac{1}{10}$ of the population in 2000. Use a calculator and round to the nearest year.

ENRICHMENT ESSAY

The Percentage of Young People Who Remain Single

Although popular stereotypes suggest that being single is more difficult for women than for men, most studies find that single women are healthier and more satisfied with their lives than single men are. Evidence suggests that women get along better without men than men get along without women. (*Source*: L. Cargan, *Singles: Myths and Realities*)

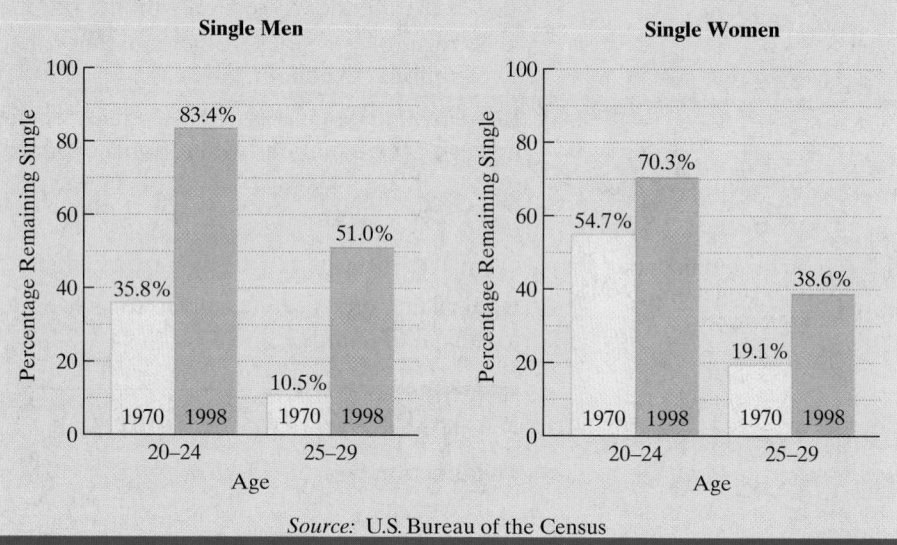

Source: U.S. Bureau of the Census

EXERCISE SET 10.6

 Practice Exercises

In Exercises 1–36, solve each radical equation.

1. $\sqrt{3x - 2} = 4$

2. $\sqrt{5x - 1} = 8$

3. $\sqrt{5x - 4} - 9 = 0$

4. $\sqrt{3x - 2} - 5 = 0$

5. $\sqrt{3x + 7} + 10 = 4$

6. $\sqrt{2x + 5} + 11 = 6$

7. $x = \sqrt{7x + 8}$

8. $x = \sqrt{6x + 7}$

9. $\sqrt{5x + 1} = x + 1$

10. $\sqrt{2x + 1} = x - 7$

11. $x = \sqrt{2x - 2} + 1$

12. $x = \sqrt{3x + 7} - 3$

13. $x - 2\sqrt{x - 3} = 3$

14. $3x - \sqrt{3x + 7} = -5$

15. $\sqrt{2x - 5} = \sqrt{x + 4}$

16. $\sqrt{6x + 2} = \sqrt{5x + 3}$

17. $\sqrt[3]{2x + 11} = 3$

18. $\sqrt[3]{6x - 3} = 3$

19. $\sqrt[3]{2x - 6} - 4 = 0$

20. $\sqrt[3]{4x - 3} - 5 = 0$

21. $\sqrt{x - 7} = 7 - \sqrt{x}$

22. $\sqrt{x - 8} = \sqrt{x} - 2$

23. $\sqrt{x + 2} + \sqrt{x - 1} = 3$

24. $\sqrt{x - 4} + \sqrt{x + 4} = 4$

25. $(2x + 3)^{\frac{1}{3}} + 4 = 6$

26. $(3x - 6)^{\frac{1}{3}} + 5 = 8$

27. $(3x + 1)^{\frac{1}{4}} + 7 = 9$

28. $(2x + 3)^{\frac{1}{4}} + 7 = 10$

29. $(x + 2)^{\frac{1}{2}} + 8 = 4$

30. $(x - 3)^{\frac{1}{2}} + 8 = 6$

31. $\sqrt{2x - 3} - \sqrt{x - 2} = 1$

32. $\sqrt{x + 2} + \sqrt{3x + 7} = 1$

33. $3x^{\frac{1}{3}} = (x^2 + 17x)^{\frac{1}{3}}$

34. $2(x - 1)^{\frac{1}{3}} = (x^2 + 2x)^{\frac{1}{3}}$

35. $(x + 8)^{\frac{1}{4}} = (2x)^{\frac{1}{4}}$

36. $(x - 2)^{\frac{1}{4}} = (3x - 8)^{\frac{1}{4}}$

Application Exercises

First the good news: The graph shows that U.S. seniors' scores in standard testing in science have improved since 1982. Now the bad news: The highest possible score is 500, and in 1970, the average test score was 304.

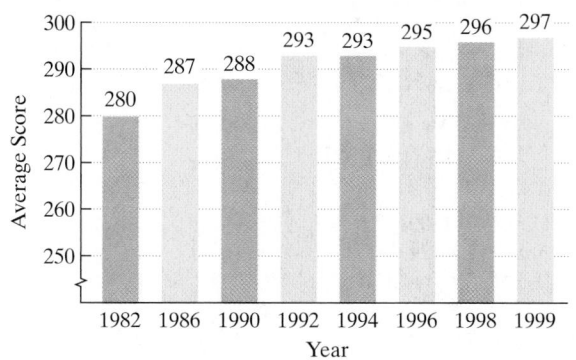

U.S. Seniors' Test Scores in Science

Source: National Assessment of Educational Progress

The function

$$f(x) = 4\sqrt{x} + 280$$

models the average science test score, f(x), x years after 1982. Use the model to solve Exercises 37–38.

37. When will the average science score return to the 1970 average of 304?

38. When will the average science test score be 300?

The function

$$f(x) = 6.75\sqrt{x} + 12$$

models the amount, f(x), in billions of dollars, of new student loans x years after 1993. Use the model to solve Exercises 39–40.

39. When is the amount loaned expected to reach $32.25 billion?

40. When is the amount loaned expected to reach $39 billion?

Out of a group of 50,000 births, the number of people, f(x), surviving to age x is modeled by the function

$$f(x) = 5000\sqrt{100 - x}.$$

The graph of the function is shown. Use the function to solve Exercises 41–42.

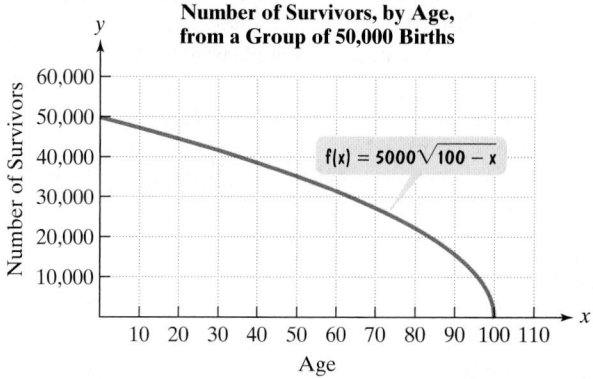

Number of Survivors, by Age, from a Group of 50,000 Births

41. To what age will 40,000 people in the group survive? Identify the solution as a point on the graph of the function.

42. To what age will 35,000 people in the group survive? Identify the solution as a point on the graph of the function.

The function

$$f(x) = 29x^{\frac{1}{3}}$$

models the number of plant species, f(x), on the various islands of the Galápagos chain in terms of the area, x, in square miles, of a particular island. Use the function to solve Exercises 43–44.

43. What is the area of a Galápagos island that has 87 species of plants?

44. What is the area of a Galápagos island that has 58 species of plants?

For each planet in our solar system, its year is the time it takes the planet to revolve once around the sun. The function

$$f(x) = 0.2x^{\frac{1}{3}}$$

models the number of Earth days in a planet's year, $f(x)$, where x is the average distance of the planet from the sun, in millions of kilometers. Use the function to solve Exercises 45–46.

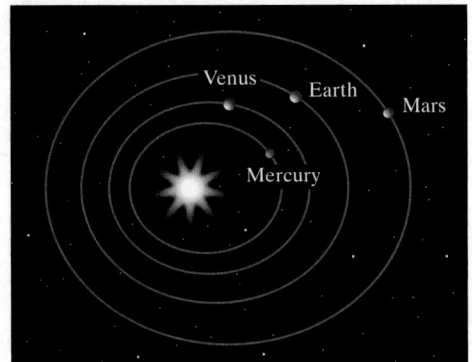

45. We, of course, have 365 Earth days in our year. What is the average distance of Earth from the sun? Use a calculator and round to the nearest million kilometers.

46. There are approximately 88 Earth days in the year of the planet Mercury. What is the average distance of Mercury from the sun? Use a calculator and round to the nearest million kilometers.

Writing in Mathematics

47. What is a radical equation?

48. In solving $\sqrt{2x - 1} + 2 = x$, why is it a good idea to isolate the radical term? What if we don't do this and simply square each side? Describe what happens.

49. What is an extraneous solution to a radical equation?

50. Explain why $\sqrt{x} = -1$ has no solution.

51. Explain how to solve a radical equation with rational exponents.

52. The radical function in Example 6 on page 699 shows a gradual year-by-year increase in the number of Americans living alone. What explanations can you offer for this trend? Describe an event that might occur in the future that could cause this trend to change.

53. Describe the trend shown by the graph of f in Exercises 41–42. When is the rate of decrease most rapid? What does this mean about survival rate by age?

Critical Thinking Exercises

54. Which one of the following is true?

    a. The first step in solving $\sqrt{x + 6} = x + 2$ is to square both sides, obtaining $x + 6 = x^2 + 4$.

b. The equations $\sqrt{x + 4} = -5$ and $x + 4 = 25$ have the same solution.

c. If $T = 2\pi\sqrt{\dfrac{L}{32}}$, then $L = \dfrac{8T^2}{\pi^2}$.

d. The equation $\sqrt{x^2 + 9x + 3} = -x$ has no solution because a principal square root is always nonnegative.

55. Find the length of the three sides of the right triangle shown in the figure.

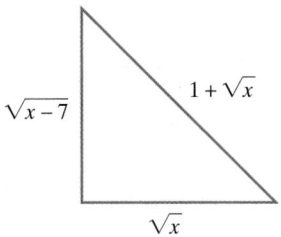

In Exercises 56–58, solve each equation.

56. $\sqrt[3]{x\sqrt{x}} = 9$

57. $\sqrt{\sqrt{x} + \sqrt{x + 9}} = 3$

58. $(x - 4)^{\frac{2}{3}} = 25$

Technology Exercises

In Exercises 59–63, use a graphing utility to solve each radical equation. Graph each side of the equation in the given viewing rectangle. The equation's solution is given by the x-coordinate of the point(s) of intersection. Check by substitution.

59. $\sqrt{2x + 2} = \sqrt{3x - 5}$

    $[-1, 10, 1]$ by $[-1, 5, 1]$

60. $\sqrt{x + 3} = 5$

    $[-1, 6, 1]$ by $[-1, 6, 1]$

61. $\sqrt{x^2 + 3} = x + 1$

    $[-1, 6, 1]$ by $[-1, 6, 1]$

62. $4\sqrt{x} = x + 3$

    $[-1, 10, 1]$ by $[-1, 14, 1]$

63. $\sqrt{x + 4} = 2$

    $[-2, 18, 1]$ by $[0, 10, 1]$

Review Exercises

64. Divide using synthetic division:
$(4x^4 - 3x^3 + 2x^2 - x - 1) \div (x + 3)$.
(Section 5.6, Example 4)

65. Divide:

$$\frac{3x^2 - 12}{x^2 + 2x - 8} \div \frac{6x + 18}{x + 4}.$$

(Section 7.2, Example 6)

66. Factor completely: $64x^3 - x$.
(Section 6.5, Example 3)

▶ SECTION 10.7 *Complex Numbers*

Objectives

1. Express square roots of negative numbers in terms of i.

2. Add and subtract complex numbers.

3. Multiply complex numbers.

4. Divide complex numbers.

5. Simplify powers of i.

SSM
PH Tutor   CD-   Video
Center   ROM

Who is this kid warning us about our eyeballs turning black if we attempt to find the square root of -9? Don't believe what you hear on the street. Although square roots of negative numbers are not real numbers, they do play a significant role in algebra. In this section, we move beyond real numbers and discuss square roots with negative radicands.

The Imaginary Unit i

1 Express square roots of negative numbers in terms of i.

In Chapter 11, we will study equations whose solutions involve the square roots of negative numbers. Because the square of a real number is never negative, there is no real number x such that $x^2 = -1$. To provide a setting in which such equations have solutions, mathematicians invented an expanded system of numbers, the complex numbers. The *imaginary number i*, defined to be a solution to the equation $x^2 = -1$, is the basis of this new set.

> **The Imaginary Unit i**
> The **imaginary unit** i is defined as
> $$i = \sqrt{-1}, \quad \text{where} \quad i^2 = -1.$$

Using the imaginary unit i, we can express the square root of any negative number as a real multiple of i. For example,

$$\sqrt{-25} = \sqrt{25(-1)} = \sqrt{25}\sqrt{-1} = 5i.$$

We can check that $\sqrt{-25} = 5i$ by squaring $5i$ and obtaining -25.
$$(5i)^2 = 5^2 i^2 = 25(-1) = -25$$

The Square Root of a Negative Number

If b is a positive real number, then
$$\sqrt{-b} = \sqrt{b(-1)} = \sqrt{b}\sqrt{-1} = \sqrt{b}i \quad \text{or} \quad i\sqrt{b}.$$

EXAMPLE 1   Expressing Square Roots of Negative Numbers as Multiples of i

Write as a multiple of i: **a.** $\sqrt{-9}$ **b.** $\sqrt{-7}$ **c.** $\sqrt{-8}$.

Solution

    a. $\sqrt{-9} = \sqrt{9(-1)} = \sqrt{9}\sqrt{-1} = 3i$

    b. $\sqrt{-7} = \sqrt{7(-1)} = \sqrt{7}\sqrt{-1} = \sqrt{7}i$

> Be sure not to write i under the radical.

    c. $\sqrt{-8} = \sqrt{8(-1)} = \sqrt{8}\sqrt{-1} = \sqrt{4 \cdot 2}\sqrt{-1} = 2\sqrt{2}i$ ■

✔ **CHECK POINT 1**   Write as a multiple of i:

    a. $\sqrt{-16}$ **b.** $\sqrt{-5}$ **c.** $\sqrt{-50}$.

      A new system of numbers, called **complex numbers**, is based on adding multiples of i, such as $5i$, to the real numbers.

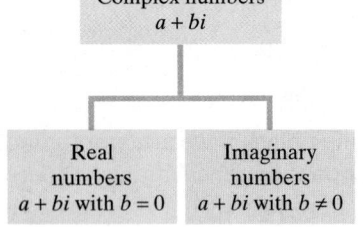

Figure 10.8 The complex number system

Complex Numbers and Imaginary Numbers

The set of all numbers in the form
$$a + bi$$
with real numbers a and b, and i, the imaginary unit, is called the set of **complex numbers**. The real number a is called the **real part**, and the real number b is called the **imaginary part**, of the complex number $a + bi$. If $b \neq 0$, then the complex number is called an **imaginary number** (Figure 10.8).

      Here are some examples of complex numbers. Each number can be written in the form $a + bi$.

$$-4 + 6i \qquad\qquad 2i = 0 + 2i \qquad\qquad 3 = 3 + 0i$$

| a, the real part, is -4. | b, the imaginary part, is 6. | a, the real part, is 0. | b, the imaginary part, is 2. | a, the real part, is 3. | b, the imaginary part, is 0. |

Can you see that b, the imaginary part, is not zero in the first two complex numbers? Because $b \neq 0$, these complex numbers are imaginary numbers. By contrast, the imaginary part of the complex number on the right is zero. This complex number is not an imaginary number. The number 3, or $3 + 0i$, is a real number.

Adding and Subtracting Complex Numbers

2 Add and subtract complex numbers.

The form of a complex number $a + bi$ is like the binomial $a + bx$. Consequently we can add, subtract, and multiply complex numbers using the same methods we used for binomials, remembering that $i^2 = -1$.

> **Adding and Subtracting Complex Numbers**
>
> **1.** $(a + bi) + (c + di) = (a + c) + (b + d)i$
>
> In words, this says that you add complex numbers by adding their real parts, adding their imaginary parts, and expressing the sum as a complex number.
>
> **2.** $(a + bi) - (c + di) = (a - c) + (b - d)i$
>
> In words, this says that you subtract complex numbers by subtracting their real parts, subtracting their imaginary parts, and expressing the difference as a complex number.

EXAMPLE 2 Adding and Subtracting Complex Numbers

Perform the indicated operations, writing the result in the form $a + bi$:

a. $(5 - 11i) + (7 + 4i)$ **b.** $(-5 + 7i) - (-11 - 6i)$.

Solution

a. $(5 - 11i) + (7 + 4i)$

$= 5 - 11i + 7 + 4i$ Remove the parentheses.

$= 5 + 7 - 11i + 4i$ Group real and imaginary terms.

$= (5 + 7) + (-11 + 4)i$ Use the distributive property.

$= 12 - 7i$ Add real parts and add imaginary parts.

b. $(-5 + 7i) - (-11 - 6i)$

$= -5 + 7i + 11 + 6i$ Remove the parentheses.

$= -5 + 11 + 7i + 6i$ Group real and imaginary terms.

$= (-5 + 11) + (7 + 6)i$

$= 6 + 13i$ ∎

✔ **CHECK POINT 2** Add or subtract as indicated:

a. $(5 - 2i) + (3 + 3i)$

b. $(2 + 6i) - (12 - 4i)$.

Study Tip

The following examples, using the same integers as in Example 2, show how operations with complex numbers are just like operations with polynomials.

a. $(5 - 11x) + (7 + 4x)$

$= 12 - 7x$

b. $(-5 + 7x) - (-11 - 6x)$

$= -5 + 7x + 11 + 6x$

$= 6 + 13x$

3 Multiply complex numbers.

Multiplying Complex Numbers

Multiplication of complex numbers is performed the same way as multiplication of polynomials, using the distributive property and the FOIL method. After completing the multiplication, we replace i^2 with -1. This idea is illustrated in the next example.

EXAMPLE 3 Multiplying Complex Numbers

Find the products: **a.** $4i(3 - 5i)$ **b.** $(7 - 3i)(-2 - 5i)$.

Solution

a. $4i(3 - 5i)$

$= 4i \cdot 3 - 4i \cdot 5i$ Distribute $4i$ throughout the parentheses.

$= 12i - 20i^2$ Multiply.

$= 12i - 20(-1)$ Replace i^2 with -1.

$= 20 + 12i$ Simplify to $12i + 20$ and write in $a + bi$ form.

b. $(7 - 3i)(-2 - 5i)$

$$\underset{\text{F}}{\quad}\underset{\text{O}}{\quad}\underset{\text{I}}{\quad}\underset{\text{L}}{\quad}$$

$$= -14 - 35i + 6i + 15i^2 \qquad \text{Use the FOIL method.}$$
$$= -14 - 35i + 6i + 15(-1) \qquad i^2 = -1$$
$$= -14 - 15 - 35i + 6i \qquad \text{Group real and imaginary terms.}$$
$$= -29 - 29i \qquad \text{Combine real and imaginary terms.} \qquad \blacksquare$$

✔ **CHECK POINT 3**   Find the products:

a. $7i(2 - 9i)$ **b.** $(5 + 4i)(6 - 7i)$.

Consider the multiplication problem

$$5i \cdot 2i = 10i^2 = 10(-1) = -10.$$

This problem can also be given in terms of square roots of negative numbers:

$$\sqrt{-25} \cdot \sqrt{-4}.$$

Because the product rule for radicals only applies to real numbers, multiplying radicands is incorrect. **When performing operations with square roots of negative numbers, begin by expressing all square roots in terms of i.** Then perform the indicated operation.

CORRECT:

$$\sqrt{-25} \cdot \sqrt{-4} = \sqrt{25}\,\sqrt{-1} \cdot \sqrt{4}\,\sqrt{-1}$$
$$= 5i \cdot 2i$$
$$= 10i^2 = 10(-1) = -10$$

INCORRECT:

$$\sqrt{-25} \cdot \sqrt{-4} = \sqrt{(-25)(-4)}$$
$$= \sqrt{100}$$
$$= 10$$

EXAMPLE 4   Multiplying Square Roots of Negative Numbers

Multiply: $\sqrt{-3} \cdot \sqrt{-5}$.

Solution

$$\sqrt{-3} \cdot \sqrt{-5} = \sqrt{3}\,\sqrt{-1} \cdot \sqrt{5}\,\sqrt{-1}$$
$$= \sqrt{3}i \cdot \sqrt{5}i \qquad \text{Express square roots in terms of } i.$$
$$= \sqrt{15}i^2 \qquad \sqrt{3} \cdot \sqrt{5} = \sqrt{15} \text{ and } i \cdot i = i^2.$$
$$= \sqrt{15}\,(-1) \qquad i^2 = -1$$
$$= -\sqrt{15} \qquad\qquad\qquad \blacksquare$$

✔ **CHECK POINT 4**   Multiply: $\sqrt{-5} \cdot \sqrt{-7}$.

4 Divide complex numbers.

Conjugates and Division

It is possible to multiply imaginary numbers and obtain a real number. Here is an example:

$$\underset{\text{F}}{\quad}\underset{\text{O}}{\quad}\underset{\text{I}}{\quad}\underset{\text{L}}{\quad}$$

$$(4 + 7i)(4 - 7i) = 16 - 28i + 28i - 49i^2$$

$$= 16 - 49i^2 = 16 - 49(-1) = 65.$$

Replace i^2 with -1.

Complex Numbers on a Postage Stamp

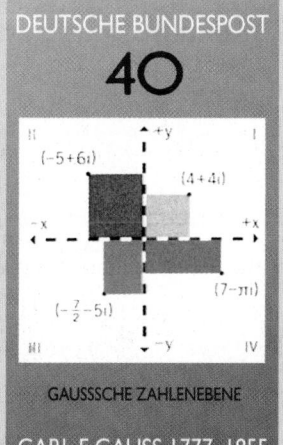

DEUTSCHE BUNDESPOST

40

$(-5+6i)$

$+y$

$(4+4i)$

$-x$ $+x$

$(7-\pi i)$

$(-\tfrac{7}{2}-5i)$

$-y$

III IV

GAUSSSCHE ZAHLENEBENE

CARL F. GAUSS 1777–1855
1977

This stamp honors the work done by the German mathematician Carl Friedrich Gauss (1777–1855) with complex numbers. Gauss represented complex numbers as points in the plane. (Stamp from the private collection of Professor C.M. Lang, photography by Gary J. Shulfer, University of Wisconsin, Stevens Point. "Germany: #5"; Scott Standard Postage Stamp Catalogue.)

You can also perform the multiplication using the formula
$$(A + B)(A - B) = A^2 - B^2.$$
A real number is obtained even faster:
$$(4 + 7i)(4 - 7i) = 4^2 - (7i)^2 = 16 - 49i^2 = 16 - 49(-1) = 65.$$

The **conjugate** of the complex number $a + bi$ is $a - bi$. The **conjugate** of the complex number $a - bi$ is $a + bi$. The multiplication problem that we just performed involved conjugates. The multiplication of conjugates always results in a real number:
$$(a + bi)(a - bi) = a^2 - (bi)^2 = a^2 - b^2i^2 = a^2 - b^2(-1) = a^2 + b^2$$

> The product eliminates i.

Conjugates are used to divide complex numbers. By multiplying the numerator and the denominator of the division by the conjugate of the denominator, you will obtain a real number in the denominator. Here are two examples of such divisions:

• $\dfrac{7 + 4i}{2 - 5i}$

• $\dfrac{5i - 4}{3i}$ or $\dfrac{5i - 4}{0 + 3i}$.

> The conjugate of the denominator is $2 + 5i$.

> The conjugate of the denominator is $0 - 3i$, or $-3i$.

The procedure for dividing complex numbers, illustrated in Examples 5 and 6, should remind you of rationalizing denominators.

EXAMPLE 5   Using Conjugates to Divide Complex Numbers

Divide and simplify to the form $a + bi$:
$$\frac{7 + 4i}{2 - 5i}.$$

Solution   The conjugate of the denominator is $2 + 5i$. Multiplication of both the numerator and denominator by $2 + 5i$ will eliminate i from the denominator.

$$\frac{7 + 4i}{2 - 5i} = \frac{7 + 4i}{2 - 5i} \cdot \frac{2 + 5i}{2 + 5i} \qquad \text{Multiply by 1.}$$

$$= \frac{\overset{F}{14} + \overset{O}{35i} + \overset{I}{8i} + \overset{L}{20i^2}}{2^2 - (5i)^2} \qquad \text{Use FOIL in the numerator and } (A - B)(A + B) = A^2 - B^2 \text{ in the denominator.}$$

$$= \frac{14 + 43i + 20i^2}{4 - 25i^2} \qquad \text{Simplify.}$$

$$= \frac{14 + 43i + 20(-1)}{4 - 25(-1)} \qquad i^2 = -1$$

$$= \frac{14 + 43i - 20}{4 + 25} \qquad \text{Perform the multiplications involving } -1.$$

$$= \frac{-6 + 43i}{29} \qquad \text{Combine real terms in the numerator and denominator.}$$

$$= -\frac{6}{29} + \frac{43}{29}i \qquad \text{Express the answer in the form } a + bi.$$

∎

✔ **CHECK POINT 5** Divide and simplify to the form $a + bi$:

$$\frac{6 + 2i}{4 - 3i}.$$

EXAMPLE 6 Using Conjugates to Divide Complex Numbers

Divide and simplify to the form $a + bi$:

$$\frac{5i - 4}{3i}.$$

Solution The conjugate of the denominator, $0 + 3i$, is $0 - 3i$. Multiplication of both the numerator and denominator by $-3i$ will eliminate i from the denominator.

$$\frac{5i - 4}{3i} = \frac{5i - 4}{3i} \cdot \frac{-3i}{-3i} \qquad \text{Multiply by 1.}$$

$$= \frac{-15i^2 + 12i}{-9i^2} \qquad \text{Multiply. Use the distributive property in the numerator.}$$

$$= \frac{-15(-1) + 12i}{-9(-1)} \qquad i^2 = -1$$

$$= \frac{15 + 12i}{9} \qquad \text{Perform the multiplications involving } -1.$$

$$= \frac{15}{9} + \frac{12}{9}i \qquad \text{Express the division in the form } a + bi.$$

$$= \frac{5}{3} + \frac{4}{3}i \qquad \text{Simplify real and imaginary parts.} \qquad \blacksquare$$

✔ **CHECK POINT 6** Divide and simplify to the form $a + bi$:

$$\frac{3 - 2i}{4i}.$$

5 Simplify powers of i.

Powers of i

Cycles govern many aspects of life —heartbeats, sleep patterns, seasons, and tides all follow regular, predictable cycles. Surprisingly, so do powers of i. To see how this occurs, use the fact that $i^2 = -1$ and express each power of i in terms of i^2:

$$i$$
$$i^2 = -1$$
$$i^3 = i^2 \cdot i = (-1)i = -i$$
$$i^4 = (i^2)^2 = (-1)^2 = 1$$
$$i^5 = i^4 \cdot i = (i^2)^2 \cdot i = (-1)^2 \cdot i = i$$
$$i^6 = (i^2)^3 = (-1)^3 = -1.$$

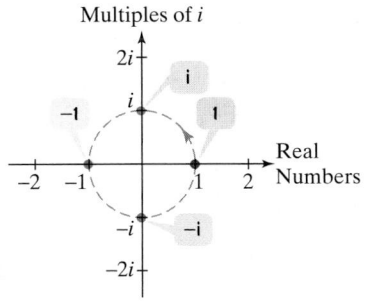

Multiples of i

Figure 10.9

Can you see that the powers of i are cycling through the values $i, -1, -i,$ and 1? The cycle is illustrated in Figure 10.9. In the figure, tick marks on the horizontal axis represent real numbers and tick marks along the vertical axis represent multiples of i. Using this representation, powers of i are equally spaced at $90°$ intervals on a circle having radius 1.

Here is a procedure for simplifying powers of i:

Simplifying Powers of i

1. Express the given power of i in terms of i^2.
2. Replace i^2 by -1 and simplify. Use the fact that -1 to an even power is 1, and -1 to an odd power is -1.

EXAMPLE 7   Simplifying Powers of i

Simplify: **a.** i^{12} **b.** i^{39} **c.** i^{50}.

Solution

    a. $i^{12} = (i^2)^6 = (-1)^6 = 1$

    b. $i^{39} = i^{38}i = (i^2)^{19}i = (-1)^{19}i = (-1)i = -i$

    c. $i^{50} = (i^2)^{25} = (-1)^{25} = -1$ ∎

 CHECK POINT 7   Simplify: **a.** i^{16} **b.** i^{25} **c.** i^{35}.

ENRICHMENT ESSAY

The Patterns of Chaos

R.F. Voss "29-Fold M-set Seahorse" computer-generated image, ©1990 R.F. Voss/IBM

One of the new frontiers of mathematics suggests that there is an underlying order in things that appear to be random, such as the hiss and crackle of background noises as you tune a radio. Irregularities in the heartbeat, some of them severe enough to cause a heart attack, or irregularities in our sleeping patterns, such as insomnia, are example of chaotic behavior. Chaos in the mathematical sense does not mean a complete lack of form or arrangement. In mathematics, chaos is used to describe something that appears to be random but is not actually random. The patterns of chaos appear in images like the one on the left, called the Mandelbrot set. Magnified portions of this image yield repetitions of the original structure, as well as new and unexpected patterns. The Mandelbrot set transforms the hidden structure of chaotic events into a source of wonder and inspiration.

The Mandelbrot set is made possible by opening up graphing to include complex numbers in the form $a + bi$. Although the details are beyond the scope of this text, the coordinate system that is used is shown in Figure 10.9. Plot certain complex numbers in this system, add color to the magnified boundary of the graph, and the patterns of chaos begin to appear.

EXERCISE SET 10.7

 Practice Exercises

In Exercises 1–16, express each number in terms of i and simplify, if possible.

1. $\sqrt{-49}$

2. $\sqrt{-36}$

3. $\sqrt{-17}$

4. $\sqrt{-11}$

5. $\sqrt{-75}$

6. $\sqrt{-20}$

7. $\sqrt{-28}$

8. $\sqrt{-45}$

9. $-\sqrt{-150}$

10. $-\sqrt{-700}$

11. $7 + \sqrt{-16}$

12. $9 + \sqrt{-4}$

13. $5 + \sqrt{-5}$

14. $10 + \sqrt{-3}$

15. $6 - \sqrt{-18}$

16. $6 - \sqrt{-98}$

In Exercises 17–32, add or subtract as indicated. Write the result in the form a + bi.

17. $(3 + 2i) + (5 + i)$

18. $(6 + 5i) + (4 + 3i)$

19. $(7 + 2i) + (1 - 4i)$

20. $(-2 + 6i) + (4 - i)$

21. $(10 + 7i) - (5 + 4i)$

22. $(11 + 8i) - (2 + 5i)$

23. $(9 - 4i) - (10 + 3i)$

24. $(8 - 5i) - (6 + 2i)$

25. $(3 + 2i) - (5 - 7i)$

26. $(-7 + 5i) - (9 - 11i)$

27. $(-5 + 4i) - (-13 - 11i)$

28. $(-9 + 2i) - (-17 - 6i)$

29. $8i - (14 - 9i)$

30. $15i - (12 - 11i)$

31. $(2 + \sqrt{3}i) + (7 + 4\sqrt{3}i)$

32. $(4 + \sqrt{5}i) + (8 + 6\sqrt{5}i)$

In Exercises 33–62, find each product. Write the result in the form a + bi.

33. $2i(5 + 3i)$

34. $5i(4 + 7i)$

35. $3i(7i - 5)$

36. $8i(4i - 3)$

37. $-7i(2 - 5i)$

38. $-6i(3 - 5i)$

39. $(3 + i)(4 + 5i)$

40. $(4 + i)(5 + 6i)$

41. $(7 - 5i)(2 - 3i)$

42. $(8 - 4i)(3 - 2i)$

43. $(6 - 3i)(-2 + 5i)$

44. $(7 - 2i)(-3 + 6i)$

45. $(3 + 5i)(3 - 5i)$

46. $(2 + 7i)(2 - 7i)$

47. $(-5 + 3i)(-5 - 3i)$

48. $(-4 + 2i)(-4 - 2i)$

49. $(3 - \sqrt{2}i)(3 + \sqrt{2}i)$

50. $(5 - \sqrt{3}i)(5 + \sqrt{3}i)$

51. $(2 + 3i)^2$

52. $(3 + 2i)^2$

53. $(5 - 2i)^2$

54. $(5 - 3i)^2$

55. $\sqrt{-7} \cdot \sqrt{-2}$

56. $\sqrt{-7} \cdot \sqrt{-3}$

57. $\sqrt{-9} \cdot \sqrt{-4}$

58. $\sqrt{-16} \cdot \sqrt{-4}$

59. $\sqrt{-7} \cdot \sqrt{-25}$

60. $\sqrt{-3} \cdot \sqrt{-36}$

61. $\sqrt{-8} \cdot \sqrt{-3}$

62. $\sqrt{-9} \cdot \sqrt{-5}$

In Exercises 63–84, divide and simplify to the form a + bi.

63. $\dfrac{2}{3 + i}$

64. $\dfrac{3}{4 + i}$

65. $\dfrac{2i}{1 + i}$

66. $\dfrac{5i}{2 + i}$

67. $\dfrac{7}{4 - 3i}$

68. $\dfrac{9}{1 - 2i}$

69. $\dfrac{6i}{3-2i}$

70. $\dfrac{5i}{2-3i}$

71. $\dfrac{1+i}{1-i}$

72. $\dfrac{1-i}{1+i}$

73. $\dfrac{2-3i}{3+i}$

74. $\dfrac{2+3i}{3-i}$

75. $\dfrac{5-2i}{3+2i}$

76. $\dfrac{6-3i}{4+2i}$

77. $\dfrac{4+5i}{3-7i}$

78. $\dfrac{5-i}{3-2i}$

79. $\dfrac{7}{3i}$

80. $\dfrac{5}{7i}$

81. $\dfrac{8-5i}{2i}$

82. $\dfrac{3+4i}{5i}$

83. $\dfrac{4+7i}{-3i}$

84. $\dfrac{5+i}{-4i}$

In Exercises 85–100, simplify each expression.

85. i^{10}

86. i^{14}

87. i^{11}

88. i^{15}

89. i^{22}

90. i^{46}

91. i^{200}

92. i^{400}

93. i^{17}

94. i^{21}

95. $(-i)^4$

96. $(-i)^6$

97. $(-i)^9$

98. $(-i)^{13}$

99. $i^{24}+i^2$

100. $i^{28}+i^{30}$

Application Exercises

Complex numbers are used in electronics to describe the current in an electric circuit. Ohm's law relates the current in a circuit, I, in amperes, the voltage of the circuit, E, in volts, and the resistance of the circuit, R, in ohms, by the formula $E = IR$. Use this formula to solve Exercises 101–102.

101. Find E, the voltage of a circuit, if $I = (4-5i)$ amperes and $R = (3+7i)$ ohms.

102. Find E, the voltage of a circuit, if $I = (2-3i)$ amperes and $R = (3+5i)$ ohms.

103. The mathematician Girolamo Cardano is credited with the first use (in 1545) of negative square roots in solving the now-famous problem, "Find two numbers whose sum is 10 and whose product is 40." Show that the complex numbers $5 + \sqrt{15}i$ and $5 - \sqrt{15}i$ satisfy the conditions of the problem. (Cardano did not use the symbolism $\sqrt{15}i$ or even $\sqrt{-15}$. He wrote R.m 15 for $\sqrt{-15}$, meaning "radix minus 15." He regarded the numbers $5 + $ R.m 15 and $5 - $ R.m 15 as "fictitious" or "ghost numbers," and considered the problem "manifestly impossible." But in a mathematically adventurous spirit, he exclaimed, "Nevertheless, we will operate.")

Writing in Mathematics

104. What is the imaginary unit i?

105. Explain how to write $\sqrt{-64}$ as a multiple of i.

106. What is a complex number? Explain when a complex number is a real number and when it is an imaginary number. Provide examples with your explanation.

107. Explain how to add complex numbers. Give an example.

108. Explain how to subtract complex numbers. Give an example.

109. Explain how to find the product of $2i$ and $5 + 3i$.

110. Explain how to find the product of $2i + 3$ and $5 + 3i$.

111. Explain how to find the product of $2i + 3$ and $2i - 3$.

112. Explain how to find the product of $\sqrt{-1}$ and $\sqrt{-4}$. Describe a common error in the multiplication that needs to be avoided.

113. What is the conjugate of $2 + 3i$? What happens when you multiply this complex number by its conjugate?

114. Explain how to divide complex numbers. Provide an example with your explanation.

115. Explain each of the three jokes in the cartoon on page 703.

116. A stand-up comedian uses algebra in some jokes, including one about a telephone recording that announces "You have just reached an imaginary number. Please multiply by i and dial again." Explain the joke.

Explain the error in Exercises 117–118.

117. $\sqrt{-9} + \sqrt{-16} = \sqrt{-25} = \sqrt{25}i = 5i$

118. $\left(\sqrt{-9}\right)^2 = \sqrt{-9}\cdot\sqrt{-9} = \sqrt{81} = 9$

Critical Thinking Exercises

119. Which one of the following is true?

  a. Some irrational numbers are not complex numbers.

  b. $(3 + 7i)(3 - 7i)$ is an imaginary number.

  c. $\dfrac{7 + 3i}{5 + 3i} = \dfrac{7}{5}$

  d. In the complex number system, $x^2 + y^2$ (the sum of two squares) can be factored as $(x + yi)(x - yi)$.

In Exercises 120–122, perform the indicated operations and write the result in the form a + bi.

120. $(8 + 9i)(2 - i) - (1 - i)(1 + i)$

121. $\dfrac{4}{(2 + i)(3 - i)}$

122. $\dfrac{1 + i}{1 + 2i} + \dfrac{1 - i}{1 - 2i}$

123. Evaluate $x^2 - 2x + 2$ for $x = 1 + i$.

Review Exercises

124. Solve: $2x - 1 = x^2 - 4x + 4$.
(Section 6.6, Example 4)

125. Simplify: $\left(2x^2\right)^{-3}$. (Section 5.7, Example 6)

126. Multiply: $\dfrac{x^2 - 6x + 9}{12} \cdot \dfrac{3}{x^2 - 9}$. (Section 7.2, Example 3)

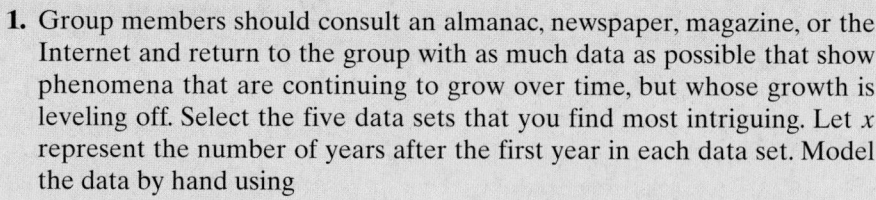

CHAPTER 10 GROUP PROJECTS

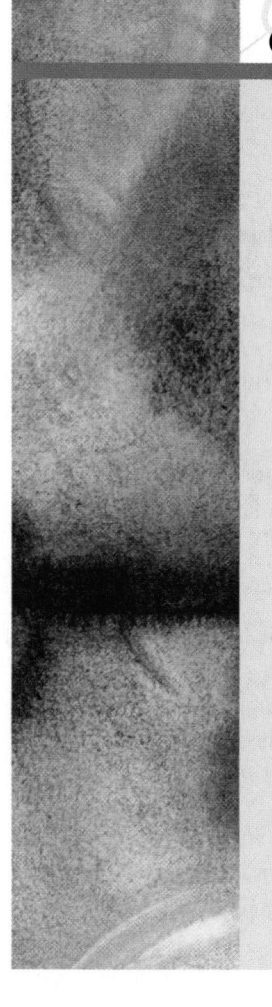

1. Group members should consult an almanac, newspaper, magazine, or the Internet and return to the group with as much data as possible that show phenomena that are continuing to grow over time, but whose growth is leveling off. Select the five data sets that you find most intriguing. Let x represent the number of years after the first year in each data set. Model the data by hand using

$$f(x) = a\sqrt{x} + b.$$

Use the first and last data points to find values for a and b. The first data point corresponds to $x = 0$. Its second coordinate gives the value of b. To find a, substitute the second data point into $f(x) = a\sqrt{x} + b$, with the value that you obtained for b. Now solve the equation and find a. Substitute a and b into $f(x) = a\sqrt{x} + b$ to obtain a square root function that models each data set. Then use the function to make predictions about what might occur in the future. Are there circumstances that might affect the accuracy of the predictions? List some of these circumstances.

2. Group members should prepare and present a seminar on chaos. Include one or more of the following topics in your presentation: fractal images, the role of complex numbers in generating fractal images, algorithms, iterations, iteration number, and fractals in nature. Be sure to include visual images that will intrigue your audience.

3. The math lab has asked the group to prepare a pamphlet for students having difficulties with the 30-item Chapter 10 test on pages 718–719. The pamphlet's title should be "How To Ace the Test on Radicals, Radical Functions, and Rational Exponents." Offer guidelines for avoiding errors, as well as hints and suggestions for succeeding with each of the test items.

CHAPTER SUMMARY, REVIEW, AND TEST

SUMMARY

DEFINITIONS AND CONCEPTS	EXAMPLES

Section 10.1 Radical Expressions and Functions

If $b^2 = a$, then b is a square root of a. The principal square root of a, designated $\sqrt{a}$, is the nonnegative number satisfying $b^2 = a$. The negative square root of a is written $-\sqrt{a}$. A square root of a negative number is not a real number. A radical function in x is a function defined by an expression containing a root of x. The domain of a square root function is the set of real numbers for which the radicand is nonnegative.	Let $f(x) = \sqrt{6 - 2x}$. $f(-15) = \sqrt{6 - 2(-15)} = \sqrt{6 + 30} = \sqrt{36} = 6$ $f(5) = \sqrt{6 - 2 \cdot 5} = \sqrt{6 - 10} = \sqrt{-4}$, not a real number Domain of f: Set the radicand greater than or equal to zero. $6 - 2x \geq 0$ $-2x \geq -6$ $x \leq 3$ Domain of $f = \{x \mid x \leq 3\}$ or $(-\infty, 3]$.
The cube root of a real number a is written $\sqrt[3]{a}$. $\quad \sqrt[3]{a} = b \quad$ means that $\quad b^3 = a$. The nth root of a real number a is written $\sqrt[n]{a}$. The number n is the index. Every real number has one root when n is odd. The odd nth root of a, $\sqrt[n]{a}$, is the number b for which $b^n = a$. Every positive real number has two real roots when n is even. An even root of a negative number is not a real number. If n is even, then $\sqrt[n]{a^n} = \lvert a \rvert$. If n is odd, then $\sqrt[n]{a^n} = a$.	• $\sqrt[3]{-8} = -2$ because $(-2)^3 = -8$. • $\sqrt[4]{-16}$ is not a real number. • $\sqrt{x^2 - 14x + 49} = \sqrt{(x-7)^2} = \lvert x - 7 \rvert$ • $\sqrt[3]{125(x+6)^3} = 5(x + 6)$

Section 10.2 Rational Exponents

• $a^{\frac{1}{n}} = \sqrt[n]{a}$ • $a^{\frac{m}{n}} = \left(\sqrt[n]{a} \right)^m$ or $\sqrt[n]{a^m}$ • $a^{-\frac{m}{n}} = \dfrac{1}{a^{\frac{m}{n}}}$	• $16^{\frac{1}{2}} = \sqrt{16} = 4$ • $27^{\frac{1}{3}} = \sqrt[3]{27} = 3$ • $8^{\frac{5}{3}} = \left(\sqrt[3]{8} \right)^5 = 2^5 = 32$ • $81^{-\frac{3}{4}} = \dfrac{1}{81^{\frac{3}{4}}} = \dfrac{1}{\left(\sqrt[4]{81} \right)^3} = \dfrac{1}{3^3} = \dfrac{1}{27}$ • $\left(\sqrt[3]{3xy} \right)^4 = (3xy)^{\frac{4}{3}}$
Properties of integer exponents are true for rational exponents. An expression with rational exponents is simplified when no parentheses appear, no powers are raised to powers, each base occurs once, and no negative or zero exponents appear.	Simplify: $\left(8x^{\frac{1}{3}} y^{-\frac{1}{2}} \right)^{\frac{1}{3}}$. $= 8^{\frac{1}{3}} \left(x^{\frac{1}{3}} \right)^{\frac{1}{3}} \left(y^{-\frac{1}{2}} \right)^{\frac{1}{3}}$ $= 2x^{\frac{1}{9}} y^{-\frac{1}{6}} = \dfrac{2x^{\frac{1}{9}}}{y^{\frac{1}{6}}}$

DEFINITIONS AND CONCEPTS	EXAMPLES

Section 10.2 Rational Exponents (continued)

Some radical expressions can be simplifed using rational exponents. Rewrite the expression using rational exponents, simplify, and rewrite in radical notation if rational exponents still appear.

- $\sqrt[9]{x^3} = x^{\frac{3}{9}} = x^{\frac{1}{3}} = \sqrt[3]{x}$

- $\sqrt[5]{x^2} \cdot \sqrt[4]{x} = x^{\frac{2}{5}} \cdot x^{\frac{1}{4}} = x^{\frac{2}{5}+\frac{1}{4}}$

  $= x^{\frac{8}{20}+\frac{5}{20}} = x^{\frac{13}{20}} = \sqrt[20]{x^{13}}$

Section 10.3 Multiplying and Simplifying Radical Expressions

The product rule for radicals can be used to multiply radicals with the same indices:

$$\sqrt[n]{a} \cdot \sqrt[n]{b} = \sqrt[n]{ab}.$$

$\sqrt[3]{7x} \cdot \sqrt[3]{10y^2} = \sqrt[3]{7x \cdot 10y^2} = \sqrt[3]{70xy^2}$

The product rule for radicals can be used to simplify radicals:

$$\sqrt[n]{ab} = \sqrt[n]{a} \cdot \sqrt[n]{b}.$$

A radical expression with index n is simplified when its radicand has no factors that are perfect nth powers. To simplify, write the radicand as the product of two factors, one of which is the largest perfect nth power. Then use the product rule to take the nth root of each factor. If all variables in a radicand are positive, then

$$\sqrt[n]{a^n} = a.$$

Some radicals can be simplified after the multiplication is performed.

- Simplify: $\sqrt[3]{54x^7y^{11}}$.

  $= \sqrt[3]{27 \cdot 2 \cdot x^6 \cdot x \cdot y^9 \cdot y^2}$

  $= \sqrt[3]{(27x^6y^9)(2xy^2)}$

  $= \sqrt[3]{27x^6y^9} \cdot \sqrt[3]{2xy^2} = 3x^2y^3\sqrt[3]{2xy^2}$

- Assuming positive variables, multiply and simplify: $\sqrt[4]{4x^2y} \cdot \sqrt[4]{4xy^3}$.

  $= \sqrt[4]{4x^2y \cdot 4xy^3} = \sqrt[4]{16x^3y^4}$

  $= \sqrt[4]{16y^4} \cdot \sqrt[4]{x^3} = 2y\sqrt[4]{x^3}$

Section 10.4 Adding, Subtracting, and Dividing Radical Expressions

Like radicals have the same indices and radicands. Like radicals can be added or subtracted using the distributive property. In some cases, radicals can be combined once they have been simplified.

$4\sqrt{18} - 6\sqrt{50}$

$= 4\sqrt{9 \cdot 2} - 6\sqrt{25 \cdot 2} = 4 \cdot 3\sqrt{2} - 6 \cdot 5\sqrt{2}$

$= 12\sqrt{2} - 30\sqrt{2} = -18\sqrt{2}$

The quotient rule for radicals can be used to simplify radicals:

$$\sqrt[n]{\frac{a}{b}} = \frac{\sqrt[n]{a}}{\sqrt[n]{b}}.$$

$\sqrt[3]{-\frac{8}{x^{12}}} = \frac{\sqrt[3]{-8}}{\sqrt[3]{x^{12}}} = -\frac{2}{x^4}$

$\sqrt[3]{x^{12}} = (x^{12})^{\frac{1}{3}} = x^4$

The quotient rule for radicals can be used to divide radicals with the same indices:

$$\frac{\sqrt[n]{a}}{\sqrt[n]{b}} = \sqrt[n]{\frac{a}{b}}.$$

Some radicals can be simplified after the division is performed.

Assuming a positive variable, divide and simplify:

$\dfrac{\sqrt[4]{64x^5}}{\sqrt[4]{2x^{-2}}} = \sqrt[4]{32x^{5-(-2)}} = \sqrt[4]{32x^7}$

$= \sqrt[4]{16 \cdot 2 \cdot x^4 \cdot x^3} = \sqrt[4]{16x^4} \cdot \sqrt[4]{2x^3}$

$= 2x\sqrt[4]{2x^3}.$

DEFINITIONS AND CONCEPTS	EXAMPLES

Section 10.5 Multiplying with More Than One Term and Rationalizing Denominators

Radical expressions with more than one term are multiplied in much the same way that polynomials with more than one term are multiplied.

- $\sqrt{5}(2\sqrt{6} - \sqrt{3}) = 2\sqrt{30} - \sqrt{15}$
- $(4\sqrt{3} - 2\sqrt{2})(\sqrt{3} + \sqrt{2})$

$$= 4\sqrt{3} \cdot \sqrt{3} + 4\sqrt{3} \cdot \sqrt{2} - 2\sqrt{2} \cdot \sqrt{3} - 2\sqrt{2} \cdot \sqrt{2}$$
$$= 4 \cdot 3 + 4\sqrt{6} - 2\sqrt{6} - 2 \cdot 2$$
$$= 12 + 4\sqrt{6} - 2\sqrt{6} - 4 = 8 + 2\sqrt{6}$$

Radical expressions that involve the sum and difference of the same two terms are called conjugates. Use
$$(A + B)(A - B) = A^2 - B^2$$
to multiply conjugates.

$$(8 + 2\sqrt{5})(8 - 2\sqrt{5})$$
$$= 8^2 - (2\sqrt{5})^2 = 64 - 4 \cdot 5$$
$$= 64 - 20 = 44$$

The process of rewriting a radical expression as an equivalent expression without a radical in the denominator is called rationalizing the denominator. When the denominator contains a single radical with an nth root, multiply the numerator and denominator by a radical of index n that produces a perfect nth power in the denominator's radicand.

Rationalize the denominator: $\dfrac{7}{\sqrt[3]{2x}}$.

$$= \frac{7}{\sqrt[3]{2x}} \cdot \frac{\sqrt[3]{4x^2}}{\sqrt[3]{4x^2}} = \frac{7\sqrt[3]{4x^2}}{\sqrt[3]{8x^3}} = \frac{7\sqrt[3]{4x^2}}{2x}$$

If the denominator contains two terms, rationalize the denominator by multiplying the numerator and denominator by the conjugate of the denominator.

$$\frac{9}{7 - \sqrt{5}} = \frac{9}{7 - \sqrt{5}} \cdot \frac{7 + \sqrt{5}}{7 + \sqrt{5}}$$
$$= \frac{9(7 + \sqrt{5})}{7^2 - (\sqrt{5})^2}$$
$$= \frac{9(7 + \sqrt{5})}{49 - 5} = \frac{9(7 + \sqrt{5})}{44}$$

Section 10.6 Radical Equations

A radical equation is an equation in which the variable occurs in a radicand.

Solving Radical Equations Containing nth Roots

1. Isolate the radical.
2. Raise both sides to the nth power.
3. Solve the resulting equation.
4. Check proposed solutions in the original equation. Solutions of an equation to an even power that is radical-free, but not the original equation, are called extraneous solutions.

Solve: $\sqrt{2x + 1} - x = -7$.

$\sqrt{2x + 1} = x - 7$ Isolate the radical.
$(\sqrt{2x + 1})^2 = (x - 7)^2$ Square both sides.
$2x + 1 = x^2 - 14x + 49$
$0 = x^2 - 16x + 48$ Subtract $2x + 1$ from both sides.
$0 = (x - 12)(x - 4)$ Factor.
$x - 12 = 0$ or $x - 4 = 0$
$x = 12$ $x = 4$

Check both proposed solutions. 12 checks, but 4 is extraneous. The solution is 12 and the solution set is $\{12\}$.

DEFINITIONS AND CONCEPTS	EXAMPLES
Section 10.7 Complex Numbers	

The imaginary unit i is defined as $$i = \sqrt{-1}, \text{ where } i^2 = -1.$$ The set of numbers in the form $a + bi$ is called the set of complex numbers; a is the real part and b is the imaginary part. If $b = 0$, the complex number is a real number. If $b \neq 0$, the complex number is an imaginary number.	• $\sqrt{-36} = \sqrt{36(-1)} = \sqrt{36}\sqrt{-1} = 6i$ • $\sqrt{-50} = \sqrt{50(-1)} = \sqrt{25 \cdot 2}\sqrt{-1} = 5\sqrt{2}i$
To add or subtract complex numbers, add or subtract their real parts and add or subtract their imaginary parts.	$(2 - 4i) - (7 - 10i)$ $= 2 - 4i - 7 + 10i$ $= (2 - 7) + (-4 + 10)i = -5 + 6i$
To multiply complex numbers, multiply as if they are polynomials. After completing the multiplication, replace i^2 with -1. When performing operations with square roots of negative numbers, begin by expressing all square roots in terms of i. Then multiply.	F O I L • $(2 - 3i)(4 + 5i) = 8 + 10i - 12i - 15i^2$ $= 8 + 10i - 12i - 15(-1)$ $= 23 - 2i$ • $\sqrt{-36} \cdot \sqrt{-100} = \sqrt{36(-1)} \cdot \sqrt{100(-1)}$ $= 6i \cdot 10i = 60i^2 = 60(-1) = -60$
The complex numbers $a + bi$ and $a - bi$ are conjugates. Conjugates can be multiplied using the formula $$(A + B)(A - B) = A^2 - B^2.$$ The multiplication of conjugates results in a real number.	$(3 + 5i)(3 - 5i)$ $= 3^2 - (5i)^2$ $= 9 - 25i^2$ $= 9 - 25(-1) = 34$
To divide complex numbers, multiply the numerator and the denominator by the conjugate of the denominator.	$\dfrac{5 + 2i}{4 - i} = \dfrac{5 + 2i}{4 - i} \cdot \dfrac{4 + i}{4 + i} = \dfrac{20 + 5i + 8i + 2i^2}{16 - i^2}$ $= \dfrac{20 + 13i + 2(-1)}{16 - (-1)}$ $= \dfrac{20 + 13i - 2}{16 + 1}$ $= \dfrac{18 + 13i}{17} = \dfrac{18}{17} + \dfrac{13}{17}i$
To simplify powers of i, rewrite the expression in terms of i^2. Then replace i^2 by -1 and simplify.	Simplify: i^{27}. $i^{27} = i^{26} \cdot i = (i^2)^{13}i$ $= (-1)^{13}i = (-1)i = -i$

Review Exercises

10.1 *In Exercises 1–5, find the indicated root, or state that the expression is not a real number.*

1. $\sqrt{81}$

2. $-\sqrt{\dfrac{1}{100}}$

3. $\sqrt[3]{-27}$

4. $\sqrt[4]{-16}$

5. $\sqrt[5]{-32}$

In Exercises 6–7, find the indicated function values for each function. If necessary, round to two decimal places. If the function value is not a real number and does not exist, so state.

6. $f(x) = \sqrt{2x - 5}$; $f(15), f(4), f\left(\frac{5}{2}\right), f(1)$

7. $g(x) = \sqrt[3]{4x - 8}$; $g(4), g(0), g(-14)$

In Exercises 8–9, find the domain of each square root function.

8. $f(x) = \sqrt{x - 2}$

9. $g(x) = \sqrt{100 - 4x}$

In Exercises 10–15, simplify each expression. Assume that each variable can represent any real number, so include absolute value bars where necessary.

10. $\sqrt{25x^2}$

11. $\sqrt{(x + 14)^2}$

12. $\sqrt{x^2 - 8x + 16}$

13. $\sqrt[3]{64x^3}$

14. $\sqrt[4]{16x^4}$

15. $\sqrt[5]{-32(x + 7)^5}$

10.2 In Exercises 16–18, use radical notation to rewrite each expression. Simplify, if possible.

16. $(5xy)^{\frac{1}{3}}$

17. $16^{\frac{3}{2}}$

18. $32^{\frac{4}{5}}$

In Exercises 19–20, rewrite each expression with rational exponents.

19. $\sqrt{7x}$

20. $\left(\sqrt[3]{19xy}\right)^5$

In Exercises 21–22, rewrite each expression with a positive rational exponent. Simplify, if possible.

21. $8^{-\frac{2}{3}}$

22. $3x(ab)^{-\frac{4}{5}}$

In Exercises 23–26, use properties of rational exponents to simplify each expression.

23. $x^{\frac{1}{3}} \cdot x^{\frac{1}{4}}$

24. $\dfrac{5^{\frac{1}{2}}}{5^{\frac{1}{3}}}$

25. $\left(8x^6y^3\right)^{\frac{1}{3}}$

26. $\left(x^{-\frac{2}{3}}y^{\frac{1}{4}}\right)^{\frac{1}{2}}$

In Exercises 27–31, use rational exponents to simplify each expression. If rational exponents appear after simplifying, write the answer in radical notation.

27. $\sqrt[3]{x^9y^{12}}$

28. $\sqrt[9]{x^3y^9}$

29. $\sqrt{x} \cdot \sqrt[3]{x}$

30. $\dfrac{\sqrt[3]{x^2}}{\sqrt[4]{x^2}}$

31. $\sqrt[5]{\sqrt[3]{x}}$

32. The function $f(x) = 350x^{\frac{2}{3}}$ models the expenditures, $f(x)$, in millions of dollars, for the U.S. National Park Service x years after 1985. According to this model, what will expenditures be in 2012?

10.3 In Exercises 33–35, use the product rule to multiply.

33. $\sqrt{3x} \cdot \sqrt{7y}$

34. $\sqrt[5]{7x^2} \cdot \sqrt[5]{11x}$

35. $\sqrt[6]{x - 5} \cdot \sqrt[6]{(x - 5)^4}$

36. If $f(x) = \sqrt{7x^2 - 14x + 7}$, express the function, f, in simplified form. Assume that x can be any real number.

In Exercises 37–39, simplify by factoring. Assume that all variables in a radicand represent positive real numbers.

37. $\sqrt{20x^3}$

38. $\sqrt[3]{54x^8y^6}$

39. $\sqrt[4]{32x^3y^{11}}$

In Exercises 40–43, multiply and simplify, if possible. Assume that all variables in a radicand represent positive real numbers.

40. $\sqrt{6x^3} \cdot \sqrt{4x^2}$

41. $\sqrt[3]{4x^2y} \cdot \sqrt[3]{4xy^4}$

42. $\sqrt[5]{2x^4y^3} \cdot \sqrt[5]{8xy^6}$

43. $\sqrt{x + 1} \cdot \sqrt{x - 1}$

10.4 Assume that all variables represent positive real numbers.

In Exercises 44–47, add or subtract as indicated.

44. $6\sqrt[3]{3} + 2\sqrt[3]{3}$

45. $5\sqrt{18} - 3\sqrt{8}$

46. $\sqrt[3]{27x^4} + \sqrt[3]{xy^6}$

47. $2\sqrt[3]{6} - 5\sqrt[3]{48}$

In Exercises 48–50, simplify using the quotient rule.

48. $\sqrt[3]{\dfrac{16}{125}}$

49. $\sqrt{\dfrac{x^3}{100y^4}}$

50. $\sqrt[4]{\dfrac{3y^5}{16x^{20}}}$

In Exercises 51–54, divide and, if possible, simplify.

51. $\dfrac{\sqrt{48}}{\sqrt{2}}$

52. $\dfrac{\sqrt[3]{32}}{\sqrt[3]{2}}$

53. $\dfrac{\sqrt[4]{64x^7}}{\sqrt[4]{2x^2}}$

54. $\dfrac{\sqrt{200x^3y^2}}{\sqrt{2x^{-2}y}}$

10.5 Assume that all variables represent positive real numbers.

In Exercises 55–62, multiply as indicated. If possible, simplify any radical expressions that appear in the product.

55. $\sqrt{3}(2\sqrt{6} + 4\sqrt{15})$

56. $\sqrt[3]{5}(\sqrt[3]{50} - \sqrt[3]{2})$

57. $\left(\sqrt{7} - 3\sqrt{5}\right)\left(\sqrt{7} + 6\sqrt{5}\right)$

58. $\left(\sqrt{x} - \sqrt{11}\right)\left(\sqrt{y} - \sqrt{11}\right)$

59. $\left(\sqrt{5} + \sqrt{8}\right)^2$

60. $\left(2\sqrt{3} - \sqrt{10}\right)^2$

61. $\left(\sqrt{7} + \sqrt{13}\right)\left(\sqrt{7} - \sqrt{13}\right)$

62. $\left(7 - 3\sqrt{5}\right)\left(7 + 3\sqrt{5}\right)$

In Exercises 63–75, rationalize each denominator. Simplify, if possible.

63. $\dfrac{4}{\sqrt{6}}$

64. $\sqrt{\dfrac{2}{7}}$

65. $\dfrac{12}{\sqrt[3]{9}}$

66. $\sqrt{\dfrac{2x}{5y}}$

67. $\dfrac{14}{\sqrt[3]{2x^2}}$

68. $\sqrt[4]{\dfrac{7}{3x}}$

69. $\dfrac{5}{\sqrt[5]{32x^4y}}$

70. $\dfrac{6}{\sqrt{3} - 1}$

71. $\dfrac{\sqrt{7}}{\sqrt{5} + \sqrt{3}}$

72. $\dfrac{10}{2\sqrt{5} - 3\sqrt{2}}$

73. $\dfrac{\sqrt{x} + 5}{\sqrt{x} - 3}$

74. $\dfrac{\sqrt{7} + \sqrt{3}}{\sqrt{7} - \sqrt{3}}$

75. $\dfrac{2\sqrt{3} + \sqrt{6}}{2\sqrt{6} + \sqrt{3}}$

In Exercises 76–79, rationalize each numerator. Simplify, if possible.

76. $\sqrt{\dfrac{2}{7}}$

77. $\dfrac{\sqrt[3]{3x}}{\sqrt[3]{y}}$

78. $\dfrac{\sqrt{7}}{\sqrt{5} + \sqrt{3}}$

79. $\dfrac{\sqrt{7} + \sqrt{3}}{\sqrt{7} - \sqrt{3}}$

10.6 *In Exercises 80–84, solve each radical equation.*

80. $\sqrt{2x + 4} = 6$

81. $\sqrt{x - 5} + 9 = 4$

82. $\sqrt{2x - 3} + x = 3$

83. $\sqrt{x - 4} + \sqrt{x + 1} = 5$

84. $(x^2 + 6x)^{\frac{1}{3}} + 2 = 0$

85. The time, $f(x)$, in seconds, for a free-falling object to fall x feet is modeled by the function

$$f(x) = \sqrt{\dfrac{x}{16}}.$$

If a worker accidently drops a hammer from a building and it hits the ground after 4 seconds, from what height was the hammer dropped?

86. Out of a group of 50,000 births, the number of people, $f(x)$, surviving to age x is modeled by the function

$$f(x) = 5000\sqrt{100 - x}.$$

To what age will 20,000 people in the group survive?

10.7 *In Exercises 87–89, express each number in terms of i and simplify, if possible.*

87. $\sqrt{-81}$

88. $\sqrt{-63}$

89. $-\sqrt{-8}$

In Exercises 90–99, perform the indicated operation. Write the result in the form a + bi.

90. $(7 + 12i) + (5 - 10i)$

91. $(8 - 3i) - (17 - 7i)$

92. $4i(3i - 2)$

93. $(7 - 5i)(2 + 3i)$

94. $(3 - 4i)^2$

95. $(7 + 8i)(7 - 8i)$

96. $\sqrt{-8} \cdot \sqrt{-3}$

97. $\dfrac{6}{5 + i}$

98. $\dfrac{3 + 4i}{4 - 2i}$

99. $\dfrac{5 + i}{3i}$

In Exercises 100–101, simplify each expression.

100. i^{16}

101. i^{23}

Chapter 10 Test

1. Let $f(x) = \sqrt{8 - 2x}$.

   a. Find $f(-14)$.

   b. Find the domain of f.

2. Evaluate: $27^{-\frac{4}{3}}$.

3. Simplify: $\left(25x^{-\frac{1}{2}}y^4\right)^{\frac{1}{2}}$.

In Exercises 4–5, use rational exponents to simplify each expression. If rational exponents appear after simplifying, write the answer in radical notation.

4. $\sqrt[8]{x^4}$

5. $\sqrt[4]{x} \cdot \sqrt[5]{x}$

In Exercises 6–9, simplify each expression. Assume that each variable can represent any real number.

6. $\sqrt{75x^2}$

7. $\sqrt{x^2 - 10x + 25}$

8. $\sqrt[3]{16x^4y^8}$

9. $\sqrt[5]{-\dfrac{32}{x^{10}}}$

In Exercises 10–17, perform the indicated operation and, if possible, simplify. Assume that all variables represent positive real numbers.

10. $\sqrt[3]{5x^2} \cdot \sqrt[3]{10y}$

11. $\sqrt[4]{8x^3y} \cdot \sqrt[4]{4xy^2}$

12. $3\sqrt{18} - 4\sqrt{32}$

13. $\sqrt[3]{8x^4} + \sqrt[3]{xy^6}$

14. $\dfrac{\sqrt[3]{16x^8}}{\sqrt[3]{2x^4}}$

15. $\sqrt{3}(4\sqrt{6} - \sqrt{5})$

16. $(5\sqrt{6} - 2\sqrt{2})(\sqrt{6} + \sqrt{2})$

17. $(7 - \sqrt{3})^2$

In Exercises 18–20, rationalize each denominator. Simplify, if possible. Assume all variables represent positive real numbers.

18. $\sqrt{\dfrac{5}{x}}$

19. $\dfrac{5}{\sqrt[3]{5x^2}}$

20. $\dfrac{\sqrt{2} - \sqrt{3}}{\sqrt{2} + \sqrt{3}}$

In Exercises 21–23, solve each radical equation.

21. $3 + \sqrt{2x - 3} = x$

22. $\sqrt{x + 9} - \sqrt{x - 7} = 2$

23. $(11x + 6)^{\frac{1}{3}} + 3 = 0$

24. The function

$$f(x) = 2.9\sqrt{x} + 20.1$$

models the average height, in inches, of boys who are x months of age, $0 \le x \le 60$. Find the age at which the average height of boys is 40.4 inches.

25. Express in terms of i and simplify: $\sqrt{-75}$.

In Exercises 26–29, perform the indicated operation. Write the result in the form $a + bi$.

26. $(5 - 3i) - (6 - 9i)$

27. $(3 - 4i)(2 + 5i)$

28. $\sqrt{-9} \cdot \sqrt{-4}$

29. $\dfrac{3 + i}{1 - 2i}$

30. Simplify: i^{35}.

Cumulative Review Exercises (Chapters 1–10)

In Exercises 1–5, solve each equation, inequality, or system.

1. $\begin{aligned} 2x - y + z &= -5 \\ x - 2y - 3z &= 6 \\ x + y - 2z &= 1 \end{aligned}$

2. $3x^2 - 11x = 4$

3. $2(x + 4) < 5x + 3(x + 2)$

4. $\dfrac{1}{x + 2} + \dfrac{15}{x^2 - 4} = \dfrac{5}{x - 2}$

5. $\sqrt{x + 2} - \sqrt{x + 1} = 1$

6. Graph the solution set of the system:

$$\begin{aligned} x + 2y &< 2 \\ 2y - x &> 4. \end{aligned}$$

In Exercises 7–15, perform the indicated operations.

7. $\dfrac{8x^2}{3x^2 - 12} \div \dfrac{40}{x - 2}$

8. $\dfrac{x + \dfrac{1}{y}}{y + \dfrac{1}{x}}$

9. $(2x - 3)(4x^2 - 5x - 2)$

10. $\dfrac{7x}{x^2 - 2x - 15} - \dfrac{2}{x - 5}$

11. $7(8 - 10)^3 - 7 + 3 \div (-3)$

12. $\sqrt{80x} - 5\sqrt{20x} + 2\sqrt{45x}$

13. $\dfrac{\sqrt{3} - 2}{2\sqrt{3} + 5}$

14. $(2x^3 - 3x^2 + 3x - 4) \div (x - 2)$

15. $(2\sqrt{3} + 5\sqrt{2})(\sqrt{3} - 4\sqrt{2})$

In Exercises 16–17, factor completely.

16. $24x^2 + 10x - 4$

17. $16x^4 - 1$

18. The amount of light provided by a light bulb varies inversely as the square of the distance from the bulb. The illumination provided is 120 lumens at a distance of 10 feet. How many lumens are provided at a distance of 15 feet?

19. You invested $6000 in two accounts paying 7% and 9% annual interest, respectively. At the end of the year, the total interest from these investments was $510. How much was invested at each rate?

20. Although there are 2332 students enrolled in the college, this is 12% fewer students than there were last year. How many students were enrolled last year?

Thanks to the efficiency of U.S. farms, most Americans enjoy an abundance of inexpensive foods. Nearly one in five U.S. kids ages 9 through 13 say they've already been on a weight-loss diet. By contrast, worldwide, every three seconds a child dies of hunger. If all the world's undernourished people were gathered into a new country, it would be the third most populous nation, just behind China and India.

Source: The State of Food Insecurity in the World, 2000; 2000 BBC News

Because we tend to take our food abundance for granted, it might surprise you to know that the number of U.S. farms has declined since the 1920s, as individually owned family farms have been swallowed up by huge agribusinesses owned by corporations. Currently, there are about 1.8 acres of cropland to grow food for each American. If the current trends in population growth and loss of farmland continue, there will be only 0.6 acre per American in the year 2050. This is the rate that currently exists worldwide.

Like the number of farms and the amount of cropland in the United States, many phenomena follow trends that involve growth and decline, or vice versa. In this chapter, you will learn to solve equations and graph functions that provide new ways of looking at and understanding these phenomena.

Chapter 11

Quadratic Equations and Functions

▶ SECTION 11.1   *The Square Root Property and Completing the Square*

Objectives

1 Solve quadratic equations using the square root property.

2 Complete the square of a binomial.

3 Solve quadratic equations by completing the square.

4 Solve problems using the square root property.

SSM
PH Tutor CD- Video
Center ROM

Sandy Skoglund
"Radioactive
Cats" cibachrome,
30 × 40 inches.

One of the most famous formulas in the world is $E = mc^2$, formulated by Albert Einstein. Eistein showed that any form of energy has mass and that mass itself is a form of energy. In this formula, E represents energy, in ergs, m represents mass, in grams, and c represents the speed of light. Because light travels at 30 billion centimeters per second, the formula indicates that 1 gram of mass will produce 900 billion ergs of energy. The mass of a golf ball could provide the daily energy needs of a large metropolitan area.

Numerous problems stand in the way of nuclear power expansion, not the least of which is what to do with nuclear waste. So far, scientists have not found a safe way to dispose of the dangerous radioactive wastes produced by nuclear reactors. By 2000, more than 25,000 tons of waste had accumulated around the 104 nuclear power plants in the United States.

The formula $E = mc^2$ shows that mass and energy are equivalent, and the transformation of even a tiny amount of mass releases an enormous amount of energy. Is there a way to work with the formula $E = mc^2$ and solve for c, thereby expressing the speed of light, c, in terms of energy, E, and mass, m? In this section, by learning how this is done, you will develop two new ways for solving quadratic equations. These new methods are called the *square root property* and *completing the square*.

Study Tip

Here is a summary of what we already know about quadratic equations and quadratic functions.

1. A **quadratic equation** in x can be written in the standard form
$$ax^2 + bx + c = 0, \quad a \neq 0.$$

2. Some quadratic equations can be solved by factoring.

Solve: $2x^2 + 7x - 4 = 0.$
$$(2x - 1)(x + 4) = 0$$
$$2x - 1 = 0 \quad \text{or} \quad x + 4 = 0$$
$$2x = 1 \qquad\qquad x = -4$$
$$x = \tfrac{1}{2}$$

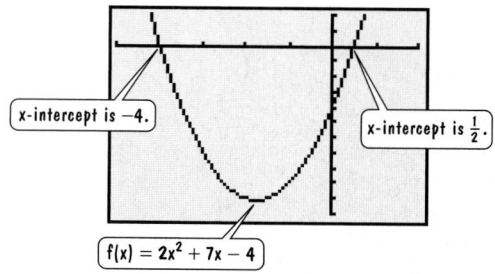

x-intercept is −4.

x-intercept is $\tfrac{1}{2}$.

$f(x) = 2x^2 + 7x - 4$

Figure 11.1

The solutions are -4 and $\tfrac{1}{2}$, and the solution set is $\left\{-4, \tfrac{1}{2}\right\}$.

3. A polynomial function of the form
$$f(x) = ax^2 + bx + c, \quad a \neq 0$$
is a **quadratic function**. Graphs of quadratic functions have cuplike shapes.

4. The real solutions of $ax^2 + bx + c = 0$ correspond to the x-intercepts for the graph of the quadratic function $f(x) = ax^2 + bx + c$. For example, the solutions of the equation $2x^2 + 7x - 4 = 0$ are -4 and $\tfrac{1}{2}$. Figure 11.1 shows that the solutions appear as x-intercepts on the graph of the quadratic function.

Now that we've summarized what we know, let's look at where we go. How do we solve a quadratic equation, $ax^2 + bx + c = 0$, if the trinomial $ax^2 + bx + c$ cannot be factored? Methods other than factoring are needed. In this section, we look at other ways of solving quadratic equations.

1 Solve quadratic equations using the square root property.

The Square Root Property

Let's begin with a relatively simple quadratic equation:

$$x^2 = 9.$$

The value of x must be a number whose square is 9. There are two numbers whose square is 9:

$$x = \sqrt{9} = 3 \quad \text{or} \quad x = -\sqrt{9} = -3.$$

Thus, the solutions of $x^2 = 9$ are 3 and −3. This is an example of the **square root property**.

Discover for Yourself

Solve $x^2 = 9$, or

$$x^2 - 9 = 0,$$

by factoring. What is the advantage of using the square root property?

The Square Root Property

If u is an algebraic expression and d is a nonzero real number, then $u^2 = d$ has exactly two solutions:

$$\text{If } u^2 = d, \quad \text{then } u = \sqrt{d} \text{ or } u = -\sqrt{d}.$$

Equivalently,

$$\text{If } u^2 = d, \quad \text{then } u = \pm\sqrt{d}.$$

Notice that $u = \pm\sqrt{d}$ is a shorthand notation to indicate that $u = \sqrt{d}$ or $u = -\sqrt{d}$. Although we usually read $u = \pm\sqrt{d}$ as "u equals plus or minus the square root of d," we actually mean that u is the positive square root of d or the negative square root of d.

EXAMPLE 1 **Solving a Quadratic Equation by the Square Root Property**

Solve: $4x^2 = 20$.

Solution To apply the square root property, we need a squared expression by itself on one side of the equation.

$$4x^2 = 20$$

We want x^2 by itself.

We can get x^2 by itself if we divide both sides by 4.

$$4x^2 = 20 \qquad \text{This is the original equation.}$$

$$\frac{4x^2}{4} = \frac{20}{4} \qquad \text{Divide both sides by 4.}$$

$$x^2 = 5 \qquad \text{Simplify.}$$

$$x = \sqrt{5} \quad \text{or} \quad x = -\sqrt{5} \qquad \text{Apply the square root property.}$$

Now let's check these proposed solutions in the original equation.

$$\textbf{Check } \sqrt{5}: \qquad\qquad \textbf{Check } -\sqrt{5}:$$

$$4x^2 = 20 \qquad\qquad\qquad 4x^2 = 20$$

$$4\left(\sqrt{5}\right)^2 \overset{?}{=} 20 \qquad\qquad 4\left(-\sqrt{5}\right)^2 \overset{?}{=} 20$$

$$4 \cdot 5 \overset{?}{=} 20 \qquad\qquad\qquad 4 \cdot 5 \overset{?}{=} 20$$

$$20 = 20, \text{ true} \qquad\qquad 20 = 20, \text{ true}$$

The solutions are $-\sqrt{5}$ and $\sqrt{5}$. The solution set is $\{-\sqrt{5}, \sqrt{5}\}$ or $\{\pm\sqrt{5}\}$. ∎

✔ **CHECK POINT 1** Solve: $5x^2 = 15$.

In this section, we will express irrational solutions in simplified radical form, rationalizing denominators when possible.

EXAMPLE 2 Solving a Quadratic Equation by the Square Root Property

Solve: $2x^2 - 7 = 0$.

Solution To solve by the square root property, we place the squared expression by itself on one side of the equation.

$$2x^2 - 7 = 0$$

 We want x^2 by itself.

$$2x^2 - 7 = 0 \qquad \text{This is the original equation.}$$

$$2x^2 = 7 \qquad \text{Add 7 to both sides.}$$

$$x^2 = \frac{7}{2} \qquad \text{Divide both sides by 2.}$$

$$x = \sqrt{\frac{7}{2}} \quad \text{or} \quad x = -\sqrt{\frac{7}{2}} \qquad \text{Apply the square root property.}$$

Because the proposed solutions are opposites, we can rationalize both denominators at once:

$$\pm\sqrt{\frac{7}{2}} = \pm\frac{\sqrt{7}}{\sqrt{2}} \cdot \frac{\sqrt{2}}{\sqrt{2}} = \pm\frac{\sqrt{14}}{2}.$$

Substitute these values in the original equation and verify that the solutions are $-\dfrac{\sqrt{14}}{2}$ and $\dfrac{\sqrt{14}}{2}$. The solution set is $\left\{-\dfrac{\sqrt{14}}{2}, \dfrac{\sqrt{14}}{2}\right\}$ or $\left\{\pm\dfrac{\sqrt{14}}{2}\right\}$. ∎

✔ **CHECK POINT 2** Solve: $2x^2 - 5 = 0$.

Some quadratic equations have solutions that are imaginary numbers.

EXAMPLE 3 Solving a Quadratic Equation by the Square Root Property

Solve: $9x^2 + 25 = 0$.

Solution

$$9x^2 + 25 = 0 \qquad \text{This is the original equation.}$$

We need to isolate x^2.

$$9x^2 = -25 \qquad \text{Subtract 25 from both sides.}$$

$$x^2 = -\frac{25}{9} \qquad \text{Divide both sides by 9.}$$

$$x = \sqrt{-\frac{25}{9}} \quad \text{or} \quad x = -\sqrt{-\frac{25}{9}} \qquad \text{Apply the square root property.}$$

$$x = \sqrt{\frac{25}{9}}\,\sqrt{-1} \qquad x = -\sqrt{\frac{25}{9}}\,\sqrt{-1}$$

$$x = \frac{5}{3}i \qquad\qquad x = -\frac{5}{3}i \qquad \sqrt{-1} = i$$

Because the equation has an x^2-term and no x-term, we can check both proposed solutions, $\pm\dfrac{5}{3}i$, at once.

Check $\dfrac{5}{3}i$ and $-\dfrac{5}{3}i$:

$$9x^2 + 25 = 0$$

$$9\left(\pm\frac{5}{3}i\right)^2 + 25 \overset{?}{=} 0$$

$$9\left(\frac{25}{9}i^2\right) + 25 \overset{?}{=} 0$$

$$25i^2 + 25 \overset{?}{=} 0$$

$i^2 = -1$

$$25(-1) + 25 \overset{?}{=} 0$$

$$0 = 0, \text{ true}$$

Using Technology

The graph of

$$f(x) = 9x^2 + 25$$

has no x-intercepts. This shows that

$$9x^2 + 25 = 0$$

has no real solutions. Example 3 algebraically establishes that the solutions are imaginary numbers.

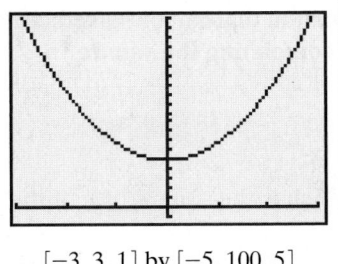

$[-3, 3, 1]$ by $[-5, 100, 5]$

The solutions are $-\dfrac{5}{3}i$ and $\dfrac{5}{3}i$. The solution set is $\left\{-\dfrac{5}{3}i, \dfrac{5}{3}i\right\}$ or $\left\{\pm\dfrac{5}{3}i\right\}$. ∎

✔ **CHECK POINT 3** Solve: $4x^2 + 9 = 0$.

Can we solve an equation such as $(x - 1)^2 = 5$ using the square root property? Yes. The equation is in the form $u^2 = d$, where u^2, the squared expression, is by itself on the left side.

$$(x - 1)^2 \qquad = \qquad 5$$

This is u^2 in $u^2 = d$ with $u = x - 1$.

This is d in $u^2 = d$ with $d = 5$.

Discover for Yourself

Try solving

$$(x - 1)^2 = 5$$

by writing the equation in standard form and factoring. What problem do you encounter?

EXAMPLE 4   Solving a Quadratic Equation by the Square Root Property

Solve by the square root property: $(x - 1)^2 = 5$.

Solution

$$(x - 1)^2 = 5 \qquad \text{This is the original equation.}$$

$$x - 1 = \sqrt{5} \quad \text{or} \quad x - 1 = -\sqrt{5} \qquad \text{Apply the square root property.}$$

$$x = 1 + \sqrt{5} \qquad x = 1 - \sqrt{5} \qquad \text{Add 1 to both sides in each equation.}$$

Check $1 + \sqrt{5}$:	**Check $1 - \sqrt{5}$:**
$(x - 1)^2 = 5$	$(x - 1)^2 = 5$
$(1 + \sqrt{5} - 1)^2 \overset{?}{=} 5$	$(1 - \sqrt{5} - 1)^2 \overset{?}{=} 5$
$(\sqrt{5})^2 \overset{?}{=} 5$	$(-\sqrt{5})^2 \overset{?}{=} 5$
$5 = 5$, true	$5 = 5$, true

The solutions are $1 \pm \sqrt{5}$, and the solution set is $\{1 + \sqrt{5}, 1 - \sqrt{5}\}$ or $\{1 \pm \sqrt{5}\}$. ∎

✔ **CHECK POINT 4**   Solve: $(x - 2)^2 = 7$.

2 Complete the square of a binomial.

Completing the Square

We return to the question that opened this section: How do we solve a quadratic equation, $ax^2 + bx + c = 0$, if the trinomial $ax^2 + bx + c$ cannot be factored? We can convert the equation into an equivalent equation that can be solved using the square root property. This is accomplished by **completing the square**.

> **Completing the Square**
>
> If $x^2 + bx$ is a binomial, then by adding $\left(\dfrac{b}{2}\right)^2$, which is the square of half the coefficient of x, a perfect square trinomial will result. That is,
>
> $$x^2 + bx + \left(\frac{b}{2}\right)^2 = \left(x + \frac{b}{2}\right)^2.$$

EXAMPLE 5   Completing the Square

Complete the square for each binomial. Then factor the resulting perfect square trinomial:

a. $x^2 + 8x$ **b.** $x^2 - 7x$ **c.** $x^2 + \dfrac{3}{5}x$.

Visualizing Completing the Square

This figure, with area $x^2 + 8x$, is not a complete square. The bottom-right corner is missing.

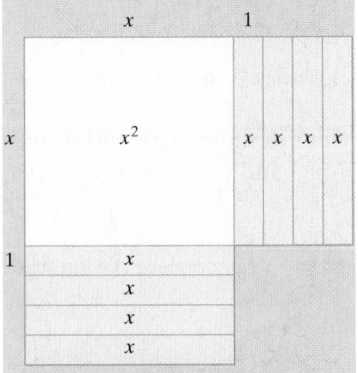

Area: $x^2 + 8x$

Add 16 square units to the missing portion and you, literally, complete the square.

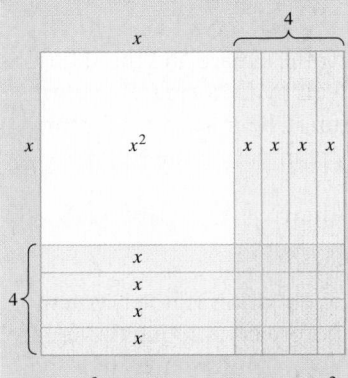

Area: $x^2 + 8x + 16 = (x + 4)^2$

Solution   To complete the square, we must add a term to each binomial. The term that should be added is the square of half the coefficient of x.

$$x^2 + 8x \qquad x^2 - 7x \qquad x^2 + \frac{3}{5}x$$

Add $\left(\frac{8}{2}\right)^2 = 4^2$. Add 16 to complete the square.

Add $\left(\frac{-7}{2}\right)^2$, or $\frac{49}{4}$, to complete the square.

Add $\left(\frac{1}{2} \cdot \frac{3}{5}\right)^2 = \left(\frac{3}{10}\right)^2$. Add $\frac{9}{100}$ to complete the square.

a. The coefficient of the x-term in $x^2 + 8x$ is 8. Half of 8 is 4, and $4^2 = 16$. Add 16.

$$x^2 + 8x + 16 = (x + 4)^2$$

b. The coefficient of the x-term in $x^2 - 7x$ is -7. Half of -7 is $-\frac{7}{2}$, and $\left(-\frac{7}{2}\right)^2 = \frac{49}{4}$. Add $\frac{49}{4}$.

$$x^2 - 7x + \frac{49}{4} = \left(x - \frac{7}{2}\right)^2$$

c. The coefficient of the x-term in $x^2 + \frac{3}{5}x$ is $\frac{3}{5}$. Half of $\frac{3}{5}$ is $\frac{1}{2} \cdot \frac{3}{5}$, or $\frac{3}{10}$, and $\left(\frac{3}{10}\right)^2 = \frac{9}{100}$. Add $\frac{9}{100}$.

$$x^2 + \frac{3}{5}x + \frac{9}{100} = \left(x + \frac{3}{10}\right)^2 \qquad\blacksquare$$

Study Tip

You may not be accustomed to factoring perfect square trinomials in which fractions are involved. The constant in the factorization is always half the coefficient of x.

$$x^2 - 7x + \frac{49}{4} = \left(x - \frac{7}{2}\right)^2 \qquad\qquad x^2 + \frac{3}{5}x + \frac{9}{100} = \left(x + \frac{3}{10}\right)^2$$

Half the coefficient of x, -7, is $-\frac{7}{2}$.

Half the coefficient of x, $\frac{3}{5}$, is $\frac{3}{10}$.

✔ **CHECK POINT 5**   Complete the square for each binomial. Then factor the resulting perfect square trinomial:

a. $x^2 + 10x$

b. $x^2 - 3x$

c. $x^2 + \frac{3}{4}x$.

3 Solve quadratic equations by completing the square.

Solving Quadratic Equations by Completing the Square

We can solve any quadratic equation by completing the square. If the coefficient of the x^2-term is one, we add the square of half the coefficient of x to both sides of the equation. **When you add a constant term to one side of the equation to complete the square, be certain to add the same constant to the other side of the equation.** These ideas are illustrated in Example 6.

EXAMPLE 6   Solving a Quadratic Equation by Completing the Square

Solve by completing the square: $x^2 - 6x + 2 = 0$.

Solution   We begin by subtracting 2 from both sides. This is done to isolate the binomial $x^2 - 6x$ so that we can complete the square.

$$x^2 - 6x + 2 = 0 \qquad \text{This is the original equation.}$$

$$x^2 - 6x = -2 \qquad \text{Subtract 2 from both sides.}$$

Next, we complete the square. Find half the coefficient of the x-term and square it. The coefficient of the x-term is -6. Half of -6 is -3 and $(-3)^2 = 9$. Thus, we add 9 to both sides of the equation.

$$x^2 - 6x + 9 = -2 + 9 \qquad \text{Add 9 to both sides to complete the square.}$$

$$(x - 3)^2 = 7 \qquad \text{Factor and simplify.}$$

$$x - 3 = \sqrt{7} \quad \text{or} \quad x - 3 = -\sqrt{7} \qquad \text{Apply the square root property.}$$

$$x = 3 + \sqrt{7} \qquad x = 3 - \sqrt{7} \qquad \text{Add 3 to both sides in each equation.}$$

The solutions are $3 \pm \sqrt{7}$, and the solution set is $\{3 + \sqrt{7}, 3 - \sqrt{7}\}$ or $\{3 \pm \sqrt{7}\}$. ∎

If you solve a quadratic equation by completing the square and the solutions are rational numbers, the equation can also be solved by factoring. By contrast, quadratic equations with irrational solutions cannot be solved by factoring. However, all quadratic equations can be solved by completing the square.

✔ **CHECK POINT 6**   Solve by completing the square: $x^2 - 10x + 18 = 0$.

If the coefficient of the x^2-term in a quadratic equation is not 1, you must divide each side of the equation by this coefficient before completing the square. For example, to solve $2x^2 + 5x - 4 = 0$ by completing the square, first divide every term by 2:

$$\frac{2x^2}{2} + \frac{5x}{2} - \frac{4}{2} = \frac{0}{2}$$

$$x^2 + \frac{5}{2}x - 2 = 0.$$

Now that the coefficient of the x^2-term is 1, we can solve by completing the square.

EXAMPLE 7   Solving a Quadratic Equation by Completing the Square

Solve by completing the square: $2x^2 + 5x - 4 = 0$.

Solution

$$2x^2 + 5x - 4 = 0$$ This is the original equation.

$$x^2 + \frac{5}{2}x - 2 = 0$$ Divide both sides by 2.

$$x^2 + \frac{5}{2}x = 2$$ Add 2 to both sides to isolate the binomial.

$$x^2 + \frac{5}{2}x + \frac{25}{16} = 2 + \frac{25}{16}$$ Complete the square: Half of $\frac{5}{2}$ is $\frac{5}{4}$ and $\left(\frac{5}{4}\right)^2 = \frac{25}{16}$.

$$\left(x + \frac{5}{4}\right)^2 = \frac{57}{16}$$ Factor and simplify. On the right: $2 + \frac{25}{16} = \frac{32}{16} + \frac{25}{16} = \frac{57}{16}$.

$$x + \frac{5}{4} = \sqrt{\frac{57}{16}} \quad \text{or} \quad x + \frac{5}{4} = -\sqrt{\frac{57}{16}}$$ Apply the square root property.

$$x + \frac{5}{4} = \frac{\sqrt{57}}{4} \quad \text{or} \quad x + \frac{5}{4} = -\frac{\sqrt{57}}{4}$$ $\sqrt{\frac{57}{16}} = \frac{\sqrt{57}}{\sqrt{16}} = \frac{\sqrt{57}}{4}$

$$x = -\frac{5}{4} + \frac{\sqrt{57}}{4} \quad \text{or} \quad x = -\frac{5}{4} - \frac{\sqrt{57}}{4}$$ Solve the equations, subtracting $\frac{5}{4}$ from both sides.

$$x = \frac{-5 + \sqrt{57}}{4} \quad \text{or} \quad x = \frac{-5 - \sqrt{57}}{4}$$ Express solutions with a common denominator.

The solutions are $\dfrac{-5 \pm \sqrt{57}}{4}$, and the solution set is $\left\{\dfrac{-5 \pm \sqrt{57}}{4}\right\}$.   ■

✔ **CHECK POINT 7**   Solve by completing the square: $2x^2 - 10x - 1 = 0$.

Using Technology

Obtain a decimal approximation for each solution of $2x^2 + 5x - 4 = 0$, the equation in Example 7.

$$\frac{-5 + \sqrt{57}}{4} \approx 0.6$$

$$\frac{-5 - \sqrt{57}}{4} \approx -3.1$$

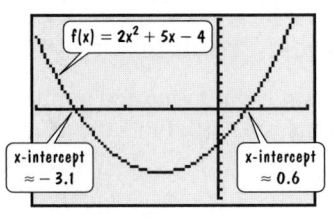

$f(x) = 2x^2 + 5x - 4$

x-intercept ≈ −3.1   x-intercept ≈ 0.6

$[-4, 2, 1]$ by $[-10, 10, 1]$

The x-intercepts of $f(x) = 2x^2 + 5x - 4$ verify the solutions.

4   Solve problems using the square root property.

Applications

We all want a wonderful life with fulfilling work, good health, and loving relationships. And let's be honest: Financial security wouldn't hurt! Achieving this goal depends on understanding how money in a savings account grows in remarkable ways as a result of *compound interest*. **Compound interest** is interest computed on your original investment as well as on any accumulated interest. For example, suppose you deposit $2000, the principal, in a savings account at a rate of 6%. The interest at the end of the first year is

$$I = Pr = (\$2000)(0.06) = \$120.$$

Interest earned is the principal times the interest rate.

Assuming that you do not withdraw the $120 interest, the amount in the account during year two is $2000 + $120, or $2120. The interest paid at the end of the second year is computed using $2120:

$$I = Pr = (\$2120)(0.06) = \$127.20.$$

The bank now adds $127.20 to the $2120 in your account. Thus, during year 3 the account contains $2120 + $127.20, or $2247.20. The interest paid at the end of the third year is computed using $2247.20:

$$I = Pr = (\$2247.20)(0.06) \approx \$134.83.$$

The amount in the savings account at the end of year three is $2247.20 + $134.83, or $2382.03.

A faster way to determine the amount, A, in an account subject to compound interest is to use the following formula:

A Formula for Compound Interest

Suppose that an amount of money, P, is invested at interest rate r, compounded annually. In t years, the amount, A, or balance, in the account is given by the formula

$$A = P(1 + r)^t.$$

Some compound interest problems can be solved using quadratic equations.

EXAMPLE 8 Solving a Compound Interest Problem

You invested $1000 in a savings account whose interest is compounded annually. After 2 years, the amount, or balance, in the account is $1210. Find the annual interest rate.

Solution We are given that

P (the amount invested) = $1000

t (the time of the investment) = 2 years

A (the amount, or balance, in the account) = $1210.

We are asked to find the annual interest rate, r. We do this by substituting the three given values in the compound interest formula and solving for r.

$$A = P(1 + r)^t \qquad \text{Use the compound interest formula.}$$

$$1210 = 1000(1 + r)^2 \qquad \text{Substitute the given values.}$$

$$\frac{1210}{1000} = (1 + r)^2 \qquad \text{Divide both sides by 1000.}$$

$$\frac{121}{100} = (1 + r)^2 \qquad \text{Simplify the fraction.}$$

$$1 + r = \sqrt{\frac{121}{100}} \quad \text{or} \quad 1 + r = -\sqrt{\frac{121}{100}} \qquad \text{Apply the square root property.}$$

$$1 + r = \frac{11}{10} \quad \text{or} \quad 1 + r = -\frac{11}{10} \qquad \sqrt{\frac{121}{100}} = \frac{\sqrt{121}}{\sqrt{100}} = \frac{11}{10}$$

$$r = \frac{11}{10} - 1 \quad \text{or} \quad r = -\frac{11}{10} - 1 \qquad \text{Subtract 1 from both sides.}$$

$$r = \frac{1}{10} \quad \text{or} \quad r = -\frac{21}{10} \qquad \frac{11}{10} - 1 = \frac{11}{10} - \frac{10}{10} = \frac{1}{10} \text{ and}$$
$$-\frac{11}{10} - 1 = -\frac{11}{10} - \frac{10}{10} = -\frac{21}{10}.$$

Because the interest rate cannot be negative, we reject $-\dfrac{21}{10}$. Thus, the annual interest rate is $\dfrac{1}{10} = 0.10 = 10\%$.

We can check this answer using the formula $A = P(1 + r)^t$. If \$1000 is invested for 2 years at 10% interest, compounded annually, the balance in the account is

$$A = \$1000(1 + 0.10)^2 = \$1000(1.10)^2 = \$1210.$$

Because this is precisely the amount given by the problem's conditions, the annual interest rate is, indeed, 10% compounded annually. ∎

✔ **CHECK POINT 8** You invested \$3000 in an account whose interest is compounded annually. After 2 years, the amount, or balance, in the account is \$4320. Find the annual interest rate.

The Pythagorean Theorem and the Square Root Property

The ancient Greek philosopher and mathematician Pythagoras (approximately 582–500 B.C.) founded a school whose motto was "All is number." Pythagoras is best remembered for his work with the **right triangle**, a triangle with one angle measuring 90°. The side opposite the 90° angle is called the **hypotenuse**. The other sides are called **legs**. Pythagoras found that if he constructed squares on each of the legs, as well as a larger square on the hypotenuse, the sum of the areas of the smaller squares is equal to the area of the larger square. This is illustrated in Figure 11.2.

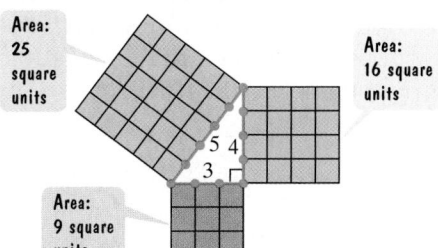

Area: 25 square units

Area: 16 square units

Area: 9 square units

Figure 11.2 The area of the large square equals the sum of the areas of the smaller squares.

This relationship is usually stated in terms of the lengths of the three sides of a right triangle and is called the **Pythagorean Theorem**.

The Pythagorean Theorem

The sum of the squares of the lengths of the legs of a right triangle equals the square of the length of the hypotenuse.

If the legs have lengths a and b, and the hypotenuse has length c, then

$$a^2 + b^2 = c^2.$$

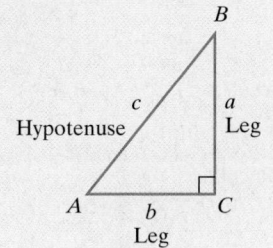

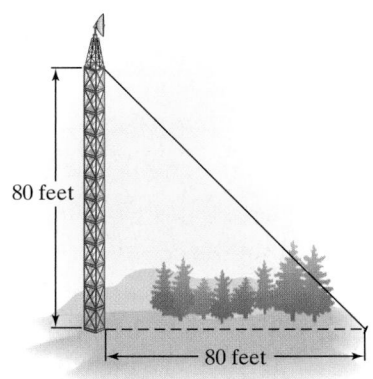

80 feet

80 feet

Figure 11.3

EXAMPLE 9 Using the Pythagorean Theorem and the Square Root Property

A supporting wire is to be attached to the top of an 80-foot antenna, as shown in Figure 11.3. Because of surrounding trees, the wire must be anchored 80 feet from the base of the antenna. What length of wire is required?

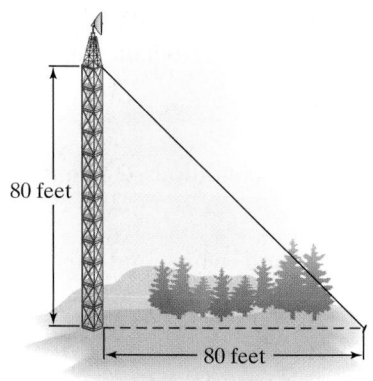

80 feet

80 feet

Figure 11.3, repeated

Solution Let w = the wire's length. We can find w in the right triangle using the Pythagorean Theorem.

(Leg)²	plus	(Leg)²	equals	(Hypotenuse)².
80^2	$+$	80^2	$=$	w^2.

We solve this equation using the square root property.

$$6400 + 6400 = w^2 \qquad \text{Square 80.}$$

$$12{,}800 = w^2 \qquad \text{Add.}$$

$$w = \sqrt{12{,}800} \quad \text{or} \quad w = -\sqrt{12{,}800} \qquad \text{Apply the square root property.}$$

$$w = \sqrt{6400}\,\sqrt{2} \quad \text{or} \quad w = -\sqrt{6400}\,\sqrt{2} \qquad \text{6400 is the largest perfect square factor of 12,800.}$$

$$w = 80\sqrt{2} \quad \text{or} \quad w = -80\sqrt{2}$$

Because w represents the wire's length, we reject the negative value. Thus, a wire of length $80\sqrt{2}$ feet is needed to support the antenna. Using a calculator, this is approximately 113.1 feet of wire. ∎

Take a second look at the right triangle in Figure 11.3. Can you see that both legs are the same length? If the lengths of both legs of a right triangle are the same, the triangle is called an **isosceles right triangle**. Figure 11.4 shows such a triangle. In Example 9, the wire's length was the length of a leg, 80 feet, times $\sqrt{2}$. We can use Figure 11.4 to show that the length of the hypotenuse in any isosceles right triangle is the length of a leg times $\sqrt{2}$.

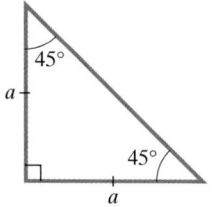

Figure 11.4 An isosceles right triangle has legs that are the same length and acute angles each measuring 45°.

$$c^2 = a^2 + a^2 \qquad \text{Apply the Pythagorean Theorem to the isosceles right triangle in Figure 11.4.}$$

$$c^2 = 2a^2 \qquad \text{Combine like terms.}$$

$$c = \sqrt{2a^2} \quad \text{or} \quad c = -\sqrt{2a^2} \qquad \text{Apply the square root property.}$$

> Reject because c cannot be negative.

$$c = \sqrt{2}a \quad \text{or} \quad a\sqrt{2} \qquad \text{Simplify: } \sqrt{2a^2} = \sqrt{2}\,\sqrt{a^2} = \sqrt{2}a.$$

Lengths within Isosceles Right Triangles

The length of the hypotenuse in an isosceles right triangle is the length of a leg times $\sqrt{2}$.

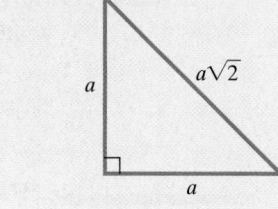

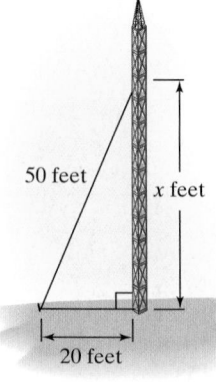

50 feet

x feet

20 feet

✔ **CHECK POINT 9** A 50-foot supporting wire is to be attached to an antenna. The wire is anchored 20 feet from the base of the antenna. How high up the antenna is the wire attached? Express the answer in simplified radical form. Then find a decimal approximation to the nearest tenth of a foot.

EXERCISE SET 11.1

Practice Exercises

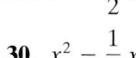

In Exercises 1–22, solve each equation by the square root property. If possible, simplify radicals or rationalize denominators. Express imaginary solutions in the form $a + bi$.

1. $3x^2 = 75$ **2.** $5x^2 = 20$

3. $7x^2 = 42$ **4.** $8x^2 = 40$

5. $16x^2 = 25$ **6.** $4x^2 = 49$

7. $3x^2 - 2 = 0$

8. $3x^2 - 5 = 0$

9. $25x^2 + 16 = 0$

10. $4x^2 + 49 = 0$

11. $(x + 7)^2 = 9$

12. $(x + 3)^2 = 64$

13. $(x - 3)^2 = 5$

14. $(x - 4)^2 = 3$

15. $(x + 2)^2 = 8$

16. $(x + 2)^2 = 12$

17. $(x - 5)^2 = -9$

18. $(x - 5)^2 = -4$

19. $\left(x + \dfrac{3}{4}\right)^2 = \dfrac{11}{16}$

20. $\left(x + \dfrac{2}{5}\right)^2 = \dfrac{7}{25}$

21. $x^2 - 6x + 9 = 36$

22. $x^2 - 6x + 9 = 49$

In Exercises 23–34, complete the square for each binomial. Then factor the resulting perfect square trinomial.

23. $x^2 + 2x$

24. $x^2 + 4x$

25. $x^2 - 14x$

26. $x^2 - 10x$

27. $x^2 + 7x$

28. $x^2 + 9x$

29. $x^2 - \dfrac{1}{2}x$

30. $x^2 - \dfrac{1}{3}x$

31. $x^2 + \dfrac{4}{3}x$

32. $x^2 + \dfrac{4}{5}x$

33. $x^2 - \dfrac{9}{4}x$

34. $x^2 - \dfrac{9}{5}x$

In Exercises 35–56, solve each quadratic equation by completing the square.

35. $x^2 + 6x = -8$

36. $x^2 + 6x = 7$

37. $x^2 + 6x = -2$

38. $x^2 + 2x = 5$

39. $x^2 + 4x + 1 = 0$

40. $x^2 + 6x - 5 = 0$

41. $x^2 + 2x + 2 = 0$

42. $x^2 - 4x + 8 = 0$

43. $x^2 + 3x - 1 = 0$

44. $x^2 - 3x - 5 = 0$

45. $x^2 = 7x - 3$

46. $x^2 = 5x - 3$

47. $x^2 + x - 1 = 0$

48. $x^2 + 3x - 1 = 0$

49. $2x^2 - 3x + 1 = 0$

50. $2x^2 - x - 1 = 0$

51. $2x^2 + 10x + 11 = 0$

52. $2x^2 + 8x + 5 = 0$

53. $3x^2 - 2x - 4 = 0$

54. $4x^2 - 2x - 3 = 0$

55. $8x^2 - 4x + 1 = 0$

56. $9x^2 - 6x + 5 = 0$

Application Exercises

In Exercises 57–60, use the compound interest formula

$$A = P(1 + r)^t$$

to find the annual interest rate, r.

57. In 2 years, an investment of \$2000 grows to \$2880.

58. In 2 years, an investment of \$2000 grows to \$2420.

59. In 2 years, an investment of \$3125 grows to \$3360.

60. In 2 years, an investment of \$1280 grows to \$1445.

Cable-TV modems make it possible to quickly access the Internet and download files. The graph shows the number, in millions, of Internet users in the United States with this new technology. The data can be modeled by the function

$$f(x) = 0.4x^2 + 0.5,$$

where $f(x)$ represents millions of people in the United States using cable-TV modems x years after 1996. Use this function and the square root property to solve Exercises 61–62.

Number of People in the United States Using Cable-TV Modems

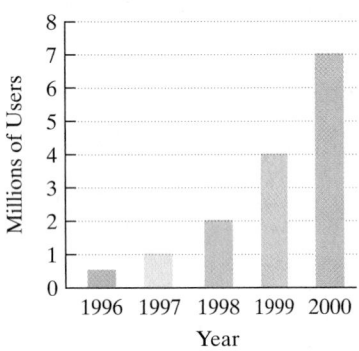

Source: The New York Times

61. In which year will 20 million Americans use cable-TV modems?

62. In which year did 4 million Americans use cable-TV modems? How well does the function model the actual number of users for that year shown in the graph?

The function $s(t) = 16t^2$ models the distance, s(t), in feet, that an object falls in t seconds. Use this function and the square root property to solve Exercises 63–64. Express answers in simplified radical form. Then use your calculator to find a decimal approximation to the nearest tenth of a second.

63. A sky diver jumps from an airplane and falls for 4800 feet before opening a parachute. For how many seconds was the diver in a free fall?

64. A sky diver jumps from an airplane and falls for 3200 feet before opening a parachute. For how many seconds was the diver in a free fall?

Use the Pythagorean Theorem and the square root property to solve Exercises 65–70. Express answers in simplified radical form. Then find a decimal approximation to the nearest tenth.

65. A rectangular park is 6 miles long and 3 miles wide. How long is a pedestrian route that runs diagonally across the park?

66. A rectangular park is 4 miles long and 2 miles wide. How long is a pedestrian route that runs diagonally across the park?

67. The base of a 20-foot ladder is 15 feet from the house. How far up the house does the ladder reach?

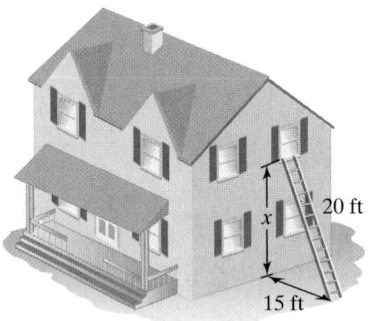

68. A baseball diamond is actually a square with 90-foot sides. What is the distance from home plate to second base?

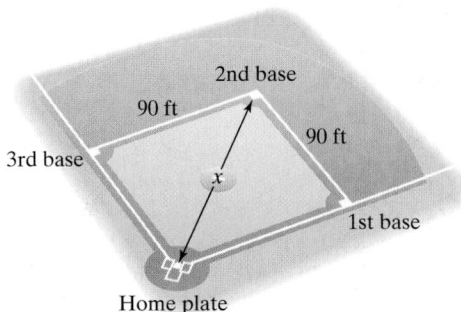

69. A supporting wire is to be attached to the top of a 50-foot antenna. If the wire must be anchored 50 feet from the base of the antenna, what length of wire is required?

70. A supporting wire is to be attached to the top of a 70-foot antenna. If the wire must be anchored 70 feet from the base of the antenna, what length of wire is required?

71. In the section opener, we looked at Einstein's famous formula

$$E = mc^2.$$

The formula implies that a mere pound of any kind of matter contains enough energy to send a large vessel on at least 100 ocean voyages. Energy, E, is equal to mass, m, multiplied by the square of the speed of light, c. Solve for c to express the speed of light in terms of energy and mass.

72. The surface area, A, of a sphere with radius r is given by the formula

$$A = 4\pi r^2.$$

Solve for r to express the radius in terms of the area.

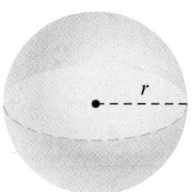

73. A square flower bed is to be enlarged by adding 2 meters on each side. If the larger square has an area of 144 square meters, what is the length of a side of the original square?

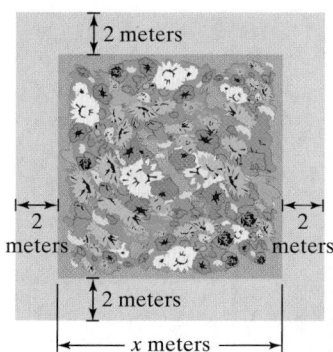

2 meters

2 meters | 2 meters

2 meters

x meters

74. A square flower bed is to be enlarged by adding 3 feet on each side. If the larger square has an area of 169 square feet, what is the length of a side of the original square?

Writing in Mathematics

75. What is the square root property?

76. Explain how to solve $(x - 1)^2 = 16$ using the square root property.

77. Explain how to complete the square for a binomial. Use $x^2 + 6x$ to illustrate your explanation.

78. Explain how to solve $x^2 + 6x + 8 = 0$ by completing the square.

79. What is compound interest?

80. In your own words, describe the compound interest formula

$$A = P(1 + r)^t.$$

81. What is an isosceles right triangle? What is the relationship between the length of its hypotenuse and the length of a leg?

82. A basketball player's hang time is the time spent in the air when shooting a basket. The function $f(x) = 4x^2$ models the player's vertical leap, $f(x)$, in feet, in terms of the hang time, x, measured in seconds. Use this function to write a problem that can be solved by the square root property. Then solve the problem. Be realistic: It is unlikely that hang time can exceed 1 second.

83. In the section opener, we mentioned the possibility of the mass of a golf ball providing the daily energy needs of a large metropolitan area. Discuss some of the problems involved with gradual and controlled release of energy through nuclear power. What problems stand in the way of nuclear power expansion? Consult the Internet to help answer the question.

Critical Thinking Exercises

84. Which one of the following is true?
   a. The equation $(x - 5)^2 = 12$ is equivalent to $x - 5 = 2\sqrt{3}$.
   b. In completing the square for $2x^2 - 6x = 5$, we should add 9 to both sides.
   c. Although not every quadratic equation can be solved by completing the square, they can all be solved by factoring.
   d. The graph of $y = (x - 2)^2 + 3$ cannot have x-intercepts.

85. Solve for y: $\dfrac{x^2}{a^2} + \dfrac{y^2}{b^2} = 1.$

86. Solve by completing the square:

$$x^2 + x + c = 0.$$

87. Solve by completing the square:

$$x^2 + bx + c = 0.$$

88. Solve: $x^4 - 8x^2 + 15 = 0.$

Technology Exercises

89. Use a graphing utility to solve $4 - (x + 1)^2 = 0$. Graph $y = 4 - (x + 1)^2$ in a $[-5, 5, 1]$ by $[-5, 5, 1]$ viewing rectangle. The equation's solutions are the graph's x-intercepts. Check by substitution in the given equation.

90. Use a graphing utility to solve $(x - 1)^2 - 9 = 0$. Graph $y = (x - 1)^2 - 9$ in a $[-5, 5, 1]$ by $[-9, 3, 1]$ viewing rectangle. The equation's solutions are the graph's x-intercepts. Check by substitution in the given equation.

91. Use a graphing utility and x-intercepts to verify any of the real solutions that you obtained for five of the quadratic equations in Exercises 35–56.

Review Exercises

92. Simplify: $4x - 2 - 3[4 - 2(3 - x)]$. (Section 1.8, Example 11)

93. Factor: $1 - 8x^3$. (Section 6.4, Example 8)

94. Divide: $(x^4 - 5x^3 + 2x^2 - 6) \div (x - 3)$. (Section 5.6, Example 2 or Example 4)

▶ SECTION 11.2   The Quadratic Formula

Objectives

1 Solve quadratic equations using the quadratic formula.

2 Use the discriminant to determine number and type of solutions.

3 Determine the most efficient method to use when solving a quadratic equation.

4 Write quadratic equations from solutions.

5 Use the quadratic formula to solve problems.

SSM
PH Tutor CD-   Video
Center   ROM

Serpico, 1973, starring Al Pacino, was a movie about police corruption.

In 2000, a police scandal shocked Los Angeles. A police officer who had been convicted of stealing cocaine held as evidence described how members of his unit behaved in ways that resembled the gangs they were targeting, assaulting and framing innocent people.

Is police corruption on the rise? How many police officers are convicted of felonies each year, and is there a trend to the data? In this section, we use a quadratic function to model this upsetting phenomenon. We begin by deriving a formula that will enable you to solve quadratic equations more quickly than the method of completing the square. In the exercise set, you will use this formula to work with the function that models data for police officers convicted as felons.

Solve quadratic equations using the quadratic formula.

Solving Quadratic Equations Using the Quadratic Formula

We can use the method of completing the square to derive a formula that can be used to solve all quadratic equations. The derivation given here also shows a particular quadratic equation, $3x^2 - 2x - 4 = 0$, to specifically illustrate each of the steps.

Deriving the Quadratic Formula

Standard Form of a Quadratic Equation	Comment	A Specific Example
$ax^2 + bx + c = 0, a > 0$	This is the given equation.	$3x^2 - 2x - 4 = 0$
$x^2 + \dfrac{b}{a}x + \dfrac{c}{a} = 0$	Divide both sides by the coefficient of x^2.	$x^2 - \dfrac{2}{3}x - \dfrac{4}{3} = 0$
$x^2 + \dfrac{b}{a}x = -\dfrac{c}{a}$	Isolate the binomial by adding $-\dfrac{c}{a}$ on both sides.	$x^2 - \dfrac{2}{3}x = \dfrac{4}{3}$
$x^2 + \dfrac{b}{a}x + \left(\dfrac{b}{2a}\right)^2 = -\dfrac{c}{a} + \left(\dfrac{b}{2a}\right)^2$ $\underbrace{\qquad}_{(\text{half})^2}$	Complete the square. Add the square of half the coefficient of x to both sides.	$x^2 - \dfrac{2}{3}x + \left(-\dfrac{1}{3}\right)^2 = \dfrac{4}{3} + \left(-\dfrac{1}{3}\right)^2$ $\underbrace{\qquad}_{(\text{half})^2}$
$x^2 + \dfrac{b}{a}x + \dfrac{b^2}{4a^2} = -\dfrac{c}{a} + \dfrac{b^2}{4a^2}$		$x^2 - \dfrac{2}{3}x + \dfrac{1}{9} = \dfrac{4}{3} + \dfrac{1}{9}$
$\left(x + \dfrac{b}{2a}\right)^2 = -\dfrac{c}{a} \cdot \dfrac{4a}{4a} + \dfrac{b^2}{4a^2}$	Factor on the left side and obtain a common denominator on the right side.	$\left(x - \dfrac{1}{3}\right)^2 = \dfrac{4}{3} \cdot \dfrac{3}{3} + \dfrac{1}{9}$
$\left(x + \dfrac{b}{2a}\right)^2 = \dfrac{-4ac + b^2}{4a^2}$	Add fractions on the right.	$\left(x - \dfrac{1}{3}\right)^2 = \dfrac{12 + 1}{9}$
$\left(x + \dfrac{b}{2a}\right)^2 = \dfrac{b^2 - 4ac}{4a^2}$		$\left(x - \dfrac{1}{3}\right)^2 = \dfrac{13}{9}$
$x + \dfrac{b}{2a} = \pm\sqrt{\dfrac{b^2 - 4ac}{4a^2}}$	Apply the square root property.	$x - \dfrac{1}{3} = \pm\sqrt{\dfrac{13}{9}}$
$x + \dfrac{b}{2a} = \pm\dfrac{\sqrt{b^2 - 4ac}}{2a}$	Take the square root of the quotient, simplifying the denominator.	$x - \dfrac{1}{3} = \pm\dfrac{\sqrt{13}}{3}$
$x = \dfrac{-b}{2a} \pm \dfrac{\sqrt{b^2 - 4ac}}{2a}$	Solve for x by subtracting $\dfrac{b}{2a}$ from both sides.	$x = \dfrac{1}{3} \pm \dfrac{\sqrt{13}}{3}$
$x = \dfrac{-b \pm \sqrt{b^2 - 4ac}}{2a}$	Combine fractions on the right side.	$x = \dfrac{1 \pm \sqrt{13}}{3}$

The formula shown at the bottom of the left column is called the **quadratic formula**. A similar proof shows that the same formula can be used to solve quadratic equations if a, the coefficient of the x^2-term, is negative.

The Quadratic Formula

The solutions of a quadratic equation in standard form $ax^2 + bx + c = 0$, with $a \neq 0$, are given by the **quadratic formula**

$$x = \frac{-b \pm \sqrt{b^2 - 4ac}}{2a}.$$

x equals negative b, plus or minus the square root of $b^2 - 4ac$, all divided by 2a.

To use the quadratic formula, write the quadratic equation in standard form if necessary. Then determine the numerical values for a (the coefficient of the squared term), b (the coefficient of the x-term), and c (the constant term). Substitute the values of a, b, and c in the quadratic formula and evaluate the expression. The $\pm$ sign indicates that there are two solutions of the equation.

EXAMPLE 1   Solving a Quadratic Equation Using the Quadratic Formula

Solve using the quadratic formula: $8x^2 + 2x - 1 = 0$.

Solution   The given equation is in standard form. Begin by identifying the values for a, b, and c.

$$8x^2 + 2x - 1 = 0$$

$a = 8 \qquad b = 2 \qquad c = -1$

Substituting these values into the quadratic formula and simplifying gives the equation's solutions.

$$x = \frac{-b \pm \sqrt{b^2 - 4ac}}{2a} \qquad \text{Use the quadratic formula.}$$

$$x = \frac{-2 \pm \sqrt{2^2 - 4(8)(-1)}}{2(8)} \qquad \text{Substitute the values for } a, b, \text{ and } c: a = 8, b = 2, \text{ and } c = -1.$$

$$= \frac{-2 \pm \sqrt{4 - (-32)}}{16} \qquad 2^2 - 4(8)(-1) = 4 - (-32)$$

$$= \frac{-2 \pm \sqrt{36}}{16} \qquad 4 - (-32) = 4 + 32 = 36$$

$$= \frac{-2 \pm 6}{16} \qquad \sqrt{36} = 6$$

Now we will evaluate this expression in two different ways to obtain the two solutions. At left, we will *add* 6 to -2. At right, we will *subtract* 6 from -2.

$$x = \frac{-2 + 6}{16} \quad \text{or} \quad x = \frac{-2 - 6}{16}$$

$$= \frac{4}{16} = \frac{1}{4} \qquad\qquad = \frac{-8}{16} = -\frac{1}{2}$$

The solutions are $-\frac{1}{2}$ and $\frac{1}{4}$, and the solution set is $\left\{-\frac{1}{2}, \frac{1}{4}\right\}$.   ∎

In Example 1, the solutions of $8x^2 + 2x - 1 = 0$ are rational numbers. This means that the equation can also be solved by factoring. The reason that the solutions are rational numbers is that $b^2 - 4ac$, the radicand in the quadratic formula, is 36, which is a perfect square. If a, b, and c are rational numbers, all quadratic equations for which $b^2 - 4ac$ is a perfect square have rational solutions.

✔ **CHECK POINT 1**   Solve using the quadratic formula: $2x^2 + 9x - 5 = 0$.

Using Technology

The graph of the quadratic function

$$y = 8x^2 + 2x - 1$$

has x-intercepts at $-\frac{1}{2}$ and $\frac{1}{4}$. This verifies that $\left\{-\frac{1}{2}, \frac{1}{4}\right\}$ is the solution set of the quadratic equation

$$8x^2 + 2x - 1 = 0.$$

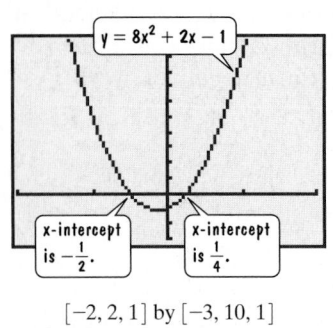

$y = 8x^2 + 2x - 1$

x-intercept is $-\frac{1}{2}$.

x-intercept is $\frac{1}{4}$.

$[-2, 2, 1]$ by $[-3, 10, 1]$

Can the equations
$$7x^5 + 12x^3 - 9x + 4 = 0$$
and
$$8x^6 - 7x^5 + 4x^3 - 19 = 0$$
be solved using a formula similar to the quadratic formula? The first equation has five solutions and the second has six solutions, but they cannot be found using a formula. How do we know? In 1832, a 20-year-old Frenchman, Evariste Galois, wrote down a proof showing that there is no general formula to solve equations when the exponent on the variable is 5 or greater. Galois was jailed as a political activist several times while still a teenager. The day after his brilliant proof he fought a duel over a woman. The duel was a political setup. As he lay dying, Galois told his brother, Alfred, of the manuscript that contained his proof: "Mathematical manuscripts are in my room. On the table. Take care of my work. Make it known. Important. Don't cry, Alfred. I need all my courage—to die at twenty." (Our source is Leopold Infeld's biography of Galois, *Whom the Gods Love*. Some historians, however, dispute the story of Galois's ironic death the very day after his algebraic proof. Mathematical truths seem more reliable than historical ones!)

EXAMPLE 2   Solving a Quadratic Equation Using the Quadratic Formula

Solve using the quadratic formula:

$$2x^2 = 4x + 1.$$

Solution   The quadratic equation must be in standard form to identify the values for a, b, and c. To move all terms to one side and obtain zero on the right, we subtract $4x + 1$ from both sides. Then we can identify the values for a, b, and c.

$$2x^2 = 4x + 1 \qquad \text{This is the given equation.}$$

$$2x^2 - 4x - 1 = 0 \qquad \text{Subtract } 4x + 1 \text{ from both sides.}$$

$$a = 2 \quad b = -4 \quad c = -1$$

Substituting these values into the quadratic formula and simplifying gives the equation's solutions.

$$x = \frac{-b \pm \sqrt{b^2 - 4ac}}{2a} \qquad \text{Use the quadratic formula.}$$

$$x = \frac{-(-4) \pm \sqrt{(-4)^2 - 4(2)(-1)}}{2(2)} \qquad \text{Substitute the values for } a, b, \text{ and } c: a = 2, b = -4, \text{ and } c = -1.$$

$$= \frac{4 \pm \sqrt{16 - (-8)}}{4} \qquad (-4)^2 - 4(2)(-1) = 16 - (-8)$$

$$= \frac{4 \pm \sqrt{24}}{4} \qquad 16 - (-8) = 16 + 8 = 24$$

$$= \frac{4 \pm 2\sqrt{6}}{4} \qquad \sqrt{24} = \sqrt{4 \cdot 6} = \sqrt{4} \cdot \sqrt{6} = 2\sqrt{6}$$

$$= \frac{2(2 \pm \sqrt{6})}{4} \qquad \text{Factor out 2 from the numerator.}$$

$$= \frac{2 \pm \sqrt{6}}{2} \qquad \text{Divide the numerator and denominator by 2.}$$

The solutions are $\dfrac{2 \pm \sqrt{6}}{2}$, and the solution set is $\left\{ \dfrac{2 + \sqrt{6}}{2}, \dfrac{2 - \sqrt{6}}{2} \right\}$ or $\left\{ \dfrac{2 \pm \sqrt{6}}{2} \right\}$.

In Example 2, the solutions of $2x^2 = 4x + 1$ are irrational numbers. This means that the equation cannot be solved by factoring. The reason that the solutions are irrational numbers is that $b^2 - 4ac$, the radicand in the quadratic formula, is 24, which is not a perfect square. Notice, too, that the solutions, $\dfrac{2 + \sqrt{6}}{2}$ and $\dfrac{2 - \sqrt{6}}{2}$, are conjugates.

Study Tip

Many students use the quadratic formula correctly until the last step, where they make an error in simplifying the solutions. Be sure to factor the numerator before dividing the numerator and denominator by the greatest common factor.

$$\frac{4 \pm 2\sqrt{6}}{4} = \frac{2(2 \pm \sqrt{6})}{4} = \frac{\overset{1}{\cancel{2}}(2 \pm \sqrt{6})}{\underset{2}{\cancel{4}}} = \frac{2 \pm \sqrt{6}}{2}$$

You cannot divide just one term in the numerator and the denominator by their greatest common factor.

INCORRECT

$$\frac{\overset{1}{\cancel{4}} \pm 2\sqrt{6}}{\underset{1}{\cancel{4}}} = 1 \pm 2\sqrt{6} \qquad \frac{4 \pm \overset{1}{\cancel{2}}\sqrt{6}}{\underset{2}{\cancel{4}}} = \frac{4 \pm \sqrt{6}}{2}$$

Can all irrational solutions of quadratic equations be simplified? No. The following solutions cannot be simplified:

$$\frac{5 \pm 2\sqrt{7}}{2} \qquad \boxed{\text{Other than 1, terms in each numerator have no common factor.}} \qquad \frac{4 \pm 3\sqrt{7}}{2}.$$

✔ **CHECK POINT 2** Solve using the quadratic formula: $2x^2 = 6x - 1$.

EXAMPLE 3 **Solving a Quadratic Equation Using the Quadratic Formula**

Solve using the quadratic formula:

$$3x^2 + 2 = -4x.$$

Solution Begin by writing the quadratic equation in standard form.

$$3x^2 + 2 = -4x \qquad \text{This is the given equation.}$$

$$3x^2 + 4x + 2 = 0 \qquad \text{Add 4x to both sides.}$$

$$\boxed{a = 3} \quad \boxed{b = 4} \quad \boxed{c = 2}$$

Substituting these values into the quadratic formula and simplifying gives the equation's solutions.

Using Technology

The graph of the quadratic function

$$y = 3x^2 + 4x + 2$$

has no x-intercepts. This verifies that the equation in Example 3

$$3x^2 + 2 = -4x, \quad \text{or}$$
$$3x^2 + 4x + 2 = 0$$

has imaginary solutions.

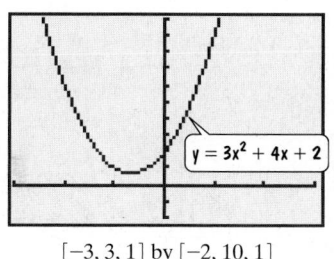

$[-3, 3, 1]$ by $[-2, 10, 1]$

$$x = \frac{-b \pm \sqrt{b^2 - 4ac}}{2a} \qquad \text{Use the quadratic formula.}$$

$$x = \frac{-4 \pm \sqrt{4^2 - 4 \cdot 3 \cdot 2}}{2 \cdot 3} \qquad \begin{array}{l} \text{Substitute the values for } a, b, \\ \text{and } c: a = 3, b = 4, \text{ and } c = 2. \end{array}$$

$$= \frac{-4 \pm \sqrt{16 - 24}}{6} \qquad \text{Multiply under the radical.}$$

$$= \frac{-4 \pm \sqrt{-8}}{6} \qquad \text{Subtract under the radical.}$$

$$= \frac{-4 \pm 2\sqrt{2}i}{6} \qquad \begin{array}{l} \sqrt{-8} = \sqrt{8(-1)} = \sqrt{8}\sqrt{-1} = \sqrt{8}i \\ = \sqrt{4 \cdot 2}i = 2\sqrt{2}i \end{array}$$

$$= \frac{2(-2 \pm \sqrt{2}i)}{6} \qquad \text{Factor out 2 from the numerator.}$$

$$= \frac{-2 \pm \sqrt{2}i}{3} \qquad \begin{array}{l} \text{Divide the numerator and} \\ \text{denominator by 2.} \end{array}$$

$$= -\frac{2}{3} \pm \frac{\sqrt{2}}{3}i \qquad \text{Write in the form } a + bi.$$

The solutions are $-\dfrac{2}{3} \pm \dfrac{\sqrt{2}}{3}i$, and the solution set is $\left\{ -\dfrac{2}{3} + \dfrac{\sqrt{2}}{3}i, -\dfrac{2}{3} - \dfrac{\sqrt{2}}{3}i \right\}$ or $\left\{ -\dfrac{2}{3} \pm \dfrac{\sqrt{2}}{3}i \right\}$. ∎

In Example 3, the solutions of $3x^2 + 2 = -4x$ are imaginary numbers. This means that the equation cannot be solved using factoring. The reason that the solutions are imaginary numbers is that $b^2 - 4ac$, the radicand in the quadratic formula, is -8, which is negative. Notice, too, that the solutions are complex conjugates.

✔ **CHECK POINT 3** Solve using the quadratic formula: $3x^2 + 5 = -6x$.

Some rational equations can be solved using the quadratic formula. For example, consider the equation

$$3 + \frac{4}{x} = -\frac{2}{x^2}.$$

The denominators are x and x^2. The least common denominator is x^2. We clear fractions by multiplying both sides of the equation by x^2. Notice that x cannot equal zero.

$$x^2\left(3 + \frac{4}{x}\right) = x^2\left(-\frac{2}{x^2}\right), \quad x \neq 0$$

$$3x^2 + \frac{4}{x} \cdot x^2 = x^2\left(-\frac{2}{x^2}\right) \qquad \text{Use the distributive property.}$$

$$3x^2 + 4x = -2 \qquad \text{Simplify.}$$

By adding 2 to both sides, we obtain the standard form of the quadratic equation:

$$3x^2 + 4x + 2 = 0.$$

This is the equation that we solved in Example 3. The two imaginary solutions are not part of the restriction that $x \neq 0$.

Use the discriminant to determine the number and type of solutions.

The Discriminant

The quantity $b^2 - 4ac$, which appears under the radical sign in the quadratic formula, is called the **discriminant**. In Example 1, the discriminant was 36, a positive number that is a perfect square. The equation had two solutions that were rational numbers. In Example 2, the discriminant was 24, a positive number that is not a perfect square. The equation had two solutions that were irrational numbers. Finally, in Example 3, the discriminant was -8, a negative number. The equation had solutions that were imaginary numbers involving i. In this case, the graph of the corresponding quadratic function had no x-intercepts.

These observations are generalized in Table 11.1.

Table 11.1   The Discriminant and the Kinds of Solutions to $ax^2 + bx + c = 0$

Discriminant $b^2 - 4ac$	Kinds of Solutions to $ax^2 + bx + c = 0$	Graph of $y = ax^2 + bx + c$
$b^2 - 4ac > 0$	Two unequal real solutions; if a, b, and c are rational numbers and the discriminant is a perfect square, the solutions are rational. If the discriminant is not a perfect square, the solutions are irrational conjugates.	Two x-intercepts
$b^2 - 4ac = 0$	One real solution (a repeated solution) that is a real number; If a, b, and c are rational numbers, the repeated solution is also a rational number.	One x-intercept
$b^2 - 4ac < 0$	No real solution; two imaginary solutions; The solutions are complex conjugates.	No x-intercepts

EXAMPLE 4   Using the Discriminant

For each equation, compute the discriminant. Then determine the number and type of solutions:

a. $3x^2 + 4x - 5 = 0$ **b.** $9x^2 - 6x + 1 = 0$ **c.** $3x^2 - 8x + 7 = 0$.

Solution Begin by identifying the values for a, b, and c in each equation. Then compute $b^2 - 4ac$, the discriminant.

a. $3x^2 + 4x - 5 = 0$

$a = 3$ $b = 4$ $c = -5$

Substitute and compute the discriminant:

$$b^2 - 4ac = 4^2 - 4 \cdot 3(-5) = 16 - (-60) = 16 + 60 = 76.$$

The discriminant, 76, is a positive number that is not a perfect square. Thus, there are two irrational solutions. (These solutions are conjugates of each other.)

b. $9x^2 - 6x + 1 = 0$

$a = 9$ $b = -6$ $c = 1$

Substitute and compute the discriminant:

$$b^2 - 4ac = (-6)^2 - 4 \cdot 9 \cdot 1 = 36 - 36 = 0.$$

The discriminant, 0, shows that there is only one solution. This solution is a rational number.

c. $3x^2 - 8x + 7 = 0$

$a = 3$ $b = -8$ $c = 7$

$$b^2 - 4ac = (-8)^2 - 4 \cdot 3 \cdot 7 = 64 - 84 = -20$$

The negative discriminant, -20, shows that there are two imaginary solutions. (These solutions are complex conjugates of each other.) ∎

✔ **CHECK POINT 4** For each equation, compute the discriminant. Then determine the number and type of solutions:

a. $x^2 + 6x + 9 = 0$

b. $2x^2 - 7x - 4 = 0$

c. $3x^2 - 2x + 4 = 0.$

3 Determine the most efficient method to use when solving a quadratic equation.

Determining Which Method to Use

All quadratic equations can be solved by the quadratic formula. However, if an equation is in the form $u^2 = d$, such as $x^2 = 5$ or $(2x + 3)^2 = 8$, it is faster to use the square root property, taking the square root of both sides. If the equation is not in the form $u^2 = d$, write the quadratic equation in standard form $(ax^2 + bx + c = 0)$. Try to solve the equation by the factoring method. If $ax^2 + bx + c$ cannot be factored, then solve the quadratic equation by the quadratic formula.

Because we used the method of completing the square to derive the quadratic formula, we no longer need it for solving quadratic equations. However, we will use completing the square in Chapter 13 to help graph certain kinds of equations.

Table 11.2 summarizes our observations about which technique to use when solving a quadratic equation.

Table 11.2   Determining the Most Efficient Technique to Use When Solving a Quadratic Equation

Description and Form of the Quadratic Equation	Most Efficient Solution Method	Example
$ax^2 + bx + c = 0$ and $ax^2 + bx + c$ can be factored easily.	Factor and use the zero-product principle.	$3x^2 + 5x - 2 = 0$ $(3x - 1)(x + 2) = 0$ $3x - 1 = 0$ or $x + 2 = 0$ $x = \dfrac{1}{3}$ $x = -2$
$ax^2 + c = 0$ The quadratic equation has no x-term. $(b = 0)$	Solve for x^2 and apply the square root property.	$4x^2 - 7 = 0$ $4x^2 = 7$ $x^2 = \dfrac{7}{4}$ $x = \pm\dfrac{\sqrt{7}}{2}$
$u^2 = d$; u is a first-degree polynomial.	Use the square root property.	$(x + 4)^2 = 5$ $x + 4 = \pm\sqrt{5}$ $x = -4 \pm \sqrt{5}$
$ax^2 + bx + c = 0$ and $ax^2 + bx + c$ cannot be factored or the factoring is too difficult.	Use the quadratic formula: $x = \dfrac{-b \pm \sqrt{b^2 - 4ac}}{2a}$.	$x^2 - 2x - 6 = 0$ $\boxed{a = 1}\ \boxed{b = -2}\ \boxed{c = -6}$ $x = \dfrac{-(-2) \pm \sqrt{(-2)^2 - 4(1)(-6)}}{2(1)}$ $= \dfrac{2 \pm \sqrt{4 - 4(1)(-6)}}{2(1)}$ $= \dfrac{2 \pm \sqrt{28}}{2} = \dfrac{2 \pm \sqrt{4}\sqrt{7}}{2}$ $= \dfrac{2 \pm 2\sqrt{7}}{2} = \dfrac{2(1 \pm \sqrt{7})}{2}$ $= 1 \pm \sqrt{7}$

4 Write quadratic equations from solutions.

Writing Quadratic Equations from Solutions

Using the zero-product principle, the equation $(x - 3)(x + 5) = 0$ has two solutions, 3 and -5. By applying the zero-product principle in reverse, we can find a quadratic equation that contains two given solutions.

The Zero-Product Principle in Reverse
If $A = 0$ or $B = 0$, then $AB = 0$.

EXAMPLE 5   Writing Equations from Solutions

Write a quadratic equation with the given solution set:

a. $\left\{-\dfrac{5}{3}, \dfrac{1}{2}\right\}$ **b.** $\{-5i, 5i\}$.

Solution

Jasper Johns, "Zero". © Jasper Johns/Licensed by VAGA, New York, NY.

The special properties of zero make it possible to write a quadratic equation from its solutions.

a. Because the solution set is $\left\{-\dfrac{5}{3}, \dfrac{1}{2}\right\}$, then

$$x = -\frac{5}{3} \quad \text{or} \quad x = \frac{1}{2}.$$

$$x + \frac{5}{3} = 0 \quad \text{or} \quad x - \frac{1}{2} = 0 \qquad \text{Obtain zero on one side of each equation.}$$

$$3x + 5 = 0 \quad \text{or} \quad 2x - 1 = 0 \qquad \text{Clear fractions, multiplying by 3 and 2, respectively.}$$

$$(3x + 5)(2x - 1) = 0 \qquad \text{Use the zero-product principle in reverse: If A = 0 or B = 0, then AB = 0.}$$

$$6x^2 - 3x + 10x - 5 = 0 \qquad \text{Use the FOIL method to multiply.}$$

$$6x^2 + 7x - 5 = 0. \qquad \text{Combine like terms.}$$

Thus, the equation is $6x^2 + 7x - 5 = 0$. Many other quadratic equations have $\left\{-\dfrac{5}{3}, \dfrac{1}{2}\right\}$ for their solution sets. These equations can be obtained by multiplying both sides of $6x^2 + 7x - 5 = 0$ by any nonzero real number.

b. Because the solution set is $\{-5i, 5i\}$, then

$$x = -5i \quad \text{or} \quad x = 5i$$

$$x + 5i = 0 \quad \text{or} \quad x - 5i = 0 \qquad \text{Obtain zero on one side of each equation.}$$

$$(x + 5i)(x - 5i) = 0 \qquad \text{Use the zero-product principle in reverse: If A = 0 or B = 0, then AB = 0.}$$

$$x^2 - (5i)^2 = 0 \qquad \text{Multiply conjugates using } (A + B)(A - B) = A^2 - B^2.$$

$$x^2 - 25i^2 = 0 \qquad (5i)^2 = 5^2 i^2 = 25i^2$$

$$x^2 - 25(-1) = 0 \qquad i^2 = -1$$

$$x^2 + 25 = 0. \qquad \text{This is the required equation.} \qquad \blacksquare$$

✔ **CHECK POINT 5** Write a quadratic equation with the given solution set:

a. $\left\{-\dfrac{3}{5}, \dfrac{1}{4}\right\}$

b. $\{-7i, 7i\}$.

5 Use the quadratic formula to solve problems.

Applications

Quadratic equations can be solved to answer questions about variables contained in quadratic functions.

EXAMPLE 6   AIDS Deaths in the United States

The bar graph in Figure 11.5 shows the cumulative number of deaths from AIDS in the United States for each year from 1990 through 1999. The cumulative number of U.S. AIDS deaths, $f(x)$, in thousands, x years after 1990 can be modeled by the quadratic function

$$f(x) = -1.65x^2 + 51.8x + 111.44.$$

According to the model, in which year will the total number of U.S. AIDS deaths reach 500 thousand?

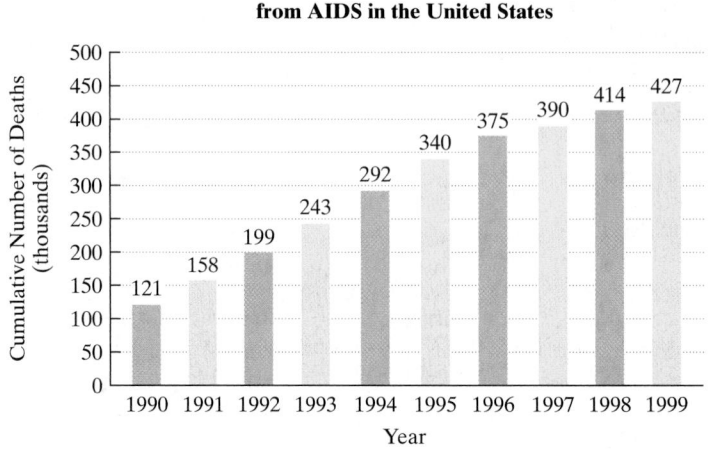

Cumulative Number of Deaths from AIDS in the United States

Figure 11.5   *Source*: Centers for Disease Control

Solution   Because we are interested in when the number of deaths will reach 500 thousand, we substitute 500 for $f(x)$ in the given function. Then we solve for x, the number of years after 1990.

$f(x) = -1.65x^2 + 51.8x + 111.44$	This is the given function.
$500 = -1.65x^2 + 51.8x + 111.44$	Substitute 500 for $f(x)$.
$0 = -1.65x^2 + 51.8x - 388.56$	Subtract 500 from both sides and write the quadratic equation in standard form.

$a = -1.65$ $b = 51.8$ $c = -388.56$

Because the trinomial on the right side of the equation is not easily factored, if it can be factored at all, we solve using the quadratic formula.

$$x = \frac{-b \pm \sqrt{b^2 - 4ac}}{2a}$$ Use the quadratic formula.

$$= \frac{-51.8 \pm \sqrt{(51.8)^2 - 4(-1.65)(-388.56)}}{2(-1.65)}$$ Substitute the values for a, b, and c: $a = -1.65$, $b = 51.8$, $c = -388.56$.

$$= \frac{-51.8 \pm \sqrt{118.744}}{-3.3}$$ Use a calculator to simplify the radicand.

$$x = \frac{-51.8 + \sqrt{118.744}}{-3.3} \quad \text{or} \quad x = \frac{-51.8 - \sqrt{118.744}}{-3.3}$$

$$x \approx 12 \qquad\qquad\qquad x \approx 19 \qquad\qquad \text{Use a calculator.}$$

According to the model, approximately 12 years after 1990, in 2002, cumulative U.S. AIDS deaths will reach 500 thousand. We must be careful not to extend the model too far into the future. The approximate solution, 19, makes no sense. The only way that cumulative deaths can reach 500 thousand in 2002 and again in $1990 + 19$, or in 2009, is that no AIDS deaths occur between 2002 and 2009. This seems unlikely, indicating that model breakdown has occurred. ∎

✔ **CHECK POINT 6** According to the model in Example 6, in which year did the cumulative number of deaths from AIDS in the United States reach 330 thousand? What was the actual number for that year? How well does the function describe the situation for that year?

EXERCISE SET 11.2

Practice Exercises

In Exercises 1–16, solve each equation using the quadratic formula. Simplify solutions, if possible.

1. $x^2 + 8x + 12 = 0$

2. $x^2 + 8x + 15 = 0$

3. $2x^2 - 7x = -5$

4. $5x^2 + 8x = -3$

5. $x^2 + 3x - 20 = 0$

6. $x^2 + 5x - 10 = 0$

7. $3x^2 - 7x = 3$

8. $4x^2 + 3x = 2$

9. $6x^2 = 2x + 1$

10. $2x^2 = -4x + 5$

11. $4x^2 - 3x = -6$

12. $9x^2 + x = -2$

13. $x^2 - 4x + 8 = 0$

14. $x^2 + 6x + 13 = 0$

15. $3x^2 = 8x - 7$

16. $3x^2 = 4x - 6$

In Exercises 17–28, compute the discriminant. Then determine the number and type of solutions for the given equation.

17. $x^2 + 8x + 3 = 0$

18. $x^2 + 7x + 4 = 0$

19. $x^2 + 6x + 8 = 0$

20. $x^2 + 2x - 3 = 0$

21. $2x^2 + x + 3 = 0$

22. $2x^2 - 4x + 3 = 0$

23. $2x^2 + 6x = 0$

24. $3x^2 - 5x = 0$

25. $5x^2 + 3 = 0$

26. $5x^2 + 4 = 0$

27. $9x^2 = 12x - 4$

28. $4x^2 = 20x - 25$

In Exercises 29–44, solve each equation by the method of your choice. Simplify solutions, if possible.

29. $3x^2 - 4x = 4$

30. $2x^2 - x = 1$

31. $x^2 - 2x = 1$

32. $2x^2 + 3x = 1$

33. $(2x - 5)(x + 1) = 2$

34. $(2x + 3)(x + 4) = 1$

35. $(3x - 4)^2 = 16$

36. $(2x + 7)^2 = 25$

37. $\dfrac{x^2}{2} + 2x + \dfrac{2}{3} = 0$

38. $\dfrac{x^2}{3} - x - \dfrac{1}{6} = 0$

39. $(3x - 2)^2 = 10$

40. $(4x - 1)^2 = 15$

41. $\dfrac{1}{x} + \dfrac{1}{x + 2} = \dfrac{1}{3}$

42. $\dfrac{1}{x} + \dfrac{1}{x + 3} = \dfrac{1}{4}$

43. $(2x - 6)(x + 2) = 5(x - 1) - 12$

44. $7x(x - 2) = 3 - 2(x + 4)$

In Exercises 45–58, write a quadratic equation in standard form with the given solution set.

45. $\{-3, 5\}$

46. $\{-2, 6\}$

47. $\left\{-\dfrac{2}{3}, \dfrac{1}{4}\right\}$

48. $\left\{-\dfrac{5}{6}, \dfrac{1}{3}\right\}$

49. $\{-6i, 6i\}$

50. $\{-8i, 8i\}$

51. $\{-\sqrt{2}, \sqrt{2}\}$

52. $\{-\sqrt{3}, \sqrt{3}\}$

53. $\{-2\sqrt{5}, 2\sqrt{5}\}$

54. $\{-3\sqrt{5}, 3\sqrt{5}\}$

55. $\{1 + i, 1 - i\}$

56. $\{2 + i, 2 - i\}$

57. $\{1 + \sqrt{2}, 1 - \sqrt{2}\}$

58. $\{1 + \sqrt{3}, 1 - \sqrt{3}\}$

Application Exercises

The bar graph shows the number of convictions of police officers throughout the United States for seven years. The number of police officers convicted of felonies, $f(x)$, x years after 1990 can be modeled by the quadratic function

$$f(x) = 23.4x^2 - 259.1x + 815.8.$$

Use the function to solve Exercises 59–60.

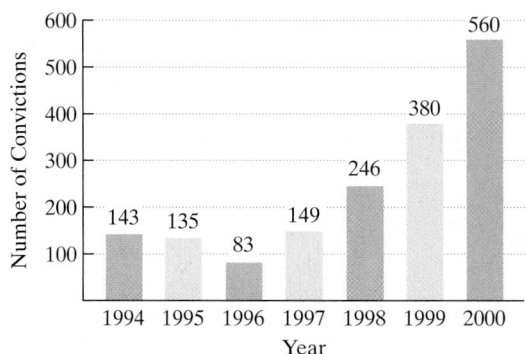

Convictions of Police Officers

Source: F.B.I.

59. In which year will 1000 police officers be convicted of felonies?

60. In which year after 1994 were 250 police officers convicted of felonies? How well does the function model the actual number of convictions for that year shown in the bar graph?

A driver's age has something to do with his or her chance of getting into a fatal car crash. The bar graph shows the number of fatal vehicle crashes per 100 million miles driven for drivers of various age groups. For example, 25-year-old drivers are involved in 4.1 fatal crashes per 100 million miles driven. Thus, when a group of 25-year-old Americans have driven a total of 100 million miles, approximately 4 have been in accidents in which someone died.

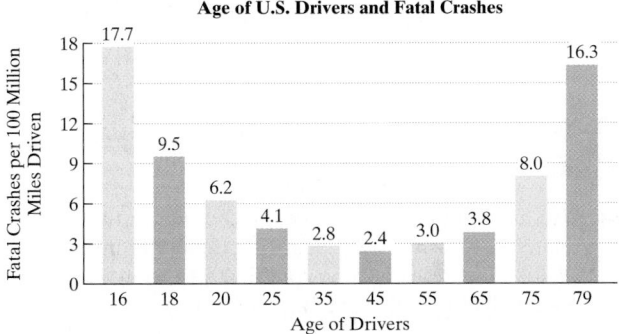

Age of U.S. Drivers and Fatal Crashes

Source: Insurance Institute for Highway Safety

The number of fatal vehicle crashes per 100 million miles, $f(x)$, for drivers of age x can be modeled by the quadratic function

$$f(x) = 0.013x^2 - 1.19x + 28.24.$$

Use the function to solve Exercises 61–62.

61. What age groups are expected to be involved in 10 fatal crashes per 100 million miles driven? How well does the function model the trend in the actual data shown in the bar graph?

62. What age groups are expected to be involved in 3 fatal crashes per 100 million miles driven? How well does the function model the trend in the actual data shown in the bar graph?

63. A football is kicked straight up from a height of 4 feet with an initial speed of 60 feet per second. The function

$$f(t) = -16t^2 + 60t + 4$$

models the ball's height above the ground, $f(t)$, in feet, t seconds after it was kicked. How long will it take for the football to hit the ground? Use a calculator and round to the nearest tenth of a second.

64. Standing on a platform 50 feet high, a person accidentally fires a gun straight into the air. The function

$$f(t) = -16t^2 + 100t + 50$$

models the bullet's height above the ground, $f(t)$, in feet, t seconds after the gun was fired. How long will it take

for the bullet to hit the ground? Use a calculator and round to the nearest tenth of a second.

65. The length of a rectangle is 3 meters longer than the width. If the area is 36 square meters, find the rectangle's dimensions. Round to the nearest tenth of a meter.

66. The length of a rectangle is 2 meters longer than the width. If the area is 10 square meters, find the rectangle's dimensions. Round to the nearest tenth of a meter.

67. The hypotenuse of a right triangle is 4 feet long. One leg is 1 foot longer than the other. Find the lengths of the legs. Round to the nearest tenth of a foot.

68. The hypotenuse of a right triangle is 6 feet long. One leg is 1 foot shorter than the other. Find the lengths of the legs. Round to the nearest tenth of a foot.

69. A rain gutter is made from sheets of aluminum that are 20 inches wide. As shown in the figure, the edges are turned up to form right angles. Determine the depth of the gutter that will allow a cross-sectional area of 13 square inches. Show that there are two different solutions to the problem. Round to the nearest tenth of an inch.

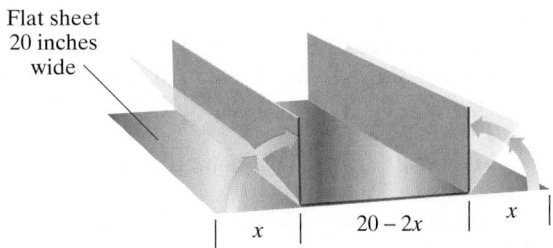

Flat sheet
20 inches
wide

x $20 - 2x$ x

70. A piece of wire is 8 inches long. The wire is cut into two pieces and then each piece is bent into a square. Find the length of each piece if the sum of the areas of these squares is to be 2 square inches.

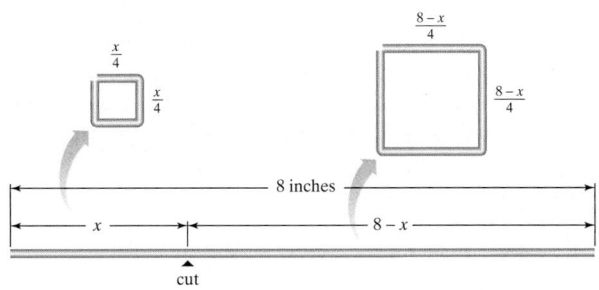

$\frac{x}{4}$

$\frac{x}{4}$

$\frac{8-x}{4}$

$\frac{8-x}{4}$

8 inches

x $8 - x$

cut

Writing in Mathematics

71. What is the quadratic formula and why is it useful?

72. Without going into specific details for every step, describe how the quadratic formula is derived.

73. Explain how to solve $x^2 + 6x + 8 = 0$ using the quadratic formula.

74. If a quadratic equation has imaginary solutions, how is this shown on the graph of the corresponding quadratic function?

75. What is the discriminant, and what information does it provide about a quadratic equation?

76. If you are given a quadratic equation, how do you determine which method to use to solve it?

77. Explain how to write a quadratic equation from its solution set. Give an example with your explanation.

78. Describe the trend shown by the data for the convictions of police officers in the graph for Exercises 59–60. Do you believe that this trend is likely to continue or might something occur that would make it impossible to extend the model into the future? Explain your answer.

Critical Thinking Exercises

79. Which one of the following is true?
   a. The quadratic formula is developed by applying factoring and the zero-product principle to the quadratic equation $ax^2 + bx + c = 0$.
   b. In using the quadratic formula to solve the quadratic equation $5x^2 = 2x - 7$, we have $a = 5, b = 2$, and $c = -7$.
   c. The quadratic formula cannot be used to solve the equation $x^2 - 9 = 0$.
   d. Any quadratic equation that can be solved by completing the square can be solved by the quadratic formula.

80. Solve: $x^2 - 3\sqrt{2}x = -2$.

81. Solve: $10^{-4}x^2 + 2 \cdot 10^{-3}x + 10^{-2} = 0$.

82. A rectangular swimming pool is 12 meters long and 8 meters wide. A tile border of uniform width is to be built around the pool using 120 square meters of tile. The tile is from a discontinued stock (so no additional materials are available), and all 120 square meters are to be used. How wide should the border be? Round to the nearest tenth of a meter. If zoning laws require at least a 2-meter-wide border around the pool, can this be done with the available tile?

83. An owner of a large diving boat that can carry as many as 70 people charges $10 per passenger to groups of between 15 and 20 people. If more than 20 divers charter the boat, the fee per passenger is decreased by 10 cents times the number of people exceeding 20. How many people must there be in a diving group to generate an income of $323.90?

Technology Exercises

84. Use a graphing utility to graph the quadratic function related to any five of the quadratic equations in Exercises 17–28. How does each graph illustrate what you determined algebraically using the discriminant?

85. Use a graphing utility to obtain a graph of the quadratic function in Exercise 63 or Exercise 64. $\boxed{\text{TRACE}}$ along the curve and verify what you determined algebraically about when the football or bullet hits the ground.

86. Reread Exercise 69. The cross-sectional area of the gutter is given by the quadratic function

$$f(x) = x(20 - 2x).$$

Graph the function in a $[0, 10, 1]$ by $[0, 60, 5]$ viewing rectangle. Then $\boxed{\text{TRACE}}$ along the curve or use the maximum function feature to determine the depth of the gutter that will maximize its cross-sectional area and allow the greatest amount of water to flow. What is the maximum area? Does the situation described in Exercise 69 take full advantage of the sheets of aluminum?

Review Exercises

87. Solve: $|5x + 2| = |4 - 3x|$. (Section 9.3, Example 2)

88. Solve: $\sqrt{2x - 5} - \sqrt{x - 3} = 1$. (Section 10.6, Example 4)

89. Rationalize the denominator: $\dfrac{5}{\sqrt{3} + x}$. (Section 10.5, Example 5)

▶ SECTION 11.3   *Quadratic Functions and Their Graphs*

Objectives

1 Recognize characteristics of parabolas.

2 Graph parabolas in the form $f(x) = a(x - h)^2 + k$.

3 Graph parabolas in the form $f(x) = ax^2 + bx + c$.

4 Solve problems involving minimizing or maximizing quadratic functions.

SSM
PH Tutor   CD-   Video
Center   ROM

The wage gap is used to desribe women's earnings relative to men's. The wage gap is expressed as a percent and is calculated by dividing the median annual earnings for women by the median annual earnings for men. In 1999, the wage gap was approximately 77%. Thus, women earned approximately 77% as much as men did. Median weekly earnings of female full-time workers were \$473, compared to \$618 for their male counterparts.

The quadratic function

$$f(x) = 0.022x^2 - 0.4x + 60.07$$

models the wage gap, $f(x)$, x years after 1960. By graphing the function, we can get a good idea of the trend for the wage gap, including the year in which women's earnings relative to men's were the lowest. In this section, we study quadratic functions and their graphs.

1 Recognize characteristics of parabolas.

Graphs of Quadratic Functions

The graph of the quadratic function

$$f(x) = ax^2 + bx + c, \quad a \neq 0,$$

is called a **parabola**. Parabolas are shaped like cups, as shown in Figure 11.6. If the coefficient of x^2 (the value of a in $ax^2 + bx + c$) is positive, the parabola opens upward. If the coefficient of x^2 is negative, the parabola opens downward. The **vertex** (or turning point) of the parabola is the lowest point on the graph when it opens upward and the highest point on the graph when it opens downward.

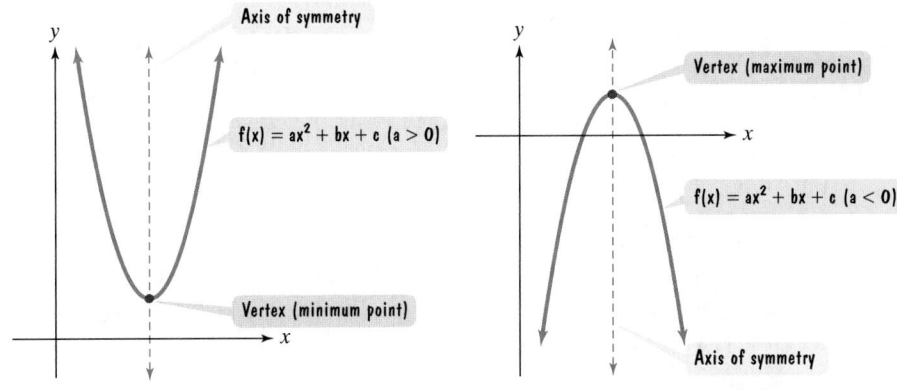

$a > 0$: Parabola opens upward. $a < 0$: Parabola opens downward.

Figure 11.6 Characteristics of graphs of quadratic functions

Look at the unusual image of the word "mirror" shown below. The artist, Scott Kim, has created the image so that the two halves of the whole are mirror images of each other. A parabola shares this kind of symmetry, in which a line through the vertex divides the figure in half. Parabolas are symmetric with respect to this line, called the **axis of symmetry**. If a parabola is folded along its axis of symmetry, the two halves match exactly.

2 Graph parabolas in the form $f(x) = a(x - h)^2 + k$.

Graphing Quadratic Functions in the Form $f(x) = a(x - h)^2 + k$

One way to obtain the graph of a quadratic function is to use point plotting. Let's begin by graphing the functions $f(x) = x^2$, $g(x) = 2x^2$, and $h(x) = \frac{1}{2}x^2$ in the same rectangular coordinate system. Select integers for x, starting with -3 and ending with 3. The partial table of coordinates for each function is shown below. The three parabolas are shown in Figure 11.7.

x	$f(x) = x^2$	(x, y) or $(x, f(x))$
-3	$f(-3) = (-3)^2 = 9$	$(-3, 9)$
-2	$f(-2) = (-2)^2 = 4$	$(-2, 4)$
-1	$f(-1) = (-1)^2 = 1$	$(-1, 1)$
0	$f(0) = 0^2 = 0$	$(0, 0)$
1	$f(1) = 1^2 = 1$	$(1, 1)$
2	$f(2) = 2^2 = 4$	$(2, 4)$
3	$f(3) = 3^2 = 9$	$(3, 9)$

x	$g(x) = 2x^2$	(x, y) or $(x, g(x))$
-3	$g(-3) = 2(-3)^2 = 18$	$(-3, 18)$
-2	$g(-2) = 2(-2)^2 = 8$	$(-2, 8)$
-1	$g(-1) = 2(-1)^2 = 2$	$(-1, 2)$
0	$g(0) = 2 \cdot 0^2 = 0$	$(0, 0)$
1	$g(1) = 2 \cdot 1^2 = 2$	$(1, 2)$
2	$g(2) = 2 \cdot 2^2 = 8$	$(2, 8)$
3	$g(3) = 2 \cdot 3^2 = 18$	$(3, 18)$

x	$h(x) = \dfrac{1}{2}x^2$	(x, y) or $(x, h(x))$
-3	$h(-3) = \dfrac{1}{2}(-3)^2 = \dfrac{9}{2}$	$\left(-3, \dfrac{9}{2}\right)$
-2	$h(-2) = \dfrac{1}{2}(-2)^2 = 2$	$(-2, 2)$
-1	$h(-1) = \dfrac{1}{2}(-1)^2 = \dfrac{1}{2}$	$\left(-1, \dfrac{1}{2}\right)$
0	$h(0) = \dfrac{1}{2} \cdot 0^2 = 0$	$(0, 0)$
1	$h(1) = \dfrac{1}{2} \cdot 1^2 = \dfrac{1}{2}$	$\left(1, \dfrac{1}{2}\right)$
2	$h(2) = \dfrac{1}{2} \cdot 2^2 = 2$	$(2, 2)$
3	$h(3) = \dfrac{1}{2} \cdot 3^2 = \dfrac{9}{2}$	$\left(3, \dfrac{9}{2}\right)$

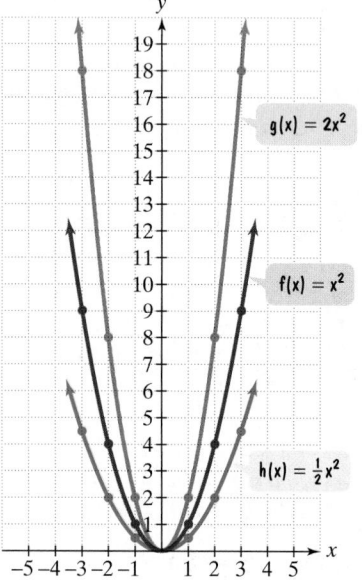

Figure 11.7

Can you see that the graphs of f, g, and h all have the same vertex, $(0, 0)$? They also have the same axis of symmetry, the y-axis, or $x = 0$. This is true for all graphs of the form $f(x) = ax^2$. However, the blue graph of $g(x) = 2x^2$ is a narrower parabola than the red graph of $f(x) = x^2$. By contrast, the green graph of $h(x) = \frac{1}{2}x^2$ is a flatter parabola than the red graph of $f(x) = x^2$.

Is there a more efficient method than point plotting to obtain the graph of a quadratic function? The answer is yes. The method is based on comparing graphs of the form $g(x) = a(x - h)^2 + k$ to those of the form $f(x) = ax^2$.

In Figure 11.8(a), the graph of $f(x) = ax^2$ for $a > 0$ is shown in black. The parabola's vertex is $(0, 0)$ and it opens upward. In Figure 11.8(b), the graph of $f(x) = ax^2$ for $a < 0$ is shown in black. The parabola's vertex is $(0, 0)$ and it opens downward.

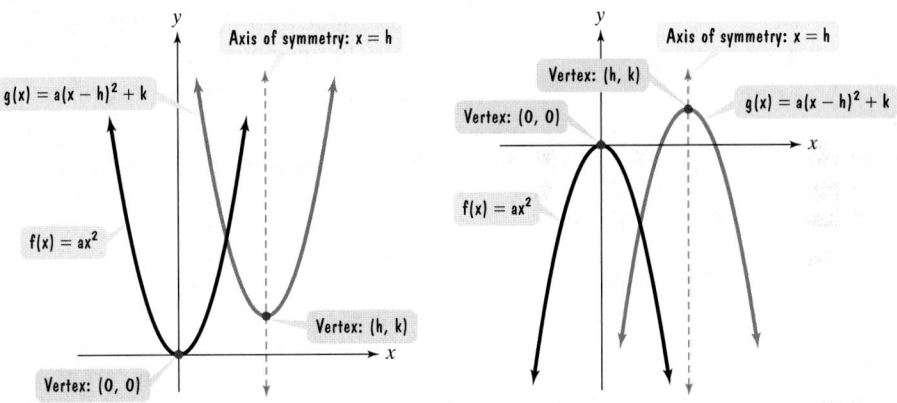

(a) $a > 0$: Parabola opens upward.   (b) $a < 0$: Parabola opens downward.

Figure 11.8 Moving, or shifting, the graph of $f(x) = ax^2$

Figure 11.8 also shows the graph of $g(x) = a(x - h)^2 + k$ in blue. Compare these graphs to those of $f(x) = ax^2$. Observe that h causes a horizontal move, or shift, and k causes a vertical shift of the graph of $f(x) = ax^2$. Consequently, the vertex $(0, 0)$ on the graph of $f(x) = ax^2$ moves to the point (h, k) on the graph of $g(x) = a(x - h)^2 + k$. The axis of symmetry is the vertical line whose equation is $x = h$.

The form of the expression for g is convenient because it immediately identifies the vertex of the parabola as (h, k). The sign of a in $g(x) = a(x - h)^2 + k$ determines whether the parabola opens upward or downward. Furthermore, if $|a|$ is small, the parabola opens more flatly than if $|a|$ is large.

Graphing Quadratic Functions with Equations in the Form $f(x) = a(x - h)^2 + k$

To graph $f(x) = a(x - h)^2 + k$:

1. Determine whether the parabola opens upward or downward. If $a > 0$, it opens upward. If $a < 0$, it opens downward.
2. Determine the vertex of the parabola. The vertex is (h, k).
3. Find any x-intercepts by replacing $f(x)$ with 0. Solve the resulting quadratic equation for x.
4. Find the y-intercept by replacing x with 0.
5. Plot the intercepts and vertex and additional points as necessary. Connect these points with a smooth curve that is shaped like a cup.

EXAMPLE 1   Graphing a Quadratic Function in the Form $f(x) = a(x - h)^2 + k$

Graph the quadratic function $f(x) = -2(x - 3)^2 + 8$.

Solution   We can graph this function by following the steps in the preceding box. We begin by identifying values for a, h, and k.

$$f(x) = a(x - h)^2 + k$$

$$a = -2 \quad h = 3 \quad k = 8$$

$$f(x) = -2(x - 3)^2 + 8$$

Step 1 Determine how the parabola opens. Note that a, the coefficient of x^2, is -2. Thus, $a < 0$; this negative value tells us that the parabola opens downward.

Step 2 Find the vertex. The vertex of the parabola is at (h, k). Because $h = 3$ and $k = 8$, the parabola has its vertex at $(3, 8)$.

Step 3 Find the x-intercepts. Replace $f(x)$ with 0 in $f(x) = -2(x - 3)^2 + 8$.

$0 = -2(x - 3)^2 + 8$ Find x-intercepts, setting f(x) equal to 0.

$2(x - 3)^2 = 8$ Solve for x. Add $2(x - 3)^2$ to both sides of the equation.

$(x - 3)^2 = 4$ Divide both sides by 2.

$x - 3 = \sqrt{4}$ or $x - 3 = -\sqrt{4}$ Apply the square root property.

$x - 3 = 2$ $x - 3 = -2$ $\sqrt{4} = 2$

$x = 5$ $x = 1$ Add 3 to both sides in each equation.

The x-intercepts are 5 and 1. The parabola passes through $(5, 0)$ and $(1, 0)$.

Step 4 Find the y-intercept. Replace x with 0 in $f(x) = -2(x - 3)^2 + 8$.

$$f(0) = -2(0 - 3)^2 + 8 = -2(-3)^2 + 8 = -2(9) + 8 = -10$$

The y-intercept is -10. The parabola passes through $(0, -10)$.

Step 5 Graph the parabola. With a vertex at $(3, 8)$, x-intercepts at 5 and 1, and a y-intercept at -10, the graph of f is shown in Figure 11.9. The axis of symmetry is the vertical line whose equation is $x = 3$. ∎

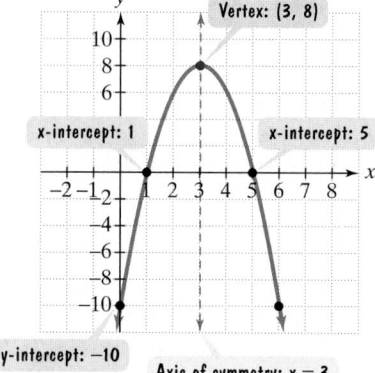

Figure 11.9 The graph of $f(x) = -2(x - 3)^2 + 8$

✔ **CHECK POINT 1**   Graph the quadratic function $f(x) = -(x - 1)^2 + 4$.

EXAMPLE 2   Graphing a Quadratic Function in the Form $f(x) = a(x - h)^2 + k$

Graph the quadratic function $f(x) = (x + 3)^2 + 1$.

Solution   We begin by finding values for a, h, and k.

$$f(x) = a(x - h)^2 + k$$ Form of given equation

$$f(x) = (x + 3)^2 + 1$$ Given equation

or $$f(x) = 1(x - (-3))^2 + 1$$

$$a = 1 \quad h = -3 \quad k = 1$$

Step 1 Determine how the parabola opens. Note that a, the coefficient of x^2, is 1. Thus, $a > 0$; this positive value tells us that the parabola opens upward.

Step 2 Find the vertex. The vertex of the parabola is at (h, k). Because $h = -3$ and $k = 1$, the parabola has its vertex at $(-3, 1)$.

Step 3 Find the x-intercepts. Replace $f(x)$ with 0 in $f(x) = (x + 3)^2 + 1$. Because the vertex is at $(-3, 1)$, which lies above the x-axis, and the parabola opens upward, it appears that this parabola has no x-intercepts. We can verify this observation algebraically.

$$0 = (x + 3)^2 + 1 \qquad \text{Find possible } x\text{-intercepts, setting } f(x) \text{ equal to 0.}$$

$$-1 = (x + 3)^2 \qquad \text{Solve for } x. \text{ Subtract 1 from both sides.}$$

$$x + 3 = \sqrt{-1} \quad \text{or} \quad x + 3 = -\sqrt{-1} \qquad \text{Apply the square root property.}$$

$$x + 3 = i \qquad\qquad x + 3 = -i \qquad \sqrt{-1} = i$$

$$x = -3 + i \qquad\qquad x = -3 - i \qquad \text{The solutions are } -3 \pm i.$$

Because this equation has no real solutions, the parabola has no x-intercepts.

Step 4 Find the y-intercept. Replace x with 0 in $f(x) = (x + 3)^2 + 1$.

$$f(0) = (0 + 3)^2 + 1 = 3^2 + 1 = 9 + 1 = 10$$

The y-intercept is 10. The parabola passes through $(0, 10)$.

Step 5 Graph the parabola. With a vertex at $(-3, 1)$, no x-intercepts, and a y-intercept at 10, the graph of f is shown in Figure 11.10. The axis of symmetry is the vertical line whose equation is $x = -3$. ∎

✔ **CHECK POINT 2** Graph the quadratic function $f(x) = (x - 2)^2 + 1$.

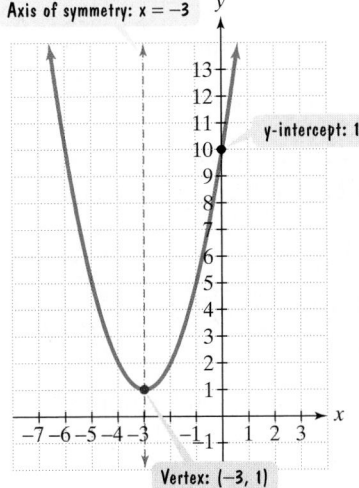

Axis of symmetry: $x = -3$

y-intercept: 10

Vertex: $(-3, 1)$

Figure 11.10 The graph of $f(x) = (x + 3)^2 + 1$

3 Graph parabolas in the form $f(x) = ax^2 + bx + c$.

Graphing Quadratic Functions in the Form $f(x) = ax^2 + bx + c$

Quadratic functions are frequently expressed in the form $f(x) = ax^2 + bx + c$. How can we identify the vertex of a parabola whose equation is in this form? Completing the square provides the answer to this question.

$$f(x) = ax^2 + bx + c$$

$$= a\left(x^2 + \frac{b}{a}x\right) + c \qquad \text{Factor out } a \text{ from } ax^2 + bx.$$

$$= a\left(x^2 + \frac{b}{a}x + \frac{b^2}{4a^2}\right) + c - a\left(\frac{b^2}{4a^2}\right)$$

Complete the square by adding the square of half the coefficient of x.

By completing the square, we added $a \cdot \dfrac{b^2}{4a^2}$. To avoid changing the function's equation, we must subtract this term.

$$= a\left(x + \frac{b}{2a}\right)^2 + c - \frac{b^2}{4a}$$

Write the trinomial as the square of a binomial and simplify the constant term.

On the next page, we compare the form of this equation with a quadratic function in the form $f(x) = a(x - h)^2 + k$.

The form we know how to graph $\quad f(x) = a(x - h)^2 + k$

$$h = -\frac{b}{2a} \qquad k = c - \frac{b^2}{4a}$$

Equation under discussion $\quad f(x) = a\left(x - \left(-\frac{b}{2a}\right)\right)^2 + c - \frac{b^2}{4a}$

The important part of this observation is that h, the x-coordinate of the vertex, is $-\dfrac{b}{2a}$. The y-coordinate can be found by evaluating the function at $-\dfrac{b}{2a}$.

The Vertex of a Parabola Whose Equation Is $f(x) = ax^2 + bx + c$

Consider the parabola defined by the quadratic function $f(x) = ax^2 + bx + c$. The parabola's vertex is at $\left(-\dfrac{b}{2a}, f\left(-\dfrac{b}{2a}\right)\right)$.

We can apply our five-step procedure and graph parabolas in the form $f(x) = ax^2 + bx + c$. The only step that is different is how we determine the vertex.

EXAMPLE 3 Graphing a Quadratic Function in the Form $f(x) = ax^2 + bx + c$

Graph the quadratic function $f(x) = -x^2 + 4x - 1$.

Solution

Step 1 Determine how the parabola opens. Note that a, the coefficient of x^2, is -1. Thus, $a < 0$; this negative value tells us that the parabola opens downward.

Step 2 Find the vertex. We know that the x-coordinate of the vertex is $x = -\dfrac{b}{2a}$. We identify a, b, and c in $f(x) = ax^2 + bx + c$.

$$f(x) = -x^2 + 4x - 1$$

$$a = -1 \quad b = 4 \quad c = -1$$

Substitute the values of a and b into the equation for the x-coordinate:

$$x = -\frac{b}{2a} = -\frac{4}{2(-1)} = \frac{-4}{-2} = 2.$$

The x-coordinate of the vertex is 2. We substitute 2 for x in the equation of the function to find the y-coordinate:

$$f(2) = -2^2 + 4 \cdot 2 - 1 = -4 + 8 - 1 = 3.$$

The vertex is at $(2, 3)$.

Step 3 Find the x-intercepts. Replace $f(x)$ with 0 in $f(x) = -x^2 + 4x - 1$. We obtain $0 = -x^2 + 4x - 1$ or $-x^2 + 4x - 1 = 0$. This equation cannot be solved by factoring. We will use the quadratic formula to solve it.

$$a = -1, \qquad b = 4, \qquad c = -1$$

$$x = \frac{-b \pm \sqrt{b^2 - 4ac}}{2a} = \frac{-4 \pm \sqrt{4^2 - 4(-1)(-1)}}{2(-1)} = \frac{-4 \pm \sqrt{16 - 4}}{-2}$$

$$x = \frac{-4 + \sqrt{12}}{-2} \approx 0.3 \quad \text{or} \quad x = \frac{-4 - \sqrt{12}}{-2} \approx 3.7$$

The x-intercepts are approximately 0.3 and 3.7. The parabola passes through $(0.3, 0)$ and $(3.7, 0)$.

Step 4 Find the y-intercept. Replace x with 0 in $f(x) = -x^2 + 4x - 1$.

$$f(0) = -0^2 + 4 \cdot 0 - 1 = -1$$

The y-intercept is -1. The parabola passes through $(0, -1)$.

Step 5 Graph the parabola. With a vertex at $(2, 3)$, x-intercepts at 0.3 and 3.7, and a y-intercept at -1, the graph of f is shown in Figure 11.11. The axis of symmetry is the vertical line whose equation is $x = 2$.

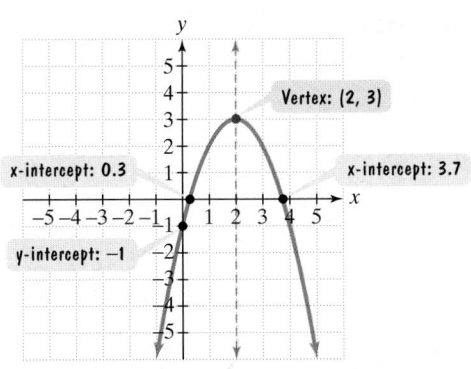

Figure 11.11 The graph of $f(x) = -x^2 + 4x - 1$

✔ **CHECK POINT 3**   Graph the quadratic function $f(x) = x^2 - 2x - 3$.

4 Solve problems involving minimizing or maximizing quadratic functions.

Applications of Quadratic Functions

When were women's earnings as a percentage of men's at the lowest? What is the age of a driver having the least number of car accidents? How many units of a product should a business manufacture to maximize its profits? The answers to these questions involve finding the maximum or minimum value of quadratic functions.

Consider the quadratic function $f(x) = ax^2 + bx + c$. If $a > 0$, the parabola opens upward and the vertex is its lowest point. If $a < 0$, the parabola opens downward and the vertex is its highest point. The x-coordinate of the vertex is $-\dfrac{b}{2a}$. Thus, we can find the minimum or maximum value of f by evaluating the quadratic function at $x = -\dfrac{b}{2a}$.

Minimum and Maximum: Quadratic Functions

Consider $f(x) = ax^2 + bx + c$.

1. If $a > 0$, then f has a minimum that occurs at $x = -\dfrac{b}{2a}$.

2. If $a < 0$, then f has a maximum that occurs at $x = -\dfrac{b}{2a}$.

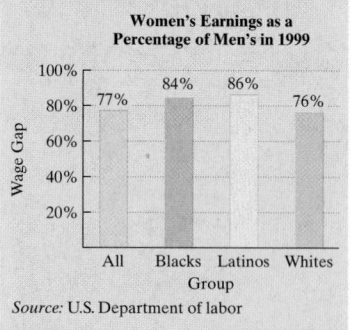
EXAMPLE 4   An Application: The Wage Gap

The function

$$f(x) = 0.022x^2 - 0.4x + 60.07$$

models women's earnings as a percentage of men's x years after 1960. In which year was this percentage at a minimum? What was the percentage for that year?

Solution   The quadratic function is in the form $f(x) = ax^2 + bx + c$ with $a = 0.022$ and $b = -0.4$. With $a > 0$, the function has a minimum when $x = -\dfrac{b}{2a}$.

$$x = -\frac{b}{2a} = -\frac{(-0.4)}{2(0.022)} \approx 9$$

This means that women's earnings as a percentage of men's were at their lowest approximately 9 years after 1960, or in 1969. The percentage for that year was

$$f(9) = 0.022(9)^2 - 0.4(9) + 60.07 \approx 58.$$

In 1969, women earned approximately 58% as much as men.   ∎

✔ CHECK POINT 4   The function $f(x) = 0.4x^2 - 36x + 1000$ models the number of accidents, $f(x)$, per 50 million miles driven, as a function of a driver's age, x, in years, where $16 \le x \le 74$. What is the age of a driver having the least number of car accidents? What is the minimum number of car accidents per 50 million miles driven?

Quadratic functions can also be formed from verbal conditions. Once we have obtained a quadratic function, we can then use the x-coordinate of the vertex to determine its maximum or minimum value. Here is a step-by-step strategy for solving these kinds of problems:

Strategy for Solving Problems Involving Maximizing or Minimizing Quadratic Functions

1. Read the problem carefully and decide which quantity is to be maximized or minimized.
2. Use the conditions of the problem to express the quantity as a function in one variable.
3. Rewrite the function in the form $f(x) = ax^2 + bx + c$.
4. If $a > 0$, f has a minimum at $x = -\dfrac{b}{2a}$. If $a < 0$, f has a maximum at $x = -\dfrac{b}{2a}$.
5. Answer the question posed in the problem.

EXAMPLE 5   Maximizing Area

You have 100 yards of fencing to enclose a rectangular region. Find the dimensions of the rectangle that maximize the enclosed area. What is the maximum area?

Solution

Step 1 Decide what must be maximized or minimized. We must maximize area. What we do not know are the rectangle's dimensions, x and y.

Step 2 Express this quantity as a function in one variable. Because we must maximize area, we have $A = xy$. We need to transform this into a function in which A is represented by one variable. Because you have 100 yards of fencing, the perimeter of the rectangle is 100 yards. This means that

$$2x + 2y = 100.$$

We can solve this equation for y in terms of x, substitute the result into $A = xy$, and obtain A as a function in one variable. We begin by solving for y.

$$2y = 100 - 2x \qquad \text{Subtract 2x from both sides.}$$

$$y = \frac{100 - 2x}{2} \qquad \text{Divide both sides by 2.}$$

$$y = 50 - x \qquad \text{Divide each term in the numerator by 2.}$$

Now we substitute $50 - x$ for y in $A = xy$.

$$A = xy = x(50 - x)$$

The rectangle and its dimensions are illustrated in Figure 11.12. Because A is now a function of x, we can write

$$A(x) = x(50 - x).$$

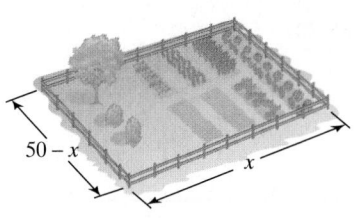

Figure 11.12 What value of x will maximize the rectangle's area?

50 − x

x

Step 3 Write the function in the form $f(x) = ax^2 + bx + c$. Applying the distributive property and rearranging terms, we obtain

$$A(x) = -x^2 + 50x.$$

Using Technology

The graph of the area function

$$A(x) = x(50 - x)$$

was obtained with a graphing utility using a [0, 50, 2] by [0, 700, 25] viewing rectangle. The maximum function feature verifies that a maximum area of 625 square feet occurs when one of the dimensions is 25 feet.

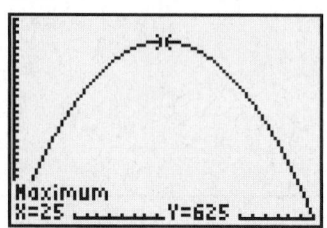

Step 4 If $a < 0$, the function has a maximum at $x = -\dfrac{b}{2a}$. Using our formula for $A(x)$, we see that $a = -1$ and $b = 50$.

$$A(x) = -x^2 + 50x$$

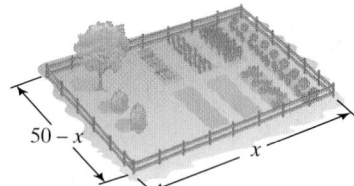

The function has a maximum at $x = -\dfrac{b}{2a}$.

$$x = -\frac{b}{2a} = -\frac{50}{2(-1)} = 25$$

Step 5 Answer the question posed by the problem. We found that $x = 25$. Figure 11.12 shows that the rectangle's other dimension is $50 - x = 50 - 25 = 25$. The dimensions that maximize area are 25 feet by 25 feet. The rectangle that gives the maximum area is actually a square with an area of 25 feet · 25 feet, or 625 square feet. ∎

Figure 11.12, repeated

✔ **CHECK POINT 5** You have 120 feet of fencing to enclose a rectangular region. Find the dimensions of the rectangle that maximize the enclosed area. What is the maximum area?

Using Technology

We've come a long way from the small nation of "embattled farmers" who launched the American Revolution. In the early days of our Republic, 95% of the population was involved in farming. The graph in Figure 11.13 shows the number of farms in the United States from 1850 through 2010 (projected). Because the graph is shaped like a cup, with an increasing number of farms from 1850 to 1910 and a decreasing number of farms from 1910 to 2010, a quadratic function is an appropriate model for the data. You can use the statistical menu of a graphing utility to enter the data in Figure 11.13. We entered the data using (number of decades after 1850, millions of U.S. farms). The data are shown to the right of Figure 11.13.

Number of U. S. Farms, 1850–2010

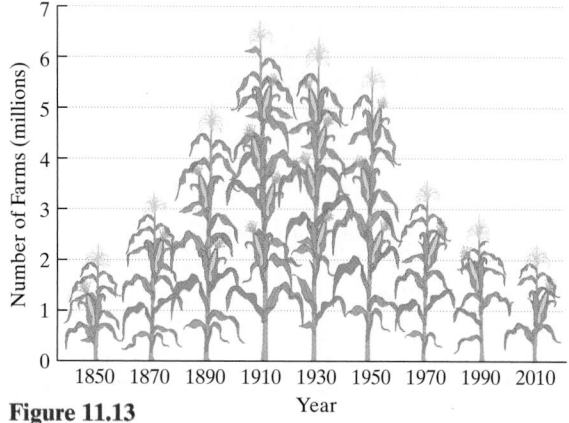

Figure 11.13

Source: U. S. Bureau of the Census

Data:
(0, 2.3), (2, 3.3), (4, 5.1),
(6, 6.7), (8, 6.4), (10, 5.8),
(12, 3.6), (14, 2.9), (16, 2.3)

```
QuadReg
 y=ax²+bx+c
 a=-.0643668831
 b=.9873701299
 c=2.203636364
```

Upon entering the QUADratic REGression program, we obtain the results shown in the screen. Thus, the quadratic function of best fit is

$$f(x) = -0.064x^2 + 0.99x + 2.2$$

where x represents the number of decades after 1850 and $f(x)$ represents the number of U.S. farms, in millions.

EXERCISE SET 11.3

Practice Exercises

In Exercises 1–4, the graph of a quadratic function is given. Write the function's equation, selecting from the following options:

$$f(x) = (x + 1)^2 - 1, \quad g(x) = (x + 1)^2 + 1,$$
$$h(x) = (x - 1)^2 + 1, \quad j(x) = (x - 1)^2 - 1.$$

In Exercises 5–8, the graph of a quadratic function is given. Write the function's equation, selecting from the following options:

$$f(x) = x^2 + 2x + 1, \quad g(x) = x^2 - 2x + 1,$$
$$h(x) = x^2 - 1, \quad j(x) = -x^2 - 1.$$

1.

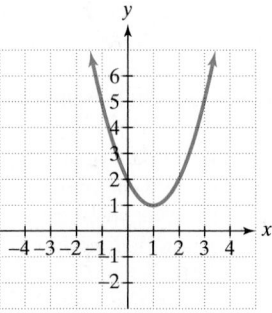

2.

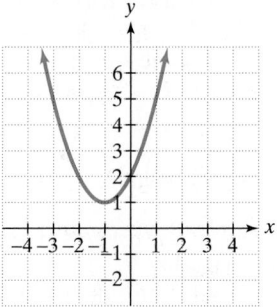

3.

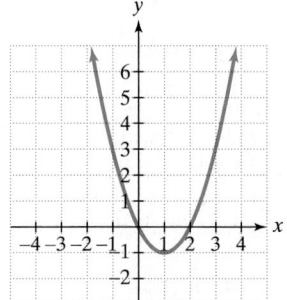

4.

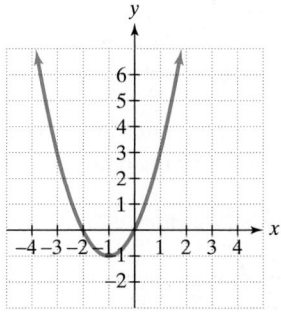

5.

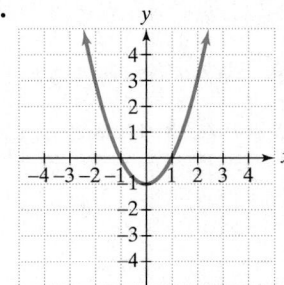

6.

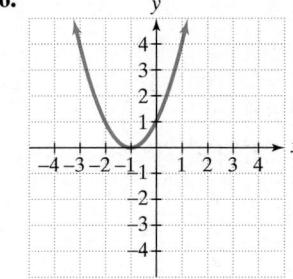

7.

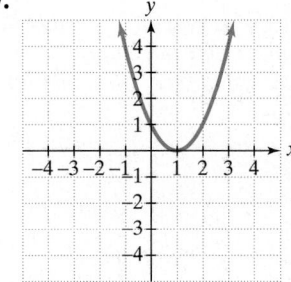

8.

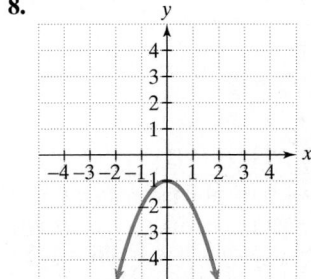

In Exercises 9–16, find the coordinates of the vertex for the parabola defined by the given quadratic function.

9. $f(x) = 2(x - 3)^2 + 1$
10. $f(x) = -3(x - 2)^2 + 12$
11. $f(x) = -2(x + 1)^2 + 5$
12. $f(x) = -2(x + 4)^2 - 8$
13. $f(x) = 2x^2 - 8x + 3$
14. $f(x) = 3x^2 - 12x + 1$
15. $f(x) = -x^2 - 2x + 8$
16. $f(x) = -2x^2 + 8x - 1$

In Exercises 17–34, use the vertex and intercepts to sketch the graph of each quadratic function. Give the equation for the parabola's axis of symmetry.

17. $f(x) = (x - 4)^2 - 1$
18. $f(x) = (x - 1)^2 - 2$
19. $f(x) = (x - 1)^2 + 2$
20. $f(x) = (x - 3)^2 + 2$
21. $y - 1 = (x - 3)^2$
22. $y - 3 = (x - 1)^2$
23. $f(x) = 2(x + 2)^2 - 1$
24. $f(x) = \frac{5}{4} - \left(x - \frac{1}{2}\right)^2$
25. $f(x) = 4 - (x - 1)^2$
26. $f(x) = 1 - (x - 3)^2$
27. $f(x) = x^2 + 2x - 3$
28. $f(x) = x^2 - 2x - 15$
29. $f(x) = x^2 + 3x - 10$
30. $f(x) = 2x^2 - 7x - 4$
31. $f(x) = 2x - x^2 + 3$
32. $f(x) = 5 - 4x - x^2$
33. $f(x) = 2x - x^2 - 2$
34. $f(x) = 6 - 4x + x^2$

In Exercises 35–40, determine, without graphing, whether the given quadratic function has a minimum value or a maximum value. Then find the coordinates of the minimum or the maximum point.

35. $f(x) = 3x^2 - 12x - 1$
36. $f(x) = 2x^2 - 8x - 3$
37. $f(x) = -4x^2 + 8x - 3$
38. $f(x) = -2x^2 - 12x + 3$
39. $f(x) = 5x^2 - 5x$
40. $f(x) = 6x^2 - 6x$

Application Exercises

41. The function
$$f(x) = -3.1x^2 + 51.4x + 4024.5$$
models the average annual per capita consumption of cigarettes, $f(x)$, by Americans 18 and older x years after 1960. According to this model, in which year did cigarette consumption per capita reach a maximum? What was the consumption for that year?

42. The function
$$f(x) = -0.0065x^2 + 0.23x + 8.47$$
models the American marriage rate, $f(x)$ (the number of marriages per 1000 people), x years after 1960. According to this model, in which year was the marriage rate the highest? What was the marriage rate for that year?

43. The function
$$f(x) = 104.5x^2 - 1501.5x + 6016$$
models the death rate per year per 100,000 males, $f(x)$, for U.S. men who average x hours of sleep each night. How many hours of sleep, to the nearest tenth of an hour, corresponds to the minimum death rate? What is this minimum death rate, to the nearest whole number?

44. There is a relationship between the amount of one's annual income, x, in thousands of dollars, and the percentage of this income, $P(x)$, that one contributes to charities. This relationship is modeled by the quadratic function
$$P(x) = 0.0014x^2 - 0.1529x + 5.855,$$
where $5 \le x \le 100$. What annual income corresponds to the minimum percentage given to charity? What is the minimum percentage?

45. A person standing close to the edge on the top of a 160-foot building throws a baseball vertically upward. The quadratic function
$$s(t) = -16t^2 + 64t + 160$$
models the ball's height above the ground, $s(t)$, in feet, t seconds after it was thrown.
    a. After how many seconds does the ball reach its maximum height? What is the maximum height?
    b. How many seconds does it take until the ball finally hits the ground? Round to the nearest tenth of a second.
    c. Find $s(0)$ and describe what this means.
    d. Use your results from parts (a) through (c) to graph the quadratic function. Begin the graph with $t = 0$ and end with the value of t for which the ball hits the ground.

46. A person standing close to the edge on the top of a 200-foot building throws a baseball vertically upward. The quadratic function
$$s(t) = -16t^2 + 64t + 200$$
models the ball's height above the ground, $s(t)$, in feet, t seconds after it was thrown.
    a. After how many seconds does the ball reach its maximum height? What is the maximum height?
    b. How many seconds does it take until the ball finally hits the ground? Round to the nearest tenth of a second.
    c. Find $s(0)$ and describe what this means.
    d. Use your results from parts (a) through (c) to graph the quadratic function. Begin the graph with $t = 0$ and end with the value of t for which the ball hits the ground.

47. Prices, as measured by the U.S. Consumer Price Index, have risen since World War II. What cost $100 in 1967, the reference year, cost $60 in 1920, $116.30 in 1970, and $511.50 by 2000. Suppose that a quadratic function is used to model the data shown in the graph with ordered pairs representing (number of years after 1920, Consumer Price Index).

Consumer Price Index, 1920–2000

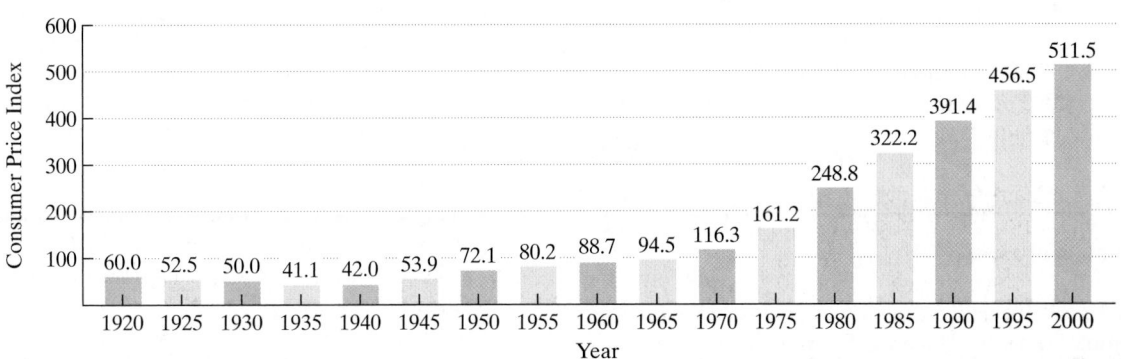

Source: U.S. Department of Labor

Determine, without obtaining an actual quadratic function that models the data, the coordinates of the vertex for the function's graph. Describe what this means in practical terms.

48. Suppose that a quadratic function is used to model the data shown in the graph with ordered pairs representing (number of years after 1971, millions of students enrolled in U.S. schools).

Millions of Students Enrolled in U.S. Schools

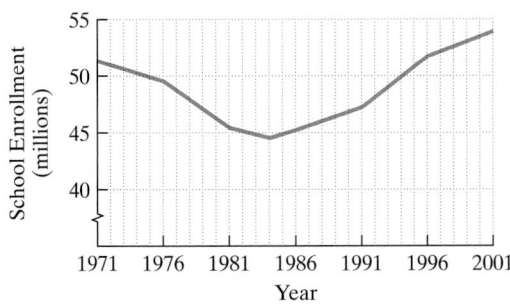

Source: National Education Association

Determine, without obtaining an actual quadratic function that models the data, the approximate coordinates of the vertex for the function's graph.

49. You have 120 feet of fencing to enclose a rectangular plot that borders on a river. If you do not fence the side along the river, find the length and width of the plot that will maximize the area. What is the largest area that can be enclosed?

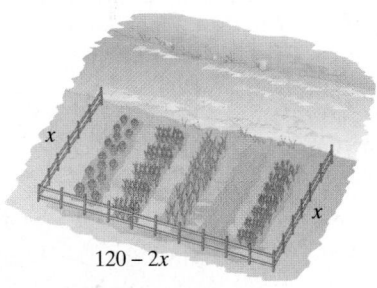

50. You have 600 feet of fencing to enclose a rectangular plot that borders on a river. If you do not fence the side along the river, find the length and width of the plot that will maximize the area. What is the largest area that can be enclosed?

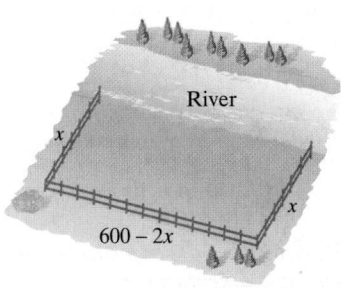

51. Among all pairs of numbers whose sum is 16, find a pair whose product is as large as possible. What is the maximum product?

52. Among all pairs of numbers whose sum is 20, find a pair whose product is as large as possible. What is the maximum product?

53. Among all pairs of numbers whose difference is 10, find a pair whose product is as small as possible. What is the minimum product?

54. Among all pairs of numbers whose difference is 8, find a pair whose product is as small as possible. What is the minimum product?

55. You have 50 yards of fencing to enclose a rectangular region. Find the dimensions of the rectangle that maximize the enclosed area. What is the maximum area?

56. You have 80 yards of fencing to enclose a rectangular region. Find the dimensions of the rectangle that maximize the enclosed area. What is the maximum area?

57. A rain gutter is made from sheets of aluminum that are 20 inches wide by turning up the edges to form right angles. Determine the depth of the gutter that will maximize its cross-sectional area and allow the greatest amount of water to flow. What is the maximum cross-sectional area?

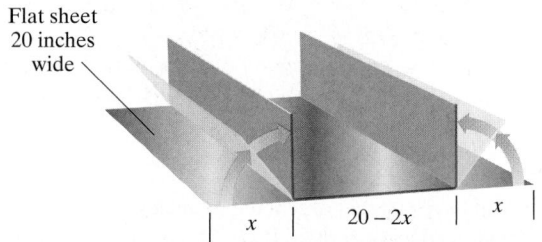

Flat sheet
20 inches
wide

x $20 - 2x$ x

58. A rain gutter is made from sheets of aluminum that are 12 inches wide by turning up the edges to form right angles. Determine the depth of the gutter that will maximize its cross-sectional area and allow the greatest amount of water to flow. What is the maximum cross-sectional area?

In Chapter 3, we saw that the profit, $P(x)$, generated after producing and selling x units of a product is given by the function

$$P(x) = R(x) - C(x)$$

where R and C are the revenue and cost functions, respectively. Use these functions to solve Exercises 59–60.

59. Hunky Beef, a local sandwich store, has a fixed weekly cost of $525.00, and variable costs for making a roast beef sandwich are $0.55.
a. Let x represent the number of roast beef sandwiches made and sold each week. Write the weekly cost function, C, for Hunky Beef.
b. The function $R(x) = -0.001x^2 + 3x$ describes the money that Hunky Beef takes in each week from the sale of x roast beef sandwiches. Use this revenue function and the cost function from part (a) to write the store's weekly profit function, P.

c. Use the store's profit function to determine the number of roast beef sandwiches it should make and sell each week to maximize profit. What is the maximum weekly profit?

60. Virtual Fido is a company that makes electronic virtual pets. The fixed weekly cost is $3000, and variable costs for each pet are $20.
a. Let x represent the number of virtual pets made and sold each week. Write the weekly cost function, C, for Virtual Fido.
b. The function $R(x) = -x^2 + 1000x$ describes the money that Virtual Fido takes in each week from the sale of x virtual pets. Use this revenue function and the cost function from part (a) to write the weekly profit function, P.

c. Use the profit function to determine the number of virtual pets that should be made and sold each week to maximize profit. What is the maximum weekly profit?

Writing in Mathematics

61. What is a parabola? Describe its shape.

62. Explain how to decide whether a parabola opens upward or downward.

63. Describe how to find a parabola's vertex if its equation is in the form $f(x) = a(x - h)^2 + k$. Give an example.

64. Describe how to find a parabola's vertex if its equation is in the form $f(x) = ax^2 + bx + c$. Use $f(x) = x^2 - 6x + 8$ as an example.

65. A parabola that opens upward has its vertex at $(1, 2)$. Describe as much as you can about the parabola based on this information. Include in your discussion the number of x-intercepts (if any) for the parabola.

66. Use the Consumer Price Index shown in Exercise 47 to answer this question: Does a worker who earned $40,000 in 1990 and $50,000 in 2000 have more buying power in 2000 than in 1990? Explain your answer.

67. The quadratic function

$$f(x) = -0.018x^2 + 1.93x - 25.34$$

describes the miles per gallon of a Ford Taurus driven at x miles per hour. Suppose that you own a Ford Taurus. Describe how you can use this function to save money.

Critical Thinking Exercises

68. Which one of the following is true?
  a. No quadratic functions have a range of $(-\infty, \infty)$.
  b. The vertex of the parabola described by $f(x) = 2(x - 5)^2 - 1$ is at $(5, 1)$.
  c. The graph of $f(x) = -2(x + 4)^2 - 8$ has one y-intercept and two x-intercepts.
  d. The maximum value of y for the quadratic function $f(x) = -x^2 + x + 1$ is 1.

In Exercises 69–70, find the axis of symmetry for each parabola whose equation is given. Use the axis of symmetry to find a second point on the parabola whose y-coordinate is the same as the given point.

69. $f(x) = 3(x + 2)^2 - 5; \quad (-1, -2)$

70. $f(x) = (x - 3)^2 + 2; \quad (6, 11)$

71. A rancher has 1000 feet of fencing to construct six corrals, as shown in the figure. Find the dimensions that maximize the enclosed area. What is the maximum area?

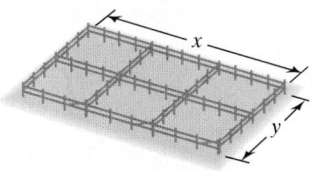

72. The annual yield per lemon tree is fairly constant at 320 pounds when the number of trees per acre is 50 or fewer. For each additional tree over 50, the annual yield per tree for all trees on the acre decreases by 4 pounds due to overcrowding. Find the number of trees that should be planted on an acre to produce the maximum yield. How many pounds is the maximum yield?

Technology Exercises

73. Use a graphing utility to verify any five of your hand-drawn graphs in Exercises 17–34.

74. a. Use a graphing utility to graph $y = 2x^2 - 82x + 720$ in a standard viewing rectangle. What do you observe?

  b. Find the coordinates of the vertex for the given quadratic function.
  c. The answer to part (b) is $(20.5, -120.5)$. Because the leading coefficient, 2, of the given function is positive, the vertex is a minimum point on the graph. Use this fact to help find a viewing rectangle that will give a relatively complete picture of the parabola. With an axis of symmetry at $x = 20.5$, the setting for x should extend past this, so try Xmin = 0 and Xmax = 30. The setting for y should include (and probably go below) the y-coordinate of the graph's minimum point, so try Ymin = -130. Experiment with Ymax until your utility shows the parabola's major features.
  d. In general, explain how knowing the coordinates of a parabola's vertex can help determine a reasonable viewing rectangle on a graphing utility for obtaining a complete picture of the parabola.

In Exercises 75–78, find the vertex for each parabola. Then determine a reasonable viewing rectangle on your graphing utility and use it to graph the parabola.

75. $y = -0.25x^2 + 40x$

76. $y = -4x^2 + 20x + 160$

77. $y = 5x^2 + 40x + 600$

78. $y = 0.01x^2 + 0.6x + 100$

79. The quadratic function $f(x) = 0.013x^2 - 0.96x + 25.4$ describes the average yearly consumption of whole milk per person, $f(x)$, in the United States x years after 1970. The linear function $g(x) = 0.41x + 6.03$ describes the average yearly consumption of low-fat milk per person, $g(x)$, in the United States x years after 1970.
  a. Use a graphing utility to graph each function in the same viewing rectangle for the years 1970 through 2000.
  b. Use the graphs to describe the trend in consumption for both types of milk. What possible explanations are there for these consumption patterns?

80. The function $y = 0.011x^2 - 0.097x + 4.1$ models the number of people in the United States, y, in millions, holding more than one job x years after 1970. Use a graphing utility to graph the function in a $[0, 20, 1]$ by $[3, 6, 1]$ viewing rectangle. TRACE along the curve or use your utility's minimum value feature to approximate the coordinates of the parabola's vertex. Describe what this represents in practical terms.

81. The following data show fuel efficiency, in miles per gallon, for all U.S. automobiles in the indicated year.

x (Years after 1940)	y (Average Number of Miles per Gallon for U.S. Automobiles)
1940: 0	14.8
1950: 10	13.9
1960: 20	13.4
1970: 30	13.5
1980: 40	15.9
1990: 50	20.2
1998: 58	21.8

Source: U.S. Department of Transportation

a. Use a graphing utility to draw a scatter plot of the data. Explain why a quadratic function is appropriate for modeling these data.

b. Use the quadratic regression feature to find the quadratic function that best fits the data.

c. Use the equation in part (b) to determine the worst year for automobile fuel efficiency. What was the average number of miles per gallon for that year?

d. Use a graphing utility to draw a scatter plot of the data and graph the quadratic function of best fit on the scatter plot.

Review Exercises

82. Solve: $\dfrac{2}{x + 5} + \dfrac{1}{x - 5} = \dfrac{16}{x^2 - 25}$. (Section 7.6, Example 4)

83. Divide: $\dfrac{x^2 - x - 6}{3x - 3} \div \dfrac{x^2 - 4}{x - 1}$. (Section 7.2, Example 6)

84. Solve using determinants (Cramer's Rule):
$$2x + 3y = 6$$
$$x - 4y = 14.$$
(Section 8.5, Example 2)

▶ **SECTION 11.4**   *Equations Quadratic in Form*

Objective

1 Solve equations that are quadratic in form.

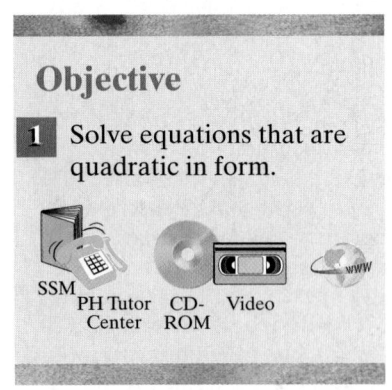

SSM   PH Tutor   CD-   Video
     Center   ROM

"My husband asked me if we have any cheese puffs. Like he can't go and lift the couch cushion up himself."
—Roseanne

How important is it for you to have a clean house? The percentage of people who find this to be quite important varies by age. In the exercise set, you will work with a function that models this phenomenon. Your work will be based on equations that are not quadratic, but that can be written as quadratic equations using an appropriate substitution. Here are some examples:

Given Equation	Substitution	New Equation
$x^4 - 10x^2 + 9 = 0$ or $\left(x^2\right)^2 - 10x^2 + 9 = 0$	$t = x^2$	$t^2 - 10t + 9 = 0$
$5x^{\frac{2}{3}} + 11x^{\frac{1}{3}} + 2 = 0$ or $5\left(x^{\frac{1}{3}}\right)^2 + 11x^{\frac{1}{3}} + 2 = 0$	$t = x^{\frac{1}{3}}$	$5t^2 + 11t + 2 = 0$

An equation that is **quadratic in form** is one that can be expressed as a quadratic equation using an appropriate substitution. Both of the preceding given equations are quadratic in form.

Equations that are quadratic in form contain an expression to a power, the same expression to that power squared, and a constant term. By letting t equal the expression to the power, a quadratic equation in t will result. Now it's easy. Solve this quadratic equation for t. Finally, use your substitution to find the values for the variable in the given equation. Example 1 shows how this is done.

> **1** Solve equations that are quadratic in form.

EXAMPLE 1 Solving an Equation Quadratic in Form

Solve: $x^4 - 10x^2 + 9 = 0$.

Solution Notice that the equation contains an expression to a power, x^2, the same expression to that power squared, x^4 or $(x^2)^2$, and a constant term, 9. We let t equal the expression to the power. Thus,

$$\text{let } t = x^2.$$

Now we write the given equation as a quadratic equation in t and solve for t.

$x^4 - 10x^2 + 9 = 0$	This is the given equation.
$(x^2)^2 - 10x^2 + 9 = 0$	The given equation contains x^2 and x^2 squared.
$t^2 - 10t + 9 = 0$	Replace x^2 with t.
$(t - 9)(t - 1) = 0$	Factor.
$t - 9 = 0 \quad$ or $\quad t - 1 = 0$	Set each factor equal to zero.
$t = 9 \quad$ or $\quad t = 1$	Solve for t.

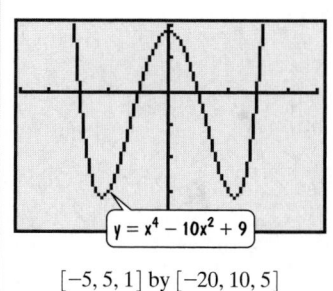
We're not done! Why not? We were asked to solve for x and we have values for t. We use the original substitution, $t = x^2$, to solve for x. Replace t with x^2 in each equation shown.

$$x^2 = 9 \quad \text{or} \quad x^2 = 1$$
$$x = \pm\sqrt{9} \qquad x = \pm\sqrt{1} \quad \text{Apply the square root property.}$$
$$x = \pm 3 \qquad x = \pm 1$$

Substitute these values into the given equation and verify that the solutions are $-3, 3, -1$, and 1. The solution set is $\{-3, 3, -1, 1\}$. You may prefer to express the solution set in numerical order as $\{-3, -1, 1, 3\}$. ∎

✔ **CHECK POINT 1** Solve: $x^4 - 17x^2 + 16 = 0$.

If checking proposed solutions is not overly cumbersome, you should do so either algebraically or with a graphing utility. The Using Technology box shows a check of the four solutions in Example 1. Are there situations when solving equations quadratic in form where a check is essential? Yes. **If at any point in the solution process both sides of an equation are raised to an even power, a check is required.** Extraneous solutions that are not solutions of the given equations may have been introduced.

EXAMPLE 2   Solving an Equation Quadratic in Form

Solve: $2x - \sqrt{x} - 10 = 0$.

Solution   In order to identify exponents on the terms, let's rewrite $\sqrt{x}$ as $x^{\frac{1}{2}}$. The equation can be expressed as

$$2x^1 - x^{\frac{1}{2}} - 10 = 0.$$

Notice that equation contains an expression to a power, $x^{\frac{1}{2}}$, the same expression to that power squared, x^1 or $\left(x^{\frac{1}{2}}\right)^2$, and a constant term, -10. We let t equal the expression to the power. Thus,

$$\text{let } t = x^{\frac{1}{2}}.$$

Now we write the given equation as a quadratic equation in t and solve for t.

$2x^1 - x^{\frac{1}{2}} - 10 = 0$	This is the given equation in exponential form.
$2\left(x^{\frac{1}{2}}\right)^2 - x^{\frac{1}{2}} - 10 = 0$	The equation contains $x^{\frac{1}{2}}$ and $x^{\frac{1}{2}}$ squared.
$2t^2 - t - 10 = 0$	Let $t = x^{\frac{1}{2}}$.
$(2t - 5)(t + 2) = 0$	Factor.
$2t - 5 = 0$ or $t + 2 = 0$	Set each factor equal to zero.
$t = \dfrac{5}{2}$ or $t = -2$	Solve for t.

Use the original substitution, $t = x^{\frac{1}{2}}$, to solve for x. Replace t with $x^{\frac{1}{2}}$ in each of the preceding equations.

$x^{\frac{1}{2}} = \dfrac{5}{2}$ or	$x^{\frac{1}{2}} = -2$	Replace t with $x^{\frac{1}{2}}$.
$\left(x^{\frac{1}{2}}\right)^2 = \left(\dfrac{5}{2}\right)^2$	$\left(x^{\frac{1}{2}}\right)^2 = (-2)^2$	Solve for x by squaring both sides of each equation.

Both sides are raised to even powers. We must check.

$x = \dfrac{25}{4}$	$x = 4$	Square $\dfrac{5}{2}$ and -2.

It is essential to check both proposed solutions in the original equation.

Check $\dfrac{25}{4}$:

$$2x - \sqrt{x} - 10 = 0$$

$$2 \cdot \frac{25}{4} - \sqrt{\frac{25}{4}} - 10 \stackrel{?}{=} 0$$

$$\frac{25}{2} - \frac{5}{2} - 10 \stackrel{?}{=} 0$$

$$\frac{20}{2} - 10 \stackrel{?}{=} 0$$

$$0 = 0, \text{ true}$$

Check 4:

$$2x - \sqrt{x} - 10 = 0$$

$$2 \cdot 4 - \sqrt{4} - 10 \stackrel{?}{=} 0$$

$$8 - 2 - 10 \stackrel{?}{=} 0$$

$$6 - 10 \stackrel{?}{=} 0$$

$$-4 = 0, \text{ false}$$

The check indicates that 4 is not a solution. It is an extraneous solution brought about by squaring each side of the equation. The only solution is $\dfrac{25}{4}$, and the solution set is $\left\{\dfrac{25}{4}\right\}$.

✔ **CHECK POINT 2** Solve: $x - 2\sqrt{x} - 8 = 0$.

The equations in Examples 1 and 2 can be solved by methods other than using substitutions.

$$x^4 - 10x^2 + 9 = 0 \qquad\qquad 2x - \sqrt{x} - 10 = 0$$

This equation can be solved
directly by factoring:
$(x^2 - 9)(x^2 - 1) = 0.$

This equation can be solved by
isolating the radical term:
$2x - 10 = \sqrt{x}.$
Then square both sides.

In the examples that follow, solving the equations by methods other than first introducing a substitution becomes increasingly difficult.

EXAMPLE 3 Solving an Equation Quadratic in Form

Solve: $(x^2 - 5)^2 + 3(x^2 - 5) - 10 = 0$.

Solution This equation contains $x^2 - 5$ and $x^2 - 5$ squared. We let

$$t = x^2 - 5.$$

$(x^2 - 5)^2 + 3(x^2 - 5) - 10 = 0$	This is the given equation.
$t^2 + 3t - 10 = 0$	Let $t = x^2 - 5$.
$(t + 5)(t - 2) = 0$	Factor.
$t + 5 = 0 \qquad$ or $\qquad t - 2 = 0$	Set each factor equal to zero.
$t = -5 \qquad$ or $\qquad t = 2$	Solve for t.

Use the original substitution, $t = x^2 - 5$, to solve for x. Replace t with $x^2 - 5$ in each of the preceding equations.

$x^2 - 5 = -5 \qquad$ or $\qquad x^2 - 5 = 2$	Replace t with $x^2 - 5$.
$x^2 = 0 \qquad\qquad\qquad x^2 = 7$	Solve for x by isolating x^2.
$x = 0 \qquad\qquad\qquad x = \pm\sqrt{7}$	Apply the square root property.

Although we did not raise both sides of an equation to an even power, checking the three proposed solutions in the original equation is a good idea. Do this now and verify that the solutions are $-\sqrt{7}$, 0, and $\sqrt{7}$, and the solution set is $\{-\sqrt{7}, 0, \sqrt{7}\}$.

✔ **CHECK POINT 3** Solve: $(x^2 - 4)^2 + (x^2 - 4) - 6 = 0$.

EXAMPLE 4 Solving an Equation Quadratic in Form

Solve: $10x^{-2} + 7x^{-1} + 1 = 0$.

Solution Notice that the equation contains an expression to a power, x^{-1}, the same expression to that power squared, x^{-2} or $(x^{-1})^2$, and a constant term, 1. We let t equal the expression to the power. Thus,

$$\text{let } t = x^{-1}.$$

Now we write the given equation as a quadratic equation in t and solve for t.

$$10x^{-2} + 7x^{-1} + 1 = 0 \qquad \text{This is the given equation.}$$

$$10(x^{-1})^2 + 7x^{-1} + 1 = 0 \qquad \text{The equation contains } x^{-1} \text{ and } x^{-1} \text{ squared.}$$

$$10t^2 + 7t + 1 = 0 \qquad \text{Let } t = x^{-1}.$$

$$(5t + 1)(2t + 1) = 0 \qquad \text{Factor.}$$

$$5t + 1 = 0 \quad \text{or} \quad 2t + 1 = 0 \qquad \text{Set each factor equal to zero.}$$

$$5t = -1 \quad \text{or} \quad 2t = -1$$

$$t = -\frac{1}{5} \quad \text{or} \quad t = -\frac{1}{2} \qquad \text{Solve for } t.$$

Use the original substitution, $t = x^{-1}$, to solve for x. Replace t with x^{-1} in each of the preceding equations.

$$x^{-1} = -\frac{1}{5} \qquad \text{or} \qquad x^{-1} = -\frac{1}{2} \qquad \text{Replace } t \text{ with } x^{-1}.$$

$$(x^{-1})^{-1} = \left(-\frac{1}{5}\right)^{-1} \qquad (x^{-1})^{-1} = \left(-\frac{1}{2}\right)^{-1} \qquad \begin{array}{l}\text{Solve for } x \text{ by raising both sides}\\ \text{of each equation to the } -1 \text{ power.}\end{array}$$

$$x = -5 \qquad\qquad x = -2$$

$$\left(-\frac{1}{5}\right)^{-1} = \frac{1}{-\frac{1}{5}} = -5 \qquad \left(-\frac{1}{2}\right)^{-1} = \frac{1}{-\frac{1}{2}} = -2$$

We did not raise both sides of an equation to an even power. A check will show that both -5 and -2 are solutions of the original equation. The solution set is $\{-5, -2\}$. ∎

✔ **CHECK POINT 4** Solve: $2x^{-2} + x^{-1} - 1 = 0$.

EXAMPLE 5   Solving an Equation Quadratic in Form

Solve: $5x^{\frac{2}{3}} + 11x^{\frac{1}{3}} + 2 = 0$.

Solution Notice that the equation contains an expression to a power, $x^{\frac{1}{3}}$, the same expression to that power squared, $x^{\frac{2}{3}}$ or $(x^{\frac{1}{3}})^2$, and a constant term, 1. We let t equal the expression to the power. Thus,

$$\text{let } t = x^{\frac{1}{3}}.$$

Now we write the given equation as a quadratic equation in t and solve for t.

$$5x^{\frac{2}{3}} + 11x^{\frac{1}{3}} + 2 = 0 \qquad \text{This is the given equation.}$$

$$5(x^{\frac{1}{3}})^2 + 11(x^{\frac{1}{3}}) + 2 = 0 \qquad \begin{array}{l}\text{The given equation contains } x^{\frac{1}{3}}\\ \text{and } x^{\frac{1}{3}} \text{ squared.}\end{array}$$

$$5t^2 + 11t + 2 = 0 \qquad \text{Let } t = x^{\frac{1}{3}}.$$

$$(5t + 1)(t + 2) = 0 \qquad \text{Factor.}$$

$$5t + 1 = 0 \quad \text{or} \quad t + 2 = 0 \qquad \text{Set each factor equal to 0.}$$

$$t = -\tfrac{1}{5} \quad \text{or} \quad t = -2 \qquad \text{Solve for } t.$$

Use the original substitution, $t = x^{\frac{1}{3}}$, to solve for x. Replace t with $x^{\frac{1}{3}}$ in each of the preceding equations.

$$x^{\frac{1}{3}} = -\frac{1}{5} \qquad\qquad x^{\frac{1}{3}} = -2 \qquad \text{Replace } t \text{ with } x^{\frac{1}{3}}.$$

$$\left(x^{\frac{1}{3}}\right)^3 = \left(-\frac{1}{5}\right)^3 \qquad \left(x^{\frac{1}{3}}\right)^3 = (-2)^3 \qquad \begin{array}{l}\text{Solve for } x \text{ by cubing both}\\ \text{sides of each equation.}\end{array}$$

$$x = -\frac{1}{125} \qquad\qquad x = -8$$

We did not raise both sides of an equation to an even power. A check will show that both -8 and $-\frac{1}{125}$ are solutions of the original equation. The solution set is $\left\{-8, -\frac{1}{125}\right\}$. ∎

 CHECK POINT 5 Solve: $3x^{\frac{2}{3}} - 11x^{\frac{1}{3}} - 4 = 0$.

EXERCISE SET 11.4

 Practice Exercises

In Exercises 1–32, solve each equation by making an appropriate substitution. If at any point in the solution process both sides of an equation are raised to an even power, a check is required.

1. $x^4 - 5x^2 + 4 = 0$

2. $x^4 - 13x^2 + 36 = 0$

3. $x^4 - 11x^2 + 18 = 0$

4. $x^4 - 9x^2 + 20 = 0$

5. $x^4 + 2x^2 = 8$

6. $x^4 + 4x^2 = 5$

7. $x + \sqrt{x} - 2 = 0$

8. $x + \sqrt{x} - 6 = 0$

9. $x - 4x^{\frac{1}{2}} - 21 = 0$

10. $x - 6x^{\frac{1}{2}} + 8 = 0$

11. $x - 13\sqrt{x} + 40 = 0$

12. $2x - 7\sqrt{x} - 30 = 0$

13. $(x - 5)^2 - 4(x - 5) - 21 = 0$

14. $(x + 3)^2 + 7(x + 3) - 18 = 0$

15. $(x^2 - 1)^2 - (x^2 - 1) = 2$

16. $(x^2 - 2)^2 - (x^2 - 2) = 6$

17. $(x^2 + 3x)^2 - 8(x^2 + 3x) - 20 = 0$

18. $(x^2 - 2x)^2 - 11(x^2 - 2x) + 24 = 0$

19. $x^{-2} - x^{-1} - 20 = 0$

20. $x^{-2} - x^{-1} - 6 = 0$

21. $2x^{-2} - 7x^{-1} + 3 = 0$

22. $20x^{-2} + 9x^{-1} + 1 = 0$

23. $x^{-2} - 4x^{-1} = 3$

24. $x^{-2} - 6x^{-1} = -4$

25. $x^{\frac{2}{3}} - x^{\frac{1}{3}} - 6 = 0$

26. $x^{\frac{2}{3}} + 2x^{\frac{1}{3}} - 3 = 0$

27. $x^{\frac{2}{5}} + x^{\frac{1}{5}} - 6 = 0$

28. $x^{\frac{2}{5}} + x^{\frac{1}{5}} - 2 = 0$

29. $2x^{\frac{1}{2}} - x^{\frac{1}{4}} = 1$

30. $2x^{\frac{1}{2}} - 5x^{\frac{1}{4}} = 3$

31. $\left(x - \dfrac{8}{x}\right)^2 + 5\left(x - \dfrac{8}{x}\right) - 14 = 0$

32. $\left(x - \dfrac{10}{x}\right)^2 + 6\left(x - \dfrac{10}{x}\right) - 27 = 0$

In Exercises 33–38, find the x-intercepts of the given function, f. Then use the x-intercepts to match each function with its graph. (The graphs are labeled (a) through (f).)

33. $f(x) = x^4 - 5x^2 + 4$

34. $f(x) = x^4 - 13x^2 + 36$

35. $f(x) = x^{\frac{1}{3}} + 2x^{\frac{1}{6}} - 3$

36. $f(x) = x^{-2} - x^{-1} - 6$

37. $f(x) = (x + 2)^2 - 9(x + 2) + 20$

38. $f(x) = 2(x + 2)^2 + 5(x + 2) - 3$

a.
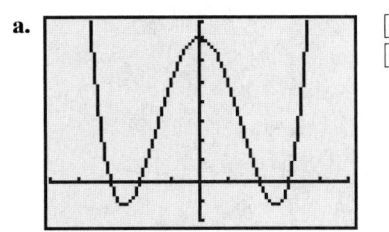
$[-5, 5, 1]$ by $[-10, 40, 5]$

b.
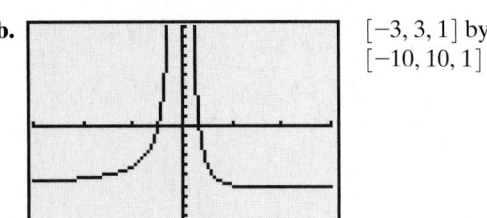
$[-3, 3, 1]$ by $[-10, 10, 1]$

c.
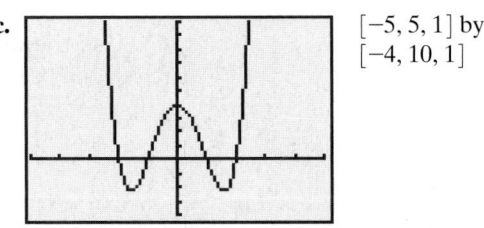
$[-5, 5, 1]$ by $[-4, 10, 1]$

d.
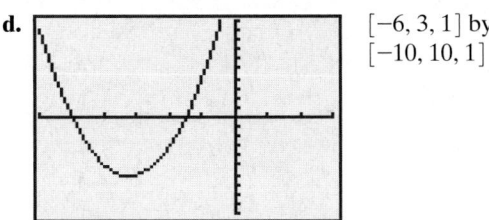
$[-6, 3, 1]$ by $[-10, 10, 1]$

e.
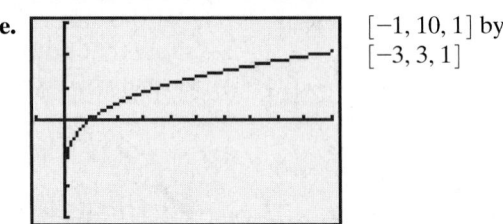
$[-1, 10, 1]$ by $[-3, 3, 1]$

f.
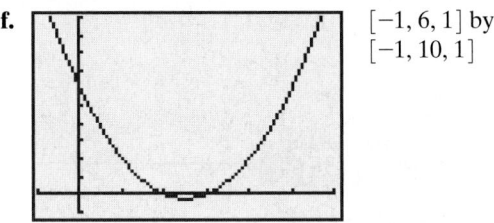
$[-1, 6, 1]$ by $[-1, 10, 1]$

Application Exercises

How important is it for you to have a clean house? The bar graph indicates that the percentage of people who find this to be quite important varies by age. The percentage, P(x), who find having a clean house very important can be modeled by the function

$$P(x) = 0.04(x + 40)^2 - 3(x + 40) + 104$$

where x is the number of years a person's age is above or below 40. Thus, x is positive for people over 40 and negative for people under 40. Use the function to solve Exercises 39–40.

The Importance of Having a Clean House, by Age

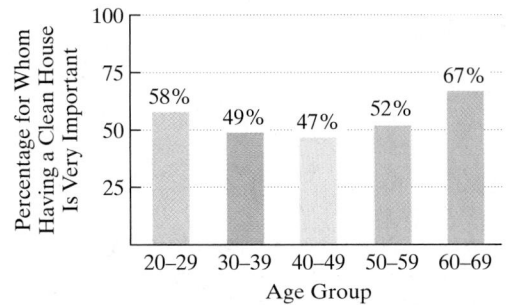

Source: Soap and Detergent Association

39. According to the model, at which ages do 60% of us feel that having a clean house is very important? Substitute 60 for $P(x)$ and solve the quadratic-in-form equation. How well does the function model the data shown in the bar graph?

40. According to the model, at which ages do 50% of us feel that having a clean house is very important? Substitute 50 for $P(x)$ and solve the quadratic-in-form equation. How well does the function model the data shown in the bar graph?

Writing in Mathematics

41. Explain how to recognize an equation that is quadratic in form. Provide two original examples with your explanation.

42. Describe two methods for solving this equation:
$$x - 5\sqrt{x} + 4 = 0.$$

43. In the twenty-first century, collecting and presenting data are big business. Because statisticians both record and influence our behavior, you should always ask yourself if the person or group presenting the data has any special case to make for or against the displayed information. Using this criterion, describe your impressions of the data in the bar graph for Exercises 39–40.

Critical Thinking Exercises

44. Which one of the following is true?
  a. If an equation is quadratic in form, there is only one method that can be used to obtain its solution.
  b. An equation that is quadratic in form must have a variable factor in one term that is the square of the variable factor in another term.
  c. Because x^6 is the square of x^3, the equation $x^6 - 5x^3 + 6x = 0$ is quadratic in form.
  d. To solve $x - 9\sqrt{x} + 14 = 0$, we let $\sqrt{t} = x$.

In Exercises 45–47, use a substitution to solve each equation.

45. $x^4 - 5x^2 - 2 = 0$

46. $5x^6 + x^3 = 18$

47. $\sqrt{\dfrac{x + 4}{x - 1}} + \sqrt{\dfrac{x - 1}{x + 4}} = \dfrac{5}{2}$ $\left(\text{Let } t = \sqrt{\dfrac{x + 4}{x - 1}}. \right)$

Technology Exercises

48. Use a graphing utility to verify the solutions of any five equations in Exercises 1–32 that you solved algebraically. The real solutions should appear as x-intercepts on the graph of the function related to the given equation.

Use a graphing utility to solve the equations in Exercises 49–56. Check by direct substitution.

49. $x^6 - 7x^3 - 8 = 0$

50. $3(x - 2)^{-2} - 4(x - 2)^{-1} + 1 = 0$

51. $x^4 - 10x^2 + 9 = 0$

52. $2x + 6\sqrt{x} = 8$

53. $2(x + 1)^2 = 5(x + 1) + 3$

54. $(x^2 - 3x)^2 + 2(x^2 - 3x) - 24 = 0$

55. $x^{\frac{1}{2}} + 4x^{\frac{1}{4}} = 5$

56. $x^{\frac{2}{3}} - 3x^{\frac{1}{3}} + 2 = 0$

Review Exercises

57. Simplify:
$$\frac{2x^2}{10x^3 - 2x^2}.$$
(Section 7.1, Example 3)

58. Divide: $\dfrac{2 + i}{1 - i}$. (Section 10.7, Example 5)

59. Solve using matrices:
$$\begin{aligned} 2x + y &= 6 \\ x - 2y &= 8. \end{aligned}$$
(Section 8.4, Example 3)

▶ SECTION 11.5 *Quadratic and Rational Inequalities*

Objectives

1 Solve quadratic inequalities.

2 Solve rational inequalities.

3 Solve problems using nonlinear inequalities.

SSM
PH Tutor   CD-   Video
Center   ROM

People are going to live longer in the twenty-first century. This will put added pressure on the Social Security and Medicare systems. The bar graph in Figure 11.14 shows the cost of Medicare, in billions of dollars, projected through 2005.

Medicare Spending

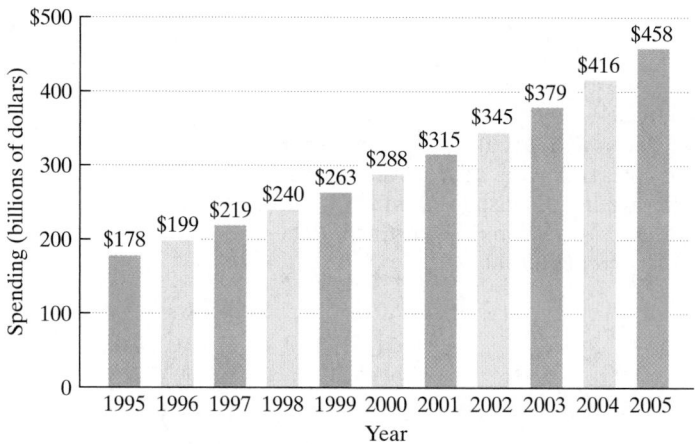

Figure 11.14   *Source*: Congressional Budget Office

Medicare spending, $f(x)$, in billions of dollars, x years after 1995 can be modeled by the quadratic function

$$f(x) = 1.2x^2 + 15.2x + 181.4.$$

To determine in which years Medicare spending will exceed $500 billion, we must solve the inequality

$$1.2x^2 + 15.2x + 181.4 > 500.$$

Medicare spending   exceeds $500 billion.

We begin by subtracting 500 from both sides. This will give us zero on the right:

$$1.2x^2 + 15.2x + 181.4 - 500 > 500 - 500$$

$$1.2x^2 + 15.2x - 318.6 > 0.$$

The form of this inequality is $ax^2 + bx + c > 0$. Such an inequality is called a *quadratic inequality*.

Definition of a Quadratic Inequality

A **quadratic inequality** is any inequality that can be put in one of the forms

$$ax^2 + bx + c < 0 \qquad ax^2 + bx + c > 0$$
$$ax^2 + bx + c \le 0 \qquad ax^2 + bx + c \ge 0$$

where a, b, and c are real numbers and $a \ne 0$.

In this section we establish the basic techniques for solving quadratic inequalities. We will use these techniques to solve inequalities containing quotients, called **rational inequalities**.

Solving Quadratic Inequalities

1 Solve quadratic inequalities

Graphs can help us to estimate the solutions of quadratic inequalities. The graph of $y = x^2 - 7x + 10$ is shown in Figure 11.15. The x-intercepts, 2 and 5, are **boundary points** between where the graph lies above the x-axis, shown in blue, and where the graph lies below the x-axis, shown in red. These boundary points play a critical role in solving quadratic inequalities.

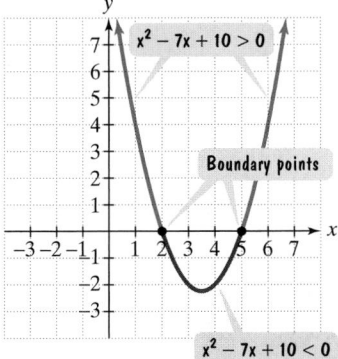

Figure 11.15 The graph of
$$y = x^2 - 7x + 10.$$

The blue parts of the graph lie above the x-axis:
$$x^2 - 7x + 10 > 0.$$
By contrast, the red part lies below the x-axis:
$$x^2 - 7x + 10 < 0.$$

Procedure for Solving Quadratic Inequalities

1. Express the inequality in the standard form
$$ax^2 + bx + c > 0 \quad \text{or} \quad ax^2 + bx + c < 0.$$

2. Solve the equation $ax^2 + bx + c = 0$. The real solutions are the **boundary points**.

3. Locate these boundary points on a number line, thereby dividing the number line into **test intervals**.

4. Choose one representative number within each test interval. If substituting that value into the original inequality produces a true statement, then all real numbers in the test interval belong to the solution set. If substituting that value into the original inequality produces a false statement, then no real number in the test interval belongs to the solution set.

5. Write the solution set, selecting the interval(s) that produced a true statement. The graph of the solution set on a number line usually appears as

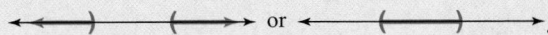

This procedure is valid if $<$ is replaced by $\le$ and $>$ is replaced by $\ge$.

EXAMPLE 1   Solving a Quadratic Inequality

Solve and graph the solution set on a real number line: $x^2 - 7x + 10 < 0$.

Solution

Step 1 Write the inequality in standard form. The inequality is given in this form, so this step has been done for us.

Step 2 Solve the related quadratic equation. This equation is obtained by replacing the inequality sign in $x^2 - 7x + 10 < 0$ by an equal sign. Thus, we will solve $x^2 - 7x + 10 = 0$.

$$x^2 - 7x + 10 = 0 \quad \text{This is the related quadratic equation.}$$

$$(x - 2)(x - 5) = 0 \quad \text{Factor.}$$

$$x - 2 = 0 \quad \text{or} \quad x - 5 = 0 \quad \text{Set each factor equal to 0.}$$

$$x = 2 \quad \text{or} \quad x = 5 \quad \text{Solve for x.}$$

The boundary points are 2 and 5.

Step 3 Locate the boundary points on a number line. The number line with the boundary points is shown as follows:

The boundary points divide the number line into three test intervals, namely $(-\infty, 2)$, $(2, 5)$, and $(5, \infty)$.

Step 4 Take one representative number within each test interval and substitute that number into the original inequality.

Test Interval	Representative Number	Substitute into $x^2 - 7x + 10 < 0$	Conclusion
$(-\infty, 2)$	0	$0^2 - 7 \cdot 0 + 10 \overset{?}{<} 0$ $10 < 0$, false	$(-\infty, 2)$ does not belong to the solution set.
$(2, 5)$	3	$3^2 - 7 \cdot 3 + 10 \overset{?}{<} 0$ $9 - 21 + 10 \overset{?}{<} 0$ $-2 < 0$, true	$(2, 5)$ belongs to the solution set.
$(5, \infty)$	6	$6^2 - 7 \cdot 6 + 10 \overset{?}{<} 0$ $36 - 42 + 10 \overset{?}{<} 0$ $4 < 0$, false	$(5, \infty)$ does not belong to the solution set.

Step 5 The solution set is the interval that produced a true statement. Our analysis shows that the solution set is the interval $(2, 5)$, or $\{x \mid 2 < x < 5\}$. The graph in Figure 11.16 confirms that $x^2 - 7x + 10 < 0$ (lies below the x-axis) in this interval. The graph of the solution set on a number line is shown as follows:

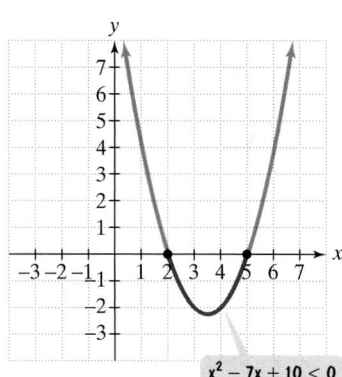

$x^2 - 7x + 10 < 0$

Figure 11.16 The graph lies below the x-axis between the boundary points 2 and 5, in the interval $(2, 5)$.

✔ **CHECK POINT 1** Solve and graph the solution set:

$$x^2 + 2x - 3 < 0.$$

EXAMPLE 2   Solving a Quadratic Inequality

Solve and graph the solution set on a real number line: $2x^2 + x \geq 15$.

Solution

Step 1 Write the inequality in standard form. We can write $2x^2 + x \geq 15$ in standard form by subtracting 15 from both sides. This will give us zero on the right.

$$2x^2 + x - 15 \geq 15 - 15$$

$$2x^2 + x - 15 \geq 0$$

Step 2 Solve the related quadratic equation. This equation is obtained by replacing the inequality sign in $2x^2 + x - 15 \geq 0$ by an equal sign. Thus, we will solve $2x^2 + x - 15 = 0$.

$2x^2 + x - 15 = 0$	This is the related quadratic equation.
$(2x - 5)(x + 3) = 0$	Factor.
$2x - 5 = 0$ or $x + 3 = 0$	Set each factor equal to 0.
$x = \frac{5}{2}$ or $x = -3$	Solve for x.

The boundary points are -3 and $\frac{5}{2}$.

Step 3 Locate the boundary points on a number line. The number line with the boundary points is shown as follows:

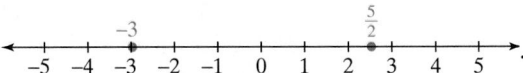

The boundary points divide the number line into three test intervals. Including the boundary points (because of the given greater than or *equal to* sign), the intervals are $(-\infty, -3]$, $[-3, \frac{5}{2}]$, and $[\frac{5}{2}, \infty)$.

Step 4 Take one representative number within each test interval and substitute that number into the original inequality.

Test Interval	Representative Number	Substitute into $2x^2 + x \geq 15$	Conclusion
$(-\infty, -3]$	-4	$2(-4)^2 + (-4) \overset{?}{\geq} 15$ $28 \geq 15$, true	$(-\infty, -3]$ belongs to the solution set.
$\left[-3, \dfrac{5}{2}\right]$	0	$2 \cdot 0^2 + 0 \overset{?}{\geq} 15$ $0 \geq 15$, false	$\left[-3, \dfrac{5}{2}\right]$ does not belong to the solution set.
$\left[\dfrac{5}{2}, \infty\right)$	3	$2 \cdot 3^2 + 3 \overset{?}{\geq} 15$ $21 \geq 15$, true	$\left[\dfrac{5}{2}, \infty\right)$ belongs to the solution set.

Using Technology

The solution set for

$$2x^2 + x \geq 15$$

or, equivalently,

$$2x^2 + x - 15 \geq 0$$

can be verified with a graphing utility. The graph of $y = 2x^2 + x - 15$ was obtained using a $[-10, 10, 1]$ by $[-16, 6, 1]$ viewing rectangle. The graph lies above or on the x-axis, representing $\geq$, for all x in $(-\infty, -3] \cup [\frac{5}{2}, \infty)$.

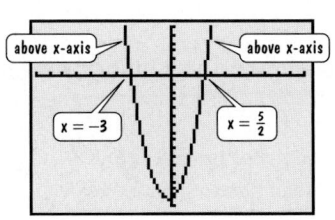

Step 5 **The solution set is the union of intervals that produced a true statement.** Our analysis on page 777 shows that the solution set is

$$(-\infty, -3] \cup \left[\frac{5}{2}, \infty\right)$$

$$\text{or } \left\{x \mid x \leq -3 \text{ or } x \geq \frac{5}{2}\right\}.$$

The graph of the solution set on a number line is shown as follows:

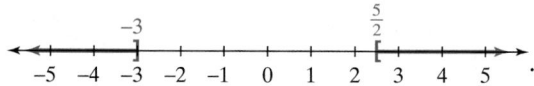

✔ **CHECK POINT 2**   Solve and graph the solution set: $x^2 - x \geq 20$.

---

2 Solve rational inequalities.

Solving Rational Inequalities

Inequalities that involve quotients can be solved in the same manner as quadratic inequalities. For example, the inequalities

$$(x + 3)(x - 7) > 0 \quad \text{and} \quad \frac{x + 3}{x - 7} > 0$$

are similar in that both are positive under the same conditions. To be positive, each of these inequalities must have two positive linear expressions

$$x + 3 > 0 \quad \text{and} \quad x - 7 > 0$$

or two negative linear expressions

$$x + 3 < 0 \quad \text{and} \quad x - 7 < 0.$$

Consequently, we solve $\dfrac{x + 3}{x - 7} > 0$ using boundary points to divide the number line into test intervals. Then we select one representative number in each interval to determine whether that interval belongs to the solution set. Example 3 illustrates how this is done.

Study Tip

Many students want to solve

$$\frac{x + 3}{x - 7} > 0$$

by first multiplying both sides by $x - 7$ to clear fractions. This is incorrect. The problem is that $x - 7$ contains a variable and can be positive or negative, depending on the value of x. Thus, we do not know whether or not to reverse the sense of the inequality.

EXAMPLE 3   Using Test Numbers to Solve a Rational Inequality

Solve and graph the solution set: $\dfrac{x + 3}{x - 7} > 0.$

Solution   We begin by finding values of x that make the numerator and denominator 0.

$$x + 3 = 0 \qquad x - 7 = 0 \qquad \text{Set the numerator and denominator equal to 0.}$$

$$x = -3 \qquad x = 7 \qquad \text{Solve.}$$

The boundary points are -3 and 7. We locate these numbers on a number line as follows:

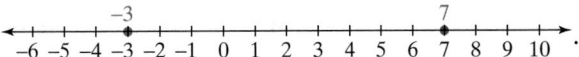

These boundary points divide the number line into three test intervals, namely $(-\infty, -3)$, $(-3, 7)$, and $(7, \infty)$. Now, we take one representative number from each test interval and substitute that number into the original inequality.

Test Interval	Representative Number	Substitute into $\dfrac{x + 3}{x - 7} > 0$	Conclusion
$(-\infty, -3)$	-4	$\dfrac{-4 + 3}{-4 - 7} \overset{?}{>} 0$ $\dfrac{-1}{-11} \overset{?}{>} 0$ $\dfrac{1}{11} > 0, \text{ true}$	$(-\infty, -3)$ belongs to the solution set.
$(-3, 7)$	0	$\dfrac{0 + 3}{0 - 7} \overset{?}{>} 0$ $-\dfrac{3}{7} > 0, \text{false}$	$(-3, 7)$ does not belong to the solution set.
$(7, \infty)$	8	$\dfrac{8 + 3}{8 - 7} \overset{?}{>} 0$ $11 > 0, \text{true}$	$(7, \infty)$ belongs to the solution set.

Our analysis shows that the solution set is

$$(-\infty, -3) \cup (7, \infty)$$

or $\{x \mid x < -3 \text{ or } x > 7\}$.

The graph of the solution set on a number line is shown as follows:

✔ **CHECK POINT 3** Solve and graph the solution set: $\dfrac{x - 5}{x + 2} > 0.$

The first step in solving a rational inequality is to bring all terms to one side, obtaining zero on the other side. Then express the nonzero side as a single quotient. At this point, we follow the same procedure as in Example 3, finding values of the variable that make the numerator and denominator 0.

Study Tip

Do not begin solving

$$\frac{x + 1}{x + 3} \le 2$$

by multiplying both sides by $x + 3$. We do not know if $x + 3$ is positive or negative. Thus, we do not know whether or not to reverse the sense of the inequality.

EXAMPLE 4   Solving a Rational Inequality

Solve and graph the solution set: $\dfrac{x + 1}{x + 3} \le 2$.

Solution

Step 1 Express the inequality so that one side is zero and the other side is a single quotient. We subtract 2 from both sides to obtain zero on the right.

$$\frac{x + 1}{x + 3} \le 2 \qquad \text{This is the given inequality.}$$

$$\frac{x + 1}{x + 3} - 2 \le 0 \qquad \text{Subtract 2 from both sides, obtaining 0 on the right.}$$

$$\frac{x + 1}{x + 3} - \frac{2(x + 3)}{x + 3} \le 0 \qquad \text{The least common denominator is } x + 3. \text{ Express 2 in terms of this denominator.}$$

$$\frac{x + 1 - 2(x + 3)}{x + 3} \le 0 \qquad \text{Subtract rational expressions.}$$

$$\frac{x + 1 - 2x - 6}{x + 3} \le 0 \qquad \text{Apply the distributive property.}$$

$$\frac{-x - 5}{x + 3} \le 0 \qquad \text{Simplify.}$$

Step 2 Find boundary points by setting the numerator and the denominator equal to zero.

$$-x - 5 = 0 \qquad x + 3 = 0 \qquad \text{Set the numerator and denominator equal to 0. These are the values that make the previous quotient zero or undefined.}$$

$$x = -5 \qquad x = -3 \qquad \text{Solve for } x.$$

The boundary points are -5 and -3. Because equality is included in the given less-than-or-equal-to symbol, we include the value of x that causes the quotient $\dfrac{-x - 5}{x + 3}$ to be zero. Thus, -5 is included in the solution set. By contrast, we do not include -3 in the solution set because -3 makes the denominator zero.

Step 3 Locate boundary points on a number line. The number line, with the boundary points, is shown as follows:

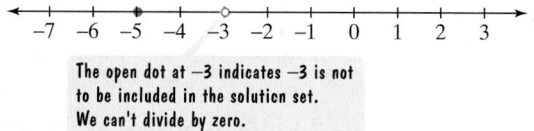

The open dot at -3 indicates -3 is not to be included in the solution set. We can't divide by zero.

The boundary points divide the number line into three test intervals, namely $(-\infty, -5]$, $[-5, -3)$, and $(-3, \infty)$.

Step 4 Take one representative number within each test interval and substitute that number into the original inequality.

Test Interval	Representative Number	Substitute into $\dfrac{x+1}{x+3} \le 2$	Conclusion
$(-\infty, -5]$	-6	$\dfrac{-6+1}{-6+3} \overset{?}{\le} 2$ $\dfrac{5}{3} \le 2$, true	$(-\infty, -5]$ belongs to the solution set.
$[-5, -3)$	-4	$\dfrac{-4+1}{-4+3} \overset{?}{\le} 2$ $3 \le 2$, false	$[-5, -3)$ does not belong to the solution set.
$(-3, \infty)$	0	$\dfrac{0+1}{0+3} \overset{?}{\le} 2$ $\dfrac{1}{3} \le 2$, true	$(-3, \infty)$ belongs to the solution set.

Step 5 The solution set consists of the intervals that produced a true statement. Our analysis shows that the solution set is

$$(-\infty, -5] \cup (-3, \infty)$$

or $\{x \,|\, x \le -5 \text{ or } x > -3\}$.

The graph of the solution set on a number line is shown as follows:

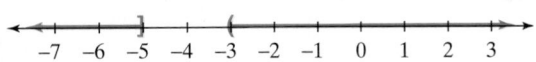

■

Discover for Yourself

Because $(x+3)^2$ is positive, it is possible so solve

$$\frac{x+1}{x+3} \le 2$$

by first multiplying both sides by $(x+3)^2$ (where $x \ne -3$). This will not reverse the sense of the inequality and will clear the fraction. Try using this solution method and compare it to the solution on pages 780–781.

✔ **CHECK POINT 4**   Solve and graph the solution set: $\dfrac{2x}{x+1} \le 1$.

3 Solve problems using nonlinear inequalities.

Applications

Quadratic inequalities can be solved to answer questions about variables contained in quadratic functions.

EXAMPLE 5   Modeling the Position of a Free-Falling Object

The Leaning Tower of Pisa is 176 feet high. Figure 11.17 on page 782 shows that a ball is thrown vertically upward from the top of the tower. The function

$$s(t) = -16t^2 + 96t + 176$$

models the ball's height above the ground, $s(t)$, in feet, t seconds after it was thrown. During which time period will the ball's height exceed that of the tower?

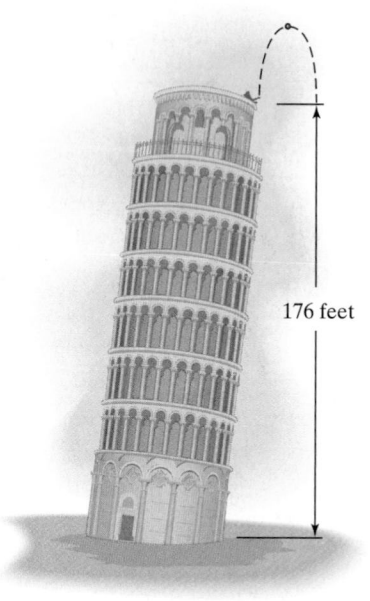

176 feet

Figure 11.17 Throwing a ball from the top of the Leaning Tower of Pisa

Solution Using the problem's question and the given model for the ball's height, $s(t) = -16t^2 + 96t + 176$, we obtain a quadratic inequality.

When will the ball's height   exceed that   of the tower?

$$-16t^2 + 96t + 176 > 176$$

$-16t^2 + 96t + 176 > 176$	This is the inequality implied by the problem's question. We must find t.
$-16t^2 + 96t > 0$	Subtract 176 from both sides.
$-16t^2 + 96t = 0$	Solve the related quadratic equation.
$-16t(t - 6) = 0$	Factor.
$-16t = 0$ or $t - 6 = 0$	Set each factor equal to 0.
$t = 0$ $t = 6$	Solve for t. The boundary points are 0 and 6.

Locate these values on a number line, with $t \geq 0$.

The intervals are $(0, 6)$ and $(6, \infty)$, although the time interval should not extend to infinity but rather to the value of t when the ball hits the ground. (By setting $-16t^2 + 96t + 176$ equal to zero, we find $t \approx 7.47$; the ball hits the ground after approximately 7.47 seconds.)

We use $(0, 6)$ and $(6, 7.47)$ for our test intervals.

Test Interval	Representative Number	Substitute into $-16t^2 + 96t + 176 > 176$	Conclusion
$(0, 6)$	1	$-16 \cdot 1^2 + 96 \cdot 1 + 176 \overset{?}{>} 176$ $256 > 176$, true	$(0, 6)$ belongs to the solution set.
$(6, 7.47)$	7	$-16 \cdot 7^2 + 96 \cdot 7 + 176 \overset{?}{>} 176$ $64 > 176$, false	$(6, 7.47)$ does not belong to the solution set.

The ball's height exceeds that of the tower between 0 and 6 seconds, excluding $t = 0$ and $t = 6$. ∎

✔ **CHECK POINT 5** An object is propelled straight up from ground level with an initial velocity of 80 feet per second. Its height at time t is modeled by

$$s(t) = -16t^2 + 80t,$$

where the height, $s(t)$, is measured in feet and the time, t, is measured in seconds. In which time interval will the object be more than 64 feet above the ground?

EXERCISE SET 11.5

Practice Exercises

Solve each quadratic inequality in Exercises 1–26, and graph the solution set on a real number line.

1. $(x - 4)(x + 2) > 0$

2. $(x + 3)(x - 5) > 0$

3. $(x - 7)(x + 3) \leq 0$
4. $(x + 1)(x - 7) \leq 0$
5. $x^2 - 5x + 4 > 0$

6. $x^2 - 4x + 3 < 0$
7. $x^2 + 5x + 4 > 0$

8. $x^2 + x - 6 > 0$

9. $x^2 - 6x + 8 \leq 0$
10. $x^2 - 2x - 3 \geq 0$

11. $3x^2 + 10x - 8 \leq 0$
12. $9x^2 + 3x - 2 \geq 0$

13. $2x^2 + x < 15$
14. $6x^2 + x > 1$

15. $4x^2 + 7x < -3$
16. $3x^2 + 16x < -5$
17. $x^2 - 4x \geq 0$
18. $x^2 + 2x < 0$
19. $2x^2 + 3x > 0$

20. $3x^2 - 5x \leq 0$
21. $-x^2 + x \geq 0$
22. $-x^2 + 2x \geq 0$
23. $x^2 \leq 4x - 2$

24. $x^2 \leq 2x + 2$

25. $x^2 - 6x + 9 < 0$
26. $4x^2 - 4x + 1 \geq 0$

Solve each rational inequality in Exercises 27–42, and graph the solution set on a real number line.

27. $\dfrac{x - 4}{x + 3} > 0$

28. $\dfrac{x + 5}{x - 2} > 0$

29. $\dfrac{x + 3}{x + 4} < 0$

30. $\dfrac{x + 5}{x + 2} < 0$

31. $\dfrac{-x + 2}{x - 4} \geq 0$

32. $\dfrac{-x - 3}{x + 2} \leq 0$

33. $\dfrac{4 - 2x}{3x + 4} \leq 0$

34. $\dfrac{3x + 5}{6 - 2x} \geq 0$

35. $\dfrac{x}{x - 3} > 0$

36. $\dfrac{x + 4}{x} > 0$

37. $\dfrac{x + 1}{x + 3} < 2$

38. $\dfrac{x}{x - 1} > 2$

39. $\dfrac{x + 4}{2x - 1} \leq 3$

40. $\dfrac{1}{x - 3} < 1$

41. $\dfrac{x - 2}{x + 2} \leq 2$

42. $\dfrac{x}{x + 2} \geq 2$

 Application Exercises

43. You throw a ball straight up from a rooftop 160 feet high with an initial speed of 48 feet per second. The function

$$s(t) = -16t^2 + 48t + 160$$

models the ball's height above the ground, $s(t)$, in feet, t seconds after it was thrown. During which time period will the ball's height exceed that of the rooftop?

44. A model rocket is launched from the top of a cliff 80 feet above sea level. The function

$$s(t) = -16t^2 + 64t + 80$$

models the rocket's height above the water, $s(t)$, in feet, t seconds after it was launched. During which time period will the rocket's height exceed that of the cliff?

The bar graph in Figure 11.14 on page 774 shows the cost of Medicare, in billions of dollars, projected through 2005. Using the regression feature of a graphing utility, these data can be modeled by

 a linear function , $f(x) = 27x + 163$;

 a quadratic function , $g(x) = 1.2x^2 + 15.2x + 181.4$.

In each function, x represents the number of years after 1995. Use these functions to solve Exercises 45–48.

45. The graph indicates that Medicare spending will reach $379 billion in 2003. Find the amount predicted by each of the functions, f and g, for that year. How well do the functions model the value in the graph?

46. The graph indicates that Medicare spending will reach $458 billion in 2005. Find the amount predicted by each of the functions, f and g, for that year. How well do the functions model the value in the graph? Which function serves as a better model for that year?

47. For which years does the quadratic model indicate that Medicare spending will exceed $536.6 billion?

48. For which years does the quadratic model indicate that Medicare spending will exceed $629.4 billion?

The United States has more people in prison, as well as more people in prison per capita, than any other western industrialized nation. The bar graph shows the number of inmates in U.S. state and federal prisons in seven selected years from 1980 through 1999. The data can be modeled by the function

$$f(x) = 1.3x^2 + 32x + 303,$$

in which $f(x)$ represents the number of inmates, in thousands, x years after 1980. Use this function to solve Exercises 49–52.

Number of Inmates in U.S. State and Federal Prisons

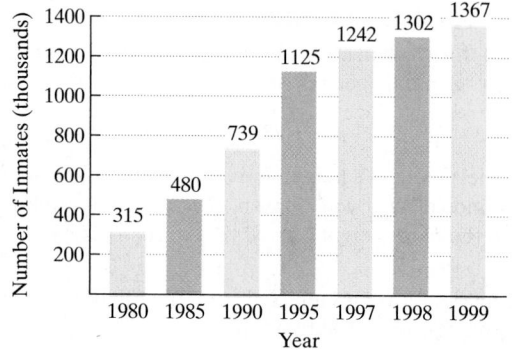

Source: U.S. Justice Department

49. Find the number of inmates in 1998 using the given function. How well does the function model the actual number shown for that year?

50. Find the number of inmates in 1999 using the given function. How well does the function model the actual number shown for that year?

51. For which years does the model indicate that U.S. inmate population will exceed 2433?

52. For which years does the model indicate that U.S. inmate population was no more than 1463 thousand?

A company manufactures wheelchairs. The average cost function, $\bar{C}$, of producing x wheelchairs per month is given by

$$\bar{C}(x) = \frac{500{,}000 + 400x}{x}.$$

The graph of the rational function is shown. Use the function to solve Exercises 53–54.

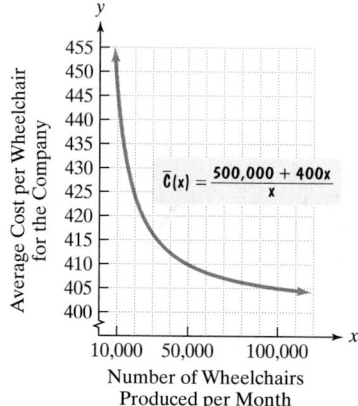

Number of Wheelchairs
Produced per Month

53. Describe the company's production level so that the average cost of producing each wheelchair does not exceed $425. Use a rational inequality to solve the problem. Then explain how your solution is shown on the graph.

54. Describe the company's production level so that the average cost of producing each wheelchair does not exceed $410. Use a rational inequality to solve the problem. Then explain how your solution is shown on the graph.

Writing in Mathematics

55. What is a quadratic inequality?

56. What is a rational inequality?

57. Describe similarities and differences between the solutions of

$$(x - 2)(x + 5) \ge 0 \quad \text{and} \quad \frac{x - 2}{x + 5} \ge 0.$$

58. What explanations can you offer for the trend shown in the bar graph for Exercises 49–52? Why do you think that the United States has more people in prison than any other industrialized nation?

Critical Thinking Exercises

59. Which one of the following is true?

   a. The solution set to $x^2 > 25$ is $(5, \infty)$.

   b. The inequality $\dfrac{x - 2}{x + 3} < 2$ can be solved by multiplying both sides by $x + 3$, resulting in the equivalent inequality $x - 2 < 2(x + 3)$.

   c. $(x + 3)(x - 1) \geq 0$ and $\dfrac{x + 3}{x - 1} \geq 0$ have the same solution set.

   d. None of the above statements is true.

60. Write a quadratic inequality whose solution set is $[-3, 5]$.

61. Write a rational inequality whose solution set is $(-\infty, -4) \cup [3, \infty)$.

In Exercises 62–65, use inspection to describe each inequality's solution set. Do not solve any of the inequalities.

62. $(x - 2)^2 > 0$

63. $(x - 2)^2 \leq 0$ **64.** $(x - 2)^2 < -1$

65. $\dfrac{1}{(x - 2)^2} > 0$

66. The graphing calculator screen shows the graph of $y = 4x^2 - 8x + 7$.

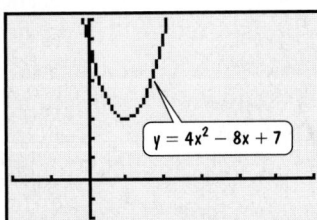

   a. Use the graph to describe the solution set for $4x^2 - 8x + 7 > 0$.

   b. Use the graph to describe the solution set for $4x^2 - 8x + 7 < 0$.

   c. Use an algebraic approach to verify each of your descriptions in parts (a) and (b).

67. The graphing calculator screen shows the graph of $y = \sqrt{27 - 3x^2}$. Write and solve a quadratic inequality that explains why the graph only appears for $-3 \leq x \leq 3$.

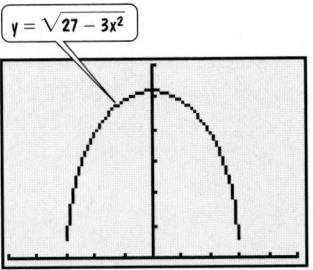

Technology Exercises

Solve each inequality in Exercises 68–73 using a graphing utility.

68. $x^2 + 3x - 10 > 0$

69. $2x^2 + 5x - 3 \leq 0$

70. $\dfrac{x - 4}{x - 1} \leq 0$

71. $\dfrac{x + 2}{x - 3} \leq 2$

72. $\dfrac{1}{x + 1} \leq \dfrac{2}{x + 4}$

73. $x^3 + 2x^2 - 5x - 6 > 0$

Review Exercises

74. Solve: $\left| \dfrac{x - 5}{3} \right| < 8$. (Section 9.3, Example 3)

75. Divide:
$$\dfrac{2x + 6}{x^2 + 8x + 16} \div \dfrac{x^2 - 9}{x^2 + 3x - 4}.$$
(Section 7.2, Example 6)

76. Factor completely: $x^4 - 16y^4$. (Section 6.5, Example 7)

CHAPTER 11 GROUP PROJECTS

1. Each group member should consult an almanac, newspaper, magazine, or the Internet to find data that can be modeled by a quadratic function. Group members should select the two sets of data that are most interesting and relevant. For each data set selected:

   a. Use the quadratic regression feature of a graphing utility to find the quadratic function that best fits the data.

(continues on next page)

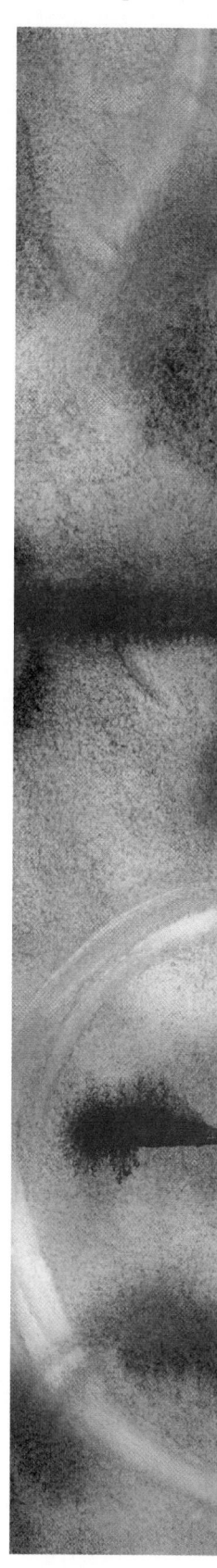

b. Use the equation of the quadratic function to make a prediction from the data. What circumstances might affect the accuracy of your prediction?

c. Use the equation of the quadratic function to write and solve a problem involving maximizing or minimizing the function.

2. Throughout the chapter, we have considered functions that model the position of free-falling objects. Any object that is falling, or vertically projected into the air, has its height above the ground, $s(t)$, in feet, after t seconds in motion, modeled by the quadratic function

$$s(t) = -16t^2 + v_o t + s_o,$$

where v_o is the original velocity (initial velocity) of the object, in feet per second, and s_o is the original height (initial height) of the object, in feet, above the ground. In this exercise, group members will be working with this position function. The exercise is appropriate for groups of three to five people.

a. Drop ball from a height of 3 feet, 6 feet, and 12 feet. Record the number of seconds it takes for the ball to hit the ground.

b. For each of the three initial positions, use the position function to determine the time required for the ball to hit the ground.

c. What factors might result in differences between the times that you recorded and the times indicated by the function?

d. What appears to be happening to the time required for a free-falling object to hit the ground as its initial height is doubled? Verify this observation algebraically and with a graphing utility.

e. Repeat part (a) using a sheet of paper rather than a ball. What differences do you observe? What factor seems to be ignored in the position function?

f. What is meant by the acceleration of gravity and how does this number appear in the position function for a free-falling object?

3. Members of your group act as the marketing research division for a large corporation. Your division estimates that at a price of x dollars per unit, the weekly cost, C, and revenue, R, in thousands of dollars, will be given by the functions

$$C(x) = 14 - x$$
$$\text{and} \quad R(x) = 8x - x^2.$$

Begin by finding a price range for x that describes when a profit will result. (A profit will result if revenue is greater than cost.) Once you find the price range, decide on the kind of product manufactured and sold by the corporation that will be consistent with this price range. Once the product is determined, present a written report to the corporation, heading the report as "Profit Analysis." Include the following in your report:

a. The price range for x, describing when a profit will result

b. A price that will result in the maximum profit

c. Graphs that the corporation might find helpful

d. Factors that might change the cost and revenue functions in the future (This will depend on the product chosen.)

e. A discussion of the possible limitations of the given functions

f. Anything else that you think might be helpful.

CHAPTER SUMMARY, REVIEW, AND TEST

SUMMARY

DEFINITIONS AND CONCEPTS	EXAMPLES

Section 11.1 The Square Root Property and Completing the Square

The Square Root Property

If u is an algebraic expression and d is a real number, then

If $u^2 = d$, then $u = \sqrt{d}$ or $u = -\sqrt{d}$.

Equivalently,

If $u^2 = d$, then $u = \pm\sqrt{d}$.

Solve: $(x - 6)^2 = 50$.

$$x - 6 = \pm\sqrt{50}$$
$$x - 6 = \pm\sqrt{25 \cdot 2}$$
$$x - 6 = \pm 5\sqrt{2}$$
$$x = 6 \pm 5\sqrt{2}$$

The solutions are $6 \pm 5\sqrt{2}$ and the solution set is $\{6 \pm 5\sqrt{2}\}$.

Completing the Square

If $x^2 + bx$ is a binomial, then by adding $\left(\dfrac{b}{2}\right)^2$, the square of half the coefficient of x, a perfect square trinomial will result. That is,

$$x^2 + bx + \left(\frac{b}{2}\right)^2 = \left(x + \frac{b}{2}\right)^2.$$

Complete the square: $x^2 + \dfrac{2}{7}x$.

> Half of $\frac{2}{7}$ is $\frac{1}{2} \cdot \frac{2}{7} = \frac{1}{7}$ and $\left(\frac{1}{7}\right)^2 = \frac{1}{49}$.

$$x^2 + \frac{2}{7}x + \frac{1}{49} = \left(x + \frac{1}{7}\right)^2$$

Solving Quadratic Equations by Completing the Square

1. If the coefficient of x^2 is not 1, divide both sides by this coefficient.
2. Isolate variable terms on one side.
3. Complete the square by adding the square of half the coefficient of x to both sides.
4. Factor the perfect square trinomial.
5. Solve by applying the square root property.

Solve by completing the square:

$$2x^2 + 12x - 4 = 0.$$
$$\frac{2x^2}{2} + \frac{12x}{2} - \frac{4}{2} = \frac{0}{2} \quad \text{Divide by 2.}$$
$$x^2 + 6x - 2 = 0 \quad \text{Simplify.}$$
$$x^2 + 6x = 2 \quad \text{Add 2.}$$

The coefficient of x is 6. Half of 6 is 3 and $3^2 = 9$.
Add 9 to both sides.

$$x^2 + 6x + 9 = 2 + 9$$
$$(x + 3)^2 = 11$$
$$x + 3 = \pm\sqrt{11}$$
$$x = -3 \pm \sqrt{11}$$

Section 11.2 The Quadratic Formula

The solutions of a quadratic equation in standard form

$$ax^2 + bx + c = 0, \quad a \neq 0,$$

are given by the quadratic formula

$$x = \frac{-b \pm \sqrt{b^2 - 4ac}}{2a}.$$

Solve using the quadratic formula:

$$2x^2 + 4x = 5.$$

First write the equation in standard form by subtracting 5 from both sides.

$$2x^2 + 4x - 5 = 0$$

> $a = 2$ $b = 4$ $c = -5$

$$x = \frac{-4 \pm \sqrt{4^2 - 4 \cdot 2(-5)}}{2 \cdot 2} = \frac{-4 \pm \sqrt{16 - (-40)}}{4}$$

$$= \frac{-4 \pm \sqrt{56}}{4} = \frac{-4 \pm \sqrt{4 \cdot 14}}{4} = \frac{-4 \pm 2\sqrt{14}}{4}$$

$$= \frac{2(-2 \pm \sqrt{14})}{2 \cdot 2} = \frac{-2 \pm \sqrt{14}}{2}$$

DEFINITIONS AND CONCEPTS	EXAMPLES

Section 11.2 The Quadratic Formula (continued)

The Discriminant

The discriminant, $b^2 - 4ac$, of the quadratic equation $ax + bx + c = 0$ determines the number and type of solutions.

Discriminant	Solutions
Positive perfect square, with a, b, and c rational numbers	2 rational solutions
Positive and not a perfect square	2 irrational solutions
Zero, with a, b, and c rational numbers	1 rational solution
Negative	2 imaginary solutions

• $2x^2 - 7x - 4 = 0$

$a = 2 \quad b = -7 \quad c = -4$

$$b^2 - 4ac = (-7)^2 - 4(2)(-4)$$

$$= 49 - (-32) = 49 + 32 = 81$$

Positive perfect square

The equations has 2 rational solutions.

Writing Quadratic Equations from Solutions

The zero-product principle in reverse makes it possible to write a quadratic equation from solutions:

If $A = 0$ or $B = 0$, then $AB = 0$.

Write a quadratic equation with the solution set $\{-2\sqrt{3}, 2\sqrt{3}\}$.

$$x = -2\sqrt{3} \quad \text{or} \quad x = 2\sqrt{3}$$

$$x + 2\sqrt{3} = 0 \quad \text{or} \quad x - 2\sqrt{3} = 0$$

$$(x + 2\sqrt{3})(x - 2\sqrt{3}) = 0$$

$$x^2 - (2\sqrt{3})^2 = 0$$

$$x^2 - 12 = 0$$

Section 11.3 Quadratic Functions and Their Graphs

The graph of the quadratic function
$$f(x) = a(x - h)^2 + k, \quad a \neq 0,$$
is called a parabola. The vertex, or turning point, is (h, k). The graph opens upward if a is positive and downward if a is negative. The axis of symmetry is a vertical line passing through the vertex. The graph can be obtained using the vertex, x-intercepts, if any, (set $f(x)$ equal to zero), and the y-intercept (set $x = 0$).

Graph: $f(x) = -(x + 3)^2 + 1$.
$$f(x) = -1(x - (-3))^2 + 1$$

$a = -1 \quad h = -3 \quad k = 1$

• Vertex: $(h, k) = (-3, 1)$.
• Opens downward because $a < 0$
• x-intercepts: Set $f(x) = 0$.

$$0 = -(x + 3)^2 + 1$$

$$(x + 3)^2 = 1$$

$$x + 3 = \pm\sqrt{1}$$

$$x + 3 = 1 \quad \text{or} \quad x + 3 = -1$$

$$x = -2 \qquad\qquad x = -4$$

• y-intercept: Set $x = 0$.

$$f(0) = -(0 + 3)^2 + 1 = -9 + 1 = -8$$

Vertex: (–3, 1)
(–4, 0)   (–2, 0)
(0, –8)
Axis of symmetry: x = –3

DEFINITIONS AND CONCEPTS	EXAMPLES

Section 11.3 Quadratic Functions and Their Graphs (continued)

A parabola whose equation is in the form

$$f(x) = ax^2 + bx + c, \quad a \neq 0,$$

has its vertex at

$$\left(-\frac{b}{2a}, f\left(-\frac{b}{2a} \right) \right).$$

The parabola is graphed as described in the left column at the bottom of page 788. The only difference is how we determine the vertex. If $a > 0$, then f has a minimum that occurs at $x = -\dfrac{b}{2a}$. If $a < 0$, then f has a maximum that occurs at $x = -\dfrac{b}{2a}$.

Graph: $f(x) = x^2 - 6x + 5$.

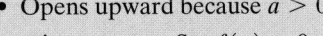
$a = 1$ $b = -6$ $c = 5$

- Vertex: $x = -\dfrac{b}{2a} = -\dfrac{-6}{2 \cdot 1} = 3$
  $$f(3) = 3^2 - 6 \cdot 3 + 5 = -4$$
  Vertex is at $(3, -4)$.
- Opens upward because $a > 0$.
- x-intercepts: Set $f(x) = 0$.
  $$x^2 - 6x + 5 = 0$$
  $$(x - 1)(x - 5) = 0$$
  $$x = 1 \quad \text{or} \quad x = 5$$
- y-intercept: Set $x = 0$.
  $$f(0) = 0^2 - 6 \cdot 0 + 5 = 5$$

Axis of symmetry: $x = 3$

$(0, 5)$

$(1, 0)$ $(5, 0)$

Vertex: $(3, -4)$

Section 11.4 Equations Quadratic in Form

An equation that is quadratic in form is one that can be expressed as a quadratic equation using an appropriate substitution. These equations contain an expression to a power, the same expression to that power squared, and a constant term. Let $t =$ the expression to the power. If at any point in the solution process both sides of an equation are raised to an even power, a check is required.

Solve: $x^{\frac{2}{3}} - 3x^{\frac{1}{3}} + 2 = 0$.
$$\left(x^{\frac{1}{3}} \right)^2 - 3x^{\frac{1}{3}} + 2 = 0$$

Let $t = x^{\frac{1}{3}}$.
$$t^2 - 3t + 2 = 0$$
$$(t - 1)(t - 2) = 0$$
$$t - 1 = 0 \quad \text{or} \quad t - 2 = 0$$
$$t = 1 \qquad\qquad t = 2$$
$$x^{\frac{1}{3}} = 1 \qquad\qquad x^{\frac{1}{3}} = 2$$
$$\left(x^{\frac{1}{3}} \right)^3 = 1^3 \qquad \left(x^{\frac{1}{3}} \right)^3 = 2^3$$
$$x = 1 \qquad\qquad x = 8$$

The solutions are 1 and 8, and the solution set is $\{1, 8\}$.

Section 11.5 Quadratic and Rational Inequalities

Solving Quadratic Inequalities

1. Write the inequality in standard form:
   $$ax^2 + bx + c > 0 \quad \text{or} \quad ax^2 + bx + c < 0.$$
2. Solve the related quadratic equation.
3. Use solutions to locate boundary points on a number line, dividing the line into test intervals.
4. Substitute a test point from each interval into the inequality. If a true statement results, the interval belongs to the solution set.
5. The solution set is the union of all such intervals.

This procedure is valid if $<$ is replaced by $\leq$ and $>$ is replaced by $\geq$. In these cases, include the boundary points in the solution set.

Solve: $2x^2 + x - 6 > 0$.
$$2x^2 + x - 6 = 0$$
$$(2x - 3)(x + 2) = 0$$
$$2x - 3 = 0 \quad \text{or} \quad x + 2 = 0$$
$$x = \frac{3}{2} \qquad\qquad x = -2$$

-2 $\dfrac{3}{2}$

$x = -3$ makes the inequality true,
$x = 0$ makes it false, and
$x = 2$ makes it true.
The solution set is $\left\{ x \mid x < -2 \text{ or } x > \frac{3}{2} \right\}$ or $(-\infty, -2) \cup \left(\frac{3}{2}, \infty \right)$.

DEFINITIONS AND CONCEPTS	EXAMPLES

Section 11.5 Quadratic and Rational Inequalities (continued)

Solving Rational Inequalities

1. Express the inequality so that one side is zero and the other side is a single quotient.
2. Find boundary points by setting the numerator and denominator equal to zero.
3. Locate boundary points on a number line, dividing the line into test intervals.
4. Substitute a test point from each interval into the inequality. If a true statement results, the interval belongs to the solution set.
5. The solution set is the union of all such intervals. Exclude any values that make the denominator zero.

Solve: $\dfrac{x}{x+4} \geq 2$.

$$\frac{x}{x+4} - \frac{2(x+4)}{x+4} \geq 0$$

$$\frac{-x-8}{x+4} \geq 0$$

$$\begin{array}{ll} -x - 8 = 0 & x + 4 = 0 \\ -8 = x & x = -4 \end{array}$$

$x = -9$ makes the original inequality false,

$x = -7$ makes it true, and

$x = -3$ makes it false.

The solution set is $\{x \mid -8 \leq x < -4\}$ or $[-8, -4)$.

Review Exercises

11.1 *In Exercises 1–5, solve each equation by the square root property. If possible, simplify radicals or rationalize denominators. Express imaginary solutions in the form $a + bi$.*

1. $2x^2 - 3 = 125$
2. $3x^2 - 150 = 0$

3. $3x^2 - 2 = 0$

4. $(x - 4)^2 = 18$
5. $(x + 7)^2 = -36$

In Exercises 6–7, complete the square for each binomial. Then factor the resulting perfect square trinomial.

6. $x^2 + 20x$

7. $x^2 - 3x$

In Exercises 8–10, solve each quadratic equation by completing the square.

8. $x^2 - 12x + 27 = 0$

9. $x^2 - 7x - 1 = 0$

10. $2x^2 + 3x - 4 = 0$

11. In 2 years, an investment of \$2500 grows to \$2916. Use the compound interest formula

$$A = P(1 + r)^t$$

to find the annual interest rate, r.

12. The function $W(t) = 3t^2$ models the weight of a human fetus, $W(t)$, in grams, after t weeks, where $0 \leq t \leq 39$. After how many weeks does the fetus weigh 1200 grams?

13. A building casts a shadow that is double the length of the building's height. If the distance from the end of the shadow to the top of the building is 300 meters, how high is the building? Express the answer in simplified radical form. Then find a decimal approximation to the nearest tenth of a meter.

11.2 *In Exercises 14–16, solve each equation using the quadratic formula. Simplify solutions, if possible.*

14. $x^2 = 2x + 4$
15. $x^2 - 2x + 19 = 0$

16. $2x^2 = 3 - 4x$

In Exercises 17–19, compute the discriminant. Then determine the number and type of solutions for the given equation.

17. $x^2 - 4x + 13 = 0$
18. $9x^2 = 2 - 3x$
19. $2x^2 + 4x = 3$

In Exercises 20–26, solve each equation by the method of your choice. Simplify solutions, if possible.

20. $2x^2 - 11x + 5 = 0$
21. $(3x + 5)(x - 3) = 5$

22. $3x^2 - 7x + 1 = 0$

23. $x^2 - 9 = 0$

24. $(x - 3)^2 - 8 = 0$

25. $3x^2 - x + 2 = 0$

26. $\dfrac{5}{x + 1} + \dfrac{x - 1}{4} = 2$

In Exercises 27–29, write a quadratic equation in standard form with the given solution set.

27. $\left\{-\dfrac{1}{3}, \dfrac{3}{5}\right\}$

28. $\{-9i, 9i\}$ **29.** $\{-4\sqrt{3}, 4\sqrt{3}\}$

30. As the use of the Internet increases, so has the number of computer infections from viruses. The function

$$N(x) = 0.2x^2 - 1.2x + 2$$

models the number of infections per month for every 1000 PCs, $N(x)$, x years after 1990. In which year will the infection rate be 13 PCs per month for every 1000 PCs?

31. A baseball is hit by a batter. The function

$$s(t) = -16t^2 + 140t + 3$$

models the ball's height above the ground, $s(t)$, in feet, t seconds after it was hit. How long will it take for the ball to strike the ground? Round to the nearest tenth of a second.

11.3 *In Exercises 32–35, use the vertex and intercepts to sketch the graph of each quadratic function. Give the equation for the parabola's axis of symmetry.*

32. $f(x) = -(x + 1)^2 + 4$

33. $f(x) = (x + 4)^2 - 2$

34. $f(x) = -x^2 + 2x + 3$

35. $f(x) = 2x^2 - 4x - 6$

36. The function

$$f(x) = -0.02x^2 + x + 1$$

models the yearly growth of a young redwood tree, $f(x)$, in inches, with x inches of rainfall per year. How many inches of rainfall per year results in maximum tree growth? What is the maximum yearly growth?

37. A model rocket is launched upward from a platform 30 feet above the ground. The quadratic function

$$s(t) = -16t^2 + 400t + 40$$

models the rocket's height above the ground, $s(t)$, in feet, t seconds after it was launched. After how many seconds does the rocket reach its maximum height? What is the maximum height?

38. Suppose that a quadratic function is used to model the data shown in the graph using

(number of years after 1960, divorce rate per 1000 population).

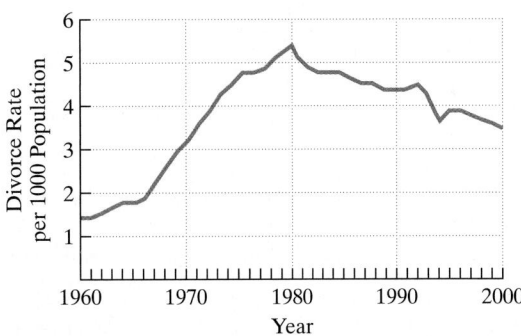

U.S. Divorce Rate

Source: U.S. Department of Health and Human Services

Determine, without obtaining an actual quadratic function that models the data, the approximate coordinates of the vertex for the function's graph. Describe what this means in practical terms.

39. A field bordering a straight stream is to be enclosed. The side bordering the stream is not to be fenced. If 1000 yards of fencing material is to be used, what are the dimensions of the largest rectangular field that can be fenced? What is the maximum area?

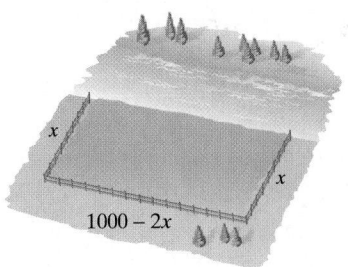

40. Among all pairs of numbers whose difference is 14, find a pair whose product is as small as possible. What is the minimum product?

11.4 *In Exercises 41–46, solve each equation by making an appropriate substitution. When necessary, check proposed solutions.*

41. $x^4 - 6x^2 + 8 = 0$

42. $x + 7\sqrt{x} - 8 = 0$

43. $(x^2 + 2x)^2 - 14(x^2 + 2x) = 15$

44. $x^{-2} + x^{-1} - 56 = 0$

45. $x^{\frac{2}{3}} - x^{\frac{1}{3}} - 12 = 0$

46. $x^{\frac{1}{2}} + 3x^{\frac{1}{4}} - 10 = 0$

11.5 *In Exercises 47–50, solve each inequality and graph the solution set on a real number line.*

47. $2x^2 + 5x - 3 < 0$

48. $2x^2 + 9x + 4 \geq 0$

49. $\dfrac{x - 6}{x + 2} > 0$

50. $\dfrac{x + 3}{x - 4} \leq 5$

51. A model rocket is launched from ground level. The function

$$s(t) = -16t^2 + 48t$$

models the rocket's height above the ground, $s(t)$, in feet, t seconds after it was launched. During which time period will the rocket's height exceed 32 feet?

52. The function

$$H(x) = \frac{15}{8}x^2 - 30x + 200$$

models heart rate, $H(x)$, in beats per minute, x minutes after a strenuous workout.

a. What is the heart rate immediately following the workout?

b. According to the model, during which intervals of time after the strenuous workout does the heart rate exceed 110 beats per minute? For which of these intervals has model breakdown occurred? Which interval provides a more realistic answer? How did you determine this?

Chapter 11 Test

Express solutions to all equations in simplified form. Rationalize denominators, if possible.

In Exercises 1–2, solve each equation by the square root property.

1. $2x^2 - 5 = 0$

2. $(x - 3)^2 = 20$

In Exercises 3–4, complete the square for each binomial. Then factor the resulting perfect square trinomial.

3. $x^2 - 16x$

4. $x^2 + \dfrac{2}{5}x$

5. Solve by completing the square: $x^2 - 6x + 7 = 0$.

6. Use the measurements determined by the surveyor to find the width of the pond. Express the answer in simplified radical form.

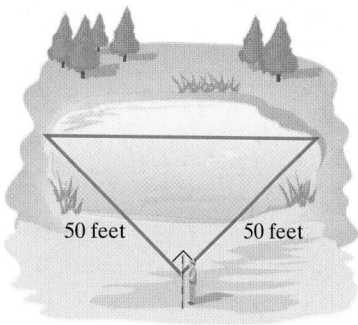

In Exercises 7–8, compute the discriminant. Then determine the number and type of solutions for the given equation.

7. $3x^2 + 4x - 2 = 0$

8. $x^2 = 4x - 8$

In Exercises 9–12, solve each equation by the method of your choice.

9. $2x^2 + 9x = 5$

10. $x^2 + 8x + 5 = 0$

11. $(x + 2)^2 + 25 = 0$

12. $2x^2 - 6x + 5 = 0$

In Exercises 13–14, write a quadratic equation in standard form with the given solution set.

13. $\{-3, 7\}$

14. $\{-10i, 10i\}$

15. The function

$$f(x) = -0.5x^2 + 4x + 19$$

models the number of people in the United States, $f(x)$, in millions, receiving food stamps x years after 1990. In which year(s) were 20 million people receiving food stamps? Use a calculator and round to the nearest year(s).

In Exercises 16–17, use the vertex and intercepts to sketch the graph of each quadratic function. Give the equation for the parabola's axis of symmetry.

16. $f(x) = (x + 1)^2 + 4$

17. $f(x) = x^2 - 2x - 3$

A baseball player hits a pop fly into the air. The function

$$s(t) = -16t^2 + 64t + 5$$

models the ball's height above the ground, $s(t)$, in feet, t seconds after it was hit. Use the function to solve Exercises 18–19.

18. When does the baseball reach its maximum height? What is that height?

19. After how many seconds does the baseball hit the ground? Round to the nearest tenth of a second.

20. The function $f(x) = -x^2 + 46x - 360$ models the daily profit, $f(x)$, in hundreds of dollars, for a company that manufactures x computers daily. How many computers should be manufactured each day to maximize profit? What is the maximum daily profit?

In Exercises 21–23, solve each equation by making an appropriate substitution. When necessary, check proposed solutions.

21. $(2x - 5)^2 + 4(2x - 5) + 3 = 0$

22. $x^4 - 13x^2 + 36 = 0$

23. $x^{\frac{2}{3}} - 9x^{\frac{1}{3}} + 8 = 0$

In Exercises 24–25, solve each inequality and graph the solution set on a real number line.

24. $x^2 - x - 12 < 0$

25. $\dfrac{2x + 1}{x - 3} \le 3$

Cumulative Review Exercises (Chapters 1–11)

In Exercises 1–7, solve each equation, inequality, or system.

1. $8 - (4x - 5) = x - 7$

2. $5x + 4y = 22$
$3x - 8y = -18$

3. $-3x + 2y + 4z = 6$
$7x - y + 3z = 23$
$2x + 3y + z = 7$

4. $|x - 1| > 3$

5. $\sqrt{x + 4} - \sqrt{x - 4} = 2$

6. $x - 4 \ge 0$ and $-3x \le -6$

7. $2x^2 = 3x - 2$

In Exercises 8–12, graph each function, equation, or inequality in a rectangular coordinate system.

8. $3x = 15 + 5y$

9. $2x - 3y > 6$

10. $f(x) = -\dfrac{1}{2}x + 1$

11. $f(x) = x^2 + 6x + 8$

12. $f(x) = (x - 3)^2 - 4$

13. Evaluate:
$$\begin{vmatrix} 3 & 1 & 0 \\ 0 & 5 & -6 \\ -2 & -1 & 0 \end{vmatrix}.$$

14. Simplify: $\dfrac{x - \frac{1}{3}}{3 - \frac{1}{x}}.$

15. Write the equation of the linear function whose graph contains the point $(-2, 4)$ and is perpendicular to the line whose equation is $2x + y = 10$.

In Exercises 16–20, perform the indicated operations and simplify, if possible.

16. $\dfrac{-5x^3y^7}{15x^4y^{-2}}$

17. $(4x^2 - 5y)^2$

18. $(5x^3 - 24x^2 + 9) \div (5x + 1)$

19. $\dfrac{\sqrt[3]{32xy^{10}}}{\sqrt[3]{2xy^2}}$

20. $\dfrac{x + 2}{x^2 - 6x + 8} + \dfrac{3x - 8}{x^2 - 5x + 6}$

In Exercises 21–22, factor completely.

21. $x^4 - 4x^3 + 8x - 32$

22. $2x^2 + 12xy + 18y^2$

23. The length of a rectangular carpet is 4 feet greater than twice its width. If the area is 48 square feet, find the carpet's length and width.

24. Working alone, you can mow the lawn in 2 hours, and your sister can do it in 3 hours. How long will it take you to do the job if you work together?

25. Your motorboat can travel 15 miles per hour in still water. Traveling with the river's current, the boat can cover 20 miles in the same time it takes to go 10 miles against the current. Find the rate of the current.

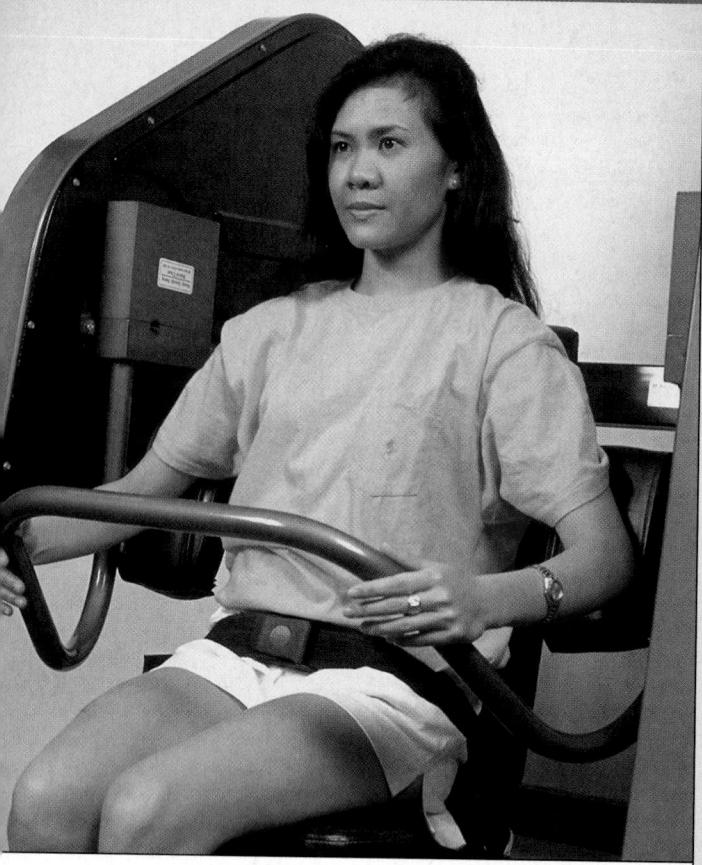

You've recently taken up weightlifting, recording the maximum number of pounds you can lift at the end of each week. At first your weight limit increases rapidly, but now you notice that this growth is beginning to level off. You wonder about a function that would serve as a mathematical model to predict the number of pounds you can lift as you continue the sport.

W hat went wrong on the space shuttle *Challenger*? Will population growth lead to a future without comfort or individual choice? Can I put aside a small amount of money and have millions for early retirement? Why did I feel I was walking too slowly on my visit to New York City? Why are people in California at far greater risk from drunk drivers than from earthquakes? What is the difference between earthquakes measuring 6 and 7 on the Richter scale? And what can I hope to accomplish in weightlifting?

The functions that you will be learning about in this chapter will provide you with the mathematics for answering these questions. You will see how these remarkable functions enable us to predict the future and rediscover the past.

Chapter 12

Exponential and
Logarithmic Functions

▶ SECTION 12.1   *Exponential Functions*

Objectives

1 Evaluate exponential functions.

2 Graph exponential functions.

3 Evaluate functions with base e.

4 Use compound interest formulas.

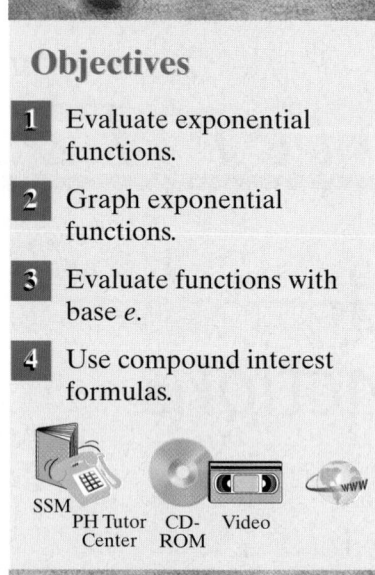

SSM
PH Tutor   CD-   Video
Center   ROM

The space shuttle *Challenger* exploded approximately 73 seconds into flight on January 28, 1986. The tragedy involved damage to ○-rings, which were used to seal the connections between different sections of the shuttle engines. The number of ○-rings damaged increases dramatically as Fahrenheit temperature falls.

The function

$$f(x) = 13.49(0.967)^x - 1$$

models the number of ○-rings expected to fail when the temperature is $x°$F. Can you see how this function is different from polynomial functions? The variable x is in the exponent. Functions whose equations contain a variable in the exponent are called **exponential functions**. Many real-life situations, including population growth, growth of epidemics, radioactive decay, and other changes that involve rapid increase or decrease can be described using exponential functions.

Definition of the Exponential Function

The **exponential function f with base b** is defined by

$$f(x) = b^x \quad \text{or} \quad y = b^x$$

where b is a positive constant other than $1 (b > 0$ and $b \neq 1)$ and x is any real number.

Here are some examples of exponential functions.

$$f(x) = 2^x \qquad g(x) = 10^x \qquad h(x) = 3^{x+1}$$

> Base is 2.   Base is 10.   Base is 3.

Each of these functions has a constant base and a variable exponent. By contrast the following functions are not exponential.

$$F(x) = x^2 \qquad G(x) = 1^x \qquad H(x) = x^x$$

> Variable is the base and not exponent.   The base of an exponential function must be a positive constant other than 1.   Variable is both the base and the exponent.

Why is $G(x) = 1^x$ not classified as an exponential function? The number 1 raised to any power is 1. Thus, the function G can be written as $G(x) = 1$, which is a constant function.

1 Evaluate exponential functions.

You will need a calculator to evaluate exponential expressions. Most scientific calculators have an $\boxed{x^y}$ key. Graphing calculators have a $\boxed{\wedge}$ key. To evaluate expressions of the form b^x, enter the base b, press $\boxed{x^y}$ or $\boxed{\wedge}$, enter the exponent x, and finally press $\boxed{=}$ or $\boxed{\text{ENTER}}$.

EXAMPLE 1   Evaluating an Exponential Function

The exponential function $f(x) = 13.49(0.967)^x - 1$ describes the number of O-rings expected to fail, $f(x)$, when the temperature is $x°$F. On the morning the *Challenger* was launched, the temperature was 31°F, colder than any previous experience. Find the number of O-rings expected to fail at this temperature.

Solution   Because the temperature was 31°F, substitute 31 for x and evaluate the function at 31.

$$f(x) = 13.49(0.967)^x - 1 \quad \text{This is the given function.}$$
$$f(31) = 13.49(0.967)^{31} - 1 \quad \text{Substitute 31 for x.}$$

Use a scientific or graphing calculator to evaluate $(0.967)^{31}$. Press the following keys on your calculator to do this:

Scientific calculator: .967 $\boxed{x^y}$ 31 $\boxed{=}$

Graphing calculator: .967 $\boxed{\wedge}$ 31 $\boxed{\text{ENTER}}$.

The display should be approximately .353362693426. Multiplying this number by 13.49 and subtracting 1, we obtain

$$f(31) = 13.49(0.967)^{31} - 1 \approx 4.$$

Thus, four O-rings are expected to fail at a temperature of 31°F.   ∎

✔ **CHECK POINT 1**   Use the function in Example 1 to find the number of O-rings expected to fail at a temperature of 60°F. Round to the nearest whole number.

2 Graph exponential functions.

Graphing Exponential Functions

We are familiar with expressions involving b^x where x is a rational number. For example,

$$b^{1.7} = b^{\frac{17}{10}} = \sqrt[10]{b^{17}} \quad \text{and} \quad b^{1.73} = b^{\frac{173}{100}} = \sqrt[100]{b^{173}}.$$

However, note that the definition of $f(x) = b^x$ includes all real numbers for the domain x. You may wonder what b^x means when x is an irrational number, such as $b^{\sqrt{3}}$ or b^π. Using the nonrepeating and nonterminating approximation 1.73205 for $\sqrt{3}$, we can think of $b^{\sqrt{3}}$ as the value that has the successively closer approximations

$$b^{1.7}, \ b^{1.73}, \ b^{1.732}, \ b^{1.73205}, \ldots.$$

In this way, we can graph the exponential function with no holes, or points of discontinuity, at the irrational domain values.

EXAMPLE 2   Graphing an Exponential Function

Graph: $f(x) = 2^x$.

Solution   We begin by setting up a table of coordinates.

x	$f(x) = 2^x$
-3	$f(-3) = 2^{-3} = \frac{1}{8}$
-2	$f(-2) = 2^{-2} = \frac{1}{4}$
-1	$f(-1) = 2^{-1} = \frac{1}{2}$
0	$f(0) = 2^0 = 1$
1	$f(1) = 2^1 = 2$
2	$f(2) = 2^2 = 4$
3	$f(3) = 2^3 = 8$

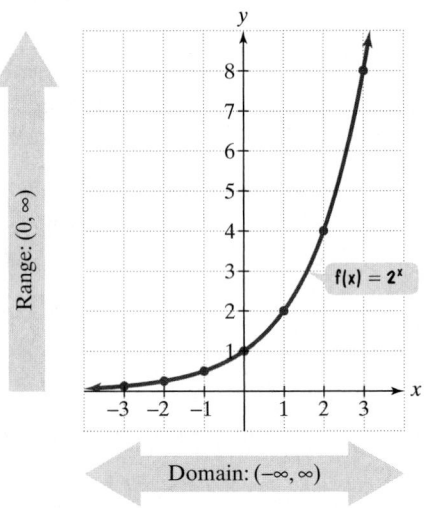

Figure 12.1 The graph of $f(x) = 2^x$

We plot these points, connecting them with a continuous curve. Figure 12.1 shows the graph of $f(x) = 2^x$. Observe that the graph approaches but never touches the negative portion of the x-axis. Thus, the x-axis is a horizontal asymptote. The range is the set of all positive real numbers. Although we used integers for x in our table of coordinates, you can use a calculator to find additional points. For example, $f(0.3) = 2^{0.3} \approx 1.231$, $f(0.95) = 2^{0.95} \approx 1.932$. The points $(0.3, 1.231)$ and $(0.95, 1.932)$ fit the graph. ∎

✔ **CHECK POINT 2**   Graph: $f(x) = 3^x$.

EXAMPLE 3   Graphing an Exponential Function

Graph: $f(x) = \left(\dfrac{1}{2}\right)^x$.

Solution   We begin by setting up a table of coordinates. We compute the function values by noting that

$$f(x) = \left(\frac{1}{2}\right)^x = \left(2^{-1}\right)^x = 2^{-x}.$$

We plot these points, connecting them with a continuous curve. Figure 12.2 on page 799 shows the graph of $f(x) = \left(\frac{1}{2}\right)^x$. This time the graph approaches but never touches the *positive* portion of the x-axis. Once again, the x-axis is a horizontal asymptote. The range consists of the set of all positive real numbers.

x	$f(x) = \left(\dfrac{1}{2}\right)^x$ or 2^{-x}
-3	$f(-3) = 2^{-(-3)} = 2^3 = 8$
-2	$f(-2) = 2^{-(-2)} = 2^2 = 4$
-1	$f(-1) = 2^{-(-1)} = 2^1 = 2$
0	$f(0) = 2^{-0} = 1$
1	$f(1) = 2^{-1} = \dfrac{1}{2^1} = \dfrac{1}{2}$
2	$f(2) = 2^{-2} = \dfrac{1}{2^2} = \dfrac{1}{4}$
3	$f(3) = 2^{-3} = \dfrac{1}{2^3} = \dfrac{1}{8}$

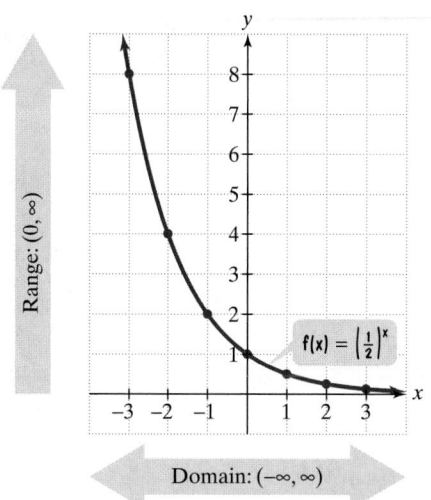

Figure 12.2 The graph of $f(x) = \left(\dfrac{1}{2}\right)^x$

Do you notice a relationship between the graphs of $f(x) = 2^x$ and $f(x) = \left(\frac{1}{2}\right)^x$ in Figures 12.1 and 12.2? The graph of $f(x) = \left(\frac{1}{2}\right)^x$ is a mirror image, or reflection, of the graph of $f(x) = 2^x$ about the y-axis.

✔ **CHECK POINT 3** Graph: $f(x) = \left(\frac{1}{3}\right)^x$. Note that $f(x) = \left(\frac{1}{3}\right)^x = \left(3^{-1}\right)^x = 3^{-x}$.

Four exponential functions have been graphed in Figure 12.3. Compare the black and green graphs, where $b > 1$, to those in blue and red, where $b < 1$. When $b > 1$, the value of y increases as the value of x increases. When $b < 1$, the value of y decreases as the value of x increases. Notice that all four graphs pass through $(0, 1)$.

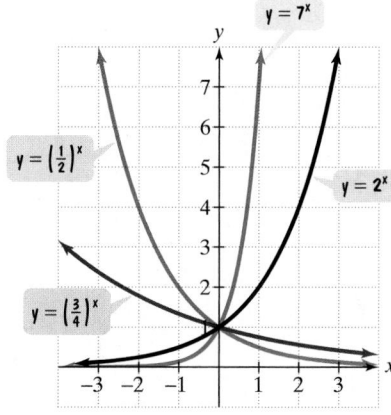

Figure 12.3 Graphs of four exponential functions

These graphs illustrate the following general characteristics of exponential functions, listed in the box on page 800.

Characteristics of Exponential Functions of the Form $f(x) = b^x$

1. The domain of $f(x) = b^x$ consists of all real numbers. The range of $f(x) = b^x$ consists of all positive real numbers.

2. The graphs of all exponential functions of the form $f(x) = b^x$ pass through the point $(0, 1)$ because $f(0) = b^0 = 1$ $(b \neq 0)$. The y-intercept is 1.

3. If $b > 1$, $f(x) = b^x$ has a graph that goes up to the right and is an increasing function. The greater the value of b, the steeper the increase.

4. If $0 < b < 1$, $f(x) = b^x$ has a graph that goes down to the right and is a decreasing function. The smaller the value of b, the steeper the decrease.

5. The graph of $f(x) = b^x$ approaches, but does not cross, the x-axis. The x-axis, or $y = 0$, is a horizontal asymptote.

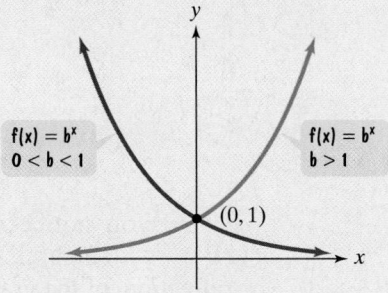

EXAMPLE 4   Graphing Exponential Functions

Graph $f(x) = 3^x$ and $g(x) = 3^{x+1}$ in the same rectangular coordinate system. How is the graph of g related to the graph of f?

Solution   We begin by setting up a table showing some of the coordinates for f and g, selecting integers from -2 to 2 for x. Notice that $x + 1$ is the exponent for $g(x) = 3^{x+1}$.

x	$f(x) = 3^x$	$g(x) = 3^{x+1}$
-2	$f(-2) = 3^{-2} = \frac{1}{9}$	$g(-2) = 3^{-2+1} = 3^{-1} = \frac{1}{3}$
-1	$f(-1) = 3^{-1} = \frac{1}{3}$	$g(-1) = 3^{-1+1} = 3^0 = 1$
0	$f(0) = 3^0 = 1$	$g(0) = 3^{0+1} = 3^1 = 3$
1	$f(1) = 3^1 = 3$	$g(1) = 3^{1+1} = 3^2 = 9$
2	$f(2) = 3^2 = 9$	$g(2) = 3^{2+1} = 3^3 = 27$

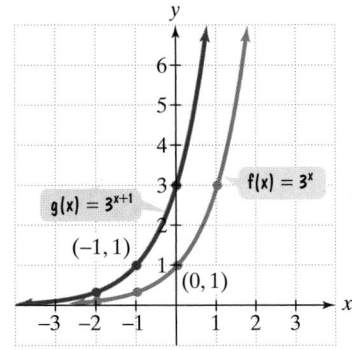

Figure 12.4

We plot the points for each function and connect them with a smooth curve. Because of the scale on the y-axis, the points on each function corresponding to $x = 2$ are not shown. Figure 12.4 shows the graphs of $f(x) = 3^x$ and $g(x) = 3^{x+1}$. The graph of g is the graph of f shifted one unit to the left. ∎

✔ **CHECK POINT 4** Graph $f(x) = 3^x$ and $g(x) = 3^{x-1}$ in the same rectangular coordinate system. Select integers from -2 to 2 for x. How is the graph of g related to the graph of f?

EXAMPLE 5   Graphing Exponential Functions

Graph $f(x) = 2^x$ and $g(x) = 2^x - 3$ in the same rectangular coordinate system. How is the graph of g related to the graph of f?

Solution   We begin by setting up a table showing some of the coordinates for f and g, selecting integers from -2 to 2 for x.

x	$f(x) = 2^x$	$g(x) = 2^x - 3$
-2	$f(-2) = 2^{-2} = \frac{1}{4}$	$g(-2) = 2^{-2} - 3 = \frac{1}{4} - 3 = -2\frac{3}{4}$
-1	$f(-1) = 2^{-1} = \frac{1}{2}$	$g(-1) = 2^{-1} - 3 = \frac{1}{2} - 3 = -2\frac{1}{2}$
0	$f(0) = 2^0 = 1$	$g(0) = 2^0 - 3 = 1 - 3 = -2$
1	$f(1) = 2^1 = 2$	$g(1) = 2^1 - 3 = 2 - 3 = -1$
2	$f(2) = 2^2 = 4$	$g(2) = 2^2 - 3 = 4 - 3 = 1$

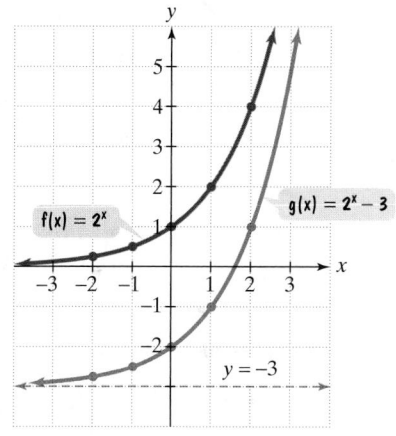

Figure 12.5

We plot the points for each function and connect them with a smooth curve. Figure 12.5 shows the graphs of $f(x) = 2^x$ and $g(x) = 2^x - 3$. The graph of g is the graph of f shifted down three units. As a result, $y = -3$ is the horizontal asymptote for g.   ∎

✔ **CHECK POINT 5**   Graph $f(x) = 2^x$ and $g(x) = 2^x + 3$ in the same rectangular coordinate system. Select integers from -2 to 2 for x. How is the graph of g related to the graph of f?

3 Evaluate functions with base e.

The Natural Base e

An irrational number, symbolized by the letter e, appears as the base in many applied exponential functions. This irrational number is approximately equal to 2.72. More accurately,

$$e \approx 2.71828\ldots.$$

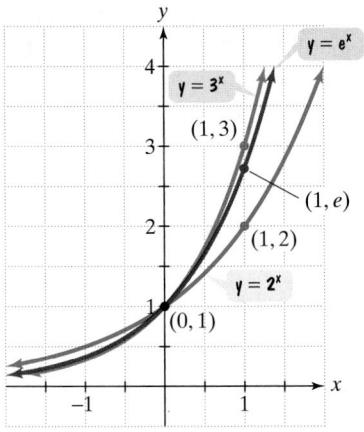

Figure 12.6 Graphs of three exponential functions

The number e is called the **natural base**. The function $f(x) = e^x$ is called the **natural exponential function**.

Use a scientific or graphing calculator with an $\boxed{e^x}$ key to evaluate e to various powers. For example, to find e^2, press the following keys on most calculators:

Scientific calculator: 2 $\boxed{e^x}$

Graphing calculator: $\boxed{e^x}$ 2 $\boxed{\text{ENTER}}$.

The display is approximately 7.389.

$$e^2 \approx 7.389$$

The number e lies between 2 and 3. Because $2^2 = 4$ and $3^2 = 9$, it makes sense that e^2, approximately 7.389, lies between 4 and 9.

Because $2 < e < 3$, the graph of $y = e^x$ is between the graphs of $y = 2^x$ and $y = 3^x$, shown in Figure 12.6.

EXAMPLE 6 World Population

In a report entitled *Resources and Man*, the U.S. National Academy of Sciences concluded that a world population of 10 billion "is close to (if not above) the maximum that an intensely managed world might hope to support with some degree of comfort and individual choice." At the time the report was issued in 1969, the world population was approximately 3.6 billion, with a growth rate of 2% per year. The function

$$f(x) = 3.6e^{0.02x}$$

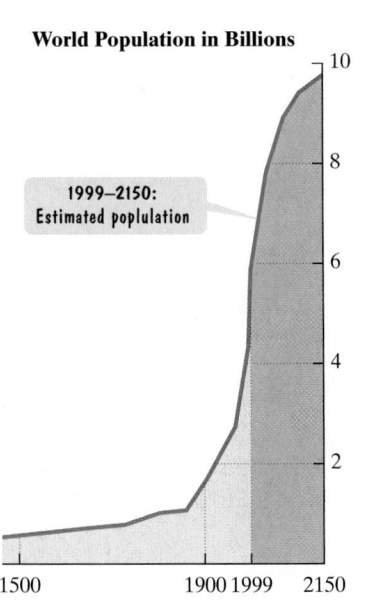

Source: U.N. Population Division

Growth rate in world population has slowed down since 1969. Current models project that world population will not be close to 10 billion until 2150.

describes world population, $f(x)$, in billions, x years after 1969. Use the function to find world population in the year 2020. Is there cause for alarm?

Solution Because 2020 is 51 years after 1969, we substitute 51 for x:

$$f(51) = 3.6e^{0.02(51)}.$$

Although this computation can be done on your calculator in one step, we will break it down into smaller steps so that you can clearly see how we use the $\boxed{e^x}$ key. First find 0.02(51):

$$0.02(51) = 1.02.$$

Now find $e^{1.02}$:

Scientific calculator: 1.02 $\boxed{e^x}$

Graphing calculator: $\boxed{e^x}$ 1.02 $\boxed{\text{ENTER}}$.

The display is approximately 2.7731948. Multiplying this number by 3.6, we obtain

$$f(51) = 3.6e^{0.02(51)} = 3.6e^{1.02} \approx 9.98.$$

This indicates that world population in the year 2020 will be approximately 9.98 billion. Because this number is quite close to 10 billion, the given function suggests that there may be cause for alarm. ■

World population in 2000 was approximately 6 billion, but the growth rate was no longer 2%. It had slowed down to 1.3%. Using this current growth rate, exponential functions now predict a world population of 7.6 billion in the year 2020. Experts think the population may stabilize at 10 billion after 2200 if the deceleration in growth rate continues.

✔ **CHECK POINT 6** The function $f(x) = 6e^{0.013x}$ describes world population, $f(x)$, in billions, x years after 2000 subject to a growth rate of 1.3% annually. Use the function to find world population in 2050.

Compound Interest

In Chapter 11, we saw that the amount of money, A, that a principal, P, will be worth after t years at interest rate r, compounded annually, is given by the formula

$$A = P(1 + r)^t.$$

Most savings institutions have plans in which interest is paid more than once a year. If compound interest is paid twice a year, the compounding period is six months. We say that the interest is **compounded semiannually**. When compound interest is paid four times a year, the compounding period is three months and the interest is said to be **compounded quarterly**. Some plans allow for monthly compounding or daily compounding.

In general, when compound interest is paid n times a year, we say that there are **n compounding periods per year**. The formula $A = P(1 + r)^t$ can be adjusted to take into account the number of compounding periods in a year. If there are n compounding periods per year, the formula becomes

$$A = P\left(1 + \frac{r}{n}\right)^{nt}.$$

Some banks use **continuous compounding**, where the number of compounding periods increases infinitely (compounding interest every trillionth of a second, every quadrillionth of a second, etc.). As n, the number of compounding periods in a year, increases without bound, the expression $\left(1 + \frac{1}{n}\right)^n$ approaches e. As a result, the formula for continuous compounding is $A = Pe^{rt}$. Although continuous compounding sounds terrific, it yields only a fraction of a percent more interest over a year than daily compounding.

n	$\left(1 + \frac{1}{n}\right)^n$
1	2
2	2.25
5	2.48832
10	2.59374246
100	2.704813829
1000	2.716923932
10,000	2.718145927
100,000	2.718268237
1,000,000	2.718280469
1,000,000,000	2.718281827

As n takes on increasingly large values, the expression $\left(1 + \frac{1}{n}\right)^n$ approaches e.

4 Use compound interest formulas.

Formulas for Compound Interest

After t years, the balance, A, in an account with principal P and annual interest rate r (in decimal form) is given by the following formulas:

1. For n compoundings per year: $A = P\left(1 + \frac{r}{n}\right)^{nt}$

2. For continuous compounding: $A = Pe^{rt}$.

Using Technology

The graphs illustrate that as x increases, $\left(1 + \dfrac{1}{x}\right)^x$ approaches e.

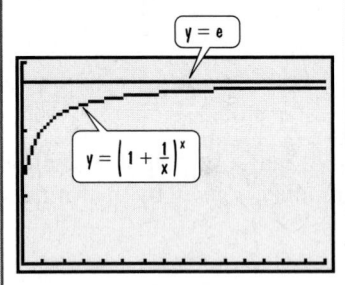

EXAMPLE 7 Choosing between Investments

You want to invest $8000 for 6 years, and you have a choice between two accounts. The first pays 7% per year, compounded monthly. The second pays 6.85% per year, compounded continuously. Which is the better investment?

Solution The better investment is the one with the greater balance in the account after 6 years. Let's begin with the account with monthly compounding. We use the compound interest model with $P = 8000$, $r = 7\% = 0.07$, $n = 12$ (monthly compounding means 12 compoundings per year), and $t = 6$.

$$A = P\left(1 + \frac{r}{n}\right)^{nt} = 8000\left(1 + \frac{0.07}{12}\right)^{12\cdot6} \approx 12{,}160.84$$

The balance in this account after 6 years is $12,160.84. For the second investment option, we use the model for continuous compounding with $P = 8000$, $r = 6.85\% = 0.0685$, and $t = 6$.

$$A = Pe^{rt} = 8000e^{0.0685(6)} \approx 12{,}066.60$$

The balance in this account after 6 years is $12,066.60, slightly less than the previous amount. Thus, the better investment is the 7% monthly compounding option. ∎

✔ **CHECK POINT 7** A sum of $10,000 is invested at an annual rate of 8%. Find the balance in the account after 5 years subject to **a**. quarterly compounding and **b**. continuous compounding.

EXERCISE SET 12.1

Practice Exercises

In Exercises 1–10, approximate each number using a calculator. Round your answer to three decimal places.

1. $2^{3.4}$ **2.** $3^{2.4}$

3. $3^{\sqrt{5}}$ **4.** $5^{\sqrt{3}}$

5. $4^{-1.5}$ **6.** $6^{-1.2}$

7. $e^{2.3}$ **8.** $e^{3.4}$

9. $e^{-0.95}$ **10.** $e^{-0.75}$

In Exercises 11–16, set up a table of coordinates for each function. Select integers from −2 to 2 for x. Then use the table of coordinates to match the function with its graph. [The graphs are labeled (a) through (f).]

11. $f(x) = 3^x$ **12.** $f(x) = 3^{x-1}$

13. $f(x) = 3^x - 1$ **14.** $f(x) = -3^x$

15. $f(x) = 3^{-x}$ **16.** $f(x) = -3^{-x}$

(a)

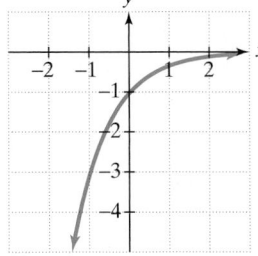

(b)

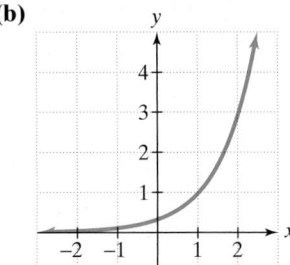

(c)

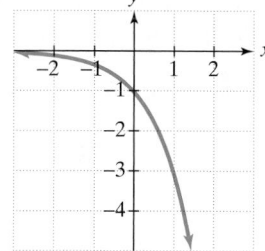

(d)

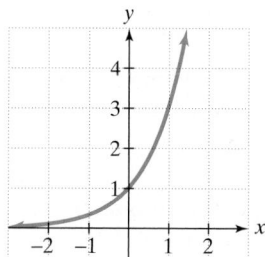

(e)

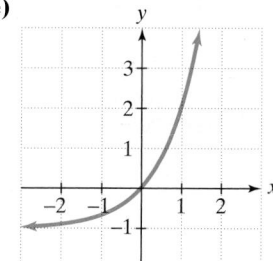

(f)

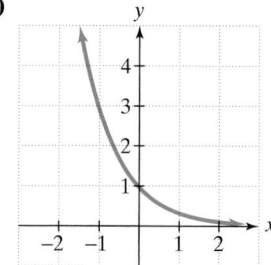

In Exercises 17–24, graph each function by making a table of coordinates. If applicable, use a graphing utility to confirm your hand-drawn graph.

17. $f(x) = 4^x$ **18.** $f(x) = 5^x$

19. $g(x) = \left(\dfrac{3}{2}\right)^x$ **20.** $g(x) = \left(\dfrac{4}{3}\right)^x$

21. $h(x) = \left(\dfrac{1}{2}\right)^x$ **22.** $h(x) = \left(\dfrac{1}{3}\right)^x$

23. $f(x) = (0.6)^x$ **24.** $f(x) = (0.8)^x$

In Exercises 25–38, graph functions f and g in the same rectangular coordinate system. Select integers from −2 to 2 for x. Then describe how the graph of g is related to the graph of f. If applicable, use a graphing utility to confirm your hand-drawn graphs.

25. $f(x) = 2^x$ and $g(x) = 2^{x+1}$

26. $f(x) = 2^x$ and $g(x) = 2^{x+2}$

27. $f(x) = 2^x$ and $g(x) = 2^{x-2}$

28. $f(x) = 2^x$ and $g(x) = 2^{x-1}$

29. $f(x) = 2^x$ and $g(x) = 2^x + 1$

30. $f(x) = 2^x$ and $g(x) = 2^x + 2$

31. $f(x) = 2^x$ and $g(x) = 2^x - 2$

32. $f(x) = 2^x$ and $g(x) = 2^x - 1$

33. $f(x) = 3^x$ and $g(x) = -3^x$

34. $f(x) = 3^x$ and $g(x) = 3^{-x}$

35. $f(x) = 2^x$ and $g(x) = 2^{x+1} - 1$

36. $f(x) = 2^x$ and $g(x) = 2^{x+1} - 2$

37. $f(x) = 3^x$ and $g(x) = \frac{1}{3} \cdot 3^x$

38. $f(x) = 3^x$ and $g(x) = 3 \cdot 3^x$

Use the compound interest formulas, $A = P\left(1 + \dfrac{r}{n}\right)^{nt}$ and $A = Pe^{rt}$, to solve Exercises 39–42. Round answers to the nearest cent.

39. Find the accumulated value of an investment of $10,000 for 5 years at an interest rate of 5.5% if the money is **a.** compounded semiannually; **b.** compounded monthly; **c.** compounded continuously.

40. Find the accumulated value of an investment of $5000 for 10 years at an interest rate of 6.5% if the money is **a.** compounded semiannually; **b.** compounded monthly; **c.** compounded continuously.

41. Suppose that you have $12,000 to invest. Which investment yields the greatest return over 3 years: 7% compounded monthly or 6.85% compounded continuously?

42. Suppose that you have $6000 to invest. Which investment yields the greatest return over 4 years: 8.25% compounded quarterly or 8.3% compounded semiannually?

Application Exercises

Use a calculator with an $\boxed{x^y}$ key or a $\boxed{\wedge}$ key to solve Exercises 43–48.

43. The exponential function $f(x) = 67.38(1.026)^x$ describes the population of Mexico, $f(x)$, in millions, x years after 1980.
  a. Substitute 0 for x and, without using a calculator, find Mexico's population in 1980.

  b. Substitute 27 for x and use your calculator to find Mexico's population in the year 2007 as predicted by this function.

  c. Find Mexico's population in the year 2034 as predicted by this function.

  d. Find Mexico's population in the year 2061 as predicted by this function.

  e. What appears to be happening to Mexico's population every 27 years?

44. The 1986 explosion at the Chernobyl nuclear power plant in the former Soviet Union sent about 1000 kilograms of radioactive cesium-137 into the atmosphere. The function $f(x) = 1000(0.5)^{\frac{x}{30}}$ describes the amount, $f(x)$, in kilograms, of cesium-137 remaining in Chernobyl x years after 1986. If even 100 kilograms of cesium-137 remain in Chernobyl's atmosphere, the area is considered unsafe for human habitation. Find $f(80)$ and determine if Chernobyl will be safe for human habitation by 2066.

The formula $S = C(1 + r)^t$ models inflation, where $C =$ the value today, $r =$ the annual inflation rate, and $S =$ the inflated value t years from now. Use this formula to solve Exercises 45–46.

45. If the inflation rate is 6%, how much will a house now worth $65,000 be worth in 10 years?

46. If the inflation rate is 3%, how much will a house now worth $110,000 be worth in 5 years?

47. A decimal approximation for $\sqrt{3}$ is 1.7320508. Use a calculator to find $2^{1.7}$, $2^{1.73}$, $2^{1.732}$, $2^{1.73205}$, and $2^{1.7320508}$. Now find $2^{\sqrt{3}}$. What do you observe?

48. A decimal approximation for π is 3.141593. Use a calculator to find 2^3, $2^{3.1}$, $2^{3.14}$, $2^{3.141}$, $2^{3.1415}$, $2^{3.14159}$, and $2^{3.141593}$. Now find 2^{π}. What do you observe?

Use a calculator with an $\boxed{e^x}$ key to solve Exercises 49–55. The function $f(x) = 24,000e^{0.21x}$ describes the number of AIDS cases in the United States among intravenous drug users x years after 1989. Use this function to solve Exercises 49–50.

49. Evaluate $f(11)$ and describe what this means in practical terms.

50. Evaluate $f(31)$ and describe what this means in practical terms.

51. In college, we study large volumes of information—information that, unfortunately, we do not often retain for very long. The function
$$f(x) = 80e^{-0.5x} + 20$$
describes the percentage of information, $f(x)$, that a particular person remembers x weeks after learning the information.
  a. Substitute 0 for x and, without using a calculator, find the percentage of information remembered at the moment it is first learned.

  b. Substitute 1 for x and find the percentage of information that is remembered after 1 week.

  c. Find the percentage of information that is remembered after 4 weeks.

  d. Find the percentage of information that is remembered after one year (52 weeks).

52. In 1626, Peter Minuit convinced the Wappinger Indians to sell him Manhattan Island for $24. If the Native Americans had put the $24 into a bank account paying 5% interest, how much would the investment have been worth in the year 2000 if interest were compounded
  a. monthly?
  b. continuously?

The function
$$f(x) = \frac{90}{1 + 270e^{-0.122x}}$$
models the percentage, $f(x)$ of people x years old with some coronary heart disease. Use this function to solve Exercises 53–54. Round answers to the nearest tenth of a percent.

53. Evaluate $f(30)$ and describe what this means in practical terms.

54. Evaluate $f(70)$ and describe what this means in practical terms.

55. The function

$$N(t) = \frac{30,000}{1 + 20e^{-1.5t}}$$

describes the number of people, $N(t)$, who become ill with influenza t weeks after its initial outbreak in a town with 30,000 inhabitants. The horizontal asymptote in the graph indicates that there is a limit to the epidemic's growth.

a. How many people became ill with the flu when the epidemic began? (When the epidemic began, $t = 0$.)

b. How many people were ill by the end of the third week?

c. Why can't the spread of an epidemic simply grow indefinitely? What does the horizontal asymptote shown in the graph indicate about the limiting size of the population that becomes ill?

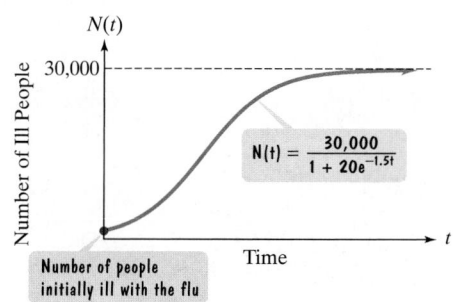

Writing in Mathematics

56. What is an exponential function?

57. What is the natural exponential function?

58. Use a calculator to evaluate $\left(1 + \dfrac{1}{x}\right)^x$ for $x = 10, 100,$ 1000, 10,000, 100,000, and 1,000,000. Describe what happens to the expression as x increases.

59. Write an example similar to Example 7 on page 804 in which continuous compounding at a slightly lower yearly interest rate is a better investment than compounding n times per year.

60. Describe how you could use the graph of $f(x) = 2^x$ to obtain a decimal approximation for $\sqrt{2}$.

61. In 2000, world population was 6 billion with an annual growth rate of 1.3%. Discuss two factors that would cause this growth rate to slow down over the next ten years.

Critical Thinking Exercises

62. Which one of the following is true?

a. As the number of compounding periods increases on a fixed investment, the amount of money in the account over a fixed interval of time will increase without bound.

b. The functions $f(x) = 3^{-x}$ and $g(x) = -3^x$ have the same graph.

c. $e = 2.718$

d. The functions $f(x) = \left(\frac{1}{3}\right)^x$ and $g(x) = 3^{-x}$ have the same graph.

63. The graphs labeled (a)–(d) in the figure represent $y = 3^x$, $y = 5^x$, $y = \left(\frac{1}{3}\right)^x$, and $y = \left(\frac{1}{5}\right)^x$, but not necessarily in that order. Which is which? Describe the process that enables you to make this decision.

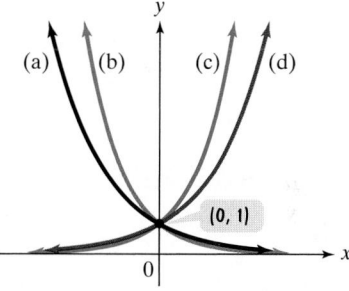

64. The hyperbolic cosine and hyperbolic sine functions are defined by

$$\cosh x = \frac{e^x + e^{-x}}{2} \quad \text{and} \quad \sinh x = \frac{e^x - e^{-x}}{2}.$$

Prove that $(\cosh x)^2 - (\sinh x)^2 = 1$.

Technology Exercises

65. Graph $y = 13.49(0.967)^x - 1$, the function for the number of O-rings expected to fail at $x°$F, in a $[0, 90, 10]$ by $[0, 20, 5]$ viewing rectangle. If NASA engineers had used this function and its graph, is it likely they would have allowed the *Challenger* to be launched when the temperature was $31°$F? Explain.

66. The student–teacher ratio in U.S. elementary and secondary schools can be modeled by $y = 25.34(0.987)^x$, where x represents the number of years after 1959 and y represents the student–teacher ratio. Graph the function in a $[1, 40, 1]$ by $[0, 26, 1]$ viewing rectangle. When did the student–teacher ratio become less than 21 students per teacher?

67. You have $10,000 to invest. One bank pays 5% interest compounded quarterly and the other pays 4.5% interest compounded monthly.

a. Use the formula for compound interest to write a function for the balance in each account at any time t, in years.

b. Use a graphing utility to graph both functions in an appropriate viewing rectangle. Based on the graphs, which bank offers the better return on your money?

68. a. Graph $y = e^x$ and $y = 1 + x + \dfrac{x^2}{2}$ in the same viewing rectangle.

b. Graph $y = e^x$ and $y = 1 + x + \dfrac{x^2}{2} + \dfrac{x^3}{6}$ in the same viewing rectangle.

c. Graph $y = e^x$ and $y = 1 + x + \dfrac{x^2}{2} + \dfrac{x^3}{6} + \dfrac{x^4}{24}$ in the same viewing rectangle.

d. Describe what you observe in parts (a)–(c). Try generalizing this observation.

Review Exercises

69. Subtract: $\dfrac{2x + 3}{x^2 - 7x + 12} - \dfrac{2}{x - 3}$.
(Section 7.4, Example 7)

70. Evaluate: $\begin{vmatrix} 3 & -2 \\ 7 & -5 \end{vmatrix}$.
(Section 8.5, Example 1)

71. Solve: $x(x - 3) = 10$.
(Section 11.2, Example 1)

▶ SECTION 12.2 *Composite and Inverse Functions*

Objectives

1 Form composite functions.

2 Verify inverse functions.

3 Find the inverse of a function.

4 Use the horizontal line test to determine if a function has an inverse function.

5 Use the graph of a one-to-one function to graph its inverse function.

SSM PH Tutor CD- Video
Center ROM

You are an over-40 student returning to college. Through consistent hard work, you've earned a near-perfect grade point average, feeling more empowered by knowledge with each academic success. Unfortunately, when it comes to technology, you are at a disadvantage compared to younger students. There were no computers in the classroom during your high school years, and you glaze over hearing students chatting about browsers, megabytes, RAM, and ROM. It's time to turn things around and become computer literate.

Luckily, your local computer store is having a sale. The models that are on sale cost either $300 less than the regular price or 85% of the regular price. If x represents the computer's regular price, both discounts can be described with the following functions.

$$f(x) = x - 300 \qquad\qquad g(x) = 0.85x$$

| The computer is on sale for $300 less than its regular price. | The computer is on sale for 85% of its regular price. |

At the store, you bargain with the salesperson. Eventually, she makes an offer you can't refuse: The sale price is 85% of the regular price followed by a $300 reduction:

$$0.85x - 300.$$

| 85% of the regular price | followed by a $300 reduction |

In terms of functions f and g, this offer can be obtained by taking the output of $g(x) = 0.85x$, namely $0.85x$, and using it as the input of f:

$$f(x) = x - 300$$

Replace x with 0.85 x, the output of g(x) = 0.85 x.

$$f(0.85x) = 0.85x - 300.$$

Because $0.85x$ is $g(x)$, we can write this last equation as

$$f(g(x)) = 0.85x - 300.$$

We read this equation as "f of g of x is equal to $0.85x - 300$." We call $f(g(x))$ the **composition of the function f with g**, or a **composite function**. This composite function is written $f \circ g$. Thus,

$$(f \circ g)(x) = f(g(x)) = 0.85x - 300.$$

Like all functions, we can evaluate $f \circ g$ for a specified value of x in the function's domain. For example, here's how to find the value of this function at 1400:

$$(f \circ g)(x) = 0.85x - 300$$

This composite function describes the offer you cannot refuse.

Replace x with 1400.

$$(f \circ g)(1400) = 0.85(1400) - 300 = 1190 - 300 = 890.$$

This means that a computer that regularly sells for \$1400 is on sale for \$890 subject to both discounts.

In this section, we will focus on the composition of two functions. We will also study functions whose compositions have a special relationship.

Before you run out to buy a computer, let's generalize our discussion of the computer's double discount and define the composition of any two functions.

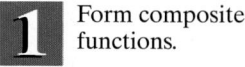
Form composite functions.

The Composition of Functions

The **composition of the function f with g** is denoted by $f \circ g$ and is defined by the equation

$$(f \circ g)(x) = f(g(x)).$$

The domain of the **composite function $f \circ g$** is the set of all x such that

1. x is in the domain of g and
2. $g(x)$ is in the domain of f.

The composition of f with g, $f \circ g$, is pictured as a machine with inputs and outputs in Figure 12.7. The diagram indicates that the output of g, or $g(x)$, becomes the input for "machine" f. If $g(x)$ is not in the domain of f, it cannot be input into machine f, and so $g(x)$ must be discarded.

Inputs, x

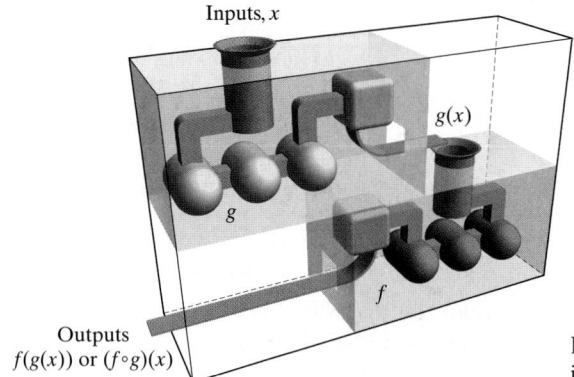

Outputs
$f(g(x))$ or $(f \circ g)(x)$

Figure 12.7 Inputting one function into a second function

EXAMPLE 1   Forming Composite Functions

Given $f(x) = 3x - 4$ and $g(x) = x^2 + 6$, find:

a. $(f \circ g)(x)$ **b.** $(g \circ f)(x)$.

Solution

a. We begin with $(f \circ g)(x)$, the composition of f with g. Because $(f \circ g)(x)$ means $f(g(x))$, we must replace each occurrence of x in the equation for f with $g(x)$.

$$f(x) = 3x - 4 \qquad \text{This is the given equation for } f.$$

Replace x with g(x).

$$(f \circ g)(x) = f(g(x)) = 3g(x) - 4$$
$$= 3(x^2 + 6) - 4 \qquad \text{Because } g(x) = x^2 + 6, \text{ replace } g(x)$$
$$\text{with } x^2 + 6.$$
$$= 3x^2 + 18 - 4 \qquad \text{Use the distributive property.}$$
$$= 3x^2 + 14 \qquad \text{Simplify.}$$

Thus, $(f \circ g)(x) = 3x^2 + 14$.

b. Next, we find $(g \circ f)(x)$, the composition of g with f. Because $(g \circ f)(x)$ means $g(f(x))$, we must replace each occurrence of x in the equation for g with $f(x)$.

$$g(x) = x^2 + 6 \qquad \text{This is the given equation for } g.$$

Replace x with f(x).

$$(g \circ f)(x) = g(f(x)) = (f(x))^2 + 6$$
$$= (3x - 4)^2 + 6 \qquad \text{Because } f(x) = 3x - 4, \text{ replace } f(x)$$
$$\text{with } 3x - 4.$$
$$= 9x^2 - 24x + 16 + 6 \qquad \text{Use } (A - B)^2 = A^2 - 2AB + B^2 \text{ to}$$
$$\text{square } 3x - 4.$$
$$= 9x^2 - 24x + 22 \qquad \text{Simplify.}$$

Thus, $(g \circ f)(x) = 9x^2 - 24x + 22$. Notice that $f \circ g$ is not the same function as $g \circ f$. ∎

✔ **CHECK POINT 1**   Given $f(x) = 5x + 6$ and $g(x) = x^2 - 1$, find:

a. $(f \circ g)(x)$ **b.** $(g \circ f)(x)$.

Inverse Functions

Here are two functions that describe situations related to the price, x, of a computer:

$$f(x) = x - 300 \qquad g(x) = x + 300.$$

Function f subtracts \$300 from the computer's price and function g adds \$300 to the computer's price. Let's see what $f(g(x))$ does. Put $g(x)$ into f:

$$f(x) = x - 300 \qquad \text{This is the given equation for } f.$$

Replace x with $g(x)$.

$$f(g(x)) = g(x) - 300$$

$$= x + 300 - 300 \qquad \text{Because } g(x) = x + 300, \text{ replace } g(x) \text{ with } x + 300.$$

$$= x. \qquad \text{This is the computer's original price.}$$

Using $f(g(x))$, the computer's price, x, went through two changes: the first, an increase; the second, a decrease:

$$x + 300 - 300.$$

The final price of the computer, x, is identical to its starting price, x.

In general, if the changes made to x by function g are undone by the changes made by function f, then

$$f(g(x)) = x.$$

Assume, also, that this "undoing" takes place in the other direction:

$$g(f(x)) = x.$$

Under these conditions, we say that each function is the **inverse function** of the other. The fact that g is the inverse of f is expressed by renaming g as f^{-1}, read "f-inverse." For example, the inverse functions

$$f(x) = x - 300 \qquad g(x) = x + 300$$

are usually named as follows:

$$f(x) = x - 300 \qquad f^{-1}(x) = x + 300.$$

With these ideas in mind, we present the formal definition of the inverse of a function.

Definition of the Inverse of a Function

Let f and g be two functions such that

$$f(g(x)) = x \qquad \text{for every } x \text{ in the domain of } g$$

and

$$g(f(x)) = x \qquad \text{for every } x \text{ in the domain of } f.$$

The function g is the **inverse** of the function f, and is denoted by f^{-1} (read "f-inverse"). Thus, $f(f^{-1}(x)) = x$ and $f^{-1}(f(x)) = x$. The domain of f is equal to the range of f^{-1}, and vice versa.

2 Verify inverse functions.

EXAMPLE 2   Verifying Inverse Functions

Show that each function is an inverse of the other:

$$f(x) = 5x \qquad \text{and} \qquad g(x) = \frac{x}{5}.$$

Solution   To show that f and g are inverses of each other, we must show that $f(g(x)) = x$ and $g(f(x)) = x$. We begin with $f(g(x))$.

$$f(x) = 5x \qquad\qquad \text{This is the given equation for f.}$$

Replace x with g(x).

$$f(g(x)) = 5g(x) = 5\left(\frac{x}{5}\right) = x$$

Next, we find $g(f(x))$.

$$g(x) = \frac{x}{5} \qquad\qquad \text{This is the given equation for g.}$$

Replace x with f(x).

$$g(f(x)) = \frac{f(x)}{5} = \frac{5x}{5} = x$$

Because g is the inverse of f (and vice versa), we can use inverse notation and write

$$f(x) = 5x \qquad \text{and} \qquad f^{-1}(x) = \frac{x}{5}.$$

Notice how f^{-1} undoes the change produced by f: f changes x by multiplying by 5 and f^{-1} undoes this by dividing by 5.   ∎

✔ **CHECK POINT 2**   Show that each function is an inverse of the other:

$$f(x) = 7x \qquad \text{and} \qquad g(x) = \frac{x}{7}.$$

EXAMPLE 3   Verifying Inverse Functions

Show that each function is an inverse of the other:

$$f(x) = 3x + 2 \qquad \text{and} \qquad g(x) = \frac{x-2}{3}.$$

Solution   To show that f and g are inverses of each other, we must show that $f(g(x)) = x$ and $g(f(x)) = x$. We begin with $f(g(x))$.

$$f(x) = 3x + 2 \qquad \text{This is the equation for f.}$$

Replace x with g(x).

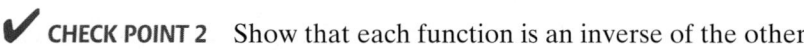

$$f(g(x)) = 3g(x) + 2 = 3\left(\frac{x-2}{3}\right) + 2 = x - 2 + 2 = x$$

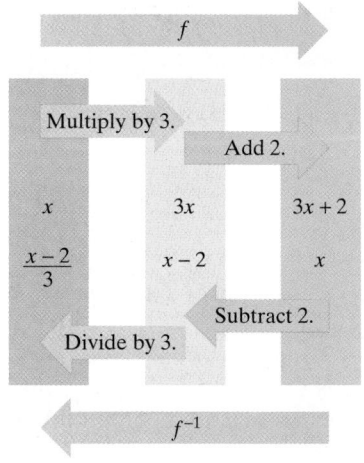

Figure 12.8 f^{-1} undoes the changes produced by f.

Next, we find $g(f(x))$.

$$g(x) = \frac{x - 2}{3}$$ *This is the equation for g.*

Replace x with f(x).

$$g(f(x)) = \frac{f(x) - 2}{3} = \frac{(3x + 2) - 2}{3} = \frac{3x}{3} = x$$

Because g is the inverse of f (and vice versa), we can use inverse notation and write

$$f(x) = 3x + 2 \quad \text{and} \quad f^{-1}(x) = \frac{x - 2}{3}.$$

Notice how f^{-1} undoes the changes produced by f: f changes x by *multiplying* by 3 and *adding* 2, and f^{-1} undoes this by *subtracting* 2 and *dividing* by 3. This "undoing" process is illustrated in Figure 12.8. ∎

✔ **CHECK POINT 3** Show that each function is an inverse of the other:

$$f(x) = 4x - 7 \quad \text{and} \quad g(x) = \frac{x + 7}{4}.$$

3 Find the inverse of a function.

Finding the Inverse of a Function

The definition of the inverse of a function tells us that the domain of f is equal to the range of f^{-1}, and vice versa. This means that if the function f is the set of ordered pairs (x, y), then the inverse of f is the set of ordered pairs (y, x). If a function is defined by an equation, we can obtain the equation for f^{-1}, the inverse of f, by interchanging the role of x and y in the equation for the function f.

Study Tip

The procedure for finding a function's inverse uses a *switch-and-solve* strategy. Switch x and y, then solve for y.

> **Finding the Inverse of a Function**
>
> The equation for the inverse of a function f can be found as follows:
>
> 1. Replace $f(x)$ with y in the equation for $f(x)$.
> 2. Interchange x and y.
> 3. Solve for y. If this equation does not define y as a function of x, the function f does not have an inverse function and this procedure ends. If this equation does define y as a function of x, the function f has an inverse function.
> 4. If f has an inverse function, replace y in step 3 with $f^{-1}(x)$. We can verify our result by showing that $f(f^{-1}(x)) = x$ and $f^{-1}(f(x)) = x$.

EXAMPLE 4 Finding the Inverse of a Function

Find the inverse of $f(x) = 7x - 5$.

Solution

Step 1 Replace $f(x)$ with y:

$$y = 7x - 5.$$

Step 2 Interchange x and y:

$$x = 7y - 5 \qquad \text{This is the inverse function.}$$

Step 3 Solve for y:

$$x + 5 = 7y \qquad \text{Add 5 to both sides.}$$

$$\frac{x + 5}{7} = y \qquad \text{Divide both sides by 7.}$$

Step 4 Replace y with $f^{-1}(x)$:

$$f^{-1}(x) = \frac{x + 5}{7} \qquad \text{The equation is written with } f^{-1} \text{ on the left.}$$

Thus, the inverse of $f(x) = 7x - 5$ is $f^{-1}(x) = \dfrac{x + 5}{7}$.

The inverse function, f^{-1}, undoes the changes produced by f. f changes x by multiplying by 7 and subtracting 5. f^{-1} undoes this by adding 5 and dividing by 7. ∎

✔ **CHECK POINT 4** Find the inverse of $f(x) = 2x + 7$.

EXAMPLE 5 Finding the Equation of the Inverse

Find the inverse of $f(x) = x^3 + 1$.

Solution

Step 1 Replace $f(x)$ with y: $y = x^3 + 1$.
Step 2 Interchange x and y: $x = y^3 + 1$.
Step 3 Solve for y.

$$x - 1 = y^3$$
$$\sqrt[3]{x - 1} = \sqrt[3]{y^3}$$
$$\sqrt[3]{x - 1} = y.$$

Step 4 Replace y with $f^{-1}(x)$: $f^{-1}(x) = \sqrt[3]{x - 1}$.

Thus, the inverse of $f(x) = x^3 + 1$ is $f^{-1}(x) = \sqrt[3]{x - 1}$. ∎

✔ **CHECK POINT 5** Find the inverse of $f(x) = 4x^3 - 1$.

4 Use the horizontal line test to determine if a function has an inverse function.

The Horizontal Line Test and One-to-One Functions

Let's see what happens if we try to find the inverse of the quadratic function $f(x) = x^2$.

Step 1 Replace $f(x)$ with y: $y = x^2$.
Step 2 Interchange x and y: $x = y^2$.
Step 3 Solve for y: We apply the square root property to solve $y^2 = x$ for y. We obtain

$$y = \pm\sqrt{x}.$$

The $\pm$ in this last equation shows that for certain values of x (all positive real numbers), there are two values of y. Because this equation does not represent y as a function of x, the quadratic function $f(x) = x^2$ does not have an inverse function.

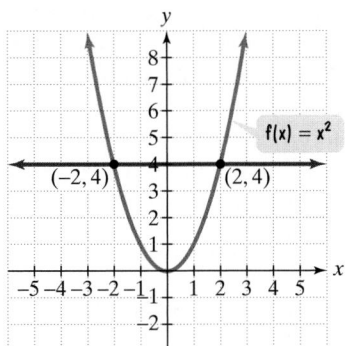

Figure 12.9 The horizontal line intersects the graph twice.

Discover for Yourself

How might you restrict the domain of $f(x) = x^2$, graphed in Figure 12.9, so that the remaining portion of the graph passes the horizontal line test?

Can we look at the graph of a function and tell if it represents a function with an inverse? Yes. The graph of the quadratic function $f(x) = x^2$ is shown in Figure 12.9. Four units above the x-axis, a horizontal line is drawn. This line intersects the graph at two of its points, $(-2, 4)$ and $(2, 4)$. Because inverse functions have ordered pairs with the coordinates reversed, let's see what happens if we reverse these coordinates. We obtain $(4, -2)$ and $(4, 2)$. A function provides exactly one output for each input. However, the input 4 is associated with two outputs, -2 and 2. The points $(4, -2)$ and $(4, 2)$ do not define a function.

If any horizontal line, such as the one in Figure 12.9, intersects a graph at two or more points, the set of these points will not define a function when their coordinates are reversed. This suggests the **horizontal line test** for inverse functions.

The Horizontal Line Test For Inverse Functions

A function f has an inverse that is a function, f^{-1}, if there is no horizontal line that intersects the graph of the function f at more than one point.

EXAMPLE 6   Applying the Horizontal Line Test

Which of the following graphs represent functions that have inverse functions?

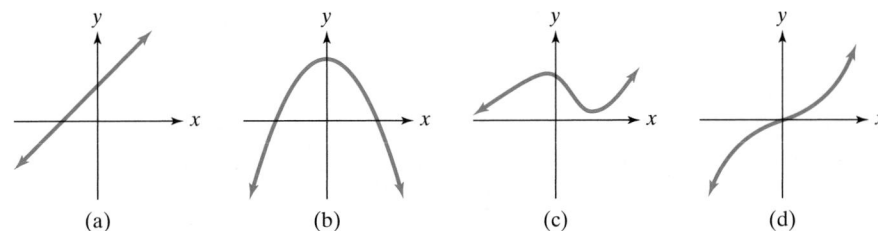

Solution   Notice that horizontal lines can be drawn in parts (b) and (c) that intersect the graphs more than once. These graphs do not pass the horizontal line test. These are not the graphs of functions with inverse functions. By contrast, no horizontal line can be drawn in parts (a) and (d) that intersects the graphs more than once. These graphs pass the horizontal line test. Thus, the graphs in parts (a) and (d) represent functions that have inverse functions.

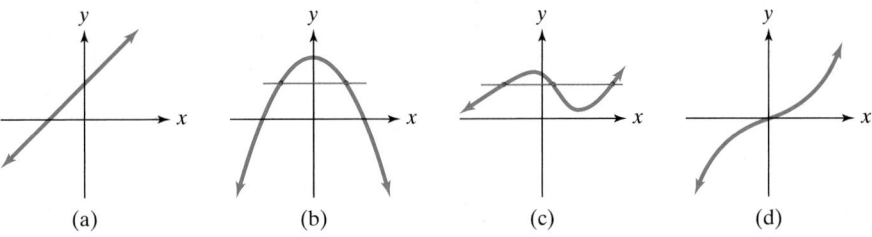

Has an inverse function   No inverse function   No inverse function   Has an inverse function ■

✔ **CHECK POINT 6**   Which of the following graphs represent functions that have inverse functions?

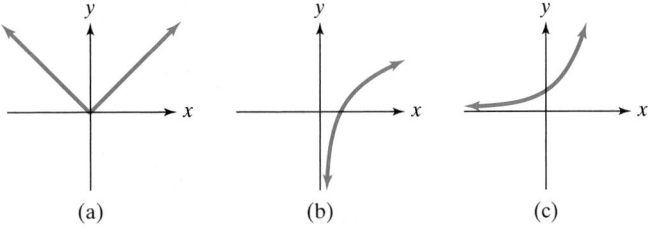

A function passes the horizontal line test when no two different ordered pairs have the same second component. This means that if $x_1 \neq x_2$, then $f(x_1) \neq f(x_2)$. Such a function is called a **one-to-one function**. Thus, a one-to-one function is a function in which no two different ordered pairs have the same second component. Only one-to-one functions have inverse functions. Any function that passes the horizontal line test is a one-to-one function. Any one-to-one function has a graph that passes the horizontal line test.

5 Use the graph of a one-to-one function to graph its inverse function.

Graphs of f and f^{-1}

There is a relationship between the graph of a one-to-one function, f, and its inverse, f^{-1}. Because inverse functions have ordered pairs with the coordinates reversed, if the point (a, b) is on the graph of f, then the point (b, a) is on the graph of f^{-1}. The points (a, b) and (b, a) are symmetric with respect to the line $y = x$. Thus, **the graph of f^{-1} is a reflection of the graph of f about the line $y = x$.** This is illustrated in Figure 12.10.

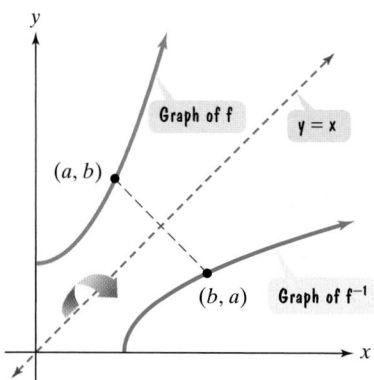

Figure 12.10 The graph of f^{-1} is a reflection of f about $y = x$.

EXAMPLE 7   Graphing the Inverse Function

Use the graph of f in Figure 12.11 to draw the graph of its inverse function.

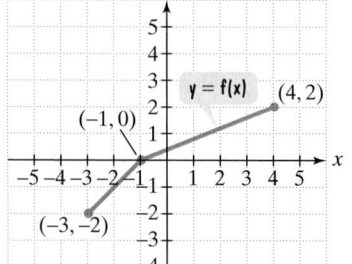

Figure 12.11

Solution   We begin by noting that no horizontal line intersects the graph of f at more than one point, so f does have an inverse function. Because the points $(-3, -2)$, $(-1, 0)$, and $(4, 2)$ are on the graph of f, the graph of the inverse function, f^{-1}, has points with these ordered pairs reversed. Thus, $(-2, -3)$, $(0, -1)$, and $(2, 4)$ are on the graph of f^{-1}. We can use these points to graph f^{-1}. The graph of f^{-1} is shown in green in Figure 12.12. Note that the green graph of f^{-1} is the reflection of the blue graph of f about the line $y = x$.

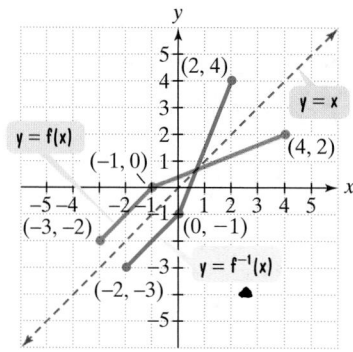

Figure 12.12 The graphs of f and f^{-1}

✔ **CHECK POINT 7** Use the graph of f in the figure below to draw the graph of its inverse function.

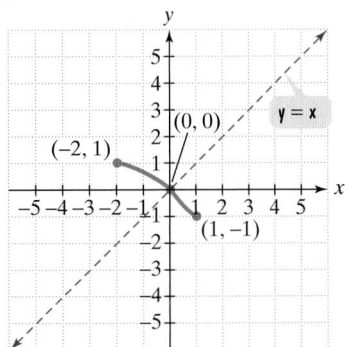

 EXERCISE SET 12.2

Practice Exercises

In Exercises 1–14, find

a. $(f \circ g)(x)$;
b. $(g \circ f)(x)$;
c. $(f \circ g)(2)$.

1. $f(x) = 2x$, $g(x) = x + 7$

2. $f(x) = 3x$, $g(x) = x - 5$

3. $f(x) = x + 4$, $g(x) = 2x + 1$

4. $f(x) = 5x + 2$, $g(x) = 3x - 4$

5. $f(x) = 4x - 3$, $g(x) = 5x^2 - 2$

6. $f(x) = 7x + 1$, $g(x) = 2x^2 - 9$

7. $f(x) = x^2 + 2$, $g(x) = x^2 - 2$

8. $f(x) = x^2 + 1$, $g(x) = x^2 - 3$

9. $f(x) = \sqrt{x}$, $g(x) = x - 1$

10. $f(x) = \sqrt{x}$, $g(x) = x + 2$

11. $f(x) = 2x - 3$, $g(x) = \dfrac{x + 3}{2}$

12. $f(x) = 6x - 3$, $g(x) = \dfrac{x + 3}{6}$

13. $f(x) = \dfrac{1}{x}$, $g(x) = \dfrac{1}{x}$

14. $f(x) = \dfrac{1}{x}$, $g(x) = \dfrac{2}{x}$

In Exercises 15–24, find $f(g(x))$ and $g(f(x))$ and determine whether each pair of functions f and g are inverses of each other.

15. $f(x) = 4x$ and $g(x) = \dfrac{x}{4}$

16. $f(x) = 6x$ and $g(x) = \dfrac{x}{6}$

17. $f(x) = 3x + 8$ and $g(x) = \dfrac{x - 8}{3}$

18. $f(x) = 4x + 9$ and $g(x) = \dfrac{x - 9}{4}$

19. $f(x) = 5x - 9$ and $g(x) = \dfrac{x + 5}{9}$

20. $f(x) = 3x - 7$ and $g(x) = \dfrac{x + 3}{7}$

21. $f(x) = \dfrac{3}{x - 4}$ and $g(x) = \dfrac{3}{x} + 4$

22. $f(x) = \dfrac{2}{x - 5}$ and $g(x) = \dfrac{2}{x} + 5$

23. $f(x) = -x$ and $g(x) = -x$

24. $f(x) = \sqrt[3]{x - 4}$ and $g(x) = x^3 + 4$

The functions in Exercises 25–44 are all one-to-one. For each function:

a. *Find an equation for $f^{-1}(x)$, the inverse function.*
b. *Verify that your equation is correct by showing that $f(f^{-1}(x)) = x$ and $f^{-1}(f(x)) = x$.*

25. $f(x) = x + 3$

26. $f(x) = x + 5$

27. $f(x) = 2x$

28. $f(x) = 4x$

29. $f(x) = 2x + 3$

30. $f(x) = 3x - 1$

31. $f(x) = x^3 + 2$

32. $f(x) = x^3 - 1$

33. $f(x) = (x + 2)^3$

34. $f(x) = (x - 1)^3$

35. $f(x) = \dfrac{1}{x}$

36. $f(x) = \dfrac{2}{x}$

37. $f(x) = \sqrt{x}$

38. $f(x) = \sqrt[3]{x}$

39. $f(x) = x^2 + 1, \quad$ for $x \geq 0$

40. $f(x) = x^2 - 1, \quad$ for $x \geq 0$

41. $f(x) = \dfrac{2x + 1}{x - 3}$

42. $f(x) = \dfrac{2x - 3}{x + 1}$

43. $f(x) = \sqrt[3]{x - 4} + 3$

44. $f(x) = x^{\frac{3}{5}}$

Which graphs in Exercises 45–50 represent functions that have inverse functions?

45.

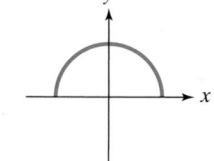

46.

47.

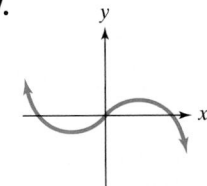

48.

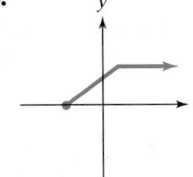

49.

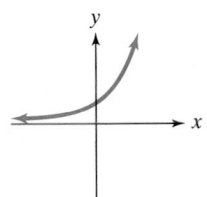

50.

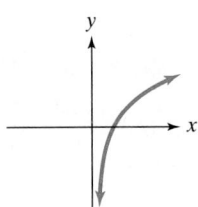

In Exercises 51–54, use the graph of f to draw the graph of its inverse function.

51.

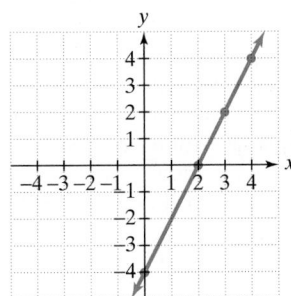

52.

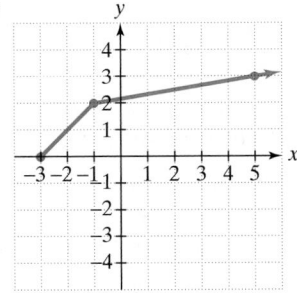

53.

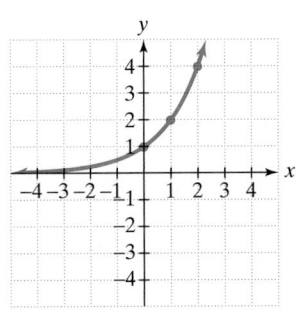

54.

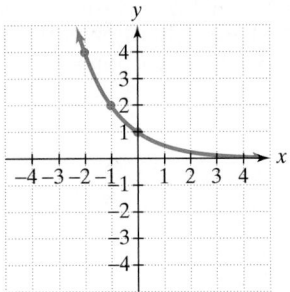

 Application Exercises

55. The regular price of a computer is x dollars. Let $f(x) = x - 400$ and $g(x) = 0.75x$.

a. Describe what the functions f and g model in terms of the price of the computer.

b. Find $(f \circ g)(x)$ and describe what this models in terms of the price of the computer.

c. Repeat part (b) for $(g \circ f)(x)$.

d. Which composite function models the greater discount on the computer, $f \circ g$ or $g \circ f$? Explain.

e. Find f^{-1} and describe what this models in terms of the price of the computer.

56. The regular price of a pair of jeans is x dollars. Let $f(x) = x - 5$ and $g(x) = 0.6x$.

a. Describe what functions f and g model in terms of the price of the jeans.

b. Find $(f \circ g)(x)$ and describe what this models in terms of the price of the jeans.

c. Repeat part (b) for $(g \circ f)(x)$.

d. Which composite function models the greater discount on the jeans, $f \circ g$ or $g \circ f$? Explain.

e. Find f^{-1} and describe what this models in terms of the price of the jeans.

57. The graph represents the probability of two people in the same room sharing a birthday as a function of the number of people in the room. Call the function f.

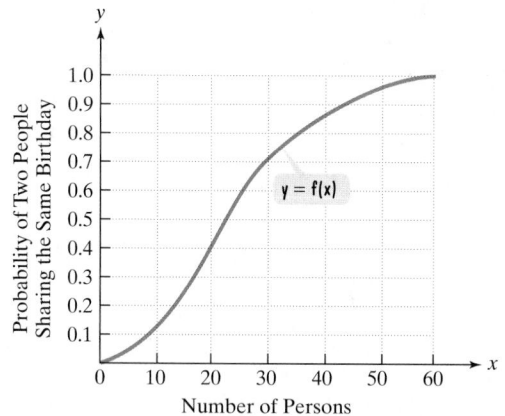

a. Explain why f has an inverse that is a function.

b. Describe in practical terms the meanings of $f^{-1}(0.25)$, $f^{-1}(0.5)$, and $f^{-1}(0.7)$.

58. The line graph shown below is based on data from the World Health Organization.
a. Explain why f has an inverse that is a function.

b. Describe in practical terms the meaning of $f^{-1}(20)$.

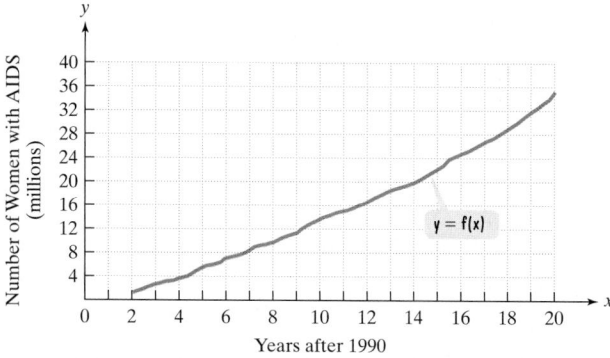

Source: Boston Globe

The graph shows the average age at which women in the United States married for the first time over a 110-year period. Use the graph to solve Exercises 59–60.

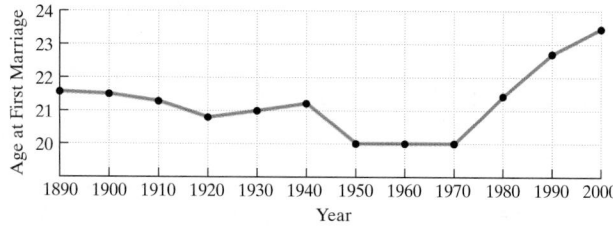

Source: U.S. Census Bureau

59. Does this graph have an inverse that is a function? What does this mean about the average age at which U.S. women married for the first time during the period shown?

60. Identify two or more years in which U.S. women married for the first time at the same average age. What is this average age?

Writing in Mathematics

61. Describe a procedure for finding $(f \circ g)(x)$.
62. Explain how to determine if two functions are inverses of each other.
63. Describe how to find the inverse of a one-to-one function.
64. What is the horizontal line test and what does it indicate?
65. Describe how to use the graph of a one-to-one function to draw the graph of its inverse function.
66. How can a graphing utility be used to visually determine if two functions are inverses of each other?
67. Consider the following function:

{(The Beatles, 20), (Elvis Presley, 18),(Michael Jackson, 13) (Mariah Carey, 13), (The Supremes, 12)}.

The domain is the set of the five recording artists with the most number 1 singles in the United States. The range is the set of the number of #1 singles for each artist. Reverse each of the five ordered pairs. Is the resulting relation a function? Describe what this means in terms of whether or not the given function is one-to-one. (*Source*: *The Popular Music Database*)

Critical Thinking Exercises

68. Which one of the following is true?
a. The inverse of $\{(1, 4), (2, 7)\}$ is $\{(2, 7), (1, 4)\}$.
b. The function $f(x) = 5$ is one-to-one.
c. If $f(x) = 3x$, then $f^{-1}(x) = \dfrac{1}{3x}$.
d. The domain of f is the same as the range of f^{-1}.
69. If $h(x) = \sqrt{3x^2 + 5}$, find functions f and g so that $h(x) = (f \circ g)(x)$.
70. If $f(x) = 3x$ and $g(x) = x + 5$, find $(f \circ g)^{-1}(x)$ and $(g^{-1} \circ f^{-1})(x)$.

71. Show that
$$f(x) = \frac{3x - 2}{5x - 3}$$
is its own inverse.
72. Consider the two functions defined by $f(x) = m_1 x + b_1$ and $g(x) = m_2 x + b_2$. Prove that the slope of the composite function of f with g is equal to the product of the slopes of the two functions.

Technology Exercises

In Exercises 73–80, use a graphing utility to graph the function. Use the graph to determine whether the function has an inverse that is a function (that is, whether the function is one-to-one).

73. $f(x) = x^2 - 1$

74. $f(x) = \sqrt[3]{2 - x}$

75. $f(x) = \dfrac{x^3}{2}$

76. $f(x) = \dfrac{x^4}{4}$

77. $f(x) = |x - 2|$

78. $f(x) = (x - 1)^3$

79. $f(x) = -\sqrt{16 - x^2}$

80. $f(x) = x^3 + x + 1$

In Exercises 81–83, use a graphing utility to graph f and g in the same viewing rectangle. In addition, graph the line $y = x$ and visually determine if f and g are inverses.

81. $f(x) = 4x + 4$, $g(x) = 0.25x - 1$

82. $f(x) = \dfrac{1}{x} + 2$, $g(x) = \dfrac{1}{x - 2}$

83. $f(x) = \sqrt[3]{x} - 2$, $g(x) = (x + 2)^3$

Review Exercises

84. Divide and write the quotient in scientific notation:
$$\frac{4.3 \times 10^5}{8.6 \times 10^{-4}}.$$
(Section 5.7, Example 9)

85. Graph: $f(x) = x^2 - 4x + 3$.
(Section 11.3, Example 3)

86. Solve: $\sqrt{x + 4} - \sqrt{x - 1} = 1$.
(Section 10.6, Example 4)

▶ **SECTION 12.3**   *Logarithmic Functions*

Objectives

1 Change from logarithmic to exponential form.

2 Change from exponential to logarithmic form.

3 Evaluate logarithms.

4 Use basic logarithmic properties.

5 Graph logarithmic functions.

6 Find the domain of a logarithmic function.

7 Use common logarithms.

8 Use natural logarithms.

SSM PH Tutor CD- Video
      Center ROM

The earthquake that ripped through northern California on October 17, 1989, measured 7.1 on the Richter scale, killed more than 60 people, and injured more than 2400. Shown here is San Francisco's Marina district, where shock waves tossed houses off their foundations and into the street.

A higher measure on the Richter scale is more devastating than it seems because for each increase in one unit on the scale, there is a tenfold increase in the intensity of an earthquake. In this section our focus is on the inverse of the exponential function, called the logarithmic function. The logarithmic function will help you to understand diverse phenomena, including earthquake intensity, human memory, and the pace of life in large cities.

The Definition of Logarithmic Functions

No horizontal line can be drawn that intersects the graph of an exponential function at more than one point. This means that the exponential function is one-to-one and has an inverse. The inverse function of the exponential function with base b is called the **logarithmic function with base b**.

Definition of the Logarithmic Function

For $x > 0$ and $b > 0$, $b \neq 1$,

$$y = \log_b x \text{ is equivalent to } b^y = x.$$

The function $f(x) = \log_b x$ is the **logarithmic function with base b**.

The equations

$$y = \log_b x \quad \text{and} \quad b^y = x$$

are different ways of expressing the same thing. The first equation is in **logarithmic form** and the second equivalent equation is in **exponential form**.

Notice that a **logarithm, y, is an exponent**. You should learn the location of the base and exponent in each form.

Location of Base and Exponent in Exponential and Logarithmic Forms

Exponent

Exponent

Logarithmic Form: $y = \log_b x$ Exponential Form: $b^y = x$

Base

Base

EXAMPLE 1 Changing from Logarithmic to Exponential Form

1 Change from logarithmic to exponential form.

Write each equation in its equivalent exponential form:

a. $2 = \log_5 x$ **b.** $3 = \log_b 64$ **c.** $\log_3 7 = y$.

Solution We use the fact that $y = \log_b x$ means $b^y = x$.

a. $2 = \log_5 x$ means $5^2 = x$.   **b.** $3 = \log_b 64$ means $b^3 = 64$.

Logarithms are exponents.

Logarithms are exponents.

c. $\log_3 7 = y$ or $y = \log_3 7$ means $3^y = 7$.   ∎

✔ **CHECK POINT 1** Write each equation in its equivalent exponential form:

a. $3 = \log_7 x$ **b.** $2 = \log_b 25$ **c.** $\log_4 26 = y$.

---

2 Change from exponential to logarithmic form.

EXAMPLE 2 Changing From Exponential to Logarithmic Form

Write each equation in its equivalent logarithmic form:

a. $12^2 = x$ **b.** $b^3 = 8$ **c.** $e^y = 9$.

Solution We use the fact that $b^y = x$ means $y = \log_b x$.

a. $12^2 = x$ means $2 = \log_{12} x$.   **b.** $b^3 = 8$ means $3 = \log_b 8$.

Logarithms are exponents.

Logarithms are exponents.

c. $e^y = 9$ means $y = \log_e 9$.   ∎

CHECK POINT 2 Write each equation in its equivalent logarithmic form:

   a. $2^5 = x$ **b.** $b^3 = 27$ **c.** $e^y = 33$.

3 Evaluate logarithms.

Remembering that logarithms are exponents makes it possible to evaluate some logarithms by inspection. The logarithm of x with base b, $\log_b x$, is the exponent to which b must be raised to get x. For example, suppose we want to evaluate $\log_2 32$. We ask, 2 to what power gives 32? Because $2^5 = 32$, $\log_2 32 = 5$.

EXAMPLE 3 Evaluating Logarithms

Evaluate:

   a. $\log_2 16$ **b.** $\log_3 9$ **c.** $\log_{25} 5$.

Solution

Logarithmic Expression	Question Needed for Evaluation	Logarithmic Expression Evaluated
a. $\log_2 16$	2 to what power gives 16?	$\log_2 16 = 4$ because $2^4 = 16$.
b. $\log_3 9$	3 to what power gives 9?	$\log_3 9 = 2$ because $3^2 = 9$.
c. $\log_{25} 5$	25 to what power gives 5?	$\log_{25} 5 = \frac{1}{2}$ because $25^{\frac{1}{2}} = \sqrt{25} = 5$.

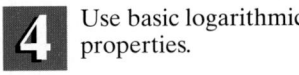 **CHECK POINT 3** Evaluate:

   a. $\log_{10} 100$ **b.** $\log_3 3$ **c.** $\log_{36} 6$.

4 Use basic logarithmic properties.

Basic Logarithmic Properties

Because logarithms are exponents, they have properties that can be verified using properties of exponents.

> **Basic Logarithmic Properties Involving One**
>
> **1.** $\log_b b = 1$ because 1 is the exponent to which b must be raised to obtain b. $(b^1 = b)$
>
> **2.** $\log_b 1 = 0$ because 0 is the exponent to which b must be raised to obtain 1. $(b^0 = 1)$

EXAMPLE 4 Using Properties of Logarithms

Evaluate:

   a. $\log_7 7$ **b.** $\log_5 1$.

Solution

   a. Because $\log_b b = 1$, we conclude $\log_7 7 = 1$.
   b. Because $\log_b 1 = 0$, we conclude $\log_5 1 = 0$.

✔ **CHECK POINT 4** Evaluate:

a. $\log_9 9$ **b.** $\log_8 1$.

The inverse of the exponential function is the logarithmic function. Thus, if $f(x) = b^x$, then $f^{-1}(x) = \log_b x$. In Section 12.2, we saw how inverse functions "undo" one another. In particular,

$$f(f^{-1}(x)) = x \quad \text{and} \quad f^{-1}(f(x)) = x.$$

Applying these relationships to exponential and logarithmic functions, we obtain the following **inverse properties of logarithms**:

Inverse Properties of Logarithms
For $b > 0$ and $b \neq 1$,

$$\log_b b^x = x \qquad \text{The logarithm with base } b \text{ of } b \text{ raised to a power equals that power.}$$

$$b^{\log_b x} = x \qquad b \text{ raised to the logarithm with base } b \text{ of a number equals that number.}$$

EXAMPLE 5 Using Inverse Properties of Logarithms

Evaluate:

a. $\log_4 4^5$ **b.** $6^{\log_6 9}$.

Solution

a. Because $\log_b b^x = x$, we conclude $\log_4 4^5 = 5$.
b. Because $b^{\log_b x} = x$, we conclude $6^{\log_6 9} = 9$. ■

✔ **CHECK POINT 5** Evaluate:

a. $\log_7 7^8$ **b.** $3^{\log_3 17}$.

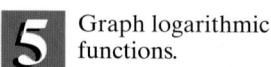 Graph logarithmic functions.

Graphs of Logarithmic Functions

How do we graph logarithmic functions? We use the fact that the logarithmic function is the inverse of the exponential function. This means that the logarithmic function reverses the coordinates of the exponential function. It also means that the graph of the logarithmic function is a reflection of the graph of the exponential function about the line $y = x$.

EXAMPLE 6 Graphs of Exponential and Logarithmic Functions

Graph $f(x) = 2^x$ and $g(x) = \log_2 x$ in the same rectangular coordinate system.

Solution We first set up a table of coordinates for $f(x) = 2^x$. Reversing these coordinates gives the coordinates for the inverse function $g(x) = \log_2 x$.

x	-2	-1	0	1	2	3
$f(x) = 2^x$	$\frac{1}{4}$	$\frac{1}{2}$	1	2	4	8

Reverse coordinates.

x	$\frac{1}{4}$	$\frac{1}{2}$	1	2	4	8
$g(x) = \log_2 x$	-2	-1	0	1	2	3

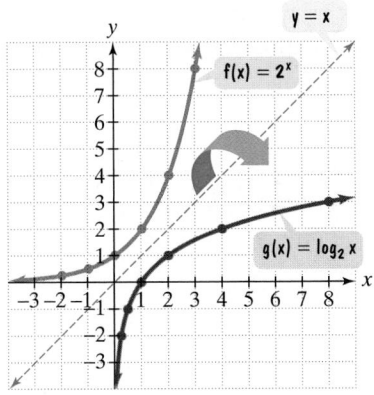

Figure 12.13 The graphs of $f(x) = 2^x$ and its inverse function

We now plot the ordered pairs in both tables, connecting them with smooth curves. Figure 12.13 shows the graphs of $f(x) = 2^x$ and its inverse function $g(x) = \log_2 x$. The graph of the inverse can also be drawn by reflecting the graph of $f(x) = 2^x$ about the line $y = x$. ∎

✔ **CHECK POINT 6** Graph $f(x) = 3^x$ and $g(x) = \log_3 x$ in the same rectangular coordinate system.

Figure 12.14 illustrates the relationship between the graph of the exponential function, shown in blue and its inverse, the logarithmic function, shown in red, for bases greater than 1 and for bases between 0 and 1.

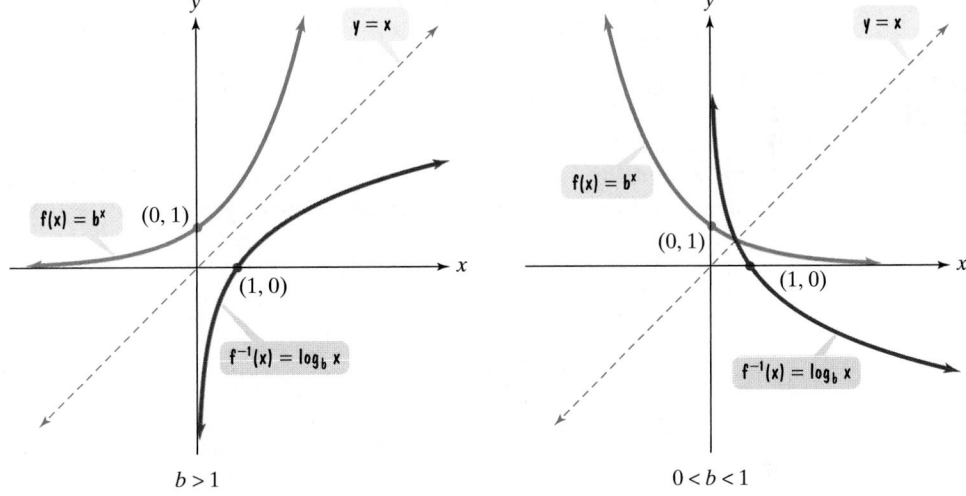

Figure 12.14 Graphs of exponential and logarithmic functions

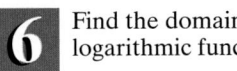
Characteristics of the Graphs of Logarithmic Functions of the Form $f(x) = \log_b x$

- The x-intercept is 1. There is no y-intercept.
- The y-axis is a vertical asymptote.
- If $b > 1$, the function is increasing. If $0 < b < 1$, the function is decreasing.
- The graph is smooth and continuous. It has no sharp corners or gaps.

6 Find the domain of a logarithmic function.

The Domain of a Logarithmic Function

In Section 12.1, we learned that the domain of an exponential function of the form $f(x) = b^x$ includes all real numbers and its range is the set of positive real numbers. Because the logarithmic function reverses the domain and the range of the exponential function, the **domain of a logarithmic function of the form $f(x) = \log_b x$ is the set of all positive real numbers**. Thus, $\log_2 8$ is defined because the value of x in the logarithmic expression, 8, is greater than zero and therefore is included in the domain of the logarithmic function $f(x) = \log_2 x$. However, $\log_2 0$ and $\log_2(-8)$ are not defined because 0 and -8 are not positive real numbers and therefore are excluded from the domain of the logarithmic function $f(x) = \log_2 x$. In general, the domain of $f(x) = \log_b(x + c)$ consists of all x for which $x + c > 0$.

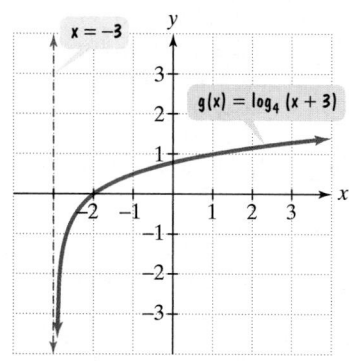

Figure 12.15 The domain of $g(x) = \log_4(x + 3)$ is $(-3, \infty)$.

7 Use common logarithms.

EXAMPLE 7   Finding the Domain of a Logarithmic Function

Find the domain of $g(x) = \log_4(x + 3)$.

Solution   The domain of g consists of all x for which $x + 3 > 0$. Solving this inequality for x, we obtain $x > -3$. Thus, the domain of g is $\{x|x > -3\}$ or $(-3, \infty)$. This is illustrated in Figure 12.15. The vertical asymptote is $x = -3$, and all points on the graph of g have x-coordinates that are greater than -3.   ∎

✔ **CHECK POINT 7**   Find the domain of $h(x) = \log_4(x - 5)$.

Common Logarithms

The logarithmic function with base 10 is called the **common logarithmic function**. The function $f(x) = \log_{10} x$ is usually expressed as $f(x) = \log x$. A calculator with a $\boxed{\text{LOG}}$ key can be used to evaluate common logarithms. Here are some examples:

Logarithm	Most Scientific Calculator Keystrokes	Most Graphing Calculator Keystrokes	Display (or Approximate Display)
$\log 1000$	$1000 \boxed{\text{LOG}}$	$\boxed{\text{LOG}}\ 1000\ \boxed{\text{ENTER}}$	3
$\log \dfrac{5}{2}$	$\boxed{(}\ 5\ \boxed{\div}\ 2\ \boxed{)}\ \boxed{\text{LOG}}$	$\boxed{\text{LOG}}\ \boxed{(}\ 5\ \boxed{\div}\ 2\ \boxed{)}\ \boxed{\text{ENTER}}$	0.39794
$\dfrac{\log 5}{\log 2}$	$5\ \boxed{\text{LOG}}\ \boxed{\div}\ 2\ \boxed{\text{LOG}}\ \boxed{=}$	$\boxed{\text{LOG}}\ 5\ \boxed{\div}\ \boxed{\text{LOG}}\ 2\ \boxed{\text{ENTER}}$	2.32192
$\log(-3)$	$3\ \boxed{+/-}\ \boxed{\text{LOG}}$	$\boxed{\text{LOG}}\ \boxed{(-)}\ 3\ \boxed{\text{ENTER}}$	$\boxed{\text{ERROR}}$

The error message given by many calculators for $\log(-3)$ is a reminder that the domain of every logarithmic function, including the common logarithmic function, is the set of positive real numbers.

Many real-life phenomena start with rapid growth, and then the growth begins to level off. This type of behavior can be modeled by logarithmic functions.

EXAMPLE 8   Modeling Height of Children

The percentage of adult height attained by a boy who is x years old can be modeled by

$$f(x) = 29 + 48.8 \log(x + 1)$$

where x represents the boy's age and $f(x)$ represents the percentage of his adult height. Approximately what percentage of his adult height is a boy at age eight?

Solution   We substitute the boy's age, 8, for x and evaluate the function at 8.

$$f(x) = 29 + 48.8 \log(x + 1) \qquad \text{This is the given function.}$$
$$f(8) = 29 + 48.8 \log(8 + 1) \qquad \text{Substitute 8 for } x.$$
$$= 29 + 48.8 \log 9 \qquad \text{Graphing calculator keystrokes:}$$
$$\approx 76 \qquad 29\ \boxed{+}\ 48.8\ \boxed{\times}\ \boxed{\text{LOG}}\ 9\ \boxed{\text{ENTER}}$$

Thus, an 8-year-old boy is approximately 76% of his adult height.   ∎

✔ **CHECK POINT 8** Use the function in Example 8 to answer this question: Approximately what percentage of his adult height is a boy at age ten?

The basic properties of logarithms that were listed earlier in this section can be applied to common logarithms.

Properties of Common Logarithms

General Properties	Common Logarithms
1. $\log_b 1 = 0$	**1.** $\log 1 = 0$
2. $\log_b b = 1$	**2.** $\log 10 = 1$
3. $\log_b b^x = x$	**3.** $\log 10^x = x$
4. $b^{\log_b x} = x$	**4.** $10^{\log x} = x$

The property $\log 10^x = x$ can be used to evaluate common logarithms involving powers of 10. For example,

$$\log 100 = \log 10^2 = 2, \quad \log 1000 = \log 10^3 = 3, \text{ and } \log 10^{7.1} = 7.1.$$

EXAMPLE 9   Earthquake Intensity

The magnitude, R, on the Richter scale of an earthquake of intensity I is given by

$$R = \log \frac{I}{I_0}$$

where I_0 is the intensity of a barely felt zero-level earthquake. The earthquake that destroyed San Francisco in 1906 was $10^{8.3}$ times as intense as a zero-level earthquake. What was its magnitude on the Richter scale?

Solution   Because the earthquake was $10^{8.3}$ times as intense as a zero-level earthquake, the intensity, I, is $10^{8.3}I_0$.

$$R = \log \frac{I}{I_0} \qquad \text{This is the formula for magnitude on the Richter scale.}$$

$$R = \log \frac{10^{8.3}I_0}{I_0} \qquad \text{Substitute } 10^{8.3}I_0 \text{ for } I.$$

$$= \log 10^{8.3} \qquad \text{Simplify.}$$

$$= 8.3 \qquad \text{Use the property } \log 10^x = x.$$

San Francisco's 1906 earthquake registered 8.3 on the Richter scale.   ∎

✔ **CHECK POINT 9** Use the formula in Example 9 to solve this problem. If an earthquake is 10,000 times as intense as a zero-level quake $(I = 10,000I_0)$, what is its magnitude on the Richter scale?

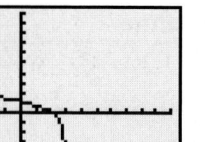

8 Use natural logarithms.

Natural Logarithms

The logarithmic function with base e is called the **natural logarithmic function**. The function $f(x) = \log_e x$ is usually expressed as $f(x) = \ln x$, read "el en of x." A calculator with an $\boxed{\text{LN}}$ key can be used to evaluate natural logarithms.

Like the domain of all logarithmic functions, the domain of the natural logarithmic function is the set of all positive real numbers. Thus, the domain of $f(x) = \ln(x + c)$ consists of all x for which $x + c > 0$.

EXAMPLE 10 Finding Domains of Natural Logarithmic Functions

Find the domain of each function:

a. $f(x) = \ln(3 - x)$ **b.** $g(x) = \ln(x - 3)^2$.

Solution

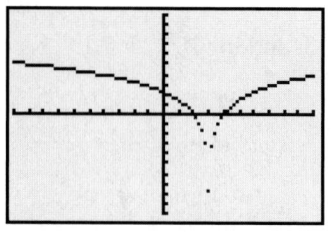

Figure 12.16 The domain of $f(x) = \ln(3 - x)$ is $(-\infty, 3)$.

a. The domain of f consists of all x for which $3 - x > 0$. Solving this inequality for x, we obtain $x < 3$. Thus, the domain of f is $\{x | x < 3\}$ or $(-\infty, 3)$. This is verified by the graph in Figure 12.16.

b. The domain of g consists of all x for which $(x - 3)^2 > 0$. It follows that the domain of g is all real numbers except 3. Thus, the domain of g is $\{x | x \neq 3\}$ or $(-\infty, 3) \cup (3, \infty)$. This is shown by the graph in Figure 12.17. To make it more obvious that 3 is excluded from the domain, we used a $\boxed{\text{DOT}}$ format. ∎

Figure 12.17 3 is excluded from the domain of $g(x) = \ln(x - 3)^2$.

✔ **CHECK POINT 10** Find the domain of each function:

a. $f(x) = \ln(4 - x)$
b. $g(x) = \ln x^2$.

The basic properties of logarithms that were listed earlier in this section can be applied to natural logarithms.

Properties of Natural Logarithms	
General Properties	**Natural Logarithms**
1. $\log_b 1 = 0$	**1.** $\ln 1 = 0$
2. $\log_b b = 1$	**2.** $\ln e = 1$
3. $\log_b b^x = x$	**3.** $\ln e^x = x$
4. $b^{\log_b x} = x$	**4.** $e^{\ln x} = x$

The property $\ln e^x = x$ can be used to evaluate natural logarithms involving powers of e. For example,

$$\ln e^2 = 2, \quad \ln e^3 = 3, \quad \ln e^{7.1} = 7.1, \quad \text{and} \quad \ln \frac{1}{e} = \ln e^{-1} = -1.$$

EXAMPLE 11 Using Inverse Properties

Use inverse properties to simplify:

a. $\ln e^{7x}$ **b.** $e^{\ln 4x^2}$.

Solution

a. Because $\ln e^x = x$, we conclude that $\ln e^{7x} = 7x$.

b. Because $e^{\ln x} = x$, we conclude that $e^{\ln 4x^2} = 4x^2$. ∎

 CHECK POINT 11 Use inverse properties to simplify:

a. $\ln e^{25x}$ **b.** $e^{\ln \sqrt{x}}$.

EXAMPLE 12   Walking Speed and City Population

As the population of a city increases, the pace of life also increases. The formula

$$W = 0.35 \ln P + 2.74$$

models average walking speed, W, in feet per second, for a resident of a city whose population is P thousand. Find the average walking speed for people living in New York City with a population of 7323 thousand.

Solution   We use the formula and substitute 7323 for P, the population in thousands.

$W = 0.35 \ln P + 2.74$	This is the given formula.
$W = 0.35 \ln 7323 + 2.74$	Substitute 7323 for P.
≈ 5.9	Graphing calculator keystrokes: $0.35 \boxed{\times} \boxed{} 7323 \boxed{+} 2.74 \boxed{\text{ENTER}}$. (On some calculators, a parenthesis is needed after 7323.)

The average walking speed in New York City is approximately 5.9 feet per second. ∎

 CHECK POINT 12 Use the formula $W = 0.35 \ln P + 2.74$ to find the average walking speed in Jackson, Mississippi with a population of 197 thousand.

EXERCISE SET 12.3

 Practice Exercises

In Exercises 1–8, write each equation in its equivalent exponential form.

1. $4 = \log_2 16$ **2.** $6 = \log_2 64$

3. $2 = \log_3 x$ **4.** $2 = \log_9 x$

5. $5 = \log_b 32$ **6.** $3 = \log_b 27$

7. $\log_6 216 = y$ **8.** $\log_5 125 = y$

In Exercises 9–20, write each equation in its equivalent logarithmic form.

9. $2^3 = 8$ **10.** $5^4 = 625$

11. $2^{-4} = \frac{1}{16}$ **12.** $5^{-3} = \frac{1}{125}$

13. $\sqrt[3]{8} = 2$ **14.** $\sqrt[3]{64} = 4$

15. $13^2 = x$ **16.** $15^2 = x$

17. $b^3 = 1000$ **18.** $b^3 = 343$

19. $7^y = 200$ **20.** $8^y = 300$

In Exercises 21–38, evaluate each expression without using a calculator.

21. $\log_4 16$ **22.** $\log_7 49$

23. $\log_2 64$ **24.** $\log_3 27$

25. $\log_7 \sqrt{7}$ **26.** $\log_6 \sqrt{6}$

27. $\log_2 \frac{1}{8}$ **28.** $\log_3 \frac{1}{9}$

29. $\log_{64} 8$ **30.** $\log_{81} 9$

31. $\log_5 5$ **32.** $\log_{11} 11$

33. $\log_4 1$ **34.** $\log_6 1$

35. $\log_5 5^7$ **36.** $\log_4 4^6$

37. $8^{\log_8 19}$ **38.** $7^{\log_7 23}$

39. Graph $f(x) = 4^x$ and $g(x) = \log_4 x$ in the same rectangular coordinate system.

40. Graph $f(x) = 5^x$ and $g(x) = \log_5 x$ in the same rectangular coordinate system.

41. Graph $f(x) = \left(\frac{1}{2}\right)^x$ and $g(x) = \log_{\frac{1}{2}} x$ in the same rectangular coordinate system.

42. Graph $f(x) = \left(\frac{1}{4}\right)^x$ and $g(x) = \log_{\frac{1}{4}} x$ in the same rectangular coordinate system.

In Exercises 43–48, find the domain of each logarithmic function.

43. $f(x) = \log_5(x + 4)$

44. $f(x) = \log_5(x + 6)$

45. $f(x) = \log(2 - x)$

46. $f(x) = \log(7 - x)$

47. $f(x) = \ln(x - 2)^2$

48. $f(x) = \ln(x - 7)^2$

In Exercises 49–62, evaluate each expression without using a calculator.

49. $\log 100$

50. $\log 1000$

51. $\log 10^7$

52. $\log 10^8$

53. $10^{\log 33}$

54. $10^{\log 53}$

55. $\ln 1$

56. $\ln e$

57. $\ln e^6$

58. $\ln e^7$

59. $\ln \dfrac{1}{e^6}$

60. $\ln \dfrac{1}{e^7}$

61. $e^{\ln 125}$

62. $e^{\ln 300}$

In Exercises 75–80, use inverse properties of logarithms to simplify each expression.

63. $\ln e^{9x}$

64. $\ln e^{13x}$

65. $e^{\ln 5x^2}$

66. $e^{\ln 7x^2}$

67. $10^{\log \sqrt{x}}$

68. $10^{\log \sqrt[3]{x}}$

 Application Exercises

The percentage of adult height attained by a girl who is x years old can be modeled by

$$f(x) = 62 + 35 \log(x - 4)$$

where x represents the girl's age (from 5 to 15) and f(x) represents the percentage of her adult height. Use the formula to solve Exercises 69–70.

69. Approximately what percentage of her adult height is a girl at age 13?

70. Approximately what percentage of her adult height is a girl at age ten?

71. The annual amount that we spend to attend sporting events can be modeled by

$$f(x) = 2.05 + 1.3 \ln x$$

where x represents the number of years after 1984 and $f(x)$ represents the total annual expenditures for admission to spectator sports, in billions of dollars. In 2000, approximately how much was spent on admission to spectator sports?

72. The percentage of U.S. households with cable television can be modeled by

$$f(x) = 18.32 + 15.94 \ln x$$

where x represents the number of years after 1979 and $f(x)$ represents the percentage of U.S. households with cable television. What percentage of U.S. households had cable television in 1990?

The loudness level of a sound, D, in decibels, is given by the formula

$$D = 10 \log(10^{12} I)$$

where I is the intensity of the sound, in watts per meter². Decibel levels range from 0, a barely audible sound, to 160, a sound resulting in a ruptured eardrum. Use the formula to solve Exercises 73–74.

73. The sound of a blue whale can be heard 500 miles away, reaching an intensity of 6.3×10^6 watts per meter². Determine the decibel level of this sound. At close range, can the sound of a blue whale rupture the human eardrum?

74. What is the decibel level of a normal conversation, 3.2×10^{-6} watt per meter²?

75. Students in a psychology class took a final examination. As part of an experiment to see how much of the course content they remembered over time, they took equivalent forms of the exam in monthly intervals thereafter. The average score for the group, $f(t)$, after t months was modeled by the function

$$f(t) = 88 - 15 \ln(t + 1), \qquad 0 \le t \le 12.$$

a. What was the average score on the original exam?

b. What was the average score after 2 months? 4 months? 6 months? 8 months? 10 months? one year?

c. Sketch the graph of f (either by hand or with a graphing utility). Describe what the graph indicates in terms of the material retained by the students.

Writing in Mathematics

76. Describe the relationship between an equation in logarithmic form and an equivalent equation in exponential form.

77. What question can be asked to help evaluate $\log_3 81$?

78. Explain why the logarithm of 1 with base b is 0.

79. Describe the following property using words: $\log_b b^x = x$.

80. Explain how to use the graph of $f(x) = 2^x$ to obtain the graph of $g(x) = \log_2 x$.

81. Explain how to find the domain of a logarithmic function.

82. New York City is one of the world's great walking cities. Use the formula in Example 12 on page 829 to describe what frequently happens to tourists exploring the city by foot.

83. Logarithmic models are well suited to phenomena in which growth is initially rapid but then begins to level off. Describe something that is changing over time that can be modeled using a logarithmic function.

84. Suppose that a girl is 4 feet 6 inches at age 10. Explain how to use the function in Exercises 69–70 to determine how all she can expect to be as an adult.

Critical Thinking Exercises

85. Which one of the following is true?
  a. $\dfrac{\log_2 8}{\log_2 4} = \dfrac{8}{4}$
  b. $\log(-100) = -2$.
  c. The domain of $f(x) = \log_2 x$ is $(-\infty, \infty)$.
  d. $\log_b x$ is the exponent to which b must be raised to obtain x.

86. Without using a calculator, find the exact value of
$$\frac{\log_3 81 - \log_\pi 1}{\log_{2\sqrt{2}} 8 - \log 0.001}.$$

87. Solve for x: $\log_4\left[\log_3(\log_2 x)\right] = 0$.

88. Without using a calculator, determine which is the greater number: $\log_4 60$ or $\log_3 40$.

Technology Exercises

In Exercises 89–92, graph f and g in the same viewing rectangle. Then describe the relationship of the graph of g to the graph of f.

89. $f(x) = \ln x, g(x) = \ln(x + 3)$

90. $f(x) = \ln x, g(x) = \ln x + 3$

91. $f(x) = \log x, g(x) = -\log x$

92. $f(x) = \log x, g(x) = \log(x - 2) + 1$

93. Students in a mathematics class took a final examination. They took equivalent forms of the exam in monthly intervals thereafter. The average score, $f(t)$, for the group after t months was modeled by the human memory function $f(t) = 75 - 10\log(t + 1)$, where $0 \le t \le 12$. Use a graphing utility to graph the function. Then determine how many months will elapse before the average score falls below 65.

94. Graph f and g in the same viewing rectangle.
  a. $f(x) = \ln(3x), g(x) = \ln 3 + \ln x$
  b. $f(x) = \log(5x^2), g(x) = \log 5 + \log x^2$
  c. $f(x) = \ln(2x^3), g(x) = \ln 2 + \ln x^3$
  d. Describe what you observe in parts (a)–(c). Generalize this observation by writing an equivalent expression. In each case, the graphs of f and g are the same for $\log_b(MN)$, where $M > 0$ and $N > 0$.

  e. Complete this statement: The logarithm of a product is equal to _____.

95. Graph each of the following functions in the same viewing rectangle and then place the functions in order from the one that increases most slowly to the one that increases most rapidly.
$$y = x, y = \sqrt{x}, y = e^x, y = \ln x, y = x^x, y = x^2$$

Review Exercises

96. Solve the system:
$$2x = 11 - 5y$$
$$3x - 2y = -12.$$

(Section 4.3, Example 4)

97. Factor completely:
$$6x^2 - 8xy + 2y^2.$$

(Section 6.5, Example 8)

98. Solve: $x + 3 \le -4$ or $2 - 7x \le 16$.

(Section 9.2, Example 6)

▶ SECTION 12.4   *Properties of Logarithms*

Objectives

1. Use the product rule.
2. Use the quotient rule.
3. Use the power rule.
4. Expand logarithmic expressions.
5. Condense logarithmic expressions.
6. Use the change-of-base property.

SSM
PH Tutor   CD-   Video
Center   ROM

We all learn new things in different ways. In this section, we consider important properties of logarithms. What would be the most effective way for you to learn about these properties? Would it be helpful to use your graphing utility and discover one of these properties for yourself? To do so, work Exercise 94 in Exercise Set 12.3 before continuing. Would the properties become more meaningful if you could see exactly where they come from? If so, you will find details of the proofs of many of these properties in Appendix B on page A4. The remainder of our work in this chapter will be based on the properties of logarithms that you learn in this section.

1   Use the product rule.

The Product Rule

Properties of exponents correspond to properties of logarithms. For example, when we multiply expressions with the same base, we add exponents:

$$b^M \cdot b^N = b^{M+N}.$$

This property of exponents, coupled with an awareness that a logarithm is an exponent, suggests the following property, called the **product rule**:

The Product Rule

Let b, M, and N be positive real numbers with $b \neq 1$.

$$\log_b(MN) = \log_b M + \log_b N$$

The logarithm of a product is the sum of the logarithms.

When we use the product rule to write a single logarithm as the sum of two logarithms, we say that we are **expanding a logarithmic expression**. For example, we can use the product rule to expand $\ln(7x)$:

$$\ln(7x) = \ln 7 + \ln x.$$

The logarithm of a product   is   the sum of the logarithms.

EXAMPLE 1   Using the Product Rule

Use the product rule to expand each logarithmic expression:

a. $\log_4(7 \cdot 5)$ **b.** $\log(10x)$.

Solution

a. $\log_4(7 \cdot 5) = \log_4 7 + \log_4 5$ The logarithm of a product is the sum of the logarithms.

b. $\log(10x) = \log 10 + \log x$ The logarithm of a product is the sum of the logarithms. These are common logarithms with base 10 understood.

$ = 1 + \log x$ Because $\log_b b = 1$, then $\log_{10} 10 = 1$. ∎

✔ **CHECK POINT 1**   Use the product rule to expand each logarithmic expression:

a. $\log_6(7 \cdot 11)$ **b.** $\log(100x)$.

2   Use the quotient rule.

The Quotient Rule

When we divide expressions with the same base, we subtract exponents:

$$\frac{b^M}{b^N} = b^{M-N}.$$

This property suggests the following property of logarithms, called the **quotient rule**:

The Quotient Rule

Let b, M, and N be positive real numbers with $b \neq 1$.

$$\log_b\left(\frac{M}{N}\right) = \log_b M - \log_b N$$

The logarithm of a quotient is the difference of the logarithms.

When we use the quotient rule to write a single logarithm as the difference of two logarithms, we say that we are **expanding a logarithmic expression**. For example, we can use the quotient rule to expand $\log \frac{x}{2}$:

$$\log \frac{x}{2} = \log x - \log 2.$$

The logarithm of a quotient   is   the difference of the logarithms.

EXAMPLE 2   Using the Quotient Rule

Use the quotient rule to expand each logarithmic expression:

a. $\log_7\left(\frac{19}{x}\right)$ **b.** $\ln\left(\frac{e^3}{7}\right)$.

Solution

a. $\log_7\left(\dfrac{19}{x}\right) = \log_7 19 - \log_7 x$ The logarithm of a quotient is the difference of the logarithms.

b. $\ln\left(\dfrac{e^3}{7}\right) = \ln e^3 - \ln 7$ The logarithm of a quotient is the difference of the logarithms. These are natural logarithms with base e understood.

$\qquad\qquad\quad = 3 - \ln 7$ Because $\ln e^x = x$, then $\ln e^3 = 3$.   ■

✔ **CHECK POINT 2**   Use the quotient rule to expand each logarithmic expression:

a. $\log_8\left(\dfrac{23}{x}\right)$ **b.** $\ln\left(\dfrac{e^5}{11}\right)$.

The Power Rule

3   Use the power rule.

When an exponential expression is raised to a power, we multiply exponents:
$$\left(b^M\right)^p = b^{Mp}.$$

This property suggests the following property of logarithms, called the **power rule**:

The Power Rule

Let b, and M be positive real numbers with $b \neq 1$, and let p be any real number.

$$\log_b M^p = p \log_b M$$

The logarithm of a number with an exponent is the product of the exponent and the logarithm of that number.

When we use the power rule to "pull the exponent to the front," we say that we are **expanding a logarithmic expression**. For example, we can use the power rule to expand $\ln x^2$:

$$\ln x^2 = 2 \ln x.$$

| The logarithm of a number with an exponent | is | the product of the exponent and the logarithm of that number. |

Figure 12.18 shows the graphs of $y = \ln x^2$ and $y = 2 \ln x$. Are $\ln x^2$ and $2 \ln x$ the same? The graphs illustrate that $y = \ln x^2$ and $y = 2 \ln x$ have different domains. The graphs are only the same if $x > 0$. Thus, we should write

$$\ln x^2 = 2 \ln x \quad \text{for} \quad x > 0.$$

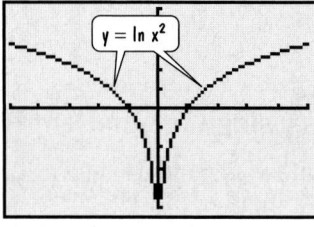

Domain: $(-\infty, 0) \cup (0, \infty)$

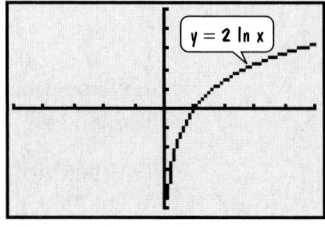

Domain: $(0, \infty)$

Figure 12.18 $\ln x^2$ and $2 \ln x$ have different domains.

When expanding a logarithmic expression, you might want to determine whether the rewriting has changed the domain of the expression. For the rest of this section, assume that all variables and variable expressions represent positive numbers.

4 Expand logarithmic expressions.

EXAMPLE 3   Using the Power Rule

Use the power rule to expand each logarithmic expression:

a. $\log_5 7^4$ **b.** $\ln \sqrt{x}$.

Solution

a. $\log_5 7^4 = 4 \log_5 7$ The logarithm of a number with an exponent is the exponent times the logarithm of that number.

b. $\ln \sqrt{x} = \ln x^{\frac{1}{2}}$ Rewrite the radical using a rational exponent.

$\quad\quad = \dfrac{1}{2} \ln x$ Use the power rule to bring the exponent to the front. ■

 CHECK POINT 3   Use the power rule to expand each logarithmic expression:

a. $\log_6 8^9$ **b.** $\ln \sqrt[3]{x}$.

Expanding Logarithmic Expressions

It is sometimes necessary to use more than one property of logarithms when you expand a logarithmic expression. Properties for expanding logarithmic expressions are as follows:

Properties for Expanding Logarithmic Expressions

1. $\log_b (MN) = \log_b M + \log_b N$ Product rule

2. $\log_b \left(\dfrac{M}{N} \right) = \log_b M - \log_b N$ Quotient rule

3. $\log_b M^p = p \log_b M$ Power rule

In all cases, $M > 0$ and $N > 0$.

EXAMPLE 4   Expanding Logarithmic Expressions

Use logarithmic properties to expand each expression as much as possible:

a. $\log_b x^2 \sqrt{y}$ **b.** $\log_6 \left(\dfrac{\sqrt[3]{x}}{36 y^4} \right)$.

Solution   We will have to use two or more of the properties for expanding logarithms in each part of this example.

a. $\log_b x^2 \sqrt{y} = \log_b x^2 y^{\frac{1}{2}}$ Use exponential notation.

$\quad\quad = \log_b x^2 + \log_b y^{\frac{1}{2}}$ Use the product rule.

$\quad\quad = 2 \log_b x + \dfrac{1}{2} \log_b y$ Use the power rule.

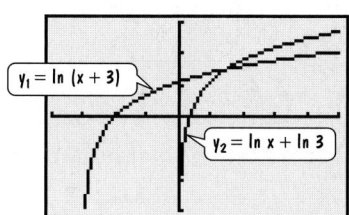

$$\textbf{b. } \log_6\left(\frac{\sqrt[3]{x}}{36y^4}\right) = \log_6 \frac{x^{\frac{1}{3}}}{36y^4} \qquad \text{Use exponential notation.}$$

$$= \log_6 x^{\frac{1}{3}} - \log_6 36y^4 \qquad \text{Use the quotient rule.}$$

$$= \log_6 x^{\frac{1}{3}} - \left(\log_6 36 + \log_6 y^4\right) \qquad \text{Use the product rule on } \log_6 36y^4.$$

$$= \frac{1}{3}\log_6 x - \left(\log_6 36 + 4\log_6 y\right) \qquad \text{Use the power rule.}$$

$$= \frac{1}{3}\log_6 x - \log_6 36 - 4\log_6 y \qquad \text{Apply the distributive property.}$$

$$= \frac{1}{3}\log_6 x - 2 - 4\log_6 y \qquad \log_6 36 = 2 \text{ because 2 is the power to which we must raise 6 to get 36. } (6^2 = 36) \ \blacksquare$$

✔ **CHECK POINT 4** Use logarithmic properties to expand each expression as much as possible:

a. $\log_b x^4 \sqrt[3]{y}$ 　　　　　　　　**b.** $\log_5\left(\frac{\sqrt{x}}{25y^3}\right)$.

⑤ Condense logarithmic expressions.

Condensing Logarithmic Expressions

To **condense a logarithmic expression**, we write the sum or difference of two or more logarithmic expressions as a single logarithmic expression. We use the properties of logarithms to do so.

Properties for Condensing Logarithmic Expressions

1. $\log_b M + \log_b N = \log_b(MN)$ 　　Product rule

2. $\log_b M - \log_b N = \log_b\left(\dfrac{M}{N}\right)$ 　　Quotient rule

3. $p\log_b M = \log_b M^p$ 　　Power rule

In all cases, $M > 0$ and $N > 0$.

EXAMPLE 5 Condensing Logarithmic Expressions

Write as a single logarithm:

a. $\log_4 2 + \log_4 32$ 　　　　　　　　**b.** $\log(4x - 3) - \log x$.

Solution

a. $\log_4 2 + \log_4 32 = \log_4(2 \cdot 32)$ 　　Use the product rule.

$$= \log_4 64 \qquad \text{We now have a single logarithm. However, we can simplify.}$$

$$= 3 \qquad \log_4 64 = 3 \text{ because } 4^3 = 64.$$

b. $\log(4x - 3) - \log x = \log\left(\dfrac{4x - 3}{x}\right)$ Use the quotient rule. 　　■

✔ **CHECK POINT 5** Write as a single logarithm:

a. $\log 25 + \log 4$ 　　　　　　　　**b.** $\log(7x + 6) - \log x$.

Coefficients of logarithms must be 1 before you can condense them using the product and quotient rules. For example, to condense

$$2 \ln x + \ln(x + 1),$$

the coefficient of the first term must be 1. We use the power rule to rewrite the coefficient as an exponent:

> 1. Make the number in front an exponent.

$$2 \ln x + \ln(x + 1) = \ln x^2 + \ln(x + 1) = \ln x^2(x + 1).$$

> 2. Use the product rule. The sum of logarithms with coefficients 1 is the logarithm of the product.

EXAMPLE 6 Condensing Logarithmic Expressions

Write as a single logarithm:

a. $\dfrac{1}{2}\log x + 4\log(x - 1)$ **b.** $3\ln(x + 7) - \ln x.$

Solution

a. $\dfrac{1}{2}\log x + 4\log(x - 1)$

$= \log x^{\frac{1}{2}} + \log(x - 1)^4$ *Use the power rule so that all coefficients are 1.*

$= \log x^{\frac{1}{2}}(x - 1)^4$ *Use the product rule.*

$= \log \sqrt{x}(x - 1)^4$

b. $3\ln(x + 7) - \ln x$

$= \ln(x + 7)^3 - \ln x$ *Use the power rule so that all coefficients are 1.*

$= \ln \dfrac{(x + 7)^3}{x}$ *Use the quotient rule.*   ■

✔ **CHECK POINT 6**   Write as a single logarithm:

a. $2\ln x + \dfrac{1}{3}\ln(x + 5)$ **b.** $2\log(x - 3) - \log x.$

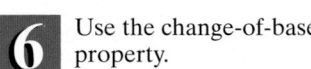

6   Use the change-of-base property.

The Change-of-Base Property

We have seen that calculators give the values of both common logarithms (base 10) and natural logarithms (base e). To find a logarithm with any other base, we can use the following change-of-base property:

The Change-of-Base Property

For any logarithmic bases a and b, and any positive number M,

$$\log_b M = \frac{\log_a M}{\log_a b}.$$

The logarithm of M with base b is equal to the logarithm of M with any new base divided by the logarithm of b with that new base.

In the change-of-base property, base b is the base of the original logarithm. Base a is a new base that we introduce. Thus, the change-of-base property allows us to change from base b to *any* new base a, as long as the newly introduced base is a positive number not equal to 1.

The change-of-base property is used to write a logarithm in terms of quantities that can be evaluated with a calculator. Because calculators contain keys for common (base 10) and natural (base e) logarithms, we will frequently introduce base 10 or base e.

Change-of-Base Property

$$\log_b M = \frac{\log_a M}{\log_a b}$$

a is the new introduced base.

Introducing Common Logarithms

$$\log_b M = \frac{\log_{10} M}{\log_{10} b}$$

10 is the new introduced base.

Introducing Natural Logarithms

$$\log_b M = \frac{\log_e M}{\log_e b}$$

e is the new introduced base.

Using the notations for common logarithms and natural logarithms, we have the following results:

The Change-of-Base Property: Introducing Common and Natural Logarithms

Introducing Common Logarithms

$$\log_b M = \frac{\log M}{\log b}$$

Introducing Natural Logarithms

$$\log_b M = \frac{\ln M}{\ln b}$$

EXAMPLE 7   Changing Base to Common Logarithms

Use common logarithms to evaluate $\log_5 140$.

Solution   Because $\log_b M = \dfrac{\log M}{\log b}$,

$$\log_5 140 = \frac{\log 140}{\log 5}$$

$$\approx 3.07.$$

Use a calculator: 140 LOG ÷ 5 LOG = or LOG 140 ÷ LOG 5 ENTER.

This means that $\log_5 140 \approx 3.07$.

✔ CHECK POINT 7   Use common logarithms to evaluate $\log_7 2506$.

EXAMPLE 8   Changing Base to Natural Logarithms

Use natural logarithms to evaluate $\log_5 140$.

Solution   Because $\log_b M = \dfrac{\ln M}{\ln b}$,

$$\log_5 140 = \frac{\ln 140}{\ln 5}$$

$$\approx 3.07.$$

Use a calculator: 140 LN ÷ 5 LN = or LN 140 ÷ LN 5 ENTER.

We have again shown that $\log_5 140 \approx 3.07$.

Discover for Yourself

Find a reasonable estimate of $\log_5 140$ to the nearest whole number: 5 to what power is 140? Compare your estimate to the value obtained in Example 7.

✔ **CHECK POINT 8** Use natural logarithms to evaluate $\log_7 2506$.

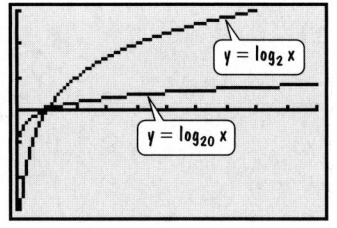

Figure 12.19 Using the change-of-base property to graph logarithmic functions

We can use the change-of-base property to graph logarithmic functions with bases other than 10 or e on a graphing utility. For example, Figure 12.19 shows the graphs of

$$y = \log_2 x \quad \text{and} \quad y = \log_{20} x$$

in a $[0, 10, 1]$ by $[-3, 3, 1]$ viewing rectangle. Because $\log_2 x = \dfrac{\ln x}{\ln 2}$ and $\log_{20} x = \dfrac{\ln x}{\ln 20}$ the functions are entered as

$$y_1 = \boxed{\text{LN}}\, x \boxed{\div} \boxed{\text{LN}}\, 2$$

$$\text{and} \quad y_2 = \boxed{\text{LN}}\, x \boxed{\div} \boxed{\text{LN}}\, 20.$$

EXERCISE SET 12.4

 Practice Exercises

In all exercises, assume that all variables and variable expressions represent positive numbers.

In Exercises 1–32, use properties of logarithms to expand each logarithmic expression as much as possible. Where possible, evaluate logarithmic expressions without using a calculator.

1. $\log_5 (7 \cdot 3)$

2. $\log_8 (13 \cdot 7)$

3. $\log_7 (7x)$

4. $\log_9 (9x)$

5. $\log (1000x)$

6. $\log (10{,}000x)$

7. $\log_7 \left(\dfrac{7}{x} \right)$

8. $\log_9 \left(\dfrac{9}{x} \right)$

9. $\log \left(\dfrac{x}{100} \right)$

10. $\log \left(\dfrac{x}{1000} \right)$

11. $\log_4 \left(\dfrac{64}{y} \right)$

12. $\log_5 \left(\dfrac{125}{y} \right)$

13. $\ln \left(\dfrac{e^2}{5} \right)$

14. $\ln \left(\dfrac{e^4}{8} \right)$

15. $\log_b x^3$

16. $\log_b x^7$

17. $\log N^{-6}$

18. $\log M^{-8}$

19. $\ln \sqrt[5]{x}$

20. $\ln \sqrt[7]{x}$

21. $\log_b x^2 y$

22. $\log_b xy^3$

23. $\log_4 \left(\dfrac{\sqrt{x}}{64} \right)$

24. $\log_5 \left(\dfrac{\sqrt{x}}{25} \right)$

25. $\log_6 \left(\dfrac{36}{\sqrt{x+1}} \right)$

26. $\log_8 \left(\dfrac{64}{\sqrt{x+1}} \right)$

27. $\log_b \left(\dfrac{x^2 y}{z^2} \right)$

28. $\log_b \left(\dfrac{x^3 y}{z^2} \right)$

29. $\log \sqrt{100x}$

30. $\ln \sqrt{ex}$

31. $\log \sqrt[3]{\dfrac{x}{y}}$

32. $\log \sqrt[5]{\dfrac{x}{y}}$

In Exercises 33–52, use properties of logarithms to condense each logarithmic expression. Write the expression as a single logarithm whose coefficient is 1. Where possible, evaluate logarithmic expressions.

33. $\log 5 + \log 2$

34. $\log 250 + \log 4$

35. $\ln x + \ln 7$

36. $\ln x + \ln 3$

37. $\log_2 96 - \log_2 3$

38. $\log_3 405 - \log_3 5$

39. $\log(2x + 5) - \log x$

40. $\log(3x + 7) - \log x$

41. $\log x + 3 \log y$

42. $\log x + 7 \log y$

43. $\frac{1}{2} \ln x + \ln y$

44. $\frac{1}{3} \ln x + \ln y$

45. $2 \log_b x + 3 \log_b y$

46. $5 \log_b x + 6 \log_b y$

47. $5 \ln x - 2 \ln y$

48. $7 \ln x - 3 \ln y$

49. $3 \ln x - \frac{1}{3} \ln y$

50. $2 \ln x - \frac{1}{2} \ln y$

51. $4 \ln(x + 6) - 3 \ln x$

52. $8 \ln(x + 9) - 4 \ln x$

In Exercises 53–60, use common logarithms or natural logarithms and a calculator to evaluate to four decimal places

53. $\log_5 13$

54. $\log_6 17$

55. $\log_{14} 87.5$

56. $\log_{16} 57.2$

57. $\log_{0.1} 17$

58. $\log_{0.3} 19$

59. $\log_\pi 63$

60. $\log_\pi 400$

 Application Exercises

61. The loudness level of a sound can be expressed by comparing the sound's intensity to the intensity of a sound barely audible to the human ear. The formula

$$D = 10(\log I - \log I_0)$$

describes the loudness level of a sound, D, in decibels, where I is the intensity of the sound, in watts per meter^2, and I_0 is the intensity of a sound barely audible to the human ear.

a. Express the formula so that the expression in parentheses is written as a single logarithm.

b. Use the form of the formula from part (a) to answer this question. If a sound has an intensity 100 times the intensity of a softer sound, how much larger on the decibel scale is the loudness level of the more intense sound?

62. The formula

$$t = \frac{1}{c}\left[\ln A - \ln(A - N)\right]$$

describes the time, t, in weeks, that it takes to achieve mastery of a portion of a task, where A is the maximum learning possible, N is the portion of the learning that is to be achieved, and c is a constant used to measure an individual's learning style.

a. Express the formula so that the expression in brackets is written as a single logarithm.

b. The formula is also used to determine how long it will take chimpanzees and apes to master a task. For example, a typical chimpanzee learning sign language can master a maximum of 65 signs. Use the form of the formula from part (a) to answer this question. How many weeks will it take a chimpanzee to master 30 signs if c for that chimp is 0.03?

Writing in Mathematics

63. Describe the product rule for logarithms and give an example.

64. Describe the quotient rule for logarithms and give an example.

65. Describe the power rule for logarithms and give an example.

66. Without showing the details, explain how to condense $\ln x - 2\ln(x + 1)$.

67. Describe the change-of-base property and give an example.

68. Explain how to use your calculator to find $\log_{14} 283$.

69. You overhear a student talking about a property of logarithms in which division becomes subtraction. Explain what the student means by this.

70. Find $\ln 2$ using a calculator. Then calculate each of the following: $1 - \frac{1}{2}$; $\quad 1 - \frac{1}{2} + \frac{1}{3}$; $\quad 1 - \frac{1}{2} + \frac{1}{3} - \frac{1}{4}$; $1 - \frac{1}{2} + \frac{1}{3} - \frac{1}{4} + \frac{1}{5}$; .... Describe what you observe.

Critical Thinking Exercises

71. Which one of the following is true?

a. $\dfrac{\log_7 49}{\log_7 7} = \log_7 49 - \log_7 7$

b. $\log_b(x^3 + y^3) = 3\log_b x + 3\log_b y$

c. $\log_b(xy)^5 = (\log_b x + \log_b y)^5$

d. $\ln\sqrt{2} = \dfrac{\ln 2}{2}$

72. Use the change-of-base property to prove that

$$\log e = \frac{1}{\ln 10}.$$

73. If $\log 3 = A$ and $\log 7 = B$, find $\log_7 9$ in terms of A and B.

74. Write as a single term that does not contain a logarithm:

$$e^{\ln 8x^5 - \ln 2x^2}.$$

Technology Exercises

75. a. Use a graphing utility (and the change-of-base property) to graph $y = \log_3 x$.

b. Graph $y = 2 + \log_3 x$, $y = \log_3(x + 2)$, and $y = -\log_3 x$ in the same viewing rectangle as $y = \log_3 x$. Then describe the change or changes that need to be made to the graph of $y = \log_3 x$ to obtain each of these three graphs.

76. Graph $y = \log x$, $y = \log(10x)$, and $y = \log(0.1x)$ in the same viewing rectangle. Describe the relationship among the three graphs. What logarithmic property accounts for this relationship?

77. Use a graphing utility and the change-of-base property to graph $y = \log_3 x$, $y = \log_{25} x$, and $y = \log_{100} x$ in the same viewing rectangle.

a. Which graph is on the top in the interval $(0, 1)$? Which is on the bottom?

b. Which graph is on the top in the interval $(1, \infty)$? Which is on the bottom?

c. Generalize by writing a statement about which graph is on top, which is on the bottom, and in which intervals, using $y = \log_b x$ where $b > 1$.

Disprove each statement in Exercises 78–82 by

a. *letting y equal a positive constant of your choice.*

b. *using a graphing utility to graph the function on each side of the equal sign. The two functions should have different graphs, showing that the equation is not true in general.*

78. $\log(x + y) = \log x + \log y$

79. $\log \dfrac{x}{y} = \dfrac{\log x}{\log y}$

80. $\ln(x - y) = \ln x - \ln y$

81. $\ln(xy) = (\ln x)(\ln y)$

82. $\dfrac{\ln x}{\ln y} = \ln x - \ln y$

Review Exercises

83. Graph: $5x - 2y > 10$. (Section 9.4, Example 1)

84. Solve: $x - 2(3x - 2) > 2x - 3$. (Section 9.1, Example 3)

85. Divide and simplify: $\dfrac{\sqrt[3]{40x^2y^6}}{\sqrt[3]{5xy}}$. (Section 10.4, Example 5)

▶ SECTION 12.5   *Exponential and Logarithmic Equations*

Objectives

1 Solve exponential equations.

2 Solve logarithmic equations.

3 Solve applied problems involving exponential and logarithmic equations.

SSM
PH Tutor   CD-   Video
Center   ROM

Is an early retirement awaiting you?

You inherited $30,000. You'd like to put aside $25,000 and eventually have over half a million dollars for early retirement. Is this possible? In this section you will see how techniques for solving equations with variable exponents provide an answer to this question.

1 Solve exponential equations.

Exponential Equations

An **exponential equation** is an equation containing a variable in an exponent. Examples of exponential equations include

$$2^{3x-8} = 16, \quad 4^x = 15, \quad \text{and} \quad 40e^{0.6x} = 240.$$

Some exponential equations can be solved by expressing each side of the equation as a power of the same base. All exponential functions are one-to-one—that is, no two different ordered pairs have the same second component. Thus, if b is a positive number other than 1 and $b^M = b^N$, then $M = N$.

EXAMPLE 1   Solving Exponential Equations

Solve: **a.** $2^{3x-8} = 16$ **b.** $16^x = 64$.

Solution   In each equation, express both sides as a power of the same base. Then set the exponents equal to each other.

a. Because 16 is 2^4, we express each side in terms of base 2.

$$2^{3x-8} = 16 \qquad \text{This is the given equation.}$$

$$2^{3x-8} = 2^4 \qquad \text{Write each side as a power of the same base.}$$

$$3x - 8 = 4 \qquad \text{If } b^M = b^N, b > 0 \text{ and } b \neq 1, \text{ then } M = N.$$

$$3x = 12 \qquad \text{Add 8 to both sides.}$$

$$x = 4 \qquad \text{Divide both sides by 3.}$$

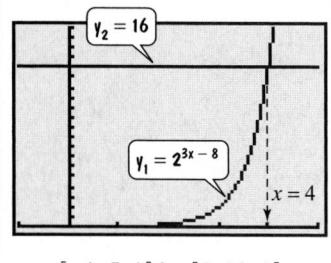

Using Technology

The graphs of

$$y_1 = 2^{3x-8}$$

and $y_2 = 16$ have an intersection point whose x-coordinate is 4. This verifies that $\{4\}$ is the solution set for $2^{3x-8} = 16$.

$y_2 = 16$

$y_1 = 2^{3x-8}$

$x = 4$

$[-1, 5, 1]$ by $[0, 20, 1]$

CHECK 4:

$$2^{3x-8} = 16$$

$$2^{3 \cdot 4 - 8} \overset{?}{=} 16$$

$$2^4 \overset{?}{=} 16$$

$$16 = 16, \quad \text{true}$$

The solution is 4 and the solution set is $\{4\}$.

b. Because $16 = 4^2$ and $64 = 4^3$, we express each side in terms of base 4.

$16^x = 64$	This is the given equation.
$(4^2)^x = 4^3$	Write each side as a power of the same base.
$4^{2x} = 4^3$	When an exponential expression is raised to a power, multiply exponents.
$2x = 3$	If two powers of the same base are equal, then the exponents are equal.
$x = \dfrac{3}{2}$	Divide both sides by 2.

CHECK $\dfrac{3}{2}$:

$$16^x = 64$$

$$16^{\frac{3}{2}} \overset{?}{=} 64$$

$$(\sqrt{16})^3 \overset{?}{=} 64 \qquad\qquad b^{\frac{m}{n}} = (\sqrt[n]{b})^m$$

$$4^3 \overset{?}{=} 64$$

$$64 = 64, \quad \text{true}$$

The solution is $\frac{3}{2}$ and the solution set is $\{\frac{3}{2}\}$. ∎

Discover for Yourself

The equation $16^x = 64$ can also be solved by writing each side in terms of base 2. Do this. Which solution method do you prefer?

✔ **CHECK POINT 1** Solve:

a. $5^{3x-6} = 125$ **b.** $4^x = 32$.

When both sides of an exponential equation cannot be written as a power of the same base, logarithms are used in solving the equation. The solution begins with isolating the exponential expression and taking the natural logarithm on both sides. Why can we do this? All logarithmic functions are one-to-one—that is, no two different ordered pairs have the same second component. Thus, if M and N are positive real numbers and $M = N$, then $\log_b M = \log_b N$.

Using Natural Logarithms to Solve Exponential Equations

1. Isolate the exponential expression.
2. Take the natural logarithm on both sides of the equation.
3. Simplify using one of the following properties:

$$\ln b^x = x \ln b \quad \text{or} \quad \ln e^x = x.$$

4. Solve for the variable.

EXAMPLE 2   Solving an Exponential Equation

Solve: $4^x = 15$.

Solution   Because the exponential expression, 4^x, is already isolated on the left, we begin by taking the natural logarithm on both sides of the equation.

$4^x = 15$	This is the given equation.
$\ln 4^x = \ln 15$	Take the natural logarithm on both sides.
$x \ln 4 = \ln 15$	Use the power rule and bring the variable exponent to the front: $\ln b^x = x \ln b.$
$x = \dfrac{\ln 15}{\ln 4}$	Solve for x by dividing both sides by $\ln 4$.

We now have an exact value for x. We use the exact value for x in the equation's solution set. Thus, the equation's solution is $\dfrac{\ln 15}{\ln 4}$ and the solution set is $\left\{ \dfrac{\ln 15}{\ln 4} \right\}$. We can obtain a decimal approximation by using a calculator: $x \approx 1.95$. Because $4^2 = 16$, it seems reasonable that the solution to $4^x = 15$ is approximately 1.95.   ∎

✔ **CHECK POINT 2**   Solve: $5^x = 134$. Find the solution, and then use a calculator to obtain a decimal approximation to two decimal places for the solution.

EXAMPLE 3   Solving an Exponential Equation

Solve: $40e^{0.6x} = 240$.

Solution

Solution   We begin by dividing both sides by 40 to isolate the exponential expression, $e^{0.6x}$. Then we take the natural logarithm on both sides of the equation.

$40e^{0.6x} = 240$	This is the given equation.
$e^{0.6x} = 6$	Isolate the exponential factor by dividing both sides by 40.
$\ln e^{0.6x} = \ln 6$	Take the natural logarithm on both sides.
$0.6x = \ln 6$	Use the inverse property $\ln e^x = x$ on the left.
$x = \dfrac{\ln 6}{0.6} \approx 2.99$	Divide both sides by 0.6.

Thus, the solution of the equation is $\dfrac{\ln 6}{0.6} \approx 2.99$. Try checking this approximate solution in the original equation, verifying that $\left\{ \dfrac{\ln 6}{0.6} \approx 2.99 \right\}$ is the solution set. ∎

✔ **CHECK POINT 3**   Solve: $7e^{2x} = 63$. Find the solution set, and then use a calculator to obtain a decimal approximation to two decimal places for the solution.

2 Solve logarithmic equations.

Logarithmic Equations

A **logarithmic equation** is an equation containing a variable in a logarithmic expression. Examples of logarithmic equations include

$$\log_4(x + 3) = 2 \quad \text{and} \quad \ln 2x = 3.$$

If a logarithmic equation is in the form $\log_b x = c$, we can solve the equation by rewriting it in its equivalent exponential form $b^c = x$. Example 4 illustrates how this is done.

EXAMPLE 4 Solving a Logarithmic Equation

Solve: $\log_4(x + 3) = 2$.

Solution We first rewrite the equation as an equivalent equation in exponential form using the fact that $\log_b x = c$ means $b^c = x$.

$$\log_4(x + 3) = 2 \qquad \text{means} \qquad 4^2 = x + 3.$$

Logarithms are exponents.

Now we solve the equivalent equation for x.

$4^2 = x + 3$	This is the equivalent equation.
$16 = x + 3$	Square 4.
$13 = x$	Subtract 3 from both sides.

CHECK 13:

$\log_4(x + 3) = 2$	This is the given logarithmic equation.
$\log_4(13 + 3) \stackrel{?}{=} 2$	Substitute 13 for x.
$\log_4 16 \stackrel{?}{=} 2$	
$2 = 2, \quad \text{true}$	$\log_4 16 = 2$ because $4^2 = 16$.

This true statement indicates that the solution is 13 and the solution set is {13}. ∎

✔ CHECK POINT 4 Solve: $\log_2(x - 4) = 3$.

Logarithmic expressions are defined only for logarithms of positive real numbers. Always check proposed solutions of a logarithmic equation in the original equation. Exclude from the solution set any proposed solution that produces the logarithm of a negative number or the logarithm of 0.

To rewrite the logarithmic equation $\log_b x = c$ in the equivalent exponential form $b^c = x$, we need a single logarithm whose coefficient is one. It is sometimes necessary to use properties of logarithms to condense logarithms into a single logarithm. In the next example, we use the product rule for logarithms to obtain a single logarithmic expression on the left side.

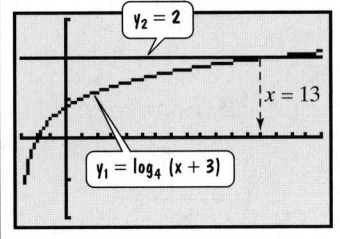

EXAMPLE 5   Using the Product Rule to Solve a Logarithmic Equation

Solve: $\log_2 x + \log_2(x - 7) = 3$.

Solution

$\log_2 x + \log_2(x - 7) = 3$	This is the given equation.
$\log_2 x(x - 7) = 3$	Use the product rule to obtain a single logarithm: $\log_b M + \log_b N = \log_b(MN)$.
$2^3 = x(x - 7)$	$\log_b x = c$ means $b^c = x$.
$8 = x^2 - 7x$	Apply the distributive property on the right.
$0 = x^2 - 7x - 8$	Set the equation equal to 0.
$0 = (x - 8)(x + 1)$	Factor.
$x - 8 = 0$ or $x + 1 = 0$	Set each factor equal to 0.
$x = 8$ $x = -1$	Solve for x.

CHECK 8:

$\log_2 x + \log_2(x - 7) = 3$

$\log_2 8 + \log_2(8 - 7) \overset{?}{=} 3$

$\log_2 8 + \log_2 1 \overset{?}{=} 3$

$3 + 0 \overset{?}{=} 3$

$3 = 3$,   true

CHECK −1:

$\log_2 x + \log_2(x - 7) = 3$

$\log_2(-1) + \log_2(-1 - 7) \overset{?}{=} 3$

The number −1 does not check. Negative numbers do not have logarithms.

The solution is 8 and the solution set is $\{8\}$.   ∎

✔ **CHECK POINT 5**   Solve: $\log x + \log(x - 3) = 1$.

EXAMPLE 6   Using the Quotient Rule to Solve a Logarithmic Equation

Solve: $\log_4(x + 3) - \log_4 x = 2$.

Solution

$\log_4(x + 3) - \log_4 x = 2$	This is the given equation.
$\log_4 \dfrac{x + 3}{x} = 2$	Use the quotient rule to obtain a single logarithm: $\log_b M - \log_b N = \log_b \dfrac{M}{N}$.
$4^2 = \dfrac{x + 3}{x}$	$\log_b x = c$ means $b^c = x$.
$16 = \dfrac{x + 3}{x}$	Square 4: $4^2 = 16$.
$16x = x + 3$	Multiply both sides by x.
$15x = 3$	Subtract x from both sides.
$x = \dfrac{3}{15} = \dfrac{1}{5}$	Divide both sides by 15.

Check by substituting $\frac{1}{5}$ in the original equation. In this equation, both x and $x + 3$ must be positive. The proposed solution, $\frac{1}{5}$, satisfies these conditions, and the solution set is $\left\{\frac{1}{5}\right\}$. ∎

✔ **CHECK POINT 6**   Solve: $\log_5(x + 1) - \log_5 x = 2$

Equations involving natural logarithms can be solved using the inverse property $e^{\ln x} = x$. For example, to solve

$$\ln x = 5$$

we write both sides of the equation as exponents on base e:

$$e^{\ln x} = e^5.$$

This is called **exponentiating both sides** of the equation. Using the inverse property $e^{\ln x} = x$, we simplify the left side of the equation and obtain the solution:

$$x = e^5.$$

EXAMPLE 7   Solving an Equation with a Natural Logarithm

Solve: $3 \ln 2x = 12$.

Solution

$3 \ln 2x = 12$	This is the given equation.
$\ln 2x = 4$	Divide both sides by 3.
$e^{\ln 2x} = e^4$	Exponentiate both sides.
$2x = e^4$	Use the inverse property to simplify the left side: $e^{\ln x} = x$.
$x = \dfrac{e^4}{2} \approx 27.30$	Divide both sides by 2.

$\mathbf{C}\text{HECK}\ \dfrac{e^4}{2}$:

$3 \ln 2x = 12$	This is the given logarithmic equation.
$3 \ln 2\left(\dfrac{e^4}{2}\right) \overset{?}{=} 12$	Substitute $\dfrac{e^4}{2}$ for x.
$3 \ln e^4 \overset{?}{=} 12$	Simplify: $\dfrac{2}{1} \cdot \dfrac{e^4}{2} = e^4$.
$3 \cdot 4 \overset{?}{=} 12$	Because $\ln e^x = x$, we conclude $\ln e^4 = 4$.
$12 = 12$, true	

This true statement indicates that the solution is $\dfrac{e^4}{2}$ and the solution set is $\left\{\dfrac{e^4}{2}\right\}$. ∎

✔ **CHECK POINT 7**   Solve: $4 \ln 3x = 8$.

3 Solve applied problems involving exponential and logarithmic equations.

Applications

Our first applied example provides a mathematical perspective on the old slogan "Alcohol and driving don't mix." In California, where 38% of fatal traffic crashes involve drinking drivers, it is illegal to drive with a blood alcohol concentration of 0.08 or higher. At these levels, drivers may be arrested and charged with driving under the influence.

EXAMPLE 8 Alcohol and Risk of a Car Accident

Medical research indicates that the risk of having a car accident increases exponentially as the concentration of alcohol in the blood increases. The risk is modeled by

$$R = 6e^{12.77x}$$

where x is the blood alcohol concentration and R, given as a percent, is the risk of having a car accident. What blood alcohol concentration corresponds to a 17% risk of a car accident?

Solution For a risk of 17%, we let $R = 17$ in the equation and solve for x, the blood alcohol concentration.

$$R = 6e^{12.77x}$$ This is the given equation.

$$6e^{12.77x} = 17$$ Substitute 17 for R and (optional) reverse the two sides of the equation.

$$e^{12.77x} = \frac{17}{6}$$ Isolate the exponential factor by dividing both sides by 6.

$$\ln e^{12.77x} = \ln\left(\frac{17}{6}\right)$$ Take the natural logarithm on both sides.

$$12.77x = \ln\left(\frac{17}{6}\right)$$ Use the inverse property $\ln e^x = x$ on the left side.

$$x = \frac{\ln\left(\frac{17}{6}\right)}{12.77} \approx 0.08$$ Divide both sides by 12.77.

For a blood alcohol concentration of 0.08, the risk of a car accident is 17%. In many states, it is illegal to drive at this blood alcohol concentration. ∎

✔ **CHECK POINT 8** Use the formula in Example 8 to solve this problem. What blood alcohol concentration corresponds to a 7% risk of a car accident? (In many states, drivers under the age of 21 can lose their license for driving at this level.)

Suppose that you inherit $30,000. Is it possible to invest $25,000 and have over half a million dollars for early retirement? Our next example illustrates the power of compound interest.

EXAMPLE 9 Revisiting the Formula for Compound Interest

The formula

$$A = P\left(1 + \frac{r}{n}\right)^{nt}$$

describes the accumulated value, A, of a sum of money, P, the principal, after t years at annual percentage rate r (in decimal form) compounded n times a year. How long will it take $25,000 to grow to $500,000 at 9% annual interest compounded monthly?

Playing Doubles: Interest Rates and Doubling Time

One way to calculate what your savings will be worth at some point in the future is to consider doubling time. The following table shows how long it takes for your money to double at different annual interest rates subject to continuous compounding.

Annual Interest Rate	Years to Double
5%	13.9 years
7%	9.9 years
9%	7.7 years
11%	6.3 years

Of course, the first problem is collecting some money to invest. The second problem is finding a reasonably safe investment with a return of 9% or more.

Solution

$$A = P\left(1 + \frac{r}{n}\right)^{nt}$$

This is the given formula.

$$500,000 = 25,000\left(1 + \frac{0.09}{12}\right)^{12t}$$

A(the desired accumulated value) = 500,000,
P(the principal) = 25,000,
r(the interest rate) = 9% = 0.09, and n = 12 (monthly compounding).

Our goal is to solve the equation for t. Let's reverse the two sides of the equation and then simplify within parentheses.

$$25,000\left(1 + \frac{0.09}{12}\right)^{12t} = 500,000$$

$$25,000(1 + 0.0075)^{12t} = 500,000$$

Divide within parentheses: $\frac{0.09}{12} = 0.0075$.

$$25,000(1.0075)^{12t} = 500,000$$

Add within parentheses.

$$(1.0075)^{12t} = 20$$

Divide both sides by 25,000.

$$\ln(1.0075)^{12t} = \ln 20$$

Take the natural logarithm on both sides.

$$12t \ln(1.0075) = \ln 20$$

Use the power rule to bring the exponent to the front: $\ln b^x = x \ln b$.

$$t = \frac{\ln 20}{12 \ln 1.0075}$$

Solve for t, dividing both sides by 12 $\ln 1.0075$.

$$\approx 33.4$$

Use a calculator.

After approximately 33.4 years, the $25,000 will grow to an accumulated value of $500,000. If you set aside the money at age 20, you can begin enjoying a life of leisure at about age 53. ∎

✔ **CHECK POINT 9** How long, to the nearest tenth of a year, will it take $1000 to grow to $3600 at 8% annual interest compounded quarterly?

EXAMPLE 10   The Growth of the Environmental Industry

The formula

$$N = 461.87 + 299.4 \ln x$$

models the thousands of workers, N, in the environmental industry in the United States x years after 1979. By which year will there be 1,500,000, or 1500 thousand, U.S. workers in the environmental industry?

Solution   We substitute 1500 for N and solve for x, the number of years after 1979.

$$N = 461.87 + 299.4 \ln x$$ This is the given formula.

$$461.87 + 299.4 \ln x = 1500$$ Substitute 1500 for N and reverse the two sides of the equation.

Our goal is to isolate $\ln x$. We can then find x by exponentiating both sides of the equation, using the inverse property $e^{\ln x} = x$.

$$299.4 \ln x = 1038.13$$ Subtract 461.87 from both sides.

$$\ln x = \frac{1038.13}{299.4}$$ Divide both sides by 299.4.

$$e^{\ln x} = e^{1038.13/299.4}$$ Exponentiate both sides.

$$x = e^{1038.13/299.4}$$ $e^{\ln x} = x$

$$\approx 32$$ Use a calculator.

Approximately 32 years after 1979, in the year 2011, there will be 1.5 million U.S. workers in the environmental industry. ■

✔ **CHECK POINT 10**   Use the formula in Example 10 to find by which year there will be two million, or 2000 thousand, U.S. workers in the environmental industry.

EXERCISE SET 12.5

Practice Exercises

Solve each exponential equation in Exercises 1–12 by expressing each side as a power of the same base and then equating exponents.

1. $2^x = 64$

2. $3^x = 81$

3. $5^x = 125$

4. $5^x = 625$

5. $2^{2x-1} = 32$

6. $3^{2x+1} = 27$

7. $4^{2x-1} = 64$

8. $5^{3x-1} = 125$

9. $32^x = 8$

10. $4^x = 32$

11. $9^x = 27$

12. $125^x = 625$

Solve each exponential equation in Exercises 13–34 by taking the natural logarithm on both sides. Express the solution set in terms of natural logarithms. Then use a calculator to obtain a decimal approximation, correct to two decimal places, for the solution.

13. $10^x = 3.91$

14. $10^x = 8.07$

15. $e^x = 5.7$

16. $e^x = 0.83$

17. $5^x = 17$

18. $19^x = 143$

19. $5e^x = 25$

20. $9e^x = 99$

21. $3e^{5x} = 1977$

22. $4e^{7x} = 10{,}273$

23. $e^{0.7x} = 13$

24. $e^{0.08x} = 4$

25. $1250e^{0.055x} = 3750$

26. $1250e^{0.065x} = 6250$

27. $30 - (1.4)^x = 0$

28. $135 - (4.7)^x = 0$

29. $e^{1-5x} = 793$

30. $e^{1-8x} = 7957$

31. $7^{x+2} = 410$

32. $5^{x-3} = 137$

33. $2^{x+1} = 5^x$

34. $4^{x+1} = 9^x$

In Exercises 35–58, solve each logarithmic equation. Be sure to reject any value of x that produces the logarithm of a negative number or the logarithm of zero in the original equation.

35. $\log_3 x = 4$

36. $\log_5 x = 3$

37. $\log_2 x = -4$

38. $\log_2 x = -5$

39. $\log_9 x = \dfrac{1}{2}$

40. $\log_{25} x = \dfrac{1}{2}$

41. $\log x = 2$

42. $\log x = 3$

43. $\ln x = -3$

44. $\ln x = -4$

45. $\log_4(x + 5) = 3$

46. $\log_5(x - 7) = 2$

47. $\log_3(x - 4) = -3$

48. $\log_7(x + 2) = -2$

49. $\log_4(3x + 2) = 3$

50. $\log_2(4x + 1) = 5$

51. $\log_5 x + \log_5(4x - 1) = 1$

52. $\log_6(x + 5) + \log_6 x = 2$

53. $\log_3(x - 5) + \log_3(x + 3) = 2$

54. $\log_2(x - 1) + \log_2(x + 1) = 3$

55. $\log_2(x + 2) - \log_2(x - 5) = 3$

56. $\log_4(x + 2) - \log_4(x - 1) = 1$

57. $\log(3x - 5) - \log 5x = 2$

58. $\log(2x - 1) - \log x = 2$

Exercises 59–66 involve equations with natural logarithms. Solve each equation by isolating the natural logarithm and exponentiating both sides. Express the answer in terms of e. Then use a calculator to obtain a decimal approximation, correct to two decimal places, for the solution.

59. $\ln x = 2$

60. $\ln x = 3$

61. $5 \ln 2x = 20$

62. $6 \ln 2x = 30$

63. $6 + 2 \ln x = 5$

64. $7 + 3 \ln x = 6$

65. $\ln \sqrt{x + 3} = 1$

66. $\ln \sqrt{x + 4} = 1$

Application Exercises

Use the formula $R = 6e^{12.77x}$, where x is the blood alcohol concentration and R, given as a percent, is the risk of having a car accident, to solve Exercises 67–68.

67. What blood alcohol concentration corresponds to certainty, or a 100% risk, of a car accident?

68. What blood alcohol concentration corresponds to a 50% risk of a car accident?

69. The function $f(t) = 18.2e^{0.001t}$ models the population of New York State, $f(t)$, in millions, t years after 1994.
  a. What was the population of New York in 1994?

  b. When will the population of New York reach 18.5 million?

70. The function $f(t) = 14e^{0.168t}$ models the population of Florida, $f(t)$, in millions, t years after 1994.
  a. What was the population of Florida in 1994?

  b. When will the population of Florida reach 18.5 million?

In Exercises 71–74, complete the table for a savings account subject to n compoundings yearly $\left(A = P\left(1 + \dfrac{r}{n}\right)^{nt}\right)$. *Round answers to one decimal place.*

	Amount Invested	Number of Compounding Periods	Annual Interest Rate	Accumulated Amount	Time t in Years
71.	$12,500	4	5.75%	$20,000	
72.	$7250	12	6.5%	$15,000	
73.	$1000	360		$1400	2
74.	$5000	360		$9000	4

In Exercises 75–78, complete the table for a savings account subject to continuous compounding $\left(A = Pe^{rt}\right)$. *Round answers to one decimal place.*

	Amount Invested	Annual Interest Rate	Accumulated Amount	Time t in Years
75.	$8000	8%	Double the amount invested	
76.	$8000		$12,000	2
77.	$2350		Triple the amount invested	7
78.	$17,425	4.25%	$25,000	

79. The function $f(x) = 15{,}557 + 5259 \ln x$ models the average cost of a new car, $f(x)$, in dollars, x years after 1989. When was the average cost of a new car $25,000?

80. The function $f(x) = 68.41 + 1.75 \ln x$ models the life expectancy, $f(x)$, in years, for African-American females born x years after 1969. In which birth year was life expectancy 73.7 years? Round to the nearest year.

The function $P(x) = 95 - 30 \log_2 x$ *models the percentage, $P(x)$, of students who could recall the important features of a classroom lecture as a function of time, where x represents the number of days that have elapsed since the lecture was given. The figure shows the graph of the function. Use this information to solve Exercises 81–82. Round answers to one decimal place.*

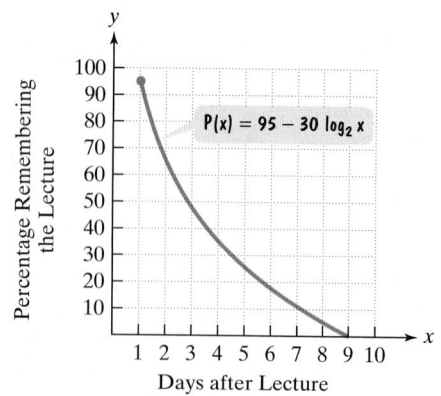

Days after Lecture

81. After how many days do only half the students recall the important features of the classroom lecture? (Let $P(x) = 50$ and solve for x.) Locate the point on the graph that conveys this information.

82. After how many days have all students forgotten the important features of the classroom lecture? (Let $P(x) = 0$ and solve for x.) Locate the point on the graph that conveys this information.

The pH of a solution ranges from 0 to 14. An acid solution has a pH less than 7. Pure water is neutral and has a pH of 7. Normal, unpolluted rain has a pH of about 5.6. The pH of a solution is given by

$$pH = -\log x$$

where x represents the concentration of the hydrogen ions in the solution, in moles per liter. Use the formula to solve Exercises 83–84.

83. An environmental concern involves the destructive effects of acid rain. The most acidic rainfall ever had a pH of 2.4. What was the hydrogen ion concentration? Express the answer as a power of 10, and then round to the nearest thousandth.

84. The figure at the top of the next page shows very acidic rain in the northeast United States. What is the hydrogen ion concentration of rainfall with a pH of 4.2? Express the answer as a power of 10, and then round to the nearest hundred-thousandth.

Acid Rain Over Canada and the United States

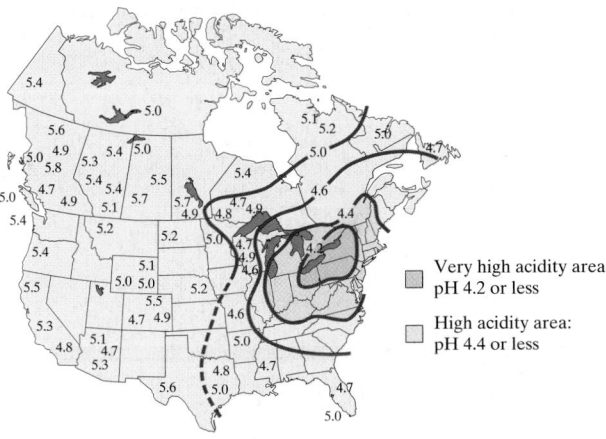

Source: National Atmospheric Program

Very high acidity area:
pH 4.2 or less

High acidity area:
pH 4.4 or less

Writing in Mathematics

85. What is an exponential equation?

86. Explain how to solve an exponential equation when both sides can be written as a power of the same base.

87. Explain how to solve an exponential equation when both sides cannot be written as a power of the same base. Use $3^x = 140$ in your explanation.

88. What is a logarithmic equation?

89. Explain how to solve a logarithmic equation. Use $\log_3(x - 1) = 4$ in your explanation.

90. In many states, a 17% risk of a car accident with a blood alcohol concentration of 0.08 is the lowest level for charging a motorist with driving under the influence. Do you agree with the 17% risk as a cutoff percentage, or do you feel that the percentage should be lower or higher? Explain your answer. What blood alcohol concentration corresponds to what you believe is an appropriate percentage?

91. Have you purchased a new or used car recently? If so, describe if the function in Exercise 79 accurately models what you paid for your car. If there is a big difference between the figure given by the function and the amount that you paid, how can you explain this difference?

Critical Thinking Exercises

92. Which one of the following is true?
   a. If $\log(x + 3) = 2$, then $e^2 = x + 3$.
   b. If $\log(7x + 3) - \log(2x + 5) = 4$, then in exponential form $10^4 = (7x + 3) - (2x + 5)$.
   c. If $x = \dfrac{1}{k} \ln y$, then $y = e^{kx}$.

d. Examples of exponential equations include $10^x = 5.71$, $e^x = 0.72$, and $x^{10} = 5.71$.

93. If $4000 is deposited into an account paying 3% interest compounded annually and at the same time $2000 is deposited into an account paying 5% interest compounded annually, after how long will the two accounts have the same balance? Round to the nearest year.

Solve each equation in Exercises 94–96. Check each proposed solution by direct substitution or with a graphing utility.

94. $(\ln x)^2 = \ln x^2$

95. $(\log x)(2 \log x + 1) = 6$

96. $\ln(\ln x) = 0$

Technology Exercises

In Exercises 97–104, use your graphing utility to graph each side of the equation in the same viewing rectangle. Then use the x-coordinate of the intersection point to find the equation's solution set. Verify this value by direct substitution into the equation.

97. $2^{x+1} = 8$

98. $3^{x+1} = 9$

99. $\log_3(4x - 7) = 2$

100. $\log_3(3x - 2) = 2$

101. $\log(x + 3) + \log x = 1$

102. $\log(x - 15) + \log x = 2$

103. $3^x = 2x + 3$

104. $5^x = 3x + 4$

Hurricanes are one of nature's most destructive forces. These low-pressure areas often have diameters of over 500 miles. The function $f(x) = 0.48 \ln(x + 1) + 27$ models the barometric air pressure, $f(x)$, in inches of mercury, at a distance of x miles from the eye of a hurricane. Use this function to solve Exercises 105–106.

105 Graph the function in a $[0, 500, 50]$ by $[27, 30, 1]$ viewing rectangle. What does the shape of the graph indicate about barometric air pressure as the distance from the eye increases?

106. Use an equation to answer this question: How far from the eye of a hurricane is the barometric air pressure 29 inches of mercury? Use the TRACE and ZOOM features or the intersect command of your graphing utility to verify your answer.

107. The function $P(t) = 145e^{-0.092t}$ models a runner's pulse, $P(t)$, in beats per minute, t minutes after a race, where $0 \le t \le 15$. Graph the function using a graphing utility. $\boxed{\text{TRACE}}$ along the graph and determine after how many minutes the runner's pulse will be 70 beats per minute. Round to the nearest tenth of a minute. Verify your observation algebraically.

108. The function $W(t) = 2600(1 - 0.51e^{-0.075t})^3$ models the weight, $W(t)$, in kilograms, of a female African elephant at age t years. (1 kilogram $\approx$ 2.2 pounds) Use a graphing utility to graph the function. Then $\boxed{\text{TRACE}}$ along the curve to estimate the age of an adult female elephant weighing 1800 kilograms.

Review Exercises

109. Solve: $\sqrt{x + 4} - \sqrt{x - 1} = 1$

(Section 10.6, Example 4)

110. Solve:

$$\frac{3}{x + 1} - \frac{5}{x} = \frac{19}{x^2 + x}.$$

(Section 7.6, Example 4)

111. Multiply: $(x - y)(x^2 + xy + y^2)$.

(Section 5.2, Example 7)

▶ **SECTION 12.6**   *Exponential Growth and Decay; Modeling Data*

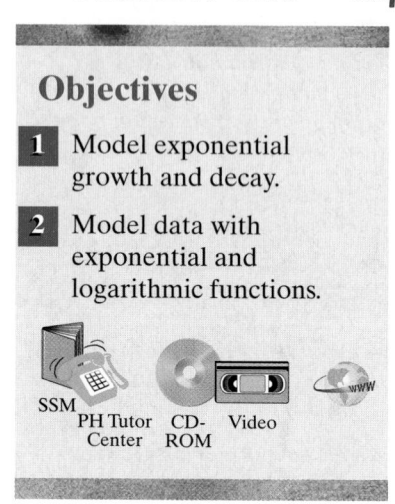

Objectives

1 Model exponential growth and decay.

2 Model data with exponential and logarithmic functions.

SSM   PH Tutor   CD-   Video   www
Center   ROM

The most casual cruise on the Internet shows how people disagree when it comes to making predictions about the effects of the world's growing population. Some argue that there is a recent slowdown in the growth rate, economies remain robust, and famines in Biafra and Ethiopia are aberrations rather than signs of the future. Others say that the 6 billion people on Earth is twice as many as can be supported in middle-class comfort, and the world is running out of arable land and fresh water. Debates about entities that are growing exponentially can be approached mathematically: We can create functions that model data and use these functions to make predictions. In this section we will show you how this is done.

1   Model exponential growth and decay.

Exponential Growth and Decay

One of algebra's many applications is to predict the behavior of variables. This can be done with **exponential growth** and **decay models**. With exponential growth and decay, quantities grow or decay at a rate directly proportional to their size. Populations that are growing exponentially grow extremely rapidly as they get larger because there are more adults to have offspring. For example, the **growth rate** for world population is 1.3%, or 0.013. This means that each year world population is 1.3% more than what it was in the previous year. In 1999, world population was 6 billion. Thus, we compute the world population in 2000 as follows:

$$6 \text{ billion} + 1.3\% \text{ of } 6 \text{ billion} = 6 + (0.013)(6) = 6.078.$$

This computation suggests that 6.078 billion people populated the world in 2000. The 0.078 billion represents an increase of 78 million people from 1999 to 2000, the equivalent of the population of Germany. Using 1.3% as the annual growth rate, world population for 2001 is found similarly:

$$6.078 \text{ billion} + 1.3\% \text{ of } 6.078 \text{ billion} = 6.078 + (0.013)(6.078) \approx 6.157.$$

This computation suggests that approximately 6.157 billion people populated the world in 2001.

The explosive growth of world population may remind you of the growth of money in an account subject to compound interest. Just as the growth rate for world population is multiplied by the population plus any increase in the population, a compound interest rate is multiplied by your original investment plus any accumulated interest. The balance in an account subject to continuous compounding and world population are special cases of an *exponential growth model*.

Exponential Growth and Decay Models

The mathematical model for **exponential growth** or **decay** is given by

$$f(t) = A_0 e^{kt} \quad \text{or} \quad A = A_0 e^{kt}.$$

- **If $k > 0$, the function models the amount or size of a *growing* entity.** A_0 is the original amount or size of the growing entity at time $t = 0$, A is the amount at time t, and k is a constant representing the growth rate.

- **If $k < 0$ the function models the amount or size of a *decaying* entity.** A_0 is the original amount or size of the decaying entity at time $t = 0$, A is the amount at time, t, and k is a constant representing the decay rate.

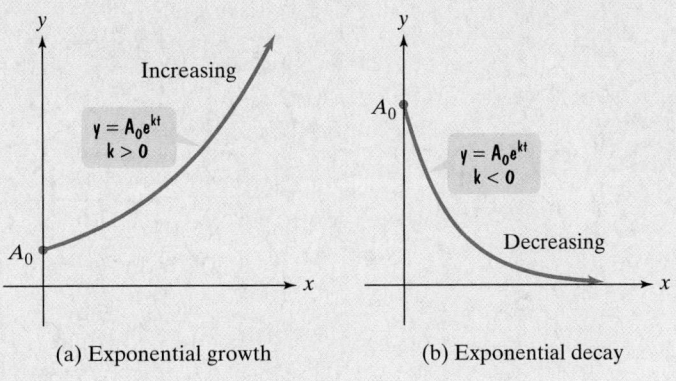

(a) Exponential growth   (b) Exponential decay

Sometimes we need to use given data to determine k, the rate of growth or decay. After we compute the value of k, we can use the formula $A = A_0 e^{kt}$ to make predictions. This idea is illustrated in our first two examples.

EXAMPLE 1   Modeling the Growth of the Minimum Wage

The graph in Figure 12.20 shows the growth of the minimum wage from 1970 through 2000. In 1970, the minimum wage was $1.60 per hour. By 2000, it had grown to $5.15 per hour.

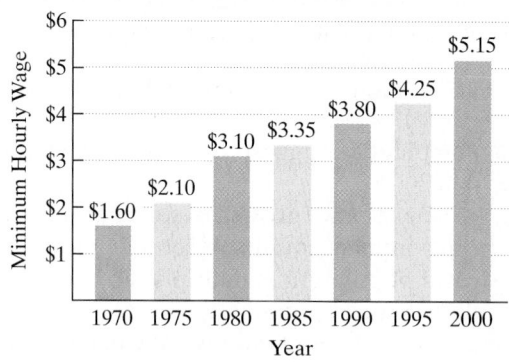

Federal Minimum Wages, 1970–2000

Figure 12.20
Source: U.S. Employment Standards Administration

a. Find the exponential growth function that models the data for 1970 through 2000.

b. By which year will the minimum wage reach $7.50 per hour?

Solution

a. We use the exponential growth model

$$A = A_0 e^{kt}$$

in which t is the number of years after 1970. This means that 1970 corresponds to $t = 0$. At that time the minimum wage was $1.60, so we substitute 1.6 for A_0 in the growth model:

$$A = 1.6 e^{kt}.$$

We are given that $5.15 is the minimum wage in 2000. Because 2000 is 30 years after 1970, when $t = 30$ the value of A is 5.15. Substituting these numbers into the growth model will enable us to find k, the growth rate. We know that $k > 0$ because the problem involves growth.

$A = 1.6 e^{kt}$	Use the growth model with $A_0 = 1.6$.
$5.15 = 1.6 e^{k \cdot 30}$	When $t = 30, A = 5.15$. Substitute these numbers into the model.
$e^{30k} = \dfrac{5.15}{1.6}$	Isolate the exponential factor by dividing both sides by 1.6. We also reversed the sides.
$\ln e^{30k} = \ln \dfrac{5.15}{1.6}$	Take the natural logarithm on both sides.
$30k = \ln \dfrac{5.15}{1.6}$	Simplify the left side using $\ln e^x = x$.
$k = \dfrac{\ln \dfrac{5.16}{1.6}}{30} \approx 0.039$	

We substitute 0.039 for k in the growth model to obtain the exponential growth function for the minimum wage. It is

$$A = 1.6 e^{0.039t}$$

where t is measured in years after 1970.

b. To find the year in which the minimum wage will reach $7.50 per hour, we substitute 7.5 for A in the model from part (a) and solve for t.

$$A = 1.6e^{0.039t}$$ This is the model from part (a).

$$7.5 = 1.6e^{0.039t}$$ Substitute 7.5 for A.

$$e^{0.039t} = \frac{7.5}{1.6}$$ Divide both sides by 1.6.

$$\ln e^{0.039t} = \ln \frac{7.5}{1.6}$$ Take the natural logarithm on both sides.

$$0.039t = \ln \frac{7.5}{1.6}$$ Simplify on the left using $\ln e^x = x$.

$$t = \frac{\ln \dfrac{7.5}{1.6}}{0.039} \approx 40$$ Solve for t by dividing both sides by 0.039.

Because 40 is the number of years after 1970, the model indicates that the minimum wage will reach $7.50 by 1970 + 40, or in the year 2010. ∎

ENRICHMENT ESSAY

Lying with Statistics

Benjamin Disraeli, Queen Victoria's prime minister, stated that there are "lies, damned lies, and statistics." The problem is not that data lie, but rather that liars use data. For example, the data in Example 1 create the impression that wages are on the rise and workers are better off each year. The graph in Figure 12.21 is more effective in creating an accurate picture. Why? It is adjusted for inflation and measured in constant 1996 dollars. Something else to think about: In predicting a minimum wage of $7.50 by 2010, are we using the best possible model for the data? We return to this issue in the exercise set.

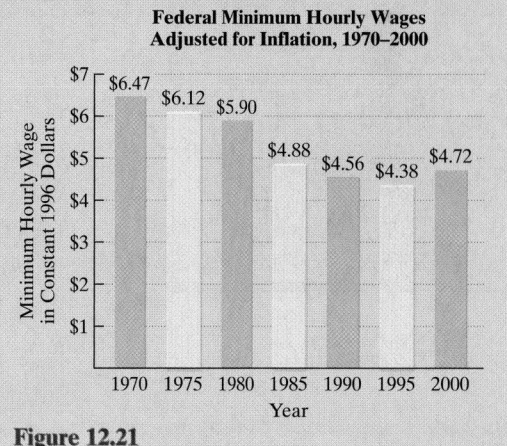

Figure 12.21

Source: U.S. Employment Standards Administration

✔ **CHECK POINT 1** In 1980, the population of Africa was 491 million and by 1990 it had grown to 643 million.

a. Use the exponential growth model $A = A_0 e^{kt}$, in which t is the number of years after 1980, to find the exponential growth function that models the data.

b. By which year will Africa's population reach 1000 million, or one billion?

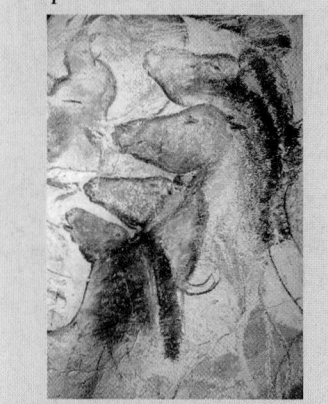
Our next example involves exponential decay and its use in determining the age of fossils and artifacts. The method is based on considering the percentage of carbon-14 remaining in the fossil or artifact. Carbon-14 decays exponentially with a *half-life* of approximately 5715 years. The **half-life** of a substance is the time required for half of a given sample to disintegrate. Thus, after 5715 years a given amount of carbon-14 will have decayed to half the original amount. Carbon dating is useful for artifacts or fossils up to 80,000 years old. Older objects do not have enough carbon-14 left to date age accurately.

EXAMPLE 2   Carbon-14 Dating: The Dead Sea Scrolls

a. Use the fact that after 5715 years a given amount of carbon-14 will have decayed to half the original amount to find the exponential decay model for carbon-14.

b. In 1947, earthenware jars containing what are known as the Dead Sea Scrolls were found by an Arab Bedouin herdsman. Analysis indicated that the scroll wrappings contained 76% of their original carbon-14. Estimate the age of the Dead Sea Scrolls.

Solution   We begin with the exponential decay model $A = A_0 e^{kt}$. We know that $k < 0$ because the problem involves the decay of carbon-14. After 5715 years ($t = 5715$), the amount of carbon-14 present, A, is half the original amount, A_0. Thus, we can substitute $\dfrac{A_0}{2}$ for A in the exponential decay model. This will enable us to find k, the decay rate.

a.

$A = A_0 e^{kt}$ Begin with the exponential decay model.

$\dfrac{A_0}{2} = A_0 e^{k5715}$ After 5715 years ($t = 5715$), $A = \dfrac{A_0}{2}$ (because the amount present, A, is half the original amount, A_0).

$\dfrac{1}{2} = e^{5715k}$ Divide both sides of the equation by A_0.

$\ln \dfrac{1}{2} = \ln e^{5715k}$ Take the natural logarithm on both sides.

$\ln \dfrac{1}{2} = 5715k$ $\ln e^x = x$

$k = \dfrac{\ln \dfrac{1}{2}}{5715} \approx -0.000121$ Solve for k.

Substituting for k in the decay model, the model for carbon-14 is $A = A_0 e^{-0.000121t}$.

b.

$A = A_0 e^{-0.000121t}$ This is the decay model for carbon-14.

$0.76A_0 = A_0 e^{-0.000121t}$ A, the amount present, is 76% of the original amount, so $A = 0.76A_0$.

$0.76 = e^{-0.000121t}$ Divide both sides of the equation by A_0.

$\ln 0.76 = \ln e^{-0.000121t}$ Take the natural logarithm on both sides.

$\ln 0.76 = -0.000121t$ $\ln e^x = x$

$t = \dfrac{\ln 0.76}{-0.000121} \approx 2268$ Solve for t.

The Dead Sea Scrolls are approximately 2268 years old plus the number of years between 1947 and the current year. ∎

✔ **CHECK POINT 2** Strontium-90 is a waste product from nuclear reactors. As a consequence of fallout from atmospheric nuclear tests, we all have a measurable amount of strontium-90 in our bones.

    a. Use the fact that after 28 years a given amount of strontium-90 will have decayed to half the original amount to find the exponential decay model for strontium-90.

    b. Suppose that a nuclear accident occurs and releases 60 grams of strontium-90 into the atmosphere. How long will it take for strontium-90 to decay to a level of 10 grams?

The Art of Modeling

2 Model data with exponential and logarithmic functions.

Throughout this chapter, we have been working with models that were given. However, we can create functions that model data by observing patterns in scatter plots. Figure 12.22 shows scatter plots for data that are exponential or logarithmic.

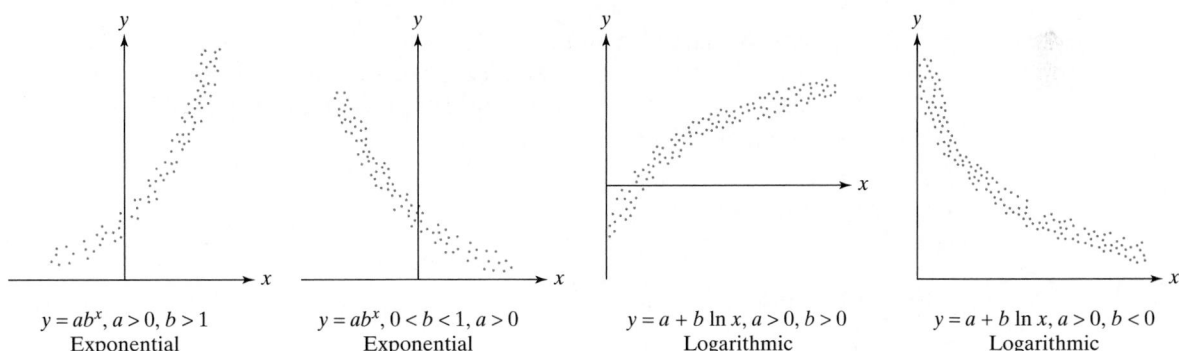

| $y = ab^x, a > 0, b > 1$ Exponential | $y = ab^x, 0 < b < 1, a > 0$ Exponential | $y = a + b \ln x, a > 0, b > 0$ Logarithmic | $y = a + b \ln x, a > 0, b < 0$ Logarithmic |

Figure 12.22 Scatter plots for exponential or logarithmic models

Graphing utilities can be used to find the equation of a function that is derived from data. For example, earlier in the chapter we encountered a function that modeled the size of a city and the average walking speed, in feet per second, of pedestrians. The function was derived from the data in Table 12.1. The scatter plot is shown in Figure 12.23.

Table 12.1

x, Population (thousands)	y, Walking Speed (feet per second)
5.5	3.3
14	3.7
71	4.3
138	4.4
342	4.8

Source: Mark and Helen Bornstein, "The Pace of Life"

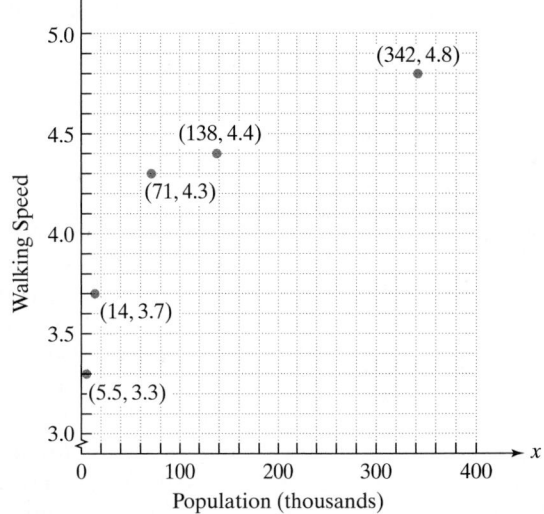

Figure 12.23 Scatter plot for data in Table 12.1

Because the data in this scatter plot increase rapidly at first and then begin to level off a bit, the shape suggests that a logarithmic model might be a good choice.

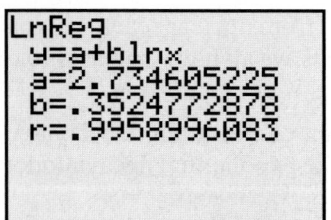

Figure 12.24 A logarithmic model for the data in Table 12.1 on page 859

A graphing utility fits the data in Table 12.1 on the previous page to a logarithmic model of the form $y = a + b \ln x$ by using the logarithmic REGression option (see Figure 12.24). From the figure, we see that the logarithmic model of the data, with numbers rounded to three decimal places, is

$$y = 2.735 + 0.352 \ln x.$$

The number r that appears in Figure 12.24 is called the **correlation coefficient** and is a measure of how well the model fits the data. The value of r is such that $-1 \leq r \leq 1$. A positive r means that as the x-values increase, so do the y-values. A negative r means that as the x-values increase, the y-values decrease. **The closer that r is to -1 or 1, the better the model fits the data.** Because r is approximately 0.996, the model

$$y = 2.735 + 0.352 \ln x$$

fits the data very well.

Now let's look at data whose scatter plot suggests an exponential model. The data in Table 12.2 indicate world population for six years. The scatter plot is shown in Figure 12.25.

Table 12.2

x, Year	y, World Population (billions)
1950	2.6
1960	3.1
1970	3.7
1980	4.5
1989	5.3
1999	6.0

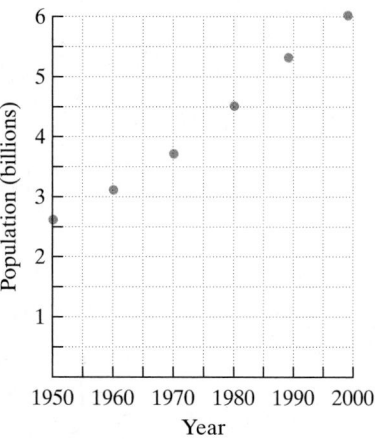

Figure 12.25 A scatter plot for the data in Table 9.2

Because the data in this scatter plot have a rapidly increasing pattern, the shape suggests that an exponential model might be a good choice. (You might also want to try a linear model.) If you select the exponential option, you will use a graphing utility's Exponential REGression option. With this feature, a graphing utility fits the data to an exponential model of the form $y = ab^x$.

When computing an exponential model of the form $y = ab^x$, a graphing utility rewrites the equation using logarithms. Because the domain of the logarithmic function is the set of positive numbers, **zero must not be a value for x.** What does this mean for our data for world population that starts in the year 1950? We must start values of x after 0. Thus, we'll assign x to represent the number of years after 1949. This gives us the data shown in Table 12.3. Using the Exponential REGression option, we obtain the equation in Figure 12.26.

Table 12.3

x, Numbers of Years after 1949		*y*, World Population (billions)
1	(1950)	2.6
11	(1960)	3.1
21	(1970)	3.7
31	(1980)	4.5
40	(1989)	5.3
50	(1999)	6.0

Figure 12.26 An exponential model for the data in Table 12.3

From Figure 12.26, we see that the exponential model of the data for world population *x* years after 1949, with numbers rounded to three decimal places, is

$$y = 2.570(1.018)^x.$$

The correlation coefficient, *r*, is close to 1, indicating that the model fits the data very well.

When using a graphing utility to model data, begin with a scatter plot, drawn either by hand or with the graphing utility, to obtain a general picture for the shape of the data. It might be difficult to determine what model best fits the data—linear, logarithmic, exponential, quadratic, or something else. If necessary, use your graphing utility to fit several models to the data. The best model is the one that yields the value *r*, the correlation coefficient, closest to 1 or −1. Finding a proper fit for data can be almost as much art as it is mathematics. In this era of technology, the process of creating models that best fit data is one that involves more decision making than computation.

ENRICHMENT ESSAY

The Future of World Population

In the future, a new world order looms. The percentage of world population in less developed countries will increase, North America's will remain relatively stable, and Europe's will decrease.

Most Populous Countries, 2050

1. India	1529 million
2. China	1478 million
3. U.S.	349 million
4. Pakistan	345 million
5. Indonesia	312 million
6. Nigeria	244 million
7. Brazil	244 million
8. Bangladesh	212 million
9. Ethiopia	169 million
10. Congo	160 million

Source: United Nations Population Fund

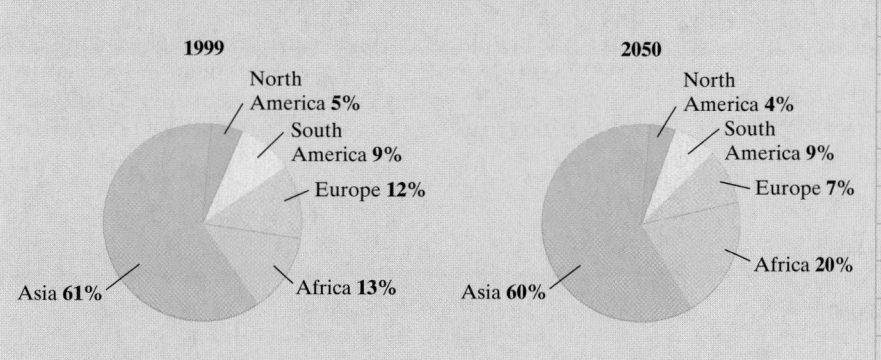

EXERCISE SET 12.6

Practice and Application Exercises

The exponential growth model $A = 208e^{0.008t}$ describes the population of the United States, in millions, t years after 1970. Use this model to solve Exercises 1–4.

1. What was the population of the United States in 1970?

2. By what percentage is the population of the United States increasing each year?

3. When will the U.S. population be 300 million?

4. When will the U.S. population be 350 million?

India is currently one of the world's fastest-growing countries. By 2040, the population of India will be larger than the population of China; by 2050, nearly one-third of the world's population will live in these two countries alone. The exponential growth model $A = 574e^{0.026t}$ describes the population of India, in millions, t years after 1974. Use this model to solve Exercises 5–8.

5. By what percentage is the population of India increasing each year?

6. What was the population of India in 1974?

7. When will India's population be 1624 million?

8. When will India's population be 2732 million?

The value of houses in a neighborhood follows a pattern of exponential growth. In the year 2000, you purchased a house in this neighborhood. The value of your house, in thousands of dollars, t years after 2000 is given by the exponential growth model $V = 140e^{0.068t}$. Use this model to solve Exercises 9–12.

9. What did you pay for your house?

10. By what percentage is the price of houses in your neighborhood increasing each year?

11. When will your house be worth $200,000?

12. When will your house be worth $300,000

13. Through the end of 1991, 200,000 cases of AIDS had been reported to the Centers for Disease Control in the United States. By the end of 1998, the number had grown to 680,000. The exponential growth function $A = 200e^{kt}$ describes the thousands of AIDS cases in the United States t years after 1991. Use the fact that 7 years after 1991 there were 680 thousand cases to find k to three decimal places. Then write the exponential growth function. According to your model, by what percentage is the number of AIDS cases in the United States increasing each year?

14. In 1980, China's population was 983 million; in 1990, it was 1154 million. The exponential growth function $A = 983e^{kt}$ describes the population of China, in millions, t years after 1980. Use the fact that 10 years after 1980 the population was 1154 million to find k to three decimal places. Then write the exponential growth function. According to your model, by what percentage is the population of China increasing each year?

An artifact originally had 16 grams of carbon-14 present. The decay model $A = 16e^{-0.000121t}$ describes the amount of carbon-14 present after t years. Use this model to solve Exercises 15–16.

15. How many grams of carbon-14 will be present in 5715 years?

16. How many grams of carbon-14 will be present in 11,430 years?

17. The half-life of the radioactive element krypton-91 is 10 seconds. If 16 grams of krypton-91 are initially present, how many grams are present after 10 seconds? 20 seconds? 30 seconds? 40 seconds? 50 seconds?

18. The half-life of the radioactive element plutonium-239 is 25,000 years. If 16 grams of plutonium-239 are initially present, how many grams are present after 25,000 years? 50,000 years? 75,000 years? 100,000 years? 125,000 years?

Use the exponential decay model for carbon-14, $A = A_0e^{-0.000121t}$, to solve Exercises 19–20.

19. Prehistoric cave paintings were discovered in a cave in France. The paint contained 15% of the original carbon-14. Estimate the age of the paintings.

20. Skeletons were found at a construction site in San Francisco in 1989. The skeletons contained 88% of the expected amount of carbon-14 found in a living person. In 1989, how old were the skeletons?

21. The August 1978 issue of *National Geographic* described the 1964 find of dinosaur bones of a newly discovered dinosaur weighing 170 pounds, measuring 9 feet, with a 6-inch claw on one toe of each hind foot. The age of the dinosaur was estimated using potassium-40 dating of rocks surrounding the bones.
   a. Potassium-40 decays exponentially with a half-life of approximately 1.31 billion years. Use the fact that after 1.31 billion years a given amount of potassium-40 will have decayed to half the original amount to show that the decay model for potassium-40 is given by $A = A_0 e^{-0.52912t}$, where t is in billions of years.

   b. Analysis of the rocks surrounding the dinosaur bones indicated that 94.5% of the original amount of potassium-40 was still present. Let $A = 0.945 A_0$ in the model in part (a) and estimate the age of the bones of the dinosaur.

22. A bird species in danger of extinction has a population that is decreasing exponentially ($A = A_0 e^{kt}$). Five years ago the population was at 1400 and today only 1000 of the birds are alive. Once the population drops below 100, the situation will be irreversible. When will this happen?

23. Use the exponential growth model, $A = A_0 e^{kt}$, to show that the time it takes a population to double (to grow from A_0 to $2A_0$) is given by $t = \dfrac{\ln 2}{k}$.

24. Use the exponential growth model, $A = A_0 e^{kt}$, to show that the time it takes a population to triple (to grow from A_0 to $3A_0$) is given by $t = \dfrac{\ln 3}{k}$.

Use the formula $t = \dfrac{\ln 2}{k}$ that gives the time for a population with a growth rate k to double to solve Exercises 25–26. Express each answer to the nearest whole year.

25. China is growing at a rate of 1.1% per year. How long will it take China to double its population?

26. Japan is growing at a rate of 0.3% per year. How long will it take Japan to double its population?

Writing in Mathematics

27. Nigeria has a growth rate of 0.031 or 3.1%. Describe what this means.

28. How can you tell if an exponential model describes exponential growth or exponential decay?

29. Suppose that a population that is growing exponentially increases from 800,000 people in 1997 to 1,000,000 people in 2000. Without showing the details, describe how to obtain the exponential growth function that models this data.

30. What is the half-life of a substance?

31. Describe the shape of a scatter plot that suggests modeling the data with an exponential function.

32. You take up weightlifting and record the maximum number of pounds you can lift at the end of each week. You start off with rapid growth in terms of the weight you can lift from week to week, but then the growth begins to level off. Describe how to obtain a function that models the number of pounds you can lift at the end of each week. How can you use this function to predict what might happen if you continue the sport?

33. Would you prefer that your salary be modeled exponentially or logarithmically? Explain your answer.

34. One problem with all exponential growth models is that nothing can grow exponentially forever. Describe factors that might limit the size of a population.

Critical Thinking Exercises

35. The World Health Organization makes predictions about the number of AIDS cases based on a compromise between a linear model and an exponential growth model. Explain why the World Health Organization does this.

36. a. Show that $y = ab^x$ is equivalent to $y = ae^{(\ln b) \cdot x}$. *Hint*: Use the fact that $b = e^{\ln b}$.

   b. In this section, we used the Exponential REGression feature of a graphing utility and obtained an exponential model for world population. World population, y, in billions, x years after 1949 is modeled by

$$y = 2.57(1.018)^x.$$

Use the equivalence from part (a) and express the model as

$$y = 2.57e^{0.018x}.$$

c. Express the model for world population in part (b) so that it has the same letters as those in the exponential growth model $A = A_0e^{kt}$. According to the model, what is the growth rate for world population?

d. The actual growth rate for world population is now 1.3%. Describe what might happen if you use the model in part (c) to make projections about world population in the future.

Technology Exercises

In Example 1 on page 856, we used two data points and an exponential function to model federal minimum wages that were not adjusted for inflation from 1970 through 2000. The data are shown again in the table. Use all seven data points to solve Exercises 37–40.

x, Number of Years after 1969	y, Federal Minimum Wage
1	1.60
6	2.10
11	3.10
16	3.35
21	3.80
26	4.25
31	5.15

37. Use your graphing utility's Exponential REGression option to obtain a model of the form $y = ab^x$ that fits the data. How well does the correlation coefficient, r, indicate that the model fits the data?

38. Use your graphing utility's logarithmic REGression option to obtain a model of the form $y = a + b \ln x$ that fits the data. How well does the correlation coefficient, r, indicate that the model fits the data?

39. Use your graphing utility's Linear REGression option to obtain a model of the form $y = ax + b$ that fits the data. How well does the correlation coefficient, r, indicate that the model fits the data?

40. Use your graphing utility's Power REGression option to obtain a model of the form $y = ax^b$ that fits the data.

How well does the correlation coefficient, r, indicate that the model fits the data?

41. Use the values of r in Exercises 37–40 to select the model of best fit. Use this model to predict by which year the minimum wage will reach $7.50. How does this answer compare to the year we found in Example 1, namely 2010? If you obtained a different year, how do you account for this difference?

42. The figure shows the number of people in the United States age 65 and over, with projected figures for the year 2000 and beyond.

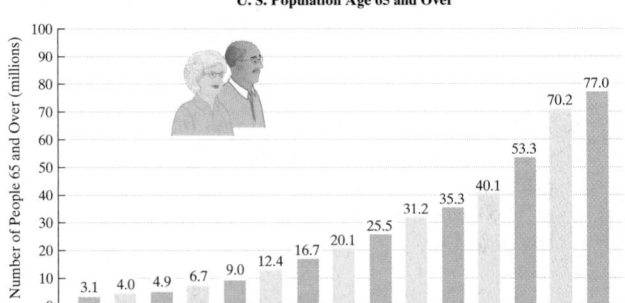

U. S. Population Age 65 and Over

Source: U. S. Bureau of the Census

Let x represent the number of years after 1899 and let y represent the U.S. population, in millions. Use your graphing utility to find the model that best fits the data in the bar graph. Then use the model to find the U.S. population age 65 and over in 2050.

Review Exercises

43. Divide:

$$\frac{x^2 - 9}{2x^2 + 7x + 3} \div \frac{x^2 - 3x}{2x^2 + 11x + 5}.$$

(Section 7.2, Example 6)

44. Solve: $x^{\frac{2}{3}} + 2x^{\frac{1}{3}} - 3 = 0$.
(Section 11.4, Example 5)

45. Simplify: $6\sqrt{2} - 2\sqrt{50} + 3\sqrt{98}$.
(Section 10.4, Example 2)

CHAPTER 12 GROUP PROJECTS

1. This group exercise involves exploring the way we grow. Group members should create a graph for the function that models the percentage of adult height attained by a boy who is x years old, $f(x) = 29 + 48.8 \log(x + 1)$. Let $x = 1, 2, 3, \ldots, 12$, find function values, and connect the resulting points with a smooth curve. Then create a graph for the function that models the percentage of adult height attained by a girl who is x years old, $g(x) = 62 + 35 \log(x - 4)$. Let $x = 5, 6, 7, \ldots, 15$, find function values, and connect the resulting points by a smooth curve. Group members should then discuss similarities and differences in the growth patterns for boys and girls based on the graphs.

2. Research applications of logarithmic functions as mathematical models and plan a seminar based on your group's research. Each group member should research one of the following areas or any other area of interest: pH (acidity of solutions), intensity of sound (decibels), brightness of stars, consumption of natural resources, human memory, progress over time in a sport, profit over time. For the area that you select, explain how logarithmic functions are used and provide examples.

3. This activity is intended for three or four people who would like to take up weightlifting. Each person in the group should record the maximum number of pounds that he or she can lift at the end of each week for the first 10 consecutive weeks. Use the logarithmic REGression option of a graphing utility to obtain a model showing the amount of weight that group members can lift from week 1 through week 10. Graph each of the models in the same viewing rectangle to observe similarities and differences among weight-growth patterns of each member. Use the functions to predict the amount of weight that group members will be able to lift in the future. If the group continues to work out together, check the accuracy of these predictions.

4. Each group member should consult an almanac, newspaper, magazine, or the Internet to find data that can be modeled by exponential or logarithmic functions. Group members should select the two sets of data that are most interesting and relevant. For each data set selected, find a model that best fits the data. Each group member should make one prediction based on the model and then discuss a consequence of this prediction. What factors might change the accuracy of each prediction?

5. In Tom Stoppard's play *Arcadia*, the characters dream and talk about mathematics, including ideas involving graphing, composite functions, symmetry, and lack of symmetry in things that are tangled, mysterious, and unpredictable. Group members should rent and view the movie. Present a report on the ideas discussed by the characters that are related to concepts involving functions, their graphs, and composite functions. Bring in a copy of the video and show appropriate excerpts.

CHAPTER SUMMARY, REVIEW, AND TEST

SUMMARY

DEFINITIONS AND CONCEPTS	EXAMPLES

Section 12.1 Exponential Functions

The exponential function with base b is defined by $f(x) = b^x$, where $b > 0$ and $b \neq 1$. The graph contains the point $(0, 1)$. When $b > 1$, the graph rises from left to right. When $0 < b < 1$, the graph falls from left to right. The x-axis is a horizontal asymptote. The domain is $(-\infty, \infty)$; the range is $(0, \infty)$. The natural exponential function is $f(x) = e^x$, $e \approx 2.71828$.

Graph $f(x) = 2^x$ and $g(x) = 2^{x-1}$.

x	$f(x) = 2^x$	$g(x) = 2^{x-1}$
-2	$2^{-2} = \frac{1}{4}$	$2^{-3} = \frac{1}{8}$
-1	$2^{-1} = \frac{1}{2}$	$2^{-2} = \frac{1}{4}$
0	$2^0 = 1$	$2^{-1} = \frac{1}{2}$
1	$2^1 = 2$	$2^0 = 1$
2	$2^2 = 4$	$2^1 = 2$

The graph of g is the graph of f shifted one unit to the right.

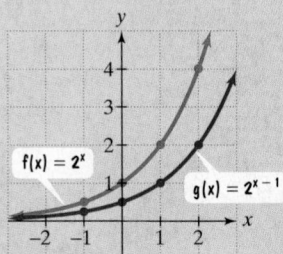

Formulas for Compound Interest
After t years, the balance, A, in an account with principal P and annual interest rate r is given by the following formulas:

1. For n compoundings per year: $A = P\left(1 + \dfrac{r}{n}\right)^{nt}$

2. For continuous compounding: $A = Pe^{rt}$

Select the better investment for $4000 over 6 years:

• 6% compounded semiannually
$$A = P\left(1 + \frac{r}{n}\right)^{nt}$$
$$= 4000\left(1 + \frac{0.06}{2}\right)^{2 \cdot 6} \approx \$5703$$

• 5.9% compounded continuously
$$A = Pe^{rt} = 4000e^{0.059(6)} \approx \$5699$$
The first investment is better.

Section 12.2 Composite and Inverse Functions

Composite Functions
The composite function $f \circ g$ is defined by
$$(f \circ g)(x) = f(g(x)).$$
The composite function $g \circ f$ is defined by
$$(g \circ f)(x) = g(f(x)).$$

Let $f(x) = x^2 + x$ and $g(x) = 2x + 1$.

• $(f \circ g)(x) = f(g(x)) = (g(x))^2 + g(x)$

Replace x by g(x).

$$= (2x + 1)^2 + (2x + 1) = 4x^2 + 4x + 1 + 2x + 1$$
$$= 4x^2 + 6x + 2$$

• $(g \circ f)(x) = g(f(x)) = 2f(x) + 1$

Replace x by f(x).

$$= 2(x^2 + x) + 1 = 2x^2 + 2x + 1$$

DEFINITIONS AND CONCEPTS	EXAMPLES

Section 12.2 Composite and Inverse Functions (continued)

Inverse Functions

If $f(g(x)) = x$ and $g(f(x)) = x$, function g is the inverse of function f, denoted f^{-1} and read "f inverse." The procedure for finding a function's inverse uses a switch-and-solve strategy. Switch x and y, then solve for y.

If $f(x) = 2x - 5$, find $f^{-1}(x)$.

$$y = 2x - 5 \qquad \text{Replace } f(x) \text{ with } y.$$

$$x = 2y - 5 \qquad \text{Exchange } x \text{ and } y.$$

$$x + 5 = 2y \qquad \text{Solve for } y.$$

$$\frac{x + 5}{2} = y$$

$$f^{-1}(x) = \frac{x + 5}{2} \qquad \text{Replace } y \text{ with } f^{-1}(x).$$

The Horizontal Line Test for Inverse Functions

A function, f, has an inverse that is a function, f^{-1}, if there is no horizontal line that intersects the graph of f at more than one point. A one-to-one function is one in which no two different ordered pairs have the same second component. Only one-to-one functions have inverse functions. If the point (a, b) is on the graph of f, then the point (b, a) is on the graph of f^{-1}. The graph of f^{-1} is a reflection of the graph of f about the line $y = x$.

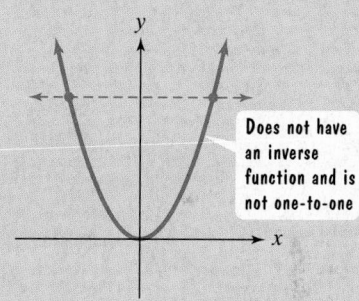

Does not have an inverse function and is not one-to-one

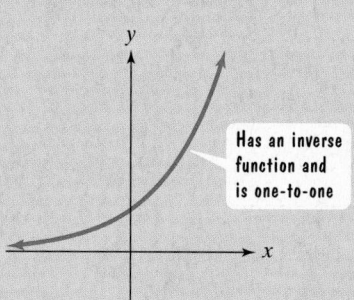

Has an inverse function and is one-to-one

Section 12.3 Logarithmic Functions

Definition of the logarithmic function: For $x > 0$ and $b > 0, b \neq 1$, $y = \log_b x$ is equivalent to $b^y = x$. The function $f(x) = \log_b x$ is the logarithmic function with base b. This function is the inverse function of the exponential function with base b.

- Write $\log_2 32 = 5$ in exponential form.

$$2^5 = 32 \qquad y = \log_b x \text{ means } b^y = x.$$

- Write $\sqrt{49} = 7$, or $49^{\frac{1}{2}} = 7$, in logarithmic form.

$$\frac{1}{2} = \log_{49} 7 \qquad b^y = x \text{ means } y = \log_b x.$$

DEFINITIONS AND CONCEPTS	EXAMPLES

Section 12.3 Logarithmic Functions (continued)

The graph of $f(x) = \log_b x$ can be obtained from $f(x) = b^x$ by reversing coordinates. The graph of $f(x) = \log_b x$ contains the point $(1, 0)$. If $b > 1$, the graph rises from left to right. If $0 < b < 1$, the graph falls from left to right. The y-axis is a vertical asymptote. The domain is $(0, \infty)$; the range is $(-\infty, \infty)$. $f(x) = \log x$ means $f(x) = \log_{10} x$ and is the common logarithmic function. $f(x) = \ln x$ means $f(x) = \log_e x$ and is the natural logarithmic function. The domain of $f(x) = \log_b(x + c)$ consists of all x for which $x + c > 0$.

- Graph $f(x) = \log_3 x$.

x	-1	0	1	2
$y = 3^x$	$\frac{1}{3}$	1	3	9

x	$\frac{1}{3}$	1	3	9
$f(x) = \log_3 x$	-1	0	1	2

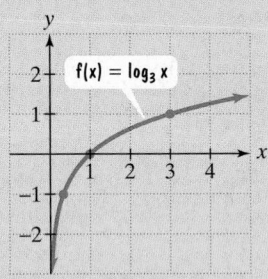

- Find the domain: $f(x) = \log_6(4 - x)$.

$$4 - x > 0$$
$$4 > x \quad (\text{or } x < 4)$$

The domain is $\{x | x < 4\}$ or $(-\infty, 4)$.

Basic Logarithmic Properties

Base b ($b > 0, b \neq 1$)	Base 10 (Common Logarithms)	Base e (Natural Logarithms)
$\log_b 1 = 0$	$\log 1 = 0$	$\ln 1 = 0$
$\log_b b = 1$	$\log 10 = 1$	$\ln e = 1$
$\log_b b^x = x$	$\log 10^x = x$	$\ln e^x = x$
$b^{\log_b x} = x$	$10^{\log x} = x$	$e^{\ln x} = x$

- $\log_8 1 = 0$ because $\log_b 1 = 0$.
- $\log_4 4 = 1$ because $\log_b b = 1$.
- $\ln e^{8x} = 8x$ because $\ln e^x = x$.
- $e^{\ln \sqrt[3]{x}} = \sqrt[3]{x}$ because $e^{\ln x} = x$.
- $\log_t t^{25} = 25$ because $\log_b b^x = x$.

Section 12.4 Properties of Logarithms

Properties of Logarithms

1. *The Product Rule*: $\log_b(MN) = \log_b M + \log_b N$
2. *The Quotient Rule*: $\log_b\left(\dfrac{M}{N}\right) = \log_b M - \log_b N$
3. *The Power Rule*: $\log_b M^p = p \log_b M$
4. *The Change-of Base Property*:

The General Property	Introducing Common Logarithms	Introducing Natural Logarithms
$\log_b M = \dfrac{\log_a M}{\log_a b}$	$\log_b M = \dfrac{\log M}{\log b}$	$\log_b M = \dfrac{\ln M}{\ln b}$

- Expand: $\log_3(81x^7)$.

$$= \log_3 81 + \log_3 x^7$$
$$= 4 + 7 \log_3 x$$

- Write as a single logarithm:

$$7 \ln x - 4 \ln y.$$

$$= \ln x^7 - \ln y^4 = \ln \frac{x^7}{y^4}$$

- Evaluate: $\log_6 92$.

$$\log_6 92 = \frac{\ln 92}{\ln 6} \approx 2.5237$$

DEFINITIONS AND CONCEPTS	**EXAMPLES**

Section 12.5 Exponential and Logarithmic Equations

An exponential equation is an equation containing a variable in an exponent. Some exponential equations can be solved by expressing both sides as a power of the same base. Then set the exponents equal to each other.	Solve: $4^{2x-1} = 64$. $$4^{2x-1} = 4^3$$ $$2x - 1 = 3$$ $$2x = 4$$ $$x = 2$$ The solution is 2 and the solution set is $\{2\}$.
If both sides of an exponential equation cannot be expressed as a power of the same base, isolate the exponential expression and take the natural logarithm on both sides. Simplify using $$\ln b^x = x \ln b \quad \text{or} \quad \ln e^x = x.$$	Solve: $7^x = 103$. $$\ln 7^x = \ln 103$$ $$x \ln 7 = \ln 103$$ $$x = \frac{\ln 103}{\ln 7}$$ The solution is $\dfrac{\ln 103}{\ln 7}$ and the solution set is $\left\{\dfrac{\ln 103}{\ln 7}\right\}$.
A logarithmic equation is an equation containing a variable in a logarithmic expression. Logarithmic equations in the form $\log_b x = c$ can be solved by rewriting as $b^c = x$. When checking logarithmic equations, reject proposed solutions that produce the logarithm of a negative number or the logarithm of zero in the original equation.	Solve: $\log_2(3x - 1) = 5$. $$2^5 = 3x - 1$$ $$32 = 3x - 1$$ $$33 = 3x$$ $$11 = x$$ The solution is 11 and the solution set is $\{11\}$.
Equations involving natural logarithms are solved by isolating the natural logarithm with coefficient 1 on one side and exponentiating both sides. Simplify using $e^{\ln x} = x$.	Solve: $3 \ln 2x = 15$. $$\ln 2x = 5$$ $$e^{\ln 2x} = e^5$$ $$2x = e^5$$ $$x = \frac{e^5}{2}$$ The solution is $\dfrac{e^5}{2}$ and the solution set is $\left\{\dfrac{e^5}{2}\right\}$.

Section 12.6 Exponential Growth and Decay; Modeling Data

Exponential growth and decay models are given by $A = A_0 e^{kt}$ in which t represents time, A_0 is the amount present at $t = 0$, and A is the amount present at time t. If $k > 0$, the model describes growth and k is the growth rate. If $k < 0$, the model describes decay and k is the decay rate. Scatter plots for exponential and logarithmic models are shown in Figure 12.22 on page 859. When using a graphing utility to model data, the closer that the correlation coefficient r is to -1 or 1, the better the model fits the data.	The 1970 population of Mexico City was 9.4 million; in 1990, it was 20.2 million. Write the exponential growth function that describes the population, in millions, t years after 1970. Begin with $A = A_0 e^{kt}$. $A = 9.4e^{kt}$ In 1970 ($t = 0$), population was 9.4 million. $20.2 = 9.4e^{k \cdot 20}$ When $t = 20$ (in 1990), $A = 20.2$. $$e^{20k} = \frac{20.2}{9.4}$$ $$\ln e^{20k} = \ln \frac{20.2}{9.4}$$ $$20k = \ln \frac{20.2}{9.4} \quad \text{and} \quad k = \frac{\ln \frac{20.2}{9.4}}{20} \approx 0.038$$ Growth function is $A = 9.4e^{0.038t}$. Growth rate is 0.038 or 3.8%.

Review Exercises

12.1 *In Exercises 1–4, set up a table of coordinates for each function. Select integers from −2 to 2 for x. Then use the table of coordinates to match the function with its graph. [The graphs are labeled (a) through (d).]*

1. $f(x) = 4^x$ **2.** $f(x) = 4^{-x}$
3. $f(x) = -4^{-x}$ **4.** $f(x) = -4^{-x} + 3$

a.
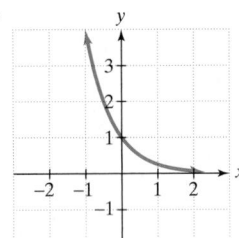

b.
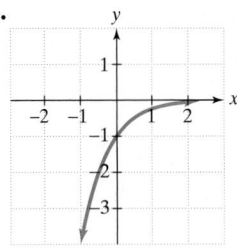

c.
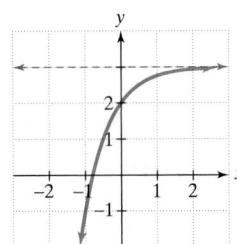

d.
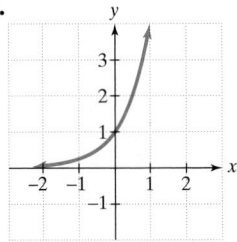

In Exercises 5–8, graph functions f and g in the same rectangular coordinate system. Select integers from −2 to 2 for x. Then describe how the graph of g is related to the graph of f. If applicable, use a graphing utility to confirm your hand-drawn graphs.

5. $f(x) = 2^x$ and $g(x) = 2^{x-1}$

6. $f(x) = 2^x$ and $g(x) = \left(\dfrac{1}{2}\right)^x$

7. $f(x) = 3^x$ and $g(x) = 3^x - 1$

8. $f(x) = 3^x$ and $g(x) = -3^x$

Use the compound interest formulas

$$A = P\left(1 + \frac{r}{n}\right)^{nt} \quad \text{and} \quad A = Pe^{rt}$$

to solve Exercises 9–10.

9. Suppose that you have $5000 to invest. Which investment yields the greater return over 5 years: 5.5% compounded semiannually or 5.25% compounded monthly?

10. Suppose that you have $14,000 to invest. Which investment yields the greater return over 10 years: 7% compounded monthly or 6.85% compounded continuously?

11. A cup of coffee is taken out of a microwave oven and placed in a room. The temperature, T, in degrees Fahrenheit, of the coffee after t minutes is modeled by the function $T = 70 + 130e^{-0.04855t}$. The graph of the function is shown in the figure. Use the graph to answer each of the following questions.

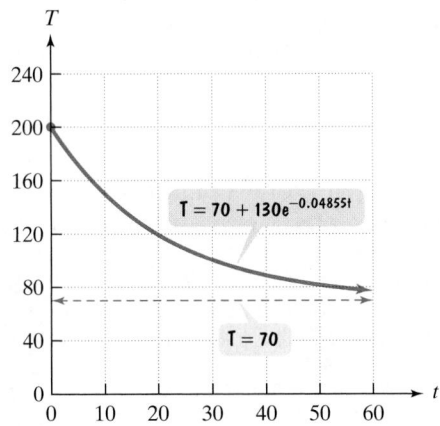

a. What was the temperature of the coffee when it was first taken out of the microwave?

b. What is a reasonable estimate of the temperature of the coffee after 20 minutes? Use your calculator to verify this estimate.

c. What is the limit of the temperature to which the coffee will cool? What does this tell you about the temperature of the room?

12.2 *In Exercises 12–13, find **a.** $(f \circ g)(x)$; **b.** $(g \circ f)(x)$; **c.** $(f \circ g)(3)$.*

12. $f(x) = x^2 + 3, \quad g(x) = 4x - 1$

13. $f(x) = \sqrt{x}, \quad g(x) = x + 1$

In Exercises 14–15, find $f\big(g(x)\big)$ and $g\big(f(x)\big)$ and determine whether each pair of functions f and g are inverses of each other.

14. $f(x) = \dfrac{3}{5}x + \dfrac{1}{2}$ and $g(x) = \dfrac{5}{3}x - 2$

15. $f(x) = 2 - 5x$ and $g(x) = \dfrac{2 - x}{5}$

The functions in Exercises 16–18 are all one-to-one. For each function:

   a. *Find an equation of $f^{-1}(x)$, the inverse function.*
   b. *Verify that your equation is correct by showing that*
    $f\big(f^{-1}(x)\big) = x$ *and* $f^{-1}\big(f(x)\big) = x.$

16. $f(x) = 4x - 3$

17. $f(x) = \sqrt{x + 2}$

18. $f(x) = 8x^3 + 1$

Which graphs in Exercises 19–22 represent functions that have inverse functions?

19.

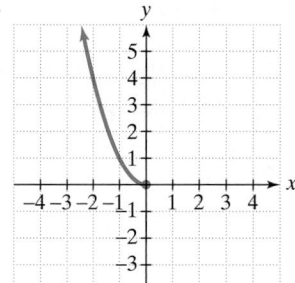

20.

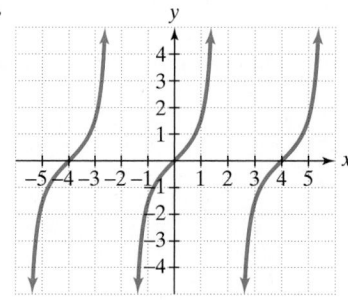

21.

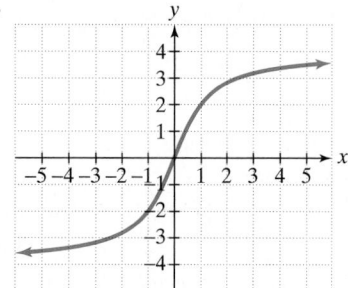

22.

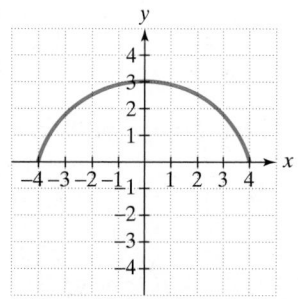

23. Use the graph of f in the figure shown to draw the graph of its inverse function.

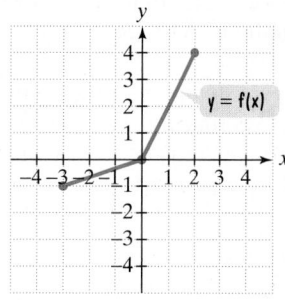

12.3 *In Exercises 24–26, write each equation in its equivalent exponential form.*

24. $\dfrac{1}{2} = \log_{49} 7$ **25.** $3 = \log_4 x$

26. $\log_3 81 = y$

In Exercises 27–29, write each equation in its equivalent logarithmic form.

27. $6^3 = 216$ **28.** $b^4 = 625$

29. $13^y = 874$

In Exercises 30–37, evaluate each expression without using a calculator. If evaluation is not possible, state the reason.

30. $\log_4 64$ **31.** $\log_5 \frac{1}{25}$

32. $\log_3(-9)$

33. $\log_{16} 4$

34. $\log_{17} 17$

35. $\log_3 3^8$

36. $\ln e^5$

37. $\log_3(\log_8 8)$

38. Graph $f(x) = 2^x$ and $g(x) = \log_2 x$ in the same rectangular coordinate system.

39. Graph $f(x) = \left(\frac{1}{3}\right)^x$ and $g(x) = \log_{\frac{1}{3}} x$ in the same rectangular coordinate system.

In Exercises 40–42, find the domain of each logarithmic function.

40. $f(x) = \log_8(x + 5)$

41. $f(x) = \log(3 - x)$

42. $f(x) = \ln(x - 1)^2$

In Exercises 43–45, use inverse properties of logarithms to simplify each expression.

43. $\ln e^{6x}$

44. $e^{\ln \sqrt{x}}$

45. $10^{\log 4x^2}$

46. On the Richter scale, the magnitude, R, of an earthquake of intensity I is given by $R = \log \dfrac{I}{I_0}$, where I_0 is the intensity of a barely felt zero-level earthquake. If the intensity of an earthquake is $1000I_0$, what is its magnitude on the Richter scale?

47. Students in a psychology class took a final examination. As part of an experiment to see how much of the course content they remembered over time, they took equivalent forms of the exam in monthly intervals thereafter. The average score, $f(t)$, for the group after t months is modeled by the function $f(t) = 76 - 18 \log(t + 1)$, where $0 \le t \le 12$.

    a. What was the average score when the exam was first given?

    b. What was the average score after 2 months? 4 months? 6 months? 8 months? one year?

    c. Use the results from parts (a) and (b) to graph f. Describe what the shape of the graph indicates in terms of the material retained by the students.

48. The formula

$$t = \frac{1}{c} \ln\left(\frac{A}{A - N}\right)$$

describes the time, t, in weeks, that it takes to achieve mastery of a portion of a task. In the formula, A represents maximum learning possible, N is the portion of the learning that is to be achieved, and c is a constant used to measure an individual's learning style. A 50-year-old man decides to start running as a way to maintain good health. He feels that the maximum rate he could ever hope to achieve is 12 miles per hour. How many weeks will it take before the man can run 5 miles per hour if $c = 0.06$ for this person?

12.4 *In Exercises 49–52, use properties of logarithms to expand each logarithmic expression as much as possible. Where possible, evaluate logarithmic expressions without using a calculator. Assume that all variables represent positive numbers.*

49. $\log_6(36x^3)$

50. $\log_4 \dfrac{\sqrt{x}}{64}$

51. $\log_2\left(\dfrac{xy^2}{64}\right)$

52. $\ln \sqrt[3]{\dfrac{x}{e}}$

In Exercises 53–56, use properties of logarithms to condense each logarithmic expression. Write the expression as a single logarithm whose coefficient is 1.

53. $\log_b 7 + \log_b 3$

54. $\log 3 - 3 \log x$

55. $3 \ln x + 4 \ln y$

56. $\dfrac{1}{2} \ln x - \ln y$

In Exercises 57–58, use common logarithms or natural logarithms and a calculator to evaluate to four decimal places.

57. $\log_6 72{,}348$

58. $\log_4 0.863$

12.5 *Solve each exponential equation in Exercises 59–61 by expressing each side as a power of the same base and then equating exponents.*

59. $2^{4x-2} = 64$

60. $125^x = 25$

61. $9^x = \dfrac{1}{27}$

Solve each exponential equation in Exercises 62–64 by taking the natural logarithm on both sides. Express the solution in terms of natural logarithms. Then use a calculator to obtain a decimal approximation, correct to two decimal places, for the solution.

62. $8^x = 12{,}143$

63. $9e^{5x} = 1269$

64. $30e^{0.045x} = 90$

In Exercises 65–71, solve each logarithmic equation.

65. $\log_5 x = -3$

66. $\log x = 2$

67. $\log_4(3x - 5) = 3$

68. $\log_2(x + 3) + \log_2(x - 3) = 4$

69. $\log_3(x - 1) - \log_3(x + 2) = 2$

70. $\ln x = -1$

71. $3 + 4\ln 2x = 15$

72. The function $f(t) = 10.1e^{0.005t}$ models the population, $f(t)$, of Los Angeles, California, in millions, t years after 1992. If the growth rate continues into the future, when will the population reach 13 million?

73. The amount of carbon dioxide in the atmosphere, measured in parts per million, has been increasing as a result of the burning of oil and coal. The buildup of gases and particles traps heat and raises the planet's temperature, a phenomenon called the greenhouse effect. Carbon dioxide accounts for about half of the warming. The function $f(t) = 364(1.005)^t$ projects carbon dioxide concentration, $f(t)$, in parts per million, t years after 2000. Using the projections given by the function, when will the carbon dioxide concentration be double the preindustrial level of 280 parts per million?

74. The function $\overline{C}(x) = 15{,}557 + 5259\ln x$ models the average cost of a new car, $\overline{C}(x)$, x years after 1989. When will the average cost of a new car be $30,000?

75. Use the compound interest formula

$$A = P\left(1 + \frac{r}{n}\right)^{nt}$$

to solve this problem. How long, to the nearest tenth of a year, will it take $12,500 to grow to $20,000 at 6.5% annual interest compounded quarterly?

Use the compound interest formula

$$A = Pe^{rt}$$

to solve Exercises 76–77.

76. How long, to the nearest tenth of a year, will it take $50,000 to triple in value at 7.5% annual interest compounded continuously?

77. What interest rate is required for an investment subject to continuous compounding to triple in 5 years?

12.6

78. According to the U.S. Bureau of the Census, in 1980 there were 14.6 million residents of Hispanic origin living in the United States. By 1997, the number had increased to 29.3 million. The exponential growth function $A = 14.6e^{kt}$ describes the U.S. Hispanic population, in millions, t years after 1980.

a. Find k, correct to three decimal places.

b. Use the resulting model to project the Hispanic resident population in 2005.

c. In which year will the Hispanic resident population reach 50 million?

79. Use the exponential decay model for carbon-14, $A = A_0 e^{-0.000121t}$, to solve this exercise. Prehistoric cave paintings were discovered in the Lascaux cave in France. The paint contained 15% of the original carbon-14. Estimate the age of the paintings at the time of the discovery.

80. The figure shows world population projections through the year 2150. The data are from the United Nations Family Planning Program and are based on optimistic or pessimistic expectations for successful control of human population growth. Suppose that you are interested in modeling these data using exponential, logarithmic, linear, and quadratic functions. Which function would you use to model each of the projections? Explain your choices. For the choice corresponding to a quadratic model, would your formula involve one with a positive or negative leading coefficient? Explain.

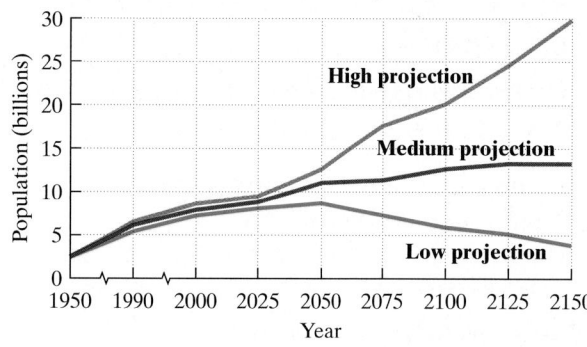

Chapter 12 Test

1. Graph $f(x) = 2^x$ and $g(x) = 2^{x+1}$ in the same rectangular coordinate system.

2. Use $A = P\left(1 + \dfrac{r}{n}\right)^{nt}$ and $A = Pe^{rt}$ to solve this problem. Suppose you have \$3000 to invest. Which investment yields the greater return over 10 years: 6.5% compounded semiannually or 6% compounded continuously? How much more (to the nearest dollar) is yielded by the better investment?

3. If $f(x) = x^2 + x$ and $g(x) = 3x - 1$, find $(f \circ g)(x)$ and $(g \circ f)(x)$.

4. If $f(x) = 5x - 7$, find $f^{-1}(x)$.

5. A function f models the amount given to charity as a function of income. The graph of f is shown in the figure.

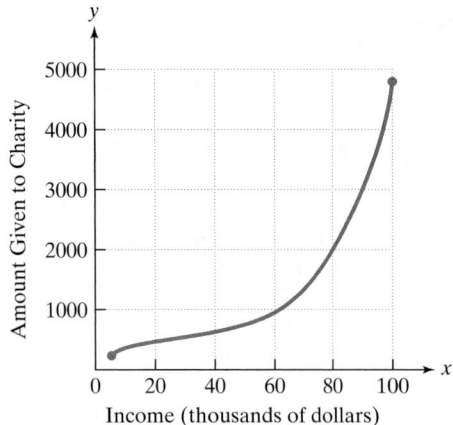

   a. Explain why f has an inverse that is a function.

   b. Find $f(80)$.

   c. Describe in practical terms the meaning of $f^{-1}(2000)$.

6. Write in exponential form: $\log_5 125 = 3$.
7. Write in logarithmic form: $\sqrt{36} = 6$.
8. Graph $f(x) = 3^x$ and $g(x) = \log_3 x$ in the same rectangular coordinate system.

In Exercises 9–11, simplify each expression.

9. $\ln e^{5x}$

10. $\log_b b$

11. $\log_6 1$

12. Find the domain: $f(x) = \log_5(x - 7)$.

13. On the decibel scale, the loudness of a sound, in decibels, is given by $D = 10 \log \dfrac{I}{I_0}$, where I is the intensity of the sound, in watts per meter^2, and I_0 is the intensity of a sound barely audible to the human ear. If the intensity of a sound is $10^{12}I_0$, what is its loudness in decibels? (Such a sound is potentially damaging to the ear.)

In Exercises 14–15, use properties of logarithms to expand each logarithmic expression as much as possible. Where possible, evaluate logarithmic expressions without using a calculator.

14. $\log_4(64x^5)$

15. $\log_3 \dfrac{\sqrt[3]{x}}{81}$

In Exercises 16–17, write each expression as a single logarithm.

16. $6 \log x + 2 \log y$

17. $\ln 7 - 3 \ln x$

18. Use a calculator to evaluate $\log_{15} 71$ to four decimal places.

In Exercises 19–25, solve each equation.

19. $3^{x-2} = 81$

20. $5^x = 1.4$

21. $400e^{0.005x} = 1600$

22. $\log_{25} x = \dfrac{1}{2}$

23. $\log_6(4x - 1) = 3$

24. $\log x + \log(x + 15) = 2$

25. $2 \ln 3x = 8$

26. The function
$$P(t) = 89.18e^{-0.004t}$$
models the percentage, $P(t)$, of married men in the United States who were employed t years after 1959.
   a. What percentage of married men were employed in 1959?
   b. Is the percentage of married men who are employed increasing or decreasing? Explain.
   c. In which year were 77% of U.S. married men employed?

Use the formulas

$$A = P\left(1 + \frac{r}{n}\right)^{nt} \quad \text{and} \quad A = Pe^{rt}$$

to solve Exercises 27–28.

27. How long, to the nearest tenth of a year, will it take $4000 to grow to $8000 at 5% annual interest compounded quarterly?

28. What interest rate is required for an investment subject to continuous compounding to double in 10 years?

29. The 1980 population of Europe was 484 million; in 1990, it was 509 million. Write the exponential growth function that describes the population of Europe, in millions, t years after 1980.

30. Use the exponential decay model for carbon-14, $A = A_0 e^{-0.000121t}$, to solve this exercise. Bones of a pre-historic man were discovered and contained 5% of the original amount of carbon-14. How long ago did the man die?

Cumulative Review Exercises (Chapters 1–12)

In Exercises 1–8, solve each equation, inequality, or system.

1. $\sqrt{2x + 5} - \sqrt{x + 3} = 2$

2. $(x - 5)^2 = -49$

3. $x^2 + x > 6$

4. $6x - 3(5x + 2) = 4(1 - x)$

5. $\dfrac{2}{x - 3} - \dfrac{3}{x + 3} = \dfrac{12}{x^2 - 9}$

6. $3x + 2 < 4 \quad \text{and} \quad 4 - x > 1$

7. $3x - 2y + z = 7$
$2x + 3y - z = 13$
$x - y + 2z = -6$

8. $\log_9 x + \log_9 (x - 8) = 1$

In Exercises 9–11, graph each function, equation, or inequality in a rectangular coordinate system.

9. $f(x) = (x + 2)^2 - 4$ **10.** $y < -3x + 5$

11. $f(x) = 3^{x-2}$

In Exercises 12–15, perform the indicated operations, and simplify, if possible.

12. $\dfrac{2x + 1}{x - 5} - \dfrac{4}{x^2 - 3x - 10}$

13. $\dfrac{\dfrac{1}{x - 1} + 1}{\dfrac{1}{x + 1} - 1}$

14. $\dfrac{6}{\sqrt{5} - \sqrt{2}}$

15. $8\sqrt{45} + 2\sqrt{5} - 7\sqrt{20}$

16. Rationalize the denominator: $\dfrac{5}{\sqrt[3]{2x^2 y}}$.

17. Factor completely: $5ax + 5ay - 4bx - 4by$.

18. Write as a single logarithm: $5 \log x - \dfrac{1}{2} \log y$.

19. Write 0.00397 in scientific notation.

In Exercises 20–22, find the domain of each function.

20. $f(x) = \dfrac{2}{x^2 + 2x - 15}$

21. $f(x) = \sqrt{2x - 6}$

22. $f(x) = \ln(1 - x)$

23. The length of a rectangular garden is 2 feet more than twice its width. If 22 feet of fencing is needed to enclose the garden, what are its dimensions?

24. With a 6% raise, you will earn $19,610 annually. What is your salary before this raise?

25. The function $F(t) = 1 - k \ln(t + 1)$ models the fraction of people, $F(t)$, who remember all the words in a list of nonsense words t hours after memorizing the list. After 3 hours, only half the people could remember all the words. Determine the value of k and then predict the fraction of people in the group who will remember all the words after 6 hours. Round to three decimal places and then express the fraction with a denominator of 1000.

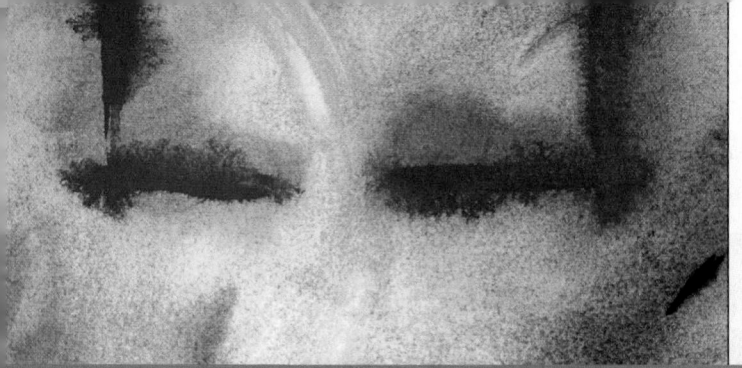

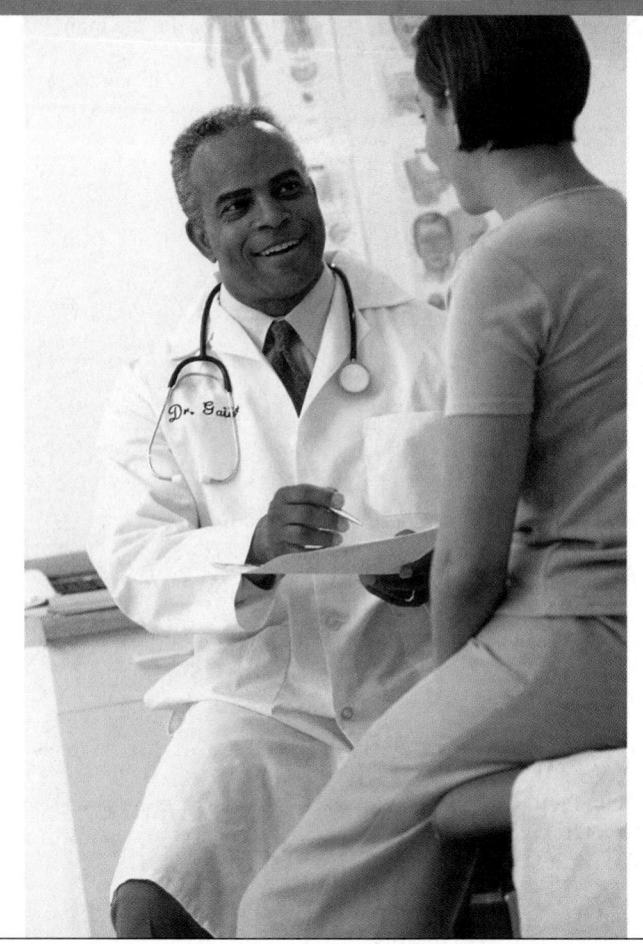

One minute you're in class, enjoying the lecture. Then a sharp pain radiates down your side. The next minute you're being diagnosed with, of all things, a kidney stone. It took your cousin six weeks to recover from kidney stone surgery, but your doctor assures you there is nothing to worry about. A new procedure, based on a curve that looks like the cross section of a football, will dissolve the stone painlessly and let you return to class in a day or two. How can this be?

From ripples in water to the path on which humanity journeys through space, certain curves occur naturally throughout the universe. Over 2000 years ago the ancient Greeks studied these curves, called *conic sections*, without regard to their immediate usefulness simply because the study elicited ideas that were exciting, challenging, and interesting. The ancient Greeks could not have imagined the applications of these curves in the twenty-first century. Overwhelmed by the choices on satellite television? Blame it on a conic section! In this chapter, we use the rectangular coordinate system to study the conic sections and the mathematics behind their surprising applications.

Chapter 13

Conic Sections and Systems of Nonlinear Equations

▶ SECTION 13.1   *Distance and Midpoint Formulas; Circles*

Objectives

1 Find the distance between two points.

2 Find the midpoint of a line segment.

3 Write the standard form of a circle's equation.

4 Give the center and radius of a circle whose equation is in standard form.

5 Convert the general form of a circle's equation to standard form.

SSM
PH Tutor CD- Video
Center ROM

It's a good idea to know your way around a circle. Clocks, angles, maps, and compasses are based on circles. Circles occur everywhere in nature: in ripples on water, patterns on a butterfly's wings, and cross sections of trees. Some consider the circle to be the most pleasing of all shapes.

The rectangular coordinate system gives us a unique way of knowing a circle. It enables us to translate a circle's geometric definition into an algebraic equation. To do this, we must first develop a formula for the distance between any two points in rectangular coordinates.

1 Find the distance between two points.

The Distance Formula

Using the Pythagorean Theorem, we can find the distance between the two points $P_1(x_1, y_1)$ and $P_2(x_2, y_2)$ in the rectangular coordinate system. The two points are illustrated in Figure 13.1.

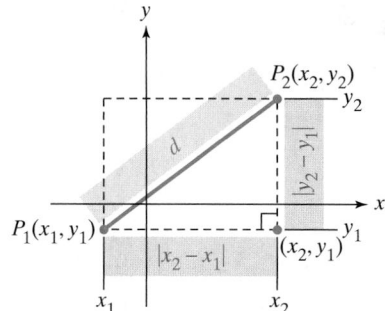

Figure 13.1

The distance that we need to find is represented by d and shown in blue. Notice that the distance between two points on the dashed horizontal line is the absolute value of the difference between the x-coordinates of the two points. This distance, $|x_2 - x_1|$, is shown in pink. Similarly, the distance between two points on the dashed vertical line is the absolute value of the difference between the y-coordinates of the two points. This distance, $|y_2 - y_1|$, is also shown in pink.

Because the dashed lines are horizontal and vertical, a right triangle is formed. Thus, we can use the Pythagorean Theorem to find distance d. By the Pythagorean Theorem,

$$d^2 = |x_2 - x_1|^2 + |y_2 - y_1|^2$$
$$d = \sqrt{|x_2 - x_1|^2 + |y_2 - y_1|^2}$$
$$d = \sqrt{(x_2 - x_1)^2 + (y_2 - y_1)^2}.$$

This result is called the **distance formula**.

The Distance Formula

The distance, d, between the points (x_1, y_1) and (x_2, y_2) in the rectangular coordinate system is

$$d = \sqrt{(x_2 - x_1)^2 + (y_2 - y_1)^2}.$$

When using the distance formula, it does not matter which point you call (x_1, y_1) and which you call (x_2, y_2).

EXAMPLE 1 Using the Distance Formula

Find the distance between $(-1, -3)$ and $(2, 3)$.

Solution Letting $(x_1, y_1) = (-1, -3)$ and $(x_2, y_2) = (2, 3)$, we obtain

$$d = \sqrt{(x_2 - x_1)^2 + (y_2 - y_1)^2} \qquad \text{Use the distance formula.}$$

$$= \sqrt{[2 - (-1)]^2 + [3 - (-3)]^2} \qquad \text{Substitute the given values.}$$

$$= \sqrt{(2 + 1)^2 + (3 + 3)^2} \qquad \text{Perform subtractions within the grouping symbols.}$$

$$= \sqrt{3^2 + 6^2}$$

$$= \sqrt{9 + 36} \qquad \text{Square 3 and 6.}$$

$$= \sqrt{45} \qquad \text{Add.}$$

$$= 3\sqrt{5} \approx 6.71. \qquad \sqrt{45} = \sqrt{9 \cdot 5} = \sqrt{9}\sqrt{5} = 3\sqrt{5}$$

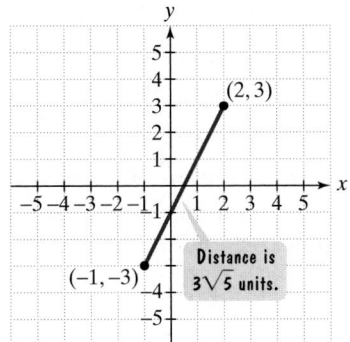

Figure 13.2 Finding the distance between two points

The distance between the given points is $3\sqrt{5}$ units, or approximately 6.71 units. The situation is illustrated in Figure 13.2. ∎

✔ **CHECK POINT 1** Find the distance between $(2, -2)$ and $(5, 2)$.

The Midpoint Formula

The distance formula can be used to derive a formula for finding the midpoint of a line segment between two given points. The formula is given in the box on page 880.

2 Find the midpoint of a line segment.

The Midpoint Formula

Consider a line segment whose endpoints are (x_1, y_1) and (x_2, y_2). The coordinates of the segment's midpoint are

$$\left(\frac{x_1 + x_2}{2}, \frac{y_1 + y_2}{2} \right).$$

To find the midpoint, take the average of the two x-coordinates and of the two y-coordinates.

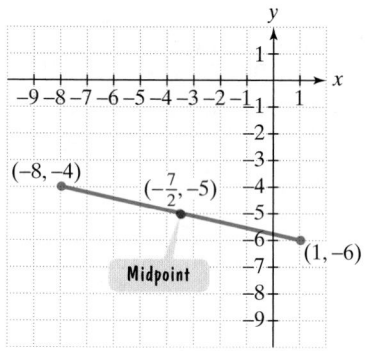

Figure 13.3 Finding a line segment's midpoint

EXAMPLE 2   Using the Midpoint Formula

Find the midpoint of the line segment with endpoints $(1, -6)$ and $(-8, -4)$.

Solution To find the coordinates of the midpoint, we average the coordinates of the endpoints.

$$\text{Midpoint} = \left(\frac{1 + (-8)}{2}, \frac{-6 + (-4)}{2} \right) = \left(\frac{-7}{2}, \frac{-10}{2} \right) = \left(-\frac{7}{2}, -5 \right)$$

Average the x-coordinates.   Average the y-coordinates.

Figure 13.3 illustrates that the point $\left(-\frac{7}{2}, -5\right)$ is midway between the points $(1, -6)$ and $(-8, -4)$. ∎

✔ **CHECK POINT 2**   Find the midpoint of the line segment with endpoints $(1, 2)$ and $(7, -3)$.

Circles

Our goal is to translate a circle's geometric definition into an equation. We begin with this geometric definition.

Definition of a Circle

A **circle** is the set of all points in a plane that are equidistant from a fixed point, called the **center**. The fixed distance from the circle's center to any point on the circle is called the **radius**.

Figure 13.4 is our starting point for obtaining a circle's equation. We've placed the circle into a rectangular coordinate system. The circle's center is (h, k) and its radius is r. We let (x, y) represent the coordinates of any point on the circle.

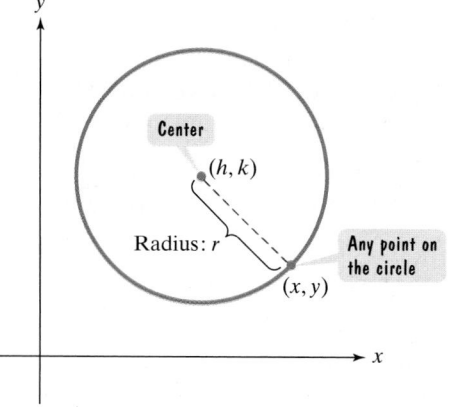

Figure 13.4 A circle centered at (h, k) with radius r

What does the geometric definition of a circle tell us about the point (x, y) in Figure 13.4? The point is on the circle if and only if its distance from the center is r. We can use the distance formula to express this idea algebraically:

The distance between (x, y) and (h, k) is always r.

$$\sqrt{(x - h)^2 + (y - k)^2} = r.$$

Squaring both sides of this equation yields the **standard form of the equation of a circle**.

The Standard Form of the Equation of a Circle

The **standard form of the equation of a circle** with center (h, k) and radius r is

$$(x - h)^2 + (y - k)^2 = r^2.$$

 Write the standard form of a circle's equation.

EXAMPLE 3   Finding the Standard Form of a Circle's Equation

Write the standard form of the equation of the circle with center $(0, 0)$ and radius 2. Graph the circle.

Solution   The center is $(0, 0)$. Because the center is represented as (h, k) in the standard form of the equation, $h = 0$ and $k = 0$. The radius is 2, so we will let $r = 2$ in the equation.

$$(x - h)^2 + (y - k)^2 = r^2 \qquad \text{This is the standard form of a circle's equation.}$$

$$(x - 0)^2 + (y - 0)^2 = 2^2 \qquad \text{Substitute 0 for } h, \text{ 0 for } k, \text{ and 2 for } r.$$

$$x^2 + y^2 = 4 \qquad \text{Simplify.}$$

The standard form of the equation of the circle is $x^2 + y^2 = 4$. Figure 13.5 shows the graph. ∎

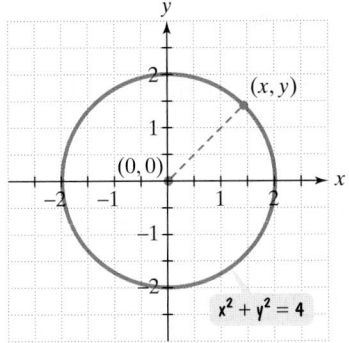

Figure 13.5 The graph of $x^2 + y^2 = 4$

✔ **CHECK POINT 3**   Write the standard form of the equation of the circle with center $(0, 0)$ and radius 4.

Using Technology

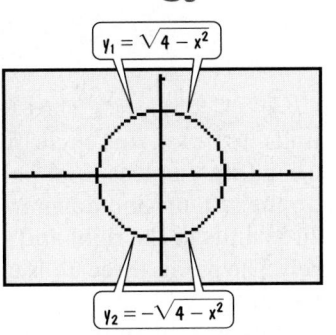

$y_1 = \sqrt{4 - x^2}$

$y_2 = -\sqrt{4 - x^2}$

To graph a circle with a graphing utility, first solve the equation for y.

$$x^2 + y^2 = 4$$

$$y^2 = 4 - x^2$$

$$y = \pm\sqrt{4 - x^2}$$

Graph the two equations

$$y_1 = \sqrt{4 - x^2} \quad \text{and} \quad y_2 = -\sqrt{4 - x^2}$$

in the same viewing rectangle. The graph of $y_1 = \sqrt{4 - x^2}$ is the top semicircle because y is always positive. The graph of $y_2 = -\sqrt{4 - x^2}$ is the bottom semicircle because y is always negative. Use a ZOOM SQUARE setting so that the circle looks like a circle. (Many graphing utilities have problems connecting the two semicircles because the segments directly across from the center become nearly vertical.)

Example 3 and Check Point 3 involved circles centered at the origin. The standard form of the equation of all such circles is $x^2 + y^2 = r^2$, where r is the circle's radius. Now, let's consider a circle whose center is not at the origin.

EXAMPLE 4 Finding the Standard Form of a Circle's Equation

Write the standard form of the equation of the circle with center $(-2, 3)$ and radius 4.

Solution   The center is $(-2, 3)$. Because the center is represented as (h, k) in the standard form of the equation, $h = -2$ and $k = 3$. The radius is 4, so we will let $r = 4$ in the equation.

$$(x - h)^2 + (y - k)^2 = r^2 \qquad \text{This is the standard form of a circle's equation.}$$

$$[x - (-2)]^2 + (y - 3)^2 = 4^2 \qquad \text{Substitute } -2 \text{ for } h, 3 \text{ for } k, \text{ and } 4 \text{ for } r.$$

$$(x + 2)^2 + (y - 3)^2 = 16 \qquad \text{Simplify.}$$

The standard form of the equation of the circle is $(x + 2)^2 + (y - 3)^2 = 16$. ∎

✔ **CHECK POINT 4**   Write the standard form of the equation of the circle with center $(5, -6)$ and radius 10.

---

4 Give the center and radius of a circle whose equation is in standard form.

EXAMPLE 5 Using the Standard Form of a Circle's Equation to Graph the Circle

Find the center and radius of the circle whose equation is

$$(x - 2)^2 + (y + 4)^2 = 9$$

and graph the equation.

Solution   In order to graph the circle, we need to know its center, (h, k), and its radius, r. We can find the values for h, k, and r by comparing the given equation to the standard form of the equation of a circle.

$$(x - 2)^2 + (y + 4)^2 = 9$$

$$(x - 2)^2 + (y - (-4))^2 = 3^2$$

| This is $(x - h)^2$, with $h = 2$. | This is $(y - k)^2$, with $k = -4$. | This is r^2, with $r = 3$. |

We see that $h = 2$, $k = -4$, and $r = 3$. Thus, the circle has center $(h, k) = (2, -4)$ and a radius of 3 units. To graph this circle, first plot the center $(2, -4)$. Because the radius is 3, you can locate at least four points on the circle by going out three units to the right, to the left, up, and down from the center.

The points three units to the right and to the left of $(2, -4)$ are $(5, -4)$ and $(-1, -4)$, respectively. The points three units up and down from $(2, -4)$ are $(2, -1)$ and $(2, -7)$, respectively.

Using these points, we obtain the graph in Figure 13.6. ∎

✔ **CHECK POINT 5**   Find the center and radius of the circle whose equation is

$$(x + 3)^2 + (y - 1)^2 = 4$$

and graph the equation.

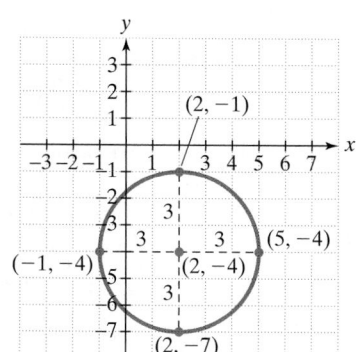

Figure 13.6 The graph of $(x - 2)^2 + (y + 4)^2 = 9$

If we square $x - 2$ and $y + 4$ in the standard form of the equation of Example 5, we obtain another form for the circle's equation.

$$(x - 2)^2 + (y + 4)^2 = 9 \qquad \text{This is the standard form of the equation from Example 5.}$$

$$x^2 - 4x + 4 + y^2 + 8y + 16 = 9 \qquad \text{Square } x - 2 \text{ and } y + 4.$$

$$x^2 + y^2 - 4x + 8y + 20 = 9 \qquad \text{Combine numerical terms and rearrange terms.}$$

$$x^2 + y^2 - 4x + 8y + 11 = 0 \qquad \text{Subtract 9 from both sides.}$$

This result suggests that an equation in the form $x^2 + y^2 + Dx + Ey + F = 0$ can represent a circle. This is called the **general form of the equation of a circle**.

> ### The General Form of the Equation of a Circle
> The **general form of the equation of a circle** is
> $$x^2 + y^2 + Dx + Ey + F = 0.$$

5 Convert the general form of a circle's equation to standard form.

We can convert the general form of the equation of a circle to the standard form $(x - h)^2 + (y - k)^2 = r^2$. We do so by completing the square on x and y. Let's see how this is done.

EXAMPLE 6   Converting the General Form of a Circle's Equation to Standard Form and Graphing the Circle

Write in standard form and graph: $x^2 + y^2 + 4x - 6y - 23 = 0$.

Solution   Because we plan to complete the square on both x and y, let's rearrange terms so that x-terms are arranged in descending order, y-terms are arranged in descending order, and the constant term appears on the right.

$$x^2 + y^2 + 4x - 6y - 23 = 0 \qquad \text{This is the given equation.}$$

$$\left(x^2 + 4x \quad\right) + \left(y^2 - 6y \quad\right) = 23 \qquad \text{Rewrite in anticipation of completing the square.}$$

$$\left(x^2 + 4x + 4\right) + \left(y^2 - 6y + 9\right) = 23 + 4 + 9 \qquad \text{Complete the square on } x: \tfrac{1}{2} \cdot 4 = 2 \text{ and } 2^2 = 4, \text{ so add 4 to both sides. Complete the square on } y: \tfrac{1}{2}(-6) = -3 \text{ and } (-3)^2 = 9, \text{ so add 9 to both sides.}$$

Remember that numbers added on the left side must also be added on the right side.

$$(x + 2)^2 + (y - 3)^2 = 36 \qquad \text{Factor on the left and add on the right.}$$

This last equation is in standard form. We can identify the circle's center and radius by comparing this equation to the standard form of the equation of a circle, $(x - h)^2 + (y - k)^2 = r^2$.

$$(x + 2)^2 + (y - 3)^2 = 36$$

$$\left(x - (-2)\right)^2 + (y - 3)^2 = 6^2$$

This is $(x - h)^2$, with $h = -2$.   This is $(y - k)^2$, with $k = 3$.   This is r^2, with $r = 6$.

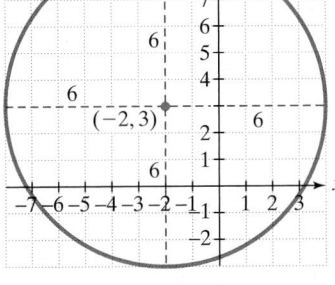

Figure 13.7 The graph of $(x + 2)^2 + (y - 3)^2 = 36$

We use the center, $(h, k) = (-2, 3)$, and the radius, $r = 6$, to graph the circle. The graph is shown in Figure 13.7.   ■

Using Technology

To graph $x^2 + y^2 + 4x - 6y - 23 = 0$, rewrite the equation as a quadratic equation in y.

$$y^2 - 6y + (x^2 + 4x - 23) = 0$$

Now solve for y using the quadratic formula, with $a = 1$, $b = -6$, and $c = x^2 + 4x - 23$.

$$y = \frac{-b \pm \sqrt{b^2 - 4ac}}{2a} = \frac{-(-6) \pm \sqrt{(-6)^2 - 4 \cdot 1(x^2 + 4x - 23)}}{2 \cdot 1} = \frac{6 \pm \sqrt{36 - 4(x^2 + 4x - 23)}}{2}$$

Because we will enter these equations, there is no need to simplify further. Enter

$$y_1 = \frac{6 + \sqrt{36 - 4(x^2 + 4x - 23)}}{2}$$

and

$$y_2 = \frac{6 - \sqrt{36 - 4(x^2 + 4x - 23)}}{2}.$$

Use a $\boxed{\text{ZOOM SQUARE}}$ setting. The graph is shown on the right.

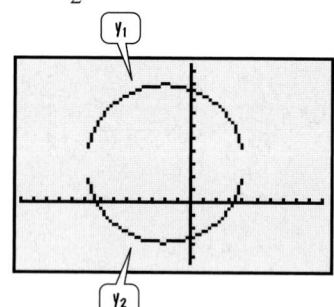

✔ **CHECK POINT 6** Write in standard form and graph:

$$x^2 + y^2 + 4x - 4y - 1 = 0.$$

EXERCISE SET 13.1

Practice Exercises

In Exercises 1–18, find the distance between each pair of points. If necessary, round answers to two decimal places.

1. $(2, 3)$ and $(14, 8)$
2. $(5, 1)$ and $(8, 5)$
3. $(4, 1)$ and $(6, 3)$
4. $(2, 3)$ and $(3, 5)$
5. $(0, 0)$ and $(-3, 4)$
6. $(0, 0)$ and $(3, -4)$
7. $(-2, -6)$ and $(3, -4)$
8. $(-4, -1)$ and $(2, -3)$
9. $(0, -3)$ and $(4, 1)$
10. $(0, -2)$ and $(4, 3)$
11. $(3.5, 8.2)$ and $(-0.5, 6.2)$
12. $(2.6, 1.3)$ and $(1.6, -5.7)$
13. $(0, -\sqrt{3})$ and $(\sqrt{5}, 0)$
14. $(0, -\sqrt{2})$ and $(\sqrt{7}, 0)$
15. $(3\sqrt{3}, \sqrt{5})$ and $(-\sqrt{3}, 4\sqrt{5})$
16. $(2\sqrt{3}, \sqrt{6})$ and $(-\sqrt{3}, 5\sqrt{6})$
17. $\left(\frac{7}{3}, \frac{1}{5}\right)$ and $\left(\frac{1}{3}, \frac{6}{5}\right)$
18. $\left(-\frac{1}{4}, -\frac{1}{7}\right)$ and $\left(\frac{3}{4}, \frac{6}{7}\right)$

In Exercises 19–30, find the midpoint of each line segment with the given endpoints.

19. $(6, 8)$ and $(2, 4)$
20. $(10, 4)$ and $(2, 6)$
21. $(-2, -8)$ and $(-6, -2)$
22. $(-4, -7)$ and $(-1, -3)$
23. $(-3, -4)$ and $(6, -8)$
24. $(-2, -1)$ and $(-8, 6)$
25. $\left(-\frac{7}{2}, \frac{3}{2}\right)$ and $\left(-\frac{5}{2}, -\frac{11}{2}\right)$
26. $\left(-\frac{2}{5}, \frac{7}{15}\right)$ and $\left(-\frac{2}{5}, -\frac{4}{15}\right)$
27. $(8, 3\sqrt{5})$ and $(-6, 7\sqrt{5})$
28. $(7\sqrt{3}, -6)$ and $(3\sqrt{3}, -2)$
29. $(\sqrt{18}, -4)$ and $(\sqrt{2}, 4)$
30. $(\sqrt{50}, -6)$ and $(\sqrt{2}, 6)$

In Exercises 31–40, write the standard form of the equation of the circle with the given center and radius.

31. Center $(0, 0)$, $r = 7$
32. Center $(0, 0)$, $r = 8$
33. Center $(3, 2)$, $r = 5$
34. Center $(2, -1)$, $r = 4$
35. Center $(-1, 4)$, $r = 2$
36. Center $(-3, 5)$, $r = 3$
37. Center $(-3, -1)$, $r = \sqrt{3}$
38. Center $(-5, -3)$, $r = \sqrt{5}$
39. Center $(-4, 0)$, $r = 10$
40. Center $(-2, 0)$, $r = 6$

In Exercises 41–48, give the center and radius of the circle described by the equation and graph each equation.

41. $x^2 + y^2 = 16$
42. $x^2 + y^2 = 49$
43. $(x - 3)^2 + (y - 1)^2 = 36$
44. $(x - 2)^2 + (y - 3)^2 = 16$
45. $(x + 3)^2 + (y - 2)^2 = 4$
46. $(x + 1)^2 + (y - 4)^2 = 25$
47. $(x + 2)^2 + (y + 2)^2 = 4$
48. $(x + 4)^2 + (y + 5)^2 = 36$

In Exercises 49–56, complete the square and write the equation in standard form. Then give the center and radius of each circle and graph the equation.

49. $x^2 + y^2 + 6x + 2y + 6 = 0$

50. $x^2 + y^2 + 8x + 4y + 16 = 0$

51. $x^2 + y^2 - 10x - 6y - 30 = 0$

52. $x^2 + y^2 - 4x - 12y - 9 = 0$

53. $x^2 + y^2 + 8x - 2y - 8 = 0$

54. $x^2 + y^2 + 12x - 6y - 4 = 0$

55. $x^2 - 2x + y^2 - 15 = 0$

56. $x^2 + y^2 - 6y - 7 = 0$

Application Exercises

57. A rectangular coordinate system with coordinates in miles is placed on the map in the figure shown. Bangkok has coordinates $(-115, 170)$ and Phnom Penh has coordinates $(65, 70)$. How long will it take a plane averaging 400 miles per hour to fly directly from one city to the other? Round to the nearest tenth of an hour. Approximately how many minutes is the flight?

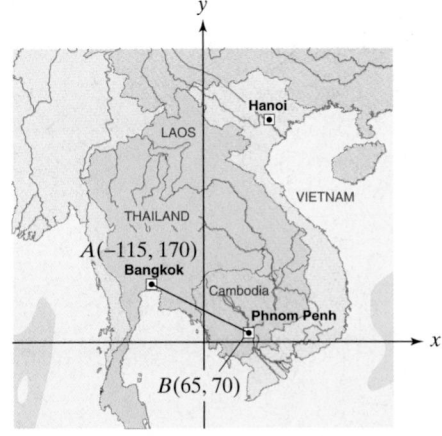

58. We refer to the driveway in the figure shown as being *circular*, meaning that it is bounded by two circles. The figure indicates that the radius of the larger circle is 52 feet and the radius of the smaller circle is 38 feet.

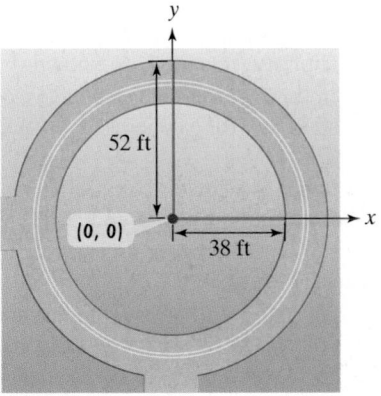

a. Use the coordinate system shown to write the equation of the smaller circle.
b. Use the coordinate system shown to write the equation of the larger circle.

59. The ferris wheel in the figure has a radius of 68 feet. The clearance between the wheel and the ground is 14 feet. The rectangular coordinate system shown has its origin on the ground directly below the center of the wheel. Use the coordinate system to write the equation of the circular wheel.

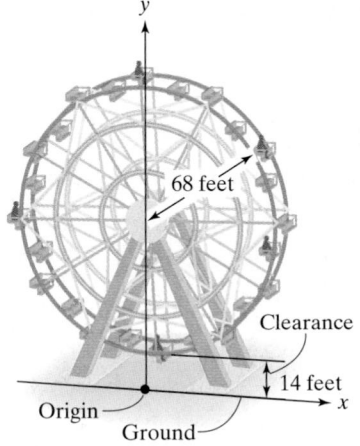

60. The circle formed by the middle lane of a circular running track can be described algebraically by $x^2 + y^2 = 4$, where all measurements are in miles. If you run around the track's middle lane twice, approximately how many miles have you covered?

Writing in Mathematics

61. In your own words, describe how to find the distance between two points in the rectangular coordinate system.

62. In your own words, describe how to find the midpoint of a line segment if its endpoints are known.

63. What is a circle? Without using variables, describe how the definition of a circle can be used to obtain a form of its equation.

64. Give an example of a circle's equation in standard form. Describe how to find the center and radius for this circle.

65. How is the standard form of a circle's equation obtained from its general form?

66. Does $(x - 3)^2 + (y - 5)^2 = 0$ represent the equation of a circle? If not, describe the graph of this equation.

67. Does $(x - 3)^2 + (y - 5)^2 = -25$ represent the equation of a circle? What sort of set is the graph of this equation?

Critical Thinking Exercises

68. Which one of the following is true?
  a. The equation of the circle whose center is at the origin with radius 16 is $x^2 + y^2 = 16$.
  b. The graph of $(x - 3)^2 + (y + 5)^2 = 36$ is a circle with a radius 6 centered at $(-3, 5)$.
  c. The graph of $(x - 4) + (y + 6) = 25$ is a circle with a radius 5 centered at $(4, -6)$.
  d. None of the above is true.

69. Show that the points $A(1, 1 + d)$, $B(3, 3 + d)$, and $C(6, 6 + d)$ are collinear (lie along a straight line) by showing that the distance from A to B plus the distance from B to C equals the distance from A to C.

70. Prove the midpoint formula by using the following procedure.

  a. Show that the distance between (x_1, y_1) and $\left(\dfrac{x_1 + x_2}{2}, \dfrac{y_1 + y_2}{2}\right)$ is equal to the distance between (x_2, y_2) and $\left(\dfrac{x_1 + x_2}{2}, \dfrac{y_1 + y_2}{2}\right)$.

b. Use the procedure from Exercise 69 and the distances from part (a) to show that the points (x_1, y_1), $\left(\dfrac{x_1 + x_2}{2}, \dfrac{y_1 + y_2}{2}\right)$, and (x_2, y_2) are collinear.

In Exercises 71–72, write the standard form and the general form of the equation of each circle.

71. Center at $(3, -5)$ and passing through the point $(-2, 1)$

72. Passing through $(-7, 2)$ and $(1, 2)$; these points lie on the circle and also on the line that passes through the circle's center.

73. Find the area of the donut-shaped region bounded by the graphs of $(x - 2)^2 + (y + 3)^2 = 25$ and $(x - 2)^2 + (y + 3)^2 = 36$.

74. A **tangent line** to a circle is a line that intersects the circle at exactly one point. The tangent line is perpendicular to the radius of the circle at this point of contact. Write the point-slope equation of a line tangent to the circle whose equation is $x^2 + y^2 = 25$ at the point $(3, -4)$.

Technology Exercises

In Exercises 75–77, use a graphing utility to graph each circle whose equation is given.

75. $x^2 + y^2 = 25$

76. $(y + 1)^2 = 36 - (x - 3)^2$

77. $x^2 + 10x + y^2 - 4y - 20 = 0$

Review Exercises

78. If $f(x) = x^2 - 2$ and $g(x) = 3x + 4$, find $f(g(x))$ and $g(f(x))$. (Section 12.2, Example 1)

79. Solve: $2x = \sqrt{7x - 3} + 3$. (Section 10.6, Example 3)

80. Solve: $|2x - 5| < 10$. (Section 9.3, Example 3)

▶ **SECTION 13.2** *The Ellipse*

Objectives

1 Graph ellipses centered at the origin.

2 Graph ellipses not centered at the origin.

3 Solve applied problems involving ellipses.

SSM PH Tutor CD- Video
      Center ROM

Tunnel cut in rock wall. © Kevin Fleming/CORBIS.

You took on a summer job driving a truck, delivering books that were ordered online. You're an avid reader, so just being around books sounded appealing. However, now you're feeling a bit shaky driving the truck for the first time. It's 10 feet wide and 9 feet high; compared to your compact car, it feels like you're behind the wheel of a tank. Up ahead you see a sign at the semielliptical entrance to a tunnel: Caution! Tunnel is 10 Feet High at Center Peak. Then you see another sign: Caution! Tunnel is 40 Feet Wide. Will your truck clear the opening of the tunnel's archway?

The mathematics of your world is present in the movements of planets, bridge and tunnel construction, navigational systems used to keep track of a ship's location, manufacture of lenses for telescopes, and even a procedure for disintegrating kidney stones. The mathematics behind these applications involves conic sections. **Conic sections** are curves that result from the intersection of a right circular cone and a plane. Figure 13.8 illustrates the four conic sections: the circle, the ellipse, the parabola, and the hyperbola.

Circle   Ellipse   Parabola   Hyperbola

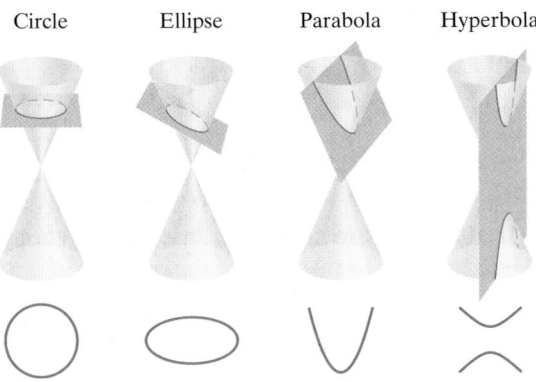

Figure 13.8 Obtaining the conic sections by intersecting a plane and a cone

In this section, we study the symmetric oval-shaped curve known as the ellipse. We will use a geometric definition for an ellipse to derive its equation. With this equation, we will determine if your delivery truck will clear the tunnel's entrance.

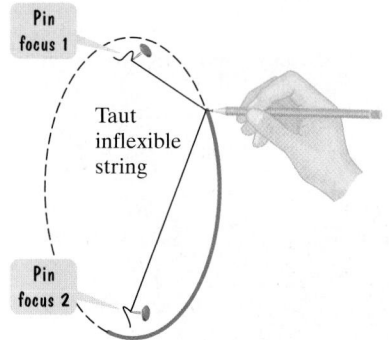

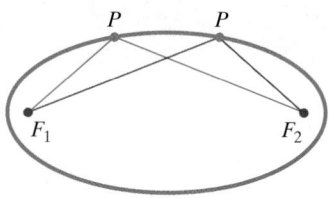

Figure 13.9 Drawing an ellipse

Definition of an Ellipse

Figure 13.9 illustrates how to draw an ellipse. Place pins at two fixed points, each of which is called a focus (plural: foci). If the ends of a fixed length of string are fastened to the pins and we draw the string taut with a pencil, the path traced by the pencil will be an ellipse. Notice that the sum of the distances of the pencil point from the foci remains constant because the length of the string is fixed. This procedure for drawing an ellipse illustrates its geometric definition.

> ### Definition of an Ellipse
>
> An **ellipse** is the set of all points, P, in a plane the sum of whose distances from two fixed points, F_1 and F_2, is constant (see Figure 13.10). These two fixed points are called the **foci** (plural of **focus**). The midpoint of the segment connecting the foci is the **center** of the ellipse.

Figure 13.11 illustrates that an ellipse can be elongated horizontally or vertically. The line through the foci intersects the ellipse at two points, called the **vertices** (singular: **vertex**). The line segment that joins the vertices is the **major axis**. Notice that the midpoint of the major axis is the center of the ellipse. The line segment whose endpoints are on the ellipse and that is perpendicular to the major axis at the center is called the **minor axis** of the ellipse.

Figure 13.10

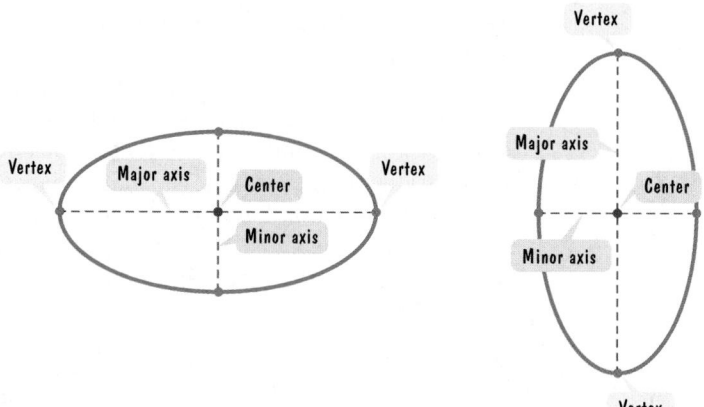

Figure 13.11 Horizontal and vertical elongations of an ellipse

Standard Form of the Equation of an Ellipse

The rectangular coordinate system gives us a unique way of describing an ellipse. It enables us to translate an ellipse's geometric definition into an algebraic equation.

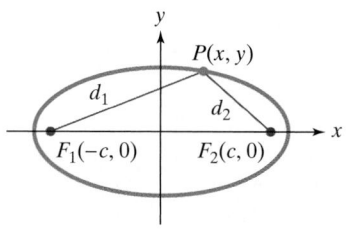

Figure 13.12

We start with Figure 13.12 to obtain an ellipse's equation. We've placed an ellipse that is elongated horizontally into a rectangular coordinate system. The foci are on the x-axis at $(-c, 0)$ and $(c, 0)$, as in Figure 13.12. In this way, the center of the ellipse is at the origin. We let (x, y) represent the coordinates of any point, P, on the ellipse.

What does the definition of an ellipse tell us about the point (x, y) in Figure 13.12? For any point (x, y) on the ellipse, the sum of the distances to the two foci, $d_1 + d_2$, must be constant. We denote this constant by $2a$. Thus, the point (x, y) is on the ellipse if and only if

$$d_1 + d_2 = 2a.$$

$$\sqrt{(x + c)^2 + y^2} + \sqrt{(x - c)^2 + y^2} = 2a \qquad \text{Use the distance formula.}$$

After eliminating radicals and simplifying, we obtain

$$(a^2 - c^2)x^2 + a^2y^2 = a^2(a^2 - c^2).$$

In order to simplify this equation, let $b^2 = a^2 - c^2$. Substituting b^2 for $a^2 - c^2$, we obtain

$$b^2x^2 + a^2y^2 = a^2b^2$$

$$\frac{b^2x^2}{a^2b^2} + \frac{a^2y^2}{a^2b^2} = \frac{a^2b^2}{a^2b^2} \qquad \text{Divide both sides by } a^2b^2.$$

$$\frac{x^2}{a^2} + \frac{y^2}{b^2} = 1 \qquad \text{Simplify.}$$

This last equation is the **standard form of the equation of an ellipse**. There are two such equations, one for a horizontal major axis and one for a vertical major axis.

Standard Forms of the Equations of an Ellipse

The **standard form of the equation of an ellipse** with center at the origin, and major and minor axes of lengths $2a$ and $2b$ (where a and b are positive, and $a^2 > b^2$) is

$$\frac{x^2}{a^2} + \frac{y^2}{b^2} = 1 \qquad \text{or} \qquad \frac{x^2}{b^2} + \frac{y^2}{a^2} = 1.$$

Figure 13.13 illustrates that the vertices are on the major axis, a units from the center. The foci are are on the major axis, c units from the center.

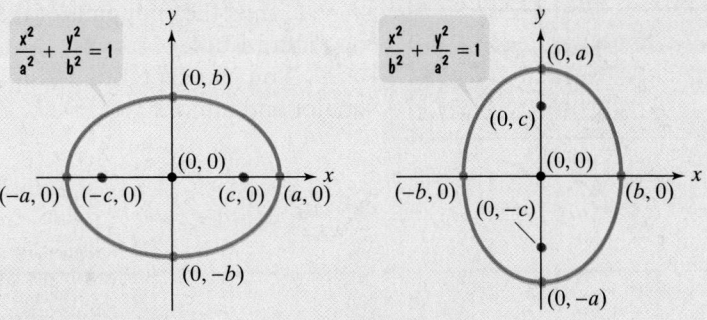

Figure 13.13 **(a)** Major axis is horizontal with length $2a$. **(b)** Major axis is vertical with length $2a$.

Using the Standard Form of the Equation of an Ellipse

We can use the standard form of an ellipse's equation to graph the ellipse.

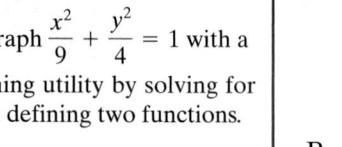 **1** Graph ellipses centered at the origin.

EXAMPLE 1 Graphing an Ellipse Centered at the Origin

Graph the ellipse: $\dfrac{x^2}{9} + \dfrac{y^2}{4} = 1$.

Solution The given equation is the standard form of an ellipse's equation with $a^2 = 9$ and $b^2 = 4$.

$$\dfrac{x^2}{9} + \dfrac{y^2}{4} = 1$$

$a^2 = 9$. This is the larger of the two denominators. $b^2 = 4$. This is the smaller of the two denominators.

Because the denominator of the x^2-term is greater than the denominator of the y^2-term, the major axis is horizontal. Based on the standard form of the equation, we know the vertices are $(-a, 0)$ and $(a, 0)$. Because $a^2 = 9$, $a = 3$. Thus, the vertices are $(-3, 0)$ and $(3, 0)$, shown in Figure 13.14.

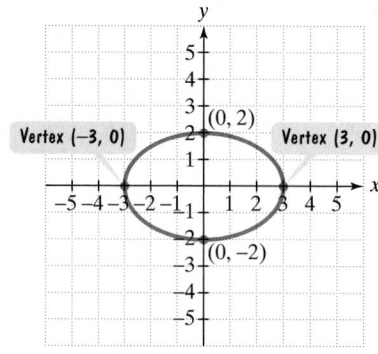

Figure 13.14 The graph of $\dfrac{x^2}{9} + \dfrac{y^2}{4} = 1$

Now let us find the endpoints of the vertical minor axis. According to the standard form of the equation, these endpoints are $(0, -b)$ and $(0, b)$. Because $b^2 = 4$, $b = 2$. Thus, the endpoints of the minor axis are $(0, -2)$ and $(0, 2)$. They are shown in Figure 13.14.

You can sketch the ellipse in Figure 13.14 by locating endpoints on the major and minor axes.

$$\dfrac{x^2}{3^2} + \dfrac{y^2}{2^2} = 1$$

Endpoints of the major axis are 3 units to the left and right of the center.   Endpoints of the minor axis are 2 units up and down from the center.

✔ **CHECK POINT 1** Graph the ellipse: $\dfrac{x^2}{36} + \dfrac{y^2}{9} = 1$.

Using Technology

We graph $\dfrac{x^2}{9} + \dfrac{y^2}{4} = 1$ with a graphing utility by solving for y and defining two functions.

$$\dfrac{y^2}{4} = 1 - \dfrac{x^2}{9}$$

$$y^2 = 4\left(1 - \dfrac{x^2}{9}\right)$$

$$y = \pm 2\sqrt{1 - \dfrac{x^2}{9}}$$

Enter

$y_1 = 2 \boxed{\sqrt{\ }} (1 \boxed{-} x \boxed{\wedge} 2 \boxed{\div} 9)$

and

$$y_2 = -y_1.$$

To see the true shape of the ellipse, use the $\boxed{\text{ZOOM SQUARE}}$ feature so that one unit on the x-axis is the same length as one unit on the y-axis.

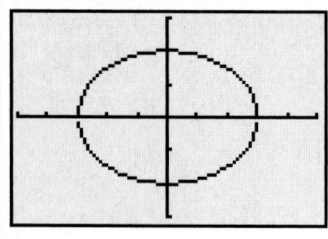

EXAMPLE 2   Graphing an Ellipse Centered at the Origin

Graph the ellipse: $25x^2 + 16y^2 = 400$.

Solution   We begin by expressing the equation in standard form. Because we want 1 on the right side, we divide both sides by 400.

$$\frac{25x^2}{400} + \frac{16y^2}{400} = \frac{400}{400}$$

$$\frac{x^2}{16} + \frac{y^2}{25} = 1$$

$b^2 = 16$. This is the smaller of the two denominators.	$a^2 = 25$. This is the larger of the two denominators.

The equation is the standard form of an ellipse's equation with $a^2 = 25$ and $b^2 = 16$. Because the denominator of the y^2-term is greater than the denominator of the x^2-term, the major axis is vertical. Based on the standard form of the equation, we know the vertices are $(0, -a)$ and $(0, a)$. Because $a^2 = 25$, $a = 5$. Thus, the vertices are $(0, -5)$ and $(0, 5)$, shown in Figure 13.15.

Now let us find the endpoints of the horizontal minor axis. According to the standard form of the equation, these endpoints are $(-b, 0)$ and $(b, 0)$. Because $b^2 = 16$, $b = 4$. Thus, the endpoints of the minor axis are $(-4, 0)$ and $(4, 0)$. They are shown in Figure 13.15.

You can sketch the ellipse in Figure 13.15 by locating endpoints on the major and minor axes:

$$\frac{x^2}{4^2} + \frac{y^2}{5^2} = 1.$$

Endpoints of the minor axis are 4 units to the left and right of the center.	Endpoints of the major axis are 5 units up and down from the center.

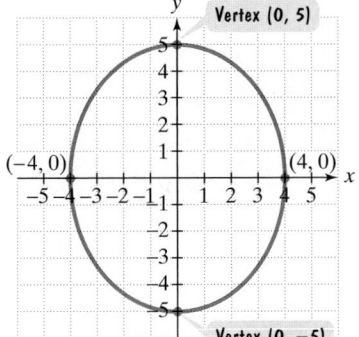

Figure 13.15 The graph of $\frac{x^2}{16} + \frac{y^2}{25} = 1$

✔ **CHECK POINT 2**   Graph the ellipse: $16x^2 + 9y^2 = 144$.

2 Graph ellipses not centered at the origin.

Ellipses Centered at (h, k)

Horizontal and vertical translations can be used to graph ellipses that are not centered at the origin. Figure 13.16 illustrates that the graphs of

$$\frac{(x - h)^2}{a^2} + \frac{(y - k)^2}{b^2} = 1 \quad \text{and} \quad \frac{x^2}{a^2} + \frac{y^2}{b^2} = 1$$

have the same size and shape. However, the graph of the first equation is centered at (h, k) rather than at the origin.

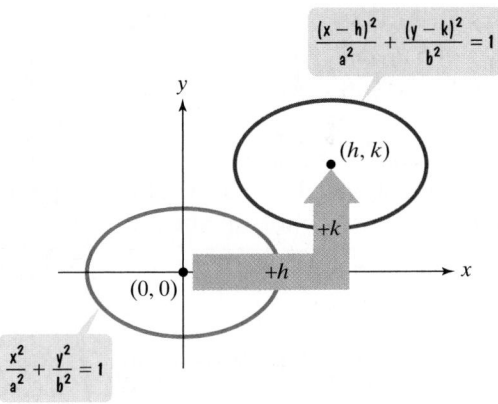

Figure 13.16 Translating an ellipse's graph

Table 13.1 gives the standard forms of equations of ellipses centered at (h, k). Figure 13.17 shows their graphs.

Table 13.1   Standard Forms of Equations of Ellipses Centered at (h, k)

Equation	Center	Major Axis	Vertices
$\dfrac{(x-h)^2}{a^2} + \dfrac{(y-k)^2}{b^2} = 1,$ $a^2 > b^2$	(h, k)	Parallel to the x-axis, horizontal	$(h - a, k)$ $(h + a, k)$
$\dfrac{(x-h)^2}{b^2} + \dfrac{(y-k)^2}{a^2} = 1,$ $a^2 > b^2$	(h, k)	Parallel to the y-axis, vertical	$(h, k - a)$ $(h, k + a)$

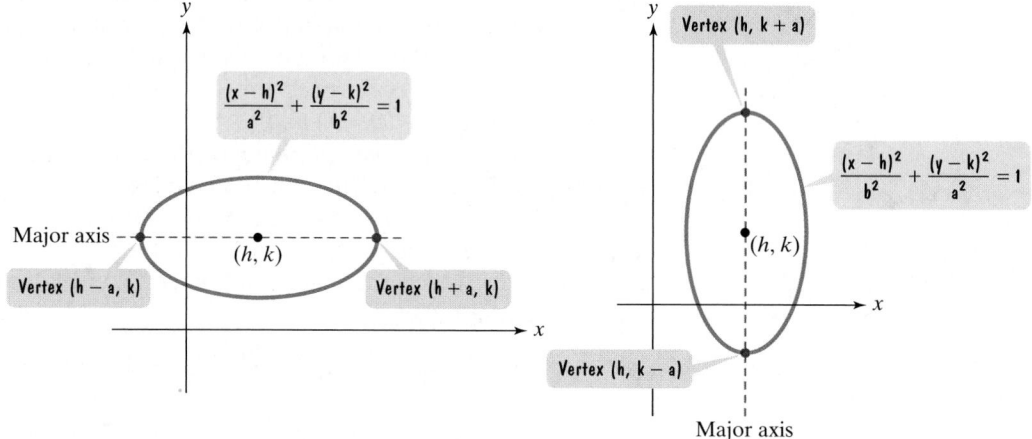

Figure 13.17 Graphs of ellipses centered at (h, k).

EXAMPLE 3   Graphing an Ellipse Centered at (h, k)

Graph the ellipse: $\dfrac{(x-1)^2}{4} + \dfrac{(y+2)^2}{9} = 1$.

Solution   In order to graph the ellipse, we need to know its center, (h, k). In the standard forms of equations centered at (h, k), h is the number subtracted from x and k is the number subtracted from y.

$$\underbrace{\frac{(x-1)^2}{4}}_{\substack{\text{This is } (x-h)^2 \\ \text{with } h = 1}} + \underbrace{\frac{(y-(-2))^2}{9}}_{\substack{\text{This is } (y-k)^2 \\ \text{with } k = -2.}} = 1$$

We see that $h = 1$ and $k = -2$. Thus, the center of the ellipse, (h, k), is $(1, -2)$. We can graph the ellipse by locating endpoints on the major and minor axes. To do this, we must identify a^2 and b^2.

$$\frac{(x-1)^2}{\underset{\substack{b^2 = 4. \text{ This is the smaller} \\ \text{of the two denominators.}}{4}} + \frac{(y+2)^2}{\underset{\substack{a^2 = 9. \text{ This is the larger} \\ \text{of the two denominators.}}{9}} = 1$$

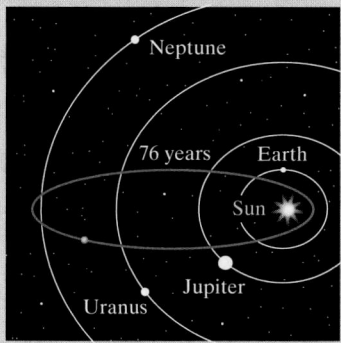
In Figure 13.20, we've constructed a coordinate system with the *x*-axis on the ground and the origin at the center of the archway. Also shown is the truck, whose height is 9 feet.

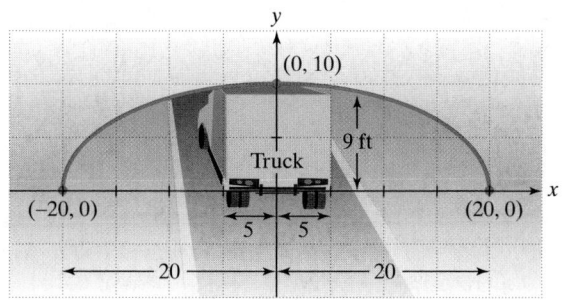

Figure 13.20

Using the equation $\dfrac{x^2}{a^2} + \dfrac{y^2}{b^2} = 1$, we can express the equation of the blue archway in Figure 13.20 as $\dfrac{x^2}{20^2} + \dfrac{y^2}{10^2} = 1$ or $\dfrac{x^2}{400} + \dfrac{y^2}{100} = 1$.

As shown in Figure 13.20, the edge of the 10-foot-wide truck corresponds to $x = 5$. We find the height of the archway 5 feet from the center by substituting 5 for *x* and solving for *y*.

$$\frac{5^2}{400} + \frac{y^2}{100} = 1 \qquad \text{Substitute 5 for x.}$$

$$\frac{25}{400} + \frac{y^2}{100} = 1 \qquad \text{Square 5.}$$

$$400\left(\frac{25}{400} + \frac{y^2}{100}\right) = 400(1) \qquad \text{Clear fractions by multiplying both sides by 400.}$$

$$25 + 4y^2 = 400 \qquad \text{Use the distributive property and simplify.}$$

$$4y^2 = 375 \qquad \text{Subtract 25 from both sides.}$$

$$y^2 = \frac{375}{4} \qquad \text{Divide both sides by 4.}$$

$$y = \sqrt{\frac{375}{4}} \qquad \text{Take only the positive square root. The archway is above the x-axis, so y is nonnegative.}$$

$$\approx 9.68$$

Thus, the height of the archway 5 feet from the center is approximately 9.68 feet. Because your truck's height is 9 feet, there is enough room for the truck to clear the archway. ∎

✔ **CHECK POINT 4**   Will a truck that is 12 feet wide and has a height of 9 feet clear the opening of the archway described in Example 4?

EXERCISE SET 13.2

 Practice Exercises

In Exercises 1–16, graph each ellipse.

1. $\dfrac{x^2}{16} + \dfrac{y^2}{4} = 1$ **2.** $\dfrac{x^2}{25} + \dfrac{y^2}{16} = 1$

3. $\dfrac{x^2}{9} + \dfrac{y^2}{36} = 1$ **4.** $\dfrac{x^2}{16} + \dfrac{y^2}{49} = 1$

5. $\dfrac{x^2}{25} + \dfrac{y^2}{64} = 1$ **6.** $\dfrac{x^2}{49} + \dfrac{y^2}{36} = 1$

7. $\dfrac{x^2}{49} + \dfrac{y^2}{81} = 1$ **8.** $\dfrac{x^2}{64} + \dfrac{y^2}{100} = 1$

9. $25x^2 + 4y^2 = 100$ **10.** $9x^2 + 4y^2 = 36$

11. $4x^2 + 16y^2 = 64$ **12.** $16x^2 + 9y^2 = 144$

13. $25x^2 + 9y^2 = 225$ **14.** $4x^2 + 25y^2 = 100$

15. $x^2 + 2y^2 = 8$ **16.** $12x^2 + 4y^2 = 36$

In Exercises 17–20, find the standard form of the equation of each ellipse.

17.

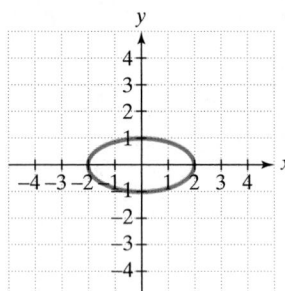

18.

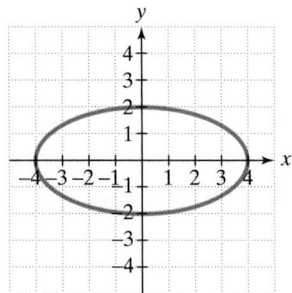

19.

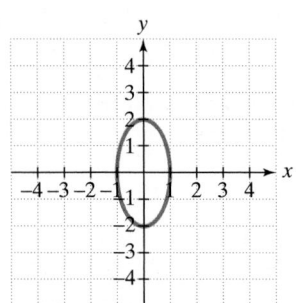

20.

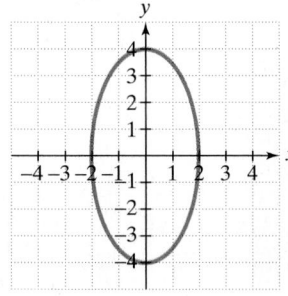

In Exercises 21–32, graph each ellipse.

21. $\dfrac{(x - 2)^2}{9} + \dfrac{(y - 1)^2}{4} = 1$

22. $\dfrac{(x - 1)^2}{16} + \dfrac{(y + 2)^2}{9} = 1$

23. $(x + 3)^2 + 4(y - 2)^2 = 16$

24. $(x - 3)^2 + 9(y + 2)^2 = 18$

25. $\dfrac{(x - 4)^2}{9} + \dfrac{(y + 2)^2}{25} = 1$

26. $\dfrac{(x - 3)^2}{9} + \dfrac{(y + 1)^2}{16} = 1$

27. $\dfrac{x^2}{25} + \dfrac{(y - 2)^2}{36} = 1$

28. $\dfrac{(x - 4)^2}{4} + \dfrac{y^2}{25} = 1$

29. $\dfrac{(x + 3)^2}{9} + (y - 2)^2 = 1$

30. $\dfrac{(x + 2)^2}{16} + (y - 3)^2 = 1$

31. $9(x - 1)^2 + 4(y + 3)^2 = 36$

32. $36(x + 4)^2 + (y + 3)^2 = 36$

 Application Exercises

33. Will a truck that is 8 feet wide carrying a load that reaches 7 feet above the ground clear the semielliptical arch on the one-way road that passes under the bridge shown in the figure?

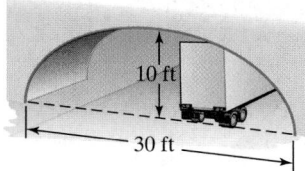

34. A semielliptic archway has a height of 20 feet and a width of 50 feet, as shown in the figure. Can a truck 14 feet high and 10 feet wide drive under the archway without going into the other lane?

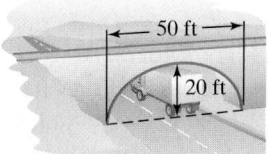

35. The elliptical ceiling in Statuary Hall in the U.S. Capitol Building is 96 feet long and 23 feet tall.

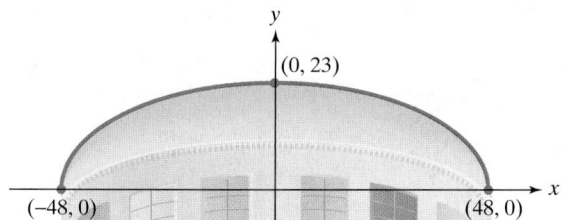

a. Using the rectangular coordinate system in the figure shown, write the standard form of the equation of the elliptical ceiling.

b. John Quincy Adams discovered that he could overhear the conversations of opposing party leaders near the left side of the chamber if he situated his desk at the focus at the right side of the chamber. How far from the center of the ellipse along the major axis did Adams situate his desk? (Round to the nearest foot.)

36. If an elliptical whispering room has a height of 30 feet and a width of 100 feet, where should two people stand if they would like to whisper back and forth and be heard?

Writing in Mathematics

37. What is an ellipse?

38. Describe how to graph $\dfrac{x^2}{25} + \dfrac{y^2}{16} = 1$.

39. Describe one similarity and one difference between the graphs of $\dfrac{x^2}{25} + \dfrac{y^2}{16} = 1$ and $\dfrac{x^2}{16} + \dfrac{y^2}{25} = 1$.

40. Describe one similarity and one difference between the graphs of $\dfrac{x^2}{25} + \dfrac{y^2}{16} = 1$ and $\dfrac{(x-1)^2}{25} + \dfrac{(y-1)^2}{16} = 1$.

41. An elliptipool is an elliptical pool table with only one pocket. A pool shark places a ball on the table, hits it in what appears to be a random direction, and yet it bounces off the edge, falling directly into the pocket. Explain why this happens.

Critical Thinking Exercises

42. Which one of the following is true?
  a. The graphs of $x^2 + y^2 = 16$ and $\dfrac{x^2}{4} + \dfrac{y^2}{9} = 1$ do not intersect.
  b. Some ellipses have equations that define y as a function of x.
  c. The graph of $\dfrac{x^2}{9} + \dfrac{y^2}{4} = 0$ is an ellipse.
  d. None of the above is true.

43. Find the standard form of the equation of an ellipse with vertices at $(0, -6)$ and $(0, 6)$, passing through $(2, -4)$.

In Exercises 44–45, convert each equation to standard form by completing the square on x and y. Then graph the ellipse.

44. $9x^2 + 25y^2 - 36x + 50y - 164 = 0$

45. $4x^2 + 9y^2 - 32x + 36y + 64 = 0$

46. An Earth satellite has an elliptical orbit described by

$$\frac{x^2}{(5000)^2} + \frac{y^2}{(4750)^2} = 1.$$

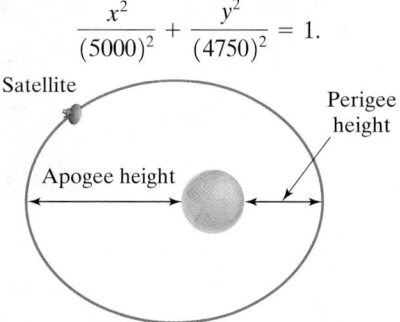

(All units are in miles.) The coordinates of the center of Earth are $(16, 0)$.

a. The perigee of the satellite's orbit is the point that is nearest Earth's center. If the radius of Earth is approximately 4000 miles, find the distance of the perigee above Earth's surface.

b. The apogee of the satellite's orbit is the point that is the greatest distance from Earth's center. Find the distance of the apogee above Earth's surface.

47. The equation of the red ellipse shown below is

$$\frac{x^2}{25} + \frac{y^2}{9} = 1.$$

Write the equation for each circle shown in the figure.

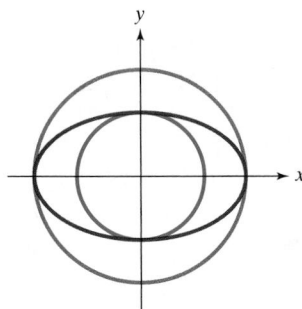

48. What happens to the shape of the graph of $\frac{x^2}{a^2} + \frac{y^2}{b^2} = 1$ as $\frac{c}{a}$ is close to zero?

Technology Exercises

49. Use a graphing utility to graph any five of the ellipses that you graphed by hand in Exercises 1–16.

50. Use a graphing utility to graph any three of the ellipses that you graphed by hand in Exercises 21–32. First solve the given equation for y by using the square root method. Enter each of the two resulting equations to produce each half of the ellipse.

Review Exercises

51. Factor completely: $x^3 + 2x^2 - 4x - 8$. (Section 6.5, Example 4)

52. Simplify: $\sqrt[3]{40x^4y^7}$. (Section 10.3, Example 5)

53. Solve: $\frac{2}{x+2} + \frac{4}{x-2} = \frac{x-1}{x^2-4}$. (Section 7.6, Example 4)

▶ **SECTION 13.3   The Hyperbola**

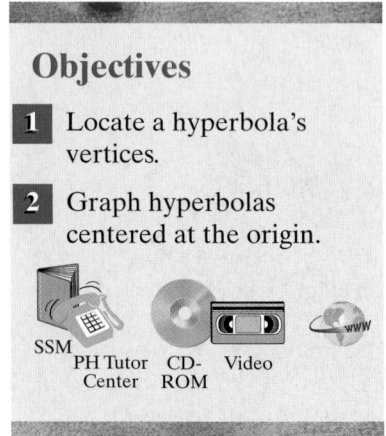

Objectives

1 Locate a hyperbola's vertices.

2 Graph hyperbolas centered at the origin.

SSM
PH Tutor   CD-   Video
Center   ROM

St. Mary's Cathedral

Conic sections are often used to create unusual architectural designs. The top of St. Mary's Cathedral in San Francisco is a 2135-cubic-foot dome with walls rising 200 feet above the floor and supported by four massive concrete pylons that extend 94 feet into the ground. Cross sections of the roof are parabolas and hyperbolas. In this section, we study the curve with two parts known as the hyperbola.

Figure 13.21 Casting hyperbolic shadows

Definition of a Hyperbola

Figure 13.21 shows a cylindrical lampshade casting two shadows on a wall. These shadows indicate the distinguishing feature of hyperbolas: Their graphs contain two disjoint parts, called **branches**. Although each branch might look like a parabola, its shape is actually quite different.

The definition of a hyperbola is similar to that of the ellipse. For the ellipse, the *sum* of the distances to the foci is a constant. By contrast, for a hyperbola the *difference* of the distances to the foci is a constant.

Definition of a Hyperbola

A **hyperbola** is the set of points in a plane the difference of whose distances from two fixed points (called foci) is a constant.

Figure 13.22 illustrates the two branches of a hyperbola. The line through the foci intersects the hyperbola at two points, called the **vertices**. The line segment that joins the vertices is the **transverse axis**. The midpoint of the transverse axis is the **center** of the hyperbola. Notice that the center lies midway between the vertices, as well as midway between the foci.

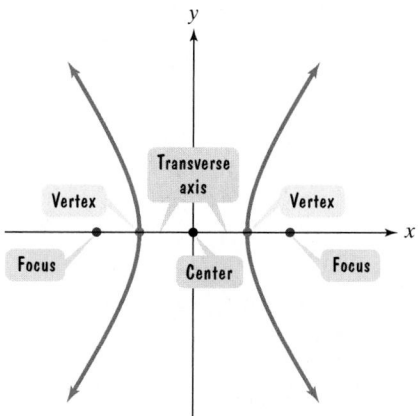

Figure 13.22 The two branches of a hyperbola

Standard Form of the Equation of a Hyperbola

The rectangular coordinate system enables us to translate a hyperbola's geometric definition into an algebraic equation. Figure 13.23 is our starting point for obtaining an equation. We place the foci, F_1 and F_2, on the x-axis at the points $(-c, 0)$ and $(c, 0)$. Note that the center of this hyperbola is at the origin. We let (x, y) represent the coordinates of any point, P, on the hyperbola.

What does the definition of a hyperbola tell us about the point (x, y) in Figure 13.23? For any point (x, y) on the hyperbola, the absolute value of the difference of the distances from the two foci, $|d_2 - d_1|$, must be constant. We denote this constant by $2a$, just as we did for the ellipse. Thus, the point (x, y) is on the hyperbola if and only if

$$|d_2 - d_1| = 2a$$

$$\left| \sqrt{(x + c)^2 + y^2} - \sqrt{(x - c)^2 + y^2} \right| = 2a \qquad \text{Use the distance formula.}$$

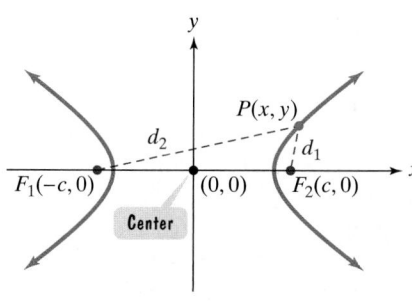

Figure 13.23

After eliminating radicals and simplifying, we obtain

$$(c^2 - a^2)x^2 - a^2y^2 = a^2(c^2 - a^2).$$

For convenience, let $b^2 = c^2 - a^2$. Substituting b^2 for $c^2 - a^2$ in the preceding equation, we obtain

$$b^2x^2 - a^2y^2 = a^2b^2.$$

$$\frac{b^2x^2}{a^2b^2} - \frac{a^2y^2}{a^2b^2} = \frac{a^2b^2}{a^2b^2} \qquad \text{Divide both sides by } a^2b^2.$$

$$\frac{x^2}{a^2} - \frac{y^2}{b^2} = 1. \qquad \text{Simplify.}$$

This last equation is called the **standard form of the equation of a hyperbola**. There are two such equations. The first is for a hyperbola in which the transverse axis lies on the x-axis. The second is for a hyperbola in which the transverse axis lies on the y-axis.

> ### Standard Forms of the Equations of a Hyperbola
> The **standard form of the equation of a hyperbola** with center at the origin is
>
> $$\frac{x^2}{a^2} - \frac{y^2}{b^2} = 1 \qquad \text{or} \qquad \frac{y^2}{a^2} - \frac{x^2}{b^2} = 1.$$
>
> Figure 13.24 illustrates that for the equation on the left, the transverse axis lies on the x-axis. For the equation on the right, the transverse axis lies on the y-axis. The vertices are a units from the center and the foci are c units from the center.

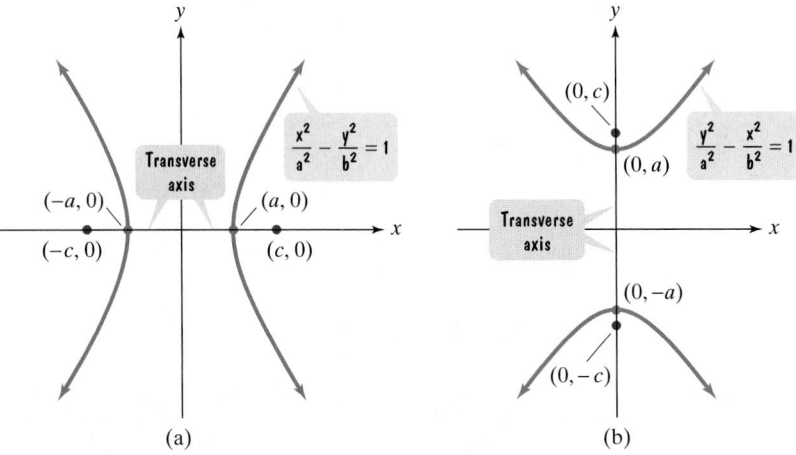

Figure 13.24 **(a)** Transverse axis lies on the x-axis. **(b)** Transverse axis lies on the y-axis.

(a)   (b)

1 Locate a hyperbola's vertices.

Using the Standard Form of the Equation of a Hyperbola

We can use the standard form of the equation of a hyperbola to find its vertices. Because the vertices are a units from the center, begin by identifying a^2 in the equation. In the standard form of a hyperbola's equation, a^2 **is the number under the variable whose term is preceded by a plus sign** (+). If the x^2-term is preceded by a plus sign, the transverse axis lies along the x-axis. Thus, the vertices are a units to the left and right of the origin. If the y^2-term is preceded by a plus sign, the transverse axis lies along the y-axis. Thus, the vertices are a units above and below the origin.

EXAMPLE 1   Finding Vertices from a Hyperbola's Equation

Find the vertices for each of the following hyperbolas with the given equation:

a. $\dfrac{x^2}{16} - \dfrac{y^2}{9} = 1$ **b.** $\dfrac{y^2}{9} - \dfrac{x^2}{16} = 1.$

Solution   Both equations are in standard form. We begin by identifying a^2 and b^2 in each equation.

a. The first equation is in the form $\dfrac{x^2}{a^2} - \dfrac{y^2}{b^2} = 1.$

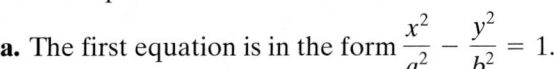

$$\frac{x^2}{16} - \frac{y^2}{9} = 1$$

$a^2 = 16.$ This is the number in the denominator of the term preceded by a plus sign.

$b^2 = 9.$ This is the number in the denominator of the term preceded by a minus sign.

Because the x^2-term is preceded by a plus sign, the transverse axis lies along the x-axis. Thus, the vertices are a units to the *left* and *right* of the origin. Based on the standard form of the equation, we know the vertices are $(-a, 0)$ and $(a, 0)$. Because $a^2 = 16$, $a = 4$. Thus, the vertices are $(-4, 0)$ and $(4, 0)$, shown in Figure 13.25.

b. The second given equation is in the form $\dfrac{y^2}{a^2} - \dfrac{x^2}{b^2} = 1.$

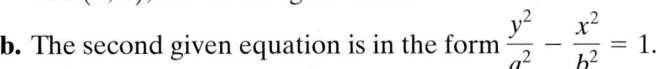

$$\frac{y^2}{9} - \frac{x^2}{16} = 1$$

$a^2 = 9.$ This is the number in the denominator of the term preceded by a plus sign.

$b^2 = 16.$ This is the number in the denominator of the term preceded by a minus sign.

Because the y^2-term is preceded by a plus sign, the transverse axis lies along the y-axis. Thus, the vertices are a units *above* and *below* the origin. Based on the standard form of the equation, we know the vertices are $(0, -a)$ and $(0, a)$. Because $a^2 = 9$, $a = 3$. Thus, the vertices are $(0, -3)$ and $(0, 3)$, shown in Figure 13.26. ∎

✔ **CHECK POINT 1**   Find the vertices for each of the following hyperbolas with the given equation:

a. $\dfrac{x^2}{25} - \dfrac{y^2}{16} = 1$

b. $\dfrac{y^2}{25} - \dfrac{x^2}{16} = 1.$

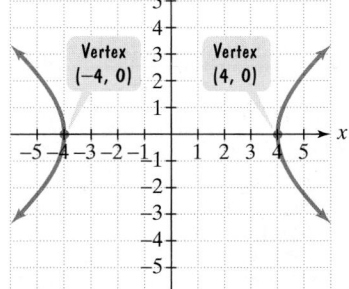

Figure 13.25 The graph of $\dfrac{x^2}{16} - \dfrac{y^2}{9} = 1$

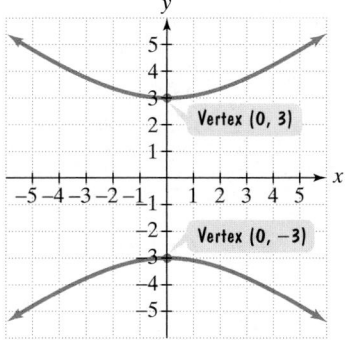

Figure 13.26 The graph of $\dfrac{y^2}{9} - \dfrac{x^2}{16} = 1$

The Asymptotes of a Hyperbola

As x and y get larger, the two branches of the graph of a hyperbola approach a pair of intersecting lines, called **asymptotes**. The asymptotes pass through the center of the hyperbola and are helpful in graphing hyperbolas.

Figure 13.27 shows the asymptotes for the graphs of hyperbolas centered at the origin. The asymptotes pass through the corners of a rectangle. Note that the dimensions of this rectangle are 2a by 2b.

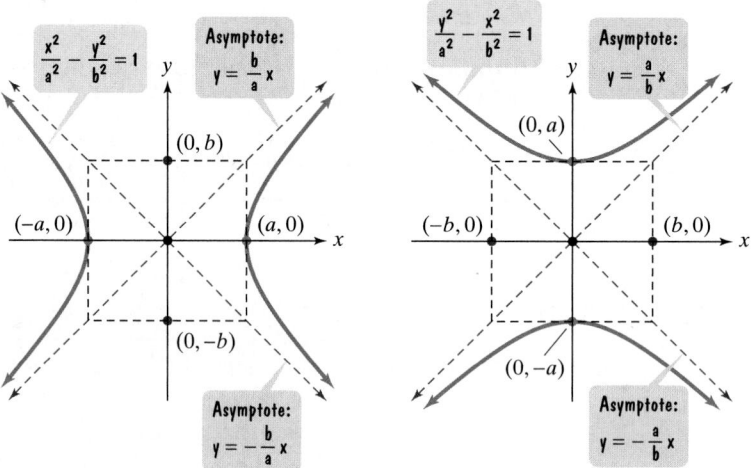

Figure 13.27 Asymptotes of a hyperbola

Why are $y = \pm \dfrac{b}{a} x$ the asymptotes for a hyperbola whose transverse axis is horizontal? The proof can be found in the appendix.

2 Graph hyperbolas centered at the origin.

Graphing Hyperbolas Centered at the Origin

Hyperbolas are graphed using vertices and asymptotes.

Graphing Hyperbolas

1. Locate the vertices.
2. Draw the rectangle centered at the origin with sides parallel to the axes, crossing one axis at $\pm a$ and the other at $\pm b$.
3. Draw the diagonals of this rectangle and extend them to obtain the asymptotes.
4. Draw the two branches of the hyperbola by starting at each vertex and approaching the asymptotes.

The rectangle in step 2 and the asymptotes in step 3 are drawn using dashed lines to show that they are not part of the hyperbola.

EXAMPLE 2   Graphing a Hyperbola

Graph the hyperbola: $\dfrac{x^2}{25} - \dfrac{y^2}{16} = 1.$

Solution

Step 1 Locate the vertices. The given equation is in the form $\frac{x^2}{a^2} - \frac{y^2}{b^2} = 1$, with $a^2 = 25$ and $b^2 = 16$.

$$\frac{x^2}{25} - \frac{y^2}{16} = 1$$

$$a^2 = 25 \qquad b^2 = 16$$

Based on the standard form of the equation with the transverse axis on the x-axis, we know that the vertices are $(-a, 0)$ and $(a, 0)$. Because $a^2 = 25$, $a = 5$. Thus, the vertices are $(-5, 0)$ and $(5, 0)$, shown in Figure 13.28.

Step 2 Draw a rectangle. Because $a^2 = 25$ and $b^2 = 16$, $a = 5$ and $b = 4$. We construct a rectangle to find the asymptotes, using -5 and 5 on the x-axis (the vertices are located here) and -4 and 4 on the y-axis. The rectangle passes through these four points, shown using dashed lines in Figure 13.28.

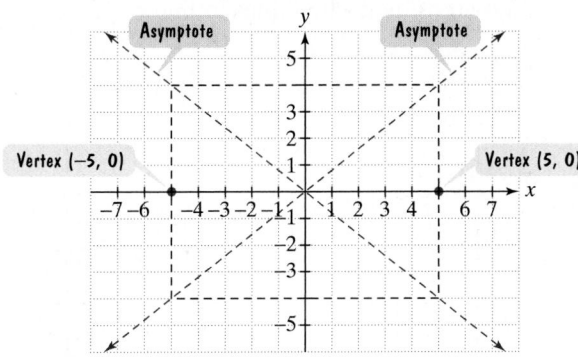

Figure 13.28 Preparing to graph $\frac{x^2}{25} - \frac{y^2}{16} = 1$

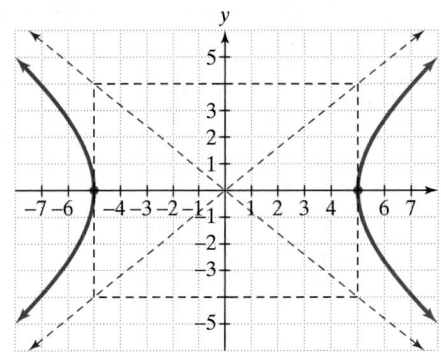

Figure 13.29 The graph of $\frac{x^2}{25} - \frac{y^2}{16} = 1$

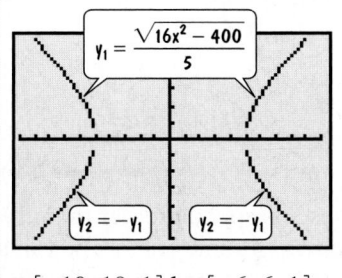
Step 3 Draw extended diagonals for the rectangle to obtain the asymptotes. We draw dashed lines through the opposite corners of the rectangle, shown in Figure 13.28, to obtain the graph of the asymptotes.

Step 4 Draw the two branches of the hyperbola by starting at each vertex and approaching the asymptotes. The hyperbola is shown in Figure 13.29. ∎

✔ **CHECK POINT 2** Graph the hyperbola: $\frac{x^2}{36} - \frac{y^2}{9} = 1$.

EXAMPLE 3 Graphing a Hyperbola

Graph the hyperbola: $9y^2 - 4x^2 = 36$.

Solution We begin by writing the equation in standard form. The right side should be 1, so we divide both sides by 36.

$$\frac{9y^2}{36} - \frac{4x^2}{36} = \frac{36}{36}$$

$$\frac{y^2}{4} - \frac{x^2}{9} = 1 \qquad \text{Simplify. The right side is now 1.}$$

Now we are ready to use our four-step procedure for graphing hyperbolas.

Step 1 Locate the vertices. The equation that we obtained is in the form $\frac{y^2}{a^2} - \frac{x^2}{b^2} = 1$, with $a^2 = 4$ and $b^2 = 9$.

$$\frac{y^2}{4} - \frac{x^2}{9} = 1$$

$$\boxed{a^2 = 4} \quad \boxed{b^2 = 9}$$

Based on the standard form of the equation with the transverse axis on the y-axis, we know that the vertices are $(0, -a)$ and $(0, a)$. Because $a^2 = 4$, $a = 2$. Thus, the vertices are $(0, -2)$ and $(0, 2)$, shown in Figure 13.30.

Step 2 Draw a rectangle. Because $a^2 = 4$ and $b^2 = 9$, $a = 2$ and $b = 3$. We construct a rectangle to find the asymptotes, using -2 and 2 on the y-axis (the vertices are located here) and -3 and 3 on the x-axis. The rectangle passes through these four points, shown using dashed lines in Figure 13.30.

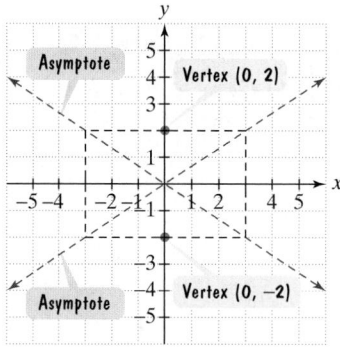

Figure 13.30 Preparing to graph $\frac{y^2}{4} - \frac{x^2}{9} = 1$

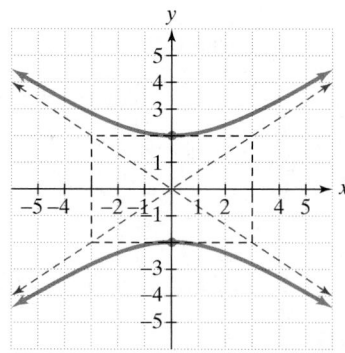

Figure 13.31 The graph of $\frac{y^2}{4} - \frac{x^2}{9} = 1$

Step 3 Draw extended diagonals of the rectangle to obtain the asymptotes. We draw dashed lines through the opposite corners of the rectangle, shown in Figure 13.30, to obtain the graph of the asymptotes.

Step 4 Draw the two branches of the hyperbola by starting at each vertex and approaching the asymptotes. The hyperbola is shown in Figure 13.31. ∎

✔ **CHECK POINT 3** Graph the hyperbola: $y^2 - 4x^2 = 4$.

Applications

Hyperbolas have many applications. When a jet flies at a speed greater than the speed of sound, the shock wave that is created is heard as a sonic boom. The wave has the shape of a cone. The shape formed as the cone hits the ground is one branch of a hyperbola.

Where Exactly Am I?

The hyperbola is the basis for the navigational system LORAN (for long-range navigation), used by a ship or aircraft to determine its location. The measured time-of-arrival difference between signals transmitted from two ground stations determines the hyperbola on which the ship or aircraft is located. The process is then repeated by taking a similar time-difference reading from a second pair of stations, determining a second hyperbola. The point of intersection of the two hyperbolas is the location of the ship or aircraft.

LORAN will eventually be replaced by the Global Positioning System. Using 24 satellites that orbit at 11,000 miles above Earth, the system is able to show you your exact position on Earth anytime, in any weather, anywhere.

Halley's Comet, a permanent part of our solar system, travels around the sun in an elliptical orbit. Other comets pass through the solar system only once, following a hyperbolic path with the sun as a focus.

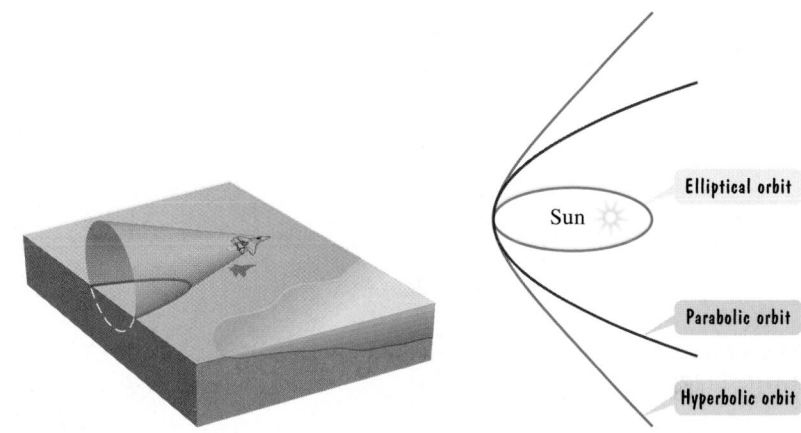

Hyperbolas are of practical importance in fields ranging from architecture to navigation. Cooling towers used in the design for nuclear power plants have cross sections that are both ellipses and hyperbolas. Three-dimensional solids whose cross sections are hyperbolas are used in some rather unique architectural creations, including the TWA building at Kennedy Airport and the St. Louis Science Center Planetarium.

EXERCISE SET 13.3

Practice Exercises

In Exercises 1–4, find the vertices of each hyperbola with the given equation. Then match each equation to one of the graphs that are shown and labeled **a–d.**

1. $\dfrac{x^2}{4} - \dfrac{y^2}{1} = 1$

2. $\dfrac{x^2}{1} - \dfrac{y^2}{4} = 1$

3. $\dfrac{y^2}{4} - \dfrac{x^2}{1} = 1$

4. $\dfrac{y^2}{1} - \dfrac{x^2}{4} = 1$

a.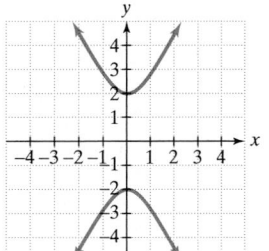

b.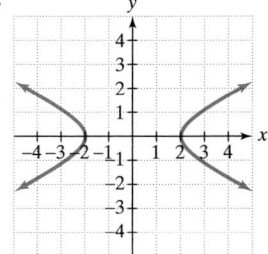

c.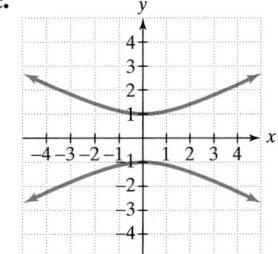

d.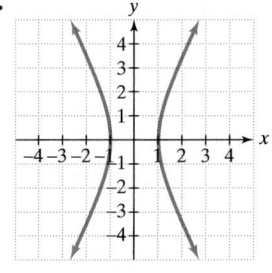

In Exercises 5–18, use vertices and asymptotes to graph each hyperbola.

5. $\dfrac{x^2}{9} - \dfrac{y^2}{25} = 1$

6. $\dfrac{x^2}{16} - \dfrac{y^2}{25} = 1$

7. $\dfrac{x^2}{100} - \dfrac{y^2}{64} = 1$

8. $\dfrac{x^2}{144} - \dfrac{y^2}{81} = 1$

9. $\dfrac{y^2}{16} - \dfrac{x^2}{36} = 1$

10. $\dfrac{y^2}{25} - \dfrac{x^2}{64} = 1$

11. $\dfrac{y^2}{36} - \dfrac{x^2}{25} = 1$

12. $\dfrac{y^2}{100} - \dfrac{x^2}{49} = 1$

13. $9x^2 - 4y^2 = 36$

14. $4x^2 - 25y^2 = 100$

15. $9y^2 - 25x^2 = 225$

16. $16y^2 - 9x^2 = 144$

17. $4x^2 = 4 + y^2$

18. $25y^2 = 225 + 9x^2$

In Exercises 19–22, find the standard form of the equation of each hyperbola.

19.

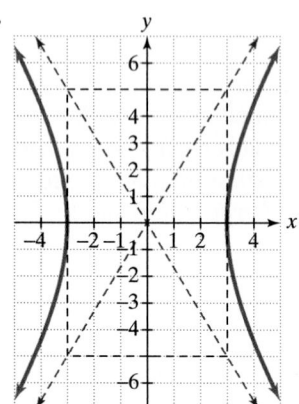

20.

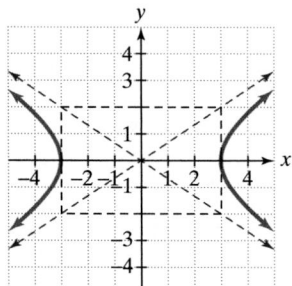

21.

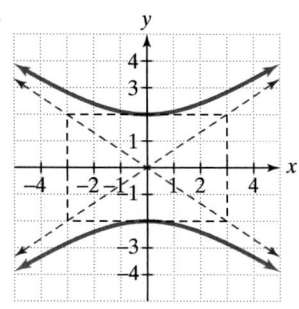

22.

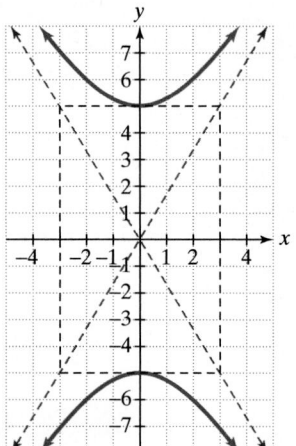

Application Exercises

23. An architect designs two houses that are shaped and positioned like a part of the branches of the hyperbola whose equation is $625y^2 - 400x^2 = 250{,}000$, where x and y are in yards. How far apart are the houses at their closest point?

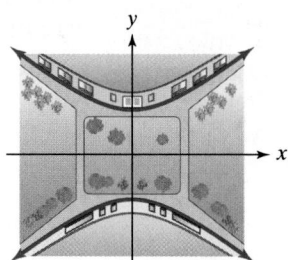

24. Scattering experiments, in which moving particles are deflected by various forces, led to the concept of the nucleus of an atom. In 1911, the physicist Ernest Rutherford (1871–1937) discovered that when alpha particles are directed toward the nuclei of gold atoms, they are eventually deflected along hyperbolic paths, illustrated in the figure. If a particle gets as close as 3 units to the nucleus along a hyperbolic path with an asymptote given by $y = \frac{1}{2}x$, what is the equation of its path?

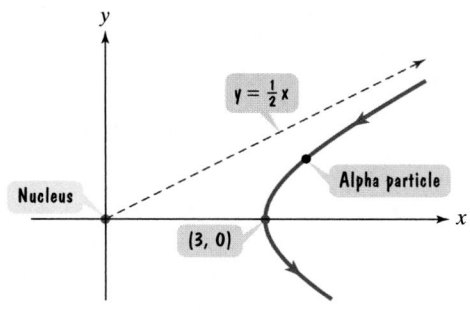

Writing in Mathematics

25. What is a hyperbola?

26. Describe how to graph $\dfrac{x^2}{9} - \dfrac{y^2}{1} = 1$.

27. Describe one similarity and one difference between the graphs of $\dfrac{x^2}{9} - \dfrac{y^2}{1} = 1$ and $\dfrac{y^2}{9} - \dfrac{x^2}{1} = 1$.

28. How can you distinguish an ellipse from a hyperbola by looking at their equations?

29. In 1992, a NASA team began a project called Spaceguard Survey, calling for an international watch for comets that might collide with Earth. Why is it more difficult to detect a possible "doomsday comet" with a hyperbolic orbit than one with an elliptical orbit?

Critical Thinking Exercises

30. Which one of the following is true?
  a. If one branch of a hyperbola is removed from a graph, then the branch that remains must define y as a function of x.
  b. All points on the asymptotes of a hyperbola also satisfy the hyperbola's equation.
  c. The graph of $\dfrac{x^2}{9} - \dfrac{y^2}{4} = 1$ does not intersect the line $y = -\dfrac{2}{3}x$.
  d. Two different hyperbolas can never share the same asymptotes.

The graph of

$$\frac{(x-h)^2}{a^2} - \frac{(y-k)^2}{b^2} = 1$$

is the same as the graph of $\dfrac{x^2}{a^2} - \dfrac{y^2}{b^2} = 1$, *except the center is at* (h, k) *rather than at the origin. Use this information to graph the hyperbolas in Exercises 31–34.*

31. $\dfrac{(x-2)^2}{16} - \dfrac{(y-3)^2}{9} = 1$

32. $\dfrac{(x+2)^2}{9} - \dfrac{(y-1)^2}{25} = 1$

33. $(x-3)^2 - 4(y+3)^2 = 4$

34. $x^2 - y^2 - 2x - 4y - 4 = 0$

In Exercises 35–36, find the standard form of the equation of each hyperbola satisfying the given conditions.

35. vertices: $(6, 0), (-6, 0)$; asymptotes: $y = 4x, y = -4x$

36. vertices: $(0, 7), (0, -7)$; asymptotes: $y = 5x, y = -5x$

Technology Exercises

37. Use a graphing utility to graph any five of the hyperbolas that you graphed by hand in Exercises 5–18.

38. Use a graphing utility to graph $\dfrac{x^2}{4} - \dfrac{y^2}{9} = 0$. Is the graph a hyperbola? In general, what is the graph of $\dfrac{x^2}{a^2} - \dfrac{y^2}{b^2} = 0$?

39. Graph $\dfrac{x^2}{a^2} - \dfrac{y^2}{b^2} = 1$ and $\dfrac{x^2}{a^2} - \dfrac{y^2}{b^2} = -1$ in the same viewing rectangle for values of a^2 and b^2 of your choice. Describe the relationship between the two graphs.

Review Exercises

40. Use intercepts and the vertex to graph the quadratic function: $y = -x^2 - 4x + 5$. (Section 11.3, Example 3)

41. Solve: $3x^2 - 11x - 4 \geq 0$. (Section 11.5, Example 2)

42. Solve: $\log_4(3x + 1) = 3$. (Section 12.5, Example 4)

▶ **SECTION 13.4** *The Parabola; Identifying Conic Sections*

Objectives

1 Graph horizontal parabolas.

2 Identify conic sections by their equations.

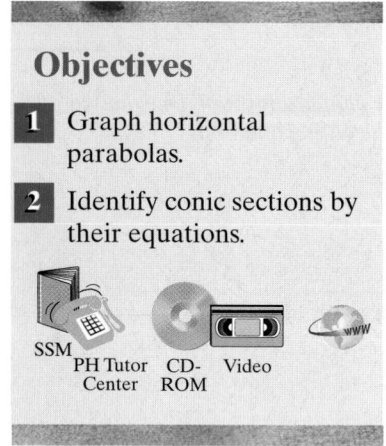

SSM
PH Tutor   CD-   Video
Center   ROM

At first glance, this image looks like columns of smoke rising from a fire into a starry sky. Those are, indeed, stars in the background, but you are not looking at ordinary smoke columns. These stand almost 6 trillion miles high and are 7000 light-years from Earth—more than 400 million times as far away as the sun.

This NASA photograph is one of a series of stunning images captured from the ends of the universe by the Hubble Space Telescope. The image shows infant star systems the size of our solar system emerging from the gas and dust that shrouded their creation. Using a parabolic mirror that is 94.5 inches in diameter, the Hubble is providing answers to many of the profound mysteries of the cosmos: How big and how old is the universe? How did the galaxies come to exist? Do other Earth-like planets orbit other sun-like stars? In this section, we study parabolas and their applications, including parabolic shapes that gather distant rays of light and focus them into spectacular images.

Definition of a Parabola

In Chapter 11, we studied parabolas, viewing them as graphs of quadratic functions in the form

$$y = a(x - h)^2 + k \quad \text{or} \quad y = ax^2 + bx + c.$$

Study Tip

Here is a summary of what you should already know about graphing parabolas.

Graphing $y = a(x - h)^2 + k$ and $y = ax^2 + bx + c$

1. If $a > 0$, the graph opens upward. If $a < 0$, the graph opens downward.
2. The vertex of $y = a(x - h)^2 + k$ is (h, k).
3. The x-coordinate of the vertex of $y = ax^2 + bx + c$ is $x = -\dfrac{b}{2a}$.

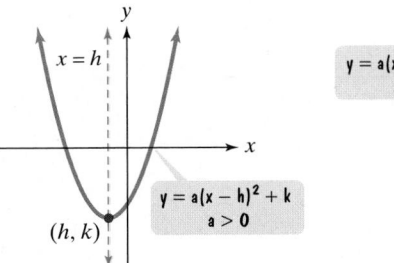
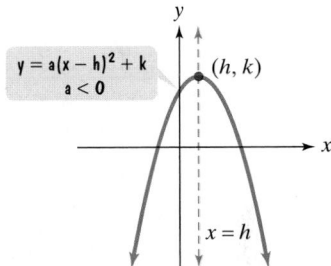

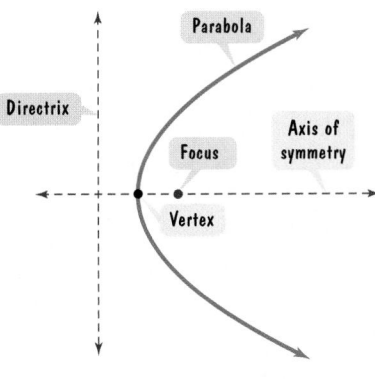

Figure 13.32

Parabolas can be given a geometric definition that enables us to include graphs that open to the left or to the right. The definitions of ellipses and hyperbolas involved two fixed points, the foci. By contrast, the definition of a parabola is based on one point and a line.

Definition of a Parabola

A **parabola** is the set of all points in a plane that are equidistant from a fixed line (the **directrix**) and a fixed point (the **focus**) that is not on the line (see Figure 13.32).

In Figure 13.32, the parabola is shown opening to the right. Find the line passing through the focus and perpendicular to the directrix. This is the **axis of symmetry** of the parabola. The point of intersection of the parabola with its axis of symmetry is the parabola's **vertex**. Notice that the vertex is midway between the focus and the directrix.

Parabolas can open to the left, right, upward, or downward. Figure 13.33 shows a basic "family" of four parabolas and their equations. Notice that the parabolas that open left or right are not functions of x because they fail the vertical line test —that is, it is possible to draw vertical lines that intersect these graphs at more than one point.

The equation $x = y^2$ interchanges the variables in the equation $y = x^2$. By interchanging x and y in the two forms of a parabola's equation,

$$y = a(x - h)^2 + k \quad \text{and} \quad y = ax^2 + bx + c,$$

we can obtain the forms of the equations of parabolas that open to the right or to the left.

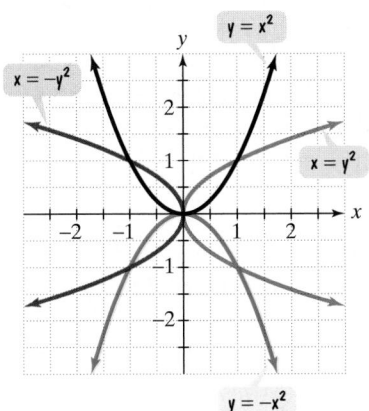

Figure 13.33

Parabolas Opening to the Left or to the Right

The graphs of

$$x = a(y - k)^2 + h \quad \text{and} \quad x = ay^2 + by + c$$

are parabolas opening to the left or to the right.

1. If $a > 0$, the graph opens to the right. If $a < 0$, the graph opens to the left.
2. The vertex of $x = a(y - k)^2 + h$ is (h, k).

3. The y-coordinate of the vertex of $x = ay^2 + by + c$ is $y = -\dfrac{b}{2a}$.

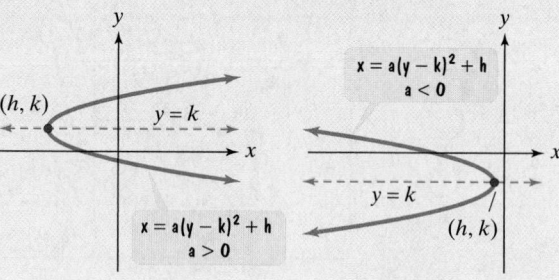

1 Graph horizontal parabolas.

Graphing Parabolas Opening to the Right or the Left

Here is a procedure for graphing horizontal parabolas that are not functions. Notice how this procedure is similar to the one that we used in Chapter 11 for graphing vertical parabolas.

Graphing Horizontal Parabolas

To graph $x = a(y - k)^2 + h$ or $x = ay^2 + by + c$:

1. Determine whether the parabola opens to the left or to the right. If $a > 0$, it opens to the right. If $a < 0$, it opens to the left.

2. Determine the vertex of the parabola. The vertex of $x = a(y - k)^2 + h$ is (h, k). The y-coordinate of the vertex of $x = ay^2 + by + c$ is $y = -\dfrac{b}{2a}$. Substitute to find the x-coordinate.

3. Find the x-intercept by replacing y with 0.

4. Find any y-intercepts by replacing x with 0. Solve the resulting quadratic equation for y.

5. Plot the intercepts and vertex. Connect these points with a smooth curve. If additional points are needed to obtain a more accurate graph, select values for y on each side of the axis of symmetry and compute values for x.

EXAMPLE 1 Graphing a Horizontal Parabola in the Form $x = a(y - k)^2 + h$

Graph: $x = -3(y - 1)^2 + 2$.

Solution   We can graph this equation by following the steps in the preceding box. We begin by identifying values for a, k, and h.

$$x = a(y - k)^2 + h$$

$a = -3$ $k = 1$ $h = 2$

$$x = -3(y - 1)^2 + 2$$

Step 1 Determine how the parabola opens. Note that a, the coefficient of y^2, is -3. Thus, $a < 0$; this negative value tells us that the parabola opens to the left.

Step 2 Find the vertex. The vertex of the parabola is at (h, k). Because $k = 1$ and $h = 2$, the parabola has its vertex at $(2, 1)$.

Step 3 Find the x-intercept. Replace y with 0 in $x = -3(y - 1)^2 + 2$.

$$x = -3(0 - 1)^2 + 2 = -3(-1)^2 + 2 = -3(1) + 2 = -1$$

The x-intercept is -1. The parabola passes through $(-1, 0)$.

Step 4 Find the y-intercepts. Replace x with 0 in the given equation.

$x = -3(y - 1)^2 + 2$	This is the given equation.
$0 = -3(y - 1)^2 + 2$	Replace x with 0.
$3(y - 1)^2 = 2$	Solve for y. Add $3(y - 1)^2$ to both sides of the equation
$(y - 1)^2 = \dfrac{2}{3}$	Divide both sides by 3.
$y - 1 = \sqrt{\dfrac{2}{3}}$ or $y - 1 = -\sqrt{\dfrac{2}{3}}$	Apply the square root property.
$y = 1 + \sqrt{\dfrac{2}{3}}$ $y = 1 - \sqrt{\dfrac{2}{3}}$	Add 1 to both sides in each equation.
$y \approx 1.8$ $y \approx 0.2$	Use a calculator.

The y-intercepts are approximately 1.8 and 0.2. The parabola passes through approximately $(0, 1.8)$ and $(0, 0.2)$.

Step 5 Graph the parabola. With a vertex at $(2, 1)$, an x-intercept at -1, and y-intercepts at approximately 1.8 and 0.2, the graph of the parabola is shown in Figure 13.34. The axis of symmetry is the horizontal line whose equation is $y = 1$.

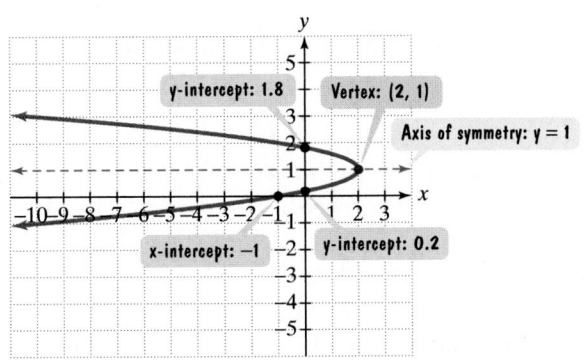

Figure 13.34 The graph of $x = -3(y - 1)^2 + 2$

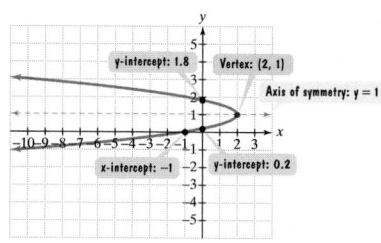

Figure 13.34, repeated

We can possibly improve our graph in Figure 13.34 by finding additional points on the parabola. Choose values of y on each side of the axis of symmetry, $y = 1$. We use $y = 2$ and $y = -1$. Then we compute values for x. The values in the table of coordinates show that $(-1, 2)$ and $(-10, -1)$ are points on the parabola. Locate each point in Figure 13.34.

x	y
-1	2
-10	-1

Choose values for y.

Compute values for x using

$x = -3(y - 1)^2 + 2$:

$x = -3(2 - 1)^2 + 2 = -3 \cdot 1^2 + 2 = -1$

$x = -3(-1 - 1)^2 + 2 = -3(-2)^2 + 2 = -10.$

∎

✔ **CHECK POINT 1**   Graph: $x = -(y - 2)^2 + 1$.

EXAMPLE 2   Graphing a Horizontal Parabola in the Form $x = ay^2 + by + c$

Graph: $x = y^2 + 4y - 5$.

Solution

Step 1 Determine how the parabola opens. Note that a, the coefficient of y^2, is 1. Thus $a > 0$; this positive value tells us that the parabola opens to the right.

Step 2 Find the vertex. We know that the y-coordinate of the vertex is $y = -\dfrac{b}{2a}$. We identify a, b, and c in $x = ay^2 + by + c$.

$$x = y^2 + 4y - 5$$

$a = 1$ $b = 4$ $c = -5$

Substitute the values of a and b into the equation for the y-coordinate:

$$y = -\frac{b}{2a} = -\frac{4}{2 \cdot 1} = -2.$$

The y-coordinate of the vertex is -2. We substitute -2 for y in the parabola's equation, $x = y^2 + 4y - 5$, to find the x-coordinate:

$$x = (-2)^2 + 4(-2) - 5 = 4 - 8 - 5 = -9.$$

The vertex is at $(-9, -2)$.

Step 3 Find the x-intercept. Replace y with 0 in $x = y^2 + 4y - 5$.

$$x = 0^2 + 4 \cdot 0 - 5 = -5$$

The x-intercept is -5. The parabola passes through $(-5, 0)$.

Step 4 Find the y-intercepts. Replace x with 0 in the given equation.

$x = y^2 + 4y - 5$	This is the given equation.
$0 = y^2 + 4y - 5$	Replace x with 0.
$0 = (y - 1)(y + 5)$	Use factoring to solve the quadratic equation.
$y - 1 = 0$ or $y + 5 = 0$	Set each factor equal to 0.
$y = 1$ $y = -5$	Solve.

The y-intercepts are 1 and -5. The parabola passes through $(0, 1)$ and $(0, -5)$.

Using Technology

To graph $x = y^2 + 4y - 5$ using a graphing utility, rewrite the equation as a quadratic equation in y.

$$y^2 + 4y + (-x - 5) = 0$$

$a = 1$ $b = 4$ $c = -x - 5$

Use the quadratic formula to solve for y and enter the resulting equations.

$$y_1 = \frac{-4 + \sqrt{16 - 4(-x - 5)}}{2}$$

$$y_2 = \frac{-4 - \sqrt{16 - 4(-x - 5)}}{2}$$

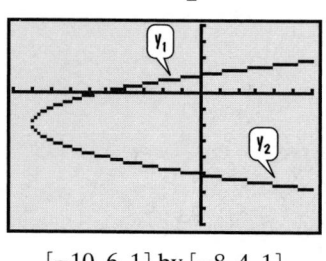

$[-10, 6, 1]$ by $[-8, 4, 1]$

Step 5 Graph the parabola. With a vertex at $(-9, -2)$, an x-intercept at -5, and y-intercepts at 1 and -5, the graph of the parabola is shown in Figure 13.35. The axis of symmetry is the horizontal line whose equation is $y = -2$.

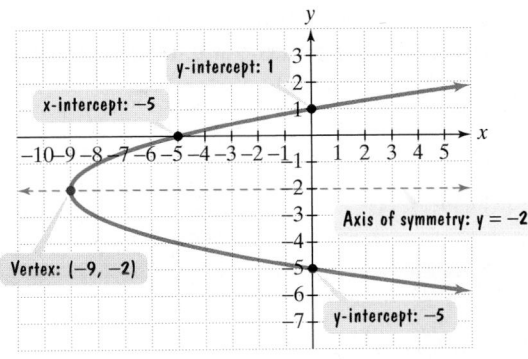

Figure 13.35 The graph of $x = y^2 + 4y - 5$

✔ **CHECK POINT 2** Graph: $x = y^2 + 8y + 7$.

Applications

Parabolas have many applications. Cables hung between structures to form suspension bridges form parabolas. Arches constructed of steel and concrete, whose main purpose is strength, are usually parabolic in shape.

Suspension bridge   Arch bridge

We have seen that comets in our solar system travel in orbits that are ellipses and hyperbolas. Some comets also follow parabolic paths. Only comets with elliptical orbits, such as Halley's Comet, return to our part of the galaxy.

A projectile, such as a baseball thrown directly upward, moves along a parabolic path, illustrated in Figure 13.36.

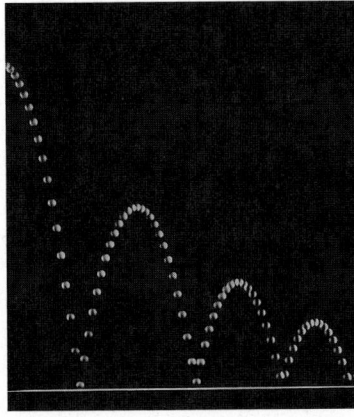

Figure 13.36 Multiflash photo showing the parabolic path of a ball thrown into the air ("Bouncing Ball" Cambridge, Massachusetts. © Berenice Abbot/Commerce Graphics, Ltd., Inc.)

If a parabola is rotated about its axis of symmetry, a parabolic surface is formed. Figure 13.37(a) shows how a parabolic surface can be used to reflect light. Light originates at the focus. Note how the light is reflected by the parabolic surface, so that the outgoing light is parallel to the axis of symmetry. The reflective properties of parabolic surfaces are used in the design of searchlights (see Figure 13.37(b)), automobile headlights, and parabolic microphones.

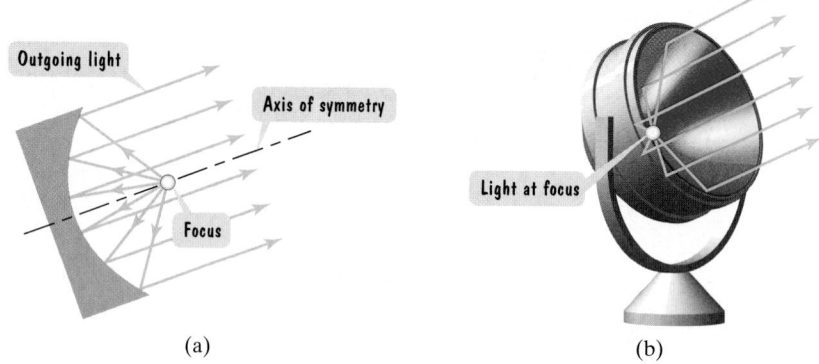

(a)   (b)

Figure 13.37 (a) Parabolic surface reflecting light   **(b)** Light from the focus is reflected parallel to the axis of symmetry.

Figure 13.38(a) shows how a parabolic surface can be used to reflect *incoming* light. Note that light rays strike the surface and are reflected *to the focus*. This principle is used in the design of reflecting telescopes, radar, and television satellite dishes. Reflecting telescopes magnify the light from distant stars by reflecting the light from these bodies to the focus of a parabolic mirror (see Figure 13.38(b)).

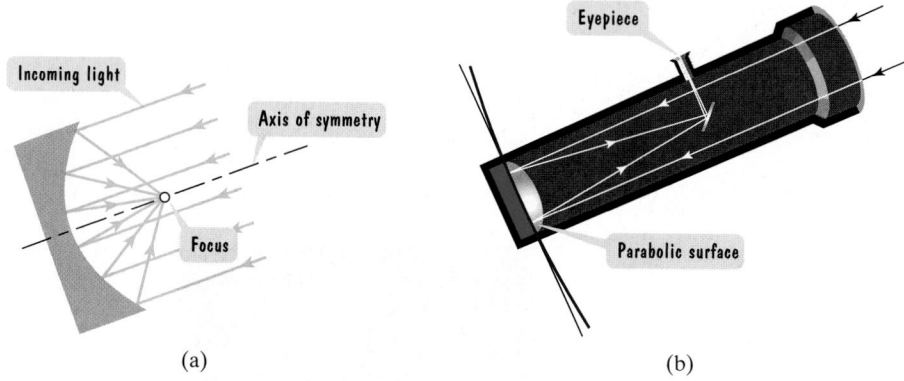

(a)   (b)

Figure 13.38 (a) Parabolic surface reflecting incoming light   **(b)** Incoming light rays are reflected to the focus.

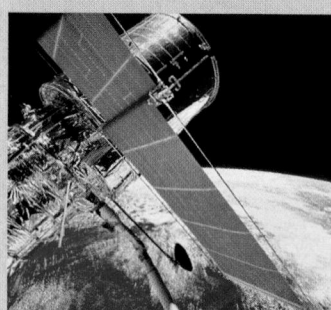
2   Identify conic sections by their equations.

Identifying Conic Sections by their Equations

What do the equations of the conic sections in this chapter have in common? They contain x^2-terms, y^2-terms, or both. Furthermore, they do not contain terms with exponents greater than 2. Here is how to identify conic sections by their equations:

Recognizing Conic Sections from Equations

Conic Section	How to Identify the Equation	Example
Circle	When x^2- and y^2-terms are on the same side, they have the same coefficient.	$x^2 + y^2 = 16$
Ellipse	When x^2- and y^2-terms are on the same side, they have different coefficients of the same sign.	$4x^2 + 16y^2 = 64$ or (dividing by 64) $\dfrac{x^2}{16} + \dfrac{y^2}{4} = 1$
Hyperbola	When x^2- and y^2-terms are on the same side, they have coefficients with opposite signs.	$9y^2 - 4x^2 = 36$ or (dividing by 36) $\dfrac{y^2}{4} - \dfrac{x^2}{9} = 1$
Parabola	Only one of the variables is squared.	$x = y^2 + 4y - 5$

EXAMPLE 3 Recognizing Equations of Conic Sections

Indicate whether the graph of each equation is a circle, an ellipse, a hyperbola, or a parabola:

a. $4y^2 = 16 - 4x^2$ **b.** $x^2 = y^2 + 9$

c. $x + 7 - 6y = y^2$ **d.** $x^2 = 16 - 16y^2$.

Solution If both variables are squared, the graph of the equation is not a parabola. In this case, we get the x^2- and y^2-terms on the same side of the equation.

a. $4y^2 = 16 - 4x^2$

Both variables, x and y, are squared.

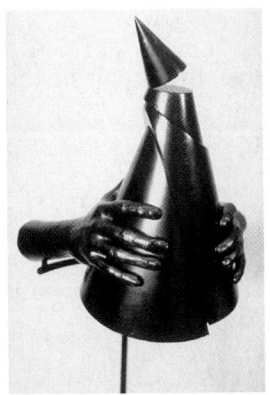

Richard E. Prince "The Cone of Apollonius" (detail), fiberglass, steel, paint, graphite. 51 x 18 x 14". Collection: Vancouver Art Gallery, Vancouver, Canada. Photo courtesy of Equinox Gallery, Vancouver, Canada.

The graph is a circle, an ellipse, or a hyperbola. To see which one it is, we get the x^2- and y^2-terms on the same side. Add $4x^2$ to both sides. We obtain

$$4x^2 + 4y^2 = 16.$$

Because the coefficients of x^2 and y^2 are the same, namely 4, the equation's graph is a circle. This becomes more obvious if we divide both sides by 4.

$$x^2 + y^2 = 4$$

$$(x - 0)^2 + (y - 0)^2 = 2^2 \qquad \text{The equation of the circle with center } (h, k) \text{ and radius } r \text{ is } (x - h)^2 + (y - k)^2 = r^2.$$

The graph is a circle with center at the origin and radius 2.

b. $x^2 = y^2 + 9$

> Both variables, x and y, are squared.

The graph cannot be a parabola. To see if it is a circle, an ellipse, or a hyperbola, we get the x^2- and y^2-terms on the same side. Subtract y^2 from both sides. We obtain

$$x^2 - y^2 = 9.$$

Because x^2- and y^2-terms have coefficients with opposite signs, the equation's graph is a hyperbola. This becomes more obvious if we divide both sides by 9 to obtain 1 on the right.

$$\frac{x^2}{9} - \frac{y^2}{9} = 1$$

> $a^2 = 9$ $b^2 = 9$

The hyperbola's vertices are $(a, 0)$ and $(-a, 0)$, namely $(3, 0)$ and $(-3, 0)$.

c. $x + 7 - 6y = y^2$

> Only one variable, y, is squared.

Because only one variable is squared, the graph of the equation is a parabola. We can express the equation of the horizontal parabola in the form $x = ay^2 + by + c$ by isolating x on the left. We obtain

$$x = y^2 + 6y - 7.$$

Because the coefficient of the y^2-term, 1, is positive, the graph of the horizontal parabola opens to the right.

d. $x^2 = 16 - 16y^2$

> Both variables, x and y, are squared.

The graph cannot be a parabola. To see if it is a circle, an ellipse, or a hyperbola, we get the x^2- and y^2-terms on the same side. Add $16y^2$ to both sides. We obtain

$$x^2 + 16y^2 = 16.$$

Because the x^2- and y^2-terms have different coefficients of the same sign, namely 1 and 16, the equation's graph is an ellipse. This becomes more obvious if we divide both sides by 16 to obtain 1 on the right.

$$\frac{x^2}{16} + \frac{y^2}{1} = 1$$

> $a^2 = 16$ $b^2 = 1$

An equation in the form $\frac{x^2}{a^2} + \frac{y^2}{b^2} = 1$ is an ellipse.

The vertices are $(4, 0)$ and $(-4, 0)$. ∎

✔ **CHECK POINT 3** Identify whether the graph of each equation is a circle, an ellipse, a hyperbola, or a parabola:

   a. $x^2 = 4y^2 + 16$

   b. $x^2 = 16 - 4y^2$

   c. $4x^2 = 16 - 4y^2$

   d. $x = -4y^2 + 16y.$

EXERCISE SET 13.4

Practice Exercises

In Exercises 1–6, the equation of a horizontal parabola is given. For each equation:
Determine how the parabola opens. Find the parabola's vertex. Use your results to identify the equation's graph. [The graphs are labeled (a) through (f).]

1. $x = (y - 2)^2 - 1$

2. $x = (y + 2)^2 - 1$

3. $x = (y + 2)^2 + 1$

4. $x = (y - 2)^2 + 1$

5. $x = -(y - 2)^2 + 1$

6. $x = -(y + 2)^2 + 1$

a.

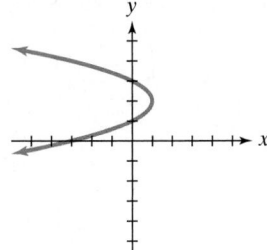

b.

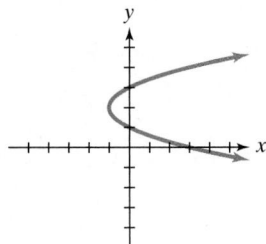

c.

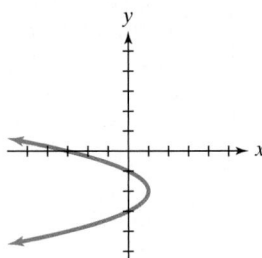

d.

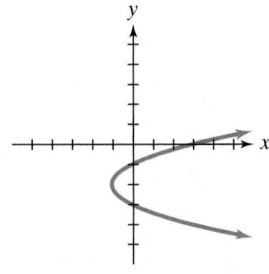

e.

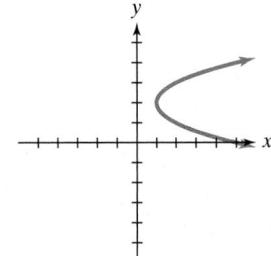

f.

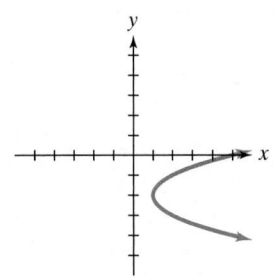

In Exercises 7–18, find the coordinates of the vertex for the horizontal parabola defined by the given equation.

7. $x = 2y^2$

8. $x = 4y^2$

9. $x = (y - 2)^2 + 3$

10. $x = (y - 3)^2 + 4$

11. $x = -4(y + 2)^2 - 1$

12. $x = -2(y + 5)^2 - 1$

13. $x = 2(y - 6)^2$

14. $x = 3(y - 7)^2$

15. $x = y^2 - 6y + 6$

16. $x = y^2 + 6y + 8$

17. $x = 3y^2 + 6y + 7$

18. $x = -2y^2 + 4y + 6$

In Exercises 19–42, use the vertex and intercepts to sketch the graph of each equation. Give the equation for the parabola's axis of symmetry. If needed, find additional points on the parabola by choosing values of y on each side of the axis of symmetry.

19. $x = (y - 2)^2 - 4$

20. $x = (y - 3)^2 - 4$

21. $x = (y - 3)^2 - 5$

22. $x = (y + 2)^2 - 3$

23. $x = -(y - 5)^2 + 4$

24. $x = -(y - 3)^2 + 4$

25. $x = (y - 4)^2 + 1$

26. $x = (y - 2)^2 + 3$

27. $x = -3(y - 5)^2 + 3$

28. $x = -2(y + 6)^2 + 2$

29. $x = -2(y + 3)^2 - 1$

30. $x = -3(y + 1)^2 - 2$

31. $x = \frac{1}{2}(y + 2)^2 + 1$

32. $x = \frac{1}{2}(y + 1)^2 + 2$

33. $x = y^2 + 2y - 3$

34. $x = y^2 - 6y + 8$

35. $x = -y^2 - 4y + 5$

36. $x = -y^2 - 6y + 7$

37. $x = y^2 + 6y$

38. $x = y^2 + 4y$

39. $x = -2y^2 - 4y$

40. $x = -3y^2 - 6y$

41. $x = -2y^2 - 4y + 1$

42. $x = -2y^2 + 4y - 3$

In Exercises 43–54, the equation of a parabola is given. Determine:

    a. *if the parabola is horizontal or vertical.*
    b. *the way the parabola opens.*
    c. *the vertex.*

43. $x = 2(y - 1)^2 + 2$

44. $x = 2(y - 3)^2 + 1$

45. $y = 2(x - 1)^2 + 2$

46. $y = 2(x - 3)^2 + 1$

47. $y = -(x + 3)^2 + 4$

48. $y = -(x + 1)^2 + 4$

49. $x = -(y + 3)^2 + 4$

50. $x = -(y + 1)^2 + 4$

51. $y = x^2 - 4x - 1$

52. $y = x^2 + 6x + 10$

53. $x = -y^2 + 4y + 1$

54. $x = -y^2 - 6y - 10$

In Exercises 55–64, indicate whether the graph of each equation is a circle, an ellipse, a hyperbola, or a parabola.

55. $x - 7 - 8y = y^2$

56. $x - 3 - 4y = 6y^2$

57. $4x^2 = 36 - y^2$

58. $4x^2 = 36 + y^2$

59. $x^2 = 36 + 4y^2$

60. $x^2 = 36 - 4y^2$

61. $3x^2 = 12 - 3y^2$

62. $3x^2 = 27 - 3y^2$

63. $3x^2 = 12 + 3y^2$

64. $3x^2 = 27 + 3y^2$

In Exercises 65–74, indicate whether the graph of each equation is a circle, an ellipse, a hyperbola, or a parabola. Then graph the conic section.

65. $x^2 - 4y^2 = 16$

66. $7x^2 - 7y^2 = 28$

67. $4x^2 + 4y^2 = 16$

68. $7x^2 + 7y^2 = 28$

69. $x^2 + 4y^2 = 16$

70. $4x^2 + y^2 = 16$

71. $x = (y - 1)^2 - 4$

72. $x = (y - 4)^2 - 1$

73. $(x - 2)^2 + (y + 1)^2 = 16$

74. $(x - 1)^2 + (y + 2)^2 = 16$

Application Exercises

75. The George Washington Bridge spans the Hudson River from New York to New Jersey. Its two towers are 3500 feet apart and rise 316 feet above the road. The cable between the towers has the shape of a parabola, and the cable just touches the sides of the road midway between the towers. The parabola is positioned in a rectangular coordinate system with its vertex at the origin. The point (1750, 316) lies on the parabola, as shown.

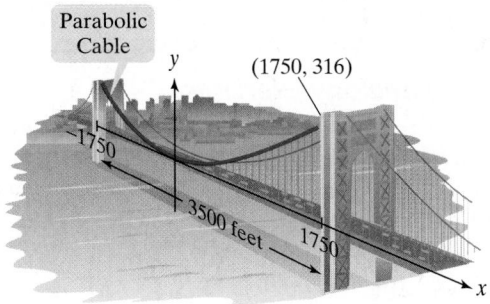

a. Write an equation in the form $y = ax^2$ for the parabolic cable. Do this by substituting 1750 for x and 316 for y and determining the value of a.

b. Use the equation in part (a) to find the height of the cable 1000 feet from a tower. Round to the nearest foot.

76. The towers of the Golden Gate Bridge connecting San Francisco to Marin County are 1280 meters apart and rise 160 meters above the road. The cable between the towers has the shape of a parabola, and the cable just touches the sides of the road midway between the towers. The parabola is positioned in a rectangular coordinate system with its vertex at the origin. The point (640, 140) lies on the parabola, as shown.

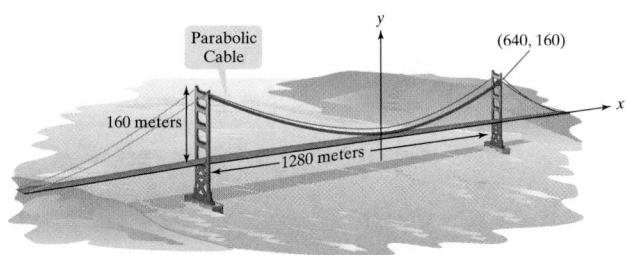

a. Write an equation in the form $y = ax^2$ for the parabolic cable. Do this by substituting 640 for x and 140 for y and determining the value of a.

b. Use the equation in part (a) to find the height of the cable 200 meters from a tower. Round to the nearest meter.

77. A satellite dish is in the shape of a parabolic surface. Signals coming from a satellite strike the surface of the dish and are reflected to the focus, where the receiver is located. The satellite dish shown has a diameter of 12 feet and a depth of 2 feet. The parabola is positioned in a rectangular coordinate system with its vertex at the origin.

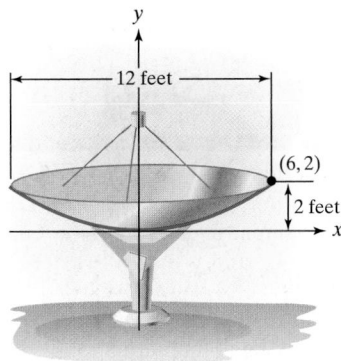

a. Write an equation in the form $y = ax^2$ for the parabola used to shape the dish.

b. The received should be placed at the focus $(0, p)$. The value of p is given by the equation $a = \dfrac{1}{4p}$. How far from the base of the dish should the received be placed?

78. An engineer is designing a flashlight using a parabolic reflecting mirror and a light source, as shown. The casting has a diameter of 4 inches and a depth of 2 inches. The parabola is positioned in a rectangular coordinate system with its vertex at the origin.

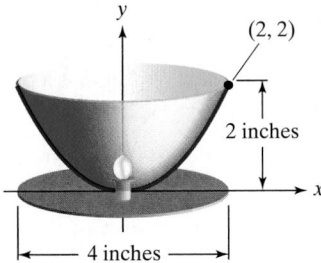

a. Write an equation in the form $y = ax^2$ for the parabola used to shape the mirror.

b. The light source should be placed at the focus $(0, p)$. The value of p is given by the equation $a = \dfrac{1}{4p}$. Where should the light source be placed relative to the mirror's vertex?

Writing in Mathematics

79. What is a parabola?

80. If you are given an equation of a parabola, explain how to determine if the parabola opens to the right, left, upward, or downward.

81. Explain how to use $x = 2(y + 3)^2 - 5$ to find the parabolas's vertex.

82. Explain how to use $x = y^2 + 8y + 9$ to find the parabolas's vertex.

83. Describe one similarity and one difference between the graphs of $x = 4y^2$ and $x = 4(y - 1)^2 + 2$.

84. How can you distinguish parabolas from other conic sections by looking at their equations?

85. How can you distinguish ellipses from hyperbolas by looking at their equations?

86. How can you distinguish ellipses from circles by looking at their equations?

Critical Thinking Exercises

87. Which one of the following is true?
   a. The parabola whose equation is $x = 2y - y^2 + 5$ opens to the right.
   b. If the parabola whose equation is $x = ay^2 + by + c$ has its vertex at $(3, 2)$ and $a > 0$, then it has no y-intercepts.
   c. Some parabolas that open to the right have equations that define y as a function of x.
   d. The graph of $x = a(y - k) + h$ is a parabola with vertex at (h, k).

88. Look at the satellite dish shown in Exercise 77. Why must the receiver for a shallow dish be farther from the base of the dish than for a deeper dish of the same diameter?

89. The parabolic arch shown in the figure is 50 feet above the water at the center and 200 feet wide at the base. Will a boat that is 30 feet tall clear the arch 30 feet from the center?

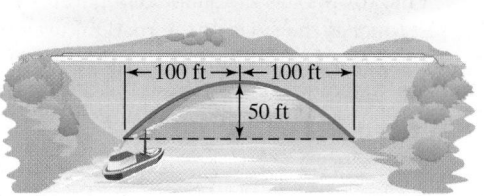

Technology Exercises

Use a graphing utility to graph the parabolas in Exercises 90–91. Write the given equation as a quadratic equation in y and use the quadratic formula to solve for y. Enter each of the equations to produce the complete graph.

90. $y^2 + 2y - 6x + 13 = 0$

91. $y^2 + 10y - x + 25 = 0$

92. Use a graphing utility to graph any three of the parabolas that you graphed by hand in Exercises 19–42. First solve the given equation for y, possibly using the square root method. Enter each of the two resulting equations to produce the complete graph.

Review Exercises

93. Graph: $f(x) = 2^{1-x}$.
(Section 12.1, Example 4)

94. If $f(x) = \frac{1}{3}x - 5$, find $f^{-1}(x)$.
(Section 12.2, Example 4)

95. Simplify:

$$\frac{x - \dfrac{1}{3}}{3 - \dfrac{1}{x}}.$$

(Section 7.5, Example 6)

▶ SECTION 13.5   Systems of Nonlinear Equations in Two Variables

Objectives

1 Recognize systems of nonlinear equations in two variables.

2 Solve nonlinear systems by substitution.

3 Solve nonlinear systems by addition.

4 Solve problems using systems of nonlinear equations.

SSM   PH Tutor   CD-   Video
      Center   ROM

Scientists debate the probability that a "doomsday rock" will collide with Earth. It has been estimated that an asteroid, a tiny planet that revolves around the sun, crashes into Earth about once every 250,000 years, and that such a collision would have disastrous results. In 1908, a small fragment struck Siberia, leveling thousands of acres of trees. One theory about the extinction of dinosaurs 65 million years ago involves Earth's collision with a large asteroid and the resulting drastic changes in Earth's climate.

Understanding the path of Earth and the path of a comet is essential to detecting threatening space debris. Orbits about the sun are not described by linear equations in the form $Ax + By = C$. The ability to solve systems that do not contain linear equations provides NASA scientists watching for troublesome asteroids with a possible collision point with Earth's orbit.

1 Recognize systems of nonlinear equations in two variables.

Systems of Nonlinear Equations and their Solutions

A **system of** two **nonlinear equations** in two variables contains at least one equation that cannot be expressed in the form $Ax + By = C$. Here are two examples:

$$x^2 = 2y + 10$$
$$3x - y = 9$$

Not in the form $Ax + By = C$. The term x^2 is not linear.

$$y = x^2 + 3$$
$$x^2 + y^2 = 9.$$

Neither equation is in the form $Ax + By = C$. The terms x^2 and y^2 are not linear.

A **solution** to a nonlinear system in two variables is an ordered pair of real numbers that satisfies all equations in the system. The **solution set** to the system is the set of all such ordered pairs. As with linear systems in two variables, the solution to a nonlinear system (if there is one) corresponds to the intersection point(s) of the graphs of the equations in the system. Unlike linear systems, the graphs can be circles, ellipses, hyperbolas, parabolas, or anything other than two lines. We will solve nonlinear systems using the substitution method and the addition method.

2 Solve nonlinear systems by substitution.

Eliminating a Variable Using the Substitution Method

The substitution method involves converting a nonlinear system to one equation in one variable by an appropriate substitution. The steps in the solution process are exactly the same as those used to solve a linear system by substitution. However, when you obtain an equation in one variable, this equation will not be linear. In our first example, this equation is quadratic.

EXAMPLE 1   Solving a Nonlinear System by the Substitution Method

Solve by the substitution method:

$$x^2 = 2y + 10 \qquad \text{The graph is a parabola.}$$
$$3x - y = 9. \qquad \text{The graph is a line.}$$

Solution

Step 1 Solve one of the equations for one variable in terms of the other. We begin by isolating one of the variables raised to the first power in either of the equations. By solving for y in the second equation, which has a coefficient of -1, we can avoid fractions.

$$3x - y = 9 \qquad \text{This is the second equation in the given system.}$$

$$3x = y + 9 \qquad \text{Add } y \text{ to both sides.}$$

$$3x - 9 = y \qquad \text{Subtract 9 from both sides.}$$

Step 2 Substitute the expression from step 1 into the other equation. We substitute $3x - 9$ for y in the first equation.

$$y = \boxed{3x - 9} \qquad x^2 = 2\boxed{y} + 10$$

This gives us an equation in one variable, namely

$$x^2 = 2(3x - 9) + 10.$$

The variable y has been eliminated.

Step 3 Solve the resulting equation containing one variable.

$$x^2 = 2(3x - 9) + 10 \qquad \text{This is the equation containing one variable.}$$

$$x^2 = 6x - 18 + 10 \qquad \text{Use the distributive property.}$$

$$x^2 = 6x - 8 \qquad \text{Combine numerical terms on the right.}$$

$$x^2 - 6x + 8 = 0 \qquad \text{Move all terms to one side and set the quadratic equation equal to 0.}$$

$$(x - 4)(x - 2) = 0 \qquad \text{Factor.}$$

$$x - 4 = 0 \quad \text{or} \quad x - 2 = 0 \qquad \text{Set each factor equal to 0.}$$

$$x = 4 \quad \text{or} \quad x = 2 \qquad \text{Solve for x.}$$

Step 4 Back-substitute the obtained values into the equation from step 1. Now that we have the x-coordinates of the solutions, we back-substitute 4 for x and 2 for x in the equation $y = 3x - 9$.

If x is 4, $y = 3(4) - 9 = 3$,   so $(4, 3)$ is a solution.

If x is 2, $y = 3(2) - 9 = -3$,   so $(2, -3)$ is a solution.

Step 5 Check the proposed solutions in both of the system's given equations. We begin by checking $(4, 3)$. Replace x with 4 and y with 3.

$x^2 = 2y + 10$	$3x - y = 9$	These are the given equations.
$4^2 \overset{?}{=} 2(3) + 10$	$3(4) - 3 \overset{?}{=} 9$	Let x = 4 and y = 3.
$16 \overset{?}{=} 6 + 10$	$12 - 3 \overset{?}{=} 9$	Simplify.
$16 = 16$, true	$9 = 9$, true	True statements result.

The ordered pair $(4, 3)$ satisfies both equations. Thus, $(4, 3)$ is a solution to the system.

Now let's check $(2, -3)$. Replace x with 2 and y with -3 in both given equations.

$x^2 = 2y + 10$	$3x - y = 9$	These are the given equations.
$2^2 \overset{?}{=} 2(-3) + 10$	$3(2) - (-3) \overset{?}{=} 9$	Let x = 2 and y = -3.
$4 \overset{?}{=} -6 + 10$	$6 + 3 \overset{?}{=} 9$	Simplify.
$4 = 4$, true	$9 = 9$, true	True statements result.

The ordered pair $(2, -3)$ also satisfies both equations and is a solution to the system. The solutions are $(4, 3)$ and $(2, -3)$, and the solution set is $\{(4, 3), (2, -3)\}$.

Figure 13.39 shows the graphs of the equations in the system and the solutions as intersection points. ∎

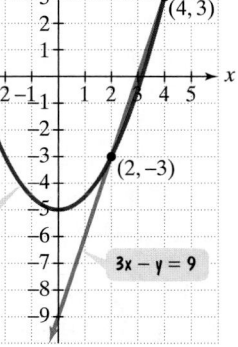

Figure 13.39 Points of intersection illustrate the nonlinear system's solutions.

✔ **CHECK POINT 1** Solve by the substitution method:

$$x^2 = y - 1$$
$$4x - y = -1.$$

EXAMPLE 2 Solving a Nonlinear System by the Substitution Method

Solve by the substitution method:

$x - y = 3$	The graph is a line.
$(x - 2)^2 + (y + 3)^2 = 4.$	The graph is a circle.

Solution Graphically, we are finding the intersection of a line and a circle with center $(2, -3)$ and radius 2.

Step 1 Solve one of the equations for one variable in terms of the other. We will solve for x in the linear equation – that is, the first equation. (We could also solve for y).

$x - y = 3$	This is the first equation in the given system.
$x = y + 3$	Add y to both sides.

Step 2 Substitute the expression from step 1 into the other equation. We substitute $y + 3$ for x in the second equation.

$$x = \boxed{y + 3} \qquad (\boxed{x} - 2)^2 + (y + 3)^2 = 4$$

This gives an equation in one variable, namely

$$(y + 3 - 2)^2 + (y + 3)^2 = 4.$$

The variable x has been eliminated.

Step 3 Solve the resulting equation containing one variable.

$(y + 3 - 2)^2 + (y + 3)^2 = 4$	This is the equation containing one variable.
$(y + 1)^2 + (y + 3)^2 = 4$	Combine numerical terms in the first parentheses.
$y^2 + 2y + 1 + y^2 + 6y + 9 = 4$	Use the formula $(A + B)^2 = A^2 + 2AB + B^2$ to square $y + 1$ and $y + 3$.
$2y^2 + 8y + 10 = 4$	Combine like terms on the left.
$2y^2 + 8y + 6 = 0$	Subtract 4 from both sides and set the quadratic equation equal to 0.
$y^2 + 4y + 3 = 0$	Simplify by dividing both sides by 2.
$(y + 3)(y + 1) = 0$	Factor.
$y + 3 = 0$ or $y + 1 = 0$	Set each factor equal to 0.
$y = -3$ or $y = -1$	Solve for y.

Step 4 Back-substitute the obtained values into the equation from step 1. Now that we have the y-coordinates of the solutions, we back-substitute -3 for y and -1 for y in the equation $x = y + 3$.

If $y = -3$: $x = -3 + 3 = 0$,   so $(0, -3)$ is a solution.

If $y = -1$: $x = -1 + 3 = 2$,   so $(2, -1)$ is a solution.

Step 5 Check the proposed solution in both of the system's given equations. Take a moment to show that each ordered pair satisfies both equations. The solution are $(0, -3)$ and $(2, -1)$, and the solution set of the given system is $\{(0, -3), (2, -1)\}$.

Figure 13.40 shows the graphs of the equations in the system and the solutions as intersection points. ∎

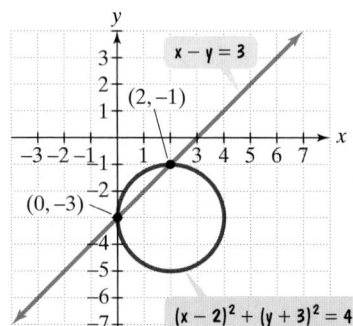

Figure 13.40 Points of intersection illustrate the nonlinear system's solutions.

✔ **CHECK POINT 2**   Solve by the substitution method:

$$x + 2y = 0$$
$$(x - 1)^2 + (y - 1)^2 = 5.$$

3 Solve nonlinear systems by addition.

Eliminating a Variable Using the Addition Method

In solving linear systems with two variables, we learned that the addition method works well when each equation is in the form $Ax + By = C$. For nonlinear systems, the addition method can be used when each equation is in the form $Ax^2 + By^2 = C$. If necessary, we will multiply either equation or both equations by appropriate numbers so that the coefficients of x^2 or y^2 will have a sum of 0. We then add equations. The sum will be an equation in one variable.

EXAMPLE 3   Solving a Nonlinear System by the Addition Method

Solve the system:

$4x^2 + y^2 = 13$	Equation 1 (The graph is an ellipse.)
$x^2 + y^2 = 10.$	Equation 2 (The graph is a circle.)

The given system:

$$4x^2 + y^2 = 13$$
$$x^2 + y^2 = 10,$$

repeated from page 924.

Solution We can use the same steps that we did when we solved linear systems by the addition method.

Step 1 Write both equations in the form $Ax^2 + By^2 = C$. Both equations are already in this form, so we can skip this step.

Step 2 If necessary, multiply either equation or both equations by appropriate numbers so that the sum of the x^2-coefficients or the sum of the y^2-coefficients is 0. We can eliminate y^2 by multiplying Equation 2 by -1.

$$4x^2 + y^2 = 13 \xrightarrow{\text{No change}} 4x^2 + y^2 = 13$$
$$x^2 + y^2 = 10 \xrightarrow{\text{Multiply by } -1.} -x^2 - y^2 = -10$$

Steps 3 and 4 Add equations and solve for the remaining variable.

$$
\begin{aligned}
4x^2 + y^2 &= 13 \\
-x^2 - y^2 &= -10 \\
\hline
3x^2 &= 3 \qquad \text{Add equations.} \\
x^2 &= 1 \qquad \text{Divide both sides by 3.} \\
x &= \pm 1 \qquad \text{Use the square root method: If } x^2 = c, \text{ then } x = \pm\sqrt{c}.
\end{aligned}
$$

Step 5 Back-substitute and find the values for the other variables. We must back-substitute each value of x into either one of the original equations. Let's use $x^2 + y^2 = 10$, Equation 2. If $x = 1$,

$$1^2 + y^2 = 10 \qquad \text{Replace x with 1 in Equation 2.}$$
$$y^2 = 9 \qquad \text{Subtract 1 from both sides.}$$
$$y = \pm 3 \qquad \text{Apply the square root method.}$$

$(1, 3)$ and $(1, -3)$ are solutions. If $x = -1$,

$$(-1)^2 + y^2 = 10 \qquad \text{Replace x with } -1 \text{ in Equation 2.}$$
$$y^2 = 9 \qquad \text{The steps are the same as before.}$$
$$y = \pm 3$$

$(-1, 3)$ and $(-1, -3)$ are solutions.

Step 6 Check. Take a moment to show that each of the four ordered pairs satisfies Equation 1 and Equation 2. The solutions are $(1, 3)$, $(1, -3)$, $(-1, 3)$, and $(-1, -3)$, and the solution set of the given system is $\{(1, 3), (1, -3), (-1, 3), (-1, -3)\}$.

Figure 13.41 shows the graphs of the equations in the system, an ellipse and a circle, and the solutions as intersection points. ∎

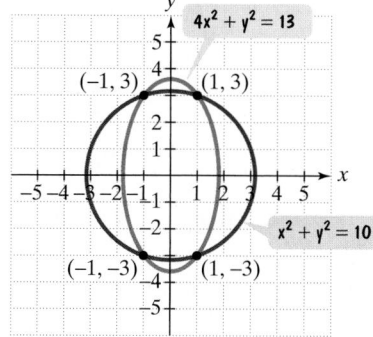

Figure 13.41 A system with four solutions

✔ **CHECK POINT 3** Solve the system:

$$3x^2 + 2y^2 = 35$$
$$4x^2 + 3y^2 = 48.$$

In solving nonlinear systems, we will include only ordered pairs with real numbers in the solution set. We have seen that each of these ordered pairs corresponds to a point of intersection of the system's graphs.

EXAMPLE 4 Solving a Nonlinear System by the Addition Method

Solve the system:

$$y = x^2 + 3 \qquad \text{Equation 1 (The graph is a parabola.)}$$
$$x^2 + y^2 = 9. \qquad \text{Equation 2 (The graph is a circle.)}$$

Solution We could use substitution because Equation 1 has y expressed in terms of x, but this would result in a fourth-degree equation. However, we can rewrite Equation 1 by subtracting x^2 from both sides and adding the equations to eliminate the x^2-terms.

$$
\begin{array}{rll}
-x^2 + y & = 3 & \text{Subtract } x^2 \text{ from both sides of Equation 1.} \\
x^2 \quad\;\; + y^2 & = 9 & \text{This is Equation 2.} \\
\hline
y + y^2 & = 12 & \text{Add the equations.}
\end{array}
$$

We now solve this quadratic equation.

$$
\begin{array}{ll}
y + y^2 = 12 & \text{This is the equation containing one variable.} \\
y^2 + y - 12 = 0 & \text{Subtract 12 from both sides and set the} \\
& \text{quadratic equation equal to 0.} \\
(y + 4)(y - 3) = 0 & \text{Factor.} \\
y + 4 = 0 \quad \text{or} \quad y - 3 = 0 & \text{Set each factor equal to 0.} \\
y = -4 \quad \text{or} \quad y = 3 & \text{Solve for } y.
\end{array}
$$

To complete the solution, we must back-substitute each value of y into either one of the original equations. We will use $y = x^2 + 3$, Equation 1. First, we substitute -4 for y.

$$
\begin{array}{ll}
-4 = x^2 + 3 & \\
-7 = x^2 & \text{Subtract 3 from both sides.}
\end{array}
$$

Because the square of a real number cannot be negative, the equation $x^2 = -7$ does not have real-number solutions. We will not include the imaginary solutions, $x = \pm\sqrt{-7}$, or $\sqrt{7}i$ and $-\sqrt{7}i$, in the ordered pairs that make up the solution set. Thus, we move on to our other value for y, 3, and substitute this value into Equation 1.

$$
\begin{array}{ll}
y = x^2 + 3 & \text{This is Equation 1.} \\
3 = x^2 + 3 & \text{Back-substitute 3 for } y. \\
0 = x^2 & \text{Subtract 3 from both sides.} \\
0 = x & \text{Solve for } x.
\end{array}
$$

We showed that if $y = 3$, then $x = 0$. Thus, $(0, 3)$ is the solution with a real ordered pair. Take a moment to show that $(0, 3)$ satisfies Equation 1 and Equation 2. The solution set of the given system is $\{(0, 3)\}$. Figure 13.42 shows the system's graphs and the solution as an intersection point. ∎

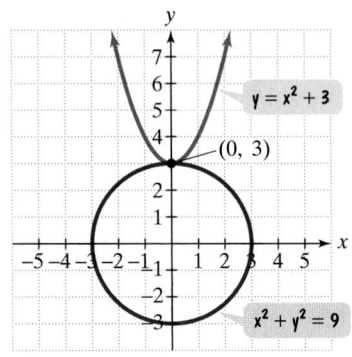

Figure 13.42 A system with one real solution

✔ **CHECK POINT 4** Solve the system:

$$y = x^2 + 5$$
$$x^2 + y^2 = 25.$$

4 Solve problems using systems of nonlinear equations.

Applications

Many geometric problems can be modeled and solved by the use of nonlinear systems of equations. We will use our step-by-step strategy for solving problems using mathematical models that are created from verbal models.

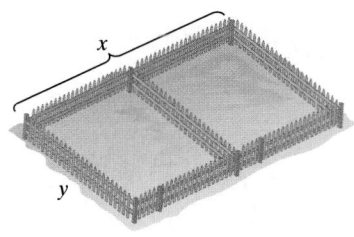

Figure 13.43 Building an enclosure

EXAMPLE 5   An Application of a Nonlinear System

You have 36 yards of fencing to build the enclosure in Figure 13.43. Some of this fencing is to be used to build an internal divider. If you'd like to enclose 54 square yards, what are the dimensions of the enclosure?

Solution

Step 1 Use variables to represent unknown quantities. Let x = the enclosure's length and y = the enclosure's width. These variables are shown in Figure 13.43.

Step 2 Write a system of equations describing the problem's conditions. The first condition is that you have 36 yards of fencing.

Fencing along both lengths	plus	fencing along both widths	plus	fencing for the internal divider	equals	36 yards.
$2x$	$+$	$2y$	$+$	y	$=$	36

Adding like terms, we can express the equation that models the verbal conditions for the fencing as $2x + 3y = 36$.

The second condition is that you'd like to enclose 54 square yards. The rectangle's area, the product of its length and its width, must be 54 square yards.

Length	times	width	is	54 square yards.
x	$\cdot$	y	$=$	54

Step 3 Solve the system and answer the problem's question. We must solve the system

$$2x + 3y = 36 \qquad \text{Equation 1}$$
$$xy = 54. \qquad \text{Equation 2}$$

We will use substitution. Because Equation 1 has no coefficients of 1 or -1, we will solve Equation 2 for y. Dividing both sides of $xy = 54$ by x, we obtain

$$y = \frac{54}{x}.$$

Now we substitute $\dfrac{54}{x}$ for y in Equation 1 and solve for x.

$$2x + 3y = 36 \qquad \text{This is Equation 1.}$$

$$2x + 3 \cdot \frac{54}{x} = 36 \qquad \text{Substitute } \frac{54}{x} \text{ for } y.$$

$$2x + \frac{162}{x} = 36 \qquad \text{Multiply.}$$

$$x\left(2x + \frac{162}{x} \right) = 36 \cdot x \qquad \text{Clear fractions by multiplying both sides by } x.$$

$$2x^2 + 162 = 36x \qquad \text{Use the distributive property on the left side.}$$

$$2x^2 - 36x + 162 = 0 \qquad \text{Subtract } 36x \text{ from both sides and set the quadratic equation equal to 0.}$$

$$x^2 - 18x + 81 = 0 \qquad \text{Simplify by dividing both sides by 2.}$$

$$(x - 9)^2 = 0 \qquad \text{Factor using } A^2 - 2AB + B^2 = (A - B)^2.$$

$$x - 9 = 0 \qquad \text{Set the repeated factor equal to zero.}$$

$$x = 9 \qquad \text{Solve for } x.$$

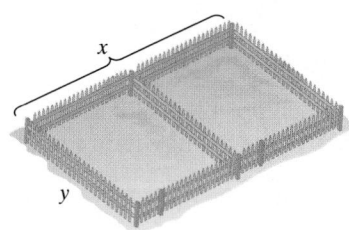

Figure 13.43, repeated

We back-substitute this value of x into $y = \dfrac{54}{x}$.

$$\text{If } x = 9, \quad y = \frac{54}{9} = 6.$$

This means that the dimensions of the enclosure in Figure 13.43 are 9 yards by 6 yards.

Step 4 Check the proposed solution in the original wording of the problem. With a length of 9 yards and a width of 6 yards, take a moment to check that this results in 36 yards of fencing and an area of 54 square yards. ∎

✔ **CHECK POINT 5** Find the length and width of a rectangle whose perimeter is 20 feet and whose area is 21 square feet.

EXERCISE SET 13.5

Practice Exercises

In Exercises 1–18, solve each system by the substitution method.

1. $x + y = 2$
$y = x^2 - 4$

2. $x - y = -1$
$y = x^2 + 1$

3. $x + y = 2$
$y = x^2 - 4x + 4$

4. $2x + y = -5$
$y = x^2 + 6x + 7$

5. $y = x^2 - 4x - 10$
$y = -x^2 - 2x + 14$

6. $y = x^2 + 4x + 5$
$y = x^2 + 2x - 1$

7. $x^2 + y^2 = 25$
$x - y = 1$

8. $x^2 + y^2 = 5$
$3x - y = 5$

9. $xy = 6$
$2x - y = 1$

10. $xy = -12$
$x - 2y + 14 = 0$

11. $y^2 = x^2 - 9$
$2y = x - 3$

12. $x^2 + y = 4$
$2x + y = 1$

13. $xy = 3$
$x^2 + y^2 = 10$

14. $xy = 4$
$x^2 + y^2 = 8$

15. $x + y = 1$
$x^2 + xy - y^2 = -5$

16. $x + y = -3$
$x^2 + 2y^2 = 12y + 18$

17. $x + y = 1$
$(x - 1)^2 + (y + 2)^2 = 10$

18. $2x + y = 4$
$(x + 1)^2 + (y - 2)^2 = 4$

In Exercises 19–28, solve each system by the addition method.

19. $x^2 + y^2 = 13$
$x^2 - y^2 = 5$

20. $4x^2 - y^2 = 4$
$4x^2 + y^2 = 4$

21. $x^2 - 4y^2 = -7$
$3x^2 + y^2 = 31$

22. $3x^2 - 2y^2 = -5$
$2x^2 - y^2 = -2$

23. $3x^2 + 4y^2 - 16 = 0$
$2x^2 - 3y^2 - 5 = 0$

24. $16x^2 - 4y^2 - 72 = 0$
$x^2 - y^2 - 3 = 0$

25. $x^2 + y^2 = 25$
$(x - 8)^2 + y^2 = 41$

26. $x^2 + y^2 = 4$
$x^2 + (y - 3)^2 = 9$

27. $y^2 - x = 4$
$x^2 + y^2 = 4$

28. $x^2 - 2y = 8$
$x^2 + y^2 = 16$

In Exercises 29–42, solve each system by the method of your choice.

29. $3x^2 + 4y^2 = 16$
$2x^2 - 3y^2 = 5$

30. $x + y^2 = 4$
$x^2 + y^2 = 16$

31. $2x^2 + y^2 = 18$
$xy = 4$

32. $x^2 + 4y^2 = 20$
$xy = 4$

33. $x^2 + 4y^2 = 20$
$x + 2y = 6$

34. $3x^2 - 2y^2 = 1$
$4x - y = 3$

35. $x^3 + y = 0$
$x^2 - y = 0$

36. $x^3 + y = 0$
$2x^2 - y = 0$

37. $x^2 + (y - 2)^2 = 4$
$x^2 - 2y = 0$

38. $x^2 - y^2 - 4x + 6y - 4 = 0$
$x^2 + y^2 - 4x - 6y + 12 = 0$

39. $y = (x + 3)^2$
$x + 2y = -2$

40. $(x - 1)^2 + (y + 1)^2 = 5$
$2x - y = 3$

41. $x^2 + y^2 + 3y = 22$
$2x + y = -1$

42. $x - 3y = -5$
$x^2 + y^2 - 25 = 0$

In Exercises 43–46, let x represent one number and let y represent the other number. Use the given conditions to write a system of nonlinear equations. Solve the system and find the numbers.

43. The sum of two numbers is 10 and their product is 24. Find the numbers.

44. The sum of two numbers is 20 and their product is 96. Find the numbers.

45. The difference between the squares of two numbers is 3. Twice the square of the first number increased by the square of the second number is 9. Find the numbers.

46. The difference between the squares of two numbers is 5. Twice the square of the second number subtracted from three times the square of the first number is 19. Find the numbers.

Application Exercises

47. A planet's orbit follows an elliptical path described by $16x^2 + 4y^2 = 64$. A comet follows the parabolic path $y = x^2 - 4$. Where might the comet intersect the orbiting planet?

48. A system for tracking ships indicates that a ship lies on a hyperbolic path described by $2y^2 - x^2 = 1$. The process is repeated and the ship is found to lie on a hyperbolic path described by $2x^2 - y^2 = 1$. If it is known that the ship is located in the first quadrant of the coordinate system, determine its exact location.

49. Find the length and width of a rectangle whose perimeter is 36 feet and whose area is 77 square feet.

50. Find the length and width of a rectangle whose perimeter is 40 feet and whose area is 96 square feet.

Use the formula for the area of a rectangle and the Pythagorean Theorem to solve Exercises 51–52.

51. A small television has a picture with a diagonal measure of 10 inches and a viewing area of 48 square inches. Find the length and width of the screen.

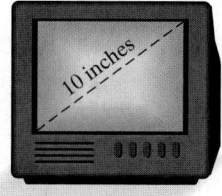

52. The area of a rug is 108 square feet, and the length of its diagonal is 15 feet. Find the length and width of the rug.

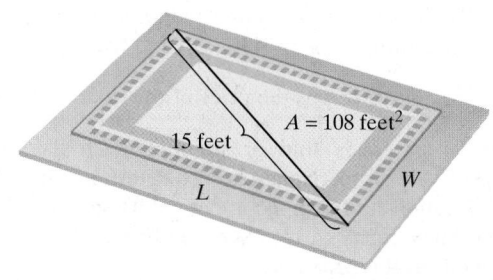

53. The figure shows a square floor plan with a smaller square area that will accommodate a combination fountain and pool. The floor with the fountain-pool area removed has an area of 21 square meters and a perimeter of 24 meters. Find the dimensions of the floor and the dimensions of the square that will accommodate the fountain and pool.

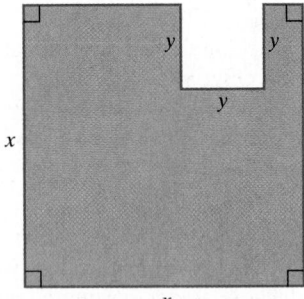

54. The area of the rectangular piece of cardboard shown on the left is 216 square inches. The cardboard is used to make an open box by cutting a 2-inch square from each corner and turning up the sides. If the box is to have a volume of 224 cubic inches, find the length and width of the cardboard that must be used.

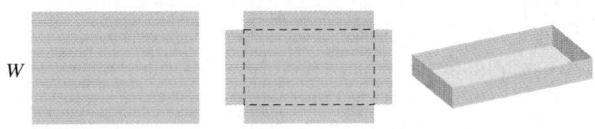

Writing in Mathematics

55. What is a system of nonlinear equations? Provide an example with your description.

56. Explain how to solve a nonlinear system using the substitution method. Use $x^2 + y^2 = 9$ and $2x - y = 3$ to illustrate your explanation.

57. Explain how to solve a nonlinear system using the addition method. Use $x^2 - y^2 = 5$ and $3x^2 - 2y^2 = 19$ to illustrate your explanation.

58. The daily demand and supply models for a carrot cake supplied by a bakery to a convenience store are given by the demand model $N = 40 - 3p$ and the supply model $N = \dfrac{p^2}{10}$, in which p is the price of the cake and N is the number of cakes sold or supplied each day to the convenience store. Explain how to determine the price at which supply and demand are equal. Then describe how to find how many carrot cakes can be supplied and sold each day at this price.

Critical Thinking Exercises

59. Which one of the following is true?
  a. A system of two equations in two variables whose graphs are a circle and a line can have four real ordered-pair solutions.
  b. A system of two equations in two variables whose graphs are a parabola and a circle can have four real ordered-pair solutions.
  c. A system of two equations in two variables whose graphs are two circles must have at least two real ordered-pair solutions.
  d. A system of two equations in two variables whose graphs are a parabola and a circle cannot have only one real ordered-pair solution.

60. Find a and b in this figure.

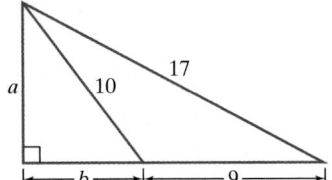

Solve the systems in Exercises 61–62.

61. $\log_y x = 3$
   $\log_y(4x) = 5$

62. $\log x^2 = y + 3$
   $\log x = y - 1$

Technology Exercises

63. Verify your solutions to any five exercises from 1 through 42 by using a graphing utility to graph the two equations in the system in the same viewing rectangle. Then use the trace or intersection feature to verify the solutions.

64. Write a system of equations, one equation whose graph is a line and the other whose graph is a parabola, that has no ordered pairs comprised of real numbers in its solution set. Graph the equations using a graphing utility and verify that you are correct.

Review Exercises

65. Graph: $3x - 2y \le 6$.
   (Section 9.4, Example 1)

66. Find the slope of the line passing through $(-2, -3)$ and $(1, 5)$.
   (Section 3.3, Example 1)

67. Multiply: $(3x - 2)(2x^2 - 4x + 3)$.
   (Section 5.2, Example 7)

CHAPTER 13 GROUP PROJECTS

MODELING PLANETARY MOTION

1. Polish astronomer Nicolaus Copernicus (1473–1543) was correct in stating that planets in our solar system revolve around the sun and not the Earth. However, he incorrectly believed that celestial orbits move in perfect circles, calling his system "the ballet of the planets."

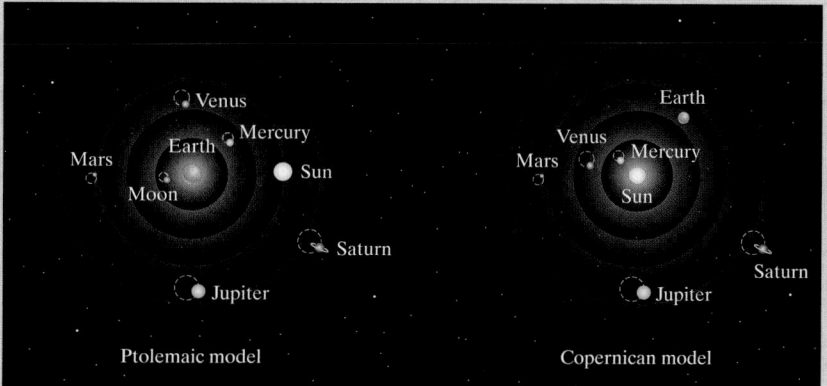

German scientist and mathematician Johannes Kepler (1571–1630) discovered that planets move in elliptical orbits with the sun at one focus. In this exercise, group members will write equations for two of these orbits and use a graphing utility to see what they look like. Use the following information:

Earth's orbit: Length of major axis: 186 million miles
                    Length of minor axis: 185.8 million miles
Mars's orbit: Length of major axis: 283.5 million miles
                    Length of minor axis: 278.5 million miles.

a. Group members should write equations in the form $\dfrac{x^2}{a^2} + \dfrac{y^2}{b^2} = 1$ for the elliptical orbits of Earth and Mars.

b. Use a graphing utility to graph the two ellipses in the same $[-300, 300, 50]$ by $[-200, 200, 50]$ viewing rectangle. Based on these graphs, explain why early astronomers incorrectly used the Copernican model to describe planetary motion.

c. Describing planetary orbits, Kepler wrote, "The heavenly motions are nothing but a continuous song for several voices, to be perceived by the intellect, not by the ear." Group members should discuss what Kepler meant by this statement.

2. Consult the research department of your library or the Internet to find an example of architecture that incorporates one or more conic sections in its design. Share this example with other group members. Explain precisely how conic sections are used. Do conic sections enhance the appeal of the architecture? In what ways?

3. Many public and private organizations and schools provide educational materials and information for the blind and visually impaired. Using your library, resources on the Worldwide Web, or local organizations, investigate how your group or college could make a contribution to enhance the study of conic sections for the blind and visually impaired. Group members should discuss how to create graphs in tactile, or touchable, form that show blind students the visual structure of each of the conic sections.

CHAPTER SUMMARY, REVIEW, AND TEST

SUMMARY

DEFINITIONS AND CONCEPTS	EXAMPLES

Section 13.1 Distance and Midpoint Formulas; Circles

The Distance Formula
The distance, d, between the points (x_1, y_1) and (x_2, y_2) is given by
$$d = \sqrt{(x_2 - x_1)^2 + (y_2 - y_1)^2}.$$

Find the distance between $(-3, -5)$ and $(6, -2)$.
$$d = \sqrt{[6 - (-3)]^2 + [-2 - (-5)]^2}$$
$$= \sqrt{9^2 + 3^2} = \sqrt{81 + 9} = \sqrt{90} = 3\sqrt{10} \approx 9.49$$

The Midpoint Formula
The midpoint of the line segment whose endpoints are (x_1, y_1) and (x_2, y_2) is the point with coordinates
$$\left(\frac{x_1 + x_2}{2}, \frac{y_1 + y_2}{2}\right).$$

Find the midpoint of the line segment whose endpoints are $(-3, 6)$ and $(4, 1)$.
$$\text{midpoint} = \left(\frac{-3 + 4}{2}, \frac{1 + 6}{2}\right) = \left(\frac{1}{2}, \frac{7}{2}\right)$$

A circle is the set of all points in a plane that are equidistant from a fixed point, the center. The distance from the center to any point on the circle is the radius.

Standard Form of the Equation of a Circle
The graph of $(x - h)^2 + (y - k)^2 = r^2$ is a circle with center (h, k) and radius r.

General Form of the Equation of a Circle
$$x^2 + y^2 + Dx + Ey + F = 0$$
Convert from the general form to the standard form by completing the square on x and y.

Find the center and radius and graph:
$$(x - 3)^2 + (y + 4)^2 = 16.$$
$$(x - 3)^2 + (y - (-4))^2 = 4^2$$
$h = 3$ $k = -4$ $r = 4$

The center, (h, k), is $(3, -4)$ and the radius, r, is 4.

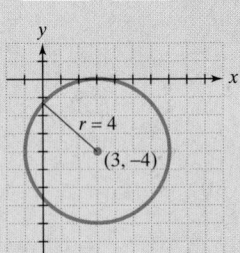

Section 13.2 The Ellipse

An ellipse is the set of all points in a plane the sum of whose distances from two fixed points, the foci, is constant. The midpoint of the segment connecting the foci is the center of the ellipse.

Standard Forms of the Equations of an Ellipse Centered at the Origin $(a^2 > b^2)$

Horizontal with vertices $(a, 0)$ and $(-a, 0)$
$$\frac{x^2}{a^2} + \frac{y^2}{b^2} = 1$$
Endpoints of major axis a units left and right of center; minor axis b units up and down from center.

Vertical with vertices $(0, a)$ and $(0, -a)$
$$\frac{x^2}{b^2} + \frac{y^2}{a^2} = 1$$
Endpoints of major axis a units up and down from center, minor axis b units left and right of center.

The equations $\dfrac{(x - h)^2}{a^2} + \dfrac{(y - k)^2}{b^2}$ and $\dfrac{(x - h)^2}{b^2} + \dfrac{(y - k)^2}{a^2} = 1$ $(a^2 > b^2)$ represent ellipses centered at (h, k).

Graph: $\dfrac{(x + 2)^2}{9} + \dfrac{(y + 4)^2}{25} = 1.$
$$\frac{(x - (-2))^2}{9} + \frac{(y - (-4))^2}{25} = 1$$
$b^2 = 9$ $a^2 = 25$

Center, (h, k), is $(-2, -4)$. With $a^2 = 25$, vertices are 5 units above and below the center. With $b^2 = 9$, endpoints of minor axis are 3 units to the left and right of the center.

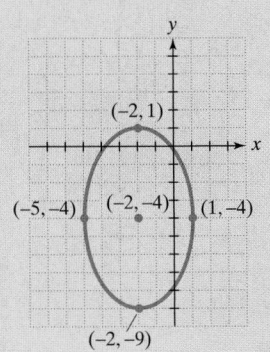

DEFINITIONS AND CONCEPTS	EXAMPLES

Section 13.3 The Hyperbola

A hyperbola is the set of all points in a plane the difference of whose distances from two fixed points, the foci, is constant.

Standard Forms of the Equations of a Hyperbola Centered at the Origin

$$\frac{x^2}{a^2} - \frac{y^2}{b^2} = 1 \qquad \frac{y^2}{a^2} - \frac{x^2}{b^2} = 1$$

Vertices are $(a, 0)$ and $(-a, 0)$.

Vertices are $(0, a)$ and $(0, -a)$.

As x and y get larger, the two branches of a hyperbola approach a pair of intersecting lines, called asymptotes. Draw the rectangle centered at the origin with sides parallel to the axes, crossing one axis at $\pm a$ and the other at $\pm b$. Draw the diagonals of this rectangle and extend them to obtain the asymptotes. Draw the two branches of the hyperbola by starting at each vertex and approaching the asymptotes.

Graph: $4x^2 - 9y^2 = 36$.

$$\frac{4x^2}{36} - \frac{9y^2}{36} = \frac{36}{36}$$

$$\frac{x^2}{9} - \frac{y^2}{4} = 1$$

$a^2 = 9$ $b^2 = 4$

Vertices are $(3, 0)$ and $(-3, 0)$. Draw a rectangle using -3 and 3 on the x-axis and -2 and 2 on the y-axis. Its extended diagonals are the asymptotes.

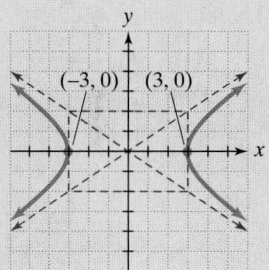

Section 13.4 The Parabola; Identifying Conic Sections

A parabola is the set of all points in a plane that are equidistant from a fixed line, the directrix, and a fixed point, the focus, that is not on the line. The line passing through the focus and perpendicular to the directrix is the axis of symmetry. The point of intersection of the parabola with its axis of symmetry is the vertex.

Equations of Horizontal Parabolas

$x = a(y - k)^2 + h$ $x = ay^2 + by + c$

   Vertex is (h, k). y-coordinate of

                                  vertex is $y = -\dfrac{b}{2a}$.

If $a > 0$, parabola opens to the right. If $a < 0$, parabola opens to the left.

Equations of Vertical Parabolas

$y = a(x - h)^2 + k$ $y = ax^2 + bx + c$

   Vertex is (h, k). x-coordinate of

                                  vertex is $x = -\dfrac{b}{2a}$.

If $a > 0$, the parabola opens upward. If $a < 0$, the parabola opens downward.

Find the vertex and graph:

$$x = -(y + 3)^2 + 4.$$
$$x = -1(y - (-3))^2 + 4$$

$a = -1$ $k = -3$ $h = 4$

Parabola opens to the left. Vertex, (h, k), is $(4, -3)$.
x-intercept: Let $y = 0$.

$$x = -(0 + 3)^2 + 4 = -9 + 4 = -5$$

y-intercepts: Let $x = 0$.

$$0 = -(y + 3)^2 + 4$$
$$(y + 3)^2 = 4$$
$$y + 3 = \pm\sqrt{4}$$
$$y = -1 \qquad \text{or} \qquad y = -5$$

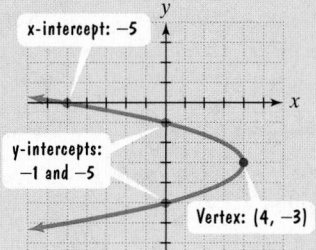

DEFINITIONS AND CONCEPTS	EXAMPLES

Section 13.4 The Parabola; Identifying Conic Sections (continued)

Recognizing Conic Sections from Equations

Conic sections result from the intersection of a cone and a plane. Their graphs contain x^2-terms, y^2-terms, or both. For parabolas, only one variable is squared. For circles, ellipses, and hyperbolas, both variables are squared. Put x^2- and y^2-terms on the same side of the equation to identify the graph:

Circle: x^2 and y^2 have the same coefficient of the same sign.

Ellipse: x^2 and y^2 have different coefficients of the same sign.

Hyperbola: x^2 and y^2 have coefficients with opposite signs.

- $9x^2 + 4y^2 = 36$

  Different coefficients of the same sign; ellipse

- $9y^2 = 25x^2 + 225$

  $9y^2 - 25x^2 = 225$

  Coefficients with opposite signs; hyperbola

- $x = -y^2 + 4y$

  Only one variable, y, is squared; parabola

- $\dfrac{x^2}{9} + \dfrac{y^2}{9} = 1$

  $x^2 + y^2 = 9$

  Same coefficient; circle

Section 13.5 Systems of Nonlinear Equations in Two Variables

A system of two nonlinear equations in two variables, x and y, contains at least one equation that cannot be expressed in the form $Ax + By = C$. Systems can be solved by the substitution method or the addition method. Each solution corresponds to a point of intersection of the system's graphs.

Solve: $\quad x^2 + y^2 = 25$

$\qquad\qquad x - 3y = -5.$

Using substitution: $\quad x = 3y - 5$

$$(3y - 5)^2 + y^2 = 25$$
$$9y^2 - 30y + 25 + y^2 = 25$$
$$10y^2 - 30y = 0$$
$$10y(y - 3) = 0$$
$$y = 0 \quad \text{or} \quad y = 3$$

If $y = 0$: $\quad x = 3y - 5 = 3 \cdot 0 - 5 = -5$

If $y = 3$: $\quad x = 3y - 5 = 3 \cdot 3 - 5 = 4$

The solution are $(-5, 0)$ and $(4, 3)$, and the solution set is $\{(-5, 0), (4, 3)\}$.

Review Exercises

13.1 *In Exercises 1–2, find the distance between each pair of points. If necessary, round answers to two decimal places.*

1. $(-2, -3)$ and $(3, 9)$

2. $(-4, 3)$ and $(-2, 5)$

In Exercises 3–4, find the midpoint of each line segment with the given endpoints.

3. $(2, 6)$ and $(-12, 4)$

4. $(4, -6)$ and $(-15, 2)$

In Exercises 5–6, write the standard form of the equation of the circle with the given center and radius.

5. Center $(0, 0)$, $r = 3$

6. Center $(-2, 4)$, $r = 6$

In Exercises 7–10, give the center and radius of each circle and graph its equation.

7. $x^2 + y^2 = 1$

8. $(x + 2)^2 + (y - 3)^2 = 9$

9. $x^2 + y^2 - 4x + 2y - 4 = 0$

10. $x^2 + y^2 - 4y = 0$

13.2 *In Exercises 11–16, graph each ellipse.*

11. $\dfrac{x^2}{36} + \dfrac{y^2}{25} = 1$ **12.** $\dfrac{x^2}{25} + \dfrac{y^2}{16} = 1$

13. $4x^2 + y^2 = 16$

14. $4x^2 + 9y^2 = 36$

15. $\dfrac{(x - 1)^2}{16} + \dfrac{(y + 2)^2}{9} = 1$

16. $\dfrac{(x + 1)^2}{9} + \dfrac{(y - 2)^2}{16} = 1$

17. A semielliptic archway has a height of 15 feet at the center and a width of 50 feet, as shown in the figure. The 50-foot width consists of a two-lane road. Can a truck that is 12 feet high and 14 feet wide drive under the archway without going into the other lane?

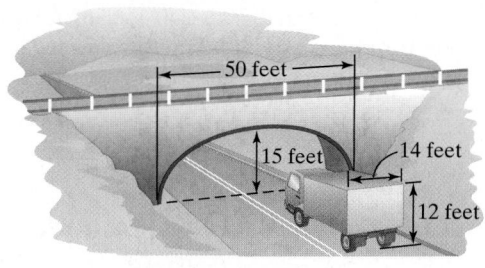

13.3 *In Exercises 18–21, use vertices and asymptotes to graph each hyperbola.*

18. $\dfrac{x^2}{16} - y^2 = 1$ **19.** $\dfrac{y^2}{16} - x^2 = 1$

20. $9x^2 - 16y^2 = 144$ **21.** $4y^2 - x^2 = 16$

13.4 *In Exercises 22–25, use the vertex and intercepts to sketch the graph of each equation. Give the equation for the horizontal parabola's axis of symmetry. If needed, find additional points on the parabola by choosing values of y on each side of the axis of symmetry.*

22. $x = (y - 3)^2 - 4$

23. $x = -2(y + 3)^2 + 2$

24. $x = y^2 - 8y + 12$

25. $x = -y^2 - 4y + 6$

In Exercises 26–32, indicate whether the graph of each equation is a circle, an ellipse, a hyperbola, or a parabola.

26. $x + 8y = y^2 + 10$

27. $16x^2 = 32 - y^2$

28. $x^2 = 25 + 25y^2$

29. $x^2 = 4 - y^2$

30. $36y^2 = 576 + 16x^2$

31. $\dfrac{(x + 3)^2}{9} + \dfrac{(y - 4)^2}{25} = 1$

32. $y = x^2 + 6x + 9$

In Exercises 33–41, indicate whether the graph of each equation is a circle, an ellipse, a hyperbola, or a parabola. Then graph the conic section.

33. $5x^2 + 5y^2 = 180$

34. $4x^2 + 9y^2 = 36$

35. $4x^2 - 9y^2 = 36$

36. $\dfrac{x^2}{25} + \dfrac{y^2}{1} = 1$

37. $x + 3 = -y^2 + 2y$

38. $y - 3 = x^2 - 2x$

39. $\dfrac{(x + 2)^2}{16} + \dfrac{(y - 5)^2}{4} = 1$

40. $(x - 3)^2 + (y + 2)^2 = 4$

41. $x^2 + y^2 + 6x - 2y + 6 = 0$

42. An engineer is designing headlight units for automobiles. The unit has a parabolic surface with a diameter of 12 inches and a depth of 3 inches. The situation is illustrated in the figure, where a coordinate system has been superimposed.

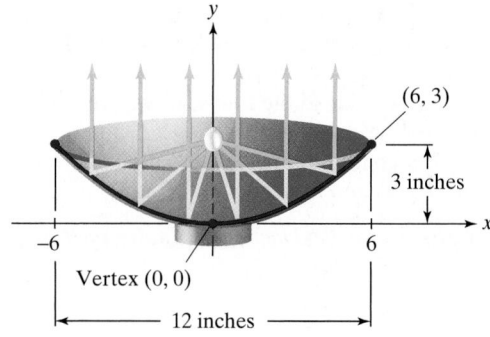

a. Use the point $(6, 3)$ to write an equation in the form $y = ax^2$ for the parabola used to design the headlight.

b. The light source should be placed at the focus $(0, p)$. The value of p is given by the equation $a = \dfrac{1}{4p}$. Where should the light source be placed? Describe this placement relative to the vertex.

13.5 *In Exercises 43–53, solve each system by the method of your choice.*

43. $5y = x^2 - 1$
$x - y = 1$

44. $y = x^2 + 2x + 1$
$x + y = 1$

45. $x^2 + y^2 = 2$
$x + y = 0$

46. $2x^2 + y^2 = 24$
$x^2 + y^2 = 15$

47. $xy - 4 = 0$
$y - x = 0$

48. $y^2 = 4x$
$x - 2y + 3 = 0$

49. $x^2 + y^2 = 10$
$y = x + 2$

50. $xy = 1$
$y = 2x + 1$

51. $x + y + 1 = 0$
$x^2 + y^2 + 6y - x = -5$

52. $x^2 + y^2 = 13$
$x^2 - y = 7$

53. $2x^2 + 3y^2 = 21$
$3x^2 - 4y^2 = 23$

54. The perimeter of a rectangle is 26 meters, and its area is 40 square meters. Find its dimensions.

55. Find the coordinates of all points (x, y) that lie on the line whose equation is $2x + y = 8$, so that the area of the rectangle shown in the figure is 6 square units.

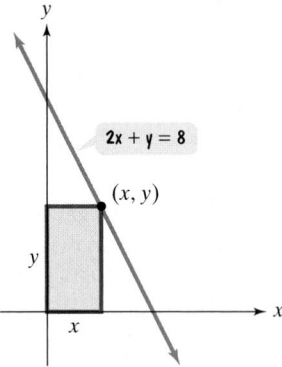

56. Two adjoining square fields with an area of 2900 square feet are to be enclosed with 240 feet of fencing. The situation is represented in the figure. Find the length of each side where a variable appears.

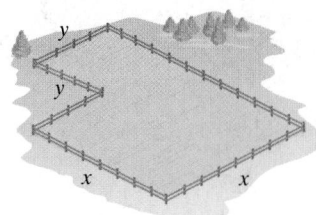

Chapter 13 Test

1. Find the distance between $(-1, 5)$ and $(2, -3)$. If necessary, round the answer to two decimal places.

2. Find the midpoint of the line segment whose endpoints are $(-5, -2)$ and $(12, -6)$.

3. Write the standard form of the equation of the circle with center $(3, -2)$ and radius 5.

In Exercises 4–5, give the center and radius of each circle.

4. $(x - 5)^2 + (y + 3)^2 = 49$

5. $x^2 + y^2 + 4x - 6y - 3 = 0$

In Exercises 6–7, give the coordinates of the vertex for each parabola.

6. $x = -2(y + 3)^2 + 7$

7. $x = y^2 + 10y + 23$

In Exercises 8–16, indicate whether the graph of each equation is a circle, an ellipse, a hyperbola, or a parabola. Then graph the conic section.

8. $\dfrac{x^2}{4} - \dfrac{y^2}{9} = 1$

9. $4x^2 + 9y^2 = 36$

10. $x = (y + 1)^2 - 4$

11. $16x^2 + y^2 = 16$

12. $25y^2 = 9x^2 + 225$

13. $x = -y^2 + 6y$

14. $\dfrac{(x - 2)^2}{16} + \dfrac{(y + 3)^2}{9} = 1$

15. $(x + 1)^2 + (y + 2)^2 = 9$

16. $\dfrac{x^2}{4} + \dfrac{y^2}{4} = 1$

In Exercises 17–18, solve each system.

17. $x^2 + y^2 = 25$
$x + y = 1$

18. $2x^2 - 5y^2 = -2$
$3x^2 + 2y^2 = 35$

19. The rectangular plot of land shown in the Figure is to be fenced along three sides using 39 feet of fencing. No fencing is to be placed along the river's edge. The area of the plot is 180 square feet. What are its dimensions?

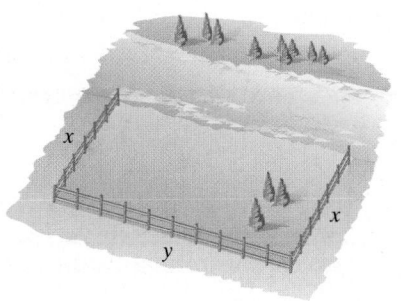

20. A rectangle has a diagonal of 5 feet and a perimeter of 14 feet. Find the rectangle's dimensions.

Cumulative Review Exercises (Chapters 1–13)

In Exercises 1–7, solve each equation, inequality, or system.

1. $3x + 7 > 4$ or $6 - x < 1$

2. $x(2x - 7) = 4$

3. $\dfrac{5}{x - 3} = 1 + \dfrac{30}{x^2 - 9}$

4. $3x^2 + 8x + 5 < 0$

5. $3^{2x-1} = 81$

6. $30e^{0.7x} = 240$

7. $3x^2 + 4y^2 = 39$
$5x^2 - 2y^2 = -13$

In Exercises 8–11, graph each function, equation, or inequality in a rectangular coordinate system.

8. $f(x) = -\dfrac{2}{3}x + 4$ **9.** $3x - y > 6$

10. $x^2 + y^2 + 4x - 6y + 9 = 0$ **11.** $9x^2 - 4y^2 = 36$

In Exercises 12–15, perform the indicated operations, and simplify, if possible.

12. $-2(3^2 - 12)^3 - 45 \div 9 - 3$

13. $(3x^3 - 19x^2 + 17x + 4) \div (3x - 4)$

14. $\sqrt[3]{4x^2y^5} \cdot \sqrt[3]{4xy^2}$

15. $(2 + 3i)(4 - i)$

In Exercises 16–17, factor completely.

16. $12x^3 - 36x^2 + 27x$

17. $x^3 - 2x^2 - 9x + 18$

18. Find the domain: $f(x) = \sqrt{6 - 3x}$.

19. Rationalize the denominator: $\dfrac{1 - \sqrt{x}}{1 + \sqrt{x}}$.

20. Write as a single logarithm: $\dfrac{1}{3}\ln x + 7\ln y$.

21. Divide using synthetic division:

$(3x^3 - 5x^2 + 2x - 1) \div (x - 2)$.

22. Write a quadratic equation whose solution set is $\{-2\sqrt{3}, 2\sqrt{3}\}$.

23. Two cars leave from the same place at the same time, traveling in opposite directions. The rate of the faster car exceeds that of the slower car by 10 miles per hour. After 2 hours, the cars are 180 miles apart. Find the rate of each car.

24. Rent-a-Truck charges a daily rental rate of $39 plus $0.16 per mile. A competing agency, Ace Truck Rentals, charges $25 a day plus $0.24 per mile for the same truck. How many miles must be driven in a day to make the daily cost of both agencies the same? What will be the cost?

25. Three apples and two bananas provide 354 calories, and two apples and three bananas provide 381 calories. Find the number of calories in one apple and one banana.

We often save for the future by investing small amounts at periodic intervals. To understand how our savings accumulate, we need to understand properties of lists of numbers that are related to each other by a rule. Such lists are called *sequences*. Learning about properties of sequences will show you how to make your financial goals a reality. Your knowledge of sequences will enable you to inform your college roommate of the best of the three appealing offers.

Something incredible has happened. Your college roommate, a gifted athlete, has been given a six-year contract with a professional baseball team. He will be playing against the likes of Mark McGwire and Barry Bonds. Management offers him three options. One is a beginning salary of $1,700,000 with annual increases of $70,000 per year starting in the second year. A second option is $1,700,000 the first year with an annual increase of 2% per year beginning in the second year. The third offer involves less money the first year—$1,500,000—but there is an annual increase of 9% yearly after that. Which option offers the most money over the six-year contract?

Chapter 14

Sequences, Series, and Probability

939

▶ **SECTION 14.1** *Sequences and Summation Notation*

Objectives

1 Find particular terms of a sequence from the general term.

2 Use factorial notation.

3 Use summation notation.

SSM PH Tutor CD- Video
Center ROM

ENRICHMENT ESSAY

Fibonacci Numbers on the Piano Keyboard

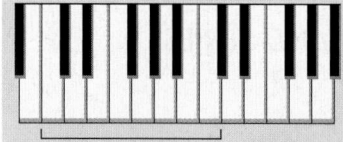

One Octave

Numbers in the Fibonacci sequence can be found in an octave on the piano keyboard. The octave contains 2 black keys in one cluster, 3 black keys in another cluster, 5 black keys, 8 white keys, and a total of 13 keys altogether. The numbers 2, 3, 5, 8, and 13 are the third through seventh terms of the Fibonacci sequence.

Sequences

Many creations in nature involve intricate mathematical designs, including a variety of spirals. For example, the arrangement of the individual florets in the head of a sunflower forms spirals. In some species, there are 21 spirals in the clockwise direction and 34 in the counterclockwise direction. The precise numbers depend on the species of sunflower: 21 and 34, or 34 and 55, or 55 and 89, or even 89 and 144.

This observation becomes even more interesting when we consider a sequence of numbers investigated by Leonardo of Pisa, also known as Fibonacci, an Italian mathematician of the thirteenth century. The **Fibonacci sequence** of numbers is an infinite sequence that begins as follows:

$$1, 1, 2, 3, 5, 8, 13, 21, 34, 55, 89, 144, 233 \ldots.$$

The first two terms are 1. Every term thereafter is the sum of the two preceding terms. For example, the third term, 2, is the sum of the first and second terms: $1 + 1 = 2$. The fourth term, 3, is the sum of the second and third terms: $1 + 2 = 3$, and so on. Did you know that the number of spirals in a daisy or a sunflower, 21 and 34, are two Fibonacci numbers? The number of spirals in a pinecone, 8 and 13, and a pineapple, 8 and 13, are also Fibonacci numbers.

We can think of the Fibonacci sequence as a function. The terms of the sequence

$$1, 1, 2, 3, 5, 8, 13, 21, 34, 55, 89, 144, 233, \ldots$$

are the range values for a function whose domain is the set of positive integers.

Domain: 1,   2,   3,   4,   5,   6,   7,   ...
          ↓   ↓   ↓   ↓   ↓   ↓   ↓
Range: 1,   1,   2,   3,   5,   8,   13,   ...

Thus, $f(1) = 1, f(2) = 1, f(3) = 2, f(4) = 3, f(5) = 5, f(6) = 8, f(7) = 13$, and so on.

The letter a with a subscript is used to represent function values of a sequence, rather than the usual function notation. The subscripts make up the domain of the sequence, and they identify the location of a term. Thus, a_1 represents the first term of the sequence, a_2 represents the second term, a_3 the third term, and so on. This notation is shown for the first six terms of the Fibonacci sequence:

1,   1,   2,   3,   5,   8.

$a_1 = 1$ $a_2 = 1$ $a_3 = 2$ $a_4 = 3$ $a_5 = 5$ $a_6 = 8$

The notation a_n represents the nth term, or **general term**, of a sequence. The entire sequence is represented by $\{a_n\}$.

> **Definition of a Sequence**
>
> An **infinite sequence** $\{a_n\}$ is a function whose domain is the set of positive integers. The function values, or **terms**, of the sequence are represented by
>
> $$a_1, a_2, a_3, a_4, \ldots, a_n, \ldots.$$
>
> Sequences whose domains consist only of the first n positive integers are called **finite sequences**.

1 Find particular terms of a sequence from the general term.

EXAMPLE 1 Writing Terms of a Sequence from the General Term

Write the first four terms of the sequence whose nth term, or general term, is given:

a. $a_n = 3n + 4$
b. $a_n = \dfrac{(-1)^n}{3^n - 1}$.

Solution

a. We need to find the first four terms of the sequence whose general term is $a_n = 3n + 4$. To do so, we replace n in the formula by $1, 2, 3,$ and 4.

a_1, 1st term $3 \cdot 1 + 4 = 3 + 4 = 7$

a_2, 2nd term $3 \cdot 2 + 4 = 6 + 4 = 10$

a_3, 3rd term $3 \cdot 3 + 4 = 9 + 4 = 13$

a_4, 4th term $3 \cdot 4 + 4 = 12 + 4 = 16$

The first four terms are 7, 10, 13, and 16. The sequence defined by $a_n = 3n + 4$ can be written as

$$7, 10, 13, \ldots, 3n + 4, \ldots.$$

b. We need to find the first four terms of the sequence whose general term is $a_n = \dfrac{(-1)^n}{3^n - 1}$. To do so, we replace each occurrence of n in the formula by $1, 2, 3,$ and 4.

a_1, 1st term $\dfrac{(-1)^1}{3^1 - 1} = \dfrac{-1}{3 - 1} = -\dfrac{1}{2}$

a_2, 2nd term $\dfrac{(-1)^2}{3^2 - 1} = \dfrac{1}{9 - 1} = \dfrac{1}{8}$

a_3, 3rd term $\dfrac{(-1)^3}{3^3 - 1} = \dfrac{-1}{27 - 1} = -\dfrac{1}{26}$

a_4, 4th term $\dfrac{(-1)^4}{3^4 - 1} = \dfrac{1}{81 - 1} = \dfrac{1}{80}$

The first four terms are $-\frac{1}{2}, \frac{1}{8}, -\frac{1}{26},$ and $\frac{1}{80}$. The sequence defined by $\dfrac{(-1)^n}{3^n - 1}$ can be written as

$$-\frac{1}{2}, \frac{1}{8}, -\frac{1}{26}, \ldots, \frac{(-1)^n}{3^n - 1}, \ldots.$$

Study Tip

The factor $(-1)^n$ in the general term of a sequence causes the signs of the terms to alternate between positive and negative, depending on whether n is even or odd.

✔ **CHECK POINT 1** Write the first four terms of the sequence whose nth term, or general term, is given:

a. $a_n = 2n + 5$
b. $a_n = \dfrac{(-1)^n}{2^n + 1}$.

Although sequences are usually named with the letter a, any lowercase letter can be used. For example, the first four terms of the sequence $\{b_n\} = \{(\frac{1}{2})^n\}$ are $b_1 = \frac{1}{2}$, $b_2 = \frac{1}{4}$, $b_3 = \frac{1}{8}$, and $b_4 = \frac{1}{16}$.

Because a sequence is a function whose domain is the set of positive integers, the **graph of a sequence** is a set of discrete points. For example, consider the sequence whose general term is $a_n = \frac{1}{n}$. How does the graph of this sequence differ from the graph of the rational function $f(x) = \frac{1}{x}$? The graph of $f(x) = \frac{1}{x}$ is shown in Figure 14.1(a) for positive values of x. To obtain the graph of the sequence $\{a_n\} = \{\frac{1}{n}\}$, remove all the points from the graph of f except those whose x-coordinates are positive integers. Thus, we remove all points except $(1, 1), (2, \frac{1}{2}), (3, \frac{1}{3}), (4, \frac{1}{4})$, and so on. The remaining points are the graph of the sequence $\{a_n\} = \{\frac{1}{n}\}$, shown in Figure 14.1(b). Notice that the horizontal axis is labeled n and the vertical axis a_n.

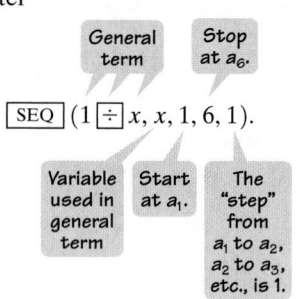

Using Technology

Graphing utilities can write the terms of a sequence and graph them. For example, to find the first six terms of $\{a_n\} = \left\{\frac{1}{n}\right\}$, enter

General term | Stop at a_6.

SEQ $(1 \div x, x, 1, 6, 1)$.

Variable used in general term | Start at a_1. | The "step" from a_1 to a_2, a_2 to a_3, etc., is 1.

The first few terms of the sequence are shown in the viewing rectangle. By pressing the right arrow key to scroll right, you can see the remaining terms.

```
seq(1/X,X,1,6,1)
{1 .5 .33333333…
Ans►Frac
{1 1/2 1/3 1/4 …
```

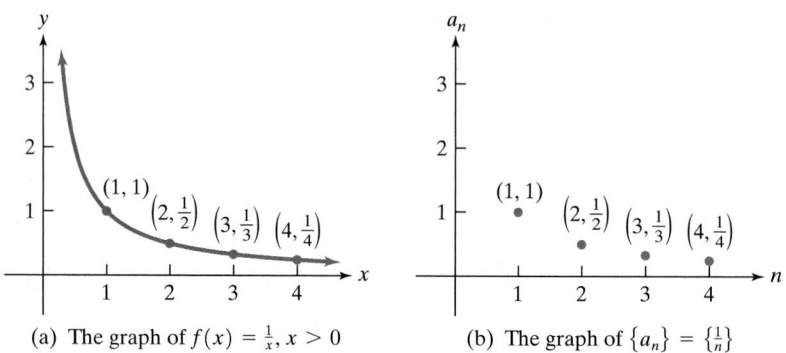

(a) The graph of $f(x) = \frac{1}{x}$, $x > 0$

(b) The graph of $\{a_n\} = \{\frac{1}{n}\}$

Figure 14.1 Comparing a continuous graph to the graph of a sequence

Factorial Notation

Products of consecutive positive integers occur quite often in sequences. These products can be expressed in a special notation, called **factorial notation**.

2 Use factorial notation.

Factorial Notation

If n is a positive integer, the notation $n!$ (read "n factorial") is the product of all positive integers from n down through 1.

$$n! = n(n - 1)(n - 2) \cdots (3)(2)(1)$$

$0!$ (zero factorial), by definition, is 1.

$$0! = 1$$

The values of $n!$ for the first six positive integers are

$$1! = 1$$
$$2! = 2 \cdot 1 = 2$$
$$3! = 3 \cdot 2 \cdot 1 = 6$$
$$4! = 4 \cdot 3 \cdot 2 \cdot 1 = 24$$
$$5! = 5 \cdot 4 \cdot 3 \cdot 2 \cdot 1 = 120$$
$$6! = 6 \cdot 5 \cdot 4 \cdot 3 \cdot 2 \cdot 1 = 720.$$

Factorials affect only the number or variable that they follow unless grouping symbols appear. For example,

$$2 \cdot 3! = 2(3 \cdot 2 \cdot 1) = 2 \cdot 6 = 12$$

whereas

$$(2 \cdot 3)! = 6! = 6 \cdot 5 \cdot 4 \cdot 3 \cdot 2 \cdot 1 = 720.$$

In this sense, factorials are similar to exponents.

EXAMPLE 2 Finding Terms of a Sequence Involving Factorials

Write the first four terms of the sequence whose nth term is

$$a_n = \frac{2^n}{(n-1)!}.$$

Solution We need to find the first four terms of the sequence. To do so, we replace each n in the formula by 1, 2, 3, and 4.

a_1, 1st term $\dfrac{2^1}{(1-1)!} = \dfrac{2}{0!} = \dfrac{2}{1} = 2$

a_2, 2nd term $\dfrac{2^2}{(2-1)!} = \dfrac{4}{1!} = \dfrac{4}{1} = 4$

a_3, 3rd term $\dfrac{2^3}{(3-1)!} = \dfrac{8}{2!} = \dfrac{8}{2 \cdot 1} = 4$

a_4, 4th term $\dfrac{2^4}{(4-1)!} = \dfrac{16}{3!} = \dfrac{16}{3 \cdot 2 \cdot 1} = \dfrac{16}{6} = \dfrac{8}{3}$

The first four terms are 2, 4, 4, and $\frac{8}{3}$. ∎

✔ **CHECK POINT 2** Write the first four terms of the sequence whose nth term is

$$a_n = \frac{20}{(n+1)!}.$$

3 Use summation notation.

Summation Notation

It is sometimes useful to find the sum of the first n terms of a sequence. For example, consider the number of AIDS cases diagnosed in the United States for each year from 1991 through 1999, shown in Table 14.1.

Table 14.1   AIDS Cases Diagnosed in the United States, 1991–1999

Year	1991	1992	1993	1994	1995	1996	1997	1998	1999
Cases Diagnosed	60,472	79,477	79,752	72,684	69,172	59,832	47,439	38,587	25,434

Source: U.S. Department of Health and Human Services

We can let a_n represent the number of AIDS cases diagnosed in year n, where $n = 1$ corresponds to 1991, $n = 2$ to 1992, $n = 3$ to 1993, and so on. The terms of the finite sequence in Table 14.1 are given as follows:

60,472, 79,477, 79,752, 72,684, 69,172, 59,832, 47,439, 38,587, 25,434.

a_1 a_2 a_3 a_4 a_5 a_6 a_7 a_8 a_9

Why might we want to add the terms of this sequence? We do this to find the total number of AIDS cases diagnosed from 1991 through 1999. Thus,

$$a_1 + a_2 + a_3 + a_4 + a_5 + a_6 + a_7 + a_8 + a_9$$

$$= 60{,}472 + 79{,}477 + 79{,}752 + 72{,}684 + 69{,}172 + 59{,}832 + 47{,}439 + 38{,}587 + 25{,}434$$

$$= 532{,}849$$

We see that there were 532, 849 AIDS cases diagnosed in the United States from 1991 through 1999.

There is a compact notation for expressing the sum of the first n terms of a sequence. For example, rather than write

$$a_1 + a_2 + a_3 + a_4 + a_5 + a_6 + a_7 + a_8 + a_9,$$

we can use **summation notation** to express the sum as

$$a_1 + a_2 + a_3 + a_4 + a_5 + a_6 + a_7 + a_8 + a_9 = \sum_{i=1}^{9} a_i.$$

We read the expression on the right as "the sum as i goes from 1 to 9 of a_i." The letter i is called the **index of summation** and is not related to the use of i to represent $\sqrt{-1}$.

You can think of the symbol Σ (the uppercase Greek letter sigma) as an instruction to add up terms of a sequence.

Summation Notation

The sum of the first n terms of a sequence is represented by the **summation notation**

$$\sum_{i=1}^{n} a_i = a_1 + a_2 + a_3 + a_4 + \cdots + a_n$$

where i is the **index of summation**, n is the **upper limit of summation**, and 1 is the **lower limit of summation**.

Any letter can be used for the index of summation. The letters i, j, and k are used commonly. Furthermore, the lower limit of summation can be an integer other than 1.

When we write out a sum that is given in summation notation, we are **expanding the summation notation**. Example 3 shows how to do this.

EXAMPLE 3   Using Summation Notation

Expand and evaluate the sum:

a. $\displaystyle\sum_{i=1}^{6} (i^2 + 1)$ **b.** $\displaystyle\sum_{k=4}^{7} [(-2)^k - 5]$ **c.** $\displaystyle\sum_{i=1}^{5} 3.$

Solution

a. We must replace i in the expression $i^2 + 1$ with all consecutive integers from 1 to 6 inclusively. Then we add.

$$\sum_{i=1}^{6} (i^2 + 1) = (1^2 + 1) + (2^2 + 1) + (3^2 + 1) + (4^2 + 1)$$
$$+ (5^2 + 1) + (6^2 + 1)$$
$$= 2 + 5 + 10 + 17 + 26 + 37$$
$$= 97$$

b. This time the index of summation is k. First we evaluate $(-2)^k - 5$ for all consecutive integers from 4 through 7 inclusively. Then we add.

$$\sum_{k=4}^{7} \left[(-2)^k - 5\right] = \left[(-2)^4 - 5\right] + \left[(-2)^5 - 5\right]$$
$$+ \left[(-2)^6 - 5\right] + \left[(-2)^7 - 5\right]$$
$$= (16 - 5) + (-32 - 5) + (64 - 5) + (-128 - 5)$$
$$= 11 + (-37) + 59 + (-133)$$
$$= -100$$

c. To find $\displaystyle\sum_{i=1}^{5} 3$, we observe that every term of the sum is 3. The notation $i = 1$ through 5 indicates that we must add the first five 3s from a sequence in which every term is 3.

$$\sum_{i=1}^{5} 3 = 3 + 3 + 3 + 3 + 3 = 15$$ ∎

✔ **CHECK POINT 3** Expand and evaluate the sum:

a. $\displaystyle\sum_{i=1}^{6} 2i^2$

b. $\displaystyle\sum_{k=3}^{5} (2^k - 3)$

c. $\displaystyle\sum_{i=1}^{5} 4$.

For a given sum, we can vary the upper and lower limits of summation as well as the letter used for the index of summation. By doing so, we can produce different-looking summation notations for the same sum. For example, the sum of the squares of the first four positive integers, $1^2 + 2^2 + 3^2 + 4^2$, can be expressed in a number of equivalent ways:

$$\sum_{i=1}^{4} i^2 = 1^2 + 2^2 + 3^2 + 4^2 = 30$$

$$\sum_{i=0}^{3} (i + 1)^2 = (0 + 1)^2 + (1 + 1)^2 + (2 + 1)^2 + (3 + 1)^2$$
$$= 1^2 + 2^2 + 3^2 + 4^2 = 30$$

$$\sum_{k=2}^{5} (k - 1)^2 = (2 - 1)^2 + (3 - 1)^2 + (4 - 1)^2 + (5 - 1)^2$$
$$= 1^2 + 2^2 + 3^2 + 4^2 = 30.$$

EXAMPLE 4   Writing Sums in Summation Notation

Express each sum using summation notation:

a. $1^3 + 2^3 + 3^3 + \cdots + 7^3$

b. $1 + \dfrac{1}{3} + \dfrac{1}{9} + \dfrac{1}{27} + \cdots + \dfrac{1}{3^{n-1}}.$

Solution   In each case, we will use 1 as the lower limit of summation and i for the index of summation.

a. The sum $1^3 + 2^3 + 3^3 + \cdots + 7^3$ has seven terms, each of the form i^3, starting at $i = 1$ and ending at $i = 7$. Thus,

$$1^3 + 2^3 + 3^3 + \cdots + 7^3 = \sum_{i=1}^{7} i^3.$$

b. The sum

$$1 + \frac{1}{3} + \frac{1}{9} + \frac{1}{27} + \cdots + \frac{1}{3^{n-1}}$$

has n terms, each of the form $\dfrac{1}{3^{i-1}}$, starting at $i = 1$ and ending at $i = n$. Thus,

$$1 + \frac{1}{3} + \frac{1}{9} + \frac{1}{27} + \cdots + \frac{1}{3^{n-1}} = \sum_{i=1}^{n} \frac{1}{3^{i-1}}.$$ ∎

✔ **CHECK POINT 4**   Express each sum using summation notation:

a. $1^2 + 2^2 + 3^2 + \cdots + 9^2$

b. $1 + \dfrac{1}{2} + \dfrac{1}{4} + \dfrac{1}{8} + \cdots + \dfrac{1}{2^{n-1}}.$

EXERCISE SET 14.1

Practice Exercises

In Exercises 1–16, write the first four terms of each sequence whose general term is given.

1. $a_n = 3n + 2$

2. $a_n = 4n - 1$

3. $a_n = 3^n$

4. $a_n = \left(\dfrac{1}{3}\right)^n$

5. $a_n = (-3)^n$

6. $a_n = \left(-\dfrac{1}{3}\right)^n$

7. $a_n = (-1)^n(n + 3)$

8. $a_n = (-1)^{n+1}(n + 4)$

9. $a_n = \dfrac{2n}{n + 4}$

10. $a_n = \dfrac{3n}{n + 5}$

11. $a_n = \dfrac{(-1)^{n+1}}{2^n - 1}$

12. $a_n = \dfrac{(-1)^{n+1}}{2^n + 1}$

13. $a_n = \dfrac{n^2}{n!}$

14. $a_n = \dfrac{(n + 1)!}{n^2}$

15. $a_n = 2(n + 1)!$

16. $a_n = -2(n - 1)!$

In Exercises 17–30, find each indicated sum.

17. $\displaystyle\sum_{i=1}^{6} 5i$

18. $\displaystyle\sum_{i=1}^{6} 7i$

19. $\displaystyle\sum_{i=1}^{4} 2i^2$

20. $\displaystyle\sum_{i=1}^{5} i^3$

21. $\displaystyle\sum_{k=1}^{5} k(k + 4)$

22. $\displaystyle\sum_{k=1}^{4} (k - 3)(k + 2)$

23. $\sum_{i=1}^{4} \left(-\frac{1}{2}\right)^i$

24. $\sum_{i=2}^{4} \left(-\frac{1}{3}\right)^i$

25. $\sum_{i=5}^{9} 11$

26. $\sum_{i=3}^{7} 12$

27. $\sum_{i=0}^{4} \frac{(-1)^i}{i!}$

28. $\sum_{i=0}^{4} \frac{(-1)^{i+1}}{(i+1)!}$

29. $\sum_{i=1}^{5} \frac{i!}{(i-1)!}$

30. $\sum_{i=1}^{5} \frac{(i+2)!}{i!}$

In Exercises 31–42, express each sum using summation notation. Use 1 as the lower limit of summation and i for the index of summation.

31. $1^2 + 2^2 + 3^2 + \cdots + 15^2$

32. $1^4 + 2^4 + 3^4 + \cdots + 12^4$

33. $2 + 2^2 + 2^3 + \cdots + 2^{11}$

34. $5 + 5^2 + 5^3 + \cdots + 5^{12}$

35. $1 + 2 + 3 + \cdots + 30$

36. $1 + 2 + 3 + \cdots + 40$

37. $\frac{1}{2} + \frac{2}{3} + \frac{3}{4} + \cdots + \frac{14}{14 + 1}$

38. $\frac{1}{3} + \frac{2}{4} + \frac{3}{5} + \cdots + \frac{16}{16 + 2}$

39. $4 + \frac{4^2}{2} + \frac{4^3}{3} + \cdots + \frac{4^n}{n}$

40. $\frac{1}{9} + \frac{2}{9^2} + \frac{3}{9^3} + \cdots + \frac{n}{9^n}$

41. $1 + 3 + 5 + \cdots + (2n - 1)$

42. $a + ar + ar^2 + \cdots + ar^{n-1}$

In Exercises 43–48, express each sum using summation notation. Use a lower limit of summation of your choice and k for the index of summation.

43. $5 + 7 + 9 + 11 + \cdots + 31$

44. $6 + 8 + 10 + 12 + \cdots + 32$

45. $a + ar + ar^2 + \cdots + ar^{12}$

46. $a + ar + ar^2 + \cdots + ar^{14}$

47. $a + (a + d) + (a + 2d) + \cdots + (a + nd)$

48. $(a + d) + (a + d^2) + \cdots + (a + d^n)$

Application Exercises

49. The bar graph shows the number of compact discs (CDs) sold in the United States. Let a_n represent the number of CDs sold, in millions, in year n, where $n = 1$ corresponds to 1991, $n = 2$ to 1992, and so on.

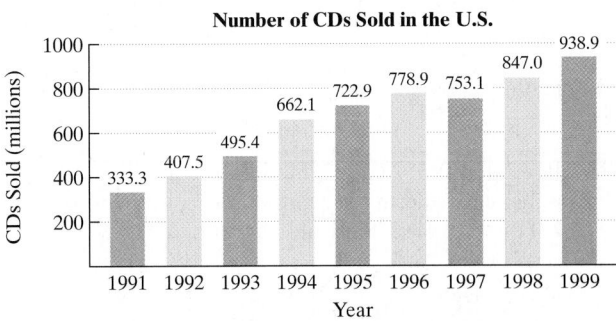

Number of CDs Sold in the U.S.

Source: Recording Industry Association of America

a. Find $\sum_{i=1}^{9} a_i$. What does this represent?

b. Find $\frac{1}{9} \sum_{i=1}^{9} a_i$. What does this represent?

50. The bar graph shows the number of vinyl long-playing records (LPs) sold in the United States. Let a_n represent the number of LPs sold, in millions, in year n, where $n = 1$ corresponds to 1991, $n = 2$ to 1992, and so on.

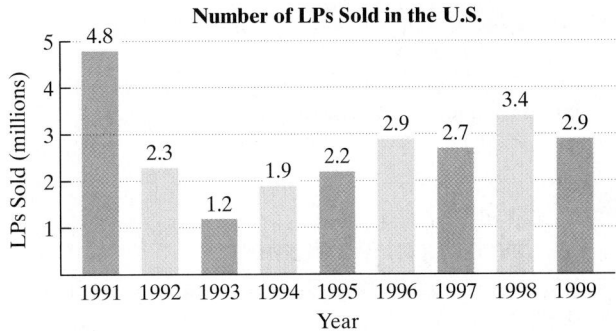

Number of LPs Sold in the U.S.

Source: Recording Industry Association of America

a. Find $\sum_{i=1}^{9} a_i$. What does this represent?

b. Find $\frac{1}{9} \sum_{i=1}^{9} a_i$. What does this represent?

51. The finite sequence whose general term is

$$a_n = 0.16n^2 - 1.04n + 7.39$$

where $n = 1, 2, 3, \ldots, 9$ models the total number of dollars, in billions, that Americans spent on recreational boating from 1991 through 1999. Find and interpret

$$\sum_{i=1}^{5} a_i.$$

52. The finite sequence whose general term is

$$a_n = 2.54e^{-0.09n}$$

where $n = 0, 1, 2, \ldots, 9$ models the number of new foreign cars sold in the United States, in millions, from 1990 through 1999. Find and interpret

$$\sum_{i=0}^{4} a_i.$$

Round to the nearest tenth of a million.

53. A deposit of $6000 is placed in an account that earns 6% interest compounded quarterly. The balance in the account after n quarters is given by the sequence

$$a_n = 6000\left(1 + \frac{0.06}{4}\right)^n, \quad n = 1, 2, 3, \ldots.$$

Find the balance in the account after five years by computing a_{20}. Round to the nearest cent.

54. A deposit of $10,000 is placed in an account that earns 8% interest compounded quarterly. The balance in the account after n quarters is given by the sequence

$$a_n = 10,000\left(1 + \frac{0.08}{4}\right)^n, \quad n = 1, 2, 3, \ldots.$$

Find the balance in the account after six years by computing a_{24}. Round to the nearest cent.

Writing in Mathematics

55. What is a sequence? Give an example with your description.

56. Explain how to write terms of a sequence if the formula for the general term is given.

57. What does the graph of a sequence look like? How is it obtained?

58. Explain how to find $n!$ if n is a positive integer.

59. What is the meaning of the symbol Σ? Give an example with your description.

60. You buy a new car for $24,000. At the end of n years, the value of your car is given by the sequence

$$a_n = 24{,}000\left(\frac{3}{4}\right)^n, \quad n = 1, 2, 3, \ldots.$$

Find a_5 and write a sentence explaining what this value represents. Describe the nth term of the sequence in terms of the value of your car at the end of each year.

Critical Thinking Exercises

61. Which one of the following is true?

a. $\displaystyle\sum_{i=1}^{2} (-1)^i 2^i = 0$

b. $\displaystyle\sum_{i=1}^{2} a_i b_i = \sum_{i=1}^{2} a_i \sum_{i=1}^{2} b_i$

c. $\displaystyle\sum_{i=1}^{4} 3i + \sum_{i=1}^{4} 4i = \sum_{i=1}^{4} 7i$

d. $\displaystyle\sum_{i=0}^{6} (-1)^i (i + 1)^2 = \sum_{j=1}^{7} (-1)^j j^2$

In Exercises 62–63, expand and write the answer as a single logarithm with a coefficient of 1.

62. $\displaystyle\sum_{i=1}^{4} \log 2i$

63. $\displaystyle\sum_{i=2}^{4} 2i \log x$

64. If $a_1 = 7$ and $a_n = a_{n-1} + 5$ for $n \geq 2$, write the first four terms of the sequence.

65. Evaluate without using a calculator: $\dfrac{600!}{599!}$.

Technology Exercises

66. Use the ⎡SEQ⎤ (sequence) capability of a graphing utility to verify the terms of the sequences you obtained for any five sequences from Exercises 1–16.

67. Use the ⎡SUM⎤ ⎡SEQ⎤ (sum of the sequence) capability of a graphing utility to verify any five of the sums you obtained in Exercises 17–30.

68. As n increases, the terms of the sequence

$$a_n = \left(1 + \frac{1}{n}\right)^n$$

get closer and closer to the number e (where $e \approx 2.7183$). Use a calculator to find a_{10}, a_{100}, a_{1000}, $a_{10,000}$, and $a_{100,000}$, comparing these terms to the decimal approximation for e.

Many graphing utilities have a sequence-graphing mode that plots the terms of a sequence as points on a rectangular coordinate system. Consult your manual; if your graphing utility has this capability, use it to graph each of the sequences in Exercises 69–72. What appears to be happening to the terms of each sequence as n gets larger?

69. $a_n = \dfrac{n}{n + 1}; \quad n:[0, 10, 1]$ by $a_n:[0, 1, 0.1]$

70. $a_n = \dfrac{100}{n}$; $n : [0, 1000, 100]$ by $a_n : [0, 1, 0.1]$

71. $a_n = \dfrac{2n^2 + 5n - 7}{n^3}$; $n : [0, 10, 1]$ by $a_n : [0, 2, 0.2]$

72. $a_n = \dfrac{3n^4 + n - 1}{5n^4 + 2n^2 + 1}$; $n : [0, 10, 1]$ by $a_n : [0, 1, 0.1]$

Review Exercises

73. Simplify: $\sqrt[3]{40x^4y^7}$.
(Section 10.3, Example 5)

74. Factor: $27x^3 - 8$.
(Section 6.4, Example 8)

75. Solve: $\dfrac{6}{x} + \dfrac{6}{x + 2} = \dfrac{5}{2}$.
(Section 7.6, Example 3)

▶ SECTION 14.2   *Arithmetic Sequences*

Objectives

1 Find the common difference for an arithmetic sequence.

2 Write terms of an arithmetic sequence.

3 Use the formula for the general term of an arithmetic sequence.

4 Use the formula for the sum of the first *n* terms of an arithmetic sequence.

SSM
PH Tutor   CD-   Video
Center   ROM

Your grandmother and her financial counselor are looking at options in case nursing home care is needed in the future. The good news is that your grandmother's total assets are $350,000. The bad news is that yearly nursing home costs average $49,730, increasing by $1800 each year. In this section, we will see how sequences can be used to describe your grandmother's situation and help her to identify realistic options.

Arithmetic Sequences

A mathematical model for the average annual salaries of major league baseball players generates the following data.

Year	1991	1992	1993	1994	1995	1996	1997	1998	1999
Salary	801,000	892,000	983,000	1,074,000	1,165,000	1,256,000	1,347,000	1,438,000	1,529,000

From 1991 to 1992, salaries increased by $892,000 - $801,000 = $91,000. From 1992 to 1993, salaries increased by $983,000 - $892,000 = $91,000. If we make these computations for each year, we find that the yearly salary increase is $91,000. The sequence of annual salaries shows that each term after the first, 801,000, differs from the preceding term by a constant amount, namely 91,000. The sequence of annual salaries

$$801,000, \ 892,000, \ 983,000, \ 1,074,000, \ 1,165,000, \ 1,256,000, \dots$$

is an example of an *arithmetic sequence*.

Definition of an Arithmetic Sequence

An **arithmetic sequence** is a sequence in which each term after the first differs from the preceding term by a constant amount. The difference between consecutive terms is called the **common difference** of the sequence.

① Find the common difference for an arithmetic sequence.

The common difference, d, is found by subtracting any term from the term that directly follows it. In the following examples, the common difference is found by subtracting the first term from the second term, $a_2 - a_1$.

ARITHMETIC SEQUENCE	COMMON DIFFERENCE
801,000, 892,000, 983,000, 1,074,000, ...	$d = 892{,}000 - 801{,}000 = 91{,}000$
2, 6, 10, 14, 18, ...	$d = 6 - 2 = 4$
$-2, -7, -12, -17, \ldots$	$d = -7 - (-2) = -5$

If the first term of an arithmetic sequence is a_1, each term after the first is obtained by adding d, the common difference, to the previous term.

② Write terms of an arithmetic sequence.

EXAMPLE 1 Writing the Terms of an Arithmetic Sequence Using the First Term and the Common Difference

Write the first six terms of the arithmetic sequence with first term 6 and common difference -2.

Solution To find the second term, we add -2 to the first term, 6, giving 4. For the next term, we add -2 to 4, and so on.

$$a_1 \text{ (first term)} = 6$$
$$a_2 \text{ (second term)} = 6 + (-2) = 4$$
$$a_3 \text{ (third term)} = 4 + (-2) = 2$$
$$a_4 \text{ (fourth term)} = 2 + (-2) = 0$$
$$a_5 \text{ (fifth term)} = 0 + (-2) = -2$$
$$a_6 \text{ (sixth term)} = -2 + (-2) = -4$$

The first six terms are

$$6, 4, 2, 0, -2, \text{ and } -4.$$ ■

✔ **CHECK POINT 1** Write the first six terms of the arithmetic sequence with first term 100 and common difference -30.

③ Use the formula for the general term of an arithmetic sequence.

The General Term of an Arithmetic Sequence

Consider an arithmetic sequence whose first term is a_1 and whose common difference is d. We are looking for a formula for the general term, a_n. Let's begin by writing the first six terms. The first term is a_1. The second term is $a_1 + d$. The third term is $a_1 + d + d$, or $a_1 + 2d$. Thus, we start with a_1 and add d to each successive term. The first six terms are

$$a_1, \quad a_1 + d, \quad a_1 + 2d, \quad a_1 + 3d, \quad a_1 + 4d, \quad a_1 + 5d.$$

a_1, first term	a_2, second term	a_3, third term	a_4, fourth term	a_5, fifth term	a_6, sixth term

Compare the coefficient of d and the subscript of a denoting the term number. Can you see that the coefficient of d is 1 less than the subscript of a denoting the term number?

$$a_3 : \text{third term} = a_1 + 2d \qquad\qquad a_4 : \text{fourth term} = a_1 + 3d$$

| 2 is one less than 3. | | 3 is one less than 4. |

Thus, the formula for the nth term is

$$a_n : n\text{th term} = a_1 + (n - 1)d.$$

| $n - 1$ is one less than n. |

General Term of an Arithmetic Sequence

The nth term (the general term) of an arithmetic sequence with first term a_1 and common difference d is

$$a_n = a_1 + (n - 1)d.$$

EXAMPLE 2 Using the Formula for the General Term of an Arithmetic Sequence

Find the eighth term of the arithmetic sequence whose first term is 4 and whose common difference is -7.

Solution To find the eighth term, a_8, we replace n in the formula with 8, a_1 with 4, and d with -7.

$$a_n = a_1 + (n - 1)d$$

$$a_8 = 4 + (8 - 1)(-7) = 4 + 7(-7) = 4 + (-49) = -45$$

The eighth term is -45. We can check this result by writing the first eight terms of the sequence:

$$4, -3, -10, -17, -24, -31, -38, -45. \qquad\qquad \blacksquare$$

✔ **CHECK POINT 2** Find the ninth term of the arithmetic sequence whose first term is 6 and whose common difference is -5.

EXAMPLE 3 Using an Arithmetic Sequence to Model Teachers' Earnings

According to the National Education Association, teachers in the United States earned an average of $21,700 in 1984. This amount has increased by approximately $1472 per year.

a. Write a formula for the nth term of the arithmetic sequence that describes teachers' average earnings n years after 1983.

b. How much will U.S. teachers earn by the year 2005?

Solution

a. We can express teachers' earnings by the following arithmetic sequence:

21,700,   23,172,   24,644,   26,116,....

| a_1: earnings in 1984, 1 year after 1983 | a_2: earnings in 1985, 2 years after 1983 | a_3: earnings in 1986, 3 years after 1983 | a_4: earnings in 1987, 4 years after 1983 |

In this sequence a_1, the first term, represents the amount teachers earned in 1984. Each subsequent year this amount increases by \$1472, so $d = 1472$. We use the formula for the general term of an arithmetic sequence to write the nth term of the sequence that describes teachers' earnings n years after 1983.

$$a_n = a_1 + (n - 1)d \qquad \text{This is the formula for the general term of an arithmetic sequence.}$$

$$a_n = 21{,}700 + (n - 1)1472 \qquad a_1 = 21{,}700 \text{ and } d = 1472.$$

$$a_n = 21{,}700 + 1472n - 1472 \qquad \text{Distribute 1472 to each term in parentheses.}$$

$$a_n = 1472n + 20{,}228 \qquad \text{Simplify.}$$

Thus, teachers' earnings n years after 1983 can be described by $a_n = 1472n + 20{,}228$.

b. Now we need to find teachers' earnings in 2005. The year 2005 is 22 years after 1983: That is, $2005 - 1983 = 22$. Thus, $n = 22$. We substitute 22 for n in $a_n = 1472n + 20{,}228$.

$$a_{22} = 1472 \cdot 22 + 20{,}228 = 52{,}612$$

The 22nd term of the sequence is 52,612. Therefore, U.S. teachers are predicted to earn an average of \$52,612 by the year 2005. ∎

✔ **CHECK POINT 3**   According to the U.S. Bureau of Economic Analysis, U.S. travelers spent \$12,808 million in other countries in 1984. This amount has increased by approximately \$2350 million yearly.

a. Write a formula for the nth term of the arithmetic sequence that describes what U.S. travelers spend in other countries n years after 1983.

b. How much will U.S. travelers spend in other countries by the year 2010?

4   Use the formula for the sum of the first n terms of an arithmetic sequence.

The Sum of the First n Terms of an Arithmetic Sequence

The sum of the first n terms of an arithmetic sequence, denoted by S_n, and called the **nth partial sum**, can be found without having to add up all the terms. Let

$$S_n = a_1 + a_2 + a_3 + \cdots + a_n$$

be the sum of the first n terms of an arithmetic sequence. Because d is the common difference between terms, S_n can be written forward and backward as follows.

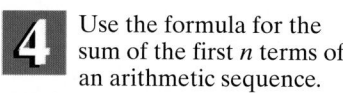
Forward: Start with the first term. Keep adding d.

Backward: Start with the last term. Keep subtracting d.

$$S_n = a_1 \qquad\qquad + (a_1 + d) + (a_1 + 2d) + \cdots + a_n$$

$$\underline{S_n = a_n \qquad\qquad + (a_n - d) + (a_n - 2d) + \cdots + a_1}$$

$$2S_n = (a_1 + a_n) \quad + (a_1 + a_n) + (a_1 + a_n) + \cdots + (a_1 + a_n) \qquad \text{Add the two equations.}$$

Because there are n sums of $(a_1 + a_n)$ on the right side, we can express this side as $n(a_1 + a_n)$. Thus, the last equation can be simplified:

$$2S_n = n(a_1 + a_n).$$

$$S_n = \frac{n}{2}(a_1 + a_n) \qquad \text{Solve for } S_n, \text{ dividing both sides by 2.}$$

We have proved the following result:

The Sum of the First n Terms of an Arithmetic Sequence

The sum, S_n, of the first n terms of an arithmetic sequence is given by

$$S_n = \frac{n}{2}(a_1 + a_n)$$

in which a_1 is the first term and a_n is the nth term.

To find the sum of the terms of an arithmetic sequence, we need to know the first term, a_1, the last term, a_n, and the number of terms, n. The following examples illustrate how to use this formula.

EXAMPLE 4   Finding the Sum of n Terms of an Arithmetic Sequence

Find the sum of the first 100 terms of the arithmetic sequence: $1, 3, 5, 7, \ldots$.

Solution   We are finding the sum of the first 100 odd numbers. To find the sum of the first 100 terms, S_{100}, we replace n in the formula with 100.

$$S_n = \frac{n}{2}(a_1 + a_n)$$

$$S_{100} = \frac{100}{2}(a_1 + a_{100})$$

The first term, a_1, is 1.   We must find a_{100}, the 100th term.

We use the formula for the general term of a sequence to find a_{100}. The common difference, d, of $1, 3, 5, 7, \ldots$, is 2.

$$a_n = a_1 + (n - 1)d \qquad \text{This is the formula for the nth term of an arithmetic sequence. Use it to find the 100th term.}$$

$$a_{100} = 1 + (100 - 1) \cdot 2 \qquad \text{Substitute 100 for } n, \text{ 2 for } d, \text{ and 1 (the first term) for } a_1.$$

$$= 1 + 99 \cdot 2$$
$$= 1 + 198 = 199$$

Now we are ready to find the sum of the 100 terms $1, 3, 5, 7, \ldots, 199$.

$$S_n = \frac{n}{2}(a_1 + a_n) \qquad \text{Use the formula for the sum of the first n terms of an arithmetic sequence. Let } n = 100, a_1 = 1,$$

$$S_{100} = \frac{100}{2}(1 + 199) = 50(200) = 10,000 \qquad \text{and } a_{100} = 199.$$

The sum of the first 100 odd numbers is 10,000. Equivalently, the 100th partial sum of the sequence $1, 3, 5, 7, \ldots$ is 10,000.   ∎

✔ **CHECK POINT 4** Find the sum of the first 15 terms of the arithmetic sequence: $3, 6, 9, 12, \ldots$.

EXAMPLE 5 Using S_n to Evaluate a Summation

Find the following sum: $\displaystyle\sum_{i=1}^{25} (5i - 9)$.

Solution

$$\sum_{i=1}^{25} (5i - 9) = (5 \cdot 1 - 9) + (5 \cdot 2 - 9) + (5 \cdot 3 - 9) + \cdots + (5 \cdot 25 - 9)$$

$$= -4 \qquad + 1 \qquad + 6 \qquad + \cdots + 116$$

By evaluating the first three terms and the last term, we see that $a_1 = -4$; d, the common difference, is $1 - (-4)$ or 5; and a_{25}, the last term, is 116.

$$S_n = \frac{n}{2}(a_1 + a_n)$$
  Use the formula for the sum of the first n terms of an arithmetic sequence. Let $n = 25$, $a_1 = -4$, and $a_{25} = 116$.

$$S_{25} = \frac{25}{2}(-4 + 116) = \frac{25}{2}(112) = 1400$$

Thus,

$$\sum_{i=1}^{25} (5i - 9) = 1400.$$
■

✔ **CHECK POINT 5** Find the following sum: $\displaystyle\sum_{i=1}^{30} (6i - 11)$.

EXAMPLE 6 Modeling Total Nursing Home Costs over a Six-Year Period

Your grandmother has assets of \$350,000. One option that she is considering involves nursing home care for a six-year period beginning in 2001. The model

$$a_n = 1800n + 49{,}730$$

describes yearly nursing home costs n years after 2000. Does your grandmother have enough to pay for the facility?

Solution We must find the sum of an arithmetic sequence. The first term of the sequence corresponds to nursing home costs in the year 2001. The last term corresponds to nursing home costs in the year 2006. Because the model describes costs n years after 2000, $n = 1$ describes the year 2001 and $n = 6$ describes the year 2006.

$$a_n = 1800n + 49{,}730$$
  This is the given formula for the general term of the sequence.

$$a_1 = 1800 \cdot 1 + 49{,}730 = 51{,}530$$
  Find a_1 by replacing n with 1.

$$a_6 = 1800 \cdot 6 + 49{,}730 = 60{,}530$$
  Find a_6 by replacing n with 6.

The first year the facility will cost $51,530. By year six, the facility will cost $60,530. Now we must find the sum of the costs for all six years. We focus on the sum of the first six terms of the arithmetic sequence

$$51{,}530, \quad 53{,}330, \dots, \quad 60{,}530.$$

$$a_1 \qquad\qquad a_2 \qquad\qquad a_6$$

We find this sum using the formula for the sum of the first n terms of an arithmetic sequence. We are adding 6 terms: $n = 6$. The first term is 51,530: $a_1 = 51{,}530$. The last term—that is, the sixth term—is 60,530: $a_6 = 60{,}530$.

$$S_n = \frac{n}{2}(a_1 + a_n)$$

$$S_6 = \frac{6}{2}(51{,}530 + 60{,}530) = 3(112{,}060) = 336{,}180$$

Total nursing home costs for your grandmother are predicted to be $336,180. Because your grandmother's assets are $350,000, she has enough to pay for the facility. ∎

✔ **CHECK POINT 6** In Example 6, how much would it cost for nursing home care for a ten-year period beginning in 2001?

EXERCISE SET 14.2

Practice Exercises

In Exercises 1–6, find the common difference for each arithmetic sequence.

1. $2, 6, 10, 14, \dots$ **2.** $3, 8, 13, 18, \dots$

3. $-7, -2, 3, 8, \dots$ **4.** $-10, -4, 2, 8, \dots$

5. $714, 711, 708, 705, \dots$ **6.** $611, 606, 601, 596, \dots$

In Exercises 7–16, write the first six terms of each arithmetic sequence with the given first term, a_1, and common difference, d.

7. $a_1 = 200, d = 20$

8. $a_1 = 300, d = 50$

9. $a_1 = -7, d = 4$

10. $a_1 = -8, d = 5$

11. $a_1 = 300, d = -90$

12. $a_1 = 200, d = -60$

13. $a_1 = \dfrac{5}{2}, d = -\dfrac{1}{2}$

14. $a_1 = \dfrac{3}{4}, d = -\dfrac{1}{4}$

15. $a_1 = -0.4, d = -1.6$

16. $a_1 = -0.3, d = -1.7$

In Exercises 17–24, use the formula for the general term (the nth term) of an arithmetic sequence to find the indicated term of each sequence with the given first term, a_1, and common difference, d.

17. Find a_6 when $a_1 = 13, d = 4$.

18. Find a_{16} when $a_1 = 9, d = 2$.

19. Find a_{50} when $a_1 = 7, d = 5$.

20. Find a_{60} when $a_1 = 8, d = 6$.

21. Find a_{200} when $a_1 = -40, d = 5$.

22. Find a_{150} when $a_1 = -60, d = 5$.

23. Find a_{60} when $a_1 = 35, d = -3$.

24. Find a_{70} when $a_1 = -32, d = 4$.

In Exercises 25–34, write a formula for the general term (the nth term) of each arithmetic sequence. Then use the formula for a_n to find a_{20}, the 20th term of the sequence.

25. $1, 5, 9, 13, \dots$

26. $2, 7, 12, 17, \dots$

27. $7, 3, -1, -5, \ldots$

28. $6, 1, -4, -9, \ldots$

29. $-20, -24, -28, -32, \ldots$

30. $-70, -75, -80, -85, \ldots$

31. $a_1 = -\dfrac{1}{3}, d = \dfrac{1}{3}$

32. $a_1 = -\dfrac{1}{4}, d = \dfrac{1}{4}$

33. $a_1 = 4, d = -0.3$

34. $a_1 = 5, d = -0.2$

35. Find the sum of the first 20 terms of the arithmetic sequence: $4, 10, 16, 22, \ldots$.

36. Find the sum of the first 25 terms of the arithmetic sequence: $7, 19, 31, 43, \ldots$.

37. Find the sum of the first 50 terms of the arithmetic sequence: $-10, -6, -2, 2, \ldots$.

38. Find the sum of the first 50 terms of the arithmetic sequence: $-15, -9, -3, 3, \ldots$.

39. Find $1 + 2 + 3 + 4 + \ldots + 100$, the sum of the first 100 natural numbers.

40. Find $2 + 4 + 6 + 8 + \ldots + 200$, the sum of the first 100 positive even integers.

41. Find the sum of the first 60 positive even integers.

42. Find the sum of the first 80 positive even integers.

43. Find the sum of the even integers between 21 and 45.

44. Find the sum of the odd integers between 30 and 54.

For Exercises 45–50, write out the first three terms and the last term. Then use the formula for the sum of the first n terms of an arithmetic sequence to find the indicated sum.

45. $\displaystyle\sum_{i=1}^{17} (5i + 3)$

46. $\displaystyle\sum_{i=1}^{20} (6i - 4)$

47. $\displaystyle\sum_{i=1}^{30} (-3i + 5)$

48. $\displaystyle\sum_{i=1}^{40} (-2i + 6)$

49. $\displaystyle\sum_{i=1}^{100} 4i$

50. $\displaystyle\sum_{i=1}^{50} (-4i)$

 Application Exercises

51. According to the U.S. Bureau of Labor Statistics, in 1990 there were 126,424 thousand employees in the United States. This number has increased by approximately 1265 thousand employees per year.
   a. Write the general term for the arithmetic sequence modeling the thousands of employees in the United States n years after 1989.
   b. How many thousands of employees will there be by the year 2005?

52. According to the National Center for Education Statistics, the total enrollment in U.S. public elementary and secondary schools in 1985 was 39.05 million. Enrollment has increased by approximately 0.45 million per year.
   a. Write the general term for the arithmetic sequence modeling the millions of students enrolled in U.S. public elementary and secondary schools n years after 1984.
   b. How many millions of students will be enrolled by the year 2005?

53. Company A pays $24,000 yearly with raises of $1600 per year. Company B pays $28,000 yearly with raises of $1000 per year. Which company will pay more in year 10? How much more?

54. Company A pays $23,000 yearly with raises of $1200 per year. Company B pays $26,000 yearly with raises of $800 per year. Which company will pay more in year 10? How much more?

55. According to the Environmental Protection Agency, in 1960 the United States recovered 3.78 million tons of solid waste. Due primarily to recycling programs, this amount has increased by approximately 0.576 million ton per year.
   a. Write the general term for the arithmetic sequence modeling the amount of solid waste recovered in the United States n years after 1959.
   b. What is the total amount of solid waste recovered from 1960 through 2000?

56. According to the Environmental Protection Agency, in 1960 the United States generated 87.1 million tons of solid waste. This amount has increased by approximately 3.14 million tons per year.
   a. Write the general term for the arithmetic sequence modeling the amount of solid waste generated in the United States n years after 1959.
   b. What is the total amount of solid waste generated from 1960 through 2000?

57. A company offers a starting yearly salary of $33,000 with raises of $2500 per year. Find the total salary over a ten-year period.

58. You are considering two job offers. Company A will start you at $19,000 per year and guarantee you a raise of $2600 per year. Company B will start you at a higher salary, $27,000 per year, but will only guarantee a raise of $1200 per year. Find the total salary that each company will pay you over a ten-year period. Which company pays the greater total amount?

59. A theater has 30 seats in the first row, 32 seats in the second row, increasing by 2 seats each row for a total of 26 rows. How many seats are there in the theater?

60. A section in a stadium has 20 seats in the first row, 23 seats in the second row, increasing by 3 seats each row for a total of 38 rows. How many seats are in this section of the stadium?

Writing in Mathematics

61. What is an arithmetic sequence? Give an example with your explanation.

62. What is the common difference in an arithmetic sequence?

63. Explain how to find the general term of an arithmetic sequence.

64. Explain how to find the sum of the first n terms of an arithmetic sequence without having to add up all the terms.

65. Teachers' earnings n years after 1983 can be described by $a_n = 1472n + 20{,}228$. According to this model, what will teachers earn in 2083? Describe two possible circumstances that would render this predicted salary incorrect.

Critical Thinking Exercises

66. Give examples of two different arithmetic sequences whose fourth term, a_4, is 10.

67. In the sequence 21,700, 23,172, 24,644, 26,116,..., which term is 314,628?

68. A *degree-day* is a unit used to measure the fuel requirements of buildings. By definition, each degree that the average daily temperature is below 65°F is 1 degree-day. For example, a temperature of 42°F constitutes 23 degree-days. If the average temperature on January 1 was 42°F and fell 2°F for each subsequent day up to and including January 10, how many degree-days are included from January 1 to January 10?

69. Show that the sum of the first n positive odd integers,

$$1 + 3 + 5 + \cdots + (2n - 1),$$

is n^2.

Technology Exercises

70. Use the ⬚SEQ⬚ (sequence) capability of a graphing utility and the formula you obtained for a_n to verify the value you found for a_{20} in any five exercises from Exercises 25–34.

71. Use the capability of a graphing utility to calculate the sum of a sequence to verify any five of your answers to Exercises 45–50.

Review Exercises

72. Solve: $\log(x^2 - 5) - \log(x + 5) = 3$. (Section 12.5, Example 6)

73. Solve: $x^2 + 3x \le 10$. (Section 11.5, Example 1)

74. Simplify: $\dfrac{x^2 + 7x + 12}{x^2 - 16}$. (Section 7.1, Example 4)

▶ SECTION 14.3   *Geometric Sequences and Series*

Objectives

1 Find the common ratio of a geometric sequence.

2 Write terms of a geometric sequence.

3 Use the formula for the general term of a geometric sequence.

4 Use the formula for the sum of the first *n* terms of a geometric sequence.

5 Find the value of an annuity.

6 Use the formula for the sum of an infinite geometric series.

SSM   PH Tutor CD-   Video
      Center   ROM

Here we are at the closing moments of a job interview. You're shaking hands with the manager. You managed to answer all the tough questions without losing your poise, and now you've been offered a job. As a matter of fact, your qualifications are so terrific that you've been offered two jobs—one just the day before, with a rival company in the same field! One company offers $30,000 the first year, with increases of 6% per year for four years after that. The other offers $32,000 the first year, with annual increases of 3% per year after that. Over a five-year period, which is the better offer?

If salary raises amount to a certain percent each year, the yearly salaries over time form a geometric sequence. In this section, we investigate geometric sequences and their properties. After studying the section, you will be in a position to decide which job offer to accept: you will know which company will pay you more over five years.

Geometric Sequences

Figure 14.2 shows a sequence in which the number of squares is increasing. From left to right, the number of squares is 1, 5, 25, 125, and 625. In this sequence, each term after the first, 1, is obtained by multiplying the preceding term by a constant amount, namely 5. This sequence of increasing number of squares is an example of a *geometric sequence*.

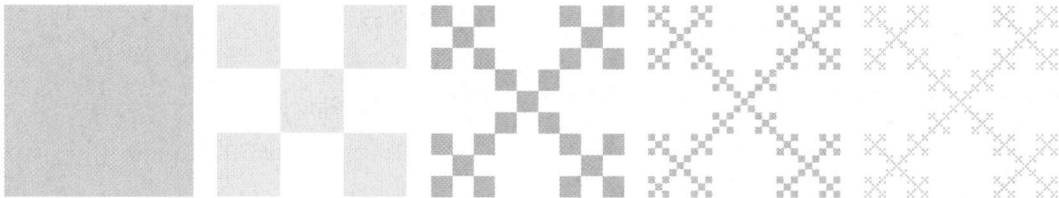

Figure 14.2 A geometric sequence of squares

Definition of a Geometric Sequence

A **geometric sequence** is a sequence in which each term after the first is obtained by multiplying the preceding term by a fixed nonzero constant. The amount by which we multiply each time is called the **common ratio** of the sequence.

1 Find the common ratio of a geometric sequence.

The common ratio, r, is found by dividing any term after the first term by the term that directly precedes it. In the following examples, the common ratio is found by dividing the second term by the first term, $\frac{a_2}{a_1}$.

GEOMETRIC SEQUENCE	**COMMON RATIO**
$1, 5, 25, 125, 625, \ldots$	$r = \dfrac{5}{1} = 5$
$4, 8, 16, 32, 64, \ldots$	$r = \dfrac{8}{4} = 2$
$6, -12, 24, -48, 96, \ldots$	$r = \dfrac{-12}{6} = -2$
$9, -3, 1, -\dfrac{1}{3}, \dfrac{1}{9}, \ldots$	$r = \dfrac{-3}{9} = -\dfrac{1}{3}$

Study Tip

When the common ratio of a geometric sequence is negative, the signs of the terms alternate.

2 Write terms of a geometric sequence.

How do we write out the terms of a geometric sequence when the first term and the common ratio are known? We multiply the first term by the common ratio to get the second term, multiply the second term by the common ratio to get the third term, and so on.

EXAMPLE 1   Writing the Terms of a Geometric Sequence

Write the first six terms of the geometric sequence with first term 6 and common ratio $\frac{1}{3}$.

Solution   The first term is 6. The second term is $6 \cdot \frac{1}{3}$, or 2. The third term is $2 \cdot \frac{1}{3}$, or $\frac{2}{3}$. The fourth term is $\frac{2}{3} \cdot \frac{1}{3}$, or $\frac{2}{9}$, and so on. The first six terms are

$$6, 2, \frac{2}{3}, \frac{2}{9}, \frac{2}{27}, \frac{2}{81}.$$

■

✔ **CHECK POINT 1**   Write the first six terms of the geometric sequence with first term 12 and common ratio $\frac{1}{2}$.

3 Use the formula for the general term of a geometric sequence.

The General Term of a Geometric Sequence

Consider a geometric sequence whose first term is a_1, and whose common ratio is r. We are looking for a formula for the general term, a_n. Let's begin by writing the first six terms. The first term is a_1. The second term is $a_1 r$. The third term is $a_1 r \cdot r$, or $a_1 r^2$. The fourth term is $a_1 r^2 \cdot r$, or $a_1 r^3$, and so on. Starting with a_1 and multiplying each successive term by r, the first six terms are

$$a_1, \qquad a_1 r, \qquad a_1 r^2, \qquad a_1 r^3, \qquad a_1 r^4, \qquad a_1 r^5.$$

a_1, first term	a_2, second term	a_3, third term	a_4, fourth term	a_5, fifth term	a_6, sixth term

Compare the exponent on r and the subscript of a denoting the term number. Can you see that the exponent on r is 1 less than the subscript of a denoting the term number?

a_3: third term $= a_1r^2$ a_4: fourth term $= a_1r^3$

2 is one less than 3.   3 is one less than 4.

Thus, the formula for the nth term is

$$a_n = a_1r^{n-1}.$$

$n - 1$ is one less than n.

General Term of a Geometric Sequence

The nth term (the general term) of a geometric sequence with first term a_1 and common ratio r is

$$a_n = a_1r^{n-1}.$$

Study Tip

Be careful with the order of operations when evaluating

$$a_1r^{n-1}.$$

First find r^{n-1}. Then multiply the result by a_1.

EXAMPLE 2   Using the Formula for the General Term of a Geometric Sequence

Find the eighth term of the geometric sequence whose first term is -4 and whose common ratio is -2.

Solution   To find the eighth term, a_8, we replace n in the formula with 8, a_1 with -4, and r with -2.

$$a_n = a_1r^{n-1}$$
$$a_8 = -4(-2)^{8-1} = -4(-2)^7 = -4(-128) = 512$$

The eighth term is 512. We can check this result by writing the first eight terms of the sequence:

$$-4, 8, -16, 32, -64, 128, -256, 512. \qquad \blacksquare$$

✔ **CHECK POINT 2**   Find the seventh term of the geometric sequence whose first term is 5 and whose common ratio is -3.

In Chapter 12, we studied exponential functions of the form $f(x) = b^x$ and the explosive exponential growth of world population. In our next example, we consider Florida's geometric population growth. Because **a geometric sequence is an exponential function whose domain is the set of positive integers**, geometric and exponential growth mean the same thing. (By contrast, an arithmetic sequence is a *linear function* whose domain is the set of positive integers.)

EXAMPLE 3   Geometric Population Growth

The population of Florida from 1980 through 1987 is shown in the following table.

Year	1980	1981	1982	1983	1984	1985	1986	1987
Population (in millions)	9.75	10.03	10.32	10.62	10.93	11.25	11.58	11.92

a. Show that the population is increasing geometrically.

b. Write the general term for the geometric sequence describing population growth for Florida n years after 1979.

c. Estimate Florida's population, in millions, for the year 2000.

Solution

a. First, we divide the population for each year by the population in the preceding year.

$$\frac{10.03}{9.75} \approx 1.029, \quad \frac{10.32}{10.03} \approx 1.029, \quad \frac{10.62}{10.32} \approx 1.029$$

Continuing in this manner, we will keep getting approximately 1.029. This means that the population is increasing geometrically with $r \approx 1.029$. In this situation, the common ratio is the growth rate, indicating that the population of Florida in any year shown in the table is approximately 1.029 times the population the year before.

b. The sequence of Florida's population growth is

$$9.75, 10.03, 10.32, 10.62, 10.93, 11.25, 11.58, 11.92, \ldots .$$

Because the population is increasing geometrically, we can find the general term of this sequence using

$$a_n = a_1 r^{n-1}.$$

In this sequence, $a_1 = 9.75$ and r [from part (a)] ≈ 1.029. We substitute these values into the formula for the general term. This gives the general term for the geometric sequence describing Florida's population n years after 1979.

$$a_n = 9.75(1.029)^{n-1}$$

c. We can use the formula for the general term, a_n, in part (b) to estimate Florida's population for the year 2000. The year 2000 is 21 years after 1979—that is, $2000 - 1979 = 21$. Thus, $n = 21$. We substitute 21 for n in $a_n = 9.75(1.029)^{n-1}$.

$$a_{21} = 9.75(1.029)^{21-1} = 9.75(1.029)^{20} \approx 17.27$$

The formula indicates that Florida had a population of approximately 17.27 million in the year 2000. ∎

✔ **CHECK POINT 3** Write the general term for the geometric sequence

$$3, 6, 12, 24, 48, \ldots .$$

Then use the formula for the general term to find the eighth term.

4 Use the formula for the sum of the first n terms of a geometric sequence.

The Sum of the First n Terms of a Geometric Sequence

The sum of the first n terms of a geometric sequence, denoted by S_n, and called the **nth partial sum**, can be found without having to add up all the terms. Recall that the first n terms of a geometric sequence are

$$a_1, a_1 r, a_1 r^2, \ldots, a_1 r^{n-2}, a_1 r^{n-1}.$$

We proceed as follows:

$$S_n = a_1 + a_1 r + a_1 r^2 + \cdots + a_1 r^{n-2} + a_1 r^{n-1}$$

S_n is the sum of the first n terms of the sequence.

$$r S_n = a_1 r + a_1 r^2 + a_1 r^3 + \cdots + a_1 r^{n-1} + a_1 r^n$$

Multiply both sides of the equation by r.

$$S_n - r S_n = a_1 - a_1 r^n$$

Subtract the second equation from the first equation.

$$S_n(1 - r) = a_1(1 - r^n)$$

Factor out S_n on the left and a_1 on the right.

$$S_n = \frac{a_1(1 - r^n)}{1 - r}.$$

Solve for S_n by dividing both sides by $1 - r$ (assuming that $r \neq 1$).

We have proved the following result:

The Sum of the First n Terms of a Geometric Sequence

The sum, S_n, of the first n terms of a geometric sequence is given by

$$S_n = \frac{a_1(1 - r^n)}{1 - r}$$

in which a_1 is the first term and r is the common ratio ($r \neq 1$).

To find the sum of the terms of a geometric sequence, we need to know the first term, a_1, the common ratio, r, and the number of terms, n. The following examples illustrate how to use this formula.

EXAMPLE 4   Finding the Sum of the First n Terms of a Geometric Sequence

Find the sum of the first 18 terms of the geometric sequence: $2, -8, 32, -128, \ldots$.

Solution   To find the sum of the first 18 terms, S_{18}, we replace n in the formula with 18.

$$S_n = \frac{a_1(1 - r^n)}{1 - r}$$

$$S_{18} = \frac{a_1(1 - r^{18})}{1 - r}$$

The first term, a_1, is 2.

We must find r, the common ratio.

We can find the common ratio by dividing the second term by the first term.

$$r = \frac{a_2}{a_1} = \frac{-8}{2} = -4$$

Now we are ready to find the sum of the first 18 terms of $2, -8, 32, -128, \ldots$.

$$S_n = \frac{a_1(1 - r^n)}{1 - r}$$ Use the formula for the sum of the first n terms of a geometric sequence.

$$S_{18} = \frac{2(1 - (-4)^{18})}{1 - (-4)}$$ a_1 (the first term) $= 2$, $r = -4$, and $n = 18$ because we want the sum of the first 18 terms.

$$= -27{,}487{,}790{,}694$$ Use a calculator.

The sum of the first 18 terms is $-27{,}487{,}790{,}694$. Equivalently, this number is the 18th partial sum of the sequence $2, -8, 32, -128, \ldots$. ∎

✔ **CHECK POINT 4**   Find the sum of the first nine terms of the geometric sequence: $2, -6, 18, -54, \ldots$.

Using Technology

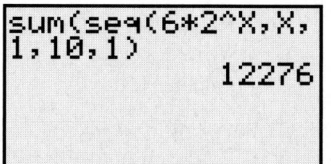

To find

$$\sum_{i=1}^{10} 6 \cdot 2^i$$

on a graphing utility, enter

$\boxed{\text{SUM}}\ \boxed{\text{SEQ}}\ (6 \times 2^x, x, 1, 10, 1)$.

Then press $\boxed{\text{ENTER}}$.

```
sum(seq(6*2^X,X,
1,10,1)
             12276
```

EXAMPLE 5   Using S_n to Evaluate a Summation

Find the following sum: $\displaystyle\sum_{i=1}^{10} 6 \cdot 2^i$.

Solution   Let's write out a few terms in the sum.

$$\sum_{i=1}^{10} 6 \cdot 2^i = 6 \cdot 2 + 6 \cdot 2^2 + 6 \cdot 2^3 + \cdots + 6 \cdot 2^{10}$$

Can you see that each term after the first is obtained by multiplying the preceding term by 2? To find the sum of the 10 terms ($n = 10$), we need to know the first term, a_1, and the common ratio, r. The first term is $6 \cdot 2$ or 12: $a_1 = 12$. The common ratio is 2.

$$S_n = \frac{a_1(1 - r^n)}{1 - r}$$ Use the formula for the sum of the first n terms of a geometric sequence.

$$S_{10} = \frac{12(1 - 2^{10})}{1 - 2}$$ a_1 (the first term) $= 12$, $r = 2$, and $n = 10$ because we are adding ten terms.

$$= 12{,}276$$ Use a calculator.

Thus,

$$\sum_{i=1}^{10} 6 \cdot 2^i = 12{,}276.$$ ∎

✔ **CHECK POINT 5**   Find the following sum: $\displaystyle\sum_{i=1}^{8} 2 \cdot 3^i$.

Some of the exercises in the previous exercise set involved situations in which salaries increase by a fixed amount each year. A more realistic situation is one in which salary raises increase by a certain percent each year. Example 6 shows how such a situation can be described using a geometric series.

EXAMPLE 6   Computing a Lifetime Salary

A union contract specifies that each worker will receive a 5% pay increase each year for the next 30 years. One worker is paid $20,000 the first year. What is this person's total lifetime salary over a 30-year period?

Solution   The salary for the first year is $20,000. With a 5% raise, the second-year salary is computed as follows:

$$\text{Salary for year 2} = 20{,}000 + 20{,}000(0.05) = 20{,}000(1.05).$$

Each year, the salary is 1.05 times what it was in the previous year. Thus, the salary for year 3 is 1.05 times $20{,}000(1.05)$, or $20{,}000(1.05)^2$. The salaries for the first five years are given in the table.

Yearly Salaries					
Year 1	**Year 2**	**Year 3**	**Year 4**	**Year 5**	...
20,000	20,000(1.05)	$20{,}000(1.05)^2$	$20{,}000(1.05)^3$	$20{,}000(1.05)^4$	...

The numbers in the bottom row form a geometric sequence with $a_1 = 20{,}000$ and $r = 1.05$. To find the total salary over 30 years, we use the formula for the sum of the first n terms of a geometric sequence, with $n = 30$.

$$S_n = \frac{a_1(1 - r^n)}{1 - r}$$

$$S_{30} = \frac{20{,}000(1 - (1.05)^{30})}{1 - 1.05}$$

Total salary over 30 years

$$= \frac{20{,}000(1 - (1.05)^{30})}{-0.05}$$

$$\approx 1{,}328{,}777 \qquad \textit{Use a calculator.}$$

The total salary over the 30-year period is approximately $1,328,777.   ∎

✔ **CHECK POINT 6**   A job pays a salary of $30,000 the first year. During the next 29 years, the salary increases by 6% each year. What is the total lifetime salary over the 30-year period?

5 Find the value of an annuity.

Annuities

The compound interest formula

$$A = P(1 + r)^t$$

gives the future value, A, after t years, when a fixed amount of money, P, the principal, is deposited in an account that pays an annual interest rate r (in decimal form) compounded once a year. However, money is often invested in small amounts at periodic intervals. For example, to save for retirement, you might decide to place $1000 into an Individual Retirement Account (IRA) at the end of each year until you retire. An **annuity** is a sequence of equal payments made at equal time periods. An IRA is an example of an annuity.

Suppose P dollars is deposited into an account at the end of each year. The account pays an annual interest rate, r, compounded annually. At the end of the first year, the account contains P dollars. At the end of the second year, P dollars is deposited again. At the time of this deposit, the first deposit has received interest earned during the second year. The **value of the annuity** is the sum of all deposits made plus all interest paid. Thus, the value of the annuity after two years is

$$P + P(1 + r).$$

Deposit of P dollars at end of second year	First-year deposit of P dollars with interest earned for a year

The value of the annuity after three years is

$$P \quad + \quad P(1 + r) \quad + \quad P(1 + r)^2.$$

Deposit of P dollars at end of third year	Second-year deposit of P dollars with interest earned for a year	First-year deposit of P dollars with interest earned over two years

The value of the annuity after t years is

$$P + P(1 + r) + P(1 + r)^2 + P(1 + r)^3 + \cdots + P(1 + r)^{t-1}.$$

Deposit of P dollars at end of year t	First-year deposit of P dollars with interest earned over $t - 1$ years

This is the sum of the terms of a geometric sequence with first term P and common ratio $1 + r$. We use the formula

$$S_n = \frac{a_1(1 - r^n)}{1 - r}$$

to find the sum of the terms:

$$S_n = \frac{P(1 - (1 + r)^t)}{1 - (1 + r)} = \frac{P(1 - (1 + r)^t)}{-r} = P\frac{(1 + r)^t - 1}{r}.$$

This formula gives the value of an annuity after t years if interest is compounded once a year. We can adjust the formula to find the value of an annuity if equal payments are made at the end of each of n yearly compounding periods.

Value of an Annuity: Interest Compounded n Times per Year

If P is the deposit made at the end of each compounding period for an annuity at r percent annual interest compounded n times per year, the value, A, of the annuity after t years is

$$A = P\frac{\left(1 + \dfrac{r}{n}\right)^{nt} - 1}{\dfrac{r}{n}}.$$

EXAMPLE 7   Determining the Value of an Annuity

To save for retirement, you decide to deposit $1000 into an IRA at the end of each year for the next 30 years. If the interest rate is 10% per year compounded annually, find the value of the IRA after 30 years.

Solution   The annuity involves 30 year-end deposits of $P = \$1000$. The interest rate is 10%: $r = 0.10$. Because the deposits are made once a year and the interest is compounded once a year, $n = 1$. The number of years is 30: $t = 30$. We replace the variables in the formula for the value of an annuity with these numbers.

$$A = P\,\frac{\left(1 + \dfrac{r}{n}\right)^{nt} - 1}{\dfrac{r}{n}}$$

$$A = 1000\,\frac{\left(1 + \dfrac{0.10}{1}\right)^{1 \cdot 30} - 1}{\dfrac{0.10}{1}} \approx 164{,}494$$

The value of the IRA at the end of 30 years is approximately $164,494.   ■

✔ **CHECK POINT 7**   If $3000 is deposited into an IRA at the end of each year for 40 years and the interest rate is 10% per year compounded annually, find the value of the IRA after 40 years.

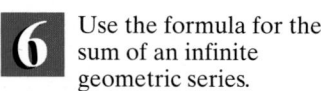

Use the formula for the sum of an infinite geometric series.

Geometric Series

An infinite sum of the form

$$a_1 + a_1 r + a_1 r^2 + a_1 r^3 + \cdots + a_1 r^{n-1} + \cdots$$

with first term a_1 and common ratio r is called an **infinite geometric series**. How can we determine which infinite geometric series have sums and which do not? We look at what happens to r^n as n gets larger in the formula for the sum of the first n terms of this series, namely

$$S_n = \frac{a_1(1 - r^n)}{1 - r}.$$

If r is any number between -1 and 1, that is, $-1 < r < 1$, the term r^n approaches 0 as n gets larger. For example, consider what happens to r^n for $r = \frac{1}{2}$:

$$\left(\frac{1}{2}\right)^1 = \frac{1}{2} \quad \left(\frac{1}{2}\right)^2 = \frac{1}{4} \quad \left(\frac{1}{2}\right)^3 = \frac{1}{8} \quad \left(\frac{1}{2}\right)^4 = \frac{1}{16} \quad \left(\frac{1}{2}\right)^5 = \frac{1}{32} \quad \left(\frac{1}{2}\right)^6 = \frac{1}{64}.$$

These numbers are approaching 0 as n gets larger.

Take another look at the formula for the sum of the first n terms of a geometric sequence.

$$S_n = \frac{a_1(1 - r^n)}{1 - r}$$ **If $-1 < r < 1$, r^n approaches 0 as n gets larger.**

Let us replace r^n with 0 in the formula for S_n. This change gives us a formula for the sum of infinite geometric series with common ratios between -1 and 1.

The Sum of an Infinite Geometric Series

If $-1 < r < 1$ (equivalently, $|r| < 1$), then the sum of the infinite geometric series

$$a_1 + a_1 r + a_1 r^2 + a_1 r^3 + \cdots$$

in which a_1 is the first term and r is the common ratio is given by

$$S = \frac{a_1}{1 - r}.$$

If $|r| \geq 1$, the infinite series does not have a sum.

To use the formula for the sum of an infinite geometric series, we need to know the first term and the common ratio. For example, consider

First term, a_1, is $\frac{1}{2}$.

$$\frac{1}{2} + \frac{1}{4} + \frac{1}{8} + \frac{1}{16} + \frac{1}{32} + \cdots.$$

Common ratio, r, is $\frac{a_2}{a_1}$.

$$r = \frac{1}{4} \div \frac{1}{2} = \frac{1}{4} \cdot 2 = \frac{1}{2}$$

With $r = \frac{1}{2}$, the condition that $|r| < 1$ is met, so the infinite geometric series has a sum given by $S = \frac{a_1}{1 - r}$. The sum of the series is found as follows:

$$\frac{1}{2} + \frac{1}{4} + \frac{1}{8} + \frac{1}{16} + \frac{1}{32} + \cdots = \frac{a_1}{1 - r} = \frac{\frac{1}{2}}{1 - \frac{1}{2}} = \frac{\frac{1}{2}}{\frac{1}{2}} = 1.$$

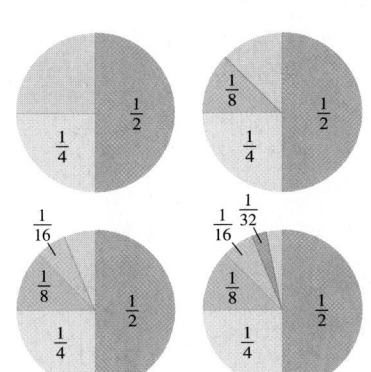

Figure 14.3 The sum $\frac{1}{2} + \frac{1}{4} + \frac{1}{8} + \frac{1}{16} + \frac{1}{32} + \cdots$ is approaching 1.

Thus, the sum of the infinite geometric series is 1. Notice how this is illustrated in Figure 14.3. As more terms are included, the sum is approaching the area of one complete circle.

EXAMPLE 8   Finding the Sum of an Infinite Geometric Series

Find the sum of the infinite geometric series: $\frac{3}{8} - \frac{3}{16} + \frac{3}{32} - \frac{3}{64} + \cdots$.

Solution   Before finding the sum, we must find the common ratio.

$$r = \frac{a_2}{a_1} = \frac{-\frac{3}{16}}{\frac{3}{8}} = -\frac{3}{16} \cdot \frac{8}{3} = -\frac{1}{2}$$

The infinite geometric series

$$\frac{3}{8} - \frac{3}{16} + \frac{3}{32} - \frac{3}{64} + \cdots,$$

repeated

Because $r = -\frac{1}{2}$, the condition that $|r| < 1$ is met. Thus, the infinite geometric series has a sum.

$$S = \frac{a_1}{1 - r} \qquad \text{This is the formula for the sum of an infinite geometric series. Let } a_1 = \frac{3}{8} \text{ and } r = -\frac{1}{2}.$$

$$= \frac{\frac{3}{8}}{1 - \left(-\frac{1}{2}\right)} = \frac{\frac{3}{8}}{\frac{3}{2}} = \frac{3}{8} \cdot \frac{2}{3} = \frac{1}{4}$$

Thus, the sum of this infinite geometric series is $\frac{1}{4}$. Put in an informal way, as we continue to add more and more terms, the sum is approximately $\frac{1}{4}$. ∎

✔ **CHECK POINT 8** Find the sum of the infinite geometric series: $3 + 2 + \frac{4}{3} + \frac{8}{9} + \cdots$.

We can use the formula for the sum of an infinite series to express a repeating decimal as a fraction in lowest terms.

EXAMPLE 9 Writing a Repeating Decimal as a Fraction

Express $0.\overline{7}$ as a fraction in lowest terms.

Solution

$$0.\overline{7} = 0.7777\ldots = \frac{7}{10} + \frac{7}{100} + \frac{7}{1000} + \frac{7}{10,000} + \cdots$$

Observe that $0.\overline{7}$ is an infinite geometric series with first term $\frac{7}{10}$ and common ratio $\frac{1}{10}$. Because $r = \frac{1}{10}$, the condition that $|r| < 1$ is met. Thus, we can use our formula to find the sum. Therefore,

$$0.\overline{7} = \frac{a_1}{1 - r} = \frac{\frac{7}{10}}{1 - \frac{1}{10}} = \frac{\frac{7}{10}}{\frac{9}{10}} = \frac{7}{10} \cdot \frac{10}{9} = \frac{7}{9}.$$

An equivalent fraction for $0.\overline{7}$ is $\frac{7}{9}$. ∎

✔ **CHECK POINT 9** Express $0.\overline{9}$ as a fraction in lowest terms.

Infinite geometric series have many applications, as illustrated in Example 10.

EXAMPLE 10 Tax Rebates and the Multiplier Effect

A tax rebate that returns a certain amount of money to taxpayers can have a total effect on the economy that is many times this amount. In economics, this phenomenon is called the **multiplier effect**. Suppose, for example, that the government reduces taxes so that each consumer has $2000 more income. The government assumes that each person will spend 70% of this (= $1400). The individuals and businesses receiving this $1400 in turn spend 70% of it (= $980), creating extra income for other people to spend, and so on. Determine the total amount spent on consumer goods from the initial $2000 tax rebate.

Solution   The total amount spent is given by the infinite geometric series

$$1400 + 980 + 686 + \cdots.$$

70% of 1400	70% of 980

The first term is 1400: $a_1 = 1400$. The common ratio is 70%, or 0.7: $r = 0.7$. Because $r = 0.7$, the condition that $|r| < 1$ is met. Thus, we can use our formula to find the sum. Therefore,

$$1400 + 980 + 686 + \cdots = \frac{a_1}{1 - r} = \frac{1400}{1 - 0.7} \approx 4667.$$

This means that the total amount spent on consumer goods from the initial $2000 rebate is approximately $4667. ∎

✔ **CHECK POINT 10**   Rework Example 10 and determine the total amount spent on consumer goods with a $1000 tax rebate and 80% spending down the line.

$1400

70% is spent.

$980

70% is spent.

$686

EXERCISE SET 14.3

 Practice Exercises

In Exercises 1–8, find the common ratio for each geometric sequence.

1. $5, 15, 45, 135, \ldots$ **2.** $5, 10, 20, 40, \ldots$

3. $-15, 30, -60, 120, \ldots$

4. $-2, 6, -18, 54, \ldots$

5. $3, \dfrac{9}{2}, \dfrac{27}{4}, \dfrac{81}{8}, \ldots$

6. $4, \dfrac{8}{3}, \dfrac{16}{9}, \dfrac{32}{27}, \ldots$

7. $4, -0.4, 0.04, -0.004, \ldots$

8. $7, -0.7, 0.07, -0.007, \ldots$

In Exercises 9–16, write the first five terms of each geometric sequence with the given first term, a_1, and common ratio, r.

9. $a_1 = 2, r = 3$

10. $a_1 = 2, r = 4$

11. $a_1 = 20, r = \dfrac{1}{2}$

12. $a_1 = 24, r = \dfrac{1}{3}$

13. $a_1 = -4, r = -10$

14. $a_1 = -3, r = -10$

15. $a_1 = -\dfrac{1}{4}, r = -2$

16. $a_1 = -\dfrac{1}{16}, r = -4$

In Exercises 17–24, use the formula for the general term (the nth term) of a geometric sequence to find the indicated term of each sequence with the given first term, a_1, and common ratio, r.

17. Find a_8 when $a_1 = 6, r = 2$.

18. Find a_8 when $a_1 = 5, r = 3$.

19. Find a_{12} when $a_1 = 5, r = -2$.

20. Find a_{12} when $a_1 = 4, r = -2$.

21. Find a_6 when $a_1 = 6400, r = -\dfrac{1}{2}$.

22. Find a_6 when $a_1 = 8000, r = -\dfrac{1}{2}$.

23. Find a_8 when $a_1 = 1{,}000{,}000, r = 0.1$.

24. Find a_8 when $a_1 = 40{,}000, r = 0.1$.

In Exercises 25–32, write a formula for the general term (the nth term) of each geometric sequence. Then use the formula for a_n to find a_7, the seventh term of the sequence.

25. $3, 12, 48, 192, \ldots$

26. $3, 15, 75, 375, \ldots$

27. $18, 6, 2, \dfrac{2}{3}, \ldots.$

28. $12, 6, 3, \dfrac{3}{2}, \ldots.$

29. $1.5, -3, 6, -12, \ldots$

30. $5, -1, \dfrac{1}{5}, -\dfrac{1}{25}, \ldots.$

31. $0.0004, -0.004, 0.04, -0.4. \ldots$

32. $0.0007, -0.007, 0.07, -0.7, \ldots$

Use the formula for the sum of the first n terms of a geometric sequence to solve Exercises 33–38.

33. Find the sum of the first 12 terms of the geometric sequence: $2, 6, 18, 54 \ldots$.

34. Find the sum of the first 12 terms of the geometric sequence: $3, 6, 12, 24, \ldots$.

35. Find the sum of the first 11 terms of the geometric sequence: $3, -6, 12, -24, \ldots$.

36. Find the sum of the first 11 terms of the geometric sequence: $4, -12, 36, -108, \ldots$.

37. Find the sum of the first 14 terms of the geometric sequence: $-\frac{3}{2}, 3, -6, 12, \ldots$.

38. Find the sum of the first 14 terms of the geometric sequence: $-\frac{1}{24}, \frac{1}{12}, -\frac{1}{6}, \frac{1}{3}, \ldots$.

In Exercises 39–44, find the indicated sum. Use the formula for the sum of the first n terms of a geometric sequence.

39. $\sum_{i=1}^{8} 3^i$

40. $\sum_{i=1}^{6} 4^i$

41. $\sum_{i=1}^{10} 5 \cdot 2^i$

42. $\sum_{i=1}^{7} 4(-3)^i$

43. $\sum_{i=1}^{6} \left(\frac{1}{2}\right)^{i+1}$

44. $\sum_{i=1}^{6} \left(\frac{1}{3}\right)^{i+1}$

In Exercises 45–52, find the sum of each infinite geometric series.

45. $1 + \frac{1}{3} + \frac{1}{9} + \frac{1}{27} + \cdots$

46. $1 + \frac{1}{4} + \frac{1}{16} + \frac{1}{64} + \cdots$

47. $3 + \frac{3}{4} + \frac{3}{4^2} + \frac{3}{4^3} + \cdots$

48. $5 + \frac{5}{6} + \frac{5}{6^2} + \frac{5}{6^3} + \cdots$

49. $1 - \frac{1}{2} + \frac{1}{4} - \frac{1}{8} + \cdots$

50. $3 - 1 + \frac{1}{3} - \frac{1}{9} + \cdots$

51. $\sum_{i=1}^{\infty} 26(-0.3)^{i-1}$

52. $\sum_{i=1}^{\infty} 51(-0.7)^{i-1}$

In Exercises 53–58, express each repeating decimal as a fraction in lowest terms.

53. $0.\overline{5} = \frac{5}{10} + \frac{5}{100} + \frac{5}{1000} + \frac{5}{10,000} + \cdots$

54. $0.\overline{1} = \frac{1}{10} + \frac{1}{100} + \frac{1}{1000} + \frac{1}{10,000} + \cdots$

55. $0.\overline{47} = \frac{47}{100} + \frac{47}{10,000} + \frac{47}{1,000,000} + \cdots$

56. $0.\overline{83} = \frac{83}{100} + \frac{83}{10,000} + \frac{83}{1,000,000} + \cdots$

57. $0.\overline{257}$

58. $0.\overline{529}$

In Exercises 59–64, the general term of a sequence is given. Determine whether the sequence is arithmetic, geometric, or neither. If the sequence is arithmetic, find the common difference; if it is geometric, find the common ratio.

59. $a_n = n + 5$

60. $a_n = n - 3$

61. $a_n = 2^n$

62. $a_n = \left(\frac{1}{2}\right)^n$

63. $a_n = n^2 + 5$

64. $a_n = n^2 - 3$

Application Exercises

Use the formula for the general term (the nth term) of a geometric sequence to solve Exercises 65–68.

In Exercises 65–66, suppose you save $1 the first day of a month, $2 the second day, $4 the third day, and so on. That is, each day you save twice as much as you did the day before.

65. What will you put aside for savings on the fifteenth day of the month?

66. What will you put aside for savings on the thirtieth day of the month?

67. A professional baseball player signs a contract with a beginning salary of $3,000,000 for the first year with an annual increase of 4% per year beginning in the second year. That is, beginning in year 2, the athlete's salary will be 1.04 times what it was in the previous year. What is the athlete's salary for year 7 of the contract? Round to the nearest dollar.

68. You are offered a job that pays $30,000 for the first year with an annual increase of 5% per year beginning in the second year. That is, beginning in year 2, your salary will be 1.05 times what it was in the previous year. What can you expect to earn in your sixth year on the job? Round to the nearest dollar.

69. The population of Iraq from 1995 through 1998 is shown in the following table.

Year	1995	1996	1997	1998
Population in millions	20.60	21.36	22.19	23.02

Source: U.N. Population Division

a. Divide the population for each year by the population in the preceding year. Round to two decimal places and show that Iraq's population is increasing geometrically.

b. Write the general term of the geometric sequence describing population growth for Iraq n years after 1994.

c. Estimate Iraq's population, in millions, for the year 2005. Round to two decimal places.

70. The population of China from 1995 through 1998 is shown in the following table.

Year	1995	1996	1997	1998
Population in millions	1218.80	1232.21	1245.76	1259.46

Source: U.N. Population Division

a. Divide the population for each year by the population in the preceding year. Round to two decimal places and show that China's population is increasing geometrically.

b. Write the general term of the geometric sequence describing population growth for China n years after 1994.

c. Estimate China's population, in millions, for the year 2005. Round to two decimal places.

Use the formula for the sum of the first n terms of a geometric sequence to solve Exercises 71–76.

In Exercises 71–72, you save $1 the first day of a month, $2 the second day, $4 the third day, continuing to double your savings each day.

71. What will your total savings be for the first 15 days?

72. What will your total savings be for the first 30 days?

73. A job pays a salary of $24,000 the first year. During the next 19 years, the salary increases by 5% each year. What is the total lifetime salary over the 20-year period? Round to the nearest dollar.

74. You are investigating two employment opportunities. Company A offers $30,000 the first year. During the next four years, the salary is guaranteed to increase by 6% per year. Company B offers $32,000 the first year, with guaranteed annual increases of 3% per year after that. Which company offers the better total salary for a five-year contract? By how much? Round to the nearest dollar.

75. A pendulum swings through an arc of 20 inches. On each successive swing, the length of the arc is 90% of the previous length.

$$20, \quad 0.9(20), \quad 0.9^2(20), \quad 0.9^3(20), \quad \dots$$

1st swing   2nd swing   3rd swing   4th swing

After 10 swings, what is the total length of the distance the pendulum has swung? Round to the nearest hundredth of an inch.

76. A pendulum swings through an arc of 16 inches. On each successive swing, the length of the arc is 96% of the previous length.

$$16, \quad 0.96(16), \quad (0.96)^2(16), \quad (0.96)^3(16), \quad \dots$$

1st swing   2nd swing   3rd swing   4th swing

After 10 swings, what is the total length of the distance the pendulum has swung? Round to the nearest hundredth of an inch.

Use the formula for the value of an annuity to solve Exercises 77–80. Round answers to the nearest dollar.

77. To save for retirement, you decide to deposit $2500 into an IRA at the end of each year for the next 40 years. If the interest rate is 9% per year compounded annually, find the value of the IRA after 40 years.

78. You decide to deposit $100 at the end of each month into an account paying 8% interest compounded monthly to save for your child's education. How much will you save over 16 years?

79. You contribute $600 at the end of each quarter to a Tax Sheltered Annuity (TSA) paying 8% annual interest compounded quarterly. Find the value of the TSA after 18 years.

80. To save for a new home, you invest $500 per month in a mutual fund with an annual rate of return of 10% compounded monthly. How much will you have saved after four years?

Use the formula for the sum of an infinite geometric series to solve Exercises 81–83.

81. A new factory in a small town has an annual payroll of $6 million. It is expected that 60% of this money will be spent in the town by factory personnel. The people in the town who receive this money are expected to spend 60% of what they receive in the town, and so on. What is the total of all this spending, called the *total economic impact* of the factory, on the town each year?

82. How much additional spending will be generated by a $10 billion tax rebate if 60% of all income is spent?

83. If the shading process shown in the figure is continued indefinitely, what fractional part of the largest square is eventually shaded?

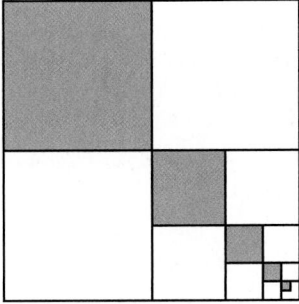

Writing in Mathematics

84. What is a geometric sequence? Give an example with your explanation.

85. What is the common ratio in a geometric sequence?

86. Explain how to find the general term of a geometric sequence.

87. Explain how to find the sum of the first n terms of a geometric sequence without having to add up all the terms.

88. What is an annuity?

89. What is the difference between a geometric sequence and an infinite geometric series?

90. How do you determine if an infinite geometric series has a sum? Explain how to find the sum of an infinite geometric series.

91. Would you rather have $10,000,000 and a brand new BMW or 1¢ today, 2¢ tomorrow, 4¢ on day 3, 8¢ on day 4, 16¢ on day 5, and so on, for 30 days? Explain.

92. For the first 30 days of a flu outbreak, the number of students on your campus who become ill is increasing. Which is worse: the number of students with the flu is increasing arithmetically or is increasing geometrically? Explain your answer.

Critical Thinking Exercises

93. Which one of the following is true?
  a. The sequence 2, 6, 24, 120, ... is an example of a geometric sequence.
  b. The sum of the geometric series $\frac{1}{2} + \frac{1}{4} + \frac{1}{8} + \cdots + \frac{1}{512}$ can only be estimated without knowing precisely what terms occur between $\frac{1}{8}$ and $\frac{1}{512}$.
  c. $10 - 5 + \frac{5}{2} - \frac{5}{4} + \cdots = \dfrac{10}{1 - \frac{1}{2}}$
  d. If the nth term of a geometric sequence is $a_n = 3(0.5)^{n-1}$, the common ratio is $\frac{1}{2}$.

94. In a pest-eradication program, sterilized male flies are released into the general population each day. Ninety percent of those flies will survive a given day. How many flies should be released each day if the long-range goal of the program is to keep 20,000 sterilized flies in the population?

95. You are now 25 years old and would like to retire at age 55 with a retirement fund of $1,000,000. How much should you deposit at the end of each month for the next 30 years in an IRA paying 10% annual interest compounded monthly to achieve your goal? Round to the nearest dollar.

Technology Exercises

96. Use the $\boxed{\text{SEQ}}$ (sequence) capability of a graphing utility and the formula you obtained for a_n to verify the value you found for a_7 in any three exercises from Exercises 25–32.

97. Use the capability of a graphing utility to calculate the sum of a sequence to verify any three of your answers to Exercises 39–44.

In Exercises 98–99, use a graphing utility to graph the function. Determine the horizontal asymptote for the graph of f and discuss its relationship to the sum of the given series.

98. Function
$$f(x) = \frac{2\left[1 - \left(\frac{1}{3}\right)^x\right]}{1 - \frac{1}{3}}$$
Series
$$2 + 2\left(\frac{1}{3}\right) + 2\left(\frac{1}{3}\right)^2 + 2\left(\frac{1}{3}\right)^3 + \cdots$$

99. Function
$$f(x) = \frac{4\left[1 - (0.6)^x\right]}{1 - 0.6}$$
Series
$$4 + 4(0.6) + 4(0.6)^2 + 4(0.6)^3 + \cdots$$

Review Exercises

100. Simplify: $\sqrt{28} - 3\sqrt{7} + \sqrt{63}$.
(Section 10.4, Example 2)

101. Solve: $2x^2 = 4 - x$.
(Section 11.2, Example 2)

102. Rationalize the denominator: $\dfrac{6}{\sqrt{3} - \sqrt{5}}$.
(Section 10.5, Example 5)

▶ **SECTION 14.4   *The Binomial Theorem***

Objectives

1 Recognize patterns in binomial expansions.

2 Evaluate a binomial coefficient.

3 Expand a binomial raised to a power.

4 Find a particular term in a binomial expansion.

SSM
PH Tutor   CD-   Video
Center   ROM

Galaxies are groupings of billions of stars bound together gravitationally. Some galaxies, such as the Centaurus galaxy shown here, are elliptical in shape.

Is mathematics discovered or invented? For example, planets revolve in elliptical orbits. Does that mean that the ellipse is out there, waiting for the mind to discover it? Or do people create the definition of an ellipse just as they compose a song? And is it possible for the same mathematics to be discovered/invented by independent researchers separated by time, place, and culture? This is precisely what occurred when mathematicians attempted to find efficient methods for raising binomials to higher and higher powers, such as

$$(x + 2)^3, (x + 2)^4, (x + 2)^5, (x + 2)^6,$$

and so on. In this section, we study higher powers of binomials and a method first discovered/invented by great minds in Eastern and Western culture working independently.

Patterns in Binomial Expansions

1 Recognize patterns in binomial expansions.

When we write out the *binomial expression* $(a + b)^n$, where n is a positive integer, several patterns begin to appear.

$$(a + b)^1 = a + b$$
$$(a + b)^2 = a^2 + 2ab + b^2$$
$$(a + b)^3 = a^3 + 3a^2b + 3ab^2 + b^3$$
$$(a + b)^4 = a^4 + 4a^3b + 6a^2b^2 + 4ab^3 + b^4$$
$$(a + b)^5 = a^5 + 5a^4b + 10a^3b^2 + 10a^2b^3 + 5ab^4 + b^5$$

Discover for Yourself

Each expanded form of the binomial expression is a polynomial. Study the five polynomials and answer the following questions:

1. For each polynomial, describe the pattern for the exponents on *a*. What is the largest exponent on *a*? What happens to the exponent on *a* from term to term?

2. Describe the pattern for the exponents on *b*. What is the exponent on *b* in the first term? What is the exponent on *b* in the second term? What happens to the exponent on *b* from term to term?

3. Find the sum of the exponents on the variables in each term for the polynomials in the five rows. Describe the pattern.

4. How many terms are there in the polynomials on the right in relation to the power of the binomial?

Expansions of binomials

$(a + b)^1 = a + b$

$(a + b)^2 = a^2 + 2ab + b^2$

$(a + b)^3 = a^3 + 3a^2b + 3ab^2 + b^3$

$(a + b)^4 = a^4 + 4a^3b + 6a^2b^2 + 4ab^3 + b^4$

$(a + b)^5 = a^5 + 5a^4b + 10a^3b^2 + 10a^2b^3 + 5ab^4 + b^5,$

repeated from page 973.

How many of the following patterns were you able to discover?

1. The first term is a^n. The exponent on a decreases by 1 in each successive term.
2. The exponents on b increase by 1 in each successive term. In the first term, the exponent on b is 0. (Because $b^0 = 1$, b is not shown in the first term.) The last term is b^n.
3. The sum of the exponents on the variables in any term is equal to n, the exponent on $(a + b)^n$.
4. There is one more term in the polynomial expansion than there is in the power of the binomial, n. There are $n + 1$ terms in the expanded form of $(a + b)^n$.

Using these observations, the variable parts of the expansion of $(a + b)^6$ are

$$a^6, \quad a^5b, \quad a^4b^2, \quad a^3b^3, \quad a^2b^4, \quad ab^5, \quad b^6.$$

The first term is a^6, with the exponent on a decreasing by 1 in each successive term. The exponents on b increase from 0 to 6, with the last term being b^6. The sum of the exponents in each term is equal to 6.

We can generalize from these observations to obtain the variable parts of the expansion of $(a + b)^n$. They are

$$a^n, \quad a^{n-1}b, \quad a^{n-2}b^2, \quad a^{n-3}b^3, \ldots, \quad ab^{n-1}, \quad b^n.$$

Exponents on a are decreasing by 1. Exponents on b are increasing by 1.

Sum of exponents: $n - 1 + 1 = n$

Sum of exponents: $n - 3 + 3 = n$

Sum of exponents: $1 + n - 1 = n$

Let's now establish a pattern for the coefficients of the terms in the binomial expansion. Notice that each row in the figure shown below begins and ends with 1. Any other number in the row can be obtained by adding the two numbers immediately above it.

Coefficients for $(a + b)^1$.   1   1

Coefficients for $(a + b)^2$.   1   2   1

Coefficients for $(a + b)^3$.   1   3   3   1

Coefficients for $(a + b)^4$.   1   4   6   4   1

Coefficients for $(a + b)^5$.   1   5   10   10   5   1

This triangular array of coefficients is called **Pascal's triangle**. If we continue with the sixth row, the first and last numbers are 1. Each of the other numbers is obtained by finding the sum of the two closest numbers above it in the fifth row.

1   1

1   2   1

1   3   3   1

1   4   6   4   1

1   5   10   10   5   1

$1 + 5$ $5 + 10$ $10 + 10$ $10 + 5$ $5 + 1$

1   6   15   20   15   6   1

We can use the numbers in the sixth row and the variable parts we found to write the expansion for $(a + b)^6$. It is

$$(a + b)^6 = a^6 + 6a^5b + 15a^4b^2 + 20a^3b^3 + 15a^2b^4 + 6ab^5 + b^6.$$

Study Tip

We have not shown the number in the top row of Pascal's triangle on the right. The top row is *row zero* because it corresponds to $(a + b)^0 = 1$. With row zero, the triangle appears as

1

1   1

1   2   1

1   3   3   1

1   4   6   4   1

etc.

2 Evaluate a binomial coefficient.

Binomial Coefficients

Pascal's triangle becomes cumbersome when a binomial is raised to a relatively large power. Therefore, the coefficients in a binomial expansion are instead given in terms of factorials. The coefficients are written in a special notation, which we define next.

Definition of a Binomial Coefficient $\dbinom{n}{r}$

For nonnegative integers n and r, with $n \geq r$, the expression $\dbinom{n}{r}$ (read "n above r") is called a **binomial coefficient** and is defined by

$$\binom{n}{r} = \frac{n!}{r!(n-r)!}.$$

The symbol $_nC_r$ is often used in place of $\dbinom{n}{r}$ to denote binomial coefficients.

Can you see that the definition of a binomial coefficient involves a fraction with factorials in the numerator and the denominator? When evaluating such an expression, try to reduce the fraction before performing the multiplications. For example, consider $\frac{26!}{21!}$. Rather than write out 26! as the product of all integers from 26 down to 1, we can express 26! as

$$26! = 26 \cdot 25 \cdot 24 \cdot 23 \cdot 22 \cdot 21!.$$

In this way, we can divide both the numerator and the denominator by the common factor, 21!.

$$\frac{26!}{21!} = \frac{26 \cdot 25 \cdot 24 \cdot 23 \cdot 22 \cdot \cancel{21!}}{\cancel{21!}} = 26 \cdot 25 \cdot 24 \cdot 23 \cdot 22 = 7,893,600$$

EXAMPLE 1   Evaluating Binomial Coefficients

Evaluate: **a.** $\dbinom{6}{2}$ **b.** $\dbinom{3}{0}$ **c.** $\dbinom{9}{3}$ **d.** $\dbinom{4}{4}$.

Solution   In each case, we apply the definition of the binomial coefficient.

a. $\dbinom{6}{2} = \dfrac{6!}{2!(6-2)!} = \dfrac{6!}{2!4!} = \dfrac{6 \cdot 5 \cdot \cancel{4!}}{2 \cdot 1 \cdot \cancel{4!}} = 15$

b. $\dbinom{3}{0} = \dfrac{3!}{0!(3-0)!} = \dfrac{\cancel{3!}}{0!\,\cancel{3!}} = \dfrac{1}{1} = 1$

Remember that $0! = 1$.

c. $\dbinom{9}{3} = \dfrac{9!}{3!(9-3)!} = \dfrac{9!}{3!6!} = \dfrac{9 \cdot 8 \cdot 7 \cdot \cancel{6!}}{3 \cdot 2 \cdot 1 \cdot \cancel{6!}} = 84$

d. $\dbinom{4}{4} = \dfrac{4!}{4!(4-4)!} = \dfrac{\cancel{4!}}{\cancel{4!}\,0!} = \dfrac{1}{1} = 1$ ∎

Using Technology

Graphing utilities can compute binomial coefficients. For example, to find $\dbinom{6}{2}$, many utilities require the sequence

6 [nCr] 2 [ENTER].

The graphing utility will display 15. Consult your manual and verify the other evaluations in Example 1.

✔ **CHECK POINT 1** Evaluate:

a. $\begin{pmatrix} 6 \\ 3 \end{pmatrix}$ b. $\begin{pmatrix} 6 \\ 0 \end{pmatrix}$

c. $\begin{pmatrix} 8 \\ 2 \end{pmatrix}$ d. $\begin{pmatrix} 3 \\ 3 \end{pmatrix}$.

3 Expand a binomial raised to a power.

The Binomial Theorem

If we use binomial coefficients and the pattern for the variable part of each term, a formula called the **Binomial Theorem** can be used to expand any positive integral power of a binomial.

A Formula for Expanding Binomials: The Binomial Theorem

For any positive integer n,

$$(a + b)^n = \binom{n}{0}a^n + \binom{n}{1}a^{n-1}b + \binom{n}{2}a^{n-2}b^2 + \binom{n}{3}a^{n-3}b^3 + \cdots + \binom{n}{n}b^n.$$

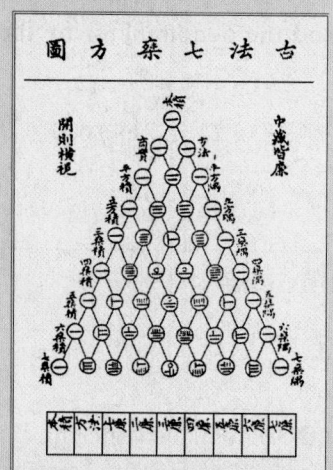
EXAMPLE 2 Using the Binomial Theorem

Expand: $(x + 2)^4$.

Solution We use the Binomial Theorem

$$(a + b)^n = \binom{n}{0}a^n + \binom{n}{1}a^{n-1}b + \binom{n}{2}a^{n-2}b^2 + \binom{n}{3}a^{n-3}b^3 + \cdots + \binom{n}{n}b^n$$

to expand $(x + 2)^4$. In $(x + 2)^4$, $a = x$, $b = 2$, and $n = 4$.

$$(x + 2)^4 = \binom{4}{0}x^4 + \binom{4}{1}x^3 \cdot 2 + \binom{4}{2}x^2 \cdot 2^2 + \binom{4}{3}x \cdot 2^3 + \binom{4}{4}2^4$$

These binomial coefficients are evaluated using $\binom{n}{r} = \dfrac{n!}{r!(n-r)!}$.

$$= \frac{4!}{0!4!}x^4 + \frac{4!}{1!3!}x^3 \cdot 2 + \frac{4!}{2!2!}x^2 \cdot 4 + \frac{4!}{3!1!}x \cdot 8 + \frac{4!}{4!0!} \cdot 16$$

$$\frac{4!}{2!2!} = \frac{4 \cdot 3 \cdot 2!}{2! \cdot 2 \cdot 1} = \frac{12}{2} = 6$$

Take a few minutes to verify the other factorial evaluations.

$$= 1 \cdot x^4 + 4x^3 \cdot 2 + 6x^2 \cdot 4 + 4x \cdot 8 + 1 \cdot 16$$

$$= x^4 + 8x^3 + 24x^2 + 32x + 16 \qquad \blacksquare$$

✔ **CHECK POINT 2** Expand: $(x + 1)^4$.

EXAMPLE 3   Using the Binomial Theorem

Expand: $(2x - y)^5$.

Solution   Because the Binomial Theorem involves the addition of two terms raised to a power, we rewrite $(2x - y)^5$ as $[2x + (-y)]^5$. We use the Binomial Theorem

$$(a + b)^n = \binom{n}{0}a^n + \binom{n}{1}a^{n-1}b + \binom{n}{2}a^{n-2}b^2 + \binom{n}{3}a^{n-3}b^3 + \cdots + \binom{n}{n}b^n$$

to expand $[2x + (-y)]^5$. In $[2x + (-y)]^5$, $a = 2x$, $b = -y$, and $n = 5$.

$$(2x - y)^5 = [2x + (-y)]^5$$

$$= \binom{5}{0}(2x)^5 + \binom{5}{1}(2x)^4(-y) + \binom{5}{2}(2x)^3(-y)^2 + \binom{5}{3}(2x)^2(-y)^3 + \binom{5}{4}(2x)(-y)^4 + \binom{5}{5}(-y)^5$$

Evaluate binomial coefficients using $\binom{n}{r} = \dfrac{n!}{r!(n - r)!}$.

$$= \frac{5!}{0!5!}(2x)^5 + \frac{5!}{1!4!}(2x)^4(-y) + \frac{5!}{2!3!}(2x)^3(-y)^2 + \frac{5!}{3!2!}(2x)^2(-y)^3 + \frac{5!}{4!1!}(2x)(-y)^4 + \frac{5!}{5!0!}(-y)^5$$

$$\frac{5!}{2!3!} = \frac{5 \cdot 4 \cdot \cancel{3!}}{2 \cdot 1 \cdot \cancel{3!}} = 10$$

Take a few minutes to verify the other factorial evaluations.

$$= 1(2x)^5 + 5(2x)^4(-y) + 10(2x)^3(-y)^2 + 10(2x)^2(-y)^3 + 5(2x)(-y)^4 + 1(-y)^5$$

Raise both factors in these parentheses to the indicated powers.

$$= 1(32x^5) + 5(16x^4)(-y) + 10(8x^3)(-y)^2 + 10(4x^2)(-y)^3 + 5(2x)(-y)^4 + 1(-y)^5$$

Now raise $-y$ to the indicated powers.

$$= 1(32x^5) + 5(16x^4)(-y) + 10(8x^3)y^2 + 10(4x^2)(-y^3) + 5(2x)y^4 + 1(-y^5)$$

Multiplying factors in each of the six terms gives us the desired expansion:

$$(2x - y)^5 = 32x^5 - 80x^4y + 80x^3y^2 - 40x^2y^3 + 10xy^4 - y^5. \qquad \blacksquare$$

✔ **CHECK POINT 3**   Expand: $(x - 2y)^5$.

4 | Find a particular term in a binomial expansion.

Finding a Particular Term in a Binomial Expansion

In the formula for expanding binomials, $\binom{n}{0}a^n b^0$ is the first term, $\binom{n}{1}a^{n-1}b$ is the second term, $\binom{n}{2}a^{n-2}b^2$ is the third term, and so on. Generalizing these observations provides a formula for finding a particular term without writing the entire expansion.

> ### Finding a Particular Term in a Binomial Expansion
> The rth term of the expansion of $(a + b)^n$ is
> $$\binom{n}{r-1}a^{n-r+1}b^{r-1}.$$

EXAMPLE 4   Finding a Single Term of a Binomial Expansion

Find the fourth term in the expansion of $(3x + 2y)^7$.

Solution   We will use the formula for the rth term of the expansion of $(a + b)^n$,

$$\binom{n}{r-1}a^{n-r+1}b^{r-1}$$

to find the fourth term of $(3x + 2y)^7$. For the fourth term, $n = 7, r = 4, a = 3x$, and $b = 2y$. Thus, the fourth term is

$$\binom{7}{4-1}(3x)^{7-4+1}(2y)^{4-1} = \binom{7}{3}(3x)^4(2y)^3 = \frac{7!}{3!(7-3)!}(3x)^4(2y)^3.$$

We use $\binom{n}{r} = \dfrac{n!}{r!(n-r)!}$ to evaluate $\binom{7}{3}$.

Now we need to evaluate the factorial expression and raise $3x$ and $2y$ to the indicated powers. We obtain

$$\frac{7!}{3!4!}(81x^4)(8y^3) = \frac{7 \cdot 6 \cdot 5 \cdot \cancel{4!}}{3 \cdot 2 \cdot 1 \cdot \cancel{4!}}(81x^4)(8y^3) = 35(81x^4)(8y^3) = 22{,}680x^4y^3.$$

The fourth term of $(3x + 2y)^7$ is $22{,}680x^4y^3$.   ∎

✔ **CHECK POINT 4**   Find the fifth term in the expansion of $(2x + y)^9$.

EXERCISE SET 14.4

Practice Exercises

In Exercises 1–8, evaluate the given binomial coefficient.

1. $\binom{8}{3}$

2. $\binom{7}{2}$

3. $\binom{12}{1}$

4. $\binom{11}{1}$

5. $\binom{6}{6}$

6. $\binom{15}{2}$

7. $\binom{100}{2}$

8. $\binom{100}{98}$

In Exercises 9–30, use the Binomial Theorem to expand each binomial and express the result in simplified form.

9. $(x + 2)^3$

10. $(x + 4)^3$

11. $(3x + y)^3$

12. $(x + 3y)^3$

13. $(5x - 1)^3$

14. $(4x - 1)^3$

15. $(2x + 1)^4$

16. $(3x + 1)^4$

17. $(x^2 + 2y)^4$

18. $(x^2 + y)^4$

19. $(y - 3)^4$

20. $(y - 4)^4$

21. $(2x^3 - 1)^4$

22. $(2x^5 - 1)^4$

23. $(c + 2)^5$

24. $(c + 3)^5$

25. $(x - 1)^5$

26. $(x - 2)^5$

27. $(3x - y)^5$

28. $(x - 3y)^5$

29. $(2a + b)^6$

30. $(a + 2b)^6$

In Exercises 31–38, write the first three terms in each binomial expansion, expressing the result in simplified form.

31. $(x + 2)^8$

32. $(x + 3)^8$

33. $(x - 2y)^{10}$

34. $(x - 2y)^9$

35. $(x^2 + 1)^{16}$

36. $(x^2 + 1)^{17}$

37. $(y^3 - 1)^{20}$

38. $(y^3 - 1)^{21}$

In Exercises 39–46, find the term indicated in each expansion.

39. $(2x + y)^6$; third term

40. $(x + 2y)^6$; third term

41. $(x - 1)^9$; fifth term

42. $(x - 1)^{10}$; fifth term

43. $(x^2 + y^3)^8$; sixth term

44. $(x^3 + y^2)^8$; sixth term

45. $(x - \frac{1}{2})^9$; fourth term

46. $(x + \frac{1}{2})^8$; fourth term

Application Exercises

47. The percentage of people taking the SAT whose intended college major is engineering, $f(t)$, can be modeled by

$$f(t) = 0.002t^3 - 0.9t^2 + 1.27t + 6.76, \qquad 0 \le t \le 20,$$

where $t = 0$ represents 1975. How can we adjust this model so that $t = 0$ corresponds to 1985 rather than 1975? We shift the graph of f ten units to the left. We obtain $g(t) = f(t + 10)$. Use the Binomial Theorem to express $g(t)$ in descending powers of t.

48. The personal income per capita in the United States, $f(t)$, in constant 1992 dollars, can be modeled by

$$f(t) = 3.75t^3 - 115.23t^2 + 1229.81t + 16{,}025.65,$$
$$0 \le t \le 15,$$

where $t = 0$ represents 1979. How can we adjust this model so that $t = 0$ corresponds to 1989 rather than 1979? We shift the graph of f ten units to the left. We obtain $g(t) = f(t + 10)$. Use the Binomial Theorem to express $g(t)$ in descending powers of t.

Writing in Mathematics

49. Describe the pattern on the exponents on a in the expansion of $(a + b)^n$.

50. Describe the pattern on the exponents on b in the expansion of $(a + b)^n$.

51. What is true about the sum of the exponents on a and b in any term in the expansion of $(a + b)^n$?

52. How do you determine how many terms there are in a binomial expansion?

53. What is Pascal's triangle? How do you find the numbers in any row of the triangle?

54. Explain how to evaluate $\binom{n}{r}$. Provide an example with your explanation.

55. Explain how to use the Binomial Theorem to expand a binomial. Provide an example with your explanation.

56. Explain how to find a particular term in a binomial expansion without having to write out the entire expansion.

57. Are there situations in which it is easier to use Pascal's triangle than binomial coefficients? Describe these situations.

Critical Thinking Exercises

58. Which one of the following is true?
   a. The binomial expansion for $(a + b)^n$ contains n terms.
   b. The Binomial Theorem can be written in condensed form as $(a + b)^n = \sum_{r=0}^{n} \binom{n}{r} a^{n-r} b^r$.
   c. The sum of the binomial coefficients in $(a + b)^n$ cannot be 2^n.
   d. There are no values of a and b such that $(a + b)^4 = a^4 + b^4$.

59. Use the Binomial Theorem to expand and then simplify the result: $(x^2 + x + 1)^3$. [*Hint*: Write $x^2 + x + 1$ as $x^2 + (x + 1)$].

60. Find the term in the expansion of $(x^2 + y^2)^5$ containing x^4 as a factor.

Technology Exercises

61. Use the $\boxed{\text{nCr}}$ key on a graphing utility to verify your answers in Exercises 1–8.

In Exercises 62–63, graph each of the functions in the same viewing rectangle. Describe how the graphs illustrate the Binomial Theorem.

62. $f_1(x) = (x + 2)^3$
   $f_2(x) = x^3$
   $f_3(x) = x^3 + 6x^2$
   $f_4(x) = x^3 + 6x^2 + 12x$
   $f_5(x) = x^3 + 6x^2 + 12x + 8$
   Use a $[-10, 10, 1]$ by $[-30, 30, 10]$ viewing rectangle.

63. $f_1(x) = (x + 1)^4$
   $f_2(x) = x^4$
   $f_3(x) = x^4 + 4x^3$
   $f_4(x) = x^4 + 4x^3 + 6x^2$
   $f_5(x) = x^4 + 4x^3 + 6x^2 + 4x$
   $f_6(x) = x^4 + 4x^3 + 6x^2 + 4x + 1$
   Use a $[-5, 5, 1]$ by $[-30, 30, 10]$ viewing rectangle.

In Exercises 64–66, use the Binomial Theorem to find a polynomial expansion for each function. Then use a graphing utility and an approach similar to the one in Exercises 62 and 63 to verify the expansion.

64. $f_1(x) = (x - 1)^3$

65. $f_1(x) = (x - 2)^4$

66. $f_1(x) = (x + 2)^6$

67. Graphing utilities capable of symbolic manipulation, such as the TI-92, will expand binomials. On the TI-92, to expand $(3a - 5b)^{12}$, input the following:

$\boxed{\text{EXPAND}}\, ((3a \boxed{-} 5b) \boxed{\wedge} 12)\, \boxed{\text{ENTER}}$.

Use a graphing utility with this capability to verify any five of the expansions you performed by hand in Exercises 9–30.

Review Exercises

68. If $f(x) = x^2 + 2x + 3$, find $f(a + 1)$.
   (Section 8.1, Example 3)

69. If $f(x) = x^2 + 5x$ and $g(x) = 2x - 3$, find $f(g(x))$ and $g(f(x))$.
   (Section 12.2, Example 1)

70. Subtract: $\dfrac{x}{x + 3} - \dfrac{x + 1}{2x^2 - 2x - 24}$.
   (Section 7.4, Example 7)

▶ SECTION 14.5   *Counting Principles, Permutations, and Combinations*

Objectives

1 Use the Fundamental Counting Principle.

2 Use the permutations formula.

3 Distinguish between permutation problems and combination problems.

4 Use the combinations formula.

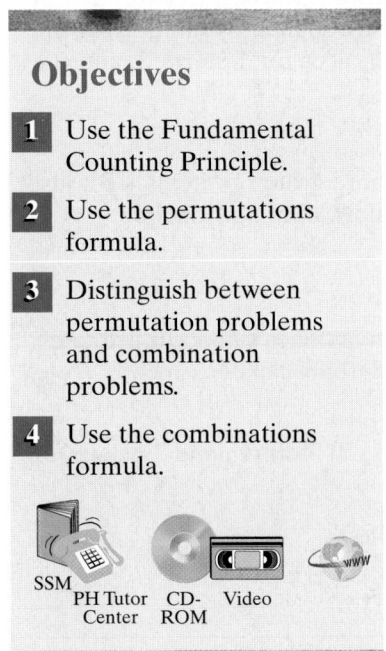

SSM
PH Tutor Center   CD-ROM   Video   www

Have you ever imagined what your life would be like if you won the lottery? What changes would you make? Before you fantasize about becoming a person of leisure with a staff of obedient elves, think about this: The probability of winning top prize in the lottery is about the same as the probability of being struck by lightning. There are millions of possible number combinations in lottery games, and only one way of winning the grand prize. Determining the probability of winning involves calculating the chance of getting the winning combination from all possible outcomes. In this section, we begin preparing for the surprising world of probability by looking at methods for counting possible outcomes.

1 Use the Fundamental Counting Principle.

The Fundamental Counting Principle

It's early morning, you're groggy, and you have to select something to wear for your 8 A.M. class. (What *were* you thinking of when you signed up for a class at that hour?!) Fortunately, your "lecture wardrobe" is rather limited—just two pairs of jeans to choose from (one blue, one black), three T-shirts to choose from (one beige, one yellow, and one blue), and two pairs of sneakers to select from (one black, one red). Your possible outfits are shown in Figure 14.4.

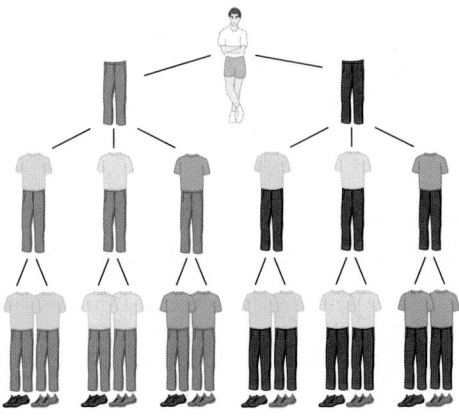

Figure 14.4 Selecting a wardrobe

The number of possible ways of playing the first four moves on each side in a game of chess is 318,979,564,000.

The **tree diagram**, so named because of its branches, shows that you can form 12 outfits from your two pairs of jeans, three T-shirts, and two pairs of sneakers. Notice that the number of outfits can be obtained by multiplying the number of choices for jeans, 2, the number of choices for T-shirts, 3, and the number of choices for sneakers, 2:

$$2 \cdot 3 \cdot 2 = 12.$$

We can generalize this idea to any two or more groups of items—not just jeans, T-shirts, and sneakers—with the **Fundamental Counting Principle**:

The Fundamental Counting Principle

The number of ways in which a series of successive things can occur is found by multiplying the number of ways in which each thing can occur.

For example, if you own 30 pairs of jeans, 20 T-shirts, and 12 pairs of sneakers, you have

$$30 \cdot 20 \cdot 12 = 7200$$

choices for your wardrobe!

EXAMPLE 1   Options in Planning a Course Schedule

Next semester you are planning to take three courses—math, English, and humanities. Based on time blocks and highly recommended professors, there are 8 sections of math, 5 of English, and 4 of humanities that you find suitable. Assuming no scheduling conflicts, how many different three-course schedules are possible?

Solution   This situation involves making choices with three groups of items.

MATH	ENGLISH	HUMANITIES
8 choices	5 choices	4 choices

We use the Fundamental Counting Principle to find the number of three-course schedules. Multiply the number of choices for each of the three groups:

$$8 \cdot 5 \cdot 4 = 160.$$

Thus, there are 160 different three-course schedules.   ∎

✔ **CHECK POINT 1**   A pizza can be ordered with three choices of size (small, medium, or large), four choices of crust (thin, thick, crispy, or regular), and six choices of toppings (ground beef, sausage, pepperoni, bacon, mushrooms, or onions). How many different one-topping pizzas can be ordered?

EXAMPLE 2   A Multiple-Choice Test

You are taking a multiple-choice test that has ten questions. Each of the questions has four choices, with one correct choice per question. If you select one of these options per question and leave nothing blank, in how many ways can you answer the questions?

ENRICHMENT ESSAY

Permutations and Rubik's Cube

First developed in Hungary in the 1970s by Erno Rubik, a Rubik's cube contains 26 small cubes. The square faces of the cubes are colored in six different colors. The cubes can be twisted horizontally or vertically. When first purchased, the cube is arranged so that each face shows a single color. To do the puzzle, you first turn columns and rows in a random way until all of the six faces are multicolored. To solve the puzzle, you must return the cube to its original state—that is, a single color on each of the six faces. With 115,880,067,072,000 arrangements, this is no easy task! If it takes one-half second for each of these arrangements, it would require over 1,800,000 years to move the cube into all possible arrangements.

Solution   We use the Fundamental Counting Principle to determine the number of ways you can answer the test. Multiply the number of choices, 4, for each of the ten questions:

$$4 \cdot 4 \cdot 4 \cdot 4 \cdot 4 \cdot 4 \cdot 4 \cdot 4 \cdot 4 \cdot 4 = 4^{10} = 1,048,576.$$

Thus, you can answer the questions in 1,048,576 different ways.   ■

Are you surprised that there are over one million ways of answering a ten-question multiple-choice test? Of course, there is only one way to answer the test and receive a perfect score. The probability of guessing your way into a perfect score involves calculating the chance of getting a perfect score, just one way, from all 1,048,576 possible outcomes. In short, prepare for the test and do not rely on guessing!

✔ **CHECK POINT 2**   You are taking a multiple-choice test that has six questions. Each of the questions has three choices, with one correct choice per question. If you select one of these options per question and leave nothing blank, in how many ways can you answer the questions?

EXAMPLE 3   Telephone Numbers in the United States

Telephone numbers in the United States begin with three-digit area codes followed by seven-digit local telephone numbers. Area codes and local telephone numbers cannot begin with 0 or 1. How many different telephone numbers are possible?

Solution   This situation involves making choices with ten groups of items.

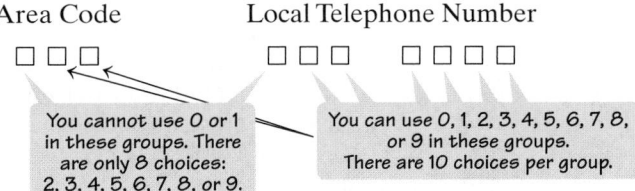

We use the Fundamental Counting Principle to determine the number of different telephone numbers that are possible. The total number of telephone numbers possible is

$$8 \cdot 10 \cdot 10 \cdot 8 \cdot 10 \cdot 10 \cdot 10 \cdot 10 \cdot 10 \cdot 10 = 6,400,000,000.$$

There are six billion four hundred million different telephone numbers that are possible.   ■

✔ **CHECK POINT 3**   License plates in a particular state display two letters followed by three numbers, such as AT-887 or BB-013. How many different license plates can be manufactured?

Permutations

You are the coach of a little league baseball team. There are 13 players on the team (and lots of parents hovering in the background, dreaming of stardom for their little "Mark McGwire"). You need to choose a batting order having 9

2   Use the permutations formula.

players. The order makes a difference, because, for instance, if bases are loaded and "Little Mark" is fourth or fifth at bat, his possible home run will drive in three additional runs. How many batting orders can you form?

You can choose any of 13 players for the first person at bat. Then you will have 12 players from which to choose the second batter, then 11 from which to choose the third batter, and so on. The situation can be shown as follows:

Batter 1	Batter 2	Batter 3	Batter 4	Batter 5	Batter 6	Batter 7	Batter 8	Batter 9
13 choices	12 choices	11 choices	10 choices	9 choices	8 choices	7 choices	6 choices	5 choices

We use the Fundamental Counting Principle to find the number of batting orders. The total number of batting orders is

$$13 \cdot 12 \cdot 11 \cdot 10 \cdot 9 \cdot 8 \cdot 7 \cdot 6 \cdot 5 = 259,459,200.$$

Nearly 260 million batting orders are possible for your 13-player little league team. Each batting order is called a **permutation** of 13 players taken 9 at a time. The number of permutations of 13 players taken 9 at a time is 259,459,200. A permutation is an ordered arrangement of items that occurs when

* No item is used more than once. (Each of the 9 players in the batting order bats exactly once.)
* The order of arrangement makes a difference.

We can obtain a formula for finding the number of permutations by rewriting our computation:

$13 \cdot 12 \cdot 11 \cdot 10 \cdot 9 \cdot 8 \cdot 7 \cdot 6 \cdot 5$

$$= \frac{13 \cdot 12 \cdot 11 \cdot 10 \cdot 9 \cdot 8 \cdot 7 \cdot 6 \cdot 5 \cdot \boxed{4 \cdot 3 \cdot 2 \cdot 1}}{\boxed{4 \cdot 3 \cdot 2 \cdot 1}} = \frac{13!}{4!} = \frac{13!}{(13-9)!}.$$

Thus, the number of permutations of 13 things taken 9 at a time is $\frac{13!}{(13-9)!}$. The special notation $_{13}P_9$ is used to replace the phrase "the number of permutations of 13 things taken 9 at a time." Using this new notation, we can write

$$_{13}P_9 = \frac{13!}{(13-9)!}.$$

The numerator of this expression is the number of items, 13 team members, expressed as a factorial: 13! The denominator is also a factorial. It is the factorial of the difference between the number of items, 13, and the number of items in each permutation, 9 batters: $(13 - 9)!$.

The notation $_nP_r$ means the **number of permutations of n things taken r at a time**. We can generalize from the situation in which 9 batters were taken from 13 players. By generalizing, we obtain the following formula for the number of permutations if r items are taken from n items.

Permutations of n Things Taken r at a Time

The number of possible permutations if r items are taken from n items is

$$_nP_r = \frac{n!}{(n-r)!}.$$

Because all permutation problems are also Fundamental Counting problems, they can be solved using the formula for $_nP_r$, or using the Fundamental Counting Principle.

Using Technology

Graphing utilities have a key for calculating permuations, usually labeled $\boxed{_nP_r}$. For example, to find $_{20}P_3$, the keystrokes are

$20 \boxed{_nP_r} 3 \boxed{ENTER}$.

If you are using a scientific calculator, check your manual for the location of the key for calculating permutations and the required keystrokes.

EXAMPLE 4   Using the Formula for Permutations

You and 19 of your friends have decided to form an Internet marketing consulting firm. The group needs to choose three officers—a CEO, an operating manager, and a treasurer. In how many ways can those offices be filled?

Solution   Your group is choosing $r = 3$ officers from a group of $n = 20$ people (you and 19 friends). The order in which the officers are chosen matters because the CEO, the operating manager, and the treasurer each have different responsibilities. Thus, we are looking for the number of permutations of 20 things taken 3 at a time. We use the formula

$$_nP_r = \frac{n!}{(n-r)!}$$

with $n = 20$ and $r = 3$.

$$_{20}P_3 = \frac{20!}{(20-3)!} = \frac{20!}{17!} = \frac{20 \cdot 19 \cdot 18 \cdot 17!}{17!} = \frac{20 \cdot 19 \cdot 18 \cdot \cancel{17!}}{\cancel{17!}} = 20 \cdot 19 \cdot 18 = 6840$$

Thus, there are 6840 different ways of filling the three offices. ∎

✔ **CHECK POINT 4**   A corporation has seven members on its board of directors. In how many different ways can it elect a president, vice-president, secretary, and treasurer?

EXAMPLE 5   Using the Formula for Permutations

You need to arrange seven of your favorite books along a small shelf. How many different ways can you arrange the books, assuming that the order of the books makes a difference to you?

Solution   Because you are using all seven of your books in every possible arrangement, you are arranging $r = 7$ books from a group of $n = 7$ books. Thus, we are looking for the number of permutations of 7 things taken 7 at a time. We use the formula

$$_nP_r = \frac{n!}{(n-r)!}$$

with $n = 7$ and $r = 7$.

$$_7P_7 = \frac{7!}{(7-7)!} = \frac{7!}{0!} = \frac{7!}{1} = 5040$$

Thus, you can arrange the books in 5040 ways. There are 5040 different possible permutations. ∎

✔ **CHECK POINT 5**   In how many ways can 6 books be lined up along a shelf?

ENRICHMENT ESSAY

How to Pass the Time for $2\frac{1}{2}$ Million Years

If you were to arrange 15 different books on a shelf and it took you one minute for each permutation, the entire task would take 2,487,965 years.

Source: Isaac Asimov's Book of Facts.

3 Distinguish between permutation problems and combination problems.

Marilyn Monroe, actress (1927–1962)

Combinations

As the twentieth century drew to a close, *Time* magazine presented a series of special issues on the most influential people of the century. In their issue on heroes and icons (June 14, 1999), they discussed a number of people whose careers became more profitable after their tragic deaths, including Marilyn Monroe, James Dean, Jim Morrison, Kurt Cobain, and Selena.

Imagine that you ask your friends the following question: "Of these five people, which three would you select to be included in a documentary featuring the best of their work?" You are not asking your friends to rank their three favorite artists in any kind of order—they should merely select the three to be included in the documentary.

One friend answers, "Jim Morrison, Kurt Cobain, and Selena." Another responds, "Selena, Kurt Cobain, and Jim Morrison." These two people have the same artists in their group of selections, even if they are named in a different order. We are interested *in which artists are named*, not the order in which they are named for the documentary. Because the items are taken without regard to order, this is not a permutation problem. No ranking of any sort is involved.

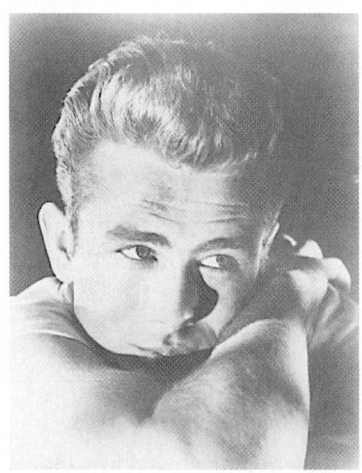

James Dean, actor (1931–1955)

Jim Morrison, musician and lead singer of the Doors (1943–1971)

Kurt Cobain, musician and front man for Nirvana (1967–1994)

Later on, you ask your roommate which three artists she would select for the documentary. She names Marilyn Monroe, James Dean, and Selena. Her selection is different from those of your two other friends because different entertainers are cited.

Mathematicians describe the group of artists given by your roommate as a combination. A **combination** of items occurs when

- The items are selected from the same group (the five stars who died young and tragically).
- No item is used more than once. (You may adore Selena, but your three selections cannot be Selena, Selena, and Selena).
- The order of items makes no difference. (Morrison, Cobain, Selena is the same group in the documentary as Selena, Cobain, Morrison.)

Do you see the difference between a permutation and a combination? A permutation is an ordered arrangement of a given group of items. A combination

Selena, musician of Tejano music (1971–1995)

is a group of items taken without regard to their order. **Permutation** problems involve situations in which **order matters**. **Combination** problems involve situations in which the **order** of items **makes no difference**.

EXAMPLE 6   Distinguishing between Permutations and Combinations

For each of the following problems, explain whether the problem is one involving permutations or combinations. (It is not necessary to solve the problem.)

  a. Six candidates are running for president, chief technology officer, and director of marketing of an Internet company. The candidate with the greatest number of votes becomes the president, the second biggest vote-getter becomes chief technology officer, and the candidate who gets the third largest number of votes will be director of marketing. How many different outcomes are possible for these three positions?

  b. From the six candidates who desire to hold office in an Internet company, a three-person committee is formed to study ways of finding new investors. How many different committees could be formed?

Solution

  a. Voters are choosing three officers from six candidates. The order in which the officers are chosen makes a difference because each of the offices (president, chief technology officer, and director of marketing) is different. Order matters. This is a problem involving permutations. (How many permutations are possible if three candidates are elected from six candidates?)

  b. A three-person committee is to be formed from the six candidates. The order in which the three people are selected does not matter because they are not filling different roles on the committee. Because order makes no difference, this is a problem involving combinations. (How many different combinations of three people can be chosen from a group of six people?)   ■

✔ **CHECK POINT 6**   For each of the following problems, explain if the problem is one involving permutations or combinations. (It is not necessary to solve the problem.)

  a. How many ways can you select 6 free videos from a list of 200 videos?

  b. In a race in which there are 50 runners and no ties, in how many ways can the first three finishers come in?

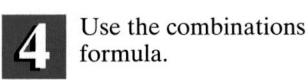
Use the combinations formula.

The notation $_nC_r$ means the **number of combinations of n things taken r at a time**. In general, there are $r!$ times as many permutations of n things taken r at a time as there are combinations of n things taken r at a time. Thus, we find the number of combinations of n things taken r at a time by dividing the number of permutations of n things taken r at a time by $r!$.

$$_nC_r = \frac{_nP_r}{r!} = \frac{\dfrac{n!}{(n-r)!}}{r!} = \frac{n!}{(n-r)!r!}$$

> **Combinations of n Things Taken r at a Time**
>
> The number of possible combinations if r items are taken from n items is
>
> $$_nC_r = \frac{n!}{(n-r)!r!}.$$

Notice that the formula for $_nC_r$ is the same as the formula for the binomial coefficient $\begin{pmatrix} n \\ r \end{pmatrix}$.

We cannot find the number of combinations if r items are taken from n items using the Fundamental Counting Principle. We must use the formula shown in the box to do so.

EXAMPLE 7   Using the Formula for Combinations

A three-person committee is needed to study ways of improving public transportation. How many committees could be formed from the eight people on the board of supervisors?

Solution   The order in which the three people are selected does not matter. This is a problem of selecting $r = 3$ people from a group of $n = 8$ people. We are looking for the number of combinations of eight things taken three at a time. We use the formula

$$_nC_r = \frac{n!}{(n-r)!r!}$$

with $n = 8$ and $r = 3$.

$$_8C_3 = \frac{8!}{(8-3)!3!} = \frac{8!}{5!3!} = \frac{8 \cdot 7 \cdot 6 \cdot 5!}{5! \cdot 3 \cdot 2 \cdot 1} = \frac{8 \cdot 7 \cdot 6 \cdot \cancel{5!}}{\cancel{5!} \cdot 3 \cdot 2 \cdot 1} = 56$$

Thus, 56 committees of three people each can be formed from the eight people on the board of supervisors. ■

✔ **CHECK POINT 7**   From a group of 10 physicians, in how many ways can four people be selected to attend a conference on acupuncture?

EXAMPLE 8   Using the Formula for Combinations

In poker, a person is dealt 5 cards from a standard 52-card deck. The order in which you are dealt the 5 cards does not matter. How many different 5-card poker hands are possible?

Solution   Because the order in which the 5 cards are dealt does not matter, this is a problem involving combinations. We are looking for the number of combinations of $n = 52$ cards drawn $r = 5$ at a time. We use the formula

$$_nC_r = \frac{n!}{(n-r)!r!}$$

with $n = 52$ and $r = 5$.

$$_{52}C_5 = \frac{52!}{(52-5)!5!} = \frac{52!}{47!5!} = \frac{52 \cdot 51 \cdot 50 \cdot 49 \cdot 48 \cdot \cancel{47!}}{\cancel{47!} \cdot 5 \cdot 4 \cdot 3 \cdot 2 \cdot 1} = 2{,}598{,}960$$

Thus, there are 2,598,960 different 5-card poker hands possible. It surprises many people that more than 2.5 million 5-card hands can be dealt from a mere 52 cards. ■

Figure 14.5 A royal flush

If you are a card player, it does not get any better than to be dealt the 5-card poker hand shown in Figure 14.5. This hand is called a *royal flush*. It consists of an ace, king, queen, jack, and 10, all of the same suit: all hearts, all diamonds, all clubs, or all spades. The probability of being dealt a royal flush involves calculating the number of ways of being dealt such a hand: just 4 of all 2,598,960 possible hands. In the next section, we move from counting possibilities to computing probabilities.

✔ **CHECK POINT 8** How many different 4-card hands can be dealt from a deck that has 16 different cards?

EXERCISE SET 14.5

 Practice Exercises

In Exercises 1–8, use the formula for $_nP_r$ to evaluate each expression.

1. $_9P_4$ **2.** $_7P_3$
3. $_8P_5$ **4.** $_{10}P_4$
5. $_6P_6$ **6.** $_9P_9$
7. $_8P_0$ **8.** $_6P_0$

In Exercises 9–16, use the formula for $_nC_r$ to evaluate each expression.

9. $_9C_5$ **10.** $_{10}C_6$
11. $_{11}C_4$ **12.** $_{12}C_5$
13. $_7C_7$ **14.** $_4C_4$
15. $_5C_0$ **16.** $_6C_0$

In Exercises 17–20, does the problem involve permutations or combinations? Explain your answer. (It is not necessary to solve the problem.)

17. A medical researcher needs 6 people to test the effectiveness of an experimental drug. If 13 people have volunteered for the test, in how many ways can 6 people be selected?

18. Fifty people purchase raffle tickets. Three winning tickets are selected at random. If first prize is $1000, second prize is $500, and third prize is $100, in how many different ways can the prizes be awarded?

19. How many different four-letter passwords can be formed from the letters A, B, C, D, E, F, and G if no repetition of letters is allowed?

20. Fifty people purchase raffle tickets. Three winning tickets are selected at random. If each prize is $500, in how many different ways can the prizes be awarded?

 Application Exercises

Use the Fundamental Counting Principle to solve Exercises 21–32.

21. The model of the car you are thinking of buying is available in nine different colors and three different styles (hatchback, sedan, or station wagon). In how many ways can you order the car?

22. A popular brand of pen is available in three colors (red, green, or blue) and four writing tips (bold, medium, fine, or micro). How many different choices of pens do you have with this brand?

23. An ice cream store sells two drinks (sodas or milk shakes), in four sizes (small, medium, large, or jumbo), and five flavors (vanilla, strawberry, chocolate, coffee, or pistachio). In how many ways can a customer order a drink?

24. A restaurant offers the following lunch menu.

Main Course	Vegetables	Beverages	Desserts
Ham	Potatoes	Coffee	Cake
Chicken	Peas	Tea	Pie
Fish	Green beans	Milk	Ice cream
Beef		Soda	

If one item is selected from each of the four groups, in how many ways can a meal be ordered? Describe two such orders.

25. You are taking a multiple-choice test that has five questions. Each of the questions has three choices, with one correct choice per question. If you select one of these options per question and leave nothing blank, in how many ways can you answer the questions?

26. You are taking a multiple-choice test that has eight questions. Each of the questions has three choices, with one correct choice per question. If you select one of these options per question and leave nothing blank, in how many ways can you answer the questions?

27. In the original plan for area codes in 1945, the first digit could be any number from 2 through 9, the second digit was either 0 or 1, and the third digit could be any number except 0. With this plan, how many different area codes were possible?

28. How many different four-letter radio station call letters can be formed if the first letter must be W or K?

29. Six performers are to present their comedy acts on a weekend evening at a comedy club. One of the performers insists on being the last stand-up comic of the evening. If this performer's request is granted, how many different ways are there to schedule the appearances?

30. Five singers are to perform at a night club. One of the singers insists on being the last performer of the evening. If this singer's request is granted, how many different ways are there to schedule the appearances?

31. In the *Cambridge Encyclopedia of Language* (Cambridge University Press, 1987), author David Crystal presents five sentences that make a reasonable paragraph regardless of their order. The sentences are:

> Mark had told him about the foxes.
> John looked out the window.
> Could it be a fox?
> However, nobody had seen one for months.
> He thought he saw a shape in the bushes.

How many different five-sentence paragraphs can be formed if the paragraph begins with "He thought he saw a shape in the bushes" and ends with "John looked out of the window"?

32. A television programmer is arranging the order that five movies will be seen between the hours of 6 P.M. and 4 A.M. Two of the movies have a G rating, and they are to be shown in the first two time blocks. One of the movies is rated NC-17, and it is to be shown in the last of the time blocks, from 2 A.M. until 4 A.M. Given these restrictions, in how many ways can the five movies be arranged during the indicated time blocks?

Use the formula for $_nP_r$ to solve Exercises 33–40.

33. A club with ten members is to choose three officers—president, vice-president, and secretary-treasurer. If each office is to be held by one person and no person can hold more than one office, in how many ways can those offices be filled?

34. A corporation has ten members on its board of directors. In how many different ways can it elect a president, vice-president, secretary, and treasurer?

35. For a segment of a radio show, a disc jockey can play 7 records. If there are 13 records to select from, in how many ways can the program for this segment be arranged?

36. Suppose you are asked to list, in order of preference, the three best movies you have seen this year. If you saw 20 movies during the year, in how many ways can the three best be chosen and ranked?

37. In a race in which six automobiles are entered and there are no ties, in how many ways can the first three finishers come in?

38. In a production of *West Side Story*, eight actors are considered for the male roles of Tony, Riff, and Bernardo. In how many ways can the director cast the male roles?

39. Nine bands have volunteered to perform at a benefit concert, but there is only enough time for five of the bands to play. How many lineups are possible?

40. How many arrangements can be made using four of the letters of the word COMBINE if no letter is to be used more than once?

Use the formula for $_nC_r$ to solve Exercises 41–48.

41. An election ballot asks voters to select three city commissioners from a group of six candidates. In how many ways can this be done?

42. A four-person committee is to be elected from an organization's membership of 11 people. How many different committees are possible?

43. Of 12 possible books, you plan to take 4 with you on vacation. How many different collections of 4 books can you take?

44. There are 14 standbys who hope to get seats on a flight, but only 6 seats are available on the plane. How many different ways can the 6 people be selected?

45. You volunteer to help drive children at a charity event to the zoo, but you can fit only 8 of the 17 children present in your van. How many different groups of 8 children can you drive?

46. Of the 100 people in the U.S. Senate, 18 serve on the Foreign Relations Committee. How many ways are there to select Senate members for this committee (assuming party affiliation is not a factor in selection)?

47. To win at LOTTO in the state of Florida, one must correctly select 6 numbers from a collection of 49 numbers (1 through 49). The order in which the selection is made does not matter. How many different selections are possible?

48. To win in the New York State lottery, one must correctly select 6 numbers from 54 numbers. The order in which the selection is made does not matter. How many different selections are possible?

In Exercises 49–58, solve by the method of your choice.

49. In a race in which six automobiles are entered and there are no ties, in how many ways can the first four finishers come in?

50. A book club offers a choice of 8 books from a list of 40. In how many ways can a member make a selection?

51. A medical researcher needs 6 people to test the effectiveness of an experimental drug. If 13 people have volunteered for the test, in how many ways can 6 people be selected?

52. Fifty people purchase raffle tickets. Three winning tickets are selected at random. If first prize is $1000, second prize is $500, and third prize is $100, in how many different ways can the prizes be awarded?

53. From a club of 20 people, in how many ways can a group of three members be selected to attend a conference?

54. Fifty people purchase raffle tickets. Three winning tickets are selected at random. If each prize is $500, in how many different ways can the prizes be awarded?

55. How many different four-letter passwords can be formed from the letters A, B, C, D, E, F, and G if no repetition of letters is allowed?

56. Nine comedy acts will perform over two evenings. Five of the acts will perform on the first evening and the order in which the acts perform is important. How many ways can the schedule for the first evening be made?

57. Using 15 flavors of ice cream, how many cones with three different flavors can you create if it is important to you which flavor goes on the top, middle, and bottom?

58. Baskin-Robbins offers 31 different flavors of ice cream. One of their items is a bowl consisting of three scoops of ice cream, each a different flavor. How many such bowls are possible?

Writing in Mathematics

59. Explain the Fundamental Counting Principle.

60. Write an original problem that can be solved using the Fundamental Counting Principle. Then solve the problem.

61. What is a permutation?

62. Describe what $_nP_r$ represents.

63. Write a word problem that can be solved by evaluating $_7P_3$.

64. What is a combination?

65. Explain how to distinguish between permutation and combination problems.

66. Write a word problem that can be solved by evaluating $_7C_3$.

Critical Thinking Exercises

67. Which one of the following is true?
   a. The number of ways to choose four questions out of ten questions on an essay test is $_{10}P_4$.
   b. If $r > 1$, $_nP_r$ is less than $_nC_r$.
   c. $_7P_3 = 3!\,_7C_3$
   d. The number of ways to pick a winner and first runner-up in a piano recital with 20 contestants is $_{20}C_2$.

68. Five men and five women line up at a checkout counter in a store. In how many ways can they line up if the first person in line is a woman, and the people in line alternate woman, man, woman, man, and so on?

69. How many four-digit odd numbers less than 6000 can be formed using the digits 2, 4, 6, 7, 8, and 9?

70. If a collection of n objects has n_1 identical objects of the same type, n_2 identical objects of a second kind, n_3 of a third kind, and so on for a total of $n = n_1 + n_2 + \cdots + n_k$ objects, the number of distinguishable permutations of the n objects is given by

$$\frac{n!}{n_1!\,n_2!\,n_3!\cdots n_k!}.$$

Use this formula to find the number of different signals consisting of eight flags that can be made using three white flags, four red flags and one blue flag.

Technology Exercises

71. Use a graphing utility with an $\boxed{_nP_r}$ key to verify your answers in Exercises 1–8.

72. Use a graphing utility with an $\boxed{_nC_r}$ key to verify your answers in Exercises 9–16.

Review Exercises

73. If $f(x) = x^2 + 2x - 5$ and $g(x) = 4x - 1$, find $(f \circ g)(x)$.
   (Section 12.2, Example 1)

74. Solve: $|2x - 5| > 3$.
   (Section 9.3, Example 4)

75. Give the center and radius. Then graph the equation:
$$x^2 + y^2 - 2x + 4y - 4 = 0.$$
   (Section 13.1, Example 6)

▶ SECTION 14.6   *Probability*

Objectives

1 Compute empirical probability.

2 Compute theoretical probability.

3 Find the probability that an event will not occur.

4 Find the probability of one event or a second event occurring.

5 Find the probability of one event and a second event occurring.

SSM   PH Tutor   CD-   Video
       Center   ROM

Table 14.2   Number of Americans and the Hours of Sleep They Get on a Typical Night

Hours of Sleep	Number of Americans, in millions
4 or less	11
5	24.75
6	68.75
7	82.5
8	74.25
9	8.25
10 or more	5.5
Total:	275

Source: Discovery Health Media

 Compute empirical probability.

How many hours of sleep do you typically get each night? Table 14.2 indicates that 11 million out of 275 million Americans are getting four hours or less sleep on a typical night. The *probability* of an American getting four hours or less sleep on a typical night is $\frac{11}{275}$. This fraction can be reduced to $\frac{1}{25}$, or expressed as 0.04 or 4%. Thus, 4% of Americans get four hours or less sleep each night.

We find a probability by dividing one number by another. Probabilities are assigned to an event, such as getting four hours or less sleep on a typical night. Events that are certain to occur are assigned probabilities of 1, or 100%. For example, the probability that a given individual will eventually die is 1. Regrettably, taxes and death are always certain! By contrast, if an event cannot occur, its probability is 0. For example, the probability that Elvis will return from the dead and serenade us with one final reprise of "Heartbreak Hotel" is 0.

Probabilities of events are expressed as numbers ranging from 0 to 1, or 0% to 100%. The closer the probability of a given event is to 1, the more likely it is that the event will occur. The closer the probability of a given event is to 0, the less likely it is that the event will occur.

Empirical Probability

Empirical probability applies to situations in which we observe how frequently an event occurs. We use the following formula to compute the empirical probability of an event:

Computing Empirical Probability

The **empirical probability** of event E is

$$P(E) = \frac{\text{observed number of times } E \text{ occurs}}{\text{total number of observed occurrences}}.$$

EXAMPLE 1   Computing Empirical Probability

An American is randomly selected. Use Table 14.2 to find the probability of that person getting eight hours sleep on a typical night.

Solution The probability of getting eight hours sleep is the observed number of Americans who do this, 74.25 million, divided by the total number of Americans, 275 million.

$$P(\text{eight hours sleep}) = \frac{\text{number of Americans who sleep 8 hours}}{\text{total numbers of Americans}}$$

$$= \frac{74.25}{275} = \frac{297}{1100} = 0.27$$

The empirical probability of randomly selecting an American who gets eight hours sleep on a typical night is $\frac{297}{1100}$, or 0.27. ∎

✔ **CHECK POINT 1** Use Table 14.2 to find the probability of randomly selecting an American who gets seven hours sleep on a typical night.

2 Compute theoretical probability.

Theoretical Probability

You toss a coin. Although it is equally likely to land either heads up, denoted by H, or tails up, denoted by T, the actual outcome is uncertain. Any occurrence for which the outcome is uncertain is called an **experiment**. Thus, tossing a coin is an example of an **experiment**. The set of all possible equally likely outcomes of an experiment is the **sample space** of the experiment, denoted by S. The sample space for the coin-tossing experiment is

$$S = \{H, T\}.$$

lands heads up   lands tails up

We can define an event more formally using these concepts. An **event**, denoted by E, is any subcollection, or subset, of a sample space. For example, the subset $E = \{T\}$ is the event of landing tails up when a coin is tossed.

Theoretical probability applies to situations like this, in which the sample space of all equally likely outcomes is known. To calculate the theoretical probability of an event, we divide the number of outcomes in the event by the number of outcomes in the sample space.

Computing Theoretical Probability

If an event E has $n(E)$ equally likely outcomes and its sample space S has $n(S)$ equally likely outcomes, the theoretical probability of event E, denoted by $P(E)$, is

$$P(E) = \frac{\text{number of outcomes in event } E}{\text{number of outcomes in sample space } S} = \frac{n(E)}{n(S)}.$$

The sum of the theoretical probabilities of all possible outcomes in the sample space is 1.

How can we use this formula to compute the probability of a coin landing tails up? We use the following sets:

$$E = \{T\} \qquad\qquad S = \{H, T\}.$$

This is the event of landing tails up.   This is the sample space with all equally possible outcomes.

The probability of a coin landing tails up is

$$P(E) = \frac{n(E)}{n(S)} = \frac{1}{2}.$$

Theoretical probability applies to many games of chance, including dice rolling, lotteries, card games, and roulette. The next example deals with the experiment of rolling a die. Figure 14.6 illustrates that when a die is rolled, there are six equally likely outcomes. The sample space can be shown as

$$S = \{1, 2, 3, 4, 5, 6\}.$$

Figure 14.6 Outcomes when a die is rolled

EXAMPLE 2   Computing Theoretical Probability

A die is rolled. Find the probability of getting a number less than 5.

Solution   The sample space of equally likely outcomes is $S = \{1, 2, 3, 4, 5, 6\}$. There are six outcomes in the sample space, so $n(S) = 6$.

We are interested in the probability of getting a number less than 5. The event of getting a number less than 5 can be represented by

$$E = \{1, 2, 3, 4\}.$$

There are four outcomes in this event, so $n(E) = 4$.

The probability of rolling a number less than 5 is

$$P(E) = \frac{n(E)}{n(S)} = \frac{4}{6} = \frac{2}{3}.$$ ∎

✔ **CHECK POINT 2**   A die is rolled. Find the probability of getting a number greater than 4.

EXAMPLE 3   Computing Theoretical Probability

Two ordinary six-sided dice are rolled. What is the probability of getting a sum of 8?

Solution   Each die has six equally likely outcomes. By the Fundamental Counting Principle, there are $6 \cdot 6$, or 36, equally likely outcomes in the sample space. That is, $n(S) = 36$. The 36 outcomes are shown here as ordered pairs. The five ways of rolling a sum of 8 appear in the highlighted diagonal as follows.

		Second Die					
		⚀	⚁	⚂	⚃	⚄	⚅
First Die	⚀	(1,1)	(1,2)	(1,3)	(1,4)	(1,5)	(1,6)
	⚁	(2,1)	(2,2)	(2,3)	(2,4)	(2,5)	(2,6)
	⚂	(3,1)	(3,2)	(3,3)	(3,4)	(3,5)	(3,6)
	⚃	(4,1)	(4,2)	(4,3)	(4,4)	(4,5)	(4,6)
	⚄	(5,1)	(5,2)	(5,3)	(5,4)	(5,5)	(5,6)
	⚅	(6,1)	(6,2)	(6,3)	(6,4)	(6,5)	(6,6)

$$S = \{(1, 1), (1, 2), (1, 3), (1, 4),$$
$$(1, 5), (1, 6), (2, 1), (2, 2),$$
$$(2, 3), (2, 4), (2, 5), (2, 6),$$
$$(3, 1), (3, 2), (3, 3), (3, 4),$$
$$(3, 5), (3, 6), (4, 1), (4, 2),$$
$$(4, 3), (4, 4), (4, 5), (4, 6),$$
$$(5, 1), (5, 2), (5, 3), (5, 4),$$
$$(5, 5), (5, 6), (6, 1), (6, 2),$$
$$(6, 3), (6, 4), (6, 5), (6, 6)\}$$

The phrase "getting a sum of 8" describes the event

$$E = \{(6, 2), (5, 3), (4, 4), (3, 5), (2, 6)\}.$$

This event has 5 outcomes, so $n(E) = 5$. Thus, the probability of getting a sum of 8 is

$$P(E) = \frac{n(E)}{n(S)} = \frac{5}{36}.$$

■

✔ **CHECK POINT 3**   What is the probability of getting a sum of 5 when two six-sided dice are rolled?

Computing Theoretical Probability Without Listing an Event and the Sample Space

In some situations, we can compute theoretical probability without having to write out each event and each sample space. For example, suppose you are dealt one card from a standard 52-card deck, illustrated in Figure 14.7. The deck has four suits: Hearts and diamonds are red, and clubs and spades are black. Each suit has 13 different face values—A(ace), 2, 3, 4, 5, 6, 7, 8, 9, 10, J(jack), Q(queen), and K(king). Jacks, queens, and kings are called **picture cards** or **face cards**.

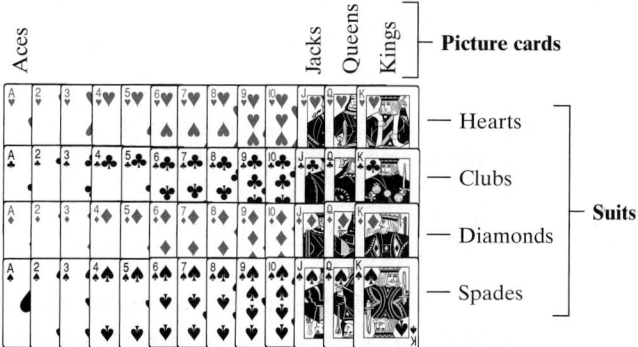

Figure 14.7 A standard 52-card bridge deck

━━━━

EXAMPLE 4   Probability and a Deck of 52 Cards

You are dealt one card from a standard 52-card deck. Find the probability of being dealt a heart.

Solution   Let E be the event of being dealt a heart. Because there are 13 hearts in the deck, the event of being dealt a heart can occur in 13 ways. The number of outcomes in event E is 13: $n(E) = 13$. With 52 cards in the deck, the total number of possible ways of being dealt a single card is 52. The number of outcomes in the sample space is 52: $n(S) = 52$. The probability of being dealt a heart is

$$P(E) = \frac{n(E)}{n(S)} = \frac{13}{52} = \frac{1}{4}.$$

■

✔ **CHECK POINT 4**   If you are dealt one card from a standard 52-card deck, find the probability of being dealt a king.

If your state has a lottery drawing each week, the probability that someone will win the top prize is relatively high. If there is no winner this week, it is virtually

State lotteries keep 50 cents on the dollar, resulting in $10 billion a year for public funding.

certain that eventually someone will be graced with millions of dollars. So how come you are unlucky compared to this undisclosed someone? In Example 5, we provide an answer to this question, using the counting principles discussed in Section 14.5.

EXAMPLE 5 Probability and Combinations: Winning the Lottery

Florida's lottery game, LOTTO, is set up so that each player chooses six different numbers from 1 to 53. If the six numbers chosen match the six numbers drawn randomly twice weekly, the player wins (or shares) the top cash prize. (As of this writing, the top cash prize has ranged from $7 million to $106.5 million.) With one LOTTO ticket, what is the probability of winning this prize?

Solution Because the order of the six numbers does not matter, this is a situation involving combinations. Let E be the event of winning the lottery with one ticket. With one LOTTO ticket, there is only one way of winning. Thus, $n(E) = 1$. The sample space is the set of all possible six-number combinations. We can use the combinations formula

$$_nC_r = \frac{n!}{(n - r)!\,r!}$$

to find the number of outcomes in the sample space. We are selecting $r = 6$ numbers from a collection of $n = 53$ numbers.

$$_{53}C_6 = \frac{53!}{(53 - 6)!6!} = \frac{53!}{47!6!} = \frac{53 \cdot 52 \cdot 51 \cdot 50 \cdot 49 \cdot 48 \cdot \cancel{47!}}{\cancel{47!} \cdot 6 \cdot 5 \cdot 4 \cdot 3 \cdot 2 \cdot 1} = 22,957,480$$

There are nearly 23 million number combinations possible in LOTTO. If a person buys one LOTTO ticket, the probability of winning is

$$P(E) = \frac{n(E)}{n(S)} = \frac{1}{22,957,480} \approx 0.0000000436.$$

The probability of winning the top prize with one LOTTO ticket is $\frac{1}{22,957,480}$, or about 1 in 23 million. ∎

In 2000, Americans spent nearly 20 billion dollars on lotteries set up by revenue-hungry states. If a pigeon, er, person, buys, say 5000 different tickets in Florida's LOTTO, that person has selected 5000 different combinations of the six numbers. The probability of winning is

$$\frac{5000}{22,957,480} \approx 0.000218.$$

The chances of winning top prize are about 218 in a million. At $1 per LOTTO ticket, it is highly probable that Mr. or Ms. Pigeon will be $5000 poorer.

✔ **CHECK POINT 5** In a state lottery, a player chooses five different numbers from 1 to 30. If the five numbers chosen match the five numbers drawn each week, the player wins (or shares) the top cash prize. With one lottery ticket, what is the probability of winning this prize?

3 Find the probability that an event will not occur.

Probability of an Event Not Occurring

A survey (*Source:* Penn, Schoen, and Berland, 1999) asked 500 Americans to rate their health. Of those surveyed, 270 rated their health as good/excellent. This means that $500 - 270$, or 230, people surveyed did not rate their health as good/excellent. Notice that

$$P(\text{good/excellent}) + P(\text{not good/excellent}) = \frac{270}{500} + \frac{230}{500} = \frac{500}{500} = 1.$$

In general, because the sum of the probabilities of all possible outcomes in any situation is 1,

$$P(E) + P(\text{not } E) = 1.$$

We now solve this equation for $P(\text{not } E)$, the probability that event E will not occur, by subtracting $P(E)$ from both sides. The resulting formula is given in the following box:

> **The Probability of an Event Not Occurring**
>
> The probability that an event E will not occur is equal to one minus the probability that it will occur.
>
> $$P(\text{not } E) = 1 - P(E)$$

EXAMPLE 6   The Probability of Not Winning the Lottery

We have seen that the probability of winning Florida's LOTTO with one ticket is $\frac{1}{22,957,480}$. What is the probability of not winning?

Solution

$$P(\text{not winning}) = 1 - P(\text{winning})$$

$$= 1 - \frac{1}{22,957,480} = \frac{22,957,480}{22,957,480} - \frac{1}{22,957,480}$$

$$= \frac{22,957,479}{22,957,480} \approx 0.99999996$$

The probability of not winning is close to 1. It is almost certain that with one LOTTO ticket, a person will not win top prize. ∎

✔ **CHECK POINT 6**   With one lottery ticket, what is the probability of not winning the lottery described in Check Point 5?

4 Find the probability of one event or a second event occurring.

Or Probabilities with Mutually Exclusive Events

Suppose that you randomly select one card from a deck of 52 cards. Let A be the event of selecting a king and B be the event of selecting a queen. Only one card is selected, so it is impossible to get both a king and a queen. The outcomes of selecting a king and a queen cannot occur simultaneously. They are called

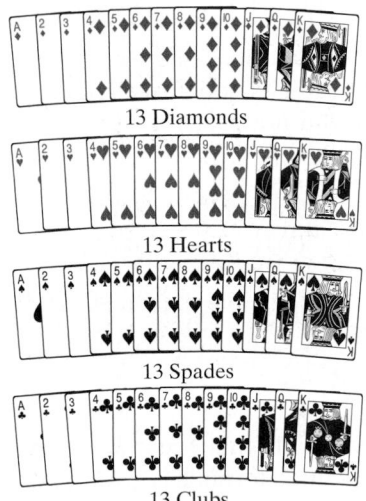

13 Diamonds

13 Hearts

13 Spades

13 Clubs

Figure 14.8 A deck of 52 cards

mutually exclusive events. If it is impossible for any two events, A and B, to occur simultaneously, they are said to be **mutually exclusive**. If A and B are mutually exclusive events, the probability that either A or B will occur is determined by adding their individual probabilities.

Or Probabilities with Mutually Exclusive Events

If A and B are mutually exclusive events, then

$$P(A \text{ or } B) = P(A) + P(B).$$

Using set notation, $P(A \cup B) = P(A) + P(B)$.

EXAMPLE 7 The Probability of Either of Two Mutually Exclusive Events Occurring

If one card is randomly selected from a deck of cards, what is the probability of selecting a king or a queen?

Solution We find the probability that either of these mutually exclusive events will occur by adding their individual probabilities.

$$P(\text{king or queen}) = P(\text{king}) + P(\text{queen}) = \frac{4}{52} + \frac{4}{52} = \frac{8}{52} = \frac{2}{13}$$

The probability of selecting a king or a queen is $\frac{2}{13}$. ∎

✔ **CHECK POINT 7** If you roll a single, six-sided die, what is the probability of getting either a 4 or a 5?

Or Probabilities with Events That Are Not Mutually Exclusive

Consider the deck of 52 cards shown in Figure 14.8. Suppose that these cards are shuffled and you randomly select one card from the deck. What is the probability of selecting a diamond or a picture card (jack, queen, king)? Begin by adding their individual probabilities:

$$P(\text{diamond}) + P(\text{picture card}) = \frac{13}{52} + \frac{12}{52}.$$

There are 13 diamonds in the deck of 52 cards.   There are 12 picture cards in the deck of 52 cards.

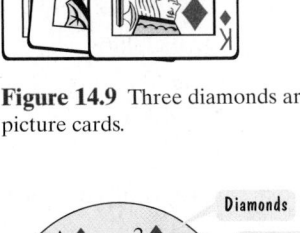

Figure 14.9 Three diamonds are picture cards.

However, this is not the probability of selecting a diamond or a picture card. The problem is that there are three cards that are simultaneously diamonds and picture cards, shown in Figure 14.9. The events of selecting a diamond and selecting a picture card are not mutually exclusive. It is possible to select a card that is both a diamond and a picture card.

The situation is illustrated in the diagram in Figure 14.10. Why can't we find the probability of selecting a diamond or a picture card by adding their individual probabilities? The diagram shows that three of the cards, the three diamonds that are picture cards, get counted twice when we add the individual probabilities. First the three cards get counted as diamonds, and then they get counted as picture cards. In order to avoid the error of counting the three cards twice, we need to subtract the probability of getting a diamond and a picture card, $\frac{3}{52}$, as follows:

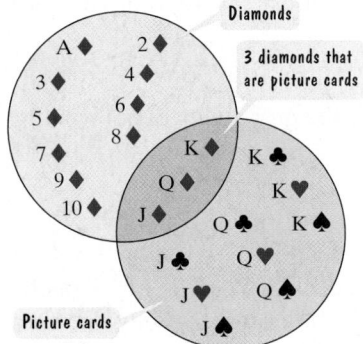

Figure 14.10

P(diamond or picture card)

$$= P(\text{diamond}) + P(\text{picture card}) - P(\text{diamond and picture card})$$

$$= \frac{13}{52} + \frac{12}{52} - \frac{3}{52} = \frac{13 + 12 - 3}{52} = \frac{22}{52} = \frac{11}{26}.$$

Thus, the probability of selecting a diamond or a picture card is $\frac{11}{26}$.

In general, if A and B are events that are not mutually exclusive, the probability that A or B will occur is determined by adding their individual probabilities and then subtracting the probability that A and B occur simultaneously.

Or Probabilities with Events That Are Not Mutually Exclusive

If A and B are not mutually exclusive events, then

$$P(A \text{ or } B) = P(A) + P(B) - P(A \text{ and } B).$$

Using set notation,

$$P(A \cup B) = P(A) + P(B) - P(A \cap B).$$

EXAMPLE 8   An Or Probability with Events That Are Not Mutually Exclusive

Figure 14.11 illustrates a spinner. It is equally probable that the pointer will land on any one of the eight regions, numbered 1 through 8. If the pointer lands on a borderline, spin again. Find the probability that the pointer will stop on an even number or a number greater than 5.

Solution   It is possible for the pointer to land on a number that is even and greater than 5. Two of the numbers, 6 and 8, are even and greater than 5. These events are not mutually exclusive. The probability of landing on a number that is even or greater than 5 is

Figure 14.11 It is equally probable that the pointer will land on any one of the eight regions.

$$P\left(\begin{array}{c}\text{even or} \\ \text{greater than 5}\end{array}\right) = P(\text{even}) + P(\text{greater than 5}) - P\left(\begin{array}{c}\text{even and} \\ \text{greater than 5}\end{array}\right)$$

$$= \quad \frac{4}{8} \quad + \quad \frac{3}{8} \quad - \quad \frac{2}{8}$$

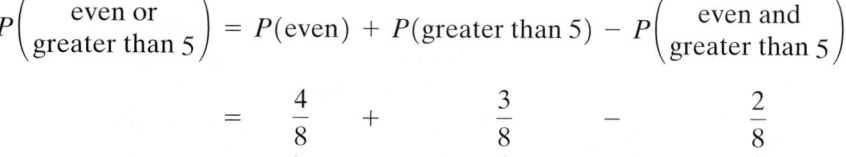

Four of the eight numbers, 2, 4, 6, and 8, are even.   Three of the eight numbers, 6, 7, and 8, are greater than 5.   Two of the eight numbers, 6, and 8, are even and greater than 5.

$$= \quad \frac{4 + 3 - 2}{8} = \frac{5}{8}.$$

The probability that the pointer will stop on an even number or a number greater than 5 is $\frac{5}{8}$. ∎

✔ **CHECK POINT 8**   Use Figure 14.11 to find the probability that the pointer will stop on an odd number or a number less than 5.

EXAMPLE 9   An *Or* Probability with Events That Are Not Mutually Exclusive

A group of people is comprised of 15 U.S. men, 20 U.S. women, 10 Canadian men, and 5 Canadian women. If a person is selected at random from the group, find the probability that the selected person is a man or a Canadian.

Solution   The group is comprised of $15 + 20 + 10 + 5$, or 50 people. It is possible to select a man who is Canadian. We are given that there are 10 Canadian men, so these events are not mutually exclusive.

$$P(\text{man or Canadian}) = P(\text{man}) + P(\text{Canadian}) - P(\text{man and Canadian})$$

$$= \frac{25}{50} + \frac{15}{50} - \frac{10}{50}$$

> Of the 50 people, 25 are men—15 U.S. men and 10 Canadian men.

> Of the 50 people, 15 are Canadian—10 Canadian men and 5 Canadian women.

> Of the 50 people, 10 are Canadian men.

$$= \frac{25 + 15 - 10}{50} = \frac{30}{50} = \frac{3}{5}$$

The probability of selecting a man or a Canadian is $\frac{3}{5}$. ∎

✔ **CHECK POINT 9**   In a group of 25 baboons, 18 enjoy picking fleas off their neighbors, 16 enjoy screeching wildly, while 10 enjoy picking fleas off their neighbors and screeching wildly. If one baboon is selected at random from the group, find the probability that it enjoys picking fleas off its neighbors or screeching wildly.

5 Find the probability of one event and a second event occurring.

And Probabilities with Independent Events

Suppose that you toss a fair coin two times in succession. The outcome of the first toss, heads or tails, does not affect what happens when you toss the coin a second time. For example, the occurrence of tails on the first toss does not make tails more likely or less likely to occur on the second toss. The repeated toss of a coin produces **independent events** because the outcome of one toss does not affect the outcome of others. Two events are *independent* if the occurrence of either of them has no effect on the probability of the other.

If two events are independent, we can calculate the probability of the first occurring and the second occurring by multiplying their probabilities.

> ### *And* Probabilities with Independent Events
>
> If A and B are independent events, then
>
> $$P(A \text{ and } B) = P(A) \cdot P(B).$$

EXAMPLE 10   Independent Events on a Roulette Wheel

Figure 14.12 shows a U.S. roulette wheel that has 38 numbered slots (1 through 36, 0, and 00). Of the 38 compartments, 18 are black, 18 are red, and 2 are green. Each play consists of spinning the wheel and a small ball in opposite directions.

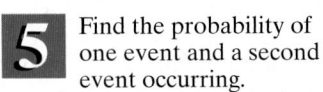

Figure 14.12 A U.S. roulette wheel

As the ball slows to a stop, it can land with equal probability on any one of the 38 numbered slots. Find the probability of red occurring on two consecutive plays.

Solution   The wheel has 38 equally likely outcomes and 18 are red. Thus, the probability of red occurring on a play is $\frac{18}{38}$, or $\frac{9}{19}$. The result that occurs on each play is independent of all previous results. Thus,

$$P(\text{red and red}) = P(\text{red}) \cdot P(\text{red}) = \frac{9}{19} \cdot \frac{9}{19} = \frac{81}{361} \approx 0.224.$$

The probability of red occurring on two consecutive plays is $\frac{81}{361}$. ∎

Some roulette players incorrectly believe that if red occurs on two consecutive plays, then another color is "due." Because the events are independent, the outcomes of previous spins have no effect on any other spins.

✔ **CHECK POINT 10**   Find the probability of green occurring on two consecutive plays on a roulette wheel.

The *and* rule for independent events can be extended to cover three or more events. Thus, if A, B, and C are independent events, then

$$P(A \text{ and } B \text{ and } C) = P(A) \cdot P(B) \cdot P(C).$$

EXAMPLE 11   Independent Events in a Family

The picture in the margin shows a family that has had nine girls in a row. Find the probability of this occurrence.

Solution   If two or more events are independent, we can find the probability of them all occurring by multiplying the probabilities. The probability of a baby girl is $\frac{1}{2}$, so the probability of nine girls in a row is $\frac{1}{2}$ used as a factor nine times.

$$P(\text{nine girls in a row}) = \frac{1}{2} \cdot \frac{1}{2} \cdot \frac{1}{2} \cdot \frac{1}{2} \cdot \frac{1}{2} \cdot \frac{1}{2} \cdot \frac{1}{2} \cdot \frac{1}{2} \cdot \frac{1}{2}$$

$$= \left(\frac{1}{2}\right)^9 = \frac{1}{512}$$

The probability of a run of nine girls in a row is $\frac{1}{512}$. (If another child is born into the family, this event is independent of the other nine, and the probability of a girl is still $\frac{1}{2}$.)

✔ **CHECK POINT 11**   Find the probability of a family having four boys in a row.

EXERCISE SET 14.6

  Practice and Application Exercises

Exercises 1–4 involve empirical probability. Use the empirical probability formula to solve each exercise. Express answers as fractions. Then use a calculator to express probabilities as decimals, rounded to the nearest thousandth.

Use the table showing U.S. family size to solve Exercises 1–2.

U.S. Families (includes only a householder and his/her relatives) by Size

Total: 70,241,000 Families	
Size	**Number of Families**
2 people	29,780,000
3 people	16,239,000
4 people	14,602,000
5 people	6,326,000
6 people	2,108,000
7 people or more	1,186,000

Source: U.S. Bureau of the Census

Find the probability that a U.S. family has:

1. 2 people. **2.** 3 people.

Use the table showing world population for selected regions to solve Exercises 3–4.

Populations of Selected Regions of the World

Total World Population: 5926 million	
Region	**Population in millions**
Africa	761
Near East	165
Asia	3363
Latin America	508
Europe	799
North America	301

Source: U.S. Bureau of the Census

If one person is randomly selected from all people on planet Earth, find the probability of selecting a person from:

3. Africa.

4. North America.

Exercises 5–20 involve theoretical probability. Use the theoretical probability formula to solve each exercise. Express each probability as a fraction reduced to lowest terms.

In Exercises 5–10, a die is rolled. The sample space of equally likely outcomes is {1, 2, 3, 4, 5, 6}. Find the probability of getting:

5. a 4. **6.** a 5.

7. an odd number.

8. a number greater than 3.

9. a number greater than 4.

10. a number greater than 7.

In Exercises 11–14, you are dealt one card from a standard 52 card deck. Find the probability of being dealt:

11. a queen. **12.** a diamond.

13. a picture card.

14. a card greater than 3 and less than 7.

In Exercises 15–16, a fair coin is tossed two times in succession. The sample space of equally likely outcomes is $\{HH, HT, TH, TT\}$. Find the probability of getting:

15. two heads.

16. the same outcome on each toss.

In Exercises 17–18, you select a family with three children. If M represents a male child and F a female child, the sample space of equally likely outcomes is $\{MMM, MMF, MFM, MFF, FMM, FMF, FFM, FFF\}$. Find the probability of selecting a family with:

17. at least one male child.

18. at least two female children.

In Exercises 19–20, a single die is rolled twice. The 36 equally likely outcomes are shown as follows:

		Second Roll					
		⚀	⚁	⚂	⚃	⚄	⚅
First Roll	⚀	(1, 1)	(1, 2)	(1, 3)	(1, 4)	(1, 5)	(1, 6)
	⚁	(2, 1)	(2, 2)	(2, 3)	(2, 4)	(2, 5)	(2, 6)
	⚂	(3, 1)	(3, 2)	(3, 3)	(3, 4)	(3, 5)	(3, 6)
	⚃	(4, 1)	(4, 2)	(4, 3)	(4, 4)	(4, 5)	(4, 6)
	⚄	(5, 1)	(5, 2)	(5, 3)	(5, 4)	(5, 5)	(5, 6)
	⚅	(6, 1)	(6, 2)	(6, 3)	(6, 4)	(6, 5)	(6, 6)

Find the probability of getting:

19. two numbers whose sum is 4.

20. two numbers whose sum is 6.

21. To play the California lottery, a person has to correctly select 6 out of 51 numbers, paying $1 for each six-number selection. If you pick six numbers that are the same as the ones drawn by the lottery, you win mountains of money. What is the probability that a person with one combination of six numbers will win? What is the probability of winning if 100 different lottery tickets are purchased?

22. A state lottery is designed so that a player chooses six numbers from 1 to 30 on one lottery ticket. What is the probability that a player with one lottery ticket will win? What is the probability of winning if 100 different lottery tickets are purchased?

23. A poker hand consists of five cards.
    a. Find the total number of possible five-card poker hands that can be dealt from a deck of 52 cards.

b. A diamond flush consists of a five-card hand containing all diamonds. Find the number of possible five-card diamond flushes.

c. Find the probability of being dealt a diamond flush.

24. A committee of five people is to be formed from six lawyers and seven teachers. Find the probability that all are lawyers.

Use these figures for the U.S. population in 2000 to answer Exercises 25–30.

Total U.S. Population: 274,634 Thousand

Age	under 5	5–13	14–17	18–24	25–34	35–44	45–64	65–84	85 and older
Population (in thousands)	18,987	36,043	15,752	26,258	37,233	44,659	60,992	30,378	4332

Source: U.S. Bureau of the Census

If a U.S. citizen is chosen at random, find the probability that this person is not:

25. under 5.

26. in the 18–24 age group.

27. in the 25–34 age group.

28. 85 and older.

Exercises 29–32 involve or probabilities with mutually exclusive events.

29. If a U.S. citizen is chosen at random, find the probability that this person is in the 14–17 or 18–24 age group.

30. If a U.S. citizen is chosen at random, find the probability that this person is in the 25–34 or 35–44 age group.

If one card is randomly selected from a standard 52-card deck, find the probability of selecting:

31. a 2 or a 3.   **32.** a red 7 or a black 8.

Exercises 33–40 involve or probabilities with events that are not mutually exclusive.

In Exercises 33–34, a single die is rolled. Find the probability of getting:

33. an even number or a number less than 5.

34. an odd number or a number less than 4.

In Exercises 35–36, you are dealt one card from a 52-card deck. Find the probability that you are dealt

35. a 7 or a red card.   **36.** a 5 or a black card.

In Exercises 37–38, it is equally probable that the pointer on the spinner shown will land on any one of the eight regions, numbered 1 through 8. If the pointer lands on a borderline, spin again.

Find the probability that the pointer will stop on:

37. an odd number or a number less than 6.

38. an odd number or a number greater than 3.

Use this information to solve Exercises 39–40. The mathematics department of a college has 8 male professors, 11 female professors, 14 male teaching assistants, and 7 female teaching assistants. If a person is selected at random from the group, find the probability that the selected person is:

39. a professor or a male.   **40.** a professor or a female.

Exercises 41–46 involve and probabilities with independent events.

In Exercises 41–44, a single die is rolled twice. Find the probability of getting:

41. a 2 the first time and a 3 the second time.

42. a 5 the first time and a 1 the second time.

43. an even number the first time and a number greater than 2 the second time.

44. an odd number the first time and a number less than 3 the second time.

45. If you toss a fair coin six times, what is the probability of getting all heads?

46. If you toss a fair coin seven times, what is the probability of getting all tails?

47. The probability that South Florida will be hit by a major hurricane (category 4 or 5) in any single year is $\frac{1}{16}$. (*Source:* National Hurricane Center)

    a. What is the probability that South Florida will be hit by a major hurricane two years in a row?

    b. What is the probability that South Florida will be hit by a major hurricane in three consecutive years?

    c. What is the probability that South Florida will not be hit by a major hurricane in the next ten years?

    d. What is the probability that South Florida will be hit by a major hurricane at least once in the next ten years?

Writing in Mathematics

48. Describe the difference between theoretical probability and empirical probability.

49. Give an example of an event whose probability must be determined empirically rather than theoretically.

50. Write a probability word problem whose answer is one of the following fractions: $\frac{1}{6}$ or $\frac{1}{4}$ or $\frac{1}{3}$.

51. Explain how to find the probability of an event not occurring. Give an example.

52. What are mutually exclusive events? Give an example of two events that are mutually exclusive.

53. Explain how to find or probabilities with mutually exclusive events. Give an example.

54. Give an example of two events that are not mutually exclusive.

55. Explain how to find *or* probabilities with events that are not mutually exclusive. Give an example.

56. Explain how to find *and* probabilities with independent events. Give an example.

57. The president of a large company with 10,000 employees is considering mandatory cocaine testing for every employee. The test that would be used is 90% accurate, meaning that it will detect 90% of the cocaine users who are tested, and that 90% of the nonusers will test negative. This also means that the test gives 10% false positive. Suppose that 1% of the employees actually use cocaine. Find the probability that someone who tests positive for cocaine use is, indeed, a user. (See the hint at the top of the next column.)

Hint: Find the following probability fraction:

$$\frac{\text{the number of employees who test positive and are cocaine users}}{\text{the number of employees who test positive}}.$$

This fraction is given by:

$$\frac{90\% \text{ of } 1\% \text{ of } 10,000}{\substack{\text{the number who test positive who actually use} \\ \text{cocaine plus the number who test positive} \\ \text{who do not use cocaine.}}}.$$

What does this probability indicate in terms of the percentage of employees who test positive who are not actually users? Discuss these numbers in terms of the issue of mandatory drug testing. Write a paper either in favor of or against mandatory drug testing, incorporating the actual percentage accuracy for such tests.

Critical Thinking Exercises

58. The target in the figure shown contains four squares. If a dart thrown at random hits the target, find the probability that it will land in a colored region.

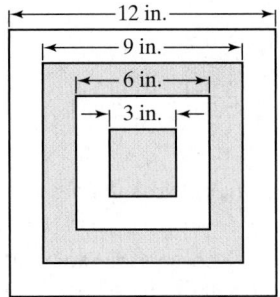

59. Suppose that it is a week in which the cash prize in Florida's LOTTO is promised to exceed $50 million. If a person purchases 13,983,816 tickets in LOTTO at $1 per ticket (all possible combinations), isn't this a guarantee of winning the lottery? Because the probability in this situation is 1, what's wrong with doing this?

60. **a.** If two people are selected at random, the probability that they do not have the same birthday (day and month) is $\frac{365}{365} \cdot \frac{364}{365}$. Explain why this is so. (Ignore leap years and assume 365 days in a year.)

    b. If three people are selected at random, find the probability that they all have different birthdays.

    c. If three people are selected at random, find the probability that at least two of them have the same birthday.

    d. If 20 people are selected at random, find the probability that at least 2 of them have the same birthday.

    e. How large a group is needed to give a 0.5 chance of at least two people having the same birthday?

Review Exercises

61. Graph: $4x^2 + 25y^2 = 100$. (Section 13.2, Example 2)

62. Solve: $\log_2(x + 5) + \log_2(x - 1) = 4$. (Section 12.5, Example 5)

63. Divide $x^3 + 5x^2 + 3x - 10$ by $x + 2$. (Section 5.6, Example 4)

CHAPTER 14 GROUP PROJECTS

1. Group members serve as a financial team analyzing the three options given to the professional baseball player described in the chapter opener on page 938. As a group, determine which option provides the most amount of money over the six-year contract and which provides the least. Describe one advantage and one disadvantage to each option.

2. Enough curiosities involving the Fibonacci sequence exist to warrant a flourishing Fibonacci Association, which publishes a quarterly journal. Do some research on the Fibonacci sequence by consulting the Internet or the research department of your library, and find one property that interests you. After doing this research, get together with your group to share these intriguing properties.

3. Members of your group have been hired by the Environmental Protection Agency to write a report on whether we are making significant progress in recovering solid waste. Use the models from Exercises 55 and 56 in Exercise Set 14.2 on page 956 as the basis for your report. A graph of each model from 1960 through the present would be helpful. What percentage of solid waste generated is actually recovered on a year-to-year basis? Be as creative as you want in your report and then draw conclusions. The group should write up the report and perhaps even include suggestions as to how we might improve recycling progress.

4. The group should select real-world situations where the Fundamental Counting Principle can be applied. These could involve the number of possible student ID numbers on your campus, the number of possible phone numbers in your community, the number of meal options at a local restaurant, the number of ways a person in the group can select outfits for class, the number of ways a condominium can be purchased in a nearby community, and so on. Once situations have been selected, group members should determine in how many ways each part of the task can be done. Group members will need to obtain menus, find out about telephone-digit requirements in the community, count shirts, pants, shoes in closets, visit condominium sales offices, and so on. Once the group reassembles, apply the Fundamental Counting Principle to determine the number of available options in each situation. Because these numbers may be quite large, use a calculator.

5. Research and present a group report on state lotteries. Include answers to some or all of the following questions: Which states do not have lotteries? Why not? How much is spent per capita on lotteries? What are some of the lottery games? What is the probability of winning top prize in these games? What income groups spend the greatest amount of money on lotteries? If your state has a lottery, what does it do with the money it makes? Is the way the money is spent what was promised when the lottery first began?

CHAPTER SUMMARY, REVIEW, AND TEST

SUMMARY

DEFINITIONS AND CONCEPTS	EXAMPLES

Section 14.1 Sequences and Summation Notation

An infinite sequence $\{a_n\}$ is a function whose domain is the set of positive integers. The function values, or terms, are represented by

$$a_1, a_2, a_3, a_4, \ldots, a_n, \ldots.$$

General term: $a_n = \dfrac{(-1)^n}{n^3}$.

$$a_1 = \frac{(-1)^1}{1^3} = -1, \quad a_2 = \frac{(-1)^2}{2^3} = \frac{1}{8},$$

$$a_3 = \frac{(-1)^3}{3^3} = -\frac{1}{27}, \quad a_4 = \frac{(-1)^4}{4^3} = \frac{1}{64}$$

The first four terms are $-1, \frac{1}{8}, -\frac{1}{27},$ and $\frac{1}{64}$.

Factorial Notation

$$n! = n(n-1)(n-2)\cdots(3)(2)(1) \quad \text{and} \quad 0! = 1$$

$$6! = 6 \cdot 5 \cdot 4 \cdot 3 \cdot 2 \cdot 1 = 720$$

$$3! = 3 \cdot 2 \cdot 1 = 6$$

Summation Notation

$$\sum_{i=1}^{n} a_i = a_1 + a_2 + a_3 + a_4 + \cdots + a_n$$

In the summation shown here, i is the index of summation, n is the upper limit of summation, and 1 is the lower limit of summation.

$$\sum_{i=3}^{7} (i^2 - 4)$$

$$= (3^2 - 4) + (4^2 - 4) + (5^2 - 4) + (6^2 - 4) + (7^2 - 4)$$

$$= (9 - 4) + (16 - 4) + (25 - 4) + (36 - 4) + (49 - 4)$$

$$= 5 + 12 + 21 + 32 + 45$$

$$= 115$$

Section 14.2 Arithmetic Sequences

In an arithmetic sequence, each term after the first differs from the preceding term by a constant, the common difference. Subtract any term from the term that directly follows to find the common difference.

General Term of an Arithmetic Sequence

The nth term (the general term) of an arithmetic sequence with first term a_1 and common difference d is

$$a_n = a_1 + (n-1)d.$$

Find the general term and the tenth term:

$$3, 7, 11, 15, \ldots.$$

$a_1 = 3 \quad d = 7 - 3 = 4$

Using $a_n = a_1 + (n-1)d$:

$$a_n = 3 + (n-1)4 = 3 + 4n - 4 = 4n - 1.$$

The general term is $a_n = 4n - 1$.
The tenth term is $a_{10} = 4 \cdot 10 - 1 = 39$.

The Sum of the First n Terms of an Arithmetic Sequence

The sum, S_n, of the first n terms of an arithmetic sequence is given by

$$S_n = \frac{n}{2}(a_1 + a_n)$$

in which a_1 is the first term and a_n is the nth term.

Find the sum of the first ten terms:

$$2, 5, 8, 11, \ldots.$$

$a_1 = 2 \quad d = 5 - 2 = 3$

First find a_{10}, the 10th term. Using $a_n = a_1 + (n-1)d$,

$$a_{10} = 2 + (10 - 1) \cdot 3 = 2 + 9 \cdot 3 = 29.$$

Find the sum of the first ten terms using

$$S_n = \frac{n}{2}(a_1 + a_n).$$

$$S_{10} = \frac{10}{2}(a_1 + a_{10}) = 5(2 + 29) = 5(31) = 155$$

DEFINITIONS AND CONCEPTS	EXAMPLES

Section 14.3 Geometric Sequences and Series

In a geometric sequence, each term after the first is obtained by multiplying the preceding term by a nonzero constant, the common ratio. Divide any term after the first by the term that directly precedes it to find the common ratio.

General Term of a Geometric Sequence
The nth term (the general term) of a geometric sequence with first term a_1 and common ratio r is

$$a_n = a_1 r^{n-1}.$$

Find the general term and the ninth term:

$$12, -6, 3, -\frac{3}{2}, \ldots.$$

$$a_1 = 12 \qquad r = \frac{-6}{12} = -\frac{1}{2}$$

Using $a_n = a_1 r^{n-1}$:

$$a_n = 12\left(-\frac{1}{2}\right)^{n-1} \text{ is the general term.}$$

The ninth term is

$$a_9 = 12\left(-\frac{1}{2}\right)^{9-1} = 12\left(-\frac{1}{2}\right)^8 = \frac{12}{256} = \frac{3}{64}.$$

The Sum of the First n Terms of a Geometric Sequence
The sum, S_n, of the first n terms of a geometric sequence is given by

$$S_n = \frac{a_1(1 - r^n)}{1 - r}$$

in which a_1 is the first term and r is the common ratio ($r \neq 1$).

Find $\sum_{i=1}^{8} 4 \cdot 3^i$

$$= 4 \cdot 3 + 4 \cdot 3^2 + 4 \cdot 3^3 + \cdots + 4 \cdot 3^8.$$

$$a_1 = 12 \qquad r = \frac{4 \cdot 3^2}{4 \cdot 3} = 3$$

Using $S_n = \dfrac{a_1(1 - r^n)}{1 - r}$:

$$S_8 = \frac{12(1 - 3^8)}{1 - 3} = 39{,}360.$$

The Sum of an Infinite Geometric Series
If $-1 < r < 1$ (equivalently, $|r| < 1$), then the sum of the infinite geometric series

$$a_1 + a_1 r + a_1 r^2 + a_1 r^3 + \cdots$$

in which a_1 is the first term and r is the common ratio is given by

$$S = \frac{a_1}{1 - r}.$$

If $|r| \geq 1$, the infinite series does not have a sum.

Find the sum:

$$6 + \frac{6}{3} + \frac{6}{3^2} + \frac{6}{3^3} + \cdots.$$

$$a_1 = 6 \qquad r = \frac{1}{3}$$

Using $S = \dfrac{a_1}{1 - r}$, the sum is

$$S = \frac{6}{1 - \frac{1}{3}} = \frac{6}{\frac{2}{3}} = 6 \cdot \frac{3}{2} = 9.$$

Section 14.4 The Binomial Theorem

Definition of a Binomial Coefficient

$$\binom{n}{r} = \frac{n!}{r!(n - r)!}$$

$$\binom{8}{3} = \frac{8!}{3!(8 - 3)!} = \frac{8!}{3!5!}$$

$$= \frac{8 \cdot 7 \cdot 6 \cdot 5!}{3 \cdot 2 \cdot 1 \cdot 5!} = 56$$

DEFINITIONS AND CONCEPTS	EXAMPLES

Section 14.4 The Binomial Theorem (continued)

A Formula for Expanding Binomials:
The Binomial Theorem
For any positive integer n,

$$(a + b)^n = \binom{n}{0}a^n + \binom{n}{1}a^{n-1}b +$$

$$\binom{n}{2}a^{n-2}b^2 + \binom{n}{3}a^{n-3}b^3 + \cdots + \binom{n}{n}b^n.$$

Expand: $(3x - y)^4 = [3x + (-y)]^4.$

$$= \binom{4}{0}(3x)^4 + \binom{4}{1}(3x)^3(-y)$$

$$+ \binom{4}{2}(3x)^2(-y)^2 + \binom{4}{3}(3x)^1(-y)^3 + \binom{4}{4}(-y)^4$$

$$= 1 \cdot 81x^4 + 4 \cdot 27x^3(-y) + 6 \cdot 9x^2y^2 + 4 \cdot 3x(-y^3) + 1 \cdot y^4$$

$$= 81x^4 - 108x^3y + 54x^2y^2 - 12xy^3 + y^4$$

Finding a Particular Term in a Binomial Expansion
The rth term of the expansion of $(a + b)^n$ is

$$\binom{n}{r-1}a^{n-r+1}b^{r-1}.$$

The 8th term of $(x + 2y)^{10}$ is

$$\binom{10}{8-1}x^{10-8+1}(2y)^{8-1}$$

$$= \binom{10}{7}x^3(2y)^7 = 120x^3 \cdot 128y^7$$

$$= 15,360x^3y.$$

Section 14.5 Counting Principles, Permutations, and Combinations

The Fundamental Counting Principle
The number of ways in which a series of successive things can occur is found by multiplying the number of ways in which each thing can occur.

How many ways can 6 applicants fill three different positions?

$$6 \cdot 5 \cdot 4 = 120 \text{ ways}$$

Permutations
A permutation from a group of items occurs when no item is used more than once and the order of arrangement makes a difference.
Permutations Formula: The number of possible permutations if r items are taken from n items is

$$_nP_r = \frac{n!}{(n-r)!}.$$

How many ways can 6 applicants fill three different positions?

$$_6P_3 = \frac{6!}{(6-3)!} = \frac{6!}{3!} = \frac{6 \cdot 5 \cdot 4 \cdot 3!}{3!}$$

$$= 6 \cdot 5 \cdot 4 = 120 \text{ ways}$$

Combinations
A combination from a group of items occurs when no item is used more than once and the order of items makes no difference.
Combinations Formula: The number of possible combinations if r items are taken from n items is

$$_nC_r = \frac{n!}{(n-r)!r!}.$$

How many different sets of 3 books can be selected from 6 books?

$$_6C_3 = \frac{6!}{(6-3)!3!} = \frac{6!}{3!3!}$$

$$= \frac{6 \cdot 5 \cdot 4 \cdot 3!}{3 \cdot 2 \cdot 1 \cdot 3!} = \frac{6 \cdot 5 \cdot 4}{3 \cdot 2 \cdot 1}$$

$$= 20 \text{ ways}$$

DEFINITIONS AND CONCEPTS	EXAMPLES

Section 14.6 Probability

Empirical Probability

Empirical probability applies to situations in which we observe the frequency of occurrence of an event. The empirical probability of event E is

$$P(E) = \frac{\text{observed number of times } E \text{ occurs}}{\text{total number of observed occurrences}}.$$

Teachers in U.S. Catholic High Schools

Total	Religious	Lay
47,730	4149	43,581

Source: National Catholic Education Association

The probability that a randomly selected U.S. Catholic high school teacher belongs to a religious order is

$$\frac{4149}{47,730} \approx 0.087.$$

Theoretical Probability

Theoretical probability applies to situations in which the sample space of all equally likely outcomes is known. The theoretical probability of event E is

$$P(E) = \frac{\text{number of outcomes in event } E}{\text{number of outcomes in sample space } S} = \frac{n(E)}{n(S)}.$$

Probability of an event not occurring:
$P(\text{not } E) = 1 - P(E)$.

A die is rolled.
$$S = \{1, 2, 3, 4, 5, 6\}$$

Probability of getting a number greater than 4
$\left(E = \{5, 6\}\right)$ is $\frac{2}{6} = \frac{1}{3}$.

Probability of not getting a number greater than 4 is
$1 - \frac{1}{3} = \frac{2}{3}$.

Or Probabilities

If it is impossible for events A and B to occur simultaneously, the events are mutually exclusive. If A and B are mutually exclusive events, then $P(A \text{ or } B) = P(A) + P(B)$.
If A and B are not mutually exclusive events, then $P(A \text{ or } B) = P(A) + P(B) - P(A \text{ and } B)$.

A die is rolled: $S = \{1, 2, 3, 4, 5, 6\}$.

Probability (2 or 5)

$$= P(2) + P(5) = \frac{1}{6} + \frac{1}{6} = \frac{2}{6} = \frac{1}{3}.$$

Probability (even or greater than 3)

$$= P(\text{even}) + P(>3) - P(\text{even and} >3)$$

$$= \frac{3}{6} + \frac{3}{6} - \frac{2}{6} = \frac{4}{6} = \frac{2}{3}.$$ This event is {4, 6}.

And Probabilities

Two events are independent if the occurrence of either of them has no effect on the probability of the other. If A and B are independent events, then

$$P(A \text{ and } B) = P(A) \cdot P(B).$$

The probability of a succession of independent events is the product of each of their probabilities.

A quiz contains six multiple-choice questions. Each question has four available options ($a, b, c,$ or d). If you guess at every question, the probability of answering all correctly is

$$\frac{1}{4} \cdot \frac{1}{4} \cdot \frac{1}{4} \cdot \frac{1}{4} = \frac{1}{256}.$$

Review Exercises

14.1 *In Exercises 1–4, write the first four terms of each sequence whose general term is given.*

1. $a_n = 7n - 4$

2. $a_n = (-1)^n \dfrac{n + 2}{n + 1}$

3. $a_n = \dfrac{1}{(n - 1)!}$

4. $a_n = \dfrac{(-1)^{n+1}}{2^n}$

In Exercises 5–6, find each indicated sum.

5. $\displaystyle\sum_{i=1}^{5} (2i^2 - 3)$

6. $\displaystyle\sum_{i=0}^{4} (-1)^{i+1} i!$

In Exercises 7–8, express each sum using summation notation. Use i for the index of summation.

7. $\dfrac{1}{3} + \dfrac{2}{4} + \dfrac{3}{5} + \cdots + \dfrac{15}{17}$

8. $4^3 + 5^3 + 6^3 + \cdots + 13^3$

14.2 *In Exercises 9–11, write the first six terms of each arithmetic sequence.*

9. $a_1 = 7, d = 4$

10. $a_1 = -4, d = -5$

11. $a_1 = \dfrac{3}{2}, d = -\dfrac{1}{2}$

In Exercises 12–14, use the formula for the general term (the nth term) of an arithmetic sequence to find the indicated term of each sequence.

12. Find a_6 when $a_1 = 5, d = 3$.

13. Find a_{12} when $a_1 = -8, d = -2$.

14. Find a_{14} when $a_1 = 14, d = -4$.

In Exercises 15–18, write a formula for the general term (the nth term) of each arithmetic sequence. Then use the formula for a_n to find a_{20}, the 20th term of the sequence.

15. $-7, -3, 1, 5, \ldots$

16. $a_1 = 200, d = -20$

17. $a_1 = -12, d = -\dfrac{1}{2}$

18. $15, 8, 1, -6, \ldots$

19. Find the sum of the first 22 terms of the arithmetic sequence: $5, 12, 19, 26, \ldots$.

20. Find the sum of the first 15 terms of the arithmetic sequence: $-6, -3, 0, 3, \ldots$.

21. Find $3 + 6 + 9 + \cdots + 300$, the sum of the first 100 positive multiples of 3.

In Exercises 22–24, use the formula for the sum of the first n terms of an arithmetic sequence to find the indicated sum.

22. $\displaystyle\sum_{i=1}^{16} (3i + 2)$

23. $\displaystyle\sum_{i=1}^{25} (-2i + 6)$

24. $\displaystyle\sum_{i=1}^{30} (-5i)$

25. In 1911, the world record for the men's mile run was 1043.04 seconds. The world record has decreased by approximately 0.4118 second each year since then.
  a. Write the general term for the arithmetic sequence modeling record times for the men's mile run n years after 1910.
  b. Use the model to predict the record time for the men's mile run for the year 2010.

26. A company offers a starting salary of $31,500 with raises of $2300 per year. Find the total salary over a ten-year period.

27. A theater has 25 seats in the first row and 35 rows in all. Each successive row contains one additional seat. How many seats are in the theater?

14.3 *In Exercises 28–31, write the first five terms of each geometric sequence.*

28. $a_1 = 3, r = 2$

29. $a_1 = \dfrac{1}{2}, r = \dfrac{1}{2}$

30. $a_1 = 16, r = -\dfrac{1}{4}$

31. $a_1 = -5, r = -1$

In Exercises 32–34, use the formula for the general term (the nth term) of a geometric sequence to find the indicated term of each sequence.

32. Find a_7 when $a_1 = 2, r = 3$.

33. Find a_6 when $a_1 = 16, r = \frac{1}{2}$.

34. Find a_5 when $a_1 = -3, r = 2$.

In Exercises 35–37, write a formula for the general term (the nth term) of each geometric sequence. Then use the formula for a_n to find a_8, the eighth term of the sequence.

35. $1, 2, 4, 8, \ldots$

36. $100, 10, 1, \frac{1}{10}, \ldots$

37. $12, -4, \frac{4}{3}, -\frac{4}{9}, \ldots$

38. Find the sum of the first 15 terms of the geometric sequence: $5, -15, 45, -135, \ldots$.

39. Find the sum of the first 7 terms of the geometric sequence: $8, 4, 2, 1, \ldots$.

In Exercises 40–42, use the formula for the sum of the first n terms of a geometric sequence to find the indicated sum.

40. $\displaystyle\sum_{i=1}^{6} 5^i$

41. $\displaystyle\sum_{i=1}^{7} 3(-2)^i$

42. $\displaystyle\sum_{i=1}^{5} 2\left(\frac{1}{4}\right)^{i-1}$

In Exercises 43–46, find the sum of each infinite geometric series.

43. $9 + 3 + 1 + \dfrac{1}{3} + \cdots$

44. $2 - 1 + \dfrac{1}{2} - \dfrac{1}{4} + \cdots$

45. $-6 + 4 - \dfrac{8}{3} + \dfrac{16}{9} - \cdots$

46. $\displaystyle\sum_{i=1}^{\infty} 5(0.8)^i$

In Exercises 47–48, express each repeating decimal as a fraction in lowest terms.

47. $0.\overline{6}$

48. $0.\overline{47}$

49. A job pays \$32,000 for the first year with an annual increase of 6% per year beginning in the second year. What is the salary in the sixth year? What is the total salary paid over this six-year period? Round answers to the nearest dollar.

50. You decide to deposit \$200 at the end of each month into an account paying 10% interest compounded monthly to save for your child's education. How much will you save over 18 years? Round to the nearest dollar.

51. A factory in an isolated town has an annual payroll of \$4 million. It is estimated that 70% of this money is spent within the town, that people in the town receiving this money will again spend 70% of what they receive in the town, and so on. What is the total of all this spending in the town each year?

14.4 *In Exercises 52–53, evaluate the given binomial coefficient.*

52. $\dbinom{11}{8}$

53. $\dbinom{90}{2}$

In Exercises 54–57, use the Binomial Theorem to expand each binomial and express the result in simplified form.

54. $(2x + 1)^3$

55. $(x^2 - 1)^4$

56. $(x + 2y)^5$

57. $(x - 2)^6$

In Exercises 58–59, write the first three terms in each binomial expansion, expressing the result in simplified form.

58. $(x^2 + 3)^8$

59. $(x - 3)^9$

In Exercises 60–61, find the term indicated in each expansion.

60. $(x + 2)^5$; fourth term

61. $(2x - 3)^6$; fifth term

14.5 *In Exercises 62–65, evaluate each expression.*

62. $_8P_3$

63. $_9P_5$

64. $_8C_3$

65. $_{13}C_{11}$

In Exercises 66–72, solve by the method of your choice.

66. A popular brand of pen comes in red, green, blue, or black ink. The writing tip can be chosen from extra bold, bold, regular, fine, or micro. How many different choices of pens do you have with this brand?

67. A stock can go up, go down, or stay unchanged. How many possibilities are there if you own five stocks?

68. A club with 15 members is to choose four officers—president, vice-president, secretary, and treasurer. In how many ways can these offices be filled?

69. How many different ways can a director select 4 actors from a group of 20 actors to attend a workshop on performing in rock musicals?

70. From the 20 CDs that you've bought during the past year, you plan to take 3 with you on vacation. How many different sets of three CDs can you take?

71. How many different ways can a director select from 20 male actors and cast the roles of Mark, Roger, Angel, and Collins in the musical *Rent*?

72. In how many ways can five airplanes line up for departure on a runway?

14.6 *Exercises 73–74 involve empirical probabilities. Express each probability as a fraction. Then use a calculator to express the probability in decimal form, rounded to the nearest thousandth. The table shows the two states with the largest Hispanic populations.*

Largest Hispanic Population

State	Total Population	Hispanic Population
California	31,878,234	9,630,188
Texas	19,128,261	5,503,372

Source: Bureau of the Census

Find the probability that:

73. a person randomly selected from California is Hispanic.

74. a person randomly selected from Texas is Hispanic.

In Exercises 75–76, a die is rolled. Find the probability of:

75. getting a number less than 5.
76. getting a number less than 3 or greater than 4.

In Exercises 77–78, you are dealt one card from a 52-card deck. Find the probability of:

77. getting an ace or a king.
78. getting a queen or a red card.

In Exercises 79–80, it is equally probable that the pointer on the spinner shown in the next column will land on any one of the six regions, numbered 1 through 6, and colored as shown. If the pointer lands on a borderline, spin again. Find the probability of:

79. not stopping on yellow.

80. stopping on red or a number greater than 3.

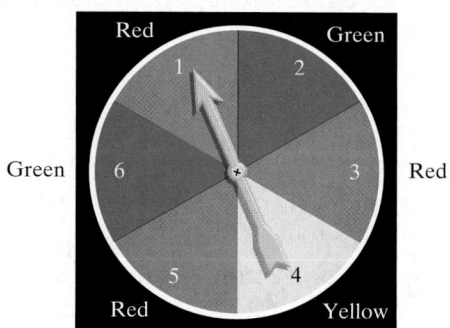

81. A lottery game is set up so that each player chooses five different numbers from 1 to 20. If the five numbers match the five numbers drawn in the lottery, the player wins (or shares) the top cash prize. What is the probability of winning the prize:

a. with one lottery ticket?

b. with 100 different lottery tickets?

Use this information to solve Exercises 82–83. At a workshop on police work and the black community, there are 50 black male police officers, 20 black female police officers, 90 white male police officers, and 40 white female police officers. If one police officer is selected at random from the people at the workshop, find the probability that the selected person is:

82. black or male.

83. female or white.

84. What is the probability of a family having five boys born in a row?

85. The probability of a flood in any given year in a region prone to floods is 0.2.

a. What is the probability of a flood two years in a row?

b. What is the probability of a flood for three consecutive years?

c. What is the probability of no flooding for four consecutive years?

Chapter 14 Test

1. Write the first five terms of the sequence whose general term is $a_n = \dfrac{(-1)^{n+1}}{n^2}$.

2. Find the indicated sum: $\displaystyle\sum_{i=1}^{5} (i^2 + 10)$.

3. Express the sum using summation notation. Use i for the index of summation.

$$\frac{2}{3} + \frac{3}{4} + \frac{4}{5} + \cdots + \frac{21}{22}$$

In Exercises 4–5, write a formula for the general term (the nth term) of each sequence. Then use the formula to find the twelfth term of the sequence.

4. $4, 9, 14, 19, \ldots$

5. $16, 4, 1, \frac{1}{4}, \ldots$

In Exercises 6–7, use the formula for the sum of the first n terms of an arithmetic sequence.

6. Find the sum of the first ten terms of the arithmetic sequence: $-7, -14, -21, -28, \ldots$.

7. Find $\displaystyle\sum_{i=1}^{20} (3i - 4)$.

In Exercises 8–9, use the formula for the sum of the first n terms of a geometric sequence.

8. Find the sum of the first ten terms of the geometric sequence: $7, -14, 28, -56, \ldots$.

9. Find $\displaystyle\sum_{i=1}^{15} (-2)^i$.

10. Find the sum of the infinite geometric series:

$$4 + \frac{4}{2} + \frac{4}{2^2} + \frac{4}{2^3} + \cdots.$$

11. Express $0.\overline{73}$ in fractional notation.

12. A job pays \$30,000 for the first year with an annual increase of 4% per year beginning in the second year. What is the total salary paid over an eight-year period? Round to the nearest dollar.

13. Evaluate: $\dbinom{9}{2}$.

14. Use the Binomial Theorem to expand and simplify: $(x^2 - 1)^5$.

15. Use the Binomial Theorem to write the first three terms in the expansion and simplify: $(x + y^2)^8$.

16. A human resource manager has 11 applicants to fill three different positions. Assuming that all applicants are equally qualified for any of the three positions, in how many ways can this be done?

17. From the ten books that you've recently bought but not read, you plan to take four with you on vacation. How many different sets of four books can you take?

18. How many seven-digit local telephone numbers can be formed if the first three digits are 279?

19. What is the difference between a permutation and a combination?

20. A lottery game is set up so that each player chooses six different numbers from 1 to 15. If the six numbers match the six numbers drawn in the lottery, the player wins (or shares) the top cash prize. What is the probability of winning the prize with 50 different lottery tickets?

21. One card is randomly selected from a deck of 52 cards. Find the probability of selecting a black card or a picture card.

22. A group of students consists of 10 male freshmen, 15 female freshmen, 20 male sophomores, and 5 female sophomores. If one person is randomly selected from the group, find the probability of selecting a freshman or a female.

23. A quiz consisting of four multiple-choice questions has four available options ($a, b, c,$ or d) for each question. If a person guesses at every question, what is the probability of answering all questions correctly?

24. If the spinner shown is spun twice, find the probability that the pointer lands on red on the first spin and blue on the second spin.

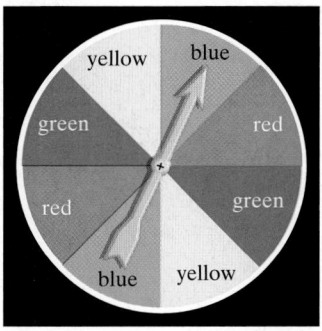

Cumulative Review Exercises (Covering the Entire Book)

In Exercises 1–17, solve each equation, inequality, or system.

1. $9(x - 1) = 1 + 3(x - 2)$

2. $3x + 4y = -7$
  $x - 2y = -9$

3. $x - y + 3z = -9$
  $2x + 3y - z = 16$
  $5x + 2y - z = 15$

4. $7x + 18 \leq 9x - 2$

5. $4x - 3 < 13$ and $-3x - 4 \geq 8$

6. $2x + 4 > 8$ or $x - 7 \geq 3$

7. $|2x - 1| < 5$

8. $\left|\frac{2}{3}x - 4\right| = 2$

9. $\dfrac{4}{x-3} - \dfrac{6}{x+3} = \dfrac{24}{x^2-9}$

10. $\sqrt{x+4} - \sqrt{x-3} = 1$

11. $2x^2 = 5 - 4x$

12. $x^{\frac{2}{3}} - 5x^{\frac{1}{3}} + 6 = 0$

13. $2x^2 + x - 6 \le 0$

14. $\log_8 x + \log_8(x+2) = 1$

15. $5^{2x+3} = 125$

16. $2x^2 - 3y^2 = 5$
$3x^2 + 4y^2 = 16$

17. $2x^2 - y^2 = -8$
$x - y = 6$

In Exercises 18–24, graph each function, equation, or inequality in a rectangular coordinate system.

18. $x - 3y = 6$

19. $f(x) = \dfrac{1}{2}x - 1$

20. $3x - 2y > -6$

21. $f(x) = -2(x-3)^2 + 2$

22. $\dfrac{x^2}{16} + \dfrac{y^2}{4} = 1$

23. $y = \log_2 x$

24. $x^2 - y^2 = 9$

In Exercises 25–35, perform the indicated operations, and simplify, if possible.

25. $4[2x - 6(x - y)]$

26. $(-5x^3y^2)(4x^4y^{-6})$

27. $(8x^2 - 9xy - 11y^2) - (7x^2 - 4xy + 5y^2)$

28. $(3x - 1)(2x + 5)$

29. $(3x^2 - 4y)^2$

30. $\dfrac{3x}{x+5} - \dfrac{2}{x^2 + 7x + 10}$

31. $\dfrac{1 - \dfrac{9}{x^2}}{1 + \dfrac{3}{x}}$

32. $\dfrac{x^2 - 6x + 8}{3x + 9} \div \dfrac{x^2 - 4}{x + 3}$

33. $\sqrt{5xy} \cdot \sqrt{10x^2y}$

34. $4\sqrt{72} - 3\sqrt{50}$

35. $(5 + 3i)(7 - 3i)$

In Exercises 36–38, factor completely.

36. $81x^4 - 1$

37. $24x^3 - 22x^2 + 4x$

38. $x^3 + 27y^3$

In Exercises 39–42, let $f(x) = x^2 + 3x - 15$ and $g(x) = x - 2$. Find each indicated expression.

39. $(f - g)(x)$ and $(f - g)(5)$

40. $\left(\dfrac{f}{g}\right)(x)$ and the domain of $\dfrac{f}{g}$

41. $f(g(x))$

42. $g(f(x))$

43. If $f(x) = 7x - 3$, find $f^{-1}(x)$.

44. Find the distance between $(6, -1)$ and $(-3, -4)$. Round to two decimal places.

45. Divide using synthetic division:

$(3x^3 - x^2 + 4x + 8) \div (x + 2)$.

46. Solve for t: $A = \dfrac{5r + 2}{t}$.

47. Write the slope-intercept form of the equation of the line through $(-2, 5)$ and parallel to the line whose equation is $3x + y = 9$.

48. Evaluate the determinant: $\begin{vmatrix} -2 & -4 \\ 5 & 7 \end{vmatrix}$.

49. Write as a single logarithm whose coefficient is 1:

$2\ln x - \dfrac{1}{2}\ln y$.

50. Find the indicated sum: $\displaystyle\sum_{i=2}^{5}(i^3 - 4)$.

51. Find the sum of the first 30 terms of the arithmetic sequence: $2, 6, 10, 14, \ldots$.

52. Express $0.\overline{3}$ as a fraction in lowest terms.

53. Use the Binomial Theorem to expand and simplify: $(2x - y^3)^4$.

In Exercises 54–55, find the domain of each function.

54. $f(x) = \dfrac{x - 2}{x^2 - 3x + 2}$

55. $f(x) = \ln(2x - 8)$

56. The price of a computer is reduced by 30% to $434. What was the original price?

57. The area of a rectangle is 52 square yards. The length of the rectangle is 1 yard longer than 3 times its width. Find the rectangle's dimensions.

58. You invested $4000 in two stocks paying 12% and 14% annual interest, respectively. At the end of the year, the total interest from these investments was $508. How much was invested at each rate?

59. Use the formula for continuous compounding, $A = Pe^{rt}$, to solve this problem. What interest rate is required for an investment of $6000 subject to continuous compounding to grow to $18,000 in 10 years?

60. The current, I, in amperes, flowing in an electrical circuit varies inversely as the resistance, R, in ohms, in the circuit. When the resistance of an electric percolator is 22 ohms, it draws 5 amperes of current. How much current is needed when the resistance is 10 ohms?

APPENDIX A

Are You Prepared for Intermediate Algebra?

Algebra is cumulative. This means that **your performance in intermediate algebra depends heavily on the skills you acquired in introductory algebra.** Do you need a quick review of introductory algebra topics before starting the intermediate algebra portion of this book? This appendix provides a fast way to review and practice the prerequisite skills needed in intermediate algebra.

Study Tip

> You can quickly review the major topics of introductory algebra by studying the review grids for each of the first seven chapter in this book. Each chart summarizes the definitions and concepts in every section of the chapter. Examples that illustrate the key concepts are also included in the chart. The review charts, with illustrative examples, for each of the book's first seven chapters begin on pages 90, 173, 236, 287, 364, 424, and 506.

A Diagnostic Test for Your Introductory Algebra Skills. You can use the 36 exercises that begin on the next page to test your understanding of introductory algebra topics. These exercises cover the fundamental algebra skills upon which the intermediate algebra portion of this book is based. Here are some suggestions for using these exercises as a diagnostic test:

1. Work through all 36 items at your own pace.

2. Use the answer section in the back of the book to check your work.

3. If your answer differs from that in the answer section or if you are not certain how to proceed with a particular item, turn to the section and the illustrative example given in parentheses at the end of each exercise. Study the step-by-step solution of the example that parallels the exercise and then try working the exercise again. If you feel that you need more assistance, study the entire section in which the example appears and work on a selected group of exercises in the exercise set for that section.

In Exercises 1–7, solve each equation, inequality, or system of equations.

1. $2 - 4(x + 2) = 5 - 3(2x + 1)$
(Section 2.3, Example 3)

2. $\dfrac{x}{2} - 3 = \dfrac{x}{5}$ (Section 2.3, Example 4)

3. $3x + 9 \geq 5(x - 1)$ (Section 2.7, Example 7)

4. $2x + 3y = 6$
$x + 2y = 5$
(Section 4.3, Example 2)

5. $3x - 2y = 1$
$y = 10 - 2x$
(Section 4.2, Example 1)

6. $\dfrac{3}{x + 5} - 1 = \dfrac{4 - x}{2x + 10}$
(Section 7.6, Example 4)

7. $x + \dfrac{6}{x} = -5$
(Section 7.6, Example 3)

In Exercises 8–16, perform the indicated operations. If possible, simplify the answer.

8. $\dfrac{12x^3}{3x^{12}}$ (Section 5.7, Example 4)

9. $4 \cdot 6 \div 2 \cdot 3 + (-5)$ (Section 1.8, Example 4)

10. $(6x^2 - 8x + 3) - (-4x^2 + x - 1)$
(Section 5.1, Example 3)

11. $(7x + 4)(3x - 5)$
(Section 5.3, Example 2)

12. $(5x - 2)^2$ (Section 5.3, Example 6)

13. $(x + y)(x^2 - xy + y^2)$
(Section 5.4, Example 8)

14. $\dfrac{x^2 + 6x + 8}{x^2} \div (3x^2 + 6x)$

(Section 7.2, Example 7)

15. $\dfrac{x}{x^2 + 2x - 3} - \dfrac{x}{x^2 - 5x + 4}$

(Section 7.4, Example 7)

16. $\dfrac{x - \dfrac{1}{5}}{5 - \dfrac{1}{x}}$ (Section 7.5, Examples 2 and 5)

In Exercises 17–22, factor completely.

17. $4x^2 - 49$ (Section 6.4, Example 1)

18. $x^3 + 3x^2 - x - 3$
(Section 6.5, Example 4)

19. $2x^2 + 8x - 42$
(Section 6.5, Example 2)

20. $x^5 - 16x$
(Section 6.4, Example 4)

21. $x^3 - 10x^2 + 25x$
(Section 6.4, Example 5)

22. $x^3 - 8$
(Section 6.4, Example 8)

In Exercises 23–25, graph each equation in a rectangular coordinate system.

23. $y = \dfrac{1}{3}x - 1$ (Section 3.4, Example 3)

24. $3x + 2y = -6$ (Section 3.2, Example 4)

25. $y = -2$ (Section 3.2, Example 7)

26. Find the slope of the line passing through the points $(-1, 3)$ and $(2, -3)$.   (Section 3.3, Example 1)

27. Write the point-slope form of the equation of the line passing through the points $(1, 2)$ and $(3, 6)$. Then use the point-slope equation to write the slope-intercept form of the line's equation.   (Section 3.5, Example 2)

In Exercises 28–36, use an equation or a system of equations to solve each problem.

28. Seven subtracted from five times a number is 208. Find the number.   (Section 2.5, Example 2)

29. After a 20% reduction, a digital camera sold for $256. What was the price before the reduction?
(Section 2.5, Example 7)

30. A rectangular field is three times as long as it is wide. If the perimeter of the field is 400 yards, what are the field's dimensions?   (Section 2.5, Example 6)

31. You invest $20,000 in two accounts paying 7% and 9% annual interest, respectively. If the total interest earned for the year is $1550, how much was invested at each rate? (Section 4.4, Example 4)

32. A chemist needs to mix a 40% acid solution with a 70% acid solution to obtain 12 liters of a 50% acid solution. How many liters of each solution should be used?   (Section 4.4, Example 5)

33. A sailboat has a triangular sail with an area of 120 square feet and a base that is 15 feet long. Find the height of the sail. (Section 2.6, Example 1)

34. In a triangle, the measure of the first angle is 10° more than the measure of the second angle. The measure of the third angle is 20° more than four times that of the second angle. What is the measure of each angle? (Section 2.6, Example 6)

35. A salesperson works in the TV and stereo department of an electronics store. One day she sold 3 TVs and 4 stereos for $2530. The next day, she sold 4 of the same TVs and 3 of the same stereos for $2510. Find the price of a TV and a stereo. (Section 4.4, Example 1)

36. The length of a rectangle is 6 meters more than the width. The area is 55 square meters. Find the rectangle's dimensions. (Section 6.6, Example 8)

APPENDIX B

Where Did That Come From? Selected Proofs

▶ **SECTION 12.4** **Properties of Logarithms**

The Product Rule

Let b, and N be positive real numbers with $b \neq 1$.

$$\log_b(MN) = \log_b M + \log_b N$$

Proof

We begin by letting $\log_b M = R$ and $\log_b N = S$.
    Now we write each logarithm in exponential form.

$$\log_b M = R \quad \text{means} \quad b^R = M.$$

$$\log_b N = S \quad \text{means} \quad b^S = N.$$

By substituting and using a property of exponents, we see that

$$MN = b^R b^S = b^{R+S}$$

Now we change $MN = b^{R+S}$ to logarithmic form.

$$MN = b^{R+S} \quad \text{means} \quad \log_b(MN) = R + S.$$

Finally, substituting $\log_b M$ for R and $\log_b N$ for S gives us

$$\log_b(MN) = \log_b M + \log_b N,$$

the property that we wanted to prove.

The quotient and power rules for logarithms are proved using similar procedures.

The Change-of-Base Property

For any logarithmic bases a and b, and any positive number M,

$$\log_b M = \frac{\log_a M}{\log_a b}.$$

Proof

To prove the change-of-base property, we let x equal the logarithm on the left side:

$$\log_b M = x.$$

Now we rewrite this logarithm in exponential form.

$$\log_b M = x \quad \text{means} \quad b^x = M.$$

Because b^x and M are equal, the logarithms with base a for each of these expressions must be equal. This means that

$$\log_a b^x = \log_a M$$

$$x \log_a b = \log_a M \qquad \text{Apply the power rule for logarithms on the left side.}$$

$$x = \frac{\log_a M}{\log_a b} \qquad \text{Solve for x by dividing both sides by } \log_a b.$$

In our first step we let x equal $\log_b M$. Replacing x on the left side by $\log_b M$ gives us

$$\log_b M = \frac{\log_a M}{\log_a b},$$

which is the change-of-base property.

▶ **SECTION 13.3**   *The Hyperbola*

The Asymptotes of a Hyperbola Centered at the Origin

The hyperbola

$$\frac{x^2}{a^2} - \frac{y^2}{b^2} = 1$$

with a horizontal transverse axis has the two asymptotes

$$y = \frac{b}{a}x \quad \text{and} \quad y = -\frac{b}{a}x.$$

Proof

Begin by solving the hyperbola's equation for y.

$$\frac{x^2}{a^2} - \frac{y^2}{b^2} = 1 \qquad \text{This is the standard form of the equation of a hyperbola.}$$

$$\frac{y^2}{b^2} = \frac{x^2}{a^2} - 1 \qquad \text{We isolate the term involving } y^2 \text{ to solve for y.}$$

$$y^2 = \frac{b^2 x^2}{a^2} - b^2 \qquad \text{Multiply both sides by } b^2.$$

$$y^2 = \frac{b^2 x^2}{a^2}\left(1 - \frac{a^2}{x^2}\right) \qquad \text{Factor out } \frac{b^2 x^2}{a^2} \text{ on the right. Verify that this result is correct by multiplying using the distributive property and obtaining the previous step.}$$

$$y = \pm\sqrt{\frac{b^2 x^2}{a^2}\left(1 - \frac{a^2}{x^2}\right)} \qquad \text{Solve for y using the square root property: If } u^2 = d, u = \pm\sqrt{d}.$$

$$y = \pm\frac{b}{a}x\sqrt{1 - \frac{a^2}{x^2}} \qquad \text{Simplify.}$$

As x increases or decreases without bound, the value of $\dfrac{a^2}{x^2}$ approaches 0. Consequently, the value of y can be approximated by

$$y = \pm \frac{b}{a}x.$$

This means that the lines whose equations are $y = \dfrac{b}{a}x$ and $y = -\dfrac{b}{a}x$ are asymptotes for the graph of the hyperbola.

Answers to Selected Exercises

CHAPTER 1

Section 1.1

Check Point Exercises

1. a. $\frac{2}{3}$ **b.** $\frac{7}{4}$ **c.** $\frac{13}{15}$ **d.** $\frac{1}{5}$ **2. a.** $\frac{8}{33}$ **b.** $\frac{18}{5}$ **c.** $\frac{2}{7}$ **3. a.** $\frac{10}{3}$ **b.** $\frac{2}{9}$ **4. a.** $\frac{5}{11}$ **b.** $\frac{2}{3}$ **5.** $\frac{14}{21}$

6. a. $\frac{11}{10}$ **b.** $\frac{7}{12}$ **c.** $\frac{1}{2}$ **7.** $\frac{1}{6}$

Exercise Set 1.1

1. $\frac{5}{8}$ **3.** $\frac{5}{6}$ **5.** $\frac{7}{10}$ **7.** $\frac{2}{5}$ **9.** $\frac{22}{25}$ **11.** $\frac{60}{43}$ **13.** $\frac{2}{15}$ **15.** $\frac{21}{88}$ **17.** $\frac{36}{7}$ **19.** $\frac{1}{12}$ **21.** $\frac{15}{14}$ **23.** $\frac{15}{16}$ **25.** $\frac{9}{5}$

27. $\frac{5}{9}$ **29.** 3 **31.** $\frac{7}{10}$ **33.** $\frac{1}{2}$ **35.** $\frac{6}{11}$ **37.** $\frac{2}{3}$ **39.** $\frac{5}{4}$ **41.** $\frac{1}{6}$ **43.** 2 **45.** $\frac{7}{10}$ **47.** $\frac{9}{10}$ **49.** $\frac{19}{24}$ **51.** $\frac{7}{18}$

53. $\frac{7}{12}$ **55.** $\frac{41}{80}$ **57.** $\frac{13}{4} + \frac{13}{9} = \frac{169}{36}$ and $\frac{13}{4} \times \frac{13}{9} = \frac{169}{36}$ **59.** $\frac{16}{25}$ **61.** $\frac{19}{50}$ **63.** 285 black teenagers **65.** $\frac{3}{8}$ c

67. $\frac{1}{3}$ of the business **69.** $588.00 **71–75.** Answers will vary. **77.** d

79.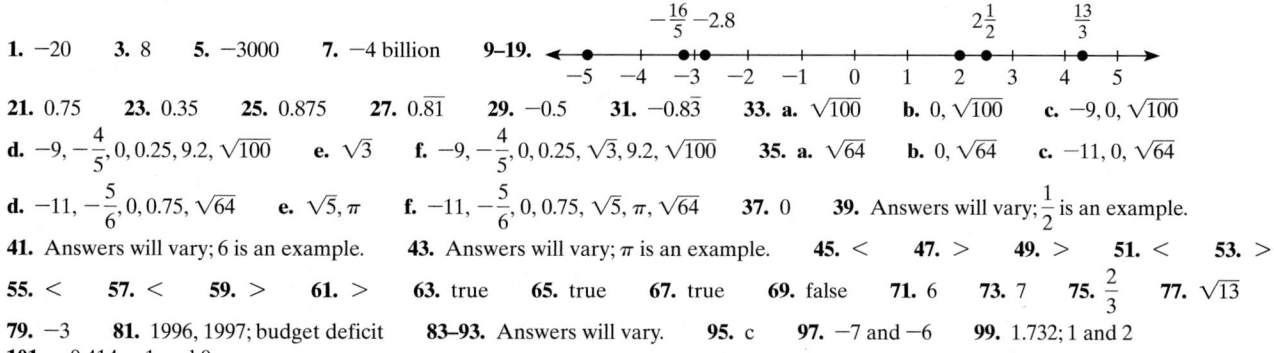

say does that Star-span-gled Ban-ner yet wave O'er the

81. $\frac{53}{120}$ **83.** $\frac{9}{40}$

Section 1.2

Check Point Exercises

1. a. -500 **b.** -282 **2.**  **3.**

4. a. 0.375 **b.** $0.\overline{45}$ **5. a.** $\sqrt{9}$ **b.** $0, \sqrt{9}$ **c.** $-9, 0, \sqrt{9}$ **d.** $-9, -1.3, 0, 0.\overline{3}, \sqrt{9}$ **e.** $\frac{\pi}{2}, \sqrt{10}$ **f.** $-9, -1.3, 0, 0.\overline{3}, \frac{\pi}{2}, \sqrt{9}, \sqrt{10}$

6. a. $>$ **b.** $<$ **c.** $<$ **d.** $<$ **7. a.** true **b.** true **c.** false **8. a.** 4 **b.** 6 **c.** $\sqrt{2}$

Exercise Set 1.2

1. -20 **3.** 8 **5.** -3000 **7.** -4 billion **9–19.**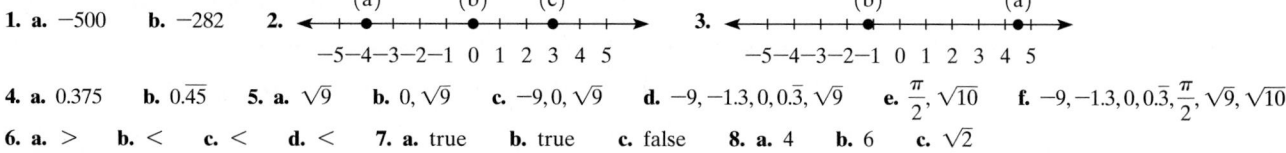

21. 0.75 **23.** 0.35 **25.** 0.875 **27.** $0.\overline{81}$ **29.** -0.5 **31.** $-0.8\overline{3}$ **33. a.** $\sqrt{100}$ **b.** $0, \sqrt{100}$ **c.** $-9, 0, \sqrt{100}$

d. $-9, -\frac{4}{5}, 0, 0.25, 9.2, \sqrt{100}$ **e.** $\sqrt{3}$ **f.** $-9, -\frac{4}{5}, 0, 0.25, \sqrt{3}, 9.2, \sqrt{100}$ **35. a.** $\sqrt{64}$ **b.** $0, \sqrt{64}$ **c.** $-11, 0, \sqrt{64}$

d. $-11, -\frac{5}{6}, 0, 0.75, \sqrt{64}$ **e.** $\sqrt{5}, \pi$ **f.** $-11, -\frac{5}{6}, 0, 0.75, \sqrt{5}, \pi, \sqrt{64}$ **37.** 0 **39.** Answers will vary; $\frac{1}{2}$ is an example.

41. Answers will vary; 6 is an example. **43.** Answers will vary; π is an example. **45.** $<$ **47.** $>$ **49.** $>$ **51.** $<$ **53.** $>$

55. $<$ **57.** $<$ **59.** $>$ **61.** $>$ **63.** true **65.** true **67.** true **69.** false **71.** 6 **73.** 7 **75.** $\frac{2}{3}$ **77.** $\sqrt{13}$

79. -3 **81.** 1996, 1997; budget deficit **83–93.** Answers will vary. **95.** c **97.** -7 and -6 **99.** 1.732; 1 and 2

101. $-0.414; -1$ and 0

Section 1.3

Check Point Exercises

1.

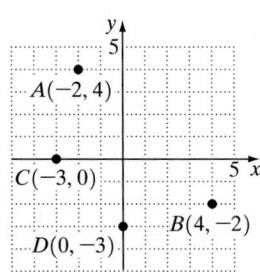

2. $E(-4, -2), F(-2, 0), G(6, 0)$

3. $B(8, 200)$; After 8 sec, the watermelon is 200 ft above ground.

4. $D(8.8, 0)$; After 8.8 sec, the watermelon is 0 ft above ground.
Equivalently, the watermelon splatters on the ground after 8.8 sec.

5. 20 years old; 1950, 1960, or 1970

6. a. 14%
   b. Tommy Hilfiger, Calvin Klein, and Nike

7. a. 13.6% of the U.S. population in 2050 will be black.
   b. 32,000,000

Exercise Set 1.3

1.

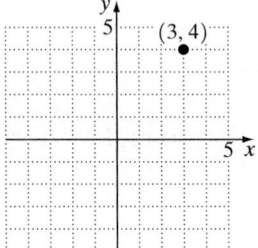

; I

3.

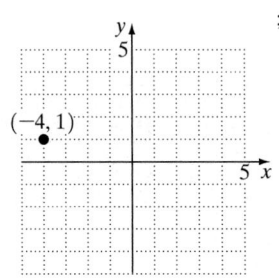

; II

5.

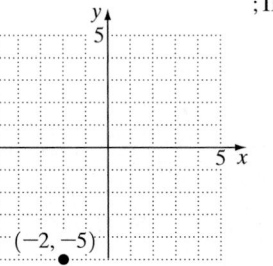

; III

7.

; IV

9–23.

25. $(5, 2)$ **27.** $(-6, 5)$
29. $(-2, -3)$ **31.** $(5, -3)$
33. $(2, 7)$; The football is 7 ft above ground when it is 2 yd from the quarterback.
35. $(6, 9.25)$ **37.** 12 ft; 15 yd
39. $(1970, 61)$; In 1970 there were approximately 61 million people under 16.
41. $(1990, 60)$; In 1990 there were approximately 60 million people under 16.
43. 5% **45.** 1982; 9.9% **47.** 33%

49. historical places and museums, outdoor recreation, and shopping **51.** 48 yr **53.** approx 14 yr **55.** $470.00

57. $20.00 **59.** 160,000,000 **61.** 6000 **63–71.** Answers will vary. **73.** c **75.** $\dfrac{23}{20}$ **76.** $<$ **77.** 5.83

Section 1.4

Check Point Exercises

1. 226.5; In 1980, the population of the United States was 226.5 million. **2. a.** 3 terms **b.** 6 **c.** 11 **d.** $6x$ and $2x$
3. a. $14 + x$ **b.** $y7$ **4. a.** $17 + 5x$ **b.** $x5 + 17$ **5. a.** $20 + x$ or $x + 20$ **b.** $30x$ **6.** $12 + x$ or $x + 12$ **7.** $5x + 15$
8. $24y + 42$ **9. a.** $10x$ **b.** $5a$ **10. a.** $18x + 10$ **b.** $4x + 4y$ **11.** $3x - 21$ **12.** $38x + 19y$ **13.** 108 beats/min

Exercise Set 1.4

1. 18 **3.** 70 **5.** 27 **7.** 20 **9.** 25 **11. a.** 2 **b.** 3 **c.** 5 **d.** no **13. a.** 3 **b.** 1 **c.** 2 **d.** yes; x and $5x$
15. a. 3 **b.** 4 **c.** 1 **d.** no **17.** $4 + y$ **19.** $3x + 5$ **21.** $5y + 4x$ **23.** $5(3 + x)$ **25.** $x9$ **27.** $x + 6y$
29. $x7 + 23$ **31.** $(x + 3)5$ **33.** $(7 + 5) + x = 12 + x$ **35.** $(7 \cdot 4)x = 28x$ **37.** $3x + 15$ **39.** $16x + 24$ **41.** $4 + 2r$
43. $5x + 5y$ **45.** $3x - 6$ **47.** $8x - 10$ **49.** $\dfrac{5}{2}x - 6$ **51.** $8x + 28$ **53.** $6x + 18 + 12y$ **55.** $15x - 10 + 20y$
57. $17x$ **59.** $8a$ **61.** $14 + x$ **63.** $11y - 3$ **65.** $9x + 1$ **67.** $8a + 10$ **69.** $15x + 6$ **71.** $15x + 2$ **73.** $41a + 4b$
75. 300; You can stay in the sun for 300 minutes without burning with a number 15 lotion.
77. 46,560; In 2000 the average yearly earnings for elementary and secondary teachers in the United States was $46,560.
79. about 108; The proper dose for a 12 year-old is approximately 108 mg. **81–91.** Answers will vary. **93.** c

95. a. $50.50; $5.50; $1.00 **b.** No; When producing 2000 clocks, the average cost is $3, which is more than the maximum price at which he can sell the clocks. **96.** $0.\overline{4}$ **97.**

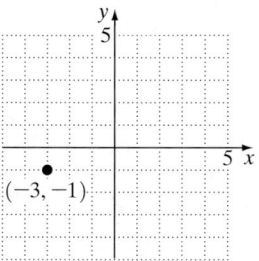

98. $\dfrac{1}{5}$

Section 1.5

Check Point Exercises

1. -3

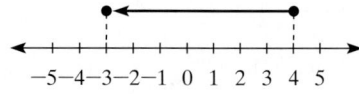

2. a. -4

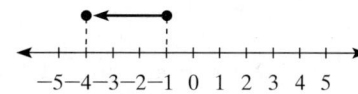

b. -2

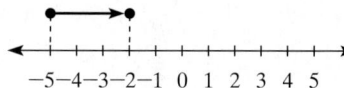

3. a. -35 **b.** -1.5 **c.** $-\dfrac{5}{6}$ **4. a.** -13 **b.** 1.2 **c.** $-\dfrac{1}{2}$ **5. a.** $-17x$ **b.** $-7y + 6z$ **c.** $20 - 20x$

6. The water level had fallen 3 ft at the end of 5 mo.

Exercise Set 1.5

1. -4

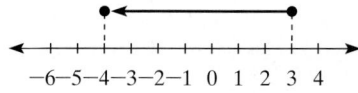

3. -9

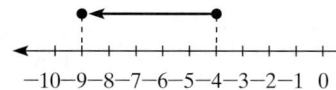

5. -6

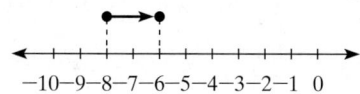

7. $2 + (-2) = 0$

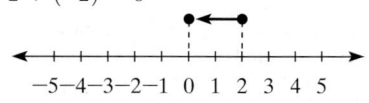

9. -4 **11.** 0 **13.** -18 **15.** -12 **17.** -1.3 **19.** -1 **21.** -5 **23.** 4
25. -3 **27.** -1.5 **29.** -5.7 **31.** $\dfrac{3}{10}$ **33.** $\dfrac{1}{8}$ **35.** $-\dfrac{43}{35}$ **37.** -8 **39.** 62
41. 8 **43.** -21 **45.** 22.1 **47.** $-3x$ **49.** $3y$ **51.** $-17a$ **53.** $1 - 6x$
55. $-4 + 6b$ **57.** $-2x - 3y$ **59.** $20x - 6$ **61.** $24 + 3y$ **63.** $47 - 33a$ **65.** $44°$ **67.** 600 ft below sea level
69. $3°F$ **71.** 25-yard line **73.** 4400 women athletes **75–81.** Answers will vary. **83.** d **85.** $-18y$ **87.** Answers will vary.
89. 5.0283 **90.** true **91. a.** $\sqrt{4}$ **b.** $0, \sqrt{4}$ **c.** $-6, 0, \sqrt{4}$ **d.** $-6, 0, 0.\overline{7}, \sqrt{4}$ **e.** $-\pi, \sqrt{3}$ **f.** $-6, -\pi, 0, 0.\overline{7}, \sqrt{3}, \sqrt{4}$
92. ; IV

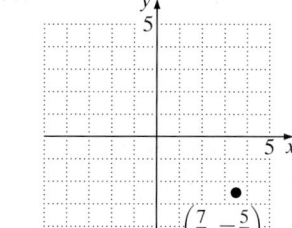

Section 1.6

Check Point Exercises

1. a. -8 **b.** 9 **c.** -5 **2. a.** 9.2 **b.** $-\dfrac{14}{15}$ **c.** 7π **3.** 15 **4.** $-6, 4a, -7ab$ **5. a.** $4 - 7x$ **b.** $-9x + 4y$
6. $19,763$ m

Exercise Set 1.6

1. a. -12 **b.** $5 + (-12)$ **3. a.** 7 **b.** $5 + 7$ **5.** 5 **7.** -7 **9.** 14 **11.** 11 **13.** -9 **15.** -28 **17.** 0 **19.** 0
21. 14 **23.** -8 **25.** 3 **27.** $-\dfrac{2}{7}$ **29.** $\dfrac{4}{5}$ **31.** -1 **33.** $-\dfrac{3}{5}$ **35.** $\dfrac{3}{4}$ **37.** $\dfrac{1}{4}$ **39.** 7.6 **41.** -2 **43.** 2.6 **45.** 0

47. 3π **49.** 13π **51.** 19 **53.** -3 **55.** -15 **57.** 0 **59.** -52 **61.** -187 **63.** $\dfrac{7}{6}$ **65.** -4.49 **67.** $-\dfrac{3}{8}$
69. $-3x, -8y$ **71.** $12x, -5xy, -4$ **73.** $-6x$ **75.** $4 - 10y$ **77.** $5 - 7a$ **79.** $-4 - 9b$ **81.** $24 + 11x$ **83.** $3y - 8x$
85. 19,757 ft **87.** \$8 billion **89.** 21 °F **91.** 3 °F **93.** 0.05 mg per 100 ml; during the 3rd hour **95.** 0.015 mg per 100 ml
97. from 0 to 3 hr **99.** 632 thousand jobs **101–105.** Answers will vary. **107.** 711 yr **109.** Answers will vary. **111.** 16.7429

113.
$$-4.5$$
$$-8\ -7\ -6\ -5\ -4\ -3\ -2\ -1\ \ 0\ \ 1\ \ 2$$
114. $10(4 + a)$ **115.** Answers will vary; -7 is an example.

Section 1.7

Check Point Exercises

1. a. -40 **b.** $-\dfrac{4}{21}$ **c.** 36 **d.** 5.5 **e.** 0 **2. a.** 24 **b.** -30 **3. a.** $\dfrac{1}{7}$ **b.** 8 **c.** $-\dfrac{1}{6}$ **d.** $-\dfrac{13}{7}$

4. a. -4 **b.** 8 **5. a.** 8 **b.** $-\dfrac{8}{15}$ **c.** -7.3 **d.** 0 **6. a.** $-20x$ **b.** $10x$ **c.** $-b$ **d.** $-21x + 28$ **e.** $-7y + 6$

7. $-y - 26$ **8. a.** \$330 **b.** \$60 **c.** \$33

Exercise Set 1.7

1. -54 **3.** 21 **5.** -12 **7.** 13 **9.** 0 **11.** -7 **13.** 15 **15.** $\dfrac{12}{35}$ **17.** $-\dfrac{14}{27}$ **19.** -3.6 **21.** 0.12 **23.** 30

25. -72 **27.** 24 **29.** -27 **31.** 90 **33.** 0 **35.** $\dfrac{1}{4}$ **37.** 5 **39.** $-\dfrac{1}{10}$ **41.** $-\dfrac{5}{2}$ **43. a.** $-32 \cdot \left(\dfrac{1}{4}\right)$ **b.** -8

45. a. $-60 \cdot \left(-\dfrac{1}{5}\right)$ **b.** 12 **47.** -3 **49.** -7 **51.** 30 **53.** 0 **55.** undefined **57.** -5 **59.** -12 **61.** 6 **63.** 0

65. undefined **67.** -4.3 **69.** $\dfrac{5}{6}$ **71.** $-\dfrac{16}{9}$ **73.** -1 **75.** -15 **77.** $-10x$ **79.** $3y$ **81.** $9x$ **83.** $-4x$ **85.** $-b$

87. $3y$ **89.** $-8x + 12$ **91.** $6x - 12$ **93.** $-2y + 5$ **95.** $y - 14$ **97.** 233 thousand liposuctions **99. a.** 11 Latin words
b. 11 Latin words **101. a.** \$2,000,000 (or \$200 tens of thousands) **b.** \$8,000,000 (or \$800 tens of thousands) **c.** Cost increases.
103–109. Answers will vary. **111.** b **113.** $5x$ **115.** Answers will vary. **117.** $1.144x + 2.5$ **119.** -9 **120.** -3 **121.** 2

Section 1.8

Check Point Exercises

1. a. 36 **b.** -64 **c.** 1 **d.** -1 **2. a.** $21x^2$ **b.** $8x^3$ **c.** cannot be simplified **3.** 15 **4.** 32 **5. a.** 36 **b.** 12

6. 82 **7.** -40 **8.** -31 **9.** $\dfrac{5}{7}$ **10.** -5 **11.** $7x^2 + 15$ **12.** 200 accidents per 50 million miles driven **13.** 30 °C

Exercise Set 1.8

1. 81 **3.** 64 **5.** 16 **7.** -64 **9.** 625 **11.** -625 **13.** -100 **15.** $17x^2$ **17.** $13x^3$ **19.** $8x^4$ **21.** $-x^2$ **23.** x^3
25. cannot be simplified **27.** 0 **29.** 25 **31.** 27 **33.** 12 **35.** 5 **37.** 45 **39.** -24 **41.** 300 **43.** 0 **45.** -32

47. 64 **49.** 30 **51.** $\dfrac{4}{3}$ **53.** 2 **55.** 2 **57.** 3 **59.** 88 **61.** -60 **63.** -36 **65.** 14 **67.** 24 **69.** 28 **71.** 9

73. -7 **75.** $15x - 27$ **77.** $15 - 3y$ **79.** $16y - 25$ **81.** $-2x^2 - 9$ **83.** 13 lb; $(4, 13)$ **85.** 135 beats/min; $(40, 135)$

87. 6.9 million Americans **89.** \$287.4 billion **91.** 20 °C **93.** -30 °C **95–99.** Answers will vary. **101.** $-\dfrac{79}{4}$

103. $\left(2 \cdot 5 - \dfrac{1}{2} \cdot 10\right) \cdot 9 = 45$ **105.** 443.99 thousands (or 443,990) **107.** 6 **108.** -24 **109.** Answers will vary; -3 is an example.

Review Exercises

1. $\frac{5}{11}$ **2.** $\frac{8}{15}$ **3.** $\frac{21}{50}$ **4.** $\frac{8}{3}$ **5.** $\frac{2}{3}$ **6.** $\frac{29}{18}$ **7.** $\frac{37}{60}$ **8.** $\frac{5}{12}$ **9.**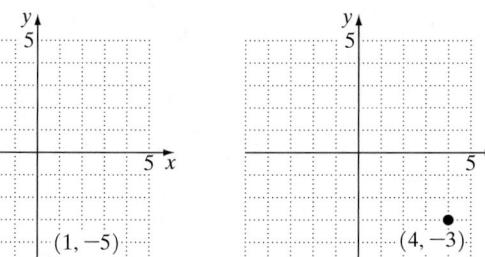

10. **11.** 0.625 **12.** $0.\overline{27}$ **13. a.** $\sqrt{81}$ **b.** $0, \sqrt{81}$ **c.** $-17, 0, \sqrt{81}$

d. $-17, -\frac{9}{13}, 0, 0.75, \sqrt{81}$ **e.** $\sqrt{2}, \pi$ **f.** $-17, -\frac{9}{13}, 0, 0.75, \sqrt{2}, \pi, \sqrt{81}$ **14.** Answers will vary; -2 is an example.

15. Answers will vary; $\frac{1}{2}$ is an example. **16.** Answers will vary; π is an example. **17.** $<$ **18.** $>$ **19.** $>$ **20.** $<$

21. false **22.** true **23.** 58 **24.** 2.75
25. IV **26.** IV **27.** I **28.** II

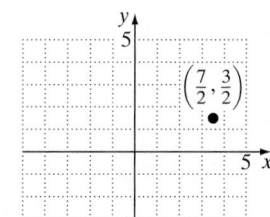

29. $A(5, 6); B(-3, 0); C(-5, 2); D(-4, -2); E(0, -5); F(3, -1)$ **30.** 7 murders **31.** 1980; 10 murders **32.** 90%
33. camcorder and satellite dish **34.** 10.8 million **35.** 73 **36.** 40 **37.** $13y + 7$ **38.** $(x + 7)9$
39. $(6 + 4) + y; 10 + y$ **40.** $(7 \cdot 10)x; 70x$ **41.** $24x - 12 + 30y$ **42.** $7a + 2$ **43.** $28x + 19$

44. 1800; The sale price at 25% off is $1800 for a $2400 computer. **45.** ; 2 **46.** -3

47. $-\frac{11}{20}$ **48.** -7 **49.** $5y - 4x$ **50.** $40 - 2y$ **51.** 800 ft below sea level **52.** 23 ft **53.** $9 + (-13)$ **54.** 4

55. $-\frac{6}{5}$ **56.** -1.5 **57.** -7 **58.** -3 **59.** $-5 - 8a$ **60.** 27,150 ft **61.** 84 **62.** $-\frac{3}{11}$ **63.** -120 **64.** -9
65. undefined **66.** 2 **67.** $3x$ **68.** $-x - 1$ **69.** 36 **70.** -36 **71.** -32 **72.** $6x^3$ **73.** cannot be simplified
74. -16 **75.** -16 **76.** 10 **77.** -2 **78.** 17 **79.** -88 **80.** 14 **81.** 6 **82.** 10 **83.** $28a - 20$ **84.** $6y - 12$
85. 5.5 million **86.** 4.1 million **87.** 740 thousand (or 740,000) **88.** As the years increase, so do the number of inmates.

Chapter 1 Test

1. 4 **2.** -11 **3.** -51 **4.** $\frac{1}{5}$ **5.** -5 **6.** 1 **7.** -4 **8.** -32 **9.** 1 **10.** $4x + 4$ **11.** $13x - 19y$ **12.** $10 - 6x$

13. $-7, -\frac{4}{5}, 0, 0.25, \sqrt{4}, \frac{22}{7}$ **14.** $>$ **15.** 12.8 **16.** II;

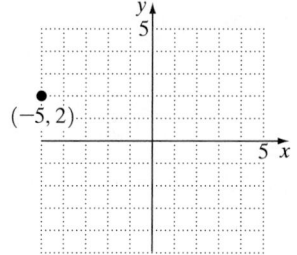

17. $(-5, -2)$ **18.** -15 **19.** 150
20. $2(3 + x)$
21. $(-6 \cdot 4)x = -24x$
22. $35x - 7 + 14y$
23. $(30, 200)$; After 30 years there are 200 elk.
24. 50 **25.** 9.7 million
26. 6.3 million acres **27.** 16 sec
28. $95 thousand (or $95,000)
29. $93 thousand (or $93,000)
30. 17,030 ft

CHAPTER 2

Section 2.1

Check Point Exercises

1. a. not a solution   **b.** solution   **2.** 17 or {17}   **3.** 2.29 or {2.29}   **4.** $\frac{1}{4}$ or $\left\{\frac{1}{4}\right\}$ **5.** 13 or {13}   **6.** 12 or {12}

7. 11 or {11}   **8.** 2100 words

Exercise Set 2.1

1. 20 or {20}   **3.** −17 or {−17}   **5.** −17 or {−17}   **7.** −13 or {−13}   **9.** 6 or {6}   **11.** −14 or {14}   **13.** 2 or {2}

15. $-\frac{17}{12}$ or $\left\{-\frac{17}{12}\right\}$ **17.** $\frac{21}{4}$ or $\left\{\frac{21}{4}\right\}$ **19.** $-\frac{11}{20}$ or $\left\{-\frac{11}{20}\right\}$ **21.** 4.3 or {4.3}   **23.** $-\frac{21}{4}$ or $\left\{-\frac{21}{4}\right\}$ **25.** 18 or {18}

27. $\frac{9}{10}$ or $\left\{\frac{9}{10}\right\}$ **29.** −310 or {−310}   **31.** 4.3 or {4.3}   **33.** 0 or {0}   **35.** 11 or {11}   **37.** 5 or {5}   **39.** −13 or {−13}

41. 6 or {6}   **43.** $212 billion   **45.** $1700   **47.** 525,000 deaths   **49–53.** Answers will vary.   **55.** −6 and 6, or {−6, 6}

57. 2.7529 or {2.7529}   **58.** II;   **59.** −12   **60.** $6 - 9x$

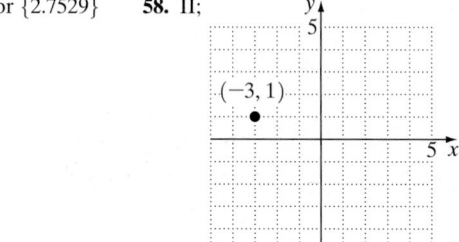

Section 2.2

Check Point Exercises

1. 36 or {36}   **2. a.** 21 or {21}   **b.** −4 or {−4}   **c.** −3.1 or {−3.1}   **3. a.** 24 or {24}   **b.** −16 or {−16}

4. a. −5 or {−5}   **b.** 3 or {3}   **5.** 6 or {6}   **6.** −10 or {−10}   **7.** 6 or {6}   **8.** 100 g

Exercise Set 2.2

1. 12 or {12}   **3.** −55 or {−55}   **5.** 9 or {9}   **7.** −8 or {−8}   **9.** −3 or {−3}   **11.** 5 or {5}   **13.** $-\frac{1}{4}$ or $\left\{-\frac{1}{4}\right\}$

15. 0 or {0}   **17.** 12 or {12}   **19.** −6 or {−6}   **21.** −7 or {−7}   **23.** 15 or {15}   **25.** 50 or {50}   **27.** −4 or {−4}

29. 5 or {5}   **31.** 6 or {6}   **33.** −1 or {−1}   **35.** −2 or {−2}   **37.** $\frac{9}{4}$ or $\left\{\frac{9}{4}\right\}$ **39.** −6 or {−6}   **41.** −3 or {−3}

43. −3 or {−3}   **45.** 4 or {4}   **47.** $-\frac{3}{2}$ or $\left\{-\frac{3}{2}\right\}$ **49.** 2 or {2}   **51.** −4 or {−4}   **53.** −6 or {−6}   **55.** 10 sec

57. 1502.2 mph   **59.** 60 yd   **61.** If $a = b$ and $c \neq 0$, then $ac = bc$; Answers will vary.   **63.** Answers will vary.   **65.** d

67. Answers will vary.   **69.** 6.5 or {6.5}   **70.** 100   **71.** −100   **72.** 3

Section 2.3

Check Point Exercises

1. 6 or {6}   **2.** 2 or {2}   **3.** 5 or {5}   **4.** −2 or {−2}   **5.** no solution or {} or ∅   **6.** all real numbers or ℝ   **7.** 124 lb

Exercise Set 2.3

1. 3 or {3}   **3.** 10 or {10}   **5.** 4 or {4}   **7.** 6 or {6}   **9.** 2 or {2}   **11.** $\frac{1}{3}$ or $\left\{\frac{1}{3}\right\}$ **13.** 6 or {6}   **15.** 8 or {8}

17. 17 or {17}   **19.** 1 or {1}   **21.** −4 or {−4}   **23.** 5 or {5}   **25.** 6 or {6}   **27.** 1 or {1}   **29.** −57 or {−57}

31. −10 or {−10}   **33.** 18 or {18}   **35.** $\frac{7}{4}$ or $\left\{\frac{7}{4}\right\}$ **37.** 1 or {1}   **39.** 24 or {24}   **41.** −6 or {−6}   **43.** 20 or {20}

45. 5 or {5}   **47.** no solution or {} or ∅   **49.** all real numbers or ℝ   **51.** all real numbers or ℝ   **53.** no solution or {} or ∅

55. no solution or {} or ∅   **57.** 0 or {0}   **59.** 85 mph   **61.** 120 chirps/min   **63.** 409.2 ft   **65–67.** Answers will vary.

69. c   **71.** 3 or {3}   **73.** −4.2 or {−4.2}   **75.** <   **76.** <   **77.** −10

Section 2.4

Check Point Exercises

1. $l = \dfrac{A}{w}$ **2.** $l = \dfrac{P - 2w}{2}$ **3.** $m = \dfrac{T - D}{p}$ **4.** $x = 15 + 12y$ **5.** 2.3%   **6. a.** 0.67   **b.** 2.5   **7.** 4.5   **8.** 15
9. 36%   **10.** 80%

Exercise Set 2.4

1. $r = \dfrac{d}{t}$ **3.** $P = \dfrac{I}{rt}$ **5.** $r = \dfrac{C}{2\pi}$ **7.** $m = \dfrac{E}{c^2}$ **9.** $m = \dfrac{y - b}{x}$ **11.** $p = \dfrac{T - D}{m}$ **13.** $b = \dfrac{2A}{h}$ **15.** $n = 5M$

17. $c = 4F - 160$ **19.** $a = 2A - b$ **21.** $r = \dfrac{S - P}{Pt}$ **23.** $b = \dfrac{2A}{h} - a$ **25.** $x = \dfrac{C - By}{A}$ **27.** 59%   **29.** 0.3%

31. 287%   **33.** 10,000%   **35.** 0.72   **37.** 0.436   **39.** 1.3   **41.** 0.02   **43.** 0.625   **45.** 6   **47.** 7.2   **49.** 5   **51.** 170

53. 20%   **55.** 12%   **57. a.** $z = 3A - x - y$ **b.** 96%   **59. a.** $t = \dfrac{d}{r}$ **b.** 2.5 hr   **61.** 23%   **63.** 34%   **65.** 359 women

67. 12.5%   **69.** $9   **71. a.** $1008   **b.** $17,808   **73. a.** $103.20   **b.** $756.80   **75–79.** Answers will vary.
81. d   **83.** 1.5 sec; 60 ft   **84.** 12 or $\{12\}$ **85.** 20 or $\{20\}$ **86.** $0.7x$

Section 2.5

Check Point Exercises

1. a. $4x + 6$ **b.** $\dfrac{x - 4}{9}$ **2.** 12   **3.** *Jagged Little Pill*: 16 million albums; *Saturday Night Fever*: 11 million albums
4. pages 96 and 97   **5.** 32; 4 mi   **6.** 40 ft by 120 ft   **7.** $940

Exercise Set 2.5

1. $x + 9$ **3.** $20 - x$ **5.** $8 - 5x$ **7.** $\dfrac{15}{x}$ **9.** $2x + 20$ **11.** $7x - 30$ **13.** $4(x + 12)$ **15.** $x + 40 = 450; 410$

17. $x - 13 = 123; 136$ **19.** $7x = 91; 13$ **21.** $\dfrac{x}{18} = 6; 108$ **23.** $4 + 2x = 36; 16$ **25.** $5x - 7 = 123; 26$ **27.** $x + 5 = 2x; 5$

29. $2(x + 4) = 36; 14$ **31.** $9x = 30 + 3x; 5$ **33.** $\dfrac{3x}{5} + 4 = 34; 50$ **35.** Titanic: $200 million; Waterworld: $160 million

37. Los Angeles: 82 hr; Miami: 57 hr   **39.** pages 314 and 315   **41.** 19 and 20   **43.** 32 and 34   **45.** 800 mi   **47.** 6 mo
49. 50 yd by 200 yd   **51.** width: 160 ft; length: 360 ft   **53.** width: 12 ft; length: 4 ft   **55.** $400   **57.** $28   **59.** $14,500

61. 11 hr   **63–65.** Answers will vary.   **67.** d   **69.** 17 min   **71.** 7 oz   **72.** -20 or $\{-20\}$ **73.** 0 or $\{0\}$ **74.** $w = \dfrac{3V}{lh}$

Section 2.6

Check Point Exercises

1. 12 ft   **2.** 400π ft$^2 \approx 1256$ ft^2; 40π ft ≈ 126 ft   **3.** large pizza   **4.** 2 times   **5.** No, about 32 more cubic inches are needed.
6. $120°, 40°, 20°$ **7.** $60°$

Exercise Set 2.6

1. 18 m; 18 m^2 **3.** 56 in.^{2} **5.** 91 m^2 **7.** 50 ft   **9.** 8 ft   **11.** 50 cm   **13.** 16π cm$^2 \approx 50$ cm^2; 8π cm ≈ 25 cm
15. 36π yd$^2 \approx 113$ yd^2; 12π yd ≈ 38 yd   **17.** 7 in.; 14 in.   **19.** 36 in.^{3} **21.** 150π cm$^3 \approx 471$ cm^3 **23.** 972π cm$^3 \approx 3052$ cm^3

25. 48π m$^3 \approx 151$ m^3 **27.** $h = \dfrac{V}{\pi r^2}$ **29.** 9 times   **31.** $x = 50°; x + 30 = 80°$ **33.** $4x = 76°; 3x + 4 = 61°; 2x + 5 = 43°$
35. $40°, 80°, 60°$ **37.** $42°$ **39.** $1°$ **41.** $69°$ **43.** $90°$ **45.** $75°$ **47.** $135°$ **49.** $50°$ **51.** $698.18   **53.** large pizza
55. $2260.80   **57.** approx 19.7 ft   **59.** 21,000 yd^3 **61.** the can with diameter of 6 in. and height of 5 in.
63. Yes, the water tank is a little over one cubic foot too small.   **65–73.** Answers will vary.   **75.** 2.25 times

77. Volume increases 8 times.   **79.** $35°$ **80.** $s = \dfrac{P - b}{2}$ **81.** 8 or $\{8\}$ **82.** 0

Section 2.7

Check Point Exercises

1. a.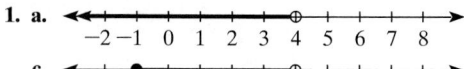

b.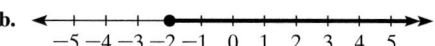

c.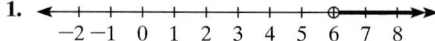

2. $\{x \mid x < 3\}$

3. $\{x \mid x \geq -2\}$

4. a. $\{x \mid x < 8\}$

b. $\{x \mid x > -3\}$

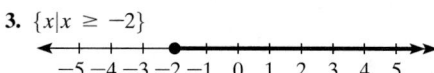

5. $\{y \mid y \geq 4\}$

6. $\{x \mid x \geq 1\}$

7. $\{x \mid x \geq 1\}$

8. no solution or {} or ∅
9. $\{x \mid x$ is a real number$\}$ or $\mathbb{R}$
10. at least 83%

Exercise Set 2.7

1.

3.

5.

7.

9.

11.

13. $\{x \mid x > -2\}$ **15.** $\{x \mid x \geq 4\}$ **17.** $\{x \mid x \geq 3\}$

19. $\{x \mid x > 5\}$

21. $\{x \mid x \leq 5\}$

23. $\{y \mid y < 3\}$

25. $\{x \mid x \leq 3\}$

27. $\{x \mid x < 16\}$

29. $\{x \mid x > 4\}$

31. $\left\{x \,\middle|\, x > \dfrac{7}{6}\right\}$

33. $\left\{y \,\middle|\, y \leq -\dfrac{3}{8}\right\}$

35. $\{y \mid y > 0\}$

37. $\{x \mid x < 8\}$

39. $\{x \mid x > -6\}$

41. $\{x \mid x < 5\}$

43. $\{x \mid x \geq -7\}$

45. $\{x \mid x > -5\}$

47. $\{x \mid x \leq -5\}$

49. $\{x \mid x < 3\}$

51. $\left\{y \mid y \geq -\dfrac{1}{8}\right\}$

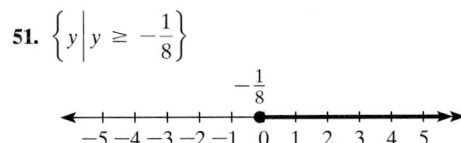

53. $\{x \mid x > -4\}$

55. $\{x \mid x > 5\}$

57. $\{x \mid x < 5\}$

59. $\{x \mid x \geq -2\}$

61. $\{x \mid x > -3\}$

63. $\{x \mid x \geq 4\}$

65. $\left\{x \mid x > \dfrac{11}{3}\right\}$

67. $\{y \mid y > 2\}$

69. $\{y \mid y < 2\}$

71. $\{x \mid x < 3\}$

73. $\left\{x \mid x > \dfrac{5}{3}\right\}$

75. $\{x \mid x \geq 9\}$

77. $\{x \mid x < -6\}$

79. no solution or {} or $\varnothing$ **81.** $\{x \mid x$ is a real number$\}$ or $\mathbb{R}$ **83.** no solution or {} or $\varnothing$ **85.** $\{x \mid x$ is a real number$\}$ or $\mathbb{R}$
87. $\{x \mid x \leq 0\}$ **89.** Raleigh, NC; Seattle, WA; San Francisco, CA; Austin, TX   **91.** San Diego, CA
93. Washington, DC; Lexington-Fayette, KY; Minneapolis, MN; Boston, MA; Arlington, TX   **95.** antisocial personality; schizophrenia
97. 20 yr; from 2009 onward   **99. a.** at least 96%   **b.** less than 66%   **101.** 1280 mi   **103.** at most 29 bags of cement
105. An open dot indicates an endpoint that is not a solution, and a closed dot indicates an endpoint that is a solution.
107–109. Answers will vary.   **111.** more than 720 mi   **113.** $\{x \mid x < 0.4\}$ **115.** 20   **116.** length: 11 in.; width: 6 in.
117. 4 or {4}

Review Exercises

1. 42 or {42}   **2.** -20 or $\{-20\}$ **3.** 12 or {12}   **4.** -22 or $\{-22\}$ **5.** 5 or {5}   **6.** 70 or {70}   **7.** -32 or $\{-32\}$
8. 11 or {11}   **9.** 4 or {4}   **10.** -15 or $\{-15\}$ **11.** -12 or $\{-12\}$ **12.** -14 or $\{-14\}$ **13.** 3 or {3}   **14.** 6 or {6}
15. -5 or $\{-5\}$ **16.** -10 or $\{-10\}$ **17.** 2 or {2}   **18.** 1 or {1}   **19.** 30 yr; 2014   **20.** -1 or $\{-1\}$ **21.** 12 or {12}
22. -13 or $\{-13\}$ **23.** -3 or $\{-3\}$ **24.** -10 or $\{-10\}$ **25.** 2 or {2}   **26.** 2 or {2}   **27.** no solution or {} or $\varnothing$
28. all real numbers or $\mathbb{R}$ **29.** 20 years old   **30.** $r = \dfrac{I}{P}$ **31.** $h = \dfrac{3V}{B}$ **32.** $w = \dfrac{P - 2l}{2}$ **33.** $B = 2A - C$ **34.** $m = \dfrac{T - D}{p}$
35. 72%   **36.** 0.35%   **37.** 0.65   **38.** 1.5   **39.** 0.03   **40.** 9.6   **41.** 200   **42.** 48%   **43. a.** $h = 7r$ **b.** 5 ft 3 in.
44. 8000 Americans   **45.** 10   **46.** New York: 55 days; Los Angeles: 213 days   **47.** pages 46 and 47
48. Streisand: 49 albums; Madonna: 47 albums   **49.** 9 yr; 2009   **50.** 18 checks   **51.** length: 150 yd; width: 50 yd   **52.** $240
53. 32.5 ft^2 **54.** 50 cm^2 **55.** 135 yd^2 **56.** 20π m $\approx$ 63 m; 100π m^2 $\approx$ 314 m^2 **57.** 6 ft   **58.** 156 ft^2 **59.** $1890
60. medium pizza   **61.** 60 cm^3 **62.** 128π yd^3 $\approx$ 402 yd^3 **63.** 288π m^3 $\approx$ 904 m^3 **64.** 4800 m^3 **65.** 16 fish
66. $x = 30°, 3x = 90°, 2x = 60°$ **67.** $85°, 35°, 60°$ **68.** $33°$ **69.** $105°$ **70.** $57.5°$ **71.** $45°$ and $135°$
72.   **73.**   **74.** $\{x \mid x > 4\}$ **75.** $\{x \mid x \leq -3\}$

76. $\{x \mid x < 4\}$ **77.** $\{x \mid x > -8\}$

78. $\{x \mid x \geq -3\}$ **79.** $\{x \mid x > 6\}$

80. $\{x|x \geq 4\}$

81. $\{x|x \leq 2\}$

82. $\{x|x \text{ is a real number}\}$ or $\mathbb{R}$ **83.** no solution or $\{\}$ or $\varnothing$ **84.** at least 64 **85.** at most 99 min

Chapter 2 Test

1. $\frac{9}{2}$ or $\left\{\frac{9}{2}\right\}$ **2.** -5 or $\{-5\}$ **3.** $-\frac{4}{3}$ or $\left\{-\frac{4}{3}\right\}$ **4.** 2 or $\{2\}$ **5.** -20 or $\{-20\}$ **6.** $-\frac{5}{3}$ or $\left\{-\frac{5}{3}\right\}$ **7.** 60 yr; 2020

8. $h = \dfrac{V}{\pi r^2}$ **9.** $w = \dfrac{P - 2l}{2}$ **10.** 8.4 **11.** 150 **12.** 5% **13.** 63.8 **14.** Reagan: 69 years old; Buchanan: 65 years old

15. 600 min **16.** length: 150 yd; width: 75 yd **17.** $35 **18.** 517 m^2 **19.** 525 in.^{2} **20.** 18 in.^{3}

21. 175π cm$^3 \approx 550$ cm^3 **22.** $650 **23.** 14 ft **24.** $126°, 42°, 12°$ **25.** $53°$

26.

27.

28. $\{x|x \leq -1\}$

29. $\{x|x < -6\}$

30. $\{x|x \leq -3\}$

31. $\{x|x > 7\}$

32. at least 92

33. widths greater than 8 in.

Cumulative Review Exercises (Chapters 1–2)

1. -4 **2.** -2 **3.** -128 **4.** $-103 - 20x$ **5.** $-4, -\dfrac{1}{3}, 0, \sqrt{4}, 1063$

6. III;

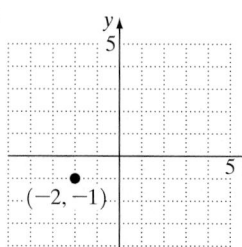

7. $<$ **8.** $24x - 6 - 30y$ **9.** 2000; 4% **10.** 1992; 7.8% **11.** 1 or $\{1\}$

12. -15 or $\{-15\}$ **13.** $A = \dfrac{3V}{h}$ **14.** 160 **15.** length: 130 yd; width: 70 yd

16. 75,000 gal **17.**

18. $\{x|x < -3\}$

19. $\{x|x \geq -6\}$

20. more than $47,500

CHAPTER 3

Section 3.1

Check Point Exercises

1. a. solution **b.** not a solution **2.** $(-2, -4), (-1, -1), (0, 2), (1, 5),$ and $(2, 8)$

3.

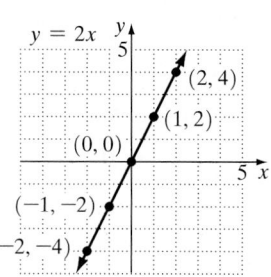

4.

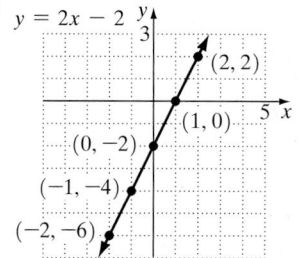

5.

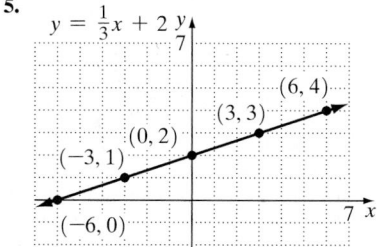

6.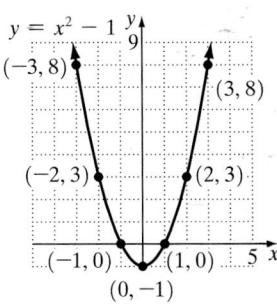

$y = x^2 - 1$

7. a.

$y = 2x$			$y = 10 + x$	
x	(x, y)		x	(x, y)
0	$(0, 0)$		0	$(0, 10)$
2	$(2, 4)$		2	$(2, 12)$
4	$(4, 8)$		4	$(4, 14)$
6	$(6, 12)$		6	$(6, 16)$
8	$(8, 16)$		8	$(8, 18)$
10	$(10, 20)$		10	$(10, 20)$
12	$(12, 24)$		12	$(12, 22)$

c. $(10, 20)$; If the bridge is used 10 times in a month, the total monthly cost without the coupon book is the same as the monthly cost with the coupon book, namely $20.

b.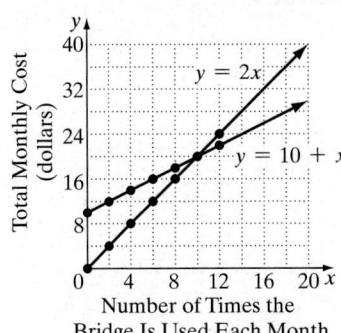

Exercise Set 3.1

1. $(2, 3)$ and $(3, 2)$ are not solutions; $(-4, -12)$ is a solution.
3. $(-5, -20)$ is not a solution; $(0, 0)$ and $(9, -36)$ are solutions.
5. $(2, -2)$ is not a solution; $(0, 6)$ and $(-3, 0)$ are solutions.
7. $(0, 5)$ is not a solution; $(-5, 6)$ and $(10, -3)$ are solutions.
9. $\left(1, \dfrac{1}{3}\right)$ is not a solution, $(0, 0)$ and $\left(2, -\dfrac{2}{3}\right)$ are solutions.
11. $(3, 4)$ and $(0, -4)$ are not solutions; $(4, 7)$ is a solution.

13.

x	(x, y)
-2	$(-2, -20)$
-1	$(-1, -10)$
0	$(0, 0)$
1	$(1, 10)$
2	$(2, 20)$

15.

x	(x, y)
-2	$(-2, 12)$
-1	$(-1, 6)$
0	$(0, 0)$
1	$(1, -6)$
2	$(2, -12)$

17.

x	(x, y)
-2	$(-2, -18)$
-1	$(-1, -13)$
0	$(0, -8)$
1	$(1, -3)$
2	$(2, 2)$

19.

x	(x, y)
-2	$(-2, 17)$
-1	$(-1, 10)$
0	$(0, 3)$
1	$(1, -4)$
2	$(2, -11)$

21.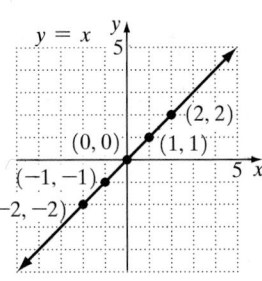

$y = x$

23.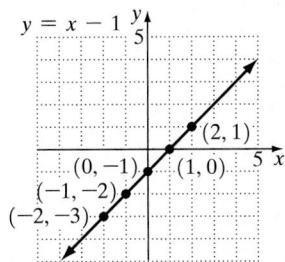

$y = x - 1$

25.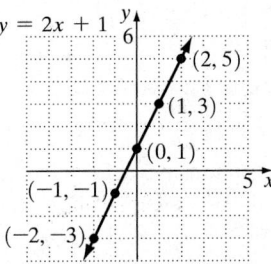

$y = 2x + 1$

27.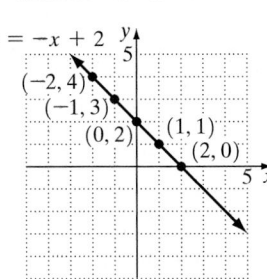

$y = -x + 2$

29.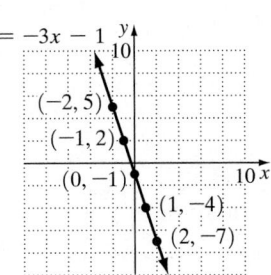

$y = -3x - 1$

31.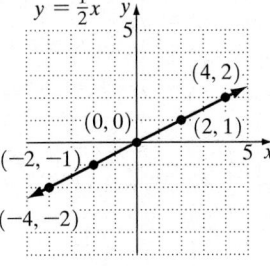

$y = \dfrac{1}{2}x$

33.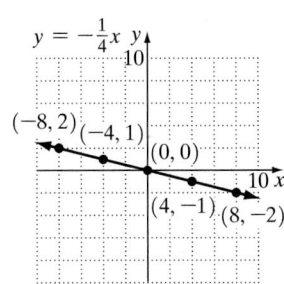

$y = -\dfrac{1}{4}x$

35.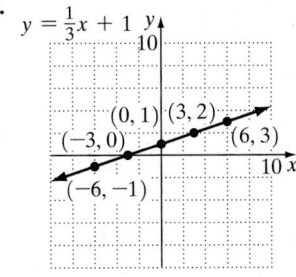

$y = \dfrac{1}{3}x + 1$

37.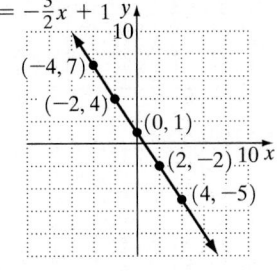

$y = -\dfrac{3}{2}x + 1$

39.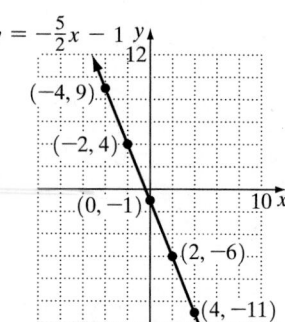

41.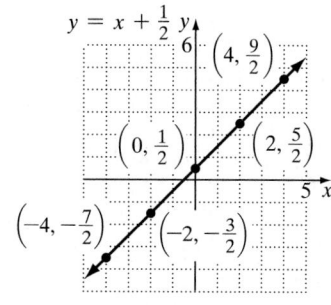

43.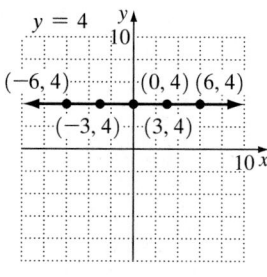

45.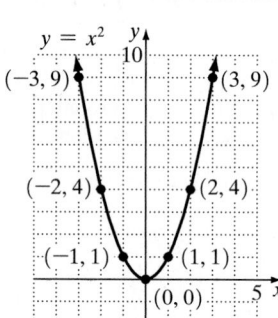

47.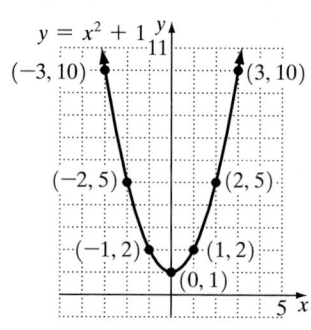

49.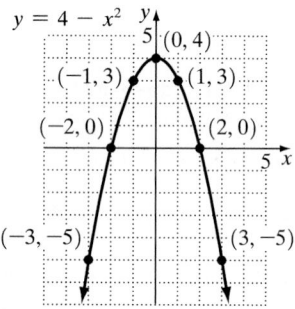

51. a.

x	(x, y)
40	$(40, 276)$
50	$(50, 300)$
60	$(60, 324)$

b. Answers will vary.

53. a.

x	(x, y)
11.6	$(11.6, 13.91)$
8.3	$(8.3, 11.105)$

b. Answers will vary.

55. a. 40; Both rental companies have the same cost when the truck is used for 40 miles.
b. 55
c. 54; Both rental companies have the same cost of $54 when the truck is used for 40 miles.

57. a.

x	(x, y)
0	$(0, 30{,}000)$
10	$(10, 30{,}500)$
20	$(20, 31{,}000)$
30	$(30, 31{,}500)$
40	$(40, 32{,}000)$

b.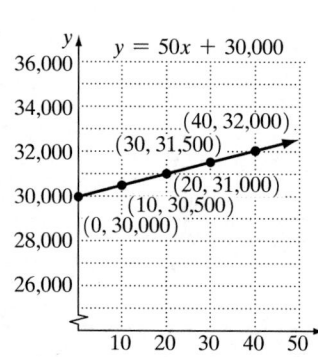

59–61. Answers will vary.

63.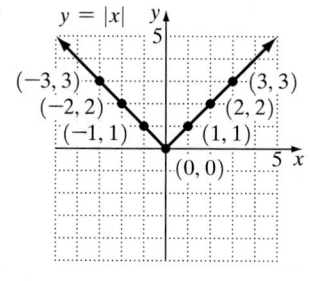

65. a. $(0, 0.6), (1, 0.3), (2, 0.2), (3, 0.3), (4, 0.6), (5, 1.1)$ **b.** between 10:00 A.M. and noon

67.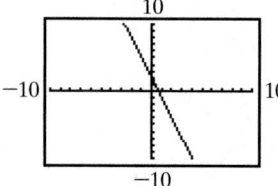

Answers will vary.

69.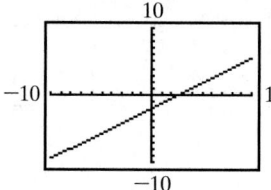

Answers will vary.

71.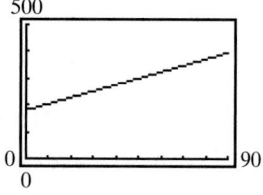

U.S. population increases.

72. 2 or $\{2\}$ **73.** 1 **74.** $h = \dfrac{3V}{A}$

Section 3.2

Check Point Exercises

1. a. x-intercept: -3; y-intercept: 5   **b.** y-intercept: 4; no x-intercept   **c.** x-intercept: 0; y-intercept: 0   **2.** 3

3. -4 **4.**

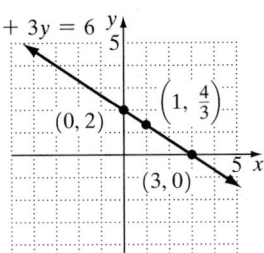

5.

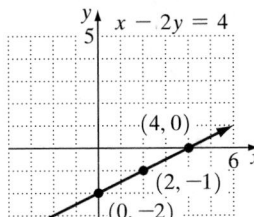

6.

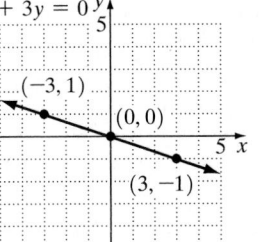

7.

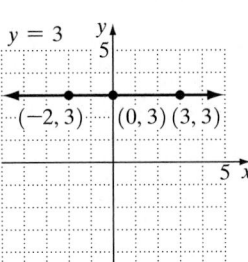

8.

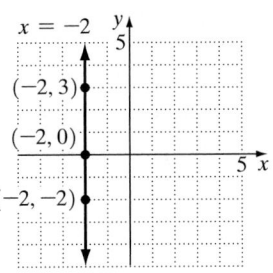

Exercise Set 3.2

1. a. 3   **b.** 4   **3. a.** -4 **b.** -2 **5. a.** 0   **b.** 0   **7. a.** no x-intercept   **b.** -2 **9.** x-intercept: 5; y-intercept: 4

11. x-intercept: 6; y-intercept: -14 **13.** x-intercept: 8; y-intercept: -2 **15.** x-intercept: 0; y-intercept: 0

17. x-intercept: -3; y-intercept: 2

19.

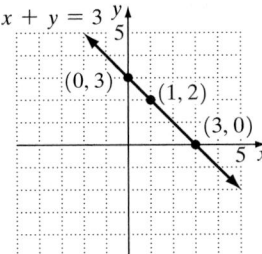

21.

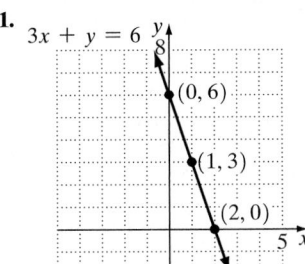

23.

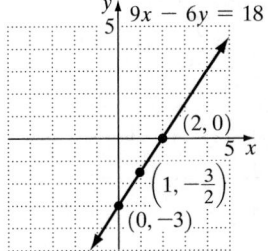

25.

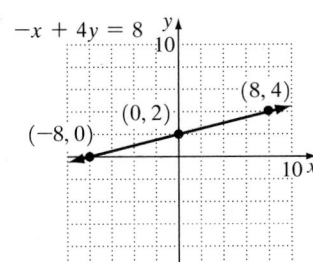

27.

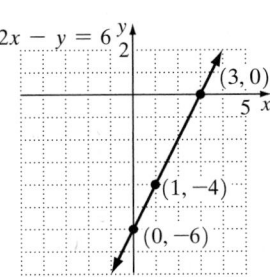

29.

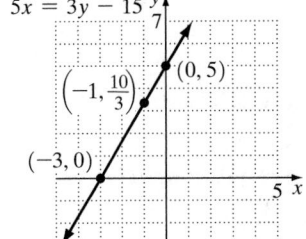

31.

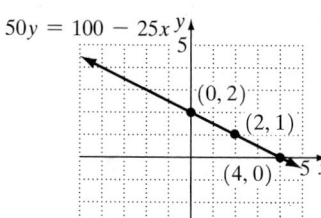

$50y = 100 - 25x$
$(0, 2)$
$(2, 1)$
$(4, 0)$

33.

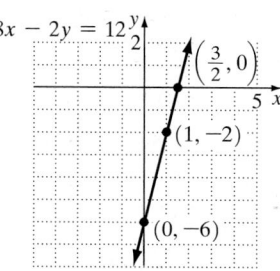

$8x - 2y = 12$
$\left(\frac{3}{2}, 0\right)$
$(1, -2)$
$(0, -6)$

35.

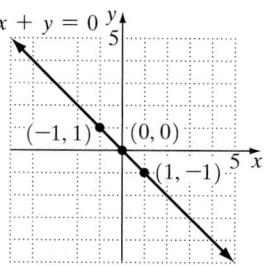

$x + y = 0$
$(-1, 1)$
$(0, 0)$
$(1, -1)$

37.

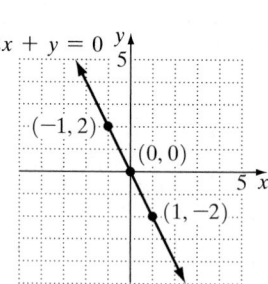

$2x + y = 0$
$(-1, 2)$
$(0, 0)$
$(1, -2)$

39.

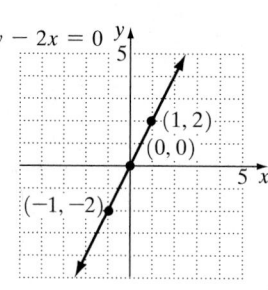

$y - 2x = 0$
$(1, 2)$
$(0, 0)$
$(-1, -2)$

41. $y = 3$
43. $x = -3$
45. $y = 0$

47.

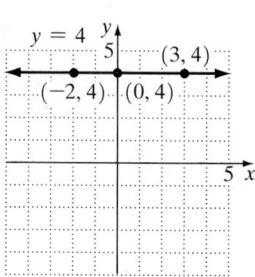

$y = 4$
$(3, 4)$
$(-2, 4)$
$(0, 4)$

49.

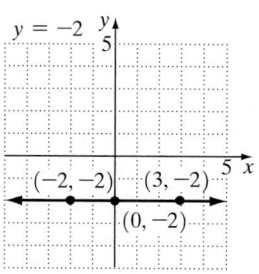

$y = -2$
$(-2, -2)$
$(3, -2)$
$(0, -2)$

51.

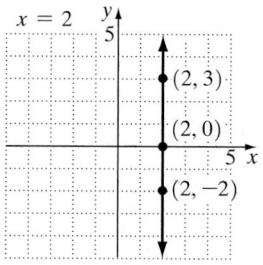

$x = 2$
$(2, 3)$
$(2, 0)$
$(2, -2)$

53.

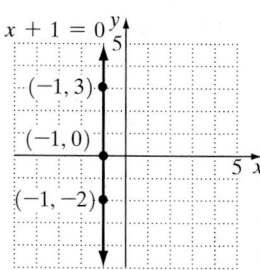

$x + 1 = 0$
$(-1, 3)$
$(-1, 0)$
$(-1, -2)$

55.

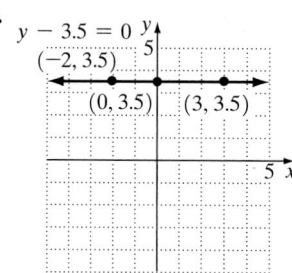

$y - 3.5 = 0$
$(-2, 3.5)$
$(0, 3.5)$
$(3, 3.5)$

57.

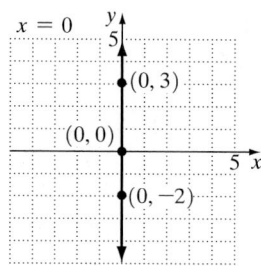

$x = 0$
$(0, 3)$
$(0, 0)$
$(0, -2)$

59.

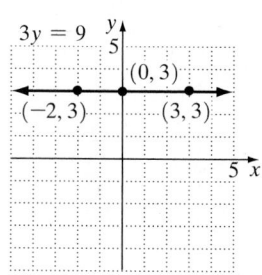

$3y = 9$
$(0, 3)$
$(-2, 3)$
$(3, 3)$

61.

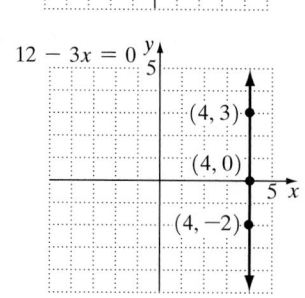

$12 - 3x = 0$
$(4, 3)$
$(4, 0)$
$(4, -2)$

63. from 3 to 12 sec
65. 45; The vulture was 45 meters above the ground when the observation started.
67. 12, 13, 14, 15, 16; The vulture is on the ground at this time.
69. 8:00 A.M. to 11:00 A.M.
71. 11:00 A.M. to 1:00 P.M.
73. $y = 80$　　**75–83.** Answers will vary.
85. $2x + 5y = 10$
87. Answers will vary.

89. x-intercept: 3; y-intercept: -9;

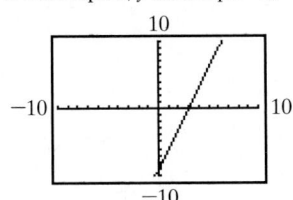

91. x-intercept: -10; y-intercept: 20;

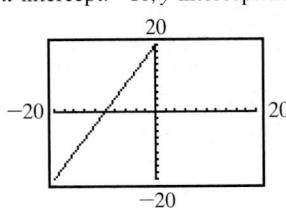

92. 13.4
93. $4x + 5$

94.

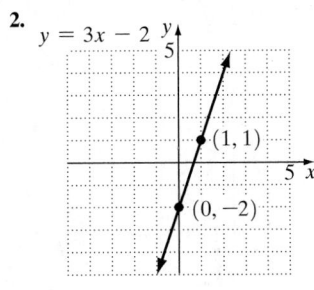

Section 3.3

Check Point Exercises

1. a. 6   **b.** $-\dfrac{7}{5}$ **2. a.** 0   **b.** undefined   **3.** Both slopes equal 2, so the lines are parallel.   **4.** $\dfrac{2}{15} \approx 0.13$; The slope indicates that the number of U.S. men living alone is projected to increase by 0.13 million each year. The rate of change is 0.13 million men per year.

Exercise Set 3.3

1. $\dfrac{3}{4}$; rises   **3.** $\dfrac{1}{4}$; rises   **5.** 0; horizontal   **7.** -5; falls   **9.** undefined; vertical   **11.** $\dfrac{1}{2}$ **13.** $-\dfrac{1}{3}$ **15.** $-\dfrac{1}{2}$ **17.** $-\dfrac{2}{3}$
19. 0   **21.** undefined   **23.** parallel   **25.** not parallel   **27.** 250; The amount spent online per U.S. online household increased by $250 each year from 1999 to 2001.   **29.** $-\dfrac{112}{3} \approx -37.33$; The federal budget surplus will decrease at a projected rate of $37.33 billion each year from 2001 to 2010.   **31.** 0; The number of books sold each year has not changed.   **33.** $0.40 per mi
35. $\dfrac{1}{3}$ **37.** 8.3%   **39–43.** Answers will vary.   **45.** b   **47.** b_2, b_1, b_4, b_3 **49.** -3 **51.** $\dfrac{3}{4}$ **53.** 12 in. and 24 in.
54. -42 **55.** $\{x \mid x \le 4\}$;

Section 3.4

Check Point Exercises

1. a. $5; -3$ **b.** $\dfrac{2}{3}; 4$ **c.** $-7; 6$
2.

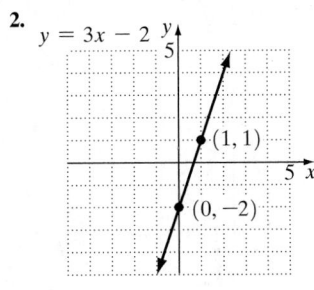

3.

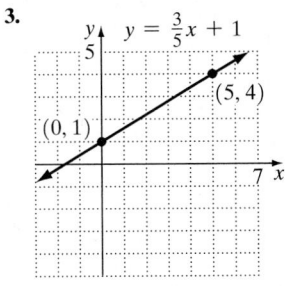

4.

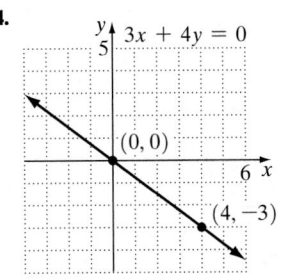

Exercise Set 3.4

1. $3; 2$ **3.** $3; -5$ **5.** $-\dfrac{1}{2}; 5$ **7.** $7; 0$ **9.** $0; 10$ **11.** $-1; 4$ **13.** $y = 5x + 7; 5; 7$ **15.** $y = -x + 6; -1; 6$

17. $y = -6x; -6; 0$ **19.** $y = 2x; 2; 0$ **21.** $y = -\dfrac{2}{7}x; -\dfrac{2}{7}; 0$ **23.** $y = -\dfrac{3}{2}x + \dfrac{3}{2}; -\dfrac{3}{2}; \dfrac{3}{2}$ **25.** $y = \dfrac{3}{4}x - 3; \dfrac{3}{4}; -3$

27.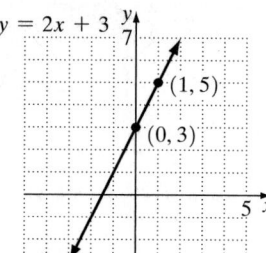

29.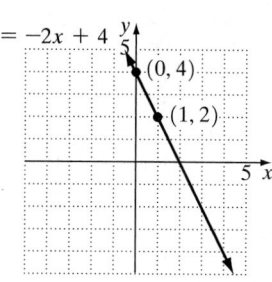

31.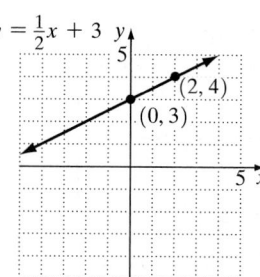

33.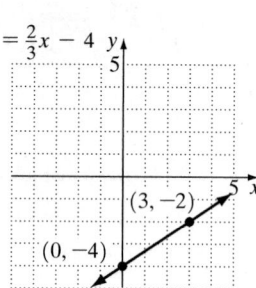

35.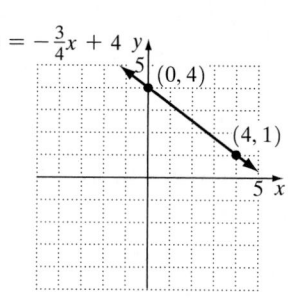

37.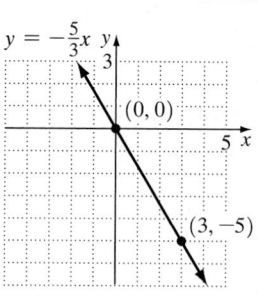

39. a. $y = -3x$ **b.** $-3; 0$

41. a. $y = \frac{4}{3}x$ **b.** $\frac{4}{3}; 0$

43. a. $y = -2x + 3$ **b.** $-2; 3$

c.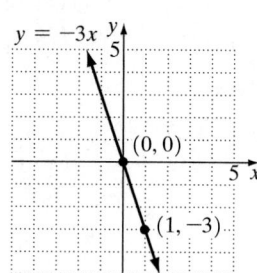

c.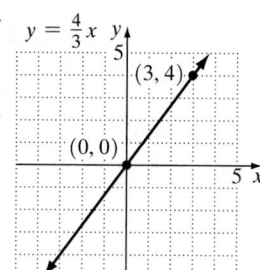

c.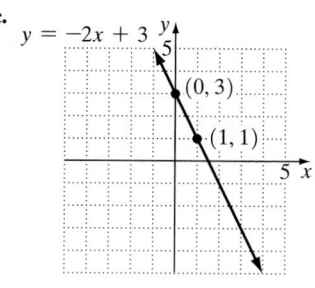

45. a. $y = -\frac{7}{2}x + 7$ **b.** $-\frac{7}{2}; 7$

c.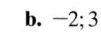

47.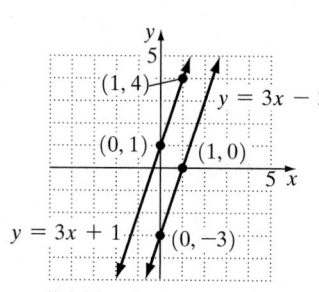

parallel; The slopes are equal.

49.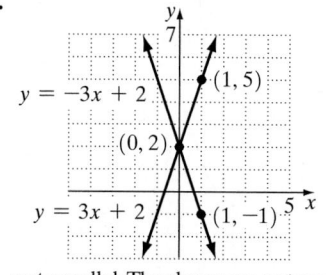

not parallel; The slopes are not equal.

51. a. 38%; 37.6%; 37.2%; 36.8%; 34%; 30%   **b.** -0.4; The percentage of U.S. men smoking is decreasing by 0.4 each year.
c. 38; In 1980 there were 38% of U.S. men smoking.
53. a. 24; In 1991 the cost was $24.   **b.** 2; The cost increased at a rate of $2 each year from 1991 to 2000.   **c.** $y = 2x + 24$ **d.** $52
55–59. Answers will vary.   **61.** 8 or {8}   **62.** 0   **63.** 56

Section 3.5

Check Point Exercises

1. $y + 5 = 6(x - 2); y = 6x - 17$ **2. a.** $y + 1 = -5(x + 2)$ or $y + 6 = -5(x + 1)$ **b.** $y = -5x - 11$
3. $y - 5 = 3(x + 2); y = 3x + 11$ **4.** 3   **5.** $y = 2.32x + 180.1; 319.3$ million

Exercise Set 4.5

1. $y - 5 = 2(x - 3); y = 2x - 1$ **3.** $y - 5 = 6(x + 2); y = 6x + 17$ **5.** $y + 3 = -3(x + 2); y = -3x - 9$

7. $y - 0 = -4(x + 4); y = -4x - 16$ **9.** $y + 2 = -1\left(x + \frac{1}{2}\right); y = -x - \frac{5}{2}$ **11.** $y - 0 = \frac{1}{2}(x - 0); y = \frac{1}{2}x$

13. $y + 2 = -\frac{2}{3}(x - 6); y = -\frac{2}{3}x + 2$ **15.** $y - 2 = 2(x - 1)$ or $y - 10 = 2(x - 5); y = 2x$

17. $y - 0 = 1(x + 3)$ or $y - 3 = 1(x - 0); y = x + 3$ **19.** $y + 1 = 1(x + 3)$ or $y - 4 = 1(x - 2); y = x + 2$

21. $y + 2 = \frac{4}{3}(x + 3)$ or $y - 6 = \frac{4}{3}(x - 3); y = \frac{4}{3}x + 2$ **23.** $y + 1 = 0(x + 3)$ or $y + 1 = 0(x - 4); y = -1$

25. $y - 4 = 1(x - 2)$ or $y - 0 = 1(x + 2); y = x + 2$ **27.** $y - 0 = 8\left(x + \frac{1}{2}\right)$ or $y - 4 = 8(x - 0); y = 8x + 4$

29. a. 5   **b.** $-\frac{1}{5}$ **31. a.** -7 **b.** $\frac{1}{7}$ **33. a.** $\frac{1}{2}$ **b.** -2 **35. a.** $-\frac{2}{5}$ **b.** $\frac{5}{2}$ **37. a.** -4 **b.** $\frac{1}{4}$ **39. a.** $-\frac{1}{2}$ **b.** 2

41. a. $\frac{2}{3}$ **b.** $-\frac{3}{2}$ **43. a.** undefined   **b.** 0   **45.** $y - 2 = 2(x - 4); y = 2x - 6$ **47.** $y - 4 = -\frac{1}{2}(x - 2); y = -\frac{1}{2}x + 5$

49. $y + 10 = -4(x + 8); y = -4x - 42$ **51.** $y + 3 = -5(x - 2); y = -5x + 7$ **53.** $y - 2 = \frac{2}{3}(x + 2); y = \frac{2}{3}x + \frac{10}{3}$

55. $y + 7 = -2(x - 4); y = -2x + 1$ **57. a.** $y - 162 = 1(x - 2)$ or $y - 168 = 1(x - 8)$ **b.** $y = x + 160$ **c.** 175 lb

59. $y - 3 = -\frac{2}{3}(x - 12); y = -\frac{2}{3}x + 11; 6.3$ **61. a–c.** Answers will vary.   **63.** Answers will vary.   **65. c**

67. $E = 2.4M - 20$ **69.**

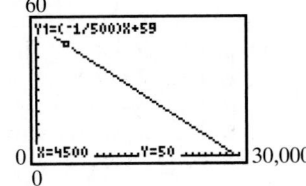

71. at most 12 sheets of paper
72. $1, \sqrt{4}$
73.

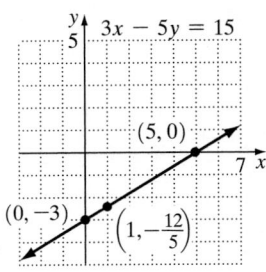

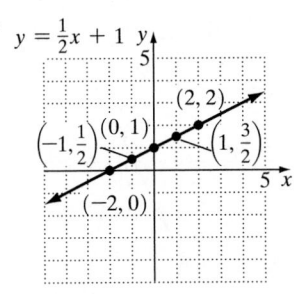

Review Exercises

1. $(-3, 3)$ is not a solution; $(0, 6)$ and $(1, 9)$ are solutions.   **2.** $(0, 4)$ and $(-1, 15)$ are not solutions; $(4, 0)$ is a solution.

3. a.

x	(x, y)
-2	$(-2, -7)$
-1	$(-1, -5)$
0	$(0, -3)$
1	$(1, -1)$
2	$(2, 1)$

b.

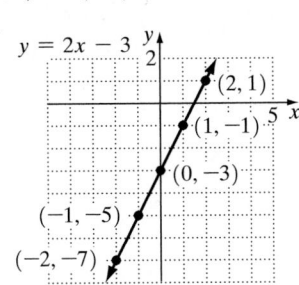

4. a.

x	(x, y)
-2	$(-2, 0)$
-1	$\left(-1, \frac{1}{2}\right)$
0	$(0, 1)$
1	$\left(1, \frac{3}{2}\right)$
2	$(2, 2)$

b.

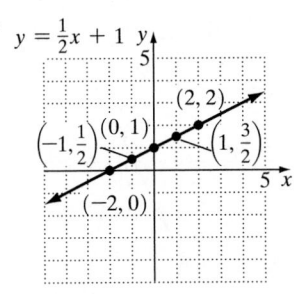

5.

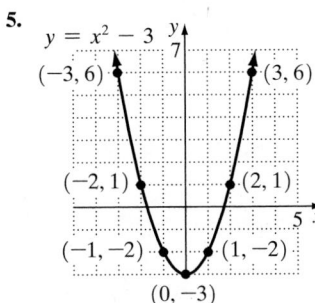

6. a.

x	(x, y)
10	$(10, 9)$
12	$(12, 19)$
14	$(14, 29)$
16	$(16, 39)$

b. Answers will vary.

7. a. -2
b. -4
8. a. no x-intercept
b. 2
9. a. 0
b. 0

10.

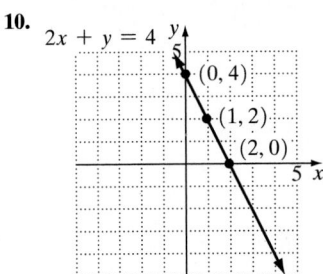

11.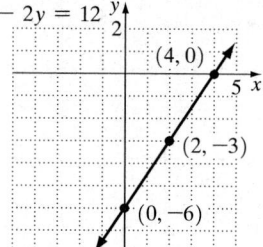

$3x - 2y = 12$
$(4, 0)$
$(2, -3)$
$(0, -6)$

12.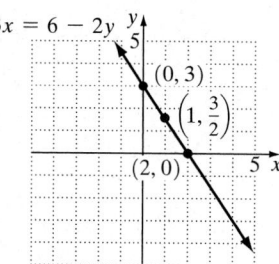

$3x = 6 - 2y$
$(0, 3)$
$\left(1, \dfrac{3}{2}\right)$
$(2, 0)$

13.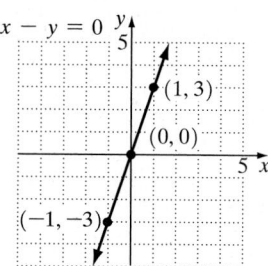

$3x - y = 0$
$(1, 3)$
$(0, 0)$
$(-1, -3)$

14.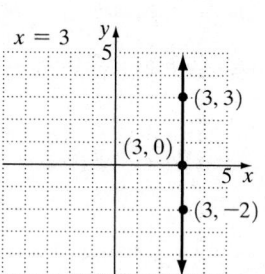

$x = 3$
$(3, 3)$
$(3, 0)$
$(3, -2)$

15.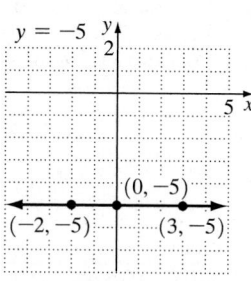

$y = -5$
$(0, -5)$
$(-2, -5)$
$(3, -5)$

16.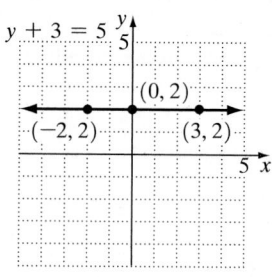

$y + 3 = 5$
$(0, 2)$
$(-2, 2)$
$(3, 2)$

17.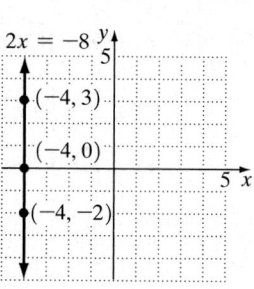

$2x = -8$
$(-4, 3)$
$(-4, 0)$
$(-4, -2)$

18. a. 5:00 P.M.; $-4\,°F$ **b.** 8:00 P.M.; $16\,°F$
c. 4 and 6; At 4:00 P.M. and 6:00 P.M., the temperature was $0\,°F$.
d. 12; At noon the temperature was $12\,°F$.
e. The temperature stayed the same, $12\,°F$.

19. $-\dfrac{1}{2}$; falls **20.** 3; rises **21.** 0; horizontal

22. undefined; vertical **23.** $\dfrac{3}{5}$ **24.** undefined

25. $-\dfrac{1}{3}$ **26.** 0 **27.** not parallel **28.** parallel

29. a. 26; The number of lawyers increased at a rate of 26 thousand per year from 1974 to 2000.
b. 6.25; The number of lawyers increased at a rate of 6.25 thousand per year from 1950 to 1974.

30. 5; -7 **31.** -4; 6 **32.** 0; 3 **33.** $-\dfrac{2}{3}$; 2

34.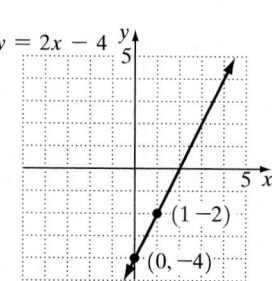

$y = 2x - 4$
$(1, -2)$
$(0, -4)$

35.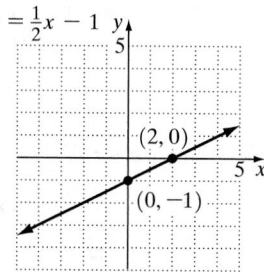

$y = \dfrac{1}{2}x - 1$
$(2, 0)$
$(0, -1)$

36.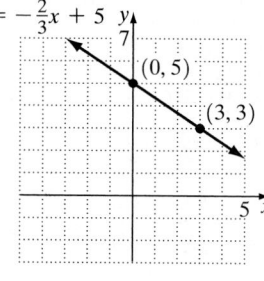

$y = -\dfrac{2}{3}x + 5$
$(0, 5)$
$(3, 3)$

37.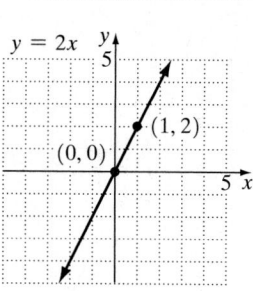

$y = 2x$
$(1, 2)$
$(0, 0)$

38.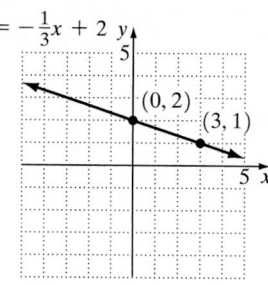

$y = -\dfrac{1}{3}x + 2$
$(0, 2)$
$(3, 1)$

39.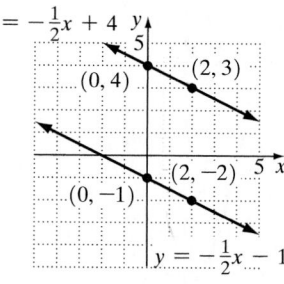

$y = -\dfrac{1}{2}x + 4$
$(0, 4)$
$(2, 3)$
$(2, -2)$
$(0, -1)$
$y = -\dfrac{1}{2}x - 1$

Yes, they are parallel since both have a slope of $-\dfrac{1}{2}$.

40. a. 25; In 1990 the average age of U.S. Hispanics was 25.
b. 0.3; The average age for U.S. whites increased at a rate of about 0.3 per year from 1990 to 2000. **c.** $y = 0.3x + 35$ **d.** 41 years old
41. $y - 7 = 6(x + 4); y = 6x + 31$ **42.** $y - 4 = 3(x - 3)$ or $y - 1 = 3(x - 2); y = 3x - 5$
43. $y + 7 = -3(x - 4); y = -3x + 5$ **44.** $y - 6 = -3(x + 2); y = -3x$
45. a. $y - 16 = -0.13(x - 0)$ or $y - 12.1 = -0.13(x - 30)$ **b.** $y = -0.13x + 16$ **c.** 6.9 ft in 1970 and 5.6 ft in 1980
d. Answers will vary.

Chapter 3 Test

1. $(-2, 1)$ is not a solution; $(0, -5)$ and $(4, 3)$ are solutions.

2.

x	(x, y)
-2	$(-2, -5)$
-1	$(-1, -2)$
0	$(0, 1)$
1	$(1, 4)$
2	$(2, 7)$

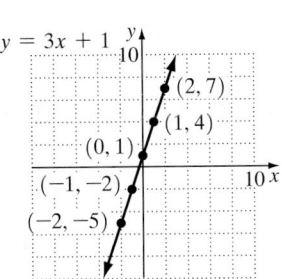

3.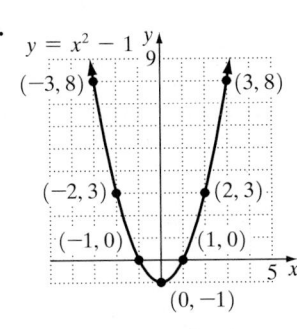

4. a. 2 **b.** -3

5.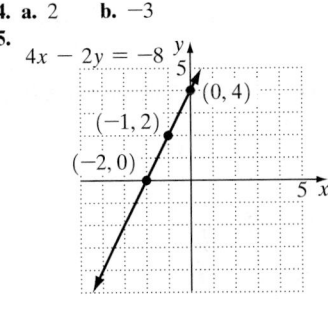

6.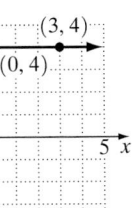

7. 3; rises **8.** undefined; vertical **9.** $\dfrac{3}{2}$ **10.** parallel **11.** $-1; 10$ **12.** $-2; 6$

13.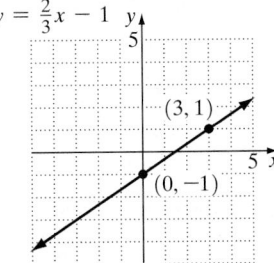

14.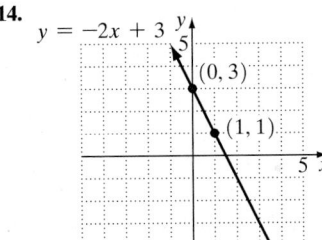

15. $y - 4 = -2(x + 1); y = -2x + 2$ **16.** $y - 1 = 3(x - 2)$ or $y + 8 = 3(x + 1); y = 3x - 5$
17. $y - 3 = 2(x + 2); y = 2x + 7$ **18.** 106; Spending per pupil increased at a rate of about \$106 per year.
19. a. $y - 320 = 42(x - 0)$ or $y - 530 = 42(x - 5)$ **b.** $y = 42x + 320$ **c.** \$866

Cumulative Review Exercises (Chapters 1–3)

1. 2 **2.** $4 - 6x$ **3.** $\sqrt{5}$ **4.** 19 or $\{19\}$ **5.** $\dfrac{5}{4}$ or $\left\{\dfrac{5}{4}\right\}$ **6.** $x = \dfrac{y - b}{m}$ **7.** 800 **8.** 40 mph

9. $\{x | x \le -2\}$

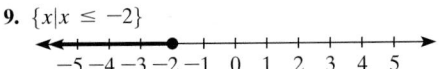

10. $\{x | x < 0\}$

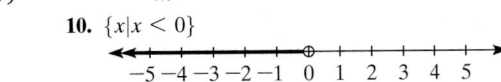

11. 39 **12.** $<$ **13.** $6°F$ **14.** 2015 **15.** width: 53 m; length: 120 m **16.** 200 lb **17.** 6 hr **18.** $40°, 60°, 80°$

19.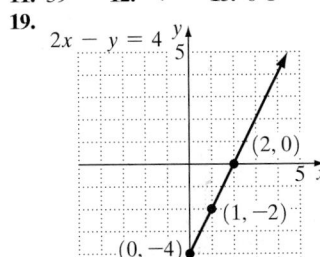

20.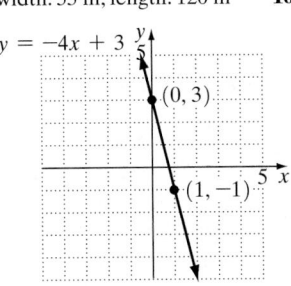

CHAPTER 4

Section 4.1

Check Point Exercises

1. a. solution   **b.** not a solution   **2.** $(1, 4)$ or $\{(1, 4)\}$ **3.** $(3, 3)$ or $\{(3, 3)\}$ **4.** no solution or $\varnothing$
5. infinitely many solutions; $\{(x, y)|x + y = 3\}$ or $\{(x, y)|2x + 2y = 6\}$

Exercise Set 4.1

1. solution   **3.** solution   **5.** not a solution   **7.** solution   **9.** not a solution   **11.** $(4, 2)$ or $\{(4, 2)\}$ **13.** $(-1, 2)$ or $\{(-1, 2)\}$
15. $(3, 0)$ or $\{(3, 0)\}$ **17.** $(1, 0)$ or $\{(1, 0)\}$ **19.** $(-1, 4)$ or $\{(-1, 4)\}$ **21.** $(2, 4)$ or $\{(2, 4)\}$ **23.** $(2, -1)$ or $\{(2, -1)\}$
25. no solution or $\varnothing$ **27.** $(-2, 6)$ or $\{(-2, 6)\}$ **29.** infinitely many solutions; $\{(x, y)|x - 2y = 4\}$ or $\{(x, y)|2x - 4y = 8\}$
31. $(2, 3)$ or $\{(2, 3)\}$ **33.** no solution or $\varnothing$ **35.** infinitely many solutions; $\{(x, y)|x - y = 0\}$ or $\{(x, y)|y = x\}$
37. $(2, 4)$ or $\{(2, 4)\}$ **39.** no solution or $\varnothing$ **41.** no solution or $\varnothing$ **43. a.** $(1996, 41)$; Mothers 30 years and older had 41 thousand
births in 1996.   **b.** There are more births to mothers 30 years and older than to mothers under 30 years old.
45–51. Answers will vary.   **53.** c   **55–57.** Answers will vary.   **59.** $(6, -1)$ or $\{(6, -1)\}$ **61.** $(3, 0)$ or $\{(3, 0)\}$
63. $(2, -1)$ or $\{(2, -1)\}$ **65.** $(-4, 4)$ or $\{(-4, 4)\}$ **66.** -12 **67.** 6   **68.** 27

Section 4.2

Check Point Exercises

1. $(3, 2)$ or $\{(3, 2)\}$ **2.** $(1, -2)$ or $\{(1, -2)\}$ **3.** no solution or $\varnothing$
4. infinitely many solutions; $\{(x, y)|y = 3x - 4\}$ or $\{(x, y)|9x - 3y = 12\}$ **5.** $\$30; 400$ units

Exercise Set 4.2

1. $(1, 3)$ or $\{(1, 3)\}$ **3.** $(5, 1)$ or $\{(5, 1)\}$ **5.** $(2, 1)$ or $\{(2, 1)\}$ **7.** $(-1, 3)$ or $\{(-1, 3)\}$ **9.** $(4, 5)$ or $\{(4, 5)\}$
11. $\left(-4, \dfrac{5}{4}\right)$ or $\left\{\left(-4, \dfrac{5}{4}\right)\right\}$ **13.** no solution or $\varnothing$ **15.** infinitely many solutions; $\{(x, y)|y = 3x - 5\}$ or $\{(x, y)|21x - 35 = 7y\}$
17. $(0, 0)$ or $\{(0, 0)\}$ **19.** $(3, -2)$ or $\{(3, -2)\}$ **21.** no solution or $\varnothing$ **23.** $(-22, -5)$ or $\{(-22, -5)\}$
25. $(5, 2)$ or $\{(5, 2)\}$ **27.** $(7, 3)$ or $\{(7, 3)\}$ **29.** $(-1, -1)$ or $\{(-1, -1)\}$ **31.** $(5, 4)$ or $\{(5, 4)\}$
33. a. 6500 sold; 6200 supplied   **b.** $\$50; 6250$ tickets   **35.** 2700 gal
37. 2032; about 0.6 deaths per 1000 live births or less than one death in 1000 live births   **39–43.** Answers will vary.
45. $x = 1, y = -3, z = 5$ **47.**
48. 12 or $\{12\}$ **49.** $-73, 0, \dfrac{3}{1}$

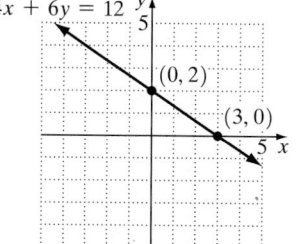
$4x + 6y = 12$
$(0, 2)$
$(3, 0)$

Section 4.3

Check Point Exercises

1. $(7, -2)$ or $\{(7, -2)\}$ **2.** $(6, 2)$ or $\{(6, 2)\}$ **3.** $(2, -1)$ or $\{(2, -1)\}$ **4.** $\left(\dfrac{60}{17}, -\dfrac{11}{17}\right)$ or $\left\{\left(\dfrac{60}{17}, -\dfrac{11}{17}\right)\right\}$ **5.** no solution or $\varnothing$
6. infinitely many solutions; $\{(x, y)|x - 5y = 7\}$ or $\{(x, y)|3x - 15y = 21\}$

Exercise Set 4.3

1. $(2, -1)$ or $\{(2, -1)\}$ **3.** $(3, 0)$ or $\{(3, 0)\}$ **5.** $(-3, 5)$ or $\{(-3, 5)\}$ **7.** $(3, 1)$ or $\{(3, 1)\}$ **9.** $(-4, 3)$ or $\{(-4, 3)\}$
11. $(-6, -2)$ or $\{(-6, -2)\}$ **13.** $(4, -1)$ or $\{(4, -1)\}$ **15.** $(3, 1)$ or $\{(3, 1)\}$ **17.** $(1, -2)$ or $\{(1, -2)\}$ **19.** $(-1, 1)$ or $\{(-1, 1)\}$
21. $(3, 1)$ or $\{(3, 1)\}$ **23.** $(-5, -2)$ or $\{(-5, -2)\}$ **25.** $\left(\dfrac{11}{12}, -\dfrac{7}{6}\right)$ or $\left\{\left(\dfrac{11}{12}, -\dfrac{7}{6}\right)\right\}$ **27.** $\left(\dfrac{23}{16}, \dfrac{3}{8}\right)$ or $\left\{\left(\dfrac{23}{16}, \dfrac{3}{8}\right)\right\}$
29. no solution or $\varnothing$ **31.** infinitely many solutions; $\{(x, y)|x + 3y = 2\}$ or $\{(x, y)|3x + 9y = 6\}$ **33.** no solution or $\varnothing$
35. $\left(\dfrac{1}{2}, -\dfrac{1}{2}\right)$ or $\left\{\left(\dfrac{1}{2}, -\dfrac{1}{2}\right)\right\}$ **37.** infinitely many solutions; $\{(x, y)|x = 5 - 3y\}$ or $\{(x, y)|2x + 6y = 10\}$

39. $\left(\frac{1}{3}, 1\right)$ or $\left\{\left(\frac{1}{3}, 1\right)\right\}$ **41.** $(-10, 21)$ or $\{(-10, 21)\}$ **43.** $(0, 1)$ or $\{(0, 1)\}$ **45.** $(2, -1)$ or $\{(2, -1)\}$
47. $(1, -3)$ or $\{(1, -3)\}$ **49.** $(4, 3)$ or $\{(4, 3)\}$ **51.** no solution or $\varnothing$
53. infinitely many solutions; $\{(x, y)|2(x + 2y) = 6\}$ or $\{(x, y)|3(x + 2y - 3) = 0\}$ **55.** $(3, 2)$ or $\{(3, 2)\}$
57–61. Answers will vary.   **63.** $(-1, 2)$ or $\{(-1, 2)\}$ **65.** $(1, 0)$ or $\{(1, 0)\}$ **67.** Answers will vary.   **69.** $(5, 1)$ or $\{(5, 1)\}$
71. 10   **72.** II   **73.** 26 or $\{26\}$

Section 4.4

Check Point Exercises

1. A bustard weighs 46 lb; a condor weighs 27 lb.   **2.** A Quarter Pounder has 420 cal; a Whopper with cheese has 589 cal.
3. 17.5 yr; $24,250   **4.** $5000 at 6%; $11,000 at 8%   **5.** 4 l of 18% solution; 8 l of 45% solution   **6.** boat: 35 mph; current: 7 mph

Exercise Set 4.4

1. 3 and 4   **3.** 2 and 5   **5.** 5.8 million lb of potato chips; 4.6 million lb of tortilla chips
7. pan pizza: 1120 calories; beef burrito: 430 calories   **9.** scrambled eggs: 366 mg; Double Beef Whopper: 175 mg
11. sweater: $12; shirt: $10   **13. a.** 300 min; $35   **b.** plan B; Answers will vary.   **15.** $600 of merchandise; $580
17. 26 yr; 2011; college grads: $1158; high school grads: $579   **19.** 2 servings of macaroni and 4 servings of broccoli
21. $12,500 at 7%; $7500 at 9%   **23.** $220,000 at 8%; $30,000 at 18%   **25.** $6000 at 12%; $2000 at a 5% loss
27. 12 g of 45% fat content; 18 g of 20% fat content   **29.** north campus: 300 students; south campus: 900 students
31. 36 lb of the $6-per-pound tea; 108 lb of the $8-per-pound tea   **33.** plane: 130 mph; wind: 30 mph
35. crew: 6 km/hr; current: 2 km/hr   **37.** speed in still water: 4.5 mph; current: 1.5 mph   **39–41.** Answers will vary.
43. 10 birds and 20 lions   **45.** There are 5 people downstairs and 7 people upstairs.   **47.** Answers will vary.
48. $-\frac{1}{4}$ **49.** $-\frac{11}{20}$ **50.**

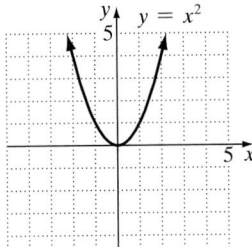

Review Exercises

1. solution   **2.** not a solution   **3.** no   **4.** $(4, -2)$ or $\{(4, -2)\}$ **5.** $(3, -2)$ or $\{(3, -2)\}$
6. $(2, 0)$ or $\{(2, 0)\}$ **7.** $(2, 1)$ or $\{(2, 1)\}$ **8.** $(4, -1)$ or $\{(4, -1)\}$ **9.** no solution or $\varnothing$
10. infinitely many solutions; $\{(x, y)|2x - 4y = 8\}$ or $\{(x, y)|x - 2y = 4\}$ **11.** no solution or $\varnothing$
12. $(-2, -6)$ or $\{(-2, -6)\}$ **13.** $(2, 5)$ or $\{(2, 5)\}$ **14.** no solution or $\varnothing$ **15.** $(2, -1)$ or $\{(2, -1)\}$
16. $(5, 4)$ or $\{(5, 4)\}$ **17.** $(-2, -1)$ or $\{(-2, -1)\}$ **18.** $(1, -4)$ or $\{(1, -4)\}$ **19.** $(20, -21)$ or $\{(20, -21)\}$
20. infinitely many solutions; $\{(x, y)|4x + y = 5\}$ or $\{(x, y)|12x + 3y = 15\}$ **21.** no solution or $\varnothing$
22. $(4, 18)$ or $\{(4, 18)\}$ **23.** $\left(-1, -\frac{1}{2}\right)$ or $\left\{\left(-1, -\frac{1}{2}\right)\right\}$ **24.** $12.50; 250 copies   **25.** $(2, 4)$ or $\{(2, 4)\}$
26. $(-1, -1)$ or $\{(-1, -1)\}$ **27.** $(2, -1)$ or $\{(2, -1)\}$ **28.** $(3, 2)$ or $\{(3, 2)\}$ **29.** $(2, 1)$ or $\{(2, 1)\}$
30. $(0, 0)$ or $\{(0, 0)\}$ **31.** $\left(\frac{17}{7}, -\frac{15}{7}\right)$ or $\left\{\left(\frac{17}{7}, -\frac{15}{7}\right)\right\}$ **32.** no solution or $\varnothing$
33. infinitely many solutions; $\{(x, y)|3x - 4y = -1\}$ or $\{(x, y)|-6x + 8y = 2\}$ **34.** $(4, -2)$ or $\{(4, -2)\}$
35. $(-8, -6)$ or $\{(-8, -6)\}$ **36.** $(-4, 1)$ or $\{(-4, 1)\}$ **37.** $\left(\frac{5}{2}, 3\right)$ or $\left\{\left(\frac{5}{2}, 3\right)\right\}$ **38.** $(3, 2)$ or $\{(3, 2)\}$
39. $\left(\frac{1}{2}, -2\right)$ or $\left\{\left(\frac{1}{2}, -2\right)\right\}$ **40.** no solution or $\varnothing$ **41.** $(3, 7)$ or $\{(3, 7)\}$ **42.** horses: 20 yr; lion: 15 yr
43. gorilla: 485 lb; orangutan: 165 lb   **44.** shrimp: 42 mg; scallops: 15 mg   **45.** 9 ft by 5 ft   **46.** room: $80; car: $60
47. 200 min; $25   **48.** $3000 at 8%; $7000 at 10%   **49.** 10 ml of 34%; 90 ml of 4%   **50.** plane: 630 mph; wind: 90 mph

Chapter 4 Test

1. solution   **2.** not a solution   **3.** $(2, 4)$ or $\{(2, 4)\}$ **4.** $(2, 4)$ or $\{(2, 4)\}$ **5.** $(1, -3)$ or $\{(1, -3)\}$
6. $(2, -3)$ or $\{(2, -3)\}$ **7.** no solution or $\varnothing$ **8.** $(-1, 4)$ or $\{(-1, 4)\}$ **9.** $(-4, 3)$ or $\{(-4, 3)\}$

10. infinitely many solutions; $\{(x, y)|3x - 2y = 2\}$ or $\{(x, y)|-9x + 6y = -6\}$
11. World War II: \$310 billion; Vietnam War: \$190 billion   **12.** 500 min; \$40   **13.** \$4000 at 9%; \$2000 at 6%
14. 40 oz of 20%; 20 oz of 50%   **15.** plane: 725 km/hr; wind: 75 km/hr

Cumulative Review Exercises (Chapters 1–4)

1. -36 **2.** $17x - 11$ **3.** -6 or $\{-6\}$ **4.** 20 or $\{20\}$ **5.** $t = \dfrac{A - P}{Pr}$

6. $\{x|x > 2\}$;

7.

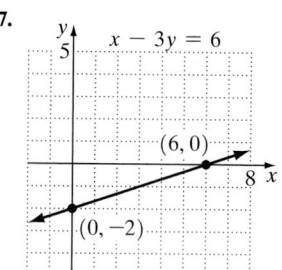

8.

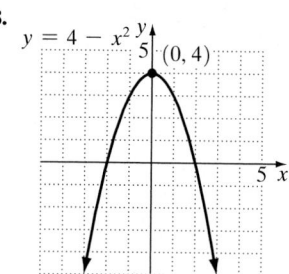

9.

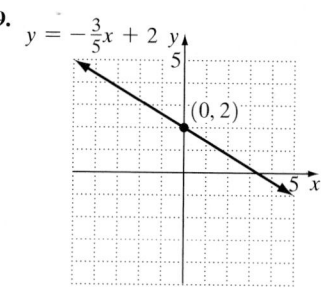

10. $(0, -2)$ or $\{(0, -2)\}$ **11.** $\left(\dfrac{3}{2}, -2\right)$ or $\left\{\left(\dfrac{3}{2}, -2\right)\right\}$ **12.** 1   **13.** $y - 6 = -4(x + 1)$; $y = -4x + 2$ **14.** 10 ft

15. pen: \$0.80; pad: \$1.20   **16.** $-93, 0, \dfrac{7}{1}, \sqrt{100}$ **17.** 20%   **18.** one computer   **19.** 7 yr; 2004   **20.** 2012

CHAPTER 5

Section 5.1

Check Point Exercises

1. $5x^3 + 4x^2 - 8x - 20$ **2.** $5x^3 + 4x^2 - 8x - 20$ **3.** $7x^2 + 11x + 4$ **4.** $7x^3 + 3x^2 + 12x - 8$ **5.** $3y^3 - 10y^2 - 11y - 8$
6. approx 4 per thousand; approx $(40, 4)$

Exercise Set 5.1

1. binomial, 1   **3.** binomial, 3   **5.** monomial, 2   **7.** monomial, 0   **9.** trinomial, 2   **11.** trinomial, 4   **13.** binomial, 3
15. monomial, 23   **17.** $-3x + 10$ **19.** $10x^2 + 15x - 11$ **21.** $7x^2 - 4x$ **23.** $4x^2 - x + 18$ **25.** $4y^3 + 10y^2 + y - 2$
27. $3x^3 + 2x^2 - 9x + 7$ **29.** $-2y^3 + 4y^2 + 13y + 13$ **31.** $-3y^6 + 8y^4 + y^2$ **33.** $10x^3 + 1$ **35.** $-\dfrac{2}{5}x^4 + x^3 - \dfrac{1}{8}x^2$
37. $0.01x^5 + x^4 - 0.1x^3 + 0.3x + 0.33$ **39.** $11y^3 - 3y^2$ **41.** $-2x^2 - x + 1$ **43.** $-\dfrac{1}{4}x^4 - \dfrac{7}{15}x^3 - 0.3$
45. $-y^3 + 8y^2 - 3y - 14$ **47.** $-5x^3 - 6x^2 + x - 4$ **49.** $7x^4 - 2x^3 + 4x - 2$ **51.** $8x^2 + 7x - 5$ **53.** $9x^3 - 4.9x^2 + 11.1$
55. $-2x - 10$ **57.** $-5x^2 - 9x - 12$ **59.** $-5x^2 - x$ **61.** $-4x^2 - 4x - 6$ **63.** $-2y - 6$ **65.** $6y^3 + y^2 + 7y - 20$
67. $n^3 + 2$ **69.** $y^6 - y^3 - y^2 + y$ **71.** $26x^4 + 9x^2 + 6x$ **73.** $\dfrac{5}{7}x^3 - \dfrac{9}{20}x$ **75.** $4x + 6$ **77.** $10x^2 - 7$
79. $-4y^2 - 7y + 5$ **81.** $9x^3 + 11x^2 - 8$ **83.** $-y^3 + 8y^2 + y + 14$ **85.** $7x^4 - 2x^3 + 3x^2 - x + 2$
87. $0.05x^3 + 0.02x^2 + 1.02x$ **89.** 54, 72 and 54; Answers will vary.   **91.** 79.3%   **93.** 1120.5 cigarettes; Answers will vary.
95. 42 human years   **97.** 3 dog years   **99–107.** Answers will vary.   **109.** $-3x^2 - x - 2$ **111.** Answers will vary.
112. -10 **113.** 5.6   **114.** -4 or $\{-4\}$

Section 5.2

Check Point Exercises

1. a. 2^6 or 64   **b.** x^{10} **c.** y^8 **d.** y^9 **2. a.** 3^{20} **b.** x^{90} **c.** $(-5)^{21}$ **3. a.** $16x^4$ **b.** $-64y^6$ **4. a.** $70x^3$ **b.** $-20x^9$
5. a. $3x^2 + 15x$ **b.** $30x^5 - 12x^3 + 18x^2$ **6. a.** $x^2 + 9x + 20$ **b.** $10x^2 - 29x - 21$ **7.** $5x^3 - 18x^2 + 7x + 6$
8. $6x^5 - 19x^4 + 22x^3 - 8x^2$

Exercise Set 5.2

1. x^{15} **3.** y^8 **5.** x^{11} **7.** 3^{19} **9.** 3^{90} **11.** x^{20} **13.** $(-2)^9$ **15.** $8x^3$ **17.** $25x^2$ **19.** $16x^6$ **21.** $16y^{24}$ **23.** $-32x^{35}$

25. $14x^2$ **27.** $24x^3$ **29.** $-15y^7$ **31.** $\frac{1}{8}a^5$ **33.** $-48x^7$ **35.** $4x^2 + 12x$ **37.** $x^2 - 3x$ **39.** $2x^2 - 12x$ **41.** $-12y^2 - 20y$

43. $4x^3 + 8x^2$ **45.** $2y^4 + 6y^3$ **47.** $6y^4 - 8y^3 + 14y^2$ **49.** $6x^4 + 8x^3$ **51.** $-2x^3 - 10x^2 + 6x$ **53.** $12x^4 - 3x^3 + 15x^2$

55. $x^2 + 8x + 15$ **57.** $2x^2 + 9x + 4$ **59.** $x^2 - 2x - 15$ **61.** $x^2 - 2x - 99$ **63.** $2x^2 + 3x - 20$ **65.** $\frac{3}{16}x^2 + \frac{11}{4}x - 4$

67. $x^3 + 3x^2 + 5x + 3$ **69.** $y^3 - 6y^2 + 13y - 12$ **71.** $2a^3 - 9a^2 + 19a - 15$ **73.** $x^4 + 3x^3 + 5x^2 + 7x + 4$

75. $4x^4 - 4x^3 + 6x^2 - \frac{17}{2}x + 3$ **77.** $x^4 + x^3 + x^2 + 3x + 2$ **79.** $x^3 + 3x^2 - 37x + 24$ **81.** $2x^3 - 9x^2 + 27x - 27$

83. $2x^4 + 9x^3 + 6x^2 + 11x + 12$ **85.** $12z^4 - 14z^3 + 19z^2 - 22z + 8$ **87.** $21x^5 - 43x^4 + 38x^3 - 24x^2$

89. $4y^6 - 2y^5 - 6y^4 + 5y^3 - 5y^2 + 8y - 3$ **91.** $x^4 + 6x^3 - 11x^2 - 4x + 3$ **93.** $2x^2 + 7x - 15$ ft^2 **95. a.** $(2x + 1)(x + 2)$

b. $2x^2 + 5x + 2$ **c.** $(2x + 1)(x + 2) = 2x^2 + 5x + 2$ **97–105.** Answers will vary. **107.** $8x + 16$ **109.** $-8x^4$

110. $\{x | x < -1\}$ **111.** **112.** $-\dfrac{2}{3}$

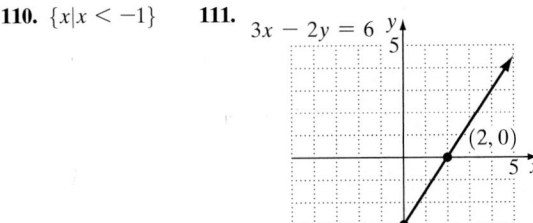

Section 5.3

Check Point Exercises

1. $x^2 + 11x + 30$ **2.** $28x^2 - x - 15$ **3.** $6x^2 - 22x + 20$ **4. a.** $49y^2 - 64$ **b.** $16x^2 - 25$ **c.** $4a^6 - 9$
5. a. $x^2 + 20x + 100$ **b.** $25x^2 + 40x + 16$ **6. a.** $x^2 - 18x + 81$ **b.** $49x^2 - 42x + 9$

Exercise Set 5.3

1. $x^2 + 8x + 15$ **3.** $y^2 - 2y - 15$ **5.** $2x^2 + 3x - 2$ **7.** $2y^2 - y - 3$ **9.** $10x^2 - 9x - 9$ **11.** $12y^2 - 43y + 35$
13. $-15x^2 - 32x + 7$ **15.** $6y^2 - 28y + 30$ **17.** $15x^4 - 47x^2 + 28$ **19.** $-6x^2 + 17x - 10$ **21.** $x^3 + 5x^2 + 3x + 15$

23. $8x^5 + 40x^3 + 3x^2 + 15$ **25.** $x^2 - 9$ **27.** $9x^2 - 4$ **29.** $9r^2 - 16$ **31.** $9 - r^2$ **33.** $25 - 49x^2$ **35.** $4x^2 - \frac{1}{4}$

37. $y^4 - 1$ **39.** $r^6 - 4$ **41.** $1 - y^8$ **43.** $x^{20} - 25$ **45.** $x^2 + 4x + 4$ **47.** $4x^2 + 20x + 25$ **49.** $x^2 - 6x + 9$

51. $9y^2 - 24y + 16$ **53.** $16x^4 - 8x^2 + 1$ **55.** $49 - 28x + 4x^2$ **57.** $4x^2 + 2x + \frac{1}{4}$ **59.** $16y^2 - 2y + \frac{1}{16}$

61. $x^{16} + 6x^8 + 9$ **63.** $x^3 - 1$ **65.** $x^2 - 2x + 1$ **67.** $9y^2 - 49$ **69.** $12x^4 + 3x^3 + 27x^2$ **71.** $70y^2 + 2y - 12$

73. $x^4 + 2x^2 + 1$ **75.** $x^4 + 3x^2 + 2$ **77.** $x^4 - 16$ **79.** $4 - 12x^5 + 9x^{10}$ **81.** $\frac{3}{16}x^4 + 7x^2 - 96$ **83.** $x^2 + 2x + 1$

85. $4x^2 - 9$ **87.** $6x + 22$ **89.** $(x + 1)(x + 2)$ yd^2 **91.** 56 yd^2; $(6, 56)$ **93.** $(x^2 + 4x + 4)$ in^2 **95–99.** Answers will vary.
101. Answers will vary. An example is $(x - 10)$ and $(x + 2)$. **103.** $x^2 + 2x$ **105.** Change $x^2 + 2x + 4$ to $x^2 + 4x + 4$.
107. Graphs coincide. **108.** $(2, -1)$ or $\{(2, -1)\}$ **109.** $(1, 1)$ or $\{(1, 1)\}$ **110.**

Section 5.4

Check Point Exercises

1. -9 **2.** polynomial degree: 9;

Term	Coefficient	Degree
$8x^4y^5$	8	9
$-7x^3y^2$	-7	5
$-x^2y$	-1	3
$-5x$	-5	1
11	11	0

3. $2x^2y + 2xy - 4$ **4.** $3x^3 + 2x^2y + 5xy^2 - 10$
5. $60x^5y^5$ **6.** $60x^5y^7 - 12x^3y^3 + 18xy^2$
7. a. $21x^2 - 25xy + 6y^2$ **b.** $4x^2 + 16xy + 16y^2$
8. a. $36x^2y^4 - 25x^2$ **b.** $x^3 - y^3$

Exercise Set 5.4

1. 1   **3.** -47 **5.** -6

7. polynomial degree: 9;

Term	Coefficient	Degree
x^3y^2	1	5
$-5x^2y^7$	-5	9
$6y^2$	6	2
-3	-3	0

9. $7x^2y - 4xy$ **11.** $2x^2y + 13xy + 13$ **13.** $-11x^4y^2 - 11x^2y^2 + 2xy$
15. $-5x^3 + 8xy - 9y^2$ **17.** $x^4y^2 + 8x^3y + y - 6x$
19. $5x^3 + x^2y - xy^2 - 4y^3$ **21.** $-3x^2y^2 + xy^2 + 5y^2$
23. $8a^2b^4 + 3ab^2 + 8ab$ **25.** $-30x + 37y$ **27.** $18x^3y^2$
29. $-14x^5y^9$ **31.** $10x^2y + 15xy^2$ **33.** $18x^3y^2 - 6xy^3$
35. $18a^3b^5 + 15a^2b^3$ **37.** $-a^2b + ab^2 - b^3$ **39.** $7x^2 + 38xy + 15y^2$ **41.** $2x^2 + xy - 21y^2$ **43.** $15x^2y^2 + xy - 2$
45. $4x^2 + 12xy + 9y^2$ **47.** $x^2y^2 - 6xy + 9$ **49.** $x^4 + 2x^2y^2 + y^4$ **51.** $x^4 - 4x^2y^2 + 4y^4$ **53.** $9x^2 - y^2$ **55.** $a^2b^2 - 1$
57. $x^2 - y^4$ **59.** $9a^4b^2 - a^2$ **61.** $9x^2y^4 - 16y^2$ **63.** $a^3 - ab^2 + a^2b - b^3$ **65.** $x^3 + 4x^2y + 4xy^2 + y^3$
67. $x^3 - 4x^2y + 4xy^2 - y^3$ **69.** $x^2y^2 - a^2b^2$ **71.** $x^6y + x^4y + x^4 + 2x^2 + 1$ **73.** $x^4y^4 - 6x^2y^2 + 9$ **75.** $x^2 + 2xy + y^2 - 1$
77. $3x^2 + 8xy + 5y^2$ **79.** $2xy + y^2$ **81.** no; need 120 more board feet   **83.** 192 ft   **85.** 0 ft; The ball has hit the ground.
87. 2.5 to 6 sec   **89.** $(2, 192)$ **91.** 2.5 sec; 196 ft   **93.** Answers will vary.   **95.** c   **97.** $8y^2 - 2x^2$ **99.** Answers will vary.
100. $W = \dfrac{2R - L}{3}$ **101.** 3.8   **102.** $y - 5 = 3(x + 2); y = 3x + 11$

Section 5.5

Check Point Exercises

1. a. 5^8 **b.** x^7 **c.** y^{19} **2. a.** 1   **b.** 1   **c.** -1 **d.** 20   **e.** 1   **3. a.** $\dfrac{x^2}{25}$ **b.** $\dfrac{x^{12}}{8}$ **c.** $\dfrac{16a^{40}}{b^{12}}$ **4. a.** $-2x^8$ **b.** $\dfrac{1}{5}$

c. $3x^5y^3$ **5.** $-5x^7 + 2x^3 - 3x$ **6.** $5x^6 - \dfrac{7}{5}x + 2$ **7.** $3x^6y^4 - xy + 10$

Exercise Set 5.5

1. 3^{15} **3.** x^4 **5.** y^8 **7.** $5^3 \cdot 2^4$ **9.** $x^{75}y^{40}$ **11.** 1   **13.** 1   **15.** -1 **17.** 100   **19.** 1   **21.** 0   **23.** -2 **25.** $\dfrac{x^2}{9}$
27. $\dfrac{x^6}{64}$ **29.** $\dfrac{4x^6}{25}$ **31.** $-\dfrac{64}{27a^9}$ **33.** $-\dfrac{32a^{35}}{b^{20}}$ **35.** $\dfrac{x^8y^{12}}{16z^4}$ **37.** $3x^5$ **39.** $-2x^{20}$ **41.** $-\dfrac{1}{2}y^3$ **43.** $\dfrac{7}{5}y^{12}$ **45.** $6x^5y^4$
47. $-\dfrac{1}{2}x^{12}$ **49.** $\dfrac{9}{7}$ **51.** $-\dfrac{1}{10}x^8y^9z^4$ **53.** $3x^4 + x^3$ **55.** $3x^3 - x^2$ **57.** $y^4 - 3y + 1$ **59.** $-5x^2 + 8x$
61. $6x^3 + 2x^2 + 3x$ **63.** $3x^3 - 2x^2 + 10x$ **65.** $4x - 6$ **67.** $-6z^2 - 2z$ **69.** $4x^2 + 3x - 1$ **71.** $5x^4 - 3x^2 - x$
73. $-9x^3 + \dfrac{9}{2}x^2 - 10x + 5$ **75.** $4xy + 2x - 5y$ **77.** $-4x^5y^3 + 3xy + 2$ **79–85.** Answers will vary.   **87.** $18x^8 - 27x^6 + 36x^4$
89. $\dfrac{3x^{14} - 6x^{12} - ?x^7}{?x^?} = \dfrac{3x^{14} - 6x^{12} - 9x^7}{-3x^7}$ **90.** 20.3   **91.** 0.875   **92.**

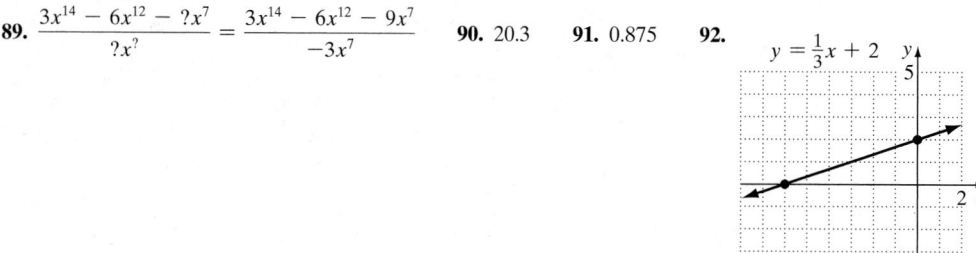

Section 5.6

Check Point Exercises

1. $x + 5$ **2.** $4x - 3 - \dfrac{3}{2x + 3}$ **3.** $x^2 + x + 1$ **4.** $x^2 - 2x - 3$

Exercise Set 5.6

1. $x + 4$ **3.** $2x + 5$ **5.** $x - 2$ **7.** $2y + 1$ **9.** $x - 2 + \dfrac{2}{x - 3}$ **11.** $y + 3 + \dfrac{4}{y + 2}$ **13.** $x^2 - 5x + 2$ **15.** $6y - 1$

17. $2a + 3$ **19.** $y^2 - y + 2$ **21.** $x - 6 + \dfrac{26}{2x + 3}$ **23.** $x^2 + 2x + 8 + \dfrac{13}{x - 2}$ **25.** $2y^2 + y + 1 + \dfrac{6}{2y + 3}$

27. $2y^2 - 3y + 2 + \dfrac{1}{3y + 2}$ **29.** $9x^2 + 3x + 1$ **31.** $y^3 - 9y^2 + 27y - 27$ **33.** $2y + 4 + \dfrac{4}{2y - 1}$

35. $y^3 + y^2 - y - 1 + \dfrac{4}{y - 1}$ **37.** $2x + 5$ **39.** $3x - 8 + \dfrac{20}{x + 5}$ **41.** $4x^2 + x + 4 + \dfrac{3}{x - 1}$

43. $6x^4 + 12x^3 + 22x^2 + 48x + 93 + \dfrac{187}{x - 2}$ **45.** $x^3 - 10x^2 + 51x - 260 + \dfrac{1300}{5 + x}$ **47.** $3x^2 + 3x - 3$

49. $x^4 + x^3 + 2x^2 + 2x + 2$ **51.** $x^3 + 4x^2 + 16x + 64$ **53.** $2x^4 - 7x^3 + 15x^2 - 31x + 64 - \dfrac{129}{x + 2}$ **55.** $x + 3$ in.

57. a. $1 + \dfrac{5}{x + 20}$ **b.** $\dfrac{5}{4}, \dfrac{6}{5}, \dfrac{7}{6}, \dfrac{10}{9}, \dfrac{15}{14}, \dfrac{20}{19}$ **c.** As x increases the ratio is decreasing and approaching 1.   **59–61.** Answers will vary.

63. b   **65.** -3 **67.** Graphs coincide.   **69.** $2x + 3$ should be $2x + 23 + \dfrac{130}{x - 5}$.   **71.** $x^2 - 2x + 3$ should be $x^2 + 2x + 3$.

72. $(5, 3)$ or $\{(5, 3)\}$ **73.** 1.2 **74.** -6 or $\{-6\}$

Section 5.7

Check Point Exercises

1. a. $\dfrac{1}{6^2} = \dfrac{1}{36}$ **b.** $\dfrac{1}{5^3} = \dfrac{1}{125}$ **c.** $\dfrac{1}{(-4)^3} = -\dfrac{1}{64}$ **d.** $\dfrac{1}{8^1} = \dfrac{1}{8}$ **2. a.** $\dfrac{7^2}{2^3} = \dfrac{49}{8}$ **b.** $\dfrac{5^2}{4^2} = \dfrac{25}{16}$ **c.** $\dfrac{y^2}{7}$ **d.** $\dfrac{y^8}{x^1} = \dfrac{y^8}{x}$ **3.** $\dfrac{1}{x^{10}}$

4. a. $\dfrac{1}{x^8}$ **b.** $\dfrac{15}{x^6}$ **c.** $-\dfrac{2}{y^6}$ **5.** $\dfrac{36}{x^3}$ **6.** $\dfrac{1}{x^{20}}$ **7. a.** $7,400,000,000$ **b.** 0.000003017 **8. a.** 7.41×10^9 **b.** 9.2×10^{-8}

9. a. 6×10^{10} **b.** 2.1×10^{11} **c.** 6.4×10^{-5} **10.** 5.2×10^5 mi

Exercise Set 6.7

1. $\dfrac{1}{5^2} = \dfrac{1}{25}$ **3.** $\dfrac{1}{2^3} = \dfrac{1}{8}$ **5.** $\dfrac{1}{(-2)^2} = \dfrac{1}{4}$ **7.** $\dfrac{1}{(-3)^3} = -\dfrac{1}{27}$ **9.** $\dfrac{1}{4^1} = \dfrac{1}{4}$ **11.** $\dfrac{1}{2^1} + \dfrac{1}{3^1} = \dfrac{1}{2} + \dfrac{1}{3} = \dfrac{5}{6}$ **13.** $3^2 = 9$

15. $(-3)^2 = 9$ **17.** $\dfrac{8^2}{2^3} = 8$ **19.** $\dfrac{4^2}{1^2} = 16$ **21.** $\dfrac{5^3}{3^3} = \dfrac{125}{27}$ **23.** $\dfrac{x^5}{6}$ **25.** $\dfrac{y^1}{x^8} = \dfrac{y}{x^8}$ **27.** $3 \cdot (-5)^3 = -375$ **29.** $\dfrac{1}{x^5}$

31. $\dfrac{8}{x^3}$ **33.** $\dfrac{1}{x^6}$ **35.** $\dfrac{1}{y^{99}}$ **37.** $\dfrac{3}{z^5}$ **39.** $-\dfrac{4}{x^4}$ **41.** $-\dfrac{1}{3a^3}$ **43.** $\dfrac{7}{5w^8}$ **45.** $\dfrac{1}{x^5}$ **47.** $\dfrac{1}{y^{11}}$ **49.** $\dfrac{16}{x^2}$ **51.** $216y^{17}$ **53.** $\dfrac{1}{x^6}$

55. $\dfrac{1}{16x^{12}}$ **57.** $\dfrac{x^2}{9}$ **59.** $-\dfrac{y^3}{8}$ **61.** $\dfrac{2x^6}{5}$ **63.** x^8 **65.** $16y^6$ **67.** $\dfrac{1}{y^2}$ **69.** $\dfrac{1}{y^{50}}$ **71.** $\dfrac{1}{a^{12}b^{15}}$ **73.** $\dfrac{a^8}{b^{24}}$ **75.** $\dfrac{4}{x^4}$ **77.** $\dfrac{y^9}{x^6}$

79. 270 **81.** $912,000$ **83.** 3.4 **85.** 0.79 **87.** 0.0215 **89.** 0.000786 **91.** 3.24×10^4 **93.** 2.2×10^8 **95.** 7.13×10^2

97. 6.751×10^3 **99.** 2.7×10^{-3} **101.** 2.02×10^{-5} **103.** 5×10^{-3} **105.** 3.14159×10^0 **107.** 6×10^5 **109.** 1.6×10^9

111. 3×10^4 **113.** 3×10^6 **115.** 3×10^{-6} **117.** 9×10^4 **119.** 2.5×10^6 **121.** 1.25×10^8 **123.** 8.1×10^{-7}

125. 2.5×10^{-7} **127.** 9.2×10^3 **129.** 2.5×10^{-16} **131.** 1.6943×10^{12} **133.** 6×10^{10} **135.** $\$4 \times 10^{11}$ **137.** $\$3.24 \times 10^{10}$

139–143. Answers will vary.   **145.** b   **147–151.** Answers will vary.   **153.** $\{x | x < 2\}$ **154.** 5 **155.** $0, \sqrt{16}$

Review Exercises

1. binomial, 4   **2.** trinomial, 2   **3.** monomial, 1   **4.** $8x^3 + 10x^2 - 20x - 4$ **5.** $13y^3 - 8y^2 + 7y - 5$ **6.** $11y^2 - 4y - 4$

7. $8x^4 - 5x^3 + 6$ **8.** $-14x^4 - 13x^2 + 16x$ **9.** $7y^4 - 5y^3 + 3y^2 - y - 4$ **10.** $3x^2 - 7x + 9$ **11.** $10x^3 - 9x^2 + 2x + 11$

12. 1451; 1451 men per 100,000 men is the death rate for men averaging 10 hr of sleep each night.   **13.** x^{23} **14.** y^{14} **15.** x^{100}

16. $100y^2$ **17.** $-64x^{30}$ **18.** $50x^4$ **19.** $-36y^{11}$ **20.** $30x^{12}$ **21.** $21x^3 + 63x$ **22.** $20x^5 - 55x^4$ **23.** $-21y^4 + 9y^3 - 18y^2$

24. $16y^8 - 20y^7 + 2y^5$ **25.** $x^3 - 2x^2 - 13x + 6$ **26.** $12y^3 + y^2 - 21y + 10$ **27.** $3y^3 - 17y^2 + 41y - 35$

28. $8x^4 + 8x^3 - 18x^2 - 20x - 3$ **29.** $x^2 + 8x + 12$ **30.** $6y^2 - 7y - 5$ **31.** $4x^4 - 14x^2 + 6$ **32.** $25x^2 - 16$

33. $49 - 4y^2$ **34.** $y^4 - 1$ **35.** $x^2 + 6x + 9$ **36.** $9y^2 + 24y + 16$ **37.** $y^2 - 2y + 1$ **38.** $25y^2 - 20y + 4$
39. $x^4 + 8x^2 + 16$ **40.** $x^4 - 16$ **41.** $x^4 - x^2 - 20$ **42.** $x^2 + 7x + 12$ **43.** $x^2 + 50x + 600 \text{ yd}^2$ **44.** 28
45. polynomial degree: 5;

Term	Coefficient	Degree
$4x^2y$	4	3
$9x^3y^2$	9	5
$-17x^4$	-17	4
-12	-12	0

46. $-x^2 - 17xy + 5y^2$ **47.** $2x^3y^2 + x^2y - 6x^2 - 4$
48. $-35x^6y^9$ **49.** $15a^3b^5 - 20a^2b^3$ **50.** $3x^2 + 16xy - 35y^2$
51. $36x^2y^2 - 31xy + 3$ **52.** $9x^2 + 30xy + 25y^2$
53. $x^2y^2 - 14xy + 49$ **54.** $49x^2 - 16y^2$ **55.** $a^3 - b^3$
56. 6^{30} **57.** x^{15} **58.** 1 **59.** -1 **60.** 400

61. $\dfrac{x^{12}}{8}$ **62.** $\dfrac{81}{16y^{24}}$ **63.** $-5y^6$ **64.** $8x^7y^3$ **65.** $3x^3 - 2x + 6$ **66.** $-6x^5 + 5x^4 + 8x^2$ **67.** $9x^2y - 3x - 6y$

68. $2x + 7$ **69.** $x^2 - 3x + 5$ **70.** $x^2 + 5x + 2 + \dfrac{7}{x-7}$ **71.** $y^2 + 3y + 9$ **72.** $4x^2 - 7x + 5 - \dfrac{4}{x+1}$

73. $3x^3 + 6x^2 + 10x + 10$ **74.** $x^3 - 4x^2 + 16x - 64 + \dfrac{272}{x+4}$ **75.** $\dfrac{1}{7^2} = \dfrac{1}{49}$ **76.** $\dfrac{1}{(-4)^3} = -\dfrac{1}{64}$

77. $\dfrac{1}{2^1} + \dfrac{1}{4^1} = \dfrac{1}{2} + \dfrac{1}{4} = \dfrac{3}{4}$ **78.** $5^2 = 25$ **79.** $\dfrac{5^3}{2^3} = \dfrac{125}{8}$ **80.** $\dfrac{1}{x^6}$ **81.** $\dfrac{6}{y^2}$ **82.** $\dfrac{30}{x^5}$ **83.** x^8 **84.** $81y^2$ **85.** $\dfrac{1}{y^{19}}$

86. $\dfrac{x^3}{8}$ **87.** $\dfrac{1}{x^6}$ **88.** y^{20} **89.** 23,000 **90.** 0.00176 **91.** 0.9 **92.** 7.39×10^7 **93.** 6.2×10^{-4} **94.** 3.8×10^{-1}
95. 3.8×10^0 **96.** 9×10^3 **97.** 5×10^4 **98.** 1.6×10^{-3} **99.** 1000 nanosec **100.** 1.22×10^{10} people

Chapter 6 Test

1. trinomial, 2 **2.** $13x^3 + x^2 - x - 24$ **3.** $5x^3 + 2x^2 + 2x - 9$ **4.** $-35x^{11}$ **5.** $48x^5 - 30x^3 - 12x^2$
6. $3x^3 - 10x^2 - 17x - 6$ **7.** $6y^2 - 13y - 63$ **8.** $49x^2 - 25$ **9.** $x^4 + 6x^2 + 9$ **10.** $25x^2 - 30x + 9$ **11.** 30
12. $2x^2y^3 + 3xy + 12y^2$ **13.** $12a^2 - 13ab - 35b^2$ **14.** $4x^2 + 12xy + 9y^2$ **15.** $-5x^{12}$ **16.** $3x^3 - 2x^2 + 5x$
17. $x^2 - 2x + 3 + \dfrac{1}{2x+1}$ **18.** $3x^3 - 4x^2 + 7$ **19.** $\dfrac{1}{10^2} = \dfrac{1}{100}$ **20.** $4^3 = 64$ **21.** $-27x^6$ **22.** $\dfrac{4}{x^5}$ **23.** $-\dfrac{21}{x^6}$

24. $16y^4$ **25.** $\dfrac{x^8}{25}$ **26.** $\dfrac{1}{x^{15}}$ **27.** 0.00037 **28.** 7.6×10^6 **29.** 1.23×10^{-2} **30.** 2.1×10^8

Cumulative Review Exercises (Chapters 1–5)

1. $\dfrac{35}{9}$ **2.** -128 **3.** 15,050 ft **4.** $-\dfrac{2}{3}$ or $\left\{-\dfrac{2}{3}\right\}$ **5.** $-\dfrac{5}{3}$ or $\left\{-\dfrac{5}{3}\right\}$ **6.** 4 ft by 10 ft
7. $\{x | x \geq 6\}$; 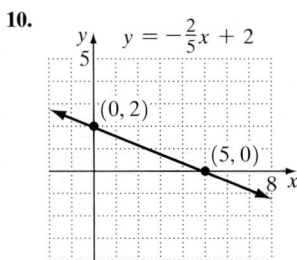 **8.** $3400 at 12%; $2600 at 14%
9. 15 liters of 70% antifreeze solution; 5 liters of 30% antifreeze solution

10.
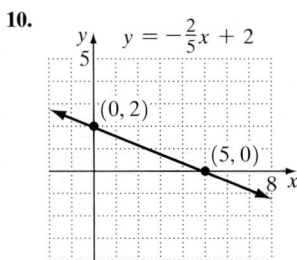
$y = -\dfrac{2}{5}x + 2$

11.
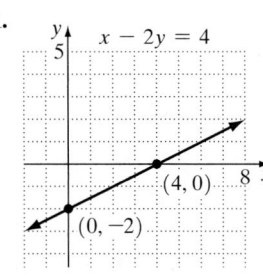
$x - 2y = 4$

12. $-\dfrac{6}{5}$; falling
13. $y + 1 = -2(x - 3); y = -2x + 5$
14. $(0, 5)$ or $\{(0, 5)\}$ **15.** $(3, -4)$ or $\{(3, -4)\}$
16. 500 min; $40 **17.** $3x^5 - 6x^3 + 9x + 2$
18. $x^2 + 2x + 3$ **19.** $\dfrac{81}{x^2}$ **20.** 0.0024

CHAPTER 6

Section 6.1

Check Point Exercises

1. a. $3x^2$ **b.** $4x^2$ **c.** x^2y **2.** $6(x^2 + 3)$ **3.** $5x^2(5 + 7x)$ **4.** $3x^3(5x^2 + 4x - 9)$ **5.** $2xy(4x^2y - 7x + 1)$
6. a. $(x + 1)(x^2 + 7)$ **b.** $(y + 4)(x - 7)$ **7.** $(x + 5)(x^2 + 2)$ **8.** $(y + 3)(x - 5)$

Exercise Set 6.1

1. Answers will vary; 3 examples are: $(2x)(4x^2), (4x)(2x^2),$ and $(8x)(x^2)$. **3.** Answers will vary; 3 examples are: $(-4x^3)(3x^2), (2x^2)(-6x^3),$ and $(-3)(4x^5)$. **5.** Answers will vary; 3 examples are: $(6x^2)(6x^2), (-2x)(-18x^3),$ and $(4x^3)(9x)$. **7.** 4 **9.** $4x$ **11.** $2x^3$ **13.** $3y$ **15.** xy **17.** $4x^4y^3$ **19.** $5(x + 1)$ **21.** $3(y - 1)$ **23.** $8(x + 2)$ **25.** $5(5x - 2)$ **27.** $x(x + 1)$ **29.** $6(3y^2 + 4)$ **31.** $12x^2(3x + 2)$ **33.** $y(25y - 13)$ **35.** $9y^4(1 + 3y^2)$ **37.** $4x^2(2 - x^2)$ **39.** $4(3y^2 + 4y - 2)$ **41.** $3x^2(3x^2 + 6x + 2)$ **43.** $50y^2(2y^3 - y + 2)$ **45.** $5x(2 - 4x + x^2)$ **47.** cannot be factored **49.** $3xy(2x^2y + 3)$ **51.** $10xy(3xy^2 - y + 2)$ **53.** $8x^2y(4xy - 3x - 2)$ **55.** $(x + 5)(x + 3)$ **57.** $(x + 2)(x - 4)$ **59.** $(y + 6)(x - 7)$ **61.** $(x + y)(3x - 1)$ **63.** $(3x + 1)(4x + 1)$ **65.** $(5x + 4)(7x^2 + 1)$ **67.** $(x + 2)(x + 4)$ **69.** $(x + 3)(x - 5)$ **71.** $(x - 2)(x^2 + 5)$ **73.** $(x - 1)(x^2 + 2)$ **75.** $(y + 5)(x + 9)$ **77.** $(y - 1)(x + 5)$ **79.** $(x - 2y)(3x + 5y)$ **81.** $(3x - 2)(x^2 - 2)$ **83.** $(x - a)(x - b)$ **85. a.** 3488 thousand or 3,488,000 students **b.** $4(2x^2 + 5x + 622)$ **c.** 3488 thousand or 3,488,000 students; Answers will vary. **87.** $x^3 - 2$ **89–93.** Answers will vary. **95.** d **97.** $x^{2n}(x^{2n} + 1 + x^n)$ **99.** Answers will vary. **101.** $(x - 2)(x - 5)$ should be $(x - 2)(x + 5)$. **103.** $x^2 + 17x + 70$ **104.** $(-3, -2)$ or $\{(-3, -2)\}$ **105.** $y - 2 = 1(x + 7)$ or $y - 5 = 1(x + 4); y = x + 9$

Section 6.2

Check Point Exercises

1. $(x + 2)(x + 3)$ **2.** $(x - 2)(x - 4)$ **3.** $(x + 5)(x - 2)$ **4.** $(y - 9)(y + 3)$ **5.** prime **6.** $(x - 3y)(x - y)$ **7.** $2x(x - 4)(x + 7)$

Exercise Set 6.2

1. $(x + 5)(x + 1)$ **3.** $(x + 3)(x + 5)$ **5.** $(x + 1)(x + 11)$ **7.** $(x - 5)(x - 3)$ **9.** $(x - 7)(x - 7)$ **11.** $(y - 3)(y - 12)$ **13.** $(x + 5)(x - 2)$ **15.** $(y + 13)(y - 3)$ **17.** $(x - 5)(x + 3)$ **19.** $(x - 4)(x + 2)$ **21.** prime **23.** $(y - 4)(y - 12)$ **25.** prime **27.** $(w - 32)(w + 2)$ **29.** $(y - 5)(y - 13)$ **31.** $(r + 3)(r + 9)$ **33.** prime **35.** $(x + 6y)(x + y)$ **37.** $(x - 3y)(x - 5y)$ **39.** $(x - 6y)(x + 3y)$ **41.** $(a - 15b)(a - 3b)$ **43.** $3(x + 2)(x + 3)$ **45.** $4(y - 2)(y + 1)$ **47.** $10(x - 10)(x + 6)$ **49.** $3(x - 2)(x - 9)$ **51.** $2r(r + 2)(r + 1)$ **53.** $4x(x + 6)(x - 3)$ **55.** $2r(r + 8)(r - 4)$ **57.** $y^2(y + 10)(y - 8)$ **59.** $x^2(x - 5)(x + 2)$ **61.** $2w^2(w - 16)(w + 3)$ **63.** $-16(t - 2)(t + 1)$ **65–67.** Answers will vary. **69.** c **71.** 3, 4 **73.** $4x(x - 3)(x - 4)$ **75.** $2(x - 3)(x + 2)$ should be $2(x + 3)(x - 2)$. **77.** $(x + 3)(x + 1)$ should be $2(x + 3)(x + 1)$. **78.** $2x^2 - x - 6$ **79.** $9x^2 + 15x + 4$ **80.** 13 or $\{13\}$

Section 6.3

Check Point Exercises

1. $(5x - 4)(x - 2)$ **2.** $(3x - 1)(2x + 7)$ **3.** $(3x - y)(x - 4y)$ **4.** $(3x + 5)(x - 2)$ **5.** $(2x - 1)(4x - 3)$ **6.** $y^2(5y + 3)(y + 2)$

Exercise Set 6.3

1. $(2x + 1)(x + 3)$ **3.** $(3x + 2)(x + 2)$ **5.** $(2x + 5)(x + 4)$ **7.** $(5y - 3)(y - 1)$ **9.** $(3y + 2)(y - 1)$ **11.** $(3x - 5)(x + 2)$ **13.** $(3x - 1)(x - 7)$ **15.** $(5y - 1)(y - 3)$ **17.** $(3x - 2)(x - 5)$ **19.** $(3w - 4)(2w - 1)$ **21.** $(8x + 1)(x + 4)$ **23.** $(5x - 2)(x + 7)$ **25.** $(7y - 3)(2y + 3)$ **27.** prime **29.** $(5z - 3)(5z - 3)$ **31.** $(3y + 1)(5y - 2)$ **33.** prime **35.** $(5y - 1)(2y + 9)$ **37.** $(4x + 1)(2x - 1)$ **39.** $(3y - 1)(3y - 2)$ **41.** $(5x + 8)(4x - 1)$ **43.** $(2x + y)(x + y)$ **45.** $(3x + 2y)(x + y)$ **47.** $(2x - 3y)(x - 3y)$ **49.** $(2x - 3y)(3x + 2y)$ **51.** $(3x - 2y)(5x + 7y)$ **53.** $(2a + 5b)(a + b)$ **55.** $(3a - 2b)(5a + 3b)$ **57.** $(3x - 4y)(4x - 3y)$ **59.** $2(2x + 3)(x + 5)$ **61.** $3(3x + 4)(x - 2)$ **63.** $2(2y - 5)(y + 3)$ **65.** $3(3y - 4)(y + 5)$ **67.** $x(3x + 1)(x + 1)$ **69.** $x(2x - 5)(x + 1)$ **71.** $3y(3y - 1)(y - 4)$ **73.** $5z(6z + 1)(2z + 1)$ **75.** $3x^2(5x - 3)(x - 2)$ **77.** $x^3(2x - 3)(5x - 1)$ **79.** $3(2x + 3y)(x - 2y)$ **81.** $2(2x - y)(3x + 4y)$ **83.** $2y(4x - 7)(x + 6)$ **85.** $2b(2a - 7b)(3a - b)$ **87. a.** $x^2 + 3x + 2$ **b.** $(x + 2)(x + 1)$ **c.** $x^2 + 3x + 2 = (x + 2)(x + 1)$ **89–91.** Answers will vary. **93.** a **95.** $5, 7, -5, -7$ **97.** $(2x^n + 1)(x^n - 4)$ **98.** $81x^2 - 100$ **99.** $16x^2 + 40xy + 25y^2$ **100.** $x^3 + 8$

Section 6.4

Check Point Exercises

1. a. $(x + 9)(x - 9)$ **b.** $(6x + 5)(6x - 5)$ **2. a.** $(5 + 2x^5)(5 - 2x^5)$ **b.** $(10x + 3y)(10x - 3y)$ **3. a.** $2x(3x + 1)(3x - 1)$ **b.** $18(2 + x)(2 - x)$ **4.** $(9x^2 + 4)(3x + 2)(3x - 2)$ **5. a.** $(x + 7)^2$ **b.** $(x - 3)^2$ **c.** $(4x - 7)^2$ **6.** $(2x + 3y)^2$ **7.** $(x + 3)(x^2 - 3x + 9)$ **8.** $(1 - y)(1 + y + y^2)$ **9.** $(5x + 2)(25x^2 - 10x + 4)$

Exercise Set 6.4

1. $(x + 5)(x - 5)$　**3.** $(y + 1)(y - 1)$　**5.** $(2x + 3)(2x - 3)$　**7.** $(5 + x)(5 - x)$　**9.** $(1 + 7x)(1 - 7x)$
11. $(3 + 5y)(3 - 5y)$　**13.** $(x^2 + 3)(x^2 - 3)$　**15.** $(7y^2 + 4)(7y^2 - 4)$　**17.** $(x^5 + 3)(x^5 - 3)$　**19.** $(5x + 4y)(5x - 4y)$
21. $(x^2 + y^5)(x^2 - y^5)$　**23.** $(x^2 + 4)(x + 2)(x - 2)$　**25.** $(4x^2 + 9)(2x + 3)(2x - 3)$　**27.** $2(x + 3)(x - 3)$
29. $2x(x + 6)(x - 6)$　**31.** prime　**33.** $2x(x^2 + 36)$　**35.** $2(5 + y)(5 - y)$　**37.** $2y(2y + 1)(2y - 1)$
39. $2x(x + 1)(x - 1)$　**41.** $(x + 1)^2$　**43.** $(x - 7)^2$　**45.** $(x - 1)^2$　**47.** $(x + 11)^2$　**49.** $(2x + 1)^2$　**51.** $(5y - 1)^2$
53. prime　**55.** $(x + 7y)^2$　**57.** $(x - 6y)^2$　**59.** prime　**61.** $(4x - 5y)^2$　**63.** $3(2x - 1)^2$　**65.** $x(3x + 1)^2$
67. $2(y - 1)^2$　**69.** $2y(y + 7)^2$　**71.** $(x + 1)(x^2 - x + 1)$　**73.** $(x - 3)(x^2 + 3x + 9)$　**75.** $(2y - 1)(4y^2 + 2y + 1)$
77. $(3x + 2)(9x^2 - 6x + 4)$　**79.** $(xy - 4)(x^2y^2 + 4xy + 16)$　**81.** $y(3y + 2)(9y^2 - 6y + 4)$　**83.** $2(3 - 2y)(9 + 6y + 4y^2)$
85. $(4x + 3y)(16x^2 - 12xy + 9y^2)$　**87.** $(5x - 4y)(25x^2 + 20xy + 16y^2)$　**89.** $x^2 - 25 = (x + 5)(x - 5)$
91. $x^2 - 16 = (x + 4)(x - 4)$　**93–95.** Answers will vary.　**97.** b　**99.** $(x + 6)(x - 4)$　**101.** $(2x^n + 3)^2$
103. $[(x + 3) - 1]^2$ or $(x + 2)^2$　**105.** 1　**107.** correctly factored　**109.** $(x - 1)(x^2 - x + 1)$ should be $(x - 1)(x^2 + x + 1)$.
110. $80x^9y^{14}$　**111.** $-4x^2 + 3$　**112.** $2x + 5$

Section 6.5

Check Point Exercises

1. $5x^2(x + 3)(x - 3)$　**2.** $4(x - 6)(x + 2)$　**3.** $4x(x^2 + 4)(x + 2)(x - 2)$　**4.** $(x - 4)(x + 3)(x - 3)$　**5.** $3x(x - 5)^2$
6. $2x^2(x + 3)(x^2 - 3x + 9)$　**7.** $3y(x^2 + 4y^2)(x + 2y)(x - 2y)$　**8.** $3x(2x + 3y)^2$

Exercise Set 6.5

1. $3x(x + 1)(x - 1)$　**3.** $3x(x^2 + 1)$　**5.** $4(x - 3)(x + 2)$　**7.** $2(x^2 + 9)(x + 3)(x - 3)$　**9.** $(x + 2)(x + 3)(x - 3)$
11. $3x(x - 4)^2$　**13.** $2x^2(x + 1)(x^2 - x + 1)$　**15.** $2x(3x + 4)$　**17.** $2(y - 8)(y + 7)$　**19.** $7y^2(y + 1)^2$　**21.** prime
23. $2(4y + 1)(2y - 1)$　**25.** $r(r - 25)$　**27.** $(2w + 5)(2w - 1)$　**29.** $x(x + 2)(x - 2)$　**31.** prime　**33.** $(9y + 4)(y + 1)$
35. $(y + 2)(y + 2)(y - 2)$　**37.** $(3y + 4)^2$　**39.** $5y(y - 7)(y - 2)$　**41.** $y(y^2 + 9)(y + 3)(y - 3)$　**43.** $5a^2(2a + 3)(2a - 3)$
45. $(4y - 1)(3y - 2)$　**47.** $(3y + 8)(3y - 8)$　**49.** prime　**51.** $(2y + 3)(y + 5)(y - 5)$　**53.** $2r(r + 17)(r - 2)$
55. $2x^3(2x + 1)(2x - 1)$　**57.** $3(x^2 + 81)$　**59.** $x(x + 2)(x^2 - 2x + 4)$　**61.** $2y^2(y - 1)(y^2 + y + 1)$　**63.** $2x(3x + 4y)$
65. $(y - 7)(x + 3)$　**67.** $(x - 4y)(x + y)$　**69.** $12a^2(6ab^2 + 1 - 2a^2b^2)$　**71.** $3(a + 6b)(a + 3b)$　**73.** $3x^2y(4x + 1)(4x - 1)$
75. $b(3a + 2)(2a - 1)$　**77.** $7xy(x^2 + y^2)(x + y)(x - y)$　**79.** $2xy(5x - 2y)(x - y)$　**81.** $2b(x + 11)^2$
83. $(5a + 7b)(3a - 2b)$　**85.** $2xy(9x - 2y)(2x - 3y)$　**87.** $(y - x)(a + b)(a - b)$　**89.** $ax(3x + 7)(3x - 2)$
91. $y(9x^2 + y^2)(3x + y)(3x - y)$　**93.** $16(4 + t)(4 - t)$　**95.** $\pi b^2 - \pi a^2; \pi(b + a)(b - a)$　**97.** Answers will vary.
99. d　**101.** $5y^2(y - 1)(y + 2)(y - 2)$　**103.** $(x - 5)^2$　**105.** $(4x - 3)^2$ should be $(2x - 3)^2$.　**107.** correctly factored
109. $2(x + 5)(x^2 + 1)$ should be $2(x + 5)(x + 1)(x - 1)$.　**110.** $(3x + 4)(3x - 4)$
111.

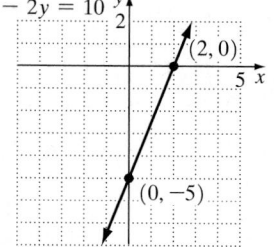

$5x - 2y = 10$, points $(2, 0)$ and $(0, -5)$

112. $20°, 60°, 100°$

Section 6.6

Check Point Exercises

1. $-\dfrac{1}{2}$ and 4, or $\left\{-\dfrac{1}{2}, 4\right\}$　**2.** 1 and 5, or $\{1, 5\}$　**3.** 0 and $\dfrac{1}{2}$, or $\left\{0, \dfrac{1}{2}\right\}$　**4.** 5 or $\{5\}$　**5.** $-\dfrac{5}{4}$ and $\dfrac{5}{4}$, or $\left\{-\dfrac{5}{4}, \dfrac{5}{4}\right\}$
6. -2 and 9, or $\{-2, 9\}$　**7.** 1 sec and 2 sec; $(1, 192)$ and $(2, 192)$　**8.** length: 9 ft; width: 6 ft

Exercise Set 6.6

1. -3 and 0, or $\{-3, 0\}$　**3.** -5 and 8, or $\{-5, 8\}$　**5.** $-\dfrac{5}{4}$ and 2, or $\left\{-\dfrac{5}{4}, 2\right\}$　**7.** $-\dfrac{7}{2}$ and 3, or $\left\{-\dfrac{7}{2}, 3\right\}$
9. -5 and -3, or $\{-5, -3\}$　**11.** -3 and 5, or $\{-3, 5\}$　**13.** -3 and 7, or $\{-3, 7\}$　**15.** -8 and -1, or $\{-8, -1\}$
17. -4 and 0, or $\{-4, 0\}$　**19.** 0 and 5, or $\{0, 5\}$　**21.** 0 and 4, or $\{0, 4\}$　**23.** 0 and $\dfrac{5}{2}$, or $\left\{0, \dfrac{5}{2}\right\}$　**25.** $-\dfrac{5}{3}$ and 0, or $\left\{-\dfrac{5}{3}, 0\right\}$

27. -2 or $\{-2\}$ **29.** 6 or $\{6\}$ **31.** $\dfrac{3}{2}$ or $\left\{\dfrac{3}{2}\right\}$ **33.** $-\dfrac{1}{2}$ and 4, or $\left\{-\dfrac{1}{2}, 4\right\}$ **35.** -2 and $\dfrac{9}{5}$, or $\left\{-2, \dfrac{9}{5}\right\}$

37. -7 and 7, or $\{-7, 7\}$ **39.** $-\dfrac{5}{2}$ and $\dfrac{5}{2}$, or $\left\{-\dfrac{5}{2}, \dfrac{5}{2}\right\}$ **41.** $-\dfrac{5}{9}$ and $\dfrac{5}{9}$, or $\left\{-\dfrac{5}{9}, \dfrac{5}{9}\right\}$ **43.** -3 and 7, or $\{-3, 7\}$

45. $-\dfrac{5}{2}$ and $\dfrac{3}{2}$, or $\left\{-\dfrac{5}{2}, \dfrac{3}{2}\right\}$ **47.** -6 and 3, or $\{-6, 3\}$ **49.** -2 and $-\dfrac{3}{2}$, or $\left\{-2, -\dfrac{3}{2}\right\}$ **51.** 4 or $\{4\}$ **53.** $-\dfrac{5}{2}$ or $\left\{-\dfrac{5}{2}\right\}$

55. $\dfrac{3}{8}$ or $\left\{\dfrac{3}{8}\right\}$ **57.** 5 sec; Each tick represents one second.   **59.** 2 sec; $(2, 276)$ **61.** $\dfrac{1}{2}$ sec and 4 sec

63. 1995; $(15, 1100)$ **65.** 10 yr   **67.** $(10, 7250)$ **69.** 10 teams   **71.** length: 15 yd; width: 12 yd   **73.** base: 5 cm; height: 6 cm
75. a. $4x^2 + 44x$ **b.** 3 ft   **77–79.** Answers will vary.   **81.** $x^2 - 2x - 15 = 0$ **83.** $-5, 0,$ and 2, or $\{-5, 0, 2\}$
85. 4 and 1, or $\{4, 1\}$ **87. a**   **89. b**   **91.** -3 and 2, or $\{-3, 2\}$ **93.** 1 or $\{1\}$ **95.** Answers will vary.

96.

$y = -\dfrac{2}{3}x + 1$

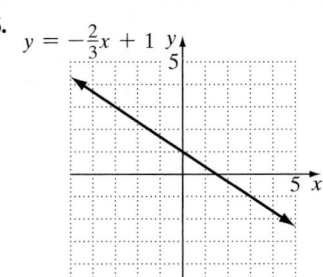

97. $\dfrac{4}{x^6}$ **98.** -2

Review Exercises

1. $15(2x - 3)$ **2.** $4x(3x^2 + 4x - 100)$ **3.** $5x^2y(6x^2 + 3x + 1)$ **4.** $5(x + 3)$ **5.** $(7x^2 - 1)(x + y)$ **6.** $(x^2 + 2)(x + 3)$
7. $(x + 1)(y + 4)$ **8.** $(x^2 + 5)(x + 1)$ **9.** $(x - 2)(y + 4)$ **10.** $(x - 2)(x - 1)$ **11.** $(x - 5)(x + 4)$ **12.** $(x + 3)(x + 16)$
13. $(x - 4y)(x - 2y)$ **14.** prime   **15.** $(x + 17y)(x - y)$ **16.** $3(x + 4)(x - 2)$ **17.** $3x(x - 11)(x - 1)$
18. $(x + 5)(3x + 2)$ **19.** $(y - 3)(5y - 2)$ **20.** $(2x + 5)(2x - 3)$ **21.** prime   **22.** $2(2x + 3)(2x - 1)$
23. $x(2x - 9)(x + 8)$ **24.** $4y(3y + 1)(y + 2)$ **25.** $(2x - y)(x - 3y)$ **26.** $(5x + 4y)(x - 2y)$ **27.** $(2x + 1)(2x - 1)$
28. $(9 + 10y)(9 - 10y)$ **29.** $(5a + 7b)(5a - 7b)$ **30.** $(z^2 + 4)(z + 2)(z - 2)$ **31.** $2(x + 3)(x - 3)$ **32.** prime
33. $x(3x + 1)(3x - 1)$ **34.** $2x(3y + 2)(3y - 2)$ **35.** $(x + 11)^2$ **36.** $(x - 8)^2$ **37.** $(3y + 8)^2$ **38.** $(4x - 5)^2$
39. prime   **40.** $(6x + 5y)^2$ **41.** $(5x - 4y)^2$ **42.** $(x - 3)(x^2 + 3x + 9)$ **43.** $(4x + 1)(16x^2 - 4x + 1)$
44. $2(3x - 2y)(9x^2 + 6xy + 4y^2)$ **45.** $y(3x + 2)(9x^2 - 6x + 4)$ **46.** $(a + 3)(a - 3)$ **47.** $(a + 2b)(a - 2b)$
48. $A^2 + 2A + 1 = (A + 1)^2$ **49.** $x(x - 7)(x - 1)$ **50.** $(5y + 2)(2y + 1)$ **51.** $2(8 + y)(8 - y)$ **52.** $(3x + 1)^2$
53. $4x^3(5x^4 - 9)$ **54.** $(x - 3)^2(x + 3)$ **55.** prime   **56.** $x(2x + 5)(x + 7)$ **57.** $3x(x - 5)^2$ **58.** $3x^2(x - 2)(x^2 + 2x + 4)$
59. $4y^2(y + 3)(y - 3)$ **60.** $5(x + 7)(x - 3)$ **61.** prime   **62.** $2x^3(5x - 2)(x - 4)$ **63.** $(10y + 7)(10y - 7)$
64. $9x^4(x - 2)$ **65.** $(x^2 + 1)(x + 1)(x - 1)$ **66.** $2(y - 2)(y^2 + 2y + 4)$ **67.** $(x + 4)(x^2 - 4x + 16)$ **68.** $(3x - 2)(2x + 5)$
69. $3x^2(x + 2)(x - 2)$ **70.** $(x - 10)(x + 9)$ **71.** $(5x + 2y)(5x + 3y)$ **72.** $x(x + 5)(x^2 - 5x + 25)$
73. $2y(4y + 3)(4y + 1)$ **74.** $2(y - 4)^2$ **75.** $(x + 5y)(x - 7y)$ **76.** $(x + y)(x + 7)$ **77.** $(3x + 4y)^2$
78. $2x^2y(x + 1)(x - 1)$ **79.** $(10y + 7z)(10y - 7z)$ **80.** prime   **81.** $3x^2y^2(x + 2y)(x - 2y)$ **82.** 0 and 12, or $\{0, 12\}$
83. 7 and $-\dfrac{9}{4}$, or $\left\{7, -\dfrac{9}{4}\right\}$ **84.** -7 and 2, or $\{-7, 2\}$ **85.** -4 and 0, or $\{-4, 0\}$ **86.** -8 and $\dfrac{1}{2}$, or $\left\{-8, \dfrac{1}{2}\right\}$

87. -4 and 8, or $\{-4, 8\}$ **88.** -8 and 7, or $\{-8, 7\}$ **89.** 7 or $\{7\}$ **90.** $-\dfrac{10}{3}$ and $\dfrac{10}{3}$, or $\left\{-\dfrac{10}{3}, \dfrac{10}{3}\right\}$

91. -5 and -2, or $\{-5, -2\}$ **92.** $\dfrac{1}{3}$ and 7, or $\left\{\dfrac{1}{3}, 7\right\}$ **93.** 2 sec   **94.** width: 5 ft; length: 8 ft   **95.** 11 m by 11 m

Chapter 6 Test

1. $(x - 3)(x - 6)$ **2.** $(x - 7)^2$ **3.** $5y^2(3y - 1)(y - 2)$ **4.** $(x^2 + 3)(x + 2)$ **5.** $x(x - 9)$ **6.** $x(x + 7)(x - 1)$
7. $2(7x - 3)(x + 5)$ **8.** $(5x + 3)(5x - 3)$ **9.** $(x + 2)(x^2 - 2x + 4)$ **10.** $(x + 3)(x - 7)$ **11.** prime
12. $3y(2y + 1)(y + 1)$ **13.** $4(y + 3)(y - 3)$ **14.** $4(2x + 3)^2$ **15.** $2(x^2 + 4)(x + 2)(x - 2)$ **16.** $(6x - 7)^2$
17. $(7x - 1)(x - 7)$ **18.** $(x^2 - 5)(x + 2)$ **19.** $3y(2y + 3)(2y - 5)$ **20.** $(y - 5)(y^2 + 5y + 25)$ **21.** $5(x - 3y)(x + 2y)$
22. -6 and 4, or $\{-6, 4\}$ **23.** $-\dfrac{1}{3}$ and 2, or $\left\{-\dfrac{1}{3}, 2\right\}$ **24.** -2 and 8, or $\{-2, 8\}$ **25.** 0 and $\dfrac{7}{2}$, or $\left\{0, \dfrac{7}{2}\right\}$
26. $-\dfrac{9}{4}$ and $\dfrac{9}{4}$, or $\left\{-\dfrac{9}{4}, \dfrac{9}{4}\right\}$ **27.** -1 and $\dfrac{6}{5}$, or $\left\{-1, \dfrac{6}{5}\right\}$ **28.** $x^2 - 4 = (x + 2)(x - 2)$ **29.** 6 sec   **30.** width: 5 ft; length: 11 ft

Cumulative Review Exercises (Chapters 1–6)

1. -48 **2.** 0 or {0} **3.** 12 or {12} **4.** $\{x \mid x < -2\}$; 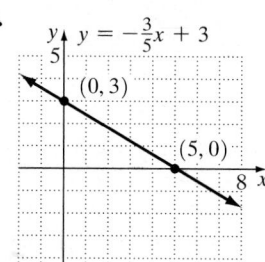 **5.** $38°, 38°, 104°$ **6.** $150

7.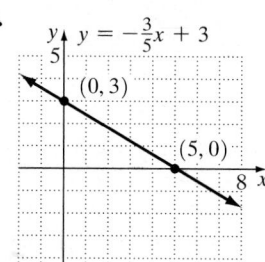

$y = -\frac{3}{5}x + 3$

$(0, 3)$ $(5, 0)$

8. $y + 4 = 5(x - 2)$ or $y - 1 = 5(x - 3); y = 5x - 14$

9.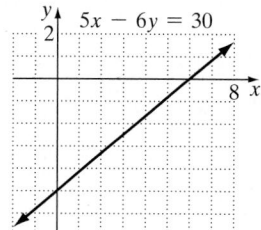

$5x - 6y = 30$

10. $(3, -1)$ or $\{(3, -1)\}$

11. $\left(-\frac{2}{13}, \frac{23}{13}\right)$ or $\left\{\left(-\frac{2}{13}, \frac{23}{13}\right)\right\}$ **12.** $-\frac{13}{40}$

13. $2x^4 - x^3 + 3x + 9$ **14.** $6x^2 + 17xy - 45y^2$

15. $2x^2 + 5x - 3 - \frac{2}{3x - 5}$ **16.** 7.1×10^{-3}

17. $(3x + 2)(x + 3)$ **18.** $y(y^2 + 4)(y + 2)(y - 2)$

19. $(2x + 3)^2$ **20.** width: 4 ft; length: 6 ft

CHAPTER 7

Section 7.1

Check Point Exercises

1. a. $x = 5$ **b.** $x = -7$ and $x = 4$ **2.** $\frac{x + 4}{3x}$ **3.** $\frac{x^2}{7}$ **4.** $\frac{x - 1}{x + 1}$ **5.** $-\frac{3x + 7}{4}$

Exercise Set 7.1

1. $x = 0$ **3.** $x = 7$ **5.** $x = 3$ **7.** $x = -7$ and $x = 3$ **9.** $x = 5$ and $x = -2$ **11.** $x = -4$ and $x = 3$

13. defined for all real numbers **15.** $y = -1$ and $y = \frac{3}{4}$ **17.** $y = -5$ and $y = 5$ **19.** defined for all real numbers **21.** $2x$

23. $\frac{x - 3}{5}$ **25.** $\frac{x - 4}{2x}$ **27.** $\frac{1}{x - 3}$ **29.** $-\frac{5}{x - 3}$ **31.** 3 **33.** $\frac{1}{x - 5}$ **35.** $\frac{2}{3}$ **37.** $\frac{1}{x - 3}$ **39.** $\frac{4}{x - 2}$

41. $\frac{y - 1}{y + 9}$ **43.** $\frac{y - 3}{y - 2}$ **45.** cannot be simplified **47.** $\frac{x + 6}{x - 6}$ **49.** $x^2 + 1$ **51.** $x^2 + 2x + 4$ **53.** $\frac{x - 4}{x + 4}$

55. cannot be simplified **57.** cannot be simplified **59.** -1 **61.** -1 **63.** cannot be simplified **65.** -2

67. $-\frac{2}{x}$ **69.** $-x - 1$ **71.** $-y - 3$ **73.** $-\frac{1}{x}$ **75.** $\frac{x - y}{2x - y}$ **77.** $\frac{6t^4 - 207t^3 + 2128t^2 - 6622t + 15{,}220}{28t^4 - 711t^3 + 5963t^2 - 1695t + 27{,}424}$

79. a. It costs $86.67 million to inoculate 40% of the population. It costs $520 million to inoculate 80% of the population. It costs $1170 million to inoculate 90% of the population. **b.** $x = 100$ **c.** The cost keeps rising, no amount of money will be enough to inoculate 100% of the population. **81.** 400 mg **83. a.** $300 **b.** $125 **c.** decrease **85.** 1.5 mg per liter; (3, 1.5)

87. after 1 hr; 2.5 mg per liter **89–95.** Answers will vary. **97.** $\frac{x^2 - x - 6}{x + 2}$ **99.** correctly simplified **101.** $x^2 - 1$ should be $x - 1$.

103. $\frac{3}{10}$ **104.** $\frac{1}{6}$ **105.** $(4, 2)$ or $\{(4, 2)\}$

Section 7.2

Check Point Exercises

1. $\frac{9x - 45}{2x + 8}$ **2.** $\frac{3}{8}$ **3.** $\frac{x + 2}{9}$ **4.** $-\frac{5(x + 1)}{7x(2x - 3)}$ **5.** $\frac{(x + 3)(x + 7)}{x - 4}$ **6.** $\frac{x + 3}{x - 5}$ **7.** $\frac{y + 1}{5y(y^2 + 1)}$

Exercise Set 7.2

1. $\frac{5x - 15}{7x + 14}$ **3.** $\frac{2x}{x + 1}$ **5.** $\frac{2}{3}$ **7.** $\frac{2}{5}$ **9.** 1 **11.** $\frac{x + 5}{x}$ **13.** $\frac{2}{y}$ **15.** $\frac{y + 2}{y + 4}$ **17.** $4(y + 3)$ **19.** $\frac{x - 1}{x + 2}$

21. $\frac{x^2 + 2x + 4}{3x}$ **23.** $\frac{x - 2}{x - 1}$ **25.** $-\frac{2}{x(x + 1)}$ **27.** $-\frac{y - 10}{y - 7}$ **29.** $(x + y)(x - y)$ **31.** $\frac{4(x + y)}{3(x - y)}$ **33.** $\frac{3x}{35}$ **35.** $\frac{1}{4}$

37. 10 **39.** $\frac{7}{9}$ **41.** $\frac{3}{4}$ **43.** $\frac{(x - 2)^2}{x}$ **45.** $\frac{(y - 4)(y^2 + 4)}{y - 1}$ **47.** $\frac{y}{3}$ **49.** $\frac{2(x + 3)}{3}$ **51.** $\frac{x - 5}{2}$ **53.** $\frac{y^2 + 1}{y^2}$

55. $\dfrac{y+1}{y-7}$ **57.** 6 **59.** $\dfrac{2}{3}$ **61.** $\dfrac{(x+y)^2}{32(x-y)^2}$ **63.** $\dfrac{y(x+y)}{(x+1)^2}$ **65.** $\dfrac{125x}{100-x}$ **67–69.** Answers will vary.

71. numerator: x; denominator: $x-4$ **73.** $\dfrac{y+3}{y-3}$ **75.** correct answer **77.** $2x-1$ should be $2x+1$. **78.** $\{x|x>18\}$

79. $3(x-7)(x+2)$ **80.** -5 and $\dfrac{1}{2}$, or $\left\{-5,\dfrac{1}{2}\right\}$

Section 7.3

Check Point Exercises

1. $x+2$ **2.** $\dfrac{x-5}{x+5}$ **3. a.** $\dfrac{3x+5}{x+7}$ **b.** $3x-4$ **4.** $\dfrac{3y+2}{y-4}$ **5.** $x+3$ **6.** $\dfrac{-4x^2+12x}{x^2-2x-9}$

Exercise Set 7.3

1. $\dfrac{7x}{9}$ **3.** $\dfrac{2x}{3}$ **5.** $\dfrac{x+1}{2}$ **7.** $\dfrac{8}{x}$ **9.** $\dfrac{4}{3x}$ **11.** $\dfrac{9}{x+3}$ **13.** $\dfrac{5x+5}{x-3}$ **15.** 2 **17.** $\dfrac{2y+3}{y-5}$ **19.** $\dfrac{7}{5y}$ **21.** $\dfrac{2}{x+2}$

23. $\dfrac{x-2}{x-3}$ **25.** $\dfrac{3x-4}{5x-4}$ **27.** 1 **29.** 1 **31.** $\dfrac{x}{2x-1}$ **33.** $-\dfrac{1}{y}$ **35.** $\dfrac{y+3}{3y+8}$ **37.** $\dfrac{3y+2}{y-3}$ **39.** $\dfrac{2}{x-3}$ **41.** $\dfrac{3x+7}{x-6}$

43. $\dfrac{3x+1}{3x-4}$ **45.** $x+2$ **47.** 0 **49.** $\dfrac{11}{x-1}$ **51.** $\dfrac{12}{x+3}$ **53.** $\dfrac{y+1}{y-1}$ **55.** $\dfrac{x-2}{x-7}$ **57.** $\dfrac{2x-4}{x^2-25}$ **59.** 1 **61.** $\dfrac{2}{x+y}$

63. $\dfrac{x-3}{x-1}$ **65. a.** $\dfrac{100\,W}{L}$ **b.** round **67.** 10 m **69–71.** Answers will vary. **73.** d **75.** $\dfrac{x+1}{x+2}$ **77.** $2x+1$ **79.** -7

81. $-4x-1$ **83.** $x-2$ should be $x-3$. **85.** $\dfrac{31}{45}$ **86.** $(9x^2+1)(3x+1)(3x-1)$ **87.** $3x^2-7x-5$

Section 7.4

Check Point Exercises

1. $30x^2$ **2.** $(x+3)(x-3)$ **3.** $7x(x+4)(x+4)$ or $7x(x+4)^2$ **4.** $\dfrac{9+14x}{30x^2}$ **5.** $\dfrac{6x+6}{(x+3)(x-3)}$ **6.** $-\dfrac{5}{x+5}$

7. $-\dfrac{5+y}{5y}$ **8.** $\dfrac{x-15}{(x+5)(x-5)}$

Exercise Set 7.4

1. $120x^2$ **3.** $30x^5$ **5.** $(x-3)(x+1)$ **7.** $7y(y+2)$ **9.** $(x+3)(x-3)$ **11.** $y(y+2)(y-2)$

13. $(y+5)(y-5)(y-5)$ **15.** $(x-5)(x+4)(2x-1)$ **17.** $\dfrac{3x+5}{x^2}$ **19.** $\dfrac{37}{18x}$ **21.** $\dfrac{8x+7}{2x^2}$ **23.** $\dfrac{x+1}{x}$ **25.** $\dfrac{3+5x}{x}$

27. $\dfrac{x+1}{2}$ **29.** $\dfrac{7x-20}{x(x-5)}$ **31.** $\dfrac{5x+1}{(x-1)(x+2)}$ **33.** $\dfrac{11y+15}{4y(y+5)}$ **35.** $-\dfrac{7}{x+7}$ **37.** $\dfrac{3x-55}{(x+5)(x-5)}$

39. $\dfrac{x^2+6x}{(x+4)(x-4)}$ **41.** $\dfrac{y+12}{(y+3)(y-3)}$ **43.** $\dfrac{7x-10}{(x-1)(x-1)}$ **45.** $\dfrac{9y}{4(y-5)}$ **47.** $\dfrac{8y+16}{y(y+4)}$ **49.** $\dfrac{17}{(x-3)(x-4)}$

51. $\dfrac{7x-1}{(x+1)(x+1)(x-1)}$ **53.** $\dfrac{x^2-x}{(x+5)(x-2)(x+3)}$ **55.** $\dfrac{y^2+8y+4}{(y+1)(y+1)(y+4)}$ **57.** $\dfrac{2x^2-4x+34}{(x+3)(x-5)}$

59. $\dfrac{5-3y}{2y(y-1)}$ **61.** $-\dfrac{x^2}{(x+3)(x-3)}$ **63.** $\dfrac{2}{y+5}$ **65.** $\dfrac{4x-11}{x-3}$ **67.** $\dfrac{3}{y+1}$ **69.** $\dfrac{-x^2+7x+3}{(x-3)(x+2)}$

71. $\dfrac{x^2+2x-1}{(x+2)(x-1)(x+1)}$ **73.** $\dfrac{-y^2+8y+9}{15y^2}$ **75.** $\dfrac{x^2-2x-6}{3(x+2)(x-2)}$ **77.** $\dfrac{-y-2}{(y+1)(y-1)}$ **79.** $\dfrac{x+2xy-y}{xy}$

81. $\dfrac{5x+2y}{(x+y)(x-y)}$ **83.** $\dfrac{D}{5}$ **85.** $\dfrac{D}{24}$ **87.** no **89.** 5 yr **91.** $\dfrac{4x^2+14x}{(x+3)(x+4)}$ **93.** Answers will vary.

95. $\dfrac{3}{x+5}$ should be $\dfrac{5+2x}{5x}$. **97.** Answers will vary. **99.** $\dfrac{-x+6}{(x+2)(x-2)}$ **101.** $\dfrac{y^3+y^2-10y-2}{(y-1)(y-1)(y+3)}$ **103.** $\dfrac{2}{x+1}$

104. $6x^2 - 11x - 35$ **105.**   **106.** $y = x - 1$

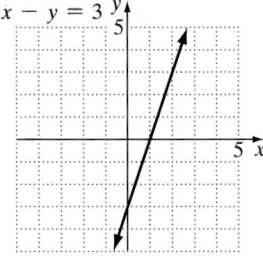

$3x - y = 3$

Section 7.5

Check Point Exercises

1. $\dfrac{11}{5}$ **2.** $\dfrac{2x - 1}{2x + 1}$ **3.** $y - x$ **4.** $\dfrac{11}{5}$ **5.** $\dfrac{2x - 1}{2x + 1}$ **6.** $y - x$

Exercise Set 7.5

1. $\dfrac{9}{10}$ **3.** $\dfrac{14}{15}$ **5.** $-\dfrac{4}{5}$ **7.** $\dfrac{3 - 4x}{3 + 4x}$ **9.** $\dfrac{5x - 2}{3x + 1}$ **11.** $\dfrac{2y + 3}{y - 7}$ **13.** $\dfrac{4 - 6y}{4 + 3y}$ **15.** $x - 5$ **17.** $\dfrac{x}{x - 1}$ **19.** $-\dfrac{1}{y}$

21. $\dfrac{xy + 1}{x}$ **23.** $\dfrac{y + x}{x^2 y^2}$ **25.** $\dfrac{x^2 + y}{y(y + 1)}$ **27.** $\dfrac{1}{y}$ **29.** $\dfrac{4 - x}{5x - 3}$ **31.** $\dfrac{2y}{y - 3}$ **33.** $\dfrac{1}{x + 3}$ **35.** $\dfrac{x^2 - 5x + 3}{x^2 - 7x + 2}$

37. $\dfrac{3xy + 2y^2}{y^2 + 2x}$ **39.** $-\dfrac{6}{5}$ **41.** $\dfrac{2r_1 r_2}{r_1 + r_2}; 34\frac{2}{7}$ mph   **43–45.** Answers will vary.   **47.** It cubes x.   **49.** $\dfrac{5y + 3}{(y - 3)(y + 1)}$

51. 2 should be $1 + x$.   **53.** $2x(x - 5)^2$ **54.** -2 or $\{-2\}$ **55.** $x^3 + y^3$

Section 7.6

Check Point Exercises

1. 4 or $\{4\}$ **2.** 3 or $\{3\}$ **3.** -3 and -2, or $\{-3, -2\}$ **4.** 24 or $\{24\}$ **5.** no solution or $\varnothing$ **6.** 75%

Exercise Set 7.6

1. 12 or $\{12\}$ **3.** 0 or $\{0\}$ **5.** -2 or $\{-2\}$ **7.** 2 or $\{2\}$ **9.** -2 or $\{-2\}$ **11.** 2 or $\{2\}$ **13.** 15 or $\{15\}$
15. 4 or $\{4\}$ **17.** -6 and -1, or $\{-6, -1\}$ **19.** -5 and 5, or $\{-5, 5\}$ **21.** -3 and 3, or $\{-3, 3\}$ **23.** -8 and 1, or $\{-8, 1\}$

25. -1 and 16, or $\{-1, 16\}$ **27.** 3 or $\{3\}$ **29.** no solution or $\varnothing$ **31.** $\dfrac{2}{3}$ and 4, or $\left\{\dfrac{2}{3}, 4\right\}$ **33.** no solution or $\varnothing$ **35.** 1 or $\{1\}$

37. $\dfrac{1}{5}$ or $\left\{\dfrac{1}{5}\right\}$ **39.** -3 or $\{-3\}$ **41.** no solution or $\varnothing$ **43.** -6 and $-\dfrac{1}{2}$, or $\left\{-6, -\dfrac{1}{2}\right\}$ **45.** 10,000 wheelchairs   **47.** 50%

49. 5 years old   **51.** either 50 or approx 67 cases; $(50, 350)$ or $\left(66\dfrac{2}{3}, 350\right)$ **53.** 10 more hits   **55–59.** Answers will vary.   **61.** b

63. $f_2 = -\dfrac{ff_1}{f - f_1}$ or $f_2 = \dfrac{ff_1}{f_1 - f}$ **65.** no solution or $\varnothing$ **67.** 8 or $\{8\}$ **69.** -3 and -2, or $\{-3, -2\}$ **70.** $(x^3 - 3)(x + 2)$

71. $-\dfrac{12}{x^8}$ **72.** $-20x + 55$

Section 7.7

Check Point Exercises

1. 2 mph   **2.** $2\dfrac{2}{3}$ hr or 2 hr 40 min   **3.** $1500   **4.** 720 deer   **5.** 32 in.

Exercise Set 7.7

1. Walking rate is 6 mph; car rate is 9 mph.   **3.** Downhill rate is 10 mph; uphill rate is 6 mph.   **5.** Water current rate is 5 mph.
7. Walking rate is 3 mph; jogging rate is 6 mph.   **9.** Still water rate is 10 mph.   **11.** It will take about 8.6 min which is enough time.
13. It will take about 171.4 hr, which is not enough time.   **15.** It will take 2.4 hr or 2 hr 24 min.   **17.** $1800   **19.** 20,490 fur seal pups
21. $950   **23.** 154.1 in.   **25.** 5 in.   **27.** 6 m   **29.** 16 in.   **31.** 16 ft   **33–41.** Answers will vary.
43. It will take the experienced carpenter 8 hr and the apprentice 24 hr.

45. Either change the rear to 12 teeth or change the front to 100 teeth.　**46.** $(5x + 9)(5x - 9)$　**47.** 6 or $\{6\}$

48.

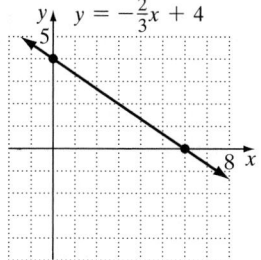

$y = -\frac{2}{3}x + 4$

Section 7.8

Check Point Exercises

1. 137.5 lb per in.^{2}　**2.** about 4.36 lb per in.^{2}　**3.** 24 min　**4.** 96π cubic feet

Exercise Set 7.8

1. 84　**3.** 25　**5.** $\frac{5}{6}$　**7.** 240　**9. a.** $G = kW$　**b.** $G = 0.02W$　**c.** 1.04 in.　**11.** \$60.00　**13.** 2442 mph　**15.** about 607 lb
17. 0.5 hr　**19.** 6.4 lb　**21.** 32; not in the desirable range　**23.** about 11.1 foot-candles　**25.** 72 ergs　**27.** 3983 calls
29. Yes, the wind will exert a force of 360 pounds on the window.　**31–35.** Answers will vary.
37. z varies jointly as the square of x and the square root of y.　**39.** If wind speed doubles, the wind pressure is 4 times more destructive.
41. To triple the amount of heat, the resistance should be divided by 3.　**43.** Answers will vary.
44. $\frac{16}{3}$ or $\left\{\frac{16}{3}\right\}$　**45.** $9x^2 - 6x + 4 - \frac{16}{3x + 2}$　**46.** $6x(x - 5)(x + 4)$

Review Exercises

1. $x = 4$　**2.** $x = 2$ and $x = -5$　**3.** $x = 1$ and $x = 2$　**4.** defined for all real numbers　**5.** $\frac{4x}{3}$　**6.** $x + 2$　**7.** x^2
8. $\frac{x - 3}{x - 6}$　**9.** $\frac{x - 5}{x + 7}$　**10.** $\frac{y}{y + 2}$　**11.** cannot be simplified　**12.** $-2(x + 3y)$　**13.** $\frac{x - 2}{4}$　**14.** $\frac{5}{2}$　**15.** $\frac{x + 3}{x + 2}$
16. $\frac{2y - 1}{5}$　**17.** $\frac{-3(y + 1)}{5y(2y - 3)}$　**18.** $\frac{x - 1}{4}$　**19.** $\frac{2}{x(x + 1)}$　**20.** $\frac{1}{7(y + 3)}$　**21.** $(y - 3)(y - 6)$　**22.** $\frac{8}{x - y}$　**23.** 4
24. 4　**25.** $3x - 5$　**26.** $3y + 4$　**27.** $\frac{4}{x - 2}$　**28.** $\frac{2x + 5}{x - 3}$　**29.** $36x^3$　**30.** $x^2(x - 1)^2$　**31.** $(x + 3)(x + 1)(x + 7)$
32. $\frac{14x + 15}{6x^2}$　**33.** $\frac{7x + 2}{x(x + 1)}$　**34.** $\frac{7x + 25}{(x + 3)^2}$　**35.** $\frac{3}{y - 2}$　**36.** $-\frac{y}{y - 1}$　**37.** $\frac{x^2 + y^2}{xy}$　**38.** $\frac{3x^2 - x}{(x + 1)^2(x - 1)}$
39. $\frac{5x^2 - 3x}{(x + 1)(x - 1)}$　**40.** $\frac{4}{(x + 2)(x - 2)(x - 3)}$　**41.** $\frac{2x + 13}{x + 3}$　**42.** $\frac{y - 4}{3y}$　**43.** $\frac{7}{2}$　**44.** $\frac{1}{x - 1}$　**45.** $x + y$　**46.** $\frac{3}{x}$
47. $\frac{3x}{x - 4}$　**48.** 12 or $\{12\}$　**49.** -1 or $\{-1\}$　**50.** -6 and 1, or $\{-6, 1\}$　**51.** no solution or $\varnothing$　**52.** 5 or $\{5\}$
53. -6 and 2, or $\{-6, 2\}$　**54.** 5 or $\{5\}$　**55.** -5 or $\{-5\}$　**56.** 3 yr　**57.** 30%　**58.** 4 mph
59. Slower car rate is 20 mph; faster car rate is 30 mph.　**60.** 4 hr　**61.** 324 teachers　**62.** 287 trout
63. 5 ft　**64.** \$154　**65.** 5 hr　**66.** 112 decibels　**67.** 16 hr　**68.** 800 cubic feet

Chapter 7 Test

1. $x = -9$ and $x = 4$　**2.** $\frac{x + 3}{x - 2}$　**3.** $\frac{4x}{x + 1}$　**4.** $\frac{x - 4}{2}$　**5.** $\frac{x}{x + 3}$　**6.** $\frac{2x + 6}{x + 1}$　**7.** $\frac{5}{(y - 1)(y - 3)}$　**8.** $2y$　**9.** $\frac{2}{y + 3}$
10. $\frac{x^2 + 2x + 15}{(x + 3)(x - 3)}$　**11.** $\frac{8x - 14}{(x + 2)(x - 1)(x - 3)}$　**12.** $-\frac{x + 1}{x - 3}$　**13.** $\frac{x + 2}{x - 1}$　**14.** $\frac{11}{(x - 3)(x - 4)}$　**15.** $\frac{4}{y + 4}$
16. $\frac{3x - 3y}{x}$　**17.** $\frac{5x + 5}{2x + 1}$　**18.** $\frac{y - x}{y}$　**19.** 6 or $\{6\}$　**20.** -8 or $\{-8\}$　**21.** 0 and 2, or $\{0, 2\}$
22. Water current rate is 2 mph.　**23.** 12 min　**24.** 6000 tule elk　**25.** 3.2 in.　**26.** 52.5 amp

Cumulative Review Exercises (Chapters 1–7)

1. 2 or {2}   **2.** $\{x | x < 2\}$ **3.** -6 and 3, or $\{-6, 3\}$ **4.** -6 or $\{-6\}$ **5.** $(3, 3)$ or $\{(3, 3)\}$ **6.** $(-2, 2)$ or $\{(-2, 2)\}$

7. $3x - 2y = 6$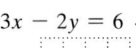

8. $y = -2x + 3$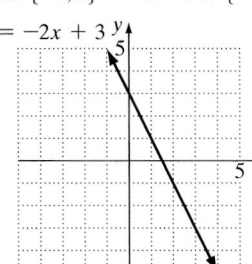

9. $y = -3$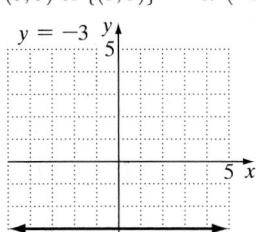

10. -19
11. $8x^9$
12. $\dfrac{1 - 2x}{4x - 1}$
13. $(4x - 1)(x - 3)$
14. $(2x - 5)^2$
15. $3(x + 5)(x - 5)$
16. $-x^2 + 4x + 8$
17. $-2x^4 + 3x^2 - 1$

18. $\dfrac{3x^2 + 6x + 16}{(x - 2)(x + 3)}$ **19.** $1225 at 5\% and \$2775 at 9\%   **20.** 17 in. and 51 in.

CHAPTER 8

Section 8.1

Check Point Exercises

1. domain: $\{5, 10, 15, 20, 25\}$; range: $\{12.8, 16.2, 18.9, 20.7, 21.8\}$ **2. a.** not a function   **b.** function   **3. a.** 29   **b.** 65   **c.** 46
d. $6a + 6h + 9$ **4. a.** function   **b.** function   **c.** not a function   **5. a.** 16   **b.** 8

Exercise Set 8.1

1. function; domain: $\{1, 3, 5\}$; range: $\{2, 4, 5\}$ **3.** not a function; domain: $\{3, 4\}$; range: $\{4, 5\}$ **5.** function; domain: $\{-3, -2, -1, 0\}$;
range: $\{-3, -2, -1, 0\}$ **7.** not a function; domain: $\{1\}$; range: $\{4, 5, 6\}$ **9. a.** 1   **b.** 6   **c.** -7 **d.** $2a + 1$ **e.** $a + 3$
11. a. -2 **b.** -17 **c.** 0   **d.** $12b - 2$ **e.** $3b + 10$ **13. a.** 5   **b.** 8   **c.** 53   **d.** 32   **e.** $48b^2 + 5$
15. a. -1 **b.** 26   **c.** 19   **d.** $2b^2 + 3b - 1$ **e.** $50a^2 + 15a - 1$ **17. a.** $\dfrac{3}{4}$ **b.** -3 **c.** $\dfrac{11}{8}$ **d.** $\dfrac{13}{9}$ **e.** $\dfrac{2a + 2h - 3}{a + h - 4}$
f. Denominator would be zero.   **19.** function   **21.** function   **23.** not a function   **25.** function   **27.** -4 **29.** 4   **31.** 0
33. 2   **35.** 2   **37.** -2 **39.** $(1, 31), (2, 53), (3, 70), (4, 86), (5, 86)\}$; domain: $\{1, 2, 3, 4, 5\}$; range: $\{31, 53, 70, 86\}$; function;
Answers will vary.   **41.** 5.22; In 2000 there were 5.22 million women enrolled in U.S. colleges.; $(2000, 5.22)$
43. 1.4; In 2004 there will be 1.4 million more women than men enrolled in U.S. colleges.
45. 440; For 20-year-old people there are 440 accidents per 50 million miles driven.; $(20, 440)$
47. $x = 45$; $y = 190$; The minimum number of accidents is 190 per 50 million miles driven and is attributed to 45-year-old drivers.
49. 3.1; In 1960 about 3.1% of the U.S. population was Jewish American.
51. 19 and 64; In 1919 and in 1962 3% of the U.S. population was Jewish American.   **53.** 1940; 3.7%
55. Each year corresponds to only one percentage.   **57.** 0.76; It costs \$0.76 to mail a 3-ounce first-class letter.   **59.** \$0.55
61–67. Answers will vary.   **69.** 3   **71.** $f(2) = 6; f(3) = 9; f(4) = 12$; no   **73.** 0   **75.** $\dfrac{y^{10}}{9x^4}$ **76.** -15 or $\{-15\}$

Section 8.2

Check Point Exercises

1. a. $\{x | x$ is a real number$\}$ **b.** $\{x | x$ is a real number and $x \neq -5\}$ **2. a.** $3x^2 + 6x + 6$ **b.** 78   **3. a.** $\dfrac{5}{x} + \dfrac{7}{x - 8}$

b. $\{x | x$ is a real number and $x \neq 0$ and $x \neq 8\}$ **4. a.** 23   **b.** $x^2 - 3x - 3; 1$ **c.** $x^3 + x^2 - 6x; -24$ **d.** $\dfrac{x^2 - 2x}{x + 3}; \dfrac{7}{2}$

Exercise Set 8.2

1. $\{x | x$ is a real number$\}$ **3.** $\{x | x$ is a real number and $x \neq -4\}$ **5.** $\{x | x$ is a real number and $x \neq 3\}$
7. $\{x | x$ is a real number and $x \neq 5\}$ **9.** $\{x | x$ is a real number and $x \neq -7$ and $x \neq 9\}$ **11. a.** $5x - 5$ **b.** 20
13. a. $3x^2 + x - 5$ **b.** 75   **15. a.** $2x^2 - 2$ **b.** 48   **17.** $\{x | x$ is a real number$\}$ **19.** $\{x | x$ is a real number and $x \neq 5\}$
21. $\{x | x$ is a real number and $x \neq 0$ and $x \neq 5\}$ **23.** $\{x | x$ is a real number and $x \neq 2$ and $x \neq -3\}$ **25.** $\{x | x$ is a real number and $x \neq 2\}$
27. $\{x | x$ is a real number$\}$ **29.** $x^2 + 3x + 2; 20$ **31.** 0   **33.** $x^2 + 5x - 2; 48$ **35.** -8 **37.** $-x^3 - 2x^2 + 8x; 0$ **39.** -135
41. $\dfrac{x^2 + 4x}{2 - x}; 5$ **43.** -1 **45.** $\{x | x$ is a real number$\}$ **47.** $\{x | x$ is a real number and $x \neq 2\}$ **49.** 20; In 2000 veterinary costs in
the U.S. for dogs and cats were about \$20 billion.   **51.** $\{1983, 1987, 1991, 1996, 2000\}$ **53.** $h(x)$ or the total world population in year x

55. 5.9 billion   **57.** −200,000; 0; 200,000; The company has profits of −200,000, 0, and 200,000 dollars when 20,000, 30,000 and 40,000 radios are produced, respectively.   **59–63.** Answers will vary.

65.

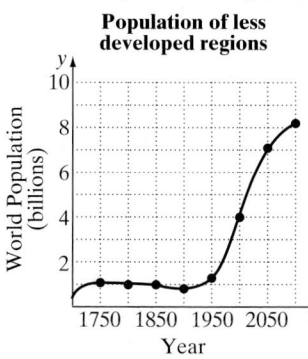

67.

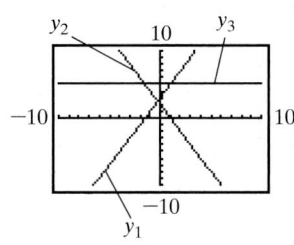

69.

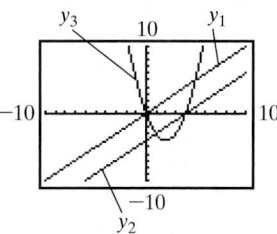

71. No y-value is displayed.; y_3 is undefined at $x = 0$.   **72.** $(−1, 2)$ or $\{(−1, 2)\}$ **73.** 7 or $\{7\}$ **74.** $6b + 8$

Section 8.3

Check Point Exercises

1. $(−1) − 2(−4) + 3(5) = 22; 2(−1) − 3(−4) − 5 = 5; 3(−1) + (−4) − 5(5) = −32$ **2.** $(1, 4, −3)$ or $\{(1, 4, −3)\}$
3. $(4, 5, 3)$ or $\{(4, 5, 3)\}$ **4.** $y = 3x^2 − 12x + 13$ or $f(x) = 3x^2 − 12x + 13$

Exercise Set 8.3

1. not a solution   **3.** solution   **5.** $(2, 3, 3)$ or $\{(2, 3, 3)\}$ **7.** $(2, −1, 1)$ or $\{(2, −1, 1)\}$ **9.** $(1, 2, 3)$ or $\{(1, 2, 3)\}$
11. $(3, 1, 5)$ or $\{(3, 1, 5)\}$ **13.** $(1, 0, −3)$ or $\{(1, 0, −3)\}$ **15.** $(1, −5, −6)$ or $\{(1, −5, −6)\}$ **17.** no solution or ∅
19. infinitely many solutions; dependent equations   **21.** $\left(\frac{1}{2}, \frac{1}{3}, −1\right)$ or $\left\{\left(\frac{1}{2}, \frac{1}{3}, −1\right)\right\}$
23. $y = 2x^2 − x + 3$ **25.** $y = 2x^2 + x − 5$ **27.** 7, 4, and 5
29. a. $(0, 5.4), (10, 4.7), (20, 6.2)$ **b.** $0a + 0b + c = 5.4; 100a + 10b + c = 4.7; 400a + 20b + c = 6.2$
31. a. $y = −16x^2 + 40x + 200$ **b.** 0; After 5 seconds, the ball hits the ground.
33. chemical: $42,758; mechanical: $39,852; electrical: $38,811   **35.** $1200 at 8%; $2000 at 10%; $3500 at 12%
37. 200 $8 tickets; 150 $10 tickets; 50 $12 tickets   **39.** 4 oz of food A; 0.5 oz of food B; 1 oz of food C
41–45. Answers will vary.   **47.** c   **49.** 13 triangles, 21 rectangles, and 6 pentagons

53.

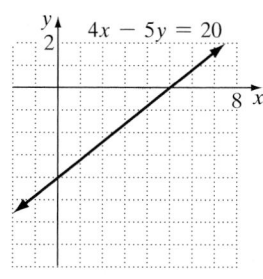

54. $\{x|x < −8\}$;

```
◄─┼──┼──⊕──┼──┼──┼──┼──┼──┼──┼──┼──►
 −10−9 −8 −7 −6 −5 −4 −3 −2 −1  0
```

55. $\dfrac{x + 9}{x − 15}$

Section 8.4

Check Point Exercises

1. $(4, −3, 1)$ or $\{(4, −3, 1)\}$ **2. a.** $\begin{bmatrix} 1 & 6 & −3 & | & 7 \\ 4 & 12 & −20 & | & 8 \\ −3 & −2 & 1 & | & −9 \end{bmatrix}$ **b.** $\begin{bmatrix} 1 & 3 & −5 & | & 2 \\ 1 & 6 & −3 & | & 7 \\ −3 & −2 & 1 & | & −9 \end{bmatrix}$ **c.** $\begin{bmatrix} 4 & 12 & −20 & | & 8 \\ 1 & 6 & −3 & | & 7 \\ 0 & 16 & −8 & | & 12 \end{bmatrix}$
3. $(−1, 2)$ or $\{(−1, 2)\}$ **4.** $(5, 2, 3)$ or $\{(5, 2, 3)\}$

Exercise Set 8.4

1. $x - 3y = 11; y = -3; (2, -3)$ or $\{(2, -3)\}$ **3.** $x - 3y = 1; y = -1; (-2, -1)$ or $\{(-2, 1)\}$ **5.** $x - 4z = 5; y - 12z = 13;$

$z = -\dfrac{1}{2}; \left(3, 7, -\dfrac{1}{2}\right)$ or $\left\{\left(3, 7, -\dfrac{1}{2}\right)\right\}$ **7.** $x + \dfrac{1}{2}y + z = \dfrac{11}{2}; y + \dfrac{3}{2}z = 7; z = 4; (1, 1, 4)$ or $\{(1, 1, 4)\}$ **9.** $\begin{bmatrix} 1 & -\dfrac{3}{2} & 5 \\ 2 & 2 & 5 \end{bmatrix}$

11. $\begin{bmatrix} 1 & -\dfrac{4}{3} & 2 \\ 3 & 5 & -2 \end{bmatrix}$ **13.** $\begin{bmatrix} 1 & -3 & 5 \\ 0 & 12 & -6 \end{bmatrix}$ **15.** $\begin{bmatrix} 1 & -\dfrac{3}{2} & \dfrac{7}{2} \\ 0 & \dfrac{17}{2} & -\dfrac{17}{2} \end{bmatrix}$ **17.** $\begin{bmatrix} 1 & -3 & 2 & 5 \\ 1 & 5 & -5 & 0 \\ 3 & 0 & 4 & 7 \end{bmatrix}$ **19.** $\begin{bmatrix} 1 & -3 & 2 & 0 \\ 0 & 10 & -7 & 7 \\ 2 & -2 & 1 & 3 \end{bmatrix}$

21. $\begin{bmatrix} 1 & 1 & -1 & 6 \\ 0 & -3 & 3 & -15 \\ 0 & -4 & 2 & -14 \end{bmatrix}$ **23.** $(4, 2)$ or $\{(4, 2)\}$ **25.** $(3, -3)$ or $\{(3, -3)\}$ **27.** $(2, -5)$ or $\{(2, -5)\}$ **29.** no solution or $\varnothing$

31. infinitely many solutions; dependent equations   **33.** $(1, -1, 2)$ or $\{(1, -1, 2)\}$ **35.** $(3, -1, -1)$ or $\{(3, -1, -1)\}$
37. $(1, 2, -1)$ or $\{(1, 2, -1)\}$ **39.** $(1, 2, 3)$ or $\{(1, 2, 3)\}$ **41.** no solution or $\varnothing$ **43.** infinitely many solutions; dependent equations
45. $(-1, 2, -2)$ or $\{(-1, 2, -2)\}$ **47. a.** $y = 2x^2 + 22x + 320$ **b.** 2780 thousand, or 2,780,000, inmates   **c.** Answers will vary.
49. 35% are under 30, 49% are 30−49, and 16% are 50 and over.   **51–55.** Answers will vary.   **57.** d
59. a. Matrices show that the system of equations for this problem is inconsistent.
b. Matrices show that the system has infinitely many solutions. There are many combinations of foods that satisfy the new requirements.
62. $-6a + 13$ **63.** $6x^2 - 9x; 15$ **64.** $2x(x - 3)(x - 5)$

Section 8.5

Check Point Exercises

1. a. -4 **b.** -17 **2.** $(4, -2)$ or $\{(4, -2)\}$ **3.** 80   **4.** $(2, -3, 4)$ or $\{(2, -3, 4)\}$

Exercise Set 8.5

1. 1   **3.** -29 **5.** 0   **7.** 33   **9.** $-\dfrac{7}{16}$ **11.** $(5, 2)$ or $\{(5, 2)\}$ **13.** $(2, -3)$ or $\{(2, -3)\}$ **15.** $(3, -1)$ or $\{(3, -1)\}$

17. $(4, 0)$ or $\{(4, 0)\}$ **19.** $(4, 2)$ or $\{(4, 2)\}$ **21.** $(7, 4)$ or $\{(7, 4)\}$ **23.** inconsistent; no solution or $\varnothing$
25. infinitely many solutions; dependent equations   **27.** 72   **29.** -75 **31.** 0   **33.** $(-5, -2, 7)$ or $\{(-5, -2, 7)\}$
35. $(2, -3, 4)$ or $\{(2, -3, 4)\}$ **37.** $(3, -1, 2)$ or $\{(3, -1, 2)\}$ **39.** $(2, 3, 1)$ or $\{(2, 3, 1)\}$ **41.** 28 sq units   **43.** yes

45. $\begin{vmatrix} x & y & 1 \\ 3 & -5 & 1 \\ -2 & 6 & 1 \end{vmatrix} = 0; y = -\dfrac{11}{5}x + \dfrac{8}{5}$ **47–53.** Answers will vary.   **55.** d   **57.** The value is multiplied by -1.

59–61. Answers will vary.   **62.** 1 and 5, or $\{1, 5\}$ **63.** $\dfrac{1}{8x^6}$ **64.** $\dfrac{x - 3}{4(x + 3)}$

Review Exercises

1. function; domain: $\{3, 4, 5\}$; range: $\{10\}$ **2.** function; domain: $\{1, 2, 3, 4\}$; range: $\{12, 100, \pi, -6\}$
3. not a function; domain: $\{13, 15\}$; range: $\{14, 16, 17\}$ **4. a.** -5 **b.** 16   **c.** -75 **d.** $14a - 5$ **e.** $7a + 9$
5. a. 2   **b.** 52   **c.** 70   **d.** $3b^2 - 5b + 2$ **e.** $48a^2 - 20a + 2$ **6.** not a function   **7.** function   **8.** function
9. not a function   **10.** -3 **11.** -2 **12.** 3   **13. a.** For each time, there is only one height.
b. 0; The vulture was on the ground after 15 seconds.   **c.** 45 m   **d.** 7 and 22; After 7 and 22 seconds the vulture's height is 20 meters.
e. Answers will vary.   **14.** $\{x \mid x \text{ is a real number}\}$ **15.** $\{x \mid x \text{ is a real number and } x \neq -8\}$ **16.** $\{x \mid x \text{ is a real number and } x \neq 5\}$
17. a. $6x - 4$ **b.** 14   **18. a.** $5x^2 + 1$ **b.** 46   **19.** $\{x \mid x \text{ is a real number and } x \neq 4\}$
20. $\{x \mid x \text{ is a real number and } x \neq -6 \text{ and } x \neq -1\}$ **21.** $x^2 - x - 5; 1$ **22.** 1   **23.** $x^2 - 3x + 5; 3$ **24.** 9
25. $x^3 - 7x^2 + 10x; -120$ **26.** $\dfrac{x^2 - 2x}{x - 5}; -8$ **27.** $\{x \mid x \text{ is a real number}\}$ **28.** $\{x \mid x \text{ is a real number and } x \neq 5\}$
29. no   **30.** $(0, 1, 2)$ or $\{(0, 1, 2)\}$ **31.** $(2, 1, -1)$ or $\{(2, 1, -1)\}$ **32.** infinitely many solutions; dependent equations
33. $y = 3x^2 - 4x + 5$ **34.** Japan: 16%; Germany: 15%; France: 14%   **35.** $2x + 3y = -10; y = -6; (4, -6)$ or $\{(4, -6)\}$
36. $x + y + 3z = 12; y - 2z = -4; z = 3; (1, 2, 3)$ or $\{(1, 2, 3)\}$ **37.** $\begin{bmatrix} 1 & -8 & 3 \\ 0 & 1 & -2 \end{bmatrix}$ **38.** $\begin{bmatrix} 1 & -3 & 1 \\ 0 & 7 & -7 \end{bmatrix}$

39. $\begin{bmatrix} 1 & -1 & \dfrac{1}{2} & -\dfrac{1}{2} \\ 1 & 2 & -1 & 2 \\ 6 & 4 & 3 & 5 \end{bmatrix}$ **40.** $\begin{bmatrix} 1 & 2 & 2 & 2 \\ 0 & 1 & -1 & 2 \\ 0 & 0 & 9 & -9 \end{bmatrix}$ **41.** $(-5, 3)$ or $\{(-5, 3)\}$ **42.** inconsistent; no solution or $\varnothing$

43. $(1, 3, -4)$ or $\{(1, 3, -4)\}$ **44.** $(-2, -1, 0)$ or $\{(-2, -1, 0)\}$ **45.** 17 **46.** 4 **47.** -86 **48.** -236

49. $\left(\dfrac{7}{4}, -\dfrac{25}{8}\right)$ or $\left\{\left(\dfrac{7}{4}, -\dfrac{25}{8}\right)\right\}$ **50.** $(2, -7)$ or $\{(2, -7)\}$ **51.** $(23, -12, 3)$ or $\{(23, -12, 3)\}$ **52.** $(-3, 2, 1)$ or $\{(-3, 2, 1)\}$

53. $y = \dfrac{5}{8}x^2 - 50x + 1150$; 30-year-olds are involved in 212.5 accidents daily, and 50-year-olds are involved in 212.5 accidents daily.

Chapter 8 Test

1. function; domain: $\{1, 3, 5, 6\}$; range: $\{2, 4, 6\}$ **2.** not a function; domain: $\{2, 4, 6\}$; range: $\{1, 3, 5, 6\}$ **3.** $3a + 10$ **4.** 28
5. function **6.** not a function **7.** -3 **8.** -5 **9.** $\{x | x$ is a real number and $x \neq 10\}$ **10.** $x^2 + 5x + 2; 26$

11. $x^2 + 3x - 2; -4$ **12.** $x^3 + 6x^2 + 8x; -15$ **13.** $\dfrac{x^2 + 4x}{x + 2}; 3$ **14.** $\{x | x$ is a real number and $x \neq -2\}$

15. $(1, 3, 2)$ or $\{(1, 3, 2)\}$ **16.** $\begin{bmatrix} 1 & 0 & -4 & | & 5 \\ 0 & -1 & 26 & | & -20 \\ 2 & -1 & 4 & | & -3 \end{bmatrix}$ **17.** $(4, -2)$ or $\{(4, -2)\}$ **18.** $(-1, 2, 2)$ or $\{(-1, 2, 2)\}$

19. 17 **20.** -10 **21.** $(-1, -6)$ or $\{(-1, -6)\}$ **22.** $(4, -3, 3)$ or $\{(4, -3, 3)\}$

Cumulative Review Exercises (Chapters 1–8)

1. -3 or $\{-3\}$ **2.** -4 and $\dfrac{3}{2}$, or $\left\{-4, \dfrac{3}{2}\right\}$ **3.** $(-3, -4)$ or $\{(-3, -4)\}$ **4.** $\dfrac{9}{7}$ or $\left\{\dfrac{9}{7}\right\}$ **5.** -5 or $\{-5\}$ **6.** $x^2 - 3x - 1; -1$

7. $-\dfrac{2}{x^4}$ **8.** 4 **9.** $\dfrac{3}{x}$ **10.** $-\dfrac{2 + x}{3x + 1}$ **11.** 5 **12.** $(x - 7)(x - 11)$ **13.** $x(x + 5)(x - 5)$ **14.** $6x^2 - 7x + 2$

15. $8x^3 - 27$ **16.** $\dfrac{1}{x - 1}$ **17.** $\dfrac{5x - 1}{4x - 3}$ **18.** $(3, 2, 4)$ or $\{(3, 2, 4)\}$

19.
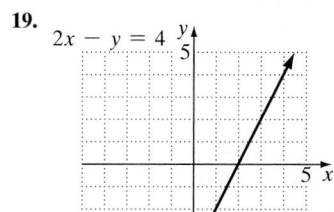
$2x - y = 4$

20.
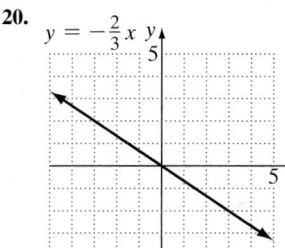
$y = -\dfrac{2}{3}x$

21. 23 **22.** $-\dfrac{8}{3}$ **23.** $y + 3 = 5(x + 2); y = 5x + 7$
24. $(3, -2, 1)$ or $\{(3, -2, 1)\}$ **25.** $(-3, 2)$ or $\{(-3, 2)\}$

CHAPTER 9

Section 9.1

Check Point Exercises

1. a.
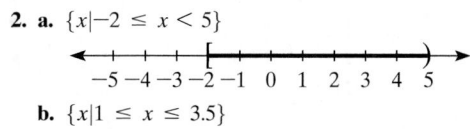

b.

c.

2. a. $\{x | -2 \leq x < 5\}$

b. $\{x | 1 \leq x \leq 3.5\}$

c. $\{x | x < -1\}$
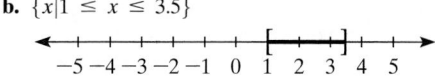
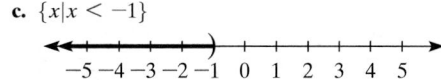
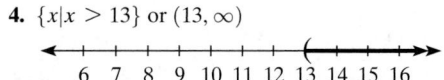

3. $\{x | x \geq -9\}$ or $[-9, \infty)$

4. $\{x | x > 13\}$ or $(13, \infty)$

5. a. $P(x) = 125x - 160,000$ **b.** more than 1280 units **6. a.** $C(x) = 300,000 + 30x$ **b.** $R(x) = 80x$
c. $P(x) = 50x - 300,000$ **d.** more than 6000 pairs

Exercise Set 9.1

1. $\{x | 1 < x \le 6\}$

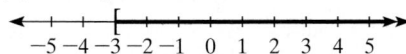

3. $\{x | -5 \le x < 2\}$

5. $\{x | -3 \le x \le 1\}$

7. $\{x | x > 2\}$

9. $\{x | x \ge -3\}$

11. $\{x | x < 3\}$

13. $\{x | x < 5.5\}$

15. $\{x | x < 3\}$ or $(-\infty, 3)$

17. $\{x | x \ge 7\}$ or $[7, \infty)$

19. $\{x | x \le -4\}$ or $(-\infty, -4]$

21. $\left\{x \,\middle|\, x \le -\dfrac{2}{5}\right\}$ or $\left(-\infty, -\dfrac{2}{5}\right]$

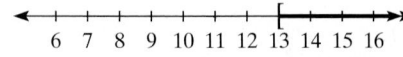

23. $\{x | x \ge 0\}$ or $[0, \infty)$

25. $\{x | x < 1\}$ or $(-\infty, 1)$

27. $\{x | x \ge 6\}$ or $[6, \infty)$

29. $\{x | x \ge -6\}$ or $[-6, \infty)$

31. $\{x | x < -6\}$ or $(-\infty, -6)$

33. $\{x | x \ge 13\}$ or $[13, \infty)$

35. a. $P(x) = 17x - 25{,}500$ **b.** more than 1500 units
37. a. $P(x) = 140x - 70{,}000$ **b.** more than 500 units
39. playing sports and sports events

41. exercise, movies, gardening, and amusement parks   **43.** movies and gardening
45. gardening, amusement parks, and home improvement   **47.** after the year 2005
49. $\{t > 175\}$ or $(175, \infty)$; The women's speed skating times will be less than the men's after the year 2075.
51. a. $C(x) = 18{,}000 + 20x$ **b.** $R(x) = 80x$ **c.** $P(x) = 60x - 18{,}000$ **d.** more than 300 canoes
53. a. $C(x) = 30{,}000 + 2500x$ **b.** $R(x) = 3125x$ **c.** $P(x) = 625x - 30{,}000$ **d.** more than 48 sold-out performances
55. more than 6250 tapes   **57.** more than 300 minutes   **59–63.** Answers will vary.   **65.** d   **67.** $\{x | -10 < x < -8\}$ or $(-10, -8)$
69.
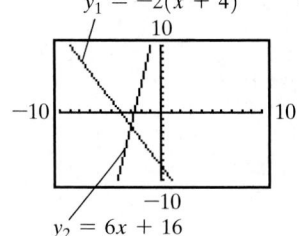

$y_1 = -2(x + 4)$

$y_2 = 6x + 16$

; $\{x | x < -3\}$ or $(-\infty, -3)$ **71.** more than 6   **73.** 29
74. $(-1, -1, 2)$ or $\{(-1, -1, 2)\}$ **75.** $(5x + 9)(5x - 9)$

Section 9.2

Check Point Exercises

1. $\{3, 7\}$ **2.** $\{x | x < 1\}$ or $(-\infty, 1)$ **3.** $\varnothing$ **4.** $\{x | -1 \le x < 4\}$ or $[-1, 4)$;

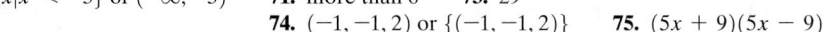

5. $\{3, 4, 5, 6, 7, 8, 9\}$ **6.** $\{x | x \le 1 \text{ or } x > 3\}$ or $(-\infty, 1] \cup (3, \infty)$ **7.** $\{x | x \text{ is a real number}\}$ or $\mathbb{R}$ or $(-\infty, \infty)$

Exercise Set 9.2

1. $\{2, 4\}$　　**3.** $\varnothing$

5. $\{x|x > 6\}; (6, \infty)$

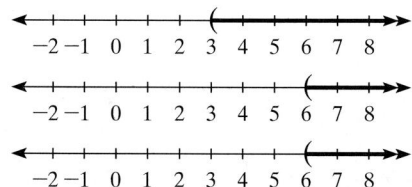

7. $\{x|x \leq 1\}; (-\infty, 1]$

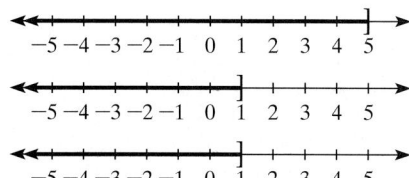

9. $\{x|-1 \leq x < 2\}; [-1, 2)$

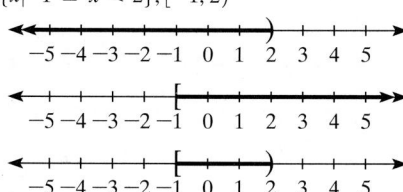

11. $\varnothing$

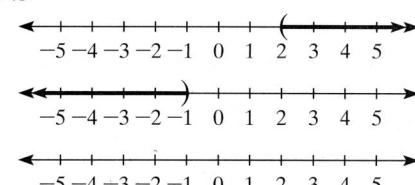

13. $\{x|-6 < x < -4\}; (-6, -4)$

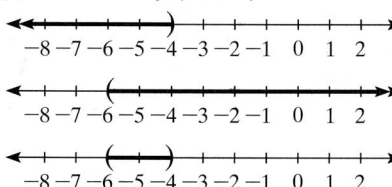

15. $\{x|-3 < x \leq 6\}; (-3, 6]$

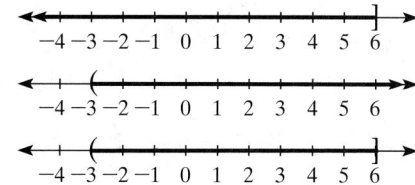

17. $\{x|2 < x < 5\}; (2, 5)$

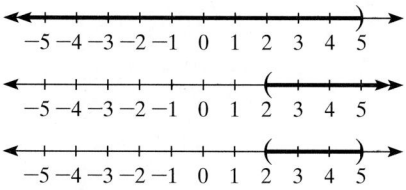

19. $\varnothing$

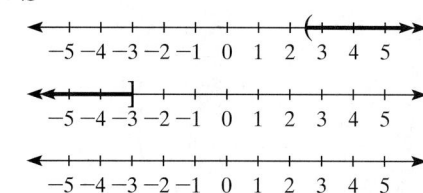

21. $\{x|0 \leq x < 2\}; [0, 2)$

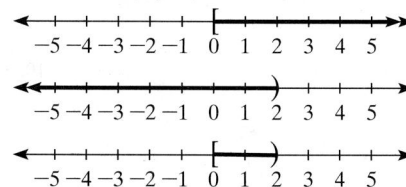

23. $\{x|3 < x < 5\}; (3, 5)$

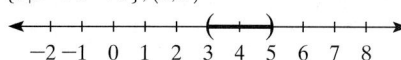

25. $\{x|-1 \leq x < 3\}; [-1, 3)$

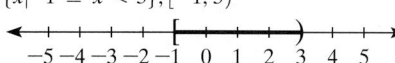

27. $\{x|-5 < x \leq -2\}; (-5, -2]$

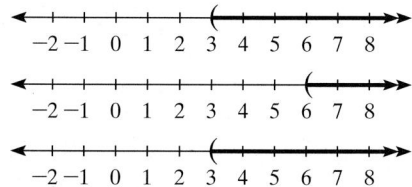

29. $\{x|3 \leq x < 6\}; [3, 6)$

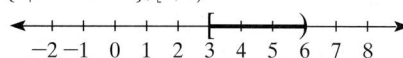

31. $\{1, 2, 3, 4, 5\}$　　**33.** $\{1, 2, 3, 4, 5, 6, 7, 8, 10\}$

35. $\{x|x > 3\}; (3, \infty)$

37. $\{x|x \leq 5\}; (-\infty, 5]$

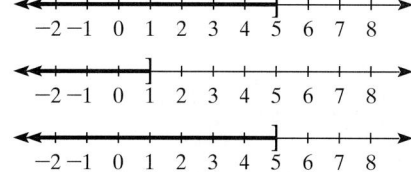

39. $\{x|x \text{ is a real number}\}; (-\infty, \infty)$

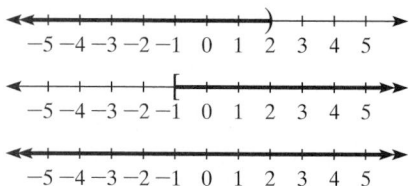

41. $\{x|x < -1 \text{ or } x \geq 2\}; (-\infty, -1) \cup [2, \infty)$

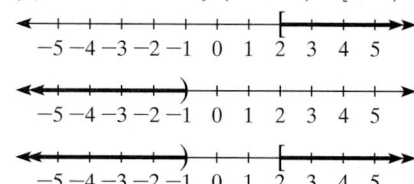

43. $\{x|x < -3 \text{ or } x > 4\}; (-\infty, -3) \cup (4, \infty)$

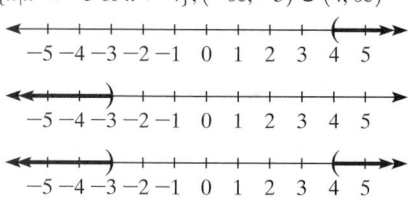

45. $\{x|x \leq 1 \text{ or } x \geq 3\}; (-\infty, 1] \cup [3, \infty)$

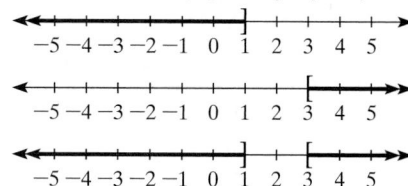

47. $\{x|x \text{ is a real number}\}; (-\infty, \infty)$

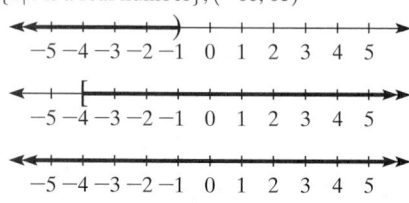

49. $\{x|x < 2\}; (-\infty, 2)$

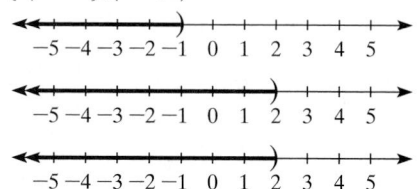

51. spatial-temporal toys, sports equipment, and toy cars and trucks **53.** dollhouses, spatial-temporal toys, sports equipment, and toy cars and trucks **55.** none of the toys **57.** from 1905 to 1955 **59.** $[76, 126]$; If the highest grade is 100, then $[76, 100]$.
61. more than 3 and less than 15 crossings per 3-month period **63–69.** Answers will vary. **71.** $(-1, 5]$
73.

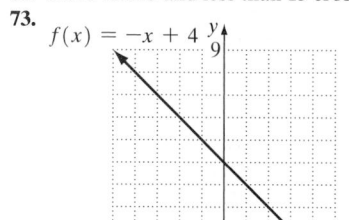

$f(x) = -x + 4$

75. $(-\infty, 4]$ **77.** $[-1, 4]$ **79.** least: 4; greatest: 7
81. $\{x|-6 < x < 2\}$ or $(-6, 2)$

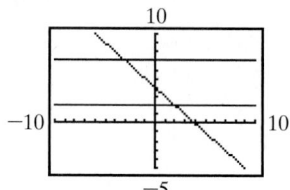

83. $\{x|-3 \leq x \leq 2\}$ or $[-3, 2]$

$\{x|-3 < x < 2\}$ or $(-3, 2)$
84. $-x^2 + 5x - 9; -15$ **85.**

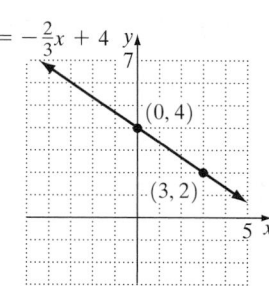

$y = -\frac{2}{3}x + 4$

$(0, 4)$

$(3, 2)$

86. $17 - 2x$

Section 9.3

Check Point Exercises

1. -2 and 3, or $\{-2, 3\}$ **2.** $\frac{4}{3}$ and 10, or $\left\{\frac{4}{3}, 10\right\}$

3. $\{x|-3 < x < 7\}$ or $(-3, 7)$

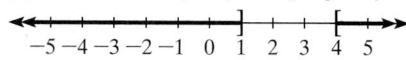

4. $\{x|x \leq 1 \text{ or } x \geq 4\}$ or $(-\infty, 1] \cup [4, \infty)$

5. $\{x|8.1 \leq x \leq 13.9\}$ or $[8.1, 13.9]$; The percentage of children in the population who think that not being able to do anything they want is a bad thing is between a low of 8.1% and a high of 13.9%.

Exercise Set 9.3

1. $\{-8, 8\}$ **3.** $\{-5, 9\}$ **5.** $\{-3, 4\}$ **7.** $\{-1, 2\}$ **9.** $\varnothing$ **11.** $\{-3\}$ **13.** $\{-11, -1\}$ **15.** $\{-3, 4\}$

17. $\left\{-\dfrac{13}{3}, 5\right\}$ **19.** $\left\{-\dfrac{2}{5}, \dfrac{2}{5}\right\}$ **21.** $\varnothing$ **23.** $\varnothing$ **25.** $\left\{\dfrac{1}{2}\right\}$ **27.** $\left\{\dfrac{3}{4}, 5\right\}$ **29.** $\left\{\dfrac{5}{3}, 3\right\}$ **31.** $\{0\}$ **33.** $\{4\}$

35. $\{4\}$ **37.** $\{-1, 15\}$

39. $\{x|-3 < x < 3\}$ or $(-3, 3)$

41. $\{x|1 < x < 3\}$ or $(1, 3)$

43. $\{x|-3 \le x \le -1\}$ or $[-3, -1]$

45. $\{x|-1 < x < 7\}$ or $(-1, 7)$

47. $\{x|x < -3 \text{ or } x > 3\}$ or $(-\infty, -3) \cup (3, \infty)$

49. $\{x|x < -4 \text{ or } x > -2\}$ or $(-\infty, -4) \cup (-2, \infty)$

51. $\{x|x \le 2 \text{ or } x \ge 6\}$ or $(-\infty, 2] \cup [6, \infty)$

53. $\left\{x \,\middle|\, x < \dfrac{1}{3} \text{ or } x > 5\right\}$ or $\left(-\infty, \dfrac{1}{3}\right) \cup (5, \infty)$

55. $\{x|-5 \le x \le 3\}$ or $[-5, 3]$

57. $\{x|-6 < x < 0\}$ or $(-6, 0)$

59. $\{x|x \le -5 \text{ or } x \ge 3\}$ or $(-\infty, -5] \cup [3, \infty)$

61. $\{x|x < -3 \text{ or } x > 12\}$ or $(-\infty, -3) \cup (12, \infty)$

63. $\varnothing$

65. $\{x|x \text{ is a real number}\}$ or $\mathbb{R}$ or $(-\infty, \infty)$

67. $\{x|-9 \le x \le 5\}$ or $[-9, 5]$

69. $\{x|x < 1 \text{ or } x > 2\}$ or $(-\infty, 1) \cup (2, \infty)$

71. $\{x|58.6 \le x \le 61.8\}$ or $[58.6, 61.8]$; The percentage of the U.S. population that watched M*A*S*H is between a low of 58.6% and a high of 61.8%.; 1.6%

73. $\{T|50 \le T \le 64\}$ or $[50, 64]$; The monthly average temperature for San Francisco, CA is between a low of 50°F and a high of 64°F.

75. $\{x|8.59 \le x \le 8.61\}$ or $[8.59, 8.61]$; A machine part that is supposed to be 8.6 centimeters is acceptable between a low of 8.59 and a high of 8.61 centimeters.   **77.** If the number of outcomes that result in heads is 41 or less or 59 or more, then the coin is unfair.

79–85. Answers will vary.   **87.** 3 and 5, or $\{3, 5\}$ **89.** The cost for refunds will be between a low of $500 and a high of $2500.

91.

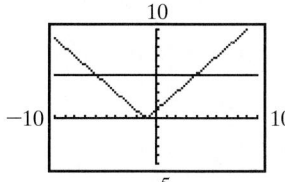

−6 and 4, or $\{-6, 4\}$

93.

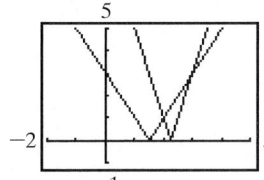

2 and 3, or $\{2, 3\}$

95.

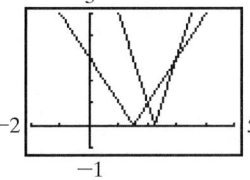

$\{x|-2 < x < -3\}$ or $(-2, 3)$

97.

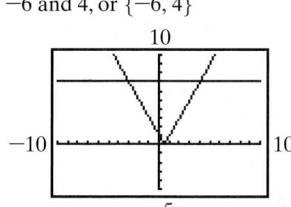

$\{x|x < -3 \text{ or } x > 4\}$ or $(-\infty, -3) \cup (4, \infty)$

99.

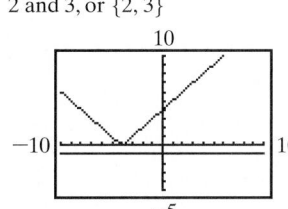

$\{x|x \text{ is a real number}\}$ or $(-\infty, \infty)$

101. 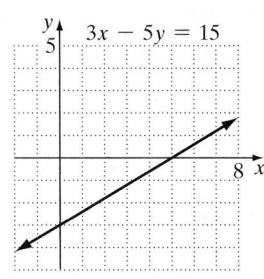 $3x - 5y = 15$

102. 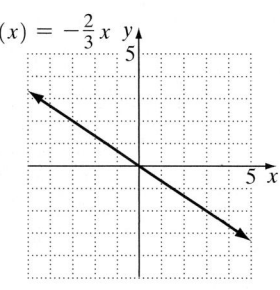 $f(x) = -\frac{2}{3}x$

103. 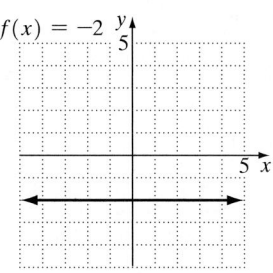 $f(x) = -2$

Section 9.4

Check Point Exercises

1. $2x - 4y \geq 8$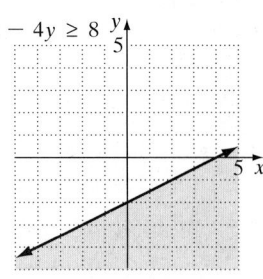

2. $y > -\frac{3}{4}x$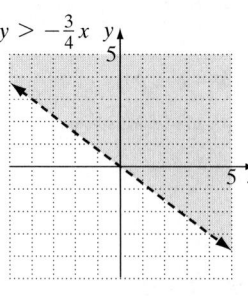

3. $x + 2y > 4$ $2x - 3y \leq -6$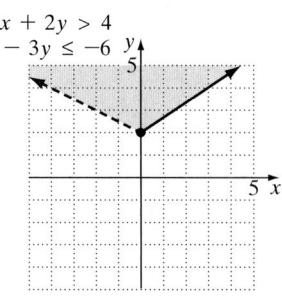

4. $x + y < 2$ $-2 \leq x < 1$ $y > -3$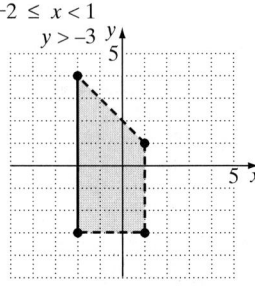

5. $B = (60, 20)$; Using $T = 60$ and $P = 20$, each of the three inequalities for grasslands is true: $60 \geq 5$, true; $5(60) - 7(20) \geq 70$, true; $3(60) - 35(20) \leq -140$, true.

Exercise Set 9.4

1. $x + y \geq 3$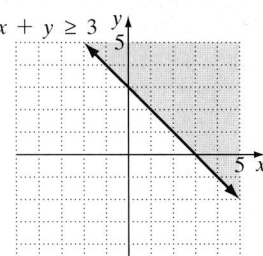

3. $x - y < 5$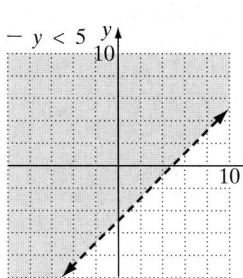

5. $x + 2y > 4$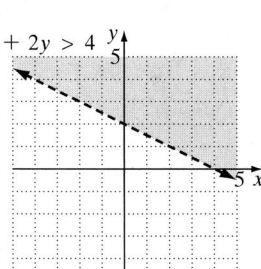

7. $3x - y \leq 6$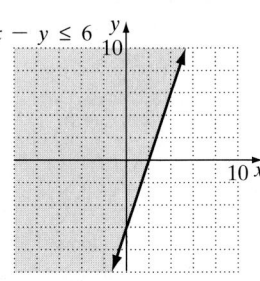

9. $\frac{x}{2} + \frac{y}{3} < 1$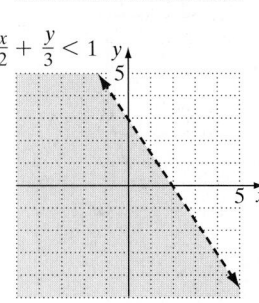

11. $y > \frac{1}{3}x$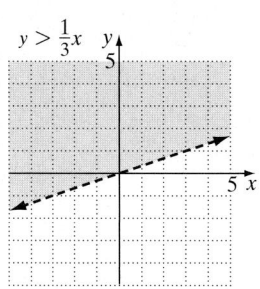

13. $y \leq 3x + 2$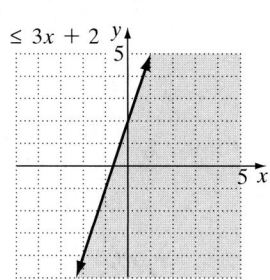

15. $y < -\frac{1}{4}x$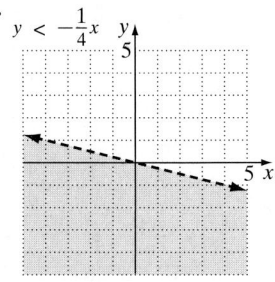

17. $x \le 2$

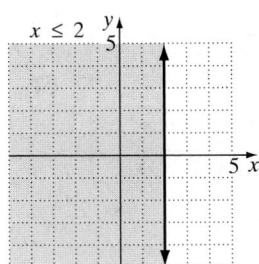

19. $y > -4$

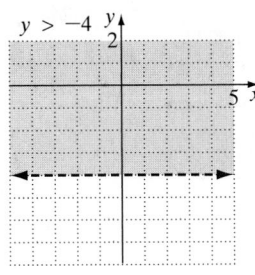

21. $y \ge 0$

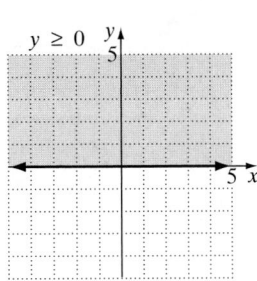

23. $3x + 6y \le 6$
$2x + y \le 8$

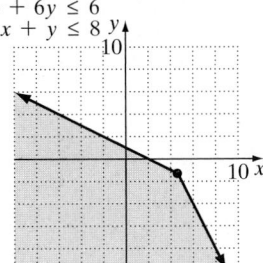

25. $2x - 5y \le 10$
$3x - 2y > 6$

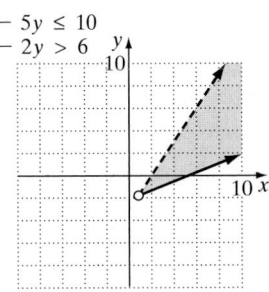

27. $y > 2x - 3$
$y < -x + 6$

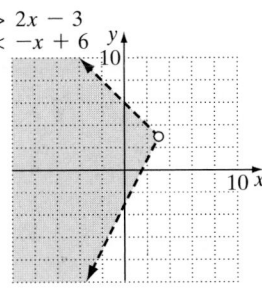

29. $x + 2y \le 4$
$y \ge x - 3$

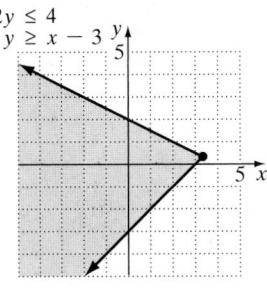

31. $x \le 2$
$y \ge -1$

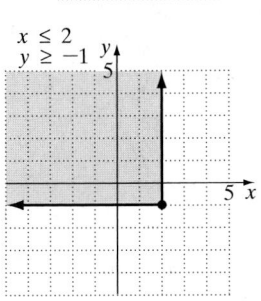

33. $-2 \le x < 5$

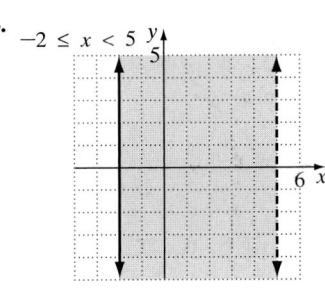

35. $x - y \le 1$
$x \ge 2$

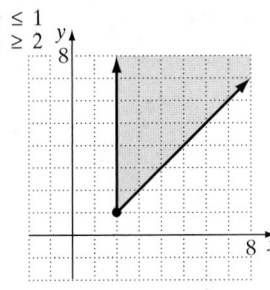

37. $\varnothing$

39. $x + y > 4$
$x + y > -1$

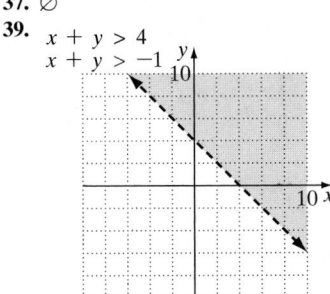

41. $x - y \le 2$
$x \ge -2$
$y \le 3$

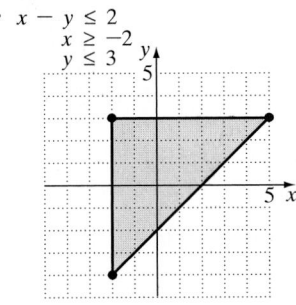

43.
$x \ge 0$
$y \ge 0$
$2x + 5y \le 10$
$3x + 4y \le 12$

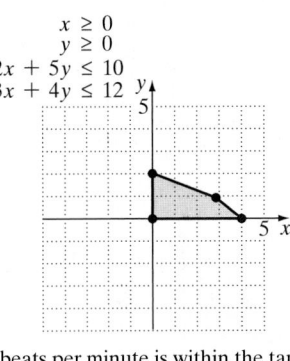

45. $3x + y \le 6$
$2x - y \le -1$
$x \ge -2$
$y \le 4$

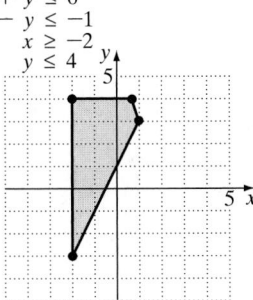

47. a. $(20, 160)$; 20 year-old with a pulse rate of 160 beats per minute is within the target zone.
   b. $10 \le 20 \le 70$, true; $2(20) + 3(160) \ge 450$, true; $20 + 160 \le 190$, true.
49. Answers will vary.
51. 190 lb
53. $7w - 25h < -800$; Using $(100, 70)$ then $7(100) - 25(70) < -800$ is true.
55. $7w - 25h \ge -800, w - 5h \le -170$; Using $(150, 70)$ then
   $7(150) - 25(70) \ge -800$ is true and $150 - 5(70) \le -170$ is true.
57–65. Answers will vary.   **67.** $y - 2x > -4$

69. $x =$ amount invested in high-risk;
$y =$ amount invested in low-risk;
$x + y \leq 15{,}000, x \geq 2000, y \geq 3x,$
$x \geq 0, y \geq 0$

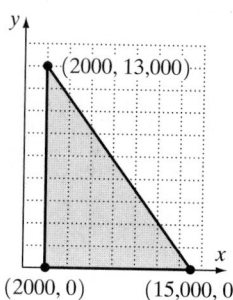

71.

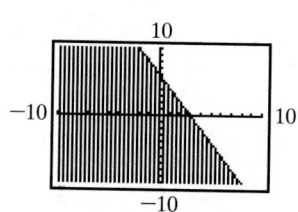

73.

75–77. Answers will vary. **78.** $(3, 1)$ or $\{(3, 1)\}$ **79.** $(2, 4)$ or $\{(2, 4)\}$ **80.** 165

Section 9.5

Check Point Exercises

1. $z = 25x + 55y$ **2.** $x + y \leq 80$ **3.** $30 \leq x \leq 80, 10 \leq y \leq 30; z = 25x + 55y, x + y \leq 80, 30 \leq x \leq 80, 10 \leq y \leq 30$
4. 50 bookshelves and 30 desks; $2900 **5.** 30

Exercise Set 9.5

1. $(1, 2)$: 17; $(2, 10)$: 70; $(7, 5)$: 65; $(8, 3)$: 58; maximum: 70; minimum: 17
3. $(0, 0)$: 0; $(0, 8)$: 400; $(4, 9)$: 610; $(8, 0)$: 320; maximum: 610; minimum: 0
5. a.

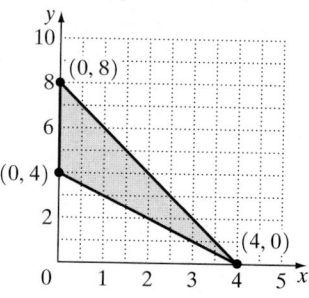

  b. $(0, 4)$: 8; $(0, 8)$: 16; $(4, 0)$: 12
  c. maximum: 16; at $(0, 8)$

7. a.

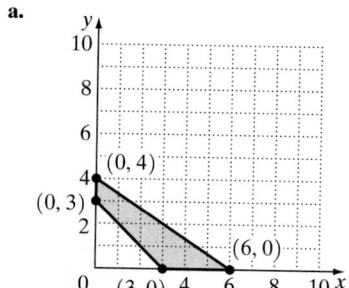

  b. $(0, 3)$: 3; $(0, 4)$: 4; $(6, 0)$: 24; $(3, 0)$: 12
  c. maximum: 24; at $(6, 0)$

9. a.

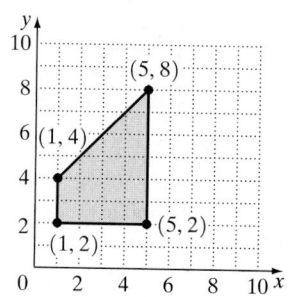

  b. $(1, 2)$: -1; $(1, 4)$: -5; $(5, 8)$: -1; $(5, 2)$: 11

  c. maximum: 11; at $(5, 2)$

11. a.

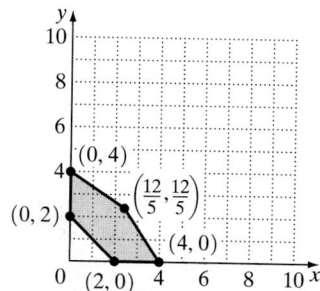

  b. $(0, 2)$: 4; $(0, 4)$: 8; $\left(\dfrac{12}{5}, \dfrac{12}{5}\right)$: $\dfrac{72}{5} = 14.4$; $(4, 0)$: 16; $(2, 0)$: 8

  c. maximum: 16; at $(4, 0)$

13. a.

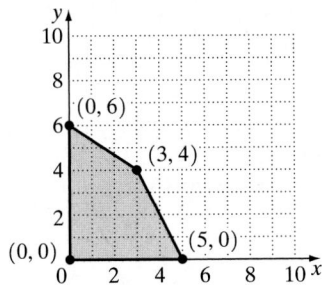

b. $(0, 0)$: 0; $(0, 6)$: 72; $(3, 4)$: 78; $(5, 0)$: 50
c. maximum: 78; at $(3, 4)$

15. a. $z = 125x + 200y$
b. $x \leq 450; y \leq 200; 600x + 900y \leq 360,000$
c.

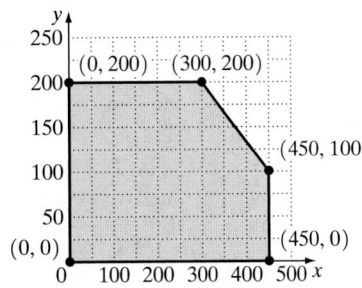

d. 0; 40,000; 77,500; 76,250; 56,250
e. 300; 200; 77,500

17. 40 of model A and 0 of model B   **19.** 300 boxes of food and 200 boxes of clothing   **21.** 50 parents and 100 students
23. 10 Boeing 727s and 42 Falcon 20s   **25–27.** Answers will vary.   **29.** $5000 in stocks and $5000 in bonds   **31.** 6 or {6}
32. $\dfrac{x + 9}{x - 15}$ **33.** 10

Review Exercises

1. $\{x|{-2} < x \leq 3\}$

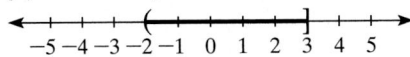

2. $\{x|{-1.5} \leq x \leq 2\}$

3. $\{x|x > -1\}$

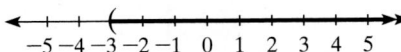

4. $\{x|x \geq -2\}; [-2, \infty)$

5. $\left\{x \middle| x \geq \dfrac{3}{5}\right\}; \left[\dfrac{3}{5}, \infty\right)$

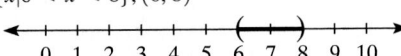

6. $\left\{x \middle| x < -\dfrac{21}{2}\right\}; \left(-\infty, -\dfrac{21}{2}\right)$

7. $\{x|x > -3\}; (-3, \infty)$

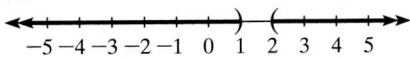
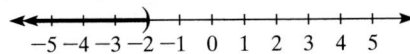

8. $\{x|x \leq -2\}; (-\infty, -2]$

9. a. $P(x) = 85x - 357,000$ **b.** more than 42000   **10.** $C(x) = 360,000 + 850x$ **11.** $R(x) = 1150x$
12. $P(x) = 300x - 360,000$ **13.** more than 1200   **14.** more than 50 checks   **15.** $\{a, c\}$ **16.** $\{a\}$
17. $\{a, b, c, d, e\}$ **18.** $\{a, b, c, d, f, g\}$
19. $\{x|x \leq 3\}; (-\infty, 3]$

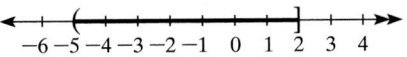

20. $\{x|x < 6\}; (-\infty, 6)$

21. $\{x|6 < x < 8\}; (6, 8)$

22. $\{x|x \leq 1\}; (-\infty, 1]$

23. $\varnothing$ **24.** $\{x|x < 1 \text{ or } x > 2\}; (-\infty, 1) \cup (2, \infty)$

25. $\{x|x \leq -4 \text{ or } x > 2\}; (-\infty, -4] \cup (2, \infty)$

26. $\{x|x < -2\}; (-\infty, -2)$

27. $\{x|x \text{ is a real number}\}; (-\infty, \infty)$

28. $\{x|{-5} < x \leq 2\}; (-5, 2]$

29. $\left\{x \middle| -\dfrac{3}{4} \leq x \leq 1\right\}; \left[-\dfrac{3}{4}, 1\right]$

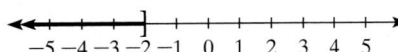
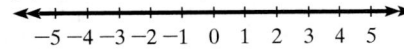

30. $[81, 131)$; If the highest grade is 100, then $[81, 100]$.　**31.** $\{-4, 3\}$　**32.** $\varnothing$　**33.** $\left\{-\dfrac{11}{2}, \dfrac{23}{2}\right\}$　**34.** $\left\{-4, -\dfrac{6}{11}\right\}$

35. $\{x | -9 \leq x \leq 6\}$; $[-9, 6]$

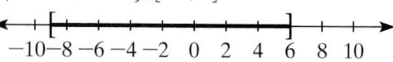

36. $\{x | x < -6 \text{ or } x > 0\}$; $(-\infty, -6) \cup (0, \infty)$

37. $\{x | -3 < x < -2\}$; $(-3, -2)$

38. $\{x | x \text{ is a real number}\}$; $(-\infty, \infty)$

39. Approximately 90% of the population sleeps between 5.5 hours and 7.5 hours daily.

40.
$3x - 4y > 12$

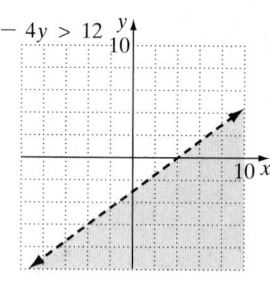

41.
$x - 3y \leq 6$

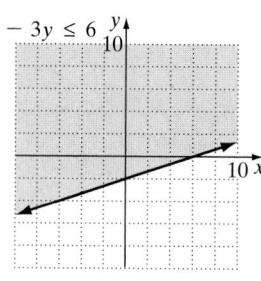

42.
$y \leq -\dfrac{1}{2}x + 2$

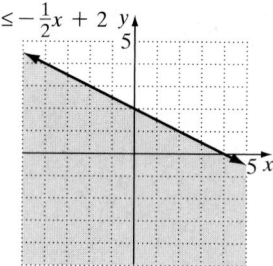

43.
$y > \dfrac{3}{5}x$

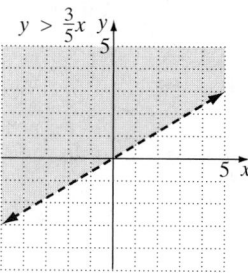

44.
$x \leq 2$

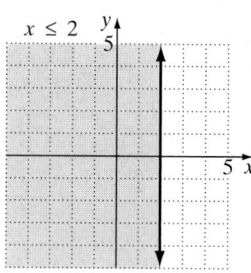

45.
$y > -3$

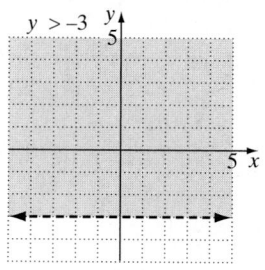

46.
$3x - y \leq 6$
$x + y \geq 2$

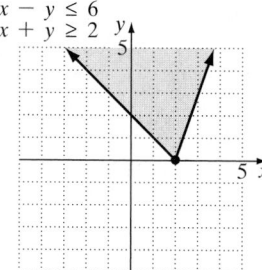

47.
$y < -x + 4$
$y > x - 4$

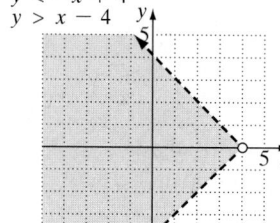

48.
$-3 \leq x < 5$

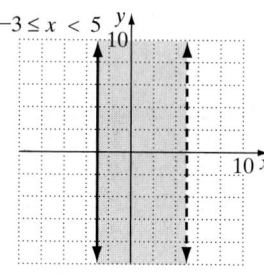

49.
$-2 < y \leq 6$

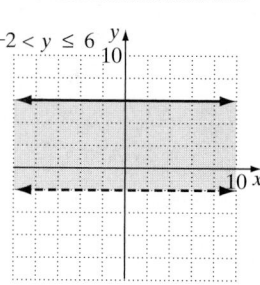

50.
$x \geq 3$
$y \leq 0$

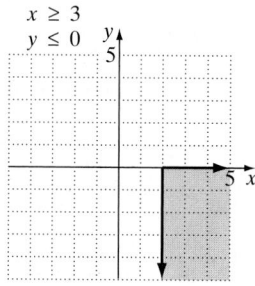

51.
$2x - y > -4$
$x \geq 0$

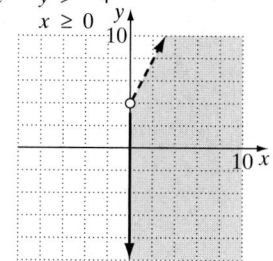

52. $x + y \leq 6$
$y \geq 2x - 3$

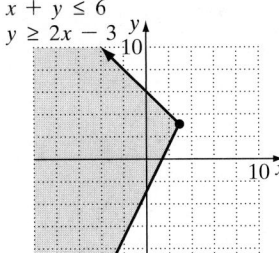

53. $3x + 2y \geq 4$
$x - y \leq 3$
$x \geq 0, y \geq 0$

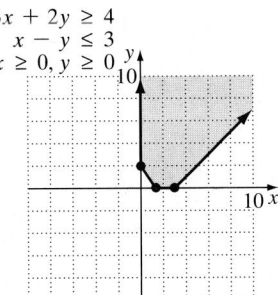

54. $\varnothing$

55. $\left(\dfrac{1}{2}, \dfrac{1}{2}\right): \dfrac{5}{2}; (2, 2): 10; (4, 0): 8; (1, 0): 2;$
maximum: 10; minimum: 2

56.

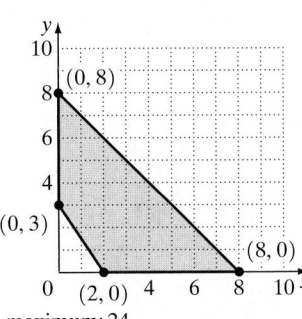

maximum: 24

57.

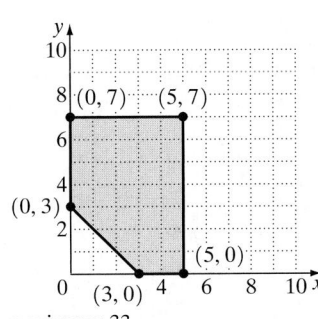

maximum: 33

58.

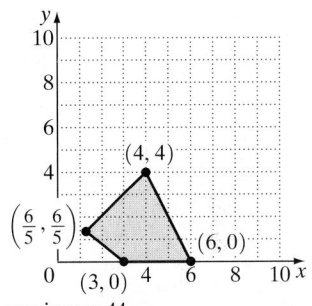

maximum: 44

59. a. $z = 500x + 350y$ **b.** $x + y \leq 200; x \geq 10; y \geq 80$
c.

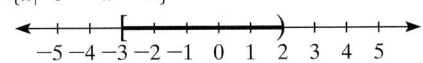

d. $(10, 80): 33,000; (10, 190): 71,500;$
$(120, 80): 88,000$
e. $120; 80; 88,000$

60. 480 of model A and 240 of model B

Chapter 9 Test

1. $\{x | -3 \leq x < 2\}$

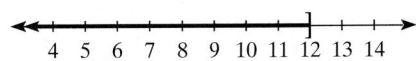

2. $\{x | x \leq -1\}$

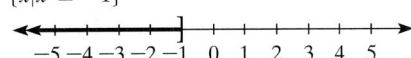

3. $\{x | x \leq 12\}; (-\infty, 12]$

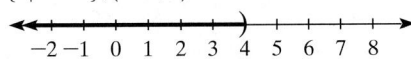

4. $\left\{x \,\middle|\, x \geq \dfrac{21}{8}\right\}; \left[\dfrac{21}{8}, \infty\right)$

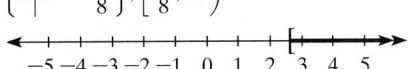

5. a. $C(x) = 60,000 + 200x$ **b.** $R(x) = 450x$ **c.** $P(x) = 250x - 60,000$ **d.** more than 240
6. $\{4, 6\}$ **7.** $\{2, 4, 6, 8, 10, 12, 14\}$
8. $\{x | -2 < x < -1\}; (-2, -1)$

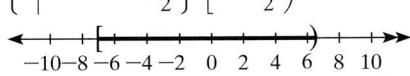

9. $\{x | x \geq -2\}; [-2, \infty)$

10. $\{x | x < 4\}; (-\infty, 4)$

11. $\{x | x \leq -4 \text{ or } x > 2\}; (-\infty, -4] \cup (2, \infty)$

12. $\left\{x \,\middle|\, -7 \leq x < \dfrac{13}{2}\right\}; \left[-7, \dfrac{13}{2}\right)$

13. $\left\{-2, \dfrac{4}{5}\right\}$

14. $\left\{-\dfrac{8}{5}, 7\right\}$

15. $\{x|-3 < x < 4\}; (-3, 4)$

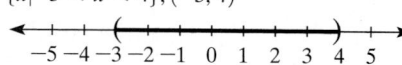

16. $\{x|x \le -1 \text{ or } x \ge 4\}; (-\infty, -1] \cup [4, \infty)$

17. $\{T|66 \le T \le 82\}$ or $[66, 82]$; The monthly average temperature for Miami, FL is between a low of $66\,°F$ and a high of $82\,°F$.

18.
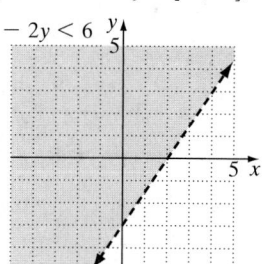
$3x - 2y < 6$

19.
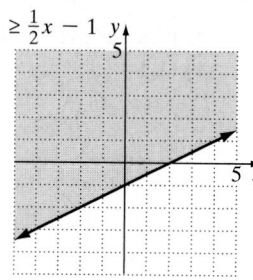
$y \ge \frac{1}{2}x - 1$

20.
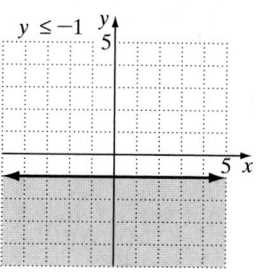
$y \le -1$

21.
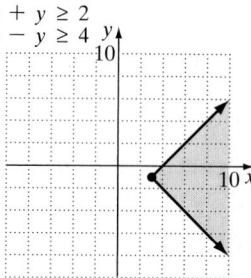
$x + y \ge 2$
$x - y \ge 4$

22.
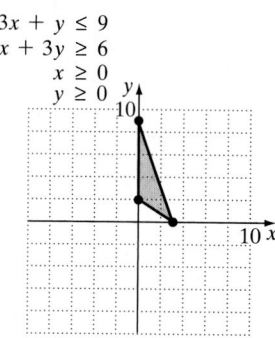
$3x + y \le 9$
$2x + 3y \ge 6$
$x \ge 0$
$y \ge 0$

23.
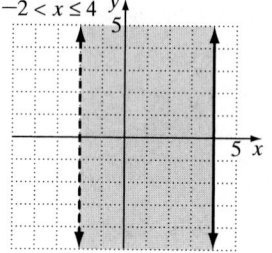
$-2 < x \le 4$

24. maximum: 26   **25.** 50 regular and 100 deluxe; $35,000

Cumulative Review Exercises (Chapters 1–9)

1. -1 or $\{-1\}$ **2.** 8 or $\{8\}$ **3.** $-\dfrac{2y^7}{3x^5}$ **4.** $22; 4a^2 - 6a + 4$ **5.** $2x^2 + x + 2; 12$ **6.** $f(x) = -\dfrac{1}{2}x + 4$

7.
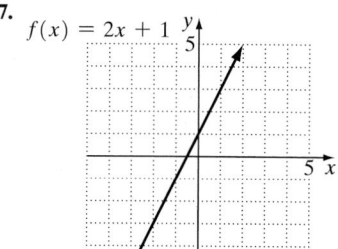
$f(x) = 2x + 1$

8.
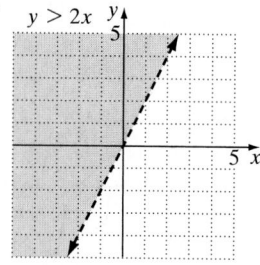
$y > 2x$

9.
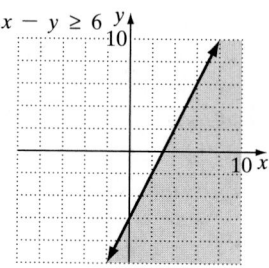
$2x - y \ge 6$

10.
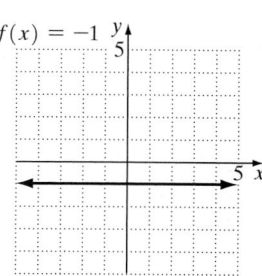
$f(x) = -1$

11. $(-4, 2, -1)$ or $\{(-4, 2, -1)\}$ **12.** $(-1, 2)$ or $\{(-1, 2)\}$ **13.** -17

14. 46 rooms with kitchen facilities and 14 without kitchen facilities

15. a. and b. are functions.

16. $\{x|x \ge -7\}; [-7, \infty)$;

17. $\{x|x < -6\}; (-\infty, -6)$;

18. $\{x|x \le 3 \text{ or } x \ge 5\}; (-\infty, 3] \cup [5, \infty)$;

19. $\{x|-10 \le x \le 7\}; [-10, 7]$

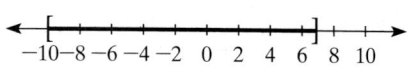

20. $\left\{x \,\middle|\, x < \dfrac{1}{3} \text{ or } x > 5\right\}; \left(-\infty, \dfrac{1}{3}\right) \cup (5, \infty)$

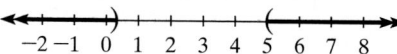

CHAPTER 10

Section 10.1

Check Point Exercises

1. a. 8 **b.** -7 **c.** $\dfrac{4}{5}$ **d.** 0.09 **e.** 5 **f.** 7 **2. a.** 4 **b.** $-\sqrt{24} \approx -4.90$ **3.** $\{x \mid x \geq 3\}$ or $[3, \infty)$

4. \$25.5 billion; The model describes the data well. **5. a.** 7 **b.** $|x + 8|$ **c.** $|7x^5|$ **d.** $|x - 3|$ **6. a.** 3 **b.** -2 **7.** $-3x$

8. a. 2 **b.** -2 **c.** not a real number **d.** -1 **9. a.** $|x + 6|$ **b.** $3x - 2$ **c.** 8

Exercise Set 10.1

1. 6 **3.** -6 **5.** not a real number **7.** $\dfrac{1}{5}$ **9.** -0.2 **11.** 3 **13.** 1 **15.** not a real number **17.** 4; 1; 0; not a real number

19. -5; $-\sqrt{5} \approx -2.24$; -1; not a real number **21.** 4; 2; 1; 6 **23.** $\{x \mid x \geq 3\}$ or $[3, \infty)$; c **25.** $\{x \mid x \geq -5\}$ or $[-5, \infty)$; d

27. $\{x \mid x \leq 3\}$ or $(-\infty, 3]$; e **29.** 5 **31.** 4 **33.** $|x - 1|$ **35.** $|6x^2|$ or $6x^2$ **37.** $-|10x^3|$ or $-10|x^3|$ **39.** $|x + 6|$

41. $-|x - 4|$ **43.** 3 **45.** -3 **47.** $\dfrac{1}{5}$ **49.** 3; 2; -1; -4 **51.** -2; 0; 2 **53.** 1 **55.** 2 **57.** -2 **59.** not a real number

61. -1 **63.** not a real number **65.** -4 **67.** 2 **69.** -2 **71.** x **73.** $|y|$ **75.** $-2x$ **77.** -5 **79.** 5 **81.** $|x + 3|$

83. $-2(x - 1)$ **85.** 40.2 in.; The model describes actual data well. **87.** 70 mph; The officer should not believe the motorist.;

Answers will vary. **89–99.** Answers will vary. **101.** Answers will vary; an example is $f(x) = \sqrt{15 - 3x}$. **103.** $|(2x + 3)^5|$

105. The graph of h is the graph of f shifted left 3 units.

107. 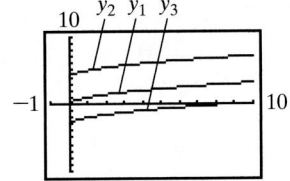 ; Answers will vary. **100.** 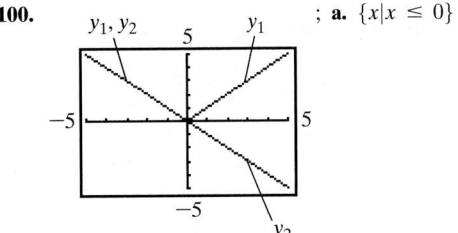 ; **a.** $\{x \mid x \leq 0\}$ **b.** $\{x \mid x > 0\}$

110. $7x + 30$ **111.** $\dfrac{x^8}{9y^6}$ **112.** $\left\{ x \,\middle|\, x < -\dfrac{7}{3} \text{ or } x > 5 \right\}$ or $\left(-\infty, -\dfrac{7}{3} \right) \cup (5, \infty)$

Section 10.2

Check Point Exercises

1. a. $\sqrt{25} = 5$ **b.** $\sqrt[3]{-8} = -2$ **c.** $\sqrt[4]{5xy^2}$ **2. a.** $(5xy)^{1/4}$ **b.** $\left(\dfrac{a^3 b}{2} \right)^{1/5}$ **3. a.** $(\sqrt[3]{27})^4 = 81$ **b.** $(\sqrt{4})^3 = 8$

c. $-(\sqrt[4]{16})^3 = -8$ **4. a.** $6^{4/3}$ **b.** $(2xy)^{7/5}$ **5. a.** $\dfrac{1}{25^{1/2}} = \dfrac{1}{5}$ **b.** $\dfrac{1}{64^{1/3}} = \dfrac{1}{4}$ **c.** $\dfrac{1}{32^{4/5}} = \dfrac{1}{16}$ **d.** $\dfrac{1}{(3xy)^{5/9}}$

6. \$126.5 thousand or \$126,500 **7. a.** $7^{5/6}$ **b.** $\dfrac{5}{x}$ **c.** $9.1^{3/10}$ **d.** $\dfrac{y^{1/12}}{x^{1/5}}$ **8. a.** $\sqrt{x}$ **b.** $2a^4$ **c.** $\sqrt[6]{x^2 y}$ **d.** $\sqrt[6]{x}$ **e.** $\sqrt[6]{x}$

Exercise Set 10.2

1. $\sqrt{49} = 7$ **3.** $\sqrt[3]{-27} = -3$ **5.** $-\sqrt[4]{16} = -2$ **7.** $\sqrt[3]{xy}$ **9.** $\sqrt[5]{2xy^3}$ **11.** $(\sqrt{81})^3 = 729$ **13.** $(\sqrt[3]{125})^2 = 25$

15. $(\sqrt[5]{-32})^3 = -8$ **17.** $(\sqrt[3]{27})^2 + (\sqrt[4]{16})^3 = 17$ **19.** $\sqrt{(xy)^4}$ **21.** $7^{1/2}$ **23.** $5^{1/3}$ **25.** $(11x)^{1/5}$ **27.** $x^{3/2}$ **29.** $x^{3/5}$

31. $(x^2 y)^{1/5}$ **33.** $(19xy)^{3/2}$ **35.** $(7xy^2)^{5/6}$ **37.** $2xy^{2/3}$ **39.** $\dfrac{1}{49^{1/2}} = \dfrac{1}{7}$ **41.** $\dfrac{1}{27^{1/3}} = \dfrac{1}{3}$ **43.** $\dfrac{1}{16^{3/4}} = \dfrac{1}{8}$ **45.** $\dfrac{1}{8^{2/3}} = \dfrac{1}{4}$

47. $\left(\dfrac{27}{8} \right)^{1/3} = \dfrac{3}{2}$ **49.** $\dfrac{1}{(-64)^{2/3}} = \dfrac{1}{16}$ **51.** $\dfrac{1}{(2xy)^{7/10}}$ **53.** $\dfrac{5x}{z^{1/3}}$ **55.** 3 **57.** 4 **59.** $x^{5/6}$ **61.** $x^{3/5}$ **63.** $\dfrac{1}{x^{5/12}}$ **65.** 25

67. $\dfrac{1}{y^{1/6}}$ **69.** $32x$ **71.** $5x^2 y^3$ **73.** $\dfrac{x^{1/4}}{y^{3/10}}$ **75.** 3 **77.** $27y^{2/3}$ **79.** $\sqrt[4]{x}$ **81.** $2a^2$ **83.** $x^2 y^3$ **85.** $x^6 y^6$ **87.** $\sqrt[5]{3y}$

89. $\sqrt[3]{4a^2}$ **91.** $\sqrt[3]{x^2 y}$ **93.** $\sqrt[6]{2^5}$ or $\sqrt[6]{32}$ **95.** $\sqrt[20]{x}$ **97.** $\sqrt{y}$ **99.** $\sqrt[8]{x}$ **101.** $\sqrt[4]{x^2 y}$ **103.** $\sqrt[12]{2x}$ **105.** $x^9 y^{15}$

107. 58 species of plants **109.** 1.9 hours **111.** 4.25% **113.** about 1872 calories per day

115. a. $L + 1.25S^{1/2} - 9.8D^{1/3} \leq 16.296$ **b.** eligible **117–125.** Answers will vary. **127.** 8 years old **129.** 61

131. Answers will vary. **133. a.** On some calculators only half the graph, for $x \geq 0$, is displayed. Other calculators display the entire graph. **b.** The entire graph is displayed. **135.** $y = -2x + 11$ or $f(x) = -2x + 11$

136.

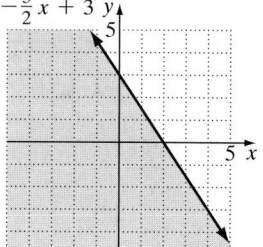

$y \leq -\dfrac{3}{2}x + 3$

137. $(3, 4)$ or $\{(3, 4)\}$

Section 10.3

Check Point Exercises

1. a. $\sqrt{55}$ **b.** $\sqrt{x^2 - 16}$ **c.** $\sqrt[3]{60}$ **d.** $\sqrt[3]{12x^4}$ **2. a.** $4\sqrt{5}$ **b.** $2\sqrt[3]{5}$ **c.** $2\sqrt[4]{2}$ **d.** $10|x|\sqrt{2y}$ **3.** $f(x) = \sqrt{3}|x - 2|$
4. $x^4y^5z\sqrt{xyz}$ **5.** $2x^3y^4\sqrt[3]{5xy^2}$ **6.** $2x^2z\sqrt[5]{x^2y^2z^3}$ **7. a.** $2\sqrt{3}$ **b.** $100\sqrt[3]{4}$ **c.** $2x^2y\sqrt[4]{2}$

Exercise Set 10.3

1. $\sqrt{15}$ **3.** $\sqrt[3]{18}$ **5.** $\sqrt[4]{33}$ **7.** $\sqrt{33xy}$ **9.** $\sqrt[5]{24x^4}$ **11.** $\sqrt{x^2 - 9}$ **13.** $\sqrt[6]{(x - 4)^5}$ **15.** $\sqrt{x}$ **17.** $\sqrt[4]{\dfrac{3x}{7y}}$
19. $\sqrt[3]{77x^5y^3}$ **21.** $5\sqrt{2}$ **23.** $3\sqrt{5}$ **25.** $5\sqrt{3x}$ **27.** $2\sqrt[3]{2}$ **29.** $3x$ **31.** $-2y\sqrt[3]{2x^2}$ **33.** $6|x + 2|$ **35.** $2(x + 2)\sqrt[3]{4}$
37. $|x - 1|\sqrt{3}$ **39.** $x^3\sqrt{x}$ **41.** $x^4y^4\sqrt{y}$ **43.** $4x\sqrt{3x}$ **45.** $y^2\sqrt[3]{y^2}$ **47.** $x^4y\sqrt[3]{x^2z}$ **49.** $3x^2y^2\sqrt[3]{3x^2}$ **51.** $(x + y)\sqrt[3]{(x + y)^2}$
53. $y^3\sqrt[5]{y^2}$ **55.** $2xy^3\sqrt[5]{2xy^2}$ **57.** $2x^2\sqrt[4]{5x^2}$ **59.** $(x - 3)^2\sqrt[4]{(x - 3)^2}$ or $(x - 3)^2\sqrt{x - 3}$ **61.** $2\sqrt{6}$ **63.** $5\sqrt{2xy}$
65. $6x$ **67.** $10xy\sqrt{2y}$ **69.** $60\sqrt{2}$ **71.** $2\sqrt[3]{6}$ **73.** $2x^2\sqrt{10x}$ **75.** $5xy^4\sqrt[3]{x^2y^2}$ **77.** $2xyz\sqrt[4]{x^2z^3}$ **79.** $2xy^2\sqrt[5]{2y^3}$
81. $(x - y)^2\sqrt[3]{(x - y)^2}$ **83.** $8\sqrt{2}$ mph; 11.3 mph **85.** $2\sqrt{5}$ sec; 4.5 sec **87. a.** $\dfrac{7.644}{2\sqrt{2}} = \dfrac{3.822}{\sqrt[4]{2}}$

b. 3.21 liters of blood per minute per square meter; $(32, 3.21)$ **89–93.** Answers will vary. **95.** Its square root is multiplied by $\sqrt{3}$.
97. $g(x) = \sqrt[3]{4x^2}$ **99.** Answers wil vary. **101.** Graphs are not the same; $\sqrt{8x^2} = 2|x|\sqrt{2}$.
103. Graphs are not the same; $\sqrt[3]{2x} \cdot \sqrt[3]{4x^2} = 2x$. **104.** $\{x|5 \leq x \leq 11\}$ or $[5, 11]$ **105.** $(2, -4)$ or $\{(2, -4)\}$
106. $(4x - 3)(16x^2 + 12x + 9)$

Section 10.4

Check Point Exercises

1. a. $10\sqrt{13}$ **b.** $(21 - 6x)\sqrt[3]{7}$ **c.** $5\sqrt[4]{3x} + 2\sqrt[3]{3x}$ **2. a.** $21\sqrt{5}$ **b.** $-12\sqrt{3x}$ **c.** cannot be simplified
3. a. $-9\sqrt[3]{3}$ **b.** $(5 + 3xy)\sqrt[3]{x^2y}$ **4. a.** $\dfrac{2\sqrt[3]{3}}{5}$ **b.** $\dfrac{3x\sqrt{x}}{y^5}$ **c.** $\dfrac{2y^2\sqrt[3]{y}}{x^4}$ **5. a.** $2x^2\sqrt{5}$ **b.** $\dfrac{5\sqrt{xy}}{2}$ **c.** $2x^2y$

Exercise Set 10.4

1. $11\sqrt{5}$ **3.** $7\sqrt[3]{6}$ **5.** $2\sqrt[5]{2}$ **7.** $\sqrt{13} + 2\sqrt{5}$ **9.** $7\sqrt{5} + 2\sqrt[3]{x}$ **11.** $4\sqrt{3}$ **13.** $19\sqrt{3}$ **15.** $6\sqrt{2x}$ **17.** $13\sqrt[3]{2}$
19. $(9x + 1)\sqrt{5x}$ **21.** $7y\sqrt[3]{2x}$ **23.** $(3x - 2)\sqrt[3]{2x}$ **25.** $4\sqrt{x - 2}$ **27.** $5x\sqrt[3]{xy^2}$ **29.** $\dfrac{\sqrt{11}}{2}$ **31.** $\dfrac{\sqrt[3]{19}}{3}$ **33.** $\dfrac{x}{6y^4}$
35. $\dfrac{2x\sqrt{2x}}{5y^3}$ **37.** $\dfrac{x\sqrt[3]{x}}{2y}$ **39.** $\dfrac{x^2\sqrt[3]{50x^2}}{3y^4}$ **41.** $\dfrac{y\sqrt[4]{9y^2}}{x^2}$ **43.** $\dfrac{2x^2\sqrt[5]{2x^3}}{y^4}$ **45.** $2\sqrt{2}$ **47.** 2 **49.** $3x$ **51.** x^2y
53. $2x^2\sqrt{5}$ **55.** $3\sqrt{xy}$ **57.** $2xy$ **59.** $2x^2y^2\sqrt[4]{y^2}$ **61.** $\sqrt[3]{x + 3}$
63. a. $R_f\dfrac{\sqrt{c^2 - v^2}}{\sqrt{c^2}} = R_f\sqrt{\dfrac{c^2 - v^2}{c^2}} = R_f\sqrt{\dfrac{c^2}{c^2} - \dfrac{v^2}{c^2}} = R_f\sqrt{1 - \dfrac{v^2}{c^2}} = R_f\sqrt{1 - \left(\dfrac{v}{c}\right)^2}$

b. $R_f\sqrt{1 - \left(\dfrac{c}{c}\right)^2} = 0$; Aging rate is 0.; A person moving at the speed of light does not age relative to a friend on earth.

65. $P = 18\sqrt{5}$ ft; $A = 100$ sq ft **67–71.** Answers will vary. **73.** d **75.** $-\dfrac{31\sqrt{5}}{12}$ **77.** Answers will vary.

79. Graphs are not the same.; $\sqrt{16x} - \sqrt{9x} = \sqrt{x}$ **81.** 0 or $\{0\}$ **82.** $(x - 2y)(x - 6y)$ **83.** $\dfrac{3x^2 + 8x + 6}{(x + 3)^2(x + 2)}$

Section 10.5

Check Point Exercises

1. a. $x\sqrt{6} + 2\sqrt{15}$ **b.** $y - \sqrt[3]{7y}$ **c.** $36 - 18\sqrt{10}$ **2. a.** $11 + 2\sqrt{30}$ **b.** 1 **c.** $a - 7$ **3. a.** $\dfrac{\sqrt{21}}{7}$ **b.** $\dfrac{\sqrt[3]{6}}{3}$

4. a. $\dfrac{\sqrt{14xy}}{7y}$ **b.** $\dfrac{\sqrt[3]{3xy^2}}{3y}$ **c.** $\dfrac{3\sqrt[5]{4x^3y}}{y}$ **5.** $12\sqrt{3} - 18$ **6.** $\dfrac{3\sqrt{5} + 3\sqrt{2} + \sqrt{35} + \sqrt{14}}{3}$ **7.** $\dfrac{1}{\sqrt{x+3} + \sqrt{x}}$

Exercise Set 10.5

1. $x\sqrt{2} + \sqrt{14}$ **3.** $7\sqrt{6} - 6$ **5.** $12\sqrt{2} - 6$ **7.** $\sqrt[3]{12} + 4\sqrt[3]{10}$ **9.** $2x\sqrt[3]{2} - \sqrt[3]{x^2}$ **11.** $32 + 11\sqrt{2}$ **13.** $34 - 15\sqrt{5}$
15. $117 - 36\sqrt{7}$ **17.** $\sqrt{6} + \sqrt{10} + \sqrt{21} + \sqrt{35}$ **19.** $\sqrt{6} - \sqrt{10} - \sqrt{21} + \sqrt{35}$ **21.** $-48 + 7\sqrt{6}$ **23.** $8 + 2\sqrt{15}$
25. $3x - 2\sqrt{3xy} + y$ **27.** -44 **29.** -71 **31.** 6 **33.** $6 - 5\sqrt{x} + x$ **35.** $\sqrt[3]{x^2} + \sqrt[3]{x} - 20$ **37.** $2x^2 + x\sqrt[3]{y^2} - y\sqrt[3]{y}$

39. $\dfrac{\sqrt{10}}{5}$ **41.** $\dfrac{\sqrt{11x}}{x}$ **43.** $\dfrac{3\sqrt[3]{3y}}{y}$ **45.** $\dfrac{\sqrt[3]{4}}{2}$ **47.** $3\sqrt[3]{2}$ **49.** $\dfrac{\sqrt[3]{18}}{3}$ **51.** $\dfrac{4\sqrt[3]{x^2}}{x}$ **53.** $\dfrac{\sqrt[3]{2y}}{y}$ **55.** $\dfrac{7\sqrt[3]{4x}}{2x}$ **57.** $\dfrac{\sqrt[3]{2x^2y}}{xy}$

59. $\dfrac{3\sqrt[4]{x^3}}{x}$ **61.** $\dfrac{3\sqrt[5]{4x^2}}{x}$ **63.** $x\sqrt[5]{8x^3y}$ **65.** $8\sqrt{5} - 16$ **67.** $\dfrac{13\sqrt{11} + 39}{2}$ **69.** $3\sqrt{5} - 3\sqrt{3}$ **71.** $\dfrac{a + \sqrt{ab}}{a - b}$

73. $25\sqrt{2} + 15\sqrt{5}$ **75.** $4 + \sqrt{15}$ **77.** $\dfrac{x - 2\sqrt{x} - 3}{x - 9}$ **79.** $\dfrac{3\sqrt{6} + 4}{2}$ **81.** $\dfrac{4\sqrt{xy} + 4x + y}{y - 4x}$ **83.** $\dfrac{3}{\sqrt{6}}$ **85.** $\dfrac{2x}{\sqrt[3]{2x^2y}}$

87. $\dfrac{x - 9}{x - 3\sqrt{x}}$ **89.** $\dfrac{a - b}{a - 2\sqrt{ab} + b}$ **91.** $\dfrac{1}{\sqrt{x+5} + \sqrt{x}}$ **93.** $\dfrac{1}{(x + y)(\sqrt{x} - \sqrt{y})}$
95. 33; In 2001, 33% of U.S. households were online.; Answer models data well. **97.** 3.5%

99. 3.4%; This answer models the actual yearly increase well. **101. a.** $\dfrac{6.85}{\sqrt{t+h} + \sqrt{t}}$ **b.** $\dfrac{6.85}{2\sqrt{t}}$ or $\dfrac{3.425}{\sqrt{t}}$ **c.** 1.7%

103. a. 18% **b.** $p(x) = \dfrac{13\sqrt{x} + x}{5}$ **105.** $P = 8\sqrt{2}$ in.; $A = 7$ sq in. **107.** $\dfrac{7\sqrt{2 \cdot 2 \cdot 3}}{6} = \dfrac{7 \cdot 2\sqrt{3}}{6} = \dfrac{14\sqrt{3}}{6} = \dfrac{7\sqrt{3}}{3} = \dfrac{7}{3}\sqrt{3}$

109–117. Answers will vary. **119.** c **121.** 6 **123.** $3\left(\dfrac{3 + \sqrt{3}}{3}\right)^2 + 2 = 6\left(\dfrac{3 + \sqrt{3}}{3}\right)$ Substitute $\dfrac{3 + \sqrt{3}}{3}$ for x.

$$\dfrac{(3 + \sqrt{3})^2}{3} + 2 = 2(3 + \sqrt{3})$$
$$(3 + \sqrt{3})^2 + 6 = 6(3 + \sqrt{3})$$
$$(9 + 6\sqrt{3} + 3) + 6 = 18 + 6\sqrt{3}$$
$$18 + 6\sqrt{3} = 18 + 6\sqrt{3} \qquad \text{True}$$

125. Graphs are not the same.; $(\sqrt{x} + 2)(\sqrt{x} - 2) = x - 4$ for $x \geq 0$

127. Graphs are the same.; Simplification is correct. **128.** $(6, -2)$ or $\{(6, -2)\}$ **129.** $\dfrac{3}{x - 2}$ **130.** $2x(x - 3)(x - 5)$

Section 10.6

Check Point Exercises

1. 20 or {20} **2.** no solution or $\varnothing$ **3.** -1 and 3, or $\{-1, 3\}$ **4.** 4 or {4} **5.** -12 or $\{-12\}$ **6.** 2013

Exercise Set 10.6

1. 6 or {6} **3.** 17 or {17} **5.** no solution or $\varnothing$ **7.** 8 or {8} **9.** 0 and 3, or {0, 3} **11.** 1 and 3, or {1, 3}

13. 3 and 7, or {3, 7} **15.** 9 or {9} **17.** 8 or {8} **19.** 35 or {35} **21.** 16 or {16} **23.** 2 or {2} **25.** $\dfrac{5}{2}$ or $\left\{\dfrac{5}{2}\right\}$

27. 5 or {5} **29.** no solution or $\varnothing$ **31.** 2 and 6, or {2, 6} **33.** 0 and 10, or {0, 10} **35.** 8 or {8} **37.** 2018 **39.** 2002
41. 36 years old; (36, 40,000) **43.** 27 sq mi **45.** 6078 million km **47–53.** Answers will vary.
55. $\sqrt{x - 7} = 3$; $\sqrt{x} = 4$; $1 + \sqrt{x} = 5$ **57.** 16 or {16}
59. **61.** **63.**

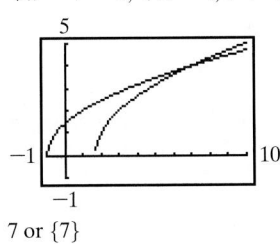

7 or {7}

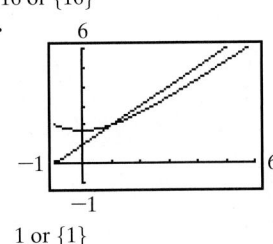

1 or {1}

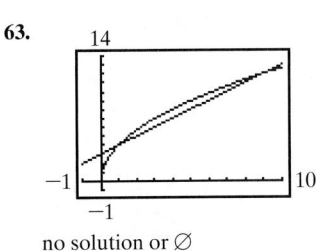

no solution or $\varnothing$

64. $4x^3 - 15x^2 + 47x - 142 + \dfrac{425}{x+3}$ **65.** $\dfrac{x+2}{2(x+3)}$ **66.** $x(8x+1)(8x-1)$

Section 10.7

Check Point Exercises

1. a. $4i$ **b.** $\sqrt{5}i$ **c.** $5\sqrt{2}i$ **2. a.** $8+i$ **b.** $-10+10i$ **3. a.** $63+14i$ **b.** $58-11i$ **4.** $-\sqrt{35}$
5. $\dfrac{18}{25} + \dfrac{26}{25}i$ **6.** $-\dfrac{1}{2} - \dfrac{3}{4}i$ **7. a.** 1 **b.** i **c.** $-i$

Exercise Set 10.7

1. $7i$ **3.** $\sqrt{17}i$ **5.** $5\sqrt{3}i$ **7.** $2\sqrt{7}i$ **9.** $-5\sqrt{6}i$ **11.** $7+4i$ **13.** $5+\sqrt{5}i$ **15.** $6-3\sqrt{2}i$ **17.** $8+3i$ **19.** $8-2i$
21. $5+3i$ **23.** $-1-7i$ **25.** $-2+9i$ **27.** $8+15i$ **29.** $-14+17i$ **31.** $9+5\sqrt{3}i$ **33.** $-6+10i$ **35.** $-21-15i$
37. $-35-14i$ **39.** $7+19i$ **41.** $-1-31i$ **43.** $3+36i$ **45.** 34 **47.** 34 **49.** 11 **51.** $-5+12i$ **53.** $21-20i$
55. $-\sqrt{14}$ **57.** -6 **59.** $-5\sqrt{7}$ **61.** $-2\sqrt{6}$ **63.** $\dfrac{3}{5} - \dfrac{1}{5}i$ **65.** $1+i$ **67.** $\dfrac{28}{25} + \dfrac{21}{25}i$ **69.** $-\dfrac{12}{13} + \dfrac{18}{13}i$ **71.** $0+i$ or i
73. $\dfrac{3}{10} - \dfrac{11}{10}i$ **75.** $\dfrac{11}{13} - \dfrac{16}{13}i$ **77.** $-\dfrac{23}{58} + \dfrac{43}{58}i$ **79.** $0 - \dfrac{7}{3}i$ or $-\dfrac{7}{3}i$ **81.** $-\dfrac{5}{2} - 4i$ **83.** $-\dfrac{7}{3} + \dfrac{4}{3}i$ **85.** -1 **87.** $-i$
89. -1 **91.** 1 **93.** i **95.** 1 **97.** $-i$ **99.** 0 **101.** $(47+13i)$ volts   **103.** $(5+\sqrt{15}i)+(5-\sqrt{15}i) = 10$;
$(5+\sqrt{15}i)(5-\sqrt{15}i) = 25 - 15i^2 = 25 + 15 = 40$ **105–115.** Answers will vary.   **117.** $\sqrt{-9} + \sqrt{-16} = 3i + 4i = 7i$
119. d   **121.** $\dfrac{14}{25} - \dfrac{2}{25}i$ **123.** 0 **124.** 1 and 5, or $\{1, 5\}$ **125.** $\dfrac{1}{8x^6}$ **126.** $\dfrac{x-3}{4(x+3)}$

Review Exercises

1. 9 **2.** $-\dfrac{1}{10}$ **3.** -3 **4.** not a real number   **5.** -2 **6.** $5; 1.73; 0;$ not a real number   **7.** $2; -2; -4$
8. $\{x \mid x \geq 2\}$ or $[2, \infty)$ **9.** $\{x \mid x \leq 25\}$ or $(-\infty, 25]$ **10.** $5|x|$ **11.** $|x+14|$ **12.** $|x-4|$ **13.** $4x$ **14.** $2|x|$
15. $-2(x+7)$ **16.** $\sqrt[3]{5xy}$ **17.** $(\sqrt{16})^3 = 64$ **18.** $(\sqrt[5]{32})^4 = 16$ **19.** $(7x)^{1/2}$ **20.** $(19xy)^{5/3}$ **21.** $\dfrac{1}{8^{2/3}} = \dfrac{1}{4}$
22. $\dfrac{3x}{(ab)^{4/5}} = \dfrac{3x}{\sqrt[5]{(ab)^4}}$ **23.** $x^{7/12}$ **24.** $5^{1/6}$ **25.** $2x^2y$ **26.** $\dfrac{y^{1/8}}{x^{1/3}}$ **27.** x^3y^4 **28.** $y\sqrt[3]{x}$ **29.** $\sqrt[6]{x^5}$ **30.** $\sqrt[6]{x}$ **31.** $\sqrt[15]{x}$
32. $\$3150$ million   **33.** $\sqrt{21xy}$ **34.** $\sqrt[5]{77x^3}$ **35.** $\sqrt[6]{(x-5)^5}$ **36.** $f(x) = \sqrt{7}|x-1|$ **37.** $2x\sqrt{5x}$ **38.** $3x^2y^2\sqrt[3]{2x^2}$
39. $2y^2\sqrt[4]{2x^3y^3}$ **40.** $2x^2\sqrt[3]{6x}$ **41.** $2xy\sqrt[3]{2y^2}$ **42.** $xy\sqrt[5]{16y^4}$ **43.** $\sqrt{x^2-1}$ **44.** $8\sqrt[3]{3}$ **45.** $9\sqrt{2}$ **46.** $(3x+y^2)\sqrt[3]{x}$
47. $-8\sqrt[3]{6}$ **48.** $\dfrac{2}{5}\sqrt[3]{2}$ **49.** $\dfrac{x\sqrt{x}}{10y^2}$ **50.** $\dfrac{y\sqrt[4]{3y}}{2x^5}$ **51.** $2\sqrt{6}$ **52.** $2\sqrt[3]{2}$ **53.** $2x\sqrt[4]{2x}$ **54.** $10x^2\sqrt{xy}$ **55.** $6\sqrt{2} + 12\sqrt{5}$
56. $5\sqrt[3]{2} - \sqrt[3]{10}$ **57.** $-83 + 3\sqrt{35}$ **58.** $\sqrt{xy} - \sqrt{11x} - \sqrt{11y} + 11$ **59.** $13 + 4\sqrt{10}$ **60.** $22 - 4\sqrt{30}$ **61.** -6 **62.** 4
63. $\dfrac{2\sqrt{6}}{3}$ **64.** $\dfrac{\sqrt{14}}{7}$ **65.** $4\sqrt[3]{3}$ **66.** $\dfrac{\sqrt{10xy}}{5y}$ **67.** $\dfrac{7\sqrt[3]{4x}}{x}$ **68.** $\dfrac{\sqrt[4]{189x^3}}{3x}$ **69.** $\dfrac{5\sqrt[5]{xy^4}}{2xy}$ **70.** $3\sqrt{3} + 3$ **71.** $\dfrac{\sqrt{35} - \sqrt{21}}{2}$
72. $10\sqrt{5} + 15\sqrt{2}$ **73.** $\dfrac{x + 8\sqrt{x} + 15}{x-9}$ **74.** $\dfrac{5 + \sqrt{21}}{2}$ **75.** $\dfrac{3\sqrt{2} + 2}{7}$ **76.** $\dfrac{2}{\sqrt{14}}$ **77.** $\dfrac{3x}{\sqrt[3]{9x^2y}}$ **78.** $\dfrac{7}{\sqrt{35} + \sqrt{21}}$
79. $\dfrac{2}{5 - \sqrt{21}}$ **80.** 16 or $\{16\}$ **81.** no solution or $\varnothing$ **82.** 2 or $\{2\}$ **83.** 8 or $\{8\}$ **84.** -4 and -2, or $\{-4, -2\}$ **85.** 256 ft
86. 84 years old   **87.** $9i$ **88.** $3\sqrt{7}i$ **89.** $-2\sqrt{2}i$ **90.** $12 + 2i$ **91.** $-9 + 4i$ **92.** $-12 - 8i$ **93.** $29 + 11i$
94. $-7 - 24i$ **95.** $113 + 0i$ or 113 **96.** $-2\sqrt{6}$ **97.** $\dfrac{15}{13} - \dfrac{3}{13}i$ **98.** $\dfrac{1}{5} + \dfrac{11}{10}i$ **99.** $\dfrac{1}{3} - \dfrac{5}{3}i$ **100.** 1 **101.** $-i$

Chapter 10 Test

1. a. 6 **b.** $\{x \mid x \leq 4\}$ or $(-\infty, 4]$ **2.** $\dfrac{1}{81}$ **3.** $\dfrac{5\sqrt[8]{y}}{\sqrt[4]{x}}$ **4.** $\sqrt{x}$ **5.** $\sqrt[20]{x^9}$ **6.** $5|x|\sqrt{3}$ **7.** $|x-5|$ **8.** $2xy^2\sqrt[3]{2xy^2}$
9. $-\dfrac{2}{x^2}$ **10.** $\sqrt[3]{50x^2y}$ **11.** $2x\sqrt[4]{2y^3}$ **12.** $-7\sqrt{2}$ **13.** $(2x + y^2)\sqrt[3]{x}$ **14.** $2x\sqrt[3]{x}$ **15.** $12\sqrt{2} - \sqrt{15}$ **16.** $26 + 6\sqrt{3}$
17. $52 - 14\sqrt{3}$ **18.** $\dfrac{\sqrt{5x}}{x}$ **19.** $\dfrac{\sqrt[3]{25x}}{x}$ **20.** $-5 + 2\sqrt{6}$ **21.** 6 or $\{6\}$ **22.** 16 or $\{16\}$ **23.** -3 or $\{-3\}$ **24.** 49 months
25. $5\sqrt{3}i$ **26.** $-1 + 6i$ **27.** $26 + 7i$ **28.** $-6 + 0i$ or -6 **29.** $\dfrac{1}{5} + \dfrac{7}{5}i$ **30.** $-i$

Cumulative Review Exercises (Chapters 1–10)

1. $(-2, -1, -2)$ or $\{(-2, -1, -2)\}$ **2.** $-\dfrac{1}{3}$ and 4, or $\left\{-\dfrac{1}{3}, 4\right\}$ **3.** $\left\{x \mid x > \dfrac{1}{3}\right\}$ or $\left(\dfrac{1}{3}, \infty\right)$ **4.** $\dfrac{3}{4}$ or $\left\{\dfrac{3}{4}\right\}$ **5.** -1 or $\{-1\}$

6. $x + 2y < 2$
$2y - x > 4$

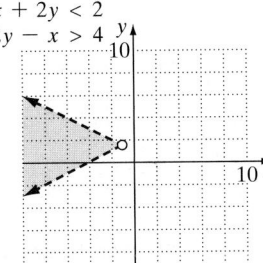

7. $\dfrac{x^2}{15(x+2)}$ **8.** $\dfrac{x}{y}$ **9.** $8x^3 - 22x^2 + 11x + 6$

10. $\dfrac{5x - 6}{(x - 5)(x + 3)}$ **11.** -64 **12.** 0

13. $-\dfrac{16 - 9\sqrt{3}}{13}$ **14.** $2x^2 + x + 5 + \dfrac{6}{x - 2}$ **15.** $-34 - 3\sqrt{6}$

16. $2(3x + 2)(4x - 1)$ **17.** $(4x^2 + 1)(2x + 1)(2x - 1)$

18. about 53 lumens   **19.** \$1500 at 7% and \$4500 at 9%   **20.** 2650 students

CHAPTER 11

Section 11.1

Check Point Exercises

1. $\pm\sqrt{3}$ or $\{\pm\sqrt{3}\}$ **2.** $\pm\dfrac{\sqrt{10}}{2}$ or $\left\{\pm\dfrac{\sqrt{10}}{2}\right\}$ **3.** $\pm\dfrac{3}{2}i$ or $\left\{\pm\dfrac{3}{2}i\right\}$ **4.** $2 \pm \sqrt{7}$ or $\{2 \pm \sqrt{7}\}$ **5. a.** $x^2 + 10x + 25 = (x + 5)^2$

b. $x^2 - 3x + \dfrac{9}{4} = \left(x - \dfrac{3}{2}\right)^2$ **c.** $x^2 + \dfrac{3}{4}x + \dfrac{9}{64} = \left(x + \dfrac{3}{8}\right)^2$ **6.** $5 \pm \sqrt{7}$ or $\{5 \pm \sqrt{7}\}$ **7.** $\dfrac{5 \pm 3\sqrt{3}}{2}$ or $\left\{\dfrac{5 \pm 3\sqrt{3}}{2}\right\}$

8. 20%   **9.** $10\sqrt{21}$ ft; 45.8 ft

Exercise Set 11.1

1. ± 5 or $\{\pm 5\}$ **3.** $\pm\sqrt{6}$ or $\{\pm\sqrt{6}\}$ **5.** $\pm\dfrac{5}{4}$ or $\left\{\pm\dfrac{5}{4}\right\}$ **7.** $\pm\dfrac{\sqrt{6}}{3}$ or $\left\{\pm\dfrac{\sqrt{6}}{3}\right\}$ **9.** $\pm\dfrac{4}{5}i$ or $\left\{\pm\dfrac{4}{5}i\right\}$

11. -10 and -4, or $\{-10, -4\}$ **13.** $3 \pm \sqrt{5}$ or $\{3 \pm \sqrt{5}\}$ **15.** $-2 \pm 2\sqrt{2}$ or $\{-2 \pm 2\sqrt{2}\}$ **17.** $5 \pm 3i$ or $\{5 \pm 3i\}$

19. $\dfrac{-3 \pm \sqrt{11}}{4}$ or $\left\{\dfrac{-3 \pm \sqrt{11}}{4}\right\}$ **21.** -3 and 9, or $\{-3, 9\}$ **23.** $x^2 + 2x + 1 = (x + 1)^2$ **25.** $x^2 - 14x + 49 = (x - 7)^2$

27. $x^2 + 7x + \dfrac{49}{4} = \left(x + \dfrac{7}{2}\right)^2$ **29.** $x^2 - \dfrac{1}{2}x + \dfrac{1}{16} = \left(x - \dfrac{1}{4}\right)^2$ **31.** $x^2 + \dfrac{4}{3}x + \dfrac{4}{9} = \left(x + \dfrac{2}{3}\right)^2$

33. $x^2 - \dfrac{9}{4}x + \dfrac{81}{64} = \left(x - \dfrac{9}{8}\right)^2$ **35.** -4 and -2, or $\{-4, -2\}$ **37.** $-3 \pm \sqrt{7}$ or $\{-3 \pm \sqrt{7}\}$ **39.** $-2 \pm \sqrt{3}$ or $\{-2 \pm \sqrt{3}\}$

41. $-1 \pm i$ or $\{-1 \pm i\}$ **43.** $\dfrac{-3 \pm \sqrt{13}}{2}$ or $\left\{\dfrac{-3 \pm \sqrt{13}}{2}\right\}$ **45.** $\dfrac{7 \pm \sqrt{37}}{2}$ or $\left\{\dfrac{7 \pm \sqrt{37}}{2}\right\}$ **47.** $\dfrac{-1 \pm \sqrt{5}}{2}$ or $\left\{\dfrac{-1 \pm \sqrt{5}}{2}\right\}$

49. $\dfrac{1}{2}$ and 1, or $\left\{\dfrac{1}{2}, 1\right\}$ **51.** $\dfrac{-5 \pm \sqrt{3}}{2}$ or $\left\{\dfrac{-5 \pm \sqrt{3}}{2}\right\}$ **53.** $\dfrac{1 \pm \sqrt{13}}{3}$ or $\left\{\dfrac{1 \pm \sqrt{13}}{3}\right\}$ **55.** $\dfrac{1}{4} \pm \dfrac{1}{4}i$ or $\left\{\dfrac{1}{4} \pm \dfrac{1}{4}i\right\}$

57. 20%   **59.** about 3.69%   **61.** 2003   **63.** $10\sqrt{3}$ sec; 17.3 sec   **65.** $3\sqrt{5}$ mi; 6.7 mi   **67.** $5\sqrt{7}$ ft; 13.2 ft

69. $50\sqrt{2}$ ft; 70.7 ft   **71.** $c = \sqrt{\dfrac{E}{m}}$ or $c = \dfrac{\sqrt{mE}}{m}$ **73.** 8 m   **75–83.** Answers will vary.

85. $y = \pm\dfrac{b\sqrt{a^2 - x^2}}{a}$ **87.** $\dfrac{-b \pm \sqrt{b^2 - 4c}}{2}$ or $\left\{\dfrac{-b \pm \sqrt{b^2 - 4c}}{2}\right\}$

89.

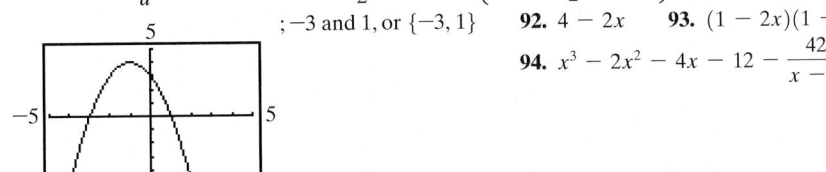

$; -3$ and 1, or $\{-3, 1\}$ **92.** $4 - 2x$ **93.** $(1 - 2x)(1 + 2x + 4x^2)$

94. $x^3 - 2x^2 - 4x - 12 - \dfrac{42}{x - 3}$

Section 11.2

Check Point Exercises

1. -5 and $\frac{1}{2}$, or $\left\{-5, \frac{1}{2}\right\}$ **2.** $\frac{3 \pm \sqrt{7}}{2}$ or $\left\{\frac{3 \pm \sqrt{7}}{2}\right\}$ **3.** $-1 \pm \frac{\sqrt{6}}{3}i$ or $\left\{-1 \pm \frac{\sqrt{6}}{3}i\right\}$ **4. a.** 0; one real solution

b. 81; two rational solutions   **c.** -44; two imaginary solutions   **5. a.** $20x^2 + 7x - 3 = 0$ **b.** $x^2 + 49 = 0$

6. 1995; actual number: 340 thousand; The function describes the situation well for that year.

Exercise Set 11.2

1. -6 and -2, or $\{-6, -2\}$ **3.** 1 and $\frac{5}{2}$, or $\left\{1, \frac{5}{2}\right\}$ **5.** $\frac{-3 \pm \sqrt{89}}{2}$ or $\left\{\frac{-3 \pm \sqrt{89}}{2}\right\}$ **7.** $\frac{7 \pm \sqrt{85}}{6}$ or $\left\{\frac{7 \pm \sqrt{85}}{6}\right\}$

9. $\frac{1 \pm \sqrt{7}}{6}$ or $\left\{\frac{1 \pm \sqrt{7}}{6}\right\}$ **11.** $\frac{3}{8} \pm \frac{\sqrt{87}}{8}i$ or $\left\{\frac{3}{8} \pm \frac{\sqrt{87}}{8}i\right\}$ **13.** $2 \pm 2i$ or $\{2 \pm 2i\}$ **15.** $\frac{4}{3} \pm \frac{\sqrt{5}}{3}i$ or $\left\{\frac{4}{3} \pm \frac{\sqrt{5}}{3}i\right\}$

17. 52; two irrational solutions   **19.** 4; two rational solutions   **21.** -23; two imaginary solutions   **23.** 36; two rational solutions

25. -60; two imaginary solutions   **27.** 0; one (repeated) rational solution   **29.** $-\frac{2}{3}$ and 2, or $\left\{-\frac{2}{3}, 2\right\}$

31. $1 \pm \sqrt{2}$ or $\{1 \pm \sqrt{2}\}$ **33.** $\frac{3 \pm \sqrt{65}}{4}$ or $\left\{\frac{3 \pm \sqrt{65}}{4}\right\}$ **35.** 0 and $\frac{8}{3}$, or $\left\{0, \frac{8}{3}\right\}$ **37.** $\frac{-6 \pm 2\sqrt{6}}{3}$ or $\left\{\frac{-6 \pm 2\sqrt{6}}{3}\right\}$

39. $\frac{2 \pm \sqrt{10}}{3}$ or $\left\{\frac{2 \pm \sqrt{10}}{3}\right\}$ **41.** $2 \pm \sqrt{10}$ or $\{2 \pm \sqrt{10}\}$ **43.** 1 and $\frac{5}{2}$, or $\left\{1, \frac{5}{2}\right\}$ **45.** $x^2 - 2x - 15 = 0$

47. $12x^2 + 5x - 2 = 0$ **49.** $x^2 + 36 = 0$ **51.** $x^2 - 2 = 0$ **53.** $x^2 - 20 = 0$ **55.** $x^2 - 2x + 2 = 0$

57. $x^2 - 2x - 1 = 0$ **59.** about 2002   **61.** 19 year olds and 72 year olds; Answers will vary.   **63.** 3.8 sec

65. 4.7 m by 7.7 m   **67.** 2.3 ft and 3.3 ft   **69.** 9.3 in. and 0.7 in.   **71–77.** Answers will vary.   **79.** d   **81.** -10 or $\{-10\}$

83. 41 people   **87.** -3 and $\frac{1}{4}$, or $\left\{-3, \frac{1}{4}\right\}$ **88.** 3 and 7, or $\{3, 7\}$ **89.** $\frac{5\sqrt{3} - 5x}{3 - x^2}$

Section 11.3

Check Point Exercises

1.

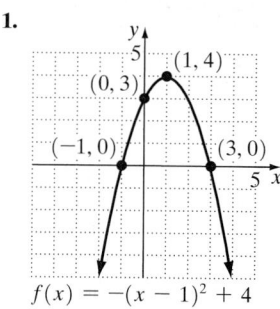

$f(x) = -(x - 1)^2 + 4$

2.

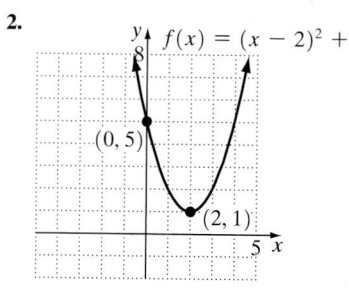

$f(x) = (x - 2)^2 + 1$

3.

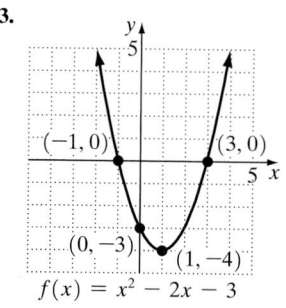

$f(x) = x^2 - 2x - 3$

4. 45 years old; 190 car accidents per 50 million miles driven   **5.** 30 ft by 30 ft; 90 sq ft

Exercise Set 11.3

1. $h(x) = (x - 1)^2 + 1$ **3.** $j(x) = (x - 1)^2 - 1$ **5.** $h(x) = x^2 - 1$ **7.** $g(x) = x^2 - 2x + 1$ **9.** $(3, 1)$ **11.** $(-1, 5)$

13. $(2, -5)$ **15.** $(-1, 9)$

17.

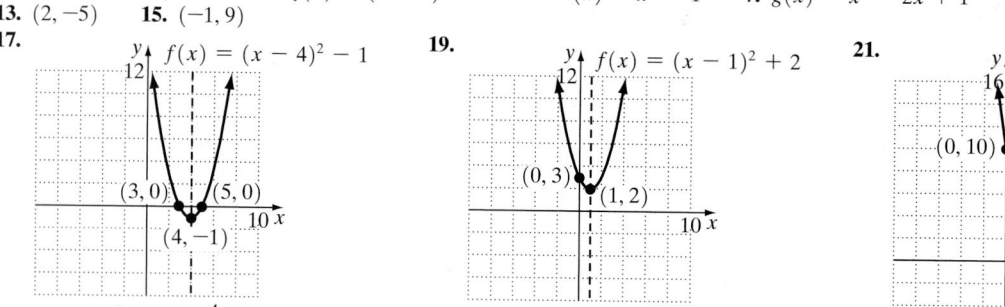

$f(x) = (x - 4)^2 - 1$

$x = 4$

19.

$f(x) = (x - 1)^2 + 2$

$x = 1$

21.

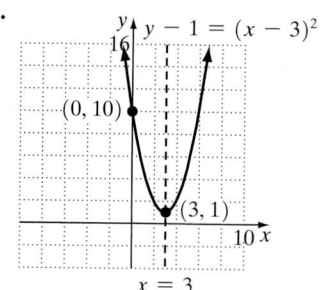

$y - 1 = (x - 3)^2$

$x = 3$

23.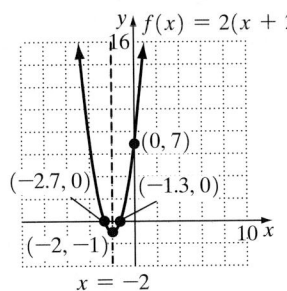

25.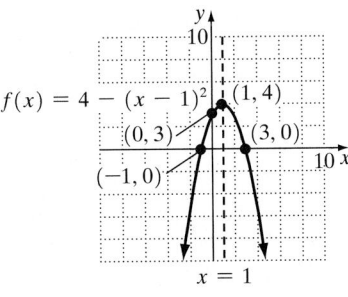

27.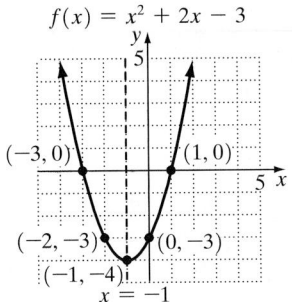

29.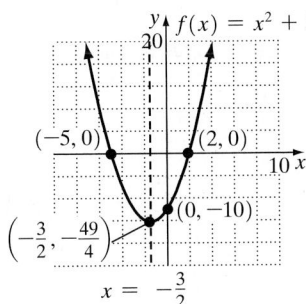

31.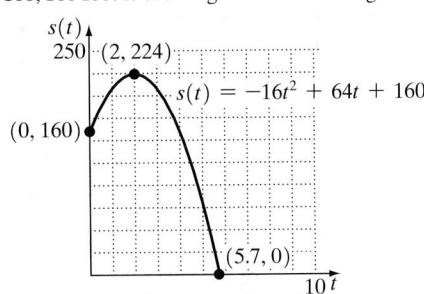

33.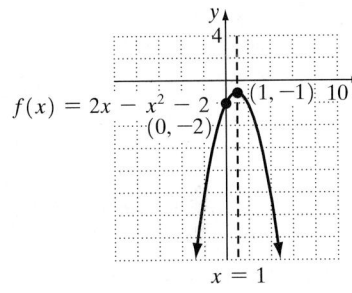

35. minimum; $(2, -13)$ **37.** maximum; $(1, 1)$ **39.** minimum; $\left(\dfrac{1}{2}, -\dfrac{5}{4}\right)$ **41.** 1968; 4237 cigarettes per person

43. 7.2 hr; 622 per 100,000 males

45. a. 2 sec; 224 ft

    b. 5.7 sec

    c. 160; 160 feet is the height of the building.

    d.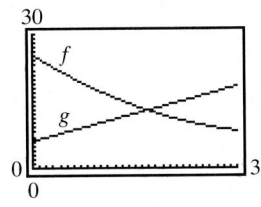

47. $(15, 41.1)$; What cost $100 in 1967 cost $41.10 in 1935.

49. length: 60 ft; width: 30 ft; 1800 sq ft

51. 8 and 8; 64   **53.** 5 and -5; -25

55. 12.5 yd by 12.5 yd; 156.25 sq yd   **57.** 5 in.; 50 sq in.

59. a. $C(x) = 525 + 0.55x$

    b. $P(x) = -0.001x^2 + 2.45x - 525$

    c. 1225 sandwiches; $975.63

61–67. Answers will vary.   **69.** $x = -2$; $(-3, -2)$

71. 125 ft by about 166.7 ft; about 20,833 sq ft

75. $(80, 1600)$;   **77.** $(-4, 520)$

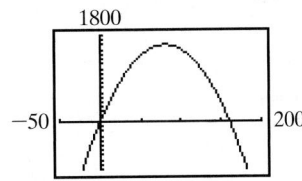   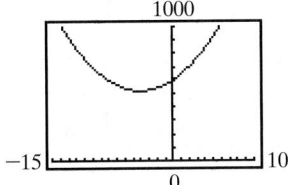

79. a.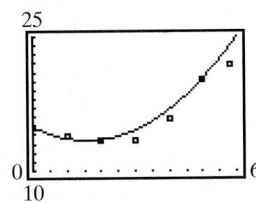

    b. The average yearly consumption of whole milk per person has been decreasing, while the average early consumption of low-fat milk has been increasing in the United States since 1970.

81. a. & d.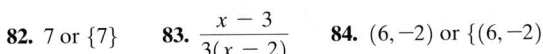

    b. $y = 0.006x^2 - 0.192x + 14.929$

    c. 1956; 13.393 miles per gallon

82. 7 or $\{7\}$ **83.** $\dfrac{x - 3}{3(x - 2)}$ **84.** $(6, -2)$ or $\{(6, -2)\}$

Section 11.4

Check Point Exercises

1. $-4, 4, -1,$ and $1,$ or $\{-4, 4, -1, 1\}$ **2.** 16 or $\{16\}$ **3.** $-\sqrt{6}, -1, 1,$ and $\sqrt{6},$ or $\{-\sqrt{6}, -1, 1, \sqrt{6}\}$ **4.** -1 and 2, or $\{-1, 2\}$
5. $-\dfrac{1}{27}$ and 64, or $\left\{-\dfrac{1}{27}, 64\right\}$

Exercise Set 11.4

1. $-2, 2, -1,$ and $1,$ or $\{-2, 2, -1, 1\}$ **3.** $-3, 3, -\sqrt{2},$ and $\sqrt{2},$ or $\{-3, 3, -\sqrt{2}, \sqrt{2}\}$ **5.** $-2i, 2i, -\sqrt{2},$ and $\sqrt{2},$ or $\{-2i, 2i, -\sqrt{2}, \sqrt{2}\}$
7. 1 or $\{1\}$ **9.** 49 or $\{49\}$ **11.** 25 and 64, or $\{25, 64\}$ **13.** 2 and 12, or $\{2, 12\}$ **15.** $-\sqrt{3}, \sqrt{3},$ and 0, or $\{-\sqrt{3}, \sqrt{3}, 0\}$
17. $-5, -2, -1,$ and 2, or $\{-5, -2, -1, 2\}$ **19.** $-\dfrac{1}{4}$ and $\dfrac{1}{5},$ or $\left\{-\dfrac{1}{4}, \dfrac{1}{5}\right\}$ **21.** $\dfrac{1}{3}$ and 2, or $\left\{\dfrac{1}{3}, 2\right\}$ **23.** $\dfrac{-2 \pm \sqrt{7}}{3}$ or $\left\{\dfrac{-2 \pm \sqrt{7}}{3}\right\}$
25. -8 and 27, or $\{-8, 27\}$ **27.** -243 and 32, or $\{-243, 32\}$ **29.** 1 or $\{1\}$ **31.** $-8, -2, 1,$ and 4, or $\{-8, -2, 1, 4\}$
33. $-2, 2, -1,$ and 1; c   **35.** 1; e   **37.** 2 and 3; f   **39.** ages 20 and 55; The function models the data well.   **41–43.** Answers will vary.
45. $\pm\dfrac{\sqrt{10 + 2\sqrt{33}}}{2}$ and $\pm\dfrac{\sqrt{10 - 2\sqrt{33}}}{2},$ or $\left\{\pm\dfrac{\sqrt{10 + 2\sqrt{33}}}{2}, \pm\dfrac{\sqrt{10 - 2\sqrt{33}}}{2}\right\}$ **47.** $-\dfrac{17}{3}$ and $\dfrac{8}{3},$ or $\left\{-\dfrac{17}{3}, \dfrac{8}{3}\right\}$
49. -1 and 2, or $\{-1, 2\}$ **51.** $-1, 1, -3,$ and 3, or $\{-1, 1, -3, 3\}$ **53.** $-\dfrac{3}{2}$ and 2, or $\left\{-\dfrac{3}{2}, 2\right\}$ **55.** 1 or $\{1\}$ **57.** $\dfrac{1}{5x - 1}$
58. $\dfrac{1}{2} + \dfrac{3}{2}i$ **59.** $(4, -2)$ or $\{(4, -2)\}$

Section 11.5

Check Point Exercises

1. $\{x | -3 < x < 1\}$ or $(-3, 1)$

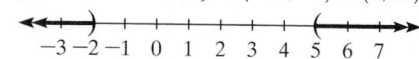

2. $\{x | x \le -4 \text{ or } x \ge 5\}$ or $(-\infty, -4] \cup [5, \infty)$

3. $\{x | x < -2 \text{ or } x > 5\}$ or $(-\infty, -2) \cup (5, \infty)$

4. $\{x | -1 < x \le 1\}$ or $(-1, 1]$

5. between 1 and 4 seconds, excluding $t = 1$ and $t = 4$

Exercise Set 11.5

1. $\{x | x < -2 \text{ or } x > 4\}$ or $(-\infty, -2) \cup (4, \infty)$

3. $\{x | -3 \le x \le 7\}$ or $[-3, 7]$

5. $\{x | x < 1 \text{ or } x > 4\}$ or $(-\infty, 1) \cup (4, \infty)$

7. $\{x | x < -4 \text{ or } x > -1\}$ or $(-\infty, -4) \cup (-1, \infty)$

9. $\{x | 2 \le x \le 4\}$ or $[2, 4]$

11. $\left\{x \middle| -4 \le x \le \dfrac{2}{3}\right\}$ or $\left[-4, \dfrac{2}{3}\right]$

13. $\left\{x \middle| -3 < x < \dfrac{5}{2}\right\}$ or $\left(-3, \dfrac{5}{2}\right)$

15. $\left\{x \middle| -1 < x < -\dfrac{3}{4}\right\}$ or $\left(-1, -\dfrac{3}{4}\right)$

17. $\{x | x \le 0 \text{ or } x \ge 4\}$ or $(-\infty, 0] \cup [4, \infty)$

19. $\left\{x \middle| x < -\dfrac{3}{2} \text{ or } x > 0\right\}$ or $\left(-\infty, -\dfrac{3}{2}\right) \cup (0, \infty)$

21. $\{x | 0 \le x \le 1\}$ or $[0, 1]$

23. $\{x | 2 - \sqrt{2} \le x \le 2 + \sqrt{2}\}$ or $[2 - \sqrt{2}, 2 + \sqrt{2}]$

25. no solution or $\varnothing$

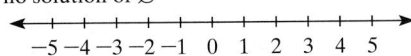

number line: $-5\ -4\ -3\ -2\ -1\ \ 0\ \ 1\ \ 2\ \ 3\ \ 4\ \ 5$

27. $\{x|x < -3 \text{ or } x > 4\}$ or $(-\infty, -3) \cup (4, \infty)$

number line: $-5\ -4\ -3\ -2\ -1\ \ 0\ \ 1\ \ 2\ \ 3\ \ 4\ \ 5$

29. $\{x|-4 < x < -3\}$ or $(-4, -3)$

number line: $-5\ -4\ -3\ -2\ -1\ \ 0\ \ 1\ \ 2\ \ 3\ \ 4\ \ 5$

31. $\{x|2 \le x < 4\}$ or $[2, 4)$

number line: $-5\ -4\ -3\ -2\ -1\ \ 0\ \ 1\ \ 2\ \ 3\ \ 4\ \ 5$

33. $\left\{x \mid x < -\dfrac{4}{3} \text{ or } x \ge 2\right\}$ or $\left(-\infty, -\dfrac{4}{3}\right) \cup [2, \infty)$

number line: $-5\ -4\ -3\ -2\ -1\ \ 0\ \ 1\ \ 2\ \ 3\ \ 4\ \ 5$

35. $\{x|x < 0 \text{ or } x > 3\}$ or $(-\infty, 0) \cup (3, \infty)$

number line: $-3\ -2\ -1\ \ 0\ \ 1\ \ 2\ \ 3\ \ 4\ \ 5\ \ 6\ \ 7$

37. $\{x|x < -5 \text{ or } x > -3\}$ or $(-\infty, -5) \cup (-3, \infty)$

number line: $-6\ -5\ -4\ -3\ -2\ -1\ \ 0\ \ 1\ \ 2\ \ 3\ \ 4$

39. $\left\{x \mid x < \dfrac{1}{2} \text{ or } x \ge \dfrac{7}{5}\right\}$ or $\left(-\infty, \dfrac{1}{2}\right) \cup \left[\dfrac{7}{5}, \infty\right)$

number line: $-5\ -4\ -3\ -2\ -1\ \ 0\ \ 1\ \ 2\ \ 3\ \ 4\ \ 5$

41. $\{x|x \le -6 \text{ or } x > -2\}$ or $(-\infty, -6] \cup (-2, \infty)$

number line: $-8\ -7\ -6\ -5\ -4\ -3\ -2\ -1\ \ 0\ \ 1\ \ 2$

43. between 0 and 3 seconds, excluding $t = 0$ and $t = 3$

45. f: \$379 billion; g: \$379.8 billion; The functions model the value in the graph well.

47. after 2007 **49.** 1300.2 thousand or 1,300,200; The function models the actual number for 1998 well. **51.** after 2010

53. The company's production level must be at least 20,000 wheelchairs per month. For values of x greater than or equal to 20,000, the graph lies on or below the line $y = 425$. **55–57.** Answers will vary. **59.** d **61.** Answers will vary; an example is: $\dfrac{x - 3}{x + 4} \ge 0$.

63. $\{2\}$ **65.** $\{x|x < 2 \text{ or } x > 2\}$ or $(-\infty, 2) \cup (2, \infty)$ **67.** $27 - 3x^2 \ge 0$; $\{x|-3 \le x \le 3\}$ or $[-3, 3]$

69. $\left\{x \mid -3 \le x \le \dfrac{1}{2}\right\}$ or $\left[-3, \dfrac{1}{2}\right]$ **71.** $\{x|x < 3 \text{ or } x \ge 8\}$ or $(-\infty, 3) \cup [8, \infty)$

73. $\{x|-3 < x < -1 \text{ or } x > 2\}$ or $(-3, -1) \cup (2, \infty)$ **74.** $\{x|-19 < x < 29\}$ or $(-19, 29)$ **75.** $\dfrac{2(x - 1)}{(x + 4)(x - 3)}$

76. $(x^2 + 4y^2)(x + 2y)(x - 2y)$

Review Exercises

1. ± 8 or $\{\pm 8\}$ **2.** $\pm 5\sqrt{2}$ or $\{\pm 5\sqrt{2}\}$ **3.** $\pm \dfrac{\sqrt{6}}{3}$ or $\left\{\pm \dfrac{\sqrt{6}}{3}\right\}$ **4.** $4 \pm 3\sqrt{2}$ or $\{4 \pm 3\sqrt{2}\}$ **5.** $-7 \pm 6i$ or $\{-7 \pm 6i\}$

6. $x^2 + 20x + 100 = (x + 10)^2$ **7.** $x^2 - 3x + \dfrac{9}{4} = \left(x - \dfrac{3}{2}\right)^2$

8. 3 and 9, or $\{3, 9\}$ **9.** $\dfrac{7 \pm \sqrt{53}}{2}$ or $\left\{\dfrac{7 \pm \sqrt{53}}{2}\right\}$ **10.** $\dfrac{-3 \pm \sqrt{41}}{4}$ or $\left\{\dfrac{-3 \pm \sqrt{41}}{4}\right\}$ **11.** 8% **12.** 20 weeks

13. $60\sqrt{5}$ m; 134.2 m **14.** $1 \pm \sqrt{5}$ or $\{1 \pm \sqrt{5}\}$ **15.** $1 \pm 3\sqrt{2}i$ or $\{1 \pm 3\sqrt{2}i\}$ **16.** $\dfrac{-2 \pm \sqrt{10}}{2}$ or $\left\{\dfrac{-2 \pm \sqrt{10}}{2}\right\}$

17. -36; two imaginary solutions **18.** 81; two rational solutions **19.** 40; two irrational solutions **20.** $\dfrac{1}{2}$ and 5, or $\left\{\dfrac{1}{2}, 5\right\}$

21. -2 and $\dfrac{10}{3}$, or $\left\{-2, \dfrac{10}{3}\right\}$ **22.** $\dfrac{7 \pm \sqrt{37}}{6}$ or $\left\{\dfrac{7 \pm \sqrt{37}}{6}\right\}$ **23.** -3 and 3, or $\{-3, 3\}$ **24.** $3 \pm 2\sqrt{2}$ or $\{3 \pm 2\sqrt{2}\}$

25. $\dfrac{1}{6} \pm \dfrac{\sqrt{23}}{6}i$ or $\left\{\dfrac{1}{6} \pm \dfrac{\sqrt{23}}{6}i\right\}$ **26.** $4 \pm \sqrt{5}$ or $\{4 \pm \sqrt{5}\}$ **27.** $15x^2 - 4x - 3 = 0$ **28.** $x^2 + 81 = 0$

29. $x^2 - 48 = 0$ **30.** 2001 **31.** 8.8 sec

32.

graph: $f(x) = -(x + 1)^2 + 4$, vertex $(-1, 4)$, points $(0, 3)$, $(-3, 0)$, $(1, 0)$, axis $x = -1$

33.

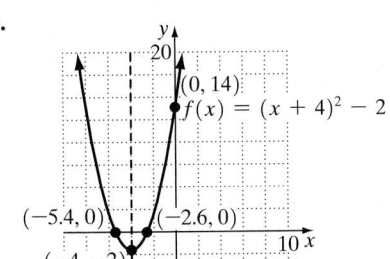

graph: $f(x) = (x + 4)^2 - 2$, point $(0, 14)$, points $(-5.4, 0)$, $(-2.6, 0)$, vertex $(-4, -2)$, axis $x = -4$

34.

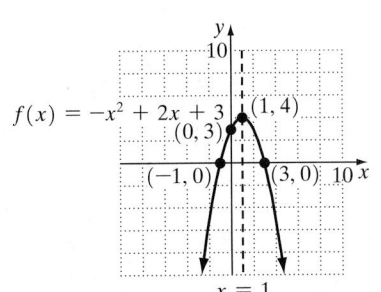

graph: $f(x) = -x^2 + 2x + 3$, vertex $(1, 4)$, point $(0, 3)$, points $(-1, 0)$, $(3, 0)$, axis $x = 1$

35.

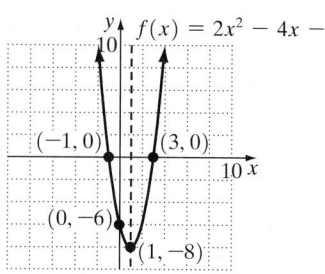

$f(x) = 2x^2 - 4x - 6$

36. 25 in. of rainfall per year; 13.5 in. of growth
37. 12.5 sec; 2540 feet
38. (20, 5.4); The maximum divorce rate of about 5.4 per 1000 population was in 1980.
39. 250 yd by 500 yd; 125,000 sq yd
40. -7 and 7; -49 **41.** $-\sqrt{2}, \sqrt{2}, -2,$ and 2, or $\{-\sqrt{2}, \sqrt{2}, -2, 2\}$ **42.** 1 or $\{1\}$
43. $-5, -1,$ and 3, or $\{-5, -1, 3\}$ **44.** $-\dfrac{1}{8}$ and $\dfrac{1}{7}$, or $\left\{-\dfrac{1}{8}, \dfrac{1}{7}\right\}$
45. -27 and 64, or $\{-27, 64\}$ **46.** 16 or $\{16\}$
47. $\left\{x \middle| -3 < x < \dfrac{1}{2}\right\}$ or $\left(-3, \dfrac{1}{2}\right)$;

48. $\left\{x \middle| x \le -4 \text{ or } x \ge -\dfrac{1}{2}\right\}$ or $(-\infty, -4] \cup \left[-\dfrac{1}{2}, \infty\right)$;

49. $\{x \mid x < -2 \text{ or } x > 6\}$ or $(-\infty, -2) \cup (6, \infty)$;

50. $\left\{x \middle| x < 4 \text{ or } x \ge \dfrac{23}{4}\right\}$ or $(-\infty, 4) \cup \left[\dfrac{23}{4}, \infty\right)$;

51. between 1 and 2 seconds, excluding $t = 1$ and $t = 2$ **52. a.** 200 beats per minute
b. between 0 and 4 minutes and more than 12 minutes after the workout; between 0 and 4 minutes; Answers will vary.

Chapter 11 Test

1. $\pm\dfrac{\sqrt{10}}{2}$ or $\left\{\pm\dfrac{\sqrt{10}}{2}\right\}$ **2.** $3 \pm 2\sqrt{5}$ or $\{3 \pm 2\sqrt{5}\}$ **3.** $x^2 - 16x + 64 = (x - 8)^2$ **4.** $x^2 + \dfrac{2}{5}x + \dfrac{1}{25} = \left(x + \dfrac{1}{5}\right)^2$
5. $3 \pm \sqrt{2}$ or $\{3 \pm \sqrt{2}\}$
6. $50\sqrt{2}$ ft   **7.** 40; two irrational solutions   **8.** -16; two imaginary solutions
9. -5 and $\dfrac{1}{2}$, or $\left\{-5, \dfrac{1}{2}\right\}$ **10.** $-4 \pm \sqrt{11}$ or $\{-4 \pm \sqrt{11}\}$ **11.** $-2 \pm 5i$ or $\{-2 \pm 5i\}$ **12.** $\dfrac{3}{2} \pm \dfrac{1}{2}i$ or $\left\{\dfrac{3}{2} \pm \dfrac{1}{2}i\right\}$
13. $x^2 - 4x - 21 = 0$ **14.** $x^2 + 100 = 0$
15. 1990 and 1998
16.

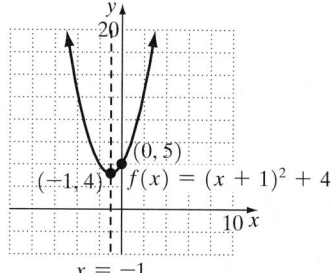

$f(x) = (x + 1)^2 + 4$

17.

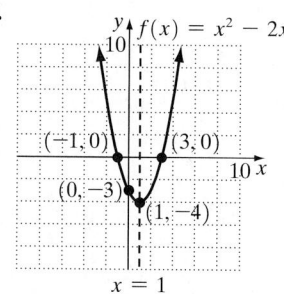

$f(x) = x^2 - 2x - 3$

18. after 2 sec; 69 ft
19. 4.1 sec
20. 23 computers; $169 hundreds or $16,900
21. 1 and 2, or $\{1, 2\}$
22. $-3, 3, -2,$ and 2, or $\{-3, 3, -2, 2\}$
23. 1 and 512, or $\{1, 512\}$

24. $\{x \mid -3 < x < 4\}$ or $(-3, 4)$

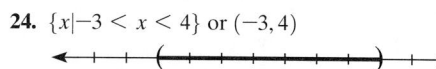

25. $\{x \mid x < 3 \text{ or } x \ge 10\}$ or $(-\infty, 3) \cup [10, \infty)$

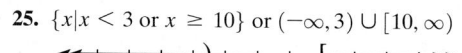

Cumulative Review Exercises (Chapters 1–11)

1. 4 or $\{4\}$ **2.** (2, 3) or $\{(2, 3)\}$ **3.** (2, 0, 3) or $\{(2, 0, 3)\}$ **4.** $\{x \mid x < -2 \text{ or } x > 4\}$ or $(-\infty, -2) \cup (4, \infty)$ **5.** 5 or $\{5\}$
6. $\{x \mid x \ge 4\}$ or $[4, \infty)$ **7.** $\dfrac{3}{4} \pm \dfrac{\sqrt{7}}{4}i$ or $\left\{\dfrac{3}{4} \pm \dfrac{\sqrt{7}}{4}i\right\}$

8.

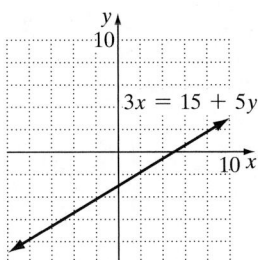

$3x = 15 + 5y$

9.

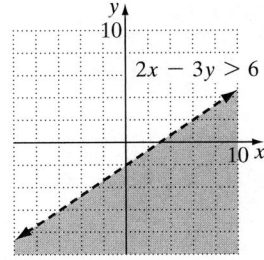

$2x - 3y > 6$

10.

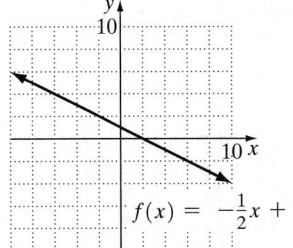

$f(x) = -\frac{1}{2}x + 1$

11.

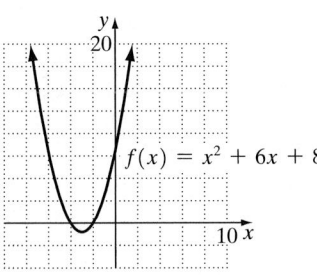

$f(x) = x^2 + 6x + 8$

12.

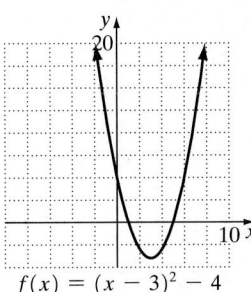

$f(x) = (x - 3)^2 - 4$

13. -6 **14.** $\frac{x}{3}$

15. $y = \frac{1}{2}x + 5$ or $f(x) = \frac{1}{2}x + 5$

16. $-\dfrac{y^9}{3x}$ **17.** $16x^4 - 40x^2y + 25y^2$

18. $x^2 - 5x + 1 + \dfrac{8}{5x + 1}$

19. $2y^2\sqrt[3]{2y^2}$ **20.** $\dfrac{4x - 13}{(x - 3)(x - 4)}$

21. $(x - 4)(x + 2)(x^2 - 2x + 4)$

22. $2(x + 3y)^2$ **23.** length: 12 ft; width: 4 ft   **24.** $\frac{6}{5}$ hr or 1 hr and 12 min   **25.** 5 mph

CHAPTER 12

Section 12.1

Check Point Exercises

1. one

2.

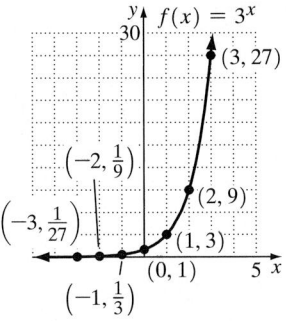

$f(x) = 3^x$

3.

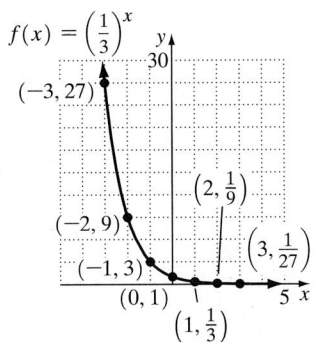

$f(x) = \left(\dfrac{1}{3}\right)^x$

4.

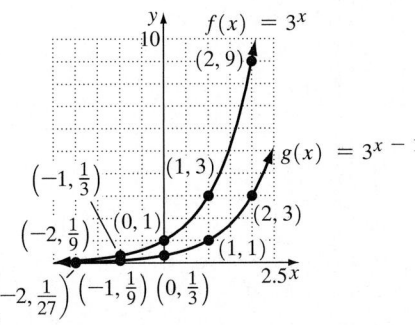

$f(x) = 3^x$

$g(x) = 3^{x-1}$

The graph of g is the graph of f shifted 1 unit to the right.

5.

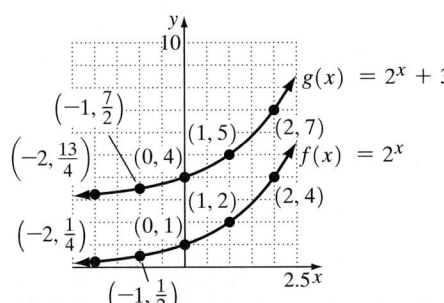

$g(x) = 2^x + 3$

$f(x) = 2^x$

The graph of g is the graph of f shifted up 3 units.

6. approximately 11.49 billion

7. a. \$14,859.47

   b. \$14,918.25

Exercise Set 12.1

1. 10.556   **3.** 11.665   **5.** 0.125   **7.** 9.974   **9.** 0.387

11.

x	$f(x)$
-2	$\dfrac{1}{9}$
-1	$\dfrac{1}{3}$
0	1
1	3
2	9

;d

13.

x	$f(x)$
-2	$-\dfrac{8}{9}$
-1	$-\dfrac{2}{3}$
0	0
1	2
2	8

;e

15.

x	$h(x)$
-2	9
-1	3
0	1
1	$\dfrac{1}{3}$
2	$\dfrac{1}{9}$

;f

17.

x	$f(x)$
-2	$\dfrac{1}{16}$
-1	$\dfrac{1}{4}$
0	1
1	4
2	16

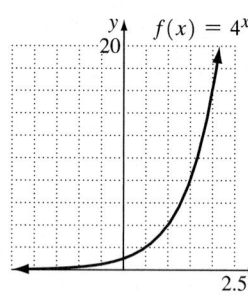

19.

x	$g(x)$
-2	$\dfrac{4}{9}$
-1	$\dfrac{2}{3}$
0	1
1	$\dfrac{3}{2}$
2	$\dfrac{9}{4}$

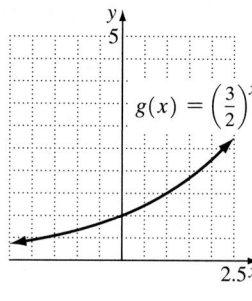

21.

x	$h(x)$
-2	4
-1	2
0	1
1	$\dfrac{1}{2}$
2	$\dfrac{1}{4}$

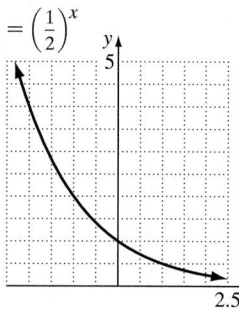

23.

x	$f(x)$
-2	2.78
-1	1.67
0	1
1	0.6
2	0.36

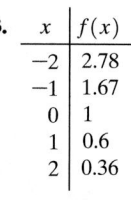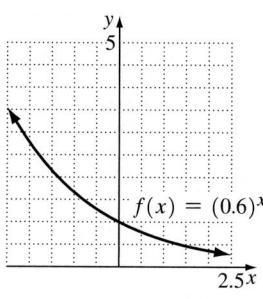

25.

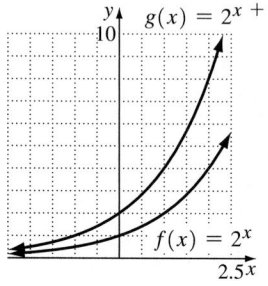

The graph of g is the graph of f shifted 1 unit to the left.

27.

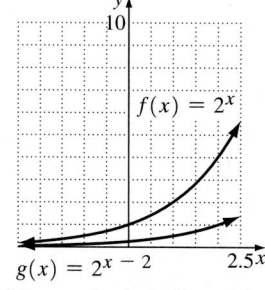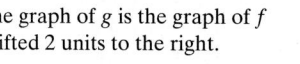

The graph of g is the graph of f shifted 2 units to the right.

29.

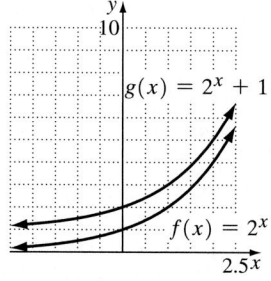

The graph of g is the graph of f shifted up 1 unit.

31.

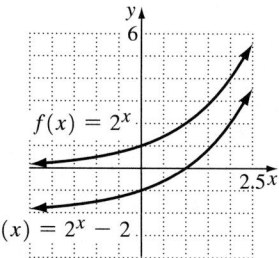

The graph of g is the graph of f shifted down 2 units.

33.

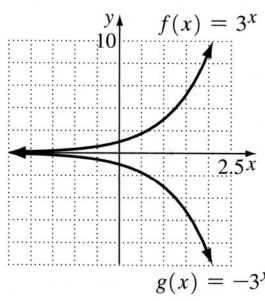

The graph of g is a reflection of the graph of f across the x-axis.

35.

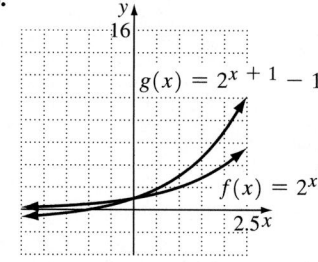

The graph of g is the graph of f shifted 1 unit to the left and 1 unit down.

37.

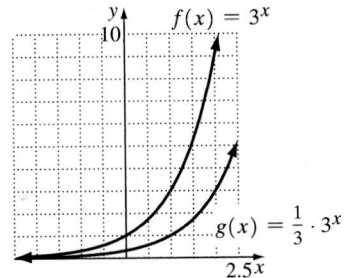

$f(x) = 3^x$

$g(x) = \frac{1}{3} \cdot 3^x$

2.5^x

The graph of g is the graph of f stretched vertically by a factor of $\frac{1}{3}$.

57–61. Answers will vary.

63. a. $y = \left(\frac{1}{3}\right)^x$ **b.** $y = \left(\frac{1}{5}\right)^x$ **c.** $y = 5^x$ **d.** $y = 3^x$; Answers will vary.

65.

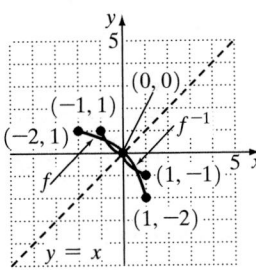

20

0   90
0

; Answers will vary.

39. a. $13,116.51 **b.** $13,157.04 **c.** $13,165.31
41. 7% compounded monthly
43. a. 67.38 million **b.** 134.74 million **c.** 269.46 million
d. 538.85 million **e.** It appears to double.
45. $116,405.10
47. $2^{1.7} \approx 3.249010$; $2^{1.73} \approx 3.317278$; $2^{1.732} \approx 3.321880$; $2^{1.73205} \approx 3.321995$;
$2^{1.7320508} \approx 3.321997$; $2^{\sqrt{3}} \approx 3.321997$; Answers will vary.
49. 241,786; There were about 241,786 AIDS cases in the United States among intravenous drug users in 2000.
51. a. 100% **b.** about 68.5% **c.** about 30.8% **d.** about 20%
53. 11.3; About 11.3% of 30-year-olds have some coronary heart disease.
55. a. about 1429 people **b.** about 24,546 people
c. The number of people cannot exceed the population.; The asymptote indicates that the number of ill people will not exceed 30,000, the population of the town.

67. a. $f(t) = 10,000\left(1 + \frac{0.05}{4}\right)^{4t}$; $f(t) = 10,000\left(1 + \frac{0.045}{12}\right)^{12t}$

b. 5% compounded quarterly; the bank that pays 5% interest compounded quarterly

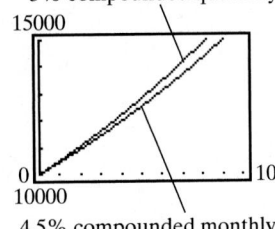

15000

0   10
10000

4.5% compounded monthly

69. $\dfrac{11}{(x - 3)(x - 4)}$ **70.** -1 **71.** -2 and 5, or $\{-2, 5\}$

Section 12.2

Check Point Exercises

1. a. $5x^2 + 1$ **b.** $25x^2 + 60x + 35$ **2.** $f(g(x)) = 7\left(\frac{x}{7}\right) = x$; $g(f(x)) = \frac{7x}{7} = x$

3. $f(g(x)) = 4\left(\frac{x + 7}{4}\right) - 7 = x + 7 - 7 = x$; $g(f(x)) = \frac{(4x - 7) + 7}{4} = \frac{4x}{4} = x$ **4.** $f^{-1}(x) = \frac{x - 7}{2}$ **5.** $f^{-1}(x) = \sqrt[3]{\frac{x + 1}{4}}$

6. (b) and (c) **7.**

y
5

(0, 0)
(−1, 1)
(−2, 1) f^{-1}
f   (1, −1) 5 x
(1, −2)
$y = x$

Exercise Set 12.2

1. a. $2x + 14$ **b.** $2x + 7$ **c.** 18 **3. a.** $2x + 5$ **b.** $2x + 9$ **c.** 9 **5. a.** $20x^2 - 11$ **b.** $80x^2 - 120x + 43$ **c.** 69
7. a. $x^4 - 4x^2 + 6$ **b.** $x^4 + 4x^2 + 2$ **c.** 6 **9. a.** $\sqrt{x - 1}$ **b.** $\sqrt{x} - 1$ **c.** 1 **11. a.** x **b.** x **c.** 2
13. a. x **b.** x **c.** 2 **15.** $f(g(x)) = x$; $g(f(x)) = x$; inverses **17.** $f(g(x)) = x$; $g(f(x)) = x$; inverses
19. $f(g(x)) = \dfrac{5x - 4}{9}$; $g(f(x)) = \dfrac{5x - 56}{9}$; not inverses **21.** $f(g(x)) = x$; $g(f(x)) = x$; inverses **23.** $f(g(x)) = x$; $g(f(x)) = x$;
inverses **25. a.** $f^{-1}(x) = x - 3$ **b.** $f(f^{-1}(x)) = (x - 3) + 3 = x$ and $f^{-1}(f(x)) = (x + 3) - 3 = x$
27. a. $f^{-1}(x) = \dfrac{x}{2}$ **b.** $f(f^{-1}(x)) = 2\left(\dfrac{x}{2}\right) = x$ and $f^{-1}(f(x)) = \dfrac{2x}{2} = x$

29. a. $f^{-1}(x) = \dfrac{x-3}{2}$ **b.** $f(f^{-1}(x)) = 2\left(\dfrac{x-3}{2}\right) + 3 = x$ and $f^{-1}(f(x)) = \dfrac{(2x+3)-3}{2} = x$

31. a. $f^{-1}(x) = \sqrt[3]{x-2}$ **b.** $f(f^{-1}(x)) = (\sqrt[3]{x-2})^3 + 2 = x$ and $f^{-1}(f(x)) = \sqrt[3]{(x^3+2)-2} = x$

33. a. $f^{-1}(x) = \sqrt[3]{x} - 2$ **b.** $f(f^{-1}(x)) = ((\sqrt[3]{x} - 2) + 2)^3 = x$ and $f^{-1}(f(x)) = \sqrt[3]{(x+2)^3} - 2 = x$

35. a. $f^{-1}(x) = \dfrac{1}{x}$ **b.** $f(f^{-1}(x)) = \dfrac{1}{\frac{1}{x}} = x$ and $f^{-1}(f(x)) = \dfrac{1}{\frac{1}{x}} = x$

37. a. $f^{-1}(x) = x^2, x \geq 0$ **b.** $f(f^{-1}(x)) = \sqrt{x^2} = x$ and $f^{-1}(f(x)) = (\sqrt{x})^2 = x$

39. a. $f^{-1}(x) = \sqrt{x-1}$ **b.** $f(f^{-1}(x)) = (\sqrt{x-1})^2 + 1 = x$ and $f^{-1}(f(x)) = \sqrt{(x^2+1)-1} = x$

41. a. $f^{-1}(x) = \dfrac{3x+1}{x-2}$ **b.** $f(f^{-1}(x)) = \dfrac{2\left(\dfrac{3x+1}{x-2}\right)+1}{\left(\dfrac{3x+1}{x-2}\right)-3} = x$ and $f^{-1}(f(x)) = \dfrac{3\left(\dfrac{2x+1}{x-3}\right)+1}{\left(\dfrac{2x+1}{x-3}\right)-2} = x$

43. a. $f^{-1}(x) = (x-3)^3 + 4$ **b.** $f(f^{-1}(x)) = \sqrt[3]{((x-3)^3+4)-4} + 3 = x$ and $f^{-1}(f(x)) = ((\sqrt[3]{x-4}+3)-3)^3 + 4 = x$

45. no inverse **47.** no inverse **49.** inverse function

51.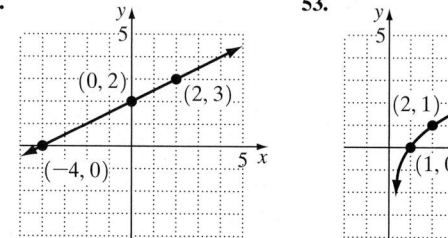

53. (see graph)

55. a. f represents the price after a \$400 discount, and g represents the price after a 25% discount (75% of the regular price).
b. $0.75x - 400; f \circ g$ represents an additional \$400 discount on a price that has already been reduced by 25%.
c. $0.75(x - 400) = 0.75x - 300; g \circ f$ represents an additional 25% discount on a price that has already been reduced \$400.
d. $f \circ g; 0.75x - 400 < 0.75x - 300;$ so $f \circ g$ represents the lower price after the two discounts.
e. $f^{-1}(x) = x + 400; f^{-1}$ represents the regular price, since the value of x here is the price after a \$400 discount.

57. a. No horizontal line intersects the graph of f in more than one point. **b.** $f^{-1}(0.25)$, or 15, represents the number of people who would have to be in the room so that the probability of two sharing a birthday would be 0.25; $f^{-1}(0.5)$, or 23, represents the number of people so that the probability would be 0.5; $f^{-1}(0.7)$, or 30, represents the number of people so that the probability would be 0.7.
59. no; There are at least two years for which the average age at which women married is the same. **61–67.** Answers will vary.
69. Answers will vary; an example is: $f(x) = \sqrt{x+5}$ and $g(x) = 3x^2$.

71. $f(f(x)) = \dfrac{3\left(\dfrac{3x-2}{5x-3}\right)-2}{5\left(\dfrac{3x-2}{5x-3}\right)-3} = \dfrac{3(3x-2)-2(5x-3)}{5(3x-2)-3(5x-3)} = \dfrac{9x-6-10x+6}{15x-10-15x+9} = \dfrac{-x}{-1} = x$

73.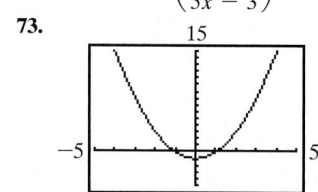

no inverse

75.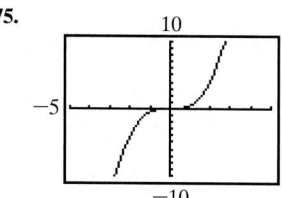

inverse

77.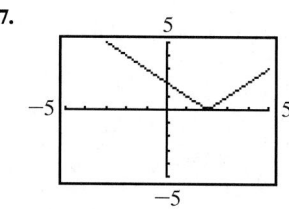

no inverse

79.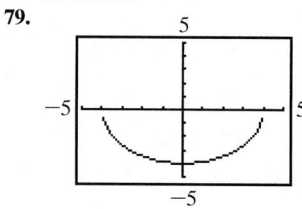

no inverse

81.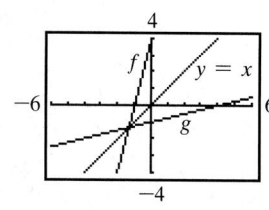

inverses

83.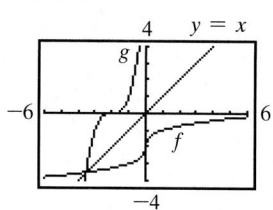

inverses

84. 5×10^8 **85.**

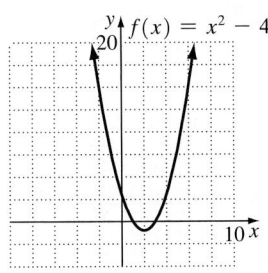

86. 5 or $\{5\}$

Section 12.3

Check Point Exercises

1. a. $7^3 = x$ **b.** $b^2 = 25$ **c.** $4^y = 26$ **2. a.** $5 = \log_2 x$ **b.** $3 = \log_b 27$ **c.** $y = \log_e 33$ **3.** 2 **b.** 1 **b.** $\frac{1}{2}$
4. a. 1 **b.** 0 **5. a.** 8 **b.** 17
6.

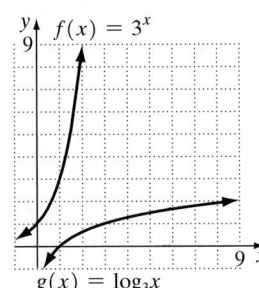

7. $\{x | x > 5\}$ or $(5, \infty)$
8. approximately 80%
9. 4
10. a. $\{x | x < 4\}$ or $(-\infty, 4)$
  b. $\{x | x \text{ is a real number and } x \neq 0\}$ or $(-\infty, 0) \cup (0, \infty)$
11. a. $25x$
  b. $\sqrt{x}$
12. approximately 4.6 feet per second

Exercise Set 12.3

1. $2^4 = 16$ **3.** $3^2 = x$ **5.** $b^5 = 32$ **7.** $6^y = 216$ **9.** $\log_2 8 = 3$ **11.** $\log_2 \frac{1}{16} = -4$ **13.** $\log_8 2 = \frac{1}{3}$ **15.** $\log_{13} x = 2$

17. $\log_b 1000 = 3$ **19.** $\log_7 200 = y$ **21.** 2 **23.** 6 **25.** $\frac{1}{2}$ **27.** -3 **29.** $\frac{1}{2}$ **31.** 1 **33.** 0 **35.** 7 **37.** 19

39.

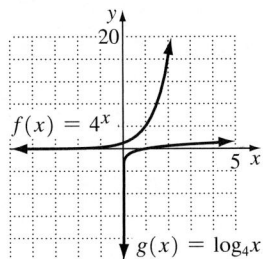

41.

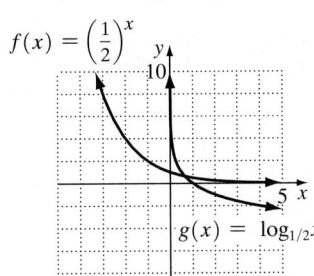

43. $\{x | x > -4\}$ or $(-4, \infty)$
45. $\{x | x < 2\}$ or $(-\infty, 2)$
47. $\{x | x \text{ is a real number and } x \neq 2\}$ or $(-\infty, 2) \cup (2, \infty)$
49. 2 **51.** 7 **53.** 33 **55.** 0
57. 6 **59.** -6 **61.** 125
63. $9x$ **65.** $5x^2$ **67.** $\sqrt{x}$
69. approximately 95.4%
71. approximately $5.7 billion
73. approximately 188 decibels; yes

75. a. 88
  b. 71.5; 63.9; 58.8; 55.0; 52.0; 49.5
  c.

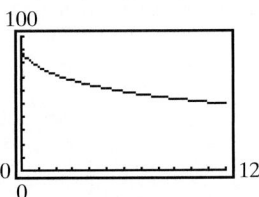

The students remembered less of the material over time.

77–83. Answers will vary.
85. d **87.** 8 or $\{8\}$
89.

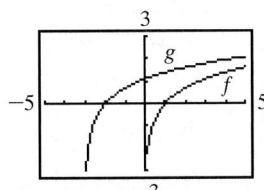

The graph of g is the graph of f shifted 3 units to the left.

91.

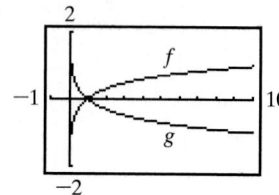

93.

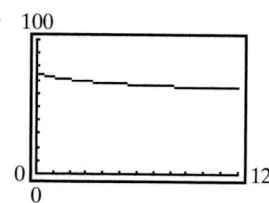

95.

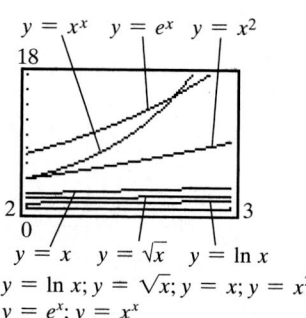

The graph of g is the reflection of the graph of f across the x-axis.

9 months

$y = \ln x; y = \sqrt{x}; y = x; y = x^2;$
$y = e^x; y = x^x$

96. $(-2, 3)$ or $\{(-2, 3)\}$ **97.** $2(3x - y)(x - y)$ **98.** $\{x | x \le -7 \text{ or } x \ge -2\}$ or $(-\infty, -7] \cup [-2, \infty)$

Section 12.4

Check Point Exercises

1. a. $\log_6 7 + \log_6 11$ **b.** $2 + \log x$ **2. a.** $\log_8 23 - \log_8 x$ **b.** $5 - \ln 11$ **3. a.** $9 \log_6 8$ **b.** $\frac{1}{3} \ln x$

4. a. $4 \log_b x + \frac{1}{3} \log_b y$ **b.** $\frac{1}{2} \log_5 x - 2 - 3 \log_5 y$ **5. a.** $\log 100 = 2$ **b.** $\log \frac{7x + 6}{x}$

6. a. $\ln x^2 \sqrt[3]{x + 5}$ **b.** $\log \frac{(x - 3)^2}{x}$ **7.** $\frac{\log 2506}{\log 7} \approx 4.02$ **8.** $\frac{\ln 2506}{\ln 7} \approx 4.02$

Exercise Set 12.4

1. $\log_5 7 + \log_5 3$ **3.** $1 + \log_7 x$ **5.** $3 + \log x$ **7.** $1 - \log_7 x$ **9.** $\log x - 2$ **11.** $3 - \log_4 y$ **13.** $2 - \ln 5$

15. $3 \log_b x$ **17.** $-6 \log N$ **19.** $\frac{1}{5} \ln x$ **21.** $2 \log_b x + \log_b y$ **23.** $\frac{1}{2} \log_4 x - 3$ **25.** $2 - \frac{1}{2} \log_6(x + 1)$

27. $2 \log_b x - \log_b y - 2 \log_b z$ **29.** $1 + \frac{1}{2} \log x$ **31.** $\frac{1}{3} \log x - \frac{1}{3} \log y$ **33.** $\log 10 = 1$ **35.** $\ln 7x$ **37.** $\log_2 32 = 5$

39. $\log \frac{2x + 5}{x}$ **41.** $\log xy^3$ **43.** $\ln y \sqrt{x}$ **45.** $\log_b x^2 y^3$ **47.** $\ln \frac{x^5}{y^2}$ **49.** $\ln \frac{x^3}{\sqrt[3]{y}}$ **51.** $\ln \frac{(x + 6)^4}{x^3}$ **53.** 1.5937

55. 1.6944 **57.** -1.2304 **59.** 3.6193 **61. a.** $D = 10 \log \frac{I}{I_0}$ **b.** 20 decibels   **63–69.** Answers will vary.   **71.** d   **73.** $\frac{2A}{B}$

75. a. & b. $y = 2 + \log_3 x$ $y = \log_3(x + 2)$; $y = \log_3 x$ is shifted up 2 units to obtain $y = 2 + \log_3 x$, $y = \log_3 x$ is shifted to the left 2 units to obtain $y = \log_3(x + 2)$, and $y = \log_3 x$ is reflected across the x-axis to obtain $y = -\log_3 x$.

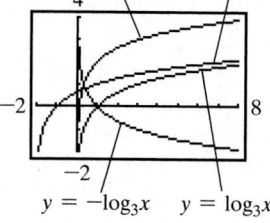

$y = -\log_3 x$ $y = \log_3 x$

77.

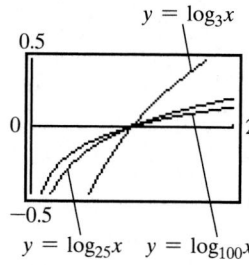

$y = \log_3 x$

$y = \log_{25} x$ $y = \log_{100} x$

a. $y = \log_{100} x$ is on the top, and $y = \log_3 x$ is on the bottom.
b. $y = \log_3 x$ is on the top, and $y = \log_{100} x$ is on the bottom.
c. If $y = \log_b x$ is graphed for two different values of b, the graph of the one with the larger base will be on top in the interval $(0, 1)$ and the one with the smaller base will be on top in the interval $(0, \infty)$.

83.

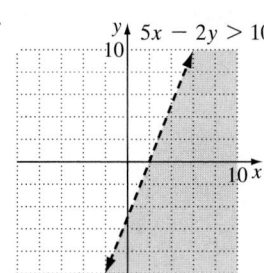

$5x - 2y > 10$

84. $\{x | x < 1\}$ or $(-\infty, 1)$ **85.** $2y\sqrt[3]{xy^2}$

Section 12.5

Check Point Exercises

1. a. 3 or $\{3\}$ **b.** $\frac{5}{2}$ or $\left\{\frac{5}{2}\right\}$ **2.** $\frac{\ln 134}{5} \approx 3.04$ or $\left\{\frac{\ln 134}{5} \approx 3.04\right\}$ **3.** $\frac{\ln 9}{2} = \ln 3 \approx 1.10$ or $\{\ln 3 \approx 1.10\}$ **4.** 12 or $\{12\}$

5. 5 or $\{5\}$ **6.** $\frac{1}{24}$ or $\left\{\frac{1}{24}\right\}$ **7.** $\frac{e^2}{3}$ or $\left\{\frac{e^2}{3}\right\}$ **8.** blood alcohol concentration of 0.01   **9.** 16.2 years   **10.** 2149

Exercise Set 12.5

1. 6 or $\{6\}$ **3.** 3 or $\{3\}$ **5.** 3 or $\{3\}$ **7.** 2 or $\{2\}$ **9.** $\frac{3}{5}$ or $\left\{\frac{3}{5}\right\}$ **11.** $\frac{3}{2}$ or $\left\{\frac{3}{2}\right\}$ **13.** $\frac{\ln 3.91}{\ln 10} \approx 0.59$ or $\left\{\frac{\ln 3.91}{\ln 10} \approx 0.59\right\}$

15. $\ln 5.7 \approx 1.74$ or $\{\ln 5.7 \approx 1.74\}$ **17.** $\frac{\ln 17}{\ln 5} \approx 1.76$ or $\left\{\frac{\ln 17}{\ln 5} \approx 1.76\right\}$ **19.** $\ln 5 \approx 1.61$ or $\{\ln 5 \approx 1.61\}$

21. $\frac{\ln 659}{5} \approx 1.30$ or $\left\{\frac{\ln 659}{5} \approx 1.30\right\}$ **23.** $\frac{\ln 13}{0.7} \approx 3.66$ or $\left\{\frac{\ln 13}{0.7} \approx 3.66\right\}$ **25.** $\frac{\ln 3}{0.055} \approx 19.97$ or $\left\{\frac{\ln 3}{0.055} \approx 19.97\right\}$

27. $\frac{\ln 30}{\ln 1.4} \approx 10.11$ or $\left\{\frac{\ln 30}{\ln 1.4} \approx 10.11\right\}$ **29.** $\frac{1 - \ln 793}{5} \approx -1.14$ or $\left\{\frac{1 - \ln 793}{5} \approx -1.14\right\}$

31. $\frac{\ln 410}{\ln 7} - 2 \approx 1.09$ or $\left\{\frac{\ln 410}{\ln 7} - 2 \approx 1.09\right\}$ **33.** $\frac{\ln 2}{\ln 5 - \ln 2} \approx 0.76$ or $\left\{\frac{\ln 2}{\ln 5 - \ln 2} \approx 0.76\right\}$ **35.** 81 or $\{81\}$

37. $\frac{1}{16}$ or $\left\{\frac{1}{16}\right\}$ **39.** 3 or $\{3\}$ **41.** 100 or $\{100\}$ **43.** $\frac{1}{e^3}$ or $\left\{\frac{1}{e^3}\right\}$ **45.** 59 or $\{59\}$ **47.** $\frac{109}{27}$ or $\left\{\frac{109}{27}\right\}$ **49.** $\frac{62}{3}$ or $\left\{\frac{62}{3}\right\}$

51. $\frac{5}{4}$ or $\left\{\frac{5}{4}\right\}$ **53.** 6 or $\{6\}$ **55.** 6 or $\{6\}$ **57.** no solution or $\varnothing$ **59.** $e^2 \approx 7.39$ or $\{e^2 \approx 7.39\}$

61. $\frac{e^4}{2} \approx 27.30$ or $\left\{\frac{e^4}{2} \approx 27.30\right\}$ **63.** $e^{-1/2} \approx 0.61$ or $\{e^{-1/2} \approx 0.61\}$ **65.** $e^2 - 3 \approx 4.39$ or $\{e^2 - 3 \approx 4.39\}$

67. a blood alcohol concentration of 0.22   **69. a.** 18.2 million   **b.** 2010   **71.** 8.2   **73.** 16.8%   **75.** 8.7
77. 15.7%   **79.** 1995   **81.** about 2.8 days; (2.8, 50)   **83.** $10^{-2.4}$ or 0.004 mole per liter   **85–91.** Answers will vary.
93. about 36 years   **95.** 10^{-2} and $10^{3/2}$, or $\{10^{-2}, 10^{3/2}\}$

97.

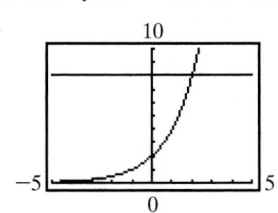

; 2 or $\{2\}$

99.

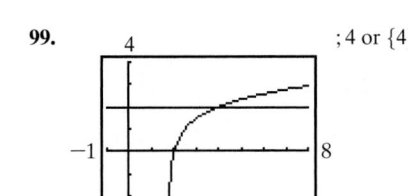

; 4 or $\{4\}$

101.

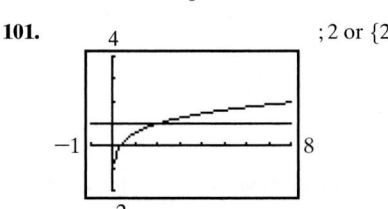

; 2 or $\{2\}$

103.

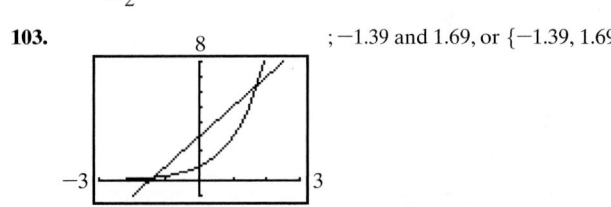

; -1.39 and 1.69, or $\{-1.39, 1.69\}$

105.

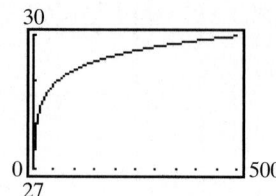

The barometric air pressure increases as the distance from the eye increases.

107.

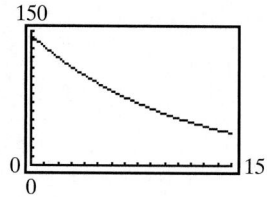

The runner's pulse will be 70 beats per minute after about 7.9 minutes

109. 5 or {5}   **110.** −12 or {−12}   **111.** $x^3 - y^3$

Section 12.6

Check Point Exercises

1. a. $A = 491e^{0.027t}$ **b.** 2006   **2. a.** $A = A_0 e^{-0.0248t}$ **b.** about 72.2 years

Exercise Set 12.6

1. 208 million   **3.** 2016   **5.** 2.6% per year   **7.** 2014   **9.** $140 thousand or $140,000   **11.** about 2005
13. $k = 0.175$; $A = 200e^{0.175t}$; 17.5% per year   **15.** approximately 8 grams   **17.** 8 grams after 10 seconds; 4 grams after 20 seconds;
2 grams after 30 seconds; 1 gram after 40 seconds; 0.5 grams after 50 seconds   **19.** approximately 15,679 years old

21. a. $\dfrac{1}{2} = e^{1.31k}$ yields $k = \dfrac{\ln \frac{1}{2}}{1.31} \approx -0.52912$.   **b.** about 0.1069 billion or 106,900,000 years old

23. $2A_0 = A_0 e^{kt}$
$2 = e^{kt}$
$\ln 2 = \ln e^{kt}$
$\ln 2 = kt$
$\dfrac{\ln 2}{k} = t$

25. about 63 years
27–35. Answers will vary.
37. $y = 1.74(1.037)^x$; $r \approx 0.97$; Since r is close to 1, the model fits the data well.
39. $y = 1.547 + 0.112x$; $r \approx 0.99$; Since r is close to 1, the model fits the data well.
41. $y = 1.547 + 0.112x$; 2022; Answers will vary.

43. $\dfrac{x + 5}{x}$ **44.** −27 and 1, or {−27, 1}   **45.** $17\sqrt{2}$

Review Exercises

1.

x	$f(x)$; d
−2	$\dfrac{1}{16}$
−1	$\dfrac{1}{4}$
0	1
1	4
2	16

2.

x	$f(x)$; a
−2	16
−1	4
0	1
1	$\dfrac{1}{4}$
2	$\dfrac{1}{16}$

3.

x	$f(x)$; b
−2	−16
−1	−4
0	−1
1	$-\dfrac{1}{4}$
2	$-\dfrac{1}{16}$

4.

x	$f(x)$; c
−2	−13
−1	−1
0	2
1	$\dfrac{11}{4}$
2	$\dfrac{47}{16}$

5.

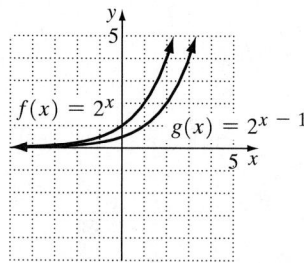

The graph of g is the graph of f shifted to the right 1 unit.

6.

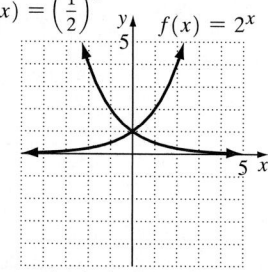

The graph of g is the reflection of f across the y-axis.

7.

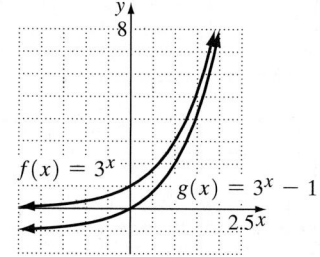

The graph of g is the graph of f shifted down 1 unit.

8.

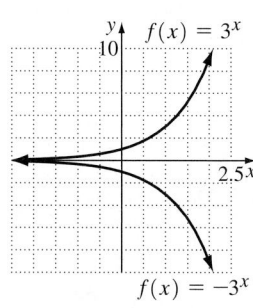

$f(x) = 3^x$

$f(x) = -3^x$

The graph of g is the reflection of f across the x-axis.

9. 5.5% compounded semiannually

10. 7% compounded monthly

11. a. 200°F
b. 119°F
c. 70°F; The temperature of the room is 70°F.

12. a. $16x^2 - 8x + 4$ **b.** $4x^2 + 11$ **c.** 124

13. a. $\sqrt{x+1}$ **b.** $\sqrt{x} + 1$ **c.** 2

14. $f(g(x)) = x - \dfrac{7}{10}; g(f(x)) = x - \dfrac{7}{6};$ not inverses

15. $f(g(x)) = x; g(f(x)) = x;$ inverses

16. a. $f^{-1}(x) = \dfrac{x+3}{4}$ **b.** $f(f^{-1}(x)) = 4\left(\dfrac{x+3}{4}\right) - 3 = x$ and $f^{-1}(f(x)) = \dfrac{(4x-3)+3}{4} = x$

17. a. $f^{-1}(x) = x^2 - 2, x \geq 0$ **b.** $f(f^{-1}(x)) = \sqrt{(x^2-2)+2} = x$ and $f^{-1}(f(x)) = (\sqrt{x+2})^2 - 2 = x$

18. a. $f^{-1}(x) = \dfrac{\sqrt[3]{x-1}}{2}$ **b.** $f(f^{-1}(x)) = 8\left(\dfrac{\sqrt[3]{x-1}}{2}\right)^3 + 1 = x$ and $f^{-1}(f(x)) = \dfrac{\sqrt[3]{(8x^3+1)-1}}{2} = x$

19. inverse **20.** no inverse **21.** inverse **22.** no inverse

23.

(4, 2)

(0, 0)

(−1, −3)

24. $49^{1/2} = 7$ **25.** $4^3 = x$ **26.** $3^y = 81$ **27.** $\log_6 216 = 3$ **28.** $\log_b 625 = 4$

29. $\log_{13} 874 = y$ **30.** 3 **31.** −2 **32.** −9 is not in the domain of $y = \log_3 x$.

33. $\dfrac{1}{2}$ **34.** 1 **35.** 8 **36.** 5 **37.** 0

38.

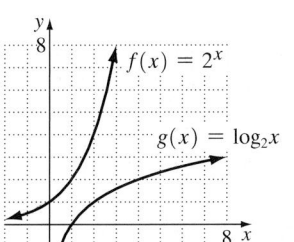

$f(x) = 2^x$

$g(x) = \log_2 x$

39.

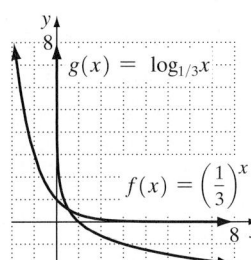

$g(x) = \log_{1/3} x$

$f(x) = \left(\dfrac{1}{3}\right)^x$

40. $\{x | x > -5\}$ or $(-5, \infty)$
41. $\{x | x < 3\}$ or $(-\infty, 3)$
42. $\{x | x < 1 \text{ or } x > 1\}$ or $(-\infty, 1) \cup (1, \infty)$
43. $6x$
44. $\sqrt{x}$ **45.** $4x^2$ **46.** 3

47. a. 76
b. 67.4 after 2 months; 63.4 after 4 months; 60.8 after 6 months; 58.8 after 8 months; 55.9 after one year
c.

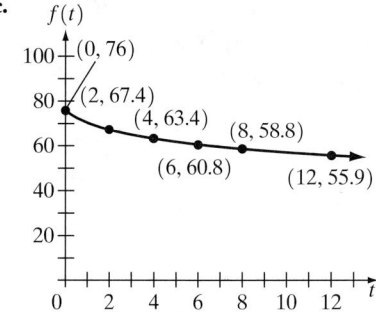

As time increases the amount of material retained by the students decreases.

48. about 9 weeks **49.** $2 + 3\log_6 x$ **50.** $\dfrac{1}{2}\log_4 x - 3$

51. $\log_2 x + 2\log_2 y - 6$ **52.** $\dfrac{1}{3}\ln x - \dfrac{1}{3}$ **53.** $\log_b 21$

54. $\log \dfrac{3}{x^3}$ **55.** $\ln(x^3 y^4)$ **56.** $\ln \dfrac{\sqrt{x}}{y}$ **57.** 6.2448

58. −0.1063 **59.** 2 or $\{2\}$ **60.** $\dfrac{2}{3}$ or $\left\{\dfrac{2}{3}\right\}$ **61.** $-\dfrac{3}{2}$ or $\left\{-\dfrac{3}{2}\right\}$

62. $\dfrac{\ln 12{,}143}{\ln 8} \approx 4.52$ or $\left\{\dfrac{\ln 12{,}143}{\ln 8} \approx 4.52\right\}$

63. $\dfrac{\ln 141}{5} \approx 0.99$ or $\left\{\dfrac{\ln 141}{5} \approx 0.99\right\}$

64. $\dfrac{\ln 3}{0.045} \approx 24.41$ or $\left\{\dfrac{\ln 3}{0.045} \approx 24.41\right\}$ **65.** $\dfrac{1}{125}$ or $\left\{\dfrac{1}{125}\right\}$

66. 100 or $\{100\}$ **67.** 23 or $\{23\}$ **68.** 5 or $\{5\}$
69. no solution or $\varnothing$

70. $\dfrac{1}{e}$ or $\left\{\dfrac{1}{e}\right\}$ **71.** $\dfrac{e^3}{2}$ or $\left\{\dfrac{e^3}{2}\right\}$ **72.** 2042 **73.** approximately 2086 **74.** 2005 **75.** 7.3 years **76.** 14.6 years **77.** 22%

78. a. $k = 0.041$ **b.** about 40.7 million **c.** 2010 **79.** 15,679 years old **80.** Answers will vary.

Chapter 12 Test

1.

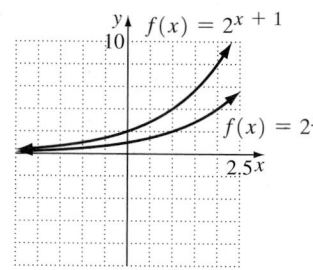

2. 6.5% compounded semiannually; $221

3. $(f \circ g)(x) = 9x^2 - 3x; (g \circ f)(x) = 3x^2 + 3x - 1$

4. $f^{-1}(x) = \dfrac{x + 7}{5}$

5. a. No horizontal line intersects the graph of f in more than one point. **b.** 2000
c. $f^{-1}(2000)$ represents the income, $80 thousand, of a family that gives $2000 to charity.

6. $5^3 = 125$ **7.** $\log_{36} 6 = \dfrac{1}{2}$

8.

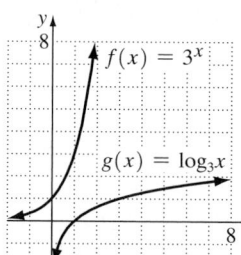

9. $5x$ **10.** 1 **11.** 0 **12.** $\{x | x > 7\}$ or $(7, \infty)$ **13.** 120 decibels **14.** $3 + 5 \log_4 x$

15. $\dfrac{1}{3} \log_3 x - 4$ **16.** $\log x^6 y^2$ **17.** $\ln \dfrac{7}{x^3}$ **18.** 1.5741 **19.** 6 or $\{6\}$ **20.** $\dfrac{\ln 1.4}{\ln 5}$ or $\left\{ \dfrac{\ln 1.4}{\ln 5} \right\}$

21. $\dfrac{\ln 4}{0.005}$ or $\left\{ \dfrac{\ln 4}{0.005} \right\}$ **22.** 5 or $\{5\}$ **23.** $\dfrac{217}{4}$ or $\left\{ \dfrac{217}{4} \right\}$ **24.** 5 or $\{5\}$ **25.** $\dfrac{e^4}{3}$ or $\left\{ \dfrac{e^4}{3} \right\}$

26. a. 89.18% **b.** decreasing; The growth rate, -0.004, is negative. **c.** 1996
27. 13.9 years **28.** about 6.9% **29.** $A = 484e^{0.005t}$ **30.** 24,758 years ago

Cumulative Review Exercises (Chapters 1–12)

1. 22 or $\{22\}$ **2.** $5 \pm 7i$ or $\{5 \pm 7i\}$ **3.** $\{x | x < -3$ or $x > 2\}$ or $(-\infty, -3) \cup (2, \infty)$ **4.** -2 or $\{-2\}$ **5.** no solution or $\varnothing$

6. $\left\{ x \middle| x < \dfrac{2}{3} \right\}$ or $\left(-\infty, \dfrac{2}{3} \right)$ **7.** $(4, 0, -5)$ or $\{(4, 0, -5)\}$ **8.** 9 or $\{9\}$

9.

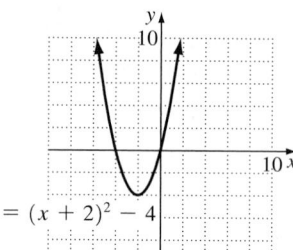

10.

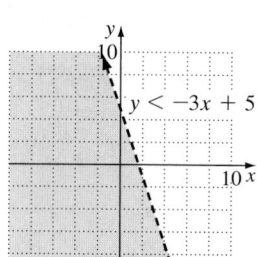

11.

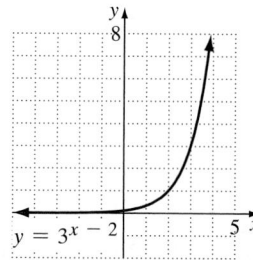

12. $\dfrac{2x^2 + 5x - 2}{(x - 5)(x + 2)}$ **13.** $-\dfrac{x + 1}{x - 1}$ **14.** $2\sqrt{5} + 2\sqrt{2}$ **15.** $12\sqrt{5}$ **16.** $\dfrac{5\sqrt[3]{4xy^2}}{2xy}$ **17.** $(x + y)(5a - 4b)$ **18.** $\log \dfrac{x^5}{\sqrt{y}}$

19. 3.97×10^{-3} **20.** $\{x | x$ is a real number and $x \neq -5$ and $x \neq 3\}$ or $(-\infty, -5) \cup (-5, 3) \cup (3, \infty)$ **21.** $\{x | x \geq 3\}$ or $[3, \infty)$

22. $\{x | x < 1\}$ or $(-\infty, 1)$ **23.** 8 ft by 3 ft **24.** $18,500 **25.** $k = \dfrac{1}{2 \ln 4} \approx 0.3607$; about 0.298 or $\dfrac{298}{1000}$

CHAPTER 13

Section 13.1

Check Point Exercises

1. 5 units **2.** $\left(4, -\dfrac{1}{2} \right)$ **3.** $x^2 + y^2 = 16$ **4.** $(x - 5)^2 + (y + 6)^2 = 100$

5. center: $(-3, 1)$; radius: 2 units

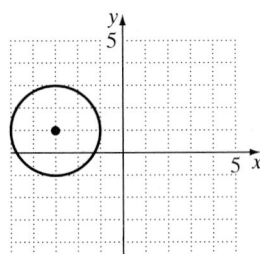

6. $(x + 2)^2 + (y - 2)^2 = 9$; center: $(-2, 2)$; $r = 3$

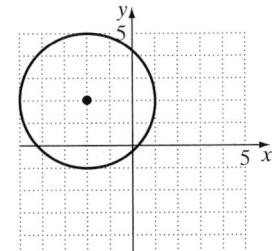

Exercise Set 13.1

1. 13 units   **3.** $2\sqrt{2}$ or 2.83 units   **5.** 5 units   **7.** $\sqrt{29}$ or 5.39 units   **9.** $4\sqrt{2}$ or 5.66 units   **11.** $2\sqrt{5}$ or 4.47 units

13. $2\sqrt{2}$ or 2.83 units   **15.** $\sqrt{93}$ or 9.64 units   **17.** $\sqrt{5}$ or 2.24 units   **19.** $(4, 6)$ **21.** $(-4, -5)$ **23.** $\left(\dfrac{3}{2}, -6\right)$

25. $(-3, -2)$ **27.** $(1, 5\sqrt{5})$ **29.** $(2\sqrt{2}, 0)$ **31.** $x^2 + y^2 = 49$ **33.** $(x - 3)^2 + (y - 2)^2 = 25$

35. $(x + 1)^2 + (y - 4)^2 = 4$ **37.** $(x + 3)^2 + (y + 1)^2 = 3$ **39.** $(x + 4)^2 + y^2 = 100$

41. center: $(0, 0)$; $r = 4$ **43.** center: $(3, 1)$; $r = 6$ **45.** center: $(-3, 2)$; $r = 2$ **47.** center: $(-2, -2)$; $r = 2$

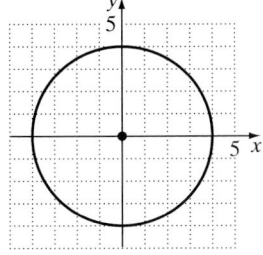
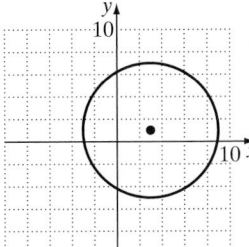
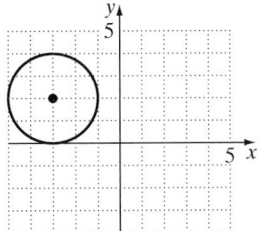
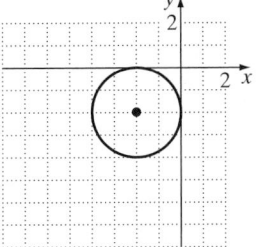

49. $(x + 3)^2 + (y + 1)^2 = 4$; center: $(-3, -1)$; $r = 2$

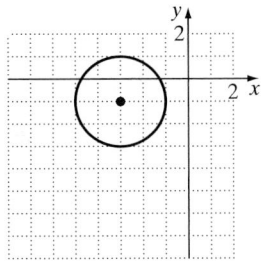

51. $(x - 5)^2 + (y - 3)^2 = 64$; center: $(5, 3)$; $r = 8$

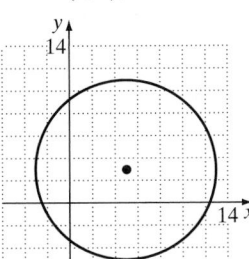

53. $(x + 4)^2 + (y - 1)^2 = 25$; center: $(-4, 1)$; $r = 5$

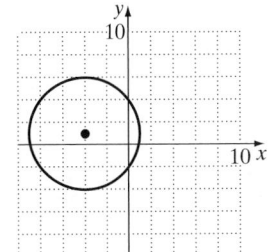

55. $(x - 1)^2 + y^2 = 16$; center: $(1, 0)$; $r = 4$

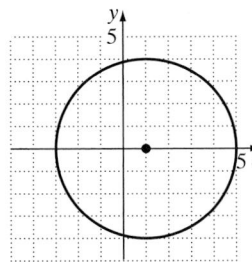

57. 0.5 hour; 30 minutes

59. $x^2 + (y - 82)^2 = 4624$ **61–67.** Answers will vary.

69. The distance from A to B is $2\sqrt{2}$ and the distance from B to C is $3\sqrt{2}$. The distance from A to C is $5\sqrt{2}$, which is equal to $2\sqrt{2} + 3\sqrt{2}$.

71. $(x - 3)^2 + (y + 5)^2 = 61$; $x^2 + y^2 - 6x + 10x - 27 = 0$

73. 11π square units, or approximately 35 square units

75.

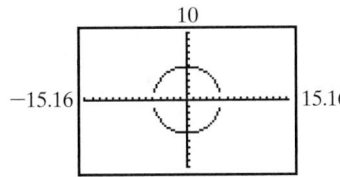

77.

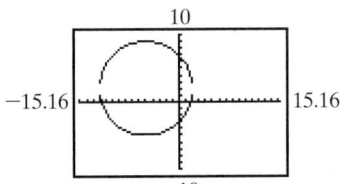

78. $f(g(x)) = 9x^2 + 24x + 14$; $g(f(x)) = 3x^2 - 2$ **79.** 4 or $\{4\}$ **80.** $\left\{x \middle| -\dfrac{5}{2} < x < \dfrac{15}{2}\right\}$ or $\left(-\dfrac{5}{2}, \dfrac{15}{2}\right)$

Section 13.2

Check Point Exercises

1.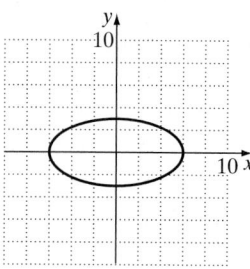

2.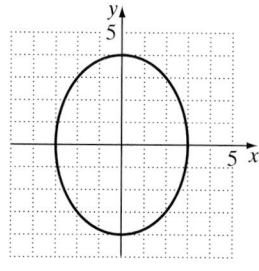

3.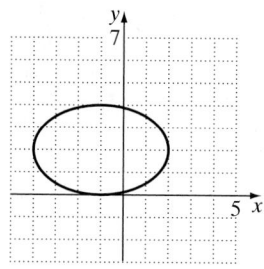

4. Yes, the height of the archway 6 feet from the center is approximately 9.54 feet.

Exercise Set 13.2

1.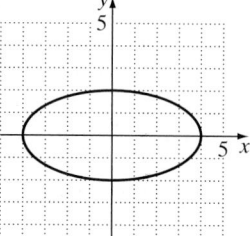

3.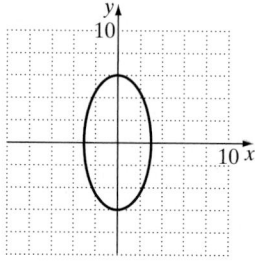

5.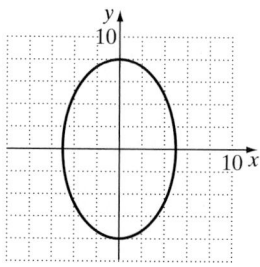

7.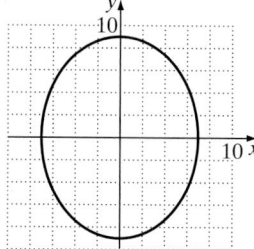

9.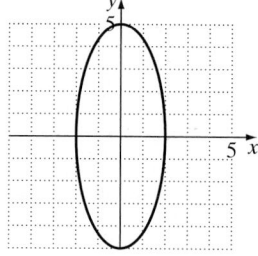

11.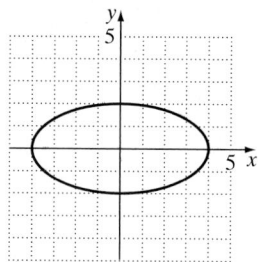

13.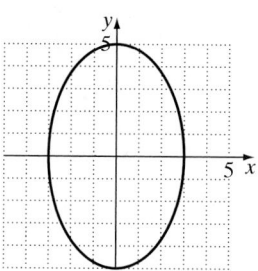

15.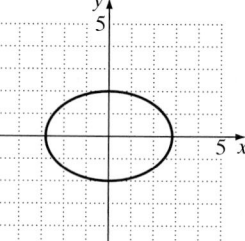

17. $\dfrac{x^2}{4} + \dfrac{y^2}{1} = 1$ or $\dfrac{x^2}{4} + y^2 = 1$

19. $\dfrac{x^2}{1} + \dfrac{y^2}{4} = 1$ or $x^2 + \dfrac{y^2}{4} = 1$

21.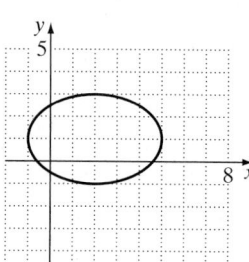

23.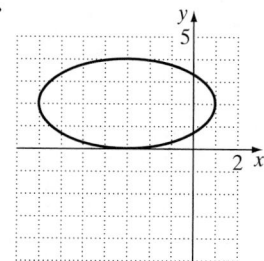

25.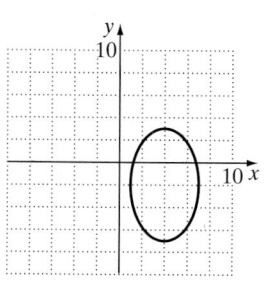

27.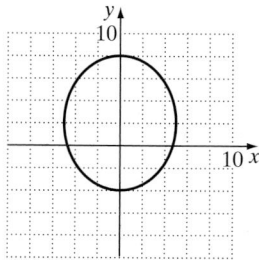

29.
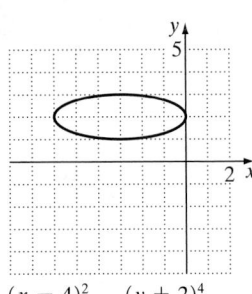

31.
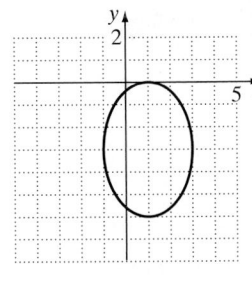

33. Yes, the height of the archway 4 feet from the center is approximately 9.64 feet.

35. a. $\dfrac{x^2}{48^2} + \dfrac{y^2}{23^2} = 1$ or $\dfrac{x^2}{2304} + \dfrac{y^2}{529} = 1$

b. 42 feet

37–41. Answers will vary.

43. $\dfrac{x^2}{\frac{36}{5}} + \dfrac{y^2}{36} = 1$

45. $\dfrac{(x-4)^2}{9} + \dfrac{(y+2)^4}{4} = 1$

47. small circle: $x^2 + y^2 = 9$; large circle: $x^2 + y^2 = 25$

51. $(x+2)(x+2)(x-2)$

52. $2xy^2 \sqrt[3]{5xy}$

53. -1 or $\{-1\}$

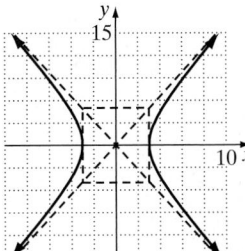

Section 13.3

Check Point Exercises

1. a. $(-5, 0)$ and $(5, 0)$ **b.** $(0, -5)$ and $(0, 5)$

2.
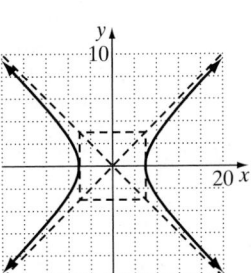

3.
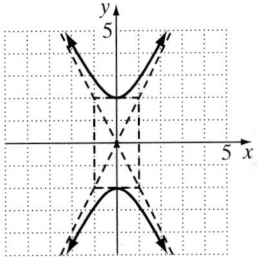

Exercise Set 13.3

1. $(-2, 0)$ and $(2, 0)$; b   **3.** $(0, -2)$ and $(0, 2)$; a

5.
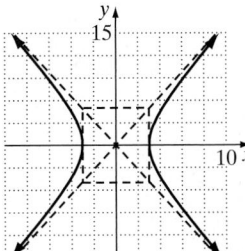

7.
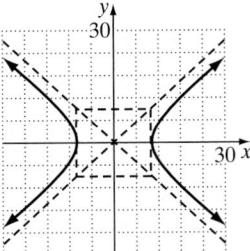

9.
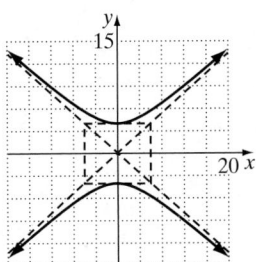

11.
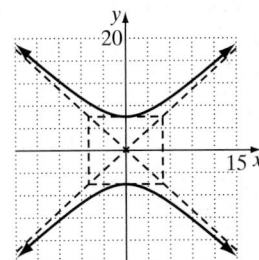

13.
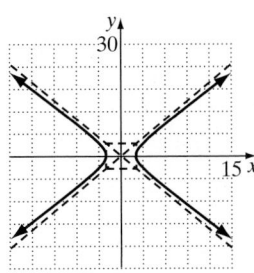

15.
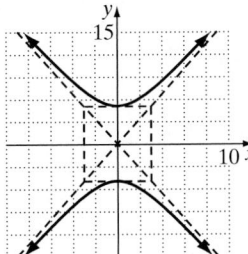

17.
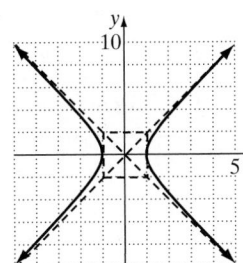

19. $\dfrac{x^2}{9} - \dfrac{y^2}{25} = 1$

21. $\dfrac{y^2}{4} - \dfrac{x^2}{9} = 1$

23. 40 yards

25–29. Answers will vary.

31.

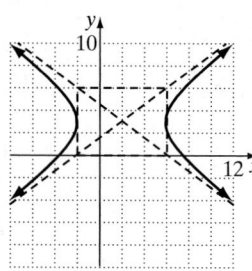

33.

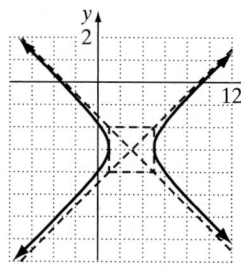

35. $\dfrac{x^2}{36} - \dfrac{y^2}{576} = 1$

39. Answers will vary.

40.

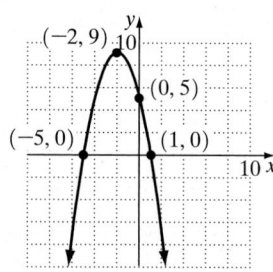

41. $\left\{ x \,\middle|\, x \le -\dfrac{1}{3} \text{ or } x \ge 4 \right\}$ or $\left(-\infty, -\dfrac{1}{3} \right] \cup [4, \infty)$ **42.** 21 or $\{21\}$

Section 13.4

Check Point Exercises

1.

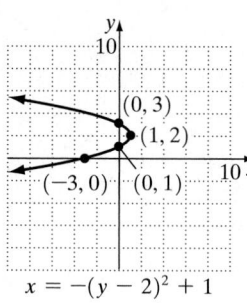

$x = -(y - 2)^2 + 1$

2.

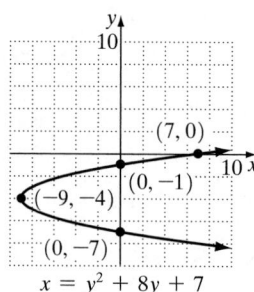

$x = y^2 + 8y + 7$

3. a. hyperbola
   b. ellipse
   c. circle
   d. parabola

Exercise Set 13.4

1. opens to right; $(-1, 2)$; b   **3.** opens to right; $(1, -2)$; f   **5.** opens to left; $(1, 2)$; a
7. $(0, 0)$ **9.** $(3, 2)$ **11.** $(-1, -2)$ **13.** $(0, 6)$ **15.** $(-3, 3)$ **17.** $(4, -1)$
19.

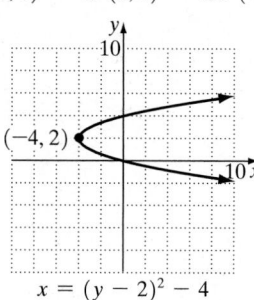

$x = (y - 2)^2 - 4$
axis of symmetry: $y = 2$

21.

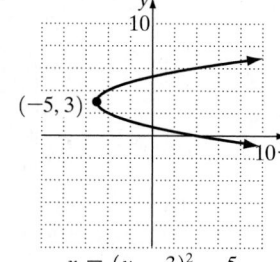

$x = (y - 3)^2 - 5$
axis of symmetry: $y = 3$

23.

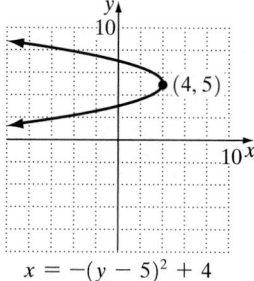

$x = -(y - 5)^2 + 4$
axis of symmetry: $y = 5$

25.

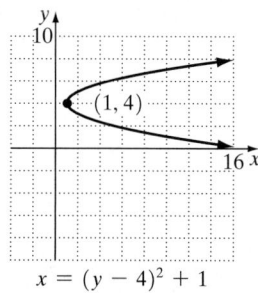

$x = (y - 4)^2 + 1$
axis of symmetry: $y = 4$

27.

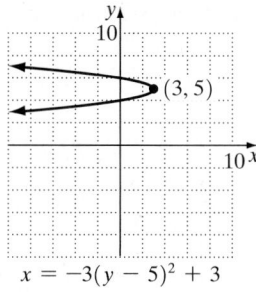

$x = -3(y - 5)^2 + 3$
axis of symmetry: $y = 5$

29.

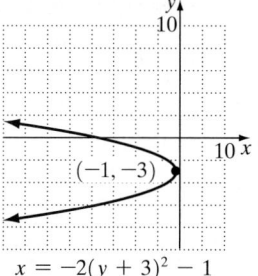

$x = -2(y + 3)^2 - 1$
axis of symmetry: $y = -3$

31.

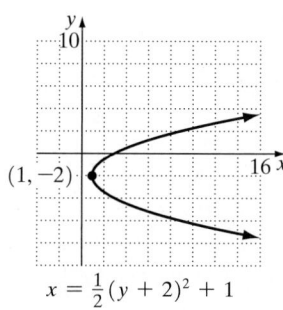

$x = \frac{1}{2}(y + 2)^2 + 1$
axis of symmetry: $y = -2$

33.

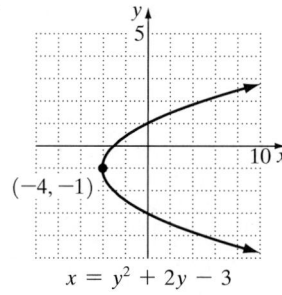

$x = y^2 + 2y - 3$
axis of symmetry: $y = -1$

35.

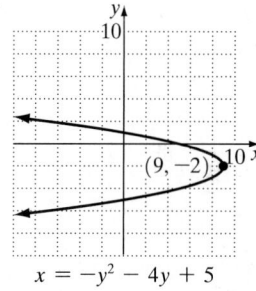

$x = -y^2 - 4y + 5$
axis of symmetry: $y = -2$

37.

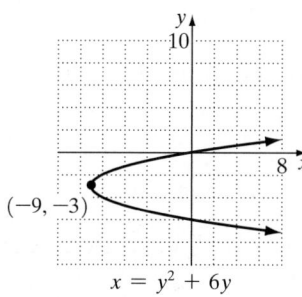

$x = y^2 + 6y$
axis of symmetry: $y = -3$

39.

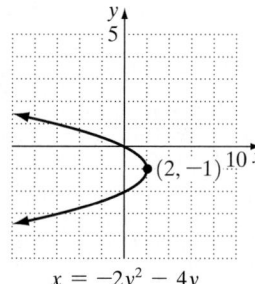

$x = -2y^2 - 4y$
axis of symmetry: $y = -1$

41.

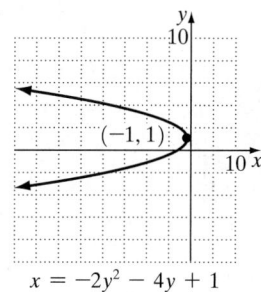

$x = -2y^2 - 4y + 1$
axis of symmetry: $y = -1$

43. a. horizontal   **b.** to the right   **c.** $(2, 1)$ **45. a.** vertical   **b.** upward   **c.** $(1, 2)$ **47. a.** vertical   **b.** downward
c. $(-3, 4)$ **49. a.** horizontal   **b.** to the left   **c.** $(4, -3)$ **51. a.** vertical   **b.** upward   **c.** $(2, -5)$ **53.** horizontal
b. to the left   **c.** $(5, 2)$ **55.** parabola   **57.** ellipse   **59.** hyperbola   **61.** circle   **63.** hyperbola

65. hyperbola   **67.** circle   **69.** ellipse   **71.** parabola

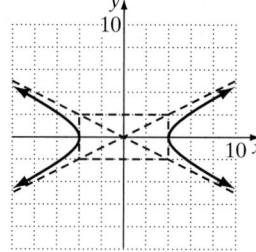

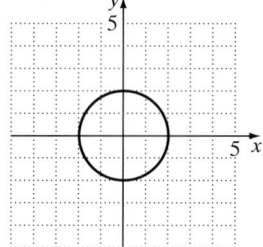

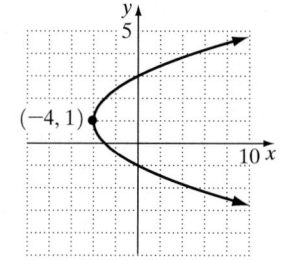

73. circle

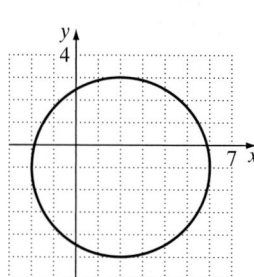

75. a. $y = 0.0001032x^2$ **b.** 58 ft   **77. a.** $y = \dfrac{1}{18}x^2$ **b.** 4.5 ft

79–85. Answers will vary.   **87.** b

89. Yes; the height of the arch 30 feet from the center is 45.5 feet.

91. $y^2 + 10y + (-x + 25) = 0$;   **93.**
$y = -5 \pm \sqrt{x}$

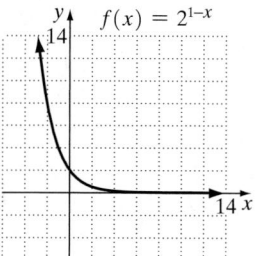

94. $f^{-1}(x) = 3x + 15$ **95.** $\dfrac{x}{3}$

Section 13.5

Check Point Exercises

1. $(0, 1)$ and $(4, 17)$, or $\{(0, 1), (4, 17)\}$ **2.** $(2, -1)$ and $\left(-\dfrac{6}{5}, \dfrac{3}{5}\right)$, or $\left\{(2, -1), \left(-\dfrac{6}{5}, \dfrac{3}{5}\right)\right\}$

3. $(3, 2), (-3, 2), (3, -2),$ and $(-3, -2)$, or $\{(3, 2), (-3, 2), (3, -2), (-3, -2)\}$ **4.** $(0, 5)$ or $\{(0, 5)\}$ **5.** length: 7 ft; width: 3 ft

Exercise Set 13.5

1. $(2, 0)$ and $(-3, 5)$, or $\{(2, 0), (-3, 5)\}$ **3.** $(2, 0)$ and $(1, 1)$, or $\{(2, 0), (1, 1)\}$ **5.** $(-3, 11)$ and $(4, -10)$, or $\{(-3, 11), (4, -10)\}$

7. $(-3, -4)$ and $(4, 3)$, or $\{(-3, -4), (4, 3)\}$ **9.** $(2, 3)$ and $\left(-\dfrac{3}{2}, -4\right)$, or $\left\{(2, 3), \left(-\dfrac{3}{2}, -4\right)\right\}$ **11.** $(3, 0)$ and $(-5, -4)$, or

$\{(3, 0), (-5, -4)\}$ **13.** $(3, 1), (-1, -3), (1, 3),$ and $(-3, -1)$, or $\{(3, 1), (-1, -3), (1, 3), (-3, -1)\}$ **15.** $(4, -3)$ and $(-1, 2)$, or

$\{(4, -3), (-1, 2)\}$ **17.** $(4, -3)$ and $(0, 1)$, or $\{(4, -3), (0, 1)\}$ **19.** $(3, 2), (-3, 2), (3, -2),$ and $(-3, -2)$, or

$\{(3, 2), (-3, 2), (3, -2), (-3, -2)\}$ **21.** $(3, 2), (-3, 2), (3, -2),$ and $(-3, -2)$, or $\{(3, 2), (-3, 2), (3, -2), (-3, -2)\}$

23. $(2, 1), (-2, 1), (2, -1),$ and $(-2, -1)$ or $\{(2, 1), (-2, 1), (2, -1), (-2, -1)\}$ **25.** $(3, 4)$ and $(3, -4)$, or $\{(3, 4), (3, -4)\}$

27. $(0, 2), (0, -2), (-1, \sqrt{3}),$ and $(-1, -\sqrt{3})$, or $\{(0, 2), (0, -2), (-1, \sqrt{3}), (-1, -\sqrt{3})\}$ **29.** $(2, 1), (-2, 1), (2, -1),$ and $(-2, -1)$, or

$\{(2, 1), (-2, 1), (2, -1), (-2, -1)\}$ **31.** $(-1, -4), (1, 4), (2\sqrt{2}, \sqrt{2}),$ and $(-2\sqrt{2}, -\sqrt{2})$, or $\{(-1, -4), (1, 4), (2\sqrt{2}, \sqrt{2}), (-2\sqrt{2}, -\sqrt{2})\}$

33. $(4, 1)$ and $(2, 2)$, or $\{(4, 1), (2, 2)\}$ **35.** $(0, 0)$ and $(-1, 1)$, or $\{(0, 0), (-1, 1)\}$ **37.** $(0, 0), (2, 2),$ and $(-2, 2)$, or

$\{(0, 0), (2, 2), (-2, 2)\}$ **39.** $(-4, 1)$ and $\left(-\dfrac{5}{2}, \dfrac{1}{4}\right)$, or $\left\{(-4, 1), \left(-\dfrac{5}{2}, \dfrac{1}{4}\right)\right\}$ **41.** $(-2, 3)$ and $\left(\dfrac{12}{5}, -\dfrac{29}{5}\right)$, or

$\left\{(-2, 3), \left(\dfrac{12}{5}, -\dfrac{29}{5}\right)\right\}$ **43.** $x + y = 10; xy = 24;$ 6 and 4   **45.** $x^2 - y^2 = 3; 2x^2 + y^2 = 9;$ 1 and 2, -2 and 1, -2 and -1 or 2 and -1

47. $(0, -4), (2, 0), (-2, 0)$ **49.** length: 11 ft; width: 7 ft   **51.** length: 8 in.; width: 6 in.

53. large square: 5 m by 5 m; small square: 2 m by 2 m   **55–57.** Answers will vary.   **59.** b   **61.** $(8, 2)$ or $\{(8, 2)\}$

63. Answers will vary.   **65.**

$3x - 2y \le 6$

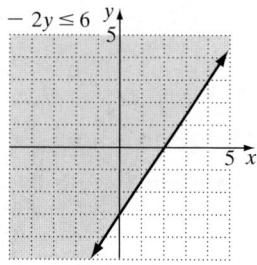

66. $m = \dfrac{8}{3}$ **67.** $6x^3 - 16x^2 + 17x - 6$

Review Exercises

1. 13 units **2.** $2\sqrt{2} \approx 2.83$ units **3.** $(-5, 5)$ **4.** $\left(-\dfrac{11}{2}, -2\right)$ **5.** $x^2 + y^2 = 9$ **6.** $(x + 2)^2 + (y - 4)^2 = 36$

7. center: $(0, 0)$; $r = 1$ **8.** center: $(-2, 3)$; $r = 3$ **9.** center: $(2, -1)$; $r = 3$ **10.** center: $(0, 2)$; $r = 2$

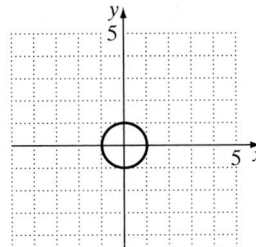 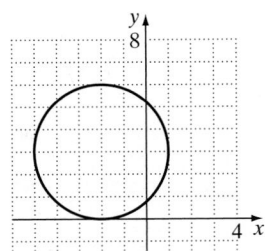 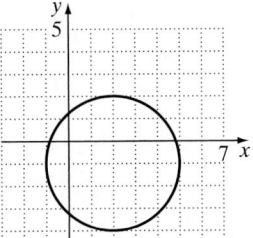 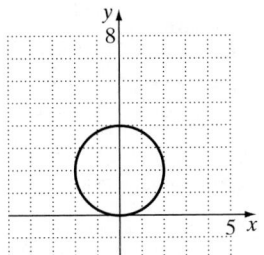

11. **12.** **13.** **14.**

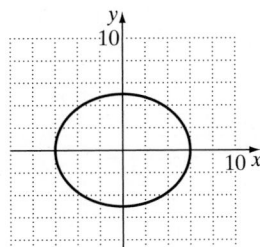 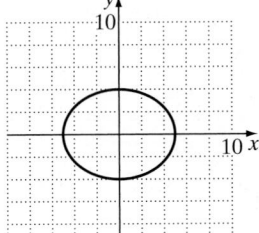 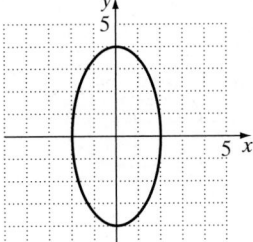 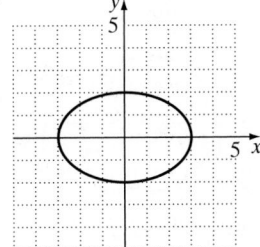

15. **16.**

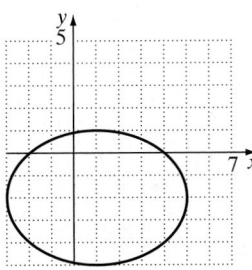 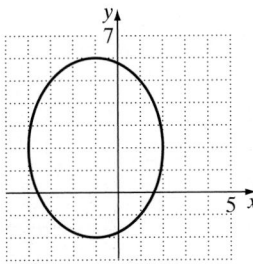

17. Yes, the height of the archway 14 feet from the center is approximately 12.43 feet.

18. **19.**

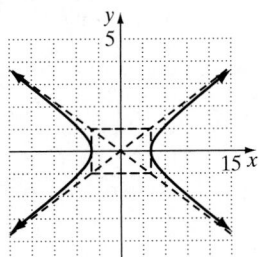 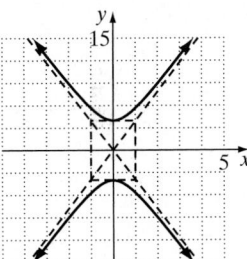

20. **21.** **22.**

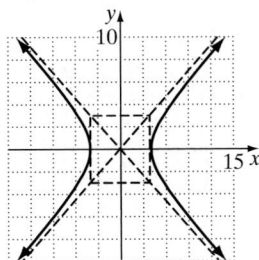 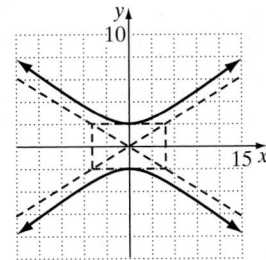 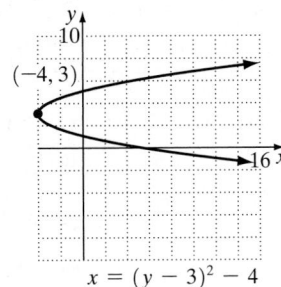

$x = (y - 3)^2 - 4$

axis of symmetry: $y = 3$

23.

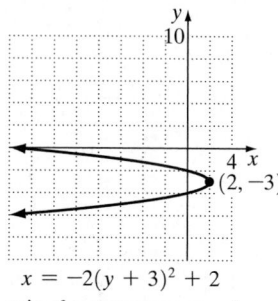

$x = -2(y + 3)^2 + 2$
axis of symmetry: $y = -3$

24.

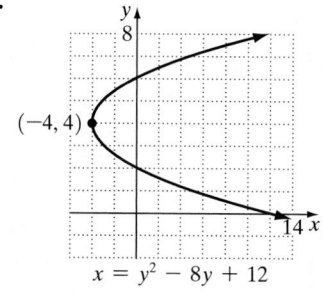

$x = y^2 - 8y + 12$
axis of symmetry: $y = 4$

25.

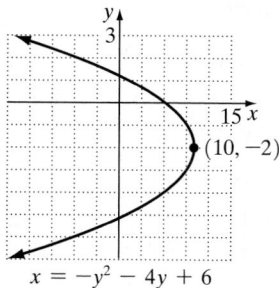

$x = -y^2 - 4y + 6$
axis of symmetry: $y = -2$

26. parabola   **27.** ellipse   **28.** hyperbola   **29.** circle   **30.** hyperbola   **31.** ellipse   **32.** parabola

33. circle   **34.** ellipse   **35.** hyperbola   **36.** ellipse

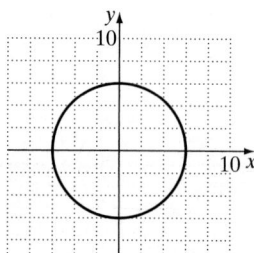
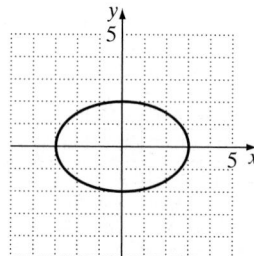
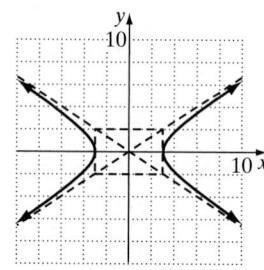
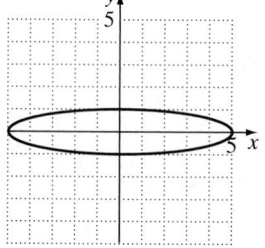

37. parabola   **38.** parabola   **39.** ellipse   **40.** circle

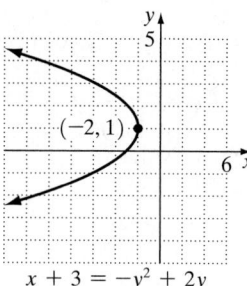

$x + 3 = -y^2 + 2y$

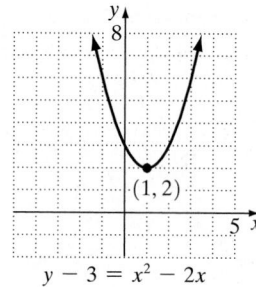

$y - 3 = x^2 - 2x$

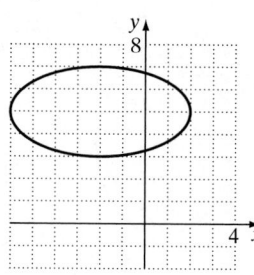

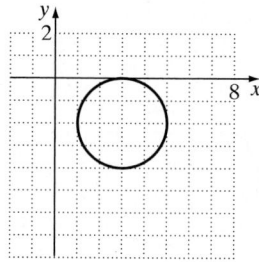

41. circle

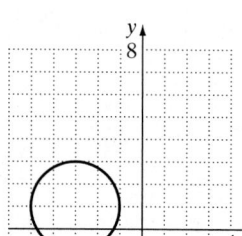

42. a. $y = \frac{1}{12}x^2$ **b.** $(0, 3)$; 3 inches above the vertex

43. $(1, 0)$ and $(4, 3)$, or $\{(1, 0), (4, 3)\}$ **44.** $(0, 1)$ and $(-3, 4)$, or $\{(0, 1), (-3, 4)\}$

45. $(-1, 1)$ and $(1, -1)$, or $\{(-1, 1), (1, -1)\}$

46. $(3, \sqrt{6}), (-3, \sqrt{6}), (3, -\sqrt{6})$, and $(-3, -\sqrt{6})$, or $\{(3, \sqrt{6}), (-3, \sqrt{6}), (3, -\sqrt{6}), (-3, -\sqrt{6})\}$

47. $(2, 2)$ and $(-2, -2)$, or $\{(2, 2), (-2, -2)\}$ **48.** $(1, 2)$ and $(9, 6)$, or $\{(1, 2), (9, 6)\}$

49. $(-3, -1)$ and $(1, 3)$, or $\{(-3, -1), (1, 3)\}$ **50.** $(-1, -1)$ and $\left(\frac{1}{2}, 2\right)$, or $\left\{(-1, -1), \left(\frac{1}{2}, 2\right)\right\}$

51. $(0, -1)$ and $\left(\frac{5}{2}, -\frac{7}{2}\right)$, or $\left\{(0, -1), \left(\frac{5}{2}, -\frac{7}{2}\right)\right\}$

52. $(3, 2), (-3, 2), (2, -3)$, and $(-2, -3)$, or $\{(3, 2), (-3, 2), (2, -3), (-2, -3)\}$

53. $(3, 1), (-3, 1), (3, -1)$, and $(-3, -1)$ or $\{(3, 1), (-3, 1), (3, -1), (-3, -1)\}$ **54.** 8 m by 5 m   **55.** $(1, 6)$ and $(3, 2)$

56. x: 50 ft; y: 20 ft

Chapter 13 Test

1. $\sqrt{73} \approx 8.54$ units **2.** $\left(\dfrac{7}{2}, -4\right)$ **3.** $(x - 3)^2 + (y + 2)^2 = 25$ **4.** center: $(5, -3)$; $r = 7$ **5.** center: $(-2, 3)$; $r = 4$

6. center: $(7, -3)$ **7.** $(-2, -5)$

8. hyperbola

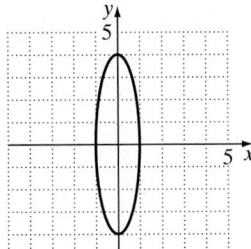

9. ellipse

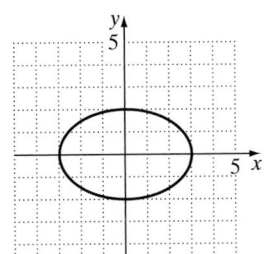

10. parabola

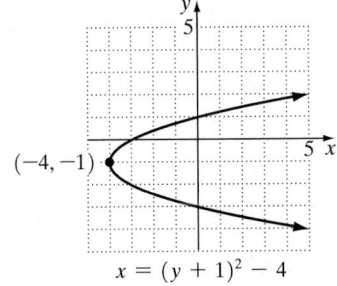

$(-4, -1)$

$x = (y + 1)^2 - 4$

11. ellipse

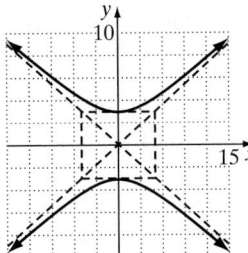

12. hyperbola

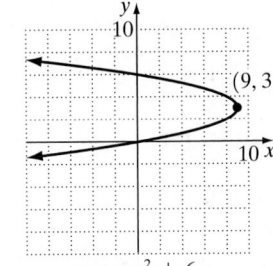

13. parabola

$(9, 3)$

$x = -y^2 + 6y$

14. ellipse

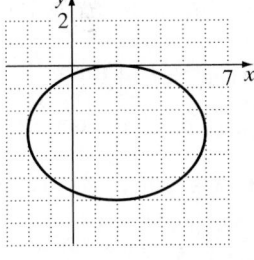

15. circle

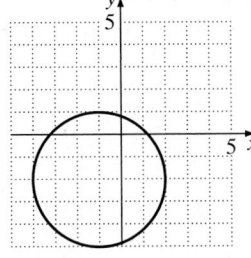

16. circle

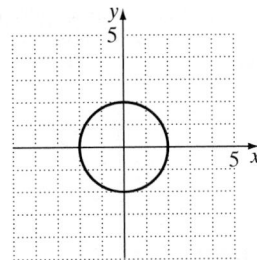

17. $(4, -3)$ and $(-3, 4)$, or $\{(4, -3), (-3, 4)\}$
18. $(3, 2), (-3, 2), (3, -2)$, and $(-3, -2)$, or $\{(3, 2), (-3, 2), (3, -2), (-3, -2)\}$
19. 15 ft by 12 ft or 24 ft by 7.5 ft
20. 4 ft by 3 ft

Cumulative Review Exercises (Chapters 1–13)

1. $\{x \mid x > -1\}$ or $(-1, \infty)$ **2.** $-\dfrac{1}{2}$ and 4, or $\left\{-\dfrac{1}{2}, 4\right\}$ **3.** 2 or $\{2\}$ **4.** $\left\{x \mid -\dfrac{5}{3} < x < -1\right\}$ or $\left(-\dfrac{5}{3}, -1\right)$ **5.** $\dfrac{5}{2}$ or $\left\{\dfrac{5}{2}\right\}$

6. $\dfrac{\ln 8}{0.7} \approx 2.97$ or $\left\{\dfrac{\ln 8}{0.7} \approx 2.97\right\}$ **7.** $(1, 3), (-1, 3), (1, -3)$, and $(-1, -3)$, or $\{(1, 3), (-1, 3), (1, -3), (-1, -3)\}$

8.

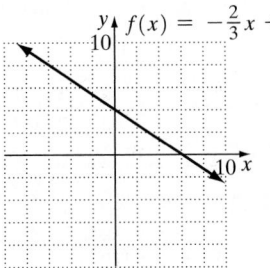

$f(x) = -\dfrac{2}{3}x + 4$

9. $3x - y > 6$

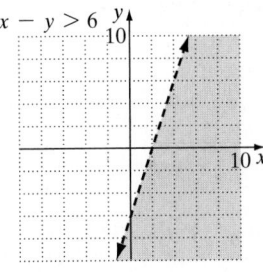

10.

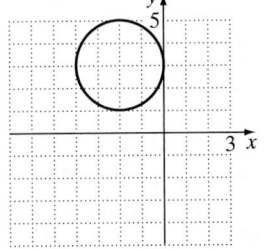

11.

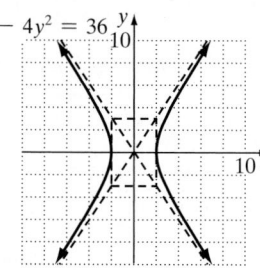

$9x^2 - 4y^2 = 36$

12. 46 **13.** $x^2 - 5x - 1$ **14.** $2xy^2\sqrt[3]{2y}$ **15.** $11 + 10i$
16. $3x(2x - 3)^2$ **17.** $(x - 2)(x + 3)(x - 3)$

18. $\{x | x \leq 2\}$ or $(-\infty, 2]$ **19.** $\dfrac{(1 - \sqrt{x})^2}{1 - x}$ or $\dfrac{1 - 2\sqrt{x} + x}{1 - x}$

20. $\ln x^{1/3} y^7$ **21.** $3x^2 + x + 4 + \dfrac{7}{x - 2}$ **22.** $x^2 - 12$

23. faster car: 50 mph; slower car: 40 mph
24. 175 miles; $67 **25.** apple: 60 calories; banana: 87 calories

CHAPTER 14

Section 14.1

Check Point Exercises

1. a. $7, 9, 11, 13$ **b.** $-\dfrac{1}{3}, \dfrac{1}{5}, -\dfrac{1}{9}, \dfrac{1}{17}$ **2.** $10, \dfrac{10}{3}, \dfrac{5}{6}, \dfrac{1}{6}$ **3. a.** $2(1)^2 + 2(2)^2 + 2(3)^2 + 2(4)^2 + 2(5)^2 + 2(6)^2 = 182$

b. $(2^3 - 3) + (2^4 - 3) + (2^5 - 3) = 47$ **c.** $4 + 4 + 4 + 4 + 4 = 20$ **4. a.** $\displaystyle\sum_{i=1}^{9} i^2$ **b.** $\displaystyle\sum_{i=1}^{n} \dfrac{1}{2^{i-1}}$

Exercise Set 14.1

1. $5, 8, 11, 14$ **3.** $3, 9, 27, 81$ **5.** $-3, 9, -27, 81$ **7.** $-4, 5, -6, 7$ **9.** $\dfrac{2}{5}, \dfrac{2}{3}, \dfrac{6}{7}, 1$ **11.** $1, -\dfrac{1}{3}, \dfrac{1}{7}, -\dfrac{1}{15}$ **13.** $1, 2, \dfrac{3}{2}, \dfrac{2}{3}$

15. $4, 12, 48, 240$ **17.** 105 **19.** 60 **21.** 115 **23.** $-\dfrac{5}{16}$ **25.** 55 **27.** $\dfrac{3}{8}$ **29.** 15 **31.** $\displaystyle\sum_{i=1}^{15} i^2$ **33.** $\displaystyle\sum_{i=1}^{11} 2^i$ **35.** $\displaystyle\sum_{i=1}^{30} i$

37. $\displaystyle\sum_{i=1}^{14} \dfrac{i}{i + 1}$ **39.** $\displaystyle\sum_{i=1}^{n} \dfrac{4^i}{i}$ **41.** $\displaystyle\sum_{i=1}^{n} (2i - 1)$ **43.** Answers will vary; examples are: $\displaystyle\sum_{k=1}^{14} (2k + 3)$ or $\displaystyle\sum_{k=2}^{15} (2k + 1)$.

45. Answers will vary; an example is: $\displaystyle\sum_{k=0}^{12} ar^k$. **47.** Answers will vary; an example is: $\displaystyle\sum_{k=0}^{n} (a + kd)$. **49. a.** 5939.1; From 1991 to 1999, a total of 5939.1 million CDs were sold. **b.** 659.9; From 1991 to 1999, an average of 659.9 million CDs were sold each year.
51. 30.15; From 1991 to 1995, Americans spent a total of $30.15 billion on recreational boating. **53.** $8081.13 **55–59.** Answers will vary.
61. c **63.** $4 \log x + 6 \log x + 8 \log x = \log x^{18}$ **65.** 600

69.

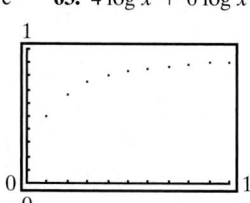

71.

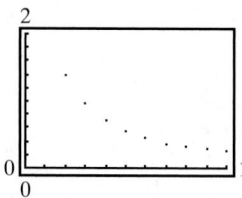

73. $2xy^2\sqrt[3]{5xy}$
74. $(3x - 2)(9x^2 + 6x + 4)$
75. $-\dfrac{6}{5}$ and 4, or $\left\{-\dfrac{6}{5}, 4\right\}$

As n gets larger, the terms get closer to 1. As n gets larger, the terms get closer to 0.

Section 14.2

Check Point Exercises

1. $100, 70, 40, 10, -20, -50$ **2.** -34 **3. a.** $a_n = 2350n + 10,458$ **b.** $73,908 million or $73,908,000,000 **4.** 360 **5.** 2460

Exercise Set 14.2

1. 4 **3.** 5 **5.** -3 **7.** $200, 220, 240, 260, 280, 300$ **9.** $-7, -3, 1, 5, 9, 13$ **11.** $300, 210, 120, 30, -60, -150$ **13.** $\dfrac{5}{2}, 2, \dfrac{3}{2}, 1, \dfrac{1}{2}, 0$

15. $-0.4, -2, -3.6, -5.2, -6.8, -8.4$ **17.** 33 **19.** 252 **21.** 955 **23.** -142 **25.** $a_n = 4n - 3; a_{20} = 77$

27. $a_n = 11 - 4n; a_{20} = -69$ **29.** $a_n = -4n - 16; a_{20} = -96$ **31.** $a_n = \dfrac{1}{3}n - \dfrac{2}{3}; a_{20} = 6$ **33.** $a_n = 4.3 - 0.3n; a_{20} = -1.7$

35. 1220 **37.** 4400 **39.** 5050 **41.** 3660 **43.** 396 **45.** $8 + 13 + 18 + \cdots + 88 = 816$
47. $2 + (-1) + (-4) + \cdots + (-85) = -1245$ **49.** $4 + 8 + 12 + \cdots + 400 = 20,200$ **51. a.** $a_n = 1265n + 125,159$
b. 145,399 thousand or 145,399,000 employees **53.** company A; $1400

55. a. $a_n = 0.576n + 3.204$ **b.** 627.3 million or 627,300,000 tons   **57.** \$442,500   **59.** 1430 seats   **61–63.** Answers will vary.

65. \$167,428; Answers will vary.   **67.** 200th term   **69.** $S_n = \dfrac{n}{2}[1 + (2n - 1)] = \dfrac{n}{2}(2n) = n^2$ **72.** 1005 or \{1005\}

73. $\{x | -5 \le x \le 2\}$ or $[-5, 2]$ **74.** $\dfrac{x + 3}{x - 4}$

Section 14.3

Check Point Exercises

1. $12, 6, 3, \dfrac{3}{2}, \dfrac{3}{4}, \dfrac{3}{8}$ **2.** 3645   **3.** $a_n = 3(2)^{n-1}; a_8 = 384$ **4.** 9842   **5.** 19,680   **6.** \$2,371,746   **7.** \$1,327,778   **8.** 9
9. 1   **10.** \$4000

Exercise Set 14.3

1. $r = 3$ **3.** $r = -2$ **5.** $r = \dfrac{3}{2}$ **7.** $r = -0.1$ **9.** 2, 6, 18, 54, 162   **11.** $20, 10, 5, \dfrac{5}{2}, \dfrac{5}{4}$ **13.** $-4, 40, -400, 4000, -40,000$

15. $-\dfrac{1}{4}, \dfrac{1}{2}, -1, 2, -4$ **17.** $a_8 = 768$ **19.** $a_{12} = -10,240$ **21.** $a_6 = -200$ **23.** $a_8 = 0.1$ **25.** $a_n = 3(4)^{n-1}; a_7 = 12,288$

27. $a_n = 18\left(\dfrac{1}{3}\right)^{n-1}; a_7 = \dfrac{2}{81}$ **29.** $a_n = 1.5(-2)^{n-1}; a_7 = 96$ **31.** $a_n = 0.0004(-10)^{n-1}; a_7 = 400$ **33.** 531,440   **35.** 2049

37. $\dfrac{16,383}{2}$ or 8191.5   **39.** 9840   **41.** 10,230   **43.** $\dfrac{63}{128}$ **45.** $\dfrac{3}{2}$ **47.** 4   **49.** $\dfrac{2}{3}$ **51.** 20   **53.** $\dfrac{5}{9}$ **55.** $\dfrac{47}{99}$ **57.** $\dfrac{257}{999}$
59. arithmetic; $d = 1$ **61.** geometric; $r = 2$ **63.** neither   **65.** \$16,384   **67.** \$3,795,957   **69. a.** 1.04; 1.04; 1.04; $r = 1.04$
b. $a_n = 20.60(1.04)^{n-1}$ **c.** 30.49 million or 30,490,000   **71.** \$32,767   **73.** \$793,583   **75.** 130.26 in.   **77.** \$844,706

79. \$94,834   **81.** \$9 million   **83.** $\dfrac{1}{3}$ **85–91.** Answers will vary.   **93.** d   **95.** \$442

99.
; horizontal asymptote: $y = 10$; sum of series: 10   **100.** $2\sqrt{7}$

101. $\dfrac{-1 \pm \sqrt{33}}{4}$ or $\left\{\dfrac{-1 \pm \sqrt{33}}{4}\right\}$
102. $-3(\sqrt{3} + \sqrt{5})$

Section 14.4

Check Point Exercises

1. a. 20   **b.** 1   **c.** 28   **d.** 1   **2.** $x^4 + 4x^3 + 6x^2 + 4x + 1$ **3.** $x^5 - 10x^4y + 40x^3y^2 - 80x^2y^3 + 80xy^4 - 32y^5$
4. $4032x^5y^4$

Exercise Set 14.4

1. 56   **3.** 12   **5.** 1   **7.** 4950   **9.** $x^3 + 6x^2 + 12x + 8$ **11.** $27x^3 + 27x^2y + 9xy^2 + y^3$ **13.** $125x^3 - 75x^2 + 15x - 1$
15. $16x^4 + 32x^3 + 24x^2 + 8x + 1$ **17.** $x^8 + 8x^6y + 24x^4y^2 + 32x^2y^3 + 16y^4$ **19.** $y^4 - 12y^3 + 54y^2 - 108y + 81$
21. $16x^{12} - 32x^9 + 24x^6 - 8x^3 + 1$ **23.** $c^5 + 10c^4 + 40c^3 + 80c^2 + 80c + 32$ **25.** $x^5 - 5x^4 + 10x^3 - 10x^2 + 5x - 1$
27. $243x^5 - 405x^4y + 270x^3y^2 - 90x^2y^3 + 15xy^4 - y^5$ **29.** $64a^6 + 192a^5b + 240a^4b^2 + 160a^3b^3 + 60a^2b^4 + 12ab^5 + b^6$
31. $x^8 + 16x^7 + 112x^6$ **33.** $x^{10} - 20x^9y + 180x^8y^2$ **35.** $x^{32} + 16x^{30} + 120x^{28}$ **37.** $y^{60} - 20y^{57} + 190y^{54}$ **39.** $240x^4y^2$

41. $126x^5$ **43.** $56x^6y^{15}$ **45.** $-\dfrac{21}{2}x^6$ **47.** $g(t) = 0.002t^3 - 0.84t^2 - 16.13t - 68.54$ **49–57.** Answers will vary.

59. $x^6 + 3x^5 + 6x^4 + 7x^3 + 6x^2 + 3x + 1$
63.
; f_2, f_3, f_4, and f_5 are approaching $f_1 = f_6$.   **65.** $x^3 - 3x^2 + 3x - 1$
68. $a^2 + 4a + 6$
69. $f(g(x)) = 4x^2 - 2x - 6; g(f(x)) = 2x^2 + 10x - 3$
70. $\dfrac{2x^2 - 9x - 1}{2(x - 4)(x + 3)}$

Section 14.5

Check Point Exercises

1. 72   **2.** 729   **3.** 676,000   **4.** 840   **5.** 720   **6. a.** combinations   **c.** permutations   **7.** 210   **8.** 1820

Exercise Set 14.5

1. 3024   **3.** 6720   **5.** 720   **7.** 1   **9.** 126   **11.** 330   **13.** 1   **15.** 1   **17.** combinations   **19.** permutations
21. 27   **23.** 40   **25.** 243   **27.** 144   **29.** 120   **31.** 6   **33.** 720   **35.** 8,648,640   **37.** 120   **39.** 15,120
41. 20   **43.** 495   **45.** 24,310   **47.** 13,983,816   **49.** 360   **51.** 1716   **53.** 1140   **55.** 840   **57.** 2730
59–65. Answers will vary.   **67.** c   **69.** 144   **73.** $(f \circ g)(x) = 16x^2 - 6$ **74.** $\{x \mid x < 1 \text{ or } x > 4\}$ or $(-\infty, 1) \cup (4, \infty)$
75. center: $(1, -2)$; $r = 3$;

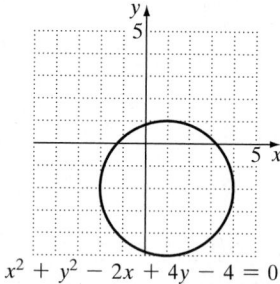

$$x^2 + y^2 - 2x + 4y - 4 = 0$$

Section 14.6

Check Point Exercises

1. $\frac{3}{10}$ or 0.3   **2.** $\frac{1}{3}$ **3.** $\frac{1}{9}$ **4.** $\frac{1}{13}$ **5.** $\frac{1}{142,506} \approx 0.00000702$ **6.** $\frac{142,505}{142,506} \approx 0.99997$ **7.** $\frac{1}{3}$ **8.** $\frac{3}{4}$ **9.** $\frac{24}{25}$
10. $\frac{1}{361} \approx 0.00277$ **11.** $\frac{1}{16}$

Exercise Set 14.6

1. $\frac{29,780}{70,241}$; 0.424   **3.** $\frac{761}{5926}$; 0.128   **5.** $\frac{1}{6}$ **7.** $\frac{1}{2}$ **9.** $\frac{1}{3}$ **11.** $\frac{1}{13}$ **13.** $\frac{3}{13}$ **15.** $\frac{1}{4}$ **17.** $\frac{7}{8}$ **19.** $\frac{1}{12}$
21. $\frac{1}{18,009,460} \approx 0.0000000555$; $\frac{5}{900,473} \approx 0.00000555$ **23. a.** 2,598,960   **b.** 1287   **c.** $\frac{33}{66,640} \approx 0.000495$ **25.** $\frac{255,647}{274,634} \approx 0.931$
27. $\frac{237,401}{274,634} \approx 0.864$ **29.** $\frac{21,005}{137,317} \approx 0.153$ **31.** $\frac{2}{13}$ **33.** $\frac{5}{6}$ **35.** $\frac{7}{13}$ **37.** $\frac{3}{4}$ **39.** $\frac{33}{40}$ **41.** $\frac{1}{36}$ **43.** $\frac{1}{3}$ **45.** $\frac{1}{64}$
47. a. $\frac{1}{256}$ **b.** $\frac{1}{4096}$ **c.** $\left(\frac{15}{16}\right)^{10} \approx 0.524$ **49–55.** Answers will vary.   **57.** $\frac{1}{12}$; Answers will vary.   **59.** Answers will vary.

61.

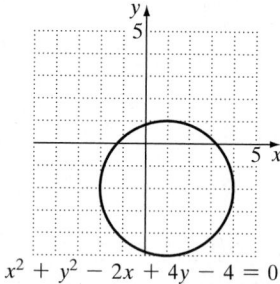

   62. 3 or $\{3\}$ **63.** $x^2 + 3x - 3 - \frac{4}{x + 2}$

Review Exercises

1. 3, 10, 17, 24   **2.** $-\frac{3}{2}, \frac{4}{3}, -\frac{5}{4}, \frac{6}{5}$ **3.** $1, 1, \frac{1}{2}, \frac{1}{6}$ **4.** $\frac{1}{2}, -\frac{1}{4}, \frac{1}{8}, -\frac{1}{16}$ **5.** 95   **6.** -20 **7.** Answers will vary; an example is:
$\sum\limits_{i=1}^{15} \frac{i}{i+2}$.   **8.** Answers will vary; examples are: $\sum\limits_{i=4}^{13} i^3$ or $\sum\limits_{i=1}^{10} (i+3)^3$.   **9.** 7, 11, 15, 19, 23, 27   **10.** $-4, -9, -14, -19, -24, -29$
11. $\frac{3}{2}, 1, \frac{1}{2}, 0, -\frac{1}{2}, -1$ **12.** 20   **13.** -30 **14.** -38 **15.** $a_n = 4n - 11$; $a_{20} = 69$ **16.** $a_n = 220 - 20n$; $a_{20} = -180$

17. $a_n = -\dfrac{23}{2} - \dfrac{1}{2}n; a_{20} = -\dfrac{43}{2}$ **18.** $a_n = 22 - 7n; a_{20} = -118$ **19.** 1727 **20.** 225 **21.** 15,150 **22.** 440 **23.** -500

24. -2325 **25. a.** $1043.4518 - 0.4118n$ **b.** 1002.2718 sec **26.** \$418,500 **27.** 1470 seats **28.** $3, 6, 12, 24, 48$

29. $\dfrac{1}{2}, \dfrac{1}{4}, \dfrac{1}{8}, \dfrac{1}{16}, \dfrac{1}{32}$ **30.** $16, -4, 1, -\dfrac{1}{4}, \dfrac{1}{16}$ **31.** $-5, 5, -5, 5, -5$ **32.** $a_7 = 1458$ **33.** $a_6 = \dfrac{1}{2}$ **34.** $a_5 = -48$

35. $a_n = 1(2)^{n-1}$ or $a_n = 2^{n-1}; a_8 = 128$ **36.** $a_n = 100\left(\dfrac{1}{10}\right)^{n-1}; a_8 = \dfrac{1}{100,000} = 0.00001$ **37.** $a_n = 12\left(-\dfrac{1}{3}\right)^{n-1}; a_8 = -\dfrac{4}{729}$

38. 17,936,135 **39.** $\dfrac{127}{8}$ or 15.875 **40.** 19,530 **41.** -258 **42.** $\dfrac{341}{128}$ **43.** $\dfrac{27}{2}$ **44.** $\dfrac{4}{3}$ **45.** $-\dfrac{18}{15}$ **46.** 20 **47.** $\dfrac{2}{3}$

48. $\dfrac{47}{99}$ **49.** \$42,823; \$223,210 **50.** \$120,113 **51.** $\$9\dfrac{1}{3}$ million **52.** 165 **53.** 4005 **54.** $8x^3 + 12x^2 + 6x + 1$

55. $x^8 - 4x^6 + 6x^4 - 4x^2 + 1$ **56.** $x^5 + 10x^4y + 40x^3y^2 + 80x^2y^3 + 80xy^4 + 32y^5$

57. $x^6 - 12x^5 + 60x^4 - 160x^3 + 240x^2 - 192x + 64$ **58.** $x^{16} + 24x^{14} + 252x^{12}$ **59.** $x^9 - 27x^8 + 324x^7$ **60.** $80x^2$ **61.** $4860x^2$

62. 336 **67.** 15,120 **63.** 56 **64.** 78 **65.** 20 **66.** 243 **67.** 32,760 **68.** 4845 **69.** 1140 **70.** 116,280 **71.** 120

72. $\dfrac{9,630,188}{31,878,234} \approx 0.302$ **73.** $\dfrac{5,503,372}{19,128,261} \approx 0.288$ **74.** $\dfrac{2}{3}$ **75.** $\dfrac{2}{3}$ **76.** $\dfrac{2}{13}$ **77.** $\dfrac{7}{13}$ **78.** $\dfrac{5}{6}$ **79.** $\dfrac{5}{6}$

80. a. $\dfrac{1}{15,504} \approx 0.0000645$ **b.** $\dfrac{25}{3876} \approx 0.00645$ **81.** $\dfrac{4}{5}$ **82.** $\dfrac{3}{4}$ **83.** $\dfrac{1}{32}$ **84. a.** 0.04 **b.** 0.008 **c.** 0.4096

Chapter 14 Test

1. $1, -\dfrac{1}{4}, \dfrac{1}{9}, -\dfrac{1}{16}, \dfrac{1}{25}$ **2.** 105 **3.** Answers will vary; examples are: $\displaystyle\sum_{i=2}^{21} \dfrac{i}{i+1}$ or $\displaystyle\sum_{i=1}^{20} \dfrac{i+1}{i+2}$. **4.** $a_n = 5n - 1; a_{12} = 59$

5. $a_n = 16\left(\dfrac{1}{4}\right)^{n-1}; a_{12} = \dfrac{1}{262,144}$ **6.** -385 **7.** 550 **8.** -2387 **9.** $-21,846$ **10.** 8 **11.** $\dfrac{73}{99}$ **12.** \$276,427 **13.** 36

14. $x^{10} - 5x^8 + 10x^6 - 10x^4 + 5x^2 - 1$ **15.** $x^8 + 8x^7y^2 + 28x^6y^4$ **16.** 990 **17.** 210 **18.** 10,000 **19.** Answers will vary.

20. $\dfrac{10}{1001}$ **21.** $\dfrac{8}{13}$ **22.** $\dfrac{3}{5}$ **23.** $\dfrac{1}{256}$ **24.** $\dfrac{1}{16}$

Cumulative Review Exercises (Covering the Entire Book)

1. $\dfrac{2}{3}$ or $\left\{\dfrac{2}{3}\right\}$ **2.** $(-5, 2)$ or $\{(-5, 2)\}$ **3.** $(1, 4, -2)$ or $\{(1, 4, -2)\}$ **4.** $\{x | x \geq 10\}$ or $[10, \infty)$ **5.** $\{x | x \leq -4\}$ or $(-\infty. -4]$

6. $\{x | x > 2\}$ or $(2, \infty)$ **7.** $\{x | -2 < x < 3\}$ or $(-2, 3)$ **8.** 3 and 9, or $\{3, 9\}$ **9.** no solution or $\varnothing$ **10.** 12 or $\{12\}$

11. $\dfrac{-2 \pm \sqrt{14}}{2}$ or $\left\{\dfrac{-2 \pm \sqrt{14}}{2}\right\}$ **12.** 8 and 27, or $\{8, 27\}$ **13.** $\left\{x \bigg| -2 \leq x \leq \dfrac{3}{2}\right\}$ or $\left[-2, \dfrac{3}{2}\right]$ **14.** 2 or $\{2\}$

15. 0 or $\{0\}$ **16.** $(2, 1), (-2, 1), (2, -1),$ and $(-2, -1),$ or $\{(2, 1), (-2, 1), (2, -1), (-2, -1)\}$

17. $(-14, -20)$ and $(2, -4),$ or $\{(-14, -20), (2, -4)\}$

18.

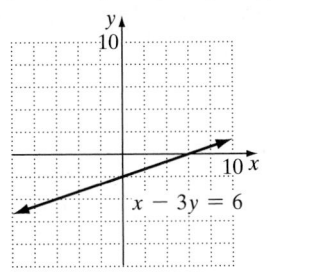

19. $f(x) = \dfrac{1}{2}x - 1$

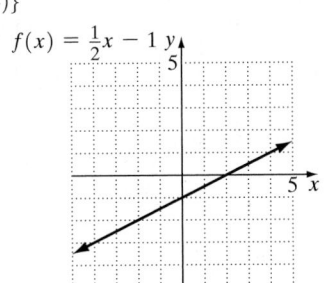

20. $3x - 2y > -6$

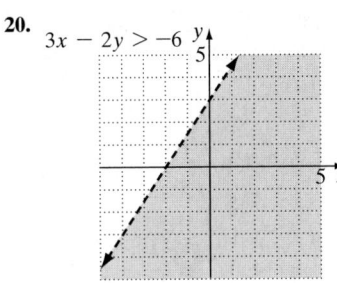

21. $f(x) = -2(x - 3)^2 + 2$

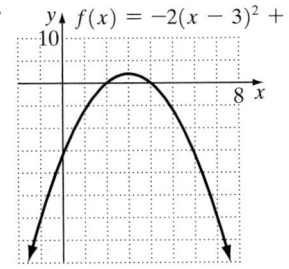

22. $\dfrac{x^2}{16} + \dfrac{y^2}{4} = 1$

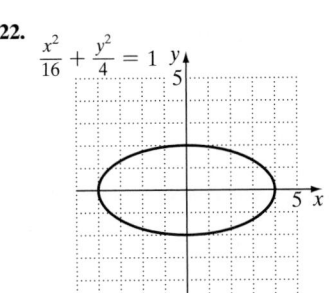

23. $y = \log_2 x$

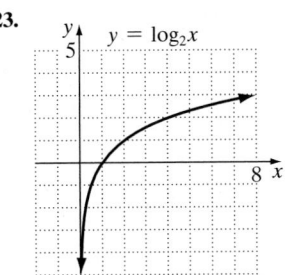

24.

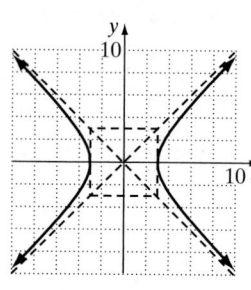

25. $-16x + 24y$ **26.** $-\dfrac{20x^7}{y^4}$ **27.** $x^2 - 5xy - 16y^2$ **28.** $6x^2 + 13x - 5$

29. $9x^4 - 24x^2y + 16y^2$ **30.** $\dfrac{3x^2 + 6x - 2}{(x + 5)(x + 2)}$ **31.** $\dfrac{x - 3}{x}$ **32.** $\dfrac{x - 4}{3x + 6}$

33. $5xy\sqrt{2x}$ **34.** $9\sqrt{2}$ **35.** $44 + 6i$ **36.** $(9x^2 + 1)(3x + 1)(3x - 1)$

37. $2x(4x - 1)(3x - 2)$ **38.** $(x + 3y)(x^2 - 3xy + 9y^2)$ **39.** $x^2 + 2x - 13; 22$

40. $x + 5$; $\{x \mid x$ is a real number and $x \neq 2\}$ or $(-\infty, 2) \cup (2, \infty)$

41. $x^2 - x - 17$ **42.** $x^2 + 3x - 17$ **43.** $f^{-1}(x) = \dfrac{x + 3}{7}$

44. $3\sqrt{10} \approx 9.49$ units **45.** $3x^2 - 7x + 18 - \dfrac{28}{x + 2}$ **46.** $t = \dfrac{5r + 2}{A}$ **47.** $y = -3x - 1$ or $f(x) = -3x - 1$

48. 6 **49.** $\ln \dfrac{x^2}{\sqrt{y}}$ **50.** 208 **51.** 1800 **52.** $\dfrac{1}{3}$ **53.** $16x^4 - 32x^3y^3 + 24x^2y^6 - 8xy^9 + y^{12}$

54. $\{x \mid x$ is a real number and $x \neq 1$ and $x \neq 2\}$ or $(-\infty, 1) \cup (1, 2) \cup (2, \infty)$ **55.** $\{x \mid x > 4\}$ or $(4, \infty)$ **56.** \$620

57. 13 yd by 4 yd **58.** \$2600 at 12% and \$1400 at 14% **59.** approximately 11% **60.** 11 amps

APPENDIX A

1. 4 or $\{4\}$ **2.** 10 or $\{10\}$ **3.** $\{x \mid x \leq 7\}$ **4.** $(-3, 4)$ or $\{(-3, 4)\}$ **5.** $(3, 4)$ or $\{(3, 4)\}$ **6.** -8 or $\{-8\}$

7. -3 and -2 or $\{-3, -2\}$ **8.** $\dfrac{4}{x^9}$ **9.** 31 **10.** $10x^2 - 9x + 4$ **11.** $21x^2 - 23x - 20$ **12.** $25x^2 - 20x + 4$ **13.** $x^3 + y^3$

14. $\dfrac{x + 4}{3x^3}$ **15.** $\dfrac{-7x}{(x + 3)(x - 1)(x - 4)}$ **16.** $\dfrac{x}{5}$ **17.** $(2x + 7)(2x - 7)$ **18.** $(x + 3)(x + 1)(x - 1)$

19. $2(x - 3)(x + 7)$ **20.** $x(x^2 + 4)(x + 2)(x - 2)$ **21.** $x(x - 5)^2$ **22.** $(x - 2)(x^2 + 2x + 4)$

23. $y = \dfrac{1}{3}x - 1$

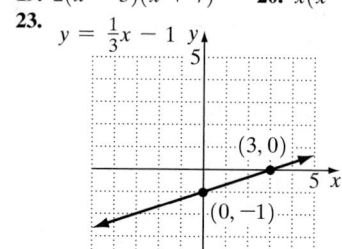

24. $3x + 2y = -6$

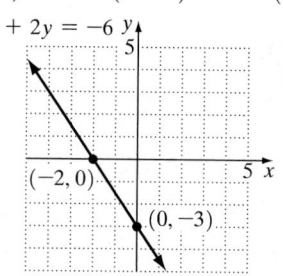

25. $y = -2$

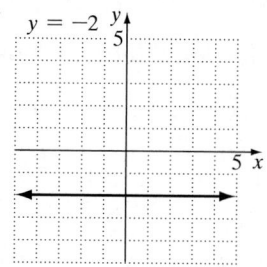

26. -2 **27.** $y - 2 = 2(x - 1)$ or $y - 6 = 2(x - 3)$; $y = 2x$ **28.** 43 **29.** \$320 **30.** length: 150 yd; width: 50 yd

31. \$12,500 at 7% and \$7500 at 9% **32.** 8 liters of 40% and 4 liters of 70% **33.** 16 ft **34.** 1st is 35°; 2nd is 25°; 3rd is 120°

35. TV: \$350; stereo: \$370 **36.** length: 11 m; width: 5 m

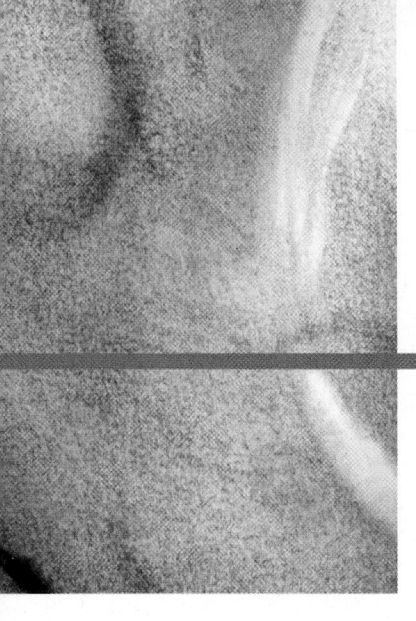

Subject Index

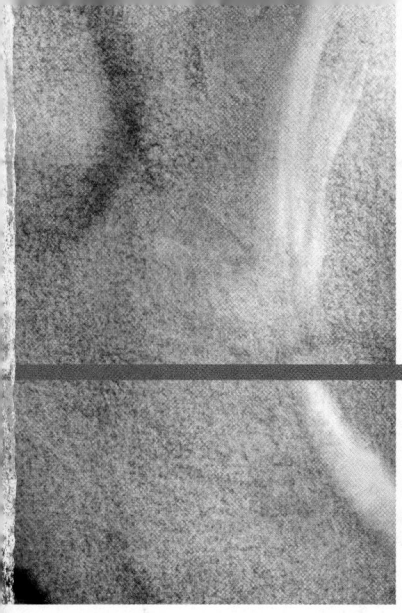

Photo Credits

Exponential and Logarithmic Functions

1. Exponential Function: $f(x) = b^x, b > 0, b \neq 1$

Graphs:

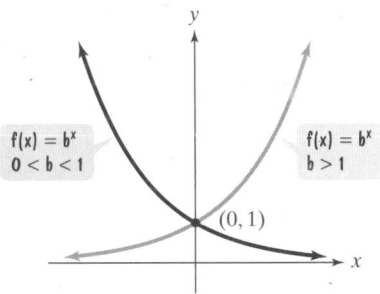

2. Logarithmic Function: $f(x) = \log_b x, b > 0, b \neq 1$

$y = \log_b x$ is equivalent to $x = b^y$.

Graphs:

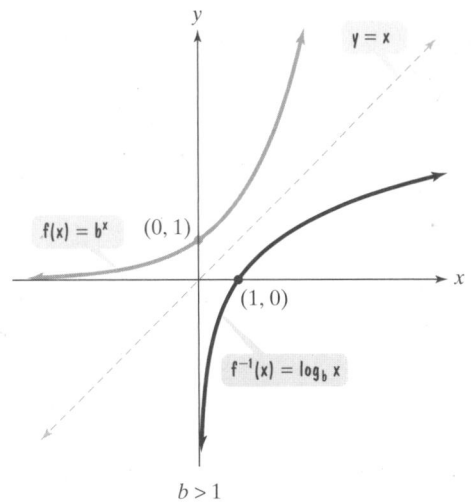

3. Properties of Logarithms

   a. $\log_b(MN) = \log_b M + \log_b N$

   b. $\log_b\left(\dfrac{M}{N}\right) = \log_b M - \log_b N$

   c. $\log_b M^p = p \log_b M$

   d. $\log_b M = \dfrac{\log_a M}{\log_a b} = \dfrac{\ln M}{\ln b} = \dfrac{\log M}{\log b}$

   e. $\log_b b^x = x; \log 10^x = x; \ln e^x = x$

   f. $b^{\log_b x} = x; 10^{\log x} = x; e^{\ln x} = x$

Distance and Midpoint Formulas

1. The distance from (x_1, y_1) to (x_2, y_2) is

$$\sqrt{(x_2 - x_1)^2 + (y_2 - y_1)^2}.$$

2. The midpoint of the line segment with endpoints (x_1, y_1) and (x_2, y_2) is

$$\left(\frac{x_1 + x_2}{2}, \frac{y_1 + y_2}{2}\right).$$

Conic Sections
Circle

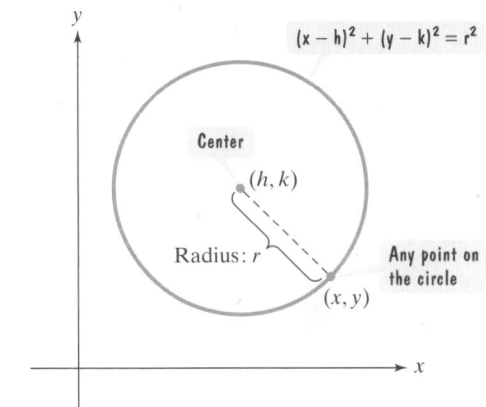

Ellipse

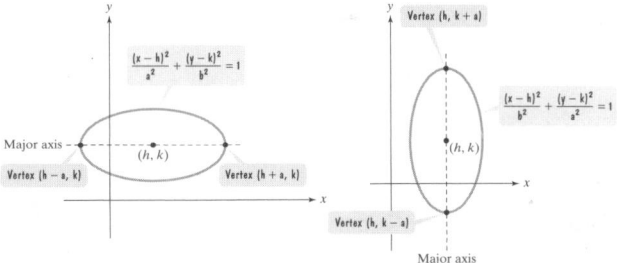

Hyperbola

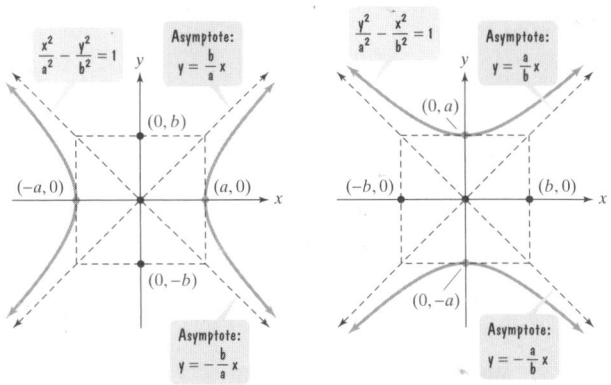